HANDBOOK OF
NANOPHYSICS

Handbook of Nanophysics

HANDBOOK OF NANOPHYSICS

Principles and Methods

Edited by

Klaus D. Sattler

CRC Press
Taylor & Francis Group
Boca Raton   London   New York

CRC Press is an imprint of the
Taylor & Francis Group, an **informa** business

CRC Press
Taylor & Francis Group
6000 Broken Sound Parkway NW, Suite 300
Boca Raton, FL 33487-2742

First issued in paperback 2020

© 2011 by Taylor and Francis Group, LLC
CRC Press is an imprint of Taylor & Francis Group, an Informa business

No claim to original U.S. Government works

ISBN 13: 978-1-138-11785-3 (pbk)
ISBN 13: 978-1-4200-7540-3 (hbk)

Library of Congress Cataloging-in-Publication Data

Handbook of nanophysics. Principles and methods / editor, Klaus D. Sattler.
    p. cm.
  "A CRC title."
  Includes bibliographical references and index.
  ISBN 978-1-4200-7540-3 (alk. paper)
  1. Microphysics--Handbooks, manuals, etc. 2. Nanotechnology--Handbooks, manuals, etc. 3. Nanoscience--Handbooks, manuals, etc. I. Sattler, Klaus D.

QC173.4.M5H358 2009
620'.5--dc22   2009047134

Visit the Taylor & Francis Web site at
http://www.taylorandfrancis.com

and the CRC Press Web site at
http://www.crcpress.com

Contents

PART I   Design and Theory

PART II   Nanoscale Systems

PART VI Nanoscale Methods

Preface

The *Handbook of Nanophysics* is the first comprehensive reference to consider both fundamental and applied aspects of nanophysics. As a unique feature of this work, we requested contributions to be submitted in a tutorial style, which means that state-of-the-art scientific content is enriched with fundamental equations and illustrations in order to facilitate wider access to the material. In this way, the handbook should be of value to a broad readership, from scientifically interested general readers to students and professionals in materials science, solid-state physics, electrical engineering, mechanical engineering, computer science, chemistry, pharmaceutical science, biotechnology, molecular biology, biomedicine, metallurgy, and environmental engineering.

What Is Nanophysics?

Modern physical methods whose fundamentals are developed in physics laboratories have become critically important in nanoscience. Nanophysics brings together multiple disciplines, using theoretical and experimental methods to determine the physical properties of materials in the nanoscale size range (measured by millionths of a millimeter). Interesting properties include the structural, electronic, optical, and thermal behavior of nanomaterials; electrical and thermal conductivity; the forces between nanoscale objects; and the transition between classical and quantum behavior. Nanophysics has now become an independent branch of physics, simultaneously expanding into many new areas and playing a vital role in fields that were once the domain of engineering, chemical, or life sciences.

This handbook was initiated based on the idea that breakthroughs in nanotechnology require a firm grounding in the principles of nanophysics. It is intended to fulfill a dual purpose. On the one hand, it is designed to give an introduction to established fundamentals in the field of nanophysics. On the other hand, it leads the reader to the most significant recent developments in research. It provides a broad and in-depth coverage of the physics of nanoscale materials and applications. In each chapter, the aim is to offer a didactic treatment of the physics underlying the applications alongside detailed experimental results, rather than focusing on particular applications themselves.

The handbook also encourages communication across borders, aiming to connect scientists with disparate interests to begin interdisciplinary projects and incorporate the theory and methodology of other fields into their work. It is intended for readers from diverse backgrounds, from math and physics to chemistry, biology, and engineering.

The introduction to each chapter should be comprehensible to general readers. However, further reading may require familiarity with basic classical, atomic, and quantum physics. For students, there is no getting around the mathematical background necessary to learn nanophysics. You should know calculus, how to solve ordinary and partial differential equations, and have some exposure to matrices/linear algebra, complex variables, and vectors.

External Review

All chapters were extensively peer reviewed by senior scientists working in nanophysics and related areas of nanoscience. Specialists reviewed the scientific content and nonspecialists ensured that the contributions were at an appropriate technical level. For example, a physicist may have been asked to review a chapter on a biological application and a biochemist to review one on nanoelectronics.

Organization

The *Handbook of Nanophysics* consists of seven books. Chapters in the first four books (*Principles and Methods, Clusters and Fullerenes, Nanoparticles and Quantum Dots,* and *Nanotubes and Nanowires*) describe theory and methods as well as the fundamental physics of nanoscale materials and structures. Although some topics may appear somewhat specialized, they have been included given their potential to lead to better technologies. The last three books (*Functional Nanomaterials, Nanoelectronics and Nanophotonics,* and *Nanomedicine and Nanorobotics*) deal with the technological applications of nanophysics. The chapters are written by authors from various fields of nanoscience in order to encourage new ideas for future fundamental research.

After the first book, which covers the general principles of theory and measurements of nanoscale systems, the organization roughly follows the historical development of nanoscience. *Cluster* scientists pioneered the field in the 1980s, followed by extensive work on *fullerenes*, *nanoparticles*, and *quantum dots* in the 1990s. Research on *nanotubes* and *nanowires* intensified in subsequent years. After much basic research, the interest in applications such as the *functions of nanomaterials* has grown. Many bottom-up and top-down techniques for nanomaterial and nanostructure generation were developed and made possible the development of *nanoelectronics* and *nanophotonics*. In recent years, real applications for *nanomedicine* and *nanorobotics* have been discovered.

For MATLAB® and Simulink® product information, please contact

The MathWorks, Inc.
3 Apple Hill Drive
Natick, MA, 01760-2098 USA
Tel: 508-647-7000
Fax: 508-647-7001
E-mail: info@mathworks.com
Web: www.mathworks.com

Acknowledgments

Many people have contributed to this book. I would like to thank the authors whose research results and ideas are presented here. I am indebted to them for many fruitful and stimulating discussions. I would also like to thank individuals and publishers who have allowed the reproduction of their figures. For their critical reading, suggestions, and constructive criticism, I thank the referees. Many people have shared their expertise and have commented on the manuscript at various stages. I consider myself very fortunate to have been supported by Luna Han, senior editor of the Taylor & Francis Group, in the setup and progress of this work. I am also grateful to Jessica Vakili, Jill Jurgensen, Joette Lynch, and Glenon Butler for their patience and skill with handling technical issues related to publication. Finally, I would like to thank the many unnamed editorial and production staff members of Taylor & Francis for their expert work.

Klaus D. Sattler
Honolulu, Hawaii

Editor

Klaus D. Sattler pursued his undergraduate and master's courses at the University of Karlsruhe in Germany. He received his PhD under the guidance of Professors G. Busch and H.C. Siegmann at the Swiss Federal Institute of Technology (ETH) in Zurich, where he was among the first to study spin-polarized photo-electron emission. In 1976, he began a group for atomic cluster research at the University of Konstanz in Germany, where he built the first source for atomic clusters and led his team to pioneering discoveries such as "magic numbers" and "Coulomb explosion." He was at the University of California, Berkeley, for three years as a Heisenberg Fellow, where he initiated the first studies of atomic clusters on surfaces with a scanning tunneling microscope.

Dr. Sattler accepted a position as professor of physics at the University of Hawaii, Honolulu, in 1988. There, he initiated a research group for nanophysics, which, using scanning probe microscopy, obtained the first atomic-scale images of carbon nanotubes directly confirming the graphene network. In 1994, his group produced the first carbon nanocones. He has also studied the formation of polycyclic aromatic hydrocarbons (PAHs) and nanoparticles in hydrocarbon flames in collaboration with ETH Zurich. Other research has involved the nanopatterning of nanoparticle films, charge density waves on rotated graphene sheets, band gap studies of quantum dots, and graphene foldings. His current work focuses on novel nanomaterials and solar photocatalysis with nanoparticles for the purification of water.

Among his many accomplishments, Dr. Sattler was awarded the prestigious Walter Schottky Prize from the German Physical Society in 1983. At the University of Hawaii, he teaches courses in general physics, solid-state physics, and quantum mechanics.

In his private time, he has worked as a musical director at an avant-garde theater in Zurich, composed music for theatrical plays, and conducted several critically acclaimed musicals. He has also studied the philosophy of Vedanta. He loves to play the piano (classical, rock, and jazz) and enjoys spending time at the ocean, and with his family.

Contributors

Yves Acremann
PULSE Institute
SLAC National Accelerator Laboratory
Stanford, California

Jurij Avsec
Faculty of Energy Technology
University of Maribor
Krško, Slovenia

Renaud Bachelot
Laboratoire de Nanotechnologie
    et d'Instrumentation Optique
Institut Charles Delaunay
Université de Technologie de Troyes
Troyes, France

Supriyo Bandyopadhyay
Department of Electrical and Computer
    Engineering
Virginia Commonwealth University
Richmond, Virginia

Rodion V. Belosludov
Institute for Materials Research
Tohoku University
Sendai, Japan

Vladimir R. Belosludov
Institute for Materials Research
Tohoku University
Sendai, Japan

and

Nikolaev Institute of Inorganic
    Chemistry
Novosibirsk, Russia

R. Stephen Berry
Department of Chemistry
The James Franck Institute
The University of Chicago
Chicago, Illinois

Alessandro di Bona
CNR-Institute of Nanoscience-Center S3
Modena, Italy

Jérôme Bürki
Department of Physics
    and Astronomy
California State University
Sacramento, California

Javier Cervera
Faculty of Physics
University of Valencia
Valencia, Spain

Sajeev Chacko
School of Information Science
Jawaharlal Nehru University
New Delhi, India

Elisabeth Charlaix
Laboratoire de Physique de la Matière
    Condensée et Nanostructures
Université Claude Bernard Lyon 1
Villeurbanne, France

James R. Chelikowsky
Center for Computational Materials
Institute for Computational Engineering
    and Sciences
and
Department of Physics and Chemical
    Engineering
University of Texas
Austin, Texas

Matteo Ciccotti
Laboratoire des Colloïdes, Verres
    et Nanomatériaux
Université Montpellier 2
Montpellier, France

Marvin L. Cohen
Department of Physics
University of California, Berkeley
and
Materials Sciences Division
Lawrence Berkeley National Laboratory
Berkeley, California

Mihail D. Croitoru
Department of Physics
University of Antwerp
Antwerp, Belgium

Arnaud Devos
Institut d'Électronique, de
    Microélectronique et de
    Nanotechnologie
Unité Mixte de Recherche
Centre national de la recherche
    scientifique
Villeneuve d'Ascq, France

Siegfried Dietrich
Max-Planck-Institut für Metallforschung
and
Institut für Theoretische und
    Angewandte Physik
Universität Stuttgart
Stuttgart, Germany

Vladimir García-Morales
Physik Department
Technische Universität München
Munich, Germany

Jacob L. Gavartin
Accelrys Ltd.
Cambridge, United Kingdom

Gian Carlo Gazzadi
CNR-Institute of Nanoscience-Center S3
Modena, Italy

James K. Gimzewski
California NanoSystems Institute
and
Department of Chemistry
    and Biochemistry
University of California, Los Angeles
Los Angeles, California

and

Material Nanoarchitectonics
National Institute for Materials Science
Tsukuba-Shi, Japan

Hermann Grabert
Physikalisches Institut
and
Freiburg Institute for Advanced Studies
Albert-Ludwigs-Universität
Freiburg, Germany

Ida Grundberg
Department of Genetics
    and Pathology
Rudbeck Laboratory
Uppsala University
Uppsala, Sweden

Alexander N. Guz
Timoshenko Institute of Mechanics
Kiev, Ukraine

Igor A. Guz
Centre for Micro- and Nanomechanics
University of Aberdeen
Scotland, United Kingdom

John D. Head
Department of Chemistry
University of Hawaii
Honolulu, Hawaii

Hirokazu Hori
Interdisciplinary Graduate School
    of Medicine and Engineering
University of Yamanashi
Kofu, Japan

Tetsuya Inoue
Department of Electronics
Yamanashi Industrial Technology
    College
Kosyu, Japan

Qing Jiang
Key Laboratory of Automobile Materials
Ministry of Education
and
Department of Materials Science
    and Engineering
Jilin University
Changchun, China

Bhargava Kanchibotla
Department of Electrical and Computer
    Engineering
Virginia Commonwealth University
Richmond, Virginia

Dilip Govind Kanhere
Department of Physics
University of Pune
Pune, India

Eduard G. Karpov
Department of Civil & Materials
    Engineering
University of Illinois at Chicago
Chicago, Illinois

Yoshiyuki Kawazoe
Institute for Materials Research
Tohoku University
Sendai, Japan

Jörg Kröger
Institut für Experimentelle und
    Angewandte Physik
Christian-Albrechts-Universität zu Kiel
Kiel, Germany

Vijay Kumar
Dr. Vijay Kumar Foundation
Haryana, India

Ulf Landegren
Department of Genetics
    and Pathology
Rudbeck Laboratory
Uppsala University
Uppsala, Sweden

Pierre Letellier
Laboratoire Interfaces et Systèmes
    Electrochimiques
Université Pierre et Marie Curie-Paris 6
Paris, France

Shuang Li
Key Laboratory of Automobile Materials
Ministry of Education
and
Department of Materials Science
    and Engineering
Jilin University
Changchun, China

Wing Kam Liu
Department of Mechanical Engineering
Northwestern University
Evanston, Illinois

Yaling Liu
Department of Mechanical & Aerospace
    Engineering
University of Texas at Arlington
Arlington, Texas

Stergios Logothetidis
Laboratory for Thin Films—
    Nanosystems and Nanometrology
Department of Physics
Aristotle University of Thessaloniki
Thessaloniki, Greece

Joachim Maier
Max Planck Institute for Solid State
    Research
Stuttgart, Germany

Tatiana Makarova
Department of Physics
Umeå University
Umeå, Sweden

and

Ioffe Physico-Technical Institute
Saint Petersburg, Russia

José A. Manzanares
Faculty of Physics
University of Valencia
Valencia, Spain

Milan Marčič
Faculty of Mechanical Engineering
University of Maribor
Maribor, Slovenia

Alain Mayaffre
Laboratoire Interfaces et Systèmes
  Electrochimiques
Université Pierre et Marie Curie-Paris 6
Paris, France

Hiroshi Mizuseki
Institute for Materials Research
Tohoku University
Sendai, Japan

Günter Möbus
Department of Engineering Materials
University of Sheffield
Sheffield, United Kingdom

Peder C. F. Møller
The Niels Bohr Institute
University of Copenhagen
Copenhagen, Denmark

Greg F. Naterer
Institute of Technology
University of Ontario
Oshawa, Ontario, Canada

Lene B. Oddershede
The Niels Bohr Institute
University of Copenhagen
Copenhagen, Denmark

Francois M. Peeters
Department of Physics
University of Antwerp
Antwerp, Belgium

Jérôme Plain
Laboratoire de Nanotechnologie
  et d'Instrumentation Optique
Institut Charles Delaunay
Université de Technologie de Troyes
Troyes, France

Sandipan Pramanik
Department of Electrical and Computer
  Engineering
University of Alberta
Edmonton, Alberta, Canada

Nicola M. Pugno
Dipartimento di Ingegneria Strutturale
  e Geotecnica
Politecnico di Torino
Turin, Italy

Markus Rauscher
Max-Planck-Institut für Metallforschung
and
Institut für Theoretische und
  Angewandte Physik
Universität Stuttgart
Stuttgart, Germany

Jeremiah J. Rushchitsky
Timoshenko Institute of Mechanics
Kiev, Ukraine

Zineb Saghi
Department of Engineering Materials
University of Sheffield
Sheffield, United Kingdom

Ryoji Sahara
Institute for Materials Research
Tohoku University
Sendai, Japan

Andreas Scherz
Stanford Institute for Material
  and Energy Science
SLAC National Accelerator Laboratory
Menlo Park, California

Werner Schindler
Department of Physics
Technische Universität München
Munich, Germany

Andreas Schmidt-Ott
Faculty of Applied Sciences
Delft University of Technology
Delft, the Netherlands

Kurt Schönhammer
Institute for Theoretical Physics
Georg-August University
Goettingen, Germany

Arkady A. Shanenko
Department of Physics
University of Antwerp
Antwerp, Belgium

Hidemi Shigekawa
Institute of Applied Physics
University of Tsukuba
Tsukuba, Japan

Tsumoru Shintake
RIKEN SPring-8 Center
Sayo, Hyogo, Japan

Hans Christoph Siegmann (deceased)
PULSE Institute
SLAC National Accelerator Laboratory
Stanford, California

Olivier Soppera
Département de Photochimie Générale
Centre national de la recherche
  scientifique
Mulhouse, France

Charles A. Stafford
Department of Physics
University of Arizona
Tucson, Arizona

Martin Stark
Center for Nanoscience
Ludwig-Maximilians-Universität
  München
Munich, Germany

Robert W. Stark
Center for Nanoscience
Ludwig-Maximilians-Universität
  München
Munich, Germany

Adam Z. Stieg
California NanoSystems Institute
Los Angeles, California

and

Material Nanoarchitectonics
National Institute for Materials Science
Tsukuba-Shi, Japan

A. Marshall Stoneham
London Centre for Nanotechnology
Department of Physics
  and Astronomy
University College London
London, United Kingdom

Oleg S. Subbotin
Institute for Materials Research
Tohoku University
Sendai, Japan

and

Nikolaev Institute of Inorganic
    Chemistry
Novosibirsk, Russia

Osamu Takeuchi
Institute of Applied Physics
University of Tsukuba
Tsukuba, Japan

Yasuhiko Terada
Institute of Applied Physics
University of Tsukuba
Tsukuba, Japan

Mireille Turmine
Laboratoire Interfaces et Systèmes
    Electrochimiques
Université Pierre et Marie Curie-Paris 6
Paris, France

Daniel F. Urban
Physikalisches Institut
Albert-Ludwigs-Universität
Freiburg, Germany

Sergio Valeri
Department of Physics
University of Modena and Reggio
    Emilia
and
CNR-Institute of Nanoscience-Center S3
Modena, Italy

Natarajan S. Venkataramanan
Institute for Materials Research
Tohoku University
Sendai, Japan

Irene Weibrecht
Department of Genetics
    and Pathology
Rudbeck Laboratory
Uppsala University
Uppsala, Sweden

Shoji Yoshida
Institute of Applied Physics
University of Tsukuba
Tsukuba, Japan

I

Design and Theory

1

The Quantum Nature of Nanoscience

Marvin L. Cohen
University of California, Berkeley

and

Lawrence Berkeley National Laboratory

1.1 Introduction

Although research on nanoscale-sized objects has been ongoing for a century or more, over the last few decades, there has been a collective effort in bringing together researchers to this area from different disciplines to form a new discipline "nanoscience" that focuses on the properties of nanostructures (Saito and Zettl 2008). The extension to nonstructures from studies of molecules and clusters is natural. There are also other conceptual paths from studies of periodic systems, which allow the application of concepts and experimental techniques originally designed for macroscopic solids (Cohen 2005). As a result, scientists and engineers have found common ground in the field of nanoscience. The physical, chemical, computational, and biological sciences have overlapping interests associated with the nanoscale and its associated energy scale. The same holds true for electrical, mechanical, and computer engineering fields.

Structures built from atoms measuring around one-tenth of a nanometer (nm) with bonds between them of the order of 0.3 nm allow the building of molecular structures of the order of several nanometers and much larger. This is how the structures of the C_{60} molecule (Kroto et al. 1985), nanotubes (Iijima 1991, Rubio et al. 1994, Chopra et al. 1995), DNA, viruses, among others, are made. Understanding electronic behavior is essential for these systems. The size scale fixes the confinement lengths of the electrons; hence, it also sets the energy scale. The theoretical tool applied to understand the size and energy domain of nanostructures is quantum mechanics. The wave nature of the particles has to be considered to explain electronic, structural, mechanical, and other properties of the nanostructures of interest.

1.2 Conceptual Models

Building from the bottom up is an obvious methodology for constructing models to ascertain the properties of nanostructures. This is the usual approach in quantum chemistry. Since we know the constituent atoms of interest and sometimes their structural arrangements, and as there are tested procedures for computing many properties of molecules and clusters, this is a valuable approach for dealing with nanostructures. Theoretical chemists and physicists may vary in their choices of specific methods, but the general approach is to arrange the positive atomic cores, each consisting of the atomic nucleus and core electrons, into a given structure, and then to treat the negative valence electrons and cores as the primary particles. In this model, it is usually assumed that the core electrons are little disturbed from their normal configurations in an isolated atom. The core–core interactions are often represented by considering Coulomb interactions between point-like particles. The treatments of the electron–core and electron–electron interactions vary. Here we use the term electron to refer to a valence electron.

The basis states for the electrons can be considered to be atomic-like assuming that the change in electronic states for free atoms is relatively small when the nanostructure is formed. Chemists often take this approach.

In contrast to the bottom-up approach, condensed matter physicists often take an almost opposite view and treat the electrons within a nearly free electron model. This model views the electrons as itinerant, and a free electron basis set is used. Both methods have their domains in which they are used with ease. In general, when fully implemented and large computers are used, both approaches can be successful. Often, tight binding, which parameterizes local orbital models, and nearly free electron models serve as approximate methods to explain particular types of data.

Hence, theoretical tools are available. The ones (Cohen 1982, 2006) that worked for bulk solids, surfaces, interfaces, clusters, and molecules can be extended to nanostructures.

1.3 Nanotubes, Fullerenes, and Graphene

If we view a carbon or boron nitride nanotube as a rolled up sheet of graphene or its boron nitride (BN) graphene equivalent, then some properties of these tubes can be easily predicted. For example, for a carbon nanotube (CNT), depending on how the graphene is rolled into a tube, the resulting system can be a semiconductor or a metal (Saito 2008). The linear dispersion of energy versus wave vector $E(k)$ found for graphene, commonly called "Dirac-like" because of its similarity to the relativistic dispersion found in the Dirac theory (Mele and Kane 2008), is altered. For an undoped boron nitride nanotube (BNNT), because of the ionic character of the BN bond, these systems are always semiconductors. The ionic potential opens a band gap at the Fermi level (Blasé et al. 1994). This system, when doped with electrons, is expected to exhibit interesting conduction with electron transport along the center of the tube.

A common feature of CNTs and BNNTs is that they can be multiwalled (MW). This cylinder-within-a-cylinder geometry allows interesting applications. For example, it is possible to pull out an inner tube of a CNT and attach a stator to it. Thus, a linear or a rotational bearing can be constructed where the inner cylinder either moves back and forth within the outer cylinder (or cylinders) or rotates (Fennimore et al. 2003). Because the bonding within the CNT is covalent, the tubes are rigid and strong; however, the bonding between the tubes within a MWCNT is van der Waals-like and very weak. Hence, these bearings have little friction, and when motors are constructed using these bearings, they are relatively friction-free. The friction mechanisms on the nanoscale are of theoretical interest (Tangney et al. 2004). It is possible to make many of these motors, which are each smaller than a virus.

The fact that CNTs have different electronic properties depending on their chiral properties, i.e., how they were "rolled up," paves the way for developing devices with interesting properties. For example, it was predicted using theoretical calculations that a junction between a semiconductor CNT and a metal CNT could be formed if the interface contained a fivefold ring of bonds next to a sevenfold ring replacing two sixfold rings. This interface, which involves only a small number of atoms, becomes a Schottky barrier (Chico et al. 1996) similar to what is found when a conventional semiconductor is put in contact with a conventional metal. The resulting device is predicted to act as a rectifier, and this property was verified experimentally (Yao et al. 1999) along with the existence of this geometrical configuration.

The Schottky barrier is only one example of a possible device. Heterojunctions formed from two semiconducting tubes are expected to have properties similar to the usual heterojunction devices made of macroscopic semiconductor materials. Hence, there is considerable excitement about the possibility of shrinking electronic devices even further using nanotubes. And since the thermal conductivities of nanotubes are very large, this may allow the relative packing densities of the electronic devices to be increased from present values. Even single nanotubes can be used as devices because of their sizes. As an example, a nanotube can be used as a sensor (Jhi et al. 2000) since its resistivity is sensitive to molecules or atoms that attach to the tubes. This is because the impurities disturb the electronic wavefunctions, which may extend over large fractions of the tube. Another example of the unusual thermal and electronic interplay is the thermoelectric power of nanotubes (Hone et al. 1998). The quantum nature of the electronic processes in NTs often leads to unexpected results. For example, we expect that two metals put in contact to allow current to flow between them. However, for some metallic tubes, this does not happen. In such cases, conductivity depends on the chirality of the tubes. So, symmetry plays an important role in addition to confinement and reduced dimensionality when dealing with the electronic properties of nanostructures.

Hence confinement and lower dimensionality have important effects. A quantum dot can be viewed as a zero-dimensional object, a graphene sheet is two-dimensional, and an NT can have some one-dimensional properties. Often, confinement leads to discrete energy levels as depicted by the classic "particle in a box" quantum mechanics problem. The size of the "box" and the dimensionality determine the detailed energy structure. For nuclei, the scale is of the order of MeVs; for alkali metal clusters with a similar potential shape, it is of the order of eVs (Knight et al. 1984); for carbon clusters, the energies are in the same range (Lonfat et al. 1999); for atoms, it varies from several eVs to keVs. So confinement, dimensionality, and symmetry all contribute to making the nanoscale interesting theoretically.

The bonding geometries are also critical in determining electronic properties as one would expect. For example, it is well known that the dramatic difference between graphite and diamond can be traced to the fourfold versus threefold bonding coordination and the sp^2 and sp^3 nature of the bonds. The coordination and lengths of the bonds are important for determining macroscopic properties. The low compressibility of diamond is associated with its short bond (Cohen 1985). However, even though the sp^2 bonds in graphite are shorter, they exist in the graphene planes. If one could pull a graphene sheet from opposite sides, it would be very strong. Hence, when a graphene sheet is rolled into a tube to produce a CNT, in a sense, this operation can be performed by pulling on the ends of the tube. The Young's moduli for CNTs and BNNTs are among the largest available for any material (Chopra and Zettl 1998, Hayashida et al. 2002). This property and the high aspect ratio for NTs make them useful for structural materials and for nanosized probes.

1.4 Some Properties and Applications

Studies of fullerenes, CNTs, BNNTs, and graphene have revealed novel properties. As a result, many applications of these systems have been proposed. A few will be discussed in this section to illustrate how the considerations described above lead to unusual and potentially useful applications.

The strength of the sp^2 covalent bond and the resulting large Young's moduli for NTs were discussed above. These properties have been exploited in applications for developing strong fibers, composites, and nanoelectromechanical systems (NEMS) in analogy with the microelectromechanical systems (MEMS) that are in use in the industry. The related structural properties have led to suggestions for using nanostructures as templates in material synthesis. Another property of NTs is their high aspect ratio. As a result, NTs have been used to probe biological systems and as tips in scanning electron and atomic force microscopy (STM, AFM) instruments. Because they are hollow, they are applications in chemical storage, molecular transport, and filtering. Another use of the hollow nature of NTs is their use to produce so-called peapods. Single-walled or MWNTs are used as confining cylinders for C_{60} molecules, which are absorbed internally. This "peapod" type geometry allows the formation of crystal structures composed of these molecules not found in nature. Both CNTs and BNNTs have been used in this fashion (Smith et al. 1998, Mickelson et al. 2003). The interaction between the BNNTs and the C_{60}s is smaller than for the CNTs, resulting in less charge transfer. This makes it easier to model the resulting structure in terms of spheres in cylinders. For spheres with diameters slightly less than the tube diameters, a linear array of spheres is expected. As the tube diameter gets larger for a fixed sphere size, a zigzag pattern results; for larger diameters, helical patterns emerge; and in some case, a new hollow center appears. Mathematicians have studied these geometries, and for BNNT, peapod structures of this kind are found. Several applications in electronics were described in Section 1.3. When used as electrical conductors, NTs behave as quantum wires with novel electronic properties and sensing ability. Often the NTs are functionalized so that they are particularly sensitive to certain adsorbates.

Although there have been observations of superconducting behavior in NTs (Tang et al. 2001), the superconductors most studied in this area are the alkali metal–doped crystals of C_{60} (Hebard et al. 1991). For example, K_3C_{60} is viewed as a metallic system in which the K outermost valence electron is donated to a sea of conduction electrons. The current consensus on the theory of the underlying mechanism for the superconducting behavior is rooted in the BCS theory of superconductivity. In these studies, electron pairing is caused by phonons as in the case of conventional superconductors. The specific phonons that appear to be dominant in the pairing interaction are associated with intramolecular vibrations. This picture of relatively itinerant electrons paired by phonons associated with local molecular motions is consistent with the isotope effect (Burk et al. 1994), photoemission studies, and a host of other experimental measurements. The superconducting transition temperature for this class of materials is fairly high with a maximum at this time of about 40 K. There are interesting suggestions for achieving higher transition temperatures. Many of these involve models designed to increase the density of electronic states at the Fermi level by increasing the lattice constant or by using metal atoms with d-electron states (Umemoto and Saito 2001). The C36 molecule appears to be promising for higher superconducting transition temperatures (Côté et al. 1998) because the higher curvature of this molecule compared to C_{60} suggests stronger electron–phonon couplings.

Considerable research has been done on the optical properties of NTs. Because of the lower dimensionality of these systems, sharp structure can appear in the absorption or reflectivity spectra. These are usually associated with van Hove singularities in the joint density of states, which are often greatly enhanced by excitonic effects (Spataru et al. 2004, Wang et al. 2007). Raman effect studies (Dresselhaus et al. 2008) have contributed to obtaining considerable information about both the electronic and vibrational properties of CNTs.

Some of the unusual optical properties seen in carbon and BN nanotubes arise because they can be thought of as rolled up sheets. Hence, it is expected that similar effects, such as unusual excitonic behavior should be observed for the sheets themselves. Recent studies on nanoribbons predict unusual electronic structure (Son et al. 2006) and excitonic effects (Yang et al. 2007). In fact, these systems are predicted to display features not expected in nanotubes, such as electric field–dependent spin effects (Son et al. 2007), anisotropic electron–phonon coupling (Park et al. 2008a), and supercollimation of electron transport (Park et al. 2008b). This area of research on graphene and graphene ribbons is expected to yield many new and unusual material properties.

Acknowledgments

This work was supported by National Science Foundation Grant No. DMR07-05941 and by the Director, Office of Science, Office of Basic Energy Sciences, Materials Sciences and Engineering Division, U.S. Department of Energy under Contract No. DE- AC02-05CH11231.

References

Blasé, X., Rubio, A., Louie, S. G., and Cohen, M. L. 1994. Stability and band gap constancy of boron nitride nanotubes. *Europhys. Lett.* 28: 335.

Burk, B., Crespi, V. H., Zettl, A., and Cohen, M. L. 1994. Rubidium isotope effect in superconducting Rb_3C_{60}. *Phys. Rev. Lett.* 72: 3706.

Chico, L., Crespi, V. H., Benedict, L. X., Louie, S. G., and Cohen, M. L. 1996. Pure carbon nanoscale devices: Nanotube heterojunctions. *Phys. Rev. Lett.* 76: 971.

Chopra, N. G. and Zettl, A. 1998. Measurement of the elastic modulus of a multi-wall boron nitride nanotube. *Solid State Commun.* 105: 297.

Chopra, N. G., Luyken, R. J., Cherrey, K. et al. 1995. Boron nitride nanotubes. *Science* 269: 966.

Cohen, M. L. 1982. Pseudopotentials and total energy calculations. *Phys. Scripta* T1: 5.

Cohen, M. L. 1985. Calculation of bulk moduli of diamond and zincblende solids. *Phys. Rev. B* 32: 7988.

Cohen, M. L. 2005. Nanoscience: The quantum frontier. *Physica E* 29: 447.

Cohen, M. L. 2006. Overview: A standard model of solids. In *Conceptual Foundations of Materials: A Standard Model for Ground- and Excited-State Properties*, S. G. Louie and M. L. Cohen (Eds.), p. 1. Amsterdam, the Netherlands: Elsevier.

Côté, M., Grossman, J. C., Cohen, M. L., and Louie, S. G. 1998. Electron–phonon interactions in solid C_{36}. *Phys. Rev. Lett.* 81: 697.

Dresselhaus, M. S., Dresselhaus, G., Saito, R., and Jorio, A. 2008. Raman spectroscopy of carbon nanotubes. In *Carbon Nanotubes: Quantum Cylinders of Graphene*, S. Saito and A. Zettl (Eds.), pp. 83–108. Amsterdam, the Netherlands: Elsevier.

Fennimore, A. M., Yuzvinsky, T. D., Han, W.-Q. et al. 2003. Rotational actuators based on carbon nanotubes. *Nature* 424: 408.

Hayashida, T., Pan, L., and Nakayama, Y. 2002. Mechanical and electrical properties of carbon tubule nanocoils. *Phys. B: Condens. Matter* 323: 352.

Hebard, A. F., Rosseinsky, M. J., Hadden, R. C. et al. 1991. Superconductivity at 18 K in potassium-doped C_{60}. *Nature* 350: 600.

Hone, J., Ellwood, I., Muno, M. et al. 1998. Thermoelectric power of single-walled carbon nanotubes. *Phys. Rev. Lett.* 80: 1042.

Iijima, S. 1991. Helical microtubules of graphitic carbon. *Nature* 354: 56.

Jhi, S.-H., Louie, S. G., and Cohen, M. L. 2000. Electronic properties of oxidized carbon nanotubes. *Phys. Rev. Lett.* 85: 1710.

Knight, W. D., Clemenger, K., de Heer, W. A., Saunders, W. A., Chou, M. Y., and Cohen, M. L. 1984. Electronic shell structure and abundances of sodium clusters. *Phys. Rev. Lett.* 5: 2141 [Erratum: *Phys. Rev. Lett.* 53: 510 (1984)].

Kroto, H., Heath, J. R., O'Brien, S. C., Curl, R. F., and Smalley, R. E. 1985. C_{60}: Bulkminster fullerene. *Nature* 318: 162.

Lonfat, M., Marsen, B., and Sattler, K. 1999. The energy gap of carbon clusters studied by scanning tunneling spectroscopy. *Chem. Phys. Lett.* 313: 539.

Mele, E. J. and Kane, C. L. 2008. Low-energy electronic structure of graphene and its Dirac theory. In *Carbon Nanotubes: Quantum Cylinders of Graphene*, S. Saito and A. Zettl (Eds.), pp. 171–197. Amsterdam, the Netherlands: Elsevier.

Mickelson, W., Aloni, S., Han, W.-Q., Cumings, J., and Zettl, A. 2003. Packing C_{60} in boron nitride nanotubes. *Science* 300: 467.

Park, C.-H., Giustino, F., McChesney, J. L. et al. 2008a. Van Hove singularity and apparent anisotropy in the electron-phonon interaction in graphene. *Phys. Rev. B* 77: 113410.

Park, C.-H., Son, Y.-W., Yang, L., Cohen, M. L., and Louie, S. G. 2008b. Electron beam supercollimation in graphene superlattices. *Nano Lett.* 8: 2920.

Rubio, A., Corkill, J. L., and Cohen, M. L. 1994. Theory of graphite boron nitride nanotubes. *Phys. Rev. B* 49: 5081.

Saito, S. 2008. Quantum theories for carbon nanotubes. In *Carbon Nanotubes: Quantum Cylinders of Graphene*, S. Saito and A. Zettl (Eds.), pp. 29–48. Amsterdam, the Netherlands: Elsevier.

Saito, S. and Zettl, A. (Eds.) 2008. *Carbon Nanotubes: Quantum Cylinders of Graphene*. Amsterdam, the Netherlands: Elsevier.

Smith, B. W., Monthioux, M., and Luzzi, D. E. 1998. Encapsulated C_{60} in carbon nanotubes. *Nature* 396: 323.

Son, Y.-W., Cohen, M. L., and Louie, S. G. 2006. Energy gaps in graphene nanoribbons. *Phys. Rev. Lett.* 97: 216803 [Erratum: *Phys. Rev. Lett.* 98: 089901 (2007)].

Son, Y.-W., Cohen, M. L., and Louie, S. G. 2007. Electric field effects on spin transport in defective metallic carbon nanotubes. *Nano Lett.* 7: 3518.

Spataru, C. D., Ismail-Beigi, S., Benedict, L. X., and Louie, S. G. 2004. Excitonic effects and optical spectra of single-walled carbon nanotubes. *Phys. Rev. Lett.* 92: 077402.

Tang, Z. K., Zhang, L., Wang, N. et al. 2001. Superconductivity in 4 Angstrom single-walled carbon nanotubes. *Science* 292: 2462.

Tangney, P., Louie, S. G., and Cohen, M. L. 2004. Dynamic sliding friction between concentric carbon nanotubes. *Phys. Rev. Lett.* 93: 065503.

Umemoto, K. and Saito, S. 2001. Electronic structure of Ba_4C_{60} and Cs_4C_{60}. *AIP Conf. Proc.* 590: 305.

Wang, F., Cho, D. J., Kessler, B. et al. 2007. Observation of excitons in one-dimensional metallic single-walled carbon nanotubes. *Phys. Rev. Lett.* 99: 227401.

Yang, L., Cohen, M. L., and Louie, S. G. 2007. Excitonic effects in the optical spectra of graphene nanoribbons. *Nano Lett.* 7: 3112.

Yao, Z., Postma, H., Balents, L., and Dekker, C. 1999. Carbon nanotube intramolecular junctions. *Nature* 402: 273.

2

Theories for Nanomaterials to Realize a Sustainable Future

Rodion V. Belosludov
Tohoku University

Natarajan S.
Venkataramanan
Tohoku University

Hiroshi Mizuseki
Tohoku University

Oleg S. Subbotin
Tohoku University

and

*Nikolaev Institute of
Inorganic Chemistry*

Ryoji Sahara
Tohoku University

Vladimir R. Belosludov
Tohoku University

and

*Nikolaev Institute of
Inorganic Chemistry*

Yoshiyuki Kawazoe
Tohoku University

2.1 Introduction

The present environmental factors and limited energy resources have led to a profound evolution in the way we view the generation, storage, and supply of energy. Although fossil fuel and nuclear energy will remain the most important sources of energy for many more years, flexible technological solutions that involve alternative means of energy supply and storage are in urgent need of development. The search for cleaner, cheaper, smaller, and more efficient energy technologies has been driven by recent developments in materials science and engineering (Lubitz and Tumas 2007). To meet the storage challenge, basic research is needed to identify new materials and to address a host of associated performance- and system-related issues. These issues include operating pressure and temperature; the durability of the storage material; the requirements for hydrogen purity imposed by the fuel cell; the

reversibility of hydrogen uptake and release; the refueling conditions of rate and time; the hydrogen delivery pressure; and overall safety, toxicity, and system efficiency and cost. No material available today comes close to meeting all the requirements for the onboard storage of hydrogen for supplying hydrogen as a fuel for a fuel cell/ electric vehicle (Schlapbach and Züttel 2004). There are several candidate groups for storage materials, each with positive and negative attributes. The traditional hydrides have excellent H-volume storage capacity, good and tunable kinetics and reversibility, but poor H-storage by weight. Highly porous carbon and hybrid materials have the capability of high mass storage capacity, but since molecular hydrogen is required to be stored, they can only work at cryogenic conditions. The light metal alloys have the required mass density but poor kinetics and high absorption temperatures/pressures. The complex hydrides undergo chemical reactions during desorption/adsorption, thus limiting kinetics and reversibility of storage. Hence, research on adequate H-storage materials remains a challenge, in particular for the vehicle transportation sector.

The host–guest or inclusion compound in which the lattice framework with porous (host) can accommodate the guest atoms or molecules is probably one of the most suitable hydrogen storage media. This type of material belongs to the field of supramolecular chemistry, which can be defined as a chemistry beyond the molecule, referring to the organized entities of higher complexity that result from the association of two or more chemical species held together by intermolecular forces (Lehn 1995). At the present time, the role of the supramolecular organization in the design and synthesis of new materials is well recognized and assumes an increasingly important position in the design of modern materials. The combination of nanomaterials as solid supports and supramolecular concepts has led to the development of hybrid materials with improved functionalities. This "heterosupramolecular" combination provides a means of bridging the gap between molecular chemistry, material science, and nanotechnology.

A number of terms are used in the literature to describe these supermolecules: host–guest compound, inclusion compound, clathrate, molecular complex, intercalate, carcerand, cavitand, crow, cryptand, podand, spherand, and so on. Many detailed schemes have been proposed for the classification of these substances according to the nomenclature given above. Thus, the term "clathrate," which is derived from the Latin word *clathratus* meaning "enclosed by bars of a grating," was used to describe a three-dimensional host lattice with cavities for accommodating guest species. The term *Einschlussverbindung* (inclusion compounds), introduced by Schlek in 1950, seems to be the most suitable for all inclusion-type systems considering some characteristic features of the host–guest association, such as no-covalent bond between the host and guest and/or the dissociation–association equilibrium in solution (Cramer 1954).

The history of inclusion compounds dates back to 1823 when Michael Faraday reported the preparation of clathrate hydrate of chlorine. However, for a long period of time, inclusion compounds were the results of discovery by chance without any importance for practical uses (Mandelcorn 1964, Davies et al. 1983). Only since the third postwar period of chemistry, known as the supramolecular era, which has bloomed since the 1970s (Vögtle 1991), have the inclusion compounds and similar co-crystalline constructions grown rapidly in importance (Atwood et al. 1984). In the mid-1990s, they became the focus for applications such as those involving separation, encapsulation, and many other applications in high-technology fields (Weber 1995).

The metal–organic framework (MOF) material is one of the inclusion compounds that may be identified as a single supramolecule host framework in which guest molecules reside completely within the host. The recent advent of MOFs as new functional adsorbents has attracted the attention of chemists due to scientific interest in the creation of unprecedented regular nanosized spaces and in the finding of novel phenomena, as well as commercial interest in their application for storage, for separation, and in heterogeneous catalysis (Kitagawa et al. 2004). In the area of MOFs, the structural versatility of molecular chemistry has allowed the

rational design and assembly of materials having novel topologies and exceptional host–guest properties, which are important for immediate industrial applications including storage of hydrogen (Rowsell and Yaghi 2005). Nowadays several hundred different types of MOFs are known and experimentally synthesized. Despite of the importance of these materials, the number of publications related to computational modeling studies is still limited in many cases due to complexity of their crystal structures. In particular, the number of atoms involved makes simulations prohibitively time-consuming. Therefore, the accurate systematic simulation of their properties including host–guest interaction, guest dynamics in MOFs, thermodynamic stability of empty host, and their structural transformations will be indispensable in providing future directions for material optimization. Using powerful computers and highly accurate methods, scientists can accelerate the realization of novel MOFs and propose these materials for different applications.

The clathrate hydrate is another type of material that has a potential application as a hydrogen storage material. This is a special class of inclusion compounds consisting of water and small guest molecules, which form a variety of hydrogen-bonded structures. These compounds are formed when water molecules arrange themselves in a cage-like structure around guest molecules. Recently, the interest in hydrogen clathrate hydrates as potential hydrogen storage materials has risen after a report that the clathrate hydrate of structure II (CS-II) can store around 4.96 wt% of hydrogen at 220 MPa and 234 K (Mao et al. 2002). However, the extreme pressure required to stabilize this material makes its application in hydrogen storage impractical. It is well known that there are several types of gas hydrate structures with different cage shapes, and some of these hydrate structures can hypothetically store more hydrogen than the hydrate of structure CS-II. Therefore, for practical application of gas clathrates as hydrogen storage materials, it is important to know the region of stability of these compounds as well as the hydrogen concentration at various pressures and temperatures. Our group developed a model that accurately predicts the phase diagram of the clathrate hydrates at the molecular level. This model significantly improves the well-known van der Waals and Platteeuw theory and will be discussed in Section 2.2.

In this chapter, we study the physical and chemical properties of hydrogen clathrate and selected MOF structures and show how the theoretical and computational techniques can provide important information for experimentalists in order to help them develop hydrogen storage materials based on MOF materials and clathrate hydrates with desired storage characteristics.

2.2 Molecular Level Description of Thermodynamics of Clathrate Systems

2.2.1 Introduction

At the present time, analytical theories of clathrate compounds, which allow the construction of the T–P diagram of gas hydrates, are based on the pioneering work of van der Waals and Platteeuw (van der Waals and Platteeuw 1959). This theory and all of its subsequent variations are based on four main assumptions. The first three are (a) cages contain at most one guest; (b) guest molecules do not interact with each other; and (c) the host lattice is unaffected by the nature as well as by the number of encaged guest molecules. These are clearly violated in the case of hydrogen clathrates, which have multiple occupancy. However, it has been shown how a nonideal solution theory can be formulated to account for guest–guest interaction (Dyadin and Belosludov 1996). A generalization of the van der Waals–Platteeuw (vdW–P) statistical thermodynamic model of clathrate hydrates, applicable for arbitrary multiple filling the cages, was formulated by Tanaka et al. (2004). However, these developments do not go far enough, and a much more comprehensive theory is desperately needed.

In this section, we discuss a theoretical model based on the solid solution theory of van der Waals and Platteeuw. Our modifications include multiple occupancies, host relaxation, and the accurate description of the behavior of guest molecule in the cavities. We used quasiharmonic lattice dynamics (QLD) method to estimate the free energies, equations of state, and chemical potentials (Belosludov et al. 2007). This is important in order to know the region of stability of the inclusion compounds as well as the guest concentration at various pressures and temperatures for practical application of these materials as storage medium. The method has been used for gas hydrate clathrates. However, our approach is general and can be applied equally well to other inclusion compounds with the same type of composition (clathrate silicon, zeolites, MOF materials, inclusion compounds of semiconductor elements, etc.). Using this approach, one can not only characterize and predict the hydrogen storage ability of known hydrogen storage materials with weak guest–host interactions but also estimate these properties for structures that have not yet been realized by experiment.

2.2.2 Thermodynamics Model of Clathrate Structures with Multiple Degree of Occupation

The following development of the model is based only on one of the assumptions of vdW–P theory: the contribution of guest molecules to the free energy is independent of mode of occupation of the cavities at a designated number of guest molecules (van der Waals and Platteeuw 1959). This assumption allows us to separate the entropy part of free energy:

$$F = F_1(V, T, y_{11}^1, \ldots, y_{nm}^k) + kT \sum_{t=1}^{m} N_t \left[\left(1 - \sum_{l=1}^{n} \sum_{i=1}^{k} y_{lt}^i \right) \right.$$

$$\left. \times \ln \left(1 - \sum_{l=1}^{n} \sum_{i=1}^{k} y_{lt}^i \right) + \sum_{l=1}^{n} \sum_{i=1}^{k} y_{lt}^i \ln \frac{y_{lt}^i}{i!} \right] \quad (2.1)$$

where F_1 is the part of the free energy of clathrate hydrate for the cases where several types of cavities and guest molecules exist, and a cavity can hold more than one guest molecule. The second term is the entropy part of free energy of a guest system, $y_{lt}^i = N_{lt}^i / N_t$ is the degree of filling of t-type cavities by i cluster

of l-type guest molecules; N_t is the number of t-type cavities; N_{lt}^i is the number of l-type guest molecules that are located in t-type cavities (Belosludov et al. 2007).

For a given arrangement $\{y_{11}^1, \ldots, y_{nm}^k\}$ of the clusters of guest molecules in the cavities, the free energy $F_1(V, T, y_{11}^1, \ldots, y_{nm}^k)$ of the crystal can be calculated within the f ramework of a lattice dynamics approach in the quasiharmonic approximation (Leifried and Ludwig 1961, Belosludov et al. 1994) as

$$F_1(V, T, y_{11}^1, \ldots, y_{nm}^k) = U + F_{vib} \quad (2.2)$$

where

U is the potential energy
F_{vib} is the vibrational contribution

$$F_{vib} = \frac{1}{2} \sum_{j\vec{q}} \hbar \omega_j(\vec{q}) + k_B T \sum_{j\vec{q}} \ln(1 - \exp(-\hbar \omega_j(\vec{q}) k_B T)) \quad (2.3)$$

where

$\omega_j(\vec{q})$ is the jth frequency of crystal vibration
$\vec{q}$ is the wave vector

The eigenfrequencies $\omega_j(\vec{q})$ of molecular crystal vibrations are determined by solving numerically the following system of equations:

$$m_k \omega^2(\vec{q}) U_\alpha^t(k, \vec{q}) = \sum_{k', \beta} \left[\widetilde{D}_{\alpha\beta}^{tt}(\vec{q}, kk') U_\beta^t(k', \vec{q}) + \widetilde{D}_{\alpha\beta}^{tr}(\vec{q}, kk') U_\beta^r(k', \vec{q}) \right] \quad (2.4)$$

$$\sum_{\beta} I_{\alpha\beta}(k) \omega^2(\vec{q}) U_\beta^r(k, \vec{q})$$

$$= \sum_{k', \beta} \left[\widetilde{D}_{\alpha\beta}^{rt}(\vec{q}, kk') U_\beta^t(k', \vec{q}) + \widetilde{D}_{\alpha\beta}^{rr}(\vec{q}, kk') U_\beta^r(k', \vec{q}) \right] \quad (2.5)$$

where $\widetilde{D}_{\alpha\beta}^{ii'}(\vec{q}, kk')$ ($\alpha, \beta = x, y, z$) are translational ($i, i' = t$), rotational ($i, i' = r$), and mixed ($i = t, i' = r$ or $i = r, i' = t$) elements of the dynamical matrix in the case of molecular crystals, the expressions for which are presented in the literature (Belosludov et al. 1988, 1994), $U_\alpha^{i'}(k, \vec{q})$ ($\alpha, \beta = x, y, z$) is the amplitude of vibration, m_k and $I_{\alpha\beta}(k)$ are the mass and inertia tensor of kth molecule in the unit cell.

In the quasiharmonic approximation, the free energy of crystal has the same form as in the harmonic approximation but the structural parameters at fixed volume depend on temperature. This dependence is determined self-consistently by the calculation of the system's free energy. Equation of state can be found by numerical differentiation of free energy:

$$P(V, T) = -\left(\frac{\partial F(V, T, y_{11}^1, \ldots, y_{nm}^k)}{\partial V} \right)_0 \quad (2.6)$$

The "zero" index mean constancy of all thermodynamic parameters except the ones which differentiation execute.

After obtaining the free energy values, we can calculate the chemical potentials, μ_{lt}^i, of i-cluster of l-type guest molecules, which are located on t-type cavities:

$$\mu_{lt}^i(P,T,y_{11}^1,\ldots,y_{nm}^k) = \left(\frac{\partial F(V(P),T,y_{11}^1,\ldots,y_{nm}^k)}{\partial N_{lt}^i}\right)_0$$

$$= \widetilde{\mu}_{lt}^i + kT\ln\frac{y_{lt}^i}{i!\left(1-\sum\limits_{l'i'}^{n,k}y_{l't}^{i'}\right)} \qquad (2.7)$$

$$\widetilde{\mu}_{lt}^i = \left(\frac{\partial F_1(V(P),T,y_{11}^1,\ldots,y_{nm}^k)}{\partial N_{lt}^i}\right)_0 \qquad (2.8)$$

The last derivative can be found by numerical calculation using the following approximation:

$$\widetilde{\mu}_{lt}^i \cong \frac{F_1\left(V(P),T,N_{11}^1,..N_{lt}^i,...N_{nm}^k\right) - F_1\left(V(P),T,N_{11}^1,..N_{lt}^i - N_{lt}^i n_{lt}^i,...N_{nm}^k\right)}{N_{lt}^i n_{lt}^i}$$

$$(2.9)$$

where $N_{lt}^i n_{lt}^i$ is number of clusters of guest molecules removed from clathrate hydrate.

If the Helmholtz free energy F and equation of state of the system are known, then the Gibbs energy $\Phi(P,T,y_{11}^1,\ldots,y_{nm}^k)$ expressed in terms of chemical potentials of host and guests is found from the following thermodynamic relation:

$$\Phi(P,T,y_{11}^1,\ldots,y_{nm}^k) = N_Q\mu_Q + \sum_{t=1}^m N_t \sum_{l=1}^n \sum_{i-1}^k y_{lt}^i \mu_{lt}^i$$

$$= F(V(P),T,y_{11}^1,\ldots,y_{nm}^k) + PV(P) \qquad (2.10)$$

Substituting the expression (2.1) for F into (2.10) allows one to obtain the chemical potential of host molecules μ_Q:

$$\mu_Q(P,T,y_{11}^1,\ldots,y_{nm}^k) = \widetilde{\mu}_Q + kT\sum_{t=l}^m \nu_t\left[\ln\left(1-\sum_{l=1}^n\sum_{i=1}^k y_{lt}^i\right)\right] \qquad (2.11)$$

$$\widetilde{\mu}_Q \equiv \frac{PV(P)}{N_Q} + \frac{1}{N_Q}F_1(V(P),T,y_{11}^1,\ldots,y_{nm}^k) - \sum_{t=1}^m\nu_t\sum_{l=1}^n\sum_{i-1}^k y_{lt}^i\widetilde{\mu}_{lt}^i \qquad (2.12)$$

$\nu_t = N_t/N_Q$, N_Q is the number of host molecules.

For the case of multiple filling of cages, we derived expressions (2.7) for chemical potentials of clusters μ_{lt}^i of guest molecules in cages. For description of phase equilibrium, we need to derive expressions for chemical potentials μ_{lt} of single guest molecules in these clusters. As a first approximation, the chemical potentials of single guest molecules inside cluster are equal and chemical potential of a cluster is equal to the sum of chemical

potentials of molecules in cluster: $\mu_{lt}^i \cong i\mu_{lt} - U_{lt}$ (U_{lt} is the interaction of guest molecules inside cluster in cavities).

The curve $P(T)$ of monovariant equilibrium can be found from the equality of the chemical potentials. In the case of gas hydrates, it can be written as

$$\mu_{lt}^i(P,T,y_{11}^1,\ldots,y_{nm}^k) = i\mu_l^{gas}(P,T) - U_{lt} \qquad (2.13)$$

$$\mu_Q(P,T,y_{11}^1,\ldots,y_{nm}^k) = \mu_Q^{ice}(P,T) \qquad (2.14)$$

where

μ_l^{gas} is the chemical potential of guest molecules in the gas phase

$\mu_Q^{ice}(P,T)$ is the chemical potentials of water molecules in ice

The following divariant equilibria lines "gas phase–hydrate" are defined by Equation 2.13. Here we assume that the ideal gas laws govern the gas phase, and then the expressions for chemical potentials of mixture components will be as follows:

$$\mu_l^{gas}(P,T) = kT\ln[x_l P/kT\Phi_l] = kT\ln\left[x_l\frac{P}{kT}\left(\frac{2\pi\hbar^2}{m_l kT}\right)^{3/2}\right] \qquad (2.15)$$

where x_l is the mole fraction of the l-type guest in the gas phase.

2.2.3 Conclusions

We have presented a general formalism for calculating the thermodynamical properties of inclusion compounds. Deviating from the well-known theory of van der Waals and Platteeuw, our model accounted the influence of guest molecules on the host lattice and guest–guest interaction. The validity of the proposed approach was checked for argon, methane, and xenon hydrates, and the results were in agreement with known experimental data (Belosludov et al. 2007). As mentioned before, the method is quite general and can be applied to the various nonstoichiometric inclusion compounds with weak guest–host interactions. However, it is significant that the present model of inclusion compounds allows the calculation thermodynamic functions starting from well-defined potentials of intermolecular guest–host and guest–guest interactions. Thus, it is important to estimate these interactions using the highly accurate first-principles methods. The applications of this model in the case of gas hydrates, including hydrogen clathrate, are presented in Section 2.3.

2.3 Gas Hydrates as Potential Nano-Storage Media

2.3.1 Introduction

Clathrate hydrates are one type of crystalline inclusion compounds in which the host framework of water molecules was linked by hydrogen bonds and formed a cage-like structure around the guest atoms or molecules. As a result, many of their

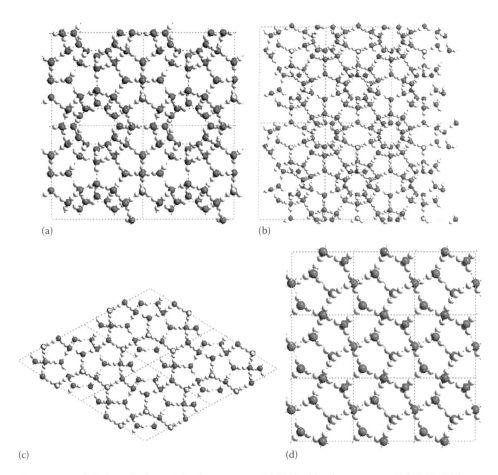

FIGURE 2.1 Crystal structure of clathrate hydrate: (a) cubic structure I (CS-I), (b) cubic structure II (CS-II), (c) hexagonal structure (sH), and (d) tetragonal structure.

physical and chemical properties are different from ice (Sloan and Koh 2007). At the present time, most of the recognized gas hydrates have one of well-known three types of structures (see Figure 2.1a–c). The cubic structures I (CS-I) and II (CS-II) of gas hydrates were first identified by Von Stackelberg and Miller (1954), Claussen (1951) and Pauling and Marsh (1952). The third one, the hexagonal structure (sH) was determined (Ripmeester et al. 1987, Udachin et al. 1997). According to general rule, CS-I hydrates are formed by molecules with van der Waals diameters of up to about 5.8 Å while CS-II hydrates are formed by large molecules up to about 7.0 Å in size. The exceptions of this rule are molecules with small van der Waals diameters up to about 4.3 Å, which form CS-II hydrates (Davidson et al. 1984, 1986). It was proved that the gas hydrates are sensitive to pressure variation due to relatively weak binding energy between water molecules and hence the friable packing of host framework. The sequential change of hydrate phase in different gas–water systems was observed by increasing pressure up to 15 kbar (Dyadin et al. 1997a,b).

Recently, it was established that these compounds are a potential source of energy in the future since natural gas hydrates occur in large amounts under conditions of high pressure and low temperature in the permafrost regions or below the ocean floor. As example, methane in gas hydrates represents one of the largest sources of hydrocarbons on earth. Moreover, the possible releases of methane

from clathrate hydrates have raised serious questions about its possible role in climate change. Among many potential applications of clathrate hydrates, these compounds can also be used as gas (such as CO, CO_2, O_2, or H_2) storage materials. Therefore, a good understanding of the chemical and physical properties of clathrate hydrates with a multiple occupation, such as electronic properties, structure, dynamics, and stability, is essential for practical manipulation of this class of inclusion compounds.

2.3.2 Argon Clathrate Hydrates with Multiple Degree of Occupation

We discuss the physical and chemical properties of argon hydrates of different structures and their stability depending on cage occupations. The phase diagram of argon hydrate at different pressures has been studied by several experimental groups (Dyadin et al. 1997b, Lotz and Schouten 1999). The possibility of double occupation of the large cages in CS-II by argon was also examined by molecular dynamics calculations (Itoh et al. 2001). Thus, it was predicted that the double-occupied argon hydrate can be stabilized by high external pressure. Moreover, the phase diagram of argon–water system was studied at high pressure and the formation of several hydrate structures was established (Manakov et al. 2001). Powder neutron diffraction study showed

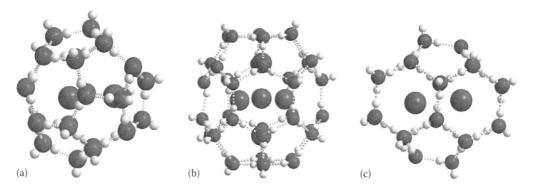

FIGURE 2.2 Large cages: (a) hexakaidecahedron ($5^{12}6^4$), (b) icosahedron ($5^{12}6^8$), and (c) tetradecahedral ($4^25^86^4$) with two, five, and two argon atoms, respectively.

that in argon–water system, CS-II hydrate exists from ambient pressure up to 4.5 kbar. Upon increasing the pressure, a phase transition occurs and the argon hydrate with hexagonal structure is formed up to 7.6 kbar. In the pressure range of 7.6–10 kbar, an argon hydrate of previously unknown type was obtained. A hydrate with tetragonal crystal structure (see Figure 2.1d) and one type of cavity (Manakov et al. 2001, 2002) was proposed.

The electronic, structural, dynamic, and thermodynamic properties of structure II, H, and tetragonal Ar clathrate hydrates with multiple filling of large cages were investigated and their stability was examined using first-principles and lattice dynamics calculations (Inerbaev et al. 2004). The geometry optimization and vibrational analyses of selected cage-like structures of water clusters with and without enclathrated argon molecules were performed at the Hartree–Fock (HF) level using the Gaussian 98 package (Frisch et al. 1998). The 6-31 + G(d) basis set was used. The inclusion of diffusion functions in the basis set is necessary for a good description of the structure and the energetics of these hydrogen-bonded complexes (Frisch et al. 1986). The redundant internal coordinate procedure was used during optimizations (Peng et al. 1996) and the vibrational frequencies were calculated from the second derivative of the total energy with respect to atomic displacement about the equilibrium geometry. These structures are indeed (at least local) minima if all frequencies are real. The difference between the total cluster energy and the energies of separated empty water cages and guest atoms at an infinite distance was considered as the stabilization energy (SE).

The multiple occupations for argon hydrate was proposed only for the large cages, the hexakaidecahedron ($5^{12}6^4$), icosahedron ($5^{12}6^8$), and tetradecahedral ($4^25^86^4$), as shown in Figure 2.2. These cavities with and without encapsulation of argon atoms were optimized using first-principles calculations. The energy values for all the structures investigated are listed in Table 2.1. For calculating the H-bond energy (HBE), we assumed that the binding energy of the water cluster is solely due to H-bonding. The value of HBE was determined as the binding energy, which is the difference between the total cluster energy without guest atoms and the separate water monomers at infinite distance, divided by the number of H-bonds. The interaction between one Ar and the hexakaidecahedron cage ($H_2O)_{28}$ is equal to −0.41 kcal/mol. In the case of the icosahedron cage

TABLE 2.1 Stabilization Energy (SE) and H-Bond Energy (HBE) for the Large Cages from Different Hydrate Structures Depend on the Number of Encapsulated Argon Atoms

Type of Water Cage	Number of Ar Atoms	SE (kcal/mol)	HBE (kcal/mol)
Hexakaidecahedron	0		−6.06
	1	−0.41	−6.06
	2	2.14	−6.06
	3	9.25	−5.99
Tetradecahedral	0		−6.17
	1	0.99	−6.17
	2	2.50	−6.16
	3	26.81	−5.87
Icosahedron	0		−5.97
	1	−0.29	−5.96
	2	−0.54	−5.96
	3	−0.04	−5.95
	4	3.54	−5.94
	5	5.68	−5.94
	6	11.85	−5.90

($H_2O)_{36}$, the interaction between one argon atom and the cage is equal to −0.29 kcal/mol. The negative value of SE means that argon has a positive stabilization effect on these cages and hence single occupations can be achieved without the applications of high external pressure. The analysis of calculated frequencies has shown that the translation of argon atom in these two cages is characterized by imaginary frequency. It was found that in these cases, the HBE values are for the respective empty cages. Moreover, the structural features of water cavities (distances, angles, etc.) are very similar to those existing in the empty cages, and hence, represent the cage structures with no distortion.

In the case of double occupancy, the negative value of SE is obtained only in the case of $5^{12}6^8$ cage. The positive values of SE are found for hexakaidecahedron and tetradecahedral cavities. However, these energies are very small and hence the double occupancy may be possible in the case of CS-II and tetragonal structures after applying external pressure. Moreover, the HBE as well as shape of cages are not changed in the presence of two argon atoms. Addition of one more argon atoms leads to significant increase in the SE and a decrease in the HBE values for

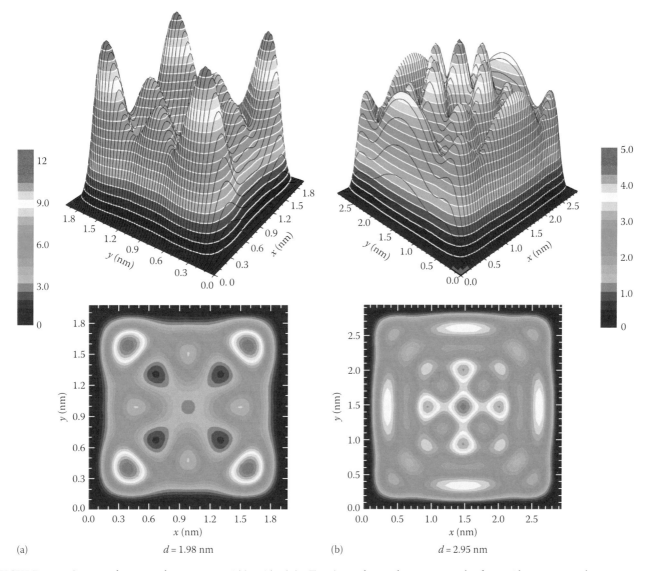

FIGURE 9.14 Superconducting order parameter $\Delta(\mathbf{r}) = \Delta(x, y)$ (at $T = 0$) together with its contour plot for an Al nanowire with square cross section for the resonant widths: (a) $d = 1.98$ nm, (b) $d = 2.95$ nm (the both points are resonances).

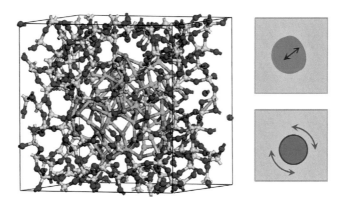

FIGURE 13.5 Atomistic model of the $(ZnS)_{47}$ cluster embedded into a-SiO_2 matrix. Detailed examination of the dynamics identifies six low-frequency modes (three rotational and three cage modes, schematically shown) with energies below 5 meV. (Based on Stoneham, A.M. and Gavartin, J.L., *Mater. Sci. Eng. C*, 27, 972, 2007.)

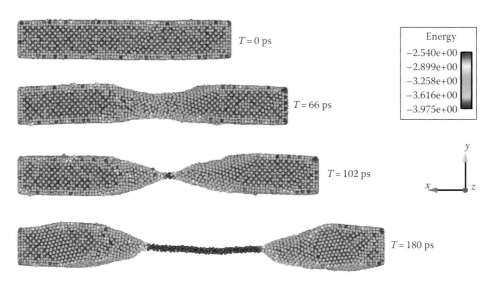

FIGURE 22.5 MD simulation of the tensile failure of a gold nanowire using an EAM potential.

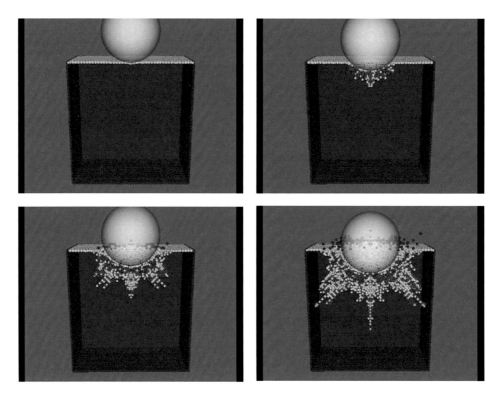

FIGURE 22.10 3D nanoindentation of a crystalline metallic substrate.

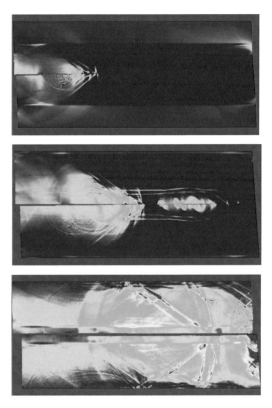

FIGURE 22.13 MD simulation of a shear-dominant crack propagation process; the middle snapshot shows the formation of a "daughter" crack ahead of the main crack tip.

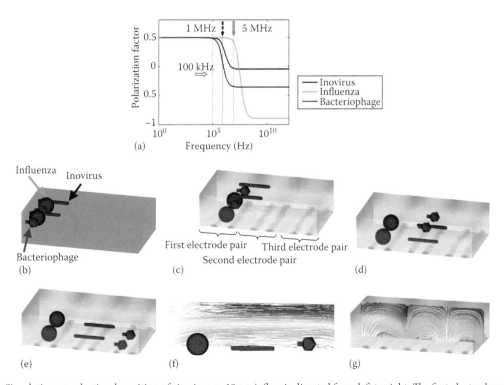

FIGURE 22.21 Simulation on selective deposition of six viruses. 18 μm/s flow is directed from left to right. The first electrode pair is operating at 5 MHz, the second pair at 1 MHz, and the third pair at 100 kHz. Influenza, inovirus, and bacteriophage are selectively deposited on the first, second, and third electrode pairs consequently. (a) Polarization factor of viruses at different frequencies. (b) Time = 0 ms. (c) Time = 15 ms. (d) Time = 30 ms. (e) Time = 45 ms. (f) Velocity profile in the channel. (g) Electric field across three electrode pairs.

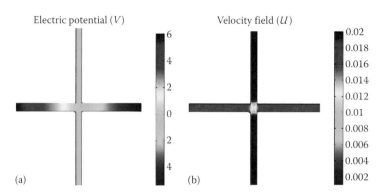

Electric potential (V)   Velocity field (U)

(a)   (b)

FIGURE 22.28 Simulation of electroosmotic flow in a cross-sectional fluid channel. (a) The distribution of electric field potential. (b) Flow velocity in the channel.

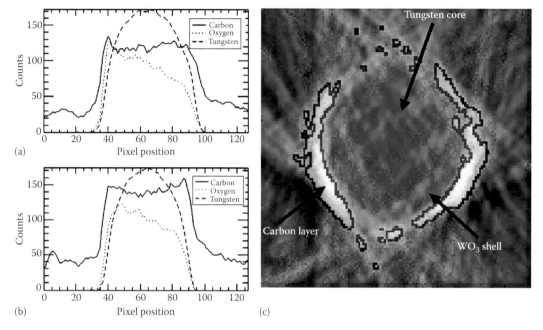

(a)

(b)

(c)

FIGURE 31.15 Energy-dispersive x-ray (EDX) spectroscopic line scan across a carbon and oxide coated tungsten tip. Two members of a tilt series of line scans using C-K, O-K, and W-Lα signals are shown at (a) −70° and (b) +50° (c) RGB-map of the tomographic reconstruction of the chemical distribution of W-Lα (blue), O-K (red), and C-K (green). (Modified from Saghi, Z. et al., *Microsc. Microanal.*, 13, 438, 2007.)

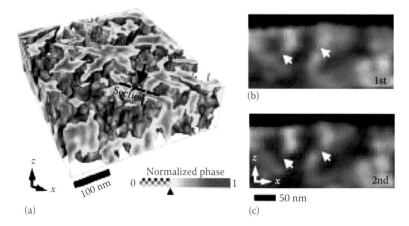

(a)

(b)

(c)

FIGURE 31.17 3D reconstruction of a volume (512 × 512 × 14 voxels) of elastomeric polypropylene (ePP). (a) Isosurface representation of the boundary between two intensity levels corresponding to amorphous and crystalline polymer phase. (b) and (c) are cross-sections in x–z-orientation with the AFM-height and AFM cantilever phase signal respectively. (From Dietz, C. et al., *Appl. Phys. Lett.*, 92, 143107, 2008. With permission.)

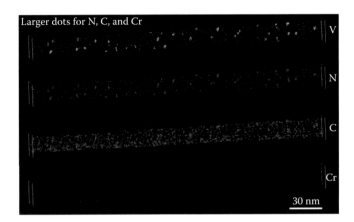

Larger dots for N, C, and Cr

V

N

C

Cr

30 nm

FIGURE 31.18 Sub-15 nm particles for strengthening of high-strength low-alloy steels. 3D reconstruction of a nanoscale volume from atom probe data. Vanadium and nitrogen are mainly partitioned into the particles, while carbon and chromium have a minor enrichment in the particles. (From Craven, A.J. et al., *Mater. Sci. Technol.*, 24, 641, 2008. With permission.)

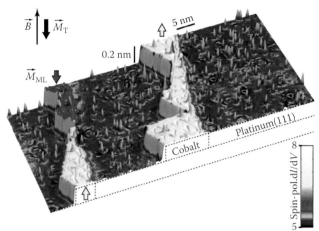

FIGURE 32.4 Using magnetic probe tips and spin-polarized tunneling effect has enabled the detection of magnetization in individual atoms. A spin-polarized dI/dV map of cobalt atoms adsorbed onto a nonmagnetic platinum surface, Pt (111), shown here in blue, demonstrates a dependence on the external magnetic field. (Reprinted from Meier, F. et al., *Science*, 320, 82, 2008. With permission.)

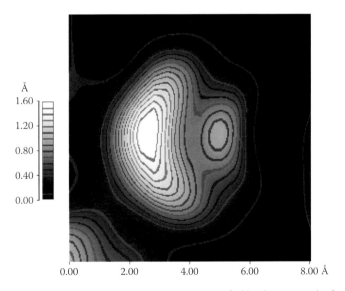

FIGURE 32.5 The current limits of resolution by local probe methods are personified by this image, the first experiment to resolve subatomic features. In this case, two lobes of charge density are detected within the single atom located at the apex of the probe tip through the angular dependence of tip–sample forces resulting in real-space characterization of a single atomic orbital. (Reprinted from Giessibl, F.J. et al., *Annalen Der Physik*, 10, 887, 2001. With permission.)

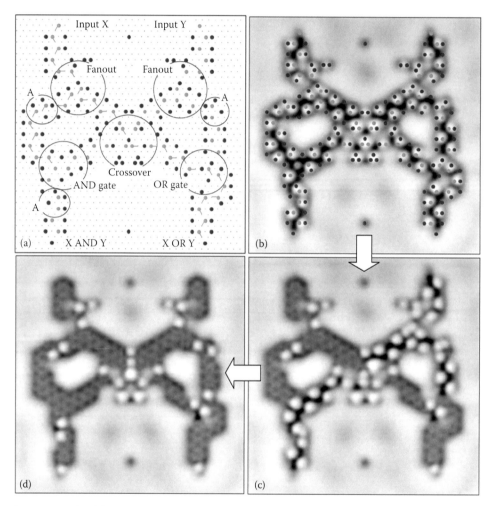

FIGURE 32.9 An elegant example of design-oriented molecular manipulation for computational logic applications. Fabrication of a two-input sorter was carried out through STM manipulation of individual carbon monoxide molecules into predetermined, atomically specific locations. "Molecule cascades" capable of computing the AND/OR logic were produced by inducing motion of one molecule with the STM tip. (Reprinted from Heinrich, A.J. et al., *Science*, 298, 1381, 2002. With permission.)

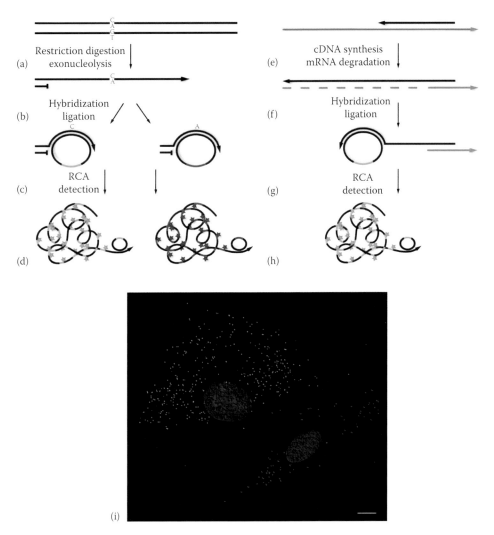

FIGURE 40.7 Padlock probes *in situ*. Detection and genotyping of DNA and mRNA *in situ* using padlock probes. (a) Double-stranded DNA to be investigated for the presence of a C or an A residue in a particular position is made single-stranded (b) to permit probe hybridization by restriction digestion and exonucleolysis to remove the non-target strand. (c) Two padlock probes, designed to have the appropriate G or T nucleotides at one end in order to recognize the two genotypes, and distinct tag sequences (green or red) are hybridized to their relative targets and the open circles are closed upon ligation. (d) Localized amplification products containing numerous copies of the original padlock probes are produced by RCA and detected by the hybridization of fluorescence-labeled oligonucleotides (green and red stars) for detection in a fluorescence microscope. (e) mRNA molecules can be detected in a similar manner by first annealing a primer to the target mRNA (gray) and generating a cDNA molecule (black) by reverse transcription. (f) Next, the copied mRNA sequence is degraded by treatment with RNase H. (g) A padlock probe is allowed to hybridize to the new cDNA strand. Upon ligation the probe is subjected to RCA creating a bundle of DNA representing a concatemer of complements of the original padlock probe. (h) Finally, the amplification product is visualized in a fluorescence microscope by hybridizing fluorescence-labeled detection probes. (i) Detection of β-actin transcripts in human and mouse fibroblasts using padlock probes. The target sequences are similar except for a single-nucleotide difference, giving rise to red RCA products in the human fibroblasts and green signals in the mouse cells. The DNA in the nuclei is stained in blue using DAPI. The scale bar represents 10 μm.

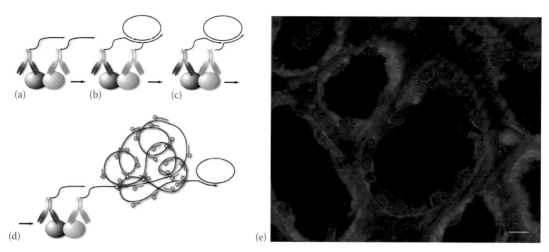

FIGURE 40.9 *In situ* PLA. Visualization of protein interactions using the *in situ* PLA. (a) Upon binding of two proximity probes to the same target molecule complex, the oligonucleotides of the proximity probes are brought into proximity. (b) Next, two oligonucleotides are added that can hybridize to the antibody-bound oligonucleotides. (c) The added oligonucleotides are ligated into a circular DNA-molecule, guided by the antibody-bound oligonucleotides. (d) Subsequently, one of the oligonucleotides on the proximity probes can prime an RCA, allowing amplification products to be detected, e.g., through hybridization of fluorescence-labeled oligonucleotides (red stars). Since the RCA product remains attached to one of the proximity probes, each amplification product gives rise to a distinct localized signal per detected protein–protein interaction where the antibodies bound their target complex. (e) An example of the visualization of the interaction of the oncoprotein c-Myc and its dimerization partner Max as red fluorescent dots by *in situ* PLA in fresh frozen human colon tissue. The cytoplasms were counterstained using a FITC-labeled anti-actin antibody (green) and the DNA in the nuclei was stained with Hoechst 33342 (blue). The scale bar represents 10 μm.

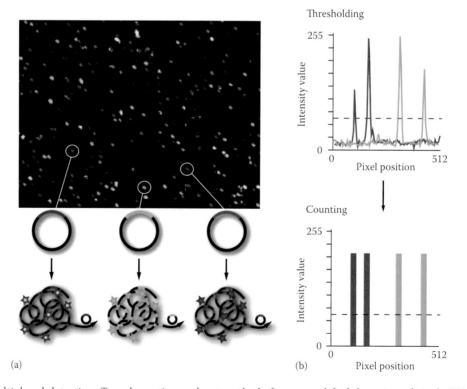

FIGURE 40.10 Multiplexed detection. Two alternative read-out methods for an amplified detection of single DNA molecules using padlock probes. (a) Random array: Padlock probes are hybridized to complementary DNA molecular targets, and after RCA, the amplification products are randomly distributed on microscope slides. Individual amplification products are visualized in a fluorescence microscope as bright objects by hybridization of short oligonucleotide detection probes carrying different fluorophores (blue, green and red stars). (b) Detection in microfabricated channels: Individual RCA products are pumped through a microchannel and detected using a confocal fluorescence microscope in line-scan mode. Amplification products representing individual-detected target molecules are easily separated from the background by applying a fluorescence threshold. Individual amplification products are recorded as digital objects.

hexakaidecahedron and tetradecahedral cages because of the distortion of the water cages. Moreover, imaginary frequencies are found. The analysis of these frequencies shows that the argon clusters interact strongly with the water cages since there is a strong coupling between the vibrations of guest and host molecules. In the case of sH hydrate, the Ar_n (n up to 5) clusters can be stabilized in an icosahedron cavity. The frequency calculations show that all frequencies are real and hence these structures are in local minima. The SE value is increased up to 5.68 kcal/mol and HBE is not significantly changed (see Table 2.1). The stabilization of the Ar_6 cluster inside this cage is energetically unfavorable since the value of SE is twice as larger as for the Ar_5 cluster.

The present calculation results are consistent with the experimental data (Manakov et al. 2002). Thus, the double occupancy of the large cage of CS-II hydrates can be achieved using external pressure. Following an increase in pressure, the stoichiometry changes from $Ar \cdot 4.25H_2O$ (double and single occupancy of large and small cages in CS-II hydrate, respectively) to $Ar \cdot 3.4H_2O$ (triple and single occupancy of large and small cages in CS-II hydrate). The same stoichiometry ($Ar \cdot 3.4H_2O$) can be achieved in sH hydrate with fivefold and single occupancy of large and both medium and small cages, respectively. The fivefold occupation of icosahedron is more energetically favorable than triple occupation in the case of hexakaidecahedron cavity (see Table 2.1). This phase transition of the hydrate with formation of sH hydrates having fivefold occupancy of icosahedron

cages was observed in the experiment. When further pressure is applied, increasing the formation of high-dense tetragonal phase with stoichiometry $Ar \cdot 3H_2O$ (double occupancy) is preferred (Inerbaev et al. 2004).

The dynamic properties and thermodynamic functions $P(V)$ of three types of hydrate structures was estimated using the lattice dynamics (LD) method. The calculations of phonon density of states (DOS) of the CS-II hydrate were performed for various fillings: empty host lattice, single occupancy of the large and small cages, and double occupancy of the large cages and single occupancy of the small cages. The results are shown in Figure 2.3a. The feature of this plot is a gap of about 240 cm^{-1} which divides the low- and high-frequency vibrations of lattice. For empty host lattice, the analysis of the eigenvectors derived from the LD method revealed that the low-frequency region (0–300 cm^{-1}) consists of translation modes of water and the high-frequency region (520–1000 cm^{-1}) consists of libration modes of water host framework. Argon atoms influence the vibrations of the host water framework only slightly and guest vibrations are located in the vicinity of the peaks of phonon DOS at 0–40 cm^{-1}, which is close to value obtained by ab initio calculations. The peak in the negative region, in the case of the single occupancy of both types of cages, corresponds to the motions of argon in large cages with imaginary frequencies, which is in agreement with the values obtained using HF methods. This means that the argon atoms are not localized in potential minima and can be freely moved inside the large

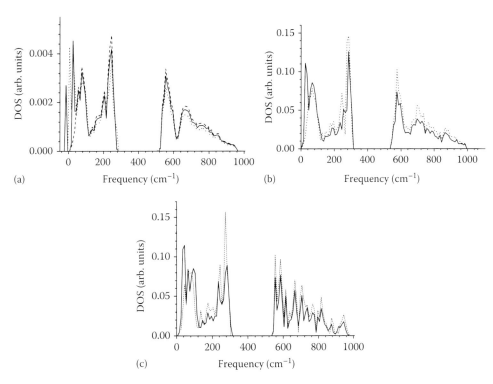

FIGURE 2.3 Phonon DOS of Ar hydrates: (a) CS-II structure, empty host lattice (solid line), double occupancy of the large cages (dashed line) and single occupancy of the large cages (dotted line); (b) sH structure, empty host lattice (dotted line), fivefold occupancy of the large cages (solid line); and (c) tetragonal structure empty host lattice (dotted line) and doubly occupancy (solid line). In the case of CS-II and sH, only single occupancy for other cages is considered. (Reproduced from Inerbaev, T.M. et al., *J. Incl. Phen. Macrocycl. Chem.*, 48, 55, 2004. With permission.)

cages. However, the all frequencies of water framework are positive in all the cases and hence both the single and the double occupancies do not disrupt the dynamical stability of the host lattice (Inerbaev et al. 2004).

In the case of sH hydrate, the dynamical properties of empty host lattice and argon hydrate of sH with maximum experimentally predicted (Manakov et al. 2002) number of guest atoms (fivefold occupancy of the large cages) was estimated. The DOS calculations were done using the experimental values of cell parameters at $T = 293$ K. The results are shown in Figure 2.3b. The large intensive peak at 20 cm^{-1} corresponds to translation of the guest atoms as in the case of CS-II hydrate. After inclusion of argon atoms, the vibrational spectrum of host lattice has practically same features as in the case of the empty hydrate structure and hence the dynamical stability of H hydrate is not significantly changed even with a fivefold occupancy of the large cages.

The unit cell for the argon hydrate with the tetragonal structure contains 12 water molecules, forming one cavity ($4^2 5^8 6^4$) (Manakov et al. 2002). The DOS calculations were performed both for empty host lattice and double occupancy of the cages using the experimental values of cell parameters at $T = 293$ K. The results are shown in Figure 2.3c. The empty host lattice is dynamically stable because all frequencies of water framework are positive. Dynamical stability of water lattice is preserved

even after the inclusion of two guest atoms in each cage. The density of vibrational states of the empty tetragonal hydrate has same features as the density of vibrational states of ice Ih (Tse 1994), hydrates of CS-I (Belosludov et al. 1990), CS-II, and sH. The frequency region of molecular vibrations is divided into two zones. In the lower zone (0–315 cm^{-1}), water molecules mainly undergo translational vibrations, whereas in the upper one (540–980 cm^{-1}), the vibrations are mostly librational. In comparison with hydrates of CS-I, the frequency spectrum of tetragonal argon hydrate is shifted toward higher frequencies, which may be explained by greater density of a new tetragonal crystal argon hydrate compared to hydrates of CS-I. Vibrational frequencies of argon atoms in the cavities lie in the region 20–45 and 60–110 cm^{-1}. The guest atoms influence the phonon spectrum of host framework, diminishing the DOS in the upper zone of translational vibrations and the DOS of librational vibrations (Inerbaev et al. 2004).

The equation of state $P(V)$ was calculated at 293 K. It was found that the studied hydrates are thermodynamically stable in selected range of pressure. These results were compared with experimental $P(V)$ data for argon hydrate of three structural types (Manakov et al. 2002). In the case of argon hydrate of CS-II type, the calculated $P(V)$ for single occupancy of the large and small cages is most closely correlated with experimental points as shown in Figure 2.4a. The largest difference between theory

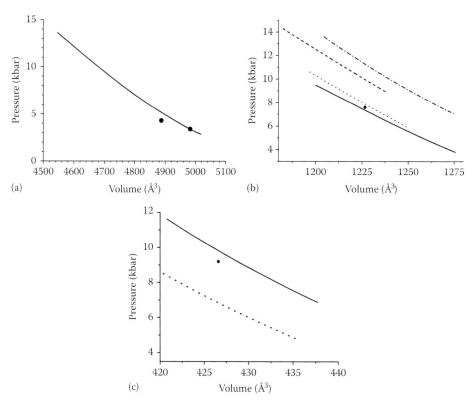

FIGURE 2.4 Equation of state of Ar hydrates at $T = 293$ K: (a) CS-II structure with single occupancy (solid line); (b) sH structure, fivefold (dotted and dashed line), threefold (dashed line), double (dotted line), and single (solid line) occupancy of the large cages (in all the cases, the occupancy of the small and medium cages are single); and (c) tetragonal structure single (dash line) and double (solid line) occupancy. (Reproduced from Inerbaev, T.M. et al., *J. Incl. Phen. Macrocycl. Chem.*, 48, 55, 2004. With permission.)

and experiment was obtained for H-hydrate using the experimentally proposed multiple (5 Ar atoms) occupation of large cages (Manakov et al. 2002) (see Figure 2.4b). In this case, the $P(V)$ function of sH Ar hydrate with $Ar \cdot 4.87H_2O$ stoichiometry, for which the occupancy of the large cages is double, is closer to experiment. A good agreement with experimental data was observed in the case of double occupancy of Ar atoms in the hydrate cages of tetragonal structure. Figure 2.4c shows that at the experimentally determined lattice parameters ($a = 6.342\,\text{Å}$, $c = 10.610\,\text{Å}$), the calculated value of pressure is $P = 9.8\,\text{kbar}$, which correlates well with the experimental value ($P = 9.2\,\text{kbar}$) (Manakov et al. 2002).

The disagreement between the LD results and experimental data on $P(V)$ diagram of CS-II and sH argon hydrates can be explained by fact that in the LD calculations, all the cages were filled. However, in practice, it is difficult to realize the full occupation of hydrate cavities. Moreover, it is experimentally known that it is not possible to occupy all cavities by guest molecules and even a small number of guest molecules are sufficient to form the clathrate structure (Sloan and Koh 2007). Therefore, the multiple occupation for sH, which was predicted experimentally, may be realized by only in a limited number of large cavities.

It can be summarized that in the studied hydrates, multiple occupancies of the large cages are possible. Moreover, the stability of Ar clusters in the large cages is correlated well with experimental phase transition from CS-II to a new tetragonal hydrate structure (Inerbaev et al. 2004).

2.3.3 Hydrogen Clathrate Hydrate

The anomalous behavior of H_2O–H_2 system was found and the formation of clathrate phase of hydrogen hydrate was proposed at hydrogen pressures of 100–360 MPa and temperature range 263–283 K (Dyadin et al. 1999a,b). The structure of the hydrogen hydrate formed at this range of pressure $P = 200$–300 MPa and lower temperature range $T = 240$–249 K was determined in 2002 (Mao et al. 2002). It was shown that hydrogen hydrates may be used as compounds for hydrogen storage because hydrogen content was 50 g/L, which corresponds to 4.96 wt% (Mao and Mao 2004). They also showed that after high pressure formation it is possible to maintain the hydrogen hydrate at ambient pressures and liquid nitrogen temperatures $T = 77$ K and decomposed them with hydrogen emission at heating to 140 K.

The large cages of clathrate hydrate structure of CS-II hold an individual large guest molecule; however, these cavities are too large for a single H_2 molecule and therefore the existence of hydrogen hydrate with single occupancy of large cavities was not suspected. Experimentally, it was shown that large ($5^{12}6^4$) cavity of CS-II hydrate can include three to four hydrogen molecules and small (5^{12}) cavities one to two hydrogen molecules. The maximum quantity of hydrogen in a clathrate hydrate was reached at high pressures in hydrogen hydrates CS-II at fourfold filling of large cages and twofold filling of small cages by hydrogen molecules.

Along with experimental studies, theoretical investigations by different methods have also been conducted. Density functional theory (DFT) (Sluiter et al. 2003, 2004), quantum chemical (Patchkovskii and Tse 2003, Patchkovskii and Yurchenko 2004, Alavi et al. 2005a), statistical thermodynamics (Inerbaev et al. 2006), Monte Carlo (Katsumasa et al. 2007), and classical molecular dynamics (Alavi et al. 2005b, 2006) modeling have been used to study pure and mixed hydrogen clathrate hydrates. The possibility of filling of large cages by clusters of hydrogen molecules was shown using these models. Various conclusions have been drawn from these studies regarding the H_2 occupancy of small cages of the host lattice.

Despite numerous experimental and theoretical investigations, the problem of the possible existence of hydrates with hydrogen content exceeding that in the hydrogen CS-II hydrate (4.96 wt%) still remained. Hydrogen storage in hydrates (gas-to-solids technology) is an alternative technology to liquefied hydrogen at cryogenic temperatures or compressed hydrogen at high pressures. Now hydrate technology for hydrogen storage as well for storage and transportation of natural gas is developing, and further investigations are needed to find a stable phase of hydrogen clathrate hydrates under moderate pressure and room temperature. This includes studies of the stability conditions under which a hydrate with higher hydrogen storage capacity can be formed at these conditions.

We use our model described in Section 2.2 to calculate the curves of monovariant three-phase equilibrium gas–hydrate–ice Ih and the degree of filling of the large and small cavities for pure hydrogen of cubic CS-II structure in a wide range of pressure and at low temperatures. The calculations were performed using the 128 water molecules in supercell of ice Ih. For calculations of free energy, the molecular coordinates of ice Ih and hydrogen hydrates (the centers of mass positions and orientation of molecules in the unit cell) were optimized by the conjugate gradient method. For clathrate hydrate of CS-II, the initial configuration for calculations was a single unit cell with 136 water molecules and different numbers of hydrogen in large (L) and small (S) cavities. The initial positions of water oxygen atoms of the hydrate lattice were taken from the x-ray analysis of the double hydrate of tetrahydrofuran and hydrogen sulfide performed by Mak and McMullan (1965).

Considering ice Ih and clathrate hydrate phases, the modified simple point charge-extended (SPCE) water–water interaction potential was used to describe the interaction between water molecules in the hydrate. The parameters describing short-range interaction between the oxygen atoms $\sigma = 3.17$ Å and the energy parameter $\varepsilon = 0.64977\,\text{kJ/mol}$ of Lennard-Jones potential of SPCE potential (Berendsen et al. 1987) were changed and taken to be $\sigma = 3.1556$ Å; $\varepsilon = 0.65063\,\text{kJ/mol}$. The charges on hydrogen ($q_H = +0.4238|e|$) and on oxygen ($q_O = -0.8476|e|$) of SPCE model were not changed. The modified SPCE potential significantly improves the agreement between the calculated cell parameters for ice Ih and methane hydrate with the experimental values. The protons were placed according to the Bernal–Fowler ice rules and the water molecules were oriented such that the total dipole moment of the unit cell of the hydrate vanishes. The long-range electrostatic interactions were computed by the Ewald method.

The guests were considered as spherically symmetric particles and their interaction potential was formulated as

$$U_{H_2-H_2} = \frac{D_0}{\zeta - 6}\left[6\exp\left(\frac{1-r}{\rho}\zeta\right) - \zeta\left(\frac{\rho}{r}\right)^6\right] \quad (2.16)$$

where

r is intermolecular separation
D_0 is the potential well depth
ρ is the intermolecular distance at the minimum of the potential (i.e., where $U_{H_2-H_2} = -D_0$)
ζ is a dimensionless "steepness factor"

The exponential-6 ("exp-6") potential used here is a more realistic potential as it gives the correct functional form at small intermolecular separations (Ross and Ree 1980).

For modeling the H_2O–H_2 interaction, the LJ part of the SPCE potential has been fitted by an exp-6 potential and host–guest interaction potential parameters are represented as

$$D_{H_2O-H_2} = \sqrt{D_{H_2O-H_2O}D_{H_2-H_2}} \quad (2.17)$$

$$\zeta_{H_2O-H_2} = \sqrt{\zeta_{H_2O-H_2O}\zeta_{H_2-H_2}} \quad (2.18)$$

$$\rho_{H_2O-H_2} = \frac{\rho_{H_2O-H_2} + \rho_{H_2-H_2}}{2} \quad (2.19)$$

The H_2–H_2 interaction potential parameters: The long-distance dispersion interaction part was taken from Murata et al. (2002) and the short-range repulsion part was estimated with the density functional calculations (Sluiter et al. 2003, 2004) using the all-electron mixed basis (TOMBO) method (Bahramy et al. 2006). Potential parameters of the fitted H_2–H_2 and fitted LJ part of host–host interaction are listed in Table 2.2. The intermolecular distance at the minimum of the potential $\rho = 2.967\,\text{Å}$ is shorter than the minimum obtained for the H_2–H_2 potential surface (3.45 Å) by ab initio calculations for the dimer (Carmichael et al. 2004). The difference is due to the fact that we did not include quadrupole–quadrupole interaction of hydrogen molecules in dispersion region. However, the present empirical potential better describes the interaction between hydrogen molecules in water cavities, because the experimental determined distance between the four tetrahedrally arranged D_2 molecules in the large cage of CS-II clathrate hydrate is found to be 2.93 Å (Lokshin et al. 2004).

TABLE 2.2 Potential Parameters of the "exp-6" Potential as Described in the Text

Parameter	D_0 (kJ/mol)	ρ (nm)	ς
H_2–H_2	0.7295	0.2967	10.92
H_2O–H_2O	0.5206	0.323	14

The calculated pressure dependence of the chemical potentials $\tilde{\mu}_Q$ of water molecules of hydrogen hydrates of CS-II with single, double, triple, and quadruple occupation of $5^{12}6^4$ cages by H_2 at $T = 200\,\text{K}$ without entropy part and μ_Q^0 of water molecules of empty host lattice of C-II hydrate are displayed in Figure 2.5. In all cases, the 5^{12} cage is occupied by only one hydrogen molecule. Figure 2.6 shows the changes of chemical potential of empty host lattice, $\Delta\mu_Q(n_L) = \tilde{\mu}_Q - \mu_Q^0$, under influence of hydrogen molecules. Based on these results, the following linear approximation, $\Delta\mu_Q(n_L) = 0.093 \times n_L$ kJ/mol, was determined in order to estimate the change of chemical potential in the cases when the occupation of large cage does not equal to integer. As can be seen from Figures 2.5 and 2.6, the change in chemical potential of the empty host lattice under influence of guest molecules was significant. The difference $\Delta\mu_Q = \tilde{\mu}_Q - \mu_Q^{ice}$ between chemical potentials of ice and host lattice is increased and reached the values

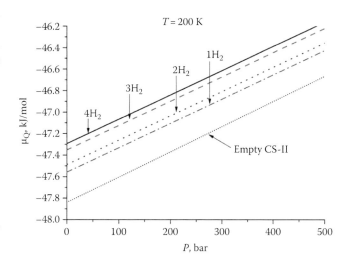

FIGURE 2.5 Chemical potentials of water molecules (host lattice) for clathrate hydrates with different occupation of hydrogen molecules in large cages.

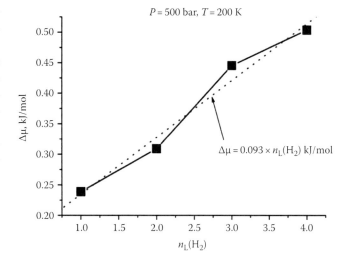

FIGURE 2.6 Changes of chemical potential of host lattice as a function of the hydrogen molecule occupation of large cage.

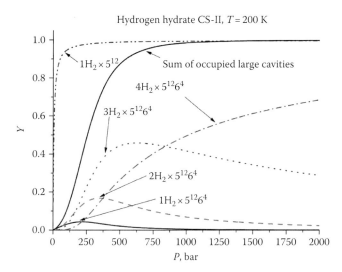

FIGURE 2.7 Degree of filling of the small and large cavities of H_2 hydrate of CS-II at 200 K.

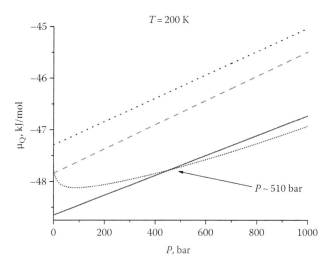

FIGURE 2.9 Chemical potentials of water molecules μ_Q of ice Ih (solid line), empty host lattice of hydrate of CS-II (dashed line), and hydrogen hydrate (CS-II) with (short dot line) and without (dot line) entropy part taken into account.

close to those that are used in construction of phase diagrams within framework of the van der Waals and Platteeuw theory. It was shown that at low temperatures with increasing pressure, the filling of large cages steadily increases from one to four hydrogen molecules (Figure 2.7). As follows from the results of calculations, the filling degrees differ notably from integer at low pressure and tend to whole number with pressure increasing.

The temperature dependence of filling degree of both large and small cages has been estimated at pressure 2 kbar and compared with experimental results (Lokshin et al. 2004) as shown in Figure 2.8. The calculated results obtained within our approach are in good agreement with experimental data. The calculated pressure dependences of the chemical potentials of water molecules of ice Ih, empty host lattice of CS-II hydrate,

of hydrogen hydrates of CS-II with and without entropy term at temperature $T = 200$ K are displayed in Figure 2.9. Intersection of chemical potential curves of ice Ih and of host lattice with the account of entropic contribution defines the pressure of monovariant equilibrium at a given temperature. At this temperature, it was found that the hydrogen hydrate exists in a metastable phase at low pressure and is stable at pressures greater than $P = 510$ bar. The calculated curves $P(T)$ of monovariant equilibrium of the gas phase, ice Ih, and H_2 hydrates CS-II are displayed in Figure 2.10. The calculated curves for the hydrogen hydrate agree well with the experiment (Barkalov et al. 2005, Lokshin and Zhao 2006).

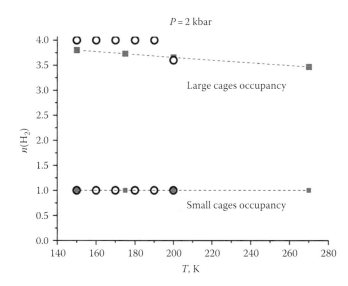

FIGURE 2.8 Number of hydrogen molecules included in small and large cages. Open black circles are experimental data taken from Lokshin et al. (2004).

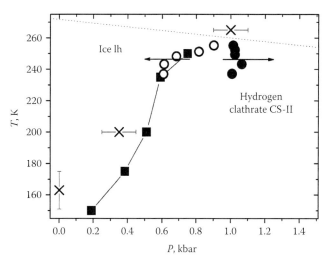

FIGURE 2.10 Calculated and experimental curve $P(T)$ of monovariant equilibrium of the gas phase, ice Ih and hydrogen hydrate. Experimental data was taken from Lokshin and Zhao (2006) and Barkalov et al. (2005) (open and filled circles, respectively). Dotted line presents ice Ih–liquid water equilibrium phase transition.

2.3.4 Guest–Guest and Guest–Host Interactions in Hydrogen Clathrate

In the case of first-principles methods, the calculations of electronic and structural properties of gas clathrate involve two steps (Patchkovskii and Tse 2003). The first step is an optimization procedure carried out using HF or DFT levels. In the second step, single point energy calculations on optimized HF or DFT structures are performed using the second-order Møller-Plesset (MP2) level, respectively. This scheme works when the guest molecule fits snugly in the cage and locates at the cage center. However, when the van der Waals volume of guest is smaller than the cavity diameter, the determination of equilibrium position of guest inside cage becomes difficult. It is well known that HF and DFT do not well reproduce the dispersive interaction, which is probably important for the proper description of guest–host and guest–guest interactions in gas clathrate systems.

Therefore, in order to avoid this problem, we have used the MP2 method for both optimization and electronic structure calculations of guest molecules inside of a large water cluster

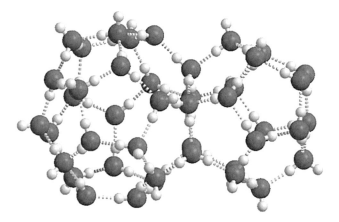

FIGURE 2.11 Initial structure of the $(H_2O)_{43}$ water cluster.

represented the fragment of CS-II hydrate structure in which the two fundamental cages (5^{12} and $5^{12}6^4$) connected directly, as shown in Figure 2.11. All calculations reported in this study were carried out using the Gaussian 03 package (Frisch et al. 2004). Full geometry optimization of selected cage-like structures of water clusters with and without enclathrated guests were performed at the MP2 level. A large yet computationally manageable basis set, 6-31 + G(d) including polarization and diffuse functions, was used. The inclusion of diffusion functions in the basis set is necessary for a better description of the structure and energetic of hydrogen-bonded complexes (Frisch et al. 1986). The optimizations were performed using the redundant internal coordinate procedure (Peng et al. 1996).

In the first step, we estimate the difference in equilibrium position of methane molecule inside 5^{12} and $5^{12}6^2$ cages obtained by HF, DFT and MP2 levels with the same basis sets. The results of calculations at different levels of theory have shown that in the case of the dodecahedral cage $(H_2O)_{20}$, the equilibrium position of methane molecule remains at the center of cavity. However, in the case of tetrakaidecahedral cage $(H_2O)_{24}$, the optimal geometry of guest is dependent on the method of calculation. Only when the MP2/6-31 + G(d) method for optimization scheme is used, the guest position moved off-center (see Figure 2.12). As a result, the guest–host interaction energy for $(H_2O)_{24}$ water cage also varies with the calculation method. Thus, the MP2 interaction energy between the CH_4 molecule and dodecahedral cage is equal to −6.81 kcal/mol. The situation is not significantly changed in the case of $(H_2O)_{24}$ cluster. The interaction energy has a value of −6.18 kcal/mol, which is close to the interaction energy of CH_4 with the $(H_2O)_{20}$ cage. When the HF/6-31 + G(d) method is used, the interaction energy as a result of interaction of methane with the dodecahedral cage $(H_2O)_{20}$ is equal to −6.63 kcal/mol, which is very close to the same energy value obtained at MP2 level. However, in the case of the large cage ($5^{12}6^2$), the HF interaction energy between the methane molecule and tetrakaidecahedral

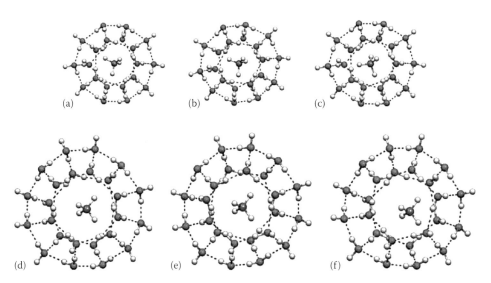

FIGURE 2.12 Optimized structures of CH_4 molecules inside 5^{12} cage. (a) HF/6−31+G(d), (b) B3LYP/6−31+G(d), (c) MP2/6−31+G(d), and $5^{12}6^2$ cage (d) HF/6−31+G(d), (e) B3LYP/6−31+G(d), and (f) MP2/6−31+G(d).

cage is equal to −0.23 kcal/mol, which is significantly different from results obtained at MP2 level of theory. This shows that the MP2 method, which captures dispersion interactions much better than DFT, is able to more accurately estimate the equilibrium position of small guests inside large cages.

Since the clathrate hydrate of cubic structure II consists of two fundamental cages (5^{12} and $5^{12}6^4$), the combination of these cages connected directly (see Figure 2.11) both without and with insertion of hydrogen molecules has been selected. In our study, three possibilities have been considered. First, there are total four hydrogen molecules located in one large cage; second, there are totally two molecules occupying one small cage; and third, there are totally six molecules with four and two in large and small cages, respectively.

Figure 2.13a shows the optimized structures of the fused cages with four H_2 molecules enclathrated in a large cage. The interaction energy between hydrogen molecules and the $(H_2O)_{43}$ cluster is equal to −0.972 kcal/mol. The distances between H_2 molecules inside the cage was found to be in a region between 2.8 and 3.1 Å, which correlated well with experimental value (2.93 Å) reported in Lokshin et al. (2004). The H–H bond lengths are slightly elongated by 0.001 Å as compared with the bond length of free molecule. Moreover, the water cages are almost undistorted. Due to the large size of the void, the hydrogen molecules, by moving closer to the cage wall, interacts with the water molecules. This leads to a small charge transfer (0.01 e) from water to hydrogen molecules and as shown in Figure 2.14a. The different results have been observed in the case of double occupancy of small cage (see Figure 2.13b). The structural properties of hydrogen dimer are different as compared to the previous case. Thus, it has been found the shorter distance (2.71 Å) between H_2 molecules that indicates the repulsion interaction between guest molecules. As in the case of water cluster with four molecules in large cage, the interaction with the cavity again leads to a charge transfer (0.014 e) from water to hydrogen molecule (see Figure 2.14b), which is larger than in the case of large cage filling. Therefore, in this case the interaction energy between hydrogen and water molecules is equal to +2.37 kcal/mol. The positive value indicates the instability of hydrogen cluster inside a small cage. This instability results from two factors. First, if we remove hydrogen cluster from water cavity and fix the geometry of the cluster, the value of interaction between hydrogen molecules is equal to +0.75 kcal/mol and hence it indicates repulsion. Second, the filling of large cavity is necessary for stabilization of the water cluster network.

Thus, in the case of filling both large and small cages by four and two hydrogen molecules, respectively, the interaction energy between hydrogen and water molecules is equal to −1.74 kcal/mol. In this case, the encapsulation of hydrogen molecules in a large cage has positive effect not only for stabilization of the water structure but also on the equilibrium position of hydrogen molecules inside a small cage, as shown in Figure 2.13c. The distances between H_2 molecules inside the small cage are found to be 2.8 Å.

Recently, it was found that the formation pressure of hydrogen clathrate can be significantly reduced by adding second guest

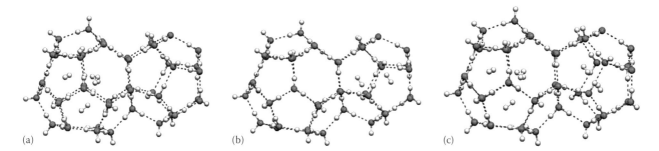

(a)   (b)   (c)

FIGURE 2.13 Optimized structures of H_2 molecules inside the $(H_2O)_{43}$ water cluster using MP2/6–31+G(d) method: (a) four guests in large cages; (b) two in small cages; (c) four and two molecules in large and small cages, respectively.

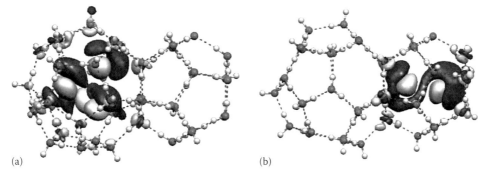

(a)   (b)

FIGURE 2.14 Charge density isosurface: (a) four hydrogen molecules in large cage; (b) two hydrogen molecules in small cage. White color is accumulation of electrons and gray is depletion of electrons.

molecule, such as tetrahydrofuran (THF) (Florusse et al. 2004). Our calculations also indicated that the encapsulation of THF in large cage is strongly affected on stability of selected water cluster. As a result, the value of interaction energy between THF and $(H_2O)_{43}$ water cluster is equal to -14.48 kcal/mol. This value is significantly larger than that in the case of four hydrogen molecules. Moreover, it is also larger than the interaction between methane molecule and water cavities.

It is found that four hydrogen molecules in a large cage energetically stabilize the large fused cluster but it is not possible by accommodating two hydrogen molecules in a small cage. The charge density distribution shows that there exists a weak interaction between hydrogen molecules and water cages. Interactions of the THF molecule with the host are larger than the interaction of hydrogen with the host, meaning that "help gas" molecules play a more significant role in the stabilization of hydrogen hydrate. These results also indicate that the interaction between the guest and the host is essential and should be accurately estimated in the calculation of the phase diagram of hydrogen hydrate.

2.3.5 Mixed Methane–Hydrogen Clathrate Hydrate

As mentioned before, hydrogen can be stored at low pressures within the clathrate hydrate lattice by stabilizing the large cages of water host framework with a second guest molecule, tetrahydrofuran (Florusse et al. 2004). Moreover, the hydrogen storage capacity in THF-containing binary clathrate hydrates can be increased at modest pressures by tuning their composition to allow the hydrogen guests to enter both the large and the small cages, while retaining low-pressure stability (Lee et al. 2005). It can be expected that mixed hydrogen-containing hydrates with a second guest molecule with smaller molecules, which are able to stabilize any cavity of hydrate and may in definite thermodynamic conditions, also to decrease hydrate formation pressure. In this case, the filling of small cavities by the second guest allows hydrogen molecules to occupy the large ones and hence increases the hydrogen storage density. In such mixed hydrates, hydrogen mass content would be more than in mixed hydrates with large second-component molecules obtained so far. The possibility of mixed hydrates formation at equilibrium with gas

phase was confirmed experimentally for gas mixtures of hydrogen with methane (Struzhkin et al. 2007). In hydrogen–methane–water system at low temperatures, $T = 250$ K and relatively high pressure $P = 3$ kbar, a new solid phase is formed. Raman spectroscopy shows hydrogen enclathration in this solid phase. Recently, the mixed $H_2 + CO_2$ CS-I hydrate at 20% CO_2 in the gas phase was obtained (Kim and Lee 2005). It was shown that clusters of two hydrogen molecules are included in the small cages of the formed hydrate.

Usually, the large cages of CS-I, CS-II, and sH clathrate hydrates fit around a single large guest molecule, but these cages are too large for a single H_2 molecule to stabilize them. For this reason, the existence of hydrogen hydrate with single occupancy of large cavities was not expected. There are pentagonal dodecahedron (a polyhedron with 12 pentagonal faces 5^{12}) cages in all hydrate structures as the basic small cavity. The basic 5^{12} cavities combined with $5^{12}6^2$ cavities form CS-I, with $5^{12}6^4$ cavities the CS-II, and with $4^35^66^3$ and $5^{12}6^8$ cages the sH structure (Figure 2.15).

At equilibrium of gas phase mixture of guest molecules with water (ice Ih) at definite conditions, the formation of a mixed hydrate can be expected. In this case, increasing of the hydrate composition may become possible by means of component ratio variation in the gas phase. Therefore, we study the thermodynamic properties of binary hydrate systems using our proposed model. In order to investigate effect of stabilization by methane molecules, we have considered CS-II and sH clathrate hydrates. Composition and fields of thermodynamic stability for these hydrates with multiple filling of large cages have been estimated and the chemical potentials of host lattice molecules were found. The conditions (pressure and temperature) for the formation of hydrogen + methane mixed CS-II and sH clathrate hydrates are determined (Belosludov et al. 2009).

The composition of mixed methane–hydrogen hydrates formed from gas mixture depends on temperature, pressure, and the composition of the gas phase. Methane molecules can fill both small and large cages of sH and CS-II hydrates while hydrogen molecules fill the remaining cages. Relationships of small and large cage fillings are determined by interaction energies of methane molecules in cages with the water host lattice and by configurational entropy part of free energy. This corresponds to the stabilization effect, which is determined by methane concentration in the gas phase.

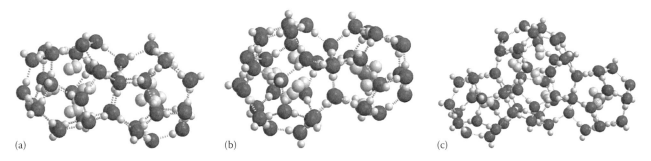

(a)   (b)   (c)

FIGURE 2.15 Types of small and large cages in gas hydrates of (a) CS-I (5^{12} and $5^{12}6^2$, respectively), (b) CS-II (5^{12} and $5^{12}6^4$, respectively), and (c) sH (5^{12}, $4^35^66^3$, and $5^{12}6^8$, respectively). In all cases, methane molecule is in 5^{12} cage and hydrogen molecules are in others.

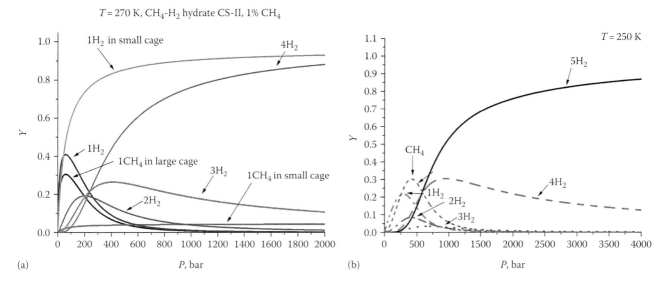

FIGURE 2.16 Degree of filling of large cages by methane and hydrogen (a) in CS-II hydrate at 10% and (b) in sH hydrate at 50% of methane in the gas phase. (Reproduced from Belosludov, V.R. et al., *Int. J. Nanosci.*, 8, 57, 2009. With permission.)

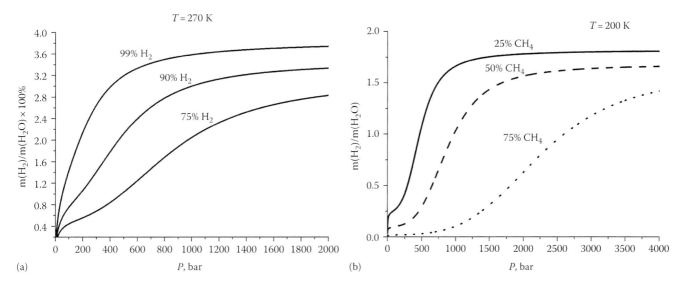

FIGURE 2.17 (a) Mass percentage at 270 K of hydrogen abundance in CS-II hydrates at 25%, 10%, and 1% of methane in the gas phase; (b) mass percentage at 200 K of hydrogen abundance in sH hydrate: at 75%, 50% and 25% of methane in gas phase. (Reproduced from Belosludov, V.R. et al., *Int. J. Nanosci.*, 8, 57, 2009. With permission.)

Our model permits to find composition of hydrates for the given *T*, *P*, and the gas phase component ratio. At increasing pressure, the large cages are first filled preferentially by methane molecules and then they are gradually expelled by hydrogen molecules, as shown in Figure 2.16a and b. Filling of small cages by methane and hydrogen molecules grows with increasing pressure. It has been seen that in the case of 1% methane concentration in gas phase, the hydrogen molecules occupied mostly 90% of small cages in CS-II hydrate in the high-pressure region. The total filling of cavities can be done by a mass of guests per mass of water in hydrate (Figure 2.17a and b). Hydrogen content continues to increase slightly due to multiple filling of large cages. Mass percentage of hydrogen in mixed hydrogen–methane

CS-II hydrate can amount up to 4 wt% at lower concentration of methane in gas phase and higher pressure, for sH hydrate, it can reach 1.5 wt%. The pressure of monovariant equilibrium "ice Ih–gas phase–mixed CS-II hydrate" decreases in comparison with the pressure of pure hydrogen hydrate formation with increasing methane concentration in the gas phase, as shown in Figure 2.18. This result indicated that it is possible to form mixture hydrogen–methane hydrate of structure CS-II at low (30 bar) pressure and 25% concentration of methane in the gas phase. This pressure is about 40 times lower that one needed to form pure hydrogen clathrate of CS-II structure. After that, the hydrogen concentration can be increased from 0.4 up to 2.8 wt% by increasing pressure.

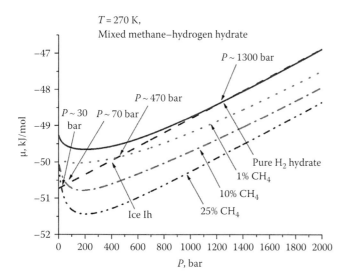

FIGURE 2.18 Chemical potentials of water molecules μ_Q for ice Ih, mixed H_2-CH_4 clathrate CS-II hydrate at 0, 1%, 10%, and 25% of methane in the gas phase. (Reproduced from Belosludov, V.R. et al., *Int. J. Nanosci.*, 8, 57, 2009. With permission.)

Monovariant equilibrium "ice Ih–gas phase–mixed sH hydrate" at $T = 200\,K$ is shown in Figure 2.19. It can be seen that the pure hydrogen clathrate of structure sH is a metastable phase at this temperature. However, sH hydrogen clathrate can be stabilized at $P = 400$ bar and at 75% methane content in the gas phase (see Figure 2.19). In this case, the maximum amount of hydrogen storage can be achieved at the value of 1.5 wt% at high pressure.

In conclusion, it was found that at increasing pressure, the large cages are filled preferentially by methane molecules, content of which in hydrate reaches its maximum and then these

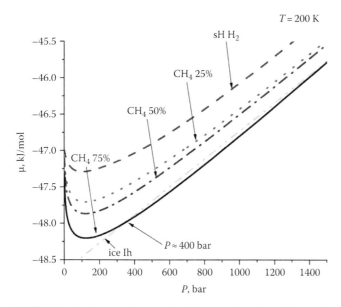

FIGURE 2.19 Chemical potentials of water molecules μ_Q for ice Ih, pure hydrogen hydrate sH and mixed methane–hydrogen sH hydrates at 25%, 50%, and 75% of methane in the gas phase. (Reproduced from Belosludov, V.R. et al., *Int. J. Nanosci.*, 8, 57, 2009. With permission.)

molecules start gradually to be expelled by hydrogen molecules. At the same time, filling of small cages by methane and hydrogen molecules grows with increasing pressure. The pressure of monovariant equilibrium "ice Ih–gas phase–mixed CS-II hydrate" lowers in comparison with the formation pressure of pure hydrogen hydrate sII and mixed H_2-CH_4 hydrate sH with increasing methane concentration in the gas phase.

2.4 Metal–Organic Framework Materials

2.4.1 Introduction

The storage of gas in solids is a technology that is currently attracting great attention because of its many important applications (Gupta 2008, Hordeski 2008, Thallapally et al. 2008). Perhaps the most well-known current area of research centers on the storage of hydrogen for energy applications, with viable energy storage for hydrogen economy as the ultimate goal (Langmi and McGrady 2007, Züttel et al. 2008, Graetz 2009, Hamilton et al. 2009). There are several reasons why one might want to store hydrogen inside a solid, rather than in a tank or a cylinder. First, it is relatively common for more gas to be stored in a given volume of solid than one can store in a cylinder, leading to an increase in storage density of the gas. Second, there may be safety advantages associated with storage inside solids, especially if high external gas pressures can then be avoided. As a good example, the high sorption ability for acetylene from acetylene/carbon dioxide gas mixture on metal–organic microporous material [$Cu_2(pzdc)_2(pyz)$] (pzdc = pyrazine-2,3-dicarboxylate, pyz = pyrazine) was recently determined at low pressure, using both extensive first-principles calculations and different experimental measurements. It was found that in the nanochannel, only acetylene molecules are indeed oriented to basic oxygen atoms and form a one-dimensional chain structure aligned to the host channel structure (see Figure 2.20). It was shown that the concept using designable regular metal–organic microporous material could be applicable to a highly stable, selective adsorption system (Matsuda

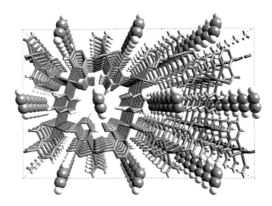

FIGURE 2.20 The stable configuration of acetylene molecules into the metal–organic microporous material ($Cu_2(pzdc)_2(pyz)$).

et al. 2005). Finally, small amounts of gases are actually easier to handle when stored in a small amount of solids. There exist several approaches to store hydrogen gas on solids (Sakintuna et al. 2007, Varin et al. 2009). One important strategy involves the storage of hydrogen reversibly on substances as in chemical hydrides (Orimo et al. 2007). Another involves the adsorption of the hydrogen gas inside a porous material, in which the adsorption occurs by a physical means (Yaghi et al. 2003, Morris and Wheatley 2008).

2.4.2 Metal Organic Frameworks as Hydrogen Storage Materials

Among the materials that are promising for the physisorption of hydrogen gas are MOFs, crystalline microporous solids comprised of metal building units and organic bridging ligands (Kitagawa et al. 2004, Rowsell and Yaghi 2004). MOFs have many advantages. Their structural versatility has allowed the rational design and assembly of materials having novel topologies and with exceptional host–guest properties important for much-needed industrial applications. They can be made from low-cost starting materials, their synthesis occur under mild conditions, and the manufacturing yields are high. They are completely regular, have high porosity, and highly designable frameworks. MOFs contain two central component connectors and linkers as shown in Figure 2.21. Transition-metal ions are versatile connectors because, depending on the metal and oxidation state (range from 2–7), they give rise to geometries such as linear, T- or Y-shaped, tetrahedral, square-planar, square-pyramidal, and so on. Linkers provide a wide variety of linking sites with tunable binding strength and directionality. Inorganic, neutral organic, anionic organic, and cationic organic ligands can act as linkers. Various combinations of the connectors and linkers afford various specific structural motifs. The important features of MOF are the ability to absorb of various gases (such

as N_2, O_2, CO_2, CH_4) as well as different organic molecules at ambient temperature, which is important for storage, catalytic, separation, and transport applications. However, because of their typically weak interaction with hydrogen, these materials function best only at very low temperature and their use as storage media in vehicles would require cryogenic cooling (Rosi et al. 2003, Rowsell et al. 2004, Chen et al. 2005, Wong-Foy et al. 2006). A very recent development in the usage of MOFs is the metal ion impregnation, which provides the necessary binding energy and also increases the storage capacity (Liu et al. 2006, Mulfort and Hupp 2007, 2008, Yang et al. 2008).

In this section, we have discussed the adsorption of hydrogen molecule on the MOFs by using DFT. We would like to caution the readers that as mentioned before, the standard DFT functionals cannot quantitatively describe the dispersion part of van der Waals interaction. However, since our system involves huge size and more than 100 atoms, to reduce the computation time, we have used DFT methods in our calculations to explain the difference in the adsorption of hydrogen molecules on the pure and Li-doped MOFs. Moreover, we show that Li cations strongly adsorbed on the organic linkers and each Li can hold up to three hydrogen molecules in quasimolecular form. Further, to understand how to control of the structure on doping Li, we have studied the isoreticular MOFs (IRMOF) and their adsorption ability toward hydrogen.

A considerable number of computational studies regarding hydrogen molecules interacting with IRMOFs have recently been published. The interaction energies and the corresponding geometries have been calculated at diverse levels of theory. In the past, the interaction energy of H_2 with the organic linkers has been determined to be 0.03–0.05 eV (Mulder et al. 2005, Mueller and Ceder 2005, Buda and Dunietz 2006, Gao and Zeng 2007). This energy is very similar to the theoretically calculated energy of interaction of a hydrogen molecule with benzene (Kolmann et al. 2008, Mavrandonakis and Klopper 2008). The metal–organic framework-5 (MOF-5) is the composition $Zn_4O(BDC)_3$ (BDC = 1,4-benzenedicarboxylate) with a cubic three-dimensional extended porous structure (Rosi et al. 2003). To model the structure of MOF-5, primitive cell containing 106 atoms were used. The atomic positions of optimized structure have good agreement with the experimental values, which provides us confidence on our theoretical method. There exist four possible sites where hydrogen could be physisorbed on the MOF unit. The sites identified are shown in Figure 2.22. The H_2 adsorption sites near to Zn atom were found to have the highest binding energy. The strongest interaction is found for H_2 perpendicular to the central Zn–O bond that is parallel to the cluster surface, as shown in Figure 2.22a. The distance between the molecular center of H_2 and the closest Zn atom is ~3.21 Å, which is in agreement with the previous theoretical prediction based on cluster models (Kuc et al. 2008a). According to the neutron powder diffraction results on the Zn-MOF system, the adsorption of hydrogen occurs near the O4Zn tetrahedral site (Rowsell et al. 2005). However, recent theoretical calculations show also that the H_2 molecule could be strongly adsorbed

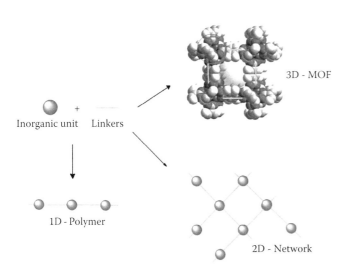

FIGURE 2.21 Schematic presentation of the basic principles of formation MOFs and their possible topologies.

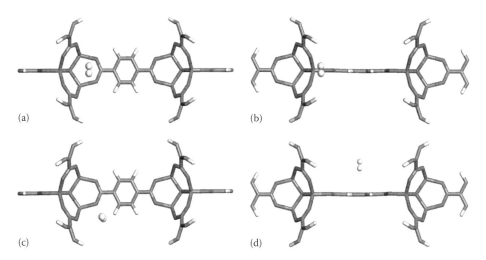

FIGURE 2.22 The main four sites for the hydrogen adsorption on MOF-5. The hydrogen molecule is indicated with a ball and stick representation.

on the linker with an interaction energy of 0.035 eV and an intermolecular distance of 3.2 Å (Kuc et al. 2008b).

Since the interaction energies are very small, a lot of efforts were made to increase their interaction energies and hence storage capacity both by experimental and theoretical studies. Recently, metal impregnation was found to increase the storage capacity of these materials (Mulfort et al. 2009). Here, we focused interest on the Li functionalization on the MOFs for the study of hydrogen adsorption (Venkataramanan et al. 2009). There exist several advantages in using Li as a dopant. First, is its small size and low weight. Second, Li can be easily ionized and the ionized molecules can hold a large amount of hydrogen by an electrostatic charge-quadruple and charge-induced dipole interactions (Niu et al. 1992). Third, Li cations can hold hydrogen

molecules in a quasimolecular form, unlike the transition metal cations that bind hydrogen covalently.

The structure of the primitive cell representing the unit cell was fully optimized without any geometrical constraints. A comparison between the optimized structural parameters with experimental value shows a good agreement, which provides confidence in our computational method. We first attempt to understand the best form of Li that could be doped on MOFs. We found that the Li as cation has the highest adsorption energy of all those studied units (cation, anion, neutral). Then we proceeded to interact one to four molecular H_2 on this Li-functionalized MOF-5 by placing them near the Li cation. The optimized structures are shown in Figure 2.23, while the selected geometrical parameters are listed in Table 2.3.

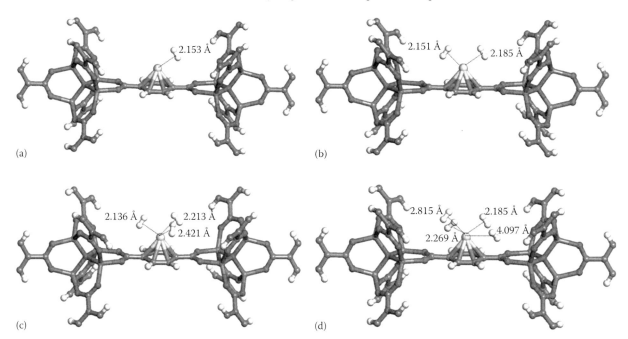

FIGURE 2.23 Optimized geometries of adsorbed hydrogen molecules on Li functionalized Zn-MOF-5 with one (a), two (b), three (c), and four (d) hydrogen molecules. Note how successive H_2 form a cluster around the Li atom sitting above the center of C_6 ring.

TABLE 2.3 Selected Bond Parameters (Å) and Binding Energy per Hydrogen Molecule (eV) for the Adsorption of Hydrogen on Li Functionalized Zn-MOF

No. of H_2	Avg. Benzene–Li (Å)	Li–H_2 (Å)	Avg. H–H (Å)	ΔE_b (eV)
0	2.206	—	0.750	—
1	2.223	2.096	0.760	0.213
2	2.241	2.124	0.759	0.209
3	2.257	2.315	0.755	0.196
4	2.252	2.379 (4.036)	0.755 (0.751)	0.163

The hydrogen interaction or the binding energy per hydrogen molecule (ΔE_b) can be defined as

$$\Delta E_b = E_T(\text{Li-MOF}) + E_T(H_2) - E_T(\text{Li-MOF} + H_2)/nH_2$$

where $E_T(\text{Li-MOF} + H_2)$ and $E_T(\text{Li-MOF})$ refer to the total energy of the Li- functionalized Zn-MOF with and without hydrogen molecule, respectively, while the $E_T(H_2)$ refers to the total energy of the free hydrogen molecule and n is the number of hydrogen molecules.

For the first H_2, the interaction energy is 0.213 eV, with an intermolecular distance of 2.153 Å. The orientation of hydrogen is in a T-shape configuration with lithium (see Figure 2.23). The H–H bond distance of 0.760 Å corresponds to a very small change compared to the 0.750 Å bond length in a free hydrogen molecule. This indicates that the Li cation holds the H_2 molecules by a charge-quadruple and charge-induced dipole interaction. When the second hydrogen is introduced, the Li–H_2 distance increases to 2.124 Å and the binding energy per H_2 molecule gets reduced to 0.209 eV. To know the number of hydrogen molecules a Li cation can hold, we doped the third and fourth H_2 near the Li cation. With the introduction of third H_2, the Li–H_2 distance increases along with a decrease in the interaction energy value. A noticeable feature is the H–H distance, which decreases with the increase in the number of H_2 molecules. When the fourth hydrogen molecule is introduced near the Li cation, three of them remain in place near the Li atom and one H_2 molecule moves away to a nonbonding distance of 4.036 Å. Thus, each Li cation can hold up to three hydrogen molecules.

To investigate the temperature effect and capability of Li ions to remain attached to the linker in Li-functionalized MOF-5, Ahuja and coworkers carried out an ab initio molecular dynamics simulations at 20, 50, 100, 200, and 300 K (Blomqvist et al. 2007). The obtained pair distribution function (PDF) shows that Li binds to the linker firmly throughout the temperature range studied, while with hydrogen molecules stay close to Li up to 200 K. It is noteworthy that MOF-5 has a hydrogen uptake of 2.9 wt% at 200 K and 2.0 wt% at 300 K. These are the highest reported uptake under comparable thermodynamic conditions.

To determine the possibility of extending the Li doping to other IRMOF-5, we studied the adsorption of Li cation on IRMOF-5. We have replaced Zn atoms by metals (M = Fe, Co, Ni, and Cu) from the same row of the periodic table. Upon doping with lithium, we found that the linker unit benzene remains

practically unchanged in all the compounds (Venkataramanan et al. 2009). However, a considerable change in the metal–metal distance, metal–linker distance, and bond angles was observed. The metal–metal distance deviation is shorter in the case of Zn, and very high reduction in distance was observed for iron system. Further, the calculated adsorption energies of Li cation for these compounds do not show any regular trend with the metals. These results suggest that Li cation doping cannot be extended to all metal systems and Li doping has a great influence on the structural and volume change on these systems.

2.4.3 Organic Materials as Hydrogen Storage Media

To achieve the U.S. Department of Energy's (DOE) target, the storage materials chosen should consist of light elements like C, B, N, O, Li, and Al. Hence, attempts were made to use pristine carbon nanotube (CNTs), graphene sheets, and BN materials as storage materials (Yang 2003, Niemann et al. 2008, Zhang et al. 2008). However, their interactions with H_2 are very weak. Some theoretical and experimental studies have shown that doping of transition metal (TM) atoms can increase the hydrogen uptake because of the enhanced adsorption energy of H_2 to the metal atoms (Yoo et al. 2004, Sun et al. 2006, Wen et al. 2008). However, later studies indicate that TM atoms tend to form clusters on the surface due to their large cohesive energy (Krasnov et al. 2007). Recently, the use of organic molecules decorated with alkali metal was found to provide high storage capacity with adsorption energy for hydrogen molecule to be in the range of 0.1–0.3 eV (Huang et al. 2008).

In the field of organic chemistry, several compounds exist that can act as a host for organic molecules and gases. Among the compounds, *p-tert*-butylcalix[4]arene (TBC) has very high stability and a high sublimation temperature. The packing mode obtained for the crystals shows the existence of each cup-shaped host molecule facing another host molecule in the adjacent layer in order to form a relatively large intramolecular cavity of 270 Å³ (see Figure

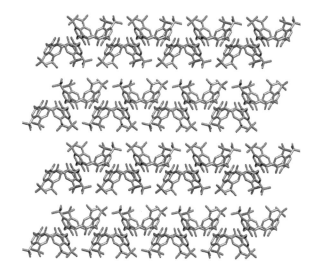

FIGURE 2.24 Packing mode of TBC molecules with inter and intra molecular cavity. Hydrogen atoms are omitted for clarity.

2.24). An intermolecular cavity was found in these crystals that can accommodate gas molecules. In addition, the cavity possesses π-rich character defined by the four aromatic rings, sufficient for storing guest molecules. We have considered only the intramolecular cavity for the storage of H_2 molecules in the TBC functionalized with the Li atom (Venkataramanan et al. 2008).

The structure of TBC was fully optimized without any geometrical constraints. The calculated bond lengths are in good agreement with the experimental values (Enrigh et al. 2003). We first studied the hydrogen molecule uptake for TBC molecules. The first hydrogen molecule stays 4.75 Å from the bottom and center of the four phenoxy units and is oriented parallel to the phenoxy group. Another important feature is that the minimum-energy structure has one *t*-butyl group oriented in the direction of the phenoxy hydrogen. Very recently, a reversible phase transition was detected following the inclusion of gaseous molecules in TBC (Thallapally et al. 2008). Upon doping additional hydrogen, the first hydrogen molecule is pulled inside and resides at a distance of 4.58 Å, and the second hydrogen molecule is partially placed inside the calixarene cavity at a distance of 6.67 Å. The hydrogen molecules in the optimized structures have a bond distance of 0.750 Å, which is the same as that obtained for isolated molecular hydrogen optimized with the PW91 GGA method. This reflects the absence of any interaction between the TBC and hydrogen molecules.

To increase the storage capacity, we doped TBC molecules by Li atoms. Lithium absorption on the TBC can take place at four different sites: by replacing the hydrogen on phenol to form an alkoxy salt (O-Li), as a cation at the center of the four phenoxy groups, and on the walls of the benzene ring, making a Li-benzene π-complex. The preferred position of Li is found to be on the inside wall of benzene in cationic mode, with a binding energy of 0.714 eV that was calculated from the energy difference between the total energy of Li-functionalized calixarene and TBC. The energy of the neutral compound was about 0.04 eV higher in energy, whereas the least stable structure was the one in which the Li atom is bonded to the outside wall of the benzene ring. In our further studies, we consider the structure with the Li atom bonded on the inside wall complex (LTBC) alone because the Li atom doped will be rigid, and the anions can occupy the pore spaces. It is worth to specify here that these anions present at the intermolecular cavity site can also hold H_2 by a charge transfer process. In addition, doping of Li cation at the intermolecular cavity would avoid Li clustering in these systems.

We then studied the interaction of LTBC with one to four hydrogen molecules by introducing them into the cavity. The optimized structures of LTBC with hydrogen molecules are shown in Figure 2.25, and their binding energies and bond lengths for the hydrogen molecules are provided in Table 2.4. The preferred position for the first H_2 molecule is found to be 2.085 Å away from the Li atom at a distance of 4.16 Å from the bottom and the center of the phenoxy unit. The H_2 binding energy per hydrogen molecule was calculated using the expression

$$\mathrm{BE}/H_2 = (E(\mathrm{LTBC} + nH_2) - E(\mathrm{LTBC}) - E(nH_2))/n,$$

where $E(\mathrm{LTBC} + H_2)$ was the total energy of Li-doped TBC, containing n number of hydrogen and $E(\mathrm{LTBC})$ total energy of the Li-doped TBC.

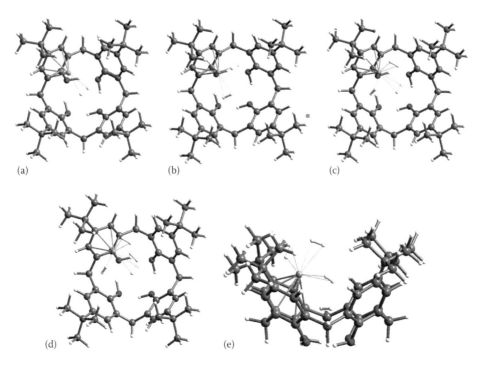

(a)   (b)   (c)

(d)   (e)

FIGURE 2.25   Optimized geometries for the Li-functionalized calixarene (a) on top view of LTBC with one hydrogen molecule, (b) LTBC with two hydrogen molecules, (c) LTBC with three hydrogen molecules, (d) LTBC with four hydrogen molecules, (e) side view of LTBC with four hydrogen molecules. (Reproduced from Venkataramanan, N.S. et al., *J. Phys. Chem. C*, 112, 19676, 2008. With permission.)

TABLE 2.4 Binding Energies (BE) for One to Four Hydrogen Molecules Absorbed on TBC and LTBC and Average H–H Bond Distance

No. of Hydrogen Molecule	TBC		LTBC	
	BE/H$_2$ (eV)	H–H (Å)	BE/H$_2$ (eV)	H–H (Å)
1 H$_2$	0.168	0.750	0.292	0.758
2 H$_2$	0.132	0.750	0.248	0.756
3 H$_2$	—	—	0.232	0.754
4 H$_2$	—	—	0.215	0.751

Source: Reproduced from Venkataramanan, N.S. et al., *J. Phys. Chem. C*, 112, 19676, 2008. With permission.

The binding energy calculated for the hydrogen molecules is provided in Table 2.4. In the ground-state configuration, hydrogen atoms are bound in nearly molecular form with a H–H bond length of 0.758 Å and a hydrogen molecule occupying the position parallel to the Li atom. Upon introduction of the second and third hydrogen molecules, the binding energy per molecule decreased to 0.292 and 0.248 eV for the second and third hydrogen molecules, respectively. Another noticeable feature is the increase in the distance between the Li atom and hydrogen with the addition of successive hydrogen. Upon adding of the fourth hydrogen, one hydrogen molecule was found to move away from the Li atom to a distance of 2.97 Å and was inside the cavity of the LTBC. Thus, the maximum number of hydrogen molecules bound by an Li atom is three, whereas the LTBC can hold four hydrogen atoms inside its cavity, yielding an approximate gravimetric density of hydrogen of 9.52 wt%.

To investigate the stability of Li-functionalized calixarene, the PDF, which is the mean distance between the benzene ring (on which the Li atom is functionalized) and the Li atom, over a temperature range of 20–300 K was calculated using ab initio molecular dynamics. The system was allowed to reach 1500 steps of 1 fs, after which the coordinates were analyzed. Lithium atom stays close to the benzene ring over the entire temperature range. Following this, LTBC stability with hydrogen molecules was also measured by calculating the mean distance between the Li atom and the center of the hydrogen molecules for the system with four H$_2$ molecules inside its cavity. Lithium was in contact with the benzene ring until 200 K. Further increases in temperature resulted in the decomposition of the complex. Therefore, in the case of the Li–H$_2$ system, we have calculated the stability until 200 K. Our simulation results show that the Li–H$_2$ system is stable up to 100 K.

Based on these results, we emphasize that the Li-functionalized calixarene is promising material for hydrogen storage applications. Taking into account both the possibility of calixarene structural versatility, which allowed to reduce the weigh of host framework as comparable with *p-tert*-butylcalix[4]arene and the possibility of the lithium adsorption on different sites, it may be likely to archive the DOE goal for hydrogen storage. However, in order to validate this hypothesis, detailed investigations, especially theoretical ones, are necessary to be carried out in the nearest future.

2.5 Conclusions

In this chapter, the computational modeling of several nanoporous materials, which show promise as hydrogen storage media, was performed. It was shown that in order to achieve the desired hydrogen storage ability of selected inclusion compounds, the computer-aided design is a useful tool. Using powerful computers and highly accurate methods, we can not only understand the physical and chemical properties of already known materials but also try to propose and optimize the way of practical realization of novel compositions by giving important information for experimentalists. Starting from accurate first-principles estimation (for example, all-electron mixed basis approach) of the guest–host interaction, the thermodynamics properties and the hydrogen concentration at various pressures can be evaluated using proposed approach. Moreover, the proposed approach can be also applied to design inclusion compounds that target other molecules, such as nitrogen oxide (NO$_x$) and sulfur oxide (SO$_x$) gas molecules, which pollute the environment, resulting in an improvement of the society in which we all live.

Acknowledgments

The authors would like to express their sincere thanks to the staff of the Center for Computational Materials Science of the Institute for Materials Research, Tohoku University, for their continuous support of the SR11000-K2/51 supercomputing facilities. We also thank Prof. Michael R. Philpott for critically reading the chapter and his valuable comments. This work has been supported by New Energy and Industrial Technology Development Organization (NEDO) under "Advanced Fundamental Research Project on Hydrogen Storage Materials."

References

Alavi, S., Ripmeester, J. A., and Klug, D. D. 2005a. NMR shielding constants for hydrogen guest molecules in structure II clathrates. *J. Chem. Phys.* 123: 024507-1–024507-4.

Alavi, S., Ripmeester, J. A., and Klug, D. D. 2005b. Molecular-dynamics study of structure II hydrogen clathrates. *J. Chem. Phys.* 123: 051107-1–051107-7.

Alavi, S., Ripmeester, J. A., and Klug, D. D. 2006. Molecular-dynamics simulations of binary structure II hydrogen and tetrahydrofurane clathrates. *J. Chem. Phys.* 124: 014704-1–014704-6.

Atwood, J. L., Davies, J. E. D., and MacNicol, D. D. (Eds.) 1984. *Inclusion Compounds*, Vols. 1–3. London, U.K.: Academic Press.

Bahramy, M. S., Sluiter, M. H. F., and Kawazoe, Y. 2006. First-principles calculations of hyperfine parameters with the all-electron mixed-basis method. *Phys. Rev B* 73: 045111-1-045111-21.

Barkalov, O. I., Klyamkin, S. N., Efimchenko, V. S., and Antonov, V. E. 2005. Formation and composition of the clathrate phase in the H$_2$O-H$_2$ system at pressures to 1.8 kbar. *JETP Lett.* 82: 413–415.

Belosludov, V. R., Lavrentiev, M. Y., and Syskin, S. A. 1988. Dynamical properties of the molecular-crystals with electrostatic interaction taken into account—low-pressure ice phases (Ih and Ic). *Phys. Status Solidi (b)* 149: 133–142.

Belosludov, V. R., Lavrentiev, M. Y., Dyadin, Y. A., and Syskin, S. A. 1990. Dynamic and thermodynamic properties of clathrate hydrates. *J. Inc. Phen. Mol. Recog. Chem.* 8: 59–69.

Belosludov, R. V., Igumenov, I. K., Belosludov, V. R., and Shpakov, V. P. 1994. Dynamical and thermodynamical properties of the acetylacetones of copper, aluminum, indium, and rhodium. *Mol. Phys.* 82: 51–66.

Belosludov, V. R., Subbotin, O. S., Krupskii, D. S. et al. 2007. Physical and chemical properties of gas hydrates: Theoretical aspects of energy storage application. *Mater. Trans.* 48: 704–710.

Belosludov, V. R., Subbotin, O. S., Belosludov, R. V., Mizuseki, H., Kawazoe, Y., and Kudoh, J. 2009. Thermodynamics and hydrogen storage ability of binary hydrogen + help gas clathrate hydrate. *Int. J. Nanosci.* 8: 57–63.

Berendsen, H. J. C., Grigera, J. R., and Straatsma, T. P. J. 1987. The missing term in effective pair potentials. *Phys. Chem.* 91: 6269–6271.

Blomqvist, A., Araújo, C. M., Srepusharawoot, P., and Ahuja, R. 2007. Li-decorated metal-organic framework 5: A route to achieving a suitable hydrogen storage medium. *Proc. Natl. Acad. Sci. USA* 104: 20173–20176.

Buda, C. and Dunietz, B. D. 2006. Hydrogen physisorption on the organic linker in metal organic frameworks: Ab initio computational study. *J. Phys. Chem. B* 110: 10479–10484.

Carmichael, M., Chenoweth, K., and Dykstra, C. E. 2004. Hydrogen molecule clusters. *J. Phys. Chem. A* 108: 3143–3152.

Chen, B. L., Ockwig, N. W., Millward, A. R., Contreras, D. S., and Yaghi, O. M. 2005. High H₂ adsorption in a microporous metal-organic framework with open metal sites. *Angew. Chem. Int. Ed.* 44: 4745–4749.

Claussen, W. F. 1951. A 2nd water structure for inert gas hydrates. *J. Chem. Phys.* 19: 1425–1426.

Cramer, F. 1954. *Einschlussverbindungen.* Berlin, Germany: Springer.

Davidson, D. W., Handa, Y. P., Ratcliffe, C. I., and Tse, J. S. 1984. The ability of small molecules to form clathrate hydrates of structure-II. *Nature* 311: 142–143.

Davidson, D. W., Handa, Y. P., Ratcliffe, C. I. et al. 1986. Crystallographic studies of clathrate hydrates. 1. *Mol. Cryst. Liq. Cryst.* 141: 141–149.

Davies, J. E., Kemula, W., Powell, H. M., and Smith, N. 1983. Inclusion compounds—Past, present, and future. *J. Incl. Phen.* 1: 3–44.

Dyadin, Y. A. and Belosludov, V. R. 1996. Stoichiometry and thermodynamics of clathrate hydrate. In *Comprehensive Supramolecular Chemistry*, J.-M. Lehn, J. L. Atwood, J. E. D. Davies, D. D. Macnicol, and F. Vögtle, (Eds.) Vol. 6, pp. 789–824. Oxford, U.K.: Elsevier Science.

Dyadin, Y. A., Larionov, E. G., Mikina, T. V., and Starostina, L. I. 1997a. Clathrate formation in Kr-H₂O and Xe-H₂O systems under pressures up to 15 kbar. *Mendeleev Commun.* 2: 74–76.

Dyadin, Y. A., Larionov, E. G., Mirinski, D. S., Mikina, T. V., and Starostina, L. I. 1997b. Clathrate formation in the Ar-H₂O system under pressures up to 15,000 bar. *Mendeleev Commun.* 1: 32–34.

Dyadin, Y. A., Larionov, E. G., Manakov, A. Y. et al. 1999a. Clathrate hydrates of hydrogen and neon. *Mendeleev Commun.* 5: 209–210.

Dyadin, Y. A., Aladko, E. Y., Manakov Y. A. et al. 1999b. Clathrate formation in water-noble gas (hydrogen) systems at high pressures. *J. Struct. Chem.* 40: 790–795.

Enrigh, G. D., Udachin, K. A., Moudrakovski, I. L., and Ripmeester, J. 2003. Thermally programmable gas storage and release in single crystals of an organic van der Waals host. *J. Am. Chem. Soc.* 125: 9896–9897.

Florusse, L. J., Peters, C. J., Schoonman, J. et al. 2004. Stable low-pressure hydrogen clusters stored in a binary clathrate hydrate. *Science* 306: 469–471.

Frisch, M. J., Del Bene, J. E., Binkley, J. S., and Schaefer III, H. F. 1986. Extensive theoretical-studies of the hydrogen-bonded complexes (H₂O)₂, (H₂O)₂H⁺, (HF)₂, (HF)₂H⁺, F₂H⁻, and (NH₃)₂. *J. Chem. Phys.* 84: 2279–2289.

Frisch, M. J., Trucks, G. W., Schlegel, H. B. et al. 1998. *Gaussian 98, Revision A. 9*, Pittsburgh, PA: Gaussian, Inc.

Frisch, M. J., Trucks, G. W., Schlegel, H. B. et al. 2004. *Gaussian 03, Revision D. 01*, Pittsburgh, PA: Gaussian, Inc.

Gao, Y. and Zeng, X. C. 2007. Ab initio study of hydrogen adsorption on benzenoid linkers in metal-organic framework materials. *J. Phys.: Condens. Matter* 19: 386220.

Graetz, J. 2009. New approaches to hydrogen storage. *Chem. Soc. Rev.* 38: 73–82.

Gupta, R. B. (Ed.) 2008. *Hydrogen Fuel: Production, Transport and Storage.* Boca Raton, FL: CRC Press.

Hamilton, C. W., Baker, R. T., Staubitz, A., and Manners, I. 2009. B-N compounds for chemical hydrogen storage. *Chem. Soc. Rev.* 38: 279–293.

Hordeski, M. F. 2008. *Alternative Fuels—The Future of Hydrogen.* Boca Raton, FL: Taylor & Francis.

Huang, B., Lee, H., Duan, W., and Ihm, J. 2008. Hydrogen storage in alkali-metal-decorated organic molecules. *Appl. Phys. Lett.* 93: 063107-1-063107-3.

Inerbaev, T. M., Belosludov, V. R., Belosludov, R. V., Sluiter, M., Kawazoe, Y., and Kudoh, J. I. 2004. Theoretical study of clathrate hydrates with multiple occupation. *J. Incl. Phen. Macrocycl. Chem.* 48: 55–60.

Inerbaev, T. M., Belosludov, V. R., Belosludov, R. V., Sluiter, M., and Kawazoe, Y. 2006. Dynamics and equation of state of hydrogen clathrate hydrate as a function of cage occupation. *Comput. Mater. Sci.* 36: 229–233.

Itoh, H., Tse, J. S., and Kawamura, K. 2001. The structure and dynamics of doubly occupied Ar hydrate. *J. Chem. Phys.* 115: 9414–9420.

Katsumasa, K., Koga, K., and Tanaka, H. 2007. On the thermodynamic stability of hydrogen clathrate hydrates. *J. Chem. Phys.* 127: 044509-1-044509-7.

Kolmann, S. J., Chan, B., and Jordan, M. J. T. 2008. Modelling the interaction of molecular hydrogen with lithium-doped hydrogen storage materials. *Chem. Phys. Lett.* 467: 126–130.

Kim, D. Y. and Lee, H. 2005. Spectroscopic identification of the mixed hydrogen and carbon dioxide clathrate hydrate. *J. Am. Chem. Soc.* 127: 9996–9997.

Kitagawa, S., Kitaura, R., and Noro, S. 2004. Functional porous coordination polymers. *Angew. Chem. Int. Ed.* 43: 2334–2375.

Krasnov, P. O., Ding, F., Singh, A. K., and Yakobson, B. I. 2007. Clustering of Sc on SWNT and reduction of hydrogen uptake: Ab-initio all-electron calculations. *J. Phys. Chem. C* 111: 17977–17980.

Kuc, A., Heine, T., Seifert, G., and Duarte, H. A. 2008a. On the nature of the interaction between H_2 and metal-organic frameworks. *Theor. Chem. Account.* 120: 543–550.

Kuc, A., Heine, T., Seifert, G., and Duarte, H. A. 2008b. H_2 adsorption in metal-organic frameworks: Dispersion or electrostatic interactions? *Chem. Eur. J.* 14: 6597–6600.

Langmi, H. W. and McGrady, G. S. 2007. Non-hydride systems of the main group elements as hydrogen storage materials. *Coord. Chem. Revs.* 251: 925–935.

Lee, H., Lee, J. W., Kim, D. Y. et al. 2005. Tuning clathrate hydrates for hydrogen storage. *Nature* 434: 743–746.

Lehn, J.-M. 1995. *Supramolecular Chemistry*. Weinheim, Germany: Wiley-VCH.

Leifried, G. and Ludwig, W. 1961. *Theory of Anharmonic Effects in Crystal*. New York: Academic Press.

Liu, Y., Kravtsov, V. C., Larsen, R., and Eddaoudi, M. 2006. Molecular building blocks approach to the assembly of zeolite-like metal-organic frameworks (ZMOFs) with extralarge cavities. *Chem. Commun.* 14: 1488–1490.

Lokshin, K. A. and Zhao, Y. S. 2006. Fast synthesis method and phase diagram of hydrogen clathrate hydrate. *Appl. Phys. Lett.* 88: 131909-1–131909-3.

Lokshin, K. A., Zhao, Y. S., He, D. et al. 2004. Structure and dynamics of hydrogen molecules in the novel clathrate hydrate by high pressure neutron diffraction. *Phys. Rev. Lett.* 93: 125503-1–125503-4.

Lotz, H. T. and Schouten, J. A. 1999. Clatrate hydrates in system H_2O–Ar at pressures and temperature up to 30 kbar and 140°C. *J. Chem. Phys.* 111: 10242–10247.

Lubitz, W. and Tumas, W. 2007. Hydrogen: An overview. *Chem. Rev.* 107: 3900–3903.

Mak, T. C. and McMullan, R. K. 1965. Polyhedral clathrate hydrates. X. Structure of double hydrate of tetrahydrofuran and hydrogen sulfide. *J. Chem. Phys.* 42: 2732–2737.

Manakov, A. Y., Voronin, V. I., Kurnosov, A. V., Teplykh, A. E., Larionov, E. G., and Dyadin, Y. A. 2001. Argon hydrates: Structural studies at high pressures. *Dokl. Phys. Chem.* 378: 148–151.

Manakov, A. Y., Voronin, V. I., Teplykh, A. E. et al. 2002. Structural and spectroscopic investigations of gas hydrates at high pressures. *Proceedings of the Fourth International Conference on Gas Hydrates*, Yokohama, Japan, pp. 630–635.

Mandelcorn, L. (Ed.) 1964. *Nonstoichiometric Compounds*. New York: Academic Press.

Mao, W. L. and Mao, H. K. 2004. Hydrogen storage in molecular compounds. *Proc. Nat. Acad. Sci.* 101: 708–710.

Mao, W. L., Mao, H., Goncharov, A. F. et al. 2002. Hydrogen clusters in clathrate hydrate. *Science* 297: 2247–2249.

Matsuda, R., Kitaura, R., Kitagawa, S. et al. 2005. Highly controlled acetylene accommodation in a metal-organic microporous materials. *Nature* 436: 238–241.

Mavrandonakis, A. and Klopper, W. 2008. First-principles study of single and multiple dihydrogen interaction with lithium containing benzene molecules. *J. Phys. Chem. C* 112: 11580–11585.

Morris, R. E. and Wheatley, P. S. 2008. Gas storage in nanoporous materials. *Angew. Chem. Int. Ed.* 47: 4966–4981.

Mueller, T. and Ceder, G. 2005. A density functional theory study of hydrogen adsorption in MOF-5. *J. Phys. Chem. B* 109: 17974–17983.

Mulder, F. M., Dingemans, T. J., Wagemaker, M., and Kearley, G. J. 2005. Modelling of hydrogen adsorption in the metal organic framework MOF5. *Chem. Phys.* 317: 113–118.

Mulfort, K. L. and Hupp, J. T. 2007. Chemical reduction of metal-organic framework materials as a method to enhance gas uptake and binding. *J. Am. Chem. Soc.* 129: 9604–9605.

Mulfort, K. L. and Hupp, J. T. 2008. Alkali metal cation effects on hydrogen uptake and binding in metal-organic frameworks. *Inorg. Chem.* 47: 7936–7938.

Mulfort, K. L., Wilson, T. M., Wasielewski, M. R., and Hupp, J. T. 2009. Framework reduction and alkali-metal doping of a triply catenating metal-organic framework enhances and then diminishes H_2 uptake. *Langmuir* 25: 503–508.

Murata, K., Kaneko, K., Kanoh, H. et al. 2002. Adsorption mechanism of supercritical hydrogen in internal and interstitial nanospaces of single-wall carbon nanohorn assembly. *J. Phys. Chem. B* 106: 11132–11138.

Niemann, M. U., Srinivasan, S. S., Phani, A. R., Kumar, A., Goswami, D. Y., and Stefanakos, E. K. 2008. Nanomaterials for hydrogen storage applications: A review. *J. Nanomater.* 2008: 950967.

Niu, J., Rao, B. K., and Jena, P. 1992. Binding of hydrogen molecules by a transition-metal ion. *Phys. Rev. Lett.* 68: 2277–2280.

Orimo, S. I., Nakamori, Y., Eliseo, J. R., Zuttel, A., and Jensen, C. M. 2007. Complex hydrides for hydrogen storage. *Chem. Rev.* 107: 4111–4132.

Patchkovskii, S. and Tse, J. S. 2003. Thermodynamic stability of hydrogen clathrates. *Proc. Natl. Acad. Sci.* 100: 14645–14650.

Patchkovskii, S. and Yurchenko, S. N. 2004. Quantum and classical equilibrium properties for exactly solvable models of weakly interacting systems. *Phys. Chem. Chem. Phys.* 6: 4152–4155.

Pauling, L. and Marsh, R. E. 1952. The structure of chlorine hydrate. *Proc. Natl. Acad. Sci. USA* 38: 112–118.

Peng, C., Ayala, P. Y., Schlegel, H. B., and Frisch, M. J. 1996. Using redundant internal coordinates to optimize equilibrium geometries and transition states. *J. Comp. Chem.* 17: 49–56.

Ripmeester, J. A., Tse, J. S., Ratcliffe, C. I., and Powell, B. M. 1987. A new clathrate hydrate structure. *Nature* 325: 135–136.

Rosi, N. L., Eckert, J., Eddaoudi, M. et al. 2003. Hydrogen storage in microporous metal-organic frameworks. *Science* 300: 1127–1129.

Ross, M. and Ree, F. H. 1980. Repulsive forces of simple molecules and mixtures at high-density and temperature. *J. Chem. Phys.* 73: 6146–6152.

Rowsell, J. L. C. and Yaghi, O. M. 2004. Metal-organic frameworks: A new class of porous materials. *Micropor. Mesopor. Mater.* 73: 3–14.

Rowsell, J. L. C. and Yaghi, O. M. 2005. Strategies for hydrogen storage in metal-organic frameworks. *Angew. Chem. Int. Ed.* 44: 4670–4679.

Rowsell, J. L. C., Millward, A. R., Park, K. S., and Yaghi, O. M. 2004. Hydrogen sorption in functionalized metal-organic frameworks. *J. Am. Chem. Soc.* 126: 5666–5667.

Rowsell, J. L. C., Eckert, J., and Yaghi, O. M. 2005. Characterization of H_2 binding sites in prototypical metal-organic frameworks by inelastic neutron scattering. *J. Am. Chem. Soc.* 127: 14904–14910.

Sakintuna, B., Lamari-Darkrim, F., and Hirscher, M. 2007. Metal hydride materials for solid hydrogen storage: A review. *Int. J. Hydrog. Energy* 32: 1121–1140.

Schlapbach, L. and Züttel, A. 2004. Hydrogen-storage materials for mobile applications. *Nature* 414: 353–358.

Sloan, E. D. and Koh, C. A. 2007. *Clathrate Hydrates of Natural Gases*, 3rd edn. Boca Raton, FL: Taylor & Francis.

Sluiter, M. H. F., Belosludov, R. V., Jain, A. et al. 2003. Ab initio study of hydrogen hydrate clathrates for hydrogen storage within the ITBL environment. *Lect. Notes Comput. Sci.* 2858: 330–341.

Sluiter, M. H. F., Adachi, H., Belosludov, R. V., Belosludov, V. R., and Kawazoe, Y. 2004. Ab initio study of hydrogen storage in hydrogen hydrate clathrates. *Mater. Trans.* 45: 1452–1454.

Struzhkin, V. V., Militzer, B., Mao, W. L., Mao, H. K., and Hemley, R. J. 2007. Hydrogen storage in molecular clathrates. *Chem. Rev.* 107: 4133–4151.

Sun, Q., Jena, P., Wang, Q., and Marquez, M. 2006. First-principles study of hydrogen storage on $Li_{12}C_{60}$. *J. Am. Chem. Soc.* 128: 9741–9745.

Tanaka, H., Nakatsuka, T., and Koga, K. 2004. On the thermodynamic stability of clathrate hydrates IV: Double occupancy of cages. *J. Chem. Phys.* 121: 5488–5493.

Thallapally, P. K., McGrail, B. P., Dalgarno, S. J., Schaef, H. T., Tian, J., and Atwood, J. L. 2008. Gas-induced transformation and expansion of a non-porous organic solid. *Nat. Mater.* 7: 146–150.

Tse, J. S. 1994. Dynamical properties and stability of clathrate hydrates. *Ann. N.Y. Acad. Sci.* 715: 187–206.

Udachin, K. A., Ratcliffe, C. I., Enright, G. D., and Ripmeester, J. A. 1997. Structure H hydrate: A single crystal diffraction study of 2,2-dimethylpentane center dot 5(Xe,H_2S)center dot 34H_2O. *Supramol. Chem.* 8: 173–176.

Varin, R. A., Czujko, T., and Wronski, Z. S. 2009. *Nanomaterials for Solid State Hydrogen Storage*. New York: Springer.

van der Waals, J. H. and Platteeuw, J. C. 1959. Clathrate solutions. *Adv. Chem. Phys.* 2: 1–57.

Venkataramanan, N. S., Sahara, R., Mizuseki, H., and Kawazoe, Y. 2008. Hydrogen adsorption on lithium-functionalized calixarenes: A computational study. *J. Phys. Chem. C* 112: 19676–19679.

Venkataramanan, N. S., Sahara, R., Mizuseki, H., and Kawazoe, Y. 2009. Probing the structure, stability and hydrogen adsorption of lithium functionalized isoreticular MOF-5 (Fe, Cu, Co, Ni and Zn) by density functional theory. *Int. J. Mol. Sci.* 10: 1601–1608.

Von Stackelberg, M. and Miller, H. R. 1954. Feste gas hydrate. II. Structure und raumchemie. *Z. Elektrochem.* 58: 25–28.

Vögtle, F. 1991. *Supramolecular Chemistry: An Introduction.* Chichester, U.K.: Wiley.

Weber, E. 1995. Inclusion compounds. In *Kirk-Othmer Encyclopedia of Chemical Technology*, 4th edn., Vol. 14, J. L. Kroschwitz, (Ed.) New York: Wiley.

Wen, S. H., Deng, W. Q., and Han, K. L. 2008. Endohedral BN metallofullerene M@$B_{36}N_{36}$ complex as promising hydrogen storage materials. *J. Phys. Chem. C* 112: 12195–12200.

Wong-Foy, A. G., Matzger, A. J., and Yaghi, O. M. 2006. Exceptional H_2 saturation uptake in microporous metal-organic frameworks. *J. Am. Chem. Soc.* 128: 3494–3495.

Yaghi, O. M., O'keeffe, M., Ockwig, N. W., Chae, H. K., Eddaoudi, M., and Kim, J. 2003. Reticular synthesis and the design of new materials. *Nature* 423: 705–714.

Yang, R. T. 2003. *Adsorbents: Fundamentals and Applications.* Hoboken, NJ: Wiley-Interscience.

Yang, S., Lin, X., Blake, A. J. et al. 2008. Enhancement of H_2 adsorption in Li^+-exchanged co-ordination framework materials. *Chem. Commun.* 46: 6108–6110.

Yoo, E., Gao, L., Komatsu, T. et al. 2004. Atomic hydrogen storage in carbon nanotubes promoted by metal catalysts. *J. Phys. Chem. B* 108: 18903–18907.

Zhang, T., Mubeen, S., Myung, N. V., and Deshusses, M. A. 2008. Recent progress in carbon nanotube-based gas sensors. *Nanotechnology* 19: 332001.

Züttel, A., Borgschulte, A., and Schlapbach, L. (Ed.) 2008. *Hydrogen as Future Energy Carrier.* Weinheim, Germany: Wiley-VCH.

3

Tools for Predicting the Properties of Nanomaterials

James R. Chelikowsky
University of Texas at Austin

3.1 Introduction

Materials at the nanoscale have been the subject of intensive study owing to their unusual electrical, magnetic, and optical properties [1–9]. The combination of new synthesis techniques offers unprecedented opportunities to tailor systems without resort to changing their chemical makeup. In particular, the physical properties of the material can be modified by confinement. If we confine a material in three, two, or one dimensions, we create a nanocrystal, a nanowire, or a nanofilm, respectively. Such confined systems often possess dramatically different properties than their macroscopic counterparts. As an example, consider the optical properties of the semiconductor cadmium selenide. By creating nanocrystals of different sizes, typically ~5 nm in diameter, the entire optical spectrum can be spanned. Likewise, silicon can be changed from an optically inactive material at the macroscopic scale to an optically active nanocrystal.

The physical confinement of a material changes its properties when the confinement length is comparable to the quantum length scale. This can easily be seen by considering the uncertainty principle. At best, the uncertainty of a particle's momentum, Δp, and its position, Δx, must be such that $\Delta p \Delta x \sim h$, where h is Planck's constant divided by 2π. For a particle in a box of length scale, Δx, the kinetic energy of the particle will scale as $\sim 1/\Delta x^2$ and rapidly increase for small values of Δx. If the electron confinement energy becomes comparable to the total energy of the electron, then its physical properties will clearly change, i.e., once the confining dimension approaches the delocalization length of an electron in a solid, "quantum confinement" occurs.

We can restate this notion in a similar context by estimating a confinement length, which can vary from solid to solid, and by considering the wave properties of the electron. For example, the de Broglie wavelength of an electron is given by $\lambda = h/p$. If λ is in the order of Δx, then we expect confinement and quantum effects to be important. Of course, this is essentially the same criterion from the uncertainty principle as $\lambda \sim \Delta x$ would give $\Delta p \Delta x \sim h$ with $p \sim \Delta p$. For a simple metal, we can estimate the momentum from a free electron gas: $p = \hbar k_f$. k_f is the wave vector such that the Fermi energy, $E_f = \hbar^2 k_f^2 / 2m$. This yields $k_f = (3\pi^2 n)^{1/3}$ where n is the electron density. The value of k_f for a typical simple metal such as Na is $\sim 10^{10}$ m^{-1}. This would give a value of $\lambda = h/p = 2\pi/k_f$ or roughly ~1 nm. We would expect quantum confinement to occur on a length scale of a nanometer for this example.

To observe the role of quantum confinement in real materials, we need to be able to construct materials routinely on the nanometer scale. This is the case for many materials. Nanocrystals of many materials can be made, although sometimes it is difficult to determine whether crystallinity is preserved. Such nanostructures provide a unique opportunity to study the properties at nanometer scales and to reveal the underlying physics occurring at reduced dimensionality. From a practical point of view, nanostructures are promising building blocks in nanotechnologies, e.g., the smallest nominal length scale in a modern CPU chip is in the order of 100 nm or less. At length scales less than this, the "band structure" of a material may no longer appear to be quasi-continuous. Rather the electronic energy levels may be described by discrete quantum energy levels, which change with the size of the system.

An understanding of the physics of confinement is necessary to provide the fundamental science for the development of nano optical, magnetic, and electronic device applications. This understanding can best be obtained by utilizing *ab initio* approaches [10]. These approaches can provide valuable insights into nanoscale phenomena without empirical parameters or adjustments extrapolated from bulk properties [11,12]. As recently as 15 years ago, it was declared that *ab initio* approaches would not be useful to systems with more than a hundred atoms or so [11,12]. Of course, hardware advances have occurred since the mid-1990s, but more significant advances have occurred in the area of algorithms and new ideas. These ideas have allowed one to progress at a much faster rate than suggested by Moore's law. We will review some of these advances and illustrate their application to nanocrystals. We will focus on two examples: a silicon nanocrystal and an iron nanocrystal. These two examples will illustrate the behavior of quantum confinement on the optical gap in a semiconductor and on the magnetic moment of a ferromagnetic metal.

3.2 The Quantum Problem

The spatial and energetic distributions of electrons with the quantum theory of materials can be described by a solution of the Kohn–Sham eigenvalue equation [13], which can be justified using density functional theory [13,14]:

$$\left[\frac{-\hbar^2 \nabla^2}{2m} + V_{\text{ext}}(\vec{r}) + V_{\text{H}}(\vec{r}) + V_{\text{xc}}(\vec{r}) \right] \Psi_n(\vec{r}) = E_n \Psi_n(\vec{r}) \quad (3.1)$$

where m is the electron mass. The eigenvalues correspond to energy levels, E_n, and the eigenfunctions or wave functions are given by Ψ_n; a solution of the Kohn–Sham equation gives the energetic and spatial distribution of the electrons. The external potential, V_{ext}, is a potential that does not depend on the electronic solution. The external potential can be taken as a linear superposition of atomic potentials corresponding to the Coulomb potential produced by the nuclear charge. In the case of an isolated hydrogen atom, $V_{\text{ext}} = -e^2/r$. The potential arising from the electron–electron interactions can be divided into two parts. One part represents the "classical" electrostatic terms and is called the "Hartree" or "Coulomb" potential:

$$\nabla^2 V_{\text{H}}(\vec{r}) = -4\pi e \rho(\vec{r}) \quad (3.2)$$

where ρ is the electron charge density; it is obtained by summing up the square of the occupied eigenfunctions and gives the probability of finding an electron at the point $\vec{r}$

$$\rho(\vec{r}) = e \sum_{n,\text{occup}} \left| \Psi_n(\vec{r}) \right|^2 \quad (3.3)$$

The second part of the screening potential, the "exchange-correlation" part of the potential, V_{xc}, is quantum mechanical in nature and effectively contains the physics of the Pauli exclusion principle. A common approximation for this part of the potential arises from the *local density approximation*, i.e.,

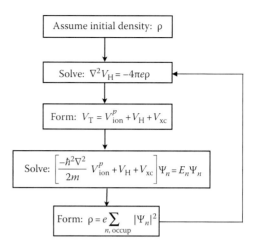

FIGURE 3.1 Self-consistent field loop. The loop is repeated until the "input" and "output" charge densities are equal to within some specified tolerance.

the potential depends only on the charge density at the point of interest, $V_{\text{xc}}(\vec{r}) = V_{\text{xc}}[\rho(\vec{r})]$.

In principle, the density functional theory is exact, provided one is given an exact functional for V_{xc}. This is an outstanding research problem. It is commonly assumed that the functional extracted for a homogeneous electron gas [15] is "universal" and can be approximated by resort to the inhomogeneous gas problem. The procedure for generating a self-consistent field (SCF) potential is given in Figure 3.1. The SCF cycle is initiated with a potential constructed by a superposition of atomic densities for a nanostructure of interest. (Charge densities are easy to obtain for an atom. Under the assumption of a spherically symmetric atom, the Kohn–Sham equation becomes one dimensional and can be solved by doing a radial integration.) The atomic densities are used to solve a Poisson equation for the Hartree potential, and a density functional is used to obtain the exchange-correlation potential. A screening potential composed of the Hartree and exchange-correlation potentials is then added to the fixed external potential, after which the Kohn–Sham equation is solved. The resulting wave functions from this solution are then employed to construct a new potential and the cycle is repeated. In practice, the "output" and "input" potentials are mixed using a scheme that accounts for the history of the previous iterations [16,17].

This procedure is difficult because the eigenvalues can span a large range of energies and the corresponding eigenfunctions span disparate length scales. Consider a heavy element such as Pb. Electrons in the 1*s* state of Pb possess relativistic energies and are strongly localized around the nucleus. In contrast, the Pb 6*s* electrons are loosely bound and delocalized. Attempting to describe the energies and wave functions for these states is not trivial and cannot easily be accomplished using simple basis functions such as plane waves. Moreover, the tightly bound core electrons in atoms are not chemically active and can be removed from the Kohn–Sham equation without significant loss of accuracy by using the *pseudopotential* model of materials.

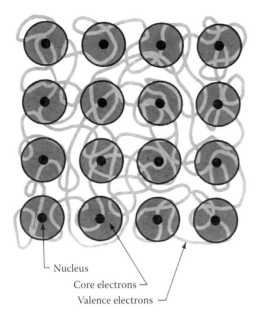

Nucleus

Core electrons

Valence electrons

FIGURE 3.2 Pseudopotential model of a solid.

The pseudopotential model is quite general and reflects the physical content of the periodic table. In Figure 3.2, the pseudopotential model is illustrated for a crystal. In the pseudopotential model of a material, the electron states are decomposed into core states and valence states, e.g., in silicon the $1s^2 2s^2 2p^6$ states represent the core states, and the $3s^2 3p^2$ states represent the valence states. The pseudopotential represents the potential arising from a combination of core states and the nuclear charge: the so-called ion-core pseudopotential. The ion-core pseudopotential is assumed to be completely transferable from the atom to a cluster or to a nanostructure.

By replacing the external potential in the Kohn–Sham equation with an ion-core pseudopotential, we can avoid considering the core states altogether. The solution of the Kohn–Sham equation using pseudopotentials will yield only the valence states. The energy and length scales are then set by the valence states; it becomes no more difficult to solve for the electronic states of a heavy element such as Pb when compared to a light element such as C.

3.2.1 Constructing Pseudopotentials from Density Functional Theory

Here we will focus on recipes for creating ion-core pseudopotentials within the density functional theory, although pseudopotentials can also be constructed from experimental data [18]. The construction of ion-core pseudopotentials has become an active area of electronic structure theory. Methods for constructing such potentials have centered on *ab initio* or "first-principles" pseudopotentials; i.e., the informational content on which the pseudopotential is based does not involve any experimental input.

The first step in the construction process is to consider an electronic structure calculation for a free *atom*. For example, in the case of a silicon atom the Kohn–Sham equation [13] can be solved for the eigenvalues and wave functions. Knowing the

valence wave functions, i.e., $3s^2$ and $3p^2$, states and corresponding eigenvalues, the pseudo wave functions can be constructed. Solving the Kohn–Sham problem for an atom is an easy numerical calculation as the atomic densities are assumed to possess spherical symmetry and the problem reduces to a one-dimensional radial integration. Once we know the solution for an "all-electron" potential, we can *invert* the Kohn–Sham equation and find the total pseudopotential. We can "unscreen" the total potential and extract the ion-core pseudopotential. This ion-core potential, which arises from tightly bound core electrons and the nuclear charge, is not expected to change from one environment to another. The issue of this "transferability" is one that must be addressed according to the system of interest. The immediate issue is how to define pseudo-wave functions that can be used to define the corresponding pseudopotential.

Suppose we insist that the pseudo-wave function be identical to the all-electron wave function outside of the core region. For example, let us consider the 3s state for a silicon atom. We want the pseudo-wave function to be identical to the all-electron state outside the core region:

$$\phi_{3s}^p(r) = \psi_{3s}(r) \quad r > r_c \tag{3.4}$$

where

ϕ_{3s}^p is a pseudo-wave function for the 3s state

r_c defines the core size

This assignment will guarantee that the pseudo-wave function will possess properties identical to the all-electron wave function, ψ_{3s}, in the region away from the ion core.

For $r < r_c$, we alter the all-electron wave function. We are free to do this as we do not expect the valence wave function within the core region to alter the chemical properties of the system. We choose to make the pseudo-wave function smooth and nodeless in the core region. This initiative will provide rapid convergence with simple basis functions. One other criterion is mandated. Namely, the integral of the pseudocharge density within the core should be equal to the integral of the all-electron charge density. Without this condition, the pseudo-wave function differs by a scaling factor from the all-electron wave function. Pseudopotentials constructed with this constraint are called "norm conserving" [19]. Since we expect the bonding in a solid to be highly dependent on the tails of the valence wave functions, it is imperative that the normalized pseudo-wave function be identical to the all-electron wave functions.

There are many ways of constructing "norm-conserving" pseudopotentials as within the core the pseudo-wave function is not unique. One of the most straightforward construction procedures is from Kerker [20] and was later extended by Troullier and Martins [21].

$$\phi_l^p(r) = \begin{cases} r^l \exp(p(r)) & r \le r_c \\ \psi_l(r) & r > r_c \end{cases} \tag{3.5}$$

$p(r)$ is taken to be a polynomial of the form

$$p(r) = c_0 + \sum_{n=1}^{6} c_{2n} r^{2n} \tag{3.6}$$

This form assures us that the pseudo-wave function is nodeless and by taking even powers there is no cusp associated with the pseudo-wave function. The parameters, c_{2n}, are fixed by the following criteria: (a) The all-electron and pseudo-wave functions have the same valence eigenvalue. (b) The pseudo-wave function is nodeless and be identical to the all-electron wave function for $r > r_c$. (c) The pseudo-wave function must be continuous as well as the first four derivatives at r_c. (d) The pseudopotential has zero curvature at the origin. This construction is easy to implement and extend to include other constraints. An example of an atomic pseudo-wave function for Si is given in Figure 3.3 where it is compared to an all-electron wave function. Unlike the $3s$ all-electron wave function, the pseudo-wave function is nodeless. The pseudo-wave function is much easier to express as a Fourier transform or a combination of Gaussian orbitals than the all-electron wave function.

Once the pseudo-wave function is constructed, the Kohn–Sham equation can be inverted to arrive at the ion-core pseudopotential

$$V_{\text{ion},l}^{p}(r) = \frac{\hbar^2 \nabla^2 \phi_l^{p}}{2m \phi_l^{p}} - E_{n,l} - V_{\text{H}}(r) - V_{\text{xc}}\left[\rho(r)\right] \tag{3.7}$$

The ion-core pseudopotential is well behaved as ϕ_l^{p} has no nodes; however, the resulting ion-core pseudopotential is both state dependent and energy dependent. The energy dependence is usually weak. For example, the $4s$ state in silicon computed by the pseudopotential constructed from the $3s$ state is usually accurate.

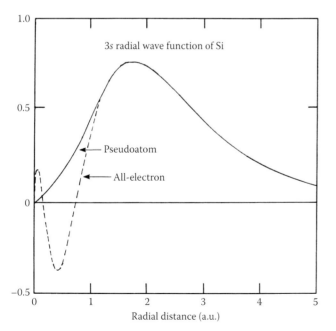

FIGURE 3.3 An all-electron and a pseudo-wave function for the silicon $3s$ radial wave function.

Physically this happens because the $4s$ state is extended and experiences the potential in a region where the ion-core potential has assumed a simple $-Z_v e^2/r$ behavior where Z_v is the number of valence electrons. However, the state dependence through l is an issue, the difference between a potential generated via a $3s$ state and a $3p$ can be an issue. In particular, for first-row elements such as C or O, the nonlocality is quite large as there are no p states within the core region. For the first-row transition elements such as Fe or Cu, this is also an issue as again there are no d-states within the core. This state dependence complicates the use of pseudopotentials. The state dependence or the nonlocal character of the ion-core pseudopotential for an atom can be expressed as

$$V_{\text{ion}}^{p}(\vec{r}) = \sum_{l=0}^{\infty} \mathcal{P}_l^{\dagger} V_{l,\text{ion}}^{p}(r)\mathcal{P}_l = \mathcal{P}_s^{\dagger} V_{s,\text{ion}}^{p}(r)\mathcal{P}_s$$
$$+ \mathcal{P}_p^{\dagger} V_{p,\text{ion}}^{p}(r)\mathcal{P}_p + \mathcal{P}_d^{\dagger} V_{d,\text{ion}}^{p}(r)\mathcal{P}_d + \cdots \tag{3.8}$$

$\mathcal{P}_l$ is an operator that projects out the lth-component. The ion-core pseudopotential is often termed "semi-local" as the potential is radial local, but possesses an angular dependence that is not local.

An additional advantage of the norm-conserving potential concerns the logarithmic derivative of the pseudo-wave function [22]. An identity exists:

$$-2\pi\left((r\phi)^2 \frac{d^2 \ln \phi}{dE\, dr}\right)_R = 4\pi \int_0^R \phi^2 r^2 dr = Q(R) \tag{3.9}$$

The energy derivative of the logarithmic derivative of the pseudo-wave function is fixed by the amount of charge within a radius, R. The radial derivative of the wave function, ϕ, is related to the scattering phase shift from elementary quantum mechanics. For a norm-conserving pseudopotential, the scattering phase shift at $R = r_c$ and at the eigenvalue of interest is identical to the all-electron case as $Q_{\text{all elect}}(r_c) = Q_{\text{pseudo}}(r_c)$. The scattering properties of the pseudopotential and the all-electron potential have the same energy variation to first order when transferred to other systems.

There is some flexibility in constructing pseudopotentials; the pseudo-wave functions are not unique. This aspect of the pseudo-wave function was recognized early in its inception, i.e., there are a variety of ways to construct the wave functions in the core. The non-uniqueness of the pseudo-wave function and the pseudopotential can be exploited to optimize the convergence of the pseudopotentials for the basis of interest. Much effort has been made to construct "soft" pseudopotentials. By "soft," one means a "rapidly" convergent calculation using plane waves as a basis. Typically, soft potentials are characterized by a "large" core size, i.e., a larger value for r_c. However, as the core becomes larger, the "goodness" of the pseudo-wave function can be compromised as the transferability of the pseudopotential becomes more limited.

A schematic illustration of the difference between an "all-electron" potential and a pseudopotential is given in Figure 3.4.

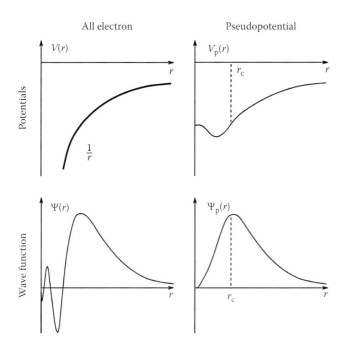

All electron | Pseudopotential

FIGURE 3.4 Schematic all-electron potential and pseudopotential. Outside of the core radius, r_c, the potentials and wave functions are identical.

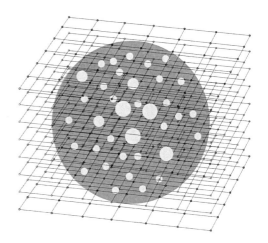

FIGURE 3.5 Real space geometry for a confined system. The grid is uniform and the wave function is taken to vanish outside the domain of interest.

$$\left(\frac{\partial^2 \psi}{\partial x^2}\right)_{x_0} \approx \sum_{n=-N}^{N} A_n \psi(x_0 + nh, y, z), \qquad (3.10)$$

where

    h is the grid spacing

    N is the number of nearest grid points

    A_n are the coefficients for evaluating the required derivatives [36]

The error scales as $O(h^{2N+2})$. Typically, $N \approx 6 - 8$ is used and there is a trade off between using a higher value for N and a coarser grid, or a smaller value for N and a finer grid. Because pseudopotentials are used in this implementation, the wave function structure is "smooth" and a higher finite difference expression for the kinetic energy converges quickly for a fine grid.

In real space, we can easily incorporate the nonlocal nature of the ion-core pseudopotential [24]. The Kleinman–Bylander form [37] can be expressed in real space as

$$\Delta V_l^{\mathrm{KB}}(x,y,z)\phi_l^p(x,y,z) = \sum_{lm} G_{lm} u_{lm}^p(x,y,z)\Delta V_l(x,y,z)$$

$$G_{lm} = \frac{\int u_{lm}^p \Delta V_l \phi_l^p \, dx\, dy\, dz}{\int u_{lm}^p \Delta V_l u_{lm}^p \, dx\, dy\, dz} \qquad (3.11)$$

where u_{lm}^p are the reference atomic pseudo-wave functions. The nonlocal nature of the pseudopotential is apparent from the definition of G_{lm}, the value of these coefficients are dependent on the pseudo-wave function, ϕ_l^p, acted on by the operator ΔV_l. This is very similar in spirit to the pseudopotential defined by Phillips and Kleinman [38].

Once the secular equation is created, the eigenvalue problem can be solved using iterative methods [32,39,40]. Typically, a method such as a preconditioned Davidson method can be used [32]. This is a robust and efficient method, which never requires one to store explicitly the Hamiltonian matrix.

3.2.2 Algorithms for Solving the Kohn–Sham Equation

The Kohn–Sham equation as cast in Equation 3.1 can be solved using a variety of techniques. Often the wave functions can be expanded in a basis such as plane waves or Gaussians and the resulting secular equations can be solved using standard diagonalization packages such as those found in the widely used code: VASP [23]. VASP is a particularly robust code, with a wide following, but it was not constructed with a parallel computing environment in mind.

Here we focus on a different approach that is particularly targeted at highly parallel computing platforms. We solve the Kohn–Sham equation without resort to an explicit basis [24–32]. We solve for the wave functions on a uniform grid within a fixed domain. The wave functions outside of the domain are required to vanish for confined systems or we can assume periodic boundary conditions for systems with translational symmetry [33–35]. In contrast to methods employing an explicit basis, such boundary conditions are easily incorporated. In particular, real space methods do not require the use of supercells for localized systems. As such, charged systems can easily be examined without considering any electrostatic divergences. The problem is typically solved on a uniform grid as indicated in Figure 3.5.

Within a "real space" approach, one can solve the eigenvalue problem using a finite element or finite difference approach [24–32]. We use a higher order finite difference approach owing to its simplicity in implementation. The Laplacian operator can be expressed using

Recent work avoids an explicit diagonalization and instead improves the wave functions by filtering approximate wave functions using a damped *Chebyshev polynomial filtered subspace iteration* [32]. In this approach, only the initial iteration necessitates solving an eigenvalue problem, which can be handled by means of any efficient eigensolver. This step is used to provide a good initial subspace (or good initial approximation to the wave functions). Because the subspace dimension is slightly larger than the number of wanted eigenvalues, the method does not utilize as much memory as standard restarted eigensolvers such as ARPACK and TRLan (Thick—Restart, Lanczos) [41,42]. Moreover, the cost of orthogonalization is much reduced as the filtering approach only requires a subspace with dimension slightly larger than the number of occupied states and orthogonalization is performed only once per SCF iteration. In contrast, standard eigensolvers using restart usually require a subspace at least twice as large and the orthogonalization and other costs related to updating the eigenvectors are much higher.

The essential idea of the filtering method is to start with an approximate initial eigenbasis, $\{\psi_n\}$, corresponding to occupied states of the initial Hamiltonian, and then to improve adaptively the subspace by polynomial filtering. That is, at a given self-consistent step, a polynomial filter, $P_m(t)$, of order m is constructed for the current Hamiltonian $\mathcal{H}$. As the eigen-basis is updated, the polynomial will be different at each SCF step since $\mathcal{H}$ will change. The goal of the filter is to make the subspace spanned by $\{\hat{\psi}_n\} = P_m(\mathcal{H})\{\psi_n\}$ approximate the eigen subspace corresponding to the occupied states of $\mathcal{H}$. There is no need to make the new subspace, $\{\hat{\psi}_n\}$, approximate the wanted eigen subspace of $\mathcal{H}$ to high accuracy at intermediate steps. Instead, the filtering is designed so that the new subspace obtained at each self-consistent iteration step will progressively approximate the wanted eigen space of the final Hamiltonian when self-consistency is reached.

This can be efficiently achieved by exploiting the Chebyshev polynomials, C_m, for the polynomials P_m. In principle, any set of polynomials would work where the value of the polynomial is large over the interval of interest and damped elsewhere. Specifically, we wish to exploit the fast growth property of Chebyshev polynomials outside of the [−1, 1] interval. All that is required to obtain a good filter at a given SCF step, is to provide a lower bound and an upper bound of an interval of the spectrum of the current Hamiltonian $\mathcal{H}$. The lower bound can be readily obtained from the Ritz values computed from the previous step, and the upper bound can be inexpensively obtained by a very small number of (e.g., 4 or 5) Lanczos steps [32]. The main cost of the filtering at each iteration is in performing the products of the polynomial of the Hamiltonian by the basis vectors; this operation can be simplified by utilizing recursion relations. To construct a "damped" Chebyshev polynomial on the interval [a, b] to the interval [−1, 1], one can use an affine mapping such that

$$l(t) = \frac{t - (a+b)/2}{(b-a)/2}.$$ (3.12)

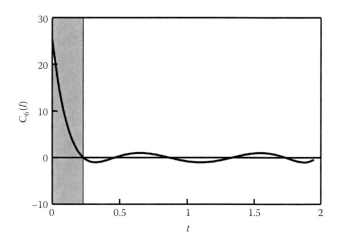

FIGURE 3.6 A damped Chebyshev polynomial, C_6. The shaded area corresponds to eigenvalue spectrum regime that will be enhanced by the filtering operation (see text).

The interval is chosen to encompass the energy interval containing the eigen space to be filtered. The filtering operation can then be expressed as

$$\{\hat{\psi}_n\} = C_m(l(\mathcal{H}))\{\psi_n\}.$$ (3.13)

This computation is accomplished by exploiting the convenient three-term recursion property of Chebyshev polynomials:

$$
\begin{aligned}
C_0(t) &= 1, \quad C_1(t) = t, \\
C_{m+1}(t) &= 2t\,C_m(t) - C_{m-1}(t)
\end{aligned}
$$ (3.14)

An example of a damped Chebyshev polynomial as defined by Equations 3.12 and 3.14 is given in Figure 3.6 where we have taken the lower bound as $a = 0.2$ and the upper bound as $b = 2$. In this schematic example, the filtering would enhance the eigenvalue components in the shaded region.

The filtering procedure for the self-consistent cycle is illustrated in Figure 3.7. Unlike traditional methods, the cycle only requires one explicit diagonalization step. Instead of repeating the diagonalization step within the self-consistent loop, a filtering operation is used to create a new basis in which the desired eigen subspace is enhanced. After the new basis, $\{\hat{\psi}_n\}$, is formed, the basis is orthogonalized. The orthogonalization step scales as the cube of the number of occupied states and as such this method is not an "order-n" method. However, the prefactor is sufficiently small that the method is much faster than previous implementations of real space methods [32]. The cycle is repeated until the "input" and "output" density is unchanged within some specified tolerance, e.g., the eigenvalues must not change by ~0.001 eV, or the like.

In Table 3.1, we compare the timings using the Chebyshev filtering method along with explicit diagonalization solvers using the TRLan and the ARPACK. These timings are for a modest-sized nanocrystal: $Si_{525}H_{276}$. The Hamiltonian size is 292,584 × 292,584 and 1194 eigenvalues were determined. The numerical runs

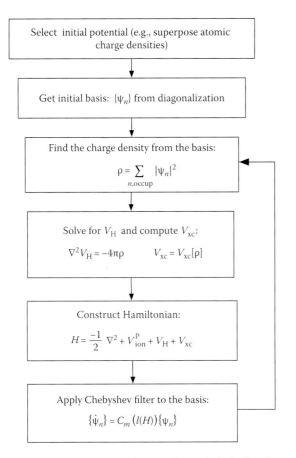

FIGURE 3.7 Self-consistent cycle using damped Chebyshev filtering. Atomic units ($e = \hbar = m$) are used here.

TABLE 3.1 Comparison of Computational Timings for Various Methods for a Nanocrystal: $Si_{525}H_{276}$

Method	SCF Its	CPU(s)
Filtering	11	5,947
ARPACK	10	62,026
TRLan	10	26,853

Note: While the number of SCF iterations is comparable for all three methods, the total time with filtering methods can be dramatically reduced.

were performed on the SGI Altix 3700 cluster at the Minnesota Supercomputing Institute. The CPU type is a 1.3 GHz Intel Madison processor. Although the number of matrix-vector products and SCF iterations is similar, the total time with filtering is over an order of magnitude faster compared to ARPACK and a factor of better than four versus the TRLan. The scaling of the algorithm with the number of processors is shown in Figure 3.8. Such improved timings are not limited to this particular example. Our focus here is on silicon nanocrystals, our method is not limited to semiconducting or insulating systems, we have also used this method to examine metallic systems such as liquid lead [43] and magnetic systems such as iron clusters [8].

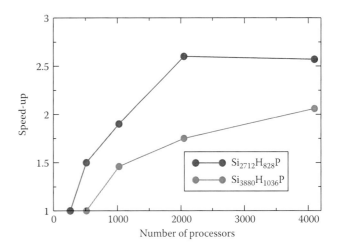

FIGURE 3.8 Examples of performance scaling for the real space pseudopotential code.

3.3 Applications

3.3.1 Silicon Nanocrystals

3.3.1.1 Intrinsic Properties

If we wish to examine the intrinsic properties of nanocrystals, we need to deal with the crystal surface, which at the nanoscale becomes increasingly important. Experimentally, this issue is often handled by passivating the intrinsic surface with surfactants or hydrogenating the surface. We chose to use the latter approach by capping surface dangling bonds with hydrogen atoms [44]. The largest nanocrystal we examined contained over ten thousand atoms: $Si_{9041}H_{1860}$, which is approximately 7 nm in diameter [32,45]. A ball and stick model of a typical nanocrystal is shown in Figure 3.9.

FIGURE 3.9 The ball and stick model of a hydrogenated silicon quantum dot. The interior consists of a diamond fragment. The surface of the fragment is capped with hydrogen atoms.

A solution of the Kohn–Sham equations yields the distribution of eigenvalues. For sufficiently large nanocrystals, one expects the distribution to approach that of crystalline silicon, i.e., the distribution of states should approach that of the crystalline density of states. The range of sizes for these nanocrystals allows us to make comparisons with the bulk crystal. This comparison is made in Figure 3.10. The "density of states" (DOS) for the eigenvalue spectrum for the nanocrystal shares similar structure as in the bulk crystal. For example, the sharp peak at about 4 eV below the top of the valence band arises from an M_1 critical point along the [110] direction in the bulk [18]. This feature is clearly present in the nanocrystal eigenvalue spectra. The only notable differences between the crystal and the nanocrystal occur because of the presence of the Si–H bonds and the lack of completely evolved bands.

We can also examine the evolution of the ionization potentials (I) and the electron affinities (A) for the nanocrystal.

$$I = E(N-1) - E(N)$$
$$A = E(N) - E(N+1) \tag{3.15}$$

In principle, these are ground state properties and, if the correct functional were known, these quantities would be accurately predicted by density functional theory. For atoms and molecules, one typically extracts accurate values for I and A, e.g., for the first-row atoms, the error is typically less than ~5%. The difference between the ionization potential and the electron affinity can be associated with the quasi-particle gap: $E_{qp} = I - A$. If the exciton (electron–hole) interaction is small, this gap can

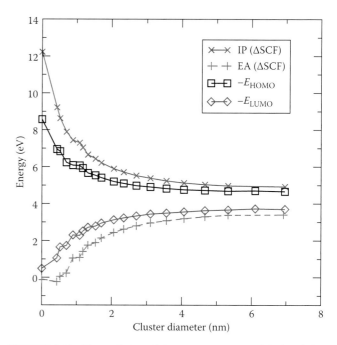

FIGURE 3.11 The evolution of the ionization potential (IP) and electron affinity (EA) with quantum dot size. Also shown are the eigenvalue levels for the highest occupied molecular orbital (HOMO) and the lowest unoccupied molecular orbital (LUMO).

be compared to the optical gap. However, for silicon nanocrystals the exciton energy is believed to be on the order of ~1 eV for nanocrystals of less than ~1 nm [46].

We can examine the scaling of the ionization potential and affinity by assuming a simple scaling and fitting to the calculated values (shown in Figure 3.11):

$$I(D) = I_\infty + A/D^\alpha$$
$$A(D) = A_\infty + B/D^\beta \tag{3.16}$$

where D is the dot diameter. A fit of these quantities results in $I_\infty = 4.5\,\text{eV}$, $A_\infty = 3.9\,\text{eV}$, $\alpha = 1.1$ and $\beta = 1.08$. The fit gives a quasi-particle gap of $E_{qp}(D \to \infty) = I_\infty - A_\infty = 0.6\,\text{eV}$ in the limit of an infinitely large dot. This value is in good agreement with the gap found for crystalline silicon using the local density approximation [47,48]. The gap is not in good agreement with experiment owing to the failure of the local density approximation to describe band gaps of bulk semiconductors in general.

A key aspect of our study is that we can examine the scaling of the ionization potential and electron affinity for nanocrystals ranging from silane (SiH_4) to systems containing thousands of atoms. We not only verify the limiting value of the quasi-particle gap, we can ascertain how this limit is reached, i.e., how the ionization potential and electron affinity scale with the size of the dot and what the relationship is between these quantities and the highest occupied and lowest empty energy levels. Since values of I and A from LDA are reasonably accurate for atoms and molecules, one can ask how the size of the nanocrystal affects the accuracy of LDA for predicting I and A. Unfortunately,

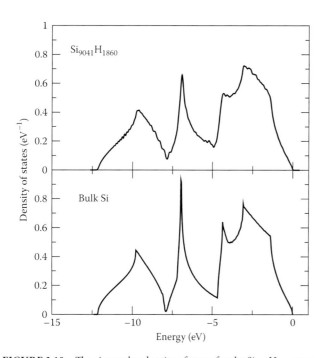

FIGURE 3.10 The eigenvalue density of states for the $Si_{9041}H_{1860}$ nanocrystal (top panel) and the electronic density of states for crystalline silicon (bottom panel). The highest occupied state in both panels is taken to be the zero energy reference.

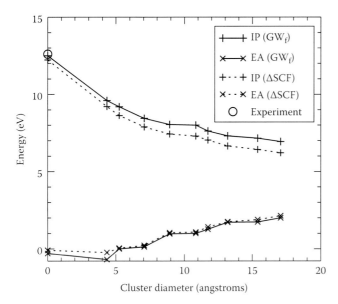

FIGURE 3.12 First ionization potential and electron affinity of passivated silicon clusters, calculated within the GW approximation [49] solid lines and ΔSCF dotted lines. Experimental data from Ref. [104] SCF results include spin-polarization effects.

experimental values for the ionization potentials and electron affinities are not known for hydrogenated silicon clusters and nanocrystals, a notable exception being silane (where the electron affinity is negative) [49]. However, there is some theoretical evidence that the issue involves errors in the ionization energies as opposed to the electron affinities. In Figure 3.12, we display the ionization potentials and electron affinities from the GW approximation [49–51] and compare to calculations using the LDA approximation. Within GW, the electron self-energy is obtained by calculating the lowest order diagram in the dynamically screened Coulomb interaction [49]. It is expected that the values for (IP,EA) are more accurately replicated by GW than the LDA work. The LDA and GW values for the EA are nearly identical. However, GW yields slightly larger IP values than LDA. As a function of size, the difference between the IP and GW values is small for the smallest nanocrystals and appears to saturate for the larger ones.

Given the energy as a function of position, we can also examine structural and vibrational properties of nanocrystals. Vibrational properties are easier to describe as they converge more rapidly to the bulk values than do electronic states; however, because they involve small changes in energy with position, the wave functions need to be better converged.

Vibrational modes in nanocrystals can play an important role within the context of the photovoltaic applications. In particular, vibrational properties are directly related to the phonon-assisted optical transitions. This is an important consideration for silicon: in the bulk limit, the lowest optical transitions are indirect and must be phonon assisted. It is expected that the vibrational dynamics in the presence of a free carrier will be different from the case of an intrinsic nanocrystal. Strains induced by a free uncompensated carrier might break the symmetry of

the vibrational modes, thereby increasing their participation probability in the optical transitions.

Owing to the localized nature of nanocrystals, it is feasible to predict vibrational modes calculations by the direct force-constant method [52]. The dynamical matrix of the system is constructed by displacing all atoms one by one from their equilibrium positions along the Cartesian directions and finding the forces induced on the other atoms of the nanocrystal. We determine the forces using the Hellmann–Feynman theorem in real space [33] and employ a symmetrized form of the dynamical matrix expression [53]. The elements of the dynamical matrix, $D_{ij}^{\alpha\beta}$, are given by

$$
D_{ij}^{\alpha\beta} = -\frac{1}{2}\left[\frac{F_i^{\alpha}(\{\mathbf{R}\}+d_j^{\beta})-F_i^{\alpha}(\{\mathbf{R}\}-d_j^{\beta})}{2d_j^{\beta}}\right.
$$
$$
\left.+\frac{F_j^{\beta}(\{\mathbf{R}\}+d_i^{\alpha})-F_j^{\beta}(\{\mathbf{R}\}-d_i^{\alpha})}{2d_i^{\alpha}}\right] \quad (3.17)
$$

where
F_i^{α} is the force on atom α in the direction i
$\{\mathbf{R}\}+d_j^{\beta}$ is the atomic configuration where only the atom β is displaced along j from its equilibrium position

The value of displacement was chosen to be 0.015 Å. The equilibrium structure was relaxed so that the maximum residual forces were less than 10^{-3} eV/Å. For this accuracy, the grid spacing h is reduced to 0.4 a.u. (1 a.u. = 0.529 Å) as compared to a value of about 0.7 a.u., typically used for electronic properties. The Chebyshev filtering algorithm is especially well suited for this procedure as the initial diagonalization need not be repeated when the geometry changes are small.

The vibrational modes frequencies and corresponding eigenvectors can be obtained from the dynamical equation

$$
\sum_{\beta,k}\left[\omega^2\delta_{\alpha\beta}\delta_{ik}-\frac{D_{ik}^{\alpha\beta}}{\sqrt{M_\alpha M_\beta}}\right]A_k^{\beta}=0 \quad (3.18)
$$

where M_α is the mass of atom labeled by α.

We apply this procedure to a Si nanocrystal: $Si_{29}H_{36}$. The surface of this nanocrystal is passivated by hydrogen atoms in order to electronically passivate the nanocrystal, i.e., remove any dangling bond states from the gap [54]. (Si clusters without hydrogen passivation has been examined previously, assuming notable surface reconstructions [55]). The choice of the number of silicon atoms is dictated so that the outermost Si atoms are passivated by no more than two hydrogen atoms. Our nanocrystal has a bulk-like geometry with the central atom having the T_d symmetry. The bonding distance to the four equivalent nearest neighbors is 2.31 Å.

We also consider cation nanocrystals by removing an electron from the system: $(Si_{29}H_{36})^+$. A relaxation of this nanocrystal leads to a distortion owing to a Jahn–Teller effect structure, i.e., there is a partial filling of the highest occupied state. This symmetry breaking leads to the central atom of the charged nanocrystal

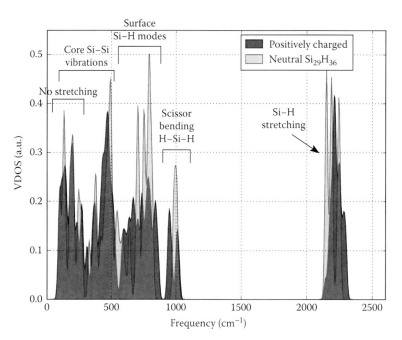

FIGURE 3.13 Vibrational density of states of $Si_{29}H_{36}$.

being bonded to four atoms with different bond lengths. Two bond lengths become slightly shorter (2.30 Å) and two others lengthen (2.35 Å). This distortion propagates throughout the nanocrystal.

Vibrational densities of states of the two nanocrystals are shown in Figure 3.13. The DOS for the nanocrystal can be divided into four distinct regions. (Vibration spectra of various silicon clusters were studied in a number of earlier papers [56,57] using model potential calculations [58]). The lowest energy vibrations involve only the Si atoms. We will assign them to "core modes." In the core modes all Si atoms are involved, while the passivating hydrogens remain static. The lowest frequency modes in this region (below ~250 cm⁻¹) correspond to modes with no bond stretching, i.e., bond bending modes dominate. For modes above ~250 cm⁻¹ stretching dynamics become more important. In the Si–Si "stretching–bending" part of the spectrum, the important part is the highest peak right below 500 cm⁻¹. The nature of this nanocrystal peak can be ascribed to modes that are characteristic transverse optical (TO) mode of crystalline silicon. The TO mode is extensively discussed in the literature as it is Raman active. The mode is sensitive to the nanocrystal size [59,60]. The next region of the vibration modes, just above 500 cm⁻¹, is related to the surface vibrations of Si and H atoms. The fourth distinct region is located around 1000 cm⁻¹ and is related to the H atoms scissor-bending modes; the Si atoms do not move in this case. The highest energy vibrations, above 2000 cm⁻¹, are the Si–H stretching modes.

The cation nanocrystal, $(Si_{29}H_{36})^+$, has a similar vibrational spectrum. However, in the charged system we see that hydrogen actively participates in the vibrations well below 500 cm⁻¹ (Figure 3.14). The peak at 500 cm⁻¹ is red shifted in the case of the changed system, which is associated with a much larger contribution from the surface atoms. The vibrational DOS spectrum of the hydrogen, shown in Figure 3.14, demonstrates that hydrogen

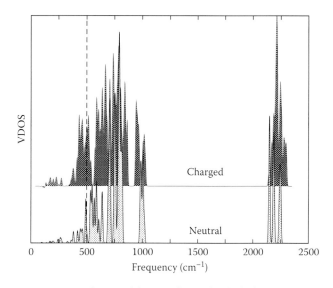

FIGURE 3.14 Vibrational density of states for the hydrogen contributions in $Si_{29}H_{36}$.

atoms vibrate at frequencies all the way down to 400 cm⁻¹. This is rather unusual, and occurs owing to an enhanced charge transfer on the surface associated with the hole charge. As noted in Ref. [55], the distribution of the hole density affects both lattice relaxations and the corresponding vibrational spectra. In order to quantify the location of the hole charge in the $(Si_{29}H_{36})^+$ nanocrystal, we calculated the "electron localization function" (ELF) for both neutral and charged structures [61]. Figure 3.15 presents two orientations of the ELF plots. On the left-hand side plots (a) and (c), the neutral nanocrystals is shown; on the right-hand side, plots (b) and (d), the charged nanocrystal is shown. Flat cuts of the ELF at the surface of the charged nanocrystal show where the hole is located.

Neutral   Charged

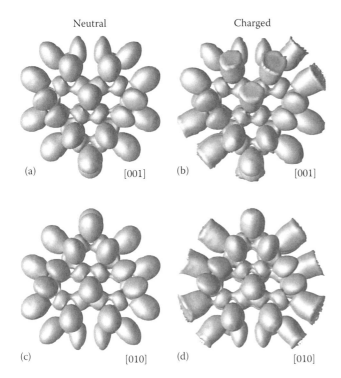

(a)   [001]   (b)   [001]

(c)   [010]   (d)   [010]

FIGURE 3.15 Electron localization function: (a,c) the neutral $Si_{29}H_{36}$ nanocrystal shown in two orientations; (b,d) the $Si_{29}H_{36}^{1+}$ cation, also in two orientations.

3.3.1.2 Extrinsic Properties

Doping a small percentage of foreign atoms in bulk semiconductors can profoundly change their electronic properties and makes possible the creation of modern electronic devices [62]. Phosphorus-doped crystalline Si introduces defect energy states close to the conduction band. For such shallow donors, electrons can be easily thermally excited, greatly enhancing the conductivity of the original pure semiconductor by orders of magnitude at room temperature.

The evolution of the semiconductor industry requires continued miniaturization. The industry is maintaining exponential gains in the performance of electronic circuits by designing devices ever smaller than the previous generation. This device miniaturization is now approaching the nanometer-scale. As a consequence, it is of the utmost importance to understand how doping operates at this length scale as quantum confinement is expected to alter the electronic properties of doped Si nanocrystals [63]. Also, doped Si nanowires have been synthesized and it has been demonstrated experimentally that they can be used as interconnects in electronic circuits or building blocks for semiconductor nanodevices [64,65]. Important questions arise as to whether the defect energy levels are shallow or not, e.g., at what length scale will device construction based on macroscopic laws be altered by quantum confinement?

Phosphorus-doped silicon nanocrystals represent *the prototypical* system for studying impurities in quantum dots. Recent experiments, designed to study this system, have utilized photoluminescence [66,67] and electron spin resonance measurements

[68–70]. Electron spin resonance experiments probe the defect energy levels through hyperfine interaction. Hyperfine splitting (HFS) arises from the interaction between the electron spin of the defect level and the spin of the nucleus, which is directly related to the probability of finding a dopant electron localized on the impurity site [71]. A HFS much higher than the bulk value of 42 G has been observed for P-doped Si nanocrystals with radii of 10 nm [68]. A size dependence of the HFS of P atoms was also observed in Si nanocrystals [69,70].

Unfortunately, theoretical studies of shallow impurities in quantum dots are computationally challenging. Owing to the large number of atoms and to the low symmetry of the system involved, most total energy calculations have been limited to studying nanocrystals that are much smaller than the size synthesized in experiment [72–75]. While empirical studies have been performed for impurities in large quantum dots, they often utilize parameters that are *ad hoc* extrapolations of bulk-like values [76–78].

The same methods used to examine intrinsic silicon can be used for extrinsic silicon [45]. It is fairly routine to examine P-doped Si nanocrystals up to a diameter of 6 nm, which spans the entire range of experimental measurements [70]. The HFS size dependence is a consequence of strong quantum confinement, which also leads to the higher binding energy of the dopant electron. Hence, P is not a shallow donor in Si if the nanocrystals are less than 20 nm in diameter. In addition, we find that there is "critical" nanocrystal size below which the P donor is not stable against migration to the surface.

As for intrinsic properties, our calculations are based on density functional theory in the local density approximation [13,14]. However, the grid spacing is chosen to be 0.55 a.u., as a finer grid must be employed to converge the system owing to the presence of the P dopant [10,45]. The geometry of Si nanocrystal is taken to be bulk-like and roughly spherical in shape, in accord with experimental observation [79]. Again, the dangling bonds on the surface of the nanocrystal are passivated by H. The experimentally synthesized Si nanocrystals are usually embedded in an amorphous silicon dioxide matrix. The Si/SiO_2 interface is in general not the same as H passivation. Nevertheless, both serve the role of satisfying the dangling bonds on the surface. The Si nanocrystals are then doped with one P atom, which substitutes a Si atom in the nanocrystal.

In Figure 3.16a, the defect state charge density along the [100] direction is illustrated. As the nanocrystal size increases, the defect wave function becomes more delocalized. This role of quantum confinement is observed in both experiments [70] and theoretical calculations [72,80]. The maxima of the charge density at around 0.2 nm correspond to the bond length between the P at the origin and its first Si neighbors. We can smooth out these atomic details by spherically averaging the defect wave function as shown in Figure 3.16b. We find that the defect wave function decays exponentially from the origin. This corresponds very well to the conventional understanding of defects in semiconductors: the defect ion and the defect electron form a hydrogen-like system with the wave function described as $\psi \sim \exp(-r/a_B^{\text{eff}})$.

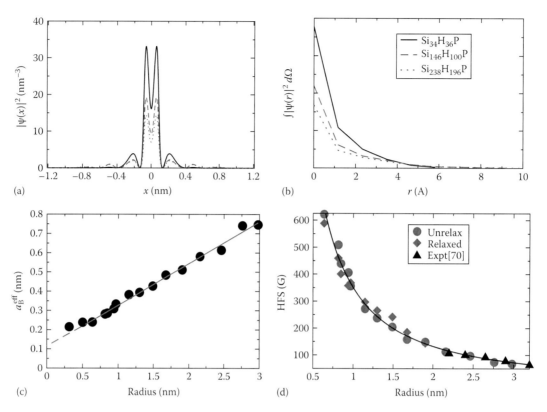

FIGURE 3.16 (a) Charge density for the dopant electron along the [100] direction for three P-doped Si nanocrystals with different radius. x is the coordinate along that direction. (b) The corresponding spherically averaged charge densities. (c) The effective Bohr radius a_B^{eff} corresponding to the dopant electron as a function of nanocrystal radius. (d) The calculated HFS of P-doped Si nanocrystals as a function of nanocrystal radius together with experimental data (▲) from Ref. [70]. Theoretical values for both the unrelaxed bulk geometries (●) and the fully relaxed structures (◆) are shown.

From the decay of the defect wave function, we can obtain an effective Bohr radius a_B^{eff} and its dependence on nanocrystal size as plotted in Figure 3.16c. We find that the effective Bohr radius varies nearly linearly with nanocrystal radius R up to 3 nm where R is approximately five times a_B^{eff}. Nonlinearity can be observed for very small Si nanocrystals, and a_B^{eff} appears to converge to ~0.2 nm as $R \to 0$. The limit trends to the size of a phosphorus atom while retaining its sp^3 hybridization. In the bulk limit, a_B^{eff} is ~2.3 nm assuming a dielectric constant of 11.4 and effective electron mass of 0.26 m_e. The dependence on R should trend to this bulk limit when the diameter of the nanocrystal is sufficiently large.

From the defect wave functions, we can evaluate the isotropic HFS as well [81,82]. Our calculated HFS is plotted in Figure 3.16d for a P atom located at the center of Si nanocrystal. There is very good agreement between experimental data [70] and our theoretical calculations. For nanocrystals with radius between 2–3 nm, the HFS is around 100 G, which considerably exceeds the bulk value of 42 G. Moreover, the HFS continues to increase as the nanocrystal size decreases. This is a consequence of the quantum confinement as illustrated in Figure 3.16a and b. The defect wave function becomes more localized at the P site as the radius decreases, leading to higher amplitude of the wave function at the P core. Only Si nanocrystals containing less than

1500 Si atoms are structurally optimized to a local energy minimum. For Si nanocrystals with radius larger than 1.5 nm, the effect of relaxation diminishes and the difference between unrelaxed and relaxed results is within ~10%.

The experimental methodology used to measure the HFS ensures that only nanocrystals with exactly one impurity atom are probed. However, experiment has no control over the location of the impurity within the nanocrystal. The spatial distribution of impurities cannot be inferred from experimental data alone. Therefore, we considered the energetics of the doped nanocrystal by varying the P position along the [100] direction as illustrated in Figure 3.17. We avoid substituting the Si atoms on the surface of the nanocrystal by P. Our results for five of the small nanocrystals after relaxation to local energy minimum are shown in Figure 3.17. For Si nanocrystals with a diameter smaller than ~2 nm, P tends to substitute Si near the surface. Otherwise, there is a bistable behavior in which both the center and the surface of the nanocrystal are energetically stable positions. This suggests that a "critical size" exists for nanocrystals. Below this size, P atoms will always be energetically expelled toward the surface.

We also calculated the binding energy E_B of the defect electron as the P position changes. The binding energy is a measure of how strongly the defect electron interacts with the P atom, and

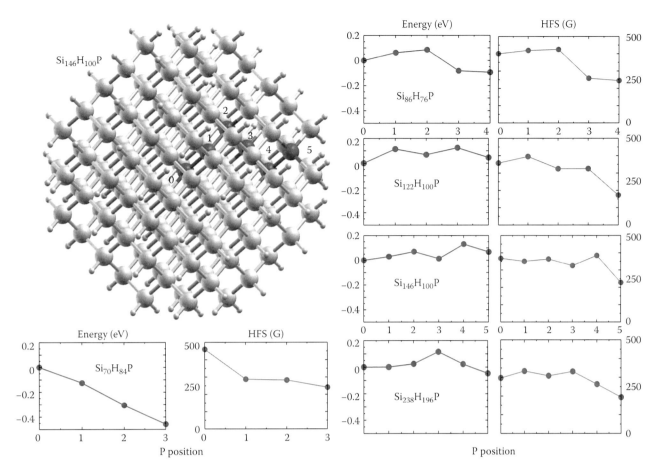

FIGURE 3.17 Difference in energy and HFS as the P atom moves away from the center of the Si nanocrystal. The energies are with respect to the energy of the Si nanocrystal with P at the center. The *x*-axis measures the distance of P atom away from the origin in the unit of Si bond length as illustrated in the perspective view of a Si nanocrystal.

is calculated by the energy required to ionize a P-doped Si nanocrystal by removing an electron (I_d) minus the energy gained by adding the electron to a pure Si nanocrystal (A_p). Figure 3.18a illustrates the typical situation for Si nanocrystals larger than 2 nm in diameter: the binding energy tends to decrease as the P moves toward the surface. This decrease occurs because the defect wave function becomes more distorted and less localized around P, leading to a loss in the Coulomb energy between the P ion and the defect electron. The change in the defect wave function explains why the center of the nanocrystal is energetically favorable. However, since the doped nanocrystal can relieve its stress by expelling the P atom toward the surface where there is more room for relaxation, positions close to the surface are always locally stable as depicted in Figure 3.17. The binding energy and P-induced stress compete with each other in determining the defect position within the Si nanocrystal.

For Si nanocrystals less than 2 nm in diameter, the binding energy is higher close to the surface of the Si nanocrystal as shown in Figure 3.18b and c. A comparison of the binding energy between relaxed and unrelaxed structures suggests that the reversal in the trend is caused by relaxation. From Figure 3.18e, the P atom is found to relax toward the center of the nanocrystal, leading to a more localized defect wave function as in

Figure 3.18d with better confinement inside the nanocrystal, and therefore higher binding energy. The relaxation of P atom toward the center of the nanocrystal causes an expansion of the Si nanocrystal in the perpendicular direction creating strain throughout the nanocrystal. This trade off between the binding energy and stress is only feasible for small Si nanocrystals as depicted in Figure 3.18f. As the diameter of the Si nanocrystal increases, the relaxation trends to a bulk-like geometry. This interplay between the binding energy and stress for small Si nanocrystals stabilizes the P atom close to the surface of the nanocrystal.

The HFS evaluated for different P positions inside the Si nanocrystal is also shown in Figure 3.17. There is a general trend for the HFS to drop drastically as the P atom approaches the surface of the nanocrystal. Smaller variations in the HFS are found for P positions away from the surface. However, for a large Si nanocrystal, the HFS varies within ~10% of the value with P at the center position. From studies of the energetics and comparison to the experimental results in Figure 3.16d, it is possible for the synthesized Si nanocrystals to have P located close to the center of the nanocrystals.

An analysis of the defect wave function in Figure 3.16 can be fitted to an effective mass model [45]. Motivated by the defect wave functions having an approximate form of $\exp(-r/a_B)$, we consider

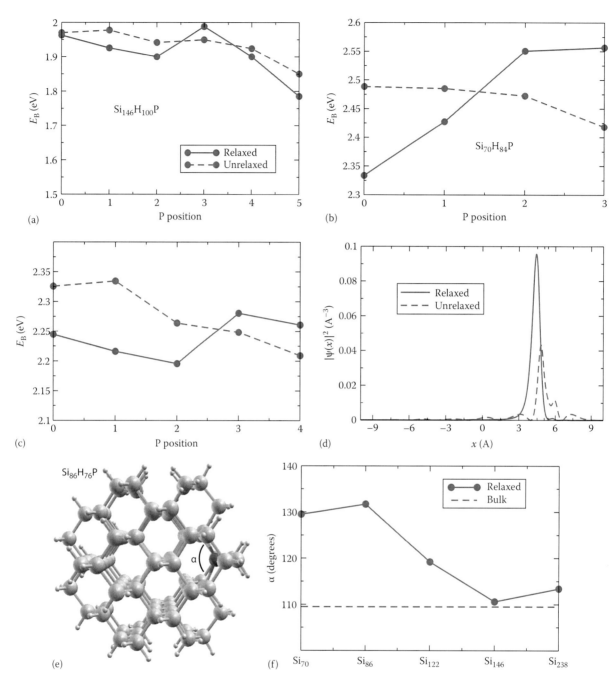

FIGURE 3.18 (a–c) Changes in binding energy as the P atom moves away from the center of three Si nanocrystals with different sizes. The same *x*-axis is used as in Figure 3.17. (d) The charge density of the dopant electron along the [100] direction (*x* direction) for the $Si_{86}H_{76}P$ cluster with the P atom located close to the surface as illustrated in (e). (f) A plot of the angle subtended by the P atom located close to the surface for five different Si nanocrystals. Only the number of Si atoms is used to label the *x*-axis for clarity. The solid line represents the results after relaxation, while the dashed line corresponds to unrelaxed bulk-like geometries in the figure.

a hydrogen-like atom in a "dielectric box." The potential that the electron experiences in atomic units is

$$V(r) = \frac{-1}{\varepsilon(R)r} + V_0 \quad \text{for } r \leq R; \quad V(r) = \frac{-1}{\varepsilon(R)r} \quad \text{for } r > R \quad (3.19)$$

where

R is the radius of the well
V_0 the well depth

The dielectric constant $\varepsilon(R)$ depends on nanocrystal size [45] and is assumed to follow Penn's model $\varepsilon(R) = 1 + ((11.4 - 1)/(1 + (\alpha/R)^n))$ [83]. The dielectric constant converges to 11.4 in the bulk limit. α and n will be used as fitting parameters. An approximate solution is obtained to the Schrodinger equation for the defect electron with an effective mass $m^* = 0.26\, m_e$ under this potential by using a trial wave function $\psi = \sqrt{\pi/a_B^3} \exp(-r/a_B)$. By applying the variational principle, the energy can be minimized with

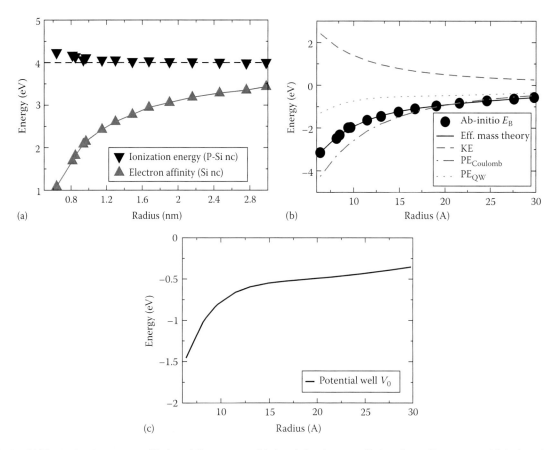

FIGURE 3.19 (a) The ionization energy of P-doped Si nanocrystal (▼) and the electron affinity of pure Si nanocrystal (▲) plotted as a function of nanocrystal radius. (b) The calculated binding energy E_B using effective mass theory plotted together with our *ab initio* results. From effective mass theory, the binding energy has contributions from the kinetic energy (KE), the Coulomb interaction between the dopant ion and its electron ($PE_{Coulomb}$), and the potential energy (PE_{QW}) due to the quantum well. (c) The potential well depth V_0 as a function of nanocrystal radius from effective mass theory based on a hydrogen atom in a box.

respect to the effective Bohr radius a_B. Hence, a_B can be found as a function of the well depth V_0 and radius R. Alternatively, V_0 can be inferred since we know a_B from our first-principles results. In fact, the energy calculated from this model corresponds to the binding energy E_B of the dopant electron. Therefore, we can fit the α and n parameters in Penn's model by calculating the binding energy. Our calculated size dependence of the binding energy ($E_B = I_d - A_p$ as defined above) is plotted in Figure 3.19a and b. A detailed explanation of the trend can be found in Refs. [45,72].

Figure 3.19 illustrates the results from the fitting to the effective mass model. By using $\alpha = 5.4$ nm and $n = 1.6$, we find that our model can reproduce almost exactly the binding energies from the first-principles calculations. The binding energy E_B scales as $R^{-1.1}$ where R the nanocrystal radius. For nanocrystals up to 6 nm in diameter, the binding energy is significantly larger than $k_B T$ at room temperature. An extrapolation of our results shows that a nanocrystal diameter of at least 20 nm is needed in order for P to be a shallow donor. Interestingly, the depth of the potential well V_0 depends on the nanocrystal radius R rather than being infinite or a constant. The finite V_0 explains why the dependence of HFS on R scales with an exponent smaller than three, which is a consequence of an infinitely deep quantum well

[84,85]. The well represents the effect of quantum confinement on the wave function, which for very large nanocrystals diminishes, corresponding to a vanishing quantum well in the bulk limit. As the nanocrystal size decreases, the quantum well becomes deeper such that it can confine the defect electron more effectively as its kinetic energy increases.

3.3.2 Iron Nanocrystals

The existence of spontaneous magnetization in metallic systems is an intriguing problem because of the extensive technological applications of magnetic phenomena and an incomplete theory of its fundamental mechanisms. In this scenario, clusters of metallic atoms serve as a bridge between the atomic limit and the bulk, and can form a basis for understanding the emergence of magnetization as a function of size. Several phenomena such as ferromagnetism, metallic behavior, and ferroelectricity have been intensely explored in bulk metals, but the way they manifest themselves in clusters is an open topic of debate. At the atomic level, ferromagnetism is associated with partially filled $3d$ orbitals. In solids, ferromagnetism may be understood in terms of the itinerant electron model [86], which assumes a partial delocalization of the $3d$ orbitals. In clusters of iron atoms, delocalization is weaker owing to the presence of a surface, whose shape affects the magnetic properties of the cluster. Because of their small size, iron clusters containing a few tens to hundreds of atoms are superparamagnetic: The entire cluster serves as a single magnetic domain, with no internal grain boundaries [87]. Consequently, these clusters have strong magnetic moments, but exhibit no hysteresis.

The magnetic moment of nano-sized clusters has been measured as a function of temperature and size [88–90], and several aspects of the experiment have not been fully clarified, despite the intense work on the subject [91–96]. One intriguing experimental observation is that the specific heat of such clusters is lower than the Dulong–Petit value, which may be due to a magnetic phase transition [89]. In addition, the magnetic moment per atom does not decay monotonically as a function of the number of atoms and for fixed temperature. Possible explanations for this behavior are structural phase transitions, a strong dependence of magnetization with the shape of the cluster, or coupling with vibrational modes [89]. One difficulty is that the structure of such clusters is not well known. First-principles and model calculations have shown that clusters with up to 10 or 20 atoms assume a variety of exotic shapes in their lowest-energy configuration [97,98]. For larger clusters, there is indication for a stable body-centered cubic (BCC) structure, which is identical to ferromagnetic bulk iron [91].

The evolution of magnetic moment as a function of cluster size has attracted considerable attention from researchers in the field [88–98]. A key question to be resolved is: What drives the suppression of magnetic moment as clusters grow in size? In the iron atom, the permanent magnetic moment arises from exchange splitting: the $3d_\uparrow$ orbitals (majority spin) are low in energy and completely occupied with five electrons, while the $3d_\downarrow$ orbitals (minority spin) are partially occupied with one electron, resulting in a magnetic moment of $4\,\mu_B$, μ_B being the Bohr magneton. When atoms are assembled in a crystal, atomic orbitals hybridize and form energy bands: $4s$ orbitals create a wide band that remains partially filled, in contrast with the completely filled $4s$ orbital in the atom; while the $3d_\downarrow$ and $3d_\uparrow$ orbitals create narrower bands. Orbital hybridization together with the different bandwidths of the various $3d$ and $4s$ bands result in weaker magnetization, equivalent to $2.2\,\mu_B$/atom in bulk iron.

In atomic clusters, orbital hybridization is not as strong because atoms on the surface of the cluster have fewer neighbors. The strength of hybridization can be quantified by the effective coordination number. A theoretical analysis of magnetization in clusters and thin slabs indicates that the dependence of the magnetic moment with the effective coordination number is approximately linear [93,95,96]. But the suppression of magnetic moment from orbital hybridization is not isotropic [99]. If we consider a layer of atoms for instance, the $3d$ orbitals oriented in the plane of atoms will hybridize more effectively than orbitals oriented normal to the plane. As a consequence, clusters with faceted surfaces are expected to have magnetic properties different from clusters with irregular surfaces, even

if they have the same effective coordination number [99]. This effect is likely responsible for a nonmonotonic suppression of magnetic moment as a function of cluster size. In order to analyze the role of surface faceting more deeply, we have performed first-principles calculations of the magnetic moment of iron clusters with various geometries and with sizes ranging from 20 to 400 atoms.

The Kohn–Sham equation can be applied to this problem using a spin-density functional. We used the generalized gradient approximation (GGA) [100] and the computational details are as outlined elsewhere [8,24,31,101]. Obtaining an accurate description of the electronic and magnetic structures of iron clusters is more difficult than for simple metal clusters. Of course, the existence of a magnetic moment means an additional degree of freedom enters the problem. In principle, we could consider non-collinear magnetism and associate a magnetic vector at every point in space. Here we assume a collinear description owing to the high symmetry of the clusters considered. In either case, we need to consider a much larger configuration space for the electronic degrees of freedom. Another issue is the relatively localized nature of the $3d$ electronic states. For a real space approach, to obtain a fully converged solution, we need to employ a much finer grid spacing than for simple metals, typically 0.3 a.u. In contrast, for silicon one might use a spacing of 0.7 a.u. This finer grid required for iron results in a much larger Hamiltonian matrix and a corresponding increase in the computational load. As a consequence, while we can consider nanocrystals of silicon with over 10,000 atoms, nanocrystals of iron of this size are problematic.

The geometry of the iron clusters introduces a number of degrees of freedom. It is not currently possible to determine the *definitive* ground state for systems with dozens of atoms as myriads of clusters can exist with nearly degenerate energies. However, in most cases, it is not necessary to know the ground state. We are more interested in determining what structures are "reasonable" and representative of the observed ensemble, i.e., if two structures are within a few meV, these structures are not distinguishable.

We considered topologically distinct clusters, e.g., clusters of both icosahedral and BCC symmetry were explored in our work. In order to investigate the role of surface faceting, we constructed clusters with faceted and non-faceted surfaces. Faceted clusters are constructed by adding successive atomic layers around a nucleation point. Small faceted icosahedral clusters exist with sizes 13, 55, 147, and 309. Faceted BCC clusters are constructed with BCC local coordination and, differently from icosahedral ones, they do not need to be centered on an atom site.

We consider two families of cubic clusters: atom-centered or bridge-centered, respectively for clusters with nucleation point at an atom site or on the bridge between two neighboring atoms. The lattice parameter is equal to the bulk value, 2.87 Å. Non-faceted clusters are built by adding shells of atoms around a nucleation point so that their distance to the nucleation point is less than a specified value. As a result, non-faceted clusters usually have narrow steps over otherwise planar surfaces and the overall shape is almost spherical. By construction, non-faceted

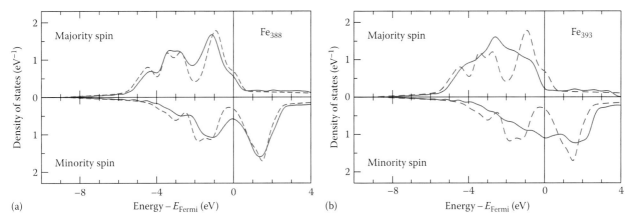

FIGURE 3.20 Density of states in the clusters Fe₃₈₈ (a) and Fe₃₉₃ (b), majority spin (upper panel), and minority spin(lower panel). For reference, the density of states in bulk iron is shown in dashed lines. The Fermi energy is chosen as energy reference.

clusters have well-defined point-group symmetries: I_h or T_h for the icosahedral family, O_h for the atom-centered family, and D_{4h} for the bridge-centered family. Clusters constructed in that manner show low tension on the surface, making surface reconstruction less likely.

As clusters grow in size, their properties approach the properties of bulk iron. Figure 3.20a shows the DOS for Fe₃₈₈, with local BCC coordination. At this size range, the DOS assumes a shape typical of bulk iron, with a three-fold partition of the $3d$ bands. In addition, the cohesive energy of this cluster is only 77 meV lower than in bulk. This evidence suggests that interesting size effects will be predominantly observed in clusters smaller than Fe₃₈₈. Figure 3.20b shows the DOS for Fe₃₉₃, which belongs to the icosahedral family. This cluster has a very smooth DOS, with not much structure compared to Fe₃₈₈ and bulk BCC iron. This is due to the icosahedral-like arrangement of atoms in Fe₃₉₃. The overall dispersion of the $3d$ peak (4 eV for $3d\uparrow$ and 6 eV for $3d\downarrow$) is nevertheless similar in all the calculated DOS.

The magnetic moment is calculated as the expectation value of the total angular momentum:

$$M = \frac{\mu_B}{\hbar}\left[g_s \langle S_z \rangle + \langle L_z \rangle \right] = \mu_B \left[\frac{g_s}{2}(n_\uparrow - n_\downarrow) + \frac{1}{\hbar}\langle L_z \rangle \right] \quad (3.20)$$

where $g_s = 2$ is the electron gyromagnetic ratio. Figure 3.21 illustrates the approximately linear dependence between the magnetic moment and the spin moment, $<S_z>$, throughout the whole size range. This results in an effective gyromagnetic ratio $g_{eff} = 2.04\,\mu_B/\hbar$, which is somewhat smaller than the gyromagnetic ratio in bulk BCC iron, 2.09 $\mu_B/\hbar$. This is probably due to an underestimation in the orbital contribution, $<L_z>$. In the absence of an external magnetic field, orbital magnetization arises from the spin-orbit interaction, which is included in the theory as a model potential,

$$V_{so} = -\xi \mathbf{L} \cdot \mathbf{S} \quad (3.21)$$

where $\xi = 80$ meV/$\hbar^2$ [93].

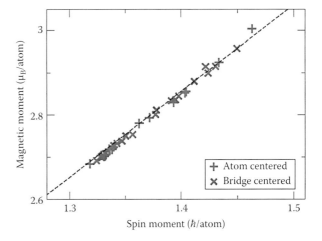

FIGURE 3.21 Magnetic moment versus spin moment calculated for the atom-centered BCC ("plus" signs) and bridge-centered BCC (crosses) iron clusters. The approximate ratio is $M/<S_z> = g_{eff} = 2.04\,\mu_B/\hbar$.

Figure 3.22 shows the magnetic moment of several clusters belonging to the three families studied: atom-centered BCC (top panel), bridge-centered BCC (middle panel), and icosahedral (bottom panel). Experimental data obtained by Billas and collaborators [88] are also shown. The suppression of magnetic moment as a function of size is readily observed. Also, clusters with faceted surfaces are predicted to have magnetic moments lower than other clusters with similar sizes. This is attributed to the more effective hybridization of d orbitals along the plane of the facets.

The non-monotonic behavior of the measured magnetic moment with size cluster can be attributed to the shape of the surface. Under this assumption, islands of low magnetic moment (observed at sizes 45, 85, and 188) are associated to clusters with faceted surfaces. In the icosahedral family, the islands of low magnetic moment are located around faceted clusters containing 55, 147, and 309 atoms. The first island is displaced by 10 units from the measured location. For the atom-centered and bridge-centered families, we found islands at (65, 175) and (92, 173) respectively, as indicated in Figure 3.22a and b. The first

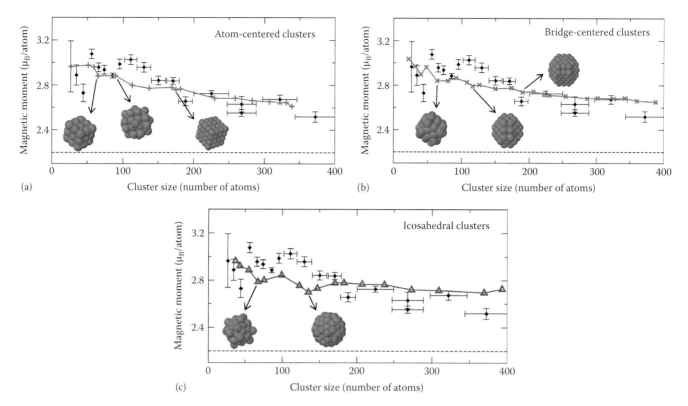

FIGURE 3.22 Calculated magnetic moments for clusters in the atom-centered ("plus" signs, a), bridge-centered (crosses, b), and icosahedral (triangles, c) families. Experimental data [88] is shown in black diamonds with error bars. Some of the faceted and non-faceted clusters are depicted next to their corresponding data points. The dashed lines indicate the value of magnetic moment per atom in bulk iron.

two islands are also close to the measured islands at 85 and 188. Clearly, there is no exact superposition in the location of calculated islands and measured islands. The magnetic moment was measured in clusters at 120 K [88,90]. At that temperature, vibrational modes or the occurrence of metastable configurations can shift the islands of low magnetic moment or make them more diffuse. Assuming that the non-monotonic decay of magnetic moment is dictated by the cluster shape, we also conclude that clusters with local structures different from the ones we discuss here (such as cobalt clusters with hexagonal-close packed coordination, or nickel clusters with face-centered cubic coordination) should have islands of low magnetic moment located at different "magic numbers," according to the local atomic coordination.

3.4 Conclusions

Here we reviewed tools for describing the electronic and vibrational properties of nanocrystals. The algorithm illustrated in this chapter replaces explicit eigenvalue calculations by an approximation of the wanted invariant subspace, obtained with the use of Chebyshev polynomial filters [32]. In this approach, only the initial self-consistent-field iteration requires solving an eigenvalue problem in order to provide a good initial subspace. In the remaining iterations, no iterative eigensolvers are involved. Instead, Chebyshev polynomials are used to refine the subspace. The subspace iteration at each step is easily one or more orders of magnitude faster than solving a corresponding eigenproblem

by the most efficient eigen-algorithms. Moreover, the subspace iteration reaches self-consistency within roughly the same number of steps as an eigensolver-based approach.

We illustrated this algorithm by applying it to hydrogenated silicon nanocrystals for both electronic and vibrational modes. The largest dot we examined contained over 10,000 atoms and was ~7 nm ($Si_{9041}H_{1860}$) in diameter. We examined the evolution of the electronic properties in these nanocrystals, which we found to assume a bulk-like configuration for dots larger than ~5 nm. In addition, we obtained scaling relations for the ionization potential, the electron affinity, and the quasi-particle gap over the size regime of interest. We found the quasi-particle gap to approach the known bulk limit within density functional theory and suggested the remaining errors in the bulk limit occurred in the ionization potential. We did a similar calculation for vibrational modes, although the size of the nanocrystals is not as large, owing to computational issues, i.e., the need for more accurate forces, and that the vibrational modes converge to the bulk values more rapidly than the electronic states. We also examined the role of a residual charge on the vibrational modes.

We also examined the role of doping and the behavior of defect wave functions in P-doped Si nanocrystals with a diameter up to 6 nm by first-principles calculations. Our calculated HFS has very good agreement with experimental data. We found that the defect wave function has a functional form similar to the hydrogen 1s orbital. A model calculation of a hydrogen atom in a quantum well can be used to describe the defect electron.

In addition, our study on the energetics of P location in the Si nanocrystals indicates that the P atom will be expelled toward the surface of the nanocrystal with diameter below a critical value of ~2 nm.

The computational tools we outlined in this chapter are not restricted to insulating materials as is often done in computational methods targeted at large systems [102]. The filtering method we employ can equally well be applied to metallic systems and we have done so for liquid Pb [43] and for iron nanocrystals [8]. We reviewed our results for the nanocrystals of iron by examining the evolution of the magnetic moment in iron clusters containing 20 to 400 atoms using our real space pseudopotential method with damped Chebyshev filtering. Three families of clusters were studied. They were characterized by the arrangement of atoms: icosahedral, BCC centered on an atom site, and BCC centered on the bridge between two neighboring atoms. We found an overall decrease of magnetic moment as the clusters grow in size toward the bulk limit. Clusters with faceted surfaces are predicted to have magnetic moment lower than other clusters with similar size. As a result, the magnetic moments is observed to decrease as a function of size in a nonmonotonic manner, which explains measurements performed at low temperature.

The utility of this numerical approach should be widely applied to a variety of problems at the nanoscale. The method is sufficiently powerful that it can be applied to systems sufficiently large that the entire nano-regime can be examined from an isolated atom to a bulk crystal. Moreover, the method has recently been extended to include systems with partial periodicity, e.g., nanowires where the system is periodic along the axis of the nanowire [35,103].

Acknowledgments

This work was supported in part by the National Science Foundation under DMR-0551195 and by the U.S. Department of Energy under DE-FG02-06ER46286 and DE-FG02-06ER15760. Computational support is acknowledged from the Texas Advanced Computing Center (TACC) and the DOE National Energy Research Scientific Computing Center (NERSC).

References

1. L. Brus, *J. Phys. Chem.* **98**, 3575 (1994).
2. A. P. Alivisatos, *Science* **271**, 933 (1996).
3. A. P. Alivisatos, *J. Phys. Chem.* **100**, 13226 (1996).
4. X. G. Peng, L. Manna, W. D. Yang, J. Wickham, E. Scher, A. Kadavanich, and A. P. Alivisatos, *Nature* **404**, 59 (2000).
5. M. Law, J. Goldberger, and P. Yang, *Ann. Rev. Mater. Res.* **34**, 83 (2004).
6. C. Burda, X. B. Chen, R. Narayanan, and M. A. El-Sayed, *Chem. Rev.* **105**, 1025 (2005).
7. R. Walters, G. Bourianoff, and H. Atwater, *Nat. Mater.* **4**, 143 (2005).
8. M. Tiago, Y. Zhou, M. M. G. Alemany, Y. Saad, and J. R. Chelikowsky, *Phys. Rev. Lett.* **97**, 147201 (2006).
9. G. Rollmann, M. E. Gruner, A. Hucht, P. Entel, M. L. Tiago, and J. R. Chelikowsky, *Phys. Rev. Lett.* **99**, 083402 (2007).
10. J. R. Chelikowsky, *J. Phys. D: Appl. Phys.* **33**, R33 (2000).
11. L. W. Wang and A. Zunger, *J. Chem. Phys.* **100**, 2394 (1994).
12. L. W. Wang and A. Zunger, *J. Phys. Chem.* **98**, 2158 (1994).
13. W. Kohn and L. J. Sham, *Phys. Rev.* **140**, A1133 (1965).
14. P. Hohenberg and W. Kohn, *Phys. Rev.* **136**, B864 (1964).
15. D. M. Ceperley and B. J. Alder, *Phys. Rev. Lett.* **45**, 566 (1980).
16. C. G. Broyden, *Math. Comp.* **19**, 577 (1965), ISSN 00255718, URL http://www.jstor. org/stable/2003941.
17. J. R. Chelikowsky and M. L. Cohen, in *Handbook of Semiconductors*, edited by T. S. Moss and P. T. Landsberg (Elsevier, Amsterdam, the Netherlands, 1992), p. 59.
18. M. L. Cohen and J. R. Chelikowsky, *Electronic Structure and Optical Properties of Semiconductors*, 2nd edn. (Springer-Verlag, Berlin, Germany, 1989).
19. D. R. Hamann, M. Schlüter, and C. Chiang, *Phys. Rev. Lett.* **43**, 1494 (1979).
20. G. P. Kerker, *J. Phys. C* **13**, L189 (1980).
21. N. Troullier and J. Martins, *Phys. Rev. B* **43**, 1993 (1991).
22. G. Bachelet, D. R. Hamann, and M. Schlüter, *Phys. Rev. B* **26**, 4199 (1982).
23. G. Kresse and J. Furthmüller, *Phys. Rev. B* **54**, 11169 (1996).
24. J. R. Chelikowsky, N. Troullier, and Y. Saad, *Phys. Rev. Lett.* **72**, 1240 (1994).
25. E. L. Briggs, D. J. Sullivan, and J. Bernholc, *Phys. Rev. B* **52**, R5471 (1995).
26. G. Zumbach, N. A. Modine, and E. Kaxiras, *Solid State Commun.* **99**, 57 (1996).
27. J.-L. Fattebert and J. Bernholc, *Phys. Rev. B* **62**, 1713 (2000).
28. T. L. Beck, *Rev. Mod. Phys.* **74**, 1041 (2000).
29. M. Heikanen, T. Torsti, M. J. Puska, and R. M. Nieminen, *Phys. Rev. B* **63**, 245106 (2001).
30. T. Torsti, T. Eirola, J. Enkovaara, T. Hakala, P. Havu, V. Havu, T. Höynälänmaa, J. Ignatius, M. Lyly, I. Makkonen et al., *Physica Status Solidi* (b) **243**, 1016 (2006).
31. L. Kronik, A. Makmal, M. L. Tiago, M. M. G. Alemany, M. Jain, X. Huang, Y. Saad, and J. R. Chelikowsky, *Physica Status Solidi* (b) **243**, 1063 (2006).
32. Y. Zhou, Y. Saad, M. L. Tiago, and J. R. Chelikowsky, *Phys. Rev. E* **74**, 066704 (2006).
33. M. M. G. Alemany, M. Jain, J. R. Chelikowsky, and L. Kronik, *Phys. Rev. B* **69**, 075101 (2004).
34. M. Alemany, M. Jain, M. L. Tiago, Y. Zhou, Y. Saad, and J. R. Chelikowsky, *Comp. Phys. Commun.* **177**, 339 (2007).
35. A. Natan, A. Mor, D. Naveh, L. Kronik, M. L. Tiago, S. P. Beckman, and J. R. Chelikowsky, *Phys. Rev. B* **78**, 075109 (2008).
36. B. Fornberg and D. M. Sloan, *Acta Numerica* **94**, 203 (1994).
37. L. Kleinman and D. M. Bylander, *Phys. Rev. Lett.* **48**, 1425 (1982).
38. J. C. Phillips and L. Kleinman, *Phys. Rev.* **116**, 287 (1959).

39. A. Stathopoulos, S. Öğüt, Y. Saad, J. Chelikowsky, and H. Kim, *Comput. Sci. Eng.* **2**, 19 (2000).

40. C. Bekas, Y. Saad, M. L. Tiago, and J. R. Chelikowsky, *Comp. Phys. Commun.* **171**, 175 (2005).

41. R. Lehoucq, D. C. Sorensen, and C. Yang, *ARPACK Users' Guide: Solution of Large Scale Eigenvalue Problems by Implicitly Restarted Arnoldi Methods* (SIAM, Philadelphia, PA, 1998).

42. K. Wu, A. Canning, H. D. Simon, and L.-W. Wang, *J. Comp. Phys.* **154**, 156 (1999).

43. M. M. G. Alemany, R. C. Longo, L. J. Gallego, D. J. González, L. E. González, M. L. Tiago, and J. R. Chelikowsky, *Phys. Rev. B* **76**, 214203 (2007).

44. S. Furukawa and T. Miyasato, *Phys. Rev. B* **38**, 5726 (1988).

45. T.-L. Chan, M. L. Tiago, E. Kaxiras, and J. R. Chelikowsky, *Nano Lett.* **8**, 596 (2008).

46. S. Ogut, J. R. Chelikowsky, and S. G. Louie, *Phys. Rev. Lett.* **79**, 1770 (1997).

47. J. P. Perdew and M. Levy, *Phys. Rev. Lett.* **51**, 1884 (1983).

48. L. J. Sham and M. Schlüter, *Phys. Rev. Lett.* **51**, 1888 (1983).

49. M. L. Tiago and J. R. Chelikowsky, *Phys. Rev. B* **73**, 205334 (2006).

50. M. S. Hybertsen and S. G. Louie, *Phys. Rev. B* **34**, 5390 (1986).

51. M. S. Hybertsen and S. G. Louie, *Phys. Rev. B* **35**, 5585 (1987).

52. K. Parlinski, Z. Q. Li, and Y. Kawazoe, *Phys. Rev. Lett.* **78**, 4063 (1997).

53. A. V. Postnikov, O. Pages, and J. Hugel, *Phys. Rev. B* **71**, 115206 (2005).

54. X. Huang, E. Lindgren, and J. R. Chelikowsky, *Phys. Rev. B* **71**, 165328 (2005).

55. J. Song, S. E. Ulloa, and D. A. Drabold, *Phys. Rev. B* **53**, 8042 (1996).

56. X. Jing, N. Troullier, J. R. Chelikowsky, K. Wu, and Y. Saad, *Solid State Commun.* **96**, 231 (1995).

57. M. R. Pederson, K. Jackson, D. V. Porezag, Z. Hajnal, and T. Frauenheim, *Phys. Rev. B* **65**, 2863 (1996).

58. A. Valentin, J. Sée, S. Galdin-Retailleau, and P. Dollfus, *J. Phys.* **92**, 1 (2007).

59. C. Meier, S. Lüttjohann, V. G. Kravets, H. Nienhaus, A. Lorke, and H. Wiggers, *Physica E* **32**, 155 (2006).

60. X.-S. Zhao, Y.-R. Ge, and X. Zhao, *J. Mater. Sci.* **33**, 4267 (1998).

61. B. Silvi and A. Savin, *Nature* **371**, 683 (1994).

62. B. G. Streetman and S. Banerjee, *Solid State Electronic Devices*, 5th edn. (Prentice Hall, Englewood Cliffs, NJ, 2000).

63. A. D. Yoffe, *Adv. Phys.* **50**, 1 (2001).

64. Y. Cui and C. M. Lieber, *Science* **291**, 851 (2001).

65. D. Appell, *Nature* **419**, 553 (2002).

66. A. Mimura, M. Fujii, S. Hayashi, D. Kovalev, and F. Koch, *Phys. Rev. B* **62**, 12625 (2000).

67. G. Mauckner, W. Rebitzer, K. Thonke, and R. Sauer, *Physica Status Solidi (b)* **215**, 871 (1999).

68. J. Müller, F. Finger, R. Carius, and H. Wagner, *Phys. Rev. B* **60**, 11666 (1999).

69. B. J. Pawlak, T. Gregorkiewicz, C. A. J. Ammerlaan, and P. F. A. Alkemade, *Phys. Rev. B* **64**, 115308 (2001).

70. M. Fujii, A. Mimura, S. Hayashhi, Y. Yamamoto, and K. Murakami, *Phys. Rev. Lett.* **89**, 206805 (2002).

71. G. Feher, *Phys. Rev.* **114**, 1219 (1959).

72. D. Melnikov and J. R. Chelikowsky, *Phys. Rev. Lett.* **92**, 046802 (2004).

73. Z. Zhou, M. L. Steigerwald, R. A. Friesner, L. Brus, and M. S. Hybertsen, *Phys. Rev. B* **71**, 245308 (2005).

74. G. Cantele, E. Degoli, E. Luppi, R. Magri, D. Ninno, G. Iadonisi, and S. Ossicini, *Phys. Rev. B* **72**, 113303 (2005).

75. S. Ossicini, E. Degoli, F. Iori, E. Luppi, R. Magri, G. Cantele, F. Trani, and D. Ninno, *Appl. Phys. Lett.* **87**, 173120 (2005).

76. C. Y. Fong, H. Zhong, B. M. Klein, and J. S. Nelson, *Phys. Rev. B* **49**, 7466 (1994).

77. I. H. Lee, K. H. Ahn, Y. H. Kim, R. M. Martin, and J. P. Leburton, *Phys. Rev. B* **60**, 13720 (1999).

78. M. Lannoo, C. Delerue, and G. Allan, *Phys. Rev. Lett.* **74**, 3415 (1995).

79. M. Fujii, K. Toshikiyo, Y. Takase, Y. Yamaguchi, and S. Hayashi, *J. Appl. Phys.* **94**, 1990 (2003).

80. G. Allan, C. Delerue, M. Lannoo, and E. Martin, *Phys. Rev. B* **52**, 11982 (1995).

81. J. A. Weil and J. R. Bolton, *Electron Paramagnetic Resonance: Elementary Theory and Practical Applications*, 2nd edn. (Wiley, Hoboken NJ, 2007).

82. C. G. V. de Walle and P. E. Blöchl, *Phys. Rev. B* **47**, 4244 (1993).

83. D. R. Penn, *Phys. Rev. B* **128**, 2093 (1962).

84. L. E. Brus, *J. Chem. Phys.* **79**, 4403 (1983).

85. L. E. Brus, *J. Chem. Phys.* **79**, 5566 (1983).

86. D. Mattis, *The Theory of Magnetism*, 2nd edn. (Springer-Verlag, Berlin, Germany, 1988).

87. C. Bean and J. Livingston, *J. Appl. Phys.* **30**, 120s (1959).

88. I. Billas, J. Becker, A. Châtelain, and W. de Heer, *Phys. Rev. Lett.* **71**, 4067 (1993).

89. D. Gerion, A. Hirt, I. Billas, A. Châtelain, and W. de Heer, *Phys. Rev. B* **62**, 7491 (2000).

90. I. Billas, A. Châtelain, and W. de Heer, *Science* **265**, 1682 (1994).

91. J. Franco, A. Vega, and F. Aguilera-Granja, *Phys. Rev. B* **60**, 434 (1999).

92. A. Postnikov and P. Entel, *Phase Transitions* **77**, 149 (2004).

93. O. Šipr, M. Košuth, and H. Ebert, *Phys. Rev. B* **70**, 174423 (2004).

94. A. Postnikov, P. Entel, and J. Soler, *Eur. Phys. J. D* **25**, 261 (2003).

95. R. Félix-Medina, J. Dorantes-Dávila, and G. Pastor, *Phys. Rev. B* **67**, 094430 (2003).

96. K. Edmonds, C. Binns, S. Baker, S. Thornton, C. Norris, J. Goedkoop, M. Finazzi, and N. Brookes, *Phys. Rev. B* **60**, 472 (1999).

97. G. Pastor, J. Dorantes-Dávila, and K. Bennemann, *Phys. Rev. B* **40**, 7642 (1989).

98. O. Diéguez, M. Alemany, C. Rey, P. Ordejón, and L. Gallego, *Phys. Rev. B* **63**, 205407 (2001).

99. P. Bruno, in *Magnetismus von Festkörpern und grenzflächen*, edited by P. Dederichs, P. Grünberg, and W. Zinn (IFF-Ferienkurs, Forschungszentrum Jülich, Germany, 1993), pp. 24.1–24.27.

100. J. P. Perdew, K. Burke, and Y. Wang, *Phys. Rev. B* **54**, 16533 (1996).

101. J. R. Chelikowsky, N. Troullier, K. Wu, and Y. Saad, *Phys. Rev. B* **50**, 11355 (1994).

102. M. J. Gillan, D. R. Bower, A. S. Torralba, and T. Miyazaki, *Comp. Phys. Commun.* **177**, 14 (2007).

103. J. Han, M. L. Tiago, T.-L. Chan, and J. R. Chelikowsky, *J. Chem. Phys.* **129**, 144109 (2008).

104. U. Itoh, Y. Toyoshima, H. Onuki, N. Washida, and T. Ibuki, *J. Chem. Phys.* **85**, 4867 (1986).

4

Design of Nanomaterials by Computer Simulations

Vijay Kumar
Dr. Vijay Kumar Foundation

4.1 Introduction

In recent years, there has been tremendous surge in research on understanding the properties of nanomaterials due to manifold interest in technological developments related to seemingly diverse fields such as miniature electronic devices (currently the device size is about 30 nm), molecular electronics, chemical and biological sensors with single molecular sensitivities, drug delivery, optical and magnetic applications (information storage, sensors, LED and other optical devices, etc.), design of novel catalysts, controlling environmental pollution, and green technologies (e.g., hydrogen-based energy storage systems, fuel cells), biological systems, drug design, protective coatings, paints, and material processing using powders, as well as the desire to develop fundamental understanding at the nanoscale, which includes a wide size range of materials in between the well-studied atomic and molecular systems on the one hand and bulk systems on the other. Advances in our ability to produce, control, and manipulate material properties at the nanoscale have grown rapidly in the past two decades. New forms of nanomaterials such as cage-like fullerenes [1] and hollow nanotubes [2] have been discovered that have opened up new vistas. This has invigorated research efforts and has brought researchers in physics, chemistry,

materials science, and biology on a common platform to address problems of materials at the nanoscale. There is so much to learn and so many new possibilities to design materials at the nanoscale that have no parallel in bulk and it would require much caution to develop applications. While there is very wide scope for research, it would be important to find materials and develop technologies that would work in a controlled manner as well as to find ways and guiding principles that could reduce experimental effort and expedite discoveries. In this direction, computer simulations have become a golden tool [3,4] and these are rapidly growing as cost-effective virtual laboratories that could also save much time and effort as well as material used in real laboratories and offer insight that may not always be possible from experiments.

4.1.1 Interplay between Theory and Experiments: A Necessity at the Nanoscale

Unlike bulk materials, the properties of nanomaterials are often sensitive to size and shape as a large fraction of atoms lie on the surface. A given material may be prepared in different nano-forms such as nanoparticles, nanowires, tubular structures, thin coatings (including core-shell type), layered or ribbon forms, etc.

and in different sizes that often exhibit very different properties in contrast to bulk. Even the structure and compositions (stoichiometries) in nanoforms could be quite different from bulk. Both of these factors are very important to understand the properties of nanomaterials similar to bulk systems but unlike bulk systems in which atomic structures of even very complex systems can be determined with good accuracy, it is often challenging to know the precise atomic structure, composition, and size distribution of nanomaterials experimentally at least for systems having the size range on the order of a nanometer. Therefore, often it is necessary to perform calculations and compare the results with experimental findings to develop a proper understanding of nanomaterials. *This interplay between theory and experiment is very important at the nanoscale.*

4.1.2 Quantum Mechanical Calculations: A Need for Predictive Simulations

As we shall discuss below, it is often important to perform ab initio calculations at the nanoscale because as mentioned above, the properties are generally very different from bulk in the nonscalable regime and surfaces play a very important role. Therefore, it is desirable to use methods that could be applicable on an equal footing to bulk, surfaces, and small systems in order to be able to ascertain the differences between bulk and nanosystems in the right perspective. In this direction, there have been developments based on the density functional theory (DFT) [5] that have attained predictive capability [3]. It is becoming increasingly possible not only to understand experimental observations but also to manipulate materials behavior in a computer experiment and explore different sizes, compositions, and shapes to unveil the properties of materials and find those that may be the desirable ones. This could accelerate materials design. While lots of the experimental data are on relatively large nanoparticles and nanowires having diameters of a few to few tens of nanometers, our ability to produce smaller nanoparticles (~1 nm size) and other nanomaterials with a control on size is increasing. It is also the size range in which quantum confinement effects become very dominant. On the other hand, theoretically it is becoming a routine to perform quantum mechanical ab initio calculations on nanosystems with a dimension of ~1 nm and study selectively materials with a size of a few nanometers [6]. With increasing computer power in the near future, calculations on larger systems would also become routine and this would be exciting both from applications point of view to design promising materials as well as from the point of view of comparisons between theory and experiments. This chapter deals with the developments in this direction with some examples primarily taken from our own work.

4.2 Small Is Different: The Unfolding of Surprises

The often different electronic, optical, magnetic, thermal, transport, and mechanical properties of nanomaterials from the corresponding bulk offer possibilities of new applications of

materials at the nanoscale. A well-known example is the formation of fullerene (cage-like) and nanotubular structures of carbon that led to wide spread research on the understanding of their properties and possible new uses of these novel materials and their derivatives as well as search for such structures in other materials. Single-wall carbon nanotubes have been found to be metallic or semiconducting depending on their type (the way the opposite edges of a graphene sheet are joined) while fullerene cages with possibilities of endohedral and exohedral doping and novel structures where a group of atoms such as C_{60} is used as building block to form solids [7] instead of atoms revolutionalized our approach to new materials. Another striking example is the finding of bright luminescence from nanostructures of silicon [8,9] in the visible range due to quantum confinement although bulk silicon is an inefficient emitter of light in the near-infrared range because of its indirect band gap. The color of the emitted light can be changed by changing the size and shape of the nanoparticle that change the energy gap and therefore the excitation energy. The finding of visible luminescence in silicon nanostructures has tremendous implications for the future of optoelectronic devices and it has the potential [10] for the development of silicon-based lasers as well as optical connections in microelectronics.

In bulk, the states at the band edges of a semiconductor arise from infinitely large systems and therefore the properties of semiconductor nanoparticles and other forms can differ from bulk over a large size range. However, metallic nanoparticles tend to attain bulk-like properties for smaller sizes as the highest occupied level lies in a bunch of states, although for small clusters of metals, there could be a sizeable highest occupied molecular orbital–lowest unoccupied molecular orbital (HOMO–LUMO) gap and such metal particles could have semiconductor-like behavior. An example of how different metal particles could be at the nanoscale is the observation of different colors of gold ranging from blue to red as the size is varied [11]. Also gold clusters become a good catalyst [12] even though bulk gold is the most noble metal. Small clusters of gold with up to 13 atoms have been found [13] to have planar structures, though one would think that metal clusters should tend to have close-packed structures (most metals in bulk crystallize in some of the closest packed structures). Similarly, as we shall discuss, small clusters of Rh [14] and Pt [6] have relatively open structures with high dispersion and become magnetic similar to their 3d counterparts in the same column in the periodic table, namely, Fe and Ni, respectively, while in their bulk form, they are nonmagnetic. Another striking observation has been that homonuclear clusters, such as those of Nb, have permanent electric dipole moments [15] though normally an electric dipole is associated with charge transfer from one constituent to another such as in a water molecule. These few examples illustrate that nanoscience is full of surprises and this gives researchers an exciting opportunity to unfold them, learn fundamentals of new phenomena using ab initio calculations, and modify them as well as design systems with desired properties.

4.2.1 Each Atom Counts

There is often a size distribution of nanoparticles, nanowires, or nanotubes in experiments and because of the variation in properties with size, one would desire to achieve size-selective nanomaterials. Further, any small number of impurities could have very significant effect on their properties. The role of impurities becomes very important because of the small size and a small number of atoms. Quantification of the size distribution as well as the impurities and defects are major difficulties from an experimental point of view as well as for reproducible applications. To illustrate the point, an Al_{13} cluster has been shown [16] to behave as a halogen-like superatom with large electron affinity of about 3.7 eV similar to that of a chlorine atom while Al_{14} has been found [17] to behave like an alkaline earth atom. When an Al atom is replaced by a Si atom at the center in icosahedral Al_{13} to form $Al_{12}Si$, it becomes an electronically closed shell cluster with a large calculated HOMO–LUMO gap of about 1.8 eV [16,18] within generalized gradient approximation (GGA) or the local density approximation (LDA) in the DFT. A more striking example is that of silicon clusters. It has been found [19] that when a transition metal atom is added to small silicon clusters, there is a dramatic change in their structure. As shown in Figure 4.1, when a Zr or Ti atom is added to Si_{16}, there is the formation of very symmetric $Zr@Si_{16}$ silicon fullerene or $Ti@Si_{16}$ Frank-Kasper (FK) polyhedral cage structure of silicon, respectively. This possibility to play with the properties of nanomaterials by changing size and atomic distribution makes their study attractive and it puts demand on the detailed understanding of the atomic distribution and size dependency of properties.

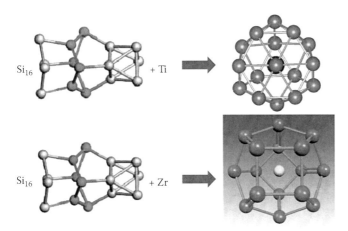

FIGURE 4.1 Transformation of a Si_{16} cluster into silicon fullerene and Frank–Kasper polyhedral structure by addition of a Zr and Ti atom, respectively. Dark balls show six-atom unit sandwiched between Si_4 and Si_6 in Si_{16} cluster. (Courtesy of X.C. Zeng.) Ti and Zr atoms are inside the silicon cage. For the fullerene structure $Zr@Si_{16}$, some Si–Si bonds are shown as double bonds due to additional pi bonding between the atoms. (Adapted from Kumar, V. and Kawazoe, Y., *Phys. Rev. Lett.*, 87, 045503, 2001; Kumar, V. et al., *Chem. Phys. Lett.*, 363, 319, 2002. With permission.)

4.2.2 Thermal Behavior

For bulk materials, a lot of thermodynamic data are available on alloys and phase diagrams (mixing, melting, etc.), and much theoretical work has been done on these problems. However, little knowledge has been accumulated on phases that may exist in the nanoform. Further, for bulk systems, empirical methods have been devised for phase diagram and thermodynamic calculations. The interatomic interactions used in such calculations are often generated from known bulk structures. However, in nanosystems, the structure itself is generally required to be determined by comparing the calculated properties with experimental observations using total energy calculations and finding the lowest energy structures. The finite size of a nanomaterial could lead to very different finite temperature behavior of atoms that lie on the surface and interior and for small clusters, the behavior of the whole system could differ very much from the bulk. Also, the unique properties of small systems may need to be appropriately protected by capping or passivation as well as nanoparticles may need functionalization for practical usage and it would be necessary to understand the effect of passivation on the properties of nanomaterials and their temperature dependence.

In the past two decades, a very large number of studies have been performed on nanomaterials and in many cases, quantitative understanding of the material properties in diverse nanoforms such as small nanoparticles, quasi-one-dimensional structures (nanowires and nanotubes), thin slabs, nanoribbons, and atomic layers has been achieved. Moreover, in some cases, it has been possible to make predictions of new nanomaterials such as silicon fullerenes by metal encapsulation [4,19,20] that have been later realized in the laboratory [21]. In the following, we first discuss briefly the methodology used in such calculations and then discuss selected results.

4.3 Method of Calculation

The DFT method within the framework of LDA or GGA or hybrid functionals for the exchange-correlation energy has been very successful to calculate the total energy of materials in different forms such as bulk, surfaces, thin slabs, layers, strips, atoms, molecules, clusters/nanoparticles, nanowires, and nanotubes and to understand the atomic and electronic structures, thermal, optical, elastic, mechanical, transport, and other properties of nanomaterials from first principles. Within the framework of DFT, different methods have been developed to solve the many-body problem of electrons and ions. These can be classified into two categories: (1) that treat all electrons such as in linearized augmented plane wave (LAPW) [22] and Gaussian methods [23] and (2) pseudopotential calculations [24] in which one treats only the valence electrons and the core is frozen. In the latter case, the calculation effort is reduced because the core electrons are not included explicitly and one often uses a plane wave basis that is convenient when ions need to be relaxed and which is often the case particularly for nanomaterials. Another way is to use a tight binding model within the linear combination of atomic orbital approach and also Gaussian method with pseudopotentials. An efficient way to perform combined electron and ion minimization was developed by Car and Parrinello [25] by which one can perform molecular dynamics with the total energy of the system calculated from DFT. In this framework, the pseudopotential method has been very widely used and the results discussed in this chapter have been obtained using such methods [26]. This approach not only allows calculations of ground state properties, but also simulated annealing can be performed to search for the low-lying structures. Finite temperature properties like diffusion, vibrational spectra, structural changes, melting, etc., can be calculated as well.

The pseudopotential method has been developed for the past nearly 50 years [24] and for systems covered in this chapter, the appropriate pseudopotentials are the so-called ab initio pseudopotentials which can treat atoms, molecules, clusters and other nanomaterials as well as bulk systems on an equal footing. These include the norm conserving pseudopotentials of Bachelet et al. [27], Troullier and Martins [28] and the like, the ultrasoft pseudopotentials [29], and the projector augmented wave (PAW) potentials [30]. These ionic pseudopotentials have been incorporated in some of the widely used programs of electronic structure calculations and it has become possible to treat a majority of elements in the periodic table.

For treating nanomaterials within the planewave-based codes, one uses a supercell approach. For clusters and nanoparticles, a large unit cell, often cubic, is considered with the cluster/nanoparticle placed at the center such that the distance between the atoms on the boundary of the cluster/nanoparticle in the cell and its periodic images is large enough to have negligible interactions and by this method, one can achieve the properties of isolated systems. In this approach, one can also obtain the properties of charged clusters by using a compensating uniform background charge so that the unit cell remains electrically neutral. In the case of quasi-one-dimensional systems such as nanowires and nanotubes, a cell that is large in two dimensions (perpendicular to the axis of the nanowire/nanotube) and has the periodicity of the infinite nanowire/nanotube along its axis is used. Infinite ribbons are treated in a similar way. Finite nanowires/nanotubes and ribbons can be treated in a way similar to those of clusters/nanoparticles. On the other hand, planar systems, such as slabs and layers, have periodicity in two dimensions and in the third dimension, the system is again made periodic by introducing vacuum space so that again the interaction between slabs/layers is negligible. When the cell dimensions are large such as in treating clusters and nanoparticles, one can often use only the gamma point for **k**-space integrations. In the past two decades, these approaches have been used quite extensively for nanomaterials and the calculated results have often been in good agreement with the available experimental data obtained from spectroscopic measurements such as photoemission, Raman or infrared, abundance spectra of clusters, ionization potentials (IPs), electron affinities (EAs), polarizabilities, magnetic moments, electric dipole moment, optical absorption, and other measurements on nanotubes/nanowires.

4.4 Clusters and Nanoparticles

Cluster and nanoparticle systems have been widely studied [31] using both experiments and theory and have contributed greatly to our understanding of nanomaterials. In the following, we present some of the developments.

4.4.1 Clusters of s-p Bonded Systems: Magic Clusters and Superatoms

Clusters of s-p bonded metals exhibit electronic shell structure similar to the shell model of nuclei and this leads to the magic behavior of clusters with 8, 20, 40, 58, 92, … valence electrons. The term "magic cluster" has been coined to represent clusters having N atoms that have high abundance in the mass spectrum while the abundance of clusters with $N + 1$ atoms is quite low. Effectively it means that N-atom clusters behave like rare gas atoms with closed electronic shells and therefore have weak interaction when one more atom is added. This has been understood and verified from ab initio calculations on clusters of elements such as alkali metals [32], aluminum [33], Ga [34], In [35], and to a certain extent noble metals [36,37] in which d electrons perturb only weakly the nearly free electron behavior by sp-d hybridization. A novel aspect of such small systems is that an aggregate of atoms could behave like an atom. Such aggregates or clusters are referred to as *superatoms*. As an example, Al_{13} has 39 valence electrons that are one electron short of the electronic shell closure at 40 valence electrons and it behaves like a halogen atom with large electron affinity of about 3.7 eV. Its interaction with an alkali atom has been shown [17] to lead to a large gain in energy and a charge transfer to the Al_{13} cluster similar to NaCl molecule. On the other hand, Al_7 cluster with 21 valence electrons behaves like a Na atom as it has one electron more than the electronic shell closing at 20 electrons. This illustrates how differently matter could behave by just changing the size. This has been confirmed from the measurement of IPs of clusters, which often vary with size [38] and this could be very important for their catalytic behavior, which often involves charge transfer to reactants. Also, experiments have shown [39] that singly negatively charged Al_{13} clusters did not react with oxygen because of their closed electronic shell.

Further similarity of superatoms to atomic behavior has been found [40] from Hund's rule that the electronic structure of atoms obeys. In the case of atomic clusters with partially filled electronic shells, often Jahn–Teller distortions remove the degeneracy of the electronic states arising due to high symmetry, leading to a lower energy of the system. However, in the case of the $Al_{12}Cu$ cluster that has an odd number of valence electrons (37 besides the 10 3d electrons on Cu atom), the electronic structure has been predicted [40] to follow the Hund's rule such that it has $3\mu_B$ magnetic moment with perfect icosahedral symmetry and Cu atom at the center of the icosahedron

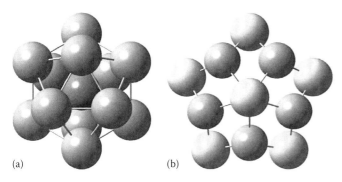

FIGURE 4.2 (a) Icosahedral structure of $Al_{12}Cu$ and $Al_{12}Si$ with Cu and Si atoms at the center and (b) the structure of $Al_{13}Li_8$ compound cluster in which ten Al atoms form a decahedron and seven Li atoms cap its faces while one Li atom is at the center. (After Kumar, V. et al., *Phys. Rev. B*, 61, 8541, 2001; Kumar, V. and Kawazoe, Y., *Phys. Rev. B*, 64, 115405, 2001; Kumar, V., *Phys. Rev. B*, 60, 2916, 1999.)

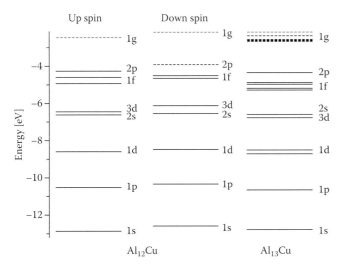

FIGURE 4.3 Electronic energy levels of an $Al_{12}Cu$ cluster (spin-polarized) show that the spin-up 2p level is fully occupied while the down-spin 2p level is empty. Interaction of this cluster with an Al atom leads to the complete filling of the 2p level and a highly stable $Al_{13}Cu$ cluster with a large HOMO–LUMO gap. (After Kumar, V. and Kawazoe, Y., *Phys. Rev. B*, 64, 115405, 2001.)

as shown in Figure 4.2a. The highest occupied up-spin state is 2p type (threefold symmetric) in the spherical jellium model and is fully occupied while the lowest unoccupied down-spin state, 2p, also a threefold degenerate state, is fully unoccupied. This is followed by a significant gap (see Figure 4.3). Therefore, this cluster behaves like an atom with half-filled p state. Accordingly, it was suggested [40] that this cluster should interact strongly with a trivalent atom. Studies on $Al_{13}Cu$ indeed showed strong stability that has also been observed [42]. In this case, the added Al atom is, however, incorporated within the shell of Al atoms and there is a large HOMO–LUMO gap (Figure 4.3). Copper atom being slightly smaller in size occupies the central position. However, doping of aluminum clusters with a bigger atom such as Sn or Pb has been shown to lead to the segregation of the impurity atom at the surface [43] as one expects on the basis of

the surface segregation theory of alloys [44] according to which a large size atom tends to segregate on the surface. Evidence for such a behavior has been obtained recently [45]. On the other hand a Si atom occupies the center of Al icosahedron and $Al_{12}Si$ is a superatom. Ordering has also predicted [41] in $Al_{10}Li_8$ cluster that has a Li atom at the centre of a decahedral cluster of Al_{10} and seven Li atoms cap this decahedron, as shown in Figure 4.2b. This cluster has 38 valence electrons and therefore its behavior is different from the known magic clusters of s-p bonded metals. The Li atoms transfer charge to Al atoms that behave like Si and are covalently bonded. The magic numbers of clusters are likely to be structure dependent. Recently, it has been shown [46] that for clusters with a shell-like structure, 32 and 90 valence electron systems act as magic clusters.

Clusters of divalent metals such as Be, Mg, Sr, and Hg are although predominantly s-p bonded, there is nonmetal to metal transition as the size grows due to the closed electronic shell structure of atoms. With increasing size, delocalization of electrons and hybridization of the occupied s states occurs with the unoccupied atomic p and d states. Such a transition was predicted [47] for clusters of Mg having about 20 atoms and it has been later confirmed from experiments [48]. In these divalent atoms, there is an interesting aspect of the electronic structure in that Be atom does not have a core p shell as compared to Mg and similarly for Ca, there is no core d-shell as compared to Sr. This leads to an interesting variation in the properties of the clusters of these elements. For Ca and Sr, the unoccupied d shell in atoms starts getting occupied, as aggregation takes place and it affects the growth behavior of the clusters. For Sr, it was shown [49] that the growth behavior is icosahedral, which is different from that of the Mg clusters. Figure 4.4 shows the calculated evolution of the electronic structure as the size grows. One can see that the d states start getting occupied as the size grows beyond about eight atoms. Close packed structures have also been obtained for Ca clusters [50]. However, large clusters of Mg ([48,51]) as well as those of alkali [52] and noble metals [53] have icosahedral structures. Often Mackay icosahedral structures with 55, 147, 309, 561, … atoms play an important role in the structures of clusters of a variety of materials [54]. Such clusters again lead to magic behavior of the clusters which is related to the completion of an atomic shell rather than an electronic shell. It has been shown [52] that alkali metal clusters with several thousand atoms could have structures different from bulk and these could be icosahedral. It has been suggested [49] that clusters of elements with large compressibility are likely to have icosahedral growth. It is because in an icosahedral structure, the center to vertex bond is about 5% shorter than the nearest neighbor vertex to vertex bond. Therefore, in order to have a compact packing, the material should be sufficiently compressible and accordingly clusters of hard material may not favor icosahedral growth. This is generally true as one finds in large clusters of alkali metals, alkaline earth metals as well as in clusters of rare gases, all of which have high compressibility.

Clusters of s-p bonded metals can be described within a jellium model and a spherical jellium model has been very helpful

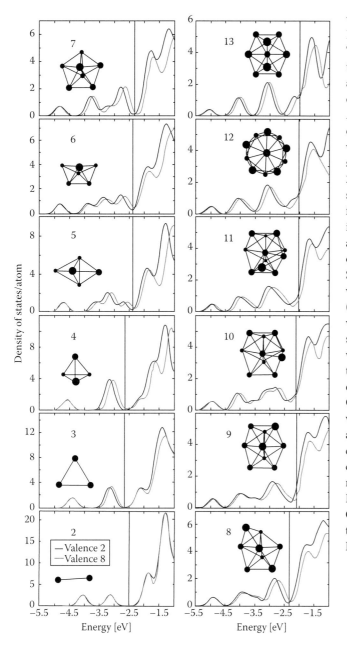

FIGURE 4.4 Evolution of the electronic structure of Sr clusters. The atomic structures of the clusters are also shown as inset. The number indicates the size of the cluster. The two curves in each figure show results for which Sr atom has been considered to have two valence electrons (only the outermost 5s electrons) and eight valence (including the six 4p semicore electrons). Vertical line shows the HOMO and the high density of states above the HOMO corresponds to the d states. (After Kumar, V. and Kawazoe, Y., *Phys. Rev. B*, 63, 075410, 2001. With permission.)

in understanding the properties of such clusters. In this model, the electronic states can be described in terms of 1s, 1p, 1d, 2s, 1f, 2p…. electronic shells. A way to design superatoms of s-p bonded metals is by considering symmetric structures such as an icosahedron and substituting another atom at the center so that a closed electronic shell is possible. The high symmetry of the structure

will lead to high degeneracy of electronic states and likely a large HOMO–LUMO gap, which is important for magic behavior. The choice of the atom at the center is done in such a way that the highest partially occupied electronic level of the substituted atom has the same angular momentum character as the HOMO of the shell of atoms. As an example, when a Si atom replaces an Al atom at the center of Al_{13} icosahedron, there is the formation of a 40 valence electron icosahedrally symmetric cluster $Al_{12}Si$ with a large HOMO–LUMO gap (Figure 4.2). The Al_{12} icosahedral shell has 36 valence electrons and the highest occupied level has 2p character in a spherical jellium model, which is a good representation for an icosahedral cluster. An Si atom with its partially filled p valence levels fits at the center of Al_{12} very well. The hybridization between the p-type states of the Al_{12} atomic shell and Si atom leads to a large HOMO–LUMO gap. Note that substitution of a Ti (also tetravalent) atom does not lead to shell closure [16]. Similarly, one can construct a large number of superatoms with 18 valence electrons such as icosahedral $Au_{12}W$ and $Au_{12}Mo$ clusters (similar structure as that of $Al_{12}Si$ in Figure 4.2) in which the 12 valence 6s electrons of Au atoms occupy 1d shell partially in the spherical jellium model (1s and 1p shells being fully occupied) and with which the d-level of W or Mo hybridizes strongly, leading to an electronically closed shell cluster with effectively 18 valence electrons and a large HOMO–LUMO gap of about 1.6 eV within GGA. Similar results have been obtained for $M@Cu_{12}$ and $M@Ag_{12}$ with M = transition metal of group 6. Other sizes of clusters can also be made by substituting different transition metals so that one can fulfill the 18 valence electron rule. Some examples of neutral clusters are: $Fe@Au_{10}$, $Ti@Au_{14}$, $Y@Au_{15}$, and $Ca@Au_{16}$. It is to be noted that high abundance of $Au_{15}Ti^+$ has been obtained [55]. In Figure 4.5, we have shown some charged clusters of Cu such as $Cu_{10}Co^+$, $Cu_{12}V^-$, $Cu_{13}Cr^+$, $Cu_{15}Ca^-$, and $Cu_{16}Sc^+$ calculated from the Gaussian method [56]. Such superatoms with a large HOMO–LUMO gap have potential for making solids just like atoms. Indeed

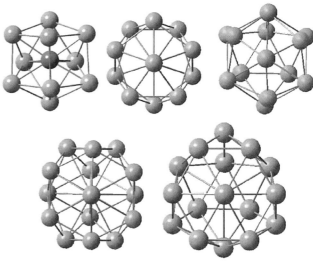

FIGURE 4.5 Atomic structures of $Cu_{10}Co^+$, $Cu_{12}V^-$, $Cu_{13}Cr^+$, $Cu_{15}Ca^-$, and $Cu_{16}Sc^+$ clusters. The transition metal atom is inside the cage of Cu atoms.

recent experiments [57] on Sc-doped copper clusters do show high stability of $Cu_{16}Sc^+$. Similarly, $Cu_{15}Sc$ has been found to be a superatom. Interestingly Cu_6Sc^+ and Cu_5Sc have also been found [58] to be very stable clusters. They correspond to effectively eight valence electron clusters, which also correspond to an electronic shell closing. Earlier experiments [59] on silver clusters doped with transition metal atoms showed high abundances of $Ag_{16}Sc^+$, $Ag_{15}Ti^+$, $Ag_{14}V^+$, $Ag_{11}Fe^+$, and $Ag_{10}Co^+$, all effectively 18 valence electron clusters and an eight valence electron cluster Ag_9Ni^+.

4.4.2 Carbon Fullerenes

Experiments on laser ablation of graphite by varying nucleation conditions in a cluster generating apparatus showed [1] special stability of C_{60} clusters, which was suggested to have an icosahedral football-shaped empty cage structure with 12 pentagons and 20 hexagons. Each carbon atom in this structure interacts with three neighboring carbon atoms. It was named "fullerene" after Buckminster Fuller, the architect of geodesic domes, and the structure was confirmed by NMR experiments [60,61]. The finding of C_{60} was considered revolutionary in the chemistry of carbon and it was thought to provide many new derivatives of C_{60} similar to benzene. Later, another exciting discovery took place in that a solid of C_{60} was formed [7]. Furthermore, when this solid phase was doped with alkali metals, superconductivity was discovered [62]. This led to feverish activities on carbon fullerene research and many other fullerenes of carbon such as C_{70} and C_{84} as well as endohedral carbon fullerenes in solid phases were produced. Further research on carbon led to the finding of nanotube structures [2]. These exciting developments have attracted the attention of a very large number of researchers around the world and a wide variety of research related to transport in nanotubes, mechanical strength, composite formation, field emission, electronic devices, support for catalysis, hydrogen storage, lubrication, among others, has taken place. Ab initio calculations have played a very important role in the understanding of the properties of these materials. Also in the larger size range, crystallites of diamond, also called nanodiamonds, have been studied. However, we focus on silicon in this section because calculations led to the discovery of new fullerene [19] and nanotube [63] structures of this technologically important material and great interest has developed in nanostructures of silicon for developing miniature devices.

Similar to carbon, silicon is a tetravalent element but it exists only in the diamond structure in bulk although a clathrate phase also exists [64] in which silicon is again tetrahedrally bonded and with doping of alkali/alkaline earth metals, it shows interesting superconducting and thermoelectric properties, whereas carbon can exist in graphite phase as well with sp^2 bonding besides the diamond structure. Carbon is versatile to form single, double, and triple bonds in a large number of molecules and in the fullerene form, the bonding is predominantly sp^2 type. The discovery of these new structures of carbon raised a question about the possibility of similar structures of silicon. Some studies were devoted to this aspect. However, silicon clusters behave quite

differently [20] from those of carbon and it was in 2001 that a fullerene of silicon was stabilized [19] by encapsulation of a metal atom using ab initio calculations. This discovery led to renewed interest in research on silicon nanoparticles and a large number of papers have been published on doping of different metals in silicon as well as other elements [4,20]. In the following, we discuss this finding and the design of a large variety of other new structures of silicon, which have shown the important role of ab initio calculations at the nanoscale in making new discoveries.

4.4.3 Bare and Hydrogenated Clusters of Silicon

When a bulk Si crystal is divided in to two pieces, two surfaces are created and some covalent bonds are broken, leading to the formation of dangling bonds on both the surfaces. In order to minimize the energy of such dangling bonds, the electronic charge density redistributes itself and often it leads to some ionic relaxation and in some cases, a reconstruction of the surface atomic structure. For nanomaterials, one divides bulk into very small pieces to create a very large surface area. Accordingly, a large fraction of atoms in nanomaterials lie on the surface, which plays a major role in the understanding of their properties as well as their applications. Because of surface reconstruction, the determination of the atomic structures of semiconductor clusters and nanoparticles is challenging and ab initio calculations have played a very important role in understanding them. It has been found [65] that Si clusters with ~1 nm diameter have structures that are very different from bulk. Clusters with up to 10 atoms have high coordination structures similar to metal clusters, while in the size range of 11–25 atoms prolate or stuffed fullerenelike structures are favored. An example of a prolate structure of Si_{16} is shown in Figure 4.1. Also a 20-atom Si cluster has prolate structure (Figure 4.6) while for Si_{25}, a 3D structure becomes favorable [66]. For larger clusters, 3D structures are lowest in energy. These are based on fullerene cage structures that are also filled with Si atoms. The presence of core atoms saturates partially the dangling bonds of the fullerene cage and this seems to be optimal for clusters with 33, 39, and 45 silicon atoms that have been found to have low reactivity and therefore a kind of magic behavior. For each cage size, there may be an optimal size of the core that would fit in the cage. Such core-cage isomers have been shown to be significantly lower in energy than other structures. Yoo and Zeng [67] have studied the optimal combinations of the core and cage units and found the carbon fullerene cages to be most favorable in the range of $N = 27$–39. Some of the most favorable combinations were reported to be $Si_3@Si_{24}$, $Si_3@Si_{28}$, $Si_3@Si_{30}$, $Si_4@Si_{32}$, $Si_4@Si_{34}$, and $Si_5@Si_{34}$. The structure of Si_{31} is shown in Figure 4.6. More recently, a spherical-shaped quantum dot of pristine Si with 600 atoms was made [68] by joining tetrahedral semiconductor fragments into an icosahedral particle. It has been shown from calculations that such icosahedral nanoparticles are more favorable than bulk fragments for diameters of less than 5 nm. These quantum dots have tetrahedral bonding and a Si_{20} fullerene at the core. Independently using molecular dynamics calculations [69], Si nanoparticles with 274, 280, and 323 atoms have been

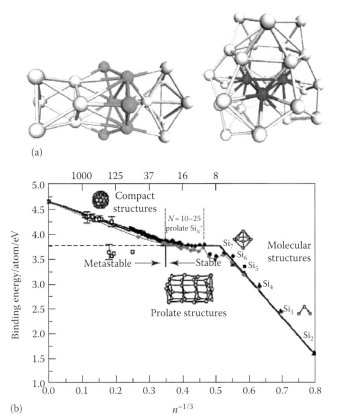

(a)

(b)

FIGURE 4.6 (a) Prolate atomic structure of Si_{20} and a stuffed fullerene structure of Si_{31}. Dark color Si_6 links Si_6 and Si_8 clusters in Si_{20} while three dark color atoms are inside a fullerene-like distorted cage of Si_{28}. (Courtesy of X.C. Zeng.) (b) The binding energy of silicon clusters as a function of $n^{-1/3}$. (After Horoi, M. and Jackson, K.A., *Chem. Phys. Lett.*, 427, 147, 2006. With permission.)

shown to form icosahedral structures as suggested by Zhao et al. The binding energy of silicon clusters increases rapidly initially and becomes nearly constant in the prolate regime and then rises again towards the bulk value when the clusters start developing again compact structures as shown in Figure 4.6b.

For larger and larger nanoparticles, the inner core would tend to have bulk diamond atomic structure while the surface region would be reconstructed. The surface of the nanoparticles can be passivated such as with hydrogen and depending upon the nanoparticle size as well as the number of H atoms, nanoparticles with different structures and different properties could be prepared. These could be nanocrystals with dangling bonds completely saturated with H, stuffed fullerene-like structures with fewer dangling bonds [71] that are saturated with H as well as empty cage Si_nH_n ($n \sim 10$–30) fullerenes [72], some of which are shown in Figure 4.7. In the latter case, each Si atom is coordinated with three Si atoms with nearly sp^3 bonding and the dangling bond on each Si atom is saturated with an H atom. Such cage structures of Si with $n = 20$, 24, and 28 are found in clathrates [73], but cages are interlinked and no hydrogen is required to saturate the dangling bonds. The HOMO–LUMO gap of these hydrogenated silicon clusters is generally large and accordingly

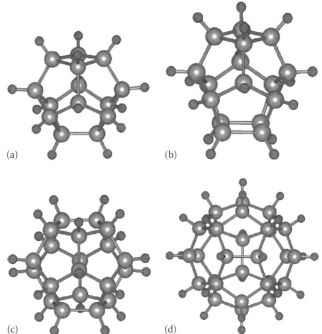

(a)   (b)

(c)   (d)

FIGURE 4.7 Hydrogenated silicon cages Si_nH_n with (a)–(d) corresponding to $n = 14$, 16, 20, and 28. (Adapted from Kumar, V. and Kawazoe, Y., *Phys. Rev. B*, 75, 155425, 2007; Kumar, V. and Kawazoe, Y., *Phys. Rev. Lett.*, 90, 055502, 2003. With permission.)

such structures have interesting optical properties as well as there could be possibilities of making derivatives and new molecules [74] and applications as sensors. Endohedral doping of such cages can be used to tailor HOMO–LUMO gap as well as to design nanomagnets with atomic-like behavior [72].

In contrast to the metal-encapsulated silicon clusters in which metal atoms interact with silicon cage very strongly, endohedral doping of these cages leads to a weak bonding of the guest atom, as shown in Figure 4.8 for a large variety of dopants in cages with

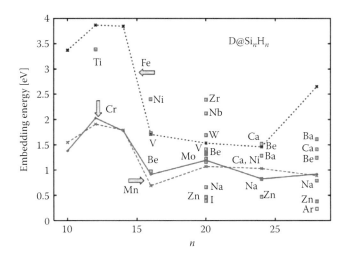

FIGURE 4.8 Binding energy of a variety of endohedral dopants D in Si_nH_n cages. (After Kumar, V. and Kawazoe, Y., *Phys. Rev. B*, 75, 155425, 2007. With permission.)

different n [72]. Cages such as with $n = 20$ are large enough to accommodate different atoms. Also the $n = 20$ cage is the most symmetric and has the highest stability. Accordingly, the guest atom can retain, to a large extent, its atomic character. For magnetic atoms Cr, Mn, and Fe, the guest atom was shown to have the same magnetic moment as in its atomic state. Such atoms were called slaved atoms [72] and hyperfine interaction was shown [75] to be a way to find endohedral doping of such cages.

The finding of bright photoluminescence in silicon was first observed [8] in a form that is known as porous silicon and which is believed to have nanoparticles of about 2 nm diameter and nanowire-like structures. Later bright red, green, and blue light emission was obtained from hydrogenated silicon nanoparticles with 3.8, 2.5, and 1.5 nm diameters [9]. The HOMO–LUMO gap of Si nanoparticles changes with size and shape due to the quantum confinement of electrons (consider a quantum mechanics text book example of a particle in a box whose size could be varied) and this makes nanoparticles of silicon attractive for applications. The nanoparticles with 1.5 nm diameter were suggested to have $Si_{29}H_{24}$ composition. Time-dependent density functional calculations [76] as well as quantum Monte Carlo calculations [9] on this cluster predicted optical gap to lie in the deep blue region, supporting the experimental findings. Calculations [77,78] on H-terminated Si clusters of varying sizes and having sp^3 bonding with diamond-like structure have shown a decreasing trend of the optical gap with increasing nanoparticle size as observed. However, in the form of Si_nH_n cages, the HOMO–LUMO gap has been found [72] to be less sensitive to size. Another way that quantum confinement can be controlled is by oxidizing silicon nanoparticles so that there is a core of silicon and a shell of SiO_2 surrounding this core. By changing the thickness of the oxide shell, the size of the core can be modified and this would affect the optical properties of silicon nanoparticles. The oxide shell also acts as protective cover on silicon nanoparticles.

4.4.4 Metal-Encapsulated Nanostructures of Silicon: Discovery of Silicon Fullerenes and Nanotubes

While much research has been done on elemental Si clusters as discussed in Section 4.4.3, no cluster size has been found to produce strikingly high abundance. For the use of silicon nanoparticles with unique properties, one would need to produce them in a controlled way in large quantities. Though there is a possibility of using hydrogen termination as discussed above, a novel way has been found to be metal encapsulation [19,20]. The first report of interaction of silicon with metal atoms appeared about two decades ago where strikingly high abundances of MSi_{15} and MSi_{16} clusters with M = Cr, Mo, and W were obtained by Beck [79,80] in experiments that were aimed to understand Schottky barrier formation in metal–silicon junctions. These metal-doped clusters were found to exist almost exclusively in this mass range and the intensities of other clusters were very small. For more than a decade, there was no theoretical work to understand this behavior though Beck speculated that the metal atom might be

surrounded by silicon atoms. In 2001, another observation was made [81] by reacting metal monomers and dimers with silane gas. This experiment produced metal–silicon–hydrogen complexes but no hydrogen was associated with WSi_{12} clusters. This led to a conclusion that WSi_{12} was a magic cluster, which did not interact with hydrogen. Using ab initio calculations a hexagonal prism structure with W atom at the center was found [81] to be of lowest energy. Henceforth, we refer this kind of endohedral structures with a cage of Si_n and M atom, inside as $M@Si_n$. Independently, a $Zr@Si_{20}$ cluster with a fullerene structure of Si_{20} was studied by Nellermore and Jackson [82] from ab initio calculations. They obtained a large gain in energy due to endohedral doping of Zr. However, Kumar and Kawazoe [19] found this cluster to deform upon optimization as Zr interacts with silicon strongly and the Si_{20} cage is too big for a Zr atom to have strong interaction with all the Si atoms. Using a shrinkage and removal of atom method akin to laser quenching, it was found that 16 Si atoms were optimal to encapsulate a Zr atom in a silicon cage, which had a structure similar to a carbon fullerene in that each Si atom had three neighboring Si atoms. The resulting silicon fullerene, $Zr@Si_{16}$, stabilized by a Zr atom (Figure 4.1), is smaller than the smallest carbon fullerene C_{20} with a dodecahedral structure (see Figure 4.9) in which all the 12 faces are regular pentagons (from Eulers theorem one needs at least 12 pentagons and an arbitrary number of hexagons to have a closed fullerene structure in which each atom has three nearest neighbors). In $Zr@Si_{16}$, there are eight pentagons (not regular) and two square faces as shown in Figure 4.1. In carbon fullerenes, pentagons are the locations of strain in bonding and there is the isolated pentagon rule. In silicon fullerenes, pentagons are favored and rhombi are the locations of strain. Accordingly, they favor an *isolated rhombus rule* [83]. This discovery showed how an unexciting Si_{16} cluster can be turned in to a very interesting structure by using a metal atom. Soon after this discovery, it was established that the size of the metal atom plays a very important role in determining the number of Si atoms that could be wrapped around it. The largest cage of silicon that can be stabilized by one M atom has

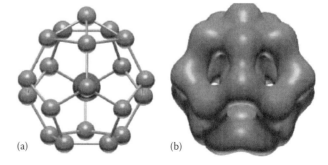

(a) (b)

FIGURE 4.9 (a) $La@Si_{20}$ dodecahedral fullerene interacting with a Cu atom. Elemental C_{20} has similar structure (without La and Cu atoms) but with icosahedral symmetry. (b) The charge density of this fullerene shows covalent bonding within the fullerene and charge transfer from Cu atom. $Th@Si_{20}$ is fully icosahedrally symmetric. (After Kumar, V. et al., *Phys. Rev. B*, 74, 125411, 2006. With permission.)

been predicted [84] to be Th@Si$_{20}$ (Figure 4.9) with a dodecahedral structure of Si$_{20}$ and icosahedral symmetry similar to that of C$_{20}$. Other rare earth atoms can also be doped and even magnetic fullerenes can be stabilized [85]. For carbon, C$_{20}$ has low stability among fullerenes as the bonding becomes nearly sp^3 type while in larger nanostructures such as C$_{60}$ fullerene and nanotubes, sp^2 bonding is more favorable. On the other hand, silicon favors sp^3 bonding and therefore Si$_{20}$ has been predicted to be an ideal structure for silicon nanoparticles as C$_{60}$ is for carbon. However, the HOMO–LUMO gap for Th@Si$_{20}$ is relatively small, though it has the largest binding energy among the silicon clusters stabilized by a metal atom. When a trivalent atom such as La is doped [85], then La@Si$_{20}$ acts like a halogen with large EA and accordingly a Na or Cu atom has ionic interaction as shown in Figure 4.9 by giving charge to the fullerene. Smaller fullerenes with Ni@Si$_{12}$ and W@Si$_{14}$ are also possible as shown in Figure 4.10. It can be noticed that in the lowest energy structure of these smaller fullerenes, the rhombi are isolated such as in W@Si$_{14}$ whereas in Ni@Si$_{21}$ some rhombi have to be nearest neighbors.

Besides these fullerene structures, a large number of other polyhedral cages M@Si$_n$ of different sizes have been stabilized by endohedral doping of a metal atom, M. Among these, a prominent structure of high symmetry (tetrahedral) is Ti@Si$_{16}$ with a Z16 Frank–Kasper polyhedron (Figure 4.1). Both Ti and Zr are isoelectronic but have slightly different atomic sizes. As the M atom in these structures is very tightly bound with the Si cage, even a small difference in the atomic size of the M atom could lead to different structures. Doping of Ti or Zr in Si$_{16}$ cage has been shown [19] to lead to a gain of ~10 eV. Therefore, optimization of this large energy gain controls the atomic structure of the Si cage. Thus the properties of such endohedral clusters can be controlled by varying M atom and the size of the cage [4]. This is also seen from the result that Zr@Ge$_{16}$ does not have a fullerene structure [86], but a Frank–Kasper polyhedral isomer as for Ti@Si$_{16}$ is the lowest in energy. Ge cage is slightly bigger than Si and a Zr atom can fit well in its Frank–Kasper cage. The stability of these high-symmetry structures can be understood from a spherical potential model [4,76] according to which the highest occupied level of the Si/Ge cage has d-character with four holes. Accordingly the d-orbitals of a Ti, Zr, or Hf atom (with four valence electrons) can interact with this cage strongly and form bonding and antibonding states, leading to full occupation

of the d-shell of the cage and large HOMO–LUMO gap. Selected cages with fullerene-like structures and Frank–Kasper-like structures are shown in Figures 4.10 and 4.11, respectively. The smallest cage has been suggested [87] to be M@Si$_{10}$ with M = Ni or Pt. Note that Si$_{10}$ is a magic cluster with a tetracapped prism structure. However, doping with a Ni or Pt atom further lowers the energy significantly and also leads to a different atomic structure. Interestingly, Pb$_{10}$Ni has been produced in bulk quantity [88]. It has a bicapped antiprism tetragonal structure of Pb$_{10}$ and Ni at its center [87].

For Zr@Si$_{16}$, the optical gap has been predicted to lie in the red region [76]. However, the Ti@Si$_{16}$ Frank–Kasper polyhedral cage has significantly larger HOMO–LUMO gap (2.35 eV) than Zr@Si$_{16}$ (1.58 eV) fullerene and it has been predicted to have the optical absorption gap in the blue region. By changing the size of the M atom, one can stabilize silicon cages with 10–20 atoms with differing properties [4]. In these silicon cages, the magnetic moment of the M atom (if any) is generally quenched due to the strong interaction of the M atom with the cage. However, it is possible to stabilize Si nanoparticles with magnetic moments by encapsulation of transition metal or rare earth atoms [85,89]. It was shown [89] that Mn@Sn$_{12}$ has 5μ_B magnetic moments in an icosahedral structure and that it behaves like a magnetic superatom. Subsequently experiments [90] did find this superatom as shown in Figure 4.12. To stabilize an icosahedral cage with Sn, Ge, or Si atoms, one needs to add two electrons to completely occupy the HOMO. Accordingly, Zn@Sn$_{12}$ is a magic structure, as one can also see in Figure 4.12. Mn has two 4s electrons in the valence shell and the 3d orbital is half filled. The two s electrons are used to stabilize the cage and the remaining five d electrons lead to its high magnetic moment. One can notice that for Cr doping, the mass spectrum does not show as prominent peak at $n = 12$ as for Mn doping, which indicates the role of the electronic configuration in the stability of the clusters. Therefore, M encapsulation provides a possibility to produce large quantities of nanoparticles with desired properties.

Subsequent to these predictions, experiments have been performed and many of the clusters such as Ti@Si$_{16}$, Zr@Si$_{16}$, isoelectronic V@Si$_{16}^+$ and Sc@Si$_{16}^-$,… have been realized in the laboratory [21]. As shown in Figure 4.13, almost exclusive abundance of Si$_{16}$Ti has been obtained in mass spectrum while Si$_{16}$Sc$^-$ and Si$_{16}$V$^+$ have high abundance compared with other sizes.

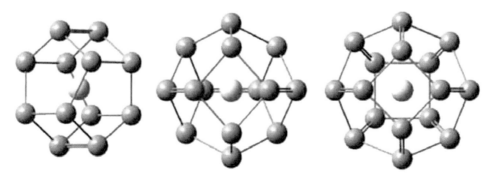

FIGURE 4.10 Atomic structures of M@Si$_N$ fullerenes with N = 12, 14, and 16.

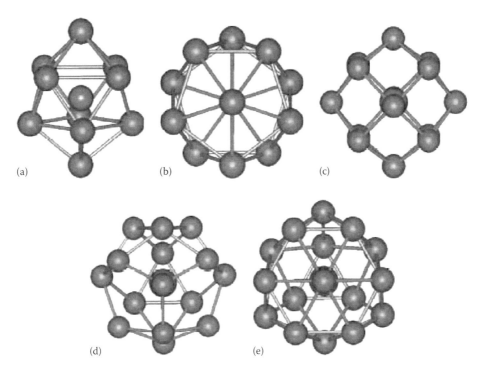

FIGURE 4.11 Frank–Kasper polyhedral structures of M@Si$_N$ clusters. (a) through (e) correspond to $N = 10$ (bicapped tetragonal antiprism), icosahederon, cubic, Z15 Frank–Kasper polyhedron, and Z16 Frank–Kasper polyhedron, respectively.

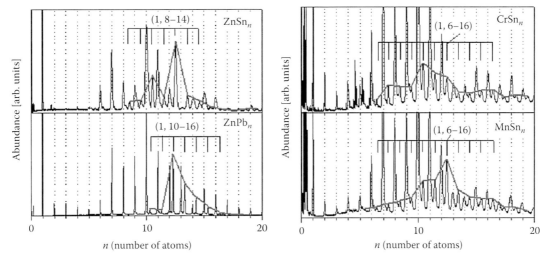

FIGURE 4.12 Mass spectra of Zn-, Cr-, and Mn-doped clusters show high abundance of ZnSn$_{12}$, ZnPb$_{12}$, and MnSn$_{12}$ clusters. (After Neukermans, S. et al., *Int. J. Mass. Spectrom.*, 252, 145, 2006. With permission.)

These results have reassured the power of ab initio calculations in the design of nanomaterials. Photoemission experiments [21] on Zr@Si$_{16}$ and Ti@Si$_{16}$ clusters have shown HOMO–LUMO gaps as predicted. Also experiments [91] on reaction of water and other molecules with M-doped silicon clusters have provided support for encapsulation of the M atom. As shown in Figure 4.14 for water on Ti@Si$_n$, small Ti-doped silicon clusters in which Ti atom is available for reaction with the molecule show higher binding energy [92] that correlates well with their higher abundance while clusters beyond a certain number of silicon atoms react very weakly as the Ti atom gets surrounded by Si atoms as

it can be seen from a sharp drop in the binding energy of a water molecule in Figure 4.14 beyond 12 silicon atoms. These results matched with experiments [91] and thus confirmed encapsulation of Ti atom in these clusters.

The idea of encapsulation to produce highly stable clusters of selected sizes is not unique to silicon and has general applicability whether it is a cluster of some metal (see discussion in Section 4.4.1) or semiconductor, and is extendable to encapsulation of a group of atoms. Indeed, encapsulated clusters of other elements such as Ge, Sn, [87,89], and some metals have been predicted as well as produced in the laboratory such as Al@Pb$_{12}$$^+$ [93].

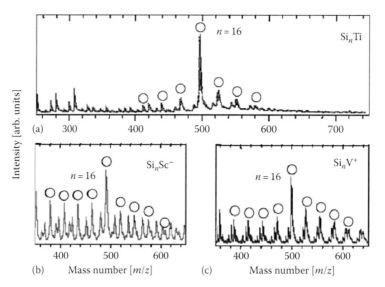

FIGURE 4.13 Mass spectra showing size-selective formation of (a) TiSi$_{16}$ neutrals, (b) ScSi$_{16}$ anions, and (c) VSi$_{16}$ cations. (Reproduced from Koyasu, K. et al., *J. Am. Chem. Soc.*, 127, 4995, 2005. With permission; Courtesy of A. Nakajima.)

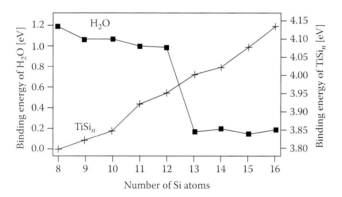

FIGURE 4.14 Binding energy (left scale) between TiSi$_N$ and a H$_2$O molecule, N = 8–16. The right scale shows the binding energy per atom of TiSi$_N$ clusters. Experiments show little abundance of TiSi$_N$ clusters with H$_2$O molecules beyond N = 12. (From Kawamura, H. et al., *Phys. Rev. B*, 70, 193402, 2004. With permission.)

As the bonding between the endohedral atom and the cage is very strong, the relative variation in the atomic sizes of the endohedral atom and the cage gives much freedom to produce a wide variety of highly stable species.

Further studies on assembly of metal-encapsulated silicon clusters showed [63,94] that one can form nanotube structures of silicon by metal encapsulation. Singh et al. [63] assembled Be@Si$_{12}$

clusters, which have some sp^3 bonding character, as shown in Figure 4.15. However, when two clusters of Be@Si$_{12}$ interact, they form hexagonal rings of silicon with Be atoms in between. This structure can be extended in the form of an infinite nanotube, which is metallic. Within the hexagonal ring, there is sp^2 bonding with two lobes of the sp^2 hybrid orbitals pointing toward each other in a silicon hexagon and one lob on each silicon atom points outward of the hexagonal ring. Such lobes on different hexagons are pi bonded in an infinite nanotube, giving rise to its metallic character. The p$_z$ orbital on each silicon atom links hexagons with a covalent bond. These studies showed that metal encapsulation can also be utilized to stabilize sp^2 bonding in silicon. Such kind of hexagonal structures are also found in nanowires of YbSi$_2$ and ErSi$_2$ silicides [95] that have been grown on substrates using anisotropy in the lattice parameters. Also following the discovery of silicon nanotubes, experiments were performed [96] on Be deposition on a Si(111) surface and the features on the surface as seen from scanning tunneling microscopy (STM) images were interpreted as representatives of the structures obtained from calculations. Several subsequent studies have been done and even magnetic nanotubes have been obtained [97] from ab initio calculations when magnetic atoms such as Mn, Fe, and Ni are doped.

Also nanowires of Zr@Si$_{16}$ fullerenes have been found [98] to be semiconducting. In an earlier work on cluster assembly, Si$_{24}$ clusters with fullerene structures were assembled to form a nanowire [99].

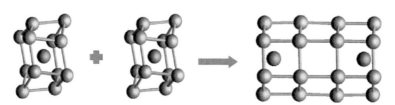

FIGURE 4.15 Assembly of two Be@Si$_{12}$ clusters leads to a tubular structure of Si with sp^2 bonding. This can be extended to form infinite nanotubes. (After Singh, A.K. et al., *Nano Lett.*, 2, 1243, 2002. With permission.)

4.4.5 Nanocoating of Gold: The Finding of a Fullerene of Al–Au

Traditionally, gold plating has been and is widely used in jewelry and other forms in bulk. However, in a recent study [100], nanocoating of gold has been studied on clusters. When gold is plated on nanoparticles, it is interesting to know what happens because in nanoform gold is quite different from bulk [13]. Using ab initio calculations, small gold clusters having up to 13 atoms have been found to have planar structures while Au_{20} is magic and has a tetrahedral structure [101] in which all atoms lie on the surface as shown in Figure 4.16. Also a fullerene structure of Au_{32} (Figure 4.16) has been found to be stable [102]. Small gold clusters have been surprisingly found to be catalytically interesting [12] although in bulk, gold is the most noble metal. Therefore, different nanoforms of gold may have interesting electronic properties and catalytic behavior. As discussed earlier in this section, an $Al_{12}Si$ or Al_{13}^- cluster has special stability in the form of a symmetric icosahedron, Kumar [100] considered coating of the 20 faces of this icosahedron with 20 gold atoms that formed a dodecahedron. Optimization of this structure and other isomers showed that Al atoms are more favorable when placed outside the dodecahedron of gold on its 12 faces such that there is a surface compound formation in which each Al atom interacts with five Au atoms and each Au atom interacts with three Al atoms as shown in Figure 4.16. As compared to Al_{13} and Au_{20}, both of which are magic structures, there is a large gain in energy of 0.55 eV/atom when the compound fullerene is formed. The binding energy of this fullerene structure (about 3 eV/atom) is much higher as compared to the value of about 2.46 eV/atom for Au_{32} also and it shows that the compound fullerene of Al–Au is energetically very stable. This finding also pointed to the possibility of such structures of other compounds. Analysis of the electronic structure of this compound fullerene showed the stability to be related to the shell closing at 58 valence electrons and

an empty cage fullerene $Al_{12}Au_{20}^{2-}$ (Figure 4.16) was suggested to be very stable with a HOMO–LUMO gap of 0.41 eV. This gap is much smaller than the values of 1.88 and 1.78 eV for Al_{13}^- and Au_{20}, respectively, and therefore the interaction of this fullerene with atoms and molecules is likely to be quite different from Al_{13} and Au_{20}. Subsequently, it has been found that a fullerene-like structure in which two Au atoms are inside the cage has lower energy. It is also possible to have endohedral fullerenes such as $Au@Al_{12}Au_{20}^-$, $Au_2@Al_{12}Au_{20}$, $Au_3@Al_{12}Au_{20}^+$, and $Al@Al_{12}Au_{20}^-$ (here the endohedral Al atom contributes only one electron to the cage), all of which have effectively 58 valence electrons.

4.4.6 Clusters and Nanostructures of Transition Metals: Designing Novel Catalysts

From technological point of view, clusters and nanoparticles of transition metals are very important in catalysis and in magnetic and optical applications. Clusters of metals such as Fe, Rh, Ru, Pd, and Pt and their alloys as well as specific atomic arrangements are of particular interest. Clusters of Fe have been observed to have high magnetic moment [103] and for a long time this has remained unresolved. On the other hand, clusters of nonmagnetic elements such as Rh, Pd, and Pt become magnetic. In recent years, much attention has been and is being paid to understand the atomic and electronic structures of these clusters and to develop new structures.

The magnetic moments in Rh clusters were observed in the early 1990s [104] but these could not be properly understood for a long time. About a decade later, detailed ab initio calculations were performed, which gave a surprising result of simple cubic structures of Rh clusters to be most favorable [14] at least up to a size of 27 atoms. These cubic structures are stabilized by eight-center bonding as shown in Figure 4.17. These results showed some interesting aspects of bonding in these clusters. Intuitively

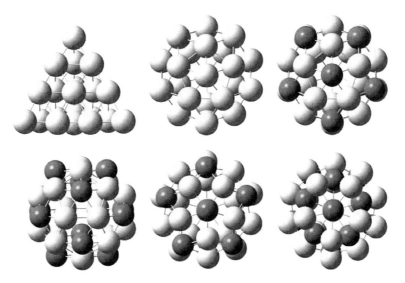

FIGURE 4.16 Atomic structures of (top from left to right) Au_{20}, Au_{32}, and $Al_{13}Au_{20}^-$, and (bottom from left to right) $Al_{12}Au_{20}^{2-}$, $Al_{12}Au_{21}^-$, and $Al_{12}Au_{21}^-$ (another view). Dark (light) balls represent Al(Au) atoms.

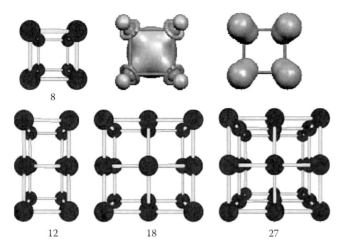

FIGURE 4.17 (Top) Cubic structure of Rh_8, eight center bonding, and spin polarization. (Bottom) cubical structures of Rh_{12}, Rh_{18}, and Rh_{27} clusters. (After Bae, Y.-C. et al., *Phys. Rev. B*, 72, 125427, 2005. With permission.)

one would consider that a lower coordination leads to a higher magnetic moment. However, in Rh cubic clusters although the coordination is significantly lower than in an icosahedron, the cubic clusters were found to have lower magnetic moments as compared to icosahedral clusters and these were in very good agreement with the experimental data as shown in Figure 4.18 and thus these studies resolved a long standing problem. The lowering of the magnetic moments is due to the specific bonding in these clusters. The magnetic moments oscillate as the size increases and become quite small for clusters with more than about 60 atoms.

The binding energy of clusters in different structures is shown in Figure 4.18 and for some sizes cuboctahedral and decahedral clusters become close in energy with cubic isomers or lie slightly lower in energy, but generally cubic clusters have lower energy

up to $n = 27$ and icosahedral clusters lie much higher in energy. The HOMO–LUMO gap of clusters with $n = 8$, 12, and 18 has local maximum value and the average nearest neighbor bond lengths are shorter in cubic clusters compared with other isomers as also shown in Figure 4.18.

Clusters of Pt are although very important catalysts and are used in fuel cells, which are currently attracting great attention, proper structures of these clusters were not known. Recently, extensive ab initio calculations [6] on clusters and nanoparticles with sizes of up to about 350 atoms have been carried out and very interesting findings have been made. Often clusters of many elements tend to attain icosahedral structure before attaining bulk atomic arrangement. However, Pt clusters have been found to attain bulk atomic structure at an early stage of about 40 atoms in the form of octahedral clusters (Figure 4.19), which have high dispersion and which is good for catalysis. The small clusters are planar with a triangle of six atoms as well as a square of nine atoms that are important building blocks. Pt_{10} is a tetrahedron with all triangular faces of six atoms while Pt_{12} is a prism made up of two Pt_6 clusters. Pt_{14} is a pyramid with a square base of Pt_9 and four triangular faces of Pt_6. Pt_{18} is again a prism with three Pt_6 clusters stacked on top of each other, forming three square faces of Pt_9. All these clusters have all atoms on the surface. Interestingly, a tetrahedron of Pt_{20} with 10-atom triangular faces is not of the lowest energy. Similarly a planar structure with a triangle of Pt_{10} is not of the lowest energy, showing the importance of Pt_6 triangle. Continuing this behavior, Pt_{22} has a decahedral cage structure with $10 Pt_6$ type faces. For Pt_{27}, a simple cubic structure with all the six square faces of Pt_9 type has been found to be of lowest energy. Beyond this size, simple cubic structures continue but soon there is a transition to octahedral structure at Pt_{44}. From calculations on octahedral clusters and icosahedral clusters, it has been concluded that icosahedral clusters have higher energy (see Figure 4.19) and therefore it has been argued that for Pt clusters and nanoparticles,

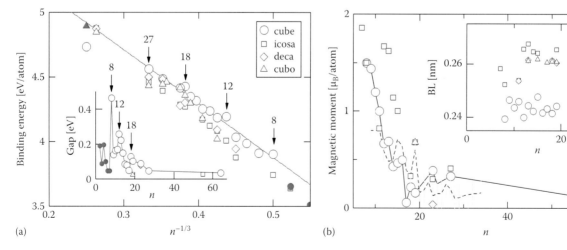

FIGURE 4.18 (a) Variation of the binding energy of different isomers of Rh_n clusters and the HOMO–LUMO gap of cubic clusters which are of the lowest energy in the small size range. (b) The magnetic moments of Rh_n clusters in cubic structures agree well with the experimental data shown by vertical lines whereas the icosahedral isomers have much higher magnetic moments. The inset also shows variation in the nearest neighbor bond lengths in different isomers. Cubic clusters have short bonds. (After Bae, Y.-C. et al., *Phys. Rev. B*, 72, 125427, 2005. With permission.)

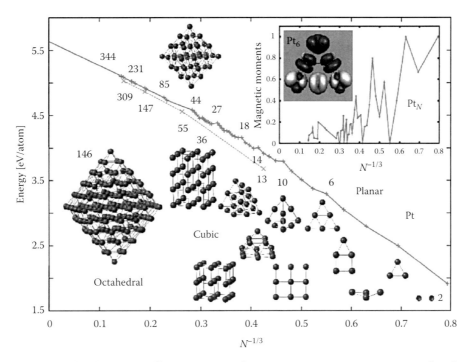

FIGURE 4.19 The variation in the binding energy of lowest energy Pt clusters. Lowest energy atomic structures for a few sizes are also shown. Also shown by crosses is the binding energy of icosahedral clusters for $N = 13, 55, 147,$ and 309, which have lower binding energy. The inset shows the variation in the magnetic moments and the spin polarization for Pt_6 in which antiferromagnetic coupling can be seen. (After Kumar, V. and Kawazoe, Y., *Phys. Rev. B*, 77, 205418, 2008.)

icosahedral growth may not occur at any size. The small clusters have magnetic moments and there is an oscillatory behavior with the variation in size, but the magnetism is weak as compared to Pd clusters [105]. Also in some cases such as Pt_6 shown in inset of Figure 4.19, the coupling is antiferromagnetic between two groups of atoms that are individually ferromagnetically coupled. In the large size range of clusters, the variation of the binding energy of Pt_N with $N^{-1/3}$ becomes nearly linear and an extrapolation of the binding energies of clusters in the limit of $N \to \infty$ gives the cohesive energy of bulk Pt. In these calculations, the ionic pseudopotential was taken from the projector-augmented wave formalism [30] with relativistic treatment, which is important for a proper description of Pt clusters. These novel structures of Pt suggest that it could be possible to manipulate and transform Pt in to other forms by alloying as well as to develop overlayers of Pt on cheaper materials. Studies are being carried out [106] on this aspect as well as on supported clusters on graphite to find if the structures of clusters remain preferred when Pt is deposited on graphite as well as the interaction of the supported structures with atoms and molecules.

In contrast to Pt in which relativistic effects are more important and the kinetic energy of the electrons could be lowered by having open structures, clusters of Pd [105] and Ni [107] have close packed structures. Extensive calculations on Pd clusters with up to 147 atoms showed [105] icosahedral growth to be favorable. Pd clusters attain magnetic moments that are distributed over the whole cluster. Interaction with H and O was shown to quench the magnetic moment of Pd_{13}. The total

density of states of icosahedral Pd_{147} with (111) type of faces was found to show significant differences from the density of states of a (111) surface of bulk Pd. Similar deviations of the total density of states have been obtained for Pt_{344} nanoparticles as shown in Figure 4.20. In this case, all the faces are of fcc (111) type as also in an icosahedron. The density of states starts developing some features as in the density of states of bulk (111) surface, but particularly near the HOMO, the deviations are significant and accordingly it has been suggested that the catalytic behavior of such clusters could differ from that of a bulk surface.

Stern Gerlach experiments [103] on clusters of iron at 120 K showed high magnetic moments of about $3\mu_B$/atom for clusters with up to about 120 atoms (bulk value $2.2\mu_B$/atom) while for clusters of Co and Ni (temperature 78 K), the magnetic moments decrease more rapidly toward the bulk value of 1.72 and $0.6\mu_B$/atom, respectively, as the cluster size increases. No proper understanding of the structures and high magnetic moments of Fe clusters could be obtained for a long time, though for Ni and Co clusters, icosahedral structures were inferred from photoemission [108] and chemical reactivity [109] experiments. Recently extensive calculations [110] on Fe clusters showed the lowest energy structures to have high average magnetic moments of $\sim 3\mu_B$/atom in a wide range of sizes in agreement with experiments.

In an interesting finding, Nb clusters were found [15] to have permanent electric dipole moments. These dipole moments vanished beyond a certain temperature that was dependent on size. This behavior of Nb clusters was interpreted to be related

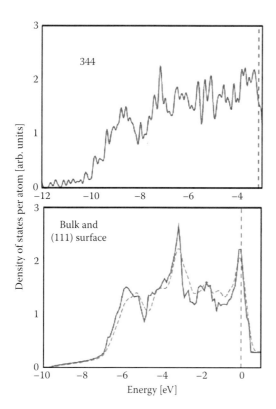

FIGURE 4.20 The electronic density of states of an octahedral Pt_{344} cluster. Also shown are the density of states of bulk Pt and (111) surface of bulk Pt (broken curve). The vertical line shows HOMO for Pt_{344} and the Fermi Energy for bulk. (After Kumar, V. and Kawazoe, Y., *Phys. Rev. B*, 77, 205418, 2008.)

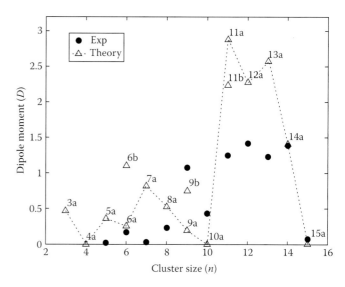

FIGURE 4.21 Calculated and experimental values of the electric dipole moments in Nb clusters. a and b show different isomers. (After Kumar, V. and Kawazoe, Y., *Phys. Rev. B*, 65, 125403, 2002; Andersen, K.E. et al., *Phys. Rev. Lett.*, 93, 246105, 2004; Andersen, K.E. et al., *Phys. Rev. Lett.*, 95, 089901, 2005; Andersen, K.E. et al., *Phys. Rev. B*, 73, 125418, 2008. With permission.)

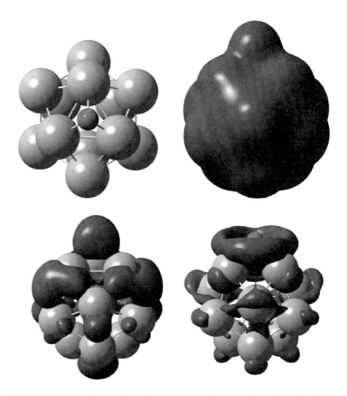

FIGURE 4.22 Interaction of an oxygen atom on a face of an icosahedral Ta_{12} (top left). Top right shows the total electronic charge density while the bottom left (right) shows excess (depletion) of charge compared with the sum of the charge densities of Ta_{12} and an oxygen atom at the same positions as in $Ta_{12}O$. One can see excess of charge around O atom from predominantly neighboring Ta atoms.

to nascent superconductivity in Nb clusters with a transition temperature, which increased when the cluster size was reduced. Ab initio calculations [111] on Nb clusters showed the permanent dipole moments to be associated with the asymmetry in atomic structures. The calculated electric dipole moments on Nb clusters compare well with those obtained from experiments as shown in Figure 4.21. Some deviations in the values from experimental data are accounted from the fact that new isomers have been found [112] that are lower in energy and their electric dipole moments are closer to the measured values. As one encounters asymmetric structures often in clusters, this phenomenon is not associated with only Nb clusters and is more general [113]. Nb clusters were also found to exhibit significant variation in their reactivity depending upon the charged state of the clusters [114] and in some cases such as Nb_{12}, the possibility of the existence of isomers was concluded. For certain sizes, the dipole moments were found to be zero, indicating the possibility of a symmetric structure. For Nb_{12}, ab initio calculations [115] suggested an icosahedral isomer to be lowest in energy and it behaves like a superatom. Similar results have been obtained for Ta_{12}. These clusters interact exohedrally with an oxygen atom (Figure 4.22) and endohedrally with Fe, Ru and Os as shown in Figure 4.23. Particularly superatoms $M@X_{12}$ have been shown to be formed with M = Fe, Ru, and Os and X = Nb and Ta. In the case of oxygen interaction, an oxygen

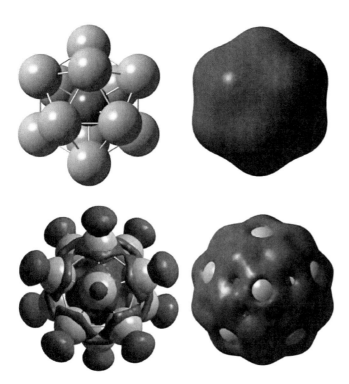

FIGURE 4.23 Interaction of an Os atom at the center of an icosahedral Ta_{12} (top left). Top right shows the total electronic charge density while the bottom left (right) shows excess (depletion) of charge compared with the sum of the charge densities of Ta_{12} and an Os atom at the same positions as in $Ta_{12}Os$. One can see excess of charge around Os atom and to the d_z^2 orbitals on Ta atoms from the icosahedral shell of Ta atoms.

atom was placed inside an icosahedron, but it came out of the cage. There is charge transfer to oxygen from predominantly neighboring Ta atoms. On the other hand, an Os atom inside the cage leads to a highly symmetric icosahedral superatom. There is charge transfer to Os atom and to the d_z^2 type orbitals on Ta atoms from the Ta cage. These results have shown that it is also possible to have superatoms of transition metals with large HOMO–LUMO gaps though the atoms have open d-shells. Similar studies have been made on $M@Pd_{12}$ and $M@Pt_{12}$ as well as on large core-shell structures to find different structures of Pd and Pt and to understand their reactivity.

4.5 Nanostructures of Compounds

Mixing of elements is an important way to modify properties of materials and there are a large variety of compounds such as oxides, sulfides, borides, nitrides, and silicides and II–VI and III–V compounds that are technologically important. Nanomaterials of compounds may have stoichiometries and structures that are different from bulk and therefore it would be important to understand them and know the behavior and properties of such materials when they are made small. Here we discuss some examples that demonstrate the point.

4.5.1 Novel Structures of CdSe Nanoparticles

Clusters of II–VI compounds have attracted great interest because of their interesting optical properties and possibilities of many applications in devices. Several studies on CdSe nanoparticles reported magic nature of clusters of about 1.5 nm size and they were considered [116] to be fragments of bulk CdSe but precise composition was lacking. It was in 2004 that nanoparticles of CdSe with precise composition of $(CdSe)_{33}$ and $(CdSe)_{34}$ were produced in macroscopic quantities [117] and characterized using mass spectrometry. High abundance of ZnS clusters was also reported independently by Martin [118] in this size range. Other clusters that were observed in the mass spectrum [117] of CdSe were $(CdSe)_{13}$ and $(CdSe)_{19}$ with much less abundance. There was almost no abundance of other clusters/nanoparticles. This identification of the number of Cd and Se atoms was used to understand the atomic structures of these clusters. Extensive calculations on a wide range of cluster sizes showed preference for novel cage structures of $(CdSe)_{13}$ and $(CdSe)_{34}$ in which one CdSe molecule is incorporated in a cubic cage of $(CdSe)_{12}$ analogous to that of $(BN)_{12}$ with a Se atom at the center and a $(CdSe)_6$ cluster is encapsulated within a $(CdSe)_{28}$ cubic cage so that there is covalent bonding not only on the cage as in BN cages but also within the cage (see Figure 4.24). Similar to silicon, as one goes down in the periodic table, empty cage structures become less favorable in these compounds unlike that of carbon or BN and stuffed cages become lowest in energy. These novel structures have lower energy as compared to those obtained from optimization of bulk fragments [119], some of which attain features similar to those of the lowest energy structure. The endohedral filling of the cages weakens the dangling bonds on the surface. Subsequent studies [120] on the optical properties of these clusters have supported these structures. These findings have opened a new way in the understanding of the properties of nanoparticles of II–VI compounds. While in these experiments bulk stoichiometric clusters were obtained, it is quite possible that under different conditions, other stoichiometries may also exist.

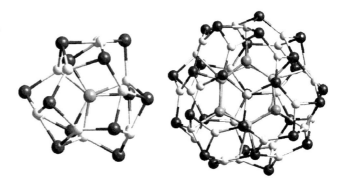

FIGURE 4.24 Lowest energy structures of (left) $(CdSe)_{13}$ (right) $(CdSe)_{34}$. Dark (on the cage) and light (inside the cage) atoms represent Se while white (on the cage) and light dark (inside the cage) balls represent Cd atoms. (After Kasuya, A. et al., *Nat. Mater.*, 3, 99, 2004. With permission.)

4.5.2 Nanostructures of Mo–S: Clusters, Platelets, and Nanowires

Bulk MoS_2 has layered structure similar to graphite and is useful as a lubricant. There has been interest in the possibilities of fullerene structures of MoS_2 and WS_2 and onion-shaped fullerenes [121] as well as multiwall tubular structures [122] have been obtained. In bulk form, Mo–S is very versatile and compounds with stoichiometries of MoS_2, Mo_2S_3, and MoS_3 are known to exist. Also a large number of cluster compounds have been synthesized such as Chevrel phases [123] with Mo_6S_8 clusters and other compounds in which such clusters condense to form finite and infinite chains. Also LiMoSe bulk structure [124] was discovered long ago. In this phase, infinite nanowires of MoSe condense to form a bulk phase in which Li intercalates between the nanowires. Such a phase has been considered interesting for Li ion battery applications. In recent years, another interesting application of nanoparticles of Mo–S system in the form of platelets has been found [125] in the removal of sulfur in petroleum refining. Recently, Murugan et al. [126] have studied bulk fragments of MoS_2 and found a tendency for clustering of Mo atoms, which form a core while S atoms cap this core. The optimized structures of bulk fragments were found to lie significantly higher in energy than other isomers in which S atoms prefer to cap a face or an edge of a Mo polyhedron and the remaining atoms occupy terminal positions on the vertices.

For two and three Mo atoms, edge capping is favored but for larger clusters, face capping becomes generally more favorable. The stability of these structures was understood from d-s-p hybrids, which prefer bond angles of 73°09′ and 133°37′ and which are also found in bulk MoS_2 (angles 82° and 136°) as well as from s-d hybrids, which favor bond angles 116°34′ and 63°26′. Extensive calculations [127] on Mo_nS_m clusters showed a hill-shaped structure of the energy versus sulfur concentration as shown in Figure 4.25 for Mo_4S_m series so that there is an optimal composition of S:Mo that has the highest binding energy. Accordingly, it was found that the optimal stoichiometry of small Mo–S clusters has S:Mo ratio of less than 2 (except for $n = 1$ and 2) such as in Mo_3S_5, Mo_4S_6, and Mo_6S_8. Among these, it was found that Mo_6S_8 has special stability and this finding goes well with the occurrence of Mo_6S_8 cluster compounds. Further calculations on larger clusters showed that nanowire-like structures in which such Mo_6S_8 clusters condense, such as in Mo_9S_{11} and $Mo_{12}S_{14}$, (Figure 4.26) were more favorable than other 3D structures [128]. Such finite chains are also found in bulk compounds. In the limit of n tending to infinity, one obtains an infinite MoS nanowire, which is metallic. Such nanowires have been formed in the laboratory [129] and their properties have been manipulated by I doping. In contrast to carbon nanotubes, which are produced with different diameters and chiralities that lead to different properties, MoS nanowires can be produced uniquely with ease and therefore these one-dimensional conductors could be useful in developing contacts in devices at the nanoscale. Hexagonal

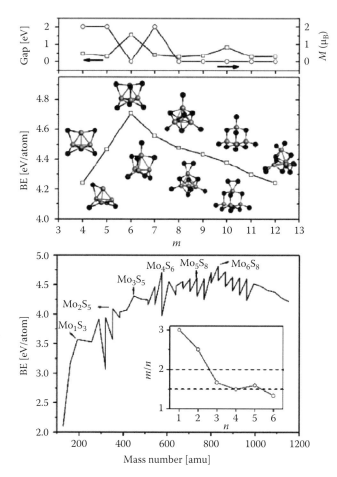

FIGURE 4.25 (Top) The variation in the binding energy of Mo_4S_m clusters as m is varied. The highest binding energy is achieved for Mo_4S_6, which is a magic structure and has the largest HOMO–LUMO gap in this series. The magnetic moment on this cluster is zero, but some of the other clusters become magnetic. (Bottom) The variation in the binding energy as the mass of the clusters is varied representing different combinations. Some of the cluster stoichiometries have been listed in the figure corresponding to locally high binding energies, the highest value being for Mo_6S_8. The optimal stoichiometry m/n changes from about 3 to 1.5 as the cluster size grows and therefore in the small size range the optimal stoichiometries are different from the bulk phases. (After Murugan, P. et al., *J. Phys. Chem. A*, 111, 2778, 2007. With permission.)

assemblies (Figure 4.27) of such nanowires have been shown [130] to be weakly bonded. An interesting finding is that such assemblies can be intercalated with Li and this leads to further metallization of this assembly. It has been suggested that such assemblies could be interesting one-dimensional electron and ion conductors and may have potential application as cathode material for 1.5 V Li ion batteries in which the change in the structure of the material when Li goes in and out could be much smaller as compared to the presently used $LiCoO_2$ cathode material. Also a new metallic phase of the $Li_3Mo_6S_6$ nanowire assemblies was predicted from ab initio calculations with monoclinic structure.

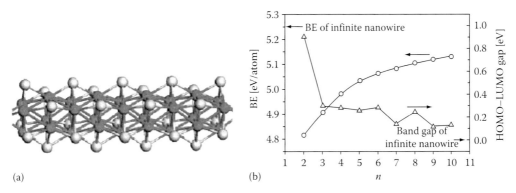

(a)

(b)

FIGURE 4.26 (a) Part of an infinite MoS nanowire. Dark (light) balls represent Mo (S) atoms. (b) Plot of the binding energy and HOMO–LUMO gap as the number n of $(MoS)_3$ is increased. (After Murugan, P. et al., *Nano Lett.*, **7**, 2214, 2007. With permission.)

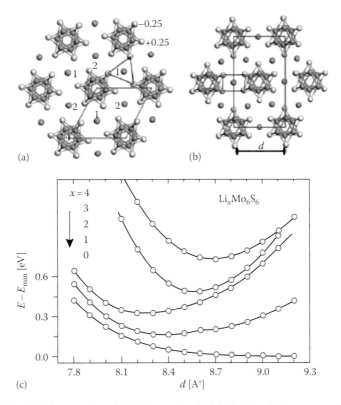

(a)

(b)

(c)

FIGURE 4.27 A hexagonal assembly of MoS nanowires which is intercalated with Li. (a) and (b) correspond to Li_2MoS and Li_3MoS intercalated assemblies. (c) The change in the binding energy of pristine MoS nanowire assemblies as well as those of Li-doped for varying Li concentration x as the internanowire spacing is changed. For $x = 0$, the variation is very flat due to very weak van der Waals bonding which is not well described within GGA. However, with increasing x, the system develops more metallic bonding. (After Murugan, P. et al., *Appl. Phys. Lett.*, **92**, 203112, 2008. With permission.)

4.6 Summary

In summary, we have presented results of some recent developments related to clusters and nanoparticles of metals and semiconductors where calculations played a very important role such as the discovery of silicon fullerenes and other novel polyhedral caged structures by metal encapsulation, some of which have been subsequently realized in laboratory. Metal encapsulation provides a novel way to produce size-selected species in high abundance, which are required for their applications.

The properties can be tailored by choosing the right combination of the metal atom. This idea is not specific to Si and can be applied to other semiconductors and metals to produce size-selected species with specific properties. Indeed 18 valence electron magic clusters of Cu and other coinage metals stabilized with transition metal atoms are examples of metal-encapsulated clusters of metals. Also, we presented results of high-symmetry icosahedral magic clusters of transition metals such as $M@Nb_{12}$ and $M@Ta_{12}$ with M = Fe, Ru, and Os that are promising for cluster assembly. As we go down in the periodic table in a column, some interesting trends are found. For elements such as C and compound BN, empty cage fullerene structures are favored while for Si and II–VI compounds such as CdSe stuffed cage structures become more favorable as we discussed for $(CdSe)_{13}$ and $(CdSe)_{34}$. So there is tendency to have tetrahedral bonding not only on the cage but also inside the cage. On the other hand, for metals such as Ni, Pd, and Pt or Cu, Ag, and Au, or Co and Rh, as one goes down in a column in the periodic table, small clusters have a tendency to change from close-packed icosahedral structures (e.g., for Co, Ni, Cu) to relatively open structures such as we discussed for Rh, Pt, and Au. Also by alloying of Au with Al, empty fullerene cage structures have been found and this development could further act as a catalyst to look for such novel structures of other compounds. Nanogold is a good catalyst and by alloying, one can tailor its structure and properties. Further nanostructures of Rh and Pt develop magnetism though bulk Rh and Pt are nonmagnetic. The properties of these catalytically important clusters can be further tailored by doping that can make drastic changes in the structure and properties similar to Si. This field has not been well explored and ab initio calculations are expected to contribute to such developments and to their applications. The hydrogenated cages of Si offer interesting possibilities of endohedral doping of atoms to produce slaved atoms with free atom-like behavior and also to have derivatives by replacement of H atoms.

We also discussed nanoclusters of Mo–S that have different stoichiometries from bulk and a tendency to have Mo polyhedral core and S atoms outside the core. Among these, Mo_6S_8 is very special and one can have assemblies in the form of finite and infinite nanowires that have potential as conductors in nanodevices. Also assemblies of such nanowires are potential electron and ion conductors such as with Li doping. We also discussed the developments related to the finding of nanotube structures of Si, which are also metallic. These results have shown the possibility of stabilizing sp^2-bonded structures of Si and with the current excitement in research on graphene, such sp^2-bonded Si nanostructures and silicene, an analog of graphene, could become very interesting. Some of the examples discussed here demonstrate the great potential of computer simulations in developing new materials and structures for nanotechnology applications and we hope that exciting discoveries would be made in the near future using computer simulations and that would expedite material development.

Acknowledgments

I would like to express my gratitude to Y. Kawazoe for all the support and cooperation at the Institute for Materials Research (IMR), Tohoku University. I am thankful to Y.-C. Bae, R.V. Belosludov, T.M. Briere, M. Itoh, A. Kasuya, H. Kawamura, C. Majumder, P. Murugan, N. Ota, F. Pichierri, A.K. Singh, M.H.F. Sluiter, and V. Sundararajan for fruitful collaborations and many discussions. I gratefully acknowledge the support of the staff of the Center for Computational Materials Science of IMR-Tohoku University for the use of SR8000/H64 supercomputer facilities, the computing facilities at RICS, AIST, Tsukuba, the support of K. Terakura and T. Ikeshoji as well as the staff of the Centre for the Development of Advanced Computing, Bangalore for allowing the use of their supercomputing facilities and excellent support. I acknowledge with sincere thanks the financial support from the Asian Office of Aerospace Research and Development.

References

1. H.W. Kroto, J.R. Heath, S.C. O'Brien, R.F. Curl, and R.E. Smalley, *Nature (London)* **318**, 162 (1985).
2. S. Iijima, *Nature* **354**, 56 (1991).
3. V. Kumar, *Sadhana* **28**, 815 (2003).
4. V. Kumar, *Comput. Mater. Sci.* **36**, 1 (2006).
5. P. Hohenberg and W. Kohn, *Phys. Rev.* **136**, B864 (1964).
6. V. Kumar and Y. Kawazoe, *Phys. Rev. B* **77**, 205418 (2008).
7. W. Krätschmer, L.D. Lamb, K. Fostiropoulos, and D.R. Huffman, *Nature (London)* **347**, 354 (1990).
8. L.T. Canham, *Appl. Phys. Lett.* **57**, 1046 (1990).
9. G. Belomoin, J. Therrien, A. Smith, S. Rao, R. Twesten, S. Chaieb, M.H. Nayfeh, L. Wagner, and L. Mitas, *Appl. Phys. Lett.* **80**, 841 (2002).
10. L. Pavesi, S. Gaponenko, and L. Dal Negro (Eds.), *Towards the First Silicon Laser*, NATO Science Series, Vol. **93**, Kluwer Academic Press, Dordrecht, the Netherlands (2003); A. Polman, *Nat. Mater.* **1**, 10 (2002).
11. P. Mulvaney, *MRS Bull.* 1009, December (2001).
12. M. Haruta, *Catal. Today* **36**, 153 (1997); M. Haruta, *Now and Future* **7**, 13 (1992); M. Haruta, *The Chem. Rec.* **3**, 75 (2003).
13. X.P. Xing, B. Yoon, U. Landman, and J.H. Parks, *Phys. Rev. B* **74**, 165423 (2006).
14. Y.-C. Bae, V. Kumar, H. Osanai, and Y. Kawazoe, *Phys. Rev. B* **72**, 125427 (2005).
15. R. Moro, X. Xu, S. Yin, and W.A. de Heer, *Science* **300**, 1265 (2003).
16. X.-G. Gong and V. Kumar, *Phys. Rev. Lett.* **70**, 2078 (1993).
17. V. Kumar, *Phys. Rev. B* **57**, 8827 (1998).
18. V. Kumar, S. Bhattacharjee, and Y. Kawazoe, *Phys. Rev. B* **61**, 8541 (2001).
19. V. Kumar and Y. Kawazoe, *Phys. Rev. Lett.* **87**, 045503 (2001); V. Kumar and Y. Kawazoe, *Phys. Rev. Lett.* **91**, 199901 (E) (2003); V. Kumar, C. Majumder, and Y. Kawazoe, *Chem. Phys. Lett.* **363**, 319 (2002).
20. V. Kumar, in *Nanosilicon*, V. Kumar (Ed.), Elsevier, Oxford, U.K. (2008).
21. K. Koyasu, M. Akutsu, M. Masaaki, and A. Nakajima, *J. Am. Chem. Soc.* **127**, 4995 (2005); K. Koyasu, J. Atobe, M. Akutsu, M. Mitsui, and A. Nakajima, *J. Phys. Chem. A* **111**, 42 (2007).
22. P. Blaha, K. Schwarz, G.K.H. Marsden, D. Kvasnicka, and J. Luitz, *Wien2k*, ISBN 3-9501031-1-2.
23. M.J. Frisch, G.W. Trucks, H.B. Schlegel, G.E. Scuseria, M.A. Robb, J.R. Cheeseman, G. Scalmani et al., Gaussian 09, Revision A.1, Gaussian, Inc., Wallingford, CT (2009).

24. R.M. Martin, *Electronic Structure: Basic Theory and Practical Methods*, Cambridge University Press, Cambridge, U.K. (2004).

25. R. Car and M. Parrinello, *Phys. Rev. Lett.* **55**, 2471 (1985).

26. G. Kresse and J. Furthmüller, *Phys. Rev. B* **54**, 11169 (1996); G. Kresse and D. Joubert, *Phys. Rev. B* **59**, 1758 (1999).

27. G.B. Bachelet, D.R. Hamann, and M. Schlüter, *Phys. Rev. B* **26**, 4199 (1982).

28. N. Troullier and J.L. Martins, *Phys. Rev. B* **43**, 1993 (1991).

29. D. Vanderbilt, *Phys. Rev. B* **41**, 7892 (1990); G. Kresse and J. Hafner, *J. Phys.: Condens. Matter* **6**, 8245 (1994).

30. P.E. Blöchl, *Phys. Rev. B* **50**, 17953 (1994).

31. W.A. de Heer, *Rev. Mod.* **65**, 611 (1993); V. Kumar, K. Esfarjani, and Y. Kawazoe, in *Clusters and Nanomaterials*, Springer Series in Cluster Physics, Y. Kawazoe, T. Kondow, and K. Ohno (Eds.), Springer, Heidelberg, Germany, p. 9 (2002).

32. W.D. Knight, K. Clemenger, W.A. de Heer, W.A. Saunders, M.Y. Chou, and M.L. Cohen, *Phys. Rev. Lett.* **52**, 2141 (1984); M. Itoh, V. Kumar, and Y. Kawazoe, *Int. J. Mod. Phys. B* **19**, 2421 (2005); M. Itoh, V. Kumar, and Y. Kawazoe, *Phys. Rev. B* **73**, 035425 (2006); G. Wrigge, M. Astruc Hoffmann, and B. v. Issendorff, *Phys. Rev. A* **65**, 063201 (2002).

33. X. Li, H. Wu, X.-B. Wang, and L.-S. Wang, *Phys. Rev. Lett.* **81**, 1909 (1998); J. Akola, M. Manninen, H. Häkkinen, U. Landman, X. Li, and L.-S. Wang, *Phys. Rev. B* **62**, 13216 (2000).

34. M. Pellarin, B. Baguenard, C. Bordas, M. Broyer, J. Lermé, and J.L. Vialle, *Phys. Rev. B* **48**, 17645 (1993).

35. B. Baguenard, M. Pellarin, C. Bordas, J. Lerme, J.L. Vialle, and M. Broyer, *Chem. Phys. Lett.* **205**, 13 (1993).

36. J. Zheng, P.R. Nicovich, and R.M. Dickson, *Annu. Rev. Phys. Chem.* **58**, 409 (2007).

37. I. Katakuse, Y. Ichihara, Y. Fujita, T. Matsuo, T. Sakurai, and H. Matsuda, *Int. J. Mass Spectrom. Ion Proc.* **74**, 22 (1986).

38. K.E. Schriver, J.L. Persson, E.C. Honea, and R.L. Whetten, *Phys. Rev. Lett.* **64**, 2539 (1990).

39. R.E. Leuchtner, A.C. Harms, and A.W. Castleman Jr., *J. Chem. Phys.* **91**, 2753 (1989); R.E. Leuchtner, A.C. Harms, and A.W. Castleman Jr., *J. Chem. Phys.* **94**, 1093 (1991).

40. V. Kumar and Y. Kawazoe, *Phys. Rev. B* **64**, 115405 (2001).

41. V. Kumar, *Phys. Rev. B* **60**, 2916 (1999).

42. O.C. Thomas, W. Zheng, and K.H. Bowen Jr., *J. Chem. Phys.* **114**, 5514 (2001).

43. V. Kumar and V. Sundararajan, *Phys. Rev. B* **57**, 4939 (1998).

44. V. Kumar, *Phys. Rev. B* **23**, 3756 (1981).

45. X. Li and L.-S. Wang, *Phys. Rev. B* **65**, 153404 (2002).

46. W.-J. Yin, X. Gu, and X.-G. Gong, *Solid State Commun.* **147**, 323 (2008).

47. V. Kumar and R. Car, *Phys. Rev. B* **44**, 8243 (1991).

48. T. Diederich, T. Döppner, J. Braune, J. Tiggesbäumker, and K.-H. Meiwes-Broer, *Phys. Rev. Lett.* **86**, 4807 (2001); O.C. Thomas, W. Zheng, S. Xu, and K.H. Bowen Jr., *Phys. Rev. Lett.* **89**, 213403 (2002).

49. V. Kumar and Y. Kawazoe, *Phys. Rev. B* **63**, 075410 (2001).

50. X. Dong, G.M. Wang, and E. Blaisten-Barojas, *Phys. Rev. B* **70**, 205409 (2004).

51. V. Kumar and Y. Kawazoe, to be published.

52. O. Kostko, B. Huber, M. Moseler, and B. von Issendorff, *Phys. Rev. Lett.* **98**, 043401 (2007); T.P. Martin, T. Bergmann, H. Göhlich, and T. Lange, *Chem. Phys. Lett.* **172**, 209 (1990).

53. O. Lopez-Acevedo, J. Akola, R.L. Whetten, H. Grönbeck, and H. Häkkinen, *J. Phys. Chem. C (Letter)* **113**, 5035 (2009).

54. V. Kumar, *Prog. Crystal Growth Charac.* **34**, 95 (1997).

55. S. Neukermans, E. Janssens, H. Tanaka, R.E. Silverans, and P. Lievens, *Phys. Rev. Lett.* **90**, 033401 (2003).

56. V. Kumar, unpublished.

57. T. Höltzl, N. Veldeman, J. De Haeck, T. Veszprémi, P. Lievens, and M.T. Nguyen, *Chem. – A Eur. J.* **15**, 3970 (2009); N. Veldeman, T. Höltzl, S. Neukermans, T. Veszprémi, M.T. Nguyen, and P. Lievens, *Phys. Rev. A* **76**, 011201 (2007).

58. T. Höltzl, N. Veldeman, T. Veszpremi, P. Lievens, and M.T. Nguyen, *Chem. Phys. Lett.* **469**, 304 (2009).

59. E. Janssens, S. Neukermans, H.M.T. Nguyen, M.T. Nguyen, and P. Lievens, *Phys. Rev. Lett.* **94**, 113401 (2005).

60. R.D. Johnson, G. Meijer, and D.S. Bethune, *J. Am. Chem. Soc.* **112**, 8983 (1990).

61. W. Krätschmer, K. Fostiropoulos, and D.R. Huffman, *Chem. Phys. Lett.* **170**, 167 (1990).

62. A.F. Hebard, M.J. Rosseinsky, R.C. Haddon, D.W. Murphy, S.H. Glarum, T.T.M. Palstra, A.P. Ramirez, and A.R. Kortan, *Nature* **350**, 600 (1991).

63. A.K. Singh, V. Kumar, T.M. Briere, and Y. Kawazoe, *Nano Lett.* **2**, 1243 (2002).

64. H. Kawaji, H. Horie, S. Yamanaka, and M. Ishikawa, *Phys. Rev. Lett.* **74**, 1427 (1995).

65. K.-M. Ho, A.A. Shvartsburg, B. Pan, Z.-Y. Lu, C.-Z. Wang, J.G. Wacker, J.L. Fye, and M.F. Jarrold, *Nature (London)* **392**, 582 (1998).

66. L. Mitas, J.C. Grossman, I. Stich, and J. Tobik, *Phys. Rev. Lett.* **84**, 1479 (2000).

67. S. Yoo and X.C. Zeng, *J. Chem. Phys.* **123**, 164303 (2005); S. Yoo, X.C. Zeng, X. Zhu, and J. Bai, *J. Am. Chem. Soc.* **125**, 13318 (2003); S. Yoo, N. Shao, C. Koehler, T. Fraunhaum, and X.C. Zeng, *J. Chem. Phys.* **124**, 164311 (2006) and references therein.

68. Y. Zhao, Y.-H. Kim, M.-H. Du, and S.B. Zhang, *Phys. Rev. Lett.* **93**, 015502 (2004).

69. K. Nishio, T. Morishita, W. Shinoda, and M. Mikami, *Phys. Rev. B* **72**, 245321 (2005).

70. M. Horoi and K.A. Jackson, *Chem. Phys. Lett.* **427**, 147 (2006).

71. L. Mitas, J. Therrien, R. Twesten, G. Belomoin, and M. Nayfeh, *Appl. Phys. Lett.* **78**, 1918 (2001); M. Nayfeh and L. Mitas, in *Nanosilicon*, V. Kumar (Ed.), Elsevier, Amsterdam, the Netherlands (2007); A.D. Zdetsis, *Phys. Rev. B* **79**, 195437 (2009).

72. V. Kumar and Y. Kawazoe, *Phys. Rev. B* **75**, 155425 (2007); V. Kumar and Y. Kawazoe, *Phys. Rev. Lett.* **90**, 055502 (2003).

73. T. Kume, H. Fukuoka, T. Koda, S. Sasaki, H. Shimizu, and S. Yamanaka, *Phys. Rev. Lett.* **90**, 155503 (2003) and references therein.

74. F. Pichierri, V. Kumar, and Y. Kawazoe, *Chem. Phys. Lett.* **383**, 544 (2004).

75. M.S. Bahramy, V. Kumar, and Y. Kawazoe, *Phys. Rev. B* **79**, 235443 (2009).

76. V. Kumar, T.M. Briere, and Y. Kawazoe, *Phys. Rev. B* **68**, 155412 (2003).

77. A. Puzder, A.J. Williamson, J.C. Grossman, and G. Galli, *Phys. Rev. Lett.* **88**, 097401 (2002).

78. S. Öğüt, J.R. Chelikowsky, and S.G. Louie, *Phys. Rev. Lett.* **79**, 1770 (1997); M. Rohlfing and S.G. Louie, *Phys. Rev. Lett.* **80**, 3323 (1998).

79. S.M. Beck, *J. Chem. Phys.* **87**, 4233 (1987).

80. S.M. Beck, *J. Chem. Phys.* **90**, 6306 (1989).

81. H. Hiura, T. Miyazaki, and T. Kanayama, *Phys. Rev. Lett.* **86**, 1733 (2001).

82. K. Jackson and B. Nellermoe, *Chem. Phys. Lett.* **254**, 249 (1996).

83. V. Kumar, *Comput. Mater. Sci.* **30**, 260 (2004).

84. A.K. Singh, V. Kumar, and Y. Kawazoe, *Phys. Rev. B* **71**, 115429 (2005).

85. V. Kumar, A.K. Singh, and Y. Kawazoe, *Phys. Rev. B* **74**, 125411 (2006).

86. V. Kumar and Y. Kawazoe, *Phys. Rev. Lett.* **88**, 235417 (2002).

87. V. Kumar, A.K. Singh, and Y. Kawazoe, *Nano Lett.* **4**, 677 (2004).

88. E.N. Esenturk, J. Fettinger, and B. Eichhorn, *Chem. Commun.* 247, (2005).

89. V. Kumar and Y. Kawazoe, *Appl. Phys. Lett.* **83**, 2677 (2003); V. Kumar and Y. Kawazoe, *Appl. Phys. Lett.* **80**, 859 (2002).

90. S. Neukermans, X. Wang, N. Veldeman, E. Janssens, R.E. Silverans, and P. Lievens, *Int. J. Mass. Spectrom.* **252**, 145 (2006).

91. M. Ohara, K. Koyasu, A. Nakajima, and K. Kaya, *Chem. Phys. Lett.* **371**, 490 (2003).

92. H. Kawamura, V. Kumar, and Y. Kawazoe, *Phys. Rev. B* **71**, 075423 (2005); H. Kawamura, V. Kumar, and Y. Kawazoe, *Phys. Rev. B* **70**, 193402 (2004).

93. S. Neukermans, E. Janssens, Z.F. Chen, R.E. Silverans, P.v.R. Schleyer, and P. Lievens, *Phys. Rev. Lett.* **92**, 163401 (2004).

94. A.K. Singh, V. Kumar, and Y. Kawazoe, *J. Mater. Chem.* **14**, 555 (2004).

95. Y. Chen, D.A.A. Ohlberg, and R.S. Williams, *J. Appl. Phys.* **91**, 3213 (2002); B.Z. Liu and J. Nogami, *J. Appl. Phys.* **93**, 593 (2003); J. Nogami, B.Z. Liu, M.V. Katkov, C. Ohbuchi, and N.O. Birge, *Phys. Rev. B* **63**, 233305 (2001); C. Preinesberger, S. Vandré, T. Kalka, and M. Dähne-Prietsch, *J. Phys. D* **31**, L43 (1998); C. Preinesberger, S.K. Becker, S. Vandré, T. Kalka, and M. Dähne, *J. Appl. Phys.* **91**, 1695 (2002); Y. Chen, D.A.A. Ohlberg, G. Medeiros-Ribeiro, Y.A. Chang, and R.S. Williams, *Appl. Phys. Lett.* **76**, 4004 (2000); N. Gonzalez Szwacki and B.I. Yakobson, *Phys. Rev. B* **75**, 035406 (2007).

96. A.A. Saranin, A.V. Zotov, V.G. Kotlyar, T.V. Kasyanova, O.A. Utas, H. Okado, M. Katayama, and K. Oura, *Nano Lett.* **4**, 1469 (2004).

97. A.K. Singh, T.M. Briere, V. Kumar, and Y. Kawazoe, *Phys. Rev. Lett.* **91**, 146802 (2003).

98. V. Kumar and Y. Kawazoe, unpublished.

99. B. Marsen and K. Sattler, *Phys. Rev. B* **60**, 11593 (1999).

100. V. Kumar, *Phys. Rev. B* **79**, 085423 (2009).

101. J. Li, X. Li, H.J. Zhai, and L.S. Wang, *Science* **299**, 864 (2003).

102. M.P. Johansson, D. Sundholm, and J. Vaara, *Angew. Chem. Int. Ed.* **43**, 2678 (2004); X. Gu, M. Ji, S.H. Wei, and X.G. Gong, *Phys. Rev. B* **70**, 205401 (2004); M. Ji, X. Gu, X. Li, X.G. Gong, J. Li, and L.-S. Wang, *Angew. Chem., Int. Ed.* **44**, 7119 (2005).

103. I.M.L. Billas, A. Chhtelain, and W.A. de Heer, *Science* **265**, 1682 (1994); I.M.L. Billas, A. Chhtelain, and W.A. de Heer, *J. Magn. Magn. Mater.* **168**, 64 (1997).

104. A.J. Cox, J.G. Louderback, and L.A. Bloomfield, *Phys. Rev. Lett.* **71**, 923 (1993); A.J. Cox, J.G. Louderback, S.E. Apsel, and L.A. Bloomfield, *Phys. Rev. B* **49**, 12295 (1994).

105. V. Kumar and Y. Kawazoe, *Phys. Rev. B* **66**, 144413 (2002).

106. V. Kumar, unpublished.

107. V. Kumar and Y. Kawazoe, unpublished.

108. M. Pellarin, B. Baguenard, J.L. Vialle, J. Lerme, M. Broyer, J. Miller, and A. Perez, *Chem. Phys. Lett.* **217**, 349 (1994).

109. E.K. Parks, B.J. Winter, T.D. Klots, and S.J. Riley, *J. Chem. Phys.* **94**, 1882 (1991); E.K. Parks, T.D. Klots, B.J. Winter, and S.J. Riley, *J. Chem. Phys.* **99**, 5831 (1993).

110. V. Kumar and Y. Kawazoe, unpublished.

111. V. Kumar and Y. Kawazoe, *Phys. Rev. B* **65**, 125403 (2002); K.E. Andersen, V. Kumar, Y. Kawazoe, and W.E. Pickett, *Phys. Rev. Lett.* **93**, 246105 (2004); K.E. Andersen, V. Kumar, Y. Kawazoe, and W.E. Pickett, *Phys. Rev. Lett.* **95**, 089901 (2005); K.E. Andersen, V. Kumar, Y. Kawazoe, and W.E. Pickett, *Phys. Rev. B* **73**, 125418 (2008).

112. V. Kumar and Y. Kawazoe, to be published.

113. V. Kumar, *Comput. Mater. Sci* **35**, 375 (2006).

114. P.J. Brucat, C.L. Pettiette, S. Yang, L.-S. Zheng, M.J. Craycraft, and R.E. Smalley, *J. Chem. Phys.* **85**, 4747 (1986); M.R. Zakin, R.O. Brickman, D.M. Cox, and A. Kaldor, *J. Chem. Phys.* **88**, 3555 (1988).

115. V. Kumar, to be published.

116. C.B. Murray, D.J. Norris, and M.G. Bawendi, *J. Am. Chem. Soc.* **115**, 8706 (1993).

117. A. Kasuya, R. Sivamohan, Y. Barnakov, I. Dmitruk, T. Nirasawa, V. Romanyuk, V. Kumar, S.V. Mamykin, K. Tohji, B. Jeyadevan, K. Shinoda, T. Kudo, O. Terasaki, Z. Liu, R.V. Belosludov, V. Sundararajan, and Y. Kawazoe, *Nat. Mater.* **3**, 99 (2004).

118. T.P. Martin, *Phys. Rep.* **272**, 199 (1996).

119. A. Puzder, A.J. Williamson, F. Gygi, and G. Galli, *Phys. Rev. Lett.* **92**, 217401 (2004).

120. S. Botti and M.A.L. Marques, *Phys. Rev. B* **75**, 035311 (2007).

121. Y.Q. Zhu, T. Sekine, Y.H. Li, W.X. Wang, M.W. Fay, H. Edwards, P.D. Brown, N. Fleischer, and R. Tenne, *Adv. Mater.* **17**, 1500 (2005).

122. R. Tenne, L. Margulis, M. Genut, and G. Hodes, *Nature (London)* **360**, 444 (1992).

123. A. Simon, *Angew. Chem. Int. Ed. Engl.* **20**, 1 (1981).

124. L. Venkataraman and C.M. Lieber, *Phys. Rev. Lett.* **83**, 5334 (1999).

125. H. Topsøe, B.S. Clausen, and F.E. Massoth, *Hydrotreating Catalysis*, J.R. Anderson and M. Boudart (Eds.), Springer Verlag, Berlin-Heidelberg, Germany (1996); R. Prins, *Adv. Catal.* **46**, 399 (2002); J.V. Lauritsen, J. Kibsgaard, S. Helveg, H. Topsøe, B.S. Clausen, E. Lægsgaard, and F. Besenbacher, *Nat. Nanotechnol.* **2**, 53 (2007).

126. P. Murugan, V. Kumar, Y. Kawazoe, and N. Ota, *Phys. Rev. A* **71**, 063203 (2005).

127. P. Murugan, V. Kumar, Y. Kawazoe, and N. Ota, *J. Phys. Chem. A* **111**, 2778 (2007).

128. P. Murugan, V. Kumar, Y. Kawazoe, and N. Ota, *Nano Lett.* **7**, 2214 (2007).

129. V. Nicolosi, P.D. Nellist, S. Sanvito, E.C. Cosgriff, S. Krishnamurthy, W.J. Blau, M.L.H. Green et al., *Adv. Mater.* **19**, 543 (2007); M. Remskar, A. Mrzel, Z. Skraba, A. Jesih, M. Ceh, J. Demsar, P. Stadelmann, F. Levy, and D. Mihailovic, *Science*, **292**, 479 (2001); A. Zimina, S. Eisebitt, M. Freiwald, S. Cramm, W. Eberhardt, A. Mrzel, and D. Mihailovic, *Nano Lett.* **4**, 1749 (2004).

130. P. Murugan, V. Kumar, Y. Kawazoe, and N. Ota, *Appl. Phys. Lett.* **92**, 203112 (2008).

Predicting Nanocluster Structures

John D. Head
University of Hawaii

5.1 Introduction

The chances of systematically making new nanoclusters with interesting and technologically useful chemical and physical properties will be greatly improved if detailed knowledge of the arrangement of atoms forming specific nanoclusters is readily available. The aim of this chapter is to provide an overview of different theoretical approaches currently being used to predict the structural properties of nanoclusters. Generally, good quality ab initio techniques, such as density functional theory (DFT) calculations, are able to produce structural models consistent with experimentally measured structures [1,2]. The stable cluster models predicted by theory are often an essential ingredient to interpreting the experimental data obtained by various experimental methods used to analyze the structure of nanoclusters. Since several chapters in this series provide an overview of the many different experimental methods such as powder diffraction, various microscopies, and spectroscopic techniques available to probe the structural properties of clusters, we do not discuss experimental methods in this chapter.

The different theoretical methods to predict the nanocluster structure described in this chapter all involve finding a local minimum on a potential energy surface (PES) expressed as a function of the component atom positions. We equate the most stable nanocluster to be the structure that has the lowest energy and corresponds to the global minimum (GM) on the PES. The PES is obtained by assuming the Born–Oppenheimer approximation enables a valid separation between the electronic and nuclear motions. A common theme to this chapter is that the challenge to predicting nanocluster structures is due to the PES, even for a relatively small molecule or cluster, being able to accommodate a huge number of local minima. For instance, different isomers of a simple organic molecule, such as ethanol

C_2H_5OH and dimethyl ether CH_3OCH_3, are an example of the many different local minima present on a PES for a specific chemical composition. Organic chemists are able to isolate and characterize the two isomers separately because of the large energy barrier on the PES that causes the kinetics of their interconversion to be slow. Similarly, a molecule, such as n-propanol, $CH_3CH_2CH_2OH$, does not spontaneously convert to an ether because of the large energy barrier separating the two isomers. However, the different conformers formed as a result of rotations about the different C–C bonds in n-propanol produce a region in the PES consisting of several local minima similar in energy but with only small energy barriers separating the different minima. These low-energy barriers enable the different n-propanol conformers to establish a thermodynamic equilibrium where the structures are populated according to a Boltzmann distribution. The topology of how the local minima are distributed on a PES will influence the theoretical method used to predict the stable structure for a nanocluster.

The PES of nanoclusters will have similar features as that for the organic molecules described above. We need a combination of both a local minimizer and the global optimizer to predict the most stable structure for a cluster. In this chapter, we focus on methods that predict the structure of lowest energy cluster on the PES since this will correspond to the most stable cluster structure at low temperature. Determining this most stable cluster structure necessitates computing the total energy of the cluster at many different geometries. The computer time needed to perform all of these cluster calculations with a good quality DFT calculations is prohibitive. Consequently, less rigorous energy calculations are typically used to prescreen the different cluster structures for possible low-energy structures before performing good quality DFT calculations on a subset of the candidate cluster structures, which are expected to contain the ab initio GM structure. The remainder of this chapter describes the steps for predicting the most stable cluster structure in more detail. Section 5.2 summarizes the properties of the PES and describes how local minima and the GM relate to the most stable structure of a cluster. This section also outlines how to perform a local geometry optimization and introduces the idea of a catchment region around each local minimum. Section 5.3 gives an overview of the different potentials functions available for calculating the cluster energies and their relative computational demands. Then, we present three of the common strategies for performing a global optimization to find the most stable cluster structure in Section 5.4. Examples of predicting nanocluster structures are given in Section 5.5 where we also describe specifically some of the work by our group on ligand-passivated silicon nanoclusters. Summary and concluding remarks are contained in the last section of the chapter.

5.2 Cluster Structural Features on the Potential Energy Surface

Within the Born–Oppenheimer approximation, the PES $E(\mathbf{x})$ is expressed as a function of the positions of the N nuclei making up the cluster where $\mathbf{x}$ is a column vector containing the $3N$ coordinates.

$$\mathbf{x}^{\dagger} = (x_1\ y_1\ z_1 \ldots x_N\ y_N\ z_N)$$

Figure 5.1 shows a hypothetical PES for a cluster where the cluster energy E is plotted against some reaction coordinate Q, which would have a complex functional dependence on the cluster coordinates $\mathbf{x}$. The energy of a cluster with geometry $\mathbf{x}_2$ on the

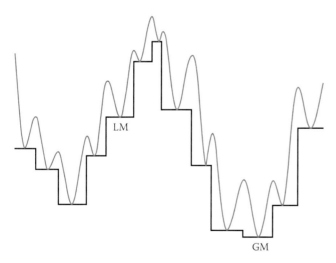

FIGURE 5.1 Schematic PES (in gray) for a cluster showing the global minimum (GM) and the kth lowest energy local minimum (LM$_k$). The transformed PES (in black) is used by the basin hopping and genetic algorithm approach to global optimization. The extent of the catchment region around each local minimum is indicated by the horizontal lines in the transformed PES.

PES can be related to the energy at a different position $\mathbf{x}_1$ by the Taylor series expansion

$$E(\mathbf{x}_2) = E(\mathbf{x}_1) + \mathbf{g}(\mathbf{x}_1)^{\dagger}\Delta + 1/2\Delta^{\dagger}\mathbf{G}(\mathbf{x}_1)\Delta + \text{higher order terms}$$

where
  $\Delta = \mathbf{x}_2 - \mathbf{x}_1$ is the displacement vector
  $\mathbf{g}(\mathbf{x}_1)$ is the gradient vector
  $\mathbf{G}(\mathbf{x}_1)$ is the second derivative or Hessian matrix both evaluated at $\mathbf{x}_1$

The elements of the gradient vector and Hessian matrix are

$$\mathbf{g}(\mathbf{x}_1)_i = \left.\frac{\partial E}{\partial x_i}\right|_{\mathbf{x}_1}$$

$$\mathbf{G}(\mathbf{x}_1)_{ij} = \left.\frac{\partial E}{\partial x_i \partial x_j}\right|_{\mathbf{x}_1}$$

A stationary point on the PES is defined as a position $\mathbf{x}^{\star}$ where the gradient vector is zero.

$$\mathbf{g}(\mathbf{x}^{\star}) = 0$$

Each stationary point is further characterized by the eigenvalues of the Hessian matrix $\mathbf{G}$. For a nonlinear cluster, six of the Hessian matrix eigenvalues can be associated with the three translational and three rotational degrees of freedom, and the remaining $3N - 6$ eigenvalues correspond to vibrational motions of the cluster and are used to classify the nature of the stationary point. A local minimum has $3N - 6$ positive eigenvalues in the Hessian, which give rise to $3N - 6$ real vibrational frequencies. A first-order stationary point with a maximum in one direction is a proper transition state and the Hessian matrix has one negative eigenvalue, which gives rise to one imaginary and $3N - 7$ real vibrational frequencies. Second and higher order stationary points with maxima in several directions contain several negative eigenvalues in the Hessian matrix and are not usually chemically important. The energy at transition state geometries provide the lowest energy pathway connecting two different local minima. There is never a reaction pathway passing over a second or higher order stationary point since there is always a lower energy pathway involving a transition state structure available instead. The schematic cluster PES in Figure 5.1 illustrates a collection of many different local minima separated from each other by the lowest energy maximum along the reaction coordinate Q corresponding to a transition state or first-order stationary point. Figure 5.1 shows the PES has only one GM and this is the local minimum with lowest energy. Apart from being at the lowest energy on the PES, the GM has no special properties which distinguish it from the other local minima. The computational difficulty with correctly locating the GM arises because the number of stationary points on a PES grows exponentially with cluster size [5,6].

Finding the local minimum on a PES is now a fairly straightforward task [1,2]. Typically a gradient-based quasi-Newton method is used to find step directions Δ toward a local minimum

$$\Delta = -\mathbf{H}\mathbf{g}(\mathbf{x}_{old})$$

where an approximation to the inverse Hessian matrix $\mathbf{H} \approx \mathbf{G}^{-1}$ is formed via an update formula using the $\mathbf{g}$ vectors calculated at previous cluster geometries used in the earlier geometry optimization cycles. The limited memory L-BFGS update formula is usually the one of choice because $\mathbf{H}$ remains positive definite for each update at the different geometries used in the local optimization [3]. The new geometry $\mathbf{x}_{new}$ used in the energy and gradient calculation is obtained via

$$\mathbf{x}_{new} = \mathbf{x}_{old} + \alpha\Delta$$

with α usually chosen to be unity unless Δ causes unphysically large coordinate changes.

An important concept to appreciate is that the PES can be divided into catchment regions around each of the various local minima [4]. The horizontal lines in Figure 5.1 depict the range of the catchment region associated with each local minimum. If the kth local minimum LM_k has the geometry $\mathbf{x}_k^\star$, then a local minimization starting from some initial geometry $\mathbf{x}_{init}$, which is inside the catchment region associated with the kth local minimum, will optimize to the $\mathbf{x}_k^\star$ geometry. This means that if one assumes a specific structural motif in an initial cluster structure, then the optimized structure would be a local minimum containing the same structural motif. A related consequence is that the point group symmetry of an initial cluster geometry is conserved throughout the optimization cycles; this can cause the geometry of the optimized cluster at the stationary point to have extra symmetry and not be a proper local minimum with $3N-6$ real vibrational frequencies. The presence of a catchment region around each local minimum means that searching for the lowest energy, or GM, cluster structure is more complicated than a local geometry optimization, where now the energies of many different local minima need to be computed and compared.

5.3 Considerations in Cluster Energy Calculations

Predicting the most stable structure for a nanocluster requires that a large number of total energy calculations are performed at different cluster geometries. The goal of the quantum chemist is to perform these calculations at a suitable level of theory to produce a stable structure, which is expected to match reasonably well with an experimentally observed structure data. However, as soon as the size of the cluster exceeds 20 or so atoms, it becomes computationally impractical to use high-quality quantum chemical methods, such as DFT, to perform local geometry optimizations on a large number of cluster structures. Consequently,

a lot of cluster structure studies have been performed using approximate or empirical potentials enabling the total energy of a cluster geometry to be evaluated in a fraction of a second rather than on the order of many minutes required by a DFT calculation. The GM cluster structures obtained by using approximate energy potentials can be intrinsically interesting as they can give valuable insights into the structural trends a nanocluster can adopt with increasing cluster size. The approximate energy potentials are also a useful tool for evaluating the effectiveness of different global optimization algorithms. Alternatively, an approximate energy potential can be used as prescreening tool, which identifies several structures as candidate clusters having low DFT energies, thereby enabling the GM at the DFT level to be identified without performing a huge number of high-quality quantum chemistry calculations.

The Lennard-Jones potential is probably the simplest potential

$$V_{LJ}(\mathbf{x}) = 4\epsilon \sum_{i<j}^{N} \left[\left(\frac{\sigma}{r_{ij}} \right)^{12} - \left(\frac{\sigma}{r_{ij}} \right)^{6} \right]$$

where ϵ and $2^{1/6}\,\sigma$ are the pair well depth and equilibrium separation [7]. The GM of $(LJ)_n$ clusters, where the cluster energy is evaluated using the LJ potential, has been found for clusters with up to 1000 atoms [8,9]. To more realistically model the structure of metal and alloy clusters, potentials that go beyond the simple pairwise interactions of an LJ potential are needed. The potentials in the embedded atom method (EAM) [10,11] and the second-moment approximation to tight binding (SMATB) methods [12,13] have been used for these types of systems. Alternatively, for silicon clusters, where there is an extensive network of covalent bonding, several groups [14–17] have used the semiempirical density functional tight binding (DFTB) method [18–20]. In our own work on passivated Si clusters, we found the semiempirical AM1 method [21,22] as another very fast way for calculating the total energy of a cluster.

The main advantage of the approximate energy calculations is that they allow many different cluster total energy calculations to be performed. The topology of the PES derived from an empirical potential is also likely to be smoother than a PES produced by DFT calculations, and this smoother PES aids in making the search for the cluster GM easier [23,24]. However, it is well known that the global minima theoretically predicted for a cluster will depend on the energy function used in the calculation [25,26]. This is further illustrated by the recent global optimization studies of Au_{20} clusters using DFT calculations directly which find the DFT calculations to consistently give the most stable structure that is quite different from those previously predicted by using various empirical Au potentials [27]. Using the lowest energy empirical structures as candidates for starting structures in DFT calculations is also troublesome. For example, performing DFT local minimizations on the 100 lowest energy clusters obtained via the empirical potential calculations produced the lowest energy Au_{20} cluster that was still 1.68 eV above the DFT GM cluster energy [27]. One approach around this problem, which depends on the flexibility of parameters in the empirical potential, is to fine-tune the parameters in the empirical potential to produce low-energy structures that are more consistent with the DFT GM.

We have performed this type of parameter tuning in our global optimization studies of Si_xH_y clusters. We originally used the standard Si and H AM1 parameters [21,22] and developed a genetic algorithm (GA) to globally optimize different $Si_{10}H_y$ clusters with $y = 4, 8, 12, 16,$ and 20 [28]. We originally picked the AM1 method because we felt it gave better optimized structures and energies for the various Si_xH_y clusters than other semiempirical methods such as PM3. Our global optimization calculations on $Si_{10}H_{16}$ using AM1 gave a very different GM to what we now find with both MP2 and DFT calculations [30,31]. AM1 produces the ab initio GM as the 10th lowest energy structure. As we performed MP2 and DFT calculations on the different low-energy AM1 structures, we realized we were generating a library of optimized structures and energies, which could be used to reparametrize the AM1 method. This lead to our reparametrized GAM1 method, where we kept the same AM1 equations but modified the parameters to predict GM structures just like those we find from ab initio calculations [30,31]. Truhlar and coworkers have used the term specific reaction (or range) parameters (SRP) for this type of approach where the parameters for an approximate method are adjusted to better reproduce the energies for a specific system [32].

5.4 Computational Approach to Finding a Global Minimum

As noted previously, the major difficulty in finding the GM structure for a nanocluster is due to the exponential growth of the number of stationary points on the PES with cluster size [5,6]. A second problem is due to the GM not having any special properties, apart from being at the lowest energy, which can be used to distinguish it readily from all the other local minima on the PES. One common approach to finding the GM is to simply use chemical intuition to guess at the lowest energy structure for a cluster. A set containing several initial cluster structures are selected on the basis that they are expected to be chemically reasonable. Out of this set of structures, the GM is taken to be the lowest energy structure found after performing a local geometry optimization on each of the initial guess structures. The obvious drawback to this approach is that if you do not guess at an initial structure, which is in the catchment region around the GM, then the locally optimized cluster structure found to have the lowest energy will not be the GM. An exhaustive search method tries to systemize this approach by generating a set of initial structures, which includes every structural possibility. For example, to find the lowest energy conformer for a *n*-alkane one might assume that staggered arrangements of the alkyl groups attached to each C–C bond should give rise to three local minima. The exhaustive search would then require the local geometry optimization of 3^m different initial structures where m is the number of C–C

bonds with R groups attached. Unfortunately, the exhaustive search approach can easily lead to an impractical huge number of different possible cluster structures; although for small clusters, point group and permutational symmetry may help to establish how many distinct initial structures need to be generated. Even when an exhaustive search is practical, the lowest energy structure obtained by the exhaustive search may still not be the correct GM if the underlying assumptions used to build the initial structures are flawed and fail to generate an initial structure inside the catchment region of the GM. Furthermore, all of the above methods are obviously going to fail if the most stable nanocluster is a consequence of some new unanticipated chemistry.

A more appealing approach to predicting cluster structures is to use an unbiased search method, which does not depend on using any prior chemical notions of the stable structure. The simplest example of this approach is to perform local geometry optimizations using randomly generated initial structures. In the simplest implementation, a random number generator is used to generate the three Cartesian coordinates for each atom in a cube with volume a^3 where a controls the density of the cluster formed. Any tendency to form cubic clusters can be avoided by randomly putting atoms in a sphere, as described by Press et al. [39]. One drawback to these randomly generated structures is that the probability of generating a new atom position unphysically close to an existing atom position increases with the number of atoms already selected to be in the cluster. However, since many different random structures for small clusters can rapidly be generated on a computer, problem structures can be ignored from further consideration when either two atoms are unphysically close together or the calculation of the initial cluster's energy fails to converge. While performing local geometry optimization on these random initial clusters avoids introducing any chemical bias into the search, again the major drawback is that many energy calculations are needed on many different cluster geometries in order to have confidence that the true GM is being correctly located.

A more efficient approach that requires fewer cluster energy calculations is to have an algorithm that modifies the initial guesses at a cluster's low-energy structures and then explores whether these modified structures have lower energy. Such algorithms should be independent of any chemical bias. In the remainder of this section, we describe simulated annealing, basin hopping, and genetic algorithms as three examples of such methods. We refer the reader to the book by Wales [33] and the review articles by Hartke [34], Johnston [35], and Springborg [36] for discussion of other promising global optimization methods. One useful feature of all three methods is that they all can be started by using either the randomly generated clusters described in the previous paragraph, or they could be applied to a series of cluster structures initially generated using chemical intuition, or by using a combination of both structural types.

Simulated annealing (SA) is an algorithm that is analogous to the physical process taking place when a liquid cools to form a crystalline solid [37]. After picking an initial cluster geometry $\mathbf{x}_1$

and computing the cluster energy $E_1 = E(\mathbf{x}_1)$, the following iterative steps are performed for several cycles at temperature T:

1. Make a random move to a new geometry $\mathbf{x}_2 = \mathbf{x}_1 + \Delta$
2. Evaluate new cluster energy $E_2 = E(\mathbf{x}_2)$
3. Accept new geometry and set $\mathbf{x}_1$ to $\mathbf{x}_2$ if $p < \exp[-(E_2 - E_1)/kT]$ where $p \in (0, 1)$ is a random probability of acceptance

A new geometry is always accepted when $E_2 < E_1$ and the initial temperature needs to be high enough to allow the geometry changes Δ to cross between catchment regions of different local minima. After several successful coordinate moves, the temperature T should be lowered. Cycling through random coordinate moves and lowering the temperature are repeated until changes in the cluster energy become small. It is both the number of needed coordinate moves and the rate of temperature lowering that determine whether the SA algorithm can actually find the lowest energy structure of a cluster. Too fast a cooling can cause a cluster structure, which has high energy barrier separating it from other lower energy structures, to become trapped in a local minimum [33,38]. Press et al. have also pointed out that the algorithm could be more efficient if the random move Δ took into account the availability of any downhill moves [39]. With such an approach, SA becomes very similar to the basin hopping algorithm discussed next.

Basin hopping (BH) is similar to SA except that it eliminates the energy barriers separating local minimum by transforming the PES into a collection of constant energy plateaus or basins defined by the energies of the different local minima [40,41]. Figure 5.1 illustrates this idea and the extent of the constant energy plateau is set by the size of the catchment region around each local minimum. The transformed cluster energy at position $\mathbf{x}$ is

$$\tilde{E}(\mathbf{x}) = \min[E(\mathbf{x})] = E(\mathbf{x}^\star)$$

where $\mathbf{x}^\star$ are the optimized coordinates of the local minimum and $\min[E(\mathbf{x})]$ signifies getting the energy from a local optimization, which is started with the initial coordinates $\mathbf{x}$. Typically, once the structure $\mathbf{x}^\star$ of the local minimum is found, the initial coordinates $\mathbf{x}$ are reset to this geometry.

Some initial cluster structure $\mathbf{x}_1$ is selected and locally optimized to give

$$E_1 = \tilde{E}(\mathbf{x}_1)$$

and $\mathbf{x}_1 = \mathbf{x}_1^\star$. Then different cluster structures are iteratively examined by

1. Making a random move to a new geometry $\mathbf{x}_2 = \mathbf{x}_1 + \Delta$
2. Evaluating the new transformed cluster energy $E_2 = \tilde{E}(\mathbf{x}_2)$ by performing a local minimization and resetting $\mathbf{x}_2$ to the optimized structure $\mathbf{x}_2^\star$
3. Accepting new geometry and setting $\mathbf{x}_1$ to $\mathbf{x}_2$ if $p < \exp[-(E_2 - E_i)/kT]$ where $p \in (0, 1)$ is a random probability of acceptance

The temperature controls the probability of allowing E_2 to be greater than E_1 and is dynamically adjusted to allow some prescribed acceptance ratio [41]. In order to hop between different basins, Δ needs to be relatively large with maximum displacements being one-third of the average near neighbor distance. The local minimization is best performed using a quasi-Newton method such as the L-BFGS algorithm [3] and requires several energy and gradient evaluations before $\mathbf{x}_2^\star$ is located.

Genetic algorithm (GA) is a global optimization strategy inspired by the Darwinian evolution process. The GA works by randomly selecting and mating the more fit individuals in a generation to produce the next generation of offspring, where the fitness is some measure of the energetic stability for an individual cluster structure. The GM is eventually located because some of the new cluster conformations created by the GA have lower energies than the structures in previous generations. A good mating operator causes good structural features in a cluster to be passed to the next generation while maintaining structural diversity in the overall population.

A first problem is how to encode the cluster representation into a form usable by a computer. The early GAs performed the genetic operations of mating and mutation on binary strings, which map in some way to the cluster geometry. In 1995, Deaven and Ho [42] introduced crossover and mutation operators, which worked directly on the clusters in real coordinate space. What we call the cut-and-paste operator in our work is the Deaven and Ho crossover method, which is shown in Figure 5.2, where two parent clusters are cut into halves along randomly oriented planes and the two halves from different parents are combined

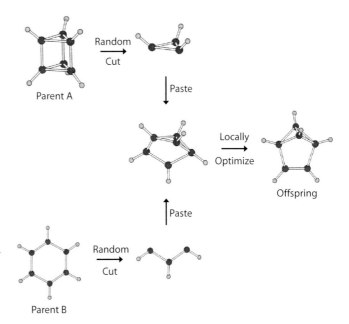

FIGURE 5.2 Example of the Deavon and Ho [42] cut-and-paste GA operator applied to Si_6H_6 clusters. The two parents A and B are randomly orientated and cut in half. The cluster halves are pasted together to produce a new cluster, which is then locally optimized to produce the offspring cluster for use in the next generation of the GA.

to give an offspring structure. Some version of the Deaven and Ho crossover operator is now used in most of the current cluster GA methods [35]. Another important idea from Deaven and Ho was to perform a gradient-driven local geometry optimization of each new cluster generated after crossover or mating [42]. The local geometry optimization is similar to the idea used in basin hopping methods and provides the ability to jump between different catchment regions around local minima without needing the energy to pass over any intervening barriers [40].

Typically, a cluster GA starts with N randomly generated clusters whose geometries are locally geometry optimized to produce the initial ancestor population, where N depends on the size of the cluster. The clusters' total energy is then used to assign a fitness to each cluster, with the lower energy clusters being assigned higher fitness. In tournament selection, several parents are initially selected randomly, and from this population subset, the two fittest parents are mated to produce offspring. The next generation of clusters is then formed by performing a local geometry optimization on each offspring and then verifying that the optimized offspring is not identical with any of the clusters already included in the new population. A simple approach for deciding if the new offspring is identical with a previous individual is to compare whether their total energies and the three principal moments of inertia all agree within some threshold [30]. The GA is then run to produce several generations of cluster populations. Judging when the GA is converged, and the GM found can be challenging with the simplest approach being that no new cluster with lower energy than the GM is found for several generations. Several different GA runs using different seeds in the random number generator and different initial random cluster geometries also serve as a consistency check that a unique GM structure is found for the cluster.

Typically, the mating methods used in the GA should keep the good features of the better (lower energy) parents while some mutations are performed to help maintain the variety of different cluster structural types in the population. For instance, in addition to the Deaven and Ho cut-and-paste operator for combining two parents, Rata et al. [17] in the global optimization of Si-only clusters mutated a cluster by applying the cut-and-paste operator on a single parent. Alternatively, a simple coordinate averaging operator that takes the arithmetic mean of the Cartesian coordinates from two parent clusters can be used to generate an offspring with a new random geometry. Although this new cluster geometry may or may not have a low energy and be physically important, the coordinate averaging operator helps to maintain the structural diversity for the population of clusters in the GA.

The advantage of the above mating operators is that they are simple to implement and avoid introducing any chemical bonding biases into the GM determination. However, such general GA operators may not be very efficient at forming a structure which resembles the GM. For instance, we found the search to find the Si_6H_6 and $Si_{14}H_{20}$ GM using general GA operators to be slow. We were able to make the GM search much faster by developing the genetic operators, which essentially mimic chemical transformations, such as the shift of H atoms between other vacant Si sites, or cause a surface SiH_3 group to be removed from or inserted into a Si–Si bond ring [31]. The GA enables these chemical transformations to take place without being concerned that there might be large energy barriers, which would normally prohibit the reaction occurring in the real chemical system. These operations can be thought of as analogues to the add/etch operations which Wolf and Landman used in their global optimization studies of large Lennard-Jones clusters [68]. The covalent bonding network ubiquitous to Si_xH_y clusters makes implementing the add/etch operations more difficult and requires that each Si atom be assigned a functional group identification, which we simply determine through connectivity information obtained from the internuclear separation matrix.

While SA is still used in a number of applications [47], it does suffer from the drawback of sometimes getting trapped in a local minimum above the correct GM when there are large energy barriers separating the different local minima. Both BH and GA circumvent this problem by accepting coordinate displacements based on energies evaluated at the local minimum rather than the energies determined in the region associated with the high energy barrier. BH is probably the easier method to implement for a general nanocluster, such as A_xB_y, which is composed of several different atom types. Our longer discussion of the GA method illustrates that implementing the genetic operators in a GA is more dependent on the type of cluster being globally optimized. For instance, the GA cut-and-paste operator to two different parent A_xB_y clusters requires an implementation where a $A_{x-m}B_{y-n}$ fragment from one parent is combined with a A_mB_n fragment from the other parent. However, one attractive feature of the GA is being able include a high degree of structural diversity in the parent cluster population thereby ensuring the energies of broad range of different local minima are compared. For instance, the third lowest energy local minimum, LM_3, on the hypothetical cluster PES, shown on the left-hand side of Figure 5.1, appears to be well separated from GM at the right-hand side of Figure 5.1. The BH algorithm is likely to spend many cycles exploring different geometries around LM_3 while the population of cluster structures in the GA can be in the vicinity of the GM and LM_3. Several groups have analyzed the topological characteristics on a PES such as the distribution of local minima and the heights of the energy barriers separating them to serve as a guide in determining the most efficient global optimization strategy [33,48].

5.5 Example Applications: Predicting Structures of Passivated Si Clusters

Theoretical prediction of the most stable structure has now been applied to many different nanocluster systems. In the next paragraph, we present a brief overview of some of the recent review articles and the Cambridge Cluster Database, which describe example applications of global optimization methods on various nanocluster systems. In the remainder of this section, we try to illustrate the utility of global optimization methods by describing our work aimed at finding the DFT GM of ligand-passivated silicon Si_xL_y clusters [28–31,43,44].

The Cambridge Cluster Database [8] and the recent book by Wales [33] both contain an extensive compendium of many different globally optimized nanoclusters. These two sources list the GM for a large collection of different-size elemental clusters, water clusters, alkali halide clusters, and fullerenes and silicon clusters where the GMs are determined using several different potential functions. Baletto and Ferrando provide an extensive review of various metal atom clusters, including Au, Ag, Cu, Pt, Pd, and Ni atoms in their recent review of structural properties of nanoclusters: energetic, thermodynamic, and kinetic effects [45]. Recently, Ferrando, Jellinek, and Johnston extensively reviewed the stable structures of bimetallic alloy clusters of the type A_xB_y where A and B are two different metal atoms and even discussed an example of a trimetallic onion-like cluster $Au_{core}Ag_{shell}Cu_{shell}$ [46]. Determining the stable structures for a nanoalloy cluster is made difficult because one needs to predict the preferred mixing pattern between the constituent atoms in the alloy. Figure 5.3 shows the four main mixing patterns found in nanoalloys and the search for the nanocluster GM is extremely challenging since a completely different cluster structure can be simply generated by permuting the positions of an A atom with a B atom.

We have been developing a GA-based global optimization method to predict the stable structures of ligand-passivated silicon nanoclusters Si_xL_y. We expect this structure to have a Si_x core encased in a shell of ligands L_y similar to the nanoalloy mixing pattern shown in Figure 5.3a. Our interest in passivated silicon clusters stems from the observation that nanometer-sized silicon clusters exhibit an intense photoluminescence (PL) and have the potential of being developed into a practical optoelectronic device [49,50]. The nanoparticles contrast with bulk silicon, which shows a low-intensity PL since it is an indirect gap semiconductor. Quantum confinement [51,52] and surface effect [53] theories have been proposed to explain the mechanism of the PL in silicon clusters. Presumably, in a practical optoelectronic device, the nanometer-sized Si clusters need to be passivated by some air stable ligand. Several groups have used quantum chemistry calculations to calculate the optoelectronic properties for relatively large atomic and molecular clusters. For instance, Zhou, Friesner, and Brus (ZFB) recently used density functional theory (DFT) to calculate the electronic structure for 1–2 nm diameter silicon clusters as large as $Si_{87}H_{76}$ [54,55]. ZFB treated several different-sized Si clusters passivated by either H, oxide, OH, hydrocarbon, or F ligands. The calculations enable the characterization of single Si nanoparticles rather than the ensemble of particles with uncertain size ranges present in most real samples. In order to do these calculations, ZFB appear to make the chemically reasonable assumption that the most stable nanoclusters consist of a Si core with the same diamond-lattice-like structure as found in bulk Si. ZFB then passivated dangling surface Si bonds of the diamond-lattice-like core with a ligand of interest and performed a full local optimization to get the geometry of specific clusters. Degoli et al. make a similar diamond-lattice-like assumption in their ground and excited state calculations on clusters ranging from Si_5H_{12} to $Si_{35}H_{36}$ [56]. Perhaps, some support for the likelihood of the Si nanocluster favoring a bulk Si-like core is provided by the recent report of the experimental synthesis and structure determination of sila-adamantane [57]: a molecule which contains a Si_{10} cluster core with a bulk Si-like structure capped by 12 methyl and 4 trimethyl silyl groups. However, by examining the structure of Si_xL_y clusters outside the catchment region for the bulk-like Si core, we are able to show that the most stable geometric arrangement

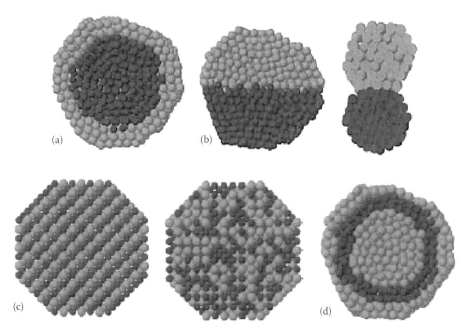

FIGURE 5.3 Cross sections of possible atom mixing patterns in nanoalloys: (a) core-shell, (b) subcluster segregated, (c) mixed, and (d) three shell. (Adapted from Ferrando, R. et al., *Chem. Rev.*, 108, 845, 2008. With permission.)

of Si atoms in a nanocluster is dependent on the ligand that is used to passivate the cluster surface [43,44]. This is an important observation because the optical properties of the Si nanocluster should be dependent on the Si core structure.

A number of different groups have attempted to find the GM for various small Si_xH_y clusters using either approximate or high-quality ab initio cluster energy calculations. The different locally optimized structures using high-level ab initio calculations used to identify the global minima for Si_2H_3 and Si_2H_4 are summarized in the recent work by Sari et al. [58] and Sillars et al. [59]. Chambreau et al. used local geometry optimizations and MP2 calculations to predict the geometries and relative stabilities of three Si_6H and five Si_6H_2 clusters by starting from geometries built by adding one or two H atoms onto the stable Si_6 cluster with D_{4h} symmetry [60]. Meleshko et al. have also attempted to identify global minima for various silicon hydride clusters using simulated annealing and the semiempirical MINDO/3 method to evaluate the cluster energy [61]. Miyazaki et al. have used density functional theory–based methods to study the stable structures of Si_6H_{2n} ($n = 1$–7) [62]. Their search for the local minima was started by using a combination of simulated annealing with pseudopotentials for Si and H atoms, along with optimizing structures initially built using chemical intuition, and by considering structures previously proposed by others. Three different groups have attempted to predict the GM for the Si_nH with $n = 4$–10 series of clusters. Prasad and coworkers developed a GA-based global optimization method, which they applied to the Si_nH clusters with $n = 4$–8 [63,64]. They evaluated the Si_nH cluster energy using a tight binding Hamiltonian and did not attempt to find the ab initio global minima for their structures. Yang et al. have performed an extensive series of DFT calculations on the

same Si_nH ($n = 4$–10) clusters and their anions but they did not explain how the initial Si_nH cluster structures were selected [65]. Presumably they used a chemical intuition approach perhaps being guided by the low-energy structures previously found by the Prasad group and they did find lowest energy DFT structures that generally agreed with the Prasad results apart from when $n = 6, 8, 9$. More recently Ona et al. have developed a GA method that uses DFT calculations directly to globally optimize the Si_nH ($n = 4$–10) clusters [66]. Their lowest energy structures agree well with the previous structures obtained by Prasad and coworkers [63,64] and Yang et al. [65], but they did find a new Si_7H GM and noted that the GA consistently produced several new low-energy isomers not found by the other two groups for the larger clusters. Ona et al. made a concluding comment on the importance of performing global optimization searches in order to correctly find the low-energy structures greatly increases with the size of the clusters [66].

We have also investigated the DFT GM for Si_xH_y clusters containing only a few H atoms using GA-based GM optimization strategy [28,29]. However, because of our interest in passivated Si nanoclusters and whether the Si_x core wants to adopt the bulk diamond-like structure in this chapter, we only focus on Si_xH_y clusters where $y > x$ [28,30,31]. Figure 5.4 illustrates the DFT GM we have found for $Si_{10}H_{16}$, $Si_{14}H_{20}$, $Si_{18}H_{24}$, $Si_{10}H_{14}$, $Si_{14}H_{18}$, and $Si_{18}H_{22}$ using the GA strategy described in Section 5.4. In our initial GA implementation [28] we used the original AM1 semiempirical method [21,22] to prescreen the different low-energy cluster structures generated by the GA. The $Si_{10}H_{16}$ cluster, shown in Figure 5.4, had the lowest DFT energy but was ranked as the 10th lowest energy structure by the AM1 calculations [28]. From these calculations, we realized the limitation to the AM1

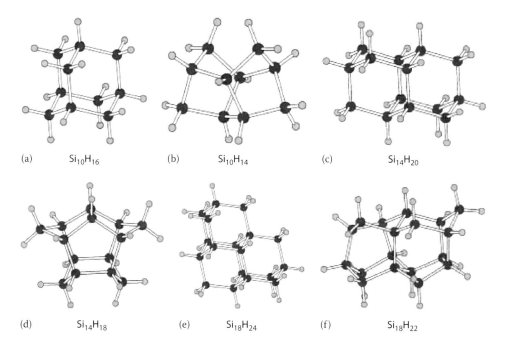

(a) $Si_{10}H_{16}$ (b) $Si_{10}H_{14}$ (c) $Si_{14}H_{20}$

(d) $Si_{14}H_{18}$ (e) $Si_{18}H_{24}$ (f) $Si_{18}H_{22}$

FIGURE 5.4 GM found for different Si_xH_y clusters. (From Ge, Y. and Head, J.D., *J. Phys. Chem. B*, 108, 6025, 2004; Ge, Y. and Head, J.D., *Chem. Phys. Lett.*, 398, 107, 2004.)

parameters is that they were designed to handle a Si atom in a variety of different bonding situations [22]. We needed a fast cluster calculation method to better reproduce the energies of just high H coverage Si clusters and this led us to develop the GAM1 method where the AM1 parameters were adjusted to this specific type of Si_xH_y clusters [29,30]. After some implementation changes to the GA [31] coupled with using the reparametrized GAM1 semiempirical method for the cluster, prescreening the efficiency of our GM search strategy was significantly improved enabling us to better find the DFT GM for the larger $Si_{18}H_{22}$ and $Si_{18}H_{24}$ clusters [30,31].

The GM in Figure 5.4 shows that, as expected by chemical intuition, when there are enough passivating H atoms, such as with the $Si_{10}H_{16}$, $Si_{14}H_{20}$, and $Si_{18}H_{24}$ stoichiometries, the Si core does favor forming a fragment of the bulk Si diamond lattice. However, it is important to emphasize these GM were obtained by starting from several randomly generated Si_xH_y clusters in the initial GA population and the GA operators eventually produced the GM structures shown in Figure 5.4. For the slightly under passivated clusters $Si_{10}H_{14}$, $Si_{14}H_{18}$, and $Si_{18}H_{22}$, which each have two less H atoms than corresponding fully H passivated cluster, the GM shown in Figure 5.4 are more difficult to predict by using chemical intuition alone since they do not appear to have any structural resemblance to the corresponding $Si_{10}H_{16}$, $Si_{14}H_{20}$, and $Si_{18}H_{24}$ GM. Furthermore, the $Si_{10}H_{14}$, $Si_{14}H_{18}$, and $Si_{18}H_{22}$ GM do not exhibit any obvious common structural trends. The $Si_{18}H_{22}$ GM does appear to retain more

of the diamond-lattice-like structure than in the smaller $Si_{14}H_{18}$ and $Si_{10}H_{14}$ clusters. Perhaps, as might be expected, our results indicate that in larger under H passivated Si clusters more of the bulk Si structure will be retained, but in the region of the cluster with incomplete H passivation, there will be a structure corresponding essentially to a Si lattice defect. An important conclusion from these studies is that the lowest energy lattice defect structure in Si_xH_{y-2} cannot be simply predicted by removing two H atoms from the diamond-lattice-like Si_xH_y GM and then performing a local geometry optimization. Instead, a more complete global optimization method is needed.

We have also considered the influence of different ligands on passivated Si nanoclusters by theoretically investigating $Si_{10}L_{16}$ clusters with the ligands L = H, CH_3, OH, and F [43,44]. Unfortunately, we were not able to use the AM1-like semiempirical calculations to prescreen the cluster energies for even the F atom ligand [67]. The chemical intuition approach of finding the $Si_{10}F_{16}$ GM by performing local geometry optimizations on different $Si_{10}F_{16}$ clusters built by replacing H with F in the low-energy clusters previously found for $Si_{10}H_{16}$ was also found not to be very reliable. Eventually an empirical correction formula was developed to calculate $Si_{10}F_{16}$ cluster energies using a GAM1 parameter set obtained by fitting to 14 Si_7F_{14} isomers where the F atoms were represented by pseudo H atoms. This enabled us to obtain the $Si_{10}F_{16}$ DFT GM shown in Figure 5.5 [43,67]. Our calculations suggested that this new structural type is preferred because the highly electronegative F atoms like to form terminal

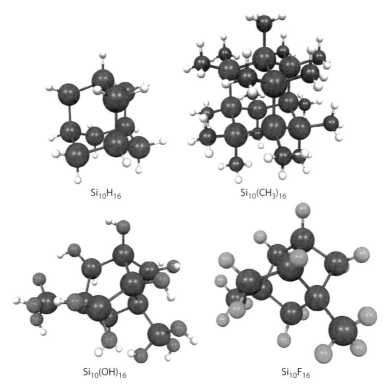

$$Si_{10}H_{16}$$

$$Si_{10}(CH_3)_{16}$$

$$Si_{10}(OH)_{16}$$

$$Si_{10}F_{16}$$

FIGURE 5.5 Low-energy $Si_{10}L_{16}$ cluster structures. The $Si_{10}H_{16}$ and $Si_{10}F_{16}$ were found by a global optimization search. The lowest energy $Si_{10}(CH_3)_{16}$ and $Si_{10}(OH)_{16}$ structures were obtained by replacing H and F atoms on the low-energy $Si_{10}H_{16}$ and $Si_{10}F_{16}$ cluster. (From Shiraishi, Y. et al., *J. Phys. Chem. C*, 112, 1819, 2008.)

SiF_3 groups rather than be evenly distributed over the available surface sites on the Si_{10} core. Based on an electronegativity argument, a CH_3 passivating ligand should be expected to produce clusters with Si core structures similar to those found for the Si_xH_y clusters, whereas the higher O atom electronegativity means the OH ligand should favor low-energy structures with Si cores similar to those found for the $Si_{10}F_{16}$ clusters.

To test the influence of different ligands on the Si_{10} core, we computed the optimized structures and relative energies by starting from the previously optimized low-energy $Si_{10}H_{16}$ or $Si_{10}F_{16}$ structure [30,31,43] and replacing either all the H or F atoms with a new ligand L = H, CH_3, OH, F and performing a local geometry optimization using B3LYP/6-31G(d) DFT calculations. Finding the lowest energy structures for the $Si_{10}(OH)_{16}$ and $Si_{10}(CH_3)_{16}$ clusters is more challenging than for $Si_{10}H_{16}$ and $Si_{10}F_{16}$ owing to the OH and CH_3 ligands being able to form several different conformers on the same Si_{10} core framework. For this reason, we used a variation of the random cluster approach described in Section 5.4: five initial geometries for the $Si_{10}(OH)_{16}$ and $Si_{10}(CH_3)_{16}$ clusters were generated for each Si_{10} core type by replacing either all the H or F atoms in the optimized $Si_{10}H_{16}$ and $Si_{10}F_{16}$ clusters with OH or CH_3 groups where the SiOH or one of SiCH planes were orientated at a randomly selected angle relative to the cluster framework. Figure 5.5 shows the locally optimized structure with lowest DFT energy for the four different $Si_{10}L_{16}$ clusters [44]. The $Si_{10}H_{16}$ and $Si_{10}F_{16}$ clusters are just the GM structures obtained previously [30,31,43], whereas a more extended GM search might find lower energy $Si_{10}(CH_3)_{16}$ and $Si_{10}(OH)_{16}$ clusters. These $Si_{10}F_{16}$ calculations illustrate that the lowest energy structure of passivated Si nanoclusters is sensitive to the type of ligands used to passivate the particle. Ligands with electronegativities similar to that of Si give rise to low-energy structures where the ligand is uniformly dispersed over the Si core surface. Providing there are enough ligands, the Si core in the lowest energy structure resembles a bulk Si-like fragment. However, more bulky low electronegativity ligands, such as CH_3 may experience static crowding on a Si_{10} core resulting in structures containing $Si(CH_3)_3$ groups to be only a few kcal/mol higher in energy than the cluster with a bulk Si-like core. The more electronegative ligands have a strong preference for forming SiL_3 groups and this tendency eliminates the likelihood of the Si atoms at the nanocluster surface to have a bulk Si-like arrangement [44].

5.6 Summary

Predicting the lowest energy nanocluster structure with high-quality ab initio calculations is still very challenging. One needs to examine many different cluster structures at their locally optimized geometries. Currently, the basin hopping and genetic algorithm–based methods appear to be the best methods for producing and evaluating the different local minima without introducing any chemical bias into the final identification of a cluster's lowest energy or GM structure. The efficiency of the GM search will be improved as better approximate cluster energy

methods, with faster evaluation times and closer matching of the cluster structure energy rankings with the high-quality ab initio methods are developed.

Acknowledgments

The author is grateful for the computing resources provided by the Dell Cluster at the University of Hawaii and the Maui High Performance Computing Center. The author also thanks his former graduate student, Prof. Yingbin Ge at the University of Central Washington, for the many stimulating discussions on genetic algorithms.

References

1. F. Jensen, *Introduction to Computational Chemistry*, 2nd edn., Wiley, Chichester, U.K., 2007.
2. C. J. Cremer, *Essential of Computational Chemistry*, Wiley, Chichester, U.K., 2002.
3. C. Zhu, R. H. Byrd, and J. Nocedal, L-BFGS-B: Algorithm 778: L-BFGS-B, FORTRAN routines for large scale bound constrained optimization, *ACM Trans. Math. Software* **23**, 550 (1997).
4. P. G. Mezey, Catchment region partitioning of energy hypersurfaces, I, *Theor. Chim. Acta* **58**, 309 (1981).
5. C. J. Tsai and K. D. Jordan, Use of the histogram and jump-walking methods for overcoming slow barrier crossing behavior in Monte Carlo simulations: Applications to the phase transitions in the $(Ar)_{13}$ and $(H_2O)_8$ clusters, *J. Chem. Phys.* **99**, 6957 (1993).
6. F. H. Stillinger, Exponential multiplicity of inherent structures, *Phys. Rev. E* **59**, 48 (1999).
7. J. E. Jones and A. E. Ingham, On the calculation of certain crystal potential constants, and on the cubic crystal of least potential energy, *Proc. R. Soc. Lond. A* **107**, 636 (1925).
8. D. J. Wales, J. P. K. Doye, A. Dullweber, M. P. Hodges, F. Y. Naumkin, and F. Calvo, The Cambridge Cluster Database, URL:http://www-wales.ch.cam.ac.uk/CCD.html.
9. Y. Xiang, H. Jiang, W. Cai, and X. Shao, An efficient method based on lattice construction and the genetic algorithm for optimization of large Lennard-Jones clusters, *J. Phys. Chem. A* **108**, 3586 (2004).
10. M. S. Daw and M. I. Baskes, Embedded-atom method: Derivation and application to impurities, surfaces, and other defects in metals, *Phys. Rev. B* **29**, 6443 (1974).
11. A. F. Voter, The embedded atom method in *Intermetallic Compounds*, eds. J. H. Westbrook and R. L. Fleischer, Wiley, New York, 1995.
12. R. P. Gupta, Lattice relaxation at a metal surface, *Phys. Rev. B* **23**, 6265 (1981).
13. A. P. Sutton and J. Chen, Long-range Finnis-Sinclair potentials, *Philos. Mag. Lett.* **61**, 139 (1990).
14. F. H. Stillinger and T. A. Weber, Computer simulation of local order in condensed phases of silicon, *Phys. Rev. B* **31**, 5262 (1985).

15. X. G. Gong, Empirical-potential studies on the structural properties of small silicon clusters, *Phys. Rev. B* **47**, 2329 (1993).

16. B. C. Bolding and H. C. Andersen, Interatomic potential for silicon clusters, crystals, and surfaces, *Phys. Rev. B* **41**, 10568 (1990).

17. I. Rata, A. A. Shvartsburg, M. Horoi, T. Frauenheim, K. W. M. Siu, and K. A. Jackson, Single-parent evolution algorithm and the optimization of Si clusters, *Phys. Rev. Lett.* **85**, 546 (2000).

18. D. Porezag, T. Frauenheim, T. Khler, G. Seifert, and R. Kaschner, Construction of tight-binding-like potentials on the basis of density-functional theory: Application to carbon, *Phys. Rev. B* **51**, 12947 (1995).

19. G. Seifert, D. Porezag, and T. Frauenheim, Calculations of molecules, clusters, and solids with a simplified LCAO-DFT-LDA scheme, *Int. J. Quantum Chem.* **58**, 185 (1996).

20. M. Elstner, D. Porezag, G. Jungnickel, J. Elsner, M. Haugk, T. Frauenheim, S. Suhai, and G. Seifert, Self-consistent-charge density-functional tight-binding method for simulations of complex materials properties, *Phys. Rev. B* **58**, 7260 (1998).

21. M. J. S. Dewar, E. G. Zoebisch, E. F. Healy, and J. J. P. Stewart, AM1: A new general purpose quantum mechanical molecular model, *J. Am. Chem. Soc.* **107**, 3902 (1985).

22. M. J. S. Dewar and C. Jie, AM1 calculations for compounds containing silicon, *Organometallics* **6**, 1486 (1987).

23. J. P. K. Doye and D. J. Wales, Structural consequences on the range of the interatomic potential. A menagerie of clusters, *J. Chem. Soc. Faraday Trans.* **93**, 4233 (1997).

24. C. Roberts, R. L. Johnston, and N. T. Wilson, A genetic algorithm for the structural optimization of Morse clusters, *Theor. Chem. Acc.* **104**, 123 (2000).

25. B. Hartke, Global geometry optimization of clusters guided by *N*-dependent model potentials, *Chem. Phys. Lett.* **258**, 144 (1996).

26. B. Hartke, Global geometry optimization of small silicon clusters at the level of density functional theory, *Theor. Chem. Acc.* **99**, 241 (1998).

27. E. Apra, R. Ferrando, and A. Fortunelli, Density-functional global optimization of gold nanoclusters, *Phys. Rev. B* **73**, 205414 (2006).

28. Y. Ge and J. D. Head, Global optimization of H-passivated Si clusters with a genetic algorithm, *J. Phys. Chem. B* **106**, 6997 (2002).

29. Y. Ge and J. D. Head, Global optimization of Si_xH_y at the ab initio level via an iteratively parametrized semiempirical method, *Int. J. Quantum Chem.* **95**, 617 (2003).

30. Y. Ge and J. D. Head, Global optimization of H-passivated Si clusters at the ab initio level via the GAM1 semiempirical method, *J. Phys. Chem. B* **108**, 6025 (2004).

31. Y. Ge and J. D. Head, Fast global optimization of Si_xH_y clusters: New mutation operators in the cluster genetic algorithm, *Chem. Phys. Lett.* **398**, 107 (2004).

32. I. Rossi and D. G. Truhlar, Parameterization of NDDO wavefunctions using genetic algorithms. An evolutionary approach to parameterizing potential energy surfaces and direct dynamics calculations for organic reactions, *Chem. Phys. Lett.* **233**, 231 (1995).

33. D. J. Wales, *Energy Landscapes*, Cambridge University Press, Cambridge, U.K., 2004.

34. B. Hartke, Application of evolutionary algorithms to global cluster geometry optimization, *Struct. Bonding* (*Berlin*), ed. R. L. Johnston, **110**, 33 (2004).

35. R. L. Johnston, Evolving better nanoparticles: Genetic algorithms for optimising cluster geometries, *J. Chem. Soc. Dalton Trans.* **2003**, 4193 (2003).

36. M. Springborg, Determination of structure in electronic structure calculations, *Chem. Model.* **4**, 249 (2006).

37. S. Kirkpatrick, C. D. Gelatt Jr., and M. P. Vecchi, Optimization by simulated annealing, *Science* **220**, 671 (1983).

38. R. S. Judson, M. E. Colvin, J. C. Meza, A. Huffer, and D. Gutierrez, Do intelligent configuration search techniques outperform random search for large molecules? *Int. J. Quantum Chem.* **44**, 277 (1992).

39. W. H. Press, S. A. Teukolsky, W. T. Vetterling, and B. P. Flannery, *Numerical Recipes*, 3rd edn., Cambridge University Press, Cambridge, U.K., 2007.

40. D. J. Wales and H. A. Scheraga, Global optimization of clusters, crystals, and biomolecules, *Science* **285**, 1368 (1999).

41. D. J. Wales and J. P. K. Doye, Global optimization by basin-hopping and the lowest energy structures of Lennard-Jones clusters containing up to 110 atoms, *J. Phys. Chem. A* **101**, 5111 (1997).

42. D. M. Deaven and K. M. Ho, Molecular geometry optimization with a genetic algorithm, *Phys. Rev. Lett.* **75**, 288 (1995).

43. Y. Ge and J. D. Head, Ligand effects on Si_xL_y cluster structures with L = H and F, *Mol. Phys.* **103**, 1035 (2005).

44. Y. Shiraishi, D. Robinson, Y. Ge, and J. D. Head, Low energy structures of ligand passivated Si nanoclusters: Theoretical investigation of Si_2L_4 and $Si_{10}L_{16}$ (L = H, CH_3, OH and F), *J. Phys. Chem. C* **112**, 1819 (2008).

45. F. Baletto and R. Ferrando, Structural properties of nanoclusters: Energetic, thermodynamic and kinetic effects, *Rev. Mod. Phys.* **77**, 371 (2005).

46. R. Ferrando, J. Jellinek, and R. L. Johnston, Nanoalloys: From theory to applications of alloy clusters and nanoparticles, *Chem. Rev.* **108**, 845 (2008).

47. F. Ruette and C. Gonzalez, The importance of global minimization and adequate theoretical tools for cluster optimization: The Ni_6 cluster case, *Chem. Phys. Lett.* **359**, 428 (2002).

48. O. M. Becker and M. Karplus, The topology of multidimensional potential energy surfaces: Theory and application to peptide structure and kinetics, *J. Chem. Phys.* **106**, 1495 (1997).

49. L. T. Canham, Silicon quantum wire array fabrication by electrochemical and chemical dissolution of wafers, *Appl. Phys. Lett.* **57**, 1046 (1990).

50. W. L. Wilson, P. F. Szajowski, and L. E. Brus, Quantum confinement in size-selected surface-oxidized silicon nanocrystals, *Science* **262**, 1242 (1993).

51. V. Lehman and U. Goesele, Porous silicon formation: A quantum wire effect, *Appl. Phys. Lett.* **58**, 856 (1991).

52. J. P. Proot, C. Delerue, and G. Allan, Electronic structure and optical properties of silicon crystallites: Application to porous silicon, *Appl. Phys. Lett.* **61**, 1948 (1992).

53. M. V. Wolkin, J. Jorne, P. M. Fauchet, G. Allan, and C. Delerue, Electronic states and luminescence in porous silicon quantum dots: The role of oxygen, *Phys. Rev. Lett.* **82**, 197 (1999).

54. Z. Zhou, R. A. Friesner, and L. Brus, Electronic structure of 1 to 2 nm diameter silicon core/shell nanocrystals: Surface chemistry, optical spectra, charge transfer, and doping, *J. Am. Chem. Soc.* **125**, 15599 (2003).

55. Z. Zhou, L. Brus, and R. A. Friesner, Electronic structure and luminescence of 1.1- and 1.4-nm silicon nanocrystals: Oxide shell versus hydrogen passivation, *Nano Lett.* **3**, 163 (2003).

56. E. Degoli, G. Cantele, E. Luppi, R. Magri, D. Ninno, O. Bisi, and S. Ossicini, Ab initio structural and electronic properties of hydrogenated silicon nanoclusters in the ground and excited state, *Phys. Rev. B* **69**, 155411 (2004).

57. J. Fischer, J. Baumgartner, and C. Marschner, Synthesis and structure of sila-adamantane, *Science* **310**, 825 (2005).

58. L. Sari, M. C. McCarthy, H. F. Schaefer, and P. Thaddeus, Mono- and dibridged isomers of Si_2H_3 and Si_2H_4: The true ground state global minima. Theory and experiment in concert, *J. Am. Chem. Soc.* **125**, 11409 (2003).

59. D. Sillars, C. J. Bennett, Y. Osamura, and R. I. Kaiser, Infrared spectroscopic detection of the disilenyl (Si_2H_3) and d3-disilenyl (Si_2D_3) radicals in silane and d4-silane matrices, *Chem. Phys. Lett.* **392**, 541 (2004).

60. S. D. Chambreau, L. Wang, and J. Zhang, Highly unsaturated hydrogenated silicon clusters, Si_nH_x ($n = 3 - 10$, $x = 0 - 3$), in flash pyrolysis of silane and disilane, *J. Phys. Chem. A* **106**, 5081 (2002).

61. V. Meleshko, Y. Morokov, and V. Schweigert, Structure of small hydrogenated silicon clusters: Global search of low-energy states, *Chem. Phys. Lett.* **300**, 118 (1999).

62. T. Miyazaki, T. Uda, I. Stich, and K. Terakura, Hydrogenation-induced structural evolution of small silicon clusters: The case of $Si_6H_x^+$, *Chem. Phys. Lett.* **284**, 12 (1998).

63. N. Chakraboti, P. S. De, and R. Prasad, Genetic algorithms based structure calculations for hydrogenated silicon clusters. *Mater. Lett.* **55**, 20 (2002).

64. D. Balamurugan and R. Prasad, Effect of hydrogen on ground-state structures of small silicon clusters, *Phys. Rev. B* **64**, 205406 (2001).

65. J. Yang, X. Bai, C. Li, and W. Xu, Silicon monohydride clusters Si_nH ($n = 4 - 10$) and their anions: Structures, thermochemistry, and electron affinities, *J. Phys. Chem. A* **109**, 5717 (2005).

66. O. Ona, V. E. Bazterra, M. C. Caputo, M. B. Ferraro, and J. C. Facelli, Ab initio global optimization of the structures of Si_nH, $n = 4 - 10$, using parallel genetic algorithms, *Phys. Rev. A* **72**, 053205 (2005).

67. Y. Ge, Global optimization of passivated Si clusters at the ab initio level via semiempirical methods, PhD thesis, University of Hawaii, Honolulu, HI (December 2004).

68. M. D. Wolf and U. Landman, Genetic algorithms for structural cluster optimization, *J. Phys. Chem. A* **102**, 6129 (1998).

II

Nanoscale Systems

6

The Nanoscale Free-Electron Model

Daniel F. Urban
Albert-Ludwigs-Universität

Jérôme Bürki
California State University,
Sacramento

Charles A. Stafford
University of Arizona

Hermann Grabert
Albert-Ludwigs-Universität

6.1 Introduction

The past decades have seen an accelerating miniaturization of both mechanical and electrical devices; therefore, a better understanding of properties of ultrasmall systems is required in increasing detail. The first measurements of conductance quantization in the late 1980s (van Wees et al. 1988, Wharam et al. 1988) in constrictions of two-dimensional electron gases formed by means of gates have demonstrated the importance of quantum confinement effects in these systems and opened a wide field of research. A major step has been the discovery of conductance quantization in metallic nanocontacts (Agraït et al. 1993, Brandbyge et al. 1995, Krans et al. 1995): The conductance measured during the elongation of a metal nanowire is a steplike function where the typical step height is frequently near a multiple of the conductance quantum $G_0 = 2e^2/h$, where e is the electron charge and h Planck's constant. Surprisingly, this was initially not interpreted as a quantum effect but rather as a consequence of abrupt atomic rearrangements and elastic deformation stages. This interpretation, supported by a series of molecular dynamics simulations (Landman et al. 1990, Todorov and Sutton 1993), was claimed to be confirmed by another pioneering experiment (Rubio et al. 1996, Stalder and Dürig 1996) measuring simultaneously the conductance and the cohesive force of gold nanowires with diameters ranging from several Ångstroms to several nanometers. As the contact was pulled apart, oscillations in the force of order 1 nN were observed in perfect correlation with the conductance steps.

It came as a surprise when Stafford et al. (1997) introduced the free-electron model of a nanocontact—referred to as the nanoscale free-electron model (NFEM) henceforth—and showed that this comparatively simple model, which emphasizes the quantum confinement effects of the metallic electrons, is able to reproduce quantitatively the main features of the experimental observations. In this approach, the nanowire is understood to act as a quantum waveguide for the conduction electrons (which are responsible for both conduction and cohesion in simple metals): Each quantized mode transmitted through the contact contributes G_0 to the conductance and a force of order E_F/λ_F to the cohesion, where E_F and λ_F are the Fermi energy and wavelength, respectively. Conductance channels act as delocalized bonds whose stretching and breaking are responsible for the observed force oscillations, thus explaining straightforwardly their correlations with the conductance steps.

Since then, free-standing metal nanowires, suspended from electrical contacts at their ends, have been fabricated by a number of different techniques. Metal wires down to a single atom thick were extruded using a scanning tunneling microscope tip (Rubio et al. 1996, Untiedt et al. 1997). Metal nanobridges were shown to "self-assemble" under electron-beam irradiation of thin metal films (Kondo and Takayanagi 1997, 2000, Rodrigues et al. 2000), leading to nearly perfect cylinders down to four atoms in diameter, with lengths up to 15 nm. In particular, the mechanically controllable break junction technique, introduced by Moreland and Ekin (1985) and refined by Ruitenbeek and coworkers (Muller et al. 1992), has allowed for systematic studies of nanowire properties for a variety of materials. For a survey, see the review by Agraït et al. (2003).

A remarkable feature of metal nanowires is that they are stable. Most atoms in such a thin wire are at the surface, with small coordination numbers, so that *surface effects* play a key role in their energetics. Indeed, macroscopic arguments comparing the surface-induced stress to the yield strength indicate a minimum radius for solidity of order 10 nm (Zhang et al. 2003). Below this critical radius without any stabilizing mechanism, plastic flow would lead to a Rayleigh instability (Chandrasekhar 1981), breaking the wire apart into clusters. Already in the nineteenth century, Plateau (1873) realized that this surface-tension-driven instability is unavoidable if cohesion is due solely to classical pairwise interactions between atoms. The experimental evidence accumulated over the past decade on the remarkable stability of nanowires considerably thinner than the above estimate clearly shows that electronic effects emphasized by the NFEM dominate over atomistic effects for sufficiently small radii.

A series of experiments on alkali metal nanocontacts (Yanson et al. 1999, 2001) identified *electron-shell effects*, which represent the semiclassical limit of the quantum-size effects discussed above, as a key mechanism influencing nanowire stability. Energetically favorable structures were revealed as peaks in conductance histograms, periodic in the nanowire radius, analogous to the electron-shell structure previously observed in metal clusters (de Heer 1993). A supershell structure was also observed (Yanson et al. 2000) in the form of a periodic modulation of the peak heights. Recently, such electron-shell effects have also been observed, even at room temperature, for the noble metals gold, copper, and silver (Díaz et al. 2003, Mares et al. 2004, Mares and van Ruitenbeek 2005) as well as for aluminum (Mares et al. 2007).

Soon after the first experimental evidence for electron-shell effects in metal nanowires, a theoretical analysis using the NFEM found that nanowire stability can be explained by a competition of the two key factors: surface tension and electron-shell effects (Kassubek et al. 2001). Both linear (Urban and Grabert 2003, Zhang et al. 2003) and nonlinear (Bürki et al. 2003, 2005) stability analyses of axially symmetric nanowires found that the surface-tension-driven instability can be completely suppressed in the vicinity of certain "magic radii." However, the restriction to axial symmetry implies characteristic gaps in the sequence of stable nanowires, which is not fully consistent with the experimentally observed nearly perfect periodicity of the conductance peak positions. A Jahn–Teller deformation breaking the symmetry can lead to more stable, deformed configurations. Recently, the linear stability analysis was extended to wires with arbitrary cross sections (Urban et al. 2004a, 2006). This general analysis confirms the existence of a sequence of magic cylindrical wires of exceptional stability, which represent roughly 75% of the main structures observed in conductance histograms. The remaining 25% are deformed and predominantly of elliptical or quadrupolar shapes. This result allows for a consistent interpretation of experimental conductance histograms for alkali and noble metals, including both the electronic shell and supershell structures (Urban et al. 2004b).

This chapter is intended to give an introduction to the NFEM. Section 6.2 summarizes the assumptions and features of the model while the general formalism is described in Section 6.3. In the following sections, two applications of the NFEM will be discussed: First, we give a unified explanation of electrical transport and cohesion in metal nanocontacts (Section 6.4) and second, the linear stability analysis for straight metal nanowires will be presented (Section 6.5). The latter will include cylindrical wires as well as wires with broken axial symmetry, thereby discussing the Jahn–Teller effect.

6.2 Assumptions and Limitations of the NFEM

Guided by the importance of conduction electrons in the cohesion of metals, and by the success of the jellium model in describing metal clusters (Brack 1993, de Heer 1993), the NFEM replaces the metal ions by a uniform, positively charged background that provides a confining potential for the electrons. The electron motion is free along the wire and confined in the transverse directions. Usually an infinite confinement potential (hard-wall boundary conditions) for the electrons is chosen. This is motivated by the fact that the effective potential confining the electrons to the wire will be short ranged due to the strong screening in good metals.

In a first approximation, electron–electron interactions are neglected, which is reasonable due to the excellent screening (Kassubek et al. 1999) in metal wires with $G > G_0$. It is known from cluster physics that a free-electron model gives qualitative agreement and certainly describes the essential physics involved. Interaction, exchange, and correlation effects as well as a realistic confinement potential have to be taken into account, however, for quantitative agreement.[*] From this, we infer that the same is true for metal nanowires, where similar confinement effects are important. Remarkably, the electron-shell effects crucial to the stabilization of long wires are described with quantitative accuracy by the simple free-electron model, as discussed below.

In addition, the NFEM assumes that the positive background behaves like an incompressible fluid when deforming the nanowire. This takes into account, to lowest order, the hard-core repulsion of core electrons as well as the exchange energy of conduction electrons. When using a hard-wall confinement, the Fermi energy E_F (or equivalently the Fermi wavelength λ_F) is the only parameter entering the NFEM. As E_F is material dependent and experimentally accessible, there is no adjustable parameter. This pleasant feature needs to be abandoned in order to model different materials more realistically. Different kinds of appropriate surface boundary conditions are imaginable in order to model the behavior of an incompressible fluid and to fit the surface properties of various metals. This will be discussed in detail in Section 6.3.5.

---

[*] Note, however, that the error introduced by using hard-walls instead of a more realistic soft-wall confining potential can be essentially corrected for by placing the hard-wall a finite distance outside the wire surface, thus compensating for the over-confinement (García-Martin et al. 1996).

A more refined model of a nanocontact would consider effects of scattering from disorder (Bürki and Stafford 1999, Bürki et al. 1999) and electron–electron interaction via a Hartree approximation (Stafford et al. 2000a, Zhang et al. 2005). The inclusion of disorder in particular leads to a better quantitative agreement with transport measurements, but does not change the cohesive properties qualitatively in any significant way, while electron–electron interactions are found to be a small correction in most cases. As a result, efforts to make the NFEM more realistic do not improve it significantly, while removing one of its main strengths, namely the absence of any adjustable parameters.

The major shortcoming of the NFEM is that its applicability is limited to good metals having a nearly spherical Fermi surface. It is best suited for the (highly reactive) s-orbital alkali metals, providing a theoretical understanding of the important physics in nanowires. The NFEM has also been proven to qualitatively (and often semiquantitatively) describe noble metal nanowires, and in particular, gold. Lately, it has been shown that the NFEM can even be applied (within a certain parameter range) to describe the multivalent metal aluminum, since Al shows an almost spherical Fermi surface in the extended-zone scheme. The NFEM is especially suitable to describe shell effects due to the conduction-band s-electrons, and the experimental observation of a crossover from atomic-shell to electron-shell effects with decreasing radius in both metal clusters (Martin 1996) and nanowires (Yanson et al. 2001) justifies *a posteriori* the use of the NFEM in the later regime. Naturally, the NFEM does not capture effects originating from the directionality of bonding, such as the effect of surface reconstruction observed for Au. For this reason, it cannot be used to model atomic chains of Au atoms, which are currently extensively studied experimentally. Keeping these limitations in mind, the NFEM is applicable within a certain range of radius, capturing nanowires with only very few atoms in cross section up to wires of several nanometers in thickness, depending on the material under consideration.

6.3 Formalism of the NFEM

6.3.1 Scattering Matrix Formalism

A metal nanowire represents an open system connected to metallic electrodes at each end. These macroscopic electrodes act as ideal electron reservoirs in thermal equilibrium with a well-defined temperature and chemical potential. When treating an open system, the Schrödinger equation is most naturally formulated as a scattering problem. The basic idea of the scattering approach is to relate physical properties of the wire with transmission and reflection amplitudes for electrons being injected from the leads.*

The fundamental quantity describing the properties of the system is the energy-dependent unitary scattering matrix $S(E)$

connecting incoming and outgoing asymptotic states of conduction electrons in the electrodes. For a quantum wire, $S(E)$ can be decomposed into four submatrices $S_{\alpha\beta}(E)$, α, $\beta = 1, 2$, where 1 (2) indicates the left (right) lead. Each submatrix $S_{\alpha\beta}(E)$ determines how an incoming eigenmode of lead β is scattered into a linear combination of outgoing eigenmodes of lead α. The eigenmodes of the leads are also referred to as scattering channels.

The formulation of electrical transport in terms of the scattering matrix was developed by Landauer and Büttiker: The (linear response) electrical conductance G can be expressed as a function of the submatrix S_{21}, which describes transmission from the source electrode 1 to the drain electrode 2 and is given by (Datta 1995)

$$G = \frac{2e^2}{h} \int dE \frac{-\partial f(E)}{\partial E} \text{Tr}_1 \left\{ S_{21}^{\dagger}(E) S_{21}(E) \right\}. \quad (6.1)$$

Here, $f(E) = \{\exp[\beta(E - \mu)] + 1\}^{-1}$ is the Fermi distribution function for electrons in the reservoirs, $\beta = (k_B T)^{-1}$ is the inverse temperature, and μ is the electron chemical potential, specified by the macroscopic electrodes. The trace Tr_1 sums over all eigenmodes of the source.

The appropriate thermodynamic potential to describe the energetics of an open system is the grand canonical potential

$$\Omega = -\frac{1}{\beta} \int dE \, D(E) \ln\left[1 + e^{-\beta(E-\mu)}\right], \quad (6.2)$$

where $D(E)$ is the electronic density of states (DOS) of the nanowire. Notably, the DOS of an open system may also be expressed in terms of the scattering matrix as (Dashen et al. 1969)

$$D(E) = \frac{1}{2\pi i} \text{Tr} \left\{ S^{\dagger}(E) \frac{\partial S}{\partial E} - \frac{\partial S^{\dagger}}{\partial E} S(E) \right\}, \quad (6.3)$$

where Tr sums over the states of both electrodes. This formula is also known as Wigner delay. Note that Equations 6.1 through 6.3 include a factor of 2 for spin degeneracy.

Thus, once the electronic scattering problem for the nanowire is solved, both transport and energetic quantities can be readily calculated.

6.3.2 WKB Approximation

For an axially symmetric constriction aligned along the z-axis, as depicted in Figure 6.1, its geometry is characterized by the z-dependent radius $R(z)$. Outside the constriction, the solutions of the Schrödinger equation decompose into plane waves along the wire and discrete eigenmodes of a circular billiard in the transverse direction. The eigenenergies $E_{\mu\nu}$ of a circular billiard are given by

$$E_{\mu\nu} = \frac{\hbar^2}{2m_e} \frac{\gamma_{\mu\nu}^2}{R_0^2}, \quad (6.4)$$

---

* Phase coherence is assumed to be preserved in the wire (a good approximation given the size of the system compared to the inelastic mean-free path) and inelastic scattering is restricted to the electron reservoirs only.

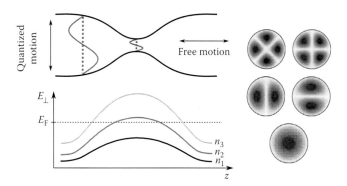

FIGURE 6.1 Upper-left part: Sketch of a nanoconstriction. Within the adiabatic approximation, transverse and longitudinal motions are separable: the motion in the transverse direction is quantized, while in the longitudinal direction the electrons move in a potential created by the transverse energies (see Equation 6.6). Lower-left part: Sketch of transverse energies for different transverse channels n_1, n_2, and n_3 as a function of the z-coordinate. Channel n_1 is transmitted through the constriction as its maximum transverse energy is smaller than the Fermi energy, channel n_2 is partly transmitted, and channel n_3 is almost totally reflected. Right part: Density plots of $|\Psi_n(r, \varphi)|^2$ for the three eigenmodes depicted on the lower-left part, corresponding to five states due to degeneracies of energies E_{n_2} and E_{n_3}.

where the quantum number $\gamma_{\mu\nu}$ is the νth root of the Bessel function $\mathcal{J}_\mu$ of order μ and R_0 is the radius of the wire outside the constriction.

In cylindrical coordinates r, φ, and z, the asymptotic scattering states read

$$\Psi_{\mu\nu}(r,\varphi,z) \sim e^{\pm ik_{\mu\nu}z + i\mu\varphi} \mathcal{J}_\mu(\gamma_{\mu\nu}r/R_0), \qquad (6.5)$$

where $k_{\mu\nu}(E) = \sqrt{2m_e(E - E_{\mu\nu})/\hbar^2}$ is the longitudinal wavevector. In the following, we use multi-indices $n = (\mu\nu)$ in order to simplify the notation.

If the constriction is smooth, i.e., $|\partial R/\partial z| \ll 1$, one may use an adiabatic approximation. In the adiabatic limit, the transverse motion is separable from the motion parallel to the z-axis even in the region of the constriction, and the channel index n of an incoming electron is preserved throughout the wire. Accordingly, Equations 6.4 and 6.5 remain valid in the region of the constriction, with R_0 replaced by $R(z)$. The channel energies become functions of z, $E_n(z) = \hbar^2\gamma_n^2/2m_eR(z)^2$, as is sketched in the lower part of Figure 6.1, and act as a potential barrier for the effective one-dimensional scattering problem in channel n. The corresponding Schrödinger equation for the longitudinal part Φ of the wave function reads

$$\frac{\partial^2}{\partial z^2}\Phi_n(z) + \frac{2m_e}{\hbar^2}\left[E - E_n(z)\right]\Phi_n(z) = 0, \qquad (6.6)$$

and is solved within the WKB approximation (see, e.g., Messiah 1999) by

$$\Phi_n(r,\varphi,z) \sim \frac{1}{\sqrt{k_n(E,z)}}\exp\left[\pm i\int_0^z k_n(E,z')dz'\right]. \qquad (6.7)$$

For a constriction of length L, the transmission amplitude in channel n is then given by the familiar WKB barrier transmission factor

$$t_n(E) = \exp\left[i\int_0^L dz\, k_n(E,z)\right] \equiv \sqrt{\mathcal{T}_n(E)}e^{i\Theta_n(E)}. \qquad (6.8)$$

Here $\mathcal{T}_n$ is the transmission coefficient of channel n and Θ_n is the corresponding phase shift. The transmission amplitude gets exponentially damped in regions where the transverse energy is larger than the state total energy.[*]

The full S-matrix is now found to be of the form

$$S = \begin{pmatrix} i\sqrt{1 - \mathcal{T}}\,e^{i\Theta} & \sqrt{\mathcal{T}}\,e^{i\Theta} \\ \sqrt{\mathcal{T}}\,e^{i\Theta} & i\sqrt{1 - \mathcal{T}}\,e^{i\Theta} \end{pmatrix}, \qquad (6.9)$$

where for simplicity of notation, we have suppressed the channel indices and each of the entries is understood to be a diagonal matrix in the channels. Using the formulas of Section 6.3.1, we may proceed to determine physical quantities. From Equation 6.1, we deduce that the electrical conductance at zero temperature reads

$$\begin{aligned} G &= \frac{2e^2}{h}\sum_n \mathcal{T}_n(E_F) \\ &= \frac{2e^2}{h}\sum_n \exp\left[-2\int_0^L dz\, \theta(E_n(z) - E_F)\sqrt{\frac{2m_e}{\hbar^2}(E_n(z) - E_F)}\right], \end{aligned}$$
$$(6.10)$$

where the second line is obtained by using Equation 6.8. Here, $\theta(x)$ denotes the Heaviside step function ($\theta(x) = 1$ for $x > 0$, 0 otherwise.) The DOS is found to be connected with the phase shift Θ_n,

$$D(E) = \frac{2}{\pi}\sum_n \frac{\partial\Theta_n(E)}{\partial E} \qquad (6.11)$$

$$= \frac{1}{\pi}\sqrt{\frac{2m_e}{\hbar^2}}\sum_n \int_0^L dz\, \frac{\theta(E - E_n(z))}{\sqrt{E - E_n(z)}}. \qquad (6.12)$$

From the DOS, one gets the grand canonical potential in the limit of zero temperature as

---

[*] This simplest WKB treatment does not correctly describe above-barrier reflection; a better approximation including this effect is described by Brandbyge et al. (1995) and by Glazman et al. (1988).

$$\Omega \overset{T\to 0}{=} -\frac{8E_{\mathrm{F}}}{3\lambda_{\mathrm{F}}} \int_0^L dz \sum_n \theta\big(E_{\mathrm{F}} - E_n(z)\big)\left(1 - \frac{E_n(z)}{E_{\mathrm{F}}}\right)^{3/2}, \qquad (6.13)$$

which can then be used to calculate the tensile force and stability of the nanowire, as discussed in the following sections.

6.3.3 WKB Approximation for Non-Axisymmetric Wires

The formalism presented in the previous subsection can be readily extended to non-axisymmetric wires. In general, the surface of the wire is given by the radius function $r = R(\varphi, z)$, which may be decomposed into a multipole expansion

$$R(\varphi, z) = \rho(z)\left\{\sqrt{1 - \sum_m \frac{\lambda_m(z)^2}{2}} + \sum_m \lambda_m(z)\cos\big[m\big(\varphi - \varphi_m(z)\big)\big]\right\},$$

$$(6.14)$$

where the sums run over positive integers. The parameterization is chosen in such a way that $\pi\rho(z)^2$ is the cross-sectional area at position z. The parameter functions $\lambda_m(z)$ and $\varphi_m(z)$ compose a vector $\Lambda(z)$, characterizing the cross-sectional shape of the wire.

The transverse problem at fixed longitudinal position z now takes the form

$$\left(\frac{\partial^2}{\partial r^2} + \frac{1}{r}\frac{\partial}{\partial r} + \frac{1}{r^2}\frac{\partial^2}{\partial\varphi^2} + \frac{2m_e}{\hbar^2}E_n(z)\right)\chi_n(r, \varphi; z) = 0, \qquad (6.15)$$

with boundary condition $\chi_n(R(\varphi, z), \varphi; z) = 0$ for all $\varphi \in [0, 2\pi]$. This determines the transverse eigenenergies $E_n(z) = E_n(\rho(z), \Lambda(z))$, which now depend on the cross-sectional shape through the boundary condition. With the cross-section parametrization (Equation 6.14), their dependence on geometry can be written as

$$E_n(\rho, \Lambda) = \frac{\hbar^2}{2m_e}\left(\frac{\gamma_n(\Lambda)}{\rho}\right)^2, \qquad (6.16)$$

where the shape-dependent functions $\gamma_n(\Lambda)$ remain to be determined. In general, and in particular for non-integrable cross sections, this has to be done by solving Equation 6.15 numerically (Urban et al. 2006).

The adiabatic approximation (long-wavelength limit) implies the decoupling of transverse and longitudinal motions. One starts with the ansatz $\Psi(r, \varphi, z) = \chi(r, \varphi; z)\, \Phi(z)$ and neglects all z-derivatives of the transverse wavefunction χ. Again, one is left with a series of effective one-dimensional scattering problems (Equation 6.6) for the longitudinal wave functions $\Phi_n(z)$, in which the transverse eigenenergies $E_n(\rho(z), \Lambda(z))$ act as additional potentials for the motion along the wire. These scattering problems can again be solved using the WKB approximation and Equations 6.11 and 6.13 apply.

6.3.4 Weyl Expansion

Semiclassical approximations often give an intuitive picture of the important physics and, due to their simplicity, allow for a better understanding of some general features. A very early analysis of the density of eigenmodes of a cavity with reflecting walls goes back to Weyl (1911) who proposed an expression in terms of the volume and surface area of the cavity. His formula was later rigorously proved and further terms in the expansion were calculated. Quite generally, we can express any extensive thermodynamic quantity as the sum of such a semiclassical Weyl expansion, which depends on geometrical quantities such as the system volume $\mathcal{V}$, surface area $\mathcal{S}$, and integrated mean curvature $\mathcal{C}$, as well as an oscillatory shell correction due to quantum-size effects (Brack and Bhaduri 1997). In particular, the grand canonical potential (Equation 6.2) can be written as

$$\Omega = -\omega\mathcal{V} + \sigma_s\mathcal{S} - \gamma_s\mathcal{C} + \delta\Omega, \qquad (6.17)$$

where the energy density ω, surface tension coefficient σ_s, and curvature energy γ_s are, in general, material- and temperature-dependent coefficients. On the other hand, the shell correction $\delta\Omega$ can be shown, based on very general arguments (Strutinsky 1968, Zhang et al. 2005), to be a single-particle effect, which is well described by the NFEM.

6.3.5 Material Dependence

Within the NFEM, there is only one parameter entering the calculation apart from the contact geometry: the Fermi energy E_{F}, which is material dependent and in general well known (see Table 6.1). Nevertheless, the energy cost of a deformation due to surface and curvature energy, which can vary significantly for different materials, plays a crucial role in determining the stability of a nanowire. Obviously, when working with a free-electron model, contributions of correlation and exchange energy are not included, while they are known to play an essential role in a correct treatment of the surface energy (Lang 1973). Using the NFEM *a priori* implies the macroscopic free energy density $\omega = 2E_{\mathrm{F}}k_{\mathrm{F}}^3/15\pi^2$, the macroscopic surface energy $\sigma_s = E_{\mathrm{F}}k_{\mathrm{F}}^2/16\pi$, and the macroscopic curvature energy $\gamma_s = 2E_{\mathrm{F}}k_{\mathrm{F}}/9\pi^2$. When drawing conclusions for metals having surface tensions and curvature energies that are rather different from these values, one has to think of an appropriate way to include these material-specific properties in the calculation.

A convenient way of modeling the material properties without losing the pleasant features of the NFEM is via the implementation of an appropriate surface boundary condition. Any atom-conserving deformation of the structure is subject to a constraint of the form

$$\mathcal{N} \equiv k_{\mathrm{F}}^3\mathcal{V} - \eta_s\, k_{\mathrm{F}}^2\mathcal{S} + \eta_c\, k_{\mathrm{F}}\mathcal{C} = \text{const.} \qquad (6.18)$$

This constraint on deformations of the nanowire interpolates between incompressibility and electroneutrality as side

TABLE 6.1 Material Parameters (Ashcroft and Mermin 1976, Perdew et al. 1991) of Several Monovalent Metals

Element	Li	Na	K	Cu	Ag	Au	Al
E_F [eV]	4.74	3.24	2.12	7.00	5.49	5.53	11.7
k_F [nm^{-1}]	11.2	9.2	7.5	13.6	12.0	12.1	17.5
σ_s [meV/Å^2]	27.2	13.6	7.58	93.3	64.9	78.5	59.2
σ_s [$E_F k_F^2$]	0.0046	0.0050	0.0064	0.0072	0.0082	0.0097	0.0017
η_s	1.135	1.105	1.001	0.939	0.866	0.755	1.146
γ_s [meV/Å]	62.0	24.6	14.9	119	96.4	161	121
γ_s [$E_F k_F$]	0.0117	0.0082	0.0094	0.0125	0.0146	0.0240	0.0059
η_c	0.802	1.06	0.971	0.741	0.583	−0.111	1.229

Source: Adapted from Urban, D.F. et al., *Phys. Rev. B*, 74, 245414, 2006.

Note: Fermi energy E_F, Fermi wavevector k_F, surface tension σ_s, and curvature energy γ_s, along with the corresponding values of η_s and η_c. The last column gives the corresponding values for the multivalent metal Al (see discussion in Section 6.5.7).

conditions, that is between volume conservation ($\eta_s = \eta_c = 0$) and treating the semiclassical expectation value for the charge Q_{Weyl} (Brack and Bhaduri 1997) as an invariant ($\eta_s = 3\pi/8$, $\eta_c = 1$).

The grand canonical potential of a free-electron gas confined within a given geometry by hard-wall boundaries, as given by Equation 6.17, changes under a deformation by

$$\Delta\Omega = -\omega\Delta\mathcal{V} + \sigma_s\Delta\mathcal{S} - \gamma_s\Delta\mathcal{C} + \Delta\big[\delta\Omega\big]$$
$$= -\frac{\omega}{k_F^3}\Delta\mathcal{N} + \left(\sigma_s - \frac{\omega}{k_F}\eta_s\right)\Delta\mathcal{S} - \left(\gamma_s - \frac{\omega}{k_F^2}\eta_c\right)\Delta\mathcal{C} + \Delta\big[\delta\Omega\big],$$
(6.19)

where the constraint (6.18) was used to eliminate $\mathcal{V}$. Now the prefactors of the change in surface $\Delta\mathcal{S}$ and the change in integrated mean curvature $\Delta\mathcal{C}$ can be identified as effective surface tension and curvature energy, respectively. They can be adjusted to fit a specific material's properties by an appropriate choice of the parameters η_s and η_c (see Table 6.1).

6.4 Conductance and Force

The formalism presented in Section 6.3 can now be applied to a specific wire geometry (Stafford et al. 1997), namely, a cosine constriction,

$$R(z) = \frac{R_0 + R_{min}}{2} + \frac{R_0 - R_{min}}{2}\cos\left(\frac{2\pi z}{L}\right),$$
(6.20)

of a cylindrical wire. One is interested in the mechanical properties of this metallic nanoconstriction in the regime of conductance quantization. The necessary condition to have well-defined conductance plateaus in a three-dimensional constriction was shown (Torres et al. 1994) to be $(\partial R/\partial z)^2 \ll 1$. In this limit, one may employ the adiabatic and WKB approximations and evaluate the expressions obtained in Section 6.3.2.

6.4.1 Conductance

The conductance is obtained from Equation 6.10. As the transmission amplitudes $\mathcal{T}_n$ vary exponentially from 1 to 0 when the transverse energy of the respective channel at the neck of the constriction traverses the Fermi energy, this results in a step-like behavior of the conductance with almost flat plateaus in between. This is the phenomenon of conductance quantization, which is observable even at room temperature for noble metal nanowires due to the large spacing of transverse energies (of order 1 eV for Au, to compare to $k_B T \simeq 10^{-3}$ eV at room temperature). The upper panel of Figure 6.2a shows the conductance obtained with an improved variant of the WKB approximation (Glazman et al. 1988, Brandbyge et al. 1995) for the geometry (6.20). The conductance as a function of elongation shows the expected steplike structure and the step heights are $2e^2/h$ and integer multiples thereof (the multiplicity depends on the degeneracy of the transverse modes). An ideal plastic deformation was assumed, i.e., the volume of the constriction was held constant during elongation.[*]

6.4.2 Force

If the wire elongation is slow enough, the electron gas has time to adjust to the wire shape changes during the deformation, and is thus always in equilibrium.[†] Under these conditions, the tensile force can be computed from the grand canonical potential, given in the WKB approximation by Equation 6.13, as $F = -\partial\Omega/\partial L|_{\mathcal{N}}$.

The lower panel of Figure 6.2a shows the tensile force for the cosine constriction (Equation 6.20). The correlations between

---

[*] Note the different abscissae for the theoretical and experimental graphs in Figure 6.2. While both contact diameter and elongation are a measure of the deformation of the contact, the latter are more easily accessible to experiments. The contact diameter on the other hand is the natural independent geometric variable that is used for the theoretical graph, since its relation to elongation is very model dependent.

[†] The experimental elongation speed is of order 1 nm/s, which is compatible with this assumption.

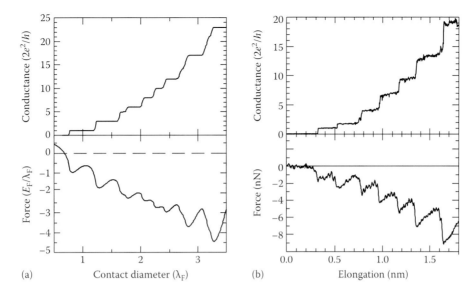

FIGURE 6.2 Electrical conductance G and tensile force F as function of elongation of a nanowire (a) calculated in a WKB approximation for a cosine constriction (Adapted from Stafford, C.A. et al., *Phys. Rev. Lett.*, 79, 2863, 1997.) and (b) measured by Rubio et al. (1996) in an experiment on gold. (Courtesy of N. Agraït.)

the force and conductance are striking: $|F|$ increases along the conductance plateaus and decreases sharply when the conductance drops. The constriction becomes unstable when the last conductance channel is cut off, which is indicated by a positive tensile force. Some transverse channels are quite closely spaced, and in these cases, the individual conductance plateaus [e.g., $G/G_0 = 14, 15, 19, 21$] and force oscillations are difficult to resolve. The force oscillations are found to have an amplitude E_F/λ_F (i.e., ~1.7 nN for gold, consistent with experimental observations) independent of the chosen geometry (circular and quadratic wires and cosine and parabolic constrictions were tested) and to persist to very large conductances.

These results can be understood within the intuitive picture of a conductance channel as a delocalized metallic bond. The increase of $|F|$ along the conductance plateaus and the rapid decrease at the conductance steps can then be interpreted as stretching and breaking of these bonds. Note that within the NFEM, the correlation between conductance changes and force oscillations comes from a pure quantum-size effect and not from atomic rearrangements.

The comparison of theoretical predictions with experimental data by Rubio et al. (1996), plotted in Figure 6.2b, shows very good qualitative agreement and underlines that, although the NFEM is simple, it gives a good qualitative description of the experimental findings. Extensions of the NFEM including structural dynamics of the wire (Bürki et al. 2003, Bürki 2007a,b), which are able to calculate the shape of the wire at all steps of elongation, find that instabilities accelerate the conductance and force jumps, making the theory even more similar to the experiment.

In addition, it is easy to show that in the NFEM, the tensile force is invariant under a stretching of the geometry $R(z) \rightarrow R(\lambda z)$, so that $F = (\varepsilon_F/\lambda_F) f(\Delta L/L_0, k_F R)$, where $f(x, y)$ is a dimensionless

function, i.e., the force oscillations are universal, and thus do not depend on the details of the wire shape, or precisely how it deforms. Nonuniversal corrections to F occur in very short constrictions, for which the adiabatic approximation breaks down.

6.5 Linear Stability Analysis

Metal nanowires are of great interest for nanotechnology since they may serve as conductors in future nanocircuits. In particular, one would like to know whether a nanowire of given length and radius remains stable at a given temperature.

At first sight, an atomistic approach seems to be more "realistic" than the NFEM and well suited to answer this question. But molecular dynamics (MD) simulations conceptually are not able to avoid the surface-tension-driven Rayleigh instability of long nanowires. Since quantum-size effects from the electron confinement are not properly taken into account, MD simulations fail to give an explanation for the electronic shell and supershell effects. On the other hand, atomistic quantum calculations using, e.g., the local-density approximation, are restricted to such small systems that their results cannot really be disentangled from finite-size effects (Stafford et al. 2000b). Therefore, to date, a stability analysis within the framework of the NFEM (and generalizations thereof) is the only approach able to correctly include the effects of electron-shell filling and thereby shed light on the puzzling stability of long metal nanowires.

The geometry of a wire of uniform cross section aligned along the z-axis is characterized by the cross-sectional area $A = \pi \rho^2$, and a set of dimensionless parameters determining the shape, which compose a vector Λ (cf. Equation 6.14). A small z-dependent perturbation of a wire of length L and initial cross section $(\bar{\rho}, \bar{\Lambda})$ can be written in terms of a Fourier series as

$$\rho(z) = \overline{\rho} + \varepsilon \delta\rho(z) = \overline{\rho} + \varepsilon \sum_q \rho_q e^{iqz},$$

$$\Lambda(z) = \overline{\Lambda} + \varepsilon \delta\Lambda(z) = \overline{\Lambda} + \varepsilon \sum_q \Lambda_q e^{iqz}, \qquad (6.21)$$

where the dimensionless small parameter ε sets the size of the perturbation.[*]

The energetic cost of a small deformation of the wire can be calculated by expanding the grand canonical potential as a series in the parameter ε,

$$\Omega = \Omega^{(0)} + \varepsilon\Omega^{(1)} + \varepsilon^2\Omega^{(2)} + \mathcal{O}(\varepsilon^3). \qquad (6.22)$$

A nanowire with initial cross section $(\overline{\rho}, \overline{\Lambda})$ is energetically stable at temperature T if and only if $\Omega^{(1)}(\overline{\rho}, \overline{\Lambda}, T) = 0$ and $\Omega^{(2)}(\overline{\rho}, \overline{\Lambda}, T) > 0$ for every possible deformation $(\delta\rho, \delta\Lambda)$ satisfying the constraint (6.18).

6.5.1 Rayleigh Instability

It is instructive to forget about quantum-size effects for a moment and to perform a stability analysis in the classical limit. For simplicity, one can restrict oneself to axial symmetry (i.e., $\Lambda \equiv 0$). In the classical limit, the grand canonical potential is given by the leading order terms of the Weyl approximation, $\Omega_{\text{Weyl}} = -\omega\mathcal{V} + \sigma_s\mathcal{S}$, and changes under the perturbation (6.21) by

$$\frac{\delta\Omega_{\text{Weyl}}}{L} = -2\pi(\overline{\rho}\omega - \sigma_s)\rho_0\varepsilon + \pi\sum_{q\neq 0}|\rho_q|^2\left[-\omega + q^2\overline{\rho}\sigma_s\right]\varepsilon^2. \qquad (6.23)$$

Because of the constraint (6.18) on possible deformations, ρ_0 can be expressed in terms of the other Fourier coefficients. Volume conservation, e.g., implies $\rho_0 = -(\varepsilon/2\overline{\rho})\sum_{q\neq 0}|\rho_q|^2$ and

$$\frac{\delta\Omega_{\text{Weyl}}(q)}{L} = \frac{\pi\sigma_s}{\overline{\rho}}\sum_{q\neq 0}|\rho_q|^2(\overline{\rho}^2q^2 - 1)\varepsilon^2, \qquad (6.24)$$

which has to be positive in order to ensure stability. Since q is restricted to integer multiples of $2\pi/L$, stability requires $L < 2\pi\overline{\rho}$. This is just the criterion of the classical Rayleigh instability (Chandrasekhar 1981): A wire longer than its circumference is unstable and likely to break up into clusters due to surface tension.

6.5.2 Quantum-Mechanical Stability Analysis

The crucial ingredient to the stabilization of metal nanowires is the oscillatory shell correction $\delta\Omega$ to the grand canonical potential (Equation 6.17), which is due to quantum-size effects.

This shell correction can be accounted for by a quantum-mechanical stability analysis based on the WKB approximation introduced in Section 6.3.2. The use of this approximation can be justified *a posteriori* by a full quantum calculation (Urban and Grabert 2003, Urban et al. 2007), which shows that the structural stability of metal nanowires is indeed governed by their response to long-wavelength perturbations. The response to short-wavelength perturbations on the other hand controls a Peierls-type instability characterized by the opening of a gap in the electronic energy dispersion relation. This quantum-mechanical instability, which is missing in the semiclassical WKB approximation, in fact limits the maximal length of stable nanowires. Nevertheless, if the wires are short enough, and/or the temperature is not too low, the full quantum calculation essentially confirms the semiclassical results.

A systematic expansion of Equation 6.13 yields

$$\frac{\Omega^{(1)}}{L/\lambda_F} = 4\sum_n\sqrt{\frac{E_F - \overline{E}_n}{E_F}}\left(\Lambda_0 \cdot \overline{E}'_n - 2\overline{E}_n\frac{\rho_0}{\overline{\rho}}\right), \qquad (6.25)$$

$$\frac{\Omega^{(2)}}{L/\lambda_F} = E_F\sum_q\begin{pmatrix}\rho_q/\overline{\rho}\\\Lambda_q\end{pmatrix}^\dagger\begin{pmatrix}A_{\rho\rho} & A_{\rho\Lambda}\\A_{\Lambda\rho} & A_{\Lambda\Lambda}\end{pmatrix}\begin{pmatrix}\rho_q/\overline{\rho}\\\Lambda_q\end{pmatrix}, \qquad (6.26)$$

where the elements of the matrix A in Equation 6.26 are given by

$$A_{\rho\rho} = \sum_n\frac{4\overline{E}_n}{E_F^{3/2}}\left[3\sqrt{E_F - \overline{E}_n} - \frac{\overline{E}_n}{\sqrt{E_F - \overline{E}_n}}\right],$$

$$A_{\Lambda\rho} = -\sum_n\frac{4\overline{E}'_n}{E_F^{3/2}}\left[\sqrt{E_F - \overline{E}_n} - \frac{\overline{E}_n}{2\sqrt{E_F - \overline{E}_n}}\right],$$

$$\qquad\qquad\qquad\qquad\qquad\qquad\qquad\qquad\qquad (6.27)$$

$$A_{\Lambda\Lambda} = \sum_n\frac{1}{E_F^{3/2}}\left[2\overline{E}''_n\sqrt{E_F - \overline{E}_n} - \frac{\overline{E}'_n \cdot (\overline{E}'_n)^\dagger}{\sqrt{E_F - \overline{E}_n}}\right] \quad\text{and}$$

$$A_{\rho,\Lambda} = A_{\Lambda,\rho}.$$

Here, $\overline{E}'_n$ denotes the gradient of E_n with respect to Λ, and $\overline{E}''_n$ is the matrix of second derivatives. The bar indicates evaluation at $(\overline{\rho}, \overline{\Lambda})$.

The number of independent Fourier coefficients in Equation 6.21 is restricted through the constraint (6.18) on allowed deformations. Hence, after evaluating the change of the geometric quantities $\mathcal{V}$, $\mathcal{S}$, and $\mathcal{C}$ due to the deformation, we can use Equation 6.18 to express ρ_0 in terms of the other Fourier coefficients, yielding an expansion $\rho_0 = \rho_0^{(0)} + \varepsilon\rho_0^{(1)} + \mathcal{O}(\varepsilon^2)$. This expansion then needs to be inserted in Equations 6.25 and 6.26, thereby modifying the first-order change of the energy $\Omega^{(1)}$ and the stability matrix A (Urban et al. 2006).

Stability requires that the resulting modified stability matrix $\tilde{A}$ be positive definite. Results at finite temperature are obtained essentially in a similar fashion, by integrating Equation 6.2 numerically.

---

[*] Assuming periodic boundary conditions, the perturbation wave vectors q must be integer multiples of $2\pi/L$. In order to ensure that $\rho(z)$ and $\Lambda(z)$ are real, we have $\rho_{-q} = \rho_q^\star$ and $\Lambda_{-q} = \Lambda_q^\star$.

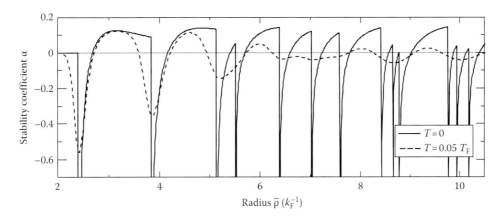

FIGURE 6.3 WKB stability coefficient, calculated using a constant-volume constraint. The sharp negative peaks at the opening of new channels (i.e., when $k_F \bar\rho = \gamma_n$) are smeared out with increasing temperature T.

6.5.3 Axial Symmetry

A straightforward application of the method outlined above is the stability analysis of cylindrical wires with respect to axisymmetric volume conserving perturbations. In this specific case, $\Lambda(z) \equiv 0$ and $\rho_0 = -(\varepsilon/2\bar\rho)\Sigma_{q\neq0}|\rho_q|^2$. Therefore, Equation 6.25 takes the form $\Omega^{(1)} \equiv 0$ and Equation 6.26 simplifies to read

$$\frac{\Omega^{(2)}}{L/\lambda_F} = E_F \sum_{q\neq0}\left|\rho_q/\bar\rho\right|^2 \alpha(\bar\rho), \tag{6.28}$$

where the *stability coefficient* $\alpha(\bar\rho) \equiv \tilde{A}_{\rho\rho}$ reads (Urban et al. 2006)

$$\alpha(\bar\rho) = \sum_n \theta(k_F\bar\rho - \gamma_n)\frac{4\gamma_n^2}{(k_F\bar\rho)^2}\left[4\sqrt{1-\frac{\gamma_n^2}{(k_F\bar\rho)^2}} - \frac{1}{\sqrt{(k_F\bar\rho)^2 - \gamma_n^2}}\right]. \tag{6.29}$$

Axial symmetry implies the use of the transverse eigenenergies $E_n/E_F = (\gamma_n/k_F\bar\rho)^2$, cf. Equation 6.4. This result, valid for zero temperature, is plotted as a function of radius in Figure 6.3 together with a numerical result at finite temperature. Sharp negative peaks at the subband thresholds, i.e., when $\bar\rho k_F = \gamma_n$ indicate strong instabilities whenever a new channel opens. On the other hand, α is positive in the regions between these thresholds giving rise to intervals of stability that decrease with increasing temperature. These islands of stability can be identified with the "magic radii" found in experiments. As will be shown below, one has to go beyond axial symmetry in order to give a full explanation of the observed conductance histograms of metal nanowires.

6.5.4 Breaking Axial Symmetry

It is well known in the physics of crystals and molecules that a Jahn–Teller deformation breaking the symmetry of the system can be energetically favorable. In metal clusters, Jahn–Teller deformations are also very common, and most of the observed structures show a broken spherical symmetry. By analogy, it is

natural to assume that for nanowires, too, a breaking of axial symmetry can be energetically favorable, and lead to more stable deformed geometries.

Canonical candidates for such stable non-axisymmetric wires are wires with a $\cos(m\varphi)$-deformed cross section (i.e., having m-fold symmetry), a special case of Equation 6.14 with only one nonzero λ_m. The quadrupolar deformation ($m = 2$) is expected to be the energetically most favorable of the multipole deformations[*] since deformations with $m > 2$ become increasingly costly with increasing m, their surface energy scaling as m^2.

The results of Section 6.5.2 can straightforwardly be used to determine stable quadrupolar configurations by intersection of the stationary curves, $\Omega^{(1)}(\bar\rho, \bar\lambda_2)|_{\mathcal{N}} = 0$, and the convex regions, $\Omega^{(2)}(\bar\rho, \bar\lambda_2)|_{\mathcal{N}} > 0$. The result is a so-called *stability diagram*, which shows the stable geometries (at a given temperature) in configuration space, that is a function of the geometric parameters $\bar\rho$ and $\bar\lambda_2$. An example of such a stability diagram is shown later in Figure 6.6 for the case of aluminum, discussed in Section 6.5.7. Results for all temperatures can then be combined, thus adding a third axis (i.e., temperature) to the stability diagram. Finally, the most stable configurations can be extracted, defined as those geometries that persist up to the highest temperature compared to their neighboring configurations.

Table 6.2 lists the most stable deformed sodium wires with quadrupolar cross section, obtained by the procedure described above. The deformation of the stable structures is characterized by the parameter λ_2 or equivalently by the aspect ratio

$$a = \frac{\sqrt{1-\lambda_2^2/2}+\lambda_2}{\sqrt{1-\lambda_2^2/2}-\lambda_2}. \tag{6.30}$$

Clearly, nanowires with highly deformed cross sections are only stable at small conductance. The maximum temperature up to which the wires remain stable, given in the last column of Table 6.2,

---

[*] The dipole deformation ($m = 1$) corresponds, in leading order, to a simple translation, plus higher order multipole deformations. Therefore, the analysis can be restricted to $m > 1$.

TABLE 6.2 Most Stable Deformed Wires with Quadrupolar Cross Sections

G/G_0	a	λ_2	T_{max}/T_ρ
2	1.72	0.26	0.50
5	1.33	0.14	0.49
9	1.22	0.10	0.50
29	1.13	0.06	0.54
59	1.11	0.05	0.49
72	1.08	0.04	0.39
117	1.06	0.03	0.55
172	1.06	0.03	0.50

Source: Adapted from Urban, D.F. et al., *Phys. Rev. B*, 74, 245414, 2006.

Note: The first column gives the quantized conductance of the corresponding wire. Both the aspect ratio a and the value of the deformation parameter λ_2 are given. The maximum temperature of stability T_{max} is given for each wire. In all cases, the surface tension was set to 0.22 N/m, corresponding to Na.

is expressed in units of $T_\rho := T_F/(k_F\overline{\rho})$. The use of this characteristic temperature reflects the temperature dependence of the shell correction to the wire energy (Urban et al. 2006).

Deformations with higher m cost more and more surface energy. Compared to the quadrupolar wires, the number of stable configurations with three-, four-, five-, and sixfold symmetry, their maximum temperature of stability, and their size of the deformations involved all decrease rapidly with increasing order m of the deformation. For $m > 6$, no stable geometries are known. All this reflects the increase in surface energy with increasing order m of the deformation.

6.5.5 General Stability of Cylinders

It is possible to derive the complete stability diagram for cylinders, i.e., to determine the radii of cylindrical wires that are stable with respect to *arbitrary* small, long-wavelength deformations (Urban et al. 2006). At first sight, considering arbitrary deformations, and therefore theoretically an infinite number of perturbation parameters seems a formidable task. Fortunately, the stability matrix $\tilde{A}$ for cylinders is found to be diagonal, and therefore the different Fourier contributions of the deformation decouple. This simplifies the problem considerably, since it allows to determine the stability of cylindrical wires with respect to arbitrary deformations through the study of a set of pure m-deformations, i.e., deformations as given by Equation 6.14 with only one nonzero λ_m.

Figure 6.4 shows the stable cylindrical wires (in dark gray) as a function of temperature. The surface tension was fixed at the value for Na, see Table 6.1. The stability diagram was obtained by intersecting a set of individual stability diagrams allowing $\cos(m\varphi)$ deformations with $m \leq 6$. This analysis confirms the extraordinary stability of a set of wires with so-called magic radii. They exhibit conductance values $G/G_0 = 1, 3, 6, 12, 17, 23, 34, 42, 51, \dots$. It is noteworthy that some wires that are stable at low temperatures when considering only axisymmetric perturbations, e.g., $G/G_0 = 5, 10, 14$, are found to be unstable when allowing more general, symmetry-breaking deformations.

The heights of the dominant stability peaks in Figure 6.4 exhibit a periodic modulation, with minima occurring near $G/G_0 = 9, 29, 59, 117, \dots$. The positions of these minima are in perfect agreement with the observed supershell structure in conductance histograms of alkali metal nanowires (Yanson et al. 2000). Interestingly, the nodes of the supershell structure, where the shell effect for a cylinder is suppressed, are precisely where the most stable deformed nanowires are predicted to occur (see Section 6.5.4). Thus, symmetry-breaking distortions and the supershell effect are inextricably linked.

Linear stability is a necessary—but not a sufficient—condition for a nanostructure to be observed experimentally. The linearly stable nanocylinders revealed in the above analysis are in fact *metastable* structures, and an analysis of their lifetime has been

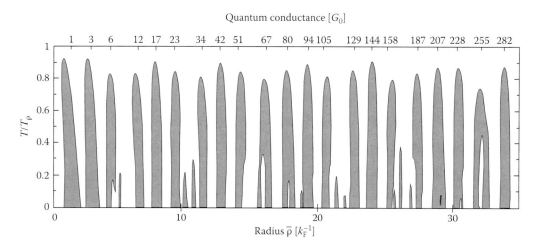

Quantum conductance [G_0]

1 3 6 12 17 23 34 42 51 67 80 94 105 129 144 158 187 207 228 255 282

FIGURE 6.4 Stability of metal nanocylinders versus electrical conductance and temperature. Dark gray areas indicate stability with respect to arbitrary small deformations. Temperature is displayed in units of $T_\rho = T_F/k_F\overline{\rho}$ (see text). The surface tension was taken as 0.22 N/m, corresponding to Na. (Adapted from Urban, D.F. et al., *Phys. Rev. B*, 74, 245414, 2006.)

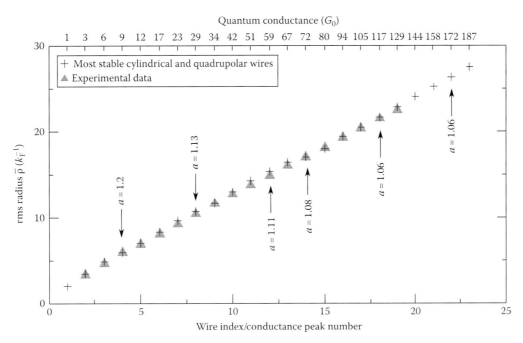

FIGURE 6.5 Comparison of the experimental shell structure for Na, taken from Yanson et al. (1999), with the theoretical predictions of the most stable Na nanowires. Non-axisymmetric wires are labeled with the corresponding aspect ratio *a*. (Adapted from Urban, D.F. et al., *Phys. Rev. Lett.*, 93, 186403, 2004a.)

carried out within an axisymmetric stochastic field theory by Bürki and Stafford (2005). There is a strong correlation between the height of the stable fingers in the linear stability analysis and the size of the activation barriers ΔE, which determines the nanowire lifetime τ through the Kramers formula $\tau = \tau_0 \exp(\Delta E/k_B T)$. This suggests that the linear stability analysis, with temperature expressed in units of $T_\rho = T_F/(k_F\bar{\rho})$, provides a good measure of the total stability of metal nanowires. In particular, the "universal" stability of the most stable cylinders is reproduced, wherein the absolute stability of the magic cylinders is essentially independent of radius (aside from the small supershell oscillations).

6.5.6 Comparison with Experiments

A detailed comparison between the theoretically most stable structures and experimental data for sodium is provided in Figure 6.5. For each stable finger in the linear stability analysis, its mean conductance is extracted and plotted as a function of its index number, together with experimental data by Yanson et al. (1999). This comparison shows that there is a one-to-one relation between observed conductance peaks and theoretically stable geometries which in particular allows for a prediction of the cross-sectional shape of the wires. This striking fit is only possible when including non-axisymmetric wires, which represent roughly 25% of the most stable structures and which are labeled by the corresponding aspect ratios *a*, as shown in Figure 6.5. The remaining 75% of the principal structures correspond to the magic cylinders. The role of symmetry in the stability of metal nanowires is thus fundamentally different from the case of atomic nuclei or metal clusters, where the vast majority of

stable structures have broken symmetry. The crucial difference between the stability of metal nanowires and metal clusters is not the shell effect, which is similar in both cases, but rather the surface energy, which favors the sphere, but abhors the cylinder.

Besides the geometries entering the comparison above, the stability analysis also reveals two highly deformed quadrupolar nanowires with conductance values of $2G_0$ and $5G_0$, cf. Table 6.2. They are expected to appear more rarely due to their reduced stability relative to the neighboring peaks, and their large aspect ratio *a* that renders them rather isolated in configuration space.* Nevertheless they can be identified by a detailed analysis of conductance histograms of the alkali metals (Urban et al. 2004b).

6.5.7 Material Dependence

Results for different metals are similar in respect to the number of stable configurations and the conductance of the wires. On the other hand, the deviations from axial symmetry and the relative stability of Jahn–Teller deformed wires is sensitive to the material-specific surface tension and Fermi temperature. The relative stability of the highly deformed wires decreases with increasing surface tension, $\sigma_s/(E_F k_F^2)$, measured in intrinsic units, and this decrease becomes stronger with increasing order *m* of the deformation. Therefore, for the simple s-orbital metals under consideration (Table 6.1), deformed Li wires have the highest and Au wires have the lowest relative stability compared to

---

* A nanowire produced by pulling apart an axisymmetric contact has a smaller probability to transform into a highly deformed configuration than into a neighboring cylindrical configuration.

cylinders of "magic radii."* Notable in this respect is aluminum with $\sigma_s = 0.0017 E_F k_F^2$, some five times smaller than the value for Au. Aluminum is a trivalent metal, but the Fermi surface of bulk Al resembles a free-electron Fermi sphere in the extended-zone scheme. This suggests the applicability of the NFEM to Al nanowires, although the continuum approximation is more severe than for monovalent metals.

Recent experiments (Mares et al. 2007) have found evidence for the fact that the stability of aluminum nanowires also is governed by shell-filling effects. Two magic series of stable structures have been observed with a crossover at $G \simeq 40G_0$ and the exceptionally stable structures have been related to electronic and atomic shell effects, respectively. Concerning the former, the NFEM can quantitatively explain the conductance and geometry of the stable structures for wires with $G > 12G_0$ and there is a perfect one-to-one correspondence of the predicted stable Al nanowires and the experimental electron-shell structure. Moreover, an experimentally observed third sequence of stable structures with conductance $G/G_0 \simeq 5, 14, 22$ provides intriguing evidence for the existence of "superdeformed" nanowires whose cross sections have an aspect ratio near 2:1. Theoretically, these wires are quite stable compared to other highly deformed structures and, more importantly, are very isolated in configuration space, as illustrated in the stability diagram shown in Figure 6.6. This favors their experimental detection if the initial structure of the nanocontact formed in the break junction is rather planar with a large aspect ratio since it is likely that the aspect ratio is maintained as the wire necks down elastically. Aluminum is unique in this respect and evidence of superdeformation has not been reported in any of the previous experiments on alkali and

noble metals, presumably because highly deformed structures are intrinsically less stable than nearly axisymmetric structures due to their larger surface energy.

6.6 Summary and Discussion

In this chapter, we have given an overview of the NFEM, treating a metal nanowire as a noninteracting electron gas confined to a given geometry by hard-wall boundary conditions. At first sight, the NFEM seems to be an overly simple model, but closer study reveals that it contains very rich and complex features. Since its first introduction in 1997, it has repeatedly shown that it captures the important physics and is able to explain qualitatively, when not quantitatively, many of the experimentally observed properties of alkali and noble metal nanowires. Its strengths compared to other approaches are, in particular, the absence of any free parameters and the treatment of electrical and mechanical properties on an equal footing. Moreover, the advantage of obtaining analytical results allows the possibility to gain some detailed understanding of the underlying mechanisms governing the stability and structural dynamics of metal nanowires.

The NFEM correctly describes electronic quantum-size effects, which play an essential role in the stability of nanowires. A linear stability analysis shows that the classical Rayleigh instability of a long wire under surface tension can be completely suppressed by electronic shell effects, leading to a sequence of certain stable "magic" wire geometries. The derived sequence of stable, cylindrical, and quadrupolar wires explains the experimentally observed shell and supershell structures for the alkali and noble metals as well as for aluminum. The most stable wires with broken axial symmetry are found at the nodes of the supershell structure, indicating that the Jahn–Teller distortions and the supershell effect are inextricably linked. In addition, a series of superdeformed aluminum nanowires with an aspect ratio near 2:1 is found, which has lately been identified experimentally. A more elaborate quantum-mechanical analysis within the NFEM reveals an interplay between Rayleigh and Peierls-type instabilities. The latter is length dependent and limits the maximal length of stable nanowires but other than that confirms the results obtained by the long-wavelength expansion discussed above. Remarkably, certain gold nanowires are predicted to remain stable even at room temperature up to a maximal length in the micrometer range, sufficient for future nanotechnological applications.

The NFEM can be expanded by including the structural dynamics of the wire in terms of a continuum model of the surface diffusion of the ions. Furthermore, defects and structural fluctuations may also be accounted for. These extensions improve the agreement with experiments but do not alter the main conclusions. However, the NFEM does not address the discrete atomic structure of metal nanowires. With increasing thickness of the wire, the effects of surface tension decrease and there is a crossover from plastic flow of ions to crystalline order, the latter implying atomic shell effects observed for thicker nanowires.

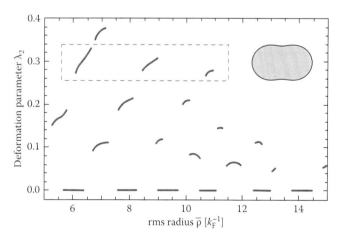

FIGURE 6.6 Stability diagram for Al wires at fixed temperature $T = 0.45\,T_\rho$. Thick lines mark stable wires in the configuration space of rms radius $\bar{\rho}$ and deformation parameter λ_2. The dashed box emphasizes a series of very stable superdeformed wires, whose peanut-shaped cross section is shown as an inset. This sequence was recently identified experimentally. (From Mares, A.I. et al., *Nanotechnology*, 18, 265403, 2007.)

---

* Concerning the absolute stability, we have to consider that the lifetime of a metastable nanowire also depends on the surface tension (Bürki and Stafford 2005).

Therefore, the NFEM applies to a window of conductance values between a few G_0 and about $100G_0$, depending on the material under consideration.

Promising extensions of the NFEM in view of current research activities are directed, e.g., toward the study of metal nanowires in nanoelectromechanical systems (NEMS) which couple nanoscale mechanical resonators to electronic devices of similar dimensions. The NFEM is ideally suited for the investigation of such systems since it naturally comprises electrical as well as mechanical properties. It is hoped that the generic behavior of metal nanostructures elucidated by the NFEM can guide the exploration of more elaborate, material-specific models in the same way that the free-electron model provides an important theoretical reference point from which we can understand the complex properties of real bulk metals.

References

Agraït, N., J. G. Rodrigo, and S. Vieira. 1993. Conductance steps and quantization in atomic-size contacts. *Phys. Rev. B* 47: 12345.

Agraït, N., A. Levy Yeyati, and J. M. van Ruitenbeek. 2003. Quantum properties of atomic-sized conductors. *Phys. Rep.* 377: 81.

Ashcroft, N. W. and N. D. Mermin. 1976. *Solid State Physics*. Saunders College Publishing, Philadelphia, PA.

Brack, M. 1993. The physics of simple metal clusters: Self-consistent jellium model and semiclassical approaches. *Rev. Mod. Phys.* 65: 677.

Brack, M. and R. K. Bhaduri. 1997. *Semiclassical Physics*, volume 96 of *Frontiers in Physics*. Addison-Wesley, Reading, MA.

Brandbyge, M., J. Schiøtz, M. R. Sørensen et al. 1995. Quantized conductance in atom-sized wires between two metals. *Phys. Rev. B* 52: 8499.

Bürki, J. 2007a. Discrete thinning dynamics in a continuum model of metallic nanowires. *Phys. Rev. B* 75: 205435.

Bürki, J. 2007b. Front propagation into unstable metal nanowires. *Phys. Rev. E* 76: 026317.

Bürki, J. and C. A. Stafford. 1999. Comment on "Quantum suppression of shot noise in atomic size metallic contacts". *Phys. Rev. Lett.* 83: 3342.

Bürki, J. and C. A. Stafford. 2005. On the stability and structural dynamics of metal nanowires. *Appl. Phys. A* 81: 1519.

Bürki, J., C. A. Stafford, X. Zotos, and D. Baeriswyl. 1999. Cohesion and conductance of disordered metallic point contacts. *Phys. Rev. B* 60: 5000. ibid. *Phys. Rev. B* 62: 2956 (2000) (Erratum).

Bürki, J., R. E. Goldstein, and C. A. Stafford. 2003. Quantum necking in stressed metallic nanowires. *Phys. Rev. Lett.* 91: 254501.

Bürki, J., C. A. Stafford, and D. L. Stein. 2005. Theory of metastability in simple metal nanowires. *Phys. Rev. Lett.* 95: 090601.

Chandrasekhar, S. 1981. *Hydrodynamic and Hydromagnetic Stability*. Dover Publishing Company, New York.

Dashen, R., S.-K. Ma, and H. J. Bernstein. 1969. S-matrix formulation of statistical mechanics. *Phys. Rev.* 187: 345.

Datta, S. 1995. *Electronic Transport in Mesoscopic Systems*. Cambridge University Press, Cambridge, U.K., pp. 48–170.

de Heer, W. A. 1993. The physics of simple metal clusters: Experimental aspects and simple models. *Rev. Mod. Phys.* 65: 611.

Díaz, M., J. L. Costa-Krämer, E. Medina, A. Hasmy, and P. A. Serena. 2003. Evidence of shell structures in Au nanowires at room temperature. *Nanotechnology* 14: 113.

García-Martin, A., J. A. Torres, and J. J. Sáenz. 1996. Finite size corrections to the conductance of ballistic wires. *Phys. Rev. B* 54: 13448.

Glazman, L. I., G. B. Lesovik, D. E. Khmel'nitskii, and R. I. Shekter. 1988. Reflectionless quantum transport and fundamental ballistic-resistance steps in microscopic constrictions. *JETP Lett.* 48: 239.

Kassubek, F., C. A. Stafford, and H. Grabert. 1999. Force, charge, and conductance of an ideal metallic nanowire. *Phys. Rev. B* 59: 7560.

Kassubek, F., C. A. Stafford, H. Grabert, and R. E. Goldstein. 2001. Quantum suppression of the Rayleigh instability in nanowires. *Nonlinearity* 14: 167.

Kondo, Y. and K. Takayanagi. 1997. Gold nanobridge stabilized by surface structure. *Phys. Rev. Lett.* 79: 3455.

Kondo, Y. and K. Takayanagi. 2000. Synthesis and characterization of helical multi-shell gold nanowires. *Science* 289: 606.

Krans, J. M., J. M. van Ruitenbeek, V. V. Fisun, I. K. Yanson, and L. J. de Jongh. 1995. The signature of conductance quantization in metallic point contacts. *Nature* 375: 767.

Landman, U., W. D. Luedtke, N. A. Burnham, and R. J. Colton. 1990. Atomistic mechanisms and dynamics of adhesion, nanoindentation, and fracture. *Science* 248: 454.

Lang, N. D. 1973. The density-functional formalism and the electronic structure of metal surfaces. *Solid State Phys.* 28: 225.

Mares, A. I. and J. M. van Ruitenbeek. 2005. Observation of shell effects in nanowires for the noble metals Cu, Ag, and Au. *Phys. Rev. B* 72: 205402.

Mares, A. I., A. F. Otte, L. G. Soukiassian, R. H. M. Smit, and J. M. van Ruitenbeek. 2004. Observation of electronic and atomic shell effects in gold nanowires. *Phys. Rev. B* 70: 073401.

Mares, A. I., D. F. Urban, J. Bürki, H. Grabert, C. A. Stafford, and J. M. van Ruitenbeek. 2007. Electronic and atomic shell structure in aluminum nanowires. *Nanotechnology* 18: 265403.

Martin, T. P. 1996. Shells of atoms. *Phys. Rep.* 273: 199.

Messiah, A. 1999. *Quantum Mechanics*. Dover Publishing Company, New York.

Moreland, J. and J. W. Ekin. 1985. Electron tunneling experiments using Nb-Sn break junctions. *J. Appl. Phys.* 58: 3888.

Muller, C. J., J. M. van Ruitenbeek, and L. J. de Jongh. 1992. Conductance and supercurrent discontinuities in atomic-scale metallic constrictions of variable width. *Phys. Rev. Lett.* 69: 140.

Perdew, J. P., Y. Wang, and E. Engel. 1991. Liquid-drop model for crystalline metals: Vacancy-formation, cohesive and face-dependent surface energies. *Phys. Rev. Lett.* 66: 508.

Plateau, J. 1873. *Statique Expérimentale et Théorique des Liquides Soumis aux Seules Forces Moléculaires.* Gauthier-Villars, Paris, France.

Rodrigues, V., T. Fuhrer, and D. Ugarte. 2000. Signature of atomic structure in the quantum conductance of gold nanowires. *Phys. Rev. Lett.* 85: 4124.

Rubio, G., N. Agraït, and S. Vieira. 1996. Atomic-sized metallic contacts: Mechanical properties and electronic transport. *Phys. Rev. Lett.* 76: 2302.

Stafford, C. A., D. Baeriswyl, and J. Bürki. 1997. Jellium model of metallic nanocohesion. *Phys. Rev. Lett.* 79: 2863.

Stafford, C. A., F. Kassubek, J. Bürki, H. Grabert, and D. Baeriswyl. 2000a. Cohesion, conductance, and charging effects in a metallic nanocontact, in *Quantum Physics at the Mesoscopic Scale.* EDP Sciences, Les Ulis, Paris, France, pp. 49–53.

Stafford, C. A., J. Bürki, and D. Baeriswyl. 2000b. Comment on "Density functional simulation of a breaking nanowire". *Phys. Rev. Lett.* 84: 2548.

Stalder, A. and U. Dürig. 1996. Study of yielding mechanics in nanometer-sized Au contacts. *Appl. Phys. Lett.* 68: 637.

Strutinsky, V. M. 1968. Shells in deformed nuclei. *Nucl. Phys. A* 122: 1.

Todorov, T. N. and A. P. Sutton. 1993. Jumps in electronic conductance due to mechanical instabilities. *Phys. Rev. Lett.* 70: 2138.

Torres, J. A., J. I. Pascual, and J. J. Sáenz. 1994. Theory of conduction through narrow constrictions in a three dimensional electron gas. *Phys. Rev. B* 49: 16581.

Untiedt, C., G. Rubio, S. Vieira, and N. Agraït. 1997. Fabrication and characterization of metallic nanowires. *Phys. Rev. B* 56: 2154.

Urban, D. F. and H. Grabert. 2003. Interplay of Rayleigh and Peierls instabilities in metallic nanowires. *Phys. Rev. Lett.* 91: 256803.

Urban, D. F., J. Bürki, C. H. Zhang, C. A. Stafford, and H. Grabert. 2004a. Jahn-Teller distortions and the supershell effect in metal nanowires. *Phys. Rev. Lett.* 93: 186403.

Urban, D. F., J. Bürki, A. I. Yanson, I. K. Yanson, C. A. Stafford, J. M. van Ruitenbeek, and H. Grabert. 2004b. Electronic shell effects and the stability of alkali nanowires. *Solid State Commun.* 131: 609.

Urban, D. F., J. Bürki, C. A. Stafford, and H. Grabert. 2006. Stability and symmetry breaking in metal nanowires: The nanoscale free-electron model. *Phys. Rev. B* 74: 245414.

Urban, D. F., C. A. Stafford, and H. Grabert. 2007. Scaling theory of the Peierls charge density wave in metal nanowires. *Phys. Rev. B* 75: 205428.

van Wees, B. J., H. van Houten, C. W. J. Beenakker et al. 1988. Quantized conductance of point contacts in a two-dimensional electron gas. *Phys. Rev. Lett.* 60: 848.

Weyl, H. 1911. Über die asymptotische Verteilung der Eigenwerte. *Nachr. akad. Wiss. Göttingen* 110–117.

Wharam, D. A., T. J. Thornton, R. Newbury et al. 1988. One-dimensional transport and the quantisation of the ballistic resistance. *J. Phys. C Solid State Phys.* 21: L209.

Yanson, A. I., I. K. Yanson, and J. M. van Ruitenbeek. 1999. Observation of shell structure in sodium nanowires. *Nature* 400: 144.

Yanson, A. I., I. K. Yanson, and J. M. van Ruitenbeek. 2000. Supershell structure in alkali metal nanowires. *Phys. Rev. Lett.* 84: 5832.

Yanson, A. I., I. K. Yanson, and J. M. van Ruitenbeek. 2001. Shell effects in alkali metal nanowires. *Fizika Nizkikh Temperatur* 27: 1092.

Zhang, C. H., F. Kassubek, and C. A. Stafford. 2003. Surface fluctuations and the stability of metal nanowires. *Phys. Rev. B* 68: 165414.

Zhang, C. H., J. Bürki, and C. A. Stafford. 2005. Stability of metal nanowires at ultrahigh current densities. *Phys. Rev. B* 71: 235404.

7

Small-Scale Nonequilibrium Systems

Peder C. F. Møller
University of Copenhagen

Lene B. Oddershede
University of Copenhagen

7.1 Introduction

Thermodynamics have proven very successful in describing physical systems in equilibrium. However, most of the systems surrounding us are, in fact, not in equilibrium. Systems out of equilibrium are difficult to describe from a physics point of view, and only little theory on such systems exists in comparison to the wealth of knowledge about thermodynamics of systems in equilibrium. One of the remaining tasks of thermodynamics is to extract thermodynamical values from systems undergoing out-of-equilibrium irreversible processes. For many systems surrounding us, the relaxation toward equilibrium is so slow that we do not experience it and do not care about it. One example is diamond, whose lattice at room temperature and normal pressure decays slowly toward graphite, the main substance of a pencil. However, the decay of a diamond is so slow in comparison to a human lifetime that they are highly valued in spite of their transient nature (Figure 7.1). Another example of a system that relaxes slowly is a mixture of oxygen and hydrogen. When a spark is added, the mixture burns rapidly and forms water, but if left alone the mixture relaxes toward water at a very slow rate. And for any explosive, it is paramount that the energy-releasing reaction does not happen spontaneously, but only when requested.

With the advances of nanoscopic techniques, it is possible to monitor systems on length scales down to nanometers, and at timescales below microseconds. At these distances and timescales, thermal fluctuations cannot be ignored. Macroscopic objects typically contain on the order of 10^{23} entities, and fluctuations around the average value are neglectable. But, when observing on the nanometer level, one typically observes a

single entity, e.g., a single functioning biological molecule with a high temporal resolution, and fluctuations are significant. If the nano-technique in addition exerts a force on the system studied, it is likely that the system is evolving through a nonequilibrium process with a corresponding energy dissipation.

The aim of this chapter is to present a theoretical framework and practical examples of how to deal with small-scale nonequilibrium systems with focus on how to extract thermodynamic values, such as the free energy, from a nonequilibrium behavior. Section 7.2 presents the most important knowledge about systems in equilibrium, introducing thermodynamic quantities and terminology that are needed also to describe the nonequilibrium systems. Section 7.3 goes through the most commonly used and easily applicable theories, which apply to small systems out of equilibrium. These encompass Kramers equations, fluctuation–dissipation relations, and finally Jarzynski's equality and Crooks fluctuation theorem, the latter two having the powerful property that they apply to systems arbitrarily far from equilibrium. Every nonequilibrium theory presented is accompanied by examples of how to apply the particular theory to a small-scale system.

7.2 Systems in Equilibrium

To clarify how the nonequilibrium dynamics, which is the subject of the subsequent sections, differ from equilibrium statistical mechanics, we here give a brief reminder of some of the basic assumptions and main results of equilibrium systems. Consider the staggering information needed to characterize the state of one liter of a noble gas at ambient temperature and pressure: It contains about 2×10^{22} atoms, giving 6×10^{22} position coordinates

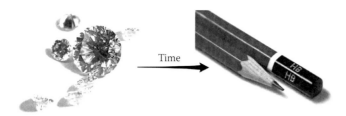

FIGURE 7.1 Diamonds are thermodynamically unstable at ambient conditions and spontaneously transform into the less spectacular graphite which is the main component of the everyday pencil. The fact that this process does not happen in spite of being thermodynamically favorable tells us, that not only is it interesting to know which reactions occur, but also the rate with which they occur.

and 6×10^{22} velocity coordinates, or 10^{23} different parameters! And if the gas is not a noble gas, more parameters are needed to characterize the different rotations and vibrations of each molecule. In spite of the practically infinite amount of information needed to describe the gas, it is completely characterized even if only a very few parameters of the system are known, for example, the volume, the temperature, and the pressure. Thinking about it, this condensation of information from 10^{23} parameters to only three is truly spectacular! And the one thing that allows us to reduce the relevant information by such a terrific extent is the implicit assumption that the system is *in equilibrium*. It is the great success of statistical mechanics that using only very few and reasonable assumptions along with the assumption of equilibrium allows us to deduce macroscopic properties of systems of seemingly untreatable complexity.

7.2.1 Fundamental Laws of Thermodynamics

7.2.1.1 The First Law of Thermodynamics

The first law of thermodynamics states that the change of the internal energy of a closed system, ΔE, is the sum of the heat added to the system, Q, and the work done on the system, W (see Figure 7.2):

$$\Delta E = Q + W. \tag{7.1}$$

Thus, the first law of thermodynamics is simply stating that energy is conserved, and that heat is a form of energy. Consider now an isolated system with energy, E, volume, V, and number of particles, N. In equilibrium, the system is completely characterized by (E, V, N), but for nonequilibrium macrostates, an additional variable, α, is needed to specify how the system differs

from the equilibrium state. Examples of α could be a parameter designating the local temperature in a system out of thermal equilibrium, or specifying the nonequilibrium transport of particles resulting from an external force (say, ions in an electric field). For given values of (E, V, N, α), there will be a huge *number of possible microscopic arrangements of the system consistent with those macroscopic constraints*. This number of microstates consistent with (E, V, N, α) is denoted by $\Omega(E, V, N, \alpha)$ and called *the statistical weight of state* (E, V, N, α). $\Omega(E, V, N, \alpha)$ is used to define the *entropy S* of a system in the macrostate (E, V, N, α) by

$$S(E, V, N, \alpha) \equiv k_B \ln(\Omega(E, V, N, \alpha)), \tag{7.2}$$

where $k_B = 1.381 \cdot 10^{-23}$ J/K is Boltzmann's constant.

7.2.1.2 The Second Law of Thermodynamics

The second law of thermodynamics states that for any process in an isolated system the entropy always increases (or is unchanged if the process is reversible), and that the equilibrium state is the state where the entropy attains its maximum under the constraints (E, V, N). This means that $S(E, V, N) \geq S(E, V, N, \alpha)$ for any α, and that α evolves in time so that if $t_2 \geq t_1$, then $S(E, V, N, \alpha(t_2)) \geq S(E, V, N, \alpha(t_1))$. Following the definition of entropy, this means that systems move toward macrostates compatible with a higher number of microstates until the macrostate consistent with the maximum number of microstates is achieved—the equilibrium state. This also means that as the system evolves, the macroscopic constraints $(E, V, N, \alpha(t))$ gives less and less information about the microstate of the system since more microstates are compatible with $(E, V, N, \alpha(t))$. This is why the second law of thermodynamics is sometimes popularly interpreted as saying that "the degree of disorder of an isolated system increases with time." As an example, consider the situation where gas particles are confined into one-half of a system by a wall that is suddenly removed (Figure 7.3). Then $\alpha(t)$ is initially specifying that one-half on the system has a particle density of zero. As time progresses, particles moving toward the region with many particles will move only a small distance before their direction is randomized by impact with other particles, while particles moving toward the region less dense in particles will move a longer distance before their direction is randomized. After a short time, the density of particles changes gradually (and no longer abruptly) from a high value in the left

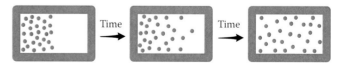

FIGURE 7.3 In a system where gas particles were initially concentrated in one-half of the system, the system is specified by (E, V, N, α) where α specifies how the system deviates from the equilibrium state of a system with the constraints (E, V, N). As time progresses the density of particles will become increasingly uniform (which $\alpha(t)$ will describe) until the equilibrium state of uniform density which maximizes the entropy of the system is achieved.

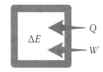

FIGURE 7.2 The energy of an isolated system increases by $\Delta E = Q + W$ where Q is the heat added to the system and W is the work done on the system.

end of the system to a very low value in the right end of the system, and $\alpha(t)$ is describing this gradual transition. It is clear that this change has increased $\Omega(E, V, N, \alpha(t))$, and hence $S(E, V, N, \alpha(t))$, and also increased the degree of disorder of the system. When a longer time has passed, the density will be uniform in the entire system and $\alpha(t)$ is no longer needed to describe the system since this state of course corresponds to the equilibrium state—the state where $S(E, V, N, \alpha)$ is maximized.

7.2.2 Thermodynamics of a System in a Heat Bath

Consider an isolated system divided into two subsystems (with parameters (E_1, V_1, N_1) and (E_2, V_2, N_2)) that can exchange only heat. Then E_1 and E_2 can vary as long as $E = E_1 + E_2$ is constant, implying that $\Delta E_1 = -\Delta E_2$ (see Figure 7.4).

According to the second law of thermodynamics, the system is in thermal equilibrium at the value of E_1 that maximizes the total entropy $S(E_1, E_2)$. Since for a given value of E_1 the arrangement of the one subsystem does not affect the other: $\Omega(E_1, E_2) = \Omega_1(E_1)\Omega_2(E_2)$, and hence $S(E_1, E_2) = S_1(E_1) + S_2(E_2)$, so the requirement for equilibrium becomes

$$\Delta S = \frac{dS_1}{dE_1}\Delta E_1 + \frac{dS_2}{dE_2}\Delta E_2 = \left[\frac{dS_1}{dE_1} - \frac{dS_2}{dE_2}\right]\Delta E_1 = 0. \quad (7.3)$$

This means that in thermal equilibrium when $T_1 = T_2$, $dS_1/dE_1 = dS_2/dE_2$. Therefore, dS/dE is somehow a measure of the temperature of a system. This fact is used to define the absolute temperature of a system as $1/T = dS/dE$ (this definition of temperature is of course identical to the one normally used). This means that $\Delta S = \Delta E(dS/dE) = \Delta E/T$ for any system at temperature T, and combining this with the first law of thermodynamics ($\Delta E = Q + W$) and the fact that no work was done on the system, gives $Q = T\Delta S$ so that the heat added to a system at temperature T is equal to the entropy added times the temperature.

If system 2 is much bigger than system 1, energy exchange between the two systems does not affect the temperature of system 2, which is then called a *heat bath* or *heat reservoir*. It is not true that the processes which can happen in system 1 are simply the processes for which $\Delta S_1 \geq 0$, since the second law of thermodynamics is stated only for an isolated system (which does not exchange heat with its surroundings). But for system 1 plus system 2, the second law can be used to show

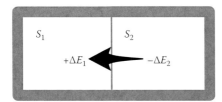

FIGURE 7.4 If energy can be transferred between two subsystems of an isolated system, $\Delta E_1 = -\Delta E_2$.

$$\Delta S = \Delta S_1 + \Delta S_2 = \Delta S_1 + \frac{\Delta E_2}{T} = \Delta S_1 - \frac{\Delta E_1}{T} \geq 0. \quad (7.4)$$

Therefore, for a system that can exchange only energy with a heat bath (not volume or particles), the processes that are allowed according to the second law of thermodynamics are the processes that obey $\Delta E - T\Delta S \leq 0$. Combining this relation with the definition of *Helmholtz free energy*, $F \equiv E - TS$, the spontaneous processes are seen to obey $\Delta F = \Delta E - T\Delta S - S\Delta T \leq -S\Delta T$, which is zero if the temperature is constant. So at constant (T, V, N) spontaneous processes obey

$$\Delta F \leq 0, \quad F \equiv E - TS. \quad (7.5)$$

For a system that can exchange both energy and volume (and hence work) with its surroundings, but is kept at constant temperature, pressure, p, and particle number, a similar relation can be derived for the *Gibbs free energy*, $G \equiv E - TS + pV$. At constant (T, p, N) spontaneous processes obey

$$\Delta G \leq 0, \quad G \equiv E - TS + pV. \quad (7.6)$$

Since most of the processes around us happen at ambient temperature and pressure, the Gibbs free energy change of a process is the quantity that usually tells us which processes occur spontaneously and which do not. Even if an exothermal process temporarily increases the temperature of a real system in thermal contact with its surroundings, the change from before the reaction to after the reaction and the overall temperature equilibration is a process where $\Delta T = 0$ and the whole process only occurs spontaneously if the total change in Gibbs free energy is negative.

At ambient temperature and pressure, the Gibbs free energy of a carbon atom in a graphite structure is about 3 kJ/mol lower than that of a carbon atom in a diamond structure. Hence, diamonds spontaneously transform into graphite. The fact that diamonds seem to be forever in spite of the process $C_{diamond} \rightarrow C_{graphite}$ having a change in Gibbs free energy of $\Delta G = -3$ kJ/mol tells us that apart from knowing in which direction a system out of equilibrium will evolve, it is also very valuable to know how fast a nonequilibrium system evolves toward the equilibrium state. One might think that the reason why diamonds appear stable is that ΔG of the process is not very large, but it is not as simple as that. For instance, a sodium and chloride ion in a crystal (NaCl) lower their total ΔG by only 9 kJ/mol when they dissolve in water, yet this process is very fast compared to the decay of diamonds. Also, the burning of a mixture of hydrogen and oxygen into water (a process we know releases a lot of energy and where $\Delta G = -237$ kJ/mol) does not happen before you strike a match; the mixture appears stable even though the reaction releases a lot of energy. So even energetically very favorable reactions do not necessarily progress rapidly, and reactions that release only very little energy can readily occur. The key to understanding this is to know the probability, $p(s)$, that a molecule is in a state, s, with energy, E_s, as function of temperature. This probability distribution is known as the Boltzmann distribution.

7.2.3 Statistical Mechanics and the Boltzmann Distribution

The assumption that forms the basis of all of statistical mechanics is the postulate of equal a priori probability which claims that *for an isolated system with a given set of constraints—say (E, V, N, α)—each microstate consistent with the macroscopic constraints is equally likely to occur.* This means, for example, that if someone hid a coin under one of three cups at random, it is equally likely to be found under cup 1, cup 2, and cup 3, and since the coin is under one of the cups, the probability of these three outcomes must sum to one. So the probability of finding the coin under any one cup is 1/3. To take a less trivial example, consider the situation where a system (system 1) can exchange energy with a heat bath (system 2). Then the probability, $p(s)$, of finding the system in one specific microstate, s, with energy, E_s, is proportional to the statistical weight of the heat bath having energy $E_{tot} - E_s$ where E_{tot} is the total energy of the two subsystems. This is because there is one way of preparing the system in state s, and $\Omega_2(E_{tot} - E_s)$ ways of preparing the heat bath in a state which takes up the remainder of the total energy of the two systems. So the total number of ways to prepare the system and the heat bath in a state where the system is in state s is $1 \cdot \Omega_2(E_{tot} - E_s)$. This gives

$$p(s) \propto \Omega_2(E_{tot} - E_s) = \exp\left[\frac{S_2(E_{tot} - E_s)}{k_B}\right] \quad (7.7)$$

according to the definition of entropy (Equation 7.2). And if $E_s \ll E_{tot}$

$$S_2(E_{tot} - E_s) = S_2(E_{tot}) - E_s\frac{dS_2(E_{tot})}{dE_{tot}} = S_2(E_{tot}) - \frac{E_s}{T}, \quad (7.8)$$

which when inserted in Equation 7.7 gives

$$p(s) \propto \exp\left(\frac{S_2(E_{tot})}{k_B} - \frac{E_s}{k_B T}\right) = \exp\left(\frac{S_2(E_{tot})}{k_B}\right)\exp\left(\frac{-E_s}{k_B T}\right) \propto \exp\left(\frac{-E_s}{k_B T}\right), \quad (7.9)$$

so that finally $p(s) = \exp(-E_s/k_B T)/Z$ where $1/Z$ is a constant of proportionality. This constant is found by the requirement that the probability of finding the system in any state must be 1: $\sum_s p(s)/Z = 1$. This gives the celebrated *Boltzmann distribution* which provides information about the probability of finding the system in a specific state, s, with energy E_s:

$$p(s) = \frac{\exp(-E_s/k_B T)}{Z}, \quad (7.10)$$

where $Z \equiv \sum_s \exp\left(\frac{-E_s}{k_B T}\right)$ is a constant of the system.

If there are $g(E_s)$ different states with energy E_s, the probability, $p(E_s)$, of finding the system with a specific energy, E_s, is $p(E_s) = p(s)g(E_s)$. If the energy levels of the system form a continuum

rather than a discrete spectrum (which is the case for all systems where quantum mechanical effects are neglectable), $g(s)$ is replaced by *the number of states with energy between E and E + dE*, $\rho(E)dE$, where $\rho(E)$ is the density of states with energy E. Hence, the Boltzmann distribution gives the probability $P(E) dE$ for finding the system in the energy range between E and $E + dE$:

$$P(E)dE = \exp\left(\frac{-E}{k_B T}\right)\rho(E)\frac{dE}{Z}, \quad \text{where } Z \equiv \int \exp\left(\frac{-E}{k_B T}\right)\rho(E)dE. \quad (7.11)$$

The Boltzmann distribution is very useful when it is applied to a large collection of identical particles in thermal contact with each other. For instance, proteins in a solvent, atoms or molecules in a gas, or even solvent molecules themselves—any large number of identical particles that interact (possibly through an intermediary). In that case, each individual entity can be considered as "the system," and the rest constitute the heat bath. Multiplying the Boltzmann distribution with the total number of systems (i.e., molecules, proteins, etc.), N, gives the total number of systems in state s: $n(s) = Np(s) \propto \exp(-E_s/k_B T)$. The ratio of the number of systems in state B to the number of systems in state A is simply

$$\frac{n(B)}{n(A)} = \frac{\exp(-E_B/k_B T)}{\exp(-E_A/k_B T)} = \exp\left(-\frac{E_B - E_A}{k_B T}\right). \quad (7.12)$$

If the collection of systems can in addition to energy also exchange work with its surroundings it is at constant temperature and pressure, and the internal energy E must be replaced by the Gibbs free energy G (Equation 7.6):

$$\frac{n(B)}{n(A)} = \exp\left(-\frac{G_B - G_A}{k_B T}\right), \quad (7.13)$$

where

$n(i)$ is the number of systems in state i
G_i is the Gibbs free energy of a system in state i.

It is hard to exaggerate the usefulness of this equation if one knows the energy difference between two states of a system. For instance, this equation can tell us how many atoms are in an excited state compared to the ground state, or how big a fraction of proteins are denatured at a temperature T. All assuming of course, that the system is in equilibrium.

7.2.3.1 The Equipartition Theorem

The Boltzmann distribution is all the physics needed to derive the famous equipartition theorem (see Ref. [1] for a derivation). The *equipartition theorem* states that *for a system in equilibrium at temperature T, each generalized coordinate that appears only as a quadratic term in the expression for the energy of the system contributes an average energy of $1/2 k_B T$ to the total energy.* A system of N particles can be described by Cartesian coordinates,

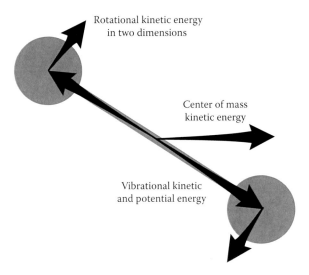

FIGURE 7.5 The total energy of a molecule is on average $1/2k_BT$ for each generalized coordinate contributing with a quadratic term to the system energy. For a diatomic molecule this becomes: $3/2k_BT$ for the kinetic energy of the center of mass (in directions x, y, and z), $2/2k_BT$ for the kinetic energy of rotation around the center of mass (along the in-plane and out-of-plane directions), $1/2k_BT$ for the kinetic energy of vibration of the interatomic bond, and finally $1/2k_BT$ for the potential energy of vibrations. So the total energy of the molecule is $7/2k_BT$.

specifying x, y, z coordinates and velocities of each particle ($6N$ coordinates in total), but alternatively in terms of *generalized coordinates* where each coordinate is some linear combination of the Cartesian coordinates. Often, by a change to generalized coordinates, the energy of the system can (to first order) be expressed as $E = \sum_i a_i q_i^2 + \sum_j b_j p_j^2$, where a_i and b_j are constants, q_i a generalized position coordinate, and p_i a generalized momentum coordinate. In this case, the system is simply described as a number of independent harmonic oscillators. In mathematical terms, one makes a change of basis to one

consisting of eigenvectors (to first order) of the Hamilton operator. In the case of a diatomic molecule (12 coordinates in total; see Figure 7.5), the generalized coordinates describe the translation of the center of motion ((x, y, z) and ($\dot{x}$, $\dot{y}$, $\dot{z}$), where the dot designates the time derivative), the vibration of the intermolecular bond (the atom–atom distance and its time derivative), and the rotation of the molecule around its center of mass—two angles and their time derivatives. According to the equipartitioning theorem, the total energy of the molecule is in equilibrium $1/2k_BT$ for each generalized coordinate that contributes with a quadratic term to the energy, i.e., $E = 7/2k_BT$ independent of the values of the prefactors a_i and b_i! However, the energy of a system thus predicted by the equipartition theorem deviates from the true value at both low and high temperatures. At high temperatures because second-order nonharmonic effects begin to play a role, and at low temperature because degrees of freedom begin to "freeze out" quantum mechanically since k_BT becomes comparable to the spacing between energy levels. For carbon monoxide, CO, at room temperature, for example, the vibrational degree of freedom is frozen out, so that the energy is only $E = 5/2k_BT$.

7.2.3.2 The Transition State Theory of Reactions

As a final application of the Boltzmann's distribution, and as our first step toward the dynamics on nonequilibrium systems, we will consider the *transition state theory* for chemical reactions, which explains why diamonds linger despite the fact that they are thermodynamically unstable. The basis of this theory is the following: A system moves from state A to state B along a *reaction coordinate* (which for the reaction *diatomic molecule → two free atoms*, for example, is simply the distance between the two atoms, see Figure 7.6a), and the two states A and B correspond to local minima of the free energy along the reaction coordinate. For instance, state A may correspond to the diamond organization of carbon atoms with four single C–C bonds and state B to the graphite organization with three stronger bonds. Both are

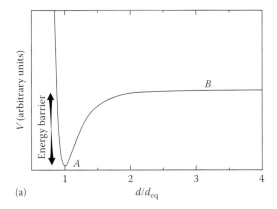

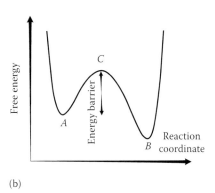

FIGURE 7.6 (a) The potential energy of a diatomic molecule as function of the distance, d, between the two atoms. The equilibrium distance, d_{eq}, is at the local minimum of the energy landscape. For the reaction that splits a diatomic molecule (state A) up into two separate atoms (state B), the reaction coordinate is simply d. When $d \approx 3d_{eq}$ there is effectively no attraction between the atoms and they are free. (b) The free energy of a system along a reaction coordinate with two (meta)stable states A and B, and an unstable transition state C. Since B has a lower energy than A, only B is a stable state, but if the energy barrier between states A and C which must be overcome for the system to move into state B is large compared to k_BT, the metastable state A may be very long lived—even compared to the age of the universe.

(meta)stable states of the system and thus correspond to local minima of the energy landscape along the reaction coordinate, and in between the two minima there is some local maximum of the free energy. State C which corresponds to this maximum is called the *transition state* since a system has to pass through this intermediate state to go from state A to B or vice versa. Since the temperature is nonzero, a system in state A is vibrating around the local minimum with some frequency, ω. Each time the system crosses A and moves toward the transition state, it has some probability, p, of having enough energy to cross the energy barrier given by the transition state (state C) and move on into state B. This reaction rate is then given by $r_{A \rightarrow B} = \omega p$. From the Boltzmann distribution, the probability of finding a system with kinetic energy E_C divided by the probability of finding the system with energy E_A is $\exp(-(E_C - E_A)/k_B T)$. If the ensemble of systems started out in state A (e.g., for a collection of carbon atoms initially in a diamond lattice) and since $E_C \gg E_A$ the probability of finding the system in state A is nearly 1. This means that $p \approx \exp(-(E_C - E_A)/k_B T)$, so the reaction rate of the process $A \rightarrow B$ is given by

$$r_{A \rightarrow B} = \omega \exp\left(-\frac{E_C - E_A}{k_B T}\right), \tag{7.14}$$

where ω is the frequency with which the system oscillates around state A and where the energy difference $E_C - E_A$ is called the *energy barrier* toward the reaction $A \rightarrow B$. This formula is immensely helpful for getting a rough understanding of chemical and biological reaction rates. Even though the frequency ω does vary from molecule to molecule, the effect this variation has on the reaction rate is often small compared to a change in the energy barrier (changing ω by a factor of 1000 has a smaller effect on the reaction rate than changing the energy barrier by a factor of 7). Equation 7.14 shows that the rate of a reaction does not depend on how much energy is released during the reaction, but rather on the energy barrier that must be overcome. And it allows us to understand why diamonds are practically stable and sodium chloride crystals in water are not: For an ion in a NaCl crystal to go into solution, it suffices to simply move into the solvent so one expects the energy barrier for this reaction to be very low and the reaction to occur rapidly. But for a carbon atom to change from having four bonds in diamond to three (stronger) bonds in graphite, the energy barrier should be on the order of the energy of the single C–C bond that must be broken. This is roughly $E_C - E_A = 346\,\mathrm{kJ/mol} = 138\,k_B T$ *per bond* at a temperature of 25°C. A reasonable value for ω is about $10^{14}\,\mathrm{s}^{-1}$ so the probability for a diamond carbon atom to become a graphite carbon atom is about $p \approx \omega \exp(-E_{\mathrm{barrier}}/k_B T) \approx 10^{14} \exp(-138)\,\mathrm{s}^{-1} \approx 10^{-46}\,\mathrm{s}^{-1}$. So one has to wait around 10^{46} s for a diamond to turn into graphite, and since the age of the universe is $\sim 10^{17}$ s the decay of diamonds into graphite really is not something to worry about!

Another thing that can be seen from Equation 7.14 is that if the temperature is increased, the effective energy barrier $E_{\mathrm{barrier}}/k_B T$ becomes smaller and reactions that have some reaction barrier will occur more rapidly. If the temperature is doubled, the effective reaction barrier is cut in half and the reaction rate is increased by a factor of $\exp(2) \approx 7$. A factor of 7 is not bad, but it is not making a huge difference either. But it is also seen that if the energy barrier can be somehow reduced, this can potentially have a much bigger affect on the reaction rate than an increase of the temperature. And this is exactly what an inorganic catalyst or an enzyme does by changing the energy of the system along the reaction coordinate. This typically happens by the catalyst/enzyme somehow binding to the intermediate state, C, thus lowering its free energy and hence the energy barrier. This is why catalysts and enzymes can facilitate reactions without being spent themselves (they do not use any energy either); they simply lower the energy of a transitory state thus making a reaction that is thermodynamically favorable happen faster by lowering the energy barrier for the reaction. These changes in the energy barrier can be quite impressive and since the reaction rate depends on the exponential of the energy barrier the change in the reaction rate can be huge. For example, an inorganic catalyst can easily increase the reaction rate by a factor of 10^4, and enzymes can easily increase reaction rates by an impressive factor of 10^{23}. This means that enzymes can make reactions that would not happen in the lifetime of the universe happen in a fraction of a second!

This terminates the section on equilibrium thermodynamics and statistical mechanics, and we now have the tools necessary to consider nonequilibrium systems. A good undergrad textbook regarding statistical mechanics of equilibrium systems is *Statistical Physics* by F. Mandl [1] while a good graduate text is *Introduction to Modern Statistical Mechanics* by D. Chandler [2].

7.3 Nonequilibrium Systems

In Section 7.2, the systems were assumed to be in equilibrium, which was necessary for the derivations. Historically, the first nonequilibrium theories concerned systems that in some sense are close to being in equilibrium, and knowledge about the equilibrium state was used to make predictions about the close-to-equilibrium behavior. Here "being close to equilibrium" means that the force driving the system out of equilibrium generates a linear response, that is, doubling the force results in a doubling of the response. We have already seen one such example when using the transition state theory to predict the rates with which species, initially in the metastable state A, escapes through the transition state, C, to the stable state B. There, it was assumed that the ratio of systems in states A and C was the same as in the equilibrium situation (i.e., given by the Boltzmann distribution), in spite of this not being the case for the ratio of systems in states A and B. If one makes absolutely no assumptions about anything being in equilibrium, it is not possible to make predictions unless all parameters of all individual particles (typically on the order of 10^{23} parameters) are known, so any successful theory must make some equilibrium assumption. The early theories such as Kramers formula and the fluctuation–dissipation relations assume that the system is being driven close to the equilibrium, but the more recent theories of Jarzynski and Crook assume only that the initial and final states of the process are in equilibrium, the process itself can be driven arbitrarily far from equilibrium!

7.3.1 Kramers Formula

In the transition state theory for reactions treated above, it was implicitly assumed that the frictional forces in the system are very low compared to inertial forces. It was assumed that a system that passes state A with a kinetic energy high enough to overcome the energy barrier will keep this energy and actually pass through state C into state B. But if there is friction in the system, not all the kinetic energy of the system will be turned into potential energy needed for climbing the energy barrier. Some of the kinetic energy will be dissipated through friction. In 1940, H. A. Kramers made a more detailed treatment of the rate of a reaction with an energy barrier where he included the role of friction [3]. This result is of particular interest to nanoscale systems where frictional forces are generally much bigger than inertial forces. That is, the motion of a nanoscale object is Brownian and not ballistic.

Consider a large ensemble of identical systems with free energy landscapes, as shown in Figure 7.6b, and with each system characterized by the position, q, and the momentum, p, along the reaction coordinate. The ensemble of systems is then completely described by $\rho(p, q, t)$ = *the probability density that a system has momentum p and position q at time t*. If the ensemble is in equilibrium, this probability distribution is given by the Boltzmann distribution

$$\rho(p,q,t) \propto \exp\left(-\frac{E_{\text{tot}}}{k_{\text{B}}T}\right) = \exp\left(-\frac{E_{\text{pot}}(q,t)}{k_{\text{B}}T}\right)\exp\left(-\frac{p^2}{2mk_{\text{B}}T}\right), \quad (7.15)$$

since the total energy is a sum of the potential and kinetic energies, $E_{\text{tot}} = E_{\text{pot}} + E_{\text{kin}} = E_{\text{pot}} + p^2/2m$ (where m is the mass of the system moving along q). But, if all systems start out in state A, the ensemble is not in equilibrium and the probability is not Boltzmann distributed. However, if the frictional force, $v\zeta$ (where v is the velocity and ζ is the frictional constant), is much bigger than the potential force, $F(q) = -\mathrm{d}E_{\text{pot}}(q)/\mathrm{d}q$, frictional forces will dominate over potential forces, so the momentum distribution will be independent of q and be distributed according to the Boltzmann distribution so that $\rho(p, q, t) \approx \sigma(q, t) \exp(-p^2/2k_{\text{B}}T)$, where $\sigma(q, t)$ is the probability density that a system is at position q at time t, i.e., the momentum distribution is in equilibrium, but the position distribution is not. So, the momentum distribution is constant in time while the probability density, σ, undergoes a slow diffusion process where the flux of probability density, j, has two components:

$$j = \frac{\sigma F}{\zeta} - D\frac{\mathrm{d}\sigma}{\mathrm{d}q}. \quad (7.16)$$

The first term is the probability density times the drift speed, F/ζ, and results from the force F. The second term is simply

diffusion, D being the diffusion constant, which according to the well-known Einstein relation, is given by $D = k_{\text{B}}T/\zeta$. Since $F = -\mathrm{d}E_{\text{pot}}/\mathrm{d}q$, j can be rewritten as

$$j = -\frac{k_{\text{B}}T}{\zeta}e^{-E_{\text{pot}}/k_{\text{B}}T}\frac{\mathrm{d}}{\mathrm{d}q}(\sigma e^{E_{\text{pot}}/k_{\text{B}}T}), \quad (7.17)$$

and by rearranging and integrating both sides between A and B one obtains

$$\int_A^B \zeta e^{E_{\text{pot}}/k_{\text{B}}T} j\,\mathrm{d}q = -k_{\text{B}}T \int_A^B \frac{\mathrm{d}}{\mathrm{d}q}(\sigma e^{E_{\text{pot}}/k_{\text{B}}T})\,\mathrm{d}q = k_{\text{B}}T\left[\sigma e^{E_{\text{pot}}/k_{\text{B}}T}\right]_B^A. \quad (7.18)$$

In a stationary state, the flux, j, is constant at each point between A and B since otherwise there would be a net accumulation or loss of density somewhere. So, finally

$$j = \frac{k_{\text{B}}T\left[\sigma e^{E_{\text{pot}}/k_{\text{B}}T}\right]_B^A}{\int_A^B \zeta e^{E_{\text{pot}}/k_{\text{B}}T}\,\mathrm{d}q} \approx \frac{k_{\text{B}}T\sigma_A}{\int_A^B \zeta e^{E_{\text{pot}}/k_{\text{B}}T}\,\mathrm{d}q}, \quad \sigma_A = (\sigma e^{E_{\text{pot}}/k_{\text{B}}T})_{\text{near } A} \quad (7.19)$$

where the last equality is true since $\sigma_A \gg \sigma_B$ because the system started out entirely in state A. Equation 7.19 gives the flux of density from state A to state B. The reaction rate for the transition $A \to B$, $r_{A \to B}$, is the fraction of systems in state A that changes to state B per unit time, so $r_{A \to B} = j/n_A$ where n_A is the density of particles near A. If the potential is harmonic near A, $E_{\text{pot}} = K_A q^2/2$ and n_A is given by

$$n_A \approx \int_{-\infty}^{\infty} \sigma_A \exp\left(-\frac{K_A q^2}{2k_{\text{B}}T}\right)\mathrm{d}q = \sigma_A\sqrt{2\pi K_A k_{\text{B}}T} \quad (7.20)$$

where the limits of integration is taken at $\pm\infty$ because the integrant is nonzero only near A. Thus

$$r_{A \to B} = \frac{j}{n_A} \approx \frac{\sqrt{k_{\text{B}}T}}{\sqrt{K_A}\,\zeta\int_A^B e^{E_{\text{pot}}/k_{\text{B}}T}\,\mathrm{d}q}. \quad (7.21)$$

The main contribution to the integral $\int_A^B e^{E_{\text{pot}}/k_{\text{B}}T}\,\mathrm{d}q$ comes from the region near C where $e^{E_{\text{pot}}/k_{\text{B}}T}$ is large. Here, the potential energy is approximately harmonic: $E_{\text{pot}} \approx E_{\text{barrier}} - K_B q^2/2$, so that

$$\int_A^B \exp\left(\frac{E_{\text{pot}}}{k_{\text{B}}T}\right)\mathrm{d}q \approx \int_{-\infty}^{\infty} \exp\left(\left(E_{\text{barrier}} - \frac{K_B q^2}{2}\right)k_{\text{B}}T\right)\mathrm{d}q$$

$$= \exp\left(\frac{E_{\text{barrier}}}{k_{\text{B}}T}\right)\sqrt{2\pi K_B k_{\text{B}}T}. \quad (7.22)$$

Finally, one obtains Kramers equation for the reaction rate

$$r_{A \to B} = \frac{\sqrt{K_A K_B}}{2\pi\zeta} e^{-E_{barrier}/k_B T}. \tag{7.23}$$

When this result is compared to the reaction rate for a non-diffusive reactions (Equation 7.14), it can be seen that the dependence on the energy barrier is the same for the two expressions, but in the diffusive reaction, the reaction rate depends on both the curvature of the energy landscape near A and C, rather than just on the dynamics near A as is the case for the non-diffusive reactions.

7.3.1.1 Application of Kramers Formula to Small-Scale Nonequilibrium Systems

Optical tweezers is the name for a technique which can trap a small bead with an index of refraction higher than the surrounding medium by focused laser light [4]. The very high spatial resolution of this technique (on the order of nanometers) together with

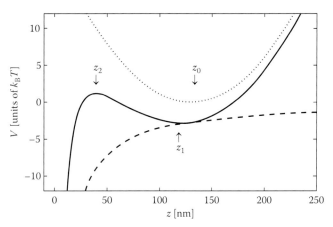

FIGURE 7.7 The total potential of an optically trapped bead as function of distance, z, from the wall is a sum of the harmonic interaction with the trap (centered at z_0) and the van der Waals interaction with the wall. The potential will always be lowest at the wall, but if z_0 is above some critical value, there will be an energy barrier for the bead to overcome before it can escape to the wall. z_1 denotes the local minimum of the total potential, and z_2 denotes the local maximum. (Reprinted from Dreyer, J.K. et al., *Phys. Rev. E*, 73, 051110, 2006, Figure 2. With permission.)

the range in which the spring constant of the harmonic trapping potential can be varied (10^{-2}–10^2 pN/nm) makes this tool practical for investigating small-scale systems, including cells and their components. Trapping a small bead by optical tweezers and moving it near a solid wall has been used to experimentally verify Kramers formula and serves as a nice model system for probing the dynamics of nanoscale systems [5]. Besides the harmonic force from the optical trap, the bead also feels the van der Waals attraction from the wall, $F_{vdW} = -AR/6z$, where A is the Hamaker constant of the van der Waals interaction between bead and wall, R is the radius of the bead, and z is the distance from the wall. So the total potential for a bead in an optical trap near a wall is given by

$$V_{tot}(z) = V_{harm}(z) + V_{vdW}(z) = \frac{K_1}{2}(z - z_0)^2 - \frac{AR}{6z} \tag{7.24}$$

where
 z_0 is the center of the optical trap
 K_1 is the spring constant of the trap

If z_0 is above some critical value, z_c, the potential has both a local minimum (at z_1) and a local maximum (at z_2) and the total potential looks as shown in Figure 7.7. If z is below z_c, there is no energy barrier at all for the bead to move to the wall which it will do "instantaneously."

Since the optical trapping force is weak compared to the Brownian forces on the bead, the escape rate of such a bead is given by Kramers equation (Equation 7.23): $r_{trap \to wall} = (\sqrt{K_1 K_2}/2\pi\zeta)\exp(-\Delta V)$ where $\Delta V = [V(z_2) - V(z_1)]/k_B T$ is the energy barrier in units of $k_B T$. In Ref. [5], the center of the trap was moved toward the wall with constant speed, $v = -dz_0/dt$, and the bead position was recorded when it escaped the trap and jumped to the wall. The outcome of an experiment where the bead jumps a distance of 157 nm is shown in Figure 7.8a. Figure 7.8b shows how the average jump length decreases as the approach velocity is increased, demonstrating that the bead escapes from the trap by a nonequilibrium process.

The fact that the jump length depends on the approach speed shows that thermal noise is important and that the system is driven far from equilibrium. From Kramers formula, an equation for the probability distribution of the jump lengths can be derived:

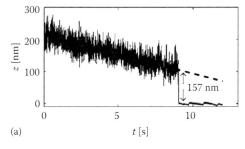

(a)

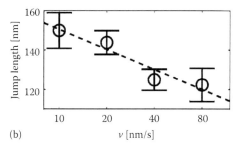

(b)

FIGURE 7.8 (a) As the center of an optical trap with the bead is approaching a wall with speed $v = -dz_0/dt$, the trapped bead position fluctuates around the local minimum that moves toward the wall. At some distance the bead jumps and sticks to the wall (no further fluctuations). In this particular experiment the jump length was 157 nm. (b) The average jump length is seen to decrease as the approach speed is increased, demonstrating that the escape process is nonequilibrium. (Reprinted from Dreyer, J.K. et al., *Phys. Rev. E*, 73, 051110, 2006, Figure 1. With permission.)

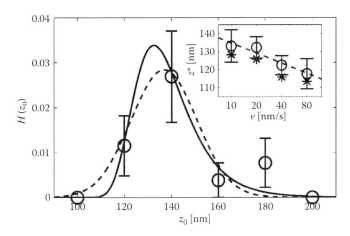

FIGURE 7.9 The experimentally obtained probability distribution for the trap position at which the bead jumps to the wall using an approach speed of 20 nm/s. The full line is a fit by Equation 7.25 which is seen to agree well with the data. The dashed line shows a Gaussian fit for comparison. The inset shows how the fitting parameter $z^\star$ changes with the approach speed. (Reprinted from Dreyer, J.K. et al., *Phys. Rev. E*, 73, 051110, 2006, Figure 4. With permission.)

$$H(z_0) \approx \alpha \exp\left[-\alpha(z_0 - z^\star) - e^{-\alpha(z_0 - z^\star)}\right] \qquad (7.25)$$

where the most likely jump distance, $z^\star$, and the local slope of the total potential, $\alpha = \dfrac{d\Delta V}{dz_0}\bigg|_{z_0 = z^\star}$, are fitting parameters that both depend on the approach speed. In Figure 7.9, the experimentally obtained jump lengths for an approach speed of 20 nm/s are fitted with Equation 7.25 and the agreement is seen to be quite good, confirming that Kramers formula can be successfully applied to thermal systems out of equilibrium, and that thermal noise is important for such processes.

7.3.2 Fluctuation–Dissipation Relations

If a system is in equilibrium, any free parameter will fluctuate around its average value. For instance the instantaneous velocity, v, of a Brownian particle fluctuates around the expectation value $\langle v \rangle = 0$. This is what gives rise to Brownian motion. If a system is driven through a nonequilibrium process, energy is irreversibly dissipated. If a force is applied to a Brownian particle for instance, it will diffuse in the direction of the force and dissipate an energy of $E_{\text{diss}} = Fx$ where F is the force on the particle and x the distance traveled. A fluctuation–dissipation relation is a relation between the spontaneous equilibrium fluctuation on the one side and the dissipation resulting from a nonequilibrium process on the other side. Such relations are possible because the processes that cause fluctuations in the equilibrium state are the same processes that dissipate energy during nonequilibrium processes. By far, the most well-known example of a fluctuation–dissipation relation is the Einstein equation for the self-diffusion of a Brownian particle [6] which will be derived here.

7.3.2.1 The Einstein Relation

A small particle in suspension is constantly experiencing random impacts by solvent molecules, and each impact changes the velocity of the particle slightly. After some time, the velocity of the particle is completely uncorrelated with its initial velocity. In other words, the particle travels for some time in one direction and at some point starts traveling in an uncorrelated direction. With the average time between randomizations denoted by $\tau = \langle t \rangle_{\text{br}}$ and the average root-mean-square distance traveled between randomizations by $L = \langle x^2 \rangle_{\text{br}}^{1/2}$, the motion of a particle is described by a *random walk*: each time step, τ, a particle moves a distance, L, in a random direction. Consider, for simplicity, a particle moving only in one dimension so that its position after N steps is given by $x = \sum_{i=1}^{N} k_i L$ where $k_i = -1$ if step i moved the particle to the left and $k_i = 1$ if step i moved the particle to the right. Since the probability of moving to the left and the right are equal $\langle x_N \rangle = L\left\langle \sum_{i=1}^{N} k_i \right\rangle = 0$, but the mean square of the particle displacement is not:

$$\langle x_N^2 \rangle = \langle (x_{N-1} + k_N L)^2 \rangle = \langle x_{N-1}^2 \rangle$$
$$+ 2L\langle x_{N-1} k_N \rangle + L^2 \langle k_N^2 \rangle = \langle x_{N-1}^2 \rangle + L^2 \langle k_N^2 \rangle \qquad (7.26)$$

where the last step uses that $2L \langle x_{N-1} k_N \rangle = 0$ since the probability for jumping left or right is independent of the particle position. Then, by iteration $\langle x_N^2 \rangle = NL^2$, and since the particle takes $N = t/\tau$ steps in a time t

$$\langle x_N^2 \rangle = 2Dt, \quad \text{where } D \equiv L^2/2\tau. \qquad (7.27)$$

For a random walk in three dimensions $\langle r^2 \rangle = \langle x^2 + y^2 + z^2 \rangle = 3\langle x^2 \rangle$ so

$$\langle r_N^2 \rangle = 6Dt, \quad \text{where } D \equiv L^2/2\tau. \qquad (7.28)$$

Hence, the random kicks a particle gets do cause it to displace itself from its initial position even if this displacement does not have any preferred direction ($\langle x \rangle = 0$). Consider now a random walker in one dimension also under influence of an external force, F. Now, τ is the average time between collisions but also the average time since the velocity of a particle was last randomized. So the average velocity of a particle *changes* from $\langle v_x \rangle = 0$ to $\langle v_x \rangle = \langle tF/m \rangle = \tau F/m$ where m is the mass of a particle. This velocity resulting from the external force is called the *drift speed, $v_d = \langle v_x \rangle$*, and the ratio between force and drift speed can be measured macroscopically and is called the friction coefficient, $\zeta = F/v_d = m/\tau$. Multiplying the friction coefficient with the diffusion coefficient, D, gives $\zeta D = (m/\tau)(L^2/2\tau) = mL^2/\tau^2/2$. By definition, $L^2 = \langle x^2 \rangle_{\text{br}} = \langle v_x^2 t^2 \rangle_{\text{br}} = \langle v_x^2 \rangle_{\text{br}} \langle t^2 \rangle_{\text{br}}$, and by the equipartition theorem $\langle v_x^2 \rangle_{\text{br}} = \langle v_x^2 \rangle = k_B T/m$, so that ζD can be written

$$\zeta D = \frac{k_B T \langle t^2 \rangle_{br}}{2\tau^2}. \tag{7.29}$$

To find a relation between $\langle t^2 \rangle_{br}$ and $\tau = \langle t \rangle_{br}$, note that $v_d \tau = \langle x \rangle_{br} = \langle Ft^2/2m \rangle_{br} = F \langle t^2 \rangle_{br}/2m$ since a particle just after a collision on average starts out with zero velocity in the x direction and then undergoes acceleration, $a = F/m$, until the next randomization. But since $v_d = \tau F/m$, $\tau^2 F/m = v_d \tau = F \langle t^2 \rangle_{br}/2m$ so that $\langle t^2 \rangle_{br} = 2\tau^2$. Inserting this into Equation 7.29 and rearranging gives

$$D = \frac{k_B T}{\zeta} = \frac{\langle x_t^2 \rangle}{2t}. \tag{7.30}$$

This relation which was originally obtained by Einstein is remarkable. Historically, it is very interesting since this relation allowed scientists to determine exactly how many carbon-12 atoms are in 12 g of carbon-12, namely, $6.02 \cdot 10^{23}$. This number is known as the Avogadro number, N_A, and was unknown until the arrival of this equation by Einstein and the subsequent experimental determination of how fast a Brownian particle in water diffuses. The experiment allowed for a determination of the Boltzmann constant, k_B, by measuring ζ, T, and $\langle x_t^2 \rangle/t$ and using Einstein's relation. This in turn allowed for a determination of the Avogadro number via $N_A = R/k_B$, since the gas constant, R, was already known from experiments on ideal gases. Equation 7.30 is also very interesting for a more fundamental reason. It is a relation between equilibrium fluctuations of a Brownian particle and the energy dissipated during a nonequilibrium process (an external force pulling the particle through a fluid). To see this more clearly, the relation $x(t) = \int_0^t v(t')dt'$ can be used to rewrite $D = \lim_{t \to \infty} \langle x_t^2 \rangle/2t = \int_0^\infty \langle v(0)v(t) \rangle dt$, where the ensemble average replaces the average over initial times. Hence, the Einstein relation can be rewritten as

$$v_d = \frac{F}{\zeta} = \frac{F}{k_B T} D = \frac{F}{k_B T} \int_0^\infty \langle v(0)v(t) \rangle dt. \tag{7.31}$$

From this equation, it is clearly seen how the nonequilibrium response, v_d, (and hence the energy dissipated per unit time, $E_{diss\ pr.\ unit\ time} = v_d F$) of v to an external force depends on the equilibrium fluctuations of the same quantity. So Einstein's relation from 1905 is a manifestation of the general fluctuation–dissipation relation proved in 1951 [7].

7.3.2.2 The Variance of Dissipated Work

Another manifestation of the fluctuation–dissipation relation connects the thermal fluctuation to the variation of the work dissipated when driving a system out of equilibrium. When driving a system in a nonequilibrium manner, work is inevitably dissipated. Consider, for instance, a sphere surrounded by fluid and connected to a spring with spring constant K. To keep the spring at a displacement x_1 from the potential minimum

position, $x = 0$, an external force of $F = Kx_1$ is needed. But the work needed to move the sphere to x_1 depends on the way in which it is moved there. If the sphere is initially at $x = 0$ and F is suddenly imposed, it will move toward x_1 where it will finally come to rest. The total work is then $W = \int_0^{x_1} F dx = \int_0^{x_1} Kx_1 dx = Kx_1^2$. But if the force had been increased sufficiently slowly from 0 to Kx_1, the position would at all times be given by $x = F/K$ and the total work would have been $W = \int_0^{x_1} F dx = \int_0^{x_1} Kx dx = Kx_1^2/2$, which is only half the work done when immediately imposing $F = Kx_1$. At the same time, the potential energy of the spring has been increased by $\Delta E_{pot} = Fx_1/2$ so when the process is driven in equilibrium (sufficiently slowly) all the work done on the system is stored as potential energy, but when the process is driven in a nonequilibrium manner, work is dissipated: $W_{diss} = W_{tot} - \Delta E_{pot}$. If the thermal fluctuations have a significant impact on the sphere velocity, the work dissipated when driving the system out of equilibrium will not be exactly the same each time the experiment is done. Sometimes, the thermal kicks the sphere gets will resist the motion a bit more than the average, sometimes a bit less. This variation of the dissipated work can be related to the thermal noise of the system.

Let λ denote a generalized coordinate along which a system is driven and let λ_0 designate the equilibrium position as function of time. If the system is driven from $\lambda_0 = 0$ to $\lambda_0 = 1$ with a constant rate, $d\lambda_0/dt$, then the variance of the dissipated work can be computed as follows [8]. Thermal fluctuations mean that the instantaneous position, λ_1, of the system is fluctuating around the equilibrium value, $\lambda_0(t)$, but if the system is not driven too far from equilibrium, the fluctuations will not be too large and the instantaneous potential energy of these fluctuations is given by

$$U(\lambda_0, \lambda_1) \approx U(\lambda_0) + \frac{K}{2}(\lambda_1 - \lambda_0)^2. \tag{7.32}$$

If the probability density, $\rho(\lambda_1)$, is not in equilibrium, it will change in time. As in Equation 7.16, the flux of probability density is a combination of diffusion and the force resulting from the derivative of the local potential, $U(\lambda_0, \lambda_1)$. Because the distribution moves with speed $d\lambda_0/dt$ without changing shape (the system is close to equilibrium), the flux is not zero, but equal to $\rho(\lambda_1)(d\lambda_0/dt)$:

$$-\frac{k_B T}{\zeta} \frac{d\rho(\lambda_1)}{d\lambda_1} - \frac{1}{\zeta} \frac{dU(\lambda_0, \lambda_1)}{d\lambda_1} \rho(\lambda_1) = \rho(\lambda_1) \frac{d\lambda_0}{dt}. \tag{7.33}$$

This differential equation is integrated to give the steady state distribution

$$\rho(\lambda_1) = \exp\left(\frac{-K\left[\lambda_1 - \lambda_0 + \dfrac{\zeta}{K} \dfrac{d\lambda_0}{dt}\right]^2}{2k_B T}\right). \tag{7.34}$$

The shape of the distribution is not changed in time, but its center, $\langle \lambda_1 \rangle$, lags behind the shifting value of λ_0 by an amount $(\zeta/K)(d\lambda_0/dt)$. The total work done on the system is given by $W = \int_0^1 (dU(\lambda_0, \lambda_1)/d\lambda_0) d\lambda_0$, and the variance of the work is defined as $\sigma^2(W) = \overline{W^2} - \overline{W}^2$ where $\bar{A}$ denotes the nonequilibrium ensemble average of A (which depends on the nonequilibrium probability distribution $\rho(\lambda_1)$). If the process lasts much longer than the relaxation time for internal fluctuations, $\tau = \zeta/K$, the work integral can be approximated as a sum:

$$W = \int_0^1 \frac{dU(\lambda_0, \lambda_1)}{d\lambda_0} d\lambda_0 \approx \sum_{i=1}^N \frac{dU(\lambda_0, \lambda_1)}{d\lambda_0} \frac{1}{N}, \quad \text{with } N\tau \frac{d\lambda_0}{dt} = 1.$$

This means that

$$\sigma^2(W) = \sigma^2 \left(\frac{1}{N} \sum_{i=1}^N \frac{dU(\lambda_0, \lambda_1)}{d\lambda_0} \right) = \frac{2}{N} \sigma^2 \left(\frac{dU(\lambda_0, \lambda_1)}{d\lambda_0} \right), \quad (7.35)$$

where the last equality is true because the variance of $(dU(\lambda_0, \lambda_1)/d\lambda_0)$ does not depend on λ_0, because $(dU(\lambda_0, \lambda_1)/d\lambda_0)$ at times t and $t + \tau$ are nearly but not quite uncorrelated, and because of the statistical relation $\sigma^2 \left(\sum_{i=1}^N X_i/N \right) = \sigma^2(X)/N$ which holds when the stochastic variables X_i all have variance $\sigma^2(X)$ and are completely uncorrelated. Since the nonequilibrium probability distribution is only shifted compared to the equilibrium distribution and their shapes are identical, the variance of $(dU(\lambda_0, \lambda_1)/d\lambda_0)$ under a shift speed is equal to the equilibrium variance of $(dU(\lambda_0, \lambda_1)/d\lambda_0)$ when $d\lambda_0/dt = 0$ so that

$$\sigma^2 \left(\frac{dU(\lambda_0, \lambda_1)}{d\lambda_0} \right) = \sigma^2 \left(\frac{dU(\lambda_0, \lambda_1)}{d\lambda_0} \bigg|_{\frac{d\lambda_0}{dt}=0} \right) \quad (7.36)$$

$$= \left\langle \left[\frac{dU(\lambda_0)}{d\lambda_0} + K(\lambda_1 - \lambda_0) \right]^2 \right\rangle - \left\langle \frac{dU(\lambda_0)}{d\lambda_0} + K(\lambda_1 - \lambda_0) \right\rangle^2 \quad (7.37)$$

$$= \left\langle K^2(\lambda_1 - \lambda_0)^2 \right\rangle, \quad (7.38)$$

where Equation 7.32 and the fact that $\langle (\lambda_1 - \lambda_0) \rangle = 0$ have been used. From the equipartition theorem, $\langle K^2(\lambda_1 - \lambda_0)^2 \rangle = 2K\langle K(\lambda_1 - \lambda_0)^2/2 \rangle = Kk_BT$. Finally, the equation for the variance of work for the process becomes

$$\sigma^2(W) = \frac{2}{N} Kk_BT = 2\zeta \frac{d\lambda_0}{dt} k_BT = 2k_BTW_{\text{diss}} \quad (7.39)$$

since the dissipated work is $W_{\text{diss}} = -\int_0^1 F_{\text{friction}} d\lambda_0 = \int_0^1 \zeta(d\lambda_0/dt) d\lambda_0 = \zeta(d\lambda_0/dt)$. This is yet another fluctuation–dissipation relation. It relates the dissipated work for driving a nonequilibrium process to the equilibrium fluctuations. As

is the case for all fluctuation–dissipation relations, Equation 7.39 is valid only if the system is driven close to the equilibrium state. In Section 7.3.3, an exact equation for the dissipated work, valid arbitrarily far from equilibrium, will be derived from Jarzynski's equality, and Equation 7.39 will be seen to be included in that result.

The general fluctuation–dissipation relation can be formulated in many ways. One of the most useful ones is in terms of power spectra, Fourier transforms, and impedances [9].

7.3.3 Jarzynski's Equality

Most nonequilibrium theories deal with systems that are close to an equilibrium state. One exception is Jarzynski's equality, which was published in 1997 [10]. Jarzynski's equality deals with the case where a system can switch irreversibly between two states, e.g., a closed state (A) and an open state (B). The energy needed to switch states, e.g., to switch from A to B, is the sum of energy needed to perform the opening reversibly plus the dissipated energy:

$$W_{\text{total}} = W_{\text{reversible}} + W_{\text{dissipated}}. \quad (7.40)$$

The reversible work, $W_{\text{reversible}}$, equals the equilibrium free energy difference, ΔF. The dissipated work, $W_{\text{dissipated}}$, is associated with the increase of entropy during the irreversible process. The following inequality is true for large N:

$$\left\langle W_{\text{total}} \right\rangle_N \geq W_{\text{reversible}}, \quad \text{for large } N \quad (7.41)$$

where $\langle \cdot \rangle_N$ denotes the average over N different switching processes. The achievement of Jarzynski was to turn this inequality into an equality, from which the thermodynamical parameter ΔG, the change in Gibbs free energy associated with the transition between the two states, could be extracted. Jarzynski's equality states

$$\exp\left[-\frac{\Delta G(z)}{k_BT} \right] = \lim_{N \to \infty} \left\langle \exp\left[-\frac{W_i(z,r)}{k_BT} \right] \right\rangle_N, \quad (7.42)$$

where
- $\Delta G(z)$ is the Gibbs free energy difference while switching from state A to B
- z is the reaction coordinate
- $W_i(r, z)$ is the irreversible work measured while switching from A to B and is dependent both on z and on the switching rate r

In general, the faster the switching rate, r, the more energy is irreversibly dissipated. The number of measurements, N, needs to be large enough for the expression on the right-hand side of Equation 7.42 to converge. How this is done in practice is shown in an example below. Two impressive things to notice about

Equation 7.42 are (1) it relates a well-defined thermodynamic quantity, ΔG, to work values measured in an irreversible process and (2) it is really an equality, not just an approximation. Here, Jarzynski's equality is stated in terms of Gibbs free energy. Often, it is formulated in terms of Helmholtz free energy, but if the experimental conditions are constant temperature and pressure, then the appropriate thermodynamic variable is Gibbs free energy.

Upon closer inspection of Jarzynski's theorem, Equation 7.42, it might be difficult to apply practically: If the fluctuations in W_i from one measurement to the next are significantly larger than $k_B T$, then the right-hand side of Equation 7.42 converges very slowly. Also, it is necessary to really measure every single work individually. Therefore, the only systems to which Jarzynski's theorem in practice applies are nanoscale systems in which the thermal fluctuations of the work, W_i, are not significantly larger than $k_B T$.

7.3.3.1 Application of Jarzynski's Equality

As stated above, to apply Jarzynski's equality, it is important that the noise of the system is only on the order of $k_B T$. In the work described in Ref. [12], they applied Jarzynski's equality to a process where a single RNA molecule is mechanically switched between two conformations. An RNA molecule consists of a string of nucleotides, where every single nucleotide preferably forms a hydrogen bond to another specific nucleotide. This creates secondary RNA structures, of which one is the so-called hairpin and is schematically shown in Figure 7.10. The RNA hairpin is held between two microscopic beads of which one is firmly held by a micropipette and the other by an optical trap. By moving the two beads away from each other, the hairpin is mechanically unfolded, and through the optical trap, the corresponding forces applied and distances moved can be controlled and measured. By varying the bead separation velocity, the unfolding can happen at different *loading rates, r* (=increase in applied external force per unit time).

To apply Jarzynski's equality, Equation 7.42, one has to define a proper reaction coordinate, z, and find the work done to switch the system from A to B, W_i, a large number of times, N. In the

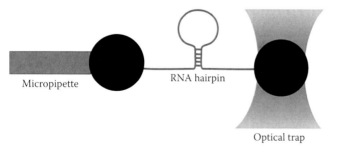

FIGURE 7.10 Schematic drawing of the experimental setup with an RNA hairpin structure mounted between two beads, one of which is held by a glass micropipette and the other held by an optical trap, capable of measuring corresponding values of forces and extensions of the RNA secondary structure. The drawing is simplified and not to scale.

particular experiment of unfolding an RNA structure, both the change in distance between the two beads, z, as well as the external force needed to unfold the structure, $F(z, r)$, are known. Notice that F depends strongly on the force loading rate, r. Hence, the work done can be found by

$$W_i(z,r) \simeq \int_0^z F_i(z',r)\,\mathrm{d}z' \quad \text{or} \quad W_i = \sum_{j=1}^N F_j \Delta x_j, \qquad (7.43)$$

where the second equation is for a discrete situation such as an experimental data set where N is the number of intervals used in the sum and F_j is the force acting on the system in the infinitesimal interval Δx_j.

Figure 7.11 shows typical force–extension traces from unfolding and refolding an RNA hairpin. The work is found as the area under the curve as the unfolding or refolding takes place. It is clear from this figure, that at high force-loading rates there is a substantial hysteresis between the unfolding and refolding curves. Typically, the unfolding happens at a larger force than the refolding. Also, the higher the loading rate, the larger is the force needed to unfold the structure. In other words, more of the work put into the system is irreversibly dissipated at high loading rates. If the applied force-loading rate is low, the unfolding takes place during a reversible process, with an equality sign in Equation 7.41.

The individually measured works, W_i, are substituted into Jarzynski's equality, Equation 7.42, and the averaging is done. As the average is over exponential terms, $\exp[-W_i/k_B T]$, the lower the value of a particular W_i, the higher its weight in the average. W_i can be negative too. Hence, the application of Jarzynski's equality can be considered as sampling the rare trajectories in the lower tails of the work distributions [15]. One important question to address is how many experiments, N, must be performed in order to give a reliable estimate of ΔG from Jarzynski's equality? The answer is basically that N must be large enough for Equation 7.42 to converge [15]. In general, the more work dissipated the larger N needs to be in order for Jarzynski's equality to converge.

In the study of the unfolding of RNA hairpins, Ref. [12], ΔG for the process was found in three different ways, all giving consistent values. One way was to conduct an unfolding of the hairpin using a loading rate low enough for the hairpin to be unfolded in a nearly reversible manner (as, e.g., shown in Figure 7.11a left trace). Then, they investigated the convergence of the Jarzynski equality as a function of the number of experiments performed, N. Figure 7.12 shows how the Jarzynski estimate of the free energy difference, ΔG, converges toward the true value as a function of distance along the reaction coordinate, z, and number of pulls, N. After approximately 40 pulls, the difference between the two are less than the experimental errors.

Jarzynski's equality can be rewritten as

$$\Delta G = -k_B T \ln \left\langle \exp\left[-\frac{W_i}{k_B T}\right] \right\rangle. \qquad (7.44)$$

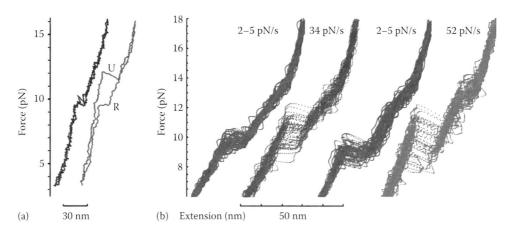

FIGURE 7.11 Force–extension traces from unfolding of RNA hairpins. (a) The left trace shows a typical force–extension trace from unfolding and refolding an RNA hairpin using a low force-loading rate, 2–5 pN/s. At this low rate the folding is nearly reversible. The unfolding and refolding takes place using a force of approximately 10 pN, the trace before and after the unfolding/folding event shows the elongation of the DNA/RNA handles holding the hairpin. The right trace shows unfolding (U) and refolding (R) of the hairpin using a higher loading rate, 52 pN/s, where the unfolding takes place at around 12 pN and the refolding at a lower force. (b) Unfolding and refolding of two RNA hairpins at fast and slow rates. (Reprinted from Liphardt, J. et al., *Science*, 296, 1832, 2002, Figure 2. With permission.)

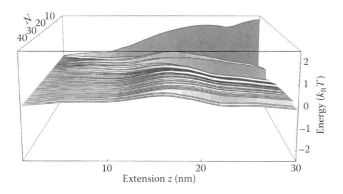

FIGURE 7.12 Convergence of Jarzynski's equality as a function of number, N of unfolding/refolding of RNA hairpins. The plot shows the numerical difference between ΔG estimated from Jarzynski's equality and the true ΔG as a function of extension along the reaction coordinate, z, and of number of pulling cycles, N. After approximately 40 pulls the ΔG estimated by Jarzynski's equality equals the true ΔG within their experimental error. Additional trajectories would further improve the convergence. (Reprinted from Liphardt, J. et al., *Science*, 296, 1832, 2002, Figure 4. With permission.)

The right-hand side of this equation can be expanded as a sum of cumulants [10] to give

$$\Delta G = \sum_{n=1}^{\infty} \left(-\frac{1}{k_B T} \right)^{n-1} \frac{\omega_n}{n!} = \omega_1 - \frac{\omega_2}{2k_B T} + \frac{\omega_3}{6(k_B T)^2} - \cdots \quad (7.45)$$

where ω_n is the nth cumulant of the distribution of works. This expansion makes it fairly easy to compare Jarzynski's equality to the work done and to the estimate from the fluctuation–dissipation theorem (see Section 7.3.2). The first term on the right-hand side of Equation 7.45 is simply the total work done, which is a good estimate of ΔG if the process happens reversibly.

The first two terms of the expansion, Equation 7.45, can be rewritten as

$$\Delta G \approx \langle W \rangle_N - \frac{\sigma^2}{2k_B T}. \quad (7.46)$$

which corresponds to the case where the work distribution is Gaussian as predicted by the fluctuation–dissipation theorem, Equation 7.39.

In Ref. [12], the performance of the fluctuation–dissipation theorem is compared to the performance of the Jarzynski equality for the unfolding of an RNA hairpin. Both at low and high loading rates, the Jarzynski equality performed better. The fluctuation–dissipation theorem did increasingly worse the higher the loading rate. This is reasonable, because the theorem is only valid at near-equilibrium conditions. Consistently, the fluctuation–dissipation theorem underestimated ΔG with respect to its true value. This is quite apparent from Equation 7.45 because the application of the fluctuation–dissipation theorem corresponds to only using the first two terms of the full Jarzynski expression, in particular, the third term is always positive, thus adding to the total estimate of ΔG.

7.3.4 Crooks Fluctuation Theorem

Another expression relating nonequilibrium measurable quantities to equilibrium thermodynamics that is valid for systems driven arbitrarily far from equilibrium is Crooks fluctuation theorem, published in 1999 [11]. It is a generalized version of the fluctuation theorem for stochastic reversible dynamics, and the Jarzynski equality is contained within Crooks fluctuation theorem. Crooks fluctuation theorem predicts a certain symmetry relation between the fluctuations in work associated with forward and backward nonequilibrium processes. Let us consider a structure, which can be either opened (O) or closed (C), the two processes being the reverse of each other. Let $P_O(W)$ denote the probability distribution of the work performed on the structure to open it during an infinite number of experiments, and let $P_C(W)$ denote the probability distribution of work performed by the structure on the surrounding system as it is closing. A requirement for Crooks fluctuation theorem to apply to the process is that the opening and closing processes need to be related by time reversal symmetry. Also, the structure needs to start in an equilibrium state and reach a well-defined end state (the "start" can be either the open or closed structure). The Crooks fluctuation theorem relates the work distributions to Gibbs free energy, ΔG, of the process:

$$\frac{P_O(W)}{P_C(W)} = \exp\left(\frac{W - \Delta G}{k_B T}\right). \qquad (7.47)$$

The theorem applies to systems driven arbitrarily far away from equilibrium. As will be shown in the following section (Section 7.3.4.1), it is fairly easy to apply to small-scale nonequilibrium systems.

At some particular value of W, the two distributions $P_O(W)$ and $P_C(W)$ might cross each other. In this case

$$P_O(W) = P_C(W) \Rightarrow \Delta G = W. \qquad (7.48)$$

In other words, it is very easy to determine ΔG of the reversible process from a plot where $P_O(W)$ and $P_C(W)$ are overlaid simply as the value of the work, W, where the two distributions cross each other.

7.3.4.1 Application of Crooks Fluctuation Theorem

For practical use, Crooks fluctuation theorem has some advantages over Jarzynski's equality [13]; due to the experimental averaging Jarzynski's equality is sensitively dependent on the experimental probing of rare events (low or even negative W values). Also, spatial drift makes it difficult to conduct reliable experimental measurements at low unfolding rates. Moreover, if one increases the loading rate, the irreversible loss is also increased and the equality converges more slowly. It seems that Crooks theorem is a bit more robust and converges more rapidly than Jarzynski's equation. One drawback, however, is the requirement of symmetric reversible events, and the question of whether one is able to measure the work associated with both the forward and backward events experimentally.

7.3.4.1.1 Application of Crooks Theorem to RNA Hairpins

Crooks theorem has been cleverly and clearly verified and presented in a form accessible to a larger audience through the work described in Ref. [13]. The model system and the setup was basically the same as depicted in Figure 7.10 and was also used by the same group to test Jarzynski's equality. This is a near ideal system because of its small size, its accessibility, the possible symmetric

operations to open and close the structure using a constant loading rate (a requirement for Crooks fluctuation theorem to hold), and finally because the folding and refolding work distributions actually do overlap over a sufficiently large range to find the situation given by Equation 7.48 from which ΔG can be found. Another important issue is that the group had alternative ways to determine the true value of ΔG which could then be compared to the value obtained from Crooks fluctuation theorem.

In order to use Crooks fluctuation theorem, the work required to mechanically open/unfold the structure must be found. This can be done, e.g., using the expressions from Equation 7.43. Figure 7.13 shows typical force–extension relationships for the unfolding (orange curves) and refolding (blue curves) of an RNA hairpin. Finding the work of a particular folding/unfolding event amounts to finding the area underneath the curve during the unfolding/refolding event. During unfolding, work must be done by the optical tweezers apparatus on the RNA hairpin. During refolding, work is done by the RNA hairpin on the optical tweezers apparatus (blue area on Figure 7.13). However, for this experiment, the found value of W must also be corrected for the work going into stretching/relaxing the backbone of the structure and the handles.

When a sufficient number of such folding and unfolding traces have been analyzed and the corresponding work found, histograms of the folding/unfolding work distributions give information about the Gibbs free energy difference of the process.

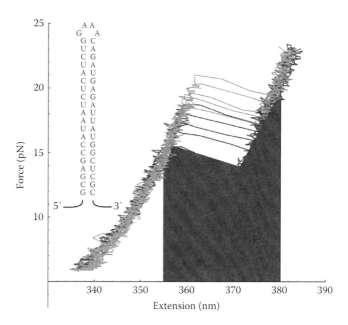

FIGURE 7.13 Force versus extension during the mechanical unfolding of an RNA hairpin mediated by an optical trap. The exact nucleotide sequence of the hairpin is also shown. The gray curves originates from unfolding of the hairpin, the black curves from refolding. The loading rate was 7.5 pN/s. The black area under the curve corresponds to the work returned to the optical trapping setup as the molecule is refolded. (Reprinted from Collin, C. et al., *Nature*, 437, 231, 2005, Figure 1. With permission.)

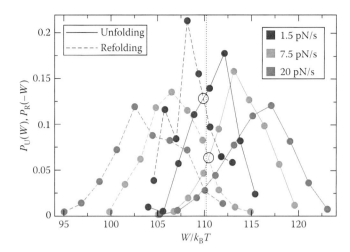

FIGURE 7.14 Probability distributions $P_O(W)$ and $P_C(W)$ for folding and refolding of an RNA hairpin at three different loading rates. The full lines are from unfolding the structure, $P_O(W)$. The dashed lines are from refolding the structure, $P_C(W)$. The black distributions are obtained with a loading rate of 1.5 pN/s, light gray distributions with 7.5 pN/s, and dark gray distributions with 20 pN/s. (Reprinted from Collin, C. et al., *Nature*, 437, 231, 2005, Figure 2. With permission.)

Figure 7.14 shows work distributions for RNA hairpin unfolding and refolding; it is the numerical value of the obtained work, which has been used. The full lines show the unfolding work distributions, the dashed lines the refolding work distributions. ΔG for the process can easily be read off the graph as the work where the two distributions are equal, i.e., 110 k_BT. A couple of other interesting observations are possible from Figure 7.14: (1) The slower the loading rate, the closer the folding and refolding traces are. If the folding/refolding is done at a zero loading rate, the process is reversible and the two curves fully overlap. (2) The higher the loading rate, the less Gaussian the work distributions are. The Gaussian appearance of the work distribution at low loading rates is because the fluctuations are not too far from equilibrium and the process is well described by the "normal" fluctuation–dissipation relation, Equation 7.39. This example also shows that the work distributions need not be Gaussian in order for Crooks fluctuation theorem to apply.

7.3.4.1.2 Application of Crooks Fluctuation Theorem to RNA Pseudoknots

Another more complex system to which Crooks theorem has been applied is the unfolding of RNA pseudoknots. An RNA pseudoknot is a tertiary RNA structure, which can be viewed as an RNA hairpin where the nucleotides of the loop have performed basepairings to the backbone. Hence, the structure has two stems, which have to be ruptured upon unfolding of the structure. Figure 7.15 shows a schematic drawing of an RNA pseudoknot and of the optical tweezers and micropipette setup used to perform a mechanical unfolding of the structure.

In Ref. [14], it was reported that the mechanical strength needed to unfold two particular pseudoknots derived from avian

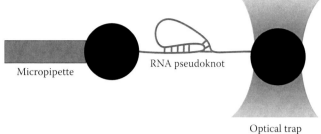

FIGURE 7.15 Simplified sketch of the setup for mechanical unfolding of an mRNA pseudoknot, the details about the settings are given in Ref. [14].

infectious bronchitis correlated with the frameshifting efficiency of the pseudoknot. In the majority of the unfolding and refolding traces, the pseudoknot unfolded in what appeared to be a single event, and Crooks theorem was applied to find the Gibbs free energy difference of the unfolding process. Figure 7.16 shows the distributions of the work needed to unfold the pseudoknot and of the work liberated by refolding the structure.

Occasionally intermediate states were observed in the force–extension curves resulting from unfolding the RNA pseudoknot, the intermediate state probably being a configuration where one but not both stems had unfolded. As it is a prerequisite for applying Crooks theorem that both the initial and final states are well defined and that the process is completely reversible, it is not clear that the theorem can be applied to this more complex tertiary structure if intermediate states are possibly present in the folding pathway. Hence, this example points toward future theoretical efforts, which encompass the challenge of dealing with structures going through metastable intermediate states.

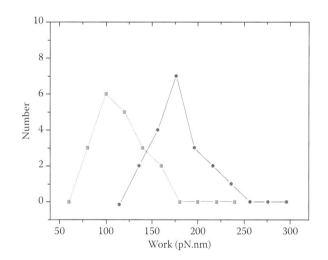

FIGURE 7.16 Supplementary information from [14]. Probability distributions $P_O(W)$ and $P_C(W)$ for folding and refolding of a RNA pseudoknot at a loading rate of 10 pN/s. The left gray line shows the work distribution from unfolding the structure, $P_O(W)$. The right black line is the work distribution from refolding the structure, $P_C(W)$. (From Hansen, T.M. et al., *PNAS*, 104, 5830, 2007. With permission.)

7.4 Conclusion and Outlook

Despite the abundantness of nonequilibrium processes and systems in our surroundings, only limited knowledge exists on their dynamics, the timescales involved, and on the switching processes between equilibrium states. In 1940, Kramer put forward his theory, which provides very useful information regarding the energy barriers to be crossed and the timescales involved. The fluctuation–dissipation theorem from 1951 explains how the fluctuations around equilibrium relates to the energy dissipated, and is very useful if the system is not too far from equilibrium. A major step forward in understanding nonequilibrium dynamics came with the publication of Jarzynski's equality in 1997. The true strength of this expression is that it is an equality and that it applies arbitrarily far from equilibrium. A more general form, containing Jarzynski's equality, was put forward in 1999 by G.E. Crooks, this generalized fluctuation relation also holding true arbitrary far from equilibrium. In practice, as shown in this chapter, using the unfolding of RNA hairpins as an example, Crooks theorem might be slightly easier to apply to small-scale nonequilibrium systems than the Jarzynski equality. However, one requirement for applying Crooks theorem is that it is possible to exactly reverse the process and to accurately measure the work put into the system both during the process and its reverse. Future challenges will include how to correctly describe and understand larger nonequilibrium systems as well as how to deal with possible intermediate states. Jarzynski's equality and Crooks theorem significantly advance thermodynamics and make thermodynamics applicable to systems from which it was previously not possible to extract equilibrium information from.

References

1. F. Mandl, *Statistical Physics*, John Wiley & Sons, Chichester, U.K., 1971.
2. D. Chandler, *Introduction to Modern Statistical Mechanics*, Oxford University Press, New York, 1987.
3. H.A. Kramers, Brownian motion in a field of force and the diffusion model for chemical reactions, *Physica VII*, **4** 285–304, 1940.
4. K.C. Neuman and S.M. Block, Optical trapping, *Review of Scientific Instruments*, **75** 2787–2809, 2004.
5. J.K. Dreyer, K. Berg-Sørensen, and L. Oddershede, Quantitative approach to small-scale nonequilibrium systems, *Physical Review E*, **73** 051110, 2006.
6. A. Einstein, On the movement of small particles suspended in stationary liquids required by the molecular-kinetic theory of heat, *Annalen der Physik*, **17** 549–560, 1905.
7. H.B. Callen and T.A. Welton, Irreversibility and generalized noise, *Physical Review*, **83** 34–40, 1951.
8. J. Hermans, Simple analysis of noise and hysteresis in (slow-growth) free energy simulations, *Journal of Physical Chemistry*, **95** 9029–9032, 1991.
9. R. Kubo, The fluctuation–dissipation theorem, *Reports on Progress in Physics*, **29** 255–284, 1966.
10. C. Jarzynski, Nonequilibrium equality for free energy differences, *Physical Review Letters*, **78** 2690–2693, 1997.
11. G.E. Crooks, Entro production fluctuation theorem and the nonequilibrium work relation for free energy differences, *Physical Review E*, **60** 2721–2726, 1999.
12. J. Liphardt, S. Dumont, S.B. Smith, I. Tinoco Jr., and C. Bustamante, Equilibrium information from nonequilibrium measurements in an experimental test of Jarzynski's equality, *Science*, **296** 1832–1835, 2002.
13. C. Collin, F. Ritort, C. Jarzynski, S.B. Smith, I. Tinoco Jr., and C. Bustamante, Verification of the Crooks fluctuation theorem and recovery of RNA folding free energies, *Nature*, **437** 231–234, 2005.
14. T.M. Hansen, S.N.S. Reihani, M.A. Sørensen, and L.B. Oddershede, Correlation between mechanical strength of messenger RNA pseudoknots and ribosomal frameshifting, *PNAS*, **104** 5830–5835, 2007.
15. H. Oberhofer, C. Dellago, and P.L. Geissler, Biased sampling of nonequilibrium trajectories: Can fast switching simulations outperform conventional free energy calculation methods? *The Journal of Physical Chemistry B*, **109** 6902–6915, 2005.

Nanoionics

Joachim Maier
*Max Planck Institute for
Solid State Research*

8.1 Introduction: Significance of Ion Conduction

While nanoelectronics refers to electronic transport and storage phenomena on the nanoscale, "nanoionics" refers—on that same scale—to ionic transport and storage phenomena. At interfaces, and particular in confined systems, exciting ionic phenomena are observed that indeed justify the use of this term (Maier 2005b).

Ion motion is no less significant for processes in nature and technology than motion of electrons. Well-known is the role of ion transport in liquid or semiliquid systems, a striking example being offered by biology in terms of nerve propagation. But also as far as solids are concerned, the role of ion transport is of paramount significance. All mass transport phenomena in ionically bound solids require ion transport, usually in the form of simultaneous transport of ions and electrons or of different ions. Beyond that, there is a whole class of applications, typically energy-related applications, for which the mobile ions are indispensable and their role cannot be taken over by electronics. Such devices include batteries and fuel cells with the help of which electrical energy can be stored or just converted to chemical energy. Related applications are various types of chemical sensors, chemical filters, or recently described resistive switches. In this case, basically chemical information is transformed into physical information.

Let us discuss some electrochemical applications based on oxygen ion conductors such as Y-doped zirconia. Two different oxygen partial pressures on the two sides of this oxide ceramic generate a cell voltage that can be used to detect oxygen partial pressures once the value on one side is known, or even to control that partial pressure as it is done in modern automobiles. If one works with very reducing gases (e.g., hydrogen) on one side (what corresponds to a low oxygen partial pressure), while having, for example, air on the outer side, the principle of a fuel cell

is realized; this case is indicated in Figure 8.1a. In this way, H_2 is electrochemically converted to H_2O. As the electrical energy can be directly used without thermal detour, Carnot's efficiency does not apply and high theoretical efficiencies can be expected. As the solid electrolytes can typically be used at high temperatures, gases such as hydrocarbons can be converted quickly enough.

If material problems that are connected with high temperatures are to be avoided, it is well advised to use proton conductors, which are mobile enough at moderate temperatures. Then the direction of the mass transport is changed and H_2O is produced on the air side where it does not pollute the fuel.

Rather than just conversion, cation conductors such as Li-conductors allow for energy storage. In the same way as the cell voltage in the previous examples is given by the difference in the chemical potential of oxygen or hydrogen (partial pressure of oxygen or hydrogen), the Li-potential difference is exploited here. Unlike in previous examples, the decisive component (Li) is accommodated in a solid phase (see Figure 8.1b). This provides the possibility to efficiently store electrical energy while in the above examples, it was rather the transformation of energy that was important. The low weight of Li and its high electronegativity guarantee high energy per weight.

As indicated in Figure 8.1b, Li storage (i.e., Li^+ and e^-) requires both ionic and electronic conductivities. Mass transport enabled by this mode is also exploited in chemical filters. Oxygen, for example, selectively permeates through oxides, which exhibit both O^{2-} and e^- conductivities. Very related applications are gas storage applications such as for H_2 provided by polar hydrides (here it is the simultaneous motion of hydrogen ions and electrons) or equilibrium conductivity sensors (conductivity effects on varied stoichiometries as response to varied partial pressures). This possibility of having both types of conductivities in the solid state enables gas filtering by mixed conducting permeation membranes.

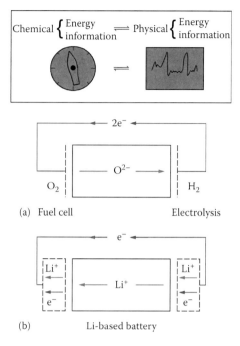

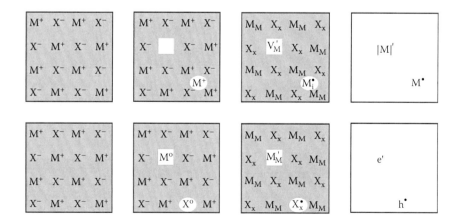

FIGURE 8.1 A few selected electrochemical applications (cf. text).

Moreover, the mixed conductor represents the general case of an electrical conductor from which the pure ion conductor and the pure electron conductor follow as special cases, making it a master material of fundamental importance. We will refer to the mixed conductor quite frequently in the following discussion.

8.2 Ionic Charge Carriers: Concentrations and Mobilities

The top row of Figure 8.2 identifies typical ionic charge carriers in crystals, namely, particles in interstitial sites or vacancies (Wagner and Schottky 1930), while the bottom row refers to electronic carriers (in a localized picture).

The ionic excitation of an ion from its regular site to an interstitial site leaving behind a vacancy (Frenkel disorder) is analogous to an excitation of an electron from the valence band to the conduction band, leaving a hole in the valence band. In many cases, the electronic carriers can be connected with different valence states (in the absence of substantial hybridization). For example, in a component such as silver chloride, a neutral silver on an Ag^+ site is equivalent to an electron in the conduction band while a missing electron on a Cl^- (i.e., neutral Cl) corresponds to a hole in the conduction band. Charge delocalization (without mixing Cl and Ag orbitals) means that the varied valence states do refer to the ensemble of cations or anions rather than to a specific identifiable ion.

If we refer to a tiny energy interval and consider the statistics in a sample of constant number of lattice sites, we have to refer to Fermi–Dirac statistics in both cases. In the case of electrons, it is Pauli's principle that excludes two electrons in the same state, and in the case of ions, it is the restriction that two ions cannot occupy the same lattice site (which after all is, of course, also a quantum mechanical effect) (Wagner and Schottky 1930; Kirchheim 1988; Maier 2004b, 2005a).

Even though the density of states is very different—in the case of electrons we may have delocalization and parabolic density of states, while in the case of ions we typically face sharp energies for interstitial sites—the statistics concerning the entire energy range of interest is very similar as long as the gap (standard free energy of formation) is sufficiently large. The chemical potential of ions and electrons will then follow a Boltzmann distribution. Figure 8.3 shows how a Boltzmann distribution results for different cases irrespective of the details concerning nature and concentration of energy levels.

Figure 8.4 refers to the ionic and electronic excitations in the energy level picture. In all cases, these energy levels are standard electrochemical potentials, while Fermi/Frenkel levels are full electrochemical potentials including also configurational terms. (One may use the term Frenkel level to emphasize the parallelity of the

FIGURE 8.2 Perfect (left) and defective crystal situations for the compound M^+X^-. A specific example may be Ag^+Cl^-. Top row: ionic defects. Bottom row: electronic defects. First and second columns: structure elements in absolute notation. Third column: structure elements in relative notation. Fourth column: building elements.

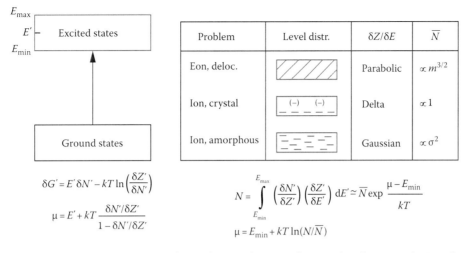

FIGURE 8.3 Whenever the gap between ground states and excited states is large, a Boltzmann distribution results. Details on the density of states enter the constant term (also the concentration measure). Eon stands for electron. If the reader is interested in more details, they are referred to Maier (2005a) and Kirchheim (1988).

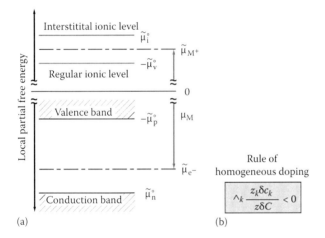

FIGURE 8.4 (a) Defect thermodynamics of a mixed conductor with Frenkel disorder (Maier 2003a). (b) Rule of homogeneous doping. (Reprinted from Maier, J., *Solid State Ionics*, 157, 327, 2003a. With permission.)

concepts.) Hence the distances between Frenkel (Fermi) levels and energy levels (band edges) are measures of interstitial and vacancy (hole and electron) concentrations.

Figure 8.4 also displays the coupling between both pictures that is provided by the relation

$$\tilde{\mu}_{M^+} + \tilde{\mu}_{e^-} = \mu_{M^+} + \mu_{e^-} = \mu_M, \tag{8.1}$$

where μ and $\tilde{\mu}$ are the chemical and electrochemical potentials (the latter include electrical potentials).

As $\mu_M = \partial G / \partial n_M$ measures the increase of Gibbs energy if neutral M is added to the master compound MX (n = mole number) at constant n_X and hence refers to compositional variations, this shows how important stoichiometry is (adding M to MX leads to variations in the M/X ratio). The stoichiometry is

determined through μ_M (=const + $RT \ln P_M$) and hence by the partial pressure of M (i.e., P_M). In a binary oxide, it is convenient and sufficient to refer to P_{O_2} (that is coupled to P_M via the phase equilibrium conditions).

Further important parameters for the equilibrium defect chemistry are temperature and doping content (C). Our energy level picture is only able to give an account of simple situations. A complex set of processes is better handled by the language of chemical thermodynamics, i.e., by explicitly writing down interaction equilibria to which then mass action laws apply. In not too complex cases, it is easy to show that the concentration for a charge carrier k, i.e., $c_k(P, T, C)$, is given by

$$c_k(P,T,C) = \alpha_k^{\beta_k} P^{N_k} C^{M_k} \Pi_r K_r^{\gamma_{rk}}(T), \tag{8.2}$$

where

$\alpha, \beta, N, M, \gamma$ are simple rational numbers
$K_r(T)$ is the mass action constant of the defect reaction r

In the simplest case, see Figures 8.3 and 8.4, we only have to refer to a single r as in the case of our energy level diagram. This equation looks slightly more complicated if more components are involved. If these additional components, i are also in equilibrium with k then P^N has to be replaced by a product of terms ($\Pi_i P_i^{N_i}$). If the other components (j with charge number z_j) are frozen however, then the respective defect concentrations (c_j) appear in the C-term as dopants ($\Sigma_j z_j c_j$). Let us briefly consider the influence of these parameters on a qualitative level (see Maier (2004b) for a quantitative treatment):

1. Temperature increase typically (but not in all cases) leads to higher defect concentrations as the formation processes are all thermally activated. Of special significance here are high defect concentrations, where interactions occur.

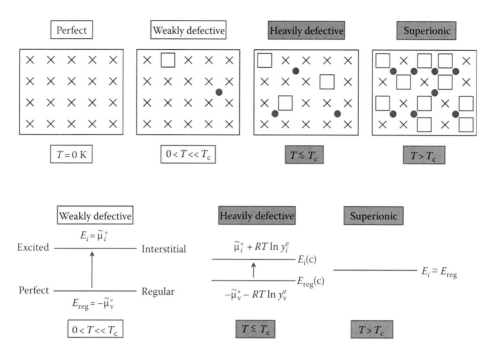

FIGURE 8.5 Thermal energy leads to Frenkel disorder. Further increase leads to attractive interactions. The "positive feedback" leads to super-ionic phase transformation. (Bottom: energy level description). (Reprinted from Maier, J. and Münch, W., *Z. Anorg. Allg. Chem.*, 626, 264, 2000. With permission.)

Interactions between intrinsic majority effects are usually attractive and can lead to the transition of the superionic state, in which all ions are defective, as it were. This is shown in Figure 8.5. In this case, however, we are dealing already with deviations from Boltzmann distribution and from Equation 8.2.

2. Figure 8.6 shows the drastic variation in ionic and electronic charge carrier concentration upon tiny changes in the partial pressure.

3. Figure 8.7 illustrates the effect of doping. An effectively negatively charged dopant leads to increase of all the positively charged defects (here $V_O^{\bullet\bullet}, h^\bullet$) and to a decrease of all negative charged defects (here O_i'', e') individually. This can be summarized by the rule of homogeneous doping (already indicated in Figure 8.4)

$$\frac{z_k \delta c_k}{z \delta C} < 0. \qquad (8.3)$$

The presence of point defects is of course not sufficient for having a perceptible conductivity (σ). For this, we need also a sufficient mobility (u) as $\sigma_k \propto u_k c_k$.

Figure 8.8 displays the three basic ionic jump mechanisms: the transport of a vacancy by allowing neighboring regular ions to occupy it (top), and the transport of interstitials that either jump directly to the next equivalent site (center) or indirectly by substituting a regular neighbor that is to occupy another interstitial site (bottom). Obviously, the respective ionic mobilities are all strongly thermally activated because of the pronounced migration barrier. As stoichiometric or dopant

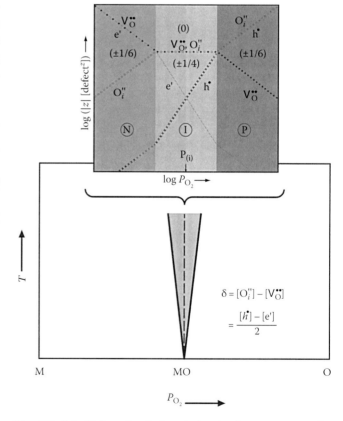

FIGURE 8.6 Defect chemical variation in the oxygen superlattice disordered oxide MO by P_{O_2} variation. (Reprinted from Maier, J., *Modern Aspects of Electrochemistry*, Conway, B.E. et al. (eds.), Vol. 38, Kluwer Academic/Plenum Publishers, New York, 2005c, 1–173. With permission.)

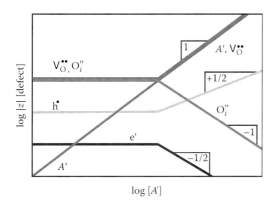

FIGURE 8.7 Acceptor doping of an oxide with oxygen sublattice disorder leads to an increase of oxygen vacancy and hole concentrations (while the concentrations of interstitial oxygen ions and electrons are depressed).

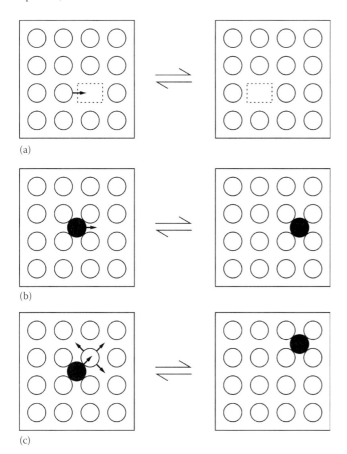

FIGURE 8.8 Three basic ion transport mechanism (a) vacancy, (b) direct interstitial, and (c) indirect interstitial. (Reprinted from Maier, J., *Physical Chemistry of Ionic Materials. Ions and Electrons in Solids*, John Wiley & Sons, Ltd., Chichester, U.K., 2004b. With permission.)

variations affect the structure only at high concentrations, the mobility is taken as independent of P and C in the following if not stated otherwise.

With these points in mind, the situation at interfaces can be tackled.

8.3 Ionic Charge Carrier Distribution at Interfaces and Conductivity Effects

As conductivity is proportional to the product of mobility and concentration, conductivity variations at the interfaces can be due to both factors. Directly in the interfacial core, these two parameters are naturally very different from the bulk. This influence is particularly decisive if we consider materials with low charge carrier conductivities in the bulk. A significant effect whose importance can be hardly overestimated is the charge carrier redistribution in the vicinity of the interfaces. Naturally, one may also meet variations of mobility due to structural variations that are expected to be minor compared to the interfacial core structure (see below). Such mobility variations are particularly important if the charge carrier concentration is already very high and bulk mobilities low, while concentration changes particularly matter for materials with high carrier mobilities and low bulk concentrations. Under these conditions, space charge effects may be order of magnitude effects. We will largely dwell on such space charge effects (Maier 1995) owing to their importance and generality of the phenomenon.

For a phenomenological description, let us consider (see Figure 8.9) the general thermodynamic situation in a mixed conductor at the interface in the approximation of an abrupt contact (i.e. we assume that the structure varies in a step-function way at $x = 0$ toward the bulk structure).

The electronic levels have to be bent and so do the levels of the mobile ions. This is equivalent to having concentration changes of both ions and electrons (Figure 8.10). If we introduce conductivity effects owing to the introduction of appropriate second phases, we term this "heterogeneous doping." The rule of heterogeneous doping, which proposes how a given surface charge (Σ) qualitatively affects the concentration of a given defect k is analogous to Equation 8.3 (see also Figure 8.9), namely,

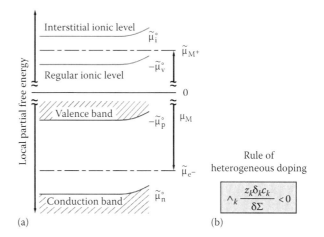

FIGURE 8.9 (a) General contact of a Frenkel disordered mixed conductor to a second phase, resulting in a negative surface charge. (b) Rule of heterogeneous doping. (Reprinted from Maier, J., *Modern Aspects of Electrochemistry*, Conway, B.E. et al. (eds.), Vol. 38, Kluwer Academic/Plenum Publishers, New York, 2005c, 1–173. With permission.)

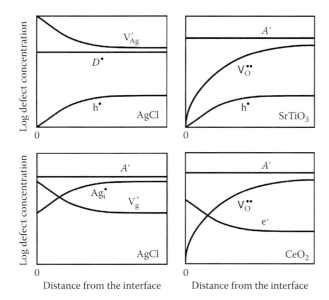

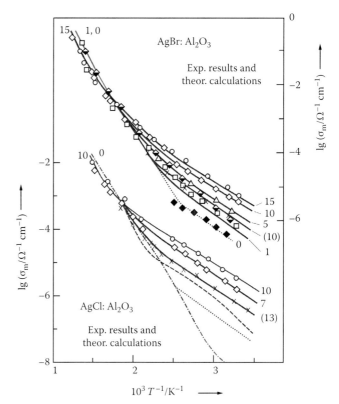

FIGURE 8.10 Four boundary situations at a positively charged contact (see text). The influence of charge number (+2 for oxygen vacancies, ±1 for e′, h•, Ag+-defects) is seen in the bulk balance as well as in the steepness of variation at the boundary.

$$\frac{z_k \delta c_k}{\delta \Sigma} < 0. \qquad (8.4)$$

Figure 8.10 shows some realistic examples for a positively charged surface. On a qualitative level, all these results can be understood on the basis of Equation 8.4. Figure 8.10 (top, left) refers to an accumulation of vacancies and an increased ion conductivity in a donor doped sample (example: $AgCl:Al_2O_3$ or AgCl:AgCl, a typical donor being Cd^{2+}) (Maier 1995), Figure 8.10 (top, right) refers to a depletion of both vacancies and holes as met in acceptor doped $SrTiO_3$ ceramics (a typical acceptor is Fe^{3+}) (Vollmann and Waser 1994; Noll et al. 1996; Vollmann et al. 1997; De Souza et al. 2003). Also inversion effects can be realized, e.g., in acceptor (S^{2-})-doped AgCl, where in the bulk we meet predominant interstitial conductivity but at the boundary vacancy conductivity, Figure 8.10 (bottom, left). In Figure 8.10 (bottom, right) we face even an inversion from ion (vacancy) to electron (n-type) conductivity in weakly acceptor (Gd)-doped ceria (Chiang et al. 1996; Tschöpe 2001; Kim and Maier 2002).

Figures 8.11 and 8.12 give experimental examples. Figure 8.11 shows a significant enhancement that is achieved by adding Al_2O_3 particles (with volume fraction φ and surface-to-volume ratio Ω) to AgCl, which leads to an adsorption of Ag^+ ions at the alumina's surface connected with an accumulation of carriers. In percolative composites, the conductivity can be derived to be (Maier 1995)

$$\sigma_m = (1-\varphi)\sigma_\infty + \beta_L \Omega \varphi F(2\lambda) c_\infty 2\vartheta_v \left[\frac{u_v}{1-\vartheta_v} - \frac{u_v}{1+\vartheta_v}\right] \quad (8.5)$$

where

　λ is the Debye length
　ϑ is the degree of influence
　β_L is the percolativity

FIGURE 8.11 Heterogeneous doping of AgCl and AgBr with Al_2O_3 and detailed description by Equation 8.5 (according to Equation 8.5 $\vartheta \to 1$). (Reprinted from Maier, J., *Prog. Solid State Chem.*, 23, 171, 1995. With permission.)

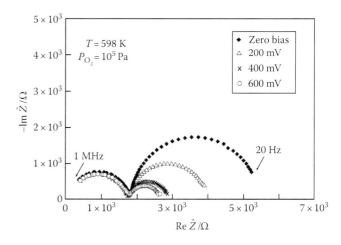

FIGURE 8.12 Bias-dependent impedance spectra of a $SrTiO_3$ bicrystal. The bicircles on the right reflect hole depletion. (Reprinted from Denk, I. et al., *J. Electrochem. Soc.*, 144, 3526, 1997. With permission.)

Figure 8.11 shows how accurately this equation describes the experimental results. If the AgCl is pure or S^{2-}-doped, the bulk conductivity is of interstitial type; then we have realized an interstitial-vacancy junction by heterogeneous doping. If we consider TlCl, which intrinsically conducts via Cl^- vacancies, then heterogeneous doping by Al_2O_3 leads to a predominant cationic

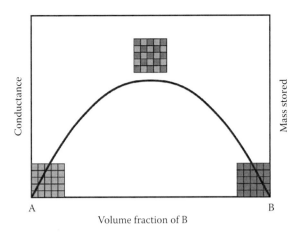

FIGURE 8.13 Conductance and mass storage anomaly in an A:B two-phase mixture owing to ion redistribution. (Reprinted from Zhukovskii, Y.F. et al., *Phys. Rev. Lett.*, 96, 058302, 2006. With permission.)

conductivity (via cation vacancies) (Maier 1995). This effect of heterogeneous doping is qualitatively sketched in Figure 8.13.

The depletion effect on the conductivity in the primarily electronically conducting (acceptor-doped) $SrTiO_3$ is directly reflected by the low frequency semicircle of a bicrystal impedance experiment (Figure 8.12) (Noll et al. 1996; Merkle and Maier 2008). The fact that also and even more strongly oxygen vacancies ($V_O^{\bullet\bullet}$) are depleted is shown by measurements in which the electronic conductivity has been blocked. (The stronger depletion of $V_O^{\bullet\bullet}$ as compared with $h^\bullet$ leads to peculiar polarization phenomena in multilayers or polycrystalline material under dc bias (Jamnik et al. 2003).)

An even more striking situation is met for a vacancy conducting oxide with excess electrons as major electronic carriers. In CeO_2 (weakly doped by Gd_2O_3) clearly a transition from ionic to electronic conductivity occurs on introducing appropriate interfaces.

A key point is the mechanism that determines the space charge potential. This obviously lies in the chemistry/crystallography of the contact (Maier 1995). Understanding this in principle would open the possibility of tuning the space charge potential. For a quantitative description, the reader is referred to the literature. One example in which a purposeful tuning succeeded refers to polycrystalline or thin films of CaF_2, which is a fluoride ion conductor. At interfaces, fluoride vacancy accumulation has been observed that could purposefully be augmented by contaminating the interfaces with an F^- attractor (SbF_5 or BF_3) (Saito and Maier 1995). Figure 8.14 shows the enormous conductivity enhancement of nanocrystalline CaF_2 (Puin et al. 2000), where it is the enormous proportion of interfaces that leads to these huge effects. Figure 8.15 gives results on thin films of BaF_2 on Al_2O_3-substrate, which leads to a fluoride vacancy ($V_F^\bullet$) accumulation (Guo and Maier 2004).

In all the examples considered, the quantitative explanation relied on an interfacial situation in which the carrier profile reached bulk values; in other words, the local situation was semi-infinite. In the following, we will treat even more exciting effects resulting from interfacial overlap. Before we do so, let us mention

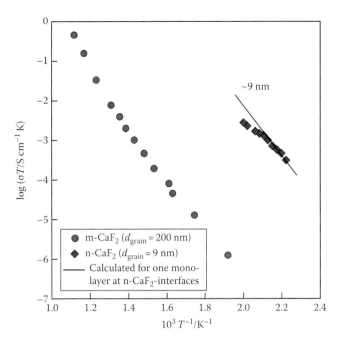

FIGURE 8.14 Fluoride conductivity of nano- and macrocrystalline CaF_2. (Reprinted from Puin, W. et al., *Solid State Ionics*, 131, 159, 2000. With permission.)

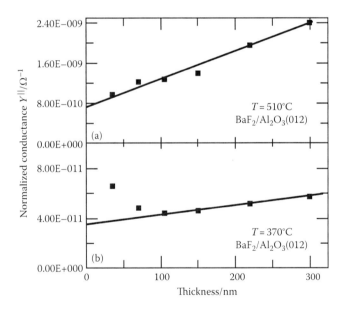

FIGURE 8.15 Parallel conductance of BaF_2 films on Al_2O_3 substrates. The explanation is based on F^--adsorption. At 370°C strain effects contribution. (Reprinted from Guo, X.X. and Maier, J., *Surf. Sci.*, 549, 211, 2004. With permission.)

one more point that is interesting for semiconductor physics and follows from the fact that majority defects typically control the space charge behavior. Also in typical semiconductors, we meet ionic point defects, very often in much greater concentrations than electrons. Owing to their low mobility, in particular at the low temperatures, they do not actually contribute to the conduction process, but act as dopants. Their concentration is typically determined by the prehistory. This is very relevant for the boundary effects in particular when pretreatment at high temperatures was performed at which the ions are locally mobile. Then very often, the ionic boundary behavior determines the space charge potential that also has to be obeyed by the electrons as minority species. This "fellow traveler effect" (Maier 1995) can explain space charge effects of electronic carriers that are not compatible with or even seemingly in contradiction with the expected polarity. It is not hopeless to use this effect for tuning in future high-temperature superconduction.

Before we deal with mesoscopic phenomena, let us consider Table 8.1, which gives results for the conductivity effect of ions and/or electrons depending on the direction of measurements (Kim et al. 2003). It is clear that in a measurement direction along the interface, the highly conductive regions are seen most sensitively, whereas it is the most resistive regions that are dominant in a measurement perpendicular to the interface.

8.4 Mesoscopic Effects

1. A well-suited model system that allows studying nano-ionic effects in detail are molecular beam epitaxially grown heterolayers consisting of CaF_2 and BaF_2 (Sata et al. 2000; Guo et al. 2007; Guo and Maier 2009). The thickness of the individual layers can be varied from almost ~1 nm to ~1 μm. The misfit stress is absorbed by misfit dislocations. It was shown that the conductivity effects are purely ionic. Impedance spectroscopy is applied to measure parallel as well as perpendicular conductivities using appropriate substrates. All the effects can be consistently explained by a fluoride ion transfer from BaF_2 to CaF_2, hence increasing the vacancy concentration in BaF_2 and the interstitial concentration in CaF_2. A fully quantitative explanation of thickness and temperature variation dependences succeeds if one takes account of background

TABLE 8.1 Effective Conductivities and Resistivities for Experiments Parallel and Perpendicular to an Interface of a Mixed Conductor (Example: Weakly Acceptor Doped CeO_2 and Positively Charged Interface; Here the Parallel Conductivity is Dominated by Electrons (n) and the Perpendicular Resistivity by the Vacancies (v))

Model	Concentration Profile	Effective Parallel Conductivity, $\sigma_m^{\parallel}$ and Perpendicular Resistivity, $\rho_m^{\perp}$
Schottky–Mott	A' e' $V_O^{\bullet\bullet}$ λ^*	$\sigma_{m,n}^{\parallel} \propto \lambda^* \dfrac{c_{n0}}{2\ln(c_{n0}/c_{n\infty})}$ $\rho_{m,v}^{\perp} \propto \lambda^* \dfrac{1}{2c_{v0}\ln(c_{v0}/c_{v\infty})}$
Gouy–Chapman	e' $V_O^{\bullet\bullet}$ λ	$\sigma_{m,n}^{\parallel} \propto (2\lambda)\sqrt{c_{n0}c_{n\infty}} \propto \sqrt{c_{n0}}$ $\rho_{m,v}^{\perp} \propto (2\lambda)\dfrac{1}{\sqrt{c_{v0}c_{v\infty}}} \propto \dfrac{1}{\sqrt{c_{v0}c_{v\infty}}}$
	A' e' $V_O^{\bullet\bullet}$ λ	$\sigma_{m,n}^{\parallel} \propto (2\lambda)\sqrt{c_{n0}c_{n\infty}} \propto \dfrac{1}{\sqrt{c_{v\infty}}}\sqrt{c_{n0}c_{n\infty}}$ $\rho_{m,v}^{\perp} \propto (2\lambda)\dfrac{1}{\sqrt{c_{v0}c_{v\infty}}} \propto \dfrac{1}{\sqrt{c_{v0}c_{v\infty}}}$
Combined	e' A' $V_O^{\bullet\bullet}$ λ	$\sigma_{m,n}^{\parallel} \propto (2\lambda)\sqrt{c_{n0}c_{v\infty}} \propto \sqrt{c_{n0}}$ $\rho_{m,v}^{\perp} \propto (2\lambda)\dfrac{1}{\sqrt{c_{v0}c_{n\infty}}} \propto \dfrac{1}{\sqrt{c_{v0}}}\dfrac{1}{\sqrt{c_{v\infty}c_{n\infty}}}$

$\infty \qquad\qquad 0$

Source: Reprinted from Kim, S. et al., *Phys. Chem. Chem. Phys.*, 5, 2268, 2003. With permission.

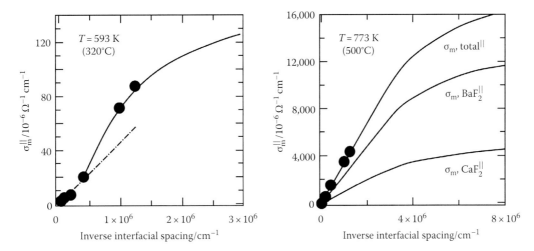

FIGURE 8.16 Parallel conductivity of CaF_2 and BaF_2 multilayers as a function of inverse spacing. The characteristic varies from Mott–Schottky type to Gouy–Chapman type if temperature is increased from 320°C to 500°C. (Reprinted from Guo, X.X. and Maier, J., *Adv. Funct. Mater.*, 19, 96, 2009. With permission.)

impurities. The treatment is even able to explain annealing effects during preparation. Let us focus on the effective parallel conductivity and its dependence on the density of the interfaces ($\propto$(thickness)$^{-1}$). Initially $\sigma_m^{\parallel}$ varies linearly with L^{-1} while the nonlinear increase can be nicely connected with space charge overlap (Figure 8.16). (The upward bending occurs in a Mott–Schottky situation but disappears at high temperature where intrinsic disorder overwhelms and a Gouy–Chapman layer is established.) There are also a variety of studies of oxide heterolayers, the most careful of which were performed by Korte et al. (2008, 2009). They use Y-ZrO_2, which due to its high disorder is not expected to show an accumulation layer that could give rise to great concentration enhancement. Here it is anticipated that mobility effects are observable. Indeed Korte et al. could show that the moderate interfacial effect scales with the degree of misorientation. Tensile strain increases the mobility of ZrO_2, while compressive strain decreases it in agreement with the migration-volume of vacancy transport. The samples investigated are not in the regime of interfacial overlap. Mesoscopic strain effects are involved in the low temperature example of Figure 8.15.

2. Overlap of depletion zones is detected in nanocrystalline ceramics of $SrTiO_3$. The bulk semicircle still seen in Figure 8.12 now has disappeared. More detailed modeling shows that this is not a matter of resolution rather the depletion zone extends throughout. Interestingly, the respective capacitance that was the bias-dependent interfacial capacitance for macrocrystalline $SrTiO_3$ is now bias independent as the surface charge becomes invariant (Balaya et al. 2006).

3. Space charge effects also can be relevant for storage: a composite of α and β can store a compound A^+B^- even if none of the phases α and β can do this individually (Figure 8.13). If α can store A^+ (not B^-) and β can store B^- (not A^+)

a synergetic storage is possible that can be appreciable in amount and speed. In the mesoscopic regime, this "job-sharing" provides the bridge between an electrostatic capacitance and a battery capacitance (Zhukovskii et al. 2006).

A particularly well-suited material in this respect is a nanocomposite of Li_2O and Ru. Owing to selective storage ability of Li^+ in Li_2O and e^- in Ru, nanocomposites can store a lot of Li with a pseudocapacitive charge/discharge behavior. The storage is also quick as Li^+ will take the Li_2O route and the electron metal route. Figure 8.17 gives the chemical potential for such a storage situation, showing that unlike bulk storage, Li storage is in this case not connected with a local variation in μ_{Li} (Maier 2007a).

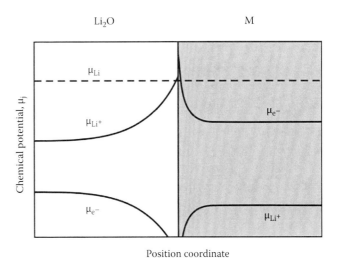

FIGURE 8.17 Excess storage is—as it occurs heterogeneously—not connected with a variation in the chemical potential of lithium (μ_{Li}). (Reprinted from Maier, J., *Faraday Discuss.*, 134, 51, 2007a. With permission.)

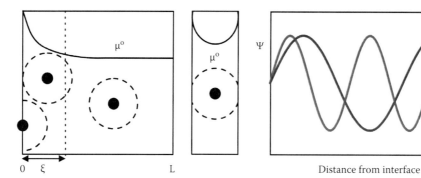

FIGURE 8.18 Confinement of localized point defects and of delocalized electrons (see text). (Reprinted from Maier, J., *Modern Aspects of Electrochemistry*, Conway, B.E. et al. (eds.), Vol. 38, Kluwer Academic/Plenum Publishers, New York, 2005c, 1–173. With permission.)

4. So far, we assumed the energy level (i.e., the local partial free energy) of the point defect to be undisturbed. Clearly, a variation will occur at tiny sizes (Figure 8.18). The analogous effect for electrons is well known for the electrons in the context of the particle-in-the-box-problem.

5. Here we have to essentially deal with classical electrostatic polarization effects that can be of longer range than the dimension of the sheer point defect. If this perturbation zone is of the order of the confinement, we expect a variation of μ° (Maier 2003b, 2004a, 2005a). As electrons and ions perceive confinement differently, the bending need not be parallel any longer.

6. So far we excluded structural variations other than the abrupt junction. If elastic effects lead to structural variations inside the phase under consideration, this is naturally also perceived by the defects to be formed within this ground structure.

7. Another effect that is known from semiconductor physics is that electrons and holes cannot evade each other in sufficiently confined systems. Then, owing to Coulomb attractions, the effective band gap decreases. The same phenomenon is expected for interstitials and vacancies in ion conductors. As such interactions are connected with phase transitions to the superionic state, also transition temperatures may be affected (Maier 2003b, 2004a, 2005a).

8. Furthermore configurational effects are expected at minute size. As this is more important for tiny crystallites, we will come back to this point in the following section where we take explicit account of curvature effects.

8.5 Consequences of Curvature for Nanoionics

Unlike films or macroparticles, nanoparticles are heavily curved (radius: r) and therefore exhibit a perceptibly increased internal (capillary) pressure. Ignoring stress effects, the chemical potential is increased by (Defay et al. 1960)

$$\mu_k(r) = \mu_k(\infty) + 2\omega V_k. \tag{8.6}$$

As now no characteristic layers can be defined like the Debye length for space charges, the curvature influence escapes the distinction between trivial and nontrivial size effects.

If stress effects are negligible and morphological equilibrium is fully established, ω is the Wulff ratio γ_i / r_i (γ_i: surface tension of surface plane with orientation i, r: the length of the normal vector pointing from the centre of the crystal to the surface plane), which is for a Wulff-shaped crystal the same for any orientation i and can also be replaced by area averaged quantities, i.e., by $\bar{\gamma}/\bar{r}$. Owing to the coherency between surface and bulk, stress effects are, however, important. Then the surface stress tensor becomes a pertinent capillary parameter (Rusanov 1995; Kramer and Weissmüller 2007). Moreover, deviations from the Wulff shape necessarily occur and the interpretation of $\bar{\gamma}$ becomes complicated. At any rate, capillary effects lead to drastic consequences such as depression of the melting point of a small nanoparticle. Capillarity also leads to a nonzero open circuit voltage for the symmetrical cell (Schroeder et al. 2004):

$$\text{Ag(nano)|Ag}^+\text{-conductor|Ag(macro)}.$$

A capillarity term also varies the chemical potential of a defect (V_k then being the partial molar volume of the defect), which is expected to lead (i) to varied defect concentrations at the same control parameters, (ii) to a charging at the contact of two differently curved identical particles, in particular if a small nanocrystal is in contact with a film of the same chemistry (Figure 8.19) (Maier 2002, 2003a,b, 2004a).

There is an important difference in the stability of nanoparticles depending on the fact whether elementary normal or binary (ternary etc.) particles are concerned. If M particles that do not undergo normal Ostwald ripening are contacted with an M$^+$ electrolyte, electrochemical Ostwald ripening can take place for kinetic reasons (Schröder et al. 2006). This is different if additional immobile elements constitute the compounds.

Finally it should be mentioned that phase transition characteristics look different on the nanoscale: (i) If in a two-phase regime the overall composition is varied, in macrocrystalline cases, the chemical potential of an exchangeable compound stays constant, as only the relative proportion of the two coexisting phases is

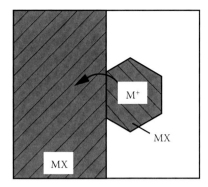

FIGURE 8.19 Capillary effects demand a space charge effect just because of different curvature (Maier 2003b). (Reprinted from Maier, J., *Z. Phys. Chem.*, 219, 35, 2005a. With permission.)

changed but not concentrations therein. This may be different on the nanoscale as sizes may change. (ii) Let us consider a first order transition (at T_c) in a macroscopic system. Below T_c the total solid is in the lower energy state, while above T_c the solid is in the upper level. This is different in molecular systems, in which always a fraction even below T_c may occupy the higher level; this is because the configurational entropy is not negligible compared to the energy difference. Nanocrystals if sufficiently small should behave as intermediates. It has however to be borne in mind that the configurational entropy rapidly decreases with increasing size (Hill 1963; Maier 2009).

How statistics have to be performed over a nanosized ensemble is generally a fascinating but not fully explored subject. All the points addressed become very different if we leave the regime of solid state physics and chemistry where each size requires a thermodynamically different equilibrium structure. Then finally we have entered the regime of cluster physics and chemistry and left our topic "nanoionics."

8.6 Conclusions

Various examples have been considered that showed the significance of small size for ion transport and mass storage properties. A variety of basic scientific problems are connected with this. Though the area of "nanoionics" is just in its infancy, already the recent research did not only reveal conductivity and storage phenomena that are order-of-magnitude effects, but also a variety of implications have been identified that are of direct importance for applications, in particular for Li-battery technology (Maier 2007b).

References

Balaya, P., Jamnik, J., Fleig, J., and Maier, J. 2006. Mesoscopic electrical conduction in nanocrystalline SrTiO₃. *Appl. Phys. Lett.* 88: 062109.

Chiang, Y. M., Lavik, E. B., Kosacki, I., Tuller, H. L., and Ying, J. Y. 1996. Defect and transport properties of nanocrystalline CeO₂₋ₓ. *Appl. Phys. Lett.* 69: 185–187.

De Souza, R. A., Fleig, J., Maier, J., Kienzle, O., Zhang, Z., Sigle, W., and Rühle, R. 2003. Electrical and structural characterization of a low-angle tilt grain boundary in iron-doped strontium titanate. *J. Am. Ceram. Soc.* 86: 922–928.

Defay, R., Prigogine, I., Bellemans, A., and Everett, H. 1960. *Surface Tension and Adsorption.* New York: John Wiley & Sons, Ltd.

Denk, I., Claus, J., and Maier, J. 1997. Electrochemical investigations of SrTiO₃ boundaries. *J. Electrochem. Soc.* 144: 3526–3536.

Guo, X. X. and Maier, J. 2004. Ionic conductivity of epitactic MBE-grown BaF₂ films. *Surf. Sci.* 549: 211–216.

Guo, X. X. and Maier, J. 2009. Comprehensive modeling of ion conduction of nanosized CaF₂/BaF₂ multilayer heterostructures. *Adv. Funct. Mater.* 19: 96–101.

Guo, X. X., Matei, I., Jamnik, J., Lee, J. S., and Maier, J. 2007. Defect chemical modelling of mesoscopic ion conduction in nanosized CaF₂/BaF₂ multilayer heterostructures. *Phys. Rev. B* 76: 125429 (1–7).

Hill, L. 1963. *Thermodynamics of Small Systems.* New York: W. A. Benjamin Inc. Publ.

Jamnik, J., Guo, X., and Maier, J. 2003. Field-induced relaxation of bulk composition due to internal boundaries. *Appl. Phys. Lett.* 82: 2820–2822.

Kim, S. and Maier, J. 2002. On the conductivity mechanism of nanocrystalline ceria. *J. Electrochem. Soc.* 149: J73–J83.

Kim, S., Fleig, J., and Maier, J. 2003. Space charge conduction: Simple analytical solutions for ionic and mixed conductors and application to nanocrystalline ceria. *Phys. Chem. Chem. Phys.* 5: 2268–2273.

Kirchheim, R. 1988. Hydrogen solubility and diffusivity in defective and amorphous metals. *Prog. Mater. Sci.* 32: 261–325.

Korte, C., Peters, A., Janek, J., Hesse, D., and Zakharov, N. 2008. Ionic conductivity and activation energy for oxygen ion transport in superlattices—the semicoherent multilayer system YSZ (ZrO₂ + 9.5 mol% Y₂O₃)/Y₂O₃. *Phys. Chem. Chem. Phys.* 10: 4623–4635.

Korte, C., Schichte, N., Hesse, D., and Janek, J. 2009. Influence of interface structure on mass transport in phase boundaries between different ionic materials. Experimental studies and formal considerations. *Chem. Monthly* 140: 1068–1080.

Kramer, D. and Weissmüller, J. 2007. A note on surface stress and surface tension and their interrelation via Shuttleworth's equation and the Lippmann equation. *Surf. Sci.* 601: 3042–3051.

Maier, J. 1995. Ionic conduction in space charge regions. *Prog. Solid State Chem.* 23: 171–263.

Maier, J. 2002. Thermodynamic aspects and morphology of nanostructured ion conductors. Aspects of nano-ionics. Part I. *Solid State Ionics* 154–155: 291–301.

Maier, J. 2003a. Defect chemistry and ion transport in nanostructured materials. Part II. Aspects of nanoionics. *Solid State Ionics* 157: 327–334.

Maier, J. 2003b. Nano-Ionics: Trivial and non-trivial size effects on ion conduction in solids. *Z. Phys. Chem.* 217: 415–436.

Maier, J. 2004a. Ionic transport in nano-sized systems. *Solid State Ionics* 175: 7–12.

Maier, J. 2004b. *Physical Chemistry of Ionic Materials. Ions and Electrons in Solids.* Chichester, U.K.: John Wiley & Sons, Ltd.

Maier, J. 2005a. Chemical potential of charge carriers in solids. *Z. Phys. Chem.* 219: 35–46.

Maier, J. 2005b. Nanoionics: Ion transport and electrochemical storage in confined systems. *Nat. Mater.* 4: 805–815.

Maier, J. 2005c. Solid state electrochemistry I: Thermodynamics and kinetics of charge carriers in solids. In *Modern Aspects of Electrochemistry*, B. E. Conway, C. G. Vayenas, and R. E. White (Eds.), vol. 38, pp. 1–173. New York: Kluwer Academic/Plenum Publishers.

Maier, J. 2007a. Mass storage in space charge regions of nano-sized systems (Nano-ionics Part V). *Faraday Discuss.* 134: 51–56.

Maier, J. 2007b. Size effects on mass transport and storage in lithium batteries. *J. Power Sources* 174: 569–574.

Maier, J. 2009. Thermodynamics of nanosystems with a special view to charge carriers. *Adv. Mater.* 21: 2571–2585.

Maier, J. and Münch, W. 2000. Thermal destiny of an ionic crystal. *Z. Anorg. Allg. Chem.* 626: 264–269.

Merkle, R. and Maier, J. 2008. How is oxygen incorporated into oxides? A comprehensive kinetic study of a simple solid-state reaction with $SrTiO_3$ as a model material. *Angew. Chem. Int. Ed. Engl.* 47: 3874–3894.

Noll, F., Münch, W., Denk, I., and Maier, J. 1996. $SrTiO_3$ as a prototype of a mixed conductor: Conductivities, oxygen diffusion, and boundary effects. *Solid State Ionics* 86–88: 711–717.

Puin, W., Rodewald, S., Ramlau, R., Heitjans, P., and Maier, J. 2000. Local and overall ionic conductivity in nanocrystalline CaF_2. *Solid State Ionics* 131: 159–164.

Rusanov, A. I. 1995. Interfacial thermodynamics: Development for last decades. *Solid State Ionics* 75: 275–279.

Saito, Y. and Maier, J. 1995. Ionic conductivity enhancement of the fluoride conductor CaF_2 by grain boundary activation using Lewis acids. *J. Electrochem. Soc.* 142: 3078–3083.

Sata, N., Eberman, K., Eberl, K., and Maier, J. 2000. Mesoscopic fast ion conduction in nanometre-scale planar heterostructures. *Nature* 408: 946–949.

Schroeder, A., Fleig, J., Drings, H., Wuerschum, R., Maier, J., and Sitte, W. 2004. Excess free enthalpy of nanocrystalline silver, determined in a solid electrolyte cell. *Solid State Ionics* 173: 95–101.

Schröder, A., Fleig, J., Gryaznov, D., Maier, J., and Sitte, W. 2006. Quantitative model of electrochemical Ostwald ripening and its application to the time-dependent electrode potential of nanocrystalline metals. *J. Phys. Chem. B* 110: 12274–12280.

Tschöpe, A. 2001. Grain size-dependent electrical conductivity of polycrystalline cerium oxide II: Space charge model. *Solid State Ionics* 139: 267–280.

Vollmann, M. and Waser, R. 1994. Grain boundary defect chemistry of acceptor-doped titanates: Space charge layer width. *J. Am. Ceram. Soc.* 77: 235–243.

Vollmann, M., Hagenbeck, R., and Waser, R. 1997. Grain-boundary defect chemistry of acceptor-doped titanates: Inversion layer and low-field conduction. *J. Am. Ceram. Soc.* 80: 2301–2314.

Wagner, C. and Schottky, W. 1930. Theorie der geordneten Mischphasen. *Z. Phys. Chem. B* 11: 163–210.

Zhukovskii, Y. F., Balaya, P., Kotomin, E. A., and Maier, J. 2006. Evidence for interfacial-storage anomaly in nanocomposites for lithium batteries from first-principles simulations. *Phys. Rev. Lett.* 96: 058302 (1–4).

9

Nanoscale Superconductivity

Francois M. Peeters
University of Antwerp

Arkady A. Shanenko
University of Antwerp

Mihail D. Croitoru
University of Antwerp

9.1 Introduction

Recent advances in nanofabrication technology resulted in *high-quality metallic nanoscale structures* like single-crystalline films with thickness down to a few monolayers (Guo et al., 2004; Bao et al., 2005; Zhang et al., 2005; Eom et al., 2006; Özer et al., 2006a,b, 2007) and wires (both single crystalline and made of strongly coupled grains) with width down to 5–10 nm (Savolainen et al., 2004; Tian et al., 2005; Zgirski et al., 2005; Altomare et al., 2006; Janković et al., 2006; Tombros et al., 2008; Wang et al., 2008). In most fabricated samples, the electron mean-free path ℓ was found to scale with the nanowire/nanofilm thickness d (Savolainen et al., 2004; Zgirski et al., 2005; Altomare et al., 2006; Özer et al., 2006b, 2007). In particular, Matthiessen-like approximation $1/\ell = 1/2d + 1/\ell_{imp}$ was used by Özer et al. (2007), where $1/2d$ represents the contribution of the boundaries (d is the sample thickness) and $\ell_{imp} \gg d$ stands for the mean-free path due to impurities. Based on experimental results reported in this chapter, one can find that $\ell_{imp} \sim 200$–400 nm in Pb single-crystalline atomically uniform nanoislands with thickness about 30–40 Å and with area $\sim$400 nm $\times$ 400 nm. In high-quality nanowires, disorder is more significant but due to the scaling $\ell \propto d$, one is also able to expect that the boundary scattering controls ℓ and, hence, $\ell_{imp} \gg d$. So, the present-day technologies make it possible to fabricate superconducting nanostructures with minor disorder, i.e., $\ell_{imp}/d \gg 1$ and $k_F\ell_{imp} \gg 1$, where k_F stands for the three-dimensional (3D) Fermi wavevector. An important consequence is that the discrete transverse electron spectrum is not smeared by impurity scattering in this case, and, as a result, the conduction band splits up into *a series of single-electron subbands*, which move in energy with changing d. As was established

half a century ago (Abrikosov and Gor'kov, 1958a,b; Anderson, 1959), the superconducting equilibrium properties are not very sensitive to minor disorder, when $k_F\ell_{imp} \gg 1$ [for instance, in Pb single-crystalline nanofilms fabricated by Özer et al. (2007), $k_F\ell_{imp} \sim 10^3$]. Thus, to first approximation, such high-quality superconducting nanofilms and nanowires can be treated in the clean limit.

Most of previous studies concerned *high-resistivity* films and wires being strongly disordered or granular. In the presence of strong disorder, the Anderson theorem is violated due to the enhancement of the Coulomb repulsion (Altshuler et al., 1980) and because of mesoscopic fluctuations (Altshuler and Spivak, 1987; Beenakker, 1991; Spivak and Zhou, 1995; Skvortsov and Feigel'man, 2005; about the mesoscopic fluctuations in normal metals, see Altshuler, 1985; Lee and Stone, 1985; Zyuzin and Spivak, 1990). In particular, the Coulomb interaction effects in the presence of the diffusive character of the electron motion lead to the suppression of the critical temperature with increasing disorder (Finkel'stein, 1987, 1994; Oreg and Finkel'stein, 1999). In granular electronic systems, superconductivity is mainly governed by the average tunneling conductance G between neighboring grains (Beloborodov et al., 2007). For the high-resistivity regime, $g = G/(2e^2/\hbar) \ll 1$, the superconducting state is usually suppressed (see, e.g., Beloborodov et al., 2007 and references therein). While the intergrain coupling is large enough ($g \gtrsim 1$), arrays of metallic nanoparticles can turn into the superconducting state, with the properties not very different from bulk.

In contrast, the present review is focused on recently fabricated *low-resistivity* specimens with minimal disorder, e.g., Pb single-crystalline nanofilms mentioned above. Notice that such ultrathin high-quality nanofilms show no significant indications of defect- or phase-driven suppression of superconductivity (see, e.g., Eom et al., 2006; Özer et al., 2006b). In high-quality metallic nanowires, clear signatures of the superconducting state are observed even for thicknesses down to 5–10 nm (Zgirski et al., 2005; Altomare et al., 2006). As is expected (see, e.g., Arutyunov et al., 2008), the quantum phase fluctuations will destroy superconductivity in nanowires with thickness ≲ 10 nm. In agreement with this expectation, the sharp crossover from the superconducting state to the normal one due to quantum fluctuations was observed in aluminum high-quality nanowires when the nanowire thickness reduces down to 8 nm (Zgirski et al., 2005). However, no signature of such a crossover was found in similar aluminum nanowires with cross section 5.2 nm × 6.1 nm and a larger length of about 100 μm, where a nonvanishing resistance due to the phase slippage was less than 1% of the corresponding normal resistance at temperatures $T < 0.9\ T_c$, with T_c the critical temperature.

Increasing T_c, critical current density j_c, and critical magnetic field H_c of a superconductor has been a major challenge since the discovery of superconductivity. On the one hand, we can look for chemically complex materials exhibiting higher critical parameters. Such a search has been very successful over the last 20 years, and new high-temperature superconducting materials have been developed. On the other hand, micro- and nanostructuring of a superconductor is a different and very promising way, which can result in the enhancement of the critical superconducting parameters. In earlier works on microstructuring of superconductors in the mesoscopic regime, enhancement of j_c was found due to trapping of vortices (Harada et al., 1996; Moshchalkov et al., 1998). Similarly, a significant increase of H_c was realized through mesoscopic structuring (Moshchalkov et al., 1995; Deo et al., 1997; Schweigert and Peeters, 1998; Geim et al., 2000), which is a consequence of enhanced surface superconductivity. However, in both situations, T_c at zero magnetic field was unaltered.

Contrary to the mesoscopic regime, T_c of high-quality superconducting nanowires and nanofilms can be enhanced due to quantum-size effects. This expectation dates back to the pioneering paper of Blatt and Thompson (1963), where the energy superconducting gap of a clean ultrathin slab was shown to exhibit thickness-dependent oscillations accompanied by pronounced resonant enhancements. The physics behind such oscillations can be outlined as follows. Due to the transverse quantization of the electron motion, the conduction band in nanowires and nanofilms splits up into a series of subbands [see, for instance, Datta (2005)], and superconductivity is supported by a set of quantum channels. Such single-electron subbands move in energy when changing the nanowire/nanofilm thickness. When the bottom of a subband passes through the Fermi surface, we have an abrupt increase in the density of single-electron states at the Fermi level and, in turn, an enhancement of the superconducting order parameter and other basic superconducting characteristics [i.e., *the superconducting resonance*, see Blatt and Thompson (1963)]. This results in *quantum-size oscillations* of superconducting properties with thickness. In the subsequent

40 years after the paper by Blatt and Thompson, very few experimental papers reported the observation of possible signatures of such a behavior (Komnik et al., 1970; Orr et al., 1984; Pfennigstorf et al., 2002), but structural defects were a serious obstacle preventing a definite conclusion on the existence of such quantum-size oscillations. For decades, atomic nuclei were the only system where the interplay of quantum confinement with pairing of fermions was studied experimentally and where the predictions of Blatt and Thompson were confirmed for nucleons [see, e.g., Hilaire et al. (2002) and references therein]. Modern developments in nanofabrication have finally resulted in high-quality nanostructures, and recently, quantum-size oscillations of the critical superconducting temperature and critical magnetic field were observed in single-crystalline Pb nanofilms (Guo et al., 2004; Bao et al., 2005; Zhang et al., 2005; Eom et al., 2006) at a high level of experimental precision and sophistication.

Below, we present an overview of recent experimental and theoretical results on superconducting nanofilms and nanowires in the quantum-size regime. Due to transverse quantization of the electron spectrum, the translational invariance is broken, and the superconducting order parameters appears to be position dependent, i.e., $\Delta = \Delta(\mathbf{r})$. In such a situation, one cannot use the Bardeen–Cooper–Schrieffer (BCS) ansatz [see, e.g., de Gennes (1966)] for the bulk many-body ground-state wave function. Moreover, the Ginzburg–Landau (GL) theory convenient and useful for mesoscopic superconductors cannot be applied anymore. From the classical Gor'kov study (Gor'kov, 1958; Abrikosov et al., 1965; de Gennes, 1966; Fetter and Walecka, 2003), it is known that the GL equations follow from the integral representation of the Gor'kov equations under the assumption of small Δ. Such an assumption results in a nonlinear integral equation for the order parameter. The simpler differential structure of the GL equations requires the additional and separate assumption: the order parameter and vector potential vary slowly with respect to the relevant kernels of the above integral equation. This criterion can hardly be satisfied for superconductors with nanoscale dimensions. In such a situation, a description based on details of the confined single-electron wave functions is required. This can be incorporated within the Bogoliubov–de Gennes (BdG) equations (Bogoliubov, 1958, 1959; de Gennes, 1966). Therefore, the theoretical study in this chapter is based on a numerical solution of the BdG equations for superconducting nanowires and nanofilms in the clean limit and with quantum confinement for the transverse electron motion.

9.2 Theoretical Formalism

In this section, the BdG formalism is outlined together with features important for the nanoscale regime. Generally, the BdG equations for nanoscale superconductors cannot be solved analytically. The use of a numerical solution requires its correct interpretation. This is why Anderson's prescription (Anderson, 1959) for an approximate semi-analytical solution to the BdG equations is also described below.

9.2.1 Bogoliubov–de Gennes Equations for Nanoscale Superconductors

9.2.1.1 General Formulation and Quantum Confinement

The BdG equations are a powerful theoretical framework for a microscopic treatment of a situation with position-dependent superconducting order parameter $\Delta(\mathbf{r})$ (Bogoliubov, 1958, 1959; de Gennes, 1966). These equations can be written as follows:

$$E_v \left| u_v \right\rangle = \hat{H}_e \left| u_v \right\rangle + \hat{\Delta} \left| v_v \right\rangle, \tag{9.1a}$$

$$E_v \left| v_v \right\rangle = \hat{\Delta}^\star \left| u_v \right\rangle - \hat{H}_e^\star \left| v_v \right\rangle, \tag{9.1b}$$

where

$\left| u_v \right\rangle$ and $\left| v_v \right\rangle$ are the particle-like and hole-like ket vectors
E_v is the quasiparticle energy
$\hat{H}_e$ stands for the single-electron Hamiltonian
$\hat{\Delta} = \Delta(\hat{\mathbf{r}})$ with $\hat{\mathbf{r}}$ the electron position operator

In the clean limit, $\hat{H}_e$ (measured from the Fermi level E_F) is written as

$$\hat{H}_e = \frac{\hat{\mathbf{P}}^2}{2m_e} + V_{\mathrm{conf}}(\hat{\mathbf{r}}) - E_F, \tag{9.2}$$

where

m_e is the band mass (set to the free electron mass)
$V_{\mathrm{conf}}(\mathbf{r})$ the confining potential
$\hat{\mathbf{P}}$ in Equation 9.2 is given by

$$\hat{\mathbf{P}} = \hat{\mathbf{p}} + m_e \hat{\mathbf{v}}_s, \tag{9.3}$$

where

$\hat{\mathbf{p}}$ is the momentum operator
$\hat{\mathbf{v}}_s = \mathbf{v}_s(\hat{\mathbf{r}})$ with $\mathbf{v}_s$ the superfluid velocity

It is taken in the gauge invariant form, absorbing the vector potential $\mathbf{A}$ (Swidzinsky, 1982; Zagoskin, 1998), i.e.,

$$\mathbf{v}_s = \nabla \chi - \frac{e}{m_e c} \mathbf{A}, \tag{9.4}$$

with χ the phase of the order parameter. When $\mathbf{v}_s$ is position independent, use of Equation 9.2 means working in the framework co-moving with the pair condensate. This results from the Galileo transformation

$$a_{\mathbf{k},\sigma} \to a_{\mathbf{k}-\mathbf{q},\sigma}, \quad a_{\mathbf{k},\sigma}^\dagger \to a_{\mathbf{k}-\mathbf{q},\sigma}^\dagger, \tag{9.5}$$

with

$a_{\mathbf{k},\sigma}(a_{\mathbf{k},\sigma}^\dagger)$ the annihilation (creation) electron operators
$\mathbf{q} = (m_e/\hbar)\mathbf{v}_s$

However, in general, the superfluid velocity is spatially dependent, and Equation 9.2 cannot be interpreted in terms of such a Galileo transformation. In such a situation, representing $\hat{H}_e$ in the form of Equation 9.2 means that the order parameter is taken as a real quantity (Swidzinsky, 1982). Strictly speaking, in this case $\Delta(\mathbf{r})$ stands for the absolute value of the order parameter in the BdG equations (Swidzinsky, 1982; Zagoskin, 1998), and $\hat{H}_e$ includes the order-parameter phase χ and, so, only "plays the role" of the single-electron Hamiltonian (this is why two different single-electron energies, ξ_v and $\tilde{\xi}_v$, are introduced below). Asterisk in Equations 9.1a and b means the complex conjugate

$$\hat{H}_e^\star = \frac{(\hat{\mathbf{P}}^\star)^2}{2m_e} + V_{\mathrm{conf}}(\hat{\mathbf{r}}) - E_F, \tag{9.6}$$

with

$$\hat{\mathbf{P}}^\star = -\hat{\mathbf{p}} + m_e \hat{\mathbf{v}}_s. \tag{9.7}$$

In most cases, the superfluid velocity is small enough such that we can work within *the linear approximation* in $\mathbf{v}_s$ given by

$$\hat{H}_e \approx \hat{H}_\mathbf{p} + \frac{\hat{\mathbf{p}}\hat{\mathbf{v}}_s + \hat{\mathbf{v}}_s\hat{\mathbf{p}}}{2}, \quad \hat{H}_e^\star \approx \hat{H}_\mathbf{p} - \frac{\hat{\mathbf{p}}\hat{\mathbf{v}}_s + \hat{\mathbf{v}}_s\hat{\mathbf{p}}}{2} \left(\text{with } \hat{H}_\mathbf{p} = \frac{\hat{\mathbf{p}}^2}{2m_e} - E_F \right). \tag{9.8}$$

We remark that the normal mean-field interaction (the Hartree–Fock term) should be, in general, included in $\hat{H}_e$ (de Gennes, 1966). However, it has little effect on the results and can be ignored.

Due to the grand-canonical formulation of the BdG equations, E_F (the same as the chemical potential for films and wires) appears in the relevant expressions. Given the mean electron density n_e, the Fermi level can be calculated from

$$n_e = \frac{2}{V} \sum_v \left[f_v \left\langle u_v \mid u_v \right\rangle + (1 - f_v)\left\langle v_v \mid v_v \right\rangle \right], \tag{9.9}$$

where V is the system volume. For the usual elemental superconductors, e.g., aluminum, tin, and lead, the superconducting energy gap (the spatially averaged order parameter) is typically much smaller than E_F. Hence, changes in the chemical potential due to the superconducting order can be ignored (Fetter and Walecka, 2003) when passing from the normal state to the superconducting one. This is why it is easier to solve Equation 9.9 for $\Delta(\mathbf{r}) = 0$, before the main procedure of numerically investigating the superconducting solution of the BdG equations.

As seen from Equation 9.2, the nanoscale BdG equations differ from those for bulk and mesoscopic samples due to the presence of the confining interaction. Here, it is assumed that $V_{\mathrm{conf}}(\mathbf{r})$ is zero inside the superconducting sample and infinite outside. This means that *the quantum-confinement boundary conditions*

$$\langle \mathbf{r} | u_\nu \rangle \big|_{\mathbf{r} \in S} = \langle \mathbf{r} | v_\nu \rangle \big|_{\mathbf{r} \in S} = 0, \qquad (9.10)$$

are imposed at the sample surface, i.e., $\mathbf{r} \in S$. For nanofilms and nanowires (we will focus on these nanostructured superconductors in the present review), the quantum-confinement boundary conditions should be applied for the transverse electron motion. For the direction parallel to the nanofilm/nanowire, periodic boundary conditions can be used with a unit cell of length L which can be considered to be typically several microns. It is interesting to get a feeling for what thickness the transverse quantization of the electron motion can significantly influence the superconducting solution. As already discussed in Section 9.1, in high-quality superconducting nanofilms/nanowires the conduction band splits up into a series of single-electron subbands with *the subband energy-spacing parameter*

$$\delta_{\text{sub}} = \frac{\hbar^2}{2m_e} \frac{\pi^2}{d^2}, \qquad (9.11)$$

where d is the nanofilm/nanowire thickness. In fact, this is the lower bound for the energy spacing between neighboring single-electron subbands. One can expect that effects of the transverse quantization have a serious impact on the superconducting characteristics when

$$\delta_{\text{sub}} \gtrsim \Delta_{\text{bulk}}, \qquad (9.12)$$

where Δ_{bulk} is the bulk order parameter (at zero temperature). Typically, $\Delta_{\text{bulk}} \sim 0.3$ to $1.3\,\text{meV}$ for elemental bulk superconductors (de Gennes, 1966; Fetter and Walecka, 2003). Hence, one can expect that the transverse quantization in high-quality metallic superconducting nanofilms/nanowires is of importance for

$$d \lesssim \sqrt{\frac{\hbar^2}{2m_e} \frac{\pi^2}{\Delta_{\text{bulk}}}} = 20-40\,\text{nm}. \qquad (9.13)$$

As is mentioned in the Introduction, superconducting nanofilms and nanowires with thickness down to several nanometers are the subject of intensive experimental study for the last decade. Thus, effects of the transverse superconducting modes in nanofilms/nanowires are an interesting and important issue.

9.2.1.2 Ultraviolet Divergence, Cutoff, and Physical Solution

As a mean-field approach, the BdG equations should be solved in a self-consistent manner, together with the relation

$$\Delta(\mathbf{r}) = g \sum_\nu \langle \mathbf{r} | u_\nu \rangle \langle v_\nu | \mathbf{r} \rangle \big[1 - 2 f_\nu \big], \qquad (9.14)$$

where $f_\nu = 1/(e^{\beta E_\nu} + 1)$ is the Fermi function [$\beta = 1/(k_B T)$ with k_B the Boltzmann constant]. When incorporating this self-consistency relation, we face a well-known problem:

one should keep in mind that the sum in Equation 9.14 diverges at large energies. This is a consequence of the fact that the electron–electron effective attraction mediated by phonons is approximated in Equations 9.1a and b by (Abrikosov et al., 1965; de Gennes, 1966; Abrikosov, 1988; Fetter and Walecka, 2003)

$$\Phi(\mathbf{r} - \mathbf{r}') = -g\,\delta(\mathbf{r} - \mathbf{r}'), \qquad (9.15)$$

where

$g > 0$ the effective coupling
$\delta(\mathbf{x})$ the Dirac 3D delta function

Equation 9.15 is actually a low-momentum approximation [see, e.g., Abrikosov (1988)] and, so, is a kind of pseudopotential. The cost for the use of this pseudopotential is *an ultraviolet divergence* [see, e.g., Fetter and Walecka (2003)]. For the case of our interest, such a divergence appears in Equation 9.14 (the sum in Equation 9.9 is convergent). The usual treatment of this problem for phonon-mediated superconductivity is to use *the energy cutoff*

$$|\xi_\nu| < \hbar\omega_D \quad \left(\xi_\nu = \langle u_\nu | \hat{H}_{\mathbf{p}} | u_\nu \rangle + \langle v_\nu | \hat{H}_{\mathbf{p}} | v_\nu \rangle \right), \qquad (9.16)$$

where

ξ_ν is the single-electron energy associated with $\hat{H}_{\mathbf{p}}$ (see Equation 9.8)
ω_D is the Debye frequency

Equation 9.16 introduces a cutoff in the momentum $\mathbf{p}$, which mimics a momentum dependence of the Fourier transform of the effective electron–electron interaction and, thus, remedies the low-momentum approximation of Equation 9.15. Notice that Equation 9.16 is the ordinary version of the cutoff [see, for example, derivation and discussion of the local approximation to the Gor'kov equations in Swidzinsky (1982) and Zagoskin (1998)]. It is convenient and elegant when working with analytical expressions but requires some additional calculations in the case of numerical investigations. This is why its simplified version is often used as well, namely, $E_\nu < E_{\text{lim}} \sim \hbar\omega_D$ (Tanaka et al., 2002; Grigorenko et al., 2008). This simplification is time-saving for numerical analysis but is not good in the presence of quantum-size oscillations because in such a situation E_{lim} should be size dependent. The question arises about possible corrections when going beyond Equation 9.15. In bulk, these corrections are known to be insignificant (de Gennes, 1966). In the nanoscale regime, this is still an open question. However, one can expect that the confining interaction is the major mechanism influencing the superconducting properties in high-quality nanostructures rather than some features of the electron–electron attraction not captured by Equation 9.15.

Another important limitation concerns the choice of the physical solution. The BdG equations are structured in such a way that for any relevant set of quantum numbers there exist two possible solutions: one is positive while another is negative (de Gennes, 1966; Swidzinsky, 1982). The well-known prescription used in the absence of a superfluid motion is to keep only positive bogolon energies, i.e., $E_v > 0$. Indeed, the bogolon vacuum is the ground state in this case, and the excitations above the ground state should be, of course, positive. However, in the presence of a strong magnetic field or a sufficiently large supercurrent, the bogolon vacuum can lose the status of ground state, which is a manifestation of the depairing reconstruction. To take into consideration such a feature, one needs to account for quasiparticles with negative energies. As a result, *the criterion for the physical solution* is of the following form:

$$\lim_{\mathbf{v}_s \to 0} E_v > 0. \qquad (9.17)$$

This means that the quasiparticle states should be selected whose energy becomes positive when a superfluid motion is, say, "switched" off. The criterion given by Equation 9.17 allows one to obtain the same results for both the BdG and Gor'kov equations. The usual expression for the ground-state energy in the Bogoliubov theory (neglecting the term $E_F N_e$ with $N_e = n_e V$ the total number of electrons) is given by (Swidzinsky, 1982; Zagoskin, 1998)

$$E = \int d^3 r \frac{|\Delta(\mathbf{r})|^2}{g} - 2 \sum_v E_v \langle v_v | v_v \rangle. \qquad (9.18)$$

When bogolons with negative energies appear, they survive at zero temperature and, so, give rise to corrections to the ground-state energy, i.e., $E \to E + \Delta E$ with [$\theta()$ is the Heaviside step function]

$$\Delta E = 2 \sum_v \theta(-E_v) E_v, \qquad (9.19)$$

a feature typical of *gapless superconductivity* (de Gennes, 1966; Swidzinsky, 1982; Zagoskin, 1998).

It is of importance to discuss the selection given by Equation 9.17 in the context of *Landau's criterion for superfluidity*. According to this criterion (Landau and Lifshitz, 1958; Pitaevskii, 2004), a superfluid can sustain nondissipative flow when quasiparticles cannot be created spontaneously. So, quasiparticles with negative energies seem to be ruled out in the superfluid state. However, Landau's criterion was originally formulated for the Bose liquid, where quasiparticles are bosons. For the Fermi case, quasiparticles are fermions, and Landau's criterion controls the gap-to-gapless transition (Zagoskin, 1998). The point is that when bogolons with negative energies appear, a superconducting solution can still survive due to the Pauli exclusion principle. A domain of the phase space with negative bogolon energies is bound and, due to the Pauli principle, can contain only a limited number of quasiparticles (contrary to the Bose case). This prevents the avalanche of converting the superfluid/superconducting electrons into the normal component.

Thus, the appearance of negative bogolon energies does not necessarily result in the immediate total decomposition of the Cooper pairs.

9.2.1.3 Superfluid Motion and Breakdown of Time-Reversal Symmetry

It is convenient, for future purposes, to introduce *the time-reversal operator K* given (in the absence of the spin indices) by the definition (for an arbitrary ket-vector $|\alpha\rangle$)

$$\hat{K}|\alpha\rangle = |\bar{\alpha}\rangle, \quad \text{with } \langle \alpha | \mathbf{r} \rangle = \langle \mathbf{r} | \bar{\alpha} \rangle, \qquad (9.20)$$

where $|\alpha\rangle$ and $|\bar{\alpha}\rangle$ are called *the time-reversed states*. Notice that $\hat{H}_e = \hat{H}_e^*$ means that the single-electron Hamiltonian is invariant under time reversal. However, in the presence of the superfluid motion, $\hat{H}_e \neq \hat{H}_e^*$, as seen from Equation 9.2, and we have *the breakdown of time-reversal symmetry*. In general, the situation with broken time reversal is more difficult to solve numerically, but on the other hand it is more interesting and promising in the context of possible applications. Notice that K defined by Equation 9.20 does not change the sign of the magnetic field. This convention is quite typical for books and articles on superconductivity due to the structure of the BdG equations (see, for instance, Chapter 8 in de Gennes, 1966). In addition, in the presence of spin indices, one needs to generalize the definition of K in such a way that a change of the sign of the spin projection is included.

A necessary supplementary to the BdG equations is the relation for the superfluid velocity (Swidzinsky, 1982):

$$\nabla \times \mathbf{v}_s = -\frac{e}{m_e c} \mathbf{H}, \qquad (9.21)$$

where

    c is the speed of light
    $\mathbf{H}$ is the magnetic field

Below, two particular situations with $\mathbf{v}_s \neq 0$ are considered in more detail: a cylindrical superconducting nanowire in a parallel magnetic field and with a longitudinal supercurrent. For the cylindrical geometry $\Delta(\mathbf{r}) = \Delta(\rho)$ (ρ, φ, and z are the cylindrical coordinates) and

$$\langle \mathbf{r} | u_{jmk} \rangle = u_{jmk}(\rho) \frac{e^{im\varphi}}{\sqrt{2\pi}} \frac{e^{ikz}}{\sqrt{L}}, \quad \langle \mathbf{r} | v_{jmk} \rangle = v_{jmk}(\rho) \frac{e^{im\varphi}}{\sqrt{2\pi}} \frac{e^{ikz}}{\sqrt{L}}, \qquad (9.22)$$

where

    $v = \{j, m, k\}$, with j the radial quantum number
    m is the azimuthal quantum number
    k is the wave vector of the quasi-free electron motion parallel to the nanowire

Due to Equation 9.10, one should set $u_{jmk}(\rho)|_{\rho=R} = v_{jmk}(\rho)|_{\rho=R} = 0$, with $R = d/2$ the nanocylinder radius.

1. A nanocylinder in a parallel external magnetic field: Here $\mathbf{v}_s = -e/m_e c\mathbf{A}$ and $\mathbf{H} = \nabla \times \mathbf{A}$. Because the nanowire diameter is typically much smaller than the magnetic penetration depth, i.e., $d \ll \lambda$, it is justified to neglect any diamagnetic screening of the external field. Hence, $\mathbf{H} = -H_{\|}\mathbf{e}_z$, where $\mathbf{e}_z$ is the unity vector in the z direction, and the vector potential can be chosen in the Coulomb gauge:

$$\mathbf{A} = -\frac{1}{2}H_{\|}\rho\,\mathbf{e}_\varphi, \qquad (9.23)$$

where $\mathbf{e}_\varphi$ is the azimuthal unit vector.

2. A nanocylinder with a longitudinal supercurrent: In this case, $\mathbf{v}_s = v_s\mathbf{e}_z$ ($\mathbf{q} = q\mathbf{e}_z$). The main problem here is to calculate how the supercurrent depends on the superfluid velocity v_s. The supercurrent density is written as

$$\mathbf{j}_s = \frac{2e}{m_e\pi R^2 L}\sum_{jmk}\left[f_{jmk}\left\langle u_{jmk}\left|\hat{\mathbf{P}}\right|u_{jmk}\right\rangle + \left(1 - f_{jmk}\right)\left\langle v_{jmk}\left|\hat{\mathbf{P}}^\star\right|v_{jmk}\right\rangle\right]. \qquad (9.24)$$

The sum in Equation 9.24 is over the physical states [over the branch (v, +)] and convergent. Using Equation 9.22 and the cylindrical-coordinate representation of the nabla operator ($\mathbf{e}_\rho$ is the unit vector in the transverse direction),

$$\nabla = \mathbf{e}_\rho\partial_\rho + \frac{\mathbf{e}_\varphi}{\rho}\partial_\varphi + \mathbf{e}_z\partial_z, \qquad (9.25)$$

we can rewrite Equation 9.24 as [$\mathbf{j}_s = (0, 0, -j_s)$]

$$j_s = \frac{|e|\hbar}{m_e\pi^2 R^2}\sum_{jm}\int_{-\infty}^{+\infty}\mathrm{d}k\left[\left(k+q\right)\mathcal{U}_{jmk}^2 f_{jmk} - \left(k-q\right)\mathcal{V}_{jmk}^2\left(1 - f_{jmk}\right)\right], \qquad (9.26)$$

with $\mathcal{U}_{jmk}$ and $\mathcal{V}_{jmk}$ given by

$$\mathcal{U}_{jmk}^2 = \left\langle u_{jmk} \mid u_{jmk}\right\rangle = \int_0^R \mathrm{d}\rho\,\rho\,u_{jmk}^2(\rho), \qquad (9.27a)$$

$$\mathcal{V}_{jmk}^2 = \left\langle v_{jmk} \mid v_{jmk}\right\rangle = \int_0^R \mathrm{d}\rho\,\rho\,v_{jmk}^2(\rho). \qquad (9.27b)$$

We remark that $\mathcal{U}_{jmk}$ and $\mathcal{V}_{jmk}$ are taken to be real and obey the usual constraint $\mathcal{U}_{jmk}^2 + \mathcal{V}_{jmk}^2 = 1$ [see, e.g., de Gennes (1966) and Swidzinsky (1982)]. In general, in the presence of a longitudinal supercurrent, $f_{jmk} \neq f_{j,m,-k}$, and, therefore, terms proportional to k should be kept in Equation 9.26.

There is one more helpful representation for the supercurrent density, namely,

$$j_s = |e|n_e v_s - j_n = |e|n_s v_s, \qquad (9.28)$$

where n_s is the density of the superfluid component, and with j_n the contribution of the normal component given by

$$j_n = \frac{e\hbar}{m_e\pi^2 R^2}\sum_{jm}\int_{-\infty}^{+\infty}\mathrm{d}k\,k\left[f_{jmk} - \mathcal{V}_{jmk}^2\right]. \qquad (9.29)$$

According to Equation 9.28, the supercurrent is the current of all electrons moving with the superfluid velocity v_s minus the term due to the normal component (bogolons). The minus sign appears because a normal current is dissipative in equilibrium, and, so, the normal component cannot participate in the superfluid motion. Since $\mathcal{U}_{jmk}$ and $\mathcal{V}_{jmk}$ do not depend on the sign of k (see, for more details, the Section 9.2.2 about Anderson's recipe), Equation 9.29 can be simplified to

$$j_n = \frac{e\hbar}{m_e\pi^2 R^2}\sum_{jm}\int_{-\infty}^{+\infty}\mathrm{d}k\,k f_{jmk}, \qquad (9.30)$$

and, thus

$$n_s = n_e + \frac{\hbar}{m_e\pi^2 R^2 v_s}\sum_{jm}\int_{-\infty}^{+\infty}\mathrm{d}k\,k\,f_{jmk}, \qquad (9.31)$$

with the integral in the right-hand side being negative for $v_s \neq 0$ ($q \neq 0$).

9.2.1.4 Numerical Procedure

Any procedure to solve the BdG equations (Equations 9.1a and b) numerically is based on converting these equations into a matrix form. Then, the numerical problem can be solved by diagonalizing the relevant matrix and invoking iterations in order to account for the self-consistency relation given by Equation 9.14. For instance, one can invoke the finite-difference scheme, in order to construct the matrix equation [see Grigorenko et al. (2008)]. Such a scheme (together with finite elements) is good for a rather complex confining geometry. For nanofilms and nanowires with a simple cross-sectional geometry, i.e., a nanocylinder or a nanorod with a rectangular cross section, it is more convenient to use an expansion in terms of the eigenfunctions of $\hat{\mathbf{p}}^2/2m_e$ which are known analytically. Below, we focus on numerical results for such idealized nanostructured superconductors without any fluctuations in thickness. However, possible effects of such fluctuations are discussed.

Our numerical solution of the BdG equations involves the expansion in terms of the single-electron wave functions associated with $\hat{\mathbf{p}}^2/2m_e$. The numerical procedure can be outlined as follows. First, a bulk value of the order parameter is used as the initial guess, and the BdG equations are solved by substituting this value for $\Delta(\mathbf{r})$ in Equations 9.1a and b. Second, the obtained particle- and hole-like wave functions, i.e., $\langle\mathbf{r}|u_v\rangle$ and $\langle\mathbf{r}|v_v\rangle$ taken together with the corresponding bogolon energy E_v, are inserted into Equation 9.14, in order to obtain a new approximate value for the order parameter. Third, the BdG equations are solved with $\Delta(\mathbf{r})$ found at the previous step. The second and third steps are repeated to satisfy an adopted accuracy. When increasing the thickness of the sample, the order parameter found for a smaller thickness can be used as an initial guess. The same is related to changing other relevant parameters such as a magnetic field or a supercurrent. This decreases the required computer time and allows us to track the superconducting state in the metastable region with possible hysteretic behavior. When far from the superconductor-to-normal transition, our algorithm is quite stable and rapidly convergent with no significant dependence on the initial guess. In the critical region, the algorithm becomes very sensitive to the initial guess and to the numerical viscosity. Here, the number of required iterations increases rapidly before convergence is reached.

When expanding $\langle \mathbf{r} \,|\, u_v \rangle$ and $\langle \mathbf{r} \,|\, v_v \rangle$ in the eigenfunctions of $\hat{\mathbf{p}}^2 / 2m_e$, their choice and number are of crucial importance. The criterion given by Equation 9.16 dictates that one should take at least all those eigenfunctions whose eigenvalues are located in the Debye window. Numerical analysis of the BdG equations for nanofilms/nanowires shows that such a choice leads, as a rule, to errors less than a few percent.

9.2.2 Anderson's Solution

After numerically solving the BdG equations, solid arguments are needed to interpret numerical results. In the case of interest, such arguments can be borrowed from *Anderson's semi-analytical approximate solution* (Anderson, 1959). An important point to be mentioned is that the results of Anderson's prescription are sensitive to whether or not time-reversal symmetry is broken.

9.2.2.1 Time-Reversal Symmetry

Anderson's superconducting solution is usually introduced on the basis of a reduced BCS-like Hamiltonian constructed with the help of the eigenvalues and eigenfunctions of $\hat{H}_e$ [see, for instance, Tanaka and Marsiglio (2000)]. Here, we prefer to proceed in another manner that makes it possible to check a useful link between the BdG equations and Anderson's prescription. First we show how to obtain Anderson's solution from the BdG equations. Then, we discuss the link to the Hamiltonian approach.

In the framework of the BdG equations Anderson's prescription means that $|u_v\rangle$ and $|v_v\rangle$ are approximated as

$$|u_v\rangle = \mathcal{U}_v |v\rangle, \quad |v_v\rangle = \mathcal{V}_v |v\rangle, \tag{9.32}$$

with $|v\rangle$ given by

$$\hat{H}_e |v\rangle = \tilde{\xi}_v |v\rangle. \tag{9.33}$$

Notice that in general, $\tilde{\xi}_v$, the single-electron energy associated with $\hat{H}_e$, should be distinguished from ξ_v, the single-electron energy associated with $\hat{H}_p$. As seen, Equation 9.32 suggests

seeking for the minimum of the relevant thermodynamic potential in the subspace where the particle- and hole-like ket vectors are proportional to $|v\rangle$.

In the presence of time-reversal symmetry (no superfluid motion, i.e., $\mathbf{v}_s = 0$), $\hat{H}_e = \hat{H}_e^* = \hat{H}_p$ and, so, $\tilde{\xi}_v = \xi_v$. It is convenient to introduce the set of transverse quantum numbers i in such a way that $v = \{i, k\}$, where k represents the components of the wavevector controlling the quasi-free motion of the electrons parallel to the nanofilm/nanowire. Hence, due to the transverse quantization, the conduction band splits up into a series of single-electron subbands controlled by the set i. Below, the x direction is taken to be the transverse direction for nanofilms, and, so, $k = \{k_y, k_z\}$ in this case. For nanowires, the z direction is longitudinal, i.e., $k = k_z$. In particular, for nanofilms,

$$\langle \mathbf{r} \,|\, v \rangle = \sqrt{\frac{2}{d}} \sin\left[\frac{\pi(i+1)z}{d} \right] \frac{e^{ik_y y}}{\sqrt{L}} \frac{e^{ik_z z}}{\sqrt{L}}, \tag{9.34}$$

with $v = \{i, k_y, k_z\}$ and

$$\xi_v = \frac{\hbar^2}{2m_e} \left(\frac{\pi^2 (i+1)^2}{d^2} + k_y^2 + k_z^2 \right) - E_F \quad (i = 0, 1, 2, \ldots), \tag{9.35}$$

which represents a sequence of two-dimensional (2D) parabolic subbands. For a nanowire with square cross section,

$$\langle \mathbf{r} \,|\, v \rangle = \sqrt{\frac{2}{d}} \sin\left[\frac{\pi(i_x+1)x}{d} \right] \sqrt{\frac{2}{d}} \sin\left[\frac{\pi(i_y+1)y}{d} \right] \frac{e^{ikz}}{\sqrt{L}}, \tag{9.36}$$

with $v = \{i_x, i_y, k\}$ and

$$\xi_v = \frac{\hbar^2}{2m_e} \left(\frac{\pi^2 (i_x+1)^2}{d^2} + \frac{\pi^2 (i_y+1)^2}{d^2} + k^2 \right) - E_F \quad (i_x, i_y = 0, 1, 2, \ldots) \tag{9.37}$$

Here we have 1D parabolic subbands, and some of them are degenerate. In turn, for a cylindrical nanowire, the transverse quantization dictates that

$$\langle \mathbf{r} \,|\, v \rangle = \vartheta_{jm}(\rho) \frac{e^{im\varphi}}{\sqrt{2\pi}} \frac{e^{ikz}}{\sqrt{L}} \quad (j = 0, 1, 2, \ldots \text{ and } m = 0, \pm 1, \pm 2, \ldots), \tag{9.38}$$

with $v = \{j, m, k\}$, and

$$\vartheta_{jm}(\rho) = \frac{\sqrt{2}}{R \mathcal{J}_{m+1}(\alpha_{jm})} \mathcal{J}_m \left(\alpha_{jm} \frac{\rho}{R} \right), \tag{9.39}$$

where

$\mathcal{J}_m(x)$ is the Bessel function of the first kind of the mth order

α_{jm} is the jth zero of this function

The corresponding single-electron spectrum is given by

$$\xi_v = \frac{\hbar^2}{2m_e}\left(\frac{\alpha_{jm}^2}{R^2} + k^2\right) - E_F. \tag{9.40}$$

Notice that $\mathcal{U}_v$ and $\mathcal{V}_v$ in Equation 9.32 can generally be taken real, and the Bogoliubov–Valatin canonical transformation (de Gennes, 1966; Fetter and Walecka, 2003) requires

$$\mathcal{U}_v^2 + \mathcal{V}_v^2 = 1. \tag{9.41}$$

We also remark that $\mathcal{U}_{jmk}$ and $\mathcal{V}_{jmk}$ introduced above in Section 9.2.1.3 are the same as in Equation 9.32.

When inserting Equation 9.32 into Equations 9.1a and b, the BdG equations can be recast into

$$E_v\,\mathcal{U}_v|v\rangle = \xi_v\,\mathcal{U}_v|v\rangle + \mathcal{V}_v\hat{\Delta}|v\rangle, \tag{9.42a}$$

$$E_v\,\mathcal{V}_v|v\rangle = \mathcal{U}_v\hat{\Delta}^*|v\rangle - \xi_v\,\mathcal{V}_v|v\rangle. \tag{9.42b}$$

Multiplying these equations on the left by $\langle v|$, one gets

$$E_v\mathcal{U}_v = \xi_v\mathcal{U}_v + \mathcal{V}_v\Delta_v, \tag{9.43a}$$

$$E_v\mathcal{V}_v = \mathcal{U}_v\Delta_v - \xi_v\mathcal{V}_v, \tag{9.43b}$$

with

$$\Delta_v = \left\langle v\left|\hat{\Delta}\right|v\right\rangle = \left\langle v\left|\hat{\Delta}^*\right|v\right\rangle. \tag{9.44}$$

Equations 9.43a and b have a nontrivial solution only when the relevant determinant is zero, i.e.,

$$\begin{vmatrix} E_v - \xi_v & -\Delta_v \\ -\Delta_v & E_v - \xi_v \end{vmatrix} = 0,$$

which yields

$$E_v = \pm\sqrt{\xi_v^2 + \Delta_v^2}. \tag{9.45}$$

According to Equation 9.17, the + sign should be taken for the physical solution in Equation 9.45. Now, Equations 9.43a and b taken together with Equation 9.41 yield (for the physical solution)

$$\mathcal{V}_v^2 = \frac{1}{2}\left(1 - \frac{\xi_v}{\sqrt{\xi_v^2 + \Delta_v^2}}\right), \tag{9.46a}$$

$$\mathcal{U}_v\mathcal{V}_v = \frac{\Delta_v}{2\sqrt{\xi_v^2 + \Delta_v^2}}. \tag{9.46b}$$

Based on Equations 9.32 and 9.46b, one can merely rewrite Equation 9.14 in the form of the BCS-like self-consistency equation,

$$\Delta_v = -\sum_{v'} V_{vv'}\frac{\Delta_{v'}}{2E_{v'}}\tanh\left(\frac{\beta E_{v'}}{2}\right), \tag{9.47}$$

with the interaction-matrix element (real) given by

$$V_{vv'} = -g\int d^3r\left|\langle \mathbf{r}|v\rangle\right|^2\left|\langle \mathbf{r}|v'\rangle\right|^2. \tag{9.48}$$

The summation in Equation 9.47 is over the states with $\xi_{v'}$ located in the Debye window (see Equation 9.16). It is necessary to stress that Equation 9.32 is not the exact solution to the BdG equations (Equations 9.1a and b). Indeed, multiplying Equations 9.42a and b on the left by $\langle v'|$ we get (for any $v \neq v'$)

$$\left\langle v'\left|\hat{\Delta}\right|v\right\rangle = \int d^3r\langle v'|\mathbf{r}\rangle\Delta(\mathbf{r})\langle \mathbf{r}|v\rangle = 0. \tag{9.49}$$

The trivial situation to satisfy Equation 9.49 is to work with a position-independent order parameter (e.g., in bulk). However, for broken translational symmetry, the order parameter is always position dependent and Equation 9.49 can hardly be exact. Thus, Anderson's prescription is a way to construct a reasonable approximate solution of the BdG equations [see, e.g., Tanaka and Marsiglio (2000), where results of Anderson's solution are checked for the attractive Hubbard model in the presence of a surface or a nonmagnetic impurity].

Notice that Equations 9.45 through 9.47 can also be derived from the reduced BCS-like many-electron Hamiltonian given by

$$\hat{\mathcal{H}} = \sum_v\sum_\sigma\tilde{\xi}_v\,a_{v,\sigma}^\dagger a_{v,\sigma} + \sum_{vv'}\sum_\sigma\frac{V_{vv'}}{2}a_{v,\sigma}^\dagger a_{\bar{v},-\sigma}^\dagger a_{\bar{v}',-\sigma}a_{v',\sigma}, \tag{9.50}$$

with $a_{v,\sigma}^\dagger$ and $a_{v,\sigma}$ creation and annihilation operators for the single-electron state (v,σ), and with $(\bar{v},-\sigma)$ the time-reversed state for (v,σ) (recall that $\langle \mathbf{r}|\bar{v}\rangle = \langle v|\mathbf{r}\rangle$). To check this, the kinetic term in Equation 9.50 should be first rearranged as

$$\sum_v\sum_\sigma\tilde{\xi}_v a_{v,\sigma}^\dagger a_{v,\sigma} = \sum_v\tilde{\xi}_v\left(a_{v\uparrow}^\dagger a_{v\uparrow} + a_{\bar{v}\downarrow}^\dagger a_{\bar{v}\downarrow}\right). \tag{9.51}$$

where $\tilde{\xi}_v = \tilde{\xi}_{\bar{v}}$ is involved with $\hat{H}_e|\bar{v}\rangle = \xi_{\bar{v}}|v\rangle$. The next step is to invoke the mean-field approximation (ignoring the Hartree–Fock term, as discussed above), which results in ($\tilde{\xi}_v = \xi_v$)

$$\hat{\mathcal{H}}_{MF} = \sum_v\xi_v\left(a_{v\uparrow}^\dagger a_{v\uparrow} + a_{\bar{v}\downarrow}^\dagger a_{\bar{v}\downarrow}\right) + \sum_{vv'}V_{vv'}\left[\left\langle a_{v\uparrow}^\dagger a_{\bar{v}\downarrow}^\dagger\right\rangle a_{\bar{v}'\downarrow}a_{v'\uparrow} \right.$$
$$\left. + a_{v\uparrow}^\dagger a_{\bar{v}\downarrow}^\dagger\left\langle a_{\bar{v}'\downarrow}a_{v'\uparrow}\right\rangle - \left\langle a_{v\uparrow}^\dagger a_{\bar{v}\downarrow}^\dagger\right\rangle\left\langle a_{\bar{v}'\downarrow}a_{v'\uparrow}\right\rangle\right]. \tag{9.52}$$

Then, based on Equation 9.52 and using the Bogoliubov–Valatin canonical transformation

$$a_{v\uparrow} = \mathcal{U}_v \gamma_{v\uparrow} - \mathcal{V}_v \gamma_{\bar{v}\downarrow}^\dagger, \quad a_{\bar{v}\downarrow} = \mathcal{U}_v \gamma_{\bar{v}\downarrow} + \mathcal{V}_v \gamma_{v\uparrow}^\dagger, \qquad (9.53)$$

with $\gamma_{v\sigma}^\dagger$ and $\gamma_{v\sigma}$ the bogolon operators, one can arrive at Equations 9.45 through 9.47. Notice that

$$\Delta_v = -\sum_{v'} V_{vv'} \left\langle a_{v'\uparrow} a_{\bar{v}'\downarrow} \right\rangle. \qquad (9.54)$$

The pairing of the time-reversed states is the key assumption when passing from the exact many-electron Hamiltonian to the reduced BCS-like Hamiltonian (see Equations 9.50 and 9.52). In addition, an approximate character of Anderson's prescription is also clear when keeping in mind Equation 9.49. Thus, the question arises to what extent Anderson's solution is good to be used in the presence of quantum confinement. To figure it out, let us take a cylindrical nanowire. Inserting Equation 9.38 into Equation 9.44, we can find that ($v = \{j, m, k\}$)

$$\Delta_v = \Delta_{jm} = \int_0^R d\rho\, \rho\, \vartheta_{jm}^2(\rho) \Delta(\rho), \quad E_v = \sqrt{\xi_{jmk}^2 + \Delta_{jm}^2}, \qquad (9.55)$$

with $\vartheta_{jm}(\rho)$ and ξ_{jmk} given by Equations 9.39 and 9.40, respectively. As seen, $\Delta_v = \Delta_{jm}$ does not depend on wavevector for longitudinal motion k. By definition (see Equations 9.43a and b), Δ_{jm} can be interpreted as the mean value of the superconducting order parameter as "watched" by electrons in the corresponding single-electron subband. This is in agreement with the structure of the quasiparticle spectrum given by Equation 9.55, where Δ_{jm} appears as the subband-dependent energy gap, as also seen from the sketch in Figure 9.1. It is necessary to remark that Δ_{jm} can be treated as the energy gap only in the presence of time-reversal

symmetry (see Section 9.2.2.2). As is discussed in Section 9.4, such a subband dependence of Δ_{jm} is related to the formation of new Andreev-type states induced by quantum confinement (Shanenko et al., 2007b, 2008b).

Based on the above reasoning, one can easily understand that a similar picture occurs for nanofilms (and for square cross-section nanowires as well). Thus, we in general have

$$\Delta_v = \Delta_i, \quad E_v = \sqrt{\xi_{ik}^2 + \Delta_i^2}, \qquad (9.56)$$

where $v = \{i, k\}$ with i the set of transverse quantum numbers. Given Equation 9.56, we can see that Anderson's prescription means that the Cooper pairs are assumed to be formed by electrons from the same single-electron subband (or from the subbands with the same energy like in the case of a pair of electrons with the opposite azimuthal quantum numbers for the cylindrical geometry). The pairing of electrons from two different subbands is ignored entirely. It is possible to expect that such an approximation is good enough when the subband-spacing energy parameter δ_{sub} (defined by Equation 9.11) is much larger than any of Δ_{jm}. This suggests that the criterion given by Equation 9.12 would also control the accuracy of Anderson's prescription for the quantized transverse spectrum of electrons.

Concluding this section, we remark that our numerical solution of the BdG equations (Equations 9.1a and b) for nanofilms/nanowires at $\mathbf{v}_s = 0$ shows that Anderson's prescription results in errors less than a few percent for diameters $d < 10$–20 nm. A systematic study of this issue for larger diameters was not yet performed due to time-consuming calculations.

9.2.2.2 Broken Time-Reversal Symmetry

When $\mathbf{v}_s \neq 0$, we get $\hat{H}_e \neq \hat{H}_e^*$ and time-reversal symmetry is broken. What about Anderson's recipe in this case? It is clear that broken time-reversal symmetry prevents us from using

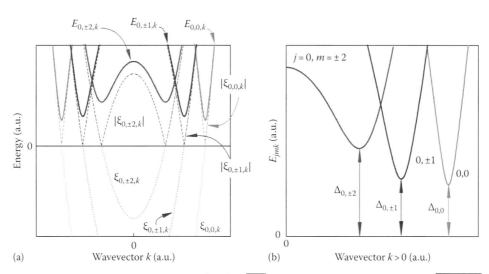

FIGURE 9.1 (a) Single-electron energy ξ_{jmk}, its absolute value $\left|\xi_{jmk}\right| = \sqrt{\xi_{jmk}^2}$ and the quasiparticle energy $E_{jmk} = \sqrt{\xi_{jmk}^2 + \Delta_{jm}^2}$ sketched versus the wavevector k for motion parallel to the nanocylinder for j, $|m| = (0, 0)$, $(0, 1)$ and $(0, 2)$. (b) The quasiparticle energy for the same single-electron subbands versus $k > 0$.

Equation 9.51 because $\tilde{\xi}_v \neq \tilde{\xi}_{\bar{v}}$. Moreover, $|\bar{v}\rangle$ is not in general an eigenvector of $\hat{H}_e$ anymore and, hence, the combination of $a_{v,\sigma}$ and $a_{\bar{v},-\sigma}$ in the interaction four-operator term in Equation 9.50 looks questionable because $a_{v,\sigma}$ and $a_{\bar{v},-\sigma}$ represent different families of creation–annihilation operators associated with the different single-electron operators $\hat{H}_e^*$ and $\hat{H}_e$. However, there is a particular situation when Anderson's prescription is still perfect even in the presence of broken time-reversal symmetry. When $\hat{H}_e$ and $\hat{H}_e^*$ are not equal but commute, it is possible to choose the single-electron basis in such a way that $|\bar{v}\rangle$ would be an eigenvector of $\hat{H}_e$. Notice that this important particular case covers both the cylindrical nanowire in a parallel magnetic field and with a longitudinal supercurrent. Numerical investigations of both situations (see Section 9.4.2) show that Anderson's solution is here as accurate as for nanofilms and nanowires in the case of time-reversal symmetry.

As mentioned above, for broken time reversal, Equation 9.51 is not valid. However, provided that $\hat{H}_e$ commutes with $\hat{H}_e^*$, it is possible to employ another transformation of the kinetic term, i.e.,

$$\sum_v \sum_\sigma \tilde{\xi}_v a_{v,\sigma}^\dagger a_{v,\sigma} = \sum_v \left(\tilde{\xi}_v a_{v\uparrow}^\dagger a_{v\uparrow} + \tilde{\xi}_{\bar{v}} a_{\bar{v}\downarrow}^\dagger a_{\bar{v}\downarrow} \right). \tag{9.57}$$

Inserting Equation 9.57 into Equation 9.50, we can rewrite the mean-field BCS-like Hamiltonian in the form

$$\hat{\mathcal{H}}_{MF} = \sum_v \left(\tilde{\xi}_v a_{v\uparrow}^\dagger a_{v\uparrow} + \tilde{\xi}_{\bar{v}} a_{\bar{v}\downarrow}^\dagger a_{\bar{v}\downarrow} \right) + \sum_v \Delta_v (a_{v\uparrow}^\dagger a_{\bar{v}\downarrow}^\dagger + a_{\bar{v}\downarrow} a_{v\uparrow})$$

$$- \sum_v \Delta_v \langle a_{v\uparrow}^\dagger a_{\bar{v}\downarrow}^\dagger \rangle, \tag{9.58}$$

where Δ_v is given by Equation 9.54. We remark again that the basis $|v\rangle$ is chosen so that $|\bar{v}\rangle$ is an eigenvector of $\hat{H}_e$, and it is possible to write $\hat{H}_e|\bar{v}\rangle = \tilde{\xi}_{\bar{v}}|v\rangle$. In spite of the difference between Equations 9.58 and 9.52, the structure of Equation 9.58 makes it possible to use the Bogoliubov–Valatin transformation (9.53) in order to get a Hamiltonian of noninteracting quasiparticles. This results in helpful generalizations of Equations 9.45 and 9.47 to the case of broken time-reversal symmetry. In particular, the bogolon spectrum now reads

$$E_v = \pm \sqrt{\left(\frac{\tilde{\xi}_v + \tilde{\xi}_{\bar{v}}}{2} \right)^2 + \Delta_v^2} + \frac{\tilde{\xi}_v - \tilde{\xi}_{\bar{v}}}{2}. \tag{9.59}$$

As seen, there is one more term, i.e., $(\tilde{\xi}_v + \tilde{\xi}_{\bar{v}})/2$, in addition to the square root in Equation 9.59. It disappears when $\mathbf{v}_s \to 0$ and, hence, according to the criterion given by Equation 9.17, the physical solution is specified by the + sign in Equation 9.59, similar to Equation 9.45. Within the linear approximation in $\mathbf{v}_s$, i.e., Equation 9.8, we have $\xi_v = (\tilde{\xi}_v + \tilde{\xi}_{\bar{v}})/2$ and E_v can be approximated as

$$E_v = \sqrt{\xi_v^2 + \Delta_v^2} + \frac{\tilde{\xi}_v - \tilde{\xi}_{\bar{v}}}{2}, \tag{9.60}$$

where Δ_v obeys the self-consistency equation

$$\Delta_v = -\sum_{v'} V_{vv'} \frac{\Delta_{v'}}{2\sqrt{\xi_{v'}^2 + \Delta_{v'}^2}} \tanh\left[\frac{\beta}{2} \left(\sqrt{\xi_v^2 + \Delta_v^2} + \frac{\tilde{\xi}_v - \tilde{\xi}_{\bar{v}}}{2} \right) \right], \tag{9.61}$$

with $V_{vv'}$ being the same as in Equation 9.48. To avoid the divergence, the sum in Equation 9.61 is cut off according to Equation 9.16.

All the above equations for broken time-reversal symmetry can also be derived by substituting Equation 9.32 into the BdG equations, which results in

$$E_v \mathcal{U}_v = \tilde{\xi}_v \mathcal{U}_v + \mathcal{V}_v \Delta_v, \tag{9.62a}$$

$$E_v \mathcal{V}_v = \mathcal{U}_v \Delta_v - \tilde{\xi}_{\bar{v}} \mathcal{V}_v. \tag{9.62b}$$

A nontrivial solution of these equations exists provided that the corresponding determinant is zero, i.e.,

$$\begin{vmatrix} E_v - \tilde{\xi}_v & -\Delta_v \\ -\Delta_v & E_v + \tilde{\xi}_{\bar{v}} \end{vmatrix} = 0,$$

and we arrive immediately at Equation 9.59. Notice that $\Delta_v = \Delta_{\bar{v}}$, $\mathcal{U}_v = \mathcal{U}_{\bar{v}}$, $\mathcal{V}_v = \mathcal{V}_{\bar{v}}$ and $\xi_v = \xi_{\bar{v}}$. On the contrary, $\tilde{\xi}_v \neq \tilde{\xi}_{\bar{v}}$ and, so, $E_v \neq E_{\bar{v}}$.

Let us have a closer look at Equations 9.60 and 9.61. For a cylindrical nanowire in a parallel magnetic field (see Section 9.2.1.3), $v = \{j, m, k\}$ and the eigenfunctions of $\hat{H}_p$ and $\hat{H}_e$ can be chosen the same (when neglecting the term $\propto \mathbf{A}^2$ in $\hat{H}_e$). Using Equations 9.23 and 9.38, we obtain

$$\frac{\hat{\mathbf{p}}\hat{\mathbf{v}}_s + \hat{\mathbf{v}}_s\hat{\mathbf{p}}}{2} |jmk\rangle = -\mu_B m H_\parallel |jmk\rangle \tag{9.63}$$

and $[\bar{v} = \{j, -m, -k\}]$

$$\tilde{\xi}_{jmk} = \xi_{jmk} - \mu_B m H_\parallel, \tag{9.64a}$$

$$\tilde{\xi}_{j,-m,-k} = \xi_{jmk} + \mu_B m H_\parallel, \tag{9.64b}$$

with ξ_{jmk} given by Equation 9.40 and $\mu_B = |e|\hbar/2m_e c$ the Bohr magneton. Equations 9.64a and b can be combined with Equation 9.60 to give

$$E_{jmk} = \sqrt{\xi_{jmk}^2 + \Delta_{jm}^2} - \mu_B m H_\parallel. \tag{9.65}$$

When "switching on" a magnetic field, the quasiparticle energies for the states with positive (negative) m are systematically

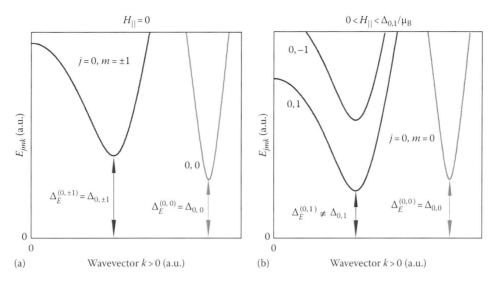

FIGURE 9.2 Quasiparticle spectrum for cylindrical nanowires in a parallel magnetic field: E_{jmk} versus k at zero temperature for (a) $H_\parallel = 0$ and (b) $0 < H_\parallel < \Delta_{0,1}/\mu_B$ (three quasiparticle branches are given, $(j, m) = (0, 0)$, $(0, \pm 1)$). Notice that the subband energy gap $\Delta_E^{(jm)}$ coincides with Δ_{jm} for all j and m at $H_\parallel = 0$. On the contrary, $\Delta_E^{(0,1)}$ and $\Delta_E^{(0,-1)}$ are not the same as $\Delta_{0,1}$ and $\Delta_{0,-1}$, respectively, when $H_\parallel > 0$ (see Equation 9.65).

shifted down (up) due to the term $-\mu_B m H_\parallel$. In particular, when $T = 0$ and all quasiparticle energies are positive, Δ_{jm} is not sensitive to $H_\parallel$ at all [it is clear from Equation 9.61 where $\tanh(\beta E_v/2) = 1$ in this case]. Therefore, E_{jmk} appears to be linear in $H_\parallel$ (with the slope $\mu_B m$) for relatively small magnetic fields. Such a behavior is illustrated by the sketch given in Figure 9.2 in the presence of the three quasiparticle branches with $(j, m) = (0, 0)$ and $(0, \pm 1)$ (this small number of branches is chosen for the sake of simplicity and corresponds to an extremely narrow nanocylinder with $R < 1$ nm). As seen from Figure 9.2b, in addition to Δ_{jm}, we need to introduce $\Delta_E^{(jm)}$, the energy gap for the electrons in the single-electron subband (j, m) (and, in turn, the total energy gap $\Delta_E = \min_{jm} \Delta_E^{(jm)}$). In the presence of time-reversal symmetry (i.e., for $H_\parallel = 0$) Δ_E^{jm} is equal to Δ_{jm}, see Figure 9.2a. However, this is not generally true for $H_\parallel \neq 0$. When $H_\parallel < \Delta_{0,1}/\mu_B$, all quasiparticle energies are positive in Figure 9.2b and, so, $\Delta_{0,\pm 1}$ does not change with $H_\parallel$. However, $\Delta_E^{(0,1)}$ exhibits a linear decrease with $H_\parallel$ while $\Delta_E^{(0,-1)}$ increases. At $H_\parallel > \Delta_{jm}/\mu_B$ some of E_{jmk} become negative (see the discussion before Equation 9.17). As quasiparticles with negative energies survive in the system at zero temperature, we face a reconstruction of the ground state, i.e., decay of the Cooper pairs induced by the magnetic field. Such a decay does not occur simultaneously in all single-electron subbands. Therefore, in the presence of many quasiparticle branches (or, in other words, for many single-electron subbands), a cascade structure of the superconductor-to-normal transition induced by the magnetic field appears [see Section 9.4.2 and Shanenko et al. (2008a)]. It is worth noticing that $\Delta_E^{(jm)}$ is always zero when there exists a sector of negative quasiparticle energies for the corresponding (j, m). However, it does not necessarily mean that $\Delta_{jm} = 0$.

Similar physics occurs for the cylindrical nanowire with a longitudinal supercurrent, see Section 9.2.1.3. In this case, we

again have $v = \{j, m, k\}$ and $\bar{v} = \{j, -m, -k\}$. Equations 9.8 and 9.38 yield ($v_s = \hbar q/m_e$)

$$\tilde{\xi}_{jmk} = \xi_{jmk} + \frac{\hbar^2}{m_e} qk, \tag{9.66a}$$

$$\tilde{\xi}_{j,-m,-k} = \xi_{jmk} - \frac{\hbar^2}{m_e} qk. \tag{9.66b}$$

Inserting Equations 9.66a and b into Equation 9.60, one gets

$$E_{jmk} = \sqrt{\xi_{jmk}^2 + \Delta_{jmk}^2} + \frac{\hbar^2}{m_e} qk. \tag{9.67}$$

The difference to a cylindrical nanowire in a parallel magnetic field is that the square root in Equation 9.67 is now accompanied by a term depending on the wave vector of the longitudinal electron motion rather than on the azimuthal quantum number (in fact, both terms are due to the Doppler shift). Therefore, the larger the absolute value of k, the more significant is the shift of the bogolon energy (it is up for positive k, and down for negative) in the presence of a longitudinal superflow. This is illustrated in Figure 9.3 for the case of the six quasiparticle branches with $(j, m) = (0, 0)$, $(0, \pm 1)$, $(0, \pm 2)$ and $(1, 0)$ (all the parameters are chosen arbitrary). It is convenient to rewrite E_{jmk} in the form

$$E_{jmk} = \sqrt{\xi_{jmk}^2 + \Delta_{jmk}^2} + \text{sgn}(k)\frac{\hbar^2}{m_e} qk_{jm}, \tag{9.68}$$

where $\text{sgn}(x)$ is the sign function, and k_{jm} is the positive root of

$$\xi_{jmk}\big|_{k=k_{jm}} = 0. \tag{9.69}$$

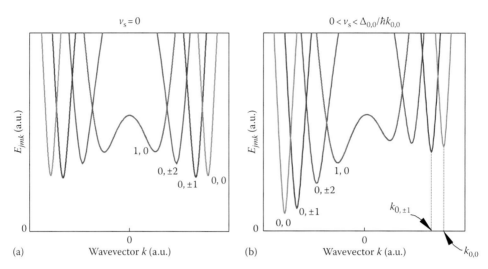

FIGURE 9.3 Quasiparticle energies for a cylindrical nanowire with a longitudinal superfluid flow: E_{jmk} versus k for (a) $v_s = 0$ and (b) $0 < v_s < \Delta_{0,0}/\hbar k_{0,0}$ at zero temperature [for the six branches $(j, m) = (0, 0)$, $(0, \pm 1)$, $(0, \pm 2)$, and $(1, 0)$].

This is a quite reasonable approximation for the single-electron states located in the Debye window $|\xi_{jmk}| < \hbar\omega_D$. Indeed, for k being in close vicinity to $\pm k_{jm}$, the single-electron energy ξ_{jmk} changes so quickly that we can set $k = \text{sgn}(k) k_{jm}$ for all the states in the Debye window. In the gap regime, similar to the situation considered in the previous paragraph, Δ_{jm} (and, hence, the order parameter) are not dependent on v_s and, so, shifts of the quasi-particle energies due to the superfluid flow are linear in v_s. In this case, the energy gap Δ_E taken as a function of v_s is a polygonal chain. In particular, for the sketch of Figure 9.3b, Δ_E includes two line segments. The first segment is determined by $k_{0,\pm 1}$ (because $\Delta_{0,0} > \Delta_{0,\pm 1}$). The slope of the second line segment is $\hbar k_{0,0}$ so that $\Delta_E = 0$ at $v_s = \Delta_{0,0}/\hbar k_{0,0}$. Thus, the gap regime is realized when $v_s < \Delta_{0,0}/\hbar k_{0,0}$ (see Figure 9.3b). For $v_s > \Delta_{0,0}/\hbar k_{0,0}$, we work with a more complex situation of gapless superconductivity.

Concluding this section, we would like to make several remarks about the most crude situation when time-reversal symmetry is broken and, in addition, $\hat{H}_e$ does not commute with $\hat{H}_e^*$. Here the procedure based on constructing a reduced BCS-like Hamiltonian is not clear because $a_{v,\sigma}$ and $a_{\bar{v},-\sigma}$ are from two different sets of creation–annihilation operators generated by $\hat{H}_e$ and $\hat{H}_e^*$, respectively. However, it is still possible to work with Equation 9.32 to have an idea about this regime. Inserting Equation 9.32 into the BdG equations, we get

$$E_v \mathcal{U}_v |v\rangle = \tilde{\xi}_v \mathcal{U}_v |v\rangle + \mathcal{V}_v \hat{\Delta} |v\rangle, \tag{9.70a}$$

$$E_v \mathcal{V}_v |v\rangle = \mathcal{U}_v \hat{\Delta}^* |v\rangle + \mathcal{V}_v \hat{H}_e^* |v\rangle. \tag{9.70b}$$

Following the procedure used earlier for Equations 9.42a and b, i.e., multiplying both BdG equations on the left by $\langle v|$, we arrive at Equations 9.59 and 9.61 but with $\tilde{\xi}_{\bar{v}}$ replaced by $\langle v|\hat{H}_e^*|v\rangle = \langle \bar{v}|\hat{H}_e|\bar{v}\rangle$. This is a reflection of the fact that $|\bar{v}\rangle$ is not an eigenvector of $\hat{H}_e$. However, this replacement is not the

only thing that differs in the present situation from the case of commuting $\hat{H}_e$ and $\hat{H}_e^*$. Indeed, multiplying Equations 9.70a and b on the left by $\langle v'|$ yields, in addition to Equation 9.49, one more requirement (for any $v \neq v'$)

$$\langle v'|\hat{H}_e^*|v\rangle = \langle \bar{v}'|\hat{H}_e|\bar{v}\rangle = 0, \tag{9.71}$$

which can only be satisfied when $|v\rangle$ represents the set of the eigenvectors of $\hat{H}_e^*$. Due to this additional requirement, one can expect that Anderson's solution is now less accurate than in the previous situations. A simple example for which $\hat{H}_e$ does not commute with $\hat{H}_e^*$, is a nanowire with square cross section placed in a parallel magnetic field (for both the Coulomb and Landau gauges, $\mathbf{A} = \frac{1}{2}H_\parallel(x\mathbf{e}_y - y\mathbf{e}_x)$ and $\mathbf{A} = H_\parallel x\mathbf{e}_y$, with $\mathbf{e}_x$ and $\mathbf{e}_y$ the unit vectors in the x and y directions). Our numerical investigations of the BdG equations in such a situation does show that Equation 9.32 is less accurate in the case of a cylindrical geometry because errors of Anderson's solution become more than 10%.

9.3 Quantum-Size Oscillations

9.3.1 Physics behind Quantum-Size Oscillations

It is known (de Gennes, 1966; Fetter and Walecka, 2003) (see also Equation 9.16) that the superconducting order parameter depends on N_D, the number of single-electron states (per spin projection) situated in the Debye window around the Fermi level. More precisely, the mean energy density of these states taken per unit volume $n_D = N_D/(2\hbar\omega_D V)$ is the key quantity. We stress that n_D is the mean density here because the density of single-electron states can diverge as is the case for the quasi-one-dimensional (quasi-1D) situation. In high-quality nanofilms and nanowires, effects of the transverse quantization are not shadowed by (nonmagnetic) impurity scattering and the conduction band splits up into a series of single-electron subbands. While thickness increases (decreases), these subbands move down (up) in energy. Note that the position of the bottom of any subband (i.e., the transverse single-electron spectrum) scales as $1/d^2$, with d, recall, the nanofilm/nanowire thickness. Each time when the bottom of a parabolic subband passes through the Fermi surface, n_D increases abruptly. This results in quantum-size oscillations of superconducting properties accompanied by substantial enhancements of superconductivity, i.e., the size-dependent superconducting resonances (Blatt and Thompson, 1963). In Figure 9.4, single-electron energies ξ_v are schematically shown versus the wavevector of the quasi-free electron motion parallel to a nanofilm/nanowire for three different thicknesses (d increases from Figure 9.4a through c). Different curves represent different single-electron subbands (1, 2, 3, 4...). The sectors of the curves corresponding to the single-electron states located in the Debye window $\xi_v \in [\hbar\omega_D, \hbar\omega_D]$ are marked with broken lines. In Figure 9.4a, the bottoms of all subbands are outside the Debye window. This is the off-resonant regime. When d increases,

the subbands move down in energy, and the bottom of subband 2 enters the Debye window (see Figure 9.4b). In this case, the number of single-electron states contributing to the superconducting characteristics sharply increases, which results in the superconducting resonance. The next size-dependent resonance arises when subband 3 enters the Debye window, Figure 9.4c. This leads to a sequence of peaks in n_D as a function of d and, as a consequence, any superconducting quantity exhibits quantum-size oscillations with remarkable resonant enhancements. Such superconducting resonances are significant in nanoscale samples but smoothed out with increasing d (see Figure 9.4d), when n_D slowly approaches its bulk limit $N(0) = m_e k_F/(2\pi^2\hbar^2)$, with k_F the 3D Fermi wavevector.

The quantum-size oscillations in the relevant mean density of single-electron states are shown in Figure 9.5a for superconducting nanofilms: the solid curve represents n_D calculated when E_F is taken to be independent of the film thickness whereas the dashed curve is obtained when the Fermi level is allowed to vary with the film thickness but the electron density

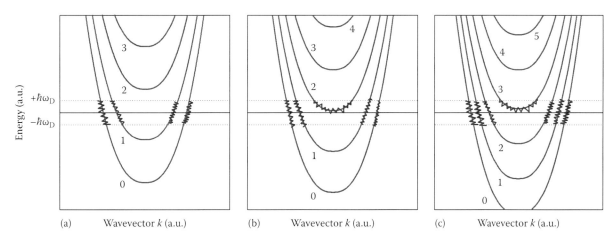

FIGURE 9.4 Single-electron energies $\xi_{i,k}$ ($i = 0, 1, 2, 3,...$) versus the wave vector k of the quasi-free electron motion parallel to the nanofilm/nanowire: (a) the off-resonant regime, the bottoms of all subbands are situated beyond the Debye window $|\xi_v| < \hbar\omega_D$; (b) and (c) correspond to the size-dependent superconducting resonances associated with single-electron subbands 2 and 3.

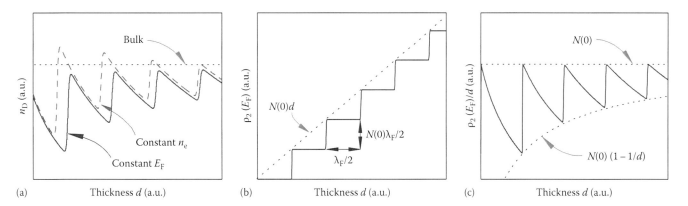

FIGURE 9.5 (a) The mean density of single-electron states in the Debye window $(|\xi_v| < \hbar\omega_D)$ versus the film thickness d: the solid curve represents results calculated for E_F independent of the film thickness; the dashed curve corresponds to n_e kept constant (the bulk limit is chosen the same in both cases). (b) The 2D density of states $\rho_2(E)$ at $E = E_F$ versus d for a thickness-independent E_F. (c) $n_D = \rho_2(E_F)/d$ as a function of d for E_F taken constant.

is kept constant. To have a feeling about n_D in the quasi-two-dimensional (quasi-2D) case, one can use a simplified but reasonable approximation, i.e.,

$$n_D \approx \frac{\rho_2(E_F)}{d}, \quad \rho_2(\varepsilon) = \frac{m_e}{2\pi\hbar^2} \sum_{i=0}^{\infty} \theta\left(\varepsilon - \frac{\hbar^2\pi^2(i+1)^2}{2m_e d^2}\right), \tag{9.72}$$

where the factor 1/2 appears in the expression for the 2D density of states $\rho_2(\varepsilon)$ due to neglecting the electron spin. In Figure 9.5b, $\rho_2(E_F)$ is sketched for the case of a constant Fermi level. As seen from Equation 9.72 and Figure 9.5b, $\rho_2(E_F)$ exhibits a sequence of jumps of amplitude $N(0)\lambda_F/2$ ($\lambda_F = 2\pi/k_F$ is the 3D Fermi wavelength). These steps are exactly regular and comes with period $\lambda_F/2 = \pi/k_F$. However, for constant electron density n_e such steps are not regular for small d due to a shift-up of E_F. This explains the difference between the solid and dashed curves given in panel (a). One can construct a more elaborate guess for n_D based on $\rho_2(E_F)$: Equation 9.72 can be replaced by the average value of $\rho_2(\varepsilon)/d$ over the Debye window, i.e., $n_D = 1/2\hbar\omega_D d \int d\varepsilon \rho_2(\varepsilon)$. In this way, we are able to understand why n_D as a function of thickness exhibits smoothed jumps at the resonant points while $\rho_2(E_F)$ exhibits a discontinuous behavior. Another consequence of this averaging is that all peaks in n_D (for the constant E_F) appear to be somewhat below the bulk limit $N(0)$. This is opposite to the peaks in $\rho_2(E_F)/d$, where $\rho_2(E_F)/d$ reaches exactly $N(0)$, see Figure 9.5c. Notice that n_D, in general, appears to be smaller than its bulk value in Figure 9.5a. However, as shown further, the corresponding superconducting characteristics, e.g., the order parameter, the critical temperature, and energy gap are enhanced at the resonant points as compared to their bulk values (even when keeping g the same as in bulk). A reason is that the single-electron wave functions are no longer 3D plane waves, and this has a significant effect on the matrix element given by Equation 9.48.

Quantum-size oscillations of n_D for nanowires are similar but, nevertheless, with a few important exceptions. First, the resonant enhancements are not equidistant in thickness and appear almost irregular, even for E_F kept independent of the nanowire thickness d. The transverse electron spectrum ε_i is now more complex (we remind that i stands for the set of quantum numbers for nanowires) than for nanofilms, where $\varepsilon_i \propto [(i+1)/d]^2$. Exactly such a dependence of the transverse single-electron energy on the ratio $(i+1)/d$ for nanofilms was a reason for the periodicity in the appearance of the superconducting resonances. Another point is that n_D as a function of d now exhibits more sharp peaks due to the van-Hove-type singularities in $\rho_1(E_F)$, i.e., the 1D-density of states at $\varepsilon = E_F$, given by

$$\rho_1(E_F) = \sqrt{\frac{m_e/2}{\pi\hbar}} \sum_{i=0}^{\infty} \frac{\theta(E_F - \varepsilon_i)}{\sqrt{E_F - \varepsilon_i}}. \tag{9.73}$$

This quantity is divergent when $E_F = E_i$, as illustrated in Figure 9.6a. Following the same scheme as before for nanofilms, we can approximate n_D by the integral of $\rho_1(E)/(2\hbar\omega_D S_d)$ taken over the Debye window (with $S_d \propto d^2$ the nanowire cross section). According to this approximation, the van-Hove-type singularities will be smoothed into sharp peaks in n_d. Such peaks associated with the van-Hove-type singularities in $\rho_1(E_F)$, are shown in Figure 9.6b, with n_D given by solid squares. The Fermi level is chosen pinned to its bulk value here. Nevertheless, the peaks in n_D exceed the bulk limit $N(0)$, contrary to the situation with E_F kept constant for nanofilms.

It is necessary to note that, to obtain the correct period for the quantum-size oscillations of the physical properties in nanosuperconductors within the simplified parabolic band approximation (based on the band mass m_e), one should use the effective Fermi level E_F rather than the real one [a detailed discussion on the effective Fermi level was presented in Shanenko

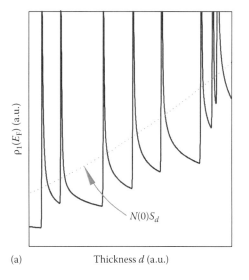
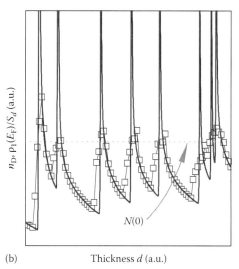

FIGURE 9.6 (a) The 1D density of states $\rho_1(E)$ taken at $E = E_F$ as a function of the nanowire thickness and (b) the mean density of single-electron states n_D (open squares) versus d together with $\rho_1(E_F)/S_d$, with $S_d \propto d^2$ the nanowire cross section.

TABLE 9.1   The Effective Fermi Energy

	Al	Sn	Pb
E_F (eV)	0.90	2.3	0.95–1.4

et al. (2007a)]. In general, the effective Fermi level depends on the complicated interplay between the crystal and confinement directions. Possible values of E_F (when ignoring the shift-up in E_F at extremely small d) are given in Table 9.1. E_F for Al is justified from a good agreement with the experimental data found for aluminum superconducting nanowires (Shanenko and Croitoru, 2006; Shanenko et al., 2006a). The data for Pb and Sn are extracted from the experimental results for superconducting nanofilms (Orr et al., 1984; Wei and Chou, 2002).

What about experimental observations of the superconducting quantum-size oscillations? To observe such oscillations with a high level of precision and sophistication excluding any possible doubts, experimentalists try to grow ultrathin high-quality films with *atomic-scale uniformity in thickness*. In this case, the value of d is limited to integer numbers of monolayers ML (atomic layers). Fabricating such films with a thickness of a few monolayers is a rather challenging task. The matter is that metal atoms prefer to form clusters and grains instead of uniformly covering an insulator substrate (Yazdani, 2006). This results in granular or amorphous nanofilms, where superconductivity is suppressed due to disorder and, so, cannot survive in ultrathin samples (Chiang, 2004). A nice exception is Pb which can, under special conditions, form smooth *single-crystalline* superconducting nanofilms on semiconductor surfaces, beginning from a few monolayers. Recently, such thin single-crystalline films were fabricated by several experimental groups (Guo et al., 2004;

Bao et al., 2005; Zhang et al., 2005; Eom et al., 2006; Özer et al., 2006a,b, 2007). The most striking is that such extremely narrow nanofilms with crystalline perfection are able to carry significant dissipationless currents (Guo et al., 2004; Bao et al., 2005; Zhang et al., 2005; Eom et al., 2006; Özer et al., 2006a,b, 2007). It means that superconductivity survives in high-quality Pb nanofilms down to extreme thicknesses of about 1 nm. Photoelectron spectroscopy demonstrates that in Pb films epitaxially grown on silicon (111), the period of the quantum-size variations in the density of single-electron states at the Fermi level is about 2 ML (Wei and Chou, 2002; Guo et al., 2004; Bao et al., 2005; Zhang et al., 2005; Eom et al., 2006). In other words, the bottoms of the relevant single-electron subbands formed due to the transverse quantization pass through the Fermi surface with a period of about 2 ML. This results in spectacular oscillations (Czoschke et al., 2004; Guo et al., 2004; Upton et al., 2004; Bao et al., 2005; Zhang et al., 2005; Eom et al., 2006; Li et al., 2006) of the physical properties between films with even and odd number of monolayers (the even–odd oscillations/staggering). Recent experimental results [see Guo et al. (2004); Bao et al. (2005)] for such even–odd oscillations of the critical superconducting temperature T_c in Pb(111) nanofilms are shown in Figure 9.7a. To make sure that these thickness-dependent oscillations well correlate with the confined electron structure, Figure 9.7b demonstrates the corresponding even–odd oscillations in the experimental results for ρ_n, the normal-state resistivity of the same films measures at temperature 8 K [see Bao et al. (2005)]. Similar even–odd staggering is also observed for the critical perpendicular magnetic field $H_{c2\perp}$ (Bao et al., 2005), as illustrated in Figure 9.8. Note that the quantum-size oscillations in $H_{c2\perp}$ are π out of phase as compared to T_c.

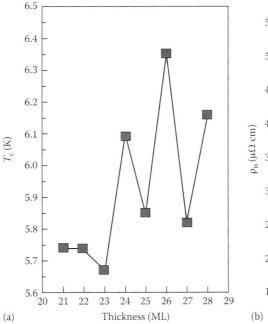
(a)

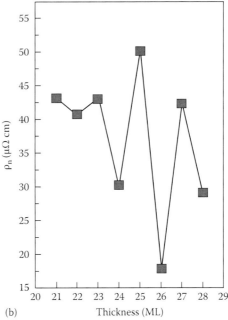
(b)

FIGURE 9.7   (a) Even–odd oscillations of the superconducting critical temperature T_c in Pb nanofilms on silicon (111), and (b) the corresponding oscillations in the normal resistivity ρ_n of the same films. (From Bao, X.-Y. et al., *Phys. Rev. Lett.*, 95, 247005-1, 2005. With permission.)

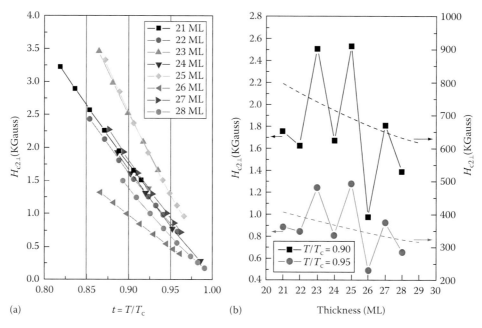

FIGURE 9.8 Pb(111) atomically uniform superconducting nanofilms: (a) the perpendicular critical magnetic field $H_{c2\perp}$ as dependent on the relative temperature T/T_c for various film thicknesses, and (b) the even–odd quantum-size oscillations in $H_{2c\perp}$ for $T = 0.9$ T_c and 0.95 T_c. (From Bao, X.-Y. et al., *Phys. Rev. Lett.*, 95, 247005-1, 2005. With permission.)

Another set of experimental results for Pb(111) single-crystalline superconducting nanofilms is illustrated by Figures 9.9 and 9.10 [from Eom et al. (2006)]. In Figure 9.7a, one can see the STM image of a Pb island on silicon (111) containing perfectly flat terraces with thickness 5, 6, and 7 ML. In panel (b) of the same figure, the tunneling current and corresponding differential conductance are shown for nanofilms with 14 and 15 ML. Peaks in the differential conductance are signatures of the formation of the single-electron subbands due to the transverse quantization of the electron motion. Study of the peak positions makes it possible to find the corresponding transverse electron spectrum given in Figure 9.7c. In particular, from the panel (c) one can learn that between 9 and 15 ML there is a clear pattern of even–odd staggering in the position of the transverse electron energies with respect to E_F. For an odd number of monolayers, one of the bottoms of the single-electron subbands is situated near the Fermi level while it is not the case for an even number. For $d > 15$ ML, this behavior is washed out, and the even–odd oscillations are changed: Now, even numbers of monolayers are characterized by an increase in the density of states near the Fermi level. This 7 ML beating is due to the fact that the period of the quantum-size oscillations is not exactly but only ~2 ML. It is instructive to compare Figure 9.9c with Figure 9.10 where the corresponding oscillations of T_c are given [from Eom et al. (2006)]. One can be convinced that enhancements of T_c are in a good correlation with the appearance of the transverse electron energies situated near (just below) the Fermi level. Notice that for Pb(111) we have 1 ML = 0.286 nm (Wei and Chou, 2002). Working within the parabolic band approximation, one can get for the period of the quantum-size oscillations in nanofilms $\lambda_F/2$ (Blatt and Thompson, 1963) (it results immediately from Equation 9.35). Hence, we are able to extract k_F,

and, for Pb(111) with the period about 0.572 nm, this yields $k_F = 5.49$ nm^{-1} and $E_F = \hbar^2 k_F^2/2m_e = 1.15$ eV. This estimate is significantly smaller than the typical value of the Fermi level in bulk lead and explains the possible values for E_F given in Table 9.1 for Pb nanofilms [the upper bound corresponds to the period 1.8 ML from Zhang et al. (2005), and the lower bound is for 2.2 ML from Wei and Chou (2002)].

Pb is a strong-coupling superconductor with a significant bulk condensation energy. This is why we cannot expect significant enhancements of superconductivity at the resonant points. Such enhancements are the result of an interplay between the confining and condensation energy: the larger the bulk gap, the less pronounced the superconducting resonances. Hence, it is natural to expect that the superconducting resonances should be substantial in weak-coupling superconducting metals, as, e.g., aluminum. Films and wires made of aluminum usually have the polycrystalline structure. However, recent advances in nanotechnology resulted in high-quality aluminum nanofilms and nanowires made of strongly coupled grains with practically no tunnel barrier between them. In this case, most of the physical properties are practically independent of the grain size and, to a great extent, such nanofilms and nanowires can be treated as high-quality samples (the electron mean-free path was estimated to be larger than the sample thickness (Savolainen et al., 2004; Zgirski et al., 2005; Altomare et al., 2006). In particular, there is no indication that the grain size has a strong influence on the energy gap measured in very recent experiments with Al superconducting films (Court et al., 2008). In addition, there are interesting experiments with aluminum superconducting nanowires (Savolainen et al., 2004; Zgirski et al., 2005; Altomare et al., 2006). It is worth noting that the superconducting gap and critical temperature of aluminum

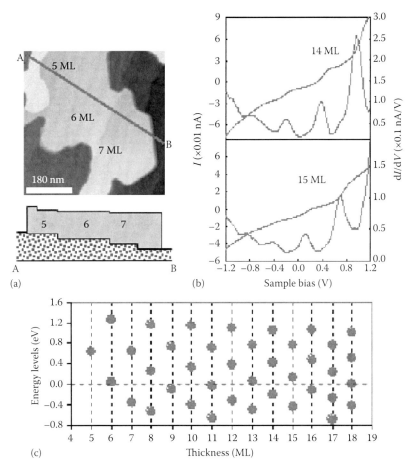

(a)   (b)   (c)

FIGURE 9.9 (a) STM image of Pb on silicon (111) containing flat terraces with 5, 6, and 7 ML height; (b) tunneling current and corresponding differential conductance for 14 and 15 ML; and (c) locations of the transverse electron energies (measured from the Fermi level) extracted from the tunneling experiments with Pb(111) islands. (From Eom, D. et al., *Phys. Rev. Lett.*, 96, 027005-1, 2006. With permission.)

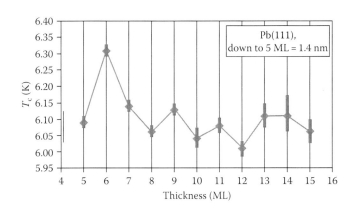

FIGURE 9.10 Quantum-size oscillations in T_c of Pb nanofilms epitaxially grown on silicon (111). (From Eom, D. et al., *Phys. Rev. Lett.*, 96, 027005-1, 2006. With permission.)

superconducting nanowires and nanofilms are enhanced as compared to bulk: by a factor of 1.5 in nanowires with thickness about $d = 10$ nm (Savolainen et al., 2004; Zgirski et al., 2005; Altomare et al., 2006; Zgirski and Arutyunov, 2007) and by a factor of 1.4–1.5 in nanofilms with $d = 5$ nm (Court et al., 2008).

Is it possible to observe, even indirectly, the quantum-size effects in such polycrystalline nanostructured samples with no atomical uniformity in thickness (this is also related to single-crystalline nanowires, where d can fluctuate significantly as compared to single-crystalline nanofilms). The quantum-size oscillations will be, of course, smoothed by thickness fluctuations. However, one can expect that some tracks of the superconducting resonances can be caught. This is owing to the proximity effect between the domains with enhanced superconductivity (the fluctuating thickness reaches a resonant point here) and domains with relatively weak superconducting properties (d takes an off-resonant value). This suggests that in the presence of thickness fluctuations, the superconducting characteristics of nanofilms and nanowires are mainly determined by the resonant enhancements that fall into the range of thicknesses covered by fluctuations. Below, in Section 9.3.2 it is shown that the quantum-size superconducting resonances provide a solid explanation of the enhancement (as compared to bulk) of the basic superconducting quantities recently found in high-quality aluminum nanowires and nanofilms. The same is related to the increase of the superconducting temperature observed in single-crystalline tin nanowires (Tian et al., 2005; Janković et al., 2006; Tombros et al., 2008).

9.3.2 Quantum-Size Oscillations and Superconducting Resonances in Nanofilms

As already mentioned in the previous section, for clear observations of the quantum-size superconducting oscillations, very narrow Pb films with thickness down to ~10 ML are fabricated. In this case, a substrate and possible protecting coverage can change the electron–phonon interaction due to interface effects (Luh et al., 2002). For instance, in ultrathin films ($d < 12$–16 ML) of Ag on V(100) and Fe(100) substrates the electron–electron coupling due to phonons was found to be significantly larger than in bulk and decreases down to its bulk value when the film thickness increases (Valla et al., 2000; Luh et al., 2002). Deviations of the coupling constant from its bulk limit follow approximately an overall $1/d$ dependence (Luh et al., 2002) and exhibit damped quantum-size oscillations (Valla et al., 2000; Luh et al., 2002). Such $1/d$ dependence can be understood [see, also, Chiang (2004)] as due to the fact that the relative number of film atoms at the film interface is proportional to $1/d$. In the case of Pb nanofilms epitaxially grown on silicon (111), g is expected to gradually increase (with quantum-size oscillations) toward its bulk value as the film gets thicker [see Zhang et al. (2005)]. The quantum-size oscillations in g can be related to a change in the transverse distribution of electrons at the interface when a new single-electron subband passes through the Fermi surface. Keeping in mind these features, one can expect that for Pb films grown on silicon (111), the dependence of g on the film thickness can be approximated as

$$g = g_0 - g_1(2k_F d)\frac{a}{d}, \qquad (9.74)$$

where

$a = 0.286$ nm is the thickness of one monolayer
$g_0 N(0) = 0.39$ provides the correct bulk limit for Pb
$g_1(x)$ is a function oscillating with period 2π

Provided that the period of the quantum-size oscillations is taken $\pi/k_F = 2a (2\,\text{ML})$.

$$g_1(2k_F d) = g_1(2k_F aN) = g_1(\pi N),$$

with $N = d/a$ the number of monolayers. Hence, for odd numbers, we have $g_1(k_F d) = g_1(\pi)$, and for even numbers $g_1(k_F d) = g_1(2\pi)$. Then, when investigating the even–odd superconducting oscillations in Pb(111) (neglecting the beating pattern of about 7 ML), one should work with two thickness-dependent couplings, i.e.,

$$g_{\text{odd}} = g_0 - \frac{g_1(\pi)}{N}, \quad g_{\text{even}} = g_0 - \frac{g_1(2\pi)}{N}, \qquad (9.75)$$

for odd and even numbers of monolayers, respectively. Notice that the well-known phonon-softening mechanism [see, e.g., Dickey

and Paskin (1968), Naugle et al. (1973), and Leavens and Fenton (1980)] due to surface phonons in nanostructured materials can be incorporated in the same manner. Indeed, here deviations of g from its bulk value can also be expected to follow the $1/d$ dependence (with possible damped quantum-size oscillations as well) because the relative number of atoms at the sample surface is proportional to S/V with S the surface area, and $S/V \propto 1/d$. Therefore, Equation 9.74 can be thought to include both, substrate and surface effects, which are opposite in the case of Pb on silicon. We also should keep in mind the presence of Au protective layer on Pb(111) films in the experiments by Guo et al. (2004), Bao et al. (2005), and Zhang et al. (2005).

Analysis of the experimental results reported by Guo et al. (2004) and Bao et al. (2005) with the help of a numerical solution of the BdG equations [see Shanenko et al. (2006b, 2007a)] shows that $g_1(\pi)N(0) \sim 0.2$ and $g_1(2\pi)N(0) \sim 0.2$–0.3. Hence, deviations in g from its bulk value can play a role when Pb(111) nanofilms on silicon are about or less than 10–20 ML (3–6 nm) in thickness. Here, it is of interest to notice that according to the comprehensive experimental study by Naugle et al. (1973), the surface phonon softening becomes of no importance in aluminum and tin films with $d > 5$ nm. A comparison of a numerical solution of the BdG equations [from Shanenko et al. (2006b, 2007a)] with the experimental results from Bao et al. (2005) is presented in Figure 9.11. As seen, the experimental data can be well understood in the framework of the BdG equations. Theoretical results given in Figure 9.11 utilize, for the sake of simplicity, the period of the quantum-size oscillations being exactly 2 ML and, so, neglect any beating pattern of the even–odd oscillations due to deviations of this period from 2 ML (see discussion in the previous section). Thickness 22 ML appears to be a node of such beating oscillations (Guo et al., 2004). In particular, below 22 ML the even-layered Pb(111) nanofilms become unstable and, as known [see, e.g., Guo et al. (2004)], the quantum-size effects strongly regulate stability and growth of such nanofilms. This explains why the theoretical results for the even–odd oscillations deviate from the experimental data in panel (c) when approaching 21–22 ML. It is interesting to notice that experimental results for Pb(111) films on silicon by another group, i.e., Eom et al. (2006), show persistent quantum oscillations of T_c down to 5 ML, without any sign of the T_c-suppression (see Figure 9.9). This suggests that parameters of the thickness-dependent g given by Equation 9.74 are sensitive to particular experimental details (e.g., on the presence/absence of a protective cap and the cap material). In addition, we note that a suppression of the critical temperature in Pb(111) nanofilms on silicon, when thickness goes down to 5–8 ML, was also observed in Özer et al., 2006a,b, as well. The authors attributed such a decrease in T_c to the surface effects and treated it by incorporating the surface-energy term into the GL free-energy functional. However, such a treatment is an oversimplification because the GL equations cannot be used for nanoscale superconductors (see our discussion in the Introduction).

Another interesting example is superconducting nanofilms made of grains with practically no tunnel barrier between them. In this case the electrons are not confined in the grains, and

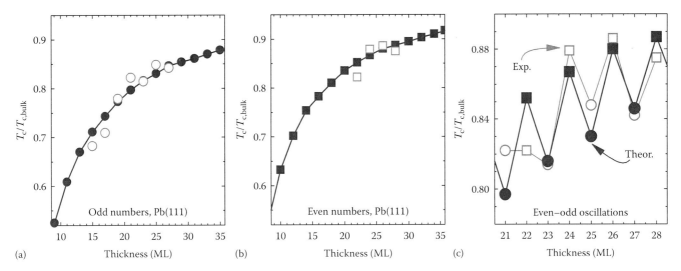

FIGURE 9.11 Critical temperature T_c in units of the bulk superconducting temperature $T_{c,\text{bulk}}$ for Pb nanofilms on silicon (111) versus the thickness: (a) films with odd number of monolayers (open circles are results of the BdG equations with $g_1(\pi)N(0) = 0.204$, solid circles are experimental data from Bao et al. (2005); (b) even number of monolayers (solid squares are the theory with $g_1(2\pi)N(0) = 0.26$, open circles are the experiment); (c) the data from panels (a) and (b) plotted together to highlight the even–odd oscillations. (From Shanenko, A.A. et al., *Phys. Rev. B*, 75, 014519, 2007a.)

the electron wave functions "feel" the sample boundaries. Such nanofilms cannot be atomically uniform in thickness, and the thickness fluctuations result in washing out any quantum-size oscillations. However, as discussed in the previous section, one can expect that the superconducting characteristics of a nonuniform nanofilm are mainly determined by size-dependent superconducting resonances in a uniform nanoslab: We should take the resonant thicknesses from the interval covered by the fluctuations. This is illustrated in Figure 9.12, where results of numerical solution of the BdG equations for energy gap Δ_E [from Peeters et al. (2008)] are compared with the corresponding experimental data for aluminum nanofilms [from Court et al. (2008)]. Panel (a)

shows theoretical data calculated with Equations 9.1a and b for Al nanofilms uniform in thickness. As well as the experimental points are for $d > 5\,\text{nm}$, the coupling constant g is taken the same as for bulk [for the bulk parameters of metal superconductors see, e.g., de Gennes (1966), Fetter and Walecka (2003), and Arutyunov et al. (2008)]. The effective Fermi level for Al is given in Table 9.1. As seen from Figure 9.12a, Δ_E (calculated at $T = 0$) exhibits significant resonant enhancements, as compared to bulk (an advantageous feature of weak-coupling superconductors). The thickness-dependent general trend of the resonant enhancements [$\Delta_E/\Delta_{\text{bulk}} = 1 + (D/d)^{4/3}$, with some characteristic length $D = 3.19\,\text{nm}$] is plotted as a function of d in Figure 9.12b

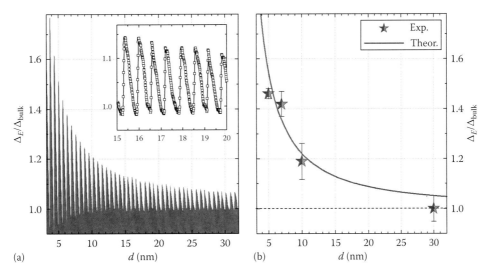

FIGURE 9.12 Zero-temperature energy gap Δ_E as function of the thickness d of aluminum nanofilm: (a) theoretical results for uniform thickness calculated from the BdG equations. (From Peeters, F.M. et al., *Physica C*, 468, 326, 2008. With permission.) and (b) experimental data (stars) versus the general trend of the superconducting enhancements from panel (a). (From Court, N.A. et al., *Supercond. Sci. Technol.*, 21, 015013-1, 2008. With permission.)

together with the four experimental points. As seen, the experimental results perfectly follow the general trend of the superconducting resonances of uniform aluminum films. Notice that the energy gap in the thickest 30 nm film was taken as Δ_{bulk} for the experimental set in Figure 9.12b.

We would like to summarize that numerical results of the BdG equations are found to be in good agreement with first recent experiments on high-quality superconducting Pb and Al nanofilms. Such a good understanding of these experimental results can be very helpful for optimization of possible future superconducting nanodevices.

9.3.3 Superconducting Resonances in Nanowires

Inevitable cross-sectional fluctuations are a specific feature of all high-quality superconducting metallic nanowires, which is similar to polycrystalline nanofilms considered in the previous section. To investigate the effect of the transverse quantization of the electron spectrum on superconductivity in this case, we need to check

a general trend of the resonant enhancements in superconducting nanowires with a uniform cross section. In Figure 9.13, results of numerically solving the BdG equations for Al and Sn nanocylinders are compared with recent experimental data for high-quality nanowires. Panels (a) and (c) show amplitudes of the resonant enhancements in Al and Sn nanocylinders, respectively, as calculated from Equations 9.1a and b (Shanenko et al., 2006a). In Figure 9.13b, the experimental points for Al nanowires (Zgirski and Arutyunov, 2007) are compared with the thickness-dependent trend of the resonant enhancements given in Figure 9.13a, and the same for Sn in Figure 9.13d [the experimental points are taken here from Tian et al. (2005) and Janković et al. (2006)]. Notice that similar to nanofilms, the trend of the resonant enhancements in superconducting nanowires can be approximated as $T_c/T_{c,bulk} = 1 + (D/d)^{4/3}$, with $D_{Al} = 7.23$ nm (compare with $D = 3.19$ nm for aluminum nanofilms in Figure 9.12b) and $D_{Sn} = 4.07$ nm. The horizontal error bars of the experimental data are due to fluctuations in the wire cross section. All the results for T_c are shown in units of the bulk critical temperature $T_{c,bulk}$. For the Al experimental set, the critical temperature of the thickest fabricated wire with

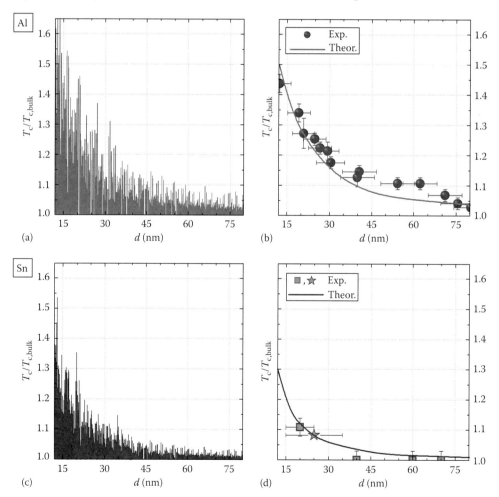

FIGURE 9.13 The thickness-dependent enhancement of T_c in metallic superconducting nanowires: (a) theoretical results from the BdG equations for Al nanocylinder (Shanenko et al., 2006a), and (b) experimental results from Zgirki and Arutyunov (2007) (solid circles) versus the general trend of the resonant enhancements in the previous panel (solid curve); (c) and (d), the same but for Sn [squares and the star are experimental results from Tian et al. (2005) and Janković et al. (2006), respectively].

$d \approx 116\,\text{nm}$ is taken as $T_{c,\text{bulk}} = 1.23\,\text{K}$ (Zgirski and Arutyunov, 2007). For Sn, the experimental points are plotted with $T_{c,\text{bulk}} = 3.7\,\text{K}$, the critical temperature of a 100 nm thick wire investigated in Tian et al. (2005). The theoretical results in Figure 9.13 were calculated with the coupling constant g being the same as in bulk. According to the analysis in the previous section, one can expect that g is not altered by surface effects (i.e., due to a substrate, protective cover, and phonon softening) in high-quality nanowires with $d \geq 10\,\text{nm}$. As to the effective Fermi level E_F (see Table 9.1), for Sn it is extracted from the experimental data of Orr et al. (1984), where the thickness-dependent oscillations of the critical temperature in Sn superconducting films were observed. However, those films were not smooth and uniform in thickness, and, moreover, oscillations of the normal-state resistivity with thickness were not observed. For Al, we take the same value of the effective Fermi level as in Han and Crespi (2004). It is of importance to note that the amplitudes of the resonant enhancements in superconducting nanowires are weakly dependent on E_F. For example, when E_F rises from 0.9 eV to $E_F = 2.6\,\text{eV}$ [this value is taken from Blatt and Thompson (1963)], the most pronounced enhancements in aluminum nanocylinder are reduced by 10%–15% while the change in the "ordinary" resonant deviations from the bulk value is practically negligible [see discussion in Shanenko and Croitoru (2006)]. However, the mean distance between two neighboring resonances is determined by λ_F and, so, is quite sensitive to E_F. For instance, instead of 1–2 resonances per 0.2 nm for $E_F = 0.9\,\text{eV}$, there appear 3–4 resonances for $E_F = 2.6\,\text{eV}$ (Shanenko and Croitoru, 2006). Thus, as seen from Figure 9.13, there is a good agreement between the experimental data and numerical results of the BdG equations (Equations 9.1a and b). This strongly suggests that the thickness-dependent increase of T_c found recently in high-quality Al and Sn superconducting nanowires is due to the quantum-size superconducting resonances.

As discussed in Section 9.2, in the presence of quantum confinement, the translational invariance is broken and the spatial distribution of the superconducting order parameter is not uniform any more. In particular, any profound resonant enhancement is accompanied by significant spatial variations of $\Delta(\mathbf{r})$. For example, this can be seen from Figure 9.14, where the superconducting order parameter for an Al nanowire with uniform square cross section is shown as calculated from the BdG equations for the resonant thicknesses $d = 1.98\,\text{nm}$ (a) and 2.95 nm (b) (Croitoru et al., 2007). For illustrative purposes, these calculations were performed for extreme narrow nanowires. In this case, the spatial variations of $\Delta(\mathbf{r})$ are much more distinctive and substantial. The coupling constant g was taken, for simplicity, as in bulk. Notice that when taking the typical aluminum parameters, i.e., $gN(0) = 0.18$ and $\hbar\omega_D = 32.3\,\text{meV}$ (de Gennes, 1966; Fetter and Walecka, 2003), the BdG equations result in the bulk energy gap $\Delta_{\text{bulk}} = 0.25\,\text{meV}$ (at $T = 0$). As seen from Figure 9.14, $\Delta(\mathbf{r})$ exhibits substantial local enhancements, i.e., by a factor of about 20–40 times larger than Δ_{bulk}. At the same time, there exist local minima in the spatial distribution of the pair condensate, where $\Delta(\mathbf{r})$ can drop down to 2–3Δ_{bulk} (see, for instance, point $x = 0.7$, $y = 0.7\,\text{nm}$ in the

contour plot for $d = 1.98\,\text{nm}$). Thus, one can say that the spatial distribution of the pair condensate is not simply nonuniform but often extremely nonuniform, especially, in the presence of resonant enhancements.

At a resonant thickness, the major contribution to $\Delta(\mathbf{r})$ is due to the subband(s) whose bottom is situated in the Debye window. In addition, such a subband is mainly responsible for the spatial variations in $\Delta(\mathbf{r})$. To have a feeling about these variations, let us switch to Anderson's solution. For nanowires with square cross section, the transverse single-electron spectrum is given by Equation 9.37, where $k = 0$ is assumed. So, when the bottom of the subband with some transverse quantum numbers i_x and i_y passes through the Fermi surface, we have

$$\frac{\pi^2}{d^2}\left[(i_x + 1)^2 + (i_y + 1)^2\right] = k_F^2. \quad (9.76)$$

Introducing the rational factor α given by $(i_y + 1) = \alpha\,(i_x + 1)$, one can find from the above expression that

$$\frac{\pi(i_x + 1)x}{d} = \frac{1}{\sqrt{1+\alpha^2}}k_F x, \quad \frac{\pi(i_y + 1)y}{d} = \frac{\alpha}{\sqrt{1+\alpha^2}}k_F y. \quad (9.77)$$

Inserting Equation 9.77 into Equation 9.36 and using Anderson's approximation given by Equation 9.32, we can find that the contribution of the resonant subband to $\Delta(x, y)$ is proportional to

$$\sin^2\left(\frac{1}{\sqrt{1+\alpha^2}}k_F x\right)\sin^2\left(\frac{\alpha}{\sqrt{1+\alpha^2}}k_F y\right).$$

So, the transverse spatial variations of the order parameter (we remind that it is uniform in the longitudinal direction) at a resonant point are governed by $\lambda_F/2$ (the shift-up in E_F for small thicknesses can slightly change λ_F). However, due to the factor α, these variations can be significantly different for the x and y directions. The situation is more complicated in the presence of many single-electron subbands with bottoms situated in the Debye window, which is typical of large thicknesses. In this case, the spatial variations of the order parameter become smoothed: the larger the thickness, the less pronounced are the variations. Notice that for nanofilms, there is only one transverse coordinate z, and the resonant subband with the transverse quantum number i (see Equation 9.35) makes a contribution to the order parameter proportional to $\sin^2(k_F z)$.

When a resonance decays (with an increase or decrease in thickness), the superconducting order parameter drops down so that its spatially averaged value becomes close to the bulk one. In the off-resonance regime, the spatial variations of the pair condensate become less extreme but still pronounced, as seen from Figure 9.15. Here, the decay of the superconducting resonance arising at $d = 2.95\,\text{nm}$ (the same as in Figure 9.14b) is shown in panel (a), where the diagonal order parameter $\Delta(x, y)|_{x=y}$ is given versus x for $d = 2.95$ (the resonant point), 3.0, and 3.1 nm (the off-resonance regime). Figure 9.15b demonstrates the same but for the resonance appearing at $d = 3.15\,\text{nm}$. Notice that in

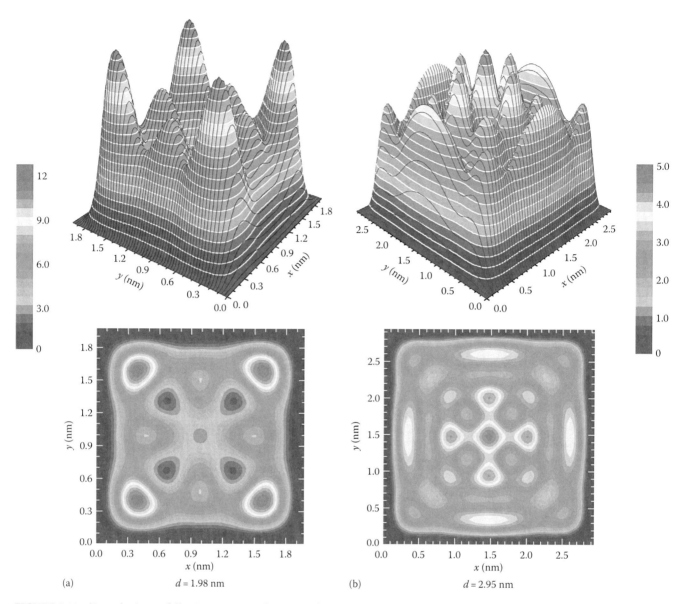

FIGURE 9.14 **(See color insert following page 25-16.)** Superconducting order parameter $\Delta(\mathbf{r}) = \Delta(x, y)$ (at $T = 0$) together with its contour plot for an Al nanowire with square cross section for the resonant widths: (a) $d = 1.98$ nm, (b) $d = 2.95$ nm (both points are resonances).

general, the smaller the nanowire width, the more profound is the increase of the order parameter (the critical temperature) at a resonant point. For mesoscopic nanowires with thickness more than 100 nm, the quantum-size oscillations are washed out and superconducting resonances disappear (see Figure 9.13). However, the spatial variations of the order parameter can be still significant in the areas close to the boundary. The reason for such a boundary effect is similar to the well-known Friedel oscillations induced by a point charge placed into an electron gas. Investigations of Anderson's solution (numerically solving of the BdG equations for $d > 100$ nm is rather time consuming) show that such Friedel-like oscillations show no tendency to disappear, as can be expected.

Concluding, in this section we discussed the quantum-size oscillations and superconducting size–dependent resonances

in nanofilms and nanowires. Such superconducting oscillations were only recently observed, thanks to progress in nanofabrication. However, it is interesting to note that during decades the quantum-size pairing-gap oscillations were investigated in nuclear physics, where the nucleon–nucleon attraction results in the superfluid pair correlations [see Hilaire et al. (2002) and references therein]. Hence, the quantum-size superconducting/superfluid oscillations are already a long-standing theoretical and experimental problem. However, this is not the only interesting point related to the transverse quantization in nanofilms and nanowires. Quantized transverse superconducting modes suggest a wealth of new interesting phenomena not investigated yet in detail. Some of them typical for nanoscale superconductivity are discussed in the following section.

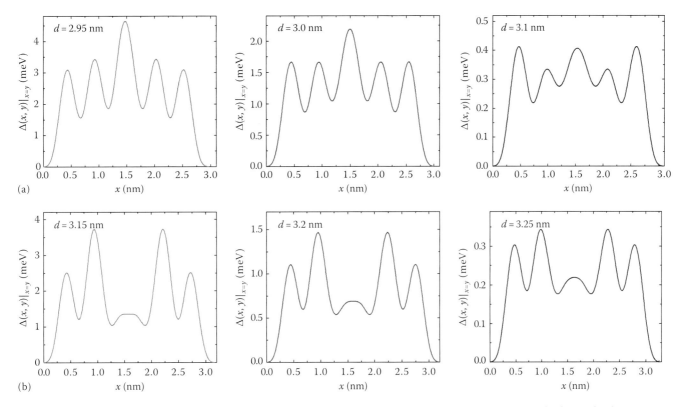

FIGURE 9.15 Decay of superconducting resonance with increasing thickness of square cross-section nanowire: (a) the diagonal order parameter $\Delta(x, y)|_{x=y}$ versus x for $d = 2.95$ (the resonant width), 3.0, and 3.1 nm (the off-resonance point); (b) the same but for $d = 3.15$ (the resonant width), 3.2, and 3.25 nm (the off-resonance regime).

9.4 Nanoscale Superconductivity in Quantum-Size Regime

9.4.1 New Andreev-Type States due to Quantum Confinement

It is well known that quasiparticles "feel" the spatial variation of the superconducting order parameter as a kind of potential barrier. This physical mechanism is the basis for Andreev quantization, and can be referred to as the *Andreev mechanism*. Andreev states were investigated previously: (1) for an isolated normal region of the intermediate state of a type-I superconductor (Andreev, 1964) or for a similar situation of superconductor–normal–superconductor contacts (de Gennes and Saint-James, 1963), and (2) in the core of a single vortex for the mixed state of a type-II superconductor (Caroli et al., 1964). It turns out that the Andreev mechanism can also manifest itself on the nanoscale through the formation of new Andreev-type states appearing due to spatial inhomogeneity of the superconducting condensate (Shanenko et al., 2007b). Such new Andreev-type states induced by quantum confinement are mainly located beyond the regions where the order parameter is enhanced. We remark that they cannot be localized in the domains where the order parameter is significantly suppressed because the characteristic length for spatial variations of the order parameter in the case of interest is of about the sample thickness $d \ll \xi$, with ξ the GL coherence length [for more detail about Andreev states, see, e.g., Duncan and Györffy (2002) and Deutscher (2005)].

The probability density to locate a quasiparticle at the point **r** is given by

$$\zeta_v(\mathbf{r}) = \left| \langle \mathbf{r} | u_v \rangle \right|^2 + \left| \langle \mathbf{r} | v_v \rangle \right|^2, \quad \int d^3 r \zeta_v(\mathbf{r}) = 1. \tag{9.78}$$

Then, the successive quasiparticles that avoid local enhancements of the superconducting condensate can be specified by the following characteristic parameter (Shanenko et al., 2008b):

$$\left\langle \hat{\Delta} \right\rangle_v = \left\langle u_v \left| \hat{\Delta} \right| u_v \right\rangle + \left\langle v_v \left| \hat{\Delta} \right| v_v \right\rangle = \int d^3 r \Delta(\mathbf{r}) \zeta_v(\mathbf{r}). \tag{9.79}$$

To proceed, Anderson's solution can be involved. When inserting Equation 9.32 into Equation 9.79, we conclude that

$$\left\langle \hat{\Delta} \right\rangle_v = \Delta_v, \tag{9.80}$$

with Δ_v given by Equation 9.44. As seen, the successive quasiparticles, i.e., the new Andreev-type states induced by quantum confinement, have generally lower excitation energies E_v. For example, in the absence of a superfluid motion, Δ_v controls the energy gap of the corresponding quasiparticles (see Section 9.2.2.1).

How can one experimentally track the formation of such Andreev-type states? A promising possibility (see Shanenko et al., 2007b) is to check the ratio $\Delta_E/k_B T_c$. The main point is that the Andreev-type states can significantly reduce the excitation energy

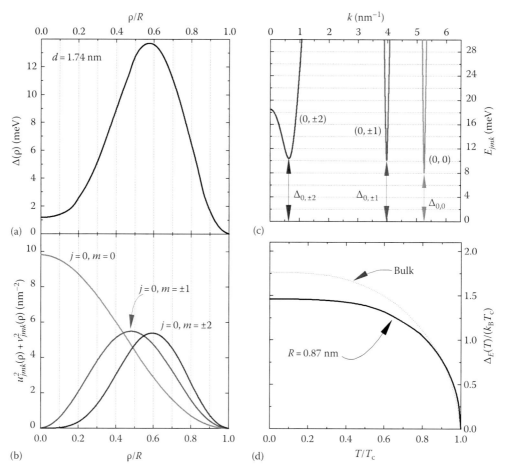

FIGURE 9.16 Aluminum nanocylinder for the resonant diameter $d = 1.74$ nm. (a) The order parameter versus ρ/R at $T = 0$; (b) the corresponding quasiparticle transverse distributions for the five relevant subbands $(j, m) = (0, 0)$, $(0, \pm1)$, and $(0, \pm2)$ (dependence on k is negligible); (c) E_{jmk} versus k for the same subbands $(T = 0)$; (d) the temperature-dependent energy gap $\Delta_E(T)$ in units of $k_B T_c$ versus the relative temperature T/T_c, the bulk results is given by the dotted curve. (From Shanenko, A.A. et al., *Phys. Rev. Lett.*, 99, 067007-1, 2007.)

gap Δ_E while having no serious effect on T_c. Indeed, let us take a look at Figure 9.16, where numerical results of the BdG equations are given for an aluminum nanocylinder with $d = 1.74$ nm. Here, for simplicity, the diameter is taken extremely small, and the coupling constant g is assumed to be the same as in bulk. The transverse profile of the order parameter $\Delta(\rho)$ is shown in Figure 9.16a, where a large enhancement is seen at $\rho/R \approx 0.6$. The spatial transverse distribution of bogolons is specified by $u_{jmk}^2(\rho) + v_{jmk}^2(\rho) = 2\pi L \zeta_{jmk}(\rho)$. This quantity is shown versus the radial coordinate ρ in panel (b) for the five relevant single-electron subbands with $(j, m) = (0, 0)$, $(0, \pm1)$, and $(0, \pm2)$ (only the states from these five subbands make a contribution to the order parameter). We remark that $u_{jmk}(\rho)$ and $v_{jmk}(\rho)$ are practically independent of k (see Section 9.2.2.1). The two subbands with $(j, m) = (0, \pm2)$ are responsible for the resonance that develops in the system at $R = 0.97$ nm. They mainly control the local enhancement of $\Delta(\rho)$, as seen from comparing the profile of $u_{jmk}^2(\rho) + v_{jmk}^2(\rho)$ with that of $\Delta(\rho)$. This is why $\Delta_{0,\pm2}$ is the largest subband gap, as seen from Figure 9.16c, where E_{jmk} is given versus k. Superconducting gaps for the other relevant subbands, i.e., $\Delta_{0,\pm1}$

and $\Delta_{0,0}$, are reduced to some extent as compared to $\Delta_{0,\pm2}$. The most pronounced reduction is for $(j, m) = (0, 0)$: $\Delta_{0,0}$ is smaller by 20%. These quasiparticles are most successful in avoiding the enhancement of the pair condensate at $\rho/R \approx 0.6$. The quasiparticles with $(j, m) = (0, \pm1)$ have practically the same transverse distribution as the bogolons with $(0, \pm2)$. As a result, $\Delta_{0,\pm1}$ is nearly the same as $\Delta_{0,\pm2}$. Thus, Figure 9.16 makes it possible to arrive at the following conclusions. Quasiparticles associated with the resonant single-electron subband(s) are responsible for significant local enhancements in $\Delta(\rho)$. The other quasiparticles "feel" such enhancements as a kind of potential barrier, which gives rise to the formation of the new Andreev-type states. As a rule, these states corresponds to single-electron subbands whose bottoms are situated quite below the Fermi level. Hence, the contribution of the Andreev-type states to the superconducting order parameter (and the critical temperature) is minor. However, they result in a pronounced reduction of the total energy gap Δ_E. This leads to a drop in the ratio $\Delta_E/k_B T_c$ as compared to bulk, see Figure 9.16d.

Clear signatures of such a drop in $\Delta_E/k_B T_c$ due to the formation of the thickness-dependent superconducting resonances are found

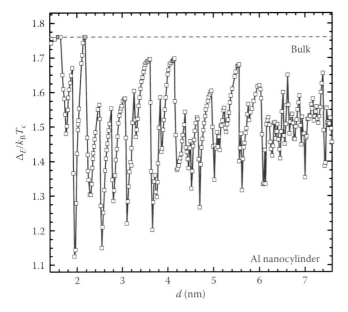

FIGURE 9.17 The zero-temperature energy gap Δ_E in units of $k_B T_c$ versus d for an aluminum superconducting nanocylinder. (From Shanenko, A.A. et al., *Phys. Rev. Lett.*, 99, 067007-1, 2007b. With permission.)

not only for extremely small diameters but survive up to diameters $d = 10–15\,\text{nm}$. In Figure 9.17, the ratio $\Delta_E/k_B T_c$ (the energy gap is taken at zero temperature) exhibits substantial quantum-size oscillations accompanied by significant drops at the resonant points. It is worth noting that even for the off-resonance points, this ratio is still below the BCS bulk value 1.763. In the presence of inevitable cross-section fluctuations, we expect that the quantum-size oscillations in $\Delta_E/k_B T_c$ will be smoothed into an overall decrease (as compared to bulk) with decreasing thickness. For the case of $d \approx 6–10\,\text{nm}$, corresponding to the narrowest high-quality superconducting nanowires fabricated in recent experiments (Savolainen et al., 2004; Tian et al., 2005; Zgirski et al., 2005; Altomare et al., 2006; Janković et al., 2006; Tombros et al., 2008; Wang et al., 2008), this ratio still deviates, on average, from the bulk value by about 10%. Thus, it can be probed through, e.g., tunneling experiments.

Concluding our discussion on the new Andreev-type states induced by quantum confinement, we notice that they play no serious role in superconducting metallic nanofilms, as seen from the calculations presented in Shanenko et al. (2008b) (however, this can be different for nanofilms fabricated of novel materials such as superconducting semiconductors). The reason is that the spatial distribution of the pair condensate in films is much more uniform as compared to nanowires. This conclusion is in agreement with the experimental results for Pb single-crystalline nanofilms where the ratio $\Delta_E(T)/k_B T_c$ was found to be close to the bulk one (Eom et al., 2006).

9.4.2 Quantum-Size Cascades

One more interesting aspect of nanoscale superconductivity concerns the quantum-size cascade transitions from the superconducting state to the normal one induced by a magnetic field

or a supercurrent. Below, such cascades are considered for two particular situations mentioned in Section 9.2.1.3, i.e., for a cylindrical nanowire in parallel magnetic field and in the presence of a longitudinal supercurrent.

9.4.2.1 Cascades Induced by a Magnetic Field

Let us consider a superconducting cylindrical nanowire in a parallel magnetic field, neglecting vortex formation (vortices can hardly appear in nanowires). According to the GL theory (Silin, 1951; Lutes, 1957), (1) the critical magnetic field increases as $1/d$ in mesoscopic wires, and (2) the superconductor-to-normal transition driven by a parallel magnetic field is of second order for narrow mesoscopic wires while of first order in bulk type-I superconductors (de Gennes, 1966). Recent calculations within the BdG equations for wires with $d = 20 \div 200\,\text{nm}$ (Han and Crespi, 2004) confirmed the GL conclusion about the second-order transition, which is in agreement with recent experimental data for Sn (Janković et al., 2006; Sorop and Jongh, 2007) and Zn (Kurtz et al., 2007) with $d > 20\,\text{nm}$.

On the nanoscale, the situation changes in a qualitative way. Based on a numerical solution of the BdG equations (Equations 9.1a and b) for cylindrical nanowires in a parallel magnetic field, it was recently shown (Shanenko et al., 2008a) that (1) the superconductor-to-normal transition occurs as a cascade of jumps in the order parameter as a function of $H_\parallel$ for diameters $d < 10–15\,\text{nm}$ at zero temperature, (2) the critical magnetic field is strongly enhanced and exhibits pronounced quantum-size oscillations. The Pauli paramagnetism was found to be significant for smaller diameters $d < 5\,\text{nm}$. In Figure 9.18, the spatially averaged order parameter

$$\bar{\Delta} = \frac{1}{\pi R^2 L} \int d^3 r \Delta(\mathbf{r}) = \frac{2}{R^2} \int_0^R d\rho \rho \Delta(\rho),$$

as calculated from the BdG equations for an aluminum nanocylinder at $T = 0$ (Shanenko et al., 2008a), is plotted versus $H_\parallel$ for the resonant diameters $d = 3.1$, 5.85, and $10.86\,\text{nm}$ (the Pauli paramagnetism is not included, the coupling constant g is taken the same as in bulk). $\bar{\Delta}$ as a function of $H_\parallel$ exhibits two distinctive regimes. For small magnetic fields, the order parameter is independent of $H_\parallel$. From the above discussion about Anderson's solution (see Section 9.2.2.2), we know that this is a signature of the gap regime, i.e., $\Delta_E > 0$. The quasiparticle energies with positive m are systematically shifted down by $H_\parallel$ (see Equation 9.65). As a result, there appear bogolons with negative energies and, so, we arrive at the gapless regime, i.e., $\Delta_E = 0$. This regime is characterized by a decrease of $\bar{\Delta}$ accompanied by a sequence of jumps. Such a *cascade of jumps* appears due to the transverse quantization of the electron motion: a drop in $\bar{\Delta}$ arises each time when a quasiparticle branch touches zero, and the system ground state is modified due to the break up of electron pairs in the corresponding single-electron subband. If such a subband controls a resonant enhancement, then we have, as a rule, a large jump down to zero, i.e., the complete decay of the superconducting

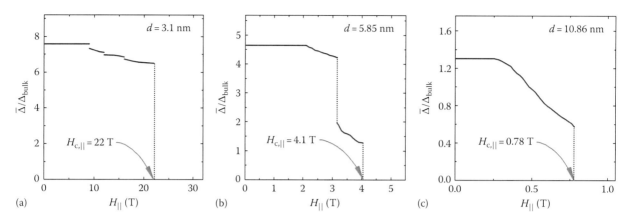

FIGURE 9.18 The spatially averaged order parameter $\bar{\Delta}$ (in units of Δ_{bulk}) as a function of the parallel magnetic field $H_\parallel$ for zero-temperature cylindrical Al nanowire for the three resonant diameters $d = 3.1$ (a), 5.85 (b), and 10.86 nm (c).

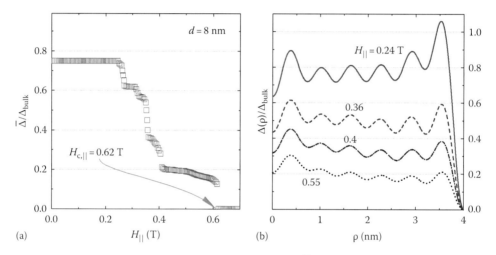

FIGURE 9.19 Aluminum nanocylinder in the off-resonance regime ($d = 8$ nm): (a) $\bar{\Delta}/\Delta_{bulk}$ versus $H_\parallel$ and (b) $\Delta(\rho)/\Delta_{bulk}$ given for $H_\parallel = 0.24$, 0.36, 0.4, and 0.55 T.

order. It is of importance to note that clear signatures of the cascades of jumps are present even in the off-resonance regime, as seen from Figure 9.19a, where $\bar{\Delta}$ is plotted versus $H_\parallel$ for the off-resonant diameter $d = 8$ nm. Figure 9.19b shows the dependence of the superconducting order parameter $\Delta(\rho)$ on the transverse coordinate ρ for different magnetic fields at the same diameter. As seen, diamagnetic current "eats" the order parameter with an increase in $H_\parallel$. This process is more intensive near the wire surface, where the density of diamagnetic current is larger.

Notice that $H_{c,\parallel}$ has been measured in Sn (Jankovič et al., 2006) and Zn (Sorop and Jongh, 2007) wires with diameters down to 20 nm. These wires were found to be still in the mesoscopic regime. It is expected that data on $H_{c,\parallel}$ for $D < 20$ nm will be available soon.

For such narrow nanowires, we can expect that the superconductor-to-normal transition driven by $H_\parallel$ occurs as a cascade of jumps (al low temperatures) and the critical magnetic field $H_{c,\parallel}$ exhibits an oscillating (but smoothed due to inevitable cross-section fluctuations) enhancement of $H_{c,\parallel}$ with decreasing thickness.

Concluding this section, we remark that a similar cascade should exist for the superconductor-to-normal transition driven

by a magnetic field perpendicular to a nanowire. As for high-quality metallic nanofilms, the situation can be more complex due to the formation of vortices.

9.4.2.2 Cascades in the Presence of a Supercurrent

Now we investigate how the superconducting condensate is suppressed in a cylindrical nanowire by a longitudinal supercurrent. One can expect that quantum confinement has a strong impact on the superconductor-to-normal transition driven by a longitudinal supercurrent, as well.

Figure 9.20 displays the spatially averaged order parameter $\bar{\Delta}$ as calculated from the BdG equations at zero temperature for an aluminum nanocylinder in the presence of a longitudinal supercurrent (all the parameters are the same as in previous example with parallel magnetic field). $\bar{\Delta}$ versus the superfluid velocity v_s is shown for the resonant diameters $d = 10.12$, 11.56, 12.94, and 13.66 nm in panel (a). The superfluid velocity is given in units of the Landau bulk velocity defined by $v_L = \Delta_{bulk}/\hbar k_F$ (see, e.g., Landau and Lifshitz, 1958; Zagoskin, 1998). As follows from Figure 9.20a, the decay of the superconducting state occurs in a way similar to the previous case of parallel magnetic field,

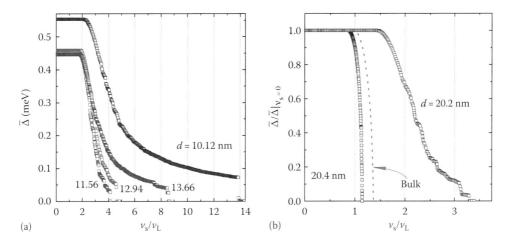

FIGURE 9.20 Superconductor-to-normal cascade transition in aluminum nanocylinder with a longitudinal supercurrent: (a) the spatially averaged order parameter $\bar{\Delta}$ as function of the relative superfluid velocity v_s/v_L for the resonant diameters $d = 10.12$, 11.56, 12.94, and 13.66 nm; (b) $\bar{\Delta}$ versus v_s/v_L for diameters $d = 20.2$ nm (resonant point) and 20.4 nm (off-resonance diameter), the dotted line represents the bulk behavior.

i.e., through a cascade of jumps in $\bar{\Delta}$ as a function of the super-fluid velocity. Jumps survive up to diameters of about 15–20 nm. However, clear signatures of the cascades of jumps smoothed into continuous steplike drops can be observed even for $d = 20$–40 nm (see Figure 9.20b). The physics behind the jumps in $\bar{\Delta}$ as a function of v_s is the same as in the situation with parallel magnetic field. However, the mechanism of the suppression of the pair condensate by a longitudinal supercurrent has some peculiarities. Figure 9.21 shows results for the resonant diameter $d = 4.2$ nm (at $T = 0$). The dependence of $\bar{\Delta}$ on v_s is given in panel (a). The quasiparticle energies for the two subbands with $(j, m) = (0, \pm 7)$ and $(1, \pm 4)$ are plotted versus k for $v_s = 54.2\ v_L$ (b) and $v_s = 79.3\ v_L$ (c). As seen from Figure 9.21a, $v_s = 54.2\ v_L$ and $v_s = 79.3\ v_L$ are exactly the points of the two most pronounced jumps in $\bar{\Delta}$. The single-electron subbands with $(j, m) = (0, \pm 7)$ and $(1, \pm 4)$ control the resonant enhancement at $d = 4.2$ nm (the bottoms of these subbands are situated in the Debye window). In the presence of a longitudinal supercurrent, the quasiparticle states with $k < 0$ and $k > 0$ are shifted down and up, respectively (see Section 9.2.2.2). The larger the superfluid velocity, the stronger

the shift. Each time when a quasiparticle branch touches zero, a jump in the dependence of $\bar{\Delta}$ on v_s appears. Such a jump is almost insignificant if the corresponding single-electron subband makes a minor contribution to the order parameter. For the case of a resonant subband, $\bar{\Delta}$ exhibits a pronounced jump like the drops at $v_s = 54.2\ v_L$ and $v_s = 79.3\ v_L$ in Figure 9.21a. The resonant sub-bands are situated within close proximity to the point $k = 0$ and, so, have small Doppler shifts as compared to other relevant sub-bands. This is why pronounced drops in $\bar{\Delta}$ are preceded by less important jumps, which results in the formation of spectacular quantum-size cascades. Notice that the states from the resonant subbands are usually specified by long longitudinal wavelength $2\pi/k \approx 20$–30 nm. Due to this fact, we can expect that the pronounced jumps in $\bar{\Delta}$ are not very sensitive to surface roughness.

The critical current density j_c is defined as the maximum value of the supercurrent density taken as a function of v_s. In bulk, such a maximal supercurrent density is only 2% above the supercurrent corresponding to the Landau velocity (Bardeen, 1962). The critical superfluid velocity v_c usually means that there is no superconducting solution at $v_s > v_c$.

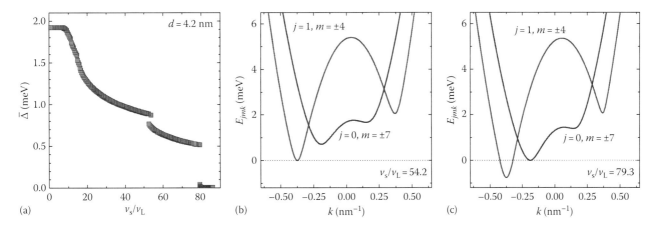

FIGURE 9.21 (a) The spatially averaged order parameter $\bar{\Delta}$ versus v_s/v_L at the resonant diameter $d = 4.2$ nm. The corresponding quasiparticle energies as function of k for (b) $v_s/v_L = 54.2$ and (c) $v_s/v_L = 79.3$ (only the two resonant subbands $(j, m) = 0, \pm 7; 1, \pm 4$ are given).

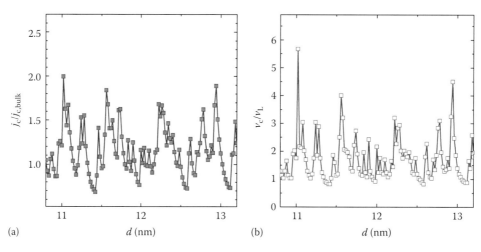

FIGURE 9.22 Aluminum superconducting nanocylinder: quantum-size oscillations of the critical current density (a) and critical superfluid velocity (b) as calculated from the BdG equations at zero temperature.

In bulk (see Figure 9.20b), $v_c \approx 1.36\, v_L$ (see, e.g., Swidzinsky, 1982; Zagoskin, 1998). Figure 9.22 shows numerical results of the BdG equations for the critical current density and critical superfluid velocity as function of the diameter of an aluminum nanocylinder. This result shows that quantum-size effects can play a very important role in the size-dependent enhancement of the critical current density. Real nanowires exhibit inevitable cross-section fluctuations that will smooth the quantum-size oscillations in an overall enhancement occurring with a decrease in thickness. This is due to the fact that the resonant enhancements in j_c and v_c are as a rule more profound than the off-resonance drops.

9.5 Conclusion

In conclusion, quantum confinement is the major mechanism governing physical properties of recently fabricated high-quality nanoscale superconductors. Based on a numerical solution of the BdG equations in the clean limit, we investigated how the superconducting properties can be modified due to quantum confinement in nanofilms and nanowires. In the case of interest, the conduction band splits up into a series of subbands, which results in quantum-size oscillations of the superconducting characteristics with profound resonant enhancements. At a resonant point, the superconducting order parameter (the energy gap, the critical temperature, and critical magnetic field) can be increased by more than an order of magnitude as compared to bulk. In nanowires, the spatial distribution of the pair condensate is very nonuniform, which can lead to the formation of new Andreev-type states. This significantly modifies the ratio of the energy gap to the critical temperature. In nanofilms, the spatial distribution of the order parameter is much more uniform, and new Andreev-type states induced by quantum confinement play no significant role. The superconducting-to-normal transition driven by a magnetic field parallel to a nanowire occurs as a cascade of jumps in the order parameter as a function of the magnetic field, contrary to the second-order transition found in the mesoscopic regime.

Similar cascade quantum transitions appear in superconducting nanowires with a longitudinal supercurrent.

Acknowledgments

This work was supported by the Flemish Science Foundation (FWO-Vl), the Interuniversity Attraction Poles Programme—Belgian State—Belgian Science Policy (IAP), the ESF-vortex and ESF-AQDJJ networks.

References

Abrikosov, A. A. 1988. *Fundamentals of the Theory of Metals*. Amsterdam, the Netherlands: Elsevier Science.

Abrikosov, A. A. and L. P. Gor'kov. 1958a. Theory for the superconductivity of alloys. 1.Electrodynamics of alloys at absolute zero. *Zh. Eksp. Teor. Fiz.* 35: 1558–1571 (in Russian) [English translation *Sov. Phys. JETP* 8: 1090 (1959)].

Abrikosov, A. A. and L. P. Gor'kov. 1958b. Superconducting alloys at temperatures above absolute zero. *Zh. Eksp. Teor. Fiz.* 36: 319–320 (in Russian) [English translation *Sov. Phys. JETP* 9: 220 (1959)].

Abrikosov, A. A., L. P. Gor'kov, and I. E. Dzyaloshinski. 1965. *Quantum Field Theoretical Methods in Statistical Physics*. Oxford, New York: Pergamon Press.

Altomare, F., A. M. Chang, M. R. Melloch, Y. Hong, and C. W. Tu. 2006. Evidence for macroscopic quantum tunneling of phase slips in long one-dimensional superconducting *Al* wires. *Phys. Rev. Lett.* 97: 017001-1–017001-4.

Altshuler, B. L. 1985. Fluctuations in the extrinsic conductivity of disordered conductors. *JETP Lett.* 41: 648–651.

Altshuler, B. L. and B. Z. Spivak. 1987. Mesoscopic fluctuations in a superconductor—Normal metal—Superconductor junctions. *Sov. Phys. JETP* 65: 343–347.

Altshuler, B. L., A. G. Aronov, and P. A. Lee. 1980. Interaction effects in disordered Fermi systems in two dimensions. *Phys. Rev. Lett.* 44: 1288–1291.

Anderson, P. W. 1959. Theory of dirty superconductors. *J. Phys. Chem. Solids* 11: 26–30.

Andreev, A. F. 1964. The thermal conductivity of the intermediate state in superconductors. *Sov. Phys. JETP* 19: 1228–1231.

Arutyunov, K. Yu., D. S. Golubev, and A. D. Zaikin. 2008. Superconductivity in one dimension. *Phys. Rep.* 464: 1–70.

Bao, X.-Y., Y.-F. Zhang, Y. Wang et al. 2005. Quantum size effects on the perpendicular upper critical field in ultrathin lead films. *Phys. Rev. Lett.* 95: 247005-1–247005-4.

Bardeen, J. 1962. Critical fields and currents in superconductors. *Rev. Mod. Phys.* 34: 667–681.

Beenakker, C. W. J. 1991. Universal limit of critical-current fluctuations in mesoscopic Josephson junctions. *Phys. Rev. Lett.* 67: 3836–3839.

Beloborodov, I. S., A. V. Lopatin, V. M. Vinokur, and K. B. Efetov. 2007. Granular electronic systems. *Rev. Mod. Phys.* 79: 469–517.

Blatt, J. M. and C. J. Thompson. 1963. Shape resonances in superconducting thin films. *Phys. Rev. Lett.* 10: 332–334.

Bogoliubov, N. N. 1958. On a new method in the theory of superconductivity. *Nuovo Cim.* 7: 794–805.

Bogoliubov, N. N. 1959. On principle of compensation and method of self-consistent field. *Sov. Phys. Usp.* 67: 549–580 (in Russian).

Caroli, C., P. G. de Gennes, and J. Matricon. 1964. Bound fermion states on a vortex line in a type II superconductor. *Phys. Lett.* 4: 307–309.

Chiang, T.-C. 2004. Superconductivity in thin films. *Science* 306: 1900–1901.

Court, N. A., A. J. Ferguson, and R. G. Clark. 2008. Energy gap measurement of nanostructured aluminium thin films for single Cooper-pair devices. *Supercond. Sci. Technol.* 21: 015013-1–015013-5.

Croitoru, M. D., A. A. Shanenko, and F. M. Peeters. 2007. Dependence of superconducting properties on the size and shape of a nanoscale superconductor: From nanowire to film. *Phys. Rev. B* 76: 024511-1–024511-6.

Czoschke, P., H. Hong, L. Basile, and T.-C. Chiang. 2004. Quantum beating patterns observed in the energetics of Pb film nanostructures. *Phys. Rev. Lett.* 93: 036103-1–036103-4.

Datta, S. 2005. *Quantum Transport: Atom to Transistor*. Cambridge, U.K.: Cambridge University Press.

Deo, P. S., V. A. Schweigert, and F. M. Peeters. 1997. Magnetization of mesoscopic superconducting disks. *Phys. Rev. Lett.* 79: 4653–4656.

Deutscher, G. 2005. Andreev–Saint-James reflections: A probe of cuprate superconductors. *Rev. Mod. Phys.* 77: 110–133.

Dickey, J. M. and A. Paskin. 1968. Phonon spectrum changes in small particles and their implications for superconductivity. *Phys. Rev. Lett.* 21: 1441–1443.

Duncan, K. P. and B. L. Györffy. 2002. Semiclassical theory of quasiparticles in the superconducting state. *Ann. Phys.* 298: 273–333.

Eom, D., S. Qin, M.-Y. Chou, and C. K. Shih. 2006. Persistent superconductivity in ultrathin Pb films: A scanning tunneling spectroscopy study. *Phys. Rev. Lett.* 96: 027005-1–027005-4.

Fetter, A. L. and J. D. Walecka. 2003. *Quantum Theory of Many-Particle Systems*. New York: Dover.

Finkel'stein, A. M. 1987. Superconducting transition temperature in amorphous films. *JETP Lett.* 45: 46–49.

Finkel'stein, A. M. 1994. Suppression of superconductivity in homogeneous disordered systems. *Physica B* 197: 636–648.

Geim, A. K., S. V. Dubonos, I. V. Grigorieva et al. 2000. Non-quantized penetration of magnetic field in the vortex state of superconductors. *Nature (London)* 407: 55–57.

de Gennes, P. G. 1966. *Superconductivity of Metals and Alloys*. New York: W. A. Benjamin.

de Gennes, P. G. and D. Saint-James. 1963. Elementary excitations in the vicinity of a normal metal-superconducting metal contact. *Phys. Lett.* 4: 151–152.

Gor'kov, L. P. 1958. On the energy spectrum of superconductors. *Sov. Phys. JETP* 7: 505–508.

Grigorenko, I., J. X. Zhu, and A. Balatsky. 2008. Optimization of the design of superconducting inhomogeneous nanowires. *J. Phys.: Condens. Matter* 20: 195204-1–195204-7.

Guo, Y., Y.-F. Zhang, X.-Y. Bao et al. 2004. Superconductivity modulated by quantum size effects. *Science* 306: 1915–1917.

Han, J. E. and V. H. Crespi. 2004. Discrete transverse superconducting modes in nanocylinders. *Phys. Rev. B* 69: 214526-1–214526-9.

Harada, K., O. Kamimura, H. Kasai, T. Matsuda, A. Tonomura, and V. V. Moshchalkov. 1996. Direct observation of vortex dynamics in superconducting films with regular arrays of defects. *Science* 274: 1167–1170.

Hilaire, S., J.-F. Berger, M. Girod, W. Satula, and P. Schuck. 2002. Mass number dependence of nuclear pairing. *Phys. Lett. B* 531: 61–66.

Janković, L., D. Gournis, P. N. Trikalitis et al. 2006. Carbon nanotubes encapsulating superconducting single-crystalline tin nanowires. *Nano Lett.* 6: 1131–1135.

Komnik, Yu. F., E. I. Bukhshtab, and K. K. Man'kovskii. 1970. Quantum size effect in superconducting tin films. *Sov. Phys. JETP* 30: 807–812.

Kurtz, J. S., R. R. Johnson, M. Tian et al. 2007. Specific heat of superconducting Zn nanowires. *Phys. Rev. Lett.* 98: 247001-1–247001-4.

Landau, L. D. and E. M. Lifshitz. 1958. *Statistical Mechanics*. London, U.K.: Pergamon Press.

Leavens, C. R. and E. W. Fenton. 1980. Superconductivity of small particles. *Phys. Rev. B* 24: 5086–5092.

Lee, P. A. and A. D. Stone. 1985. Universal conductance fluctuations in metals. *Phys. Rev. Lett.* 55: 1622–1625.

Li, S.-C., X. Ma, J.-F. Jia et al. 2006. Influence of quantum size effects on Pb island growth and diffusion barrier oscillations. *Phys. Rev. B* 74: 075410-1–075410-5.

Luh, D. A., T. Miller, J. J. Paggel, and T.-C. Chiang. 2002. Large electron-phonon coupling at an interface. *Phys. Rev. Lett.* 88: 256802-1–256802-4.

Lutes, O. S. 1957. Superconductivity of microscopic tin filaments. *Phys. Rev.* 105: 1451–1458.

Moshchalkov, V. V., L. Gielen, C. Strunk et al. 1995. Effect of sample topology on the critical fields of mesoscopic superconductors. *Nature* 373: 319–322.

Moshchalkov, V. V., M. Baert, V. V. Metlushko et al. 1998. Pinning by an antidot lattice: The problem of the optimum antidot size. *Phys. Rev. B* 57: 3615–3622.

Naugle, D. G., J. W. Baker, and R. E. Allen. 1973. Evidence for a surface-phonon contribution to thin-film superconductivity: Depression of T_c by noble-gas overlayers. *Phys. Rev. B* 7: 3028–3037.

Oreg, Y. and A. M. Finkel'stein. 1999. Suppression of T_c in superconducting amorphous wires. *Phys. Rev. Lett.* 83: 191–194.

Orr, B. G., H. M. Jaeger, and A. M. Goldman. 1984. Transition-temperature oscillations in thin superconducting films. *Phys. Rev. Lett.* 53: 2046–2049.

Özer, M. M., J. R. Thompson, and H. H. Weitering. 2006a. Hard superconductivity of a soft metal in the quantum regime. *Nat. Phys.* 2: 173–176.

Özer, M. M., J. R. Thompson, and H. H. Weitering. 2006b. Robust superconductivity in quantum-confined Pb: Equilibrium and irreversible superconductive properties. *Phys. Rev. B* 74: 235427-1–235427-11.

Özer, M. M., Y. Jia, Z. Zhang, J. R. Thompson, and H. H. Weitering. 2007. Tuning the quantum stability and superconductivity of ultrathin metal alloys. *Science* 316: 1594–1597.

Peeters, F. M., M. D. Croitoru, and A. A. Shanenko. 2008. Nanowires and nanofilms: Superconductivity in quantum-size regime. *Physica C* 468: 326–330.

Pfennigstorf, O., A. Petkova, H. L. Guenter, and M. Henzler. 2002. Conduction mechanism in ultrathin metallic films. *Phys. Rev. B* 65: 045412–045419.

Pitaevskii, L. P. 2004. 50 years of Landau's theory of superfluidity. *J. Low Temp. Phys.* 87: 127–135.

Savolainen, M., V. Touboltsev, P. Koppinen, K.-P. Riikonen, and K. Arutyunov. 2004. Ion beam sputtering for progressive reduction of nanostructures dimensions. *Appl. Phys. A* 79: 1769–1773.

Schweigert, V. A. and F. M. Peeters. 1998. Phase transitions in thin mesoscopic superconducting disks. *Phys. Rev. B* 57: 13817–13832.

Shanenko, A. A. and M. D. Croitoru. 2006. Shape resonances in the superconducting order parameter of ultrathin nanowires. *Phys. Rev. B* 73: 012510-1–012510-4.

Shanenko, A. A., M. D. Croitoru, M. Zgirski, F. M. Peeters, and K. Arutyunov. 2006a. Size-dependent enhancement of superconductivity in Al and Sn nanowires: Shape-resonance effect. *Phys. Rev. B* 74: 052502-1–052502-4.

Shanenko, A. A., M. D. Croitoru, and F. M. Peeters. 2006b. Quantum-size effects on T_c in superconducting nanofilms. *Europhys. Lett.* 76: 498–504.

Shanenko, A. A., M. D. Croitoru, and F. M. Peeters. 2007a. Oscillations of the superconducting temperature induced by quantum well states in thin metallic films: Numerical solution of the Bogoliubovde Gennes equations. *Phys. Rev. B* 75: 014519-1–014519-9.

Shanenko, A. A., M. D. Croitoru, R. G. Mints, and F. M. Peeters. 2007b. New Andreev-type states in superconducting nanowires. *Phys. Rev. Lett.* 99: 067007-1–067007-4.

Shanenko, A. A., M. D. Croitoru, and F. M. Peeters. 2008a. Magnetic-field induced quantum-size cascades in superconducting nanowires. *Phys. Rev. B* 78: 024505-1–024505-9.

Shanenko, A. A., M. D. Croitoru, and F. M. Peeters. 2008b. Superconducting nanofilms: Andreev-type states induced by quantum confinement. *Phys. Rev. B* 78: 054505-1–054505-8.

Silin, V. P. 1951. Superconducting cylinders and spheres in a magnetic field. *Zh. Eksp. Teor. Fiz.* 21: 1330–1336 (in Russian).

Skvortsov, M. A. and M. V. Feigel'man. 2005. Superconductivity in disordered thin films: Giant mesoscopic fluctuations. *Phys. Rev. Lett.* 95: 057002-1–057002-4.

Sorop, T. G. and L. J. de Jongh. 2007. Size-dependent anisotropic diamagnetic screening in superconducting Sn nanowires. *Phys. Rev. B* 75: 014510-1–014510-5.

Spivak, P. and F. Zhou. 1995. Mesoscopic effects in disordered superconductors near H_{c2}. *Phys. Rev. Lett.* 74: 2800–2803.

Swidzinsky, A. V. 1982. *Spatially Inhomogeneous Problems in the Theory of Superconductivity*. Moscow, Russia: Nauka (in Russian).

Tanaka, K. and F. Marsiglio. 2000. Anderson prescription for surfaces and impurities. *Phys. Rev. B* 62: 5345–5348.

Tanaka, K., I. Robel, and B. Janko. 2002. Electronic structure of multiquantum giant vortex states in mesoscopic superconducting disks. *Proc. Natl. Acad. Sci. U.S.A.* 99: 5233–5236.

Tian, M. L., J. G. Wang, J. S. Kurtz et al. 2005. Dissipation in quasi-one-dimensional superconducting single-crystal Sn nanowires. *Phys. Rev. B* 71: 104521-1–104521-7.

Tombros, N., L. Buit, I. Arfaoui et al. 2008. Charge transport in a single superconducting tin nanowire encapsulated in a multiwalled carbon nanotube. *Nano Lett.* 8: 3060–3064.

Upton, M. H., C. M. Wei, M. Y. Chou, T. Miller, and T.-C. Chiang. 2004. Thermal stability and electronic structure of atomically uniform Pb films on Si(111). *Phys. Rev. Lett.* 93: 026802-1–026802-4.

Valla, T., M. Kralj, A. Šiber et al. 2000. Oscillatory electron-phonon coupling in ultra-thin silver films on V(100). *J. Phys.: Condens. Matter* 12: L477–L482.

Wang, J., X.-C. Ma, L. Lu et al. 2008. Anomalous magnetoresistance oscillations and enhanced superconductivity in single-crystal Pb nanobelts. *Appl. Phys. Lett.* 92: 233119-1–233119-3.

Wei, C. M. and M. Y. Chou. 2002. Theory of quantum size effects in thin Pb(111) films. *Phys. Rev. B* 66: 233408-1–233408-4.

Yazdani, A. 2006. Lean and mean superconductivity. *Nat. Phys.* 2: 151–152.

Zagoskin, A. M. 1998. *Quantum Theory of Many–Body Systems: Techniques and Applications.* New York: Springer-Verlag.

Zgirski, M. and K. Yu. Arutyunov. 2007. Experimental limits of the observation of thermally activated phase-slip mechanism in superconducting nanowires. *Phys. Rev. B* 75: 172509-1–172509-4.

Zgirski, M., K.-P. Riikonen, V. Touboltsev, and K. Arutyunov. 2005. Size dependent breakdown of superconductivity in ultranarrow nanowires. *Nano Lett.* 5: 1029–1033.

Zhang, Y.-F., J.-F. Jia, T.-Z. Han et al. 2005. Band structure and oscillatory electron-phonon coupling of Pb thin films determined by atomic-layer-resolved quantum-well states. *Phys. Rev. Lett.* 95: 096802-1–096802-4.

Zyuzin, A. Yu. and B. Z. Spivak. 1990. Mesoscopic fluctuations of the resistance of point contact. *Sov. Phys. JETP* 71: 563–566.

10

One-Dimensional Quantum Liquids

Kurt Schönhammer
Georg-August-University
Goettingen

10.1 Introduction

The low-temperature thermodynamic properties of simple metals, like the alkalis, can be qualitatively understood in terms of the Sommerfeld model [1], which treats the conduction electrons as noninteracting fermions in a box. Typical results obtained using the model are a specific heat linear in temperature and a constant spin susceptibility which are in qualitative agreement with experiments. It took almost 30 years to understand why the low-temperature properties of interacting fermions are qualitatively the same. Landau's phenomenological Fermi liquid theory [12] rests on the assumption of quasiparticles, which are in a one-to-one correspondence to noninteracting fermions. This leads to a linear specific heat and a constant spin susceptibility but involves renormalized quantities like the effective mass and the quasiparticle interaction parameters [12], which are difficult to calculate microscopically. The consistency of the approach was shown using perturbation theory to infinite order and more recently by renormalization group techniques [23]. Landau's paper [12] marks the beginning of the general theory of (normal) quantum liquids [17], which are many-body systems in which the indistinguishability of the elementary constituents is important. The description of interacting bosons (liquid Helium 4, or the Bose alkali gases) is another theoretical challenge. These systems can undergo the phenomenon of Bose condensation. The related effect of Cooper pairing can occur in interacting Fermion systems (liquid Helium 3, electrons in the superconducting state).

The problem of interacting fermions (and bosons) simplifies in one dimension. This chapter focuses on interacting fermions. Comments on interacting bosons are presented at the end. In a pioneering paper [25] Tomonaga treated the case of a two-body interaction with a range much larger than the mean particle distance. He showed that the low-energy excitations of the noninteracting as well as the interacting system can be described

in terms of noninteracting bosons [19]. The important idea to solve the case of interacting fermions was the observation that a long-range interaction in real space is short ranged in momentum space and therefore only particles and holes in the vicinity of the Fermi points are involved in the interacting ground state and states with a low excitation energy. To obtain his results, Tomonaga linearized the energy dispersion of the noninteracting system around the two Fermi points $\pm k_F$. Luttinger [14] later used a model with strictly linear energy dispersion and presented the exact result for the mean occupation numbers. The complete solution for the Luttinger model was presented by Mattis and Lieb [15]. A very elegant method to exactly calculate correlation functions for the model is the bosonization of the fermion field operator [4,7,8,10,13]. The exponents of the power law decay of various correlation functions are determined by the anomalous dimension, which can be calculated explicitly for the Tomonaga–Luttinger (TL) model [13–15]. It was an important observation of Haldane [10] that the low-energy physics of the exactly solvable TL model provides the generic scenario for one-dimensional fermions with repulsive interactions. Like in the Landau Fermi liquid picture [12], a few parameters completely determine the low-energy physics. Generally they are as difficult to calculate as the Landau parameters. In contrast to the higher dimensional case there are additional exactly solvable models for which Haldane's Luttinger liquid scenario can be tested and the parameters like the anomalous dimension can be calculated using the Bethe-ansatz technique [22].

Fortunately even with only little background in the quantum many-body problem, the emergence of the Luttinger liquid (LL) picture can be understood. What is needed is a fresh look at the statistical mechanics of noninteracting fermions in one dimension and a comparison with the physics of the harmonic chain. These simple systems are treated in textbooks on solid state physics and statistical mechanics [1,18].

10.2 Noninteracting Fermions and the Harmonic Chain

A simple textbook example in quantum mechanics is a single fermion in one dimension in a box of length L with fixed boundary conditions $\phi(0) = \phi(L) = 0$. The single-particle eigenfunctions and energies are given by

$$\phi_n(x) = \sqrt{2/L} \sin(k_n x), \tag{10.1}$$

and

$$\epsilon_n \equiv \epsilon(k_n) = \frac{\hbar^2 k_n^2}{2m}, \tag{10.2}$$

with $k_n = n\pi/L$ and $n \in \mathbb{N}$. In Section 10.3, periodic boundary conditions are considered. Then the spacing of the k-values is doubled and negative as well as positive values occur. In the present section, the spin of the fermions is neglected for simplicity ("spinless fermions"). The ground state of N fermions is given by filling the lowest states up to the Fermi wavevector $k_F = n_F \pi/L$ where $n_F = N$, and the excited states can be classified by the fermionic occupation numbers $n_{k_n}^F$, which are 0 or 1. For fixed particle number N the sum of the occupation numbers has to be equal to N. This constraint makes the evaluation of the canonical partition function difficult and most textbooks only use the grand canonical ensemble.

The grand canonical results can be found in standard textbooks [1,18]. The total energy E_{gc} as a function of the temperature T and the chemical potential μ is given by

$$E_{gc} = \sum_n \frac{\epsilon_n}{e^{\beta[\epsilon_n - \mu]} + 1}, \tag{10.3}$$

where
$\beta = 1/(k_B T)$
k_B is the Boltzmann constant

The chemical potential μ becomes a function of temperature if the average particle number is assumed to be given by the fixed value N

$$N = \sum_n \frac{1}{e^{\beta[\epsilon_n - \mu(T)]} + 1}. \tag{10.4}$$

In the thermodynamic limit $L \to \infty$, keeping N/L fixed, the low-temperature results for the energy, specific heat, and other thermodynamic quantities can be evaluated explicitly using the Sommerfeld technique [1,18]. They are determined by the one-particle density of states per unit length $\rho_0(\epsilon) = 1/\pi\hbar v(\epsilon)$ in the *vicinity* of the Fermi energy $\epsilon_F \equiv \epsilon_{n_F}$, where ($\hbar$ times) the velocity $v(\epsilon)$ is given by $d\epsilon(k)/dk$ as a function of ϵ. The result for the low-temperature specific heat was first presented by Pauli. For spinless fermions in $d = 1$ the result for the heat capacity per unit length is

$$c_L^{\text{Pauli}} = \frac{\pi}{3} k_B \left(\frac{k_B T}{\hbar v_F} \right), \tag{10.5}$$

where $v_F = \hbar k_F/m = \hbar\pi(N/L)/m$ is the Fermi velocity.

Another simple many-body system discussed in textbooks is the harmonic chain. Consider N atoms on a $1d$ lattice with lattice constant a. In the harmonic approximation which corresponds to a coupling by springs the eigenstates of the system can be expressed by the occupation numbers n_j of the eigenmodes [1,18]. Like for harmonic oscillators they are $n_j \in \{0, 1, 2, \ldots \infty\}$ and the mode frequencies are denoted by ω_j. The energy of the corresponding eigenstate is given by

$$E(N, \{n\}) = \sum_j \hbar\omega_j \left(n_j + \frac{1}{2} \right). \tag{10.6}$$

Averaging with the canonical ensemble yields for the mean energy at temperature T

$$E(N, T) = \sum_j \hbar\omega_j \left(\frac{1}{e^{\beta\hbar\omega_j} - 1} + \frac{1}{2} \right). \tag{10.7}$$

As first realized by Debye for the three-dimensional case, the phonon dispersion can be linearized for the calculation of low-temperature properties ($k_B T \ll \hbar\omega_{\text{max}}$). For a chain with fixed ends one obtains

$$\hbar\omega_j \approx \hbar c_s \frac{j\pi}{L} \equiv j\Delta_B, \tag{10.8}$$

where
c_s is the sound velocity
$L = Na$
$j \in 1, 2, 3, \ldots$

The mode energies are multiples of Δ_B. Taking the limit $L \to \infty$ the sum in Equation 10.7 can be converted to an integral which can be evaluated analytically after the upper integration limit is assumed to be infinity, which does not alter the low-energy properties. This leads to the $d = 1$ version of Debye's famous $T^3(=T^d)$ law

$$c_L^{\text{Debye}} = \frac{\pi}{3} k_B \left(\frac{k_B T}{\hbar c_s} \right), \tag{10.9}$$

With the replacement $c_s \leftrightarrow v_F$, the results for the harmonic chain and the noninteracting fermions are identical. This suggests that apart from a scale factor, the (low energy) excitation energies and the degeneracies in the two types of systems are identical. By the use of the grand canonical ensemble in the fermionic calculation the deeper reason for this "agreement" remains mysterious. A deeper understanding involves two steps:

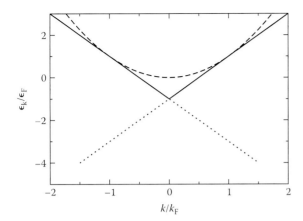

FIGURE 10.1 Energy dispersion as a function of momentum. The dashed curve shows the usual quadratic "nonrelativistic" dispersion and the full curve the linearized version. For fixed boundary conditions, only the $k > 0$ branch exists. The dot-dashed parts are the additional states (discussed in the text) for $k_0 = -1.5k_F$. The model introduced by Luttinger and used for the following classification of states corresponds to $k_0 \to -\infty$ (added "Dirac sea"). (From Schönhammer, K., *Strong Interactions in Low Dimensions*, eds. D. Baeriswyl and L. Degiorgi, Kluwer Academic Publishers, Dordrecht, the Netherlands. With permission.)

1. *Linearization* of the kinetic energy $\epsilon_k = \hbar^2 k^2/(2m)$ of the free fermions around the Fermi point k_F for fixed boundary conditions or both Fermi points $\pm k_F$ for periodic boundary conditions. The argument is simplest for fixed boundary conditions

$$\epsilon_n^{(lin)} = \epsilon_F + \hbar v_F (n - n_F) \pi/L, \qquad (10.10)$$

which corresponds to the $k > 0$ part of Figure 10.1. Then the energy differences $\epsilon_n - \epsilon_F$ are integer multiples of $\Delta_F \equiv \hbar v_F \pi/L$.

2. Classification of the excited state of the Fermi system by the number n_j^s of upward shifts by j units *of* Δ_F with respect to the ground state [20].

As the fermions are indistinguishable the construction of the $\{n_j^s\}$ shown in Figure 10.2 completely specifies the excited state. The highest occupied level in the excited state is connected with the highest occupied level in the ground state and so forth for the second, third highest level and so on. The new quantum numbers $\{n_j^s\}$ are not fermionic quantum numbers. The only upper restriction in $n_j^s \in \{0,1,2,3....\}$ comes from the fact the total number of fermions N is finite. If, for example, all fermions are shifted by the same number of j_0 energy units the corresponding shift number is N and all other n_j^s are 0. For the discussion of low-temperature properties ($k_B T \ll \epsilon_F$) this problem can easily circumvented by realizing that in this case no electrons at the bottom of the Fermi sea are excited to states above the Fermi level. Therefore, the addition of occupied one-particle states below the bottom of the Fermi sea does not change the low-temperature results. Figure 10.1 shows the fictitious states with $k_0 < k \le 0$. In the limit $k_0 \to -\infty$, which corresponds to a system with an additional infinite "Dirac sea," there is no longer any restriction

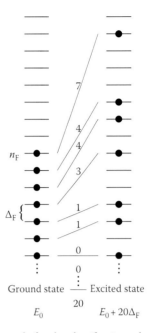

FIGURE 10.2 Example for the classification scheme for the excited states in terms of the numbers n_j^s of upward shifts by j units of Δ_F. In the example shown, the nonzero n_j^s are $n_7^s = 1$, $n_4^s = 2$, $n_3^s = 1$, and $n_1^s = 2$. (From Schönhammer, K., *Strong Interactions in Low Dimensions*, eds. D. Baeriswyl and L. Degiorgi, Kluwer Academic Publishers, Dordrecht, the Netherlands. With permission.)

on the numbers n_j^s of upward shifts and the n_j^s behave as bosonic quantum numbers $n_j^s \in \{0,1,2,...\infty\}$. The excitation energy of an eigenstate classified by the numbers of upward shifts is given by

$$E(\{n^s\}) - E_0 = \sum_{j=1}^{\infty} (j\Delta_F) n_j^s. \qquad (10.11)$$

Therefore, the canonical partition function for the system of noninteracting fermions apart from $\Delta_F \leftrightarrow \Delta_B$ has the same form as the canonical partition function for the harmonic chain with the linear phonon dispersion for all k-values after adding the infinitely deep Dirac sea. The construction of Figure 10.2 allows the deeper understanding of the "equality" of c^{Pauli} and c^{Debye}.

For the model with the linearized dispersion, Equation 10.10, also the many-particle eigenstates are equidistant $E_M - E_0 = M\Delta_F$. The state in Figure 10.2 corresponds to $M = 20$. For large M the total energy eigenvalues are highly degenerate [20]. The determination of the degeneracy is an old problem in combinatorics, the number of partitions [6] of the integer M. For a general M, a partition is defined by a set of M numbers $m_i \in \mathbb{N}_0$ such that

$$M = m_1 + 2m_2 + 3m_3 + \cdots + Mm_M = \sum_{l=1}^{M} l m_l \qquad (10.12)$$

Figure 10.2 corresponds to the partition $M = 7 + 4 + 4 + 3 + 1 + 1 = 20$. There are 626 other partitions for $M = 20$. The m_i are just the n_i^s.

The equivalence of the low-energy spectra of the noninteracting fermions and the harmonic chain can be presented in a more formal language by the method of "second quantization," which in Section 10.3 is used for the discussion of interacting fermions. Fermionic creation (annihilation) operators, which obey canonical anticommutation relations, are introduced:

$$[c_m, c_l]_+ = 0, \quad [c_m, c_l^\dagger]_+ = \delta_{m,l}. \tag{10.13}$$

Then the Hamiltonian for noninteracting fermions reads

$$H_0 = \sum_m \epsilon_m c_m^\dagger c_m \equiv \sum_m \epsilon_m \hat{n}_m, \tag{10.14}$$

where the $\hat{n}_m$ are the (fermionic) occupation number operators which have eigenvalues 0 or 1.

In order to obtain a formal "bosonic" description of the noninteracting fermions, bilinear operators in the fermionic operators are defined:

$$b_l^\dagger \equiv \frac{1}{\sqrt{l}} \sum_{n=1}^\infty c_{n+l}^\dagger c_n, \tag{10.15}$$

with $l \geq 1$. This is a sum of operators which all shift a fermion upwards by l energy units Δ_F. Using the fermionic anticommutation rules, it is easy to show that $[b_l^\dagger, b_{l'}^\dagger] = 0$ and

$$[b_l, b_l^\dagger] = \frac{1}{l} \sum_{n=1}^l c_n^\dagger c_n. \tag{10.16}$$

The operator on the right-hand side counts the number of fermions in the lowest l states at the bottom of the band divided by l. For $l \ll n_F$ and N fermion states, which have no holes at the bottom of the Fermi sea, one can replace the operator on the right-hand side of Equation 10.16 by the unit operator. One can make this an exact statement if the Dirac sea is added, i.e. the sum in Equation 10.15 does not run from 1 but from $-m_0$ with $m_0 \to \infty$. An analogous discussion for $l \neq l'$ leads to

$$[b_l, b_{l'}^\dagger] = \delta_{l,l'} \hat{1} \tag{10.17}$$

i.e., the operators obey boson commutation relations in the low-energy subspace [15].

Even without adding the Dirac sea the noninteracting Hamiltonian (shifted by a constant) can be expressed in terms of the bosonic "shift operators" [10]

$$\begin{aligned} H_0 &= \Delta_F \sum_{n=1}^\infty n c_n^\dagger c_n \\ &= \Delta_F \left[\sum_{l=1}^\infty l b_l^\dagger b_l + \frac{1}{2} \mathcal{N}(\mathcal{N}+1) \right], \end{aligned} \tag{10.18}$$

where $\mathcal{N} \equiv \sum_{n=1}^\infty c_n^\dagger c_n$ is the fermionic particle number operator. The groundstate $|E_0(N)\rangle^{(0)}$ of H_0 with $N = n_F$ electrons is obtained by occupying the lowest N states. As for all l and N

$$b_l \, | E_0(N)\rangle^{(0)} = 0 \tag{10.19}$$

holds, the particle number operator term in the second line in Equation 10.18 is necessary to yield the correct ground state energy. With the help of the boson creation operators $b_n^\dagger$, the excited states can be written as familiar from harmonic oscillator systems

$$|\{m\}, N\rangle^{(0)} = \prod_j \frac{(b_j^\dagger)^{m_j}}{\sqrt{m_j!}} \, | E_0(N)\rangle^{(0)}, \tag{10.20}$$

where the m_j are bosonic quantum numbers. If the product using the definition of the $b_j^\dagger$ is calculated, a linear combination of many energetically degenerate "Slater determinants" $|\{n^{(F)}\}, N\rangle$ is obtained, where the fermionic quantum numbers $n^{(F)}$ take the values 0 or 1 indicating which of the levels are occupied.

With the "Kronig identity" (Equation 10.18), the Hamiltonian H_0 is "bosonized." This raises the question whether the individual operators $c_n^\dagger c_n$ in the first line, or even the c_n can also be bosonized. It can in fact be done and is the key to a simple understanding of the anomalous behavior of the correlation functions of Luttinger liquids discussed in Section 10.4. Therefore, also the bosonization of the field operator is explained shortly.

After adding the Dirac sea one introduces the 2π-periodic auxiliary field operator $\tilde{\psi}(v)$,

$$\tilde{\psi}(v) \equiv \sum_{l=-\infty}^\infty e^{ilv} c_l. \tag{10.21}$$

The c_n are given by the inverse relation

$$c_n = \frac{1}{2\pi} \int_0^{2\pi} e^{-ivn} \tilde{\psi}(v) dv. \tag{10.22}$$

Now one tries to find an expression for $\tilde{\psi}(v)$ in terms of the boson operators. This is done with help of the commutation relations

$$[b_m, \tilde{\psi}(v)] = -\frac{1}{\sqrt{m}} e^{-imv} \tilde{\psi}(v), \quad [b_m^\dagger, \tilde{\psi}(v)] = -\frac{1}{\sqrt{m}} e^{imv} \tilde{\psi}(v). \tag{10.23}$$

Products of exponentials of operators linear in the boson operators

$$A_+ \equiv \sum_{n \neq 0} \lambda_n b_n^\dagger; \quad B_- \equiv \sum_{n \neq 0} \mu_n b_n \tag{10.24}$$

with arbitrary constants λ_n and μ_n obey similar commutation relations

$$\left[b_m, e^{A_+} e^{B_-} \right] = \lambda_m e^{A_+} e^{B_-}; \quad \left[b_m^\dagger, e^{A_+} e^{B_-} \right] = -\mu_m e^{A_+} e^{B_-}, \tag{10.25}$$

which follow from $[b_m, e^{\lambda b_m^\dagger}] = \lambda e^{\lambda b_m^\dagger}$. One therefore makes the ansatz

$$\tilde{\psi}(v) = \hat{O}(v)e^{i\phi^\dagger(v)}e^{i\phi(v)}, \tag{10.26}$$

where the operator $i\phi(v)$ is given by [10]

$$i\phi(v) = \sum_{n=1}^{\infty} \frac{e^{inv}}{\sqrt{n}} b_n. \tag{10.27}$$

Then the commutation relations Equation 10.23 are fulfilled if the "Klein factor" $\hat{O}(v)$ commutes with all the b_m and $b_m^\dagger$. It has the form

$$\hat{O}(v) = Ue^{iv\mathcal{N}}, \tag{10.28}$$

where the unitary operator U reduces the particle number by 1 without changing the bosonic quantum numbers. The importance of the exponential factor can be seen by calculating the matrix element of $\tilde{\psi}(v)$ between the groundstates with $N-1$ and N particles. The direct fermionic calculation using Equation 10.21 yields e^{ivN}. The same result is obtained using the bosonized expression Equations 10.26 and 10.28:

$$^{(0)}\langle E_0(N-1)\,|\,Ue^{i\phi^\dagger(v)}e^{i\phi(v)}e^{iv\mathcal{N}}\,|\,E_0(N)\rangle^{(0)}$$

$$= {}^{(0)}\langle E_0(N-1)\,|\,U\,|\,E_0(N)\rangle^{(0)}e^{ivN} = e^{ivN}. \tag{10.29}$$

Using Equation 10.19 the exponential factors with the boson operators could be replaced by unity.

The bosonization of the field operator allows, for example, the calculation of the average occupancies $\langle c_n^\dagger c_n\rangle$ in the canonical ensemble [20]. The full power of the method is shown in Section 10.4. If in Equation 10.29, the unperturbed ground states are replaced by the interacting ones, one obtains an important modification which lies at the heart of LL physics.

10.3 The Tomonaga–Luttinger Model

When a two-body interaction between the fermions is switched on, the ground state is no longer the filled Fermi sea but it has admixtures of (multiple) particle–hole pair excitations. In order to simplify the problem, Tomonaga used periodic boundary conditions [25] which corresponds to the motion on a ring of length L. He studied the high-density limit where the range of the interaction is much larger than the mean interparticle distance. Then the Fourier transform $\tilde{v}(k)$ of the two-body interaction is nonzero only for values $|k| \leq k_c$ where the cut-off k_c is much smaller than the Fermi momentum $k_c \ll k_F$. This implies that for not too strong interaction the ground state and low energy excited states have negligible admixtures of holes deep in the Fermi sea and particles with momenta $|k| - k_F \gg k_c$. In the two intermediate regions around the two Fermi points

$\pm k_F$, with particle–hole pairs present, the dispersion is linearized as shown in Figure 10.1.

$$k \approx \pm k_F: \quad \epsilon_k = \epsilon_F \pm v_F\hbar(k \mp k_F). \tag{10.30}$$

Tomonaga realized that the Fourier components of the operator of the density

$$\hat{\rho}_n = \int_{-L/2}^{L/2} \hat{\rho}(x)e^{-ik_n x}\,\mathrm{d}x = \sum_{n'} c_{n'}^\dagger c_{n'+n}, \tag{10.31}$$

where $c_n^\dagger(c_n)$ creates (annihilates) a fermion in the state with momentum $k_n = (2\pi/L)n$, play a central role in the interaction term, as well as for the kinetic energy. Apart from an additional term linear in the particle number operator, which is usually neglected, the two-body interaction is given by

$$\hat{V} = \frac{1}{2L}\sum_{n\neq 0}\tilde{v}(k_n)\hat{\rho}_n\hat{\rho}_{-n} + \frac{1}{2L}\mathcal{N}^2\tilde{v}(0), \tag{10.32}$$

where $\mathcal{N}$ is the particle number operator. Tomonaga's important step was to split $\hat{\rho}_n$ for $|n| \ll n_F$ into two parts, one containing operators of "right movers", i.e., involving fermions near the right Fermi point k_F with velocity v_F and "left movers" involving fermions near $-k_F$ with velocity $-v_F$

$$\hat{\rho}_n = \sum_{n'\geq 0} c_{n'}^\dagger c_{n'+n} + \sum_{n'<0} c_{n'}^\dagger c_{n'+n} \equiv \hat{\rho}_{n,+} + \hat{\rho}_{n,-}, \tag{10.33}$$

where the details of the splitting for small $|n'|$ are irrelevant. Apart from the square root factor, the $\hat{\rho}_{n,\alpha}$ are similar to the b_l defined in Equation 10.15. Their commutation relations in the low-energy subspace are

$$[\hat{\rho}_{m,\alpha}, \hat{\rho}_{n,\beta}] = \alpha m\delta_{\alpha\beta}\delta_{m,-n}\hat{1}. \tag{10.34}$$

If the operators are defined

$$b_n \equiv \frac{1}{\sqrt{|n|}}\begin{cases} \hat{\rho}_{n,+} & \text{for } n>0 \\ \hat{\rho}_{n,-} & \text{for } n<0 \end{cases} \tag{10.35}$$

as well as the corresponding adjoint operators $b_n^\dagger$, using $\rho_{n,\alpha}^\dagger = \rho_{-n,\alpha}$ the bosonic commutation relations can be obtained:

$$[b_n, b_m] = 0, \quad [b_n, b_m^\dagger] = \delta_{mn}\hat{1}. \tag{10.36}$$

Now the kinetic energy of the right movers as well as that of the left movers can be "bosonized" as in Equation 10.18. As $\hat{V}$ is bilinear in the $\hat{\rho}_n$ the same is true for the $\hat{\rho}_{n,\alpha}$. Therefore, apart from a constant and an additional term containing particle number operators, the Hamiltonian for the interacting fermions is a quadratic form in the boson operators

$$H = \sum_{n>0} \hbar k_n \left\{ \left(v_F + \frac{\tilde{v}(k_n)}{2\pi\hbar} \right) (b_n^\dagger b_n + b_{-n}^\dagger b_{-n}) \right.$$
$$\left. + \frac{\tilde{v}(k_n)}{2\pi\hbar} (b_n^\dagger b_{-n}^\dagger + b_{-n} b_n) \right\} + \frac{\hbar\pi}{2L} \left[v_N \mathcal{N}^2 + v_J \mathcal{J}^2 \right]$$
$$\equiv H_B + H_{\mathcal{N},\mathcal{J}}, \tag{10.37}$$

where $\mathcal{N} \equiv \mathcal{N}_+ + \mathcal{N}_-$ is the total particle number operator above the Dirac sea, $\mathcal{J} \equiv \mathcal{N}_+ - \mathcal{N}_-$ the "current operator," and the velocities are given by $v_N = v_F + \tilde{v}(0)/\pi\hbar$ and $v_J = v_F$.

Here, v_N determines the energy change for adding particles without generating bosons while v_J enters the energy change when the difference in the number of right and left movers is changed. As the particle number operators $\mathcal{N}_\alpha$ commute with the boson operators $b_m(b_m^\dagger)$, the two terms H_B and $H_{\mathcal{N},\mathcal{J}}$ in the Hamiltonian commute and can be treated separately.

Because of translational invariance, the two-body interaction only couples the modes described by $b_n^\dagger$ and b_{-n}. The modes can be decoupled by the Bogoliubov transformation $\alpha_n^\dagger = \cosh[\theta(k_n)] b_n^\dagger - \sinh[\theta(k_n)] b_{-n}$ and its inverse

$$b_n^\dagger = \cosh[\theta(k_n)] \alpha_n^\dagger + \sinh[\theta(k_n)] \alpha_{-n}. \tag{10.38}$$

The Hamiltonian H_B then takes the form [10,15,25]

$$H_B = \sum_{n \neq 0} \hbar\omega_n \alpha_n^\dagger \alpha_n + \text{const.}, \tag{10.39}$$

where the $\omega_n = v_F |k_n| \sqrt{1 + \tilde{v}(k_n)/\pi\hbar v_F}$ follow from 2×2 eigenvalue problems corresponding to the condition $[H_B, \alpha_n^\dagger] = \hbar\omega_n \alpha_n^\dagger$. The parameter $\theta(k_n)$ in the Bogoliubov transformation is given by [10]

$$e^{2\theta(k_n)} = \sqrt{\frac{\pi\hbar v_F}{\pi\hbar v_F + \tilde{v}(k_n)}} \tag{10.40}$$

For small $|k_n|$, for smooth potentials $\tilde{v}(k)$ again a linear dispersion is obtained:

$$\omega_n \approx v_c |k_n|, \tag{10.41}$$

with the charge velocity $v_c = \sqrt{v_N v_J}$, which is larger than v_F for $\tilde{v}(0) > 0$. The groundstate $|E_0(N_+, N_-)\rangle$ of the TL model is annihilated by the α_n replacing Equation 10.19 and the excited states are analogous to Equation 10.20 with the $b_n^\dagger$ replaced by $\alpha_n^\dagger$. For fixed particle numbers N_+ and N_-, the excitation energies of the interacting Fermi system are given by $\sum_j \hbar\omega_j n_j$ with integer occupation numbers $0 \leq n_j < \infty$. For small enough excitation energies, the only difference of the excitation spectrum with respect to the noninteracting case is the replacement $v_F \leftrightarrow v_c$ and the low-temperature specific heat per unit length is given by

$$c_L^{TL} = \frac{\pi}{3} k_B \left(\frac{k_B T}{\hbar v_c} \right). \tag{10.42}$$

Concerning the low-temperature thermodynamics, the situation is similar to the one in Fermi liquid theory. The results are qualitatively as in the noninteracting case but with effective parameters. For properties related to correlation functions, the scenario is different. The "stiffness constant" $K = \sqrt{v_J/v_N}$ plays a central role which also shows up in the $k_n \to 0$ limit of Equation 10.40. Before addressing this issue in Section 10.4, the inclusion of spin is presented.

Electrons are spin one-half particles and for their description it is necessary to include the spin degree of freedom in the model. For a fixed quantization axis, the two spin states are denoted by $\sigma = \uparrow, \downarrow$. The fermionic creation (annihilation) operators carry an additional spin label as well as the $\hat{\rho}_{n, \pm, \sigma}$ and the boson operators $b_{n,\sigma}$, which in a straightforward way generalize Equation 10.8. It is useful to switch to new boson operators $b_{n,a}$ with $a = c, s$

$$b_{n,c} \equiv \frac{1}{\sqrt{2}} (b_{n,\uparrow} + b_{n,\downarrow})$$
$$b_{n,s} \equiv \frac{1}{\sqrt{2}} (b_{n,\uparrow} - b_{n,\downarrow}), \tag{10.43}$$

which obey $[b_{a,n}, b_{a',n'}] = 0$ and $b_{a,n}, b_{a',n'}^\dagger = \delta_{aa'}\delta_{nn'}\hat{1}$. The kinetic energy can be expressed in terms of the "charge" (c) and "spin" (s) boson operators using $b_{n,\uparrow}^\dagger b_{n,\uparrow} + b_{n,\downarrow}^\dagger b_{n,\downarrow} = b_{n,c}^\dagger b_{n,c} + b_{n,s}^\dagger b_{n,s}$. If one defines the interaction matrix elements $\tilde{v}_c(q) \equiv 2\tilde{v}(q)$, and $v_s(q) = 0$, $\mathcal{N}_{\pm,c} \equiv (\mathcal{N}_{\pm,\uparrow} + \mathcal{N}_{\pm,\downarrow})/\sqrt{2}$ and $\mathcal{N}_{\pm,s}$ as the corresponding difference, one can write the TL-Hamiltonian $H_{TL}^{(1/2)}$ for spin one-half fermions as

$$H_{TL}^{(1/2)} = H_{TL,c} + H_{TL,s}, \tag{10.44}$$

where the $H_{TL,a}$ are of the form Equation 10.37 but the interaction matrix elements have the additional label a. The two terms on the right-hand side of Equation 10.44 commute, i.e., the "charge" and "spin" excitations are completely independent. This is usually called "spin-charge separation." The diagonalization of the two separate parts proceeds exactly as before and the low-energy excitations are "massless bosons" $\omega_{n,a} \approx v_a|k_n|$ with the charge velocity $v_c = (v_{Jc}v_{Nc})^{1/2}$ and the spin velocity $v_s = (v_{Js}v_{Ns})^{1/2} = v_F$. The corresponding two stiffness constants are given by $K_c = (v_{Jc}/v_{Nc})^{1/2}$ and $K_s = 1$. If the coupling constants in Equation 10.37 in front of the $b^\dagger b$ and the $b^\dagger b^\dagger$ terms are different, the spin velocity v_s differs from v_F. If in addition the interaction is not spin rotationally invariant, K_s differs from 1 [7]. The low-temperature thermodynamic properties of the TL model including spin can be expressed in terms of the four quantities v_c, K_c, v_s, K_s. Because of spin-charge separation, the low-temperature specific heat has two additive contributions of the same form as in the spinless case. If one denotes, as usual, the proportionality factor in the linear T-term by γ, one obtains

$$\frac{\gamma}{\gamma_0} = \frac{1}{2}\left(\frac{v_F}{v_c} + \frac{v_F}{v_s}\right), \tag{10.45}$$

where γ_0 is the value in the noninteracting limit. In the zero temperature, spin susceptibility χ_s and compressibility κ also the stiffness constants enter. For the ratios to the noninteracting case, one obtains

$$\frac{\chi_s}{\chi_{s,0}} = \frac{v_F}{v_{N_s}} = K_s\frac{v_F}{v_s}, \quad \frac{\kappa}{\kappa_0} = \frac{v_F}{v_{N_c}} = K_c\frac{v_F}{v_c}. \tag{10.46}$$

A simple manifestation of spin-charge separation occurs in the time evolution of a localized deviation of, for example, the spin-up density from its average value $\delta\langle\rho_\uparrow(x,0)\rangle = F(x)$. If the deviation involves the right movers only, the initial (charge) current density is given by $\langle j_c(x,0)\rangle = v_c F(x)$. As the Fourier components of the operator for the density are proportional to the boson operators (see Equation 10.35), the time evolution of the density can easily be calculated. If $F(x)$ is sufficiently smooth, the initial deviation is split into four parts which move with velocities $\pm v_c$ and $\pm v_s$ without changing the initial shape. Using the simple time evolution $\alpha_{n,a}(t) = \alpha_{n,a}e^{-i\omega_{n,a}t} \approx \alpha_{n,a}e^{-iv_a|k_n|t}$ for $a = c, s$ one obtains for $t > 0$

$$\delta\langle\rho_\uparrow(x,t)\rangle = \sum_a\left[\frac{1+K_a}{4}F(x - v_a t) + \frac{1-K_a}{4}F(x + v_a t)\right]. \tag{10.47}$$

For the spin rotational invariant case $K_s = 1$, there is no contribution which moves to the left with the spin velocity. Two of the three contributions move to the right with the different velocities v_c and v_s. This is a manifestation of spin-charge separation. In the spinless model, the sum in Equation 10.47 is absent and the 4 in the denominators are replaced by 2. Obviously, spin-charge separation cannot occur and the initial deviation is split into two contributions, with the "fraction" $(1 - K)/2$ moving to the left. This is sometimes called "charge fractionalization." The following comment should be made: Spin-charge separation is often described as the fact that when an electron is injected into the system its spin and charge move independently with different velocities. This is very misleading as it is a collective effect of the total system that produces expectation values like in Equation 10.47. A similar argument holds for "charge fractionalization."

10.4 Non-Fermi Liquid Properties

In order to elucidate the non-Fermi liquid character of the TL model, it is useful to first study the dynamics of states $c_{k_n}^{(\dagger)}|E_0(N)\rangle$ with an additional particle (hole) for the spinless model. Only in the noninteracting limit these states are eigenstates of the Hamiltonian and therefore have an infinite lifetime. In the (three-dimensional) Fermi liquid theory the quasi-particle(hole) concept plays a central role. Landau assumed that hole states

$c_{\vec{k}}|F\rangle$ of the noninteracting system are adiabatically connected to "quasi-hole" states of the interacting system, when the interaction is turned on. Using perturbation theory to all orders he showed that the lifetime of states $c_{\vec{k}}^\dagger|E_0(N)\rangle$ goes to infinity when the momentum approaches the Fermi surface.

If the "quasi-hole-weight" Z_F is defined for the TL model as

$$Z_F \equiv |\langle E_0(N_+ - 1, N_-)|c_{k_F}|E_0(N_+, N_-)\rangle|^2, \tag{10.48}$$

one has $Z_F = 1$ for noninteracting fermions. If Fermi liquid theory would be valid, the quasi-hole weight Z_F would go to a constant $0 < Z_F < 1$ in the limit $L \to \infty$. The exact calculation of Z_F for the TL model shows that the Fermi liquid theory expectation does not hold. This can be done using the bosonization of the field operator presented in Section 10.2. The only property of the groundstate used is $\alpha_n|E_0(N_+, N_-)\rangle = 0$. One expresses the original boson operators b_n and $b_n^\dagger$ in Equation 10.26 by the α_n and $\alpha_n^\dagger$ and brings the annihilation operators α_n to the right, i.e., performs the bosonic normal ordering with respect to the new boson operators. This is achieved with help of the Baker-Hausdorff relation $e^{A+B} = e^A e^B e^{-\frac{1}{2}[A,B]}$ which holds if the operators A and B commute with $[A,B]$. This yields for the physical field operators $\psi_\pm(x) = \tilde\psi_\pm(2\pi x/L)/\sqrt{L}$ for a system of finite length L with periodic boundary conditions [10]

$$\psi_\pm(x) = \frac{A(L)}{\sqrt{L}}\hat{O}_\pm\left(\frac{2\pi x}{L}\right)e^{i\chi_\pm^\dagger(x)}e^{i\chi_\pm(x)} \tag{10.49}$$

with

$$i\chi_\pm(x) = \sum_{m\neq 0}\frac{\Theta(\pm m)}{\sqrt{|m|}}\left(\cosh[\theta(k_n)]e^{ik_m x}\alpha_m - \sinh[\theta(k_n)]e^{-ik_m x}\alpha_{-m}\right),$$

$$A(L) = \exp\left(-\sum_{n>0}\frac{\sinh^2[\theta(k_n)]}{n}\right). \tag{10.50}$$

where $\Theta(x)$ is the unit step function. This is a very useful formula for the calculation of properties like the quasihole weight, the occupancies or spectral functions of one-dimensional interacting fermions. The quasihole weight is given by $Z_F = |A(L)|^2$ as the exponential factors involving the boson operators can be replaced by unity as in Equation 10.29. The L-dependence of $A(L)$ follows from $\sum_{n=1}^{n_c}1/n \to \log n_c + \gamma$, where γ is Euler's constant. The logarithmic divergence with $n_c = k_c L/(2\pi)$ in the exponent is converted to a power law dependence for $A(L)$ itself using $e^{a\log x} = x^a$. Therefore, in the large L limit

$$Z_F \sim \left(\frac{1}{k_c L}\right)^\alpha, \quad \alpha = 2\sinh^2[\theta(0)] = \frac{(K-1)^2}{2K}, \tag{10.51}$$

where α is called the anomalous dimension, as α also determines the anomalous slow spatial decay of the one-particle Green's function. The attempt to calculate Z_F by perturbation theory in the two-body interaction strength $\bar{v}$ leads to logarithmically diverging terms which are summed in the exact solution presented above. To summarize, in contrast to Fermi liquid theory, the quasihole weight Z_F vanishes in a power law fashion for $L \to \infty$, if $\bar{v}(0)$ is different from zero.

The appearance of power laws in the TL model was first realized by Luttinger [14]. He found that the average occupation $\left\langle n_{k,+} \right\rangle \equiv \left\langle E_0(N) \,|\, c_k^\dagger c_k \,|\, E_0(N) \right\rangle$ in the interacting ground state for $k \approx k_F$ behaves as

$$\left\langle n_{k,+} \right\rangle - \frac{1}{2} \sim \left| \frac{k - k_F}{k_c} \right|^\alpha \mathrm{sign}(k_F - k) \qquad (10.52)$$

for $0 < \alpha < 1$. This is shown in Figure 10.3 in comparison to Fermi liquid theory. The full line was calculated assuming $\sinh^2\left[\theta(k)\right] = 0.3\,e^{-2|k|/k_c}$, while the dashed curve corresponds to $\sinh^2\left[\theta(k)\right] = 0.6\left(|k|/k_c\right)e^{-2|k|/k_c}$. For $\alpha > 1$ the leading deviation of $\left\langle n_{k,+} \right\rangle - 1/2$ is *linear* in $k - k_F$. At finite temperatures T the k-derivative of $\left\langle n_{k,+} \right\rangle$ for $0 < \alpha < 1$ no longer diverges at k_F but is proportional to $T^{\alpha-1}$.

The vanishing of the quasihole (or quasiparticle) weight Z_F is the simplest hallmark of LL physics. To reach a deeper understanding, the momentum resolved one-particle spectral functions should be studied [13,16,26]. These spectral functions, which are relevant for the description of angular-resolved photoemission, are just the spectral resolutions of the hole states $c_{k,\sigma}|E_0(N)\rangle$ discussed earlier:

$$\rho_\sigma^<(k,\omega) = \sum_j |\left\langle E_j(N-1) \,|\, c_{k,\sigma} \,|\, E_0(N) \right\rangle|^2\, \delta(\omega + E_j(N-1) - E_0(N)).$$

$$(10.53)$$

The spectral function $\rho_\sigma^>(k,\omega)$ relevant for inverse photoemission involves a similar spectral resolution of $c_{k,\sigma}^\dagger |E_0(N)\rangle$ and the total spectral function $\rho_\sigma(k,\omega)$ is the sum of $\rho^<$ and $\rho^>$. For $k \to k_F$, the spinless model and the model including spin show qualitatively the same behavior. The absence of a sharp quasiparticle peak is manifest from $\rho_+(k_F,\omega) \sim \alpha\,|\omega|^{\alpha-1}\,e^{-|\omega|/k_c v_c}$, where ω is the deviation from the chemical potential. Instead of a delta function at the chemical potential a weaker power law divergence is present for $0 < \alpha < 1$.

For $k \neq k_F$, the k-resolved spectral functions for the spinless model and the model with spin-1/2 differ qualitatively. The delta peaks of the noninteracting model are broadened into one power law threshold in the model without spin and two power law singularities (see Figure 10.4) in the model including spin if the interaction is not too large [16,26]. The "peaks" disperse linearly with $k - k_F$.

For the momentum integrated spectral functions, relevant for angular integrated photoemission, $\rho_{\pm,\sigma}(\omega) \sim |\omega|^\alpha$ as in the spinless model is obtained.

It is also straightforward to calculate various response functions for the TL model [13]. The static $\pm 2k_F + Q$ density response for the spinless model diverges for repulsive interaction proportional to $|Q|^{2(K-1)}$ which has to be contrasted with the logarithmic singularity in the noninteracting case. In the model including spin, the exponent $2K - 2$ is replaced by $K_c + K_s - 2$. Results for other response functions can be found in the given literature [4,7,8,21].

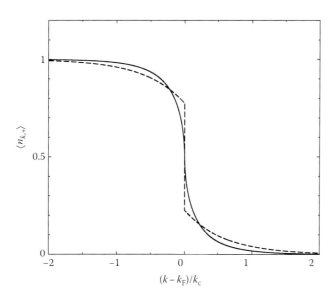

FIGURE 10.3 The full line shows the average occupation $\langle n_{k,+} \rangle$ for a TL model with $\alpha = 0.6$. The dashed line shows the expectation from Fermi liquid theory, where the discontinuity at k_F is given by Z_F. This can also be realized in a TL model with $\bar{v}(0) = 0$. The details of the interaction are specified in the text. (From Schönhammer, K., *Strong Interactions in Low Dimensions*, eds. D. Baeriswyl and L. Degiorgi, Kluwer Academic Publishers, Dordrecht, the Netherlands. With permission.)

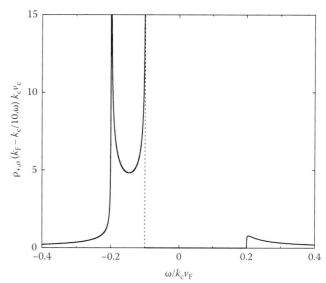

FIGURE 10.4 Spectral function $\rho_\sigma(k_F + \tilde{k},\omega) = \rho_\sigma^<(k_F + \tilde{k},\omega) + \rho_\sigma^<(k_F - \tilde{k},-\omega)$ as a function of normalized frequency for $\tilde{k} = -k_c/10$ for the TL-model with a spin independent interaction. The parameters are chosen such that $v_c = 2v_F$ and $\alpha = 1/8$. In the noninteracting limit there is a delta peak at -0.1. (From Schönhammer, K., *Strong Interactions in Low Dimensions*, eds. D. Baeriswyl and L. Degiorgi, Kluwer Academic Publishers, Dordrecht, the Netherlands. With permission.)

The exact solution of the TL model essentially depends on the fact that the numbers of right and left movers are conserved. This symmetry can be destroyed by a one-particle potential with $\pm 2k_F$ Fourier components or by interaction terms which change the individual particle numbers, like $2k_F$ "backscattering" terms or Umklapp terms for a half-filled band. With such additional terms, the model is in general no longer exactly solvable.

The unusual behavior of spinless model with an additional impurity was presented in a seminal paper by Kane and Fisher [11]. The backscattering term $\hat{V}_B$ due to the impurity is proportional to $(\psi_+^\dagger \psi_- + H.c.)$. By normal ordering $\psi_+^\dagger \psi_-$ similar to Equation 10.49, it follows that the backscattering term $\hat{V}_B$ scales as $(1/L)^K$ while the TL Hamiltonian in Equation 10.37 scales as $1/L$. Therefore, the influence of $\hat{V}_B$ depends crucially on the sign of the two-body interaction. For repulsive interactions, one has $K < 1$, which shows that $\hat{V}_B$ gets more important with increasing L, i.e., it is a relevant perturbation. For $K > 1$, i.e., an attractive interaction, $\hat{V}_B$ is irrelevant. As the strength of the impurity increases with the system size for repulsive interaction, one enters the regime of two weakly coupled semi-infinite Luttinger liquids. Near the boundary of a semi-infinite LL at zero temperature, the low-energy local spectral function behaves as $\rho(\omega) \sim |\omega|^{\alpha_B}$, with $\alpha_B = 1/K - 1$, the "boundary exponent," and at finite temperatures $\rho(0) \sim T^{\alpha_B}$. A simple "golden rule" calculation using the weak hopping between the semi-infinite LLs as the perturbation leads to a linear conductance which vanishes in the low-temperature limit like $G_{lin} \sim T^{2\alpha_B}$. This is very different from a noninteracting system, where a weak impurity only leads to a weak suppression of the ideal conductance. The unusual influence of an arbitrarily weak impurity in the LL is related to the power law divergence of the $2k_F$-density response mentioned above.

Tomonaga was well aware of the limitations of his approach for more generic two-body interactions ("In the case of force of too short range this method fails" [25]). For a short-range interaction $k_c \gg k_F$, low-energy scattering processes with momentum transfer $\approx \pm 2k_F$ are possible and have to be included in the theoretical description of the low-energy physics. The more general model including spin and terms changing right movers into left movers and vice versa is usually called the "g-ology model." It is no longer exactly solvable but spin-charge separation still holds. An important step toward the general Luttinger liquid concept came from the renormalization group (RG) study of this model [24]. It was shown that for repulsive interactions the renormalized interactions flow toward a fixed point Hamiltonian of the TL-type unless in lattice models for commensurate electron fillings strong enough interactions (for the half-filled Hubbard model this happens for arbitrarily small on-site Coulomb interaction U) destroy the metallic state by opening a Mott-Hubbard gap. These RG results as well as insight from models that allow an exact solution by the Bethe ansatz led Haldane [9,10] to propose the concept of Luttinger liquids as a replacement of Fermi liquid theory in one dimension.

As results for integrable models, which can be solved exactly by the Bethe ansatz played a central role in the emergence of the general "Luttinger liquid" concept, it is appropriate to shortly present results for the two most important lattice models of this type, the model of spinless fermions with nearest neighbor interaction and the 1d-Hubbard model. The one-dimensional single band lattice model of spinless fermions with nearest neighbor hopping matrix element $t(>0)$, and nearest neighbor interaction U (often called V in the literature) is given by

$$H = -t \sum_j (c_j^\dagger c_{j+1} + H.c.) + U \sum_j \hat{n}_j \hat{n}_{j+1} \equiv \hat{T} + \hat{U}, \qquad (10.54)$$

where
  j denotes the sites
  $\hat{n}_j = c_j^\dagger c_j$ are the local occupation number operators

In the noninteracting limit $U = 0$ for lattice constant $a = 1$, the well known dispersion $\epsilon_k = -2t \cos k$ is obtained. The interacting model $(U \neq 0)$ is here only discussed in half-filled band case $k_F = \pi/2$ with $v_F = 2t$. In contrast to the (continuum) Tomonaga model, Umklapp terms appear. They are irrelevant at the noninteracting $(U = 0)$ fixed point [23]. Therefore, the system is a Luttinger liquid for small enough values of $|U|$. The large U limit of the model is easy to understand: For $U \gg t$, charge density wave (CDW) order develops in which only every other site is occupied, thereby avoiding the "Coulomb penalty" U. For large but negative U, the fermions want to be as close as possible and phase separation occurs. For the quantitative analysis it is useful that the model in Equation 10.54 can be exactly mapped to a $S = 1/2$ Heisenberg chain with uniaxially anisotropic nearest neighbor exchange ("XXZ" model) in a magnetic field by use of the Jordan-Wigner transformation [7,8]. The point $U \equiv U_c = 2t$ corresponds to the isotropic Heisenberg model. For $U > 2t$, the Ising term dominates and the ground state is a well-defined doublet separated by a gap from the continuum and long-range antiferromagnetic order exists. For $-2t < U \leq 2t$ there is no long-range magnetic order and the spin-excitation spectrum is a gapless continuum. The mapping to the XXZ model correctly suggests that the spinless fermion model Equation 10.54 in the half-filled band case is a Luttinger liquid for $|U| < 2t$. Exact analytical results for the Luttinger liquid parameters for the half-filled model can be obtained from the Bethe Ansatz solution [9]. For the stiffness constant, for example, $K = \pi/[2 \arccos(-U/2t)]$ is obtained.

There exists a monograph [5] on the 1d-Hubbard model and an excellent earlier discussion of its LL behavior [22]. As the model includes spin, the on-site interaction between electrons of opposite spins is not forbidden by the Pauli principle. This is taken as the only interaction in the model. The 1d Hubbard Hamiltonian reads

$$H = -t \sum_{j,\sigma} (c_{j,\sigma}^\dagger c_{j+1,\sigma} + H.c.) + U \sum_j \hat{n}_{j,\uparrow} \hat{n}_{j,\downarrow}. \qquad (10.55)$$

An important difference to the spinless model Equation 10.54 shows up in the half-filled band case, which is metallic for $U = 0$.

For $U \gg t$, the "Coulomb penalty" is avoided when each site is singly occupied. Then, only the spin degrees of freedom matter. In this limit the Hubbard model can be mapped to a spin-1/2 Heisenberg antiferromagnet with an exchange coupling $J = 4t^2/U$. In the charge sector there is a large gap $\Delta_c \sim U$ while the spin excitations are gapless. The $1d$ Hubbard model can also be solved exactly using the Bethe ansatz [5] and properties like the charge gap or the ground state energy can be obtained by solving Lieb and Wu's integral equation. In contrast to the spinless model, the charge gap in the Hubbard model is finite for *all* $U > 0$. While for $U \gg t$ it is asymptotically given by U it is exponentially small $\Delta_c \approx (8t/\pi) \sqrt{U/t} \, \exp(-2\pi t/U)$ for $0 < U \ll t$. When the band is not half-filled Umklapp is not a low-energy process and the Hubbard model is a Luttinger liquid with $K_s = 1$. The LL parameters K_c and v_a can be obtained by (numerically) solving Lieb and Wu's integral equation [22].

The calculation of correlation functions not only requires excitation energies but also many-electron matrix elements that are difficult to evaluate using the Bethe ansatz. Presently no exact analytical results are available.

10.5 Additional Remarks

Strictly one-dimensional systems are a theoretical idealization. Apart from this even the coupling to an experimental probe presents a nontrivial disturbance of a Luttinger liquid. Unfortunately, the weak coupling of a 1d system to such a probe as well as the coupling between several LLs is theoretically not completely understood [7,21]. The coupling between the chains in a very anisotropic 3d compound generally, at low enough temperatures, leads to true long-range order. The order develops in the phase for which the algebraic decay of the corresponding correlation function of the single chain LL is slowest [7]. This can lead, for example, to CDW, spin density wave (SDW) order or superconductivity.

There exist several types of experimental systems where a predominantly 1d character of the low-energy electronic excitations can be hoped to lead to an (approximate) verification of the physics of Luttinger liquids in an appropriate temperature and energy window. Promising systems are highly anisotropic "quasi-one-dimensional" conductors like the Bechgaard salts, artificial quantum wires in semiconductor heterostructures or on surface substrates, and (armchair) carbon nanotubes [21]. Electrons at the edges of a two-dimensional fractional quantum Hall system can be described as a chiral Luttinger liquid. The power law tunneling density of states observable in the tunneling current–voltage characteristics shows power laws of extraordinary quality [3]. As in these chiral LLs the right- and left-movers are spatially separated, the edge state transport is quite different from the case of quantum wires.

Promising experimental techniques to verify LL behavior are, for example, high-resolution photoemission, transport, and optical properties [7,21]. There are intensive experimental activities in the attempt to verify the LL concept for interacting fermions

put forward by theoreticians. Further work on both sides is necessary to come to unambiguous conclusions.

As mentioned in the introduction, the focus of this chapter was on fermionic systems. Bosons in one dimension with repulsive interactions also behave as Luttinger liquids [7]. In contrast to fermions, there is an essential difference between the noninteracting and the interacting system. Free bosons at zero temperature all occupy the lowest one-particle state, for example, the $n = 1$ state of Equation 10.1 for a system of a box of length L with fixed boundary conditions or the $k = 0$ plane wave state for bosons on a ring. Only if a finite interaction is present the excitation spectrum is linear in $|k|$ i.e., has the typical LL form. In addition to the sound velocity, the low-energy physics is again determined by the stiffness constant K which determines again, for example, the spatial decay of correlation functions. In contrast to the fermionic system, the noninteracting system corresponds to $K \to \infty$. Very versatile experimental systems to experimentally test the LL behavior for bosonic systems are ultracold gases in strong optical lattices [2].

References

1. Ashcroft, N. W. and Mermin, N. D. 1976. *Solid State Physics*. New York: Holt, Rinehart and Winston.
2. Bloch, I., Dalibard, J., and Zwerger, W. 2008. *Rev. Mod. Phys. 80*. 885.
3. Chang, A. M., Pfeiffer, L. N., and West, K. W. 1998. *Phys. Rev. Lett. 77*. 2538.
4. Delft, J. v. and Schoeller, H. 1998. *Ann. Phys. 7*. 225.
5. Essler, F. H., Frahm, H., Göhmann, F. et al. 2005. *The One-Dimensional Hubbard Model*. Cambridge, U.K.: Cambridge University Press.
6. Euler, L. 1753. De partitione numerorum. In *Leonardi Euleri opera omnia*. Leipzig, Germany: Teubner.
7. Giamarchi, T. 2004. *Quantum Physics in One Dimension*. Oxford and New York: Oxford Science Publishers.
8. Gogolin, A. O., Nersesyan, A. A., and Tsvelik, A. M. 1998. *Bosonization and Strongly Correlated Systems*. Cambridge, U.K.: Cambridge University Press.
9. Haldane, F. D. M. 1980. *Phys. Rev. Lett. 45*. 1358.
10. Haldane, F. D. M. 1981. *J. Phys. C 14*. 2585.
11. Kane, C. L. and Fisher, M. P. A. 1992. *Phys. Rev. Lett. 68*. 1220; 1992. *Phys. Rev. B 46*. 15233.
12. Landau, L. D. 1957. *Sov. Phys. JETP 3*. 920.
13. Luther, A. and Peschel, I. 1974. *Phys. Rev. B 9*. 2911.
14. Luttinger, J. M. 1963. *J. Math. Phys. 4*. 1154.
15. Mattis, D. M. and Lieb, E. H. 1965. *J. Math. Phys. 6*. 304.
16. Meden, V. and Schönhammer, K. 1992. *Phys. Rev. B 46*. 15753.
17. Pines, D. and Nozieres, P. 1966. *The Theory of Quantum Liquids*. New York: W. A. Benjamin.
18. Reif, F. 1965. *Fundamentals of Statistical and Thermal Physics*. New York: McGraw-Hill.
19. Reprints of many of the important papers in this field together with an excellent introduction are presented in *Bosonization,* 1994. ed. M. Stone. Singapore: World Scientific.

20. Schönhammer, K. and Meden, V. 1996. *Am. J. Phys. 64.* 1168.

21. Schönhammer, K. 2004. Luttinger liquids: The basic concepts. In: *Strong Interactions in Low Dimensions.* eds. D. Baeriswyl and L. Degiorgi. Dordrecht, the Netherlands: Kluwer Academic Publishers; this review contains an extended list of important experimental papers.

22. Schulz, H. J. 1990. *Phys. Rev. Lett. 64.* 2831.

23. Shankar, R. 1994. *Rev. Mod. Phys. 66.* 129.

24. Solyom, J. 1979. *Adv. Phys. 28.* 201.

25. Tomonaga, S. 1950. *Prog. Theor. Phys. 5.* 544–569.

26. Voit, J. 1993. *Phys. Rev. B 47.* 6740.

Nanofluidics of Thin Liquid Films

Markus Rauscher
Max-Planck-Institut für Metallforschung

and

Universität Stuttgart

Siegfried Dietrich
Max-Planck-Institut für Metallforschung

and

Universität Stuttgart

11.1 Introduction

Nanofluidics deals with liquid systems in which there is movement and control over liquids in or around objects with at least one dimension of the order of 10–50 nm or below [1]. This includes the study of nanoflows in and around nanosized objects [2]. Open nanofluidic systems exhibit a free liquid–vapor or liquid–liquid interface as they occur, e.g., in liquid films with a thickness of less than 10–50 nm [3].* Nanofluidic systems are ubiquitous in biology: with a diameter of a few Angstroms ion-channels are the ultimate closed nanofluidic device [4]. But nanofluidics is also technologically interesting as the basis for further miniaturization of microfluidic devices down to the nanoscale [5].

11.1.1 From Microfluidics to Nanofluidics

Microfluidics refers not only to fluid systems on the micron scale but also to the technologically relevant lab-on-a-chip concept, which copies the extremely successful concept of microelectronics by integrating complete miniature laboratories into single chemical devices [6,7]. In many cases, the techniques used to build nanofluidic devices have been developed originally for producing microelectronic devices. In turn, nanofluidics can be even relevant to microelectronics: for example, the thickness of photoresist films used in lithographic processes is in the submicron range.

Building a microfluidic device is mainly an engineering problem. Besides the fact that gravity is negligible and flows on the micron scale are in general laminar (i.e., there is no turbulence which might promote mixing), liquids in microfluidic devices behave in the same way as liquids in macroscopic chemical apparatuses: the flow is described by the Navier-Stokes equations and wetting properties of surfaces can be characterized by the equilibrium contact angle.

11.1.2 What Is Different on the Nanoscale?

On the nanoscale, however, hydrodynamic equations reach the limit of their validity. At least they have to be modified in order to take the microscopic properties of liquids into account.

11.1.2.1 Hydrodynamic Slip

Whereas in their bulk fluids at rest are homogeneous, they exhibit significant inhomogeneities of the local number densities in the vicinity of channel walls or substrate surfaces [8]. Depending on the type of fluid (simple fluids such as liquid argon, complex fluids such as polymer melts/solutions or colloidal suspensions, or ionic solutions such as salty water) and the type of substrate (wetting or nonwetting), these influences of walls can be significant [9]. On very flat substrates and if this liquid layer near the substrate acts as to decrease the friction of the liquid with the wall, this can lead to the so-called hydrodynamic slip, i.e., to a modification of the common no-slip hydrodynamic boundary condition, which assumes that the fluid velocity vanishes at the substrate surface [10]. For Newtonian fluids, slip is quantified by the so-called slip length b, which is of the order of a few nanometers for simple liquids; but it can reach up to microns in complex

---

* Note that a nanoliter droplet has a diameter of the order of 100 μm and therefore it does not qualify as a nanofluidic system in this sense. A nanodroplet has a volume of zepto liter (zL, i.e., 10^{-21} L).

fluids. Accordingly, slip will be only relevant in nanofluidic systems with at least one spatial extension comparable with the slip length.

11.1.2.2 Range of Intermolecular Forces

In particular, in open nanofluidic systems, the finite range of intermolecular interactions has an appreciable effect. On the macroscopic scale, these interactions give rise to surface and interface tensions and therefore determine the equilibrium contact angle of a drop on a surface. On the nanoscale, they significantly distort the shape of drops, giving rise to a plethora of equilibrium wetting phenomena; moreover, they stabilize or destabilize thin liquid films [11,12]. In particular dispersion forces (van der Waals forces) play an important role. They give rise to a disjoining pressure $\Pi(z) = H/z^3$ with a Hamaker constant H of the order of 10^{-20} J. For equilibrium contact angles of 90° and surface tension coefficients of the order of 10^{-2} N/m, the disjoining pressure at the apex of a drop is of the same order as the Laplace pressure for droplet heights of a nanometer and lower. For smaller contact angles, the slopes of even larger droplets are affected.

11.1.2.3 Thermal Fluctuations

Unstable and metastable systems are particularly susceptible to thermal fluctuations. The chaotic molecular motion at finite temperatures gives rise to fluctuations in thermodynamic quantities such as volume, pressure, or particle numbers, which, in general, scale with the inverse of the square root of the system size. For example, the equilibrium thickness of wetting films is influenced by thermal fluctuations in the form of thermally excited capillary waves [13–15]: the presence of the substrate capillary wavelike fluctuations with long wavelengths, which leads to an effective, entropic repulsion of the liquid–vapor interface from the substrate surface. Moreover, not only are thermal fluctuations responsible for nucleation of holes in metastable films (this process is called homogeneous nucleation and leads to dewetting and finally to the formation of droplets), but they also accelerate spinodal dewetting of unstable films by promoting capillary wavelike fluctuations with short wavelengths [16].

11.1.2.4 Molecular Structure of the Fluid

The finite size of the constituents of the liquid becomes directly evident. The corresponding relevant length scale can range from microns for colloidal suspensions and polymer melts/solutions down to a few Angstroms for simple fluids. For example, in thin channels the above mentioned interface layers, in which the fluid structure is modified by the presence of the walls, can overlap, such that in the channel bulk-like liquid structures can no longer form. At this point, the hydrodynamic description of fluids breaks down. The same holds for ultrathin liquid films.

11.2 Theoretical Description of Open Nanofluidic Systems

As stated above, nanofluidic systems exhibit at least one length scale in between the macroscopic hydrodynamic scale and the atomistic scale. This means that they are too large for fully

atomistic models to be suitable but too small for macroscopic hydrodynamics. In order to describe nanofluidic systems, one can either follow a bottom-up approach starting with molecular dynamics or adopt a top-down approach by augmenting the hydrodynamic equations in order to extent their range of validity.

11.2.1 Bottom-Up Approaches

11.2.1.1 Molecular Dynamics

Under technologically or biologically relevant conditions fluids are, with the only exception of liquid helium, warm enough such that quantum effects are negligible in the sense that the thermal wavelength $\lambda = h/\sqrt{4\pi m k_B T}$ with the Planck constant h, the Boltzmann constant k_B, and the temperature T, is larger than the de Broglie wavelength h/p (with the momentum p) of the molecules. Therefore, one can describe the molecular dynamics of N molecules of a fluid by Newton's equations of motion [17,18]

$$m_n \frac{d^2 \mathbf{r}_n}{dt^2} = \mathbf{F}_n(\mathbf{r}_1, \mathbf{r}_2, \ldots, \mathbf{r}_N), \quad n = 1, \ldots, N, \quad (11.1)$$

where

 $\mathbf{r}_n(t)$ is the position of molecule n with mass m_n at time t and
 $\mathbf{F}_n$ is the force exerted on this molecule by all other molecules in the fluid as well as by the molecules in the channel walls or the substrate.

These equations of motion are coupled ordinary differential equations, which are solved numerically in molecular dynamics (MD) simulations [17,18].

The main challenge is the large number of equations. One mole of any substance, corresponding to a few grams, consists of $N \approx 6 \times 10^{23}$ molecules. For each molecule and in each time step of the numerical solution, the interaction force with all the $N - 1$ other molecules has to be calculated, amounting to $N(N - 1)/2$ force calculations in total. Even on modern supercomputing clusters with a peak power of a petaflop (10^{15} floating point operations per second), calculations would be limited to a really small number of molecules. Therefore, in such simulations, the range of intermolecular interactions is truncated such that each molecule interacts only with the relatively small number of neighboring molecules. As a consequence, the number of force calculations in each numerical time step is $\propto N$ (rather than N^2). However, this implies that the aforementioned effects due to the long range of the intermolecular interactions cannot be covered directly by MD simulations. And even then $N \approx 10^{12}$ is an upper limit for such MD simulations on dedicated supercomputers—keeping track of the coordinates in phase space (position and velocity) of such a large number of molecules requires a storage capacity of 43 GB (double precision).

A second major problem is the time scale. The intermolecular potentials $\Phi_{ij}(\mathbf{r})$ between pairs (i, j) of molecules together with their mass define the time scale for the chaotic molecular motion. In most actual fluids, this time scale is of the order of nanoseconds. The time steps in a MD simulation must resolve this time scale, although, in general, one is interested in phenomena that happen on the scale of seconds or even longer (such as in the case of a dewetting ultrathin fluid films). Currently, there are significant efforts to overcome these difficulties by multiscale modeling for bridging the time and length scales in the simulations [19].

Newton's equations conserve energy. For simulations of driven systems, this can pose a problem because the drive pumps energy into the system, which is dissipated into heat. In an experiment, the environment (substrate, channel walls, etc.) acts as a thermal bath that keeps the temperature under control. In MD simulations, this environment is either not taken into account at all or reduced to a small number of molecules in the first few layers of the substrate or the channel wall. In addition, in order to reduce the gap in time scales between the molecular chaos and the center of mass motion of the fluid (e.g., the speed of droplets), one often resorts to unrealistically large external forces to drive the flow. This aggravates the problem of heating. In order to overcome this problem, one therefore introduces a so-called thermostat into the numerical scheme, which keeps the temperature constant by shaping the velocity distribution of the particles, i.e., by coupling them to an external heat bath of a given temperature.

Nevertheless, MD simulations are contributing significantly to the understanding of the dynamics of nanofluidic systems.

11.2.1.2 Density Functional Theory

In general, researchers are not interested in the positions and momenta of all individual molecules in a fluid, but in the average density. For systems in thermal equilibrium, the techniques of statistical physics allow the local equilibrium number density distribution $\rho(\mathbf{r})$ to be obtained from a variational principle for the grand canonical functional $\Omega[\rho(\mathbf{r})]$ [20,21]. For monodisperse ideal gases, i.e., if the interaction between the gas molecules can be neglected, this functional is known analytically:

$$\Omega_{id}[\rho(\mathbf{r})] = \int (k_B T \rho(\mathbf{r})(\ln(\rho(\mathbf{r})\lambda^3) - 1) + (V(\mathbf{r}) + \mu)\rho(\mathbf{r}))\,d^3 r, \tag{11.2}$$

with the chemical potential μ and the external potential $V(\mathbf{r})$, which describes, e.g., the effect of channel walls on the gas. The corresponding Euler–Lagrange equation obtained from the first variation of Equation 11.2,

$$k_B T \ln(\rho(\mathbf{r})\lambda^3) + V(\mathbf{r}) + \mu = 0, \tag{11.3}$$

has the simple but important solution

$$\rho(\mathbf{r}) = \frac{1}{\lambda^3} \exp\left(-\frac{V(\mathbf{r}) + \mu}{k_B T}\right). \tag{11.4}$$

Accordingly, the particles accumulate in the minima of $V(\mathbf{r})$ and are less frequently visiting its maxima, but the particles are able to explore all spatial regions where $V(\mathbf{r})$ is finite. The higher the temperature the smaller will be the density differences in the system.

The effect of intermolecular interactions are taken into account by adding the so-called excess Helmholtz free energy $\mathcal{F}_{ex}[\rho(\mathbf{r})]$ to the ideal part in Equation 11.2

$$\Omega[\rho(\mathbf{r})] = \Omega_{id}[\rho(\mathbf{r})] + \mathcal{F}_{ex}[\rho(\mathbf{r})]. \tag{11.5}$$

For three-dimensional systems, there is no analytical expression for $\mathcal{F}_{ex}[\rho]$ available. However, intensive research has produced approximate expressions for $\mathcal{F}_{ex}[\rho]$, which yield rather accurate results for the structure and the thermodynamical properties of the systems. For soft particles with weak and bounded pair potentials Φ (e.g., star polymers), the so-called random phase approximation (RPA) can be used:

$$\mathcal{F}_{ex}[\rho(\mathbf{r})] = \frac{1}{2} \iint \rho(\mathbf{r})\Phi(\mathbf{r} - \mathbf{r}')\rho(\mathbf{r}')\,d^3 r\,d^3 r'. \tag{11.6}$$

Even for this simple functional, the Euler–Lagrange equations take the form of an integral equation, which, in general, has to be solved numerically. For fluid particles exhibiting a short-ranged repulsive interaction and a long-ranged attraction (e.g., a Lennard-Jones potential), the potential is split into a repulsive part and a smoothly varying attractive Φ_{att}. The former is mapped onto an appropriate hard-sphere reference system and the latter is treated perturbatively, giving rise to a term as in Equation 11.6 with Φ replaced by Φ_{att}. Within more sophisticated versions of density functional theory Φ_{att} is decorated with the direct correlation function (see Ref. [22] for more information on the various types of correlation functions) of the homogeneous fluid. Rather sophisticated functionals are available for the hard sphere fluids. Those hard sphere functionals are based on integral geometry and they have a rather complicated analytic form [23]. If correlations in the fluid at short distances leading to packing effects are neglected, a local approximation can be used for the hard sphere part of the density functional, leading to

$$\mathcal{F}_{ex}[\rho(\mathbf{r})] = \int \left(f_{HS}(\rho(\mathbf{r})) + \frac{1}{2}\rho(\mathbf{r}) \int \Phi_{att}(\mathbf{r} - \mathbf{r}')\rho(\mathbf{r}')\,d^3 r' \right) d^3 r, \tag{11.7}$$

with the Carnahan–Starling expression

$$f(\rho) = k_B T \rho \left(\ln \rho\lambda^3 - 1 + \frac{4\eta - 3\eta^2}{(1 - \eta)^2} \right)$$

and the packing fraction $\eta = (4\pi/3)R_{eff}^3\rho$ [22]. The suitably determined effective particle radius R_{eff} depends on the temperature and the repulsive part of the intermolecular potential.

Wetting phenomena on homogeneous or structured substrates have been described successfully with this type of functional. In this context it is particularly instructive to study the so-called effective interface approximation. In general, in a liquid film the number density of fluid molecules as a function of the normal distance from the substrate rises from zero density right at the wall and (in some cases after some damped oscillations with a period given by the molecular diameter of the fluid particles) approaches the bulk liquid density ρ_l. The liquid–gas interface has a certain width and the density changes gradually from the bulk liquid value to the bulk gas value ρ_g. Far from critical points, this width is of the order of a few molecular diameters (see Figure 11.1). Within the so-called sharp-kink approximation, the density profile of a film of thickness h is approximated by a steplike profile:

$$\rho_{\text{step}}(z,h) = \begin{cases} 0, & \text{for } z < d \\ \rho_l, & \text{for } d < z < h. \\ \rho_g, & \text{for } h < z \end{cases} \tag{11.8}$$

Within this approximation, the density profile of a film of laterally varying thickness $h(x,y)$ (see Figure 11.2) is given by $\rho_{\text{step}}(z, h(x, y))$. Accordingly, the density distribution is parameterized by the film thickness $h(x, y)$ and the density functional in Equation 11.5 reduces to a functional of the film thickness, which is called effective interface Hamiltonian $\mathcal{H}[h(x, y)]$. For interface configurations without strong curvatures (i.e., in a local approximation for the surface tension, see below) Equations 11.2, 11.5, 11.7, and 11.8 lead to

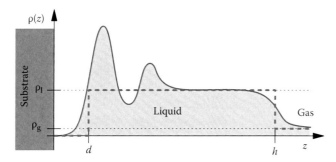

FIGURE 11.1 Schematic density profile (full blue line) of a laterally homogeneous thin liquid film on a solid substrate ($z < 0$). Within the sharp-kink approximation the profile is approximated by a steplike profile (dashed) with the value ρ_l (liquid density) inside the film and ρ_g (gas density) outside the film. For $z \lesssim d$ the repulsive substrate potential causes the liquid density to vanish.

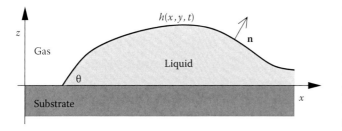

FIGURE 11.2 Cross-section at time t through a moving fluid film on a flat and homogeneous substrate. If there are no overhangs, the liquid–gas interface can be parameterized uniquely by the local thickness $z = h(x, y, t)$. The surface normal vector **n** points toward the gas phase. The liquid–gas interface and the liquid–substrate interface intersect at an angle θ, which, in general, depends on the velocity of the moving contact line.

$$\mathcal{H}[h] = \iint \left(\frac{\sigma}{2} (\nabla h)^2 + \omega(x, y, z = h(x, y)) - \Delta p h(x, y) \right) dx\, dy, \tag{11.9}$$

with the surface tension coefficient σ (which is determined independently within the same scheme [24]) and the so-called effective interface potential $\omega(x, y, z)$, which provides the effective interaction between the liquid–substrate and the liquid–vapor interface due to correlation effects and the finite range of the intermolecular potentials $V(\mathbf{r})$ and $\Phi(\mathbf{r})$ [12]. The effective potential $\omega(z)$ is known also in the case $\rho_{\text{step}}(z)$ in Equation 11.8 is replaced by realistic, smoothly varying profiles [25]. The dependence of ω on the local coordinates x and y occurs if the substrate has a geometrical or chemical inhomogeneous lateral structure (see Section 11.5). The pressure Δp in the film can be either thought of as to act as a Lagrange multiplier, which allows one to fix the volume of liquid in the film or in a grand canonical ensemble to be proportional to the deviation of the chemical potential from Equation 11.2 from its value $\mu_0(T)$ at liquid–vapor coexistence. The equilibrium profile $h(x, y)$ minimizes $\mathcal{H}$. The Euler–Lagrange equation corresponding to Equation 11.9 is

$$\sigma \nabla^2 h + \Pi(x, y, h(x, y)) + \Delta p = 0, \tag{11.10}$$

with the disjoining pressure $\Pi(x, y, z) = -\partial_z \omega(x, y, z)$.

If there are only pairwise interactions, the disjoining pressure can be expressed in terms of an integral over the substrate volume ($\rho_g \ll \rho_l$) [26]

$$\Pi(\mathbf{r}) = \int_{\text{substrate}} \rho_l^2 \Phi_{\text{att}}(\mathbf{r} - \mathbf{r}')\, d^3 r - \rho_l V(\mathbf{r}). \tag{11.11}$$

Equations 11.10 and 11.11 form the link between the microscopic description of the fluid and macroscopic thermodynamic properties, such as wetting phenomena, which will be discussed in Section 11.4.1.

Unfortunately, for nonequilibrium systems, there is no guiding principle available that is as powerful as density functional theory for thermal equilibrium. Although there are good arguments in favor of the existence of a dynamical analogue of density functional theory [27] (based on a classical version of the Runge-Gross theory for electrons [28]), it cannot be formulated explicitly even for ideal gases. The situation is somewhat more promising concerning the dynamics of suspended particles, for which inertia is negligible. Although it is demanding to take the important hydrodynamic interactions among these Brownian particles into account [29], an equation of motion for the particle density can be formulated [30]. However, only in certain, rather special cases this Brownian dynamics represents a viable model for open nanofluidic systems. Nonetheless, some of the results from equilibrium density functional theory can be incorporated into mesoscopic hydrodynamic equations (see, cf., Section 11.2.2.2), which link bottom-up and top-down approaches.

11.2.2 Top-Down Approaches

11.2.2.1 Macroscopic Description

Even in its simple form, the functional in Equation 11.7 plays the role of a bridge between the molecular and the hydrodynamic length scales. Macroscopically, fluids are described by hydrodynamic equations with the corresponding boundary conditions. In the simplest case of a viscous, incompressible, Newtonian, and nonvolatile fluid they are represented by the Navier-Stokes equation

$$\rho_l\left(\frac{\partial \mathbf{u}}{\partial t} + (\mathbf{u}\cdot\nabla)\mathbf{u}\right) = -\nabla p + \eta\nabla^2\mathbf{u} \qquad (11.12)$$

and the mass conservation equation

$$\nabla\cdot\mathbf{u} = 0 \qquad (11.13)$$

for the flow velocity $\mathbf{u}(\mathbf{r}, t)$ and the pressure $p(\mathbf{r}, t)$. ρ_l and η denote the fluid density and the viscosity, respectively [31]. In most microfluidic and nanofluidic systems, the Reynolds number $\mathrm{Re} = \rho u L/\eta$, i.e., the ratio of the dynamic and the shear pressure (u and L are the characteristic velocity and the characteristic length scale of the system, respectively) is very small, such that the inertial terms on the left-hand side of Equation 11.12 can be neglected and a much simpler, linear Stokes equation results:

$$\nabla p = \eta\nabla^2\mathbf{u}. \qquad (11.14)$$

As a consequence, flow in such systems is laminar and there is no turbulence to support mixing of, e.g., reactants, which can be a challenge for designing microfluidic devices.

Polymeric liquids play a large role in nanofluidics. However, they tend to be non-Newtonian. Equation 11.12 can be generalized to take into account viscoelasticity [32]; however, this is beyond the scope of this chapter.

In order to discuss the boundary conditions, we consider a drop or a fluid film of variable thickness on a flat and impermeable substrate (with the liquid–substrate interface located in the *x*–*y*-plane) as depicted in Figure 11.2. For simplicity, we assume in addition that the liquid–gas interface has no overhangs such that it can be parameterized by the local film thickness $z = h(x, y, t)$ (Monge parameterization). In this case, boundary conditions at the fluid–substrate and at the fluid–gas interface have to be prescribed. At the interface between the liquid and the substrate, the fluid velocity vanishes, i.e.,

$$\mathbf{u} = \mathbf{0} \quad \text{at } z = 0, \qquad (11.15)$$

and at the liquid–gas interface, the tangential component of the stress $\tau_{ij} - \delta_{ij}p$ (i.e., the anisotropic "deviatoric" part of the stress tensor $\tau_{ij} = \eta(\partial_i u_j + \partial_j u_i)$) vanishes and the normal component is balanced by the surface tension

$$\tau\cdot\mathbf{n} = (p + \sigma\kappa)\mathbf{n} \quad \text{at } z = h(x, y, z), \qquad (11.16)$$

with the surface tension coefficient σ, the local mean curvature κ, and the local surface normal vector $\mathbf{n}$ (with $|\mathbf{n}| = 1$) pointing toward the gas phase. For non-volatile fluids mass conservation implies that the local normal velocity υ of the liquid–gas interface is given by the normal component of the fluid velocity at the fluid surface, i.e.,

$$v = \mathbf{n}\cdot\mathbf{u} \quad \text{at } z = h(x, y, z). \qquad (11.17)$$

11.2.2.2 Augmenting Hydrodynamic Equations

As mentioned in Section 11.1.2, the hydrodynamic description of fluids as outlined in Section 11.2.2.1 breaks down at the nanoscale. However, hydrodynamic equations can be augmented such that they can be used down to the scale at which the molecular structure of the fluid becomes relevant—in simple liquids, the corresponding relevant scale is of the order of a nanometer. The resulting equations are often called mesoscopic hydrodynamic equations.

Hydrodynamic slip can be incorporated in a straightforward manner by replacing the noslip boundary condition (Equation 11.15) with a slip boundary condition. The most common form is the Navier slip condition for the component $\mathbf{u}_\| = (u_x, u_y)$ of the flow velocity parallel to the substrate,

$$\mathbf{u}_\| = b\partial_z\mathbf{u}_\|, \quad \text{at } z = 0 \qquad (11.18)$$

with the Navier slip length b. This form of the slip boundary condition is only applicable to Newtonian fluids, for which the stress τ is proportional to the strain rate. In a more general form, which is also applicable to viscoelastic fluids [33,34], the slip velocity is proportional to the shear stress.

Primarily, the finite range of the interaction between the fluid molecules generates an effective interaction between the liquid–substrate and the liquid–vapor interfaces, which is encoded in the effective interface potential $\omega(h)$ [12]. (If the liquid–vapor interface has a significant curvature, the local expression of the surface tension in Equation 11.16 has to be replaced by the actual nonlocal theory [24,35], or corrections have to be applied [36].) The corresponding disjoining pressure Π enters into the boundary condition in Equation 11.16 in terms of an external force field, leading to

$$\tau\cdot\mathbf{n} = (p + \sigma\kappa + \Pi)\mathbf{n} \quad \text{at } z = h(x, y, z). \qquad (11.19)$$

Thermal fluctuations can be included into the hydrodynamic equations as a randomly fluctuating stress tensor $\mathbf{S}(\mathbf{r}, t)$ [37]. For an incompressible liquid, the stochastic version of the Navier-Stokes equation (11.12) has the form

$$\rho_l\left(\frac{\partial\mathbf{u}}{\partial t} + (\mathbf{u}\cdot\nabla)\mathbf{u}\right) = -\nabla p + \eta\nabla^2\mathbf{u} + \nabla\cdot\mathbf{S}, \qquad (11.20)$$

With $\langle S \rangle = 0$ and

$$\langle S_{ij}(\mathbf{r},t)S_{lm}(\mathbf{r}',t')\rangle = 2\eta k_B T\delta(\mathbf{r}-\mathbf{r}')\delta(t-t')(\delta_{il}\delta_{jm}+\delta_{im}\delta_{jl}). \tag{11.21}$$

With $\langle \cdot \rangle$ we denote the ensemble average over all realizations of the Gaussian noise. The noise tensor also appears in the boundary conditions at the liquid–vapor interface and Equation 11.19 is replaced by

$$(\tau + S)\cdot\mathbf{n} = (p + \sigma\kappa + \Pi)\mathbf{n} \quad \text{at } z = h(x,y,z). \tag{11.22}$$

For further information on stochastic differential equations, see Refs. [38,39].

In accordance with our discussion, the density functional theory presented in Section 11.2.1.2, from which the disjoining pressure that appears in Equation 11.22 is derived, is a theory for ensemble-averaged densities, and, therefore, should already contain all effects of thermal fluctuations. From that point of view, it should not appear in an equation together with a stochastic noise term. However, although in principle fluctuations are included, all available, physically relevant functionals in three spatial dimensions are of mean field character (in particular the functional in Equation 11.7) and therefore they do not cover important aspects of fluctuations. Moreover, the effect of fluctuations on the dynamics is not covered at all.

Unfortunately, up to now it is hardly possible to take the finite size of the fluid molecules into account by augmenting hydrodynamic equations; corresponding bottom-up approaches for simple liquids are still an active field of research.

11.2.2.3 Thin Film Models

Thin liquid films are the best studied nanofluidic system, first because they are experimentally well-controllable and accessible (see Section 11.3), and second because the theoretical understanding of the dynamics of thin liquid films is developed further than that of other (open or closed) nanofluidic systems. One reason for this state of affairs is that the theory of equilibrium wetting phenomena has been developed and tested experimentally forming a solid scientific basis, such that the only remaining challenge has been to add dynamics. Conceptually, this is straightforward within the top-down approach described above.

Thin films fulfill the criterion of the separation of two length scales: the mean film thickness H (i.e., the volume of liquid on the substrate divided by the covered substrate area) and the characteristic lateral length scale L on which the film thickness varies. The ratio $\varepsilon = H/L \ll 1$ is a small dimensionless number and the solution of the augmented hydrodynamic equations can be expanded in terms of powers of ε, simplifying the equations of motion of the film significantly. The outcome of this analysis crucially depends on the slip length b: for $b \ll L$, $b \approx L$, and $b \gg L$. A weak-slip, the intermediate-slip, and the strong-slip regime, respectively, are obtained [40]. The thin film equations in the weak-slip and in the intermediate-slip regime have a rather similar form:

$$\eta\frac{\partial h}{\partial t} = -\nabla\cdot\left(\left(\frac{h^3}{3}+bh^2\right)\nabla\left(\Pi(x,y,h)+\sigma\nabla^2 h\right)\right) \quad \text{(weak-slip)} \tag{11.23}$$

and

$$\eta\frac{\partial h}{\partial t} = -\nabla\cdot\left(bh^2\nabla\left(\Pi(x,y,h)+\sigma\nabla^2 h\right)\right) \quad \text{(intermediate-slip)}, \tag{11.24}$$

respectively. These equations do not take thermal fluctuations into account. A careful comparison with Equation 11.10 shows that both equations have the same underlying structure

$$\eta\frac{\partial h}{\partial t} = -\nabla\cdot\left(M(h)\nabla\frac{\delta\mathcal{H}[h]}{\delta h}\right), \tag{11.25}$$

with a slip-regime dependent mobility factor $M(h) = (h^3/3) + bh^2$ in the weak-slip regime and $M(h) = bh^2$ in the intermediate slip regime, and with the effective interface Hamiltonian $\mathcal{H}[h]$ from Equation 11.9. Equations 11.23 and 11.24 have been used successfully to understand the dynamics of thin films and in particular their dewetting behavior (see Section 11.4.3).

The thin film equation in the strong-slip limit has a qualitatively different form. Instead of a fourth-order hyperbolic equation for the film thickness as a function of time the following relations are obtained (for low Reynolds numbers $Re \ll 1$) [41,42]:

$$\eta u_x = b\eta[\partial_x(4h\partial_x u_x + 2h\partial_y u_y)+\partial_y(h\partial_x u_y + h\partial_y u_x)]$$
$$+ bh\partial_x[\Pi(x,y,h)+\nabla^2 h], \tag{11.26a}$$

$$\eta u_y = b\eta[\partial_y(4h\partial_y u_y + 2h\partial_x u_x)+\partial_x(h\partial_y u_x + h\partial_x u_y)]$$
$$+ bh\partial_y[\Pi(x,y,h)+\nabla^2 h], \tag{11.26b}$$

and

$$\frac{\partial h}{\partial t} = -\partial_x(hu_x)-\partial_y(hu_y). \tag{11.26c}$$

Equations 11.26a and 11.26b cannot be solved for u_x and u_y, respectively, so that a closed equation for $h(x,y,t)$ does not result, as in the weak-slip and in the intermediate-slip limits. This changes the characteristics of thin film dynamics qualitatively.

Up to now, thermal fluctuations in thin film dynamics have been discussed only in the no-slip case, i.e., for $b = 0$. In this limit, the fluctuations lead to an additional multiplicative noise term in the thin film equation [43–45]:

$$\eta\frac{\partial h}{\partial t} = -\nabla\cdot\left(\frac{h^3}{3}\nabla\frac{\delta\mathcal{H}[h]}{\delta h}+\sqrt{\frac{k_B T\eta h^3}{3}}\mathbf{N}\right). \tag{11.27}$$

$N(x, y, t)$ is a stochastic current with zero mean value, $\langle N(x, y, t) \rangle = 0$, and Gaussian statistics with the correlator $\langle N_i(x, y, t) \, N_j(x', y', t') \rangle = 2\delta_{ij} \, \delta(x - x') \, \delta(y - y') \, \delta(t - t')$. From the Fokker–Planck equation corresponding to Equation 11.27, it can be inferred that the Boltzmann distribution

$$W_B[h] = \mathcal{Z}^{-1} \exp(-\mathcal{H}[h]/(k_B T)), \qquad (11.28)$$

of interface shapes $h(x, y)$ with the partition sum $\mathcal{Z}$ obeys detailed balance as required by equilibrium thermodynamics [45]. The approach for deriving Equation 11.27 from mesoscopic hydrodynamic equations as used in Ref. [45] cannot be generalized directly to the weak-slip regime. However, assuming a stochastic current $N(x, y, t)$ as in the no-slip case and following the procedure described in Ref. [44], the fluctuation-dissipation theorem dictates the form of the stochastic versions of Equations 11.23 and 11.24. With the slip-regime-dependent mobility factor $M(h)$, the following expression is obtained:

$$\eta \frac{\partial h}{\partial t} = -\nabla \cdot \left(M(h) \nabla \frac{\delta \mathcal{H}[h]}{\delta h} + \sqrt{k_B T \, \eta \, M(h)} \, N \right).$$

Currently a stochastic version of the strong-slip thin film equation is not available.

Thin film models have been developed also for viscoelastic fluids, but this topic is beyond the scope of our discussion. However, the role of viscoelasticity for the dewetting dynamics has been overestimated for some time; for further details, see Refs. [33,34,46].

11.3 Experimental Methods

Experimental analyses of open nanofluidic systems present a number of challenges. The most obvious one is spatial resolution although, in some cases, high resolution is required only in one spatial direction (e.g., the thickness), so that optical methods can play a significant role in studying open nanofluidic systems. A second challenge concerns time dependence: in particular, by using scanning techniques taking a sufficiently large image of a film can take a long time (up to minutes). The third main challenge concerns the environmental conditions. Most liquids exhibit a significant vapor pressure, such that methods that require vacuum (e.g., standard electron microscopy) cannot be used. The fourth challenge is to avoid perturbing the system too much by the measurement process: scanning probe microscope tips can deform the liquid–vapor interface and hard X-rays can damage organic molecules in the fluid. As a result, there is not a single, overall optimal experimental method to study open nanofluidic systems. In the following, we discuss the most frequently used methods.

11.3.1 Scanning Probe Microscopy

In scanning probe microscopy, a sharp tip (the probe) mounted on a soft cantilever is scanned across the sample surface and the interaction of the tip with the sample is measured locally. The first

such device was the scanning tunneling microscope (STM) developed in the early 1980s [47]. It provides atomic resolution but it can only image conducting surfaces so that it cannot be used for most fluid systems.

The most frequently used scanning probe technique is the atomic force microscope (AFM, also called scanning force microscopy [SFM], see Ref. [48]), which can be operated in three basic modes. An AFM in contact mode is reminiscent of the old-fashioned gramophone: the needle is pressed to the surface and the topography is measured via the deflection of the cantilever. Apparently this mode of operation is unsuitable for soft surfaces. In the noncontact mode, the tip hovers at a distance of 50–150 Å above the surface and the van der Waals forces between the tip and the substrate are measured [49]. Since the tip is never in direct contact with the sample surface, even fluid surfaces can be imaged [50,51]. A variant of an AFM in the noncontact mode is the scanning polarization force microscope (SPFM), which uses the electrostatic interaction between the charged tip and a dielectric sample instead of the van der Waals forces [52]. This method is particularly useful for imaging liquid films, droplets, and other weakly adsorbed materials [53]. However, for both techniques, the noncontact mode AFM as well as the SPFM, the lateral resolution is reduced as compared to the contact-mode AFM. The most frequently used mode of AFM operation is the so-called tapping mode. The cantilever is forced to oscillate (50–500 kHz) with a significant amplitude (typically larger than 20 nm) and touches the sample surface only once per cycle. By measuring the damping of the oscillation and its phase, information about the topography and the mechanical properties of the sample surface can be obtained. This method has been used successfully for surfaces of polymeric liquids [54] and for determining the tension of the three-phase contact line of a droplet residing on a solid substrate [55,56].

11.3.2 X-Ray and Neutron Scattering

X-rays as well as neutrons with thermal energies have a wavelength in the order of Angstroms up to nanometers [57], so that the diffraction limit would allows construction of microscopes with subnanometer resolution. However, there are no lenses available with sufficient performance. In contrast to light and electrons, both x-rays and neutrons typically penetrate samples significantly and therefore probe the bulk structure rather than surfaces and interfaces (or thin films). Although at first glance, they seem to be rather inadequate for analyzing open nanofluidic systems, x-rays and neutrons do play a significant role in this research field [58].

X-rays are extremely high-frequency electromagnetic waves for which the index of refraction n of most materials is almost 1 (i.e., almost identical to the index of refraction of vacuum), but slightly smaller (for visible light, the index of refraction of most materials is larger than one). As a consequence, for x-rays, total external reflection (rather than total internal reflection as for visible light) can be observed [59], which is used, e.g., in x-ray telescopes (e.g., Rosat) and for focusing the beams in synchrotron radiation sources. The critical angle is of the order of 1°, such that beams sufficiently brilliant on that scale can be supplied only by synchrotron x-ray sources. The situation for neutrons is similar, although the index of refraction for neutrons can be smaller and or larger than 1, depending on the hydrogen and deuterium content of the sample. As a consequence, under grazing angles of incidence and exit, the scattering signal stems only from a thin layer at the sample surface, which is illuminated by an exponentially decaying evanescent wave. But x-ray and neutron reflectivity of thin films reveals, e.g., layering effects of molecules in the direction normal to interfaces, the diffusively scattered intensity reports on the lateral structure of the film, such as droplet sizes or the occurrence of phase separation. This can be used to analyze the structure of thin films on the molecular scale [60], the capillary wave spectrum [61], the internal structure of droplets [62], e.g., in microphase separating liquids or the structure of liquid surfaces [63–65] and solid–liquid interfaces [66,67]. However, scattering methods rely on reciprocal space and because the phase information is lost in the detector, the real space image cannot be retrieved directly; holographic methods are only available for soft x-rays [68]. But reciprocal space techniques also have advantages: the scattering process averages over the sample surface, such that statistical information (e.g., mean droplet size etc.) can be accessed directly [69]. With real space techniques, it is often difficult to obtain sufficient statistical sampling.

Local information on the internal structure of thin films can be obtained using x-ray microbeams [70,71]. X-rays can be focused with mirrors, Fresnel zone plates, and refractive lenses or by using wave guides to spot sizes with linear extensions of a few microns. Thus, although the scattering signal provides only information in Fourier space, one can nonetheless address small localized spots on the sample surface.

While advanced x-ray sources offer a higher flux (which allows for better collimation, higher resolution, and shorter measurement times), with a kinetic energy of less than 1 eV, thermal neutrons allow investigation of the dynamics of fluids. In addition, the contrast between different material compounds can be modified by deuteration without changing the chemistry. In order to achieve sufficient contrast for x-rays, e.g., in microphase separated polymers, one often has to "stain" the sample with metal salts.

11.3.3 Optical Methods

Optical microscopy is diffraction limited to resolutions of a few hundred nanometers due to the relatively large wavelength of visible light. However, the thickness of thin and ultra-thin films often varies laterally on relatively large lengthscales (up to microns and more). Measuring the change of the polarization of light upon reflection allows inference of the structure (i.e., thickness or dielectric properties) of thin films. The method is called ellipsometry and it provides thickness resolutions in the Angstrom range [72]. Lateral spatial resolution is obtained by scanning [73] or by using CCD cameras, which provides submicron lateral resolution [74]. Ellipsometry has been used very

successfully to study the structure of the precursor film ahead of spreading droplets [75–77] (see Section 11.4.2).

The molecular orientation within interfaces can be assessed by nonlinear light scattering, in particular by second harmonic generation in an optically nonlinear medium [78,79]. In a bulk material, this process is suppressed by symmetry so that the corresponding signal stems only from interfaces. Typically an incident infrared laser beam excites vibrational modes in the molecules and it is tuned such that the second harmonic is in the visible range. This technique has been used to analyze the orientation of molecules in monolayers and of water molecules at the water surface, but also to investigate the distribution of charges at water surfaces [80].

Although confocal microscopy is limited to micrometer resolutions [81], it has been used to observe nanofluidic features in suspensions. The interface tension in phase-separating colloid–polymer mixtures is orders of magnitude lower than the surface tension of most simple liquids. Accordingly, capillary waves have a much larger wavelength, so that they can be observed in real space and in real time [82]. However, recently, it became possible to determine optically the position of fluorescent particles, which can switch between a dark and a bright state, with nanometer resolution; for a review, see Ref. [83]. However, the method is limited to fluorescent particles such that the liquid–vapor interface of a simple fluid cannot be imaged directly.

11.4 Homogeneous Substrates

When discussing liquid films and droplets, a distinction has to be made between volatile and nonvolatile substances. Volatile fluids are mostly considered in thermal equilibrium with their vapor, i.e., the temperature and the chemical potential in the film is fixed and droplets are intrinsically unstable. For nonvolatile fluids, the volume of fluid in the film or droplet is fixed and they are not in chemical equilibrium with a bath (in particular the vapor phase). As a consequence, nonvolatile fluids have a much richer phase space of equilibrium configurations. However, much of the underlying physics is the same for both volatile and nonvolatile fluids.

11.4.1 Wetting

In general, completely wetting and partially wetting substrates can be distinguished. To be more precise, fluid–substrate combinations have to be considered, and, in the more general case of a liquid drop on a substrate, which is covered by a second immiscible fluid (e.g., a water droplet beneath an oil layer) of fluid–fluid–substrate combinations. For reasons of simplicity of the presentation and since there is no conceptual difference between the case of a droplet in vacuum (or a dilute gas) and a droplet beneath a layer of a second liquid, we restrict the following discussion to the first case of a droplet exposed to vacuum or a very dilute vapor.

11.4.1.1 Contact Angle

Macroscopically, the mechanical equilibrium shape of a droplet on a substrate is determined by the interface tensions between the liquid and the gas $\gamma_{lg} = \sigma$, between the liquid and the substrate

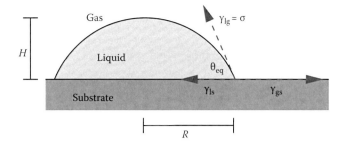

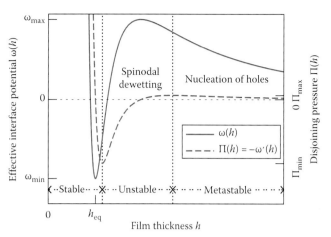

FIGURE 11.3 Macroscopic droplet on a partially wetting substrate. The equilibrium contact angle θ_{eq} at the three-phase contact line is determined by the mechanical equilibrium of the interface tensions σ, γ_{ls}, and γ_{gs}. The vertical force component is balanced by the substrate taken to be rigid.

γ_{ls}, and between the substrate and the gas γ_{gs}. At the three-phase contact line between liquid, substrate, and gas (provided that it is not pinned by defects or heterogeneities, see Section 11.5), the equilibrium contact θ_{eq} angle is fixed by Young's force balance equation [84]

$$\sigma\cos(\theta_{eq}) = \gamma_{gs} - \gamma_{ls} \tag{11.29}$$

If the substrate is passive, i.e., not in chemical equilibrium with the liquid (as it would be, e.g., in the case of salt in contact with a droplet of a concentrated salt solution), only the difference $\gamma_{gs} - \gamma_{ls}$ is an experimentally accessible quantity, such as σ and θ_{eq} [85]. In the absence of external forces such as gravity or if the external forces are negligible (i.e., if the Bond number $Bo = (\rho_l gHR)/2\sigma$, with the acceleration g, the droplet height H, and the base radius R, is small), the drop surface is a minimal surface, i.e., a surface of constant mean curvature. Therefore, on a homogeneous substrate, micron-sized droplets are spherical caps, as illustrated in Figure 11.3.

Within the effective interface model, the shape of a nanosized drop follows from Equation 11.10, and it therefore depends on the disjoining pressure. The effective interface potential of partially wetting substrates, which exhibit first-order wetting transitions, have the generic form shown in Figure 11.4 [12]. The effective interface potential has a global minimum at the thickness h_{eq} of

FIGURE 11.4 Generic dependence of the effective interface potential $\omega(h)$ (full line) and of the corresponding disjoining pressure $\Pi(h) = -\partial\omega/\partial h$ (dashed line) on the film thickness h for a partially wetting substrate. The position h_{eq} of the minimum of $\omega(h)$ corresponds to the equilibrium wetting film thickness at liquid–vapor coexistence. The depth ω_{min} determines the equilibrium contact angle of macroscopic droplets (Equation 11.30). Films with thicknesses for which $\partial_h^2\omega < 0$ are linearly unstable; otherwise the films are stable or metastable. $\omega(h)$ is defined such that $\omega(h \to \infty) = 0$.

the equilibrium wetting layer at liquid–vapor coexistence. Since $\gamma_{gs} = \gamma_{ls} + \sigma + \omega_{min}$ [12] the value of the interface potential $\omega_{min} = \omega(h_{eq})$ at this thickness determines the macroscopic equilibrium contact angle of *macroscopic* droplets:

$$\cos\theta_{eq} = 1 + \frac{\omega_{min}}{\sigma}. \tag{11.30}$$

The shape of microscopic droplets can deviate significantly from the macroscopic hemispherical shape. Figure 11.5 shows the cross-section of a nanosized droplet on a substrate with the effective interface potential shown in Figure 11.4 (with $\omega_{min}/\sigma = -1/24$) together with a circle fitted to the apex. The equilibrium contact angle calculated from Equation 11.30 is 16.6°. Since the contact line is blurred, the contact angle of a microscopic droplet is not a self-evidently defined quantity. One way to define the contact angle

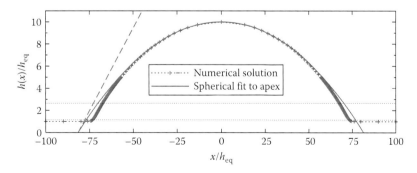

FIGURE 11.5 Profile of a nano-sized droplet on a substrate for the effective interface potential shown in Figure 11.4 with $\omega_{min}/\sigma = -1/24$ (blue symbols) calculated by numerically solving Equation 11.10. Also shown is a circle which is fitted to the apex of the droplet (full line). The contact angle of the circle with the substrate is 14.0°, i.e., smaller than the macroscopic contact angle of 16.6° (indicated by the dashed line). The thin horizontal lines mark the thickness values at which $\omega(h)$ has an inflection point (see Figure 11.4).

of a nanosized droplet is to fit a circle (or a sphere) to the apex of the droplet and to measure the angle of intersection between the circle and the substrate. For the example shown in Figure 11.5, the circle intersects the substrate surface at an angle of 14.0°, which is less than the macroscopic value 16.6°. Obviously, the value of θ obtained according to this procedure depends on which plane is chosen as the one describing the substrate surface [85]. For any choice of the location of the substrate surface, in the limit of macroscopic drops, the contact angle measured as described above converges to the one predicted by Equation 11.30.

The difference between the measured contact angle and the macroscopic contact angle can be summarized into the so-called line tension τ. The line tension is a thermodynamic concept: the free energy of a droplet is decomposed into a volume contribution (which is constant for nonvolatile fluids), an interface contribution encompassing the surface tension σ and the interface tensions γ_{ls} and γ_{gs}, and a line contribution, which is proportional to the length of the three-phase contact line multiplied by τ. As a result, the force balance (Young-Laplace) equation (11.29) for a circular droplet with base radius R (radius of the circular wetted area) turns into

$$\cos(\theta_{eq}) = \frac{\gamma_{gs} - \gamma_{ls}}{\sigma} - \frac{\tau}{\sigma R}. \tag{11.31}$$

The line tension coefficient τ can be calculated from the interface potential with the de Feijter and Vrij formula [86,87]:

$$\tau = \sqrt{2\sigma} \int\limits_{h_{eq}}^{\infty} \left(\sqrt{\omega(h) - \omega(h_{eq})} - \sqrt{-\omega(h_{eq})} \right) dh. \tag{11.32}$$

This formula has been used to actually determine the line tension of a droplet experimentally by measuring the droplet shape with an AFM [51,55,56]. In contrast to interface tensions, the line tension coefficient can be negative. In the latter case, it is tempting to expect a buckling of the three-phase contact line [88–90]. However,

the corresponding critical wavelength is comparable with the width of the three-phase contact line and a careful analysis of the effective interface model in Equation 11.9 shows that within a truly microscopic theory, this instability does not occur [91].

11.4.1.2 Wetting Transitions

Although they are part of our everyday experience, droplets as discussed in Section 11.4.1.1 are unstable if they are in chemical contact with their vapor. Mechanical and chemical equilibrium can be obtained if the vapor pressure (of the liquid which the drop consists of) around the drop is equal to the pressure inside the drop. However, the situation is unstable: moving a few molecules from the drop into the vapor reduces the drop size and therefore increases the pressure in it, such that the system is out of equilibrium and the drop evaporates. On the other hand, moving a few molecules into the drop increases the drop volume and decreases the pressure inside the drop, which leads to growth of the drop that is limited only by the capacity of the vapor reservoir. As a consequence, in thermal equilibrium, homogeneous substrates in contact with a vapor are either completely dry or covered by a laterally homogeneous film of microscopic thickness (up to a few nanometers in the case of incomplete wetting) or macroscopic thickness (in the case of complete wetting).

The thickness of this film is determined by the effective interface potential ω(h) [12], which depends on temperature and vapor pressure, i.e., the chemical potential. Figure 11.6 shows the dependence of the effective interface potential on temperature T. At low temperatures, ω(h) has a global negative minimum (corresponding to a nonzero equilibrium contact angle for non-volatile fluids) at a microscopic thickness h_0. The minimum becomes shallower as T increases and at the wetting temperature T_w the depth of the minimum becomes zero (corresponding to $\theta_{eq} = 0$ for a nonvolatile liquid). At this temperature, the equilibrium film thickness at liquid–vapor coexistence (i.e., for $\Delta p = 0$ in Equation 11.10) jumps from a microscopic value given by the position h_0 of the left minimum of ω(h), to infinity. Slightly off two-phase coexistence in

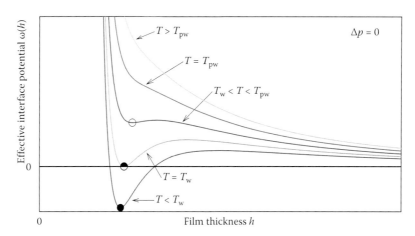

FIGURE 11.6 Generic functional form of the effective interface potential of a substrate, which exhibits a first-order wetting transition, for various temperatures (see Figure 11.8). At the wetting temperature, the left minimum of ω(h) at $h = h_0$ has depth zero. Therefore upon raising the temperature above $T = T_w$ leads to a first-order phase transition of the film thickness from $h = h_0$ to the global minimum at $h = \infty$. For $T > T_{pw}$ the graph of ω(h) has no inflection points.

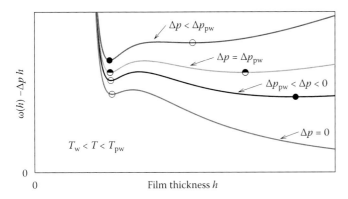

FIGURE 11.7 The equilibrium film thickness, for various values of Δp and for a temperature between T_w and T_{pw} (see Figures 11.6 and 11.8), corresponds to the global minimum (full circle) of $\omega(h) - \Delta p\, h$. For slightly negative Δp (i.e., $|\Delta p|$ small), which means slightly below the liquid–vapor coexistence line, the function $\omega(h) - \Delta p\, h$ has a global minimum (full circle) at the large but finite thickness h_m. In addition, there is a local minimum (open circle) at a microscopic film thickness h_0. For the prewetting transition value $\Delta p = \Delta p_{pw}$ both minima have equal depth (half filled circle) and for $\Delta p < \Delta p_{pw}$ the minimum at h_0 becomes the global minimum. Therefore the thick film turns into a thin one abruptly at Δp_{pw}. The prewetting transition requires that $\omega(h)$ exhibits two turning points which implies $T < T_{pw}$ (see Figure 11.6).

the bulk gas-phase region ($\Delta p < 0$ in Equation 11.10) $\omega(h) - \Delta p\, h$ has its second minimum at a large but finite value h_m (see Figure 11.7). For $0 < \Delta p < \Delta p_{pw}$ this minimum is deeper than the minimum at h_0 and the thickness of the wetting film is given by h_m. As Δp decreases, i.e., $|\Delta p|$ increases, this second minimum moves toward smaller film thicknesses and it becomes less shallow until at $\Delta p = \Delta p_{pw}$, the prewetting pressure, both minima (at h_0 and at h_m) are equally deep. At this pressure, the wetting film thickness jumps from h_m to the microscopic value h_0. For even lower Δp the film thickness remains microscopically small. This jump is a thermodynamic first-order interfacial phase transition.

Figure 11.8 shows the schematic phase diagram of a simple* liquid with an effective interface potential as shown in Figures 11.6 and 11.7. The phase boundaries separate the gas, liquid, and solid phases. Upon approaching the liquid–vapor coexistence curve along an isotherm from below at a temperature $T < T_w$, there is only a microscopically thin film (or adsorbate layer) up to liquid–vapor coexistence, above which the bulk is liquid. Increasing the pressure accordingly for $T_w < T < T_{pw}$, the thickness of the microscopically thin film jumps to a larger value as one crosses the prewetting line $P_{pw}(T)$. The height of this jump decreases for higher temperatures and vanishes at the prewetting temperature T_{pw}, at which the prewetting line ends in a critical point. Above the prewetting line, for $T > T_w$, the film thickness diverges close to the liquid–vapor coexistence line as $h(\Delta p \to 0, T) \sim |\Delta p|^{-1/3}$.

For many actual liquid–substrate combinations, the wetting temperature T_w happens to be lower than the triple point temperature T_t, so that the substrate surface is always wet. (Known

---

* Water is not a simple liquid in the sense that its liquid–solid coexistence line has a negative slope.

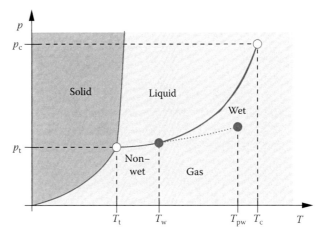

FIGURE 11.8 Schematic phase diagram of a simple liquid with phase boundaries (solid lines) separating the gas, liquid, and solid phases. (T_t, p_t) and (T_c, p_c) are the triple point and the critical point, respectively. For substrates exhibiting an effective interface potential as shown in Figure 11.6, the substrate is covered by a macroscopically thick wetting film for $T > T_w$ and pressures just below the liquid–vapor coexistence line (thick part of the coexistence line). Upon increasing the pressure at the prewetting line $P_{pw}(T)$ (dotted line) the wetting film thickness jumps from a microscopic to a macroscopic but finite value. The prewetting line meets the liquid–vapor coexistence line tangentially [92]. Assuming a positive latent heat associated with the prewetting transition, the slope of the prewetting line is positive.

exceptions are ^{4}He on Cs [93–96] or Rb [97] for which bona fide first-order wetting transitions have been identified.) In addition, there are also substrates that exhibit a second-order wetting transition. So far the only systems for which second-order wetting transitions have been found involve a liquid substrate [98–101]. Therefore we will not discuss this type of wetting transitions in this chapter.

11.4.2 Moving Three-Phase Contact Lines

In the case of a moving drop or a spreading film, the three-phase contact line between the liquid, the substrate, and the gas phase has to move. This is a longstanding problem in hydrodynamics: the stress and the viscous dissipation in the wedge-shaped volume next to the contact line diverges, such that on that basis, the contact line should not be able to move [102]. However, everyday experience tells that it does. Phenomenologically one often observes the following relation between the actual contact angle θ, the equilibrium contact angle θ_{eq}, and the contact line velocity w [103,104]:

$$w \propto \frac{\sigma}{\eta}\,\theta(\theta^2 - \theta_{eq}^2) \approx \frac{\pi}{4}R^3(t)\theta(t). \tag{11.33}$$

The stress divergence is a defect of the macroscopic description: hydrodynamics breaks down at the molecular scale, and slip as well as the existence of a precursor film regularize the stress divergence at the contact line. In addition, molecular kinetics is another important source of dissipation in the moving contact line: molecules "jumping" from the liquid onto the substrate just ahead of the contact line dissipate the energy that they gain in the process (see Ref. [105]). Although the concept of a dynamic contact angle is clearly a macroscopic one and mostly phenomenological, the problem of the moving contact line is a nanofluidic problem.

In coating processes, moving contact lines are of great technological importance. However, most of the knowledge of the contact line dynamics has been gained by studying the spreading of droplets [18,106]. Although significant progress has been made in this field, there is not yet a complete understanding of the dynamics of three-phase contact lines, in particular not on actual, i.e., inhomogeneous (structured, rough, or dirty) substrates. Since a comprehensive treatment of the subject is beyond the scope of this presentation, in the following we focus on results obtained for simple liquids on ideal homogeneous substrates.

In a spreading experiment, a droplet (usually of macroscopic size but small enough such that gravity does not play a role) is deposited on a surface with an initial contact angle, which is larger than θ_{eq}. If the spreading process is not too rapid, in good approximation, the droplet keeps the shape of a spherical cap throughout the whole spreading process such that the time-dependent contact angle $\theta(t)$ is related to the droplet base radius $R(t)$ and the fixed droplet volume V due to geometry:

$$V = \frac{\pi R^3(t)}{\sin^3 \theta(t)} \left(1 - \cos \theta(t)\right)^2 \left(2 + \cos \theta(t)\right) \approx \frac{\pi}{4} R^3(t)\theta(t), \quad (11.34)$$

where the latter relation holds for $\theta(t) \ll 1$. Assuming a power law dependence of the contact line speed on the contact angle, $dR/dt \sim \theta^m(t)$, in the limit of small contact angles, one obtains $dR/dt \sim (V/R^3(t))^m$ and therefore $R(t) \sim (V^m t)^{1/(3m+1)}$ (see Section IV in Ref. [11]). Depending on the dissipation mechanism, different spreading power laws are obtained. If viscous dissipation dominates, $m = 4$ and therefore $R(t) \sim t^{1/10}$ or $\theta(t) \sim t^{-3/10}$. This result is also known as Tanner's law [107]. Molecular dissipation as a source of dissipation was discussed by Blake and it leads to $m = 2$. Accordingly, this type of spreading is faster than in the viscous case, i.e., $R(t) \sim t^{1/7}$ and $\theta(t) \sim t^{-3/7}$ [108]. More recently, there are indications that molecular kinetic dissipation dominates at the initial stages of spreading, while viscous dissipation takes over at later stages [109,110].

Of course, the power laws mentioned above are only valid for contact angles that are larger than the equilibrium contact angle of the liquid on the substrate under study. On a completely wetting substrate, i.e., if $\theta_{eq} = 0$, a droplet is expected to spread until it is reduced to a microscopically thin film. However, in this case, one observes a so-called precursor film of microscopic thickness (i.e., a few Angstroms thick) spreading ahead of the three-phase contact line. There have been experimental reports that the radius of the precursor film increases much faster, i.e., $\sim \sqrt{t}$ [111], than that of the spreading drop. Molecular dynamics simulations [110] of spreading drops also show a spreading precursor film with the same growth law. This growth law was also found in kinetic Monte Carlo simulations of spreading liquid monolayers [112,113]. The growth rate $\sim \sqrt{t}$ is consistent with diffusive growth as well as with hydrodynamic flow in an ultrathin film, such that it is not possible to infer the underlying dynamics from the growth law alone.

In most cases, inertia does not play a role in spreading and dewetting processes. However, there are systems (e.g., gold on glass or graphite) in which the surface tensions are so large that a droplet with an initial contact angle, which is small compared with θ_{eq}, contracts so quickly that it actually jumps off the substrate [114]. The reason is that during contraction, the drop grows in height such that the center of mass of the droplet acquires a finite velocity component normal to the substrate.

11.4.3 Dewetting of Thin Films

Dewetting, i.e., the breakup of thin films into droplets, is much better understood than the spreading of droplets, mainly because in the most critical phase of the dewetting process—the film rupture—there is no three-phase contact line present [115]. Moreover, for many aspects of dewetting, the dynamics of the three-phase contact line plays a subdominant role. In the following discussion of dewetting, we focus on nonvolatile fluids; the interplay of evaporation and dewetting is much less understood.

11.4.3.1 Origin of Holes

Depending on the film thickness, a thin film can be stable, metastable, or unstable. Evidently, dewetting occurs only on partially wetting substrates, i.e., for $\theta_{eq} > 0$. The generic interface potential for this case, as shown in Figure 11.4, has two inflection points. Between these two inflection points, the curvature of $\omega(h)$ is negative and, as we shall discuss in Section 11.4.3.2, the film is linearly unstable: infinitesimal perturbations (e.g., generated by thermal fluctuations) grow exponentially and spinodal dewetting comes into play. Only very thin films are stable in the sense that they represent a thermodynamic equilibrium state. Thick films are metastable: a free energy barrier has to be overcome in order to form the nucleus of a growing hole in the film [116,117]. In general, this barrier is larger for thicker films. Depending on the source of the free energy for overcoming this barrier—thermal fluctuations or heterogeneities in the substrate (e.g., dirt or roughness) or in the film (e.g., inclusions or bubbles)—homogeneous or heterogeneous nucleation, respectively, of holes is possible. Experimentally it is extremely difficult to observe pure homogeneous nucleation because either the time scale for the formation of holes due to thermal fluctuations is too large or the films are so thin that they are extremely sensitive with respect to heterogeneities. While spinodal dewetting and nucleation can be distinguished by the resulting dewetting patterns [118], it is difficult do distinguish homogeneous and heterogeneous nucleation: in both cases, the positions of the emerging holes are random and uncorrelated. Sometimes nucleation sites can be identified by repeating a dewetting experiment on the same substrate: if a hole appears repeatedly on the same spot one can safely assume that there is a defect underneath. However, to determine whether the hole really appears at the same spot is an experimental challenge.

11.4.3.2 Spinodal Dewetting

Spinodal dewetting has been studied mostly in the no-slip regime, but in this respect the behavior of films in the no-slip, weak-slip, and intermediate-slip regime is rather similar; only the time scales differ. In order to observe spinodal dewetting in a well-controlled experiment, the film in an immobilized state should be prepared such that the starting time of the dewetting process can be defined. This can be achieved, e.g., by spincoating a polymer in a volatile solvent, which rapidly evaporates at a temperature below the glass transition temperature of the polymer. The resulting films are almost perfectly flat with some small initial roughness. In order to start the dewetting process, the system is heated above the glass transition temperature. Linearizing the corresponding thin film equation (11.25) around the initial flat film with thickness h_0 yields for the perturbation $\delta h(x, y, t)$:

$$\eta \frac{\partial \delta h}{\partial t} = M(h_0) \nabla^2 \left(\frac{\partial^2 \omega(h_0)}{\partial h^2} \delta h - \sigma \nabla^2 \delta h \right). \quad (11.35)$$

Solutions of the form $\delta h(x, y, t) = \tilde{\delta h}(k_x, k_y, t) \exp(-i k_x x - i k_y y)$ with $\tilde{\delta h}(k_x, k_y, t) = \tilde{\delta h}(k_x, k_y, 0) \exp(\Omega(k)t)$ are obtained. $\tilde{\delta h}(k_x, k_y, 0)$ is the Fourier transform of the initial roughness. The dispersion relation, i.e., the growth rate of the perturbation, is given by

$$\Omega(k) = -\frac{M(h_0)}{\eta} \left(\frac{\partial^2 \omega(h_0)}{\partial h^2} k^2 + \sigma k^4 \right), \quad (11.36)$$

with $k^2 = k_x^2 + k_y^2$. The dispersion relation is independent of the slip regime, except for the strong-slip regime. For $\partial_h^2 \omega(h_0) < 0$, there is a band of unstable (i.e., exponentially growing) modes with $\Omega(k) > 0$ between $k = 0$ and $k = \sqrt{2} Q$. $Q = \sqrt{|\partial_h^2 \omega(h_0)|/\sigma}$ is the position of the maximum of $\Omega(k)$. In Figure 11.9, $\Omega(k)$ is shown in for spinodally unstable systems as well as for stable or metastable films with $\partial_h^2 \omega(h_0) > 0$.

The dependence of Q on $\omega(h_0)$ has been exploited to experimentally determine the interface potential [119,120]. The power spectrum $S(k,t)$ of the fluctuations in the linear regime is given by

$$S(k,t) = \left\langle \left| \tilde{\delta h}(k_x, k_y, t) \right|^2 \right\rangle = S_0(k) e^{2\Omega(k)t}, \quad (11.37)$$

with the power spectrum $S_0(k)$ of the initial roughness. For sufficiently smooth initial power spectra (i.e., without a pronounced peak) the power spectrum $S(k, t)$ has its maximum at the maximum of $\Omega(k)$, i.e., at $k = Q$. The position Q of the maximum of $S(k, t)$ renders directly the curvature $\partial_h^2 \omega(h_0) = (\sigma Q)^2$. Measuring Q for many film thicknesses allows the determination of $\omega(h)$ by numerical integration. The integration constants are determined by the position and the depth of the minimum of $\omega(h)$, which have to be determined independently by measuring the wetting film thickness between the droplets after the dewetting process is finished and by measuring the equilibrium contact angle of the droplets θ_{eq}, respectively (see Equation 11.30).

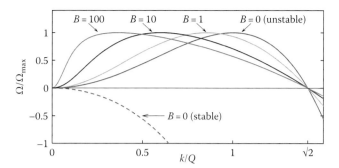

FIGURE 11.9 Dispersion relations $\Omega(k)$ (normalized by their maximum value Ω_{max}) as a function of k/Q for linearly unstable films ($\partial_h^2 \omega(h_0) < 0$, full lines); $k = Q = \sqrt{|\partial_h^2 \omega(h_0)|/\sigma}$ is the position of the maximum of $\Omega(k)$ as valid in the no-slip, weak-slip, and intermediate-slip regime (Equation 11.36). In these latter cases, $\Omega(k)/\Omega_{max}$ is given by the curve marked by $B = 0$ (unstable); $B = 4bh_0 Q^2$. In the strong slip regime Equation 11.36 is replaced by Equation 11.41 leading to a decrease of the position of the maximum upon increasing B, shown for $B = 1$, 10, and 100. All films have the same band $0 < k < \sqrt{2} Q$ of unstable modes, independently of the slip length. The dashed line shows the dispersion relation of a stable film, i.e., $\partial_h^2 \omega(h_0) > 0$. In this case, Ω is normalized by $\tilde{\Omega}_{max}$ which is obtained by calculating $\Omega(Q)$ via Equation 11.36 but with $\partial_h^2 \omega(h_0)$ replaced by $-\partial_h^2 \omega(h_0)$.

The characteristics of the power spectrum $S(k, t)$ differ from the deterministic spectrum in Equation 11.37 if thermal noise is relevant. In this case, by linearizing Equation 11.27 one obtains

$$S(k,t) = S_0(k) e^{2\Omega(k)t} + \frac{k_B T M(h_0)}{\eta} \frac{k^2}{\Omega(k)} \left(e^{2\Omega(k)t} - 1 \right). \quad (11.38)$$

The second, temperature-dependent term is generated by the thermal fluctuations and, in contrast to the deterministic spectrum in Equation 11.37, is nonzero for $t > 0$ even for an initially perfectly flat film, i.e., for $S(k, 0) = 0$. If $S(k, 0)$ depends on k only weakly, as a function of time the power spectrum develops a peak at $k_m(t) > Q$, which approaches Q from above in the limit $t \to \infty$ [43]. A shifting peak in a power spectrum is often associated with the importance of nonlinearities. Here, the reason for this noise-induced coarse graining is that initially thermal fluctuations generate mostly fluctuations with short wavelengths and only later the instability associated with longer wavelengths corresponding to the maximum of $\Omega(k)$ sets in. Recently, it has been possible to demonstrate the relevance of thermal fluctuations for spinodal dewetting of thin polymer films by analyzing the variance of the film thickness [16],

$$\langle \delta h^2 \rangle = \int_0^{k_c} \frac{k}{2\pi} S(k,t) \, dk, \quad (11.39)$$

and the variance of the slope,

$$\langle (\nabla h)^2 \rangle = \int_0^{k_c} \frac{k^3}{2\pi} S(k,t) \, dk, \quad (11.40)$$

with the microscopic cutoff k_c, which is induced by the molecular scale. However, numerical studies have also demonstrated that later stages of dewetting, when holes in the film start to coalesce, are hardly influenced by thermal fluctuations [45].

Beyond the linear regime, the thin film equation (11.25) can be solved only numerically. It is particularly difficult to cope numerically with the vanishing mobility factor $M(h \rightarrow 0) = 0$. With schemes that preserve the non-negativity of $h(x, y, t)$ [121,122], it has been possible to demonstrate [54] that spinodal dewetting patterns can be modeled quantitatively by Equation 11.25.

11.4.3.3 Dewetting on Slippery Substrates

As already mentioned above, the dispersion relation of a thin liquid film in the strong slip regime has a form different from Equation 11.36. Linearizing Equation 11.26 around a flat film with thickness h_0 yields [123]

$$\Omega(k) = -\frac{bh_0^2 k^2 \left(\partial_h^2 \omega(h_0) + \sigma k^2\right)}{\eta(1 + 4bh_0 k^2)} \qquad (11.41)$$

The strong-slip dispersion relation is also positive for $0 < k < \sqrt{2}\, Q$, but the position k_m of the maximum decreases upon increasing the dimensionless slip length $B = 4bh_0 Q^2$ (see Figure 11.9). The power spectrum of a film in the strong slip regime is also given by Equation 11.37, but with the dispersion relation given by Equation 11.41, and it also develops a peak at the position of the maximum of $\Omega(k)$. However, since this position depends on the slip length and the film thickness, the effective interface potential for films with strong slip cannot be determined from the power spectrum (as described above) without knowledge of the substrate rheology, i.e., the slip length [42].

Slip influences not only the fluctuation spectrum of spinodally dewetting films but also the dynamics and the structure of dewetting patterns of thicker, metastable films. During dewetting of a metastable film, initially separated holes, which grow in size and time, are observed. The liquid from inside the holes accumulates in a rim around the hole, which, as a consequence, grows in time. This growth rate depends on the dissipation mechanism [124]: on substrates without slip, neither the driving force for the motion of the rim nor the dissipation (which only occurs within the contact line) depends on the hole size, such that the hole radius R grows linearly in time, i.e., $R \sim t$. In the strong-slip (or plug-flow) regime, however, the dissipation happens in the liquid–substrate interface, which means that the overall dissipation rate grows with the width of the rim. In this regime, theoretically $R \sim t^{2/3}$ is obtained. A combined model, which includes both the dissipation in the contact line as well as the dissipation in the liquid–substrate interface, yields an implicit relation between R and t, which quantitatively describes the dewetting rate of polymer films with large slip [125].

However, the rim around the holes is not stable and the type of instabilities observed depends on the slip length. On substrates with weak slip, a depression is observed between the rim and the resting film, which can act as a nucleation site for satellite holes [126,127]. On substrates with strong slip, this depression is absent [128–130] and therefore it cannot serve as a nucleation site for secondary holes. Since this depression of the film thickness had been observed predominantly in Newtonian fluids and a monotonic decay toward the resting film in viscoelastic films, there were speculations that viscoelasticity would prevent the formation of the depression [131]. However, recently it has been shown that this is not the case [46]. The reason for this misinterpretation of the experimental findings was that experiments were compared that involved polymers with short (Newtonian) and long (viscoelastic) chain lengths, respectively. However, the slip length of polymers with long chain lengths tends to be much larger than that of polymers of the same type but with short chain lengths.

Stationary liquid ridges are unstable with respect to pearling [90,132]. This surface tension driven instability (related to the Plateau-Rayleigh instability which, inter alia, causes a falling stream of fluid to break up into drops) also affects the dewetting rim around holes [133,134]. Recently, it has been proposed that hydrodynamic slip should not only increase the growth rate of the rim instability by orders of magnitude, but that in the presence of slip the initial modulations of the rim become asymmetric by developing protrusions toward the hole [135]. Figures 4 and 6 in Ref. [133] indeed show such asymmetric rim undulations; however, in this reference the slip length was not assessed.

11.5 Heterogeneous Substrates

Most actual surfaces are heterogeneous. However, with microfluidic applications in mind, in the following we focus on structured substrates rather than random heterogeneities (such as roughness or dirt), although these have significant influence on the wetting properties [136,137] as well as on the dynamics of moving contact lines [138]. In general, we distinguish between topographic and chemical heterogeneities, i.e., between substrates of homogeneous chemical composition but with a nonflat surface and chemically heterogeneous substrates with a flat surface.

11.5.1 Topography

Recently, topographically structured surfaces have attracted significant attention due to their ability to change the macroscopic contact angle of droplets significantly: depending on the wetting properties of the material forming a corresponding flat substrate, roughness, and in particular topographic textures can lead to so-called superhydrophobic or superhydrophilic states. While superhydrophilic substrates are completely wetted hydrophilic substrates, in the texture or grooves of the rough surface of superhydrophobic substrates pockets of gas or vapor are trapped, which effectively increase the macroscopic equilibrium contact angle. (Since the contact line on rough surfaces is not a circle and near the substrate the drop shape deviates from a spherical cap, the contact angle has to be defined as explained in Section 11.4.1.1.) The topographical features of substrates can be generated by using advanced photolithographic techniques originally developed in the semiconductor industry. However, they are, in general, on the micron scale and we refer the reader to Refs. [139,140] for more information.

For a long time, roughness was associated with an effective no-slip boundary condition [141]. However, it turns out that the vapor trapped between the solid and the liquid in the superhydrophobic state leads to significant slip [142–144]. Since the viscosity of gases is negligible as compared to the viscosity of most fluids, the gas bubbles in the grooves of the substrate surface act as a lubricant and significantly reduce the friction of the fluid.

11.5.1.1 Wetting of Wedges and Nanosculptured Surfaces

Topographically structured surfaces exhibit wetting phenomena which are qualitatively different from the ones of flat substrates. One important aspect of this is, that the amount of adsorbed fluid on a rough surface is significantly different (in most cases larger) from the amount on corresponding smooth surfaces [26,145,146] and depends on the characteristics of the roughness [147]. The simplest topographic structure, which shows a nontrivial wetting behavior, is a wedge that can be found at the bottom of a groove. If such a structure is in chemical equilibrium with a vapor, the wedge is filled with liquid even at temperatures and pressures at which the flat surface of the same material is still nonwet. The filling of the wedge is a thermodynamic phase transition, which can be understood in terms of a simple macroscopic picture [148]: The equilibrium contact angle is a function of temperature and pressure. The liquid–vapor interface is a section of a circle with a curvature given by the pressure. Calculating the filling level of the groove follows the principles of geometry: the groove is filled/empty if $\pi/2 - \theta_{eq}$ is larger/smaller than half the opening angle of the wedge. Taking into account long-ranged intermolecular interactions leads to a very rich phase diagram [149]. On the gas side of the bulk phase diagram, a prefilling line is found, the crossing of which leads to a first-order interfacial phase transition between a small and a large but finite filling height. At bulk gas–liquid coexistence, this jump becomes macroscopically large forming a bona fide filling transition at a temperature below the wetting transition temperature of the corresponding flat surface. The prefilling line varies as function of the wedge opening angle and can even intersect the prewetting line. Generically, the order of the filling transition is the same as that of the wetting transition. Decreasing undersaturation along an isotherm above the filling transition temperature leads to a continuous complete filling transition associated with interesting critical phenomena, which have attracted considerable theoretical [150] and experimental [151] attention. Monitoring adsorption on nanosculptured surfaces [152] demonstrates that the details of the shapes of the surface structures together with the long range of the underlying dispersion forces strongly influences the macroscopic wetting properties of such surfaces [153,154].

Topographically structured surfaces do not only exhibit nontrivial wetting phenomena. Droplets of nonvolatile fluids, e.g., at steps [155] or in grooves [156], undergo a number of morphological phase transitions as a function of fluid volume and geometrical parameters. As an example, a small drop spreads along a surface step whereas a large one assumes a more compact shape [155]. The morphology also depends on the equilibrium contact angle on the corresponding flat surface: the compact shape is favored for large contact angles, while for $\theta_{eq} \to 0$ drops of any size spread along the step. θ_{eq} can be conveniently modified by applying an electrical voltage (a phenomenon called electrowetting [157]) such that the morphologies can be switched electrically [158]. For a topical review of such phenomena, see Ref. [159]. However, up to now these phenomena have not been studied on the nanoscale.

11.5.1.2 Dynamics of Thin Films and Droplets

The discussion of the moving contact line in Section 11.4.2 assumes a homogeneous substrate. If a moving contact line encounters a surface heterogeneity, such as, e.g., a receding contact line approaching a topographic step as illustrated in Figure 11.10, it can be pinned [11,140]: the local contact angle has to decrease considerably so that the contact line can overcome the obstacle. The same is true for an advancing contact angle approaching the step from above as well as for a steps with rounded edges. Within this macroscopic picture, the relevant parameter for the pinning is the slope of the steepest part of the step, which is $\pi/2$ for the example shown in Figure 11.10. A contact line that approaches a step from below will not be pinned. However, experimentally it has been shown that a step has to have a certain minimum height, which depends on the type of fluid, in order to pin a receding contact line [160]. The mechanisms for this phenomenon is not yet fully understood, but the observation that for decreasing step height the disjoining pressure in the vicinity of the step, calculated according to Equation 11.11, smoothly converges to the disjoining pressure of a flat (barrier free) substrate suggests that

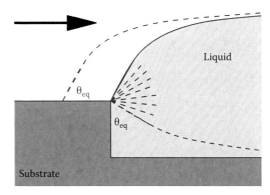

FIGURE 11.10 Macroscopic picture of a receding three-phase contact line (the direction of motion is indicated by the arrow) pinned at a topographic step. On the flat part of the substrate, the contact angle is given by the equilibrium contact angle θ_{eq} (or by the receding contact angle in the case of hysteresis). At the edge the contact angle is ill-defined. In order to overcome the step the equilibrium or receding contact angle on the vertical part of the step has to be reached.

the free energy barrier for the depinning of the contact line also vanishes continuously. This naturally explains why the step has to have a minimum height in order to effectively pin the contact line. The probability for the contact line to overcome this barrier by a thermal fluctuation decreases very strongly with the barrier height.

Nanodroplets in the vicinity of topographic steps behave qualitatively differently from macroscopic drops. If a macroscopic drop resides completely on the terrace above or below the step and if the three-phase contact line of the drop does not touch the step, it does not interact with the step and therefore does not move. Nanodrops, however, do react on the presence of the step via the long-ranged part of the intermolecular interaction potentials. Since the equilibrium contact angle θ_{eq} is determined by the interplay of long- and short-ranged forces, the direction of motion of the nanodroplets cannot be inferred from the equilibrium contact angle but only from the sign of the leading large-distance term of the disjoining pressure, i.e., the Hamaker constant. As a consequence, the direction of motion of the nanodroplets does not depend on whether they start from the top or from the base of the step: they move in step-down direction for negative and in step-up direction for positive Hamaker constants [161].

At this point it is not clear whether the alignment of condensed droplets along terrace steps of vicinal surfaces as observed in Ref. [53] is due to the migration of nanodroplets toward the step or due to an instability related to the morphological phase transition of a liquid condensate growing at the base of a step [155]. Clearly the interplay of condensation dynamics, droplet migration, and morphological instabilities is a promising field of research.

11.5.2 Chemically Inhomogeneous Substrates

Macroscopically (i.e., on length scales large compared with the structures), chemically, and topographically inhomogeneous substrates behave rather similarly: although the three-phase contact lines of droplets are not circular, an effective macroscopic equilibrium contact angle can be defined and the contact lines tend to be pinned at any kind of heterogeneity [162]. Moreover, since many techniques to chemically pattern surfaces involve coatings or grafted monolayers of large molecules such as polymers, chemical patterns are often accompanied with topographic steps [163]. However, there are methods available, such as local oxidation nanolithography [164], which allow one to create chemical nanopatterns with topographically flat surfaces. Alternatively, endgroups of grafted polymer brushes can be removed by ultraviolet light, which allows one to create patterns optically [165]. There are also coatings that change their wettability reversibly when illuminated by light [166] or when exposed to electrical fields [167].

By varying the density of coatings, surfaces with a chemical gradient leading to a position-dependent equilibrium contact angle $\theta_{eq}(x, y)$ can be realized. On such surfaces, droplets move toward regions with smaller contact angles [168–170]. The forces acting on the droplets can be strong enough to drive droplets uphill [171].

Chemical patterns have recently received significant attention due to their potential application as open microfluidic devices in which liquids are not confined to closed pipes but to chemical channels, i.e., hydrophilic stripes embedded into hydrophobic surfaces [172–174]. In such systems, flow cannot be generated by applying pressure to an inlet as in the case of a closed pipe. Several alternative means to drive flow have been discussed, ranging from gravity and shear in a covering layer of fluid [175] to substrates with dynamically switching wetting properties [176]. In contrast to what is possible for closed channel systems, manipulating the stresses at the liquid–vapor interface is a means to generate flow in an open channel system [173]. In most systems, the surface tension coefficient σ decreases as a function of temperature, such that locally heating the substrate results in a lateral variation of the surface tension, which, in turn, leads to a tangential stress in the fluid surface which generates flow. This phenomenon is called Marangoni effect.

11.5.2.1 Wetting Phenomena on Chemically Structured Substrates

While a liquid film on a homogeneous substrate, which is in thermodynamic equilibrium with its vapor, is always flat and droplets are unstable, the local thickness of wetting films on a chemically inhomogeneous substrate reflects the heterogeneity. Straight chemical steps between homogeneous half planes of different wettability and straight chemical channels (i.e., stripes of a wettability different from the surrounding homogeneous substrate) have been studied extensively as paradigmatic examples.

The thicknesses of the wetting film on the two sides of a chemical step depends on the chemical potential (i.e., the vapor pressure) and temperature and in general, they are different. Therefore the chemical step results in a lateral variation of the wetting film thickness. In the case that the wetting temperature on one half of the substrate is higher than on the other half, one can find temperatures at which a macroscopically thick liquid film on one half of the substrate crosses over into a microscopically thin wetting film on the other side. In this case, the thickness $h(x)$ on the wet side as a function of the distance x from the chemical step follows the power law $h(x) \sim \sqrt{x}$ [177].

In the case of a completely wetting chemical channel surrounded by a nonwet substrate, due to surface tension effects, the film thickness on the channel remains finite even for vanishing undersaturation [178,179]. If the effective interface potential $\omega(h)$ describing the channel has two minima at finite film thicknesses, the film thickness on the substrate surrounding the chemical channel can influence the thickness on the chemical channel and its change can induce a transition of the film thickness on the channel from the thickness corresponding to one minimum of $\omega(h)$ to the thickness corresponding to the other minimum [180]. This transition is a quasi-first-order morphological phase transition.

While the film thickness of volatile fluids on straight chemical channels is constant along the channel, one observes morphological phase transitions for nonvolatile fluids. The control parameters are the equilibrium contact angle on the stripe as well as on the surrounding substrate (which should be larger than θ_{eq} on the stripe in order to be able to confine liquids to the channel; both equilibrium contact angles depend on temperature), and V/W^3 where V is the droplet volume and W the stripe width. For small volumes and small but finite contact angles on the channel the fluid spreads in a cigar-shaped rivulet along the channel. Upon increasing the fluid volume at equilibrium contact angles on the channel below $\theta_{eq}^{(c)} \approx 39.2°$, one observes a first-order morphological phase transition to a state in which the fluid contracts into a bulge-like droplet. The values $\theta_{eq} = \theta_{eq}^{(c)}$ and $V/W^3 \approx 2.85$ form the critical end point of a line of phase transitions. On substrates with a finite contact angle, this bulge spills onto the surrounding substrate [181,182].

On a completely wetting channel ($\theta_{eq} = 0$), homogeneous rivulets can be observed only if the contact angle θ of the rivulet (which is pinned at the channel edge by the chemical step if θ is smaller than the equilibrium contact angle on the surrounding substrate) is smaller than a certain critical angle. Within the macroscopic capillary model, i.e., if only surface tensions are taken into account but not the effect of long-ranged intermolecular forces, this angle equals 90° [183]. If θ is larger the rivulet is linearly unstable and breaks up into a string of droplets. This surface tension–driven instability is similar to the Plateau-Rayleigh instability of a homogeneous cylinder of liquid: the local pressure in the fluid (determined by the Laplace pressure) decreases with the radius such that a part of the cylinder, which is only slightly thinner will inflate the rest. For the same reason, the smaller balloon inflates the thicker balloon if they are connected. In the case of the rivulet, the pressure in the fluid increases for increasing filling level, if $\theta < 90°$ but it decreases for larger θ. Indeed one can show that homogeneously filled channels are not unstable with respect to pearling if the pressure increases with the filling level even if the effect of long-ranged intermolecular forces are taken into account in terms of the effective interface potential [184]. Although a homogeneously filled straight chemical channel of macroscopic length can be linearly stable, it does not necessarily represent the state with the lowest free energy: collecting all the liquid into a single macroscopically large drop always reduces the surface area of the liquid vapor interface and, if the equilibrium contact angle on the channel is finite, also reduces the free energy. In this case, the diameter of the drop is large as compared to the channel width. If the equilibrium contact angle on the substrate surface surrounding the chemical channel is smaller than 180°, the shape of the drop resembles the hemispherical shape of a drop on a homogeneous surface with the same wetting properties, but with a small perturbation of the circular contact line at the positions where it crosses the channel. If the surrounding substrate is dry, i.e., if the equilibrium contact angle is 180°, the drop is basically spherical but still connected to the chemical channel with a narrow neck, which is extended along a portion of the channel.

Morphological transitions have been observed experimentally as well as in simulations not only on straight channels [185,186] but also on rings [187,188]: In this case, the ratio of the ring diameter and the stripe width provides an additional control parameter and one observes two transitions: At low volumes, the rivulet on the ring is cylindrically symmetric. But this configuration undergoes a surface tension–driven symmetry breaking instability toward a single bulge at very large volumes, the fluid assumes again a symmetric configuration with a big drop spanning the whole ring including the nonwetting circle in its center.

11.5.2.2 Rivulets and Droplets on Chemical Channels

The above-mentioned surface tension–driven instabilities persist in driven systems: the rivulet on a homogeneously filled chemical channel is stable or unstable with respect to pearling independent of whether the fluid flows along the channel or not [184,189]. However, the range of linear instability of the modes is shifted toward smaller wavenumbers, which leads to larger droplets. The maximal growth rate increases with the flow speed for well-filled channels with contact angles at the channel edge larger than 90° [189] but for rivulets with low height, it decreases with the flow velocity [184]. While for low flow rates the influence of the flow on the onset of instability is moderate, the coarsening dynamics of the droplets changes qualitatively. The instability leads to a string of almost equally sized droplets. Without flow, larger droplets grow at the expense of their smaller neighbors because in them the pressure is lower. But the transport of fluid between the droplets is slow because it occurs through the relatively thin fluid film connecting them. If there is flow in the chemical channel, e.g., driven by a body force like gravity or by centrifugal forces, larger droplets move faster than smaller ones because the driving force is proportional to their volume but the friction is proportional to the base area. Accordingly, the coarsening process is accelerated significantly because big drops overrun smaller ones and, as a consequence of their volume increase, move even faster [189].

Droplets or rivulets on a chemical channel have a lower free energy than droplets on the surrounding hydrophobic substrate. In the macroscopic picture, a droplet residing near a sharp chemical step between a homogeneous hydrophilic and a homogeneous hydrophobic part of the substrate does not respond to the presence of the step unless its three-phase contact line reaches the step: moving the droplet slightly in lateral directions does not change the free energy of the system. Only if the droplet spans the step, it will experience a lateral force pulling it toward the more wettable side. However, if the initial droplet has a rather low height so that its initial contact angle is much smaller than the equilibrium contact angle on either side of the substrate, it can happen that during its initial contraction process, the droplet ends up completely on the less wettable side of the step and stays there [190].

On the nanoscale, however, the droplets respond the presence of the chemical step close to them due to the long-ranged intermolecular interactions and, as a result, they will start to migrate. In analogy to the behavior of droplets in the vicinity of topographic steps, the direction of motion is given by the difference of the Hamaker constants of the substrates on the two sides of the chemical step: the droplet moves toward the side with the larger Hamaker constant [191]. Since the equilibrium contact angle is determined by the interplay of the long-ranged and the short-ranged parts of the intermolecular potentials, there are situations in which the droplet moves toward the less wettable side, i.e., in the unexpected direction. Starting on the less wettable side, it moves away from the step with a velocity that decreases rapidly as a function of the distance from the step. Starting on the other side, it moves toward the step where its advancing three-phase contact line gets pinned before crossing over to the less wettable side.

Also the behavior of nanodroplets spanning the chemical step differs qualitatively from the macroscopically expected one. Within the macroscopic picture, the driving force for droplet migration across a chemical step is the difference in equilibrium contact angles on both sides. However, microscopic droplets are driven by the integral, over the droplet surface, of the laterally varying disjoining pressure. In the limiting case of large drops one recovers the macroscopic behavior: the drops are so tall, that the integral of the disjoining pressure over the drop surface renders the effective interface potential, which is related to the equilibrium contact angle via Equation 11.30. For nanodroplets, the disjoining pressure at the droplet apex has not yet vanished so that the driving force is not given by the difference in contact angles at the two sides of the step. In fact, even the sign of the force can change with the droplet size [192] such that small droplets can migrate to the unexpected side of the chemical step.

This might have important implications for the coarsening dynamics of droplets near chemical boundaries, which is strongly influenced by the migration of droplets across chemical steps [193]. Although this has not yet been discussed in detail, chemical gradients can be expected to act in directions that differ for macroscopic and microscopic droplets.

11.6 Summary and Outlook

Nanofluidics is a wide field of research combining physics, chemistry, and engineering. It poses experimental as well as theoretical challenges with a variety of possible applications. On the experimental side, the main challenge is to probe very soft systems with sufficient resolution and contrast without perturbing them with the probe. In this context, AFM is by far the most widely used method for imaging, followed by ellipsometry and scattering techniques. The main theoretical challenge is to bridge the gap in length and time scales between the molecular motion and the collective movement of the fluid, e.g., the translation of droplets. Top-down approaches combine macroscopic hydrodynamics with equilibrium statistical physics and yield mesoscopic hydrodynamic model equations, which (partially) include the effects of boundary slip, thermal fluctuations, and the finite range of molecular interactions. Bottom-up approaches for nonequilibrium systems are only available for a small class of systems with purely diffusive dynamics. For simple liquids, one has to resort to molecular dynamics simulations.

The most intensively studied and best understood systems are dewetting thin films and droplets spreading on homogeneous substrates. However, there are still open questions concerning the intrinsic dynamics of moving three-phase contact lines and in the adjoining precursor films, in particular for liquid crystals, for which the precursor film has a distinct structure inducing peculiar instabilities of the three-phase contact line [194]. Among the greatest challenges in the field is to control the behavior of open nanofluidic systems in order to guide fluids or to pattern films on the nanoscale. Both problems are related in that chemical wettability patterns are used in order to achieve this.

Chemical patterns are used not only to guide liquids but also to control their flow [165]. While in technological applications, micron-sized channels are used, all biological cells use a large number of nanochannels to exchange ions with the extracellular medium. These ion-channels are highly sophisticated nanofluidic devices. While the selectivity to a certain type of ions is due to a combination of steric effects and electrostatics, gating, i.e., the process of opening and closing the channels, has been recently related to capillary evaporation, i.e., the formation of a vapor or gas bubble inside the channel, which blocks the ion exchange [4]. This model for gating also provides a new insight for understanding the way narcotics work.

In summary, understanding open nanofluidic systems is the key to the miniaturization of microfluidic systems down to the nanoscale as well as a prerequisite for further progress in many active areas in biology and even medicine.

References

1. Mukhopadhyay R. 2006. What does nanofluidics have to offer? *Anal. Chem.* 78:7379–7382.
2. Eijkel JCT and van den Berg A. 2005. Nanofluidics: What is it and what can we expect from it? *Microfluid. Nanofluid.* 1:249–267.
3. Rauscher M and Dietrich S. 2008. Wetting phenomena in nanofluidics. *Ann. Rev. Mater. Sci.* 38:143–172.
4. Roth R and Kroll M. 2006. Capillary evaporation in pores. *J. Phys.: Condens. Matter* 18:6517–6530.
5. Kovarik ML and Jacobson SC. 2007. Attoliter-scale dispensing in nanofluidic channels. *Anal. Chem.* 79:1655–1660.
6. Mitchell P. 2001. Microfluidics—downsizing large-scale biology. *Nat. Biotech.* 19:717–721.
7. Thorsen T, Maerkl SJ, and Quake SR. 2002. Microfluidic large scale integration. *Science* 298:580–584.
8. Evans R. 1992. Density functionals in the theory of nonuniform fluids. In *Fundamentals of Inhomogeneous Fluids*, ed. Henderson D. pp. 85–173. New York: Marcel Dekker.
9. Goel G, Krekelberg WP, Errington JR, and Truskett TM. 2008. Tuning density profiles and mobility of inhomogeneous fluids. *Phys. Rev. Lett.* 100:106001.
10. Lauga E, Brenner MP, and Stone HA. 2007. The no-slip boundary condition: A review. In *Springer Handbook of Experimental Fluid Mechanics*, eds. Tropea C, Yarin AL, and Foss JF. pp. 1219–1240. Berlin, Germany: Springer.
11. de Gennes PG. 1985. Wetting: Statistics and dynamics. *Rev. Mod. Phys.* 57:827–860.

12. Dietrich S. 1988. Wetting phenomena. In *Phase Transitions and Critical Phenomena*. Vol. 12, eds. Domb C and Lebowitz JL. pp. 1–218. London U.K.: Academic Press.

13. Binder K, Müller M, Schmid F, and Werner A. 1999. Interfacial profiles between coexisting phases in thin films: Cahn-Hilliard treatment versus capillary waves. *J. Stat. Phys.* 95:1045–1068.

14. Mecke KR. 2001. Thermal fluctuations of thin liquid films. *J. Phys.: Condens. Matter* 13:4615–4636.

15. Vorberg J, Herminghaus S, and Mecke K. 2001. Adsorption isotherms of hydrogen: The role of thermal fluctuations. *Phys. Rev. Lett.* 87:196105.

16. Fetzer R, Rauscher M, Seemann R, Jacobs K, and Mecke K. 2007. Thermal noise influences fluid flow in thin films during spinodal dewetting. *Phys. Rev. Lett.* 99:114503.

17. Frenkel D and Smit B. 2002. *Understanding Molecular Simulation*. 2nd edn. San Diego, CA: Academic Press.

18. De Coninck J and Blake TD. 2008. Wetting and molecular dynamics simulations of simple liquids. *Ann. Rev. Mater. Sci.* 38:1–22.

19. de Pablo JJ and Curtin WA. 2007. Multiscale modeling in advanced materials research: Challenges, novel methods, and emerging applications. *MRS Bull.* 32:905–911.

20. Evans R. 1979. The nature of the liquid-vapour interface and other topics in the statistical mechanics of non-uniform, classical fluids. *Adv. Phys.* 28:143–200.

21. Evans R. 1990. Microscopic theories of simple fluids and their interfaces. In *Liquids at Interfaces*, eds. Charvolin J, Joanny JF, and Zinn-Justin J. Les Houches, Session XLVIII. pp. 1–98. Amsterdam, the Netherlands: Elsevier.

22. Hansen JP and McDonald IR. 1990. *Theory of Simple Liquids*. 2nd edn. London, U.K.: Academic Press.

23. Roth R, Evans R, Lang A, and Kahl G. 2002. Fundamental measure theory for hard-sphere mixtures revisited: The White Bear version. *J. Phys.: Condens. Matter* 14:12063–12078.

24. Napiórkowski M and Dietrich S. 1993. Structure of the effective Hamiltonian for liquid-vapor interfaces. *Phys. Rev. E* 47:1836–1849.

25. Dietrich S and Napiórkowski M. 1991. Analytic results for wetting transitions in the presence of van der Waals tails. *Phys. Rev. A* 43:1861–1885.

26. Robbins MO, Andelman D, and Joanny JF. 1991. Thin liquid films on rough or heterogeneous solids. *Phys. Rev. A* 43:4344–4354.

27. Chan GKL and Finken R. 2005. Time-dependent density functional theory of classical fluids. *Phys. Rev. Lett.* 94:183001.

28. Runge E and Gross EKU. 1984. Density-functional theory for time-dependent systems. *Phys. Rev. Lett.* 52:997–1000.

29. Rex M and Löwen H. 2008. Dynamical density functional theory with hydrodynamic interactions and colloids in unstable traps. *Phys. Rev. Lett.* 101:148302–148304.

30. Archer AJ and Rauscher M. 2004. Dynamical density functional theory for interacting Brownian particles: Stochastic or deterministic? *J. Phys. A: Math. Gen.* 37:9325–9333.

31. Moffatt HK. 1977. Six lectures on general fluid dynamics and two on hydromagnetic dynamo theory. In *Fluid Dynamics*, eds. Balian R and Peube JL. pp. 149–233. London, U.K.: Gordon and Breach.

32. Bird RB, Armstrong RC, and Hassager O. 1977. *Dynamics of Polymeric Fluids*. Vol. 1. New York: John Wiley & Sons.

33. Blossey R, Münch A, Rauscher M, and Wagner B. 2006. Slip vs. viscoelasticity in dewetting thin films. *Eur. Phys. J. E* 20:267–271.

34. Münch A, Wagner B, Rauscher M, and Blossey R. 2006. A thin film model for viscoelastic fluids with large slip. *Eur. Phys. J. E* 20:365–368.

35. Edelen DGB. 1976. Nonlocal field theories. In *Continuum Physics*. Vol. IV—*Polar and Nonlocal Field Theories*, ed. Eringen AC. pp. 75–204. New York: Academic Press.

36. Napiórkowski M and Dietrich S. 1995. Curvature corrections to the capillary wave Hamiltonian. *Z. f. Physik B* 97:511–513.

37. Landau LD and Lifshitz EM. 2005. *Fluid Mechanics*. Vol. 6 of *Course of Theoretical Physics*. 2nd edn. Amsterdam, the Netherlands; Heidelberg, Germany: Elsevier Butterworth-Heinemann.

38. Gardiner CW. 1983. *Handbook of Stochastic Methods for Physics, Chemistry and the Natural Sciences*. Vol. 13 of *Springer Series in Synergetics*. 1st edn. Berlin, Germany: Springer.

39. Risken H. 1984. *The Fokker-Planck Equation*. Vol. 18 of *Springer Series in Synergetics*. Berlin, Germany: Springer.

40. Münch A, Wagner B, and Witelski T. 2005. Lubrication models with small to large slip lengths. *J. Eng. Math.* 53:359–383.

41. Münch A and Wagner B. 2008. Contact-line instability of dewetting films for large slip. In preparation.

42. Rauscher M, Blossey R, Münch A, and Wagner B. 2008. Spinodal dewetting of thin films with large interfacial slip: Implications from the dispersion relation. *Langmuir* 24: 12290–12294.

43. Mecke K and Rauscher M. 2005. On thermal fluctuations in thin film flow. *J. Phys.: Condens. Matter* 17:S3515–S3522.

44. Davidovitch B, Moro E, and Stone HA. 2005. Spreading of thin films assisted by thermal fluctuations. *Phys. Rev. Lett.* 95:244505.

45. Grün G, Mecke KR, and Rauscher M. 2006. Thin-film flow influenced by thermal noise. *J. Stat. Phys.* 122:1261–1291.

46. Rauscher M, Münch A, Wagner B, and Blossey R. 2005. A thin-film equation for viscoelastic liquids of Jeffreys type. *Eur. Phys. J. E* 17:373–379.

47. Binnig G, Rohrer H, Gerber C, and Weibel E. 1982. Surface studies by scanning tunneling microscopy. *Phys. Rev. Lett.* 49:57–61.

48. Sarid D. 1994. *Scanning Force Microscopy: With Applications to Electric, Magnetic, and Atomic Forces*. Revised edn. New York: Oxford University Press.

49. Morita S, Wiesendanger R, and Meyer E. 2002. *Noncontact Atomic Force Microscopy. Nanoscience and Technology*. Heidelberg, Germany: Springer.

50. Müller-Buschbaum P. 2003. Dewetting and pattern formation in thin polymer films as investigated in real and reciprocal space. *J. Phys.: Condens. Matter* 15:R1549–R1582.

51. Checco A, Guenoun P, and Daillant J. 2003. Nonlinear dependence of the contact angle of nanodroplets on contact line curvature. *Phys. Rev. Lett.* 91:186101.

52. Hu J, Xiao XD, and Salmeron M. 1995. Scanning polarization force microscopy: A technique for imaging liquids and weakly adsorbed layers. *Appl. Phys. Lett.* 67:476–478.

53. Hu J, Carpick RW, Salmeron M, and Xiao XD. 1996. Imaging and manipulation of nanometer-size liquid droplets by scanning polarization force microscopy. *J. Vac. Sci. Technol. B* 14:1341–1343.

54. Becker J, Grün G, Seemann R, Mantz H, Jacobs K, Mecke KR, and Blossey R. 2003. Complex dewetting scenarios captured by thin film models. *Nat. Mater.* 2:59–63.

55. Pompe T and Herminghaus S. 2000. Three-phase contact line energetics from nanoscale liquid surface topographies. *Phys. Rev. Lett.* 85:1930–1933.

56. Pompe T. 2002. Line tension behavior of a first-order wetting system. *Phys. Rev. Lett.* 89:076102.

57. Als-Nielsen J and McMorrow D. 2006. *Elements of Modern X-Ray Physics*. New York: Wiley.

58. Tolan M. 1999. *X-Ray Scattering from Soft-Matter Thin Films*. Vol. 148 of *Springer Tracts in Modern Physics*. Heidelberg, Germany: Springer.

59. Daillant J and Gibaud A. 1999. *X-Ray and Neutron Reflectivity: Principles and Applications*. Vol. 58 of *Lecture Notes in Physics/New Series M*. Berlin, Germany: Springer.

60. Salditt T and Brotons G. 2004. Biomolecular and amphiphilic films probed by surface sensitive X-ray and neutron scattering. *Anal. Bioanal. Chem.* 379:960–973.

61. Tolan M, Seeck OH, Wang J, Sinha SK, Rafailovich MH, and Sokolov J. 2000. X-ray scattering from polymer films. *Physica B* 283:22–26.

62. Müller-Buschbaum P, Bauer E, Wunnicke O, and Stamm M. 2005. The control of thin film morphology by the interplay of dewetting, phase separation and microphase separation. *J. Phys.: Condens. Matter* 17:S363–S386.

63. Mora S, Daillant J, Mecke K, Luzet D, Braslau A, Alba M, and Struth B. 2003. X-ray synchrotron study of liquid-vapor interfaces at short length scales: Effect of long-range forces and bending energies. *Phys. Rev. Lett.* 90:216101.

64. Shpyrko OG, Streitel R, Balagurusamy VSK, Grigoriev AY, Deutsch M, Ocko BM, Meron M, Lin B, and Pershan PS. 2006. Surface crystallization in a liquid AuSi alloy. *Science* 313:77–80.

65. Reichert H, Bencivenga F, Wehinger B, Krisch M, Sette F, and Dosch H. 2007. High-frequency subsurface and bulk dynamics of liquid indium. *Phys. Rev. Lett.* 98:096104.

66. Mezger M, Reichert H, Schöder S, Okasinski J, Schröder H, Dosch H, Palms D, Ralston J, and Honkimäki V. 2006. High-resolution in situ x-ray study of the hydrophobic gap at the water-octadecyl-trichlorosilane interface. *PNAS* 103:18401–18404.

67. Wolff M, Akgun B, Walz M, Magerl A, and Zabel H. 2008. Slip and depletion in a Newtonian liquid. *Europhys. Lett.* 82:36001.

68. Eisebitt S, Lüning J, Schlotter WF, Lörgen M, Hellwig O, Eberhardt W, and Stöhr J. 2004. Lensless imaging of magnetic nanostructures by X-ray spectro-holography. *Nature* 432:885–888.

69. Holý V, Pietsch U, and Baumbach T. 1999. *High-Resolution X-Ray Scattering from Thin Films and Multilayers*. Vol. 149 of *Springer Tracts in Modern Physics*. Berlin, Germany: Springer.

70. Riekel C. 2000. New avenues in x-ray microbeam experiments. *Rep. Prog. Phys.* 63:233–262.

71. Roth SV, Müller-Buschbaum P, Burghammer M, Walter H, Panagiotou P, Diethert A, and Riekel C. 2005. Microbeam grazing incidence small angle X-ray scattering-a new method to investigate heterogeneous thin films and multilayers. *Spectrochim. Acta Part B: Atomic Spectrosc.* 59:1765–1773.

72. Tompkins HG and Irene EA. 2005. *Handbook of Ellipsometry*. Norwich, NY: William Andrew.

73. Liu AH, Wayner Jr PC, and Plawsky JL. 1994. Image scanning ellipsometry for measuring nonuniform film thickness profiles. *Appl. Opt.* 33:1223–1229.

74. Zhan Q and Leger JR. 2002. High-resolution imaging ellipsometer. *Appl. Opt.* 41:4443–4450.

75. Léger L, Erman M, Guinet-Picard AM, Ausserré D, and Strazielle C. 1988. Precursor film profiles of spreading liquid drops. *Phys. Rev. Lett.* 60:2390–2393.

76. Vandyshev DI and Skakun SG. 1991. Some properties of a precursor film. *J. Eng. Phys. Thermophys.* 61:1482–1485.

77. Vouéa M, De Coninck J, Villette S, Valignat MP, and Cazabat AM. 1998. Investigation of layered microdroplets using ellipsometric techniques. *Thin Solid Films* 313–314:819–824.

78. Shen YR. 1986. A few selected applications of surface nonlinear optical spectroscopy. *PNAS* 93:12104–12111.

79. Shen YR. 1989. Surface properties probed by second-harmonic and sum-frequency generation. *Nature* 337:519–525.

80. Petersen PB and Saykally RJ. 2006. On the nature of ions at the liquid water surface. *Ann. Rev. Phys. Chem.* 57:333–364.

81. Corle TR and Kino GS. 1996. *Confocal Scanning Optical Microscopy and Related Imaging Systems*. San Diego, CA: Academic Press.

82. Aarts DGAL, Schmidt M, and Lekkerkerker HNW. 2004. Direct visual observation of thermal capillary waves. *Science* 304:847–850.

83. Hell SW. 2007. Far-field optical nanoscopy. *Science* 316:1153–1158.

84. Young T. 1805. An essay on the cohesion of fluids. *Phil. Trans. R. Soc. Lond.* 95:65–87.

85. Schimmele L, Napiórkowski M, and Dietrich S. 2007. Conceptual aspects of line tensions. *J. Chem. Phys.* 127:164715.

86. De Feijter JA and Vrij A. 1972. I. Transition regions, line tensions and contact angles in soap films. *J. Electroanal. Chem.* 37:9–22.

87. Indekeu JO. 1992. Line tension near the wetting transition: Results from an interface displacement model. *Physica A* 183:439–461.

88. Rosso R and Virga EG. 2003. General stability criterion for wetting. *Phys. Rev. E* 68:012601.

89. Rosso R and Virga EG. 2004. Sign of line tension in liquid bridge stability. *Phys. Rev. E* 70:031603.

90. Brinkmann M, Kierfeld J, and Lipowsky R. 2005. Stability of liquid channels or filaments in the presence of line tension. *J. Phys.: Condens. Matter* 17:2349–2364.

91. Mechkov S, Oshanin G, Rauscher M, Brinkmann M, Cazabat AM, and Dietrich S. 2007. Contact line stability of ridges and drops. *Europhys. Lett.* 80:66002.

92. Hauge EH and Schick M. 1983. Continuous and first-order wetting transition from the van der Waals theory of fluids. *Phys. Rev. B* 27:4288–4301.

93. Nacher PJ and Dupont-Roc J. 1991. Experimental evidence for nonwetting with super-fluid helium. *Phys. Rev. Lett.* 67:2966–2969.

94. Rutledge JE and Taborek P. 1992. Prewetting phase diagram of ^{4}He on cesium. *Phys. Rev. Lett.* 69:937–940.

95. Taborek P and Rutledge JE. 1993. Tuning the wetting transition: Prewetting and super-fluidity of ^{4}He on thin cesium substrates. *Phys. Rev. Lett.* 71:263–266.

96. Klier J, Stefanyi P, and Wyatt AFG. 1995. Contact angle of liquid ^{4}He on a Cs surface. *Phys. Rev. Lett.* 75:3709–3712.

97. Klier J and Wyatt AFG. 2002. Nonwetting of liquid ^{4}He on Rb. *Phys. Rev. B* 65:212504.

98. Ragil K, Meunier J, Broseta D, Indekeu JO, and Bonn D. 1996. Experimental observation of critical wetting. *Phys. Rev. Lett.* 77:1532–1535.

99. Shahidzadeh N, Bonn D, Ragi K, Broseta D, and Meunier J. 1998. Sequence of two wetting transitions induced by tuning the hamaker constant. *Phys. Rev. Lett.* 80:3992–3995.

100. Pfohl T and Riegler H. 1999. Critical wetting of a liquid/vapor interface by octane. *Phys. Rev. Lett.* 82:783–786.

101. Bertrand E, Dobbs H, Broseta D, Indekeu J, Bonn D, and Meunier J. 2000. First-order and critical wetting of alkanes on water. *Phys. Rev. Lett.* 85:1282–1285.

102. Dussan VEB. 1979. On the spreading of liquids on solid surfaces: Static and dynamic contact lines. *Ann. Rev. Fluid Mech.* 11:371–400.

103. de Gennes PG. 1986. Deposition of Langmuir-Blodgett layers. *Colloid Polym. Sci.* 264:463–465.

104. Léger L and Joanny JF. 1992. Liquid spreading. *Rep. Prog. Phys.* 55:431–486.

105. Blake TD. 1993. Dynamic contact angle and wetting kinetics. In *Wettability*, ed. Berg JC. pp. 252–309. New York: Marcel Dekker.

106. Ralston J, Popescu M, and Sedev R. 2008. Dynamics of wetting from an experimental point of view. *Ann. Rev. Mater. Sci.* 38:23–43.

107. Tanner LH. 1979. The spreading of silicone oil drops on horizontal surfaces. *J. Phys. D: Appl. Phys.* 12:1473–1484.

108. Blake TD and Haynes JM. 1969. Kinetics of liquid/liquid displacement. *J. Colloid Interface Sci.* 30:421–423.

109. de Ruijter MJ, Charlot M, Voué M, and De Coninck J. 2000. Experimental evidence of several time scales in drop spreading. *Langmuir* 16:2363–2368.

110. De Coninck J, de Ruijter MJ, and Voué M. 2001. Dynamics of wetting. *Curr. Opin. Colloid Interface Sci.* 6:49–53.

111. Heslot F, Fraysse N, and Cazabat AM. 1989. Molecular layering in the spreading of wetting liquid drops. *Nature* 338:640–642.

112. Popescu MN and Dietrich S. 2003. Spreading of liquid monolayers: From kinetic Monte Carlo simulations to continuum limit. In *Interface and Transport Dynamics*, eds. Emmerich H, Nestler B, and Schreckenberg M. pp. 202–207. Heidelberg, Germany: Springer.

113. Popescu MN and Dietrich S. 2004. Model for spreading of liquid monolayers. *Phys. Rev. E* 69:061602.

114. Habenicht A, Olapinski M, Burmeister F, Leiderer P, and Boneberg J. 2005. Jumping nanodroplets. *Science* 309:2043–2045.

115. Blossey R. 2008. Thin film rupture and polymer flow. *Phys. Chem. Chem. Phys.* 10:5177–5183.

116. Bausch R, Blossey R, and Burschka MA. 1994. Critical nuclei for wetting and dewetting. *J. Phys. A: Math. Gen.* 27:1405–1406.

117. Foltin G, Bausch R, and Blossey R. 1997. Critical holes in undercooled wetting layers. *J. Phys. A: Math. Gen.* 30:2937–2946.

118. Jacobs K, Seemann R, and Mecke K. 2000. Dynamics of structure formation in thin films: A special spatial analysis. In *Statistical Physics and Spatial Statistics*. Vol. 554 of *Lecture Notes in Physics*, eds. Mecke KR and Stoyan D. pp. 72–91. Berlin, Heidelberg, Germany: Springer.

119. Seemann R, Herminghaus S, and Jacobs K. 2001. Dewetting patterns and molecular forces: A reconciliation. *Phys. Rev. Lett.* 86:5534–5537.

120. Seemann R, Herminghaus S, and Jacobs K. 2001. Gaining control of pattern formation of dewetting liquid films. *J. Phys.: Condens. Matter* 13:4925–4938.

121. Grün G and Rumpf M. 2000. Nonnegativity preserving convergent schemes for the thin film equation. *Numer. Math.* 87:113–152.

122. Becker J and Grün G. 2005. The thin-film equation: Recent advances and some new perspectives. *J. Phys.: Condens. Matter* 17:S291–S307.

123. Kargupta K, Sharma A, and Khanna R. 2004. Instability, dynamics, and morphology of thin slipping films. *Langmuir* 20:244–253.

124. Brochard-Wyart F, de Gennes P-G, Hervert H, and Redon C. 1994. Wetting and slippage of polymer melts on semi-ideal surfaces. *Langmuir* 10:1566–1572.

125. Fetzer R and Jacobs K. 2007. Slippage of Newtonian liquids: Influence on the dynamics of dewetting thin films. *Langmuir* 23:11617–11622.

126. Neto C, Jacobs K, Seemann R, Blossey R, Becker J, and Grün G. 2003. Satellite hole formation during dewetting: Experiment and simulation. *J. Phys.: Condens. Matter* 15:3355–3366.

127. Neto C, Jacobs K, Seemann R, Blossey R, Becker J, and Grün G. 2003. Correlated dewetting patterns in thin polystyrene films. *J. Phys.: Condens. Matter* 15:S421–S426.

128. Fetzer R, Jacobs K, Münch A, Wagner B, and Witelski TP. 2005. New slip regimes and the shape of dewetting thin liquid films. *Phys. Rev. Lett.* 95:127801.

129. Fetzer R, Rauscher M, Münch A, Wagner BA, and Jacobs K. 2006. Slip-controlled thin film dynamics. *Europhys. Lett.* 75:638–644.

130. Fetzer R, Münch A, Wagner B, Rauscher M, and Jacobs K. 2007. Quantifying hydrodynamic slip: A comprehensive analysis of dewetting profiles. *Langmuir* 23:10559–10566.

131. Herminghaus S, Seemann R, and Jacobs K. 2002. Generic morphologies of viscoelastic dewetting fronts. *Phys. Rev. Lett.* 89:056101.

132. Davis SH. 1980. Moving contact line and rivulet instabilities. Part 1: The static rivulet. *J. Fluid Mech.* 98:225.

133. Sharma A and Reiter G. 1996. Instability of thin polymer films on coated substrates: Rupture, dewetting, and drop formation. *J. Colloid Interface Sci.* 178:383–399.

134. Herminghaus S, Seemann R, Podzimek D, and Jacobs K. 2001. Strukturbildung und Dynamik in makromolekularen Filmen. *Nachrichten a. d. Chemie* 49:1398–1404.

135. Münch A and Wagner B. 2005. Contact-line instability of dewetting thin films. *Physica D* 209:178–190.

136. Netz RR and Andelman D. 1997. Roughness-induced wetting. *Phys. Rev. E* 55:687–700.

137. Rascón C and Parry AO. 2000. Geometry-dominated fluid adsorption on sculpted solid substrates. *Nature* 407:986–989.

138. Quéré D. 2008. Wetting and roughness. *Ann. Rev. Mater. Sci.* 38:71–99.

139. Lafuma A and Quéré D. 2003. Superhydrophobic states. *Nat. Mater.* 2:457–460.

140. Quéré D. 2005. Non-sticking drops. *Rep. Prog. Phys.* 68:2495–2532.

141. Richardson S. 1973. On the no-slip boundary condition. *J. Fluid Mech.* 59:707–719.

142. Cottin-Bizonne C, Jurine S, Baudry J, Crassous J, Restagno F, and Charlaix E. 2002. Nanorheology: An investigation of the boundary condition at hydrophobic and hydrophilic interfaces. *Eur. Phys. J. E* 9:47–53.

143. Cottin-Bizonne C, Barrat JL, Bocquet L, and Charlaix E. 2003. Low friction flows of liquids at nanopatterned interfaces. *Nat. Mater.* 2:237–240.

144. Ybert C, Barentin C, Cottin-Bizonne C, Joseph P, and Bocquet L. 2007. Achieving large slip with superhydrophobic surfaces: Scaling laws for generic geometries. *Phys. Fluids* 19:123601.

145. Starov VM and Churaev NV. 1977. Thickness of wetting films on rough substrates. *Coll. J. USSR* 39:975–979.

146. Andelman D, Joanny JF, and Robbins MO. 1989. Wetting of rough solid surfaces by liquids. In *Phase Transitions in Soft Condensed Matter*, eds. Riste T and Sherrington D. pp. 161–164. New York: Plenum Press.

147. Pfeifer P, Wu YJ, Cole MW, and Krim J. 1989. Multilayer adsorption on a fractally rough surface. *Phys. Rev. Lett.* 62:1997–2000.

148. Hauge EH. 1992. Macroscopic theory of wetting in a wedge. *Phys. Rev. A* 46:4994–4998.

149. Rejmer K, Dietrich S, and Napiórkowski M. 1999. Filling transition for a wedge. *Phys. Rev. E* 60:4027–4042.

150. Parry AO, Rascón C, and Wood AJ. 2000. Critical effects at 3D wedge wetting. *Phys. Rev. Lett.* 85:345–348.

151. Bruschi L, Carlin A, and Mistura G. 2002. Complete wetting on a linear wedge. *Phys. Rev. Lett.* 89:166101.

152. Gang O, Alvine KJ, Fukuto M, Pershan PS, Black CT, and Ocko BM. 2005. Liquids on topologically nanopatterned surfaces. *Phys. Rev. Lett.* 95:217801.

153. Tasinkevych M and Dietrich S. 2007. Complete wetting of nanosculptured substrates. *Phys. Rev. Lett.* 97:106102.

154. Tasinkevych M and Dietrich S. 2007. Complete wetting of pits and grooves. *Eur. Phys. J. E* 23:117–128.

155. Brinkmann M and Blossey R. 2004. Blobs, channels and "cigars": Morphologies of liquids at a step. *Eur. Phys. J. E* 14:79–89.

156. Seemann R, Brinkmann M, Kramer EJ, Lange FF, and Lipowsky R. 2005. Wetting morphologies at microstructured surfaces. *PNAS* 102:1848–1852.

157. Mugele F and Baret JC. 2005. Electrowetting: From basics to applications. *J. Phys.: Condens. Matter* 17:R705–R774.

158. Khare K, Herminghaus S, Baret JC, Law BM, Brinkmann M, and Seemann R. 2007. Switching liquid morphologies on linear grooves. *Langmuir* 23:12997–13006.

159. Herminghaus S, Brinkmann M, and Seemann R. 2008. Wetting and dewetting of complex surface geometries. *Ann. Rev. Mater. Sci.* 38:101–121.

160. Ondarçuhu T and Piednoir A. 2005. Pinning of a contact line on nanometric steps during the dewetting of a terraced substrate. *Nano Lett.* 5:1744–1750.

161. Moosavi A, Rauscher M, and Dietrich S. 2006. Motion of nanodroplets near edges and wedges. *Phys. Rev. Lett.* 97:236101.

162. Raphaël E and de Gennes PG. 1989. Dynamics of wetting with nonideal surfaces. The single defect problem. *J. Chem. Phys.* 90:7577–7584.

163. Fukuzawa K, Deguchi T, Kawamura J, Mitsuya Y, Muramatsu T, and Zhang H. 2005. Nanoscale patterning of thin liquid films on solid surfaces. *Appl. Phys. Lett.* 87:203108.

164. Maoz R, Cohen SR, and Sagiv J. 1999. Nanoelectrochemical patterning of monolayer surfaces: Toward spatially defined self-assembly of nanostructures. *Adv. Mater.* 11:55–61.

165. Zhao B, Moore JS, and Beebe DJ. 2002. Principles of surface-directed liquid flow in microfluidic channels. *Anal. Chem.* 74:4259–4268.

166. Abbott S, Ralston J, Reynolds G, and Hayes R. 1999. Reversible wettability of photoresponsive pyrimidine-coated surfaces. *Langmuir* 15:8923–8928

167. Lahann J, Mitragotri S, Tran TN, Kaido H, Sundaram J, Choi IS, Hoffer S, Somorjai GA, and Langer R. 2003. A reversible switching surface. *Science* 299:371–374.

168. Greenspan HP. 1978. On the motion of a small viscous droplet that wets a surface. *J. Fluid Mech.* 84:125–143.

169. Moumen N, Subramanian RS, and McLaughlin JB. 2006. Experiments on the motion of drops on a horizontal solid surface due to wettability gradients. *Langmuir* 22:2682–2690.

170. Pismen LM and Thiele U. 2006. Asymptotic theory for a moving droplet driven by a wettability gradient. *Phys. Fluids* 18:042104.

171. Chaudhury MK and Whitesides GM. 1992. How to make water run uphill. *Science* 256:1539–1541.

172. Delamarche E, Juncker D, and Schmid H. 2005. Microfluidics for processing surfaces and miniaturizing biological assays. *Adv. Mater.* 17:2911–2933.

173. Darhuber AA and Troian SM. 2005. Principles of microfluidic actuation by modulation of surface stresses. *Ann. Rev. Fluid Mech.* 37:425–455.

174. Dietrich S, Popescu MN, and Rauscher M. 2005. Wetting on structured substrates. *J. Phys.: Condens. Matter* 17:S577–S593.

175. Rauscher M, Dietrich S, and Koplik J. 2007. Shear flow pumping in open microfluidic systems. *Phys. Rev. Lett.* 98:224504.

176. Huang JJ, Shu C, and Chew YT. 2008. Numerical investigation of transporting droplets by spatiotemporally controlling substrate wettability. *J. Colloid Interface Sci.* 328:124–133.

177. Bauer C and Dietrich S. 1999. Wetting films on chemically heterogeneous substrates. *Phys. Rev. E* 60:6919–6941.

178. Bauer C and Dietrich S. 1999. Quantitative study of laterally inhomogeneous wetting films. *Eur. Phys. J. B* 10:767–779.

179. Checco A, Gang O, and Ocko BM. 2006. Liquid nanostripes. *Phys. Rev. Lett.* 96:056104.

180. Bauer C, Dietrich S, and Parry AO. 1999. Morphological phase transitions of thin fluid films on chemically structured substrates. *Europhys. Lett.* 47:474–480.

181. Brinkmann M and Lipowsky R. 2002. Wetting morphologies on substrates with striped surface domains. *J. Appl. Phys.* 92:4296–4306.

182. Lipowsky R, Brinkmann M, Dimova R, Franke T, Kierfeld J, and Zhand X. 2005. Droplets, bubbles, and vesicles at chemically structured surfaces. *J. Phys.: Condens. Matter* 17:S537–S558.

183. Davis JM and Troian SM. 2005. Generalized linear stability of noninertial coating flows over topographical features. *Phys. Fluids* 17:072103.

184. Mechkov S, Rauscher M, and Dietrich S. 2008. Stability of liquid ridges on chemical micro- and nanostripes. *Phys. Rev. E* 77:061605.

185. Gau H, Herminghaus S, Lenz P, and Lipowsky R. 1999. Liquid morphologies on structured surfaces: From microchannels to microchips. *Science* 283:46–49.

186. Darhuber AA, Troian SM, Miller SM, and Wagner S. 2000. Morphology of liquid microstructures on chemically patterned surfaces. *J. Appl. Phys.* 87:7768–7775.

187. Lenz P, Fenzl W, and Lipowsky R. 2001. Wetting of ring-shaped surface domains. *Europhys. Lett.* 53:618–624.

188. Porcheron F, Monson PA, and Schoen M. 2006. Wetting of rings on a nanopatterned surface: A lattice model study. *Phys. Rev. E* 73:041603.

189. Koplik J, Lo TS, Rauscher M, and Dietrich S. 2006. Pearling instability of nanoscale fluid flow confined to a chemical channel. *Phys. Fluids* 18:032104.

190. Ondarçuhu T and Raphaël E. 1992. Étalement d'un ruban liquide à cheval entre deux substrats solides différents. *C. R. Acad. Sci. Paris, Série II* 314:453–456.

191. Moosavi A, Rauscher M, and Dietrich S. 2008. Motion of nanodroplets near chemical heterogeneities. *Langmuir* 24:734–742.

192. Moosavi A, Rauscher M, and Dietrich S. 2008. Size dependent motion of nanodroplets on chemical steps. *J. Chem. Phys.* 129:044706.

193. Jéopoldès J and Bucknall DG. 2005. Coalescence of droplets on chemical boundaries. *Europhys. Lett.* 72:597–603.

194. Poulard C and Cazabat AM. 2005. The spontaneous spreading of nematic liquid crystals. *Langmuir* 21:6270–6276.

Capillary Condensation in Confined Media

Elisabeth Charlaix
Université Claude Bernard Lyon 1

Matteo Ciccotti
Université Montpellier 2

12.1 Physics of Capillary Condensation

12.1.1 Relevance in Nanosystems

As the size of systems decreases, surface effects become increasingly important. Capillary condensation, which results from the effect of surfaces on the phase diagram of a fluid, is an ubiquitous phenomenon at the nanoscale, occurring in all confined geometries, divided media, cracks, or contacts between surfaces (Bowden and Tabor 1950). The very large capillary forces induced by highly curved menisci have strong effect on the mechanical properties of contacts. The impact of capillary forces in micro/nano electromechanical systems (MEMS & NEMS) is huge and often prevents the function of small-scale active systems under ambient condition or causes damage during the fabrication process.

Since the nanocomponents are generally very compliant and present an elevated surface/volume ratio, the capillary forces developing in the confined spaces separating the components when these are exposed to ambient condition can have a dramatic effect in deforming them and preventing their service. Stiction or adhesion between the substrate (usually silicon based) and the microstructures occurs during the isotropic wet etching of the sacrificial layer (Bhushan 2007). The capillary forces caused by the surface tension of the liquid between the microstructures (or in the gaps separating them from the substrate) during the drying of the wet etchant cause the two surfaces to adhere together. Figure 12.1 shows an example of the effect of drying after the nanofabrication of a system of lamellae of width and spacing of 200 nm and variable height. Separating

the two surfaces is often complicated due to the fragile nature of the microstructures.[*]

In divided media, capillary forces not only control the cohesion of the media but also have dramatic influence on the aging properties of materials. Since the condensation of liquid bridges is a first-order transition, it gives rise to slow activated phenomena that are responsible for long time scale variations of the cohesion forces (cf. Section 12.2). Capillary forces also have a strong effect on the friction properties of sliding nanocontacts where they are responsible for aging effects and enhanced stick-slip motion (cf. Section 12.3). Finally, the presence of capillary menisci and nanometric water films on solid surfaces has deep consequences on the surface physical and chemical properties, notably by permitting the activation of nanoscale corrosion processes, such as local dissolution and recondensation, hydration, oxidation, hydrolysis, and lixiviation. These phenomena can lead either to the long-term improvement of the mechanical properties of nanostructured materials by recrystallization of solid joints or to the failure of microstructures due to crack propagation by stress corrosion (cf. Section 12.4).

12.1.2 Physics of Capillary Condensation

Let us consider two parallel solid surfaces separated by a distance D, in contact with a reservoir of vapor at a pressure P_v and temperature T. If D is very large, the liquid–vapor

---

[*] Stiction is often circumvented by the use of a sublimating fluid, such as supercritical carbon dioxide.

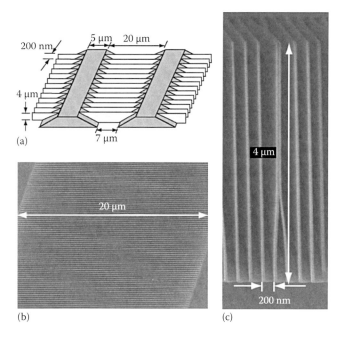

(a)

(b)   (c)

FIGURE 12.1 Stiction effect due to drying in the nanofabrication of a nano-mirror array (c). The spacing between the lamellae is 200 nm as described in (a) and imaged in (b). If the aspect ratio is larger than a critical value, the stiffness of the lamellae becomes too small to withstand the attractive action of the capillary forces induced by the meniscus in the drying process. We remark that in drying processes the meniscus curvature and capillary pressure may be quite smaller than the equilibrium values (cf. Insert B), but they still have a great impact. (After Heilmann, R. et al., SPIE Newsroom, 2008)

equilibrium occurs at the saturating pressure $P_v = P_{sat}$. For a finite D, if the surface tension γ_{sl} of the wet solid surface (see Insert A) is lower than the one γ_{sv} of the dry solid surface, the solid favors liquid condensation. One should therefore ask if the solid can successfully stabilize a liquid phase when the vapor phase is stable in the bulk, i.e., $P_v < P_{sat}$. To answer this question, one must compare the grand canonical potential (see Insert A) of two configurations: the "liquid-filled interstice," which we shall call the condensed state, and the "vapor-filled interstice," i.e., the non-condensed state, with $\mu = \mu_{sat} - \Delta\mu$ the chemical potential of the reservoir (Figure 12.2). Outside of coexistence, i.e., if $\Delta\mu \neq 0$, the pressure in the two phases is different and is given by the thermodynamic relation $\partial(P_l - P_v)/\partial\mu = \rho_l - \rho_v$ (with ρ_l, ρ_v the number of molecules per unit volume in each phase). As the liquid is usually much more dense and incompressible than the vapor, the pressure difference reduces to $(P_v - P_l)(\mu) \simeq \rho_l\Delta\mu = \rho_l k_B T \ln(P_{sat}/P_v)$ if the vapor can be considered as an ideal gas. Thus, the condensed state is favored if the confinement is smaller than the critical distance $D_c(\mu)$:

$$\rho_l\Delta\mu D_c(\mu) = 2(\gamma_{sv} - \gamma_{sl}) \tag{12.1}$$

The left-hand side of Equation 12.1 represents the free energy required to condense the unfavorable liquid state and the right-hand side, the gain in surface energy. $D_c(\mu)$ is, thus, the critical distance that balances the surface interactions and the

INSERT A: SURFACE TENSION AND CONTACT ANGLE

The surface tension of a fluid interface is defined in terms of the work required to increase its area:

$$\gamma_{lv} = \left(\frac{\partial \mathcal{F}}{\partial A_{lv}}\right)_{N_l, V_l, N_v, V_v, T}$$

Here $\mathcal{F}$ is the free energy of a liquid–vapor system, T its temperature, A_{lv} the interface area, and N_l, V_l, N_v, V_v the number of molecules and the volume of each phase respectively (Rowlinson and Widom 1982). For a solid surface, one can likewise define the difference of surface tension $\gamma_{sl} - \gamma_{sv}$ for wet and dry surfaces in terms of the work $d\mathcal{F}$ required to wet a fraction dA_{sl} of the surface initially in the dry state.

It is shown in thermodynamics that the surface tension is a grand canonical excess potential per unit area. The total grand canonical potential of a multiphase system

$$\Omega = -P_v V_v - P_l V_l - P_s V_s + \gamma_{lv} A_{lv} + (\gamma_{sl} - \gamma_{sv}) A_{sl}$$

is the potential energy for an open system. Its variation is equal to the work done on the system during a transformation, and its value is minimal at equilibrium.

On the diagram of Insert B, let us consider a horizontal translation dx of the meniscus. At equilibrium, the grand canonical potential is minimum: $d\Omega = -P_l dV_l - P_v dV_v + (\gamma_{sl} - \gamma_{sv})dA_{sl} = 0$. Thus $P_v - P_l = (\gamma_{sv} - \gamma_{sl})/D$. But, according to Laplace's law of capillarity, the pressure difference $P_v - P_l$ is also related to the curvature of the meniscus: $P_v - P_l = \gamma_{lv}/r = 2\gamma_{lv} \cos\theta/D$, where θ is the contact angle. We deduce the Young–Dupré law of partial wetting:

$$\gamma_{lv} \cos\theta = \gamma_{sv} - \gamma_{sl} \quad \text{valid if } S = \gamma_{sv} - \gamma_{sl} - \gamma_{lv} \leq 0 \quad (12.2)$$

The parameter S is the wetting parameter (de Gennes et al. 2003). The situation $S > 0$ corresponds to perfect wetting. In this case, a thin liquid layer covers the solid surface (see Insert C).

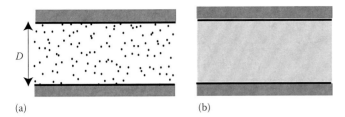

(a)   (b)

FIGURE 12.2 (a) $\Omega_{\text{non-condensed}}(\mu) = 2A\gamma_{\text{sv}} - DAP_{\text{v}}(\mu)$. (b) $\Omega_{\text{condensed}}(\mu) = 2A\gamma_{\text{sl}} - DAP_{\text{l}}(\mu)$.

bulk interactions to determine the phase diagram of the fluid (Israelachvili 1992). From the above equation, it is clear that capillary condensation can occur only if the liquid wets, at least partially, the solid surfaces.

In the case of partial wetting, the difference between the dry and the wet surface tension is related to the contact angle θ of the liquid onto the solid surface (see Insert A) and the critical distance reduces to

$$D_{\text{c}}(\mu) = \frac{2\gamma_{\text{lv}}\cos\theta}{\rho_{\text{l}}\Delta\mu} = 2r_{\text{K}}\cos\theta \qquad (12.3)$$

where r_{K} is the Kelvin's radius associated to the undersaturation $\Delta\mu$ (see Insert B).

For an estimation of the order of magnitude of the confinement at which capillary condensation occurs, consider the case of water at room temperature: $\gamma_{\text{lv}} = 72\,\text{mJ/m}^2$, $\rho_{\text{l}} = 5.5 \times$ $10^4\,\text{mol/m}^3$, and assume a contact angle $\theta = 30°$. In ambient conditions with a relative humidity of $P_{\text{v}}/P_{\text{sat}} = 40\%$, one has $r_{\text{K}} \simeq 0.6\,\text{nm}$ and $D_{\text{c}} \simeq 1\,\text{nm}$. The scale is in the nanometer range, and increases quickly with humidity: it reaches $4\,\text{nm}$ at 80% and $18\,\text{nm}$ at 95% relative humidity. Therefore, capillary condensates are ubiquitous in ambient conditions in high confinement situations.

We see from the Laplace–Kelvin equation that the pressure in capillary condensates is usually very low: taking the example of water in ambient conditions with relative humidity $P_{\text{v}}/P_{\text{sat}} = 40\%$, the pressure in the condensates is $P_{\text{l}} = -120\,\text{MPa}$, i.e., $-1200\,\text{bar}$. With these severe negative pressures, condensates exert strong attractive capillary forces on the surfaces to which they are adsorbed. Thus capillary condensation is usually associated to important mechanical aspects, such as cohesion, friction, elastic instabilities and micro-structures destruction. Furthermore, if the liquid phase wets totally the solid surfaces (see Insert A), the surfaces may be covered by a liquid film even in a nonconfined geometry (see Inserts C and D). In this case the critical distance for capillary condensation can be significantly enhanced at low humidity. In the case of water, the condensation of a liquid film has important consequences on surface chemistry as surface species can be dissolved in the liquid phase, and the capillary condensation at the level of contact between surfaces increases solute transport and is responsible for dissolution-recrystallization processes, which lead to slow temporal evolution of mechanical properties of the materials (cf. Section 12.4).

INSERT B: LAPLACE–KELVIN EQUATION

Another way to address capillary condensation is to consider the coexistence of a liquid and its vapor across a curved interface. Because of the Laplace law of capillarity the pressure in the two phases are not equal: $P_{\text{int}} - P_{\text{ext}} = \gamma_{\text{lv}}/r$, with r is the radius of mean curvature of the interface. The pressure is always higher on the concave side. Because of this pressure difference the chemical potential of coexistence is shifted:

$$\mu_{\text{v}}(P_{\text{v}}) = \mu_{\text{l}}\left(P_{\text{l}} = P_{\text{v}} - \frac{\gamma_{\text{lv}}}{r}\right) = \mu_{\text{sat}} - \Delta\mu$$

We have assumed here that the liquid is on the convex side, a configuration compatible with an undersaturation. For an ideal vapor and an uncompressible liquid:

$$\Delta\mu = k_{\text{B}}T\ln\left(\frac{P_{\text{sat}}}{P_{\text{v}}}\right) \quad P_{\text{l}}(\mu) \simeq \rho_{\text{l}}\Delta\mu + P_{\text{v}}(\mu) \qquad (12.4)$$

from where we get the Laplace–Kelvin equation for the equilibrium curvature (Thomson 1871):

$$\frac{\gamma_{\text{lv}}}{r_{\text{K}}} = P_{\text{v}}(\mu) - P_{\text{l}}(\mu) \simeq \rho_{\text{l}}\Delta\mu = \rho_{\text{l}}k_{\text{B}}T\ln\left(\frac{P_{\text{sat}}}{P_{\text{v}}}\right) \qquad (12.5)$$

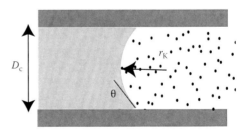

We check that in a flat slit, the critical confinement and the Kelvin's radius are related by $D_{\text{c}} = 2r_{\text{K}}\cos\theta$. The capillary condensate is thus limited by a meniscus whose curvature is equal to the Kelvin's radius. The Laplace–Kelvin law is however more general than Equation 12.3 and allows to predict the critical confinement in arbitrarily complex geometries.

INSERT C: PERFECT WETTING:
THE DISJOINING PRESSURE

When the energy of the dry solid surface γ_{sv} is larger than the sum $\gamma_{sl} + \gamma_{lv}$ of the solid–liquid and liquid–vapor interfaces ($S > 0$), the affinity of the solid for the fluid is such that it can stabilize a liquid film of thickness e in equilibrium with an undersaturated vapor without any confinement. The existence of such wetting films must be taken into account when determining the liquid–vapor equilibrium in a confined space.

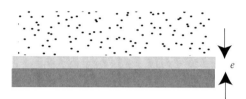

In the theory of wetting, liquid films are described by the concept of interface potential (Derjaguin 1944; de Gennes 1985). The excess potential per unit area of a solid surface covered by a wetting film does not reduce to the sum $\gamma_{sl} + \gamma_{lv}$ of the surface tensions: a further excess must be taken into account corresponding to the fact that the molecular interactions which generate the surface tension do not operate over a thickness of liquid that can be considered infinite. The excess grand canonical potential of the humid solid surface of area A is then

$$\frac{\Omega_{sv}}{A} = \tilde{\gamma}_{sv} = \gamma_{sl} + \gamma_{lv} + W_{slv}(e) - e(P_l(\mu) - P_v(\mu)) \quad (12.6)$$

where the interface potential $W_{slv}(e)$ vanishes for a macroscopic film. The excess potential Ω_{sv}/A describes the thermodynamic properties of the liquid film. It is minimum at equilibrium, so that the pressure in the liquid is not the same as in the vapor:

$$P_v(\mu) - P_l(\mu) = -\frac{dW_{slv}(e)}{de} = \Pi_d \quad (12.7)$$

The pressure difference Π_d is called the disjoining pressure. The interface potential $W_{slv}(e)$ and the wetting parameter $\tilde{\gamma}_{sv}(\Pi_d) - \gamma_{sl} - \gamma_{lv}$ are Legendre transforms of each other:

$$W_{slv}(e) = \tilde{\gamma}_{sv}(\Pi_d) - \gamma_{sl} - \gamma_{lv} - e\Pi_d \quad e = \frac{\partial \tilde{\gamma}_{sv}}{\partial \Pi_d} \quad (12.8)$$

For instance in the case of van der Waals forces, the interface potential results from dipolar interactions going as $1/r^6$ between molecules, and varies as $1/e^2$:

$$W_{slv}(e) = -\frac{A_{slv}}{12\pi e^2} \quad \Pi_d(e) = -\frac{A_{slv}}{6\pi e^3}$$

$$\tilde{\gamma}_{sv} = \gamma_{sl} + \gamma_{lv} + \left(\frac{-9A_{slv}}{16\pi}\right)^{1/3} \Pi_d^{2/3} \quad (12.9)$$

The Hamaker constant A_{slv} has the dimension of an energy (Israelachvili 1992). It lies typically between 10^{-21} and 10^{-18} J and has negative sign when the liquid wets the solid, i.e., if the interface potential is positive.

12.1.3 Mesoporous Systems

Capillary condensation has been extensively studied in relation to sorption isotherms in mesoporous media—i.e., nanomaterials with pore sizes between 2 and 50 nm—in the prospect of using those isotherms for the determination of porosity characteristics such as the specific area and the pore size distribution. Figure 12.3, for instance, shows a typical adsorption isotherm of nitrogen in a mesoporous silica at 77 K. In a first domain of low vapor pressure, the adsorption is a function of the relative vapor saturation only, and corresponds to the mono- and polylayer accumulation of nitrogen on the solid walls. This regime allows the determination of the specific area, for instance through the Brunauer–Emmett–Teller model (Brunauer et al. 1938). At a higher pressure, a massive adsorption corresponds to capillary condensation, and the porous volume is completely filled by liquid nitrogen before the saturating pressure is reached. This adsorption branch shows usually a strong hysteresis and the capillary desorption is obtained at a lower vapor pressure than the condensation. This feature underlines the first-order nature of capillary condensation. It is shown in the next paragraph that for sufficiently simple pore shapes the desorption branch is the stable one and corresponds to the liquid–vapor equilibrium through curved menisci. The desorption branch may be used to determine the pore size distribution of the medium through the Laplace–Kelvin relation using appropriate models (Barret-Joyner-Halenda, Barret et al. 1951). More can be found on the physics of phase separation in confined media in the review of Gelb et al. (1999).

INSERT D: PERFECT WETTING: THE PREWETTING TRANSITION AND CAPILLARY CONDENSATION

In a situation of perfect wetting, a liquid film condenses on a flat isolated solid surface if the humid solid surface tension $\tilde{\gamma}_{sv}$ is lower than the dry one:

$$\tilde{\gamma}_{sv} = \gamma_{sl} + \gamma_{lv} + W_{slv}(e) + e\Pi_d \leq \gamma_{sv} \qquad (12.10)$$

If the film exists, its thickness at equilibrium with the vapor is implicitly determined by the analogue of the Laplace–Kelvin equation (12.5):

$$\Pi_d(e) = -\frac{\partial W_{slv}(e)}{\partial e} = P_v(\mu) - P_l(\mu) = \rho_l k_B T \ln \frac{P_{sat}}{P_v} \qquad (12.11)$$

The thickness e^* realizing the equality in relation (12.10) is a minimum thickness for the wetting film, and the associated chemical potential μ^* and vapor pressure P_v^* correspond to a *prewetting transition*. Above the transition the thickness of the adsorbed film increases with the vapor pressure until it reaches a macroscopic value at saturation. In the case of van der Waals wetting, for instance, the vapor pressure at the prewetting transition is given by

$\Pi_d^* = \rho_l k_B T \ln(P_{sat}/P_v^*) = \sqrt{16\pi S^3/(-9A_{slv})}$ with S the wetting parameter (12.2).

In a confined geometry such as sketched in Figure 12.2, the grand canonical potential of the "noncondensed" state is shifted above the prewetting transition because the solid surface tension γ_{sv} has to be replaced by the humid value $\tilde{\gamma}_{sv}$. The modified Equation 12.1 and the Laplace–Kelvin relation $\gamma_{lv}/r_K = \Pi_d = \rho_l \Delta\mu$ give the critical distance (Derjaguin and Churaev 1976):

$$D_c = 2r_K + 2e + 2\frac{W_{slv}(e)}{\Pi_d} \qquad (12.12)$$

The difference with the partial wetting case is not simply to decrease the available interstice by twice the film thickness. In the case of van der Waals forces, for example,

$$D_c = 2r_K + 3e \quad \text{with } e = (-A_{slv}/6\pi\rho_l\Delta\mu)^{1/3} \qquad (12.13)$$

The effect of adsorbed films becomes quantitatively important for determining the critical thickness at which capillary condensation occurs in situations of perfect wetting.

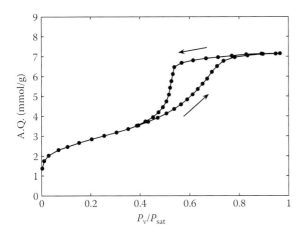

FIGURE 12.3 Sorption isotherm of nitrogen at 77 K in Vycor (A.Q., adsorbed quantity). (From Torralvo, M.J. et al., *J. Colloid Interface Sci.*, 206, 527, 1998. With permission.)

12.2 Capillary Adhesion Forces

12.2.1 Measurements by SFA

Because of their high curvature, capillary condensates exert a large attractive force on the surfaces they connect. Hence, these large forces represent a valuable tool to study the thermodynamic and mechanical properties of the condensates. Experimentally, the ideal geometry involves a contact with at least one curved surface—either a sphere on a plane, two spheres or two crossed cylinders—so that locally the topology resumes to a sphere of radius R close to a flat. Surface force apparatus (SFA) use macroscopic radius R in order to take advantage of the powerful Derjaguin approximation, which relates the interaction force $F(D)$ at distance D to the free energy per unit area (or other appropriate thermodynamic potential) of two flat parallel surfaces at the same distance D (see Insert E). It must be emphasized that the Derjaguin approximation accounts exactly for the contribution of the Laplace pressure, and more generally for all "surface terms" contributing to the force, but it does not account properly for the "perimeter terms" such as the line forces acting on the border of the meniscus, so that it neglects terms of order $\sqrt{r_K/R}$.

In a surrounding condensable vapor, the appropriate potential is the grand potential per unit area considered in Figure 12.2:

$$\Omega(D < D_c) = (\rho_l - \rho_v)\Delta\mu D + 2\gamma_{sl} + W_{sls}(D)$$

$$\Omega(D > D_c) = 2\gamma_{sv} + W_{svs}(D) \qquad (12.14)$$

INSERT E: DERJAGUIN APPROXIMATION

Consider a sphere of macroscopic radius R at a distance $D \ll R$ from a plane. If the range of the surface interactions is also smaller than R, then all surface interactions take place in the region where the sphere is almost parallel to the plane and the global interaction force $F(D)$ is

$$F(D) = -\frac{d}{dD}\left(\int_D^\infty dA(z)[U(z) - U(\infty)]\right)$$

$$= -2\pi R \frac{d}{dD}\int_D^\infty [U(z) - U(\infty)]dz$$

with $dA(z) \simeq 2\pi R dz$ the sphere area between altitude z and $z + dz$, and $U(z)$ the appropriate potential energy per unit area for flat, parallel surfaces at distance z.

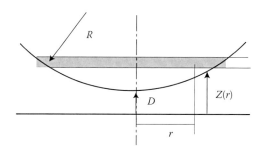

This gives the Derjaguin approximation:

$$F(D) = 2\pi R[U(D) - U(\infty)]$$

with $W_{sls}(D)$ (resp. $W_{svs}(D)$) the solid–solid interaction through the liquid phase (resp. vapor phase). This gives the interaction force in the case of partial wetting:

$$F(D < D_c) = 2\pi R\left[-2\gamma_{lv}\cos\theta + \frac{\gamma_{lv}}{r_K}D\right] + 2\pi R W_{sls}(D) \quad (12.15)$$

and the pull-off force (Fisher et al. 1981):

$$F_{\text{pull-off}} = -F(D = 0) = 4\pi R\gamma_{lv}\cos\theta + 4\pi R\gamma_{sl} = 4\pi R\gamma_{sv} \quad (12.16)$$

It should be mentioned here that the prefactor $4\pi R$ entering Equation 12.16 is valid only for rigid surfaces, i.e., if the Tabor parameter $\lambda = (\gamma_{lv}R^2/E^{*2}h_o^3)^{1/3} < 0.1$ with E^* the (mean) reduced Young modulus of the solid, and h_o the range of the interaction. In the general case, the $4\pi R$ prefactor in Equation 12.16 should be replaced by $\epsilon(\lambda)\pi R$ with $3 \leq \epsilon \leq 4$ (Maugis 1992; Maugis and Gauthier-Manuel 1994).

The pull-off force (12.16) is the sum of the capillary force $4\pi R\gamma_{lv}\cos\theta$ and of the solid–solid adhesion in the liquid phase $4\pi R\gamma_{sl}$. A remarkable feature of Equation 12.16, valid in a *partial wetting* situation as we should recall, is that the capillary force does not depend on the volume of the liquid bridge nor on its curvature, a property due to the fact that the wetted area and the capillary pressure exactly compensate each other when the size of the condensate varies. Furthermore the pull-off force does not even depend on the presence of the condensate and is identically equal to its value in dry atmosphere, $4\pi R\gamma_{sv}$. Of course these remarkable results require a perfectly smooth geometry. In practice, even minute surface roughness screens very efficiently the solid–solid interactions, so that it can be neglected only in condensable vapor close to saturation. A major experimental problem is then to follow the transition of the pull-off force from $4\pi R\gamma_{sv}$ to $4\pi R(\gamma_{lv}\cos\theta + \gamma_{sl})$ as the relative vapor pressure varies

from 0 to 1. This problems has been addressed through a surface force apparatus by using atomically smooth mica surfaces glued on crossed cylinders. However even with these physically ideal surfaces, the pull-off force measured with dry surfaces gives a solid-solid surface tension γ_{sv} significantly lower than the one obtained, for instance, from cleavage experiments or contact angle measurements (Christenson 1988). The same occurs in liquid phase, where a monolayer of liquid remains stuck between the mica surfaces and prevents the measurement of the actual solid–liquid surface tension. Therefore a full experimental verification of Equation 12.16 over the whole vapor pressure range is not available. It is found however that the pull-off force is dominated by the capillary force and that it does not depend on the vapor pressure over a significant range including saturation (cf. Figure 12.4) (Fisher et al. 1981; Christenson 1988).

As the capillary force is less sensitive to roughness than the direct solid-solid adhesion (cf. Insert H), SFA pull-off forces coupled to direct measurements of the Kelvin's radius via the observation of the liquid bridge extension have been used to investigate the validity of the classical theory of capillarity at nanoscale, and an eventual drift of the liquid surface tension for highly curved menisci (Fisher et al. 1981). The remarkable result is that the Laplace law of capillarity holds for very small radii of curvature (i.e., less than 2–3 nm) for simple liquids. Discrepancies have been reported with water, but they have been shown to depend on the type of ions covering the mica surfaces and have been attributed to the accumulation of involatile material in the water condensate (Christenson 1985, 1988).

Valuable information has been obtained by studying the full profile of the capillary force instead of only the pull-off force, as the slope of $F(D)$ gives a direct measurement of the Laplace pressure γ_{lv}/r_K. This requires, however, a SFA of large stiffness. Such analysis has been performed by Crassous et al. (1994) in a *perfect*

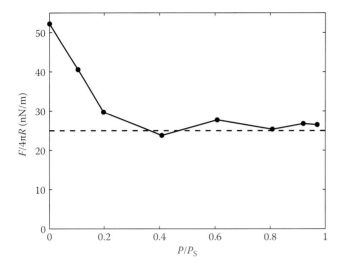

FIGURE 12.4 Measured pull-off force as a function of the relative vapor pressure of cyclohexane. The dashed line is the bulk surface tension of cyclohexane ($\gamma_{lv} = 25\,mN/m$). (From Christenson, K., *J. Colloid Interface Sci.*, 121, 170, 1988. With permission.)

wetting situation. In this case, the theoretical force is obtained from Equation 12.14 using the reference $\Omega(D = \infty) = 2\tilde{\gamma}_{sv}$ instead of $2\gamma_{sv}$ (Charlaix et al. 2005):

$$F(D < D_c) = 2\pi R\left[\frac{\gamma_{lv}}{r_K}D + 2\gamma_{sl} - 2\tilde{\gamma}_{sv}\right] + 2\pi R W_{sls}(D) \quad (12.17)$$

The first term on the right-hand side is the capillary force at liquid-vapor equilibrium. It also writes $F_{cap} = 2\pi R\Pi_d(D - D_c)$ with D_c

given by Equation 12.12. The slope of $F(D)$ provides the Laplace/disjoining pressure $\gamma_{lv}/r_K = \Pi_d$, while the offset gives access to $\tilde{\gamma}_{sv}$ and to the wetting potential with the help of Equation 12.8. Crassous et al. (1994) verified experimentally Equation 12.17 down to Kelvin's radii of 3 nm in the case of heptane condensation on platinum surfaces.

A consequence of Equation 12.17 is to give the pull-off force as a function of the vapor saturation

$$F_{pull\text{-}off} = 4\pi R\tilde{\gamma}_{sv} \quad (12.18)$$

with $\tilde{\gamma}_{sv}(\Pi_d)$ given by (12.8) and Π_d by (12.11). For instance with van der Waals forces, one expects

$$F_{pull\text{-}off} = 4\pi R\left[\gamma_{sl} + \gamma_{lv} + a\left(\ln\frac{P_{sat}}{P_v}\right)^{2/3}\right] \quad (12.19)$$

with $a = (-9A_{slv}\rho_l^2(k_BT)^2/16\pi)^{1/3}$. When the vapor pressure is increased from 0 to $P_v^\star$, the pull-off force remains equal to the dry value $4\pi R\gamma_{sv}$. Above $P_v^\star$ the pull-off force *decreases* with increasing vapor pressure, until it reaches the wet limit $4\pi R(\gamma_{sl} + \gamma_{lv})$.

A major feature of capillary condensation is the strong hysteresis in the formation of the condensates. As shown on Figure 12.5 for instance, the noncondensed state is highly metastable for $D < D_c$, and an energy barrier has to be overcome to initiate the condensation process. The kinetics of capillary condensation has been much less studied than the static properties of

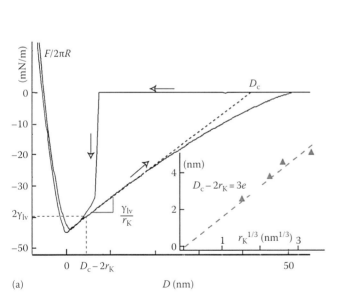

(a)

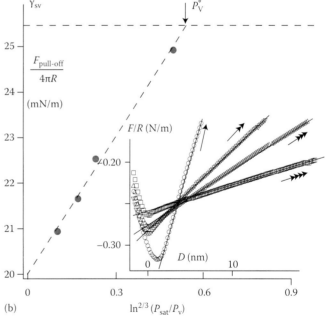

(b)

FIGURE 12.5 (a) The capillary force between platinum surfaces measured in *n*-heptane vapor. The attractive force varies linearly with the distance on a significant range where the condensate is at equilibrium with the vapor. For van der Waals forces the thickness *e* of the wetting film covering the surfaces is obtained through $F_{cap}(3e) = 4\pi R\gamma_{lv}$. It varies as the power 1/3 of the Kelvin's radius. (b) The pull-off force measured for various vapor pressures, and the comparison with the van der Waals dependence (12.19). The pull-off force is expected to saturate to the dry pull-off force at the prewetting transition. (From Crassous, J. 2005 and Crassous, J. et al., *MRS Symposium Proceedings*, 336, 33, 1995. With permission.)

the condensate. Several theoretical works have focused on the calculation of the energy barrier that depends on the confined geometry (Restagno et al. 2000; Lefevre et al. 2004). In the case of a wetting situation and of an extended slit like in SFA experiments, it has been shown that the distance of condensation D_s (see Figure 12.5) corresponds to the spinodal instability of the wetting film covering the solid surfaces (Derjaguin and Churaev 1976; Christenson 1994; Crassous et al. 1994). The growth kinetics of the capillary bridge after its initial formation has been shown by Kohonen et al. (1999) to be reasonably well described by a model based on the diffusion-limited flow of vapor toward the bridge, except for water condensates that grow significantly more slowly, which is attributed to the effect of dissolution of inorganic material from the mica surfaces.

12.2.2 Measurements by Atomic Force Microscopy

The great relevance of capillary forces between nanoscale objects has stimulated the application of contact atomic force microscopy (AFM) to provide a more quantitative measurement of the forces acting between the very sharp probe tips (with typical radius of curvature of few nanometers) and samples with variable compositions or textures. The great advantage of AFM is the elevated lateral resolution, which is comparable with the curvature radius of the terminal portion of the probe tip. The first-order effect of the probe size reduction is a proportional reduction of the adhesion forces to typical values in the nanonewton (nN) range, which can easily be measured by AFM pull-off tests, i.e., by measuring the force of contact break down during a force-distance (approach-retract) curve. However, the typical cantilever stiffness (in the range of a few N/m) prevents the full characterization of the $F(D)$ curve that can be obtained by SFA (cf. Section 12.2.1), and typical data are limited to the variation of the pull-off force with relative humidity, corresponding to $D = 0$ (Xiao and Qian 2000).

Before we engage in a deeper analysis, we should mention that the AFM force measurements are very sensitive to many ill-controlled factors, the more important being the exact details of the tip shape, including its nanoscale roughness. Although continuum-derived laws become questionable at scales smaller than 1–2 nm, we will first establish how the capillary force at a finite distance D should properly be evaluated for nanoscale probes according to classical modeling and then discuss the limitations of this approach.

Let the AFM tip be modeled as a cone terminating with a spherical tip of radius R and contact angle θ_1 and the sample as a flat surface of contact angle θ_2 (cf. Figure 12.6). The Derjaguin approximation (Insert E) leading to Equations 12.16 and 12.18 is no longer applicable since the r_K/R ratio is not negligible for nanoscale probes. Although other approximations exist, such as the circular meniscus surface model, their domain of validity is limited and their inattentive use can lead to severe errors. We thus recommend to refer to the complete analytical solutions provided by Orr et al. (1975), which describe constant mean curvature meniscus surfaces for systems amenable to a generic

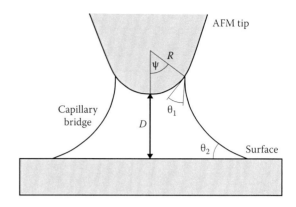

FIGURE 12.6 Graphical representation of a capillary bridge formed between an AFM tip of radius R and a flat surface separated by a distance D. ψ is the filling angle, θ_1 and θ_2 are the contact angles of the AFM tip and surface.

sphere/plane geometry (Insert F), along with a computation of capillary forces. A special attention must be paid to the case of perfect wetting, since the Orr equations do not take into account the presence of wetting films. This situation was properly treated by Crassous (1995) using both an approximated solution and numerical simulation. In the case of a vanishing distance D, the solid–solid interactions should also be accounted for.

The capillary force F_C acting on the sphere has two main contributions: a capillary pressure force F_P, which is related to the negative Laplace pressure in the liquid bridge and acts on the sphere surface, and a surface tension force F_S, which resides in the meniscus and acts on the contact line between the sphere and the liquid bridge. While this second term is negligible in the Derjaguin approximation, when R decreases to nanoscale, its contribution becomes of the same order as the Laplace pressure term, since the ratio of the perimeter to surface of the capillary meniscus contact is increased. These terms can be written without any approximation as

$$F_C = F_P + F_S \quad F_P = -2\pi\gamma HR^2 \sin^2\psi \quad F_S = 2\pi\gamma R \sin\psi \sin(\theta_1 + \psi)$$

(12.20)

The filling angle ψ should then be related to relative humidity by using the Orr et al. relations to evaluate the surface mean curvature (Insert F) and by substituting it into the Laplace–Kelvin relation (12.5).

However, the applicability of this subtle modeling to AFM pull-off force measurements is presently hampered by several technical problems. On the opposite end of SFA, the shape of typical AFM probes is generally ill-defined and subject to easy degradation and contamination. Figure 12.7 shows a typical example of pull-off force measurements as a function of humidity (Xiao and Qian 2000), along with the approximated modeling by de Lazzer et al. (1999), which aimed at showing the effect of a more general tip shape in the form of a revolution solid with radial profile $y(x)$. Even if the considered tip profiles (cf. inset in Figure 12.7) are quite similar, the modeled force curves are quite different and cannot

INSERT F: ORR SOLUTIONS FOR NEGATIVE CURVATURE MENISCI

As observed by Plateau (1864), the meridian meniscus profile of a capillary bridge is expected to step through a series of shapes, including portions of nodoids, catenoids, unduloids, cylinders, and spheres, as a function of the system's geometrical and wetting parameters. A clear report on the meniscus shapes as a function of the constant mean curvature can be found in Orr et al. (1975), generally expressed in terms of elliptic integrals, as well as a characterization of the capillary forces that are the focus here. However, only nodoids are appropriate for describing capillary bridges equilibrated with undersaturated humidity, and catenoids are limited to the case of vanishing curvature (corresponding to saturated humidity or contact with a bulk liquid reservoir). We will thus only discuss here the case of a negative curvature. The mean curvature $H = -1/2r_K < 0$ can be related to the parameters R, D, θ_1, θ_2 and ψ by

$$\frac{2HR}{\Psi} = \Theta - \frac{1}{k}\Big[E(\phi_2, k) - E(\phi_1, k) \Big] + \frac{1-k^2}{k}\Big[F(\phi_2, k) - F(\phi_1, k) \Big]$$

(12.21)

where

$$\Psi = \frac{1}{D + 1 - \cos\psi} \qquad \Theta = -\cos(\theta_1 + \psi) - \cos\theta_2$$

$$\phi_1 = -(\theta_1 + \psi) + \frac{\pi}{2} \qquad \phi_2 = \theta_2 - \frac{\pi}{2} \qquad k = \frac{1}{(1+c)^{1/2}}$$

$$c = 4H^2 R^2 \sin^2\psi - 4HR\sin\phi\sin(\theta_1 + \psi)$$

and the functions F and E are the incomplete elliptic integrals of first and second type (Abramowitz and Stegun 1972). Since the curvature H is present on both left- and right-hand terms of Equation 12.21, an iterative procedure is needed for numerical solution. The mean curvature H should then be related to the relative humidity through the Laplace–Kelvin relation (12.5). The contribution of capillary forces to the pull-off force can then be simply obtained by taking the maximum value of the total force, which is obtained for $D = 0$.

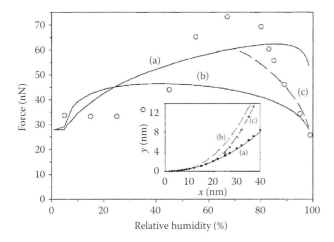

FIGURE 12.7 Adhesion force (pull-off) measurements between a Si_3N_4 AFM tip and a SiO_2 surface (circles). The theoretic approximated expressions calculated from the equations of de Lazzer et al. (1999) are also shown for the three different tip shapes with radial profile $y(x)$ represented in the inset (overlapping for $x < 10$ nm): (a) parabolic profile $y = k_1 x^2$, (b) $y = k_2 x^{2.5}$, and (c) more complex dull shape. (From Xiao, X. and Qian, L., *Langmuir*, 16, 8153, 2000. With permission.)

account for the measured data. Although this modeling is based on the circular meniscus approximation and on an approximated estimate of the solid–solid van der Waals interactions, it can be retained as a clear proof of the critical effect of the tip shape on the capillary forces (cf. also Pakarinen et al. 2005 for a less approximated modeling of the same data, leading to similar conclusions). It must be acknowledged that articles on adhesion measurements from the AFM community often do not show the same maturity as the SFA community in the subtle modeling of these phenomena, and thus often contain misleading interpretations that delay the establishment of a sound and consistent modeling.

The physical interpretation of AFM adhesion measurements in moist air still needs much work. The data in Figure 12.7 can be considered as quite typical. The presence of a plateau at low humidity was often interpreted as the consequence of the non-formation of a capillary bridge (Xiao and Qian 2000), reducing the adhesion force to the dry van der Waals solid–solid interaction. However, we showed in Section 12.2.1 that no changes in the total adhesion force should be observed in the case of a vanishing meniscus between a perfect sphere and a plane in contact: even if the capillary bridge is strictly disappearing, the increased contribution of unscreened solid–solid interactions should strictly compensate the total adhesion force. The low-humidity plateau could also be related to the effect of a small scale roughness in the tip and substrate (cf. Insert H). At low humidity, only the small contacting asperities would contribute to the pull-off force, and the rising regime is likely to be related to a transition toward a monomeniscus that may be treated according to the modeling presented in this section. More insights into the behavior at low humidity will be provided in Section 12.3

by investigating the effect of capillary forces on cohesion and friction forces. The decreasing trend at higher humidity can be explained according to Equation 12.20: for small values of R (less than 1 μm), the increase of the meniscus section with humidity is not compensating any more the decrease of the Laplace pressure in the pressure term F_P, due to the large values of the filling angle ψ. We remark that although the surface tension term F_S is significantly increasing at elevated humidity, the decrease of the pressure term has a dominant effect on the total capillary force.

A major subject of open debate is the limit of validity of the continuum modeling at nanoscale, for both the capillary forces and their combination with the other relevant interactions such as van der Waals, electrostatic interactions, hydrogen bonding and other kinds of specific chemical interactions between the contacting tip and substrate. Although the capillary forces are generally dominant in the adhesion force (due the efficient screening of other interactions), this is not the case at low humidity where the solid–solid interactions become more relevant, especially for small radius probes. A notable example is given by the strong impact of electrostatic interactions in the adhesion force in dry atmosphere (Wan et al. 1992). Although the continuum derived laws were shown to describe the adhesion forces in SFA measurements down to scales as small as 1–2 nm (Lefevre et al. 2004), the discrete nature and structure of solid and liquid molecules should become relevant at subnanometer scales, where the phase behavior of water is still debated. Several investigators have reported the complexity of the structure of thin water films, especially at low humidity where ice-like behavior is expected (cf. Verdaguer et al. 2006). Moreover, Xu et al. (1998) have shown that the contact with an AFM tip can subtly modify the structure of the water layer on a mica substrate: 2D islands are left on the surface after removal of the tip and can persist several hours before evaporating. However, a significant disagreement is still found in literature on the details of the humidity dependence of these behaviors for different kinds of substrates and AFM tips and more accurate research is needed. In particular, the delicate role of nonvolatile material in water condensates should be carefully investigated.

12.2.3 Measurements in Sharp Cracks

Sharp crack tips in very brittle materials such as glasses have recently been proven to constitute an invaluable tool to study capillary condensation (Grimaldi et al. 2008). By using a stable testing configuration such as the DCDC specimen, it is possible to obtain extremely flat crack surfaces (with roughness of the order of 0.4 nm RMS on a $1 \times 1\ \mu m^2$ region) and to induce a controlled crack opening profile by carefully loading the sample in pure mode I (Pallares et al. 2009). We should highlight that although a nanometric roughness could relevantly affect the capillary forces (cf. Section 12.3), the complementarity of the roughness of brittle crack surfaces makes the local crack opening variations negligible. The crack opening $2u$ can thus be considered as a monotonously increasing function of the distance X

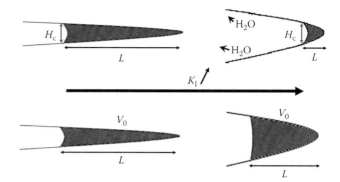

FIGURE 12.8 Sketch of the variation of the condensation length after an increase of the stress intensity factor K_I: (top) case of equilibrium; (bottom) constant volume model. NB: exaggerated vertical scale.

from the crack tip. In a close neighborhood of the crack tip, this can be expressed by a parabolic shape (cf. Figure 12.8) according to the Irwin equation:

$$2u(X, K_I) = \frac{2K_I}{E'}\left(\frac{8X}{\pi}\right)^{1/2} \tag{12.22}$$

where
K_I is the stress intensity factor
E' is an effective Young modulus

The confinement in the sharp crack tips in glass is very elevated. Typical crack tip radii are of the order of 0.5 nm and the crack opening is of the order of 10 nm at 1 μm distance from the crack tip. This can be interpreted locally as a plane pore with progressive opening and it is thus expected to be filled with equilibrium capillary condensation up to a critical opening $H_c = D_c(H_c = 2r_K \cos\theta$ for partial wetting and $H_c = 2r_K + 3e$ for perfect wetting under the assumption that the liquid-solid interactions are of the vdW type, cf. Insert D).

Since the critical distance H_c is generally in the nanometer range, the length L of the condensed phase is generally in the order of hundreds to thousands of nanometers and it has the advantage of significantly amplifying the effect of variations of H_c. Accurate measurements of H_c can be obtained by measuring the variations of the condensation length that are induced by a modulation of the crack opening profile through changes of the sample loading as shown in Figures 12.8 and 12.10. Such a measurement was recently made possible by the development of AFM phase imaging techniques (cf. Insert G) that can detect the presence of capillary condensation in the crack cavity by measuring the energy dissipated in the formation of a liquid bridge between the condensed phase and the AFM tip during the imaging in tapping mode (Figure 12.9).

The strongly negative capillary pressure in the crack cavity will apply important internal forces on the crack walls, which add to the effect of the external forces and tend to close the crack. However, the advantage of such a stiff configuration is that even huge capillary pressures will have little effect on the crack opening profile when applied over a length L of micrometer size near the crack tip. The above cited technique can thus be comfortably applied to first-order determinations of H_c with 10% accuracy by neglecting the action of capillary forces. On the other hand, these forces can become more relevant if the external load (and thus the crack opening) are reduced enough to let the condensation extend to several tens of microns. In this context, the deviations of the measured condensation length from the above cited prediction may be used to infer the value of the capillary pressure. When implemented in controlled atmosphere, this technique can be used to measure both the critical condensation distance and the capillary pressure as a function of relative humidity and thus provide important information on the surface tension of the liquid, and on its Kelvin radius. This technique provides interesting complement to the SFA

measurements since it does not suffer from the limitations of the SFA equipment to sustain the very strong negative pressures generated by the capillary meniscus at low humidity.

We remark that the values of the measured critical distance H_c for water condensation in silica glass in Figure 12.10 are quite large in relation to the prediction of Equation 12.13, which should not exceed 3 nm at 70% RH. This phenomenon of enhanced condensation is analog to other similar observations, such as the measurement of large meniscus radii in SFA measurements between mica sheets (Christenson 1985) where capillary bridges of water condensed from a non polar liquid were argued to be enriched by ionic solute. These interesting effects still need more investigation and will provide important information on the chemical effects due to glass water interactions under tensile stress (cf. Section 12.4).

INSERT G:　CAPILLARY FORCES IN AFM PHASE IMAGING (TAPPING MODE)

High-amplitude oscillating AFM techniques (tapping mode) are widely used in surface imaging due to high lateral resolution and reduced tip and substrate damage. The cantilever oscillation is stimulated by a high-frequency piezo driver at a constant frequency ω near the cantilever resonant frequency $\omega_0 (f_0 = \omega_0/2\pi \simeq 330\,\text{kHz})$ in order to have a free oscillation amplitude $A_0 \simeq 30\,\text{nm}$ at resonance. During the scan of a surface, a feedback loop acts on the vertical position of the cantilever in order to maintain the amplitude of the tip oscillation constant to a set point value $A_{sp} \simeq 20\,\text{nm}$. The vertical position of the cantilever motion provides the vertical topographic signal. Although the tip-sample interaction is nonlinear, the motion of the tip can be well approximated by an harmonic oscillation $z(t) = A\cos(\omega t + \theta)$. The phase delay θ between the stimulation and the oscillation of the cantilever can be related to the energy E_{diss}, which is dissipated during contact of tip and sample (Cleveland et al. 1998):

$$\sin\theta = \frac{\omega}{\omega_0}\frac{A_{sp}}{A_0} + \frac{QE_{\text{diss}}}{\pi k A_0 A_{sp}} \qquad (12.23)$$

where $Q \sim 600$ and $k \sim 40\,\text{N/m}$ are the quality factor and the stiffness, respectively, of typical tapping cantilevers. Phase imaging can thus provide a map of the local energy dissipation caused by the tip–sample interaction during the cantilever oscillation. This quantity contains several contributions related to viscoelastic or viscoplastic sample deformation and also from the local adhesion properties. However, for very stiff substrates such as glass or silicon, the bulk dissipation will be negligible and phase imaging will mostly be sensitive

to adhesion and in particular to the formation and rupture of capillary bridges between the tip and the sample at each oscillation. Zitzler et al. (2002) have developed a model to relate the energy dissipation to the hysteretic behavior of the capillary force–distance curve. In particular, it was shown that the formation of capillary bridges has a very sensitive influence on the position of the transition between attractive and repulsive modes in the tapping regime (Ciccotti et al. 2008), and that this effect can be used to increase the sensitivity in the measurement of the local wetting properties.

On the other hand, capillary interactions can become a significant source of perturbation and artifacts in the topographical AFM images in tapping mode. Changes in the relative humidity during imaging can alter the stability of the mode of operation and induce an intermittent mode change, which is the source of intense noise and local artifacts. In particular, in the presence of local patches with different wetting properties, the oscillation mode can systematically change between different areas of the image, inducing fake topographic mismatches. However, the topographical artifacts are of the order of the nanometers and are only relevant in very sensitive measurements on flat substrates. In such cases, an attentive monitoring of the phase imaging is required for detecting and avoiding the mode changes. Another drawback of the formation of capillary bridges is the significant loss of lateral resolution caused by the increase of the interaction area between tip and sample (cf. the smearing of the crack-tip condensate in Figure 12.9). Again, a careful tailoring of the imaging conditions can lead the repulsive interaction to be dominant and thus allows recovering a higher resolution.

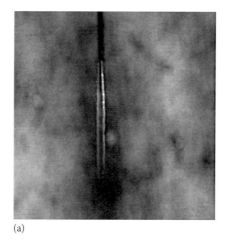   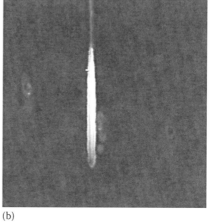

(a)   (b)

FIGURE 12.9 Typical AFM height (a) and phase (b) images of the crack tip for a fracture propagating from top to bottom of the image ($1 \times 1\,\mu m$). The linear gray scale range is respectively 10 nm and 10°. (From Grimaldi, A. et al., *Phys. Rev. Lett.*, 100, 165505, 2008. With permission.)

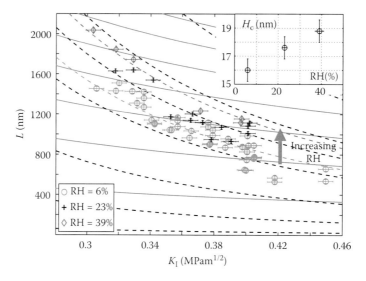

FIGURE 12.10 Plot of the condensation length L versus K_I for three different relative humidities (RH). Dashed curves represent the equilibrium model which is consistent with the data. Continuous curves represent the rejected constant volume model. The inset contains the fitted values of the critical distance H_c for each relative humidity. (From Grimaldi, A. et al., *Phys. Rev. Lett.*, 100, 165505, 2008. With permission.)

12.3 Influence on Friction Forces

12.3.1 Static Friction and Powder Cohesion

It is an everyday experience that a small amount of liquid can change drastically the physical properties of granular matter. Such effects are important not only on the beach for building sandcastles, but also in a number of industries as diverse as pharmaceutical, construction, and agriculture, and in a number of geophysical applications. The changes induced by a small amount of liquid in granular materials are primarily caused by adhesive forces associated with interstitial liquid bridges between grains (Zimon 1969; Bocquet et al. 2002).

In the absence of cohesion forces the stability of a heap of beads can be described by a modified Mohr-Coulomb analysis, thus predicting a maximum angle of stability $\theta_m = \phi$, where $\mu = \tan\phi = \tau/\sigma$ is the static friction coefficient between the beads.

The effect of an adhesive stress c between the beads can be modeled by writing a modified Coulomb yield criterion $\tau = \mu(\sigma + c)$ which provides an implicit equation for the critical angle θ_m (Halsey and Levine 1998):

$$\tan\theta_m = \tan\phi\left(1 + \frac{c}{\rho g h \cos\theta_m}\right) \qquad (12.24)$$

where ρ and h are the density and height of the granular layer. The critical angle is thus an increasing function of the adhesive stress c and it saturates to $\pi/2$ (stuck heap) when $\mu c > \rho g h$.

Several experiments have recently been investigating the applicability of this relation and the consistency of the model with capillarity-based adhesion forces. A first series of experiments has shown that the injection of a very small fixed amount of nonvolatile liquids to a granular heap can significantly increase the critical angle of stability (Hornbaker et al. 1997;

Halsey and Levine 1998). A different kind of experiments has then investigated the more subtle effect of interparticle liquid bridges formed by condensation from a humid atmosphere (Bocquet et al. 1998; Fraysse et al. 1999). While the first kind of experiments can not be related to equilibrium quantities, the second kind has allowed a deep investigation of the first order transition of capillary condensation.

The critical angle was shown to be logarithmically increasing function of the aging time t_w (after shaking) of a granular heap (Bocquet et al. 1998):

$$\tan\theta_m = \tan\theta_0 + \alpha\frac{\log_{10} t_w}{\cos\theta_m} \qquad (12.25)$$

The coefficient α is a measure of the aging behavior of friction in the granular medium and it was shown to be an increasing function of relative humidity, being substantially null in dry air. Moreover, the aging behavior was shown to be enhanced by both increasing the rest angle of the heap during the aging period and by intentionally wearing the particles (by energetic shaking) before the measurements (Restagno et al. 2002).

The logarithmic time dependency of the adhesion forces was modeled as an effect of the dynamic evolution of the total amount of condensed water related to progressive filling of the gaps induced by the particle surface roughness (cf. Insert H). The increase of the aging rate α with the rest angle before the measurements can be explained by the effect of a series of small precursor

INSERT H: EFFECT OF ROUGHNESS ON ADHESION FORCES

The first-order effect of roughness is to screen the interactions between surfaces, with an increased efficiency for the shorter range interaction. The molecular range solid–solid interactions are thus very efficiently screened by nanoscale roughness, while the capillary interactions have a more subtle behavior. Three main regimes were identified by Halsey and Levine (1998) as a function of the volume of liquid V available for the formation of a capillary bridge between two spheres: (1) the *asperity regime* prevails for small volumes, where the capillary force is dominated by the condensation around a single or a small number of asperities; (2) the *roughness regime* governs the intermediate volume range, where capillary bridges are progressively formed between a larger number of asperities and the capillary force grows linearly with V (as in Hornbaker et al. 1997); (3) the *spherical regime* where Equation 12.16 for the force caused by a single larger meniscus is recovered. The extension of the three domains is determined by the ratio between the characteristic scales of the gap distribution (height l_R and correlation length ξ) and the sphere radius R. When dealing with capillary bridges in equilibrium with undersaturated humidity, the roughness regime should be reduced to a narrow range of humidity values such that $l_R \sim r_k$, in which the capillary force jumps from a weak value to the spherical regime value (as in the experiments of Fraysse et al. 1999).

However, due to the first-order nature of capillary condensation transition, the equilibrium condition may be preceded by a long time-dependent region where the capillary bridges between asperities are progressively formed due to thermal activation. The energy barrier for the formation of a liquid bridge of volume $v_d \sim hA$ between asperities of curvature radius R_c may be expressed as (Restagno et al. 2000):

$$\Delta\Omega = v_d\rho_l\Delta\mu = v_d\rho_l k_B T\log\frac{P_{sat}}{P_v}$$

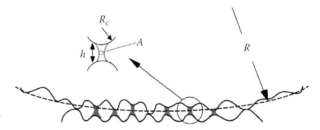

The probability that condensation occurs before a time t_w is

$$\Pi(t_w) = 1 - \exp\left(-\frac{t_w}{\tau}\right)$$

$$\tau = \tau_0\exp\left(\frac{\Delta\Omega}{k_B T}\right)$$

By integrating over a roughness dependent distribution of N_{tot} nucleation sites, each one contributing with a force $2\pi\gamma R_c$, an expression for the total force can be derived (Bocquet et al. 2002) that predicts a logarithmic increase as a function of the aging time t_w:

$$F_C = F_0 + 2\pi\gamma R_c N_{tot}\frac{k_B T}{V_0\rho_l\Delta\mu}\log\frac{t_w}{\tau_0} \qquad (12.26)$$

where V_0 is a roughness-dependent range for the distribution of the individual liquid bridge volume v_d.

beads adjustments, inducing a modification of the contacts and of the condensed bridges population. The effect of wear of the particles can be explained by accounting for both the increased roughness of the beads and the presence of a wear-induced dust consisting in small particles that significantly enhance the nucleation of further bridges (Restagno et al. 2002).

12.3.2 Time and Velocity Dependence in Nanoscale Friction Forces

Sliding friction is an everyday life issue, and its universal nature emerges from the great variety of industrial processes and natural phenomena in which it plays a central role (Persson 2000). With the miniaturization of moving components in many technological devices, such as microelectromechanical systems and hard disks, it has become of primary importance to study surface forces like friction, viscous drag and adhesion at microscales and nanoscales. The relevance of surface forces is greatly enhanced with regard to volume forces when the spatial scale is reduced to nanometers, but another major physical change comes from the increasing role of thermal fluctuations in the surface processes.

AFM has become the most efficient tool to study surface forces at the nanoscale, and the AFM tip sliding on a surface can often be considered as a model system for technologically relevant devices. The terminal apex of typical AFM tips can be roughly approximated by a sphere of radius between 10 and 100 nm, and the AFM contact imaging of a nanoscale rough surface can provide both a measurement of its roughness through the vertical deflection of the probe cantilever and a measurement of the friction forces through the dependence of the lateral deflection of the cantilever on the scan speed and on the normal applied load.

Nanoscale friction was rapidly shown not to respect the Amonton laws, being dependent on both sliding velocity and normal load. Moreover, it was shown to be strongly affected by the nanoscale roughness of the substrate and by the wetting properties of both the AFM tip and the substrate as well as by relative humidity. Nanofriction measurements have long been controversial, but recent careful measurements in controlled atmosphere have allowed defining a clearer scenario.

The sliding kinetics of an AFM tip has been shown to be determined by both the thermally activated stick-slip dynamics of the AFM tip on the substrate and the time-dependent formation of capillary bridges between the tip and the asperities of the rough substrate (cf. Insert H). When relative humidity is low, or the substrate is weakly wettable, stick-slip sliding has a dominant effect and results in a positive logarithmic dependence of the friction force on the sliding velocity (Gnecco et al. 2000; Riedo et al. 2003). The AFM tip keeps being stuck on nanoscale asperities and intermittently slips when thermal fluctuations allow to overcome a local energy barrier, which is progressively reduced by the accumulation of elastic energy due to the scanning. The dynamic friction force can be described by the following equation:

$$F_F = \mu(F_N + F_{SS}) + \mu[F_C(t)] + m\log\left(\frac{v}{v_B}\right) \quad (12.27)$$

where
- F_N is the normal force
- F_{SS} is the solid-solid adhesion force within the liquid
- $F_C(t)$ describes the eventual presence of time dependent capillary forces and the last term describes the positive velocity dependence induced by stick-slip motion
- v_B is a characteristic velocity

For more hydrophilic substrates, higher relative humidity, or increasing substrate roughness, the formation of capillary bridges and wetting films deeply modifies the friction dynamics letting the friction force be a logarithmically decreasing function of the sliding velocity (Riedo et al. 2002). This effect was successfully explained by applying the modeling developed by Bocquet et al. (1998) to account for the time-dependent thermally activated formation of capillary bridges between the nanoscale asperities of both the probe and the substrate. When the two rough surfaces are in relative sliding motion, the proximity time t_w of opposing tip and substrate asperities is a decreasing function of the sliding velocity v. The number of condensed capillary bridges (and thus the total capillary force) is thus expected to be a decreasing function of the sliding velocity according to Equation 12.26, and this trend should be reflected in the dynamic friction force according to Equation 12.27.

The study of AFM sliding friction forces has thus become an important complementary tool to study the time and load dependence of capillary forces. Notably, the friction forces were shown to present a 2/3 power law dependence on the applied load F_N (Riedo et al. 2004) and an inverse dependence on the Young modulus E of the substrate (Riedo and Brune 2003). These two effects were both explained by the increase of the nominal contact area where capillary bridges are susceptible to be formed, when either the normal load is increased, or the Young modulus of the substrate is decreased. The following equation for the capillary force during sliding was proposed in order to account for all these effects:

$$F_C = 8\pi\gamma_{lv}R(1+KF_N^{2/3})\left(\frac{1}{\rho_l l_R R_c^2}\right)\frac{\log(v_0/v)}{\log(P_{sat}/P_v)} \quad K = \frac{1}{r_K}\left(\frac{9}{16RE^2}\right)^{1/3}$$

$$(12.28)$$

the variables being defined as in the Insert H. Quantitative measurements of the AFM friction forces were thus shown to be useful in determining several physical parameters of interest, such as an estimation of the AFM tip radius and contact angle or important information on the activation energy for the capillary bridge formation (Szoskiewicz and Riedo 2005).

The formation of capillary bridges can have significant effects on the AFM imaging in contact mode due to the variations of the contact forces and consequently of the lateral forces during the scan. Thundat et al. (1993) have investigated the effect of humidity on the contrast when measuring the atomic level topography of a mica layer. The topographic contrast is shown to decrease with humidity above 20% RH due to an increase of the lateral force that acts in deforming the AFM cantilever and thus influences the measurement of the vertical deflection.

12.3.3 Friction Forces at Macroscale

Capillary forces can also affect the friction properties between macroscopic objects. In dry solid friction, aging properties have been studied on various materials, and have been related to the slow viscoplastic increase of the area of contact between asperities induced by the high values of the stress in the contact region (Baumberger et al. 1999). However, the importance of humidity has been reported by geophysicists in rock onto rock solid friction (Dieterich and Conrad 1984).

In the presence of a vapor atmosphere, the static friction coefficient was shown to increase logarithmically with the contact time, while the dynamic friction coefficient was shown do decrease with the logarithm of the sliding velocity (Dieterich and Conrad 1984; Crassous et al. 1999). The first effect is analog to the aging behavior of the maximum contact angle in granular matter as discussed in Section 12.3.1. The second effect is analogous to what observed in the sliding friction of a nanoscale contact as discussed in Section 12.3.2. However, the general behavior is strongly modified due to the greater importance of the normal stresses that induce significant plastic deformation at the contact points. The aging behavior must then be explained by the combined action of the evolution of the contact population due to plastic deformation and the evolution of the number of capillary bridges due to thermally activated condensation. This induces a more complex dependence on the normal load, since this modifies both the elastic contact area and the progressive plastic deformation of the contacts, and thus influences the residual distribution of the intersurface distances that govern the kinetics of capillary condensation.

Based on these experimental observations, Rice and Ruina (1983) have proposed a phenomenological model for non stationary friction, in which friction forces depend on both the instantaneous sliding velocity v and a state variable φ according to

$$F(v,\varphi) = F_N \left[\mu_0 + A \log\left(\frac{v}{v_0} \right) + B \log\left(\frac{\varphi v_0}{d_0} \right) \right] \quad (12.29)$$

$$\dot{\varphi} = 1 - \frac{v\varphi}{d_0} \quad (12.30)$$

where

A and B are positive constants
d_0 and v_0 are characteristic values of the sliding distance and velocity

The first term accounts for the logarithmic dependence on the sliding velocity, while the second term accounts for the logarithmic dependence on the static contact time through the evolution of the state variable φ according to Equation 12.30. This phenomenological modeling can be applied to other intermittent sliding phenomena (peeling of adhesives, shear of a granular layer, etc.) and the significance of the state variable φ is not determined a priori. However, in the case of solid friction, the state variable φ can be related to the population of microcontacts and capillary bridges (Dieterich and Kilgore 1994).

12.4 Influence on Surface Chemistry

The presence of nanometric water films on solid surfaces has a significant impact on the surface physical and chemical properties. Chemistry in these extremely confined layers is quite different than in bulk liquids due to the strong interaction with the solid surface, to the presence of the negative capillary pressure, to the reduction of transport coefficients and to the relevance of the discrete molecular structure and mobility that can hardly be represented by a continuum description. The role of thermal fluctuations and their correlation length also become more relevant.

Thin water films can have a major role in the alteration of some surface layer in the solid due to their effect on the local dissolution, hydration, oxidation, hydrolysis, and lixiviation, which are some of the basic mechanisms of the corrosion processes. Water condensation from a moist atmosphere is quite pure and it is thus initially extremely reactive toward the solid surface. However, the extreme confinement prevents the dilution of the corrosion products, leading to a rapid change in the composition and pH of the liquid film. Depending on the specific conditions this can either accelerate the reaction rates due to increased reactivity and catalytic effects, or decelerate the reaction rate due to rapid saturation of the corrosion products in the film. This condition of equilibrium between the reactions of corrosion and recondensation can lead to a progressive reorganization of the structure of the surface layer in the solid. The extreme confinement and the significance of the fluctuations can cause the generation of complex patterns related to the dissolution-recondensation process, involving inhomogeneous redeposition of different amorphous, gel or crystalline phases (Christenson and Israelachvili 1987; Watanabe et al. 1994).

The dissolution–recondensation phenomenon also happens at the capillary bridges between contacting solid grains or between the contacting asperities of two rough solid surfaces. When humidity undergoes typical ambient oscillations, capillary bridges and films are formed or swollen in moist periods, inducing an activity of differential dissolution. The subsequent redeposition under evaporation in more dry conditions is particularly effective in the more confined regions, i.e., at the borders of the solid contact areas, acting as a weld solid bridge between the contacting parts (cf. Figure 12.11). This can be responsible of a progressive increase in the cohesion of granular matters, which has important applications in the pharmaceutical and food industry, and of the progressive increase of the static friction coefficient between contacting rocks.

Another domain where the formation of capillary condensation has a determinant impact on the mechanical properties is the stress-corrosion crack propagation in moist atmosphere (cf. Ciccotti 2009 for a review). We already mentioned in Section 12.2.3 that the crack tip cavity in brittle materials like glass is so confined that significant capillary condensation can be observed at its interior. During slow subcritical crack propagation, the crack advances due to stress-enhanced chemical reactions of hydrolyzation and leaching that are deeply affected by the local

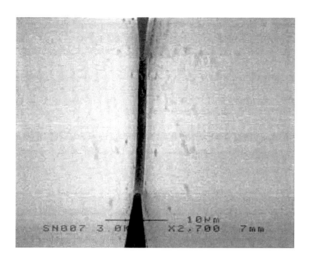

FIGURE 12.11   MEB photograph of a solid bridge between two glass beads (magnification ×2700). (From Olivi-Tran, N. et al., *Eur. Phys. J.B* 25, 217, 2002. With permission.)

crack tip environment. Capillary condensation has a fundamental impact on several levels on the kinetics of this reaction: (1) the presence of a liquid phase makes the preadsorption of water molecules near the crack tip easier; (2) the negative Laplace pressure determines the chemical activity of the water molecules in the meniscus and directly affects the reaction rate; (3) the confined nature of the condensation along with its limited volume are responsible of an evolution of the chemical composition of the condensate that has a direct and major effect on the corrosion reactions at the crack tip, especially by changes of the pH and by the enrichment in alkali species due to stress-enhanced leaching (Célarié et al. 2007).

References

Abramowitz, M. and Stegun, I. A. (eds.), 1972. *Handbook of Mathematical Functions with Formulas, Graphs, and Mathematical Tables.* New York: United States Government Printing.

Barrett, E. P., Joyner, L. G., and Halenda, P. P. 1951. The determination of pore volume and area distributions in porous substances. 1. Computation from nitrogen isotherms. *J. Am. Chem. Soc.* 73: 373–380.

Baumberger, T., Berthoud, P., and Caroli, C. 1999. Physical analysis of the state- and rate-dependent friction law. II. Dynamic friction. *Phys. Rev. B* 60: 3928–3939.

Bhushan, B. (ed.), 2007. *Springer Handbook of Nanotechnology.* 2nd edn. New York: Springer.

Bocquet, L., Charlaix, E., Ciliberto, S., and Crassous, J. 1998. Moisture-induced ageing in granular media and the kinetics of capillary condensation. *Nature* 296: 735–737.

Bocquet, L., Charlaix, E., and Restagno, F. 2002. Physics of humid granular media. *C. R. Physique* 3: 207–215.

Bowden, F. P. and Tabor, D. 1950. *Friction and Lubrication in Solids.* Oxford, U.K.: Clarendon Press.

Brunauer, S., Emmett, P. H., and Teller, E. 1938. Adsorption of gases in multimolecular layers. *J. Am. Chem. Soc.* 60: 309–319.

Célarié, F., Ciccotti, M., and Marlière, C. 2007. Stress-enhanced ion diffusion at the vicinity of a crack tip as evidenced by atomic force microscopy in silicate glasses. *J. Non-Cryst. Solids* 353: 51–68.

Charlaix, E. and Crassous, J. 2005. Adhesion forces between wetted solid surfaces. *J. Chem. Phys.* 122: 184701.

Christenson, H. K. 1985. Capillary condensation in systems of immiscible liquids. *J. Colloid Interface Sci.* 104: 234–249.

Christenson, K. 1988. Adhesion between surfaces in undersaturated vapors–A reexamination of the influence of meniscus curvature and surface forces. *J. Colloid Interface Sci.* 121: 170–178.

Christenson, H. K. 1994. Capillary condensation due to van der Waals attraction in wet slits. *Phys. Rev. Lett.* 73: 1821–1824.

Christenson, H. K. and Israelachvili, J. N. 1987. Growth of ionic crystallites on exposed surfaces. *J. Colloid Interface Sci.* 117: 576–577.

Ciccotti, M. 2009. Stress-corrosion mechanisms in silicate glasses. *J. Phys. D: Appl. Phys.* 42: 214006.

Ciccotti, M., George, M., Ranieri, V., Wondraczek, L., and Marlière, C. 2008. Dynamic condensation of water at crack tips in fused silica glass. *J. Non-Cryst. Solids* 354: 564–568.

Cleveland, J. P., Anczykowski, B., Schmid, A. E., and Elings, V. B. 1998. Energy dissipation in tapping-mode atomic force microscopy. *Appl. Phys. Lett.* 72: 2613–2615.

Crassous, J. 1995. Etude d'un pont liquide de courbure nanométrique: Propriétés statiques et dynamiques. PhD thesis. Ecole Normale Supérieure de Lyon, Lyon, France.

Crassous, J., Charlaix, E., and Loubet, J. L. 1994. Capillary condensation between high-energy surfaces: Experimental study with a surface force apparatus. *Europhys. Lett.* 28: 37–42.

Crassous, J., Loubet, J.-L., and Charlaix, E. 1995. Adhesion force between high energy surfaces in vapor atmosphere. Material Research Society Symposium Proceedings. 366: 33–38.

Crassous, J., Bocquet, L., Ciliberto, S., and Laroche, C. 1999. Humidity effect on static aging of dry friction. *Europhys. Lett.* 47: 562–567.

de Gennes, P. G. 1985. Wetting: Statics and dynamics. *Rev. Modern Phys.* 57: 827–863.

de Gennes, P. G., Brochard, F., and Quere, D. 2003. *Capillarity and Wetting Phenomena: Drops, Bubbles, Pearls, Waves.* New York: Springer.

de Lazzer, A., Dreyer, M., and Rath, H. J. 1999. Particle-surface capillary forces. *Langmuir* 15: 4551–4559.

Derjaguin, B.V. 1944. A theory of capillary condensation in the pores of sorbents and of other capillary phenomena taking into account the disjoining action of polymolecular liquid films. *Acta Physiochimica URSS* 12: 181–200.

Derjaguin, B. V. and Churaev, N. V. 1976. Polymolecular adsorption and capillary condensation in narrow slit pores. *J. Colloid Interface Sci.* 54: 157–175.

Dieterich, J. H. and Conrad, G. 1984. Effect of humidity on time-dependent and velocity-dependent friction in rocks. *J. Geophys. Res.* 89: 4196–4202.

Dieterich, J. H. and Kilgore, B. D. 1994. Direct observation of frictional contacts-new insights for state-dependent properties. *Pure Appl. Geophys.* 143: 283–302.

Fisher, L. R. and Israelachvili, J. N. 1981. Direct measurement of the effect of meniscus force on adhesion: A study of the applicability of macroscopic thermodynamics to microscopic liquid interfaces. *Coll. Surf.* 3: 303–319.

Fraysse, N., Thomé, H., and Petit, L. 1999. Humidity effects on the stability of a sandpile. *Eur. Phys. J. B* 11: 615–619.

Gelb, L. D., Gubbins, K. E., Radhakrishnan, R., and Sliwinska-Bartkowiak, M. 1999. Phase separation in confined systems. *Rep. Progr. Phys.* 62: 1573–1659.

Gnecco, E., Bennewitz, R., Gyalog, T., Loppacher, Ch., Bammerlin, M., Meyer, E., and Güntherodt, H.-J. 2000. Velocity dependence of atomic friction. *Phys. Rev. Lett.* 84: 1172–1175.

Grimaldi, A., George, M., Pallares, G., Marlière, C., and Ciccotti, M. 2008. The crack tip: A nanolab for studying confined liquids. *Phys. Rev. Lett.* 100: 165505.

Halsey, T. C. and Levine, A. J. 1998. How sandcastles fall. *Phys. Rev. Lett.* 80: 3141–3144.

Heilmann, R., Ahn, M., and Schattenburg, H. 2008. Nanomirror array for high-efficiency soft x-ray spectroscopy. SPIE Newsroom. 27 August 2008. DOI: 10.1117/2.1200808.1235.

Hornbaker, D. R., Albert, I., Barabasi, A. L., and Shiffer, P. 1997. What keeps sandcastles standing? *Nature* 387: 765–766.

Israelachvili, J. N. 1992. *Intermolecular and Surface Forces.* 2nd edn. New York: Academic Press.

Kohonen, M. M., Maeda, N., and Christenson, H. K. 1999. Kinetics of capillary condensation in a nanoscale pore. *Phys. Rev. Lett.* 82: 4667–4670.

Lefevre, B., Sauger, A., Barrat, J. L., Bocquet, L., Charlaix, E., Gobin, P. F., and Vigier, G. 2004. Intrusion and extrusion of water in hydrophobic mesopores. *J. Chem. Phys.* 120: 4927–4938.

Maugis, D. 1992. Adhesion of spheres: The JKR-DMT transition using a dugdale model. *J. Colloid Interface Sci.* 150: 243–269.

Maugis, D. and Gauthier-Manuel, B. 1994. JKR-DMT transition in the presence of a liquid meniscus. *J. Adhes. Sci. Technol.* 8: 1311–1322.

Olivi-Tran, N., Fraysse, N., Girard, P., Ramonda, M., and Chatain, D. 2002. Modeling and simulations of the behavior of glass particles in a rotating drum in heptane and water vapor atmospheres. *Eur. Phys. J. B* 25: 217–222.

Orr, F. M., Scriven, L. E., and Rivas, A. P. 1975. Pendular rings between solids: Meniscus properties and capillary force. *J. Fluid Mech.* 67: 723–742.

Pakarinen, O. H., Foster, A. S., Paajanen, M., Kalinainen, T., Katainen, J., Makkonen, I., Lahtinen, J., and Nieminen, R. M. 2005. Towards an accurate description of the capillary force in nanoparticle-surface interactions. *Model. Simul. Mater. Sci. Eng.* 13: 1175–1186.

Pallares, G., Ponson, L., Grimaldi, A., George, M., Prevot, G., and Ciccotti, M. 2009. Crack opening profile in DCDC specimen. *Int. J. Fract.* 156: 11–20.

Persson, B. N. J. 2000. *Sliding Friction: Physical Principles and Applications.* 2nd edn. Heidelberg: Springer.

Plateau, J. 1864. The figures of equilibrium of a liquid mass. In *The Annual Report of the Smithsonian Institution.* Washington D.C. pp. 338–369.

Restagno, F., Bocquet, L., and Biben, T. 2000. Metastability and nucleation in capillary condensation. *Phys. Rev. Lett.* 84: 2433–2436.

Restagno, F., Ursini, C., Gayvallet, H., and Charlaix, E. 2002. Aging in humid granular media. *Phys. Rev. E* 66: 021304.

Rice, J. R. and Ruina, A. L. 1983. Stability of steady frictional slipping. *J. Appl. Mech.* 50: 343–349.

Riedo, E., Lévy, F., and Brune, H. 2002. Kinetics of capillary condensation in nanoscopic sliding friction. *Phys. Rev. Lett.* 88: 185505.

Riedo, E. and Brune, H. 2003. Young modulus dependence of nanoscopic friction coefficient in hard coatings. *Appl. Phys. Lett.* 83: 1986–1988.

Riedo, E., Gnecco, E., Bennewitz, R., Meyer, E., and Brune, H. 2003. Interaction potential and hopping dynamics governing sliding friction. *Phys. Rev. Lett.* 91: 084502.

Riedo, E., Palaci, I., Boragno, C., and Brune, H. 2004. The 2/3 power law dependence of capillary force on normal load in nanoscopic friction. *J. Phys. Chem. B* 108: 5324–5328.

Rowlinson, J. S. and Widom, B. 1982. *Molecular Theory of Capillarity.* Oxford, U.K.: Clarendon Press.

Szoskiewicz, R. and Riedo, E. 2005. Nucleation time of nanoscale water bridges. *Phys. Rev. Lett.* 95: 135502.

Thomson, W. 1871. On the equilibrium of vapour at a curved surface of liquid. *Phil. Mag.* 42: 448–452.

Thundat, T., Zheng, X. Y., Chen, G. Y., and Warmack, R. J. 1993. Role of relative humidity in atomic force microscopy imaging. *Surf. Sci. Lett.* 294: L939–L943.

Torralvo, M. J., Grillet, Y., Llewellyn, P. L., and Rouquerol, F. 1998. Microcalorimetric study of argon, nitrogen and carbon monoxide adsorption on mesoporous Vycor glass. *J. Colloid Interface Sci.* 206: 527–531.

Verdaguer, A., Sacha, G. M., Bluhm, H., and Salmeron, M. 2006. Molecular structure of water at interfaces: Wetting at the nanometer scale. *Chem. Rev.* 106: 1478–1510.

Wan, K. T., Smith, D. T., and Lawn, B. R. 1992. Fracture and contact adhesion energies of mica-mica, silica-silica, and mica-silica interfaces in dry and moist atmospheres. *J. Am. Ceram. Soc.* 75: 667–676.

Watanabe, Y., Nakamura, Y., Dickinson, J. T., and Langford, S. C. 1994. Changes in air exposed fracture surfaces of silicate glasses observed by atomic force microscopy. *J. Non-Cryst. Solids* 177: 9–25.

Xiao, X. and Qian, L. 2000. Investigation of humidity-dependent capillary force. *Langmuir* 16: 8153–8158.

Xu, L., Lio, A., Hu, J., Ogletree, D. F., and Salmeron, M. 1998. Wetting and capillary phenomena of water on mica. *J. Phys. Chem. B* 102: 540–548.

Zimon, A. D. 1969. *Adhesion of Dust and Powder.* New York: Plenum Press.

Zitzler, L., Herminghaus, S., and Mugele, F. 2002. Capillary forces in tapping mode atomic force microscopy. *Phys. Rev. B* 66: 155436.

13

Dynamics at the Nanoscale

A. Marshall Stoneham
University College London

Jacob L. Gavartin
Accelrys Ltd.

13.1 Introduction

At the nanoscale, time-dependent behavior gains importance and standard properties of the bulk crystal are less crucial. Some of the dynamical behavior is random: fluctuations may control rate processes and thermal ratchets become possible. Dynamics is important in the transfers of energy, signals, and charge. Such transfer processes are especially efficiently controlled in biological systems. Other dynamical processes are crucial for control at the nanoscale, such as to avoid local failures in gate dielectrics, to manipulate structures using electronic excitation, or to manipulate spins as part of quantum information processing. Our aim here is to scope the wide-ranging time-dependent nanoscale phenomena.

Why does dynamics at the nanoscale deserve a special attention? The prime reason is that the *timescales* of processes and the *length scales* of a nanosystem often become interrelated, so that a range of dynamical properties show significant size dependences. These properties fall into a number of classes. A fundamental category relates to linear response effects, ultimately based on vibrational or electronic spectra. Another category of dynamical properties are nonlinear phenomena, including exciton decay mechanisms and energy dissipation. These are related to phenomena that occur when the thermodynamic limit is not reached or when thermal equilibrium is not attained. A further category includes transport phenomena: this is the mean free path or a diffusion distance l comparable to the system size L, when different factors are considered, than when l is controlled by carrier concentration. And, of course, there is the question of heterogeneity, and how behavior depends on the relative fractions of surface and bulk atoms.

A cluster of 100 atoms in thermal equilibrium at room temperature will have root mean square volume fluctuations of the order of 1%, similar to the root mean square volume fluctuation of a human breathing normally. It is true that the timescales differ by a factor of order 10^{12}, but the example emphasizes the ubiquitous nature of its dynamics. Nanoparticles also have other characteristics: they grow, restructure, and interact. Electronic excitation leads to processes on the femtosecond timescale, to relaxation processes on the picosecond timescale, and to optical and nonradiative transitions on the nano- and microsecond timescales.

Biological processes at the nanoscale are more complex and surprisingly efficient. Such processes may involve energy propagation, signal propagation, and the controlled and correlated movements of many atoms. Biological systems contain molecular motors that operate with relatively soft components. Even in living humans, coherent vibrational excitations, so-called solitons, seem to shift modest amounts of energy with minimal loss. When a large molecule meets a receptor, the initial processes might be limited by shape and size, but depend strongly on fluctuations. For very small molecules, other factors come into play: For example, for serotonin, the process may be proton transfer, and for scent molecules at olfactory receptors, inelastic electron tunneling is a strong candidate for the critical step.

Quantum computing based on condensed matter systems is inherently nanoscale, since quantum entanglement is effective only at the submicron level. Quantum information processing is also inherently dynamic, for manipulations of qubits have to be faster than decoherence (quantum dissipation) mechanisms. Strikingly, quantum information processing and life processes display commonalities in exploiting behavior far from equilibrium.

TABLE 13.1   Characteristic Timescales for Dynamics of Nanoscale Objects

Class	Phenomenon	Typical Timescale
Fast: electronic (femtoseconds to picoseconds)	Plasmon frequency	~0.1 fs
	Electron collision times	~0.1 fs
	Electronic excitation creating dynamics	~fs
	Chemically induced dynamics	~fs
	Electron moves 1 nm in a metal	1–2 fs
	Electron (1/40 eV) moves 1 nm in a semiconductor	10 fs
Fairly fast: lattice relaxation, etc. (picoseconds to nanoseconds)	Sound crosses a nanometer scale dot	~ps
	Vibrational energy loss from dot	Few ps
	Ballistic motion of electron in a nanotube	Few ps
	Intrachain movement of solitons or polarons in *trans*-polyacetylene	Few ps
	Confined vibrational modes in dots	~ns
	Spin dynamics for spintronics or spin-based quantum gates	Ideally ns
Moderately fast (nanoseconds to microseconds)	Energy transfer: soliton in α-helix	~10–100 ns
	Interchain movement of solitons and polarons in *trans*-polyacetylene	~100 ns
Moderate (microseconds to milliseconds)	Random dynamics and noise	Wide variations
	Decoherence times for "good" quantum systems	
	Diffusion and other incoherent processes	
Relatively slow: milliseconds and longer	Dynamics of equipment, like STM	Wide variations.
	Dynamics in processing, e.g., scent, sight	Our senses have
	Molecular motor systems	characteristic
	Dynamics of system failure	times of 1–100 ms

Note: There are many simplifications here, but the range and variety of behaviors is represented.

Almost all scientifically interesting systems, whether biological or physical, change with time at some scale: some time dependence is unavoidable. At the nanoscale, their functionality depends both on the object itself and on its working environment, and also on certain consistencies that have to be achieved. Table 13.1 shows some of the typical timescales. It is often useful to distinguish between natural and operational timescales (cf. the length scales discussed by Stoneham and Harding 2003). *Natural timescales* might be defined as the time taken by sound to cross a nanodot, or for spontaneous optical emission at the sum rule limit. *Operational timescales* are designed and structured to chosen criteria, often with difficulty. Thus, in state-of-the-art microelectronics devices, the structures have sizes determined partly by nature, partly by compatibility with previous generations of device (since reengineering fabrication plants is expensive), and partly by the laws of physics and the art of the possible. The choices of materials and how they are organized are intended to maximize signal speeds, delay memory decay, and keep energy dissipation under control. These choices have to be compatible and consistent with the need for many sequential steps in the fabrication process. *Biological* systems have evolved to make operational timescales seem natural. We are gradually learning how such timescales are designed. There is an opportunity to turn scientific understanding into technological advantage.

In this chapter, we discuss just a few of the many time-dependent processes taking place in nanoscale objects. Two systems at the smaller end of the nanoscale are especially interesting. Thus, II–VI

(e.g., CdSe) quantum dots of perhaps 200 ions, which show a wealth of time-dependent processes, are often considered quantally, but can largely be described classically. The other system, which is usually described classically, appears to need a quantum description: how do scent molecules (rarely, if ever, bigger than 100 atoms) provoke receptors and initiate signals that ultimately reach the brain?

13.2 Time-Dependent Behavior and the II–VI Nanodot

The term "quantum dot" is used in several different ways. There are the "large" quantum dots of silicon or III–V semiconductors, typically containing tens of thousands of atoms. These dots are central to optoelectronic devices and some variants of quantum computing. Then there are the "small" quantum dots—typically a few hundred atoms of a II–VI semiconductor—which are a couple of nanometers in diameter. These small dots show varied dynamical features, and only some of these features are understood. We discuss these features mainly to illustrate the diversity of phenomena. We emphasize the point that the nanoscale needs new ways of thinking about what is important.

13.2.1 Growth and What Stops It

Microbial synthesis offers a convenient way to produce industrial amounts of CdS nanodots (Williams et al. 1996). But why do the dots stop growing and stay nanosized, with sizes as uniform as can be achieved by sophisticated chemical methods? If ice and mushrooms can break up concrete, how can soft biomaterials constrain size when there seems to be a large thermodynamic force that can help them grow? In fact, nature has found many ways to use soft, flexible materials in ways that, at least macroscopically, are associated with stiff, rigid structures (Stoneham 2007). Examples include soft templates for the growth of an inorganic crystal with specific facets and orientations, or the growth of small nanocrystals of controlled size. How can this behavior of "soft" biological materials, including organization, be achieved? Organization has several variants. The first is organization at the atomic scale, when a particular crystal structure is selected. This *selection* step may involve a choice of chirality or a choice between structures (e.g., wurtzite versus zincblende). Such selectivity is exploited in the purification of pharmaceuticals and is also key to the creation of small II–VI quantum dots. A second type of organization is mesoscopic in scale and leads to *ordering*, often termed self-organization, though this is only a description and not an explanation. A third type of organization leads to specific *shapes* (usually external shape), primarily at the larger mesoscale or the macroscale. The structure may be relatively soft, such as in some cell structures, where topology is crucial, or it may be stiff, as in bone or shell. Complex patterns can be generated reproducibly (Meinhardt 1992, Koch and Meinhardt 1994), including periodic patterns of units that have a complex and polar substructure, such as photoreceptor cells in the *Drosophila* eye. Spatial organization has its own timescales, sometimes related to characteristic length scales, e.g., through a diffusion constant.

Restricting growth is an important phenomenon in biology, and the shapes of structures (like shells) determine their function. Sometimes a *clever control* is used by organisms to exploit DNA's capabilities. For example, protein cages are crucial in synthesizing magnetic nanoparticles, like single-domain ferrimagnetic particles of Fe_3O_4 found in magnetotactic bacteria (Klem et al. 2005). The mammalian ferritin structure uses two types of subunit (H, L) that align in antiparallel pairs to form a shell, with narrow (~3 Å) channels. One set of channels is hydrophobic and the other hydrophilic. Mineralization involves iron oxidation, hydrolysis, nucleation, and growth. The Fe ions enter the cage via the hydrophilic channels, and presumably electrostatics controls their entry and so limits their growth. The outer entrance is a region of positive potential, guiding cations into the cage until the ferritin fills the internal cavity with some precision. It is possible that similar mechanisms operate in cases like CdS nanoparticle formation in yeasts. Size alone can be controlled in other ways, such as imposing surface nucleation barriers (Frank 1952), limiting materials supply, or capping to block growth sites.

Access to a nanodot surface is important in medical applications, and it is found that chaperonin proteins form ATP-responsive barrel-like cages for nanoparticles (Ishli et al. 2003). The distribution of sizes of biologically controlled nanoparticles seems to be at least as good (perhaps 5% variance in radius)

as the best cases in solution chemistry. A 5% variance in radius amounts to a much larger variation in ion number, which is an important factor, in that when even adding a single ion can have important consequences.

13.2.2 Vibrational Spectra of Quantum Dots

The smallest ionic dots from molecular beams show crystal structures that are different from the bulk form. Partly, this variation in structures can be explained by the presence of large electric fields in such dots. So do the vibrational features of nanoclusters differ qualitatively from the bulk as well? Are there signs of discreteness in the phonon spectra (Stoneham 1965)? Could one find modes with frequencies higher than those of the corresponding bulk zone-center LO phonon (Gavartin and Stoneham 2003)? Can we associate these effects, if they exist, with the bulk or the surface, or is there an intimate mixture?

13.2.2.1 Polar Quantum Dots

All nanodots—whether ionic, covalent, organic, or metallic—should show effects of confinement. We expect some differences between the relatively close-packed ionic systems (like NaCl with sixfold coordination) and more open covalent systems (like the fourfold ZnS structure) to show trends in phonon confinement that are analogous to electron confinement in some ways. There is substantial electron confinement (band gap opening) at the appropriately terminated silicon surface, whereas surface states are found in the band gaps of MgO or NaCl. For surface phonons, ab initio calculations predict a surface band at ~4 meV above the maximum bulk frequency at the silicon (2 × 1) surface (Fritsch and Pavone 1995, Screbtii et al. 1995), but the (001) surface vibrations of NaCl and MgO do not exceed the energy of the bulk LO phonons. This can be seen from the vibrational dynamics of NaCl and ZnS nanoclusters (Figures 13.1 and 13.2).

Cubic NaCl nanocrystals with only (001) type surfaces have a vibrational density of states similar to bulk material. Faceted clusters, with less stable surfaces, show modes with frequencies up to 5 meV above the bulk maximum of 32 meV. The specific nature of the modes depend on the precise surface termination, but all faceted clusters have both surface-like and bulk-like high-frequency modes. The bulk-like modes should probably be considered as resulting from constructive interference of the surface modes localized near the opposite high index faces of a crystallite, and should eventually disappear in larger nanocrystallites.

Figure 13.2 shows the vibrational density of states for the zincblende-structured cluster $(ZnS)_{47}$ derived from harmonic analysis using density functional theory with an atomic basis set and the PBE density functional, as implemented in the DMol³ code (Delley 2000, Accelrys 2008). The mode observed around 56 meV is well above the largest bulk (LO) phonon of 47.5 meV (Tran et al. 1997), and is in line with results from earlier shell model predictions and plane wave density functional calculations (Stoneham and Gavartin 2007).

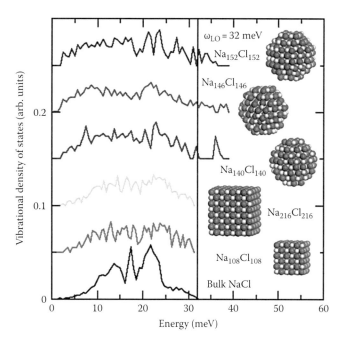

FIGURE 13.1 Vibrational densities of states of selected NaCl nanocrystals calculated using shell model as implemented in Gulp program (Gale and Rohl 2003, Accelrys 2008). Cubic nanocrystals $(NaCl)_{108}$ and $(NaCl)_{216}$ have no vibrations with frequencies higher than the bulk $\omega_{LO} = 32$ meV (vertical line), while the faceted clusters $(NaCl)_{140}$, $(NaCl)_{146}$, and $(NaCl)_{152}$ display high frequency tails. (Based on Stoneham, A.M. and Gavartin, J.L., *Mater. Sci. Eng. C*, 27, 972, 2007.)

Clearly, modes with frequencies higher than the largest (longitudinal optic) bulk frequencies exist at least in some stable nanoparticles. More work is needed to decide how universal this phenomenon is for small nanodots. More important is the unresolved question of how strongly such vibrations couple to electronic excitations. The standard view of electron–phonon coupling suggests that the dominant coupling in ZnS would be with the LO modes around 47 meV. Calculations predict that the higher frequency modes are infrared active, and may couple strongly to the electronic excitations. Further analysis reveals that high-frequency modes are strongly localized at the surface. Thus, the doubly degenerate highest frequency is associated with the in-plane optical vibration on two faces of the nanoparticle, as indicated in Figure 13.2. Intriguingly, the infrared spectra reported in the figure contrast strongly with our previous analysis of the electron–phonon interaction based on the dynamics of single-particle levels (Gavartin and Shluger 2006, Stoneham and Gavartin 2007), predicting strong electron interaction with much softer modes. Although the rigorous studies of electron–phonon coupling are in their infancy, they imply strongly that the origin of electron–phonon coupling in a nanocrystal is radically different from that in the bulk material. This would have major implications on dynamics of self-trapping, in which a carrier localizes as a result of the electron–phonon coupling.

13.2.2.2 Metallic Quantum Dots

Just as quantum dots of polar materials show features different from the bulk, so do metallic nanocrystals. Figure 13.3 shows the vibrational density of states of a Pt_{116} cluster obtained from

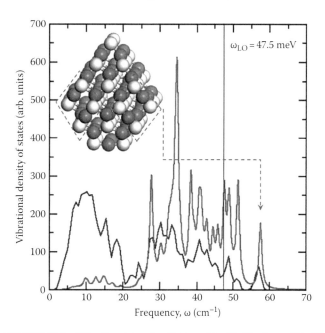

FIGURE 13.2 Vibrational density of states of the $(ZnS)_{47}$ cluster (black) obtained from the harmonic analysis using ab initio density functional theory with an atomic basis set and the PBE density functional as implemented in the DMol³ program (Delley 2000, Accelrys 2008). The same spectrum weighted with infrared intensities is shown in gray. Maximum frequency in the bulk ZnS, ω_{LO}, is indicated by a vertical line. Highlighted are two equivalent faces containing four sulfur and two Zn atoms, on which the highest frequency vibration is localized.

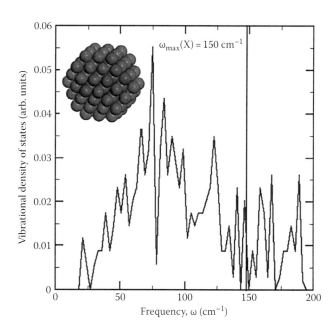

FIGURE 13.3 Vibrational density of states of the Pt_{116} cluster obtained from the harmonic analysis using embedded atomic potentials due to Sutton and Chen (1990). The vertical line corresponds to the maximum bulk frequency for the same potentials.

harmonic analysis using the embedded atomic potentials of Sutton and Chen (1990) and GULP code (Gale and Rohl 2003, Accelrys 2008). A significant number of vibrations are predicted with frequencies above the maximum bulk frequency (the X-phonon, shown as a vertical line). The maximum frequency in the bulk fcc Pt corresponds to the X-point phonon measured at ~190 cm⁻¹. Although the force field parameters used in this study predict a somewhat lower frequency of 150 cm⁻¹, similar uncertainty is expected in a cluster vibrational analysis. The highest frequency in the Pt_{116} cluster is calculated as 190 cm⁻¹ and corresponds to an isotropic breathing.

For metal nanodots, which are a major interest for solar collectors and some biomedical applications, the important property is absorption of energy from incident photons. For a dot of 100 atoms, the absorption of a 2 eV photon would give an energy equivalent to about 232°C. Similar temperature rises may occur during catalysis of exothermic reactions. So it is natural to be curious about the process and dynamics of melting. In addition, the melting of small isolated clusters reveals many aspects of nanosystems, and standard thermodynamic considerations that are used to explain them need to be modified.

Fundamentally, thermodynamic quantities represent ensemble averages, which are independent of the ensemble for very large systems. This is not generally true for the finite nanoscale systems, like isolated clusters (microcanonical ensemble) and clusters implanted in the bulk or deposited on a surface (canonical ensemble), which may behave qualitatively differently. A somewhat related issue concerns the ergodic hypothesis that thermodynamic quantities are the same whether they are obtained by the ensemble average or a time average over the evolution of a single system. Many nanosystems, especially those that can be considered isolated, are essentially nonergodic. The nature of averaging for these systems is defined by experiment design. Thus, there may be systematic differences between the same characteristics *measured* as an ensemble average and as a time average. A cluster's temperature is one of its most fundamental characteristics. For an isolated cluster, a temperature T_1 can be defined as the average kinetic energy per particle, that is,

$$\frac{3N}{2} k_B T_1 = \left\langle \sum_{i,\alpha} \frac{m_i(v_{0,\alpha}{}^2 - v_{i\alpha}{}^2)}{2} \right\rangle_t$$

where

    i runs over all particles in the cluster (N)
    α runs over three Cartesian coordinates
    $v_{0,\alpha}$ represents the velocity of the center-of-mass; an average is then taken over the time evolution of the cluster

Jellinek and Goldberg (2000) noticed that for very small isolated clusters, T_1 is different from the temperature T_2 defined thermodynamically from the following equation:

$$T_2 = \left(\frac{\partial E}{\partial S} \right)_V$$

This reflects on the general observation that measurements of cluster temperature will generally depend on the experimental probe (Makarov 2008). Some small metallic clusters have been observed to exhibit a negative heat capacity near a phase transition: they cool down when they are heated (Roduner 2006, Makarov 2008 and references therein). Technically, this violates the second law of thermodynamics as usually given, exposing limits of its applicability in small isolated systems. We have already noted the increased role of fluctuations in small systems, which indicates that thermodynamic averages do not capture all essential physics of the process, even at equilibrium.

Figure 13.4 shows a 6214 atom Pt cluster, both solid and at least partly molten as modeled using microcanonical molecular dynamics with embedded atom potentials (Sutton and Chen 1990) implemented in GULP package (Gale and Rohl 2003, Accelrys 2008). Both the cut through the center, which shows a crystalline core, and the root mean square atomic displacements show a liquid outer layer and a solid core, rather well separated by a boundary of about one atomic diameter. Melting first occurs at the surface at a lower temperature than for the bulk. There is a critical cluster size at which energy fluctuations are of the same order as the latent heat between two phases. Below this size, no phase separation is possible, though two phases may be present dynamically as a superposition. Above the critical size, the fluctuation correlation length is smaller than the cluster size, and the cluster behaves as a "big" material. In our example, this 6 nm Pt nanoparticle is still remarkably free from nano features, though both its melting temperature T_m and specific latent heat may be still lower than that of the bulk material. This cluster is also incredibly thermodynamic, even in microcanonical regime, in the sense that detailed equipartition of energy between various degrees of freedom is preserved. Liquid shell–solid core structures are common in isolated metallic clusters (Ferrando et al. 2008). However, this behavior may be reversed in molecular or semiconductor clusters and in clusters embedded in another material (Roduner 2006).

13.2.3 Dynamics and Nanocrystal Structure

Some remarkable observations (Buffat 2003) show that the electron diffraction peak for an individual dot, even one like Au, apparently switches off for periods of a few seconds or longer. Why this happens is still not known, but there are several relevant time-dependent processes.

The first model simply involves dot rotation. This seems credible in a soft matrix, like a polymer, when rotation is easy or on a surface. But it is hard to see how such rotation could work for systems like as CdS in a rigid SiO_2 matrix, even though soft rotation and cage modes below 5 meV have been predicted in the molecular dynamics modeling of ZnS in a SiO_2 matrix (Stoneham and Gavartin 2007; see Figure 13.5). Even with soft

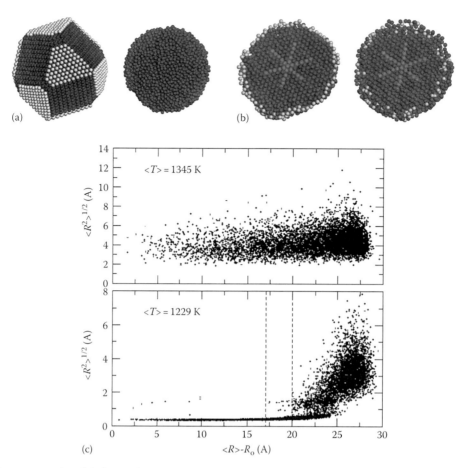

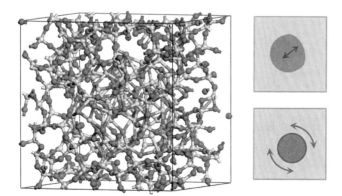

FIGURE 13.4 (a) Comparison of a solid cluster of 6214 Pt atoms at 439 and 1229 K, when it has melted, at least in part. (b) A cut through the middle of the molten cluster, indicating quasi-solid core atoms (dark gray) that keep their crystalline arrangement, whereas surface atoms (light gray) move rapidly within the melted region. (c) Root mean square atomic displacements for all particles in the cluster plotted against their average distance from the cluster's center of mass at $T = 1229$ K (solid core, liquid surface) and $T = 1345$ K (liquid cluster throughout). The solid/liquid boundary at $T = 1229$ K is just about one interatomic distance thick.

FIGURE 13.5 (See color insert following page 25-16.) Atomistic model of the $(ZnS)_{47}$ cluster embedded into a-SiO_2 matrix. Detailed examination of the dynamics identifies six low-frequency modes (three rotational and three cage modes, schematically shown) with energies below 5 meV. (Based on Stoneham, A.M. and Gavartin, J.L., *Mater. Sci. Eng. C*, 27, 972, 2007.)

rotational modes, it is difficult to identify the forces that drive rotation. Heating and thermal expansion do not readily cause rotation. It is possible to use such models to study heat transfer between a hot dot and a cool matrix, and calculations show that vibrational coherence across the boundary can be one of the factors in the energy transfer across the boundary.

The second possibility might involve photochemical effects on adsorbed species (e.g., H_2O or C oxidation) for a dot on a substrate. If there is some well-defined asymmetry, the dot may rotate, possibly because reaction products leaving the surface transfer momentum to the dot.

The third model supposes melting or quasimelting (see Section 13.2.2) in which there is melting only on an outer surface or the interfacial layer, with an unmelted core. If so, the diffraction peak should not disappear, rather it might drop in intensity. Asymmetric (local) melting might cause rotation, as in the second model, though the diffraction pattern is expected to streak before it vanishes in a liquid. Also an acoustic (thermal) mismatch between dot and host should be important: big differences between densities or elastic constants will keep the dot hot for longer, with a dependence on the geometric match of dot and matrix structures. The switch-off time should depend on the excitation rate. Melting could happen in metallic dots, as sometimes observed, as well as in semiconducting or insulating ones.

The fourth model, not relevant for metals, presumes a change of geometry driven by charge transfer. Thus, (Stoneham and Harker 1999, unpublished) the electron beam causes charge transfer within the (nonmetallic) dot, analogous to some of the charge transfers inferred from spectroscopy. This shift of charge causes ionic polarization within the dot, affecting the diffraction peak. Essentially, the charge transfer transition takes the dot into a metastable state that could survive for a reasonable amount of time. Simple molecular dynamic models show that this process should work for very small dots of a few tens of ions.

Structural changes are observed even for lower energy (sub-band gap) optical excitations (Itoh and Stoneham 2001, Stoneham 2003b). A 2.33 eV light causes an orthorhombic to cubic transformation in CdS (Yakovlev et al. 2000). When ionization of the nanoparticle occurs, as in some molecular beam experiments, a Coulomb explosion can be observed. For a typical "Coulomb explosion," the presence of two holes and less than some critical number of atoms, roughly 20 molecular units for NaI, 30 Pb atoms, or 52 Xe atoms, are observed (Sattler et al. 1981).

13.2.4 Intermittency and Luminescence

Ensembles of nanoparticles under continuous excitation behave much as expected. They exhibit steady fluorescence whose intensity decays in time as the excitation ceases. Experiments on individual nanoparticles reveal unexpected and intriguing jumps in fluorescence intensity. The particle ceases to luminesce for a period of seconds and then returns to the on-state. Such blinking is observed in single molecules, polymers, and proteins, as well as in semiconductor nanoparticles, nanorods, and nanowires (Frantsuzov et al. 2008).

The spectroscopy of II–VI quantum dots lies outside the scope of our discussion of dynamics. However, we need to recall that the small stoichiometric II–VI dots (such as the $(ZnS)_{47}$ dots discussed by Gavartin and Stoneham 2003) do have spherical symmetry: virtually all charge-neutral dots with a zincblende structure have a dipole moment. Experimental spectroscopy, however, gives insight into the dynamical processes and their rates. Many studies of II–VI dots (e.g., Delerue et al. 1995, Nirmal et al. 1995, Klimov et al. 1999) show intermittency. The dark periods have characteristic statistical recurrence periods, which are linked somehow to the statistical shifts with time of the luminescence energy. Structural intermittency (diffraction) and intermittency in luminescence seem to be two separate phenomena.

The common explanation for intermittent luminescence resides in an Auger process. Double excitation of a dot (producing 2e + 2h) with recombination of one electron–hole pair can excite an electron into the surrounding matrix, leaving

a charged dot that has different luminescence behavior from the original neutral one. The dot appears dark until an electron is captured. Looking at the stochastic energy shifts and assuming that these are associated with trapped electrons in the matrix, it seems that these trapped electrons must be very close to the dot–matrix interface. Such changes may damage the nanocrystal irreversibly (Blanton et al. 1996). The behavior can be more complex, for example, Hess et al. (2001) found evidence for a metastable dark state (possibly involving a surface transformation) on heating dots in a solution or by changing the dot environment in other ways, making recovery possible with illumination. Light possessing above the band gap energy causes a dark to bright transformation. Without such light, the dot may remain dark for months, whereas the bright to dark transformation can be fast, perhaps in a few seconds.

Heyes et al. (2007) find good experimental support for a model of Frantsuzov and Marcus (2005) in their work on CdSe/ZnS core shell dots. Frantsuzov and Marcus suggest, in line with the ideas above, that after photoinduced creation of an electron–hole pair, the hole is trapped in a deep surface state of the CdSe core, which is excited by energy from an Auger process. The key energy interval results in a stochastic diffusion, moving in and out of resonance with hole state energy gaps. Heyes et al. suggest that this model explains the power-law behavior of on/off time distributions, the observed exponential cutoff of power-law dependence at long "on" times, and the lack of dependence of blinking kinetics on shell thickness. It also explains why the overall quantum yield observed is governed by the fraction of nonemitting particles in the sample.

13.3 Cycles of Excitation and Luminescence

Optical excitation and de-excitation cycles involve several natural timescales. Optical excitation depends on the optical system and its intensity, and is largely under experimental control. There are natural timescales following excitation that determine operational timescales according to what it is we wish to do, e.g., to provide picosecond optical switch or exploit the altered refractive index (polarizability) in the excited state.

There are several distinct types of subsequent relaxation processes. First, charge redistribution on excitation changes forces on the ions. The system must relax to eliminate surface shear stresses, as vacuum cannot support shear. This (Stoneham and McKinnon 1998) takes a few picoseconds, about the time taken for an acoustic pulse to cross the particle (Itoh and Stoneham 2001 gives an alternative estimate of this timescale). A consequence of this relaxation process is a dynamic dilation: the volume change is roughly independent of dot size, so the fractional change (dilational strain) is inversely proportional to dot volume. This strain, and hence energy shifts as a result of deformation potential coupling, are inversely proportional to dot volume and can be significant. Eliminating shear stress does not need energy redistribution, but involves mainly changing the mean atomic positions about which the system oscillates. The dipole moment is *reduced* in the excited state (Stoneham and Gavartin 2007) as the electron associates with the more positive regions and the hole with the more negative regions. Energy redistribution is a second stage. "Cooling" processes (loss of energy from coherent motion in the configuration coordinate) compete with luminescence, nonradiative transitions, and further possible electronic transitions, such as those into so-called dark states.

13.3.1 Optical Excitation Cycle: Cooling

We should distinguish two types of cooling following excitation. One type of cooling establishes equilibrium among the different vibrational modes of the dot itself. The other type takes energy from the dot as a whole, as the dot equilibrates with its surrounding matrix. Slower cooling is expected for dots resting on a substrate than for those embedded in a matrix, simply because of lower thermal contact. Even if a phonon temperature is established within a dot, it may differ from that of the surroundings, as in the spatial phonon bottleneck discussed by Eisenstein (1951). As noted above, the amount of energy can be quite large: a 2 eV photon absorbed by a dot of 100 atoms can give added energy per atom up to a temperature rise of 232°C. As a result, some modes are more strongly excited (higher effective temperature) than others. Even when light is emitted, there will be some cooling in the excited state before emission, and in the ground state after emission.

In a dot, the phonon system may equilibrate only slowly, i.e., exhibit a spectral bottleneck as energy is exchanged with what is called the configuration coordinate, in analogy with color center studies. The configuration coordinate is a reaction coordinate, not usually a normal mode (see Itoh and Stoneham 2001, p. 90), and it describes the vibrational relaxation toward equilibrium associated with coupling to the excitation. This gives a second class of cooling, largely internal to the dot.

Experimentally, hot luminescence can be identified (Tittel et al. 1997, Stoneham 1999, unpublished), since the luminescence spectrum looks like a zero-phonon line of energy $\hbar\omega_0$ with sidebands. Suppose we describe the dot vibronic behavior with a configuration coordinate diagram, the ground state having a characteristic vibration frequency ω_g and the excited state a frequency ω_x. Emission occurs from excited electronic states with n_x phonons (using the word "phonon" for clarity, even though we do not strictly have a normal mode) to the ground electronic state with n_g phonons. This luminescent transition now has the energy $\hbar\omega_0 + n_x\hbar\omega_x - n_g\hbar\omega_g$. Energies lower than $\hbar\omega_0$ correspond to transitions that result in a vibrationally excited ground state; energies higher than ε_0 correspond to transitions from vibrationally excited initial states, i.e., hot luminescence. The relative importance of hot luminescence gives a measure of the transient temperature of the dot.

Analysis of data for small CdS dots (Stoneham 1999, unpublished) supports this description. Unrefined analysis of the sideband structure suggests a ground state phonon energy of ~32 meV, with the higher value ~35 meV in the excited state; both energies are fairly close to the 40 meV bulk LO phonon energy. The degree of thermal excitation at the time of luminescence is consistent with an energy input proportional to laser intensity, and with cooling at an independent rate, so the dots did not cool instantly to the matrix temperature. Nonoptimized analysis suggests temperature rises of the order 100°C. The zero phonon line has contributions from all components with $n_x = n_g$, and hence there is broadening and an energy shift as different components become important. In this case, this part of the shift would be to the blue line, as $\hbar\omega_x > \hbar\omega_g$; in addition, there is a red shift from thermal expansion, which dominates in these data. The Huang–Rhys model (Huang and Rhys 1950) also predicts changes in sideband intensities (see, e.g., Stoneham (1975) and Chapter 10 for the relevant formulae).

13.3.2 Electron–Phonon Coupling and Huang–Rhys Factors

For the data just described, the Huang–Rhys factors would be of the order 0.1–0.5, similar to other published data (Woggon 1997). Thus a typical Stokes shift might be ~0.1 eV, in line with a Huang–Rhys factor of 0.3 or so. We stress that this analysis makes no assumptions about the nature of the electronic excited state, whether effective mass or charge transfer states. There are various predictions of Huang–Rhys factor S as a function of dot radius, mostly for very simple initial and final wave functions, and bulk-like lattice vibrations and electron–phonon couplings (e.g., Fedorov and Baranov 1996). Few workers (such as Vasilevskiy's (2002) treatment of dipolar vibration modes) recognize the subtle but significant changes at the nanoscale, partly because of the boundary conditions. Fröhlich coupling to bulk-like longitudinal optic modes is an assumption, as is the neglect of deformation potential and piezoelectric couplings to acoustic modes. Simple analytical calculations can be generalized (Ridley et al. 2002, unpublished, following Ridley 2000 and Stoneham 1979). These show S to depend on the form factors of the initial and final electronic states. An important distinction arises between states for which the boundary determines the wave function dimensions (e.g., when the exciton radius exceeds the dot radius) and those for which a local interaction is dominant (e.g., a deep defect). S also depends on the wave-vector dependence of the electron–phonon coupling. In the most useful cases (unscreened piezoelectric coupling, small dot radius R), the dependences are roughly $S \sim 1/R$ (Frohlich), $1/R^2$ (deformation potential), and R^0 (piezoelectric). Depending on details, any one of these dependences can dominate. So, when resonant Raman data suggest $S \sim 1/R$ (e.g., Baranov et al. 1997) and photoluminescence data likewise (CuCl dots in glass (Itoh et al. 1995) and CuBr (Inoue et al. 1996)), there could be several interpretations. As yet, there are no calculations of Huang–Rhys factors at the level of the analysis discussed in Section 13.2.2.

The magnitude of the Huang–Rhys factor is sensitive to the boundary conditions. First-principles electronic structure calculations of the $(ZnS)_{47}$ cluster in vacuum (Figure 13.2) give rather similar ground and excited (triplet) state relaxation energies of 80 and 110 meV, respectively; hence a Stokes shift of ~190 meV. Given the bulk LO phonon energy of ~48 meV, one obtains $S \sim 2$, i.e., intermediate rather than weak coupling (Yoffe 2001). The hole component of the exciton in the $(ZnS)_{47}$ cluster of Figure 13.2 is strongly localized at the surface (Stoneham and Gavartin 2007), and so the environment surely affects exciton relaxation and the Huang–Rhys factor.

13.4 Where the Quantum Enters: Exploiting Spins and Excited States

We turn to an example of electron and spin dynamics that is intrinsically nanoscale. Our example is one route to the storage and manipulation of quantum information for quantum information processing. The underlying processes of the Stoneham–Fisher–Greenland proposal (SFG; Stoneham et al. 2003) specifically exploit properties of impurities in silicon or silicon-compatible hosts. Electron spins, used as qubits, are distributed *randomly* in space such that mutual interactions are small in the normal (ground) state when they store quantum information. In an electronic excited state, entangling interactions between qubits allow manipulation of pairs of qubits by magnetic fields and optical pulses. So what does this illustrate with regard to dynamics at the nanoscale?

First, quantum information processing needs dynamic and coherent manipulation of the spins: all the quantum manipulations must be done faster than decoherence processes. In the SFG approach, decoherence arises primarily from spontaneous emission, photoionization, spin lattice relaxation, and loss of quantum information to nonparticipating spins. Secondly, a characteristic range over which it is possible to entangle two spins exists. Donors in their ground states should be too far apart to interact, yet an excited control electron must overlap two qubits through a shaped optical pulse to give a transient interaction. These constraints give a length scale ~10 nm. However, the wavelength of light (say 1000 nm, or 1 μm) is so long that one cannot focus on just one chosen pair of qubits. The laser system can focus on (say) one square micron. To address individual gates, the use of both spatial and spectroscopic selectivity is needed. The natural disorder and spatial randomness in doped semiconductors is crucial, and even the steps of the silicon surface are useful. Simply because the spacings of the donors and control dopants are random, the excitations to manipulate qubits will have different energies from one qubit pair to another. Randomness is beneficial. These ideas put a limit on the number of qubits in one square micron that can be linked from the spectral bandwidth available. With sensible values, this would be about 20 qubits. This would enable a linking of, say, 250 qubits. Can this be done? Imagine "patches" of, say, 20 gates in a small zone (say 100 nm) of each micron-sized region. Can quantum information be transferred from one patch to another as a "flying qubit"? If practical—and there are proposals—then a linked set of say 12 patches, each containing 20 qubits, would give 240 qubits. The architecture would, however, have implications for efficient algorithms.

If there is to be widespread public use of quantum information processing, the room-temperature processor will have to work alongside conventional classical devices. Any quantum information processor will be controlled by classical microelectronic devices. So the quantum device must link well with the silicon technology that dominates current information processing. Classical silicon technology continues to evolve in a truly impressive way. It will not be replaced by quantum information processing. Instead, quantum behavior will extend its possibilities. There are strong reasons to look for silicon-based quantum information processors, like the SFG scheme. The optically controlled SFG quantum gates do not rely on small energy scales, so might function at or near room temperature, if decoherence mechanisms permit. Quantum behavior is not intrinsically a low-temperature phenomenon, as we emphasize in Section 13.5. Quantum behavior is displayed in two main ways. In quantum statistics, the quantal $\hbar$ appears in combinations like $\hbar\omega/kT$, so high temperatures make quantal effects less and less evident. But statistics relate primarily refer to equilibrium behavior. In quantum dynamics, $\hbar$ appears without T, and the quantum role may be to open new channels. Quantum information processing relies on dynamics and staying far from equilibrium. There is no *intrinsic* problem with high temperatures. Practical issues may be another matter, of course, since the rate of approach to equilibrium tends to be faster at higher temperatures.

13.5 Scent Molecule: Nasal Receptor

Nanoscience encompasses both physical and biological systems. Our example shows behavior that combines the nanoscale and quantum effects in a biological system at ambient temperatures. In many life processes, molecules interact with highly specific and selective receptors. The actuation of these receptors initiates important biophysical phenomena. The molecules might be small molecules, neurotransmitters, like NO or serotonin or steroids, or large molecules, like many enzymes. For larger molecules, shape (in some general sense, including distribution of adhesive patches) is a major factor. It is almost a mantra that there is a "lock and key" mechanism in which shape is the only significant factor in selectivity, despite lack of clarity about the activation step (the key needs to be turned in a human-scale lock). For small molecules, while shape may be necessary, it is manifestly not sufficient. Some extra feature is needed to understand what actuates the receptor once the molecule has arrived. One idea (Stoneham 2003a, unpublished) is a "swipe card" model: your human scale swipe card (credit card or keycard) has to fit well enough, but it is something other than shape (often in the magnetic strip for swipe cards) that transfers information and actuates the system. In the swipe card model, there is a natural actuation event, e.g., electron or proton transfer. At the molecular scale, it might be *proton transfer* (as seems likely for serotonin, Wallace et al. 1993), or *inelastic electron tunneling* for scent molecules, as suggested by Turin (1996). So what would be the processes in Turin's model of olfaction (Brookes et al. 2007)?

Scent molecules have to be volatile, and so are small, rarely more than 50 atoms in size, and clearly nanoscale. Each molecule may interact with a number of different receptors, from whose signals the brain discerns a particular scent. Shape alone is inadequate to explain the existence of olfactants that are structurally essentially identical, yet smell different (e.g, ferrocene smells spicy, whereas nickelocene has an oily chemical smell). Further, olfactants that smell the same can be quite different in shape. Turin realized that olfactants that smell the same, even if chemically very different (e.g., decaborane and hydrogen sulfide), have similar vibrational frequencies, and made the imaginative proposal that nasal receptors might exploit inelastic tunneling to recognize different vibrational frequencies. Thus, a receptor might have a donor component and an acceptor component (Figure 13.6a). Without the olfactant molecule, no tunneling occurs, partly because the tunneling distance is too large and partly because of energy conservation, as the energies do not match. With an olfactant present (Figure 13.6b), inelastic tunneling conserves energy by exciting an odorant vibration of definite energy (Figure 13.6c); there is still no elastic channel

that conserves energy. This simplified outline of a sequence of processes has been analyzed in a quantitative model by Brookes et al. (2007).

Some of these ideas can be checked by full-scale electronic structure calculations on the small olfactant molecule itself. Receptor structures are not known with any certainty unfortunately, certainly not to better than ~2 Å, whereas tunneling may be sensitive to changes of 0.1 Å. However, there are well-defined constraints, such as the time between exposure to an odorant and its detection, and there is also information from other biomolecules. Turin's model needs no special electronic resonances of receptor and molecule. Brookes et al. showed that there is nothing *un*physical in the model, i.e., the Turin model should work with sensible values of all parameters, and that it was robust, in the sense that there was quite a range of parameters that would work. Their detailed analysis suggests interesting features of the receptor that warrant further attention and experiments.

There are various clear challenges to the Turin theory. Shouldn't there be an isotope effect, as changing H for D would alter frequencies? This is still controversial. Some authors say there is no difference; others say that humans, dogs, and rats can discern isotope differences. There are experimental difficulties as well, since there can be isotope exchange and other isotope-related reactions in the nose, and the definitive experiment has

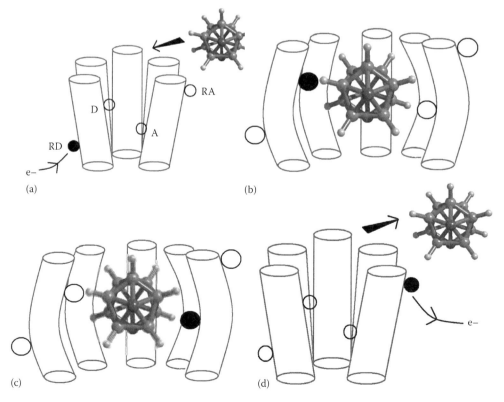

FIGURE 13.6 Schematic illustration of the Turin mechanism. (a) The olfactory receptor, a barrel structure formed from polypeptide chains. Inelastic tunneling occurs between donor D and acceptor A. The two "reservoirs" RD and RA ensure that there is an electron on D and that the electron is removed after tunneling to A. (b) The scent molecule enters the receptor, deforming it. (c) Inelastic electron tunneling occurs with the excitation of a vibration of the scent molecule. (d) The scent molecule leaves and the system re-initializes. (Based on Brookes, J.C. et al., *Phys. Rev. Lett.*, 98, 038101, 2007.)

not been done. Shouldn't enantiomers (chiral odorants with left- and right-handed forms) smell the same, since their vibrations are the same (whereas shape theories would say all should smell different)? Experimentally, the extensive Leffingwell (2001) lists suggest about half such pairs smell the same, and about half smell different. Within the Turin picture, those that smell different do so because there are different intensities from left- and right-handed forms, and these different intensities are determined in part by shape factors. A detailed analysis (Brookes et al. 2009) shows, surprisingly, that enantiomer pairs of molecules that are dynamically flexible (exploring char, boat, and twist geometries at room temperature) can be distinguished, whereas rigid ones cannot be. This result is surprising, since one might expect flexible molecules could wriggle into receptors an enable left- and right-handed forms to smell the same. The whole story is not yet clear.

Interestingly, many of the processes (electron transport, electron tunneling) are relatively slow compared with the time to discern a scent, no faster than a millisecond. But all involve dynamics at the nanoscale. Indeed, the swipe card description—which is a new paradigm for receptor processes, with possibly very wide application—is naturally dynamic and nanoscale.

13.6 Conclusions

Behavior at the nanoscale presents some generic challenges (Stoneham 2003a,b). The first challenge is to identify just what are the most important scientific ingredients. The temptation is to assume that the significant questions are the familiar questions. The second challenge is how to bring together a mix of computer-based, analytical, and statistical theories to address these key issues. The temptation for those used to macroscopic theory is to believe nanoscience is miniaturized macroscience; for those used to the atomic scale, the temptation is to believe that it suffices to extend familiar atomistic ideas. The third challenge is how to understand the link between structure and performance. The temptation is to believe that structures that look alike will actually behave alike, when even one extra atom can make a difference. But perhaps the fourth challenge is the most important: process is more significant than structure. Structures are not validated by appearance alone, but by how they perform. Knowledge of ground-state energies for idealized systems, crystal structures, and surface reconstructions is only a beginning. Dynamics is an unavoidable ingredient at the nanoscale, whether the movement is electronic, a near-equilibrium fluctuation, or a subtle biological process. Our examples have aimed to illustrate the range of dynamic phenomena and, in particular, to identify cases where there are surprises. If we were to identify themes that we regard as especially important in the next stages of nanoscale science, then we would note four personal choices. The first theme involves the ways in which living organisms exploit hard and soft matter with such ingenuity. Our example of olfaction attempts to understand one example of a remarkable biological phenomenon. A second theme might be the exploitation of selective electronic excitation: the use of spatial and spectral resolutions together for low thermal budget nanoprocessing, as well as for quantum information processing. A third theme is the significance of coherence, whether vibrational, electronic, or quantum. The final theme, and the main thrust of this paper, is the need to recognize that, at the nanoscale, dynamics rather than structure dominates behavior.

Acknowledgments

We gratefully acknowledge comments, suggestions, and practical assistance of our colleagues Gabriel Aeppli, Polina Bayvel, Ian Boyd, Jenny Brookes, Mike Burt, Andrew Fisher, Thornton Greenland, Tony Harker, Filio Hartoutsiou, Sandrine Heutz, Andrew Horsfield, Christoph Renner, Brian Ridley, and Luca Turin. This work was funded in part through EPSRC grants GR/S23506 and GR/M67865EP69 and the IRC in Nanotechnology.

References

Accelrys. 2008. *Accelrys: DMol³ and GULP Is a Part of Materials Studio Environment*. San Diego, CA: Accelrys Inc.

Baranov A V, S Yamauchi, and Y Masumoto. 1997. Exciton–LO-phonon interaction in CuCl spherical quantum dots studied by resonant hyper-Raman spectroscopy. *Phys. Rev. B* 56: 10332–10337.

Blanton S A, M A Hines, and P Guyot-Sionnest. 1996. Photoluminescence wandering in single CdSe nanocrystals. *Appl. Phys. Lett.* 69: 3905–3907.

Brookes J C, F Hartoutsiou, A Horsfield, and A M Stoneham. 2007. Can humans recognize odor by phonon assisted tunneling? *Phys. Rev. Lett.* 98: 038101.

Brookes J C, A Horsfield, and A M Stoneham. 2009. Odour character differences for enantiomers correlate with molecular flexibility. *J. R. Soc.: Interface* 5 10.1098/rsif.2008.0165 (print version 2009, 6: 75–86).

Buffat P. 2003. Dynamical behaviour of nanocrystals in transmission electron microscopy: Size, temperature or irradiation effects. *Phil. Trans. R. Soc. Lond. A* 361: 291.

Delerue C, M Lannoo, G Allan, E Martin, I Mihalcescu, J C Vial, R Romestain, F Muller, and A Bsiesy. 1995. Auger and Coulomb charging effects in semiconductor nanocrystallites. *Phys. Rev. Lett.* 75: 2228–2231.

Delley B. 2000. From molecule to solids with DMol³ approach. *J. Chem. Phys.* 113: 7756–7764.

Eisenstein J. 1951. Size and thermal conductivity effects in paramagnetic relaxation. *Phys. Rev.* 84: 548–550.

Fedorov A V and A V Baranov. 1996. *Sov. Phys. JETP* 83: 610.

Ferrando R, J Jellinek, and R L Johnston. 2008. Nanoalloys: From theory to applications of alloy clusters and nanoparticles. *Chem. Rev.* 108: 845–910.

Frank, F C. 1952. Crystal growth and dislocations. *Adv. Phys.* 1: 91–109.

Frantsuzov P and R A Marcus. 2005. Explanation of quantum dot blinking without the long-lived trap hypothesis. *Phys. Rev. B* 72: 155321.

Frantsuzov P, M Kuno, B Janko, and R A Marcus. 2008. Universal emission intermittency in quantum dots, nanorods and nanowires. *Nat. Phys.* 4: 519–522.

Fritsch J and P Pavone. 1995. Ab initio calculation of the structure, electronic states, and the phonon dispersion of the Si(100) surface. *Surf. Sci.* 344: 159–173.

Gale, J D and A L Rohl. 2003. The General Utility Lattice Program (GULP). *Mol. Simul.*, 29: 291–341.

Gavartin J L and A L Shluger. 2006. Ab initio modeling of electron-phonon coupling in high-k dielectrics. *Phys. Status Solidi (C)* 3: 3382–3385.

Gavartin J L and A M Stoneham. 2003. Quantum dots as dynamical systems. *Phil. Trans. R. Soc. Lond. A* 361: 275–290.

Hess B C, I G Okhrimenko, R C Davis, B C Stevens, Q A Schlulzke, K C Wright, C D Bass, C D Evans, and S L Summers. 2001. Surface transformation and photoinduced recovery in CdSe nanocrystals. *Phys. Rev. Lett.* 86: 3132–3135.

Heyes C D, A Yu Kobitski, V V Breus, and G U Nienhaus. 2007. Effect of the shell on the blinking statistics of core-shell quantum dots: A single-particle fluorescence study. *Phys. Rev. B* 75: 125431.

Huang K and A Rhys. 1950. Theory of light absorption and non-radiative transitions in F-centres. *Proc. R. Soc. Lond. Ser. A* 204: 406–423.

Inoue K, A Yamanaka, K Toba, A V Baranov, A A Onushchenko, and A V Fedorov. 1996. Anomalous features of resonant hyper-Raman scattering in CuBr quantum dots: Evidence of exciton-phonon-coupled states similar to molecules. *Phys. Rev. B* 54: R8321–R8324.

Ishli D, K Kinbara, Y Ishida, N Ishil, M Okochi, M Yohda, and T Alda. 2003. Chaperonin-mediated stabilization and ATP-triggered release of semiconductor nanoparticles. *Nature* 423: 629–632.

Itoh N and A M Stoneham. 2001. *Materials Modification by Electronic Excitation*. Cambridge, U.K.: Cambridge University Press.

Itoh T, M Nishijima, A I Ekimov, C Gourdon, A L Efros, and M Rosen. 1995. Polaron and exciton-phonon complexes in CuCl nanocrystals. *Phys. Rev. Lett.* 74: 1645–1648.

Jellinek, J and A Goldberg. 2000. On the temperature, equipartition, degrees of freedom, and finite size effects: Application to aluminium clusters. *J. Chem. Phys.* 113: 2570–2582.

Klem M T, M Young, and T Douglas. 2005. Biomimetic magnetic nanoparticles. *Mater. Today* 8: 28, doi:10.1016/S1369-7021(05)71078-6.

Klimov V I, Ch J Schwarz, D W McBranch, C A Leatherdale, and M G Bawendi. 1999. Ultrafast dynamics of inter- and intra-band transitions in semiconductor nanocrystals: Implications for quantum-dot lasers. *Phys. Rev. B* 60: R2177–R2180.

Koch A J and H Meinhardt. 1994. Biological pattern formation: From basic mechanisms to complex structures. *Rev. Mod. Phys.* 66: 1481–1507.

Leffingwell J C. 2001. Leffingwell Reports. *Leffingwell Reports* 5: 1. Available: http://www.leffingwell.com/

Makarov G N. 2008 Cluster temperature. Methods for its measurement and stabilization. *Physics Uspekhi* 51: 319–353.

Meinhardt H. 1992. Pattern formation in biology: A comparison of models and experiments. *Rep. Prog. Phys.* 55: 797–850.

Nirmal M, D J Norris, M Kuno, and M G Bawendi, Al L Efros, and M Rosen. 1995. Observation of the "Dark Exciton" in CdSe quantum dots. *Phys. Rev. Lett.* 75: 3728–3731.

Ridley B K. 2000. *Quantum Processes in Semiconductors*. Oxford, U.K.: Oxford University Press.

Ridley B K, A M Stoneham, and J L Gavartin. 2002. Unpublished.

Roduner E. 2006. *Size Dependent Phenomena*. Cambridge, U.K.: The Royal Society of Chemistry.

Sattler K, J Mühlbach, O Echt, P Pfau, and E Recknagel. 1981. Evidence for Coulomb explosion of doubly charged microclusters. *Phys. Rev. Lett.* 47: 160–163.

Screbtii A I, R Di Felice, CM Bertoni, and R Del Sole. 1995. *Ab initio* study of structure and dynamics of the Si(100) surface. *Phys. Rev. B* 51: 1204–11201.

Stoneham A M. 1965. Paramagnetic relaxation in small crystals. *Solid State Commun.* 3: 71–73.

Stoneham A M. 1975. *Theory of Defects in Solids*. Oxford, U.K.: Oxford University Press.

Stoneham A M. 1979. Phonon coupling and photoionisation cross sections in semiconductors. *J. Phys. C: Solid State Phys.* 12: 891–897.

Stoneham A M. 1999. Unpublished analysis.

Stoneham A M. 2003a. Unpublished.

Stoneham A M. 2003b. The challenge of nanostructures for theory. *Mater. Sci. Eng.* C23: 235–241.

Stoneham A M. 2007. How soft materials control harder ones: Routes to bio-organisation. *Rep. Prog. Phys.* 70: 1055–1097.

Stoneham A M and J L Gavartin. 2007. Dynamics at the nanoscale. *Mater. Sci. Eng. C* 27: 972–980.

Stoneham A M and J H Harding. 2003. Not too big, not too small: The appropriate scale. *Nat. Mater.* 2: 77–83.

Stoneham A M and A H Harker. 1999. Unpublished.

Stoneham A M and B McKinnon. 1998. Excitation dynamics and dephasing in quantum dots. *J. Phys.: Condens. Matter* 10: 7665–7677.

Stoneham A M, A J Fisher, and P T Greenland. 2003. Optically-driven silicon-based quantum gates with potential for high temperature operation. *J. Phys.: Condens. Matter* 15: L447–L451.

Sutton A P and J Chen. 1990. Long range Finnis Sinclair potentials. *Phil. Mag. Lett.* 61: 139–146.

Tittel J, W Gohde, F Koberling, T Basche, A Kornowski, H Weller, and A Eychmuller. 1997. Fluorescence spectroscopy on single CdS nanocrystals. *J. Phys. Chem. B* 101: 3013–3016.

Tran T K, W Park, W Tong, M M Kyi, B K Wagner, and C J Summers. 1997. Photoluminescence properties of ZnS epilayers. *J. Appl. Phys.* 81: 2803–2809.

Turin L. 1996. A spectroscopic mechanism for primary olfactory reception. *Chem. Senses* 21: 773–791.

Vasilevskiy M I. 2002. Dipolar vibrational modes in spherical semiconductor quantum dots. *Phys. Rev. B* 66: 195326–195335.

Wallace D, A M Stoneham, A Testa, A H Harker, and M M D Ramos. 1993. A new approach to the quantum modelling of biochemicals. *Mol. Simul.* 2: 385–400.

Williams P, E Keshavarz-Moore, and P Dunnill. 1996 Efficient production of microbially synthesized cadmium sulfide quantum semiconductor crystallites. *Enzyme Microb. Technol.* 19: 208–213.

Woggon U. 1997. *Optical Properties of Semiconductor Quantum Dots.* Berlin, Germany: Springer.

Yakovlev V V, V Lazarov, J Reynolds, and M Gajdardziska-Josifovska. 2000. Laser-induced phase transformations in semiconductor quantum dots. *Appl. Phys. Lett.* 76: 2050–2052.

Yoffe A D. 2001. Semiconductor quantum dots and related systems: Electronic, optical, luminescence and related properties of low dimensional systems. *Adv. Phys.* 50: 1–208.

14

Electrochemistry and Nanophysics

Werner Schindler
Technische Universität München

14.1 Introduction

The increasing miniaturization of devices and the importance of nanoscopic structures for specific functionalities in many prospective fields, such as electronics, sensorics, or catalysis, require a detailed knowledge of both basic physical and basic chemical processes on nanometer scales, as well as the availability of reliable fabrication processes suitable to prepare thoroughly tailored nanostructures. Moore's law [1], the semiconductor industry's roadmap for miniaturization of electronic devices, shows, for example, that the structure size of electronic devices approaches already the dimensions of molecules in the near future and is predicted to reach the atomic level in one or two decades (Figure 14.1). The technological effort by the semiconductor industry to reach these goals is huge. Each new generation of smaller structure sizes requires a still more expensive equipment, but fundamental and very difficult to solve problems of the patterning processes remain [2]. Therefore, it is unclear if the traditional top-down nanostructuring technology may be useful for future device generations or if bottom-up technologies are to be developed to fulfill the future demands for integration density and functionality.

Nanostructures in the range of a few nanometers are of fundamental importance for exploiting quantum size effects at room temperature [3,4] or for unprecedented sensitivity in molecular detection [5,6]. They are necessary in the fields of energy conversion and environmental issues for the development of high-efficiency catalyst materials, which are required to produce alternative (in view of fossil) fuels like hydrogen or to convert fuel into electricity making use of fuel cells. The research in this field has shown that the specific catalytic efficiency can be improved significantly when decreasing the particle size down to a few nanometers in diameter. Since many of the currently discussed applications of nanostructures utilize a solid/liquid working environment, the research close to real conditions at solid/liquid interfaces is of great importance.

Besides considerations concerning the benefits of nanostructures for certain functional applications, there are technological issues in the preparation of nanostructures that can be solved only by electrochemistry: Sputter or evaporation processes do not allow for a decent metal deposition into prestructured holes of large aspect ratio, i.e., small diameter and large height, when the diameter is smaller than a few tens of nanometers. This feature, however, is required for the fabrication of electrically conducting interconnect lines in electronic chips from one conducting layer to another through vertical interconnects (VIAs). Here, the so-called electrochemical superfilling technology offers the unique possibility to fabricate reliably the vertical interconnects in electronic devices. It utilizes additives to the electrolyte, which control precisely where the metal is deposited in the hole, and guarantee a perfect metal growth from bottom to top of the hole. Overgrowth by metal at the top of the hole, resulting in a nonconducting spot in the inner of the hole, can be completely avoided [7,8].

Another example for exploiting the unique features of electrochemistry is the fabrication of magnetic disc read/write heads and their coils, or heads consisting of materials showing a giant magnetoresistance, which are difficult to fabricate by vacuum-based evaporation or sputtering techniques. They have been the precondition to achieve the performance of today's hard disk drives found in each computer [7,8].

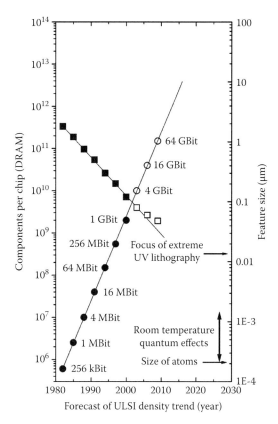

FIGURE 14.1 Prediction of the size and density of ultralarge-scale integrated (ULSI) memory devices according to Moore's law. The feature sizes of molecular electronic devices, or structures showing room temperature quantum effects, will be well below the focus of extreme UV lithography, which is required for the production of 20 nm structure sizes. (From Hugelmann et al., *Surf. Sci.*, 597, 156, 2005. With permission.)

The latter two examples from existing technologies show that electrochemistry can be applied in mass fabrication processes and is even the basis for further technological progress, since it provides possibilities that cannot be provided by vacuum-based sputter or evaporation techniques.

Another fast growing and prospective field are self-organization processes on molecular scales, which may be possibly useful for molecular electronic circuits, fabricated from bottom-up. They may allow for a functional and tailored architecture of molecules showing novel possibilities in electronics or sensorics.

Thus, it becomes clear why already in 1990 Nobel price winner Rohrer called the electrochemical interface the interface of the future [9].

A significant amount of information on electrochemical processes like corrosion or electroplating is documented already. Most of it has been derived from investigations on large, macroscopic scales. Since the invention of scanning probe techniques by Binnig and Rohrer in 1983 [10], electrochemical processes can be monitored and studied on atomic or molecular scales, which has been a great step forward toward a detailed understanding of the fundamentals of electrochemical processes. Thus, delocalized electrochemical metal deposition processes [11], underpotential deposition (UPD) of metals [12], reconstruction [13],

or adsorption [14] phenomena at solid/liquid interfaces have been well understood during the past two decades.

A further progress has been achieved in the recent years from the feasibility of studies at individual structures at atomic or molecular scales. Such nanoscale investigations eliminate statistical variations over large ensembles of structures, and avoid herewith associated errors in the measurements.

As in ultrahigh vacuum (UHV) physics, the scanning tunneling microscope (STM) plays a dominant role in nanoscale investigations at solid/liquid interfaces. It is nearly the only measurement technique that allows for a (sub-)atomic resolution of nanostructures in "real space." In this chapter, solely electrochemical processes on nanoscales are discussed, not the variety of nanostructuring processes where the scanning tunneling microscopy (STM) is used for a local modification of solid/liquid interfaces at the nanometer scale, utilizing processes different from electrochemical processes. The key point is the interaction between STM tip and substrate surface at a certain distance (gap width), which may range from ten or more nanometers for solely electrochemical processes to less than one nanometer for electron tunneling processes. A consequence of distances in the sub-nanometer range is the very difficult to solve problem of making mechanical and, hence, simultaneously electrical contact between STM tip and substrate surface when the gap is too small. Such contacts result in mechanical damage and disturb the potentials of the tip or/and the substrate surface, resulting in undefined electrochemical conditions.

Therefore, special attention must be given to the aspect whether solely electrochemical, or also other interactions between a tip or nanoelectrode and the adjacent electrode surface determine basically the experiment and its results. This discussion is not at all ridiculous, as can be seen from the history of nanostructure formation with the STM during the past two decades: Various techniques have been tried to prepare nanostructures with the STM tip on substrate surfaces, but most of them utilize mechanical, electrical, or other mainly unknown interactions between STM tip apex and substrate surface underneath [15–20], or a "jump-to-contact" deposition of metal clusters [21–23]. A clear separation of mechanisms for the nanostructuring process is difficult to achieve when the gap width is around or below 0.5 nm, although it would be crucial for defining the basic physical mechanisms involved in the nanostructure formation. So far, only methods with the tip in substantial distance to a substrate surface like the burst-like electrodeposition of metal nanostructures, utilizing a STM tip as an electrochemical nanoelectrode, are based exclusively on electrochemistry [24,25].

Upon discussing the nanoscale electrodeposition and spectroscopy with the STM at solid/liquid interfaces, at the end of this contribution a section on technical aspects of STM in a solid/liquid environment is added. The reason is that measurements on atomic scales and the investigation of modifications on nanoscales due to electrochemical processes require a very sophisticated control of the electrochemical potential at both the substrate and the STM tip and have to consider the influence of the STM tip on the measured STM images. Serious investigations at solid/liquid interfaces utilize a bipotentiostat to control tip and substrate potentials thoroughly. In the section on technical aspects (Section 14.8) particularly the importance of a small tip apex diameter and the electronic bandwidth requirements of the bipotentiostat are addressed. In contrast to UHV STM, there is always the electrochemical double layer at the STM tip and substrate surface/electrolyte interface present. The capacitance associated with this double layer results in a limitation of the potential control. These aspects are neglected in many investigations at solid/liquid interfaces found in the literature. However, they are decisive for the quality of STM investigations at solid/liquid interfaces.

14.2 Solid/Liquid Interface from a Molecular Point of View

In contrast to the solid/ultrahigh vacuum interface, the presence of a liquid in contact with a solid surface results in a variety of important changes. The chemical potential (and the electrochemical potential when a voltage is applied between the solid surface and the electrolyte) must be maintained. Unless there are special electrochemical conditions, i.e., a potential control applied, upon contact, a respective modification of electrolyte or electrode surface occurs by dissolution or deposition processes in order to establish the chemical and electrochemical equilibrium, respectively. Charges at the electrode surface are screened by ions or partially solvated ions opposite to the electrode surface, forming the so-called electrochemical double layer. The most simplified picture developed by Helmholtz is the rigid double layer consisting of charges localized in front of the electrode surface at a certain distance of a few Angstroms and adjacent opposite charges in the electrode surface. This model has been extended by Guy, Chapman, and Stern (see Ref. [26]) toward a more realistic point of view. The electrochemical interfacial layer can be treated by the Poisson equation, which correlates the local charge distribution in the interfacial region with the variation of the electrical potential across the interfacial region, similar to the theoretical treatment of semiconductor space charge regions. In all models of the electrochemical interfacial layer, positive and negative charges are located in a distance to each other, thus resulting in a capacitor arrangement that gives rise to the electrochemical double layer capacity.

The fact that there are water dipoles, ionic species, or molecules in the electrolyte and in contact with the electrode surface does not interfere with cleanliness considerations, unless these species or their concentration is unknown. Admittedly, it is easy to perform experiments in an undefined environment, i.e., in polluted electrolytes, since there is not a simple indicator like the vacuum pressure in UHV experiments. Therefore, an often discussed issue is cleanliness, which can be achieved in an electrochemical environment. In comparison to solid/ultrahigh vacuum (UHV) interfaces, the cleanliness conditions are comparable to the conditions of a vacuum pressure of 10^{-10} mbar or less [27].

Experimental and theoretical information on the detailed properties of solid/liquid interfaces exists, as summarized by various

reports [28,29]. The overwhelming majority of results have been derived from integral measurements at solid/liquid interfaces by, for example, spectroscopic methods (impedance spectroscopy) or x-ray techniques. Beyond the simple picture of an electrochemical double layer or modified interfacial layer as described above, the molecular structure of the interfacial layer in front of a solid electrode surface is much more complicated. In the case of aqueous electrolytes, water molecules (dipoles) form several discrete molecular layers in specific distances in front of a solid surface, which has been proven at first by x-ray scattering [30]. In scanning tunneling microscopy (STM) experiments, the layering of water molecules is measured as a modulation of the tunneling barrier height with the tunneling gap width (Figure 14.2) [31,32], although usually a strictly exponential dependence of the tunneling current on the distance z is assumed:

$$I_{\text{tunnel}} \propto \frac{U_{\text{bias}}}{z} \cdot \exp(-A\sqrt{\Phi} \cdot z) \tag{14.1}$$

Both mentioned techniques provide a high sub-atomic resolution perpendicular to the electrode surface and can, thus, precisely detect structures with modulations perpendicular to the electrode surface. The modulation period corresponds to the diameter of a single water molecule. Tunneling of electrons from a STM tip to an adjacent substrate surface proceeds vacuum-like at small gap widths, and across individual water molecules at gap widths larger than the diameter of a single water molecule.

The substructure in the interfacial layer is potential dependent [30,33] and may influence redox processes accordingly. The dipoles and ionic species reorientate in the interfacial layer

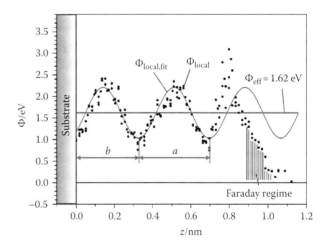

FIGURE 14.2 Dependence of the tunneling barrier height on the distance between STM tip and substrate surface. The zero point of the distance scale has been defined by the point of jump-to-contact. The data points (black dots) are from a series of 12 single experiments, showing the statistical distribution of tunneling barrier heights. Φ_{local} denotes the barrier height values as determined from the current–distance measurements according to $\Phi_{\text{local}} \approx (\text{d} \ln I/\text{d}z)^2 \, \text{Å}^{-2} = (2.302)^2(\text{d} \log I/\text{d}z)^2 \, \text{Å}^{-2}$, using Equation 14.1. Gray lines are fits to the data. Parameter a denotes the modulation period, parameter b denotes roughly the distance of the first water layer from the Au(111) surface.

with the electrode potential, depending on positive or negative electrode surface charges with respect to the point of zero charge (pzc). One may expect, that such discrete interfacial layers at solid/liquid interfaces determine or at least influence charge transfer processes at solid/liquid interfaces in general. Since the substructure in the interfacial layer is a dynamical arrangement of ionic species or water dipoles of high mobility, large statistical fluctuations are possible. Since the detection of such effects requires sophisticated experimental equipment, there are at present very few investigations reported in the literature [31,32].

14.3 Tunneling Process at Solid/Liquid Interfaces

In contrast to the tunneling process in ultrahigh vacuum, the mechanism of electron tunneling from the very last atom of the STM tip apex to the closest adjacent atom of the electrode surface has been unclear for a long time. The low tunneling barrier heights around 1.5 eV, which have been measured in many STM experiments at solid/liquid interfaces have remained basically unexplained. In non-state-of-the-art experiments, tunneling barrier heights even lower than 1 eV [34] may originate also from polluted electrolytes, which provide tunneling states at lower energy levels across the polluting molecules in the tunneling gap, which are nonexistent in clean electrolytes.

Although STM has been performed for nearly two decades, the usual interpretation of STM data is based on the assumption that the tunneling barrier height is laterally constant and independent of the gap width, as it is in ultrahigh vacuum, except for the closest distances where the barrier height decreases with the gap width. Any height variation of the STM tip in the constant current imaging mode is assumed to be caused by a corresponding change of the substrate topography. The overwhelming success of in-situ STM, for example, in the investigation of metal growth processes at solid/liquid interfaces proves the applicability of this assumption for many cases. But the more investigations are performed on molecular or atomic scales at single nanostructures or molecules, the more would be known about the impact of the solid/liquid interface on the tunneling process.

The observation of a modulation in the current–distance curves measured at Au(111) (Figure 14.3), which originates from a modulation of the tunneling barrier height with the gap width (Figure 14.2), indicates that the electronic structure of the tunneling gap varies with the gap width. It is evident that the electronic structure in the gap should be correlated with the molecules arrangement in the gap. There are water molecules, solvated cations and anions, or other molecular species in the electrolyte. From a geometrical point of view, solvated ions may not fit into small tunneling gaps. The diameter of a solvated ion is two times its Pauling radius plus the diameter of two water molecules [35,36]. The diameter of water molecules is between 0.28 and 0.34 nm, the ion showing one of the smallest Pauling radii is Na^+ with a Pauling radius of 0.1 nm. From this, the diameter of solvated Na^+ ions can be calculated to be approximately 0.76 nm. The widely used ClO_4^- or SO_4^{2-} anions in electrolyte solutions show diameters of approximately 1.05 nm

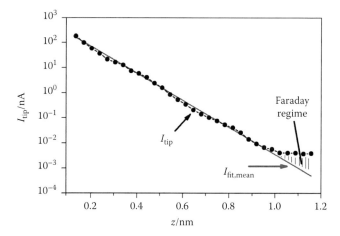

FIGURE 14.3 Modulation of the tunneling current at a Au(111) surface, as measured ba in-situ DTS using a Au STM tip. The Faraday current at the tip is below 10 pA. Black dots: Current–distance curve measured with a tip movement of 6.7 nm s^{-1}; gray line: fit of a straight line to the data according to Equation 14.1, taking only current values between 50 pA and 1000 nA into account. The mean value of an effective tunneling barrier height corresponding to the slope of the fit is $\Phi_{eff} = 1.51$ eV. $E_{WE} = 240$ mV, $E_{tip} = 340$ mV, $U_{bias} = E_{tip} - E_{WE} = 100$ mV in the range $z > 0$, electrolyte: 0.02 M $HClO_4$. Potentials are quoted with respect to the standard hydrogen electrode (SHE).

in their solvated state [35,36]. From these considerations, it can be deduced that anions will hardly fit into a tunneling gap unless it is larger than 1 nm or, correspondingly, unless the tunneling current is smaller than 10 pA, according to Figure 14.3.

Figure 14.4 shows in a sketch drawn to a realistic scale how a modulation of the molecular structure in the tunneling gap can be caused by different configurations of water molecules in the tunneling gap. The orientation of the water molecule dipoles (hydrogen or oxygen atoms pointing toward the electrode surface) depends on the electrode potential [30], but is not decisive for the discussion here. At small tunneling gaps with widths below 0.3 nm there is no space left in the gap for the smallest available molecules which are single water molecules. The result is that the electron tunneling process occurs either across water molecules outside the direct tunneling gap, or across the direct tunneling gap at closest distance of substrate surface and adjacent STM tip apex atoms (Figure 14.4a). In the latter case, tunneling occurs across a vacuum-like tunneling gap.

With increasing gap width, water molecules can penetrate into the tunneling gap. The tunneling barrier height is at its minimum when full multiples of a water layer fit into the tunneling gap (Figure 14.4b). The tunneling barrier height is at its maximum when the gap width corresponds to full multiples plus one half of a single water layer (Figure 14.4c). At these conditions, there is a vacuum-like contribution to the tunneling process across the gap. Thus, the modulation of the barrier height indicates a strong layering of interfacial water at solid/liquid interfaces.

The effect can be observed in the measurements due to the interplay of gap width and barrier height in the expression for the tunneling current (Equation 14.1). The tunneling process

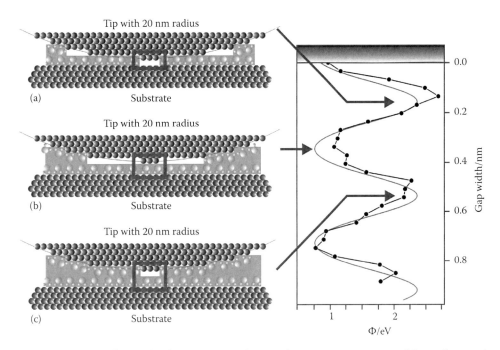

FIGURE 14.4 Schematic illustration of the molecular structure in the tunneling gap at various gap widths, explaining the modulation in the tunneling barrier height across the tunneling gap. The size of the STM tip is drawn to a realistic scale. The water molecules are oriented for electrode potentials of both substrate and STM tip negative to the point of zero charge.

is more likely across the direct gap where the gap width is minimum, even if the barrier height in the direct gap is larger than the barrier height in the areas surrounding the direct gap (Figure 14.4). As far as is known, the tunneling barrier height across water molecules is smaller than the vacuum tunneling barrier height [29], according to the results presented in Figures 14.2 and 14.4, it is approximately 0.8 eV. Considering case (a) in Figure 14.4, the gap across a single layer of water molecules would be approximately 0.2 nm (one atomic metal layer thickness) larger than the direct gap. Without considering the larger area for tunneling across the water molecules (see Figure 14.4), the ratio of the two tunneling current (density) contributions would be according to

$$\frac{I_{\text{directgap}}}{I_{\text{watergap}}} = \frac{\Delta z_{\text{watergap}}}{\Delta z_{\text{directgap}}} \cdot \exp\left(-A \cdot \left(\sqrt{\Phi_{\text{directgap}}}\right.\right.$$
$$\left.\left. \cdot \Delta z_{\text{directgap}} - \sqrt{\Phi_{\text{watergap}}} \cdot \Delta z_{\text{watergap}}\right)\right) \quad (14.2)$$

This ratio is approximately 0.5 for values of $\Delta z_{\text{directgap}} \approx 0.23$ nm, $\Phi_{\text{directgap}} \approx 5$ eV, and $\Delta z_{\text{watergap}} \approx \Delta z_{\text{directgap}} + 0.2$ nm, $\Phi_{\text{watergap}} \approx 0.8$ eV. The direct tunneling current across the vacuum gap increases on the cost of the tunneling current across water molecules outside the direct gap when the tunneling gap width decreases, and direct tunneling becomes less important when the tunneling gap becomes larger.

These considerations show that the detailed geometric configuration of the tip apex is important. Very small tip radii are required to minimize the current across the water molecules

at distances larger than the direct gap width. The parallel path for the tunneling current across the water molecules lowers the achievable lateral resolution. Highest resolution may be achieved when the direct gap is vacuum-like at small gap widths.

Theoretical studies, which have been stimulated by the experimental findings, confirm the significance of the electronic states in water molecules in the gap for the tunneling barrier height [33].

14.4 Electrochemical Processes at Nanoscale

Electrochemical growth and dissolution processes, or electrochemical reduction and oxidation processes in general, are controlled by the potential that is applied across the electrochemical double layer between an electrode (working electrode [WE] or substrate) surface and a reference electrode in the bulk electrolyte.

From a macroscopic point of view, a conducting or semiconducting electrode surface provides a uniform potential at the electrode surface, which controls the redox reactions at the electrode surface. Assuming a redox reaction

$$\text{Red} \Leftrightarrow \text{Ox}^{n+} + n \cdot e^- \quad (14.3)$$

the corresponding Nernst equation becomes

$$E^0 = E^{00} + \frac{RT}{nF} \cdot \ln \frac{a_{\text{Ox}}}{a_{\text{Red}}}. \quad (14.4)$$

It describes the equilibrium potential for the redox reaction when no net current flows across the electrochemical double layer, i.e., no macroscopic reaction occurs, and forward current equals the backward current. a_{Ox} is the activity of the Ox^{n+} ions in the electrolyte, and a_{Red} the activity of the reduced phase. The activity is equal to unity for a three-dimensional (3D) bulk phase. At potentials $E_{WE} > E^0$ (undersaturation conditions, underpotential $\Delta E = E_{WE} - E^0 > 0$), the oxidation process is favored, at potentials $E_{WE} < E^0$ (supersaturation conditions, overpotential $\eta = E_{WE} - E^0 < 0$), the reduction process is favored. The standard potential E^{00} of any electrochemical reaction can be calculated from the corresponding Gibb's free enthalpy ΔG^{00} of this reaction: $E^{00} = (-\Delta G^{00}/nF)$. Standard potentials (at standard conditions 1013 mbar and 25°C) are tabulated for a large variety of chemical reactions [37].

The electrode current is determined by the reaction kinetics and by diffusion processes in the electrolyte during the course of electrochemical reactions. Diffusion is responsible for the transport of ions to or from the electrode surface; migration effects may occur only within the electrochemical interfacial layer around the electrode surface since the potential decays within this layer as discussed above. The reaction current based on the reaction kinetics at an electrode surface is described by the Butler–Volmer equation

$$j = j_0\left(e^{\frac{\alpha_a nF}{RT}\eta} - e^{-\frac{\alpha_c nF}{RT}\eta}\right) \tag{14.5}$$

with the exchange current density j_0, the charge transfer coefficients α_a and α_c for anodic and cathodic direction, respectively, and the overpotential η. The equation describes the current voltage characteristics of an electrode when a net reaction current j flows at the overpotential η, giving rise to either a macroscopic oxidation or reduction process at this electrode surface. A detailed discussion of the current, related to kinetic and diffusion phenomena can be found in Ref. [38].

From detailed studies of electrodeposition and dissolution processes, and corrosion phenomena, in general oxidation and reduction processes at electrode surfaces, it becomes evident that redox processes at surfaces are not only determined by the electrode potential, as suggested by the above discussion and equations, but additionally determined by the specific electrode surface structure. Although the impact of surface defects on nucleation and growth has been known since a long time [11], in particular scanning probe techniques like STM, which allow for an atomically resolved imaging of surfaces in real space [39–41], resulted in a rapid increase of a detailed knowledge about the impact of electrode surface structure on the occurrence of redox processes. There is a comprehensive literature available on detailed studies on various systems, as presented, for example, in the book *Electrochemical Phase Formation and Growth* by Budevski, Lorenz, and Staikov [11].

The atomic modulation of a perfect single crystal lattice plane as the best realized electrode surface provides top, bridge, or hollow sites as positions for adsorbates, molecules, or atoms. In addition, many surfaces show reconstruction phenomena that give rise to a more or less periodic surface structure, and there are unavoidable surface defects, such as step edges, kink sites, vacancies, or screw

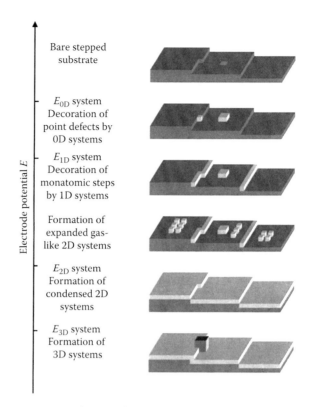

FIGURE 14.5 Substrate inhomogeneities responsible for the delocalized formation of low dimensional systems (LDSs) in the undersaturation range. The underpotential $\Delta E_{iD,\mathrm{system}} = (E - E_{iD,\mathrm{system}}) > 0$ for a phase formation process of cations, and $\Delta E_{iD,\mathrm{system}} = (E - E_{iD,\mathrm{system}}) < 0$ for a phase formation process of anions. E must be replaced on the ordinate by $-E$ for the latter case.

dislocations, to mention only a few (Figure 14.5). Each structural inhomogeneity corresponds to a certain adsorption energy for molecules or atoms, and redox processes occur preferentially at specific surface sites, although the electrode potential is uniformly applied at the electrode surface. The correlation of the nanoscale inhomogeneity in surface structure and redox processes occurring at these inhomogeneities is of greatest interest for electrodeposition processes as well as for redox processes, for example, in the field of catalysis, where precisely taylored catalyst nanoparticles may increase the mass-specific catalytic activity significantly.

A prominent example for surface structure–dependent growth is the decoration of the kink sites of the Au(111) herringbone reconstruction by metal atoms, for example, Co or Ni [42]. In model systems, the type of surface inhomogeneity–dependent growth and dissolution behavior can be observed, for example, in the electrodeposition of Pb on Ag(111) (Figure 14.6). Pb deposition onto the Ag(111) surface, or dissolution from this surface, occurs in the underpotential range at specific potentials depending on the particular sites where the deposition/dissolution process occurs: at point defect sites, step edges, or on flat terraces [43]. These types of inhomogeneities can be classified as zero-dimensional (0D), one-dimensional (1D), or two-dimensional (2D). They are energetically different sites that show a different equilibrium potential E^0 for the corresponding electrochemical

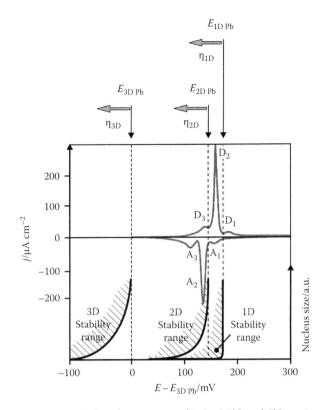

FIGURE 14.6 Cyclic voltammogram of Ag(111)/Pb^{2+} and ClO_4^- anions, showing the formation of iD metal systems ($i = 1, 2, 3$). The typical adsorption and desorption peaks are denoted by A_n and D_n ($n = 1, 2, 3$), respectively. The estimated equilibrium potentials $E_{i\mathrm{D,Pb}}$ of the corresponding iD Pb systems as well as the overpotential regions $\eta_{i\mathrm{D}}$ are indicated in the upper part of the figure. The curves in black illustrate the sizes of the respective iD nuclei as a function of overpotential. (From Lorenz, W.J. et al., *J. Electrochem. Soc.*, 149, K47, 2002. With permission.)

process. These experimental observations led to the formulation of the concept of low-dimensional systems (LDSs), which describes these experimental findings in electrodeposition processes in terms of a variation of the Nernst potential with the type or dimensionality i of particular defects or surface inhomogeneities [43]:

$$E^0_{i\mathrm{D\ system}} = E^{00}_{3\mathrm{D\ system}} + \frac{RT}{nF} \cdot \ln \frac{a_{\mathrm{Ox}}}{a_{\mathrm{Red},\,i\mathrm{D\ system}}} \qquad (14.6)$$

Among the various low-dimensional inhomogeneities, step edges on atomically flat surfaces show often a decoration by metal atoms in the underpotential range. Utilizing this knowledge, pyrolytic graphite (HOPG) surfaces can be used as substrates to grow electrochemically large arrays of metal nanowires at the HOPG step edges [44]. The large atomically flat terraces and the low-achievable defect density on the terraces allow for a nucleation of metal atoms almost exclusively at the HOPG step edges. A precise adjustment of the electrode potential (in the underpotential regime) allows then the decent growth of metal wires of different sizes.

The decoration of metal step edges by metal atoms is difficult to resolve in STM or other scanning probe measurements, since

these techniques provide usually a low contrast between different metals. Decoration effects are in this case rather observed when the charge flowing at a certain potential to or from the electrode surface is determined. Step edge decoration results in a charge peak of the order of μC or less, which is detectable in carefully performed cyclic voltammetry experiments. A nice example is the electrodeposition of Pd onto Au(111), in which the corresponding charge peak can be found [45], but where the Pd growth at Au(111) step edges cannot be seen in STM images. Thus, considering only STM images, and neglecting the influence of nanoscale inhomogeneities on the electrodeposition process, results in wrong results [46].

14.5 Localization of Electrochemical Processes

Electrochemical processes, as discussed in Section 14.4, are always localized in the sense that they depend strongly on the atomic or nanoscale structure of the electrode surface they utilize. This is a more or less passive influence of a certain electrode surface structure on a specific electrochemical redox process.

In this section, electrochemical processes are discussed, which are initiated and proceed only on specific areas of an electrode surface, thus, laterally confined. These may be redox processes at single nanostructures on foreign electrode surfaces, redox processes on lithographically structured electrode surface areas, or redox processes initiated by special electrochemical conditions only in a specific substrate area by an adjacent electrochemical nanoelectrode.

Supposing there is a single nanostructure on a foreign electrode surface, electrochemical redox processes can be initiated at this single nanostructure by a proper adjustment of the electrode potential. An example would be a catalytically active cluster on a catalytically inactive electrode surface as, e.g., Pt on C. The reaction at such a nanostructure is governed by the reaction kinetics as well as by diffusional aspects for the transport of species to and from the active cluster. There is either the possibility to measure the reaction currents integrally over the whole electrode surface, which would require a very high resolution of the measured electrode current, or to measure the reaction current locally resolved with a detector nanoelectrode, which applies the reverse redox reaction for probing the concentration of reaction products from the first redox reaction of interest. The charge transfer during a redox reaction is $k \cdot n \cdot 1.602 \cdot 10^{-19}$ A, with n the number of transferred electrons in each molecule reaction and k the reaction rate. In the first case, this may be difficult to measure for reactions showing low reaction rates since a current resolution of fA is required, assuming k values of the order of 10^4 s^{-1}. In the second case, the measured current and its time dependence are determined by the surface properties of the detector nanoelectrode, by the kinetics of the reverse redox reaction at the detector nanoelectrode, and by diffusion processes in the gap between active nanostructure on the electrode surface and the detector nanoelectrode. A substantial influence of a detector nanoelectrode on the measured results may not be excluded unless there is evidence from modeling that the disturbing influence can be estimated and accounted for in the measurements.

When a nanoelectrode is positioned above a substrate surface in a distance of $\geq 10\,\text{nm}$, the electrochemical double layers of both substrate and nanoelectrode are well separated. At ion concentrations of 10^{-3} to 10^{-1} M, as typically used for electrochemical experiments, the thickness of the electrochemical double layer ranges from less than 1 nm to a few nanometers [47]. The electrochemical processes in such a geometry are mainly determined by the potentials of the two working electrodes, i.e., nanoelectrode and substrate, and by diffusion processes of ions in the electrolyte between the two electrodes. Migration effects need not be considered since the potentials across the solid/liquid interfaces at the electrodes drop down across the electrochemical double layers. A direct local influence of potentials on the deposition process may be reasonable only in the case of overlapping electrochemical double layers from a nanoelectrode and the adjacent substrate surface area.

Due to an appropriate adjustment of over- and undersaturation conditions, electrochemical nanoelectrodes provide the possibility to change or to measure local ion concentrations [24,25,48]. This allows for a local modification of electrochemical conditions at surfaces and to perform a local electrochemistry with a lateral resolution in the nanometer range.

This has been realized in scanning electrochemical microscopy (SECM) [49–53], which utilizes solely electrochemical processes for imaging of surfaces, and can be used to study locally electrochemical processes occurring either at the substrate underneath the SECM tip or in the electrolyte in the gap between substrate and SECM tip. The achievable lateral resolution is typically in the micrometer range due to (1) the metal tip electrode diameter which is variable, but hard to downsize below 100 nm, and due to (2) the diffusion behavior of the electroactive species in the gap between substrate and tip electrode, which are separated by micrometers rather than nanometers. Such a geometry results in diffusion profiles of micrometer width at the position of the tip electrode even if there is a point source at the substrate surface.

In the STM, a tip can be either used as a nanoscale generator electrode, which can release tip material locally into or collect ions locally from the electrolyte surrounding the tip [24,25,54], or used as a local sensor for electrolyte constituents with a spatial resolution of the order of $10^{-15}\,\text{cm}^3$ [48,55]. Using the STM tip as a generator electrode, single metal clusters can be electrodeposited on metal, as well as on semiconducting substrate surfaces. This is a substantial difference to other preparation techniques that work only on specific substrate surfaces, as for example the jump-to-contact mechanism [56].

In addition to the requirements for STM, an electrochemical nanoelectrode must provide a known geometry and a clean surface. This nanoelectrode provides lateral resolution in the nanometer range, which requires nanoelectrode apex diameters also in the nanometer range. A suitable procedure to achieve such electrodes is sputtering of electrodes [57].

The electrochemical deposition of metal clusters onto substrate surfaces can be fully understood by solving the diffusion equation with appropriate boundary conditions. Modeling the STM tip apex as a hemisphere with radius a, neglecting in a first approach the other tip areas exposed to the electrolyte, which is reasonable since these do not contribute much to the ion diffusion to the substrate area opposite to the tip apex, allows for solving the diffusion equation for metal (Me) ions dissolving from the STM tip. The concentration enhancement of Me^{z+} at a particular distance R from the center of the STM tip apex hemisphere and at the time t_0 is given by [58]

$$C(R,t_0) = \frac{a\, j_{Me^{z+}}}{eR\sqrt{\pi D}} \times \int_0^{t_0} \frac{1}{\sqrt{t}} \left\{ \exp\left(-\frac{(R-a)^2}{4Dt} \right) - \exp\left(-\frac{R^2+a^2}{4Dt} \right) \right\} dt$$

(14.7)

Typical diffusion constants for ions in diluted aqueous electrolytes are $D \approx 10^{-5}\,\text{cm}^2\,\text{s}^{-1}$ [59].

The possible enhancement of ion concentration is important for the supersaturation that can be achieved. For reasonable geometries, this enhancement is of the order of 10–20, depending on the distance r.

The major parameters determining the diameter of the growing metal cluster, that is the growth area, are the (1) emission current density $j_{Me^{z+}}$, (2) the STM tip–substrate distance Δz, (3) the substrate potential E_{WE}, and (4) the STM tip apex diameter. The process is discussed in detail in Ref. [58].

This local variation of the metal ion concentration around the STM tip can be exploited to control supersaturation conditions in the volume around the STM tip, and in particular at the surface of a substrate underneath the STM tip. Since STM tips can be easily positioned above any substrate surface, the metal deposition process onto a substrate surface can be controlled laterally resolved by adjusting a particular stationary metal ion concentration in the volume around the STM tip. Thus, this technique provides the possibility to control the metal deposition process precisely by the metal ion concentration rather than by adjusting the electrode potential, which is the same in all substrate surface areas. The reverse process, the generation of undersaturation conditions around the STM tip can be correspondingly achieved by applying an appropriate potential to the STM tip where ions are electrodeposited from the electrolyte onto the STM tip.

In order to achieve a sufficient supersaturation in a small substrate area underneath the STM tip, a special potential routine must be applied at the STM tip (Figure 14.7). Since mostly noble metal STM tips are used in electrochemical environments, in a first step, the STM tip is covered with a layer of metal (Me), which is deposited from the electrolyte around the STM tip, and a nanoelectrode is formed. The second step of the procedure consists then of a dissolution of this metal from the STM tip, resulting in a concentration profile for metal ions around the STM tip and resulting finally in a sufficient supersaturation to initiate nucleation and subsequent growth of a cluster on a particular small substrate surface area underneath the STM tip. To achieve this, the substrate potential must be properly adjusted such that supersaturation conditions are achieved when the

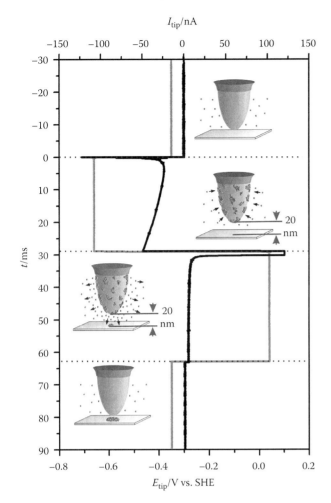

FIGURE 14.7 Potential transient as applied at the STM tip, and current transient as measured at the STM tip upon application of the potential routine during the deposition procedure. In a first step, metal is deposited electrochemically from the electrolyte onto the STM tip. The deposited charge is Q_{cat}. Localized electrodeposition is achieved in a subsequent step by a burst-like dissolution of metal ions from the STM tip, resulting in the generation of supersaturation conditions at the substrate surface underneath the STM tip. Potentials are quoted with respect to the standard hydrogen electrode (SHE). (From Lorenz, W.J. et al., *J. Electrochem. Soc.*, 149, K47, 2002. With permission.)

Me^{z+} ion concentration is increased during the second step of the deposition procedure.

An example of a single Co cluster, which is three atomic layers high, deposited onto a reconstructed Au(111) surface is shown in Figure 14.8. A sequence of STM images before and after deposition and after subsequent dissolution of a single Pb cluster on hydrogen terminated n-Si(111):H is shown in Figure 14.9.

Clusters deposited by localized electrodeposition can be dissolved upon increase of the substrate potential, as shown in Figure 14.9c. Like localized electrodeposition, localized dissolution processes can be also performed by generation of local undersaturation conditions using a STM tip. This can be observed during normal scanning in a STM, when the electrochemical conditions are adequately applied [60].

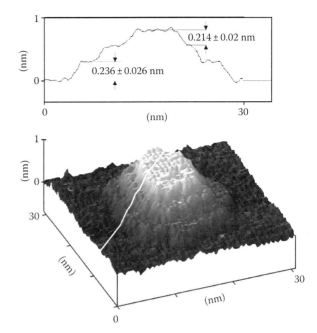

FIGURE 14.8 STM image of a single Co cluster deposited on Au(111) by localized electrodeposition. The fwhm of the cluster is 15 nm as measured by STM. The cluster shows three atomic layers which form a pyramidal shape. Cluster deposition at $E_{WE} = -460\,mV$, $\Delta z = 20\,nm$, cathodic predeposited tip charge $Q_{cat} = 1500\,pC$ which has been fully dissolved during the burst-like dissolution from the STM tip. STM image measured at $I_{tunnel} = 916\,pA$. The step heights are mean values which have been derived from a series of measurements across different clusters and various line profiles across each cluster, as exemplarily indicated by the line profile shown. Electrolyte: 0.25 M Na_2SO_4 + 1 mM $CoSO_4$. Potentials are quoted with respect to the standard hydrogen electrode (SHE). (From Schindler, W. and Hugelmann, P., *Electrocrystallization in Nanotechnology*, G. Staikov, ed., Wiley-VCH, Weinheim, Germany, 2007, p. 117. With permission.)

14.6 Beyond Electrochemical Processes: In-Situ Tunneling Spectroscopy

Spectroscopy using the tip of a STM as probe is a very suitable tool to investigate local electronic properties of surfaces, electronic states, or changes of the work function of surfaces or single particles. The principle is schematically illustrated in Figure 14.10. Due to the atomic resolution of STM, surface inhomogeneities like step edges on atomically flat surfaces, defects, or impurity atoms can be studied with respect to their electronic properties. This has been demonstrated in ultrahigh vacuum, for example, by Eigler and coworkers for the case of magnetic atomic impurities adsorbed on atomically flat surfaces [61].

In an electrochemical environment, electrochemical processes always occur at substrate or tip surface, which result in electrochemical (Faraday) currents. The total tip current measured is then the sum of Faraday and tunneling currents at the STM tip. In order to minimize Faraday currents, the tip has to be properly isolated except for the tip apex, which makes tunneling contact to the adjacent substrate surface. Various scanning

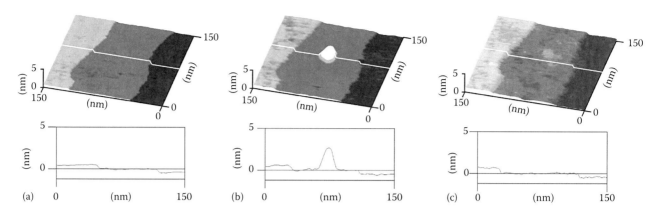

FIGURE 14.9 Localized electrodeposited Pb cluster on n-Si(111): H. (a) Bare n-Si(111): H surface before cluster deposition; (b) same n-Si(111): H surface after deposition of a Pb cluster; (c) same n-Si(111): H surface upon a change of the substrate potential E_{WE} resulting in a dissolution of the Pb cluster. Electrolyte: 0.1 M HClO$_4$ + 1 mM Pb(ClO$_4$)$_2$. Imaging conditions: E_{WE} = −240 mV, E_{tip} = +640 mV, I_{tip} = 200 pA. Cluster deposition parameters: E_{WE} = −240 mV, cathodic predeposited tip charge Q_{cat} = 2000 pC, which has been fully dissolved during the burst-like dissolution from the STM tip, Pb^{2+} ion current during the burst-like dissolution from the STM tip $I_{Pb^{2+}}$ = 120 nA, tip substrate distance during cluster deposition Δz = 20 nm. Potentials are quoted with respect to the standard hydrogen electrode (SHE).

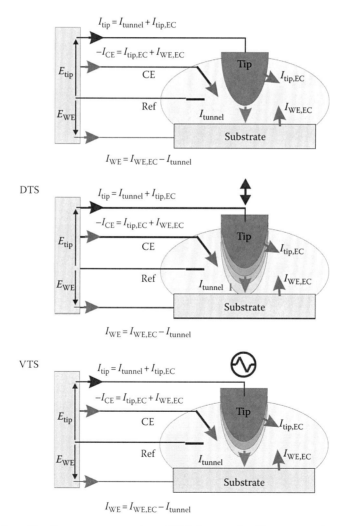

FIGURE 14.10 Schematic drawing of in-situ spectroscopy using a STM tip as probe. The currents through STM tip ($I_{tip,EC}$ and I_{tunnel}) and substrate surface ($I_{WE,EC}$ and $-I_{tunnel}$) are determined by the electrochemical interface of the electrode areas exposed to the electrolyte, and by the tunneling contact as shown in the figure. In DTS the gap width is changed at constant bias voltage, in VTS the bias voltage is changed at constant gap width.

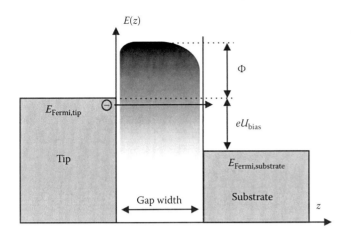

FIGURE 14.11 Idealized schematic of a tunneling contact between STM tip and substrate surface. The tunneling barrier height is Φ, the bias voltage is $U_{bias} = E_{tip} - E_{WE}$.

probe spectroscopy (SPS) investigations at solid/liquid interfaces have been reported in the last 20 years [34,62–66].

The dependence of the tunneling current on the distance between STM tip and substrate surface, that is the gap width z, can be calculated by quantum mechanics and is described for a rectangular potential barrier of height Φ and width z as shown in Figure 14.11 by the WKB relation [67]:

$$I_{tunnel} \propto \frac{U_{bias}}{z} \cdot \exp(-A\sqrt{\Phi} \cdot z) \qquad (14.8)$$

U_{bias} denotes the voltage applied across the tunneling contact, that is, between STM tip and substrate surface.
$A = 10.12 \ (eV)^{-1/2} \ nm^{-1}$ is for the case of a vacuum tunneling gap, and also used for tunneling at solid/liquid interfaces.

Equation 14.3 neglects any nonlinear bias voltage dependencies of the tunneling current, and is therefore only valid for small bias voltages. Additionally, it assumes identical Fermi levels on both sides of the tunneling barrier, in STM tip and substrate.

The exponential dependence of the tunneling current on the gap width results in the very high atomic lateral resolution of STM. This is a principal advantage of STM compared to AFM which shows only a power-law-like $1/z$ distance dependence of the atomic forces. But on the other hand, the exponential decay of the tunneling current with the gap width requires an optimum in the stability of the gap width during a spectroscopic measurement, and requires the application of fast measurements, which are discussed in more detail in Section 14.8.

There are basically two different types of spectroscopic techniques: distance tunneling spectroscopy (DTS) and voltage tunneling spectroscopy (VTS), as shown in Figure 14.10. The tunneling current is determined by the initial and final density of states on either side of the tunneling gap, but also by the tunneling barrier in the tunneling gap, as discussed in Section 14.3. The tunneling barrier height across a vacuum gap is typically of the order of the work function, which is approximately 5 eV for metals [68]. In contrast, the tunneling barrier at solid/liquid interfaces depends

on the molecular configuration in the tunneling gap, and has been found to show modulations due to the molecular arrangement in the tunneling gap as discussed in Section 14.3. The constant average values of tunneling barrier heights of approximately 1.5–1.6 eV, as reported in some publications for solid/liquid interfaces [31,65,69] are physically not relevant.

DTS denotes the measurement of the distance dependence of the tunneling current at a constant bias voltage (Figure 14.12). Both substrate potential and STM tip potential, whose difference is the bias voltage, are adjusted with respect to a reference electrode. The bias voltage is kept at a constant value, which defines the initial and final electronic states in the substrate and tip, respectively. The gap width is varied by either approaching or retracting the STM tip toward the substrate surface.

Since the potential of the substrate (first working electrode) and the potential of the STM tip (second working electrode) are kept constant throughout the measurement, there is no influence of double layer charging currents on the measurement. Thus, in the case of DTS, the most important requirement is a sufficient

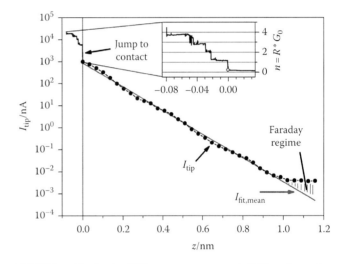

FIGURE 14.12 In-situ DTS measurement at a Au(111) surface using a Au STM tip, starting from the regime of the formation of quantized conductance channels at tip currents of several μA to the regime of Faraday tip currents below 10 pA. The complete data set has been superposed from three different DTS measurements in different overlapping current ranges, using different tip current converter modules for the measurements. This has been necessary to achieve the required signal-to-noise ratio. The zero point of the distance scale has been defined by the point of jump-to-contact. Black dots: Current–distance curve measured with a tip movement of 6.7 nm s⁻¹; gray line: fit of a straight line to the data according to Equation 14.1, taking only current values between 50 pA and 1000 nA into account. The mean value of an effective tunneling barrier height corresponding to the slope of the fit is $\Phi_{eff} = 1.51 \ eV$. $E_{WE} = 240 \ mV$ in the range $z > 0$, $E_{WE} = 175.5 \ mV$ in the range $z < 0$, $E_{tip} = 340 \ mV$ in the range $z > 0$, $E_{tip} = 240 \ mV$ in the range $z < 0$, $U_{bias} = E_{tip} - E_{WE} = 100 \ mV$ in the range $z > 0$, $U_{bias} = 64.5 \ mV$ in the range $z < 0$. Electrolyte: 0.02 M $HClO_4$. Potentials are quoted with respect to the standard hydrogen electrode (SHE). (From Schindler, W. and Hugelmann, P., *Electrocrystallization in Nanotechnology*, G. Staikov, ed., Wiley-VCH, Weinheim, Germany, 2007, p. 117. With permission.)

bandwidth of the experimental setup to allow (1) for measurements in sufficiently small times matched to the thermal drift rate of the gap width and (2) for recording changes of the tunneling current with the gap width at sufficient accuracy.

The example shown in Figure 14.12 is the current distance curve of a Au(111) substrate in 0.02 M $HClO_4$ and a Au STM tip [31,69]. This configuration ensures the same Fermi level and the same density of states on both sides of the tunneling contact, in the STM tip and in the substrate, respectively. Since the current is recorded up to values where quantized conductance channels are formed, the position of the STM tip can be also absolutely scaled. The zero point on the z-scale of the STM tip position is defined as the position where the jump-to-contact occurs. This definition of the zero point of the STM tip position shows an absolute error of approximately 0.05 nm [31,69].

Upon jump-to-contact, quantized conductance channels are formed. The tip current in this regime is described by $I_{tip} \approx U_{bias} \cdot nG_0$, where $G_0 = 2e^2/h \approx (12.9\,k\Omega)^{-1}$, and $n = 1, 2, 3\dots$ (inset of Figure 14.12) [69].

In the tunneling regime at $0 < z < 1$ nm, the tunneling current is modulated when the gap width increases. At gap widths larger than 1 nm the Faraday current at the STM tip surface exposed to the electrolyte becomes larger than the tunneling current. The Faraday current results from electrochemical processes at the unisolated tip apex surface and is basically independent of the gap width.

The modulation of the tunneling current distance curve is due to the layering of the water molecules at solid/liquid interfaces. The modulation period coincides with the diameter of single water molecules and with the thickness of a single water layer at a solid/liquid interface [31,32].

It should be mentioned that in all experiments using STM techniques at room temperature it is likely that many atomic or molecular configurations are averaged and the resulting images or data are averaged over all possible configurations. Relaxation or reorientation processes in the electrochemical double layer or in the tunneling gap may occur on much smaller time scales than the time scale of a STM measurement, though this may be performed in milliseconds. It is known that reorientation processes in the solvent occur on atomic scales within a timescale of 10^{-11} to 10^{-13} s [70–72]. Therefore, precise measurements may require multiple sweeps in order to increase the statistical significance of the data.

VTS is performed at a constant gap width and probes the dependence of the tunneling current on the bias voltage (Figure 14.13). In contrast to DTS, VTS requires to change the working electrode potential of preferentially the STM tip at a constant substrate potential in order to vary the bias voltage.

This variation of potential causes charging currents of the respective electrochemical interfacial layers at substrate electrode surface and tip apex surface, respectively. The time constants for these charging processes are typically slow, and depend on the surface area, i.e., the capacitance. Therefore, it is advantageous to vary the potential of the tip apex, since its surface area is much smaller than the electrode substrate area. The arrangement of

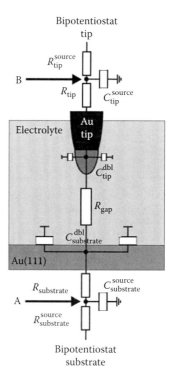

FIGURE 14.13 Electronic R/C network of the STM tip/tunneling gap/substrate surface region. The charging currents of the double layer capacitances of tip and substrate contribute to the tunneling current in a VTS measurement, and have to be considered accordingly.

electrode surface, tip apex surface, and gap forms an electrical network (Figure 14.13), which requires special attention, when measuring VTS data. Principal studies of VTS at Au(111) substrates using a Au STM tip are reported in Ref. [73].

14.7 Beyond Electrochemical Processes: In-Situ Electrical Transport Measurements at Individual Nanostructures

Besides the two spectroscopy modes DTS and VTS, there is the possibility of formation of quantized contacts between STM tip and substrate surface, or between STM tip and a nanostructure deposited onto a substrate surface, which allow to perform transport measurements (contact spectroscopy, Figure 14.14) [74–76]. The STM tip and the substrate serve as the two terminals in this two-wire measurement. The gentle approach of the STM tip, followed by the formation of a quantized contact, is the best nondestructive method to form a contact with smallest cross-sectional area to a nanostructure. Transport measurements can be performed in-situ with this method, which is a great advantage due to the sensitivity of nanostructures to environmental influences.

The quantized contacts are formed upon jump-to-contact at sufficiently small gap widths between STM tip and substrate surface. Their formation can be precisely controlled by the position of the STM tip as shown in Figure 14.15. The individual plateaus

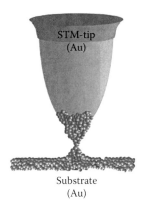

FIGURE 14.14 Schematic image of the formation of quantized contacts between a STM tip and a substrate surface or nanostructure underneath. The gentle formation of quantized conductance channels may be the best nondestructive method to contact nanostructures for transport measurements.

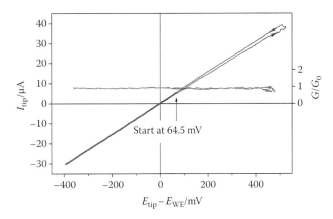

FIGURE 14.16 Ohmic behavior of quantized contacts as measured by a triangular voltage sweep across the contact as shown in Figure 14.15. (From Hugelmann et al., *Surf. Sci.*, 597, 156, 2005. With permission.)

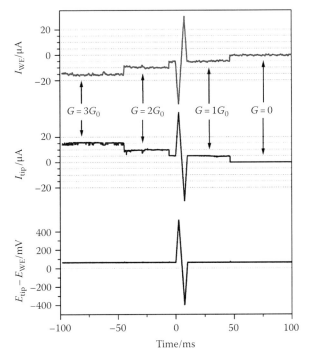

FIGURE 14.15 Au conductance channels as formed between Au STM tip and Au(111) substrate surface during variation of the distance between STM tip and substrate surface. An approach of the STM tip toward the substrate surface results in an increase of the number n of conductance channels, a retraction of the STM tip results in a decrease of the number n of conductance channels between STM tip and substrate surface. The constant bias voltage $U_{bias} = E_{tip} - E_{WE} = 64.5$ mV results in a constant height of the tip or WE current jumps according to $I_{tip} = U_{bias} \times nG_0$, with $G_0 = (12906 \ \Omega)^{-1}$ and $n = 1, 2, 3, \ldots$. The triangular bias voltage sweep applied at a contact in the state $n = 1$ results in an ohmic behavior of the quantized contact within the accuracy of the measurement. This allows to use quantized contacts to contact single nanostructures and to study their electronic properties by in-situ current–voltage transport measurements. Potentials are quoted with respect to the standard hydrogen electrode (SHE). (From Hugelmann et al., *Surf. Sci.*, 597, 156, 2005. With permission.)

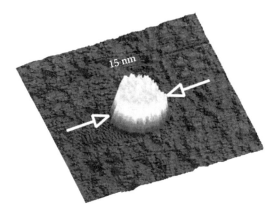

FIGURE 14.17 STM image of a 15 nm Au cluster, electrodeposited on n-Si(111), as used for the in-situ transport measurement (current–voltage characteristics) shown in Figure 14.18.

of the tip current correspond either to a single conductance quantum $G_0 = (12.9 \text{k}\Omega)^{-1}$ or to multiples of a conductance quantum ($n \times G_0$, $n = 1, 2, 3, \ldots$). While a particular conductance channel is adjusted, for example $n = 1$ in Figure 14.15, a triangular bias voltage sweep can be applied, which results in a corresponding triangular response of the STM tip current I_{tip} (Figure 14.15, midcurve). The working electrode current is inverted, $I_{WE} = -I_{tip}$. The conductance channels behave like ohmic contacts (Figure 14.16).

An example for the application of this technique at Au nanoclusters (Figure 14.17) is shown in Figure 14.18 for current–voltage measurements at Au/n-Si(111) nanodiodes with diameters of 10–20 nm. The curves represent the forward direction of the nanodiodes, and correspond to the behavior of Schottky barriers between a metal and a semiconductor. All measured nanodiodes show current-voltage characteristics, which may be compatible with the thermionic emission model [76], when the confinement of the space charge layer underneath the metal cluster in the silicon substrate is considered.

The voltage drop across quantized contacts is of the order of $U = I_{tip}/(nG_0)$. Although the resistance of conductance channels is $n \times 12.9 \text{k}\Omega$ ($n = 1, 2, 3, \ldots$) typical voltage drops are in the

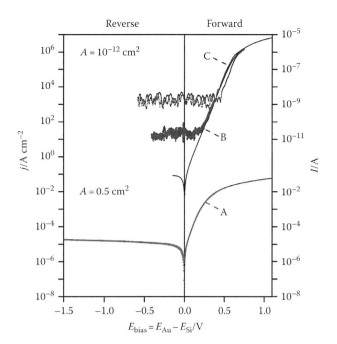

FIGURE 14.18 Current–voltage characteristics of a Au/n-Si(111) Schottky nanodiode, prepared by electrodeposition of a Au cluster with a diameter of 15 nm onto a n-Si(111): H substrate [63]. The current–voltage characteristic has been measured by contacting the Au cluster with a quantized conductance channel by STM and performing subsequently the in-situ transport measurement (curve B, C). The current–voltage characteristics of a macroscopic Au/n-Si(111) Schottky contact (curve A) is plotted for comparison.

millivolt range since currents through nanostructures of cross-sectional areas as small as 10^{-12} cm^2 can be assumed to be not larger than microamperes. The voltage drop across the quantized contact can be easily corrected in the measurements since it is linear with the current (Figure 14.16).

The size of the quantized contact applied in the measurement of Figure 14.18 is small compared to the 10–20 nm diameter of the nanodiodes. The influence of this type of contacts on the measurements may become important when the size of the nanostructure measured is comparable to the approximate 1 nm size of the contact.

14.8 Some Technical Aspects of the Application of Scanning Probe Techniques at Solid/Liquid Interfaces at Sub-Nanometer Resolution

14.8.1 Importance of Nanoelectrode Tip Shape and Surface Quality

The detailed shape and the surface morphology of electrochemically etched STM tips, as usually used in STM experiments at solid/liquid interfaces, is rather unreproducible, although a variety of STM tip-etching procedures have been published in the past two decades [77–79]. This can be deduced from numerous

studies of electrochemically etched STM tips by scanning electron microscopy (SEM). Additionally, the surface of electrochemically etched STM tips is in part electrochemically inactive due to etching residuals on the surface, unwanted adsorbates, and oxidized surface areas.

This general feature of electrochemically etched tips is usually not at all considered to be a problem if such tips are exclusively used for STM imaging. On the first view, the detailed shape and electrochemical quality of the STM tip seems to be completely irrelevant for STM imaging of surfaces, because STM imaging assumes the whole tip current measured to be the tunneling current between the two closest adjacent atoms of tip apex and substrate surface forming the tunneling gap. To achieve this, Faraday currents must be minimized by reduction of the electrochemically active tip surface exposed to the electrolyte. Typically, this is achieved by an isolation of the tips with nonconducting material, like Apiezon wax or electrophoretic lacquers, and an electrochemically inactive tip surface even helps to achieve this goal.

In fact, when extended atomically flat surfaces are imaged by blunt STM tips still atomically resolved images can be obtained. Difficulties arise when surfaces with a larger height variation shall be imaged at high lateral resolution. A prominent example for such a situation is supported clusters with diameters in the lower nanometer range and heights of several atomic layers. STM images of such clusters show usually no individual atomic layers and step edges [21–23,80], as expected, but rather a hemispherical shape, which results from the convolution of the real shape of the cluster with the actual geometry of the tip apex [81,82]. Unfortunately, this convolution changes both diameter and height of the scanned object, which results in a complicated interpretation of the corresponding STM images. Figure 14.8 shows an example of the high resolution of step edges, which can be achieved at a 15 nm diameter cluster when appropriate STM tips are used.

Severe problems arise, however, if electrochemically etched tips shall be used for more advanced purposes. They would comprise all techniques using the tip for initiation or detection of electrochemical processes on a nanometer scale, i.e., using the tip as a nanoelectrode or as a high resolution electrochemical sensor probe. Such purposes require a well-defined tip shape and an electrochemically clean tip surface, which is both usually not the case for electrochemically etched STM tips.

In order to combine the requirements for STM imaging, a very small diameter of the tip apex for high lateral resolution, with the requirements for electrochemistry at the STM tip, namely a well-defined geometrical shape and electrochemically clean surface area of the STM tip exposed to the electrolyte, the STM tip preparation can be substantially improved by applying a field emission/sputtering process subsequently to the electrochemical etching step [57]. Although the mechanisms of electron field emission and sputtering by ionized ions are known for almost 50 years [83–85], this technique has not been applied routinely for the preparation of STM tips. The basic idea is the fact, that the diameter of a metal tip can be precisely determined by the voltage for field emission of electrons. This correlation can be exploited to precisely determine the diameter of a STM tip from

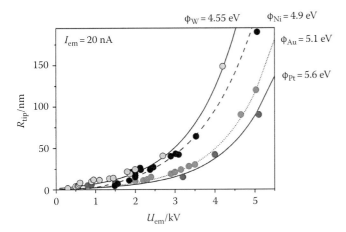

FIGURE 14.19 STM tips prepared in ultrahigh vacuum by a field emission/sputtering technique. The emission voltage for field emission in ultrahigh vacuum is a direct measure for the diameter of the STM tip apex. The curves show data at an emission current of 20 nA for Au, W, Ni, and Pt/Ir STM tips. The tip apex diameters have been determined by electron microscopy. (From Schindler, W. and Hugelmann, P., *Electrocrystallization in Nanotechnology*, G. Staikov, ed., Wiley-VCH, Weinheim, Germany, 2007, p. 117. With permission.)

its field emission voltage. Figure 14.19 shows this correlation for various STM tips prepared with this technique. It is worth mentioning that STM tips can be produced from nearly any metal wire, in particular also from Au which is often assumed to be not suitable to form stable and sharp STM tips.

Such STM tips with a defined geometrical shape can be modeled in a first approximation as a sphere of certain diameter, which is typically of the order of 5–30 nm, as proven by transmission electron microscopy (TEM) images [57]. These STM tips can be used for both purposes, STM imaging as well as localized electrochemistry at the STM tip in-situ in the same experiment.

14.8.2 STM Tip Isolation

Most STM investigations at solid/liquid interfaces have been carried out at potential control of both substrate and STM tip, and the STM tip has been only used as a sensor for the tunneling current during STM imaging of surfaces. Since a STM tip current I_{tip} consists at sufficiently small distance between tip apex and substrate surface, that is the gap width, of both tunneling current I_{tunnel} and electrochemical (Faraday) currents, STM tips are usually isolated with wax or lacquers, leaving only a small tip apex surface exposed to the electrolyte [86–88]. This ensures that at appropriate STM tip potentials almost the whole STM tip current results from electron tunneling between substrate surface and tip apex, and electrochemical (Faraday) currents are sufficiently low. It is obvious that STM imaging works the better the lower the level of Faraday currents is at the particular STM tip potential adjusted. For the purpose of STM imaging, the electrochemically most inactive tip surface is the most desirable surface, because it shows the lowest Faraday currents, and $I_{tip} \approx I_{tunnel}$ is realized to the best extent.

14.8.3 Electronic/Measurement Bandwidth Considerations

The acceptable deviation in the actual position of the STM tip, or equivalently in the actual gap width, during a spectroscopic measurement depends on the particular physical problem studied, and on the magnitude of the physical effect measured. In general, it can be supposed that the accuracy of the measured tunneling current should be better than 1%, i.e., the deviation of the actually measured tunneling current from its correct value I_{tunnel} at the correct gap width z should be less than 1%. Then, the allowable deviation Δz of the gap width z from its correct value can be calculated:

$$\ln\left(\frac{1.01 \cdot I_{tunnel}}{I_{tunnel}}\right) = \ln\left(\frac{z}{z + \Delta z}\right) - A\sqrt{\Phi} \cdot \Delta z \approx -A\sqrt{\Phi} \cdot \Delta z \tag{14.9}$$

or

$$\Delta z \approx -\frac{\ln 1.01}{A\sqrt{\Phi}} = 8 \times 10^{-4}\, nm \tag{14.10}$$

taking the value for A from Section 14.6 and a mean value of the tunneling barrier heights in aqueous electrolytes of $\Phi = 1.5\, eV$.

Typically, the gap width in a STM operating at a solid/liquid interface at room temperature drifts with a thermal drift rate of the order of $0.01\, nm\, s^{-1}$. A tolerable $\Delta z = 8 \times 10^{-4}\, nm$ during a spectroscopic measurement requires then that the measurement time of a complete spectroscopic measurement is kept below

$$\Delta t = \frac{8 \times 10^{-4}\, nm}{0.01\, nm\, s^{-1}} = 80\, ms \tag{14.11}$$

Allowing for thermal drift rates of $0.1\, nm\, s^{-1}$ which are frequently found in day-to-day experimental conditions, the measurement time must be decreased by a factor 10 to approximately 8 ms in the above example for maintaining the presumed 1% accuracy.

Spectroscopy can be performed in different operation modes of the STM, DTS, and VTS.

In VTS, usually linear or triangular bias voltage sweeps as shown in Figure 14.20 are applied across the tunneling gap. Such waveforms can be expressed by a Fourier series of sine waves:

$$I = \frac{8}{\pi^2} \sum_n \frac{(-1)^{\frac{n-1}{2}}}{n^2} \sin n\omega t \quad \text{with } n = 1, 3, 5, \ldots \tag{14.12}$$

As shown in Figure 14.20, the linear or triangular waveform is the better reproduced the more high order terms of the Fourier series are included in the signal. An accuracy of a linear voltage sweep of better than 1% (see Figure 14.20) requires $n = 33$, that is, 15 terms of the Fourier series, indicating that a bandwidth of approximately 30 times the fundamental frequency is required. This means for the above example of a total measurement time of $\Delta t = 8\, ms$, i.e., a frequency of approximately

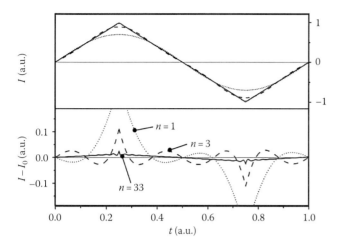

FIGURE 14.20 Ideal triangular waveform (solid curve) and various Fourier series (calculated using Equation 14.5) with different cutoff frequencies reproducing the ideal waveform (dashed line: cutoff at $n = 3$; dotted line: cutoff at $n = 33$).

$33 \times (2 \times 8\,\text{ms})^{-1} = 2062.5\,\text{Hz}$ must be applied at the STM tip electrode without a significant attenuation by the whole measurement system. This requires a 3 dB bandwidth of the measurement system of approximately 20 kHz.

The bandwidth requirements become even more demanding, when multiple bias sweeps during a single spectroscopic measurement are desirable for the reason of checking the reproducibility and increasing the statistics of data. In such cases, the limit for the required bandwidth increases proportional to the number of voltage sweeps, for example, by a factor of 10 for 10 subsequent voltage sweeps in a single measurement, resulting for the above example in an essential bandwidth of the measurement system of approximately 200 kHz at a thermal drift rate of 0.1 nm s⁻¹.

A measurement system bandwidth of 20 or 200 kHz can be compatible with the electrode capacitance of a STM tip and the time constant resulting from its double layer charging, when certain requirements are fulfilled. Typically, surfaces exposed to an electrolyte show a double layer capacity of $25\,\mu\text{C cm}^{-2}$ [26]. Unisolated STM tip apex surfaces exposed to the electrolyte show typically areas of $10^{-7}\,\text{cm}^2$, resulting in capacitances of 2.5 pF. The time constant required for charging these double layer capacitances depends on the charging resistors used. For the case of a charging resistor of 1 MΩ

$$\tau = R \cdot C = 1\,\text{M}\Omega \cdot 2.5\,\text{pF} = 2.5\,\mu\text{s}. \qquad (14.13)$$

This time constant corresponds to 400 kHz, well above the required bandwidth of 20–200 kHz. However, typical charging resistors in commercial STM I/U converters are 100 MΩ for a conversion factor of 10^8 V/A. Such high resistances result in time constants corresponding to 4 kHz for charging of the STM tip electrochemical double layer. This is much smaller than the required measurement system bandwidth for a spectroscopic measurement, and special attention must be given to this aspect in the experiments.

The desired fast charging of the double layer capacitance requires the bias voltage sweep to be applied at the STM tip electrode rather than at the substrate electrode. Substrate surfaces show usually large areas, resulting in time constants for charging of more than 100 μs, and resulting in effective bandwidths of less than 10 kHz.

So far, only the bandwidth necessary for sufficient stability and for application of the linear potential sweep of the bias voltage has been considered. However, the purpose of tunneling spectroscopy is to record current variations due to the physical properties of the studied system. An estimation of the bandwidth required for recording changes in the tunneling current, that is in the tunneling resistance across the tunneling gap, with varying bias voltage (in VTS) or with varying gap width (in DTS) can be derived from the following consideration: When a linear bias voltage scan (in VTS) with a scan rate of 100 V s⁻¹ is assumed, that is a scan range of 1 V in 10 ms, the bias voltage is changed by 1 mV in a time interval of 10 μs. When the gap width is changed (in DTS) with a scan rate of 100 nm s⁻¹, that is 1 nm in 10 ms, the gap width is varied by 1 pm in 10 μs. In case of a desired resolution of 1 mV in the I/V spectrum or 1 pm in the current–distance measurement, a jump in the current signal must be recorded within these 10 μs. Similar to linear potential sweeps, current signal jumps can be also represented by Fourier series:

$$I = \frac{8}{\pi^2} \sum_n \frac{1}{n} \sin n\omega t \quad \text{with} \quad n = 1, 3, 5, \ldots \qquad (14.14)$$

Figure 14.21 shows an ideal jump in the current signal at $t = 0$ and three Fourier series based on Equation 14.14 with $n = 5, 17,$ and 33. It can be clearly seen that the measurement system must be able to record frequency components around 10 kHz without significant attenuation, which requires a 3 dB bandwidth of the measurement system of the order of 100 kHz.

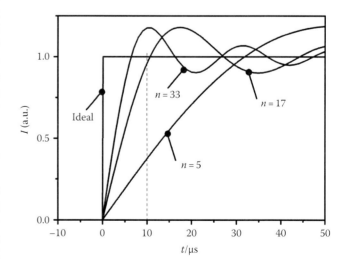

FIGURE 14.21 Ideal signal jump with infinite slope and various Fourier series (calculated using Equation 14.7) with different cutoff frequencies reproducing the ideal waveform ($n = 5, 17,$ or 33, respectively). Fundamental frequency is 1.6 kHz.

Summarizing the above discussion, the most critical parameters for tunneling spectroscopy at solid/liquid interfaces are (1) the thermal drift rate of the tunneling gap width, (2) the time constant required for the electrochemical double layer charging of the STM tip apex surface exposed to the electrolyte, and (3) the response time to changes in the tunneling current. For day-to-day experimental conditions, reliable and reproducible measurements require a 3 dB bandwidth of the measurement system of 20–200 kHz. In principle, the higher the bandwidth of the utilized experimental equipment is, the more precise are the spectroscopic data measured with this equipment.

References

1. G. Moore, *IEDM Tech. Dig.* 11, 1975.
2. J. Meindl et al., *Proceedings of the IEEE* 89, No. 3, 2001.
3. K. Matsumoto, M. Ishii, K. Segawa, Y. Oka, B. J. Vartanian, and J. S. Harris, *Appl. Phys. Lett.* 68, 34, 1996.
4. K. Matsumoto, Y. Gotoh, T. Maeda, J. A. Dagata, and J. S. Harris, *Appl. Phys. Lett.* 76, 239, 2000.
5. C. Z. Li, H. X. He, A. Bogozi, J. S. Bunch, and N. J. Tao, *Appl. Phys. Lett.* 76, 1333, 2000.
6. C. Li, D. Zhang, X. Liu, S. Han, T. Tang, J. Han, and C. Zhou, *Appl. Phys. Lett.* 82, 1613 (2003).
7. M. Datta et al., *IBM J. Res. Develop.* 42 (5) 1998.
8. P. M. Vereecken et al., *IBM J. Res. Develop.* 49 (1) 2005.
9. H. Rohrer, Nanoscale probes of the solid/liquid interface, in: *Proceedings of NATO ASI Sophia Antipolis*, France, July 10–20, 1993, A. A. Gewirth and H. Siegenthaler (Eds.), *NATO Science Series* E, Vol. 288, Kluwer, New York, 1995.
10. G. Binnig and H. Rohrer, *Rev. Mod. Phys.* 59, 615, 1987.
11. E. Budevski, G. Staikov, and W. J. Lorenz, *Electrochemical Phase Formation and Growth*, VCH, Weinheim, Germany, 1996.
12. O. M. Magnussen, J. Hotlos, R. J. Nichols, D. M. Kolb, and R. J. Behm, *Phys. Rev. Lett.* 64, 2929–2932, 1990.
13. D. M. Kolb, *Prog. Surf. Sci.* 51, 109–173, 1996.
14. M. Wilms, P. Broekmann, C. Stuhlmann, and K. Wandelt, *Surf. Sci.* 416, 121–140, 1998.
15. W. Li, J. A. Virtanen, and R. M. Penner, *Appl. Phys. Lett.* 60, 1181, 1992.
16. W. Li, J. A. Virtanen, and R. M. Penner, *J. Phys. Chem.* 96, 6529, 1996.
17. W. Li, G. S. Hsiao, D. Harris, R. M. Nyffenegger, J. A. Virtanen, and R. M. Penner, *J. Phys. Chem.* 100, 20103, 1996.
18. R. Schuster, V. Kirchner, X. H. Xia, A. M. Bittner, and G. Ertl, *Phys. Rev. Lett.* 80, 5599, 1998.
19. Y. Zhang, S. Maupai, and P. Schmuki, *Surf. Sci.* 551, L33–L39, 2004.
20. R. T. Pötzschke, G. Staikov, W. J. Lorenz, and W. Wiesbeck, *J. Electrochem. Soc.* 146, 141, 1999.
21. R. Ullmann, T. Will, and D. M. Kolb, *Chem. Phys. Lett.* 209, 238, 1993.
22. D. M. Kolb, R. Ullmann, and T. Will, *Science* 275, 1097, 1997.
23. D. M. Kolb, R. Ullmann, and J. C. Ziegler, *Electrochim. Acta* 43, 2751, 1998.
24. D. Hofmann, W. Schindler, and J. Kirschner, *Appl. Phys. Lett.* 73, 3279, 1998.
25. W. Schindler, D. Hofmann, and J. Kirschner, *J. Electrochem. Soc.* 148, C124, 2000.
26. J. O'M. Bockris and A. K. N. Reddy (Eds.), *Modern Electrochemistry*, p. 1, Plenum, New York, 1998.
27. W. Schindler and J. Kirschner, *Phys. Rev. B* 55, R1989, 1997.
28. M. A. Henderson, *Surf. Sci. Rep.* 46, 273, 2002.
29. W. Schmickler, *Surf. Sci.* 335, 416, 1995.
30. M. F. Toney, J. N. Howard, J. Richter, G. L. Borges, J. G. Gordon, O. R. Melroy, D. G. Wiesler, D. Yee, and L. B. Sorensen, *Nature* 368, 444–446, 1994.
31. M. Hugelmann and W. Schindler, *Surf. Sci. Lett.* 541, L643–L648, 2003.
32. M. Hugelmann, P. Hugelmann, and W. Schindler, *J. Electrochem. Soc.* 151, E97, 2004.
33. F. C. Simeone, D. M. Kolb, S. Venkatachalam, and T. Jacob, *Surf. Sci.* 602, 1401, 2008.
34. J. Halbritter, G. Repphuhn, S. Vinzelberg, G. Staikov, and W. J. Lorenz, *Electrochim. Acta* 40, 1385, 1995.
35. Y. Marcus, *Chem. Rev.* 88, 1475–1498, 1988.
36. Y. Marcus, *Ion Properties*, Marcel Dekker, New York, 1997.
37. A. J. Bard, R. Parsons, and J. Jordan (Eds.), *Standard Potentials in Aqueous Solution*, Marcel Dekker, New York, 1985.
38. A. J. Bard and L. R. Faulkner (Eds.), *Electrochemical Methods*, Wiley, New York, 2001.
39. R. Christoph, H. Siegenthaler, H. Rohrer, and H. Wiese, *Electrochim. Acta* 34, 1011–1022, 1989.
40. K. Itaya and E. Tomita, *Surf. Sci.* 201, L501–L512, 1988.
41. J. Wiechers, T. Twomey, D. M. Kolb, and R. J. Behm, *J. Electroanal. Chem.* 248, 451–460, 1988.
42. F. A. Möller, O. M. Magnunssen, and R. J. Behm, *Phys. Rev. Lett.* 77, 5249, 1996.
43. W. J. Lorenz, G. Staikov, W. Schindler, and W. Wiesbeck, *J. Electrochem. Soc.* 149, K47, 2002.
44. E. C. Walter, B. J. Murray, F. Favier, G. Kaltenpoth, M. Grunze, and R. M. Penner, *J. Phys. Chem.* 106, 11407, 2002.
45. J. Tang, M. Petri, L. A. Kibler, and D. M. Kolb, *Electrochim. Acta* 51, 125, 2005.
46. S. Pandelov and U. Stimming, *Electrochim. Acta* 52, 5548, 2007.
47. C. H. Hamann and W. Vielstich, *Elektrochemie I*, VCH, Weinheim, Germany, 1985.
48. J. Meier, K. A. Friedrich, and U. Stimming, *Faraday. Discuss.* 121, 365, 2002.
49. A. J. Bard, F. R. F. Fan, and J. Kwak, *Anal. Chem.* 61, 132–138, 1988.
50. A. J. Bard, *Scanning Electrochemical Microscopy*, Taylor & Francis, Oxon, U.K., 2001.
51. V. Radtke and J. Heinze, *Z. Phys. Chem.* 218, 103, 2004.
52. O. Sklyar, T. H. Treutler, N. Vlachopoulos, and G. Wittstock, *Surf. Sci.* 597, 181, 2005.

53. M. Etienne, E. C. Anderson, S. R. Evans, W. Schuhmann, and I. Fritsch, *Anal. Chem.* 78, 7317, 2006.

54. W. Schindler, P. Hugelmann, M. Hugelmann, and F. X. Kärtner, *J. Electroanal. Chem.* 522, 49–57, 2002.

55. M. Eikerling, J. Meier, and U. Stimming, *Z. Phys. Chem.* 217, 395–414, 2003.

56. D. M. Kolb and F. C. Simeone, *Electrochim. Acta* 50, 2989, 2005.

57. P. Hugelmann and W. Schindler, *J. Electroanal. Chem.* 612, 131, 2008.

58. W. Schindler and P. Hugelmann, Localized electrocrystallization of metals by STM tip nanoelectrodes, in: *Electrocrystallization in Nanotechnology*, G. Staikov (Ed.), p. 117, Wiley-VCH, Weinheim, Germany, 2007.

59. P. Vanysek, *CRC Handbook of Chemistry and Physics*, D. R. Lide and H. P. R. Frederikse (Eds.), CRC Press, Boca Raton, FL, 1993.

60. S. G. Garcia, D. R. Salinas, C. E. Mayer, W. J. Lorenz, and G. Staikov, *Electrochim. Acta* 48, 1279, 2003.

61. M. F. Crommie, C. P. Lutz, and D. M. Eigler, *Phys. Rev. B* 48, 2851–2854, 1993.

62. M. Binggeli, D. Carnal, R. Nyffenegger, H. Siegenthaler, R. Christoph, and H. Rohrer, *J. Vac. Sci. Technol. B* 9, 1985, 1991.

63. J. Pan, T. W. Jing, and S. M. Lindsay, *J. Phys. Chem.* 98, 4205, 1994.

64. A. Vaught, T. W. Jing, and S. M. Lindsay, *Chem. Phys. Lett.* 236, 306, 1995.

65. G. E. Engelmann, J. Ziegler, and D. M. Kolb, *Surf. Sci. Lett.* 401, L420, 1998.

66. G. Nagy, *J. Electroanal. Chem.* 409, 19, 1996.

67. W. Kramer, Brillouin relation, in: *Quantum Mechanics*, A. Messiah (Ed.), North-Holland Publishers, Amsterdam, the Netherlands, 1964.

68. J. K. Gimzewski and R. Möller, *Phys. Rev. B* 36, 1284–1287, 1987.

69. M. Hugelmann and W. Schindler, *J. Electrochem. Soc.* 151, E97–E101, 2004.

70. M. Buettiker and R. Landauer, *Phys. Rev. Lett.* 49, 1739, 1982.

71. K. L. Sebastian and G. Doyen, *Surf. Sci. Lett.* 290, L703, 1993.

72. K. L. Sebastian and G. Doyen, *J. Chem. Phys.* 99, 6677, 1993.

73. P. Hugelmann and W. Schindler, *J. Phys. Chem. B* 109, 6262–6267, 2005.

74. M. Hugelmann, P. Hugelmann, W. J. Lorenz, and W. Schindler, *Surf. Sci.* 597, 156–172, 2005.

75. W. Schindler, M. Hugelmann, and P. Hugelmann, *Electrochim. Acta* 50, 3077–3083, 2005.

76. M. Hugelmann and W. Schindler, *Appl. Phys. Lett.* 85, 3608–3610, 2004.

77. A. J. Melmed, *J. Vac. Sci. Technol. B* 9, 601, 1991.

78. M. C. Baykul, *Mater. Sci. Eng. B* 74, 229, 2000.

79. M. Klein and G. Schwitzgebel, *Rev. Sci. Instrum.* 68, 3099, 1997.

80. S. Maupai, A. S. Dakkouri, M. Stratmann, and P. Schmuki, *J. Electrochem. Soc.* 150, C111–C114, 2003.

81. N. Breuer, U. Stimming, and R. Vogel, *Electrochim. Acta* 40, 1401–1409, 1995.

82. W. Schindler, in: *The Electrochemical Society Proceedings*, Vol. 2003-27: Scanning Probe Techniques for Materials Characterization at Nanometer Scale, W. Schwarzacher and G. Zangari (Eds.), p. 615, The Electrochemical Society, Pennington, NJ, 2003.

83. R. Gomer, *Field Emission and Field Ionization*, Harvard University Press, Cambridge, MA, 1961.

84. E. W. Müller and T. T. Tsong, *Field Ion Microscopy*, Elsevier, New York, 1969.

85. K. M. Bowkett and D. A. Smith, *Field–Ion–Microscopy*, Elsevier, Amsterdam, the Netherlands, 1970.

86. M. J. Heben, M. M. Dovek, N. S. Lewis, and R. M. Penner, *J. Microsc.* 152, 651–661, 1988.

87. L. A. Nagahara, T. Thundat, and S. M. Lindsay, *Rev. Sci. Instrum.* 60, 3128, 1989.

88. A. A. Gewirth, D. H. Craston, and A. J. Bard, *J. Electroanal. Chem.* 261, 477–482, 1989.

III

Thermodynamics

15

Nanothermodynamics[*]

Vladimir García-Morales
Technische Universität München

Javier Cervera
University of Valencia

José A. Manzanares
University of Valencia

15.1 Introduction

Progress in the synthesis of nanoscale objects has led to the appearance of scale-related properties not seen or different from those found in microscopic/macroscopic systems. For instance, monolayer-protected Au nanoparticles with average diameter of 1.9 nm have been reported to show ferromagnetism while bulk Au is diamagnetic (Hasegawa 2007). Nanotechnology brings the opportunity of tailoring systems to specific needs, significantly modifying the physicochemical properties of a material by controlling its size at the nanoscale. Size effects can be of different types. Smooth size effects can be described in terms of a size parameter such that we recover the bulk behavior when this parameter is large. The physicochemical properties then follow relatively simple scaling laws, such as a power-law dependence, that yield a monotonous variation with size. Specific size effects, on the contrary, are not amenable to size scaling because the variation of the relevant property with the size is irregular or nonmonotonic. They are characteristic of small clusters. Finally, some properties are unique for finite systems and do not have an analog in the behavior of the corresponding bulk matter (Jortner and Rao 2002, Berry 2007).

Nanothermodynamics can be defined as the study of small systems using the methods of statistical thermodynamics. Small systems are those that exhibit nonextensive behavior and contain such a small number of particles that the thermodynamic limit cannot be applied (Gross 2001). Even though Boltzmann, presumably, did not think of nonextensive systems, his formulation of statistical thermodynamics relied neither on the use of the thermodynamic limit nor on any assumption of extensivity (Gross

2001). The same applies to Gibbs ensemble theory, which can also be used to describe the behavior of small systems. However, this is not true for classical thermodynamics, which is based on a number of assumptions that may lead to questioning its validity on the nanoscale. Care must be exercised when applying thermodynamics to nanosystems. First, quantities such as interfacial energy, which could be safely neglected for large systems, must be taken into consideration (Kondepudi 2008). These and other effects lead to nonextensive character of the thermodynamic potentials. Second, the fluctuations of thermodynamic variables about their average values may be so large in a small system that these variables no longer have a clear physical meaning (Feshbach 1988, Mafé et al. 2000, Hartmann et al. 2005). Fluctuations may also lead to violations of the second law of thermodynamics (Wang et al. 2002). Third, quantum effects may also become important (Allahverdyan et al. 2004).

In this chapter, we mostly concentrate on Hill's equilibrium nanothermodynamics. For historical and pedagogical reasons, it is convenient to start the description of size effects in nanosystems using classical equilibrium thermodynamics including interfacial contributions. Thus, in Section 15.3.1, it is shown that the smooth size dependence of many thermodynamic properties can be understood without introducing any "new theory." Similarly, Section 15.3.2 shows that the methods of traditional statistical thermodynamics can be used to describe small systems under equilibrium conditions without the need of introducing significantly new ideas. The thermodynamic behavior of small systems is somehow different from the macroscopic systems. Particularly important is the fact that fluctuations break the equivalence between the different statistical ensembles and that the ensemble that accurately describes the interaction between the small system and its surroundings

---

[*] Dedicated to Prof. Julio Pellicer on occasion of his retirement.

must be used. For some of these interactions, a completely open statistical ensemble, with no macroscopic analogue, must be used. Hill's nanothermodynamics can be considered as a rigorous formulation of the consequences of nonextensivity, due to the small number of particles composing the system, on the classical thermodynamic equations. It is shown that the new degree of freedom brought about by the nonextensivity of thermodynamic potentials can be conveniently dealt with through the definition of a new thermodynamic potential, the subdivision potential. An introduction to modern applications of Hill's nanothermodynamics is included in order to show that this theory can be applied to metastable states, to interacting small systems, and to describe microscopically heterogeneous systems like glasses and ferromagnetic solids.

Hill's theory can be related to another thermodynamic theory emphasizing nonextensivity—Tsallis' theory. This theory is discussed in Section 15.3.3, where the relation between the entropic index q and Hill's subdivision potential is presented. Similarly, Hill's theory is related in Section 15.3.4 to superstatistics, a new term coined to emphasize the existence of two different statistical probability distributions in complex systems. Finally, Section 15.3.5 describes the most recent ideas on nonequilibrium nanothermodynamics and their relation to fluctuation theorems.

15.2 Historical Background

The roots of nanothermodynamics go back to the seminal works by J. W. Gibbs and Lord Kelvin in the nineteenth century when the importance of surface contributions to the thermodynamic functions of small systems was realized. These topics have taken on a new significance due to the recent development of nanoscience.

During the first half of twentieth century, there were some interesting contributions to this field. For instance, the starting point for the kinetic interpretation of condensation phenomena in supersaturated phases, the melting point depression in small metal particles, and the size-dependent chemical potential in the droplet model developed by Becker and Döring (1935). Also, in his theory of liquids, Frenkel (1946) worked out the correction to the thermodynamic functions to extend their validity to small systems.

The major contributions have occurred during the second half of twentieth century and the current decade. The formulation of nanothermodynamics as a generalization of equilibrium thermodynamics of macroscopic systems was carried out in the early 1960s. Similar to the introduction of the chemical potential by J. W. Gibbs in 1878 to describe open systems, T. L. Hill (1962) introduced the subdivision potential to describe small systems. Hill anticipated two main classes of applications of nanothermodynamics: (1) as an aid in analyzing, classifying, and correlating equilibrium experimental data on "small systems" such as (noninteracting) colloidal particles, liquid droplets, crystallites, macromolecules, polymers, polyelectrolytes, nucleic acids, proteins, etc.; and (2) to verify, stimulate, and provide a framework for statistical thermodynamic analysis of models of finite systems.

The first computer simulations of hard sphere fluids carried out in the late 1950s and early 1960s also showed the importance of size effects. It was soon realized that the different statistical ensembles are not equivalent when applied to small systems. A key feature of small systems is that thermodynamic variables may have large fluctuations so that only the appropriate ensemble describes correctly the behavior of the system (Chamberlin 2000). Moreover, it was found that the thermodynamic functions of very small systems exhibit a variation with the number of particles that is not only due to the surface contribution and some anomalous effects may also appear (Hubbard 1971). Statistical ensembles whose natural variables are intensive were identified as the best choice to describe small systems and some specific tools were developed (Rowlinson 1987).

The development of nonextensive thermodynamics based on Tsallis entropy in the 1990s motivated a widespread interest in the modifications of classical thermodynamics for complex systems, including nanosystems. It has been shown recently, however, that Tsallis and Hill's theories can be mapped onto each other (García-Morales et al. 2005).

The interest in Hill's nanothermodynamics has grown in the 2000s after realizing that this theory can describe the behavior of a microheterogeneous material, such as a viscous liquid exhibiting complex dynamics or a ferromagnetic material, by considering it as an ensemble of small open systems. These materials have an intrinsic correlation length, which changes with the temperature. The "small systems" would then be associated to physical regions with a size related to the correlation length. These small systems would then be completely open, in the sense that they could exchange energy with the bulk material, and vary their volume and number of particles, so that they should be described using the generalized ensemble. Remarkably, Chamberlin has proved that the generalized ensemble of nanothermodynamics with unrestricted cluster sizes yields nonuniform clustering, nonexponential relaxation, and nonclassical critical scaling, similar to the behavior found near the liquid–glass and ferromagnetic transitions (Chamberlin 2003).

Besides the progress described above at equilibrium, there were some important developments on nonequilibrium statistical physics during the 1970s and 1980s coming from the modern theory of dynamical systems and the study of thermostated systems as applied to nonequilibrium fluids and molecular dynamics simulations (Evans and Morriss 2008). These developments led during the 1990s to a series of breakthroughs whose impact is today the subject of very active research at the nanoscale: the theoretical prediction (contained in the so-called fluctuation theorems), and later experimental observation, of violations of the second law of thermodynamics for small systems and short time scales, and the Jarzynski equality, which allows one to obtain equilibrium free energy differences from nonequilibrium measurements.

15.3 Presentation of State-of-the-Art

15.3.1 Surface Thermodynamics

Most modern developments of thermodynamics of nanosystems involve the introduction of new magnitudes like Hill's subdivision potential or new equations like Tsallis' entropy equation, which are sometimes controversial or difficult to accept. On the contrary, surface thermodynamics is still considered a successful framework to analyze thermodynamic properties that show a monotonous variation on the nanoscale (Delogu 2005, Rusanov 2005, Wang and Yang 2005, Jiang and Yang 2008).

15.3.1.1 Unary Systems with Interfaces

Classical macroscopic thermodynamics deals with systems composed of one or several bulk homogeneous phases. If their interfaces have definite shape and size, they can be described using the Gibbs dividing surface model, which associates zero volume and number of particles to the interface of a unary (i.e., monocomponent) system. For a homogeneous phase β, the Gibbs equation is $dU^\beta = TdS^\beta - p^\beta dV^\beta + \mu^\beta dn^\beta$ or, in terms of molar quantities $x^\beta \equiv X^\beta/n^\beta$ ($X = S, U, V$), $du^\beta = Tds^\beta - p^\beta dv^\beta$. From the intensive character of the molar quantities, it can be concluded that phase β satisfies the Gibbs–Duhem equation

$$d\mu^\beta = -s^\beta dT + v^\beta dp^\beta \qquad (15.1)$$

If another homogeneous phase α is separated from phase β by an interface σ, the Gibbs equation of the $(\alpha + \sigma)$ system is

$$dU^{\alpha+\sigma} = TdS^{\alpha+\sigma} - p^\alpha dV^\alpha + \mu^\alpha dn^\alpha + \gamma dA \qquad (15.2)$$

where A is the interfacial area and γ is the interfacial free energy. The internal energy $U^{\alpha+\sigma}$ and the entropy $S^{\alpha+\sigma}$ have a bulk contribution and an interfacial contribution. The volume V^α and the bulk contributions to internal energy and to entropy are bulk extensive variables, and therefore they are proportional to the number of moles n^α. The interfacial free energy is independent of the area A, that is, γ is an interfacial intensive variable. The interfacial contributions to internal energy and to entropy are proportional to the interfacial area A, so that they are interfacial extensive variables. The concepts of bulk and interfacial extensivity would only match if A were proportional to n^α, which is not generally the case.

The Euler and Gibbs–Duhem equations of the $(\alpha + \sigma)$ system are

$$U^{\alpha+\sigma} - TS^{\alpha+\sigma} + p^\alpha V^\alpha = \mu^\alpha n^\alpha + \gamma A \qquad (15.3)$$

$$Ad\gamma = -S^{\alpha+\sigma}dT + V^\alpha dp^\alpha - n^\alpha d\mu^\alpha \qquad (15.4)$$

In unary systems, the interfacial variables, $x^\sigma \equiv X^\sigma/A$ ($X = S, U, F$), are $s^\sigma = -d\gamma/dT$, $u^\sigma = \gamma - Td\gamma/dT$, and $\gamma = f^\sigma$, so that Equation 15.4 reduces to

$$d\mu^\alpha = -s^\alpha dT + v^\alpha dp^\alpha \qquad (15.5)$$

where $v^\alpha \equiv V^\alpha/n^\alpha$ and $s^\alpha \equiv S^\alpha/n^\alpha$ are the molar volume and entropy of phase α.

Introducing the energy $\mathsf{E} \equiv \gamma A(1 - d\ln A/d\ln n^\alpha)$ and the molar chemical potential $\mu^{\alpha+\sigma} \equiv \mu^\alpha + \gamma(dA/dn^\alpha)$, the equations of the $(\alpha + \sigma)$ system can be transformed to

$$dU^{\alpha+\sigma} = TdS^{\alpha+\sigma} - p^\alpha dV^\alpha + \mu^{\alpha+\sigma}dn^\alpha \qquad (15.6)$$

$$U^{\alpha+\sigma} - TS^{\alpha+\sigma} + p^\alpha V^\alpha = \mu^{\alpha+\sigma}n^\alpha + \mathsf{E} \qquad (15.7)$$

$$d\mathsf{E} = -S^{\alpha+\sigma}dT + V^\alpha dp^\alpha - n^\alpha d\mu^{\alpha+\sigma} \qquad (15.8)$$

It is noteworthy that the energy E only appears in the Euler and Gibbs–Duhem equations and that its value is determined by the relation between A and n^α, which depends on the system geometry. Moreover, the Gibbs potential $G^{\alpha+\sigma} = \mu^{\alpha+\sigma} n^\alpha + \mathsf{E}$ is bulk nonextensive, that is, $G^{\alpha+\sigma}$ is not proportional to n^α because neither does E.

When phase α is a spherical drop of radius r, the conditions $dV^\alpha = 4\pi r^2 dr$ and $dA = 8\pi r dr = (2/r)dV^\alpha$ are satisfied. The mechanical equilibrium condition, $(\partial F^{\alpha+\beta+\sigma}/\partial V^\alpha)_{T,n^\alpha,n^\beta} = 0$, leads to the Young–Laplace equation $p^\alpha = p^\beta + 2\gamma/r$, and the distribution equilibrium condition, $d\mu^\alpha = d\mu^\beta$ with Equations 15.1 and 15.5, requires that (DeHoff 2006)

$$(s^\beta - s^\alpha)dT - (v^\beta - v^\alpha)dp^\beta + 2\gamma v^\alpha d(1/r) = 0 \qquad (15.9)$$

Equation 15.9 allows us to evaluate the dependence of the thermodynamic properties on the curvature radius r. This dependence becomes significant for the nanoscale and practically disappears for microparticles.

15.3.1.2 Phase Diagrams

The influence of curved interfaces upon the behavior of materials systems is manifested primarily through the shift of phase boundaries on phase diagrams derived from the altered condition of mechanical equilibrium (Defay and Prigogine 1966, DeHoff 2006). For a number of substances, the metastable high-pressure phases and even some more dense packing phases do not exist in the bulk state. However, these phases are easily formed at the ambient pressure when the material size decreases to the nanoscale. For instance, in the nucleation stage of clusters from gases during chemical vapor deposition (CVD), the phase stability is quite different from that of the phase diagram that is determined at ambient pressure (Figure 15.1). The high additional internal pressure associated with the interfacial free energy through Young–Laplace equation makes it possible to observe "unusual" phases (Zhang et al. 2004, Wang and Yang 2005). Thus, nanodiamond has been found to be more stable than nanographite when the crystal size approaches the deep nanoscale (Yang and Li 2008).

15.3.1.3 Kelvin's Equation for the Vapor Pressure of a Drop

If phase β is the vapor of the condensed phase α, the integration of Equation 15.9 at constant temperature making use of the approximation $v^\alpha \ll v^\beta \approx RT/p^\beta$ leads to

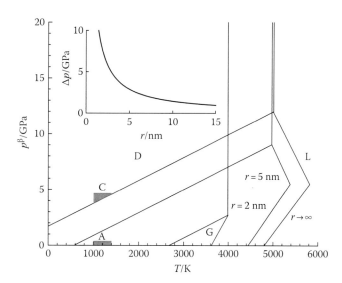

FIGURE 15.1 *T–p* phase diagrams of bulk ($r \to \infty$) and nanocrystalline (r = 2 and 5 nm) carbon: G, D, and L denote graphite, diamond, and liquid carbon. The inset shows the additional pressure ($p^\alpha - p^\beta$) as a function of the nuclei radius r. The A region means a metastable phase region of CVD diamond nucleation; the C region means a new stable phase region of CVD diamond nucleation with respect to the effect of nanosized-induced additional pressure. (Adapted from Yang, C.C. and Li, S., *J. Phys. Chem. C*, 112, 1423, 2008; Zhang, C.Y. et al., *J. Phys. Chem. B*, 108, 2589, 2004. With permission.)

$$RT \ln \frac{p^\beta(T,r)}{p^\beta(T,\infty)} = \frac{2\gamma v^\alpha}{r} \quad (15.10)$$

This equation, derived by Lord Kelvin in 1871, shows that the vapor pressure of a substance can be increased by preparing it in the form of particles of radius r. That is, the spherical particles (with a radius smaller than about 1 μm) have a greater tendency to evaporate than a flat surface of the same bulk material at the same temperature. Interestingly, the term $2\gamma v^\alpha/r$ has been shown to be proportional to the interfacial energy $u^\sigma = \gamma - T d\gamma/dT$ (Dobruskin 2006) and to the enthalpy of formation of a spherical particle from bulk material (Schäfer 2003).

15.3.1.4 Chemical Potential of a Spherical Particle

The greater tendency to evaporate commented above implies that the chemical potential $\mu^\alpha = \mu^\beta = \mu$ must increase with decreasing particle size. Using $v^\alpha \ll v^\beta$ in Equation 15.9, the variation of the chemical potential at constant temperature is $d\mu = v^\beta dp^\beta \approx 2\gamma v^\alpha d(1/r)$, and integration leads to the Gibbs–Thomson–Freundlich equation

$$\mu(T,r) = \mu(T,\infty) + \frac{2\gamma v^\alpha}{r} \quad (15.11)$$

This equation is valid when the phases α and β are under conditions of thermodynamic equilibrium, so that the vapor pressure is given by Kelvin's equation and that of the condensed phase is given by Young–Laplace equation.

15.3.1.5 Solubility

The solubility of solid particles in a liquid solution is also size-dependent. In the solid phase, the chemical potential of the dissolving component (denoted by index 1) is $\mu_1^s(T,r) = \mu_1^s(T,\infty) + 2\gamma v_1^s/r$ and in the liquid solution, it is $\mu_1^l = \mu_1^{\circ,l} + RT \ln(y_1^l c_1^l)$. If the activity coefficient y_1^l of this component in the liquid solution is not much affected by the change in particle radius, the equilibrium condition $\mu_1^s(r) = \mu_1^l$ implies that the molar concentration of the dissolving component under equilibrium conditions increases with decreasing particle size according to the Ostwald–Freundlich relation (Defay and Prigogine 1966, Kondepudi 2008)

$$c_1^l(r) = c_1^l(\infty) \exp\left[\frac{2\gamma v_1^s}{RT}\frac{1}{r}\right] \quad (15.12)$$

That is, in the same way that Equation 15.10 shows that a liquid drop is only in equilibrium with a supersaturated vapor, Equation 15.12 shows that a solid particle is only in equilibrium with a supersaturated solution. One of the most interesting consequences of the higher solubility of smaller particles is the Ostwald ripening phenomenon first described in 1896 and fundamental in modern technology for the solution synthesis of quantum dots (Vengrenovich et al. 2001).

15.3.1.6 Melting Point Depression

The distribution equilibrium condition can be written in terms of molar enthalpies and entropies as $h^\beta - h^\alpha = T(s^\beta - s^\alpha)$. At constant p^β, Equation 15.9 can then be integrated to give

$$\ln \frac{T(p^\beta, r)}{T(p^\beta, \infty)} = -\frac{2\gamma v^\alpha}{h^\beta - h^\alpha}\frac{1}{r} \quad (15.13)$$

When this equation is applied to the liquid–vapor equilibrium, $h^\beta - h^\alpha = \Delta H_{vap} > 0$ is the molar enthalpy of vaporization and therefore $T_{vap}(p^v, r) < T_{vap}(p^v, \infty)$, which is equivalent to Kelvin's equation for the vapor pressure.

When Equation 15.13 is applied to the equilibrium of solid particles in the liquid of the same component, $h^\beta - h^\alpha = \Delta H_m^b > 0$ is the bulk molar enthalpy of melting and $T_m(r) = T_m(p^l, r) < T_m(p^l, \infty) = T_m^b$, which shows that the spherical particles melt at lower temperature than the corresponding bulk phase. If we consider that this effect is not very large, Equation 15.13 can be transformed to Gibbs–Kelvin equation (Couchman and Jesser 1977)

$$1 - \frac{T_m(r)}{T_m^b} = \frac{2\gamma^{sl}v^s}{\Delta H_m^b}\frac{1}{r} \quad (15.14)$$

Many authors have attempted to predict the theoretical dependence of the melting point on the particle size using different thermodynamic approaches. Most studies consider spherical particles and thermodynamic equilibrium conditions between homogeneous bulk-like phases, and deduce equations of the form (Peters et al. 1998)

$$1 - \frac{T_m(r)}{T_m^b} = \frac{2\beta_m}{\Delta H_m^b \rho^s} \frac{1}{r} \qquad (15.15)$$

where ρ^s is the mass density of the solid and β_m is a parameter that depends on the model and is related to the interfacial free energy. Since atomistic features are missing, these models are expected to be valid only when the condensed phases contain at least several atomic layers. The thermodynamic equilibrium condition can be written in terms of the equality of the chemical potentials of the solid and melted particle at the melting point (Buffat and Borel 1976), equality of vapor pressures (Chushak and Bartell 2001), or extremal for the free energy (Vanfleet and Mochel 1995). In particular, using the latter type of approaches, Reiss et al. (1988) showed that the condition of equality of chemical potential is incorrect, and the work by Bartell and Chen (1992) added further caveats.

Although the $1/r$ dependence is widely accepted in the case of nanoparticles whose diameter is larger than a few nanometers, at very low radii, some studies have shown that the melting temperature depends nonlinearly on the reciprocal radius (Chushak and Bartell 2000). If we consider even smaller nanoparticles, i.e., atomic clusters, the study of the melting transition is necessarily more complicated not only because of experimental difficulties, but also because the very concept of melting has no meaning for atoms and molecules, and there must be a minimum size of the cluster that allows to classify the state of the atoms as solid- or liquid-like. At low temperatures, the atoms in a cluster make only small amplitude vibrations around a fixed position. At the melting temperature, the motion becomes quite anharmonic. At even higher temperatures, atoms in the cluster can visit neighboring places and start a diffusive motion (Schmidt and Haberland 2002).

15.3.1.7 Size Dependence of Interfacial Free Energy

The derivation of Equations 15.3 and 15.4 is based on the assumption that the interfacial contribution to the thermodynamic potentials is proportional to the interfacial area A. However, this area is not a good state variable when the particle size is very small and the interfacial variables $x^\sigma \equiv X^\sigma/A$ are no longer independent of A. This occurs because the approximation of size-independent interfacial free energy is not valid for very small particles of, e.g., $r < 1$ nm in the case of water and metal nuclei (Onischuk et al. 2006). The thermodynamic theory of Tolman (1949) suggests that the interfacial free energy of liquids changes with the droplet radius as

$$\gamma(r) = \frac{\gamma(\infty)}{1 + 2\delta/r} \qquad (15.16)$$

where the Tolman length δ is of the order of 0.1 nm. This size dependence of the interfacial free energy is correlated with the mechanical instability of small objects (Samsonov et al. 2003).

15.3.1.8 Nanocrystalline Solids

Most of the results of surface thermodynamics explained above were originally devised with attention to fluids, and hence isotropic behavior is assumed. However, many nanomaterials are nonisotropic crystalline solids, like nanoparticles with faceting effects and polycrystalline solids with a nanoscale grain size. Thus, for instance, since the shape of grains can be arbitrary, there is no way to relate energy A and n^α and the Gibbs–Thomson–Freundlich equation (Equation 15.11) does not apply to the "nanograins" in polycrystalline solids. The generalization of the formalism of surface thermodynamics to solids is among the achievements in thermodynamics in the twentieth century (Weissmüller 2002).

15.3.2 Hill's Nanothermodynamics

T. L. Hill (Hill 1963) expressed the belief that "The applicability of statistical mechanical ensemble theory to small systems as well as large suggests that a parallel thermodynamics should exist." In the nanoscale, the systems consist of only several tens to several hundred atoms and this casts some doubts on the statistical meaning of thermodynamic variables. Macroscopic thermodynamics should not be applied to a single small system, but it can be applied to, e.g., a solution of small systems which is considered as a Gibbs "ensemble" of independent small systems. Macroscopic thermodynamic functions are well defined for such a large sample of small systems. The thermodynamic variables of one small system should then be understood as averages over the ensemble of small systems, since it is this ensemble that we observe. This is one of the pillars of Hill's nanothermodynamics (Hill 1962, 1963, 1964), a theory whose fundamental thermodynamic equations for a small system involve average values of fluctuating extensive quantities. Nanothermodynamics provides thermodynamic functions and relations for a single small system, including, in general, variations in the system size. Allowance of these variations in size is, indeed, the important new feature of nanothermodynamics.

15.3.2.1 Subdivision Potential

Hill's theory is a generalization of classical thermodynamics that accounts for size effects via the introduction of a new thermodynamic potential called the subdivision potential E, and its conjugate variable, the number of small systems N. This potential can be positive or negative, depending on the nature of the small systems, and takes into account the energetic contributions usually negligible for macroscopic systems, such as surface effects, system rotation, etc. To understand its meaning, we should compare two composite systems ($i = 1$ and 2) with the same extensive variables S_t, V_t and N_t and differing in the number of small systems ($\mathsf{N}_1 \neq \mathsf{N}_2$). The relation between the extensive variables of a small system and those of the collection of small systems, identified with a subscript t, is $X_i = X_t/\mathsf{N}_i$, ($X = U, S, V, N$) ($i = 1, 2$). In classical thermodynamics, the Euler equations $U_1 = TS_1 - pV_1 + \mu N_1$ and $U_2 = TS_2 - pV_2 + \mu N_2$ would lead us to the conclusion that $\mathsf{N}_1 U_1 = \mathsf{N}_2 U_2$ or, equivalently, that the two composite systems have the same energy, $U_{t1}(S_t, V_t, N_t, \mathsf{N}_1) = U_{t2}(S_t, V_t, N_t, \mathsf{N}_2)$, in agreement with the fact that the Gibbs equation, $dU_t = TdS_t - pdV_t + \mu dN_t$, forbids the variation of U_t while keeping constant S_t, V_t, and N_t. In nanothermodynamics, on the contrary, it is considered that $U_{t1}(S_t, V_t, N_t, \mathsf{N}_1) \neq U_{t2}(S_t, V_t, N_t, \mathsf{N}_2)$.

The formulation of Hill's nanothermodynamics is based on the idea that the natural variables of the internal energy U_t are $(S_t, V_t, N_t, \mathsf{N})$ and that they can be varied independently. Thus, the Gibbs equation in nanothermodynamics is

$$dU_t = TdS_t - pdV_t + \mu dN_t + \mathsf{E}d\mathsf{N} \qquad (15.17)$$

where, unlike that used in Equations 15.1 through 15.8, the chemical potential $\mu \equiv (\partial U_t / \partial N_t)_{S_t, V_t, \mathsf{N}}$ is defined here per particle. The last two terms in Equation 15.17 bear some similarity, so that the subdivision potential $\mathsf{E} \equiv (\partial U_t / \partial \mathsf{N})_{S_t, V_t, N_t}$ is like the chemical potential of a small system; but the energy required to add another identical small system to the ensemble is $(\partial U_t / \partial \mathsf{N})_{S, V, N} = U \neq \mathsf{E}$. Strictly, E is the energy required to increase in one unit the number of subdivisions of the composite system, while keeping constant the total number of particles and other extensive parameters. That is, Hill's theory incorporates the possibility that processes taking place in a closed system produce or destroy small systems N (Chamberlin 2002).

In classical thermodynamics, the energy U_t is assumed to be a first-order homogeneous function of its natural variables, that is, U_t is extensive. This leads, for instance, to the Euler equation $G_t(T, p, N_t) = U_t - TS_t + pV_t = \mu N_t$. We then say that the intensive state of the system is determined by intensive variables such as T and p, and that N_t determines the size of the system. The extensive character of the thermodynamic potentials like U_t and G_t means that, for a given intensive state, they are proportional to N_t.

In nanothermodynamics the energy U_t is also assumed to be a first-order homogeneous function of all its natural variables, N included. This can be justified because the composite system is macroscopic and the small systems are noninteractive. Therefore, the Euler and Gibbs–Duhem equations take the form

$$U_t = TS_t - pV_t + \mu N_t + \mathsf{E}\mathsf{N} \qquad (15.18)$$

$$\mathsf{N}d\mathsf{E} = -S_t dT + V_t dp - N_t d\mu \qquad (15.19)$$

The Gibbs and Euler equations can also be presented as $dG_t = -S_t dT + V_t dp + \mu dN_t + \mathsf{E}d\mathsf{N}$ and $G_t(T, p, N_t, \mathsf{N}) = \mu N_t + \mathsf{E}\mathsf{N}$. Interestingly, we cannot conclude that G_t is proportional to N_t at constant T, p because N_t and N are independent variables. That is, G_t is no longer an extensive potential, and the same applies to other potentials like U_t.

Dividing Equations 15.18 and 15.19 by N, and remembering that $X = X_t/\mathsf{N}$, the thermodynamic equations for a small system are

$$U = TS - pV + \mu N + \mathsf{E} \qquad (15.20)$$

$$d\mathsf{E} = -SdT + Vdp - Nd\mu \qquad (15.21)$$

From Equations 15.20 and 15.21, the Gibbs equation for a small system is

$$dU = TdS - pdV + \mu dN \qquad (15.22)$$

which turns out to be the same as in classical thermodynamics. In classical thermodynamics, the Gibbs equation can be written in terms of intensive quantities, $x = X/N$, as $du = Tds - pdv$, and the chemical potential does not show up in this equation because N is not a natural variable of u. Similarly, the subdivision potential does not show up in Equation 15.22 because N is not a natural variable of U. On the contrary, the subdivision potential appears in Equation 15.17 because N is a natural variable of U_t, in the same way as μ appears in Equation 15.22 because N is a natural variable of U.

The Gibbs potential of a small system is $G(T, p, N) = \mu N + \mathsf{E}$ and its Gibbs equation is $dG = -SdT + Vdp + \mu dN$. The important point to be noticed is that μ and E can still vary when T and p are kept constant (see Equation 15.21) and therefore we cannot conclude that G is proportional to N. In multicomponent small systems, the Gibbs potential is $G = \sum_i \mu_i N_i + \mathsf{E}$. The presence of the subdivision potential is characteristic of small systems and evidences the nonextensive character of the Gibbs potential (Gilányi 1999).

15.3.2.2 Relation between Nano, Surface, and Nonextensive Thermodynamics

Hill's nanothermodynamics can describe interfacial contributions in a very natural way; and it can also describe nonextensive contributions of different nature. When the nonextensivity of the thermodynamic potentials arises from interfacial effects, Equations 15.20 through 15.22 would become identical to Equations 15.6 through 15.8 if we choose to define the subdivision potential as $\mathsf{E} \equiv \gamma A(1 - d \ln A / d \ln n^\alpha)$, where γ is the interfacial free energy. In relation to Equations 15.6 through 15.8, we should remember that the interfacial quantities are proportional to the interfacial area A, that the bulk extensive quantities are proportional to the number of moles n^α of the single component in bulk phase α and, more importantly, that the relation between A and n^α depends on the system geometry. If A were directly proportional to n^α then the subdivision potential $\mathsf{E} \equiv \gamma A(1 - d \ln A / d \ln n^\alpha)$ would vanish. However, this is not generally the case and the subdivision potential then accounts for the interfacial contributions to the thermodynamic potentials.

In relation to this, it can be mentioned that some authors present the Gibbs equation in nonextensive thermodynamics as

$$dU = TdS - pdV + \mu dn + \tau d\chi \qquad (15.23)$$

where n is the number of moles, μ is the chemical potential (per mole), and τ and χ are the quantities introduced to account for nonextensivity. Equation 15.23 holds a close resemblance to Equations 15.2 and 15.17, so that τ is associated with γ and χ with

A in the case of interfacial systems. Letellier et al. (2007a,b) have used Equation 15.23 to derive the Ostwald–Freundlich relation (Equation 15.12) and the Gibbs–Kelvin equation for the melting point depression (Equation 15.13). The magnitudes τ and χ could be used to describe nonextensive contributions other than interfacial, and power laws of the type $\chi \propto n^m$ with $m \neq 2/3$ can also be used. The value of *m* becomes thus a free parameter that must be determined by comparison with experiments (Letellier et al. 2007b).

15.3.2.3 Generalized or Completely Open (T, p, μ) Ensemble

The extensive character of internal energy and entropy in macroscopic thermodynamics implies that the intensive state of a monocomponent system can be characterized by two variables, e.g., *T* and *p*. If we consider a collection of macroscopic small systems, *T* and *p* take the same values for a single small system and for the collection of small systems. These variables contain no information on the size of the small systems. In nonextensive small systems, on the contrary, *T* and *p* do vary with system size and we need one additional variable to specify the state of the system. The Gibbs–Duhem equation (Equation 15.21) evidences one of the key features of nanothermodynamics: the intensive parameters *T*, *p*, and μ can be varied independently due to the additional degree of freedom brought by nonextensivity. This enables the possibility of using the completely open or generalized (T, p, μ) statistical ensemble. This ensemble describes the behavior of small systems in which the extensive variables, such as the amount of matter in the system, fluctuate under the constraint that intensive variables (T, p, μ) are fixed by the surroundings.

The equilibrium probability distribution in the completely open ensemble is

$$p_j = \frac{1}{Y} e^{-\beta(E_j + pV_j - \mu N_j)} \tag{15.24}$$

where $\beta \equiv 1/k_B T$ and $Y(T, p, \mu) \equiv \sum_j e^{-\beta(E_j + pV_j - \mu N_j)}$ is the generalized partition function and the sum extends over microstates. Introducing the absolute activity $\lambda \equiv e^{\beta\mu}$, this partition function can also be written as $Y(T, p, \lambda) \equiv \sum_N \lambda^N \Xi(T, p, N)$. The relation with the subdivision potential is

$$\mathsf{E} = -k_B T \ln Y \tag{15.25}$$

and the Gibbs–Duhem equation (Equation 15.21) allows us to obtain the extensive variables of the small system as $S = -(\partial \mathsf{E}/\partial T)_{p,\mu}$, $\langle V \rangle = (\partial \mathsf{E}/\partial p)_{T,\mu}$, $\langle N \rangle = -(\partial \mathsf{E}/\partial \mu)_{T,p}$, and $U = TS - p\langle V \rangle + \mu\langle N \rangle + \mathsf{E}$. When the small systems can be assumed to be incompressible, so that *p* and *V* are not state variables, the partition function *Y* becomes equal to the grand partition function, and the subdivision potential becomes then equal to the grand potential. The generalized ensemble is incompatible with the thermodynamic limit, $N \to \infty$.

15.3.2.4 The Incompressible, Spherical Aggregate, and the Critical Wetting Transition

In macroscopic thermodynamics, all the ensembles are equivalent and predict the same values and relations for the thermodynamics potentials and variables. In nanothermodynamics, on the contrary, this is no longer true. The state of the nanosystem is affected by the fluctuations in its thermodynamic variables and these are determined by the surroundings, so the statistical description of the nanosystem has to be done using the ensemble that correctly describes the constraints imposed to the nanosystem. We can illustrate this statement by describing a spherical aggregate under two different environmental constraints: canonical (T, N) and grand canonical (T, μ). The crystallite is assumed to be incompressible, so that *p* and *V* are not state variables; and there is no difference between the Gibbs and Helmholtz potentials, *G* and *F*, on the one hand, and the subdivision and grand potentials, E and Ω, on the other hand.

Consider first that each aggregate contains *N* particles in a volume *V*. Each particle has an intrinsic partition function $z(T) = z'(T)e^{\beta\varepsilon}$, which also includes the energy of interaction per particle, $-\varepsilon$. The canonical partition function is then

$$Z(T, N) = z^N e^{-\beta a N^{2/3}} \tag{15.26}$$

where $aN^{2/3} = \gamma A = F^\sigma$ is the surface contribution to the free energy of the crystallite. This partition sum is valid only when $N \gg 1$ (though not macroscopic) since it assumes that it is possible to distinguish a surface and a bulk in the aggregate. The subdivision potential is

$$\mathsf{E}(T, N) = G - \mu N = -k_B T \left[\ln Z - N \left(\frac{\partial \ln Z}{\partial N} \right)_\beta \right] = \frac{1}{3} a N^{2/3} \tag{15.27}$$

In the thermodynamic limit, $N \to \infty$, the surface contribution is negligible and $\mathsf{E}/N \to 0$. This result is in agreement with Equation 15.7, since $\mathsf{E} \equiv \gamma A (1 - d \ln A/d \ln N) = \gamma A/3$ for spherical particles, $A \propto N^{2/3}$. Note also that, since $A = 4\pi (3V/4\pi)^{2/3}$ and the molar volume in the condensed phase is $v^l = V N_A/N$, the relation between parameter *a* and the interfacial free energy is $a = \gamma \pi^{1/3} (6 v^l/N_A)^{2/3}$.

Consider now that the aggregates are in (distribution) equilibrium with a solution of the particles, which fixes *T* and μ. The generalized partition function $Y(T, \mu)$ or $Y(T, \lambda)$ is

$$Y(T, \lambda) = \sum_{N=0}^{\infty} Z(T, N) \lambda^N \approx \sum_{N=0}^{\infty} e^{-\beta a N^{2/3}} (z\lambda)^N \tag{15.28}$$

where $\lambda = e^{\beta\mu}$ is the absolute activity of the particles. The approximation sign is used because this form of $Z(T, N)$ is expected to be good for relatively large *N* only. The chemical potential in a bulk liquid is $\mu^\infty = -k_B T \ln z$ and the corresponding absolute activity is $\lambda^\infty = 1/z$. If $\lambda > \lambda^\infty$ the sum diverges and a finite system is not possible. In order to obtain aggregates of reasonable size we must choose $\lambda^\infty - \lambda \geq 0$, that is the aggregate must be approximately in equilibrium with the bulk liquid; the aggregates could be clusters in a saturated vapor phase in equilibrium with a liquid. But even with a saturated vapor, the aggregate will be sizeable only near the critical temperature, when the interfacial energy is very small, $a(T)N^{2/3} \ll k_{\mathrm{B}}T$.

Because we are interested only in $\langle N \rangle$ being fairly large, we can replace the sum in Y by an integral and extend the integration to $N = 0$. Even though the expression used for $Z(T, N)$ is not good when N is very small, the error introduced in this region of the integration is not serious when $\langle N \rangle$ is large enough (Hill 1964). Thus, the partition function is

$$Y(T, \lambda) \approx \int_0^\infty \mathrm{e}^{-\beta a N^{2/3}} (z\lambda)^N \mathrm{d}N \approx \int_0^\infty \mathrm{e}^{-\beta a N^{2/3}} \left[1 + N \ln(z\lambda)\right] \mathrm{d}N$$

$$= \frac{3\sqrt{\pi}}{4(\beta a)^{3/2}} + \frac{3}{(\beta a)^3} \ln(z\lambda) \tag{15.29}$$

The average number of particles in an aggregate is then

$$\langle N \rangle = \lambda \left(\frac{\partial \ln Y}{\partial \lambda} \right)_T \approx \frac{4}{\sqrt{\pi}(\beta a)^{3/2}} = \frac{2N_A}{3\pi v^1} \left(\frac{k_{\mathrm{B}}T}{\gamma} \right)^{3/2} \tag{15.30}$$

and the subdivision potential is

$$\mathsf{E}(T, \lambda) = -k_{\mathrm{B}}T \ln Y \approx -k_{\mathrm{B}}T \ln \frac{3\sqrt{\pi}}{4(\beta a)^{3/2}} = -k_{\mathrm{B}}T \ln \frac{3\pi \langle N \rangle}{16} \tag{15.31}$$

As it should be expected, the aggregate size $\langle N \rangle$ increases as the interfacial free energy decreases.

This example illustrates two interesting characteristics of nanothermodynamics. First, the state of the small system is determined by the environment and different ensembles are not equivalent. For instance, the subdivision potential is different in the two ensembles, $\mathsf{E}(T, N) = aN^{2/3}/3$ and $\mathsf{E}(T, \lambda) = -k_{\mathrm{B}}T$ $\ln(3\pi\langle N \rangle/16)$, and any magnitude related to the aggregate size fluctuation can only be described in the generalized ensemble. Second, we can determine an extensive variable, like the average aggregate size $\langle N \rangle$, from the intensive variables T and μ. In macroscopic thermodynamics, this would be impossible since, for example, the amount of water in a container cannot be determined from the values of T and μ. However, in nanothermodynamics, considering a completely open ensemble characterized by intensive parameters and evaluating the generalized partition function, all the equilibrium information on the small system is known, including the extensive parameters.

These results are found useful in analyzing the liquid–vapor equilibrium near the critical temperature and, particularly, cluster formation in metastable states. A recent successful application of nanothermodynamics to determine the critical wetting parameter is based on a modification of Equation 15.29. When

rotational and translational degrees of freedom of the aggregate are taken into account, the generalized partition function is

$$Y(T, \lambda) \propto \int_0^\infty \mathrm{e}^{-\beta a N^{2/3}} \left[1 + N \ln(z\lambda)\right] N^d \, \mathrm{d}N$$

$$\approx \frac{3[(3d + 1)/2]!}{2(\beta a)^{3(1+d)/2}} + \frac{3[(3d + 4)/2]!}{2(\beta a)^{3(1+d/2)}} \ln(z\lambda) \tag{15.32}$$

where d is related to the aggregate dynamics. The average number of particles in an aggregate is then

$$\langle N \rangle = \lambda \left(\frac{\partial \ln Y}{\partial \lambda} \right)_T \approx \frac{[(3d + 4)/2]!}{[(3d + 1)/2]!} \frac{\ln(z\lambda)}{(\beta a)^{3/2}} \tag{15.33}$$

and the critical wetting parameter is obtained as (García-Morales et al. 2003)

$$\omega_{\mathrm{c}} = 4 \frac{[(3d + 4)/2]!}{[(3d + 1)/2]!} \tag{15.34}$$

Good agreement with the experimentally determined value is obtained for $d = 2$, which corresponds to a combination of translation and vortex rotational motion of the aggregate.

15.3.2.5 Lattice Model for the Ideal Gas

As a second example, we discuss the lattice ideal gas. In this model, N particles are distributed in a lattice of sites of volume v_{s} so that every site can accommodate only one particle. In the isothermal–isobaric ensemble, the partition sum is

$$\Xi(T, p, N) = \sum_{N_{\mathrm{S}} = N}^\infty \frac{N_{\mathrm{s}}!}{N!(N_{\mathrm{s}} - N)!} x^{N_{\mathrm{s}}} z^N = \frac{(xz)^N}{(1 - x)^{N+1}} \tag{15.35}$$

where $x \equiv \mathrm{e}^{-\beta p v_{\mathrm{s}}}$ and $z(T)$ is the intrinsic partition function. The average number of sites is

$$\langle N_{\mathrm{s}} \rangle = x \left(\frac{\partial \ln \Xi}{\partial x} \right)_{T, N} = \frac{N + x}{1 - x} \tag{15.36}$$

and its relative fluctuation is

$$\frac{\sigma_{N_{\mathrm{s}}}}{\langle N_{\mathrm{s}} \rangle} = \frac{\sqrt{\langle N_{\mathrm{s}}^2 \rangle - \langle N_{\mathrm{s}} \rangle^2}}{\langle N_{\mathrm{s}} \rangle} = \frac{\sqrt{(N + 1)x}}{N + x} \propto \frac{1}{\sqrt{N}} \tag{15.37}$$

where we have used that $x \approx 1$. That is, the fluctuations in the volume vanish when the system becomes macroscopic. This is a consequence of the fact that the extensive variable N is fixed in this ensemble. The subdivision potential is $\mathsf{E}(T, p, N) = G - \mu N = k_{\mathrm{B}}T \ln(1 - x) \approx p v_s$, where $G = -k_{\mathrm{B}}T \ln \Xi$ is the Gibbs potential and

$\mu = -k_\mathrm{B}T(\partial \ln \Xi / \partial N)_{T,p}$ is the chemical potential. The chemical potential μ equals to G/N only for macroscopic systems, $N \gg 1$. The relation between T, p, and μ for the latter is $(1 + \lambda z)x = 1$, where $\lambda \equiv e^{\beta \mu}$ is the absolute activity.

Similarly, in the grand canonical ensemble the partition sum is

$$Q(T,V,\mu) = \sum_{N=0}^{N_s} \frac{N_s!}{N!(N_s - N)!}(\lambda z)^N = (1 + \lambda z)^{N_s} \tag{15.38}$$

The macroscopic equilibrium relation between T, p, and μ is again $(1 + \lambda z)x = 1$, where pressure is determined as $p = (k_\mathrm{B}T/v_s)(\partial \ln Q / \partial N_s)_{T,\mu}$. The average number of particles is

$$\langle N \rangle = \lambda z \left(\frac{\partial \ln Q}{\partial (\lambda z)} \right)_{T,V} = \frac{N_s \lambda z}{1 + \lambda z} \tag{15.39}$$

and its relative fluctuation is

$$\frac{\sigma_N}{\langle N \rangle} = \frac{\sqrt{\langle N^2 \rangle - \langle N \rangle^2}}{\langle N \rangle} = \frac{1}{\sqrt{N_s \lambda z}} \propto \frac{1}{\sqrt{\langle N \rangle}} \tag{15.40}$$

That is, the fluctuations in the number of particles is again normal and vanish when the system becomes macroscopic, which is a consequence of the fact that the extensive variable V is fixed in this ensemble. The subdivision potential is $\mathsf{E}(T,V,\mu) = G - \mu \langle N \rangle = \Omega - pV = 0$.

Finally, the generalized partition sum is

$$Y(T,p,\mu) = \sum_{N_s=0}^{\infty} \sum_{N=0}^{N_s} \frac{N_s!}{N!(N_s - N)!} x^{N_s} (\lambda z)^N$$

$$= \sum_{N_s=0}^{\infty} (1 + \lambda z)^{N_s} x^{N_s} = \frac{1}{1 - (1 + \lambda z)x} \tag{15.41}$$

where the condition $(1 + \lambda z)x < 1$, required for the convergence of the sum, determines when the small system can exist. The subdivision potential is $\mathsf{E}(T,p,\mu) = -k_\mathrm{B}T \ln Y = k_\mathrm{B}T \ln [1 - (1 + \lambda z)x]$. From this partition sum the average number of particles is

$$\langle N \rangle = \lambda \left(\frac{\partial \ln Y}{\partial \lambda} \right)_{T,x} = \frac{\lambda z x}{1 - (1 + \lambda z)x} \tag{15.42}$$

and the average number of sites is $\langle N_s \rangle = x(\partial \ln Y / \partial x)_{T,\lambda} = \langle N \rangle (1 + 1/\lambda z)$. The important result now is that the relative fluctuation is $\sigma_N / \langle N \rangle = \sqrt{(1-x)/\lambda z x} > 1$. The fact that no extensive variable is held constant in the generalized ensemble has an important consequence: since there is no fixed extensive variable that provides some restraint on the fluctuations of extensive properties, they are of a larger magnitude than in other ensembles

(Hill and Chamberlin 2002). This example also shows that different ensembles lead to different results and stress the importance of choosing the right ensemble for the problem at hand.

15.3.2.6 Micelle Formation

Hill's nanothermodynamics has been applied to study ionic (Tanaka 2004) and nonionic micelles in solution (Hall 1987), as well as to describe polymer–surfactant complex formation (Gilányi 1999). The micellar solution of surfactant is treated as a completely open ensemble of small systems (micelles or polymer–surfactant complexes) dispersed in monomeric solution. The main advantage of the nanothermodynamics approach is that it imposes no restrictions on the distribution of micelles sizes.

15.3.2.7 Reactions Inside Zeolite Cavities and Other Confined Spaces

Crystalline zeolites with well-defined cavities and pores have long been used in the chemical industry as nanospaces for catalytic reactions. During the last decades, a variety of tailor-made "nanoreactors," with confined nanospaces where selected chemical reactions can take place very efficiently in controlled environments, have been fabricated and studied. The growing research activity in this area is also justified from the observed increased reactivity in nanospaces.

Hill (1963) applied nanothermodynamics to the study of the isomerization reaction in small closed systems and found no difference in the reaction extent from the macroscopic behavior. However, Polak and Rubinovich (2008) have considered other reactions and found a universal confinement effect that explains why the equilibrium constants of exothermic reactions are significantly enhanced in confined geometries that contain a small number of reactant and product molecules. The effect is universal in the sense that it has an entropic origin associated to the fact that when the number of molecules is small the Stirling approximation cannot be used to evaluate the number of microstates, that is, it is related to the nonextensivity of the entropy in small systems.

15.3.2.8 Mean-Field Theory of Ferromagnetism

In the classical mean-field theory of ferromagnetism, the material is described as a lattice of N particles (or spins) that can have two orientations. Let $s_i = \pm 1$ be the orientation variable of spin i, l be the number of particles in the up state ($s_i = +1$), J be the strength of the exchange interaction, and c be the coordination number of the lattice. The average energy per particle within the mean-field approximation is (Chamberlin 2000)

$$\varepsilon(l,N) = -\frac{cJ}{2}\left(4\frac{l(l-1)}{N(N-1)} - 4\frac{l}{N} + 1 \right) \tag{15.43}$$

In the macroscopic limit, $N \geq l \gg 1$, this equation simplifies to $u^\infty(m) = -cJm^2/2$, where $m = (1/N)\sum_{i=1}^{N} s_i = (2l/N) - 1$ is the average value of the spin orientation variable. In this limit, the thermodynamic potentials are extensive and the canonical partition sum factorizes, $Z^\infty(T, N) = [z^\infty(T)]^N$ where $z^\infty = 2e^{-\beta cJm^2/2}\cosh[\beta cJm]$. Introducing the macroscopic critical temperature $T_c^\infty \equiv cJ/k_B$, the relation between T and m under thermodynamic equilibrium conditions is $T/T_c^\infty = m/\text{arctanh}(m)$, and the absolute activity $\lambda^\infty = e^{\beta\mu^\infty} = 1/z^\infty$ can be presented as $\lambda^\infty(m) = \exp[m\,\text{arctanh}(m)/2]\sqrt{1-m^2}/2$, and the entropy per particle is $s^\infty(m)/k_B = (u^\infty - \mu^\infty)/k_B T = -m\,\text{arctanh}(m) + \ln\left(2/\sqrt{1-m^2}\right)$. These two functions are represented together with $u^\infty(m)$ in Figures 15.2 through 15.4. Note that $m = 0$ and $\lambda^\infty = 0.5$ at $T \geq T_c^\infty$.

In the case of finite clusters, the canonical partition sum is

$$Z(T,N) = \sum_{l=0}^{N} \frac{N!}{l!(N-l)!} e^{-\beta N\varepsilon(l,N)} \qquad (15.44)$$

and the equilibrium thermodynamic properties per particle are obtained as $\lambda(T, N) = Z^{-1/N}$, $u(T, N) = -(\partial \ln Z/\partial\beta)_N/N$ and $s(T, N) = [\partial(k_B T \ln Z)/\partial T]_N/N$. These functions have also been represented in Figures 15.2 through 15.4 for $N = 20, 50$, and 100. Interestingly, the absolute activity and hence the free energy per particle is lower for finite-size clusters than for a macroscopic sample, and this effect is particularly significant for temperatures in the vicinity of T_c^∞. The reduction in free energy mostly

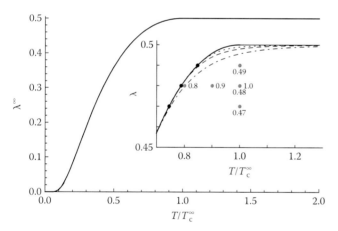

FIGURE 15.2 Temperature dependence of the absolute activity of ferromagnetic particles in the classical mean-field theory. In the case of finite-size ferromagnetic clusters, the mean-field theory predicts a decrease in the absolute activity, which is only noticeable in the vicinity of the macroscopic critical temperature T_c^∞. The inset shows the mean-field activity for restricted-size finite cluster with number of particles $N = 20$ (dot-dashed), 50 (dashed), 100 (dotted), and ∞ (solid line). The gray points in the inset describe the unrestricted-size finite clusters studied below. Some of them correspond to fixed temperature $T = T_c^\infty$ and variable activity $\lambda = 0.47, 0.48$, and 0.49; note that $\lambda^\infty(T_c^\infty) = 0.5$. The other gray points correspond to fixed activity $\lambda = 0.48$ and variable temperature $T/T_c^\infty = 1.0, 0.9$, and 0.8. Finally, the black points mark the finite-size critical temperatures for $\lambda = 0.48$ $(T_c/T_c^\infty = 0.846)$, 0.48 $(T_c/T_c^\infty = 0.788)$, and 0.47 $(T_c/T_c^\infty = 0.745)$; note that $\lambda = \lambda^\infty(T_c)$.

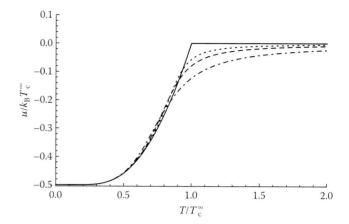

FIGURE 15.3 Internal energy per particle in the mean-field theory for restricted-size finite ferromagnetic cluster with number of particles $N = 20$ (dot-dashed), 50 (dashed), 100 (dotted), and ∞ (solid line). The classical Weiss transition at T_c^∞ is suppressed by finite-size effects.

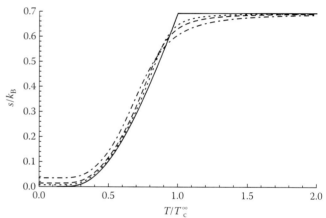

FIGURE 15.4 Entropy per particle in the mean-field theory for restricted-size finite ferromagnetic cluster with number of particles $N = 20$ (dot-dashed), 50 (dashed), 100 (dotted), and ∞ (solid line). The residual entropy $s(0) = k_B \ln 2/N$ appears because of the degeneracy associated to all-up and all-down spin states, and vanishes for macroscopic systems.

arises from the reduction in internal energy per particle due to fractionation in finite-size clusters. In particular, $m = 0$ and $u^\infty = 0$ for $T \geq T_c^\infty$ under equilibrium conditions but a macroscopic system can decrease its internal energy by subdividing into finite-size clusters, which can then become magnetized even for $T \geq T_c^\infty$.

Clusters with unrestricted sizes can be described by the generalized partition sum

$$Y(T,\lambda) = \sum_{N=2}^{\infty} \lambda^N \sum_{l=0}^{N} \frac{N!}{l!(N-l)!} e^{-\beta N\varepsilon(l,N)} \qquad (15.45)$$

where $\lambda \equiv e^{\beta\mu}$ and the sum over particles numbers start at $N = 2$ to avoid the ill-defined interaction energy of an isolated spin. The probability of finding a cluster of size N is then

$$p(N,T,\lambda) = \frac{\lambda^N}{Y(T,\lambda)} \sum_{l=0}^{N} \frac{N!}{l!(N-l)!} e^{-\beta N \varepsilon(l,N)} \quad (15.46)$$

and the average cluster size is $\langle N \rangle = \lambda(\partial \ln Y/\partial \lambda)_T = \sum_{N=2}^{\infty} Np(N,T,\lambda^N)$. The generalized partition sum only converges for $\lambda < \lambda^{\infty}(T)$, and the average cluster size becomes increasingly large as λ approaches $\lambda^{\infty}(T)$. The divergence of the average cluster size is associated to critical behavior and can be used to define a critical temperature $T_c(\lambda)$ for unrestricted finite-size clusters from the condition $\lambda^{\infty}(T_c) = \lambda$. Figure 15.5 shows $p(N, T_c^{\infty}, \lambda)$ for $\lambda^{\infty}(T_c^{\infty}) - \lambda = 0.03, 0.02,$ and 0.01. Similarly, Figure 15.6 shows $p(N, T, 0.48)$ for $T/T_c^{\infty} = 0.8, 0.9,$ and 1.0. It is observed that the probability distribution flattens and the average cluster size increases as λ approaches $\lambda^{\infty}(T_c^{\infty}) = 0.5$ in Figure 15.5 and as T approaches $T_c(0.48) = 0.788 T_c^{\infty}$ in Figure 15.6. Thus, it is predicted from this theory that $\langle N \rangle \gg 100$ near the critical temperature, so that the magnetic order parameter increases rapidly with decreasing temperature near T_c.

The subdivision potential is $\mathsf{E}(T, \lambda) = -k_B T \ln Y$ and the equilibrium thermodynamic properties of clusters with unrestricted sizes can be evaluated as $u(T, \lambda) = -(\partial \ln Y/\partial \beta)_\lambda/\langle N \rangle$ and $s(T, \lambda) = (u - \mu - \mathsf{E}/\langle N \rangle)/T$. Figures 15.7 and 15.8 show $u(T, \lambda)$ and $s(T, \lambda)$ for $\lambda^{\infty}(T_c^{\infty}) - \lambda = 0.01, 0.02,$ and 0.03. It is remarkable that, by subdividing into clusters with unrestricted sizes, a macroscopic ferromagnet can do both decrease its energy per particle and increase its entropy per particle. It should be remembered that in the case of fixed cluster size (i.e., in the canonical ensemble), the entropy per particle in the vicinity of T_c^{∞} was $s(T_c^{\infty}, N) < s^{\infty}(T_c^{\infty}) = 0$. If we denote the entropy per particle obtained from the canonical ensemble for $\langle N \rangle$ (T, λ) as $s(T, \langle N \rangle)$, it can be shown that (Hill 1964)

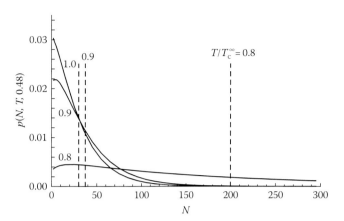

FIGURE 15.6 Probability function $p(N, T, 0.48)$ characterizing the distribution of cluster sizes for $\lambda = 0.48$. The curves correspond to temperatures $T/T_c^{\infty} = 1.0, 0.9,$ and 0.8. The size distribution broadens and the average size (marked with the dashed vertical lines) increases when the critical temperature $T_c(0.48)/T_c^{\infty} = 0.788$ is approached.

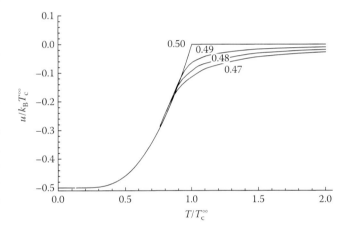

FIGURE 15.7 Internal energy per particle in the mean-field theory for unrestricted-size finite ferromagnetic cluster with absolute activity $\lambda = 0.47, 0.48,$ and 0.49. The solid line that covers the whole temperature range corresponds to the macroscopic system. The classical Weiss transition at T_c^{∞} is suppressed by finite-size effects.

$$\left[s(T,\lambda) - s(T,\langle N \rangle) \right]\langle N \rangle = -k_B \sum_{N=2}^{\infty} p(N,T,\lambda) \ln p(N,T,\lambda) \quad (15.47)$$

That is, the increased entropy in the generalized ensemble arises from the different ways in which the total number of particles can be distributed into clusters of average size $\langle N \rangle$.

In the previous paragraphs, we have considered a single cluster (or small system) with average size $\langle N \rangle$, which may become infinite at the critical point. Furthermore, the sample can increase its entropy by forming aggregates of indistinguishable clusters. These aggregates are described by the partition sum (Chamberlin 1999)

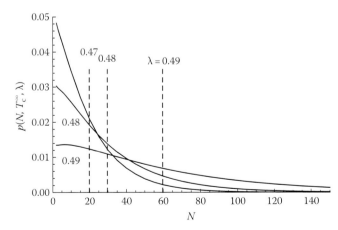

FIGURE 15.5 Probability function $p(N, T_c^{\infty}, \lambda)$ characterizing the distribution of cluster sizes at temperature $T = T_c^{\infty}$. The curves correspond to the activity values $\lambda = 0.47, 0.48,$ and 0.49. The size distribution broadens and the average size (marked with the dashed vertical lines) increases when λ approaches $\lambda^{\infty}(T_c^{\infty}) = 0.5$.

$$\Gamma(T,\lambda) = \sum_{N=0}^{\infty} \frac{1}{N!} \left[Y(T,\lambda) \right]^N \quad (15.48)$$

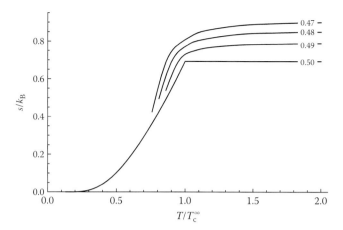

FIGURE 15.8 Entropy per particle in the mean-field theory for unrestricted-size finite ferromagnetic cluster with absolute activity $\lambda = 0.47, 0.48,$ and 0.49. The solid line that covers the whole temperature range corresponds to the macroscopic system.

and the average number of clusters per aggregate is

$$\langle N \rangle = Y \left(\frac{\partial \ln \Gamma}{\partial Y} \right)_{T,\lambda} \qquad (15.49)$$

In the case of ferromagnetic clusters in the presence of external magnetic field H, the energy per particle within the mean-field approximation is

$$\varepsilon(H,l,N) = -h \left(2\frac{l}{N} - 1 \right) - \frac{cJ}{2} \left(4\frac{l(l-1)}{N(N-1)} - 4\frac{l}{N} + 1 \right) \qquad (15.50)$$

where $h \equiv \mu_m \mu_0 H$, μ_m is the magnetic moment of the spin and μ_0 is the magnetic permeability of vacuum. The generalized partition sum is

$$Y(T,H,\lambda) = \sum_{N=2}^{\infty} \lambda^N \sum_{l=0}^{N} \frac{N!}{l!(N-l)!} e^{-\beta N \varepsilon(H,l,N)} \qquad (15.51)$$

The magnetization of the cluster is $\langle M \rangle = (\mu_m k_B T/V)(\partial \ln Y/\partial h)_{T,\lambda}$ and its magnetic susceptibility is $\chi_m = (\partial \langle M \rangle/\partial H)_{T,\mu} = (\mu_0 \mu_m^2 k_B T/V)(\partial^2 \ln Y/\partial h^2)_{T,\lambda}$. In the absence of external field $(H \to 0)$, the average magnetization vanishes above T_c^∞ and, therefore, the susceptibility reduces to

$$\chi_m^{H \to 0}(T,\lambda) = \mu_0 \mu_m^2 \frac{k_B T}{V} \frac{1}{Y} \left(\frac{\partial^2 Y}{\partial h^2} \right)_{T,\lambda}$$

$$= \frac{\mu_0 \mu_m^2}{V k_B T} \frac{1}{Y} \sum_{N=2}^{\infty} \lambda^N \sum_{l=0}^{\infty} \frac{N!(2l-N)^2}{l!(N-l)!} e^{-\beta N \varepsilon(H,l,N)} \qquad (15.52)$$

The value of the absolute activity λ is determined by comparison with experimental data and typical values are found in the range $0.0006 < \lambda^\infty(T_c^\infty) - \lambda < 0.004$ (Chamberlin 2000).

Two separate theories are often used to characterize the paramagnetic properties of ferromagnetic materials (Chamberlin 2000). Above the Weiss temperature Θ, classical mean-field theory yields the Curie–Weiss law for the magnetic susceptibility $\chi_m(T) = C/(T - \Theta)$. Close to the Curie or critical temperature T_c, however, the standard mean-field approach breaks down so that better agreement with experimental data is provided by the critical scaling theory $\chi_m(T) \propto (T - T_c)^{-\gamma}$ where γ is a scaling exponent. However, there is no known model capable of predicting either the measured values of γ or its variation among different substances.

By combining the mean-field approximation with Hill's nanothermodynamics, as explained above, the extra degrees of freedom from considering clusters with unrestricted sizes give the correct critical behavior, because the fraction of clusters with a specific amount of order diverges at T_c. At all temperatures above T_c, the model matches the measured magnetic susceptibilities of crystalline EuO, Gd, Co, and Ni, thus providing a unified picture for both the critical-scaling and Curie–Weiss regimes (Chamberlin 2000). Interestingly, Equation 15.52 gives a better agreement with experimental results for the entire paramagnetic phase with less fitting parameters and without introducing a separate transition temperature and amplitude prefactor for the scaling regime. Furthermore, when the average number of clusters per aggregate $\langle N \rangle$ is evaluated from Equation 15.49, the results are also in agreement with measurements of the correlation length in crystalline cobalt.

In conclusion, Chamberlin (2000) proved that the critical behavior of ferromagnets can be described by the mean-field theory, thereby eliminating the need for a separate scaling regime, provided that the clusters are described using the generalized or completely open ensemble because this is the only ensemble that does not artificially restrict the internal fluctuations of a bulk sample.

15.3.2.9 Supercooled Liquids and the Glass Transition

Motivated by the close similarity between the Vogel–Tamman–Fulcher (VTF) law for the characteristic relaxation time of supercooled liquids and the Curie–Weiss law of ferromagnetism, Chamberlin (1999, 2002) applied the Weiss mean-field theory to finite systems with unrestricted sizes, as explained in Section 15.3.2.8, to derive a generalized partition function for supercooled liquids. Finite-size effects broaden the transition and induce a Curie–Weiss-like energy reduction which provides an explanation for the VTF law. Moreover, the distribution of aggregate sizes derived from the generalized partition function of this nanothermodynamic theory provides an explanation for the Kohlrausch–Williams–Watt law. And standard fluctuation theory also helps to explain the measured specific heats.

15.3.3 Tsallis' Thermostatistics

Nonextensivity may appear in systems that are not in the thermodynamic limit because correlations are of the order of the system size, and this can be due to finite size effects, the presence of long-range interactions, the existence of dissipative structures, etc. Tsallis considered that the Boltzmann–Gibbs–Shannon (BGS) entropy is not appropriate to nonextensive behavior and proposed to adopt the Havrda–Charvat structural entropy inspired by the multifractal formalism. Tsallis' equation for the nonextensive entropy is (Tsallis 2001)

$$S_q \equiv k \frac{\sum_{j=1}^{W} p_j^q - 1}{1-q}, \quad \sum_{j=1}^{W} p_j = 1 \qquad (15.53)$$

where q is a real number known as the entropic index and W is the total number of microstates of the system. The entropic index q characterizes the degree of nonextensivity reflected in the following pseudoadditivity rule:

$$S_q(A+B) = S_q(A) + S_q(B) + [(1-q)/k]S_q(A)S_q(B) \qquad (15.54)$$

where A and B are two independent systems in the sense that the probabilities of $A + B$ factorize into those of A and of B. Since $S_q \geq 0$, the cases $q < 1$, $q = 1$, and $q > 1$ correspond, respectively, to superextensivity, extensivity, and subextensivity. Equation 15.53 is the only entropic form that satisfies the nonextensivity rule given in Equation 15.54, in the same way as BGS entropy is the only one that satisfies the extensivity rule $S(A + B) = S(A) + S(B)$.

The constant, k, in Equation 15.53 differs from Boltzmann's constant, k_B, but reduces to it when $q = 1$. Moreover, S_q tends to the BGS entropy $S = -k_B \sum_j p_j \ln p_j$ when $q = 1$. In the microcanonical ensemble, all microstates are equally probable, $p_j = 1/W$, and Tsallis' entropy becomes $S_q = k \ln_q W$ where $\ln_q$ is a function called the q-logarithm defined as $\ln_q x \equiv (x^{1-q} - 1)/(1 - q)$. Its inverse function is the q-exponential $e_q^x \equiv [1 + (1 - q)x]^{1/(1-q)}$ and the equation $\ln_q(e_q^x) = e_q^{\ln_q x} = x$ is satisfied. Obviously, these functions have been introduced to resemble Boltzmann's expression $S = k_B \ln W$, which is the limit of S_q when $q = 1$.

The probability distribution in Tsallis' statistics is the q-exponential distribution

$$p_j = \frac{[1 - (1-q)\beta U_j]^{1/(1-q)}}{Z_q} = \frac{e_q^{-\beta U_j}}{Z_q} \qquad (15.55)$$

where $Z_q = \sum_j e_q^{-\beta U_j}$ is a generalized canonical partition function. Equation 15.55 can be obtained by maximizing Equation 15.53 under the constraint that a generalized average energy $\langle U \rangle_q \equiv \left(\sum_j p_j^q U_j \right) / \sum_j p_j^q$ is fixed. To some extent, Equation 15.55 is responsible for the great success that Tsallis' theory has experienced since it replaces the classical Boltzmann distribution by a family of distributions with a parameter q that can be determined by fitting the experimental data (Luzzi et al. 2002).

At the same time, this widespread use of q as a fitting parameter is one of the major drawbacks of Tsallis' theory. Phenomena

characterized by the probability distribution in Equation 15.55 abound in nature. This type of statistics may arise from the convolution of the normal distribution with either a gamma or a power-law distribution, the latter being, for instance, a manifestation of the polydispersity of the system. The fact that they can be satisfactorily explained without any assumption of nonergodicity, long-range correlations, or thermodynamic nonequilibrium casts some doubts on the relevance of Tsallis formalism for many systems (Gheorghiu and Coppens 2004).

Yet, Tsallis' entropy has got a place in modern statistical mechanics, which is supported by the growing evidence of its relevance to many complex physical systems and the great success in some of its applications. For instance, the entropic index q has been shown to be intimately related to the microscopic dynamics (Cohen 2002). Furthermore, in systems with fractal phase space, the entropic index q has been shown to be equal to the fractal dimension of the available phase space (García-Morales and Pellicer 2006). This connection has allowed to interpret unequivocally the observation that q tends to vanish in the strong coupling regime found in ionic solutions, since the available phase space collapses into regions of strikingly lower dimensions when the multivalent ions of the same charge are located close to a highly charged surface and crystallize forming a Wigner crystal (García-Morales et al. 2004).

The thermodynamic equations corresponding to Tsallis' statistics have been deduced by García-Morales et al. (2005) after work by Vives and Planes (2002) and take the form

$$dS_q = \frac{1}{T} d\langle U \rangle_q + \frac{p}{T} d\langle V \rangle_q - \frac{\mu}{T} d\langle N \rangle_q \qquad (15.56)$$

$$k \frac{\Sigma_q \ln \Sigma_q}{1-q} = \frac{\langle U \rangle_q}{T} + \frac{p\langle V \rangle_q}{T} - \frac{\mu \langle N \rangle_q}{T} \qquad (15.57)$$

$$\ln \Sigma_q \, dS_q = \langle U \rangle_q \, d\frac{1}{T} + \langle V \rangle_q \, d\frac{p}{T} - \langle N \rangle_q \, d\frac{\mu}{T} \qquad (15.58)$$

where $\Sigma_q \equiv \Sigma_{j=1}^W p_j^q = 1 + (1-q)S_q/k$, and $\langle X \rangle_q \equiv (\Sigma_j p_j^q X_j)/\Sigma_q$ for $X = U, V, N$. The comparison of these equations with those obtained in Hill's nanothermodynamics shows that it is possible to connect these two nonextensive formalisms through the relation (García-Morales et al. 2005)

$$E = kT \frac{\Sigma_q \ln \Sigma_q}{1-q} - TS_q \qquad (15.59)$$

Note that when $q = 1$, the first term becomes equal to $TS_q (=TS)$ and hence $E = 0$, so that extensivity is recovered. This relation between Hill's subdivision potential E and Tsallis' entropic index q may help to clarify the physical foundations of Tsallis's entropy, and shows that, when the thermodynamic forces are properly defined, Tsallis' entropy can be used to describe the size-effects on thermodynamic magnitudes.

15.3.4 Superstatistics

In Hill's theory, the ensemble contain N-independent nanosystems, each of which is in thermal contact with the thermal bath and has a fixed temperature, that of the bath. The energy of every nanosystem u_i fluctuates, and the same applies to the mean internal energy

$$\langle u \rangle_N = \frac{\langle U \rangle_N}{N} = \frac{1}{N}\sum_{i=1}^{N} u_i \tag{15.60}$$

The probability that the nanosystem *i* is found in a microstate of energy u_i is

$$p(u_i) = \frac{e^{-\beta_0 u_i}}{Z(\beta_0)} \tag{15.61}$$

and the joint probability distribution, i.e., the probability that the ensemble is found in the microstate $\{u_1, u_2,\ldots, u_N\}$ is

$$P_N(U_N) = \prod_{i=1}^{N} p(u_i) = \frac{e^{-\beta_0 u_1}}{Z(\beta_0)}\frac{e^{-\beta_0 u_2}}{Z(\beta_0)}\cdots\frac{e^{-\beta_0 u_n}}{Z(\beta_0)} = \left[\frac{e^{-\beta_0 \langle u \rangle_N}}{Z(\beta_0)}\right]^{N} = \frac{e^{-\beta_0 \langle U \rangle_N}}{Z_N(\beta_0)} \tag{15.62}$$

which is the usual Boltzmann distribution.

An alternative approach proposed by Rajagopal et al. (2006) would be to consider that the temperature of each nanosystem *fluctuates* around the temperature of the reservoir. The Boltzmann parameter $\beta \equiv 1/k_BT$ of a nanosystem would then be a fluctuating magnitude and the thermal equilibrium of the ensemble with the bath would only ensure that the ensemble average value of β is determined by the bath, $\langle\beta\rangle = \beta_0$; (the averaging routine to calculate $\langle\beta\rangle$ is still to be defined). The origin of these fluctuations lies in the very same nanosize and thus they come to quasithermodynamic equilibrium with the reservoir. This means that the Boltzmann–Gibbs distribution has to be averaged over the temperature fluctuations induced by the reservoir. Recently, this idea has been further developed in different physical contexts using a noisy reservoir (Wilk and Wlodarczyk 2000, Beck 2002).

When temperature fluctuations are taken into account, the probability distribution that replaces that shown in Equation 15.62 can be derived by taking an integral over all possible fluctuating (inverse) temperatures. Let us work out this idea in detail starting from the concept of *superstatistics* (i.e., from the superposition of two different statistics) (Beck 2002). If all nanosystems in the ensemble had the same temperature, their probability distribution would be described by ordinary statistical mechanics, i.e., by Boltzmann factors $e^{-\beta u_i}$. However, if the nanosystems differ in temperature, we also need another statistics to describe the ensemble (the Boltzmann statistics $e^{-\beta u_i}$ and that of β), hence the name "superstatistics." One may define an average Boltzmann factor $B(u_i)$ as

$$B(u_i) \equiv \int_0^{\infty} f(\beta)e^{-\beta u_i}\,d\beta \tag{15.63}$$

where $f(\beta)$ is the probability distribution of β. The stationary long-term probability distribution is obtained by normalizing this effective Boltzmann factor as

$$p(u_i) = \frac{B(u_i)}{\int_0^{\infty} B(u_i)\,du_i} \equiv \frac{1}{Z}B(u_i) \tag{15.64}$$

which can be considered as the generalization of Equation 15.61.

It should be noticed that we have linked the concepts of Hill's ensemble of nanosystems and the description of temperature fluctuations through superstatistics to make clear the limitations of the former. However, the concepts of superstatistics and temperature fluctuations can be applied to many other situations. For example, spatiotemporal fluctuations in temperature (or in other intensive magnitudes) may arise in driven nonequilibrium system with a stationary state. The different spatial regions (cells) with different values of β would then play the role of different nanosystems.

Among all possible probability distributions $f(\beta)$, there is one that has received much attention. This is the χ^2 distribution (also called Γ distribution) and is given by

$$f(\beta) = \frac{1}{\Gamma(\gamma)}\frac{(\gamma\beta/\langle\beta\rangle)^{\gamma}\,e^{-\gamma\beta/\langle\beta\rangle}}{\beta} \tag{15.65}$$

where $\langle\beta\rangle = \int_0^{\infty}\beta\,f(\beta)d\beta$ is the average value of β. The parameter γ is a measure of the variance, $\langle\beta^2\rangle - \langle\beta\rangle^2$, of the distribution such that

$$\langle\beta^2\rangle = \int_0^{\infty}\beta^2 f(\beta)d\beta = \left(1 + \frac{1}{\gamma}\right)\langle\beta\rangle^2 \tag{15.66}$$

and

$$\frac{\langle(\beta - \langle\beta\rangle)^2\rangle}{\langle\beta\rangle^2} = \frac{\langle\beta^2\rangle - \langle\beta\rangle^2}{\langle\beta\rangle^2} = \frac{1}{\gamma} \geq 0 \tag{15.67}$$

The average Boltzmann factor $B(u_i)$ corresponding to the χ^2 distribution in Equation 15.65 is

$$B(u_i) = \int_0^{\infty} f(\beta)e^{-\beta u_i}\,d\beta = [1 + \langle\beta\rangle u_i/\gamma]^{\gamma} \tag{15.68}$$

Introducing the entropic index as $q \equiv 1 + 1/\gamma$, this factor can also be presented as $B(u_i) = e_q^{-\langle\beta\rangle u_i}$, which turns out to be the Tsallis distribution corresponding to the average temperature. Hence, it can be concluded that the entropy associated with small systems with temperature fluctuations is Tsallis' entropy. The Boltzmann–Gibbs statistics corresponds to $1/\gamma = 0$ and absence of temperature fluctuations.

In conclusion, the theory of superstatistics contains Tsallis' statistics as a particular case that corresponds to the χ^2 distribution. The validity of Tsallis' distributions observed in a large variety of physical systems, many of them in a driven stationary state far from equilibrium, can thus be justified because the χ^2 distribution naturally arises in many circumstances.

Since these or similar distributions are often observed in experiments, it seems justified to look for dynamical arguments for the occurrence of Tsallis statistics in suitable classes of non-equilibrium systems (Beck 2001). And this is indeed possible: Tsallis statistics can be generated from stochastic differential equations with fluctuating parameters. For many systems, the reason why Tsallis distributions are observed can be related to the fact that there are spatiotemporal fluctuations of an intensive parameter (e.g., the temperature). If these fluctuations evolve on a long time scale and are distributed according to a particular distribution, the χ^2 distribution, one ends up with Tsallis' statistics in a natural way.

15.3.5 Nonequilibrium Approaches

As in macroscopic systems, there exist nonequilibrium steady states with net currents flowing across small systems where physical properties do not display any observable time dependence. For example, a small system in contact with two thermal sources at different temperatures has a heat flux as current. Another example is a resistor connected to a voltage source, which has an electric current across it. Such systems require a constant input of energy to maintain their steady state because the systems constantly dissipate net energy and operate away from equilibrium. Most biological systems, including molecular machines and even whole cells, are found in nonequilibrium steady states. Out of a steady state, the most general case, one or more of the system's properties change in time. The entropy production σ is perhaps the most important fact in nonequilibrium thermodynamics, since it is totally absent in thermostatics. In macroscopic irreversible thermodynamics (de Groot and Mazur 1962), it is usual to look at it as a function of two sets of variables, the thermodynamic fluxes $\{\phi_i\}$ and forces $\{y_i\}$, defined so that the entropy production can be expressed as a sum of products of conjugates, $\sigma = \sum_i \phi_i y_i$, the fluxes being zero at equilibrium. This expression is supplemented by a set of phenomenological relations, which gives the fluxes as functions of the forces, these relations being such that the forces cancel at equilibrium. It is an experimental fact that there exists a neighborhood of equilibrium where the relations between the two sets of variables are linear, that is, $\phi_i = \sum_j L_{ij} y_j$ so that $\sigma = \sum_{i,j} L_{ij} y_j y_i$.

Onsager's result is the symmetry of the phenomenological coefficients $L_{ij} = L_{ji}$, proven on the basis of two general

hypothesis: regression of fluctuations and microscopic dynamic reversibility (de Groot and Mazur 1962). This implies that the matrix of Onsager coefficients is definite positive, and, therefore, that entropy production is always a positively defined quantity. This situation can change for a nanosystem where violations of the second law for short times have been observed experimentally (Wang et al. 2002). Dissipation and thermal properties out of equilibrium in nanosystems and small times have been the subject of intense research in the last two decades and remarkable rigorous results have been derived that have been found experimentally to hold out of equilibrium. In this section, we summarize some of these results.

15.3.5.1 Jarzynski Equality (JE)

C. Jarzynski derived an expression allowing the equilibrium free energy difference ΔF between two configurations A (initial) and B (final) of the system to be determined from finite-time measurements of the work W performed in parametrically switching from one configuration to the other. This result, which is independent of both the path γ from A to B, and the rate at which the parameters are switched along the path, is surprising: It says that we can extract equilibrium information from an ensemble of *nonequilibrium* (finite-time) measurements. Jarzynski equality reads (Jarzynski 1997a, 1997b)

$$\left\langle e^{-\beta W} \right\rangle_{\chi(t)} = e^{-\beta \Delta F} \tag{15.69}$$

where $\chi(t)$ is the time-dependent protocol specifying the switching between the two configurations and the brackets $<\cdots>$ denote an average over an ensemble of measurements of W. Each measurement is made after first allowing the system and reservoir to equilibrate at temperature T, with parameters fixed at A. (The path in parameter space γ from A to B, and the protocol at which the parameters are switched along this path, remain unchanged from one measurement to the next.) Formally, W is defined by

$$W = \int_0^{t_s} dt\, \dot{\chi} \frac{\partial H_\chi(\mathbf{z}(t))}{\partial \chi} \tag{15.70}$$

where $\mathbf{z}(t)$ is the mechanical (stochastic) trajectory followed by the system and the dynamical role of χ, a parameter that is tuned externally in the experiments, is clarified. The system' Hamiltonian H_χ depends explicitly on the latter external parameter. χ varies between 0 (at configuration A) and 1 (at configuration B) over a total switching time t_s. Now imagine an ensemble of realizations of the switching process (with γ and t_s fixed), with initial conditions for the system and reservoir generated from a canonical ensemble at temperature T. Then W may be computed for each trajectory $\mathbf{z}(t)$ in the ensemble, and the brackets in Equation 15.69 indicate an average over the distribution of values of W thus obtained. This provides a means for a numerical checking of Equation 15.69. Alternatively W defined by Equation 15.70 can be readily measured in the experiments and then brackets in Equation 15.69 denote the average obtained from the set of measurements. Equation 15.69 holds independently of the path γ. In the limiting case of an infinitely slow switching of the external parameters the system is in quasistatic equilibrium with the reservoir throughout the switching process and Equation 15.69 takes the form

$$\Delta F = \int_{0}^{1} d\chi \left\langle \frac{\partial H_{\chi}(\mathbf{z}(t))}{\partial \chi} \right\rangle \qquad (15.71)$$

In the opposite limit of infinitely fast switching ($t_s \to 0$), the switching is instantaneous and therefore $W = \Delta H = H_1 - H_0$ in Equation 15.70. Since we have a canonical distribution of initial conditions Equation 15.69 becomes

$$\Delta F = -k_B T \ln \left\langle e^{-\beta \Delta H} \right\rangle_0 \qquad (15.72)$$

Equations 15.71 and 15.72 are well known from previous work (Kirkwood 1935, Zwanzig 1954) and the JE generalizes them to *any* switching protocol χ(t). "The free energy difference between initial and final equilibrium states can be determined not just from a reversible or quasistatic process that connects those states, but also via a nonequilibrium, irreversible process that connects them" (Bustamante et al. 2005). This property makes the JE to have enormous practical importance. The exponential average appearing in Equation 15.69 implies that $\langle W \rangle \geq \Delta F$, which, for macroscopic systems, is the statement of the second law of thermodynamics in terms of free energy and work. The Clausius inequality combined with the JE allows relating mean entropy dissipation to experimental observables (Ben-Amotz and Honig 2006). The Carnot engine has been then elegantly shown to emerge as a limiting case of a family of irreversible processes arising from an interface between materials at different temperatures. The following expression for the entropy change during an irreversible process has been proposed (Ben-Amotz and Honig 2006)

$$dS = \frac{\left\langle \delta W_{dis} \right\rangle_{\chi(t)}}{T} + k_B \ln \left\langle e^{-\beta \delta W_{dis}} \right\rangle_{\chi(t)} \qquad (15.73)$$

The JE considers processes where the system is driven out of equilibrium by a mechanical external agent while remaining in contact with a thermal reservoir at a fixed temperature. Quite recently, a generalization of the JE to situations where the reservoir drives the system out of equilibrium through temperature changes has also been provided (Williams et al. 2008). The JE has also been extended to quantum systems (Mukamel 2003, Teifel and Mahler 2007).

The biophysical relevance of the JE was recently demonstrated through single-molecule experiments carried out under nonequilibrium conditions, which allowed extracting free energy differences (Hummer and Szabo 2001). The JE was also tested by

mechanically stretching a single molecule of RNA, both reversibly and irreversibly, between its folded and unfolded conformations (Liphardt et al. 2002).

15.3.5.2 Fluctuation Theorems (FTs)

The question of how reversible microscopic equations of motion can lead to irreversible macroscopic behavior has been the object of intense work in the last two decades. The fluctuation theorem (FT) was formulated heuristically in 1993 for thermostated dissipative nonequilibrium systems (Evans et al. 1993) and gives an answer to the problem of macroscopic irreversibility under reversible microscopic dynamics. The theorem, which was successfully tested in a recent experimental work (Wang et al. 2002), is entirely grounded on the postulates of causality and ergodicity at equilibrium states. Gallavotti and Cohen derived rigorously the FT in 1995 (Gallavotti and Cohen 1995) for thermostated deterministic steady-state ensembles. The authors proved the following asymptotic expression:

$$\frac{P_\tau(+\sigma)}{P_\tau(-\sigma)} = e^{\sigma \tau} \qquad (15.74)$$

Here $P_\tau(\pm \sigma)$ is the probability of observing an average entropy production σ on a trajectory of time τ. Equation 15.74 establishes that *there is a nonvanishing probability of observing a negative entropy production* (thus violating the second law of thermodynamics) which is, however, exponentially small with increasingly longer times compared to the probability of observing a positive entropy production. In small systems (and short-trajectory times), the probability of observing a violation of the second law is, however, significant. A FT for stochastic dynamics was also derived (Kurchan 1998, Lebowitz and Spohn 1999, Maes 1999). Other FTs have been reported differing in details on as whether the kinetic energy or some other variable is kept constant, and whether the system is initially prepared in equilibrium or in a nonequilibrium steady state (Evans and Searles 2002). FTs can be of crucial interest for nanosystems and especially for the development of nanoelectronics (van Zon et al. 2004, Garnier and Ciliberto 2005).

Another result that connects the FT to the JE was obtained by Crooks (1999) who derived a generalized FT for stochastic microscopic dynamics. Crooks FT, which was experimentally tested in recovering RNA folding free energies (Collin et al. 2005), provides an independent and succinct proof of the JE and has similar practical relevance as the JE. Crooks theorem has been extended to quantum systems in the microcanonical ensemble yielding interesting insights on the concept of nonequilibrium entropy in the quantum regime (Talkner et al. 2008).

15.3.5.3 Thermodynamics Based on the Principle of Least-Abbreviated Action

In invoking concepts as microscopic reversibility and deterministic or stochastic trajectories, all works mentioned above point directly to several aspects of the dynamical foundations of statistical thermodynamics. These were already the concern of Boltzmann and Clausius at the end of the nineteenth century (see Bailyn 1994) and are still open issues. Recently, a dynamical definition of nonequilibrium entropy based solely in the Hamiltonian dynamics of conservative systems has been introduced (García-Morales et al. 2008). The theory is based on the Maupertuis principle of least-abbreviated action and the definition of the entropy is of relevance to *systems of any size*, since it is grounded directly on the Hamiltonian H of the system. Finite-size effects and nonextensivity in small systems are satisfactorily captured by the formulation. The nonequilibrium entropy takes the form

$$S = k_B \ln \left(\prod_{i=1}^{N} J_i \right) \quad (15.75)$$

where the $J_i = \oint p_i dq_i$ are suitable action variables (q_i and p_i are, respectively, generalized position and momenta and the integral extends over the region that bounds each degree of freedom). When all degrees of freedom are separable in the Hamiltonian and the system is integrable, the system remains forever *out of equilibrium*, since the degrees of freedom cannot thermalize. Under this picture, macroscopic irreversibility is entirely grounded in the *nonintegrability* of the dynamics coming from complicated interactions between the degrees of freedom that lead to their thermalization. The propagation of the error in using approximate action variables to describe the nonintegrable dynamics of the system is directly linked to the entropy production (García-Morales et al. 2008), which is defined through Hamiltonian mechanics as

$$\sigma = -k_B \sum_{i=1}^{N} \frac{1}{J_i} \frac{\partial H}{\partial \theta_i} \quad (15.76)$$

where θ_i are the angle variables conjugate to the J_i, which are present in the Hamiltonian H, making the dynamics of the system nonintegrable.

15.3.5.4 Nonequilibrium Nanothermodynamics

In the entropic representation, the thermodynamic equations for the average small system in Hill´s nanothermodynamics are (García-Morales et al. 2005, Carrete et al. 2008)

$$dS = \sum_{\alpha} y_\alpha d \langle X_\alpha \rangle \quad (15.77)$$

$$S = \sum_{\alpha} y_\alpha \langle X_\alpha \rangle - \frac{\mathsf{E}}{T} \quad (15.78)$$

$$d \left(-\frac{\mathsf{E}}{T} \right) = -\sum_{\alpha} \langle X_\alpha \rangle dy_\alpha \quad (15.79)$$

The E-dependence of Equations 15.78 (Euler equation) and 15.79 (Gibbs–Duhem equation) makes entropy to be a nonhomogeneous function of the extensive variables. By using Equations 15.73 and 15.78 a statistical definition can be given to the subdivision potential change $\Delta\mathsf{E}$ of the nanosystem under an irreversible process (Carrete et al. 2008)

$$\Delta\mathsf{E}\Big|_{\chi(t)} = T \sum_{\alpha} y_\alpha \langle X_\alpha \rangle \Big|_0^{t_0} + \int_{\chi(t)} \left[\delta W_{dis} + k_B T \ln \langle e^{-\beta \delta W_{dis}} \rangle \right] \quad (15.80)$$

Nonequilibrium nanothermodynamics (Carrete et al. 2008) follows Hill's course of reasoning to establish nonequilibrium transport equations in the linear regime that generalize macroscopic irreversible thermodynamics. The key idea is to consider a macroscopic ensemble of nanosystems, with a possible gradient in their number. Assuming that linear macroscopic irreversible thermodynamics holds for the entire ensemble, transport equations can be derived for quantities regarding each nanosystem. The nanoscopic transport coefficients are also found to be symmetric (Carrete et al. 2008), ensuring that the second law of thermodynamics is obeyed by the average systems although it can be transitorily violated by a small system.

It is important to note that Hill's equilibrium nanothermodynamics is consistent with Gibbs definition of the equilibrium entropy. Out of equilibrium a link of thermodynamic properties and statistical properties of nanosystems is provided by the Gibbs' entropy postulate (Reguera et al. 2005):

$$S = S_{eq} - k_B \int P(\gamma, t) \ln[P(\gamma, t)/P_{eq}(\gamma)] d\gamma \quad (15.81)$$

where S_{eq} denotes the equilibrium Gibbs entropy when the degrees of freedom γ are at equilibrium (where the integrand of the second term in the r.h.s. cancels). The probability distribution at an equilibrium state of a given configuration in γ-space is given by

$$P_{eq}(\gamma) \approx e^{-\beta \Delta W(\gamma)} \quad (15.82)$$

where $\Delta W(\gamma)$ is the minimum reversible work to create such a state. Taking variations of Equation 15.81, we have

$$\delta S = -k_B \int \delta P(\gamma, t) \ln \left[P(\gamma, t)/P_{eq}(\gamma) \right] d\gamma \quad (15.83)$$

The evolution of the probability density in the γ-space is governed by the continuity equation

$$\frac{\partial P}{\partial t} = -\frac{\partial J}{\partial \gamma} \quad (15.84)$$

where $J(\gamma, t)$ is a current or density flux in γ-space which has to be specified. Its form can be obtained by taking the time derivative in Equation 15.83 and by using Equation 15.84. After a partial integration, one then arrives at

$$\frac{dS}{dt} = -\int \frac{\partial J_s}{\partial \gamma} \, d\gamma + \sigma \qquad (15.85)$$

where $J_s = k_B J \ln [P/P_{eq}]$ is the entropy flux and

$$\sigma = -k_B \int J(\gamma, t) \frac{\partial \ln[P(\gamma, t)/P_{eq}(\gamma)]}{\partial \gamma} \, d\gamma \qquad (15.86)$$

is the entropy production. In this scheme, the thermodynamic forces are identified as the gradients in the space of mesoscopic variables of the logarithm of the ratio of the probability density to its equilibrium value. By assuming a linear dependence between fluxes and forces and establishing a linear relationship between them we have

$$J(\gamma, t) = -k_B L(\gamma, P(\gamma)) \frac{\partial \ln[P(\gamma, t)/P_{eq}(\gamma)]}{\partial \gamma} \qquad (15.87)$$

where $L(\gamma, P(\gamma))$ is an Onsager coefficient, which, in general, depends on the state variable $P(\gamma)$ and on the mesoscopic parameters γ. To derive this expression, locality in γ-space is taken into account, and only fluxes and forces with the same value of γ become coupled. The resulting kinetic equation follows by substituting Equation 15.87 back into the continuity Equation 15.84:

$$\frac{\partial P}{\partial t} = \frac{\partial}{\partial \gamma} \left(D P_{eq} \frac{\partial}{\partial \gamma} \frac{P}{P_{eq}} \right) \qquad (15.88)$$

where the diffusion coefficient is defined as

$$D(\gamma) \equiv \frac{k_B L(\gamma, P)}{P} \qquad (15.89)$$

By using Equation 15.82, Equation 15.88 can be written as

$$\frac{\partial P}{\partial t} = \frac{\partial}{\partial \gamma} \left(D \frac{\partial P}{\partial \gamma} + \frac{D}{k_B T} \frac{\partial \Delta W}{\partial \gamma} P \right) \qquad (15.90)$$

which is the Fokker–Planck equation for the evolution of the probability density in γ-space. The dynamics of the probability distribution depends explicitly on equilibrium thermodynamic properties through the reversible work ΔW. This formalism allows to analyze the effects of entropic barriers $\Delta W = -T\Delta S$ in the nonequilibrium dynamics of the system. Entropic barriers are present in many situations, such as the motion of macromolecules through pores, protein folding, and in general in the dynamics of small confined systems (Reguera et al. 2005). As we have seen above, in mesoscopic physics, besides the diffusion processes coming from, for example, mass transport, one finds a diffusion process for the probability density of measuring certain values for experimental observables in the space of mesoscopic degrees of freedom γ.

15.4 Critical Discussion and Summary

Usually, there is some arbitrariness in all thermostatistical approaches that arise from the difficulty of relating a very limited number of macroscopic variables to an enormous number of microscopic degrees of freedom. The existence of mesoscopic degrees of freedom, like rotation and translation of mesoscopic clusters, pose additional problems since the robustness of the thermodynamic limit is lost, and efficient ways of handling a very complicated and rich dynamics coming from a sufficiently high number of degrees of freedom need to be devised. Hill's nanothermodynamics constitutes an elegant approach whose philosophy, as we have seen, is averaging over the mesoscopic degrees of freedom and over ensembles of mesoscopic systems in order to bridge the mesoscopic dynamics with the macroscopic behavior that might be expected from a huge collection of mesoscopic samples. Although Hill's nanothermodynamics is based on equilibrium statistical thermodynamics, and hence it is strictly valid only for systems in equilibrium states, it has also proved to be successful in describing metastable states in the liquid–gas phase transition (Hill and Chamberlin 1998).

The nanosystems considered in Hill's ensembles are all identical, and they are all in equilibrium with their surroundings, so that fluctuations in intensive parameters such as temperature are neglected. Fluctuations in extensive parameters such as the number of particles in the nanosystem are considered, however, and this makes it useful to describe systems close to phase transitions. Chamberlin has adapted Hill's theory to treat finite-sized thermal fluctuations inside bulk materials (Javaheri and Chamberlin 2006). Thus, for example, in the study of supercooled liquids, Chamberlin and Stangel (2006) incorporated the fact that every "small system" was in thermal contact with an ensemble of similar systems, not an infinite external bath, and this yielded a self-consistent internal temperature.

Since the direct interactions between the nanosystems are neglected in Hill's theory, some authors consider that it cannot be applied to systems where local correlations are important. Indeed, the correction terms predicted in Hill's theory do not depend on temperature, whereas it is well known that correlations become more important the lower the temperature (Hartmann et al. 2004). However, Chamberlin has proved in several systems that interactions can be satisfactorily described using the mean-field approach when the "small systems" are considered as completely open, like in the generalized ensemble. Furthermore, the use of the partition function Γ describing a collection of completely open small systems somehow also accounts for interactions among small systems, because the small systems are then allowed to vary in size due to a redistribution of the components among the small systems. This partially solves the criticism raised against Hill's theory.

Other formulations valid for nanosystems and systems exhibiting nonextensivity, like Tsallis thermostatistics, can be shown to be related to Hill's nanothermodynamics (García Morales et al. 2005) and therefore the same considerations apply.

Because of the increased importance of the specificities of the microscopic dynamics out of equilibrium, the problems to lay a general foundation of nonequilibrium nanothermodynamics are much harder than in the equilibrium case. These difficulties are partially softened in the linear branch of nonequilibrium thermodynamics, where linear relationships between fluxes and thermodynamic forces are expected. Out of this linear branch, nonlinear effects cause couplings between microscopic degrees of freedom and it is a far from a trivial task in many problems to decide which of these degrees of freedom are irrelevant to the collective dynamics or can be accounted for, for example, by means of adiabatic elimination. Despite all these problems, rigorous results of general validity have been derived in the last two decades from which we have given an overview here. These results include the Jarzynski equality and the fluctuation theorems, which can be of enormous interest for the understanding of thermal properties at the nanoscale. We have also pointed out how, in the linear regime of nonequilibrium thermodynamics, the macroscopic approach can be extended to nanosystems both from a thermodynamic and a statistical point of view.

15.5 Future Perspectives

Besides the further development of nanothermodynamics, especially of its nonequilibrium branch, there are also some other topics that might likely be of great interest in a near future. Fluctuations play a significant role in the thermodynamics of small systems, near critical points, in processes taking place at small time scales, and in nonequilibrium thermodynamics (Lebon et al. 2008). One of the consequences of fluctuations is the nonequivalence of statistical ensembles that we have shown above in small systems, and also occurs at critical points and in other systems that are mesoscopically inhomogeneous, like complex fluids. A common feature of these systems is that they possess a mesoscopic length scale, known as the correlation length which is associated with fluctuations. Finite-size scaling (Bruce and Wilding 1999) is a powerful theoretical approach that has already been applied to small systems (Anisimov 2004) and may yield more interesting results in the near future.

The development of thermodynamic concepts at the nanoscale is also of crucial interest for the development of Brownian motors. The dynamical behavior of machines based on chemical principles can be described as a random walk on a network of states. In contrast to macroscopic machines whose function is determined predominately by the connections between the elements of the machine, the function of a Brownian machine in response to an external stimulus is completely specified by the equilibrium energies of the states and of the heights of the barriers between them. The thermodynamic control of mechanisms will be crucial in the next steps of interfacing synthetic molecular machines with the macroscopic world (Astumian 2007).

Interesting thermodynamic ideas that have arisen recently in applied physics and engineering and which might be of interest for nonequilibrium nanosystems are provided by the so-called *constructal theory* (Bejan 2000, 2006). We have refrained from discussing this theory here because, until now, the applications that it has found concern purely macroscopic systems. However, an extension of these ideas to nanosystems might have great interest for the engineering of nanodevices and, specially, in the field of nanofluidics. The heart of constructal theory is contained in what might be arguably considered a new law of thermodynamics (Bejan 2000): "For a finite-size system to persist in time (to live), it must evolve in such a way that it provides easier access to the imposed currents that flow through it." This principle connects global optimization techniques employed in engineering with local constraints and has been extremely successful in providing a foundation for scaling laws found in nature as, for example, the relationship between metabolic rate and body size known as Kleiber's law, or different empirical relationships found in the locomotion of living beings. This principle also connects for the first time thermodynamics with the occurrence of definite shapes in nature: it explains, for example, why human beings have a bronchial tree with 23 levels of bifurcation. The constructal theory of the flow architecture of the lung predicts and offers an explanation for the dimensions of the alveolar sac, the total length of the airways, the total alveolar surface area and the total resistance to oxygen transport in the respiratory tree. Further research relating the constructal principle to the microscopic physical dynamics might yield valuable insight for all branches of nanoengineering.

Acknowledgments

This research was funded by the European Commission through the New and Emerging Science and Technology programme, DYNAMO STREP, project No. FP6-028669-2. Financial support from the excellence cluster NIM (Nanosystems Initiative München) is also gratefully acknowledged.

References

Allahverdyan, A. E., Balian, R., and Nieuwenhuizen, Th. M. 2004. Quantum thermodynamics: Thermodynamics at the nanoscale. *J. Modern Opt.* 51: 2703–2711.

Anisimov, M. A. 2004. Thermodynamics at the meso- and nanoscale. In *Dekker Encyclopedia of Nanoscience and Nanotechnology*, J. A. Schwarz, C. I. Contescu, and K. Putyera (Eds.), pp. 3893–3904. New York: Marcel Dekker.

Astumian, R. D. 2007. Design principles for Brownian molecular machines: How to swim in molasses and walk in a hurricane. *Phys. Chem. Chem. Phys.* 9: 5067–5083.

Bailyn, A. 1994. *A Survey of Thermodynamics*. New York: American Institute of Physics Press.

Bartell, L. S. and Chen, J. 1992. Structure and dynamics of molecular clusters. 2. Melting and freezing of carbon tetrachloride clusters. *J. Phys. Chem.* 96: 8801–8808.

Beck, C. 2001. Dynamical foundations of nonextensive statistical mechanics. *Phys. Rev. Lett.* 87: 180601.

Beck, C. 2002. Non-additivity of Tsallis entropies and fluctuations of temperature. *Europhys. Lett.* 57: 329–333.

Becker, R. and Döring, W. 1935. Kinetische Behandlung der Keimbildung in übersättingten Dämpfen. *Ann. Phys.* 24: 719–752.

Bejan, A. 2000. *Shape and Structure: From Engineering to Nature.* Cambridge, U.K.: Cambridge University Press.

Bejan, A. 2006. Constructal theory of generation of configuration in nature and engineering. *J. Appl. Phys.* 100: 041301.

Ben-Amotz, D. and Honig, J. M. 2006. Average entropy dissipation in irreversible mesoscopic processes. *Phys. Rev. Lett.* 96: 020602.

Berry, R. S. 2007. The power of the small. *Eur. Phys. J. D* 43: 5–6.

Bruce, A. D. and Wilding, N. B. 1999. Critical-point finite-size scaling in the microcanonical ensemble. *Phys. Rev. E* 60: 3748–3760.

Buffat, Ph. and Borel, J. P. 1976. Size effect on the melting temperature of gold particles. *Phys. Rev. A* 13: 2287–2298.

Bustamante, C., Liphardt, J., and Ritort, F. 2005. The nonequilibrium thermodynamics of small systems. *Phys. Today* 58: 43–48.

Carrete, J., Varela, L. M., and Gallego, L. J. 2008. Nonequilibrium nanothermodynamics. *Phys. Rev. E* 77: 022102.

Chamberlin, R.V. 1999. Mesoscopic mean-field theory for supercooled liquids and the glass transition. *Phys. Rev. Lett.* 82: 2520–2523.

Chamberlin, R. V. 2000. Mean-field cluster model for the critical behaviour of ferromagnets. *Nature* 408: 337–339.

Chamberlin, R. V. 2002. Nanoscopic heterogeneities in the thermal and dynamic properties of supercooled liquids. *ACS Symp. Ser.* 820: 228–248.

Chamberlin, R. V. 2003. Critical behavior from Landau theory in nanothermodynamic equilibrium. *Phys. Lett. A* 315: 313–318.

Chamberlin, R. V. and Stangel, K. J. 2006. Monte Carlo simulation of supercooled liquids using a self-consistent local temperature. *Phys. Lett. A* 350: 400–404.

Chushak, Y. G. and Bartell, L. S. 2000. Crystal nucleation and growth in large clusters of SeF_6 from molecular dynamics simulations. *J. Phys. Chem. A* 104: 9328–9336.

Chushak, Y. G. and Bartell, L. S. 2001. Melting and freezing of gold nanoclusters. *J. Phys. Chem. B* 105: 11605–11614.

Cohen, E. G. D. 2002. Statistics and dynamics. *Physica A* 305: 19–26.

Collin, D., Ritort, F., Jarzynski, C., Smith, S., Tinoco Jr., I., and Bustamante, C. 2005. Verification of the Crooks fluctuation theorem and recovery of RNA folding free energies. *Nature* 437: 231–234.

Couchman, P. R. and Jesser, W. A. 1977. Thermodynamic theory of size dependence of melting temperature in metals. *Nature* 269: 481–483.

Crooks, G. E. 1999. Entropy production fluctuation theorem and the nonequilibrium work relation for free-energy differences. *Phys. Rev. E* 60: 2721–2726.

de Groot, S. R. and Mazur, P. 1962. *Non-Equilibrium Thermodynamics.* Amsterdam, the Netherlands: North-Holland Publishing Co.

Defay, R. and Prigogine, I. 1966. *Surface Tension and Adsorption.* London, U.K.: Longmans.

DeHoff, R. 2006. *Thermodynamics in Materials Science.* Boca Raton, FL: Taylor & Francis.

Delogu, F. 2005. Thermodynamics on the nanoscale. *J. Phys. Chem. B* 109: 21938–21941.

Dobruskin, V. K. 2006. Size-dependent enthalpy of condensation. *J. Phys. Chem. B* 110: 19582–19585.

Evans, D. J. and Morriss, G. 2008. *Statistical Mechanics of Nonequilibrium Liquids.* Cambridge, U.K.: Cambridge University Press.

Evans, D. J. and Searles, D. J. 2002. The fluctuation theorem. *Adv. Phys.* 51: 1529–1585.

Evans, D. J., Cohen, E. G. D., and Morriss, G. P. 1993. Probability of second law violations in shearing steady states. *Phys. Rev. Lett.* 71: 2401–2404.

Feshbach, H. 1988. Small systems: When does thermodynamics apply? *IEEE J. Quant. Electron.* 24: 1320–1322.

Frenkel, J. 1946. *Kinetic Theory of Liquids.* New York: Oxford University Press.

Gallavotti, G. and Cohen, E. G. D. 1995. Dynamical ensembles in stationary states. *J. Stat. Phys.* 80: 931–970.

García-Morales, V. and Pellicer, J. 2006. Microcanonical foundation of nonextensivity and generalized thermostatistics based on the fractality of the phase space. *Physica A* 361: 161–172.

García-Morales, V., Cervera, J., and Pellicer, J. 2003. Calculation of the wetting parameter from a cluster model in the framework of nanothermodynamics. *Phys. Rev. E* 67: 062103.

García-Morales, V., Cervera, J., and Pellicer, J. 2004. Coupling theory for counterion distributions based in Tsallis statistics. *Physica A* 339: 482–490.

García-Morales, V., Cervera, J., and Pellicer, J. 2005. Correct thermodynamic forces in Tsallis thermodynamics: Connection with Hill nanothermodynamics. *Phys. Lett. A* 336: 82–88.

García-Morales, V., Pellicer, J., and Manzanares, J. A. 2008. Thermodynamics based on the principle of least abbreviated action: Entropy production in a network of coupled oscillators. *Ann. Phys. (NY)* 323: 1844–1858.

Garnier, N. and Ciliberto, S. 2005. Nonequilibrium fluctuations in a resistor. *Phys. Rev. E* 71:060101.

Gheorghiu, S. and Coppens, M. O. 2004. Heterogeneity explains features of "anomalous" thermodynamics and statistics. *Proc. Natl. Acad. Sci. USA* 101: 15852–15856.

Gilányi, T. 1999. Small systems thermodynamics of polymer-surfactant complex formation. *J. Phys. Chem. B* 103: 2085–2090.

Gross, D. H. E. 2001. *Microcanonical Thermodynamics. Phase Transitions in "Small" Systems.* Singapore: World Scientific.

Hall, D. G. 1987. Thermodynamics of micelle formation. In *Nonionic Surfactants. Physical Chemistry*, M. J. Schick (Ed.), pp. 233–296. New York: Marcel Dekker.

Hartmann, M., Mahler, G., and Hess, O. 2004. Local versus global thermal states: Correlations and the existence of local temperatures. *Phys. Rev. E* 70: 066148.

Hartmann, M., Mahler, G., and Hess, O. 2005. Nano-thermodynamics: On the minimal length scale for the existence of temperature. *Physica E* 29: 66–73.

Hasegawa, H. 2007. Non-extensive thermodynamics of transition-metal nanoclusters. *Prog. Mat. Sci.* 52: 333–351.

Hill, T. L. 1962. Thermodynamics of small systems. *J. Chem. Phys.* 36: 153–168.

Hill, T. L. 1963. *Thermodynamics of Small Systems. Part I.* New York: W.A. Benjamin.

Hill, T. L. 1964. *Thermodynamics of Small Systems. Part II.* New York: W.A. Benjamin.

Hill, T. L. and Chamberlin, R.V. 1998. Extension of the thermodynamics of small systems to open metastable states: An example. *Proc. Natl. Acad. Sci. USA*, 95: 12779–12782.

Hill, T. L. and Chamberlin, R. V. 2002. Fluctuations in energy in completely open small systems. *Nano Lett.* 2: 609–613.

Hubbard, J. 1971. On the equation of state of small systems. *J. Chem. Phys.* 55: 1382–1385, and references therein.

Hummer, G. and Szabo, A. 2001. Free energy reconstruction from nonequilibrium single-molecule pulling experiments. *Proc. Natl. Acad. Sci. USA* 98: 3658–3661.

Jarzynski, C. 1997a. Nonequilibrium equality for free energy differences. *Phys. Rev. Lett.* 78: 2690–2693.

Jarzynski, C. 1997b. Equilibrium free-energy differences from nonequilibrium measurements: A master-equation approach. *Phys. Rev. E* 56: 5018–35.

Javaheri, M. R. H. and Chamberlin, R. V. 2006. A free-energy landscape picture and Landau theory for the dynamics of disordered materials. *J. Chem. Phys.* 125: 154503.

Jiang, Q. and Yang, C. C. 2008. Size effect on the phase stability of nanostructures. *Curr. Nanosci.* 4: 179–200.

Jortner, J. and Rao, C. N. R. 2002. Nanostructured advanced materials. Perspectives and directions. *Pure Appl. Chem.* 74: 1491–1506.

Kirkwood, J. G. 1935. Statistical mechanics of fluid mixtures. *J. Chem. Phys.* 3: 300–313.

Kondepudi, D. 2008. *Introduction to Modern Thermodynamics.* New York: Wiley.

Kurchan, J. 1998. Fluctuation theorem for stochastic dynamics. *J. Phys. A: Math. Gen.* 31 3719–3729.

Lebon, G., Jou, D., and Casas-Vázquez, J. 2008. *Understanding Non-Equilibrium Thermodynamics.* Berlin, Germany: Springer-Verlag.

Lebowitz, J. L. and Spohn, H. 1999. A Gallavotti-Cohen-type symmetry in the large deviation functional for stochastic dynamics. *J. Stat. Phys.* 95: 333–365.

Letellier, P., Mayaffre, A., and Turmine, M. 2007a. Solubility of nanoparticles: Nonextensive thermodynamics approach. *J. Phys.: Condens. Matter* 19: 436229.

Letellier, P., Mayaffre, A., and Turmine, M. 2007b. Melting point depression of nanoparticles: Nonextensive thermodynamics approach. *Phys. Rev. B* 76: 045428.

Liphardt, J., Dumont, S., Smith, S. B., Tinoco Jr., I., and Bustamante, C. 2002. Equilibrium information from nonequilibrium measurements in an experimental test of Jarzynski's equality. *Science* 296: 1832–1835.

Luzzi, R., Vasconcellos, A. R., and Galvao Ramos, J. 2002. Trying to make sense out of order. *Science* 298: 1171–1172.

Maes, C. 1999. The fluctuation theorem as a Gibbs property. *J. Stat. Phys.* 95: 367–392.

Mafé, S., Manzanares, J. A., and de la Rubia, J. 2000. On the use of the statistical definition of entropy to justify Planck's form of the third law of thermodynamics. *Am. J. Phys.* 68: 932–935.

Mukamel, S. 2003. Quantum extension of the Jarzynski relation: Analogy with stochastic dephasing, *Phys. Rev. Lett.* 90:170604.

Onischuk, A. A., Purtov, P. A., Bakalnov, A. M., Karasev, V. V., and Vosel, S. K. 2006. Evaluation of surface tension and Tolman length as a function of droplet radius from experimental nucleation rate and supersaturation ratio: Metal vapor homogenous nucleation. *J. Chem. Phys.* 124: 014506.

Peters, K. F., Cohen, J. B., and Chung, Y. W. 1998. Melting of Pb nanocrystals. *Phys. Rev. B* 57: 13430–13438.

Polak, M. and Rubinovich, L. 2008. Nanochemical equilibrium involving a small number of molecules: A prediction of a distinct confinement effect. *Nano Lett.* 8: 3543–3547.

Rajagopal, A. K., Pande, C. S., and Abe, S. 2006. Nanothermodynamics—A generic approach to material properties at nanoscale. In *Nano-Scale Materials: From Science to Technology*, S. N. Sahu, R. K. Choudhury, and P. Jena (Eds.), pp. 241–248. Hauppauge, NY: Nova Science.

Reguera, D., Rubí, J. M., and Vilar, J. M. G. 2005. The mesoscopic dynamics of thermodynamic systems. *J. Phys. Chem. B* 109: 21502–21515.

Reiss, H., Mirabel, P., and Whetten, R. L. 1988. Capillary theory for the coexistence of liquid and solid clusters. *J. Phys. Chem.* 92: 7241–7246.

Rowlinson, J. S. 1987. Statistical thermodynamics of small systems. *Pure Appl. Chem.* 59: 15–24.

Rusanov, A. I. 2005. Surface thermodynamics revisited. *Surf. Sci. Rep.* 58: 111–239.

Samsonov, V. M., Sdobnyakov, N. Yu., and Bazulev, A. N. 2003. On thermodynamic stability conditions for nanosized particles. *Surf. Sci.* 532–535: 526–530.

Schäfer, R. 2003. The chemical potential of metal atoms in small particles. *Z. Phys. Chem.* 217: 989–1001.

Schmidt, M. and Haberland, H. 2002. Phase transitions in clusters. *C. R. Physique* 3: 327–340.

Talkner, P., Hänggi, P., and Morillo, M. 2008. Microcanonical quantum fluctuation theorems. *Phys. Rev. E* 77: 051131.

Tanaka, M. 2004. New interpretation of small system thermodynamics applied to ionic micelles in solution and Corrin-Harkins equation. *J. Oleo Sci.* 53: 183–196.

Teifel, J. and Mahler, G. 2007. Model studies on the quantum Jarzynski relation. *Phys. Rev. E* 76: 051126.

Tolman, R. C. 1949. The effect of droplet size on surface tension. *J. Chem. Phys.* 17: 333–337.

Tsallis, C. 2001. Nonextensive statistical mechanics and thermodynamics: Historical background and present status. In *Nonextensive Statistical Mechanics and Its Applications*, S. Abe and Y. Okamoto (Eds.), pp. 3–98. Berlin, Germany: Springer-Verlag.

van Zon, R., Ciliberto, S., and Cohen, E. G. D. 2004. Power and heat fluctuation theorems for electric circuits. *Phys. Rev. Lett.* 92: 130601.

Vanfleet, R. R. and Mochel, J. M. 1995. Thermodynamics of melting and freezing in small particles. *Surf. Sci.* 341: 40–50.

Vengrenovich, R. D., Gudyma, Yu. V., and Yarema, S. V. 2001. Ostwald ripening of quantum-dot nanostructures. *Semiconductors* 35: 1378–1382.

Vives, E. and Planes, A. 2002. Is Tsallis thermodynamics nonextensive? *Phys. Rev. Lett.* 88: 020601.

Wang, C. X. and Yang, G. W. 2005. Thermodynamics of metastable phase nucleation at the nanoscale. *Mater. Sci. Eng. R* 49: 157–202.

Wang, G. M., Sevick, E. M., Mittag, E., Searles, D. J., and Evans, D. J. 2002. Experimental demonstration of violations of the second law of thermodynamics for small systems and short time scales. *Phys. Rev. Lett.* 89: 050601.

Weissmüller, J. 2002. Thermodynamics of nanocrystalline solids. In *Nanocrystalline Metals and Oxides. Selected Properties and Applications*, P. Knauth and J. Schoonman (Eds.), pp. 1–39. Boston, MA: Kluwer.

Wilk, G. and Wlodarczyk, Z. 2000. Interpretation of the nonextensivity parameter q in some applications of Tsallis statistics and Lévy distributions. *Phys. Rev. Lett.* 84: 2770–2773.

Williams, S. R., Searles, D. J., and Evans, D. J. 2008. Nonequilibrium free-energy relations for thermal changes. *Phys. Rev. Lett.* 100: 250601.

Yang, C. C. and Li, S. 2008. Size-dependent temperature-pressure phase diagram of carbon. *J. Phys. Chem. C,* 112: 1423–1426.

Zhang, C. Y., Wang, C. X., Yang, Y. H., and Yang, G. W. 2004. A nanoscaled thermodynamic approach in nucleation of CVD diamond on nondiamond surfaces. *J. Phys. Chem. B* 108: 2589–2593.

Zwanzig, R. W. 1954. High-temperature equation of state by a perturbation method. I. Nonpolar gases. *J. Chem. Phys.* 22: 1420–1426.

16

Statistical Mechanics in Nanophysics

Jurij Avsec
University of Maribor

Greg F. Naterer
University of Ontario

Milan Marčič
University of Maribor

16.1 Introduction

Billions of years ago, when enormous quantities of energy were released after the Big Bang, the fundamental particles followed by molecules were formed into complex structures according to certain coincidental events. In the period of several billion years of development, the Earth was also shaped as one of the planets in space, after which life was created on it. Over time, humans gradually learned how to exploit substances and materials. The ability of making tools and devices distinguished humans from other living beings. Around 400,000 years ago, people were capable of making wooden spears and lances. They made tools and devices twice their own size. It has always been people's desire to make ever larger machines and devices. The reason was often simple: the leaders ruling at the time wanted to be ranked among the immortals. In Egypt, for example, pyramids were constructed in 2600 BC for the needs of the pharaohs, with the tallest being 147 m high (Keops' pyramid). In 1931, the 449 m tall Empire State Building was built in New York. Currently, the last preparations are underway in Shanghai to construct a 1000 m high housing building.

Despite the development of increasingly larger devices, many inventors and scientists wanted to reveal the smallest secrets of micro and nano processes. For centuries, only clock makers worked on a diminishing size of devices. In the seventeenth century, the invention of the microscope opened the way to the observation of microbes, plants, and animal cells. In the late

twentieth century, microdevices were technologically refined. Today, the size of transistors in integrated circuits is 0.18 μm. Transistors measuring 10 nm are already being developed in laboratories.

December 29, 1958 is cited as the date of the beginning of micromechanics and nanomechanics, when at the California Institute of Technology, the Nobel prize winner Richard P. Feynman delivered a lecture for the American Physical Association. He introduced a vision of reducing the size of machines to a nanosize. At that time, Professor Feynman could not see the economic implications of the devices made on the basis of nanotechnology. But today, nanomechanics and micromechanics are becoming increasingly important in the industry. The concepts of invisible aircrafts, pumps, and so on are now becoming a reality. At the same time, problems have arisen in advanced mechanics, not even dreamt of before. Thermodynamic and transport properties of a gas flowing through a tube with the diameter of a few nanometers are modeled completely differently due to the unusual influence of surface effects. Even classical hydromechanics is not sufficient. In addition to temperature and pressure, the Knudsen number is becoming increasingly important. Euler's equation gives inaccurate results almost over the entire range: Navier-Stokes equations at a Knudsen number of 0.1 and Burnett's equation at a Knudsen number of 10. However, in order to analyze free molecular flow in micro and nanochannels, the nonequilibrium mechanics and original Boltzmann's equation must be used. In this case, the computation of hydromechanical problems is possible over the entire range of Knudsen numbers, temperatures, and pressures [1,2].

The term "nanofluid" describes a solid–liquid mixture that consists of nanoparticles and a base liquid. This is one of the new challenges for thermosciences provided by nanotechnology. The possible application area of nanofluids is in advanced cooling systems, micro/nano electromechanical systems, and many others. The investigation of the effective thermal conductivity of liquids with nanoparticles has recently attracted much interest experimentally and theoretically. The effective thermal conductivity of nanoparticle suspensions can be much higher than for a fluid without nanoparticles.

16.2 Calculation of Thermal Conductivity

16.2.1 Calculation of Thermal Conductivity for Pure Fluids

Accurate knowledge of nonequilibrium and transport properties of pure gases and liquids is essential for the optimum design of equipment in chemical process plants and many other industrial applications [3–6]. It is needed for the determination of intermolecular potential energy functions and development of accurate theories of transport properties in dense fluids. Transport coefficients describe the process of relaxation to equilibrium from a state perturbed by the application of temperature, pressure, density, velocity, or composition gradients. The theoretical description of these phenomena constitutes a part of nonequilibrium statistical mechanics that is known as kinetic theory.

This chapter will use a Chung–Lee–Starling model (CLS) [4,5]. Equations for the thermal conductivity are developed based on kinetic gas theories and correlated with experimental data. The low-pressure transport properties are extended to fluids at high densities by introducing empirically correlated, density-dependent functions. These correlations use an acentric factor ω, dimensionless dipole moment μ_r, and empirically determined association parameters to characterize the molecular structure effects of polyatomic molecules κ, polar effects, and the hydrogen bonding effect. New constants for fluids are developed in this paper.

The dilute gas thermal conductivity for the CLS model is written as

$$\lambda = \lambda_k + \lambda_p, \qquad (16.1)$$

where

$$\lambda_k = \lambda_0 \left(\frac{1}{H_2} + B_6 Y \right). \qquad (16.2)$$

The thermal conductivity in the region of dilute gases for the CLS model is written as

$$\lambda_0 = 3119.41 \left(\frac{\eta_0}{M} \right) \psi, \qquad (16.3)$$

where ψ represents the influence of polyatomic energy contributions to the thermal conductivity. This term makes use of the Taxman theory, which includes the influence of internal degrees of freedom on the basis of Weeks, Chandler, Uhlenbeck, and de Boer (WCUB) theory [3] and the approximations provided by Mason and Monschick [3,6]. The final expression for the influence of internal degrees of freedom is represented as

$$\psi = 1 + C_{\text{int}}^* x \left\{ \frac{0.2665 + ((0.215 - 1.061\beta)/Z_{\text{coll}}) + 0.28288(C_{\text{int}}^*/Z_{\text{coll}})}{\beta + (0.6366/Z_{\text{coll}}) + (1.061\beta C_{\text{int}}^*/Z_{\text{coll}})} \right\}, \qquad (16.4)$$

where

C_{int}^* is the reduced internal heat capacity at a constant volume

β is the diffusion term

Z_{coll} is the collision number

The heat capacities are calculated by statistical thermodynamics. This chapter features all important contributions (translation, rotation, internal rotation, vibration, intermolecular potential energy, and the influence of electron and nuclei excitation). The residual part λ_p of the thermal conductivity can be represented by the following equation:

$$\lambda_p = \left(0.1272 \left(\frac{T_c}{M} \right)^{1/2} \frac{1}{V_c^{2/3}} \right) B_7 Y^2 H_2 \left(\frac{T}{T_c} \right)^{1/2}, \qquad (16.5)$$

where λ_p is in units of W/m K.

$$H_2 = \left\{ B_1 \left[1 - \exp(-B_4 Y) \right] \frac{1}{Y} + B_2 G_1 \exp(B_5 Y) + B_3 G_1 \right\}$$

$$\times \frac{1}{B_1 B_4 + B_2 + B_3}. \qquad (16.6)$$

The constants B_1–B_7 are linear functions of the acentric factor, reduced dipole moment, and the association factor:

$$B_i = b_0(i) + b_1(i)\omega + b_2(i)\mu_r^4 + b_3(i)\kappa, \quad i = 1, 10, \qquad (16.7)$$

where the coefficients b_0, b_1, b_2, and b_3 are presented in an earlier work of Chung et al. [4,5].

16.2.2 Calculation of Thermal Conductivity for Pure Solids [7–11]

16.2.2.1 Electronic Contribution to the Thermal Conductivity

The fundamental expression for the electronic contribution λ_{el} to the thermal conductivity can be calculated on the basis of the theory of thermal conductivity for a classical gas:

$$\lambda_{el} = \frac{1}{3} n c_{el} v_{el} l_{el}, \qquad (16.8)$$

where

c_{el} is the electronic heat capacity (per electron)
n is the number of conduction electrons per volume
v_{el} is the electron speed
l_{el} is the electron mean free path

In Equation 16.8, it is assumed that electrons travel the same average distance before transferring their excess thermal energy to the atoms by collisions.

We can express the mean free path in terms of the electron lifetime τ ($l_{el} = v_F \tau$):

$$\lambda_{el} = \frac{\pi^2 n k_B^2 T \tau}{3m}. \qquad (16.9)$$

Using Drude's theory [6,7], we can express thermal conductivity as the function of electrical conductivity σ_e:

$$\lambda_{el} = \sigma_e L T, \qquad (16.10)$$

where L is a temperature-dependent constant.

16.2.2.2 Phonon Contribution to the Thermal Conductivity

It is more difficult to determine thermal conductivity when there are nonfree electrons. Solids that obey this rule are called nonmetallic crystals. Because the atoms in a solid are closely coupled together, an increase in temperature will be transmitted to other parts. In modern theory, heat is considered as being transmitted by phonons, which are the quanta of energy in each mode of vibration. We can again use the following expression:

$$\lambda_{ph} = \frac{1}{3} C v l. \qquad (16.11)$$

16.2.2.3 Calculation of Electronic Contribution Using Eliashberg Transport Coupling Function

Grimwall [9] showed the following analytical expression for the electrical conductivity σ:

$$\sigma_e = \frac{ne^2}{m_b} \left\langle \tau(\varepsilon, \vec{k}) \right\rangle, \qquad (16.12)$$

where

m_b represents the electron band mass
τ is an electron lifetime that depends both on the direction of the wave vector $\vec{k}$ and the energy distance ε

The brackets $\langle\ \rangle$ describe an average over all electron states. We can also describe the electronic part of thermal conductivity with the help of Equation 16.12:

$$\lambda_{el} = \frac{n k_B T}{m_b} \left\langle \left(\frac{\varepsilon_k - E_F}{k_B T} \right)^2 \tau(\varepsilon, \vec{k}) \right\rangle. \qquad (16.13)$$

The lifetime for the scattering of electrons by phonons contains quantum-mechanical quantum matrix elements for electron–phonon interaction, and statistical Bose–Einstein and Fermi–Dirac factors for the population of phonon and electron states. A very useful magnitude in this context is the Eliashberg transport coupling function, $\alpha_{tr}^2 F(\omega)$. A detailed theoretical expression is derived by Grimwall [9,11]. The Eliashberg coupling function allows us to write the thermal conductivity in the next expression

$$\frac{1}{\lambda_{el}} = \frac{(4\pi)^2}{L_0 T \omega_{pl}^2} \int_0^{\omega_{max}} \frac{\hbar\omega/k_B T}{\left[\exp\left(\hbar\omega/k_B T - 1\right) \right]\left[1 - \exp\left(-\hbar\omega/k_B T\right) \right]}$$

$$\times \left\{ \left[1 - \frac{1}{2\pi^2} \left(\frac{\hbar\omega}{k_B T} \right)^2 \right] \alpha_{tr}^2 F(\omega) + \frac{3}{2\pi^2} \left(\frac{\hbar\omega}{k_B T} \right)^2 \alpha_{tr}^2 F(\omega) \right\} d\omega. \qquad (16.14)$$

We can describe the phonons by an Einstein model:

$$\alpha_{tr}^2 F(\omega) = A\delta(\omega - \omega_E), \qquad (16.15)$$

$$\alpha^2 F(\omega) = B\delta(\omega - \omega_E). \qquad (16.16)$$

In Equations 16.15 and 16.16, B and A are constants. With help of Equations 16.15 and 16.16, we can solve the integral in Equation 16.14 as follows:

$$\frac{1}{\lambda_{el}} = k_E C_{har} \left(\frac{T}{\theta_E} \right) \left[\frac{A}{B} + \left(\frac{\theta_E}{T} \right)^2 \frac{1}{2\pi^2} \left(3 - \frac{A}{B} \right) \right], \qquad (16.17)$$

where

k_E represents a constant
θ_E is the Einstein temperature
C_{har} represents the lattice heat capacity in the Einstein model:

$$C_{har} = 3 N k_B T \left(\frac{\theta_E}{T} \right)^2 \frac{\exp(\theta_E/T)}{\left[\exp(\theta_E/T) - 1 \right]^2}. \qquad (16.18)$$

Motokabbir and Grimwall [10] discussed Equation 16.17 with A/B as a free parameter with an assumption that $A/B \approx 1$.

16.2.2.4 Phonon Contribution to Thermal Conductivity

In an isotropic solid, we can express the thermal conductivity as an integral over ω containing the phonon density of states $F(\omega)$:

$$\lambda_{ph} = \frac{N}{3V} v_g^2 \int_0^{\omega_{max}} \tau(\omega) C(\omega) F(\omega) \, d\omega, \qquad (16.19)$$

where

v_g is an average phonon group velocity

C is the heat capacity of a single phonon mode

the ratio N/V is the number of atoms per volume

A relaxation time can be expressed as the ratio of a mean free path to a velocity, so that the thermal conductivity can be expressed as

$$\lambda_{ph} = \frac{N}{3V} v_g \int_0^{\omega_{max}} l(\omega) C(\omega) F(\omega) d\omega. \tag{16.20}$$

The crucial aspect of Equation 16.20 is the determination of relaxation time. If we consider scattering in and out of state 1, we can use quantum mechanics to describe $\tau(1)$:

$$\frac{1}{\tau(1)} = \frac{2\pi}{\hbar} \sum_{2,3} |H(1,2,3)|^2 \frac{n(2)n(3)}{n(1)}, \tag{16.21}$$

$$|H(1,2,3)|^2 = A \frac{\hbar^2 \gamma^2 \Omega_a^{1/3}}{3MN} \frac{\omega_1 \omega_2 \omega_3}{v_g^2}, \tag{16.22}$$

The evaluation of $\tau(1)$ in Equation 16.21 requires a summation over modes 2 and 3. This cannot be done analytically, so it is not possible to give a closed-form expression for the temperature dependence of thermal conductivity at all temperatures.

For the low-temperature region (where the temperature is lower than the Debye temperature θ_D), we have used the following solution:

$$\lambda_{ph} = \lambda_0 \exp\left(-\frac{\theta_D}{T}\right), \tag{16.23}$$

where λ_0 is a constant.

For the high-temperature region ($T \gg \theta_D$), the solution of Equation 16.23 gives the following result:

$$\lambda_{ph} = \frac{B}{(2\pi)^3} \frac{M\Omega_a^{1/3} k_B^3 \theta_D^3}{\hbar^3 \gamma^2 T}, \tag{16.24}$$

where

B is a dimensionless constant

Ω_a is the atomic volume

γ is the Grüneisen constant

The relation between the Einstein and Debye temperature may be written as

$$\theta_E = (0.72\ldots0.75)\theta_D. \tag{16.25}$$

16.2.3 Calculation of Thermal Conductivity for Nanoparticles [11–37]

In nanoparticle fluid mixtures, other effects such as microscopic motion of particles, particle structures, and surface properties may cause additional heat transfer in nanofluids.

Nanofluids also exhibit superior heat transfer characteristics to conventional heat transfer fluids. One of the main reasons is that suspended particles remarkably increase the thermal conductivity of nanofluids. The thermal conductivity of a nanofluid is strongly dependent on the nanoparticle volume fraction. It remains an unsolved problem to develop a precise theory to predict the thermal conductivity of nanofluids. This chapter calculates the thermal conductivity of a nanofluid analytically. Hamilton and Crosser developed a macroscopic model for the effective thermal conductivity of two-component mixtures as a function of the conductivity of the pure materials, and composition and shape of dispersed particles. The thermal conductivity can be calculated via the following expression [12–34]:

$$\lambda = \lambda_0 \left\{ \frac{\lambda_p + (n-1)\lambda_0 - (n-1)\alpha(\lambda_0 - \lambda_p)}{\lambda_p + (n-1)\lambda_0 + \alpha(\lambda_0 - \lambda_p)} \right\}, \tag{16.26}$$

where

λ is the mixture thermal conductivity

λ_0 is the liquid thermal conductivity

λ_p is the thermal conductivity of solid particles

α is the volume fraction

n is the empirical shape factor given by

$$n = \frac{3}{\psi}, \tag{16.27}$$

where ψ is sphericity, defined as the ratio of the surface area of a sphere (with a volume equal to that of a particle) to the area of the particle. The volume fraction α of the particles is defined as

$$\alpha = \frac{V_p}{V_0 + V_p} = n \frac{\pi}{6} d_p^3, \tag{16.28}$$

where

n is the number of particles per unit volume

d_p is the average diameter of particles

An alternative expression for calculating the effective thermal conductivity of solid–liquid mixtures was introduced by Wasp [34]:

$$\lambda = \lambda_0 \left\{ \frac{\lambda_p + 2\lambda_0 - 2\alpha(\lambda_0 - \lambda_p)}{\lambda_p + 2\lambda_0 + \alpha(\lambda_0 - \lambda_p)} \right\}. \tag{16.29}$$

A comparison between Equations 16.26 and 16.29 shows that the Wasp model is a special case with a sphericity of 1.0 in the Hamilton and Crosser model. From past literature [14–34], we can find some other models (Maxwell, Jeffrey, Davis, Lu-Lin) that give almost identical analytical results.

In nanofluids, many possible mechanisms explain the increased effective thermal conductivity:

- Influence of nanolayer thickness
- Hyperbolic heat conduction
- Brownian motion
- Particle driven or thermally driven natural convection
- Hyperbolic thermal natural convection

16.2.3.1 Influence of Nanolayer around Nanoparticle

The HC model gives very good results for particles larger than 13 nm. For smaller particles, the theory yields inaccurate results with a deviation more than 100% in comparison with experimental results. The theoretical models for the calculation of thermal conductivity for nanofluids are only dependent on the thermal conductivity of the solid and liquid, and their relative volume fraction, but not on particle size and the interface between particles and the fluid. For the calculation of effective thermal conductivity, we have used Xue's theory [35], based on Maxwell's theory and the average polarization theory. Because the interfacial shells existed between the nanoparticles and the liquid matrix, we can regard both the interfacial shell and nanoparticle as a complex nanoparticle. So the nanofluid system should be regarded as complex nanoparticles dispersed in the fluid. We assume that λ is the effective thermal conductivity of the nanofluid, and λ_c and λ_m are the thermal conductivities of the complex nanoparticles and the fluid, respectively. The final expression of Xue's [18] model (X) is expressed by the following equation:

$$
9\left(1-\frac{\alpha}{\lambda_r}\right)\frac{\lambda-\lambda_0}{2\lambda+\lambda_0}
$$
$$
+\frac{\alpha}{\lambda_r}\left[\frac{\lambda-\lambda_{c,x}}{\lambda+B_{2,x}(\lambda_{c,x}-\lambda_e)}+4\frac{\lambda-\lambda_{c,y}}{2\lambda+(1-B_{2,x})(\lambda_{c,y}-\lambda)}\right]=0,
$$

$$(16.30)$$

$$
\lambda_{c,j}=\lambda_1\frac{(1-B_{2,j})\lambda_1+B_{2,j}\lambda_2+(1-B_{2,j})\lambda_r(\lambda_2-\lambda_1)}{(1-B_{2,j})\lambda_1+B_{2,j}\lambda_2-B_{2,j}\lambda_r(\lambda_2-\lambda_1)}. \quad (16.31)
$$

We assume that a complex nanoparticle is composed of an elliptical nanoparticle with thermal conductivity λ_2 with half radii of (a, b, c) and an elliptical shell of thermal conductivity λ_1 with a thickness of t. In Equations 16.30 and 16.31, λ_r represents the spatial average of the heat flux component. For simplicity, we assume that all fluid particles are spherical and all nanoparticles are the same rotational ellipsoid.

We have used the model of Yu and Choi [23], wherein the nanolayer of each particle could be combined with the particle to form an equivalent particle, and that the particle volume concentration is so low that there is no overlap of equivalent particles. On this basis, we can express the effective volume fraction as follows:

$$
\alpha_e=\alpha\left(1+\frac{h}{r}\right)^3, \quad (16.32)
$$

where h represents the liquid layer thickness. We have also made the assumption that the equivalent thermal conductivity of the equivalent particles has the same value as the thermal conductivity of a particle. On the basis of these assumptions, we have derived the following new model (RHC) for the thermal conductivity of nanofluids:

$$
\lambda=\lambda_f\left\{\frac{\lambda_{pt}+(n-1)\lambda_f-(n-1)\alpha_e(\lambda_f-\lambda_{pt})}{\lambda_{pt}+(n-1)\lambda_f+\alpha_e(\lambda_f-\lambda_{pt})}\right\}. \quad (16.33)
$$

16.2.3.2 Hyperbolic Heat Conduction

Heat transport in nanoparticles occurs predominantly by electron and crystal vibrations, and it depends on the material. Macroscopic theories assume diffusive heat transport with the following Laplace equation [41]:

$$
\rho c_p\frac{\partial T}{\partial t}=\lambda\nabla^2 T+\dot{q}, \quad (16.34)
$$

where $\dot{q}$ represents the internal energy source term. From Fourier's law,

$$
\vec{J}_Q=-\lambda\vec{\nabla}T, \quad (16.35)
$$

where J_Q is the heat flux. In crystalline nanoparticles, heat is transferred by phonons. Such phonons are created in random, propagating directions, and they are scattered by each other. With the theory of Debye, the mean free path of the phonon is given by [14]

$$
l_{ph}=\frac{10aT_m}{\gamma T}, \quad (16.36)
$$

where

T_m is the melting point
a is the lattice constant
γ the Grüneisen parameter

For typical nanoparticles such as Al_2O_3 at room temperature, we obtain the result that the mean phonon free path is 35 nm [12]. For this reason, phonons cannot diffuse in the 10 nm particles, but must move ballistically across the particle.

In metals, the heat is primarily carried by electrons, which also exhibit diffusive motion at the macroscopic level. Due to the Drude formula, we can express the mean electron free path as follows [11]:

$$
l_{el}=\frac{9.2r_s^2}{\rho_{el}\left[\mu\Omega\,cm\right]}10^{-9}\,[m], \quad (16.37)
$$

where

ρ_{el} is the electrical resistivity
r_s is the dimensionless parameter

For Cu it is l_{el} 350 nm, and for Al it is l_{el} 65 nm. Due to this reason, electrons cannot diffuse in the 10 nm particles, but must move ballistically across the particle.

It is difficult to demonstrate how ballistic heat transport could be more effective than very fast diffusion transport [24,38]. Hereafter we will take into account the ballistic heat transfer phenomena in nanoparticles on the basis of Boltzmann's law [38].

The ratio between the mean free path, l, and the characteristic length, L, is called the Knudsen number [40]:

$$Kn = \frac{l}{L}.$$ (16.38)

In nanosystems, the Knudsen number becomes comparable or higher than 1. As a result, the heat transport is no longer diffusive but instead ballistic. In that situation, the usual Fourier law describing diffusive transport must be generalized to cover the transition,

$$q = \lambda \frac{\Delta T}{L} \quad \text{(diffusive transport)},$$ (16.39)

$$q = \Lambda \Delta T \quad \text{(ballistic transport)},$$ (16.40)

where

    λ represents the thermal conductivity
    Λ is the heat conduction transport coefficient

The transport phenomena can be solved in different ways:

- Direct application of universal Boltzmann equation
- Application of extended irreversible thermodynamics
- Application of dual time lag equations
- Numerical simulations of heat transport on the basis of lattice theory

In this chapter, we have focused on the extended irreversible thermodynamics theory:

$$q = \lambda \left(T, \frac{l}{L} \right) \frac{\Delta T}{L}.$$ (16.41)

The limiting behavior of this generalized conductivity to recover expressions in suitable situations should be

$$\lambda \left(T, \frac{l}{L} \right) \to \lambda(T) \quad \text{for } \frac{l}{L} \to 0,$$ (16.42)

$$\lambda \left(T, \frac{l}{L} \right) \to \frac{\lambda(T)}{a} \frac{L}{l} \equiv \Lambda(T)L \quad \text{for } \frac{l}{L} \to \infty,$$ (16.43)

where a is a constant depending on the system. On that basis, we obtain the following equation [40] for the determination of real thermal conductivity in all regimes:

$$\lambda(\omega, k) = \cfrac{\lambda(T)}{1 + i\omega\tau_1 + \cfrac{k^2 l_1^2}{1 + i\omega\tau_2 + \cfrac{k^2 l_2^2}{1 + i\omega\tau_3 + \cfrac{k^2 l_3^2}{1 + i\omega\tau_4}}}}$$ (16.44)

when we choose $l_n^2 = \alpha_{n+1} l^2$ with $\alpha_n = n^2 [(2n+1)(2n-1)]^{-1}$. This assumption is of interest because it corresponds to a detailed analysis of photon or phonon heat transport [39]. The assumption leads to the following expression:

$$\lambda \left(T, \frac{l}{L} \right) = \frac{3\lambda}{Kn^2} \left[\frac{Kn}{\tan^{-1}(Kn)} - 1 \right].$$ (16.45)

For the determination of the effective thermal conductivity due to electron heat transport and depending on Knudsen number, we use the same method as presented in Equation 16.45.

16.2.3.3 Brownian Motion

In many cases, it is postulated that the enhanced thermal conductivity of a nanofluid is mainly due to Brownian motion, which produces micromixing. Because of the small size of particles in the fluids, additional energy terms can arise from motions induced by stochastic (Brownian) and interparticle forces. The motion of particles cause microconvection that enhances heat transfer (Koo and Kleinstreuer [35,36]):

$$\lambda_{\text{Brownian}} = 5 \times 10^4 \beta \alpha \rho_l c_l \sqrt{\frac{\kappa T}{\rho_p D}} f,$$ (16.46)

$$f = (-6.04\alpha + 0.4705)T + (1722.3\alpha - 134.63).$$

Kumar et al. [32]:

$$\lambda_{\text{Brownian}} = c \frac{2k_B T}{\pi \eta d_p^2}.$$ (16.47)

The heat transfer enhancement due to Brownian motion can be estimated with a known temperature of the fluid and size of particles. The increase of thermal conductivity due to the rotational and translational motion of spherical particle is modeled by

$$\lambda_{\text{Brownian}} = \lambda_f \alpha \left(\frac{1.17(\lambda_p - \lambda_f)^2}{(\lambda_p + 2\lambda_f)^2} + 5(0.6 - 0.028) \frac{(\lambda_p - \lambda_f)}{(\lambda_p + 2\lambda_f)^2} \right) Pe_f^{1.5}$$

$$+ (0.0556 Pe_t + 0.1649 Pe_t^2 - 0.0391 Pe_t^3 + 0.0034 Pe_t^4)\lambda_f.$$

(16.48)

Prasher [33]:

$$\lambda = \lambda_0 \left\{ \frac{\lambda_p + (n-1)\lambda_0 - (n-1)\alpha(\lambda_0 - \lambda_p)}{\lambda_p + (n-1)\lambda_0 + \alpha(\lambda_0 - \lambda_p)} \right\} \left(1 + \frac{Re \, Pr}{4} \right),$$ (16.49)

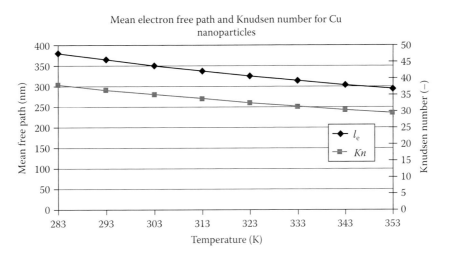

FIGURE 16.1 Mean electron free path and Knudsen number for Cu nanoparticles with an average diameter 10 nm. (From Avsec, J., *Int. J. Heat Mass Transfer*, 51, 4589, 2008.)

$$Re = \frac{1}{v}\sqrt{\frac{18 k_B T}{\pi \rho d}}, \qquad (16.50)$$

where v is the kinematic viscosity of the fluid. In our case, we have used the model for Brownian motion from Prasher [33]. We have slightly corrected the Prasher equation into the next expression:

$$\lambda = \lambda_0 \left\{ \frac{\lambda_p + (n-1)\lambda_0 - (n-1)\alpha(\lambda_0 - \lambda_p)}{\lambda_p + (n-1)\lambda_0 + \alpha(\lambda_0 - \lambda_p)} \right\}(1 + CRe^m Pr^n), \qquad (16.51)$$

where C represents the fitting parameter.

16.2.3.4 Experimental Error

A majority of experimentalists working on nanofluids have measured the thermal conductivity by using Fourier's law with the transient hot wire technique. Vadasz et al. [34] showed that the reason of thermal conductivity enhancement is due to experimental inaccuracy.

16.2.4 Results and Comparison with Experimental Data

In this section, we will show analytical computations for mixtures of copper nanoparticles and ethylene glycol, and also for a mixture between aluminum oxide (Al₂O₃) nanoparticles and water. The copper nanoparticles dispersed in the fluid are interesting for nanofluid industrial applications due to high thermal conductivity in comparison with copper or aluminum oxides. In our case, we have used experimental results from past literature [15], where copper average nanoparticle diameters are smaller than 10 nm.

Figure 16.1 shows the temperature influence of the electron mean free path and Knudsen number. Figure 16.2 shows the

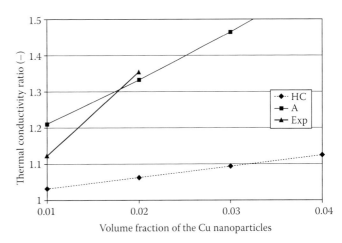

FIGURE 16.2 Thermal conductivity of mixture copper nanoparticles + ethylene glycol at various compositions at 303 K. (From Avsec, J. and Oblak, M., *Int. J. Heat Mass Transfer*, 50, 4331, October 2007.)

analytical calculation of a mixture between ethylene glycol and copper nanoparticles for the thermal conductivity ratio. The results for thermal conductivity obtained by the developed Avsec (A) model show relatively good agreement. The thermal conductivity predicted by a Hamilton–Crosser (HC) model give much lower values than experimental results (Exp). Figure 16.2 is based on the theory that the nanolayer thickness is one of the reasons for heat transfer enhancement. The theory was made on the assumption that the nanolayer thickness is one of the most important contributions [12] in very small nanoparticles ($d \ll 10$ nm). The AO model takes into account the influence on ballistic heat conduction and the influence of Brownian motion. Based on the assumption that the nanolayer thickness is the most important contribution, we have also obtained successful analytical results for viscosity and thermodynamic properties of nanofluids at room temperature in comparison with experimental data [12]. Unfortunately, in the nanofluid theory, the nanolayer thickness is the reason for heat

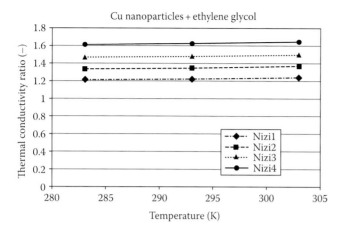

FIGURE 16.3 Analytical prediction of thermal conductivity of mixture copper nanoparticles + ethylene glycol at various composition at various temperatures (Niz1-α = 0.01, Niz2-α = 0.02, Niz3-α = 0.03, Niz4-α = 0.04).

transfer enhancement, which did not give satisfactory results for the temperature dependence. It is possible that the nanolyer thickness is temperature dependent, but till date, we did not find a successful theory or experimental results for verification. Figure 16.3 shows the analytical prediction (A model) of the temperature influence of thermal conductivity of the mixture between copper nanoparticles and ethylene glycol as the reference fluid.

Figures 16.4 through 16.7 show a comparison between our new model (A) and experimental results [26] for Al_2O_3 nanoparticles and water as the reference fluid. Figures 16.4 through 16.7 show the thermal conductivity ratio and dependence of the temperature field. The average diameter of Al_2O_3 nanoparticles is 38.4 nm. Figure 16.4 shows the predictions for Brownian motion velocity and Knudsen number, with the dependence of the temperature field [42,43].

Figure 16.8 shows the thermal conductivity and dependence of the temperature field and Knudsen number. The detailed analysis shows that the mathematical model predicts the thermal conductivity accurately in terms of the dependence of volume fraction of nanoparticles and the temperature field.

16.3 Calculation of Viscosity in Nanofluids

16.3.1 Calculation of Viscosity for Pure Fluid

The CLS [4,5] is used in this section. Equations for the viscosity and thermal conductivity are developed based on kinetic gas theories and correlated with experimental data. The low-pressure transport properties are extended to fluids at high densities by introducing empirically correlated, density-dependent functions. These correlations use an acentric factor ω, dimensionless dipole moment μ_r, and empirically determined association parameters to characterize the molecular structure effect of polyatomic molecules κ, polar effect, and the hydrogen bonding effect. In this paper, new constants for fluids are determined.

The dilute gas viscosity in our model is obtained analytically [8] with the exception of the correction factor:

$$\eta_0(T) = 26.69579 \times 10^{-1} \frac{\sqrt{MT}}{\Omega^{(2,2)^*} \sigma^2} F_c, \qquad (16.52)$$

where
η is in units of Pa s
M is the molecular mass in g mol^{-1}
T is in K
$\Omega^{(2,2)}$ is a collision integral
σ is the Lennard-Jones parameter

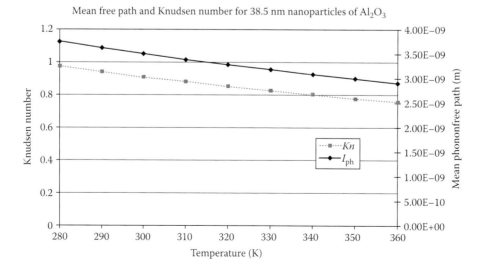

FIGURE 16.4 Knudsen number and mean free phonon path for Al_2O_3 nanoparticles with an average diameter 38.5 nm. (From Avsec, J., *Int. J. Heat Mass Transfer*, 51, 4589, 2008.)

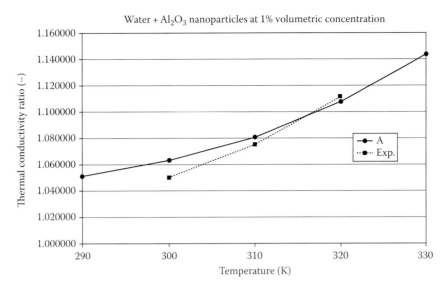

FIGURE 16.5 Thermal conductivity of water + Al$_2$O$_3$ nanoparticles with an average diameter 38.4 nm at 1% of volume concentration of nanoparticles. (From Avsec, J., *Int. J. Heat Mass Transfer*, 51, 4589, 2008.)

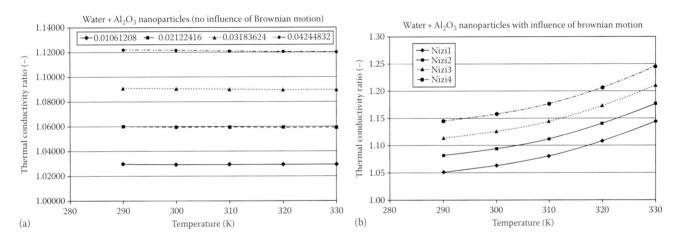

FIGURE 16.6 Analytical prediction of thermal conductivity and influence of Brownian motion in the mixture of water with Al$_2$O$_3$ nanoparticles for the A model (Niz1-α = 0.01, Niz2-α = 0.02, Niz3-α = 0.03, Niz4-α = 0.04). (From Avsec, J., *Int. J. Heat Mass Transfer*, 51, 4589, 2008.)

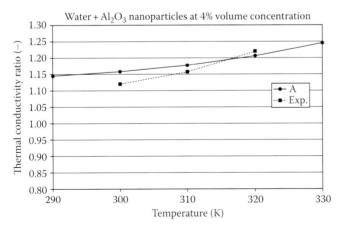

FIGURE 16.7 Thermal conductivity of water + Al$_2$O$_3$ nanoparticles with and average diameter 38.4 nm at 4% of volume concentration of nanoparticles. (From Avsec, J. and Oblak, M., *Int. J. Heat Mass Transfer*, 50, 4331, October 2007.)

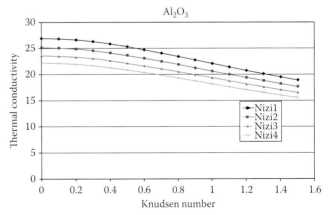

FIGURE 16.8 Thermal conductivity dependence of Knudsen number (Niz1-T = 280 K, Niz2-T = 300 K, Niz3-T = 320 K, Niz4-T = 340 K). (From Avsec, J., *Int. J. Heat Mass Transfer*, 51, 4589, 2008.)

The factor F_c has been empirically found to be

$$F_c = 1 - 0.2756\omega + 0.059035\mu_r^4 + \kappa, \qquad (16.53)$$

where

ω is the acentric factor

μ_r is the relative dipole moment

κ is a correction factor for hydrogen-bonding effects of associating substances such as alcohols, ethers, acids, and water

For dense fluids, Equation 16.52 is extended to account for the effects of temperature and pressure by developing an empirically correlated function of density and temperature as shown below:

$$\eta = \eta_k + \eta_p, \qquad (16.54)$$

$$\eta_k = \eta_0 \left(\frac{1}{G_2} + A_6 Y \right), \qquad (16.55)$$

$$\eta_p = \left[36.344 \times 10^{-6} - \frac{(MT_c)^5}{V_C^{2/3}} \right] A_7 Y^2 x$$
$$G_2 \exp\left(A_8 + \frac{A_9}{T^*} + \frac{A_{10}}{T^{*2}} \right), \qquad (16.56)$$

$$Y = \rho V_c / 6, \quad G_1 = \frac{1.0 - 0.5Y}{(1.0 - Y)^3}, \qquad (16.57)$$

$$T_c = \frac{1.2593\varepsilon}{k}, \quad V_c = (0.809\sigma(\dot{A}))^3, \qquad (16.58)$$

$$G_2 = \frac{\{ A_1(1 - \exp(-A_4 Y)) + A_2 G_1 \exp(A_5 Y) + A_3 G_1 \}}{A_1 A_4 + A_2 + A_3}. \qquad (16.59)$$

The constants A_1–A_{10} are linear functions of the acentric factor, reduced dipole moment, and the association factor:

$$A_i = a_0(i) + a_1(i)\omega + a_2(i)\mu_r^4 + a_3(i)\kappa, \quad \text{with } i = 1, 10 \qquad (16.60)$$

where the coefficients a_0, a_1, a_2, and a_3 are presented in past work of Chung et al. [4].

16.3.2 Calculation of Effective Viscosity for Nanofluids

It is well known that the earliest theoretical work on the effective viscosity was due to Einstein, whose derivation led to the effective viscosity to be linearly related to the particle concentration as follows [20]:

$$\eta_r = 1 + 2.5\alpha, \qquad (16.61)$$

where

η_r is the relative viscosity defined as the ratio of the effective viscosity of the particle fluid-mixture to the viscosity of the fluid

α is the volumetric concentration of the particles

Equation 16.61 is applicable to suspensions with low particle concentrations (less then 2%). With help of an exponential model, we obtain the following Cheng–Law [21] expression for the relative viscosity:

$$\eta_r = 1 + 2.5\alpha + \left(\frac{35}{8} + \frac{5}{4}\beta \right)\alpha^2 + \left(\frac{105}{16} + \frac{35}{8}\beta + \frac{5}{12}\beta^2 \right)\alpha^3$$
$$+ \left(\frac{1155}{128} + \frac{935}{96}\beta + \frac{235}{96}\beta^2 + \frac{5}{48}\beta^3 \right)\alpha^4 + \cdots, \qquad (16.62)$$

where β is called an exponent. If we choose β = 2, we obtain a result close to Ward (W), who suggested the following expression for spherical particles:

$$\eta_r = 1 + (2.5\alpha) + (2.5\alpha)^2 + (2.5\alpha)^3 + (2.5\alpha)^4 \ldots. \qquad (16.63)$$

The equation is fitted with experimental data for a concentration up to 35%.

The viscosity of nanofluids is strongly dependent on the nanoparticle volume fraction. A generalized theory to predict the viscosity of nanofluids is currently unavailable. This section is an attempt to calculate the thermal conductivity of the nanofluid analytically. Cheng and Law [21] developed a model for the effective thermal conductivity of two-component mixtures as a function of the viscosity of the pure fluid, composition of particles, and an exponent factor.

The Cheng, Law, and Ward models give good results for two-phase flow with particles larger than 100 nm. For smaller particles, the theory gives inaccurate results with a deviation more than 100% in comparison to experimental results. The theoretical models for calculation of the viscosity of nanofluids are only dependent on the viscosity of the liquid and their relative volume fraction, but not on particle size and the interaction between particles and the fluid.

In convection heat transfer in nanofluids, not only for the thermal conductivity but also other properties such as specific heat, dynamic viscosity, and so forth, must be calculated. Possible mechanisms for the enhancement of viscosity are discussed in the literature [17–27]: the motion of a nanoparticle, molecular level layering of the liquid at the liquid–particle interface, ballistic phenomena in nanoparticles, effects of clustering in nanoparticles, etc.

As for the case of analytical calculations of thermal conductivity, and the viscosity of nanofluids, we have made the hypothesis that the most important additional contribution is liquid layering. With the help of Equation 16.32, we express the following Ward model (RW):

$$\eta_r = 1 + (2.5\alpha_e) + (2.5\alpha_e)^2 + (2.5\alpha_e)^3 + (2.5\alpha_e)^4 \ldots. \qquad (16.64)$$

Figures 16.2 and 16.3 show a comparison between experimental results for the mixture water + TiO_2, water + Al_2O_3, and analytical results. The mean diameter of TiO_2 particles is 27 nm and Al_2O_3 particles is 13 nm. The deviation of results between W (Equation 16.45), RW model (Equation 16.46), and experimental results is high. The Ward model (RW) yields excellent results.

16.4 Calculation of Thermodynamic Properties of a Pure Fluid

To calculate thermodynamic functions of state, we applied the canonical partition [44]. Utilizing a semiclassical formulation for the purpose of the canonical ensemble for the N indistinguishable molecules, the partition function Z can be expressed as follows:

$$Z = \frac{1}{N! h^{Nf}} \int \ldots \int \exp\left(-\frac{H}{kT}\right) \cdot d\vec{r}_1 d\vec{r}_2 \ldots d\vec{r}_N d\vec{p}_1 d\vec{p}_2 \ldots d\vec{p}_N,$$

where

  f refers to the number of degrees of freedom of individual molecules
  H designates the Hamiltonian molecule system
  vectors $\vec{r}_1, \vec{r}_2 \ldots \vec{r}_N$ describe the positions of N molecules and $\vec{p}_1, \vec{p}_2 \ldots \vec{p}_N$ momenta
  k is Boltzmann's constant
  h is Planck's constant

The canonical ensemble of partition functions for the system of N molecules can be expressed as

$$Z = Z_0 Z_{\text{trans}} Z_{\text{vib}} Z_{\text{rot}} Z_{\text{ir}} Z_{\text{el}} Z_{\text{nuc}} Z_{\text{conf}}. \tag{16.65}$$

Thus, the partition function Z is a product of terms of the ground state (0), translation (trans), vibration (vib), rotation (rot), internal rotation (ir), influence of electron excitation (el), influence of nuclei excitation (nuc), and the influence of intermolecular potential energy (conf).

Utilizing the canonical theory for computing the thermodynamic functions of the state [5,6]:

$$\text{Pressure } p = kT\left(\frac{\partial \ln Z}{\partial V}\right)_T, \quad \text{Internal energy } U = kT^2\left(\frac{\partial \ln Z}{\partial T}\right)_V,$$

$$\text{Free energy } A = -kT \times \ln Z, \quad \text{Entropy } S = k\left[\ln Z + T\left(\frac{\partial \ln Z}{\partial T}\right)_V\right],$$

$$\text{Free enthalpy } G = -kT\left[\ln Z - V\left(\frac{\partial \ln Z}{\partial T}\right)_V\right], \tag{16.66}$$

$$\text{Enthalpy } H = kT\left[T\left(\frac{\partial \ln Z}{\partial T}\right)_V + V\left(\frac{\partial \ln Z}{\partial V}\right)_T\right],$$

where

  T is temperature
  V is the volume of the molecular system

The computation of individual terms of the partition function and their derivatives, except the configurational integral, is dealt with by the works of Lucas [44].

16.4.1 Revised Cotterman Model (CYJ)

The Cotterman EOS is based on hard sphere perturbation theory. The average relative deviation for pressure and internal energy in comparison with Monte-Carlo simulations are 2.17% and 2.62%, respectively, for 368 data points. The configurational free energy is given by

$$A_{\text{conf}} = A^{\text{hs}} + A^{\text{pert}}, \tag{16.67}$$

$$\frac{A^{\text{hs}}}{R_m T} = \frac{4\eta - 3\eta^2}{(1-\eta)^2}, \tag{16.68}$$

$$A^{\text{pert}} = \frac{A^{(1)}}{T^\star} + \frac{A^{(2)}}{T^{\star 2}}, \tag{16.69}$$

$$\frac{A^{(1)}}{R_m T} = \sum_{m=1}^{4} A_{1m}\left(\frac{\eta}{\tau}\right)^m, \quad \frac{A^{(2)}}{R_m T} = \sum_{m=1}^{4} A_{2m}\left(\frac{\eta}{\tau}\right)^m, \tag{16.70}$$

$$\tau = 0.7405, \quad \eta = \frac{\pi \rho D^3}{6}, \tag{16.71}$$

where

  η is the packing factor
  D is the hard-sphere diameter

Using configurational free energy, we can calculate all configurational thermodynamic properties. Expressions for the calculation of configurational entropy and internal energy are found in past literature.

16.4.2 Calculation of Thermodynamic Properties of Pure Solids

The thermodynamic system consists of N particles associated by attractive forces. Atoms in a crystal lattice are not motionless but they thermally oscillate around their positions of equilibrium. At temperatures below the melting point, the motion of atoms is approximately harmonic [1,2]. This assembly of atoms has $3N - 6$ vibration degrees of freedom. Ignore 6 vibration degrees of freedom and mark the number of vibration degrees of freedom with $3N$.

For independent harmonic oscillators, the distribution function Z can be derived as follows:

$$Z = \left[\frac{\exp(-(h\nu/2k_B T))}{1 - \exp(-(h\nu/k_B T))}\right]^{3N}. \tag{16.72}$$

In Equation 16.20, ν is the oscillation frequency of the crystal. The term $h\nu/k$ is the Einstein temperature.

In comparing the experimental data for simple crystals, a relatively good match with analytical calculations at higher temperatures is observed, whereas at lower temperatures, the discrepancies are higher. This explains why Debye corrected the Einstein model by taking into account the interactions between a number of quantized oscillators. The Debye approximation treats a solid as an isotropic elastic substance. Using the canonical distribution, the partition function [1] may be written as

$$\ln Z = -\frac{9}{8} N \frac{\theta_D}{T} - 3N \cdot \ln\left(1 - \exp\left(-\frac{\theta_D}{T}\right)\right)$$

$$+ 3N \frac{T^3}{\theta_D^3} \int_0^{\theta_D/T} \frac{\xi^3}{\exp(\xi) - 1} d\xi. \qquad (16.73)$$

In Equation 16.52, θ_D is the Debye temperature: $\theta_D = (\nu_{max} h/k)$. By developing the third term in Equation 16.73 into a series for a higher temperature range [2]:

$$\frac{\xi^3}{\exp(\xi) - 1} = \xi^2 - \frac{1}{2}\xi^3 + \frac{1}{12}\xi^4 - \frac{1}{720}\xi^6 + \cdots. \qquad (16.74)$$

Using Equation 16.53, Equation 16.52 leads to the following expression:

$$\ln Z = -\frac{9}{8} N \frac{\theta_D}{T} - 3N \cdot \ln\left(1 - \exp\left(-\frac{\theta_D}{T}\right)\right)$$

$$+ 3N \left(\frac{T}{\theta_D}\right)^3 \left[\frac{1}{3}\left(\frac{\theta_D}{T}\right)^3 - \frac{1}{8}\left(\frac{\theta_D}{T}\right)^4 + \frac{1}{60}\left(\frac{\theta_D}{T}\right)^5\right.$$

$$\left. - \frac{1}{5040}\left(\frac{\theta_D}{T}\right)^7 + \frac{1}{272160}\left(\frac{\theta_D}{T}\right)^9 - \cdots\right]. \qquad (16.75)$$

The relation between the Einstein and Debye temperature may be written as [1–2]

$$\theta_E = (0.72 \ldots 0.75)\theta_D. \qquad (16.76)$$

The Debye characteristic temperature was determined by means of the Grüneisen-independent constant γ:

$$\theta = CV^{-\gamma}, \qquad (16.77)$$

where C is a constant dependent on material.

We developed a mathematical model for the calculation of thermodynamic properties of polyatomic crystals. The derivations of the Einstein and Debye equations, outlined in the previous paragraphs, apply specifically to monoatomic solids, i.e., those belonging to the cubic system. However, experiments have shown that the Debye equation represents the values of specific

heat and other thermophysical properties of certain other monoatomic solids, such as zinc, which crystallizes in the hexagonal system. Suppose that the crystal contains N molecules, each composed of s atoms. Since there are Ns atoms, the crystal as a whole has $3Ns$ vibrational modes. A reasonable approximation is obtained by classifying the vibration into

a. $3N$ lattice vibrations, which are the normal modes discussed in the Debye treatment (acoustical modes).
b. Independent vibrations of individual molecules in which bond angles and lengths may vary. There must be $3N(s-1)$ of these (optical modes). We expressed the optical modes using the Einstein model.

16.4.2.1 Electronic Gas in Metals

Consider that electrons capable of moving in a crystal and do not belong to any individual atoms but entirely to the crystal. For example, this includes conduction electrons in metals. A number of such electrons may be called an electronic gas. Using Fermi–Dirac statistics, the configuration integral may then be calculated for temperatures lower than the Fermi temperature:

$$T_F = \frac{\varepsilon_F}{k_B}. \qquad (16.78)$$

For metals, the Fermi temperature is a few thousand kelvins. In Equation 16.26, ε_F is the Fermi energy.

$$A_{el} = \frac{3}{5} N\varepsilon_F \left(1 + \frac{5\pi^2}{12}\left(\frac{k_B T}{\varepsilon_F}\right)^2\right) - Nk_B \frac{\pi^2}{2} \frac{k_B T^2}{\varepsilon_F}. \qquad (16.79)$$

The analytical calculation of the configuration integral in solids is a difficult task. Most frequently, numerical procedures are applied in practical computations by means of the Monte Carlo method [11]. Nevertheless, the previous method requires much computer time. Another drawback is that it does not provide a functional dependence of thermodynamic properties on temperature and volume. Empirical equations [6] are frequently used as well, though mostly without any theoretical basis, built on a molecular view. In this chapter, we used the perturbation VDW theory for solids with the model of hard spheres [4] to calculate the thermodynamic properties of state. In order to calculate the mixtures of atoms of hard spheres, we obtain the configuration free energy for a certain binary crystal:

$$A_{conf0} = Nk_B T \times \left(-3\ln\left(\frac{V^\star - 1}{V^\star}\right) + 5.124 \cdot \ln V^\star - 20.78 V^\star + 9.52 V^{\star 2}\right.$$

$$\left. - 1.98 V^{\star 3} + C_0 + \psi_1 \cdot \ln\psi_1 + \psi_2 \cdot \ln\psi_2\right), \qquad (16.80)$$

$$C_0 = 15.022, \quad V^\star = \frac{V}{V_0}, \quad V_0 = \frac{N\sigma^3}{\sqrt{2}}.$$

In case of a crystal formed of atoms of the same type, the free energy can be written as

$$A_{\text{conf0}} = Nk_\text{B}T\left(-3\ln\left(\frac{V^{*}-1}{V^{*}}\right) + 5.124\cdot\ln V^{*} \right.$$
$$\left. -20.78V^{*} + 9.52V^{*2} - 1.98V^{*3} + C_0 \right). \quad (16.81)$$

To calculate the perturbation contribution, the VDW model was used. In most of the literature [4], the VDW model is treated only in relation to an atomic structure, whereas we additionally present below the temperature-dependent coefficients:

$$A_{\text{conf1}} = -\frac{a(\psi_1, \psi_2, T)}{V}. \quad (16.82)$$

The configuration integral is thus formed by the contribution of hard spheres and perturbation:

$$A_{\text{conf}} = A_{\text{conf0}} + A_{\text{conf1}}. \quad (16.83)$$

In our case, the coefficient "*a*" was determined as a temperature-dependent polynome following a comparison between experimental data and analytical results:

$$a = a_0 + a_1 T + a_2 T^2. \quad (16.84)$$

The coefficients a_0, a_1, and a_2 are obtained by numerical approximation and a comparison with thermodynamic data.

16.4.3 Thermodynamic Properties of Nanofluids

The third main parameter involved in calculating the heat transfer rate of the nanofluid is the heat capacity. For the synthesized nanoparticle-fluid suspension, we can predict the thermodynamic properties with the following expression:

$$(\rho c_p)_{\text{nf}} = (1-\varphi)(\rho c_p)_{\text{f}} + \varphi(\rho c_p)_{\text{p}}, \quad (16.85)$$

where ρ is a density of nanofluid:

$$\rho = (1-\phi)\rho_{\text{f}} + \phi\rho_{\text{p}}. \quad (16.86)$$

16.4.4 Results and Comparison with Experimental Data

Figures 16.9 and 16.10 show a comparison between experimental results for the mixture water + TiO_2, water + Al_2O_3, and analytical results. The mean diameter of TiO_2 particles is 27 nm and Al_2O_3 particles is 13 nm. The deviation of results between the W (Equation 16.63) and RW models (Equation 16.64) and experimental results is high. However, the Ward model (RW) yields excellent results. Figure 16.11 shows the calculation for

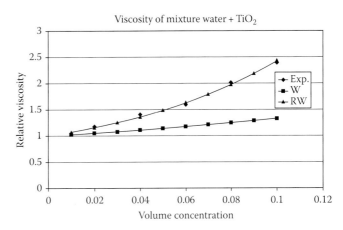

FIGURE 16.9 Relative viscosity for the mixture water + TiO_2 nanoparticles. (From Avsec, J. and Oblak, M., *Int. J. Heat Mass Transfer*, 50, 4331, October 2007.)

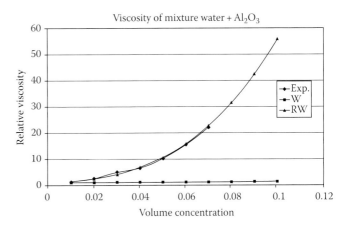

FIGURE 16.10 Relative viscosity for the mixture water + Al_2O_3 nanoparticles. (From Avsec, J. and Oblak, M., *Int. J. Heat Mass Transfer*, 50, 4331, October 2007.)

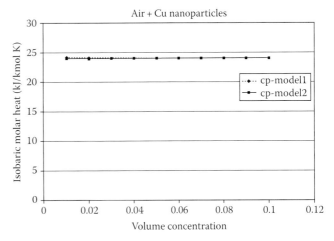

FIGURE 16.11 Thermodynamic properties for Cu nanoparticles + air. (From Avsec, J. and Oblak, M., *Int. J. Heat Mass Transfer*, 50, 4331, October 2007.)

an isobaric molar heat for copper nanoparticles + air. Figure 16.11 shows that for equilibrium thermomechanical properties of state, there is almost no enhancement with a variation of volume concentration of nanoparticles. Both models show almost the same analytical results.

16.5 Conclusions

The chapter presents a mathematical model for the computation of thermodynamic and transport properties for nanofluids. The analytical results are compared with experimental data and show relatively good agreement.

Nomenclature

c_p	heat capacity at constant pressure per mole
c_v	heat capacity at constant volume
c_{el}	electronic heat capacity
c_{har}	lattice heat capacity
CLS	Chung–Lee–Starling
CYJ	revised Cotterman model
E	Eliashberg coupling function
FPC	fluid properties calculator
HC	Hamilton–Crosser model
k_B	Boltzmann constant
Kn	Knudsen number
L	temperature-dependent constant
l_{el}	electron mean free path
LJ	Lennard-Jones
M	molecular mass
N	number of molecules in system
R_m	universal gas constant
RHC	revised Hamilton–Crosser model
RD	relative deviation
RW	revised Ward model
T	temperature
$T^{\star}$	reduced temperature
T_c	critical temperature
v_{el}	electron speed
V	volume
V_c	critical volume
W	Ward model
Z_{coll}	collision number
X	Xue model
β	diffusion term
ε	Lennard-Jones parameter
η	viscosity
θ_D	Debye temperature
θ_E	Einstein temperature
λ	thermal conductivity
Λ	heat conduction transport coefficient
μ_r	relative dipole moment
κ	correction factor for hydrogen bonding effect
κ_s	shear viscosity
ρ	density

$\rho^{\star}$	reduced density
σ	Lennard-Jones parameter
σ_e	electrical conductivity
ψ	influence of polyatomic molecules
ψ_i	molar fraction of component i
ω	acentric factor
Ω	collision integral
$\Omega^{\star}$	reduced collision integral

Superscripts and Subscripts

0	dilute gas state
c	critical condition
el	electron
ph	phonon
p	influence of high densities

References

1. Avsec, J., The combined analysis of phonon and electron heat transfer mechanism on thermal conductivity for nanofluids, *Int. J. Heat Mass Transfer*, 51(22/22), September 2008, 4589–4598.
2. Avsec, J. and Oblak, M., The calculation of thermal conductivity, viscosity and thermodynamic properties for nanofluids on the basis of statistical nanomechanics, *Int. J. Heat Mass Transfer*, 50(21/22), October 2007, 4331–4341.
3. Avsec, J. and Oblak, M., Thermal conductivity, viscosity, and thermal diffusivity calculation for binary and ternary mixtures, *J. Thermophys. Heat Transfer*, 18(3), 2004, 379–387.
4. Chung, T.-H., Lee, L.L., and Starling, K.E., Applications of kinetic gas theories and multiparameter correlation for prediction of dilute gas viscosity and thermal conductivity, *Ind. Eng. Chem. Res.*, 27(4), 1988, 671–659.
5. Chung, T.-H., Ajlan, M., Lee, L.L., and Starling, K.E., Generalized multiparameter correlation for nonpolar and polar fluid transport properties, *Ind. Eng. Chem. Fundam.*, 23(1), 1984, 8–13.
6. Avsec, J. and Oblak, M., Analytical calculation of diffusion coefficients and other transport properties in binary mixtures, *J. Thermophys. Heat Transfer*, 20(1), January–March 2006, 154–158.
7. Avsec, J. and Oblak, M., The calculation of thermal conductivity for metals. In: *43rd AIAA Aerospace Sciences Meeting and Exhibit,* AIAA 2005, January 10–13, 2005, Reno, NV, p. 758.
8. Avsec, J. and Oblak, M., The calculation of thermal conductivity for nanofluids on the basis of statistical nano-mechanics. In: *43rd AIAA Aerospace Sciences Meeting and Exhibit,* AIAA 2005, January 10–13, 2005, Reno, NV, p. 759.
9. Grimwall, G., *Thermophysical Properties of Materials*, Elsevier, New York, 1999.
10. Motakabbir, K.A. and Grimwall, G., Thermal conductivity minimum of metals, *Phys. Rev. B*, 23(2), 523–526.

11. Grimwall, G., *The Electron-Phonon Interaction in Metals*, North-Holland, Amsterdam, the Netherlands, 1980.

12. Avsec, J. and Oblak, M., The calculation of thermodynamic and transport properties for nanofluids on the basis of statistical nano-mechanics. In: *38th AIAA Thermophysics Conference,* AIAA 2005, June 6–9, 2005, Toronto, Canada, American Institute of Aeronautics and Astronautics, Reston, VA, p. 5329.

13. Xuan, Y. and Li, Q., Heat transfer enhancement of nanofluids, *Int. J. Heat Fluid Flow*, 21, 2000, 58–64.

14. Kebliski, P., Phillpot, S.R., Choi, S.U.S., and Eastman, J.A., Mechanisms of heat flow in suspensions of nano-sized particles, *Int. J. Heat Mass Transfer*, 45, 2002, 855–863.

15. Lee, S., Choi, S.U.S., Li, S., and Eastman, J.A., Measuring thermal conductivity of fluids containing oxide nanoparticles, *J. Heat Transfer, Trans. ASME*, 121, 1999, 280–289.

16. Xuan, Y. and Roetzel, W., Conceptions for heat transfer correlation of nanofluids, *Int. J. Heat Mass Transfer*, 43, 2000, 3701–3707.

17. Eastman, J.A., Choi, S.U.S., Li, S., Yu, W., and Thompson, L.J., Anomalously increased effective thermal conductivities of ethylene glycol-based nanofluids containing copper nanoparticles, *Appl. Phys. Lett.*, 78(6), 2001, 718–720.

18. Xue, Q.-Z., Model for effective thermal conductivity of nanofluids, *Phys. Lett. A*, 307, 2003, 313–317.

19. Wang, X., Xu, X., and Choi, S.U.S., Thermal conductivity of nanoparticle-fluid mixture, *J. Thermophys. Heat Transfer*, 13(4), 1999, 474–480.

20. Liu, D.-Mo., Particle packing and rheological property of highly-concentrated ceramic suspensions, *J. Mater. Sci.*, 35, 2000, 5503–5507.

21. Cheng, N.-S. and Law, A.W.-K., Exponential formula for computing effective viscosity, *Powder Technol.*, 129, 2003, 156–160.

22. Yu, W. and Choi, S.U.S., The role of interfacial layers in the enhanced thermal conductivity of nanofluids, a renovated Maxwell model, *J. Nanopart. Res.*, 5, 2003, 167–171.

23. Yu, W. and Choi, S.U.S., The role of interfacial layers in the enhanced thermal conductivity of nanofluids, a renovated Hamilton-Crosser model, *J. Nanopart. Res.*, 6, 2004, 355–361.

24. Eastman, J.A., Philpot, S.R., Choi, S.U.S., and Keblinski, P., Thermal transport in nanofluids, *Ann. Rev. Mater. Res.*, 34, 2004, 219–246.

25. Pak, B.C. and Cho, Y.I., Hydrodynamic and heat transfer study of dispersed fluids with submicron metallic oxide metallic particles, *Exp. Heat Transfer*, 11, 1998, 151–170.

26. Das, S.M., Putra, N., Thiesen, P., and Roetzel, W., Temperature dependence of thermal conductivity for nanofluids, *J. Heat Transfer*, 125(4), 2003, 567–574.

27. Xue, L., Keblinski, P., Philpot, S.R., Choi, S.U.S., and Eastman, J.A., Effect of liquid layering at the liquid solid interface on thermal transport, *Int. J. Heat Mass Transfer*, 47, 2004, 4277–4284.

28. Prasher, R., Thermal conductivity of nanoscale colloidal solutions (nanofluids), *Phys. Rev. Lett.*, 9, January 2005, 025901-1–025901-3.

29. Shackelford, J.F. and Alexander, W., *Materials Science and Engineering Handbook*, 3rd edn., CRC Press, Boca Raton, FL, 2001.

30. Gao, L. and Zhou, X.F., Differential effective medium theory for thermal conductivity in nanofluids, *Phys. Lett. A*, 348, 2006, 335–350.

31. Keblinski, P., Eastman, J.A., and Cahill, D.G., Nanofluids for thermal transport, *Mater. Today*, June 2005, 36–44.

32. Kumar, D.H., Kumar, V.R.R., Sundararajan, T., Pradeep, T., and Das, S.K., Model for heat conduction in nanofluids, *Phys. Rev. Lett.*, 93(14), 2004, 144301-1–144301-4.

33. Prasher, R., Generalized equation for phonon radiative transport, *Appl. Phys. Lett.*, 83(1), 2003, 48–50.

34. Vadasz, J.J., Govender, S., and Vadasz, P., Heat transfer enhancement in nano-fluids suspensions: Possible mechanisms and explanations, *Int. J. Heat Mass Transfer*, 48, 2005, 2673–2683.

35. Koo, J. and Kleinstreuer, C., Impact analysis of nanoparticle motion mechanisms in the thermal conductivity of nanofluids, *Int. J. Heat Mass Transfer*, 32, 2005, 1111–1118.

36. Koo, J. and Kleinstreuer, C., A new thermal conductivity model for nanofluids, *J. Nanopart. Res.*, 6, 2004, 577–588.

37. Jang, S.K. and Choi, S.U.S., Role of Brownian motion in the enhanced thermal conductivity of nanofluids, *Appl. Phys. Lett.*, 84(21), 2004, 4316–4318.

38. Majumdar, A., Microscale heat conduction in dielectric thin films, *J. Heat Transfer*, 155(2), 1993, 7–16.

39. Jou, D., Casas-Zazquez, J., Lebon, G., and Grmela, M., A phenomenological scaling approach for heat transport in nanosystems, *Appl. Math. Lett.*, 18, 2005, 963–967.

40. Chen, G., Ballistic-diffusive equations for transient heat conduction from nano to macroscales, *J. Heat Transfer*, 124(2), 2002, 320–328.

41. Incropera, F.P. and DeWitt, F.P., *Fundamentals of Heat and Mass Transfer*, 5th edn., Wiley, New York, 2002.

42. Avsec, J. and Oblak, M., The dependence of thermal conductivity for nanofluids in dependence of temperature field. In: *9th AIAA/ASME Joint Thermphysics Conference*, June 5–8, 2006, San Francisco, CA.

43. Krishnamurthy, S., Battacharya, P., Phelan, P.E., and Prasher, R.S., Enhanced massed transport in nanofluids, *Nano Lett.*, 6(3), 2006, 419–423.

44. Lucas, K., *Applied Statistical Thermodynamics*, Springer-Verlag, New York, 1992.

17

Phonons in Nanoscale Objects

Arnaud Devos
Centre national de la
recherche scientifique

17.1 Introduction

Many efforts have been devoted to the study of the impact of the size reduction on the electronic properties of nanoscale objects. In a semiconductor quantum dot, the electron levels are quantized due to a three-dimensional confinement. From this, quantum dot can be seen as artificial atoms. The optical properties have also received much attention. Metallic nanoparticles are known to present a pronounced optical resonance related to the effect of the size reduction on the surface plasmon. In a semiconductor quantum dot, the electronic confinement offers an unique way of controlling the optical properties of the nanoparticle and those of macroscopic objects built from them (color glass filter, aqueous solution).

More recently, science has examined the phonon case. A first interest in the vibrational properties of nanoscale objects has been motivated by a better knowledge of the electron–phonon interactions at nanoscale. Indeed, the emission of phonons plays a major role in carrier relaxation, which is itself of crucial importance for optoelectronic applications of nanos. The simultaneous discretization of electronic and vibrational states can strongly affect the coupling between electrons and phonons. In semiconductor quantum dots, for example, a theoretical study has predicted that due to the quantization of the electronic levels, the interaction of carriers with acoustic phonons is inhibited.

This so-called bottleneck effect has motivated numerous studies, both experimental and theoretical (Kammerer et al., 2001; Muljarov and Zimmermann, 2004; Krugel et al., 2006).

Another interest in studying acoustic phonons at nanoscale is to learn more about the object. The acoustic eigenmodes are connected to the geometry so that measuring eigenmode frequencies yield information about the object itself. Furthermore, measuring the way the acoustic energy disappears from the object provides precious information about its environment. Thus studying phonons is also a way of extracting information about the nanoscale objects.

Phonons play a major role in heat transport in insulators and semiconductors. Spatial confinement of phonons in nanostructures can strongly affect the phonon spectrum and thus the thermal properties at nanoscale. By engineering phonon dispersion in nanostructures, heat transport is expected to be controlled as the carriers are under control in heterostructures owing to band-gap engineering.

The first products that have been elaborated using the unique properties of nanoscale objects were directly issued from the properties of the individual object. For example, a color filter in the visible range based on CdS_xSe_{1-x} nanoparticles is designed from the optical properties of the nanoparticle, which are determined by its radius. Particles are embedded in a glass matrix and one obtains a macroscopic object whose properties are governed by those of the nanoobject. In a more recent approach, one elaborates new materials by ordering nanometer-sized objects in an artificial crystal. An opal that consists of spheres of silica stacked together in a lattice with dimensions of several hundred nanometers is a well-known example of such an object. It is important to note that the optical properties of the whole are governed not only by the nanoscale object itself (here the silica sphere) but also by the arrangement inside the artificial crystal. The same idea can be applied to elasticity. Novel materials that offer exceptional control over phonons, sound, and other mechanical waves can be obtained by assembling nanoscale objects in an artificial lattice. We already mentioned the interest of making a phonon engineering for controlling the thermal properties of a nanostructured material. A promising interest of these so-called (hypersonic) phononic crystals is to obtain unique physical properties issued from the fact that both phonons and photons experience periodicity effects.

In this chapter, we examine the acoustic properties of nanostuctures both isolated and organized in artificial crystals. In Section 17.1.1, we give some generalities about phonons in nanoscale objects. What are the main differences with a macroscopic object? What is the confinement? How can one calculate the resonance frequencies? In Section 17.2, we examine the experimental ways of investigation of the acoustical properties of such small objects. Two different approaches exist: the first one explores the frequency domain using Brillouin or Raman scattering and in the second approach, vibrations can also be observed directly in the time domain using ultrafast laser pulses and pump–probe schemes. This so-called ultrafast acoustic technique offers the possibility of studying phonon dynamics as well as coherently controlling the vibrational modes. Section 17.3 is devoted to the observation of individual vibrations on various nanoscale objects. If spherical nanoparticles have been the most studied objects, the literature has recently been enriched with studies on more complex objects. Beyond the fundamental interest of studying the acoustic vibrations of such small object, we also present some possible applications to industrial objects. In Section 17.4, we examine the phonons in artificial crystals. Collective modes that exist only in the artificial crystal are observed. They depend both on the dot and on their arrangement inside the crystal. The main idea of Section 17.5 is that nanos can also be used as efficient emitter of phonons. We present two improvements of the ultrafast acoustic technique based on nanos. First semiconductor quantum dots are shown to emit very high amplitude acoustic pulses when excited by a short laser pulse. Furthermore, due to the strong confinement of the emitting layer, such a system is promising for extending the ultrafast acoustic technique to the THz range. Secondly, arrays of metallic nanocubes offer the possibility of modifying the nature of the acoustic waves emitted from the optical pulse. Roughly we show that they permit the generation and detection of high-frequency surface acoustic wave emitted in various directions at the sample surface. From this information, complete mechanical characterization can be ascertained on thin layers.

17.1.1 What about Acoustic Vibrations in Nanoscale Objects?

In this first section, we give some generalities about acoustic vibration in nanoscale objects. First, we discuss eigenmode in an isolated object. The question is which specificity this problem has compared to a macroscopic object. Does it make any difference with the vibration of a macroscopic object? What is designated by the *elastic quantum confinement*? Then, we briefly describe how one can calculate the eigenmode frequencies. In the particular case of a spherical object, the formula established by Lamb at the end of the nineteenth century is still useful. In other cases, numerical methods are required. Finally, we examine the acoustic vibrations in an artificial crystal made from the ordering of nanoscale objects. Several classes of modes are identified, those coming from the nanoscale object itself and those which exist only in the crystal.

17.1.2 A Few Generalities about Individual Modes

The problem of the vibrations of an isolated nanoscale object can be addressed following two distinct ways. First, as any finite-size object, a nanoscale object can ring like a bell at specific resonance frequencies so-called eigenmodes. The only difference is concerned with the frequency domain, which is much higher due to the small size of the object. As an example, the lowest resonance frequency of a 10 nm diameter gold sphere is close to 300 GHz, a frequency slightly higher than the 29th octave above the note A given by the tuning fork commonly used by musicians. In other words, the problem of the vibrations of a nanoscale object is not a new problem, the same except the scale in space, time, or frequency domain. This is related to an important relation between frequency and size of acoustic modes. The period of an acoustical oscillation is proportional to the size of the object. This is a major point for identifying the acoustical nature of an oscillation detected in an experiment, as shown in the following. Consider an object whose size is measured by d. The period of an eigen vibration mode T can be written as

$$T = \xi \frac{d}{v}, \qquad (17.1)$$

where

v is the sound velocity of the material, which composed the object

ξ is a factor related to the geometry of the object

The frequencies of a cube and a sphere of aluminum would only differ by the ξ factor. Inversely, spheres composed of different materials would have resonance frequencies that would only differ by the v factor. It is important to note that two objects that would only differ by the size (d), the geometry and the material, being the same, the periods of the eigenmodes would only be rescaled by d.

In spite of what has been said previously, one can imagine that by reducing the object size, such a small size can be reached that the atomic nature of matter starts to play a role and then the description in terms of homogeneous elastic solid fails. Among the numerous studies that have been devoted to the observation of acoustic vibrations of nanoscale objects, such an effect has not been observed for particles larger than 2 nm.

The second approach concerning vibrations of nanoscale objects is based on confinement effect. In that case, the nanoscale object is seen as a small nanocrystal, which means a very small part of an infinite and perfect crystal. In a crystal, vibrations are described using phonons, that is, optical and acoustical phonons, which are elastic waves propagating in the crystal. By reducing the object size, new boundary conditions are applied from which the vibrational frequencies are quantized. This is particularly true for phonons whose wavelength is comparable to the dot size. This is exactly similar to what happens in a quantum well concerning the electronic state. This is the reason why confinement of the vibrational properties is important.

Optic and acoustic phonons do not experience the same quantization effects. Indeed because of the small frequency dispersion around the Brillouin zone center of the optic phonons, the discrete frequencies resulting from the finite size are not found to be sensitive to the dot size. On the contrary, acoustic phonons present a strong and linear dispersion. Furthermore, their wavelength is comparable to the size of the nanostructure so that quantization is more pronounced. These two reasons justifies the great interest that has been devoted to the so-called *confined acoustic modes*.

It is important to note that in both approaches, the same modes are focused upon. The frequencies of the eigen vibrational modes in the first approach are the quantized frequencies in the second one.

17.1.3 Calculation of Frequencies

In this section, we examine how the resonance frequency of any nanoobject can be calculated. First we discuss symmetry, which is very important for identifying the modes that can be observed experimentally. Then we focus on sphere modes that are of crucial importance for our discussion since many reports have been concerned with spherical objects. Finally, we discuss the way of reaching the eigenmode frequencies in case of complex geometry nanostructures through numerical calculations.

17.1.3.1 Symmetry

First, a few words about the impact of the symmetry. The symmetry of the object is first useful for classifying the vibrations. Assume that the solid under study is invariant under transformations of a symmetry group G. According to the Wigner theorem, each vibration eigenmode corresponds to an irreducible representation of G. For example, in the case of a sphere, the solid is unchanged by any rotation in three dimensions [the symmetry group is also known as $O(3)$]. Exactly as the electronic states of an atom, the vibrations of a solid sphere are thus classified using an angular momentum.

Symmetry can also be used to predict if a vibrational mode can be observed experimentally or not, which is called *selection rules*. From group theory considerations, Duval established that only two classes of sphere modes can be detected using Raman spectroscopy (Duval, 1992a). In calculating the normal modes and frequencies of vibration, symmetry considerations can reduce the labor of the calculations enormously. The symmetry and geometry of a molecular model can be used to determine the number and symmetry of fundamental frequencies, their degeneracies, and the selection rules for the infrared and Raman spectra.

17.1.3.2 Sphere Modes

The problem of a vibrating sphere was first addressed by Poisson, but the complete analysis of the vibrations of a free elastic sphere was obtained by Lamb (1882). He derived the analytical solutions for the free vibrations of a homogeneous spherical elastic body from the end of the 1800s. Such calculation starts with the equation of motion for a spherical elastic body under stress-free boundary conditions:

$$(c_l^2 - c_t^2)\vec{\nabla}(\vec{\nabla} \cdot \vec{u}) - c_t^2 \vec{\nabla} \times (\vec{\nabla} \times \vec{u}) = \frac{\partial^2 \vec{u}}{\partial t^2}, \qquad (17.2)$$

where $\vec{u}$, c_l, and c_t are the displacement field, the longitudinal and transverse sound velocities, respectively. The equation is written in spherical coordinates and then solved by introducing proper scalar and vector potentials. Indeed, the displacement field $\vec{u}$ can be decomposed into two displacement fields—longitudinal and transverse. The first $\vec{u}_l$ is a gradient field ($\vec{\nabla} \times \vec{u}_l = 0$). It means that a scalar field f exists, which satisfies $\vec{u}_l = \vec{\nabla}f$. The second $\vec{u}_t$ verifies $\vec{\nabla} \cdot \vec{u}_t = 0$, which means that $\vec{u}_t$ is derived from a vector potential: $\vec{u}_t = \vec{\nabla} \times \vec{A}$. Then the energies are obtained. The vibrations are first classified into spheroidal and torsional modes. In a torsional mode, the displacement is given by the curl of the vector potential. Physically the sphere volume remains unchanged in such a vibration or in other words, the radial displacement is zero. The spheroidal displacement is in general a combination of two displacements, a gradient of the scalar potential and the curl of the vector potential. For each category, modes are classified according to the spherical symmetry. As any spherical harmonics, one introduces two integer numbers l, the angular momentum, and m with $|m| \leq l$, which measures the degeneracy of the mode. It is important to note that for each (l, m) there is an infinity of overtones designated by the index n.

The simplest spheroidal modes are the radial breathing modes in which the entire spherical volume periodically expands and contracts. They present the full spherical symmetry and correspond to an angular momentum $l = 0$. The lowest spheroidal $l = 2$ mode corresponds to an uniaxial cigar-to-pancake deformation of the sphere. It is the so-called *football mode* since the extrema in this oscillation are shaped like an American football. It is important to note that the lowest frequency spheroidal mode is not the lowest breathing mode but instead the lowest $l = 2$ mode (Saviot et al., 2004b). The displacement of the sphere in these first modes is illustrated in Figure 17.1.

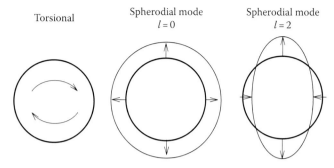

FIGURE 17.1 Simplified description of the displacement of the first acoustic modes of the solid sphere. For a first class of vibration labeled as torsional there is no volume change during the vibration. On the contrary, spheroidal modes present such a change. The simplest mode is the breathing mode in which the shape of sphere is unchanged. The third illustration concerns $l = 2$ spheroidal mode in which the sphere elongates along one direction.

Among the infinity of normal modes, only few can be observed experimentally. Indeed, in Raman spectroscopy, following the group theory considerations from Duval, it can be shown that only spheroidal $l = 0$ or $l = 2$ modes can be observed (Duval, 1992b). In time-resolved experiments, almost all the reports did concern purely radial $l = 0$ modes. This is mainly due to the fact that most studies are performed on particles in an isotropic environment. The absorption of the laser light that excites the vibration is uniform within the particle so that the acoustical energy is delivered only to purely radial modes. In most cases, the fundamental $l = 0$ mode dominates the time-domain response. Because of an optically control of the vibration, Arbouet et al. have observed the first overtone on silver nanospheres embedded in a solid matrix (Arbouet et al., 2006). Recently, Van Dijk et al. observed simultaneously the lowest $l = 0$ and $l = 2$ spheroidal modes in a pump–probe experiment performed on a single nanoparticle (van Dijk et al., 2005a). They explained the observation of the $l = 2$ mode by the presence of a substrate under the studied particle. Up to now, the torsional modes cannot be observed experimentally.

The free sphere model hardly seems applicable to realistic cases because in the investigated samples, the nanoparticles are solidly embedded in a matrix. A few extension of the Lamb's theory have been proposed by Tamura et al. (1982). In that model, the sphere is not free and the mechanical properties of the surrounding matrix are taken into account. It has two impacts on the vibrations of the sphere. First a part of the acoustical energy is radiated into the medium and the life time of modes is reduced. Of course, it strongly depends on the acoustical contrast between matrix and sphere materials. The Lamb's solutions correspond to the limiting case of an infinite contrast. The second effect of coupling the sphere to a matrix is to shift the frequencies (Tamura et al., 1982). More recently, Saviot et al. have also examined the problem of an elastically anisotropic sphere (Saviot et al., 2004a). One consequence of anisotropy they identified is that the degeneracy of the modes is broken.

17.1.3.3 Objects with Complex Geometry

The normal modes of an arbitrary shape structure can be investigated using finite element method, molecular dynamics simulation (MD) (Saviot et al., 2004a) or course-grained molecular dynamics. The finite element method is a good choice for solving partial differential equations over complex domains. It can, in particular, be applied to the determination of the eigenmodes by solving the Helmholtz equation, which describes the vibration of a solid. In MD calculation, the atomic composition of the nanostructure is described as point particles interacting through springs. The interaction is described using a potential energy or a force field, which consists of a summation of bonded forces associated with chemical bonds, bond angles, and bond dihedrals, and nonbonded forces associated with van der Waals forces and electrostatic charge. These potentials contain free parameters that are obtained by fitting against experimental physical properties such as elastic constants, lattice parameters, etc. Mechanical equations are numerically solved in the time-domain and vibrational modes are obtained after a Fourier transform. The course-grained molecular dynamics model has been developed for being able to apply MD at a larger scale, typically on micron-scale objects (Rudd and Broughton, 1998). The structure to be studied is divided into cells that may contain only one atom in important regions or a high number of atoms in other regions. This method has been successfully applied by Antonelli et al. for the determination of the normal modes in a complex nanostructure composed of copper wires embedded in a glass matrix and deposited onto a silicon substrate (Antonelli et al., 2002a).

17.1.4 Vibrations in Crystals of Nanoscale Objects

It is known that a single quantum dot behaves like an artificial atom: electronic states and vibrational states exhibit confinement effects due to finite size and a quantized energy spectrum results. If you imagine that you can build a crystal in which any atom is replaced by a quantum dot, you realize what is called an *artificial crystal*. Doing that with a molecule leads to the formation of a molecular crystal. Here we discuss the expected vibrational properties of an ensemble of nanoscale objects ordered in a crystal lattice.

First, due to periodicity, a vibration in the lattice becomes a mode that can then be labeled by a wave vector k and an angular frequency ω. From the theoretical point of view, the mechanical coupling between nanoobjects leads to the formation of vibrational bands of varying nature. We distinguish several kinds of acoustic modes in the artificial crystal. The passage from the isolated object to the artificial crystal is illustrated in Figure 17.2.

First, the highest modes arise from the individual vibrations of the nanoobject and the coupling between objects transforms the vibration into a branch of vibrations. In the case of a weak coupling, the resulting branch is flat and located at a frequency that is close to the original frequency of the isolated object. These

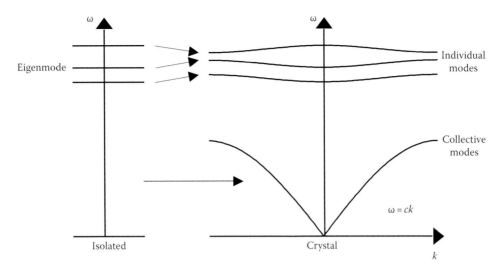

FIGURE 17.2 Illustration of the consequences of the organization of nanoscale objects in an artificial crystal. The discrete frequencies of the eigenmode of the individual object become branches due to the coupling between objects. New modes appear. They are similar to acoustic phonon in an usual crystal and are designated here as collective modes.

vibrations can therefore be approximately classified by the individual mode from which they arise. Such modes are the complete similar to *intramolecular* phonons in a molecular crystal as in a C_{60} for example (Gunnarsson, 1997).

Much lower frequency modes exist only in the crystal. They correspond to translations of the rigid pattern from the equilibrium position. Such modes are similar to the acoustic phonon branches in any crystal. Close to $k = 0$ the angular frequency is proportional to k and the slope is the sound velocity. Compared to the first type, the frequencies of these modes are lower and strongly dependent on the lattice. These modes are the analog of *intermolecular* modes in a molecular crystal.

In the case of a lattice with two dots per unit cell, modes analog to optical phonons would also exist. For optical branch and in the long wavelength limit, the two dots in the unit cell move opposite to each other. Close to $k = 0$, the frequency of such a mode is not zero similar to an individual branch. In spite of this, it is important to make a difference between both kinds of modes. An optical mode is only issued from the interaction between dots and due to that can be designated as an intermolecular mode. On the contrary, an individual mode or intramolecular mode arises from a vibration of the isolated object.

In the following, we designate the intramolecular modes as "individual" and the intermolecular modes as "collective."

17.2 Experimental Ways of Investigation

Experimental studies on the optical or acoustic phonon modes in nanostructures have been first carried out using Raman scattering. Simultaneously, some authors reported coherent oscillation related to vibrational modes by studying the ultrafast dynamics of nanoparticles. Confined acoustic phonons then became observable in the time-domain using femtosecond pump and probe measurements. We examine both kinds of experiments but focus on ultrafast acoustics since the rest of the chapter is mainly concerned with coherent phonons.

17.2.1 Frequency versus Time-Resolved Experiments

Vibrations of nanoscale objects have been first observed experimentally by low-frequency inelastic light scattering (Raman or Brillouin) (Duval et al., 1986; Champagnon et al., 1991). In Raman spectroscopy, the incident light is inelastically scattered by phonons thermally excited in the sample. A laser light is sent on a sample and the light collected from the sample is analyzed. The majority of the photons reemitted by the sample have the same frequency as the incident photons, scattering is then elastic and is called Rayleigh scattering. Photons can also have an energy that differ from the incident light due to energy exchange with phonons in matter. In the inelastic light scattering, scattered photons can have a reduced frequency, Stokes process, or a higher frequency, anti-Stokes process. Acoustic eigenmodes of nanostructure can produce Stokes and anti-Stokes processes. But as the energy shift is low, in practice, the main difficulty of applying Raman spectroscopy is separating the useful signal from the intense Rayleigh scattering. For achieving this separation, a high-resolution laser (narrow spectrum) is used and a very efficient wavelength selector is needed.

Another class of experiments provide the ability to probe vibrations of nanoscale object in the time domain. In these experiments, laser light is used both for exciting and detecting the acoustic modes. An important difference between Raman and ultrafast experiments is the fact that in the last case, the detected phonons have been excited by the laser whereas in Raman spectroscopy, light is scattered on existing phonons.

In the time domain, the detected phonons are then designated as "coherent" due to the fact that they are launched by the pump so that they are coherently detected by the probe.

17.2.2 Time-Resolved Experiments

In this section, we describe the pump–probe setup, which is commonly used for resolving the vibrational states of nanoscale objects in the time domain.

17.2.2.1 Principle of the Pump–Probe Method

No electronic device is sufficiently fast for measuring in a direct way the dynamical properties on the femtosecond time scale. In spite of this, it is possible to reach such an ultimate time scale thanks to an elegant experimental setup, the so-called pump–probe method. The general principle of pump–probe setup is the following. A schematic view of a pump–probe setup is given in Figure 17.3. A first optical pulse, the pump pulse, is sent on the sample, which excites various physical phenomena inside the sample. A second optical pulse, the probe pulse, usually arising from the same laser reaches the sample at the same place but after an adjustable time delay. One measures the transmission or reflection of the probe for which a fast detector is not required. Indeed, for each time delay, the motion is frozen and the probe signal is averaged over many pulses. By monitoring the probe signal as a function of the time delay, it is possible to reconstruct the history of the sample, before, during, and after the arrival of the pump. This so-called *optical sampling* method is apparently to stroboscopy. It is important to note that the temporal resolution is fundamentally limited only by the laser pulse duration. The wavelengths of pump and probe beam do not need to be identical. A so-called two-color pump–probe measurement, based on two synchronized sources of short pulses (e.g., fundamental- and frequency-doubled parts of a same laser or a laser and an optical parametric oscillator) offers additional capabilities in ultrafast spectroscopy.

17.2.2.2 Experimental Pump–Probe Setup

In this section, we give a detailed description of a pump–probe setup and we especially focus on high repetition rate setups since they offer an unique way of reaching the high sensitivity needed for probing acoustic phonons. First one needs a femtosecond laser with a repetition rate of several tens of MHz. A femtosecond laser is a laser that emits optical pulses with a duration below 1 ps, i.e., in the domain of femtoseconds (1 fs = 10^{-15} s). Most studies have been performed using titanium-sapphire lasers, which deliver ultrashort optical pulses (typically from 10 to 200 fs in duration) in the 700–1000 nm wavelength range. The laser output is split into two parts, the pump and the probe. Usually, the pump has a much higher power than the probe. The two beams follow different optical paths but are overlapped at the focus point on the sample. The delay between pump and probe can be adjusted by an optical delay line, which is a reflective optics (two mirrors or a retroreflector or a prisms pair) mounted on a motorized translation stage. With a 1 μm accuracy, such a translation achieves a time resolution of a few femtoseconds, which demonstrates that it is not the main limitation of the time resolution in a pump–probe setup. Typically, we are interested in measuring dynamical properties of samples during from a few 10 ps up to a few nanoseconds. As light propagates at 30 cm per ns, a stage displacement of 50 cm (1 m round trip) gives a time delay of 3.3 ns. After reflection or transmission through the sample, the intensity of the probe is monitored using a photodiode.

The detection of vibrations in nanoscale objects requires an experimental setup with a sensitivity better than 10^{-5}. To improve the signal-to-noise ratio, further refinements are needed. First the pump light is chopped, for example, by an acousto-optic modulator. The probe light reflected or transmitted by the sample is then amplified by a lock-in amplifier that has as its reference source the signal used to drive the pump chopper. This way the changes induced by the pump on the probe light can be extracted. The pump and probe beams are made cross-polarized in order to reject as much as possible the scattered pump from the photodiode using a polarizer. Another improvement concerns the use of a reference beam and a balanced photodetector. This way the fluctuations of the laser intensity are canceled out. In practice, the measurement procedure consists in recording the lock-in output as a function of the position of the mechanical stage. This procedure is repeated a number of times and the results averaged to further improve the signal-to-noise ratio.

In most experiments, the diameter of the region illuminated by the pump light is large compared to a nanoscale object. In that case, a large number of structures are simultaneously excited by the laser. That is typically the case when studying a gold colloid (nanoparticles in aqueous solution) using a 800 nm laser wavelength focused into a liquid cell by a 50 mm lens. The beams are then focused to a 20 μm diameter spot, and it can be estimated that about one million of nanoparticles are simultaneously

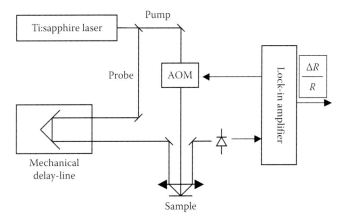

FIGURE 17.3 Schematic diagram of a pump and probe setup commonly used for time-resolving the vibrations of nanoscale objects. The pump beam is chopped at high frequency thanks to an acousto-optic modulator (AOM). The reflected probe is amplified through a lock-in scheme in order to extract the relative change in reflectivity of the probe ($\Delta R/R$) induced by the pump.

excited by the pump light. To be able to look at individual nano-scale object, a much smaller laser spot is needed. This can be achieved by focusing the beams using a microscope (van Dijk et al., 2005b). This has also been done by collecting the light with a fiber tip, which can have a diameter as small as 100 nm (Vertikov et al., 1996).

17.2.2.3 As an Example

Here we illustrate the previous description by presenting an experimental signal obtained using a pump–probe setup on a thin metallic film. As shown below, the pump pulse excites some acoustic vibrations in the film, which rings at a frequency related to the first thickness mode. The studied sample is composed of a very thin aluminum layer deposited on a dielectric film whose mechanical impedance is much lower than the metallic layer. When the pump light reaches the sample, it is mainly absorbed in the metallic film, which experiments a thermal expansion. From such an expansion, acoustic waves are excited. Due to the acoustic contrast between the metallic layer and the underlayer film, a significant part of the acoustic energy is confined within the Al film. The mechanical resonance of the Al film induces an oscillation of the optical reflectivity through the photoelastic coupling. In the Al layer, the strain field modulates the optical index, which affects the optical reflectivity of the sample. This is detected as an oscillation in the transient reflectivity as shown in Figure 17.4. From the period of the detected oscillation, we can measure the film thickness (assuming the longitudinal sound velocity). Indeed a period correspond to a round-trip in the Al film so that we can write: $T = 2e/v$, where e is the film thickness and v is the longitudinal sound velocity. The damping of the oscillation is mainly related to the transmission of a part of the acoustic energy in the underlayer.

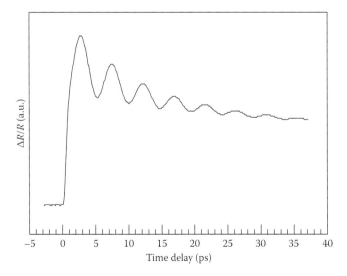

FIGURE 17.4 Transient reflectivity measured in a thin Al film. Under the laser light, the 150 Å thick Al film vibrates at a frequency close to 210 GHz. This is an illustration of the capability of a time-resolved setup to excite and to detect vibrations in nanoscale objects.

17.2.3 Generation and Detection Mechanisms

Here we discuss the various mechanisms involved in pump–probe experiments especially when devoted to the study of vibrational properties at the nanoscale. In the time-resolved technique, light is used to generate and detect vibrations in nanoscale objects. Concerning generation, we assume that the material that made up the studied structure is absorbing (metallic or semiconductor). When an ultrashort pump pulse excites a nanoscale object, the absorbed energy is first given to the conduction electrons, which collide within some tens of femto-seconds through electron–electron interactions. On a 1 ps time scale, the hot electrons thermalize with the lattice. On a larger time scale (typically 10 ps for a 10 nm particle), the whole particle cools down to ambient temperature via heat diffusion. Two mechanisms can convert a part of the pump energy into elastic oscillations. First the change in the energy distribution of the carriers sets up a stress, an electronic stress, which is proportional to the deformation potential. Second as some energy is shared with the lattice modes, vibrations can also be launched through thermal expansion (Thomsen et al., 1986).

Concerning the detection, we must identify a mechanism that explains how a vibrations can modulate the optical properties of the sample. The photoelastic detection relies on the changes induced by the mechanical strain on the optical constants. The efficiency of such a mechanism is measured by photoelastic constants, which strongly depend on the studied material and of the probe wavelength (Devos and Lerouge, 2001; Devos and LeLouarn, 2003; Devos and Côte, 2004).

In some cases, the studied systems present a plasmon resonance, which provides an alternative detection mechanism. Indeed the plasmon resonance appears due to the size reduction and a modulation of the particle size affects the plasmon resonance in position or in width. The size variations induced by the vibrations can thus be detected via shifts of the plasmon resonance.

17.3 Individual Vibrations

In this section, we explore in various systems some experimental results demonstrating that it is possible to time-resolve some acoustic vibrations in nanoscale objects. We first start with spherical object, which is the mostly studied geometry. Then we present results obtained on particles whose shape is more complicated. Finally we present measurements performed on technological objects that demonstrate that time-resolved phonon measurements can also have an applied interest.

17.3.1 Vibration of Nanospheres

Acoustic vibrations in spherical nanoparticles have been extensively studied on ensembles using ultrafast pump–probe spectroscopy. Experiments have been conducted on various systems composed of spherical nanoscale objects mostly nonorganized. In most of these experiments, coherent breathing modes have been observed in the transient reflectivity or transmitivity of the sample.

17.3.1.1 Studied Systems

We first give a brief description of the systems under study. The goal is to give an overview of the large variety of the studied systems. Some of them are made using chemistry, others using physical deposition and some others using technology facilities initially dedicated to silicon technology. The sample parameters are the nanosphere diameter (from a few to a few hundred nanometers), their composition (metal, semiconductor), and the nature of the environment (liquid, glassy matrix).

The oldest way of producing nanospheres is through the principles of metallic colloidal chemistry. A metallic colloid is a suspension of nanometric metallic spheres in water (diameter comprise between 1 and 100 nm). The most fascinating property of such a solution is its color: with 10 nm diameter spheres the solution appears to be red. Ancient Roman already used colloidal gold to color glass. In 1857, Michael Faraday prepared the first pure sample of colloidal gold by using phosphorus to reduce a solution of gold chloride (Faraday, 1847). He was the first to recognize that the color was due to the minute size of the gold particles. The complete understanding of the varied colors of colloidal gold particles suspended in water was achieved by Gustav Mie at the beginning of the twentieth century. A small metal particles present a strong surface plasmon resonance in the UV-visible region of the spectrum (Bohren and Huffman, 1998). The two most important and well-characterized metal colloids are the gold and silver colloids. They present a surface plasmon band absorption around 520 and 390 nm, respectively. Concerning gold, many authors use the recipe of Turkevitch et al. (1985). As gold colloid is frequently used in biological applications, it is now possible to buy gold colloid already made and whose size distribution is known from a transmission electron microscope observation. To date, it is possible to obtain various metal colloids, gold, silver, palladium, and gallium. Colloidal semiconductors also designated as quantum dots or nanocrystallites, such as InAs, PbS, CdS, CdSe, can also be produced.

An important limitation of this way of producing nanoparticles is the size dispersion, the standard deviation being usually slightly smaller than 10%. The size distribution of the particles is determined by transmission electronic microscopy (TEM). The time-resolved experiments can be easily performed in liquids thanks to an optical cell containing a few milliliters of colloid and by studying the transient transmitivity. In some studies, the nanoparticles first synthesized in solution are then embedded in a polymer film (Cerullo et al., 1999; van Dijk et al., 2005b).

Another way of producing nanospheres relies on heating and condensation. This way, metallic or semiconducting nanoparticles can be embedded in a solid matrix (Nisoli et al., 1997; Voisin et al., 2002). By evaporating a material in ultrahigh vacuum, it is possible to condensate some clusters on a substrate. A simple way of getting such semiconductor samples is to study the response of some commercial color filters, which are made using CdS_xSe_{1-x} nanocrystallites embedded in a borosilicate glass matrix (Verma et al., 1999).

17.3.1.2 Example of Time-Resolved Vibrations of Spherical Nanoparticles

The first observation of acoustical resonance in an ultrafast pump–probe experiment was reported by Nisoli et al. (1997). They studied tin and gallium nanoparticles whose radius was between 2 and 30 nm. They observed a damped oscillation in the transient reflectivity. The measured period is found to depend linearly on the size of the particles, which is used by the authors for interpreting it as an acoustical vibration of the spheres.

Many studies follow this first one. Vallee and Del Fatti studied metallic nanoparticles in vitreous matrix (Fatti et al., 1999; Voisin et al., 2002). At the same time, a few studies also concern metallic particles in aqueous solution Hodak et al. (1998, 1999), Devos et al. (1999), or semi-conductor quantum dots in solid matrix (Krauss and Wise, 1997; Thoen et al., 1998).

In all these reports, the authors have observed oscillations in the transient reflectivity or transmitivity. The period between a few picoseconds to a few tens of picoseconds is found to linearly depend on the particle size. This last point is used for claiming that acoustic oscillations are launched and detected by the laser pulses.

Here we illustrate this by presenting some results obtained by Devos and Perrin in a gold colloid. The studied sample was synthesized by the authors following the Turkevitch receipt. The pump and probe laser beams were focused spatially overlapped into a 5 mm thick optical sample cell. The transient transmission change $\Delta T/T$ measured in a large collection of gold nanoparticles is reproduced in Figure 17.5. The sharp peak visible close to $t = 0$ reflects the optical effect induced by the absorption of the pump energy by the electron of the system. The fast decrease visible on a 1 ps time scale is due to the electron–lattice energy transfer. On a longer time scale, the experimental signal exhibits

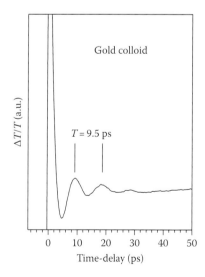

FIGURE 17.5 Relative change in transmission as a function of the time-delay measured by Devos and Perrin in a gold colloid. The oscillation visible in the first tens of picoseconds is attributed to the mechanical resonances of the spherical nanoparticles.

an oscillation with a period close to 10 ps, which is the value expected for the first breathing mode of a 30 nm gold particle. Furthermore the period of the oscillation is found to vary linearly with the particle size which confirms the acoustic nature of the observation. Each nanoparticle is heated by the pump and due to thermal expansion vibrations are launched. One should note that the oscillation is strongly damped, which is attributed to the large dispersion in particle size of the sample. Initially all the particles vibrate in-phase but due to the dispersion in particle size the various frequencies interfere destructively and the oscillation disappears. In such a case the damping is qualified as inhomogeneous.

Most of experiments performed on nanospheres have exhibited similar oscillations usually attributed to the first breathing mode. Indeed a very good agreement is found between the measured periods and those obtained using Lamb theory. This result is not so surprising since the heat deposited by the absorption of the pump is isotropic. Thus the excitation presents a spherical symmetry. In a spherical particle and in a isotropic environment, only $l = 0$ modes, so-called breathing modes, are expected to be excited.

Using two pump pulses, it is possible to stop the oscillation of the fundamental mode and thus to observe some other modes. During the observation, Arbouet et al. (2006) found a the second order mode, which is also a breathing mode but higher in frequency.

By studying a single gold nanoparticle, Van Dijk et al. (2005b) succeeded in the observation of two acoustic modes, the first being the first breathing mode, as in previous studies and the second one was detected at a lower frequency. The authors attributed this last mode to the first $l = 2$ mode (lower in frequency for gold according to Lamb theory) and related its observation to a difference in the particle environment. Indeed in most suitable samples, the particles are numerous and in an isotropic medium. On the contrary they performed experiments on a single nanoparticle whose boundary conditions are thus modified.

17.3.2 Other Geometries

17.3.2.1 Studied Systems

Most time-resolved studies have concerned roughly spherical nanoparticles, mostly because of the difficulties associated with preparing uniform samples of nonspherical nanoobjects. Recently, chemistry has offered the opportunity of producing objects with a more complex geometry. This has motivated time-resolved studies on more complex nanoscale objects. (Bonacina et al. (2006) studied aqueous solutions of triangular silver nanoparticles. Taubert et al. (2007) investigated the acoustic properties of nanoscale gold triangles deposited on various substrates. Petrova et al. (2007) time-resolved vibrational modes in silver nanocubes in solution in water. A few recent studies have also been devoted to metallic nanoshells obtained by growing metal shells on dielectric spheres (Guillon et al., 2007; Mazurenko et al., 2007).

A few other studies have been devoted to technological objects. This means that the systems under study have been realized using equipments and processes initially dedicated to the fabrication of silicon integrated circuits. Various materials can be deposited in thin layers using sputtering or evaporation. A thin film can be patterned using photoresist, lithography, and a lift-off process. In a few words, a thin photoresist layer is first spin-coated onto the film to be patterned. Then UV light or an electron beam is used to write the pattern through a mask or directly on the sample. After that, the resist is developed: the photoresist that has been exposed is removed. Then, another material can be deposited at the corresponding places or the unprotected parts of the sample can be etched. In a last step, so-called "lift-off," the photoresist is completely removed from the film using solvent.

The main advantage of e-beam lithography is that it is one of the ways to beat the diffraction limit of light and make features in the nanometer regime due to the small wavelength of the electron. It offers an unique way of realizing nanoscale objects whose shape and size can be perfectly controlled down to a few nanometers. It can also be a convenient way of organizing objects in an artificial crystals (Lin et al., 1993, Robillard et al., 2007, 2008).

17.3.2.2 Comparison with Spheres

As an illustration, we present individual vibrations measured on nanocubes (Robillard et al., 2007). The samples under study are lattices of nanocubes whose size varies between 50 and 200 nm. Cubes are composed of aluminum and are deposited on an aluminum layer. A typical experimental signal is reproduced in Figure 17.6. It has been obtained on a sample composed of 50 nm cubes. As shown, the transient reflectivity presents some oscillations. One should remark that the oscillation is not monochromatic. In the present case, the experimental signal can be very well reproduced using two frequencies. This means that two modes are simultaneously excited by the laser. This is a major difference with the sphere case.

The acoustical nature of the modes is demonstrated by studying the dependence of the period on the cube size. As shown in Figure 17.7, the period is found to vary linearly with the cube size as expected for an acoustical mode. Furthermore, the inverse of the slope gives a realistic value for a sound velocity (4060 and 3230 m/s for the low and high modes respectively).

The authors have also reported that the oscillation magnitude is proportional to the number of cubes. This is consistent with the fact that the laser excites a large collection of dots that vibrate in-phase. By varying the lattice parameter, the number of cubes that are exposed to the laser beam and also the acoustical signal varies.

17.3.2.3 Applications to Technological Objects

Maris et al. used the possibility of probing the individual vibration modes of nanostructures for a practical problem (Antonelli et al., 2002a,b). They studied arrays of copper wires

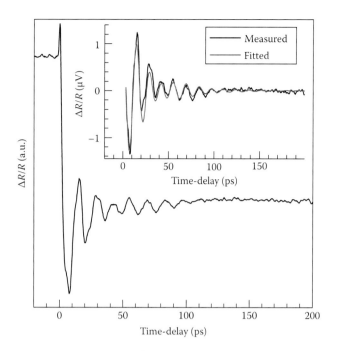

FIGURE 17.6 Transient reflectivity measured on a square lattice of 50 nm cube. Oscillations are detected in the first picoseconds. As shown in the inset, the signal can be well-reproduced by using a combination of two different frequencies (57 and 73 GHz).

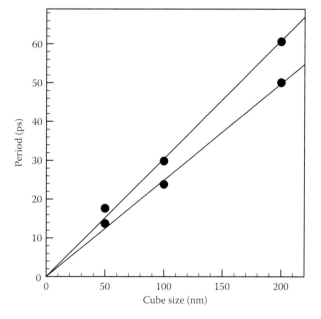

FIGURE 17.7 Period of the oscillation detected in the transient reflectivity of nanocubes lattices as a function of the dot size. One should note the linearity obtained for both modes which confirms their acoustical nature.

of cross-sectional dimension 360 nm by 355 nm embedded in SiO_2. Each wire is surrounded by a liner of 25 nm of Ta. Such copper lines are usually used in microelectronic industry in order to connect transistors in integrated circuits. Copper presents many mechanical problems at the submicron scale

and measuring elastic properties of such lines is a crucial issue for controlling the quality of the interconnect lines.

The goal of the experiment was to determine as many normal modes of the structure as possible and to compare the measured frequencies with the results of calculations. Measurements were made on the sample in a series of experiments in which the polarization and the propagation direction of the pump and probe beams were varied. It was possible to make an accurate determination of the frequency of six normal modes.

This study illustrates that studying acoustic phonons in nanoscale objects is not a pure fundamental problem but can also addressed some practical applications.

17.3.3 Possible Sources of Damping of the Acoustic Resonances

We discuss the possible sources of damping of the mechanical resonance of nanoscale objects.

The first one is the inhomogeneous damping. It is due to the dispersion in particle size of the sample. The laser excites a large collection of particles which initially vibrate in phase. But due to the size dispersion, the resonance frequencies is slightly different from one particle to another. Destructive interference between the various frequencies leads to the fast disappearance of the oscillation. This is the case of colloid as shown in Figure 17.5. This is supported by the recent study performed on one single particle in which the damping is found to be much lower (van Dijk et al., 2005b).

A second origin of the damping is the transmission of acoustic energy to the environment. Liquid or solid matrix, the nanoscale object is in contact with a medium in which a part of the acoustic energy radiates. This is illustrated in Figure 17.6 where the mechanical resonance of nanocubes are strongly damped in spite of a very low size dispersion. In that case, the cubes have been deposited on a layer composed of the same material, which is itself deposited on a substrate whose acoustic impedance is similar. Due to that a significant part of the acoustic energy initially launched in the cube is transmitted to the substrate. On the contrary, in Figure 17.4, the Al thin film is deposited on a low acoustic impedance material which permits to observe a long lived oscillation.

The last source of damping is the sound attenuation. The higher the frequency is, the stronger the acoustic energy is lost. A simple way of avoiding this effect is to work at lower temperature since sound attenuation diminishes strongly at low temperature.

17.4 Collective Acoustic Modes

Up to now, we focused on individual vibrations of nanoscale objects. As described in Section 17.1.4, when organized in an artificial crystal, new acoustical excitations are expected due to the mechanical coupling between objects. They are expected to depend both on the lattice parameter and on the dot size. If the normal modes in nanoscale objects have been observed for a long time, the observation in the time-domain of such collective modes is much more recent (Comin et al., 2006; Robillard et al., 2007). Two main reasons may justify

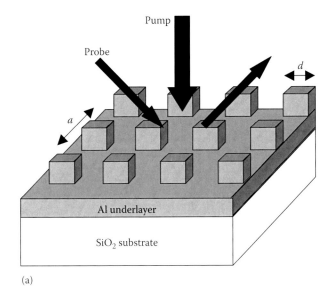

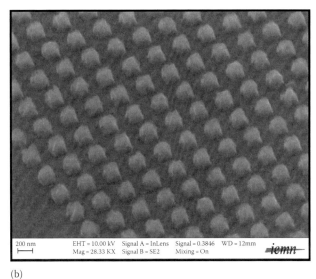

FIGURE 17.8 (a) Schematic diagram of the experiment performed on 2D artificial crystals of nanocubes. (b) SEM image of a 200 nm cubes sample, a 400 nm step size square lattice.

this point. First one must be able to build an artificial crystal by organizing nanoscale objects with a submicronic lattice step, which is not easy to achieve. Second, as shown below, some specific experimental conditions have to be satisfied for detecting the collective modes. This last point could explain why pioneer studies performed on artificial crystals did not reveal any collective mode.

17.4.1 Samples Description: Back to Nanocubes Lattices

The samples studied are artificial crystals made of arrays of nanocubes on a substrate using electron beam lithography. First, an aluminum layer is deposited by metal evaporation onto the substrate. It is the underlayer on which cubes are deposited. Its thickness, designated as *h* in the following, varies between 20, 100 or 400 nm depending on the sample. Then e-beam lithography is used to write the cubic pattern in a positive photoresist. Aluminum cubes are then built-up by a second metal evaporation and a lift-off process. The high resolution of e-beam lithography (less than 10 nm) enables a precise control of the geometrical parameters and assures a low cube width dispersion. Furthermore, it offers a simple way of realizing almost perfect lattice with various symmetry (square, rectangular or hexagonal) lattices of various cube widths (*d*). For a given cube width, underlayer thickness (*h*) and lattice symmetry, a set of crystals with various lattice parameter (*a*) is fabricated. A SEM image of a 200 nm cubes samples is shown in Figure 17.8.

17.4.2 Under a Particular Experimental Condition, Strong Oscillations Appear

In Figure 17.9 we compare the experimental signal measured on the same sample (200 nm cubes in a 400 nm step square lattice), using the same pump but different probe wavelengths. As shown,

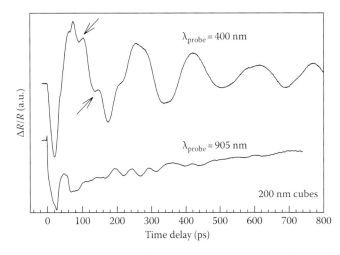

FIGURE 17.9 Transient reflectivity measured in a square lattice of 200 nm cubes at two distinct probe wavelengths. Using an infrared probe, one only detect individual vibrations in the first 10 ps. On the contrary, using a blue probe one detects strong oscillations which last several nanoseconds.

the signal obtained using a blue probe is qualitatively different from the first one. In the first 400 ps a weak oscillation whose period is close to 50 ps is visible on both curves. This component does not change with the lattice constant. It is related to some individual vibrations of the aluminum cubes as already mentioned in Section 17.3.2. The signal detected using a blue probe presents another oscillating contribution, which differs from the first one by some major points. First point to notice is that the oscillations last for at least 8 ns (note the change in the time scale). Second, the signals have a much higher amplitude and a much lower damping rate. The last point but not least, the low-frequency signal is found to depend both on the lattice constant and the cube width, which suggests its collective origin.

The second kind of modes is detected in various crystals each time the probe wavelength is lower than the lattice step a. Provided this condition is satisfied the low-frequency signal appears, but the probe wavelength does not affect its frequency content.

17.4.3 Collective Acoustic Modes

In this section, we demonstrate that the strong oscillations detected in the transient reflectivity when the probe wavelength is lower than the lattice constant are collective acoustic modes. We already mentioned that the measured periods depend on the lattice constant. This is visible in Figure 17.10 where the signals measured on the $a = 700$ nm and on the $a = 500$ nm sample are reproduced. This suggests the collective character of the signal. To go further, we perform transient reflectivity measurements on the whole set of crystals with increasing lattice constant. The use of Fourier transform reveals up to four frequencies per crystal, which are plotted as a function of the inverse of the lattice constant in Figure 17.10c. The multiple frequencies detected order regularly as branches. The continuous lines are obtained from the model developed in Section 17.4.4.

These plots have several relevant characteristics. First, the upper branches of Figure 17.10c can be deduced by scaling the lower one with remarkable factors $\sqrt{2}$, 2, and $\sqrt{5}$. Second, each branch extends linearly to zero for lower and lower $1/a$ in a similar manner as an acoustic phonon branch. This is consistent with the expected cancellation of collective modes when the dots are infinitely separated. Furthermore, the slope of the branch around zero is close a standard sound velocity value. All of this confirms the acoustic origin of the low-frequency oscillations.

17.4.4 Interpretation

In this section, we derive a theoretical description of the curves presented in Figure 17.10. We derive the model in the square lattice case, which is the simplest one. The complete derivation and generalization to any kind of two-dimensional crystal can be found in Robillard et al. (2008).

We start from an usual dispersion relation for an acoustic mode written as

$$\omega(\vec{k}) = c(a) \cdot \|\vec{k}\|, \tag{17.3}$$

where
$\vec{k}$ is the wavevector
$c(a)$ the sound velocity of the mode

It is important to note that the collective nature of the mode appears in the dependence of the sound velocity on the lattice parameter. We first show that due to periodicity only a discrete set of wavevectors is permitted, which explains that all the branches derive from the first one. Then we derive an analytical expression for the sound velocity $c(a)$ as a function of a, d, and h, which is valid for any crystal symmetry. Finally, we discuss the influence of each parameter on the frequencies.

Among all the normal modes of the structure, the initial strain launched by the pump light will distribute on the subset of modes whose displacements comply with the sample periodicity. Let $\eta(\vec{r},t)$ be the strain field of one mode at position $\vec{r}$ and at time t. Then whatever the vector $\vec{A}$ of the direct lattice is, we can write

$$\eta(\vec{r} + \vec{A},t) = \eta(\vec{r},t). \tag{17.4}$$

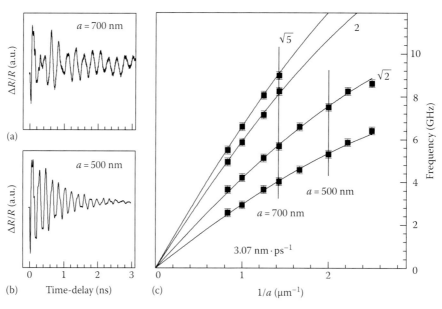

FIGURE 17.10 Experimental method: for each lattice parameter we measured the transient reflectivity as shown in (a) sample $a = 700$ nm, in (b) sample $a = 500$ nm. (c) Then we plot the measured frequencies as a function of $1/a$ (dots). The data obtained are finally fitted using an analytical model (continuous line) described in the text.

Assuming $\eta(\vec{r},t)$ to be propagative parallel to the surface, we can write

$$\eta(\vec{r},t) = \eta_0 e^{i(\omega t - \vec{k}\cdot\vec{r})}. \qquad (17.5)$$

The respect of the condition (17.4) leads to $\vec{k} \cdot \vec{A} = 2n\pi$, where n is an integer. As a consequence for any crystal, the only propagative modes excited are those whose wavevector $\vec{k}$ is a reciprocal lattice vector (Brillouin, 1953). In the case of a square lattice with a lattice constant equal to a, the permitted wavevectors $\vec{k}$ can be written as

$$\vec{k} = \frac{2\pi}{a}(i\vec{u}_x + j\vec{u}_y), \qquad (17.6)$$

where i and j are integer numbers. The corresponding mode thus propagates in the (i, j) direction and its wavenumber $k_{i,j}$ is given by

$$k_{i,j} = \sqrt{i^2 + j^2}\left(\frac{2\pi}{a}\right). \qquad (17.7)$$

By reporting this in Equation 17.3, we get for one crystal several frequencies between which we retrieve the observed remarkable factors between branches. The first fourth modes are those obtained with the directions (1, 0), (1, 1), (2, 0), (2, 1). The ratios between the corresponding frequencies are 1, $\sqrt{2}$, 2, and $\sqrt{5}$ identical to those observed experimentally in Figure 17.10c. A first conclusion is thus that for any crystal, a same nondispersive acoustic wave is excited at several discrete wavevectors whose wavenumbers vary (which leads to different frequencies) and which propagates along different crystal directions.

The discrete set of wavevectors is derived from the reciprocal lattice. For example in a hexagonal lattice the magnitude of the wavevectors $k_{i,j}$ are

$$k_{i,j} = \sqrt{i^2 - ij + j^2}\left(\frac{2}{\sqrt{3}}\right)\left(\frac{2\pi}{a}\right). \qquad (17.8)$$

In that case, the ratio between the two first branches is expected to be $\sqrt{3}$, given by the ratio between wavevectors in the (1, 0) and (2, 1) directions.

Let us describe an analytical model for the velocity of the collective mode [$c(a)$]. Since the curvature is larger for high $1/a$ values, the velocity $c(a)$ decreases with the dot density at the sample surface. As shown below, such an effect can be explained by considering the mass loading of the cubes. Since all the branches are deduced from the first branch, in the following, we focus on the lower collective mode frequency ($f_{1,0}$) and its dependence on the lattice constant a. We also assume that it arises from a vibration of a unit cell described as a harmonic oscillator of stiffness K and mass M. The unit cell is composed of a cube and a square piece of the underlayer. We can write

$$f_{1,0} = \frac{1}{2\pi}\sqrt{\frac{K}{M}} = \frac{1}{2\pi}\sqrt{\frac{K}{\rho V}}, \qquad (17.9)$$

where
ρ is the mass density of aluminum
V the unit cell volume

V can then be written as a function of a, d and h the underlayer thickness.

In the square lattice case, we can write

$$f_{1,0} = \frac{1}{2\pi}\sqrt{\frac{K}{\rho}}\,\frac{1}{\sqrt{ha^2 + d^3}}. \qquad (17.10)$$

According to Equation 17.3 the angular frequency of the lowest mode is

$$2\pi f_{1,0} = c(a)k_{1,0} = c(a)\frac{2\pi}{a}, \qquad (17.11)$$

from which we can derive the formula of $c(a)$ valid for a square lattice:

$$c(a) = \frac{1}{2\pi}\sqrt{\frac{K}{\rho}}\sqrt{\frac{a^2}{ha^2 + d^3}}. \qquad (17.12)$$

We now examine the limit of infinitely spaced dots. In Figure 17.10c we note a linear behavior of frequencies close to zero, which defines a constant sound velocity for large a, here designated by c_0. In other words

$$c_0 = \lim_{a\to\infty} c(a). \qquad (17.13)$$

By introducing c_0 in Equation 17.12, we get the sound velocity of the collective mode in the square lattice case:

$$c(a) = c_0\sqrt{\frac{a^2}{a^2 + d^{3/h}}}. \qquad (17.14)$$

c_0 is here the only adjustable parameter of the model.

This result can be generalized to any 2D lattice and to samples in which cubes and underlayer are composed of two different materials (Robillard et al., 2008).

17.4.5 Experimental Confirmations of the Model

An important point is that there is only one fitting parameter for the whole set of branches. The continuous lines visible in Figure 17.10c are obtained using a velocity $c_0 = 3.07 \pm 0.1$ nm·ps^{-1}. The agreement between experiments and model is excellent.

From this model, we can now examine the influence of the lattice symmetry, the cube width, and the underlayer thickness on

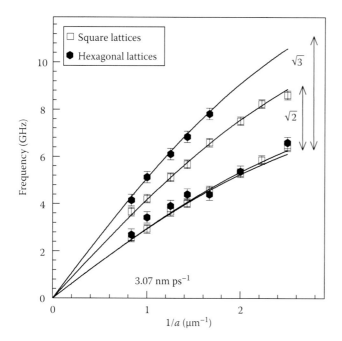

FIGURE 17.11 Collective acoustic frequencies as a function of the inverse of the lattice parameter measured on two different lattice symmetries. The lowest frequencies are very close in both cases. On the contrary, the scaling factor which permits to obtain the second branch from the first equals $\sqrt{2}$ in the square lattice case and $\sqrt{3}$ in the hexagonal case. Theses results are those expected from the theoretical description.

the lattice frequencies. The lattice symmetry plays a major role in the determination of the set of discrete wavevectors, which itself determines the ratios between branches. For example, we expect a $\sqrt{2}$ factor between the first two branches in a square lattice and $\sqrt{3}$ factor for a hexagonal one. In Figure 17.11, we compare the two first branches measured on a square lattice series and a hexagonal lattice series. The cube width and the underlayer thickness are identical in both series. The lower branches of both sets are very close and well reproduced using the previous model with the same sound velocity $c_0 = 3.07 \pm 0.1\,\mathrm{nm \cdot ps^{-1}}$. This is in accordance with the fact that for infinitely spaced cubes the surface has an intrinsic velocity, which depends only on the substrate and underlayer system. The most noticeable point is the difference in the factors needed for deducing the second from the first branch. The $\sqrt{2}$ factor needed in the square case is replaced by a $\sqrt{3}$ factor, as expected. This confirms the role of the lattice symmetry in the selective excitation of a discrete vibration spectrum in the crystal.

Second from Equation 17.14, we show that curvature is governed by the factor $\chi = d^3/h$. Indeed, for large a values, $c(a)$ in Equation 17.14 approaches c_0 since d^3/h being negligible compared to a^2. On the contrary, for small value of a, d^3/h is not negligible and the sound velocity decreases, which manifests itself as a curvature of the branches. Such an effect is expected in any phononic crystal in which a band structure appears due to the interferences between multiple reflections. The cube size d plays a major role on the branch curvature. The larger the cubes

are, the more pronounced the curvature is expected to be. We also note that the thinner the underlayer is, the larger the curvature is. Furthermore, increasing the cube mass density produces similar effects. Each of these three influences can be understood by referring to the mass loading effect induced by the cubes. The curvature is governed by the impact of the cube mass on the unit cell mass. The heavier the cube is, the more pronounced the curvature is.

To test this effect, we compared the frequencies measured on lattices differing only by the cube width (100 and 200 nm). While 200 nm cubes have a strong effect on the velocity $c(a)$, the 100 nm ones have a slight influence. The model accounts very well for this difference due to the cube width change. The initial velocity c_0 is still unchanged compared to previous results. These results confirm the role of the cubes as mass loading of the surface. We also explored the influence of the underlayer thickness. Three series of identical square lattices were built on different underlayers. The thicknesses are 20, 100, and 400 nm, respectively, while cube width is 200 nm. As predicted by the model, the thinner the underlayer is, the more pronounced the curvature is found to be.

17.4.6 Conclusion about Collective Modes

Collective acoustical modes issued from the mechanical coupling between dots are detected. We derived an analytical model, which fully reproduces the experimental data first in the square lattice case and then generalized to any 2D crystal. Numerous experimental data are fitted using only this model and one parameter, which is the sound velocity of a surface acoustic wave of the underlayer/substrate stack. From the theoretical analysis, we show that several waves propagate along the sample surface in various directions defined by the reciprocal lattice. We introduced c_0 as a limit value for the sound velocity of the collective mode in the extent of widely spaced cubes. While the perturbation of cubes gives a specific velocity to the surface mode the velocity, here c_0 does not depend on the cubes any more. As propagation acts along the surface it can also be interpreted as the surface acoustic wave velocity in the underlayer/substrate stack. Thus, by studying the impact of nanocubes on sound velocity, we get some information about acoustic propagation in the underlayer. From that, we suggest that such lattices could be used for measuring in-plane elastic properties in picosecond ultrasonics. A demonstration of this capability is given in Section 17.5.2.

An important point raised by the observation of the collective modes is their detection mechanism. Indeed, it has been shown that the photoelastic mechanism could not be involved in that case. The detection of collective modes is based on a different and more efficient mechanism compared to individual vibrations. This mechanism for which a laser-wavelength condition has been identified may involve plasmonic effects. Indeed, the nanostructured metallic surface of the studied samples can couple the incident light to plasmons propagating along the surface (Martín-Moreno et al., 2001). For that, the probe must be diffracted by the lattice, which could explain the criterion $\lambda \leq a$.

As collective modes also propagate along the surface, there is a simultaneous localization of electromagnetic and acoustic fields. This can enhance the acousto-optic interaction, which is consistent with the high efficiency of collective modes detection (Mazurenko et al., 2007).

17.5 Quantum Dots as Ultrahigh-Frequency Transducer

In this last section, we examine some applications of nanoscale objects in acoustic transduction. More precisely, we present results showing that nanoscale objects are efficient acoustic phonons emitters. First, we present some results that demonstrate that semiconductor quantum dots are a very efficient transducer of high-frequency acoustic waves. Furthermore they may constitute a possible way of reaching a THz transducer. Second, we show that arrays of nanocubes offer an unique way of emitting high-frequency acoustic wave of various nature. Indeed if ultrafast acoustics has reached an unexplored frequency range by using ultrafast laser pulses to produce and detect acoustic waves, one should keep in mind that only longitudinal acoustic phonons are launched by the laser pulse. Due to that, a part of the physical acoustics in this frequency range is still to be explored. As shown below, arrays of nanocubes also permit the generation and the detection of high-frequency surface waves using the same experimental setup. Thanks to that complete mechanical measurements can be performed on various thin films.

17.5.1 Acoustic Phonon Emission from Quantum Dot Layers

Phonon studies by time-resolved pump–probe experiments can be listed in two categories. Most of these experiments have been applied to the study of confined vibrational modes in metallic or semiconducting nanostructures. As shown in Section 17.3, the transient transmission or reflectivity reveals in the first tens of picoseconds an oscillating part, which is assigned to vibrational normal modes of the nanostructure (Krauss and Wise, 1997; Nisoli et al., 1997; Özgür et al., 2001; Bragas et al., 2004, 2006). Such a regime is obtained in dots whose mechanical properties significantly differ from those of the matrix, vibrational modes being confined. A few other time-resolved studies were reported on semiconductor heterostructures, which present a small acoustic mismatch. These studies deal mainly with quantum wells, in which the light pulses excite and detect ballistic phonon wavepackets whose propagation is monitored on a larger time scale (Baumberg et al., 1997; Liu et al., 2005; Matsuda et al., 2005). In the following we present the detection of acoustic phonon wavepackets emitted from quantum dots.

17.5.1.1 Emission from SiGe Quantum Dots Layers

Experiments were carried out on self-assembled SiGe quantum dots (QDs). The samples were grown by gas source molecular beam epitaxy on (100) Si substrates. The samples differ by the number of layers of buried QDs, which varies between 1 and 5. The time-resolved experiments were conducted using a two-color pump and probe setup. The pump pulse is directly issued from the femtosecond oscillator. At a wavelength close to 800 nm, the optical absorption length in silicon is large so that the pump light penetrates deeply the sample. On the contrary, the probe is frequency doubled to produce a blue pulse, which is absorbed near the sample surface in the Si cap. All the experiments are performed at room temperature. The choice of using a blue probe was initially motivated by the huge acoustic detection, which acts in silicon in the blue range (Devos and Côte, 2004).

In Figure 17.12, we present the experimental results obtained on two samples differing by the number of QD layers (one and two layers). The cap thickness is the same in both samples. The most noticeable point is the detection of short oscillating structures, which are acoustic echoes detected at the sample surface. Several points support such an interpretation. First the number of echoes is equal to the number of layers. Second the time of arrival of the first structure in both cases is equal to the time of flight of a longitudinal acoustic wave through 160 nm silicon, which is precisely the cap thickness. Finally, in the sample containing two layers, the two echoes are separated by a time-delay, which perfectly fits the time needed for an acoustic wave to go through the interlayer. From this observation, we can conclude that the pump pulse excites acoustic phonons in each QDs layer. The phonon wavepacket then propagates at the longitudinal sound velocity and reaches the free surface where it is detected by the blue probe. This first result is confirmed by a similar

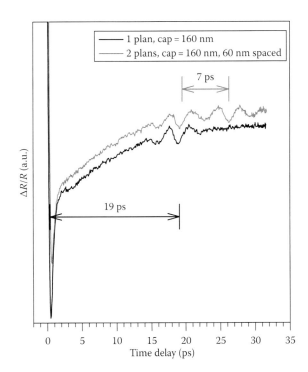

FIGURE 17.12 Coherent acoustic phonons are emitted from each quantum dot layer and detected at the sample surface using a blue probe.

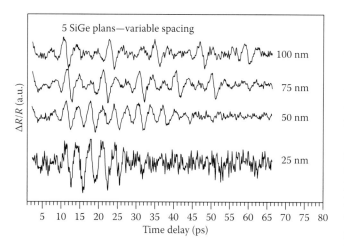

FIGURE 17.13 Coherent acoustic phonons detected in a series of samples composed of 5 QDs layers whose spacing varies between 25 and 100 nm.

observation made in a series of samples composed of five layers whose spacing varies between 25 and 100 nm. As shown in Figure 17.13, in that case we detect five structures whose time spacing varies. Here again the time-delay spacing is given by the ratio between the interlayer distance and the sound velocity in silicon. This is a first demonstration of the capability of a QDs layer to emit acoustic phonons.

17.5.1.2 Giant Phonon Emission in InAs/InP Quantum Dots

We report similar experiments but performed in another QDs system. As shown below, we also observed the emission of acoustic phonons from each QDs layer but in that case high-amplitude echoes are detected and we show that it is due to a giant emission from each QD layer.

Experiments were carried out on samples grown by gas source molecular beam epitaxy on (100) InP substrates (Paranthoen et al., 2001). Transmission electron microscopy observations have shown InAs islands of quantum size, with an average height of 2 nm and an average width of 20 nm. In the following, we present the results obtained on two samples hereafter designated as A and B. Sample A contains two QD layers separated by a 480 nm thick interlayer and the final InP cap thickness is 320 nm. Sample B contains one single QDs layer buried under a 400 nm thick InP cap.

Figure 17.14 shows the transient reflectivity obtained on samples A and B using an infrared pump and a blue probe. Due to the strong optical absorption of InP in the blue range, the probe only detects the reflectivity changes at the sample surface. The photogenerated carriers first produce an initial spike visible on ultrafast time scale (≤1 ps). The relaxation contribution, which produces a long-lived change in reflectivity, has been subtracted for clarity. One notes that in both samples

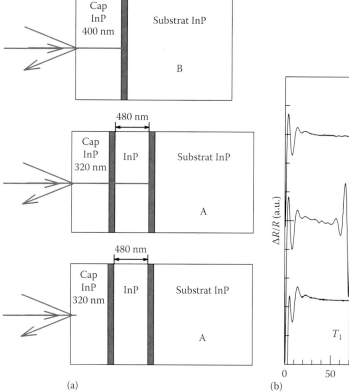

(a)

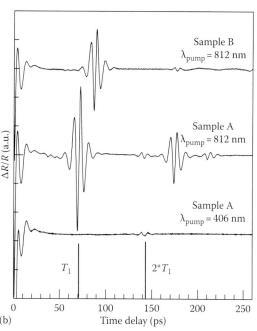

(b)

FIGURE 17.14 (a) Two-color pump and probe experiments performed on two InAs/InP quantum dots (QDs) samples. Sample A contains two QDs layers whereas sample B contains only one single QDs layer. (b) One should notice the strong echoes detected near 70 and 170 ps in sample A and near 90 ps in sample B which correspond to strong phonon wavepackets emitted from the QDs.

the number of strong detected structures (in the following designated by "echo") is equal to the QD layer number (two for sample A at T_1 and T_2 and one for sample B at T_1'). As in the SiGe case, the echoes delays correspond to the time needed for a longitudinal acoustic wave to travel from the QDs layers to the surface.

The subtraction of the relaxation contribution reveals a difference between these results and those obtained in SiGe QDs (Figure 17.12). Indeed the response of both samples displays a small oscillating part in the first picoseconds and several smaller echoes.

The oscillation results from the detection of the acoustic pulse generated at the free surface and which propagates towards the substrate. The blue pulses only probe a few tens of nanometers in the InP cap (the absorption length in InP close to 20 nm at 400 nm is much shorter than the cap thickness in both samples). The oscillating character is related to the photoelastic detection mechanism (Brillouin, 1953). This point is confirmed by measuring the oscillation period (T) as a function of the probe wavelength (λ_{probe}). It reveals a very good agreement with the theoretical result expected from a photoelastic model: $T = \lambda_{probe}/2n\upsilon$, where the optical index (n) and the longitudinal sound velocity (υ) were extracted from the literature (Madelung, 1991). In the SiGe case, such an oscillation was not detected in the first picosecond. This is related to the low efficiency of bulk silicon to emit acoustic phonons when pumped with an infrared optical pulse. On the contrary, most of the III–V semiconductors present strong deformation potential from which acoustic phonons are easily emitted with light pulses.

A small part of the acoustic pulse generated in the cap layer is reflected on the first QD layer, which leads to a small structure detected at a delay which corresponds to a round-trip in the InP cap. In sample A (resp. B) such a structure is detected at $2T_1 = 140$ ps (resp. $2T_1' = 176$ ps). These delays correspond to those expected after one round trip in 321 nm of InP (resp. 403 nm), and both values are very close to the nominal cap thicknesses (320 and 400 nm respectively).

In Figure 17.14 we also present the experimental result obtained using a blue beam for both pump and probe on sample A. Compared to the previous results, only the generation is modified due to the strong optical absorption of pump pulse in the final cap. The most noticeable point in the comparison between two-color and one-color experiments is the extinction of the strong acoustic structures detected at T_1 and T_2. Due to the wavelength change, no blue pump light can reach the QDs layers. The small echo detected at a delay corresponding to the round trip in the cap is still detected. That is consistent with the fact that this echo results from a generation and a detection near the free surface.

From all these results we conclude that coherent acoustic phonons are generated from the sample surface and from each QDs layer. Using a numerical modeling of the generation and detection processes we can examine independently the acoustic generation from the successive layers of the sample. The

numerical tool first solves the electromagnetic continuity equations in order to get the pump absorption profile in the sample which serves as a source term in the resolution of the acoustic continuity equations. The first derivation of the model can be found in Thomsen et al. (1986). The input parameters are the optical and mechanical parameters of materials, an effective generation parameter and a complex photoelastic constant for the detection.

Figure 17.15 presents numerical and experimental results obtained on sample A. We first consider the generation only acting in InP layers. By fitting the photoelastic constants, we can well reproduce intensity and phase of the experimental oscillation detected in the first picoseconds (Figure 17.15a and c). The 810 nm pump can reach up to 1 μm in the sample depth. Even in the case of a generation limited to InP, each QD layer introduces a discontinuity in the absorption and generation profiles. The resulting strain pulse also presents such discontinuities and its photoelastic detection gives echoes at T_1 and T_2 as shown in Figure 17.15c. Nevertheless, these structures cannot correspond to the measured echoes for two obvious reasons revealed in the comparison of Figure 17.15a and c. First, the phase of the numerical structures is opposite to the experimental one. Second, in this scheme, the numerical echoes cannot be higher than the first picoseconds oscillation due to the absorption profile what is unlike to the experimental observation. To reach the excellent agreement between theory and experiment presented in Figure 17.15a and b we need to introduce a strong generation from the InAs QDs. The numerical study further demonstrates the strong acoustic generation in the self-assembled InAs/InP QDs layers.

As shown here we obtained an excellent agreement between experimental and numerical signals. For doing that we needed to introduce a huge generation inside each QD layer. One can note that the contribution issued from InP also produces echoes at the expected delays. But these echoes have an opposite phase

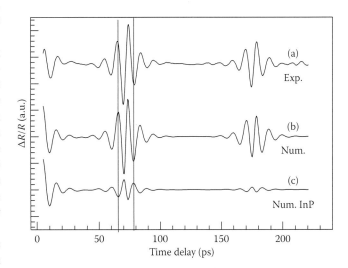

FIGURE 17.15 (a) Experimental signal measured in sample A. (b) Numerical modeling of the detected echoes which perfectly fits the signal (a). (c) Numerical modeling when acoustic generation is limited to InP.

compared to experimental echoes. This means that the acoustic pulse generated inside QD layer is even higher since they must compensate these echoes. Huge or giant means here that in the QD layer we have a generation, which is about 20 times higher than in aluminum, which is a metal known to be very efficient in ultrafast acoustics.

17.5.1.3 What about the Generation Mechanism?

These results raise a fundamental question. By which mechanism such high-amplitude acoustic pulses can be emitted inside quantum dots? In semiconductors, the dominant mechanism is the deformation potential (Gusev and Karabutov, 1993). Carriers are excited by the laser and the modification of their occupancy of the electronic band produces a strain through the deformation potential. In our case, the absorption inside the QD layer is too small for justifying such strong pulses. But in this system InAs quantum dot in InP, it is known that carriers that are photoexcited all around the QD layers are captured by the dots within less than 1 ps (Hinooda et al., 2001). Thanks to this mechanism, we can thus imagine that in a very short time a much more important carrier density appears in the QD layer. And then through the deformation potential a strong strain appears.

We performed similar experiments on various systems. In silicon germanium we observed echoes but small echoes whose amplitude can be explained with the conventional mechanisms. On the contrary, in InAs in GaAs, we measured strong echoes too. It is important to note that the first system does not present such a capture mechanism whereas the second one does. These results thus confirm the role capture may play in this observation.

17.5.1.4 Toward a THz Acoustics

As a resume, we must keep that the most noticeable point here is the high amplitude, a point that is now well established. Another point of interest is the frequency content of such pulses. Indeed, as generation acts in a QD layer, it is strongly confined. And emission from a confined region means high frequency. For example, a 3 nm height layer would excite a pulse whose Fourier content will extend up to one THz.

Unfortunately we can not prove this statement with previous results. Indeed in our experiment, we detect the acoustic pulses at the sample surface using a blue probe and a photoelastic mechanism in InP. Due to that, we limit the detection to 100 GHz, the Brillouin frequency of InP at 400 nm.

Let us examine what would happen if we could change the Brillouin frequency involved in the detection. For that, we numerically change the probe wavelength. Please note that we keep the generation unchanged and the pulse emitted from the dots is thus the same. As shown in Figure 17.16, it is possible to detect 300 GHz with a higher amplitude echo. This point confirms the intuition according to which the generation must be very high in frequency higher than in aluminum thin films. This point has to be experimentally demonstrated.

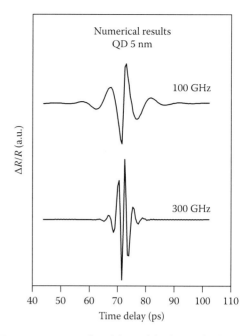

FIGURE 17.16 Numerical modeling of the detected echo in InP emitted from an InAs quantum dots (QDs) layer. The upper curve has been obtained using realistic values for the optical properties of InP. It is in excellent agreement with the experimental echo visible in Figure 17.14. The lower curve has been obtained by adjusting the optical properties of the InP cap in order to simulate a detection at a much higher frequency. In both cases the strain pulse issued from the QDs layer is the same. As shown, one expects an even higher amplitude echo at 300 GHz than at 100 GHz which confirms the high frequency content of pulse emitted by the QDs.

17.5.2 Arrays of Nanocubes as High-Frequency Surface Acoustic Wave Emitter

17.5.2.1 About the Context

Ultrafast acoustics has been widely used to measure the elastic properties of thin films and nanostructures. One implements a sonar at the submicronic scale using ultrashort optical pulses and a pump–probe scheme (the principle of such an experiment is given in Section 17.2). In the standard technique, the laser pump pulse, which generates the acoustical pulse, is focused onto the surface of an absorbing sample. Optical absorption is strong in metals so that the depth over which the laser energy is deposited is much smaller than the spot size (typically the absorption length is about 20 nm whereas the diameter of the focused spot is about 10 μm). Due to that, the conventional setup only produces longitudinal waves.

From the metrological point of view, by measuring echo time-delay or period of Brillouin oscillation, ultrafast acoustics can access the longitudinal sound velocity and mass density. An isotropic material is completely characterized from the mechanical point of view if one knows three physical parameters: the mass density ρ, the Young modulus E, and the Poisson ratio v. The longitudinal sound velocity v_l and the mass density ρ are related to E and v according to (Landau and Lifchitz, 1959):

$$v_l = \sqrt{\frac{E(1-\nu)}{\rho(1+\nu)(1-2\nu)}}. \qquad (17.15)$$

As shown, as long as ultrafast acoustics only accesses to longitudinal waves, mechanical measurements in submicron films could not be complete.

Some authors have recently proposed alternative ways of performing ultrafast acoustic experiments in order to excite transverse waves. Multiple acoustic waves can be excited by tightly focusing the laser beam (Rossignol et al., 2005) or by choosing a specific anisotropic substrate (Matsuda et al., 2004). Up to now, none of these techniques can be applied to some material or other deposited on a silicon substrate, which is the most common geometry in microelectronics. Here we show that using 2D crystals of nanoscale objects, it is possible to generate and detect high-frequency surface acoustic waves using the standard experimental setup. Combining surface and longitudinal waves, mechanical characterization becomes complete in any thin isotropic film.

17.5.2.2 Pioneer Works

A first proposal for generating high-frequency surface acoustic waves using ultrashort optical pulses was to use a 1D diffraction grating deposited onto the sample surface. Antonelli et al. used e-beam lithography for patterning aluminum thin films with polymethyl methacrylate (PMMA) bars (Antonelli et al., 2002c). The bar width and periodicity are close to 1 μm. When the pump light strikes the patterned region, the periodicity of the grating is transferred to the metallic film where a surface acoustic wave results. The wavelength of this surface wave is imposed by the grating period and it propagates along the periodicity direction.

Similar 1D gratings but with much lower dimensions (the bars width and periodicity are typically 100 and 200 nm respectively) were recently studied by Hurley et al. (2006, 2008). This way the authors demonstrate the possibility of generating 22 GHz surface acoustic wave in Si substrates.

17.5.2.3 Back to Arrays of Nanocubes

We recently studied the vibrations of nanocubes arrays by ultrafast acoustics (Robillard et al., 2007, 2008). We revealed two kinds of vibrations that were found to be individual resonances of cubes and collective modes, respectively. Collective modes, which depend on the lattice constant and the cube-width, were found to propagate along the sample surface. Here we show that this feature can be used for measuring in-plane properties on thin films.

In Section 17.4, we demonstrated that the large amplitude and low frequency oscillations detected in the transient reflectivity measured on nanocubes arrays are collective acoustical vibrations of the artificial crystals. By measuring the frequencies as a function of the lattice parameter, we build dispersion curves. All the frequencies order along a few branches, which can be derived from the first one through a scaling by remarkable factors. For example in a square lattice, the second branch can be deduced

from the first one by scaling the frequencies by $\sqrt{2}$. These factors have been identified as resulting from the excitation of a mode propagating in various directions at the sample surface. The first branch corresponds to a propagation in the (1, 0) direction and the second to a propagation in the (1, 1). The theoretical justification of this point can be found in Section 17.4.4.

The collective modes have points in common with a surface acoustic wave but are not. Indeed, as the sample surface is nanostructured the surface acoustic waves are much more complicated than usual. This is illustrated by the strong dispersion, which is found, whereas a surface acoustic wave in a infinite medium is nondispersive. On the contrary, in the limit of infinitely spaced cubes, the collective mode should increasingly resemble a surface acoustic wave. This point is supported by the linear behavior of the dispersion, which has been found in the vicinity of $k = 0$. That defines a sound velocity, which corresponds to the surface acoustic wave velocity. This way, by measuring collective mode frequencies for various artificial crystals, we can extract the velocity in the limit where no cube exists at the sample surface and thus learn more about the mechanical properties of the underlayer.

17.5.2.4 Applications to Thin Film Metrology

Here we illustrate what the metrological facility offers by the arrays of nanocubes by measuring the elastic constants of a thin silica layer deposited onto a silicon substrate. The studied sample is a 600 nm-thick SiO_2 film deposited by plasma enhanced chemical vapor deposition on a (100) silicon substrate. As the silica layer is transparent, a 28 nm-thick aluminum film is deposited at the top acting as a transducer in conventional ultrafast acoustics (pump light is absorbed in the Al film). In order to measure the elastic properties in-plane, we also realize some square lattices of Al nanocubes on a part of this transducer using e-beam lithography. Here the cube size is 200 nm and the lattice constant varies from 400 to 800 nm. The experiments were performed at two different locations: Out of the array of nanocubes, directly on the Al film as shown in Figure 17.17a; On lattices as in Figure 17.17b. Off-the-cubes, the pump beam launches a *longitudinal* pulse in the Al layer. Using a blue probe (Devos et al., 2005) we measure the longitudinal sound velocity, the mass density, and the thickness of the SiO_2 layer. In Figure 17.18a, we reproduce an off-the-cubes transient reflectivity signal. Several acoustic contributions appear. First an acousto-optic oscillation detected during the propagation of the wave inside the SiO_2 film. Such an oscillation as the so-called Brillouin oscillation results from interferences between the probe light that is reflected from the interfaces and the light partially reflected from the acoustic pulse (Devos and Côte, 2004). Its period is given by $T_1 = \lambda/(2n v_l) \simeq 24\,ps$ where $\lambda = 400\,nm$ is the probe wavelength, n is the SiO_2 refractive index at λ, and v_l is the SiO_2 longitudinal sound velocity. Assuming $n = 1.47$ (Palik, 1985) from T, we deduce $v_l = 5.85\,nm \cdot ps^{-1}$. When the pulse reaches the substrate another oscillation is detected, it corresponds to a Brillouin oscillation detected in Si. From the delay at which this high-amplitude signal starts and from

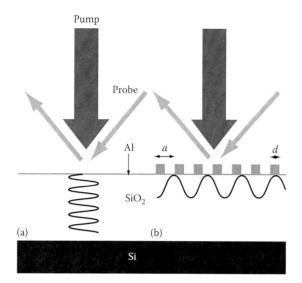

FIGURE 17.17 Two different experiments are performed for measuring all the elastic constants of a thin SiO₂ layer. (a) Off-the-cubes where the pump light is absorbed by the Al thin film deposited at the top of the sample. The laser excites longitudinal waves which propagate perpendicularly to the film plane. (b) On the cubes. In that case the laser excites collective acoustic modes in the artificial crystal of nanocubes. These modes propagate in the plane of the SiO₂ layer.

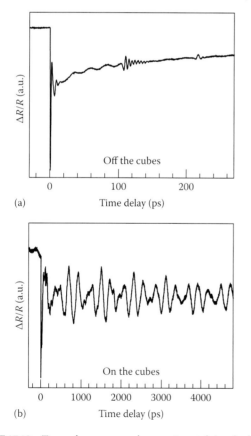

FIGURE 17.18 Two-color pump–probe experimental signals obtained using 800 nm femtosecond laser pulses. (a) Transient reflectivity measured aside of the cubes. (b) Transient reflectivity measured on the cubes (a = 800 nm).

the sound velocity we can deduce the thickness of SiO₂. The last acoustic contribution is an echo detected in Al at a longer delay. It corresponds to a round trip of the pulse inside the Al-SiO₂ stack. Then from the echo time-delay, we extract the Al layer thickness. The part of the pulse, which is reflected by the SiO₂/Si interface to the surface, also produces a low-frequency Brillouin oscillation. By comparing the magnitude of such an oscillation before and after reflection, we can measure the acoustic reflection coefficient at the interface. As the acoustic impedance of the Si substrate is well-known, we can deduce the SiO₂ acoustic impedance and then the SiO₂ mass density from the reflection coefficient (ρ = 2170 g·cm⁻³).

We now focus on the transient reflectivity measured on the cube lattices as shown in Figure 17.17b. Figure 17.18b reproduces the signal obtained on a 800 nm step-size lattice. The main difference with a classical PU experiment is the time scale. The signal detected on-the-cubes lasts at least 7 ns and also has a much higher amplitude than the signal detected off-the-cubes. As explained in Section 17.4.3, it corresponds to collective vibrations of the nanocube crystal. We measured the frequencies on various arrays whose lattice constant a varies from 400 to 800 nm. As shown in Figure 17.10, the multiple frequencies orders regularly as branches. The upper branches can be deduced by scaling the lower one with remarkable factors $\sqrt{2}, 2, \sqrt{5}, 2\sqrt{2}$. One notes that each branch extends linearly to zero for lower and lower $1/a$ in a similar manner as an acoustic phonon branch. All the measured frequencies can be reproduced by the theoretical model derived in Section 17.4.4. According to this description, the frequency of the mode, which propagates in the (i, j) direction is given by

$$f_{i,j}(a) = c\sqrt{\frac{a^2 h}{a^2 h + d^3}} \cdot \frac{\sqrt{i^2 + j^2}}{a},\qquad(17.16)$$

where

    h is the transducer thickness
    d is the cube size
    c is a sound velocity
    i and j are the integer coordinates of the wavevector in the reciprocal lattice

The excellent agreement between experimental frequencies and the theoretical model as shown in Figure 17.10 is obtained using one fitting parameter c. We obtained a sound velocity $c = 3.35$ nm·ps⁻¹. c is the slope of the lowest branch close to zero. It corresponds to the velocity of the mode in the case of infinitely spaced cubes, i.e., of a surface wave in the sample without the cubes. The accuracy of the velocity determination is greatly improved by measuring a large number of frequencies. The surface wave velocity we determine from the fit is the Rayleigh's velocity of the Al-SiO₂ stack and not SiO₂. It is possible to eliminate the influence of the Al layer on the sound velocity following a method described in Robillard et al. (2008). We then reach the Rayleigh's velocity of SiO₂, $v_R = 3.4$ nm·ps⁻¹.

From the longitudinal and Rayleigh's velocities, we can extract the Young's modulus (E) and Poisson's ratio (v). First we deduce the transverse sound velocity from the Rayleigh wave velocity using the Viktorov's approximation (Landau and Lifchitz, 1959; Royer and Dieulesaint, 1996). Then using Equation 17.15 and the relation (Landau and Lifchitz, 1959)

$$v_t = \sqrt{\frac{E}{2\rho(1+v)}}, \qquad (17.17)$$

we extract the Young modulus and the Poisson ratio. In the present case of a 600 nm thick SiO_2 layer and using $v_l = 5.85$ nm ps^{-1}, $\rho = 2170$ g cm^{-3}, and $v_R = 3.4$ nm ps^{-1}, we successively get $v_t = 3.76$ nm ps^{-1}, $E = 71.6$ GPa and $v = 0.16$, which is in very good agreement with the literature (Kim, 1996).

This study has illustrated another interest of nanos when organized in artificial crystals. We have shown that they act as an in-plane transducer for the underlayer. It is important to note that the improvement of the capability of ultrafast acoustics does not require any modification of the experimental setup. From that it becomes possible to determine the velocity of longitudinal and surface waves, and then Young's modulus and Poisson's ratio. The same technique using the same transducer can be applied to any isotropic thin film (metallic, dielectric…) thicker than 100 nm. The characterization could be extended to thinner layers provided that the lattice steps and cube width are reduced. The method could also be extended to anisotropic materials by taking advantage of the various directions of propagation simultaneously excited by the laser and which appear as several branches in Figure 17.10.

17.6 Conclusion

In this chapter, we explored multiple aspects of the phonons in various nanoscale objects. We started from some theoretical generalities concerning the acoustic vibrations in single or organized nanos. The important points to be kept in mind are the finite size leads to a quantization of the acoustical resonances, which can be designated as a phonon confinement effect; the frequency range in which the resonances fall starts typically from 100 GHz. We then examined experimental methods that permit to resolve the phonons in such small objects. We focused on time-resolved experiment that offers a unique way of studying the dynamic of phonons at nanoscale. Ultrashort laser pulses can excite mechanical resonances of nanos and those vibrations can be monitored in the time-domain using another optical pulse time-delayed with respect with the first one. We then present a review of studies dedicated to the individual vibrations of nanoscale objects. Many reports have been concerned with spherical objects for which the first breathing mode is usually detected. Similar experiments have recently been performed on much more complex objects, the main difference being the observation of several modes. If in most of the experiments a large collection of objects are simultaneously excited by the laser. Some authors have demonstrated that it is possible to study the phonons in one single nanoscale object. Such studies provide information about the object and its environment. Indeed the acoustic phonons quantization has been shown to be very dependent on the geometry and the size of the object under study. The damping of the vibration can also offer a way of analyzing the coupling between an object and its environment.

We then examined the phonons in artificial crystals made of nanoscale objects organized along a lattice. We identified two kinds of modes in such an artificial lattice. First modes arise from the individual vibrations of the nanoscale object. The organization and the coupling between objects produce a branch that is flat and centered at a frequency close to the isolated object. We also show that new acoustic modes appear in the crystal. Similar to acoustic phonons in an usual crystal, these low-frequency vibrations correspond to displacement of one nanoscale object with respect to another. These modes are found to strongly depend on the organization. By studying series of artificial crystals differing by the lattice step, we built dispersion curves. We established a model for reproducing the measured frequencies and deduced this way that collective modes propagates along the sample surface.

In a last part, we show that nanos can also be used as acoustic waves emitter. Semiconductor quantum dots and especially InAs quantum dots have been shown to be very efficient in conversion of ultrashort light pulses into high-frequency acoustic pulses. Furthermore, such systems are promising tools for realizing a THz transducer. From the analysis of the collective modes excited and detected in 2D artificial crystals, we proposed to use the lattice of nanos as a high-frequency surface acoustic wave transducer.

Beyond the fundamental interest of studying the phonons in nanoscale objects, we show that it can also have some practical applications. We mentioned the possible applications of measuring eigen modes in interconnect lines of copper in microelectronics. We also show that crystal of nanos can be useful for measuring the elastic properties of thin films.

Elastic waves are also expected to experiment such periodic effects in the so-called phononic crystals (PC). The perspective of designing an on demand material by adjusting the lattice type and parameter has recently motivated numerous studies on PC (Kushwaha et al., 1993; Page et al., 2004). Most applications of PC have concerned low-frequency ranges due to technological constraints. In order to reach the hypersonic range (i.e., beyond 1 GHz), the lattice constant must be between a few hundred nanometers and a few microns.

Phonons play a major role in heat transport in insulators and semiconductors. Spatial confinement of phonons in nanostructures can strongly affect the phonon spectrum and thus the thermal properties at nanoscale. Organization of nanos in an artificial crystal constitutes what is the so-called hypersonic phononic crystal. Any wave that propagates in a periodic structure whose lattice parameter is comparable to the wavelength experiments similar "band" effects (Brillouin, 1953). There are analogies between electronic states in a crystal or light modes in a photonic crystal, the most common being the apparition of forbidden energies also called band gaps. A first step has been demonstrated here by showing that the acoustic phonons in phononic crystals depend both on the nanoscale object and on the lattice. Similarly to what has been done in heterostructures due to band-gap engineering, one expects to be able of realizing a phonon dispersion engineering in nanostructures from which one expects to control the heat transport. Hypersonic phononic crystals also offer the unique properties of being simultaneously periodic for photons and phonons. Such crystals are both phononic and photonic from which new effects are expected especially concerning the acousto-optic interaction (Gorishnyy et al., 2005).

References

Antonelli, G. A., Maris, H. J., Malhotra, S. G. et al. 2002a. Picosecond ultrasonics study of the vibrational modes of a nanostructure. *J. Appl. Phys.*, 91(5):3261–3267.

Antonelli, G. A., Maris, H. J., Malhotra, S. G. et al. 2002b. A study of the vibrational modes of a nanostructure with picosecond ultrasonics. *Physica B*, 317:434–437.

Antonelli, G. A., Zannitto, P., and Maris, H. J. 2002c. New method for the generation of surface acoustic waves of high frequency. *Physica B*, 317:377–379.

Arbouet, A., Fatti, N. D., and Vallee, F. 2006. Optical control of the coherent acoustic vibration of metal nanoparticles. *J. Chem. Phys.*, 124(14):144701.

Baumberg, J. J., Williams, D. A., and Köhler, K. 1997. Ultrafast acoustic phonon ballistics in semiconductor heterostructures. *Phys. Rev. Lett.*, 78(17):3358–3361.

Bohren, C. and Huffman, D. 1998. *Absorption and Scattering of Light by Small Particles*. Wiley, New York.

Bonacina, L., Callegari, A., Bonati, C. et al. 2006. Time-resolved photodynamics of triangular silver nanoplates. *Nano Lett.*, 6:7.

Bragas, A. V., Aku-Leh, C., Costantino, S., Ingale, A., Zhao, J., and Merlin, R. 2004. Ultrafast optical generation of coherent phonons in CdTe$_{1-x}$Se$_x$ quantum dots. *Phys. Rev. B*, 69(20):205306.

Bragas, A. V., Aku-Leh, C., and Merlin, R. 2006. Raman and ultrafast optical spectroscopy of acoustic phonons in CdTe$_{0.68}$Se$_{0.32}$ quantum dots. *Phys. Rev. B*, 73(12):125305.

Brillouin, L. 1953. *Wave Propagation in Periodic Structures*. Dover Publications, New York.

Cerullo, G., De Silvestri, S., and Banin, U. 1999. Size-dependent dynamics of coherent acoustic phonons in nanocrystal quantum dots. *Phys. Rev. B*, 60(3):1928–1932.

Champagnon, B., Andrianasolo, B., and Duval, E. 1991. Nanocrystallites vibration modes of CdS$_x$Se$_{1-x}$ semiconductors in glasses: Size determination by Raman scattering. *J. Chem. Phys.*, 94(7):5237–5239.

Comin, A., Giannetti, C., Samoggia, G. et al. 2006. Elastic and magnetic dynamics of nanomagnet-ordered arrays impulsively excited by subpicosecond laser pulses. *Phys. Rev. Lett.*, 97(21):217201.

Devos, A. and Lerouge, C. 2001. Evidence of laser-wavelength effect in picosecond ultrasonics: Possible connection with interband transitions. *Phys. Rev. Lett.*, 86(12):2669–2672.

Devos, A. and LeLouarn, A. 2003. Strong effect of interband transitions in the picosecond ultrasonics response of metallic thin films. *Phys. Rev. B*, 68:045405.

Devos, A. and Côte, R. 2004. Strong oscillations detected by picosecond ultrasonics in silicon: Evidence for an electronic-structure effect. *Phys. Rev. B*, 70:125208.

Devos, A., Perrin, B., Bonello, B. et al. 1999. Ultrafast photoacoustics in colloids. *Photoacoustic and Photothermal Phenomena: 10th International Conference AIP Conference Proceedings*, Rome, Italy, 463:445.

Devos, A., Cote, R., Caruyer, G. et al. 2005. A different way of performing picosecond ultrasonic measurements in thin transparent films based on laser-wavelength effects. *Appl. Phys. Lett.*, 86(21):211903.

Duval, E. 1992a. Far-infrared and Raman vibrational transitions of a solid sphere: Selection rules. *Phys. Rev. B*, 46(9):5795–5797.

Duval, E. 1992b. Far-infrared and Raman vibrational transitions of a solid sphere: Selection rules. *Phys. Rev. B*, 46:5795.

Duval, E., Boukenter, A., and Champagnon, B. 1986. Vibration eigenmodes and size of micro-crystallites in glass: Observation by very-low-frequency Raman scattering. *Phys. Rev. Lett.*, 56(19):2052–2055.

Faraday, M. 1847. The Bakerian lecture: Experimental relations of gold (and other metals) to light. *Phil. Trans. R. Soc. (London)*, 147:145.

Fatti, N. D., Voisin, C., Chevy, F., Vallee, F., and Flytzanis, C. 1999. Coherent acoustic mode oscillation and damping in silver nanoparticles. *J. Chem. Phys.*, 110(23):11484–11487.

Gorishnyy, T., Ullal, C. K., Maldovan, M. et al. 2005. Hypersonic phononic crystals. *Phys. Rev. Lett.*, 94(11):115501.

Guillon, C., Langot, P., del Fatti, N. et al. 2007. Coherent acoustic vibration of metal nanoshells. *Nano Lett.*, 7:138–142.

Gunnarsson, O. 1997. Superconductivity in fullerides. *Rev. Mod. Phys.*, 69(2):575–606.

Gusev, V. and Karabutov, A. 1993. *Laser Optoacoustics*. AIP, New York.

Hinooda, S., Loualiche, S., Lambert, B. et al. 2001. Wetting layer carrier dynamics in InAs/InP quantum dots. *Appl. Phys. Lett.*, 78(20):3052–3054.

Hodak, J. H., Martini, I., and Hartland, G. V. 1998. Observation of acoustic quantum beats in nanometer sized au particles. *J. Chem. Phys.*, 108(22):9210–9213.

Hodak, J. H., Henglein, A., and Hartland, G. V. 1999. Size dependent properties of au particles: Coherent excitation and dephasing of acoustic vibrational modes. *J. Chem. Phys.*, 111(18):8613–8621.

Hurley, D. H. 2006. Optical generation and spatially distinct interferometric detection of ultrahigh frequency surface acoustic waves. *Appl. Phys. Lett.*, 88(19):191106.

Hurley, D. H., Lewis, R., Wright, O. B., and Matsuda, O. 2008. Coherent control of gigahertz surface acoustic and bulk phonons using ultrafast optical pulses. *Appl. Phys. Lett.*, 93(11):113101.

Kammerer, C., Cassabois, G., Voisin, C. et al. 2001. Efficient acoustic phonon broadening in single self-assembled InAs/GaAs quantum dots. *Phys. Rev. B*, 65(3):033313.

Kim, M. T. 1996. Influence of substrates on the elastic reaction of films for the microindentation tests. *Thin Solid Films*, 283:12–16.

Krauss, T. D. and Wise, F. W. 1997. Coherent acoustic phonons in a semiconductor quantum dot. *Phys. Rev. Lett.*, 79(25):5102–5105.

Krugel, A., Axt, V. M., Kuhn, T. et al. 2006. Back action of nonequilibrium phonons on the optically induced dynamics in semiconductor quantum dots. *Phys. Rev. B*, 73:035302 and references therein.

Kushwaha, M. S., Halevi, P., Dobrzynski, L. et al. 1993. Acoustic band structure of periodic elastic composites. *Phys. Rev. Lett.*, 71(13):2022–2025.

Lamb, H. 1882. On the vibrations of an elastic sphere. *Proc. London Math. Soc.*, 13:189.

Landau, L. and Lifchitz, E. 1959. *Theory of Elasticity*. Pergamon Press, London, U.K.

Lin, H.-N., Maris, H. J., Freund, L. B. et al. 1993. Study of vibrational modes of gold nanostructures by picosecond ultrasonics. *J. Appl. Phys.*, 73(1):37–45.

Liu, R., Sanders, G. D., Stanton, C. J. et al. 2005. Femtosecond pump–probe spectroscopy of propagating coherent acoustic phonons in $In_xGa_{1-x}N$/GaN heterostructures. *Phys. Rev. B*, 72(19):195335.

Madelung, O. 1991. *Semiconductors Groupe IV Elements and III–V Compounds*. Springer-Verlag, Berlin, Germany.

Martín-Moreno, L., Garcia-Vidal, F. J., Lezec, H. J. et al. 2001. Theory of extraordinary optical transmission through subwavelength hole arrays. *Phys. Rev. Lett.*, 86(6):1114–1117.

Matsuda, O., Wright, O. B., Hurley, D. H. et al. 2004. Coherent shear phonon generation and detection with ultrashort optical pulses. *Phys. Rev. Lett.*, 93(9):095501.

Matsuda, O., Tachizaki, T., Fukui, T. et al. 2005. Acoustic phonon generation and detection in $GaAs/Al_{0.3}Ga_{0.7}As$ quantum wells with picosecond laser pulses. *Phys. Rev. B*, 71(11):115330.

Mazurenko, D. A., Shan, X., Stiefelhagen, J. C. P. et al. 2007. Coherent vibrations of submicron spherical gold shells in a photonic crystal. *Phys. Rev. B*, 75(16):161102.

Muljarov, E. A. and Zimmermann, R. 2004. Dephasing in quantum dots: Quadratic coupling to acoustic phonons. *Phys. Rev. Lett.*, 93(23):237401.

Nisoli, M., De Silvestri, S., Cavalleri, A., Malvezzi, A. M., Stella, A., Lanzani, G., Cheyssac, P., and Kofman, R. 1997. Coherent acoustic oscillations in metallic nanoparticles generated with femtosecond optical pulses. *Phys. Rev. B*, 55(20):R13424–R13427.

Özgür, U., Lee, C.-W., and Everitt, H. O. 2001. Control of coherent acoustic phonons in semiconductor quantum wells. *Phys. Rev. Lett.*, 86(24):5604–5607.

Page, J. H., Sukhovich, A., Yang, S. et al. 2004. Phononic crystals. *Phys. Status Solidi(B)*, 241:3454.

Palik, E. D. 1985. *Handbook of Optical Constants of Solids*. Academic Press, London, U.K.

Paranthoen, C., Bertru, N., Dehaese, O. et al. 2001. Height dispersion control of InAs/InP quantum dots emitting at 1.55 mu m. *Appl. Phys. Lett.*, 78(12):1751–1753.

Petrova, H., Lin, C.-H., de Liejer, S. et al. 2007. Time-resolved spectroscopy of silver nanocubes: Observation and assignment of coherently excited vibrational modes. *J. Chem. Phys.*, 126(9):094709.

Robillard, J.-F., Devos, A., and Roch-Jeune, I. 2007. Time-resolved vibrations of two-dimensional hypersonic phononic crystals. *Phys. Rev. B (Condens. Mater. Mater. Phys.)*, 76(9):092301.

Robillard, J.-F., Devos, A., Roch-Jeune, I. et al. 2008. Collective acoustic modes in various two-dimensional crystals by ultrafast acoustics: Theory and experiment. *Phys. Rev. B (Condens. Mater. Mater. Phys.)*, 78(6):064302.

Rossignol, C., Rampnoux, J. M., Perton, M. et al. 2005. Generation and detection of shear acoustic waves in metal submicrometric films with ultrashort laser pulses. *Phys. Rev. Lett.*, 94(16):166106.

Royer, D. and Dieulesaint, E. 1996. *Ondes Elastiques dans les Solides*. Masson, Paris, France.

Rudd, R. E. and Broughton, J. Q. 1998. Coarse-grained molecular dynamics and the atomic limit of finite elements. *Phys. Rev. B*, 58(10):R5893–R5896.

Saviot, L., Murray, D. B., and Marco de Lucas, M. C. 2004a. Vibrations of free and embedded anisotropic elastic spheres: Application to low-frequency Raman scattering of silicon nanoparticles in silica. *Phys. Rev. B*, 69(11):113402.

Saviot, L., Murray, D. B., Mermet, A. et al. 2004b. Comment on testimate of the vibrational frequencies of spherical virus particles. *J. Phys. Rev. E*, 69(2):023901.

Tamura, A., Higeta, K., and Ichinokawa, T. 1982. Lattice vibrations and specific heat of a small particle. *J. Phys. C*, 15:4975.

Taubert, R., Hudert, F., Bartelsand, A., Merkt, F., Habenicht, A., Leiderer, P., and Dekorsy, T. 2007. Coherent acoustic oscillations of nanoscale Au triangles and pyramids: Influence of size and substrate. *New J. Phys.*, 9:376.

Thoen, E. R., Steinmeyer, G., Langlois, P. et al. 1998. Coherent acoustic phonons in PbTe quantum dots. *Appl. Phys. Lett.*, 73(15):2149–2151.

Thomsen, C., Grahn, H. T., Maris, H. J. et al. 1986. Surface generation and detection of phonons by picosecond light pulses. *Phys. Rev. B*, 34:4129.

Turkevitch, J. 1985. Colloidal gold. *Gold Bull.*, 18:86–91.

van Dijk, M. A., Lippitz, M., and Orrit, M. 2005a. Detection of acoustic oscillations of single gold nanospheres by time-resolved interferometry. *Phys. Rev. Lett.*, 95(26):267406.

van Dijk, M. A., Lippitz, M., and Orrit, M. 2005b. Detection of acoustic oscillations of single gold nanospheres by time-resolved interferometry. *Phys. Rev. Lett.*, 95(26):267406.

Verma, P., Cordts, W., Irmer, G., and Monecke, J. 1999. Acoustic vibrations of semiconductor nanocrystals in doped glasses. *Phys. Rev. B*, 60(8):5778–5785.

Vertikov, A., Kuball, M., Nurmikko, A. V., and Maris, H. J. 1996. Time-resolved pump-probe experiments with subwavelength lateral resolution. *Appl. Phys. Lett.*, 69(17):2465–2467.

Voisin, C., Christofilos, D., Fatti, N. D., and Valle, F. 2002. Environment effect on the acoustic vibration of metal nanoparticles. *Physica B*, 316–317:89–94.

18

Melting of Finite-Sized Systems

Dilip Govind Kanhere
University of Pune

Sajeev Chacko
Jawaharlal Nehru University

18.1 Introduction

The world we live in consists of objects that are "macroscopic," i.e., objects having large sizes. Typically, we talk about length scales in centimeters, meters, and even larger. The world at the other end of the length scale, i.e., the microscopic world, can be very fascinating. In fact, it is well known that the properties of objects such as surfaces, thin films, wires etc., having reduced dimensionality, can be very different as compared to those of the familiar three-dimensional extended solids. The reduced dimensionality of such systems can be two, one, or even zero. By zero-dimension, we mean finite-sized systems that are restricted in all the three dimensions to a few tens of nanometers. Atomic clusters and quantum dots are well-known examples of zero-dimensional systems. Materials having sizes larger than this, i.e., with length scales of up to few hundreds of nanometers, are termed as *nanomaterials*.

Fundamental to all properties is the basic electronic structure, i.e., the eigenvalues, the nature of the energy spectrum, the charge and spin densities, the molecular orbitals or the Bloch states, the density of states, the energy gaps, etc. of the interacting electrons. It is natural to expect the electronic structure to change significantly if the dimensionality is reduced. This is best illustrated by examining the density of states, a property often studied. The density of states tells us the number of energy states available within a small energy interval. In Figure 18.1, we show the density of states for the materials with different dimensionalities within the free electron approximation. The model systems are shown on top of the density of states.

The density of states of three-dimensional (bulk) systems is known to have a $\sqrt{E}$ behavior, while in two-dimension (the quantum wells, as shown) it is constant. In one-dimension, it goes as $1/\sqrt{E}$, while in zero-dimension, i.e., for finite-sized systems, the eigenvalue spectrum becomes discrete. The steps seen in a 2D case are due to the discretization in the third direction. The dramatic change in the nature of electronic structure is self-evident even from these simple jellium model–based results. Quite clearly, we should expect the properties of the zero-dimensional system to be very different from their bulk counterparts.

18.1.1 What Are Atomic Clusters?

Atomic clusters, the focus of this chapter, are experimentally realizable finite-sized systems. These clusters are stable aggregates of atoms or molecules held together under various conditions. We are familiar with molecules. But, how do clusters differ from molecules?

1. Clusters, by and large, are artificially synthesized. For example, it is possible to produce sodium clusters containing two to a few thousands of atoms. These can be size selected to carry out various measurements (de Heer, 1993).
2. Clusters are synthesized out of atoms of metallic, nonmetallic, semiconducting materials, or inert gas atoms. Thus, one can get homogeneous clusters, formed out of the same atomic species, or heterogeneous clusters, formed out of more than one type of atom giving rise to tiny alloys, impurity doped clusters, etc.

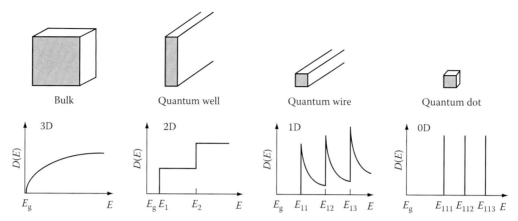

FIGURE 18.1 Density of states per unit volume and energy for a 3D semiconductor, a 2D quantum well, a 1D quantum wire, and a 0D quantum dot.

3. The thermodynamically stable state of clusters is generally their native infinite size bulk. This is clear from the fact that the total energy per atom is the lowest for the bulk. However, in some cases of heterogeneous clusters, the bulk counterpart may not exist. For example, small clusters of GaO are stable, while there is no corresponding bulk form with the same composition.

4. An interesting class of clusters is the "core-shell clusters" having a core of type A atoms and an outer layer of type B atoms. The formation of such clusters opens up an exciting possibility of designer's clusters.

Because of a great variety of sizes, shapes, and component materials, physics and chemistry of clusters turn out to be very rich. In fact, their properties are neither atomic like nor solidlike. Sometimes, clusters are termed as another state of matter synthesized artificially with tunable properties.

The study of clusters consists of investigating various properties such as vibrational, optical, magnetic, thermal, and chemical reactivities. One of the most interesting facts that distinguish clusters from the bulk is the size sensitivity of their properties. A great majority of the properties including stability, magnetism, optical spectrum, and the gap between the highest occupied and the lowest unoccupied molecular orbital are size sensitive. This was brought out in the most dramatic way by experiments reported by Knight's group (Knight et al., 1984). In Figure 18.2a, we show the abundance (mass) spectrum of sodium clusters, measured by them, showing substantially enhanced stability for certain sizes. It can be seen that the peaks corresponding to the sizes of $N = 8, 20, 40, 58$ are large. Hence, this enhanced stability is associated with the "shell closing" of the electronic spectrum. This is illustrated in Figure 18.2c, where the one-electron levels of three different potentials are schematically shown. The numbers near the levels correspond to the total number of electrons occupying the shells. The second difference in energies $\Delta^2(N) (= E_{n+1} + E_{n-1} - 2E_n)$ obtained by jellium as well as the Clemenger–Nielsson model is shown in the bottom panel of Figure 18.2b. It can be seen that the major features in stability are due to electronic shell closing. It may

be noted that such a magic behavior is characteristic of metallic clusters, where the electronic structure effects are dominant. Other types of clusters, e.g., van der Waals or covalently bonded clusters, may show effects due to geometric shell closing.

Just as clusters come in a variety of sizes and shapes, they also display a variety of bonding. A convenient classification of clusters could be on the lines used for bulk systems.

1. *Metallic clusters*: These are clusters formed by atoms of metallic elements such as Li, Na, Al, Cu, etc. Such clusters could be homogeneous or heterogeneous. These clusters are characterized by delocalized charge density. In Figure 18.3a, we show the charge density for Na_3 cluster, depicting a typical delocalized nature.

2. *Covalent clusters*: These are the clusters formed by semiconducting, nonmetallic atoms such as Si, Ge, C, B, etc. In Figure 18.3b, we show the charge density of Si_3. Clearly, the charge density is localized along the bond between the atoms. Interestingly, small clusters of aluminum also display highly directional localized bonding.

3. *Ionic clusters*: These are formed by two species, A and B, having significant difference in their electronegativities, e.g., Al–Li, Ti–O, etc. There is a substantial charge transfer from one species to another, leading to a strong ionic bond between them. In Figure 18.3c, the charge density for Ti–O dimer is shown. Note the spherically symmetric charge distribution around oxygen atom.

4. *van der Waals clusters*: These are formed mainly by inert gas atoms and typically assume compact icosahedral shapes.

In Figure 18.4a through e, we show some interesting clusters displaying different geometries. In Figure 18.4a, an interesting structure showing (perhaps) the smallest silicon cage structure formed by doping a single Zr atom in Si_{16} is shown. The resulting structure is a Frank–Kasper polyhedron. Figure 18.4b shows a fullerene-like structure formed out of 80 boron atoms. In Figure 18.4c we show a hollow cage of Au with 32 atoms, while Figure 18.4d is a core-shell cluster consisting of As–Ni–As

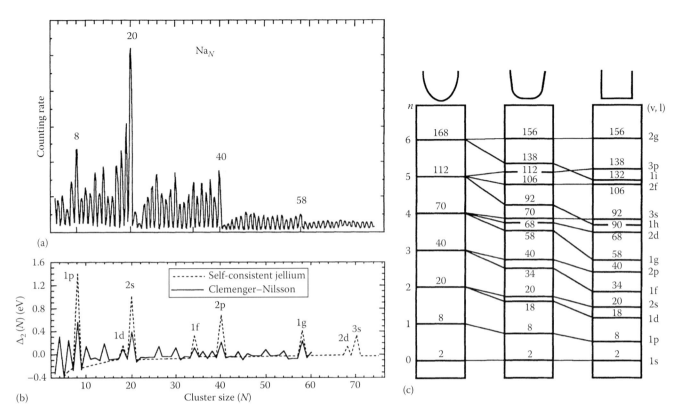

FIGURE 18.2 Abundance spectrum for sodium cluster: (a) experimental. (After Knight, W.D. et al., *Phys. Rev. Lett.*, 52, 2141, 1984.); (b) dashed-line, using Woods–Saxon potential. (After Knight, W.D. et al., *Phys. Rev. Lett.*, 52, 2141, 1984.) and solid-line, using the ellipsoidal shell (Clemenger–Nilsson) model. (After de Heer, W.A. et al., *Solid State Physics*, Vol. 40, Academic Press, New York, 1987.); and (c) energy levels and occupancy for three-dimensional harmonic, intermediate, and square-well potentials. (After de Heer, W.A. et al., *Solid State Physics*, Vol. 40, Academic Press, New York, 1987.)

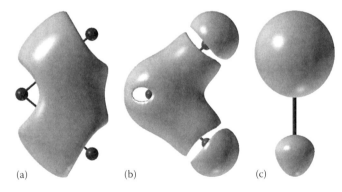

FIGURE 18.3 Isovalued surfaces of the total electronic charge density for (a) Na_3 (at half of the maximum value of 0.05 electrons/Å³), (b) Si_3 (at half of the maximum value of 0.62 electrons/Å³). In (c), Ti–O dimer (at 1/8th of the maximum value of 8.11 electrons/Å³) (c) The top atom is oxygen and the bottom atom is titanium. Note that the electron charge density of Na_3 cluster is seen in the region between all the three atoms, while that of Si_3 is localized along the bonding region between the silicon atoms. In case of Ti–O dimer, a charge transfer from titanium atom (represented by the larger atom in Figure 18.3b) to oxygen atom is clearly evident.

shells. A perfectly formed icosahedron out of 147 sodium atoms is displayed in Figure 18.4e, and Figure 18.4f shows a cluster formed out of 36 gallium atoms showing a layered structure.

In the last few decades, a substantial work on free, unsupported clusters has been reported. The advent of nanoscience has

given further impetus to the study of the physics and chemistry of atomic clusters. The reader is referred to a number of exhaustive and excellent reviews for more details (Berry, 1997; Ekardt, 1999; Haberland, 1994; Jellinek, 1999; Jena et al., 1992; Kumar et al., 2000; Wales, 2003, 2005).

18.1.2 Atomic Clusters at Finite Temperature

In the following sections, we will deal with the finite temperature properties of atomic clusters. Such properties are much more complicated to calculate, since they are beyond the ground-state equilibrium geometry. It may also be mentioned that we deal with ionic motion. The electrons essentially play a role in generating the forces on ions. Thus, the dynamics that we deal with is classical with forces on ions derived from the quantum mechanical treatment of electrons. During the last couple of decades, *ab initio* molecular dynamics has turned out to be a powerful tool to compute the ionic trajectories using appropriate ensembles, i.e., canonical or microcanonical ensemble. Such methods do not rely upon interatomic forces, either fitted or otherwise, and are more reliable in generating the accurate trajectories. Of course, the price we pay is in terms of computational expenses. *Ab initio* molecular dynamical simulations are an order of magnitude more expensive than conventional molecular dynamics. Consequently, these simulations are restricted to a smaller number of atoms and short simulation times. The development and the

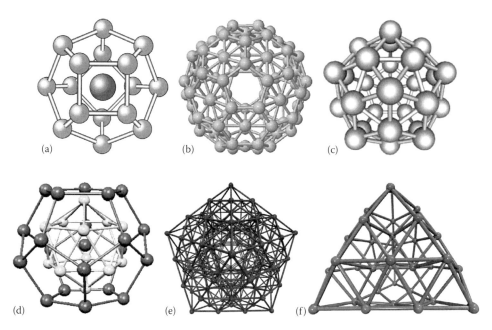

FIGURE 18.4 The geometries of: (a) Zr-Si$_{16}$—caged silicon cluster. (After Kumar, V. and Kawazoe, Y., *Phys. Rev. Lett.*, 87, 045503, 2001.) (The black circle at the center represents Zr atom.) (b) B$_{80}$ fullerene. (After Szwacki, N.Z. et al., *Phys. Rev. Lett.*, 98, 166804, 2007.) (c) Au$_{32}$ hollow cage. (From Johansson, M.P. et al., *Angew. Chem. Int. Ed.*, 43, 2678, 2004. With permission.) (d) As–Ni$_{12}$–As$_{20}$—"core–shell" onionlike cluster. (From Baruah, T. et al., *Phys. Rev. A*, 68, 241404, 2003. With permission.) (The black circles represent As atoms and the gray circles represents Ni atoms.) (e) Na$_{147}$—three-shell Mackay icosahedron. (f) Geometry of Ga$_{36}$.

achievements of *ab initio* simulation methods that have been reported in the last decade or so will be reviewed.

In Sections 18.2 and 18.3, we introduce the basic concepts of density functional theory and molecular dynamics. Density functional theory is a quantum mechanical theory used in physics and chemistry to investigate the electronic structure (principally the ground state) of many-body systems, in particular atoms, molecules, and the condensed phases. It maps a many-body problem onto an effective one-particle problem. With this theory, the properties of a many-electron system can be determined using functionals of electron density. This theory is among the most popular and versatile methods available in condensed-matter physics, computational physics, and computational chemistry. Molecular dynamics, on the other hand, is a computer simulation method in which particles are allowed to interact and move for an extended period of time under Newton's laws of motion. This method is used to generate trajectories of all the particles in order to compute the statistical averages and to understand the molecular motion on an atomic scale. In Section 18.4, we discuss in brief various tools generally used to analyze the trajectory data generated by molecular dynamics in conjunction with density-functional theory. We summarize the mathematical expressions for these tools and discuss their reliability in understanding the finite temperature behavior of atomic clusters. Finally in Section 18.5, we give a brief review of the recent experimental and theoretical work carried out in the last decade or so. We discuss three main issues observed in the finite temperature behavior of atomic clusters: (1) the irregular variations in the melting point with respect to the size of the clusters, (2) higher-than-bulk melting

temperature, and (3) the size-sensitive nature of the specific heat curves. We also discuss the effect of impurities on the finite temperature properties of cluster.

18.2 Theoretical Background

A cluster may be considered as a collection of ionic cores (ions) surrounded by valence electrons. In this section, we discuss the methods to compute thermodynamics quantities such as density of states, entropy, and specific heat of the collection of these ions. If the interaction between the atoms are modeled using some suitably chosen interatomic potentials, then the total energy of the system, the forces acting on the ions, and the trajectories of ions required for carrying out statistical averages are straightforward to compute.

However, the presence of interacting electrons complicates the situation because during the motion of ions the accompanied electronic subsystem must be kept on the so-called *Born–Oppenheimer surface*. The electronic degrees of freedom need to be treated quantum mechanically while the ionic degrees are to be treated classically. These two subsystems are weakly coupled. Fortunately, computationally tractable schemes are available that unify these two subsystems. The basic ingredients for carrying out such calculations are as follows:

1. The density-functional theory for treating interacting electronic system.
2. Classical molecular dynamics for the computation of trajectories of ions.

3. Born–Oppenheimer and Car–Parrinello schemes for combining the above two methods.
4. Tools for analyzing data generated by the molecular dynamical simulations.

18.2.1 Density Functional Theory

Density functional theory is the most preferred method in the electronic structure community. It represents a general approach for treating interacting many particle systems and has a simple conceptual framework. It is computationally tractable and has been successfully applied to a wide variety of systems. It is a natural extension of the Thomas–Fermi theory, where the basic idea was to express the energy of a many particle system as a functional of the one-particle density. The search for such a functional lacked a firm foundation until 1964, when Hohenberg and Kohn established two crucial theorems (Hohenberg and Kohn, 1964). For a thorough understanding, we refer the readers to the book by Parr and Yang (1989).

Consider a cluster consisting of electrons and ions. The Hamiltonian for such a system with N_a ions and N_e interacting electrons is given as

$$\hat{H} = \frac{-\hbar^2}{2m_e} \sum_i^{N_e} \nabla_i^2 + \sum_I^{N_a} \sum_i^{N_e} \frac{Z_I e^2}{|\mathbf{R}_I - \mathbf{r}_i|} + \sum_i^{N_e-1} \sum_{j>i}^{N_e} \frac{e^2}{|\mathbf{r}_i - \mathbf{r}_j|}$$
$$+ \sum_I^{N_a-1} \sum_{J>I}^{N_a} \frac{Z_I Z_J e^2}{|\mathbf{R}_I - \mathbf{R}_J|} \tag{18.1}$$

The first term is the kinetic energy operator of electrons, $\hat{T}$; the second term is the external potential generated by a collection of nuclei, $V_{ext} = \Sigma_i V(\mathbf{r}_i)$; the third is the electron–electron Coulomb potential, V_c; and the fourth term is the nuclear–nuclear Coulomb potential, V_{II}. The associated eigenvalue problem is $\hat{H}\Psi = E\Psi$, where E is the total energy of the system given as the expectation value of the Hamiltonian $\hat{H}$; $E = \langle \hat{H} \rangle = \langle \hat{T} \rangle + \langle V_{ext} \rangle + \langle V_c \rangle + \langle V_{II} \rangle$, and Ψ is the many-body wave function, $\Psi \equiv \Psi(\mathbf{r}_1, \mathbf{r}_2, ..., \mathbf{r}_{N_e})$. Quite clearly this is a formidable problem to solve. We assume the Born–Oppenheimer approximation[*] and therefore focus on the electronic part keeping the nuclear coordinates fixed. However, we still land up with many-electron problem, which is intractable. This problem was reduced to an effective one-electron problem by Hohenberg, Kohn, and Sham. The essence of their formulation is to replace the interacting electron system by a non-interacting electron systems having the same charge density distribution. They represented the many-body wave function as a single Slater determinant from which the electronic charge density is obtained.

18.2.1.1 Hohenberg–Kohn–Sham Formulation

In 1964, Hohenberg and Kohn proved two theorems which laid the foundation for reducing the above many-body problem to an effective one-electron problem on rigorous ground (Hohenberg and Kohn, 1964). The first theorem shows that the external potential, $V(\mathbf{r})$, is uniquely determined by the ground-state electronic charge density.[†] The second theorem gives a variational principle. It asserts that the ground-state energy regarded as the functional of the charge density attains its minimum value with respect to variation in the charge density when the system is in its ground state, subject to the normalization condition. The result is truly remarkable. It means that the ground-state properties of an interacting electron system are completely determined by the charge density, a simple function of one variable that is real and positive definite. However, so far, nothing has been said about how to calculate this quantity and the functional form relating to the total energy.

Later, in 1965, Kohn and Sham (1965) presented a formulation leading to a practical implementation based on Hohenberg–Kohn theorems. They proposed an idea of replacing the kinetic energy functional $T[\rho]$ of interacting electrons with that of a non-interacting system having the same ρ. Their formulation led to a set of self-consistent equations now known as Kohn–Sham equations. This is achieved by introducing a set of variational orbitals into the problem in such a way that the kinetic energy can be computed simply to a good accuracy, leaving small residual correction that is handled separately. In this formulation, the total energy of a system with N_e electrons can be written as

$$E[\rho] = F[\rho] + \int V_{ion}(\mathbf{r})\rho(\mathbf{r})d\mathbf{r} \tag{18.2}$$

with

$$F[\rho] = T[\rho] + E_c[\rho] + E_{xc}[\rho] \tag{18.3}$$

where

V_{ion} is the external potential due to ionic cores
$T[\rho]$ is the kinetic energy of the interacting electron having charge density $\rho(\mathbf{r})$
E_c is the Coulomb energy of electrons
E_{xc} contains all the omitted effects, i.e., the exchange and correlation effects

Apart from the ill-defined exchange–correlation energy, it is the kinetic energy functional that is unknown. Kohn–Sham

---

[*] Due to large difference in the masses of nuclei and electrons, their velocities differ by orders of magnitude. The electrons travel much faster than the nuclei and hence it is assumed that the electrons instantaneously follow the motion of the nuclei. This allows us to separate the electronic and nuclear degrees of freedom and treat them independently. This approximation is called as the Born–Oppenheimer approximation. Further, the nuclear degrees of freedom are treated classically while the electronic degrees of freedom are treated quantum mechanically.

[†] The single-particle electron density $\rho(\mathbf{r})$ is defined as $\rho(\mathbf{r}_1) = N_e \int ... \int |\Psi(\mathbf{r}_2, \mathbf{r}_3, ..., \mathbf{r}_{N_e})|^2 d^3\mathbf{r}_2 d^3\mathbf{r}_3 ..., d^3\mathbf{r}_{N_e}$, where N_e is the total number of electrons in the system. The integration is carried out on all $\mathbf{r}_i$s for $i = 2$ to N_e.

replaced this exact kinetic energy by that of non-interacting electrons having the same charge density and redefined $E_{xc}[\rho]$ as

$$E_{xc}[\rho] = E_{xc}^{exact}[\rho] + T[\rho] - T_s[\rho] \qquad (18.4)$$

In other words, they incorporate the quantum correction into the exchange–correlation energy term. With this the problem reduced to solving the non-interacting one-electron equations. Thus, the Kohn–Sham total energy functional based on density-functional theory, where each electron moves in an effective field due to the rest of the electrons, can be written as a functional of the single particle wave functions ψ_i, and ionic coordinates R_I. This energy functional is given as,

$$E\left[\{\psi_i\}, \{\mathbf{R}_I\}\right] = T_s\left[\{\psi_i\}\right] + E_c[\rho] + E_{ext}\left[\rho, \{\mathbf{R}_I\}\right] + E_{xc}[\rho] + E_{ion} \qquad (18.5)$$

where $\rho(\mathbf{r})$ is the electronic charge density given by

$$\rho(\mathbf{r}) = \sum_i^{occ} n_i \, |\, \psi_i(\mathbf{r})\,|^2 \qquad (18.6)$$

where n_i is the occupancy of the ith eigenstate, and the sum is over all occupied states.

In Equation 18.5, the first term represents the kinetic energy,

$$T_s = \sum_i n_i \int \psi_i^* \left[-\frac{\nabla^2}{2}\right] \psi_i d\mathbf{r} \qquad (18.7)$$

where T_s is exact only for the system of non-interacting electrons.

The second term is the classical coulomb energy contribution,

$$E_c = \frac{1}{2} \iint \frac{\rho(\mathbf{r})\rho(\mathbf{r}')}{|\mathbf{r} - \mathbf{r}'|} \, d\mathbf{r}\, d\mathbf{r}' \qquad (18.8)$$

The third term is the electron-ion interaction energy,

$$E_{ext} = \int V_{ion}(\mathbf{r})\rho(\mathbf{r})d\mathbf{r} \qquad (18.9)$$

The fourth term E_{xc} represents the contribution due to the exchange–correlation energy. The exact form of the exchange and correlation energies is unknown for almost all systems of interest with the exception of a few idealized models. Hence, approximations are necessary to evaluate E_{xc}. One of the most widely used approximation for the exchange-correlation functional is the *local density approximation* (LDA). Within the LDA, an approximate parameterized form of the exchange–correlation energy functional for an inhomogeneous electron gas is constructed from the knowledge of exchange–correlation energy of a homogeneous electron gas. Thus, E_{xc} is given by

$$E_{xc}^{LDA} = \int \rho(\mathbf{r})\epsilon(\rho)d\mathbf{r} \qquad (18.10)$$

where $\epsilon(\rho)$ is the exchange–correlation energy per particle of a uniform electron gas of density ρ. This approximation has been by and large successful in describing ground-state properties of systems bound by metallic, covalent, or ionic interactions; however, there a few failures too. Considerable progress has been made into improving the exchange–correlation energies by going beyond LDA. A class of potentials based on *generalized gradient approximation* (GGA) are quite popular. The E_{xc} within the GGA is given by

$$E_{xc}^{GGA} = E_{xc}^{LDA} + \int \rho(\mathbf{r})f_{xc}\left[\rho(\mathbf{r}), \nabla\rho(\mathbf{r})\right]d\mathbf{r} \qquad (18.11)$$

The choice of f_{xc} is not unique, and several different approximations have been proposed in recent years (Becke, 1996; Lee et al., 2005b; Perdew et al., 1996a,b, 1997).

The last term E_{ion} is the coulomb energy contribution from interactions among the ions:

$$E_{ion} = \sum_I \sum_{J \neq I} \frac{Z_I Z_J}{|\mathbf{R}_I - \mathbf{R}_J|} \qquad (18.12)$$

where Z_I denotes the nuclear charge on the Ith nuclei. This term contributes a constant to the total energy for fixed ionic positions but is required to be evaluated when ion dynamics is incorporated. At the minimum, the Kohn–Sham energy functional is equal to the ground-state energy of the system of electrons and ions. It is necessary to determine the set of functions ψ_i that minimize the Kohn–Sham energy functional. By following variational procedure, we get a set of equations known as *Kohn–Sham* equations:

$$\left[-\frac{\nabla^2}{2} + V_{eff}(\mathbf{r})\right]\psi_i(\mathbf{r}) = \epsilon_i\psi_i(\mathbf{r}) \qquad (18.13)$$

where ψ_i is the wave function of the electronic state i and ϵ_i is the Kohn–Sham eigenvalue, The effective potential V_{eff} is given by

$$V_{eff}(\mathbf{r}) = V_c(\mathbf{r}) + V_{ion}(\mathbf{r}) + V_{xc}(\mathbf{r}) \qquad (18.14)$$

where V_c is the coulomb potential given by

$$V_c(\mathbf{r}) = \int \frac{\rho(\mathbf{r}')}{|\mathbf{r} - \mathbf{r}'|}d\mathbf{r}' \qquad (18.15)$$

where $\rho(\mathbf{r})$ is given by Equation 18.6 and the exchange–correlation potential V_{xc} is given by

$$V_{xc}(\mathbf{r}) = \frac{\delta E_{xc}\left[\rho(\mathbf{r})\right]}{\delta\rho(\mathbf{r})} \qquad (18.16)$$

The above set of equations must be solved self-consistently. What does it mean in practice? Let us note that the Equation 18.13 is a single particle equation and in principle can always be solved if the effective potential is known. But to know V_{eff}, we need to know the charge density and that means the occupied set of orbitals ψ_i, which are the solutions of Kohn–Sham equations. Therefore, we assume some trial charge density and generate V_{eff}. Then, we solve the Kohn–Sham equations to get set $\{\epsilon_i, \psi_i\}$ and regenerate $\rho(\mathbf{r})$. If this output charge density is the same as the input charge density, then we say that *self-consistency* is achieved. Obviously, we need to iterate this process quite a few times to get the self-consistent solutions.

We must also mention that set $\{\epsilon_i, \psi_i\}$ has been introduced as parameters to get the equations. They have no physical interpretation, except in a special case (see Janak's theorem in Parr and Yang (1989)). In particular, ϵ_i are not the excitation energies of the ith single particle states. The only physical quantities are the total energy E and the charge density $\rho(\mathbf{r})$. In spite of this, all the band structure calculations do use these eigenvalues as dispersion relation or bands. In fact, it works quite well. It is possible to use the Kohn–Sham equation to rewrite the total energy in terms of eigenvalues values ϵ_i. Then, the total energy is given by

$$E[\rho] = \sum_i^{\text{occ}} n_i \epsilon_i - \frac{1}{2} \iint \frac{\rho(\mathbf{r})\rho(\mathbf{r}')}{|\mathbf{r} - \mathbf{r}'|} d\mathbf{r} d\mathbf{r}' + E_{\text{xc}} - \int v_{\text{xc}}(\mathbf{r})\rho(\mathbf{r}) d\mathbf{r} \quad (18.17)$$

This alternative form that does not involve the external potential is some times computationally more useful. Thus, the basic steps for obtaining the ground-state energy of an electron system in the Kohn–Sham formalism are as follows:

1. Starting with some trial potential or charge density, solve Equation 18.13 with appropriate boundary conditions and get ϵ_i and ψ_i for all the occupied states.
2. Calculate the charge density using Equation 18.6 and generate Hartree potential and exchange–correlation potential. The Hartree potential involves the solution of Poisson equation that may not be trivial.
3. Calculate total energy and effective potential, and iterate till convergence is obtained for say total energy, charge density, etc.

The technical complexity of the implementation of the above is dependent on a number of factors, e.g., system size, dimensionality (one, two, or three), boundary conditions, numerical schemes used, the type of basis function, etc.

It must be mentioned that the modern methods of the electronic structure, developed by Car and Parrinello, changed the complexion of the methodology. A host of new techniques using minimization are now routinely used. One of the most significant achievements is the ability to carry out ion dynamics. Before we make more comments on *ab initio* molecular dynamical methods, it is appropriate to introduce standard methods for carrying out molecular dynamics.

18.3 Molecular Dynamics

Consider a system of N interacting particles. The systems could be gases, liquids, solids, or surfaces. Let us restrict ourselves to a simple model consisting of structureless particles representing atoms. Our objective is to study the finite temperature properties of such a system using a suitable ensemble. Since we wish to simulate the behavior of the system as faithfully as possible, we will carry out atomistic dynamics, i.e., obtaining the trajectories of all the atoms evolving in time, under appropriate laws of motion with given physical conditions. We assume the particles to be classical and hence use Newton's laws to describe their motion. The physical conditions depend on the nature of the investigation and will fix parameters such as temperature, number of particles, volume, pressure, etc. Thus, the idea is very simple. Set up the physical conditions, use Newton's laws, and explore the phase space as completely as possible by recording the trajectories of all the particles. Then compute the required observables as trajectory averages. The ingredients of molecular dynamical simulations are as follows:

1. The nature of the interacting system; in the present case interacting electrons and ions.
2. The nature of the interactions, normally taken as a suitably chosen easy to evaluate two-body interaction form. Another way of treating the interactions accurately is to use *ab initio* methods, which we will describe at the end of this section.
3. The physical conditions such as temperature, volume, pressure, etc. The most common simulations are constant temperature, constant volume, and constant energy.
4. Visualization: Seeing is believing. It is a wonderful experience to visualize the dynamical behavior of the particles. Indeed, plots organize a large volume of data to bring out systematic trends and provide a lot more fun.

18.3.1 Ingredients

It is most convenient to illustrate the molecular dynamical calculations by using interatomic potentials. For the sake of simplicity, we assume a simple binary form that simplifies the calculation of total energy and the forces acting on the ions.

18.3.1.1 Total Potential Energy

We assume that the total potential energy of a system of N interacting classical particles can be written as a sum of binary interaction and chose a suitable form for the two-body interaction potential. The total potential energy (in fact it is the total binding energy) is given by

$$V(\mathbf{r}_i) = \sum_{i>j}^{N} v(r_{ij}) \quad (18.18)$$

where $r_{ij} = |\mathbf{r}_i - \mathbf{r}_j|$, which can be taken as the Lennard Jones potential,

$$v(r) = 4\epsilon \left[\left(\frac{\sigma}{r} \right)^{12} - \left(\frac{\sigma}{r} \right)^{6} \right] \qquad (18.19)$$

The first term represents the repulsive part due to the atomic cores and the second term is the long-range attractive van der Waals interaction. As is well known, this is a suitable form for representing the interactions between inert gas atoms. Note that ϵ and σ are having dimensions of energy and length respectively. These are different for different types of atoms and are obtained by suitable fitting methods. It is easy to see that $2^{1/6}\sigma$ is the dimer bond length with $-\epsilon$ as the binding energy for a two atom system.

In order to carry out dynamics, we need forces on each of the atoms. The x–component of the force on ith atom is given by

$$F_i^x = -\frac{\partial V}{\partial x_i} = \sum_{j \neq i}^{N} f_{ij}^x \qquad (18.20)$$

with

$$f_{ij}^x = \left(\frac{48\epsilon}{\sigma^2} \right) \left[\left(\frac{\sigma}{r_{ij}} \right)^{14} - \left(\frac{\sigma}{r_{ij}} \right)^{8} \right] \qquad (18.21)$$

These are $3N$ equations of motion for N particles in three-dimension.

18.3.1.2 Simulating Atomic Motion: Calculation of Trajectories

The next important ingredient is the calculation of trajectories, which must be obtained by solving equations of motion. Given the initial conditions for N atoms, namely, the coordinates and the velocities (at time $t = 0$), the trajectories can be obtained by solving Newton's laws. The solution is sought by numerical methods. The time is discretized and the successive positions as a function of time (at discrete time intervals) are obtained by a suitable algorithm. The commonly used algorithm, which turns out to be surprisingly accurate, is the Verlet algorithm. It is instructive to derive the Verlet equations by using good old Taylor series expansion.

Expanding the x–coordinate for ith atom around time t, we get

$$x_i(t + dt) = x_i(t) + \dot{x}_i dt + \ddot{x}_i \frac{dt^2}{2} + \dddot{x}_i \frac{dt^3}{3!} + O(dt^4) \qquad (18.22)$$

Similarly, for $-dt$,

$$x_i(t - dt) = x_i(t) - \dot{x}_i dt + \ddot{x}_i \frac{dt^2}{2} - \dddot{x}_i \frac{dt^3}{3!} + O(dt^4) \qquad (18.23)$$

Adding these two equations we get

$$x_i(t + dt) = 2x_i(t) - x_i(t - dt) + \frac{F_i^x}{m_i} dt^2 \qquad (18.24)$$

where we have used $F_i^x = m_i a_i^x = m_i \ddot{x}_i$.

The algorithm is very simple. If we know the positions at the last two time steps, then we obtain the coordinates for the next step and thus generate the trajectories for sufficiently long time.

A few comments are in order:

1. The error in the calculation of the coordinates is $O(dt^4)$.
2. The Verlet equations do not involve velocities explicitly and hence velocities have to be calculated from available positions using

$$v(t) = \frac{v(t + dt) - v(t - dt)}{2dt} \qquad (18.25)$$

This means that we need position at next time step to get velocity at the current time step.
3. The algorithm is simple to program, needs very little memory, and has very nice desirable properties. First, it has a time reversal symmetry; second, it shows excellent energy conservation; and third, it is area preserving.
4. In spite of this, there are always limitations due to the truncation errors. After all we are solving the second-order equations approximately with finite precision. We should be cautious in choosing time step, dt, and the number of iteration that sets a limit of the total simulation time. A drift in the energy must be avoided or at least minimized to an acceptable level.

A popular form of Verlet algorithm is velocity Verlet algorithm that we now state. This algorithm involves explicit use of velocities in computing the coordinates at the next iteration. The Velocity Verlet equations are given by

$$x_i(t + dt) = x_i(t) + v_i^x(t)dt + \frac{F_i^x(t)}{2m_i} dt^2 \qquad (18.26)$$

$$v_i(t + dt) = v_i^x(t) = \frac{F_i^x(t + dt) + F_i^x(t)}{2m_i} dt \qquad (18.27)$$

Note that we have incorporated the steps for calculating the total kinetic energy and the total potential energy.

18.3.1.3 Ensemble

There are two commonly used ensembles *viz.* microcanonical and canonical ensembles. The choice of the ensemble is dependent on the nature of the physical system and the problem at hand.

- Constant energy simulations—microcanonical ensemble

We can maintain the cluster at constant energy by not allowing it to interact with the outside world, i.e., keeping the system isolated. For example, we can start by having an arbitrary set of positions of the atoms. The cluster at time $t = 0$ will have some potential energy (nonzero—unless we are lucky to generate equilibrium position). If we let the position evolve according to Verlet equations with zero initial velocities, then at each instant of time the total energy $E = T + V$ will be conserved. Another way of simulating the isolated system, especially when it is in an equilibrium position, is to set up random initial velocities derived from the Maxwell–Boltzmann distribution. This will set the initial kinetic energy.

- Constant temperature simulation—canonical ensemble

This is a common situation, where in addition to the number of particles and volume the temperature is kept constant. In the present case of atomic clusters, we do not have any volume parameter as clusters need no enclosure. But in most of the simulations, a suitable box representing the volume of the system is required, along with the specified boundary conditions. The temperature can be calculated from the total kinetic energy using equipartition theorem:

$$T = \frac{1}{2} k_{\mathrm{B}} T \times f = \sum_i \frac{1}{2} m_i v_i^2 \qquad (18.28)$$

where f is the number of degrees of freedom. In our case, $f = 3N - 6$, since we conserve the linear momentum and the angular momentum; both set to be zero at time $t = 0$.

For the present problem, we are required to carry out three processes, namely, heating, cooling, and maintaining the system at a given temperature. We introduce a rather simple method to control the system temperature: the velocity scaling method. At any time instant t, let T_c be the current temperature and T_r the required one. We can raise or lower the temperature by scaling all the velocities uniformly by

$$v_i^r = v_i^c \sqrt{\frac{T_r}{T_c}} \qquad (18.29)$$

where v_i^c is the current velocity. The new scaled velocities v_i^r are then used in equation (Equation 18.26) for propagating the system. Of course, the scaling is carried out for all the three components of velocities of all the atoms.

We would like to note that this velocity scaling is a crude way of mimicking a heat bath. More sophisticated baths such as Nosé thermostat are available and should used if a true canonical distribution is warranted. We note some excellent books on this topic (Allen and Tildesley, 1987; Frenkel and Smit, 1996; Rapaport, 1998).

18.3.2 *Ab Initio* Molecular Dynamics: Marrying DFT and MD

It is clear from this brief introduction that the total energy and the forces on ions are crucial for an accurate computation of trajectories. For a parameter-free computation, it is necessary to solve the electronic structure problem during the time evolution of the ionic trajectory. There are two computationally tractable schemes. The first one is the well-known Car and Parrinello method, which uses a modified Lagrangian to incorporate the electronic degrees of freedom. The dynamics of electron is fictitious. The ions and the electronic degrees of freedom are evolved simultaneously. The method has revolutionized the way electronic structure calculations were carried out. The second method is known as the Born–Oppenheimer dynamics. At time $t = 0$, the electrons are assumed to be on the Born–Oppenheimer surface, i.e., in the instantaneous eigenstates of the Kohn–Sham Hamiltonian, with the external potential provided by the ions. Then the total energy is calculated within the density-functional theory, from which the forces on the ions can be evaluated as $-\partial E / \partial R_I$. This calculation is then followed by a molecular dynamical move for the ions. This move changes the external potential generated by the ions, leading to the effective potential. Now the electrons are no longer in the instantaneous eigenstate of the new Kohn–Sham Hamiltonian. Modern minimization techniques such as conjugate gradient methods may be used to bring the electrons back into the instantaneous eigenstate, i.e., on the Born–Oppenheimer surface. This procedure of moving the ions classically and bringing the electrons back on the Born–Oppenheimer surface constitutes one step of *ab initio* Born–Oppenheimer molecular dynamics. This procedure when iterated for a long time, i.e., for many steps, generates ionic trajectories, which can be used to extract the thermodynamic quantities such as density of states, entropy, specific heat, etc. Many sophisticated techniques have been successfully developed to make this procedure more efficient. It is now possible to carry out the dynamics of up to a few hundred electron system for a simulation time of the order of up to about 100 ps per trajectory[*] with modest computational power. Quite clearly, the Born–Oppenheimer dynamics is much more general than the Car–Parrinello dynamics. A full exposition of total energy and force methods is beyond the scope of this chapter. A number of excellent articles and books are available for interested readers (Kohanoff, 2006; Martin, 2004; Payne et al., 1992).

Having computed the trajectories, it is necessary to compute the relevant thermodynamic quantities. In the next section, we discuss the computation of some of these quantities. Apart from conventional ones, we also present a very useful technique to extract entropy and specific heat, i.e., the multiple-histogram method.

18.4 Data Analysis Tools

In this section, we present the basic tools of analyzing the finite temperature trajectory data that can probe the phase transformation. Since we are going to deal with the statistical mechanics, let us recall some of the basic terms. There are two commonly used ensembles in statistical mechanics: the constant energy (microcanonical) ensemble and the constant temperature (canonical) ensemble. Generally, clusters are simulated in free space. This action leads to zero pressure on it. In simulations, the interacting atoms are allowed to evolve in time, in principle exactly—in practice using Verlet or similar suitable algorithm, in order

---

[*] 100 ps corresponds to 40,000 molecular dynamical steps with a time step of 2.5 fs.

to span a reasonable region in the phase space. Then, the averages over the trajectories yield the required statistical averages. In the present case, it is more profitable to extract the entropies via the multiple-histogram method. This method differs slightly depending upon the nature of the ensemble. We will illustrate the method for canonical ensemble. In addition, we also discuss the computation of a number of other traditional parameters commonly used such as the root-mean-square bond-length-fluctuations, mean-squared displacements, etc. We begin with a mathematical function, the electron localization function, which is convenient to describe the nature of bonding in clusters.

18.4.1 Electron Localization Function

The electron localization function is not a thermodynamical indicator. However, it has been extremely useful in elucidating the nature of bonding between atoms in a cluster. This is especially useful since it has been observed that the nature of bonding in atomic clusters can be very different from that in their bulk counterpart. Further, as noted earlier, each atom in a cluster can be bonded to other atoms with different bond strengths. Analysis of electron localization function is useful in bringing out these differences in the bonding between different atoms or a group of atoms. Since the nature of bonding in clusters can affect their finite temperature properties, it may be necessary to carry out such analysis.

The electron localization function was originally introduced by Becke and Edgecombe as a simple measure of electron localization in atomic and molecular systems (Becke and Edgecombe, 1990). However, their definition was electron spin dependent. Silvi and Savin later generalized this function for any density independent of the spin (Silvi and Savin, 1994). According to their definition, for a single determinantal wave function built from, say Kohn–Sham orbitals ψ_i, the electron localization function is given by

$$\chi_{\text{ELF}}(\mathbf{r}) = \left[1 + \left(\frac{D}{D_{\text{h}}} \right)^2 \right]^{-1} \qquad (18.30)$$

where

$$D_{\text{h}} = \frac{3}{10} (3\pi^2)^{5/3} \rho^{5/3}(\mathbf{r}) \quad \text{and} \quad D = \frac{1}{2} \sum_i \left| \nabla \psi_i(\mathbf{r}) \right|^2 - \frac{1}{8} \frac{\left| \nabla \rho(\mathbf{r}) \right|^2}{\rho(\mathbf{r})} \qquad (18.31)$$

where
- $\rho(\mathbf{r})$ is the valence electron density
- D is the excess local kinetic energy density due to Pauli repulsion
- D_{h} is the Thomas–Fermi kinetic energy density for homogeneous electron gas

The numerical values of χ_{ELF} are conveniently normalized to a value between zero and unity. A value of 1 represents a perfect localization of the valence charge, while the value for uniform electron gas is 0.5. Typically, the existence of an isosurface in the

region between two atoms at a high value of χ_{ELF}, say ≥ 0.7, signifies a localized bond in that region.

Silvi and Savin also proposed a topological classification and rationalization of the electron localization function, which helps in giving the quantification of the chemical concepts associated to the function. According to their description, the molecular space is partitioned into regions or basins of localized electron pairs. At very low values of electron localization function, all the basins are connected (disynaptic basins). In other words, there is a single basin containing all the atoms. As the value of χ_{ELF} is increased, the basins begin to split, and finally, we will have as many basins as the number of atoms. The value of electron localization function at which the basins split (a disynaptic basin splits into two monosynaptic basins) is a measure of interaction between the different basins (i.e., a measure of the electron localization).

18.4.2 Traditional Indicators of Melting

It is not easy to give a precise definition of melting in a finite-sized system such as atomic cluster, which does not undergo a proper phase transition in the (discontinuous or singular) sense seen in infinite systems. Even the notion of solid state and liquid state is a little vague. It is preferable to use the terms *solidlike* and *liquidlike*. Roughly speaking, by *solidlike*, we mean that the constituent ions of the cluster vibrate about fixed points. By *liquidlike*, we mean that the ions undergo a diffusive motion, exploring the entire volume of the cluster; there is permutational equivalence between the ions. To make this discussion more rigorous, one should introduce the time scale on which such behavior is to be observed, which depends in turn on the conditions of the experiment or simulation. In a general sense, the melting of a cluster is a process by which the cluster goes from a *solidlike* state to a *liquidlike* state as the cluster is heated.

In the following, we shall discuss a number of indicators that have been used traditionally in simulations to investigate melting behavior. The multiple-histogram method for extracting the density of states and thermodynamic averages will be discussed later in Section 18.4.3.

18.4.2.1 The Caloric Curve

A direct indicator of melting is the plot of the internal energy with respect to temperature (in the ensemble of interest). For bulk systems, this curve exhibits a discontinuity at the transition temperature, and the difference in the internal energy at the temperature at which the discontinuity occurs is the latent heat of melting of the system. The derivative of the internal energy with respect to the temperature is the specific heat and shows a δ function behavior for bulk systems at the transition. In contrast, for finite-sized systems, the caloric curve does not show a sharp discontinuity.

18.4.2.2 Root Mean Square Bond-Length Fluctuations—Lindemann Criterion

As thermal energy is supplied to a cluster, the average amplitude of vibrations increases and in a *liquidlike* state particles diffuse.

The Lindemann criterion asserts that a system may no longer be considered to be solid if the average bond-length fluctuations exceed about 10% to 15% of their ground-state value. The average bond-length fluctuation for each atom in a cluster with N_a ions is defined as

$$\delta_{rms}^I = \frac{1}{N_a - 1} \sum_{J \neq I}^{N_a} \frac{\sqrt{\langle r_{IJ}^2 \rangle_t - \langle r_{IJ} \rangle_t^2}}{\langle r_{IJ} \rangle_t} \tag{18.32}$$

where

$\langle \ldots \rangle_t$ denotes either a time average (microcanonical ensemble) or a thermal average (canonical ensemble)

r_{IJ} is the distance between ions I and J

18.4.2.3 Mean-Squared Displacement and Diffusion Coefficient

The mean-squared displacement differs from the bond-length fluctuation in that one considers the motion of a single particle over time and averages over all the particles. It attempts to capture the onset of a diffusive, *liquidlike* state. The mean-squared displacement is given by

$$\langle \mathbf{r}^2(t) \rangle = \frac{1}{N_a M} \sum_{m=1}^{M} \sum_{I=1}^{N_a} \left[\mathbf{R}_I(t_{0m} + t) - \mathbf{R}_I(t_{0m}) \right]^2 \tag{18.33}$$

Here t represents a time delay, and the average is performed over the N_a ions and also over M time origins t_{0m} taken at regular intervals throughout a molecular dynamics simulation. In a *solidlike* state, $\langle \mathbf{r}^2(t) \rangle$ reaches a plateau on a time scale of order the vibrational frequency with a value of order the mean amplitude squared of the ionic vibration. In a *liquidlike* state, $\langle \mathbf{r}^2(t) \rangle$ displays a linear portion corresponding to diffusive motion with diffusion constant:

$$D = \frac{1}{6} \frac{d}{dt} \langle \mathbf{r}^2(t) \rangle \tag{18.34}$$

18.4.2.4 Velocity Autocorrelation Function and Power Spectra

The velocity autocorrelation function is defined as

$$C(t) = \frac{\left\langle \left(v(t_0 + t) - \langle v \rangle \right) \cdot \left(v(t_0) - \langle v \rangle \right) \right\rangle}{\left\langle \left(v(t_0) - \langle v \rangle \right)^2 \right\rangle} \tag{18.35}$$

where

v is a component of velocity of a given ion

t is a time delay

t_0 is a time origin

One may also average $C(t)$ over ions and different values of the time origin, as in Equation 18.33, as well as over x-, y-, and z-components.

The power (or phonon) spectrum is

$$C(\omega) = 2 \int_0^\infty C(t) \cos(\omega t) dt \tag{18.36}$$

A nonzero value at the zero frequency of the power spectrum results when the motion has a diffusive character.

18.4.2.5 Shape Analysis of the Density

Yet another possible indicator has been suggested by the work of Rytkönen et al. (1998) on the 40-atom sodium cluster. They analyzed the shapes of the density before and after the melting transition using a dimensionless shape parameter defined as

$$S_l = \sum_{m=-l}^{l} |a_{lm}|^2 = \sum_{m=-l}^{l} \frac{4\pi(2l+1)}{9r_s^{2l} N^{2l/3+2}} |Q_{lm}|^2 \tag{18.37}$$

$$Q_{lm} = \sqrt{\frac{4\pi}{2l+1}} \int d^3 r\, r^l Y_{lm}(\theta, \phi) \rho(\mathbf{r}) \tag{18.38}$$

where

Q_{lm} are the multipole moments

Y_{lm} is a spherical harmonic

$\rho(\mathbf{r})$ is the electronic density

$r_s(\mathbf{r}) = [4\pi\rho(\mathbf{r})/3]^{-1/3}$ is the Wigner–Seitz density parameter of the electron gas

The shape parameter S_l as defined by Equation 18.37 integrates out all the possible values of m and is rotationally invariant. This indicator is especially interesting since it can use the electron density—a quantity that is available through *ab initio* approaches.

18.4.2.6 Quadrupole Deformation

It has been observed in experiments that upon melting, the shape of certain clusters such as Sn_N changes from prolate to spherical. This structural deformation was measured by the change in diffusion coefficient in ion mobility experiments by Shvartsburg (Shvartsburg et al., 1998). Such a change in the geometric shape of the cluster can be analyzed more conveniently using the quadrupole deformation parameter defined below:

$$\epsilon_{def} = \frac{2Q_1}{Q_2 + Q_3} \tag{18.39}$$

where $Q_1 \geq Q_2 \geq Q_3$ are the eigenvalues of the quadrupole tensor

$$Q_{ij} = \sum_I R_{Ii} R_{Ij} \tag{18.40}$$

Here

i and j run from 1 to 3

I runs over the number of ions

R_{Ii} is the ith coordinate of ion I relative to the cluster center of mass

18.4.3 Multiple Histogram Method

All the above quantities used to characterize the melting transition have been motivated phenomenologically. It is desirable to compute thermodynamic indicators such as the entropy and the specific heat to characterize the melting transition completely. In principle, an observation of the thermodynamic properties of a system should involve a set of simulations over a range of temperatures that are closely spaced. The need to maintain each temperature for long times, coupled with the high cost of first-principles simulations means that the overall simulations can be very expensive. We need a technique that allows us to reduce the number of temperatures being simulated and to reliably interpolate the behavior of the system at temperatures in between.

The multiple-histogram method is a technique that permits a better estimation of the classical density of states and also makes reliable interpolations possible. It was developed independently by Bichara et al. (1987), by Ferrenberg and Swendsen (1988), and in the context of clusters by Labastie and Whetten (1990). It was applied initially to data collected by Monte Carlo simulations. What follows is a brief description of the technique with its theoretical basis.

18.4.3.1 Idea of the Multiple Histogram Method: Canonical Ensemble

Given a finite number of data sets of some physical quantity, e.g., N_i potential energies per temperature and τ temperatures, we would like to interpolate reliably the values over a range of temperatures including the ones for which simulations have not been performed. The contribution of a point in the phase space of a system to the statistical quantity being observed is, in the canonical ensemble, proportional to the Boltzmann factor at that temperature. Hence, given a data point from the simulations of one temperature point, its contribution at another temperature would be proportional to $\exp(\beta(E - E_0))$. Another advantage of the multiple-histogram method in the context of cluster simulations is that it permits a separate treatment of the configurational and kinetic parts of the problem. This separation is desirable because the kinetic part of the problem can be handled analytically; a numerical sampling of the phase space is only required for the configuration space.

We assume that the Hamiltonian of the cluster is separable, $H(R, P) = V(R) + K(P)$, where $R \equiv \{\mathbf{R}_I\}$ are the ionic coordinates, $P \equiv \{\mathbf{P}_I\}$ are the ionic momenta, $V(R)$ is the potential-energy surface, and $K(P) = \Sigma_I P_I^2/(2M_I)$ is the kinetic energy. The classical density of states can be expressed as a convolution integral:

$$\Omega(E) = \int_{V_0}^{E} \Omega_\mathrm{C}(W)\Omega_\mathrm{K}(E - W)\,dW \qquad (18.41)$$

where

V_0 is the global minimum of the potential-energy surface

$\Omega_\mathrm{C}(E)$ and $\Omega_\mathrm{K}(E)$ are the so-called configurational and kinetic densities of states, respectively,

$$\Omega_\mathrm{C}(E) = \int \delta(E - V(R))\,d\mathbf{R} \qquad (18.42)$$

$$\Omega_\mathrm{K}(E) = \int \delta(E - K(P))\,d\mathbf{P} \qquad (18.43)$$

Now, $\Omega_\mathrm{K}(E - W)$ is known analytically (Kittel, 1958) as $\Omega_\mathrm{K}(E-W) \sim (E - W)^{\nu/2-1}$ (neglecting unimportant constant factors), where ν is the number of independent degrees of freedom of the system. Thus, only the configurational density of states $\Omega_\mathrm{C}(E)$ is required. This can be extracted from the potential-energy values obtained from the simulations. We need not assume here that the sampling simulations follow a canonical distribution in kinetic space, but we will assume that they are canonical in configuration space, that is, that the probability for finding a potential energy V at a temperature T is given by

$$p(V, T) = \frac{\Omega_\mathrm{C}(V)\exp(-V/k_\mathrm{B}T)}{Z_\mathrm{C}(T)} \qquad (18.44)$$

where $Z_\mathrm{C}(T)$ is the configurational partition function required for normalization. The configurational density of states $\Omega_\mathrm{C}(E)$ is extracted from the simulation by comparing the expected potential-energy distribution, Equation 18.44, with that obtained numerically in the sampling runs.

The first step in extracting $\Omega_\mathrm{C}(E)$ is to construct a histogram of the potential energy at each of the τ temperatures used for the sampling runs. For this purpose, the potential-energy scale, which ranges from V_0 to some maximum observed value $V_{\max}$, is divided into N_V intervals (or *bins*) of width $\delta V = (V_{\max} - V_0)/N_V$. The same bins should be used for all temperatures. We shall denote each temperature by an index i satisfying $1 \le i \le \tau$, and each bin by an index j satisfying $1 \le j \le N_V$, with V_j the central value of the potential energy in the jth bin.

Let n_{ij} be the number of times the potential energy assumes a value lying in the jth bin at a temperature i. (In dynamical sampling methods, one would rather take n_{ij} to be the total time spent by the system in this bin.) Then, the probability that the system takes a potential energy in the jth bin at an inverse temperature $\beta_i = 1/(k_\mathrm{B}T_i)$ is estimated from the simulation as

$$p_{ij}^{\mathrm{sim}} = \frac{n_{ij}}{\displaystyle\sum_j n_{ij}} \qquad (18.45)$$

On the other hand, the theoretical probability is obtained from Equation 18.44 as

$$p_{ij}^{\mathrm{theo}} = p(V_j, T_i)\delta V = \Omega_\mathrm{C}(V_j)\delta V \frac{\exp(-\beta_i V_i)}{Z_\mathrm{C}(\beta_i)} \qquad (18.46)$$

Equating p_{ij}^{sim} and p_{ij}^{theo} and taking logarithms yield

$$S_j + \alpha_i = \beta_i V_j + \ln p_{ij}^{\mathrm{sim}} \qquad (18.47)$$

where

$$S_j = \ln[\Omega_C(V_j)\delta V] \quad (18.48)$$

$$\alpha_i = -\ln Z_C(\beta_i) \quad (18.49)$$

Now, Equation 18.47 is a system of equations whose right-hand side is known and whose left-hand side is unknown. Since we have τN_V equations (i.e., pairs ij) but only $N_V + \tau$ unknowns (i.e., the S_j and the α_i), this system is overdetermined. We therefore solve it in a least-squares sense, regarding the τN_V different values of

$$W_{ij} \equiv \beta_i V_j + \ln p_{ij}^{sim} \quad (18.50)$$

in effect as "experimental" data to which we wish to fit the $N_V + \tau$ quantities S_j and α_i. Thus, from Equation 18.47, we are led to choose S_j and α_i such as to minimize

$$\chi = \sum_{ij} \frac{(W_{ij} - S_j - \alpha_i)^2}{(\delta W_{ij})^2} \quad (18.51)$$

where δW_{ij} is the statistical error in the quantity W_{ij}. From Equation 18.50, this error satisfies

$$\delta W_{ij} = \frac{\delta p_{ij}^{sim}}{p_{ij}^{sim}} \approx c n_{ij}^{-1/2} \quad (18.52)$$

where c is an unimportant constant. The second step in Equation 18.52 follows because the value n_{ij} of each bin of the histogram follows approximately a Poisson distribution. Putting together Equations 18.50 through 18.52, the final least-squares problem is thus to minimize

$$\chi = \sum_{ij} n_{ij}(\beta_i V_j + \ln p_{ij}^{sim} - S_j - \alpha_i)^2 \quad (18.53)$$

with respect to S_j and α_i, which requires solving the linear equations $\partial\chi/\partial S_j = 0$ and $\partial\chi/\partial\alpha_i = 0$ for S_j and α_i. The S_j give us the configurational density of states, Equation 18.48, while the α_i give us the configurational partition function, Equation 18.49.

We use the Gauss elimination method to solve these linear equations. Since the equations are overdetermined, the α_i that are obtained are relative to one of them. Further, our temperatures are ordered in ascending order, choosing $\alpha_0 = 0$ corresponds to the choice of entropy reference. The parameters α_i then enable us to compute the entropy S_j, the partition function $Z(\beta_i)$, the internal energy $U(T)$, and the configurational specific heat C_v as follows:

$$S_j = k_B \frac{\sum_{i=1}^{N_\tau}(\ln n_{ij} + \beta_i V_j - \alpha_i)}{\sum_{i=1}^{N_\tau}} \quad (18.54)$$

$$Z(T) = \sum_{j=1}^{N_v} \exp\left(S_j - \frac{V_j}{T}\right) \quad (18.55)$$

$$U(T) = \frac{2T(N-1)}{2} + \frac{1}{Z}\sum_{j=1}^{N_v} \exp\left(S_j - \frac{V_j}{T}\right)V_j \quad (18.56)$$

$$C_v = \frac{2T(N-1)}{2} + \frac{1}{T^2}\left(\langle V^2\rangle - \langle V\rangle^2\right) \quad (18.57)$$

where $\langle V\rangle = U(T)$. It may be noted that due to the use of least square fit, the resulting specific heat curves are smooth.

We have glossed over one small point in this derivation: the configurational partition function $Z_C(\beta)$ is by definition related to an integral over $\Omega_C(V)$, and thus the S_j and the α_i are not independent. However, enforcing the relation between S_j and α_i converts the linear least-squares problem, Equation 18.53, to a nonlinear one. It is therefore simpler in practice to treat the S_j and α_i as independent, as we have done above, and check that the partition function is consistent with the extracted density of states at the end. This invariably turns out to be the case for data that are reasonably statistically converged.

With $\Omega_C(E)$ in hand, we can now construct the full density of states $\Omega(E)$ from Equation 18.41. This in turn gives us access to a large range of thermodynamic averages for the system using the standard statistical mechanical relations (Kittel, 1958). Note, in particular, that it is possible to evaluate averages in a variety of ensembles, not just in the ensemble that was used for the sampling runs themselves.

18.5 Atomic Clusters at Finite Temperature

Atomic clusters, as noted earlier, are finite-sized systems that are restricted in all three directions. Such restrictions lead to their unusual properties including the finite temperature characteristics. Before we discuss the phenomenon of melting of clusters, let us describe in brief the process of melting in bulk systems.

18.5.1 Melting of Bulk Systems

Melting and freezing are commonly occurring phenomena. Many substances such as ice and metals transform from solid state to liquid state upon heating; i.e., they undergo melting. This phenomenon is commonly termed as "phase transition." A phase is an equilibrium thermodynamic state of a substance over a range of thermodynamic variables. The study of such phase transition has been one of the serious activities in the area of solid-state physics for the past several decades. It is also well known that such a change of phase occurs at a well-defined temperature. In the case of melting, this temperature is called the "melting temperature" of the substance. In other words this change takes place suddenly as the temperature increases through the melting temperature. Apart from solid–liquid transition, there are many other phase transitions such as order–disorder, magnetic–nonmagnetic, superconducting state–normal state, etc.

Let us examine the physical process when a homogeneous ordered solid, say a sodium metal, is heated. At low enough temperatures, the atoms exhibit small harmonic oscillations around their mean positions. As the temperature increases, the amplitude of these oscillations also increases. Essentially, the lattice executes harmonic vibrations. At low temperatures, it is possible to identify the space, i.e., the unit cell, to which the individual atoms belong. As the solid is heated further, the amplitude of such oscillations increases so much that the atoms begin to execute anharmonic motion and eventually the bonds between them start breaking. The individual atoms now no more belong to a particular "unit cell." This leads to a diffusive motion of the atoms and, at a critical temperature, both the solid and the liquid states coexist. At this stage, pumping in more energy does not increase the temperature, i.e., the kinetic energy, of the system. If the system is heated further, the solid gets completely transformed into a liquid state. This solid–liquid phase transition involves latent heat. The phase diagram of such a transition with respect to thermodynamic variables such as pressure, volume, and temperature for many solid–liquid–vapor transition has been extensively studied (Stanley, 1987).

The phase transitions are traditionally classified by Ehrenfest's classification, which, although faulty, is quite useful. According to this classification, if G is the Gibbs free energy, then its nth derivative with respect to temperature T is given as

$$G^n = \frac{\partial^n G}{\partial T^n}\bigg|_{P,V}$$

The nth order transition is characterized by discontinuous G^n, with all the lower derivatives being continuous. Solid–liquid transition is known to be a first-order transition, since

$$S = -\frac{\partial G}{\partial T}\bigg|_{P}$$

is discontinuous, where S is the entropy. The behavior of G and S is schematically depicted in Figure 18.5.

It may be noted that a more convenient description of the first-order transition is the one which involves latent heat. It is interesting to note that it is possible to define a parameter called an "order parameter," which is zero for one phase and changes to a nonzero value across the phase transition. A solid is a periodically ordered system having long-range correlations in density. The liquid does not have any long-range correlations. As the temperature of liquid is lowered, a (mass) density wave is set up. When the temperature is lowered below the phase transition temperature, the system gets locked into one of the modes of this density wave. Therefore, $\rho(\mathbf{G})$, where ρ is density and $\mathbf{G}$ the first reciprocal lattice vector, can be identified as an order parameter. Let us recall that Helmholtz free energy F is given as,

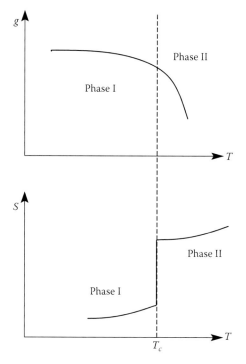

FIGURE 18.5 The schematic diagram of Gibb's free energy per particle ($g = G/N$) and entropy (S) of a N particle system as a function of temperature (T).

$F = U - TS$, U being the internal energy. At zero temperature, the phase of the system is determined by the internal energy U only. As temperature is increased entropy starts playing a significant role and at the phase transition there is a balance between the internal energy and the entropy contributions. Thus, the internal energy as well as the entropy plays a significant role in phase transitions. It may be noted that such transitions as solid–liquid are well understood by examining the caloric curve and the specific heat for the system under consideration as a function of temperature.

18.5.2 Melting of Atomic Clusters

Atomic clusters, the object of our interest, exhibit significantly different finite temperature characteristics as compared to the ordered extended systems. It may be noted that the finite size systems do not show sharp phase transitions. Hence, it is preferable to use the term "phase transformation" to describe any phase change seen in the cluster. It is of interest to ask the question: What happens when cluster is heated? Let us recall that every atom in a typical ordered homogeneous solid has a similar environment, i.e., to say, all the atoms are bonded to other atoms by similar strengths, the only exception being the surface atoms. Therefore, it is natural to expect all the atoms to respond similarly when a solid is heated (with the exception of the surface atoms). However, in the case of clusters, especially for small sizes with number of atoms up to about $N \sim 100$, the environment of each atom could be different. Each atom may be bonded to other atoms with different strengths. Therefore, we expect their dynamical response to heating to be different. Further, in a cluster the ratio of surface atoms to the total number is quite large compared to that of the extended systems. Therefore, the melting process is not expected to exhibit sharp phase transformation.

Clusters differ in another aspect from solids. They come in different shapes and sizes. In fact, clusters in this sense have individuality. Their finite temperature properties, e.g., shapes of specific heat curves and melting points, depend on the size. It turns out that in many cases, as we shall see, the finite temperature behavior is extremely size sensitive where even one atom makes a dramatic effect. We will examine two parameters: the melting temperatures and the shapes of the specific heat curves. The melting temperature of a cluster is taken to be the temperature corresponding to the peak in the specific heat curve. It is obvious that a cluster will have a number of isomers extending to high energies. The distribution of energies of these isomers affects the nature of specific heat. The behavior below melting temperature could be significantly affected by isomerization process (Bixon and Jortner, 1989). Apart from these parameters, issues such as surface melting, pre-melting, and impurity-induced effects need to be considered.

The response of a cluster to heating depends on the nature of bonding. A question of interest is: "Is the bonding in a cluster similar to that seen in its bulk counterpart?" A priori, there is no reason to believe that the nature of the bonding will change. However, as we shall see there are at least two cases experimentally established where the nature of bonding changes significantly leading to rather interesting phenomena, higher than the bulk melting temperatures for clusters in certain size range.

Since there is no sharp phase transition in finite size systems, it is common to describe a state of a cluster as *solidlike* and *liquidlike*. Although phase transformation in small systems is gradual and not sharp, as pointed out by Berry et al., there is some precision and a certain kind of sharpness to these changes (Berry, 1997). At sufficiently low temperatures, the clusters can be described as *solidlike*. Many clusters exhibit well-ordered geometries that are not necessarily compatible with periodic geometries. It is possible to identify a temperature T_f below which the cluster is *solidlike*. Between T_f and T_m, i.e., the melting temperature of the cluster, the *solidlike* and *liquidlike* phases coexist. And above T_m, it is clearly in the *liquidlike* state. Considerable amount of work has been carried out to elucidate many of the points raised above and have been discussed in a lucid article by Berry (1997).

The two common tools used were classical molecular dynamics and Monte Carlo techniques using canonical or microcanonical ensemble. Recent experiments carried out in the last 10 years or so, brought out the necessity of using *ab initio* molecular dynamics. There have been three sets of experimental work. The first set of experiments was performed on sodium clusters by Haberland and coworkers, which brought the irregular behavior of measured melting temperatures (Schmidt et al., 1997, 1998). The second intriguing phenomenon observed was the higher than the bulk melting temperatures of clusters of tin and gallium (Breaux

et al., 2003; Shvartsburg and Jarrold, 2001). Equally dramatic observations were reported on shapes of the specific heat curves for gallium and aluminum clusters where even one atom made a difference. For example, the specific heat of Ga_{31} showed an identifiable sharp peak while Ga_{30} was nearly flat displaying a near continuous phase transformation. The understanding and explanation of these observations required *ab initio* or density-functional methods. It may be noted that the dynamics of ions is still classical and governed by Newton's laws. However, the forces acting on ions as cluster evolves in time, are now determined from the instantaneous electronic structure. Thus, density-functional molecular dynamics brought in the effects of full interacting electron–ion system. It was possible to carry out such a program in the late 1990s because of three ingredients. First, the advent of Car–Parrinello molecular dynamics and Born–Oppenheimer molecular dynamics. Second, the availability of ultrasoft pseudopotentials; and third, the availability of phenomenal computing power at affordable cost. With these ingredients, it is now possible to carry out molecular dynamical simulations over a time scales of 100–150 ps per temperature, depending on the nature and size of the systems. For clusters of simple metal atoms such as sodium sizes up to about $N \sim 200$ atoms are accessible. However, for clusters of transition metal atoms, dynamics may be carried out for sizes up to about $N \sim 50$ atoms. Mixed clusters obviously will require much more computing time in order to span a reasonable configuration space. Even then such *ab initio* simulations fall significantly short of the time scales accessible to classical dynamics using parameterized interatomic potentials. If a reliable data is needed for clusters over a few hundred atoms, there is no alternative to using appropriately chosen interatomic potentials.

In what follows, we mainly focus on the work carried out in the last 10 years or so using first-principles calculation. We will begin the discussion on the clusters of sodium, followed by higher than the bulk melting system Sn and Ga. Then, we will present a review of the Ga and Al dealing with size-sensitive effect on specific heat. We end this section by reviewing work on cages of Au and impurity-induced effects.

A large body of simulation work, both using interatomic potential and first-principle methods exists. It is not possible to summarize all contributions in this chapter. The reader is referred to a recent exhaustive review (Baletto and Ferrando, 2005).

18.5.2.1 Irregular Variations in the Melting Point

Clusters of sodium are perhaps the most intensely studied. In a series of experiments the melting temperatures, latent heat, and entropy of sodium clusters in the size range of $N = 55–355$ atoms per cluster have been measured by Haberland and coworkers (Schmidt and Haberland, 2002, Schmidt et al., 1997, 1998). A brief experimental procedure is noted below. In the first step, cluster ions are produced in a gas aggregation cell and thermalized in a heat bath typically of helium gas at 70 Pa. The clusters leave the heat bath and are transferred into a high vacuum and then mass selected.

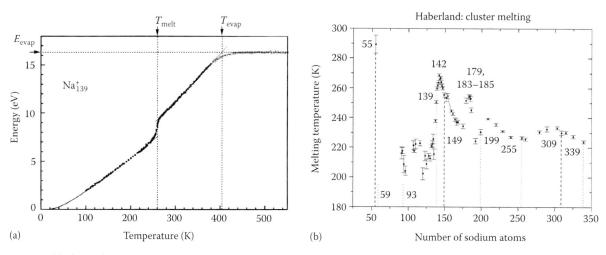

FIGURE 18.6 (a) The total energy of Na$_{139}^+$ cluster as a function of temperature. The sudden increase at $T = 260$ K is due to melting. The melting temperature is marked. (b) The melting temperatures of Na$_n^+$ clusters as a function of size (n). (After Schmidt, M. and Haberland, H., *C. R. Phys.*, 3, 327, 2002. With permission.)

The experiment consists of measuring the mass spectra as a function of temperature T of the heat bath from which caloric curve $E(T)$ is extracted. It turns out that the increase of the cluster temperature leads to the ejection of a certain number of atoms from the cluster. The same effect could be achieved when the cluster has absorbed one photon of known energy. Therefore, the increase in the internal energy due to increase in the temperature can be equated to the change in internal energy due to the absorption of one photon. An experimental caloric curve for Na$_{139}^+$ is shown in Figure 18.6a.

The most interesting result is shown in Figure 18.6b, where the observed melting temperatures in Kelvin are shown as a function of size. A number of unusual characteristics are immediately evident. The melting point varies irregularly between 230 to 290 K. The oscillatory behavior persists even in the size range of about $N \sim 350$ atoms. The reduction in the melting points, expected for the finite-sized system, does not follow the simple $1/R$ scaling, where R is the radius of the system. However, the most peculiar observation relates to the position of the maxima in the melting points. The positions of the maxima do not correlate with the electronic or geometric shell closing. Interestingly the highest melting temperature belongs to Na$_{55}^+$ having icosahedral shape. For reference, we note the bulk melting temperature as 371 K. We also note that the lowest melting temperature is observed for the cluster with size $N = 92$. Interestingly, Na$_{92}$ is an electronically closed-shell system.

In spite of extensive theoretical studies using interatomic potentials, these observations could not be understood. Density functional molecular dynamical simulations gave quantitative agreement and a clue toward the understanding of the maxima and minima observed in the experimental melting temperatures.

In a recent work (Haberland et al., 2005), the authors demonstrate that the energy and entropy differences between *liquidlike* and *solidlike* phases of clusters are relevant parameters

exhibiting pronounced maxima that correlate well with the geometric shell closing. Thus, the magic numbers displayed by sodium clusters for the melting temperatures are geometric in nature. They demonstrated that the icosahedral geometry dominates the melting phenomena in these clusters, a conclusion corroborated by photoelectron spectra. Their work brought out an interesting conclusion: this simple metal atom system displays two different kinds of magic numbers depending upon the properties investigated.

Now we turn to the *ab initio* theoretical investigations of the melting of sodium clusters.

- Theoretical investigation on the melting of sodium clusters

Aguado et al. reported first-principle isokinetic molecular dynamical simulations for the sizes between $N = 55 - 299$ for 10 representative clusters (Aguado and López, 2005). They used orbital free version of the density-functional theory and obtained very good agreement with the experimental melting temperatures. Figure 18.7a demonstrates the level of agreement. Their calculated latent heats also agree with the measured ones. They found that structural aspects can affect all the broad features. The clusters that are compact, having shortened bonds between surface and core atoms, show high melting temperatures.

Chacko et al. have also carried out one of the first successful Kohn–Sham based calculations for Na$_{55}$, Na$_{92}$, and Na$_{142}$. The calculated melting temperatures and the latent heat were again in very good agreement with the experimental values (Chacko et al., 2005). Interestingly, they found that upon melting, Na$_{55}$ deforms its shape from spherical to quadrupole. The shape deformation parameter ϵ defined in Equation 18.39 is shown in Figure 18.7b. Clearly, upon melting the shape of Na$_{55}$ changes sharply from spherical to prolate (quadrupole deformation). Similar shape deformation has been observed by Rytkönen et al. in their *ab initio* simulations on Na$_{40}$ as well

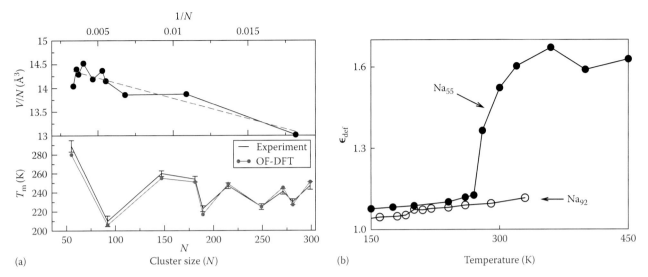

FIGURE 18.7 (a) Size variation of volume per atom (top) and melting temperature (bottom) in sodium clusters. The dashed line in the upper panel is the best linear fit to the data. (After Aguado, A. and López, J.M., *Phys. Rev. Lett.*, 94, 233401, 2005. With permission.) (b) Deformation parameter ϵ_{def} for Na_{55} and Na_{92}. (After Chacko, S. et al., *Phys. Rev. B*, 71, 155407, 2005. With permission.)

as Na_{20} and Na_{55} (Rytkönen et al., 1998), except that this magic Na_{40} cluster underwent an octupole deformation rather than a quadrupole deformation as observed for Na_{55}. This group has also investigated the thermal behavior of Na_{59}^{+} and Na_{93}^{+}. Their estimated melting temperatures were consistent with the experimental ones.

A systematic and detailed investigation of the finite temperature behavior over a broad size range of $8 \leq N \leq 55$ has been reported (Lee et al., 2005a; Zorriasatein et al., 2007a). We show calculated heat capacities for $40 \leq N \leq 55$ in Figure 18.8. The striking feature in the figures is that the shapes of specific heat are sensitive to cluster size. Although small clusters with $N < 25$ do not show any recognizable peak, the specific heat of clusters with $N \sim 50$ are rather flat. It turns out that the shapes of the specific heat curves are dependent on the nature of the ground-state geometries, *viz.* ordered or disordered. We will have an occasion to discuss the shape sensitive features while presenting the results on gallium clusters.

18.5.2.2 Higher-Than-Bulk Melting Temperature

As noted earlier, finite-sized systems are expected to melt at temperatures well below the bulk melting temperature. However, experimental measurements on clusters of tin and gallium brought out some surprises (Breaux et al., 2003, 2005; Shvartsburg and Jarrold, 2001). The melting temperatures of these clusters turned out to be substantially higher than their bulk melting point. The first experimental observation of higher than the bulk melting temperature was seen in tin clusters in the size range of $10 \leq N \leq 30$. These measurements based on the ionic mobilities showed that the clusters Sn_{10} and Sn_{20} do not melt at least about 50 K above the bulk melting temperature of 500 K.

The same group carried out extensive calorimetric measurements on a series of gallium clusters in the size range of 17–55 atoms. Their measured specific heat is shown in Figure 18.9.

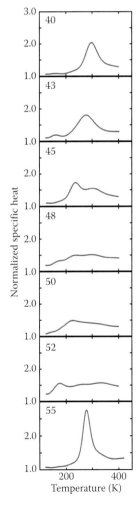

FIGURE 18.8 The normalized specific heat as a function of temperature for Na_N, $N = 40, 43, 45, 48, 50, 52,$ and 55. (After Zorriasatein, S. et al., *Phys. Rev. B*, 76, 165414, 2007b. With permission.)

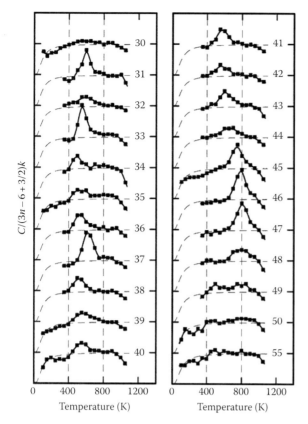

FIGURE 18.9 Heat capacities plotted against temperature for Ga_n^+ with $n = 31–50$ and 55. The points are the measured values, and the dashed lines are calculated from statistical thermodynamics. ($3n − 6 + 3/2)k_b$ is the classical (vibrational + rotational) heat capacity, where k_b is the Boltzmann constant. (After Breaux, G.A. et al., *J. Am. Chem. Soc.*, 126, 8628, 2004. With permission.)

Note that the melting temperature of bulk Ga is 305 K. Three intriguing aspects can be immediately seen from the figure. Firstly, the melting temperatures identified with the recognizable peaks are well above the bulk melting temperature. Secondly, there is an extreme size sensitivity seen in the shapes of the specific heat curves. For example, adding just one atom to Ga_{30} changes the shape of the specific heat curve dramatically. Thirdly, the variation in the melting temperature with respect to size is about 350 K. We now summarize the results and offer a plausible explanation for some of the above observations. The discussion is mainly based on simulations carried out by our group.

- Tin clusters

It is worth examining some properties such as the band gap of group IV elements to which tin belongs. These are, carbon, silicon, germanium, tin, and lead having band gaps (in electron-Volts) of 5.5, 1.17, 0.75, 0.1 (gray tin) respectively, while lead is a metal. Clearly, tin a semimetal can be viewed as a failed semiconductor at room temperature. Tin transforms into a metallic phase (white tin) at temperature about 286 K. Further, small clusters of tin are known to show similar growth pattern as silicon and germanium clusters. These observations suggest a possibility

of similar bonding in tin clusters as in silicon and germanium, i.e., covalent bonding. Our extensive density-functional calculations reveal that both Sn_{10} and Sn_{20} consist of highly stable tricapped trigonal prism (Joshi and Kanhere, 2003; Joshi et al., 2002). This unit turns out to be highly stable and enhances the melting temperature. We show the tricapped trigonal prism unit and the corresponding electron localization function in Figure 18.10 that brings out the covalent nature of bonding. This change in the bonding from their bulk counterpart and the existence of the tricapped trigonal prism unit is responsible for the higher than the bulk melting temperature. A similar result has been obtained by Chuang et al. for small clusters of tin by using *ab initio* density-functional methods (Chuang et al., 2004). However, in a later work it was observed that Sn_{20} fragments rather than becoming *liquidlike* at higher temperatures. It turns out that the fragmentation temperature (as seen in the simulations) depends on the nature of the exchange–correlation potential. Within LDA it is about 1250 K while for GGA it is about 650 K in agreement with the experimental results. Our studies also brought out the importance of long simulation times. The results with 50 ps per temperature were actually erroneous (Krishnamurty et al., 2006b).

- Gallium clusters

Now, we turn our attention to clusters of gallium for which systematic and extensive data are available. In order to understand the higher than the bulk melting temperature, it is convenient to examine the nature of bonding and the electronic structure of bulk gallium. We follow the work by Gong et al. (1991). The stable bulk structure at ambient temperature, pressure is the α-Ga and can be viewed as base centered orthorhombic with eight

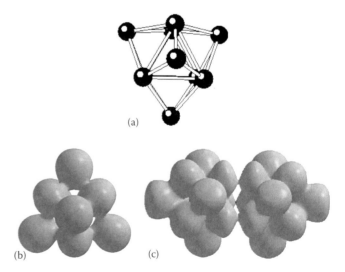

FIGURE 18.10 (a) Ground state geometry of Sn_{10} cluster depicting the tricapped trigonal prism unit, (b) the isovalued surface of the electron localization function for this structure at the value of 0.7. (After Joshi, K. et al., *Phys. Rev. B*, 67, 235413, 2003. With permission.) (c) The isovalued surface of the electron localization function for a low-lying isomeric structure of Sn_{20} at the value of 0.55. (After Joshi, K. et al., *Phys. Rev. B*, 67, 235413, 2003. With permission.)

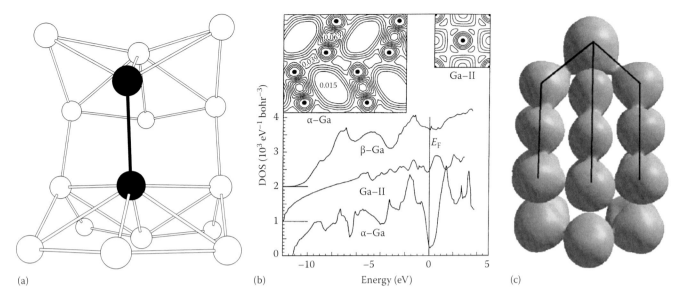

FIGURE 18.11 (a) A part of the bulk structure of α-Ga (not a unit cell). It shows a part of two buckled planes connected by a short covalent bond represented by the dark line joining the black atoms. (b) The electron density of states for α-Ga, β-Ga, and Ga II. Note the deep pseudogap of α-Ga corresponding to the strong Ga_2 covalent bond depicted in the inset. (After Gong, X.G. et al., *Phys. Rev. B*, 43, 14277, 1991. With permission.) (c) The electron localization function of G_{31} at the value of 0.68. The black lines show the connected basins. (After Krishnamurty, S. et al., *Phys. Rev. Lett.*, 86, 135703, 2006a. With permission.)

atoms per unit cell (See Figure 18.11a. The structure is peculiar in the sense that there is one nearest neighbor with a short bond of 2.44 Å. The six other neighbors are at 2.71 and 2.79 Å. The band structure of α gallium shows an interesting feature, i.e., a pseudogap related to the covalent bond between the atoms in the two buckled planes (see Figure 18.11b). Within the buckled plane, the charge density is metallic. It turns out that in the finite-sized systems, where the atoms are free to move and reconfigure, the bonding between all the atoms is mainly covalent. This can be inferred from the electron localization function analysis. This is illustrated in Figure 18.11c where we have shown the isovalued surface of Ga_{31}. Thus, both tin and gallium show higher than the bulk melting temperature due to localized bonding as opposed to delocalized bonding in their bulk counterpart.

18.5.2.3 Size Sensitivity of the Specific Heat

In this section, we discuss the most interesting aspect of size sensitivity in the specific heat curves. It may be emphasized that such a size-sensitive behavior has been observed experimentally in gallium clusters (Breaux et al., 2005) and aluminum (Breaux et al., 2005; Neal et al., 2007) and in simulations for cluster of sodium, gold atoms (Krishnamurty et al., 2006a). Thus the phenomenon is generic. It turns out that the origin of this phenomenon is certainly geometric. It depends on the nature of the ground-state geometry and also on the distribution of energies of all the isomeric structures. The second factor is not independent of the first one.

In general, the geometry of a cluster could be highly symmetric displaying rotational symmetry (e.g., Na_{55}) or it could be completely disordered. The degree of order or disorder will vary with size. Recall that in a typical homogeneous ordered

solid, each atom being equivalent is bonded to other atoms with nearly equal strength. In contrast to this, in a typical cluster each atom is individual in the sense that it may be bonded with others with different strengths. As we have seen, the first consequence of this is the broadening of the phase transition region over a range of temperatures. If the cluster is well-ordered, there is still a recognizable sharp peak. But when the cluster is disordered, each atom is likely to have a different local environment. Consequently, their dynamical behavior as a response to temperature will differ. Some of the atoms will pick up the kinetic energy at lower temperature leading to early diffusive motion while others may still be executing harmonic motion. In a given cluster if a large group of atoms are bonded together with similar strengths forming a region of local order, then it will melt at nearly the same temperature showing a recognizable peak. We illustrate these remarks for two clusters Ga_{30} and Ga_{31}. In Figure 18.12a, we show the ground-state geometries of these two clusters with two different perspectives. Clearly, the difference in the nature of the order is rather subtle. A detailed examination of the geometry and the nature of the bonding reveals that Ga_{31} is significantly more ordered than Ga_{30}. In Figure 18.12b, we show mean-squared displacement for individual atoms at 250 K. All the atoms in Ga_{31} have very small mean-squared displacements, while there is a distribution of mean-squared displacements for Ga_{30}, the largest displacement being of the order of 8 Å. As a consequence, Ga_{30}, "melts" continuously showing no recognizable peak. The calculated specific heat of these clusters are shown in Figure 18.12c. The dramatic difference seen experimentally is correctly captured by these *ab initio* simulations.

Such size-sensitive behavior is also observed in the case of sodium clusters. In Figure 18.8, we show calculated specific heat

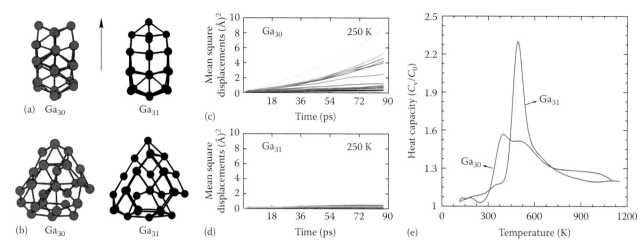

FIGURE 18.12   (a and b) The ground-state geometries of Ga_{30} and Ga_{31} with two different perspectives. Perspective (b) is rotated by 90° with respect to perspective (a) about an axis shown in (a). (c and d) The mean square displacements for individual atoms of Ga_{30} and Ga_{31} computed over 90 ps. (e) The specific heat curves of Ga_{30} and Ga_{31} computed over 90 ps. (After Krishnamurty, S. et al., *Phys. Rev. Lett.*, 86, 135703, 2006a. With permission.)

for the sizes of $N = 40$–55. Even here, the analysis of the ground-state geometries of these clusters indicate that these clusters grow from an ordered structure (Na_{40}) to another ordered structure (Na_{55}) via a disordered route; the maximum disorder is observed for the clusters around $N = 50$. This observation correlates completely with the calculated specific heat curves.

Small clusters of gold have attracted interest firstly because of their reactivity and more recently due to the discovery of hollow cages in the size range of 16–18 atom (Bulusu et al., 2006). Au_{20} happens to be another intriguing cluster showing a pyramidal structure exhibiting a large energy gap. Au_{19} is again a pyramid with missing apex atom (Krishnamurty et al., 2007). The finite temperature simulations of these clusters show some interesting feature again conforming the generic nature of size-sensitivity noted above. In Figure 18.13, we show the ground-state geometry and the respective heat capacities of Au_{19} and Au_{20}. Interestingly, the flat nature of Au_{19} arises due to the "vacancy" induced diffusive motion at low temperatures.

18.5.2.4 Impurity-Induced Effects

It is well-known in the solid-state community that impurities, even dilute ones, are capable of significantly changing many of the properties such as local structure and conductivity. Although there are no experimental results, a few simulations have shown very promising results. Such simulations are motivated by the desire to modify the finite temperature properties of the host system upon doping with impurities.

Mottet and coworkers carried out molecular dynamical simulations of alloying effects in silver clusters by introducing a single impurity in clusters containing more than a hundred atoms (Mottet et al., 2005). The impurities doped were nickel and copper atoms. They found a considerable increase in the melting temperature of the silver clusters upon doping. They correlated this upward shift in the melting temperatures to the strain relaxation induced by the small central impurity in the icosahedral silver clusters.

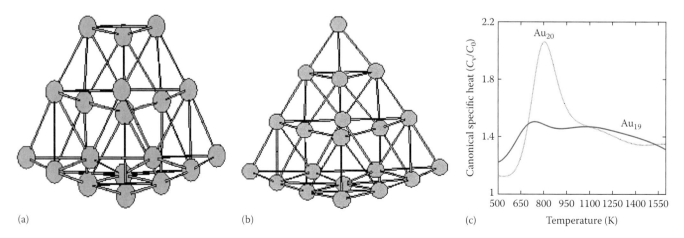

FIGURE 18.13   (a and b) The ground-state geometries of Au_{19} and Au_{20} clusters. Note that one atom at the vertex of Au_{19} is missing. (c) The specific heat curves of these clusters. (After Krishnamurty, S. et al., *J. Phys. Chem. A*, 111, 10769, 2007. With permission.)

Another interesting observation has been seen in the case of silicon clusters. Small clusters of silicon, in the range of 15–20 atoms, fragment upon heating typically above 1300 K. It turns out that a single impurity of titanium in silicon clusters converts that structure into Frank–Kasper polyhedra. Interestingly, this caged structure remains non-fragmented until up to about 2200 K (Zorriasatein et al., 2007a). Thus, impurity in these silicon clusters stabilizes their structure. In a very recent work, Chandrachud et al. examined the effect of doping a single carbon in Al_{13} and Ga_{13} (Chandrachud et al., 2007). The effect of impurity was to lower the melting temperature substantially in both cases.

18.6 Summary

Atomic clusters are considered as models of nanostructure materials. It is known that their properties are neither bulk like nor atomic like. We have examined issues related to finite temperature behavior as observed in the experiments on atomic clusters. In particular three observations need to be explained: the irregular behavior of the melting points as a function of size, the size-sensitive nature of the specific heat, and higher than bulk melting points in clusters of gallium and tin. We present methods to undertake finite temperature simulations i.e., the density-functional molecular dynamics, and we also present the tools for the analysis of the finite temperature data. It is shown that the nature of the ground state, both geometric and electronic, influences the finite temperature properties. We bring out the role of bonding and establish a strong correlation between the geometric order (or the absences of it) with the shapes of the specific heat curves. Finally, the effect of impurity doped in the host cluster has been brought out.

Acknowledgments

It is a pleasure to acknowledge discussions with S. Blundell, A. Vichare, K. Joshi, S. Krishnamurty, M.-S. Lee, S. Zorriasatein, M. Ghazi, V. Kaware, and P. Chandrachud. We gratefully acknowledge C-DAC India for providing HPC facility.

References

Aguado, A. and J. M. López, 2005, *Phys. Rev. Lett.* **94**, 233401.

Allen, M. P. and D. J. Tildesley, 1987, *Computer Simulation of Liquids* (Clarendon Press, Oxford, U.K.).

Baletto, F. and R. Ferrando, 2005, *Rev. Mod. Phys.* **77**, 371.

Baruah, T., R. Zope, S. L. Richardson, and M. R. Pederson, 2003, *Phys. Rev. A* **68**, 241404.

Becke, A. D., 1996, *J. Chem. Phys.* **104**, 1040.

Becke, A. D. and K. E. A. Edgecombe, 1990, *J. Chem. Phys.* **92**, 5397.

Berry, R. S., 1997, *Micro. Therm. Eng.* **97**, 1089.

Bichara, C., J. P. Gaspard, and J. C. Mathieu, 1987, *Phys. Lett. A* **119**, 462.

Bixon, M. and J. Jortner, 1989, *J. Chem. Phys.* **91**, 1631.

Breaux, G. A., R. C. Benirschke, T. Sugai, B. S. Kinnear, and M. F. Jarrold, 2003, *Phys. Rev. Lett.* **91**, 215508.

Breaux, G. A., D. A. Hillman, C. M. Neal, and R. C. Benirschke, 2004, *J. Am. Chem. Soc.* **126**, 8628.

Breaux, G. A., C. M. Neal, B. Cao, and M. F. Jarrold, 2005, *Phys. Rev. Lett.* **94**, 173401.

Bulusu, S., L.-S. Wang, and X. C. Zeng, 2006, *Proc. Natl. Acad. Sci.* **103**, 8326.

Chacko, S., D. G. Kanhere, and S. A. Blundell, 2005, *Phys. Rev. B* **71**, 155407.

Chandrachud, P., K. Joshi, and D. G. Kanhere, 2007, *Phys. Rev. B* **76**, 235423.

Chuang, F.-c., B. B. Wang, J. R. Chelikowsky, and K. M. Ho, 2004, *Phys. Rev. B* **69**, 165408.

Ekardt, W. (ed.), 1999, *Metal Clusters* (Wiley Blackwell, West Sussex, U.K.).

Ferrenberg, A. M. and R. H. Swendsen, 1988, *Phys. Rev. Lett.* **61**, 2635.

Frenkel, D. and B. Smit, 1996, *Understanding Molecular Simulations* (Academic Press Ltd., San Diego, CA).

Gong, X. G., G. L. Chiarotti, M. Parrinello, and E. Tosatti, 1991, *Phys. Rev. B* **43**, 14277.

Haberland, H. (ed.), 1994, *Clusters of Atoms and Molecules* (Springer-Verlag, Berlin, Germany).

Haberland, H., T. Hoppler, J. Donges, O. Kostko, M. Schmidt, and B. von Issendorf, 2005, *Phys. Rev. Lett.* **94**, 035701.

de Heer, W. A., 1993, *Rev. Mod. Phys.* **65**, 611.

de Heer, W. A., W. D. Knight, M. Y. Chou, and M. L. Cohen, 1987, *Solid State Physics*, Vol. 40 (Academic Press, New York).

Hohenberg, P. and W. Kohn, 1964, *Phys. Rev.* **136**, B864.

Jellinek, J. (ed.), 1999, *Theory of Atomic and Molecular Clusters* (Springer-Verlag, Berlin, Germany).

Jena, P., S. N. Khanna, and B. K. Rao (eds.), 1992, *Physics and Chemistry of Finite Systems: From Clusters to Crystals*, Vols. 1 and 2 (Kluwer Academic Publishers, Dordrecht, the Netherlands).

Johansson, M. P., D. Sundholm, and J. Vaara, 2004, *Angew. Chem. Int. Ed.* **43**, 2678.

Joshi, K. and D. G. Kanhere, 2003, *J. Chem. Phys.* **119**, 12301.

Joshi, K., D. G. Kanhere, and S. A. Blundell, 2002, *Phys. Rev. B* **66**, 155329.

Joshi, K., D. G. Kanhere, and S. A. Blundell, 2003, *Phys. Rev. B* **67**, 235413.

Kittel, C., 1958, *Elementary Statistical Physics* (Wiley, New York).

Knight, W. D., K. Clemenger, W. A. de Heer, W. A. S. M. Y. Chou, and M. L. Cohen, 1984, *Phys. Rev. Lett.* **52**, 2141.

Kohanoff, J., 2006, *Electronic Structure Calculations for Solids and Molecules: Theory and Computational Methods* (Cambridge University Press, Cambridge, U.K.).

Kohn, W. and L. J. Sham, 1965, *Phys. Rev.* **140**, A1133.

Krishnamurty, S., K. Joshi, and D. G. Kanhere, 2006a, *Phys. Rev. Lett.* **86**, 135703.

Krishnamurty, S., K. Joshi, D. G. Kanhere, and S. A. Bundell, 2006b, *Phys. Rev. B* **73**, 045419.

Krishnamurty, S., G. Sadatshafai, and D. G. Kanhere, 2007, *J. Phys. Chem. A* **111**, 10769.

Kumar, V. and Y. Kawazoe, 2001, *Phys. Rev. Lett.* **87**, 045503.

Kumar, V., K. Esfarjini, and Y. Kawazoe, 2000, *Advances in Cluster Science* (Springer-Verlag, Heidelberg, Germany).

Labastie, P. and R. L. Whetton, 1990, *Phys. Rev. Lett.* **65**, 1567.

Lee, M.-S., S. Chacko, and D. G. Kanhere, 2005a, *J. Chem. Phys.* **123**, 164310.

Lee, M.-S., D. G. Kanhere, and K. Joshi, 2005b, *Phys. Rev. A* **72**, 015201.

Martin, R. M., 2004, *Electronic Structure: Basic Theory and Practical Methods* (Cambridge University Press, Cambridge, U.K.).

Mottet, C., G. Rossi, F. Baletto, and R. Ferrando, 2005, *Phys. Rev. Lett.* **95**, 035501.

Neal, C. M., A. K. Starace, and M. F. Jarrold, 2007, *Phys. Rev. B* **76**, 054113.

Parr, R. and W. Yang, 1989, *Density Functional Theory of Atoms and Molecules* (Oxford University Press, New York).

Payne, M. C., M. P. Teter, D. C. Allan, T. A. Arias, and J. D. Joannopoulous, 1992, *Rev. Mod. Phys.* **64**, 1045.

Perdew, J. P., K. Burke, and M. Ernzerhof, 1996a, *Phys. Rev. Lett.* **77**, 3865.

Perdew, J. P., K. Burke, and Y. Wang, 1996b, *Phys. Rev. B* **54**, 16533.

Perdew, J. P., K. Burke, and M. Ernzerhof, 1997, *Phys. Rev. Lett.* **78**, 1396.

Rapaport, D. C., 1998, *The Art of Molecular Dynamics Simulation* (Cambridge University Press, Cambridge, U.K.).

Rytkönen, A., H. Häkkinen, and M. Manninen, 1998, *Phys. Rev. Lett.* **80**, 3940.

Schmidt, M. and H. Haberland, 2002, *C. R. Phys.* **3**, 327.

Schmidt, M., R. Kusche, W. Kronmüller, B. von Issendorff, and H. Haberland, 1997, *Phys. Rev. Lett.* **79**, 99.

Schmidt, M., R. Kusche, B. von Issendorff, and H. Haberland, 1998, *Nature* **393**, 238.

Shvartsburg, A. A. and M. F. Jarrold, 2001, *Phys. Rev. Lett.* **85**, 2530.

Shvartsburg, A. A., M. F. Jarrold, B. Liu, Z. Y. Lu, C. Z. Wang, and K.-M. Ho, 1998, *Phys. Rev. Lett.* **81**, 4616.

Silvi, B. and A. Savin, 1994, *Nature* **371**, 683.

Stanley, H. E., 1987, *Introduction to Phase Transitions and Critical Phenomena* (Oxford University Press, New York).

Szwacki, N. Z., A. Sadrzadeh, and B. I. Yakobson, 2007, *Phys. Rev. Lett.* **98**, 166804.

Wales, D. J., 2003, *Energy Landscapes: With Applications to Clusters, Biomolecules and Glasses* (Cambridge University Press, Cambridge, U.K.).

Wales, D. J. (ed.), 2005, *Intermolecular Forces and Clusters—II* (Springer, Berlin, Germany).

Zorriasatein, S., K. Joshi, and D. G. Kanhere, 2007a, *Phys. Rev. B* **75**, 045117.

Zorriasatein, S., M.-S. Lee, and D. G. Kanhere, 2007b, *Phys. Rev. B* **76**, 165414.

19

Melting Point of Nanomaterials

Pierre Letellier
*Université Pierre et
Marie Curie-Paris 6*

Alain Mayaffre
*Université Pierre et
Marie Curie-Paris 6*

Mireille Turmine
*Université Pierre et
Marie Curie-Paris 6*

19.1 Introduction

When we try to define what is a nanomaterial or a nanoparticle, the first idea that comes to our mind is to specify its size. Whether it is a sphere, a crystal, a wire, or a film, we admit that at least one of its dimensions is of the order of few nanometers. But this geometrical definition is insufficient to assess whether the system has the properties of a nanomaterial. Indeed, the "measured size" of a nanomaterial essentially depends on the method of analysis that is available. The shape of these nanoelements is often complicated and complex and they are rarely identical to one another.

Generally, we know that we are faced with a nanomaterial if the latter shows physicochemical properties and reactivity very different compared to what they exhibit on a macroscale. For instance, copper (Zong et al., 2005), which is an opaque substance in the macroscopic scale, becomes transparent in nanoscale, inert materials (such as platinum, Sasaki et al., 1999) attain catalytic properties, stable materials (as aluminum, Kuo et al., 2008) turn combustible, insulators (as silicon, Lechner et al., 2008) become conductors. At nanometric scale, noble metals like silver (Li and Zhu, 2006) and gold embedded as nanoparticles (Shi et al., 2000) in a silica matrix, react with hydrochloric acid, leading to the metal chloride with hydrogen release.

The characterization of a nanomaterial, besides the determination of its structure, implies necessarily the determination of its physicochemical properties and its reactivity. But since they are different from those of the material in the macroscopic state, it is fundamental to understand why and to wonder about the use of the traditional thermodynamics of Gibbs to describe the behavior of such systems.

19.2 Is the Gibbs Thermodynamics Adapted to Describe the Behaviors of Nanosystems?

The strict application of the Gibbs thermodynamics to nanometric systems does not allow to account properly for the physicochemical properties of these systems. Indeed, this approach predicts that for a material, a number of parameters remain constant and independent of the mass of the system. This is the case of saturated vapor pressure of liquids, their boiling and crystallization points, and the melting point of pure solids. But, experience shows that for these same systems considered at nanoscale, the value of these parameters depends on the size, the shape, and the mass of the considered systems.

In the traditional Gibbs thermodynamics, there is no way to account for this type of behavior. This is due to the fact that Gibbs thermodynamics is extensive and that the parameters previously mentioned are of intensive magnitudes; that is, they do not depend on the mass of the system. Thus, to account for the physicochemical properties of nanosystems, it is necessary to make the thermodynamics nonextensive. One of the mostly used ways to introduce the nonextensivity is to introduce into the thermodynamic equations Laplace's law. This law assumes that the pressure inside a nanoparticle is higher than the surrounding pressure and dependent on the surface tension. For a spherical particle of radius r, this difference of pressure is expressed as

$$P_{\text{int}} - P_{\text{ext}} = \frac{2\gamma}{r} \tag{19.1}$$

where

γ is the surface tension

r the radius of the particle

Then the equilibrium condition between the particle and its environment is expressed by a null nonisobare chemical affinity. Thus, for a particle i in equilibrium with its environment, one can write

$$A_i = {}^{P_{\text{int}}}\mu_{i(\text{int})} - {}^{P_{\text{ext}}}\mu_{i(\text{ext})} = 0 \qquad (19.2)$$

The chemical potential of i at the internal pressure can be expressed at the external pressure by supposing that its molar volume does not depend on the pressure,

$$\left(\frac{\partial \mu_{i(\text{int})}}{\partial P}\right)_T = V^\star_{i(\text{int})} \qquad (19.3)$$

That leads to

$${}^{P_{\text{int}}}\mu_{i(\text{int})} = {}^{P_{\text{ext}}}\mu_{i(\text{int})} + V^\star_{i(\text{int})}(P_{\text{int}} - P_{\text{ext}}) = {}^{P_{\text{ext}}}\mu_{i(\text{int})} + V^\star_{\text{int}}\frac{2\gamma}{r} = {}^{P_{\text{ext}}}\mu_{i(\text{ext})} \qquad (19.4)$$

In the case of a liquid drop of radius r in equilibrium with its vapor at the pressure P_i, one can find

$$\mu^\star_{i(\text{liq})} + V^\star_i\frac{2\gamma}{r} = \mu^\star_{i(\text{gaz})} + RT\ln\{P_i\} \qquad (19.5)$$

If $P^\star_i$ is the vapor pressure of i at equilibrium with its pure liquid in the form of unlimited phase (Defay, 1934),

$$\ln\frac{P_i}{P^\star_i} = \frac{V^\star_i}{RT}\frac{2\gamma}{r} \qquad (19.6)$$

The Kelvin's relation is then found. When r decreases, the vapor pressure, at equilibrium, increases. Obviously, other characteristic properties of pure compounds and especially the melting point of a spherical solid nanoparticle of radius r can be described in the same way. The phenomenon of melting point depression for nanosolids was first described at the beginning of the twentieth century (Pawlow, 1910) by the Gibbs–Thompson relation. This links the difference between the melting point of the particle, T_x, and the melting point of the material, T_m, in unlimited phase form (Defay, 1934) (no size effect), by

$$T_m - T_x = \Delta T = \frac{T_m V_{(T_x)}}{\Delta H_m}\frac{2\gamma^{\text{SL}}}{r} \qquad (19.7)$$

At temperature T_x, $V_{(T_x)}$ is the molar volume of the solid, γ^{SL} is the surface tension between the solid and the liquid, and ΔH_m is the melting molar enthalpy of the solid ($\Delta H_m > 0$, endothermic). This type of development has been thoroughly discussed by Defay and Prigogine in their book (Defay and Prigogine, 1966).

If the introduction of Laplace's law in the equations of thermodynamics allows reporting the variation of physicochemical properties of nanoparticles with their size, the method is fundamentally questionable, because it introduces a condition of nonextensivity into reasoning whose consistency stems from the fact that it is extensive. If it is well accepted to place the particular properties of the nanosystems in their interfacial energy, the classic way of writing it in the expression of the internal energy does not allow escaping from the extensivity:

$$dU = T\,dS - P\,dV + \sum \mu_i\,dn_i + \gamma\,dA \qquad (19.8)$$

where

A is the area

γ is the surface tension

In this expression, S, V, n_i, and A are extensive magnitudes. This implies that if the mass of the system is multiplied by λ, the values of these extents will also be multiplied by λ. It is the condition so that T, P, μ, and γ are intensive extents. When we introduce artificially Laplace's law in the equilibrium relations, the chemical potential is then dependent on the size of the system. They are no more intensive extents.

Beyond this fundamental point, the approach, which consists in introducing Laplace's law, does not allow to report correctly a number of situations. Very often, real particles of nanosolid are not, in general, perfect spheres. This is the case, for example, of mineral or organic crystals and their assemblies, of polymer aggregates, and of films. For some materials, like microporous materials, it is difficult to describe their spatial structure from the classical geometrical variables of dimension (volume, surface, length) and some authors (Jaroniec et al., 1990) have suggested using fractal approaches. In such conditions, it is unreal to define a geometrical surface; the notion of area does not have a meaning anymore. The variations of certain extents such as the melting points with the size of the particle can be in one way or in an opposite way according to the conditions of the experiment.

The conclusion of this is that it is not enough to introduce artificially a nonextensivity condition into the thermodynamics to have a tool of analysis, because the latter also has to be autoconsistent.

It is the reason for which we introduced a non extensive thermodynamics which we widely illustrated in various domains of the physics and in particular for the melting points of nanoparticles.

19.3 The Bases of Nonextensive Thermodynamics

In 2004, we introduced the bases of a nonextensive thermodynamics (NET) adapted to the description of physicochemical behaviors of complex systems (Turmine et al., 2004) for which the interfaces are geometrically ill-defined (including porous systems, interpenetrated phases, dispersed solutions, nanoparticles, and films). The contribution of the shape of the system to its properties is not a characteristic of the system's chemistry itself. In physics, there are some systems for which the state variables depend on their shape and their size: for example, the potential difference between the ends of a capacitor depends on its shape (Tsallis, 2002). This type of observation has led numerous physicists (Tsallis et al., 1998, Lavagno, 2002, Vives and Planes, 2002, Abe and Rajagopal, 2003) to consider the unicity of the thermodynamic formulations and to question the fundamental law that posits that state functions must be strictly extensive. Statistical analyses different from the standard Boltzmann formalisms indicate that a different type of thermodynamics can be constructed based on a nonextensive form of the entropy. Currently, the physics literature contains references to more than 20 different entropic forms. This fruitful reflection has not been exploited or indeed even considered in chemistry. We successfully applied our nonextensive thermodynamic approach to various systems such as the wettability phenomena (Letellier et al., 2007a), solubility of nanoparticles (Letellier et al., 2007c), redox properties (Letellier et al., 2008b), and also micellar solution properties (Letellier et al., 2008a).

The basis of NET is identical to that of classical thermodynamics, with the same functions of state, but NET supposes that these functions of state can be nonextensive. This property is introduced by means of integer or fractional *thermodynamic dimensions*. As a result, various physicochemical behaviors can be described by power laws without resorting to the concept of fractality.

To explain our approach, we will first describe some of the underlying principles of NET.

19.3.1 Conceptual Bases of Nonextensive Thermodynamics

Let us consider a system defined by its content (n_1, n_2, n_i moles) in contact with its environment by a geometrically ill-defined interface. We chose not to try to specify the exact borders of this system and to characterize its interface(s) with the environment by an interfacial energy. By convention, we call these interfaces "fuzzy interfaces" (Figure 19.1).

The description of the behavior of this system requires the usual variables, S, V, and n_i, and a variable of extensity χ. The internal energy can then be written

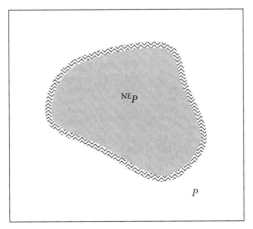

FIGURE 19.1 Scheme of fuzzy interface. ^{NE}P is the pressure of the nonextensive phase and P is the pressure of the environment.

$$dU = T\,dS - P\,dV + \sum \mu_i\,dn_i + \tau\,d\chi \tag{19.9}$$

where τ is an intensive tension extent, associated with χ. The product $\tau d\chi$ characterizes the contribution of the interfacial energy to the internal energy. The form of this relation is classical (Hill, 2001a) and in the case of interfacial systems, χ is associated with area and τ with surface tension. Classically, in thermodynamics, the variables of extensity associated to tension extents are assumed to be extensive variables, i.e., Euler's functions of the system mass of order $m = 1$.

We considered the possibility that they are not extensive ($m \neq 1$). Adopting this condition, the notion of nonextensivity is introduced in the expression of the internal energy (U). The internal energy, remaining a function of state, has all the mathematical properties associated to these functions. This implies new relations between the variables of state and in particular with χ. In a system consisting of n_1 moles of 1, n_2 moles of 2, and n_i moles of i, the extensity χ is a function of the system mass

$$\chi = \chi(n_1, n_2, ..., n_i) \tag{19.10}$$

By convention, this extent has the property of Euler's function of order m. If the system content is multiplied by λ, then

$$\chi_\lambda = \chi(\lambda n_1, \lambda n_2, ..., \lambda n_i) = \lambda^m \chi \tag{19.11}$$

The parameter m is the degree of homogeneity of the Euler's function, named by convention, *the thermodynamic dimension of the system*. In this approach, the thermodynamic dimension is defined with regard to the mass of the system. Its value can be equal to 1, in which case classical thermodynamics apply. The introduction of nonextensive thermodynamics in the extensity magnitudes implies that the functions of state of thermodynamics (U, S, etc.) are not extensive. Consequently, the tension extents associated with the extensities may not be intensive. We chose, by convention, to conserve this property for the temperature T^* and for τ.

For consistency, the chemical potentials and the pressures become nonintensive extents; this means that they vary with the systems' mass.

We have now to link the pressure variations to the variables of dimension. For this, a volume ^EV is defined for the environment and a volume ^{NE}V for the nonextensive phase. The system being at equilibrium, the sum of resulting works has to be null. Therefore

$$-P\,d^EV - {}^{NE}P\,d^{NE}V + \tau\,d\chi = 0 \tag{19.12}$$

Assume that ^{NE}V increases. Moving the "fuzzy interface" involves

$$d^{NE}V = -d^EV \tag{19.13}$$

---

* In physics, there are several developments of nonextensive thermodynamics using conventions in which the temperature is considered as a nonintensive variable, see for example, Abe et al. (2001) and Toral (2003).

and as the state of the system remains the same,

$$^{NE}P - P = \tau \frac{d\chi}{d^{NE}V} \qquad (19.14)$$

The value of the ratio $d\chi/d^{NE}V$ fixes the way in which dimensions of the system evolve in the transformation. This equation can be also explained by noting that the variables χ and ^{NE}V are homogeneous functions of different orders of the same dimension content. This implies that χ is a homogeneous function of order m with regard to ^{NE}V, which is expressed by the relationship

$$m\chi = \frac{d\chi}{d^{NE}V}^{NE}V \qquad (19.15)$$

and for the pressure difference

$$^{NE}P - P = \tau \frac{d\chi}{d^{NE}V} = m \frac{\tau\chi}{^{NE}V} \qquad (19.16)$$

This is one of the most important results of our research. This relation generalizes Laplace's relationship for nonextensive systems. It does not involve a radius of curvature or precise geometrical borders of the nonextensive system but is defined only from physicochemical parameters.

This obviously allows the solution to be found if the geometry is simple. In the case of liquid drop of interfacial tension γ^{LV}, radius r, and volume V, the pressure difference between the inside of the drop ($^{NE}P = P_d$) and the external pressure of the gaseous atmosphere is obtained by introducing the interfacial parameters into Equation 19.16, $\tau = \gamma^{LV}$, $\chi = A^{LV}$. Note that when the drop volume is multiplied by λ, the area A^{LV} is multiplied by $\lambda^{2/3}$. The area is then an extensity of dimension $m = 2/3$ toward the mass or the drop volume. *The liquid drop is a nonextensive phase with a thermodynamic dimension equal to 2/3.* We can write

$$(P_d - P) = m \frac{\tau\chi}{V} = \frac{2}{3} \frac{\gamma^{LV} A^{LV}}{V} = 2 \frac{\gamma^{LV}}{r} \qquad (19.17)$$

which corresponds to Laplace's relationship.

19.3.2 Application of the NET to Solid–Liquid Equilibrium

At first, we will apply these NET definition relations to the change of melting point of a nanosolid, when the size and the shape are modified. We will assume that the solid constitutes a nonextensive phase (Figure 19.2).

19.3.3 Melting Point

The equilibrium between the nanosolid in the form of a nonextensive phase and the liquid at the melting point T_x is expressed by writing the equality of the chemical potentials of the compound in its two forms. The solid (S) is at the pressure ^{NE}P in the nonextensive phase, and the liquid (liq) is subject to the external pressure P:

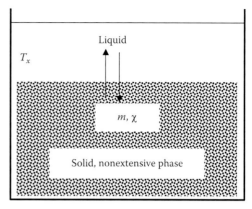

FIGURE 19.2 Scheme of the considered system. The nanosolid constitutes a nonextensive phase of dimension m and of extensity χ. This solid is in contact with the liquid phase at temperature T_x. (Reprinted from Letellier, P. et al., *Phys. Rev. B*, 76, 8, 2007b. With permission.)

$$^{NE}P\mu_{(S,T_x)} = {}^P\mu_{(liq,T_x)} \qquad (19.18)$$

At temperature T_x and constant content, the chemical potential of pure solid varies with the pressure P according to Equation 19.3. The integration of this expression leads to

$$^{NE}P\mu_{(S,T_x)} = {}^P\mu_{(S,T_x)} + V_{(T_x)}({}^{NE}P - P) \qquad (19.19)$$

$V_{(T_x)}$ is the molar volume of pure solid at temperature T_x. By considering the volume V which corresponds to the volume occupied by the nanosolid in the system, then

$$^{NE}P\mu_{(S,T_x)} = {}^P\mu_{(S,T_x)} + V_{(T_x)}m\tau\frac{\chi}{V} \qquad (19.20)$$

At the pressure P of the environment, the equilibrium condition (Equation 19.18) can be written as

$$\mu_{(liq,T_x)} = \mu_{(S,T_x)} + V_{(T_x)}m\tau\frac{\chi}{V} \qquad (19.21)$$

The pressure being constant (isobar condition), we removed P from the notation of the chemical potential for clarity. The free energy of melting at temperature T_x, $\Delta G_{m(T_x)}$, is linked to the characteristic parameters of the nonextensive phase:

$$\mu_{(liq,T_x)} - \mu_{(S,T_x)} = \Delta G_{m(T_x)} = V_{(T_x)}m\tau\frac{\chi}{V} \qquad (19.22)$$

At the melting temperature of the solid in the form of an unlimited phase (T_m), the melting free energy is null. The integration of the Gibbs–Helmholtz relation between T_m and T_x leads to

$$\frac{\Delta G_{m(T_x)}}{T_x} = -\int_{T_m}^{T_x} \Delta H_m \frac{dT}{T^2} \qquad (19.23)$$

where ΔH_m is the melting molar enthalpy of pure solid at ambient pressure P. As the melting enthalpy varies slightly in the temperature range considered for integration, one can write

$$\Delta G_{m(T_x)} = \Delta H_m \left(1 - \frac{T_x}{T_m}\right) = V_{(T_x)} m \tau \frac{\chi}{V} \qquad (19.24)$$

Then, the melting temperature T_x of the nonextensive phase can be easily linked to the melting point T_m of the unlimited phase by the following equation, which can be written in two different ways:

$$T_m - T_x = \Delta T = T_m \frac{V_{(T_x)} m \tau (\chi/V)}{\Delta H_m} = T_m \frac{M^\circ m \tau (\chi/M)}{\Delta H_m} \qquad (19.25)$$

where
  M° is the molecular weight
  M the mass of the solid

19.4 Application to the Melting Temperature of a Nonextensive Phase

Consider first the case where the geometry of the nonextensive phase is sufficiently well defined for the extensity χ to be identified with an area.

19.4.1 Extensity Is an Area

19.4.1.1 Gibbs–Thompson Law

Consider a solid particle of spherical shape of radius r in equilibrium with a molten liquid. In this case, dimension m is equal to 2/3. If the extensity is identified with the solid–liquid area, A^{SL}, the tension τ with the surface tension γ^{SL}, the volume of the nonextensive phase to that of the particle, the Gibbs–Thompson relation is then found:

$$\Delta T = \frac{T_m V_{(T_x)}}{\Delta H_m} \left(\frac{2}{3} \gamma^{SL} \frac{4\pi r^2}{(4/3)\pi r^3}\right) = \frac{T_m V_{(T_x)}}{\Delta H_m} \frac{2\gamma^{SL}}{r} \qquad (19.26)$$

Note that this relation supposes that the solid–liquid surface tension γ^{SL} is experimentally accessible, because the melting temperature variations of a material in spherical particles of known sizes can be accurately determined. The same is true for the Ostwald–Freundlich expression, which governs nanoparticle solubility. It is thus surprisingly simple to determine an extent that is generally calculated by semiempirical approaches and wettability studies (Kwok and Neumann, 1999, 2000, Graf and Riegler, 2000). However, in addition to the experimental difficulties of determining the particle size (and the particles must be spherical), the introduction of a solid–liquid interfacial tension assumes that the interface is at equilibrium and of constant curvature: this can only be an assumption in the case of a solid. The result is that the validity of the determination of γ^{SL} by this method is uncertain.

In our approach, this problem does not explicitly appear because Equation 19.25 is based on a property of the system's response to the variations of its mass (extensity); neither the interfacial area nor the use of Laplace's relation is required. Thus, it seems appropriate to express the previous equilibrium without identifying the tension τ as the interfacial tension γ^{SL} in the Gibbs–Thompson law and to write the depression melting point in the form

$$\Delta T = \frac{T_m V_{(T_x)}}{\Delta H_m} \frac{2\tau}{r} \qquad (19.27)$$

We will proceed on this basis for the following cases.

19.4.1.2 Particle Is Not Spherical

Equation 19.25 can be used to address the case of particles of diverse forms, and especially those corresponding to classical unit cells, for example, a cube of edge a and of volume a^3. Initially, the thermodynamic dimension of the system, m, is determined by multiplying its mass by a number λ and leaving its shape unchanged. In this operation, the volume will also be multiplied by λ (at constant density) whereas the surface area of the cube, $6a^2$, will be multiplied by $\lambda^{2/3}$. The thermodynamic dimension of the system is $m = 2/3$. Then

$$\Delta T = T_m \frac{V_{(T_x)}(2/3)\tau(6a^2/a^3)}{\Delta H_m} = \frac{T_m V_{(T_x)}}{\Delta H_m} \frac{4\tau}{a} \qquad (19.28)$$

The melting point depression of a cubic nanosolid is related to the length of the cube edge. The depression is greater as the edge length decreases.

Our approach has its limitations because the extensities of some structures do not always display the properties of Euler's functions. Consider, for example, a cylindrical particle of height h whose base is of diameter d (Figure 19.3). Suppose that the particle grows without any change in base area but with increasing height h. The cylinder volume is $V = (\pi d^2/4)h$ and the area $A = 2(\pi d^2/4) + \pi dh$. If the cylinder mass is multiplied by λ, only h will be multiplied by λ and consequently the surface area of cylinder will become $A_\lambda = 2(\pi d^2/4) + \lambda \pi dh$. In this case, the area is not an Euler's function of the mass and the relations of NET do not apply. For this reason, it must be systematically verified that the extensity is an Euler's function of the mass before Equation 19.25 can be applied. However, for cylindrical particles that are sufficiently long for the surface area of the base to be negligible relative to the surface area of the sides ($h \gg (d/2)$), the total surface area approximates to an Euler's function of order one of the mass: the dimension of the system is 1. Then,

$$\Delta T = T_m \frac{V_{(T_x)}\tau(4\pi dh/\pi d^2 h)}{\Delta H_m} = \frac{T_m V_{(T_x)}}{\Delta H_m} \frac{4\tau}{d} \qquad (19.29)$$

In this case, the melting point depression of particles is dependent on the cylinder diameter but is independent of its length.

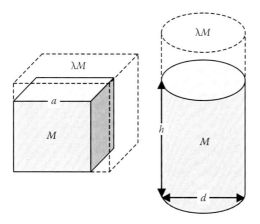

FIGURE 19.3 Two kinds of geometrically defined particles are considered. Their mass is multiplied by λ. The cube increases in size with no change in shape. The cylinder increases in size with no change in base area. (Reprinted from Letellier, P. et al., *Phys. Rev. B*, 76, 8, 2007b. With permission.)

There are many examples which can be used to illustrate the value of applying Equation 19.25 to objects of various sizes and forms. We will consider the case of real nanoparticles whose spatial structure cannot be described simply from classical dimensions (volumes, areas, lengths).

19.4.2 General Case. Power Laws: Consequence of the NET

Consider a given mass, M, of solid presumed to constitute a non-extensive phase. The extensity χ is a homogeneous function of order m of the system mass, so the ratio between χ and M is a homogeneous function of order $(m - 1)$ of the solid mass. This corresponds to

$$\frac{\chi}{M} = kM^{m-1} \qquad (19.30)$$

where k is a characteristic constant of the nonextensive phase considered. This condition combined with Equation 19.25 leads to

$$\Delta T = T_m \frac{M^\circ m \tau k}{\Delta H_m} M^{m-1} \qquad (19.31)$$

This can be written in logarithmic form

$$\ln(\Delta T) = \ln\left(T_m \frac{M^\circ m Y}{\Delta H_m}\right) + (m-1)\ln M \qquad (19.32)$$

To simplify the notation, the product τk is replaced by Y. Y is a characteristic extent of the solid. Its unit, u, depends on the value of m, $u = $ J kg^{-m}. Thus, we show that the melting point depression or elevation of a nanosolid follows a power law of the particle mass.

For the melting point of the nanosolid T_x to find its value T_m when the solid mass becomes large (unlimited phase), it is necessary that $m < 1$. Equation 19.31 implies that for positive m and τ, the melting point of the nanosolid must be depressed as the particle size decreases: this is the classical behavior. However, theory does not require this behavior. Indeed, situations in which m or $\tau(Y)$ are negative can be envisaged, leading to the opposite phenomenon, i.e., an increase in the melting point as particle size decreases. Behavior of this type has been observed, mostly for particles embedded in a matrix (Sun et al., 1997, Zhang et al., 2000b, Lu and Jin, 2001) and for Vycor glass as reported by Christenson (2001).

19.4.2.1 Case of Nanoparticles of Dimension m and of Mass M_P

Above, we considered a mass of solid without specifying its state of division We will now examine the behavior of a solid in the form of identical nanoparticles, having the same property of non-extensive phase of dimension m. The mass of a particle is M_P, its volume, V_P, and its extensity χ_P. We will assume that any addition of solid to the system will only increase the number, N, of nanoparticles. This situation is described by the following relations:

$$M = NM_P$$

$$\chi = N\chi_P$$

$$\frac{\chi}{V} = \frac{\chi_P}{V_P} = kM_P^{m-1} \qquad (19.33)$$

and then

$$\ln(\Delta T) = \ln\left(T_m \frac{M^\circ m Y}{\Delta H_m}\right) + (m-1)\ln M_P \qquad (19.34)$$

The depression or elevation of the melting point then depends on the nanoparticle mass according to a power law.

19.4.2.2 Applications to Real Systems

The validity of the previous relations is difficult to judge because the data in the literature concerning melting point depression are generally given according to the "particle size." Size may be described by a radius for presumed spherical particles, a diameter for nanowires, or a thickness for films. We thus modified the form of the previous equations so as to make them more generally applicable and take into account all measurements reported in the literature without considering particular shapes for the nanosolids.

Our reasoning is as follows. Assume that the nanosolid size is characterized by a geometrical dimension (radius, diameter, thickness), which is denoted by ω. We will suppose that the extensity, χ, and the volume are Euler's functions of ω so

$$V = V_{(\omega)} = \alpha \omega^q$$

$$\chi = \chi_{(\omega)} = \beta \omega^p \qquad (19.35)$$

The pressure difference between the nonextensive solid phase and the solution is then written as

$$^{\text{NE}}P - P = \tau \frac{d\chi}{dV} = \tau \frac{\beta}{\alpha} \frac{p}{q} \omega^{p-q} \qquad (19.36)$$

Similar reasoning leads to the melting temperature variation with the size, ω, of the particles according to a power law:

$$T_\text{m} - T_x = \Delta T = T_\text{m} \frac{V_{(T_x)}(\beta/\alpha)\tau}{\Delta H_\text{m}} \frac{p}{q} \omega^{p-q} = T_\text{m} \frac{M°Y_\omega}{\Delta H_\text{m}} \frac{p}{q} \omega^\eta \qquad (19.37)$$

By convention, we will note $p - q = \eta$ and $Y_\omega = (V_{(T_x)}/M°)(\beta/\alpha)\tau$.

The form of the Gibbs–Thompson relation can be verified by taking the radius as dimension $\omega = r$, $p = 2$ and $q = 3$. In this case, for a compound of density ρ, $Y_\omega = (3\gamma^{\text{SL}}/\rho)$ (J m kg^{-1}).

We will now test the validity of Equation 19.37 under its logarithmic form for published data:

$$\ln(\Delta T) = \ln\left(T_\text{m} \frac{M°Y_\omega}{\Delta H_\text{m}} \frac{p}{q} \right) + \eta \ln(\omega) \qquad (19.38)$$

The plot of $\ln(\Delta T)$ against the logarithm of the geometrical dimension, ω, chosen by the author of the study to characterize nanosolid size will only be a straight line if the extensies and the masses are Euler's functions of dimension, and if variations of the parameters Y_ω, p, and q with the temperature are small.

19.5 Analyses of Published Data

19.5.1 Nanoparticles

First, consider the case where the nanosolid is in the form of nanoparticles. We will examine two series of data.

Example 19.1

We examined the results reported by Sun and Simon (2007) concerning the melting behavior of Al nanoparticles having an oxide passivation layer by differential scanning calorimetry (DSC). Figure 19.4 shows $\ln(\Delta T)$ plotted against the logarithm of the particle radius (in nanometers). The line of correlation is fair, with a y-axis intercept of 4.57 and $\eta = -0.79$, which is lower (in absolute value) than that corresponding to the Gibbs–Thompson relation ($\eta = -1$). From these two parameters, we can calculate the melting point for different values of nanoparticles' radius. We compared in Figure 19.5 the experimental data and the calculated melting point from the parameters determined in Figure 19.4.

The value of η takes into account both the nonextensivity of the mass and the extensity with respect to the measured dimension. This value indicates that if the mass varies with ω (with $q = 3$) as is generally the case, then the extensity, χ, of the nanoparticles would be of power $p = 2.21$ with respect to the nanoparticle radius. This value is higher than 2, which would be characteristic of an area. The result is that the extensity, χ, increases more quickly with ω than an area would.

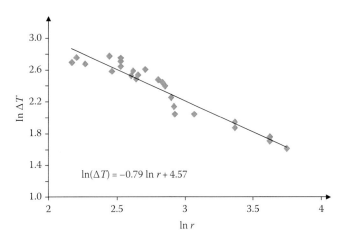

FIGURE 19.4 Plot of $\ln(\Delta T)$ against $\ln(r)$ for Pb particles according to the data of Sun and Simon (2007). The diameter, r, is expressed in nanometers.

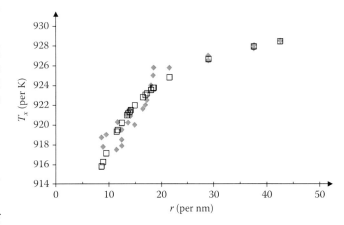

FIGURE 19.5 Experimental (dark points) and calculated (open squares) values of melting point, T_x, against the particle size (radius r) for aluminum nanoparticles (Data from Sun, J. and Simon, S.L., *Thermochimica Acta*, 463, 32, 2007).

Example 19.2

We examined the melting behavior of tin nanoparticles (Lai et al., 1996). These particles are formed by thermal evaporation. For the small amounts of Sn deposited, the films are discontinuous and form self-assembled nanometer-sized islands on the inert substrate. According to the authors, in contrast to embedding metal particles in bulk matrix, this type of sample preparation produces spherical Sn particles with high purity and free surfaces; this system is thus ideal for studies of melting of small metal particles. We plotted the experimental $\ln(\Delta T)$ against the logarithm of the particle radius (in nm) in Figure 19.6.

Once again, an excellent linear correlation is found; the y-axis intercept is 5.995 and the slope is $\eta = -1.20$, lower than -1, which corresponds to spherical particles. Contrarily to the previous case, the extensity, χ, increases less quickly than the area with the particle radius: $p = 1.8$. Thus, it seemed that in the first example the nanoparticles of Al were not spherical.

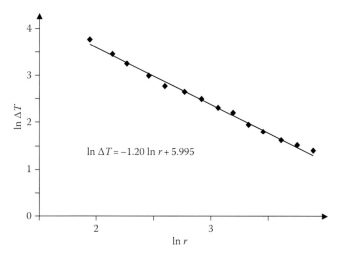

FIGURE 19.6 Plot of $\ln(\Delta T)$ against $\ln(r)$ for tin particles. Data from Lai et al. (1996). ΔT is the melting point depression in degrees Celsius and r is the particle radius in nanometers. (Reprinted from Letellier, P. et al., *Phys. Rev. B*, 76, 8, 2007b. With permission.)

19.5.2 Nanowires

Metal nanowires have attracted a great deal of research interest in recent years, because of their importance in fundamental low-dimensional physics research as well as for technological applications. The melting behavior of Zn nanowires with various diameters embedded in the holes of a porous anodic alumina membrane has been studied (Wang et al., 2006). These nanosolids are particularly interesting because they are assumed to be composed of one-dimensional nanostructures. Differential scanning calorimetry showed that the melting temperature of the Zn nanowire arrays was strongly dependent on nanowire size. We report the values of the logarithm of melting point depression against the logarithm of nanowire diameter (in nanometers) in Figure 19.7.

An excellent linear correlation is obtained with a y-axis intercept of 4.27 and a slope $\eta = -0.57$. This behavior is very different from that observed for nanoparticles; indeed if one supposes that $q = 3$ as above, then p is equal to 2.43. In this case, the extensity increases more quickly than the particle area and obviously does not follow the Gibbs–Thompson law, contrary to the expectations of the authors.

19.5.3 Films

Films of nanometric thickness can be considered alongside nanoparticles: the dimension ω is in this case the film thickness. Two series of data extracted from the literature will be analyzed.

Example 19.3

Bismuth films (Olson et al., 2005): The particles were formed by evaporating bismuth onto a silicon nitride substrate, which was then heated. The particles self-assemble into truncated spherical particles. At mean film thicknesses below 5 nm, mean particle sizes increased linearly with deposition thickness but for 10 nm-thick films, particle size increased rapidly. A plot of the logarithms of melting point depressions against film thickness is given in Figure 19.8. A reasonable linear correlation is obtained. The y-axis intercept is 2.78 and the slope is $\eta = -0.94$, which as for nanowires, is greater than -1. Supposing that $q = 3$, the extensity varies more than the area with the increase of the nanoparticle film thickness ($p = 2.06$).

Example 19.4

The melting point depression phenomenon can also be considered for organic nanosolids. Thus, the behavior of films of triglyceride nanoparticles (Figure 19.9) was studied (Unruh et al., 2001).

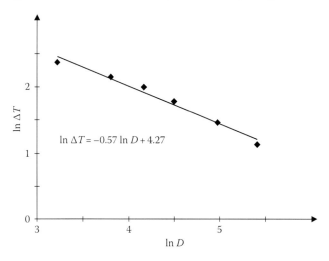

FIGURE 19.7 Plot of $\ln(\Delta T)$ against $\ln(D)$ for zinc nanowires in an alumina matrix (Data extracted from Wang et al. (2006). ΔT is the melting point depression in degrees Celsius and D is the particle diameter in nanometers. (Reprinted from Letellier, P. et al., *Phys. Rev. B*, 76, 8, 2007b. With permission.)

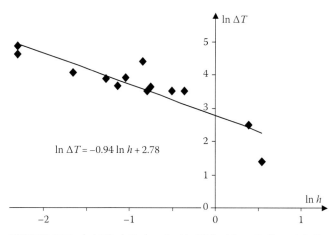

FIGURE 19.8 $\ln(\Delta T)$ plotted against $\ln(h)$ for bismuth films. h is the thickness of the film in nanometers. Data extracted from Olson et al. (2005). (Reprinted from Letellier, P. et al., *Phys. Rev. B*, 76, 8, 2007b. With permission.)

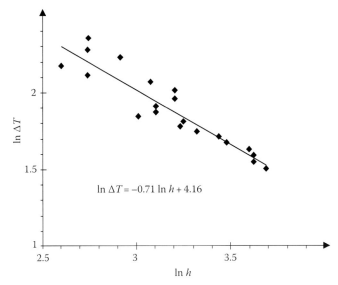

FIGURE 19.9 Plot of $\ln(\Delta T)$ against $\ln(h)$ for films of triglycerides. h is the thickness of the film in nanometers. Data extracted from Unruh et al. (2001). (Reprinted from Letellier, P. et al., *Phys. Rev. B*, 76, 8, 2007b. With permission.)

The linear correlation is satisfactory. The y-axis intercept is 4.16 and the slope is $\eta = -0.71$; this is greater than -1 as was the case for mineral films. Supposing that $q = 3$, the extensity varies more than the area with the increase of the nanoparticle film thickness ($p = 2.29$).

The description of the behavior of films is worse than those for the previous examples. This is probably because films consist of nanosolids in juxtaposition.

19.5.4 Melting Point Elevation

The equations that we have developed can also be used to consider melting point elevation. This situation is mainly found for particles that are coated or embedded in a matrix (Grabaek et al., 1990, Chattopadhyay and Goswami, 1997, Jiang et al., 2000, Zhang et al., 2000a,b, Lu and Jin, 2001). There are many explanations proposed for this phenomenon. All these explanations use the interfacial energy between the liquid compound and the solid constituting the matrix. It is certain that for embedded particles, the borders between the particle and its environment are very badly defined geometrically; we show, below, that our approach can be used to overcome this problem. We analyzed the published values (Lu and Jin, 2001) concerning the variations of melting point with the particle size for nanoparticles of In embedded in an Al matrix. Two kinds of In/Al nanogranular samples were prepared by means of melt-spinning and ball-milling. For melt-spun nanoparticles, the melting point increased as the particle size decreased, whereas for ball-milled nanoparticles the melting point decreases with particle size. We exploited these two data series. For the melting point elevation series, we changed the sign in relation 38 such that extents were positive under the logarithmic terms,

$$\ln(T_m - T_x) = \ln\left(T_m \frac{M^{\circ} Y_{\omega}}{\Delta H_m} \frac{p}{q}\right) + \eta \ln(\omega) \qquad (19.38')$$

The ln/ln correlations are excellent for both melting point elevation and depression data series. We plotted the experimental data and the calculated values (Figure 19.10) with the following parameters.

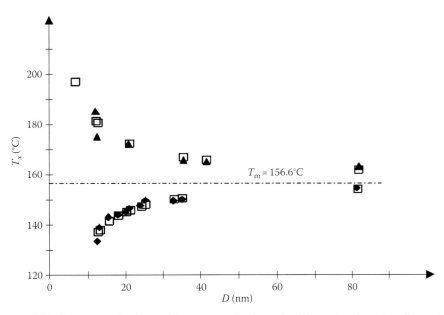

FIGURE 19.10 Experimental (black points) and calculated (open squares) values of melting point, T_x, against the particle size (diameter D) for In nanoparticles embedded in an Al matrix prepared by melt-spinning (triangles) and ball-milling (diamonds) (Lu and Jin, 2001). (Reprinted from Letellier, P. et al., *Phys. Rev. B*, 76, 8, 2007b. With permission.)

For melting point elevation, $\eta = -0.805$ and the y-axis intercept is 5.203. For melting point depression, $\eta = -1.149$ and the y-axis intercept is 5.858.

For the melting point depression, if we take $q = 3$ for the volume dimension, then $p = 1.851$. Thus, the extensity varies less quickly than the area with the particle mass. In the case of the melting point elevation, *the value of τ is negative*. p is equal to 2.195, taking, by convention, $q = 3$. Thus, the extensity varies more quickly than the area with the particle mass. This approach allows a description of the particle and its matrix. According to the matrix structure or the particle shape, one of these two behaviors is observed; the sign of τ expresses and formalizes this difference of behavior. Thus under constraint, the sign of τ can be reversed.

Remark 19.1: To analyze the behaviors of embedded particles, we considered only one extensity χ. This seems sufficient in the case examined to provide a good representation of the experimental results. However, since the growth of a system in contact with a substrate or a matrix is considered, other dimensions can intervene. This is illustrated by Auer and Frenkel in a recent article (Auer and Frenkel, 2003) in which they show the importance of the line tension in the phenomenon of aggregation. This raises the issue of the form of the relations we propose in such a situation. In fact, these relations can easily be generalized and we can use several extensities. Equation 19.16 becomes

$$^{\mathrm{NE}}P - P = \sum_i \tau_i \frac{d\chi_i}{dV} = \sum_i m_i \tau_i \frac{\chi_i}{V} \qquad (19.39)$$

∎

When one of these extensities is a triple line, then

$$^{\mathrm{NE}}P - P = \sum_{i-1} m_i \tau_i \frac{\chi_i}{V} + m_{\mathrm{line}} \frac{\tau_{\mathrm{line}}\chi_{\mathrm{line}}}{V} \qquad (19.40)$$

In this case,

the dimension m_{line} is then equal to 1/3
τ_{line} is the line tension
χ_{line} is the length of the considered line

The exploitation of this relation is possible only if the particle shape is known and if its interfaces with the matrix can be characterized geometrically. An application of this is found in a paper concerning contact angles (Letellier et al., 2007a), in which we show that for nanodrops, the term for line tension can become dominant in Equation 19.40.

19.6 Conclusion

Our conclusion is an answer to a question: "Is the nanometric the world nonextensive?" The response is obviously, yes, the world of nanomaterials is nonextensive. We could furthermore

formulate otherwise this assertion by saying that the thermodynamics is generally nonextensive, but the macroscopic world is a particular case where thermodynamics is extensive. The extension we proposed for thermodynamics allows describing from the same laws the behavior of matter between the nanoscopic and the macroscopic scale.

When the NET relations are applied to the melting of pure components, we showed that for nanomaterials, the melting point depression of nanosolids follows a power law with respect to their geometrical dimensions, as do nanoparticles, nanowires, and films. The NET relations provide a theoretical justification for these behaviors and give a meaning to the various parameters implied in these laws. Note, however, that our findings have their limitations.

The first and undoubtedly most important is that the extensions that we propose address only small particles, without nuclearity being too weak (i.e., made of few atoms or molecules). Many authors assume that for these systems there is a new state of matter, intermediate between the atom and the crystal (Belloni et al., 1982, Belloni, 2006). Kubo (Kubo, 1962) thus suggests that an isolated atom, or a few atoms linked together in a cluster for example in a molecule, should be considered to possess discrete electron levels, introducing a *quantum-size effect*. It has been shown, indeed, that the thermodynamic properties of a metallic cluster vary with the number of atoms, n, which it contains, in solutions (Henglein, 1977) or in the vapor phase (Morse, 1986, Schumacher et al., 1988).

Concerning melting points, the calorimetric measurements reported by Jarrold and co-workers (Shvartsburg and Jarrold, 2000, Breaux et al., 2003) indicate that small clusters of tin and gallium—in the size range of 17–55 atoms—have *higher than bulk* melting temperatures (T_m bulk). A striking experimental result from the same group showed extreme size sensitivity in the nature of the heat capacity of Ga clusters of 30–55 atoms (Breaux et al., 2004). Recently, Joshi et al. (2006) presented a study of extensive ab initio molecular dynamic simulations with Ga_{30} and Ga_{31}, where they attribute the origin of this size sensitivity of heat capacities to the relative order in their respective ground state geometries. It turns out that the addition of even one atom changes the heat capacity dramatically.

The relations that we propose make sense if the aggregates have sufficient nuclearity for average behaviors to appear, and this implies several hundreds of atoms, and sizes higher than 1 nm (a spherical aggregate of silver of 2 nm comprises approximately 2000 atoms).

This condition of size is not the only one that limits the application of the relations we suggest. They can apply only if

- The system is at equilibrium.
- The variation of parameters p, q, Y_{ω}, or their association is largely independent of the temperature. This property cannot be taken as a general condition.

Our analysis shows that for systems of a nanometric magnitude, the laws of thermodynamics must be reconsidered (Hill, 2001b,c).

References

Abe, S. and Rajagopal, A. K. (2003) Validity of the second law in nonextensive quantum thermodynamics. *Physical Review Letters,* 91, 120601-1–120601-3.

Abe, S. Y., Martinez, S., Pennini, F., and Plastino, A. (2001) Nonextensive thermodynamic relations. *Physics Letters A,* 281, 126–130.

Auer, S. and Frenkel, D. (2003) Line tension controls wall-induced crystal nucleation in hard-sphere colloids. *Physical Review Letters,* 91, 015703-1–015703-4.

Belloni, J. (2006) Nucleation, growth and properties of nano-clusters studied by radiation chemistry—Application to catalysis. *Catalysis Today,* 113, 141–156.

Belloni, J., Delcourt, M. O., and Leclere, C. (1982) Radiation-induced preparation of metal-catalysts—Iridium aggregates. *Nouveau Journal De Chimie-New Journal of Chemistry,* 6, 507–509.

Breaux, G. A., Benirschke, R. C., Sugai, T., Kinnear, B. S., and Jarrold, M. F. (2003) Hot and solid gallium clusters: Too small to melt. *Physical Review Letters,* 91, 215508-1–215508-4.

Breaux, G. A., Hillman, D. A., Neal, C. M., Benirschke, R. C., and Jarrold, M. F. (2004) Gallium cluster "magic melters." *Journal of the American Chemical Society,* 126, 8628–8629.

Chattopadhyay, K. and Goswami, R. (1997) Melting and super-heating of metals and alloys. *Progress in Materials Science,* 42, 287–300.

Christenson, H. K. (2001) Confinement effects on freezing and melting. *Journal of Physics: Condensed Matter,* 13, R95–R133.

Defay, R. (1934) *Etude Thermodynamique de la Tension Superficielle,* Gauthier-Villars & Cie, Paris, France.

Defay, R. and Prigogine, I. (1966) *Surface Tension and Adsorption,* Longmans, Green & Co Ltd., London, U.K.

Grabaek, L., Bohr, J., Johnson, E., Johansen, A., Sarholtkristensen, L., and Andersen, H. H. (1990) Superheating and supercooling of lead precipitates in aluminum. *Physical Review Letters,* 64, 934–937.

Graf, K. and Riegler, H. (2000) Is there a general equation of state approach for interfacial tensions? *Langmuir,* 16, 5187–5191.

Henglein, A. (1977) Reactivity of silver atoms in aqueous solutions—(Gamma-radiolysis study). *Berichte Der Bunsen-Gesellschaft-Physical Chemistry Chemical Physics,* 81, 556–561.

Hill, T. L. (2001a) A different approach to nanothermodynamics. *Nano Letters,* 1, 273–275.

Hill, T. L. (2001b) Extension of nanothermodynamics to include a one-dimensional surface excess. *Nano Letters,* 1, 159–160.

Hill, T. L. (2001c) Perspective: Nanothermodynamics. *Nano Letters,* 1, 111–112.

Jaroniec, M., LU, X. C., Madey, R., and Avnir, D. (1990) Thermodynamics of gas-adsorption on fractal surfaces of heterogeneous microporous solids. *Journal of Chemical Physics,* 92, 7589–7595.

Jiang, Q., Zhang, Z., and Li, J. C. (2000) Superheating of nanocrystals embedded in matrix. *Chemical Physics Letters,* 322, 549–552.

Joshi, K., Krishnamurty, S., and Kanhere, D. G. (2006) "Magic melters" have geometrical origin. *Physical Review Letters,* 96, 135703-1–135703-4.

Kubo, R. (1962) Electronic properties of metallic fine particles. 1. *Journal of the Physical Society of Japan,* 17, 975–986.

Kuo, Y.-C., Huang, H.-K., and Wu, H.-C. (2008) Thermal characteristics of aluminum nanoparticles and oilcloths. *Journal of Hazardous Materials,* 152, 1002–1010.

Kwok, D. Y. and Neumann, A. W. (1999) Contact angle measurement and contact angle interpretation. *Advances in Colloid and Interface Science,* 81, 167–249.

Kwok, D. Y. and Neumann, A. W. (2000) Contact angle interpretation: Re-evaluation of existing contact angle data. *Colloids and Surfaces A-Physicochemical and Engineering Aspects,* 161, 49–62.

Lai, S. L., Guo, J. Y., Petrova, V., Ramanath, G., and Allen, L. H. (1996) Size-dependent melting properties of small tin particles: Nanocalorimetric measurements. *Physical Review Letters,* 77, 99–102.

Lavagno, A. (2002) Relativistic nonextensive thermodynamics. *Physics Letters A,* 301, 13–18.

Lechner, R., Stegner, A. R., Pereira, R. N., Dietmueller, R., Brandt, M. S., Ebbers, A., Trocha, M., Wiggers, H., and Stutzmann, M. (2008) Electronic properties of doped silicon nanocrystal films. *Journal of Applied Physics,* 104, 7.

Letellier, P., Mayaffre, A., and Turmine, M. (2007a) Drop size effect on contact angle explained by nonextensive thermodynamics. Young's equation revisited. *Journal of Colloid and Interface Science,* 314, 604–614.

Letellier, P., Mayaffre, A., and Turmine, M. (2007b) Melting point depression of nanosolids: Nonextensive thermodynamics approach. *Physical Review B,* 76, 8.

Letellier, P., Mayaffre, A., and Turmine, M. (2007c) Solubility of nanoparticles: Nonextensive thermodynamics approach. *Journal of Physics: Condensed Matter,* 19, 9.

Letellier, P., Mayaffre, A., and Turmine, M. (2008a) Micellar aggregation for ionic surfactant in pure solvent and electrolyte solution: Nonextensive thermodynamics approach. *Journal of Colloid and Interface Science,* 321, 195–204.

Letellier, P., Mayaffre, A., and Turmine, M. (2008b) Redox behavior of nanoparticules: Nonextensive thermodynamics approach. *Journal of Physical Chemistry C,* 112, 12116–12121.

Li, L. and Zhu, Y. J. (2006) High chemical reactivity of silver nano-particles toward hydrochloric acid. *Journal of Colloid and Interface Science,* 303, 415–418.

Lu, K. and Jin, Z. H. (2001) Melting and superheating of low-dimensional materials. *Current Opinion in Solid State & Materials Science,* 5, 39–44.

Morse, M. D. (1986) Clusters of transition-metal atoms. *Chemical Reviews,* 86, 1049–1109.

Olson, E. A., Efremov, M. Y., Zhang, M., Zhang, Z., and Allen, L. H. (2005) Size-dependent melting of Bi nanoparticles. *Journal of Applied Physics,* 97, 034304.

Pawlow, P. (1910) The dependence of the melting point on the surface energy of a solid body. *Zeitschrift Fur Physikalische Chemie-Stochiometrie Und Verwandtschaftslehre,* 65, 1–35.

Sasaki, M., Osada, M., Higashimoto, N., Yamamoto, T., Fukuoka, A., and Ichikawa, M. (1999) Templating fabrication of platinum nanoparticles and nanowires using the confined mesoporous channels of FSM-16—Their structural characterization and catalytic performances in water gas shift reaction. *Journal of Molecular Catalysis A-Chemical,* 141, 223–240.

Schumacher, E., Blatter, F., Frey, M., Heiz, U., Rothlisberger, U., Schar, M., Vayloyan, A., and Yeretzian, C. (1988) Metal-clusters—Between atom and bulk. *Chimia,* 42, 357–376.

Shi, H. Z., Bi, H. J., Yao, B. D., and Zhang, L. D. (2000) Dissolution of Au nanoparticles in hydrochloric acid solution as studied by optical absorption. *Applied Surface Science,* 161, 276–278.

Shvartsburg, A. A. and Jarrold, M. F. (2000) Solid clusters above the bulk melting point. *Physical Review Letters,* 85, 2530–2532.

Sun, J. and Simon, S. L. (2007) The melting behavior of aluminum nanoparticles. *Thermochimica Acta,* 463, 32–40.

Sun, N. X., Lu, H., and Zhou, Y. C. (1997) Explanation of the melting behaviour of embedded particles; equilibrium melting point elevation and superheating. *Philosophical Magazine Letters,* 76, 105–109.

Toral, R. (2003) On the definition of physical temperature and pressure for nonextensive thermostatistics. *Physica A-Statistical Mechanics and Its Applications,* 317, 209–212.

Tsallis, C. (2002) Entropic nonextensivity: A possible measure of complexity. *Chaos Solitons & Fractals,* 13, 371–391.

Tsallis, C., Mendes, R. S., and Plastino, A. R. (1998) The role of constraints within generalized nonextensive statistics. *Physica A-Statistical Mechanics and Its Applications,* 261, 534–554.

Turmine, M., Mayaffre, A., and Letellier, P. (2004) Nonextensive approach to thermodynamics: Analysis and suggestions, and application to chemical reactivity. *Journal of Physical Chemistry B,* 108, 18980–18987.

Unruh, T., Bunjes, H., Westesen, K., and Koch, M. H. J. (2001) Investigations on the melting behaviour of triglyceride nanoparticles. *Colloid and Polymer Science,* 279, 398–403.

Vives, E. and Planes, A. (2002) Is Tsallis thermodynamics nonextensive? *Physical Review Letters,* 88, 020601-1–020601-4.

Wang, X. W., Fei, G. T., Zheng, K., Jin, Z., and De Zhang, L. (2006) Size-dependent melting behavior of Zn nanowire arrays. *Applied Physics Letters,* 88, 173114-1–173114-3.

Zhang, L., Jin, Z. H., Zhang, L. H., Sui, M. L., and Lu, K. (2000a) Superheating of confined Pb thin films. *Physical Review Letters,* 85, 1484–1487.

Zhang, Z., Li, J. C., and Jiang, Q. (2000b) Modelling for size-dependent and dimension-dependent melting of nanocrystals. *Journal of Physics D-Applied Physics,* 33, 2653–2656.

Zong, R. L., Zhou, J., Li, B., Fu, M., Shi, S. K., and Li, L. T. (2005) Optical properties of transparent copper nanorod and nanowire arrays embedded in anodic alumina oxide. *Journal of Chemical Physics,* 123, 094710-1–094710-5.

20

Phase Changes of Nanosystems

R. Stephen Berry
The University of Chicago

20.1 Introduction

Small nanoparticles, notably those consisting of only tens or hundreds of atoms or molecules, have a kind of behavior in their phase changes that may, at first, seem to violate the laws of thermodynamics. While it is possible to identify solid and liquid forms for many, probably most such particles, they do not obey the long-established Gibbs phase rule, which relates the number of degrees of freedom, f, the number of chemically distinct components c, and the number of phases present in thermodynamic equilibrium, p, through what is probably the simplest equation in thermodynamics, even perhaps in science, simply because it involves only addition and subtraction: $f = c - p + 2$. (The only truly subtle term in this relation is the "2.") The simulations of such clusters of atoms or molecules show clearly that there can be a solid form and a liquid form in a *dynamic* equilibrium within a range of temperatures at a fixed pressure. The phase rule would require that the solid and liquid could be in thermodynamic equilibrium at only one temperature, at a fixed pressure. That is, according to the phase rule, f is 1 if c is 1 and p is 2. In fact, simulations also show that it is possible for more than two phases to coexist in dynamic equilibrium within a range of temperatures and pressures. This chapter explores how this can be, what kinds of phases can be seen in simulations, what kinds of experimental evidence there is for such a coexistence, how one can estimate the range of temperature within which such coexistence is observable (as a function of cluster size), what special properties molecular clusters may exhibit, and what some of the open, unanswered issues are.

Let us begin here with the most basic issue, of how a finite range of coexistence for such small systems is compatible with the Gibbs phase rule. We can begin by writing an equilibrium constant for a solid in equilibrium with a liquid:

$$K_{eq} = \frac{[\text{solid}]}{[\text{liquid}]} = e^{-(F_{\text{liq}} - F_{\text{sol}})/kT} \tag{20.1}$$

but the free energies F_{liq} and F_{sol} can be written in terms of the corresponding chemical potentials μ_{liq} and μ_{sol} and the number N of particles in the system, so

$$F_{\text{liq}} - F_{\text{sol}} = N(\mu_{\text{liq}} - \mu_{\text{sol}}) = N\Delta\mu. \tag{20.2}$$

Now suppose we measure the chemical potential difference in units of kT, i.e., in the same energy units we use for temperature. If the chemical potentials and free energies of the two forms are equal, then of course the equilibrium constant is 1 and the two forms are present in equilibrium, strictly in equal amounts. Suppose now that the system is just a tiny bit away from equilibrium, say only 10^{-10} in units of kT, which is surely a very tiny deviation. This could be positive or negative. But suppose that the system is macroscopic and consists of 10^{20} particles, i.e., $N = 10^{20}$. Then $N\Delta\mu = \pm 10^{20} \cdot 10^{-10} = \pm 10^{10}$, and the equilibrium constant is the exponential of this number. Consequently the equilibrium constant is so large or so small that it is totally impossible for one to see the thermodynamically unfavored form, even this close to equilibrium. This is precisely what sets the constraint that leads to the Gibbs phase rule.

But now let us consider a system made of only 100 particles, so $N = 100$. If $\Delta\mu$ were 1% of kT, i.e., if $\Delta\mu = \pm 10^{-2}$, then the equilibrium constant would be $e^{\pm 1}$ and both solid and liquid could be present in easily observable amounts. In short, the Gibbs phase rule is not a universal rule, but a rule for moderately or very large systems, but not for very small systems.

In the next section, we review the evidence for such coexistence of two or more phases of atomic clusters, and how this coexistence can be represented in ways related to traditional phase diagrams. Section 20.3 describes the phase behavior of some molecular clusters and connects the phase behavior of molecular clusters to that of bulk systems in terms of the order of the transition. Then we show how one can estimate the range of conditions under which coexisting phases can be seen experimentally. Finally, we close with a brief description of some of the open, unanswered questions regarding the phase behavior of small systems.

That solid and liquid phases of clusters could occur at all was anticipated at least tacitly by Sir William Thomson (later Lord Kelvin) when he showed in 1871 that a small particle with positive curvature must have a higher vapor pressure than a bulk, flat surface of the same material, implying that the small system is more volatile than the bulk (Thomson, 1871). In 1909, Pawlow treated the question of the melting points of small particles explicitly to show that they should melt at lower temperatures than the bulk material (Pawlow, 1909a,b). Hill, in his first book on the thermodynamics of small systems, showed schematically that one can expect small systems to exhibit a gradual passage between one phase and another, and that the process becomes sharper as the size of the system increases (Hill, 1963). Experiments, particularly based on electron diffraction, demonstrated the existence of solid forms of clusters over ranges of temperature and pressure first in the late 1970s (Farges et al., 1975, 1977), and continued vigorously in the next decade and after (Farges et al., 1983, 1986, 1987a,b; Valente and Bartell, 1983, 1984a,b; Bartell, 1986). Similarly, liquid forms also revealed themselves in electron diffraction experiments (Bartell et al., 1988), as did transitions between solid and liquid phases (Bartell et al., 1989; Bartell and Dibble, 1990, 1991a,b; Bartell and Chen, 1992). Stace inferred from the ion fragments he produced by electron impact from argon clusters doped with a molecule of dimethyl ether that when the clusters were produced in a low-pressure jet, they were liquid and, from a high-pressure jet, they were solid (Stace, 1983). The range of coexistence appeared in calorimetric experiments with roughly 500-atom clusters of tin atoms (Bachels et al., 2000); Figure 20.1, from that work, compares (in Figure 20.1a) the internal energy of the 430-atom cluster with that of bulk tin, as functions of temperature, The second part of the figure shows the heat capacity of the cluster, the derivative of the curve of internal energy vs. temperature.

A new approach to the experimental investigation of phase changes in clusters came from the studies of ion fragments by the Freiburg group (Schmidt et al., 1997, 1998, 2001a). In this work, singly charged clusters of a single selected size equilibrated thermally at various temperatures and then were photoionized. The fragmentation pattern revealed the internal energy content. The experiments were the first to show not only the melting transition but also the vaporization. Figure 20.2 is a caloric curve taken from the third of the cited works.

Alkali halide clusters, different as they are from metal or molecular clusters, also show bands of coexisting phases (Breaux et al., 2004).

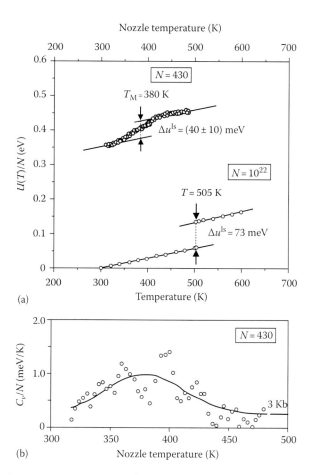

(a)

FIGURE 20.1 Temperature dependence of phase behavior of a 430-atom cluster of tin atoms; (a) internal energy, measured calorimetrically, as a function of temperature for clusters (upper curve) and for bulk solid (lower set of open circles); (b) heat capacity of the 430-atom cluster as a function of temperature, the temperature derivative of the curve in (a). (Taken from Bachels, T. et al., *Phys. Rev. Lett.*, 85, 1250, 2000. With permission.)

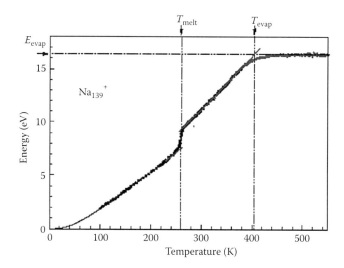

FIGURE 20.2 The experimental caloric curve for the cluster Na_{139}^+, showing both the melting region and the vaporization. (Taken from Schmidt, M. et al., *Phys. Rev. Lett.*, 87, 203402, 2001b. With permission.)

20.2 Evidence from Simulation of Bands of Coexistence of Phases of Small Nanoparticles

The phase behavior of clusters became apparent with the advent of molecular dynamics and Monte Carlo simulation of rare gas clusters. McGinty (1973), then Etters and Kaelberer (Etters and Kaelberer, 1975, 1977; Etters et al., 1977; Kaelberer and Etters, 1977), using the Monte Carlo method, and Briant and Burton (1973, 1975), with molecular dynamics, carried out their simulation studies. McGinty found liquid-like behavior at high temperatures and solid-like behavior at low temperatures, but was unable to distinguish phase changes. In contrast, the molecular dynamics results of Briant and Burton indicated that solid and liquid forms of small argon clusters could be in dynamic equilibrium, rather like chemical isomers, passing back and forth between phases on time scales of many vibrational periods. Figure 20.3a shows a group of caloric curves produced by Briant and Burton; more recent results, examples of which are in Figure 20.3b and c, sometimes differ from those early presentations, but the qualitative ideas certainly have been sustained. The simulations from the 1970s led to an investigation to establish conditions under which such coexistence behavior could occur, in terms of the densities of states of the solid and liquid

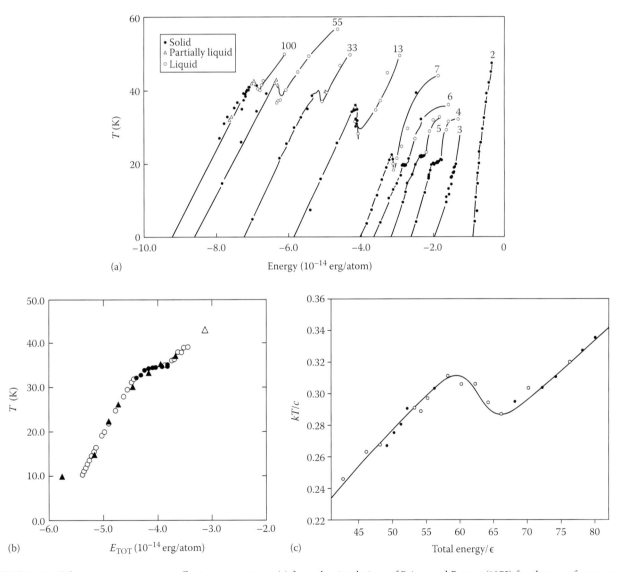

FIGURE 20.3 Caloric curves, energy vs. effective temperature; (a) from the simulations of Briant and Burton (1975) for clusters of argon atoms ranging from 2 to 100 atoms. Open circles indicate liquid, filled circles indicate solid, and triangles indicate an intermediate form sometimes called "liquid surface." (From Briant, C.L. and Burton, J.J., *J. Chem. Phys.*, 63, 2045, 1975. With permission.) (b) Later simulations of the Ar_{13} cluster, showing no negative slope; here triangles are based on Monte Carlo simulations at constant temperature and circles are based on molecular dynamics simulations at constant energy. (From Davis, H.L. et al., *J. Chem. Phys.*, 86, 6456, 1987. With permission.) (c) A later caloric curve for Ar_{55}, derived by two separate methods, clearly showing a region of negative slope. (From Doye, J.P.K. and Wales, D.J., *J. Chem. Phys.*, 102, 9659, 1995. With permission.)

forms (Natanson et al., 1983; Berry et al., 1984a,b). This, in turn, stimulated new molecular dynamics simulations of clusters of various sizes, some at constant energy (Amar and Berry, 1986; Jellinek et al., 1986; Beck et al., 1987), and some at constant temperature (Beck and Berry, 1988; Beck et al., 1988; Berry et al., 1988; Davis et al., 1988). From these simulations, a coherent picture emerged that explained the nature of the coexistence, and many of the conditions under which it can occur. It was not yet apparent at that time what conditions would make the coexistence observable, but it did seem likely that it should be. Here, we now discuss the simulations and their results. Then, in the next section, we examine the theoretical interpretation of the phenomenon.

We concentrate more on molecular dynamics simulations in this discussion than on Monte Carlo, simply because the former come as close as we know how to representing real time-dependent behavior. The simplest kind of molecular dynamics is based on maintaining constant energy. A clear example that reveals the phase behavior of a small cluster is that shown in Figure 20.4 (Jellinek et al., 1986). This gives the time dependence of the mean kinetic energy per atom for three successively higher total energies, for a cluster of 13 atoms bound by a Lennard-Jones potential, simulating an argon cluster. The total energies, per atom, for the three cases are (a) -4.20×10^{-14} erg/atom, (b) -4.16×10^{-14} erg/atom, and (c) -3.99×10^{-14} erg/atom. At the energy of (a), the system is solid-like most of the time, as indicated by the relatively high kinetic energy with small fluctuations. At still lower energies, the system remains in a state of small fluctuations at all times, i.e., is a cold solid. At the energy of (c), the highest of the three, the kinetic energy fluctuates widely, corresponding to the motion of the atoms when the cluster is liquid. Between, in (b), the cluster passes back and forth rather randomly between the high-kinetic-energy solid and the low-kinetic-energy liquid. These, of course, correspond to passage between regions of low potential energy (solid) and high potential energy (liquid). We shall return to this point and to the implication of the shapes of the caloric curves of Figure 20.3 in the discussion on heat capacities of small systems.

The essential point of what appears in Figure 20.4 as a bimodal distribution of kinetic energies is that we can distinguish (in this case) two distinct kinds of behavior. The system dwells long enough in each form that we can evaluate properties, essentially equivalent to internally equilibrated properties for each form, and, by determining the relative fractions of the total time spent in each form, we can determine a long-time average of the distribution between the two forms. The simulations can be carried out under isothermal conditions as well as under constant-energy conditions, and in many ways, the results are very similar. For example, for Ar$_{13}$, see Davis et al. (1987). The caloric curves may differ, but the bimodal (or multimodal) distributions appear and time-average distributions can reveal the same kind of behavior: a unimodal distribution corresponding to a solid phase at low temperature or low energy, a single liquid phase at sufficiently high temperatures, and a bimodal distribution between these two temperatures. Figure 20.5 shows a succession

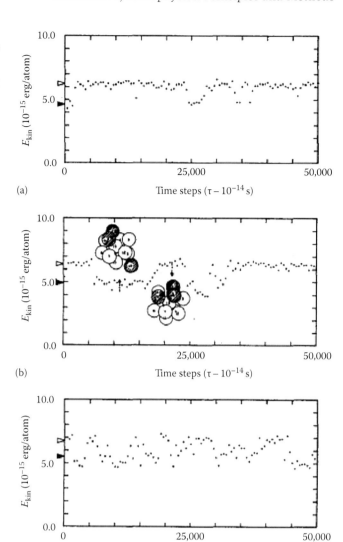

(a)
(b)
(c)

FIGURE 20.4 Time dependences of the short-time average kinetic energy of a 13-atom cluster of atoms bound by a Lennard-Jones potential, simulating a cluster of argon atoms. The time steps are each 10^{-14} s. The short-time averaging to construct each point was taken for 500 such time steps. The total energies, per atom, for the three cases are (a) -4.20×10^{-14} erg/atom, (b) -4.16×10^{-14} erg/atom, and (c) -3.99×10^{-14} erg/atom; (a), the system is mostly solid-like. At the highest energy, (c), the cluster is liquid. In (b), the cluster passes back and forth rather randomly between the high-kinetic energy solid and the low-kinetic energy liquid and one sees a dynamic equilibrium of the two phase-like forms. (From Jellinek, J. et al., *J. Chem. Phys.*, 84, 783, 1986. With permission.)

of distributions of mean potential energies—short-term means, taken over just 500 time steps—for an Ar$_{13}$ cluster at successively higher temperatures (Davis et al., 1987). The passage from only solid, through a range of temperatures (all at the same zero pressure) where the solid and liquid coexist in dynamic equilibrium, to a temperature high enough that only the liquid is stable. This is a very general phenomenon, exhibited by clusters of all sorts.

One aspect of phase coexistence that goes beyond the simple two-phase dynamic equilibrium appears with somewhat larger clusters. Nauchitel and Pertsin showed from simulations that the

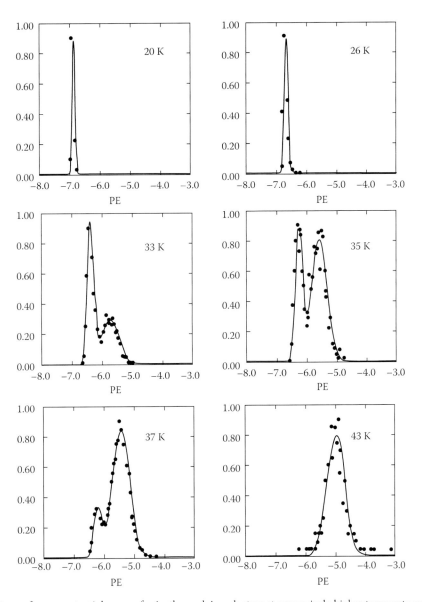

FIGURE 20.5 Distributions of mean potential energy for isothermal Ar_{13} clusters at successively higher temperatures, showing passage from a unimodal distribution corresponding to a solid phase, through a region of bimodal distributions, to a temperature high enough that only the liquid phase is present. (From Davis, H.L. et al., *J. Chem. Phys.*, 86, 6456, 1987. With permission.)

cluster of 55 atoms bound by Lennard-Jones forces, modeling the Ar_{55} cluster, exhibits what they called "surface melting." There was no question that there is a range of temperature in which the 42 atoms on the surface of this cluster are more mobile than at low temperatures at which the system has a completed icosahedral structure (Nauchitel and Pertsin, 1980). Later investigations showed that this phenomenon occurs with clusters of other sizes and kinds including metal clusters, in the range from the mid-40s through at least the 150-range (Cheng and Berry, 1992). However, this work showed, through the use of animations, that "surface melting" was perhaps a bit of a misnomer, in the sense that the phenomenon involves promotion of a few particles from the outermost cluster layer to the outer surface, on which those few particles, roughly 1 in about 30, can move about fairly freely, while the particles remaining in the outer shell undergo large-

amplitude, anharmonic motions but nonetheless motion about a well-defined polyhedral structure. The amplitudes of the anharmonic motions are large enough that instantaneous "snapshots" of the cluster suggest that the structure of the outer shell is amorphous, but the eye recognizes from animations that those atoms actually do oscillate around a well-defined polyhedral structure. This is illustrated in Figure 20.6, which shows the mean square displacement of the atoms, shell by shell, of a simulated Ar_{130} cluster. The lower curves show that the inner shells are fixed at the temperature of 35.2 K, while the uppermost curve shows the large-amplitude motion of the outer-shell atoms. Still further investigations (Kunz and Berry, 1993, 1994) of "surface melting" showed clearly that there can be a range of temperatures in which *three* phase-like forms can coexist in dynamic equilibrium. Figure 20.7 shows the short-time-averaged internal

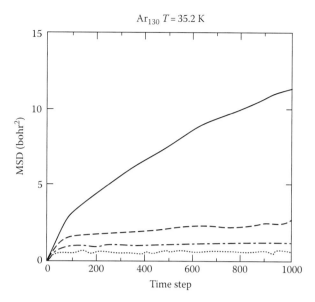

FIGURE 20.6 An example of "surface melting" as illustrated by the large and growing mean square displacement of the outermost shell (upper curve, in contrast to the essentially fixed positions of the inner-shell atoms. The cluster is composed of 130 atoms bound by Lennard-Jones forces, simulating Ar_{130} at a temperature of 35.2 K. (From Cheng, H.-P. and Berry, R.S., *Phys. Rev. A*, 45, 7969, 1992. With permission.)

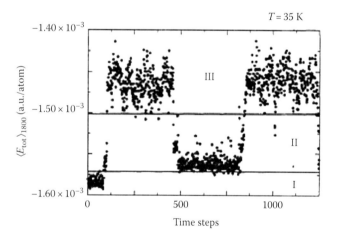

FIGURE 20.7 Results of a constant-temperature simulation of Ar_{55} in the region in which solid (I), liquid (III) and surface-melted (II) phase-like forms are in dynamic equilibrium. (From Kunz, R.E. and Berry, R.S., *Phys. Rev. E*, 49, 1895, 1994. With permission.)

energy as a function of time for an Ar_{55} cluster (Lennard-Jones potential) in a simulation at the constant temperature of 35 K (Kunz and Berry, 1994). The high-energy regions (III) are of course liquid, the lowest-energy regions (I) are solid, and the intermediate region, II, is that of the surface-melted form. We shall return to how this phenomenon can be represented in a kind of phase diagram.

Other kinds of atomic clusters also exhibit the same sort of melting behavior. Both true melting and surface melting have been seen experimentally (Kofman et al., 1989, 1990; Vlachos et al., 1992) and theoretically (Garzón and Jellinek,

1992; Bonacic-Koutecky et al., 1997; Rytkönen et al., 1998). Semiconductor clusters also show similar behavior (Dinda et al., 1994). There is, however, a remarkable situation in at least two kinds of systems that was completely unexpected. In 1871, Thomson (later Lord Kelvin) established that small particles should melt at temperatures lower than their bulk counterparts (Thomson, 1871). This has been demonstrated again and again. The results with large clusters of tin, for example (Bachels et al., 2000), show that clusters of order of 500 atoms melt at temperatures 125 K lower than does bulk tin. However, experiments showed conclusively that smaller clusters of gallium melt at temperatures far higher than bulk gallium (Shvartsburg and Jarrold, 2000; Breaux et al., 2003). Small clusters of tin show the same sort of elevated melting points (Shvartsburg and Jarrold, 1999). The interpretations of this have been presented in two formulations. Both are based on calculations using density functional theory. They attribute the high melting to covalent bonding, in contrast to the metallic bonding of the bulk (Joshi, 2002; Joshi et al. 2003, 2006; Chacko et al., 2004).

20.3 Thermodynamic Interpretation of Bands of Coexisting Phases

The concept of bands of temperature and pressure in which two or more phases could coexist in thermodynamic equilibrium seems at first sight to contradict one of the fundamental concepts of this most general of sciences. The discussion at the outset here clarifies how there is no real contradiction between the Gibbs phase rule, which is valid for macroscopic systems, and the phase behavior of small systems, which are composed of tens, hundreds or perhaps even thousands of particles. That interpretation lies entirely in the behavior of exponentials, and the difference between large and small exponents. Here, we can probe a little deeper into the thermodynamic basis of the simultaneous stability of two or more phases.

Stability of a system corresponds to its being at or in the vicinity of a minimum of a characteristic property. For a mechanical system, this typically means being at a minimum energy. For a system at a fixed temperature, it is a free energy that is a minimum when the system is stable and at rest. However, we distinguish between a local minimum and the most stable of all attainable states, the global minimum. A system may well remain for a very long time in a minimum that is not the lowest. People live in towns in the mountains that are far above the gravitational minimum, for example. Likewise, chemical species with more than one isomeric form can remain for arbitrarily long times in a stable structure that is not the form of lowest energy or free energy. This is precisely the situation when two or more phase-like forms of a nanoparticle are present either in unequal concentrations in a large ensemble or a single nanoparticle passes back and forth between forms, spending unequal amounts of time in each phase-like form.

A way to envision this kind of equilibrium lies in the concept of a kind of order parameter γ, which we can call a "nonrigidity

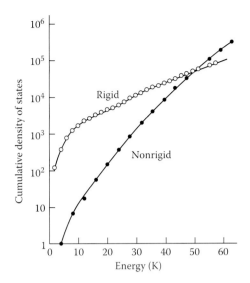

FIGURE 20.8 Schematic representation of the densities of states of a typical system of 6 or 7 inert gas atoms near their rigid and nonrigid limits, corresponding to solid and liquid forms of the same system. (From Berry, R.S. et al., *Phys. Rev. A*, 30, 919, 1984b. With permission.)

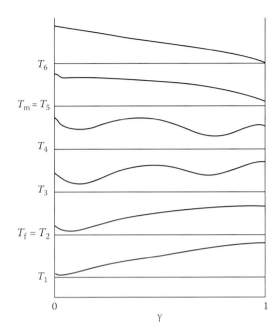

FIGURE 20.9 Schematic curves of free energies at different temperatures, as functions of an order parameter characterizing the degree of rigidity. The left limit, $\gamma = 0$, corresponds to a completely rigid solid; the right limit, $\gamma = 1$, to an extremely nonrigid system. The minima correspond to locally stable states, solid for low γ and liquid for the higher value. The temperatures increase from the lowest, T_1, to the highest, T_6. This figure describes a system with only two coexisting states. They can coexist in the regions of T_3 and T_4. (From Berry, R.S., *Theory of Atomic and Molecular Clusters*, Jellinek, J. (ed.), Springer-Verlag, Berlin, Germany, 1999, 7.)

parameter." We can think of an extreme in which $\gamma = 0$ in the limit that the particles cannot move at all, and an opposite extreme where $\gamma = 1$ and the particles are completely free as a very dilute gas. A real solid thus corresponds to a low but non-zero value of γ because the particles do exhibit some vibrational motion. Likewise, a real liquid corresponds to a much higher value of γ but to something much below 1 because the motion of the particles in a liquid is far from completely free. We can recognize the way phase changes occur first by comparing the density of states of a system near its rigid limit with the corresponding density of states for the same system but near its nonrigid limit, and especially, we want to see how these two depend on energy. Figure 20.8 shows how the cumulative density of states of the solid dominates at low energies, but is overtaken by the nonrigid, liquid states at high energies. The result is that at high energies, the entropic contribution to the free energy, whether Helmholz or Gibbs, of the liquid makes it the more thermodynamically favored form.

The free energy can, at least in some metaphoric fashion, be treated as a continuous function of an order parameter characterizing the degree of rigidity or nonrigidity. One can envision curves of free energy as functions of temperature; minima in such curves correspond to stable states. At low temperatures, we expect to see only one minimum, in the rigid region, corresponding to a solid. At sufficiently high temperatures, again the curve should show only a single minimum, that of the liquid. For small systems, the nanoscale systems we are considering here, we find a range of temperatures (for a given pressure or volume) in which the curves exhibit two significant minima. This behavior is just what the curves of Figure 20.9 show. These curves represent schematically the free energies of a system small enough to show a band of coexistence for just two phases. Were the system very large, macroscopic, then the temperature

range of coexisting minima would be so narrow that it would be unobservable; the two minima would appear in the curve at only the sharp temperature corresponding to the classical melting point.

Temperatures T_2 and T_5 of Figure 20.9 carry other designations, T_f and T_m, respectively. These two are the *freezing limit* and *melting limit* temperatures. Below T_f, there is no local minimum in the nonrigid region, meaning that there is no locally stable liquid state at temperatures below this. Likewise, the local minimum in the rigid region disappears at T_m, so there is no stability above this temperature for a solid form. The two phases thus can coexist only within sharply bounded temperature limits. Of course, T_f and T_m presumably depend on pressure, but that dependence has not been studied. Furthermore, the question is still open as to whether or how precisely the freezing and melting limits could be determined experimentally. At issue is how large the influence of fluctuations would be.

In situations such as that portrayed in Figure 20.7, where more than two phases of a cluster system coexist, it may be possible to use a single order parameter, as the nonrigidity parameter γ is being used here, to portray the multiphase coexistence. Figure 20.10 is a schematic representation of that situation. In addition to the local minima to the left or ordered side and the right or nonrigid side, at temperatures T_3 and T_4, we see a local minimum between the other two, at T_5, this has become just an inflection point. The extent to which such a portrayal is useful will depend

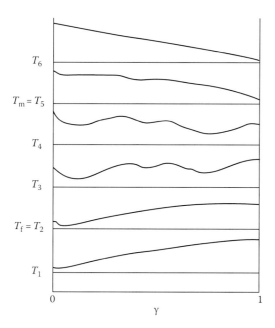

FIGURE 20.10 Schematic curves of free energies at successively higher temperatures, for a case in which three locally stable, phase-like forms may coexist. In this hypothetical example, in contrast to the surface-melting cases of clusters of 40–60 argon atoms, the curves suggest that the phase with a midrange value of γ does not appear at a temperature lower than the temperature where the liquid appears. However, the solid does disappear in this illustration at a temperature lower than that for the intermediate phase. (From Berry, R.S. and Levine, R.D., *Progress in Experimental and Theoretical Studies of Clusters*, Kondow, T. and Mafuné, F. (Eds.), World Scientific, Singapore, 2003, 22. With permission.)

on whether a single order parameter is sufficient to characterize the different phases. It may be that one will suffice in many situations, but if one of the phases differs from the others in some characteristic that is not clearly related to the first-chosen parameter, then a second may be required. For example, we can imagine a third phase in which a system becomes paramagnetic, while the other phases are diamagnetic. In such a case, we may well want to introduce a second order parameter.

20.4 Phase Diagrams for Clusters

A general and powerful tool for understanding phase coexistence of bulk materials has long been the phase diagram. In a traditional phase diagram for a single, pure substance, one typically displays the boundary curves separating regions of stability of individual phases, as functions of the relevant variables, most often pressure p and temperature T. The phase rule determines that sharp curves are the boundaries between regions of stability and only on those curves can phases coexist. The liquid–vapor boundary curve terminates at the critical point; the solid–liquid, solid–vapor and liquid–vapor curves intersect at the triple point. But we have seen that the phase rule is inadequate for describing small systems, and that their coexisting phases over bands of temperature and pressure would make the traditional phase diagram inadequate to portray the behavior of such systems. We can, however, make

a bit of an extension to the traditional phase diagram to give us just that description (Berry, 1994). We simply need to introduce one more variable. The most convenient for one kind of extended phase diagram is a variable that ranges between −1 and 1, namely a distribution we call D, which is a simple function of the equilibrium constant K_{eq} of Equation 20.1:

$$D = \frac{K_{eq} - 1}{K_{eq} + 1} \tag{20.3}$$

which is −1 if the system is all solid and +1 if the system is all liquid. Hence the extended phase diagram simply adds an axis for D to augment those of p and T, and extend the diagram into a third dimension. For a large system, the transition between $D = -1$ and $D = +1$ is too sharp and sudden to reveal any region of intermediate values. The curve of coexistence simply jumps from the plane of $D = -1$ to that of $D = +1$ where the normal, two-dimensional phase boundary is. This is the situation in Figure 20.11a. However if the system is small, then there is a region in which D takes on intermediate values, a specific value for each pressure and temperature. This behavior is what the extended phase diagram of Figure 20.11b shows. Moreover the shaded regions in the back plane indicate the temperatures and pressures of the freezing and melting limits, T_f and T_m.

It is sometimes useful to use a different kind of phase diagram for clusters and nanoscale particles, especially when one is dealing with three phases in equilibrium. This second kind of diagram is somewhat analogous to the pressure–volume curves one draws for a van der Waals system in pressure–volume space. In that situation, one has a curve of stationary points; where the curve slopes downward, the points correspond to locally stable states; where the curve slopes upward, the points correspond to unstable states, and we connect those points to the left and right branches of the downward sloping regions, corresponding to stable phases. In Figure 20.12, we have a schematic curve of the stationary points for a system that may be solid, liquid, or surface-melted (Berry, 1997). The horizontal axes are the scales of defect concentrations $\rho_{surface}$ and ρ_{core}, for the surface and the core, respectively. The vertical axis is T^{-1}, the inverse of the temperature. The heavy curve is the locus of stationary points for the system. Where the curve slopes downward, starting from the top of the diagram, the coldest temperature, the points correspond to stable states; where the curve slopes upward, the corresponding states are unstable. We see an initial downward slope where both defect densities are very small; this corresponds of course to the solid. The next downward sloping region is in a region where $\rho_{surface}$ is fairly large but ρ_{core} is still very low; this is obviously the region of stability of the surface-melted form. The next and last downward-sloping region is one where both $\rho_{surface}$ and ρ_{core} are large, the region of stability for the liquid. The temperature ranges where two or three downward slopes occur are the regions of stability for coexisting phases. The limits of these zones are shown by dashed lines. Between the top and second dashed lines, the solid and surface-melted forms can coexist;

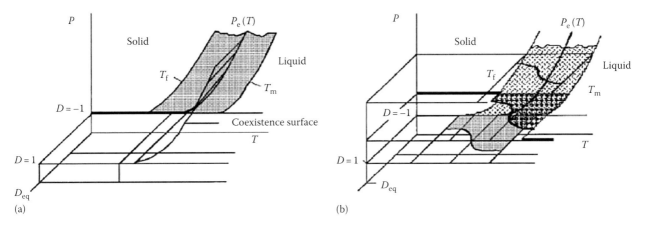

FIGURE 20.11 Extended phase diagrams that use the distribution variable D of Equation 20.3 for a third dimension; (a) a schematic solid–liquid phase diagram for a macroscopic system, in which the passage from the plane of $D = -1$ to $D = +1$ is so sharp that the third dimension is unnecessary; (b) a schematic solid–liquid phase diagram for a small cluster, in which there is a temperature–pressure *band* in which the equilibrium composition contains both phases. The shaded regions in the rear plane of $D = -1$ indicate the limits set by T_f and T_m, the freezing and melting limits. (From Berry, R.S., *Theory of Atomic and Molecular Clusters*, Jellinek, J. (ed.), Springer-Verlag, Berlin, Germany, 1999, 14.)

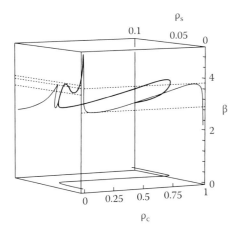

FIGURE 20.12 A second kind of extended phase diagram to illustrate coexistence of three phases. The vertical axis is T^{-1}, the inverse of the absolute temperature. The two horizontal axes, $\rho_{surface}$ and ρ_{core}, represent the densities of defects in the surface layer and in the core, respectively. The heavy curve in the three-dimensional space is the locus of stationary points. If we follow it from the top, the lowest temperature, the points on the curve are points of stability whenever the curve is dropping, and are stationary points of instability, analogous to the upward-sloping portion of curves for a van der Waals system, in a pressure-volume phase diagram. The light solid curves are projections of the heavy curve on the left side and front planes, to exhibit explicitly where the heavy curve slopes up and where it slopes down. There are three regions of downward slope: that near the region where both defect densities are nearly zero, which of course corresponds to the region of stability of the solid; then a region in which $\rho_{surface}$ is nonzero but the core is essentially defect-free, and corresponds to a stable surface-melted form; and finally, to the right, is a region corresponding to ρ_{core}, the liquid region. Between the two uppermost dashed lines, the solid and surface-melted forms can coexist; between the second and third dashed lines, the three phases can coexist; then, between the third and fourth dashed lines, the surface-melted and liquid phases can coexist. Above the top dashed line and below the lowest, only a single phase is stable. (From Berry, R.S., *Theory of Atomic and Molecular Clusters*, Jellinek, J. (ed.), Springer-Verlag, Berlin, Germany, 1999, 19.)

between the second and third dashed lines, all three phases can coexist. Then, between the third and fourth dashed lines, just the surface-melted and liquid phases are stable together. Above the top dashed line, the only stable form is the solid; below the fourth, only the liquid is stable.

20.5 Observability of Coexisting Phases

One concern arises immediately when one realizes the physical possibility of two or more phases coexisting over a band of temperatures at a given pressure. This is the question of how wide that temperature band is, and how easily it would be to observe such coexistence. Another way to phrase the question is to ask how large a system would show such coexistence over an observable temperature range. To address this question, we need to determine the temperature range within which the equilibrium constant K_{eq} is relatively near 1, or the distribution parameter γ is close to zero (Berry and Smirnov, 2009). Let us arbitrarily decide what we want "relatively close" to mean, in terms of there being an observable amount of the unfavored phase present. Specifically, let us set out to find the range of temperature such that

$$0.1 \leq K_{eq} \equiv \exp(-\Delta F) \leq 10$$

with ΔF expressed in units of temperature. The free energy change associated with the phase change is, of course,

$$\Delta F = \frac{\Delta E}{T} - \Delta S \tag{20.4}$$

which is zero at the classical melting point. Now let us define the temperature range of observability of the phases as

$$\partial F = \frac{\Delta E \partial T}{T^2} \tag{20.5}$$

and, because the range of observability of the coexistence is narrow, we can safely suppose that both ΔE and ΔS are constant within that range. But our assumption that the range of K_{eq} lies between 0.1 and 10 tells us that the free energy change ΔF lies between −2.3 and +2.3; that is, $\delta F \approx 4.6$ or, for convenience, 5. Hence we can write, as a useful approximation for the range of observable coexistence,

$$\partial T \approx \frac{5T}{\Delta S} \quad \text{or} \quad \frac{\partial T}{T_{m}} \approx \frac{5}{\Delta S} \qquad (20.6)$$

if we take as T_{m} the temperature at which $\Delta F = 0$ reduces the problem to estimating the entropy change in the change of phase.

The estimation of ΔS breaks naturally into two parts: a configurational part and a vibrational part. For metallic clusters, it may sometimes be appropriate to include an electronic contribution, but we assume here that the electronic densities of states of the solid and liquid forms (and other phases, if they are present) are approximately equal, so there is no significant contribution to the entropy change from the electrons. It is also useful to separate two cases, one in which all the phase changes occur in the outer layer of the cluster, and the other in which the entire cluster changes phase.

In the simpler case of the outer layer "melting," we know that the process involves promotion of one or more particles from that layer to become "floaters" on the surface of the system, while the vacancy left in the outer shell provides an increased volume in which the particles in that shell can move. Suppose that the radius of the initial, unexcited cluster is R_0, the radius of the atoms is r, and their mean nearest-neighbor distance in the unexcited system is d. The initial volume available to each particle is $v_0 = (4\pi/3)(d/2)^3$. Simulations indicate that roughly 1/15 of the n_s surface atoms is promoted to become a floater when "surface melting" occurs. The promoted particles move in a volume of approximately $(4\pi/3)[(R_0 + 2r)^3 − R_0^3]$. Hence the total increase in available volume for the cluster's particles following surface melting, V_{sm}, is

$$\Delta V_{sm} = \frac{n_s}{15} \frac{4\pi}{3} \left\{ \left[\left(R_0 + 2r\right)^3 − R_0^{\,3} \right] − \left(\frac{n_s}{15} − 1 \right) r^3 \right\} \qquad (20.7)$$

This makes the configurational change in the entropy

$$\Delta S_{sm} = \ln \left(\frac{\Delta V_{sm}}{v_0} \right) \qquad (20.8)$$

For surface-melting of clusters with partly filled outer shells, we can treat this entropy change slightly differently. If the outer shell has less than two-thirds of its sites filled, we can simply assume that the volume available to each of the surface-shell atoms is that empty space. For systems with two-thirds or more of the surface sites filled, we can assume that the volume made available on melting is that due to promotion plus the empty volume in the partly filled surface.

There is one other contribution to the entropy still to be estimated. This is the change in the vibrational contribution. This can be done with a rather crude approximation, specifically that the vibrational entropy per atom is a linear function of temperature. From simulations, the temperature dependence of the vibrational entropy, per atom, based on a closed-icosahedral Ar_{13} cluster, is simply $\Delta s_{13} = 2.2T + 0.13$, in the same dimensionless units based on the pair dissociation energy D. For open-shell clusters with less than two-thirds of the available sites occupied in the outer shell, the treatment above is adequate for that outer shell and one can use the vibrational entropy contribution, per atom, as derived for the Ar_{13} cluster for the next-outermost shell.

In this context, we can also estimate the energy associated with promoting an atom to be a floater. This is essentially the change in the number of nearest neighbors times the energy D of each pairwise interaction at equilibrium. For an icosahedral cluster, the number of contacts, due to promotion to become a floater, changes from 6 to 3, so, in units of D, this change is

$$\Delta E_{sm} = \frac{3n_s}{15} \qquad (20.9)$$

Returning now to our problem of finding the observable size range on the basis of Equation 20.6, we see that high-melting clusters must show wider coexistence ranges than low-melting clusters. As an example, if we make the rough estimate that the entropy change per atom in the size range of 50–100 atoms is the amount one obtains for the 55-atom cluster, namely 45/55 or 0.82 per atom, then the range of observable coexistence based on at least 10% of the minority phase is only 0.3 K, from 47.07 to 47.37 K. However, if the melting point were in the range of 270 K, as it is for the high-melting Na_{139} to Na_{147} (Schmidt et al., 1998), the predicted coexistence range would be close to 2 K. In fact, the results from this very crude estimate seem narrower (and hence more pessimistic about the observability) for these clusters than was seen in the experiments (Schmidt et al., 1998). However, the estimate used here makes no allowance for the difference in behavior of argon clusters and metal clusters, clearly a significant factor here. In practice, one can generally make more firmly based estimates of that entropy change.

20.6 Phase Changes of Molecular Clusters

There are far fewer studies of molecular clusters, and particularly, the phase changes of molecular clusters than of their atomic counterparts. However, there have been enough experimental and theoretical investigations of these to show that they can reveal new insights and new phenomena that do not appear in the behavior of bulk systems. Here we shall focus on what we have learned from clusters of octahedral molecules such as SF_6, in some ways the closest molecular parallels to atomic clusters.

Because of their nonspherical geometries, molecules in clusters have rotational degrees of freedom. As a result, clusters of molecules, and macroscopic solids as well, can have varieties of solid phases that do not occur for structures built from simple atoms. This phenomenon reveals itself in clusters of octahedral molecules. Moreover, the transitions between these solid phases exhibit the characteristics we associate with transitions of first and second order—but the characteristic order of a particular solid–solid transition in clusters of tens or hundreds of atoms may differ from the order of what we must consider the same transition in the bulk solid. Let us briefly review the difference between first- and second-order transitions. The most important difference between these for characterizing clusters is this: a first-order transition has a *latent heat*, a discrete change of energy as the system passes from one form to the other. The phases can coexist because each corresponds to a minimum in the free energy, but, as we saw, in a bulk system, the phase with the higher free energy, the unfavored phase, is present only in unobservably small quantities unless the two phases have equal free energies. A second-order transition has no latent heat; the internal energy changes continuously with the change of phase, but the *derivative, the slope* of the energy as a function of temperature is discontinuous in a traditional second-order transition. Hence the heat capacity, $\partial E/\partial T$ (for whatever conditions are specified) is effectively infinite at the point of a classical, bulk phase transition because the inverse of the heat capacity is zero there; the temperature does not change at all until the system has absorbed the full latent heat of the transition.

A finite system, a cluster, has a smooth heat capacity that remains finite through its equivalent of a first-order transition because there is a continuous change in the relative amounts of the coexisting phases as the temperature changes. Strictly, one might expect zeros in the inverse of the heat capacity of a cluster at the temperatures T_f and T_m of Figure 20.9, the temperatures at which one or the other second minimum appears. At these temperatures, there should be an effective latent heat, corresponding to an equilibrium fraction of the minority phase appearing or disappearing at the temperature at which the second, higher minimum can sustain a nonzero amount of material. That this quantity does not increase or decrease from zero is a consequence of the zero-point energy level appearing when the higher-energy minimum is just deep enough to sustain that level. Strictly, in a quantum system, the equilibrium population of the minority phase should change from zero to a finite, nonzero value when that zero-point level appears. In contrast, the heat capacity of a classical cluster would behave continuously. However, there is no experimental evidence at this time for any such discontinuity, and it is not clear how difficult it would be to observe such a thing.

As an illustrative system that has been studied both experimentally and theoretically, we examine collections of the octahedral molecule TeF_6. The bulk material exhibits several phases (Bartell et al., 1987): liquid to 233 K, a body-centered cubic (bcc) structure from 233 to about 50 K and an orthorhombic structure below that. Clusters of roughly 100 molecules or more of

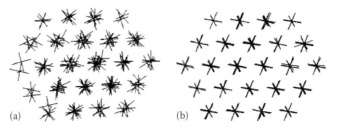

FIGURE 20.13 (a) Randomly oriented molecules of TeF_6 in an 89-molecule cluster with body-centered cubic structure; (b) the same cluster in its monoclinic form, viewed along the axis with respect to which the clusters have orientational order; along either axis perpendicular to this, the molecules have no orientational order in this phase. (From Proykova, A. et al., *J. Chem. Phys.*, 115, 8583, 2001. With permission.)

TeF_6 or SF_6 solidify first into a bcc structure (Bartell et al., 1987; Maillet et al., 1998) also show a monoclinic structure between 100 and 50 K that does not appear in the bulk systems (Bartell et al., 1989; Proykova et al., 1999, 2001). Simulations show that the cubic and monoclinic phases coexist in the same manner as the solid and liquid clusters of atomic systems. This is a clear indication that this transition has the properties we associate with it being first order. Two local minima are present for a range of temperatures and the system can stay in equilibrium in each for some time, while the two forms are in dynamic equilibrium on a long time scale. Small clusters, e.g., of 59, 89, or 137 molecules, show the expected band of coexistence. In the cubic phase, the molecules are completely randomly oriented, and appear to be able to rotate relatively freely. In the monoclinic phase that results from cooling the bcc cluster, the molecules are ordered around only one axis. Figure 20.13 compares these two structures (Proykova et al., 2001).

At a still lower temperature, 60 K in the case of the 89-molecule cluster of TeF_6, the cluster undergoes another phase transition, this time to a completely orientationally ordered monoclinic structure. In this case, one sees no coexisting phases; as the system cools, it passes, with no latent heat, from the partially ordered to the completely ordered phase. This is clearly the analogue of a classic second-order phase transition (Proykova et al., 2001). Figure 20.14 shows this behavior: a low-temperature solid–solid transition to the completely ordered state and a higher-temperature transition between the partially ordered solid and the randomly oriented solid.

The TeF_6 system is not unique; clusters of SF_6 show, in simulations, the same behavior but the barriers between different forms are lower than for the tellurium compound, so the coexistence range is narrower and coexistence is more difficult to find (Proykova et al., 2001). Nevertheless both systems show two solid–solid phase transitions, one first-order and one second-order. Similarly, clusters of different sizes exhibit this behavior. The 59-molecule cluster of TeF_6, for example, shows a first-order transition between bcc and monoclinic structures at about 76 K and a second-order transition between a partially oriented monoclinic structure and a fully oriented structure at approximately

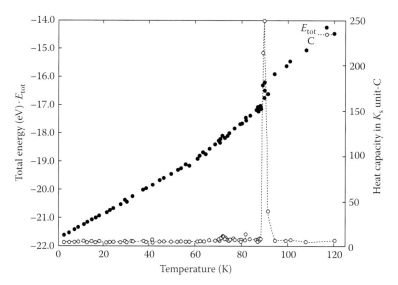

FIGURE 20.14 The caloric curve (dots) and heat capacity for $(TeF_6)_{89}$. The jump in the curve at approximately 90 K corresponds to the monoclinic-to-cubic transition; the change in slope of the caloric curve at 60 K corresponds to the transition from the fully orientationally ordered monoclinic or orthorhombic state at lower temperatures to the monoclinic state with orientational order around only one axis. (From Proykova, A. et al., *J. Chem. Phys.*, 115, 8583, 2001. With permission.)

30 K (Proykova et al., 2003). Clusters consisting of 28 or more TeF_6 molecules exhibit this low-temperature continuous, second-order-like transition.

The first-order character of the bcc-monoclinic transition arises from the change in structure and symmetry that the system undergoes there. In contrast, the second-order transition can be thought of as a simple passage between a lower symmetry to a higher symmetry in which the group of higher symmetry contains that of the lower symmetry as a subgroup. No fundamental structural change or movement of the molecules occurs there (Proykova et al., 2002).

Electron diffraction has been the most effective tool to study phase behavior in molecular clusters. For example, chlorinated hydrocarbons show liquid-to-solid transitions that this technique can recognize, and, likewise, one can see a solid-to-solid transition in SeF_6 clusters of just the kind we have been considering (Dibble and Bartell, 1992a,b).

20.7 A Surprising Phenomenon: Negative Heat Capacities

One interesting aspect of cluster behavior, but not something strictly a property of these small systems, is a striking kind of behavior that shows itself in some *noncanonical* conditions. Most frequently, this is observable in some systems constrained to be at constant energy, rather than at constant temperature. This was already evident in some of the caloric curves in Figure 20.3, such as that for Ar_{100}: There is a region in which the caloric curve of $T(E)$ derived from constant-energy simulations shows a negative slope. Even if some of those early curves (Briant and Burton, 1975) do not look quite like those computed more recently, there is no question now that some clusters, under constant-energy

conditions, show regions of negative heat capacity, precisely in the range of coexisting phases. The rationale for this, for such systems, was presented in the 1990s (Labastie and Whetten, 1990; Wales and Berry, 1994): Suppose a system can reside either in a deep, narrow potential minimum corresponding to a solid form, or in a broad, liquid-like region of high potential energy, much like a high rolling plain. The energy is constant, so the system's kinetic energy is low in the liquid region and high in the solid region. It is simplest here to use the mean kinetic energy per degree of freedom as the measure of effective temperature, although one can come to the same conclusion using a definition based on the variation of energy with respect to the microcanonical entropy, the analogue for a microcanonical system of the traditional canonical relation $T = \partial E / \partial S$ (Jellinek and Goldberg, 2000). The reasoning is thus: the density of states in the broad, high-energy liquid region increases significantly faster with energy than does the density of states in the solid region. Hence as the energy increases, more and more systems move into the liquid region, where the potential energy is high, so the kinetic energy is low, making the effective temperature low. This means that as the energy increases, the mean kinetic energy of the entire ensemble goes *down*, corresponding to a negative heat capacity (Berry, 2004). This has been seen in a variety of simulations, and in experiment as well (Schmidt et al., 2001a; Gobet et al., 2002). This is an example of a phenomenon that illustrates an often-overlooked subtlety of thermodynamics, akin in that sense to the way the phase rule does not apply to small systems.

20.8 Summary

We have surveyed phase changes of clusters, beginning with the evidence from simulations and experiment that these systems can show coexisting phases over ranges of temperature and pressure, in apparent violation of the Gibbs phase rule. We have then seen how that rule is a consequence of the large numbers of particles in macroscopic samples, and how the behavior of small systems merges into that of their bulk counterparts as the number of their constituent particles increases. We have examined the thermodynamics of the phase equilibria of atomic clusters and showed how one can use thermodynamics to estimate the range within which coexisting phases could be present in observable amounts. Clusters of many kinds of elements exhibit coexistence. However, clusters of gallium and of tin, in certain size ranges, show anomalously *high* melting ranges. Extended phase diagrams can show how up to three phases can coexist for clusters. We then turned to molecular clusters and examined the way these can exhibit both first-order and second-order transitions. Finally we saw how noncanonical ensembles of clusters may show negative heat capacities in specific temperature ranges. The overall view we have tried to convey is that the phase behavior of clusters is not inconsistent at all with thermodynamics, but that these systems can often show a rich variety of properties that may seem at first to confound conventional ideas but, with a deeper understanding, are in fact entirely reasonable and interpretable.

References

Amar, F. and Berry, R. S., 1986. The onset of nonrigid dynamics and the melting transition in Ar_7. *J. Chem. Phys.* 85: 5943–5954.

Bachels, T., Güntherodt, H.-J., and Schäfer, R., 2000. Melting of isolated tin nanoparticles. *Phys. Rev. Lett.* 85: 1250–1253.

Bartell, L. S., 1986. Diffraction studies of clusters generated in supersonic flow. *Chem. Rev.* 86: 492–505.

Bartell, L. S. and Chen, J., 1992. Structure and dynamics of molecular clusters. 2. Melting and freezing of CCl_4 clusters. *J. Phys. Chem.* 96: 8801–8808.

Bartell, L. S. and Dibble, T. S., 1990. Observation of the time evolution of phase changes in clusters. *J. Am. Chem. Soc.* 112: 890–891.

Bartell, L. S. and Dibble, T. S., 1991a. Electron diffraction studies of the kinetics of phase changes in molecular clusters. Freezing of CCl_4 in supersonic flow. *J. Phys. Chem.* 95: 1159–1167.

Bartell, L. S. and Dibble, T. S., 1991b. Kinetics of phase changes in large molecular clusters. *Z. Phys. D* 20: 255–257.

Bartell, L. S., Valente, E., and Calliat, J. C., 1987. Electron diffraction studies of supersonic jets. 8. Nucleation of various phases of SF_6, SeF_6 and TeF_6. *J. Phys. Chem.* 91: 2498–2503.

Bartell, L. S., Sharkey, L. R., and Shi, X., 1988. Electron diffraction and Monte Carlo studies of liquids. 3. Supercooled benzene. *J. Am. Chem. Soc.* 110: 7006–7013.

Bartell, L. S., Harsanyi, L., and Valente, E. J., 1989. Phases and phase changes of molecular clusters generated in supersonic flow. *J. Phys. Chem.* 93: 6201–6205.

Beck, T. L. and Berry, R. S., 1988. The interplay of structure and dynamics in the melting of small clusters. *J. Chem. Phys.* 88: 3910–3922.

Beck, T. L., Jellinek, J., and Berry, R. S., 1987. Rare gas clusters: Solids, liquids, slush and magic numbers. *J. Chem. Phys.* 87: 545–554.

Beck, T. L., Leitner, D. M., and Berry, R. S., 1988. Melting and phase space transitions in small clusters: Spectral characteristics, dimensions and K-entropy. *J. Chem. Phys.* 89: 1681–1694.

Berry, R. S., 1994. Phase transitions in clusters: A bridge to condensed matter. In *Linking Gaseous and Condensed Phases of Matter: The Behavior of Slow Electrons*, Christophorou, L. G., Illenberger, E., and Schmidt, W. F. (Eds.), pp. 231–249. New York: Plenum.

Berry, R. S., 1997. Melting and freezing phenomena. *Microsc. Thermophys. Eng.* 1: 1–18.

Berry, R. S., 1999. *Theory of Atomic and Molecular Clusters*, Jellinek, J. (Ed.), Springer-Verlag, Berlin, Germany, p. 7, 14, 19.

Berry, R. S., 2004. Remarks on the negative heat capacities of clusters. *Israel J. Chem.* 44: 211–214.

Berry, R. S. and Levine, R. D., 2003. *Progress in Experimental and Theoretical Studies of Clusters*, Kondow, T. and Mafuné, F. (Eds.), World Scientific, Singapore, p. 22.

Berry, R. S. and Smirnov, B. M., 2009. Observability of coexisting phases of clusters. *Int. J. Mass Spectrom.* 280: 204–208.

Berry, R. S., Jellinek, J., and Natanson, G., 1984a. Unequal freezing and melting temperatures for clusters. *Chem. Phys. Lett.* 107: 227–230.

Berry, R. S., Jellinek, J., and Natanson, G., 1984b. Melting of clusters and melting. *Phys. Rev. A* 30: 919–931.

Berry, R. S., Beck, T. L., Davis, H. L., and Jellinek, J., 1988. Solid-liquid phase behavior in microclusters. In *Evolution of Size Effects in Chemical Dynamics, Part 2*, Prigogine, I. and Rice, S. A. (Eds.), pp. 75–138. New York: John Wiley & Sons.

Bonacic-Koutecky, V., Jellinek, J., Wiechert, M., and Fantucci, P., 1997. Ab initio molecular dynamics study of solid-to-liquid transitions in Li_0^+, Li_{10} and Li_{11}^+ clusters. *J. Chem. Phys.* 107: 6321–6334.

Breaux, G. A., Benirschke, R. C., Sugai, T., Kinnear, B. S., and Jarrold, M. F., 2003. Hot and solid gallium clusters: Too small to melt. *Phys. Rev. Lett.* 91: 215508 (4).

Breaux, G. A., Benirschke, R. C., and Jarrold, M. F., 2004. Melting, freezing, sublimation, and phase coexistence in sodium chloride. *J. Chem. Phys.* 121: 6502–6507.

Briant, C. L. and Burton, J. J., 1973. Thermodynamics–melting of small clusters of atoms. *Nature Phys. Sci.* 243: 100–102.

Briant, C. L. and Burton, J. J., 1975. Molecular dynamics study of the structure and thermodynamic properties of argon microclusters. *J. Chem. Phys.* 63: 2045–2058.

Chacko, S., Joshi, K., Kanhere, D. G., and Blundell, S. A., 2004. Why do gallium clusters have a higher melting point than the bulk? *Phys. Rev. Lett.* 92: 135506 (4).

Cheng, H.-P. and Berry, R. S., 1992. Surface melting of clusters and implications for bulk matter. *Phys. Rev. A* 45: 7969–7980.

Davis, H. L., Jellinek, J., and Berry, R. S., 1987. Melting and freezing in isothermal Ar_{13} clusters. *J. Chem. Phys.* 86: 6456–6469.

Davis, H. L., Beck, T. L., Braier, P. A., and Berry, R. S., 1988. Time scale considerations in the characterization of melting and freezing in microclusters. In *The Time Domain in Surface and Structural Dynamics*, Long, G. J. and Grandjean, F. (Eds.), pp. 535–549. Dordrecht, the Netherlands: Kluwer.

Dibble, T. S. and Bartell, L. S., 1992a. Electron diffraction studies of the kinetics of phase changes in molecular clusters. 2. Freezing of CH_3Cl_3. *J. Phys. Chem.* 96: 2317–2322.

Dibble, T. S. and Bartell, L. S., 1992b. Electron diffraction studies of the kinetics of phase changes in molecular clusters. 3. Solid-state phase transitions in SeF_6 and $(CH_3)_3CCl$. *J. Phys. Chem.* 96: 8603–8610.

Dinda, P. T., Vlastou-Tsinganos, G., Flytzanis, N., and Mistriotis, A. D., 1994. The melting behavior of small silicon clusters. *Phys. Lett. A* 191: 339–345.

Doye, J. P. K. and Wales, D. J., 1995. Calculation of thermodynamic properties of small Lennard-Jones clusters incorporating anharmonicity. *J. Chem. Phys.* 102: 9659–9672.

Etters, R. D. and Kaelberer, J. B., 1975. Thermodynamic properties of small aggregates of rare-gas atoms. *Phys. Rev. A* 11: 1068–1079.

Etters, R. D. and Kaelberer, J. B., 1977. On the character of the melting transition in small atomic aggregates. *J. Chem. Phys.* 66: 5112–5116.

Etters, R. D., Danilowicz, R., and Kaelberer, J., 1977. Metastable states of small rare gas crystallites. *J. Chem. Phys.* 67: 4145–4148.

Farges, J., DeFeraudy, M. F., Rouault, B., and Torchet, G., 1975. Structure compacte désordonnée et agrégats moléculaires modèls polytétaédriques. *J. Phys., Paris* 36; 13–17.

Farges, J., DeFeraudy, M. F., Rouault, B., and Torchet, G., 1977. Transition dans l'ordre local des agrégats de quelques dizaines d'atomes. *J. Phys., Paris* 38: 47–51.

Farges, J., deFeraudy, M. F., Raoult, B., and Torchet, G., 1983. Noncrystalline structure of argon clusters. I. Polyicosahedral structure of Ar_N. *J. Chem. Phys.* 78: 5067–5080.

Farges, J., deFeraudy, M. F., Raoult, B., and Torchet, G. 1986. Noncrystalline structure of argon clusters. 2. Multilayer icosahedral structure of Ar_N clusters ($50 \leq N \leq 750$). *J. Chem. Phys.* 84: 3491–3501.

Farges, J., DeFeraudy, M. F., Rouault, B., and Torchet, G. 1987a. From five-fold to crystalline symmetry in large clusters. In *Large Finite Systems*, Jortner, J., Pullman, A. and Pullman, B. (Eds.), pp. 113–118. Dordrecht, the Netherlands: Reidel.

Farges, J., DeFeraudy, M. F., Roualt, B., and Torchet, G. 1987b. Atomic structure of small clusters: The why and how of the five-fold symmetry. In *Physics and Chemistry of Small Clusters*, Jena, P., Khanna, S. N., and Rao, B. K. (Eds.), pp. 15–24. New York: Plenum.

Garzón, I. L. and Jellinek, J., 1992. Melting of nickel clusters. In *Physics and Chemistry of Finite Systems: From Clusters to Crystals*, Jena, P., Khanna, S. N. and Rao, B. K. (Eds.), pp. 405–410. Dordrecht, the Netherlands: Kluwer.

Gobet, F., Farizon, B., Farizon, M. et al., 2002. Direct experimental evidence for a negative heat capacity in the liquid-to-gas phase transition in hydrogen cluster ions: Backbending of the caloric curve. *Phys. Rev. Lett.* 89: 183403 (4).

Hill, T. L., 1963. *The Thermodynamics of Small Systems, Part 1.* New York: W. A. Benjamin.

Jellinek, J. and Goldberg, A., 2000. On the temperature, equipartition, degrees of freedom and finite size effects: Application to aluminum clusters. *J. Chem. Phys.* 113: 2570–2582.

Jellinek, J., Beck, T. L., and Berry, R. S., 1986. Solid-liquid phase changes in simulated isoenergetic Ar_{13}. *J. Chem. Phys.* 84: 783–2794.

Joshi, K., Kanhere, D. G., and Blundell, S. A., 2002. Abnormally high melting temperature of the Sn_{10} cluster. *Phys. Rev. B* 66: 155329(5).

Joshi, K., Kanhere, D. G., and Blundell, S. A., 2003. Thermodynamics of tin clusters. *Phys. Rev. B* 67: 235413 (8).

Joshi, K., Krishnamurty, S., and Kanhere, D. G., 2006. "Magic Melters" have geometric origin. *Phys. Rev. Lett.* 96: 135703 (4).

Kaelberer, J. B. and Etters, R. D., 1977. Phase transitions in small clusters of atoms. *J. Chem. Phys.* 66: 3233–3239.

Kofman, R., Cheyssac, P., Garrigos, R., Lereah, Y., and Deutscher, G., 1989. Solid liquid transition of metallic clusters—Occurrence of surface melting. *Physica A* 157: 631–638.

Kofman, R., Cheyssac, P., and Garrigos, R., 1990. From the bulk to clusters: Solid-liquid phase transitions and precursor effects. *Phase Transit.* 24–26: 283–342.

Kunz, R. E. and Berry, R. S., 1993. Coexistence of multiple phases in finite systems. *Phys. Rev. Lett.* 71: 3987–3990.

Kunz, R. E. and Berry, R. S., 1994. Multiple phase coexistence in finite systems. *Phys. Rev. E* 49: 1895–1908.

Labastie, P. and Whetten, R. L., 1990. Statistical thermodynamics of the cluster solid–liquid transition. *Phys. Rev. Lett.* 65: 1567–1570.

Maillet, J. B., Boutin, A., Buttefey, S., Calvo, F., and Fuchs, A. H., 1998. From molecular clusters to bulk matter. I. Structure and thermodynamics of small CO_2, N_2, and SF_6 clusters. *J. Chem. Phys.* 109: 329–337.

McGinty, D. J., 1973. Molecular dynamics studies of the properties of small clusters of argon atoms. *J. Chem. Phys.* 58: 4733–4742.

Natanson, G., Amar, F., and Berry, R. S., 1983. Melting and surface tension in microclusters. *J. Chem. Phys.* 78: 399–408.

Nauchitel, V. V. and Pertsin, A. J., 1980. A Monte Carlo study of the structure and thermodynamic behaviour of small Lennard-Jones clusters. *Mol. Phys.* 40: 1341–1355.

Pawlow, P., 1909a. Über die Abhängigkeit des Schmelzpunkte von der Oberflächenenergie eines festen Körpers. *Z. Phys. Chem.* 65: 1–35.

Pawlow, P., 1909b. Über die Abhängigkeit des Schmelzpunkte von der Oberflächenenergie eines festen Körpers (Zusatz.). *Z. Phys. Chem.* 65: 545–548.

Proykova, A., Radev, R., Li, F.-Y., and Berry, R. S., 1999. Structural transitions in small molecular clusters. *J. Chem. Phys.* 110: 3887–3896.

Proykova, A., Pisov, S., and Berry, R. S., 2001. Dynamical coexistence of phases in molecular clusters. *J. Chem. Phys.* 115: 8583–8591.

Proykova, A., Nikolova, D., and Berry, R. S., 2002. Symmetry in order-disorder changes of molecular clusters. *Phys. Rev. B* 65: 085411 (6).

Proykova, A., Pisov, S., Radev, R., Mihailov, P., Daykov, I., and Berry, R. S., 2003. Temperature induced phase transformations of molecular nanoclusters. *Vacuum* 68: 87–95.

Rytkönen, A., Häkkinen, H., and Manninen, M., 1998. Melting and octupole deformation of Na_{40}. *Phys. Rev. Lett.* 80: 3940–3943.

Schmidt, M., Kusche, R., Kronmüller, W., von Issendorff, B., and Haberland, H., 1997. Experimental determination of the melting point and heat capacity for a free cluster of 139 sodium atoms. *Phys. Rev. Lett.* 79: 99–102.

Schmidt, M., Kusche, R., von Issendorff, B., and Haberland, H., 1998. Irregular variations in the melting point of size-selected atomic clusters. *Nature* 393: 238–240.

Schmidt, M., Kusche, R., Hippler, T., Donges, J., Kronmüller, W., von Issendorff, B., and Haberland, H., 2001a. Negative heat capacity for a cluster of 147 sodium atoms. *Phys. Rev. Lett.* 86: 1191–1194.

Schmidt, M., Hippler, T., Donges, J., Kronmüller, W., von Issendorff, B., and Haberland, H., 2001b. Caloric curve across the liquid-to-gas change for sodium clusters. *Phys. Rev. Lett.* 87: 203402 (1–4).

Shvartsburg, A. A. and Jarrold, M. F., 1999. Tin clusters adopt prolate geometries. *Phys. Rev. A* 60: 1235–1239.

Shvartsburg, A. A. and Jarrold, M. F., 2000. Solid clusters above the bulk melting point. *Phys. Rev. Lett.* 85: 2530–2532.

Stace, A. J., 1983. Experimental evidence of a phase transition in a microcluster. *Chem. Phys. Lett.* 99: 470–474.

Thomson, W., 1871. On the equilibrium of vapor at a curved surface of liquid. *Philos. Mag.* 42: 448–452.

Valente, E. J. and Bartell, L. S., 1983. Electron diffraction studies of supersonic jets. V. Low temperature crystalline forms of SF_6, SeF_6, and TeF_6. *J. Chem. Phys.* 79: 2683–2686.

Valente, E. J. and Bartell, L. S., 1984a. Electron diffraction studies of supersonic jets. VI. Microdrops of benzene. *J. Chem. Phys.* 80: 1451–1457.

Valente, E. J. and Bartell, L. S., 1984b. Electron diffraction studies of supersonic jets. VII. Liquid and plastic crystalline carbon tetrachloride. *J. Chem. Phys.* 80: 1458–1461.

Vlachos, D. G., Schmidt, L. D., and Aris, R., 1992. Structures of small metal clusters. II. Phase transitions and isomerizations. *J. Chem. Phys.* 96: 6891–6901.

Wales, D. J. and Berry, R. S., 1994. Coexistence in finite systems. *Phys. Rev. Lett.* 73: 2875–2878.

21

Thermodynamic Phase Stabilities of Nanocarbon

Qing Jiang
Jilin University

Shuang Li
Jilin University

21.1 Introduction

Carbon, as one of the most versatile, interesting, and useful elements, is abundant in the earth's crust, constituting about 0.02%. Carbon is also a unique element in the Periodic Table of Elements, which is the basis for a variety of compounds due to its propensity to form a wide range of bonding networks. Under special circumstances, the s and p orbitals in an atom combine to form hybrid sp^n orbitals, where n indicates the number of p-orbitals involved, which may have a value of 1, 2, or 3. The 3A, 4A, and 5A group elements of the Periodic Table are those which most often form these hybrids. The driving force for the formation of hybrid orbit is a lower energy state for the valence electrons. For carbon, sp, sp^2, and sp^3 hybrids may be formed. The sp^3 electronic state is of primary importance in organic and polymer chemistries. The shape of the sp^3 hybrid determines the 109° (or tetrahedral) angle found in polymer chains. Organic chemistry, which is established on the basis of the C–H bonding, owes its existence to the carbon. Moreover, carbon is also regarded as the interface between living and nonliving matters.

The solid carbon itself exhibits different structures. In bulk form, carbon shows amorphous as well as crystalline diamond (D_B) and graphite (G_B) structures.

D_B is a metastable carbon polymorph at room temperature under atmospheric pressure. Its crystalline structure is a variant of the zinc blende, in which carbon atoms occupy all positions (both Zn and S), as indicated in Figure 21.1a. Thus, each carbon bonds to four other carbons, and these bonds are totally covalent by sp^3 hybridization. This is appropriately called the diamond cubic crystal structure, which is also found for other group IVA elements in the Periodic Table.

G_B has a crystal structure (Figure 21.1b) being distinctly different from that of diamond, which is a more stable state than G_B at ambient temperature and pressure. The structure of G_B is composed of layers of hexagonally arranged carbon atoms; within the layers, each carbon atom is bonded to three coplanar neighbor atoms by strong covalent bonds (sp^2 hybridization). The fourth bonding electron participates in a weak van der Waals type of bond between the layers.

Although G_B widely exists in nature, another crystalline carbon, D_B, is even more distinguished, which is ascribed to its unique physical and chemical properties. D_B has the highest incompressibility among all elements under ambient pressure with high elastic modulus and strength-to-weight ratio (Bean et al. 1986). In addition, its high melting temperature, high Debye temperature, small intrinsic anharmonicity, and chemical inertness, qualify D_B as a special reference material for the realization of an "international practical pressure scale" (IPPS) (Aleksandrov et al. 1987, Holzapfel 1997, Tse and Holzapfel 2008). The tribological properties of D_B have been exploited in macro- and micro/nanometer scales. It is demonstrated that the D_B could be used as a coating for the seals of rotating shafts and as a monolithic atom force microscope (AFM) tip for imaging micromachining applications. The D_B films are also used as structural materials in microelectric-mechanic system (MEMS) and nanoelectric-mechanic system (NEMS) as mechanical resonators, micromechanical switches, and ink jets for corrosive liquids. Moreover, as a conformal coating, D_B has been used for the passivation of surfaces and the fabrication of scanning probe microscope (SPM) cantilevers and efficient field emitters.

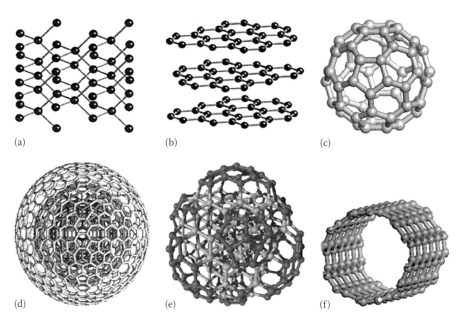

FIGURE 21.1 Schematical diagram of different structures of nanocarbon: (a) D, (b) G, (c) F, (d) O, (e) B, (f) U.

Beginning in the late 1980s, fine-grained polycrystalline films of D_B were grown for optical coatings, thin support membranes in x-ray windows and x-ray lithography masks (Butler and Sumant 2008), and optical materials to fabricate "whispering gallery" mode optical resonators, two-dimensional photonic crystals, and UV-transparent electrodes on SiC. Nanodiamond (D_n) films with grain size d in the nanometer scale have also been incorporated into Si on insulator (SOI) wafers. Thin-film transistors with covalent molecular functionalization have been demonstrated. Moreover, chemically modified D_n surfaces are proved to be an important and stable platform for chemical and DNA sensing.

One of the basic concerns in mechanics, physics, chemistry, and biomedical engineering of solids is the microstructure of a solid, which is determined by the chemical composition, which refers to the arrangement of the atoms and electrons (the atomic and electronic structures), as well as the size of a solid in one, two, or three dimensionalities (Gleiter 2000). Crystal structures reflect a complex interplay between intrinsic factors (composition, band structure, valence electrons, bonding states, structural symmetry, etc.) and extrinsic factors (temperature, pressure/stress, electric field, magnetic field etc.). A change in any of these factors may trigger a structural transition. Conventionally, studies of phase transitions in condensed matters have assumed that the pressure P, the temperature T, and the amount of the ith component or molecular number in a system N_i, are the variables in determining the stable states of a material. This has been well described by using the classical thermodynamics theory based on Gibbs free energy and the statistic mechanics.

With the progress of nanoscience and nanotechnology in recent 30 years, a diverse range of nanomaterials have been vigorously developed in order to exploit their properties for high-performance nanodevices. The traditional thermodynamics cannot explain the phase transition behavior of nanomaterials (Clark et al. 2005, Abudukelimu et al. 2006). In this special field, a new freedom degree—the material size d, plays an important role in determining the physical properties of nanomaterials (Jiang and Yang 2008). d as an intrinsic factor, which is associated with the increased surface/volume ratio, directly relates to the dimensionalities and the shapes (particles, nanowires, thin films, polyhedra, etc.) of nanocrystals (Barnard 2006a). For nanosized carbon, several new structures of fullerenes (F), nanotubes (U), and onion-like carbon (O) have been observed in experiments. Recent work also indicates that the transition from D_n to O has led to an introduction of a new intermediate phase of bucky diamond (B) with a D core encased into an O shell (Barnard 2006b).

In a chemical viewpoint, these structures of carbon are bonded by two essential forms, namely, sp^2 (trigonal) and sp^3 (tetrahedral) hybridizations. The details are listed below (1) sp^2 hybridization: G, F, O, and U consist of two-dimensional carbon layers stacked in an AB sequence with three bonds, which is linked by a weak van der Waals interaction induced by a delocalized π-orbital; (2) sp^3 hybridization: D has a three-dimensional structure in which each carbon atom is bonded by four other carbon atoms; (3) both sp^2 and sp^3 hybridizations: B has sp^2 and sp^3 structures at the surface and in the core, respectively. Figure 21.1 shows a schematic diagram for these different structures of nanocarbon, which will be illustrated in details in the following.

F exists in discrete molecular form, and consists of a hollow spherical cluster of 60 carbon atoms; a single molecule is denoted by C_{60}. Each molecule is composed of groups of carbon atoms that are bonded to one another to form both hexagon (six-carbon atom) and pentagon (five-carbon atom) geometrical configurations. One such molecule, shown in Figure 21.1c, is found to consist of 20 hexagons and 12 pentagons, which are arrayed such that no two pentagons share a common side; the molecular surface thus exhibits the symmetry of a soccer ball.

U, by virtue of its curved graphitic structure, has a small diameter (1 to ≤ 100 nm), and a high aspect ratio (Figure 21.1f).

O, which consists of concentric graphitic shells, is one of the fullerene-related materials together with C_{60} and carbon nanotubes. An ideal O is composed of up to several tens of concentric graphitic spherical shells with adjacent shell separation using a C_{60} as nucleus (Figure 21.1d). The innermost shell is formed by 60 carbon atoms, and the carbon atoms of other layers increase in turn by $60n^2$ (n denotes the number of layer). An intermediate phase between O and D is B, which is formed by a diamond-like core and an onion-like outer shell (Figure 21.1e).

It is known that G_B is the stable phase while D_B is the metastable phase in bulk form of carbon under ambient pressure P_a and room temperature T_r. However, the energetic difference between G_B and D_B is only 0.02 eV/atom, where the unit eV/atom denotes the electronic volt per atom. As a result, their relative stability could be changed easily through changing surrounding conditions, such as T, P, and d of the material. For example, as P increases, D_B becomes more stable than G_B due to higher density of the D_B. Note that $N_i = N$ is considered as a constant since the considered system consists of the unique component—carbon, where N denotes the total number of carbon atoms.

Under the condition of $P = P_a$ and $T = T_r$, D_B could also be stable by changing its d. As d decreases, D_n and the above new denser packing phases are easily formed and become stable. It is noted that the metastable state could also persist due to kinetic reasons, such as sufficiently large energetic barriers and low transition temperature.

In 1960s and 1970s, it was mentioned that D_n could be more stable than the nanosized G (G_n) based on the successful synthesis of D_n by using vapor deposition of carbon (Fedoseev et al. 1989). In this work, the G–D phase transition will be first considered, which acts as a reference for other phase transitions.

21.2 Nanothermodynamics

The classical thermodynamics on macroscopic systems has long been well established (Gibbs 1878, Rusanov 1996), which describes adequately the macroscopic behaviors of bulk systems with the change of macroscopic parameters where the astrophysical objects and nanoscaled systems are excluded (Wang and Yang 2005). The basic thermodynamic relationship for a macroscopic system at equilibrium can be expressed as (Hill 2001),

$$du = T\,dS - P\,dV \tag{21.1}$$

where
  u is the internal energy
  S is the entropy
  P denotes the pressure
  V denotes the volume

This equation connects incremental changes of internal energy, heat, and work.

In 1878, the monumental work of Gibbs first formulated a detailed thermodynamic phase equilibrium theory (Gibbs 1878) in which he transformed the previous complicated thermodynamics of cycles into the simpler thermodynamics of potentials and introduced chemical potentials. Gibbs generalized Equation 21.1 by allowing, explicitly, variations in N_i of the different components in the system (Hill 2001). As a result, Equation 21.1 became $du = T\,dS - P\,dV + \Sigma\mu_i\,dN_i$, or in a more general and modernized form,

$$dg = -S\,dT + V\,dP + \gamma dA + \sum_i \mu_i\,dN_i \tag{21.2}$$

where
  g and μ_i denote Gibbs free energy and the chemical potential of the component i, respectively
  A is the surface (interface) area

Equation 21.2 thus could predict various equilibria (chemical, phase, osmotic, surface, etc.) and examine many other topics, such as the equilibrium condition of a solid and a surrounding medium (Rusanov 1996). Equation 21.2 is much less difficult to use than to understand although it is very simple in mathematics. Although Equation 21.2 has a wide application range, it is essentially used to solve the phase equilibrium and related phenomena, such as phase diagrams.

Later, the thermodynamic definitions of γ and surface stress f are clarified to formulate surface thermodynamics (Hill 2001). This issue becomes increasingly important due to the appearance of nanotechnology. However, due to the content limitation of this chapter, size dependences of surface thermodynamic functions can be referenced elsewhere (Jiang and Lu 2008).

Note that Equation 21.2 with a statistic basis is only valid for materials being at least larger than submicron size while the parameter d in Equation 21.2 is actually a constant of bulk. Since nanomaterials and nanotechnology go into the scientific and technical worlds now, extending the validity of the macroscopic thermodynamics and statistical mechanics into the nanometer scale becomes an urgent task.

The above task can be reached by deeply analyzing the size dependence of a typical known process of the thermodynamic phase equilibrium, such as the melting (Dash 1999). There are the so-called specific and smooth size effects. The former is responsible for the existence of "magic numbers" and related irregular variation of properties in clusters, whereas the latter pertains to nanostructures in the size domain between clusters and bulk systems. Within this broad size range, mechanical, physical, and chemical properties are often seen to change according to relatively simple scaling equations involving a power-law dependence on the system size due to the energy contributions of γ to the total g of the system (Lu 1996). These property changes lead to, on one hand, an emerging interdisciplinary field involving solid state physics, chemistry, biology, and materials science to synthesize materials and/or devices with new properties by controlling their microstructures on the atomic level. To utilize the new properties of nanomaterials and to guarantee the working stability of the nanosystems, it is of consequence and exigency to develop a suitable theoretical tool to address them naturally. On the other hand, the rapid progress in the synthesis and processing of materials with the structures at the nanometer size has created a demand for better scientific understanding of the thermodynamics on nanoscale, namely thermodynamics of small systems or nanothermodynamics.

One of the most important applications of the nanothermodynamics is to establish the corresponding phase diagram where the relative stability of phases can be well discussed and understood with the change of variables of T, P, and d. This is also the case for carbon. Traditionally, in order to make use of D_n in a variety of engineering applications, T–P phase diagram of carbon has been established to understand the condition of high pressure synthesis of D_n (Bean et al. 1986, Barnard 2006b). Recently, constructing the phase diagram of nanocarbon and establishing the phase boundary between D_n and G_n by taking into account the effects of d, T, and P together have been carried out (Wang et al. 2005), which will be introduced in the next section.

21.3 Phase Equilibria and Phase Diagram of Bulk and Nanocarbon

Before the beginning of the discussion for the phase diagram of nanocarbon and the nanophase stability, we first describe the future of the phase diagram of bulk carbon.

For bulk carbon, the liquid carbon existed at higher T is metallic (Van Thiel and Ree 1993), although there is still some conjectures regarding its exact structure. The phase transition is associated with changes of the density and the structure. At $4000 < T < 6500\,K$ under high P, first-order liquid–liquid phase transitions have been undertaken where the liquid clusters are more likely to be G-like sp^2 than D-like sp^3 (Van Thiel and Ree 1993) However, some studies have predicted sp^3 liquid clusters (Galli et al. 1989), such as a simple cubic structure (Grumback and Martin 1996). Moreover, the sp bond is considered to be predominant in low-density liquid with little sp^3 character whereas the high-density liquid is mostly sp^3 bonded with little sp character (Glosli and Ree 1999). It is evident that more detailed works, especially theoretical works, are needed to understand structural characteristics of liquid carbon since they are difficult to be determined accurately in experiments at high temperatures and under high-pressure environments.

Under more moderate P and T, a wide spectrum of many metastable forms and complex hybrid carbon may be present (including various specific types of graphite) due to high activation energies for solid–solid phase transitions (Gust 1980).

In the past decades (beginning in the 1960s) (Young 1991), the P–T phase diagram (including G, D, and liquid L phases) of carbon has been continually updated on the basis of the information obtained from newly developed technologies. One of the most comprehensive carbon phase diagrams proposed by Bundy et al. (1996) (the solid lines in Figure 21.2) shows that (1) G_B undergoes a well-characterized first-order transition upon pressurization

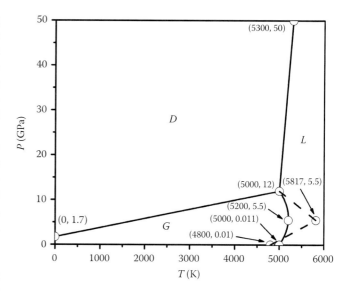

FIGURE 21.2 T–P phase diagram of bulk carbon (solid lines) (Bundy et al. 1996) where the G–L phase boundary has been corrected with recent experimental results (the dash line) (Young 1991, Bundy et al. 1996, Korobenko 2001, Korobenko and Savvatimskiy 2003). The symbols O denote characterized points. (Reproduced from Yang, C.C. and Li, S., *J. Phys. Chem. C*, 112, 1423, 2008. With permission.)

to D_B and (2) both phases melt to form L as T increases. On the other hand, recent advances provide new experimental results to improve the accuracy of G_B–L phase boundary (the dash line in Figure 21.2) (Korobenko 2001, Korobenko and Savvatimskiy 2003, Savvatimskii 2003, Basharin et al. 2004, Savvatimskiy 2005). It is found that the phase line is terminated by a critical point at 8801 K and 10.56 GPa and by a triple point on the melting line of G_B at 5133 K and 1.88 GPa.

Recently, the P–T phase diagram of carbon was reviewed (Barnard 2006b) where the underlying basis for recent studies has been summarized, extending the bulk carbon phase diagram to the nanometer scale. In general, the thermodynamic properties of carbon nanocrystals may be calculated by considering the coexistence of several phases of gases, L, and solids in chemical equilibrium with the same d.

One method of introducing size dependence into the bulk carbon phase diagram is to add the contribution of γ to g for an N-atom cluster in a given phase, and to define the phase equilibrium by equating g values for the cluster of each phase (Ree et al. 1999). A number of phase diagrams based on this principle have been proposed, each exhibiting displacement of the phase equilibrium lines for the clusters with $10^2 < N < 10^4$. In this case, a time t and P–T path dependent value for the nonequilibrium D_n fraction of the soot mixture is simulated and the approximate computing methods for computing the detonation product pressure for the kinetics derived mixture of D_n and G_n are discussed (Viecelli and Ree 2000, Viecelli et al. 2001). In addition, when $N \sim 100$, the melting temperature of small particles is lower than that of the bulk carbon (Viecelli et al. 2001).

With similar considerations, a three-dimensional phase diagram for nanocarbon has been established where the size dependence was validated and the bulk *P–T* phase diagram was shown in the horizontal plane (Verechshagin 2002). In this study, although D_n appears as the most stable phase at $d < 3$ nm, a lower limit for D_n phase stability is introduced at 1.8 nm.

A *T*-dependent transition size d_c between G_n and D_n was discussed by using a thermodynamic model with an inclusion of charge lattice energy. The d_c is calculated to be 15 nm at 0 K, 10.2 nm at T_r, 6.1 nm at 798 K, 4.8 nm at 1073 K, and 4.3 nm at 1373 K (Gamarnik 1996a,b). Therefore, d_c decreases with the increasing of *T*. Hwang et al. outlined a chemical potential model (Hwang et al. 1996a) and a charged cluster model (Hwang et al. 1996b) to describe the relative stability of G_n and D_n for the low-pressure synthesis of D_n. Around this time, using first-principles and semiempirical potentials, the phase stability of carbon nanoparticles has been investigated on the basis of the formation heat of graphene sheet and hydrogenated D_n (Ree et al. 1999, Barnard et al. 2003a). Most recently, the size dependence of G_n–D_n transition using Laplace–Yang equation by considering the effects of γ or *f* has been calculated (Jiang et al. 2000, Yang and Li 2008). Moreover, a theoretical result shows that the surface of D_n being larger than 1 nm reconstructs in a fullerene-like manner, giving rise to B (Raty et al. 2003).

The polymorphic behavior of nanocarbon in light of considering the contribution of P_{in} induced by the *f* is also investigated, where P_{in} denotes the curvature-induced internal pressure. Figure 21.3 shows a *d*-dependent *T–P* phase diagram of carbon where the phase equilibria of G_n–D_n, G_n–L_n, as well as D_n–L_n are considered individually (Yang and Li 2008). As shown in the figure, the melting temperature of G_n decreases with decreasing *d*. Moreover, the D_n/G_n/L_n triple point shifts toward lower *T* and *P* regions with decreasing *d*, resulting in large reduction of the stable region for G_n.

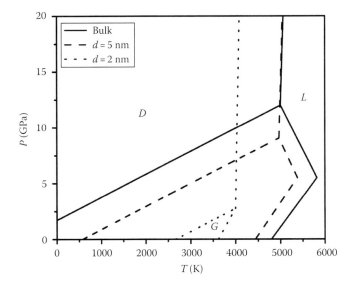

FIGURE 21.3 *T–P* phase diagrams of bulk and nanocarbon. The solid, dash, and dot lines denote the bulk $d = 5$ nm and $d = 2$ nm, respectively. (Reproduced from Yang, C.C. and Li, S., *J. Phys. Chem. C*, 112, 1423, 2008. With permission.)

The phase transition between D_n and O structures has also been previously addressed by a thermodynamic quasiequilibrium theory (Banhart and Ajayan 1996). The crossover from F to closed U has also been analyzed recently (Park et al. 2002).

Using traditional analysis, the relative stability of D_n and G_n, and the defined size regions of the stability for F, O, and B are considered (Barnard et al. 2003b). As d increases, the stability of carbon changes from F to O to B to D_n, and to G_n.

Three stability regions of nanocarbon can be outlined as follows (Tomanek and Schluter 1991): (1) $N < 20$, the most stable geometry is one-dimensional ring cluster; (2) $20 < N < 28$, the energetics of quite different types of geometries of clusters is similar; (3) for larger clusters, F should be more stable.

To better understand the above experimental, theoretical and simulation results, a systematic and standard thermodynamic description for these transitions is required to establish the relationship between *d* and stability of D_n, and to exploit the origin of the stability of D_n in the nanometer scale.

21.4 Solid Transition between D_n and G_n with the Effects of γ and *f*

To better clarify the size effect on the polymorphism of carbon with nanothermodynamics, the bulk phase diagram of carbon (Bertsch 1997) can be taken as the basis. The phase-equilibrium line function $P(T)$ of D–G phase transition in the bulk is approximately expressed as (Bundy et al. 1996)

$$P(T)(\text{Pa}) = 2.01 \times 10^6 T + 2.02 \times 10^9 \quad (21.3)$$

Nanothermodynamic analysis takes into account the capillary effect induced by the curvature of D_n and G_n with the Laplace–Young equation for spherical and quasi-isotropic nanocrystals (although the real shape of D_n should be polyhedral (Kwon and Park 2007), this consideration should result in minor error) (Zhao et al. 2002, Jiang and Chen 2006), and the P_{in} is given as

$$P_{in} = \frac{4f}{d} = \left(\frac{4}{d}\right) h \left[\frac{S_{vib} H_m}{2\kappa V_m R}\right]^{1/2} \quad (21.4)$$

where
   R is the ideal gas constant
   H_m denotes the bulk melting enthalpy of crystals
   S_{vib} is the vibrational part of the overall melting entropy S_m
   κ is the compressibility
   h is the atomic diameter

From Equation 21.4, $f = 3.54$ J/m^2 is determined for the widely studied C_{60}, which is consistent with the computer simulation result of $f = 2.36 \sim 4.02$ J/m^2 (Robertson et al. 1992), where the *f* value is transformed from eV/atom to J/m^2.

Adding P_{in} term into Equation 21.3, we have

$$P(T,d) = 2.01 \times 10^6 T + 2.02 \times 10^9 - \frac{4f}{d} \quad (21.5)$$

Under $P_a = 10^5$ Pa in equilibrium, $4f/d + 10^5 = 2.01 \times 10^6 T + 2.02 \times 10^9$, or $d = 4f/(2.01 \times 10^6 T + 2.02 \times 10^9 - 10^5)$ based on Equation 21.5. This can be used to calculate the d–T phase-equilibrium line, as shown in Figure 21.4. It is evident that the calculation results are in good agreement with experimental results.

At equilibrium, D_n and G_n should be under the same P or f where D_n and G_n have different h values and thus distinct d values. As a first order approximation, $f = (f_D + f_G)/2 = 3.6 \text{ J/m}^2$ is taken in Equation 21.5 with $f_D = 6.1 \text{ J/m}^2$ and $f_G = 1.1 \text{ J/m}^2$ (Jiang et al. 2000), where the subscripts D and G denote the related phases, respectively.

From Figure 21.4, it is found that the equilibrium d of D_n and G_n decreases from 8.5 nm at 0 K to 3 nm at 1500 K. As d decreases, the stability of D_n increases compared with that of G_n. The predictions of this model correspond to the experimental results better than that of the charge lattice model (Gamarnik 1996a,b).

As shown in the Figure 21.4, the model predictions based on $f \approx \gamma$ are also consistent with the experimental results, where $\gamma = (\gamma_D + \gamma_G)/2 = 3.485 \text{ J/m}^2$ with $\gamma_D = 3.7 \text{ J/m}^2$ and $\gamma_G = 3.27 \text{ J/m}^2$. Thus, $f \approx \gamma$ for carbon. However, this criterion is not suitable for other materials in most cases. From a thermodynamic viewpoint, the fundamental difference between a solid surface and a liquid surface is the distinction between f and γ. Essentially γ describes a reversible work per unit area to form a new surface while f denotes a reversible work per unit area to elastically stretch the surface, which corresponds to the derivative of γ with respect to the strain tangential to the surface (Jiang et al. 2000, Yang and Li 2008). For L, $\gamma = f$ while for the solid, $\gamma \neq f$. It is noted that the P_{in} for a solid differs from the Laplace pressure for a spherical L droplet surrounded by other L in the equilibrium with $P_{in} = 4\gamma/d$. In fact, f is a vital factor in promoting sp^3 bonding in the synthesis of D_n by low T and low P methods (Ree et al. 1999).

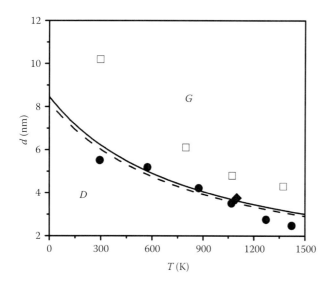

FIGURE 21.4 The d–T transition diagram of nanocarbon under $P = 0$ (solid lines) (Yang and Li 2008) where other theoretical [□ (Gamarnik 1996b)] and experimental results [◆ (Chen et al. 1999) and ● (Wang et al. 2005)] are presented for comparisons.

21.5 Relative Phase Stabilities of D_n, Compared with B, O, and F

As $d < 1.8$ nm, carbon structures of nanoparticles are abundant, such as F and O (Shenderova et al. 2002). In order to address a thermodynamic model of size-dependent phase stability and coexistence of B with D_n and F phases, the structural energy or the standard formation heat ΔH_f^0 at 298.15 K as functions of N for each carbon phase has been calculated by using the density functional theory with the generalized gradient approximation (Barnard et al. 2003a, Barnard 2006b). The technique for obtaining the ΔH_f^0 is outlined by Winter and Ree (1998) and developed by Barnard et al. (2003a). The ΔH_f^0 (G), ΔH_f^0 (D), ΔH_f^0 (F) $\approx \Delta H_f^0$ (O) (the both structures are similar) can be expressed as

$$\frac{\Delta H_f^0(G)}{N} = \frac{3}{2} E_{CC}^G + \frac{N_H}{N}\left(E_{CH}^G - \frac{1}{2} E_{CC}^G + \Delta H_f^0(H) \right)$$
$$+ \Delta H_f^0(C) + \frac{1}{2} E_{CC}^{vdw} \tag{21.6}$$

$$\frac{\Delta H_f^0(D)}{N} = 2 E_{CC}^D + \frac{N_{DB}}{N}\left(E_{DB}^D - \frac{1}{2} E_{CC}^D + \Delta H_f^0(DB) \right)$$
$$+ \Delta H_f^0(C) \tag{21.7}$$

$$\frac{\Delta H_f^0(F)}{N} = \frac{3}{2} E_{CC}^F + \Delta H_f^0(C) + \frac{E_{strain}^F}{R^2} + \frac{1}{2} E_{CC}^{vdw} \tag{21.8}$$

where

N_H is the number of terminating hydrogen atoms

E_{CC} and E_{CH} denote C–C and C–H bond energy, respectively

E_{DB} is the dangling bond energy, which is linearly dependent upon N_{DB}/N with N_{DB} being the number of surface dangling bonds (Barnard et al. 2003a)

E_{strain}^F denotes the strain energy associated with the curvature of F

E_{CC}^{vdw} is the van der Waals attraction between G sheets or O layers

There is no interlayer attraction in F, thus $E_{CC}^{vdw} = 0$.

The calculation results of ΔH_f^0 (O), ΔH_f^0 (F), ΔH_f^0 (B), and ΔH_f^0 (D_n) (as a function of d) are shown in Figure 21.5. It is found that (1) ΔH_f^0 (O) and ΔH_f^0 (F) of the sp^2-bonded O and F are indistinguishable to each other at $N < 2000$; (2) ΔH_f^0 (B) is more akin to O than D_n; (3) in the range of $500 < N < 1850$, there is a thermodynamic coexistence region of B and D_n. The region was then further broken into three subregions, as indicated in Figure 21.5. At $500 < N < 900$ or $1.4 < d < 1.7$ nm, ΔH_f^0 (B) $\approx \Delta H_f^0$ (F). Although O is the most stable structure at $900 < N < 1350$ or $1.7 < d < 2.0$ nm, B and O could coexist in this size range. Moreover, B coexists with D_n at $1350 < N < 1850$ or $2.0 < d < 2.2$ nm. It is noted that the intersection of B and O stability is very close to that of D_n and F at $N = 1100$ where a sp^3-bonded core becomes more favorable than a sp^2-bonded core, irrespective of surface structure.

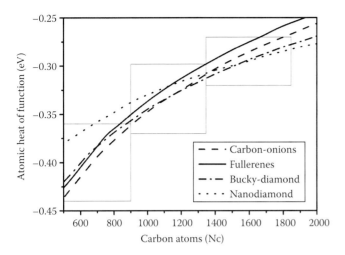

FIGURE 21.5 Atomic formation heat of carbon nanoparticles, indicating the relative subregions of coexistence of B with other phases. Uncertainties are indicated for B only.

The relative stability of D_n as a function of surface hydrogenation is considered in this size range by using the first-principles calculations (Raty and Galli 2003). It is found that B is in fact energetically preferred even over hydrogenated D_n when $d < 3\,nm$. By comparing various degrees of hydrogen coverage, the difference in ΔH_f^0 between particles with and without hydrogenated surfaces was found to decrease as d increases in the range of $29 \leq N \leq 275$. Interestingly, the results was not dependent significantly on the hydrogen chemical potential (Raty and Galli 2003). Although the calculations cannot establish exact size at which the crossover between hydrogenated and bucky surfaces occurs, the numerical results suggest that varying the hydrogen pressure (and thus its chemical potential) during synthesis may promote different types of thin films. Two ranges for the hydrogen chemical potential corresponding to two different growth conditions of D thin films are proposed: one favors the formation of D_n and the other does that of bulk diamond-like films (Raty and Galli 2003).

21.6 Graphitization Dynamics of D_n

To avoid graphitization of the full D_n, which is the most important thermal stability condition for any application of D_n, detailed kinetic condition of graphitization must be considered. In other words, even if D_n is metastable, as long as the graphitization is absent, D_n is still workable. As was experimentally reported (Chen et al. 1999a, Butenko et al. 2000, Wang et al. 2005, Osipov et al. 2006, Bi et al. 2008), graphitization occurs at the surface of D_n, or a transition from D_n to B. This is simply due to the energetic drop induced by decrease of deficit bond number where the surface bonding changes from sp³ to sp². The reaction rate was modeled as a migration rate of the interface between the developing G_n and the remaining D_n. A "reducing sphere" model was used to obtain the rates from the changes in densities (Butenko et al. 2000). The estimated kinetic parameters in an Arrhenius

expression, namely the activation energy, $E = 45 \pm 4\,kcal/mol$, and the preexponential factor, $A = 74 \pm 5\,nm/s$, allow quantitative calculations of the diamond graphitization rates in and around the indicated temperature range. The calculated graphitization rates agree well with the graphitization rates of D_n with different disparity estimated from high-resolution transmission electron microscopy data (Butenko et al. 2000).

According to T–P phase diagram of carbon shown in Figure 21.2, D_n is metastable in the region of $713 < T < 1273\,K$ and of $81 < P < 200\,MPa$ where D_n derived from gas phase during the duration of phase transition nucleates. Since d is small enough, γ plays an important role in this process (Zaiser et al. 2000). The phase transition of D_n in this metastable region differs from that of the solid phase D_n. The graphitization is an energetically favorable process, the corresponding Gibbs free energy change is expressed as (Wang et al. 2004)

$$\Delta G(d_G, P, T) = \left(\frac{\pi d_G^3 \Delta g}{6 V_{mG}} + \pi d_G^2 \gamma_G \right) f(\theta) \qquad (21.9)$$

where

V_m denotes the molar volume, the subscript G denotes the corresponding phase

Δ shows the change

θ is the contact angle

$f(\theta) = (1 - \cos\theta)^2(2 + \cos\theta)/4$ is the so-called heterogeneous factor (in the range from 0 to 1). When $\partial \Delta G(d)/\partial d = 0$, the critical diameter of graphitization nuclei $d_G^\star$ is obtained (Wang et al. 2004, Wang and Yang 2005):

$$d_G^\star = \frac{4\gamma}{\left(\frac{4\gamma_G}{d} + 1 \times 10^5 - 2.01 \times 10^6 T - 2.02 \times 10^9 \right) \frac{V_{mG}}{\Delta V}} \qquad (21.10)$$

Accordingly, we obtain the relationship curves between d_D of D_n and $\Delta G(d)$ at different annealing temperatures T_a and $f(\theta)$, where d_D is the size of the residual D_n, which is shown in Figure 21.6. It is clear that $\Delta G(d)$ increases with decreasing d_D at a given T_a. For different T_a, there are corresponding thresholds of d_D in graphitization. As a result, only the surface region of D_n particles is graphitized at a given T_a, whereas the phase is still B. Then, the graphitization proceeds inward on further elevation of T_a at the expense of D_n phase, as proved by the appearance of bands for graphitic structures in the Raman spectrum and x-ray diffraction. Thus, graphitization would display a staircase behavior with increasing T_a. For example, the threshold diameters of the residual D_n are about 3.5 and 3 nm at annealing temperatures of 1073 and 1423 K, respectively. Graphitization thus hardly proceeds further when $d_D < 3 \sim 3.5\,nm$, as the experiments have shown (Wang et al. 2004). For a comparison, a function between the graphitization energy and $f(\theta)$ is shown as inset in Figure 21.6.

One such property, fundamental to the stability of D_n, is the degree of surface hydrogenation. The analysis on the

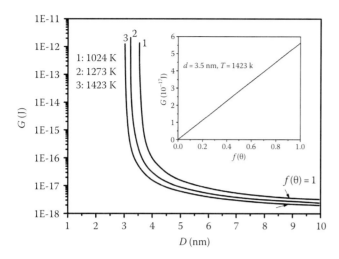

FIGURE 21.6 Relationship curves of d and $\Delta G(d)$ of D_n graphitization at different T_a and a given heterogeneous factor. The inset shows the relationship between the heterogeneous factor and the critical energy of graphitization.

size-dependent stability of D_n and G_n expressed by the T_a–d_D relationship and the critical condition for the formation of the stable and metastable phases of D_n could be predicted along with the calculation of $d_G^\star$ of the synthesis of D_n.

21.7 Summary and Prospects

From the numerous studies outlined above, it is found that advances have been made in understanding the relative stability of sp^2- and sp^3-bonded carbon particles at the nanometer scale. These studies have clearly identified the two important size regimes, where sp^2–sp^3–sp^2 phase transitions may be readily expected. In the case of larger particles, the crossover in stability between D_n and G_n may be expected at $d = 5$–10 nm, and for smaller particles, the crossover between D_n and F may be expected at $d = 1.5$–2 nm. These results are supported by theory (the majority of which are thermodynamic arguments). That is, the upper limit of D_n phase stability is $d < 5$–6 nm, while the lower one is $d = 1.4$–2.2 nm, along with the phase coexistence of D_n and F at this lower limit via the formation of B.

Although D_n can be stable in this size range, D_n as a metastable phase may also exist outside of this range. Furthermore, the identification of a coexistence region implies that (1) the phase transitions are not entirely driven by the volume thermodynamic amount and (2) other thermodynamic factors, such as γ, f, charge, and kinetic considerations, may also affect phase transitions. Therefore, a complete examination of the phase stability of nanocarbon should include not only the use of a sophisticated computational method and large cluster sizes, but also theoretical terms to describe dependencies on a variety of experimentally relevant cluster properties.

Nanocarbons are one of the most fantastical materials in the twenty-first century. The corresponding research brings out not only much understanding in science, but also many industrial

applications. The deep understanding in science and the wide application in industry will certainly benefit our society. In the following, several present main research fields on nanocarbon are summarized.

Moore's law, a scaling rule of thumb turned into self-fulfilling prophecy, has dictated the exponential growth of the semiconductor industry over the last four decades (Moore 1965). To keep the law being valid in future development, carbon based electronics offers one of the most promising options to replace Si. Great attention has been paid to carbon nanotubes due to their intriguing electronic properties (Iijima 1991, Chau et al. 2005, Lin et al. 2005). Their random orientation and spatial distribution, however, inhibit their introduction into the real applications where up to one billion devices need to be connected in a chip.

Another potential solution could be graphene, a single layer of G, or an individual sheet of sp^2-hybridized carbon bound in two dimensions. After the theoretical prediction of the peculiar electronic properties of graphene in 1947 by Wallace (1947) and the subsequent studies on its magnetic spectrum (McClure 1956), it took half a century until graphene could be manufactured experimentally (Novoselov et al. 2004) while the corresponding electronic properties have been measured only recently (Novoselov et al. 2005, Zhang et al. 2005). Graphene has excellent electronic properties, with carrier mobility between 3000 and 27,000 cm²/(V s) at T_a (Novoselov et al. 2004, Berger et al. 2006), being an extremely promising material for nanoelectronic devices. Since the graphene consists of a single layer graphite, its stability is similar to single wall U or O although graphene has no curvature on it. The corresponding thermal stability can be considered to be the same of O with minor error.

The single-wall U (Liu et al. 1999, Chen et al. 2008, Wu et al. 2008), C_{60} (Sun et al. 2006, Chandrakumar and Ghosh 2008, Pupysheva et al. 2008), and graphene (Rojas and Leiva 2007) have emerged as potential candidates for hydrogen storage. For practical applications, a moderate binding strength in the range of $-0.70 \leq E_{ad} \leq -0.20$ eV/H_2 was suggested at T_r, where E_{ad} is the adsorption energy between the H_2 molecules and the storage media (Jhi 2006, Shevlin and Guo 2006). This presents a quandary since this requires enhancing the weak binding between the H_2 and solid surfaces on one hand, which results from the strong H–H bond and the closed-shell electronic configuration (Jhi and Kwon 2004). On the other hand, an excessively strong bond is not ideal either since ultimately both storage and desorption of H_2 are necessary (Sun et al. 2006).

H_2 is either chemisorbed in atomic form or physisorbed in molecular form. The storage abilities of the nanocarbon can be largely improved via metal doping due to the enhanced charge transfer from metal to carbon. Noteworthy, although the alkali (Chen et al. 1999a,b, 2008, Sun et al. 2006, Wu et al. 2008) and transition metals (Yildirim and Ciraci 2005, Shevlin and Guo 2006, Rojas and Leiva 2007) are often utilized as dopants, a recent work predicted that Ca is the most attractive metal for functionalizing F (Yoon et al. 2008).

F, U, and D_n appear also to be valuable resources for biomedical applications (Frietas 1999, 2003). It was demonstrated that

F compounds have biological activity, and their potential as therapeutic products for the treatment of several diseases has been reported. At $d = 0.72\,nm$, C_{60} is similar in size to steroid hormones or peptide α-helices, and thus F compounds are ideal molecules to serve as ligands for enzymes and receptors (Wilson 2000). The exploration of bucky tubes in biomedical applications is also underway. In addition, U has been used for immobilization of proteins, enzymes, and oligonucleotides (Wilson 2000).

Acknowledgments

National Key Basic Research and Development Program (Grant No. 2004CB619301), and "985 Project" of Jilin University are acknowledged.

References

Abudukelimu, G., Guisbiers, G., Wautelet, M. 2006. Theoretical phase diagrams of nanowires. *J. Mater. Res.* 21: 2829–2834.

Aleksandrov, I. V., Goncharov, A. F., Zisman, A. N., Stishov, S. M. 1987. Diamond at high pressures: Raman scattering of light, equation of state, and high-pressure scale. *Sov. Phys. JETP* 66: 384–390.

Banhart, F., Ajayan, P. M. 1996. Carbon onions as nanoscopic pressure cells for diamond formation. *Nature* 382: 433–435.

Barnard, A. S. 2006a. Using theory and modelling to investigate shape at the nanoscale. *J. Mater. Chem.* 16: 813–815.

Barnard, A. S. 2006b. Theory and modeling of nanocarbon phase stability. *Diam. Relat. Mater.* 15: 285–291.

Barnard, A. S., Russo, S. P., Snook, I. K. 2003a. Size dependent phase stability of carbon nanoparticles: nanodiamond versus fullerenes. *J. Chem. Phys.* 118: 5094–5097.

Barnard, A. S., Russo, S. P., Snook, I. K. 2003b. Coexistence of bucky diamond with nanodiamond and fullerene carbon phase. *Phys. Rev. B* 68: 073406.

Basharin, A. Yu., Brykin, M. V., Marin, M. Yu., Pakhomov, I. S., Sitnikov, S. F. 2004. Methods of increasing the measurement accuracy during the experimental determination of the melting point of graphite. *High Temp.* 42: 60–67.

Bean, V. E., Akimoto, S., Bell, P. M., Block, S., Holzapfel, W. B., Manghnani, M. H., Nicol, M. F., Stishov, S. M. 1986. Another step toward an international practical pressure scale: 2nd AIRAPT IPPS task group report. *Physica B & C* 139: 52–54.

Berger, C., Song, Z., Li, X. et al. 2006. Electronic confinement and coherence in patterned epitaxial graphene. *Science* 312: 1191–1196.

Bertsch, G. 1997. Melting in clusters. *Science* 277: 1619.

Bi, H., Kou, K. C., Ostrikov, K. K., Zhang, J. Q. 2008. Graphitization of nanocrystalline carbon microcoils synthesized by catalytic chemical vapor deposition. *J. Appl. Phys.* 104: 033510.

Bundy, F. P., Bassett, W. A., Weathers, M. S., Hemley, R. J., Mao, H. K., Goncharov, A. F. 1996. Pressure–temperature phase and transformation diagram for carbon; updated through 1994. *Carbon* 34: 141–153.

Butenko, Yu. V., Kuznetsov, V. L., Chuvilin, A. L., Kolomiichuk V. N. 2000. Kinetics of the graphitization of dispersed diamonds at "low" temperatures. *J. Appl. Phys.* 88: 4380–4388.

Butler, J. E., Sumant, A. V. 2008. The CVD of nanodiamond materials. *Chem. Vap. Depos.* 14: 145–160.

Chandrakumar, K. R. S., Ghosh, S. K. 2008. Alkali-metal-induced enhancement of hydrogen adsorption in C-60 fullerene: An ab initio study. *Nano Lett.* 8: 13–19.

Chau, R., Datta, S., Doczy, M. et al. 2005. Benchmarking nanotechnology for high-performance and low-power logic transistor applications. *IEEE T. Nanotechnol.* 4: 153–158.

Chen, J., Deng, S. Z., Chen, J., Yu, Z. X., Xu, N. S. 1999a. Graphitization of nanodiamond powder annealed in argon ambient. *Appl. Phys. Lett.* 74: 3651–3653.

Chen, P., Wu, X., Lin, J., Tan, K. L. 1999b. High H_2 uptake by alkali-doped carbon nanotubes under ambient pressure and moderate temperatures. *Science* 285: 91–93.

Chen, L., Zhang, Y., Koratkar, N., Jena, P., Nayak, S. K. 2008. First-principles study of interaction of molecular hydrogen with Li-doped carbon nanotube peapod structures. *Phys. Rev. B* 77: 033405.

Clark, S. M., Prilliman, S. G., Erdonmez, C. K., Alivisatos, A. P. 2005. Size dependence of the pressure-induced gamma to alpha structural phase transition in iron oxide nanocrystals. *Nanotechonology* 16: 2813–2818. (References therein).

Dash, J. G. 1999. History of the search for continuous melting. *Rev. Mod. Phys.* 71: 1737–1743.

Fedoseev, D. V., Deryagin, B. V., Varshavskayu, I. G. 1989. The crystallization of diamond. *Surf. Coat. Tech.* 38: 1–122.

Frietas, R. A. 1999. *Nanomedicine*, Vol. I: *Basic Capabilities*. Georgetown, TX: Landes Bioscience. Available: http://www.nanomedicine.com.

Frietas, R. A. 2003. *Nanomedicine*, Vol. IIA: *Biocompatibility*. Georgetown, TX: Landes Bioscience. Available: http://www.nanomedicine.com

Galli, G., Martin, R. M., Carr, R., Parrinello, M. 1989. Structural and electronic properties of amorphous carbon. *Phys. Rev. Lett.* 62: 555–558.

Gamarnik, M. Y. 1996a. Energetical preference of diamond nanoparticles. *Phys. Rev. B* 54: 2150–2156.

Gamarnik, M. Y. 1996b. Size-related stabilization of diamond nanoparticles. *Nanostruct. Mater.* 7: 651–658.

Gibbs, J. W. 1878. On the equilibrium of heterogeneous substances. *Trans. Conn. Acad.* 3: 343–524.

Gleiter, H. 2000. Nanostructured materials: Basic concepts and microstructure. *Acta Mater.* 48: 1–29.

Glosli, J. N., Ree, F. H. 1999. Liquid-liquid phase transformation in carbon. *Phys. Rev. Lett.* 82: 4659–4662.

Grumback, M. P., Martin, R. M. 1996. Phase diagram of carbon at high pressure: Analogy to silicon. *Solid State Commun.* 100: 61–65.

Gust, W. H. 1980. Phase transition and shock-compression parameters to 120 GPa for three types of graphite and for amorphous carbon. *Phys. Rev. B* 22: 4744–4756.

Hill, T. L. 2001. A different approach to nanothermodynamics. *Nano Lett.* 1: 273–275.

Holzapfel, W. B. 1997. In: *High Pressure Techniques in Chemistry and Physics*, ed. Holzapfel, W. B. and Isaacs, N., p. 47. Oxford, U.K.: Oxford University Press.

Hwang, N. M., Hahn, J. H., Yoon, D. Y. 1996a. Chemical potential of carbon in the low pressure synthesis of diamond. *J. Cryst. Growth* 160: 87–97.

Hwang, N. M., Hahn, J. H., Yoon, D. Y. 1996b. Charged cluster model in the low pressure synthesis of diamond. *J. Cryst. Growth* 162: 55–68.

Iijima, S. 1991. Helical microtubules of graphitic carbon. *Nature* 354: 56–58

Jhi, S. H. 2006. Activated boron nitride nanotubes: A potential material for room-temperature hydrogen storage. *Phys. Rev. B* 74: 155424.

Jhi, S. H., Kwon, Y. K. 2004. Hydrogen adsorption on boron nitride nanotubes: A path to room-temperature hydrogen storage. *Phys. Rev. B* 69: 245407.

Jiang, Q., Chen, Z. P. 2006. Thermodynamic phase stabilities of nano-carbon. *Carbon* 44: 79–83.

Jiang, Q., Lu, H. M. 2008. Size dependent interface energy and its applications. *Surf. Sci. Rep.* 63: 427–464.

Jiang, Q., Yang, C. C. 2008. Size effect on phase stability of nanostructures. *Curr. Nanosci.* 4: 179–200.

Jiang, Q., Li, J. C., Wilde, G. 2000. The size dependence of the diamond-graphite transition. *J. Phys.: Condens. Matter* 12: 5623–5627.

Korobenko, V. N. 2001. Experimental investigation of liquid metals and carbon properties under high temperatures, PhD thesis, Associated Institute for High Temperatures of the Russian Academy of Sciences, Moscow, Russia.

Korobenko, V. N., Savvatimskiy, A. I. 2003. Blackbody design for high temperature (1800 to 5500 K) of metals and carbon in liquid states under fast heating, In: *Temperature: Its Measurement and Control in Science and Industry, Conference Proceedings*, ed. Ripple, D. C., Vol. 7, pp. 783–788. Melville, NY: American Institute of Physics.

Kwon, S. J., Park, J. G. 2007. Theoretical analysis of the graphitization of a nanodiamond. *J. Phys.: Condens. Matter* 19: 386215.

Lin, Y. M., Appenzeller, J., Chen, Z. H., Chen, Z. G., Cheng, H. M., Avouris, P. 2005. High-performance dual-gate carbon nanotube FETs with 40-nm gate length. *IEEE Electron Device Lett.* 26: 823–825.

Liu, C., Fan, Y. Y., Liu, M., Cong, H. T., Cheng, H. M., Dresselhaus, M. S. 1999. Hydrogen storage in single-walled carbon nanotubes at room temperature. *Science* 286: 1127–1129.

Lu, K. 1996. Nanocrystalline metals crystallized from amorphous solids: Nanocrystallization, structure, and properties. *Mater. Sci. Eng. R.* 16: 161–221.

McClure, J. W. 1956. Diamagnetism of graphite. *Phys. Rev.* 104: 666–671.

Moore, G. E. 1965. Cramming more components onto integrated circuits. *Electronics* 38: 114–117

Novoselov, K. S., Geim, A. K., Morozov, S. V., Jiang, D., Zhang, Y., Dubonos, S. V., Grigorieva, I. V., Firsov, A. A. 2004. Electric field effect in atomically thin carbon films. *Science* 306: 666–669.

Novoselov, K. S., Geim, A. K., Morozov, S. V. et al. 2005. Two-dimensional gas of massless Dirac fermions in graphene. *Nature* 438: 197–200.

Osipov, Yu. V., Enoki, T., Takai, K., Takahara, K., Endo, M., Hayashi, T., Hishiyama, Y., Kaburagi, Y., Vul, A. Y. 2006. Magnetic and high resolution TEM studies of nanographite derived from nanodiamond. *Carbon* 44: 1225–1234.

Park, N., Lee, K., Han, S. W., Yu, J. J., Ihm, J. 2002. Energetics of large carbon clusters: Crossover from fullerenes to nanotubes. *Phys. Rev. B* 65: 121405.

Pupysheva, O. V., Farajian, A. A., Yakobson, B. I. 2008. Fullerene nanocage capacity for hydrogen storage. *Nano Lett.* 8: 767–774.

Raty, J. Y., Galli, G. 2003. Ultradispersity of diamond at the nanoscale. *Nat. Mater.* 2: 792–795.

Raty, J. Y., Calli, G., Bostedt, C., Van Buuren, T. W., Terminello, L. J. 2003. Quantum confinement and fullerene like surface reconstructions in nanodiamonds. *Phys. Rev. Lett.* 90: 37401.

Ree, F. H., Winter, N. W., Glosli, J. N., Viecelli, J. A. 1999. Kinetics and thermodynamic behavior of carbon clusters under high pressure and high temperature. *Physica B* 265: 223–229.

Robertson, D. H., Brenner, D. W., Mintmire, J. W. 1992. Energetics of nanoscale graphitic tubules. *Phys. Rev. B* 45: 12592–12595.

Rojas, M. I., Leiva, E. P. M. 2007. Density functional theory study of a graphene sheet modified with titanium in contact with different adsorbates. *Phys. Rev. B* 76: 155415.

Rusanov, A. I. 1996. Thermodynamics of solid surfaces. *Surf. Sci. R.* 23: 173–247.

Savvatimskii, A. I. 2003. Melting point of graphite and liquid carbon—(Concerning the paper "Experimental investigation of the thermal properties of carbon at high temperatures and moderate pressures" by Asinovskii, E. I., Kirillin, A. V., and Kostanovskii, A. V.). *Phys. Usp.* 46: 1295–1303.

Savvatimskiy, A. I. 2005. Measurements of the melting point of graphite and the properties of liquid carbon (a review for 1963–2003). *Carbon* 43: 1115–1142.

Shenderova, O. A., Zhirnov, V. V., Brenner, D. W. 2002. Carbon nanostructures. *Crit. Rev. Solid State Mater. Sci.* 27: 227–356.

Shevlin, S. A., Guo, Z. X. 2006. Transition-metal-doping-enhanced hydrogen storage in boron nitride systems. *Appl. Phys. Lett.* 89: 153104.

Sun, Q., Jena, P., Wang, Q., Marquez, M. 2006. First-principles study of hydrogen storage on $Li_{12}C_{60}$. *J. Am. Chem. Soc.* 128: 9741–9745.

Tomanek, D., Schluter, M. A. 1991. Growth regimes of carbon clusters. *Phys. Rev. Lett.* 67: 2331–2335.

Tse, J. S., Holzapfel, W. B. 2008. Equation of state for diamond in wide ranges of pressure and temperature. *J. Appl. Phys.* 104: 043525.

Van Thiel, M., Ree, F. H. 1993. High-pressure liquid-liquid phase change in carbon *Phys. Rev. B* 48: 3591–3599.

Verechshagin, A. L. 2002. Phase diagram of ultrafine carbon. *Combust. Exp. Shock Waves* 38: 358–359.

Viecelli, J. A., Ree, F. H. 2000. Carbon particle phase transformation kinetics in detonation waves. *J. Appl. Phys.* 88: 683–690.

Viecelli, J. A., Bastea, S., Glosli, J. N., Ree, F. H. 2001. Phase transformations of nanometer size carbon particles in shocked hydrocarbons and explosives. *J. Chem. Phys.* 115: 2730–2736.

Wallace, P. R. 1947. The band theory of graphite. *Phys. Rev.* 71: 622–634.

Wang, C. X., Yang, G. W. 2005. Thermodynamics of metastable phase nucleation at the nanoscale. *Mater. Sci. Eng. R.* 49: 157–202.

Wang, C. X., Yang, Y. H., Xu, N. S., Yang, G. W. 2004. Thermodynamics of diamond nucleation on the nanoscale. *J. Am. Chem. Soc.* 126: 11303–11306.

Wang, C. X., Chen, J., Yang, G. W., Xu, N. S. 2005. Thermodynamic stability and ultrasmall-size effect of nanodiamonds. *Angew. Chem. Int. Ed.* 44: 7414–7418.

Wilson, S. R. 2000. Biological aspects of fullerenes. In: *Fullerenes: Chemistry, Physics, and Technology*, ed. Kadish, K. M. and Ruoff, R. S., pp. 437–466. New York: John Wiley & Sons.

Winter, N., Ree, F. H. 1998. Carbon particle phase stability as a function of size. *J. Comput.-Aided Mater. Des.* 5: 279–294.

Wu, X. J., Gao, Y., Zeng, X. C. 2008. Hydrogen storage in pillared Li-dispersed boron carbide nanotubes. *J. Phys. Chem. C* 112: 8458–8463.

Yang, C. C., Li, S. 2008. Size-dependent temperature-pressure phase diagram of carbon. *J. Phys. Chem. C* 112: 1423–1426.

Yildirim, T., Ciraci, S. 2005. Titanium-decorated carbon nanotubes as a potential high-capacity hydrogen storage medium. *Phys. Rev. Lett.* 94: 175501.

Yoon, M., Yang, S. Y., Hicke, C., Wang, E., Geohegan, D., Zhang, Z. Y. 2008. Calcium as the superior coating metal in functionalization of carbon fullerenes for high-capacity hydrogen storage. *Phys. Rev. Lett.* 100: 206806.

Young, D. A. 1991. *Phase Diagrams of the Elements*. Berkeley, CA: University of California Press.

Zaiser, M., Lyutovich, Y., Banhart, F. 2000. Irradiation-induced transformation of graphite to diamond: A quantitative study. *Phys. Rev. B* 62: 3058–3064.

Zhang, Y., Tan, Y. W., Stormer, H. L., Kim, P. 2005. Experimental observation of the quantum hall effect and Berry's phase in graphene. *Nature* 438: 201–204.

Zhao, D. S., Zhao, M., Jiang, Q. 2002. Size and temperature dependence of nanodiamond–nanographite transition related with surface stress. *Diam. Relat. Mater.* 11: 234–236.

IV

Nanomechanics

22

Computational Nanomechanics

Wing Kam Liu
Northwestern University

Eduard G. Karpov
University of Illinois at Chicago

Yaling Liu
University of Texas at Arlington

22.1 Introduction

Over the past three decades, we have acquired new tools and techniques to synthesize nanoscale objects and to learn their many incredible properties. The high-resolution electron microscopes that are available today allow the visualization of single atoms; furthermore, the manipulation of these individual atoms is possible using scanning probe techniques. Advanced materials synthesis provides the technology to tailor-design systems from as small as molecules to structures as large as the fuselage of a plane. We now have the technology to detect single molecules, bacteria, or virus particles. We can make protective coatings more wear-resistant than diamond and fabricate alloys and composites stronger than ever before.

Advances in the synthesis of nanoscale materials have stimulated ever-broader research activities in science and engineering devoted entirely to these materials and their applications. This is due in large part to the combination of their expected structural perfection, small size, low density, high stiffness, high strength, and excellent electronic properties. As a result, nanostructured materials may find use in a wide range of applications in material reinforcement, field emission panel display, chemical sensing, drug delivery, nanoelectronics, and tailor-designed materials. Nanoscale devices have great potential as sensors and as medical diagnostic and delivery systems.

While microscale and nanoscale systems and processes are becoming more viable for engineering applications, our knowledge of their behavior and our ability to model their performance remain limited. Continuum-based computational capabilities are obviously not applicable over the full range of operational conditions of these devices. Non-continuum behavior is observed in large deformation behavior of nanotubes, ion deposition processes, gas dynamic transport, and material mechanics as characteristic scales drop toward the micron scale. At the scales of nanodevices, interactions between thermal effects and mechanical response can become increasingly important. Furthermore, nanoscale components will be used in conjunction with components that are larger and respond at different timescales. In such hybrid systems, the interaction of different time and length scales may play a crucial role in the performance of the complete system. Single scale methods such as ab initio methods or molecular dynamics (MD) would have difficulty in analyzing such hybrid structures due to the large range of time and length scales. For the design and study of nanoscale materials and devices in microscale systems, models must span length scales from nanometers to hundreds of microns.

Computational power has doubled approximately every 18 months in accordance with Moore's law. Despite this fact and the fact that desktop computers can now routinely simulate million atom systems, simulations of realistic atomic system require at least tens of billions of atoms. In short, such systems can be modeled neither by continuum methods because they are too small nor by molecular methods because they are too large and, therefore, require usage of multiscale methods.

Multiple scale methods generally imply the utilization of information at one length scale to model the response of the material at subsequently larger length scales. These methods can be divided into two categories: hierarchical and concurrent.

Hierarchical multiple scale methods directly utilize the information at a small length scale as an input into a larger model via some type of averaging process. The Young's modulus is a good example of this; the structural material stiffness is found as a single quantity, through homogenization of all defects and microstructure at the micro and nanoscales. Concurrent multiple scale methods are those which run simultaneously; in these methods, the information at the smaller length scale is calculated and input to the larger scale model on the fly. In this chapter, we shall concentrate on the development of concurrent multiple scale methods, much of which has occurred within the past decade. We note in particular the work of Li and Liu (2004), as well as two excellent review papers that comprehensively cover the field, those of Liu et al. (2004a), and Curtin and Miller (2003).

The material presented in this chapter informs researchers and educators about specific fundamental concepts and tools in nanomechanics and materials, including solids and fluids, and their modeling via multiple scale methods and techniques. In recognizing the importance of engineering education, the material presented in this chapter is correlated with several newly developed courses taught by the authors at Northwestern University and University of Illinois, including multiscale simulations, molecular modeling, and principles of nanomechanics. Furthermore, this material was utilized as a basis for the interdisciplinary NSF-sponsored Summer Institute on Nano Mechanics and Materials, www.tam.northwestern.edu/ summerinstitute/Home.htm, which has been held at Northwestern University during years 2004–2009. This chapter, therefore, can serve as a starting point to the researchers willing to contribute to the emerging field of computational nanomechanics.

22.2 Classical Molecular Dynamics

This section is devoted to the methods of classical mechanics that allow studying the motion of gas, liquid, and solid particles as a system of interactive, dimensionless mass points. The classical dynamic equations of motion are valid for slow and heavy particles, with typical velocities $v \ll c$, c being the speed of light, and masses $m \gg m_e$, m_e being the electron mass. Therefore, only slow motion (not faster than thermal vibrations) of atoms, ions, and molecules can be considered, and the internal electronic structure is ignored. The atoms and molecules exert internal forces on each other that are determined by instantaneous values of the total potential energy of the system. The potential energy is typically considered only as a function of the system spatial configuration and is described by means of *interatomic potentials*. These potentials are considered as known input information; they are either found experimentally, or computed by averaging over the motion of the valence electrons in the ion's Coulomb field by means of quantum *ab initio* methods. During the course of the system's dynamics, the interatomic potentials are not perturbed by possible changes in the internal electronic states of the simulated particles.

Analytical solutions of the equations of particle dynamics are possible only for a limited set of interesting problems, and only for systems with a small number of degrees of freedom. Numerical methods of solving the classical equations of motion for multiparticle systems with known interatomic potentials are collectively referred to as *molecular dynamics*. MD is regarded as a major practical application of the classical particle dynamics. The subsequent computer postprocessing and visualization of the results accomplished in a dynamic manner are called the MD simulation.

22.2.1 Mechanics of a System of Particles

Classical dynamics studies the motion of mass points (ideal dimensionless particles) due to known *forces* exerted on them. These forces serve as qualitative characteristics of the interaction of particles with each other (internal forces) and with exterior bodies (external forces). The general task of dynamics consists of solving for the positions (trajectories) of all particles in a given mechanical system over the course of time. In principle, such a solution is uniquely determined by a set of initial conditions, i.e., positions and velocities of all particles at time $t = 0$, and the interaction forces. If there are no external forces applied to the system, then this system is *isolated* or *closed*, otherwise it is called *non-isolated*.

22.2.1.1 Generalized Coordinates

The spatial configuration of N dimensionless particles can be determined by N radius vectors $\mathbf{r}_1, \mathbf{r}_2, \ldots, \mathbf{r}_N$, or by $3N$ coordinates (Cartesian x_i, y_i, z_i, spherical r_i, θ_i, ϕ_i, etc.). In some cases, the motion of these particles is constrained in a specific manner, i.e., under given provisos, it cannot be absolutely arbitrary. Then we say that such a system has *mechanical constraints*, and the system itself is called *constrained*; otherwise the system is called *non-constrained*. If some mechanical constraint can be expressed as a function of the coordinates of the particles,

$$f(\mathbf{r}_1, \mathbf{r}_2 \ldots, \mathbf{r}_N) = 0 \tag{22.1}$$

we call them *holonomic*, otherwise *non-holonomic*. Example systems with holonomic constraints are a pendulum in the field of gravity, as well as diatomic and polyatomic gas molecules with rigid interatomic bonding.

In the presence of k holonomic constraints of the type (22.1), there exist only $s = 3N - k$ independent coordinates. Any s independent variables $q_1, q_2, \ldots, q_s$ (lengths, angles, etc.) that fully determine the spatial configuration of the system are referred to as the *generalized coordinates*, and their time derivatives $\dot{q}_1, \dot{q}_2, \ldots, \dot{q}_s$ – *generalized velocities*. The relationship between the radius vectors and the generalized coordinates can be expressed by the *transformation equations*

$$\mathbf{r}_i = \mathbf{r}_i(q_1, q_2, \ldots, q_s), \quad i = 1, 2, \ldots, N \tag{22.2}$$

that provide parametric representations of the old coordinates $\mathbf{r}_i$ in terms of the new coordinates q_i. The corresponding velocities are given by

$$\mathbf{v}_i = \dot{\mathbf{r}}_i = \frac{\partial \mathbf{r}_i}{\partial t} + \sum_{j=1}^{s} \frac{\partial \mathbf{r}_i}{\partial q_i} \dot{q}_i, \quad i = 1, 2, \ldots, N \qquad (22.3)$$

The transformations are assumed to be invertible, i.e., the Equations 22.2 combined with the constraint rules (22.1) can be inverted to obtain the generalized coordinates as functions of the radius vectors.

22.2.1.2 Mechanical Forces and Potential Energy

Besides being classified as internal or external, all mechanical forces in a system of particles can be classified as *conservative* or *nonconservative*. Conservative forces are those whose work depends *only* on positions of the particles, without regard for their instantaneous velocities and trajectories of passage between these positions. All other forces are called *nonconservative*. They comprise two major types: dissipative and gyroscopic. Forces of mechanical friction and viscous friction in gases and liquids are dissipative. Current magnitudes and directions of dissipative forces may depend on instantaneous velocities, and/or a time history of the atomic motion in the system. The work of these forces in a closed system is always negative, including the case of looped trajectories. Gyroscopic forces depend on instantaneous velocities of particles and act in directions that are orthogonal to these velocities. The work of these forces is always trivial over the course of motion of the particles. Examples of gyroscopic forces are the Coriolis force, felt by mass particles moving in a rotating coordinate system inwards or outwards from the axis of rotation, as well as the Lorentz force felt by charged particles moving in a magnetic field.

For a system characterized by only conservative forces, there exists a specific function of coordinates of the particles

$$U = U(\mathbf{r}_1, \mathbf{r}_2, \ldots, \mathbf{r}_N) \qquad (22.4)$$

called the *potential energy*, or simply the *potential*, of the system. At the same time, partial derivatives of U with respect to the coordinates of a particle i yield the corresponding components of the resultant force felt by this particle due to the potential U:

$$\mathbf{f}_i = -\frac{dU}{d\mathbf{r}_i} = -\left(\frac{\partial}{\partial x_i} + \frac{\partial}{\partial y_i} + \frac{\partial}{\partial z_i} \right) U \equiv -\nabla_i U, \quad i = 1, 2, \ldots, N$$
$$(22.5)$$

22.2.1.3 Newtonian Equations

For a system of N interacting monoatomic molecules, treated as individual mass points, there are no holonomic constraints, and the generalized coordinates can be chosen equivalent to the Cartesian coordinates. The equation of motion of such a system can be written in the *Newtonian form*, e.g., Goldstein (1980),

$$m_i \ddot{\mathbf{r}}_i = -\nabla_i U + \mathbf{F}_i, \quad i = 1, 2, \ldots, N \qquad (22.6)$$

where

the nabla operator is such as in Equation 22.5
U is the potential energy function (22.4)
$\mathbf{F}_i$ is the nonconservative or external force on the ith particle

The Newtonian equations are also applicable to a system of polyatomic molecules, provided that these molecules can be approximated as structureless mass points.

22.2.1.3.1 Dissipative Equations

For a nonconservative system, the right-hand side of the Equation 22.6 will include also all the available nonconservative forces. One typical example is the dissipative system, where the damping force exerted on a particle is proportional to its instantaneous velocity with the opposite sign:

$$m_i \ddot{\mathbf{r}}_i = -\nabla_i U - m_i \gamma_i \dot{\mathbf{r}}_i, \quad i = 1, 2, \ldots, N \qquad (22.7)$$

and γ is the damping constant. The damping force

$$\mathbf{F}_i^S = -m_i \gamma_i \dot{\mathbf{r}}_i \qquad (22.8)$$

is also called the viscous, or Stokes' friction, because it is similar to decelerating forces exerted on a solid particle moving in a liquid solvent. This model can be updated with a stochastic external force $\mathbf{R}_i$ that represents thermal collisions of the system particle i with the hypothetical solvent molecules:

$$m_i \ddot{\mathbf{r}}_i = -\nabla_i U - m_i \gamma_i \dot{\mathbf{r}}_i + \mathbf{R}_i(t), \quad i = 1, 2, \ldots, N \qquad (22.9)$$

Assuming that interactions/collisions between the particle i and the solvent molecules are frequent and fast, i.e., the magnitude of the random force $\mathbf{R}$ varies over much shorter timescales than the timescale over which the particle's position and velocity change, Equation 22.9 is said to represent *Brownian motion* of the particle i. Furthermore, if the stochastic force satisfies the following relationships

$$\lim_{t \to \infty} \frac{1}{t} \int_0^t \mathbf{R}(\tau) d\tau = 0 \qquad (22.10)$$

$$\lim_{t \to \infty} \frac{1}{t} \int_0^t \mathbf{R}_i(\tau) \cdot \mathbf{R}_{i'}(t_0 + \tau) d\tau = a \delta(t_0) \delta_{ii'} \qquad (22.11)$$

where

a is a constant
$\delta(t_0)$ and $\delta_{ii'}$ are the Dirac and Kronecker deltas, respectively

Equation 22.9 is often called the *Langevin equation*, and the particle i is referred to as a Langevin particle.

22.2.1.3.2 Generalized Langevin Equation

The relationship (22.10) implies that there is no directional preference for the stochastic force in the Langevin model. An integral of the type (22.11) is called the autocorrelation function of a continuous function **R**. The Dirac delta autocorrelation (22.11) implies that the function **R** does not represent any time history, so that the current value of **R** is not affected by behavior of this function at preceding times. The Kronecker delta in (22.11) means that the force exerted on the particle i is uncorrelated with the force exerted on another particles i'.

A more general form of the Equation 22.9, which is used in application to solids, gas/solid and solid–solid interfaces, is given by

$$m_i \ddot{\mathbf{r}}_i(t) = -\nabla_i U - \sum_{i'} \int_0^t \beta_{ii'}(t-\tau)\dot{\mathbf{r}}_{i'}(\tau) + \mathbf{R}_i(t), \quad i = 1, 2, \ldots, N$$

(22.12)

Here, the second term on the right-hand side represents a dissipative damping force that depends on the entire *time history* of velocities of the current atom i and a group of neighboring atoms i'; the matrix β is called the time-history damping kernel. This damping force can be alternatively defined in terms of the history of atomic positions, rather than velocities:

$$m_i \ddot{\mathbf{r}}_i(t) = -\nabla_i U - \sum_{i'} \int_0^t \theta_{ii'}(t-\tau)\mathbf{r}_{i'}(\tau) + \mathbf{R}_i(t), \quad i = 1, 2, \ldots, N$$

(22.13)

where θ is a new damping kernel. Newtonian equations of the type (22.12), or (22.13), are often called the *generalized Langevin equations*, e.g., Adelman and Doll (1976), Adelman and Garrison (1976), Karpov et al. (2005). One application of these equations is related to the motion of individual atoms in a large crystal lattice subjected to external pulse excitations. We consider this application in greater detail in Section 22.3.5.

22.2.2 Molecular Forces

As discussed in Section 22.2.1 the general forms of the governing equation of particle dynamics are given by straightforward second-order ordinary differential equations, which allow a variety of numerical solution techniques. Meanwhile, the potential function U for Equation 22.6 can be an extremely complicated object, when the accurate representation of the atomic interactions within the system under consideration is required. The nature of these interactions is due to complicated quantum effects taking place at the subatomic level that are responsible for chemical properties such as valence and bond energy. The effects are also responsible for the spatial arrangement (topology) of the interatomic bonds, their formation, and breakage. In order to obtain reliable results in MD computer

simulations, the classical interatomic potentials should accurately account for the quantum effects, even in an averaged sense. Typically, the function U is obtained from experimental observations, as well as from the quantum-scale modeling and simulation, e.g., La Paglia (1971), Mueller (2001), Ratner and Schatz (2001).

The issues related to the form of the potential function for particular types of atomic systems have been extensively discussed in the literature. The general structure of this function can be presented as

$$U(\mathbf{r}_1, \mathbf{r}_2, \ldots, \mathbf{r}_N) = \sum_i V_1(\mathbf{r}_i) + \sum_{i,j>i} V_2(\mathbf{r}_i, \mathbf{r}_j) + \sum_{i,j>i,k>j} V_3(\mathbf{r}_i, \mathbf{r}_j, \mathbf{r}_k) + \cdots$$

(22.14)

where $\mathbf{r}$'s are radius vectors of the particles, and the function V_m is called the m-body potential. The first term represents the energy due to an external force field, such as gravity or electrostatic, which the system is immersed into, or bounding fields such as potential barriers and wells. The second term shows potential energy of pair-wise interaction of the particles; the third gives the three-body components, etc. Respectively, the function V_1 is also called the external potential, V_2 is the interatomic (pair-wise), and V_m at $m > 2$ is a multi-body potential. In order to reduce the computational expense of numerical simulations, it is practical to truncate the sum (22.14) after the second term and incorporate all the multi-body effects into V_2 with some appropriate degree of accuracy; this approach is further discussed in Section 22.2.2.2.

22.2.2.1 External Fields

The effect of external fields for a particle i can be generally described as

$$V_1 = V_1(\mathbf{r}_i)$$

(22.15)

where V_1 is a function of the radius vectors of this particle. Thus, the instantaneous force exerted on this particle due to V_1 depends only on the spatial location of this particular particle, and it is independent of the positions of any other particles in the system. Simple examples include the uniform field of gravity:

$$V_G(\mathbf{r}) = V_G(y) = mgx$$

(22.16)

where x is the component of the radius vector, orthogonal to the Earth's surface; the field of a one-dimensional (1D) harmonic oscillator,

$$V_h(x) = kx^2$$

(22.17)

and a spherical oscillator,

$$V_h(r) = kr^2, \quad r^2 = x^2 + y^2 + z^2$$

(22.18)

External bounding fields and potential barriers give further examples of the one-body interaction. A 1D potential well can be expressed by

$$V_W(x) = -\Delta E e^{-\frac{(x-x_0)^s}{2R^s}} \qquad (22.19)$$

and a two-sided potential barrier,

$$V_B(x) = -V_W(x) \qquad (22.20)$$

where x_0, $2R$, s, and ΔE are the coordinate of the center, width, steepness, and total depth/height of the well/barrier, respectively. The steepness s is an even integer, and the width $2R$ corresponds to $V_W = \Delta E e^{-1/2}$. By manipulating these functions, one can also obtain 2D, 3D, cylindrical, and ellipsoid shapes. In case ΔE is increasingly large compared with the average kinetic energy of particles, a function similar to V_W is also called a wall function that models a system of particles inside a vessel with impenetrable walls. In general, potential wells and barriers are introduced to confine the spatial domain occupied by a finite system of atoms and molecules in gaseous or liquid phase.

22.2.2.2 Pair-Wise Interaction

The pair-wise function V_2 describes the dependence of the total potential energy U on the interparticle distances. Letting $\mathbf{r}_i$ and $\mathbf{r}_j$ be radius vectors of two arbitrary particles in the system, we can generally write

$$V_2(\mathbf{r}_i, \mathbf{r}_j) = V_2(r), \quad r = |\mathbf{r}_{ij}| = |\mathbf{r}_i - \mathbf{r}_j| \qquad (22.21)$$

Such a function serves as an addition to V_1, formula (22.15), which only describes separate dependence of U on the radius vectors $\mathbf{r}_i$ and $\mathbf{r}_j$. Pair-wise coordination of particles is depicted in Figure 22.1.

There are two major types of pair-wise interactions: long-range electrostatic interactions and short-range interactions between electrically neutral particles.

22.2.2.2.1 Long-Range Coulomb Interaction

One basic physical characteristic of atoms, molecules, and their elementary components is the electric charge, e. Electric charge is related to the property of particles to exert forces on each other by means of electric fields. The standard SI unit of electric charge is 1 coulomb, $[e] = 1$ C. The electric charge of a particle is always quantized, occurring as a multiple of the elementary charge, $e_0 = 1.602177 \cdot 10^{-19}$ C. An electron, proton, and neutron, the atomic components, have the charges $-e_0$, e_0, and 0, respectively. The charge of an atomic nucleus, comprised by protons and neutrons, is Ze_0, where Z is the number of protons. An atomic ion can have a charge $(Z - N_e)e_0$, where N_e is the number of available electrons. As a result, classical particle dynamics deals with positively and negatively charged atomic and molecular ions, as well as with electrically neutral particles ($e = 0$).

A pair of particles bearing electric charges e_1 and e_2 exert on each other repulsive (at $e_1/e_2 > 0$), or attractive (at $e_1/e_2 < 0$) forces, described by

$$\mathbf{F}_1 = -\mathbf{F}_2 = -\nabla_i V_C(\mathbf{r}_1, \mathbf{r}_2), \quad i = 1, 2 \qquad (22.22)$$

where $\mathbf{r}_1$ and $\mathbf{r}_2$ are radius vectors of the particles, and

$$V_C(\mathbf{r}_1, \mathbf{r}_2) = V_C(r) = \frac{1}{4\pi\epsilon_0} \frac{e_1 e_2}{r}, \quad r = |\mathbf{r}_{12}| = |\mathbf{r}_2 - \mathbf{r}_1| \qquad (22.23)$$

is the electrostatic *Coulomb potential*; $\epsilon_0 = 8.854188 \cdot 10^{-12}$ C/(Vm) is the permittivity constant in a vacuum. Equations 22.22 and 22.23 account for the convention where attractive forces are defined as negative, and repulsive forces as positive.

The absolute value of the interaction force (Equation 22.22) can be expressed as a function of the separation distance:

$$F_C(r) = -\frac{\partial V(r)}{\partial r} = \frac{1}{4\pi\epsilon_0} \frac{e_1 e_2}{r^2} \qquad (22.24)$$

For a system of N charged particles, the pair-wise interaction energy is written as

$$U_C = \sum_{i,j>i} V_C(r_{ij}), \quad r_{ij} = |\mathbf{r}_{ij}| = |\mathbf{r}_j - \mathbf{r}_i| \qquad (22.25)$$

where

V_C is the Coulomb potential (22.23)
r_{ij} is the separation distance for a pair of particles i and j, see Figure 22.1

The relevant interaction forces can be computed according to the general formula (22.5).

The Coulomb interaction is of greatest magnitude at large separations distances, because the potential (22.23) decays slowly with the growth of r. For this reason, it is called a *long-range* interaction.

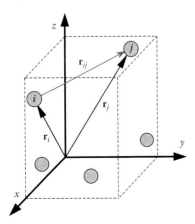

FIGURE 22.1 Pair-wise coordination of particles.

22.2.2.2.2 Short-Range Interaction

For a pair of electrically neutral atoms or molecules, the electrostatic field of the positively charged atomic nuclei or ion is neutralized by the negatively charged electron clouds surrounding the nuclei. Quantum mechanical descriptions of the electron motion involves a probabilistic framework to evaluate the probability densities at which the electrons can occupy particular spatial locations; in other words, quantum mechanics provides probability densities in the configuration space of electrons, e.g., La Paglia (1971), Ratner and Schatz (2001). The term "electron cloud" is typically used in relation to the spatial distributions of these densities. The negatively charged electron clouds, however, experience cross-atomic attraction, which grows as the distance between the nuclei decreases. On reaching some particular distance, which is referred to as the equilibrium (bond) length, this attraction is equilibrated by the repulsive force due to the positively charged atomic nuclei or ions. A further decrease in the interparticle distance results in a quick growth of the resultant repulsive force. The potential energy of such a system will be a continuous function of the separation distance, provided that internal quantum states of the electron cloud are not excited. Alternatively, these quantum states are excited and consequently relaxed at timescales that are significantly faster than the characteristic time of the ions' thermal motion. Then the "heavy" and "slow" (in the quantum sense) nuclei or ions can be considered as a classical system of particles, interacting through a time-independent potential, averaged over the electronic degrees of freedom.

There exist a number of mathematical models to adequately describe the dual attractive/repulsive character of interactions between a pair of neutral atoms or molecules. In 1924, Jones (1924a,b) proposed the following potential:

$$V_{LJ}(r) = 4\varepsilon\left(\frac{\sigma^{12}}{r^{12}} - \frac{\sigma^6}{r^6}\right) \qquad (22.26)$$

This model is currently known as the *Lennard-Jones* (LJ) *potential*, and it is used in computer simulations of a great variety of nanoscale systems and processes. Here, σ is the collision diameter, the distance at which $V_{LJ} = 0$, and ε is the dislocation energy. In a typical atomistic system, the collision diameter is equal to several angstroms (Å), $1\,\text{Å} = 10^{-10}$ m. The value ε corresponds to the minimum of function (22.26), which occurs at the equilibrium bond length $\rho = 2^{1/6}\sigma$; $V_{LJ}(\rho) = -\varepsilon$. Physically, ε represents the amount of work that needs to be done in order to move the interacting particles apart from the equilibrium distance ρ to infinity. The availability of a minimum in the LJ potential represents the possibility of bonding for two colliding particles, provided that their relative kinetic energy is less than ε.

The first term of the LJ potential represents atomic repulsion, dominating at small separation distances while the second term shows attraction (bonding) between two atoms or molecules. Since the bracket quantity is dimensionless, the choice of units for V depends on the definition of ε. Typically, $\varepsilon \sim 10^{-19} - 10^{-18}$ J,

therefore it is more convenient to use a smaller energy unit, such as the electron volt (eV),

$$1\,\text{eV} = 1.602 \times 10^{-19}\,\text{J} \qquad (22.27)$$

rather than joules. One electron volt represents the work done if an elementary charge is accelerated by an electrostatic field of a unit voltage. This is a typical atomic-scale unit; therefore, it is often used in computational nanomechanics and materials.

The absolute value of the LJ interaction force, as a function of the interparticle distance, gives

$$F_{LJ}(r) = 24\varepsilon\left(\frac{2\sigma^{12}}{r^{13}} - \frac{\sigma^6}{r^7}\right) \qquad (22.28)$$

The potential (22.26) and force (22.28) functions are plotted in Figure 22.2a in terms of dimensionless quantities. Note that $F_{LJ}(\rho) = 0$.

Another popular model for pair-wise interactions is the Morse potential shown in Figure 22.2b:

$$V_M(r) = \varepsilon(e^{2\beta(\rho-r)} - 2e^{\beta(\rho-r)}) \qquad (22.29)$$

and

$$F_M(r) = 2\varepsilon\beta(e^{2\beta(\rho-r)} - e^{\beta(\rho-r)}) \qquad (22.30)$$

This potential is commonly used for systems found in solid state at normal conditions. These include elemental metallic systems and alloys, e.g., Harrison (1988). For the solid state, the typical kinetic energy of particles is less than the dislocation energy, and the particles are restrained to move in the vicinity of some equilibrium positions that form a regular spatial pattern, the crystal lattice.

In the case of multiple particles, the total potential energy due to the LJ or Morse interaction is computed similar to Equation 22.25, and the required internal forces are found by utilizing Equation 22.5.

The interaction between neutral particles described by a LJ or Morse potential is said to be short ranged, as contrasts the long-range Coulomb interactions. As seen from Figure 22.2, short-range potentials are effectively zero if the separation r is larger than several equilibrium distances.

The LJ and Morse potentials are the most common models for short-range pair-wise interactions. They have found numerous applications in computational chemistry, physics, and nanoengineering.

22.2.2.2.3 Cutoff Radius

One important issue arising from MD computer simulations relates to the truncation of the potential functions, such as (22.26) and (22.29). Note that computing the internal forces (22.5) for the equations of motion due to only the pair-wise interaction in

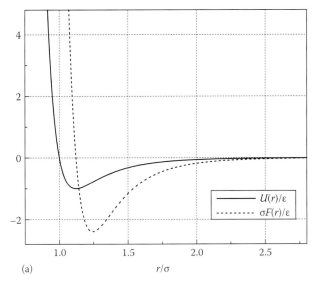
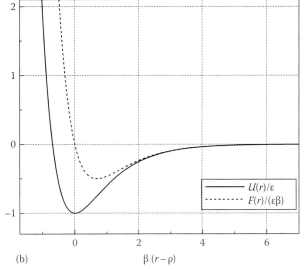

FIGURE 22.2 Short-range potentials: (a) Lennard-Jones and (b) Morse.

Equation 22.14 results in $(N^2 - N)/2$ terms, where N is the total number of particles. This value corresponds to the case when one takes into account the interaction of each current atom i with all other atoms in the system $j \neq i$; this can be computationally expensive even for considerably small systems. For short-range potentials, we can assume that the current atom only interacts with its nearest neighbors, found not further than some critical distance R. Typically, the value R is equal to several equilibrium distances ρ, and is called the *cutoff radius* of the potential. Limiting the cutoff radius to a sphere of several neighboring atoms can reduce the computational effort significantly. A truncated pair-wise potential can then be written as the following:

$$V^{(\mathrm{tr})}(r) = \begin{cases} V(r), & r \leq R \\ 0, & r > R \end{cases} \qquad (22.31)$$

If each atom interacts with only n atoms in its R-vicinity, the evaluation of the internal pair-wise forces will result in only $nN/2$ terms, which is considerably less than the $(N^2 - N)/2$ terms for a non-truncated potential.

In order to assure continuity (differentiability) of V^{tr}, a "skin" factor can be alternatively introduced for the truncated potential by means of a smooth steplike function f_c, which is referred to as the *cutoff function*. The function f_c provides a smooth and quick transition from 1 to 0, when the value of r approaches R, and it is usually chosen as a simple analytical function of the separation distance r. One example of a trigonometric cutoff function is given by Equation 22.36.

22.2.2.3 Multi-Body Interaction

The higher order terms of the potential function (22.14) ($m > 2$) can be of importance in modeling of solids and complex

molecular structures to account for chemical bond formation, their topology and spatial arrangement, as well as the chemical valence of atoms. However, the practical implementation of multi-body interactions can be extremely involved. As a result, all the multi-body effects of the order higher than three are usually ignored.

Meanwhile, the three-body potential V_3 is intended to provide contributions to the total potential energy U that depends on the value of the angle θ_{ijk} between a pair of interparticle vectors $\mathbf{r}_{ij}$ and $\mathbf{r}_{ik}$, forming a triplet of particles i, j, and k (see Figure 22.3):

$$V_3(\mathbf{r}_i, \mathbf{r}_j, \mathbf{r}_k) = V_3(\cos\theta_{ijk}), \quad \cos\theta_{ijk} = \frac{\mathbf{r}_{ij} \cdot \mathbf{r}_{ik}}{r_{ij} r_{ik}} \qquad (22.32)$$

Such a function is viewed as an addition to the two-body term (Equation 22.21), which accounts only for the absolute values of r_{ij} and r_{ik}. Three-body potentials are dedicated to reflect changes in molecular shapes and bonding geometries in atomistic structures, e.g., Stillinger and Weber (1985), Takai et al. (1985).

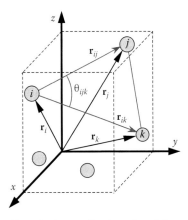

FIGURE 22.3 Three-body coordination of particles.

22.2.2.3.1 Local Environment Potentials

As a matter of fact, explicit three-body potentials, such as Equation 22.32, are impractical in terms of computer modeling. Furthermore, they have been criticized for being unable to describe the energetics of all possible bonding geometries; see Biswas and Hamann (1985, 1987), Tersoff (1988b). At the same time, four- and five-body potentials appear computationally intractable, and generally contain too many free parameters. As a result, a number of advanced two-body potentials have been proposed to efficiently account for the specifics of a local atomistic environment by incorporating some specific multi-body dependencies inside the function V_2, known as *bond order functions*, rather than introducing the multi-body functions $V_{m>2}$. Such potentials are called local environment potentials. The bond-order function is intended to implicitly describe the angular dependence of interatomic forces, while the overall pair-wise formulation is preserved. Local environment potentials are usually short ranged; therefore, cutoff functions can be utilized. Some of the most common models of this type are the Tersoff potential for a class of covalent systems, such as carbon, silicon, and germanium, the Tersoff (1986, 1988a,b), Brenner (1990), Los and Fasolino (2002), Rosenblum et al. (1999), and REBO Brenner et al. (2002) potentials for carbon and hydrocarbon molecules, and the Finnis–Sinclair potential for BCC metals, Finnis and Sinclair (1984) and Konishi et al. (1999).

Most of the existing local environment potentials feature the following common structure:

$$V_{LE}(\mathbf{r}_i, \mathbf{r}_j) = (V_R(r_{ij}) - B_{ij} V_A(r_{ij})), \quad r_{ij} = |\mathbf{r}_{ij}| \tag{22.33}$$

where V_R and V_A are pair-wise repulsive and attractive interactions, respectively and B is the bond order function, which is intended to represent multi-body effects by accounting for spatial arrangements of the bonds in a current atom's vicinity.

The silicon potential model by Tersoff (1988b) gives an example of the local environment approach:

$$V_T(r_{ij}) = f_c(r_{ij})(Ae^{-\lambda_1 r_{ij}} - B_{ij} e^{-\lambda_2 r_{ij}}) \tag{22.34}$$

where A is a constant, and

$$B_{ij} = (1 + \beta^n \zeta_{ij}^n)^{-1/2n}$$

$$\zeta_{ij} = \sum_{k \neq i,j} f_c(r_{ik}) g(\theta_{ijk}) e^{\lambda_3^3 (r_{ij} - r_{ik})^3} \tag{22.35}$$

$$g(\theta) = 1 + \frac{c^2}{d^2} - \frac{c^2}{d^2 + (h - \cos\theta)^2}$$

The cutoff function is chosen as

$$f_c(r) = \frac{1}{2} \begin{cases} 2, & r < R - D \\ 1 - \sin(\pi(r - R)/2D), & R - D < r < R + D \\ 0, & r > R + D \end{cases} \tag{22.36}$$

where the middle interval function is known as the "skin" of the potential. Note that if the local bond order is ignored, so that $B = 2A = const$, and $\lambda_1 = 2\lambda_2$, the Tersoff potential reduces to the Morse model (22.29). In other words, all deviations from a simple pair potential are ascribed to the dependence of the function B on the local atomic environment. The value of this function is determined by the number of competing bonds, the strength α of the bonds and the angles θ between them; for example, θ_{ijk} shows the angle between the bonds i–j and i–k. The function ζ in Equation 22.35 is a weighted measure of the number of bonds competing with the bond i–j, and the parameter n shows how much the closer neighbors are favored over more distant ones in the competition to form bonds.

The potentials proposed by Brenner and coworkers, e.g., Brenner (1990), Brenner et al. (2002), Los and Fasolino (2002), Rosenblum et al. (1999) are considered to be more accurate, though more involved, extensions of the Tersoff (1986, 1988a,b) models. The Brenner potentials include more detailed functions V_A, V_R, and B_{ij} to account for different types of chemical bonds that occur in the diamond and graphite phases of the carbon, as well as in hydrocarbon molecules.

22.2.2.3.2 Embedded Atom Potential

Another special form of a multi-body potential is provided by the *embedded atom method* (EAM) for metallic systems, e.g., Daw (1989), Daw et al. (1993), Johnson (1988). One appealing aspect of the EAM potential is its physical picture of metallic bonding, where each atom is embedded in a host electron gas created by all neighboring atoms. The atom–host interaction is inherently more complicated than the simple pair-wise model. This interaction is described in a cumulative way in terms of an empirical *embedding energy function*. The embedding function incorporates some important many-atom effects by providing the amount of energy (work) required to insert one atom into the electron gas of a given density. The total potential energy U includes the embedding energies G of all the atoms in the system, and an electrostatic Coulomb interaction V_C:

$$U = \sum_i G_i \left(\sum_{j \neq i} \rho_j^a(r_{ij}) \right) + \sum_{i,j>i} V_C(r_{ij}) \tag{22.37}$$

Here, ρ_j^a is the averaged electron density for a host atom j, viewed as a function of the distance between this atom and the embedded atom i. Thus, the host electron density is employed as a linear superposition of contributions from individual atoms, which in turn are assumed to be spherically symmetric. Information on the specific shapes of the functions G, ρ, and V_C for various metals and alloys can be gathered from these two references: Clementi and Roetti (1974) and Foiles et al. (1986). The embedded atom method has been applied successfully to study defects and fracture, grain boundaries, interdiffusion in alloys, liquid metals, and other metallic systems and processes; a comprehensive review of the embedded atom methodology and applications is provided in Daw et al. (1993).

22.2.3 Numerical Heat Bath Techniques

The modeling and simulation of multiparticle systems, investigation of various temperature-dependent macroscopic properties (e.g., internal energy, pressure, viscosity) often require the availability of numerical methods for maintaining the temperature of the system at a particular target/reference value. From the thermodynamical point of view, such a system can be regarded as interacting with an external thermostat that keeps the temperature at a constant level by providing or removing heat in the course of time.

One issue associated with such an approach is the way to represent the mathematical coupling between the degrees of freedom of the simulated system and the hypothetical heat bath. Standard coupling techniques include the thermostatting approach of Berendsen et al. (1984), the stochastic collisions method of Andersen (1980), and the extended systems method originated by Nosé (1984a) and extended by Hoover (1985). The Nosé–Hoover approach incorporates the external heat bath as an integral part of the system. This is achieved by assigning the reservoir an additional degree of freedom, and including it in the system Hamiltonian. The novel phonon approach (Karpov et al. 2007) that encapsulates the intrinsic mechanical properties of the crystalline lattice is more adequate for the modeling of nonmetallic solids, including carbon nanostructures. Below we review the Berendsen, Nosé–Hoover, and the phonon method in more detail.

22.2.3.1 Berendsen Thermostat

The Berendsen model corresponds physically to a system of particles that experience viscous friction and are subject to frequent collisions with light particles that form an ideal gas at temperature T_0, see Figure 22.4. Mathematically, Berendsen et al. (1984) utilize the Langevin equation (22.9) with a viscous friction force (22.8) and a stochastic external force with the properties (22.10 and 22.11). These two forces are intended to represent coupling to an external heat bath and scale the atomic velocities during the numerical simulation to add or remove energy from the system as desired. The Langevin equations are written either for the entire set of particles (in gases and liquids), or for a local group of particles corresponding to a pre-boundary region of a solid structure.

For the purpose of further discussion, we rewrite the Langevin equation (22.9) in terms of the velocity components:

$$m\dot{v}_j(t) = F_j(t) - m\gamma v_j(t) + R_j(t), \quad j = 1, 2, \ldots, 3N \quad (22.38)$$

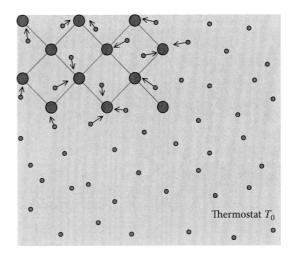

FIGURE 22.4 Physical model of the Berendsen thermostatting approach: atoms of a structure under analysis (large circles) are damped and subject to frequent collisions with light gas particles (small circles) at a target temperature T_0.

where F_j is the standard interatomic force. According to the ergodic hypothesis, the time-averaged quantities (22.10 and 22.11) for a system in thermodynamic equilibrium are equal to the corresponding ensemble average quantities; therefore,

$$\overline{R_j(t)} = \langle R(t) \rangle = 0 \quad (22.39)$$

$$\overline{R_j(t)R_j(t+t_0)} = \langle R(t)R(t+t_0) \rangle = a\delta(t_0) \quad (22.40)$$

22.2.3.1.1 Fluctuation-Dissipation Theorem

A reasonable physical assumption about the intensity of the random force R in Equation 22.40 can be made on the basis of the *fluctuation-dissipation theorem* that gives the relationship between a fluctuating force on some degree of freedom and the damping coefficient that determines dissipation in this degree of freedom:

$$\langle R(t)R(t+t_0) \rangle = 2m\gamma k_B T_0 \delta(t_0) \quad (22.41)$$

where T_0 is the equilibrium system temperature.

This theorem can be proven by utilizing the Langevin equation (22.38), where the interatomic force is omitted as not participating in the dissipation process,

$$m\dot{v}(t) = -m\gamma v(t) + R(t) \quad (22.42)$$

The general solution of this equation reads

$$v(t) = v(0)e^{-\gamma t} + \frac{1}{m}\int_0^t R(\tau)e^{-\gamma(t-\tau)}\,d\tau \quad (22.43)$$

For an equilibrium system, the mean square value of this function is equal to

$$\overline{v^2(t)} = \left\langle v^2(t) \right\rangle = \frac{k_B T_0}{m} \tag{22.44}$$

Substituting Equation 22.43 into Equation 22.44, where the ensemble average is utilized yields,

$$\left\langle v^2(t) \right\rangle = \left\langle v^2(0) \right\rangle e^{-2\gamma t} + \frac{2}{m} \int_0^t \left\langle v(0) R(\tau) \right\rangle e^{-2\gamma(t-\tau)} \, d\tau$$

$$+ \frac{2}{m^2} \int_0^t \int_0^t \left\langle R(\tau) R(\tau') \right\rangle e^{-\gamma(2t-\tau-\tau')} \, d\tau' d\tau \tag{22.45}$$

In this expression,

$$\left\langle v^2(0) \right\rangle = k_B T_0 / m \tag{22.46}$$

$$\left\langle v(0) R(\tau) \right\rangle = \overline{v(0) R(t)} = v(0) \overline{R(t)} = 0 \tag{22.47}$$

$$\left\langle R(\tau) R(\tau') \right\rangle = a\delta(\tau - \tau') \tag{22.48}$$

Therefore, relationship (22.45) can be reduced to

$$\frac{k_B T_0}{m} (1 - e^{-2\gamma t}) = \frac{a}{m^2} \int_0^t e^{-2\gamma(t-\tau)} \, d\tau \tag{22.49}$$

where

$$a = 2m\gamma k_B T_0 \tag{22.50}$$

Utilizing this value for Equation 22.40 proves the fluctuation-dissipation theorem (Equation 22.41).

As follows from Equations 22.11 and 22.50, dynamic properties of the random force R in Langevin equation (22.38) can be summarized as

$$\overline{R_j(t)} = 0, \quad \overline{R_j(t) R_{j'}(t')} = 2m\gamma k_B T_0 \delta(t - t') \delta_{jj'} \tag{22.51}$$

where T_0 is the target system temperature. These relationships can be physically interpreted as follows: (1) the function R has no directional preference, (2) R is a Gaussian random function of time with zero mean variance $2m\gamma k_B T_0$ for all degrees of freedom interacting with the heat bath; (3) the force R_j on degree of freedom j is uncorrelated with the force $R_{j'}$ on another degree of freedom j'; (4) the instantaneous value of R is not affected by its preceding values, i.e., the function R is uncorrelated with its time history.

In practice, the random force R can be sampled at each time step of a numerical simulation as a random Gaussian variable with zero mean and variance (mean square amplitude) $2m\gamma k_B T_0$. The sampling procedure is performed independently for each degree of freedom exposed to the thermal noise. Also, samples for two successive time steps are evaluated independent of each other.

22.2.3.1.2 Elimination of the Random Force

In various applications, only the global thermodynamic behavior of the system is of importance. Then, it is computationally effective to eliminate the local random noise $\mathbf{R}_i(t)$ in the Langevin equation (22.38), and to characterize the system/heat bath coupling via a *time-dependent* damping term. This damping term can be introduced on the basis of the following requirement: The new equation of motion must yield the same averaged behavior of the system's kinetic and total energy for a given target temperature T_0, as the original equation (22.9). We detail the relevant mathematical derivations below.

First, write the time derivative of kinetic energy of the system in the form

$$\dot{E}_k = \lim_{\Delta t \to 0} \sum_{j=1}^{3N} \frac{m}{2\Delta t} (v_j^2(t + \Delta t) - v_j^2(t)) \tag{22.52}$$

and rearrange it to get

$$\dot{E}_k = \lim_{\Delta t \to 0} \sum_{j=1}^{3N} \frac{m}{2\Delta t} (2v_j \Delta v_j + \Delta v_j^2) \tag{22.53}$$

According to (22.38), the change of velocity over a short time interval is

$$\Delta v_j = v_j(t + \Delta t) - v_j(t) = \frac{1}{m} \int_t^{t+\Delta t} (F_j(t') - m\gamma v_j(t') + R_j(t')) \, dt'$$

$$\simeq \frac{1}{m} (F_j(t)\Delta t - m\gamma v_j(t)\Delta t) + \frac{1}{m} \int_t^{t+\Delta t} R_j(t') \, dt' \tag{22.54}$$

Substituting Equation 22.54 into Equation 22.53 and separately considering the first term,

$$\lim_{\Delta t \to 0} \sum_{j=1}^{3N} \frac{m v_j \Delta v_j}{\Delta t} = \sum_{j=1}^{3N} (F_j v_j - m\gamma v_j^2) + \lim_{\Delta t \to 0} \sum_{j=1}^{3N} \frac{1}{\Delta t} \int_t^{t+\Delta t} v_j(t) R_j(t') \, dt'$$

$$= -\dot{U} - 2\gamma E_k + \lim_{\Delta t \to 0} \frac{3N}{\Delta t} \int_t^{t+\Delta t} \left\langle v(t) R(t') \right\rangle dt'$$

$$= -\dot{U} - 2\gamma E_k \tag{22.55}$$

For this derivation, we utilized the time derivative of potential energy,

$$\dot{U} = \frac{dU(x_1, x_2, \ldots, x_3 N)}{dt} = \sum_{j=1}^{3N} \frac{\partial U}{\partial x_j} \frac{dx_j}{dt} = -\sum_{j=1}^{3N} F_j v_j \quad (22.56)$$

current kinetic energy,

$$E_k = \sum_{j=1}^{3N} \frac{m v_j^2}{2} \quad (22.57)$$

and the ensemble average,

$$\frac{1}{3N} \sum_{j=1}^{3N} v_j(t) R_j(t') = \langle v(t) R(t') \rangle = 0 \quad (22.58)$$

as follows from Equation 22.47.

According to Equation 22.54, the second term in Equation 22.53 becomes

$$\lim_{\Delta t \to 0} \sum_{j=1}^{3N} \frac{m \Delta v_j^2}{2\Delta t} = \lim_{\Delta t \to 0} \frac{1}{2m\Delta t} \sum_{j=1}^{3N} \int_t^{t+\Delta t} R_j(t') dt' \int_t^{t+\Delta t} R_j(t'') dt''$$

$$= \lim_{\Delta t \to 0} \frac{3N}{2m\Delta t} \int_t^{t+\Delta t} \int_t^{t+\Delta t} \langle R(t') R(t'') \rangle dt'' dt'$$

$$= \lim_{\Delta t \to 0} \frac{3N\gamma k_B T_0}{\Delta t} \int_t^{t+\Delta t} \int_t^{t+\Delta t} \delta(t'-t'') dt'' dt'$$

$$= 3\gamma N k_B T_0 \lim_{\Delta t \to 0} \frac{1}{\Delta t} \int_t^{t+\Delta t} dt' = 3\gamma N k_B T_0 \quad (22.59)$$

Here, we showed only one of the six terms arising from Δv_j^2. The other five terms are trivial due to the following arguments:

$$\lim_{\Delta t \to 0} (F_j(t)\Delta t)^2 / \Delta t = 0$$

$$\lim_{\Delta t \to 0} (v_j(t)\Delta t)^2 / \Delta t = 0$$

$$\lim_{\Delta t \to 0} (F_j(t)\Delta t)(v_j(t)\Delta t) / \Delta t = 0 \quad (22.60)$$

$$\sum_j F_j(t) R_j(t') = 3N \langle F(t) R(t') \rangle = 0$$

$$\sum_j v_j(t) R_j(t') = 3N \langle v(t) R(t') \rangle = 0$$

Next we can utilize the results of Equations 22.55 and 22.59 for Equation 22.53,

$$\dot{E}_k = -\dot{U} - 2\gamma E_k + 3\gamma N K_B T_0 \quad (22.61)$$

and rearrange this relationship to obtain the energy equation,

$$\dot{H} = 3\gamma N k_B (T_0 - T) \quad (22.62)$$

where H is the system Hamiltonian. This equation must be satisfied by the sought equation of motion after elimination of the random force term R. The transformation from Equation 22.61 to Equation 22.62 employs $E_k = 3N k_B T/2$, where T is the current system temperature viewed as a function of time.

We can show that the following equation,

$$m_j \dot{v}_j = -F_j - m_j \gamma \left(1 - \frac{T_0}{T}\right) v_j, \quad j = 1, 2, \ldots, 3N \quad (22.63)$$

known as Berendsen equation of motion, satisfies Equation 22.62. In terms of the radius vectors, it can be written as

$$m_i \ddot{\mathbf{r}}_i = -\nabla_i U - m_i \gamma \left(1 - \frac{T_0}{T}\right) \dot{\mathbf{r}}_i, \quad i = 1, 2, \ldots, N \quad (22.64)$$

Pre-multiplying Equation 22.63 with v_i and summing over all the degrees of freedom gives

$$\sum_j m_j v_j \dot{v}_j = -\sum_j v_j \nabla_j U - \gamma \left(1 - \frac{T_0}{T}\right) \sum_j m_j v_j^2$$

$$\Rightarrow \dot{E}_k = -\dot{U} - 2\gamma E_k \left(1 - \frac{T_0}{T}\right)$$

$$\Rightarrow \dot{H} = -3\gamma N k_B T \left(1 - \frac{T_0}{T}\right) = 3\gamma N k_B (T_0 - T) \quad (22.65)$$

Thus, from an energetic point of view, the Berendsen equation (22.63) is equivalent to the original Langevin equation (22.38) in application to multiparticle systems. Berendsen thermostatting equations are amongst the most widely used in practical MD simulations.

Note that the Hamiltonian H represents internal energy of a thermodynamic system. Assume the internal energy and temperature are related to each other as

$$U = \frac{3}{2} N k_B T \quad (22.66)$$

Then the time rate of temperature change is represented by the first-order differential equation,

$$\dot{T} = 2\gamma (T_0 - T) \quad (22.67)$$

The value $\tau_T = (2\gamma)^{-1}$ gives the coupling constant that represents the characteristic time of equilibration of the system with the heat bath.

The time-dependent parameter

$$\zeta(t) = \gamma\left(1 - \frac{T_0}{T(t)}\right) \qquad (22.68)$$

in the Berendsen equation can be viewed as a dynamic damping parameter. This parameter turns to zero when the system temperature approaches the target value T_0.

The current system temperature T for Equation 22.63 or Equation 22.64 can be computed as

$$T = \frac{2}{k_B}\langle E_k \rangle \simeq \frac{1}{3Nk_B}\sum_{j=1}^{3N} m_j v_j^2 \qquad (22.69)$$

for each successive time step of a numerical solution. The value γ should be chosen small enough so the physical properties of the system are not violated by the artificial damping forces. On the other hand, γ has to be significantly large to ensure thermal equilibrium within a reasonable simulation time. In applications to crystal lattices, the value of γ is normally on the order of several percent of the maximum atomic lattice frequency.

22.2.3.2 Nosé–Hoover Heat Bath

The Nosé–Hoover heat bath utilizes a dynamic friction coefficient, which evolves according to a first-order differential equation:

$$\dot{\zeta}(t) = \frac{1}{Q}\left(\sum_j^{3N} m_j v_j^2 - (3N+1)k_B T_0\right) \qquad (22.70)$$

where Q is a method parameter of dimension energy × (time)². Thus, the difference between the simultaneous and target kinetic energy of the system determines the *time derivative* of the friction coefficient that contrasts with Berendsen thermostat, where the dynamic friction coefficient is simply Equation 22.68. Hoover (1985) has shown that the Berendsen equations are overdamped and they lead to a statistical ensemble not compliant with the Gibbs canonical distribution (??). Provided that such a distribution is sought for the microstates associated with solution of the Langevin equation, the friction coefficient must follow the relaxation equation (22.70). Note that the term $3N + 1$ stands for the total number of degrees of freedom in the system that includes the $3N$ coordinates of the particles, plus the variable ζ.

The relaxation equation (22.70) is solved simultaneously with the Langevin equations (22.9), or (22.38), where γ is replaced by $\zeta(t)$. The Langevin equations can be presented in the alternative Hamiltonian form:

$$\dot{\mathbf{r}}_i = \frac{\mathbf{p}_i}{m_i}, \quad \dot{\mathbf{p}}_i = -\nabla_i U - \zeta(t)\mathbf{p}_i \qquad (22.71)$$

The Hoover equations (22.70 and 22.71) serve as a practical update of the Nosé (1984a,b) thermostatting method, which also used an additional variable for the momenta rescaling,

utilized the scaled coordinates, and first reproduced canonical distribution for positions and scaled momenta. These works provided a vital background for the present method. As results, Equations 22.70 and 22.71 are usually referred to as the *Nosé–Hoover thermostat*.

Berendsen and Nosé–Hoover thermostatting approaches have been used in MD simulations of a vast range of physical and engineering systems and processes, and particularly in simulations of gases and liquids. Applications to ionic and metallic crystals, where the heat is mostly transferred by the electron gas, often yield a sufficiently accurate physical model. On the other hand, the physical concept behind these models makes them less adequate in applications to solid–solid interfaces, where the heat exchange mechanism is governed dominantly by oscillations of the lattice atoms (e.g., nonmetallic crystals, carbon nanostructures), and where the heat bath cannot be viewed as a gaseous substance surrounding the simulated crystal structure. In these instances, the phonon or configurational method by Karpov et al. (2007) provides a more accurate physical model.

22.2.4 Molecular Dynamics Applications

MD is used for the numerical solution of Lagrange or Newtonian equations (22.6) for classical multiparticle systems, as well as postprocessing and computer visualization of the time-dependent solution data.

In this section, we review applications of the Newtonian formalism to some typical MD simulations in the field of nanomechanics and materials.

22.2.4.1 Modeling Inelasticity and Failure in Gold Nanowires

A current research emphasis in nanostructured materials is the behavior of metallic nanowires. Nanowires are envisioned to have great potential as structural reinforcements, biological sensors, elements in electronic circuitry, and many other applications. Interested readers can find reviews on this comprehensive subject by Lieber (2003) and Yang (2005).

The examples shown here are MD simulations of the tensile failure of gold nanowires, as shown in Park and Zimmerman (2005). The wire size was initially 16 nm in length with a square cross section with length 2.588 nm. The wire was first quasistatically relaxed to a minimum energy configuration with free boundaries everywhere, then thermally equilibrated at a fixed length to 300 K using a Nosé–Hoover thermostat, which is described in Section 22.2.3.2. Finally, a ramp velocity was applied to the nanowire ranging from zero at one end to a maximum value at the loading end; thus, one end of the nanowire was fixed, while the other was elongated at a constant velocity each time step corresponding to an applied strain rate of $\dot{\epsilon} = 3.82 \cdot 10^9 \text{ s}^{-1}$.

As can be seen in Figure 22.5, the gold nanowire shows many of the same failure characteristics as macroscopic tension specimens, such as necking and yield. However, one very interesting quality of gold nanowires concerns their incredible ductility, which is manifested in the elongation of extremely thin nanobridges, as

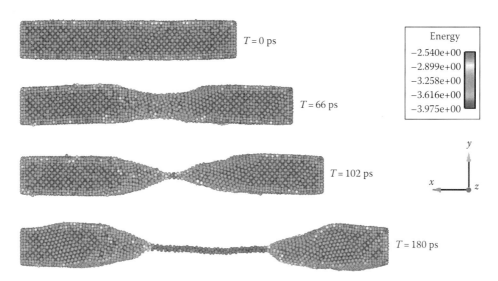

$T = 0$ ps

$T = 66$ ps

$T = 102$ ps

$T = 180$ ps

Energy
-2.540e+00
-2.899e+00
-3.258e+00
-3.616e+00
-3.975e+00

FIGURE 22.5 **(See color insert following page 25-16.)** MD simulation of the tensile failure of a gold nanowire using an EAM potential.

seen in the later snapshots in Figure 22.5. These nanobridges are extremely low coordinated chains of atoms which in first principles simulations are shown to form single atom chains. For MD simulations to capture these phenomena, it is very important to utilize an interatomic potential, which can accurately model the material stacking fault energy; this point is elucidated in Zimmerman et al. (2000).

As quantum mechanical calculations cannot yet model entire nanowires, MD simulations will continue to be a necessary tool in modeling nanostructured materials and atomic-scale plasticity.

22.2.4.2 Interaction of Nanostructures with Gas/Liquid Molecules

There has been significant effort aimed at the modeling and simulation of interactions between nanostructures and the flow of liquids and gases at the atomic scale. Particular interest arose from the interaction of carbon nanotubes (CNT), e.g., Qian et al. (2002) and Saether et al. (2003), with a surrounding liquid or gas, including the resultant deformation and vibration of the CNT, drag forces, slip boundary effects, hydrophobic/hydrophilic behavior of the nanotubes, and nanosensors applications. Example references on these topics are the following: Bolton and Gustavsson (2003), Bolton and Rosen (2002), Li et al. (2003), and Walther et al. (2001, 2004).

Snapshots of a typical gas-structure atomic-scale simulation are shown in Figure 22.6. Here, a carbon nanotube is immersed in helium at given temperature and concentration. The carbon atoms are initially at rest; however, collisions with fast helium atoms induce vibration and deflection of the nanostructure. The mathematical modeling of these types of systems, considered as a system of N spherical particles, utilizes the Newtonian equations of motion

$$m_i \ddot{\mathbf{r}}_i = -\nabla_i U, \quad i = 1, 2, \ldots, N \quad (22.72)$$

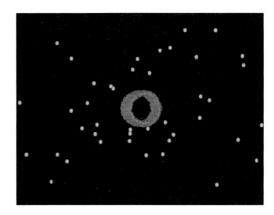

FIGURE 22.6 Molecular dynamics simulation of a carbon nanotube immersed into monoatomic helium gas at a given temperature and concentration.

where the potential function describes three types of atomic interactions present in the system,

$$U = U_{C-C} + U_{He-He} + U_{C-He} \quad (22.73)$$

The first term describes the interaction between carbon atoms, the second the interaction of helium molecules (note that helium molecules are monoatomic), and the third the potential of interaction between carbon atoms and helium molecules. For the example shown in Figure 22.6, the C–C interaction is modeled by the Tersoff potential (22.34), while the He–C and He–He interactions occur via two LJ potentials (22.26) with different sets of the parameters σ and ϵ. In general, if the system under analysis is comprised of n_f distinct phases, the total number of different components in the potential U is equal to $n_f(n_f + 1)/2$.

One convenient approach to modeling cylindrical macromolecules, such the carbon nanotube, is depicted in Figure 22.7; the modeling procedure starts by producing a flat sheet of carbon

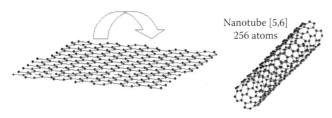

FIGURE 22.7 Approach to modeling 3D nanotube structures.

rings, on the basis of given chirality and tubule translation vectors (Qian et al. 2002), and followed by wrapping of the flat sheet into a 3D structure.

Initial radius vectors for the equations of motion (Equation 22.72) are determined by the geometry of the system, as well as physical properties of the gas, such as concentration and volume occupied. Initial velocities of the carbon and helium particles are determined by the initial temperature of the CNT and the gas, and are evaluated (sampled) from the Maxwell–Boltzmann distribution.

22.2.4.3 Nanoindentation

Atomic-scale indentation of thin films and nanostructured materials is an effective experimental technique for the analysis of material properties. This technique consists of pushing a sharp tip made of a hard material, usually diamond, into a matrix/substrate material under investigation, see Figure 22.8, and measuring the loading force as a function of indentation depth. Material properties of the matrix are then evaluated from the analysis of a resultant load-indentation curve, properties of the tip, as well as plastic behavior of the substrate material.

Numerical modeling and simulation of the nanoindentation process for a tip-substrate system comprised of N spherical atoms or molecules requires utilization of the Newtonian equations (Equation 22.72) where the gradient of function U determines the internal potential forces of atomic interaction. The function U describes the interaction between the substrate atoms, and may also include components describing interactions between matrix and indenter, and between indenter atoms. In the simplest case, displacement boundary conditions are applied

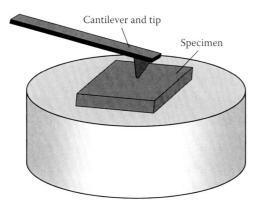

FIGURE 22.8 Nanoindentation, experimental scheme.

throughout the domain including atoms subject to the indenter load, so that a sole substrate potential is required for Equation 22.72. Note that for solid domains, this potential is such that the substrate atoms, in the absence of external forcing, cannot move freely in the domain; they are constrained to vibrate in the vicinity of some equilibrium configuration determined by the local minima of function U. Initial coordinates of the atoms utilized for solving Equation 22.72 usually correspond to one of these equilibrium configurations. As in most atomic-scale simulations, initial velocities are sampled from the Maxwell–Boltzmann distribution.

The results of a typical 2D simulation are depicted in Figure 22.8. One interesting feature of nanoindentation simulations, which is emphasized in this example, is the initiation and propagation of lattice dislocations that determine the plastic behavior of the substrate. Here, the substrate material is modeled as an initially perfect hexagonal lattice structure governed by the LJ potential (22.26) with a cutoff between the second- and third-nearest neighbors. The boundary conditions are the following: fixed y-components on the lower edge of the block and under the rectangular indenter. The speed of sound in this material is 1 km/s. The loading rate is 15 m/s.

Figure 22.9 shows a sequence of averaged contour plots of potential energy of the atoms. As seen from this figure, in the vicinity of a dislocation core, atoms are mis-coordinated and have a higher potential energy than other atoms in the material. Dislocations move away from the nucleation site with the velocity of about 0.2 of the speed of sound, or 13 times the indenter speed.

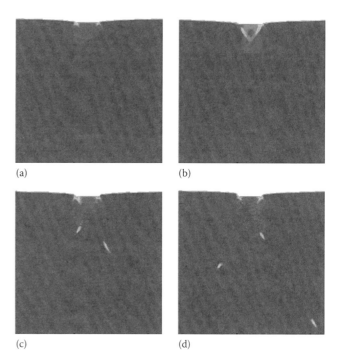

FIGURE 22.9 Dislocation dynamics in a 2D nanoindentation simulation: averaged contour plots of potential energy of the atomic system; individual atoms are not shown.

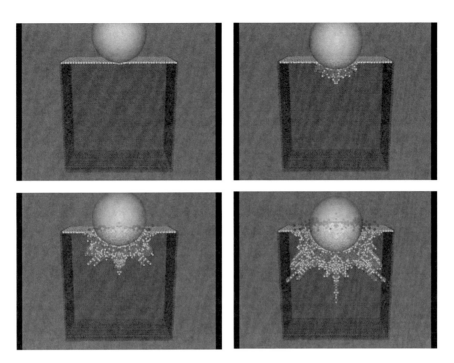

FIGURE 22.10 **(See color insert following page 25-16.)** 3D nanoindentation of a crystalline metallic substrate.

Figure 22.10 shows a cross section through a relatively small, though typical 3D nanoindentation problem. Here, the domain is a 25 nm face-centered cube with roughly 376,000 atoms which interact via a $\sigma = \epsilon = 1$ LJ potential (22.26). The interactions are nearest neighbor, and the indenter, shown as a sphere, is modeled as a repulsive pair-wise potential,

$$V_2 = V_2(|\mathbf{r}_i - \mathbf{r}_I|) \qquad (22.74)$$

where

$\mathbf{r}_i$ are radius vectors of the substrate atoms
$\mathbf{r}_I$ is the radius vector of the center of the indenter

The coloring is given in Figure 22.10 by a coordination number, which is the total number of neighboring atoms within the cutoff radius of the potential for a given atom: blue is lowest, red is highest, and all atoms with coordination number equal to 12 (perfect face-centered lattice) have not been visualized to show lattice defects more clearly. Thus, the colored atoms shown represent the region of plastic deformation in the vicinity of the indenter. In contrast to the previous example, far-reaching localized dislocations are not formed here due to the large relative size and smoothness of the indenter.

22.2.4.4 Nanodeposition

The modeling and simulation of deposition of individual vapor (gas) molecules on the surface of a solid body (substrate) is an important problem in the area of surface engineering, mechanics of thin films, and physical chemistry. The computational modeling of this process, which is called often

nanodeposition, is comprised typically of a solid structure and a gaseous domain governed by a three-component potential, similar to Equation 22.73.

A directional deposition process viewed as a sequence of depositions of individual atomic or molecular ions with known orientation and modulus of the incident velocity vector is sometimes called *ion-beam deposition*; more generally, they are referred to as a physical or chemical *vapor deposition* process. Atomic ions can be controlled by means of electromagnetic fields in order to provide the required intensity, kinetic energy, and orientation of the ion beam with respect to the surface of a substrate.

The ion-beam deposition process is illustrated in the MD simulation frame depicted in Figure 22.11. Individual carbon atoms of known mean kinetic energy and angle of incidence are deposited on the surface of a monocrystal diamond substrate to form a thin amorphous film. The situation shown corresponds to energetic ions with mean kinetic energy several times higher than the bonding energy of carbon atoms in the diamond substrate. Then, the bombarding ions destroy the crystalline structure of the substrate surface, and the growing amorphous film is comprised of cross-diffused deposited and substrate atoms. The carbon–carbon interaction can be modeled via the Tersoff (1988a) or Brenner (1990) potentials.

22.2.4.5 Crack Propagation Simulations

MD simulations have been successful in application to atomic-scale dynamic fracture processes, e.g., Abraham et al. (1997, 2002). Two snapshots of a typical MD simulation of fracture in a 2D structure under tensile load are depicted in Figure 22.12.

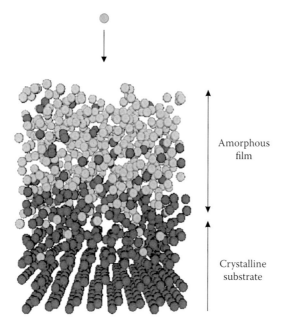

FIGURE 22.11 Deposition of an amorphous carbon film (light gray atoms) on top of a diamond substrate (dark atoms).

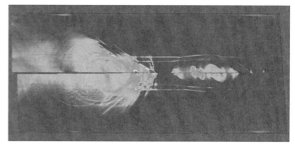

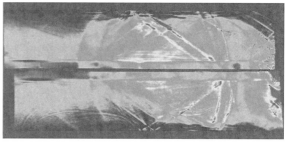

FIGURE 22.13 **(See color insert following page 25-16.)** MD simulation of a shear-dominant crack propagation process; the middle snapshot shows the formation of a "daughter" crack ahead of the main crack tip.

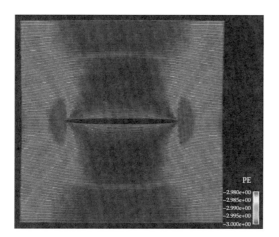

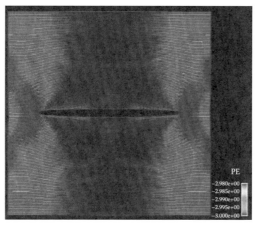

FIGURE 22.12 Averaged potential energy profiles for an MD fracture simulation at two different time steps.

Note that the wave created due to opening the fracture surfaces heads out toward the MD boundary, reflects from it, and propagates back in toward the crack.

Fracture under shear loading is presented in Figure 22.13. It shows a reproduction of the numerical experiments performed by Abraham and Gao (2000) and Gao et al. (2001) to analyze the mechanism by which a mode II (shear) dominated crack is able to accelerate past the Rayleigh wave speed, the lower theoretical max limit, over the "forbidden" velocity zone to the longitudinal speed of sound, the absolute theoretical maximum. The mechanism was seen to be the formation of a "daughter" crack ahead of the main crack tip; see the second snapshot in Figure 22.13. The system shown is a 2D hexagonal lattice with a LJ $\sigma = \epsilon = 1$ strip through which the crack propagates. The loading is mixed shear/tension, but the shear is dominant (5:1 shear:tensile strain rate ratio). The system is 1424 atoms long by 712 atoms high and the precrack is 200 atoms long in the center of the left vertical face. The model shows several interesting phenomena, namely, the presence of a distinct displacement wavefront, and corresponding boundary reflections, as well as the formation of a Mach cone upon accelerating to the speed of sound in the material. Additionally, an expanding halo can be seen behind the Mach cone as a result of the passing shock wave.

The main crack tip propagates at nearly the speed of sound, and produces shock waves that form the Mach cone. The third snapshot indicates that the resultant crack propagation speed is higher than the speed of sound, since the Mach cone is still approaching the right vertical boundary after complete separation of the structure.

In conclusion, we note that materials applications of classical particle dynamics, such as nanoindentation, deposition, atomic-scale failure, and others, often require the MD simulation to be very large, in order to diminish boundary effects. Otherwise, the waves that have been emitted by the indenter, atoms deposited, or crack tip, reflect from the boundaries and continue to incorrectly participate in the phenomena under investigation. Though the increase in computational power has made million atom MD calculations fairly commonplace, it appears to be computationally and physically unnecessary to have full atomistic resolution far from the crack, or indenter tip. The waves emitted are usually elastic in nature, and do not cause atomic lattice imperfections, i.e., plasticity, far from the localized domains of plastic deformation. Then, for example, the crack propagation can be correctly modeled using full atomistic resolution, while the propagation of the elastic waves away from the cracktip can be accurately modeled and captured using a continuum formulation. The usage of finite elements (FEs) at a sufficient distance from the crack tip would reduce the computational expense of having full atomistic resolution, while still accurately capturing the necessary physics. In greater detail, these issues are discussed in Section 22.3 summarizing the multiscale modeling methods that allow concurrent and hierarchical coupling of MD and FEM simulations.

22.3 Introduction to Multiple Scale Modeling

In recent years, thanks mainly to the constantly increasing surge in computational power, atomistic simulations have been utilized with great success in the modeling of nanoscale materials phenomena. However, despite these technological improvements, MD simulations still cannot be utilized to simulate more than a few billion atoms, or a few cubic microns in volume. Thus, while the modeling and simulation of nanoscale materials is within the realm of MD simulations, many important microdevices are too large to be simulated using MD. Furthermore, in many interesting applications, nanoscale materials will be used in conjunction with other components that are larger, and have different response times, thus operating at different time and length scales. Thus, single scale methods such as MD, e.g., Allen and Tildesley (1987), Haile (1992), or *ab initio* methods, will have difficulty in analyzing such hybrid structures due to the limitations in terms of time and length scales that each method is confined to. Therefore, the need arises to couple atomistic methods with approaches that operate at larger length scales and longer timescales.

Continuum methods have in contrast had much success in the macroscale modeling and simulation of structures. FE methods, e.g., Belytschko et al. (2000), Hughes (1987), are now the standard numerical analysis tool to study diverse problems such as the modeling of crashworthiness in automobiles, the fluid-structure interaction of submarines, plasticity in manufacturing processes, and blast and impact simulations. Therefore, the logical approach taken by many researchers in the desire to create truly multiple scale simulations that exist at disparate length and timescales has been to couple MD and FE in some manner. Unfortunately, the coupling of these methods is not straightforward for the reasons discussed next.

The major problem in multiscale simulations is that of pathological wave reflection, which occurs at the interface between the MD and FE regions. The issue is that wavelengths emitted by the MD region are considerably smaller than that which can be captured by the continuum FE region. Because of this and the fact that an energy conserving formulation is typically used, the wave must go somewhere and is thus reflected back into the MD domain. This leads to spurious heat generation in the MD region, and a contamination of the simulation. The retention of heat within the MD region can have extremely deleterious effects, particularly in instances of plasticity where heat generated within the MD region is trapped; in such an extreme situation, melting of the MD region can eventually occur.

A separate, but related issue to effective multiscale modeling is that of extending the timescale available to MD simulations. This issue still remains despite the efforts of current multiscale methods to limit the MD region to a small portion of the computational domain. Despite the reduction in the MD system size, limits still exist on the duration of time for which the MD system can be simulated. Research has been ongoing in the physics community to prolong the MD simulation time, particularly for infrequent events such as surface diffusion. Two excellent examples of the types of methods currently under investigation can be found in the works of Voter (1997) and Voter et al. (2002).

Other issues that typically arise in attempting to couple simulations that operate at disparate length and timescales are discussed below in the context of some recent methods dealing with these issues in various manners. While this chapter attempts to cover many of the existing approaches to multiple scale modeling, it is by no means complete. Two recently written review papers that comprehensively cover the field of multiple scale modeling are those of Liu et al. (2004a) and Curtin and Miller (2003).

22.3.1 MAAD

One pioneering multiscale approach was the work of Abraham et al. (1998). The idea was to concurrently link tight binding (TB), MD, and FEs together in a unified approach called MAAD (macroscopic, atomistic, *ab initio* dynamics). Concurrent linking here means that all three simulations run at the same time, and dynamically transmit necessary information to and receive information from the other simulations. In this approach, the FE mesh is graded down until the mesh size is on the order of the atomic spacing, at which point the atomic dynamics are governed via MD. Finally, at the physically most interesting point, i.e., at a crack tip, TB is used to simulate the atomic bond breaking processes.

The idea of meshing the FE region down to the atomic scale was one of the first attempts to eliminate spurious wave reflection at the MD/FE interface. The logic was that gradually increasing the FE mesh size would reduce the amount of reflection back into the MD region. However, meshing the FE region down to the atomic spacing presents two problems, one numerical and one physical. The numerical issue is that the timestep in an FE simulation is governed by the smallest element in the mesh. Thus, if the FEs are meshed down to the atomic scale, many timesteps will be wasted simulating the dynamics in these regions. Furthermore, it seems unphysical that the variables of interest in the continuum region should evolve at the same timescales as the atomistic variables.

The physical issue in meshing the FE region down to the atomic scale lies in the FE constitutive relations. The constitutive relations typically used in FE calculations, e.g., for plasticity, are constructed based on the bulk behavior of many dislocations. Once the FE mesh size approaches the atomic spacing, each FE can represent only a small number of dislocations, the bulk assumption disappears, and the constitutive relation is invalidated.

The overlapping regions (FE/MD and MD/TB) are termed "handshake" regions, and each makes a contribution to the total energy of the system. The total energy of the handshake regions is a linear combination of the energies of the relevant computational methods, with weight factors chosen depending on which computational method contributes the most energy in the handshake region. The three equations of motion (TB/FE/MD) are all integrated forward using the same timestep.

The interactions between the three distinct simulation tools are governed by conserving energy in the system as in Broughton et al. (1999)

$$H_{\text{TOT}} = H_{\text{FE}} + H_{\text{FE/MD}} + H_{\text{MD}} + H_{\text{MD/TB}} + H_{\text{TB}} \qquad (22.75)$$

More specifically, the Hamiltonian, or total energy of the MD system can be written as

$$H_{\text{MD}} = \sum_{i<j} V^{(2)}(r_{ij}) + \sum_{i,(j<k)} V^{(3)}(r_{ij}, r_{ik}, \Theta_{ijk}) + K \qquad (22.76)$$

where
the summations are over all atoms in the system
K is the kinetic energy of the system
r_{ij} and r_{ik} indicate the distance between two atoms i and j and i and k respectively
Θ_{ijk} is the bonding angle between the three atoms

The summation convention $i < j$ is performed so that each atom ignores itself in finding its nearest neighbors. Here, the potential energy is comprised of two parts. The first ($V^{(2)}$) are the two-body interactions, for example, nearest-neighbor spring interactions in 1D. The second part are the three-body interactions ($V^{(3)}$), which incorporate features such as angular bonding between atoms. The three-body interactions also make the potential energy of each atom dependent on its environment.

The FE Hamiltonian can be written as the sum of the kinetic and potential energies in the elements, i.e.,

$$H_{\text{FE}} = V_{\text{FE}} + K_{\text{FE}} \qquad (22.77)$$

Expanding these terms gives

$$V_{\text{FE}} = \frac{1}{2} \int_{\Omega} \boldsymbol{\epsilon}(\mathbf{r}) \cdot \mathbf{C} \cdot \boldsymbol{\epsilon}(\mathbf{r}) \, d\Omega \qquad (22.78)$$

$$K_{\text{FE}} = \frac{1}{2} \int_{\Omega} \rho(\mathbf{r})(\dot{\mathbf{u}})^2 \, d\Omega \qquad (22.79)$$

where
$\boldsymbol{\epsilon}$ is the strain tensor
$\mathbf{C}$ is the stiffness tensor
ρ is the material density
$\dot{\mathbf{u}}$ are the nodal velocities

The TB total energy is written as

$$V_{\text{TB}} = \sum_{n=1}^{N_{\text{occ}}} \boldsymbol{\epsilon}_n + \sum_{i<j} V^{\text{rep}}(r_{ij}) \qquad (22.80)$$

This energy can be interpreted as having contributions from an attractive part $\boldsymbol{\epsilon}_n$ and a repulsive part V^{rep}. N_{occ} are the number of occupied states. While a detailed overview of tight binding methods is beyond the scope of this work, further details can be found in Foulkes and Haydock (1989). MAAD was applied to the brittle fracture of silicon by Abraham et al. (1998).

22.3.2 Coarse-Grained Molecular Dynamics

An approach related to the TB/MD/FE approach of Abraham et al. was developed by Rudd and Broughton (1998), called coarse-grained molecular dynamics (CGMD). This approach removes the TB method from the MAAD method and instead couples only FE and MD. The basic idea in CGMD is that a coarse-grained energy approximation which converges to the exact atomic energy is utilized to derive the governing equations of motion. The coarse-grained energy from which the equations of motion are extracted is defined to be

$$E(\mathbf{u}_k, \dot{\mathbf{u}}_k) = U_{\text{int}} + \frac{1}{2} \sum_{j,k} (M_{jk}\dot{\mathbf{u}}_j \cdot \dot{\mathbf{u}}_k + \mathbf{u}_j \cdot K_{jk} \cdot \mathbf{u}_k) \quad (22.81)$$

where

the internal energy $U_{\text{int}} = 3(N - N_{\text{node}})kT$
the kinetic energy is defined as $M_{jk}\dot{\mathbf{u}}_j \cdot \dot{\mathbf{u}}_k$
the potential energy is defined as $\mathbf{u}_j \cdot K_{jk} \cdot \mathbf{u}_k$
the displacement are $\mathbf{u}$ and the velocities are $\dot{\mathbf{u}}$

The internal energy represents the thermal energy of those degrees of freedom which have been coarse grained (eliminated) out of the system; clearly, as the number of nodes approaches the number of atoms, this term disappears, and the full atomistic energy is recovered.

The stiffness matrix K_{jk} and mass matrix M_{jk} are calculated using weight functions which are similar in form to FE shape functions. Therefore, while the explicit equation of motion, which is solved in CGMD, is not found in the work of Rudd and Broughton (1998), it appears as though CGMD mimics the behavior of a FE mesh, which is graded down to the MD atomic spacing in regions of interest, and is coarsened away from the MD region. Therefore, it is expected that CGMD would suffer from the same issues as MAAD, i.e., that the mesh grading would eventually reach a point where the high-frequency MD wavelengths would not be representable in the continuum, and hence be reflected back into the MD domain.

This notion is supported by the paper of Rudd (2001), in which the notion of dissipative Langevin dynamics (Adelman and Doll 1976) is introduced into the CGMD formulation. The equation of motion is then given to be

$$M_{ij}\ddot{u}_j = -G_{ik}^{-1}u_k + \int_{-\infty}^{t} \eta_{ik}(t - \tau)\dot{u}_k(\tau)d\tau + F_i(t) \quad (22.82)$$

where

M_{ij} is a mass matrix
G_{jk} is a stiffness-like quantity
η_{ik} is a time history, or memory function
$F_i(t)$ is a random force

The addition of the dissipative terms to the equation of motion seems to clearly indicate that the original formulation of CGMD did suffer from spurious wave reflection as coarse graining of the mesh occurred.

It is interesting to note that a similar expression to (22.81) was derived by Wagner and Liu (2003) for a system involving multiple scales. In the Wagner and Liu work, the ensemble multiple scale kinetic energy behaves similar to the energy in (22.81), in that as the number of FE nodes approaches the atomic limit, the purely atomistic kinetic energy is recovered.

22.3.3 Quasicontinuum Method

A well-known quasistatic multiple scale method, the quasicontinuum method, was developed by Tadmor et al. (1996). Examples of applications and further improvements on the quasicontinuum method are the works of Miller et al. (1998) and Knap and Ortiz (2001). A recent review concentrating on the history and development of the quasicontinuum method is given by Miller and Tadmor (2003). While the quasicontinuum method is essentially an adaptive FE method, the atomistic to continuum link is achieved here by the use of the Cauchy–Born rule. The Cauchy–Born rule assumes that the continuum energy density W can be computed using an atomistic potential, with the link to the continuum being the deformation gradient $\mathbf{F}$. To briefly review continuum mechanics, the deformation gradient $\mathbf{F}$ maps an undeformed line segment $d\mathbf{X}$ in the reference configuration onto a deformed line segment $d\mathbf{x}$ in the current configuration:

$$d\mathbf{x} = \mathbf{F}d\mathbf{X} \quad (22.83)$$

In general, $\mathbf{F}$ can be written as

$$\mathbf{F} = 1 + \frac{d\mathbf{u}}{d\mathbf{X}} \quad (22.84)$$

where $\mathbf{u}$ is the displacement. If there is no displacement in the continuum, the deformation gradient is equal to unity.

The major restriction and implication of the Cauchy–Born rule is that the deformation of the lattice underlying a continuum point must be homogeneous. This results from the fact that the underlying atomistic system is forced to deform according to the continuum deformation gradient $\mathbf{F}$, as illustrated in Figure 22.14. By using the Cauchy–Born rule, Tadmor et al. (1996) were able to derive a continuum stress tensor and tangent stiffness directly from the interatomic potential, which allowed the usage of nonlinear FE techniques. This can be done by the following relations:

$$\mathcal{P} = \frac{\partial W}{\partial \mathbf{F}^{\mathrm{T}}} \quad (22.85)$$

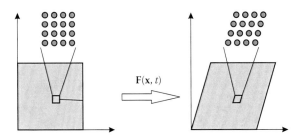

FIGURE 22.14 Illustration of the Cauchy–Born rule.

$$\mathbf{C} = \frac{\partial^2 W}{\partial \mathbf{F}^\mathrm{T} \partial \mathbf{F}^\mathrm{T}} \qquad (22.86)$$

where
  $\mathbf{C}$ is the Lagrangian tangent stiffness
  $\mathcal{P}$ is the first Piola–Kirchoff stress tensor

An updated version of the Cauchy–Born rule was proposed by Arroyo and Belytschko (2004).

Adaptivity criteria were used in regions of large deformation so that full atomic resolution could be achieved in these instances, i.e., near a dislocation. A nonlocal version of the Cauchy–Born rule was also developed so that nonhomogeneous deformations such as dislocations could be modeled. The quasicontinuum method has been applied to quasistatic problems such as nanoindentation, atomic-scale fracture, and grain boundary interactions.

22.3.4 CADD

Recently, a new method for quasistatic coupling termed CADD (coupled atomistics and discrete dislocation) was presented by Curtin and Miller (2003) and Shilkrot et al. (2002, 2004). The approach taken here is to couple molecular statics with discrete dislocation plasticity (van der Giessen and Needleman 1995). The motivation in doing so is such that defects, mainly dislocations, generated within the atomistic region are allowed to pass through the atomistic/continuum border into the continuum, where they are represented via discrete dislocation mechanics. Because discrete dislocation mechanics incorporates the elastic stress field emitted from a dislocation into the continuum stress and modulus expressions, the defects are able to be tracked once they pass into the continuum region, and also evolve in the continuum by following predefined sets of evolution laws.

As the atomistic side of the calculation relies on standard principles, we briefly discuss the discrete dislocation continuum to which the atomistic region is coupled. The continuum energy E^c is defined to be

$$E^c = \sum_\mu E_\mu(\mathbf{U}_\mathrm{I}, \mathbf{U}_\mathrm{c}, \mathbf{d}^i) - \int_{\mathrm{d}\Omega_\mathrm{T}} \mathbf{T}_0 \mathbf{u} \, \mathrm{d}A \qquad (22.87)$$

where
  $\mathbf{U}_\mathrm{I}$ are the MD/FE interface nodes
  $\mathbf{U}_\mathrm{c}$ are the continuum nodes
  $\mathbf{d}^i$ are the positions of the discrete dislocations in the continuum
  $\mathbf{T}_0$ is the prescribed traction on the continuum boundary $\mathrm{d}\Omega_\mathrm{T}$

The total stresses, strains, and displacements in the continuum can all be written as functions of the contribution from the discrete dislocations, and a correction term (we write for the displacement only):

$$\mathbf{u} = \tilde{\mathbf{u}} + \hat{\mathbf{u}} \qquad (22.88)$$

where
  $\tilde{\mathbf{u}}$ is the contribution from the discrete dislocations
  $\hat{\mathbf{u}}$ is a correction term, which is necessary due to the fact that the discrete dislocation solution is for an infinite medium

Noting that the strains and stresses can be decomposed accordingly, the continuum energy E^c can be rewritten as

$$E^c = \frac{1}{2} \int_{\Omega_\mathrm{c}} (\hat{\mathbf{u}} + \tilde{\mathbf{u}}) : (\hat{\boldsymbol{\epsilon}} + \tilde{\boldsymbol{\epsilon}}) \, \mathrm{d}V - \int_{\mathrm{d}\Omega_\mathrm{T}} \mathbf{T}_0(\tilde{\mathbf{u}} + \hat{\mathbf{u}}) \, \mathrm{d}A \qquad (22.89)$$

The equilibrium displacement $\tilde{\mathbf{u}}$ fields are obtained by minimizing (22.89), i.e.,

$$\frac{\partial E^c}{\partial \tilde{\mathbf{u}}_\mathrm{c}} = 0 \qquad (22.90)$$

After the continuum displacement fields are known, the forces $\mathbf{p}^i$ on the discrete dislocations are calculated by minimizing E^c with respect to the discrete dislocation positions

$$\mathbf{p}^i = -\frac{\partial E^c}{\partial \mathbf{d}^i} \qquad (22.91)$$

At this point, an iterative procedure involving the discrete dislocation positions, FE positions, and atomic positions is solved until all degrees of freedom are at equilibrium. We note that the atomic degrees of freedom are defined separately, but the interface FE displacements $\mathbf{U}_\mathrm{I}$ are used to prescribe boundary conditions for the atomistic iterative procedure, and vice versa.

The approach has been validated via 2D problems, including fracture, nanoindentation, and atomic-scale void growth. Current issues facing the CADD developers include the extension to dynamic problems, and the passing of dislocations from the atomistic to continuum regions in 3D, where a dislocation is a loop that can reside in both atomistic and continuum regions at the same time.

22.3.5 Bridging Scale Method

This section introduces the bridging scale concurrent method, which was recently proposed by Wagner and Liu (2003) to couple atomistic and continuum simulation methods. The fundamental idea is to decompose the total displacement field $\mathbf{u}(\mathbf{x})$ into coarse and fine scales

$$\mathbf{u}(\mathbf{x}) = \bar{\mathbf{u}}(\mathbf{x}) + \mathbf{u}'(\mathbf{x}) \qquad (22.92)$$

This decomposition has been used before in solid mechanics in the variational multiscale methods (Hughes et al. 1998), and reproducing kernel particle methods (Liu et al. 1995a,b). The coarse scale $\bar{\mathbf{u}}$ is that part of the solution which can be represented by a set of basis functions, i.e., FE shape functions. The fine scale $\mathbf{u}'$ is defined as the part of the solution whose projection onto the coarse scale basis functions is zero; this implies orthogonality of the coarse and fine scale solutions.

In order to describe the bridging scale, first imagine a body in any dimension which is described by N_a atoms. The notation used here will mirror that used by Wagner and Liu (2003). The total displacement of an atom α is written as $\mathbf{u}_\alpha$. The coarse scale displacement is a function of the initial positions $\mathbf{X}_\alpha$ of the atoms. It should be noted that the coarse scale would at first glance be thought of as a continuous field, since it can be interpolated between atoms. However, because the fine scale is defined only at atomic positions, the total displacement and thus the coarse scale are discrete functions that are defined only at atomic positions. For consistency, Greek indices (α, β, ...) will define atoms for the remainder of Section 22.3, and uppercase Roman indices (I, J,...) will define coarse scale nodes.

The coarse scale is defined to be

$$\tilde{\mathbf{u}}(\mathbf{X}_\alpha) = \sum_I N_I^\alpha \mathbf{d}_I \tag{22.93}$$

Here, $N_I^\alpha = N_I(\mathbf{X}_\alpha)$ is the shape function of node I evaluated at the initial atomic position. $\mathbf{X}_\alpha$, and $\mathbf{d}_I$ are the FE nodal displacements associated with node I.

The fine scale in the bridging scale decomposition is simply that part of the total displacement that the coarse scale cannot represent. Thus, the fine scale is defined to be the projection of the total displacement $\mathbf{u}$ onto the FE basis functions subtracted from the total solution $\mathbf{u}$. We will select this projection operator to minimize the mass-weighted square of the fine scale, which we call J and can be written as

$$J = \sum_\alpha m_\alpha \left(\mathbf{u}_\alpha - \sum_I N_I^\alpha \mathbf{w}_I \right)^2 \tag{22.94}$$

where

m_α is the atomic mass of an atom α
$\mathbf{w}_I$ are temporary nodal (coarse scale) degrees of freedom

It should be emphasized that Equation 22.94 is only one of many possible ways to define an error metric. In order to solve for $\mathbf{w}$, the error is minimized with respect to $\mathbf{w}$, yielding the following result:

$$\mathbf{w} = \mathbf{M}^{-1}\mathbf{N}^T\mathbf{M}_A\mathbf{u} \tag{22.95}$$

where the coarse scale mass matrix $\mathbf{M}$ is defined as

$$\mathbf{M} = \mathbf{N}^T\mathbf{M}_A\mathbf{N} \tag{22.96}$$

In Equations 22.95 and 22.96, $\mathbf{M}_A$ is a diagonal matrix with the atomic masses on the diagonal, and $\mathbf{N}$ is a matrix containing the values of the FE shape functions evaluated at all the atomic positions. In general, the size of $\mathbf{N}$ is $N_{a1} \times N_{n1}$, where N_{n1} is the number of FE nodes whose support contains an atomic position, and N_{a1} is the total number of atoms. The fine scale $\mathbf{u}'$ can thus be written as

$$\mathbf{u}' = \mathbf{u} - \mathbf{N}\mathbf{w} \tag{22.97}$$

or

$$\mathbf{u}' = \mathbf{u} - \mathbf{P}\mathbf{u} \tag{22.98}$$

where the projection matrix $\mathbf{P}$ is defined to be

$$\mathbf{P} = \mathbf{N}\mathbf{M}^{-1}\mathbf{N}^T\mathbf{M}_A \tag{22.99}$$

The total displacement $\mathbf{u}_\alpha$ can thus be written as the sum of the coarse and fine scales as

$$\mathbf{u} = \mathbf{N}\mathbf{d} + \mathbf{u} - \mathbf{P}\mathbf{u} \tag{22.100}$$

The final term in the above equation is called the bridging scale. It is the part of the solution that must be removed from the total displacement so that a complete separation of scales is achieved, i.e., the coarse and fine scales are orthogonal to each other. This bridging scale approach was first used by Liu et al. (1997) to enrich the FE method with meshfree shape functions. Wagner and Liu (2001) used this approach to consistently apply essential boundary conditions in meshfree simulations. Zhang et al. (2002) applied the bridging scale in fluid dynamics simulations. Qian et al. (2004) recently used the bridging scale in quasistatic simulations of carbon nanotube buckling. The bridging scale was also used in conjunction with a multiscale constitutive law to simulate strain localization by Kadowaki and Liu (2004).

Now that the details of the bridging scale have been laid out, some comments are in order. In Equation 22.94, the fact that an error measure was defined implies that $\mathbf{u}_\alpha$ is the "exact" solution to the problem. In our case, the atomistic simulation method we choose to be our "exact" solution is MD. After determining that the MD displacements shall be referred to by the variable $\mathbf{q}$, Equation 22.94 can be rewritten as

$$J = \sum_\alpha m_\alpha \left(\mathbf{q}_\alpha - \sum_I N_I^\alpha \mathbf{w}_I \right)^2 \tag{22.101}$$

where the MD displacements $\mathbf{q}$ now take the place of the total displacements $\mathbf{u}$. The equation for the fine scale $\mathbf{u}'$ can now be rewritten as

$$\mathbf{u}' = \mathbf{q} - \mathbf{P}\mathbf{q} \tag{22.102}$$

The fine scale is now clearly defined to be the difference between the MD solution and its projection onto a predetermined coarse scale basis space. This implies that the fine scale can thus be interpreted as a built-in error estimator to the quality of the coarse scale approximation. Finally, the equation for the total displacement **u** can be rewritten as

$$\mathbf{u} = \mathbf{Nd} + \mathbf{q} - \mathbf{Pq} \tag{22.103}$$

22.3.5.1 Multiscale Equations of Motion

The next step in the multiscale process is to derive the coupled MD and FE equations of motion. This is done by first constructing a Lagrangian $\mathcal{L}$, which is defined to be the kinetic energy minus the potential energy

$$\mathcal{L}(\mathbf{u}, \dot{\mathbf{u}}) = \mathcal{K}(\dot{\mathbf{u}}) - V(\mathbf{u}) \tag{22.104}$$

Ignoring external forces, Equation 22.104 can be written as

$$\mathcal{L}(\mathbf{u}, \dot{\mathbf{u}}) = \frac{1}{2} \dot{\mathbf{u}}^{\mathsf{T}} \mathbf{M}_A \dot{\mathbf{u}} - U(\mathbf{u}) \tag{22.105}$$

where $U(\mathbf{u})$ is the interatomic potential energy. Differentiating the total displacement **u** in Equation 22.103 with respect to time gives

$$\dot{\mathbf{u}} = \mathbf{N}\dot{\mathbf{d}} + \mathbf{Q}\dot{\mathbf{q}} \tag{22.106}$$

where the complimentary projection operator $\mathbf{Q} \equiv \mathbf{I} - \mathbf{P}$. Substituting Equation 22.106 into the Lagrangian equation (22.105) gives

$$\mathcal{L}(\mathbf{d}, \dot{\mathbf{d}}, \mathbf{q}, \dot{\mathbf{q}}) = \frac{1}{2} \dot{\mathbf{d}}^{\mathsf{T}} \mathbf{M} \dot{\mathbf{d}} + \frac{1}{2} \dot{\mathbf{q}}^{\mathsf{T}} \mathcal{M} \dot{\mathbf{q}} - U(\mathbf{d}, \mathbf{q}) \tag{22.107}$$

where the fine scale mass matrix $\mathcal{M}$ is defined to be $\mathcal{M} = \mathbf{Q}^{\mathsf{T}} \mathbf{M}_A$. One elegant feature of Equation 22.107 is that the total kinetic energy has been decomposed into the sum of the coarse scale kinetic energy plus the fine scale kinetic energy.

The multiscale equations of motion are obtained from the Lagrangian by following the Lagrange equations:

$$\frac{\mathrm{d}}{\mathrm{d}t}\left(\frac{\partial \mathcal{L}}{\partial \dot{\mathbf{d}}}\right) - \frac{\partial \mathcal{L}}{\partial \mathbf{d}} = 0 \tag{22.108}$$

$$\frac{\mathrm{d}}{\mathrm{d}t}\left(\frac{\partial \mathcal{L}}{\partial \dot{\mathbf{q}}}\right) - \frac{\partial \mathcal{L}}{\partial \mathbf{q}} = 0 \tag{22.109}$$

Substituting the Lagrangian (22.107) into (22.108) and (22.109) gives

$$\mathbf{M}\ddot{\mathbf{d}} = -\frac{\partial U(\mathbf{d}, \mathbf{q})}{\partial \mathbf{d}} \tag{22.110}$$

$$\mathcal{M}\ddot{\mathbf{q}} = -\frac{\partial U(\mathbf{d}, \mathbf{q})}{\partial \mathbf{q}} \tag{22.111}$$

Equations 22.110 and 22.111 are coupled through the derivative of the potential energy U, which can be expressed as functions of the interatomic force **f** as

$$\mathbf{f} = -\frac{\partial U(\mathbf{u})}{\partial \mathbf{u}} \tag{22.112}$$

Expanding the right-hand sides of Equations 22.110 and 22.111 with a chain rule and using Equation 22.112 together with Equation 22.103 gives

$$\mathbf{M}\ddot{\mathbf{d}} = \frac{\partial U}{\partial \mathbf{u}}\frac{\partial \mathbf{u}}{\partial \mathbf{d}} = \mathbf{N}^{\mathsf{T}}\mathbf{f} \tag{22.113}$$

$$\mathcal{M}\ddot{\mathbf{q}} = \frac{\partial U}{\partial \mathbf{u}}\frac{\partial \mathbf{u}}{\partial \mathbf{q}} = \mathbf{Q}^{\mathsf{T}}\mathbf{f} \tag{22.114}$$

Using the fact that $\mathcal{M} = \mathbf{Q}^{\mathsf{T}}\mathbf{M}_A$, Equation 22.114 can be rewritten as

$$\mathbf{Q}^{\mathsf{T}}\mathbf{M}_A \ddot{\mathbf{q}} = \mathbf{Q}^{\mathsf{T}}\mathbf{f} \tag{22.115}$$

Because **Q** can be proven to be a singular matrix (Wagner and Liu 2003), there are many unique solutions to Equation 22.115. However, one solution which does satisfy Equation 22.115 and is beneficial to us is (including the coarse scale equation of motion):

$$\mathbf{M}_A \ddot{\mathbf{q}} = \mathbf{f}(\mathbf{q}) \tag{22.116}$$

$$\mathbf{M}\ddot{\mathbf{d}} = \mathbf{N}^{\mathsf{T}}\mathbf{f}(\mathbf{u}) \tag{22.117}$$

Now that the coupled multiple scale equations of motion have been derived, we make some relevant comments:

1. The fine scale equation of motion (Equation 22.116) is simply the MD equation of motion. Therefore, a standard MD solver can be used to obtain the MD displacements **q**, while the MD forces **f** can be found by using any relevant potential energy function.
2. The coarse scale equation of motion (Equation 22.117) is simply the FE momentum equation. Therefore, we can use standard FE methods to find the solution to Equation 22.117, while noting that the FE mass matrix **M** is defined to be a consistent mass matrix.

3. The coupling between the two equations is through the coarse scale internal force $\mathbf{N}^{\mathrm{T}}\mathbf{f}(\mathbf{u})$, which is a direct function of the MD internal force $\mathbf{f}$. In the region in which MD exists, the coarse scale force is calculated by interpolating the MD force using the FE shape functions $\mathbf{N}$. In the region in which MD has been eliminated, the coarse scale force can be calculated in multiple ways. The elimination of unwanted MD degrees of freedom is discussed in Section 22.3.6.

4. We note that the total solution $\mathbf{u}$ satisfies the same equation of motion as $\mathbf{q}$, i.e.,

$$\mathbf{M}_A\ddot{\mathbf{u}} = \mathbf{f} \qquad (22.118)$$

This result is due to the fact that $\mathbf{q}$ and $\mathbf{u}$ satisfy the same initial conditions, and will be utilized in deriving the boundary conditions on the MD simulation in a later section.

5. Because of the equality of $\mathbf{q}$ and $\mathbf{u}$, it would appear that solving the FE equation of motion is unnecessary, since the coarse scale can be calculated directly as the projection of $\mathbf{q}$, i.e., $\mathbf{Nd} = \mathbf{Pq}$. However, because the goal is to eliminate the fine scale from large portions of the domain, the MD displacements $\mathbf{q}$ are not defined over the entire domain, and thus it is not possible to calculate the coarse scale solution everywhere via direct projection of the MD displacements. Thus, the solution of the FE equation of motion everywhere ensures a continuous coarse scale displacement field.

6. Due to the Kronecker delta property of the FE shape functions, for the case in which the FE nodal positions correspond exactly to the MD atomic positions, the FE equation of motion (Equation 22.117) converges to the MD equation of motion (Equation 22.116).

7. The FE equation of motion is redundant for the case in which the MD and FE regions both exist everywhere in the domain, because the FE equation of motion is simply an approximation to the MD equation of motion, with the quality of the approximation controlled by the FE shape functions $\mathbf{N}$. This redundancy will be removed by eliminating the fine scale from large portions of the domain.

8. We note that the right-hand side of Equation 22.116 constitutes an approximation; the internal force $\mathbf{f}$ should be a function of the total displacement $\mathbf{u}$. We utilize the MD displacements $\mathbf{q}$ for two reasons. The first reason, as stated above, is the equality of $\mathbf{q}$ and $\mathbf{u}$. The second reason relates to computational efficiency, as determining $\mathbf{u}$ at each MD time step requires the calculation of the coarse scale solution $\mathbf{Nd}$, which would defeat the purpose of keeping a coarse FE mesh over the entire domain.

Because of the redundance of the FE equation of motion, one also requires to eliminate the MD region from a large portion of the domain, such that the redundance of the FE equation of motion is removed. This elimination techniques are discussed

by Karpov et al. (2005, 2007) and Wagner et al. (2004). Once the redundancy is removed, the coarse scale exists in large portions of the domain from which the fine scale MD is eliminated. Furthermore, the coarse scale variables are allowed to influence the motion of the fine scale. Thus, a two-way coarse scale/fine scale coupling can be achieved at the MD boundary where information originating from the coarse scale can act as a *boundary condition* for the fine scale.

22.3.6 Predictive Multiresolution Continuum Method

Predicting microstructure-property relationships in engineering materials via direct simulation of the underlying micromechanics remains an elusive goal due to the massive disparities in length and timescales; huge structures such as bridges may fail after decades of operation due to nanoscale fracture events, which occur at a micro/nano second timescale. In this section, we outline a multiscale method by Liu and McVeigh (2008), McVeigh and Liu (2008), and McVeigh et al. (2006), which predicts macroscale mechanical deformation in terms of the underlying evolving microstructure. This type of approach usually involves coupling between simulations at different scales. If information is passed from one simulation (scale) to another in one direction only, it is known as a *hierarchical* approach. This approach is different to the concurrent approaches, such as bridging scale or Section 22.3.5, where the information is exchanged in both directions, and the simulations must be performed simultaneously at all scales.

An affordable multiresolution theory can be proposed that captures the key underlying micromechanics while remaining in the context of continuum mechanics. Here, deformation at a continuum point is decomposed into the homogeneous deformation and a set of inhomogeneous deformations; each inhomogeneous measure is associated with a particular characteristic scale of inhomogeneous deformation in the microstructure. This introduces a set of microstresses in the governing equations that represent a resistance to inhomogeneous deformation.

22.3.6.1 Multiresolution Stress and Deformation Measures

A multiscale, also *micromorphic*, material is one which contains discrete microstructural constituents at N scales of interest. For example, a material may contain weakly bonded microscale particles and nanoscale dislocations. In that example, three scales are of interest: the macroscale, the microscale, and the nanoscale. The material behavior of each scale will differ considerably. A model which hopes to simulate the structure–property relationship of a material must capture the micromechanics at each distinct scale. A *general multiresolution framework* is thus defined as one in which (1) the material structure and the deformation field are resolved at each scale of interest; (2) the resulting internal power is a multifield expression with contributions from the average deformation at each scale i.e., the overall properties depend on the average deformation at each scale; (3) the deformation behavior at each scale is found by examining the

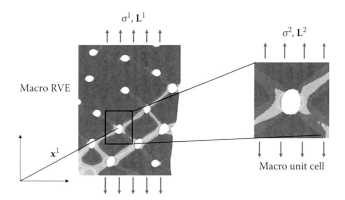

FIGURE 22.15 A macro RVE and a unit cell and their respective stress and rate of deformation measures.

micromechanics at each scale; and (4) the constitutive relations can be developed at each scale.

Two-scale material: We begin by deriving the internal power of a simple two-scale material containing a metal matrix with microscale inclusions or voids. One example is the porous material shown in Figure 22.15. In such a material, microscale deformation is controlled by the growth of voids. Void growth is a volumetric process and a function of the hydrostatic stress only. It is reasonable to assume that the volumetric part of the microde formation plays a prominent role in the microscale deformation of such materials. Here, the macroscale and microscale behavior are of interest. Therefore, the goal is to resolve the material structure and deformation to the macro- and microscales. The general equations for N scales will be discussed next.

In a conventional continuum simulation, homogenized constitutive behavior at a point is predetermined by finding the average behavior of a material sample. This can be achieved through experimental mechanical testing. Alternatively, for a multicomponent material with known individual component properties, computational simulations can be performed on a sample called a representative volume element (RVE). This is directly analogous to an experimental test.

To determine the influence of the subscales, additional deformation fields corresponding to each scale are introduced. This is achieved by mathematically decomposing the position and velocity within a representative volume element into components associated with each scale of interest. Figure 22.15 shows the mathematical decomposition for a simple two-scale micromorphic material along with a physical interpretation. For such a case, the position and velocity of a material point becomes

$$\mathbf{x} = \mathbf{x}^1 + \mathbf{x}^2, \quad \mathbf{v} = \mathbf{v}^1 + \mathbf{v}^2 \qquad (22.119)$$

where

   $\mathbf{v}^1$ is the macrovelocity
   $\mathbf{v}^2$ is the relative microvelocity

The macro RVE incorporates a microstructural feature as an *inclusion*. Thus, the unit cell is defined at each scale so that the

gradient of the relative microvelocity can be considered constant within the cell, i.e., the relative micro velocity varies linearly,

$$\mathbf{v}^2(\mathbf{x}^2) = \mathbf{L}^2 \cdot \mathbf{x}^2 \qquad (22.120)$$

The Figure 22.15 represents a macro RVE with an average macro stress, σ^1 and macro rate of deformation, $\mathbf{L}^1$. These are the average macro stress and rate of deformation over a macroscopic RVE of the material, which are constant within the micro unit cell. By zooming into a micro unit cell in the RVE, we can examine the total micro stress, σ^2 and rate of deformation, $\mathbf{L}^2$ associated with a micro unit cell. These will differ from the corresponding macro measures as long as the RVE is not homogeneous. The relative rate of microdeformation associated with a micro unit cell is defined as $(\mathbf{L}^2 - \mathbf{L}^1)$. The micro stress associated with a micro unit cell is denoted β^2.

The *internal power* of a unit cell can be provided as a resultant of the contributions from the macroscopic rate of deformation, $\mathbf{L}^1$, and the rate of relative microdeformation, $(\mathbf{L}^2 - \mathbf{L}^1)$,

$$p_{\text{int}} = \sigma^1 : \mathbf{L}^1 + \beta^2 : (\mathbf{L}^2 - \mathbf{L}^1) \qquad (22.121)$$

The micro-stress associated with a unit cell, β^2, is defined as the power conjugate of the relative rate of microdeformation. The micro-stress is therefore constant within the micro unit cell. The value β^2 is the measure of a stress field's tendency to produce a strain gradient in the averaging domain. An average value of the internal power (at a discrete position $\mathbf{x}$) is required for use within the mathematical multiscale framework. This is determined numerically by averaging the internal power over a micro domain of influence. This domain is chosen to be representative of the range of interactions between the microstructural features at the microscale. In Figure 22.16, the micro domain of influence (DOI) is chosen to include nearest-neighbor interactions between the voids.

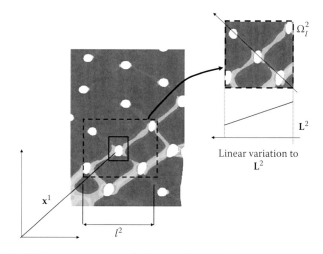

FIGURE 22.16 Domain of influence chosen at the microscale, where the rate of microdeformation varies linearly in space.

The average internal power is given by averaging over the micro DOI:

$$\bar{p}_{int}(\mathbf{x}) = \frac{1}{V_1^2} \int\limits_{\Omega_1^2} p_{int}(\mathbf{x} + \mathbf{y}) \, d\Omega_1^2 \qquad (22.122)$$

where an over bar represents an average value and $\mathbf{y}$ is simply an integration position variable within the DOI. $\bar{p}_{int}$ is given by (22.121). An expression for the variation of the rate of micro-deformation within the DOI can be obtained through a Taylor expansion. Truncating the first-order terms, we get

$$\mathbf{L}^2(\mathbf{x} + \mathbf{y}) = \bar{\mathbf{L}}^2(\mathbf{x}) + \mathbf{y} \cdot \overline{\nabla \mathbf{L}^2}(\mathbf{x}) \qquad (22.123)$$

where

$\bar{\mathbf{L}}^2(\mathbf{x})$ is the average microdeformation rate for the micro DOI

$\overline{\nabla \mathbf{L}^2}(\mathbf{x})$ is the average gradient of the rate of microdeformation over the micro DOI

The linear approximation in Equation 22.123 is justified in Figure 22.16 showing the variation in total microdeformation along a micro DOI.

The average internal power for a two-scale micromorphic material gives

$$\bar{p}_{int} = \sigma^1 : \mathbf{L}^1 + \bar{\beta}^2 : (\bar{\mathbf{L}}^2 - \bar{\mathbf{L}}^1) + \bar{\bar{\beta}}^2 : \overline{\Delta \mathbf{L}^2} \qquad (22.124)$$

where

$\bar{\beta}$ is the average of the relative micro-stress over the domain of influence

$\bar{\bar{\beta}}$ is the average of the first moment of the micro stress, i.e., $\mathbf{y} \cdot \bar{\beta}$, over the domain of influence

Equation 22.124 includes the contributions to the power density from the macro and microdeformation rates. It also includes the gradient of these rates. This gradient term arises from the variation in the microdeformation within the domain of influence. Indeed, the DOI should be chosen such that this variation can be considered linear and the first-order approximation (22.123) holds within the micro DOI. Physically, the gradient arises due to the interaction between microstructural features, as shown in Figure 22.16. Use of a DOI implicitly imbeds a length scale into the mathematical framework at the microscale, which is closely related to the size of the smallest microstructural feature at that scale. Not only does this make the model more physically realistic, it also eliminates the pathological mesh dependency associated with conventional continuum approaches.

Note that in a conventional continuum approach, where only the homogenized macroscale behavior is considered, the macro domain of influence coincides with the macro representative volume element (RVE) shown in Figure 22.15. The total micro rate

of deformation $\mathbf{L}^2$ is equal to the total macro rate of deformation $\mathbf{L}^1$ and expression (22.124) reduces to give the average internal power of a conventional homogenized continuum:

$$\bar{p}_{int} = \sigma^1 : \mathbf{L}^1 \qquad (22.125)$$

The *external power* can similarly be derived in terms of average measures. By applying the principle of virtual power and using the divergence theorem, the resulting strong form gives rise to a coupled multi-field system of governing differential equations. This forms the basis of the mathematical model used in multi-scale micromorphic FE simulations.

The *generalized stress and deformation tensors* can be defined as

$$\Sigma = \begin{bmatrix} \sigma^1 & \bar{\beta}^2 & \bar{\bar{\beta}}^2 \end{bmatrix}, \quad \Upsilon = \begin{bmatrix} \mathbf{L}^1 & [\bar{\mathbf{L}}^2 - \bar{\mathbf{L}}^1] & \overline{\nabla \mathbf{L}^2} \end{bmatrix} \qquad (22.126)$$

These are related through an elasto-plastic micromorphic multi-scale constitutive law.

N-scale material: The multiresolution internal power, stress and strain tensors have been developed and generalized to N scales of interest by Vernerey et al. (2007, 2008). The N-scale internal power density is given by

$$\bar{p}_{int} = \sigma^1 : \mathbf{L}^1 + \sum_{\alpha=2}^{N} (\bar{\beta}^\alpha : (\bar{\mathbf{L}}^\alpha - \bar{\mathbf{L}}^1) + \bar{\bar{\beta}}^\alpha : \overline{\nabla \mathbf{L}^\alpha}) \qquad (22.127)$$

where we denoted the coarsest and finest length scales as the 1st and Nth, respectively.

The generalized stress and deformation measures can be defined as

$$\Sigma = \begin{bmatrix} \sigma^1 & \bar{\beta}^2 & \bar{\bar{\beta}}^2 & \bar{\beta}^3 & \bar{\bar{\beta}}^3 & \dots & \bar{\beta}^N & \bar{\bar{\beta}}^N \end{bmatrix} \qquad (22.128)$$

$$\Upsilon = \begin{bmatrix} \mathbf{L}^1 & [\bar{\mathbf{L}}^2 - \bar{\mathbf{L}}^1] & \overline{\nabla \mathbf{L}^2} & [\bar{\mathbf{L}}^3 - \bar{\mathbf{L}}^1] & \overline{\nabla \mathbf{L}^3} \dots [\bar{\mathbf{L}}^N - \bar{\mathbf{L}}^1] & \overline{\nabla \mathbf{L}^N} \end{bmatrix} \qquad (22.129)$$

In concise notations, the internal power density (Equation 22.127) can be written as

$$\bar{p}_{int} = \Sigma \cdot \Upsilon \qquad (22.130)$$

For FE implementation, the principle of virtual power can be applied and discretized as usual. The generalized stress Σ and generalized deformation rate Υ tensors replace the standard stress and strain tensors. Hence, the multiresolution approach can be implemented in a standard FE code with increased degrees of freedom.

22.3.6.2 Multiresolution Constitutive Laws

In a multiresolution FE analysis, a generalized constitutive relation is required to relate the generalized stress to the generalized deformation. The first step is to derive individual constitutive relationships at each scale by examining the micromechanics at each scale. The individual constitutive relationships at each scale can then be combined to create a generalized constitutive relation:

$$\Sigma = \mathbf{C}^{ep} : \Upsilon \qquad (22.131)$$

where $\mathbf{C}^{ep}$ is a generalized elastic–plastic tangent modulus.

Analytical derivation of a generalized constitutive relation, and hence determination of $\mathbf{C}^{ep}$ may be possible if the mechanics are simple and well understood. This approach was demonstrated for a polycrystalline material and a granular material by McVeigh et al. (2006). For more complex problems involving more than two scales, a multiresolution cell modeling approach can be used. In a manner similar to the hierarchical approach, a constitutive relationship can be determined at the lowest scale of interest first. This relationship is used as an input when finding the constitutive behavior at the next largest scale and so on. The average behavior at each scale is determined by examining the average response over the averaging domain Ω^{α}, a domain of influence at the scale α. At the macroscale, the averaging domain is simply the RVE. A generalized constitutive law is then formed based on the scale-specific behavior. This technique is applied to design of a bio-inspired self-healing material and other advanced materials systems.

Potentially, the cell modeling approach can be enriched through incorporation of data from atomistic modeling of interfacial bonding strengths between particles of various sizes and matrix materials. We point out that one can utilize these techniques to resolve also the inverse problem, finding the way to *design the microstructure of materials on the basis of desired macroscopic properties*. The multiresolution continuum approach can predict the evolving scale of deformation due to the changing microstructure without having to perform large-scale detailed microstructure-level simulations; only limited RVE scale microstructural simulations are required to calibrate the constitutive relationships. The ability to predict the scale of deformation is crucial in order to compute the correct macroscale performance.

22.4 Manipulation of Nanoparticles and Biomolecules

This section provides an example of a hybrid (multiphysics) approach of bio-nanomechanics, which is applicable to the manipulation of nanoparticles and biomolecules by electric field and surface tension. This class of problems calls for a hybrid mathematical description of the solid, liquid, and electric field components of the system. Electric field has become one of the most widely used tools for manipulating cells, biomolecules, and nanoscale particles in microfluidic devices. Here, the 3D dynamic assembly of nanowires on various microelectrodes under dielectrophoretic force is presented with discussion on capillary action and electroosmosis effects in the manipulation. The various approaches to manipulate the small-scale materials are addressed both numerically and experimentally. For successful prediction and analysis on nanoscale, a hierarchical and multiscale scheme for modeling fluid transportation in nanochannels is suggested. The results show that the combined effects of electric field and capillary action–induced forces are crucial for precise control over nanoscale materials.

22.4.1 Introduction

Transport, alignment, and assembly of nano- and biomaterials in liquid phase are drawing more attention with the rapid development of small-scale electromechanical devices. The current trend toward high-performance microsystems can have an impact on a broad range of applications such as medical devices, fuel cells, communication systems, and biological or chemical sensors. To improve the performance of such devices, controlled manipulation of the nano- and biomaterials involved is the key issue because the materials having specific functions need to be assembled in a consistent manner with the design. The major challenge for the manipulation lies in the small length scale of nano- and biomaterials, which limits the effective observation either by naked eyes or by conventional optical microscopes. Electron microscopes are powerful tools for observation that is also limited by environmental conditions including high vacuum, material conditions, material size, etc. Furthermore, in situ monitoring of nanostructured materials during the manipulation and assembly is more challenging than observation after the assembly because of the assembly materials in liquid environment. Recently, rigorous computational methods have been proposed to model, analyze, and predict nanoscale processes and behaviors in order to augment and enhance experiments. These modeling technologies have demonstrated the ability to investigate the behavior of bulk nanostructured materials (Liu et al. 2004b, 2005). These methods enable the precise prediction of material behavior in multiple scales, which is crucial to shorten the incubation process for manufacturing nano- and biomaterials. These methods are also essential for quickly verifying a new idea on nanomaterials requiring an enormous effort in experiment. Various mechanisms have been proposed in recent years to assemble and pattern nano- and biomaterials. Among them, electric-field-induced manipulation has been demonstrated by changing the amplitude of an electric potential and its frequency (Jones 1995). The simple working principles and convenient experimental setups involving electric forces have been successfully applied to assemble nano- and biomaterials (Chen et al. 2001). The major effects induced by an electric field are dielectrophoretic force, electroosmotic flow, and electrothermal effect. The dielectrophoretic force arises from induced dipole moments on a particle embedded in a nonuniform electric field. The electroosmosis flow is driven by the electrostatic force applied to the charged double layer on the surfaces of the electrodes. The electric field can also increase the temperature of the fluid through Joule heating. However, under most conditions discussed in this chapter, the increase of temperature due to Joule heating is far less than 1 K, thus the electrothermal effect is negligible. Among the major forces induced by an electric field, the electrophoretic force gained more attention due to its successful applications to gel electrophoresis for separating nucleic acids and proteins (Brisson and Tilton 2003, Trau et al. 1997). The dielectrophoretic force was introduced in 1951 by Pohl (1978) and was theoretically formulated by Jones (1995). The dielectrophoretic force was used to attract and repel particles as designed due to the difference in permittivity. Besides the attraction or repulsion, the manipulation of flexible biopolymers or cells has been of greater interest. Zimmermann et al. used a high-amplitude AC electric field to manipulate mammalian cells (Zimmermann et al. 2000). Recently, more precise assembly methods have been proposed. Washizu et al. (2003) used the electroosmotic flow to immobilize and stretch DNA molecules. Chung and Lee (2003) and Chung et al. (2004) assembled multiwalled carbon nanotubes and DNA molecules by combining an AC with DC electric field. These technologies have allowed for the control and deposition of individual nanoscale components in micro/nano systems. During the manipulation using an electric field, most nanomaterials are processed in liquid phase, released from solution and used in liquid and gas phases. In the transition from solution to gas phase, both capillary action and surface tension inevitably affect the final stage of the nanomaterials. In fact, the force induced by the capillary action is usually a few orders of magnitude larger than that induced by electric field, especially in micro/nanoscale dimension. The capillary action caused by adhesive intermolecular force is dominant in the small dimension due to the relatively small inertia.

This explains why small insects can float on water and sometimes be trapped in a small drop. The dexterous control of the capillary action in material processing and manufacturing is critical regardless of its resultant effects. In spite of such a dominant effect, the capillary action has not been paid much attention possibly because it is not easy to be controlled in nanoscale material processing. Thus, how to utilize or minimize capillary action is the main concern in nano- and biomaterial assembly, which may also provide an ample opportunity for the generation of new structures. Wu and Whitesides (2001) and Lu et al. (2001) have utilized the surface tension due to water/air interface to assemble microscale polystyrene ball on a patterned surface. Manoharan et al. (2003), Lauga and Brenner (2004) reported the experimental and modeling results on the evaporation-induced assembly of colloidal particles. It was found that the colloidal particles absorbed at the interface of liquid droplets were arranged into a uniform pattern during evaporation. The creation of the unique packing was ascribed to geometrical constraints during the drying. Combination of an electric field and capillary action can create more interesting results. Single-walled carbon nanotube (SWCNT) bundles have been formed using both an AC electric field and capillary action by

Tang et al. (2003, 2004, 2005). The dielectrophoretic force due to the AC field was applied to attract SWCNTs near a tungsten (W) tip. In the experiment, the dielectrophoretic force directed the motion of nanoparticles along the orientation of the field. Subsequently, the capillary action formed the fibril shape of the SWCNTs. To move toward the next-generation assembly methods, a comprehensive understanding of the underlying physics is crucial, but is yet to be fully realized due to the complexity of the assembly processes at such a small scale. There have been various numerical studies in literature on motions of liquid or particles under an electric field. Ramos et al. (1998) have reviewed the motion of particles in a suspension when subjected to an AC electric field. Wan et al. (2003) have studied the dynamics of a charged particle in a fluid via lattice Boltzmann simulation. Dimaki et al. (2004) have used point dipole approximation to calculate the dielectrophoretic force and simulate the trapping process of a carbon nanotube on microelectrodes. In these studies, the solid particles are assumed to have no influence on local electric field, thus behave like a point in the electric field, which is accurate only when the particles are small enough or far from the electrodes. To overcome these limitations, Aubry and Singh (2006a,b); Kadaksham et al. (2004, 2005, 2006); and Singh et al. (2005) have used distributed Lagrangian method and Maxwell stress tensor to study the dielectrophoretic assembly of rigid spherical particles in an electric field cage. It is demonstrated that the point dipole approximation is not valid when the particle size is close to the cage size, thus has to be replaced by the general Maxwell stress tensor method. Furthermore, Liu et al. (2006a,b) have studied the dielectrophoretic assembly of nanowires and the electrodeformation of cells through coupled electrokinetics and immersed finite element method (IFEM) (Liu et al. 2006b, Wang and Liu 2004, Zhang et al. 2004). Electric field has also been used in liquid transportation in micro/nano channels. Patankar and Hu (1998) have used FE methods to calculate the continuum electroosmotic flow in a 2D channel. Qiao and Aluru (2002, 2003, 2004, 2005) have studied the electroosmotic flow in nanochannels by MD simulation. Similar molecular simulation of electroosmotic transportation in a charged nanopore has been presented by Thompson (2003). Currently, a complete theory or model is yet to be available for the three-dimensional dynamic assembly processes for multiple, arbitrary shaped nanomaterials under complex environments such as electric fields, surface tension, and moving boundaries. In this section, we discuss the fundamental physics involved in the assembly process and how ideas motivated by the immersed boundary method (Peskin 2002) can be utilized for the modeling of the manipulation process. It provides an excellent demonstration of the broad-range capabilities and applications of computational methods to analyze and predict the performance of electromechanical devices at micrometer to nanometer scale. We first give a review of the electrohydrodynamic coupling approach. This coupling scheme is applied to model nanowire assembly and viruses sorting. Next, we discuss the liquid/ion transportation in nanochannels to illustrate the consideration of both electric field and fluid flow. By comparing the analytical and computational models to experimental results, it is demonstrated that modeling can be combined with experimental work to improve the accuracy, yield, and control of the assembly process.

22.4.2 Electrohydrodynamic Coupling

Electric fields are widely used to manipulate biological objects such as cells or biofibros. It is important to incorporate the electric field into the fluid-structure interaction problem, thus solving an electrohydrodynamic problem. A typical electrohydrodynamic problem involves three components: the liquid domain, the solid structure immersed in the liquid, and an external electrostatic field. The structure is responsive to the electric field due to induction of an electric dipole moment. The liquid remains electrically neutral and does not respond to the electric field. In this section, we discuss electrohydrodynamic coupling that utilizes the IFEM formulation of Section 22.4.3.

22.4.2.1 Maxwell Equations

The electrostatic field in a continuous media or structure is governed by a system of time-independent *Maxwell equations* that can be written in the integral form as follows:

$$\oint \mathbf{E} \cdot d\mathbf{l} = 0$$

$$\oint \mathbf{D} \cdot d\mathbf{S} = 4\pi Q \qquad (22.132)$$

$$Q = \int_V \rho \, dV$$

Here, the continuum structure occupies physical volume V, and carries distributed charge of density $\rho(r)$; Q is the total volume charge of the structure. Other quantities in the above equations characterize the electric field and charge distributions: the vector $\mathbf{E}$ is a measure of intensity of the field, $\mathbf{D}$ is the vector of electric displacement, and $\mathbf{P}$ the polarization vector defined as a dipole moment, $d\mathbf{p}$, of a unit volume:

$$\mathbf{P} = \frac{d\mathbf{p}}{dV} \qquad (22.133)$$

In case of a homogeneous isotropic media, one obtains

$$\mathbf{D} = \varepsilon \mathbf{E} \qquad (22.134)$$

where ε is the electric permittivity of the media.

The vector $\mathbf{E}$ is called the intensity of the electric field, or *field strength*, and it serves as the force characteristic of the electric field. In particular, for a point charge q, the electrostatic force is given by

$$\mathbf{F} = \mathbf{E}q \qquad (22.135)$$

The field strength is determined by the electric potential φ that determines the voltage V_{AB} between two points A and B:

$$\mathbf{E} = -\nabla\varphi, \quad \varphi_A - \varphi_B \equiv V_{AB} = \int_A^B \mathbf{E} \cdot d\mathbf{l} \qquad (22.136)$$

The electric potential satisfies the Poisson equation

$$\nabla^2 \varphi(\mathbf{r}) = -\frac{4\pi}{\varepsilon} \rho(\mathbf{r}) \qquad (22.137)$$

whose solution is written in the form

$$\varphi(\mathbf{r}) = \frac{1}{\varepsilon} \int_V \frac{\rho(\mathbf{r}')}{|\mathbf{r} - \mathbf{r}'|} dV' \qquad (22.138)$$

Note that the electric potential $\varphi(\mathbf{r})$ is a special case of the general one-body potential $V_1(\mathbf{r})$ discussed in Section 22.2.2.1 of this chapter; see Equation 22.15. For simple systems, such as point charge, spherical charge, charged straight line, halfspace, etc., the function $\varphi(\mathbf{r})$ is known in closed form, e.g., Benenson et al. (2002).

The vector of electric displacements $\mathbf{D}$ represents the quantity of charge ΔQ per element ΔA displaced by electrostatic induction. The magnitude of this vector is equal to the surface charge density

$$|\mathbf{D}| = \lim_{\Delta A \to 0} \frac{\Delta Q}{\Delta A} = \frac{dQ}{dA} \qquad (22.139)$$

The second equation in (22.132) is also called *Gauss theorem*. This equation is particularly convenient in cases where the symmetry of the system allows evaluation of the surface integral for this equation in closed form. Note that if there is a finite surface S_0 at which the normal vector $d\mathbf{S}$ and vector $\mathbf{D}$ are coplanar, and $|\mathbf{D}| = \text{const}$, then the flux of vector $\mathbf{D}$ through this surface

$$\oint_{S_0} \mathbf{D} \cdot d\mathbf{S} = D_n S_0 \qquad (22.140)$$

In this case, the second equation of the system (22.132) reduces to an algebraic form allowing straightforward evaluation of the vector $\mathbf{D}$.

22.4.2.2 Electromanipulation

Experimental techniques used for the manipulation of particles and small-scale structures are generally referred to as electromanipulation, as well as *nanomanipulation*, when associated with directed motion of nanoscale objects.

A general list of factors that can be involved in electromanipulation and their applicable conditions are summarized in Table 22.1.

TABLE 22.1 Mechanisms of Electromanipulation

Mechanism	Applicable Condition
Electrophoresis (EP)	DC or low frequency AC
Dielectrophoresis (DEP)	AC
Electroosmosis flow	DC or low frequency AC
Drag force	Viscous fluid
Brownian motion	Nano/microscale particles

22.4.2.2.1 Electrophoresis (EP)

The first mechanism of electromanipulation, *electrophoresis*, is based on directed motion of suspended charged particles in a nonconductive liquid under the action of external electric fields. The driving force of electrophoresis is, in principle, the electrostatic force given by the expression (22.135).

22.4.2.2.2 Dielectrophoresis (DEP)

In contrast to electrophoresis, the *dielectrophoresis* (DEP) utilizes movement of *uncharged*, i.e., electrically neutral structures caused by a spatially nonuniform electrical field. The structure becomes polarized when immersed into the electric field. DEP only arises when the structure and the surrounding media, typically, a nonconductive gas or liquid, have different polarizabilities. If the structure is more polarizable compared to the surrounding media, it will be pulled toward regions of stronger field; this effect is called the positive DEP. Otherwise, the structure is repelled toward regions of weaker field; this is known as the negative DEP. One commonly known demonstration of the positive DEP is the attraction of small pieces of paper to a charged plastic comb or stick. The general concept of the positive DEP is illustrated in Figure 22.17.

The driving force of dielectrophoresis results from a dipole moment induced in the structure when it is immersed into an electric field. If the electric field is inhomogeneous, the field strength and thus distribution of the electrostatic force acting on each part of the structure is not uniform; that leads to a relative motion of the structure in the medium. The force exerted by an electric field $\mathbf{E}$ on a dipole with dipole moment $\mathbf{p}$ is given generally by

$$\mathbf{F}^{\text{dep}} = (\mathbf{p} \cdot \nabla)\mathbf{E} \qquad (22.141)$$

A widely used expression for the time-averaged DEP force on a particle is given as (Jones 1995)

$$\left\langle \mathbf{F}^{\text{dep}} \right\rangle = \Gamma \cdot \varepsilon_1 \operatorname{Re}\left\{ K_f \right\} \nabla \mid \mathbf{E} \mid^2. \qquad (22.142)$$

where

Γ is a parameter that depends on the particle shape and size

ε_1 is the real part of the permittivity of the medium

K_f is a factor that depends on the complex permittivities of both the particle and the medium

$\nabla|\mathbf{E}|^2$ is the gradient of the energy density of the electric field

For a sphere with a radius a, $\Gamma = 2\pi a^3$, $K_f = (\varepsilon_2^* - \varepsilon_1^*)/(\varepsilon_2^* + 2\varepsilon_1^*)$ (called the Clausius–Mossotti factor). For a cylinder with a diameter r and a length l, $\Gamma = \pi r^2 l/6$, $K_f = (\varepsilon_1^* - \varepsilon_1^*)/\varepsilon_1^*$. The frequency-dependent complex permittivities shown with the asterisk are expressed by the complex combination of conductivity σ, permittivity ε, and electric field frequency ω as $\varepsilon_1^* = \varepsilon_1 - j\sigma_1/\omega$, and $\varepsilon_1^* = \varepsilon_2 - j\sigma_2/\omega$, where $j = \sqrt{-1}$, and the indices 1 and 2 refer to the medium and the particle, respectively.

For an arbitrarily shaped solid structure, the dielectrophoresis force can be computed as a surface or volume integral,

$$\mathbf{F}^{\text{dep}} = \int_{\Gamma}(\sigma^M \cdot \mathbf{n})\,dA = \int_{\Omega^s} \nabla \cdot \sigma^M \, d\Omega \qquad (22.143)$$

where σ^M is the Maxwell stress tensor

$$\sigma^M = \varepsilon_1 \mathbf{E}\mathbf{E} - \frac{1}{2}\varepsilon_1 \mathbf{E} \cdot \mathbf{E}\mathbf{I} \qquad (22.144)$$

22.4.2.2.3 Electroosmotic Flow

The electroosmotic flow, or electroosmosis, is driven by the electric static force applied onto the charged double layer, see Figure 22.18. Electroosmosis flow can be induced by DC electric field or by low-frequency AC electric field.

Within the fluid-structure interaction approach, electroosmosis is treated as a slip boundary condition for the fluid. Also, since the Debye layer, the layer close to the wall

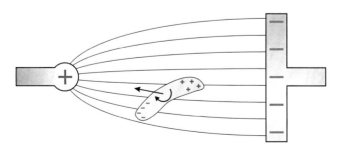

FIGURE 22.17 Positive dielectrophoresis. Positive and negative charges polarized in the structure are equal. The structure will rotate clockwise and move left to the area of stronger field.

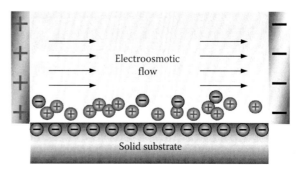

FIGURE 22.18 Illustration of electroosmosis developed within a channel.

where the velocity is varying, is only a few nanometers, only the steady velocity is taken into account, Solomentsev et al. (1997),

$$\mathbf{v} = -\frac{\varepsilon \psi_0 \mathbf{E}}{\mu} \qquad (22.145)$$

where

- ψ_0 is the zeta potential (the electric potential at the slipping plane close to the solid surface)
- μ and ε are the viscosity and permittivity of the medium, respectively

The Brownian force induced by thermal fluctuations will influence the motion of nanoscale particles. The approach described in the work by Sharma and Patankar (2004) on Brownian motion of rigid particles can be used to capture the thermal motion of complex-shaped objects without using approximations for the viscous drag on these objects.

22.4.2.2.4 Fluid-Structure Interaction: Governing Equations

The coupled electrohydrodynamic equation of motion, Liu and Iqbal (2009), can be written in the context of fluid-structure interaction to give two coupled governing equations: (1) a fluid equation with a fluid-structure interaction force and (2) a solid equation (MD, FEM, or rigid solid).

The fluid equation with fluid-structure interaction takes the form

$$\rho^f \dot{\mathbf{v}} = \nabla \cdot \sigma^f + \mathbf{f}^{FSI} \qquad (22.146)$$

The interaction force, representing the solid equation, reads

$$\mathbf{f}^{FSI} = -(\rho^s - \rho^f)\dot{\mathbf{v}}^s + \nabla \cdot (\sigma^s - \sigma^f) + \mathbf{F}^{ext} + \mathbf{F}^e + \mathbf{F}^{dep} + \mathbf{F}^c \quad (22.147)$$

where $\mathbf{F}^{ext}$, $\mathbf{F}^e$, $\mathbf{F}^{dep}$, and $\mathbf{F}^c$ are the external, electrostatic (electrophoresis), dielectrophoresis, and solid–solid interaction forces, respectively. The external force is usually the gravity force. Expressions for the electrostatic force and dielectrophoresis force are given in Equations 22.135 and 22.143. The force of interaction between two charged solid surfaces is evaluated by integrating the Coulomb potential (22.23). In this formulation, as well as in the applications considered below, thermal and Brownian motion effects are ignored.

In this way, the solid, fluid, and electrokinetic equations are coupled together. In the current simulations, the electrokinetic equation and solid/fluid motion equation are solved iteratively, i.e., a semi-static approach. Since the transition time of the electric field is much shorter than the characteristic time of the solid/fluid motion, this iterative approach is reasonable.

22.4.3 Nanostructure Assembly Driven by Electric Field and Fluid Flow

In recent years, assembly of nanoparticles has becomes a crucial step in many bio/MEMS devices. The assembled pattern largely depends on the electric field strength, electrode geometry, and electric properties of the particles. In this section, the electrohydrodynamic coupling, discussed in Section 22.4.2, is used to explore the dynamic process of nanowire assembly between microelectrodes. The assembly of nanoparticle nanowires by an electric field is mainly controlled by dielectrophoretic forces, and has been investigated both numerically and experimentally, e.g., Liu et al. (2006b, 2007, 2008). Such an assembly technique has been improved recently by introducing fluid flow in assembly, thus taking advantages from both electric field and fluid flow. As an example, an interesting phenomenon under the combined effect of fluid flow and electric field, pivoting of a nanowire during assembly, is presented here. In the experiment, Au electrodes are patterned on 130 nm-thick thermally oxidized silicon wafer by photolithography. The fluid channel is made of polydimethylsiloxane (PDMS), Dow Corning Corp. It is bonded to the electrodes by using stamp-and-stick bonding technique (Satyanarayana et al. 2005). The channel cross section is 500 μm wide and 40 μm high. SiC nanowires (Advanced Composite Materials Corporation, Greer, SC) used in the experiment are 1–100 μm in length and 300–500 nm in diameters. The concentration of nanowire solution is 5.0 μg/mL in dimethylformamide (DMF). The solution is placed in the inlet of the fluidic device and the flow is controlled by a syringe pump (Pump 11 pico plus, Harvard Apparatus). An AC field (0.5 Vpp, 5 MHz) is applied for the assembly of nanowires. The assembly process was recorded by a video camera (DXC-390, Sony Corp.) through an optical microscope. In the experiment, a nanowire is approaching the electrode by fluid flow, Figure 22.19a, and one end of the nanowire is landed on an edge of the electrode. After landing, one end of the nanowire is pivoted and the other end is rotating, Figure 22.19b and c. The simulated pivot process is shown in Figure 22.20, which agrees well with the experimental observation. The details

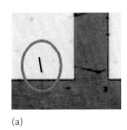 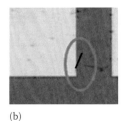  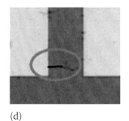

(a)   (b)   (c)   (d)

FIGURE 22.19 Alignment of a SiC nanowire (circled) with pivoting process in DMF. (a) Time = 0 s. (b) Time = 1.5 s. (c) Time = 2 s. (d) Time = 3 s. AC voltage: 500 mV at 5 MHz.

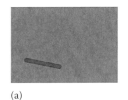 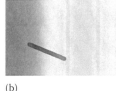 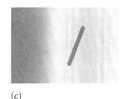 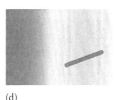

(a)   (b)   (c)   (d)

FIGURE 22.20 Pivot simulation: the flow is directed from left to right. The nanowire is attracted, pinned at one end, rotated, and oriented to the fluid flow direction on the electrodes. (a) Time = 0 s. (b) Time = 1 s. (c) Time = 2 s. (d) Time = 3 s.

TABLE 22.2 Problem Configuration for the Pivot Simulation

Fluid	29,525 nodes	157,079 elements	$\gamma_f = 0.0056\,\text{S/cm}$
$\rho = 1\,\text{g/cm}^3$	$\mu = 0.01\,\text{g cm/s}$		$\epsilon_f = 20\,\epsilon_0$
Nanowire	1,536 nodes	6,253 elements	$\gamma_s = 0.0112\,\text{S/cm}$
$\rho_s = 1\,\text{g/cm}^3$	Length = 6 m		$\epsilon_s = 100\,\epsilon_0$
$\Delta t = 0.03\,\text{s}$	Diameter = 600 nm		Cap rad. = 300 nm

TABLE 22.3 Properties of the Three Virus Types Used in the Simulation of Virus Detection

	Inovirus	Influenza	Bacteriophage P22
Permittivity (ϵ_0)	70	3	30
Conductivity (mS/m)	8	80	3
Size (nm)	200 (length)	100 (diameter)	65 (length) 20 (diameter)

of the simulation are shown in Table 22.2. It should be noted that the method we proposed is general and applicable to different types of particles and mediums. The values of permittivity and conductivity vary largely for different types of particles and mediums. Both permittivity and conductivity contribute to the DEP force calculation, which is the one of the most computationally challenging cases demonstrates the capability of this method.

22.4.3.1 Virus Deposition

In recent years, lots of so-called lab-on-a-chip devices have been proposed for point-of-care diagnostics and prognostics of various diseases. One important procedure involved in the diagnosis is to selectively separate one type of virus out of a suspension with multiple viruses. Since an electric field can be easily applied and controlled, individual virus manipulation by electric field–induced forces has attracted considerable interests. The working principle of selective deposition by electric field is that the sign of dielectrophoretic force on the particle is determined by the electric properties of the particle and the medium and the electric field frequency. By applying an AC field of a specific frequency, one type of viral particles can be selectively attracted under positive dielectrophoresis. We demonstrate the feasibility of the selection process by modeling the electric deposition of three different types of viruses. Suppose a sample solution contains the inovirus, influenza, and bacteriophage P22. The physical and electric properties of these three viruses (Patolsky et al. 2004) are listed in Table 22.3. The entire virus is assumed to be a homogeneous rigid body. The DEP force versus frequency curve for each individual virus is shown in Figure 22.21. The DEP force on a virus is calculated through the Maxwell stress tensor. The proposed strategy to sort the three kinds of viruses is as follows. First, the frequency indicated by the solid arrow (5 MHz) in Figure 22.21a is chosen to selectively attract influenza in the first array of electrodes in

our proposed device since only influenza particles experience positive DEP force. Then, the frequency indicated by the dashed arrow (about 1 MHz) is chosen to attract inovirus in the second array of electrodes. Finally, the frequency indicated by the hollow arrow (about 100 kHz) is chosen to deposit bacteriophage.

The simulations for selective deposition of six viruses (two viruses from each type) on three pairs of parallel-rectangular-shaped electrodes are shown in Figure 22.21(b–e). The six viruses are initially aligned at the same height of 0.2 μm above the electrode surface. An AC field of 5 MHz is applied to the first set of electrode pair. An inflow of 18 μ/s is applied to transport the virus. The flow profile of the cross section in the middle of the channel is shown in Figure 22.21f. The electric field distribution across the electrode pairs is shown in Figure 22.21g. From the electric field distribution, it shows that each electrode pair generates a local electric field that has little impact on the nearby electrode pairs. The inovirus and bacteriophage are under negative DEP, while the Influenza is under positive DEP at 5 MHz. Thus, only influenza is trapped exactly in the gap and the other two are transported by the flow toward the second electrode pair. On the second electrode pair, when a 1 MHz AC field is applied, the DEP force on the inovirus is switched to be positive while bacteriophage is still under a negative DEP. Thus, only inovirus is attracted on the second electrode pair, while bacteriophage is again transported by the flow. On the third electrode pair, when an AC field of 100 kHz is applied, the Bacteriophage is finally deposited under a positive DEP. Through these three steps, three different groups of viruses are selectively deposited.

22.4.4 Ion and Liquid Transportation in Nanochannels

In recent years, nanoscale fluidic channels are believed to provide a platform for single molecule detection and evaluation. The transportation of ion species, biomolecules, and chemical

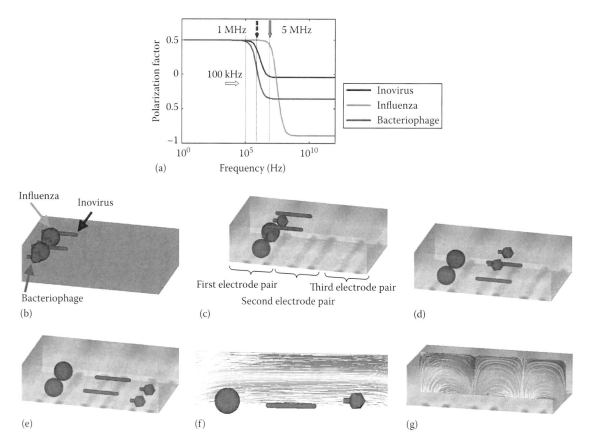

FIGURE 22.21 (See color insert following page 25-16.) Simulation on selective deposition of six viruses. 18 μm/s flow is directed from left to right. The first electrode pair is operating at 5 MHz, the second pair at 1 MHz, and the third pair at 100 kHz. Influenza, inovirus, and bacteriophage are selectively deposited on the first, second, and third electrode pairs consequently. (a) Polarization factor of viruses at different frequencies. (b) Time = 0 ms. (c) Time = 15 ms. (d) Time = 30 ms. (e) Time = 45 ms. (f) Velocity profile in the channel. (g) Electric field across three electrode pairs.

agents in nanofluidic channels provides a novel tool for controlling and separating such molecules.

The difference of nanochannels from micro/macroscale channels originates from small dimensions. At such a small scale, surface-charging effect dominates the fluid behavior. Two major factors are involved in the diffusion of liquid in a nanochannel: ion diffusion and capillary action. The diffusion of the liquid into the nanochannel can be triggered by the diffusion of ions into the channel covered by surface charges. Capillary action at air–liquid interface may also result in liquid filling. Such filling process strongly depends on liquid property, ion concentration, and the liquid–wall interface. The physics involved in the ion diffusion process and capillary action will be addressed in this section.

22.4.4.1 Capillary Action–Driven Flow

Liquid filling is a phenomenon in which fluid is introduced through a channel by surface tension or surface charges. Surface effect is dominant as the size of a channel decreases to micro or nanoscale. Two major forces in such a small channel are surface tension and an externally applied pressure or vacuum. Without the external forces, fluid flow is generated by capillary action when the contact angle is less than 90° (hydrophilic). When the

contact angle is greater than 90° (hydrophobic), the fluid has to be pressurized to induce a fluid flow. To trigger fluid flow in a hydrophobic channel, the external pressure has to be larger than the pressure jump across the interface. The detailed analysis of capillary action–driven flow has been presented by Yang et al. (2004).

The liquid–air interface is an evolving surface driven by the surface energy. The dynamics of the liquid/air interface in the nanochannel is controlled by the capillary force and viscous friction from the channel wall.

The pressure drop across the air–liquid interface is described by the Young–Laplace equation:

$$P_c = \frac{2\gamma \cos(\theta)}{r} \tag{22.148}$$

where

r is the channel radius
γ is the surface tension of the liquid
θ is the contact angle

The rate of liquid penetration into a small channel of radius r is given by the Washburn equation:

$$\frac{dl}{dt} = \frac{(P+P_c)r^2}{8\mu l} \qquad (22.149)$$

where

P is the external driving pressure
μ is the viscosity
l is the penetration length

Solving Equation 22.149 without external driving pressure gives an estimation of the kinetics of the capillary filling process:

$$l = a\sqrt{t} = \sqrt{\frac{\gamma \cos(\theta)r}{2\mu}} \cdot \sqrt{t} \qquad (22.150)$$

Equation 22.150 is based on a constant capillary pressure (Equation 22.148) driving the filling process and a linear hydraulic flow resistance. This resistance increases in proportion to the length of the liquid plug. In micro/nanoscale channel, the gravity force is much smaller than the surface tension, thus the filling length is as large as a few centimeters.

Such capillary action–based filling process is tested in an open microchannel experiment. Two kinds of microfluidic channels were fabricated to investigate capillary action: 4 and 10 μm wide channels (the depth and length are the same for both channels; 2 and 200 μm respectively). The open microchannel means that the top of the channel is open to air, which is not enclosed. When water suspending microspheres

(average diameter: 6 μm; standard derivation: 0.37 μm; Bangs Laboratories, Inc., Fisher, Indiana) was gently dropped in a reservoir of a 10 μm-wide channel, the water was driven along the wall of the mesa structure in Figure 22.22a. The channel, however, was not completely wet, since the top of the fluid channel was open to air.

In the 4 μm-wide channel, the capillary action was strong enough to make the channel wholly wet. It demonstrated that the water could be transported into an open microchannel by capillary action. Figure 22.22b and c show that microspheres were gradually transported by the flow due to capillary action. It was interestingly observed that the continuous flow was generated due to evaporation of water in the other side of the reservoir. As water evaporated, the capillary action continued to supply water to fill in the channel from the reservoir. The continuous supply of fluid was achieved by the open channel configuration without an external pressure.

Figure 22.23 shows magnified pics/images of the inset in Figure 22.22c. A small sphere (diameter ~1 μm) was transported through the channel due to the continuous flow. The experimental result showed that a particle could be isolated from larger particles (diameter 6 μm) upon its diameter due to the channel dimension and the formed meniscus. In spite of continuous flow using capillary action, the particle in Figure 22.23c stopped at the corner of the channel due to the drag force.

The experimental results presented in this section were explained by the capillary-driven filling theory. Such filling process in channels having various sizes and surface properties could be designed to control the liquid/particle transportation.

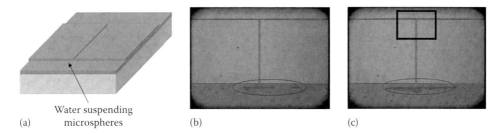

(a) Water suspending microspheres   (b)   (c)

FIGURE 22.22 Capillary flow in a 4 μm-wide open channel of length 200 μm and depth 2 μm. (a) Structure of an open fluidic channel. (b) and (c) Sequential pics/images showing the supply of water suspending spheres with a mean diameter of 6 μm and a standard deviation of 0.37 μm, and the aggregation of microspheres near the channel due to capillary force and its induced flow.

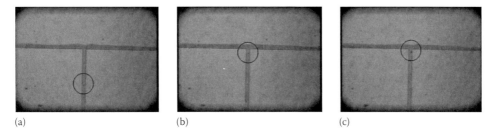

(a)   (b)   (c)

FIGURE 22.23 Transport of a 1 μm diameter microsphere in a 4 μm wide channel (inset of Figure 22.22c). (a) Microsphere marked in the circle is 1 μm in diameter, which is isolated from 6 μm diameter spheres. (b) The sphere is transported to the edge of the channel. The water is continuously supplied to the channel due to evaporation of water and capillary reaction. (c) The particle stops at the corner of the channel.

22.4.4.2 Liquid Diffusion in Nanochannels

Compared to microchannels, nanochannels are more challenging in terms of both fabrication challenge and physical complexity. Besides capillary action, charges and ions in the fluid also play an important role in fluid transportation in nanochannels. Using a developed fabrication method, an array of nanochannels was fabricated. Figure 22.24a shows the optical pics/image of the array of the fabricated nanochannels. The spacing of the nanochannels is 20 μm. As shown in SEM pics/images in Figure 22.24b and c, the entrance of the channel is 23 nm wide and the average width of the channel is 28 nm.

To study the diffusion effect in nanochannels, a drop of phosphate buffer (PB; Fisher Scientific, pH 7.2) solution was used. A droplet of the solution was gently placed on the array of nanochannels. The solution entered into the channel paths, which made the channels disappear from the vision under an optical microscope, Figure 22.24. In the figure, the channels filled with PB were not observed, but the channels unfilled with the solution were observed as bright-blue stripes.

To verify the diffusion in nanochannels, a base solution (KOH) was mixed with an acid solution (H₂SO₄) in nanochannels. For the purpose, a drop of the KOH solution was placed on nanochannels, as shown in Figure 22.25, and the solution was diffused into the channels. The diffusion length was limited to 200–300 μm. On the right side of the KOH drop, a drop of

H₂SO₄ was also placed on the nanochannels. The solution was diffused into the nanochannels at the distance of over 500 μm and reacted with KOH solution, which resulted in salt generation. In Figure 22.24c, the gray area shows the salt generated due to the reaction. Since the diffusion length of KOH solution is smaller than that of H₂SO₄, the reaction was generated near the KOH solution. In its magnified view, the salt generated by the reaction is shown in Figure 22.24d.

22.4.4.3 Modeling Liquid Diffusion in Nanochannels

The nanochannels in Section 22.4.4.2 can potentially provide a platform for various applications including bio/chemical sensors. To achieve this goal, it is important to investigate the physics involved in the diffusion process. In this section, we suggest a model to help understand the liquid diffusion problem in the nanochannels, and explore the feasibility for precise control over the diffusion length.

Fluid transportation in nanochannels can be modeled in various ways at different scales. At molecular level, MD can be used to analyze the molecular interaction of liquid/ion species with the channel surface. Unlike continuum non-slip theory, liquid flow in nanochannels usually experiences slip at liquid–wall interface, which can be revealed by MD simulation (Thompson et al. 1997). Such molecular-level simulation is applicable to nanochannels having a diameter of a few nanometers. Beyond

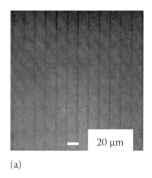
(a)

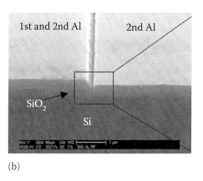
(b)

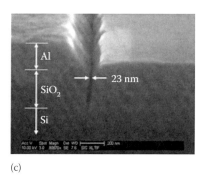
(c)

FIGURE 22.24 Fabricated nanochannels by the shadow edge lithography: (a) Optical pictures/image of an array of nanofluidic channels after RIE etching; the channels are 28 nm wide and 240 nm deep. (b) SEM pictures/image of a clearly defined nanochannel. (c) Magnified pictures/image at the nanochannel (From Bai, JG. et al., *Nanotech.*, 18(40), 405307, 2007; Bai, JG. et al., *Lab on a Chip*, 9, 449, 2009.).

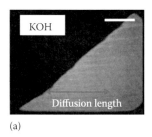
(a)

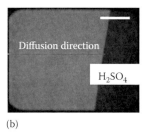
(b)

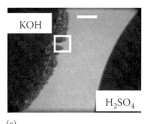
(c)

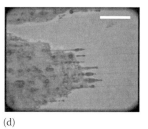
(d)

FIGURE 22.25 Diffusion experiment in open nanochannels. (a) KOH solution on nanochannels, scale bar 100 μm. (b) H₂SO₄ solution, scale bar 100 μm. (c) Reaction due to mixture of KOH and H₂SO₄, scale bar 200 μm. (d) Exploded pics/image of (c); salt produced in the reaction, scale bar 100 μm.

that, continuum approach is a more reasonable choice due to the computational cost. Continuum calculation can predict the basic features of electroosmotic flow, but lacks in accurate representation of the interaction between ions, biomolecules, electrical double layer, and surface charges on the wall. Continuum calculation assumes that the characteristic length of the channel is much larger than the Debye length. However, in the nanochannels, the thickness of the electric double layer (EDL) can be comparable to the channel width and depth. Transportation of ions and species in nanochannels involves the coupling of ions with the EDL, at which the continuum assumptions may not hold anymore. The EDL not only induces nonuniform motion of the solvent but also generates nonuniform transverse electric fields. In the case of transporting biomolecules in nanochannels, steric interactions of biomolecules with walls should be considered as well.

In this section, we introduce a hierarchical model that brings molecular-level diffusion and slip coefficient into the continuum description. The details of the EDL are analyzed by MD calculation while the macroscopic fluid flow is solved by continuum calculation.

22.4.4.3.1 Continuum Calculation

The primary variables to be solved are the electric field strength and the species concentration. The species concentration is governed by the conservation equation:

$$\frac{\partial c_k}{\partial t} + v^f \cdot \nabla c_k = \nabla \cdot \left[-\eta_k e z_k c_k E + \eta_k k_B T \nabla c_k \right], \quad k = 1, \dots, N \tag{22.151}$$

where

e is the charge on a proton

z_k is the valence of the kth species whose concentration is c_k

η_k is the mobility that describes how fast a given ion moves in a uniform electric field gradient

v^f is the fluid velocity

k_B is the Boltzmann constant

T is the temperature

$c_k = 0$ at the channel surface

The free charge density is related to the species concentration as

$$\rho^e = \sum_k e z_k c_k, \tag{22.152}$$

The electric field is calculated through the electric potential:

$$E = -\nabla \psi \tag{22.153}$$

For convenience, the electric potential can be decomposed into two parts:

$$\psi = \phi + \varphi \tag{22.154}$$

where

φ is the potential induced by the external electric field

ϕ is the zeta potential induced by the charges in the fluid or on the solid wall

The external potential should satisfy the Poisson equation:

$$\nabla^2 \varphi = 0 \tag{22.155}$$

The zeta potential is governed by the Gauss equation:

$$\nabla^2 \phi = -\frac{1}{\varepsilon_0 \varepsilon} \rho^e \tag{22.156}$$

where

ε_0 is the permittivity of free space

ε is the dielectric constant of the media

The flux due to kth species is given by Nernst–Planck equation:

$$J_k = -D_k \left(\nabla c_k + \frac{z_k e c_k}{kT} \nabla \psi \right) \tag{22.157}$$

where D_k is the diffusivity of kth ion species, which describes the capability of ions to minimize a concentration gradient.

Combining these equations together, the fluid motion can be described by the Navier–Stokes equation with the electric force from the free charges as

$$\nabla \cdot v = 0 \tag{22.158}$$

$$\rho^f \left(\frac{\partial v}{\partial t} + v \cdot \nabla v \right) = \nabla \cdot \sigma + f_e = -\nabla p + \mu \nabla^2 v + E \rho^e, \tag{22.159}$$

Equations 22.152 through 22.159 are the general governing equations to calculate the ion distribution and fluid flow in a channel.

At nanoscale channels, the non-slip boundary condition may not be applicable anymore. Thus, it is reasonable to adopt a slip boundary at the fluid–wall interface: $u = u_{slip}$, where u_{slip} is the slip velocity at the boundary that is determined by MD simulation.

It should be noted that the viscosity μ and mobility η_k in the continuum governing equations cannot be treated as a material constant due to the nature of fluid flow in nanochannels, and should be obtained through molecular-level calculation. The charge diffusion coefficient is related to the mobility through the generalized Einstein relation (see Chapter 29 of Ashcroft and Mermin 1976, Plecis et al. 2005). At low density limit, the relation is given by $D/\eta = k_B T/q$ (Einstein relation) where q is the charge of the particle.

In the following part of this section, we make approximations to obtain a solution of electroosmotic flow in a 2D rectangular channel.

First, let us assume that the electric double layer has a finite length. Based on the Debye–Huckel approximation for the charge density, the equilibrium zeta potential can be described by the meanfield Poisson–Boltzmann equation for a monovalent solution:

$$\nabla^2 \phi = -\frac{2cze}{\varepsilon_0 \varepsilon} \sinh\left(\frac{-ze\phi}{k_B T}\right) \tag{22.160}$$

If we assume further that $ze \ll k_B T$, Equation 22.160 can be approximated as

$$\nabla^2 \phi = \kappa^2 \phi, \quad \kappa = \sqrt{\frac{2cz^2 e^2}{\varepsilon_0 \varepsilon k_B T}} \tag{22.161}$$

The inverse value κ^{-1} is also called Debye length and represents the characteristic length of the electric double layer.

For a 2D channel at equilibrium, the zeta potential should be constant in the flow direction x, thus we only need to solve the zeta potential in direction y perpendicular to the wall. For a 2D rectangular-shaped channel with a width w, in case of zeta potential $\phi = 1$ at top and bottom wall ($y = 0$ and $y = w$), an analytical solution of ϕ is given as [33]

$$\phi = \frac{\cosh\left[\kappa w(y - 1/2)\right]}{\cosh(\kappa w/2)} \tag{22.162}$$

The distribution of zeta potential across the channel at different Debye lengths is shown in Figure 22.26a. The y coordinate is normalized by channel width w. When the Debye length is much smaller than the channel width, the zeta potential gradient

(electric field induced by zeta potential) is only localized in the vicinity of the wall. The potential gradient exponentially decreases as the distance from the wall increases, as shown in Figure 22.26b. However, when the Debye length is comparable to the channel width, the weak zeta potential gradient is generated across the channel. As shown in Figure 22.26a, the local electric field close to the wall decreases as the Debye length increases. When the Debye length is large enough, the EDL at the upper and lower channel wall may even overlap. This potential gradient can contribute to change of surface tension, thus influencing liquid motion in nanochannels. In the classical electrowetting (Pollack et al. 2005) phenomena in microfluidic channels, an external voltage is usually required to generate large enough electric potential gradient between electrolyte drop and electrode surface in order to induce surface tension change. In nanochannels, due to its small size, the potential difference generated by EDL itself may be sufficient to change the surface tension and lead to liquid filling. To understand this phenomenon, the electrowetting theory is briefly reviewed first.

Electrowetting usually refers to the motion of electrolyte drop induced by an applied voltage between the electrolyte and the electrode surface. The applied potential difference can partially modify the surface tension on the electrolyte drop, which induces a local pressure in the drop. It can generate the motion of electrolyte drop in order to reach a uniform pressure in the drop. This process can be described by an energy equation, which relates the total surface tension with a contribution from electric potential:

$$\gamma_{dw} = \gamma_{dw}^0 - \frac{CV^2}{2} \tag{22.163}$$

where

γ_{dw} is the total surface tension including both chemical and electrical terms

γ_{dw}^0 is the total surface tension with only chemical components

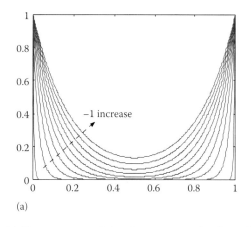

(a)

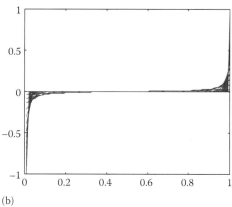

(b)

FIGURE 22.26 (a) The zeta potential distribution across the channel width at different Debye lengths. (b) Distribution of the electric field induced by the zeta potential.

$C = \varepsilon_m \varepsilon_0 / h$ is the capacitance of the interface for a uniform dielectric of thickness h and permittivity ε_m

V is the potential difference between the electrolyte drop and wall surface

The contact angle θ is given by the Young–Dupre equation as

$$\gamma_{dw} = \gamma_w - \gamma_d \cos(\theta) \qquad (22.164)$$

where

γ_d is the surface tension on the electrolyte drop
γ_w is the surface tension on the wall surface

Thus the contact angle can be expressed as

$$\theta = \cos^{-1}\left(\frac{\gamma_w - \gamma_{dw}^0 + CV^2/2}{\gamma_d} \right) \qquad (22.165)$$

Through this expression, it is found that the contact angle can be changed by increasing the electric potential gradient; thus electric field strength is changed at the interface.

The electrical properties of a surface can be modified in various ways, such as ion concentration variation, surface coating, chemical modification, and temperature change. Among these approaches, ion concentration variation has received more attention in recent years, because it is easy to apply and accurate to control. When the ion concentration is low, the electric field due to surface charges is small but spans across the channel width. When the ion concentration is high, a strong electric field is induced by surface charges close to the wall. This strong electric field localized near the wall induces the change of surface tension. The local field applied on the drop surface tends to pull the drop surface to the wall surface, which decreases the contact angle. This phenomenon in nanochannels is comparable to the electrowetting phenomenon under high voltage potential at microchannels.

Equation 22.165 gives the relationship between the contact angle and the potential difference and, thus, electric field strength, between the electrolyte and wall. Basically, the contact angle decreases with the increase of the potential difference. When the contact angle is less than 90°, the interface becomes hydrophilic and leads to liquid motion. This agrees with the experimental measurement of the diffusion time and length of liquids at different ion concentrations (Plecis et al. 2005). It shows that the diffusion time is long when the ion concentration is very low. Below a certain critical ion concentration, the diffusion time decreases exponentially as the ion concentration increases. The diffusion time finally reaches a plateau when the ion concentration is large enough.

The physics involved in this process can be explained in the proposed model. The simulated filling length by combining Equations 22.157 and 22.160 is shown in Figure 22.27. Here, the ion concentration is normalized by 0.1 M. The electric field strength and diffusion length are normalized by their maximum values. When the ion concentration is low, the potential difference induced by the EDL is small, which hardly influences the contact angle. The filling length at this stage is close to a constant, and it is dominated by the diffusion of ions in the channels. When the ion concentration is large enough, the contact angle decreases to be less than 90° and the filling length increases exponentially according to the ion concentration. As the ion concentration increases further, the potential gradient induced by the EDL finally reaches a plateau; thus the filling length reaches a maximum constant.

If we assume the electroosmotic flow reaches equilibrium and there is no pressure difference, Equation 22.159 can be simplified in the flow direction as

$$\frac{d}{dx}\left(\mu \frac{du(x)}{dx} \right) + \rho^e E_{ext} = 0 \qquad (22.166)$$

where u is the flow velocity in the channel direction. Inserting Equations 22.156 and 22.161 into Equation 22.166 leads to

$$\frac{d}{dx}\left(\mu \frac{du(x)}{dx} \right) - \varepsilon_0 \varepsilon \kappa^2 \phi E_{ext} = 0 \qquad (22.167)$$

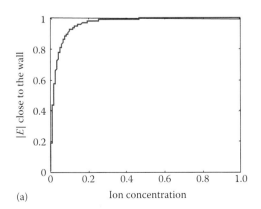

(a)

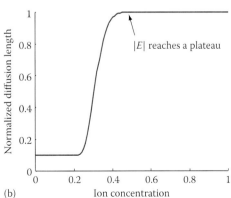

(b)

FIGURE 22.27 (a) Electric field strength near the wall as a function of ion concentration. (b) Predicted liquid diffusion length at different ion concentrations.

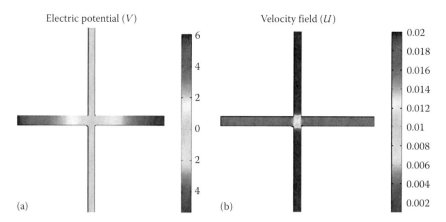

Electric potential (*V*)   Velocity field (*U*)

(a)   (b)

FIGURE 22.28 **(See color insert following page 25-16.)** Simulation of electroosmotic flow in a cross-sectional fluid channel. (a) The distribution of electric field potential. (b) Flow velocity in the channel.

Equations 22.155, 22.162, and 22.167 can be solved together to obtain the electric potential and fluid flow inside a 2D rectangular-shaped channel.

A simulation of electroosmotic flow in a 2D cross channel is shown in Figure 22.28. The electrical potential applied on the channel is shown in Figure 22.28a. Under the electric field, an electroosmotic flow is generated due to the ionic motion close to the surface of the wall, as shown in Figure 22.28b. The fluid flow is uniform in the most part of the horizontal channel except the intersection region.

The theory proposed in this section predicts that the diffusion length can be precisely controlled by changing ion concentration, applying external electric field, or modifying surface properties of the channels. In the future, we will combine the modeling work presented in this section with designed experiments in order to comprehensively understand diffusion in nanochannels.

22.4.4.3.2 Molecular Dynamics Simulation

MD simulation can be used to provide molecular-level details for continuum calculation, i.e., determining the proper coefficients and boundary conditions. Such calculation is essential for electrically charged nanochannels having a few nanometers in diameter. The characteristic length of ion double layer forming on the channel surface, the Debye length, is determined by the local balance of electromigration toward the surface and diffusion from the surface. For the nanochannels having surface charges, the electric charge density is higher at the vicinity of the surface where the ion mobility is lower. The MD simulation can be used to calculate the diffusion coefficient of the ions, i.e., through the slope of the mean square displacement. These coefficients are used to obtain the mobility of ions in the continuum calculation (Figure 22.28).

From the MD, diffusion coefficient is calculated as

$$\langle r_k^2 \rangle = 2kDt \qquad (22.168)$$

Moreover, the viscosity of fluid in nanochannels is usually substantially higher than the bulk viscosity due to the local structure of fluid near the surface. This viscosity at different heights across the channel can be obtained through MD.

The viscosity of the fluid can be calculated by a simple MD simulation as illustrated in Figure 22.29. A velocity *U* applied on the top and bottom of a solid wall generates a shear flow. The force exerted by the liquid on the solid wall is *F*, thus the viscosity of the liquid is expressed as

$$\mu = \frac{hF}{2lU} \qquad (22.169)$$

where
h is the channel width
l is the channel length

Slip boundary condition is derived from MD. Thompson and Troian (1997) have used MD simulations to quantify the slip-flow boundary condition dependence on shear rate. A generalized boundary condition that relates the degree of slip to the underlying static properties, and the dynamic interactions between the walls and the fluid have been proposed. In the formulation, the linear Navier slip boundary condition is described as

$$\Delta u\,|_\mathrm{w} = u_\mathrm{fluid} - u_\mathrm{wall} = L_\mathrm{s} \left. \frac{\partial u}{\partial y} \right|_\mathrm{w} \qquad (22.170)$$

where
L_s is a constant slip length
$\partial u/\partial y|_\mathrm{w}$ is the strain rate computed at the wall

The major goal is to determine the degree of slip at a solid–liquid interface as the interfacial parameters and the shear rate change. To achieve this, the slip length is computed as

$$L_\mathrm{s} = \frac{\Delta u\,|_\mathrm{w}}{\dot{\gamma}} = \frac{(U/\dot{\gamma} - h)}{2} \qquad (22.171)$$

where $\dot{\gamma}$ is the shear rate.

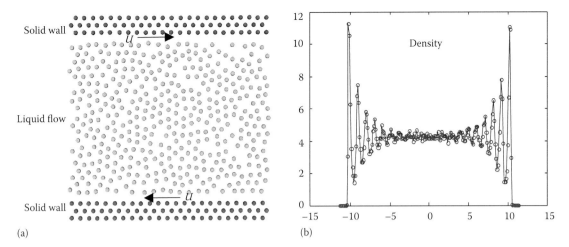

FIGURE 22.29 Molecular dynamic simulation of fluid flow in a nanochannel: (a) nanochannel flow configuration. (b) Liquid density profile across the channel.

Scaling L_s by its asymptotic limiting value L_s^0 and $\dot{\gamma}$ by its critical value $\dot{\gamma}_c$, a generalized boundary condition is expressed as

$$L_s = L_s^0 (1 - \dot{\gamma} / \dot{\gamma}_c)^{-\alpha} \qquad (22.172)$$

In this formulation, L_s^0, $\dot{\gamma}_c$, and fitting coefficient α depend on the interfacial parameters, this should be determined by MD simulation under the specific condition.

22.5 Conclusion

The specific fundamental concepts and tools in nanomechanics and materials, including solids and fluids, and their modeling via multiple scale methods and techniques have been outlined above. Meantime, this chapter should only be viewed as a starting point from which interested researchers may commence their literature review and further contribute to the emerging field of computational nanomechanics. Other interesting issues of computational nanomechanics beyond the scope of this chapter include the Monte Carlo and multiple timescale modeling approaches, a range of coupled multiphysics methods, including those with evolutionary and growth models for mesoscale material mechanics, internal structure evolution, and phase transitions. These exciting topics are in focus of the contemporary nanomechanics research.

Acknowledgments and Copyright

Wing Kam Liu would like to acknowledge the support from the National Science Foundation (CMMI-0823327), Army Research Lab, the World Class University program (R33-10079) under the Ministry of Education, Science and Technology, Republic of Korea, Office of Naval Research (ONR) and the NSF Summer Institute on Nano Mechanics and Materials. Support of Eduard G. Karpov from the NSF (CMMI-0900498) and UIC College of Engineering is gratefully acknowledged. Yaling Liu acknowledges the supports from National Science Foundation (CBET-0955214) and National Institute of Health (R03-EB009786) and thanks Jaehyun Chung (UW) for sharing experimental images of particle manipulation. Material of Sections 22.2.1 through 22.3.6 and 22.4.2 is reproduced from *Nano Mechanics and Materials: Theory, Multiscale Methods and Applications* by W.K. Liu, E.G. Karpov and H.S. Park with permission from John Wiley & Sons. Material of Sections 22.4.1, 22.4.2, and 22.4.3 is reprinted from *Computer Methods in Applied Mechanics and Engineering* 197, Y. Liu, K. Oh, John G., et. al., Manipulation of nanoparticles and biomolecules by electric field and surface tension, 2156–2172, 2008, with permission from Elsevier.

References

Abraham F and Gao H 2000 How fast can cracks propagate? *Physical Review Letters* **84**(14), 3113–3116.

Abraham F, Brodbeck D, Rudge W, and Xu X 1997 A molecular dynamics investigation of rapid fracture mechanics. *Journal of the Mechanics and Physics of Solids* **45**(9), 1595–1619.

Abraham F, Walkup R, Gao H, Duchaineau M, la Rubia TD, and Seager M 2002 Simulating materials failure by using up to one billion atoms and the world's fastest computer: Brittle fracture. *Proceedings of the National Academy of Sciences of the United States of America* **99**(9), 5777–5782.

Abraham FF, Broughton J, Bernstein N, and Kaxiras E 1998 Spanning the continuum to quantum length scales in a dynamic simulation of brittle fracture. *Europhysics Letters* **44**, 783–787.

Adelman SA and Doll JD 1976 Generalized Langevin equation approach for atom/solid-surface scattering: General formulation for classical scattering off harmonic solids. *Journal of Chemical Physics* **64**, 2375–2388.

Adelman SA and Garrison BJ 1976 Generalized Langevin theory for gas/solid processes: Dynamic solid models. *Journal of Chemical Physics* **65**, 3751–3761.

Allen MP and Tildesley D 1987 *Computer Simulation of Liquids.* Oxford University Press, New York.

Anderson HC 1980 Molecular dynamics simulations at constant pressure and/or temperature. *Journal of Chemical Physics* **72**(4), 2384–2393.

Arroyo M and Belytschko T 2004 Finite crystal elasticity of carbon nanotubes based on the exponential Cauchy-Born rule. *Physical Review B* **69**(11), 115415.

Ashcroft NW and Mermin ND 1976 *Solid State Physics.* HRW, Philadelphia, PA.

Aubry N and Singh P 2006a Control of electrostatic particle-particle interactions in dielectrophoresis. *Europhysics Letters* **4**(4), 623–629.

Aubry N and Singh P 2006b Influence of particle-particle interactions and particles rotational motion in traveling wave dielectrophoresis. *Electrophoresis* **27**(3), 703–715.

Bai JG, Chang CL, Chung JH, and Lee KH 2007 Shadow edge lithography for nanoscale patterning and manufacturing. *Nanotechnology* **18**(40), 405307–405315.

Bai JG, Yeo WH, and Chung JH 2009 Nanostructured biosensing platform shadow edge lithography for high-throughput nanofabrication. *Lab on a Chip* **9**, 449–455.

Belytschko T, Liu WK, and Moran B 2000 *Nonlinear Finite Elements for Continua and Structures.* Wiley, Chichester, U.K.

Benenson W, Harris J, Stocker H, and Lutz H 2002 *Handbook of Physics.* Springer, New York.

Berendsen HJC, Postma JPM, van Gunsteren WF, DiNola A, and Haak JR 1984 Molecular dynamics with coupling to an external bath. *Journal of Chemical Physics* **81**(8), 3684–3690.

Biswas R and Hamann DR 1985 Interatomic potential for silicon structural energies. *Physical Review Letters* **55**(19), 2001–2004.

Biswas R and Hamann DR 1987 New classical models for silicon structural energies. *Physical Review B* **36**(12), 6434–6445.

Bolton K and Gustavsson S 2003 Energy transfer mechanisms in gas-carbon nanotube collisions. *Chemical Physics* **291**(2), 161–170.

Bolton K and Rosen A 2002 Computational studies of gas-carbon nanotube collision dynamics. *Physical Chemistry Chemical Physics* **4**(18), 4481–4488.

Brenner DW 1990 Empirical potential for hydrocarbons for use in simulating the chemical vapor deposition of diamond films. *Physical Review B* **42**(15), 9458–9471.

Brenner DW, Shenderova OA, Harrison JA, Stuart SJ, Ni B, and Sinnott SB 2002 A second-generation reactive empirical bond order (rebo) potential energy expression for hydrocarbons. *Journal of Physics: Condensed Matter* **14**, 783–802.

Brisson V and Tilton R 2003 Self-assembly and two-dimensional patterning of cell arrays by electrophoretic deposition. *Biotechnology and Bioengineering* **77**(3), 290–295.

Broughton JQ, Abraham FF, Bernstein N, and Kaxiras E 1999 Concurrent coupling of length scales: Methodology and applications. *Physical Review B* **60**, 2391–2403.

Chen XQ, Saito T, Yamada H, and Matsushige K 2001 Aligning single-wall carbon nanotubes with an alternating-current electric field. *Applied Physics Letters* **78**(23), 3714–3416.

Chung J and Lee J 2003 Nanoscale gap fabrication and integration of carbon nanotubes by micromachining. *Sensors and Actuators: A. Physical* **103**(3), 229–235.

Chung J, Lee K, Lee J, and Ruoff R 2004 Toward large-scale integration of carbon nanotubes. *Langmuir* **20**(8), 3011–3017.

Clementi E and Roetti C 1974 Roothaan-hartree-fock atomic wavefunctions. *Atomic Data and Nuclear Data Tables* **14**(3–4), 177–478.

Curtin WA and Miller RE 2003 Atomistic/continuum coupling in computational materials science. *Modelling and Simulation in Materials Science and Engineering* **11**, R33–R68.

Daw MS 1989 Model of metallic cohesion: The embedded-atom method. *Physical Review B* **39**(11), 7441–7452.

Daw MS, Foiles SM, and Baskes MI 1993 The embedded-atom method: A review of theory and applications. *Materials Science Reports* **9**, 251–310.

Dimaki M and Boggild P 2004 Dielectrophoresis of carbon nanotubes using micro-electrodes: A numerical study. *Nanotechnology* **15**(8), 1095–1102.

Finnis MW and Sinclair JE 1984 A simple empirical n-body potential for transition metals. *Philosophical Magazine A* **50**(1), 45–55.

Foiles SM, Baskes ML, and Daw MS 1986 Embedded-atom-method functions for the fcc metals Cu, Ag, Au, Ni, Pd, Pt, and their alloys. *Physical Review B* **33**(12), 7893–7991.

Foulkes WMC and Haydock R 1989 Tight-binding models and density-functional theroy. *Physical Review B* **39**, 12520.

Gao H, Huang Y, and Abraham F 2001 Continuum and atomistic studies of intersonic crack propagation. *Journal of Mechanics and Physics of Solids* **49**(9), 2113–2132.

Goldstein H 1980 *Classical Mechanics.* Addison-Wesley, Reading, MA.

Haile J 1992 *Molecular Dynamics Simulations.* Wiley & Sons, New York.

Harrison DE 1988 Application of molecular dynamics simulations to the study of ion-bombarded metal surfaces. *Critical Reviews in Solid State and Materials Science* **14**(Sup.1), S1–S78.

Hoover WG 1985 Canonical dynamics: Equilibrium phase-space distributions. *Physical Review A* **31**(3), 1695–1697.

Hughes TJR 1987 *The Finite Element Method: Linear Static and Dynamic Finite Element Analysis.* Prentice-Hall, Englewood Cliffs, NJ.

Hughes TJR, Feijoo GR, Mazzei L, and Quincy JB 1998 The variational multiscale method: A paradigm for computational mechanics. *Computer Methods in Applied Mechanics and Engineering* **166**, 3–24.

Johnson RA 1988 Analytic nearest-neighbor model for fcc metals. *Physical Review B* **37**(8), 3924–3931.

Jones JE 1924a On the determination of molecular fields. i. From the variation of the viscosity of a gas with temperature. *Proceedings of the Royal Society A* **106**, 441–462.

Jones JE 1924b On the determination of molecular fields. ii. From the equation of state of a gas. *Proceedings of the Royal Society A* **106**, 463–477.

Jones TB 1995 *Electromechanics of Particles*. Cambridge University Press, New York.

Kadaksham A, Singh P, and Aubry N 2004 Dielectrophoresis of nanoparticles. *Electrophoresis* **25**(21–22), 3625–3632.

Kadaksham J, Singh P, and Aubry N 2005 Dielectrophoresis induced clustering regimes of viable yeast cells. *Electrophoresis* **26**(19), 3738–3744.

Kadaksham J, Singh P, and Aubry N 2006 Manipulation of particles using dielectrophoresis. *Mechanics Research Communications* **33**(1), 108–122.

Kadowaki H and Liu WK 2004 Bridging multi-scale method for localization problems. *Computer Methods in Applied Mechanics and Engineering* **193**, 3267–3302.

Karpov EG, Wagner GJ, and Liu WK 2005 A Green's function approach to deriving non-reflecting boundary conditions in molecular dynamics simulations. *International Journal for Numerical Methods in Engineering* **62**(9), 1250–1262.

Karpov EG, Park HS, and Liu WK 2007 A phonon heat bath approach for the atomistic and multiscale simulation of solids. *International Journal for Numerical Methods in Engineering* **70**(3), 351–378.

Knap J and Ortiz M 2001 An analysis of the quasicontinuum method. *Journal of the Mechanics and Physics of Solids* **49**(9), 1899–1923.

Konishi T, Ohsawa K, Abe H, and Kuramoto E 1999 Determination of n-body potential for Fe-Cr alloy system and its application to defect study. *Computational Materials Science* **14**(1–4), 108–113.

La Paglia SR 1971 *Introductory Quantum Chemistry*. Harper and Row Publishers, New York.

Lauga E and Brenner M 2004 Evaporation-driven assembly of colloidal particles. *Physical Review Letters* **93**(23), 238301.

Li J, Lu Y, Ye Q, Cinke M, Han J, and Meyyappan M 2003 Carbon nanotube sensors for gas and organic vapor detection. *Nano Letters* **3**(7), 929–933.

Li S and Liu WK 2004 *Meshfree Particle Methods*. Springer, Berlin, Germany.

Lieber CM 2003 Nanoscale science and technology: Building a big future from small things. *MRS Bulletin* **28**(7), 486–491.

Liu WK and McVeigh C 2008 Predictive multiscale theory for design of heterogeneous materials. *Computational Mechanics* **42**(2), 147–170.

Liu Y and Iqbal S 2009 A mesoscale model for molecular interaction in functionalized nanopores. *Applied Physics Letters* **95**(22), 223701.

Liu WK, Jun S, and Zhang YF 1995a Reproducing kernel particle methods. *International Journal for Numerical Methods in Fluids* **20**, 1081–1106.

Liu WK, Jun S, Li S, Adee J, and Belytschko T 1995b Reproducing kernel particle methods for structural dynamics. *International Journal for Numerical Methods in Engineering* **38**, 1655–1680.

Liu WK, Uras R, and Chen Y 1997 Enrichment of the finite element method with the reproducing kernel particle method. *Journal of Applied Mechanics* **64**, 861–870.

Liu WK, Karpov EG, Zhang S, and Park HS 2004a An introduction to computational nano mechanics and materials. *Computer Methods in Applied Mechanics and Engineering* **193**, 1529–1578.

Liu WK, Karpov EG, Zhang S, and Park HS 2004b An introduction to computational nanomechanics and materials. *Computer Methods in Applied Mechanics and Engineering* **193**(17–20), 1529–1578.

Liu WK, Karpov EG, and Park HS 2005 *Nano Mechanics and Materials: Theory, Multiple Scale Analysis, and Applications*. Springer, Berlin, Germany.

Liu WK, Liu Y, Farrell D, Zhang L, Wang XS, Fukui Y, Patankar N et al. 2006a Immersed finite element method and its applications to biological systems. *Computer Methods in Applied Mechanics and Engineering* **195**(13–16), 1722–1749.

Liu Y, Chung JH, Liu WK, and Ruoff RS 2006b Dielectrophoretic assembly of nanowires. *Journal of Physical Chemistry B* **110**(29), 14098–14106.

Liu Y, Liu WK, Belytschko T, Patankar N, Chung JH, and To A 2007 Immersed electrokinetic finite element method. *International Journal for Numerical Methods in Engineering* **71**, 379–405.

Liu Y, Oh K, Bai JG, Chang CL, Yeo W, Chung JH, Lee KH and Liu WK 2008 Manipulation of nanoparticles and biomolecules by electric field and surface tension. *Computer Methods in Applied Mechanics and Engineering* **197**(25–28), 2156–2172.

Los JH and Fasolino A 2002 Monte Carlo simulations of carbon-based structures based on an extended Brenner potential. *Computer Physics Communications* **147**(1–2), 178–181.

Lu Y, Yin Y, and Xia Y 2001 A self-assembly approach to the fabrication of patterned, two-dimensional arrays of microlenses of organic polymers. *Advanced Materials* **13**(1), 34–37.

Manoharan V, Elsesser M, and Pine D 2003 Dense packing and symmetry in small clusters of microspheres. *Science* **301**(5632), 483–487.

McVeigh C and Liu WK 2008 Linking microstructure and properties through a predictive multiresolution continuum. *Computer Methods in Applied Mechanics and Engineering* **197**, 3268–3290.

McVeigh C, Vernerey F, Liu WK, and Brinson LC 2006 Multiresolution analysis for material design. *Computer Methods in Applied Mechanics and Engineering* **195**, 5053–5076.

Miller RE and Tadmor EB 2003 The quasicontinuum method: Overview, applications and current directions. *Journal of Computer-Aided Materials Design* **9**, 203–239.

Miller RE, Tadmor EB, Phillips R, and Ortiz M 1998 Quasicontinuum simulation of fracture at the atomic scale. *Modelling and Simulation in Materials Science and Engineering* **6**(5), 607–638.

Mueller M 2001 *Fundamentals of Quantum Chemistry*. Kluwer Academic/Plenum Publisher, New York.

Nosé S 1984a A molecular dynamics method for simulations in the canonical ensemble. *Molecular Physics* **53**, 255–268.

Nosé S 1984b A unified formulation of the constant temperature molecular dynamics methods. *Journal of Chemical Physics* **81**(1), 511–519.

Park HS and Zimmerman JA 2005 Modeling inelasticity and failure in gold nanowires. *Physical Review B* **72**, 054106.

Patankar NA and Hu HH 1998 Numerical simulation of electroosmotic flow. *Analytical Chemistry* **70**, 1870–1881.

Patolsky F, Zheng G, Hayden O, Lakadamyali M, Zhuang X, and Lieber CM 2004 Electrical detection of single viruses. *PNAS* **101**(39), 14017–14022.

Peskin CS 2002 The immersed boundary method. *Acta Numerica* **11**(14), 479–517.

Plecis A, Schoch RB, and Renaud P 2005 Ionic transport phenomena in nanofluidics: Experimental and theoretical study of the exclusion-enrichment effect on a chip. *Nano Letters* **5**(6), 1147–1155.

Pohl H 1978 *Dielectrophoresis*. Cambridge University Press, Cambridge, U.K.

Pollack MG, Fair RB, and Shenderov AD 2005 Electrowetting-based actuation of liquid droplets for microfluidic applications. *Applied Physics Letters* **77**(11), 1725–1726.

Qian D, Wagner GJ, Liu WK, Yu MF, and Ruoff RS 2002 Mechanics of carbon nanotubes. *Applied Mechanics Reviews* **55**(6), 495–533.

Qian D, Wagner GJ, and Liu WK 2004 A multiscale projection method for the analysis of carbon nanotubes. *Computer Methods in Applied Mechanics and Engineering* **193**, 1603–1632.

Qiao R and Aluru NR 2002 A compact model for electroosmotic flows in microfluidic devices. *Journal of Micromechanics and Microengineering* **12**(5), 625–635.

Qiao R and Aluru NR 2003 Ion concentrations and velocity profiles in nanochannel electroosmotic flows. *Journal of Chemical Physics* **118**(10), 4692–4701.

Qiao R and Aluru NR 2004 Charge inversion and flow reversal in a nanochannel electro-osmotic flow. *Physical Review Letters* **92**(19), 198301.

Qiao R and Aluru NR 2005 Surface-charge-induced asymmetric electrokinetic transport in confined silicon nanochannels. *Applied Physics Letters* **86**(14), 14305.

Ramos A, Morgan H, Green NG, and Castellanos A. 1998 Ac electrokinetics: A review of forces in microelectrode structures. *Journal of Physics D: Applied Physics* **31**(18), 2338–2353.

Ratner MA and Schatz GC 2001 *Introduction to Quantum Mechanics in Chemistry*. Prentice Hall, Upper Saddle River, NJ.

Rosenblum I, Adler J, and Brandon S 1999 Multi-processor molecular dynamics using the Brenner potential: Parallelization of an implicit multi-body potential. *International Journal of Modern Physics C* **10**(1), 189–203.

Rudd RE 2001 Coarse-grained molecular dynamics: Dissipation due to internal modes. *Materials Research Society Symposium Proceedings*, San Francisco, CA.

Rudd RE and Broughton JQ 1998 Coarse-grained molecular dynamics and the atomic limit of finite elements. *Physical Review B* **58**, 5893–5896.

Saether E, Frankland S, and Pipes R 2003 Transverse mechanical properties of single-walled carbon nanotube crystals. Part I: Determination of elastic moduli. *Composites Science and Technology* **63**(11), 1543–1550.

Satyanarayana S, Karnik RN, and Majumdar A 2005 Stamp-and-stick room-temperature bonding technique for microdevices. *Journal of Microelectromechanical Systems* **14**(2), 392–399.

Sharma N and Patankar N 2004 Direct numerical simulation of Brownian motion of particles using fluctuating hydrodynamics equation. *Journal of Computational Physics* **201**(2), 466–486.

Shilkrot LE, Curtin WA, and Miller RE 2002 A coupled atomistic/continuum model of defects in solids. *Journal of the Mechanics and Physics of Solids* **50**, 2085–2106.

Shilkrot LE, Miller RE, and Curtin WA 2004 Multiscale plasticity modeling: Coupled atomistics and discrete dislocation mechanics. *Journal of the Mechanics and Physics of Solids* **52**, 755–787.

Singh P and Aubry N 2005 Trapping force on a finite-sized particle in a dielectrophoretic cage. *Physical Review E* **72**(1), 016602.

Solomentsev Y, Bohmer M, and Anderson J 1997 Particle clustering and pattern formation during electrophoretic deposition: A hydrodynamic model. *Langmuir* **13**, 6058–6068.

Stillinger FH and Weber TA 1985 Computer simulation of local order in condensed phases of silicon. *Physical Review B* **31**(8), 5262–5271.

Tadmor E, Ortiz M, and Phillips R 1996 Quasicontinuum analysis of defects in solids. *Philosophical Magazine A* **73**, 1529–1563.

Takai T, Halicioglu T, and Tiller WA 1985 Prediction for the pressure and temperature phase transitions of silicon using a semiempirical potential. *Scripta Metallurgica* **19**(6), 709–713.

Tang J, Gao B, Geng H, Velev OD, Qin L-C, and Zhou O 2003 Assembly of 1d nanostructures into sub-micrometer diameter fibrils with controlled and variable length by dielectrophoresis. *Advanced Materials* **15**(16), 1352–1355.

Tang J, Gao B, Geng H, Velev OD, Qin L-C, and Zhou O 2004 Manipulation and assembly of SWNTs by dielectrophoresis. *Abstracts of Papers of the American Chemical Society* **227**, 1273.

Tang J, Yang G, Zhang Q, Parhat A, Maynor B, Liu J, Qin L-C, and Zhou O 2005 Rapid and reproducible fabrication of carbon nanotube AFM probes by dielectrophoresis. *Nano Letters* **5**(1), 11–14.

Tersoff J 1986 New empirical model for the structural properties of silicon. *Physical Review Letters* **56**(6), 632–635.

Tersoff J 1988a Empirical interatomic potential for carbon, with applications to amorphous carbon. *Physical Review Letters* **61**(25), 2879–2882.

Tersoff J 1988b New empirical approach for the structure and energy of covalent systems. *Physical Review B* **37**(12), 6991–7000.

Thompson AP 2003 Nonequilibrium molecular dynamics simulation of electro-osmotic flow in a charged nanopore. *Journal of Chemical Physics* **119**(14), 7503–7511.

Thompson PA and Troian SM 1997 A general boundary condition for liquid flow at solid surfaces. *Nature* **389**(6649), 360–362.

Trau M, Saville DA, and Aksay I 1997 Assembly of colloidal crystals at electrode interfaces. *Langmuir* **13**(24), 6375–6381.

van der Giessen E and Needleman A 1995 Discrete dislocation plasticity: A simple shear model. *Modelling and Simulation in Materials Science and Engineering* **3**, 689–735.

Vernerey FJ, Liu WK, and Moran B 2007 Multi-scale micromorphic theory for hierarchical materials. *Journal of the Mechanics and Physics of Solids* **55**(12), 2603–2651.

Vernerey FJ, Liu WK, Moran B, and Olson G 2008 A micromorphic model for the multiple scale failure of heterogeneous materials. *Journal of the Mechanics and Physics of Solids* **56**(4), 1320–1347.

Voter AF 1997 Hyperdynamics: Accelerated molecular dynamics of infrequent events. *Physical Review Letters* **78**(20), 3908–3911.

Voter AF, Montalenti F, and Germann TC 2002 Extending the time scale in atomistic simulation of materials. *Annual Review of Materials Research* **32**, 321–346.

Wagner GJ and Liu WK 2001 Hierarchical enrichment for bridging scales and meshfree boundary conditions. *International Journal for Numerical Methods in Engineering* **50**, 507–524.

Wagner GJ and Liu WK 2003 Coupling of atomistic and continuum simulations using a bridging scale decomposition. *Journal of Computational Physics* **190**, 249–274.

Wagner GJ, Karpov EG, and Liu WK 2004 Molecular dynamics boundary conditions for regular crystal lattices. *Computer Methods in Applied Mechanics and Engineering* **193**, 1579–1601.

Walther J, Jaffe R, Halicioglu T, and Koumoutsakos P 2001 Carbon nanotubes in water: Structural characteristics and energetics. *Journal of Physical Chemistry* **105**(41), 9980–9987.

Walther J, Jaffe R, Kotsalis E, Werder T, Halicioglu T, and Koumoutsakos P 2004 Hydrophobic hydration of C-60 and carbon nanotubes in water. *Carbon* **42**(5–6), 1185–1194.

Wan R, Fang H, Lin Z, and Chen S 2003 Lattice Boltzmann simulation of a single charged particle in a Newtonian fluid. *Physical Review E* **68**(1), 011401.

Wang X and Liu WK 2004 Extended immersed boundary method using FEM and RKPM. *Computer Methods in Applied Mechanics and Engineering* **193**(12–14), 1305–1321.

Washizu M, Nikaido Y, Kurosawa O, and Kabata H 2003 Stretching yeast chromosomes using electroosmotic flow. *Journal of Electrostatics* **57**(3–4), 395–405.

Wu M and Whitesides G 2001 Fabrication of arrays of two-dimensional micropatterns using microspheres as lenses for projection photolithography. *Applied Physics Letters* **78**(16), 2273–2275.

Yang HQ, Habchi SD, and Przekwas AJ 2004 General strong conservation formulation of Navier-Stokes equations in non-orthogonal curvilinear coordinates. *AIAA Journal* **32**(5), 936–941.

Yang P 2005 The chemistry and physics of semiconductor nanowires. *MRS Bulletin* **30**(2), 85–91.

Zhang L, Gerstenberger A, Wang X, and Liu. WK 2004 Immersed finite element method. *Computer Methods in Applied Mechanics and Engineering* **193**(21–22), 2051–2067.

Zhang LT, Wagner GJ, and Liu WK 2002 A parallel meshfree method with boundary enrichment for large-scale CFD. *Journal of Computational Physics* **176**, 483–506.

Zimmerman JA, Gao H, and Abraham FF 2000 Generalized stacking fault energies for embedded atom fcc metals. *Modelling and Simulation in Materials Science and Engineering* **8**, 103–115.

Zimmermann U, Friedrich U, Mussauer H, Gessner P, Hamel K, and Sukhoruhov V 2000 Electromanipulation of mammalian cells: Fundamentals and application. *IEEE Transactions on Plasma Science* **28**(1), 72–82.

Nanomechanical Properties
of the Elements

Nicola M. Pugno
Politecnico di Torino

23.1 Introduction

Recently, the interest on the mechanical properties of materials at the nanoscale level has been remarkably growing. Just in the last few decades, material scientists have been able to make direct measurements at such a critical size scale, three orders of magnitude smaller than the more known and accessible microscale. An example is given by the exceptional mechanical properties observed in nanotubes (Ross, 1991; Treacy et al., 1996; Yakobson et al., 1996, 1997; Yakobson and Smalley, 1997) since their re-discovery (see Pugno 2008) by Sumio Iijima (1991) and other scientists (Chopra et al., 1995; Weng-Sieh et al., 1995; Loiseau et al., 1996). The tremendous mechanical properties coupled with the exceptional electronic ones lead to consider nanoscale materials as optimal candidates for innovative materials (e.g., bio-inspired), for biomechanical applications (e.g., nanorobots), or electronics (e.g., nanoelectromechanical systems) (see the review by Qian et al. 2002).

Following the increasing interest in nanomechanics, this chapter intends to review the new results reported in the study of the mechanical properties of materials at the nanoscale (Pugno et al. 2006). We have shown that only two parameters are needed to describe the nanomechanics of materials: the cohesion energy and the atomic size. The proposed simple but general model gives, as a result, a preliminary periodic table for the nanomechanical properties of elements in which a periodicity of the mechanical properties emerges. As a simple example of application for

the reader, the theory is applied to the very recent results on the measurement of the elastic properties and intrinsic strength of monolayer graphene (Lee et al., 2008).

23.2 Nonlinear Normal Stress–Strain Law

Let us consider—for the sake of simplicity—a material arranged in the simple cubic lattice with lattice spacing a. Around the equilibrium position, linear elasticity is expected whereas for larger displacements, a nonlinearity takes place. As we will show, the nonlinearity in the constitutive equation has to be considered for developing a general model including the effect of the coefficient of thermal expansion.

Anisotropy is not taken into account in our model, the aim of which is to give simple estimations of the nanomechanical properties of materials. However, different types of lattices could be treated by considering as a first approximation an equivalent simple cubic lattice (e.g., by equating the atomic volumes).

The interatomic potential U between atoms depends substantially on their chemical bonding. The atoms do not come into contact owing to Pauli's and nuclei repulsions and reach their equilibrium positions. Even if different chemical bonds imply different interatomic potentials (e.g., Lennard-Jones), we can consider a general form (representing the interaction between one atom and all the others) according to the following series expansion:

$$U(x) = \sum_{n=0}^{N} c_n x^n, \qquad (23.1)$$

where

  x is the displacement around the equilibrium position
  c_n are unknown coefficients
  N is the order of the polynomial approximation

The force F between atoms will be

$$F(x) = \frac{dU(x)}{dx}. \qquad (23.2)$$

We can assume $c_0 = 0$, the energy being defined through its differential, and we must have $c_1 = 0$, the net force being vanishing at the equilibrium point $x = 0$. The classical harmonic approximation sets $N = 2$ and gives a linear relationship between force and displacement, the so-called Hooke's law. In addition, this symmetric form for the potential energy, related to small displacements, predicts vanishing thermal expansion, in contrast to the experimental evidence. Thus, at least an additional term has to be assumed so that a value of $N = 3$ is here considered. According to the simplified hypothesis of isotropic regular lattice, a volume $a \times a \times a$ per each atom is considered (simple cubic lattice). The two unknown constants c_2 and c_3 can be obtained by imposing the definitions of Young's modulus and coefficient of thermal expansion, i.e.,

$$\lim_{x \to 0} \frac{F(x)}{xa} = E, \qquad (23.3a)$$

$$\frac{\langle x \rangle}{aT} = \alpha, \qquad (23.3b)$$

$\langle x \rangle$ being the mean value of the displacement due to the thermal vibration at temperature T. The first condition implies $c_2 = Ea/2$. On the other hand, the second one, evaluating $\langle x \rangle$ by means of the Boltzmann's distribution, i.e.,

$$\langle x \rangle = \frac{\int_{-\infty}^{+\infty} x e^{-\beta U(x)} dx}{\int_{-\infty}^{+\infty} e^{-\beta U(x)} dx} \approx \frac{\int_{-\infty}^{+\infty} x e^{-\beta c_2 x^2} (1 - \beta c_3 x^3) dx}{\int_{-\infty}^{+\infty} e^{-\beta c_2 x^2} dx}, \quad (23.4)$$

where

  $\beta = (k_B T)^{-1}$
  k_B is Boltzmann's constant, gives $c_3 = -E^2 a^3 \alpha/(3k_B)$

In terms of local stress $\sigma = F/a^2$ and strain $\varepsilon = x/a$, the result is

$$\sigma(\varepsilon) \approx E\varepsilon - \frac{E^2 a^3 \alpha}{k_B} \varepsilon^2, \quad \text{for } \varepsilon \le \varepsilon_C. \qquad (23.5)$$

Even if the considered approach, based on the interatomic potential, is very simple (Kittel, 1966), the result of Equation 23.5 is original and describes a general form for the stress–strain relationship at nanoscale: for small displacements, it recovers the

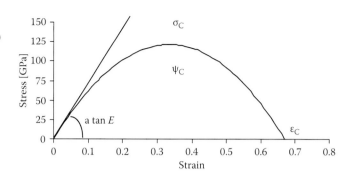

FIGURE 23.1 Nano-stress vs. nano-strain relationship (for carbon). Young's modulus E, critical stress σ_C and strain ε_C and energy density (or fragmentation energy) ψ_C. (From Pugno, N. et al., *Int. J. Solids Struct.*, 43, 5647, 2006. With permission.)

well-known Hooke's law, whereas for large displacements, a non-Hookean softening regime dominates. Note that the multibody nature of the atomic interaction is automatically taken into account in Equation 23.5 via Young's modulus. However, more sophisticated multibody potential could be easily considered (Zhang et al., 2002). The cutoff at ε_C is analogous to those classically introduced in the interatomic potentials. This is imposed by the fact that, after the critical (i.e., maximum) strain ε_C for which the stress vanishes, the approximation of Equation 23.5 loses its validity. Equation 23.5 is general in the sense that the atomic (or electronic or chemical bonding) structure of the solid is traduced in terms of global parameters.

According to Equation 23.5, the critical (i.e., maximum) stress and strain are predicted to be

$$\sigma_C \approx \frac{k_B}{4\alpha a^3}, \qquad (23.6a)$$

$$\varepsilon_C \approx \frac{k_B}{E\alpha a^3}. \qquad (23.6b)$$

Note that ε_C represents the maximum strain assuming a displacement-controlled process. Vice-versa, for a force-controlled process, the critical strain is one-half of the previous one and will be reached at the maximum stress (see Figure 23.1).

23.3 Cohesion Energies

The energy dissipated per unit volume a^3, that we could call fragmentation energy (Carpinteri and Pugno, 2002), can be calculated starting from the nonlinear relationship of Equation 23.5 as

$$\Psi_C \approx \int_0^{\varepsilon_C} \sigma(\varepsilon) d\varepsilon \approx \frac{k_B^2}{6Ea^6\alpha^2}. \qquad (23.7)$$

Thus, the energy dissipated per unit area a^2, the so-called fracture energy (Carpinteri, 1997), is

$$G_C \approx \Psi_C a \approx \frac{k_B^2}{6Ea^5\alpha^2}. \qquad (23.8)$$

The fracture energy plays a fundamental role in the prediction of the resistance against brittle crack propagation for a structural element (Pugno and Carpinteri, 2003).

On the other hand, the energy to pull-out an atom from the lattice, the so-called cohesion energy (Kittel, 1966), must be equal to

$$W_C \approx G_C 6a^2 \approx \frac{k_B^2}{Ea^3\alpha^2}, \tag{23.9}$$

where $6a^2$ is the new surface area created after the pull-out of one atom from the lattice, according to our simplified model.

23.4 Young's Modulus and Coefficient of Thermal Expansion Correlation

Young's modulus is an index of the stiffness of a material against mechanical loadings, whereas the coefficient of thermal expansion is an index of the compliance of a material under thermal variations. The aim of this section is to find, by virtue of a simplified argument, their correlation.

Let us consider the differential of the free energy u, given by (Kittel, 1966) $du = -p\,dV - S\,dT$, where the pressure p and the entropy S are defined as $p = -(\partial u/\partial V)_T$, $S = -(\partial u/\partial T)_V$, and V is the volume. At the thermodynamical equilibrium, $du = 0$, so that $p = -S(dT/dV) = (\partial u/\partial T)_V(dT/dV)$. For one atom, the contribution to the free energy due to the thermal vibrations can be classically considered equal to $\sim 3/2 k_B T$. Assuming constant pressure, the coefficient of thermal expansion being defined as $\alpha = (3V)^{-1}\,dV/dT$, the previous relationship would give $p = k_B/(2V\alpha)$. On the other hand, assuming constant temperature, differentiating and introducing the stress σ and strain ε under hydrostatic pressure p, for one atom of volume $V \approx a^3$, we have

$$dp = -d\sigma = -\frac{k_B}{2\alpha}\frac{dV}{V^2} = -\frac{k_B}{2\alpha a^3}3d\varepsilon. \tag{23.10}$$

Isotropic linear elastic constitutive laws (see Carpinteri, 1997) are expected for small strains, implying, under hydrostatic regime, the following relationship:

$$\frac{d\sigma}{d\varepsilon} = \frac{E}{1-2\nu}, \tag{23.11}$$

where ν is the Poisson's ratio of the material. The combination of the two previous equations provides the following correlation between Young's modulus and coefficient of thermal expansion as

$$E \approx \frac{3k_B(1-2\nu)}{2\alpha a^3}. \tag{23.12}$$

As expected, they are inversely related. This result coincides with Grüneisen's relation evaluated for one atom, in which the

thermal capacity is assumed to be equal to $3k_B$ (the classical value around room temperature) and Grüneisen's experimental constant is assumed to be equal to 3/2, close to its experimental value for many chemical elements (Slater, 1940). However, note that the thermal expansion coefficient and Young's modulus are functions of temperature (Jiang et al., 2004).

23.5 Nonlinear Shear Stress–Strain Law

In Equation 23.12 a new elastic constant appears, i.e., the Poisson's ratio ν that, with Young's modulus E, allows one to describe the elastic properties of isotropic materials. What is the expected value for ν at the nanoscale? To evaluate this coefficient, that thermodynamically must be comprised between -1 and $1/2$, we can alternatively evaluate the shear elastic modulus G.

For small displacements, the shear stress τ is connected with the displacement y (perpendicular to the previously introduced x coordinate) by Hooke's law, i.e.,

$$\tau = G\frac{y}{a}. \tag{23.13}$$

Due to the periodicity of the lattice with respect to shear, the relation shear stress vs. displacement can be assumed as (Frenkel, 1926):

$$\tau \approx \frac{G}{2\pi}\sin\left(2\pi\frac{y}{a}\right), \tag{23.14}$$

showing a non-Hookean region for large displacements. Obviously, for small displacements it becomes Hooke's law of Equation 23.13. As a consequence, the maximum value of the shear stress is

$$\tau_C \approx \frac{G}{2\pi}. \tag{23.15}$$

The ideal shear strength is predicted to be approximately only one order of magnitude smaller than the shear elastic modulus (Frenkel, 1926). Even if the correct coefficient of proportionality remains unknown, depending on the adopted model, this result is experimentally verified and represents an interesting tool to discriminate if the measurements on material strength are close or not to the ideal material strength. The simple approach reported in Section 23.2 can be considered the extension of this approach for the normal stress–strain relationship.

The shear strain γ is defined by $\tan\gamma = y/a$, so that the nonlinear shear stress vs. strain relationship at the nanoscale is predicted as

$$\tau \approx \frac{G}{2\pi}\sin(2\pi\tan\gamma). \tag{23.16}$$

The critical value of the shear strain γ will be reached, in a displacement-controlled process, when the shear stress vanishes, for

$$\gamma_C = a \tan 1/2 \approx 27°. \qquad (23.17)$$

On the other hand, if the process is force-controlled, then the critical value of the shear strain will be reached when the stress equals its critical value. The corresponding shear strain level is atan1/4 ≈ 14°. This parameter is very large if compared with the measured values at human size scale (of the order of the meter). In addition, it is material-independent. This means that, at nanoscale, the ductility—which is not a material property but a size-dependent parameter—seems to "universally" prevail over brittleness, independently of the considered material.

Considering the derived strength of Equation 23.6a, and replacing α by Equation 23.12, gives $\sigma_C \approx E/(6(1 − 2\nu))$. Thus, the model confirms that the ideal strength is expected as a significant fraction of Young's modulus: such result can be considered a proof of consistency for the simple model that we are proposing. Assuming the well-known tensional Tresca's or energetic von Mises' criteria (usually considered in plasticity but still applicable if a brittle collapse is assumed, see Carpinteri, 1997)

$$\sigma_C \approx \lambda_{T,vM} \tau_C, \qquad (23.18)$$

where $\lambda_T = 2$ or $\lambda_{vM} = \sqrt{3}$ for the two criteria respectively. By comparison between the normal and shear strengths, noting that $G = E/(2(1 + \nu))$, we deduce an estimation of the Poisson's ratio at the nanoscale as

$$\nu \approx \frac{3\lambda_{T,vM} − 2\pi}{6\lambda_{T,vM} + 2\pi} \approx 0. \qquad (23.19)$$

According to Tresca's criterion, the prediction is of $\nu_T = −0.015$, as well as for von Mises' criterion of $\nu_{vM} = −0.065$. Practically, both criteria suggest Poisson's ratio close to zero. A prediction of ν outside its thermodynamical domain [−1, 1/2] would show an inconsistency of our model. On the contrary and in spite of its simplicity, it appears to be *self-consistent*. Obviously, the prediction of Poisson's ratio close to zero has to be taken with caution, representing only an estimation of our simplified model. However, a surprisingly close to zero Poisson's ratio of ν ≈ 0.07 has been measured for nanotubes by means of Brillouin light scattering (Casari et al., 2001).

23.6 Nanomechanical Properties of the Elements

Eliminating the coefficient of thermal expansion in the derived nanomechanical properties, and assuming a Poisson's ratio equal to zero, gives the following estimation for the nanomechanical properties as a function of the cohesion energy W_C and of the atomic size a:

$$G_C \approx \frac{W_C}{6a^2}, \quad \Psi_C \approx \frac{G_C}{a}, \qquad (23.20a)$$

$$\nu \approx 0, \quad E \approx \frac{27}{2}\Psi_C, \qquad (23.20b)$$

$$\sigma_C \approx \frac{E}{6}, \quad \tau_C \approx \frac{E}{4\pi}, \qquad (23.20c)$$

$$\varepsilon_C \approx \frac{2}{3}, \quad \gamma_C \approx a \tan\frac{1}{2}. \qquad (23.20d)$$

Better estimations could be deduced relaxing the simplified hypothesis of ν ≈ 0. Note the large critical normal and shear strains suggest large ductility at the nanoscale, independently from the considered material. Such a result seems to be confirmed by the large ductility shown by classically brittle materials (if considered at the human size scale) like glass or carbon, e.g., glass whiskers or carbon nanotubes (Yakobson et al., 1997).

23.7 Comparison with the Literature

The most well-known prediction for the ideal strength of crystals was derived by Orowan (1948) in the following form:

$$\sigma_C^{(Orowan)} \approx \sqrt{\frac{EG_C}{2a}} \qquad (23.21)$$

A detailed comparison between the Orowan's prediction and a large number of experimental observations was reported by Macmillan (1983), demonstrating that, in spite of its simplicity, Equation 23.21 can reasonably predict the ideal strength of materials. Thus, if our approach agrees with such a prediction, we conclude that it has to be considered in agreement with the experimental observations on ideal strength of solids. Obviously, our approach as well as the Orowan's estimation have to be considered as reasonable estimations rather than as exact predictions. Rearranging Equations 23.20 we find

$$\sigma_C \approx \sqrt{\frac{27}{36}}\sqrt{\frac{EG_C}{2a}} \approx 0.9\sigma_C^{(Orowan)}. \qquad (23.22)$$

Thus, the two estimations are in reciprocal agreement.

Finally, we note that, applying quantized fracture mechanics (Pugno and Ruoff, 2004) considering the fracture quantum as coincident with the atomic size, the prediction of the ideal strength is

$$\sigma_C^{(QFM)} \approx \sqrt{\frac{4}{\pi}}\sqrt{\frac{EG_C}{2a}} \approx 1.1\sigma_C^{(Orowan)}, \qquad (23.23)$$

again in agreement with the previous model.

Now, let us focus the attention on carbon (graphitic form), for which $a \approx 1.54$ Å and $W_C \approx 7.36$ eV/atom (Kittel, 1966). Correspondingly, from Equations 23.20, we estimate the following:

1. ν ≈ 0; experiments on carbon nanotubes seem to confirm this prediction: a surprisingly close to zero value of ν ≈ 0.07 has been measured (Casari et al., 2001).
2. $E ≈ 725$ GPa; it is well-known that Young's modulus for ideal carbon nanotubes, or graphite, is expected to be of

the order of $E \approx 1\,\text{TPa}$ (Qian et al., 2002). Values close to $800\,\text{GPa}$ were measured by Yu et al. (2000).

3. $G_C \approx 8.3\,\text{N/m}$ (and $\Psi_C \approx 54\,\text{GPa}$); a reference value for carbon nanotubes is $G_C \approx 8.4\,\text{N/m}$ (Lambin et. al., 1998).

4. $\varepsilon_C \approx 67\%$ (and $\gamma_C \approx 27°$); based on molecular dynamics atomistic simulations (Yakobson et al., 1997), a value of $\varepsilon_C \approx 40\%$ is locally predicted in monoatomic chains due to high strain fracture of carbon nanotubes.

5. $\sigma_C \approx 121\,\text{GPa}$ (and $\tau_C \approx 58\,\text{GPa}$); strength of ideal carbon nanotubes, or graphite, is expected to be of the order of $\sigma_C \approx 100\,\text{GPa}$ (Qian et al., 2002). Values up to $64\,\text{GPa}$ were measured by Yu et al. (2000).

Eventually, the toughness at the nanoscale is predicted by definition as $K_{IC} = \sqrt{G_C E} \approx 2.45\,\text{MPa}\sqrt{m}$, and, from Equation 23.12, the coefficient of thermal expansion is $\alpha \approx 8 \times 10^{-6}\,\text{K}^{-1}$.

The nanoscale stress–strain relationship of carbon, Equation 23.5, is reported in Figure 23.1.

It is important to note that the predicted values substantially agree with the experimental results at the nanoscale and that they are completely different from the corresponding values at the macroscale. In fact, strong size effects on material properties are expected (Carpinteri and Pugno, 2004).

23.8 Nanomechanics Is the Borderline between Classical and Quantum Mechanics

The last considerations are on brittle crack propagation at the nanoscale. The velocity of the crack propagation, as well as of the elastic waves, is of the order of $\sqrt{E/\rho}$, ρ being the density of the considered material. According to special relativity, it must be smaller than light velocity c, so that the corresponding maximum value of Young's modulus results to be $E_{SR\,max} \approx \rho c^2$, and therefore around $10^{20}\,\text{Pa}$ for $\rho \approx 10^3\,\text{kg/m}^3$ (thus, much larger than the observed values).

A more interesting upper-bound for Young's modulus is imposed by quantum mechanics, considering fracture propagation at the nanoscale coupled with the Heisenberg's principle. In one of its forms, the principle states that $\Delta W\,\Delta t \geq \hbar$, where ΔW and Δt are, respectively, the energy and the time spent in the process, and $\hbar = h/2\pi$, where h is the Planck's constant. With reference to fracture propagation, evaluating the time as $\Delta t \approx a/\sqrt{E/\rho}$, and the energy as $\Delta W \approx G_C a^2$, we obtain

$$E_{QM\,max} \approx \frac{\rho G_C^2 a^6}{\hbar^2}, \tag{23.24}$$

which, for $\rho \approx 10^3\,\text{kg/m}^3$, $G_C \approx 10\,\text{N/m}$, and $a \approx 1\,\text{Å}$, is found to be around $10\,\text{TPa}$ and of the same order of magnitude (TPa) observed for example in carbon nanotubes. This very simple argument is intended to show that nanomechanics can be considered at the borderline between classical and quantum mechanics (and, obviously, outside the domain of special relativity). This is the reason why both classical and quantum mechanics have been successfully applied in nanomechanical treatments.

23.9 Periodic Table for the Nanomechanical Properties of Elements

According to Equations 23.22, and based on the values of the interatomic distances in the stable lattice reported (Table of Periodic Properties of The Elements by Sargent-Scientific Laboratory Equipment Catalog Number S18806) and of the cohesion energies (Kittel, 1966; referred to 0 K), the nanomechanical properties of the elements (for which both the interatomic distance and cohesion energy are known) as functions of their atomic number are depicted in Figures 23.2 through 23.8.

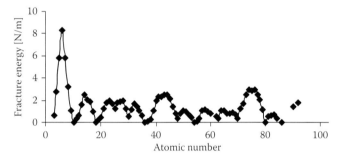

FIGURE 23.2 Nano-fracture energy G_C of elements vs. atomic number. (From Pugno, N. et al., *Int. J. Solids Struct.*, 43, 5647, 2006. With permission.)

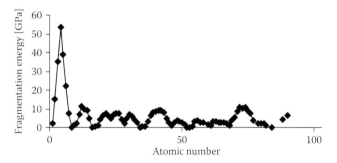

FIGURE 23.3 Nano-fragmentation energy ψ_C of elements vs. atomic number. (From Pugno, N. et al., *Int. J. Solids Struct.*, 43, 5647, 2006. With permission.)

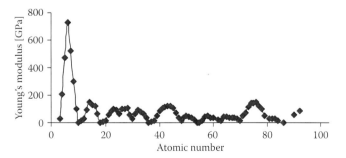

FIGURE 23.4 Nano-Young's modulus E of elements vs. atomic number. (From Pugno, N. et al., *Int. J. Solids Struct.*, 43, 5647, 2006. With permission.)

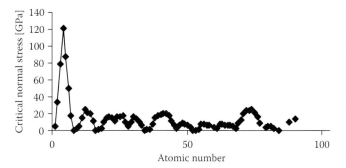

FIGURE 23.5 Nano-normal strength σ_C of elements vs. atomic number. (From Pugno, N. et al., *Int. J. Solids Struct.*, 43, 5647, 2006. With permission.)

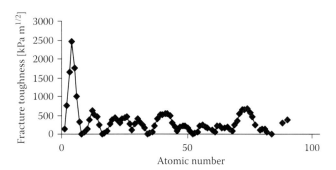

FIGURE 23.6 Nano-shear strength τ_C of elements vs. atomic number. (From Pugno, N. et al., *Int. J. Solids Struct.*, 43, 5647, 2006. With permission.)

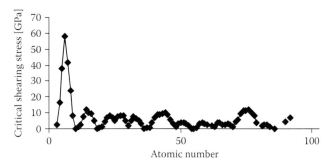

FIGURE 23.7 Nano-fracture toughness K_{IC} of elements vs. atomic number. (From Pugno, N. et al., *Int. J. Solids Struct.*, 43, 5647, 2006. With permission.)

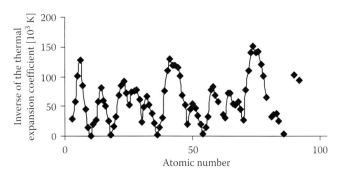

FIGURE 23.8 Inverse of the coefficient of thermal expansion α of elements vs. atomic number. (From Pugno, N. et al., *Int. J. Solids Struct.*, 43, 5647, 2006. With permission.)

A preliminary periodic table for the nanomechanical properties of elements is reported in Table 23.1. We have focused our attention on the main mechanical properties of materials, i.e., Young's modulus, the strength, and the fracture toughness. The periodicity in the nanomechanical properties appears in a very clear way. Carbon—the chemical foundation of life—appears to be the strongest element.

23.10 Example of Application: Nonlinear Elasticity and Strength of Graphene

An explosion of interest in the fabrication and characterization of graphene sheets is currently taking place (Novoselov et al., 2004; Berger et al., 2006; Stankovich et al., 2006; Dikin et al., 2007) due to their predicted fascinating mechanical (and electrical) properties and thanks to recently available new nanotechnological tools. In the paper by Lee et al. (2008), the authors measured, by atomic force nanoindentation, the elastic properties and the ideal strength of free-standing monolayer graphene sheets.

The material constitutive law was assumed to be isotropic nonlinear elastic, in the form of $\sigma = E\varepsilon + D\varepsilon^2$, where σ is the stress, ε is the strain, E is Young's modulus, and D is the third-order elastic modulus. They measured values of $E = 1.0 \pm 0.1$ TPa and $D = -2.0 \pm 0.4$ TPa. While a one terapascal Young's modulus was expected and consistent with the abundant data in the literature, the measurement of D reveals new insights regarding the nonlinear elastic behavior of graphene. In order to check the plausibility of their computed value, we may note that our model predicts $D = -E^2 a^3 \alpha / k_B$, where α is the (linear) expansion coefficient, k_B is the Boltzmann's constant, and a is here the C–C bond length (in graphene). Thus, we could indirectly estimate the thermal expansion coefficient of the tested monolayer graphene membranes, plausibly finding $\alpha \approx 1.0 \times 10^{-5}$ K^{-1}, which suggests the consistence of the reported third-order elastic modulus (or, vice versa, assuming $\alpha \approx 1.0 \times 10^{-5}$ K^{-1} one would deduce $D \approx -2.1$ TPa).

The nonlinear elasticity affected only the small region of the graphene around the point where the load was applied and not the cubic force–load displacement curve (as imposed by the large displacements involved in the stretching). Accordingly, under the atomic force tip, a stress/strain concentration took place, numerically computed by the authors (Lee et al., 2008), from which the material strength was derived. Even if the precision of an intrinsic strength measurement based on the stress–concentration concept of the continuum (which ignores energy release rate and quantization) is questionable, leading toward stress-intensifications to obvious strength overestimations, a value of $\sigma_{int} = 130 \pm 10$ GPa was deduced following a rigorous approach. The predicted huge strength suggests to have measured the ideal material intrinsic strength, expected to be of the order of one tenth of Young's modulus. Weibull moduli, one order of magnitude larger than those that we observed in carbon nanotubes, showed a more deterministic failure and seem

TABLE 23.1 Periodic Table for the Nanomechanical Properties of the Elements

The Periodic Table for the Nanomechanical Properties of the Elements

Key:
$$Z=6\,C^{121}=\sigma_C[\mathrm{GPa}]$$
$$E[\mathrm{GPa}]=725\,C$$
$$2450=K_{IC}\,[\mathrm{kPa}\sqrt{\mathrm{m}}]$$

Cell format: ^{Z}Sym$^{\sigma_C}$ with E / K_{IC} below.

1	2	3	4	5	6	7	8	9	10	11	12	13	14	15	16	17	18
^{1}H																	^{2}He
^{3}Li^{5} 31/137	^{4}Be^{34} 206/751											^{5}B^{79} 474/1653	^{6}C^{121} 725/2450	^{7}N^{87} 525/1749	^{8}O^{50} 301/989	^{9}F^{17} 101/331	^{10}Ne$^{0.07}$ 0.4/1.8
^{11}Na^{2} 14/67	^{12}Mg^{5} 31/138											^{13}Al^{15} 91/382	^{14}Si^{25} 153/619	^{15}P^{22} 130/514	^{16}S^{20} 121/471	^{17}Cl^{11} 65/249	^{18}Ar$^{0.1}$ 0.7/3.5
^{19}K^{1} 6/30	^{20}Ca^{3} 16/79	^{21}Sc^{10} 59/274	^{22}Ti^{14} 87/390	^{23}V^{17} 99/442	^{24}Cr^{15} 90/391	^{25}Mn^{11} 67/290	^{26}Fe^{16} 97/417	^{27}Co^{17} 101/435	^{28}Ni^{17} 105/449	^{29}Cu^{10} 60/271	^{30}Zn^{5} 27/119	^{31}Ga^{10} 63/270	^{32}Ge^{16} 96/408	^{33}As^{13} 80/336	^{34}Se^{10} 61/255	^{35}Br^{6} 37/152	^{36}Kr$^{0.1}$ 0.8/4.1
^{37}Rb^{1} 4/23	^{38}Sr^{2} 11/58	^{39}Y^{8} 46/227	^{40}Zr^{15} 88/411	^{41}Nb^{18} 108/502	^{42}Mo^{19} 114/518	^{43}Tc^{20} 123/550	^{44}Ru^{21} 124/551	^{45}Rh^{18} 108/479	^{46}Pd^{11} 69/310	^{47}Ag^{6} 37/177	^{48}Cd^{3} 16/75	^{49}In^{7} 39/181	^{50}Sn^{8} 50/223	^{51}Sb^{8} 46/202	^{52}Te^{6} 37/164	^{53}I^{4} 21/94	^{54}Xe$^{0.1}$ 0.8/4.4
^{55}Cs^{1} 3/19	^{56}Ba^{2} 11/58	^{57}La^{7} 42/210	^{72}Hf^{12} 72/351	^{73}Ta^{19} 117/544	^{74}W^{24} 145/658	^{75}Re^{24} 142/639	^{76}Os^{25} 149/668	^{77}Ir^{21} 124/556	^{78}Pt^{17} 100/453	^{79}Au^{8} 58/238	^{80}Hg	^{81}Tl^{4} 26/122	^{82}Pb^{5} 29/135	^{83}Bi^{5} 31/145	^{84}Po^{2} 12/63	^{85}At	^{86}Rn$^{0.2}$ 0.9/5.2
^{87}Fr	^{88}Ra	^{89}Ac															

Lanthanides:

^{58}Ce^{8} 48/236	^{59}Pr^{7} 39/193	^{60}Nd^{6} 34/168	^{61}Pm	^{62}Sm^{3} 21/103	^{63}Eu^{2} 13/67	^{64}Gd^{7} 45/218	^{65}Tb^{8} 46/223	^{66}Dy^{6} 35/168	^{67}Ho^{6} 34/166	^{68}Er^{6} 38/185	^{69}Tm^{5} 31/148	^{70}Yb^{2} 15/74	^{71}Lu^{9} 53/252
^{90}Th^{10} 59/294	^{91}Pa	^{92}U^{14} 85/340	^{93}Np	^{94}Np	^{95}Am	^{96}Cm	^{97}Bk	^{98}Cf	^{99}Es	^{100}Fm	^{101}Md	^{102}No	^{103}Lw

Based on:

$$G_C \approx \frac{W_C}{6a^2} \qquad \Psi_C \approx \frac{G_C}{a} \qquad \nu \approx 0 \qquad E \approx \frac{27}{2}\Psi_C \qquad \sigma_C \approx \frac{E}{6} \qquad \tau_C \approx \frac{E}{4\pi} \qquad \varepsilon_C \approx \frac{2}{3} \qquad \gamma_C \approx \operatorname{atan}\frac{1}{2}$$

ν = Poisson's ratio (material-independent), E = Young's modulus, σ_C = critical normal stress, ε_C = critical normal strain, τ_C = critical shear stress, γ_C = critical shear strain (material-independent), G_C = fracture energy (per unit area), Ψ_C = fragmentation energy (per unit volume), W_C = cohesion energy, a = interatomic distance, Z = atomic number – (material properties referred to 0 K).

Source: Pugno, N. et al., *Int. J. Solids Struct.*, 43, 5647–5657, 2006. With permission.

to confirm the observation of the ideal strength. Our model agrees with such a prediction for carbon.

The membranes were analyzed by scanning tunneling microscopy (STM), confirming the absence of defects over an area of hundreds of square nanometers, a size comparable to that of the highly stressed zone developed under the nanoindenter tip. Since the stress–concentration rapidly decays by increasing the distance from the point where the load is applied, moderate defects placed far from the contact zone could not prevail, as observed. However, defects are thermodynamically unavoidable. At the thermal equilibrium, the vacancy fraction, $f = n/N$, where n is the number of vacancies and N is the total number of atoms, is estimated to be $f \approx e^{-W_C/(k_B T)}$, where $W_C \approx 7$ eV is the energy to remove one carbon atom and T is the absolute temperature at which the carbon is assembled. Considering a maximum value of $T \approx 4000$ K leads to $f \approx 1.5 \times 10^{-9}$, thus to a maximum number $N \approx 6.5 \times 10^8$ of atoms in which less than one vacancy is expected (Pugno, 2007). This corresponds to a defect-free maximum surface area of the order of one square micrometer, thus again compatible with the observation. Even if atomistic defects are tediously and not easily observable by STM investigation and the defect density is usually imposed by the fabrication process rather than by the thermodynamic limit, our model agrees with the observations.

23.11 Model Limitations

The values that we have reported in the "preliminary periodic table for the nanomechanical properties of elements" (Table 23.1) are affected by different uncertainties.

In particular, we have simply assumed the Poisson's ratio equal to zero, as suggested by the considerations reported in Section 23.5; however, zero represents an intermediate value between its thermodynamic limits of −1 and 1/2. Furthermore, we have to note that the Poisson's ratio is an anisotropic parameter, depending on the crystallographic direction along which it is measured: thus, different values should be considered for each different crystallographic direction. For the sake of simplicity, to present a preliminary periodic table, we have chosen to ignore anisotropy. Note that the classical periodic table of the elements itself ignores anisotropy, reporting mean values, as for example for the electrical or thermal conductivity, as well as for the atomic radius. The "atomic radius" itself has a degree of uncertainty, which affects our predictions. In fact, different types of atomic radii can be defined through different models, e.g., Hartree–Fock approach, rigid spheres, and so on. A few of them are reported in the periodic table of elements in terms of atomic (or also covalent or ionic) radius or volume. We note that the atomic radius and volume are independent parameters, thus representing a first reason of uncertainty. In addition, as previously emphasized, we have neglected anisotropy. However, in our approach, different values for each crystallographic direction of the parameter a could allow us to roughly take into account anisotropy, as well as a "mean value" (e.g., the cubic root of the volume per atom) would allow us to consider not only simple cubic lattice. Finally, our model ignores plastic deformations.

To clarify the previous points, we can treat as a simple example the case of sodium, considering its lattice parameter (BCC, 0.42906 nm) and applying the rigid sphere model (two atoms per cell in BCC). Young's modulus of sodium usually reported in the literature is close to 10 GPa, against our preliminary prediction of 14 GPa. According to the periodic table of the elements, the atomic radius of the sodium (that we have used) is 0.190 nm, whereas the atomic radius that we calculate, remembering that the closed packed direction is [111] (diagonal), is 0.186 nm. The same value is deduced starting from the volume of the unit cell (cube of the lattice parameter for BCC) and taking into account the packing factor for BCC (0.68). Thus, considering 0.186 nm instead of 0.190 nm would yield $E \approx 14 \cdot 0.190^3/0.186^3 \approx 15$ GPa. On the other hand, removing the approximation of a vanishing Poisson's ratio, $E/(1 - 2\nu) \approx 15$ GPa, so that to capture the correct value of $E \approx 10$ GPa, a value of $\nu \approx 0.17$ is deduced.

23.12 Conclusions

We conclude that our model must be considered a basic treatment; however, in spite of its limits, the approach reported in this chapter could be of interest due to its simplicity and generality for estimating the nanomechanical properties of the elements.

References

Berger, C., Song, Z., Li, X., Wu, X., Brown, N., Naud, C., Mayou, D. et al., 2006, Electronic confinement and coherence in patterned epitaxial graphene, *Science* 312, 1191–1196.

Carpinteri, A., 1997, *Structural Mechanics—A Unified Approach*, E & FN Spon, New York.

Carpinteri, A. and Pugno, N., 2002, One-, two- and three-dimensional universal laws for fragmentation due to impact and explosion, *J. Appl. Mech.* 69, 854–856.

Carpinteri, A. and Pugno, N., 2004, Scale-effects on average and standard deviation of the mechanical properties of condensed matter: An energy based unified approach, *Int. J. Fract.* 128, 253–261.

Casari, C. S., Li Bassi, A., and Bottani, C. E., 2001, Acoustic phonon propagation and elastic properties of cluster-assembled carbon films investigated by Brillouin light scattering, *Phys. Rev. B* 64, 85417-1–85417-5.

Chopra, N. G., Luyken, R. J., Cherrey, K., Crespi, V. H., Cohen, M. L., Louie, S. G., and Zettl, A., 1995, Boron nitride nanotubes, *Science* 269, 966–968.

Dikin, D. A., Stankovich, S., Zimney, E. J., Piner, R. D., Dommett, G., Evmenenko, S., Nguyen, T. et al., 2007, Preparation and characterization of graphene oxide paper, *Nature* 448, 457–460.

Frenkel, J. Z., 1926, Zur theorie der elastizit. atsgrenze und der festigkeit krystallinischer körper, *Z. für Phys.* 37, 572–609.

Iijima, S., 1991, Helical microtubules of graphitic carbon, *Nature* 354, 56–58.

Jiang, H., Liu, B., Huang, Y., and Hwang, K. C., 2004, Thermal expansion of single wall carbon nanotubes, *J. Eng. Mater. Technol.* 126, 265–270.

Kittel, C., 1966, *Introduction to Solid State Physics,* John Wiley & Sons, New York.

Lambin, P. H., Meunier, V., and Biró, L. P., 1998, Elastic deformation of a carbon nanotube adsorbed on a stepped surface, *Carbon* 36, 701–704.

Lee, C., Wei, X., Kysar, J. W., and Hone, J., 2008, Measurement of the elastic properties and intrinsic strength of monolayer graphene, *Science* 321, 385–388.

Loiseau, A., Willaime, F., Demoncy, H., Hug, G., and Pascard, H., 1996, Boron nitride nanotubes with reduced numbers of layers synthesized by arc discharge, *Phys. Rev. Lett.* 76, 4737–4740.

Macmillan, N. H., 1983, The ideal strength of solids, in: R. Latanision and J.R. Pickens (eds.), *Atomistics of Fracture,* Plenum Press, New York, pp. 95–164.

Novoselov, K. S., Geim, A. K., Morozov, S. V., Jiang, D., Zhang, Y., Dubonos, S. V., Grigorieva, I. V. et al., 2004, Electric field effect in atomically thin carbon films, *Science* 306, 666–669.

Orowan, E., 1948, Fracture and strength of solids, *Rep. Prog. Phys.* XII, 185–232.

Pugno, N., 2007, The role of defects in the design of the space elevator cable: From nanotube to megatube, *Acta Mater.* 55, 5269–5279.

Pugno, N., 2008, A journey on the nanotube, in the top ten advances in materials, *Mater. Today* 11, 40–45.

Pugno, N. and Carpinteri, A., 2003, Tubular adhesive joints under axial load, *J. Appl. Mech.* 70, 832–839.

Pugno, N. and Ruoff, R., 2004, Quantized fracture mechanics, *Philos. Mag.* 84, 2829–2845.

Pugno, N., Marino, F., and Carpinteri, A., 2006, Towards a periodic table for the nanomechanical properties of elements, *Int. J. Solids Struct.* 43, 5647–5657.

Qian, D., Wagner, G. J., Liu, W. K., Yu, M.-F., and Ruoff, R. S., 2002, Mechanics of carbon nanotubes, *Appl. Mech. Rev.* 55, 495–532.

Ross, P. E., 1991, Buckytubes: Fullerenes may form the finest, toughest fibers yet, *Sci. Am.* 265, 24–25.

Slater, J. C., 1940, Note on Grüneisen's constant for the incompressible metals, *Phys. Rev.* 57, 744–746.

Stankovich, S., Dikin, D. A., Dommett, G. H. B., Kohlhaas, K. M., Zimney, E. J., Stach, E. A., Piner, R. D. et al., 2006, Graphene-based composite materials, *Nature* 442, 282–285.

Treacy, M. M., Ebbesen, T. W., and Gibson, J. M., 1996, Exceptionally high Young's modulus observed for individual carbon nanotubes, *Nature* 381, 678–680.

Weng-Sieh, Z., Cherrey, K., Chopra, N. G., Blase, X., Yoshiyuki Miyamoto., Angel Rubio., Cohen, M. L. et al., 1995, Synthesis of $B_xC_yN_z$ nanotubules, *Phys. Rev. B* 51, 11229–11232.

Yakobson, B. I. and Smalley, R. E., 1997, Fullerene nanotubes: C-1000000 and beyond, *Am. Sci.* 85, 324–337.

Yakobson, B. I., Brabec, C. J., and Bernholc, J., 1996, Nanomechanics of carbon nanotubes: Instabilities beyond linear response, *Phys. Rev. Lett.* 76, 2511–2514.

Yakobson, B. I., Campbell, M. P., Brabec, C. J., and Bernholc, J., 1997, High strain rate fracture and C-chain unraveling in carbon nanotubes, *Comput. Mater. Sci.* 8, 341–348.

Yu, M.-F., Lourie, O., Dyer, M. J., Moloni, K., Kelly, T. F., and Ruoff, R. S., 2000, Strength and breaking mechanism of multiwalled carbon nanotubes under tensile load, *Science* 287, 637–640.

Zhang, P., Huang, Y., Geubelle, P. H., Klein, P. A., and Hwang K. C., 2002, The elastic modulus of single-wall carbon nanotubes: A continuum analysis incorporating interatomic potentials, *Int. J. Solids Struct.* 39, 3893–3906.

24

Mechanical Models for Nanomaterials

Igor A. Guz
University of Aberdeen

Jeremiah J. Rushchitsky
Timoshenko Institute of Mechanics

Alexander N. Guz
Timoshenko Institute of Mechanics

24.1 Introduction

Nanotechnologies and nanomaterials are arguably the most actively and extensively developing research areas at the beginning of this century. The number of publications in scientific periodicals and conference proceedings, fully or partially devoted to nanotechnologies and nanomaterials, is rapidly increasing. However, the development of nanomechanical models and their application to investigation of mechanical behavior of nanomaterials in a systematic way is not happening yet. Present studies of the mechanical behavior of nanoparticles, nanoformations, and nanomaterials are still in their infancy. Only external manifestations of mechanical phenomena are detected, but their mechanisms have not been studied yet.

An attempt to formulate basic problems of nanomechanics and to suggest possible ways how to solve them using the knowledge accumulated within the solid mechanics and, more generally, continuum mechanics was undertaken by Guz et al. (2007a,b). These papers, which also contain extensive reviews of publications on the topic, revisited some of the well-known models in the mechanics of structurally heterogeneous media for the purpose of analyzing their suitability to describe properties of nanomaterials (nanoparticles and nanocomposites) and their mechanical behavior. A number of the macro-, meso- and micromechanical models were then reviewed and new nanomechanical models were suggested based on the knowledge accumulated within the realm of micromechanics of composite materials. New directions of research in nanomechanics of materials were also pointed out.

Following Guz et al. (2007b), this chapter gives a brief overview of the usage of some concepts of continuum mechanics in nanomechanics. As an example of application of the developed approaches, this chapter contains the results of prediction of the effective properties of particular nanocomposites.

24.2 Length Scales in Mechanics of Materials

24.2.1 General Classification of the Modeling Levels

In mechanics of materials, the internal structure of a material is always taken into consideration in one form or another. In the majority of studies, the internal structure is used only to characterize or identify materials (e.g., by means of tracking changes in the internal structure under loading or during manufacture). In fewer studies, the information about the internal structure is incorporated into the model of a material and is used to formulate constitutive equations. The mechanics of materials that accounts, both qualitatively and quantitatively, for the internal structure of materials while developing models of materials and solving various problems is often called "structural mechanics of materials." Its subject of research is a wide range of modern materials, including reinforced concrete, whose internal structure is defined by the reinforcement; metals, alloys, and ceramics, whose internal structure is defined by the presence of grains and other structural elements; microcomposites, whose internal structure is determined by the presence of particles, fibers, and layers; and nanomaterials (or nanocomposites)—a new class of materials, whose internal structure is determined by the presence of nanoparticles. If understood in this way, structural mechanics of materials incorporates macromechanics, mesomechanics, micromechanics, and nanomechanics, which are well-defined and widely used terms. The only necessary common requirement for the four research areas is to take the internal structure of materials into consideration when establishing the mechanical models and solving the corresponding problems.

To characterize quantitatively the internal structure of materials, it is expedient to introduce a geometrical parameter, h. For reinforced concrete, h is the average minimum cross-sectional diameter of metal reinforcement. For metals, alloys, and ceramics, h is the average minimum size of cells, grains, and other structural heterogeneities. For composites with polymer or metal matrix, h is the average minimum diameter of particles in granular materials, the average minimum cross-sectional diameters of fibers in fibrous materials, and the average minimum thickness of layers in laminated materials. For nanomaterials (nanocomposites), h is the average minimum diameter of nanoparticles.

The average minimum cross-sectional diameter of metal reinforcement in reinforced concrete may be on the order of centimeters (10^{-2} m), whereas the average minimum diameter of nanoparticles in nanomaterials may reach 0.4 nm. Thus, the structural mechanics of materials can be said to investigate materials with parameter h that varies within the following limits:

$$0.4 \times 10^{-9} \text{ m} \leq h \leq 1.0 \times 10^{-2} \text{ m} \qquad (24.1)$$

With such a wide range of h, it can be subdivided into several levels for convenience of further analysis, though such a division is rather conditional. Sih and Lui (2001) suggested considering three levels: macro (10^{-4} to 10^{-5} m), meso (10^{-5} to 10^{-7} m), and micro (10^{-7} to 10^{-8} m). In view of the ever-increasing significance of nanomaterials, the introduction of the fourth level seems pertinent. The classification of the modeling levels proposed by Guz et al. (2007b) is demonstrated in Figure 24.1. Since the atomic level, as defined by the interatomic distance in a crystal lattice, has an order of one to several Angstroms (10^{-10} m), the nanolevel in Figure 24.1 is conditionally restricted to 10^{-9} m. In the authors' opinion, the levels do not have clear borders and are overlapping.

Very often, the ideas developed for studying the materials at one level can be successfully modified or extended for application at another level, which will be demonstrated later in this chapter following the general methodology proposed by Guz et al. (2007b).

24.2.2 Examples of Research Areas with Specific Length Scales

24.2.2.1 Example 1

Classical Griffith–Irwin fracture mechanics does not account for the internal structure of the material at the crack tip. With the exception of the Neuber model, which isolates a grain at the crack tip to average local stresses within it, the classical fracture mechanics was developed for homogeneous solids (see, for example, Cherepanov 1979). This also includes the case when a crack is located at, or is extending onto the interface between dissimilar homogeneous solids. Sih and Lui (2001) concluded that the next stage in the development of fracture mechanics will account for the internal structure at the crack tip. Fracture mechanics that considers the internal structure at the crack tip at any structural level was termed fracture mesomechanics. Elements of the internal structure at various levels interact at the crack tip. If we exclude from consideration dimensions commensurable with the interatomic distance in a crystal lattice (approximately 10^{-10} m), it follows from Sih and Lui (2001) that fracture mesomechanics studies cracked materials with parameter h, varying within the following limits:

$$10^{-8} \text{ m} \leq h \leq 10^{-5} \text{ m} \qquad (24.2)$$

This interval includes several levels defined by Figure 24.1.

24.2.2.2 Example 2

The main concepts, approaches, and results of the physical mesomechanics of metals, alloys, and ceramics are given by Panin et al. (1995). To analyze plastic deformations, methods of solid mechanics and the continuum theory of dislocations were applied, focusing on the analysis of mechanisms and their interaction at various structural levels. In order to gain more insight, the concept of mesoscopic structural elements was introduced.

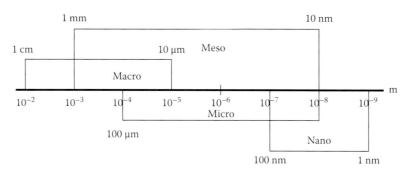

FIGURE 24.1 Length scales in structural mechanics of materials.

(According to Sih and Lui (2001), mesoscopic structural elements are cells, grains, strip structures, precipitated phases, structural heterogeneities, or fragments of structures.) Generally, physical mesomechanics (Panin et al. 1995) addresses a somewhat wider or detailed group of structures than fracture mesomechanics (Sih and Lui 2001). To determine the range of the parameter h (Equation 24.1), which characterizes the internal structure in physical mesomechanics, let us analyze the data on the size of grains in various metals and alloys. The grains can be as large as 10^{-3} m; in ultradispersed structures, the grain size is between 0.05×10^{-6} and 1×10^{-6} m. The grain size in other metals and alloys is somewhere between 10^{-3} and 5×10^{-6} m. Under a shock-wave load, some grains of special ceramics based on ZrO_2 powder may become 3 nm in size (Panin et al. 1995). Thus, the interval of variation for h in the physical mesomechanics of metals, alloys, and ceramics is

$$3 \times 10^{-9}\,\text{m} \le h \le 10^{-3}\,\text{m} \qquad (24.3)$$

The interval defined by Equation 24.3 encompasses several levels defined by Figure 24.1.

24.2.2.3 Example 3

Micromechanics of polymer and metal matrix composites (PMCs and MMCs) is arguably one of the most developed branches of the structural mechanics of materials. PMCs and MMCs can be classified according by their internal structure as follows: dispersion-strengthened composites, large-particle composites, and fiber-reinforced composites (Katz and Milewski 1978; Christensen 1979; Kelly and Zweben 2000). The dispersion-strengthened composites are composites whose filler is dispersion particles ranging in size from 0.01 to 0.1 μm, with the volume fraction ranging from 1% to 15%. The large-particle composites are composites whose filler is particles with the diameter bigger than 1 μm and the volume fraction of more than 25%. The fiber-reinforced composites are composites whose filler is fibers (from 0.1 μm to hundreds of microns in size) with a volume fraction of almost 70%. Therefore, for PMCs and MMCs the parameter h varies within the following limits:

$$10^{-8}\,\text{m} \le h \le 10^{-4}\,\text{m} \qquad (24.4)$$

Again, as in the previous two examples, his interval corresponds to several levels defined by Figure 24.1.

24.2.2.4 Example 4

Since various authors define the nanolevel somewhat differently, we will use the definition used by the National Scientific Foundation (NSF). According to this definition, the nanolevel is the following interval of variation of the structural parameter h:

$$10^{-9}\,\text{m} \le h \le 10^{-7}\,\text{m} \qquad (24.5)$$

Nanoparticles may reach from 200 to 300 nm in size or constitute tenths of a nanometer. Therefore, nanomechanics, as

a branch of the structural mechanics of materials, describes mechanical processes in materials with the internal structure defined by Equation 24.5. It should be noted that, as in the other branches of the structural mechanics of materials, nanomaterials are a group of specific materials by no means solely defined by the interval (Equation 24.5). We may consider nanomaterials as materials grouped according to the similarity between their manufacturing processes, physical and mechanical properties, and structural peculiarities.

24.2.3 "Macro," "Meso," "Micro," and "Nano": A Very Conditional Classification

The prefixes *macro, meso, micro,* and *nano* in the words "macro-mechanics," "mesomechanics," "micromechanics," and "nano-mechanics" do not at all imply that these research areas study materials, whose internal structure is necessarily defined by Figure 24.1. As a rule, an isolated research area covers several structural levels. For instance, the physical mesomechanics of metals and alloys (Example 2) covers completely the meso- and microlevels and partially the macro- and nanolevels. The micromechanics of PMCs and MMCs (Example 3) covers completely the meso- and microlevels and partially the macro- and nanolevels.

A group of materials studied by a specific branch of the structural mechanics of materials is by no means defined solely by the interval of variation for parameter h. For example, the physical mesomechanics and the micromechanics of composites have almost equal intervals for h. A group of materials studied by a specific branch of the structural mechanics of materials is also defined by the similarity in manufacturing processes, and in physical and mechanical properties of these materials. For example, the restructuring and hardening techniques for materials studied in physical mesomechanics are quite similar (Panin et al. 1995). One more example is the possibility to produce the materials addressed in the micromechanics of composites in such a way that they would be suitable for certain external loads. Composite materials are also united by the possibility of manufacturing both a material and a structural element in a single concurrent technological process.

In some cases, subdivision of the structural mechanics of materials into macro-, meso-, micro-, and nanomechanics has only a methodological meaning. To solve specific problems, the branches may employ the same formulations, and the same methods of solving the problems, though the phenomena that are being analyzed can be at different structural levels.

To illustrate this point, let us consider a composite laminate made of unidirectional plies, with fibers in neighboring layers located at different angles to each other. The reduced properties of a single unidirectional ply can be determined within the micromechanics of composite materials using various homogenization techniques. The effective properties of a multidirectional laminate can be determined using homogenization techniques within the macromechanics of composites. Here, the same formulations and the same methods of solution are applied at various structural levels (macro and micro). Another example: both the physical mesomechanics and the micromechanics of composites, as defined by Equations 24.3 and 24.4, partially cover the nanolevel.

Thus, some problems of nanomechanics associated with the upper part of the nanolevel may have been already studied within the framework of physical mesomechanics and the micromechanics of composites. The limits of applicability of the micromechanics of composites may actually be even wider than those defined by Equation 24.4. For instance, the one and only limitation for applying the micromechanics of composites developed in Guz et al. (1992) and Guz (1993–2003) is the possibility of using the solid mechanics formulations in order to describe the stress–strain relations in each reinforcing element. Whether the micromechanics of composites can be extended to some classes of nanomechanical problems will be discussed in Section 24.2.4.

24.2.4 Continuum Solid Mechanics: How Small Can It Go?

24.2.4.1 General Considerations

First it should be noted that the nanolevel interval (Figure 24.1) is fairly wide. Its upper part coincides with the lower part of the microlevel interval, while in the lower part of the nanolevel, the parameter h may be as small as several interatomic distances. Most nanomechanical problems for nanomaterials with the parameter h in the upper part of the range (Equation 24.5) can be formulated and solved using approaches and methods of the micromechanics of PMCs and MMCs.

Since all nanoparticles consist of discrete elements (molecules and atoms), the question naturally arises about the limits of applicability of the continuum solid mechanics to the description of mechanical processes in nanoparticles. Answering this question will automatically answer the question about the applicability limits for the micromechanics of composites. The rigorous and complete mathematical solution of this problem is quite difficult; therefore, we will discuss here an approximate estimate.

Along with the parameter characterizing the average minimum size of particles in the internal structure, we introduce two geometrical parameters, l and L_V. The parameter l is the average center-to-center distance between the particles in the internal structure, and the parameter L_V is a characteristic linear dimension of the minimum volume within which the material may be modeled by a homogeneous continuum. Many experts believe that L_V must be larger than l by more than one order of magnitude. Hence, to perform an approximate qualitative analysis, we assume that $L_V > 10l$. This geometrical condition does not fully define the applicability of continuum solid mechanics because the nature of studied mechanical processes is not taken into account. In this connection, we introduce a geometrical parameter, L_p, to characterize those mechanical processes. The parameter L_p can be the minimum distance at which the stress and strain fields (in the case of statics), or the wavelength (in the case of dynamics), or the buckling mode wavelength (in the case

of stability theory) change substantially. This restriction on the studied mechanical processes can be represented in the form $L_p > L_V$. Thus, the continuum solid mechanics can be applied to description of a mechanical process in a material with the internal structure, if the geometrical parameters that characterize the process, the minimum volume of the material, and its internal structure satisfy the following inequality:

$$L_p > L_V > 10l \qquad (24.6)$$

24.2.4.2 Examples

Now let us consider examples of applying the above condition to a number of materials.

Example A. Consider nanomaterials with crystalline structure. In this case, the parameter l in Equation 24.6 is the average distance between the neighboring atomic planes. Since the average interatomic distance in a crystal lattice comprises several Angstroms, it follows from Equation (24.6) that L_V must be of magnitude of several nanometers. Therefore, to analyze nanomaterials with crystalline structure, we can apply the tools of continuum solid mechanics (i.e., the homogeneous continuum model) within the minimum volume L_V of the material if the mechanical processes being studied satisfy the condition given by Equation 24.6 and change substantially at distances much larger than several nanometers.

Example B. Consider nanomaterials whose atoms form granules. Wilson et al. (2002) and Bhushan (2007) reported such nanomaterials with granules up to 100 nm in size. From this value and Equation (24.6), it follows that the mechanical processes in granules to be studied within the framework of the homogeneous continuum model must satisfy the condition $L_p > 100$ nm.

24.2.5 Main Research Areas in Nanomechanics

Bearing in mind the discussion of some aspects of nanomechanics in the Sections 24.2–24.4 let us point out three main areas of research in the mechanics of nanomaterials, as it is envisaged by the authors.

The first area comprises the studies of materials and processes, for which the applicability conditions, Equation 24.6, are satisfied. It covers a wide range of static, dynamic, and stability problems for nanoparticles and nanocomposites and those fracture problems that can be solved within the framework of continuum solid mechanics. One of the well-known approaches within this area is based on the principle of homogenization (see, for example, Guz 1993–2003; Nemat-Nasser and Hori 1999).

The second research area comprises the studies of materials and processes, for which the applicability conditions (Equation 24.6) are not generally satisfied, but an approximate approach (which may be named the principle of continualization) can be applied. By continualization, we mean constructing approximate continuum theories that describe changes of at least integral characteristics of discrete structures. The principle of continualization is widely used in various branches of physics. A classical example is the continuum theory of dislocation presented, for instance, by Cottrell (1964) and Eshelby (1956). Dislocations are discrete defects in a discrete system, a crystal lattice. Nevertheless, a continuum theory that describes the laws of propagation of dislocations was developed for such a discrete system. Certainly, the mechanics of nanoparticles requires applying the principle of continualization; after that, problems for nanocomposites can be treated within the framework of continuum representations. Due to the application of the principle of continualization, new formulations are expected to appear based on specific models of continuum solid mechanics.

The third research area comprises the studies of particularly discrete systems, for which the applicability conditions (Equation 24.6) are not satisfied. It seems obvious that the third research area includes investigations of the mechanics of nanoparticles. Here, it is difficult to obtain specific results for a wide range of structures and mechanical phenomena.

Subdivision into three research areas proposed here, although conditional, may be useful in analyzing approaches to the study of phenomena and in interpreting results. Three possible research areas mainly with reference to the mechanics of nanoparticles were discussed above. Let us now point out some issues relevant to the mechanics of nanocomposites.

The term "nanocomposite," when used in nanomechanics, may imply two different types of nanomaterials. The first type is a material consisting of a matrix reinforced with nanoparticles. These composites can be studied within the first and second research area using approaches and methods of the micromechanics of composites—the latter is comprehensively developed (see, for example, Christensen 1979; Guz et al. 1992; Guz 1993–2003; Kelly and Zweben 2000). As each nanoparticle has a rather complex internal structure, it can be considered as a nanocomposite itself with its own internal structure. Such a composite can be studied using approaches from all the three research areas. To analyze mechanical phenomena in nanocomposites at the final stage of studies within the first and second research areas, we can use continuum representations of the mechanics of materials with internal structure. With such an approach, it is possible to analyze a wide range of static, dynamic, and stability problems for nanocomposites and those fracture mechanics problems that can be solved within the framework of solid mechanics.

The detailed examples of application of micro- and macromechanical models to nanocomposites were also given by Guz et al. (2005, 2007a,b, 2008a,b), Guz and Rushchitsky (2004, 2007), Zhuk and Guz (2006, 2007).

24.3 New Type of Reinforcement— "The Bristled Nanocentipedes"

24.3.1 New Incarnation of an Old Idea

As mentioned above, very often, the ideas developed for studying the materials at one level can be successfully modified or extended for application at another level. As an example, the

procedure of determining the effective properties of a new type of nanocomposites is considered below.

Several decades ago, whiskers were looked at as a new class of materials: very promising and very expensive—much the same as we look at the nanofibers now. The whiskers are strong because they are essentially perfect crystals and their extremely small diameters allow little room for the defects which weaken larger crystals. More than 100 materials, including metals, oxides, carbides, halides, nitrides, graphite, and organic compounds can be prepared as whiskers (Katz and Milewski 1978). The whiskerization was suggested as a special form of supplementary reinforcement particular to carbon fibers. Since carbon fibers were frequently utilized with resin matrices, such composite materials exhibited low shear strength due to poor bonding. The technique of whiskerization consists in the addition of whiskers to the composite by growing them directly on the surfaces of the carbon fibers. Whiskerizing introduces an integral bond between whiskers and the carbon fibers. The three to five times increase of the interfacial shear bond for different combinations of carbon fibers and epoxy matrices was reported by Katz and Milewski (1978).

After almost four decades since the introduction of whiskerized microfibers, they found a new incarnation in the emerging world of nanomaterials. The recent paper by Wang et al. (2004) describes the CdTe nanowires coated with SiO_2 nanowires. The resulting composition is wittily named by the authors as "the bristled nanocentipedes." Two types of CdTe nanowires are reported: stabilized with MSA (mercatptosuccinic acid, $HO_2CCH_2CH(SH)CO_2H$) and TGA (thioglycolic acid, $HSCH_2CO_2H$). The former has a diameter of 4.5–6.5 nm, and the latter 3 nm. Figure 24.2 (Wang et al. 2004) shows the MSA stabilized "wire-coating" with numerous nearly parallel bristles growing perpendicular to the surface called also the brush-like composition. Evidently, the structure "CdTe nanowire–SiO_2 nanowires" consists of three components: CdTe wire forms the solid core, which is jointed continuously with coating from SiO_2 nanowires in the form of some solid shell, and then the shell is coated periodically by

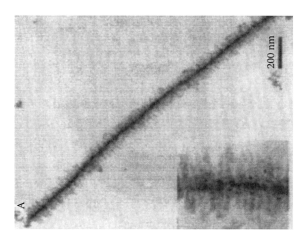

FIGURE 24.2 CdTe nanowires coated with SiO_2 nanowires: "the bristled nanocentipede." (From Wang, Y. et al., *Nano Lett.*, 4(2), 225, 2004. With permission.)

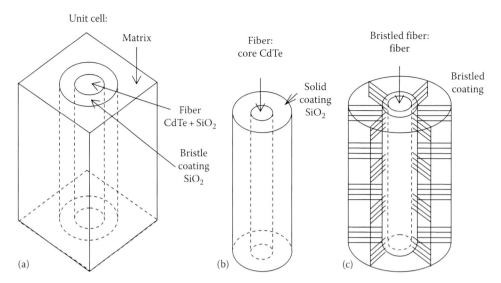

FIGURE 24.3 General schematic of the model.

bristled and smooth zones. Particularly, for the case shown in Figure 24.2, we can loosely assume that the core radius is 3 nm, the shell thickness is 15 nm, the bristled/smooth zone length is 6 nm, and the length of bristles is 32 nm.

There are striking similarities between the well-known process of whiskerization of microfibers and the recent idea of bristled nanowires. Both whiskerized microfibers and bristled nanowires exhibit similar geometrical structure and can be used for the same ultimate purpose of fabricating composite materials with improved fiber–matrix adhesion and hence the increased shear strength. If we consider the structure "CdTe nanowire–SiO_2 nanowires" placed in the matrix, the total number of components in the composition would be as high as four–the matrix being the fourth component in addition to the three components mentioned above. Therefore, in order to study the effective properties of the entire composite and the effect of reinforcement with whiskerized microfibers or bristled nanowires on the overall performance of the material, we need a four-component model. The necessity of this model was emphasized by Guz et al. (2008b), where the general methodology was outlined.

Further in our discussion, we consider the problem of modeling properties of fibrous composite materials, which are additionally reinforced either by whiskerizing the microfibers or by bristlizing the nanowires, and give the details of the method for deriving the explicit formulas for effective elastic constants of such materials.

24.3.2 Structural Model

The proposed structural model for composites reinforced by whiskerized microfibers or bristled nanowires is based on the assumption that mircofibers or nanowires (henceforth, called simply "fibers") are arranged in the matrix periodically as a quadratic or hexagonal lattice. Then the representative volume element (unit cell) consists of the matrix and a coated fiber (Figure 24.3a). At that, the coating itself has several

subcomponents (Figure 24.3b and c), which is a new feature of the model. The coated fiber is assumed to be consisting of three different parts: a solid core, a solid coating (homogeneous shell), and a "bristled coating" (composite shell). The fourth component in the model is the matrix. A segment of the model representing the core fiber with solid coating and bristles attached to the solid coating is shown in Figure 24.4. Subsequently, the following notations are used to distinguish the four components of composite: (1) for the fiber core, (2) for the fiber solid coating, (3) for the fiber bristled coating, and (4) for the matrix. For instance, $c^{(1)}$, $c^{(2)}$, $c^{(3)}$, and $c^{(4)}$ are the volume fractions of the fiber core, the fiber solid coating, the fiber bristled coating, and the matrix, respectively. The radii of the fiber core, the fiber solid coating, and the fiber bristled coating are, respectively, $r_{(1)}$, $r_{(2)}$, and $r_{(3)}$.

Three out of four components are homogeneous materials, e.g., the epoxy matrix, SiO_2 solid coating and CdTe fiber core, with certain physical properties (Young's modulus, shear modulus, Poisson's ratio, density, etc.). However, the bristled coating is itself a composite consisting of, e.g., the epoxy matrix reinforced by SiO_2 nanowires (Figures 24.3c and 24.4). The effective properties of this

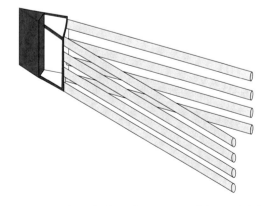

FIGURE 24.4 A closer look at the segment of the model shown in Figure 24.3c.

component are evaluated separately beforehand. The easiest way to do it is by using the classical Voigt and Reuss bounds (see, for example, Christensen 1979; Kelly and Zweben 2000). For this purpose, we would need to know the radius, r_b, and the length, l_b, of bristles, their number per unit surface area of fiber solid coating, and the radius of the fiber solid coating, r_b. Then the elemental volume of the bristled coating, $V_{(3)}$, which corresponds to the unit length along the fiber, can be expressed as

$$V_{(3)} = \pi \left((r_{(2)} + l_b)^2 - r_{(2)}^2 \right) \tag{24.7}$$

and the volume of a single bristle, V_b, as

$$V_b = \pi r_b^2 l_b \tag{24.8}$$

If there are M bristles growing over the fiber circumference and K bristles growing over the fiber unit length, then the volume fraction of bristles, c_b, in the elemental volume will be

$$c_b = MK \frac{V_b}{V_{(3)}} = MK \frac{r_b^2}{2r_{(2)} + l_b} \tag{24.9}$$

Also, in order to use the known formulas of the rule of mixture (Christensen 1979), we assume here that all the bristles are parallel to each other, i.e., that the properties of the bristled coating (3) (Figure 24.3c) do not change with the radius. This assumption, being, of course, a certain simplification, seems reasonable, since the length of SiO$_2$ nanowires used for reinforcement is rather small.

Now we can proceed with the proposed four-component model. The model is based on the recent idea presented by Guz et al. (2008b). Instead of the thin shell model for the fiber coating used in existing models, two different components are distinguished: the solid coating surrounding the fiber and the bristled coating surrounding the solid coating.

The mathematical formulation of the model is based on considering the four simple states of plane elastic equilibrium of the unit cell (a square with the side l_{cell}, Figure 24.5)—i.e., longitudinal tension, transverse tension, longitudinal shear and transverse shear—and using the Muskhelishvili complex potentials (Muskhelishvili 1953) for each domain occupied by a separate component. The model yields the explicit formulas for five effective elastic constants of the transversally isotropic medium, which represent the macroscopic properties of the considered composite.

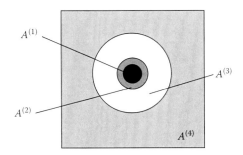

FIGURE 24.5 The cross-section of the four-component model.

24.3.3 Application of the Muskhelishvili Complex Potentials

The procedure of deriving the explicit expressions for effective elastic constants of the suggested four-component structural model is by no means a trivial mathematical exercise. For the lack of space, here it can be given only in outline for one of the constants, namely the shear modulus, G.

Let us consider one of the simple states of equilibrium mentioned in Section 24.3.2, namely, the longitudinal shear, and denote the shear moduli of the components (i.e., the fiber core, the fiber solid coating, the fiber bristled coating and the matrix) as $G^{(1)}$, $G^{(2)}$, $G^{(3)}$ and $G^{(4)}$, respectively.

On the first stage, two shear stress components, σ_{12} and σ_{13}, and one displacement component, u_1, are expressed in each domain occupied by a separate component (Figure 24.5) using the Muskhelishvili complex potentials (z is a complex coordinate in the transverse cross section, and i is imaginary unit):

In the circle $A^{(1)}$

$$\sigma_{12}^{(1)}(z) = G^{(1)} \left[\varphi'_{(1)}(z) + \overline{\varphi'_{(1)}(z)} \right];$$
$$\sigma_{13}^{(1)}(z) = iG^{(1)} \left[\varphi'_{(1)}(z) - \overline{\varphi'_{(1)}(z)} \right]; \tag{24.10}$$
$$u_1^{(1)}(z) = \varphi_{(1)}(z) + \overline{\varphi_{(1)}(z)}$$

In the ring $A^{(2)}$

$$\sigma_{12}^{(2)}(z) = G^{(2)} \left[\varphi'_{(2)}(z) + \overline{\varphi'_{(2)}(z)} \right];$$
$$\sigma_{13}^{(2)}(z) = iG^{(2)} \left[\varphi'_{(2)}(z) - \overline{\varphi'_{(2)}(z)} \right]; \tag{24.11}$$
$$u_1^{(2)}(z) = \varphi_{(2)}(z) + \overline{\varphi_{(2)}(z)}$$

In the ring $A^{(3)}$

$$\sigma_{12}^{(3)}(z) = G^{(3)} \left[\varphi'_{(3)}(z) + \overline{\varphi'_{(3)}(z)} \right];$$
$$\sigma_{13}^{(3)}(z) = iG^{(3)} \left[\varphi'_{(3)}(z) - \overline{\varphi'_{(3)}(z)} \right]; \tag{24.12}$$
$$u_1^{(3)}(z) = \varphi_{(3)}(z) + \overline{\varphi_{(3)}(z)}$$

In the domain of the matrix $A^{(4)}$

$$\sigma_{12}^{(4)}(z) = G^{(4)} \left[\varphi'_{(4)}(z) + \overline{\varphi'_{(4)}(z)} \right];$$
$$\sigma_{13}^{(4)}(z) = iG^{(4)} \left[\varphi'_{(4)}(z) - \overline{\varphi'_{(4)}(z)} \right]; \tag{24.13}$$
$$u_1^{(4)}(z) = \varphi_{(4)}(z) + \overline{\varphi_{(4)}(z)}$$

The three boundary conditions on the domain interfaces (Figure 24.5) are the conditions of perfect bonding between the components:

On the boundary between $A^{(1)}$ and $A^{(2)}$

$$\left(1+\frac{G^{(1)}}{G^{(2)}}\right)\varphi_{(1)}(z)+\left(1-\frac{G^{(1)}}{G^{(2)}}\right)\overline{\varphi_{(1)}(z)}=2\varphi_{(2)}(z) \quad (24.14)$$

On the boundary between $A^{(2)}$ and $A^{(3)}$

$$\left(1+\frac{G^{(2)}}{G^{(3)}}\right)\varphi_{(2)}(z)+\left(1-\frac{G^{(2)}}{G^{(3)}}\right)\overline{\varphi_{(2)}(z)}=2\varphi_{(3)}(z) \quad (24.15)$$

On the boundary between $A^{(3)}$ and $A^{(4)}$

$$\left(1+\frac{G^{(3)}}{G^{(4)}}\right)\varphi_{(3)}(z)+\left(1-\frac{G^{(3)}}{G^{(4)}}\right)\overline{\varphi_{(3)}(z)}=2\varphi_{(4)}(z) \quad (24.16)$$

The possible case of imperfect adhesion between the fiber core and the matrix can be taken into account by considering one of the four components, i.e., the coating layer, with the appropriately reduced properties.

The cornerstone of the analytical procedure is the representation of the Muskhelishvili potentials by

- A harmonic complex function, which is regular in the domain of fiber core (circle $A^{(1)}$ in Figure 24.5):

$$\varphi_{(1)}(z)=\sum_{k=0}^{\infty}a_{2k}^{(1)}\frac{z^{2k+1}}{2k+1} \quad (24.17)$$

- A function in the form of Laurent series, which is regular in the domain of fiber solid coating (ring $A^{(2)}$ in Figure 24.5):

$$\varphi_{(2)}(z)=\sum_{k=-\infty}^{\infty}a_{2k}^{(2)}z^{2k+1} \quad (24.18)$$

- A function in the form of Laurent series, which is regular in the domain of fiber bristle coating (ring $A^{(3)}$ in Figure 24.5):

$$\varphi_{(3)}(z)=\sum_{k=-\infty}^{\infty}a_{2k}^{(3)}z^{2k+1} \quad (24.19)$$

- A doubly-periodic function constructed utilizing the Weierstrass functions in the domain of the matrix ($A^{(4)}$ in Figure 24.5):

$$\varphi_{(4)}(z)=a_0^{(4)}z-\lambda^2 a_2^{(4)}\left(\frac{1}{z}-\sum_{n=1}^{\infty}\alpha_{n,0}\frac{z^{2n+1}}{2n+1}\right)$$

$$+\sum_{k=1}^{\infty}\sum_{n=1}^{\infty}a_{2k+2}^{(4)}\lambda^{2k+2}\alpha_{n,k}\frac{z^{2n+1}}{2n+1}$$

$$-\sum_{k=1}^{\infty}\frac{a_{2k+2}^{(4)}\lambda^{2k+2}}{(2k+1)z^{2k+1}} \quad (24.20)$$

where

$\alpha_{n,k}$ are the constants used in the theory of Weierstrass functions (Gradshteyn and Ryzhik 2000)

$\lambda=2r_{(3)}/l_{cell}$

$a_{2k}^{(1)}, a_{2k}^{(2)}, a_{2k}^{(3)}, a_{2k}^{(4)}$ are the yet unknown coefficients in the series given by Equations 24.17 through 24.20

Then the averaged stresses and strains for each of the domains are calculated using the contour integrals about closed paths:

$$\sigma_{12}^{*}-i\sigma_{13}^{*}=\frac{1}{l_{cell}^{2}}\oint(\sigma_{12}^{*}-i\sigma_{13}^{*})dx_2dx_3$$

$$=\frac{i}{l_{cell}^{2}}\left[G_{(4)}\oint_{S_4}\varphi_{(4)}(z)d\bar{z}+G_{(3)}\oint_{S_4}\varphi_{(3)}(z)d\bar{z}\right.$$

$$\left.+G_{(2)}\oint_{S_4}\varphi_{(2)}(z)d\bar{z}+G_{(1)}\oint_{S_4}\varphi_{(1)}(z)d\bar{z}\right] \quad (24.21)$$

where the closed paths for each of the contour integrals are shown in Figure 24.6. Note that S_4 consists of the outer boundary of the unit cell and the boundary between the bristled coating and the matrix. In order to make the path S_4 a closed contour, a virtual mathematical section is introduced to the area filled with the matrix (Figure 24.6). The integrals, Equation 24.21, can be taken as (Gradshteyn and Ryzhik 2000)

$$\oint_{S_1}\varphi_{(1)}(z)d\bar{z}=2il_{cell}^{2}c^{(1)}a_0^{(1)}, \quad c^{(1)}=\pi r_{(1)}^{2}l_{cell}^{-2} \quad (24.22)$$

$$\oint_{S_2}\varphi_{(2)}(z)d\bar{z}=2il_{cell}^{2}\left(c^{(1)}+c^{(2)}\right)a_0^{(1)}, \quad c^{(2)}=\pi\left(r_{(2)}^{2}-r_{(1)}^{2}\right)l_{cell}^{-2} \quad (24.23)$$

$$\oint_{S_3}\varphi_{(3)}(z)d\bar{z}=2il_{cell}^{2}\left(c^{(1)}+c^{(2)}+c^{(3)}\right)a_0^{(1)}, \quad c^{(3)}=\pi\left(r_{(3)}^{2}-r_{(2)}^{2}\right)l_{cell}^{-2}$$

$$\quad (24.24)$$

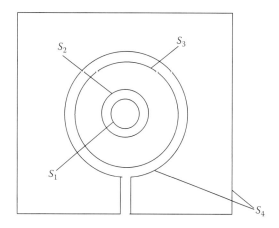

FIGURE 24.6 Paths used for computing contour integrals in the four-component model.

$$\oint_{S_4} \varphi_{(4)}(z)\overline{dz} = 2i\, l_{\text{cell}}^2 \left[-c^{(4)}a_0^{(4)} + \left(1-c^{(4)}\right)\sum_{k=1}^{\infty} a_{2k+2}^{(4)} r_{(3)}^{2k+2}\alpha_{0,k} \right],$$

$$c^{(4)} = 1 - \pi r_{(3)}^2 l_{\text{cell}}^{-2} \tag{24.25}$$

Bearing in mind that for the volume fraction of the components

$$c^{(1)} + c^{(2)} + c^{(3)} = 1 - c^{(4)} \tag{24.26}$$

the procedure results in the following expression for the average stresses, $\sigma_{12}^{\star}$ and $\sigma_{13}^{\star}$, in a unit area are

$$\sigma_{12}^{\star} - i\sigma_{13}^{\star} = 2G^{(4)}\left[-c^{(4)}a_0^{(4)} + \left(1-c^{(4)}\right)\sum_{k=1}^{\infty} a_{2k+2}^{(4)} r_{(3)}^{2k+2}\alpha_{0,k} \right]$$
$$+ \left(1-c^{(4)}\right)a_0^{(3)}G^{(3)} + \left(c^{(2)}+c^{(3)}\right)a_0^{(3)}G^{(2)} + c^{(1)}a_0^{(1)}G^{(1)} \tag{24.27}$$

Using Equations 24.14 through 24.16, coefficients $a_0^{(1)}$, $a_0^{(2)}$, $a_0^{(3)}$ and $a_0^{(4)}$ can be related as

$$a_0^{(3)} = \frac{2a_0^{(4)}}{1+\left(G^{(3)}/G^{(4)}\right)}, \quad a_0^{(2)} = \frac{2a_0^{(3)}}{[1+\left(G^{(2)}/G^{(3)}\right)]}$$
$$= \frac{4a_0^{(4)}}{[1+\left(G^{(3)}/G^{(4)}\right)][1+\left(G^{(2)}/G^{(3)}\right)]} \tag{24.28}$$
$$a_0^{(1)} = \frac{2a_0^{(2)}}{[1+\left(G^{(1)}/G^{(2)}\right)]}$$
$$= \frac{8a_0^{(4)}}{[1+\left(G^{(3)}/G^{(4)}\right)][1+\left(G^{(2)}/G^{(3)}\right)][1+\left(G^{(1)}/G^{(2)}\right)]}$$

For average shear strains, $\varepsilon_{12}^{\star}$ and $\varepsilon_{13}^{\star}$, we have

$$\varepsilon_{12}^{\star} = a_0^{(4)} + \overline{a_0^{(4)}} - c^{(3)}\left(a_2^{(4)} + \overline{a_2^{(4)}}\right) \tag{24.29}$$

$$\varepsilon_{13}^{\star} = i\left[a_0^{(4)} - \overline{a_0^{(4)}} + c^{(3)}\left(a_2^{(4)} - \overline{a_2^{(4)}}\right) \right] \tag{24.30}$$

After expressing $a_2^{(4)}$ in terms of $a_0^{(4)}$, from Equation 24.29 we get

$$\varepsilon_{12}^{\star} = 2\,\mathrm{Re}\{a_0^{(4)}\}\left[1 + \left(c^{(3)}+c^{(2)}+c^{(1)}\right)\frac{1-\left(G^{(3)}/G^{(4)}\right)}{1+\left(G^{(3)}/G^{(4)}\right)} \right] \tag{24.31}$$

Combining Equations 24.27 and 24.29 through 24.31, the following expression for the effective longitudinal shear modulus of the entire four-component composition, $G^{\star}$, can be deduced:

$$G^{\star} = G_{12}^{\star} = \frac{\sigma_{12}^{\star}}{\varepsilon_{12}^{\star}}$$
$$= G_{(4)}\left(\frac{c^{(4)} + 2c^{(3)} + c^{(4)}G_{(4)}G_{(3)}^{-1} + 4c^{(2)}\left(1+G_{(3)}G_{(2)}^{-1}\right)^{-1}}{c^{(4)} + (2-c^{(4)})G_{(4)}G_{(3)}^{-1}} \right.$$
$$\left. + \frac{8c^{(1)}\left(1+G_{(3)}G_{(2)}^{-1}\right)^{-1}\left(1+G_{(2)}G_{(1)}^{-1}\right)^{-1}}{c^{(4)} + (2-c^{(4)})G_{(4)}G_{(3)}^{-1}} \right) \tag{24.32}$$

Equation 24.32 yields the well-known formulas for two-component and three-component models (Nemat-Nasser and Hori 1999; Kelly and Zweben 2000; Guz 1993–2003; Rushchitsky 2006) as the particular cases. The three-component model will follow from Equation 24.32 if the volume fraction of bristled coating $c^{(3)} = 0$ and the shear moduli of bristled, $G_{(3)}$, and solid, $G_{(2)}$, coatings are the same ($G_{(3)} = G_{(2)}$):

$$G^{\star} = G_{(4)} \frac{c^{(4)} + 2c^{(2)} + c^{(4)}G_{(4)}G_{(2)}^{-1} + 4c^{(1)}\left(1+G_{(2)}G_{(1)}^{-1}\right)^{-1}}{c^{(4)} + (2-c^{(4)})G_{(4)}G_{(2)}^{-1}} \tag{24.33}$$

The two-component model will follow from Equation 24.33 if, additionally, the volume fraction of solid coating $c^{(2)} = 0$ and the shear moduli of solid coating, $G_{(2)}$, and fiber core, $G_{(1)}$, are the same ($G_{(2)} = G_{(1)}$):

$$G^{\star} = G_{(4)} \frac{2c^{(1)} + c^{(4)}G_{(4)}G_{(1)}^{-1}}{c^{(4)} + (2-c^{(4)})G_{(4)}G_{(1)}^{-1}} \tag{24.34}$$

Similarly, the explicit expressions for other four effective constants for the entire four-component composition are deduced.

24.3.4 Computing the Effective Constants for a Particular Composition

In this subsection, we illustrate how to use the proposed four-component model for computing the effective elastic constants for a generic composite material, which have the internal structure similar to the one given in Figure 24.3 and some typical properties of reinforcing elements (i.e., the core fibers, bristles/whiskers, etc.).

Let us consider the unidirectional fiber-reinforced composite consisting of the epoxy matrix and Thornel 300 fibers bristled by the graphite whiskers—a structure similar to that presented in Figure 24.2. The properties of the matrix, the fiber core and the fiber solid coating are given in Table 24.1 according to Katz and Milewski (1978), Lubin (1982), and Nemat-Nasser and Hori (1999). The composite is simulated by the model suggested in Section 24.3.2 and shown in Figure 24.3. Here the bristled coating is itself a composite (Figures 24.3c and 24.4). The effective properties of this component are evaluated separately beforehand, as described in Section 24.3.2. The radii of the fiber core, the fiber solid coating, and the fiber bristled coating used for

TABLE 24.1   Properties of the Matrix, the Fiber Core and the Fiber Solid Coating

Components	Density ρ, kg/m^3	Young's Modulus E, GPa	Shear Modulus G, GPa	Poisson's Ratio ν
Graphite whiskers	2250	1000	385	0.3
Fiber core Thornel 300	1750	228.0	88.00	0.3
Epoxy matrix	1210	2.68	0.96	0.4

computing are, respectively, $r_{(1)} = 4\,\mu m$, $r_{(2)} = 6\,\mu m$, $r_{(3)} = 50\,\mu m$. Three different densities of bristlization are examined: dense, with 120 bristles over the fiber circumference and 50 bristles over $100\,\mu m$ of the fiber length; medium, with, respectively, 60 and 50 bristles; and sparse, with, respectively, 30 and 50 bristles. The medium density is a limiting case for single bristles growing from the fiber surface. The dense density (two times higher then the medium density) corresponds to two bristles growing from the same nest on the fiber surface. At a distance from the fiber surface, the bristles separate with some space between them still remaining for the matrix material to fill in. According to Equation 24.9, the three cases give the volume fractions of bristles $c_b = 0.25; 0.125; 0.063$, respectively, for sparse, medium, and dense bristlization. For the case of no whiskers added to the system, $c_b = 0$.

The computed volume fractions for the four components used in the model, i.e., the fiber core, $c^{(1)}$, the fiber solid coating, $c^{(2)}$, the fiber bristled coating, $c^{(3)}$, and the matrix, $c^{(4)}$, are shown in Table 24.2. These values are not affected by variations in density of bristlization. The latter influences only the effective elastic properties of bristled coating.

The values of all five effective elastic constants representing the transversely isotropic response of the entire composite were computed by the method outlined in Section 24.3.3. The results show that the properties in the direction of fibers are the most sensitive to the density of bristles. The increase in the number of bristles per unit surface of the fibers gives a very strong rise to the value of Young's modulus. However, the shear modulus,

being the driving parameter for the strength estimation of the entire composition (Guz et al. 2005, 2008b), is significantly less sensitive to this factor. The values of shear modulus for the considered cases of sparse, G_{sparse}, medium, G_{medium}, and dense, G_{dense}, bristlization are given in Table 24.3 together with the values of shear modulus for the same composition without whiskers, G_0. In the latter case, $c_w = 0$. The difference between G_{sparse}, G_{medium}, and G_{dense} is less than 3% (Table 24.3). In the same time, the presence of bristled fibers itself—either with dense, medium, or sparse density of bristles—gives the significant increase in the shear modulus in composites if compared with the case without whiskers, i.e., with G_0. The considered case (Table 24.3) shows the increase from 1.4 times to up to 2 times, depending on the volume fraction of the matrix, $c^{(4)}$.

24.4 Discussion

The above example gives an illustration of application of the proposed four-component model. It was given primarily for the purpose of explaining the computational procedure rather than for producing a correct estimate of effective properties of real materials. For the latter, we still lack some basic information about mechanical properties of nanocomponents (CdTe core nanofibers and SiO_2 nanobristles). Therefore, designing special experiments for testing CdTe nanofibers and SiO_2 nanobristles is a necessary step for a using the proposed four-component model in engineering practice. On the next stage, after acquiring the necessary experimental data, the effective properties for particular real nanocomposites can be computed using the procedure described in Section 24.3.4.

Undoubtedly, even after verification of the predicted properties for bristled nanowires by comparing them with the results of specially designed experiments, the suggested approach can be considered as merely the first step toward modeling bristled nanowires and their application. Even a four-component model is an idealization of the complex internal structure of the considered materials. However, it can provide us with important insight into some basic relationships between the properties of constituents and the overall performance of such materials.

Ultimately, any mechanics of materials, including mechanics of nanomaterials, envisages analysis of materials for structural applications, be it on macro-, micro-, or nanoscale (Guz et al. 2007a,b; Windle 2007). It is therefore a logical conclusion that any research on nanomaterials should be followed by the analysis of nanomaterials working in various structures and devices. Micro- and nanostructural applications look like

TABLE 24.2   Computed Volume Fractions for the Four Components Used in the Model

$c^{(1)}$	$c^{(2)}$	$c^{(3)}$	$c^{(4)}$
0.00384	0.02016	0.576	0.4
0.00320	0.01680	0.480	0.5
0.00256	0.01344	0.384	0.6
0.00192	0.01008	0.288	0.7

TABLE 24.3   Values of Shear Modulus for the Cases of Sparse, Medium, and Dense Bristlization

G, GPa	$c^{(4)} = 0.7$	$c^{(4)} = 0.6$	$c^{(4)} = 0.5$	$c^{(4)} = 0.4$
G_{dense}	1.355	1.505	1.674	1.872
G_{medium}	1.374	1.528	1.701	1.905
G_{sparse}	1.391	1.544	1.715	1.913
G_0	0.9636	0.9649	0.9660	0.9671

the most natural and promising areas of the nanomaterials utilization. They do not require large industrial production of nanoparticles, which are currently rather expensive. It seems pertinent to recall a discussion on mechanical properties of new materials that took place more than 40 years ago. In the concluding remarks, Bernal (1964) said: "Here we must reconsider our objectives. We are talking about new materials but ultimately we are interested, not so much in materials themselves, but in the structures in which they have to function." The authors believe that nanomechanics faces the same challenges that micromechanics did 40 years ago, which Professor Bernal described so eloquently.

24.5 Conclusions

This chapter revisited some of the well-known models in the mechanics of structurally heterogeneous media for the purpose of analyzing their suitability to describe properties of nanomaterials and nanocomposites and their mechanical behavior. New areas of research in nanomechanics of materials were also pointed out.

As an example of application of the proposed approaches, the prediction of effective properties for a new type of nanocomposites was considered. This chapter presented a new four-component model for predicting the mechanical properties of microcomposites reinforced with whiskers and nanocomposites reinforced with bristled nanowires. The mathematical formulation of the model is based on using the Muskhelishvili complex potentials for each domain occupied by a separate component.

To illustrate the method for computing effective elastic constants within the proposed four-component model, a generic fibrous composite with three different densities of bristles growing on the core reinforcing fibers was considered. It was shown that the increase in the number of bristles per unit surface of the fibers gives a very strong rise to the value of Young's modulus (Guz et al. 2007b, 2008b). However, the shear modulus, being the driving parameter for the strength estimation of the entire composition, is less sensitive to this factor.

Acknowledgments

The authors would like to express their gratitude to Dr. M. Kashtalyan and all researchers from the Centre for Micro- and Nanomechanics (CEMINACS) at the University of Aberdeen (Scotland, United Kingdom) for the helpful discussions and suggestions.

Financial support of the part of this research by the Royal Society, the Royal Academy of Engineering, and the Engineering and Physical Sciences Research Council (EPSRC) is gratefully acknowledged.

References

Bernal, J.D. 1964. Final remarks. A discussion on new materials. *Proceedings of the Royal Society A* 282: 1388–1398.

Bhushan, B. (Ed.). 2007. *Springer Handbook of Nanotechnology*. Berlin, Germany: Springer.

Cherepanov, G.P. 1979. *Mechanics of Brittle Fracture*. New York: McGraw-Hill.

Christensen, R.M. 1979. *Mechanics of Composite Materials*. New York: John Wiley & Sons.

Cottrell, A.H. 1964. *Theory of Crystal Dislocation*. London, U.K.: Blackie & Son Ltd.

Eshelby, J.D. 1956. The continuum theory of lattice defects. In: *Progress in Solid State Physics*, vol. 3, F. Seitz and D. Turnbull (Eds.), New York: Academic Press.

Gradshteyn, I.S. and I.M. Ryzhik. 2000. *Tables of Integrals, Series, and Products*. San Diego, CA: Academic Press.

Guz, A.N. (Ed.). 1993–2003. *Mechanics of Composites*, vols. 1–12. Kiev, Ukraine: Naukova dumka (vols. 1–4), Kiev, Ukraine: A.C.K. (vols. 5–12).

Guz, A.N., Akbarov, S.D., Shulga, N.A., Babich, I.Yu., and V.N. Chekhov. 1992. Micromechanics of composite materials: Focus on Ukrainian research (Special Issue). *Applied Mechanics Review* 45(2): 13–101.

Guz, A.N., Rodger, A.A., and I.A. Guz. 2005. Developing a compressive failure theory for nanocomposites. *International Applied Mechanics* 41(3): 233–255.

Guz, A.N., Rushchitsky, J.J., and I.A. Guz. 2007a. Establishing fundamentals of the mechanics of nanocomposites. *International Applied Mechanics* 43(3): 247–271.

Guz, A.N., Rushchitsky, J.J., and I.A. Guz. 2008a. Comparative computer modeling of carbon-polymer composites with carbon or graphite microfibers or carbon nanotubes. *Computer Modeling in Engineering & Sciences (CMES)* 26(3): 139–156.

Guz, I.A. and J.J. Rushchitsky. 2004. Comparison of mechanical properties and effects in micro and nanocomposites with carbon fillers (carbon microfibres, graphite microwhiskers and carbon nanotubes). *Mechanics of Composite Materials* 40(3): 179–190.

Guz, I.A. and J.J. Rushchitsky. 2007. Computational simulation of harmonic wave propagation in fibrous micro- and nanocomposites. *Composite Science & Technology* 67(5): 861–866.

Guz, I.A., Rodger, A.A., Guz, A.N., and J.J. Rushchitsky. 2007b. Developing the mechanical models for nanomaterials. *Composites Part A* 38(4): 1234–1250.

Guz, I.A., Rodger, A.A., Guz, A.N., and J.J. Rushchitsky. 2008b. Predicting the properties of micro and nanocomposites: From the microwhiskers to bristled nano-centipedes. *The Philosophical Transactions of the Royal Society A* 366(1871): 1827–1833.

Katz, H.S. and J.V. Milewski (Eds.). 1978. *Handbook of Fillers and Reinforcements for Plastics*. New York: Van Nostrand Reinhold Company.

Kelly, A. and C. Zweben (Eds.). 2000. *Comprehensive Composite Materials*, vols. 1–6. Amsterdam, the Netherlands: Elsevier Science.

Lubin, G. (Ed.). 1982. *Handbook of Composites*. New York: Van Nostrand Reinhold Company.

Muskhelishvili, N.I. 1953. *Some Basic Problems of the Mathematical Theory of Elasticity*. Leiden, the Netherlands: Noordhoff.

Nemat-Nasser, S. and M. Hori. 1999. *Micromechanics: Overall Properties of Heterogeneous Materials.* Amsterdam, the Netherlands: North-Holland.

Panin, V.E., Egorushkin, V.E., and N.V. Makarov. 1995. *Physical Mesomechanics and Computer Design of Materials*, vols. 1–2. Novosibirsk, Russia: Siberia Publishing House of the RAS.

Rushchitsky, J.J. 2006. Sensitivity of structural models of composite material to structural length scales. *International Applied Mechanics* 42(12): 1364–1370.

Sih, G.C. and B. Lui. 2001. Mesofracture mechanics: A necessary link. *Theoretical & Applied Fracture Mechanics* 37(1–3): 371–395.

Wang, Y., Tang, Z.Y., Liang, X.R., Liz-Marzan, L.M., and N.A. Kotov. 2004. SiO_2-coated CdTe nanowires: Bristled nano centipedes. *Nano Letters* 4(2): 225–231.

Wilson, M.A., Kannangara, K., Smith, G., Simmons, M., and B. Raguse. 2002. *Nanotechnology. Basic Science and Emerging Technologies.* Boca Raton, FL: Chapman & Hall/CRC.

Windle, A.H. 2007. Two defining moments: A personal view by Prof. Alan H. Windle. *Composite Science & Technology* 67(5): 929–930.

Zhuk, Y. and I.A. Guz. 2006. Influence of prestress on the velocities of plane waves propagating normally to the layers of nanocomposites. *International Applied Mechanics* 42(7): 729–743.

Zhuk, Y.A. and I.A. Guz. 2007. Features of plane wave propagation along the layers of a prestrained nanocomposite. *International Applied Mechanics* 43(4): 361–379.

V

Nanomagnetism and Spins

25

Nanomagnetism in Otherwise Nonmagnetic Materials

Tatiana Makarova
Umeå University

25.1 Introduction

The physics of nanomagnetism is concerned with the studies of magnetic phenomena specific to nanostructured materials, i.e., materials with the size of typical structural elements from 1 to 100 nm. A porous specimen built of small particles is the simplest example of a nanostructured material. Due to dramatically enhanced surface-to-volume ratio, the magnetic properties of nanoparticles may be markedly different from those of the bulk material with the same chemical composition. Numerous experiments show that the magnetic properties of bulk ferromagnetic and antiferromagnetic materials are modified in the corresponding nanomaterials. Most surprisingly, nanostructured materials or thin films may show ferromagnetism at room temperatures (RTFM) even when the starting material is magnetically inactive.

Whether a given material shows magnetism is delicately controlled by its structural and geometrical properties, such as the interatomic distances, coordination number (i.e., the number of nearest neighbors), and symmetry. The main consequences of the structural modifications in nanostructured materials are the following (after Feng et al. 1989):

1. Coordination number decreases, and hence there is a decrease in the overlap of the nearby atomic orbitals. This leads to the sharper density of states. The magnetic moment per atom increases with the decrease of coordination number.

2. With the decrease of coordination number, the effect of vacancies on the nearest-neighbor magnetic atoms tends to enhance its magnetic moment.

3. However, surface relaxations decrease the interlayer separation, overlap increases, and magnetic moment decreases.

4. The effect of interatomic distances is larger than that due to coordination number. Thus, the monolayers of magnetic elements are likely to provide a very strong moment.

To summarize, the local environment of the atoms can be altered by introducing defects such as impurities, vacancies, and vacancy complexes. The reduced coordination number and symmetry changes are expected to narrow the electronic bands. This enhances magnetism in ferromagnetic materials and may cause magnetization in nonmagnetic materials.

Spintronics, which enables the manipulation of spin and charges in electronic devices, is a field of study that holds promise for a revolution in electronics. The materials explored today for spintronics applications are mainly dilute magnetic semiconductors (DMS), i.e., semiconductors doped with magnetic elements. A novel family of spintronic materials is emerging, i.e., nanometer-scale magnets built from nominally nonmagnetic elements. As an example, the progress article (Bogani and Wernsdorfer 2008) emphasizes the importance of systems of reduced dimensionality for enhanced information storage capacity.

25.2 What Are Nonmagnetic Materials?

Strictly speaking, all materials are magnetically active in the sense that their magnetization can be induced by an external magnetic filed: they show either negative diamagnetic susceptibility (diamagnetism) or positive magnetic susceptibility (paramagnetism). The total magnetic susceptibility, χ, of a certain material includes several terms: the diamagnetic contribution from the core electrons, the orbital diamagnetism, the paramagnetic van Vleck term originating from virtual magnetic dipole transitions between the valence and conduction bands, the Landau diamagnetism of the itinerant electrons in metals, and the Pauli spin paramagnetism of itinerant electrons in metals or the Curie paramagnetism exhibited by localized unpaired spins.

A material is called nonmagnetic if the magnetization density, local or global, is always zero in the absence of an external magnetic field. If the magnetization density in a material is finite even in the absence of the external field, the material is generally called magnetic. In some magnetic materials—antiferromagnets—the local density varies from point to point in both magnitude and sign on a microscopic scale, so the magnetization density measured in macroscopic volumes vanishes. Below, we mainly focus on the case of ferromagnets, i.e., materials in which the macroscopic magnetization density is finite. If not stated otherwise, the ferromagnetic state is implied when we use the word "magnetic."

Ferromagnetic state requires unpaired electrons and mechanisms leading to ferromagnetic coupling of unpaired electrons. Every electron is a tiny magnet, but an atom does not necessarily have a magnetic field just because its constituent electrons do. An atom can have a net magnetic field if it has unpaired electrons in one of its outer shells. If an element has unpaired electrons, a sample of that element can become magnetic if the spins in the bulk material are oriented properly. In isolated atoms, many elements exhibit magnetic moments according to Hund's rules, while in the solid state, only a few of them are magnetic. The disappearance of magnetic moments in a solid is a consequence of the delocalization of the electrons, which favors equal occupation of states having opposite projections of magnetic moments.

Light elements, say carbon, may have unpaired spins in elemental state, as shown in Figure 25.1, but in a bulk state all electrons are paired. That is why magnetism based on s and p

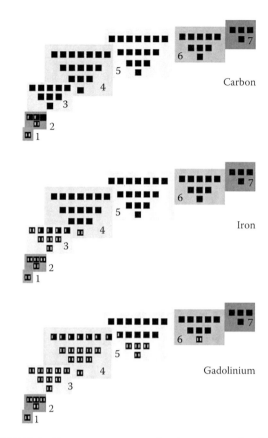

FIGURE 25.1 Electronic structure of carbon, iron, and gadolinium atoms. Plotted using www.webelements.com

electrons is a phenomenon that is not expected. "The principal quantum number for electrons responsible for the magnetism must be ≥3" (Heisedberg 1928).

Unpaired electrons arise inevitably as one moves down the periodic table toward larger atoms. Iron atom has a strong magnetic field because it has four unpaired electrons in its outer shell. If an element has many unpaired electrons, there is a large probability that in certain compounds of this element the spins are oriented.

The peculiar role of d and f electrons in magnetism can be understood from the electronic configuration. Figure 25.1 shows electronic configuration of iron atom, which contains four unpaired electrons.

However, the presence of unpaired electrons is a necessary but not a sufficient condition for ferromagnetism. Electrons with higher angular momentum (i.e., higher orbitals) have higher kinetic energy, and thus they have lower potential energy and potential energy is in turn responsible for the correlation. Therefore, the most significant contribution to the magnetic moment is from unpaired electrons closest to the nucleus. That is why heavy elements like Gd (Figure 25.1c) with its eight unpaired electrons do not have large magnetic moments.

Magnetic periodic table (Skomski and Coey 1995, Coey and Sanvito 2004) includes three islands of magnetic stability: one around 3d elements Fe, Ni, Co; one around the 3f element Gd; and one around oxygen, which is known to order

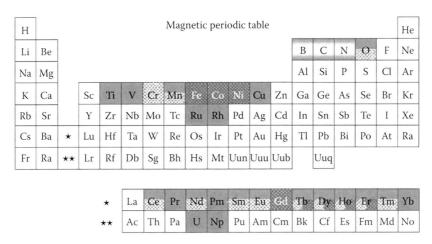

Magnetic periodic table

FIGURE 25.2 The magnetic periodic table. Shaded cells are the "islands of magnetic stability." Elements which are able to form ferromagnetic and antiferromagnetic compounds are cross-hatched black and white correspondingly. (Adapted from Coey, M. and Sanvito, S., *Phys. World*, 17, 33, 2004.)

antiferromagnetically at 22 and 30 K (Figure 25.2). Oxygen is the only element without unpaired d or f electrons that orders magnetically (Meier and Helmholdt 1984).

Whereas in usual materials, magnetism is determined mainly by the choice of constituting elements from the periodic table, in nanometer-scale materials the above statement is not true. When the sample of a material becomes small or modified in the nanometer scale, shapes or boundaries in nanostructures, defects, and vacancies play important roles in characteristics of electron states (Oshiyama and Okada 2006).

A term "d-zero ferromagnetism" (Coey 2005) was coined in response to a growing number of reports on hexaborides, metal oxides, and carbon structures displaying small ferromagnetic moments despite the absence of atoms with partially filled d or f shells. Speaking about materials that are nonmagnetic in the usual state, one can supplement the list with nonmagnetic metals like Au, Pd, Ru, and Rh, which, in nanoscale, may also exhibit magnetism, and with semiconductor nanoparticles and quantum dots GaN, CdS, CdSe. In all cases, the unexpected collective behavior, also known as emergent behavior, is related to the nanometer-scale changes of the structure.

The weakness of the ferromagnetic signal and low reproducibility are common features of these publications. They give rise to understandable doubts concerning the intrinsic origin of some of the reported data. However, in many cases, the content of metallic impurities is too low to account for the observed value of magnetization. By now, the impurity scenario is rather unlikely, since it cannot explain recurring regularities in structure–property relationships. The magnetic properties of the ferromagnetic nanostructures are highly sensitive to the conditions of synthesis or subsequent annealing. Analysis of the synthesis and annealing conditions provides the basis for the models for ferromagnetism in these systems. In several cases, the firm evidence for magnetism in nonmagnetic materials has been obtained from elementally sensitive magnetic measurements like x-ray magnetic circular dichroism. This is the case of Au, Ag, Cu nanoparticles, some oxides, and graphite.

25.3 Formation of Magnetic State through Introducing Nonmagnetic sp Elements to Nonmagnetic Matrices

Promising materials for spintronic applications are magnetic half-metals (MHMs) (see Katsnelson et al. 2008 for a review), i.e., systems that are characterized by nonzero density of carriers at the Fermi level (E_F) for only one spin direction, say, up, ($N{\uparrow}(E_F) > 0$), but there is an energy gap (FG) for the reverse spin projection ($N{\downarrow}(E_F) = 0$) (Figure 25.3). Therefore, in the ideal case, spin density polarization at the Fermi level is

$$P = \frac{N_\uparrow\left(E_F\right) - N_\downarrow\left(E_F\right)}{N_\uparrow\left(E_F\right) + N_\downarrow\left(E_F\right)} = 1$$

As a result, the electric current in magnetic half-metals is accompanied by a spin current, and MHM materials exhibit nontrivial spin-dependent transport properties (after Ivanovskii 2007).

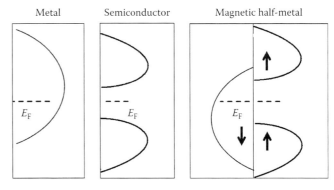

Metal   Semiconductor   Magnetic half-metal

FIGURE 25.3 Density of states for the metallic, semiconducting and metallic half metal states. (Reprinted from Edwards, D.M. and Katsnelson, M.I., *J. Phys.: Condens. Matter*, 18, 7209, 2006. With permission.)

The MHM materials are manufactured by doping semiconductor hosts with magnetic atoms, and these materials are called dilute magnetic semiconductors (DMS). The physical picture of magnetism in the DMS is described by Zener's p–d exchange mechanism and superexchange mechanism that compete to determine the magnetic states (Katayama-Yoshida et al. 2007, Dietl 2008). These materials are reviewed in several papers (see, for example, a recent review (Jungwirth et al. 2006)). Dilute magnetic systems include manganites $LaSrMnO_3$ (Salamon and Jaime 2001) and diluted magnetic semiconductors like GaMnAs (Ohno 1998). Still, the ordering temperatures in DMS are far below room temperature.

Semiconducting or insulating oxides like ZnO doped with several percent of transition metal cations were predicted (Dietl et al. 2000) and found experimentally to be ferromagnetic at room temperature. These materials are described below. However, there is growing evidence that the experimental observations are incompatible with a picture based on magnetic moments carried by the d-electrons of the transition metal cations mediated by the itinerant electrons. The main areas of discussions range from indirect d-electron exchange interaction mediated by defects (Kittilstved et al. 2006) to parasitic ferromagnetism. An unexpected fact that the oxides do not need magnetic cations to become ferromagnetic turned the discussions to recognition of a novel type of magnetism, which is tentatively called interface magnetism.

There are many theoretical and experimental confirmations that under certain conditions magnetism arises without transition metal elements. The best known example of a magnetic p-compound is molecular oxygen, which orders in an antiferromagnetic fashion. Nonmetallic impurities as well as intrinsic defects in metal-free compounds may offer a path to new ferromagnetic materials. It is shown theoretically that ferromagnetism can be induced in CaO with calcium vacancies (Elfimov et al. 2002), boron, carbon, or nitrogen-doped CaO (Kenmochi et al. 2004), calcium pnictides, i.e., CaP, CaAs, and CaSb with the zinc blende structure (the pnictides are the compounds of phosphorus, arsenic, antimony, and bismuth) (Kusakabe et al. 2004). There are persistent indications that small ferromagnetic moment in CaB_6 (Young et al. 1999), which depends on sample stoichiometry and heat treatment (Lofland et al. 2003) and is observed in impurity-free disordered thin films (Dorneles et al. 2004), is not the result of trivial sample contamination. The same picture emerges for another system with closed shell configuration, oxide films, metal nanoparticles, semiconductor quantum dots and interfaces, and carbon nanostructures. Two main reasons for nontrivial magnetism are considered:

1. Dopants that come to the spin-polarized state and order ferromagnetically.
2. Magnetism due to nonstoichiometry (Ivanovskii 2007), including increasingly observed phenomenon of "interfacial magnetism" (Hernando et al. 2006a,b), charge transfer ferromagnetism (Coey et al. 2008), and some exotic phenomena like "even-odd effects" (Lounis et al. 2008).

25.4 Ferromagnetism in Hexaborides: Discovery, Disproof, Rebuttal

25.4.1 Discovery

Young et al. (1999) reported ferromagnetism in La-doped calcium hexaboride (CaB_6) in which a few of the calcium atoms a replaced with lanthium atoms. This discovery was taken as a mark of a long-sought mechanism for ferromagnetism in metals, where the "electron gas" is susceptible to magnetic ordering at low density (Ceperley 1999); in other words, a phenomenon of high-temperature weak ferromagnetism at low-carrier concentration (HTFLCC) with no atomic localized moments. Later unusual ferromagnetism in hexaborides was reported for undoped MB_6 (M = Ca, Sr, Ba) (Ott et al. 2000; Vonlanthen et al. 2000). None of the constituent elements in these compounds possess partially filled d or f levels, and these reports were considered as the first clear evidence of magnetism in otherwise nonmagnetic materials.

Sharp decrease in saturation with the increase of doping level ruled out the effects of an accident contamination but required consideration from the viewpoint of the electronic band structure. One school of thought attributed this phenomenon to the polarization of low density electronic gas and another school of thought to the hole doped excitonic insulator, whereas several authors insisted that the mechanism for magnetism in this compound is strongly connected to defects.

At the moment of the discovery of weak ferromagnetism, alkaline-earth hexaborides were believed to be either semiconductors or semimetals, in close vicinity to the border between semimetals and small-gap semiconductors and having a peculiar configuration of the electronic excitation spectrum: the valence and conduction band are separated with a gap in all points of the Brillouin zone, except for the X points, where a weak overlap does exist (Massidda et al. 1997). Low electron doping with La^{3+} (order of 0.1%) was shown to result in an itinerant type of ferromagnetism stable up to 600 K (Young et al. 1999) and almost 1000 K (Ott et al. 2000). The saturation magnetic moment is quite sensitive to the doping level and reaches $0.07 \mu_B$/electron, the electron density being $7 \times 10^{19} \, cm^{-3}$. Taking into account the absence of localized magnetic moments, magnetic order was ascribed to the itinerant charge carriers: a ferromagnetic phase of a dilute three-dimensional electron gas (Ceperley 1999, Young et al. 1999).

A different approach is based on the formation of excitons between electrons and holes in the overlap region around the X point (Zhitomirskyi et al. 1999, Murakami et al. 2002). Electrons and holes are created as a result of the band overlap. Coulombic attraction between these electrons and holes can lead to a condensation of the bound exciton pairs. Condensation opens a gap in the quasiparticle spectrum. Excitonic insulator is thus created, which contains a condensate of a spin-triplet state of electron–hole pairs. A ferromagnet with a small magnetic moment but with a high Curie temperature can be obtained by doping an excitonic insulator. Doping provides electrons to the conduction band, change the electron–hole equilibrium, pairing becomes less favorable, and extra electrons are ferromagnetically aligned. The idea of ferromagnetic instability in the excitonic metal was further developed by adding the effect of imperfect nesting on the excitonic state (Veillette and Balents 2002).

ESR experiments give some evidence that ferromagnetism of hexaborides is not a bulk phenomenon, but spins exist only within the surface layer approximately 1.5 μm thick (Kunii 2000). Band calculations (Jarlborg 2000) demonstrated a possibility for ferromagnetism below the Stoner limit in doped hexaborides, and the phenomenon has been attributed to various defects.

Observations of anomalous NMR spin–lattice relaxation for hexaborides showed the presence of a band with a coexistence of weakly interacting localized and extended electronic states (Gavilano et al. 2001). Measurements of thermoelectric power and the thermal conductivity in the range of 5–300 K showed that it is described by scattering of electrons on acoustic phonons and ionized impurities in the conduction band, which is well separated from the valence band (Gianno et al. 2002). The explanation of the origin of weak but stable ferromagnetism with the model of a band overlap has become a problem.

The starting point of the majority of models has been a semimetallic structure with a small overlap. However, parameter-free calculations of the single-particle excitation spectrum based on the so-called GW approximation predict a rather large band gap. In this case, CaB_6 should not be considered a semimetal but as a semiconductor with a band gap of 0.8 eV (Tromp et al. 2001). Angle-resolved photoemission provides an answer for the fundamental question of whether divalent hexaborides are intrinsic semimetals or defect-doped band gap insulators: there is a gap between the valence and conduction bands in the X point, which exceeds 1 eV (Denlinger et al. 2002). Assuming that CaB_6 is a semiconductor, magnetism is considered to be due to a La-induced impurity band, arising on the metallic side of the Mott transition for the impurity band (Tromp et al. 2001).

The magnetic properties of these structures have been considered from the point of view of imperfections in the hexaboride lattice. It has been found that of all intrinsic point defects, the B vacancy bears a magnetic moment of $0.04 \mu_B$. The ordering of the moments can be understood assuming that in the presence of compensating cation vacancies, a B_6 vacancy cannot be neutral (Monnier and Delley 2001). A support for the impurity-band mechanism has been found from the electrical and magnetic measurements on several La-doped samples (Terashima et al. 2000): all the samples show metallic behavior of conductivity. Prepared at nominally identical conditions, some of the samples are paramagnetic and some are ferromagnetic, suggesting that ferromagnetic state can be spatially inhomogeneous. On the other hand, the models for a doped excitonic insulator also included spatial inhomogeneity (Balents and Varma 2000) and phase separation with appearance of a superstructure (Barzykin and Gorkov 2000).

25.4.2 Disproof

Shortly after the striking observation of high-temperature weak ferromagnetism, a discussion was opened concerning the possibility of a parasitic origin of the hexaboride ferromagnetism (Matsubayashi et al. 2002). Ferromagnetic CaB_6 and LaB_6 were created with the magnetic properties, including the Curie temperature, similar to reported earlier. These magnetic samples were washed in HCl several times, and the mass of the washed-out iron was measured. Linear dependence of magnetization reduction versus iron mass apparently left no doubts that high-temperature ferromagnetism in hexaborides is not intrinsic but that is instead due to alien phases of iron and boride, namely FeB and Fe_2B that have the Curie temperatures at 598 and 1015 K, respectively.

Replying to the claims, the authors of the pioneering work (Young et al. 2002) noted that these new findings did not contradict their picture (Fisk et al. 2002) where the magnetism is due to strongly interacting magnetically active defects in off-stoichiometric CaB_6 crystals. In single crystals of CaB_6 with intentionally added iron, no dependence of the measured ordered moment on the iron concentration was found, suggesting that alien Fe–B phases are not the source of ferromagnetism. The important point here is that not only iron, but also surface moments are removed in acid solution. There is the experimental evidence that during the sample storage in air, magnetization is enhanced and the ordering temperature progressively increases. Were the ferromagnetism due to the Fe contribution, the formation of Fe oxides in air would lead to opposite results.

The discussion finally arrived at the conclusion that iron with a concentration of about 0.1 at.% is indeed involved in the weak high-temperature ferromagnetism of CaB_6 although the exact mechanism is still unclear and probably highly nontrivial (Young et al. 2002).

The evidence that transition metals might play a role in ferromagnetism of CaB_6 raised doubts about the very existence of magnetism in otherwise nonmagnetic materials. The magnetism could be ascribed to iron impurities originating from the crucible used in the synthesis (Matsubayashi et al. 2003), from boride commercial powders, boron powder, and from aluminum metal, used in significant quantity as a flux for crystal growth (Otani and Mori 2002, 2003, Mori and Otani 2002).

The speculations that the ferromagnetism is due to the defect surface states have been refuted by the argument that the iron is concentrated in the surface region, as the Auger depth profiles demonstrated (Meegoda et al. 2003). Electron microprobe experiments reveal that Fe and Ni are found at the edges of facets and growth steps, which served as the indication of extrinsic origin of weak ferromagnetism in undoped CaB_6 (Bennett et al. 2004). These findings made two of the authors of the original report (Young et al. 1999) to state that "the weak ferromagnetism in electron-doped CaB_6 is extrinsic, due to surface contamination by ferromagnetic compounds containing Fe and Ni" (Bennett et al. 2004). The authors made however an important note that the data do not exclude the existence of an intrinsic ferromagnetic phase, which is masked by a strong extrinsic signal.

25.4.3 Rebuttal

The recognition of an unpleasant truth that ferromagnetic metals concentrate at the edges of facets and growth steps (Bennett et al. 2004) did not stop investigations in this area. There were questions that had to be answered. For instance, the presence of transition metals could not account for the dependence of magnetism on La or Ca concentration (Young et al. 2002, Cho et al. 2004) as well as the fact that the magnetization is enhanced and the ordering temperature progressively increases during the sample storage in air. Again, were ferromagnetism due to the Fe contribution, the formation of Fe oxides in air would lead to opposite results because iron oxides have lower values of both the magnetization and Curie temperature (Lofland et al. 2003). For CaB_6 samples grown from Al flux, which was deliberately contaminated with a wide Fe, no dependence of the saturation magnetization on Fe concentration was found (Young et al. 2002).

The understanding that iron is somehow involved in the ferromagnetism of CaB_6 has not stopped claims that magnetism is of a nontrivial origin. Having studied CaB_6 crystals of different purity, the experimenters analyzed the formation of the mid-gap states and suggested that the exotic ferromagnetism in CaB_6, in part, cannot be ascribed to a magnetic impurity (Cho et al. 2004).

Further development showed that the retraction of the original ideas concerning an intrinsic nature of magnetism in hexaborides (Bennett et al. 2004) was indeed premature. The experiments on thin films (Dorneles et al. 2004) have shown that Fe/Ca atomic ratio exceeding 100% would be needed to explain the sample magnetization. Huge magnetic moments were detected at the surface layers of thin films of disordered CaB_6 and SrB_6 films deposited by pulsed-laser deposition on MgO (100) or Al_2O_3 (001) substrates. The moment, which is present in films as thin as 12 nm, appears to reside in an interface layer where the magnetization density corresponds to approximately 0.4 T (Dorneles et al. 2004). Lattice defects are suggested as the origin of the high-temperature magnetism in hexaborides.

Ab initio calculations (Maiti 2008) show that a vacancy in the boron sublattice leads to the formation of an impurity band in the vicinity of the Fermi level, which exhibits finite exchange splitting and, therefore, a magnetic moment. Supporting this theoretical prediction, a photoemission study (Rhyee and Cho 2004) reveals the presence of weakly localized states in the vicinity of the Fermi level in the ferromagnetic CaB_6, whereas it is absent in the paramagnetic LaB_6 (Medicherla et al. 2007). All these results suggest that the nature of ferromagnetism is intrinsic. Its nature is different from that of diluted magnetic semiconductors because it is *related to the presence of localized states at the Fermi level* originating from the boron vacancies (Maiti et al. 2007).

Itinerant electron ferromagnetism in a narrow impurity band has been compared to the well-known case of the 3d band of transition metals (Edwards and Katsnelson 2006). Figure 25.4 shows the energy spectrum for the 3d-band bulk case (a) and for the impurity band, which contains a broad main band and a narrow impurity band. In the d-band bulk case, a weak itinerant

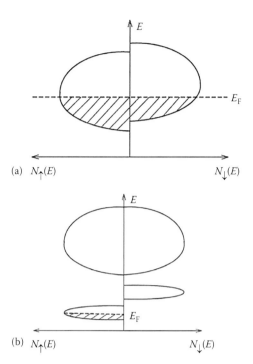

FIGURE 25.4 Schematic density of states for (a) a weak itinerant ferromagnet and (b) the ferromagnetic impurity band model. (Reprinted from Edwards, D.M. and Katsnelson, M.I., *J. Phys.: Condens. Matter*, 18, 7209, 2006.)

electron ferromagnetism is due to a small exchange splitting between majority ($\uparrow$) and minority ($\downarrow$) spin bands (Figure 25.4a). Complete spin alignment of low-density carriers takes place in an impurity band (Figure 25.4b). The Stoner criterion is satisfied due to a high density of states in the narrow impurity band, and ferromagnetism is achieved. The impurity band scenario of magnetism is qualitatively similar to the case of ferromagnetic transition metals. In the CaB_6 case, the role of the d band is played by a narrow impurity band formed from boron p orbitals, and the conduction band is formed by calcium d orbitals. The boron lattice is regarded as formed from B_2 dimers, and the nature of magnetic ordering is analogous to another second-row dimer, O_2. The spins of two p electrons in B_2 are aligned by the Hund's rule mechanism, similar to the alignment of the transition metal d band.

The model (Edwards and Katsnelson 2006) is able to explain not only ferromagnetism of undoped CaB_6, but also sheds light on the nontrivial role of iron in iron-contaminated samples: It is conceivable that Fe is a defect that leads to a partially occupied impurity band just above the valence band. Magnetism of hexaborides is associated with the impurities, but they need not to be intrinsically magnetic: an impurity band responsible for RTFM in CaB_6 is quite independent of whether or not the defects responsible for the impurity band are magnetic impurities.

In this section we have followed the evolution of the understanding of unconventional ferromagnetism phenomenon in transition metal-free hexaborides, MB_6, where M = Ca, Sr, Ba, etc. Obviously, the exotic ferromagnetism shown by the hexaborides is far from being understood, and this development is in line with the old wisdom that "For every complex problem there is an answer that is clear, simple, and wrong" (Mencken 1920).

25.5 Magnetic Semiconducting Oxides

Dilute magnetic oxides (DMO) is another example of systems where magnetism appears *eh nihilo* (Coey 2005b). A challenge of current research is to verify the very existence of intrinsic magnetism in these materials and to find out its origin. DMO were initially considered as a variety of diluted magnetic semiconductors DMS, like doped III–V compounds like (Ga, Mn) As, or II–VI materials such as (Zn, Mn)Te alloys. The intrinsic ferromagnetic order in DMS is presumably mediated by mobile charge carriers. This feature is favorable for spintronics application since it allows one to influence magnetic behavior through charge manipulation. As predicted by Dietl et al. (2000) ZnO and GaN can be the host candidate for the room-temperature ferromagnetic DMS. At present, there is no consensus on the source of ferromagnetism in DMO, but there is growing evidence that the mechanism responsible for the emergent behavior is different from DMS and stems from the interface effects (Brinkman et al. 2007, Hernando et al. 2006a,b), although the carrier mediation mechanism is still under discussion (Durst et al. 2002, Calderón and Das Sarma 2007).

25.5.1 Zink Oxide

First observations of magnetism in DMO were made on the oxides TiO_2 (Matsumoto et al. 2001) and ZnO (Ueda et al. 2001) doped with a transition metal cobalt. Immediately, the question has been raised whether the reported magnetism is an intrinsic effect or it is a trivial one, due to nanoclusters of the transition metal magnetic impurity (Norton et al. 2003), since some studies reported the phase separation and the formation of ferromagnetic clusters (see Janisch et al. 2005 for a review).

Surprisingly, the most recent results have demonstrated that oxide thin films or nanostructures do not need magnetic cations to become magnetic. Coey et al. found room-temperature ferromagnetism in non-transition-metal-doped ZnO (Coey et al. 2005a). Ferromagnetism in ZnO single crystals was triggered by the implantation of Ar ions with an energy of 100 keV (Borges et al. 2007). Experiment and theory confirmed that carbon induces ferromagnetism in nonmagnetic ZnO oxides (Pan et al. 2007). Apparently, the effect is connected with the presence of implantation-induced lattice defects.

Enhancement of ferromagnetism upon thermal annealing was found in pure ZnO (Banerjee et al. 2007) and explained by the formation of the anionic vacancy clusters where the magnetic state is achieved either through the superexchange between vacancy clusters via isolated F^+ centers or through a limited electron delocalization between vacancy clusters.

Surprisingly, such a magnetically strong element as cobalt *suppresses* ferromagnetism in ZnO. Ghoshal and Kumar (2008) were able to achieve ferromagnetic state in ZnO films without

transition metals, just by tuning the oxygen content in the film. Co doping of the intrinsically magnetic films suppressed the magnetization of the films.

The experiments of Xu et al. (2008) attribute the observed ferromagnetism in ZnO films to zinc vacancies and not to oxygen. The authors suggest that a careful control of defects in ZnO rather than doping with magnetic ions might be possibly a better method to obtain reproducible, intrinsic, and homogeneous ferromagnetism in ZnO at room temperature.

A comparative first-principles study has been done for ZnO in both pure and cobalt-doped states (Sanchez et al. 2008). A robust ferromagnetic state is predicted at the O (0001) surface even in the absence of magnetic atoms, correlated with the number of p holes in the valence band of the oxide.

ZnO nanowires prepared by oxidation of electrodeposited Zn wires show ferromagnetism at room temperature (Yi et al. 2008). A detailed study indicates that, owing to incomplete oxidation, Zn clusters embedded in the ZnO matrix may attribute to the room-temperature ferromagnetism.

Another method for triggering RTFM in ZnO films, namely, by deposition of nonmagnetic metallic clusters on the surface of ZnO film has been demonstrated (Ma et al. 2008). Both transmission electron microscopy and x-ray photoelectron spectroscopy suggest that the observed RTF is associated with the presence of the clusters of some of nonmagnetic metals. ZnO films covered with Zn, Al, and Pt do show room temperature ferromagnetism after vacuum annealing while (Ag, Au)/ZnO films do not. In addition, the ferromagnetism is normally destroyed when the metal clusters are oxidized. Even so, the magnetism, which is destroyed by oxidation of the Al/ZnO structure, survives in the case Pt/ZnO. The latter result speaks in favor of the model of the cluster-triggered ferromagnetism because Pt is stable against oxidation.

A clear evidence that RTFM in nanocrystalline ZnO is governed by defects comes from the experiments where paramagnetic ZnO becomes ferromagnetic once oxygen defects are introduced in it (Sanyal et al. 2007). Room-temperature ferromagnetism (FM) has been observed in laser-ablated ZnO thin films. The FM in this type of compound does not stem from oxygen vacancies as in the case of TiO_2 and HfO_2 films, but from defects on Zn sites, which are located mostly at the surface and/or the interface between the film and the substrate (Hong et al. 2007a,b). Size and shape of nanocrystals are important as follows from the observation of RTFM in ZnO nanorods with diameters about 10 nm and lengths of below 100 nm (Yan et al. 2008).

Ferromagnetic order can be induced in ZnO by 2p light element (N) doping (Shen et al. 2008) or by means of Fe ion implantation or just by vacuum annealing at mild temperatures without any transition metal doping (Zhou et al. 2008). Comparison of the results obtained on the samples with and without magnetic atoms speaks against the DMS model where magnetic coupling of localized d-moments of the implanted Fe, and FM properties are discussed with respect to defects in the ZnO host matrix.

In the context of the discussions of the role of magnetic impurity clustering, Cu-doped ZnO is an interesting examples because Cu atoms do not have clustering tendency and neither of Cu-based compounds is ferromagnetic. RTFM in Cu-doped ZnO was theoretically predicted (Park and Min 2003, Park et al. 2003, Wu et al. 2006, Ye et al. 2006) and observed experimentally (Ando et al. 2001, Cho et al. 2004, Chakraborti et al. 2007, Hou et al. 2007, Xing et al. 2008). However, a magnetic circular dichroism study in Cu-doped ZnO thin films did not show any significant spin polarization on the Cu 3d and O_2 p states, although the samples showed RTFM and were free of contamination (Keavney et al. 2007). Several Cu-doped oxides have been studied (Dutta et al. 2008), but only ZnO oxide demonstrates ferromagnetic behavior, and thus the CuO phase is suggested to be a paramagnet. A comparative study on Cu-doped ZnO nanowires prepared by two distinct methods and demonstrated unambiguously an enhancement of RTFM by structural inhomogeneity (Xing et al. 2008). The results suggest that RTFM is not a homogeneous bulk property, but a surface effect by nature: the alteration of the electronic structure induced by impurities and defects plays an important role in magnetism (Buchholz et al. 2005, Garcia et al. 2007, Seehra et al. 2007).

Several different explanations for the behavior of Cu-doped ZnO materials have been proposed, and the model of oxygen vacancies (Chakraborti et al. 2007) contradicts the DFT calculations (Ye et al. 2006), which show that vacancies tend to destroy ferromagnetism. In the experiments on ZnO:Cu nanowires (Shuai et al. 2007), the samples annealed in oxygen showed stronger RTFM than those annealed in Ar, and this result speaks against the oxygen vacancies. The authors believe that FM occurs due to the hybridization between partly occupied Cu 3d bands and O 2p bands: these energy levels are closely situated, and the delocalized holes induced by O 2p and Cu 3d hybridization can efficiently mediate the ferromagnetic exchange interaction (Huang et al. 2006).

XMCD studies on Co and Li doped ZnO have not revealed *any element-specific signature of ferromagnetism*. Only paramagnetic signal was recorded from cobalt. This result suggests that RTFM in doped ZnO has an intrinsic origin and is caused by the oxygen vacancies (Tietze et al. 2008).

Ferromagnetism was observed in Ga-doped ZnO films, and the following mechanism has been proposed (Bhosle and Narayan 2008):

- Ferromagnetism is attributed to oxygen vacancies.
- The vacancies act as F-centers and trap free electrons.
- The trapped electrons tend to get easily polarized under the influence of the magnetic field.
- The F-centers freeze the spin, resulting in almost temperature-independent magnetism.
- RTFM is greatly facilitated by the high concentration of free carriers.

The mechanism has been verified by the experiments on the films deposited at different technological conditions and, therefore, different vacancy and free carrier concentrations. The experiments show that RTFM decreases with the decrease of the concentrations. Supporting evidence comes from annealing, which leads to the decrease of both the F-center and free carrier concentration and RTFM quenching. Upon annealing, three effects were observed simultaneously: the redshift of the optical absorption edge, the decrease in the carrier concentration and the oxygen vacancies obtained from electrical resistivity and Hall measurements, and loss of ferromagnetism.

A suggestion has been made about the role of Ga in inducing RTFM in the ZnO system: first, Ga alters the energy levels in oxygen vacancies, and second, it provides additional carriers and thus enhances the free carrier mediation of the spin interaction of the polarized F-centers. These results clearly demonstrate the dependence of magnetic properties on the vacancy concentration in ZnGaO films.

But the model of oxygen vacancies is not the only one that is under consideration. The observed magnetism could be due to Zn vacancy rather than O vacancy, and magnetic moment arises from the unpaired 2p electrons at O sites surrounding the Zn vacancy (Wang et al. 2008a,b).

What has become clear now is that ZnO can show ferromagnetic properties from defects created by doping, implantation, or annealing. An easy way to create defect induced ferromagnetism in ZnO by means of mechanical force has been recently found (Potzger et al. 2008), and this is an easy mechanical approach using a conventional hammer and producing small "flakes" of ZnO. There is large evidence that the ferromagnetic signal comes from strain or domain boundaries in the flakes.

25.5.2 Titanium Oxide

Similar results were obtained from the experiments on titanium oxide films. Films of TiO_2, which showed ferromagnetism at room temperature, were manufactured by various deposition techniques: spin-coated TiO_2 thin films or pulsed laser ablated TiO_2 thin films. The fact that in thick films deposited under the same conditions the magnetic ordering degraded enormously give the grounds to suggest that not only defects but also the confinement effects seem to be important.

A semiconducting material, $TiO_{2-\delta}$ is ferromagnetic up to $880\,K$, without the introduction of magnetic ions (Yoon et al. 2006). The subscript $(2-\delta)$ implies oxygen deficiency in the samples, or the presence of oxygen vacancies. Magnetism in these films is controlled by anion defects. Magnetism scales with conductivity, suggesting the double exchange interaction scenario.

Unprecedentedly strong RTFM has been observed in $TiO_{2-\delta}$ nanoparticles synthesized by the sol-gel method and annealed under different reducing atmosphere (Zhao et al. 2008). In a model of oxygen vacancies, the authors (Zhao et al. 2008) explain their results by the aggregation of the oxygen vacancies. This process is more pronounced for small-size nanoparticles that have a larger surface-to-volume ratio.

Pure TiO_2 thin films produced by both spin-coating and sputter-deposition techniques on sapphire and quartz substrates demonstrated RTFM when annealed in vacuum (Sudakar et al. 2008), while the air-annealed samples showed much smaller, often negligible, magnetic moments.

Spin-coated pristine TiO_2 thin films show magnetic behavior similar to that of pulsed laser ablated TiO_2 thin films (Hassini et al. 2008). Observation of the same effect on the films obtained by different deposition techniques instills confidence in the intrinsic nature of ferromagnetism. Two other observations give an additional support: (1) annealing in the oxygen atmosphere degrades the moment and (2) thicker films deposited at the same conditions have the magnetic ordering degraded enormously. Consequently, two factors are considered: first, defects and/or oxygen vacancies and second, confinement effects seem to be important.

The set of experimental data on transition metal-free oxides raised the question: Does Mn doping play any key role in tailoring the ferromagnetic ordering of TiO_2 thin films? (Hong et al. 2006a,b) When the Mn concentration is small and does not distort the TiO_2 structure, it enhances the ferromagnetic component into the magnetic moment of the already ferromagnetic TiO_2 film base, which is already ferromagnetic, and then enhances it. But at higher concentrations, Mn drastically degrades and destroys the ferromagnetic ordering. These experiments show that Mn doping indeed does not play any key role in introducing FM in TiO_2 thin films. In other words, magnetic semiconducting oxides cannot be considered as a variety of dilute magnetic semiconductors DMS: *DMO are not DMS*.

25.5.3 Hafnium Oxide

Unexpected magnetism was observed in undoped HfO_2 thin films on sapphire or silicon substrates (Venkatesan et al. 2004). While for the films prepared by pulsed laser deposition the results were confirmed (Hong et al. 2006a,b), other groups did not observe the effect on the films grown by metalorganic chemical vapor deposition (Abraham et al. 2005) and by pulsed-laser deposition (Rao et al. 2006). While a very weak signal was observed in lightly Co-doped HfO_2 films, it came presumably from a Co-rich surface layer (Rao et al. 2006). The ferromagnetic signal measured on pulsed-laser deposited thin HfO_2 films at different oxygen pressure was ascribed to possible tweezer contamination (Hadacek et al. 2007). Similarly, very weak ferromagnetism observed in pure and Gd-doped HfO_2 films on different substrates was attributed to either impure target materials or signals from the substrates. An intrinsic magnetism was not identified either on films, or on HfO_2 powders annealed in pure hydrogen flow (Wang et al. 2006a,b).

On the theoretical side, the RTFM in the HfO_2 system is not forbidden. Isolated cation vacancies in HfO_2 may form high-spin defect states, resulting in a ferromagnetic ground state (Das Pemmaraju and Sanvito 2005). A model for vacancy-induced ferromagnetism is based on a correlated model for oxygen orbitals with random potentials representing cation vacancies (Bouzerar and Ziman 2006). For certain potentials, moments appear on oxygen sites near defects (Beltran et al. 2008). A specific nonmagnetic host doping is proposed for HfO_2 or ZrO_2.

Such subtle phenomena as magnetism in otherwise nonmagnetic materials are not immediately noticeable, and one should

excel in experiments to observe the effect. To make closed-shell-oxides magnetic, experimental methods are required, which can produce the highly nonequilibrium defect concentrations, and one of these methods is low-temperature colloidal syntheses. Colloidal HfO_2 nanorods with controllable defects have been synthesized (Tirosh and Markovich 2007). Defects have been studied by high-resolution electron microscopy and by optical absorption spectroscopy, and it was shown that nanocrystals with a high defect concentration exhibit ferromagnetism and superparamagnetic-like behavior (Shinde et al. 2004).

25.5.4 Other Nonmagnetic Oxides

Ferromagnetism in oxides can be induced by the elements of Periodic Table, which have nothing to do with magnetism in the bulk state. Ab initio study (Maca et al. 2008) of the induced magnetism in ZrO_2 shows that the substitution of the cation by an impurity from the groups IA or IIA of the Periodic Table (K and Ca) leads to opposite results: K impurity induces magnetic moment on the surrounding O atoms in the cubic ZrO_2 host while Ca impurity leads to a nonmagnetic ground state. Moreover, the authors suggest switching on/off ferromagnetism in potassium doped oxides by density of states tuning through applying the gate voltage.

Similarly, experiments on Mn-doped SnO_2 films show that a transition metal doping does not play any key role in introducing FM in the system (Hong et al. 2008). Mn doping, even if with a small content, degrades the structure of the SnO_2 host and reduces its magnetic moment. Both oxygen vacancies and confinement effects are assumed to be key factors in introducing magnetic ordering into SnO_2.

Observation of room-temperature ferromagnetism in nanoparticles of nonmagnetic oxides such as CeO_2, Al_2O_3, ZnO, In_2O_3, and SnO_2 (Sundaresan et al. 2006) lead to a conclusion that is probably adventurous: Ferromagnetism is a universal feature of nanoparticles of otherwise nonmagnetic oxides. To explain the origin of ferromagnetism in these nonmagnetic oxides, they assumed the existence of oxygen vacancies at the surfaces of these nanoparticles.

Nanoparticles of cerium oxide show distinct dependence of magnetic properties on particle size and shape. CeO_2 nanoparticles and nanocubes have been investigated both experimentally and theoretically (Ge et al. 2008), and it is found that monodisperse CeO_2 nanocubes with an average size of 5.3 nm do show ferromagnetic behavior at ambient temperature. First-principles calculations reveal that oxygen vacancies in pure CeO_2 cause spin polarization of f electrons for Ce ions surrounding oxygen vacancies, resulting in net magnetic moment for pure CeO_2 samples. The role of particle size is actually played by the surface area because an oxygen vacancy at surface induces more magnetic moments than in bulk. The role of oxygen is, however, a controversial point: size-dependent ferromagnetism in cerium oxide nanostructures was found to be independent of oxygen vacancies (Liu et al. 2008). Ferromagnetism in nanosized CeO_2 powders was studied in nanoparticles of different sizes but found only in sub-20 nm powders. Annealing studies combined with photoluminescence measurements showed that oxygen vacancies did not mediate ferromagnetism in the samples.

Remarkable room-temperature ferromagnetism was observed in undoped TiO_2, HfO_2, and In_2O_3 thin films (Hong et al. 2006a,b). On the other hand, in another study, no trace of ferromagnetism has been detected in In_2O_3 even with samples sintered under argon, except extrinsic ferromagnetism for samples with magnetic dopant concentrations exceeding the solubility limit (Berardan et al. 2008).

The room-temperature weak ferromagnetism of amorphous $HfAlO_x$ thin films has been demonstrated (Qiu et al. 2006) and it is argued that interfacial defects are one of the possible sources of the weak ferromagnetism.

Room-temperature size-dependent ferromagnetism was observed in sub-20-nm sized CeO_2 nanopowders. In order to check the role of oxygen vacancies, which have been speculated to be the cause of ferromagnetism in undoped oxides, annealing was performed in different atmospheres. This study showed that ferromagnetism is not linked to oxygen vacancies, but possibly to the changes of a cation surface defect state.

The occurrence of spin polarization at ZrO_2, Al_2O_3, and MgO surfaces is proved by means of ab initio calculations within the density functional theory (Gallego et al. 2005). Large spin moments develop at O-ended polar terminations, transforming the nonmagnetic insulator into a half-metal. The magnetic moments mainly reside in the surface oxygen atoms and their origin is related to the existence of 2p holes of well-defined spin polarization at the valence band of the ionic oxide. The direct relation between magnetization and local loss of donor charge makes it possible to extend the magnetization mechanism beyond surface properties.

The creation of collective ferromagnetism in nonmagnetic oxides by intrinsic point defects such as vacancies has been discussed (Osorio-Guillen et al. 2007). This effect is in principle possible, but the minimum concentration of vacancies is eight orders of magnitude higher than the equilibrium vacancy concentration in HfO_2 in the most favorable growth conditions. Thus, equilibrium growth cannot lead to ferromagnetism, and experimental methods are required that can produce the highly nonequilibrium defect concentrations.

25.6 Magnetism in Metal Nanoparticles

Hori et al. (1999) observed magnetism in ~3 nm gold nanoparticles with an unexpected large magnetic moment of about 20 spins per particle. Since then, several papers reported magnetism in gold nanoparticles with the emphasis on the stabilization of the particles by polymers, diameter dependence of the ferromagnetic spin moment, experimentally (Hori et al. 2004, Reich et al. 2006) and theoretically (Michael et al. 2007) and observation of spin polarization of gold by x-ray magnetic circular dichroism.

Self-assembled alkanethiol monolayers on gold surfaces have been reported to show permanent magnetism (Crespo et al. 2004). Ferromagnetic behavior observed in thiol-capped Au

nanoparticles (NP) cannot be ascribed to the presence of magnetic impurities. Au nanoparticles with similar size but stabilized by means of a surfactant, i.e., weak interaction between protective molecules and Au surface atoms, are diamagnetic, as bulk Au samples are. Gold nanoparticles (Au NPs) capped with dodecanethiol showed superparamagnetic or diamagnetic behavior depending on its size (Dutta et al. 2007).

A thiol is a compound that contains the functional group composed of a sulfur atom and a hydrogen atom (–SH). Alkanes are chemical compounds that consist only of the elements carbon (C) and hydrogen (H) (i.e., hydrocarbons), wherein these atoms are linked together by single bonds. Sulfur has particular affinity for gold, and alkanes with a thiol head group will stick to the gold surface, and alkane thiols (or alkanethiols, Figure 25.5) are well known for their ability to form monolayers on gold. The discovery of ferromagnetism of gold capped with alkanethiols has convincingly shown that magnetism of oxides does not require transition metal atoms (Crespo et al. 2008).

It has been suggested that ferromagnetism is associated with 5d localized holes generated through Au–S bonds (Crespo et al. 2004). These holes give rise to localized magnetic moments that are frozen due to the combination of the high spin–orbit coupling (1.5 eV) of gold and the symmetry reduction associated with two types of bonding: Au–Au and Au–S. Thus, the ferromagnetism stems from the charge transfer processes. According to electron circular dichroism measurements carried out on thiolated organic monolayers on gold (Vager et al. 2004), the magnetic moment originates from the orbital momentum. Highly anisotropic giant moments were also observed for self-organized organic molecules linked by thiols bonds to gold films (Carmeli et al. 2003). FM has been observed in gold capped with thiol groups, both in the form of thin films and nanoparticles. However, there is a noticeable difference in magnetic behavior: the magnetic moment reaches 10 or even $100\,\mu_B$ per atom) for films, but it is extremely low ($0.01\,\mu_B$ per atom) for nanoparticles. Probably, this phenomenon is due to the directional nature of the assembled organic layers (Figure 25.6.).

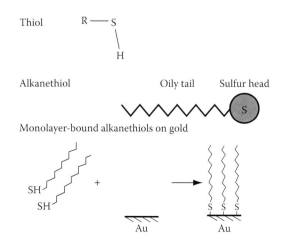

FIGURE 25.5 Schematic picture of a thiol, an alkanethiol and the formation of an alkanethiol/gold interface.

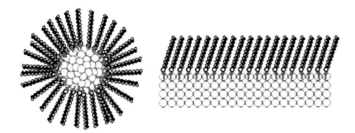

FIGURE 25.6 Scheme of the magnetic moments arising at surfaces capped with organic molecules. Due to the spin–orbital interaction, the magnetic moments are fixed in the bond direction that for NPs are uniformly distributed while for films there is a unique orientation perpendicular to the surface. (Reprinted from Quesada, A. et al., *Eur. Phys. J. B*, 59, 457, 2007. With permission.)

Self-assembled monolayers on gold of double-stranded DNA oligomers create a strong and oriented magnetic field. There is clear difference between monolayers made from single-stranded DNA and those made from double-stranded DNA despite the fact that both molecules are chiral (Ray et al. 2006).

X-ray magnetic circular dichroism experiments (Yamamoto et al. 2004) provided direct evidence for ferromagnetic spin polarization of Au nanoparticles with a mean diameter of 1.9 nm. X-ray magnetic circular dichroism is an elementally sensitive method, so in these experiments only the gold magnetization is explored. Magnetization of gold atoms as estimated by XMCD shows a good agreement with results obtained by conventional magnetometry. Further XMCD studies (Negishi et al. 2006) confirmed that the presence of localized holes created by Au–S bonding at the interface, rather than the quantum size effect, is responsible for the spin polarization of gold clusters. Magnetic moment increases with the cluster size but remains constant when calculated for a surface atom.

The available explanation of orbital ferromagnetism and giant magnetic anisotropy at the nanoscale (Hernando et al. 2006a,b) assumes the induction of orbital motion of surface electrons around ordered arrays of Au–S bonds. It is considered that electrons are pumped up from the substrate to the molecular layer; at the same time, spin–orbit interaction effects, known to be extremely important in gold surfaces, are taken into account.

The experiments with different capping agents clarified the role of adsorbing molecules. Two types of thiol-capped gold nanoparticles (NPs) with similar diameters between 2.0 and 2.5 nm and different organic molecules linked to the sulfur atom: dodecanethiol and tiopronin have been studied (Guerrero et al. 2008). The third capping agent, tetraoctyl ammonium bromide, has also been included in the investigation since it interacts only weakly with the gold surface atoms and, therefore, this system can serve as a reference sample of naked gold nanoparticles. Modifications of the electronic structure clearly reveals itself by the quenching of the surface plasmon resonance for dodecanethiol capping and its total disappearance for tiopronin-capped NPs (Figure 25.7). Regarding the magnetization, dodecanethiol-capped NPs have a ferromagnetic-like behavior, while the NPs capped with tiopronin exhibit a paramagnetic behavior, whereas

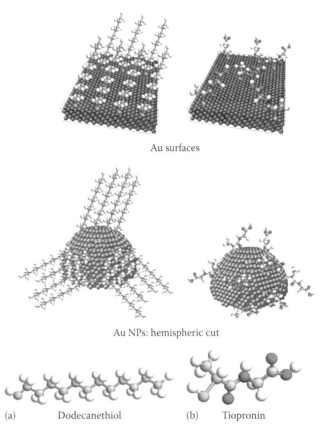

Au surfaces

Au NPs: hemispheric cut

(a)   Dodecanethiol   (b)   Tiopronin

FIGURE 25.7 Scheme of self-assembled monolayer and nonordered array formation: (a) for dodecanethiol- and (b) for tiopronin-functionalized gold surfaces and gold NPs. The structure of the capping molecules is shown using a "ball–stick" model: hydrogen (white), carbon (gray), nitrogen, oxygen and sulfur (dark-gray). (Reprinted from Guerrero, E. et al., *Nanotechnology*, 19, 17501, 2008. With permission.)

the reference samples remained diamagnetic. It is has been concluded that straight chains with a well-defined symmetry axis induce orbital momentum, which not only contributes to the magnetization but also to the local anisotropy. Due to the domain structure of the adsorbed molecules, orbital momentum is not induced for tiopronin-capped NPs.

The following scenario for the gold nanoscale magnetism has been proposed: Capping gold surfaces with certain organic molecules creates surface bonds. The bonds give rise to magnetic moments, and in the case of atomically flat gold surface the magnetic moments are giant. Due to the strong spin–orbit interaction characteristic to gold, these magnetic moments are blocked along the bond direction showing huge anisotropy (De La Venta et al. 2007).

Reversible phototuning of ferromagnetism that was observed in gold nanoparticles passivated with azobenzenederivatized ligands is believed to become the basis for developing future magneto-optical memory (Suda et al. 2008).

A direct observation of the intrinsic magnetism of Au-atoms in thiol-capped gold nanoparticles, which possess a permanent magnetization at room temperature (Garitaonandia et al. 2008), was obtained by using two element specific techniques: x-ray magnetic circular dichroism on the L edges of the Au and Au-197 Mössbauer spectroscopy. Besides, silver and copper nanoparticles synthesized by the same chemical procedure also present room-temperature permanent magnetism. The observed permanent magnetism at room temperature in Ag and Cu dodecanethiol-capped nanoparticles proves that the same physics can be applied to more elements, opening the way to new and still not-discovered applications and to new possibilities to research basic questions of magnetism. Two main questions have been answered: is the origin of the magnetism really located on the Au atoms? And if it is so, are there more elements susceptible to induce magnetism on them? XMCD spectra determined the location of the magnetic atoms in the nanoparticles and demonstrated that the effect is observable also on copper and silver (Garitaonandia et al. 2008).

Some 4d elements, nominally nonmagnetic, e.g., Ru, Rh, Pd, may also exhibit nanoscale magnetism. Magnetism in Pd nanoparticles has been ascribed to the electronic structure alteration by the twin boundaries at the nanoparticles surfaces (Sampedro et al. 2003). A Pd nanoparticle containing only single Co atom exhibits a single-domain nanomagnet behavior (Ito et al. 2008). Platinum atoms, not magnetic in the bulk, become magnetic when grouped together in small clusters (Liu et al. 2006). Platinum monatomic nanowires were predicted to spontaneously develop magnetism (Smogunov et al. 2008), and it was shown that Pd and Pt nanowires are ferromagnetic at room temperature, in contrast to their bulk form (Teng et al. 2008). Two remarkable effects are observed at the nanowires at low temperatures, namely below 40 K for Pd and 60 K for Pt. The first one is a magnetic memory effect: Hysteresis loop is not symmetric but shifts along the field's axis on a certain value HB (biased field). Another unusual effect is the temperature dependence of the coercive force: while magnetization increases at low temperatures, coercivity becomes weaker. Apparently, this is the consequence of electron localization at low temperatures, which enhances magnetic moment but quenches the exchange interactions between them.

The magnetic properties of Ru/Ta mixture drastically change by Xe atom irradiation, and the reason is the formation of small clusters in the Ru/Ta matrix under irradiation (Wang et al. 2006a,b).

More surprisingly, potassium clusters display a nontrivial magnetic behavior on the nanoscale: low T ferromagnetism when the clusters are incorporated into zeolite lattice (Nozue et al. 1992). Two models have been invoked to explain this behavior: spin-canting mechanism of antiferromagnet (Nakano et al. 2000) and N-type ferrimagnetism (Nakano et al. 2006), which is constructed of nonequivalent magnetic sublattices of K clusters the matrix. Potassium clusters that are 60 atoms on average, when accommodated in the magnetic nanographene-based porous network, become antiferromagnetic (Takai et al. 2008) due to the charge transfer with the host nanographene.

The results on metal films and nanoparticles point out the possibility to observe magnetism at nanoscale in materials without transition metals and rare earths atoms, and are of fundamental value to understand the magnetic properties of surfaces.

25.7 Magnetism in Semiconductor Nanostructures

Clear evidence of nanoscale magnetism comes from the observation of room-temperature ferromagnetic behavior in semiconductor nanoparticles and quantum dots (Jian et al. 2006), as well at in the heterostructures.

Semiconductor nanoparticles show increasing magnetization for decreasing diameter (Neeleshwar et al. 2005). RTFM in undoped GaN and CdS semiconductor nanoparticles of different sizes was observed for the particles with the average diameter in the range 10–25 nm. RT saturation magnetization is of the order of 10^{-3} emu/g, which is comparable to that observed in nanoparticles of nonmagnetic oxides. Agglomerated particles of GaN and CdS loose the FM properties: the saturation magnetic moment decreases with the increase in particles size, suggesting that ferromagnetism is due to the defects confined to the surface of the nanoparticles (Madhu et al. 2008).

Ferromagnetism has been also measured in PbS attached to the GaAs substrate (Zakrassov et al. 2008). PbS nanoparticles were attached to GaAs through organic linkers. The magnetization is anisotropic and the magnetic moment reaches saturation for a magnetic field of about 2000 Oe applied parallel to the surface, while it responds to the magnetic field almost linearly when the field is applied perpendicular to the surface. Interestingly, the magnetic anisotropy depends on the alignment of the long axis of the organic molecule linker relative to the surface normal (Figure 25.8). Anisotropy follows the orientation of the long axis of the organic molecule. When the PbS were replaced with CdSe NPs no magnetic signal could be detected.

RTFM in CdSe quantum dots (QD) capped with TOPO (tri-*n*-octylphosphine) has been observed (Seehra et al. 2008). The strength of magnetism weakens with increase in size of the QDs. This phenomenon is classified as *ex nihilo* magnetism since the effect stems from the contact of two diamagnetic materials, namely CdSe and TOPO. The magnetism here is possibly due to the charge transfer from Cd d-band to the oxygen atoms of TOPO.

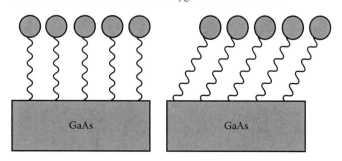

FIGURE 25.8 Scheme of the PbS nanoparticles attached to the GaAs substrate via organic molecules when the molecules are aligned parallel (left) or at some angle (right) relative to the surface normal. (Reprinted from Zakrassov, A. et al., *Adv. Mater.*, 20, 2552, 2008. With permission.)

Adsorption of monolayers of organic molecules onto the surface of ferromagnetic semiconductor heterostructures produces large, robust changes in their magnetic properties (Kreutz et al. 2003). Effects of chemisorption of polar organic molecules onto ferromagnetic GaAs/GaMnAs heterostructures has been investigated (Carmeli et al. 2006). The chemisorbed heterostructures exhibit striking anisotropic enhancement of the magnetization, while GaAs substrates physisorbed with the same molecules show no change in magnetic properties. Thus, the enhanced magnetism of the chemisorbed heterostructures reflects changes in the spin alignment that arise from the surface bonding of the organic monolayer.

The adsorption of closed packed monolayers on solid substrates enriches the surface with novel qualities. The new electronic and magnetic properties emerge from the charge transfer from the organized organic layer substrate and, in particular, from the alignment of the spin of the transferred electrons/holes (Naaman and Vager 2006).

Defect-induced ma GaN is believed to be due to a Ga vacancy defect, which can show induced local magnetic moment in N atoms (Hong 2008). Cation-vacancy-induced intrinsic magnetism in GaN and BN is investigated, and a dual role of defects is shown: First, the defects create a net magnetic moment, and second, the extended tails of defect wave functions mediate surprisingly long-range magnetic interactions between the defect-induced moments (Dev et al. 2008).

25.8 Ferromagnetism in Carbon Nanostructures

25.8.1 Pyrolytic Carbonaceous Materials

Magnetic ordering at high temperatures in carbon-based compounds has been persistently reported since 1986 when several tens of papers and patents describing ferromagnetic structures containing either pure carbon or carbon combined with first row elements were published (reviewed in Ref. Makarova 2004). The earlier results on magnetic carbon compounds obtained by pyrolysis of organic compounds at relatively low temperatures were poorly reproducible, and this fact caused natural skepticism about the observed effects.

25.8.2 Graphite

A new step in the studies of carbon magnetism started when ferromagnetic and superconducting-like magnetization hysteresis loops in highly oriented pyrolytic graphite (HOPG) samples above room temperature were reported (Kopelevich et al. 2000). Later the authors retracted from superconducting-like hysteresis loops as they had been partially influenced by an artifact produced by the SQUID current supply, but ferromagnetic behavior remained beyond doubt (Kopelevich and Esquinazi 2007). Absence of correlation between magnetic properties and impurity content was found in highly oriented pyrolytic graphite (Esquinazi et al. 2002), suggesting intrinsic ferromagnetic signal.

25.8.3 Porous Graphite

Various independent groups have reported ferromagnetism and anomalous magnetic behaviors in porous graphitic based materials. Ferromagnetic correlations have been observed in activated mesocarbon microbeads mainly composed of graphitic microcrystallites. Magnetization curves measured at 1.7 K showed a marked hysteresis, which becomes less and less visible with increasing temperature although still was present at room temperature (Ishii et al. 1995). The occurrence of high-temperature ferromagnetism has been found in microporous carbon with a three-dimensional nanoarray (zeolite) structure and was associated with the fragments with positive and/or negative curvature (Kopelevich et al. 2003). Similar behavior was described for glassy carbon (Wang et al. 2002). The carbon nanofoam produced by bombarding carbon with a high-frequency pulsed laser in an inert gas displays strong paramagnetic behavior, which is very unusual for the carbon allotropies (Rode et al. 2004). The material contains both sp^2 and sp^3 bonded carbon atoms and exhibits ferromagnetic-like behavior with a narrow hysteresis curve and a high saturation magnetization. Strong magnetic properties fade within hours at room temperature; however, at 90 K the foam's magnetism persists for up to 12 months. Magnetic behavior of the nanofoam is complex: the nonlinear component does not scale with the H/T which is expected for the superparamagnetism, and the thermal behavior is more consistent with the spin glass picture with unusually high freezing temperature (Blinc et al. 2006). The higher *g* factors are typical of amorphous carbon systems with significant sp^3 character, i.e., strongly nonplanar parts of a carbon sheet, and magnetization values at low temperatures 25 times exceed the extrinsic contribution (Arcon et al. 2006). Oxygen-eroded graphite (Mombru et al. 2005, Pardo et al. 2006) shows multilevel ferromagnetic behavior with the Curie temperature at about 350 K.

25.8.4 Carbon Nanoparticles

In the experimental studied of carbon nanoparticles magnetization values more than an order of magnitude larger than the expected saturated magnetization due to any possible transition metal impurity were reported. For carbon nanoparticles prepared in helium plasma (Akutsu and Utsushikawa 1999), the saturation magnetization increases with decreasing grain size, and the grain size of the carbon fine particles having the highest magnetization is 19 nm. More recently, ferromagnetism was found in carbon nanospheres (Caudillo et al. 2006), macrotubes (Li et al. 2007), necklace-like chains and nanorods (Parkansky et al. 2008), and highly oriented pyrolytic graphite nanospheres grown from Pb-C nanocomposites (Li et al. 2008a,b). Interestingly, magnetic carbon nanoparticles have definite shapes, mainly spherical shapes (Figure 25.9). In the experiments of Parkansky et al. (2008) the particles were magnetically separated, and magnetic particles included and nanotubes and nanorods with lengths of 50–250 nm and spheres with diameters of 20–30 nm whereas nonmagnetic stuff did not have particular shapes.

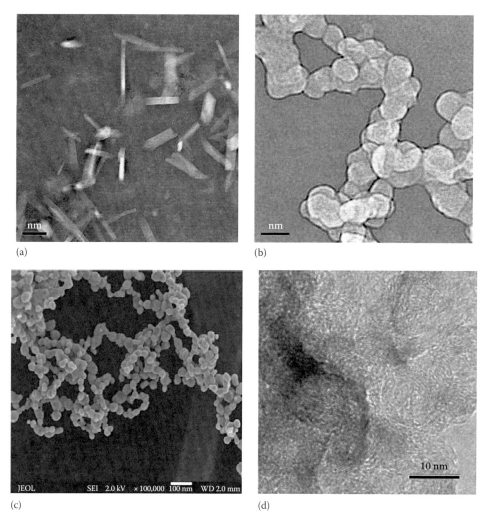

FIGURE 25.9 Magnetic carbon nanoparticles. (a) Nanotubes and nanorods with lengths of 50–250 nm and diameters of 20–30 nm. (Reprinted from Parkansky, N. et al., *Carbon*, 46, 215, 2008. With permission.) (b) Chains of 30–50 nm diameter spheres. (From Parkansky, N. et al., *Carbon*, 46, 215, 2008.) (c) Nanospheres. (Reprinted from Caudillo, R. et al., *Phys. Rev. B*, 74, 214418, 2006. With permission.) (d) Pyrolytic carbon nanospheres. (Reprinted from Li, D. et al., *J. Phys. D Appl. Phys.*, 41, 115005, 2008a. With permission.)

25.8.5 Nanographite

Various types of magnetic behavior were discovered in nanocarbon derived from graphite. There exists strong experimental evidence that the edge states in nanographite disordered network govern its magnetic properties. One example is nanographite obtained from the heat treatment of nano-diamond particles (Andersson et al. 1998). Strong antiferromagnetic coupling has been found between the spins localized on the surface of similar particles (Osipov et al. 2006). Another example is activated carbon fibers (ACF), which can be considered as a three-dimensional random network of nanographitic domains with characteristic dimensions of several nanometers (Shibayama et al. 2000). Temperature dependencies of the susceptibility taken in zero field cooled regime indicate a presence of a quenched disordered magnetic structure like a spin glass state. This effect appears in the vicinity of the metal–insulator transition, giving grounds to believe that the coexistence of the edge-state localized spins and

the conduction π-electrons causes the magnetic state in which the exchange interactions between the localized spins are mediated by the conduction electrons (Enoki and Takai 2006).

An important proof for the edge-state inherited unconventional magnetism is the magnetic switching phenomenon, which has been found in the activated carbon fibers. Physisorption of water drastically changes magnetic properties, although water itself is nonmagnetic (Sato et al. 2003). Water molecules compress the nanographite domains, reducing the interlayer distance in a stepwise manner. Physisorption leads to the enhancement of the antiferromagnetic exchange interaction of the edge-state localized spins situated at the adjacent nanographene layers (Sato et al. 2007). The physisorption of various guest materials can cause a reversible low-spin/high-spin magnetic switching phenomenon, while physisorption of oxygen molecules is responsible for the giant magnetoresistance of the nanographite network (Enoki and Takai 2008).

25.8.6 Fullerenes

Polymer–C_{60} composite with room-temperature ferromagnetism was first reported when C_{60} was ultrasonically dispersed in a dimethylformamide solution of polyvinylidenefluoride (Ata et al. 1994). Fullerene hydride $C_{60}H_{36}$ (Lobach et al. 1998) and $C_{60}H_{24}$ (Antonov et al. 2002) have been reported to be room-temperature ferromagnets. Room-temperature ferromagnetism of polymerized fullerenes was first reported when the samples were exposed to oxygen under the action of the strong visible light (Murakami and Suematsu 1996). In photopolymers saturation, magnetization progressively increases with increasing exposure time (Makarova et al. 2003). The existence of ferromagnetic phase in photolyzed C_{60} was confirmed by the three methods: (1) SQUID; (2) ferromagnetic resonance in the EPR; and (3) low-field nonresonance derivative signal (Owens et al. 2004). The experiments were done in a chamber with flowing oxygen, which exclude any possibility of penetration of metallic particles during the experiment.

The presence of a magnetically ordered phase was revealed in pressure-polymerized C_{60} (Makarova et al. 2001a,b). Later, several authors retracted from this paper on the grounds that the impurity content measured on the surface of the samples by particle-induced x-ray scattering (PIXE) was higher than that obtained by the bulk impurity analysis and that the Curie temperature was close to that of Fe_3C (Makarova and Palacio 2006). Some authors did not agree with the retraction. One of the reasons of disagreement was that the concentration of impurities within the information depth of the PIXE method (36 μm) measured on the same samples was still three times less than necessary for the observed magnetic signal (Han et al. 2003, Spemann et al. 2003). The whole set of experiments was repeated with the same team of technologists and on the same equipment (Makarova and Zakharova 2008), and for several samples the magnetization values were higher than those expected from the metallic contamination. Having followed in situ the depolymerization process through the temperature dependence of the ESR signal (Zorko et al. 2005), the authors conclude that the magnetic signal is directly connected with the polymerized fullerene phase and cannot be attributed to iron compounds. Systematic study of synthesis conditions for the production of the ferromagnetic fullerene phase was made by another team. Only samples prepared in a narrow temperature range show a ferromagnetic signal with a qualitatively similar magnetic behavior (Narozhnyi et al. 2003). A different method was used for the preparation of the ferromagnetic polymers of C_{60}: multi-anvil octupole press (Wood et al. 2002). Inelastic neutron scattering analysis of the ferromagnetic phase in the polymerized fullerene sample showed a sufficient presence of hydrogen (Chan et al. 2004).

25.8.7 Irradiated Carbon Structures

Studies of irradiated carbon structures provided convincing proof for the intrinsic origin of the effect. This is the case of the proton-irradiated HOPG where the ultimate purity of the material is proved by simultaneous measurements of the magnetic impurities (Esquinazi et al. 2003). Elementally sensitive experiments on proton bombarded graphite provided fast evidence for metal-free carbon magnetism (Ohldag et al. 2007). The temperature behavior suggests two-dimensional magnetic order (Barzola-Quiquia et al. 2007). Ferromagnetism was found in irradiated fullerenes with 250 keV Ar and 92 MeV Si ions (Kumar et al. 2006), with 10 MeV oxygen ion beam (Kumar et al. 2007), 2 MeV protons (Mathew et al. 2007). Paradoxically, if one bombards graphite with iron and hydrogen, both produce similar paramagnetic contributions. However, only protons induce ferromagnetism (Barzola-Quiquia et al. 2008; Hohne et al. 2008).

The mechanism of ferromagnetism in H^+-irradiated graphite is largely unknown and may result from the appearance of bound states due to disorder and the enhancement of the density of states (Araujo and Peres 2006), and can be induced by single carbon vacancies in a three-dimensional graphitic network (Faccio et al. 2008); magnetism decreases for both diamond and graphite with increase in vacancy density (Zhang et al. 2007). The role of hydrogen is not well understood as magnetism should survive only at low H concentrations (Boukhvalov et al. 2008). The mechanism of ferromagnetism in disordered graphite samples is considered to arise from unpaired spins at defects, induced by a change in the coordination of the carbon atoms (Guinea et al. 2006). Several works discuss theoretical models that address the effects of electron–electron interactions and disorder in graphene planes (González et al. 2001, Stauber et al. 2005).

25.8.8 Magnetic Nature of Intrinsic Carbon Defects

There are strong reasons why high-temperature ferromagnetism in carbon is hard to expect. A major requisite for magnetism in an all-carbon structure is the presence and stability of carbon radicals. The occurrence of radicals, which can introduce an unpaired spin, is cut down by the strong ability of pairing all valence electrons in covalent bonds. These reasons may explain the difficulties and poor reproducibility in preparation of magnetic carbon compounds. All known carbon allotropes are diamagnets. Diamagnetic susceptibility of bulk crystalline graphite is very large, and it yields only to superconductors in this respect. The situation changes for graphite containing certain type of defects and for nano-sized graphene layers. According to theoretical suggestions, the presence of edges in nanographene produces edge-inherited nonbonding π-electronic state (edge state) in addition to the π- and π*-bands, giving entirely different electronic structure from bulk graphite. These so-called "peculiar" states are extended along the edges but at the same time are localized at the edges (Fujita et al. 1996). Nanographite is characterized by the dependence of electronic structure on edge termination: edge states are present on variously terminated zigzag edges but are absent at the armchair edges. These states produce large electronic density of states at the Fermi level and play an important role in the unconventional nano-magnetism.

It is suggested that the basic magnetic mechanism is spin polarization in these highly degenerate orbitals or in a flat band (Kusakabe 2006).

Diamagnetism of nanographenes can be understood in terms of diamagnetic ring currents. Defects in graphite always reduce the diamagnetic signal. In a simplified picture, vacancies, adatoms, pores, and bond rotations enhance local paramagnetic ring currents and produce local magnetic moments (Lopez-Urias et al., 2000). Theory allows also magnetism in diamond structures (Cho and Choi 2008).

Several scenarios that account for (or predict) the magnetism of carbon have been suggested: bulk magnetism, induced magnetism, and atomic-scale magnetism caused by structural imperfection. Figure 25.10 illustrates the intrinsic carbon defects that may lead to the magnetic ordering in carbon structures.

25.8.8.1 Adatoms and Vacancies

Carbon adatoms (Figure 25.10a) possess a magnetic moment of about $0.5\,\mu_B$ whereas carbon vacancies in graphitic network generate a magnetic moment of about $1\,\mu_B$ (Ma et al. 2004).

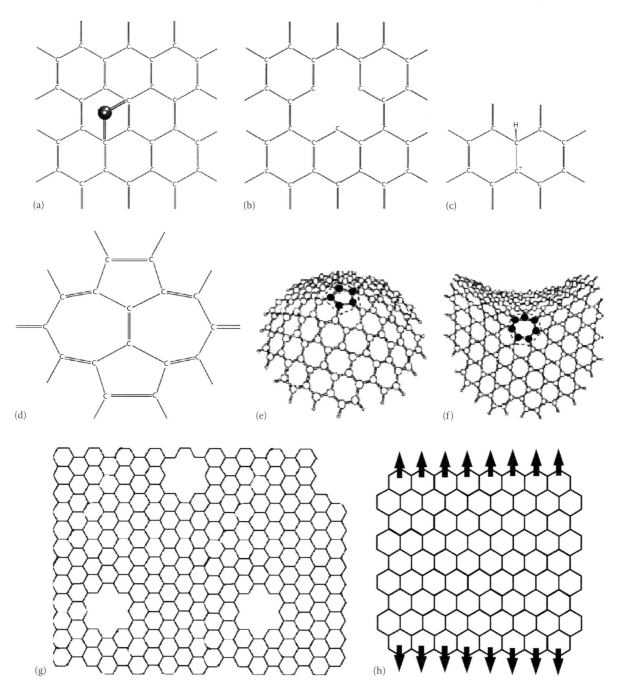

FIGURE 25.10 Intrinsic carbon defects which are thought to lead to magnetic ordering in carbon structures. (a) Adatom. (b) Vacancy. (c) Hydrogen. (d) Stone-Wales defect. (e) Positive curvature. (f) Negative curvature. (g) Porosity. (h) Zigzag edge.

Vacancies in graphite, both ordinary (Figure 25.10b), and hydrogenated (Figure 25.10c), create new states below the Fermi level. The extra π-electrons are induced to the system when vacancies are introduced. For noninteracting vacancies, these extra electrons give rise to an unpaired spin associated with the vacancy (Lehtinen et al. 2003).

25.8.8.2 First-Raw Elements

A specific case of defects is the presence of the first-raw elements, although they cannot be unambiguously classified as intrinsic carbon defects. The most important defect is hydrogen: Unsaturated valence bonds at the boundaries of graphene flakes are filled with stabilizing elements; among these stabilizers hydrogen atoms are the common ones. The entrapment of hydrogen by dangling bonds at the nanographite perimeter can induce a finite magnetization. A theoretical study of a graphene ribbon in which each carbon atom is bonded to two hydrogen atoms at one edge and to a single hydrogen atom at the other edge shows that the structure has a finite total magnetic moment (Kusakabe and Maruyama 2003). Combination of different edge structures (by means of hydrogenation, fluorination, or oxidation) is proposed as a guiding principle to design magnetic nanographite (Maruyama and Kusakabe 2004). Hydrogenation of carbon materials can induce magnetism through termination of nanographite ribbons, adsorption on the CNT external surface (Pei et al. 2006), trapping at a carbon vacancy or pinning by a carbon adatom (Ma et al. 2005).

Other elements that may strongly influence magnetic behavior of carbon are boron and nitrogen. Border states in hexagonally bonded BNC heterosheets have been predicted to lead to a ferromagnetic ground state, a manifestation of flat band ferromagnetism (Okada and Oshiyama 2001). In heterostructured nanotubes, partly filled states at the interface of carbon and boron nitride segments may acquire a permanent magnetic moment. Depending on the atomic arrangement, heterostructured C/BN nanotubes may exhibit an itinerant ferromagnetic behavior owing to the presence of localized states at the zigzag boundary of carbon and boron nitride segments (Choi et al. 2003).

25.8.8.3 Curvature

Stone–Wales defects (Figure 25.10e) are responsible for Gaussian curvature in carbon structures. Gaussian negative curvature provides a mechanism for steric protection of the unpaired spin (Park and Min 2003, Park et al. 2003). A particular case of a graphene modification is the Stone–Wales defects, or topological defects, caused by the rotation of carbon atoms, which leads to the formation of five- or sevenfold rings. A novel class of curved carbon structures, Schwarzites and Haeckelites, has been proposed theoretically (Mackay and Terrones 1991). Schwarzite is a form of carbon containing graphite-like sheets with hyperbolic curvature. So far, periodic Schwarzites have not been realized experimentally; however, there is experimental evidence that random Schwarzite structures are present in a cluster form in such carbon phases as spongy carbon (Barborini et al. 2002) and carbon nanofoam (Rode et al. 2004). In the systems with

negative Gaussian curvature, unpaired spins can be introduced by sterically protected carbon radicals. Not only negative Gaussian curvature may lead to magnetism; the same is true for the positive: carbon compounds that display an odd number of pentagons and heptagons present polarization in the ground state (Azevedo et al. 2008).

25.8.8.4 Zigzag Edges

Electronic states are strongly influenced by the existence and the shape of graphite edge, and zigzag edges favor the spin polarization with ferromagnetic alignment. Nanographite, a stack of nanosized graphene layers, is a nanosized π-electron system with open edges. The periphery of a nanographite pattern can be described as a combination of zigzag and armchair edges (Figure 25.10h). In the open-edge systems, the edges around their boundary produce distinctive electronic features, namely, the zigzag edges produce strongly spin-polarized states, which are spatially localized around the edges. The presence of these states modifies the electronic structure of nanographite as a whole: It produces edge-inherited nonbonding π-electronic state (edge state) in addition to the π- and π*-bands, giving entirely different electronic structure from bulk graphite. These so-called "peculiar" states are extended along the edges but at the same time are localized at the edges. These states produce large electronic density of states at the Fermi level and play an important role in the unconventional nanomagnetism (Enoki et al. 2007).

Interestingly, similar predictions have been made for ZnO nanoribbons with zigzag-terminated edges. A net magnetic moment was found for single- and triple-layered zigzag nanoribbons, however, for two, four, and five layers it vanishes even when in the latter case there are states in the Fermi level (Botello-Mendez et al. 2008).

25.8.8.5 Defects in Fullerenes

Ferromagnetism in fullerenes is also thought to be of defect nature. It was shown both theoretically (Okada and Oshiyama 2003) and experimentally (Boukhvalov et al. 2004) that the ideal polymerized fullerene matrix is not magnetic. The following type of defects have been considered: broken or shortened interfullerene bonds, distortion of fullerene cages, vacancies in the fullerene cages, adatoms on fullerene cages, local charge inhomogeneities, and open-cage defect structure with hydrogen atom bonded chemically to one of defect carbon atoms. Partial disruption of interfullerene bonds: linking of molecules through a single bond is preferred for multiplet states of system (Chan et al. 2004). Cage distortion of C_{60} in polymerized two-dimensional network leads to competition between diamagnetic and ferromagnetic states (Nakano et al. 2004). Certain types of vacancies in coexistence with the 2 + 2 cycloaddition bonds represent a generalized McConnell's model for high-spin ground states in the systems with mixed donor–acceptor stacks (Andriotis et al. 2005). Donor–acceptor mechanism was considered for the case of microscopic electric charge inhomogeneities introduced in a polymeric network. Two charged adjacent fullerenes interact ferromagnetically, and the ground state of a charged dimer is triplet (Kvyatkovskii et al. 2004). The C_{60} doublet radicals appear after the application of pressure, and this state has a long life state (Ribas-Arino and Novoa 2004a). The evaluation of capability of the C_{60} molecule to act as a magnetic coupling unit was made: C_{60} diradical is an excellent magnetic coupler (Ribas-Arino and Novoa 2004b). Some metastable isomer states with zigzag-type arrangement of the edge atoms of C_{60} may form during the cage opening process (Kim et al. 2003). Long-range spin coupling, which is an essential condition for the ferromagnetism, has been considered through the investigation of an infinite, periodic system of polymerized C_{60} network. Chemically bonded hydrogen plays a vital role, providing a necessary pathway for the ferromagnetic coupling of the considered defect structure.

It is well known that C_{60} molecules become magnetically active due to the spin (and charge) transfer from dopants. Magnetic transitions were reported for the TDAE-C_{60} ferromagnet, the $(NH_3)K_3C_{60}$ antiferromagnet, AC_{60} and Na_2AC_{60} polymers (A = K, Rb, Cs). Kvyatkovskii et al. (2005, 2006) consider the situation when C_{60} molecule is doped through the presence of structural defects and impurities which create stable molecular ions $C_{60}^{\pm}$ and analyze the interaction of two adjacent molecules (i.e., dimer) embedded in a two-dimensional polymeric network. The main result is that the ferromagnetic interaction is possible only in the crystals where fullerenes have specific orientation. This type of orientation is provided by (2 + 2) cycloaddition reaction, which forms a double bond (DB) between the buckyballs.

25.8.9 Magnetism of Graphene

A quickly developing topic is magnetism of graphene. The first experimental isolation of a single nanographene was obtained by electrophoretic deposition and heat treatment of diamond nanoparticles (Affoune et al. 2001). Experimentally, the observation of room-temperature graphene magnetism was claimed on the graphene material prepared from graphite oxide (Wang et al. 2009).

It has been shown theoretically (Vozmediano et al. 2006) that the interplay of disorder and interactions in a 2D graphene layer gives rise to a rich phase diagram where strong coupling phases can become stable. Local defects can lead to the magnetic ordering. The theories predict itinerant magnetism in graphene due to the defect-induced extended states (Yazyev and Helm 2007) while only *single-atom defects* can induce FM in graphene-based materials (Yazyev 2008) or short-range magnetic order peculiar to the honeycomb lattice with the vacancies (Kumazaki and Hirashima 2007a) or with hydrogen termination or a chemisorption defect (Kumazaki and Hirashima 2007b). The graphene magnetic susceptibility is temperature dependent, unlike an ordinary metal (Kumazaki and Hirashima 2007c). Spin susceptibility, which decreases with temperature without impurities, takes a finite value with impurities which may enhance the tendency to a ferromagnetic ordered state (Peres et al. 2006).

Finite graphene fragments of certain shapes, e, g, triangular or and hexagonal "nanoislands" terminated by zigzag edges (Fernandez-Rossier and Palacios 2007), or variable-shaped graphene nanoflakes (Wang et al. 2008a,b), as well as some "Star of David"-like fractal structures (Yazyev 2008) possess a high-spin ground state and behave as artificial ferrimagnetic atoms. Ferrimagnetic order emerges in rhombohedral voids with imbalance charge in graphene ribbons, and the defective graphene ribbons behave as diluted magnetic semiconductors (Palacios and Fernández-Rossier 2008). A defective graphene phase is foreseen to behave as a room temperature ferromagnetic semiconductor (Pisani et al. 2008). Both magnetic and ferroelectric orders are predicted (Fernandez-Rossier 2008).

Edge state magnetism has been studied on realistic edges of graphene and is shown that only elimination of zigzag parts with $n > 3$ will suppress local edge magnetism of graphene (Kumazaki and Hirashima 2008). The edge irregularities and defects of the bounding edges of graphene nanostructures do not destroy the edge state magnetism (Bhowmick and Shenoy 2008). However, such edge defects (vacancies) and impurities (substitutional dopants) suppress spin polarization on graphene nanoribbons, which is caused by the reduction and removal of edge states at the Fermi energy (Huang et al. 2008). Magnetic order in zigzag bilayers ribbons is also related to the properties of zigzag edges (Sahu et al. 2008). Neutral graphene bilayers are proposed to be pseudospin magnets (Min et al. 2008). In a biased bilayer graphite (Stauber et al. 2008) a tendency toward a ferromagnetic ground state is investigated and shown that the phase transition between paramagnetic and ferromagnetic phases is of the first order. Spin is confined in the superlattices of graphene ribbons, and in specific geometries magnetic ground state changes from antiferromagnetic to ferrimagnetic (Topsakal et al. 2008).

An alternative approach is connected with pentagons, dislocations, and other topological defects (Carpio et al. 2008). Single pentagons and glide dislocations made of a pentagon–heptagon pair alter the magnetic behavior, whereas the Stone–Wales defects are harmless in the flat lattice.

The combination of hydrogen-induced magnetism and changeable thermodynamics upon variation of the graphene layer spacing makes graphene a reversible magnetic system (Lei et al. 2008). In graphene magnetism survives at low H concentrations (Boukhvalov et al. 2008). A number of nanoscale spintronics devices utilizing the phenomenon of spin polarization localized at one-dimensional (1D) zigzag edges of graphene have been proposed (Yazyev and Katsnelson 2008). Some of the theories created for graphene explain the experimental observations in proton-bombarded graphite (Yazyev et al. 2008).

Carbon materials that exhibit ferromagnetic behavior have been predicted theoretically and reported experimentally in recent years (Makarova and Palacio 2006). The initial surprising experiments were confirmed by the independent groups. The fact that carbon atoms can be magnetically ordered at room temperature was confirmed by the direct experiment: an element-sensitive method x-ray magnetic circular dichroism.

25.9 Possible Traps in Search of Magnetic Order

This section critically reviews the data in the literature that report experimental observations of magnetism without transition metal cations. While only a few of the experiments that have reported the "nonmagnetic magnetism" are free from obvious or possible artifacts, those few along with the theoretical predictions or computer simulations suggest it is real.

The experimental situation in this field is unclear with many conflicting results. The source of RTFM is often controversial and contentious particularly because of the possible role of undetected ferromagnetic impurities such as Fe, Co, Ni, etc. (Janisch et al. 2005). Some of the experimental results were plagued by the precipitates of doped magnetic elements and even unintentional contaminations during sample handling. The sources of unintentional sample contamination are numerous. Even pure single crystalline sapphire substrates show a ferromagnetic behavior that partially changes after surface cleaning. The amount of magnetic impurities in the substrates was determined by particle-induced x-ray emission, and for 10 commercial substrates the iron concentration ranged from 1 to 260 ng/cm^2 (Salzer et al. 2007). This amount of impurities is enough to overshadow the intrinsic signal from thin films grown on oxide substrates.

Impurities can be introduced during various technological processes, for example by the procedure used to fix the substrates to the oven (Golmar et al. 2008). Nanoparticles of various sizes and shapes were observed as a result of hydrothermal treatment of cyanometalate polymers and the authors emphasize the extreme care that must be taken in the studies of magnetism of apparently analytically pure materials (Lefebvre et al. 2008).

Silicon—the main element of the modern electronics—has been declared as a magnetic element (Kopnov et al. 2007). RTFM observed in Co-doped ZnO grown on Si (100) has been characterized as coming from Si/SiO$_x$ interface (Yin et al. 2008). It was shown that iron from the Pyrex glassware appears on silicon substrate after etching in hot KOH in the form of well-separated ferromagnetic nanoparticles (Figure 25.11) (Grace et al. 2009).

A detailed investigation by magnetic measurements and EPR spectroscopy of the magnetic fraction of cigarettes and its variation with the smoking process shows that complex magnetic

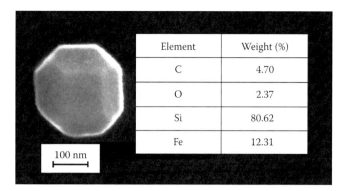

Element	Weight (%)
C	4.70
O	2.37
Si	80.62
Fe	12.31

FIGURE 25.11 Scanning electron microscope image of an etched silicon surface, with a magnified image of a magnetic particle and elemental analysis of the particle and the surrounding area showing that the particles are iron-rich. (Adapted from Grace, P.J. et al., *Adv. Mater.*, 21, 71, 2009. With permission.)

properties of both ashes and tobacco, which should be considered as a possible source of data contamination to be avoided in magnetism laboratories (Cador et al. 2008).

An important source of mistakes in the interpretation of the experimental data is underestimation of the role of iron in the measured magnetic signal. Usually the authors use the following logic: "Let us start with a rather unrealistic assumption that all iron impurities form metallic clusters and all clusters are large enough to behave ferro- or ferrimagnetically, then the maximum magnetization produced by 1 ppm of impurities is calculated using the parameters listed in Table 25.1." Indeed, in order to contribute to the total magnetization of the samples, the impurities must interact magnetically. Simple addition of transition metals does not lead to ferromagnetic properties, as no interaction pathway is provided. Table 25.1 can be used for estimating the maximum values of the parasitic signals. However, this method must be used with caution.

First, the assumption that all atoms of transition metals form clusters large enough to behave ferro- or ferrimagnetically is *not an unrealistic assumption*. If the sample was prepared at high temperatures, metallic atoms could aggregate during the cooling process with the formation of clusters (Lefebvre et al. 2008). To behave ferromagnetically, the clusters must be sufficiently

TABLE 25.1 Magnetization Values Produced by One Weight ppm of Transition Metal Impurity Provided the Impurities Interact Magnetically

Type of Impurity	Maximum Magnetization, emu/g	Curie Temperature (K)	Type of Ordering
Iron, α-Fe	0.00022	1045	Ferromagnetic
Magnetite, Fe$_3$O$_4$	0.000092	860	Ferrimagnetic
Maghemite, Fe$_2$O$_3$	0.00008	880	Ferrimagnetic
Hematite, Fe$_2$O$_3$	0.0000004	950	Canted antiferromagnetic
Iron carbide, Fe$_3$C	0.00013	483	Ferromagnetic
Nickel, Ni	0.000055	630	Ferromagnetic
Cobalt, Co	0.000161	1130	Ferromagnetic

Note that the laboratories carrying out elemental analysis issue the results not in the elemental *ppm*, but in the *weight ppm* units, i.e., 1 ppm = 1 mg/kg = 1/1,000,000 part by weight.

large. Below a certain size, the particles consist of single magnetic domains. For iron this limit is of the order of 150 Å, i.e., about 10^5 atoms per Fe cluster (Kittel 1946). Normally for these particles the superparamagnetic behavior is expected. However, describing the magnetic behavior of the small clusters, one must take into account the effects of the local environment on the electronic structure and magnetic moments, which can be different for various structural forms (Liu et al. 1989).

Second, the sample may provide special conditions for the metallic atoms to aggregate. One example is iron in nanographite matrix. Figure 25.12 illustrates an accidental matching of the Fe–Fe distance 2.866 Å with that of the C1–C4 distance ~2.842 Å of the hexagonal rings in graphite (Kosugi et al. 2004). The x-ray diffraction pattern indicates that the particles are composed of a-Fe and graphitic carbon. Due to the matching of two distances, the growth of graphitic planes from iron ionic salts has been observed (Kosugi et al. 2004), and one could imagine that an opposite effect might happen: the decoration of the armchair edges of nanographites with metallic iron. The latter example is only a suspicion of the author of this paper and does not have any experimental confirmation; on the contrary, proximity of carbon generally leads the reduced magnetization of iron (Saito et al. 1997, Host et al. 1998, Fauth et al. 2004).

Third, in bulk Fe, magnetism and structure are strongly dependent. The magnetic moments of free Fe monolayers are theoretically found to be larger than those of the surface with values equal to $3.2\,\mu_B$ for Fe(001) (Freeman and Wu 1991). Magnetic moments $\mu(N)$ of iron clusters μ ($25 \leq \underline{N} \leq 130$) is $3\,\mu_B$ per atom, decreasing to the bulk value ($2.2\,\mu_B$ per atom) near $N = 500$. For all sizes, μ decreases with increasing temperature, and is approximately constant above a temperature $T_C(N)$. For example, $T_C(130)$ is about 700 K, and $T_C(550)$ is about 550 K (T_C bulk = 1043 K) (Billas et al. 1993). This means that neither the absence of large clusters nor an unusual Curie temperature can be taken as an evidence of iron-independent magnetism.

The enhancement of magnetism at the surface can be qualitatively understood on the basis of a simple picture for the evolution of magnetism from the atom to the bulk. Iron atom has the electron configuration $[Ar]3d^64s^2$. The total spin must be maximized according to the Hund's rule. From Figure 25.12, one may think that iron has magnetic moment about $4\,\mu_B$ per atom,

but this is wrong. Magnetism is a collective phenomenon, and single atoms do not produce magnetism. When two Fe atoms are brought together, the s-electron will jump to the d-level because the spd hybridization leads to the powering of the total energy. Surface is a low-coordinated system. For the iron atom on the surface one s-electron is transferred to the d-levels whereas the d-levels corresponding to the surface are strongly localized on the atomic sites. According to the Hund's rule the spin-up d-band (i.e., the majority band) will be completely filled with five electrons and therefore the spin-down d-band (i.e., the minority band) will contain two electrons. The spin imbalance between the majority and minority spins equals to three $n(\uparrow) - n(\downarrow) = 3$ for the surface Fe atoms. As the system grows in size, the coordination number z increases, and hybridization is changed to dd- and sd-hybridization. The spin imbalance of bulk Fe finally reduces to $n(\uparrow) - n(\downarrow) = 2.2$ for bulk Fe (Fritsche et al. 1987).

In many experiments, the presence of iron in the samples does not lead to ferromagnetic behavior of iron. It is a matter of common knowledge that carbon quenches ferromagnetism of iron in the case of stainless steel. If the iron concentration is small, the absence of superparamagnetic behavior, which is typical for Fe_xC_{1-x} iron–carbon nanocomposites (Babonneau et al. 2000, Enz et al. 2006, Schwickardi et al. 2006) suggests that impurities either do not contribute to magnetic properties, or their role is far from trivial. Nontrivial origin of magnetic behavior in contaminated carbon-based materials may result from catalytic or template properties of transition metal atoms.

Small iron clusters are driven into a nonmagnetic state by the interaction to graphitic surfaces (Fauth et al. 2004). Proximity of carbon generally leads the reduced magnetization of the transition metal clusters (Saito et al. 1997, Host et al. 1998). The reduced magnetism in case of very small clusters is explained by the fact that the transition metal 4s-related density of states is strongly shifted upward in energy due to the repulsive interaction with the carbon π orbitals (Duffy and Blackman 1998).

Fe impurities weaken the ferromagnetic behavior in Au by delocating the charge from the surface of the NPs (Crespo et al. 2006). Such magnetic element as cobalt suppresses ferromagnetism in intrinsically magnetic ZnO films (Ghoshal and Kumar 2008).

A paper analyzing several uses of spurious effects (Garcia et al. 2009) must become "The Deskbook on Professional Responsibility" for everybody working with small magnetic signals. In addition to the above, the paper analyses the errors coming from such units of SQUID equipment as polyimide Kapton® tape, gelatin capsules, cotton, plastically deformed straws, anisotropy artifacts coming from irregular distributed impurities.

25.10 Nontrivial Role of Transition Metals: Charge Transfer Ferromagnetism

One of the mechanisms of nontrivial role of magnetic metals in d-zero ferromagnetism is contact-induced magnetism, which arises when nonmagnetic materials brought in the proximity

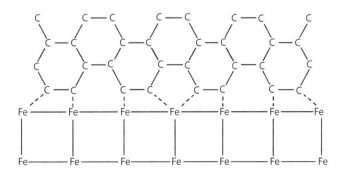

FIGURE 25.12 Accidental coincidence of carbon and iron lattice constants.

with magnetic ones. Close proximity of iron and carbon can result in magnetic moments (~0.05 μ_B/C; 1 μ_B/C $\equiv$ 465 A m^2 kg^{-1}) for atoms near the surfaces of multiwalled carbon nanotubes (Céspedes et al. 2004) or thin films in C/Fe multilayer stacks (Mertins et al. 2004).

There are theoretical models that explain the role or iron as just the role of a defect, and magnetic nature of the dopant does not play a role (Weyer et al. 2007). In case of ferromagnetism of undoped CaB$_6$ iron serves as one of the defects, which lead to a partially occupied impurity band that is responsible for RTFM (Edwards and Katsnelson 2006).

A nontrivial role of transition metal impurities in oxide ferromagnetism is proposed in Osorio-Guillen et al. (2008). The impurities introduce excess electrons in oxides and either (1) introduce resonant states inside the host conduction band and produce free electrons or (2) introduce a deep gap state that carries a magnetic moment. The second scenario leads to a ferromagnetic behavior.

The following arguments have been put forward by Coey et al. (2008) that magnetism in the dilute magnetic oxide films is not related to the transition metal cations: The oxides are not good crystalline materials, on the contrary, magnetism is governed by the defect structure; high Curie temperatures are incompatible with low concentration of magnetic dopants, and the dopants itself are paramagnetic whereas the whole sample is ferromagnetic. Also, the magnetic semiconductor effects such as Hall, Faraday, and Kerr effects are *not* observed. Thus, magnetism in these systems is not due to the ferromagnetically ordered moments of the doping cations mediated by the carriers. The role of transition metal ions is to provide a charge reservoir, and this ability is due to the fact that the cations can exist in two different charge states. "It is therefore the ability of the 3d cations to exhibit *mixed valence*, rather than their possession of a localized moment, which is the key to the magnetism" (Coey et al. 2008). The proposed model of a formation of a defect-based narrow band and tuning the position of the Fermi level by transferred charges was named charge-transfer ferromagnetism.

25.11 Interface Magnetism

There is growing experimental evidence that a new type of magnetism has been identified, namely, a magnetism related to surfaces and interfaces of nonmagnetic materials, the "interface magnetism" (Brinkman et al. 2007, Eckstein 2007). Similar effects have been described for different objects: organic molecules adsorbed on metals (Carmeli et al. 2003) HfO$_2$-coated silicon or sapphire (Venkatesan et al. 2004) silicon/silicon oxide interfaces (Kopnov et al. 2007) or PbS self-assembled nanoparticles on GaAs (Zakrassov et al. 2008).

As has been already mentioned, such elements as Au, Ru, Rh, Pd, which do not show bulk magnetization, becomes magnetic at the nanoscale. Semiconductor nanoparticles show increasing magnetization for decreasing diameter (Neeleshwar et al. 2005), strong size dependence was found for Ge quantum dots (Liou and Shen 2008). Ferromagnetism of a different nature is

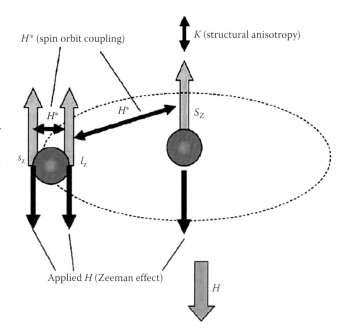

FIGURE 25.13 Scheme of the different magnetic moments and the interaction controlling their orientation. The relative orientation of S_Z, s_z, and l_z is fixed by H^*. The structural anisotropy acts only on S_Z, while the reversal magnetic applied field acts on all of them. (Reprinted from Hernando, A. et al., *Phys. Rev. B*, 74, 052403, 2006b. With permission.)

observed in thin films and nanoparticles capped with organic molecules (Crespo et al. 2004, Yamamoto et al. 2004).

The interface magnetism is characterized by

- being temperature-independent in the range of 0–400 K
- having high anisotropy
- having very large magnetic signal per atom on the surface

It is believed that the combination of these features speaks in favor of collective orbital magnetism initiated by the charge transfer between the substrate and the thin layer.

Accumulating data clearly indicate that a new type of magnetism exists, related to a cooperative effects on the surface or interfaces (Cahen et al. 2005). At present, the precise mechanism of this phenomenon is not known and theory is not yet constructed. From the analysis of the anisotropy of thiol capped gold films, a conclusion is made that the *orbital momentum induced at the surface conduction electrons* is crucial to understand the observed giant anisotropy (Hernando et al. 2006a,b).

The orbital motion is driven by spin–orbit interaction, which reaches extremely high values at the surfaces (Figure 25.13). The induced orbital moment gives rise to an effective field of the order of 1000 T, which is responsible for the giant anisotropy.

25.12 Conclusions

This chapter reviews the current status of research on nanomagnetism along with history of research developments in this field. An attempt has been made to bring together the results obtained in different areas: from metals and semiconductors to dielectric oxides, from quantum dots to graphite and graphene. Particular attention is paid to the pitfalls awaiting the researcher at every turn. Therefore, we expect that the chapter will be useful to all scientists facing with the problem "To believe or not to believe." Science is not religion and it doesn't just come down to faith; science is based upon verifiable evidence.

Science is all about testing ideas via experiment: the models that match the current experimental evidence can be disproved by new experiments. Science is not about "always being right," and any scientist meets with wrong turns in trying to understand a complicated phenomenon. A bright example is the beginning of the twentieth century, which created "the gallery of failed atomic models," from Thomson "plum pudding" model of the atom, through the dipole model of Lenard, the Saturnian model of Nagaoka, electron fluid model by Lord Rayleigh, expanding electron model by Schott, the archion model by Stark, until the Bohr orbital model appeared, which "won" being the only model capable of explaining the Rydberg formula. Somewhat similarly, magnetism of nonmagnetic materials seems to be constructing "a gallery of failed magnetic models" from ferromagnetic ground state of a dilute electron gas to a trivial parasitic ferromagnetism. There is a divergence of opinion on the role of vacancies, defects, and the carrier mediation, and novel ideas of charge-transfer ferromagnetism and that of interface magnetism are being developed. Due to the present status of researches in this field, the author does not take liberty to give preference to any of the theories. What is unambiguously clear now is that nanoscale magnetism of otherwise nonmagnetic materials is *sui generic,* i.e., "outside the family."

Due to the lack of a theoretical understanding and due to the difficulty in reproducibility, the origin of the defect-induced ferromagnetism is under intense debate, but its existence is beyond doubt. It is not easy to predict what will happen when a magnetic atom is introduced into or onto the surface of a non-magnetic host crystal because the atom "sees" the surface crystal lattice through orbital overlap of the electrons (Schneider 2008). Even more difficult is to predict the magnetic behavior of the materials that are nonmagnetic in the bulk but become magnetic at the nanoscale. Recent magnetic experiments on nanostructures are showing us new approaches that open a new world of possibilities for creating the materials for next generation spintronic devices.

References

Abraham, D. W., Frank, M. M., and Guha, S. 2005. Absence of magnetism in hafnium oxide films. *Appl. Phys. Lett.* 87: 252502.

Affoune, A. M., Prasad, B. L. V., Sato, H., Enoki, T., Kaburagi, Y., and Y. Hishiyama, Y. 2001. Experimental evidence of a single nano-graphene. *Chem. Phys. Lett.* 348: 17–20.

Akutsu, S. and Utsushikawa, Y. 1999. Magnetic properties and electron spin resonance of carbon fine particles prepared in He plasma. *Mater. Sci. Res. Int.* 5: 110–115.

Andersson, O. E., Prasad, B. L. V., and Sato, H. 1998. Structure and electronic properties of graphite nanoparticles. *Phys. Rev. B* 58: 16387–16393.

Ando, K., Saito, H., Jin, Z. W. et al. 2001. Magneto-optical properties of ZnO-based diluted magnetic semiconductors. *J. Appl. Phys.* 89: 7284–7286.

Andriotis, A. N., Menon, M. R., and Sheetz, M. 2005. Are s–p- and d-ferromagnetisms of the same origin? *J. Phys.: Condens. Matter* 17: L35–L38.

Antonov, V. E., Bashkin, I. O., Khasanov, S. S. et al. 2002. Magnetic ordering in hydrofullerite $C_{60}H_{24}$. *J. Alloys Compd.* 330: 365–368.

Araujo, M. A. N. and Peres, N. M. R. 2006. Weak ferromagnetism and spiral spin structures in honeycomb Hubbard planes. *J. Phys.: Condens. Matter* 18: 1769–1779.

Arcon, D., Jaglicic, Z., Zorko, A. et al. 2006. Origin of magnetic moments in carbon nanofoam. *Phys. Rev. B* 74: 014438.

Ata, M., Machida, M., Watanabe, H., and Seto, J. 1994. Polymer—C_{60} composite with ferromagnetism. *Jpn. J. Appl. Phys.* 33: 1865–1871.

Azevedo, S., de Paivaa, R., and Kaschny, J. R. 2008. Spin polarization in carbon nanostructures with disclinations. *Phys. Lett. A* 372: 2315–2318.

Babonneau, D., Briatico, J., Petrof, F. et al. 2000. Structural and magnetic properties of Fe_xC_{1-x} nanocomposite thin films. *J. Appl. Phys.* 87: 3432–3443.

Balents, L. and Varma, C. M. 2000. Ferromagnetism in doped excitonic insulators. *Phys. Rev. Lett.* 84: 1264–1267.

Banerjee, S., Mandal, M., Gayathri, N., and Sardar, M. 2007. Enhancement of ferromagnetism upon thermal annealing in pure ZnO. *Appl. Phys. Lett.* 91: 182501.

Barborini, E., Piseri, P., Milani, P. et al. 2002. Negatively curved spongy carbon. *Appl. Phys. Lett.* 81: 3359–3361.

Barzola-Quiquia, J., Esquinazi, P., Rothermel, M, Spemann, D., Butz, T., and Garcia, N. 2007. Experimental evidence for two-dimensional magnetic order in proton bombarded graphite. *Phys. Rev. B* 76: 161403.

Barzola-Quiquia, J., Hohne, R., Rothermel, M., Setzer, A., Esquinazi, P., and Heera, V. 2008. A comparison of the magnetic properties of proton- and iron-implanted graphite. *Eur. Phys. J. B* 61: 127–130.

Barzykin, V. and Gorkov, L. P. 2000. Ferromagnetism and superstructure in $Ca_{1-x}La_xB_6$. *Phys. Rev. Lett.* 84: 2207–2210.

Beltran, J. I., Munoz, M. C., and Hafner, J. 2008. Structural, electronic and magnetic properties of the surfaces of tetragonal and cubic HfO_2. *New J. Phys.* 10: 063031.

Bennett, M. C., van Lierop, J., Berkeley, E. M. et al. 2004. Weak ferromagnetism in CaB_6. *Phys. Rev. B* 69: 132407–132410.

Berardan, D., Guilmeau, E., and Pelloquin, D. 2008. Intrinsic magnetic properties of In_2O_3 and transition metal-doped-In_2O_3. *J. Magn. Magn. Mater.* 320: 983–989.

Bhosle, V. and Narayan, J. 2008. Observation of room temperature ferromagnetism in Ga:ZnO: A transition metal free transparent ferromagnetic conductor. *Appl. Phys. Lett.* 93: 021912.

Bhowmick, S. and Shenoy, V. B. 2008. Edge state magnetism of single layer graphene nanostructures. *J. Chem. Phys.* 128: 244717.

Billas, I. M. L., Becker, J. A., Châtelain, A., and de Heer, W. A. 1993. Magnetic moments of iron clusters with 25 to 700 atoms and their dependence on temperature. *Phys. Rev. Lett.* 71: 4067–4070.

Blinc, R., Cevc, P., Arcon, D. et al. 2006. C-13 NMR and EPR of carbon nanofoam. *Phys. Status Solidi B* 243: 3069–3072.

Bogani, L. and Wernsdorfer, W. 2008. Molecular spintronics using single-molecule magnets. *Nat. Mater.* 7: 179–186.

Borges, R. P., da Silva, R. C., Magalhaes, S., Cruz, M. M., and Godinho, M. 2007. Magnetism in Ar-implanted ZnO. *J. Phys.: Condens. Matter* 19: 476207.

Botello-Mendez, A. R., Lopez-Urias, F., Terrones, M. et al. 2008. Magnetic behavior in zinc oxide zigzag nanoribbons. *Nano Lett.* 8: 1562–1565.

Boukhvalov, D. W., Karimov, R. F., Kurmaev, E. Z. et al. 2004. Testing the magnetism of polymerized fullerene. *Phys. Rev. B* 69: 115425.

Boukhvalov, D. W., Katsnelson, M. I., and Lichtenstein, A. I. 2008. Hydrogen on graphene: Electronic structure, total energy, structural distortions and magnetism from first-principles calculations. *Phys. Rev. B* 77: 035427.

Bouzerar, G. and Ziman, T. 2006. Model for vacancy-induced d(0) ferromagnetism in oxide compounds. *Phys. Rev. Lett.* 96: 207602.

Brinkman, A., Huijben, M., Van Zalk, M. et al. 2007. Magnetic effects at the interface between non-magnetic oxides. *Nat. Mater.* 6: 493–496.

Buchholz, D. B., Chang, R. P. H., Song, J. H., and Ketterson, J. B. 2005. Room-temperature ferromagnetism in Cu-doped ZnO thin films. *Appl. Phys. Lett.* 87: 082504.

Cador, O., Caneschi, A., Rovai, D., Sangregorio, C., Sessoli, R., and Sorace, L. 2008. From multidomain particles to organic radicals: The multifaceted magnetic properties of tobacco and cigarette ash. *Inorg. Chim. Acta* 361: 3882–3886.

Cahen, D., Naaman, R., and Vager, Z. 2005. The cooperative molecular field effect. *Adv. Funct. Mater.* 15: 1571–1578.

Calderón, M. J. and Das Sarma, S. 2007. Theory of carrier mediated ferromagnetism in dilute magnetic oxides. *Ann. Phys.* 322: 2618–2634.

Carmeli, I., Leitus, G., Naaman, R., Reich, S., and Vager, Z. 2003. Magnetism induced by the organization of self-assembled monolayers. *J. Chem. Phys.* 118: 10372–10375.

Carmeli, I., Bloom, F., Gwinn, E. G. et al. 2006. Molecular enhancement of ferromagnetism in GaAs/GaMnAs heterostructures. *Appl. Phys. Lett.* 89: 112508.

Carpio, A., Bonilla, L. L., de Juan, F., and Vozmediano, M. A. H. 2008. Dislocations in graphene. *New. J. Phys.* 10: 053021.

Caudillo, R., Gao, X., Escudero, R. et al. 2006. Ferromagnetic behavior of carbon nanospheres encapsulating silver nanoparticles. *Phys. Rev. B* 74: 214418.

Ceperley, D. 1999. Return of the itinerant electron. *Nature* 397: 386–387.

Céspedes, O., Ferreira, M. S., Sanvito, S., Kociak, M., and Coey, J. M. D. 2004. Contact induced magnetism in carbon nanotubes. *J. Phys.: Condens. Matter* 16: L155–L161.

Chakraborti, D., Narayan, J., and Prater, J. T. 2007. Room temperature ferromagnetism in $Zn_{1-x}Cu_xO$ thin films. *Appl. Phys. Lett.* 90: 062504.

Chan, J. A., Montanari, B., Gale, J. D., Bennington, S. M., Taylor, J. W., and Harrison, N. M. 2004. Magnetic properties of polymerized C_{60}: The influence of defects and hydrogen. *Phys. Rev. B* 70: 041403.

Cho, J.-H. and Choi, J.-H. 2008. Antiferromagnetic ordering in one-dimensional dangling-bond wires on a hydrogen-terminated C(001) surface: A density-functional study. *Phys. Rev. B* 77: 075404.

Cho, B. K., Rhyee, J. S., Oh, B. H. et al. 2004. Formation of midgap states and ferromagnetism in semiconducting CaB_6. *Phys. Rev. B* 69.

Choi, J., Kim, Y., Chang, K. J., and Tomanek, D. 2003. Itinerant ferromagnetism in heterostructured C/BN nanotubes. *Phys. Rev. B* 67: 125421.

Coey, J. M. D. 2005. d(0) ferromagnetism. *Solid State Sci.* 7: 660–667.

Coey, M. and Sanvito, S. 2004. The magnetism of carbon. *Phys. World* 17: 33–37.

Coey, J. M. D., Venkatesan, M., and Fitzgerald, C. B. 2005a. Donor impurity band exchange in dilute ferromagnetic oxides. *Nat. Mater.* 4: 173–179.

Coey, J. M. D., Venkatesan, M., Stamenov, P., Fitzgerald, C. B., and Dorneles, L. S. 2005b. Magnetism in hafnium dioxide. *Phys. Rev. B* 72.

Coey, J. M. D., Wongsaprom, K., Alaria, J., and Venkatesan, M. 2008. Charge-transfer ferromagnetism in oxide nanoparticles. *J. Phys. D: Appl. Phys.* 41: 134012.

Crespo, P., Litran, R., Rojas, T. C. et al. 2004. Permanent magnetism, magnetic anisotropy, and hysteresis of thiol-capped gold nanoparticles. *Phys. Rev. Lett.* 93: 087204.

Crespo, P., Garcia, M. A., Pinel, E. F. et al. 2006. Fe impurities weaken the ferromagnetic behavior in Au nanoparticles. *Phys. Rev. Lett.* 97: 177203.

Crespo, P., Garcia, M. A., Fernandez-Pinel, E. et al. 2008. Permanent magnetism in thiol capped nanoparticles, gold and ZnO. *Acta Phys. Pol. A* 113: 515–520.

Das Pemmaraju, C. and Sanvito, S. 2005. Ferromagnetism driven by intrinsic point defects of HfO_2. *Phys. Rev. Lett.* 94: 217205.

De La Venta, J., Pinel, E. F., Garcia, M. A., Crespo, P., and Hernando, A. 2007. Magnetic properties of organic coated gold surfaces. *Mod. Phys. Lett. B* 21: 303–319.

Denlinger, J. D., Clack, J. A., Allen, J. W. et al. 2002. Bulk band gaps in divalent hexaborides. *Phys. Rev. Lett.* 89: 157601.

Dev, P., Xue, Y., and Zhang, P. 2008. Defect-induced intrinsic magnetism in wide-gap III nitrides. *Phys. Rev. Lett.* 100: 117204.

Dietl, T. 2008. Origin and control of ferromagnetism in dilute magnetic semiconductors and oxides (invited). *J. Appl. Phys.* 103: 07D111.

Dietl, T., Ohno, H., Matsukura, F., Cibert, J., and Ferrand, D. 2000. Zener model description of ferromagnetism in zinc-blende magnetic semiconductors. *Science* 287: 1019–1022.

Dorneles, L. S., Venkatesan, M., Moliner, M., Lunney, J. G., and Coey, J. M. D. 2004. Magnetism in thin films of CaB_6 and SrB_6. *Appl. Phys. Lett.* 85: 6377–6379.

Duffy, D. M. and Blackman, J. A. 1998. Magnetism of 3d transition-metal adatoms and dimers on graphite. *Phys. Rev. B* 58: 7443–7449.

Durst, A. C., Bhatt, R. N., and Wolff, P. A. 2002. Bound magnetic polaron interactions in insulating doped diluted magnetic semiconductors. *Phys. Rev. B* 65: 235205.

Dutta, P., Pal, S., Seehra, M. S. et al. 2007. Magnetism in dodecanethiol-capped gold nanoparticles: Role of size and capping agent. *Appl. Phys. Lett.* 90: 213102.

Dutta, P., Seehra, M. S., Zhang, Y., and Wender, I. 2008. Nature of magnetism in copper-doped oxides: ZrO_2, TiO_2, MgO, SiO_2, Al_2O_3, and ZnO. *J. Appl. Phys.* 103: 07D104.

Eckstein, J. N. 2007. Oxide interfaces—Watch out for the lack of oxygen. *Nat. Mater.* 6: 473–474.

Edwards, D. M. and Katsnelson, M. I. 2006. High-temperature ferromagnetism of sp electrons in narrow impurity bands: Application to CaB6. *J. Phys.: Condens. Matter* 18: 7209–7225.

Elfimov, I. S., Yunoki, S., and Sawatzky, G. A. 2002. Possible path to a new class of ferromagnetic and half-metallic ferromagnetic materials. *Phys. Rev. Lett.* 89.

Enoki, T. and Takai, K. 2006. Unconventional magnetic properties of nanographite, in: *Carbon Based Magnetism*, T. L. Makarova and F. Palacio (Eds.), Elsevier, Amsterdam, the Netherlands, pp. 397–416.

Enoki, T. and Takai, K. 2008. Unconventional electronic and magnetic functions of nanographene-based host-guest systems. *Dalton Trans.* 29: 3773–3781.

Enoki, T., Kobayashi, Y., and Fukui, K. I. 2007. Electronic structures of graphene edges and nanographene. *Int. Rev. Phys. Chem.* 26: 609–645.

Enz, T., Winterer, M., Stahl, B., Bhattacharya, S. et al. 2006. Structure and magnetic properties of iron nanoparticles stabilized in carbon. *J. Appl. Phys.* 99: 044306.

Esquinazi, P., Setzer, A. R., Höhne, R. et al. 2002. Ferromagnetism in oriented graphite samples. *Phys. Rev. B* 66: 24429.

Esquinazi, P., Spemann, D., Höhne, R., Setzer, A, Han, K.-H., and Butz, T. 2003. Induced magnetic ordering by proton irradiation in graphite. *Phys. Rev. Lett.* 91: 227201.

Faccio, R., Pardo, H., Denis, P. A. et al. 2008. Magnetism induced by single carbon vacancies in a three-dimensional graphitic network. *Phys. Rev. B* 77: 035416.

Fauth, K., Gold, S., Hessler, M., Schneider, N., and Schutz, G. 2004. Cluster surface interactions: Small Fe clusters driven nonmagnetic on graphite. *Chem. Phys. Lett.* 392: 498–502.

Fernandez-Rossier, J. 2008. Prediction of hidden multiferroic order in graphene zigzag ribbons. *Phys. Rev. B* 77: 075430.

Fernandez-Rossier, J. and Palacios, J. J. 2007. Magnetism in graphene nanoislands. *Phys. Rev. Lett.* 99: 177204.

Fisk, Z., Ott, H. R., Barzykin, V. et al. 2002. The emerging picture of ferromagnetism in the divalent hexaborides. *Physica B* 312: 808–810.

Freeman, A. J. and Wu, R. Q. 1991. Electronic structure theory of surface, interface and thin-film magnetism. *J. Magn. Magn. Mater.* 100: 497–514.

Fritsche, L., Noffke, J., and Eckard, H. 1987. Relativistic treatment of interacting spin-aligned electron systems: Application to ferromagnetic iron, nickel and palladium metal. *J. Phys. F: Met. Phys.* 17: 943–965.

Fujita, M., Wakabayashi, K., Nakada, K., and Kusakabe, K. 1996. Peculiar localized state at zigzag graphite edge. *J. Phys. Soc. Jpn.* 65: 1920–1923.

Gallego, S., Beltran, J. I., Cerda, J., and Munoz, M. C. 2005. Magnetism and half-metallicity at the O surfaces of ceramic oxides. *J. Phys.: Condens. Matter* 17: L451–L457.

Garcia, M. A., Merino, J. M., Pinel, E. F. et al. 2007. Magnetic properties of ZnO nanoparticles. *Nano Lett.* 7: 1489–1494.

Garcia, M. A., Pinel, E. F., de la Venta, J. et al. 2009. Sources of experimental errors in the observation of nanoscale magnetism. *J. Appl. Phys.* 105: 013925.

Garitaonandia, J. S., Insausti, M., Goikolea, E. et al. 2008. Chemically induced permanent magnetism in Au, Ag, and Cu nanoparticles: Localization of the magnetism by element selective techniques. *Nano Lett.* 8: 661–667.

Gavilano, J. L., Mushkolaj, S., Rau, D. et al. 2001. Anomalous NMR spin-lattice relaxation in SrB6 and $Ca_{1-x}La_xB_6$. *Phys. Rev. B* 63: 140410.

Ge, M. Y., Wang, H., Liu, E. Z. et al. 2008. On the origin of ferromagnetism in CeO_2 nanocubes. *Appl. Phys. Lett.* 93: 062505.

Ghoshal, S. and Kumar, P. S. A. 2008. Suppression of the magnetic moment upon Co doping in ZnO thin film with an intrinsic magnetic moment. *J. Phys.: Condens. Matter* 20: 192201.

Gianno, K., Sologubenko, A. V., Ott, H. R., Bianchi, A. D., and Fisk, Z. 2002. Low-temperature thermoelectric power of CaB_6. *J. Phys.: Condens. Matter* 14: 1035–1043.

Golmar, F., Navarro, A. M. M., Torres, C. E. R. et al. 2008. Extrinsic origin of ferromagnetism in single crystalline $LaAlO_3$ substrates and oxide films. *Appl. Phys. Lett.* 92: 262503.

González, J., Guinea, F., and Vozmediano, M. A. H. 2001. Electron-electron interactions in graphene sheets. *Phys. Rev. B* 63: 134421.

Grace, P. J., Venkatesan, M., Alaria, J., Coey, J. M. D., Kopnov, G., and Naaman, R. 2009. The origin of the magnetism of etched silicon. *Adv. Mater.* 21: 71–74.

Guerrero, E., Munoz-Marquez, M. A., Garcia, M. A. et al. 2008. Surface plasmon resonance and magnetism of thiol-capped gold nanoparticles. *Nanotechnology* 19: 17501.

Guinea, F. M., Lopez-Sancho, P., and Vozmediano, M. A. H. 2006. Interactions and disorder in 2D graphite sheets, in: *Carbon Based Magnetism*, T. L. Makarova and F. Palacio (Eds.), Elsevier, Amsterdam, the Netherlands, pp. 353–371.

Hadacek, N., Nosov, A., Ranno, L., Strobel, P., and Galera, R.-M. 2007. Magnetic properties of HfO_2 thin films. *J. Phys.: Condens. Matter* 19: 486206.

Han, K. H., Spemann, D., Hohne, R. et al. 2003. Observation of intrinsic magnetic domains in C_{60} polymers. *Carbon* 41: 785–795.

Hassini, A., Sakai, J., Lopez, J. S., and Hong, N. H. 2008. Magnetism in spin-coated pristine TiO_2 thin films. *Phys. Lett. A* 372: 3299–3302.

Heisedberg, W. 1928. Zur Theorie des Ferromagnetismus. *Z. Phys.* 49: 615–636.

Hernando, A., Crespo, P., and García, M. A. 2006a. Origin of orbital ferromagnetism and giant magnetic anisotropy at the nanoscale. *Phys. Rev. Lett.* 96: 057206.

Hernando, A., Crespo, P., García, M. A. et al. 2006b. Giant magnetic anisotropy at the nanoscale: Overcoming the superparamagnetic limit. *Phys. Rev. B* 74: 052403.

Hohne, R., Esquinazi, P., Heera, V. et al. 2008. The influence of iron, fluorine and boron implantation on the magnetic properties of graphite. *J. Magn. Magn. Mater.* 320: 966–977.

Hong, J. S. 2008. Local magnetic moment induced by Ga vacancy defect in GaN. *J. Appl. Phys.* 103: 063907.

Hong, N. H., Sakai, J., Poirot, N., and Brize, V. 2006a. Room-temperature ferromagnetism observed in undoped semiconducting and insulating oxide thin films. *Phys. Rev. B* 73 132404.

Hong, N. H., Sakai, J., Ruyter, A., and Brize, V. 2006b. Does Mn doping play any key role in tailoring the ferromagnetic ordering of TiO_2 thin films? *Appl. Phys. Lett.* 89: 252504.

Hong, N. H., Sakai, J., and Brize, V. 2007a. Observation of ferromagnetism at room temperature in ZnO thin films. *J. Phys.: Condens. Matter* 19: 036219.

Hong, N. H., Sakai, J., and Gervais, F. 2007b. Magnetism due to oxygen vacancies and/or defects in undoped semiconducting and insulating oxide thin films. *J. Magn. Magn. Mater.* 316: 214–217.

Hong, N. H., Poirot, N., and Sakai, J. 2008. Ferromagnetism observed in pristine SnO_2 thin films. *Phys. Rev. B* 77: 033205.

Hori, H., Teranishi, T., Nakae, Y. et al. 1999. Anomalous magnetic polarization effect of Pd and Au nano-particles. *Phys. Lett. A* 263: 406–410.

Hori, H., Yamamoto, Y., Iwamoto, T., Miura, T., Teranishi, T., and Miyake, M. 2004. Diameter dependence of ferromagnetic spin moment in Au nanocrystals. *Phys. Rev. B* 69: 174411.

Host, J. J., Block, J. A., Parvin, K. et al. 1998. Effect of annealing on the structure and magnetic properties of graphite encapsulated nickel and cobalt nanocrystals. *J. Appl. Phys.* 83: 793–801.

Hou, D. L., Ye, X. J., Meng, H. J. et al. 2007. Magnetic properties of n-type Cu-doped ZnO thin films. *Appl. Phys. Lett.* 90: 142502.

Huang, L. M., Rosa, A. L., and Ahuja, R. 2006. Ferromagnetism in Cu-doped ZnO from first-principles theory. *Phys. Rev. B* 74: 075206.

Huang, B., Feng, L., Wu, J., Gu, B.-L., and Duan, W. 2008. Suppression of spin polarization in graphene nanoribbons by edge defects and impurities. *Phys. Rev. B* 77: 153411.

Ishii, C., Matsumura, Y., and Kaneko, K. 1995. Ferromagnetic behaviour of superhigh surface-area carbon. *J. Phys. Chem.* 99: 5743–5745.

Ito, Y., Miyazaki A., Fukui F., Valiyaveettil S., Yokoyama T., and Enoki, T. 2008. Pd nanoparticle embedded with only one Co atom behaves as a single-particle magnet. *J. Phys. Soc. Jpn.* 77: 103701.

Ivanovskii, A. L. 2007. Magnetic effects induced by sp impurities and defects in nonmagnetic sp materials. *Phys.-Usp.* 50: 1031–1052.

Janisch, R., Gopal, P., and Spaldin, N. A. 2005. Transition metal-doped TiO2 and ZnO—present status of the field. *J. Phys.: Condens. Matter* 17: R657–R689.

Jarlborg, T. 2000. Ferromagnetism below the Stoner limit in La-doped SrB_6. *Phys. Rev. Lett.* 85: 186–189.

Jian, W. B., Lu, W. G., Fang, J., Lan, M. D., and Lin, J. J. 2006. Spontaneous magnetization and ferromagnetism in PbSe quantum dots. *J. Appl. Phys.* 99: 08N708.

Jungwirth, T., Sinova, J., Masek, J., Kucera, J., and MacDonald, A. H. 2006. Theory of ferromagnetic (III,Mn)V semiconductors. *Rev. Mod. Phys.* 78: 809–864.

Katayama-Yoshida, H., Sato, K., Fukushima, T. et al. 2007. Theory of ferromagnetic semiconductors. *Phys. Status Solidi A* 204: 15–32.

Katsnelson, M. I., Irkhin, V. Y., Chioncel, L., Lichtenstein, A. I., and de Groot, R. A. 2008. Half-metallic ferromagnets: From band structure to many-body effects. *Rev. Mod. Phys.* 80: 315–378.

Keavney, D. J., Buchholz, D. B., Ma, Q., and Chang, R. P. H. 2007. Where does the spin reside in ferromagnetic Cu-doped ZnO? *Appl. Phys. Lett.* 91: 012501.

Kenmochi, K., Seike, M., Sato, K., Yanase, A., and Katayama-Yoshida, H. 2004. New class of diluted ferromagnetic semiconductors based on CaO without transition metal elements. *Jpn. J. Appl. Phys.* 43: L934–L936.

Kim, Y.-H., Choi, J., and Chang, K. J. 2003. Defective fullerenes and nanotubes as molecular magnets: An ab initio study. *Phys. Rev. B* 68: 125420.

Kittel, C. 1946. Theory of the structure of ferromagnetic domains in films and small particles. *Phys. Rev.* 70: 965–971.

Kittilstved, K. R., Liu, W. K., and Gamelin, D. R. 2006. Electronic structure origins of polarity-dependent high-T_C ferromagnetism in oxide-diluted magnetic semiconductors. *Nat. Mater.* 5: 291–297.

Kopelevich, Y. and Esquinazi, P. 2007. Ferromagnetism and superconductivity in carbon-based systems. *J. Low Temp. Phys.* 146: 629–639.

Kopelevich, Y., Esquinazi, P., Torres, J. H. S., and Moehlecke, S. 2000. Ferromagnetic- and superconducting-like behavior of graphite. *J. Low Temp. Phys.* 119: 691–697.

Kopelevich, Y., da Silva, R. R., Torres, J. H. S., Penicaud, A., and Kyotani, T. 2003. Local ferromagnetism in microporous carbon with the structural regularity of zeolite. *Phys. Rev. B* 68: 092408.

Kopnov, G., Vager, Z., and Naaman, R. 2007. New magnetic properties of silicon/silicon oxide interfaces. *Adv. Mater.* 19: 925–928.

Kosugi, K., Bushiri, M. J., and Nishi, N. 2004. Formation of air stable carbon-skinned iron nanocrystals from FeC_2. *Appl. Phys. Lett.* 84 1753–1755.

Kreutz, T. C., Gwinn, E. G., Artzi, R., Naaman, R., Pizem, H., and Sukenik, C. N. 2003. Modification of ferromagnetism in semiconductors by molecular monolayers. *Appl. Phys. Lett.* 83: 4211–4213.

Kumar, A., Avasthi, D. K., Pivin, C., Tripathi, A., and Singh, F. 2006. Ferromagnetism induced by heavy-ion irradiation in fullerene films. *Phys. Rev. B.* 74: 153409.

Kumar, A., Avasthi, D. K., Pivin, C. et al. 2007. Magnetic force microscopy of nano-size magnetic domain ordering in heavy ion irradiated fullerene films. *J. Nanosci. Nanotechnol.* 7: 2201–2205.

Kumazaki, H. and Hirashima, D. S. 2007a. Possible vacancy-induced magnetism on a half-filled honeycomb lattice. *J. Phys. Soc. Jpn.* 76: 034707.

Kumazaki, H. and Hirashima, D. S. 2007b. Nonmagnetic-defect-induced magnetism in graphene. *J. Phys. Soc. Jpn.* 76: 064713.

Kumazaki, H. and Hirashima, D. S. 2007c. Magnetism of a two-dimensional graphite sheet. *J. Magn. Magn. Mater.* 310: 2256.

Kumazaki, H. and Hirashima, D. S. 2008. Local magnetic moment formation on edges of graphene. *J. Phys. Soc. Jpn.* 77: 044705.

Kunii, S. 2000. Surface-layer ferromagnetism and strong surface anisotropy in $Ca_{1-x}La_xB_6$ (x=0.005) evidenced by ferromagnetic resonance. *J. Phys. Soc. Jpn.* 69: 3789–3791.

Kusakabe, K. 2006. Flat-band ferromagnetism in organic crystals, in: *Carbon Based Magnetism*, T. L. Makarova and F. Palacio (Eds.), Elsevier, Amsterdam, the Netherlands, pp. 305–328.

Kusakabe, K. and Maruyama, M. 2003. Magnetic nanographite. *Phys. Rev. B* 67: 092406.

Kusakabe, K., Geshi, M., Tsukamoto, H., and Suzuki, N. 2004. Design of new ferromagnetic materials with high spin moments by first-principles calculation. *J. Phys.: Condens. Matter* 16: S5639–S5644.

Kvyatkovskii, O. E., Zakharova, I. B., Shelankov, A. L., and Makarova, T. L. 2004. Electronic properties of the $(C_{60})_2$ and $(C_{60})_2^{2-}$ fullerene dimer. *AIP Conf. Proc.* 723: 385–388.

Kvyatkovskii, O. E., Zakharova, I. B., Shelankov, A. L., and Makarova, T. L. 2005. Spin-transfer mechanism of ferromagnetism in polymerized fullerenes: Ab initio calculations *Phys. Rev. B* 72: 214426.

Kvyatkovskii, O. E., Zakharova, I. B., and Shelankov, A. L. 2006. Magnetic properties of polymerized fullerene doped with hydrogen, fluorine and oxygen. *Fuller. Nanot. Car. Nan.* 14: 373–380.

Lefebvre, J., Trudel, S., Hill, R. H. et al. 2008. A closer look: Magnetic behavior of a three-dimensional cyanometalate coordination polymer dominated by a trace amount of nanoparticle impurity. *Chem. Eur. J.* 14: 7156–7167.

Lehtinen, P. O., Foster, A. S., Ayuela, A., Krasheninnikov, A., Nordlund, K., and Nieminen, R. M. 2003. Magnetic properties and diffusion of adatoms on a graphene sheet. *Phys. Rev. Lett.* 91: 017202.

Lei, Y., Shevlin, S. A., Zhu, W., and Guo, Z. X. 2008. Hydrogen-induced magnetization and tunable hydrogen storage in graphitic structures. *Phys. Rev. B* 77: 134114.

Li, S. D., Huang, Z. G., Lue, L. Y. et al. 2007. Ferromagnetic chaoite macrotubes prepared at low temperature and pressure. *Appl. Phys. Lett.* 90: 232507.

Li, D., Han, Z., Wu, B. et al. 2008a. Ferromagnetic and spin-glass behaviour of nanosized oriented pyrolytic graphite in Pb-C nanocomposites. *J. Phys. D Appl. Phys* 41: 115005.

Li, Q. K., Wang, B., Woo, C. H., Wang, H., Zhu, Z. Y., and Wang, R. 2008b. Origin of unexpected magnetism in Cu-doped TiO_2. *Europhys. Lett.* 81: 17004.

Liou, Y. and Shen, Y. L. 2008. Magnetic properties of germanium quantum dots. *Adv. Mater.* 20: 779–783.

Liu, F., Press, M. R., Khanna, S. N., and Jena, P. 1989. Magnetism and local order: *Ab initio* tight-binding theory. *Phys. Rev. B* 39: 6914–6924.

Liu, X., Bauer, M., Bertagnolli, H., Roduner, E., van Slageren, J., and Phillipp, F. 2006. Structure and magnetization of small monodisperse platinum clusters. *Phys. Rev. Lett.* 97: 253401.

Liu, Y. L., Lockman, Z., Aziz, A., and MacManus-Driscoll, J. 2008. Size dependent ferromagnetism in cerium oxide (CeO_2) nanostructures independent of oxygen vacancies. *J. Phys.: Condens. Matter* 20: 165201.

Lobach, A. S., Shul'ga, Y. M., Roshchupkina, O. S. et al. 1998. $C_{60}H_{18}$, $C_{60}H_{36}$ and $C_{70}H_{36}$ fullerene hydrides: Study by methods of IR, NMR, XPS, EELS and magnetochemistry. *Fuller. Sci. Technol.* 6: 375–391.

Lofland, S. E., Seaman, B., Ramanujachary, K. V., Hur, N., and Cheong, S. W. 2003. Defect driven magnetism in calcium hexaboride. *Phys. Rev. B* 67: 020410.

Lopez-Urias, F., Pastor, G. M., and Bennemann, K. H. 2000. Calculation of finite temperature magnetic properties of clusters. *J. Appl. Phys.* 87: 4909–4911.

Lounis, S., Dederichs, P. H., and Bluegel, S. 2008. Magnetism of nanowires driven by novel even-odd effects. *Phys. Rev. Lett.* 101: 107204.

Ma, Y. C., Lehtinen, P. O., Foster, A. S., and Nieminen, R. M. 2004. Magnetic properties of vacancies in graphene and single-walled carbon nanotubes. *New J. Phys.* 6: 68.

Ma, Y. C., Lehtinen, P. O., Foster, A. S., and Nieminen, R. M. 2005. Hydrogen-induced magnetism in carbon nanotubes. *Phys. Rev. B* 72: 085451.

Ma, Y. W., Yi, J. B., Ding, J., Van, L. H., Zhang, H. T., and Ng, C. M. 2008. Inducing ferromagnetism in ZnO through doping of nonmagnetic elements. *Appl. Phys. Lett.* 93: 042514.

Maca, F., Kudrnovsky, J., Drchal, V., and Bouzerar, G. 2008. Magnetism without magnetic impurities in ZrO_2 oxide. *Appl. Phys. Lett.* 92: 212503.

Mackay, A. L. and Terrones, H. 1991. Diamond from graphite. *Nature* 352: 762.

Madhu, C., Sundaresan, A., and Rao, C. N. R. 2008. Room-temperature ferromagnetism in undoped GaN and CdS semiconductor nanoparticles. *Phys. Rev. B* 77: 201306.

Maiti, K. 2008. Role of vacancies and impurities in the ferromagnetism of semiconducting CaB6. *Europhys. Lett.* 82: 67006.

Maiti, K., Medicherla, V. R. R., Patil, S., and Singh, R. S. 2007. Revelation of the role of impurities and conduction electron density in the high resolution photoemission study of ferromagnetic hexaborides. *Phys. Rev. Lett.* 99: 266401.

Makarova, T. L. 2004. Magnetic properties of carbon structures. *Semiconductors* 38: 615–638.

Makarova, T. and Palacio, F. (Eds.). 2006. *Carbon Based Magnetism: An Overview of the Magnetism of Metal Free Carbon-Based Compounds and Materials*, Elsevier, Amsterdam, the Netherlands.

Makarova, T. L. and Zakharova, I. B. 2008. Separation of intrinsic and extrinsic contribution to fullerene magnetism. *Fuller. Nanot. Car. Nan.* 16: 567–573.

Makarova, T. L., Sundqvist, B., Höhne, R. et al. 2001a. Magnetic carbon. *Nature* 413: 716–719.

Makarova, T. L., Sundqvist, B., Höhne, R. et al. 2001b. Magnetic carbon. (retraction). *Nature* 440: 707.

Makarova, T. L., Han, K.-H., Esquinazi, P. et al. 2003. Magnetism in photopolymerized fullerenes. *Carbon* 41 (8) 1575–1584.

Maruyama, M. and Kusakabe, K. 2004. Theoretical prediction of synthesis methods to create magnetic nanographite. *J. Phys. Soc. Jpn.* 73: 656–663.

Massidda, S., Continenza, A., de Pascale, T. M., and Monnier, R. 1997. Electronic structure of divalent hexaborides. *Z. Phys. B: Condens. Matter* 102: 83–89.

Mathew, S., Satpati, B., Joseph B. et al. 2007. Magnetism in C$_{60}$ films induced by proton irradiation. *Phys. Rev. B* 75: 075426.

Matsubayashi, K., Maki, M., Tsuzuki, T., Nishioka, T., and Sato, N. K. 2002. Magnetic properties—Parasitic ferromagnetism in a hexaboride? *Nature* 420: 143–144.

Matsubayashi, K., Maki, M., Moriwaka, T. et al. 2003. Extrinsic origin of high-temperature ferromagnetism in CaB$_6$. *J. Phys. Soc. Jpn.* 72: 2097–2102.

Matsumoto, Y., Murakami, M., Shono, T. et al. 2001. Room-temperature ferromagnetism in transparent transition metal-doped titanium dioxide. *Science* 291: 854–856.

Medicherla, V. R. R., Patil, S., Singh, R. S., and Maiti, K. 2007. Origin of ground state anomaly in LaB$_6$ at low temperatures. *Appl. Phys. Lett.* 90: 062507.

Meegoda, C., Trenary, M., Mori, T., and Otani, S. 2003. Depth profile of iron in a CaB$_6$ crystal. *Phys. Rev. B* 67: 172410.

Meier, R. J. and Helmholdt, R. B. 1984. Neutron-diffraction study of α- and β-oxygen. *Phys. Rev. B* 29: 1387–1393.

Mencken, H. L. 1920. *Prejudices: Second Series*, Knopf, New York.

Mertins, H.-C., Valencia, S., Gudat, W., Oppeneer, P. M., Zaharko, O., and Grimmer, H. 2004. Direct observation of local ferromagnetism on carbon in C/Fe multilayers. *Europhys. Lett.* 66: 743–748.

Michael, F., Gonzalez, C., Mujica, V., Marquez, M., and Ratner, M. A. 2007. Size dependence of ferromagnetism in gold nanoparticles: Mean field results. *Phys. Rev. B* 76: 224409.

Min, H., Borghi, G., Polini, M., and MacDonald, A. H. 2008. Pseudo-spin magnetism in graphene. *Phys. Rev. B* 77: 041407(R).

Mombru, A. W., Pardo, H., Faccio, R. et al. 2005. Multilevel ferromagnetic behavior of room-temperature bulk magnetic graphite. *Phys. Rev. B* 71:100404.

Monnier, R. and Delley, B. 2001. Point defects, ferromagnetism, and transport in calcium hexaboride. *Phys. Rev. Lett.* 87: 157204.

Mori, T. and Otani, S. 2002. Ferromagnetism in lanthanum doped CaB$_6$: Is it intrinsic? *Solid State Commun.* 123: 287–290.

Murakami, Y. and Suematsu, H. 1996. Magnetism of C$_{60}$ induced by photo-assisted oxidation. *Pure Appl. Chem.* 68: 1463–1467.

Murakami, S., Shindou, R., Nagaosa, N., and Mishchenko, A. S. 2002. Theory of ferromagnetism in Ca$_{1-x}$La$_x$B$_6$. *Phys. Rev. Lett.* 88: 126404.

Naaman, R. and Vager, Z. 2006. New electronic and magnetic properties emerging from adsorption of organized organic layers. *Phys. Chem. Chem. Phys.* 8: 2217–2224.

Nakano, T., Ikemoto, Y., and Nozue, Y. 2000. Loading density dependence of ferromagnetic properties in potassium clusters arrayed in a simple cubic structure in zeolite LTA. *Physica B* 281: 688–690.

Nakano, S., Kitagawa, Y., Kawakami, T., Okumura, M., Nagao, H., and Yamaguchi, K. 2004. Theoretical studies on electronic states of Rh-C$_{60}$. Possibility of a room-temperature organic ferromagnet. *Molecules* 9: 792–807.

Nakano, T., Gotoa, K., Watanabeb, I., Prattc, F. L., Ikemotod, Y., and Nozue, Y. 2006. μSR study on ferrimagnetic properties of potassium clusters incorporated into low silica X zeolite. *Physica B* 374: 21–25.

Narozhnyi, V. N., Müller, K.-H., Eckert, D. et al. 2003. Ferromagnetic carbon with enhanced Curie temperature. *Physica B* 329: 1217–1218.

Neeleshwar, S., Chen, C. L., Tsai, C. B., Chen, Y. Y., and Chen, C. C. 2005. Size-dependent properties of CdSe quantum dots. *Phys. Rev. B* 71: 201307.

Negishi, Y., Tsunoyama, H., Suzuki, M. et al. 2006. X-ray magnetic circular dichroism of size-selected, thiolated gold clusters. *J. Am. Chem. Soc.* 128: 12034–12035.

Norton, D. P., Overberg, M. E., Pearton, S. J. et al. 2003. Ferromagnetism in cobalt-implanted ZnO. *Appl. Phys. Lett.* 83: 5488–5490.

Nozue, Y., Kodaira, T., and Goto, T. 1992. Ferromagnetism of potassium clusters incorporated into zeolite LTA. *Phys. Rev. Lett.* 68: 3789–3792.

Ohldag, H., Tyliszczak, T., Hohne, R. et al. 2007. pi-electron ferromagnetism in metal-free carbon probed by soft x-ray dichroism. *Phys. Rev. Lett.* 98: 187204.

Ohno, H. 1998. Making nonmagnetic semiconductors ferromagnetic. *Science* 281: 951–956.

Okada, S. and Oshiyama, A. 2001. Magnetic ordering in hexagonally bonded sheets with first-row elements. *Phys. Rev. Lett.* 87: 146803.

Okada, S. and Oshiyama, A. 2003. Electronic structure of metallic rhombohedral C-60 polymers. *Phys. Rev. B* 68: 235402.

Oshiyama, A. and Okada, S. 2006. Magnetism in nanometer-scale materials that contain no magnetic elements, in: *Carbon Based Magnetism*, T. Makarova and F. Palacio (Eds.), Elsevier, Amsterdam, the Netherlands.

Osipov, V., Baidakova, M., Takai, K., Enoki, T., and Vul, A. 2006. Magnetic properties of hydrogen-terminated surface layer of diamond nanoparticles. *Fuller. Nanot. Car. Nan.* 14: 565–572.

Osorio-Guillen, J., Lany, S., Barabash, S. V., and Zunger, A. 2007. Nonstoichiometry as a source of magnetism in otherwise nonmagnetic oxides: Magnetically interacting cation vacancies and their percolation. *Phys. Rev. B* 75: 184421.

Osorio-Guillen, J., Lany, S., and Zunger, A. 2008. Atomic control of conductivity versus ferromagnetism in wide-gap oxides via selective doping: V, Nb, Ta in anatase TiO_2. *Phys. Rev. Lett.* 100: 036601.

Otani, S. and Mori, T. 2002. Flux growth and magnetic properties of CaB_6 crystals. *J. Phys. Soc. Jpn.* 71: 1791–1792.

Otani, S. and Mori, T. 2003. Flux growth of CaB_6 crystals. *J. Mater. Sci.* 22: 1065–1066.

Ott, H. R., Gavilano, J. L., Ambrosini, B. et al. 2000. Unusual magnetism of hexaborides. *Physica B* 281: 423–427.

Owens, F. J., Iqbal, Z., Belova, L. et al. 2004. Evidence for high-temperature ferromagnetism in photolyzed C_{60}. *Phys. Rev. B* 69: 033403.

Palacios, J. J. and Fernández-Rossier, L. 2008. Vacancy-induced magnetism in graphene and graphene ribbons. *Phys. Rev. B* 77: 195428.

Pan, H., Yi, J. B., Shen, L., Wu, R. Q., Yang, J. H., Lin, J. Y., Feng, Y. P., Ding, J., Van, L. H., and Yin, J. H. 2007. Room-temperature ferromagnetism in carbon-doped ZnO. *Phys. Rev. Lett.* 99: 127201.

Pardo, H., Faccio, R., Araujo-Moreira, F. M. et al. 2006. Synthesis and characterization of stable room temperature bulk ferromagnetic graphite. *Carbon* 44: 565–569.

Park, M. S. and Min, B. I. 2003. Ferromagnetism in ZnO codoped with transition metals: $Zn_{1-x}(FeCo)_{(x)}O$ and $Zn_{1-x}(FeCu)_{(x)}O$. *Phys. Rev. B* 68: 224436.

Park, N., Yoon, M., Berber, S., Ihm, J., Osawa, E., and Tomanek, D. 2003. Magnetism in all-carbon nanostructures with negative Gaussian curvature. *Phys. Rev. Lett.* 91: 237204.

Parkansky, N., Alterkop, B., Boman, R. L. et al. 2008. Magnetic properties of carbon nano-particles produced by a pulsed arc submerged in ethanol. *Carbon* 46: 215–219.

Pei, X. Y., Yang, X. P., and Dong, J. M. 2006. Effects of different hydrogen distributions on the magnetic properties of hydrogenated single-walled carbon nanotubes. *Phys. Rev. B* 73: 195417.

Peres, N. M. R., Guinea, F., and Castro, A. H. 2006. Electronic properties of disordered two-dimensional carbon. *Phys. Rev. B* 73: 125411.

Pisani, L., Montanari, B., and Harrison, N. M. 2008. A defective graphene phase predicted to be a room temperature ferromagnetic semiconductor. *New J. Phys.* 10: 033002.

Potzger, K., Zhou, S. Q., Grenzer, J. et al. 2008. An easy mechanical way to create ferromagnetic defective ZnO. *Appl. Phys. Lett.* 92: 182504.

Qiu, X. Y., Liu, Q. M., Gao, F., Lu, L. Y., and Liu, J.-M. 2006. Room-temperature weak ferromagnetism of amorphous $HfAlO_x$ thin films deposited by pulsed laser deposition. *Appl. Phys. Lett.* 89: 242504.

Quesada, A., Garcia, M. A., de la Venta, J., Pinel, E. F., Merino, J. M., and Hernando, A. 2007. Ferromagnetic behaviour in semiconductors: A new magnetism in search of spintronic materials. *Eur. Phys. J. B* 59: 457–461.

Rao, M. S. R., Kundaliya, D. C., Ogale, S. B. et al. 2006. Search for ferromagnetism in undoped and cobalt-doped HfO_2-delta. *Appl. Phys. Lett.* 88: 142505.

Ray, S. G., Daube, S. S., Leitus, G., Vager, Z., and Naaman, R. 2006. Chirality-induced spin-selective properties of self-assembled monolayers of DNA on gold. *Phys. Rev. Lett.* 96: 036101.

Reich, S., Leitus, G., and Feldman, Y. 2006. Observation of magnetism in Au thin films. *Appl. Phys. Lett.* 88: 222502.

Rhyee, J. S. and Cho, B. K. 2004. The effect of boron purity on electric and magnetic properties of CaB_6. *J. Appl. Phys.* 95: 6675–6677.

Ribas-Arino, J. and Novoa, J. J. 2004a. Magnetism in compressed fullerenes. The origin of the magnetic moments in compressed crystals of polymeric C_{60}. *Angew. Chem. Int. Ed.* 43: 577–580.

Ribas-Arino, J. and Novoa, J. J. 2004b. Evaluation of the capability of C-60-fullerene to act as a magnetic coupling unit. *J. Phys. Chem. Solids* 65: 787–791.

Rode, A. V., Gamaly, E. G., Christy, A. G. et al. 2004. Unconventional magnetism in all-carbon nanofoam. *Phys. Rev. B* 70: 054407.

Sahu, B, Min, H., MacDonald, A. H., and Banerjee, S. K. 2008. Energy gaps, magnetism, and electric-field effects in bilayer graphene nanoribbons. *Phys. Rev. B* 78: 045404.

Saito, Y., Ma, J., Nakashima, J., and Masuda, M. 1997. Synthesis, crystal structures and magnetic properties of Co particles encapsulated in carbon nanocapsules. *Z. Phys. D* 40: 170–172.

Salamon, M. B. and Jaime, M. 2001. The physics of manganites: Structure and transport. *Rev. Mod. Phys.* 73: 583–628.

Salzer, R., Spemann, D., Esquinazi, P. et al. 2007. Possible pitfalls in search of magnetic order in thin films deposited on single crystalline sapphire substrates. *J. Magn. Magn. Mater.* 317: 53–60.

Sampedro, B., Crespo, P., Hernando, A. et al. 2003. Ferromagnetism in fcc twinned 2.4 nm size Pd nanoparticles. *Phys. Rev. Lett.* 91: 237203.

Sanchez, N., Gallego, S., and Munoz, M. C. 2008. Magnetic states at the oxygen surfaces of ZnO and Co-doped ZnO. *Phys. Rev. Lett.* 101: 067206.

Sanyal, D., Chakrabarti, M., Roy, T. K., and Chakrabarti, A. 2007. The origin of ferromagnetism and defect-magnetization correlation in nanocrystalline ZnO. *Phys. Lett. A* 371: 482–485.

Sato, H., Kawatsu, N., Enoki, T., Endo, M., Kobori, R., Maruyama, S., and Kaneko, K. 2003. Physisorption-induced change in the magnetism of microporous carbon. *Solid State Commun.* 125: 641.

Sato, H., Kawatsu, N., Enoki, T. et al. 2007. Physisorption-induced change in the magnetism of microporous carbon. *Carbon* 45: 214–217.

Schneider, A. M. 2008. Nanomagnetism: A matter of orientation. *Nat. Phys.* 4: 831–832.

Schwickardi, M., Olejnik, S., Salabas, E. L., Schmidt, W., and Schuth, F. 2006. Scalable synthesis of activated carbon with superparamagnetic properties. *Chem. Commun.* 38: 3987–3989.

Seehra, M. S., Dutta, P., Singh, V., Zhang, Y., and Wender, I. 2007. Evidence for room temperature ferromagnetism in $Cu_xZn_{1-x}O$ from magnetic studies in $Cu_xZn_{1-x}O/CuO$ composite. *J. Appl. Phys.* 101: 09H107.

Seehra, M. S., Dutta, P., Neeleshwar, S. et al. 2008. Size-controlled Ex-nihilo ferromagnetism in capped CdSe quantum dots. *Adv. Mater.* 20: 1656–1660.

Shen, L., Wu, R. Q., Pan, H. et al. 2008. Mechanism of ferromagnetism in nitrogen-doped ZnO: First-principle calculations. *Phys. Rev. B* 78: 073306.

Shibayama, Y., Sato, H., Enoki, T. et al. 2000. Novel electronic properties of a nano-graphite disordered network and their iodine doping effects. *J. Phys. Soc. Jpn.* 69: 754–767.

Shinde, S. R., Ogale, S. B., Higgins, J. S. et al. 2004. Co-occurrence of superparamagnetism and anomalous Hall effect in highly reduced cobalt-doped rutile $TiO_{2-\delta}$ films. *Phys. Rev. Lett.* 92: 166601.

Shuai, M., Liao, L., Lu, H. B. et al. 2007. Room-temperature ferromagnetism in Cu^+ implanted ZnO nanowires. *J. Phys. D. Appl. Phys.* 41: 135010.

Skomski, R. and Coey, J. M. D. 1995. *Permanent Magnetism.* IOP Publishing, Bristol, U.K.

Smogunov, A., Dal Corso, A., Delin, A., Weht, R., and Tosatti, E. 2008. Colossal magnetic anisotropy of monatomic free and deposited platinum nanowires. *Nat. Nanotechnol.* 3: 22–25.

Spemann, D., Han, K. H., Hohne, R. et al. 2003. Evidence for intrinsic weak ferromagnetism in a C_{60} polymer by PIXE and MFM. *Nucl. Instrum. Meth. B* 210: 531–536.

Stauber, T., Guinea, F., and Vozmediano, M. A. H. 2005. Disorder and interaction effects in two dimensional graphene sheets. *Phys. Rev. B.* 71: 041406.

Stauber, T., Castro, E. V., Silva, N. A. P. et al. 2008. First-order ferromagnetic phase transition in the low electronic density regime of biased graphene bilayers. *J. Phys.: Condens. Matter* 20: 335207.

Suda, M., Kameyama, N., Suzuki, M., Kawamura, N., and Einaga, Y. 2008. Reversible phototuning of ferromagnetism at Au-S interfaces at room temperature. *Angew. Chem. Int. Ed.* 47: 160–163.

Sudakar, C., Kharel, P., Suryanarayanan, R. et al. 2008. Room temperature ferromagnetism in vacuum-annealed TiO_2 thin films. *J. Magn. Magn. Mater.* 320: L31–L36.

Sundaresan, A., Bhargavi, R., Rangarajan, N., Siddesh, U., and Rao, C. N. R. 2006. Ferromagnetism as a universal feature of nanoparticles of the otherwise nonmagnetic oxides. *Phys. Rev. B* 74: 161306(R).

Takai, K., Eto, S., Inaguma, M., and Enoki, T. 2008. Magnetic potassium clusters in the nanographite-based nanoporous system. *J. Phys. Chem. Sol.* 69: 1182–1184.

Teng, X. W., Han, W. Q., Ku, W. et al. 2008. Synthesis of ultra-thin palladium and platinum nanowires and a study of their magnetic properties. *Angew. Chem. Int. Ed.* 47: 2055–2058.

Terashima, T., Terakura, C., Umeda, Y., Kimura, N., Aoki, H., and Kunii, S. 2000. Ferromagnetism vs paramagnetism and false quantum oscillations in lanthanum-doped CaB_6. *J. Phys. Soc. Jpn.* 69: 2423–2426.

Tietze, T., Gacic, M., Schutz, G., Jakob, G., Bruck, S., and Goering, E. 2008. XMCD studies on Co and Li doped ZnO magnetic semiconductors. *New J. Phys.* 10: 055009.

Tirosh, E. and Markovich, G. 2007. Control of defects and magnetic properties in colloidal HfO2 nanorods. *Adv. Mater.* 19: 2608–2612.

Topsakal, M., Sevinçli, H., and Ciraci1, S. 2008. Spin confinement in the superlattices of graphene ribbons. *Appl. Phys. Lett.* 92: 173118.

Tromp, H. J., van Gelderen, P., Kelly, P. J., Brocks, G., and Bobbert, P. A. 2001. CaB6: A new semiconducting material for spin electronics. *Phys. Rev. Lett.* 87: 016401.

Ueda, K., Tabata, H., and Kawai, T. 2001. Magnetic and electric properties of transition-metal-doped ZnO films. *Appl. Phys. Lett.* 79: 988–990.

Vager, Z., Carmeli, I., Leitus, G., Reich, S., and Naaman, R. 2004. Surprising electronic-magnetic properties of closed packed organized organic layers. *J. Phys. Chem. Sol.* 65: 713–717.

Veillette, M. Y. and Balents, L. 2002. Weak ferromagnetism and excitonic condensates. *Phys. Rev. B* 65: 14428.

Venkatesan, M., Fitzgerald, C. B., and Coey, J. M. D. 2004. Unexpected magnetism in a dielectric oxide. *Nature* 430: 630.

Vonlanthen, P., Felder, E., Degiorgi, L. et al. 2000. Electronic transport and thermal and optical properties of $Ca_{1-x}La_xB_6$. *Phys. Rev. B* 62: 10076–10082.

Vozmediano, M. A. H., Guinea, F., and Lopez-Sancho, M. P. 2006. Interactions, disorder and local defects in graphite. *J. Phys. Chem. Sol.* 67: 562.

Wang, X., Liu, Z. X., Zhang, Y. L., Li, F. Y., and Jin, C. Q. 2002. Evolution of magnetic behaviour in the graphitization process of glassy carbon. *J. Phys.: Condens. Matter.* 14: 10265.

Wang, W. C., Kong, Y., He, X., and Liu, B. X. 2006a. Observation of magnetism in the nanoscale amorphous ruthenium clusters prepared by ion beam mixing. *Appl. Phys. Lett.* 89: 262511.

Wang, W. D., Hong, Y. J., Yu, M. H., Rout, B., Glass, G. A., and Tang, J. K. 2006b. Structure and magnetic properties of pure and Gd-doped HfO2 thin films. *J. Appl. Phys.* 99: 08M117.

Wang, Q., Sun, Q., Chen G., Kawazoe, Y., and Jena, P. 2008a. Vacancy-induced magnetism in ZnO thin films and nanowires. *Phys. Rev. B* 77: 205411.

Wang, W. L., Meng, S., and Kaxiras, E. 2008b. Graphene nanoflakes with large spin. *Nano Lett.* 8: 241–245.

Wang, Y., Huang, Y., Song, Y. et al. 2009. Room-temperature ferromagnetism of graphene. *Nano Lett.* 9: 220–224.

Weyer, G., Gunnlaugsson, H. P., Mantovan, R. et al. 2007. Defect-related local magnetism at dilute Fe atoms in ion-implanted. *J. Appl. Phys.* 102: 113915.

Wood, R. A., Lewis, M. H., Lees, M. R. et al. 2002. Ferromagnetic fullerene. *J. Phys.: Condens. Matter* 14: L385–L391.

Wu, R. Q., Peng, G. W., Liu, L., Feng, Y. P., Huang, Z. G., and Wu, Q. Y. 2006. Cu-doped GaN: A dilute magnetic semiconductor from first-principles study. *Appl. Phys. Lett.* 89: 062505.

Xing, G. Z., Yi, J. B., Tao, J. G. et al. 2008. Comparative study of room-temperature ferromagnetism in Cu-doped ZnO nanowires enhanced by structural inhomogeneity. *Adv. Mater.* 20: 3521–3527.

Xu, Q. Y., Schmidt, H., Zhou, S. Q. et al. 2008. Room temperature ferromagnetism in ZnO films due to defects. *Appl. Phys. Lett.* 92: 082508.

Yamamoto, Y., Miura, T., Suzuki, M. et al. 2004. Direct observation of ferromagnetic spin polarization in gold nanoparticles. *Phys. Rev. Lett.* 93: 116801.

Yan, Z., Ma, Y., Wang, D. et al. 2008. Impact of annealing on morphology and ferromagnetism of ZnO nanorods. *Appl. Phys. Lett.* 92: 081911.

Yazyev, O. V. 2008. Magnetism in disordered graphene and irradiated graphite. *Phys. Rev. Lett.* 101: 037203.

Yazyev, O. V. and Helm, L. 2007. Defect-induced magnetism in graphene. *Phys. Rev. B* 75: 125408.

Yazyev, O. V. and Katsnelson, M. I. 2008. Magnetic correlations at graphene edges: Basis for novel spintronics devices. *Phys. Rev. Lett.* 100: 047209.

Yazyev, O. V., Wang, W. L., Meng, S., and Kaxiras, E. 2008. Comment on graphene nanoflakes with large spin: Broken-symmetry states. *Nano Lett.* 9: 766–767.

Ye, L.-H., Freeman, A. J., and Delley, B. 2006. Half-metallic ferromagnetism in Cu-doped ZnO: Density functional calculations. *Phys. Rev. B* 73: 033203.

Yi, J. B., Pan, H., Lin, J. Y. et al. 2008. Ferromagnetism in ZnO nanowires derived from electro-deposition on AAO template and subsequent oxidation. *Adv. Mater.* 20: 1170–1174.

Yin, Z. G., Chen, N. F., Li, Y. et al. 2008. Interface as the origin of ferromagnetism in cobalt doped ZnO film grown on silicon substrate. *Appl. Phys. Lett.* 93: 142109.

Yoon, S. D., Chen, Y., Yang, A. et al. 2006. Oxygen-defect-induced magnetism to 880 K in semiconducting anatase TiO_2-delta films. *J. Phys.: Condens. Matter* 18: L355–L361.

Young, D. P., Hall, D., Torelli, M. E. et al. 1999. High-temperature weak ferromagnetism in a low-density free-electron gas. *Nature* 397: 412–414.

Young, D. P., Fisk, Z., Thompson, J. D., Ott, H. R., Oseroff, S. B., and Goodrich, R. G. 2002. Magnetic properties—Parasitic ferromagnetism in a hexaboride? *Nature* 420: 144–144.

Zakrassov, A., Leitus, G., Cohen, S. R., and Naaman, R. 2008. Adsorption-induced magnetization of PbS self-assembled nanoparticles on GaAs. *Adv. Mater.* 20: 2552–2555.

Zhang, Y., Talapatra, S., Kar, S., Vajtai, R., Nayak, S. K., Ajayan, P. M. 2007. First-principles study of defect-induced magnetism in carbon. *Phys. Rev. Lett.* 99: 107201.

Zhao, Q., Wu, P., Li, B. L., Lu, Z. M., and Jiang, E. Y. 2008. Room-temperature ferromagnetism in semiconducting TiO_2-delta nanoparticles. *Chin. Phys. Lett.* 25: 1811–1814.

Zhitomirskyi, M. E., Rice, T. M., and Anisimov, V. I. 1999. Magnetic properties: Ferromagnetism in the hexaborides. *Nature* 402: 251–253.

Zhou, S. Q., Potzger, K., Talut, G. et al. 2008. Ferromagnetism and suppression of metallic clusters in Fe implanted ZnO-a phenomenon related to defects? *J. Phys. D.* 41: 105011.

Zorko, A., Makarova, T. L., Davydov, V. A. et al. 2005. Study of defects in polymerized C_{60}: A room-temperature ferromagnet. *AIP Conf. Proc.* 786: 21–24.

26

Laterally Confined Magnetic Nanometric Structures

Sergio Valeri
University of Modena and Reggio Emilia

and

CNR-Institute of Nanoscience-Center S3

Alessandro di Bona
CNR-Institute of Nanoscience-Center S3

Gian Carlo Gazzadi
CNR-Institute of Nanoscience-Center S3

26.1 Introduction

In modern society, applications of magnetic materials can be found almost everywhere and our daily life is intimately connected to magnetism and magnetic materials. The most noteworthy impact of magnetism occurs via information transport and data storage devices, which mostly consist of artificially nanostructured magnetic materials. Here, artificially structured materials refer to the materials which are made into either reduced dimensions such as two-dimensional (2D) ultrathin films, one-dimensional (1D) wires, and zero-dimensional (0D) dots or assemblies of these low-dimensional structures such as multilayers, wire arrays, and dot arrays. Interesting phenomena come about by the imposed spatial confinement, which is comparable in size to some internal length scale of the material used, as spin diffusion length, carrier mean free path, magnetic domain extension, and domain wall width (Osborn 1945; Daughton 1999; Shi et al. 1999; Kirk et al. 2001; Shen and Kirschner 2002). Magnetic nanostructures by virtue of their extremely small size possess very different properties from their parent bulk materials.

Nanostructured magnetic materials have been utilized in technologies with a large and growing economic impact. The magnetic recording industry is continuously pushing the technology as evidenced by the fact that the density of information storage has been steadily increasing at a compound annual rate exceeding 60% per year (McKendrick et al. 2000). This progress has been made possible by a series of scientific and technological advances, mostly marked by the synthesis of artificially structured magnetic materials. With the ever-improved knowledge of low-dimensional magnetism, particularly the correlation between the magnetic properties, electronic properties, and structural properties, we are achieving an ability to design and to fabricate low-dimensional materials with desired magnetic properties.

The ability to tailor magnetism is, therefore, strictly related to the ability to fabricate low-dimensional objects. It is highly desirable not only to fabricate ultrafine nanostructures but also to fabricate arrays of such nanostructures. Ordered arrays of magnetic nanostructures (Figure 26.1) are particularly interesting to study, as one can probe both the individual and collective behavior of the elements in a well-defined and reproducible fashion (Cowburn and Welland 2000; Valeri et al. 2006). Ordered magnetic patterns are also technologically very important in a number of applications (as magnetic memory, recording media, magnetic switches, etc.) (Chou 1997; Nordquist et al. 1997; White et al. 1997). These activities require a high degree of control on the quality of the magnetic material and on the geometry and morphology of the arrays. In particular, the control on the morphology of structures, like roughness and sharpness at the edges, is a fundamental issue when shape-related magnetic properties are investigated. A number of nanofabrication methods have been developed, including different lithographic, direct writing, and nanotemplating approaches (Martin et al. 2003). Methods based on either extended or focused ion beams (FIBs) have been proved to be effective for the preparation of extended arrays of magnetic, laterally confined nanometric structures as the physical-based approach is applicable to virtually any solid

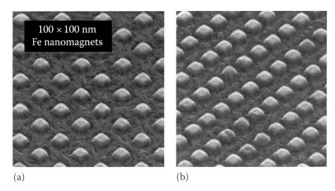

(a)   (b)

FIGURE 26.1 Nanomagnets produced by focused ion beam (FIB) milling. On the left (a) are shown the noninteracting particles, while on the right (b) the particles are interacting along one direction.

material. FIB is a versatile nanofabrication tool based on the interaction of nanosize beams of energetic Ga ions with solids (Orloff et al. 2003; Giannuzzi and Stevie 2005). Etching occurs by physical ion sputtering (Townsend et al. 1976), optionally gas-assisted to enhance material removal rates or species selectivity. With respect to state-of-the-art, competing lithographic technologies, FIB offers a comparable resolution (few tens of nanometers) with higher flexibility. On the other hand, FIB milling is a slow process compared to standard lithographic nanofabrication and its low throughput is the main limiting factor. Improved fabrication capabilities in turn call for measurement techniques that can probe and help to understand the magnetic properties at the relevant length scales of nanostructures.

26.2 Background

26.2.1 From Bulk to Surfaces and Thin Films: The Vertical Confinement

The deep past in the field of nanomagnetism is related to surface or thin-film magnetism, where the confinement was restricted just to the "vertical" dimension. This aspect started out and developed analogously as its "parent" field of surface science, both pushed by the emergence of new equipments and techniques (ultra-high-vacuum technology) that could ensure the cleanliness of surfaces and films.

Critical aspects in fabrication were the accurate control of film thickness, the stoichiometry and defectivity of the films, the sharpness of the interfaces between film and substrate or between different films in multilayers. From the magnetic point of view, phenomena controlled by the film thickness (i.e., by the vertical confinement) were mainly investigated, namely, the ferromagnetic–superparamagnetic transition in ultrathin ferromagnetic films, the thickness-dependent Néel temperature in antiferromagnetic layers, and the coupling of ferro-, antiferro-, and nonmagnetic films in multilayers (Baibich et al. 1988; Alders et al. 1998; Lang et al. 2006).

The breakthrough in the field of surface and thin-film magnetism came with the advent of giant magnetoresistance (GMR), first observed in 1988 in a metallic multilayer system consisting

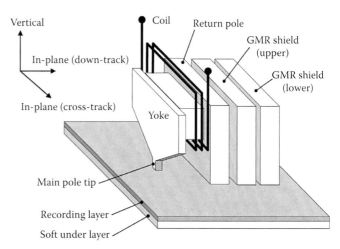

FIGURE 26.2 Schematic representation of a perpendicular head on a double layer perpendicular media and definition of applied field direction. (Reprinted from Hamaguchi, T. et al., *IEEE Trans. Magnetics*, 43, 704, 2007. With permission.)

of stacked thin films of magnetic and nonmagnetic materials (Baibich et al. 1988). GMR is a phenomenon in layered magnetic structures associated with enhanced sensitivity of the electrical resistivity to external magnetic fields: the electrical resistance reduces to minimum when the magnetization of the ferromagnetic layers are parallel, and increases to maximum when the magnetization of the two layers are antiparallel. This phenomenon was exploited for a new generation of high-performance read heads in commercial hard disk systems.

The main driving force to develop nanomagnetism was the need for magnetic storage media with increasing density. Magnetic storage media consist of homogeneous, polycrystalline magnetic films where, provided the film thickness is small enough, a uniform magnetization state may occur over an extended region (magnetic domains) separated by interfaces where the magnetization can undergo a rapid variation from one orientation to the other (domain walls).

The current method for writing/reading information is the control of magnetization in magnetic domains, each domain being an individual unit of information or *bit* (Figure 26.2) (Hamaguchi et al. 2007). Increasing the areal storage density calls not only for appropriate magnetic media that can accommodate a smaller bit size but also for the optimization of the complete disk-drive system. It includes smaller and more sensitive read/write heads, as well as progress in head/disk tribology (Brug et al. 1996; Fisher and Modlin 1996; Kryder et al. 1996).

26.2.2 From Thin Films to Wires and Dots: The Lateral Confinement

Lateral confinement represents the quest for lower dimensionality, from layers to stripes and dots, to approach 1D and "0D," respectively. Stripes and dots can be fabricated by using the controlled manipulation or the (self) assembling of elementary bricks (single atoms or clusters), in the so-called bottom-up approach, and can be organized in lateral periodic arrays. For technologically oriented applications, nanostructures are mostly fabricated by using top-down approaches, which usually involve lithographic methods in order to downsize an extended film.

Again, the driving force for this step is mainly the need to push further and further the density of magnetic storage media, moving from continuous, longitudinal media, where the bits are written onto magnetic films, to vertical and discrete media, where the bits are written onto patterned magnetic arrays, each element of the array representing a bit (Bader 2002). Longitudinal and vertical media refer to the orientation of the magnetization as either in the plane of the media or perpendicular to the plane of the media, respectively. In this field, there exist expectations for successful future of magnetic systems in lateral confinement. Especially, periodic arrays of single-domain magnetic dots attract substantial interest because they allow a much larger areal storage densities than it would be possible on the basis of the current technologies used in conventional hard disks. The dots must be a single magnetic domain and they should have a perpendicular magnetic anisotropy. Furthermore, all the dots should be spatially separated such that any existing dipolar coupling between adjacent dots is too weak in order to flip the magnetization (e.g., the stored information) of any of their neighbors (Carl and Wassermann 2002).

The quest for fabrication of ordered arrays of magnetic nanostructures was satisfied by the use of the technologies originally developed for semiconductor devices (Madou 1997). Most of the methods were directly adapted from microelectronic technologies. Different techniques currently used for the fabrication of magnetic patterns are discussed in recent review papers (Martin et al. 2003; Shen et al. 2003). Lithography technologies were, for instance, used for pattern transfer into magnetic films. In the most common optical or x-ray lithography (Thompson et al. 1994; Sheats and Smith 1998), where ultraviolet (UV) light or x-ray are used, respectively, large areas can be easily patterned. The resolution limit is ultimately determined by the radiation wavelength. The e-beam lithography technique uses an electron beam to expose an electron-sensitive resist (Fischer and Chou 1993; New et al. 1994). One of the main advantages of this technique is its versatility. Direct writing technique by dissociation of a resist containing the desired magnetic atom is also used to pattern nanostructures (Streblechenko and Scheinfein 1998). In laser interference lithography, the interference of two coherent laser beams is the mechanism used to produce periodic structures in the resist film (Farhoud et al. 2000).

Bottom-up methods appear to be potential successors/competitors to lithography. Tremendous effort has been made in the last decade toward the fabrication of nanoscale features using self-organized assembly of atoms and molecules into nanoscale dots and wires. The substrate can also be morphologically modified such that when the magnetic material is deposited it creates the desired nanostructures, e.g., by strain engineering in epitaxial films (Leroy et al. 2005). These techniques are limited by the advantages and/or disadvantages of the method utilized to modify the substrate. Moreover, the range of possible structures

is very limited and, although most of these processes are locally ordered, they usually do not have true long-range order. Deposition through nanomasks is another method recently exploited for fabrication of laterally confined magnetic nanostructures. Shadow masks with nanometric holes are placed very close to the substrate. Depositing magnetic materials through the holes creates the desired nanostructures on the substrate (Stamm et al. 1998; Marty et al. 1999). A sort of shadow mask can be also obtained by using a chemical solution containing nanometer-scale polymer spheres to coat the substrate (spheres lithography) (Li et al. 2000). This technique is capable of obtaining large patterned areas in a quick, simple, and cost-effective way.

New ways to create self-organized 0D and 1D nanostructures are suggested by the increasing knowledge of morphology evolution during bombardment of solid surfaces and films by extended ion beams (Valbusa et al. 2002; Moroni et al. 2003). During ion bombardment, atoms are removed from the surface (sputtering process): To this respect, sputtering is to a first approximation the inverse of growth. The significant amount of energy deposited by the incident ions also support surface atoms self-organization. The combined effect of these two main processes leads to relevant surface modifications that can be to some extent controlled to prepare nanometer-scale (ordered) structures. Surfaces eventually get rough during ion erosion. When ions are incident normal to a surface, the resulting pattern consists of mounds and pits. Under oblique incidence, on the contrary, the net result is the formation of ripples with the "wave vector" either parallel or perpendicular to the surface component of the beam direction. The particle size and separation can be adjusted by varying the temperature, the ion energy and flux, and the ion irradiation geometry.

Besides the use of extended ion beams, another means for nanostructuring interface regions is via FIBs (Orloff et al. 2003; Giannuzzi and Stevie 2005). The field of high-resolution FIB has advanced rapidly since pioneering applications in the early 1970s. Its evolution mainly followed the evolution in the ion source technology, from the first field emission sources (adapted from surface physics) to the liquid metal sources that definitely enabled a very high resolution to be attained. The technological "push" for improving FIB performances was the need for failure analysis and repair in integrated circuits. The present use of FIB covers a wide range of applications, including ion lithography, direct implantation, lithographic mask repair, TEM sample preparation, deposition of materials, imaging (scanning ion microscopy, SIM), and a number of nanomachining uses. In particular, FIB is an advanced technique for magnetic nanofabrication, due to its flexibility for the preparation of well-defined arbitrary element shapes and array configurations, the high spatial resolution and the possibility to join top-down (milling) and bottom-up (deposition) approaches.

Many specialized techniques were developed to probe magnetic properties. Magnetic force microscopy (MFM) is a well-established method to probe the micro-nanomagnetic properties with lateral resolution down to nanometers (Figure 26.3). The technique yields information on both the morphological as well

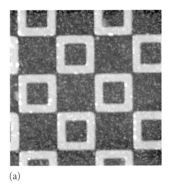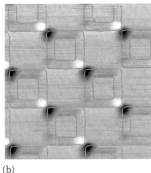

(a)   (b)

FIGURE 26.3 Topographic (a) and magnetic force (b) images of 1 μm side square rings made of permalloy. The magnetic force image has been measured after full magnetization of the structures along the top-left to bottom-right diagonal. The outline of the topographic image has been superimposed on the magnetic force image.

as the magnetic properties of surfaces (Grutter et al. 1992; Carl and Wassermann 2002). Therefore, the topology and the magnetic domain structure of a magnetic system may be correlated on the nanometer scale. MFM investigations are performed using a magnetically saturated tip attached at the end of an oscillating cantilever. The tip is then raster-scanned across the investigated surface. Each single scan line is scanned twice, but at two different distances between the tip and the surface. During the first scan at small distance, the topography (AFM image) is measured, by taking advantage of the short-range repulsive forces between the tip and the surface. During the second scan at larger distance, basically only the long-range magnetic interactions between the tip and the surface are detected, from which the MFM image is obtained.

The magneto-optic Kerr effect (MOKE) is currently used to study the magnetic properties (Bader 1991). When a beam of polarized light reflects off a magnetized surface, the plane of polarization of the light can slightly rotate. This change is caused by off-diagonal elements in the dielectric tensor of the material as a consequence of spin–orbit coupling (Bader 1991). The theoretical explanation of the Kerr effect is that any plane-polarized light can be decomposed into two circularly polarized light. The index of refraction of the right-handed circularly polarized light and the index of refraction of the left-handed circularly polarized become different when a material is magnetized. The rotation is directly related to the magnetization of the material within the probing region of the light. The technique is sometimes referred to as SMOKE, where the S stands for surface. The experimental setup normally involves a laser beam passing through a polarizer and then reflecting the light off the sample. The reflected light then passes through a polarization sensitive detector. Slight changes in the plane of polarization will thus cause variations in the detected light intensity after the second filter. MOKE is frequently used to measure the hysteresis loops in thin magnetic films, by studying the light intensity as a function of applied magnetic field, and to imaging magnetization domains by scanning the laser on the investigated surface.

26.3 State of the Art

26.3.1 Fundamental Properties of Magnetic Bodies

This section is focused on the analysis of the relevant magnetic processes that play a role in determining the magnetic state of the nanomagnet. Each physical process is characterized by an energy term whose magnitude depends on the magnetic configuration of the magnet. Some of the relevant magnetic processes (e.g., exchange and magnetocrystalline anisotropy) have a quantum mechanical origin. Unfortunately, the equations of the full quantum mechanical system have not been solved, yet, unless rough approximations or simplifying assumptions are introduced. The common approach is to use quantum mechanics to understand the physical phenomena and to find an expression for the energy of the individual magnetic process. Based on that, simplified classical expressions for significant energies are determined. In these expressions, a number of phenomenological constants are used to specify the macroscopic properties of the magnetic material.

The resulting picture is a model where the magnetic system is represented by an assembly of magnetic moments whose configuration is fully specified by a spatial varying magnetization **M**, intended as a local parameter averaged over a so-called *physically small* volume, i.e., a volume that is large with respect to the lattice constant of the material, but small with respect to the length over which **M** changes. Maxwell equations and classical physics are used for building an energy functional of the magnetic configuration. The actual magnetic state of the system is eventually determined by variational calculus, imposing that the total free energy is a minimum. This approach has been referred to as *micromagnetics* because the magnetic configuration of the system comes out as the *result* of the variational calculation as opposed to the *domain theory*, where the class of magnetic configurations of the system (i.e., magnetic domains separated by negligibly thin domain walls) is *assumed, a priori*. However, micromagnetics is not a microscopic theory, in the sense that the details of the atomic structure of matter are ignored and the material is assumed to be continuous by using a suitable averaging procedure similar to that used for discussing the macroscopic media in classical electrodynamics textbooks.

26.3.1.1 Magnetic Materials

All known materials, under the action of a magnetic field **H**, acquire a magnetic moment. The magnetization is a vector quantity **M** defined as the magnetic moment per unit volume. In most cases, **M** can be considered as a continuous vector field without considering the atomic structure of the matter. For most materials, **M** is proportional to **H**:

$$\mathbf{M} = \chi \mathbf{H} \qquad (26.1)$$

where χ is called the *magnetic susceptibility* of the material. If $\chi > 0$, the material is said *paramagnetic*, otherwise *diamagnetic*.

It is possible to define a vector $\mathbf{B} = \mu_0(\mathbf{H} + \mathbf{M})$, called the *magnetic flux density*, that satisfies Maxwell equations for the material. $\mu_0 = 1.256 \cdot 10^{-6}$ N/A² is called the *permeability of a vacuum*. If (26.1) is satisfied, then $\mathbf{B} = \mu_0(1 + \chi)\mathbf{H}$. The quantity $\mu = \mu_0(1 + \chi)$ is also referred to as the *magnetic permeability*. However, there are materials for which the relation between $\mathbf{M}$ and $\mathbf{H}$ is not linear, rather it is a multi-valued function of $\mathbf{H}$ that depends on the history of the applied field. These are called *ferromagnetic* materials. The typical magnetization curve (i.e., $\mathbf{M}$ vs. $\mathbf{H}$) for this class of materials is shown in Figure 26.4a, where the component of $\mathbf{M}$ in the direction of the applied field, M_H, is plotted as a function of the magnitude of the field.

The bold curve is the so-called *limiting hysteresis curve* that is obtained by applying a sufficiently large field for saturating the material in one direction, decreasing it to zero and then increasing it to the saturation value in the opposite direction. As the sample is saturated in both directions, the curve is precisely retraced in consecutive cycles of the applied field. The dashed curve is called the *virgin magnetization curve* and can be traced only after the demagnetization of the sample that can be obtained by heating the material at high temperatures and cooling it in zero applied field or by repeated cycles of applied field with steadily decreasing amplitudes. If the field is applied and decreased before the limiting hysteresis curve is reached, a minor loop is obtained, as shown in Figure 26.4b as a thin continuous line. There are infinite minor loops and with an appropriate sequence of applied fields, any point inside the limiting hysteresis curve can be reached. In particular, one can end a minor loop at $\mathbf{H} = 0$ with any value of M_H between $+M_r$ and $-M_r$, where M_r is called the *remanence magnetization* or *remanence*, defined as the value of M_H at zero applied field in the limiting hysteresis curve. The *coercive field* or *force* or *coercivity* is defined as the field for which $M_H = 0$ in the limiting curve. The *saturation magnetization* or *spontaneous magnetization* is an intrinsic property of the ferromagnetic material as it is defined as the value of M_H at large applied fields. It depends on the temperature as shown in Figure 26.4b. The temperature T_C at which the spontaneous magnetization becomes zero in zero applied field is called the Curie temperature.

26.3.1.2 Exchange Interaction

Exchange interaction is a quantum mechanical effect that tends to align the magnetic moment of neighboring atoms. It arises from the Coulomb coupling between the electron orbitals and from the Pauli exclusion principle. It is a strong but short-range interaction. The exchange energy operator for two localized spins is $E_{ex} = -J\mathbf{S}_1 \cdot \mathbf{S}_2$, where $\mathbf{S}_1$ and $\mathbf{S}_2$ are the spin operators and J is the so-called exchange parameter or integral. It can be calculated from first principles once the electrons wavefunctions are known, but in most practical cases, approximations have to be introduced and the resulting accuracy is inadequate. Exchange integral is positive for ferromagnetic materials. Assuming that the angle between adjacent spins is small, a semiclassical approximation for the exchange energy can be worked out, in which the classical magnetization vector is used instead of the spin operators and only nearest-neighbor interactions are considered:

$$E_{ex} = \int w_e dV$$
$$w_e = A\left[(\nabla m_x)^2 + (\nabla m_y)^2 + (\nabla m_z)^2\right]$$
(26.2)

where
w_e is the exchange energy density, and the volume integral is extended to all the body
A is a constant specific for the material, called exchange stiffness constant
$\mathbf{m} = \mathbf{M}/M_s$ is a unit vector pointing along the local magnetization $\mathbf{M}$ which is assumed to have a constant size M_s
m_x, m_y, and m_z are the Cartesian components of $\mathbf{m}$

The assumption of small angle between neighbors is justified by the fact that exchange interaction is strong at short range; therefore, it will not allow any large angle to develop. The exchange energy is positive defined, it is minimum (actually zero) in the case of aligned magnetization and it is large for large spatial variations, consistent with the fact that the exchange interaction tends to smear out such configurations. The semiclassical exchange functional is based on the assumption of a continuous

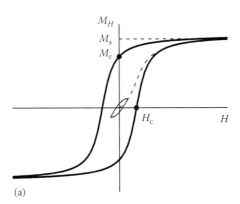

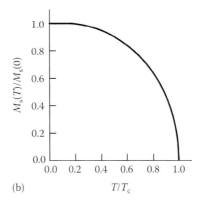

(a)   (b)

FIGURE 26.4 (a) Magnetization curves for a ferromagnetic material. The bold line is the limiting hysteresis loop, the dashed line is the virgin magnetization curve, and the thin line is one of the possible minor loops. Saturation magnetization M_s, remanence magnetization M_r, and coercive field H_c are also indicated. (b) Temperature dependence of the saturation magnetization M_s for a ferromagnetic material.

TABLE 26.1 Exchange Stiffness and Saturation Magnetization for Common Ferromagnetic Materials

Materials	A [J/m]	M_s (A/m)
Fe	$2.1 \cdot 10^{-11}$	$1.7 \cdot 10^6$
Ni	$9.0 \cdot 10^{-12}$	$4.9 \cdot 10^5$
Co	$3.0 \cdot 10^{-11}$	$1.4 \cdot 10^6$
Permalloy ($Ni_{80}Fe_{20}$)	$1.3 \cdot 10^{-11}$	$8.6 \cdot 10^5$

material. No meaningful information can be drawn beyond the validity of this approximation, i.e., at length scales below several atomic distances. Typical values for the exchange stiffness and saturation magnetization for common magnetic materials are reported in Table 26.1.

26.3.1.3 Magnetic Anisotropy

The exchange interaction is isotropic in space meaning that the energy of a given magnetic state does not depend on the direction in which the body is magnetized. It can be demonstrated that a magnetic system subject only to thermal fluctuations and to exchange interaction behaves like a paramagnet, i.e., it asymptotically magnetizes up to M_s in increasing applied field, but shows no magnetization at zero magnetic field. This contradicts the experiments where nonzero magnetization is often observed as the magnetic field is switched off. The reason for the failure is that real magnetic systems are never fully isotropic and not all the directions in space are equally probable in thermodynamic terms; therefore, the value of the magnetization does not always average to zero under the action of the thermal fluctuations.

The work needed to bring a body from the demagnetized to the fully magnetized state is called the magnetic free energy. It includes the isotropic contribution coming from the exchange interaction. But other energy terms contribute to the magnetic free energy, some of them are anisotropic, i.e., show directional dependence. An example is the magnetocrystalline anisotropy whose origin is related to the spin–orbit and orbit–lattice interaction. The shape of the body can also introduce anisotropy, but this effect is customarily included in the dipolar interaction effect (see. Section 26.3.1.6).

Magnetocrystalline anisotropy is usually small with respect to exchange, the latter being the main process that forces the spins to align, making a net magnetization to appear at macroscopic level, but the direction of the magnetization is determined only by the anisotropy, since the exchange is isotropic. The result is that the total free energy depends on the direction in which the magnetic field is applied. The directions along which the magnetization energy is minimal are called the easy magnetization axes. For a given crystal structure and atom species, a quantomechanical calculation of the magnetocrystalline anisotropy is possible, but the accuracy is generally poor. Further, whatever the microscopic origin of the anisotropy is, it is always possible to write phenomenological expressions for the anisotropy energy as a series expansion that takes into account the crystal symmetry, with coefficients taken from the experiments. The

anisotropy energy is written as a volume integral of an anisotropy energy density:

$$E_a = \int w_a dV \tag{26.3}$$

where the integral is extended over the volume of the magnetic body. The anisotropy energy density w_a has specific expressions for the different symmetries. In hexagonal crystals, uniaxial symmetry is observed:

$$\begin{aligned} w_a &= K_1(1 - m_z^2) + K_2(1 - m_z^2)^2 + \cdots \\ &= K_1 \sin^2 \theta + K_2 \sin^4 \theta + \cdots \end{aligned} \tag{26.4}$$

where the z axis is parallel to the crystalline c-axis; $K_1 > 0$ and K_2, called anisotropy constants, are the expansion coefficients specific for the material and m_x, m_y, and m_z are the Cartesian components of the reduced magnetization unit vector $\mathbf{m} = \mathbf{M}/M_s$. Although the power series expansion can be carried to higher order, in all practical cases it is not required and even K_2 is often negligible. For cubic symmetry, the anisotropy energy is

$$\begin{aligned} w_a &= K_1(m_x^2 m_y^2 + m_y^2 m_z^2 + m_z^2 m_x^2) + K_2 m_x^2 m_y^2 m_z^2 + \cdots \\ &= (K_1 + K_2 \sin^2 \theta) \cos^4 \theta \sin^2 \varphi \cos^2 \varphi + K_1 \sin^2 \theta \cos^2 \theta + \cdots \end{aligned} \tag{26.5}$$

The main anisotropy constant K_1 is positive for some cubic materials (like Fe) and negative for others (like Ni). For the former, the easy magnetization axes are along (100) and equivalent directions, i.e., the sides of the cubic cell, while for the latter the easy magnetization axes lie along (111), i.e., the long diagonals of the cubic cell. Anisotropy constants for common magnetic materials are reported in Table 26.2.

26.3.1.4 External Applied Field

As a consequence of the Lorentz force, a magnetic moment $\boldsymbol{\mu}$ in an applied field $\mathbf{H}$ is subjected to a torque $\mathbf{G} = \mu_0 \boldsymbol{\mu} \times \mathbf{H}$. This allows to define a potential energy for a magnetic moment in an applied field $U = -\mu_0 \boldsymbol{\mu} \cdot \mathbf{H}$. In the case of a magnetized material, the magnetization vector $\mathbf{M}$ represents the local magnetic moment density, i.e., the magnetic moment per unit volume at a given position. It is, therefore, possible to define an energy functional that describes the interaction of the magnetized body with the external field, also called the Zeeman energy term:

TABLE 26.2 Anisotropy Constants for Different Ferromagnetic Materials

Materials	K_1 (J/m³)	Symmetry
Fe	$4.8 \cdot 10^4$	Cubic
Ni	$-5.7 \cdot 10^3$	Cubic
Co	$5.2 \cdot 10^5$	Uniaxial
Permalloy ($Ni_{80}Fe_{20}$)	$\approx 2 \cdot 10^2$	Cubic

$$E_Z = \int w_Z dV$$

$$w_Z = -\mu_0 \mathbf{M} \cdot \mathbf{H}_{ext} = -\mu_0 M_s \, \mathbf{m} \cdot \mathbf{H}_{ext}$$

(26.6)

where the volume integral is extended to the volume of the magnetized body and $\mathbf{H}_{ext}$ is the external applied field.

26.3.1.5 Magnetic Hysteresis and Superparamagnetism

It is instructive to analyze the behavior of a magnetic system in an applied field subjected only to exchange interaction and magnetic anisotropy. In absence of any other energy terms, the magnetic free energy is minimized by the uniformly magnetized state. The magnetic state of the system is then fully specified by the angle θ between the magnetization $\mathbf{M}$ and the z-axis. Let us assume that the system has a uniaxial anisotropy with easy axis directed along the z-axis and that a magnetic field H (positive or negative) is applied along the z-axis. Since for uniformly magnetized states the exchange energy (26.2) is zero, the magnetic free energy of this system is the sum of the magnetic anisotropy energy and the Zeeman term:

$$E = V(K_1 \sin^2 \theta - \mu_0 M_s H \cos \theta)$$

(26.7)

where

V is the volume of the body
K_1 is the main uniaxial anisotropy constant (for simplicity K_2 is assumed to be zero)

This function is shown in Figure 26.5 for different values of the applied field. For $H = 0$, the curve has two equivalent minima at $\theta = 0$ and $\theta = \pi$, corresponding to the magnetic easy axes for the uniaxial anisotropy, separated by an energy barrier whose height is $K_1 V$. As the magnetic field is increased and positive, the

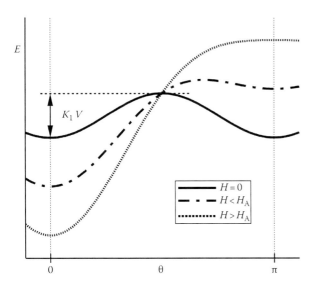

FIGURE 26.5 The magnetic free energy for a magnetic system subjected to exchange interaction and magnetic anisotropy, plotted for different values of the applied field.

minima at $\theta = 0$ become deeper than the other, and the energy barrier progressively lowers until it disappears at a critical field value, called the anisotropy field $H_A = 2K_1/M_s$. The role of the two minima are exchanged if the field is increased in the negative direction; i.e., the minima at $\theta = \pi$ becomes deeper than the other and the energy barrier will disappear at $H = -H_A$.

The energy scale is determined by the volume of the magnetic system. If the volume is large enough, the thermal energy $k_B T$ at a given temperature will be negligible with respect to the energy range of (26.7), and the magnetic system will spend most of the time in the states corresponding to the local minima of the free energy. As the magnetic field is varied, the free energy curve changes, as shown in Figure 26.5, and the system follows the local minima of the free energy. Let us suppose the system is magnetized along $-z$ direction (i.e., the system is initially in the minima at $\theta = \pi$) and a magnetic field is applied in the $+z$ direction. As long as an energy barrier exists between the two minima (i.e., for $H < H_A$), the transition toward the absolute minimum is hindered and the magnetization will remain in the initial state. For $H > H_A$, there is no barrier and the magnetization will switch abruptly to $\theta = 0$, i.e., parallel to the applied field. Once the magnetization has switched, it cannot switched back by lowering the field below H_A because a new energy barrier exists for the reverse transition and it needs a field larger than $-H_A$ for the reverse transition to occur. This phenomenon is called the *magnetic hysteresis* and explains why a ferromagnetic material, once it is immersed in a strong magnetic field, maintains its magnetization even if the magnetic field is removed. Magnetic hysteresis is a desired property in data storage applications, where the magnetic transitions between energetically equivalent states, should not occur spontaneously. Magnetic hysteresis occurs because the local minima of the magnetic free energy are separated from the absolute minimum by energy barriers.

When the system is in a local minimum, it is said that it is in a *metastable* state because, sooner or later, the thermal fluctuations will make the system to transition toward the absolute energy minimum. But how long the transition will take to occur? The relaxation time τ for a barrier-limited, thermally activated process, based on the Maxwell–Boltzmann statistics is given by the Arrhenius equation

$$\tau = \tau_0 e^{\Delta E / k_B T}$$

(26.8)

where

ΔE is the height of the energy barrier
k_B is the Boltzmann's constant
T is the absolute temperature
τ_0, also called the *pre-exponential factor*, is the characteristic attempt time of the system and depends on the details of the microscopic processes involved in the transition

For magnetization switching, typical values for τ_0 are in the 10^{-8}–10^{-12} s range, slightly dependent on the temperature and other quantities, while $\Delta E = K_1 V$. Even if the actual value of the pre-exponential term is *a priori* not precisely known, it is the

TABLE 26.3 Relaxation Time for Spherical Particles with Different Diameters

Material	Diameter (nm)	τ (s)
Co	6.5	$7.0 \cdot 10^{-2}$
	8.0	$4.2 \cdot 10^{5}$
Fe	14	$1.7 \cdot 10^{-2}$
	18	$2.4 \cdot 10^{6}$
Ni	29	$4.3 \cdot 10^{-2}$
	36	$4.0 \cdot 10^{5}$

strong dependence of the exponential factor on its argument $K_1 V / k_B T$ that determines the order of magnitude of the relaxation time. In order to demonstrate how strong this dependence is, Table 26.3 gives the relaxation time for Co and Ni spherical particles, as a function of their diameter, at room temperature ($k_B T = 4.14 \cdot 10^{-12}$ J) and assuming $\tau_0 = 10^{-9}$ s.

Within a small change in the particle diameter, the relaxation time becomes orders of magnitude larger or smaller of an arbitrarily chosen observation time τ_{exp} of 100 s. A completely different value of τ_0 will shift slightly the diameters corresponding to the same relaxation times, but it is always possible to identify a sharp transition between short and long relaxation times. Different magnetic materials, corresponding to different values of K_1, have different transition diameters, but the general behavior is that ferromagnetic materials show two distinct regimes: $\tau \ll \tau_{exp}$ or $\tau \gg \tau_{exp}$. Given a characteristic observation time τ_{exp}, the transition between the two regimes is quite sharp and depends on the diameter of the particle and on its magnetic anisotropy. If $\tau \gg \tau_{exp}$, the magnetization does not change during the observation time. This is the region of *stable ferromagnetism* where a magnetic system, once magnetized, held its magnetization for times as long as τ.

On the other hand, if $\tau \ll \tau_{exp}$, the magnetization of the system fluctuates and it must be calculated as an ensemble average using the classical Maxwell–Boltzmann distribution function:

$$M_z = M_s \langle \cos\theta \rangle = \frac{\int_0^{2\pi} \int_0^{\pi} e^{-E/k_B T} \cos\theta \sin\theta \, d\theta \, d\varphi}{\int_0^{2\pi} \int_0^{\pi} e^{-E/k_B T} \sin\theta \, d\theta \, d\varphi} \quad (26.9)$$

where E is the magnetic free energy. It can be easily verified that if the magnetic particle has no anisotropy ($K_1 = 0$), the magnetization is

$$\frac{M_z}{M_s} = L\left(\frac{\mu_0 M_s V}{k_B T} H \right) \quad (26.10)$$

where $L(x) = \coth x - 1/x$ is the so-called Langevin function, an s-shaped odd function that saturates to ± 1 for $|x| \gg 1$. This result is similar to that obtained for a system of independent magnetic atoms, also called a *paramagnetic gas*. For that, $M_s V$ must be substituted by the magnetic moment of the single atom, μ. The main difference between a paramagnetic gas and an exchange-coupled magnetic system is that for the latter saturation is obtained at

magnetic fields that are few orders of magnitudes lower, because normally $M_s V \gg \mu$ and the exchange-coupled system behaves like a huge atom with spin number in the 10^3–10^4 range, instead of 10^0 like in conventional paramagnets. As far as it concerns, the general case with $K_1 \neq 0$, the shape of the magnetization curve *vs.* H is no more a Langevin function, but the general behavior is conserved: We have zero magnetization in zero applied field (i.e., no magnetic hysteresis) and magnetic saturation for $|H| \gg k_B T / \mu_0 M_s V$.

Reducing the size of a ferromagnetic system makes the transition from $\tau \gg \tau_{exp}$ to $\tau \ll \tau_{exp}$ to occur. The correspondent loss of stable ferromagnetism is called *superparamagnetism*. For a given particle size, the transition between the two regimes occurs at the so-called blocking temperature T_B, corresponding to the condition $\tau = \tau_{exp}$. The blocking temperature is related to the measuring time, and may be very different if evaluated by different experimental techniques.

26.3.1.6 Dipolar Interaction and Shape Anisotropy

A magnetized body generates a magnetic field that extends all over the space, called the *magnetostatic* field. It is customary to call it *demagnetizing field* inside the body and *stray field* outside the body. The sources of the magnetostatic field are the magnetic dipole moments of the atoms that constitute the magnetic material. Each atom produces a dipolar field $\mathbf{h}_i^d$ centered at the atomic position r_i that extends to long range. The superposition of all these sources builds up the magnetostatic field $\mathbf{H}_d$:

$$\mathbf{H}_d(r) = \sum_i \mathbf{h}_i^d(r) = \sum_i \left[-\frac{\mu_i}{|\mathbf{r} - \mathbf{r}_i|^3} + \frac{3[\mu_i \cdot (\mathbf{r} - \mathbf{r}_i)](\mathbf{r} - \mathbf{r}_i)}{|\mathbf{r} - \mathbf{r}_i|^5} \right] \quad (26.11)$$

where μ_i is the magnetic moment of the atom i and the sum extends to all the atoms in the body. In micromagnetics, the atomic structure is ignored. The local magnetic moment is represented by the magnetization $\mathbf{M}$, i.e., the average magnetic moment per unit volume, and the average is taken over a volume that contains a large number of atoms. It can be demonstrated that, in the continuum approximation, the magnetostatic field can be calculated as the field generated by a distribution of volume and surface density of "magnetic charges":

$$\mathbf{H}_d(r) = \int_V \frac{\rho_V(\mathbf{r}')(\mathbf{r} - \mathbf{r}')}{|\mathbf{r} - \mathbf{r}'|^3} dV + \int_S \frac{\rho_S(\mathbf{r}')(\mathbf{r} - \mathbf{r}')}{|\mathbf{r} - \mathbf{r}'|^3} dS \quad (26.12)$$

where the first and the second integrals are over the volume and the surface of the body, respectively, and $\mathbf{r}'$ is the integration variable. The volume and surface magnetic charge densities have the following expressions:

$$\rho_V = \nabla \cdot \mathbf{M}$$
$$\rho_S = \mathbf{n} \cdot \mathbf{M} \quad (26.13)$$

where $\mathbf{n}$ is a unit vector normal to the surface. Magnetic charges are, therefore, generated by the spatial variations of $\mathbf{M}$, whenever they are generated in the volume by the divergence of the magnetization vector, or at the discontinuity represented by the body surface, where the magnetization must sudden drop to zero. It must be pointed out that the magnetic charges associated to the above densities are useful mathematical abstractions, but they have no physical meanings. The magnetostatic field is also called *demagnetizing field* because, inside the body, it acts against the "magnetizing field," i.e., the external applied field that induces magnetization in the body. In general, the demagnetizing field is not uniform inside the body even if the magnetization is uniform. For particular geometries, like spheres, ellipsoids, infinitely long cylinders, and thin films, the resulting demagnetizing field is uniform and reads

$$\mathbf{H}_d = -N\,\mathbf{M} \tag{26.14}$$

where N is symmetric a tensor with unit trace called the *demagnetization tensor*. If the symmetry axes of the body are directed along the Cartesian axes, the tensor is diagonal. The demagnetization tensor for few geometries is reported in Table 26.4.

A magnetostatic energy is associated to the magnetostatic field. It can be considered as the energy stored in the magnetostatic field or as the sum of the works needed to orient the individual atomic magnetic moments in the field generated by the others. In this case, it is also referred to as *magnetostatic self-energy*. The magnetostatic energy is also called *demagnetization energy* or *dipolar energy*. The magnetostatic energy is a functional of the magnetization vector field:

$$E_d = \int w_d \mathrm{d}V$$
$$w_d = -\frac{1}{2}\mu_0 \mathbf{H}_d \cdot \mathbf{M} \tag{26.15}$$

where the integral extends over the volume of the body and w_d is the magnetostatic energy density.

Using Maxwell equations and the boundary conditions at infinity on the magnetostatic potential, it can be demonstrated that the following expression is equivalent to (26.15):

$$E_d = \frac{1}{2}\mu_0 \int H_d^2 \, \mathrm{d}V \tag{26.16}$$

In this case, the integral extends over all the space and shows that the magnetostatic energy is positive definite and that it is zero only if $\mathbf{H}_d = 0$ everywhere. This is the *pole avoidance principle* that states that a magnetized body tries to avoid as much as possible the formation of magnetic charges, the latter being the sources of the demagnetizing field.

In many applications it is possible to assume that the body is uniformly magnetized. In this case in (26.13) $\rho_V = \nabla \cdot \mathbf{M} = 0$ and only surface magnetic charges are formed. The resulting demagnetizing field (26.12) and magnetostatic energy (26.15) depend only on the shape of the body and on the direction of the magnetization vector. The angular dependence of the magnetostatic energy per unit volume is also referred to as the *shape anisotropy*. An illustrative example is the uniformly magnetized thin ferromagnetic film. Using (26.14) and the appropriate entry in Table 26.4, the demagnetizing field can be calculated as

$$\mathbf{H}_d = -N\,\mathbf{M} = -\hat{z}\,M_z \tag{26.17}$$

Thus the demagnetization field has only a component directed along the surface normal. In (26.15), the expression to be integrated over the volume of the film is $\mathbf{H}_d \cdot \mathbf{M} = -M_z^2$; therefore, the magnetostatic energy per unit volume is

$$w_d = \frac{1}{2}\mu_0 M_z^2 = \frac{1}{2}\mu_0 M_s^2 \cos^2\theta \tag{26.18}$$

where θ is the angle between the magnetization vector and the film normal. This expression has the same formal form of the anisotropy energy density described in (26.4) or (26.5). The constant $\mu_0 M_s^2 / 2 > 0$ plays the role of the anisotropy constant and usually its value overwhelms that of the magnetocrystalline anisotropy of the film. In this case, the shape-induced anisotropy tends to keep the magnetization within the film plane. Similar arguments holds for elongated bodies, where the shape anisotropy tends to align the magnetization along the major axis of the body.

26.3.1.7 Magnetic Domains

The existence of *magnetic domains* was first postulated by Weiss in 1907 (and experimentally demonstrated forty years later) for explaining why a small magnetic field can induce a huge change in the magnetization state of a ferromagnetic body and why the magnetization can assume any value between $-M_s$ and $+M_s$, depending on the history of the applied field. Magnetic domains are regions within the ferromagnetic material, which are uniformly magnetized at the saturation value M_s, but the direction of the magnetization vector varies from domain to domain. Magnetic domains are separated from each other by thin regions, called *domain walls*, where the magnetizations vary. The fraction of the volume occupied by the domain walls is usually

TABLE 26.4 Components of the Demagnetization Tensor for Different Geometries of a Body

Geometry	$N_x = N_y$	N_z
Sphere	1/3	1/3
Infinite cylinder with axis along z	1/2	0
Thin film with normal along z	0	1
Prolate spheroid or egg-shaped revolution ellipsoid with long axis along z	$\dfrac{1-N_z}{2}$	$\dfrac{1}{m^2-1}\left[\dfrac{1}{2\xi}\ln\left(\dfrac{1+\xi}{1-\xi}\right)-1\right]$
Oblate spheroid or disk-shaped revolution ellipsoid with short axis along z	$\dfrac{1-N_z}{2}$	$\dfrac{1}{\xi^2}\left[1-\dfrac{1}{m\xi}\sin^{-1}\xi\right]$

Note: $m > 1$ is the ratio between the major and the minor axis of the revolution ellipsoid and $\xi = \sqrt{m^2 - 1}/m$.

negligible with respect to the volume of the body. Within each magnetic domain, the magnetization is aligned to some easy magnetization direction. The measured magnetization value is the average over the domains structure and it can assume any value between zero and M_s and any direction, depending on the number and orientation of the magnetic domains involved in the measurement. A small field applied to a ferromagnetic body causes a large effect on the domain structure because it does not have to reorder the atomic magnetic moments, which are already ordered, but only to make the domains to align to the applied field. This can be achieved by growing the size of the domains that are already aligned at the expenses of those that are not. This process is called *domain walls motion*, and energetically costs a fraction of the energy needed for the rotation of the whole domain.

Domain walls are usually classified by the angle between the magnetization directions in the adjacent domains; thus we have 90° domain walls when they separate perpendicularly magnetized domains and 180° domain walls when they separate domains of opposite magnetization. Two common types of 180° domain walls are shown in Figure 26.6. In a *Bloch domain wall* (Figure 26.6a), the magnetization rotates in a plane parallel to the domain wall while in a Néel domain wall (Figure 26.6b), the magnetization rotates in a plane perpendicular to the domain wall. Domain walls in which the magnetization is a function of only one parameter (in these cases the distance from the wall surface) are called 1D walls.

The formation of a domain wall costs energy because (a) adjacent spins are not parallel and this costs exchange energy and (b) spins are not pointing along the easy magnetization axes and this costs anisotropy energy. The formation energy for a Bloch domain wall can be readily calculated making the assumption that, in the transition region, the rotation angle of the magnetization is a linear function of the distance. Let us assume that the thickness of the transition region is L and that the magnetization is pointing along $+z$ in the $y > +L/2$ region and along $-z$ in the $y < -L/2$ region. The domain wall is centered around the $y = 0$ plane, and let us assume that the magnetization rotates towards the positive x axis. In the transition region, the magnetization unit vector is thus

$$\mathbf{m} = \frac{\mathbf{M}}{M_s} = \left(\cos\frac{\pi y}{L}, 0, \sin\frac{\pi y}{L} \right) \qquad (26.19)$$

The exchange energy density is given by (26.2) and it is constant within such a domain wall:

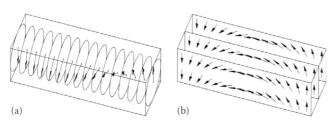

(a)   (b)

FIGURE 26.6   (a) A Bloch domain wall. (b) A Néel domain wall.

$$w_e = A\left[(\nabla m_x)^2 + (\nabla m_y)^2 + (\nabla m_z)^2 \right] = \frac{A\pi^2}{L^2} \qquad (26.20)$$

The exchange energy per unit domain wall area is thus

$$\frac{E_e}{\sigma} = \frac{A\pi^2}{L} \qquad (26.21)$$

where σ is the area of the wall. Let us assume that the ferromagnetic material has a uniaxial magnetocrystalline anisotropy with easy magnetization axis along the z axis and that $K_2 = 0$. The anisotropy energy density is thus given by (26.4):

$$w_a = K_1(1 - m_z^2) = K_1 \cos^2\frac{\pi y}{L} \qquad (26.22)$$

The anisotropy energy per unit area is therefore

$$\frac{E_a}{\sigma} = \int_{-L/2}^{+L/2} w_a \, dy = \frac{K_1 L}{2} \qquad (26.23)$$

As far as the dipolar interaction is concerned, we observe that, for this domain wall, $\nabla \cdot \mathbf{M} = 0$; therefore, we have no volume charges. We neglect, for the moment, the surface charges because we assume they are at infinity; then there are no magnetic charges (i.e., sources of demagnetizing field) and therefore magnetostatic energy is zero. The total energy per unit area of the domain wall is thus the sum of two terms:

$$\frac{E}{\sigma} = \frac{A\pi^2}{L} + \frac{K_1 L}{2} \qquad (26.24)$$

The first term is proportional to $1/L$ and tends to make the wall as large as possible, while the second term, proportional to N, tries to shrink the wall and make it as small as possible. The lowest domain wall energy corresponds to the condition $dE/dL = 0$, which leads to the following domain wall width:

$$L = \pi\sqrt{\frac{2A}{K_1}} \qquad (26.25)$$

that represents the result of the optimal balance between exchange and anisotropy energy costs. The corresponding energy cost per unit area is

$$\frac{E}{\sigma} = \pi\sqrt{2AK_1} \qquad (26.26)$$

The assumption that the rotation angle of the magnetization is a linear function is useful for the calculations, but it does not have physical basis. However, the more general hypothesis that the rotation angle is an arbitrary function of y (called the *Landau–Lifschitz's one-dimensional wall*) leads to the same conclusions: There is a finite domain wall width proportional to $\sqrt{A/K_1}$ and an energy per unit area proportional to $\sqrt{AK_1}$. The arbitrary function that results from the energy minimization leads to the following domain wall:

$$\mathbf{m} = \frac{\mathbf{M}}{M_s} = (\sin\theta, 0, \cos\theta) \qquad (26.27)$$

where $\cos\theta = \tanh y/L$.

In most cases, the formation of magnetic domains saves magnetostatic energy associated to the dipolar field. If the energy cost for the formation of the domain wall is lower than the gain in magnetostatic energy, then the magnetic state of the body *with* the domain wall has a magnetic free energy lower than the state *without* the domain wall. Magnetostatic energy is saved because surface magnetic charges are always generated at the edges of the body, where **M** starts and stops according to (26.13), second equation. Magnetic charges are sources of magnetostatic field which fills space, but this costs $\mu_0 H_d^2/2$ J/m³, as shown in (26.15). Magnetostatic energy can be saved if domain walls are formed as shown in Figure 26.7a and b.

The single-domain structure in Figure 26.7a has no domain walls and magnetic charges form at the top and bottom edges, causing a large dipolar energy. Breaking the body in two domains reduces the dipolar energy because part of the positive charge is moved from the top to the bottom surface. Charges of opposite sign attract each other, reducing the energy of configuration (b) at the energy cost of a domain wall. Configuration (c), also called the *closure domain* structure, has no magnetic charge because **M** is parallel to the edges, but introduces a number of domain walls. The actual magnetic configuration will be the result of a balance between the energy cost of the domain walls and the energy cost of the magnetostatic field.

On reducing the size of the body, the relative contributions of the magnetostatic and domain wall energy to the total magnetic free energy are changed. The former is proportional to the volume of the body, the latter to the surface of the domain wall. At small

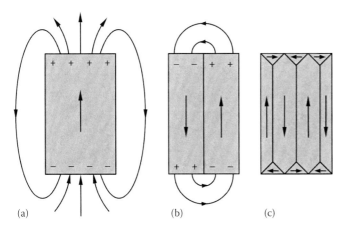

(a)   (b)   (c)

FIGURE 26.7 A ferromagnetic body which is (a) uniformly magnetized, (b) divided into two magnetic domains, and (c) an example of a closure domain structure.

sizes, there will be a point at which the surface-dependent term will prevail over the volume term and the single-domain state will be favorable energetically. In this state, the ferromagnetic body behaves like a permanent magnet, if the thermal fluctuations are low, or superparamagnetic (cf. Section 26.3.1.5), if the thermal fluctuations are larger than the energy barriers induced by the anisotropy or shape.

26.3.1.8 Micromagnetic Equations

The total magnetic free energy of a ferromagnetic body immersed in a magnetic field can be expressed as an integral over the volume of the body:

$$G(\mathbf{m}, \mathbf{H}_{ext}) = \int_{body} (w_e + w_a + w_Z + w_d)\,dV \qquad (26.28)$$

where w_e, w_a, w_Z, and w_d are defined in (26.2), (26.3), (26.6), and (26.15), respectively

$$\mathbf{m}(\mathbf{r}) = \mathbf{M}(\mathbf{r})/M_s$$

By solving a so-called *variational problem* it is possible to find all the possible configurations $\mathbf{m}(\mathbf{r})$ that are local minima of (26.28), under the constraint that $|\mathbf{m}(\mathbf{r})|=1$ everywhere in the body. They correspond to the equilibrium states of the system. It can be demonstrated that the solutions of the variational problem satisfy the condition

$$\mathbf{m} \times \mathbf{H}_{eff} = 0 \qquad (26.29)$$

where $\mathbf{H}_{eff}$ is called the *effective field*, defined as

$$\mathbf{H}_{eff} = \mathbf{H}_{ext} + \mathbf{H}_d + \frac{2A}{\mu_0 M_s}\nabla^2 \mathbf{m} - \frac{1}{\mu_0 M_s}\frac{\partial w_a}{\partial \mathbf{m}} \qquad (26.30)$$

Here $\nabla^2\mathbf{m}$ and $\partial w_a/\partial \mathbf{m}$ are compact notations for vectors whose Cartesian components are $\nabla^2 m_x$, $\nabla^2 m_y$, and $\nabla^2 m_z$ and $\partial w_a/\partial m_x$, $\partial w_a/\partial m_y$, and $\partial w_a/\partial m_z$, respectively. The condition (26.29) states that the torque exerted on the magnetization must vanish at each point $\mathbf{r}$ of the ferromagnetic body, i.e., that the magnetization is parallel to the effective field. The condition (26.29) must be solved together with the demagnetizin g field equations (26.12) and (26.13), which establish a set of nonlocal, nonlinear, integro-differential equations called *micromagnetic* or *Brown's equations* that give complete information on the stable and metastable equilibrium states for the magnetic body, corresponding to the specified external field $\mathbf{H}_{ext}$. However, the solution of such equations is definitely not straightforward and requires complex numerical calculations.

An alternative approach to Brown's equations is the dynamical micromagnetic method also known as Landau–Lifschitz–Gilbert (LLG) equation. Starting from a given magnetic configuration, the system is allowed to relax toward the equilibrium solution with a speed that depends on a phenomenological damping factor α. The LLG equation is

$$\frac{d\mathbf{m}}{dt} = -\gamma_0\,\mathbf{m} \times \mathbf{H}_{\text{eff}} - \alpha\,\mathbf{m} \times \frac{d\mathbf{m}}{dt} \qquad (26.31)$$

where

  t is the time
  $\gamma_0 = 1.76086 \cdot 10^{-11}$ rad/s/T is the electron gyromagnetic ratio
  $\mathbf{H}_{\text{eff}}$ is given by (26.30)
  α is a dimensionless in the range of 0.004 to 0.15 for most materials

In some sense, Brown static equations (26.29) and (26.30) can be considered as a particular case of LLG equation (26.31), but the latter is much easier to solve numerically.

26.3.2 Nanofabrication by Focused Ion Beam (FIB)

The use of FIB as a nanostructuring tool has much increased in the last decade as new instruments, often combining an ion beam column with a SEM in a so-called double-beam apparatus, have been developed specifically for research and academic purposes.

The versatility offered by the direct approach to material nanomachining, performed through ion sputtering at the nanoscale, and the easiness of FIB operation, which is very similar to the SEM one, have made FIB an ideal tool for a variety of applications: nanoprototyping and device modification, cross-section analysis and TEM sample preparation, and surface and thin films patterning (instead of using dedicated lithography systems). In the following, we will describe the main components and the working principle of an FIB instrument and review the most common nanofabrication types.

26.3.2.1 FIB Apparatus

A FIB apparatus (Orloff et al. 2003; Giannuzzi and Stevie 2005) shares many similarities with the well-known SEM: in essence it can be considered as a scanning ion microscope. The ion-optical

column, schematized in Figure 26.8, consists of an ion source, electrostatic lenses, and a set of mechanical apertures, devoted to the formation of a finely focused beam, and a scanning system to move the beam on the sample over a desired pattern. The column is mounted on a vacuum chamber where the sample is hosted on a multi-axis manipulator.

Much of the FIB nanomachining capability is due to the development of liquid metal ion sources (LMIS), cold field-emission sources featuring high brightness (10^7 A/cm^2 sr), i.e., a high emission current from a nanosize area with low angular divergence of the ion trajectories. The source is a metallic tip wetted by the liquid metal (usually gallium), mounted in front of an extractor electrode held at a high negative potential (10–12 keV) with respect to the tip, which is positively biased at the beam acceleration potential (5–50 keV). The liquid metal, immersed in the extractor electrical field, is deformed and takes the shape of a cone (the "Taylor cone") protruding out of the tip as a result of the balance between the attracting field force and the opposing surface-tension force. The high electrical field (10^8 V/cm) concentrated at the apex of the cone, having a radius of few nanometers, is responsible for the ion evaporation of the liquid. The source size as seen by the ion-optical system (virtual source) is around 30–50 nm and it will be demagnified to the beam spot size by the ion-optical system. The choice of Ga as a source material is motivated by the combination of a low melting point (29.8°C) with a high surface tension, which results in small energy spread and angular divergence of the beam, key parameters to minimize the contribution of lens aberrations to beam spot size.

The extracted ions are shaped into a beam and focused on the sample by the condenser and the objective lenses, respectively. These are electrostatic lenses with cylindrical symmetry and are preferred over electromagnetic ones (those employed to focus electrons in SEMs) because they are more compact and easy to realize, and their focus length is independent of the charge/mass ratio of the ion. A typical electrostatic lens consists of three

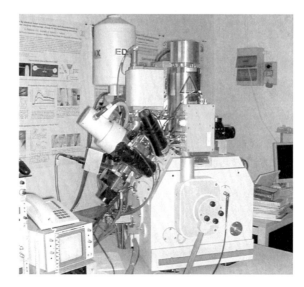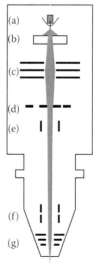

FIGURE 26.8 Double-beam (FIB-SEM) instrument and a scheme of the FIB column: (a) LMIS source, (b) extractor, (c) condenser lens, (d) apertures, (e) blanking plates, (f) stigmation and scanning octupole, and (g) objective lens.

cylinders or annular disks closely packed along the optical axis, where the outer electrodes are held at ground potential and the inner one is at the focus potential (either positive or negative). The focusing action occurs at the gap between the electrodes by the Coulomb force. This kind of lens, known as symmetric Einzel lens, is always focusing, irrespective of the sign of the inner potential, and it does not change the beam energy.

Below the condenser lens, there is a set of mechanical apertures crossing the beam path with variable diameters to select the beam portion entering the final objective lens. The beam current and probe diameter on the sample are proportional to the aperture size. Typical beam currents range between few picoamperes to tens of nanoamperes and the corresponding spot size from few to hundreds of nanometers. At this stage of the column there is also the beam blanker, a couple of plates switching off the beam rapidly (100 ns) by deviating it from the axis. The beam scanning and the adjustment of astigmatism is performed by an octupolar lens system, eight electrode plates arranged as the sides of an octagon and switched on in the proper sequence.

The vacuum chamber is equipped with detectors collecting the secondary particles (electrons or ions) generated by the ion beam interaction within the sample. Such signals can be employed to return an image of the scanned area, in strict analogy with the imaging performed with SEM.

The focused beam is controlled by a pattern generator system, which selects the pattern shape and the scanning path to cover the pattern, the dwell time spent on each pixel, and the overlap between adjacent spots. These parameters, together with beam spot size, are those relevant to determine the final quality of the nanofabrication.

26.3.2.2 Ion–Solid Interactions

The dominant energy-loss mechanism for ions in the tens of kiloelectron volts range is the nuclear energy loss, the energy and momentum transferred in binary elastic collisions from the moving ion to the atomic nucleus at rest inside the solid (Townsend et al. 1976). In a pictorial way, we could think of a billiard balls' collision, but more precisely, the ion is scattered by the repulsive potential of the nucleus screened by the electron shells (screened nuclear potential) since ion energy is not high enough to penetrate the inner electron shells and hit the nucleus (Rutherford scattering). The nuclear stopping power $S_n(E)$ [eV · cm^2] is the quantity defining the efficiency of this interaction mechanism, and the nuclear energy loss per unit length traveled by the ion inside the solid is given by $dE/dx = \rho_N S_n(E)$ [eV/cm], where ρ_N is the target atomic density. $S_n(E)$ depends on the primary ion mass (m_1), atomic number (Z_1), energy (E_1), target atomic mass (m_2), and atomic number (Z_2): it increases with m_1 and also with m_2/m_1 until the ratio is approximately unity, then it saturates; in the energy range typical of FIB (up to 50 keV), it rapidly increases with E, except for very low m_1, where it has a maximum at few kiloelectron volts and then decreases. The nuclear energy loss is strongly modulated by ρ_N, and for Ga at 30 keV, it may vary between 300 and 3000 eV/nm average values. The momentum transfer changes the particles' trajectory, and we are interested in the maximum scattering angle of the primary ion (ϕ_M): ϕ_M gradually increases to 180°, the backscattering angle, as m_2 approaches m_1. This means that ions moving inside a matrix of lighter atoms are weakly deflected, and those processes, described in the following, depending on the energy transfer in the backward direction, like sputtering, or on the lateral spread of ion trajectory, like ion damage, have a reduced efficiency.

As depicted in Figure 26.9, the energy transferred to recoil atoms by the primary ion is high enough to trigger a sequence of scattering events between each recoil and its nearest neighbors, and generate a so-called collisional cascade process. Atoms involved in the cascade are displaced from their equilibrium position, giving rise to a local disorder of the sample (structural and compositional for compound targets) referred to as ion damage. Those atoms in the surface region receiving sufficient energy to overcome the surface binding energy and a momentum transfer in the outward direction are ejected out of the sample. The surface erosion produced by atom ejection is known as ion sputtering or milling.

The efficiency of the sputtering process is quantified by the sputtering yield (Y_S), the number of atoms ejected per incident ion, which depends on the same parameters introduced for the nuclear stopping power, energy, masses and atomic numbers, and on additional ones like the surface binding energy (W) of the target and the beam incidence angle with respect to the surface (θ).

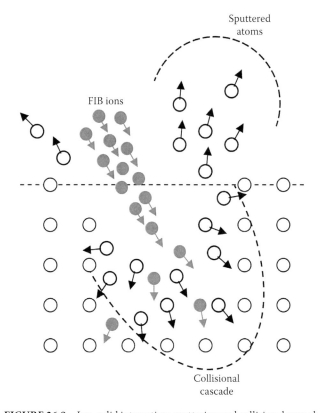

FIGURE 26.9 Ion–solid interaction: sputtering and collisional cascade effects generated by the incoming FIB ions.

Y_S increases with m_1 and with the m_2/m_1 ratio, and decreases with W. It is slowly increasing with E_1 following the $S_n(E)$ behavior. Dependence on θ reflects the efficiency of energy transfer in the surface region. Y_S increases as $1/\cos\theta$ from $0°$ to $70°-80°$, because as θ grows, more ion track is spent close to the surface; for values above $80°$ ion reflection at the surface takes place and Y_S drops. Typical Y_S values for Ga at 30 keV, impinging at $\theta = 0°$, are in the range of 1–10 atoms/ion, depending on the W value.

The collisional sequence of the primary ion ends when its kinetic energy drops below a critical value and the ion stops inside the solid becoming an implanted ion. The distance between the ion position and the surface, projected on the incidence direction, is defined as the ion projected range (R_p). Ion distribution inside the solid, plotted as a function of depth, follows a Gaussian-like profile peaked at R_p, the average ion projected range. R_p increases with E_1 and decreases with the ratio m_2/m_1; typical values for Ga at 30 keV are between 10 and 80 nm, depending on the target density.

Besides the nuclear component, the other mechanism contributing to ion energy loss is the electronic energy loss generated by the inelastic interaction with electrons of the solid. Valence or conduction band electrons are excited above the Fermi level and may escape the sample as secondary electrons, provided they are generated few nanometers below the surface. The electronic stopping power is proportional to ion velocity ($S_e(E) \propto E_1^{1/2}$) and, in FIB conditions, it is roughly a 10% of the nuclear stopping value. The ion-induced secondary electrons yield (γ_e) can be greater than unity, and it is generally higher than the one generated by primary electrons because range and energy transfer of ions are much shallower than for electrons. The secondary electrons can be collected to obtain a FIB microscopy of the scanned area, and they also play an important role in the ion beam–induced deposition (IBID) process. Here, precursor gas molecules are injected close to the surface, adsorbed on it, and decomposed by FIB irradiation. The gas molecules, e.g., metalorganic compounds with a metal atom surrounded by a cage of organic species, are dissociated by the primary beam and by the outgoing secondary particles (ions and electrons): the heavy metal component sticks on the surface, forming a deposit, and volatile fragments are evacuated. The decomposition cross section of these molecules is typically high at few electron volts, closely matching the low energies of secondary electrons; thus, these behave as very efficient bond breakers.

26.3.2.3 Nanofabrication by FIB

FIB nanofabrication can be performed using three different approaches, as sketched in Figure 26.10: material erosion (ion milling, Figure 26.10a), material structure/composition modification (ion-induced damage, mixing, and ion implantation, Figure 26.10b), or material addition (IBID, Figure 26.10c). Each one of these processes is strictly related to a particular aspect of the ion–solid interaction between the incoming ion and the atoms of the sample.

26.3.2.3.1 Nanofabrication by Ion Milling

Ion milling can produce nanostructures either with a "positive" process, where the structure is laterally defined by removing the material around it, or with a "negative" process, where the structure is the empty volume removed from the material. Due to the typical sputtering rates (tenths of $\mu m^3/nC$) and the sequential nature of the scanning process, FIB milling is suited to surface structuring, with depths limited to a few microns and areas in the range of few hundreds × hundreds μm^2.

Besides ion–solid interaction parameters, nanofabrication by ion milling is strongly influenced by ion beam parameters like the beam spot size and profile, the dwell time, and the overlap.

Though minimum spot size of 5–10 nm can be achieved with the lowest beam currents, lateral resolution of FIB milling is always larger than FIB spot size and it is limited by two effects: lateral range of ion–solid interaction and beam profile. The first effect can be quantified through the lateral ion range, whose values increases from 10% to 50% of R_p as m_2/m_1 increases, and represents the area from which atoms can be sputtered off the sample. Beam profile is the spatial distribution of ions around the beam axis. It follows a Gaussian profile in the central region, with a full width at half maximum (FWHM) corresponding to the beam spot size, but the external tails deviate from the Gaussian distribution, extending higher and longer. The problem of such a beam shape is the erosion contributed by these tails: the edges and sidewalls of the cuts become rounded and sloped, respectively, and not square-sharp as ideally desired. This worsens the lateral resolution as the milling depth increases. These effects can be minimized by selecting a small beam current spot size: Typical sidewall slopes range from $1°-2°$ to $7°-8°$ on going from few picoamperes to nanoampere beams.

Milling a pattern implies that material has to be removed down to a certain depth, and one can realize this either by milling a

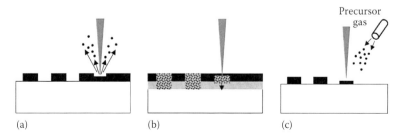

(a)   (b)   (c)

FIGURE 26.10 Types of FIB nanofabrication: (a) material erosion by ion milling, (b) structural damage and atomic mixing by ion irradiation, and (c) material addition by ion beam–induced deposition.

single frame, where each pixel has the full dwell time necessary to reach the target depth, or by repeating many frames at a fraction of the full dwell time. The first kind of approach results in a "dirty" pattern, with regions behind the beam scanning direction coated by redeposited sputtered material; conversely, repeating the frame many times helps in keeping the structures clean of redeposition and well defined. Dwell time also plays a role in the sharpness of the patterned nanostructures. It has been shown (Gazzadi et al. 2005) that higher dwell times (10, 100 μs) give sharper structures while values close to the beam blanking speed (0.1 μs) produce fuzziness and drag effects. The possible reason is that longer dwell times allow to mill well-defined grooves in the first frames, and these induce a channeling effect on the beam, helping the overall cut sharpness.

Beam overlap is defined as the crossing fraction between two adjacent spots: the higher it is, the more continuous is the beam scanning, and according to the relative magnitude between spot size and pixel separation, which depends on the magnification, this may have an influence on the structure resolution.

The aspect ratio (height/width) of structures fabricated by FIB milling is not arbitrary. In the case of "negative" processing, like milling a hole, the ratio is limited to 3–5:1 by the ejection of the sputtered material, which cannot escape from the hole beyond a certain depth and is redeposited at the sidewalls. For "positive" processing, as the milling depth increases, there is a progressive rounding of the top edges of the structure, due to beam tails erosion, and this eventually turns into a lowering of the structure.

Ion milling can be gas-assisted to enhance the removal rates of a specific element selectively. The gas is chosen so that it forms a volatile compound with the sputtered atoms: halogens (I_2, Cl_2) for removing metals, silicon, and GaAs, XeF_2 for silicon and silicon dioxide, and H_2O for carbon. Gas processing eliminates redeposition and gives cleaner structures, but it also enhances unwanted effects like the beam tail erosion.

26.3.2.3.2 Nanofabrication by Ion Irradiation

Nanofabrication by ion irradiation is performed at much lower ion doses (10^{13}–10^{15} ions/cm²) than those employed for milling and it is based either on the structural disorder produced by the collisional cascade or on the change in material composition. Since the aim is to locally alter the material properties, minimizing the sputtering effect, light primary ions are preferable for this process.

In crystalline semiconductor materials, the lattice disorder induced by ion irradiation (amorphization) lowers the material density, and amorphous regions, laterally confined by the higher density crystalline material, are pushed up producing a surface swelling effect (Gazzadi et al. 2005). In this way, a topographical nanostructuring can be achieved. Structural disorder generally worsens the electrical transport properties and localized ion irradiation has often been employed to draw blocking patterns to electrical conduction.

Compositional change has been exploited in the patterning of magnetic thin films. A sequence of ultrathin magnetic layers, crossed by a light ion beam, are subject to local atomic mixing at the films' interface, and this may generate a new phase with different magnetic properties (e.g., different magnitude or orientation of the magnetization) (Chappert et al. 1998).

Other examples of patterning can be achieved by local Ga ion implantation. Ga-implanted silicon is resistant to KOH anisotropic etching; thus, FIB irradiation can be exploited to perform nanostructuring through selective etching (Xu and Steckl 1994). Another method is to employ Ga-implanted nanoareas as templates for the localized growth of Ga-based nanodots (InGaN (Lachab et al. 2000), GaN (Gierak et al. 2004)).

26.3.2.3.3 Nanofabrication by Ion-Induced Deposition

Ion beam–induced deposition (Utke et al. 2008) is a high-resolution direct additive lithography commonly employed in the microelectronics industry for device failure analysis and photomask defects repair. In recent years, this technique and its electron-beam analogue, the electron beam–induced deposition (EBID), have been extended to nanoscience and nanotechnology applications, including electrical and mechanical connections at the nanoscale (e.g., on nanotubes), the fabrication of nanotips for scanning probe microscopes, and the fabrication of electronic and photonic nanodevices. The deposition of laterally confined nanostructures has been explored especially with EBID, as focused electron beams offer a higher lateral resolution than FIBs. A strong effort has been made to improve the minimum size (1–10 nm can be achieved) and the purity of metallic deposits, in particular, which are affected by the high content of organic species (C, O) present in the precursor molecule. The IBID approach is less resolved (50–100 nm minimum size) but gives higher metallic content as the shallower ion–solid interactions are more efficient in decomposing the molecules adsorbed on the substrate. On the other hand, the deposit contains the ion species implanted during the deposition process. The IBID process is in competition with sputtering erosion, which is always present whenever an ion hits; therefore, a delicate balance exists between the molecular and ion fluxes in order to maximize deposition over erosion. Typical values at room temperature for 30 keV Ga FIB are of 2–8 pA/μm² for 30 keV Ga ions and gas precursor pressure of 10^{-5}–10^{-6} torr.

26.4 Critical Discussion of Selected Applications

26.4.1 Magnetocrystalline and Configurational Anisotropies in Fe Nanostructures

A key issue in data storage technology is to control the magnetic switching of small magnetic elements. Many properties of such systems come about by imposing a geometric shape on the magnet. For instance, the magnetostatic energy associated to the lateral confinement is known to induce a dependence of the magnet free energy on the magnetization direction (shape magnetic anisotropy). In symmetric elements, as square magnets, the free energy of a uniformly magnetized square element do not depend on the magnetization direction due to the prefect balance of the surface magnetic charges that forms at the magnet boundaries. This means that the shape anisotropy (cf. Section 26.3.1.6) for square elements is zero. However, the uniform magnetization state cannot be set up in non-ellipsoidal magnets and at any finite applied field the actual magnetic configuration will deviate from the uniform magnetization state. As a consequence, the perfect balance of the surface magnetic charges does not take place and an anisotropic energy contribution, called configurational anisotropy, sets up (Cowburn et al. 1998; Cowburn and Welland 1998a,b), which may compete with other anisotropies such as the magnetocrystalline anisotropy. Since the magnetic properties of a magnet depend critically on anisotropy, an understanding of the overall anisotropy in nanomagnets is therefore essential, especially for technological applications.

These effects have been studied on a set of arrays of single-crystal Fe micron and submicron elements on MgO (Vavassori et al. 2005a,b). They have been fabricated by FIB and they have been magnetically characterized by means of magneto-optical Kerr effect magnetometry (MOKE) (Bader 1991; Vavassori 2000). The anisotropy field (cfr. Section 26.3.1.5) of the film has been measured by modulated field magneto-optical anisotropy (MFMA) (Cowburn et al. 1997). In detail, an epitaxial, 10 nm thick Fe film on MgO(00 1) single crystal has been grown, which has shown a good crystalline quality with its (100) axis parallel to the (110) direction of the substrate. To avoid oxidation, the film has been capped with a 10 nm MgO film. FIB has been subsequently used to remove portions of the bilayer to produce different arrays of nanostructures. The main advantage of the FIB technique is the ability to sculpt nanostructures of arbitrary shape starting from high-quality single-crystal films with elevated spatial resolution (down to 20 nm). Figure 26.11 shows scanning electron microscopy (SEM) images of three samples. The inter-element distance is large enough such that any magnetostatic interactions between nanomagnets is negligible if compared to any other energy contribution.

The MFMA characterization of the Fe continuous film shows that the film has a cubic magnetocrystalline anisotropy, with anisotropy field $H_a = 2K_1/M_s$ of about 560 Oe, where K_1 is the first-order

cubic anisotropy constant and M_s is the saturation magnetization. Assuming a value of $1.7 \cdot 10^6$ A/m for M_s, this corresponds to an anisotropy constant of $4.8 \cdot 10^4$ J/m^3, in good agreement with that of the Fe bulk. Out-of-plane component of the magnetization has not been found, as expected for thin films (cf. Section 26.3.1.6).

The easy and hard axis MOKE loops of the continuous Fe film are shown on the left-hand side of Figure 26.12. The small coercive field ($\approx$20 Oe to be compared to H_a) indicates that the magnetization reversal is determined by nucleation and expansion of reversed domains. The orientation of the film's easy axes with respect to the patterned structures is shown by the white arrows in the SEM image of Figure 26.11. The square elements have been oriented to have their diagonals parallel to the film easy axes. At first order, the configurational anisotropy in square nanomagnets was found to have an in-plane fourfold symmetry with easy directions along the square diagonals (Cowburn and Welland 1998b). The easy axes directions of configurational anisotropy being coincident with those of intrinsic magnetocrystalline anisotropy, the symmetry of the overall anisotropy of the square nanomagnets should be the same as in the continuous film. The same symmetry is expected for the circular nanomagnets of pattern 3 in this case because the magnetic in-plane configurations are energetically isotropic, as confirmed by the hysteresis loops shown on the right-hand side of Figure 26.12.

The main effect introduced by the FIB patterning is visible in the corresponding hysteresis loops. They are very different compared to the continuous film for what concerns shape and coercive field. These differences are due to the lateral confinement, which hinders domain formation during magnetization reversal. As a result, the nucleation of magnetization reversal is retarded, the coercive field increased, and the magnetization switching takes place more gradually.

26.4.2 Ion Irradiation

In addition to defining the geometry of the data bits, most of the fabrication techniques change the surface topography of the magnetic media. In order to preserve the surface flatness of the medium (a major prerequisite for high-density information storage

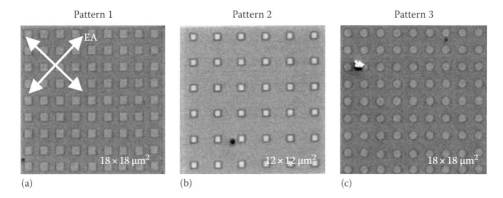

Pattern 1   Pattern 2   Pattern 3

(a)   (b)   (c)

FIGURE 26.11 Scanning electron microscope images of portions of arrays. The arrows indicate the direction of the film magnetocrystalline anisotropy easy axes. Pattern 1 is an array (pitch of 2 μm) of square elements of 1 μm side; in pattern 2, the lateral size of the square elements is 500 nm; pattern 3 is an array of circular elements of 1 μm diameter. (Reprinted from Vavassori, P. et al., *J. Magn. Magn. Mater.*, 290, 183, 2005b. With permission.)

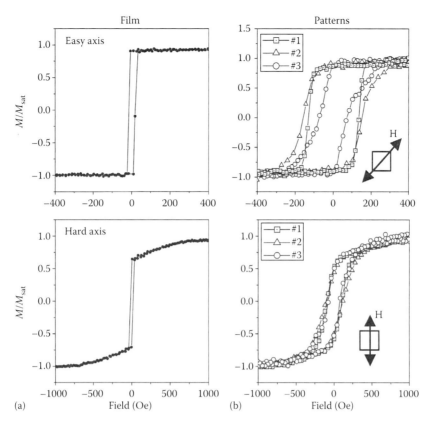

FIGURE 26.12 (a) MOKE hard and easy axis hysteresis loops of Fe film. (b) MOKE loops measured in the patterned areas, applying the field along the same direction as for the continuous film. (Reprinted from Vavassori, P. et al., *J. Magn. Magn. Mater.*, 290, 183, 2005b. With permission.)

nanotechnologies) and to avoid detrimental effects due to the worsening of surface quality, 2D magnetic patterns in continuous films were produced by exploiting ion irradiation to induce local modifications like intermixing, changes in crystallinity, chemical composition, and strain.

Several investigations have been performed on Co/Pt and Fe/Pt multilayers (Blon et al. 2004). Two main effects are produced by ion irradiation of these multilayers: alloying of the two elements resulting in the modification of the magnetic state (from ferromagnetic to paramagnetic) and intermixing at interfaces resulting in the modification of the magnetic anisotropy (from perpendicular to parallel).

Magnetic properties of Pt/Co/Pt sandwiches or $(Pt/Co)_n$/Pt multilayers (that are ferromagnetic at RT and have a perpendicular easy magnetization axis) were tailored without affecting their surface roughness by combining ion irradiation with standard electron-beam lithography (Chappert et al. 1998). Electron beam lithography was used to prepare a suitable mask on the surface of the sputter-deposited metal films. Irradiation was performed using a 30 keV He$^+$ ion beam with an increasing fluence up to 10^{16} ions/cm^2. The ion range into the sample is much higher than the metal layer thickness so that all ions are implanted in the substrate, but their energy is mainly deposited just in correspondence to the multilayers. The structural origin of the magnetic changes was investigated by diffraction methods and ascribed

to progressive modifications of the Co atoms environment near the Pt/Co interfaces due to the elastic collisions induced by ion irradiation. Under these specific irradiation conditions, the displacement of atoms at the Pt/Co interfaces is very low (typically a few interatomic distances), and they can relax to a stable position whose local surroundings differ from their initial one. The composition modulation in the growth direction is increasingly reduced (Figure 26.13). Interfacial Co atoms have a high

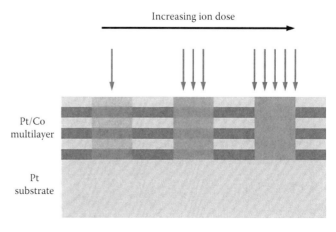

FIGURE 26.13 Progressive mixing of Pt/Co multilayer by increasing ion dose.

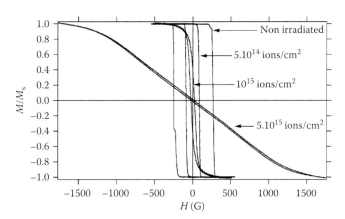

FIGURE 26.14 Room temperature polar MOKE hysteresis loops obtained on Pt–Co (1 nm) Pt after irradiation with different N^+ ion fluences (indicated in ions/cm²). (Reprinted from Blon, T. et al., *J. Magn. Magn. Mater.*, 272, E803, 2004. With permission.)

probability to experience a more Pt-rich environment, resulting in weaker magnetic character of the whole system in the irradiated region. In particular, very thin Co-layer samples (≤0.5 nm) essentially comprise only interfaces, so that almost all the Co atoms are involved. The overall result is a progressive lowering of the magnetic anisotropy energy, a reduction of the coercive force (Figure 26.14), and then a change of the magnetization axis from perpendicular to in-plane. The total magnetization is not affected, until for higher irradiation fluences the films become paramagnetic at RT. A crucial feature of the method is the low density of displaced atoms provided by light ion irradiation. No evidence of irradiation-induced surface roughness was found by atomic force microscopy (AFM) measurements.

On thicker films, attempts were made to chemically order alloys while not changing their microstructure (Devolder et al. 2001). Room temperature–prepared FePt alloy films show no chemical order, and post-annealing at 700°C is necessary to induce the appearance of an ordered phase, i.e., a stacking of Pt and Fe atomic planes. It is worth noticing that in FePt the direction and the intensity of the magnetocrystalline anisotropy are very sensitive to structural order and crystallographic orientation. Therefore, it appears to be the ideal candidate for the realization of smooth and continuous patterned perpendicular systems by locally modifying the degree of structural order by ion irradiation. Atomic displacement induced by irradiation with 130 keV He^+ ions was found to favor chemical order in the FePt disordered structure, triggering and controlling the ordering process at temperatures well below the standard ones.

Different conditions of ion irradiation were found on the contrary to destroy chemical order in FePt structures previously ordered by annealing at suitable temperatures (Albertini et al. 2008). To obtain 2D patterns of perpendicular magnetic structures based on FePt thin films, by controlling coercivity and direction of magnetization (i.e., out-of-plane versus in-plane), maskless Ga^+ irradiation with FIB has been used. Thin films of thickness 10 nm were grown on MgO. Morphological characterization was performed by atomic force microscopy (AFM). Magneto-optical Kerr effect (MOKE)

magnetometry was used to characterize the magnetic properties. Magnetic force microscopy was performed to characterize magnetic domain structure. It has been shown that continuous films with high anisotropy can be obtained. The continuous morphology also allows the presence of continuous domain patterns. Thin films were subsequently processed by 30 keV Ga^+ irradiation with doses up to $4 \cdot 10^{16}$ ions/cm², to study in detail the effects of different Ga^+ doses on structure, morphology, and magnetism. The lowest effective dose for which the complete disordering takes place was found to be $1 \cdot 10^{14}$ ions/cm². Disordering eliminates the perpendicular magnetocrystalline anisotropy that arises from the ordered structure. As a consequence, the perpendicular coercivity dramatically drops, leading to a change of easy magnetization direction from perpendicular to in-plane. At the lowest effective dose for a complete disordering of structure, the morphology was found not to be significantly affected by ion irradiation. Just a small enlargement of grains has been observed, accompanied by an increase of the surface roughness. The effects of surface erosion become pronounced after irradiation with $2 \cdot 10^{16}$ ions/cm².

By using the lowest effective dose of $1 \cdot 10^{14}$ ions/cm² 2D continuous patterns were fabricated, e.g., by alternating irradiated and nonirradiated stripes with lateral size of 1 μm. Other patterns consisted of nonirradiated dots (1 μm and 250 nm in diameter) arranged in square array with period twice their diameter, surrounded by an irradiated matrix. Due to the preservation of surface quality after irradiation, the AFM measurements are practically insensitive to ion irradiation. On the other hand, MFM, performed with a tip magnetized perpendicularly to the film is sensitive to the perpendicular stray field gradients emanating from the sample and consequently to the magnetic patterns (Porthun et al. 1998). MFM large-scale images show hard and soft zones corresponding respectively to perpendicular and parallel magnetization directions. At a lower scale, it is possible to analyze the domain structure of patterned samples (Figure 26.15a). The 1 μm-diameter nonirradiated dots in an irradiated matrix show a bi-domain structure, different from the continuous film, with concentric magnetic domains reflecting the shape

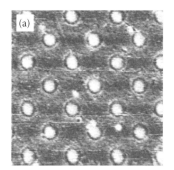
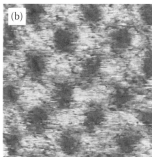

FIGURE 26.15 (a) MFM image of 2D patterned FePt films of 1 μm-diameter unirradiated dots in an irradiated matrix (dose = 1×10^{14} ions/cm²) showing concentric perpendicular magnetic domains. (b) Same as (a), but the diameter of the unirradiated areas is reduced to 250 nm. (Reprinted from Albertini, F. et al., *J. Appl. Phys.*, 104, 053907, 2008. With permission.)

of the dot (Ha et al. 2003; Komineas et al. 2005) while the 250 nm dots appear as single-domain structures (Figure 26.15b).

26.4.3 High-Density Magnetic Media Patterned by FIB

One approach to achieve magnetic storage densities of the order of 1 Tb/in.[2] is the use of patterned magnetic media (Ross 2001; Lodder 2004; Terris and Thomson 2005; Terris et al. 2007). As magnetic grain sizes in conventional media are reduced, the magnetic energy per grain becomes too small to prevent thermally activated reversals. This is the so-called superparamagnetic limit (cf. Section 26.3.1.5). A typical requirement for hard disk storage is a data retention interval of 10 years. In order to avoid thermally activated reversal during such an interval, the value of $K_1 V/k_B T$ (see Equation 26.8) must remain greater than ~35 for monodispersed grain size distribution, and shifts to about 60 if grain size dispersion is taken into account (Weller and Moser 1999). To maintain sufficient signal-to-noise ratio, it is desirable to maintain the number of grains per bit as the density is increased. Thus, the grain volume must be reduced. However, K_1 cannot be increased without bound in order to maintain $K_1 V$, as the required magnetic field to write a bit, called the anisotropy field (see Equation 26.7) increases with K_1. Nowadays, the write field is limited to ~10 kOe. This limits the diameter of thermally stable grains which can be written to around 8 nm (Moser and Weller 2001). For discrete bit media, the signal-to-noise ratio argument is different. In this case, the number of grains, or more correctly the number of magnetic switching volumes per bit is reduced to 1, and there is no statistical averaging over many grains to reduce noise. The switching volume is now defined by the disk patterning and not by the field generated by the write head and thus islands as small as 10 nm and below will be thermally stable and still writeable with current recording heads. The linear density of the bits will be limited by the media thickness in order to prevent the head from unintentionally writing neighboring islands with the tails of the head field gradient. The taller the pillar, the greater the head field a neighboring bit will experience, and hence the

more likely it is to be inadvertently written. Thus, media with a thickness on the order of the bit spacing will be required. The symmetry of hard disk recording favors media with perpendicular anisotropy, as higher head write fields can be realized. Thus, given the media thickness limitation, we conclude that it will be difficult to use tall pillars for high-density recording. This implies that the shape anisotropy cannot be used to achieve the necessary perpendicular anisotropy. The most likely perpendicular media are based on using interfacial anisotropy, such as Co/Pt or Co/Pd multilayers, or media similar to that proposed for perpendicular recording based on CoPtCr alloys grown with the c-axis normal to the substrate (Albrecht et al. 2002). To achieve single-domain islands, it will be desirable to have high exchange coupling within the individual islands, rather than the low exchange desired in conventional magnetic recording.

In order to assess the potential of patterned media, prototype nanometer-scale magnetic structures were made by patterning a layer of $Co_{70}Cr_{18}Pt_{12}$ perpendicular medium using a FIB of Ga^+ from an FEI 830XL dual-beam instrument (Albrecht et al. 2002). The magnetic film was protected by 5 nm of CN_x. Patterns were cut by scanning the beam using beam currents of ≈1 pA. While the cut depth of 6 ± 2 nm is only slightly deeper than the 5 nm overcoat, the 30 keV Ga^+ penetrates the full 20 nm depth of the media, even with an intact overcoat (Rettner et al. 2001). We conclude that the magnetic isolation is caused by a combination of media removal and disruption associated with collision cascades, with an additional contribution from the Ga and C implantation. Whatever the mechanism, the dose required for full isolation is about 0.03 nC/μm². An array of uniform square islands with a period of $p = 103$ nm corresponding to a lateral island size of about 80 nm was fabricated over a 2×4 μm area.

A qualitative comparison between the decay of a patterned structure and a continuous media of the same composition is shown in Figure 26.16 (Albrecht et al. 2002). Figure 26.16a shows an AFM image with Figure 26.16b the corresponding magnetic force microscopy (MFM) image for a section of a patterned structure (top), and area of continuous media (bottom) after applying a saturating field of 20 kOe and aging it for 5×10^5 s at room

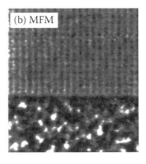
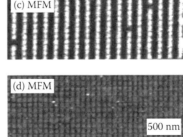

FIGURE 26.16 (a) AFM and (b) MFM images of patterned (top) and unpatterned (bottom) $Co_{70}Cr_{18}Pt_{12}$ perpendicular media after saturation and aging of the sample at room temperature for 5×10^5 s. All islands are in the original magnetized state (dark) while some areas of the unpatterned media have reversed the magnetization and appear bright. (c) MFM image of in-phase written island array with 10^3 nm pitch. A square wave pattern with a linear density of 9996 field changes per mm and a write current of 8 mA was applied. (d) AFM image of the topographic pattern. (Reprinted from Albrecht, M. et al., *J. Appl. Phys.*, 91, 6845, 2002. With permission.)

temperature in zero field. In the original state, the islands and media appear dark in the MFM image. It is clear that after aging, all islands have retained their original state of magnetization while the continuous film shows obvious decay with many local reversals of the magnetization, shown as white on the image. A total of 800 islands were monitored during the course of the experiment and out of this total not a single island reversal event was recorded. The demagnetizing field H_d plays an important role in the time decay of fully magnetized perpendicular media. For a fully magnetized film, H_d can be estimated from (26.14) and results 5.7 kOe for this medium. This substantial demagnetizing field contributes to a relatively fast decay of the continuous film. In the case of islands of $80 \times 80 \times 20$ nm in size, the demagnetizing field is reduced substantially to ≈ 0.70 (approximating the island to an oblate ellipsoid) of the value for the continuous film. This reduction leads to stability enhancement. In addition, the inability of islands below 130 nm to support domain walls (Lohau et al. 2001) may act as a further impediment to decay, since a reversal event must nucleate the switching of an entire island.

Writing and reading experiments on the patterned islands were performed using a read/write tester (Moser et al. 1999). Since the width of the write and read elements of the head (2 and 0.8 μm, respectively) were larger than the island size, the

patterned media were addressed in columns of 20 islands. Figure 26.16c through d show AFM and MFM images of a typical island array with a period of 103 nm. In order to obtain the magnetic pattern shown in the MFM image of Figure 26.16d, a square wave bit pattern was written at a linear density of 9666 field changes per mm to match the island periodicity, reversing the write field when the head is over a trench and so adjacent columns are magnetized in opposite directions. The written pattern corresponds to an areal density of about 60 Gbit/in.²

26.4.4 Sculpting by Broad Ion Beams

Nanostructuring of magnetic thin films is also performed by ion sculpting with low-energy and broad ion beams. On epitaxial films grown on crystalline surfaces, the interplay between the angular dependence of the sputtering yield and the energy barriers experienced by displaced atoms, diffusing at the step edges, builds up regular patterns of nano-ripples along specific directions. Height, lateral distance, and order of these structures can be tailored with ion irradiation parameters like ion energy, dose, incident angle, and substrate temperature. U. Valbusa and coworkers (Moroni et al. 2003) have investigated the magnetic properties of Co/Cu(100) films (Figure 26.17a) bombarded with

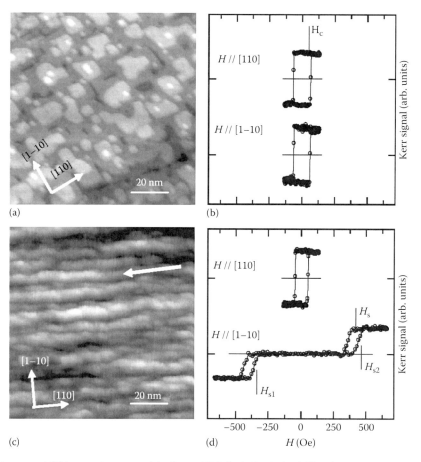

FIGURE 26.17 (a) STM image and (b) hysteresis curves of the flat 12 MLE thick Co/Cu(001) film. (c) STM image and (d) hysteresis curves of the Co film after nanostructuring with an ion dose of about 12 MLE. The arrow indicates the projection on the surface plane of the ion beam direction. (Reprinted from Moroni, R. et al., *Phys. Rev. Lett.*, 91, 167207, 2003. With permission.)

grazing-incidence (70°), low-energy (1 keV) Ar ions, imping-ing along the [110] surface direction (Figure 26.17c). The for-mation of ripples oriented along the beam incidence direction, with a periodicity in the 10 nm range, is evident. As the ion dose increases, the film is progressively eroded and Co remains only on the ripples' crest, producing parallel Co nanowires. The mag-netic properties, studied with MOKE, show a dramatic change in the symmetry of the in-plane magnetic signal. The as-depos-ited film has two equivalent easy axes (Figure 26.17b), deriving from the square-symmetry of the Co surface cell, while the rip-pled film displays a strong uniaxial anisotropy, dictated by the nanopattern symmetry (Figure 26.17d). The ripple axis behaves as an easy axis, similarly to the pristine film, while applying the field perpendicularly to the ripple axis gives a hysteresis curve split into two opposite loops. The loops are separated by a wide region around the origin where the magnetic signal is zero because in-plane magnetization is pinned along the ripple axis. Therefore, ion-patterned nanoripples introduce a strong uniax-ial anisotropy.

26.4.5 Beam-Induced Deposition

In this section, we cover examples of localized deposition of magnetic materials performed with both the ion and electron beam–induced depositions. IBID of laterally confined magnetic nanostructures stems from the quest for patterned magnetic media, where magnetic bits are stored on arrays of physically separated elements. A systematic approach has been undertaken by T. Suzuki and coworkers (Pogoryelov and Suzuki 2007). They first investigated the deposition of micron-size (2 μm in diame-ter) and nano-size (150 nm in diameter) Co dots, deposited from Co-carbonyl precursor with a Ga FIB. Magnetic characteriza-tion of the micron dots by alternated-gradient force microscopy (AGFM) showed weak ferromagnetic behavior with coercive fields of 100 Oe and saturation magnetization of 1000 emu/cm^3 in both the perpendicular and the in-plane magnetic field configurations. Nanosized dots were investigated only qualitatively with TEM Lorentz microscopy: image contrast suggested single-domain structures with in-plane magnetization. In a subsequent study (Xu et al. 2005), they performed a thorough characterization of CoPt and FePt micron-size dots, obtained from co-injection of Pt and Fe or Co precursors. Structure of the dots changed from amor-phous to crystalline fcc periodicity upon annealing at 600°C; the CoPt dots displayed higher order than the FePt dots. MFM mea-surements on the annealed samples, polarized with a perpendicu-lar magnetic field (20 kOe), are shown in Figure 26.18 (Xu et al. 2005). They reveal a concentric ring–domain pattern, particularly evident for the FePt dots, ascribed to shape anisotropy. Magnetic contrast for the CoPt dots is higher and more uniform. Contrast reversal upon reversing the magnetic field direction is observed in both cases. Continuing with the co-deposition approach, they studied micron and submicron FeCoPt dots, finding similar features as for the other deposits: weak ferromagnetism (110 Oe coercive field) and ring-domain patterns. These works indicate that IBID of magnetic materials is well feasible at the micron scale while going to the nanoscale requires some refinement.

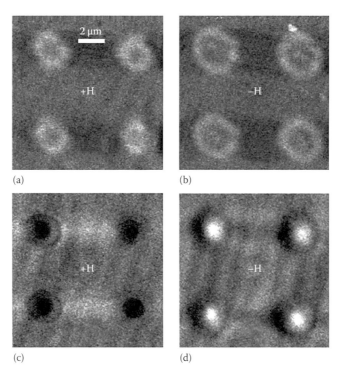

(a) (b)

(c) (d)

FIGURE 26.18 The remnant magnetic domain patterns taken by MFM after applying a 20 kOe magnetic field perpendicular to Si$_3$N$_4$ substrate with opposite directions. (a) and (b) are FePt particles annealed at 600°C for 25 h and (c) and (d) are CoPt particles annealed at 600°C for 1 h. (Reprinted from Xu, Q.Y. et al., *J. Appl. Phys.*, 97, 10K308, 2005. With permission.)

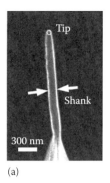

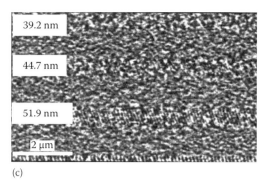

(a)   (b)   (c)

FIGURE 26.19   Magnetic tip grown on a Si AFM tip by Co EBID. (a) Tip diameter is 70 nm, shank diameter is 160 nm, and tip height is 1.7 μm. (b) The tip after magnetic cap coating by Co EBID; tip diameter improves to 50 nm. (c) MFM image of magnetic disk tracks taken with the tip in (b): track spacing down to 51.9 nm is resolved. (Reprinted from Utke, I. et al., *Appl. Phys. Lett.*, 80, 4792, 2002. With permission.)

Interesting cases of beam-induced deposition of magnetic materials can be found in the EBID literature. In one example, a magnetic "supertip" for improved MFM resolution was fabricated by depositing a Co pillar at the apex of an AFM tip (Utke et al. 2002). Deposition was obtained by focusing a 25 keV, 100 pA electron beam on the tip apex for 2 min, under Co-carbonyl precursor flux. The resulting pillar, shown in Figure 26.19a, has a high aspect ratio (~10) and a small curvature radius (35 nm). The tip was then overcoated with Co, with the same EBID procedure (Figure 26.19b), and finally tested on hard disk tracks with decreasing spacing. As shown in Figure 26.19c, the alternately magnetized tracks, appearing as black and white stripes, are resolvable down to a size of 51.9 nm. Always based on EBID of Co-carbonyl is the fabrication of a Hall effect magnetic sensor with an active area of 500 nm², realized by deposition of submicron crossing lines (Boero et al. 2005).

26.5 Outlook

Moving into the twenty-first century, emerging opportunities and exciting challenges can be envisaged in the field of laterally confined magnetic nanostructures. We foresee that the following directions will likely focus the interest for research and technology.

On the side of fabrication improvements, better definition of small structures using advanced photolithographies and electron/ion focused beam techniques is a real challenge. With this respect, the Achilles' heel of the FIB is that the gallium beam inevitably damages the surface of any milled or imaged sample. The damage consists of both gallium implantation and amorphization. This can be ameliorated by using low beam currents (at a cost of longer milling times) and/or low voltages (at a cost of worse imaging contrast). A much more powerful technique is under study to mill nanoscale features using the conventional FIB and then "polish" the sample using a broad-beam argon ion miller where the ion energy can be reduced to as low as 30 eV, so as to gently remove the gallium-implanted and amorphized layer. This would greatly reduce the depth of damage done to the sample. Problems also arise when positive ions are used for micromachining insulating materials or multilayers including

insulating layers. The target material is charged by the positive ions; as the positive charge builds up on the sample, it repels the ions and defocuses the beam. New systems are under study that uses, instead of the LMIS standard in many FIB devices, a plasma generated by radio-frequency electromagnetic fields, which separate gas molecules into their component electrons and positive ions. An ion beam and an electron beam are formed and accelerated by a suitable arrangement of electrodes. Both beams combine in a single, self-neutralizing mixed beam and are extracted by the accelerator column.

Parallel to FIB development is the continuous expansion of the family of self-assembling methods, e.g., from molecular precursor building blocks, that encompasses the domains of physics, chemistry, and biology. In this context, it is important to mention the increasing amount of fabrication techniques based on chemical synthesis methods that are very promising for production on a large and quick scale. Molecular magnets are a relevant example of this approach to magnetic nanofabrication. They consist of large-scale molecular structures where the magnetism is confined within discrete clusters within the unit cell. Most elegant in the regime of artificial structuring is the atom-by-atom assembly made possible by STM manipulation of individual atoms to "design" specific structures at the ultimate limits of miniaturization.

Since the dimensions of magnetic objects are being made smaller and smaller, magnetic imaging with atomic resolution is desirable. Scanning tunneling microscopy (STM) holds the most promise to solve this problem. In STM, a sharp tip scans across the surface and the tunneling current from the (into the) sample is recorded, enabling topographic images to be recorded with atomic spatial resolution. Spin-polarized scanning tunneling microscopy (SPSTM) is highly expected to show similar capability. The operating principle of SPSTM is close to that of the STM, but the spin-polarized component of the tunneling component is recorded in this case. The major challenge for developing the SPSTM lies with the very small signal, because the spin-polarized tunneling current is usually several percent or less of the total tunneling current.

Synchrotron techniques are increasingly emerging as powerful probes of surface magnetism. Photoelectron emission microscopy, where magnetic dichroism with electron yield detection is the method for generating the magnetic contrast at the surface, and spin-polarized photoemission, both benefit from the photon brightness available at third-generation synchrotron facilities.

Other areas of expanding interest refer to materials and phenomena. As the regime of lateral confinement becomes prime territory for the exploration of new properties, the steps and edges that surround the confined regions might be anticipated to play an important role. Atoms located in sites of reduced coordination (including surface atoms) become in fact a relevant fraction of the total number of atoms and their specific contribution, different from the contribution of "bulk," and fully coordinated atoms significantly influence the properties of nano-objects.

Recently, it has been demonstrated that a spin-current flowing directly through a nanomagnet can switch its magnetization direction by a mechanism called spin transfer. Spin transfer relies on a strong, short-range interaction between a spin current and the background magnetization of a nanomagnet. Spin transfer–induced switching, therefore, has important advantages over field induced switching and will likely form the basis for a new generation of magnetic devices called *spintronic* devices. The field of spintronics has been growing dramatically in recent years.

Besides the understanding of the intrinsic, individual properties of laterally confined magnetic nanostructures, it is also crucial to achieve a good understanding of their collective properties as they are assembled in ordered or disordered arrays. When the distance between the nano-elements becomes small enough, important interaction effects mainly due to dipolar fields are in fact observable, like changes in coercivity, presence of additional anisotropies, and modification of the magnetization dynamic.

Interaction effects between nanostructures of magnetic materials and other systems are also of emerging interest. This includes the creation of new pathways not only to create new materials and discover new properties and phenomena but also to finalize this "body of competences" to address society issues, enabling new approaches to energy production and storage, healthcare, homeland security, etc.

References

Albertini, F., L. Nasi, F. Casoli et al. 2008. Local modifications of magnetism and structure in FePt (001) epitaxial thin films by focused ion beam: Two-dimensional perpendicular patterns. *Journal of Applied Physics* 104: 053907.

Albrecht, M., S. Anders, T. Thomson et al. 2002. Thermal stability and recording properties of sub-100 nm patterned CoCrPt perpendicular media. *Journal of Applied Physics* 91: 6845–6847.

Alders, D., L. H. Tjeng, F. C. Voogt et al. 1998. Temperature and thickness dependence of magnetic moments in NiO epitaxial films. *Physical Review B* 57: 11623–11631.

Bader, S. D. 1991. SMOKE. *Journal of Magnetism and Magnetic Materials* 100: 440–454.

Bader, S. D. 2002. Magnetism in low dimensionality. *Surface Science* 500: 172–188.

Baibich, M. N., J. M. Broto, A. Fert et al. 1988. Giant magnetoresistance of (001)Fe/(001)Cr magnetic superlattices. *Physical Review Letters* 61: 2472–2475.

Blon, T., D. Chassaing, G. Ben Assayag et al. 2004. Effects of ion irradiation on cobalt thin films magnetic anisotropy. *Journal of Magnetism and Magnetic Materials* 272: E803–E805.

Boero, G., I. Utke, T. Bret et al. 2005. Submicrometer Hall devices fabricated by focused electron-beam-induced deposition. *Applied Physics Letters* 86: 042503.

Brug, J. A., L. Tran, M. Bhattacharyya, J. H. Nickel, T. C. Anthony, and A. Jander 1996. Impact of new magnetoresistive materials on magnetic recording heads. *Journal of Applied Physics* 79: 4491–4495.

Carl, A. and E. F. Wassermann 2002. Magnetic structures for future magnetic data storage: Fabrication and quantitative characterization by magnetic force Microscopy. In *Magnetic nanostructures*, Ed. H. S. Nalwa, Los Angeles, CA: American Scientific Publishers.

Chappert, C., H. Bernas, J. Ferre et al. 1998. Planar patterned magnetic media obtained by ion irradiation. *Science* 280: 1919–1922.

Chou, S. Y. 1997. Patterned magnetic nanostructures and quantized magnetic disks. *Proceedings of the IEEE* 85: 652–671.

Cowburn, R. P. and M. E. Welland 1998a. Micromagnetics of the single-domain state of square ferromagnetic nanostructures. *Physical Review B* 58: 9217–9226.

Cowburn, R. P. and M. E. Welland 1998b. Phase transitions in planar magnetic nanostructures. *Applied Physics Letters* 72: 2041–2043.

Cowburn, R. P. and M. E. Welland 2000. Room temperature magnetic quantum cellular automata. *Science* 287: 1466–1468.

Cowburn, R. P., A. Ercole, S. J. Gray, and J. A. C. Bland 1997. A new technique for measuring magnetic anisotropies in thin and ultrathin films by magneto-optics. *Journal of Applied Physics* 81: 6879–6883.

Cowburn, R. P., A. O. Adeyeye, and M. E. Welland 1998. Configurational anisotropy in nanomagnets. *Physical Review Letters* 81: 5414–5417.

Daughton, J. M. 1999. GMR applications. *Journal of Magnetism and Magnetic Materials* 192: 334–342.

Devolder, T., H. Bernas, D. Ravelosona et al. 2001. Beam-induced magnetic property modifications: Basics, nanostructure fabrication and potential applications. *Nuclear Instruments & Methods in Physics Research Section B-Beam Interactions with Materials and Atoms* 175: 375–381.

Farhoud, M., H. I. Smith, M. Hwang, and C. A. Ross 2000. The effect of aspect ratio on the magnetic anisotropy of particle arrays. *Journal of Applied Physics* 87: 5120–5122.

Fischer, P. B. and S. Y. Chou 1993. 10-nm electron-beam lithography and sub-50-nm overlay using a modified scanning electron-microscope. *Applied Physics Letters* 62: 2989–2991.

Fisher, K. D. and C. S. Modlin 1996. Signal processing for 10 GB/in(2) magnetic disk recording and beyond. *Journal of Applied Physics* 79: 4502–4507.

Gazzadi, G. C., P. Luches, S. F. Contri, A. di Bona, and S. Valeri 2005. Submicron-scale patterns on ferromagnetic-antiferromagnetic Fe/NiO layers by focused ion beam (FIB) milling. *Nuclear Instruments & Methods in Physics Research Section B-Beam Interactions with Materials and Atoms* 230: 512–517.

Giannuzzi, L. A. and F. A. Stevie 2005. *Introduction to Focused Ion Beams: Instrumentation, Theory, Techniques, and Practice.* New York: Springer.

Gierak, J., E. Bourhis, R. Jede, L. Bruchhaus, B. Beaumont, and P. Gibart 2004. FIB technology applied to the improvement of the crystal quality of GaN and to the fabrication of organised arrays of quantum dots. *Microelectronic Engineering* 73–74: 610–614.

Grutter, P., H. J. Mamin, and D. Rugar 1992. Scanning tunneling microscopy II: Further applications and related scanning techniques. In *Springer Series in Surface Sciences 28*, (Eds.) R. Wiesendanger and H. J. Güntherodt, pp. 151–207. Berlin, Germany; New York: Springer-Verlag.

Ha, J. K., R. Hertel, and J. Kirschner 2003. Concentric domains in patterned thin films with perpendicular magnetic anisotropy. *Europhysics Letters* 64: 810–815.

Hamaguchi, T., M. Mochizuki, T. Matsui, and R. Wood 2007. Perpendicular magnetic recording integration and robust design. *IEEE Transactions on Magnetics* 43: 704.

Kirk, K. J., M. R. Scheinfein, J. N. Chapman et al. 2001. Role of vortices in magnetization reversal of rectangular NiFe elements. *Journal of Physics D-Applied Physics* 34: 160–166.

Komineas, S., C. A. F. Vaz, J. A. C. Bland, and N. Papanicolaou 2005. Bubble domains in disc-shaped ferromagnetic particles. *Physical Review B* 71: 060405(R).

Kryder, M. H., W. Messner, and L. R. Carley 1996. Approaches to 10 Gbit/in(2) recording. *Journal of Applied Physics* 79: 4485–4490.

Lachab, M., M. Nozaki, J. Wang et al. 2000. Selective fabrication of InGaN nanostructures by the focused ion beam/metalorganic chemical vapor deposition process. *Journal of Applied Physics* 87: 1374–1378.

Lang, X. Y., W. T. Zheng, and Q. Jiang 2006. Size and interface effects on ferromagnetic and antiferromagnetic transition temperatures. *Physical Review B* 73: 224444.

Leroy, F., G. Renaud, A. Letoublon, R. Lazzari, C. Mottet, and J. Goniakowski 2005. Self-organized growth of nanoparticles on a surface patterned by a buried dislocation network. *Physical Review Letters* 95: 185501.

Li, S. P., W. S. Lew, Y. B. Xu et al. 2000. Magnetic nanoscale dots on colloid crystal surfaces. *Applied Physics Letters* 76: 748–750.

Lodder, J. C. 2004. Methods for preparing patterned media for high-density recording. *Journal of Magnetism and Magnetic Materials* 272–276: 1692–1697.

Lohau, J., A. Moser, C. T. Rettner, M. E. Best, and B. D. Terris 2001. Writing and reading perpendicular magnetic recording media patterned by a focused ion beam. *Applied Physics Letters* 78: 990–992.

Madou, M. J. 1997. *Fundamentals of Microfabrication.* Boca Raton, FL: CRC Press.

Martin, J. I., J. Nogues, K. Liu, J. L. Vicent, and I. K. Schuller 2003. Ordered magnetic nanostructures: Fabrication and properties. *Journal of Magnetism and Magnetic Materials* 256: 449–501.

Marty, F., A. Vaterlaus, V. Weich, C. Stamm, U. Maier, and D. Pescia 1999. Ultrathin magnetic particles. *Journal of Applied Physics* 85: 6166–6168.

McKendrick, D., R. F. Doner, and S. Haggard 2000. *From Silicon Valley to Singapore: Location and Competitive Advantage in the Hard Disk Drive Industry.* Stanford, CA: Stanford University Press.

Moroni, R., D. Sekiba, F. B. de Mongeot et al. 2003. Uniaxial magnetic anisotropy in nanostructured Co/Cu(001): From surface ripples to nanowires. *Physical Review Letters* 91: 167207.

Moser, A. and D. Weller 2001. Thermal effects in high-density recording media. In *The Physics of Ultra-High-Density Magnetic Recording*, (Eds.) M. L. Plumer, J. v. Ek, and D. Weller. Berlin, Germany; New York: Springer.

Moser, A., D. Weller, M. E. Best, and M. F. Doerner 1999. Dynamic coercivity measurements in thin film recording media using a contact write/read tester. *Journal of Applied Physics* 85: 5018–5020.

New, R. M. H., R. F. W. Pease, and R. L. White 1994. Submicron patterning of thin cobalt films for magnetic storage. *Journal of Vacuum Science & Technology B* 12: 3196–3201.

Nordquist, K., S. Pendharkar, M. Durlam et al. 1997. Process development of sub-0.5 mu m nonvolatile magnetoresistive random access memory arrays. *Journal of Vacuum Science & Technology B* 15: 2274–2278.

Orloff, J., L. Swanson, and M. W. Utlaut 2003. *High Resolution Focused Ion Beams: FIB and Its Applications: The Physics of Liquid Metal Ion Sources and Ion Optics and Their Application to Focused Ion Beam Technology.* New York: Kluwer Academic/Plenum Publishers.

Osborn, J. A. 1945. Demagnetizing factors of the general ellipsoid. *Physical Review* 67: 351.

Pogoryelov, Y. and T. Suzuki 2007. Fabrication of alloy FeCoPt particles by IBICVD and their characterization. *IEEE Transactions on Magnetics* 43: 888–890.

Porthun, S., L. Abelmann, and C. Lodder 1998. Magnetic force microscopy of thin film media for high density magnetic recording. *Journal of Magnetism and Magnetic Materials* 182: 238–273.

Rettner, C. T., M. E. Best, and B. D. Terris 2001. Patterning of granular magnetic media with a focused ion beam to produce single-domain islands at > 140 Gbit/in(2). *IEEE Transactions on Magnetics* 37: 1649–1651.

Ross, C. 2001. Patterned magnetic recording media. *Annual Review of Materials Research* 31: 203–235.

Sheats, J. R. and B. W. Smith 1998. *Microlithography: Science and Technology.* New York: Marcel Dekker.

Shen, J. and J. Kirschner 2002. Tailoring magnetism in artificially structured materials: The new frontier. *Surface Science* 500: 300–322.

Shen, J., J. P. Pierce, E. W. Plummer, and J. Kirschner 2003. The effect of spatial confinement on magnetism: Films, stripes and dots of Fe on Cu(111). *Journal of Physics: Condensed Matter* 15: R1-R30.

Shi, J., S. Tehrani, T. Zhu, Y. F. Zheng, and J. G. Zhu 1999. Magnetization vortices and anomalous switching in patterned NiFeCo submicron arrays. *Applied Physics Letters* 74: 2525–2527.

Stamm, C., F. Marty, A. Vaterlaus et al. 1998. Two-dimensional magnetic particles. *Science* 282: 449–451.

Streblechenko, D. and M. R. Scheinfein 1998. Magnetic nanostructures produced by electron beam patterning of direct write transition metal fluoride resists. *Journal of Vacuum Science & Technology a-Vacuum Surfaces and Films* 16: 1374–1379.

Terris, B. D. and T. Thomson 2005. Nanofabricated and self-assembled magnetic structures as data storage media. *Journal of Physics D-Applied Physics* 38: R199-R222.

Terris, B. D., T. Thomson, and G. Hu 2007. Patterned media for future magnetic data storage. *Microsystem Technologies-Micro-and Nanosystems-Information Storage and Processing Systems* 13: 189–196.

Thompson, L. F., C. G. Willson, and M. J. Bowden 1994. *Introduction to Microlithography*. Washington, DC: American Chemical Society.

Townsend, P. D., J. C. Kelly, and N. E. W. Hartley 1976. *Ion Implantation, Sputtering and Their Applications*. New York: Academic Press.

Utke, I., P. Hoffmann, R. Berger, and L. Scandella 2002. High-resolution magnetic Co supertips grown by a focused electron beam. *Applied Physics Letters* 80: 4792–4794.

Utke, I., P. Hoffmann, and J. Melngailis 2008. Gas-assisted focused electron beam and ion beam processing and fabrication. *Journal of Vacuum Science & Technology B* 26: 1197–1276.

Valbusa, U., C. Boragno, and F. B. de Mongeot 2002. Nanostructuring surfaces by ion sputtering. *Journal of Physics: Condensed Matter* 14: 8153–8175.

Valeri, S., A. di Bona, and P. Vavassori 2006. Magnetic anisotropies in focused ion beam sculpted arrays of submicrometric magnetic dots. In *Magnetic Properties of Laterally Confined Nanometric Structures*, (Ed.) G. Gubbiotti. Kerala, India: Transworld Research Network.

Vavassori, P. 2000. Polarization modulation technique for magneto-optical quantitative vector magnetometry. *Applied Physics Letters* 77: 1605–1607.

Vavassori, P., D. Bisero, F. Carace et al. 2005a. Interplay between magnetocrystalline and configurational anisotropies in Fe(001) square nanostructures. *Physical Review B* 72: 054405.

Vavassori, P., D. Bisero, F. Carace et al. 2005b. Magnetocrystalline and configurational anisotropies in Fe nanostructures. *Journal of Magnetism and Magnetic Materials* 290: 183–186.

Weller, D. and A. Moser 1999. Thermal effect limits in ultrahigh-density magnetic recording. *IEEE Transactions on Magnetics* 35: 4423–4439.

White, R. L., R. M. H. New, and R. F. W. Pease 1997. Patterned media: A viable route to 50 Gbit/in2 and up for magnetic recording? *IEEE Transactions on Magnetics* 33: 990–995.

Xu, J. and A. J. Steckl 1994. Fabrication of visibly photoluminescent Si microstructures by focussed ion-beam implantation and wet etching. *Applied Physics Letters* 65: 2081–2083.

Xu, Q. Y., Y. Kageyama, and T. Suzuki 2005. Ion-beam-induced chemical-vapor deposition of FePt and CoPt particles. *Journal of Applied Physics* 97: 10K308.

Nanoscale Dynamics in Magnetism

Yves Acremann*
SLAC National Accelerator Laboratory

Hans Christoph Siegmann
SLAC National Accelerator Laboratory

27.1 Introduction

In the past decades, we have learned how nanometric fabrication techniques lead to nanometer-scale magnets. It is now well established that nanomagnets can possess quite different *static* magnetic properties compared to their parent bulk material [1]. More recently, it has been found that magnetization *dynamics* is generally also quite different from what is expected from bulk studies, in particular the switching of the magnetization to the opposite direction, one of the basic operations in the application of magnetism which follows a pathway that is unique and differs quite dramatically depending on size, shape, and interface of the magnetic structure with its substrate [2]. Nanomagnets are of high current interest as they have many applications, e.g., in data storage technology, magnetic field sensing, and development of hard magnets [3]. Magnetization dynamics is known to involve a large range of time scales extending from millions of years in geomagnetism to years in magnetic storage media, milli- and microseconds in AC transformers, and nanoseconds in magnetic data writing and reading. This survey covers ultrafast magnetization dynamics ranging from nanosecond (10^{-9} s) over picosecond (10^{-12} s) to femtosecond (10^{-15} s) in nano-sized particles and magnetic structures. New tools such as pulsed lasers, x-ray sources, and polarized electron currents have extended the study of magnetism dynamics to the femtosecond time-range at nanoscale spatial resolution. Specifically, the new pulsed x-ray sources such as synchrotron-based sources and x-ray lasers combine the possibility to observe the dynamics of magnetic processes in nanoscale structures with element specificity.

Magnetism has attracted physicists since ancient times. It is the only solid-state quantum phenomenon that exists up to 1000 K, yet it is to this day still not understood satisfactorily. However, we have seen great advances in the last decades, based on the development of surface science and spectroscopy. Furthermore, experiments with polarized electron beams or currents can probe the elusive exchange interaction inducing all magnetic phenomena, ranging from static long-range magnetic order as it depends on the band structure and temperature to the plethora of nanoscopic magnetic structures and their dynamics. Polarized electrons can give insight into very fast dynamics at the femtosecond level as well, exploring particularly the promising new field of magnetization dynamics excited by spin currents.

The spin of the electron and its interaction with various stimuli in the solid-state environment underlie all magnetization dynamics. The theoretical concept of the spin has been developed mostly by Pauli, Dirac, and Heisenberg in the time period 1925–1928. However, today the electron spin has become a reality that entered everyday life and is modifying the human civilization through its utilization in magnetic devices such as computer hard drives and high-density advanced magnetic memories without which the Internet and many other modern commodities could not exist. It turns out that small magnetic structures, layered magnetic materials, and their interfaces are the building blocks of advanced information technology as we know it today. Some fundamental discoveries emerging from the art of producing such structures with well-controlled interfaces as well as new spin-based spectroscopies initiated a paradigm shift in our thinking and a technological revolution where the electron spin is now the sensor as well as the carrier of information.

After a survey of important static magnetic structures and their dynamics, we will describe new experimental techniques

---

* He is currently affiliated with Laboratory for Solid State Physics, ETH Zürich, Zürich, Switzerland.

developed to dynamically image these structures down to femtosecond temporal and nanoscale spatial resolution. It turns out that this cannot be done in full generality; one rather has to confine the discussion to few elementary cases that give first physical insight into the complex phenomena that are still under investigation in numerous research groups. The present survey attempts to lead the reader in an easy way to the frontiers of what is known today on ultrafast magnetization dynamics at the nanoscale.

27.2 Magnetic Structures in Nanoscopic Samples and Their Dynamics

27.2.1 Static Magnetic Structures

Ferromagnetic samples show a rich variety of magnetic structures depending on their size and shape. The main factor that determines the magnetic structure is the exchange interaction that tends to align spins in parallel in ferromagnets. However, this interaction, albeit strong, is short range, extending only to the nearest neighbors in most cases. In the Heisenberg model, explained in the textbooks on magnetism, e.g., in Ref. [2], one assumes that a spin s is located at each lattice site. The difference ΔE in energy in a state where two neighboring spins are parallel compared to where they enclose an angle β is given by

$$\Delta E = 2Js^2 \left[1 - \cos\beta\right] \qquad (27.1)$$

where J is the exchange integral. The smallest magnetic structure is thus given by reversing the spin at one lattice site. However, this smallest magnetic structure known as a Stoner excitation is energetically very unfavorable and therefore has a very short lifetime amounting to ≤ 1 fs in a perfect lattice. However, if the angle $\beta \ll 1$, the energy difference between two neighboring spins is $\Delta E \cong Js^2\beta^2$. Hence, the exchange excitation energy vanishes faster than linear with small angles β between the spins. This enables the ferromagnet to sustain stable nontrivial magnetic structures if the spins change direction slowly over large-enough distances. The next important fact to know is that there exists magnetic anisotropy that leads to energetically preferred directions of the spins in the magnetic body. The magnetic anisotropy energy is generally small compared to the exchange energy of the order of ~10^{-6} eV in the bulk of simple and perfect lattices. Magnetic anisotropies are generated by the electric crystal fields through spin orbit coupling by stress due to magnetostriction or by the shape of the sample. If the magnetic anisotropy was the only energy determining the direction of the magnetization, a ferromagnetic sample would still tend to be homogeneously magnetized along one of its easy directions determined by the magnetic anisotropy. However, depending on its physical shape and special magnetic anisotropies, a homogeneously magnetized sample will generate a magnetic stray field. The generation of a magnetic field outside the sample costs energy. To reduce this energy, the magnetization will split up into a variety of structures to minimize the stray field outside the sample. Well known are the

ferromagnetic domains that occur in macroscopic samples and may range in size from millimeters to nanometers. The magnetic domains have been postulated by Pierre Weiss at the beginning of the twentieth century and have been essential in understanding the hysteresis loop, i.e., the behavior of a magnetic specimen in an external magnetic field and the performance of electrical transformers, one of the cornerstones of our electromagnetic society. Yet in the presence of the homogeneously magnetized domains, there have to be transitions from one direction of the magnetization to another. These transitions occur in domain walls first postulated by Felix Bloch in the 1930s, but the formation of a domain wall costs some energy as well. The properties of a domain wall are determined by the competition between the exchange energy and the magnetic anisotropy energy. The domain wall energy increases with both the exchange energy and the anisotropy energy because both favor a collinear alignment of the moments. The domain wall width increases with the exchange energy but decreases with increasing magnetic anisotropy energy. A typical width of a domain wall, e.g., in bulk Fe is 50 nm. However, this is not the smallest dimension a magnetic structure can have. Namely, if the magnetization is forced to be perpendicular to a surface, the energy to produce the stray field will be larger compared to the crystalline anisotropy in the 3d-transition metals and their alloys. Then, a magnetic vortex may form, as explained below, where the extension of the vortex core with a magnetization perpendicular to the surface is as small as compatible with the cost of exchange energy, in practice a few nanometers only. The vortex is a delicate magnetic structure that cannot readily be imaged, e.g., by approaching a magnetic tip and measuring the force between tip and vortex, because the vortex easily gets distorted and might move. Yet beautiful pictures of a magnetic vortex in Fe have been obtained with spin-polarized scanning tunneling microscopy using a stray-field-free Cr-tip [4]. Scanning ion microscopy with polarization analysis also has produced pictures of the vortex and its close relative the antivortex in Co dots [5], but in these pictures the core is unreasonably large. We expect to encounter stable magnetic vortex structures in the nanometer-size range in common ferromagnets such as Fe, Co, Ni, and their alloys [6].

In Figure 27.1, some examples of stable magnetic states in small thin-film elements are given. Such structures are large enough

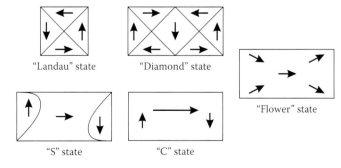

FIGURE 27.1 Examples of magnetization patterns in small thin-film elements and the corresponding names.

to be imaged with the magneto-optic Kerr-effect. However, they are composite structures, i.e., they are composed of several basic structural elements to be discussed below, such as Bloch walls or Néel walls in which magnetic vortices or antivortices might be hidden as well. The internal structure of domain walls is a fascinating and very rich topic of magnetic microscopy [7]. Detailed wall structures with attending vortices have been studied in garnets by optical techniques [8]. But with the ferromagnetic metals, both vortex and antivortex, as well as the internal structure of domain walls, are too small to be resolved by optical microscopy.

One important multidomain state of a small thin-film element is the *Landau state*, shown in Figure 27.1 together with other often-encountered magnetic structures. In the Landau state, four in-plane magnetic domains form a flux-closed structure in which the magnetostatic energy is minimized. In the center of the Landau state, a *magnetic vortex* may exist in which the magnetization points are in a direction perpendicular to the film plane. This "vortex core" costs a large magnetostatic energy and is therefore very small, of the order of ≈10 nm in a "soft," i.e., low-anisotropy magnetic material such as permalloy [9].

There are four basic vortex structures, as shown in Figure 27.2. The handedness of a vortex is determined by aligning the thumb into the direction of the out-of-plane magnetization of the core, as shown, and then matching the directions of the fingers of the right or left hand with the in-plane curl direction of the Landau state. The magnetic vortex structure is special in that it combines the basic symmetry properties of *inversion* and *time reversal*. This is seen by inspection of Figure 27.2. The four basic structures correspond to the two types of handedness, which are transformed into each other by the parity operation, and the two magnetization directions of the core, which are transformed into each other by the time-reversal operation, since the magnetization is an axial vector changing sign when time is inverted. We will see below that the handedness determines the dynamic behavior and that the vortex core can be switched by exciting the Landau state with short bursts of an external magnetic field [10].

However, the Landau state and its even more complex relative, the diamond state with two vortices [11], possess in fact a composite magnetic character in which three different magnetic substructures exist: *domains, domain walls*, and the *vortex core*. These three different substructures differ in their dynamic behavior and are mutually coupled, yet are generally simultaneously excited. This leads to an overall very complex dynamic behavior of the Landau state as will become evident below in Section 27.2.2. Within one of the four magnetic domains of the Landau state, the spontaneous magnetization $\vec{M}_s$ is constant and the dynamics are dominated by precession and damping of $\vec{M}_s$ in the anisotropy fields. The vortex itself is dominated by the exchange energy, yet its motion is necessarily strongly coupled to one of the domain walls. The walls taken by themselves are intermediate between domains and vortex with contributions from both exchange and magnetostatic or magnetocrystalline energy.

In a uniaxial thin film, i.e, a thin-film magnetized in-plane with one easy axis of magnetization, the magnetization can have two directions. These two regions are separated by a Néel domain wall. Within such a wall, $\vec{M}$ rotates in the plane of the film into the other direction. However, the sense of rotation can be different, and usually both possibilities are realized in practice. At the location where the sense of rotation changes, the magnetization is forced to be perpendicular to the plane of the film defining a Bloch point. Around such a Bloch point, a magnetic structure known as an antivortex is formed. This situation is illustrated in Figure 27.3. Thus, a domain wall is usually not a homogeneous magnetic structure. Indeed, high-resolution images of magnetic domain walls obtained with spin-resolved scanning electron microscopy known as SEMPA [12] reveal very complex structures of domain walls in metals [13].

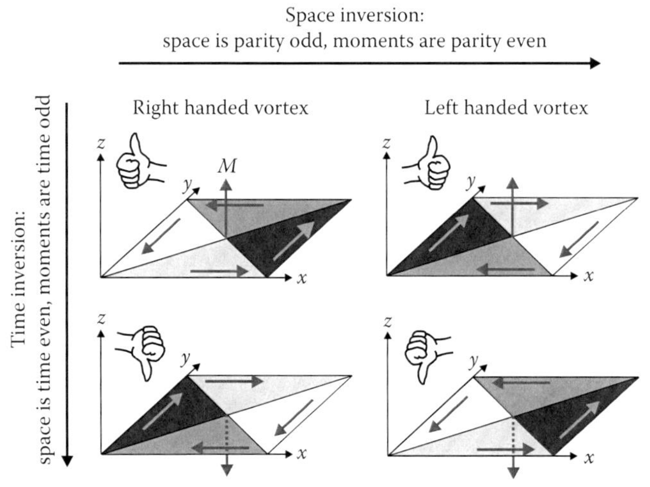

FIGURE 27.2 The four basic magnetic vortices, classified by their handedness and out-of-plane direction of the core. A magnetic vortex combines the concepts of handedness and time reversal and the four basic structures are transformed into each other by either the time reversal or inversion operations.

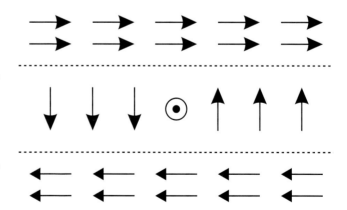

FIGURE 27.3 Néel walls with different sense of rotation are separated by a region in which $\vec{M}$ is perpendicular to the plane of the film. The direction of $\vec{M}$ is indicated only in the center of the wall. Around the point with $\vec{M}_\perp$ called a Bloch point, the magnetization rotates in the opposite sense as in magnetic vortex. Therefore, this magnetic structure is called an antivortex.

If one reduces the thickness and lateral dimensions of a magnetic film below the micrometer range but maintains a spherical shape of the sample, one encounters the vortex as a stable magnetic structure without domain walls. A "single domain" state is favored only at still smaller dimensions. For example, in self-organized Fe films on W(001), uniform single-domain, simple vortex, and distorted vortex states have been observed with spin-polarized scanning tunneling spectroscopy [14]. Magnetic vortices become stable over the single-domain state at larger film thicknesses ≥10 nm and for lateral dimensions ≥200 nm [14]. However, it has been found that magnetization states in small structures may be bistable, e.g., depending on its magnetic history, a sample may exhibit either a vortex state or a uniform magnetized state [15].

Figure 27.4 shows the distribution of $\vec{M}$ for a magnetic vortex and a magnetic antivortex. In both cases, the in-plane magnetization direction rotates by an angle of 2π in a closed loop around the core. But the sense in which $\vec{M}$ rotates over the closed loop is positive for the vortex and negative for the antivortex. The winding number n counts the number of rotations of $\vec{M}$ in units of 2π. It can generally assume every integer number except 0, but we shall limit the discussion to the relevant cases of $n = \pm 1$ shown in Figure 27.4. The winding number of a thin-film element is a topological invariant; hence, vortex and antivortex can only be created or annihilated pair wise, unless, as possible in practice, the vortex or antivortex enters through the boundary of the sample. The "skyrmion number" is a second topological invariant including the direction of $\vec{M}_\perp$ in the core. The half integer skyrmion charge is given by $q = np/2$. Thus, a vortex/antivortex pair with parallel core polarizations p has opposite skyrmion numbers adding to zero and consequently belong to the same topological sector as a uniform state. From a topological perspective, such a texture can be deformed continuously into the homogeneous ground state [16]. The detailed pathway of such an annihilation in a thin-film element has been confirmed in simulations based on the Landau–Lifshitz–Gilbert (LLG) equation explained below [17]. While an isolated vortex is stable in a circular thin film, the antivortex is not stable by itself unless the thin film has a special shape with four extensions on the edges of a square [18].

27.2.2 Dynamics of Small Magnetic Structures

The dynamics of magnetic structures is revealed when an external field is applied or when a spin-polarized electron current is injected. Let us first consider excitation with a magnetic field. The volume density of the torque exercised on $\vec{M}$ by a magnetic field $\vec{H}$ is given by

$$\vec{T} = \vec{M} \times \vec{H} \tag{27.2}$$

The motion of the magnetization is in the direction of the torque, hence perpendicular to $\vec{M}$ and $\vec{H}$.

This torque induces a precession of $\vec{M}$ about $\vec{H}$. It describes a reversible process because $\vec{T}$ does not change sign under time reversal as both $\vec{M}$ and $\vec{H}$ change sign. In precession, the angle θ between $\vec{M}$ and $\vec{H}$ does not change, and the energy $E = M.H \cos\theta$ thus remains constant. However, there is an additional torque $\vec{T}_D$ that takes care of what we all know from childhood on about magnetization dynamics: a compass needle turns into the direction of the magnetic field. This torque is a "pseudotorque" T_D called the damping torque that moves $\vec{M}$ eventually into the direction of $\vec{H}$:

$$\vec{T}_D = \frac{\alpha}{M}\left[\vec{M} \times \left(\vec{M} \times H\right)\right] \tag{27.3}$$

$\vec{T}_D$ is not invariant under time reversal like $\vec{T}$, and hence it is not a real torque, but a torque that dissipates energy in an irreversible process. α is a dimensionless empirical constant usually of the order of 10^{-4} to 10^{-2} that describes the transfer of angular momentum to the crystal lattice, called the spin lattice relaxation. With $\gamma\vec{T} = d\vec{M}/dt$, where γ is the negative gyromagnetic ratio of the electron and adding $\vec{T}$ and $\vec{T}_D$, we obtain the LLG equation that is commonly used to simulate the dynamics of magnetic structures:

$$\frac{d\vec{M}}{dt} = \gamma\left[\vec{M} \times \vec{H}\right] + \frac{\alpha\gamma}{M}\left[\vec{M} \times \left(\vec{M} \times H\right)\right] \tag{27.4}$$

With the sample shown in Figure 27.3 and applying $\vec{H}$ in the plane of the film parallel or antiparallel to the direction of $\vec{M}$ in the domains, we see that a torque exists only within the domain wall. This torque moves $\vec{M}$ out of the film plane and thus generates a demagnetizing field H_{demag} which is perpendicular to the plane of the film and lets $\vec{M}$ precess into a direction parallel to $\vec{H}$. Thus, the region with a direction of $\vec{H}$ parallel to $\vec{M}$ is enlarged. Consequently, the Néel wall of either kind parallel to the applied field moves in the direction perpendicular to $\vec{H}$. The speed with which the wall moves is limited, because the demagnetizing field has the maximum value $H_{demag} = -M/\mu_o$, and the maximum speed of the wall is given by the precession frequency of $\vec{M}$ in that field. The Bloch point, contracting to a small structure (not shown in the schematic Figure 27.3), gets easily stuck in lattice imperfections of similar spatial extension where it

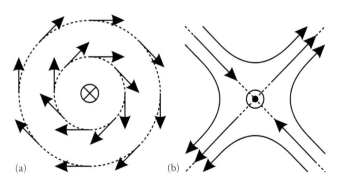

FIGURE 27.4 Two basic magnetic structures important for magnetization dynamics: the vortex (a) and the antivortex (b), distinguished by opposed winding numbers of +1 or −1 respectively. In both structures, the core is magnetized perpendicular in either up or down direction.

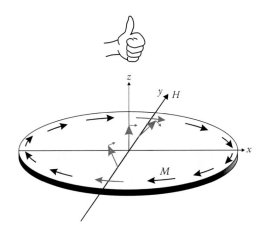

FIGURE 27.5 Left-handed magnetic vortex structure near the center of a Landau state. The magnetization $\vec{M}$ rotates out of the x–y-plane into the z direction. In a magnetic field $\vec{H} \parallel y$, the location where $\vec{M}$ is perpendicular moves toward $-y$, i.e., the vortex core moves opposite to $\vec{H}$. If the vortex is right handed, the vortex core moves into the direction $\vec{H}$.

can lower its energy. Thus, the wall jumps from imperfection to imperfection. Sometimes, the wall gets stuck and bulges out rather than moves.

In contrast to the wall motion, the motion of a magnetic vortex in response to an in-plane field is in the direction of $\vec{H}$ or against it, depending on the chirality of the vortex. To illustrate this, we show in Figure 27.5 the magnetic structure of a left-handed magnetic vortex with emphasis on the out-of-plane structure along the y direction. The direction of motion is again determined by the direction of the torque on $\vec{M}$. The magnetization $\vec{M}$ located at a negative y position moves out of the film plane, and $\vec{M}$ located at positive y position moves toward the film plane as indicated by little arrows. By inspection of the torques for the four basic vortex structures (Figure 27.2), one finds that with an in-plane magnetic field $\vec{H}$ the center of the left-handed vortex always moves against the direction of $\vec{H}$, while the center of the right-handed vortex moves into the direction of $\vec{H}$. If however the magnetic field is applied perpendicular to the film plane, the vortex core experiences no torque and hence does not move. But oscillations are excited in the in-plane closed flux magnetic structure surrounding the core. It has been shown that these modes are analogous to the eigenmodes of a membrane excited with a drumstick [2,19].

Vortices and their dynamics in metallic magnetic Landau states excited by magnetic field pulses have been extensively studied and imaged in real time by means of magneto-optical Kerr microscopy [19–21] or x-ray magnetic circular dichroism microscopy [10,22–24]. In the frequency domain, vortex excitations in microstructures have been detected by means of microwave reflection in a nearby coplanar waveguide [25].

Let us now discuss the case of exciting a magnetic structure by injection of spins. A current of spin-polarized electrons may be injected ballistically through an insulating barrier at energies of ≈1 eV above the Fermi level or diffusely through a transparent metallic contact. With the exception of [26], only currents of polarized electrons have been used, i.e., one has a flow of charge producing an Oersted magnetic field plus a flow of spins importing angular momentum. The inhomogeneous Oersted field produces a torque on $\vec{M}$ comparable in practice to the spin torque.

The occurrence of torques acting on the magnetization upon injecting spins may be viewed as a consequence of Newton's third law according to which *actio = reactio*. Once an electron is injected across the surface barrier potentials into the ferromagnet, we have a closed system with no external forces, meaning that angular momentum must be strictly conserved. However, the lattice of nanomagnets is necessarily coupled to the substrate, and angular momentum is lost by transfer to the lattice. Yet as long as the gain of angular momentum from the injected spins is much larger than the loss due to spin lattice relaxation, we can neglect the spin lattice relaxation although it is still relevant as it sets a minimum value for the current density needed to excite or switch the magnetic structure. It turns out that the injected current density must be very large, close to the electromigration limit of 10^{12} A/m^2. Such large current densities can only be produced in nanostructures, as only nanostructures have a large surface-to-volume ratio to provide adequate cooling by heat conduction through the interface with the surroundings. Yet in all spin injection experiments, Joule heating through the charge current is an issue.

To measure the spin torques, electron beams with a spin polarization vector $\vec{P}$ have been passed through the ferromagnet (Fm) [27,28]. Such polarized electron beams can be produced by optical pumping from solid photocathodes [29]. Although low-energy electron beams have a low current density due to space charge limitations and can thus not measurably excite $\vec{M}$, the electron beam experiments are very welcome to evaluate the spin torques. Namely, from the motion of $\vec{P}$ as the beam traverses the sample, we can infer the angular momentum transmitted to $\vec{M}$ of the Fm using Newton's third law. As one has to measure the direction and magnitude of $\vec{P}$ in this experiment, we refer the reader to an example of a modern high-performance electron spin polarimeter [30]. We can also reflect polarized electrons from a magnetic surface and find that $\vec{P}$ changes in reflection as well [31], hence torque is also transferred to $\vec{M}$ in reflection of electrons. In these experiments as well as in the solid-state ballistic injection experiments where the spin-polarized current densities are high enough to excite and even switch $\vec{M}$ [32], the electrons spend a time of only ≈0.3 fs/nm film thickness in the sample as they are traveling with velocities $\geq v_\mathrm{F}$ with v_F the Fermi velocity. This means that the processes transferring the angular momentum to $\vec{M}$ occur on the sub-femtosecond time scale. However, if one injects a diffusive spin current through a transparent contact, one has to calculate the interaction time from the drift velocity v_D, depending on the resistivity and the injected current density $v_\mathrm{D} \ll v_\mathrm{F}$ by one or two orders of magnitude with the extremely high current densities of spin torque experiments. This means, that slower processes like the excitation of spin waves can be important as well when using diffusive currents for excitation of the magnetic structure.

As we deal with nonrelativistic electrons in all spin injection experiments, the orbital and spin part of the wave function are separated. For simplicity we assume that the electron beam is completely polarized, i.e., $P = 1$. This is not unrealistic as it has been shown that polarized electron beams with $P = 0.92$ can be produced [33]. It is also important to inject the electrons with spin polarization $\vec{P}$ perpendicular to the magnetization, as in this initially instable configuration of $\vec{M}$ and $\vec{P}$ the motion of $\vec{P}$ is most easily observed. When entering the Fm, the spin part of the electron wave function is a coherent superposition of spin-up and spin-down wave functions with equal amplitude:

$$\psi = \frac{1}{\sqrt{2}}\left[\begin{pmatrix}1\\0\end{pmatrix}+\begin{pmatrix}0\\1\end{pmatrix}\right]e^{i\varphi} = \frac{1}{\sqrt{2}}\begin{pmatrix}1\\1\end{pmatrix}e^{i\varphi} \qquad (27.5)$$

The electrons emerge from the ferromagnet with the wave function

$$\psi' = \frac{1}{\sqrt{2}}\begin{pmatrix}\sqrt{1+A} & e^{+i\epsilon/2}\\\sqrt{1-A} & e^{-i\epsilon/2}\end{pmatrix}e^{i\varphi} \qquad (27.6)$$

and with the polarization vector $\vec{P}' = \left[\sqrt{1-A^2}\cos\epsilon, \sqrt{1-A^2}\sin\epsilon, A\right]$. This corresponds to two types of motion of the spin polarization vector, namely, a precession by an angle ϵ around M that increases linearly with film thickness, and a rotation by the angle $\Delta\theta$ into the direction of $\vec{M}$. The two motions correspond two types of torques acting on $\vec{M}$ [2].

The precession is very fast, namely 33°/nm for Fe, 19°/nm for Co, and 7°/nm for Ni, even at energies ≤5 eV above the Fermi level [28]. This fast precession reflects the generally very large, but different, magnitude of the exchange fields in these Fm. The resulting precessional torque on $\vec{M}$, also called the NEXI (non-equilibrium exchange interaction) torque, rotates in space by the above angles per nanometer film thickness. It can therefore hardly lead to a coherent excitation of the magnetization in films of several nanometer thickness; the most important effect of the NEXI-torque might be the excitation of spin waves leading to incoherent fluctuations of $\vec{M}$. But in reflection, where the electron penetration is limited to a depth of ≈1 nm, and the precessional torque is even larger due to the effects of band gaps [31]; it can exert the dominant pressure on the magnetization of the nanoscopic magnetic sample.

The rotation takes place in the plane spanned by $\vec{P}$ and $\vec{M}$. This rotation is due to spin-dependent absorption A in which the minority spin wave function is more strongly attenuated than the majority spin wave function, an effect also known as the spin filter effect and the actual cause of giant magneto-resistance (GMR). $A = (I^\uparrow - I^\downarrow)/(I^\uparrow + I^\downarrow)$, where $I^\uparrow$ and $I^\downarrow$ are the electron currents transmitted in the two spin states. A leads to a torque that pushes $\vec{M}$ toward the direction of the spin of the injected electrons, independent of the precession. The angle θ of rotation decays exponentially from its initial value of $\theta = \pi/2$ to $\theta \to 0$ with increasing film thickness. The experiments show that the decay of $\Delta\theta$ to $1/e$ of its initial value occurs in the transition

metals after the electrons travel a distance of ≈10 nm. The rotation of $\vec{P}$ into $\vec{M}$ reveals the torque $\vec{T}_I$ due to the injected spins. $\vec{T}_I$ is a pseudotorque antiparallel to the damping torque caused by the irreversible processes of electron scattering, and constitutes the main torque in spin injection, given according to [34] by

$$\vec{T}_I = g(\theta)\frac{\hbar I}{2eS_1 S_2^2}\vec{S}_2 \times\left[\vec{S}_1 \times \vec{S}_2\right] \qquad (27.7)$$

Here I is the injected current and $g(\theta) \geq 1$ a dimensionless scalar function of the angle θ between the injected spin density $\vec{S}_1$ and the spin density $\vec{S}_2$ in the Fm. $\vec{T}_I$ vanishes when $\theta = 0$ or $\theta = \pi$ but develops as soon as $\vec{S}_1$ or $\vec{S}_2$ moves out of the instable antiparallel configuration, e.g., by thermal excitation, or by the Oersted field of the charge current. The spin torque can be added to LLG yielding an equation that promises to yield the dynamics of magnetic structures under spin injection. However, simulations based on Equation 27.4 require that $\vec{M}$ changes only direction, but not magnitude, which is approximated if the volume element for which the simulation is done is sufficiently small. However, we have already seen that transient magnetic structures can be as small as one lattice diameter, and as the simulations cannot be done with very small volume elements, a priori predictions of what will happen in a specific case are not possible. Therefore, one still needs the experiments, and (27.4) can then help to understand what one sees.

The magnetic-field-induced torques as well as the superimposed effect of the spin torques [34,35] have been satisfactorily verified in a large number of experimental studies. The torques lead to magnetization reversal, motion, and reconstruction of domain walls, and to steady precession modes of $\vec{M}$. The steady-state precessional modes of nanoscale magnetic structures are due to the fact that the damping torque $\vec{T}_D$ can be compensated by the spin injection torque $\vec{T}_I$ leading to zero damping, i.e., to a steady precession. It has been proposed to use the steady-state precession as a tunable microwave generator [36] and as a nanoscale radiofrequency detector in telecommunication circuits [37]. The dominant dynamical technique to observe this has been GMR measurements, which, however, can only measure the average magnetization of a nanoscopic sample relative to a reference sample in close proximity, which is assumed to be, and to remain, homogeneously magnetized under the excitation.

27.3 Selected Experimental Results

With dynamic GMR measurements one can determine magnetic oscillation frequencies, but special magnetic structures that may be present transiently remain unidentified and must be inferred from the frequency of the oscillation. As this is not always possible and convincing, we have selected for this introductory review a number of basic experiments; yet, so far, only the last one can image the dynamical structures present transiently during the excitation.

27.3.1 Gyration and Switching of the Vortex Core

Let us first discuss the dynamics of the Landau state (Figure 27.1) excited with an external magnetic field pulse. The first work of this kind is due to Choe and collaborators [22] and Raabe and collaborators [23], using a technique know as XMCD-PEEM. XMCD stands for x-ray magnetic dichroism and is a phenomenon that is at the center of all studies of magnetism with x-rays.

To understand the XMCD-effect [2], we realize that the 3d-valence states are split by the exchange interaction, and that there are more holes available for the minority spins compared to the majority spins. The x-rays must be circularly polarized and be monochromatic with an energy of the 2p-3d transition specific for the magnetic atom under investigation. For maximum XMCD-effect, $\vec{M}$ of the sample and the photon spin, or in other words the direction of the circularly polarized x-ray beam must be parallel or antiparallel to $\vec{M}$. Then, due to optical selection rules in transitions from the $2p_{3/2}$- or $2p_{1/2}$-state to the 3d-valence state, the electrons being excited by absorption of the circularly polarized photons are spin polarized, but the intensity of the transition depends on the spin-dependent availability of a final state. In this way, the majority or minority holes in the, e.g., 3d-band can be measured, and the background can be subtracted by switching from right to left circularly polarized light or by inverting $\vec{M}$ and recording the difference. The measurement is element specific, i.e., in an alloy, the local magnetization at each element can be measured separately.

PEEM stands for photoemission electron microscopy. The photoelectron intensity emitted following XMCD excitation by decay of the 2p core hole in an Auger process and subsequent cascade formation depends on the magnitude of the XMCD-effect and is thus a measure of the magnitude of the magnetization close to the surface. The photoelectron emission from a surface is then magnified by electron optics and delivers a picture of the magnetic structure of the sample.

The magnetic field pulse is generated by a current pulse flowing in an impedance-matched strip line. A variety of sample shapes and sizes is deposited directly on the strip line. The current pulse in the strip line generating the magnetic field pulse parallel to the surface of the samples is due to the optical pump pulse which opens a photoelectric switch (known as Auston switch) synchronized with the probing with circularly polarized x-ray pulses from the synchrotron. The time resolution is limited mainly by the width of the synchrotron radiation x-ray probing pulses emitted by individual bunches in the electron storage ring. Depending on the operating conditions of the storage ring, the pulse length varies from about 30 to 100 ps. The PEEM spatial resolution is well below 100 nm.

PEEM images with XMCD magnetic contrast were recorded by pump-probe techniques as a function of delay time between the Auston-switch-triggered magnetic field pulse and the x-ray probe pulse. Figure 27.6 shows magnetic images of two adjacent $1 \times 1.5\,\mu m^2$ samples that were patterned by focused ion beam milling into a 20 nm thick CoFe film deposited onto the Cu waveguide. The images clearly show the Landau states of the two samples.

By pump-probe technique one can now record a sequence of images as a function of the delay time of the x-ray probe pulse relative to the laser pump pulse. When the Landau structures are excited by the horizontal field $\vec{H}$ in the shown direction, the initial directions of motion of the vortex centers and their ensuing gyrotropic motions are found to be in opposite directions as indicated by arrows. Following the discussion in conjunction with Figure 27.5, this indicates that the two vortices have opposite handedness. After the vortex is initially shifted from its equilibrium position, the imbalance of the in-plane magnetization creates magnetostatic fields driving the vortex into a spiraling motion back to its equilibrium position. Raabe and collaborators [23] used larger permalloy squares of $6\,\mu m$ and found in contradiction to the results shown in Figure 27.6 that the vortex core of the Landau

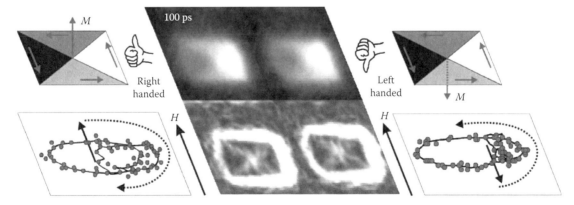

FIGURE 27.6 *Center*: Magnetic images of two adjacent $1 \times 1.5\,\mu m^2$ CoFe rectangles, patterned into a 20 nm thick film with in-plane easy axis [22]. The images were recorded by PEEM using XMCD magnetic contrast. We show images of the two patterned regions at delay times of 2.1 and 4.5 ns after field pulse excitation. The time evolution of the vortex cores is indicated by dots in the *lower left* and *right* insets. The points represent sequential vortex core positions (in 100 ps steps) up to 8 ns and the lines represent the time-averaged positions. When the magnetization in the two samples is excited by a horizontal field $\vec{H}$, the initial directions of motion of the vortex core (*straight arrows*) and the ensuing gyrotropic motions (*dotted curved arrows*) are found to be in opposite directions. This indicates that the two vortices have opposite handedness.

state moved only up and down in the direction of the applied field but did not gyrate. However, this is most likely due to the larger size of their samples in which the magnetostatic fields are weaker when the vortex with the attending domain walls moves out of the center of the sample.

The sense of the gyration of the vortex core can thus be used to determine the polarization of the core which could not be imaged directly in the PEEM experiments. However, in transmission x-ray microscopy by Stoll and collaborators [10,24], the spatial resolution is better and the core can be imaged as well. It has been found in these experiments that the sense of gyration of the Landau state can be inverted under the influence of a sinusoidal magnetic field with amplitude of only 1.5 mT at the resonance frequency of the vortex gyration of ≈250 MHz. The reversal of the gyration proves that the core polarization of the vortex has switched sign. Such sign reversal was thought to require much higher magnetic fields of 0.5 T [38].

The dynamics of the switching of the vortex core could be reproduced by simulations based on the LLG equation (27.4). As Figure 27.7 illustrates, a distortion of the out-of-plane vortex structure occurs when the vortex core is driven by the resonant excitation. After a time interval of few nanoseconds, a vortex/antivortex pair with skyrmion number 0 appears.

Such a pair can emerge continuously from the homogeneously magnetized state. The pair has the opposite core polarization as the initial vortex, and in the unification of the antivortex with the initial vortex, a displaced new vortex is left behind with the opposite core polarization. The finer details of this simulation are arguable [39] as in the simulation angles between spins can be as large as 100°, violating the basic assumption when using Equation 27.4 that $|\vec{M}|$ = constant in the volume element of the calculation. The annihilation of a vortex and an antivortex of opposite core polarizations involves, according to simulations done by other authors [17], the propagation of a Bloch point causing a burstlike emission of spin waves. According to simulations, the antivortex may analogously be switched in response to an external field pulse by adding a vortex/antivortex pair [18].

The dynamics of vortices has been studied extensively as well by injecting spin currents. However, in all cases, the experiments could not image the vortex directly. Rather, high-frequency resistivity measurements indicate a precession frequency that is lower compared to the precession frequency of a homogenous magnetization. With the help of simulations, the low precession frequency was then interpreted as due to the precession or rather gyration of a vortex. Intuitively it is not possible to understand

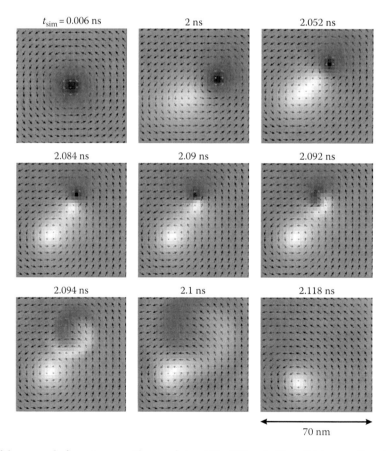

FIGURE 27.7 Simulation of the reversal of a vortex core. The sample is a 500 × 500 nm², 40 nm thick permalloy element, excited by a sinusoidal magnetic field at the resonance frequency of the vortex gyration (≈250 MHz). The arrows indicate the in-plane distribution of $\vec{M}$ while black is the negative out-of-plane and white the positive out-of-plane $\vec{M}_\perp$. The times at each panel indicate the temporal evolution of the magnetic structures. (Courtesy of Kang Wei Chou.)

why the vortex is not destroyed or driven out of the sample or at least deformed as the torque $\vec{T}_I$ in Equation 27.7 has opposite sign on each side of the vortex. Yet one finds in the simulations that nonuniform magnetic structures such as vortices, isolated within a nanoscale spin valve element similar to the one shown in Figure 27.9, can be excited into a persistent oscillation of the vortex core by a spin-polarized DC current [40]. Such highly compact spin torque oscillator devices are candidates for applications in microwave signal processing. In metallic nanocontacts, the vortex is found to execute an orbital motion around the nanocontact injecting the spin current. The Oersted field of the charge current connected to the spin current is important for the stabilization of the orbiting vortex [41]. Spin-torque-induced vortex gyration excited by in-plane spin currents is found as well in Landau states of micrometer-sized samples by x-ray microscopy [43]. Furthermore, it has been found that the vortex core can be switched by an in-plane rotating magnetic field, whether or not the vortex core switches depend on the sense of rotation of the magnetic field [42]. These and other observations have implications for the development of magnetic storage devices using spin-torque-driven magnetization switching and/or domain wall motion because vortices and antivortices appear to be responsible for the dynamics. There is a phase difference between the spin torques and the Oersted field torques that can be used, with the help of simulations, to quantify their relative influence [43].

27.3.2 Switching of a Néel Domain Wall by an Antivortex

While extensive theoretical literature existed since long on magnetic domain walls in the simple ferromagnetic metals, the experimental results seemed to be contradictory and variable until quite recently. Important progress has been made by imaging the structure of domain walls in thin wires where the magnetization is "head to head" or "tail to tail" at either side of the wall due to the shape anisotropy of a wire forcing the magnetization to be directed along the axis of the wire. If now a current flows in the wire, the electrons have to pass through the domain wall. As an electron current flows in the wire traversing the domain wall, it may change its internal structure and at the same time change its location. The wall motion and transformation is due to the magnetic field of the current as well as the spin torque exercised by the electrons polarized according to the direction of $\vec{M}$ in the part of the wire they originated. Beautiful high-resolution pictures of static wall structures in permalloy wires have been obtained by XMCD-PEEM and scanning electron microscopy with polarization analysis (SEMPA) [12]. The pictures also reveal the complex changes of the wall structures after a current pulse [13,44]. A head-to-head domain wall can be a vortex wall or a simple Néel wall. In a vortex wall, $\vec{M}$ curls around a vortex core in which $\vec{M}$ points out of the surface of the wire to reduce the overall exchange energy of the wall. If a current is passed through the wire, the wall moves stochastically in the direction of the electron flow as it gets stuck on defects and changes its structure. The theoretical treatment is complicated

by the question whether the spin of the electrons adapts to the local magnetization or whether there is a delay in the spin motion relative to the change of $\vec{M}$ in the wall. A term with this nonadiabaticity parameter has to be added into Equation 27.4 to generate some similarity of the observations with the simulation.

For the purpose of this survey, we have chosen to discuss the simple case that the wall type does not change during the motion. This can be achieved by making the ferromagnetic wire only 150 nm wide at 10 nm thickness [45]. In that case, a pure Néel wall forms with no Bloch points. These transverse walls can occur with two polarities. From a data storage perspective, transverse walls are attractive because they occur in particularly thin and narrow wires promising high storage densities, while at the same time, their location and polarity can be readily detected via the stray field they produce at the surface of the wire. Generally, domain wall memories could make possible a three-dimensional magnetic memory and logic elements with much increased information density [46]. It is important to note that the current densities flowing in these experiments are high, of the order of 2.10^{12} A/m^2. However, the temperature rise due to Joule heating is estimated to remain below 250 K, but the transverse magnetic field generated by these currents is large, favoring one type of Néel wall in which $\vec{M}$ in the wall is parallel to that field. The measurement of the magnetic structure can only be done before and after the current pulse as the wall moves in addition to its transformation and as SEMPA provides high spatial resolution of 10 nm but no time resolution. Hence at present, the dynamics of the wall transformation can only be studied by simulations.

Figure 27.8 shows that, according to the simulations, the switching of the Néel wall polarity proceeds by motion of

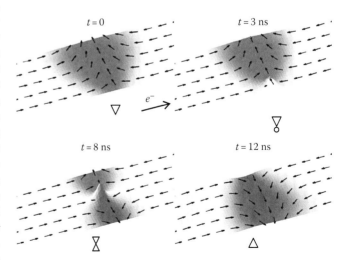

FIGURE 27.8 Transformation of the transverse domain wall structure in a uniform NiFe wire during the superimposed propagation of the wall, with electrons flowing to the right. The times and wall structures are obtained by simulation. In the end, the wall starting as a Néel wall with $\vec{M}$ up has transformed into a Néel wall with $\vec{M}$ down due to the motion of an antivortex along the wall. Triangles indicate the shape of the domain wall. It is evident that the wall width is contracted close to the Bloch point. (Adapted from Vanhaverbeke, A. et al., *Phys. Rev. Lett.*, 101, 107202, 2008. With permission.)

an antivortex across the wire. The typical time to invert the domain wall polarity is a few nanoseconds. Current pulses of this duration are thus needed to move and possibly invert the wall. With longer pulses, and if the wire is made of pure NiFe, there is a 50% probability that the wall polarity switches. This is not desirable in applications such as, e.g., in the proposed race-track memory [46]. However, if the wire is made of a double layer of 8 nm $Ni_{70}Fe_{30}$ and 2 nm Fe on top, a certain current direction always results in one particular wall polarity. Since Fe and NiFe have different electrical resistivities, the current flow through the wire is inhomogeneous, biasing a certain polarity of the Néel wall. The finding that the sign of the transverse wall polarity shown in Figure 27.8 can be set by choosing the shown direction of the electron flow is not explained correctly by the simulations. Thus, as mentioned before, the simulations are not complete and entirely trustworthy at this point. Yet the authors [45] report that the wall polarity is reliably set in their experiments by the direction of the electron flow. It therefore seems possible to use the asymmetric nature of bilayers in future device architectures.

27.3.3 Reversing the Magnetization by a Vortex

Reversing the magnetization in nanoscale magnetic specimens is one of the basic operations in magnetic recording and spintronics applications. The presently used switching by applying external magnetic fields is unfavorable as the generation of appropriate magnetic field pulses extending over the small space occupied by the specimen without crosstalk to the neighboring elements is cumbersome and wasteful. Therefore, the presently used switching by application of magnetic field pulses is envisioned to be replaced by injection of spin currents. The flow of these currents is limited to the sample and the spin torque is a short-range phenomenon, and hence neighboring elements as close as they might be are not affected. The spin switching of $\vec{M}$ in small magnetic elements has been observed in GMR studies, and spontaneous radio frequency emission has also been detected. Yet the evolution of the nanoscale magnetic structure during the process of switching remains hidden in these experiments. It can only be observed in space and time by using pump/probe x-ray microscopy yielding an actual picture of the magnetic microstructure at each point in time and space, i.e., motion pictures of the switching process can be obtained as we will show. Although the time resolution is still limited to ~100 ps and the spatial resolution to 30 nm mainly by the properties of the presently available synchrotron radiation sources and the quality of the zone plates focusing the x-rays, a striking picture of the actual switching process has emerged that is quite different from what one expects to see based on the uniform precession model of $\vec{M}$ that was thought to be universally valid once the sample is small enough to be stable in a uniform distribution of $\vec{M}$ [48]. Although initial and final states of the specimen exhibit uniform magnetization, the switching does not occur by uniform precession of $\vec{M}$ at all.

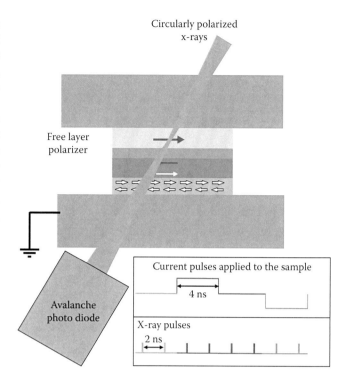

FIGURE 27.9 Schematic of a magnetic read head as used in high-density magnetic recording and principle of the time- and space-resolved imaging of the magnetic switching process in the free layer. The circularly polarized x-rays are focused in a 30 nm spot onto the free layer, and their absorption is observed in a fast photodiode. The x-ray pulses from the synchrotron are synchronized at variable time delays with a pulse generator delivering positive and negative current pulses of 4 ns duration to the pillar structure. (Adapted from Acremann, Y. et al., *Phys. Rev. Lett.*, 96, 217202, 2006.)

The sample shown in Figure 27.9 is a common spin valve structure used for detecting the magnetization of the magnetic bits in a high-density magnetic recording medium. The top and bottom of the spin valve are Cu-electrodes delivering the current pulses to the actual pillar structure. The pillar consists of the bottom of an antiferromagnet that keeps $\vec{M}$ in the adjacent ferromagnetic film aligned through exchange biasing [2]. A second ferromagnetic layer is coupled antiferromagnetically to this first layer by a Ru-film of 0.8 nm thickness. This second layer is actually the polarizer of the electron current, i.e., it produces a spin polarization of the electrons traversing it. $\vec{M}$ in the polarizer must remain fixed, yet without producing a stray field which could actually produce in turn a preferred direction of the free layer. This is achieved by the artificial antiferromagnetic structure at the bottom of the pillar which reduces the stray magnetic field of the polarizing layer. Additionally, to cut any exchange coupling of the free layer to the polarizer, a spacer layer of Cu with a thickness of 3.5 nm is inserted between polarizer and free layer. If now the electron current flows from the bottom to the top, the resistivity of the whole pillar will depend on the direction and distribution of $\vec{M}$ in the free layer. As $\vec{M}$ in the free layer responds to external magnetic fields, it is used as a sensor to read the stray field from the magnetic bits in the hard disk, establishing a readout device for the information stored in the disk. One wants a large reading current to detect GMR, but if this reading current gets too large, it will affect or even switch $\vec{M}$ in the free layer. If $\vec{M}$ has switched to the other direction determined by the shape anisotropy of the sample, it can be reset by letting the electron current flow from the top to the bottom of the pillar. In this case, the switch is done by the electrons reflected from the polarizer reentering the sensor layer. The switching threshold in the sensor layer with a coercivity of 18 mT is given by $I_{AP \to P} = 4.8$ mA but the backswitching requires a higher current $I_{P \to AP} = -8.6$ mA. P and AP refer to parallel or antiparallel magnetization in the polarizer and free layer.

The sensor layer is thus buried under a host of auxiliary structure, and there is no way to test it with anything else but x-rays. Images of $\vec{M}(x, y, t)$ where x and y are the coordinates in the plane of the sensor layer, and t is the time, are obtained by scanning transmission x-ray microscopy. The circularly polarized x-ray beam from an undulator beam line at the Advanced Light Source in Berkeley, CA, is incident at 30° from the surface normal. It is focused to ≈ 30 nm by a zone plate onto the sensor layer. Tuning the x-ray energy to the characteristic resonance in our case of the Co edge in the present $Co_{86}Fe_{14}$ soft magnetic sensor layer provides magnetic contrast through XMCD. The M_x and M_y components of $\vec{M}$ are obtained by rotating the pillar by 90° in the x–y-plane. The time-resolved pump-probe experiments were done using pulses of 4 ns width from a pulse generator. The

leading edge of the pulses was 100 ps wide, similar to the duration of the x-ray pulse of 70 ps. Hence, the temporal resolution obtained here is ≈ 100 ps.

Results of this experiment are shown in Figure 27.10. Several nominal identical samples of this size showed the same switching behavior characterized by the motion of a magnetic vortex across the sample for both directions of the electron flow, while the starting configuration is a uniformly magnetized state.

However, when smaller samples of 85 × 135 nm and 110 × 150 nm were used, full reversal of $\vec{M}$ from a homogeneous magnetic structure and back was also observed, yet there is no sign of a vortex, in the contrary, the magnetic contrast during the process of switching disappears. It is proposed in [49] that this is due to the fact that the vortex core cannot enter the sample if it is too small, it has to stay outside and pass around the sample to the other side completing the switching in this way. As there are two ways to get around the sample to the other side of it, the average switching process reduces the average magnetic contrast inside the sample to zero during the switching. Simulations based on the LLG equation (27.4) confirm this process of switching in detail.

Other switching experiments [50] seem to indicate the possibility of quasi-ballistic magnetization reversal by fine-tuning of the injection current. Quasi-ballistic reversal implies coherent motion of $\vec{M}$ in the hole sample, settling without ringing in its reversed direction. However, in this work the reversal of $\vec{M}$

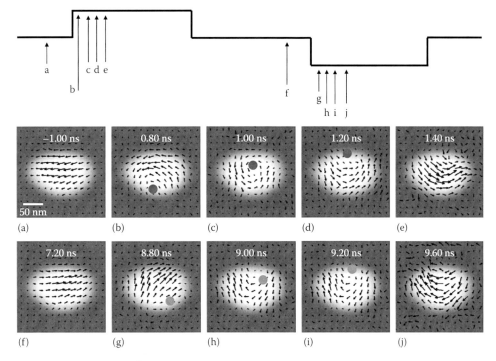

FIGURE 27.10 Evolution of the direction of $\vec{M}$ in the free layer of dimensions 110 × 180 nm during the pulse sequence shown above. The length of the arrows is $\propto$ the magnetic contrast in XMCD. The signal in the nonmagnetic region outside the sample is due to noise and the finite width of the x-ray focus. The uniform structure at the beginning (a) of a cycle transforms first in (b) into a C-state Figure 27.1. Subsequently in (c), a magnetic vortex is inside the sample. In (d), the vortex core marked by a dot moves to the edge, and in (e), the core of the vortex exits the sample leaving behind in (f) a homogeneous magnetization switched to the other direction. The sequence is reversed in (f) to (j), i.e., $\vec{M}$ is switched back by reversing the motion of the vortex core. (Adapted from Strachan, J.P. et al., *Phys. Rev. Lett.*, 100, 247201, 2008.)

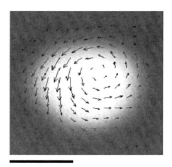

~100 nm   ~500 km

FIGURE 27.11 A magnetic vortex and a hurricane entering the Atlantic. When the hurricane passes over an island, the wind switches direction by 180° just as in magnetic switching where the passage of a vortex over a magnetic island inverts the magnetization.

was not induced by diffusive injection of the polarized electron current but rather by ballistic injection of electrons through an insulating MgO-layer. In this mode of spin injection, the charge current necessary to switch is lower, resulting in a smaller Oersted field while the degree of spin polarization is higher. Apart from the different injection mode which could generate a different switching behavior, the precession of the magnetization was detected by time-resolved resistance measurements which only give the average value of $\vec{M}$ in the sample during the process of switching; therefore, it is not clear at this point whether or not the switching is truly by homogeneous precession of $\vec{M}$ from a homogeneous initial state.

27.4 Conclusion

The first concept of the vortex has been created in 1965 by Feldtkeller and Thomas [51]. Meanwhile, the magnetic vortex and its sister structure the antivortex have assumed an ever more important role in understanding magnetization dynamics at the nanoscale. Generally, vortices are a fascinating physical phenomenon created by conservation laws [52]. Figure 27.11 compares a hurricane and a magnetic vortex. When the hurricane passes over an island, the wind direction switches by 180° just like the magnetization in a nanoscopic sample when the magnetic vortex passes.

Acknowledgments

We gratefully acknowledge support by the U.S. Department of Energy, Office of Basic Energy Sciences. We also would like to thank R. Allenspach, X. W. Yu, and J. Stöhr for helpful comments.

References

1. R.P. Cowburn, *J. Phys. D: Appl. Phys.* **33**, 1 (2000).
2. J. Stöhr and H.C. Siegmann, *Magnetism: From Fundamentals to Nanoscale Dynamics*, Springer Series in Solid-State Sciences, Springer, Berlin, Germany, 2006.
3. S.D. Bader, *Rev. Mod. Phys.* **78**, 1 (2006).
4. A. Wachowiak, J. Wiebe, M. Bode, O. Pietzsch, M. Morgenstern, and R. Wiesendanger, *Science* **298**, 577 (2002).
5. J. Li and C. Rau, *Phys. Rev. Lett.* **97**, 107201, (2006).
6. M. Bode, O. Pietzsch, A. Kubetzka, W. Wulfhegel, D. McGrouther, S. McVitie, and I.M. Chapman, *Phys. Rev. Lett.* **100**, 029703 (2008).
7. A. Hubert and R. Schäfer, *Magnetic Domains*, Springer, Heidelberg, Germany, 1998.
8. B.E. Argyle, E. Terrenzio, and J.C. Slonczewski, *Phys. Rev. Lett.* **53**, 190 (1984).
9. J. Miltat, G. Albuquerque, and A. Thiaville, in *Spin Dynamics in Confined Magnetic Structures*, eds. B. Hillebrands and K. Ounadjela, Springer, Berlin, Heidelberg, Germany; New York, 2002, p. 1.
10. B. Van Waeyenberge, A. Puzic, H. Stoll, K.W. Chou, T. Tyliszczak, R. Hertel, M. Fähnle et al., *Nature* **444**, 461 (2006).
11. K.S. Buchanan, P.E. Roy, M. Grimsditch, F.Y. Fradin, K. Yu. Guslienko, S.D. Bader, and V. Novosad, *Nat. Phys.* **1**, 172 (2005).
12. R. Allenspach, *IBM J. Res. Develop.* **44**, 553 (2000).
13. M. Kläui, P.-O. Jubert, R. Allenspach, A. Bischof, J.A.C. Bland, G. Faini, U. Rüdiger, C.A.F. Vaz, L. Vila, and C. Vouille, *Phys. Rev. Lett.* **95**, 026601 (2005).
14. A. Yamasaki, W. Wulfhekel, R. Hertel, S. Suga, and J. Kirschner, *Phys. Rev. Lett.* **91**, 127201 (2003).
15. H.F. Ding, A.K. Schmid, D. Li, K.Y. Guslienko, and S.D. Bader, *Phys. Rev. Lett.* **94**, 157202 (2005).
16. O.A. Tretiakov and O. Tchernyshov, *Phys. Rev. B* **75**, 012408 (2007).
17. R. Hertel and C.M. Schneider, *Phys. Rev. Lett.* **97**, 177202 (2006).
18. S. Gliga, M. Yan, R. Hertel, and C.M. Schneider, *Phys. Rev. B* **77**, 060404 (2008).
19. Y. Acremann, C.H. Back, M. Buess, O. Portmann, A. Vaterlaus, D. Pescia, and H. Melchior, *Science* **290**, 492 (2000).
20. W.K. Hiebert, A. Stankiewicz, and M.R. Freeman, *Phys. Rev. Lett.* **79**, 1134 (1997).
21. J.P. Park, P. Eames, D.M. Engebretson, J. Berezovsky, and P.A. Crowell, *Phys. Rev. B* **67**, 020403 (2003).
22. S.B. Choe, Y. Acremann, A. Scholl, A. Bauer, A. Doran, J. Stöhr, and H.A. Padmore, *Science* **304**, 420 (2004).
23. J. Raabe, C. Quitmann, C.H. Back, F. Nolting, S. Johnson, and C. Buehler, *Phys. Rev. Lett.* **94**, 217204 (2005).
24. H. Stoll, A. Puzic, B. van Waeyenberge, P. Fischer, J. Raabe, M. Buess, T. Haug et al., *Appl. Phys. Lett.* **84**, 3328 (2004).
25. V. Novosad, F.Y. Fradin, P.E. Roy, K.S. Buchanan, K.Y. Guslienko, and S.D. Bader, *Phys. Rev. B* **72**, 024455 (2005).
26. T. Kimura, Y. Otani, and J. Hamrle, *Phys. Rev. Lett.* **96**, 037201 (2006).
27. D. Oberli, R. Burgermeister, S. Riesen, W. Weber, and H.C. Siegmann, *Phys. Rev. Lett.* **81**, 4228 (1998).

28. W. Weber, S. Riesen, and H.C. Siegmann, *Science* **291**, 1015 (2001).

29. F. Meier and B.P. Zakharchenya, *Modern Problems in Condensed Matter Science, Vol. 8, Optical Orientation*, North Holland, Amsterdam, the Netherlands, 1984.

30. V.N. Petroff, V.V. Grebenshikov, A.N. Androniov, P.G. Gabdullin, and A.V. Maslevtcov, *Rev. Sci. Instrum.* **78**, 025102 (2007).

31. L. Joly, J.K. Ha, M. Alouani, J. Kortus, and W. Weber, *Phys. Rev. Lett.* **96**, 137206 (2006).

32. S. van Dijken, X. Jiang, and S.S.P. Parkin, *Appl. Phys. Lett.* **82**, 775 (2003).

33. Yu.A. Mamaev, L.G. Gerchikov, Yu.P. Yashin, D.A. Vasiliev, V.V. Kuzmichev, V.M. Ustinov, A.E. Zhukov, V.S. Mikhrin, and A.P. Vasiliev, *Appl. Phys. Lett.* **93**, 081114 (2008).

34. J.C. Slonczewski, *J. Magn. Magn. Mater.* **159**, L1 (1996).

35. L. Berger, *Phys. Rev. B* **54**, 9353 (1996).

36. I.N. Krivorotov, N.C. Emley, J.C. Sankey, S.I. Kiselev, D.C. Ralph, and R.A. Buhrman, *Science* **307**, 228 (2005).

37. A.A. Tulapurkar, Y. Suzuki, A. Fukushima, H. Kubota, H. Maehara, K. Tsunekawa, D.D. Djayaprawira, N. Watanabe, and S. Yuasa, *Nature* **438**, 339 (2005).

38. A. Thiaville, J. Garcia, R. Dittrich, J. Miltat, and J. Schrefl, *Phys. Rev. B* **67**, 094410 (2003).

39. K.W. Chou, Vortex dynamics studied by time-resolved x-ray microscopy, PhD dissertation, Universität Stuttgart, Stuttgart, Germany, 2007.

40. V.S. Pribiak, I.N. Krivorotov, G.D. Fuchs, P.M. Braganca, O. Ozatay, J.C. Sankey, D.C. Ralph, and R.A. Buhrman, *Nat. Phys.* **3**, 498 (2007).

41. Q. Mistral, M. van Kampen, G. Hrkac, J.-V. Kim, T. Devolder, P. Crozat, C. Chappert, L. Lagae, and T. Schrefl, *Phys. Rev. Lett.* **100**, 257201 (2008).

42. M. Curcic, B.V. Waeyenberge, A. Vansteenkiste, M. Weigand, V. Sackmann, H. Stoll, M. Fähnle et al., *Phys. Rev. Lett.* **101**, 197204 (2008).

43. M. Bolte, G. Meier, B. Krüger, A. Drews, R. Eiselt, L. Bocklage, S. Bohlens et al., *Phys. Rev. Lett.* **100**, 176601 (2008).

44. L. Heyne, M. Kläui, D. Backes, T.A. Moore, S. Krzyk, U. Rüdiger, L.J. Heyderman et al., *Phys. Rev. Lett.* **100**, 066603 (2008).

45. A. Vanhaverbeke, A. Bischof, and R. Allenspach, *Phys. Rev. Lett.* **101**, 107202, (2008).

46. S.S.P. Parkin, M. Hayashi, and L. Thomas, *Science* **320**, 190 (2008).

47. Y. Acremann, J.P. Strachan, V. Chembrolu, S.D. Andrews, T. Tyliszczak, J.A. Katine, M.J. Carey, B.M. Clemens, H.C. Siegmann, and J. Stöhr, *Phys. Rev. Lett.* **96**, 217202 (2006).

48. W. Wernsdorfer, *Adv. Chem. Phys.* **118**, 99 (2001).

49. J.P. Strachan, V. Chembrolu, Y. Acremann, X.W. Yu, A.A. Tualapurkar, T. Tyliszczak, J.A. Katine et al., *Phys. Rev. Lett.* **100**, 247201 (2008).

50. S. Serrano-Guisan, K. Rott, G. Reis, J. Langer, B. Ocker, and W. Schumacher, *Phys. Rev. Lett.* 101, 087201 (2008).

51. E. Feldtkeller and H. Thomas, *Phys. Kondens. Materie* **4**, 8 (1965).

52. J. Miltat and A. Thiaville, *Science* **298**, 555 (2002).

28

Spins in Organic Semiconductor Nanostructures

Sandipan Pramanik
University of Alberta

Bhargava Kanchibotla
Virginia Commonwealth University

Supriyo Bandyopadhyay
Virginia Commonwealth University

28.1 Introduction

Spin-polarized carrier transport in organic semiconductors has recently become a topic of immense interest since organics exhibit extremely weak spin–orbit interaction [1,2]. The weak spin–orbit interaction is a consequence of the fact that organics are mostly hydrocarbons composed of light elements and the spin–orbit interaction strength increases with the atomic number of the nucleus. Since this interaction is usually the primary cause for spin relaxation in solids, we expect organics to exhibit long spin lifetimes, which they often do [3]. The long spin lifetime and other attractive properties such as compatibility with flexible substrates and the relative ease of synthesis [4,5] have catapulted organics to the forefront of a worldwide research enterprise dedicated to the development of a technology predicated on the utilization of the spin degree of freedom of a charge carrier in an organic molecule to store, process, and communicate information. This exciting new field has motivated fundamental studies of the spin properties of electrons and holes in organic semiconductors, culminating in the birth of the subdiscipline known as "organic spintronics."

Long spin lifetimes [6] have many applications in engineering. They are the linchpin of spin-based digital computing where the spin polarization of a single electron in an organic molecule is utilized to encode binary logic bits 0 and 1. Spin–spin interaction between neighboring electrons in a molecule implements Boolean logic functionality. This is the basis of certain futuristic classical computing paradigms such as single spin logic [7]. A long spin lifetime reduces the bit error probability in such a paradigm and makes reliable, fault-tolerant computing possible. Quantum computing, too, draws on spintronics and the most advanced proposals for solid-state quantum computers advocate the use of an electron's spin for encoding a quantum bit (qubit). Thus, there is a myriad of applications of molecular spins in electronics. Spintronics has also made inroads into the world of optics. The performance of organic light-emitting devices for inexpensive and flexible display technology [8] can be improved by exploiting spin properties [9], and there too, a long spin lifetime is beneficial. Therefore, the reach of organic spintronics is vast and expanding; it has broad applications, covering nearly the entire gamut of modern device engineering. Additionally, organics possess other attractive properties such as mechanical flexibility, chemical tunability of electrical and optical properties, and amenability to low-temperature fabrication processes [10,11]. For decades, all this has spurred significant activity in organic electronics and it has ultimately spawned two immensely successful products, namely, conducting polymers and organic transistors for chemical and bio-sensing. Thus, a fusion of organic electronics with spintronics is expected to be rich in device physics as well as in commercialization prospects.

In this chapter, we discuss some recent results obtained by different groups (including ours) on *spin injection and transport and spin relaxation* in organic semiconductor nanostructures. Note that there exist several other spin-dependent effects in organics such as magnetic field–dependent electroluminescence [9,12], organic magnetoresistance (OMAR) [13], and organic molecular magnetism [14]. These topics are not directly related to spin injection and transport and are therefore not covered in this chapter.

We have organized this chapter as follows. In Section 28.2, we briefly review the main developments of organic electronics—a process which started in the 1970s and has culminated today in the formation of several industries worldwide that commercialize myriad organic electronic gadgets [18]. This is followed by Section 28.3 where we discuss the major developments in spintronics with special emphasis on spin-valve devices and spin relaxation issues. These concepts play a central role in spin transport studies in organics. Next, in Section 28.4, we discuss and analyze few earlier studies on spin transport in organic thin films. These are among the very first attempts to probe spin transport in organic semiconductors. In Section 28.5, we describe our work which, to our knowledge, is the only report (as of now) of spin transport in organic *nanowires*. In this section, we explain the motivation of this study, provide the details of the sample fabrication process, and analyze and interpret the transport measurements used to extract the spin relaxation length and time in organic nanowires. We also establish the dominant spin relaxation mechanism in organics. In Section 28.6, we introduce an intriguing topic dealing with spin dephasing in organics and make the provocative claim that organics are probably the ideal platform for spin-based scalable solid-state renditions of quantum computers. This is because spin dephasing times in organics are long enough to enable fault-tolerant quantum computing *at room temperature*, but more importantly, organic molecules offer excellent gate fidelity [15] and there exists an elegant scheme for the readout of spin-based qubits in organics, which, to our knowledge, has no analog in inorganics. We also discuss an experiment carried out in our group that hints at a possible phonon bottleneck effect in isolated few-molecule ensembles of an organic semiconductor confined in a 1–2 nm sized nanovoid. If confirmed, this would be the first observation of the phonon bottleneck effect in organics. It is fundamentally different from the analogous effect in inorganic nanostructures [17] where carriers and phonons are both quantum confined. In organic molecules, the carrier wavefunctions are typically localized over individual atoms or molecules that are much smaller than the confining space, and hence the carriers are almost never "quantum confined" by the nanostructure but the phonons may be. Hence, this effect is fundamentally different from the traditional phonon bottleneck effect that received wide attention in the context of inorganic semiconductor quantum dot structures. Finally, in Section 28.8 we summarize our findings and outline our implications.

28.2 Advent of Organics

Organic materials are primarily made of carbon, hydrogen, and oxygen. For a long time, until the 1970s, the application of organic materials in the electronics industry was limited to either sacrificial photoresists in the chip manufacturing process or passive insulators of electrical conductors. This situation began to change in the early 1970s when Shirakawa and coworkers successfully synthesized high-quality polycrystalline films of polyacetylene $(CH)_x$ and developed thermal isomerization techniques for controlling the *cis/trans* content [19,20]. These materials, as shown in Figure 28.1, have a π-conjugated structure; i.e., they consist of alternating single and double bonds and the π orbitals overlap to form a *delocalized* π electron cloud along the polymer chain. As a result, these materials are able to transport charge carriers to some extent and exhibit semiconducting property. During the late 1970s, it was discovered that these organic materials can be controllably doped in such a way that their conductivity increases by several orders of magnitude and hence these doped organics essentially behave like metals [21–23]. This new generation of semiconducting and metallic organic materials laid the foundation of a new field, which is known today as *organic electronics*.

Conjugated polymers, both semiconducting and metallic, find numerous applications in the microelectronics industry as (a) discharge layer and conducting resists in electron beam lithography, (b) printed circuit board technology, (c) electrostatic

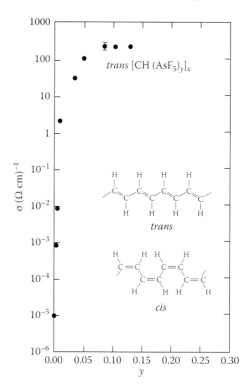

FIGURE 28.1 Electrical conductivity of *trans*-$(CH)_x$ as a function of (AsF_5) dopant concentration. The *trans* and *cis* polymer structures are shown in the inset. (Reproduced from Chiang, C.K. et al., *Phys. Rev. Lett.*, 39, 1098, 1977. With permission.)

discharge protection for packages and housing of electronic equipment, (d) corrosion protection of metals, and (e) shielding against electromagnetic interference. These applications have been reviewed in Ref. [24]. Industrial interest was further triggered when it was recognized that many of these materials are optically active [25,26] in the visible range. Organic light-emitting devices (OLED) have already made their debut in the commercial market and now the race is on to develop 32 in. TV screens from OLEDs. Other extremely active areas of technological enterprise are organic thin-film transistors and photovoltaic cells for low-cost solar energy conversion.

Organic materials offer several distinct and inherent advantages that have helped them to carve a niche in the silicon-dominated electronics market. For example, organics are (a) mechanically flexible albeit tough, allowing realization of new applications such as "roll-up" type displays and electronic newspapers, and (b) inexpensive to process—they can be deposited by low-temperature techniques such as vacuum evaporation, solution casting, ink jet printing, and stamping. Such depositions can be made on low-cost substrates such as paper, plastic, glass, or metal foils. This leads to large area coverage at a low cost.

28.3 Spin Transport Concepts

28.3.1 Introduction: The Rise of Organic Spintronics

Organic spintronics is a subfield within organic electronics. Spintronics (acronym for "spin-based electronics") is defined as the science and technology of manipulating the spin degree of freedom of a single charge carrier (electron or hole) or an ensemble of such carriers in solids to encode, store, process, and deliver information [27–29]. In conventional electronic devices (e.g., diodes or transistors), the carrier spins point along random directions in space and play no role in the performance of these devices. Spintronic devices, on the other hand, rely on the controlled transport of spin-polarized carriers. An example of a spintronic device is the read head sensor [30] in every state-of-the-art computer hard drive. The main advantage of this sensor (over its previous generation counterpart) is its higher sensitivity to the magnetic fields originating from the recorded bits on a magnetic hard disc drive (HDD). This increased sensitivity allows efficient signal detection (read operation) from densely packed bits and thus leads to enormous storage capacity exceeding $10\,GB/in.^2$ [31]. The operating principle of the read head sensor is the giant magnetoresistance effect (GMR) [32,33] observed in heterogeneous metallic systems in which magnetic (e.g., Fe, Co) and nonmagnetic layers (e.g., Cu, Ag) are alternately stacked to form a superlattice. The resistance of this multilayered structure depends on the relative magnetization orientations of the ferromagnetic layers. In particular, when the magnetizations of the ferromagnets are parallel (antiparallel), device resistance is low (high). A magnetic field switches the magnetizations of the ferromagnetic layers from "parallel" to "antiparallel" (or vice versa) and thus switches the device resistance from low to high

(or vice versa). As a result, this device acts a sensitive detector of tiny magnetic fields, and is therefore able to read the binary bits 0 and 1 stored in magnetic storage media. The change in resistance of a GMR device is approximately 100% at low temperature at a magnetic flux density of a few mT, and hence this effect is dubbed as "giant magnetoresistance" (GMR) effect as opposed to the anomalous magnetoresistance effect where the change in resistance is a meager 1%–2%. The origin of the GMR effect lies in spin-polarized electron injection from ferromagnetic layers into the paramagnet and spin-dependent scattering at the ferromagnet/paramagnet interface. A detailed review on this topic is available in Ref. [34].

Another spintronic device that has attracted considerable industrial interest is the so-called magnetic tunnel junction (MTJ), which serves as the primitive unit of the memory cell in magnetic random access memories (MRAMs). An MTJ is basically a "spin-valve" device in which two ferromagnetic electrodes are separated by a tunnel barrier. The operating principle of spin valves will be discussed later. The advantages of MRAM are as follows: It is nonvolatile and less power hungry, provides fast access time, allows dense architecture, and is durable and inexpensive [35,36]. This new class of spintronic memory therefore has the potential to replace its traditional cousins such as static RAM (SRAM), dynamic RAM (DRAM), flash memory, hard disc, etc., and emerge as a *universal memory* [37]. The commercial possibilities of MRAM have triggered the interest of leading electronics companies like Freescale (United States); Hitachi, NEC, Toshiba (Japan); and ITRI (Taiwan). As of July 2006, Freescale has released the first commercial MRAM module (MR2A16A), with 4 GB of memory, for $25 per chip.

Motivated by these initial commercial successes of metal- or metal/insulator-based spintronics [38], significant effort has been invested in the search for similar revolutionary effects in semiconductor-based systems. Early efforts in this area were focused on developing spin-based analogues of classical signal processing devices (e.g., field-effect [39,40] or bipolar junction [41–44] transistors). The motivation behind this was a tacit belief that spintronic transistors would consume less power and operate faster than their electronic brethren. A closer reexamination of these concepts reveals that as far as the signal processing functions are concerned, the spintronic transistors may not offer significant advantage over their charge-based counterparts, if any at all [45]. However, these spintronic devices can play a role in memory applications, which do not require high gain or high-frequency performance or isolation between the input and output terminals. Other unconventional areas where semiconductor spintronics may find niche applications include (a) single spin logic [7], (b) spin neurons [46], and (c) quantum computing using spin in a quantum dot to encode qubits [47,48]. These topics have been discussed in detail in a recent textbook on spintronics [49].

The commercially available spintronic devices mentioned above (i.e., GMR read head and MTJ) rely on spin injection and transport in nonmagnetic materials. This is what we discuss next. These concepts will be used later to understand spin transport in our material of interest i.e., organic semiconductors.

28.3.2 Spin Injection, Transport, and Relaxation

Spin injection and transport in solids (especially semiconductors) is a subject of much interest from the perspective of both fundamental physics and device applications. The basic principle here is as follows. We create (or inject) a population of *spin-polarized* charge carriers (electrons or holes) at one end of a paramagnetic semiconductor (say $r = r_0$) at time $t = t_0$. By "spin-polarized injection," we mean that at $r = r_0$ (and $t = t_0$), the spin polarizations of the carriers are either pointing parallel or antiparallel to a particular direction (say $\hat{\theta}_0$) in space such that

$$P_0 = \frac{n_\uparrow - n_\downarrow}{n_\uparrow + n_\downarrow} = \frac{n_\uparrow - n_\downarrow}{N} > 0 \qquad (28.1)$$

Here
  N is the total number of injected carriers
  $n_{\uparrow(\downarrow)}$ is the number of carriers with spins parallel (antiparallel) to $\hat{\theta}_0$

Thus, the quantity P_0 denotes the net spin polarization of the injected carriers (pointing along $\hat{\theta}_0$). In an all-electrical spin transport experiment, this situation is realized by using either nonideal or perfectly ideal (half-metallic) ferromagnets as spin injectors. In ferromagnets, the carriers at the Fermi level are spin polarized (in ideal half metals, they are 100% spin polarized) so that when they enter the paramagnetic semiconductor from the ferromagnet, they retain some of their initial spin polarization. This is the basis of electrical spin injection. The magnitude of P_0 in the paramagnet is, in general, not equal to the polarization in the ferromagnetic spin injector, since there is always some spin flip (and loss of spin polarization) at the injector/semiconductor interface.

After injection, the partially *spin-polarized* carriers travel through the paramagnetic semiconductor under the influence of a transport-driving electric field. During their transit, different spins interact with their environments differently and their original orientations along $\pm\hat{\theta}_0$ get changed by various amounts. Thus the ensemble spin polarization (along $\hat{\theta}_0$) decreases with time as well as distance (measured from r_0). This gradual loss of the original spin polarization P_0 is termed as *spin relaxation*. Spin relaxation length (time) is defined as the distance (duration) over which the spin polarization reduces to $1/e$ times its initial value. In order to exploit the nonzero spin polarization in spintronic applications, one always attempts to suppress spin relaxation, i.e., enhance spin relaxation length and time in the paramagnetic semiconductor.

28.3.3 Spin-Valve Devices

The spin relaxation length and time in a paramagnetic material can be experimentally determined by performing a spin-valve experiment. A spin valve is a trilayered structure, in which a paramagnetic *spacer* material is sandwiched between two ferromagnetic electrodes of different coercivities [50]. Unlike giant

magnetoresistive devices [32], these ferromagnets are generally well separated in space (i.e., the spacer layer is quite thick) and therefore the two ferromagnets are not magnetically coupled with each other. As a result, their magnetizations can be independently controlled. One of these ferromagnets acts as spin injector, i.e., under an applied electrical bias, it injects spins into the paramagnet. The function of the other ferromagnet is to act like a *spin detector* in the following way. The transmission probability (T) of a single electron through the detector is proportional to $\cos^2(\theta/2)$ where θ is the angle between the electron's spin orientation and magnetization of the detector ferromagnet. This means that when the magnetizations of the ferromagnets are parallel and there is no spin relaxation in the spacer material, the transmission coefficient is unity, which will result in a small device resistance. Similarly, when the magnetizations are antiparallel, one should observe infinite device resistance. In an actual device, there is always some spread in the value of θ since different electrons undergo different degrees of spin relaxation and this makes θ different for different electrons. As a result, ensemble averaging over all the electrons makes the resistance finite even when the magnetizations of the two ferromagnetic contacts are antiparallel. Nonetheless, the resistance is measurably larger when the contacts are in antiparallel orientation than when they are in parallel orientation.*

In a spin-valve experiment, resistance of the device is measured as a function of the applied magnetic field (H). The resistance versus magnetic field characteristic (magnetoresistance) exhibits distinct features that allow one to extract the spin diffusion length and time in the paramagnetic spacer. For the following discussion, we assume that the spin polarizations in the two ferromagnetic contacts of a spin-valve device have the same sign, meaning that majority spins in one are also majority spins in the other. We also assume that the coercivities of the ferromagnets are given by H_1 (>0) and H_2 (>0) with $H_1 < H_2$. The basic principle is that when the magnetizations of the two ferromagnets are parallel, the majority spins injected by one ferromagnet are transmitted by the other and the device resistance is low. When they are antiparallel, the majority spins injected by one are blocked by the other and the device resistance is high.

At first, the device is subjected to a strong magnetic field ($H = H_{scan} > H_2$), which magnetizes both the ferromagnets along the direction of the field. Next, the field is decreased, swept through zero, and reversed. At this stage, when the magnetic field strength just exceeds H_1 (i.e., $-H_2 < H < -H_1$), the ferromagnet with the lower coercivity (i.e., H_1) flips magnetization. Now, the two ferromagnetic contacts have their magnetizations

---

* This discussion assumes that the spin polarizations in the two ferromagnets have the same sign (e.g., in the case of cobalt and nickel). This means that majority spins in both ferromagnets point in the *same* direction when they are magnetized parallel. If the spin polarizations have opposite signs (e.g., in the case of iron and cobalt), then the majority spins in one ferromagnet and those in the other point in opposite directions when they are magnetized parallel. In that case, the spin-valve's resistance will be lower when the ferromagnetic contacts are antiparallel and higher when they are parallel.

antiparallel to each other. Hence at $H = -H_1$ a jump (increase) in the device resistance is observed. As the magnetic field is made stronger in the same (i.e., reverse) direction, the coercive field of the second ferromagnet will be reached ($H = -H_2$). At this point, the second ferromagnet also flips its magnetization direction, which once again places the two ferromagnets in a configuration where their magnetizations are parallel. Thus, the resistance drops again at $H = -H_2$. Therefore, during a single scan of magnetic field (say from H_{scan} to $-H_{scan}$), a spin-valve device shows a resistance peak between the coercive fields of the two ferromagnets (i.e., between $-H_1$ and $-H_2$). If the magnetic field is varied from $-H_{scan}$ to H_{scan}, an identical peak is observed between H_1 and H_2. These are the "spin-valve peaks." The spin-valve response is pictorially explained in Figure 28.2.

As mentioned before, the spin-valve device is the basic component of MRAM. There, it is fashioned out of a tunnel barrier sandwiched between two ferromagnets. When the magnetizations of the contacts are parallel, the device resistance is low and stores, say, the binary bit 0. When the magnetizations are antiparallel, the device resistance is high and stores the binary bit 1. One of the ferromagnets is a hard magnetic layer with high coercivity, while the other is a soft magnetic layer with low coercivity. Writing of bits is achieved by switching the magnetization of the soft layer with a small magnetic field, which does not affect

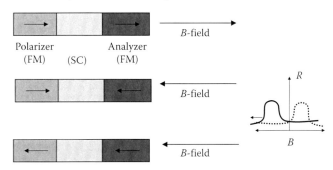

FIGURE 28.2 Explanation of the occurrence of the spin-valve peak in the magnetoresistance of a tri-layered spin-valve structure. (Top) At a high magnetic field, pointing to the right, both ferromagnets (designated spin polarizer and spin analyzer in analogy with optics) are magnetized in the same direction so that injected spins transmit easily (assuming little spin relaxation in the spacer semiconductor layer). In this case, the device resistance is low. (Middle) When the field is reversed and its strength exceeds the coercive field of one of the ferromagnets (the one with lower coercivity), this ferromagnet's magnetization flips, placing the contacts in the antiparallel configuration. Injected spins now do not transmit easily and the device resistance increases. (Bottom) As the magnetic field strength increases further in the reverse direction, it ultimately exceeds the coercive field of the second ferromagnet as well and its magnetization flips, placing the magnetizations of the two ferromagnets once again in the parallel configuration. The device resistance now drops, giving rise to a resistance peak between the two coercive fields. This is the "spin-valve peak" and its height is the strength of the spin-valve signal. The resulting magnetoresistance traces for both forward and reverse scans of the magnetic field are shown on the right. (Reproduced from Pramanik, S. et al., Electrochemical self-assembly of nanostructures: Fabrication and device applications, in Nalwa, H.S. (ed.), *Encyclopedia of Nanoscience and Nanotechnology*, 2nd edn., to appear. With permission.)

the hard layer. For "electrical writing" as opposed to "magnetic writing," one can switch the magnetization of the soft layer electrically by using the so-called spin-torque effect [51]. Reading of the stored bit is accomplished by simply measuring the device resistance (high resistance = 1; low resistance = 0).

28.3.3.1 Determining the Spin Relaxation Time and Length Using a Spin-Valve Device

The spin-valve device is exceedingly versatile. Not only can it serve as a memory element, but it is a "meter" used to measure the spin relaxation time and length in the spacer material. We now describe how this is accomplished.

The spin-valve signal is proportional to the height of the spin-valve peak and is generally defined as the ratio

$$\frac{\Delta R}{R} = \frac{R_{AP} - R_P}{R_P} \tag{28.2}$$

where R_{AP} and R_P denote the device resistances when the magnetizations of the two ferromagnetic contacts are antiparallel and parallel, respectively. The height of the spin-valve peak is ΔR. If the spacer is a semiconductor material, there exists a Schottky barrier at the ferromagnet/semiconductor interface. Under the influence of an applied bias, carriers are injected from the ferromagnet into the semiconductor via tunneling through this barrier with a surviving spin polarization P_1. As long as the barrier is thin enough, we can ignore any loss of spin polarization in traversing the barrier and assume that P_1 is approximately the spin polarization of carriers at the Fermi energy in the injecting ferromagnet. After injection, carriers drift and diffuse through the spacer with exponentially decaying spin polarization given by $P_1 \exp[-x/L_s]$ where x is the distance traveled and L_s is the spin diffusion length (or spin relaxation length) in the spacer. The exponential decay follows from the drift-diffusion model of spin transport [52]. Finally, the carriers tunnel through the Schottky barrier at the interface of the spacer and the second ferromagnet to reach the detecting contact. The conduction energy band diagram of the spin valve, along with the nature of transport in three different regions of the spacer layer, is shown in Figure 28.3.

If we apply the Jullière formula [53] at the "detecting" interface, we get

$$\frac{\Delta R}{R} = \frac{2P_1 P_2 e^{-d/L_s}}{1 - P_1 P_2 e^{-d/L_s}} \tag{28.3}$$

where

P_2 is the spin polarization of the carriers at the Fermi energy of the second (i.e., detecting) ferromagnet

d is the length of the spacer layer

If we know P_1, P_2, and d and experimentally measure $\Delta R/R$, then we can determine L_s from Equation 28.3.

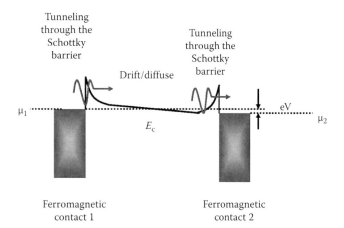

FIGURE 28.3 The transport picture that allows application of the Julliere formula to determine spin relaxation length and time. An injected carrier from the left ferromagnetic contact tunnels through the first Schottky barrier with spin polarization essentially intact, then drifts and diffuses through the paramagnet losing spin polarization and finally tunnels into the second ferromagnet through yet another Schottky barrier. We show the conduction band profile in the semiconductor under a small bias voltage V. The chemical potentials in the two (metallic) ferromagnets are μ_1 and μ_2.

One limitation of Jullière formalism is that this model ignores any possible loss of spin polarization at the interfaces between the spacer and either ferromagnetic contact. As a result, the spin relaxation length determined from Equation 28.3 is always an *underestimate*. To understand this, we replace P_1 and P_2 in Equation 28.3 by P_1/η_1 and P_2/η_2, respectively. Here η_1, $\eta_2 \geq 1$ account for the loss of spin polarizations at the interfaces. The equality holds when there is no loss at the interfaces. We can recast Equation 28.3 in the following form:

$$\frac{e^{-d/L_s}}{\eta} = C \qquad (28.4)$$

where
  $\eta = \eta_1\eta_2$
  $C = \Delta R/[P(2R + \Delta R)]$ with $P = P_1 P_2$

Clearly if we ignore spin flips at the interfaces (i.e., assume $\eta = 1$), we will obtain an underestimated value of L_s.

Spin relaxation length (L_s) is related to the spin relaxation time (τ_s) as follows:

$$L_s = \sqrt{D_s \tau_s} \qquad (28.5)$$

where D_s is the spin diffusion coefficient that is not necessarily equal to the charge diffusion coefficient [54]. Knowledge of D_s in the spacer allows us to find τ_s.

Spin-valve structures have been extensively employed for probing various features of spin-polarized carrier transport in different types of paramagnetic materials including metals [55],

tunnel barriers [56], inorganic semiconductors [57], and carbon nanotubes [58]. It is to be noted that in the case of all-metal spin valves (where all three layers are metallic), there is no Schottky barrier at the interfaces and the Jullière model is not applicable. In that case, one uses a different approach to measure the spin relaxation length and time [59], which is not discussed here since we are discussing organic *semiconductors* that inevitably form a Schottky barrier at a metal interface. In spin-valve experiments, it is customary to ignore any loss of spin polarization while traversing the Schottky barrier at the interface since $d_0/L_s \approx 0$, where d_0 is the thickness of the Schottky barrier. Since this thickness has to be small enough to allow tunneling through it, it is invariably much smaller than L_s.

28.3.3.2 Inverse Spin-Valve Effect

In some particular cases, it is possible to observe the so-called inverse spin-valve effect where the device resistance is lower when the magnetizations of the ferromagnetic contacts are antiparallel [60–63]. In this case, the spin-valve signal ΔR is negative, so that one gets a spin-valve "trough" instead of a spin-valve "peak." This can happen because of various reasons. First, if the spin polarizations of the two ferromagnets (P_1 and P_2) have opposite signs, then obviously the sign of R will be inverted (Equation 28.3). A second explanation has been provided in Refs. [61,62], which predicts that resonant tunneling through an impurity state in the paramagnetic spacer layer inverts the sign of the spin-valve response since it effectively inverts the sign of the spin polarization of the ferromagnetic contact nearer to the impurity. There is also a third explanation. A ferromagnet like cobalt or nickel has both d- and s-electrons at the Fermi level. The d-electrons are more numerous (because of the higher density of states in the d-band at the Fermi level), but also have the heavier effective mass. The majority s-spins and majority d-spins at the Fermi level are mutually antiparallel. In fact, the d-electron spins will tend to point antiparallel to an applied magnetic field (as if their Landé g-factor is negative) while the s-electron spins will tend to point parallel to the applied field (as if their Landé g-factor is positive). Even though the d-electrons are more numerous, the s-electrons are faster (because of their lower effective mass) and therefore may contribute more to the current than the d-electrons. Consider a spin valve with cobalt as the injecting contact and nickel as the detecting contact. Assume, for the sake of argument, that both s- and d-electrons have a very high degree of spin polarization at the Fermi level, so that all injected s-electrons have the same spin polarization and all injected d-electrons also have the same spin polarization, but the s- and d- spin polarizations are mutually antiparallel. If the s-electrons from cobalt contribute more to the current, then when the ferromagnets are in the parallel configuration, the s-electrons impinging on the detector find their spins to be antiparallel to those in the d-electron band (with the higher density of states) of the detector (nickel) and therefore do not transmit. This makes the device resistance high. When the magnetizations of the two contacts are antiparallel, the injected s-electrons find their spins parallel to those in the d-band of nickel and hence transmit well. This makes the device resistance low. Therefore, the spin-valve signal will be negative (i.e., we will observe a spin-valve trough instead of a spin-valve peak). If, instead, the d-electrons contribute more to current than s-electrons, then, of course, the spin-valve signal will be positive. Thus, the sign depends on whether the s- or the d-electrons are majority contributors to the current.

28.3.4 Spin Relaxation Mechanisms

There are various mechanisms that cause spin relaxation in the paramagnetic spacer layer. In case of semiconductors (as well as metals), the most dominant mechanisms [64] are (a) Elliott–Yafet mechanism [65,66], (b) D'yakonov–Perel' mechanism [67,68], (c) Bir–Aronov–Pikus mechanism [69], and (d) hyperfine interaction with nuclei [70]. Among these, the first two mechanisms accrue from spin–orbit interaction. The third one originates from exchange coupling between electron and hole spins, and the last is due to interaction between carrier spins and nuclear spins. These mechanisms are briefly described below.

28.3.4.1 Elliott–Yafet Mechanism

In the presence of spin–orbit coupling, Bloch states of a real crystal are not spin eigenstates. Therefore, these states are not pure spin states with a fixed spin quantization axis, but are either pseudo-spin-up or pseudo-spin-down, in the sense that a particular state with a given spin orientation (say ↑) has a small admixture of the opposite spin state ↓. This is an outcome of the presence of spin–orbit coupling in the crystal which mixes the pure spin-up and pure spin-down states. Therefore, we can write the Bloch states as

$$u_{\vec{k}}(\vec{r}) = a_{\vec{k}}(\vec{r})\big|\!\uparrow\big\rangle + b_{\vec{k}}(\vec{r})\big|\!\downarrow\big\rangle. \tag{28.6}$$

The degree of admixture (i.e., the ratio $a_{\vec{k}}/b_{\vec{k}}$) is a function of the electronic wavevector $\vec{k}$. As a result, when a momentum relaxing scattering event causes a transition between two states with different wavevectors, it will also reorient the spin and cause spin relaxation. Even a spin-independent scatterer, such as a nonmagnetic impurity or an acoustic phonon can cause spin relaxation by this mechanism as long as it also relaxes momentum (i.e., changes the wavevector). This is the Elliott–Yafet mechanism of spin relaxation.

From the above discussion, one naturally expects that spin relaxation rate due to Elliott–Yafet mechanism should be proportional to the momentum scattering rate. This is indeed true, and from [64], we quote a formula relating these two quantities:

$$\frac{1}{\tau_{s,EY}(E_k)} = A\left(\frac{\Delta_{so}}{E_g + \Delta_{so}}\right)^2\left(\frac{E_k}{E_g}\right)^2\frac{1}{\tau_p(E_k)} \tag{28.7}$$

This formula is valid for III-V semiconductors. Here $\tau_{p(s,EY)}(E_k)$ is the momentum relaxation time (spin relaxation time due to

Elliott–Yafet process) for electrons with wavevector k and energy E_k. The bandgap is denoted by E_g and Δ_{so} is the spin–orbit splitting of the valence band. The prefactor A, depends on the nature of the scattering mechanism. The above equation indicates that Elliott–Yafet process is significant for semiconductors with small band gap and large spin–orbit splitting. Typical example of such a semiconductor is indium arsenide (InAs). It is important to note that in case of Elliott–Yafet mechanism, the mere presence of spin–orbit interaction in the system does not cause spin relaxation. Only if the carriers are scattered during transport, spin relaxation takes place. Higher the momentum scattering rate, higher is the spin scattering rate. This observation is valid even in the case of hopping transport in disordered (noncrystalline) solids such as organics where there is no bandstructure and no Bloch states as such, but momentum relaxation still causes concomitant spin relaxation.

28.3.4.2 D'yakonov–Perel' Mechanism

The D'yakonov–Perel' mechanism of spin relaxation is dominant in solids that lack inversion symmetry. Examples of such systems are III-V semiconductors (e.g., GaAs) or II-VI semiconductors (e.g., ZnSe) where inversion symmetry is broken by the presence of two distinct atoms in the Bravais lattice. Such kind of asymmetry is known as *bulk inversion asymmetry*. Inversion symmetry can also be broken by an external or built-in electric field, which makes the conduction band energy profile inversion asymmetric along the direction of the electric field. Such asymmetry is called *structural inversion asymmetry*. Both types of asymmetries result in effective electrostatic potential gradients (or electric fields) that a charge carrier experiences. In the rest frame of a moving carrier, the electric field Lorentz transforms to an effective magnetic field $\mathbf{B}_{eff}$ whose strength depends on the electron's velocity. The interaction of an electron's spin with this effective magnetic field is the basis of spin–orbit interaction. Both bulk inversion asymmetry and structural inversion asymmetry give rise to spin–orbit interaction and associated magnetic fields $\mathbf{B}_{eff}$. The former gives rise to the Dresselhaus spin–orbit interaction [71] and the latter to the Rashba spin–orbit interaction [72]. In a disordered organic semiconductor, the Rashba interaction is overwhelmingly dominant over the Dresselhaus interaction since there is no bulk inversion asymmetry. Microscopic electric fields arising from charged impurities and surface states (e.g., dangling molecular bonds) break structural inversion symmetry locally, causing Rashba spin–orbit interaction.

A carrier's spin in a solid with Rashba and/or Dresselhaus spin–orbit interaction Larmor precesses continuously about $\mathbf{B}_{eff}$. Since the magnitude of $\mathbf{B}_{eff}$ is proportional to the magnitude of carrier velocity $\mathbf{v}$, it is different for different carriers that have different velocities owing to different scattering histories. Thus, collisions randomize $\mathbf{B}_{eff}$ and therefore the orientations of the precessing spins. As a result, the ensemble-averaged spin polarization decays with time leading to continual depolarization. This is the D'yakonov–Perel' mode of spin relaxation.

If a carrier experiences frequent momentum relaxing scattering (i.e., small mobility and small τ_p), then $\mathbf{v}$ is small implying $|\mathbf{B}_{eff}|$ is small and the spin precession frequency (which is proportional to B_{eff}) is also small. As a result, the D'yakonov–Perel' process is less effective in low-mobility samples than in high-mobility samples. It therefore stands to reason that the spin relaxation rate due to D'yakonov–Perel' process will be inversely proportional to the momentum scattering rate. The following formula is valid for bulk nondegenerate semiconductors where the carriers are in quasi-equilibrium [64]:

$$\frac{1}{\tau_{DP}} = Q\alpha^2 \frac{(kT)^3}{\hbar^2 E_g} \tau_p \qquad (28.8)$$

where

    Q is a dimensionless quantity ranging from 0.8–2.7 depending on the dominant momentum relaxation process
    E_g is the bandgap
    τ_p is the momentum relaxation time
    α is a measure of the spin–orbit coupling strength

Note that the Elliott–Yafet and the D'yakonov–Perel' mechanisms can be distinguished from each other by the opposite dependences of their spin relaxation rates on mobility. In the former mechanism, the spin relaxation rate is *inversely* proportional to the mobility and in the latter mechanism, it is *directly* proportional. We will later show how these opposite dependences can be exploited to identify the dominant spin relaxation mechanism in organics.

28.3.4.3 Bir–Aronov–Pikus Mechanism

This mechanism of spin relaxation is dominant in bipolar semiconductors. The exchange interaction between electrons and holes is described by the Hamiltonian $H = A\vec{S} \cdot \vec{J}\delta(\vec{r})$ where A is proportional to the exchange integral between the conduction and valence states, J is the angular momentum operator for holes, and $\vec{S}$ is the electron spin operator. Now, if the hole spin flips (owing to strong spin–orbit interaction in the valence band), then electron–hole coupling will make the electron spin flip as well, resulting in spin relaxation of electrons. A more detailed description is available in Ref. [64]. In case of unipolar transport, i.e., when current is carried by either electrons or holes but not both simultaneously, this mode of spin relaxation is obviously ineffective.

28.3.4.4 Hyperfine Interaction

Hyperfine interaction is the magnetic interaction between the magnetic moments of electrons and nuclei. This is the dominant spin relaxation mechanism for quasi-static carriers, i.e., when carriers are strongly localized in space and have no resultant momentum. In that case, they are virtually immune to Elliott–Yafet or D'yakonov–Perel' relaxation since those two require carrier motion. Therefore, the only remaining channel for spin relaxation is hyperfine interaction. We can view this mechanism as caused by an effective magnetic field (B_N) created by nuclear spins, which interacts with electron spins and causes dephasing.

In most organics, carrier wavefunctions are quasi-localized over individual atoms or molecules, and carrier transport is by hopping from site to site (which causes the mobility to be exceedingly poor). Because the average carrier velocity is so small (mobility so low), the D'yakonov–Perel' mechanism is almost certainly not going to be dominant. That leaves the Elliott–Yafet and hyperfine interactions as the two likely mechanisms for spin relaxation. Which one is the more dominant may depend on the specific organic. In the case of the Alq_3 molecule, we found out (see later sections) that the Elliott–Yafet mode is the dominant spin relaxation channel, at least at moderate-to-high transport-driving electric fields. This is probably because the carrier wavefunctions in this molecule are quasi-localized over carbon atoms [6] whose naturally abundant isotope ^{12}C has no net nuclear spin and hence cannot cause hyperfine interaction. However, in some other molecule where the carrier wavefunctions are spread over atoms with nonzero nuclear spin, it is entirely possible for the hyperfine interaction to outweigh the Elliott–Yafet mechanism.

28.4 Spin Transport in Organic Semiconductors: Organic Spin Valves

Study of spin transport in organic semiconductors is a relatively new area of scientific endeavor for both organic electronics and spintronics research communities. The first study in this direction [73] dates back to 2002. It reported spin injection and transport through an organic semiconductor sexithienyl (T_6). This material is a π-conjugated rigid-rod oligomer which, because of its relatively high mobility (~10^{-2} cm²V⁻¹s⁻¹), is a promising material for organic field-effect transistors [74–76]. Organic light-emitting diodes based on this material are capable of producing polarized electroluminescence [77]. The device reported in [73] had a planar spin-valve geometry in which two planar LSMO (lanthanum strontium manganate: $La_{1-x}Sr_xMnO_3$ with $0.2 < x < 0.5$) electrodes were separated by a T_6 spacer layer (see Figure 28.4). LSMO is a half-metallic ferromagnet and acts as an excellent spin injector/detector due to near 100% spin polarization at low temperatures. The overall device resistance was primarily determined by the T_6 region whose resistance was approximately six orders of magnitude higher than that of LSMO. The interface resistance did not play a significant role in this structure since the device resistance was found to scale linearly with T_6 thickness. The device resistance showed a strong dependence on magnetic field. It is to be noted that since both LSMO contacts were nominally identical, they did not allow independent switching of magnetizations as required in the case of a spin valve. However, one could still change the magnetizations from random (zero field) to parallel (at high field). In the absence of any magnetic field, i.e., when the LSMO electrodes had random magnetization orientations, the device resistance was high. When a sufficiently large (saturation) magnetic field of 3.4 kOe was applied, the LSMO electrodes acquired magnetizations that were parallel to each

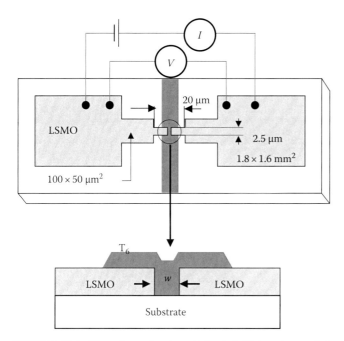

FIGURE 28.4 The schematic view of the hybrid junction and dc 4-probe electrical setup. The cross-sectional view below shows a region near the spin transport channel. (Reproduced from Dediu, V. et al., *Solid State Commun.*, 122, 181, 2002. With permission.)

other and the device resistance dropped. Maximum resistance change of ~30% was observed when the width of the T_6 spacer was ~140 nm. For larger spacer widths, the amount of resistance change decreased and disappeared beyond ~300 nm. This indicated that the spin relaxation length in T_6 was between 140 and 300 nm. Additionally T_6 did not show any intrinsic magnetoresistance effect up to 1 T. Therefore, the observed magnetoresistance in the LSMO–T_6–LSMO device originated from the spin-valve effect discussed above and confirmed spin-polarized carrier injection and transport in T_6. The spin relaxation length (L_s) and spin relaxation time τ_s (or more accurately the spin-flip time or

spin-lattice relaxation time T_1) in T_6 was estimated to be ~200 nm and ~1 μs, respectively, at room temperature. Interestingly, the resistance change was immune to the applied bias at least in the range of 0.2–0.3 MV/cm as reported in [73]. However, in other organics (e.g., Alq$_3$; see later), we found that the spin-valve signal is strongly sensitive to applied bias and falls off rapidly with increasing bias.

The next study [78] explored spin transport in the organic semiconductor Alq$_3$. The spin-valve device in this case had a vertical configuration (Figure 28.5) with cobalt and LSMO acting as spin injector and detector. The spin-valve peaks appeared between the coercive fields of LSMO (30 Oe) and cobalt (150 Oe). Interestingly, in this case, the sign of the spin-valve peak was negative i.e., the device resistance was *low* when magnetizations were antiparallel and *high* when they were parallel. This was to be expected since carriers at the Fermi energy in Co and LSMO have opposite signs of spin polarization. Unlike in the case of T_6, the Alq$_3$ spin-valve signal vanished at room temperature. This was attributed to enhanced spin relaxation rate in Alq$_3$ at elevated temperatures. From the measured data, the spin relaxation length (L_s) in Alq$_3$ at 11 K was estimated as ~45 nm (based on the Julliére formula). The spin-valve signal also decreased with increasing bias, an effect that was not observed in the case of T_6.

One important aspect of fabricating vertical spin-valve structures is that when a ferromagnet like cobalt is deposited on organics, the interface is generally ill-defined. This happens due to the softness of the organic layer that allows significant diffusion of the ferromagnetic material into the organic at high temperatures of deposition. In the device of Ref. [78], when cobalt was deposited on Alq$_3$, the diffusion of cobalt atoms extended up to a depth of 100 nm inside Alq$_3$. So, any device with Alq$_3$ thickness of 100 nm or less should not show any spin-valve response since the two ferromagnets will be electrically shorted via pinhole current paths. This has been confirmed independently in Ref. [79]. At the same time, in order to observe any spin-valve signal,

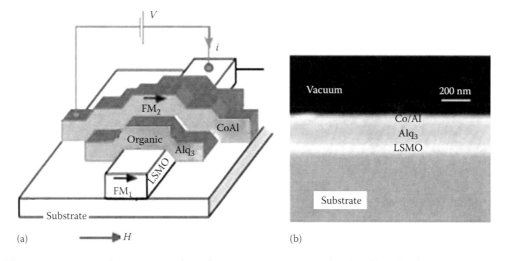

FIGURE 28.5 (a) An organic spin-valve structure where the organic spacer is an Alq$_3$ thin film. (b) Shows a cross-sectional micrograph. (Reproduced from Xiong, Z.H. et al., *Nature*, 427, 821, 2004. With permission.)

the total thickness of the Alq_3 layer must not exceed ~(100 nm + L_s) = 145 nm [78,79]. In Ref. [80], the thickness of the Alq_3 layer in a Co–Alq_3–Fe spin-valve device was always less than 100 nm, which is probably the reason why they failed to observe any measurable spin-valve signal because the Co and Fe contacts could have been electrically shorted. It is to be noted that existence of a few pinhole shorts does not necessarily guarantee ohmic current–voltage characteristics [81], but significant interdiffusion of the contact metal into the organic will surely reduce the Schottky barrier height at the metal–organic interface enough to allow strong tunneling and thermionic emission of electrons from the metal into the organic, thereby rendering the current–voltage characteristic ohmic.

Ref. [82] reported tunneling magnetoresistance in a spin-valve device where the tunnel barrier was a composite of ultrathin layers of Al_2O_3 and Alq_3. The device structure, listed in the order of deposition was as follows: Co/Al_2O_3 (0.6 nm)/Alq_3 (1–4 nm)/$Ni_{80}Fe_{20}$. In this case, the top $Ni_{80}Fe_{20}$ layer did not seem to penetrate at all inside the thin Alq_3 layer. This was in stark contrast to the interdiffusion phenomenon described earlier. This device showed tunneling magnetoresistance (TMR) peaks of height 6% at room temperature. As expected, the height of the TMR peaks decreased gradually with increasing bias. Measurement of the spin polarization of the tunneling current, using Meservey–Tedrow technique [83], showed that the existence of the interfacial alumina barrier improved spin-injection efficiency. Similar studies performed on amorphous organic semiconductor rubrene [86] revealed a spin relaxation length (L_s) of 13.3 nm in this material and provided further evidence in support of the efficacy of the alumina barrier for spin injection.

It is well known that a tunnel barrier increases spin-injection efficiency from a metallic ferromagnet into a semiconducting paramagnet since it ameliorates the notorious "conductivity mismatch problem" [84,85]. However, this is well understood only in the case of diffusive carrier transport. In organics, transport is mostly by hopping from site to site and there is no clear theory yet (to our knowledge) that can explain why a tunnel barrier might increase spin-injection efficiency from a metallic ferromagnet into an organic. This remains an open theoretical problem.

The spin-valve effect has also been demonstrated for regioregular poly(3-hexylthiophene) (RR-P3HT) [87]. There exist reports of another magnetoresistance effect that is observed in organics even when the electrodes are nonmagnetic. This effect is dubbed OMAR. The origin of this phenomenon is still unclear. Although some propositions have been advanced to explain this effect citing hyperfine interactions and other mechanisms, no widespread consensus has emerged.

28.5 Spin Transport in the Alq_3 Nanowires

The spin injection and transport experiments in organics, discussed so far, demonstrated unequivocally that it is possible to inject spin electrically from a ferromagnet into an organic. However, they did not shed any light on the nature of the dominant spin relaxation mechanism (Elliott–Yafet, Dyakonov–Perel, or hyperfine interactions) in organics. Lack of this knowledge motivated us to investigate spin transport in organics with a view to establishing which spin relaxation mechanism is dominant. Consequently, we focused on organic *nanowires* [88,89] instead of standard two-dimensional geometries since comparison between the results obtained in nanowires and thin films can offer some insight into what type of spin relaxation mechanism holds sway in organic semiconductors.

Carriers in nanowires will typically have lower mobility than in thin films because nanowires have a much larger surface-to-volume ratio and hence carriers experience more frequent scattering from charged surface states in nanowires than they do in thin films. These scatterings are *not* surface roughness scatterings but rather Coulomb scattering from the charged surface states [90]. The surface roughness scattering does not increase significantly in nanowires since the mean free path in organics is very small (fractions of a nanometer) and as long as the nanowire diameter is much larger than the mean free path, we do not expect significantly increased surface roughness scattering. However, the Coulomb scattering from surface states increases dramatically. The Coulomb scattering is long range and affects carriers that are many mean free paths from the surface. As a result, nanowires invariably exhibit significantly lower mobilities than thin films.

The mobility difference offers a handle to probe the dominant spin relaxation mechanism in organics. As explained earlier, the Elliot–Yafet spin relaxation rate is directly proportional to the momentum relaxation rate and hence inversely proportional to the mobility while the Dyakonov–Perel rate is directly proportional to mobility. Hence, if we observe an increased spin relaxation rate in nanowires compared to thin films, then we will infer that Elliot–Yafet is dominant over Dyakonov–Perel; otherwise, we will conclude that the opposite is true.

We carried out experiments that clearly showed the spin-valve effect in organic *nanowires*. From that, we were able to determine the spin relaxation length (L_s) and the spin relaxation time (τ_s) in nanowires, contrast them with the quantities measured in thin films, and thus determine the dominant spin relaxation mechanism in organics. We found that the spin relaxation time in nanowires is about an order of magnitude smaller than in thin films, which immediately suggests that the dominant spin relaxation mode is the Elliot–Yafet channel. We also showed that the spin relaxation time in Alq_3 can approach 1 s at a temperature of 100 K. This is the longest spin relaxation time reported in any nanostructure above the liquid nitrogen temperature (77 K).

In a different set of experiments, we observed a surprising correlation between the sign of the spin-valve peak and the background magnetoresistance in organic nanowires. We offered a possible explanation for this intriguing correlation and in the process showed that a magnetic field can increase the spin relaxation rate in organics, which is consistent with the Elliott–Yafet mechanism in two ways: (1) First, a magnetic field bends the electron trajectories bringing them closer to the surface of the nanowire, which decreases mobility and increases the Elliott–Yafet spin relaxation rate. (2) Second, a magnetic field causes "spin mixing," which increases the admixing of spin-up and spin-down states (recall the discussion of Elliott–Yafet mechanism) which exacerbates spin relaxation. The increase of spin relaxation rate in a magnetic field lends further support to our conclusion that the Elliott–Yafet mode is the dominant spin relaxation mechanism in organics.

In the next section, we describe some of these experiments.

28.5.1 Experimental Details

28.5.1.1 Fabrication of Nanowire Organic Spin Valves

We employ an electrochemical self-assembly technique [91] to fabricate nanowire organic spin valves. In this process, the starting material is a high-purity (99.999% pure) aluminum foil of nominal thickness ~0.1 mm. The surface of the foil, when purchased off the shelf, is typically rough (rms surface roughness ~1 μm) and is unsuitable for nanofabrication. Therefore, we electropolish this foil following a well-established procedure [92] that reduces the rms value of the surface roughness to about 3 nm. Next, the aluminum foil is anodized in 0.3 M oxalic acid with an anodization voltage of 40 V dc. For this purpose, the foil is immersed in oxalic acid and connected to the positive terminal of a dc voltage source so that it acts as the anode. A platinum mesh is used as the cathode and a voltage drop of 40 V is maintained between the two electrodes. This process creates a porous alumina (Al_2O_3) film on the surface of the foil, containing a hexagonal array of pores of nominal diameter ~50 nm and areal pore density 2×10^{10} cm^{-2}. A typical atomic force micrograph of the top surface of the alumina film is shown in Figure 28.6. The dynamics of the pore formation process has been reviewed elsewhere [93] and will not be repeated here. The anodization is carried out for 10 min to produce a 1 μm thick alumina film and therefore yields pores that are 1 μm deep. At the bottom of the pores, there is a ~20 nm thick layer of *nonporous* alumina known as the "barrier layer." As shown in Figure 28.7, this layer separates the pore bottom from the aluminum substrate and inhibits vertical conduction of electrical current through any material hosted in the pores. Consequently, it is necessary to remove this barrier layer before depositing materials inside the pores. This can be accomplished either by a "reverse-polarity etching" procedure [94] or a *pore soaking* procedure [95]. In the former method, the porous film is immersed in phosphoric acid and connected to the negative terminal of a voltage source so that it acts as a cathode. A platinum mesh acts as the anode. A constant voltage of 7 V dc is maintained between the two electrodes, which etches away the alumina from the barrier layer until it is completely removed and the aluminum underneath the barrier layer is exposed. The second method is to simply soak the film in hot chromic/phosphoric acid without passing an electrical current. This process dissolves out the alumina from the barrier layer, as well as the pore walls, gradually. Owing to the isotropic nature of the etching process, the barrier layer removal process inevitably widens the pores, so that the pore diameter becomes ~60 nm. To confirm

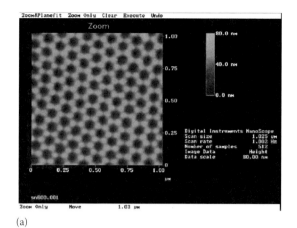

(a)

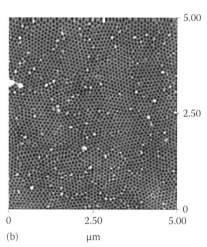

(b)   μm

FIGURE 28.6 (a) Atomic force micrograph of a nanoporous anodic alumina film produced by anodizing 99.999% pure aluminum in 0.3 M oxalic acid with a dc voltage of 40 V. The pore diameter is 50 nm. (b) The near-perfect regimentation of the pores into a hexagonal close-packed order within domains with size 0.5–1.0 μm. (Reproduced from Pramanik, S. et al., *Phys. Rev. B*, 74, 235329, 2006. With permission.)

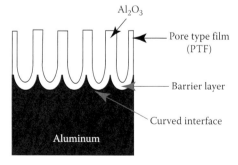

FIGURE 28.7 Schematic cross section of the porous alumina film. (Reproduced from Pramanik, S. et al., Electrochemical self-assembly of nanostructures: Fabrication and device applications, in Nalwa, H.S. (ed.), *Encyclopedia of Nanoscience and Nanotechnology*, 2nd edn., to appear. With permission.)

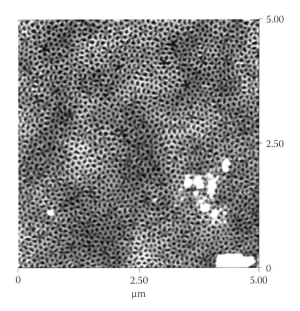

FIGURE 28.8 Atomic force micrograph of the back side of the porous alumina film after detaching it from the aluminum surface. Most of the pores have been opened up by the reverse-polarity etching procedure. (Reproduced from Pramanik, S. et al., *Phys. Rev. B*, 74, 235329, 2006. With permission.)

that the barrier layer has been indeed removed, we have stripped off the aluminum substrate in some sacrificial samples by soaking them in $HgCl_2$ solution. This releases the porous alumina film from the aluminum substrate. The released film is captured and imaged from the back side. Figure 28.8 shows an atomic force micrograph of the back side of the alumina film. Most of the pores are clearly visible in this image, indicating that the barrier layer has been removed successfully from the bottom of the majority of pores by the etching procedure. Removing the barrier layer exposes the aluminum at the pore bottom and allows dc electrodeposition of materials selectively inside the pores, since the aluminum layer conducts dc current. At the same time, we are now able to perform transport measurements on nanowires electrodeposited within the pores, since we can electrically contact them from the bottom through the aluminum and also from the top by depositing a thin layer of metal that seeps inside the pores by diffusion and attaches to the nanowires. Note from Figure 28.8 that not every pore is open at the bottom so that not every nanowire can be electrically contacted, but the vast majority of them will be contacted and electrically interrogated in any transport measurement where the top metal film acts as one electrode and the bottom aluminum film as the other electrode.

In order to fabricate nanowire spin valves, we have to sequentially deposit a ferromagnet, an organic, and a second ferromagnet within the pores to produce the trilayered structure where the organic acts as a spacer layer between two ferromagnetic contacts. For this purpose, we first electrodeposit nickel within the pores from a mild acidic solution of $NiSO_4 \cdot 6H_2O$ by applying a dc bias of 1.5 V at a platinum counter-electrode with respect to the aluminum substrate. A small deposition current ($\sim \mu A$) ensures well-controlled and slow-but-uniform electrodeposition of Ni

inside the pores. We had previously calibrated the deposition rate of Ni under these conditions by monitoring the deposition current during electrodeposition of Ni inside pores of known length. The deposition current increases drastically when the pores are completely filled and a nickel percolation layer begins to form on the surface. We stop the electrodeposition at this point. The deposition rate is determined by calculating the ratio of pore length to pore filling time. Using this (calibrated) deposition rate, we deposit 500 nm of Ni inside the pores. Transmission electron microscopy (TEM) characterization of these Ni nanowires showed that the wire lengths are very uniform and indeed conform to 500 nm. These samples are air dried and then Alq_3 is thermally evaporated on top of the Ni layer through a mask with a window of area of 1 mm² in a vacuum of 10^{-6} Torr. The deposition rate (calibrated with the help of a crystal oscillator) is in the range 0.1–0.5 nm/s. Note that during this step, a tunnel barrier of NiO may form at the interface between Ni and Alq_3. Fortunately, this unintentionally grown layer can only improve spin injection since it acts as a tunnel barrier [85]. During evaporation, Alq_3 seeps into the pores by surface diffusion and capillary action and reaches the nickel. The fact that Alq_3 is a short-stranded organic of low molecular weight is helpful in transporting it inside the pores. The thickness of the evaporated Alq_3 layer is monitored by a crystal oscillator and subsequently confirmed by TEM analysis. In this study, we prepared two sets of samples: in one set, the thickness of the Alq_3 layer is 33 nm and in another set the thickness is 26 nm (see Figure 28.9). Finally, cobalt is evaporated on the top without breaking the vacuum. The resulting structure is schematically depicted in Figure 28.10. The thickness of the cobalt layer that ends up inside the pores is also 500 nm since the total pore length is ~ 1 μm. Thus, we fabricate an array of nominally identical spin-valve nanowires. Since the cobalt contact pad has an area of 1 mm², approximately 2×10^8 nanowires are electrically contacted in parallel (the areal density of the nanowires is 2×10^{10} cm⁻²). Note that the surrounding alumina walls provide a natural encapsulation and protect the Alq_3 layer from moisture contamination. For electrical measurements, gold wires are attached to the top cobalt layer and the bottom aluminum foil with silver paste.

28.5.1.2 Control Experiments and Spin-Valve Measurements

From the measured sample conductance, we can estimate the number of nanowires that are electrically contacted from both ends and therefore contribute to the overall conductance. For example, the resistivity of an Alq_3 thin film is typically 10^5 Ω cm at room temperature [96]. When Alq_3 is confined in pores, we assume that the resistivity increases by an order of magnitude because of the increase in scattering due to charged surface states and the resulting decrease in carrier mobility. This is a typical assumption used in similar contexts [97]. Therefore, the resistivity of Alq_3 nanowires is 10^6 Ω cm. The resistivities of the ferromagnetic nanowire electrodes are $\sim 10^{-3}$ Ω cm [97]. Thus, the resistance of a single trilayered nanowire is 10^{11} Ω. Since the measured resistance of a sample is typically ~ 1 kΩ, we can conclude that 10^8 nanowires are electrically active and therefore must have contacts from

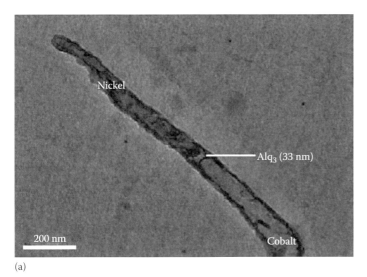

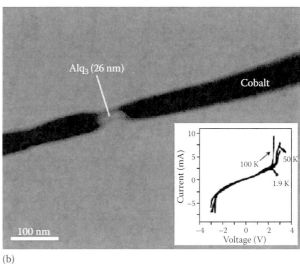

(a)

(b)

FIGURE 28.9 Cross-sectional transmission electron micrograph of two organic spin-valve nanowires: (a) spacer layer thickness is 33 nm and (b) spacer layer thickness is 26 nm. The inset shows the measured current–voltage characteristic at different temperatures for the nanowires with 26 nm spacer. (Reproduced from Pramanik, S. et al., *Nat. Nanotechnol.*, 2, 216, 2007. With permission.)

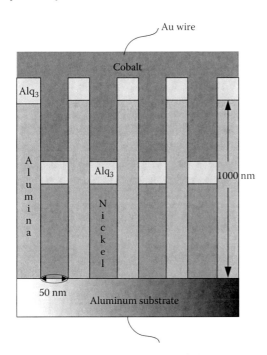

FIGURE 28.10 Schematic representation of the nanowire organic spin-valve array. The nanowires are hosted in the pores of an anodic alumina film and are electrically accessed by Au wires from both ends. The magnetic field is directed along the axis of the nanowires. (Reproduced from Pramanik, S. et al., *Phys. Rev. B*, 74, 235329, 2006. With permission.)

both ends. The contact area is 1 mm² and the density of nanowires is 2×10^{10} cm⁻². Therefore, the contacts covered 2×10^8 nanowires. This means that, on the average, 50% of the nanowires end up having contacts from both ends and become electrically active.

Note that since the resistivity of Alq₃ is nine orders of magnitude larger than the resistivities of the ferromagnets, we will always probe the resistance of the Alq₃ layer only, and not the resistance of the ferromagnetic electrodes, which are in series with the Alq₃ layer. Thus, all features in the magnetoresistance and current–voltage plots accrue from the organic layer and have nothing to do with the ferromagnetic contacts. Consequently, if there are features originating from the anisotropic magnetoresistance effects in the ferromagnets, we will never see them in our experiments. This is a fortunate happenstance since this gives us confidence that any magnetoresistance feature that we observe is associated with the organic and not the ferromagnets.

To further confirm that the contribution of the ferromagnetic layers to the resistance of the structure is indeed negligible, we fabricated a set of control samples without any Alq₃ layer. Note that a parallel array of 2×10^8 Ni/Co bilayered nanowires contacted by Al at the bottom and a thin film of Co at the top with area 1 mm² would produce a resistance of ~25 μΩ, which is below the sensitivity of our measurement apparatus. Therefore, we made control samples where we probe only ~500 nanowires. The trick employed to achieve this was to remove the barrier layer incompletely from the bottom intentionally, so that only a small fraction of the pores opened up from the bottom. We measure a resistance of ~10 Ω in the control samples at room temperature, which tells us that about ~500 nanowires are electrically probed. The magnetoresistances of the control samples were measured over a magnetic field range of 0–6 kOe and at a temperature of 1.9 K. The magnetic field was directed along the axis of the nanowires. A typical trace is shown in Figure 28.11. We observe a featureless monotonic positive magnetoresistance $\delta R(|B|)$ which accrues either from the anisotropic magnetoresistance effect associated with the ferromagnetic contacts or from the magnetoresistance of the aluminum substrate. However, the maximum value of $\delta R(|B|)$ that we observed over the entire measurement range was only ~0.08 Ω, which is more than an order of magnitude smaller than the resistance peak ΔR measured in the trilayered structures (see later). Thus, the resistance peak measured in the trilayered structures undoubtedly

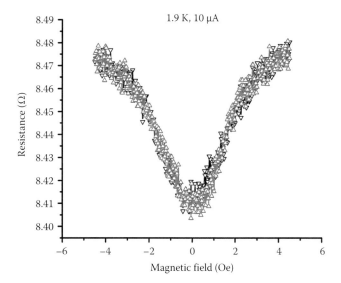

FIGURE 28.11 Magnetoresistance trace of the control sample consisting of ~500 Co–Ni bilayered nanowires (no organic spacer). No spin-valve peaks are visible in the data as expected. (Reproduced from Pramanik, S. et al., *Phys. Rev. B*, 74, 235329, 2006. With permission.)

originates from the spin-valve effect and has nothing to do with either the anisotropic magnetoresistance associated with the ferromagnetic contacts or the magnetoresistance of the aluminum substrate.

We fabricated ~90 trilayered samples using the procedure described earlier. Room-temperature resistances of these samples range from 1–10 kΩ depending on the number of nanowires that are electrically contacted from both ends (this number

varies because the process of barrier layer removal is not precisely controllable). The magnetoresistance of these samples was measured in a quantum design physical property measurement system (which has a superconducting magnet housed within a cryostat) with an ac bias current of 10 μA rms over a temperature range 1.9–100 K and over a magnetic field range of 0–6 kOe. The measured distribution of spin-valve signal $\Delta R/R$ is shown in Figure 28.12. The distribution is very broad and peaks near zero, i.e., most samples do not exhibit any measurable spin-valve signal. Among the remaining samples, some exhibit positive spin-valve signals (peaks) and others exhibit negative signals (troughs). The insets of Figure 28.12 show the magnetoresistance traces for the highest positive and negative spin-valve signals that we have measured among all samples tested. In every sample, the spin-valve peak always occurs between the coercive fields of Ni (~800 Oe) and Co (~1800 Oe) nanowires, as expected. Surprisingly, we found that the coercive fields do not vary significantly from sample to sample, indicating that the variation of coercive fields between different nanowires, and therefore different samples, is extremely small. The magnetoresistances of the devices exhibiting the inverse spin-valve effect typically saturate at low fields (~0.2 T in the figure shown), but those of devices exhibiting the normal spin-valve effect tend to saturate at much higher fields (see Figure 28.13).

28.5.2 Calculation of Spin Relaxation (or Spin Diffusion) Length L_s

Figures 28.14 and 28.15 show the positive and negative spin-valve signals, respectively, measured at different temperatures. The

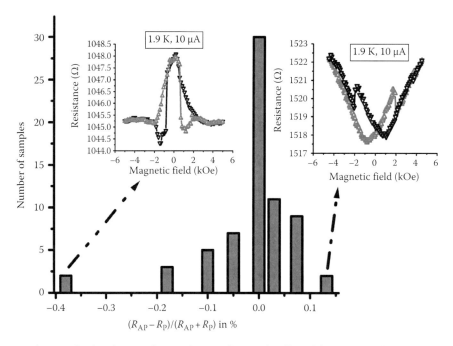

FIGURE 28.12 Histogram showing the distribution of spin-valve signal strength collected from 90 samples. Some samples show a positive and the rest a negative spin-valve signal. All data were collected with a bias current of 10 mA and at a temperature of 1.9 K. (Reproduced from Pramanik, S. et al., *Phys. Rev. B*, 74, 235329, 2006. With permission.)

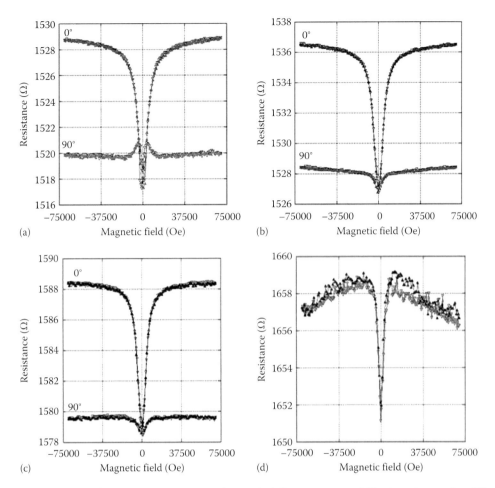

FIGURE 28.13 Magnetoresistance of organic nanowire spin valves with different magnetic field orientations and at different temperatures. (a) Magnetoresistance of Ni–Alq$_3$–Co nanowires at $T = 1.9$ K for various angles between field and nanowire axis. (b) Magnetoresistance of Ni–Alq$_3$–Co nanowires at $T = 50$ K (large field scan) for various angles between magnetic field and wire axis. (c) Magnetoresistance of Ni–Alq$_3$–Co nanowires at $T = 100$ K (large field scan) for various angles between magnetic field and the wire axis. (d) Magnetoresistance of Ni–Alq$_3$–Co nanowires at $T = 250$ K (large field scan). Here magnetic field is along the wire axis.

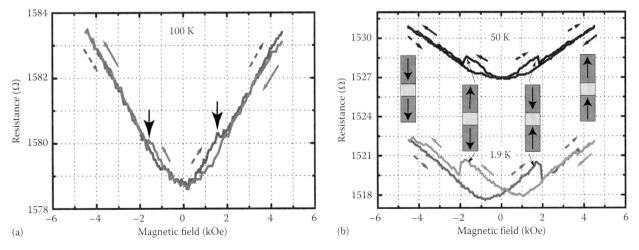

FIGURE 28.14 Magnetoresistance traces of spin-valve nanowires with 33 nm of Alq$_3$ layer. The magnetic field is directed along the axis of the nanowire. The solid and broken arrows indicate reverse and forward scans of the magnetic field. (a) For a measurement temperature of 100 K and (b) for 1.9 and 50 K. The parallel and antiparallel configurations of the ferromagnetic layers are shown within the corresponding magnetic field ranges. (Reproduced from Pramanik, S. et al., *Nat. Nanotechnol.*, 2, 216, 2007. With permission.)

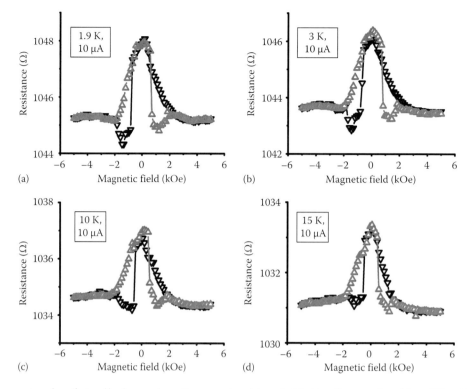

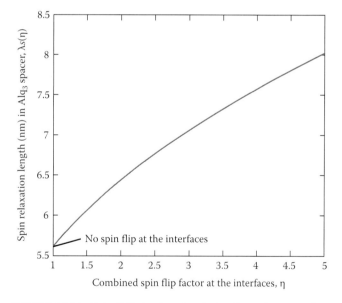

FIGURE 28.15 Inverse spin-valve effect and background negative magnetoresistance in Ni–Alq₃–Co nanowires at four different temperatures and fixed bias current (10 μA). (a) 1.5 K, (b) 3 K, (c) 10 K, and (d) 15 K. (Reproduced from Pramanik, S. et al., *Phys. Rev. B*, 74, 235329, 2006. With permission.)

bias current is kept constant at 10 μA rms in both cases. From the relative height of the spin-valve peak $\Delta R/R$, we can extract the spin diffusion length in the Alq₃ layer following the technique outlined in Section 28.3.3. We first assume that the spin polarization at the Fermi energy of the injecting ferromagnetic contact is P_1 and that there is no loss of spin polarization at the interface between Alq₃ and the injecting contact because of the so-called self-adjusting capability of the organic [98] invoked in [78]. The self-adjusting capability is a "proximity effect" in which the region of the organic in contact with a ferromagnet becomes spin polarized up to a short distance (few lattice constants) into the organic. As a result, there is no abrupt loss of spin polarization at the interface. It turns out, however, that the measured spin diffusion length is not particularly sensitive to the spin polarization of the contacts, so that even if there were any abrupt loss of spin polarization at the interface, it would not affect our result significantly. Figure 28.16 justifies this claim. Next, we assume, that there is a thin Schottky barrier at each organic/ferromagnet interface. Injected carriers tunnel through the first interface, which is too thin to cause spin randomization. After this tunneling, the carriers drift and diffuse through the remainder of the organic layer, with exponentially decaying spin polarization $P_1 \exp[-(d - d_0)/L_s(T)]$, where d is the total width of the organic layer, d_0 is the total width of the two Schottky barriers, and $L_s(T)$ is the spin diffusion length in Alq₃ at a temperature T. Finally, these carriers arrive at the detecting contact where there is a second Schottky barrier, through which they tunnel to cause the current. In our structures, $d_0 \ll d$, because

FIGURE 28.16 Spin diffusion length calculated as a function of the factor η representing spin flip at the interfaces.

d is 33 nm in one set and 26 nm in another set whereas d_0 is the width of the tunneling barrier, which must be less than 1 nm in order to produce any measurable current in the samples. In the following, we prove that $d_0 \ll d$. In that case, the loss of spin polarization in tunneling through the ultrathin potential barrier is negligible. Therefore, P_1 is approximately the spin polarization of the injecting contact. As the spin polarizations in cobalt and

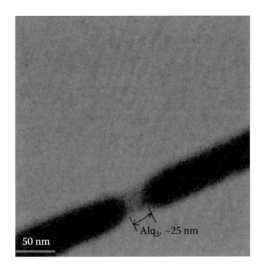

FIGURE 28.17 Transmission electron micrograph of a single nanowire showing the Alq_3 layer sandwiched between cobalt and nickel electrodes. This image was produced by releasing the nanowires from the alumina host by dissolution of alumina in dilute phosphoric acid and capturing the nanowires on TEM grids for imaging. (Reproduced from Pramanik, S. et al., *Phys. Rev. B*, 74, 235329, 2006. With permission.)

nickel at their Fermi energies are 42% and 33%, respectively [99], $P_1 = 0.42$ and $P_2 = 0.33$. In order to determine the value of d, we have carried out TEM of the nanowires. The wires were released from their alumina host by dissolution in very dilute chromic/phosphoric acid, washed, and captured on TEM grids for imaging. The TEM micrograph of a typical wire is shown in Figure 28.17. The Alq_3 layer thickness d is found to be 26 nm, which is quite close to the layer thickness estimated during fabrication (25 nm was the organic layer thickness estimated during deposition using a crystal oscillator). This agreement gives us confidence that d does not vary too much from one wire to another. We assume that it varies by ±5 nm when we calculate $L_s(T)$ from Equation 28.3 with d replaced by $d - d_0$.

Current–voltage characteristics of the nanowires are shown in the inset of Figure 28.9. These characteristics are reversible with no hysteresis. They are symmetric because of equal coupling to the contacts, but are nonlinear between −3.5 and 3.5 V at all measurement temperatures, indicating that the contacts are Schottky in nature. This means there has not been significant interdiffusion of Co or Ni into the Alq_3 layer, as that would have produced an ohmic contact.* As a result, the layer thickness d is well defined in the nanowires, which allows us to apply Equation 28.3 to estimate $L_s(T)$. In estimating $L_s(T)$ from Equation 28.3, we assume that $d - d_0 \approx d$. If this approximation is valid, then the estimated $L_s(T)$ will be independent of d. This is confirmed by our second set of samples which have slightly smaller d of

---

* This is a pleasant surprising result, which is not well understood. Evidently, the diffusion coefficients of Co and Ni in Alq_3 are much reduced when the materials are confined into nanowires. Why this happens is not known at this time.

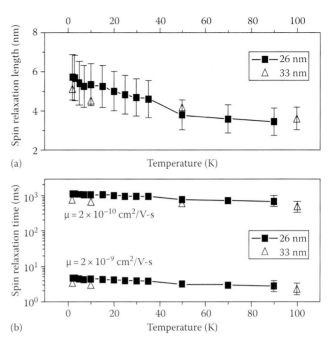

FIGURE 28.18 (a) Temperature dependence of the spin relaxation length and time in Alq_3. The data are shown for two sets of samples with spacer layer thicknesses of 26 and 33 nm. The error bars accrue from the ±5 nm uncertainty in the spacer layer thickness. The two curves in (b) correspond to the maximum and minimum values of the spin relaxation time (T_1 time) at any temperature. The wide spread between the maximum and minimum values arises entirely from the wide spread in the reported mobility of charge carriers in Alq_3. (Reproduced from Pramanik, S. et al., *Nat. Nanotechnol.*, 2, 216, 2007. With permission.)

26 nm. Figures 28.9b and 28.17 show the TEM micrographs of wires from this set where the layer thickness is found to be 26 nm. The quantity $L_s(T)$ measured from this set at any temperature agrees (to within 10%) with that measured from the other set at the same temperature. For example, at a temperature of 50 K, the value of $L_s(T)$ measured in the set with $d = 33$ nm is 4.25 ± 0.75 nm, whereas that measured in the set with $d = 26$ nm is 3.95 ± 0.75 nm. The difference between these two values is only 7.5%, which is less than the uncertainty of 19% in the measurement (accruing from the ±5 nm uncertainty in d), supporting the conclusion that $L_s(T)$ is reasonably independent of d. The values of $L_s(T)$ as a function of temperature are plotted in Figure 28.18a for both $d = 26$ nm and $d = 33$ nm.

Comparing the measured values of $L_s(T)$ with those reported in thin films of Alq_3 (45 nm at 4.2 K in [78]), we find that in a nanowire, $L_s(T)$ has been reduced by almost an order of magnitude. According to our previous discussion in Section 28.5, this indicates that the *Elliott–Yafet is the dominant mode of spin relaxation* in Alq_3.

28.5.3 Calculation of the Spin Relaxation Time τ_s

From the spin diffusion lengths, we can deduce the spin relaxation times $\tau_s(T)$ as described below. We use the formula

$$\tau_s(T) = \frac{L_s^2}{D_s(T)} = \frac{eL_s^2}{kT\mu(T)} \qquad (28.9)$$

where $\mu(T)$ is the temperature-dependent drift mobility. Since the operative spin relaxation mode is Elliott–Yafet, here we have assumed $D_s = D_c$. The mobility in our structure is calculated as follows:

The reported drift mobility in Alq$_3$ is given by the relation [100]

$$\mu(E) = \mu_0 e^{\alpha\sqrt{E}} \qquad (28.10)$$

where

μ_0 and α are temperature-dependent quantities given by $\mu_0 \approx A/kT$; $\alpha \approx B/kT$ where A and B are constants
E is the electric field

Ref. [100] reports $\mu_0 = 10^{-7} - 10^{-9}$ cm^2/V s, and $\alpha = 10^{-2}$ (cm/V)$^{1/2}$ in the *bulk* or *thin-film organic* at room temperature.

In order to determine the electric field E in the organic, we proceed as follows. The voltage over the nanowires can be estimated from the measured resistance and the current using Ohm's law: $V = IR = 10\,\mu A \times 1520\,\Omega = 15.2$ mV. Since the Alq$_3$ layer (in the first set) is nominally 33 nm wide, the average electric field across it is 15.2 mV/33 nm = 4.6 kV/cm. Using this value in Equation 28.10, we estimate the carrier mobility in the thin-film organic assuming reasonable values for the constants A and B: $A = 4.18 \times 10^{-34}$ to 4.18×10^{-36} C-m^2/sec and $B = 5 \times 10^{-7}$ eV-(m/V)$^{1/2}$. In nanowires, the mobility is expected to be lower.

Elliott has derived a relation between the spin relaxation time τ_s and the momentum relaxation time τ_p, which Yafet has shown to be temperature independent [66]:

$$\frac{\tau_p}{\tau_s} \propto \frac{\Delta}{E_g} \qquad (28.11)$$

Here Δ is the spin–orbit interaction strength in the band where the carrier resides (in our case the LUMO band) and E_g is the energy gap to the nearest band (in our case the HOMO–LUMO gap). Now one can write $\tau_s(T) = L_s^2(T)/D(T) = m^\star L_s^2/(kT\tau_p)$, where $D(T)$ is the temperature-dependent diffusion coefficient related to the mobility by the Einstein relation and $m^\star$ is the effective mass. Using the above, Equation 28.11 can be recast as $\tau_p^2 kT/m^\star L_s^2 \propto \Delta/E_g$. Since neither Δ nor E_g is affected by confinement in a nanowire, we can posit that at any temperature:

$$\frac{\tau_p^{\text{thin film}}}{\tau_p^{\text{nanowire}}} = \frac{L_s^{\text{thin film}}}{L_s^{\text{nanowire}}}$$

Since we found that the spin relaxation length is suppressed tenfold in a nanowire compared to the value in thin films, we conclude that the momentum relaxation time (and hence the mobility) must also be suppressed tenfold in a nanowire. Using

this assumption, we have calculated the spin relaxation time from Equation 28.9. These results are plotted in Figure 28.18b. The two curves give the maximum and minimum values of $\tau_s(T)$. They range from few milliseconds to over 1 s at 1.9 K (the spread in the range is entirely due to the spread in the reported value of mobility in Alq$_3$). These are among the longest spin relaxation times reported in any system. It is important to note that this estimate of τ_s is actually a worst-case estimate since we assume the bulk value of spin polarization in cobalt and nickel. Electrodeposited ferromagnetic nanowires (like Ni and Co nanowires in our case) most likely have lower spin polarization due to poor surface condition. That, in addition to the fact that we ignore any spin-flip event at the ferromagnet–organic interface, makes the 0.01–1 s estimate a lower bound. The actual lifetime may be even longer.

28.5.4 Bias Current Dependence of the Spin-Valve Signal: A Conundrum

Figure 28.19 shows the bias current dependence of the negative spin-valve signal in a typical sample at a constant temperature of 1.9 K. As the bias current is increased, the spin-valve signal decays rapidly, and at 200 μA no signal is measurable with our apparatus. Why does this happen? In Alq$_3$, the mobility should *increase* with increasing bias current (increasing electric field across the sample) according to Equation 28.10. Therefore, we expect the spin relaxation time to increase (and consequently the spin-valve signal to increase), rather than decrease, with increasing bias. Yet, we observe the opposite behavior, which seems to present a conundrum. There are many possible explanations for this anomaly. First, a high bias could cause Schottky barrier lowering and erode the tunnel barrier at the ferromagnetic/organic interface, which degrades spin-injection efficiency. Second, the high bias could open up new channels for scattering (e.g., spontaneous phonon emission) that is not captured by Equation 28.10. Third, a high bias could increasingly charge up the surface states, which exacerbates Coulomb scattering—again an effect not captured by Equation 28.10. All of these, however, are mere speculations at this point and lack confirmation. The high current does not destroy the sample irreversibly. As the current is decreased, the spin-valve signal is recovered.

28.5.5 Correlation between Sign of Spin-Valve Peak and Sign of Background Magnetoresistance

We find consistently that, whenever the spin-valve signal is negative, the background magnetoresistance is also negative, and whenever the spin-valve signal is positive, the background magnetoresistance is positive (see the insets of Figure 28.12). The background magnetoresistance has very little sensitivity to temperature (Figure 28.15) but it is extremely sensitive to bias—just like the spin-valve signal—as can be seen in Figure 28.19. It disappears at a bias current of 200 μA.

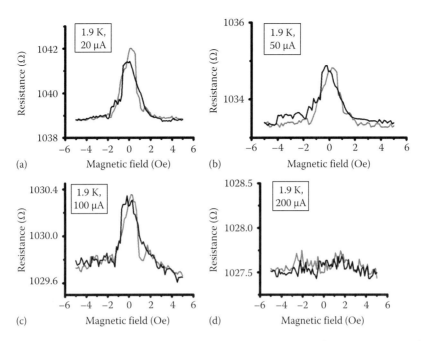

FIGURE 28.19 Inverse spin-valve effect and background negative magnetoresistance in Ni–Alq–Co nanowires at four different bias currents and at a fixed temperature of 1.9 K. (a) 20 μA, (b) 50 μA, (c) 100 μA, and (d) 200 μA. (Reproduced from Pramanik, S. et al., *Phys. Rev. B*, 74, 235329, 2006. With permission.)

The three known possible origins of the negative (inverted) spin-valve peak were discussed before. In our samples, we believe that the second mechanism is the most likely cause since the only other possible mechanism (faster s-electrons contributing more to current than the more numerous but slower d-electrons at the Fermi level of the injecting contact) is not likely since the spin polarization of the s-electrons is considerably less than that of the d-electrons at the Fermi energies in cobalt or nickel. Therefore, we do not expect much spin-polarized injection of s-electrons at all. Consequently, the inverted spin-valve effect must accrue from carriers resonantly tunneling through a localized defect or impurity state in the organic material. This requires that the carrier energy is resonant with the impurity level. In some nanowires, this indeed happens, and they exhibit a trough. In others, this does not happen so that they exhibit a peak, instead of a trough. Since each sample consists of a large number (~10^8) of nanowires, there is some cancellation between the positive and negative signals, which decreases the measured signal as a result of ensemble averaging. This is probably the reason why the distribution in Figure 28.12 peaks near zero.

We will now explain why a peak is accompanied by a positive background magnetoresistance and a trough is accompanied by a negative background magnetoresistance. At any magnetic field, except between the coercive fields of the two ferromagnets, the magnetizations of the injecting and detecting contacts are parallel. Assume also that both ferromagnets have the same sign of the spin polarization (as is indeed the case with cobalt and nickel). Now consider the case when the spin-valve peak is positive, meaning that there is no resonant tunneling through impurity states, resulting in an effective inversion of

the spin polarization of the nearest contact. In this case, an injected carrier will transmit and contribute to current if its spin does not flip within the spacer layer. In the presence of spin–orbit interaction, a magnetic field will increase the spin-flip rate by inducing spin mixing [101,102]. Thus, the probability of spin flipping increases with increasing magnetic field. If the injected carrier's spin flips, then it will be blocked by the detecting contact and the current will decrease, resulting in an increase in resistance. Thus, the resistance should increase with increasing magnetic field, resulting in a positive background monotonic magnetoresistance. This is what we observe.

In the case of negative spin-valve signal, resonant tunneling through an impurity state results in effective inversion of the spin polarization of the nearer ferromagnetic contact. In this case, spin flipping within the spacer layer will allow the flipped spin to transmit through the detector contact, which would have otherwise blocked it. Thus spin-flip events decrease the device resistance, instead of increasing it. Since a magnetic field increases the spin-flip rate, the resistance will decrease with increasing magnetic field, resulting in a negative monotonic background magnetoresistance. Again, this is exactly what we observe. These mechanisms are illustrated in Figure 28.20.

Note that the above mechanism for the background monotonic magnetoresistance does not call for phase coherence of charge carriers and therefore can persist up to high temperatures. Of course, this mechanism is spin dependent and therefore does not explain the OMAR that manifests even for nonmagnetic contacts. Our mechanism requires correlation of the signs of the spin-valve signal and the background magnetoresistance. If they turn out to be anticorrelated, then this will not be the

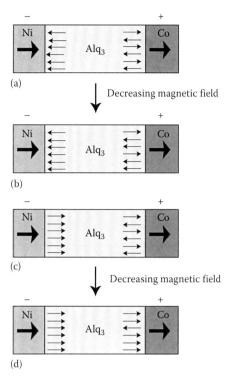

(a)

Decreasing magnetic field

(b)

(c)

Decreasing magnetic field

(d)

FIGURE 28.20 Explanation of the correlation between the sign of the background magnetoresistance and the sign of the spin-valve peak. (a) Consider the case when $H > |H_{Co}| > |H_{Ni}|$ and inversion of the injector's spin polarization has occurred owing to resonant tunneling through an impurity somewhere in the channel. The spin-valve signal is *negative*. We assume that the impurity was closer to the Ni contact so that its polarization has flipped and the spins injected by the Ni effectively have a polarization opposite to the magnetization of the Ni electrode as shown. Owing to the high magnetic field, the spin depolarization rate is high and the spins are completely depolarized by the time they reach the Co contact. Therefore, 50% of the spins will transmit through the Co contact and contribute to current. Let the device resistance be R_1. (b) When the magnetic field is decreased, the depolarization rate decreases and hence less than 50% of the spins have their polarizations aligned along the magnetization of the Co contact and transmit. Let the resistance now be R_2 which would be less than R_1. This will cause *negative* magnetoresistance. Thus, the background magnetoresistance is negative whenever the spin-valve signal is negative. (c) Again, $H > |H_{Co}| > |H_{Ni}|$ but no inversion of the spin polarization in Nickel takes place and hence spin-valve signal is *positive*. The high magnetic field depolarizes the spins and 50% conduct at the Co contact so that the device resistance is now R_1'. (d) When the field is reduced, the depolarization rate decreases so that more than 50% of the spins are aligned with the Co contact's magnetization and transmit. Let the device resistance be R_2' which is less than R_1'. This will give rise to *positive* magnetoresistance. Hence, the sign of the magnetoresistance is always the same as the sign of the spin-valve signal, which is what we experimentally observe over 90 samples. (Reproduced from Pramanik, S. et al., *Phys. Rev. B*, 74, 235329, 2006. With permission.)

cause. We have always observed correlation, and never observed anticorrelation, in all our experiments (~90 samples, multiple traces.) Therefore, we believe the mechanism suggested here is indeed the likely cause. Further experiments, of course, are needed to confirm it beyond all reasonable doubt.

28.6 The Transverse Spin Relaxation Time in Organic Molecules: Applications in Quantum Computing

There is an emerging consensus among the quantum information processing community that the ideal vehicle to host a "quantum bit" (or qubit) in a solid-state rendition of a quantum computer is an electron's or nucleus' "spin," as opposed to "charge." This is because spin is more robust than charge. The primary mechanism for corrupting a qubit and destroying the information contained therein is *decoherence* that takes place when the qubit couples to its environment. Decoherence rate is measured by the coherence time; the shorter it is, the higher is the decoherence rate. Fortunately, electron spin coherence times in solids can be much longer than the charge coherence times. Spin coherence times longer than 1 μs at room temperature have been demonstrated in nitrogen vacancies in diamond [103,104], whereas the charge coherence time saturates to only ~1 ns in solids as the temperature is lowered to the milli-Kelvin range [105]. Therefore, spin is clearly superior to charge in terms of the robustness against decoherence.

Having established that, the next important question would be as follows: What is the minimum (spin) coherence time that is required to allow fault-tolerant quantum computing? This question was answered by Knill in a seminal work [106]. Using quantum error correction theory, he demonstrated that it is possible to detect and correct qubit occurs (caused by decoherence) using quantum error correction algorithms if the error probability remains below 3%. Actually, the only errors that matter are the ones that occur within a clock period, since at the end of the clock period, the spin is intentionally rotated to execute quantum algorithms. If the clock period is T, then the probability of a qubit decohering during this period is

$$P_e = 1 - e^{-T/T_2},   (28.12)$$

where T_2 is the decoherence time. In the case of spin, this time is the so-called transverse spin relaxation time.

Setting $P_e = 0.03$ and then inverting Equation 28.12, we can see immediately that fault-tolerant quantum computing becomes possible if

$$T_2 > 33T   (28.13)$$

Thus, the transverse spin relaxation time does not have to be inordinately long; it merely has to exceed the clock period by about 33 times. The faster the clock period, the shorter can the T_2 time be and yet allow fault-tolerant quantum computing. It then behooves us to estimate the fastest clock that will be practical in spin-based quantum computing. The clock period must remain larger than the time it takes an external agency to rotate the spin by 180°. Spins are typically rotated by first splitting spin levels with a static (dc) magnetic field using the Zeeman effect, and then applying an ac magnetic field (usually from a microwave generator) whose frequency is resonant with the Zeeman splitting energy to induce Rabi oscillation [107]. This rotates the spin by an angle θ given by

$$\theta = \frac{g\mu_B B_{ac}}{\hbar}\tau, \tag{28.14}$$

where

g is the Landé g-factor
μ_B is the Bohr magneton
$\hbar$ is the reduced Planck's constant
B_{ac} is the amplitude of the ac magnetic flux density
τ is the duration for which the resonance is maintained (or the ac magnetic field is kept on)

Thus, by varying τ, one can rotate the spin through arbitrary angles. This is known as "single qubit rotation" and is one of the two ingredients necessary to form a universal quantum gate. The other ingredient is a 2-qubit "square-root-of-swap" operation [108]. Since the maximum angle that the spin needs to be rotated through is 180°, the maximum value of τ, which is the minimum clock period, is

$$T_{min} = \tau_{max} = \frac{h}{2g\mu_B B_{ac}}. \tag{28.15}$$

Time varying magnetic flux densities of amplitude 500 Gauss are available in standard electron spin resonance spectrometers [109]. If we apply a flux density of that magnitude to rotate spin using Rabi oscillations, then $T_{min} = 0.36$ ns. This is the minimum clock period.* Hence, from Equation 28.13, we need $T_2 = 11.8$ ns at least. There are many systems where the spin coherence time (or transverse relaxation time) exceeds 11.8 ns by a large margin. However, we will be particularly interested in organic molecules for two reasons: (1) First, it has been shown that organic molecules allow very high gate fidelity, as high as 98% [15]. Selective rotation of a target qubit usually requires electrical or optical gating. An example of electrical gating can be found in Ref. [110]. If we want to rotate the qubit in a targeted host, while leaving the qubits in all other hosts unaffected, we can either apply the ac magnetic field to the target host alone, or apply a global ac magnetic field but make the spin splitting in only the target host resonant with the global ac magnetic field. The former is practically impossible since confining a microwave field (with a wavelength of few cm) to a host of size ~10 nm is nearly impossible. In the latter approach, we apply a global dc magnetic field to induce a Zeeman splitting in every host, but then apply an electric field to the target host only (using "gates") to fine-tune the spin splitting in that host and make it resonant with the global ac magnetic field. The electric field increases the spin splitting energy in the target host by virtue of the Rashba spin–orbit coupling effect [111] and makes

it resonant with the global ac magnetic field. Thus, the spin is rotated in the selected host *only*. It is also possible to rotate a spin using just an ac electric field (applied through a gate) instead of an ac magnetic field. Coherent spin rotations using this approach has been demonstrated on times scales of ~50 ns [112] and also less than 1 ns [113]. However, in both cases, the application of the gate potential can disrupt the qubit. A high gate fidelity keeps this disruption to a minimum. It has been shown that some organic molecules are best suited to retain a high degree of gate fidelity [15]. (2) Second, optically active organic molecules, like Alq_3, will allow for a simple and elegant qubit readout scheme. Once the quantum computation is over, the final result has to be read out, i.e., the polarization of the electron spin has to be determined. The act of reading collapses the qubit to a classical bit, so that the final spin polarization will be either aligned along a chosen axis ("up" direction representing the bit 1) or anti-aligned ("down" direction representing the bit 0). In order to read which of these two polarizations has been assumed by the electron spin, we can use the following scheme [16]. Using a p-type dilute magnetic semiconductor like GaAs, magnetized in the "up" direction, we inject a spin-polarized hole into the Alq_3 host. In the organic, only the singlet excitons recombine radiatively while the triplet excitons are dark and do not recombine radiatively. That means a photon will be emitted only if the spin of the electron in the organic and the spin of the injected hole are antiparallel. Since the spin of the injected hole is known, we can determine the spin polarization of the electron (and hence read the collapsed qubit) by monitoring light emission from the organic. This readout scheme has no analog in inorganics.

Based on the above, we undertook a study of the prospect of using Alq_3 as a qubit host. The only requirement was that the transverse spin relaxation time (T_2 time) had to exceed 11.8 ns at a reasonable temperature, preferably room temperature. We therefore carried out experiments to determine this time as a function of temperature.

Unfortunately, it is very difficult to measure the single particle T_2 time directly in any system (including Alq_3 molecules) since it requires complicated spin echo sequences. Therefore, we have measured the ensemble-averaged T_2^* time instead, since it can be ascertained easily from the line width of electron spin resonance spectrum. This time, however, is orders of magnitude shorter than the actual T_2 time of an isolated spin because of additional decoherence caused by interactions between multiple spins in an ensemble [114,115]. It is particularly true of organics where spin–spin interaction is considered to be the major mechanism for spin decoherence [2]. Consequently, bulk samples (where numerous spins interact with each other) should behave differently from one or few molecules containing fewer interacting spins. In the rest of the chapter, we will designate the T_2^* times of bulk and few-molecule samples as T_2^b and T_2^f, respectively. We have found that they are discernibly different.

In order to prepare samples containing one or few molecules, we followed a time-honored technique adapted from Ref. [116]. We first produced a porous alumina film with 10 nm pores by anodizing an aluminum foil in 15% sulfuric acid. A two-step

---

* The clock frequency will then be 2.8 GHz.

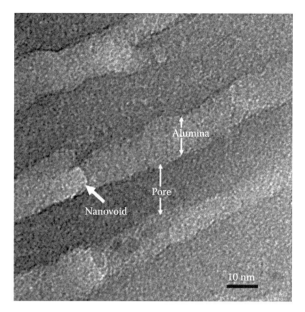

FIGURE 28.21 Cross-sectional TEM of the pores showing a nanovoid. (Reproduced from Kanchibotla, B. et al., *Phys. Rev. B*, 78, 193306, 2008. With permission.)

anodizing process was employed to improve the regimentation of the pores [117]. These porous films were then soaked in 1,2-dichloroethane ($C_2H_4Cl_2$) solution of Alq_3 for over 24 h to impregnate the pores with Alq_3 molecules. The films were subsequently rinsed several times in pure $C_2H_4Cl_2$ to remove excess Alq_3. There are cracks of size 1–2 nm in the anodic alumina film produced in sulfuric acid [116,118,119]. In Figure 28.21, we show a cross-sectional transmission electron micrograph of the porous film where this is clearly visible.

Ref. [116] claims that when the anodic alumina film is soaked in Alq_3 solution, Alq_3 molecules of 0.8 nm size diffuse into the cracks and come to rest in the nanovoids. Since the cracks are only 1–2 nm wide and the nanovoids have diameters of 1–2 nm, at best 1–2 molecules of Alq_3 can fit inside the nanovoids. Surplus molecules, not in the nanovoids, will be removed by repeated rinsing in $C_2H_4Cl_2$ [116]. $C_2H_4Cl_2$ completely dissolves out all the Alq_3 molecules, except those in the nanovoids, because the $C_2H_4Cl_2$ molecule cannot easily diffuse through the 1–2 nm wide nanocracks to reach the nanovoids. Therefore, after the repeated rinsing procedure is complete, only the nanovoids will contain isolated clusters of 1–2 molecules. The nanovoids are sufficiently far from each other that interaction between them is negligible [116]. Therefore, if we use the fabrication technique of Ref. [116], we will be confining one or two isolated molecules in nanovoids and measuring their T_2^f times. In contrast, the T_2^b times are measured in bulk Alq_3 powder containing a very large number of interacting molecules.

In electron spin resonance experiments, we apply a microwave field of fixed frequency ω to the spins while lifting the spin degeneracy by applying a dc magnetic field B_{dc}. When the resonance condition $\hbar\omega = g\mu_B B_{dc}$ is attained, the microwave is absorbed. From the absorption linewidth, we can determine the T_2^* time.

There are two magnetic fields at which the resonance condition is satisfied, which means that there are effectively two different g-factors for spins in Alq_3. These two g-factors are 2 and 4 [120]. Ref. [120] determined from the temperature dependence of the ESR intensity that the $g = 4$ resonance is associated with spins of localized electrons in Alq_3 (perhaps attached to an impurity or defect site) while the $g = 2$ resonance is associated with quasi-free (delocalized) electrons whose wavefunctions spread over multiple atoms. From the measured linewidths of these two resonances, we can estimate the T_2^f and T_2^b times for each resonance individually using the following standard formula:

$$T_2^f \text{ or } T_2^b = \frac{1}{r_e(g/2)\sqrt{3}\Delta B_{pp}} \qquad (28.16)$$

where

r_e is a constant $= 1.76 \times 10^7 (G - s)^{-1}$
g is the Landé g-factor
ΔB_{pp} is the full-width-at-half-maximum of the ESR line shape (the linewidth)

We checked that the line shape is almost strictly Lorentzian, so that the above formula can be applied with confidence [121]. Figure 28.22 shows typical magnetic field derivatives of the ESR spectrum obtained at a temperature of 10 K corresponding to $g = 2$ and $g = 4$ resonances. There are three curves in each figure corresponding to the blank alumina host, bulk Alq_3 powder, and Alq_3 in 1–2 nm voids (labeled "quantum dots"). The alumina host has an ESR peak at $g = 2$ and $g = 4$ (possibly due to oxygen vacancies) [122], but they are both much weaker than the resonance signals from Alq_3 and hence can be easily separated. Note that the g-factor of the isolated Alq_3 molecules in nanovoids is slightly larger than that of bulk powder since the resonance occurs at a slightly higher magnetic field.

In Figure 28.23, we plot the measured T_2^f and T_2^b times (associated with the resonance corresponding to $g = 2$) as functions of temperature from 4.2 to 300 K. The inequality $T_2^b < T_2^f$ is always satisfied except at one anomalous data point at 4.2 K. Two features stand out: (1) First, both T_2^f and T_2^b are relatively temperature independent over the entire range from 4.2 to 300 K. This indicates that spin–phonon interactions do not play a significant role in spin dephasing; (2) Second, both T_2^f and T_2^b times are quite long, longer than 3 ns, even at room temperature.

In Figure 28.24, we plot the measured T_2^f and T_2^b times as functions of temperature corresponding to the $g = 4$ resonance. The T_2^f time is plotted from 4.2 to 300 K, but the T_2^b time in bulk powder can only be plotted up to a temperature of 100 K. Beyond that, the intensity of the ESR signal fades below the detection limit of our equipment. The important features are as follows: (1) T_2^f and T_2^b are no longer temperature independent unlike in the case of the $g = 2$ resonance. T_2^f decreases monotonically with increasing temperature and falls by a factor of 1.7 between 4.2 and 300 K, (2) $T_2^b < T_2^f$ and the ratio T_2^f/T_2^b decreases with increasing temperature. The maximum value of the ratio T_2^f/T_2^b is 2.4,

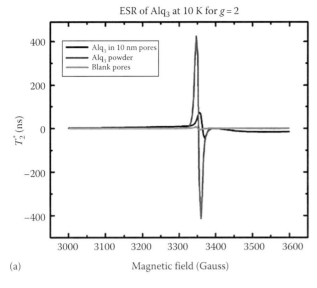

ESR of Alq₃ at 10 K for g = 2

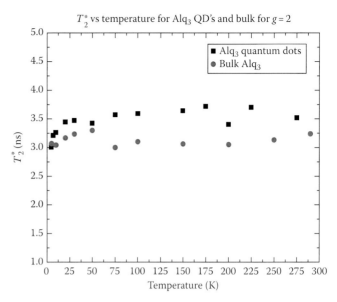

T_2^* vs temperature for Alq₃ QD's and bulk for g = 2

FIGURE 28.23 The ensemble-averaged transverse spin relaxation time as a function of temperature for the $g = 2$ resonance. (Reproduced from Kanchibotla, B. et al., *Phys. Rev. B*, 78, 193306, 2008. With permission.)

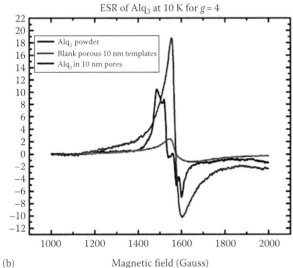

ESR of Alq₃ at 10 K for g = 4

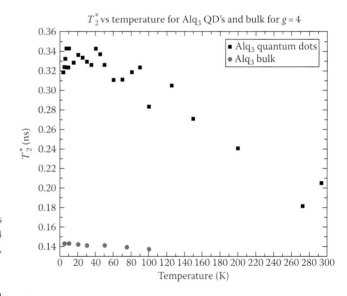

T_2^* vs temperature for Alq₃ QD's and bulk for g = 4

FIGURE 28.22 First derivative of the electron spin resonance signal as a function of the dc magnetic field. (a) for $g = 2$ resonance and (b) for $g = 4$ resonance. (Reproduced from Kanchibotla, B. et al., *Phys. Rev. B*, 78, 193306, 2008. With permission.)

FIGURE 28.24 The ensemble-averaged transverse spin relaxation time as a function of temperature for the $g = 4$ resonance. (Reproduced from Kanchibotla, B. et al., *Phys. Rev. B*, 78, 193306, 2008. With permission.)

occurring at the lowest measurement temperature of 4.2 K, and (3) both T_2^f and T_2^b times are about an order of magnitude shorter for the $g = 4$ resonance compared to the $g = 2$ resonance.

The strong temperature dependence of T_2^f and T_2^b tells us that for $g = 4$ resonance, spin–phonon coupling causes spin dephasing instead of spin–spin interaction. The spin–phonon coupling is absent or significantly suppressed for the $g = 2$ resonance, which is why T_2^f and T_2^b are an order of magnitude longer and also temperature independent for $g = 2$. Ref. [120] has ascribed the $g = 2$ resonance to quasi-free carrier spins in Alq₃ (whose wavefunctions are extended over an entire molecule) and $g = 4$ resonance to localized spins (whose wavefunctions are localized over an impurity atom). If that is the case, then it is likely that the localized spins and the delocalized spins will have very

different couplings to phonons since their wavefunctions are very different.

28.6.1 Application in Quantum Computing

We started this section by alluding to the suitability of Alq₃ molecules for quantum computing applications and mentioned that the requirement is that the single particle T_2 time should exceed 11.8 ns. The ensemble-averaged T_2^* time was measured to be about

3 ns (up to room temperature), while we could not directly measure the single particle T_2 time. However, it is entirely reasonable to assume that for a single isolated spin in Alq$_3$, T_2 should be at least an order of magnitude longer than T_2^* [114,115] particularly when spin–spin interaction is the major dephasing mechanism ($g = 2$). Therefore, we expect that the single spin T_2 time will be at least 30 ns up to room temperature. This exceeds 11.8 ns and hence meets Knill's criterion for fault-tolerant quantum computing.

We emphasize that Alq$_3$ does not have exceptionally long T_2 times, but it is still adequate for fault-tolerant quantum computing. Nitrogen vacancy NV$^-$ in diamond exhibits a much longer T_2 time of several tens of μs at room temperature [103,104]. However, quantum computing paradigms based on NV$^-$ require optical gating [123,124] or cavity dark states [125] since it would be nearly impossible to place an electrical gate on top of an atomic vacancy using any of the known fabrication methods. As a result, NV$^-$ computers are not truly miniaturizable since the gate is not miniaturizable. In contrast, the spins in Alq$_3$ are not bound to specific atomic sites. Instead, they extend over molecules of size ~1 nm, which may allow electrical gating and therefore lends itself to miniaturized quantum processors. Inorganic semiconductor qubit hosts that will also allow electrical gating and are miniaturizable, typically have a shorter T_2^* time than Alq$_3$ at room temperature [126]. Therefore, the Alq$_3$ system deserves due attention from the quantum information community. It is attractive since Alq$_3$-based quantum processors (1) are scalable, (2) are capable of fault-tolerant operation at room temperature, (3) possibly have a high degree of gate fidelity, and (4) lend themselves to an elegant qubit readout scheme. All this makes them attractive candidates for quantum computers.

28.7 A Novel Phonon Bottleneck in Organics?

The traditional *phonon bottleneck effect* [17], found in inorganic nanostructures such as quantum dots, is a consequence of quantum confinement of electrons *and* phonons. The walls of the quantum dot confines the delocalized electron wavefunction, which discretizes the electron energy, allowing the electron to exist only in very specific energy eigenstates. The allowed energy states are determined entirely by the size and shape of the quantum dot.

When an electron in the quantum dot absorbs or emits a phonon to make a transition from one energy state to another, both energy and momentum have to be conserved, which requires that the absorbed or emitted phonons have very specific energy and wavevector. Actually, in very small quantum dots, strict momentum conservation will not be required because of the Heisenberg Uncertainty Principle, but energy conservation is still required, so that

$$E_{\text{final}} - E_{\text{initial}} = \pm h\omega, \qquad (28.17)$$

where

E_{initial} and E_{final} are the energies of the initial and final energy states

$\hbar\omega$ is the phonon energy (the plus sign stands for absorption and the minus sign for emission)

Phonons of this energy must be allowed inside the dot, i.e., they must belong to allowed phonon modes. Now, if the phonons are also confined with the dot, then the phonon modes may be discretized as well, and phonons of arbitrary energy may not be available. If a phonon of energy $\hbar\omega$ is not available, then the corresponding electronic transition becomes forbidden and will not occur. This is the well-known phonon bottleneck effect that suppresses electronic transitions (including inelastic spin-dephasing transitions) in inorganic quantum dots.

In organic molecules, there is no quantum confinement effect on the electrons since the electron wavefunction is localized over individual atoms or molecules, which are typically much smaller than the size of the nanostructure housing the molecule. Therefore, the confining nanostructure (nanovoids in our case) has no influence on the allowed electronic energy states, which are the lowest unoccupied molecular orbital (LUMO) and highest occupied molecular orbital (HOMO) states of the molecule. However, any inelastic spin-dephasing transition will still require emission or absorption of a phonon of specific energy in order to conserve energy. Unlike the electrons, the phonons are delocalized over the entire nanovoid so that the walls of the nanovoid will confine the phonons and discretize the phonon modes. If the phonon of the right energy is not available because it is not contained within the allowed modes, then the spin-dephasing electronic transition will be suppressed, resulting in an increase in the spin-dephasing time or T_2 time. This too is a "phonon bottleneck effect" but somewhat different from the traditional one. It is different because it is the result of phonon confinement *alone*, without any accompanying electron confinement caused by the nanostructure.

This unconventional phonon bottleneck effect can explain why T_2^f is considerably longer than T_2^b for the $g = 4$ resonance. The bulk sample has many more interacting spins than the few-molecule sample, but if spin–spin interaction is overshadowed by spin–phonon coupling (as is the case for $g = 4$ resonance), then this should not make any difference. However, only the few-molecule samples are housed in 1–2 nm sized nanovoids and are subjected to the phonon bottleneck effect. Consequently, only they experience a suppression of the inelastic spin-dephasing transitions. That can make $T_2^f > T_2^b$, which is exactly what we observe. The bottleneck will be more severe at lower temperatures since fewer phonon modes will be occupied (Bose–Einstein statistics) so that the difference between T_2^b and T_2^f will be exacerbated at lower temperatures. This is precisely what we found. If this explanation is true, it will be the first observation of this effect in organic molecules. We raise the specter of phonon bottleneck only as a possibility, but cannot confirm it experimentally beyond all reasonable doubt since that would require showing progressive suppression of dephasing with decreasing nanovoid size, something that is experimentally not accessible. Nonetheless, we believe that there is a strong suggestion for the phonon bottleneck effect.

28.8 Conclusion

In this chapter, we have discussed some intriguing spin properties of the π-conjugated optically active organic semiconductor Alq_3. We have (1) shown that the longitudinal spin relaxation time (T_1) is exceptionally long (~1 s) and relatively temperature independent from 1.9–100 K, (2) identified the likely dominant spin relaxation mechanism in the organic as the Elliott–Yafet mechanism, (3) reported an intriguing correlation between the sign of the spin-valve signal and the background monotonic magnetoresistance, based on which we have proposed a likely origin of the background magnetoresistance, (4) demonstrated that the transverse spin relaxation time (T_2) is long enough to satisfy Knill's criterion for fault-tolerant quantum computing at *room temperature*, (5) proposed the Alq_3 molecule as a potential host for spin-based qubits, along with a simple and elegant scheme for qubit readout, and (6) showed some experimental results hinting at a possible phonon bottleneck effect in few-molecule samples of Alq_3 confined in 1–2 nm spaces.

Finally a comment on the importance of long spin relaxation time (T_1) in opto-spintronics is in order. It is often claimed that OLEDs will capture 50% of the global display market by 2015, since these are inexpensive compared to semiconductor (inorganic) LEDs, and can be produced on flexible substrates. The OLED consists of a p–n junction diode, just like an inorganic LED, with one difference: the p-type region is a hole transport layer and the n-type region is an electron transport layer. Alq_3 is an important electron transport layer used in OLEDs. In OLEDs, the electron–hole pairs form excitons, which recombine to produce photons or light. Because of the valley degeneracies in the HOMO level of organic molecules, 75% of the excitons formed are triplets and 25% are singlets. Only the singlets recombine radiatively to produce photons, while the triplets recombine non-radiatively to produce phonons and are wasted. Therefore, the maximum efficiency is limited to meager 25%. This can be changed if we inject spin-polarized carriers into the electron and hole transport layers to produce only singlets. In that case, the maximum efficiency can be 100%, resulting in brighter OLEDs. For all this to happen, it is necessary that the spin relaxation time exceed the exciton lifetime (or radiative recombination time) considerably. That will ensure that the singlets remain as singlets until they recombine. For this purpose, long spin relaxation times are very desirable. Our work shows that long spin relaxation times are indeed possible in optically active organics like Alq_3.

Acknowledgments

Much of the work described in this chapter was carried out in collaboration with the group of Professor Marc Cahay at the Department of Electrical and Computer Engineering, University of Cincinnati, Cincinnati, Ohio. Elements of this work were sponsored by the U.S. Air Force Office of Scientific Research under grant FA9550-04-1-0261 and the U.S. National Science Foundation under Grants ECCS-0608854 and CCF-0726373.

References

1. W. Naber, S. Faez, and W. van der Wiel, Organic spintronics, *Journal of Physics D: Applied Physics* **40** (2007), R205.
2. S. Sanvito, Spintronics goes plastic, *Nature Materials* **6** (2007), 803.
3. V. Krinichnyi, 2-mm Waveband electron paramagnetic resonance spectroscopy of conducting polymers, *Synthetic Metals* **108** (2000), 173.
4. D. Voss, Cheap and cheerful circuits, *Nature* **407** (2000), 442.
5. S. Forrest, P. Burrows, and M. Thompson, The dawn of organic electronics, *IEEE Spectrum* **37** (2000), 29.
6. S. Sanvito and A. Rocha, Molecular spintronics: The art of driving spin through molecules, *Journal of Computational and Theoretical Nanoscience* **3** (2006), 624–642.
7. S. Bandyopadhyay, B. Das, and A. Miller, Supercomputing with spin-polarized single electrons in a quantum coupled architecture, *Nanotechnology* **5** (1994), 113.
8. A. Davis and K. Bussman, Organic luminescent devices and magnetoelectronics, *Journal of Applied Physics* **93** (2003), 7358.
9. G. Salis, S. Alvarado, M. Tschudy, T. Brunschwiler, and R. Allenspach, Hysteretic electroluminescence in organic light-emitting diodes for spin injection, *Physical Review B* **70** (2004), 085203.
10. M. Berggren, D. Nilsson, and N. Robinson, Organic materials for printed electronics, *Nature Materials* **6** (2007), 3.
11. S. Forrest, The path to ubiquitous and low-cost organic electronic appliances on plastic, *Nature* **428** (2004), 911.
12. J. Wilkinson, A. Davis, K. Bussman, and J. Long, Evidence for charge-carrier mediated magnetic-field modulation of electroluminescence in organic light emitting diodes, *Applied Physics Letters* **86** (2005), 111109.
13. G. Veeraraghavan, T. Nguyen, Y. Sheng, O. Mermer, and M. Wohlgenannt, Magnetic field effects on current, electroluminescence and photocurrent in organic light-emitting diodes, *Journal of Physics: Condensed Matter* **19** (2007), 036209.
14. E. Carlegrim, A. Kanciurzewska, P. Nordblad, and M. Fahlman, Air-stable organic-based semiconducting room temperature thin film magnet for spintronics applications, *Applied Physics Letters* **92** (2008), 163308.
15. J. Lehmann, A. Gaita-Arino, E. Coronado, and D. Loss, Spin qubits with electrically gated polyoxometalate molecules, *Nature Nanotechnology* **2** (2007), 312.
16. B. Kanchibotla, S. Pramanik, S. Bandyopadhyay, and M. Cahay, Transverse spin relaxation time in organic molecules, *Physical Review B* **78** (2008), 193306.
17. H. Benisty, C. M. Sotomayor-Torres, and C. Weisbuch, Intrinsic mechanism for the poor luminescence properties of quantum box systems, *Physical Review B* **44** (1991), 10945.

18. http://www.icis.com/Articles/2008/07/14/9139393/market-for-organic-electronics-materials-is-growing.html

19. T. Ito, H. Shirakawa, and S. Ikeda, Simultaneous polymerization and formation of polyacetylene film on the surface of concentrated soluble Ziegler-type catalyst solution, *Journal of Polymer Science: Polymer Chemistry Edition* **12** (1974), 11.

20. H. Shirakawa, Synthesis of polyacetylene, Chapter 7, in *Handbook of Conducting Polymers*, 2nd edn., Eds. T. A. Skotheim, R. L. Elsenbaumer, and J. R. Reynolds (CRC Press, Boca Raton, FL, 1998).

21. H. Shirakawa, E. J. Louis, A. G. MacDiarmid, C. K. Chiang, and A. J. Heeger, Synthesis of electrically conducting organic polymers: Halogen derivatives of polyacetylene, (CH)$_x$, *The Journal of Chemical Society, Chemical Communications* (1977), 578.

22. C. K. Chiang, C. R. Fincher, Y. W. Park, A. J. Heeger, H. Shirakawa, E. J. Louis, S. C. Gau, and A. G. MacDiarmid, Electrical conductivity in doped polyacetylene, *Physical Review Letters* **39** (1977), 1098.

23. C. K. Chiang, M. A. Druy, S. C. Gau, A. J. Heeger, E. J. Louis, A. G. MacDiarmid, Y. W. Park, and H. Shirakawa, Synthesis of highly conducting films of derivatives of polyacetylene, (CH)$_x$, *Journal of the American Chemical Society* **100** (1978), 1013.

24. M. Angelopoulos, Conducting polymers in microelectronics, *IBM Journal of Research and Development* **45**(1) (2001), 57.

25. J. Burroughes, D. Bradley, A. Brown, R. Marks, K. Mackay, R. Friend, P. Burns, and A. Holmes, Light-emitting diodes based on conjugated polymers, *Nature* **347** (1990), 539.

26. D. Braun and A. Heeger, Visible light emission from semiconducting polymer diodes, *Applied Physics Letters* **58** (1991), 1982.

27. M. Cahay and S. Bandyopadhyay, Spintronics, *IEE Proceedings—Circuits, Devices and Systems* **152** (2005), 293.

28. D. Awschalom, M. Flatte, and N. Samarth, Spintronics, *Scientific American* **286** (2002), 66.

29. S. Wolf and D. Treger, Special issue on spintronics, *Proceedings of the IEEE* **91** (2003), 647.

30. C. Tsang, R. Fontana, T. Lin, D. Heim, B. Gurney, and M. Williams, Design, fabrication, and performance of spin-valve read heads for magnetic recording applications, *IBM Journal of Research and Development* **42** (1998), 103.

31. E. Grochowski, Emerging trends in data storage on magnetic hard disk drives, *Datatech* (1998), 11.

32. M. Baibich, J. Broto, A. Fert, F. Vandau, F. Petroff, P. Eitenne, G. Creuzet, A. Friederich, and J. Chazelas, Giant magnetoresistance of (001)Fe/(001)Cr magnetic superlattices, *Physical Review Letters* **61** (1988), 2472.

33. G. Binasch, P. Grunberg, F. Saurenbach, and W. Zinn, Enhanced magnetoresistance in layered magnetic structures with antiferromagnetic interlayer exchange, *Physical Review B* **39** (1989), 4828.

34. J. Gregg, I. Petej, E. Jouguelet, and C. Dennis, Spin electronics—A review, *Journal of Physics D—Applied Physics* **35** (2002), R121.

35. W. Gallagher and S. Parkin, Development of the magnetic tunnel junction MRAM at IBM: From first junctions to a 16-Mb MRAM demonstrator chip, *IBM Journal of Research and Development* **50** (2006), 5.

36. K. Inomata, Magnetic random access memory (MRAM) in *Giant Magnetoresistance Devices*, Eds. E. Hirota, H. Sakakima, and K. Inomata, Springer Series in Surface Sciences, Vol. 40 (2002).

37. J. Akerman, Toward a universal memory, *Science* **308** (2005), 508.

38. C. Chappert, A. Fert, and F. Van Dau, The emergence of spin electronics in data storage, *Nature Materials* **6** (2007), 813.

39. S. Datta and B. Das, Electronic analog of the electro-optic modulator, *Applied Physics Letters* **56** (1990), 665.

40. S. Bandyopadhyay and M. Cahay, Alternate spintronic analog of the electro-optic modulator, *Applied Physics Letters* **85** (2004), 1814.

41. J. Fabian, I. Zutic, and S. Das Sarma, Magnetic bipolar transistor, *Applied Physics Letters* **84** (2004), 85.

42. M. Flatte, Z. Yu, E. Johnston-Halperin, and D. Awschalom, Theory of semiconductor magnetic bipolar transistors, *Applied Physics Letters* **82** (2003), 4740.

43. M. Flatte and G. Vignale, Unipolar spin diodes and transistors, *Applied Physics Letters* **78** (2001), 1273.

44. M. Flatte and G. Vignale, Heterostructure unipolar spin transistors, *Journal of Applied Physics* **97** (2005), 104508.

45. S. Bandyopadhyay and M. Cahay, Re-examination of some spintronic field-effect device concepts, *Applied Physics Letters* **85** (2004), 1433–1435.

46. N. Wu, N. Shibata, and Y. Amemiya, Boltzmann machine neuron device using quantum-coupled single electrons, *Applied Physics Letters* **72** (1998), 3214.

47. S. Bandyopadhyay and V. Roychowdhury, Switching in a reversible spin logic gate, *Superlattices and Microstructures* **22** (1997), 411.

48. D. Loss and D. DiVincenzo, Quantum computation with quantum dots, *Physical Review A* **57** (1998), 120.

49. S. Bandyopadhyay and M. Cahay, *Introduction to Spintronics* (CRC Press, Boca Raton, FL, 2008).

50. B. Dieny, V. Speriosu, S. Parkin, B. Gurney, D. Wilhoit, and D. Mauri, Giant magnetoresistance in soft ferromagnetic multilayers, *Physical Review B* **43** (1991), 1297.

51. J. Slonczewski, Current-driven excitation of magnetic multilayers, *Journal of Magnetism and Magnetic Materials* **159** (1996), L1.

52. S. Saikin, A drift-diffusion model for spin-polarized transport in a two-dimensional non-degenerate electron gas controlled by spin-orbit interaction, *Journal of Physics: Condensed Matter* **16** (2004), 5071–5081.

53. M. Jullière, Tunneling between ferromagnetic films, *Physics Letters A* **54** (1975), 225–226.

54. S. Pramanik, S. Bandyopadhyay, and M. Cahay, The inequality of charge and spin diffusion coefficients, *Journal of Applied Physics* **104** (2008), 014304.

55. F. Jedema, A. Filip, and B. van Wees, Electrical spin injection and accumulation at room temperature in an all-metal mesoscopic spin-valve, *Nature* **410** (2001), 345.

56. J. Moodera, L. Kinder, J. Nowak, P. LeClair, and R. Meservey, Geometrically enhanced magnetoresistance in ferromagnet-insulator-ferramagnet tunnel junctions, *Applied Physics Letters* **69** (1996), 708.

57. P. Hammar and M. Johnson, Detection of spin-polarized electrons injected into a two-dimensional electron gas, *Physical Review Letters* **88** (2002), 066806.

58. K. Tsukagoshi, B. Alphenaar, and H. Ago, Coherent transport of electron spin in a ferromagnetically contacted carbon nanotube, *Nature* **401** (1999), 572.

59. S. Patibandla, S. Pramanik, S. Bandyopadhyay, and G. C. Tepper, Spin relaxation in a Germanium nanowire, *Journal of Applied Physics* **100** (2006), 044303.

60. J. De Teresa, A. Barthelemy, A. Fert, J. Contour, R. Lyonnet, F. Montaigne, P. Seneor, and A. Vaures, Inverse tunnel magnetoresistance in $Co/SrTiO_3/La_{0.7}Sr_{0.3}MnO_3$: New ideas on spin-polarized tunneling, *Physical Review Letters* **82** (1999), 4288.

61. E. Tsymbal, A. Sokolov, I. Sabirianov, and B. Doudin, Resonant inversion of tunneling magnetoresistance, *Physical Review Letters* **90** (2003), 186602.

62. A. Sokolov, I. Sabirianov, E. Tsymbal, B. Doudin, X. Li, and J. Redepenning, Resonant tunneling in magnetoresistive Ni/NiO/Co nanowire junctions, *Journal of Applied Physics* **93** (2003), 7029.

63. J. De Teresa, A. Barthelemy, A. Fert, J. Contour, F. Montaigne, and P. Seneor, Role of metal-oxide interface in determining the spin polarization of magnetic tunnel junctions, *Science* **286** (1999), 507.

64. I. Zutic, J. Fabian, and S. Das Sarma, Spintronics: Fundamentals and applications, *Reviews of Modern Physics* **76** (2004), 323.

65. R. Elliott, Theory of the effect of spin-orbit coupling on magnetic resonance in some semiconductors, *Physical Review* **96** (1954), 266.

66. Y. Yafet, in *Solid State Physics*, Vol. 14 (Academic Press, New York, 1963).

67. M. D'yakonov and V. Perel', Spin orientation of electrons associated with the interband absorption of light in semiconductors, *Soviet Physics—JETP* **33** (1971), 1053.

68. M. D'yakonov and V. Perel', Spin relaxation of conduction electrons in noncentrosymmetric semiconductors, *Soviet Physics—Solid State* **13** (1972), 3023.

69. G. Bir, A. Aronov, and G. Pikus, Spin relaxation of electrons due to scattering by holes, *Soviet Physics—JETP* **42** (1976), 705.

70. A. Abragam, *The Principles of Nuclear Magnetism* (Oxford University Press, Oxford, U.K., 1973).

71. G. Dresselhaus, Spin-orbit coupling effects in zinc blende structures, *Physical Review* **100** (1955), 580.

72. Y. A. Bychkov and E. I. Rashba, Oscillatory effects and the magnetic susceptibility of carriers in inversion layers, *Journal of Physics C: Solid State Physics* **33** (1984), 6039.

73. V. Dediu, M. Murgia, F. C. Matacotta, C. Taliani, and S. Barbanera, Room temperature spin polarized injection in organic semiconductors, *Solid State Communications* **122** (2002), 181.

74. F. Garnier, G. Horowitz, X. Peng, and D. Fichou, An all-organic "soft" thin film transistor with very high carrier mobility, *Advanced Materials* **2** (1990), 592.

75. F. Garnier, R. Hajlaoui, A. Yassar, and P. Srivastava, All-polymer field-effect transistor realized by printing techniques, *Science* **265** (1994), 1684.

76. A. Dodabalapur, L. Torsi, and H. Katz, Organic transistors: Two-dimensional transport and improved electrical characteristics, *Science* **268** (1995), 270.

77. R. Marks, F. Biscarini, R. Zamboni, and C. Taliani, Polarized electroluminescence from vacuum-grown organic light-emitting diodes, *Europhysics Letters* **32** (1995), 523.

78. Z. H. Xiong, D. Wu, Z. Valy Vardeny, and J. Shi, Giant magnetoresistance in organic spin-valves, *Nature* **427** (2004), 821.

79. H. Vinzelberg, J. Schumann, D. Elefant, R. Gangineni, J. Thomas, and B. Büchner, Low temperature tunneling magnetoresistance on (La, Sr)MnO_3/Co junctions with organic spacer layers, *Journal of Applied Physics* **103** (2008), 093720.

80. J. Jiang, J. Pearson, and S. Bader, Absence of spin transport in the organic semiconductor Alq_3, *Physical Review B* **77** (2008), 035303.

81. S. Mukhopadhyay and I. Das, Inversion of magnetoresistance in magnetic tunnel junctions: Effect of pinhole nano-contacts, *Physical Review Letters* **96** (2006), 026601.

82. T. Santos, J. Lee, P. Migdal, I. Lekshmi, B. Satpati, and J. Moodera, Room-temperature tunnel magnetoresistance and spin-polarized tunneling through an organic semiconductor barrier, *Physical Review Letters* **98** (2007), 016601.

83. R. Meservey and P. Tedrow, Spin-polarized electron tunneling, *Physics Reports* **238** (1994), 173.

84. G. Schmidt, D. Ferrand, L. W. Molenkamp, A. T. Filip, and B. J. van Wees, Fundamental obstacle for electrical spin injection from a ferromagnetic metal into a diffusive semiconductor, *Physical Review B* **72** (2000), R4790.

85. E. I. Rashba, Theory of electrical spin injection: Tunnel contacts as a solution of the conductivity mismatch problem, *Physical Review B* **62** (2000), R16267.

86. J. Shim, K. Raman, Y. Park, T. Santos, G. Miao, B. Satpati, and J. Moodera, Large spin diffusion length in an amorphous organic semiconductor, *Physical Review Letters* **100** (2008), 226603.

87. N. Morley, A. Rao, D. Dhandapani, M. Gibbs, M. Grell, and T. Richardson, Room temperature organic spintronics, *Journal of Applied Physics* **103** (2008), 07F306.

88. S. Pramanik, C. Stefanita, S. Patibandla, S. Bandyopadhyay, K. Garre, N. Harth, and M. Cahay, Observation of extremely long spin relaxation times in an organic nanowire spin-valve, *Nature Nanotechnology* **2** (2007), 216.

89. S. Pramanik, S. Bandyopadhyay, K. Garre, and M. Cahay, Normal and inverse spin-valve effect in organic semiconductor nanowires and the background magnetoresistance, *Physical Review B* **74** (2006), 235329.

90. V. Pokalyakin, S. Tereshin, A. Varfolomeev, D. Zaretsky, A. Baranov, A. Banerjee, Y. Wang, S. Ramanathan, and S. Bandyopadhyay, Proposed model for bistability in nanowire nonvolatile memory, *Journal of Applied Physics* **97** (2005), 124306.

91. B. Kanchibotla, S. Pramanik, and S. Bandyopadhyay, Self assembly of nanostructures using nanoporous alumina templates, Chapter 9 in *Nano and Molecular Electronics Handbook*, Ed. S. Lyshevski (CRC Press, Boca Raton, FL, 2007).

92. S. Bandyopadhyay, A. Miller, H. Chang, G. Banerjee, V. Yuzhakov, D. Yue, R. Ricker, S. Jones, J. Eastman, E. Baugher, and M. Chandrasekhar, Electrochemically assembled quasiperiodic quantum dot arrays, *Nanotechnology* **7** (1996), 360.

93. S. Pramanik, B. Kanchibotla, S. Sarkar, G. Tepper, and S. Bandyopadhyay, Electrochemical self-assembly of nanostructures: Fabrication and device applications, in the *Encyclopedia of Nanoscience and Nanotechnology*, 2nd edn., Ed. H. S. Nalwa (to appear).

94. O. Rabin, P. Herz, S. Cronin, Y. Lin, A. Akinwande, and M. Dresselhaus, Nanofabrication using self-assembled alumina templates, *Materials Research Society Symposium* **636** (2001), D 4.7.1.

95. T. Ohgai, X. Hoffer, L. Gravier, J. Wegrowe, and J. Ansermet, Bridging the gap between template synthesis and microelectronics: Spin-valves and multilayers in self-organized anodized aluminium nanopores, *Nanotechnology* **14** (2003), 978.

96. A. Mahapatro, R. Agrawal, and S. Ghosh, Electric-field-induced conductance transition in 8-hydroxyquinoline aluminum (Alq$_3$), *Journal of Applied Physics* **96** (2004), 3583.

97. S. Pramanik, C. Stefanita, and S. Bandyopadhyay, Spin transport in self-assembled all-metal nanowire spin-valves: A study of the pure Elliott-Yafet mechanism, *Journal of Nanoscience and Nanotechnology* **6** (2006), 1973.

98. J. Xie, K. Ahn, D. Smith, A. Bishop, and A. Saxena, Ground state properties of ferromagnetic metal/conjugated polymer interfaces, *Physical Review B* **67** (2003), 125202.

99. E. Tsymbal, O. Mryasov, and P. LeClair, Spin dependent tunneling in magnetic tunnel junctions, *Journal of Physics: Condensed Matter* **15** (2003), R109.

100. B. Chen, W. Lai, Z. Gao, C. Lee, S. Lee, and W. Gambling, Electron drift mobility and electroluminescent efficiency of tris(8-hydroxyquinolinolato) aluminum, *Applied Physics Letters* **75** (1999), 4010.

101. M. Cahay and S. Bandyopadhyay, Conductance modulation of spin interferometers, *Physical Review B* **68** (2003), 115316.

102. M. Cahay and S. Bandyopadhyay, Phase-coherent quantum mechanical spin transport in a weakly disordered quasi-one-dimensional channel, *Physical Review B* **69** (2004), 045303.

103. T. A. Kennedy, J. S. Colton, J. E. Butler, R. C. Linares, and P. J. Doering, Long coherence time at 300 K for nitrogen vacancy center spins in diamond grown by chemical vapor deposition, *Applied Physics Letters* **83** (2003), 4190.

104. F. Jelezko, T. Gaebel, I. Popa, A. Gruber, and J. Wrachtrup, Observation of coherent oscillations in a single electron spin, *Physical Review Letters* **92** (2004), 076401.

105. P. Mohanty, E. M. Q. Jariwalla, and R. A. Webb, Intrinsic decoherence in mesoscopic systems, *Physical Review Letters* **78** (1997), 3366.

106. E. Knill, Quantum computing with realistically noisy devices, *Nature (London)* **434** (2005), 39.

107. I. I. Rabi, N. F. Ramsey, and J. Schwinger, Use of rotating coordinates in magnetic resonance problems, *Reviews of Modern Physics* **26** (1954), 167.

108. G. Burkard, D. Loss, and D. P. DiVincenzo, Coupled quantum dots as quantum gates, *Physical Review B* **59** (1999), 2070.

109. A. Akhkalkasti, T. Gegechkori, G. Mamniashvili, Z. Shermadini, A. N. Pogorely, and O. M. Kuzmak, Influence of pulsed magnetic field on single- and two-pulse nuclear spin echoes in multidomain magnets, arXiv:0705.3979.

110. S. Bandyopadhyay, Self assembled nanoelectronic quantum computer based on the Rashba effect in quantum dots, *Physical Review B* **61** (2000), 13813.

111. D. Bhowmik and S. Bandyopadhyay, Gate control of the spin-splitting energy in a quantum dot: Applications in single qubit rotation, *Physica E* **41** (2009), 587.

112. K. C. Nowack, F. H. L. Koppens, Yu. V. Nazarov, and L. M. K. vandersypen, Coherent control of single electron spins with electric fields, *Science* **318** (2007), 1430.

113. Y. Kato, R. C. Meyers, D. C. Driscoll, A. C. Gossard, J. Levy, and D. D. Awschalom, Gigahertz electron spin manipulation using voltage controlled g-tensor modulation, *Science* **299** (2003), 1201.

114. R. De Sousa and S. Das Sarma, Electron spin coherence in semiconductors: Considerations for a spin based quantum computing architecture, *Physical Review B* **67** (2003), 033301.

115. X. Hu, R. De Sousa, and S. Das Sarma, Decoherence and dephasing in spin based solid state quantum computers, in *Foundations of Quantum Mechanics in the Light of New Technology*, Eds. Y. A Ono and K. Fujikawa (World Scientific, Singapore, 2003), p. 3.

116. G. S. Huang, X. L.Wu, Y. Xie, Z. Y. Zhang, G. G. Siu, and P. K. Chu, Photoluminescence from 8-hydroxy-quinolinoline aluminum embedded in porous anodic alumina membrane, *Applied Physics Letters* **87** (2005), 15190 and references therein.

117. H. Masuda and M. Satoh, Fabrication of gold nanodot array using anodic porous alumina as an evaporation mask, *Japanese Journal of Applied Physics, Part 2* **35** (1996), L126.

118. D. D. Macdonald, On the formation of voids in anodic oxide films in aluminum, *Journal of the Electrochemical Society* **140** (1993), L27.

119. S. Ono, H. Ichinose, and N. Masuko, Defects in porous anodic films formed on high purity aluminum, *Journal of the Electrochemical Society* **138** (1991), 3705.

120. M. N. Grecu, A. Mirea, C. Ghica, M. Cölle, and M. Schwoerer, Paramagnetic defect centers in crystalline Alq$_3$, *Journal of Physics: Condensed Matter* **17** (2005), 6271 and references therein.

121. P. H. Reiger, *Electron Spin Resonance: Analysis and Interpretation* (The Royal Society of Chemistry, Cambridge, U.K., 2007).

122. Y, Du, W. L. Cai, C. M. Mo, J. Chen, L. D. Zhang, and X. G. Zhu, Preparation and pho-toluminescence of alumina membranes with ordered pore arrays, *Applied Physics Letters* **74** (1999), 2951.

123. J. Wrachtrup, Ya. S. Kilin, and A. P. Nizovtsev, Quantum computation using the 13C nuclear spins near the NV defect center in diamond, *Optics and Spectroscopy* **91** (2001), 429.

124. M. V. G. Dutt, L. Childress, L. Jiang, E. Togan, J. Maze, F. Jelzko, A. S. Zibrov, P. R. Hemmer, and A. E. Craig, Quantum registers based on electronic and nuclear spin qubits in diamond, *Science* **316** (2007), 1312.

125. M. S. Shariar, J. A. Bowers, B. Demsky, P. S. Bhatia, S. Lloyd, P. R. Hemmer, and A. E. Craig, Cavity dark states for quantum computing, *Optics Communications* **195** (2001), 411.

126. S. Ghosh et al., Room temperature spin coherence in ZnO, *Applied Physics Letters* **86** (2005), 232507.

VI

Nanoscale Methods

29

Nanometrology

Stergios Logothetidis
*Aristotle University
of Thessaloniki*

29.1 Introduction

Nanometrology is the science and practice of measurement of functionally important, mostly dimensional parameters and components with at least one critical dimension, which is smaller than 100 nm. The application and use of nanomaterials in electronic and mechanical devices, optical and magnetic components, quantum computing, tissue engineering, and other biotechnologies, with smallest features widths well below 100 nm, are the economically most important parts of the nanotechnology nowadays and presumably in the near future. In parallel with the shrinking dimensions of the components and structures produced in the relative industries, the required measurement uncertainties for dimensional metrology in these important technology fields are decreasing too (Bhushan 2004, Nomura et al. 2004).

Success in nanomanufacturing of devices will rely on new nanometrologies needed to measure basic material properties including their sensitivities to environmental conditions and their variations, to control the nanofabrication processes and material functionalities and to explore failure mechanisms (Alford et al. 2007). In order to study and explore the complex nanosystems, highly sophisticated experimental, theoretical, and modeling tools are required (Alford et al. 2007, Nomura et al. 2004). Especially, the visualization, characterization, and manipulation of materials and devices require sophisticated imaging and quantitative techniques with spatial and temporal resolutions on the order of 10^{-6} and below to the molecular level. In addition, these techniques are critical for understanding the relationship and interface between nanoscopic and mesoscopic/macroscopic scales, a particularly important objective for biological and medical applications (Whitehouse 2002). Examples of important tools available at the moment include highly focused synchrotron x-ray sources and related techniques that provide detailed molecular structural information by directly probing the atomic arrangement of atoms; scanning probe microscopy that allows three-dimensional-type topographical atomic and molecular views or optical responses of nanoscale structures; in-situ optical monitoring techniques that allow the monitoring and evaluation of building block assembly and growth; optical methods, with the capability of measuring in air, vacuum, and liquid environment for the study of protein and cells adsorption on solid surfaces, that have been employed to discriminate and identify bacteria at the species level and it is very promising for analytical purposes in biochemistry and medicine (Lousinian and Logothetidis 2008, Lousinian et al. 2008, Karagkiozaki et al. 2009).

The nanometrology methods need measurements that should be performed in real time to allow simultaneous measurement of properties and imaging of material features at the nanoscale. These nanometrology techniques should be supported by physical models that allow the de-convolution of probe–sample interactions as well as to interpret subsurface and interface behaviors. Ellipsometry is a key technique meeting the aforementioned demands. It can be applied during the nanofabrication processes and provides valuable information concerning the optical, vibrational, structural, and morphological properties, the composition, as well as the thickness and the mechanisms of the specimen under growth or synthesis conditions in nanoscale (Azzam and Bashara 1977, Keldysh et al. 1989, Logothetidis 2001). Further correlation between optical and other physical properties can lead to a more complementary characterization and evaluation of materials and devices (Azzam and Bashara 1977, Keldysh et al. 1989, Logothetidis 2001).

Also, x-ray reflectivity (XRR) is a powerful tool for investigating monolithic and multilayered film structures. It is one of the few methods that, with great accuracy, allows not only information on the free surface and the interface to be extracted but also the mass density and the thickness of very thin film of the order of a few nanometers along the direction normal to the sample surface to be determined. XRR is able to offer accurate thickness determination for both homogeneous thin films and multilayers with the same precision, as well as densities, surface, and interface roughness of constituent layers. In addition, other promising nanometrology techniques include the tip-enhanced Raman spectroscopy (TERS) (Motohashi et al. 2008, Steidtner and Pettinger 2008, Yi et al. 2008). TERS combines the capabilities of Raman spectroscopy that has been used for many years for the single layer and even single molecule detection in terms of its chemical properties, with the advantages of an atomic force microscopy tip that is put close to the sample area that is illuminated by the Raman laser beam. In this way, a significant increase of the Raman signal and of the lateral resolution by up to nine orders of magnitude takes place. Thus, TERS can be used for the chemical analysis of very small areas and for the imaging of nanostructures as well as of other materials such as proteins and biomolecules (Yeo et al. 2007, Zhang et al. 2007).

Another important nanometrology method is nanoindentation that has been rapidly become the method of choice for quantitative determination of mechanical properties (as hardness and elastic modulus) of thin films and small volumes of material. Finally, scanning probe microscopes (SPMs) are standard instruments at scientific and industrial laboratories that allow imaging, modifications, and manipulations with the nanoobjects. They permit imaging of a surface topography and correlation with different physical properties within a very broad range of magnifications, from millimeter to nanometer-scale range. Atomic force microscopy (AFM) and AFM-related techniques (e.g., scanning near-field optical microscopy [SNOM]) have become sophisticated tools, not only to image surfaces of molecules, but also to measure molecular forces between molecules. This is substantially increasing our knowledge of molecular interactions.

29.2 Presentation of State of the Art

In the following paragraphs, some of the most important nanometrology methods and techniques for the study of nanomaterials, nanoparticles, thin films, biomolecules, etc., are described, together with some specific examples. Since the field of nanometrology methods and their associated applications for the study of nanomaterials, nanostructures and thin films as well as interactions of light with matter at the nanoscale is quite large, in the following paragraphs, we will focus mainly to two directions.

The first will be the study of materials used for the emerging field of organic electronics that include the flexible polymeric substrates, the hybrid (organic–inorganic) polymers, and inorganic materials that are used for the protection of the organic electronic devices against the atmospheric molecule permeation (barrier layers) and the electrode materials, such as zinc oxide (ZnO), which are used as electrodes. These applications are of significant importance since the fabrication of organic electronic devices onto flexible polymeric films by large-scale production processes offers many exciting new opportunities and will also reduce several technical limitations that characterize the production processes of conventional microelectronics. In conventional Si microelectronics, patterning is most often done using photolithography, in which the active material is deposited initially over the entire substrate area, and selected areas of it are removed by physical or chemical processes (Logothetidis 2005).

The second direction is the study of biomolecules such as blood plasma proteins adsorbed onto inorganic thin film surfaces. These materials are investigated by the use of spectroscopic ellipsometry (SE), AFM, and SNOM techniques. In addition, the study of amorphous carbon nanocoatings and transition metal nitrides as hard or biocompatible materials with the use of nanoindentation and XRR will be also presented and discussed.

29.2.1 Spectroscopic Ellipsometry

29.2.1.1 Basic Concepts and Definitions

Spectroscopic ellipsometry (SE) has become the standard technique for measuring the bulk materials and surfaces, nanomaterials, thin film, their thickness, crystallite size, and optical constants such as refractive index or dielectric function (Azzam and Bashara 1977, Keldysh et al. 1989, Logothetidis 2001). SE is used for characterization of all types of materials: dielectrics, semiconductors, metals, organics, opaque, semitransparent, even transparent, etc. This technique and its instrumentation relies on the fact that the reflection of a dielectric interface depends on the polarization of the light while the transmission of light through a transparent layer changes the phase of the incoming wave depending on the refractive index of the material (Azzam and Bashara 1977, Logothetidis 2001). An ellipsometer can be used to measure layers as thin as 0.2 nm up to layers that are several microns thick. Applications include the accurate thickness measurement of thin films, the identification of materials and thin layers, the evaluation of vibrational, compositional and nanostructural properties, and the characterization of surfaces and interfaces (Logothetidis 2001).

In order to understand how SE works, it is important to outline some basic concepts from the electromagnetic theory. Let us consider first the propagation of an electromagnetic plane wave through a nonmagnetic medium, which can be described by the vector of the electric field $\vec{E}$; in the simple case this is a plane wave given by the following expression:

$$\vec{E} = \vec{E}_i e^{i(kz - \omega t)}. \tag{29.1}$$

For oblique incidence, plane waves are typically referenced to a local coordinate system (x, y, z), where z is the direction of propagation of light k and x and y define the plane where the transverse electromagnetic wave oscillates. That is, the latter are the directions (see Figure 29.1a) parallel (p) and perpendicular (s) to the plane of incidence, respectively (these two directions are the two optical eigenaxes of the material under study) (Azzam

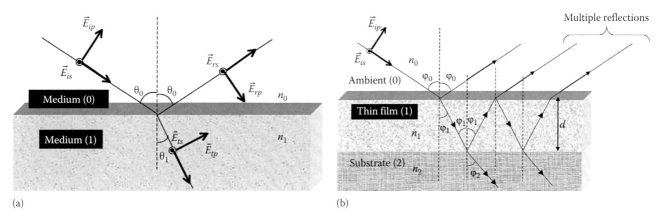

FIGURE 29.1 Oblique reflection and transmission of a plane electromagnetic wave: (a) at the sharp interface between two media 0 and 1 with refractive indexes n_0 and n_1, respectively, (b) at the surface of a thin film (medium 1) deposited onto a bulk substrate (medium 2). The electric field components E_p and E_s parallel-*p* and perpendicular-*s* to the plane of incidence, and the wave vector for the incident (*i*), reflected (*r*) and transmitted (*t*) waves are shown. θ and θ_1 are the angles of incidence and refraction.

and Bashara 1977, Logothetidis 2001). The complex electric field amplitudes, E_p and E_s, represent the projections of the plane wave $\vec{E}$ along *x* and *y* axes, respectively. Therefore, the quantity $\vec{E}_i$ in Equation 29.1 carries information not only about the amplitude of the plane wave $\vec{E}$ when it is propagated in vacuum but also about its polarization. That is,

$$\vec{E}_i = \vec{E}_{ix}\hat{x} + \vec{E}_{iy}\hat{y}. \tag{29.2}$$

In addition, the amplitude of the wave-vector *k* during the propagation of the wave in matter is in general a complex number given by the dispersion expression $k = \tilde{n}\omega/c$. ω is the photon energy and $\tilde{n}(\omega)$ is the *refractive index*, which is in general a complex quantity, and it is related to *dispersion (n)* and *absorption (k)* of the radiation by the medium:

$$\tilde{n}(\omega) = n(\omega) + i\kappa(\omega). \tag{29.3}$$

The complex dielectric function $\tilde{\varepsilon}(\omega)=(\varepsilon_1 + i\varepsilon_2)$ is actually the quantity directly related to the material properties and is connected to the refractive index through the following equation:

$$\tilde{\varepsilon}(\omega) = \varepsilon_1 + i\varepsilon_2 \equiv \tilde{n}^2(\omega) = (n+i\kappa)^2 \Rightarrow \begin{cases} \varepsilon_1 = n^2 - \kappa^2 \\ \varepsilon_2 = 2n\kappa \end{cases}. \tag{29.4}$$

For an electromagnetic wave transmitted in a dispersive and absorbing material Equation 29.1 can be rewritten as follows:

$$\vec{E}_i = \left(\vec{E}_{ix}\hat{x} + \vec{E}_{iy}\hat{y}\right) e^{i\omega n\frac{z}{c}} e^{-\frac{\omega\kappa z}{c}} e^{-i\omega t}, \tag{29.5}$$

with the quantity (absorption coefficient)

$$\alpha = \frac{2\omega\kappa}{c} \left(= \frac{4\pi\kappa}{\lambda} \right)$$

to define the penetration depth of the wave in the material and λ being the wavelength in vacuum. When the electromagnetic wave is reflected by the material smooth surface (see Figure 29.1a), the polarization of the outgoing wave can be represented as

$$\vec{E}_r = \left(\tilde{r}_p \vec{E}_{0x}\hat{x} + \tilde{r}_s \vec{E}_{0y}\hat{y}\right). \tag{29.6}$$

In this approximation, the interaction of the electromagnetic wave with the material is described by the two complex Fresnel reflection coefficients, $\tilde{r}_p$ and $\tilde{r}_s$. These reflection coefficients describe the influence of the material on the electric field components, *p* and *s*, that correspond to the directions parallel (*p*) and perpendicular (*s*) to the plane of incidence and characterize the interface between two media (e.g., the ambient—medium 0 and the material studied—medium 1) and are given by the following expressions:

$$\tilde{r}_p = \frac{\tilde{E}_{r,p}}{\tilde{E}_{i,p}} = \frac{\left|\tilde{E}_{r,p}\right| \cdot e^{i\theta_{r,p}}}{\left|\tilde{E}_{i,p}\right| \cdot e^{i\theta_{i,p}}} = \left|\frac{\tilde{E}_{r,p}}{\tilde{E}_{i,p}}\right| \cdot e^{i(\theta_{r,p}-\theta_{i,p})} = \left|\tilde{r}_p\right| e^{i\delta_p}, \tag{29.7}$$

$$\tilde{r}_s = \frac{\tilde{E}_{r,s}}{\tilde{E}_{i,s}} = \frac{\left|\tilde{E}_{r,s}\right| \cdot e^{i\theta_{r,s}}}{\left|\tilde{E}_{i,s}\right| \cdot e^{i\theta_{i,s}}} = \left|\frac{\tilde{E}_{r,s}}{\tilde{E}_{i,s}}\right| \cdot e^{i(\theta_{r,s}-\theta_{i,s})} = \left|\tilde{r}_s\right| e^{i\delta_s}. \tag{29.8}$$

The Fresnel reflection coefficients in the interface between two media *i* and *j*, e.g., medium (0) and medium (1) in Figure 29.1a, with refractive index $\tilde{n}_i$ and $\tilde{n}_j$, respectively, are given by the following expressions:

$$\tilde{r}_{ij,p} = \frac{\tilde{n}_j \cos\theta_i - \tilde{n}_i \cos\theta_j}{\tilde{n}_j \cos\theta_i + \tilde{n}_i \cos\theta_j}, \tag{29.9}$$

$$\tilde{r}_{ij,s} = \frac{\tilde{n}_i \cos\theta_i - \tilde{n}_j \cos\theta_j}{\tilde{n}_i \cos\theta_i + \tilde{n}_j \cos\theta_j}, \tag{29.10}$$

where the incident θ_i and refracted θ_j angles are correlated through the *Snell law* $\tilde{n}_i \sin \vartheta_i = \tilde{n}_j \sin \vartheta_j$. When the light beam does not penetrate the medium (1), due to either its high absorption coefficient or its infinite thickness, as shown in Figure 29.1a, we are referred to a two-phase (ambient-substrate) system or a bulk material surrounding by medium (0). In this case, the ratio of the *p*-, *s*-Fresnel reflection coefficients, namely, the *complex reflection ratio* is the quantity measured directly by SE and it is given by the following expression:

$$\tilde{\rho} = \frac{\tilde{r}_p}{\tilde{r}_s} = \left| \frac{\tilde{r}_p}{\tilde{r}_s} \right| e^{i(\delta_p - \delta_s)} = \tan \Psi \, e^{i\Delta}, \qquad (29.11)$$

that characterizes any bulk material. In this expression, Ψ and Δ are the ellipsometric angles, and for a bulk material take values $0° < \Psi < 45°$ and $0° < \Delta < 180°$. From an ellipsometric measurement, the complex reflection ratio $\tilde{\rho}$ is estimated through the calculation of amplitude ratio $\tan \Psi$ and the phase difference Δ. From these two quantities, one can extract all the other optical constants of the material. For example, the complex dielectric function of a bulk material with smooth surfaces is directly calculated by the following expression (Azzam and Bashara 1977):

$$\tilde{\varepsilon}(\omega) = \varepsilon_1 + i\varepsilon_2 = \tilde{\eta}^2(\omega) = \tilde{\varepsilon}_0 \sin^2\theta \left\{ 1 + \left[\frac{1 - \tilde{\rho}(\omega)}{1 + \tilde{\rho}(\omega)} \right]^2 \tan^2\theta \right\}, \quad (29.12)$$

where

    θ is the angle of incidence of the beam
    $\tilde{\varepsilon}_0$ is the dielectric constant of the ambient medium (for the case of air $\tilde{\varepsilon}_0 = \tilde{n}_0 = 1$)

In the case of a film deposited on a substrate (see Figure 29.1b), we have the three-phase model where the film with thickness d [medium (1)] is confined between the semi-infinite ambient [medium (0)] and the substrate [medium (2)]. The complex reflectance ratio is defined as (Azzam and Bashara 1977)

$$\tilde{\rho} = \frac{\tilde{R}_p}{\tilde{R}_s}, \qquad (29.13)$$

$$\tilde{R}_p = \frac{\tilde{r}_{01p} + \tilde{r}_{12p} e^{i2\beta}}{1 + \tilde{r}_{01p} \tilde{r}_{12p} e^{i2\beta}}, \qquad (29.14a)$$

$$\tilde{R}_s = \frac{\tilde{r}_{01s} + \tilde{r}_{12s} e^{i2\beta}}{1 + \tilde{r}_{01s} \tilde{r}_{12s} e^{i2\beta}}, \qquad (29.14b)$$

where $\tilde{r}_{01i}$ and $\tilde{r}_{12i}$ ($i = p,s$) are the Fresnel reflection coefficients for the interfaces between medium (0) and (1) and medium (1) and (2), respectively.

These depend on film thickness due to the multiple reflections of light between the media (0)–(1) and (1)–(2), through the phase angle β given by

$$\beta = 2\pi \left(\frac{d}{\lambda} \right) \left(n_1^2 - n_0^2 \sin^2 \theta \right)^{1/2}, \qquad (29.14c)$$

where

    d is the film thickness
    λ is the wavelength
    θ is the angle of incidence
    n_0, n_1 are the complex refraction indices of the ambient and the film, respectively

Thus, the measured quantity is the pseudodielectric function $\langle \varepsilon(\omega) \rangle = (\langle \varepsilon_1(\omega) \rangle + i\langle \varepsilon_2(\omega) \rangle)$, which contains information also about the substrate and the film thickness. As a result, the phase angle β diminishes in the energy region of high absorption, leading to $\tilde{R}_i = \tilde{r}_{01i} = \tilde{r}_i$, ($i = p,s$) and consequently to $\langle \rho \rangle = \rho$. Thus, at the absorption bands, we obtain information only of the optical properties of the bulk of the film.

A representative setup of a spectroscopic ellipsometer is shown in Figure 29.2. A well-collimated monochromatic beam is generated from a suitable light source (in the case of Vis–fUV spectral region is a high pressure Xe arc lamp) and it is passed through a polarizer in order to produce light of known-controlled linear polarization and a photoelastic modulator that modulates the polarization of the light beam. Then, the light beam is reflected by the surface of the sample under investigation, with a light beam spot size to be of the order 1–3 mm². After its reflection on the sample, the light beam is passed through the second polarizer (analyzer), to be analyzed with respect to the new polarization stage established under the reflection, and finally it is detected and transformed to raw data through the computing devices. This last part of the ellipsometric setup defines the accuracy and the speed of the information and the data obtained about the investigated sample. The excitation head and the optical holder with the analyzer can be easily mounted and dismounted from such an ellipsometric setup to be adapted, for example, on deposition systems for in-situ and real-time ellipsometric measurements (Logothetidis 2001).

The energy of the reflected light beam can be analyzed by a grating monochromator providing the full spectrum capability. In the simplest setup, a photomultiplier is used as detector in the wavelength range 225–830 nm. In addition, several detectors (for example, 32 photomultipliers) consisting of a multiwavelength unit can be adapted to a spectrograph through optical fibers providing the real-time measuring capability. Using digital parallel signal processing, a spectrum consisting of at least 32-photon energies between 1.5 and 6.5 eV can be recorded in less than 60 ms (Gioti et al. 2000, Logothetidis 2001, Gravalidis et al. 2004, Laskarakis et al. 2010).

Other common ellipsometer configurations include rotating analyzer, rotating polarizer, and rotating compensator configurations in order to modulate the light beam polarization before its reflection on the sample surface. The rotating element methods are generally easy to automate and can be used on a large spectral range.

In the case of rotating polarizer technique, a source with well-known polarization state is required, whereas after reflection of

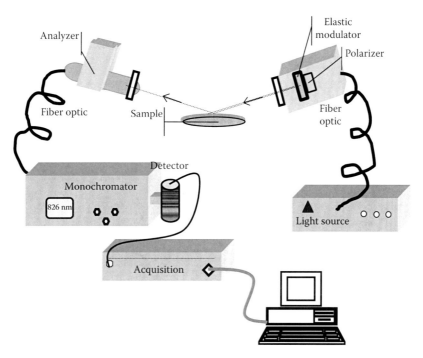

FIGURE 29.2 An optical setup of a phase modulated spectroscopic ellipsometer in the Visible–UV energy region. The light source (Xe arc lamp), the polarizer and elastic modulator consist the excitation head, whereas the analyzer, monochromator and detector consist the detection head. The acquisition board and the computing device are necessary to collect and transform the information into raw data for further analysis.

the light on the sample, the analyzer is fixed. In these systems, it is not necessary to have a detector insensitive to the polarization and the spectrometer can be located between the analyzer and the detector. In the case of the rotating analyzer technique, the detector must be insensitive to the polarization and the spectrometer must be located between the source and the polarizer. Finally, in the case of rotating compensator systems, the polarization problems at the source and detector levels can be suppressed but the spectral calibration of the compensator is difficult (Tompkins and Irene 2005).

The SE experimental setup in the infrared (IR) spectral region is quite similar to the above described setup and it is also based on the conventional polarizer-modulator-sample-analyzer ellipsometer configuration. The incident beam is provided by a conventional IR spectrometer, which contains a SiC light source and a Michelson interferometer. The parallel IR light beam coming out from the spectrometer is focused on ~1 cm² area on the sample surface by means of mirrors. Before reaching the sample, the IR beam passes through the IR grid polarizer and the ZnSe photoelastic modulator at 37 kHz (Laskarakis et al. 2001). After reflection from the sample, the beam goes through the analyzer (IR grid polarizer) and is focused on the sensitive area of a photovoltaic detector: InSb (above 1850 cm⁻¹) and HgCdTe (MCT). The spectral range of the ellipsometer is limited by the detector specifications. The resolution of the spectra can be varied from 1 to 128 cm⁻¹. A full spectrum of the dielectric function can be recorded in less than 2 s, which corresponds to a single scan of the interferometer. By increasing the integration time to a few minutes, the precision on Ψ and Δ can be improved by more than one order of magnitude. Thus a resolution better

of the order of 0.01° on both Ψ and Δ angles can be achieved (Laskarakis et al. 2001, Gravalidis et al. 2004).

SE is one of the most promising techniques for the study of inorganic, organic, and hybrid materials to be used for state-of-the-art applications, such as for organic electronic devices deposited onto flexible polymeric substrates (Logothetidis 2005, Laskarakis et al. 2009). These applications include flexible organic light emitting diodes (OLEDs), flexible organic photovoltaic cells (OPVs), and organic circuits and consist of a multilayer structure of inorganic, organic, and hybrid nanolayers. Also, SE plays an important role for the study of the optical properties of inorganic, organic, and hybrid films (e.g., ultrahard films, corrosion resistant films, optical materials, and biocompatible films) deposited onto flat and complex substrates. For the above applications, the deposition of functional thin films with nanometer precision is absolutely necessary and therefore use of nondestructive nanometrology tools is of major importance (Logothetidis 2001, Laskarakis and Logothetidis 2007).

29.2.1.2 Theoretical Models for Analysis of Ellipsometric Data

The study of the material properties by ellipsometry is performed by the use of the dielectric function $\varepsilon(\omega) = \varepsilon_1(\omega) + i\varepsilon_2(\omega)$. The real $\varepsilon_1(\omega)$ and the imaginary $\varepsilon_2(\omega)$ parts of the dielectric function are strongly related through the well-known Kramers–Kronig relation (is based on the principle of the causality (Keldysh et al. 1989)):

$$\varepsilon_1(\omega) = 1 + \frac{2}{\pi} P \int_0^\infty \frac{\omega' \varepsilon_2(\omega')}{\omega'^2 - \omega^2} d\omega', \tag{29.15a}$$

$$\varepsilon_2(\omega) = -\frac{2\omega}{\pi} P \int_0^\infty \frac{\varepsilon_1(\omega')-1}{\omega'^2-\omega^2} d\omega', \qquad (29.15b)$$

where P means the principal value of the integral around the characteristic of the material electronic resonance ($\omega' = \omega$) and

$$\varepsilon_1(\omega=0) = 1 + \frac{2}{\pi} P \int_0^\infty \frac{\varepsilon_2(\omega')}{\omega'} d\omega' \qquad (29.15c)$$

is the static dielectric function, the material strength (deviation from the strength of vacuum $\varepsilon_0 = 1$), which describes all losses in the whole electromagnetic spectrum in the material due to the electron absorption.

The optical response of the thin films can be deduced by the parameterization of the measured $\langle \varepsilon(\omega) \rangle$ by the use of appropriate theoretical models. One of these models is the damped harmonic oscillator (Lorenz model), which is described by the expression (Keldysh et al. 1989, Gioti et al. 2000, Logothetidis 2001):

$$\tilde{\varepsilon}(\omega) = 1 + \frac{f\omega_0^2}{\omega_0^2 - \omega^2 + i\Gamma\omega}, \qquad (29.16)$$

where

ω is the energy of light

ω_0 is the absorption energy of the electronic transition

The constants f and Γ denote the oscillator strength and the damping (broadening) of the specific transition, respectively

The quantity $\varepsilon_1(\omega = 0)$ is given by the following relation

$$\varepsilon_1(\omega=0) = 1 + \frac{\omega_p^2}{\omega_0^2} = f, \qquad (29.17)$$

which is the static dielectric constant and represents the contribution of the electronic transition that occurs at an energy ω_0 in the NIR–Visible–UV energy region, on the dielectric function (Laskarakis et al. 2001) and ω_p is the plasma energy.

In the case that more than one electronic transition occurs, their contribution in $\varepsilon_1(\omega = 0)$ is accounted by the summation:

$$\varepsilon_1(\omega=0) = 1 + \sum_i f_i, \qquad (29.18)$$

where $\sum_i f_i$ describes the losses in the material in the whole electromagnetic region due to the electronic transitions.

In the case of semiconducting materials, interband electronic transitions take place due to the interaction between the electromagnetic radiation (photons) and the matter (electrons). According to classical Lorentz oscillator model, the dielectric function is given by Equation 29.16. However, one of the models that are used for the modeling of the measured dielectric spectra of amorphous semiconductors is the Tauc–Lorentz (TL) model (Jellison and Modine 1996). This is based on the combination of the classical Lorentz dispersion relation and the Tauc density of states (Tauc et al. 1966) in the proximity of the fundamental optical gap E_g. This results in an asymmetrical Lorentzian lineshape for the imaginary part $\varepsilon_2(\omega)$ of the dielectric function. The TL dispersion model is described by the following relations in which the real part $\varepsilon_1(\omega)$ is determined by the imaginary part $\varepsilon_2(\omega)$ by the Kramers–Kronig integration (Jellison and Modine 1996) (see Equation 29.15a):

$$\varepsilon_2(\omega) = \begin{cases} \dfrac{AE_0C(\omega-E_g)^2}{(\omega^2-E_0^2)^2+C^2\omega^2}\dfrac{1}{\omega}, & \omega > E_g, \\ 0, & \omega \le E_g, \end{cases} \qquad (29.19)$$

where

E_g is the fundamental band gap energy

A is related to the transition probability

E_0 is the Lorentz resonant energy

C is the broadening term, which is a measure of the material disorder (Jellison and Modine 1996)

In the following, we present some representative examples of the investigation of the optical properties of various state-of-the-art materials by SE. These materials are for example, flexible polymeric films, and functional thin films used for state-of-the-art applications (as flexible organic electronic devices), as well as biocompatible inorganic surfaces and thin films and examples of protein layers adsorbed on these films.

29.2.1.3 Probing the Optical–Electronic and Anisotropic Properties of Polymers

Poly(ethylene terephthalate) (PET) and poly(ethylene naphthalate) (PEN) polymeric films have attracted significant attention due to their use in a numerous applications and industrial fields, as for the fabrication of high-performance polarization optics, data storage and recording media and optoelectronic devices, in food and pharmaceutical packaging, as well as in artificial heart valves, sutures and artificial vascular grafts (Gould et al. 1997, Wangand and Lehmann 1999, Tonelli 2002, Gioti et al. 2004, Laskarakis et al. 2004). Moreover, PET and PEN are excellent candidate materials to be used in the production of flexible organic electronic devices, such as flexible organic displays and photovoltaic cells by large scale manufacturing processes (Hamers 2001, Forrest 2004, Nomura et al. 2004), since they exhibit a combination of very important properties such as easy processing, flexibility, low cost, good mechanical properties, and reasonably high resistance to oxygen and water vapor penetration (Wang and Lehmann 1999, Yanaka et al. 2001, Fix et al. 2002, Tonelli 2002).

The monomer units of PET and PEN are schematically shown in Figure 29.3. The unit cell of PET, which is triclinic with a density of $1.455\,g/cm^3$, presents a C2h point symmetry and consists of an aromatic ring and an ester function that form the terephthalate group, and by a short aliphatic chain that constitutes the ethylene segment. PEN exhibits large similarities with PET and it has a

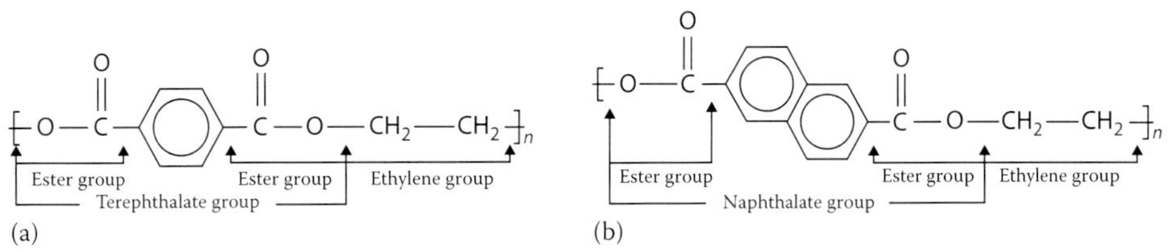

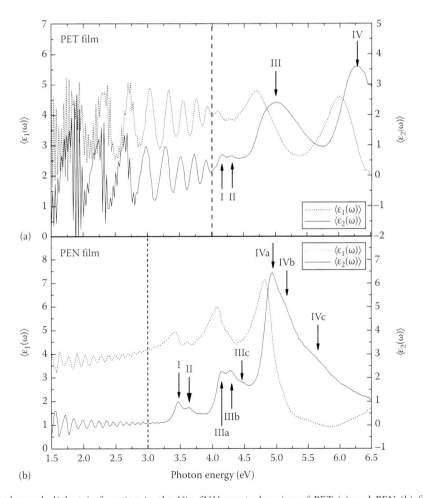

FIGURE 29.3 Monomer units of (a) PET and (b) PEN.

triclinic unit cell with the addition of a second phenyl ring, forming the naphthalene group (Laskarakis and Logothetidis 2006).

The large-scale production of these polymeric films includes a stretching process in order to achieve the necessary thickness and other desirable mechanical properties. During this process, the polymer films obtain a higher structural symmetry due to the preferred molecular orientation of the macromolecular chains leading to optical anisotropy. The optical axes are usually not perfectly oriented along the reference axis of the production line (stretching direction, referred as machine direction [MD]), introducing substantial difficulties to the investigation of their optical properties for production optimization and quality control (Laskarakis and Logothetidis 2006).

The understanding of PET and PEN optical and electronic properties and bonding structure and their relation to the degree of the macromolecular orientation is of fundamental importance and can significantly contribute toward the understanding of the mechanisms that take place during their surface functionalization or during the deposition of functional thin films on their surfaces.

Figures 29.4 and 29.5 show the measured pseudodielectric function $\langle \varepsilon(\omega) \rangle$ of the PET and PEN polymeric substrates in the Vis–fUV and IR spectral region, respectively. At energies below 4.0 (3.0) eV (Laskarakis and Logothetidis 2007) for PET (PEN), $\varepsilon(\omega)$ consists of interference fringes due to the multiple reflections of light at their back interface, as a result of

FIGURE 29.4 Measured pseudodielectric function in the Vis–fUV spectral region of PET (a) and PEN (b) films at a high symmetry orientation $\theta = 0°$.

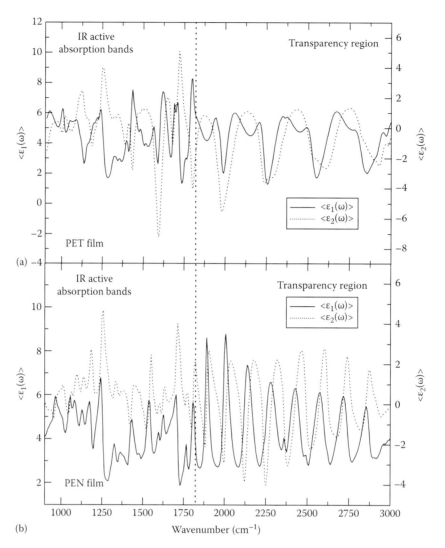

FIGURE 29.5 Real and imaginary parts of the measured $\langle\varepsilon(\omega)\rangle$ in the IR spectral region of (a) PET and (b) PEN film at azimuth angle $\theta = 0°$ and at an angle of incidence of $70°$.

their optical transparency (Gioti et al. 2004, Laskarakis and Logothetidis 2006). The existence of naphthalene group leads to the significant shift of the absorption bands of PEN to lower energies and to a characteristic split in all of them than in PET (Laskarakis et al. 2004). Peaks I and II can be attributed to the $n \rightarrow \pi^\star$ electronic transition of the nonbonded electron of the carbonyl O atom (Ouchi 1983, Ouchi et al. 2003, Laskarakis and Logothetidis 2006). The peak III (PET), which is possibly attributed to the spin-allowed, orbitally forbidden $^1A_{1g} \rightarrow {}^1B_{1u}$ transition, has been reported to be composed by two subpeaks with parallel polarization dependence (Laskarakis and Logothetidis 2006). Its higher broadening can be attributed to the break of the symmetry of the phenyl rings due to the substitution of carbon atoms (Kitano et al. 1995, Martínez-Antón 2002). In PEN, the peak III can be decomposed to the subpeaks IIIa (4.16 eV), IIIb, (4.32 eV), IIIc (4.49 eV). Finally, peak IV (PET) can be analyzed as two subpeaks with different polarizations (6.33 & 6.44 eV) after molecular orbital calculation based on π-electron approximation (Ouchi 1983, Ouchi

et al. 2003) and it can be attributed to the $^1A_{1g} \rightarrow {}^1B_{1u}$ electronic transition of the *para*-substituted benzene and naphthanene rings of the PET and PEN films with polarization rules rings plane. This peak in PEN can be analyzed to three subpeaks at 4.97(IVa), 5.2(IVb) and 5.7(IVc) eV (Ouchi 1983, Ouchi et al. 2003, Laskarakis and Logothetidis 2006).

Besides the detailed study of the electronic transitions of the PET and PEN polymeric films, it is possible to investigate the optical anisotropy of these materials by the analysis of the measured $\varepsilon(\omega)$ spectra. The detailed study of the optical anisotropy can provide important structural and morphological information regarding the arrangement of macromolecular chains. This can be achieved by the study of the dependence of the electronic transitions of PET and PEN with the angle between the plane of incidence and the MD (angle θ).

The measured $\varepsilon(\omega) = \varepsilon_1(\omega) + i\varepsilon_2(\omega)$ can be analyzed using damped harmonic Lorentzian oscillators (Equation 29.17), in combination to two-phase (air/bulk material) model as a function of the angle θ (see Figure 29.6). It is clear that the oscillator

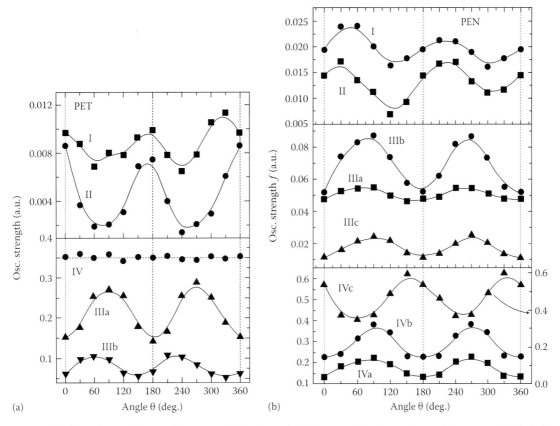

FIGURE 29.6 Angular dependence of the oscillator strength f with angle θ, determined by the analysis of the measured $\langle\varepsilon(\omega)\rangle$. The solid lines are guide to the eye.

strength f corresponding to peaks I to IV has a harmonic azimuthal dependence, with a period of half-complete rotation (180°), due to the light interaction with oriented and nonoriented regions. Also, the perpendicular azimuthal dependence of f_I and f_{II} is similar in both films, in agreement with the assignment of peaks I and II to the $n \to \pi^*$ electronic transition of the carbonyl group with selection rules perpendicular to the macromolecular chains. Moreover, all the subpeak components of peak III, attributed to the $\pi \to \pi^*$ excitation of the phenyl and naphthalene ring structures, are characterized by same parallel azimuthal dependence, as a result of the $\pi \to \pi^*$ electronic transition selection rule parallel to the MD. Finally, the azimuthal dependence of f_{IV} appears different between PET and PEN. In PET, the minor deviation of f_{IV} from the average value at ~0.3 is in agreement with the argument that peak IV consists of two overlapped subpeak components with opposite polarization dependence (Ouchi 1983, Ouchi et al. 2003). In PEN, the IVa and IVb subpeaks are characterized by similar parallel polarization dependence. However, the perpendicular polarization dependence of peak IVc could be attributed to the interaction to peak IVc of the intense absorption band that has been reported, at ~7.3 eV (characterized by perpendicular polarization) (Ouchi 1983, Ouchi et al. 2003, Laskarakis and Logothetidis 2006).

From the above, it can be supported that although the PET and PEN polymeric films are characterized by biaxial optical anisotropy, their optical response can be approximated as the

response of a uniaxial material, with its optic axis parallel to its surface. A more solid justification for this assumption can be deduced by the calculation of the refractive index n at the transparent region, or more precisely at $n^2(\omega) = \varepsilon_1(\omega \approx 0)$, shown in Figure 29.7. In the case of a uniaxial material, with its optic axis parallel to the surface, the dependence of $n(\omega = 0)$ with the angle θ is shown Figure 29.5, ($n_\parallel$ and $n_\perp$ are the principal values of the

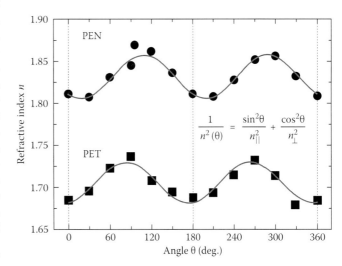

FIGURE 29.7 Refractive index n of PET (■) and PEN (●) in the transparent region, fitted with the relation shown in the inset (solid lines).

refractive index). In this analysis, we have used the formulation $\theta \rightarrow \theta + \Delta\theta$ to shift from the polymer film axis system to the optic axis system. $\Delta\theta$ represents the angle between the optic axis, (high symmetry axis corresponding to macromolecular chains), and the MD. It is clear that there is excellent agreement between the calculated values of n and the fit, justifying the approximation of PET and PEN as uniaxial materials with their optic axes parallel to their surface. Furthermore, by the azimuthal dependence of n, we obtain the angle between the optic axis and the MD. By the analysis, we obtain the following values: $n_{\perp PET} = 1.680$, $n_{\parallel PET} = 1.730$, $\Delta\theta_{(PET)} = 5.30° \pm 0.70°$, $n_{\perp PEN} = 1.805$, $n_{\parallel PEN} = 1.857$ and $\Delta\theta_{(PEN)} = 18.98° \pm 1.44°$ (Laskarakis and Logothetidis 2006).

The optical response of PET and PEN polymeric films in the IR spectral region are shown in Figure 29.5. Between 900 and 1800 cm^{-1} the strong absorption bands denote the contribution of the vibrational modes corresponding to the IR-active chemical bonds of PET and PEN. Above 1800 cm^{-1}, both films are optically transparent and their Fourier transform IR spectroscopic ellipsometry (FTIRSE) spectra are dominated by Fabry–Perot oscillations due to the multiple reflections of light at the film interfaces (Gioti et al. 2004, Laskarakis et al. 2004). Among the more intense characteristic vibration bands in the FTIRSE spectra of PET (Figure 29.6a), we observe the vibration modes at ~940 and ~971 cm^{-1} (trans) that could be attributed to the C—O stretching mode, the aromatic CH$_2$ stretching mode at ~1125 cm^{-1}, the ester mode at ~1255 cm^{-1}, the in-plane deformation of the C—H bond of the para-substituted benzene rings at ~1025 and ~1410 cm^{-1} and furthermore, the characteristic vibration band at 1720 cm^{-1} corresponding to the stretching vibration of the carbonyl C=O groups (Miller and Eichinger 1990, Cole et al. 1994, 1998). The band at 1342 cm^{-1} is attributed to the wagging mode of the ethylene glycol CH$_2$ groups of the *trans* conformations (Miller and Eichinger 1990, Cole et al. 1994, 1998). Also, we observe at 1470 cm^{-1} the characteristic peak corresponding to the CH$_2$ bending mode, whereas the C—H in plane deformation mode appears at ~1505 cm^{-1}.

Due to the existence of naphthalene ring structure in the monomer unit of PEN instead of a benzene ring structure, in PET, the FTIRSE spectra (Figure 29.5b) shows a similar IR response, however, with some additional vibration bands. These include the bands at 1098 cm^{-1} that has been attributed to the stretching and bending modes of ethylene glycol attached to the aromatic structures of the PEN monomer units. Moreover, the characteristic band at 1184 cm^{-1} corresponds to the C—C stretching modes of the naphthalene group. The complex bands at 1335 and 1374 cm^{-1} reveal the bending mode of the ethylene glycol CH$_2$ group in the *gauche* and *trans* conformations, respectively. The C=C stretching modes of the aromatic (naphthalene) ring structures of PEN can be observed at ~1635 cm^{-1}. Moreover, the stretching vibration of the carbonyl C=O group appears in lower energy in case of PEN (1713 cm^{-1}) than in PET (1720 cm^{-1}). This could be the result of the increased conjugation due to the existence of naphthalene (PEN) instead of benzene (PET) rings structures, which shifts the maximum absorbance to lower wavenumbers.

Finally, the determination of the optical response of PET and PEN films in the extended spectral region, from the IR (bonding vibration modes) to Vis–fUV (band-to-band electronic transitions), allows the calculation of the bulk dielectric function $\varepsilon(\omega)$. Figure 29.8 shows the determined $\varepsilon(\omega)$ of PET and PEN, calculated at the high symmetry orientation $\theta = 0°$ (plane of incidence parallel to the MD) (Gioti et al. 2004, Laskarakis et al. 2004). The $\varepsilon(\omega)$ has been calculated using the best-fit parameters obtained by the analysis of the measured $\langle\varepsilon(\omega)\rangle$ taking account the peaks I to IV (Vis–fUV) and the characteristic bands corresponding to the more intense bonding vibrations in the wavenumber region 900–1800 cm^{-1} (Gioti et al. 2004, Laskarakis et al. 2004).

The high-energy dielectric constant that represents the optical transitions at energies $\omega \geq 6.5$ eV, has the values of 2.39 (PET) and 2.44 (PEN), which are higher than unity (Equation 29.16). This indicates that more electronic transitions should be expected at higher energies. Indeed, it has been found at the literature, that both PET and PEN polymers show electronic transitions at energies (around 13.3 and 15.5 eV), which are characterized by optical anisotropy (Ouchi et al. 2003). These band-to-band transitions do not affect the optical absorption measured up to 6.5 eV measurement limit, however contribute to the $\varepsilon_1(\omega)$, inducing an increase from unity (see Equations 29.15c and 29.16).

From the above it is clear that the implementation of ellipsometry as a nanometrology tool for the investigation of the optical and electronic properties of flexible polymer films is of major importance and provides significant information on the understanding of their optical, structural, and vibrational properties. This information is required for the study of the active (e.g., small molecule and polymer organic semiconductors) and passive materials (electrodes, barrier layers) that can be deposited onto the polymeric films used as substrates for organic electronics applications.

29.2.1.4 Optical Studies of Inorganic–Organic Layers with Embedded SiO$_2$ Nanoparticles

One of the major challenges that have to be addressed and overcome in order for flexible organic electronic devices (such as organic light emitting diodes [OLED], electrochromic displays, flexible lighting, and flexible photovoltaic cells [OPVs]) to reveal their full potential is their encapsulation in transparent media that will provide the necessary protection against atmospheric gas molecule (H$_2$O and O$_2$) permeation (Logothetidis and Laskarakis 2008, Laskarakis et al. 2009).

The use of SiO$_x$ thin films and hybrid (inorganic–organic nanocomposite) polymers for the encapsulation of flexible electronic devices provides sufficient final protection against permeation of atmospheric gas molecules (H$_2$O and O$_2$). A SiO$_x$ thin film or nanolayer of 30–50 nm thick reduces the permeation of atmospheric gases by 50–100 times (Logothetidis and Laskarakis 2008). These are prepared by electron beam evaporation, and sputtering vacuum techniques on PET substrates and will be discussed in the next paragraph. On the other hand, the hybrid barrier materials are synthesized via the sol–gel processes from organoalkoxysilanes, and they have strong covalent or

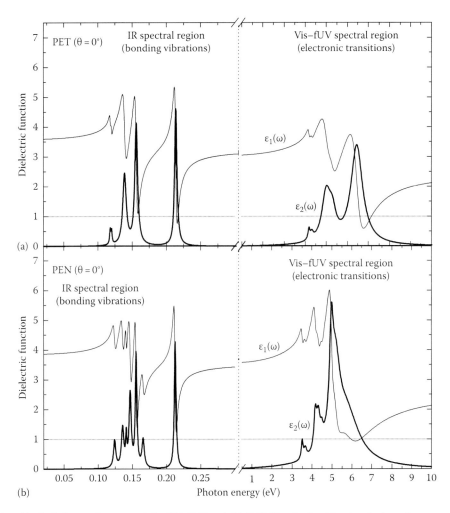

FIGURE 29.8 Calculated bulk dielectric function $\varepsilon(\omega)$ of (a) PET and (b) PEN film for $\theta = 0°$ using the best-fit parameters deduced by the SE analysis in the IR and Vis–fUV spectral regions.

ionic–covalent bonds between the inorganic and organic components. A schematic representation of these hybrid materials is shown in Figure 29.9 (Laskarakis et al. 2009).

One of the main factors that affect the barrier response of hybrid polymers is the crosslinking between the inorganic and

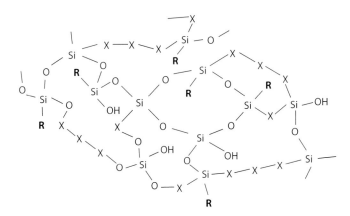

FIGURE 29.9 Schematic representation of the structure of the hybrid barrier materials.

the organic components as well as its adhesion on the substrate. Also, one promising approach in this direction and for the reduction of permeation of atmospheric gases further more by ~100 times is the inclusion of SiO_2 nanoparticles (SiO_2–NP). This can be done during the synthesis process of the hybrid polymers, with the final aim to enhance the inorganic–organic crosslinking and to realize a more cohesive bonding network between the organic and inorganic components (Laskarakis et al. 2009).

In the following, we will present some representative examples on the investigation of the optical properties of hybrid polymer used as barrier layers deposited onto SiO_x/PET substrates and the effect of the inclusion of SiO_2–NP on their microstructure, optical properties, and functionality. The optical properties were measured in an extended spectral region from the IR to the Vis–fUV spectral region in order to stimulate different light-matter mechanisms (bonding vibrations, interband transitions).

The measured pseudodielectric function $\langle\varepsilon(\omega)\rangle$ in the Vis–fUV energy region (1.5–6.5 eV) of a representative hybrid barrier material with different amounts of SiO_2 nanoparticles (SiO_2–NP) is shown in Figure 29.10. In the lower energy region, the measured $\langle\varepsilon(\omega)\rangle$ is dominated by interference fringes that are attributed to

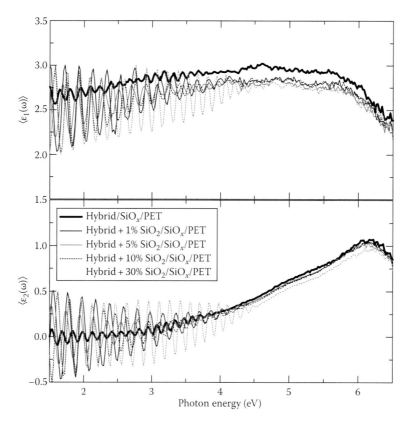

FIGURE 29.10 Measured pseudodielectric function $\langle\varepsilon(\omega)\rangle = \langle\varepsilon_1(\omega)\rangle + i\,\langle\varepsilon_2(\omega)\rangle$ in the Vis–fUV spectral region of the hybrid barrier layer deposited onto SiO$_x$/PET.

the multiple light reflections at the interfaces between the hybrid layer and the SiO$_x$ intermediate layer, as the result of their optical transparency in this energy region. At higher energies, the optical absorption of the hybrid polymer takes place. The hybrid polymers can be treated as composite materials consisting of an organic and an inorganic component. Thus, in order to extract quantitative results from the measured $\langle\varepsilon(\omega)\rangle$, this has been analyzed by the use of four-phase geometrical model (air/hybrid polymer/SiO$_x$/PET substrate) that consists of a hybrid layer (with thickness d) on top of a SiO$_x$ layers grown on top of a PET (bulk) substrate, where the ambient is air ($\varepsilon_1(\omega) = 1$, $\varepsilon_2(\omega) = 0$, for all

energy values ω). This is shown in Figure 29.11. In order to take into account the optical response of the intermediate SiO$_x$ layer deposited directly onto the PET substrate, the SiO$_x$ layer thickness and optical properties has been measured in advance. This information has been used for the analysis of the measured $\langle\varepsilon(\omega)\rangle$ from the complete hybrid/SiO$_x$/PET layer stack.

The dependence of the optical parameters deduced from the $\langle\varepsilon(\omega)\rangle$ analysis (fundamental gap E_g and electronic transition E_0) as a function of the SiO$_2$—NP content in two representative hybrid polymers (hybrid #1, hybrid #2) is shown in Figure 29.12. It is clear that the hybrid #2 is characterized by higher optical

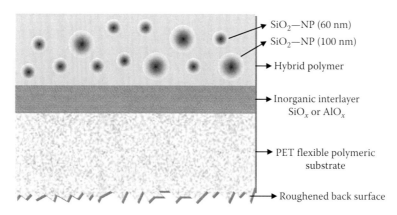

FIGURE 29.11 Theoretical structure used for the analysis of the measured $\langle\varepsilon(\omega)\rangle$ spectra of the hybrid barrier layers.

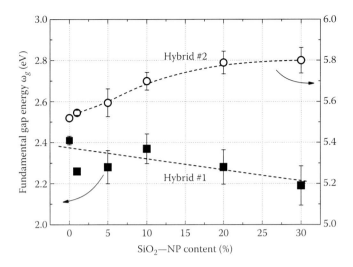

FIGURE 29.12 Dependence of the fundamental energy gap E_g of the two hybrid materials as a function of the content of SiO_2—NP.

transparency due to the higher values of the fundamental energy gap E_g, from 5.5 eV (0% SiO_2—NP) to 5.8 eV (30% SiO_2—NP). However, the hybrid #1 is characterized by lower optical transparency (E_g values in the range of 2.2–2.4 eV), and the increase of the amount of SiO_2—NP results in a slight decrease in the fundamental gap E_g (reduction of the material's transparency) as well as a reduction of E_0 values (Laskarakis et al. 2009).

In addition to the above investigations, the optical measurements of the hybrid polymers with embedded nanoparticles in the IR spectral region can provide information about the different bonding groups based in the study of their vibration modes. In Figure 29.13 the imaginary part $\langle\varepsilon_2(\omega)\rangle$ of the measured $\langle\varepsilon(\omega)\rangle$ of a representative hybrid layer with various SiO_2—NP contents deposited onto PET substrates are shown. Above 1800 cm^{-1}, the samples are optically transparent and their $\langle\varepsilon(\omega)\rangle$ spectra are dominated by Fabry–Pérot oscillations due to the multiple reflections of light at the hybrid polymer, SiO_x nanolayer, and PET interfaces. Among the more intense vibration bands, we can distinguish the glycol CH_2 wagging peak at 1340 cm^{-1} and the intense complex bands around 1240–1330 cm^{-1} that arise mainly from ester group vibrations. Also, the stretching mode of the carbonyl C=O group of PET is shown at 1720 cm^{-1}. The contribution of the Si—O bonding vibration is dominant in the wavenumber region of 1050–1100 cm^{-1}. It is clear from Figure 29.13 that the increase of the SiO_2—NP content in the hybrid layer is associated to the increase of the Si—O vibration mode in the area of 1050 cm^{-1}. The effect of the several vibration modes on the complex dielectric function can be described by Equation 29.16.

For the parameterization of the Si—O stretching vibration, the $\langle\varepsilon(\omega)\rangle$ spectra has been analyzed by the use of a four-phase model (air/(hybrid + SiO_2—NP)/SiO_x/PET substrate). The parameters that have been fitted are the thickness and the optical properties of the hybrid polymer layer by the use damped harmonic oscillator at the wavenumber region of 1070 cm^{-1} in order to describe the optical response of the Si—O bonding group, as well as the volume fraction of the SiO_2 phase (describing the embedded SiO_2—NP) (Logothetidis and Laskarakis 2008).

The oscillator strength f of the Si—O bonding group has been found to increase with increasing SiO_2—NP content. This is due to

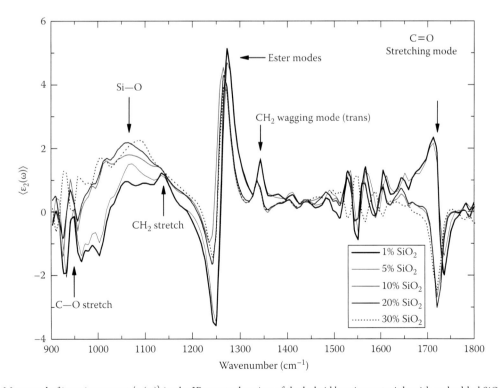

FIGURE 29.13 Measured of imaginary part $\langle\varepsilon_2(\omega)\rangle$ in the IR spectral region of the hybrid barrier materials with embedded SiO_2—NP of different percentages at an angle of incidence of 70°.

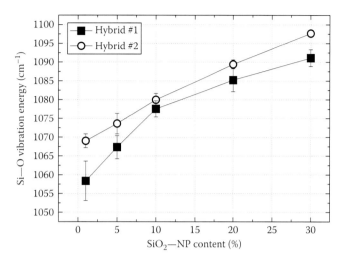

FIGURE 29.14 Dependence of the bonding vibration energy of the Si—O group as a function of the SiO$_2$—NP content in the two hybrid materials.

the enhanced contribution of the Si—O bonds from both the SiO$_2$—NPs and the inorganic component of the hybrid polymer. The calculated values of the Si—O stretching vibration energy and their dependence with the content of the SiO$_2$—NP in two representative hybrid materials are shown in Figure 29.14. It can be seen that the increase of the SiO$_2$—NP content is correlated with an increase of the Si—O vibration energy from 1058 (for 1% SiO$_2$—NP) to 1091 cm^{-1} (for 30% SiO$_2$—NP) in hybrid #1, whereas in the case of hybrid #2, this increase takes place from 1069 to 1097 cm^{-1}. Since the value of the Si—O bonding vibration energy is associated with the stoichiometry x of the material (Pai et al. 1986), the increase of the SiO$_2$—NP results to hybrid polymer materials, which are characterized by a higher amount of Si—O bonds (Logothetidis and Laskarakis 2008, Laskarakis et al. 2009).

The above findings can be combined with the results obtained from other nanometrology tools in order to provide a complete

aspect of the materials that are used for state-of-the-art applications, such as flexible organic electronics. For example, AFM can be employed for the imaging of the SiO$_2$—NPs that are embedded in the hybrid polymer materials. Figure 29.15a and b shows the surface morphology and phase image of a hybrid polymer sample, which contains 10% SiO$_2$ nanoparticles and was deposited on a PET substrate. Nanoparticles of about 60 nm in diameter are clearly observed embedded in the organic polymer matrix. The phase image shows a high contrast between the two materials. The values of the RMS surface roughness is $R_q = 2$ nm and peak-to-valley distance is 20.4 nm.

Also, from the above, it is clear that spectroscopic ellipsometry can be employed for the investigation of the optical properties of barrier layers deposited onto PET substrates and of the effect of the inclusion of SiO$_2$—NP on their microstructure, optical properties, and functionality. The optical properties were measured in an extended spectral region from the IR to the Vis–fUV spectral region in order to stimulate different light-matter mechanisms (bonding vibrations, interband transitions), that will provide significant insights on their properties. These results demonstrate the importance of optical characterization of barrier materials on the understanding of the mechanisms that dominate their functionality with the aim to optimize their optical and barrier response in order to be used for the encapsulation of flexible organic electronic devices.

29.2.1.5 Real-Time Optical and Structural Studies of Nanolayer Growth on Polymeric Films

In-situ and real-time SE is a powerful, nondestructive and surface sensitive optical technique used to monitor the deposition rate and the growth mechanisms of thin films during their deposition (Logothetidis 2001, Gravalidis et al. 2004). The implementation of real-time SE monitoring and control to large-scale production of functional thin films for numerous applications

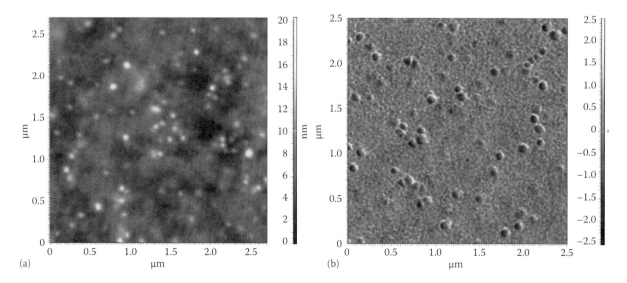

(a)   (b)

FIGURE 29.15 (a) AFM image and (b) phase image of a hybrid polymer with 10% of SiO$_2$ nanoparticles. The scan size is $2.7 \times 2.7\,\mu m^2$.

will lead to the optimization of the materials quality and increase of production yield (Logothetidis 2001, Gravalidis et al. 2004).

In the following, we will show some representative examples concerning the real-time investigation of the optical properties of the growing SiO_x thin films deposited onto PET substrates by electron beam evaporation process in an ultrahigh vacuum (UHV) deposition chamber. The investigation of the optical properties has been performed in the Vis–fUV energy region (1.5–6.5 eV or 190–826 nm) by an ultrafast multiwavelength phase-modulated spectroscopic ellipsometer that was adapted onto the UHV chamber at an angle of 70°. This system is equipped with a 32-fiber-optic array detector for simultaneous measurements at 32 different wavelengths (MWL mode), in the energy range 3–6.5 eV. The sampling time (ST) of the MWL measurements (time interval between subsequent measurements) were 200 and 500 ms, whereas the integration time for each measurement was 100 ms. In this way, it is possible to realize the real-time investigation of the growth mechanisms of the deposited inorganic thin films.

Figure 29.16 shows the evolution of the measured imaginary part $\langle \varepsilon_2(\omega) \rangle$ in the 3–6.5 eV region with the deposition time. The $\langle \varepsilon_2(\omega) \rangle$ of the PET flexible substrate is shown with black squares (■) whereas the evolution of the $\langle \varepsilon_2(\omega) \rangle$ (shown with gray lines) with time indicates the growth of the inorganic film on top of PET substrate. The last spectrum indicated by hollow circles (○) is measured by the monochromator unit of the SE system in the energy range 3–6.5 eV with a 20 meV step.

The determination of the optical and electronic properties of the SiO_x thin films has been realized by the analysis of the measured pseudodielectric function $\langle \varepsilon(\omega) \rangle$ with the TL dispersion model (see Equation 29.19) (Jellison and Modine 1996). The geometrical structure that has been used for the analysis procedure

includes a three-phase model that consists of air, the inorganic layer of thickness d, and the flexible polymeric substrate.

Based on the analysis of the $\langle \varepsilon(\omega) \rangle$ spectra, the evolution of the SiO_x films thickness and optical properties can be evaluated with the deposition time. The best-fit parameters for the evolution of the SiO_x film thickness d and the deposition rate (derivative of the thickness as a function of the deposition time) are shown in Figure 29.17. As it can be seen, the growth of the SiO_x film onto PET can be separated into three distinct stages, according to the growth mechanisms that take place (Laskarakis et al. 2010). These stages are the following:

- *Stage I:* During this stage, there is a sudden increase of the deposition rate and the film's thickness. This increase takes place for the first 1 s to the values of film thickness up to 50 Å. This is attributed to the formation of a composite material on the substrate that consists of PET and of the growing film. This is associated with the modification of the PET surface from the arrived SiO_x (charged and neutral) particles and species as well as from the ultraviolet radiation emitted by the plasma (Koidis et al. 2008).
- *Stage II:* The thickness increases rapidly with a high deposition rate of ~60 Å/s, which remains stable for the period $1 < t < 5$ s, leading to a thickness value of ~300 Å. This behavior can be attributed to the formation of separate clusters of the growing film's surface. These clusters are growing homogeneously during stage II (Koidis et al. 2008).
- *Stage III:* The coalescence of the individual clusters and the homogeneous film growth for both films takes place. The deposition rate at higher deposition times ($t > 5$ s) follows a constant but rapidly oscillating behavior. The oscillation of the deposition rate is characteristic of the island-type growth and defines the distinct growth stages (Gravalidis et al. 2004).

In addition to the study of the growing mechanisms, the evolution of the electronic and optical properties of SiO_x films can be interpreted by the parameters of TL model (Equation 29.19) and provide valuable information in terms of their compositional and nanostructural properties, as well as on the stoichiometry. Firstly, the energy position of the maximum absorption E_0 determines the electronic transitions and depends on the stoichiometry of the material. E_0 shows a significant difference for the SiO_2, and SiO, and SiO_x films, as shown in Figure 29.18. In the case of SiO_2 film, the E_0 values exhibit fluctuations between 9 and 11 eV, and they are eliminated after the completion of the first ~35 s of the deposition time, taking an almost constant value of approximately 11 eV. Similar behavior is also obtained for the E_0 of the SiO_x film, with its value to be stabilized to ~7 eV. On the contrary, for the SiO film we obtain a gradual increase in E_0, with its constant value to reach ~6.2 eV. If we take into account the E_0 of the reference bulk SiO_2 and SiO materials, i.e., 10.8 (Palik 1991)–12.0 (Herzinger et al. 1998) and 5.7 eV (Palik 1991), respectively, we can verify the respective x values for the studied SiO_2 and SiO films. Furthermore, the estimated x value for the SiO_x film should be $1 < x \ll 2$.

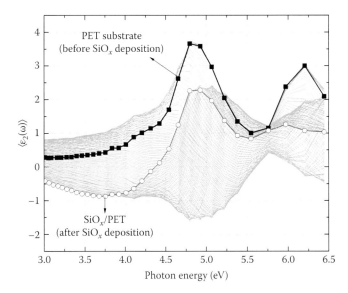

FIGURE 29.16 The time evolution of the imaginary part of the measured pseudo-dielectric function $\langle \varepsilon(\omega) \rangle = \langle \varepsilon_1(\omega) \rangle + i \langle \varepsilon_2(\omega) \rangle$ obtained for the flexible PET substrate and the deposition of the SiO_x film by e-beam evaporation.

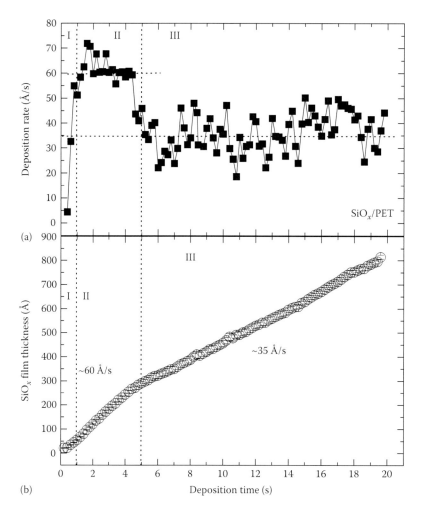

FIGURE 29.17 The calculated thickness of SiO$_x$ film grown onto PET substrate as a function of time, after the analysis of the measured $\langle \varepsilon(\omega) \rangle$ in real-time during their growth. Also, figure (a) shows the calculated deposition rate of the SiO$_x$ film. The vertical lines show the different stages of growth.

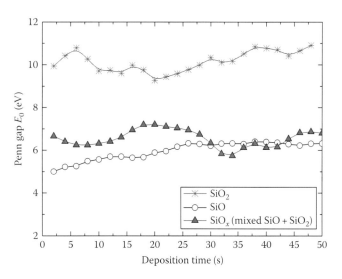

FIGURE 29.18 The evolution of E_0 of the SiO$_2$, SiO, and SiO$_x$ films during the first 50 s of their deposition.

The evolution of the fundamental energy gap E_g with the thickness for the three films is presented in Figure 29.19. The correlation of the E_g values between the SiO$_2$, SiO, and SiO$_x$ films is similar to that of E_0. As it can be seen in Figure 29.19, the film deposited from SiO$_2$ evaporation material has the highest E_g compared to the films deposited from SiO and mixed SiO + SiO$_2$ evaporation materials. This in combination to E_0 means that the excess of the O creates an insulating film, appropriate for microelectronic and optical applications meeting the demands of the optical transparency. The lower percentage of the O in the deposition of the two latter films controls their stoichiometry and leads to the formation of films with x well below 2. This means that this kind of films can be normally used as gas barriers coatings for the flexible packaging applications (Krug 1990, Gioti et al. 2009).

Real-time SE has been also implemented for the study of the deposition mechanisms of zinc oxide (ZnO) thin films onto various substrates, such as crystalline Si, PET, and PEN. ZnO is a wide direct band-gap semiconductor having the hexagonical

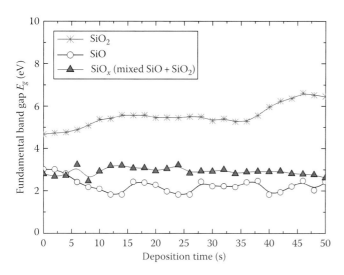

FIGURE 29.19 The evolution of E_g of the SiO_2, SiO, and SiO_x films during the first 50 s of their deposition.

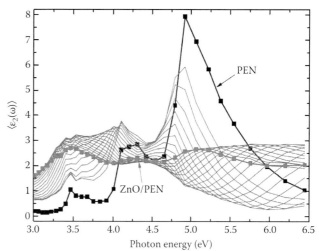

FIGURE 29.20 Evolution of the imaginary part of the measured dielectric function $\langle \varepsilon_2(\omega) \rangle$ during the deposition of ZnO onto PEN.

crystal structure of wurtzite (Koidis et al. 2008, Logothetidis and Laskarakis 2008). It is very promising material for application in flexible electronic devices (FEDs) and it has attracted large commercial and scientific interest. It combines a wide range of benefits such as high electrical conductivity, piezoelectricity, easy fabrication, low cost, nontoxicity, ultraviolet absorption behavior, and its compatibility with large-scale applications (Koidis et al. 2008, Logothetidis and Laskarakis 2008). In addition, it is of great importance to fully comprehend the growth mechanisms, the functionality, and the combination of the organic–inorganic materials as well as the effect of the deposition parameters in their optical, structural, and electronic properties in order to reveal the full potential of flexible electronic devices, such as flexible OLEDs and OPVs.

Figure 29.20 shows the evolution of the imaginary part of the measured pseudodielectric function $\langle \varepsilon_2(\omega) \rangle$, during the deposition of the ZnO film on PEN film. Figure 29.21 shows the evolution of the growing ZnO film thickness onto different substrates (c-Si, PET and PEN) as determined by the analysis of the measured $\langle \varepsilon(\omega) \rangle$ spectra. Focusing on the early growth stages (first 120 s of deposition) as shown in more detail in Figure 29.22, it is clear that the chemical structure and surface properties of the substrate affects the growth mechanism of ZnO film. The ZnO film growth can be qualitatively separated into three stages (I, II and III), according to the growth mechanisms taking place (Koidis et al. 2008). During the early stages of growth (Stage I), in the cases of ZnO/PET and ZnO/PEN, there is a significant increase of the apparent thickness at

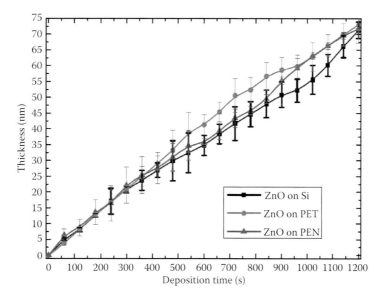

FIGURE 29.21 The evolution of the ZnO films total thickness during the deposition onto Si, PET, and PEN substrates.

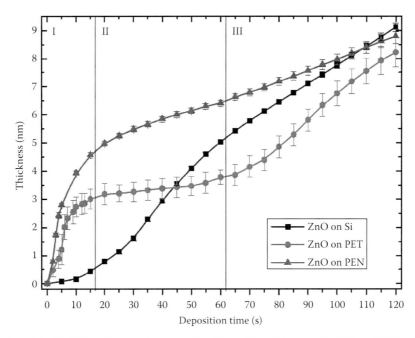

FIGURE 29.22   The evolution of the ZnO film thickness during the first 120 s of deposition onto Si, PET, and PEN substrates.

$t = 15$ s for ZnO/PET and at $t = 20$ s for ZnO/PEN, attributed to polymer surface modification by the incoming ZnO and (neutral and charged) Ar particles and the formation of an interface layer. This modification includes the changing of the surface chemical bonding and the incorporation of the arriving particles in the polymeric surface. High-resolution transmission electron microscopy (HRTEM) investigation of ZnO/PET sample revealed a composite nanocrystalline overlayer of ~10 nm on top of PET surface, including an interface layer of 1–2 nm (Koidis et al. 2009). As it can be seen in Figure 29.22, at the end of stage I, the ZnO film thicknesses are in the range of ~3 nm (ZnO/PET), ~5 nm (ZnO/PEN), and ~1 nm (ZnO/Si), whereas the deposition rates are ~0.24, ~0.37, and ~0.06 nm/s, respectively (Koidis et al. 2008).

The larger apparent thickness in the case of ZnO/PEN can be explained by taking into account the fact that the PEN surface is characterized by higher roughness values than in the case of PET and Si. More specifically, the rms surface roughness values of PEN, PET, and Si substrates, as measured by AFM, are ~1.8, ~1, and ~0.15 nm, respectively. During Stage II, in the case of ZnO/Si, it is shown from Figure 29.22 that the ZnO film is deposited directly to the aforedeposited layer, favoring the film growth (deposition rate of 0.11 nm/s). In the case of ZnO/PET, the low deposition rate (0.01 nm/s) indicates the early stages of ZnO deposition to the modified interface layer. In ZnO/PEN, the deposition rate at this stage is higher (0.04 nm/s), indicating a more effective deposition onto the modified surface. Finally, during Stage III the ZnO film growth is dominated by a layer-by-layer mechanism (deposition rates: 0.11, 0.04 and 0.07 nm/s for PET, PEN and Si, respectively), lasting until the end of deposition (1200 s), as it

is shown in Figure 29.17 (Koidis et al. 2008, 2009, Logothetidis and Laskarakis 2008).

Finally, the above two examples clearly demonstrate the capabilities of real-time SE toward the understanding of the growth mechanisms of materials during their deposition onto polymeric substrates. This nanometrology method has an enormous potential for implementation to production lines for state-of-the-art applications as a quality control tool. Also, its robustness and high flexibility for adaptation gives numerous capabilities for the improvement of the produced structures (thin films, nanomaterials, etc.) as well as for the optimization of the production process in terms of cost and waste.

29.2.1.6 IR to Far UV Ellipsometry in Nanobiology and Nanomedicine

The major clinical problem in the use of artificial materials is their blood compatibility. When these materials come into direct contact with plasma, blood coagulation, i.e., thrombosis, may occur. The mechanisms of the plasma protein adsorption and platelet adhesion on the surface of an artificial implant are very complicated. In a first approximation, the investigation and evaluation of the blood compatibility of the implant can be carried out by probing the interaction of the surface with the Human Serum Albumin (HSA) and Fibrinogen (Fib) proteins. The enhancement of HSA adhesion against to the Fib is the main target for the functionalization of hemocompatible coatings.

Ellipsometry is a critical technique for understanding the relationship and interface between artificial materials and biomolecules, a particularly important objective for biological and medical applications. With ellipsometry, it is possible to determine the optical functions of ultrathin layers, such as the protein layers (Arwin 2000, Wormeester et al. 2004) that are formed after dipping for some time the biofunctional surfaces, such as carbon-based, and amorphous boron nitride thin films, into the proteins solutions and drying them afterwards. Early measurements were done by recording spectra in air on surfaces before and after protein adsorption.

In Figures 29.23 and 29.24 ellipsometric spectra measured before and after HSA and Fib proteins adsorption, respectively,

in the carbon-based layers are shown. The difference between the two spectra contains information about the amount of protein adsorbed as well as the density of the adsorbed protein layers. Quantification is achieved by analysis using multilayer model according to which the geometrical structure of the examined specimens is composed by two-layers: the carbon-based thin film on top of c-Si substrate and a protein layer as an overlayer (see insets in Figures 29.23 and 29.24).

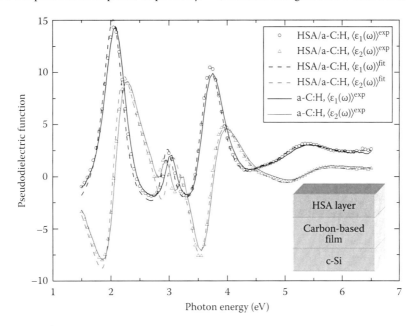

FIGURE 29.23 The measured $\langle \varepsilon(\omega) \rangle^{exp}$ before (solid lines) and after (symbols) the HSA protein adsorption on the surface of #721 a-C:H film, together with the calculated through the modeling and fitting procedures $\langle \varepsilon(\omega) \rangle^{fit}$ (dash lines). The inset shows the geometrical model used for the SE modeling/analysis.

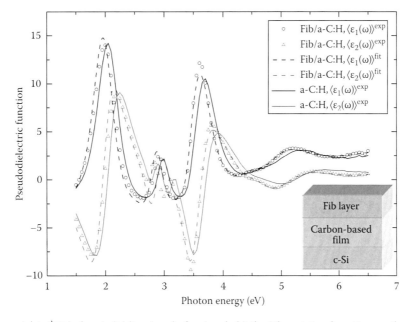

FIGURE 29.24 The measured $\langle \varepsilon(\omega) \rangle^{exp}$ before (solid lines) and after (symbols) the Fib protein adsorption on the surface of #721 a-C:H film, together with the calculated through the modeling and fitting procedures $\langle \varepsilon(\omega) \rangle^{fit}$ (dash lines). The inset shows the geometrical model used for the SE modeling/analysis.

Characteristics for most protein layers are that their optical functions are rather featureless in the Vis–UV energy range (Arwin 2000). Thus, the extension of the experimentally measured upper limit of energy to fUV range is very crucial for the accurate determination of their dielectric function or refractive index spectral dependence. The parameterization of the HSA and Fib layers' optical response is realized by using the TL model. By this, it is possible to determine the bulk $\varepsilon(\omega)$ for both HSA and Fib. A representative example on the derived results by this analysis is shown in Figure 29.25. The left axis measures the refractive index n whereas the right axis the imaginary part of the dielectric function $\varepsilon_2(\omega)$. The optical transparency is evident by the zero values of $\varepsilon_2(\omega)$ up to about 4.0 eV, and the relatively low absolute values above this energy. Also, the $n(\omega)$ exhibits a weak dispersion up to 4.0 eV with almost constant values for the two protein layers.

The size of the HSA molecule is at the order of $3 \times 8 \times 8$ nm, and that of Fib significantly larger at the order of $9 \times 7 \times 46$ nm. Their possible configuration after their adsorption on the surfaces should result considerable empty space between them, which is incorporated in the description of the optical response of the protein layer with an overall reduction of the $n(\omega)$ or the real part of the dielectric function $\varepsilon_1(\omega)$. This happens due to the optical model we applied, in which the protein layers are taken homogenous and the surface roughness are not taken into account. In a first approximation, by using this simplified model we aim to evaluate the surface concentration of HSA and Fib on various carbon-based thin films than to derive details on the arrangement of the protein molecules on the surface. The low thickness and density of these protein layers lead to small alterations between the measured SE spectra of the samples before and after proteins' adsorption (see Figures 29.23 and 29.24). Thus, the application of a simple model with the minimum possible fitting parameters is more preferable and the respective results are more reliable.

The surface concentration Γ (μg/cm^2) of the HSA and Fib can be derived by the Cuypers formula (Cuypers et al. 1983):

$$\Gamma = 0.1 \cdot d \cdot \frac{M}{A} \cdot \frac{n_f^2 - 1}{n_f^2 + 2}, \qquad (29.20)$$

where

A is the molar refractivity (cm^3/mol)
M is the molecular weight of the protein
d and n_f the thickness and the refractive index of the protein layer

The final comparison of the examined carbon films is performed through the ratio between the calculated surface concentrations for HSA and Fib protein layers (HSA/Fib ratio). The numerical data are presented in Table 29.1. The dependence of HSA/Fib ratio on the refractive index $n(\omega = 0\,\text{eV}) = [\varepsilon_1(\omega = 0\,\text{eV})]^{1/2}$ of the a-C, tetrahedral amorphous carbon (ta-C) and amorphous hydrogenated carbon (a-C:H) films is shown in Figure 29.26. The enhancement of the HSA/Fib ratio for the a-C:H films grown with a floating bias onto the substrate and a small percentage of H$_2$ partial pressure is pronounced to that developed with the higher H$_2$ concentration in plasma, appears to be the most hemocompatible film among those examined. The a-C:H films developed with a negatively biased substrate exhibit a moderate HSA/Fib ratio. On the other hand for the sputtered a-C films there is no any monotonic relation between their refractive index and the calculated HSA/Fib ratio, but a direct correlation between the sp^3 fraction and the HSA/Fib ratio can be derived. Finally, for the ta-C film it was obtained the lower HSA/Fib ratio. This can be justified due to the surface properties of this type of films (Logothetidis et al. 2005).

Also, Fourier transform IR phase modulated spectroscopic ellipsometry (FTIRSE) in the 900–3500 cm^{-1} wavenumber

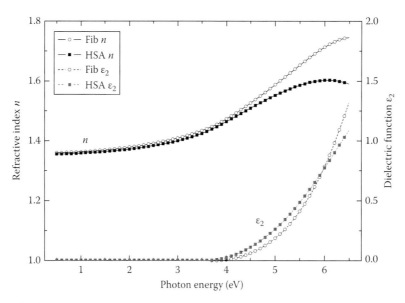

FIGURE 29.25 The calculated bulk $\varepsilon(\omega)$ using the best-fit parameters for the Fib and HSA protein layers.

TABLE 29.1 Sample Types, Deposition Parameters, Thickness, Refractive Index, and the sp³ Content for the Carbon-Based Thin Films Considered in This Work

Film Type	V_b (V)	H₂ at.% in Plasma	Thickness, d (Å)	Refractive Index, n	sp³ Fraction (%)	HSA/Fib Ratio
a-C	+10	0	320	2.023	20	0.125
a-C	−200	0	680	2.436	30	0.288
a-C	−40	0	450	1.911	45	0.592
ta-C	—	0	290	2.505	>80	0.100
a-C:H	−40	10	2.500	1.666	42–45	0.440
a-C:H	−40	5	2.500	1.676	42–45	0.271
a-C:H	+10	5	2.500	1.477	42–45	0.902
a-C:H	+10	10	2.500	1.501	42–45	1.002

Note: The calculated HSA/Fib ratio is also included.

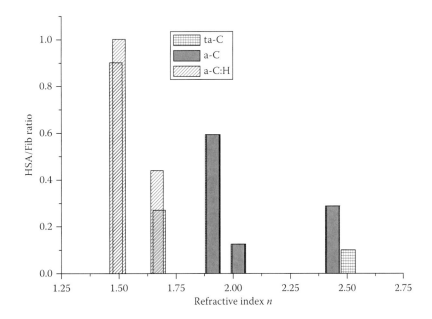

FIGURE 29.26 The calculated HSA/Fib ratio for the studied carbon-based thin films, versus the refractive index $n(\omega = 0\,\text{eV})$.

region has been used for the detailed study of the bonding structure of the absorbed proteins and platelets onto the a-C:H thin films. The angle of incidence of the IR light onto the samples was 70° and the measurements have been performed for 30 min, after 24 h incubation of the thin films into single HSA and Fib solutions and after 30 min incubation of the sample into plasma rich in platelets (PRP), in ambient environment. FTIRSE is a nondestructive optical technique for investigation of vibrational properties of samples, which provides identification of IR responses even at a monolayer level, and it can be used for the investigation of the optical properties in a variety of media (vacuum, air, transparent liquids, etc.) without special conditions for the measured materials. Its capability for data acquisition time down to 2 s, allows its application for in-situ and real-time monitoring of processes.

By using FTIRSE, the characteristic bands corresponding to the different bonding structures of the HSA and Fib proteins have been investigated. However, due to weak contribution of the characteristic vibration bands, and in order to deduce accurate results from the analysis of the protein/a-C:H IR spectra, we have used the ellipsometric optical density D defined by the relation $D = \ln(\rho_s/\rho)$, where ρ and ρ_s refer to the ellipsometric ratio of the protein/a-C:H and of the a-C:H, respectively. Figure 29.27 shows the real (Re D) and imaginary (Im D) part of the optical density D of the a-C:H thin film, on which the HSA and Fib proteins have adsorbed. As it can be seen in Figure 29.27, the contribution of the complex vibrational modes Amide I and Amide II are evident, which are characteristic of the protein secondary structures. Amide I corresponds to the absorption modes in the 1615–1700 cm⁻¹ region involving all the C=O peptide groups whereas the Amide II absorptions in the 1560–1510 cm⁻¹ range are related to CONH units (C—N stretching coupled with N—H bending modes). Also, the multiple bands at the 1000–1300 cm⁻¹ are attributed to the complex stretching vibration of C—O attached to different bonding structures of the protein backbone.

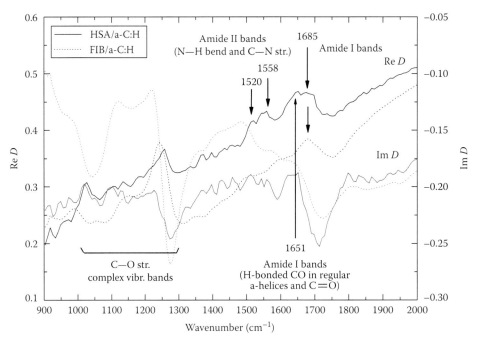

FIGURE 29.27   Real (Re D) and imaginary (Im D) part of the ellipsometric optical density D, of a-C:H thin film, with HSA and Fib proteins adsorbed onto the surface (incubation time 24 h),

FTIRSE has also been employed for the identification of the bonding structure of the adsorbed platelets onto the a-C:H thin films. Figure 29.28 shows the measured pseudodielectric function $\langle \varepsilon(\omega) \rangle = \langle \varepsilon_1(\omega) \rangle + i \langle \varepsilon_2(\omega) \rangle$ of the platelet/a-C:H sample as obtained by the FTIRSE technique. The $\langle \varepsilon(\omega) \rangle$ spectra has been acquired in $t = 30$ min, whereas the high intensity of the IR light reflected from the sample surface indicates an enhanced contribution of the IR-active bands of the plasma, which is mostly composed of platelets. As can be seen from Figure 29.28, the contribution of the different vibration bands is intense in the measured $\langle \varepsilon(\omega) \rangle$

by FTIRSE. Among the various vibration bands, we can distinguish the Amide III band at ~1242 cm⁻¹, the vs CO_2 symmetric stretching mode at 1397 cm⁻¹, the Amide II band (C—N stretch, N—H bending) at 1535 cm⁻¹, the Amide I band (C—O stretch) at ~1643 cm⁻¹, and the Amide A band (N—H stretch) at 2971 cm⁻¹. The enhanced contribution of these vibration bands and mainly of the first two ones correspond mainly to the bonding structure of the platelets and indicates their existence onto the a-C:H surface.

The phenomenon of protein adsorption can be monitored in real-time by using in-situ and real-time SE. This is a powerful,

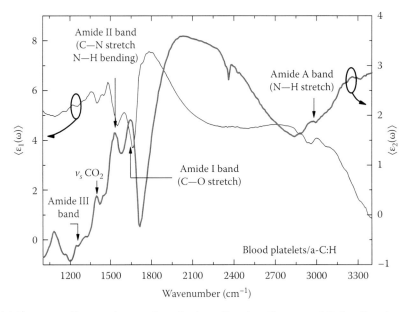

FIGURE 29.28   FTIRSE $\langle \varepsilon(\omega) \rangle$ spectra of human plasma rich in platelets adhered on Floating a-C:H thin films (incubation time 30 min).

nondestructive and surface sensitive optical technique used to monitor the deposition rate and the growth mechanisms of thin films during their deposition. The in-situ and real-time monitoring of Fib adsorption has been performed by an ultrafast multiwavelength (MWL) phase-modulated ellipsometer in the visible to far ultraviolet (Vis–fUV) energy region (1.5–4 eV). This system was equipped with a 32-fiber-optic array detector for simultaneous measurements at 32 different wavelengths (MWL mode), in the energy range 1.5–4 eV. The sampling time of measurements was ST = 500 ms and the angle of incidence was 60°. A special solid–liquid cell was used for the experiment.

In Figure 29.29, the experimental data of random in-situ measurements of a-C:H sample and Fib on a-C:H as a function of time (from 16 to 5300 s for pH 7.4) and representative fitting results (insets) are presented (Lousinian and Logothetidis 2008, Lousinian et al. 2008). An appropriate ellipsometric model (see insets of Figure 29.29) was developed for the real-time investigation of Fib adsorption, providing valuable information about the thickness and composition changes of the Fib layers formed on the a-C:H thin films during adsorption. More precisely, the protein adsorption is considered as a thermodynamic phenomenon, which takes place spontaneously whenever protein solutions contact solid surfaces, but it is not yet fully described. A simple

yet reasonable description of protein adsorption is the one of reversible attachment in an initial state followed by a subsequent change to a more strongly bound state involving greater surface contact. In the initial stages of protein adsorption, when the surface coverage is low, the protein molecules transform immediately from the solution to the adsorbed state. A slow saturation of adsorption kinetics is observed and the number of molecules in their native form is gradually decreased to zero, while the number of the protein molecules in their adsorbed form is maximized, as the equilibrium of the phenomenon is reached. Therefore, it is supposed that each moment there is a layer formed on the a-C:H film surface that is consisted of Fib molecules that are in native state (as in Fib solution) and in bound state (adsorbed Fib).

We also supposed that the volume fraction of the adsorbed Fib increases during the evolution of the protein adsorption, reaching its maximum (~100%) when the equilibrium of the phenomenon takes place. Thus, supposing that the protein layer formed on a-C:H thin film is composed by adsorbed Fib and molecules of Fib in the solution, with the volume fractions of the two phases varying through time, then Bruggeman effective medium approximation (BEMA) (Bruggeman 1935) can be used to estimate the volume fraction of Fib in adsorbed and liquid state. The BEMA is one case of the effective-medium theories (EMT) that

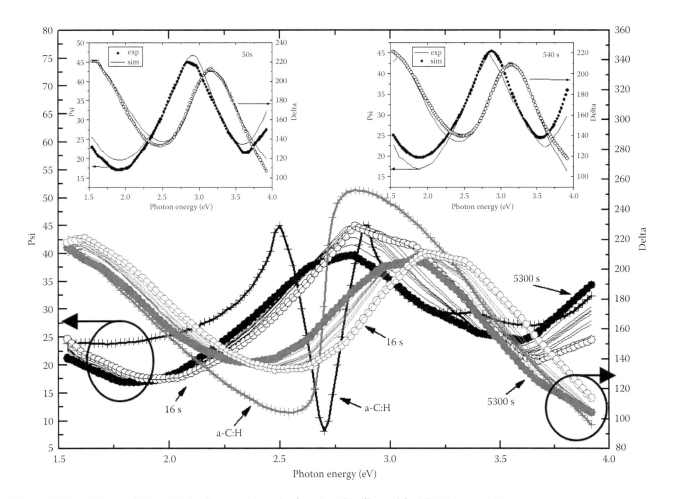

FIGURE 29.29 Measured Psi and Delta ellipsometric angles for a-C:H thin film and for Fib/a-C:H at pH 7.4.

are used to describe the macroscopic dielectric response of microscopically heterogeneous materials, such as bulk films including both compositional and shape aspects. More specifically, in the case of a composite film this has an effective dielectric function that depends on the constituent fractions shape, and size, as well as on the orientation of the individual grains. The use of EMTs allows the calculation of the macroscopic dielectric response of these heterogeneous materials from the dielectric functions of its constituents $\tilde{\varepsilon}_i$ and the wavelength-independent parameters f_i. Thus, EMTs appear as a basic tool in material characterization by optical means and can be used if the separate regions are small compared to the wavelength of the light but large enough so that the individual component dielectric functions are not distorted by size effects. All EMTs should obey in the quasistatic approximation and can be represented by the relation:

$$\frac{\tilde{\varepsilon} - \tilde{\varepsilon}_h}{\tilde{\varepsilon} + \kappa\tilde{\varepsilon}_h} = \sum_i f_i \frac{\tilde{\varepsilon}_i - \tilde{\varepsilon}_h}{\tilde{\varepsilon}_i + \kappa\tilde{\varepsilon}_h}, \quad \sum_i f_i = 1, \quad (29.21)$$

where
   κ accounts for the screening effect (for example, for spheres $\kappa = 2$)
   ε_h the host dielectric function

BEMA is self-consistent with $\varepsilon = \varepsilon_h$ and in the small sphere limit ($\kappa = 2$) of the Lorentz-Mie theory (Niklasson and Granqvist 1984) Equation 29.24, gives

$$\sum_{i=1} \frac{\tilde{\varepsilon}_i - \tilde{\varepsilon}}{\tilde{\varepsilon}_i + 2\tilde{\varepsilon}} f_i = 0, \quad \sum_i f_i = 0. \quad (29.22)$$

The use of BEMA will lead to the description of the Fib adsorption phenomenon, taking into account the effect of the liquid ambient in the suggested model. TL model (Equation 29.17) was used to parameterize the dielectric function with the optical constants on the wavelength of light for Fib solution and the adsorbed Fib dielectric functions (Lousinian and Logothetidis 2008, Lousinian et al. 2008).

In Figure 29.30a and b, the evolution of Fib thickness and the volume fraction of adsorbed Fib with time during Fib adsorption on a-C:H samples with (namely biased) and without (namely floating) the application of negative bias voltage are shown, respectively. Thickness reaches its maximum value at about 2600 s on the floating sample, which means that the equilibrium of the phenomenon is then reached, while on the biased sample the equilibrium is reached earlier, at about 1200 s (Figure 29.30a). The first molecules of Fib come in contact directly with the film surface, and then the thickness and the volume fraction of adsorbed protein are increasing extremely fast, until the whole surface is covered by the protein layer. After this stage, the interactions between the Fib molecules start to take place, and the thickness of the layer is increasing until it reaches its maximum. This is monitored and confirmed quantitatively with nanoscale precision by the use of real-time SE technique. It is observed that the Fib thickness as well as the volume fraction of adsorbed Fib is larger on the floating a-C:H thin film, especially at the initial stage of the protein adsorption. This could be attributed to the surface topography of the a-C:H thin films. As mentioned, the surface roughness R_{rms} of the floating sample (~2 nm) is one order of magnitude larger than that of the biased sample (~0.3 nm). This means that the topography of the floating a-C:H thin film offers a larger area for the Fib molecules to bind and to transform to the adsorbed form of Fib.

Another feature that differs between the two studied samples is the variation of the thickness and the volume fraction of adsorbed Fib during time (Figure 29.30a and b). A continuous increase of both thickness and volume fraction of adsorbed Fib takes place on the floating sample, while on the biased sample, variations of the values of the above parameters occur, revealing the conformational changes of the Fib molecules during the adsorption. This could be due to the different bonding configuration of

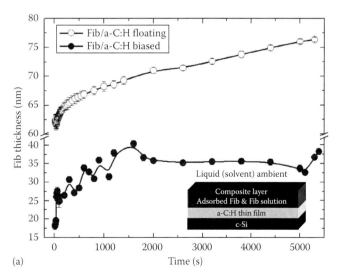

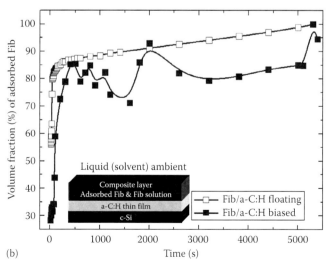

FIGURE 29.30 Evolution of (a) Fib thickness and (b) volume fractions of adsorbed Fib, on the floating and biased thin film.

the two samples and the bonds, which are formed at the interface between the a-C:H surface and the protein molecules on it (Lousinian and Logothetidis 2008, Lousinian et al. 2008).

29.2.2 Low-Angle X-Ray and Reflectivity Techniques on the Nanoscale

The most interactions taking place between electromagnetic radiation in x-ray region and a surface are: transmittance, absorption, reflection (specular reflection), and scattering (nonspecular reflection). The critical quantity in these interactions and phenomena is the refractive index n of the material under study. We can easily see from Equation 29.17 that in the x-rays energy region n is a complex quantity close to unit and is given by

$$n = 1 - \delta + i\beta, \tag{29.23}$$

where the real part corresponds to the dispersion and the imaginary to the absorption of x-rays, in the matter. The value of δ ranges between 10^{-6} and 10^{-5} and is proportional to the electron density ρ_e, whereas β is even smaller. The parameter δ is also related to the critical angle of total reflection θ_c through the expression (Daillant and Gibaud 2009)

$$\sin \theta_c^2 = 2\delta = \frac{\lambda^2}{\pi} r_e \rho_e. \tag{29.24}$$

In this expression λ is the x-rays wavelength and r_e is the electron radius. The critical angle for the most of the materials lies in the angular range between 0.2° and 0.6°. In the following paragraphs, the intensity of the reflected beam at low angles is expressed in terms of the angles of incidence, reflection, and the refractive index.

29.2.2.1 X-Rays Specular Reflection

For the case of the specular reflection, where the angle of incidence and the angle of reflection are equal, the intensity is proportional to Fresnel reflectivity. The x-rays specular reflection (XRR) is applicable in the thin film multilayer structures, such as in Figure 29.31.

In this case the reflected intensity is proportional to $|R_0|^2$, where R_0 is calculated by the recursive formula:

$$R_j = e^{-i2k_{z,j}z_j} \frac{r_{j,j+1} + R_{j+1}e^{i2k_{z,j}z_j}}{1 + r_{j,j+1} \cdot R_{j+1}e^{i2k_{z,j}z_j}},$$

$$r_{j,j+1} = \frac{k_{z,j} - k_{z,j+1}}{k_{z,j} + k_{z,j+1}}, \tag{29.25}$$

$$k_{z,j} = \frac{2\pi}{\lambda} \sqrt{\sin^2\theta - \sin^2\theta_{c,j} - i2\beta_j}.$$

In Equation 29.25 $\theta_{c,j}$ and β_j are the critical angle and the absorption of the j-layer. This means that except of the layer thickness z_j, also the layer density (Equation 29.24) can be calculated.

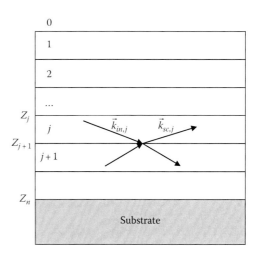

FIGURE 29.31 Illustration of the plane of incidence for a multilayer structure. Air is labeled medium 0 and the strata are identified by $i < j < n$ layers in which upwards and downwards waves travel.

In the case of a thin layer over a substrate, the reflectivity is given by the formula

$$R_0 = \frac{r_{0,1} + r_{1,2}e^{-i2k_{z,1}h}}{1 + r_{0,1} \cdot r_{1,2}e^{-i2k_{z,1}h}}, \tag{29.26}$$

where h is the film thickness.

The surface and interface roughness σ_i of the thin film can be incorporated in the reflectivity with a Debye–Waller factor in the Equation 29.25 of the Fresnel coefficient like

$$r_{j,j+1}^{\text{rough}} = \frac{k_{z,j} - k_{z,j+1}}{k_{z,j} + k_{z,j+1}} e^{-2k_{z,j}k_{z,j+1}\sigma_{j+1}^2}. \tag{29.27}$$

As an application of the utilization of the above formulations we present, in the next few paragraphs three characteristic examples of XRR measurements, each with a unique concern. At first, we present an example of a SiO/SiO_2 multilayer structure and the formation of Si nanocrystals through the annealing of the multilayer structure. Then we show a characteristic example of a Ti/TiB_2 multilayer developed on c-Si substrates and the application of an alternative method to calculate each layer and the bilayer thickness called low-angle x-ray diffraction; its principles lies in reflection though. Finally, we show an implementation of XRR technique to study the structural evolution of a-C films developed on c-Si (100) substrates.

The SiO/SiO_2 multilayer was deposited on c-Si using the e-beam evaporation technique (Gravalidis and Logothetidis 2006, Gravalidis et al. 2006). The high-temperature annealing (~1100°C) results in the phase separation of the ultrathin SiO layers and formation of Si nanoparticles surrounded by amorphous SiO_2, which is described by the equation $SiO_x \rightarrow (x/2)SiO_2 + (1 - x/2)Si$ (Yi et al. 2003). Alternative a-Si/SiO_2 systems can have the same results during annealing (Ioannou-Sougleridis et al. 2003).

The processes take place during annealing can be divided to two stages (Yi et al. 2002, 2003): (a) stage 1 (300°C–900°C): Rearrangement of SiO and SiO$_2$ components and initialization of the nc-Si nucleation; and (b) stage 2 (900°C–1100°C): the phase separation of nc-Si and SiO$_2$ has finished and Si cluster are being further crystallized. Further annealing does not have any effect on Si—O bond.

XRR takes advantage of the small wavelength ($\lambda_{CuKa} = 0.154$ nm) of x-rays, which is appropriate for the structural study of multilayered system. XRR was applied in the range from 0° to 3° and the spectra for the samples are depicted in Figure 29.32. At angles below critical angle θ_c, which is proportional to the mass density, there is the plateau indicating the region of the total reflection of x-rays and for angles above θ_c the reflectivity decays almost exponentially. Additionally, due to the layer thickness, interference fringes appear in the spectrum. The data were fitted using a model consisted of SiO$_2$/(SiO/SiO$_2$) × 3/c-Si (Gravalidis and Logothetidis 2006, Gravalidis et al. 2006) and using Parrat's formalism combined with a Nevot–Croce factor for the description of the surface and interface roughness (Parratt 1954, Nevot and Croce 1980). Thus, thickness, roughness, and density of the film can be calculated. What was found from the above analysis is that for the SiO layers the thickness is decreasing and the average density of the multilayer is increasing with the increase of annealing temperature as it is shown in Figure 29.33. This

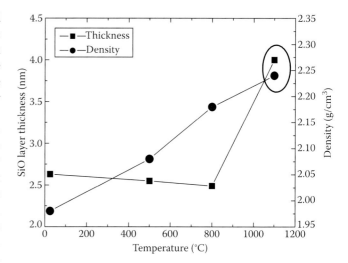

FIGURE 29.33 The average density (circles) and the thickness (rectangle) of the SiO layers and with the annealing temperature. The point in the circle corresponds to the model SiO$_2$/(Si/SiO$_2$)×3/c-Si used for the description of the data.

densification can be assigned to the dissociation of SiO$_x$, which enhances the formation of the denser SiO$_2$ and Si compounds. In the case of the sample annealed at 1100°C, the model used for the lower temperatures failed to describe the data well. This is rather expected due to the phase separation and the breakdown of the layered structure that are taking place at this temperature. To overcome this problem, we replaced the SiO layers with Si layers and the whole analysis gave a total thickness of 28 nm. The phase separation that is taking place at this temperature is one of the reasons for the increase in the thickness. This probably means the strain effects during the formation of the nc-Si in the SiO$_2$ matrix from SiO$_x$, instead of creating a single interface between the two materials, create a transition layer of stressed SiO$_2$ and Si, surrounding the nc-Si as it is reported elsewhere (Daldosso et al. 2003).

A second typical example of an XRR measurement of a Ti/TiB$_2$ multilayer film is shown Figure 29.34. The film consists of 24 alternating layers of Ti and TiB$_2$ (Kalfagiannis et al. 2009). The peaks that are observed in these patterns indicate the superimposed bilayer profile. Using Parratt's formalism we calculated the total thickness of the film at 513 nm while each layer of Ti was calculated at 9.8 nm and TiB$_2$ at 11.7 nm. Fluctuations on the layer thickness of TiB$_2$ were calculated at 0.9 nm and of Ti at 0.5 nm. The Ti/TiB$_2$ multilayer sample was prepared in a high vacuum chamber employing the unbalanced magnetron sputtering technique (Logothetidis, 2001). The multilayer structure was achieved through the rotation of the c-Si (100) substrate between two targets (Ti and TiB$_2$) with a rotation speed of 1 rpm (Kalfagiannis et al. 2009).

In order to achieve a more comprehensive structural study of the multilayer structures the implementation of XRR in higher angles can clearly determine the bilayer thickness of the films. LAXRD (low-angle XRD) technique results from the reflection

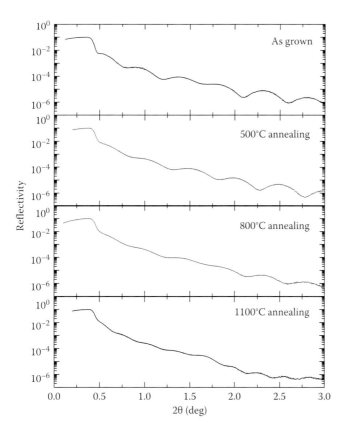

FIGURE 29.32 The XRR spectra of the samples before and after annealing.

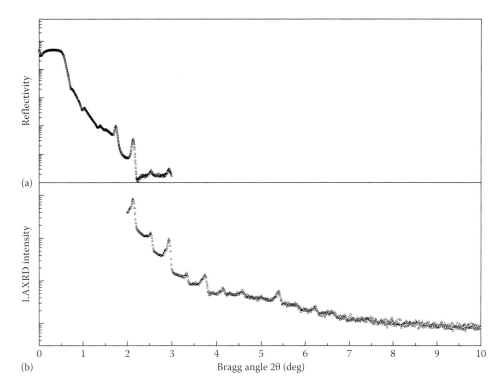

(a)

(b)

Bragg angle 2θ (deg)

FIGURE 29.34 A representative XRR and LAXRD patterns of a Ti/TiB$_2$ multilayer (Kalfagiannis and Logothetidis, unpublished data).

of x-rays by the interfaces between layers. Thus, LAXRD can be regarded as an alternating XRR method and its results are not affected by the crystalline quality within each layer.

The reflectivity of an interface between two layers in a multilayer depends on the differences in electron density of the two layers. In principle, low-angle XRD directly gives the Fourier transform of the electron density, which is related to the composition modulation. However, refraction and absorption effects become important at small angles making these patterns difficult to interpret (Yashar and Sproul 1999). An example of a low-angle XRD pattern from a Ti/TiB$_2$ multilayer is shown in Figure 29.34b. The observed peaks occur at position 2θ given by a modified form of Bragg's law:

$$\sin^2 \vartheta = \left(\frac{m\lambda}{2\Lambda} \right)^2 + 2\delta,$$

where

    2θ is the angular position of the Bragg peak
    m is the order of the reflection
    λ is the wavelength of the x-rays (1.5406 Å for CuKα light)
    Λ is the bilayer period

By plotting the sin^{2}θ versus $(m\lambda/2)^2$ one can easily calculate, from the slope of the fitted line, the bilayer period, as shown in Figure 29.35.

XRR technique is also used to identify the growth evolution of thin films and several other important nanoengineering parameters related with the density, and their morphology.

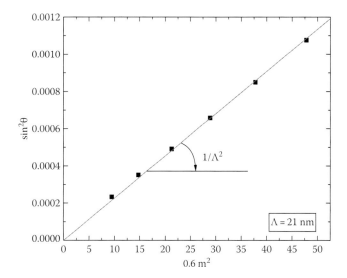

FIGURE 29.35 A graphic example of how the bilayer thickness of a multilayer can be calculated from a LAXRD pattern. The fitted line crosses through zero. This is due to the fact that δ can be neglected from modified Bragg's law, since typical values of δ are ~10–6. The slope of the line is directly related to Λ.

We will give here a representative example of magnetron sputtered amorphous carbon (a-C) thin film developed without (MS) or with ion irradiation (BMS) during deposition, by applied a negative bias V_b on the Si substrate. X-ray reflectivity scans obtained from successive ultrathin a-C layers, grown by MS on Si substrate, are shown in Figure 29.36a. Figure 29.36b also shows details from the same scans close to θ$_c$. The arrows

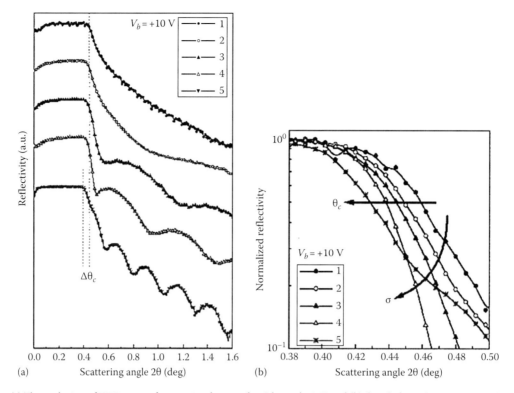

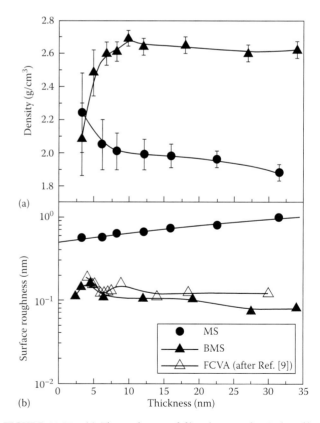

FIGURE 29.36 (a) The evolution of XRR scans of successive deposited a-C layers by MS and (b) details from the same scans close to critical angle θ_c; the arrows indicate the variation of θ_c with thickness and the effect of surface roughness to the line shape.

indicate the variation of θ_c with thickness and the effect of roughness (σ) to the line shape. A similar study has been carried out for the growth of a-C by BMS with low $V_b = -20$ V. The XRR scans have been modeled using the same Monte Carlo algorithm and taking into account only one Debye–Waller factor for the film surface; the film interface has been considered atomically sharp. This procedure was employed because for the ultrathin films the electron density variations exhibit spatial resolution comparable to the film thickness making the Monte Carlo fits with two Debye–Waller factors rather ambiguous. This is also justified by the low interface roughness of thick (~30 nm) films grown by identical conditions ($\sigma_i = 0$ nm for MS and below 0.5 nm for BMS at $V_b = -20$ V).

The results of the XRR data analysis for both cases of sputtered a-C films (MS and BMS) are presented in Figure 29.37. The density evolution of a-C thin films (Figure 29.37a) is strikingly different for the two cases. The density of the BMS film at the very initial stages of growth is very low and it is typical of a sp²-rich material; then, it gradually increases to reach the steady-state (bulk) value of 2.62 g/cm³, after thickness of 10 nm. This is in agreement with the computational results (Patsalas et al. 2005). On the other hand, the MS grown film exhibits the same density (within the experimental error) with the BMS film at the very initial stage of growth. This clearly indicates that the properties of the ultra-thin sputtered a-C films (<4 nm thick) are determined dominantly by the thermodynamics and stress relaxation of the a-C/Si interface than the ion irradiation and the growth mechanism (e.g.,

FIGURE 29.37 (a) The evolution of film density of a-C thin films grown by MS and BMS and (b) the surface roughness for films grown by MS, BMS and FCVA techniques.

subplantation), which prevail for thicker films (>10 nm). After the initial stages of MS growth the density gradually drops to 1.88 g/cm³ (Daldosso et al. 2003).

29.2.2.2 X-Rays Nonspecular Reflection

The case of the nonspecular reflection is called x-rays diffuse scattering (XDS). The intensity in nonspecular reflection is proportional to the differential scattering cross-section, which is the solution of the Schrödinger equation under the appropriate boundary conditions:

$$-\frac{\hbar^2}{2m}\left(\nabla^2\psi(\vec{r})+k^2\psi(\vec{r})\right)=V(\vec{r})\psi(\vec{r}),\qquad(29.28)$$

where

$\hbar$ is Plank constant
m is the electron mass
∇^2 the second derivative
$\psi(\vec{r})$ is the function of the electron-field
$k = 2\pi/\lambda$ (λ is the wavelength of x-rays) is the magnitude of the wave vector

Distorted-wave Born approximation is one of the solving method. Sinha et al. (1988) derived the expressions for the differential scattering cross-section, by splitting the interaction potential $V(\vec{r})$ in two parts, the unperturbed part correspond to smooth surface and the perturbation correspond to the roughness. Thus, the nonspecular (diffuse) part of the differential scattering cross section from a bare substrate is then given by

$$\frac{d\sigma_{\text{diffuse}}}{d\Omega}(q)=\frac{Ak^4}{16\pi^2}\left|(1-n^2)\right|^2\left|t_F(k_1)\right|^2\left|t_F^*(k_2)\right|^2 S(q_{tz}),\qquad(29.29)$$

where

A is the illuminated area
Ω is the solid angle of the detector
$t_F(\mathbf{k}_1)$ and $t_F^*(\mathbf{k}_2)$ are the Fresnel transmission coefficients of the incident and scattered wavevectors k_1 and k_2, respectively

Because the electric field reaches a maximum of twice the incident field at the interface when $\mathbf{k}$ makes an angle equals to the critical angle θ_c, $|t_F(\mathbf{k})|^2$ is maximum in this case. Hence, whenever θ_1 or θ_2 equals θ_c, Equation 29.29 predicts the presence of maxima in the diffuse scattering. These are known as the anomalous (Yoneda) scattering or "angel wings." The structure factor $S(q_{tz})$, with $q_t = (q_{tx},q_{ty},q_{tz}) = \mathbf{k}_{2s} - \mathbf{k}_{1s}$ being the wave vector transfer in the substrate, is written as

$$S(q_{tz})=\frac{e^{\left(-\text{Re}\left\{q_{tz}^2\right\}\sigma^2\right)}}{\left|q_{tz}\right|^2}\int_0^\infty\left(e^{\left|q_{tz}\right|^2 C(x)}-1\right)\cdot\cos(q_x x)dx,\qquad(29.30)$$

where Re denotes the real part of the complex quantity q_{tz}.

The intensity of the scattered beam is directly related to the type of scan. There are two types of scan for measuring diffuse scattering: the rocking scan and the longitudinal scan (offset or diffuse near specular scan). In the case of rocking scan the detector has a fixed position and the sample is being rocked under the condition $\theta_1 + \theta_2 = \theta_d$. The measured intensity then is normalized to the intensity at specular position:

$$I_{\text{norm}}=\frac{1-ro}{I_{\max}}\cdot\left(\frac{d\sigma}{d\Omega}\right)_{\text{diff}}+ro,\qquad(29.31)$$

where

ro is the average between the first and the last point of the spectrum
$I_{\max}$ is the maximum intensity at specular position

In the case of longitudinal scan ($\theta_i = \theta_s \pm \Delta\theta$) and for $\theta > \theta_c$ $q_{tz} \approx q_z$ the scattered intensity is given by (Thompson et al. 1994)

$$I \propto q_z^{-\left(3+\frac{1}{h}\right)}.\qquad(29.32)$$

Equations 29.31 and 29.32 are used for the analysis of the XDS spectra to obtain the surface features such as the roughness σ, the correlation length ξ and the Hurst coefficient h.

As an example of the utilization of the above formulations we present samples that were prepared by liquid spray coating of PTFE diluted in isopropyl alcohol containing 10% of solids having molecular weight equal to 30,000, density 2.2 g/cm³, and average bulk size 3.7 μm. The melting point of these solids is in the range 322°C–326°C (http://www.dupont.com). The PTFE suspension was sprayed directly on preheated (~120°C) Si wafers (Gravalidis and Logothetidis 2006). After the spraying the samples were annealed in inert atmosphere up to 350°C. The spraying process is performed as follows: on a moving sample holder, the samples are passing through a preheat process with maximum temperature ~120°C. The substrate is placed at angle ~20° with respect to the direction of the movement. Afterwards the heated samples are passing in-front of a spraying plume with specific velocity and due to the high temperature of the substrate the solvent is vaporized and a white powder of PTFE is spread all over surface. The experiments were carried out at three different velocities 1.8, 2.5, and 3.2 m/min, thus the study will be focused on the effect of the material quantity (inversely proportional to velocity) on the surface morphology. For convenience, the samples after the spraying are named as follows: A1 (1.8 m/min), A2 (2.5 m/min), A3 (3.2 m/min), and after annealing B1 (1.8 m/min), B2 (2.5 m/min), and B3 (3.2 m/min) (Gravalidis and Logothetidis 2006).

The surface features can be studied using the x-rays diffuse scattering at grazing incidence in two ways: longitudinal and

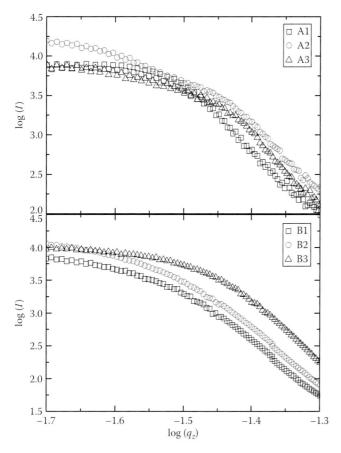

rocking scans. In the case of XDS longitudinal scan, the angle of incidence is slightly different from the scattering angle and the spectra are depicted in Figure 29.38a as $\log_{10} I$ vs $\log_{10} q_z$. For $q_z > q_c$ the $\log_{10} I$ decays linearly with $\log_{10} q_z$, meaning that the intensity depend from the q_z exponentially. From the linear fit of this region, Hurst parameter h can be calculated through Equation 29.32 (Thompson et al. 1994). This analysis gave that for the B samples the Hurst coefficient h is the same and equal to 0.15, whereas for the A samples is a bit different and equal to 0.19.

The above results show that the surface morphology of the samples after annealing become more jagged. This result comes in agreement with the fact that the annealing enhances the untwisting and the entanglements of the macromolecules and furthermore the local distortions from smoothness and regularity. Additionally, the spraying velocity does not seem to have any effect to the final morphology of the samples. The second type of scan that can give information about the surface features (correlation length and Hurst coefficient) is the rocking curve and the spectra are depicted in Figure 29.39. As can be observed from these curves, the unsintered samples have identical spectrum meaning that the surface morphology is independent from the velocity and thus the quantity of PTFE. On the other hand, the surface morphology of the sintered sample shows significant dependence on material's quantity. Taking into account the results from the off-specular XRR the reason for the different surface morphology is probably the correlation length which is related to the distribution of the height fluctuations on the surface (Gravalidis and Logothetidis 2006).

FIGURE 29.38 XDS rocking curves of both A (full points) and B (open points) samples, solid lines represent the fitted curves.

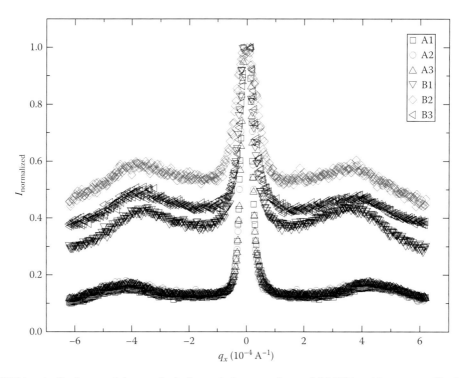

FIGURE 29.39 (a) XDS longitudinal scans of the samples before and after annealing and (b) XDS rocking curves of both A (full points) and B (open points) samples.

TABLE 29.2 Results from the XRR Data Fitting

Sample	B1	B2	B3
σ (nm)	0.5	0.5	0.5
ξ (nm)	126	268	103
h	0.15	0.15	0.15

Following Sinha's expression for the data fitting, we can calculate surface roughness and the correlation length. The Hurst coefficient that is used in the fitting procedure is already known from the off-specular XRR. The results from the fitting are summarized in the Table 29.2.

As is clear from Table 29.2 correlation length ξ depends on the spray velocity, whereas the other parameters do not. This fact means that although the three surfaces have the same roughness, the distribution of the macromolecules is related strongly to the material's quantity (through spray velocity).

The above analysis revealed the high potential of x-rays reflectivity techniques and especially XDS technique to describe polymeric surfaces grown by spray coating and sintered at high temperature in inert atmosphere. The grain size of PTFE is decreasing with the material quantity and after the annealing. Furthermore, for the annealed samples the grain size does not depend monotonically with the quantity, but has a maximum value. The XDS longitudinal scan shows that all the annealed samples have the same surface texture. Finally, the analysis on XDS rocking scans spectra showed for the correlation length that has the same dependence with the spray velocity as the grain size. Thus, the most significant conclusion from this study is that the surface characteristics do not depend from the spray velocity of PTFE, monotonically or linearly, but exhibit a maximum value for specific surface features.

29.2.3 Nanoindentation Techniques

29.2.3.1 Basic Concepts

Nanometrology in the field of nanoscale materials and devices mechanical properties is not an easy task and nanoindentation (NI) is the state-of-art experimental technique for measuring (1) the applying force, from a few μN to hundreds of mN, and (2) at the same time, the deformation with almost nanometer resolution. In nanoindentation experiments, a rigid (usually diamond) indenter is used to vertically penetrate and deform the material by controlling either the applied force or the vertical displacement of the indenter. The penetration is assumed to be frictionless and the resulting load (*P*)–displacement (*h*) curve is analyzed to estimate the hardness (*H*) and the elastic modulus (*E*) of a material and to comment on the elastic and plastic deformation and phenomena of the materials. NI is nowadays used to characterize the mechanical properties of all types of materials such as semiconductors, metals, polymers, organic materials, and biomatter (bones, dents, etc.).

29.2.3.2 Background: History and Definitions

Indentation experiments were firstly introduced in the beginning of the twentieth century by Brinnel, who used smooth ball bearings as indenters. At these conventional indentation experiments, constant load was applied by a theoretically rigid indenter, which penetrated into the sample. After the unloading of the indenter, the dimensions of the indentation imprint was measured using an optical microscope and only the hardness of the material was estimated, as the ratio of the applied load divided by the area of the residual imprint (Tabor 2000).

Contemporary nanoindentation experiments are based on the same idea, but the indenters are very sharp (with tip roundness usually <50 nm) and experimental setups (nanoindenters) can control and read both the applied load and the displacement continuously from the beginning till the end of the experiment. That is why the technique is now called "depth-sensing" nanoindentation (Oliver and Pharr 1992, 2004).

In Figure 29.40 the simplest *P*–*h* curve derived from a nanoindentation experiment is presented. The *P*–*h* curve is composed of (1) the loading part, in which the applied load (or the displacement) is increasing up to a selected load (or displacement) value (2) the hold part, in which the applied load is kept constant, and (3) the unloading part, in which the load is decreasing up to zero. Knowledge of the indenter shape function and the displacement enables calculation of the projected contact area (*A*) between the rigid indenter and the sample, which is critical for the determination of the material's hardness and the elastic modulus.

29.2.3.3 Theoretical Models for Analysis of Nanoindentation Data

In Figure 29.41 a nanoindentation *P*–*h* curve together with the critical contact depths according to the elastic contact model of Oliver–Pharr are presented (Oliver and Pharr 1992) This model concerns the elastic portion of the unloading part of the *P*–*h* curve and in Figure 29.42 a sketch of the cross-section of the

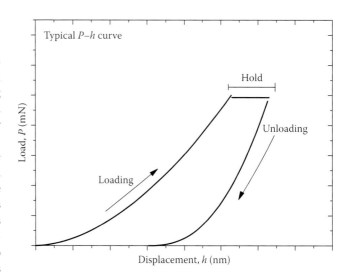

FIGURE 29.40 A typical nanoindentation load (*P*)–displacement (*h*) curve giving the loading, hold, and unloading procedure.

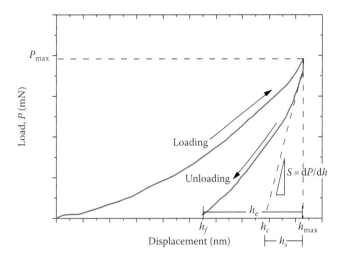

FIGURE 29.41 Nanoindentation load–displacement curve of the Oliver–Pharr elastic contact model (see also Figure 29.33).

contact between the indenter and the sample is presented. In Figures 29.40 and 29.41, h_{max} is the maximum displacement of the indenter at the maximum load (P_{max}), h_f is residual contact depth after removal of the applied load (unloading), h_c is the contact depth between the rigid indenter and the sample, h_e is the final elastic recovery of the sample's surface, and h_s is the elastic displacement of the surface, which is not in contact with the diamond indenter.

According to the Oliver–Pharr model for the elastic/plastic contact, the unloading part of the P–h curve is well approximated by power law relation:

$$P = c(h - h_f)^n, \qquad (29.33)$$

where

   P is the load
   h is the displacement
   c and n are constants related to the material and to the indenter geometry, respectively

The contact depth h_c is given by Equation 29.34 as we can also see in Figure 29.41:

$$h_c = h_{max} - h_s. \qquad (29.34)$$

h_s is estimated using the Sneddon analysis (Sneddon 1965) and we conclude Equation 29.35 for the contact depth h_c:

$$\left. \begin{aligned} h_s &= \frac{\pi - 2}{\pi}\left(h_{max} - h_f\right) \\ \varepsilon &= 2\frac{\pi - 2}{\pi} \\ h_{max} - h_f &= \frac{2P}{S} \end{aligned} \right\} \Rightarrow h_c = h_{max} - \varepsilon\frac{P_{max}}{S}, \qquad (29.35)$$

where

   ε is a coefficient, whose value depends on the indenter geometry (1 for a flat punch indenter, 0.72 for conical, and 0.75 paravoloid of revolution)
   S is the (Stiffness) given by the slope of the elastic unloading in the P–h curve:

$$S = \frac{dP}{dh}. \qquad (29.36)$$

The stiffness is directly related to the reduced elastic modulus via Equation 29.37:

$$S = \beta \cdot \left(\frac{2}{\sqrt{\pi}}\right) \cdot \sqrt{A} \cdot E_r, \qquad (29.37)$$

where

   A is the area of indenter surface in contact with the sample
   E_r is related to the elastic modulus of the sample with Equation 29.38:

$$\frac{1}{E_r} = \frac{1 - v_{sa}^2}{E_{sa}} + \frac{1 - v_i^2}{E_i}, \qquad (29.38)$$

where the subscripts sa, i correspond to the sample and the indenter, respectively and v is the Poisson ratio. In Equation 29.38, β is a dimensionless parameter that is used to account for deviations in S caused by lack of axial symmetry for pyramidal

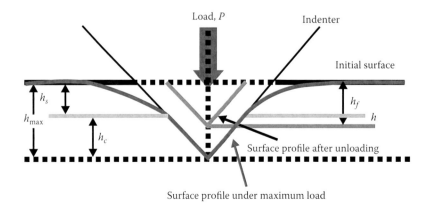

FIGURE 29.42 Sketch of the surface profile under maximum load of the Oliver–Pharr elastic contact model.

indenters (β = 1.034 for Berkovich and cubic, β = 1.012 for Vickers and Knoop indenters).

The nanohardness of a sample measured from nanoindentation P–h curves is given by Equation 29.39:

$$H = \frac{P_{max}}{A}. \tag{29.39}$$

All the above refer to the quasistatic nanoindentation test, in which the E and H values of matter is measures only at the maximum indentation depth. Last decade, the scientific community interest was attracted by dynamic nanoindentation tests (DNI) (Oliver and Pharr 1992, Lucas et al. 1998) In DNI the contact stiffness is measured continuously versus the displacement of the indenter and enables the measurement of E and H continuously. Specifically at the loading segment of the P–h curve, a small AC-driven force is superimposed to the applied force, resulting in the oscillated displacement of the indenter with the same frequency, but (usually) with phase difference (φ). The force/displacement variation is exceptionally small and does not influence the deformation of the sample. A phase-sensitive lock-in amplifier records with high precision the corresponding shift (displacement resolution 0.04 nm) that is caused by the force and the phase difference φ between the force and displacement signals. The differential equation of the forced oscillation is the following:

$$F_0 \sin \omega t - D \frac{dh}{dt} - K_s = m \frac{dh^2}{dt^2}, \tag{29.40}$$

where

F_0 is the amplitude of the oscillating force
ω the frequency of the oscillating force
dh is the instant displacement
D is the damping coefficient
K_s is the stiffness of the indenter supporting system
m is the mass of the indenter
S is the contact stiffness (see Figure 29.43)

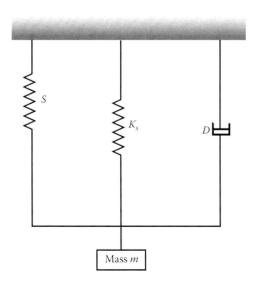

FIGURE 29.43 Nanoindentation dynamic model.

The contact stiffness can be calculated using the following equations:

$$\left| \frac{F_0}{h(\omega)} \right| = \sqrt{(K - m\omega^2)^2 + \omega^2 D^2}, \tag{29.41}$$

$$\tan \varphi = \frac{\omega D}{K - m\omega^2}, \tag{29.42}$$

where

$K = (S^{-1} + C)^{-1} + K_s$ is the reduced spring constant of the system shown in Figure 29.43
C_f is the compliance of the load frame

From the contact stiffness S, both the elastic modulus and the hardness can be calculated using Equations 29.38 and 29.39, respectively.

29.2.3.4 Mechanical Properties of Nanostructured Materials

29.2.3.4.1 Nanomechanical Properties of Amorphous Carbon Nanocoatings

The DNI is a very useful technique, especially for measuring the nanomechanical properties of few nanometers thick ($\leq$100 nm) nanostructured coatings. In Figure 29.44, the hardness versus the normalized contact depth of a hard, nanostructured amorphous carbon (a-C) coating measured using DNI is presented (d is the nanocoating thickness). The a-C nanocoating consists of thin sequential layers, rich in sp^2 (layer A) and sp^3 (layer B) hybridized carbon bonds, forming a multilayer structure of ~100 nm total thickness and was grown on c-Si (001) substrate by *rf* magnetron sputtering (Logothetidis et al. 2004).

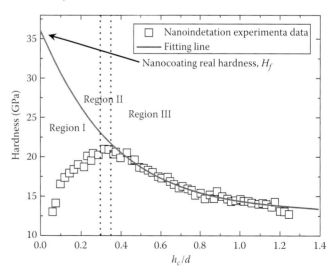

FIGURE 29.44 The hardness of a hard a-C nanocoating versus the normalized contact depth.

Three different regions of the hardness values are showed. In the first one, the hardness values are increasing continuously with the contact depth. This behavior is related to the nature of the contact (usually elastic/plastic) between the indenter and the sample and the in-depth width of this value region depends on the roundness of the indenter tip. In the second hardness value region, a "small" value plateau is present and for further increase of the contact depth (third hardness value region), the hardness values decrease approaching that of the c-Si (001) substrate (12–14 GPa). The absence of any clear H value plateau suggests that these values are substrate dependent and that the real films hardness must be even higher. This barrier for the measurement of the nanocoating real hardness is the so-called substrate effect, which is caused by the mismatch between the mechanical properties of the nanocoating and those of the substrate and is more intense as the thickness of the nanocoating decreases.

Several models, which take into account the substrate effect, have been proposed and DNI enables this kind of studies by providing the variance of H versus the nanoindentation penetration depth. In Figure 29.44 the solid line is the result of a fitting procedure to the DNI experimental data using Equation 29.43), which comes from the Bhattacharya–Nix model for the substrate effect to nanoindentation measurements (Bhattacharya and Nix 1988):

$$H = H_s + \left(H_f - H_s\right)\exp\left[-\frac{h_c}{d}\frac{H_f/H_s}{\alpha/E_s^{1/2}}\right], \qquad (29.43)$$

where subscripts f,s correspond to the nanocoating and substrate hardness and elastic modulus (E_{Silicon} = 160–180 GPa), respectively, and α is a coefficient related to the mismatch between the hardness values of the nanocoating and the substrate. By fitting the experimental results with Equation 29.42, the H_f was calculated to be ~36 GPa, instead of 21 GPa, which is the substrate affected (experimentally measured) hardness value.

Knowledge of the real hardness values enables correlation between the nanocoatings mechanical properties and the nanostructure features, such as the sp³-hybridized carbon bonds content and the bilayer thickness. Figure 29.45 shows the H_f of the multilayer nanocoatings as a function of their sp³ content, together with the single layer sp²-(denoted as layer A) and sp³-rich (denoted as layer B) a-C nanocoatings, grown at the same deposition conditions. The hardness increases linearly with the sp³ content and follows the relation $H_f \approx 0.87 \times$ (% sp³ hybridized carbon bonds content) GPa. From Figure 29.45, it is obvious that the multilayer nanocoatings appear to be much harder compared to single layer rich in sp³ hybridized carbon bonds (H_f = 24 GPa, % sp³ content ~ 45) although they have a lower content of sp³ hybridized carbon bonds. This implies that except of the bonding structure, the strengthening mechanisms in these nanomaterials depend also on other characteristics, such as the bilayer thickness (Λ), as shown in Figure 29.46. In this figure, the bilayer thickness varies by changing the thickness of the layer A (rich in sp² hybridized carbon bonds). As shown, the hardness increases linearly with $1/\Lambda$, revealing a size effect on the measured nanomechanical properties.

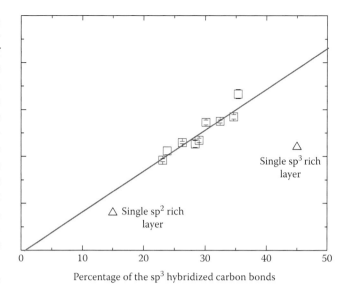

Percentage of the sp³ hybridized carbon bonds

FIGURE 29.45 Correlation of the calculated hardness with the percentage of the sp³ hybridized carbon bonds of the multilayer nanocoatings.

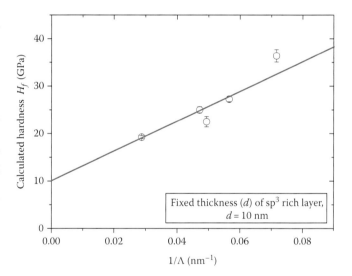

FIGURE 29.46 Correlations of the hardness with the bilayer thickness (Λ) of the multilayer nanocoatings.

29.2.3.4.2 Nanomechanical Properties of Transition Metal Nitrides

Another example of nanostructured materials with enormous technological importance, mainly due to the exceptional mechanical properties, is the transition metal nitrides. Among them TiN and CrN are the most widely investigated due to its refractory character and gold-like color the first one and to the increased resistance to severe environments, where TiN is not functional, the second. The herein presented nanoindentation results from TiN and CrN nanocoatings (≤150 nm thick) aims to reveal the correlation between the nanomechanical properties and their structure. The CrN and TiN nanocoatings were deposited on c-Si (001) substrate by unbalanced magnetron sputtering by varying the applied to the substrate bias voltage (V_b) (Patsalas

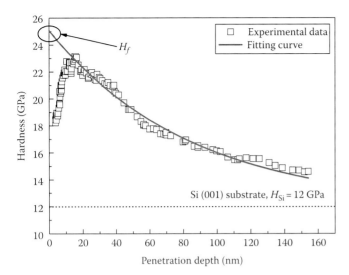

FIGURE 29.47 Hardness versus the nanoindenter penetration depth.

and Logothetidis 2001, 2003). Both CrN and TiN exhibit the rocksalt crystal structure, while their grains grow along the [111] and [100] crystallographic orientation, which are the most common for the d-metal nitrides.

In Figure 29.47, the hardness of a representative CrN thin film versus the contact depth is presented. The experimental values of hardness were fitted using Equation 29.42, in order to take into account the substrate effect and to calculate the real hardness value at the surface ($h_c = 0$). The horizontal dotted line corresponds to the mechanical properties of the c-Si (001) substrate. The hardness is a macroscopic quantity, which represents the plastic deformation of a material under specific load conditions and depends on the mobility of the dislocations across specific

slip planes within the lattice. Therefore, the hardness is affected by nonintrinsic properties of the material, such as structural defects, grain boundaries and morphological features not present in the corresponding single crystal. Figure 29.48 shows the hardness, fraction of the [111] growth orientation calculated as in (Patsalas and Logothetidis 2001), the compressive stress, and the mass density of CrN and TiN (open and closed squares respectively) nanocoatings versus V_b. This figure provides a better overview for the nanomechanical response of the above samples in terms of their structural characteristics.

In general, the nanomechanical response of the materials is affected by the crystallographic orientation. The elastic modulus along the [100] crystallographic direction is higher than this of the crystallographic direction [111] (Mathioudakis et al. 2002). Thus, the increase of the grains fraction which grow along the [100] orientation results in an increase in elastic modulus and consequently in hardness. On the other hand, the higher hardness values in TiN with increased fraction of the [111] orientation has been reported (Chen et al. 1998). This is attributed to the fact that TiN (and CrN similar) exhibits the crystal structure of NaCl. In the above structure the active slip system is the {111} ⟨110⟩ (Hultman et al. 1994). In the case in which the external force is perpendicular to the (111) plane than the resolved shear stress is to all slip directions zero. As a result, lower plastic deformation can be induced and the measured hardness is higher.

Besides the internal stress level influences the hardness. Higher internal stress level results in an increased lattice deformation. Thus, the induced plastic deformation decreases and subsequently the hardness increases. Finally, increase of the mass density results in the formation of a more compact structure with fewer grain boundaries and thus higher hardness. The above analysis is clearly illustrated in Figure 29.48. In particular, in the

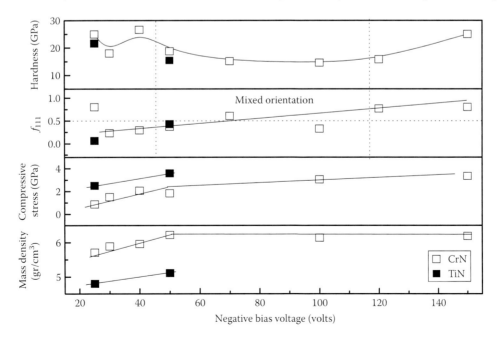

FIGURE 29.48 Hardness, the fraction of [111] orientation, the internal compressive stresses and mass density of CrN (open squares) and TiN (closed squares) nanocoatings versus substrate bias voltage. The solid lines are guides to the eye.

case of well-defined [111] grain orientation of CrN (low and high values of V_b) the hardness is maximized, while at intermediate values of V_b (mixed orientation regime in Figure 29.48) the hardness is lower. The effect of mass density and stress is more clear at low V_b values. Thus, while the increase of negative V_b from 30 to 40 V results in an increase of [111] orientation fraction, leading toward to the mixed orientation regime, the hardness increases due to the internal compressive stress, and the mass density increases. Further increase of negative V_b does not induce any increase of hardness, although it results in further increase in stress and mass density, since we are in the mixed orientation regime. Consequently, in this case of CrN, the grain orientation seems to be the dominant structural factor, which determines the nanomechanical response of the films. In the case of TiN nanocoatings, the dominant effect of the preferred orientation is more pronounced. The increase of negative V_b results in the decrease of hardness, since it causes transition from pure [100] orientation to a mixed orientation regime, although a significant increase in compressive stress and mass density takes place.

29.2.3.5 Nanoindentation Using Atomic Force Microscope

One of the inherent limitations of the commercially available nanoindenters is the lower limit in the applied force range. Specifically the applied force by a nanoindenter cannot "go" lower than 5–10 μN, and this is a limitation for the study of the early stages deformation modes of soft matter, (i.e., polymers, biomolecules, cells, etc.). In the following the atomic force microscope probe will be used as a nanoindenter, in order to plastically deform the surface of soft materials such as flexible poly(ethylene terephthalate) (PET) films and polymer-like hydrogenated amorphous carbon (a-C:H) nanocoatings (Kassavetis et al. 2007).

The surface imaging and plastic deformation (patterning) was made with a commercial SPM (SOLVER P47H, NT-MDT Co.). Both contact and semicontact scanning modes were utilized. More specifically, the initial scanning to detect a smooth enough surface was carried out in semicontact mode. Then, the deformation parameters were established (e.g., the shape of the pattern to be engraved, the applied force, etc.) and the mode was switched automatically to scan in contact mode, in order to perform the patterning, in the form of "vector" lithography (i.e., force pulses of equal magnitude and duration). The lithography results were "visualized" again by scanning in semicontact mode. A rectangular Si cantilever with relatively high spring constant (k_c = 11 N/m is the nominal value) and nominal resonance frequency f_0 = 226 kHz. The latter is directly and accurately measured by the software itself enabling a more precise estimation of the k_c value than the nominal one.

In Figure 29.49 a typical AFM force–distance (F–d) curve is presented. The DFL signal is proportional to the deflection of the cantilever. The horizontal part of the curve in Figure 29.49 refers to the condition of "tip far from sample" (ambient), when no interaction between the tip and the sample exists (no change in "DFL" signal). The sloped part of the curve represents the increase of cantilever deflection (and therefore of the applied

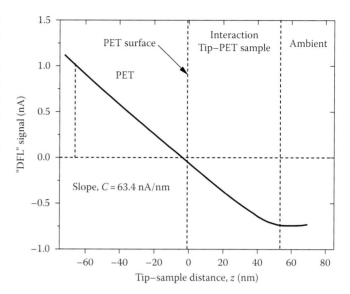

FIGURE 29.49 A representative (for the case of PET) force–distance curve.

force F_a) and, after contact with the sample surface, the penetration into the sample is started. F_a is calculated according to Hooke's law, by regarding the cantilever as an elastic spring (k_c is the spring constant) and using Equation 29.44 to transform the "DFL" signal into force values:

$$F_a \, (\text{nN}) = k_c \, (\text{N/m}) \cdot C \, (\text{nm/nA}) \cdot \text{DFL} \, (\text{nA}), \qquad (29.44)$$

where C is the slope of the curve. Thus the applied force can be controlled by adjusting accordingly the DFL signal.

From Equation 29.44, it comes out that it is necessary to define precisely the k_c value, which can be "roughly" done by using the following equation:

$$k_c = m^\star \omega_0^2 = m^\star \cdot 4\pi^2 f_0^2 = \frac{m}{4} \cdot 4\pi^2 f_0^2 = \rho \cdot l \cdot w \cdot t \cdot \pi^2 \cdot f_0^2,$$
$$(29.45)$$

where

ρ is the cantilever material mass density
l, w, and t are the length, width and thickness of the cantilever
f_0 is its resonance frequency
$m^\star$ is the "effective mass" of the cantilever–tip system equal to $m/4$ (Rabe et al. 1996)

The density is taken to be 2330 kg/m³ (Bhushan 2004) and cantilever dimensions nominal values (including error values) are; $l \pm \Delta l$ = (100 ± 5) μm, $w \pm \Delta w$ = (35 ± 3) μm, $t \pm \Delta t$ = (2.0 ± 0.3) μm. Thus, from Equation 29.45, k_c = 8.2 ± 2.3 N/m. The error in k_c propagates to the calculation of the F_a values,

In Table 29.3 the applied force, which was used in order to achieve initiation of the plastic deformation of the surface of 12 μm thick PET film and several a-C:H nanocoatings with various hardness values, is presented. In order to proceed to the

TABLE 29.3 Applied Force and Pressure by the AFM Tip on Each Sample

Samples	Applied Force, F_a (nN)	Pressure, P_a (GPa)
#1 a-C:H	2520 ± 702	4.0 ± 1.0
#2 a-C:H	1551 ± 442	2.5 ± 0.7
#3 a-C:H	1094 ± 306	1.7 ± 0.5
PET 12 μm	760 ± 255	1.2 ± 0.4

estimation of the exerted pressure (P_a), which should be $\geq H$ for occurrence of plastic deformation, from the tip to the sample surface the following assumptions were made: (1) the shape of the AFM tip edge is regarded as hemispherical (2) the tip radius R is considered 10 nm, equal to the nominal value, and (3) the contact area S is also assumed to be hemispherical, equal to $S = 2\pi d\, R^2$. P_a can be calculated using Equation 29.46:

$$P_a = \frac{F_a}{2\pi \cdot R_{tip}^2}. \qquad (29.46)$$

In Figure 29.50, the calculated exerted pressure P_a is plotted against the hardness of the samples measured using nanoindentation with a diamond Berkovich indenter. The line stands for P_a values equal to H. According to the previously mentioned condition ($P_a \geq H$) for permanent surface deformation, the P_a values should lie either on or above this line. Indeed, this is the case within the error limits, showing that the results from nanoindentation with the AFM are in accordance with those from the nanoindenter. Moreover, for the reasonable question "how such low forces, of the order of a few hundreds of nanonewtons, can deform permanently the surface of hard materials with hardness ranging from 1 to 4 GPa?", the answer lies in another size effect: the nanoscale contact area between the cantilever tip and the sample surface.

Further efforts to plastically deform samples with $H > 5$ GPa using the same cantilever were made but to no point. An upper limit of the applied force and pressure is set, firstly, by the AFM apparatus itself (and in particular, by the maximum detectable cantilever deflection symbolized by "DFL") and for this cantilever was around 5000 nN (corresponds to maximum exerted pressure $P_{max} = 7.9$ GPa). On the other hand, limitation to the maximum P_a comes from the cantilever itself and specifically from its spring constant. For cantilevers with higher k_c the same force can be achieved with less cantilever deflection.

In Figure 29.51 the image presents indents made by and AFM cantilever on the surface of the PET. The pits and the surrounding piled-up area are easily distinguished from the other surface characteristics. Analysis of the created pits in terms of their dimensions gives their real and precise geometrical characteristics, which are presented in Figures 29.52 and 29.53. It is

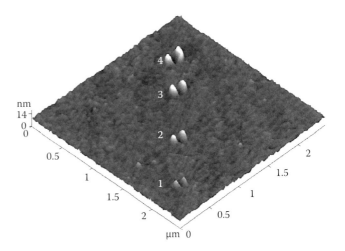

FIGURE 29.51 3D AFM image of a pits array on PET surface made by the AFM cantilever, using different force Fa for every pit.

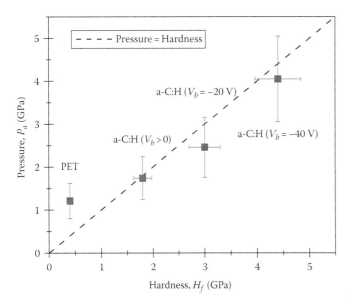

FIGURE 29.50 Pressure calculated using Equation 29.45 versus the hardness of the PET and the a-C:H nanocoatings.

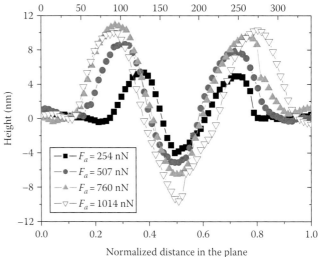

FIGURE 29.52 The profiles of the pits presented in Figure 29.44 versus the normalized distance in the plane. The upper scale shows the in plane distance of the pit using $F_a = 1041$ nN.

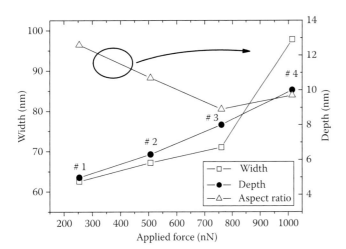

FIGURE 29.53 The width, the height and the aspect ratio of the pits presented in Figure 29.44 versus the applied force F_a.

interesting to notice that the shape of all four pits is almost the same, no matter how high the applied F_a values were, and approximately symmetrical, a fact that gives the evidence of vertical penetration. The height and the width of the piled-up area are found to increase as the F_a increases from 254 to 760 nN (Figure 29.52). Further increase of the F_a resulted only to the broadening and deepening of the pits (Figure 29.52). A clearer view of how the F_a affects the geometrical characteristics of the pits is given by the "aspect ratio," i.e., the ratio of the width over the depth (Figure 29.53). The aspect significantly decreases from 12.7 to 8.9 as the F_a increases from 254 to 760 nN. From a phenomenological point of view, this behavior correlates the decrease of the pits width expansion rate and the simultaneous formation and growth of the piled-up area, as F_a increases.

29.2.4 Scanning Probe Microscopy

AFM is a microscopy technique that belongs to the family of scanning probe microscopies in which images are obtained from the interaction between a probe and the surface of the sample and not by using light as in optical microscopy. AFM can give images from the surface of both conducting and nonconducting samples with atomic resolution (Binning et al. 1987).

In this technique, a cantilever with a very sharp tip at its end (radius of curvature on the order of 10 nm) is scanning line by line the surface of the sample. A laser beam is focused on the backside of the cantilever and is reflected onto a photodetector. During scan the tip passes above peaks or valleys causing the cantilever to bend, which in turn results in a displacement of the spot at the photodetector. AFM can work in two modes: constant height and constant force. In constant height mode, this displacement can be directly used to get the topography image of the sample. In constant force mode, the deflection of the cantilever can be used as input to a feedback system trying to keep the force constant by keeping the cantilever deflection constant, thus moving up and down the scanner in the z-axis, responding to the topography. The movement of the feedback generates the topography image.

When the tip approaches the surface, van der Waals forces start acting upon it. In ambient conditions, there is always a very thin layer of humidity on top of the surface of the sample. When the tip contacts the surface, a meniscus is formed between the tip and the sample, and the capillary force that results holds the tip in contact with the surface. Electrostatic interaction between the probe and the sample, which can be either attractive or repulsive, may appear rather often (Bhushan 2004, http://www.ntmdt.com).

AFM can work in contact, semicontact, and noncontact modes. In contact mode during scan, the tip is always in contact with the surface of the sample. The forces between the tip and the sample are repulsive as shown in Figure 29.54.

Contact mode gives easily high-resolution images, but at soft materials such as polymers or biological samples the tip can scratch and damage the sample or move material, which finally results in a distorted and unreal image. For imaging such samples, semicontact and noncontact modes that exert less pressure on the sample can be utilized. In these modes, an ac voltage is applied to a piezo, which induces oscillations to the cantilever. In semicontact mode, the cantilever-tip is oscillating near its resonance frequency and taps the surface at its lowest point of oscillation. The forces in this mode are mostly attractive and become repulsive, when the tip is getting close to the surface. During scan, the feedback system tries to keep the oscillation amplitude constant and the topography image derives by its movement. Simultaneously with topography, a phase shift between cantilever oscillations and driving ac voltage is recorded. This phase image is strongly dependent and provides useful information about the sample properties such as adhesion, elasticity, and viscoelasticity (Bhushan 2004, http://www.ntmdt.com).

Similar to the semicontact, at noncontact mode, the cantilever oscillates above the surface but never touches it. This mode is the safest for the sample but it has the lowest resolution from either contact or semicontact mode.

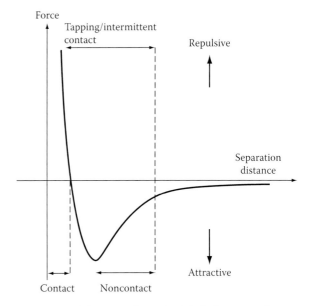

FIGURE 29.54 Schematic of van der Waals forces as a function of probe-tip–surface spacing [M54].

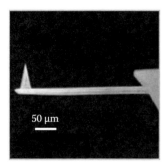 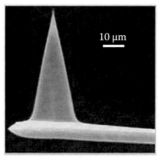

FIGURE 29.55 Silicon cantilever-tip NSG10 [NT-MDT].

The tips that are typically used are made of silicon or silicon nitride. Their geometric characteristics (see Figure 29.55) play an important role at the resolution and the quality of the AFM images.

29.2.4.1 Investigation of Polymeric Thin Films by Atomic Force Microscopy

AFM is an excellent tool for the investigation of the surface morphology for both conducting and nonconducting polymeric films. This is of great technological interest, since such films are being used in applications as OLEDs or OPVs. AFM can provide information about surface morphology and roughness of insulating polymeric films of poly(ethylene terephthalate) (PET) and poly(ethylene naphthalate) (PEN), which are the most common candidates for substrates in flexible organic electronics. In Figure 29.56, a $5 \times 5\,\mu m^2$ 3D image of a PET film is presented.

AFM can also give useful information about the morphology of organic conducting thin films. This is of great importance, since the electrical properties of these films is closely related to their morphology. A good example is the polymer complex poly(3,4-ethylenedioxythiophene): poly(styrenesulfonate) (PEDOT:PSS), which is currently being used as a hole transport layer in organic electronics. It is also one of the most promising candidates to replace indium tin oxide (ITO) and act as an anode itself (Snaith et al. 2005). PEDOT is a low-molecular-weight polymer, highly

conductive, and attached on the PSS polymer chain, which is an insulator. PEDOT:PSS spin-coated thin films are composed of gel grain-like particles, whose core is PEDOT-rich and the outer layer is PSS-rich. There is also an excess amount of PSS at the grain boundaries (Nardes et al. 2007). Conduction within the grains occurs by hopping transport from one PEDOT segment to another, while conduction between the grains occurs via tunneling as the PSS at the grain boundaries acts as a barrier to the transport of the carriers (Nardes et al. 2007).

AFM gives detailed information about the grain size, boundary thickness, and the surface roughness of the films. In Figure 29.57, AFM topography and the phase images of an PEDOT:PSS film, which was glycol treated in order to increase its conductivity are presented.

29.2.4.2 Implementation of AFM for Protein Adsorption and Conformation

Imaging is becoming an ever more important tool in the diagnosis of human diseases. Imaging at cellular, and even subcellular and molecular level, is still largely a domain of basic research. However, it is anticipated that these techniques will find their way into routine clinical use. AFM and AFM-related techniques (e.g., scanning near-field optical microscopy [SNOM]) have become sophisticated tools, not only to image surfaces of molecules or subcellular compartments, but also to measure molecular forces between molecules. This is substantially increasing our knowledge of molecular interactions. AFM has been used for the visualization of the single protein molecules as well as for the comprehension of protein adsorption mechanisms on several biocompatible surfaces. It can provide detailed information about the conformation and size of biomolecules with nanoscale precision, protein layer surface morphology, and roughness, and it can reveal protein adsorption mechanisms and cell activation, which cause morphology changes. In the following discussion, an overview of the AFM analysis related with the hemocompatibility (possibility of thrombus formation) of a surface is presented. In this context, Fib and HSA morphological

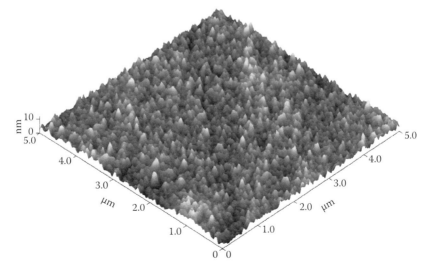

FIGURE 29.56 AFM 3D-image $5 \times 5\,\mu m^2$ of the surface of PET film (Melinex ST504).

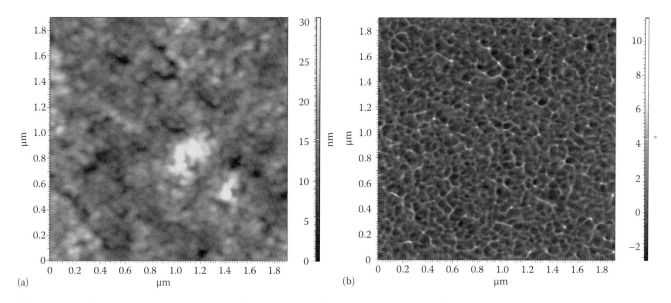

FIGURE 29.57 Topography (a) and phase image (b) of a PEDOT:PSS thin film glycol treated deposited on PET substrate by spin coating. Scan size is $1.9 \times 1.9\,\mu m^2$.

characteristics and adsorption mechanisms and platelet adhesion/activation mechanisms are investigated and presented.

29.2.4.2.1 Study of the Size and Conformation of Biomolecules with Nanoscale Precision

It is known that Fib is a linear blood plasma protein with size $48 \times 6 \times 9\,nm$ with three globular domains, with a significant role in blood coagulation and thrombus formation. In Figure 29.58, a $350 \times 350\,nm^2$ 2D image of adsorbed Fib on amorphous hydrogenated carbon (a-C:H) thin film is presented.

Protein conformation changes are extremely important for the in-depth comprehension of the protein adsorption mechanisms. These are observed through AFM technique. Causing protein denaturation, for example through heating or through pH change of the protein solution used, is a way to explore protein conformation changes in detail.

AFM technique provided information about the surface roughness of the thin films and the protein layers formed on them, as well as the change of the topography characteristics of the Fib layers with the increase in temperature. For this purpose, a special

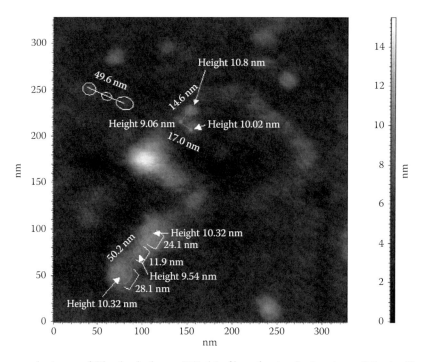

FIGURE 29.58 AFM topography image of Fib adsorbed on a-C:H thin film, after incubation time of 70 min. The length and height of Fib molecules, as well as their conformation (either linear or folded) are also indicated.

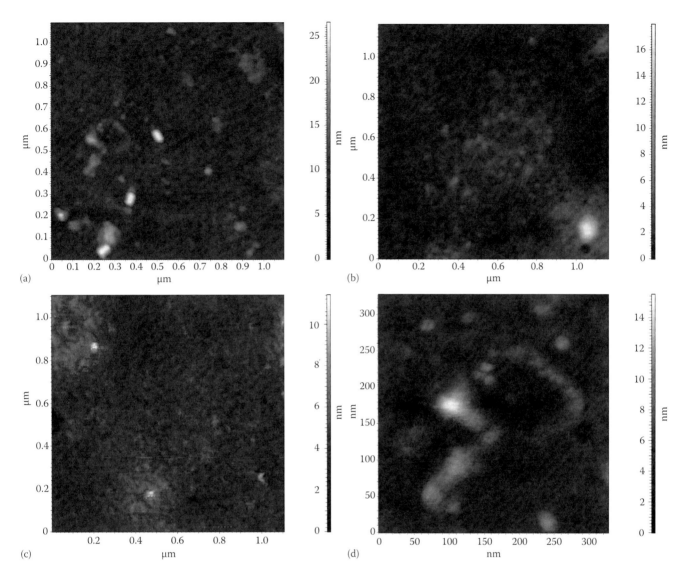

FIGURE 29.59 AFM topography image of Fib adsorbed on a-C:H thin film, (a) at room temperature (RT), (b) at 73°C, (c) at 116°C, and (d) detail in the red square of (a).

heating stage was integrated to the AFM and the samples (incubated for 70 min) were heated up, while their surface was being scanned by the AFM probe. Topography images (1 × 1 μm) of Fib on a-C:H thin film at room temperature (RT) (no denaturation), 73°C (mild denaturation), and 116°C (advanced stage of denaturation) are presented in Figure 29.59a through c, respectively. In the AFM topography image of Figure 29.59d (detail from Figure 29.59a), it is obvious that the Fib molecule preserves its trinodular shape with the three globular domains, either in a linear or in a folded conformation, on the a-C:H thin film. This is the reason why this sample was selected to be heated on the AFM heating stage. Peak-to-peak and root-mean-square roughness (R_{rms}) at the three temperatures are also presented in the graph of Figure 29.60. Peak-to-peak and R_{rms} are decreasing with increasing temperature, due to protein unfolding and dehydration. By the calculation of the diameter of randomly selected globular aggregates, it is observed that it varies from 10 to 25 nm, independent

of temperature. However, the distance between them decreases when temperature increases (data not shown) and the typical protein conformation is not easily distinct. This is caused by a possible change of the molecule from a linear to a folded conformation due to the temperature effect (Lousinian et al. 2007a,b).

29.2.4.2.2 Study of Protein Adsorption Mechanisms

The mechanisms of protein adsorption can be revealed through the implementation of AFM. For example, in the case of human serum albumin, a heart-shaped blood plasma globular protein with dimensions 8 × 8 × 3 nm, which plays an important role in inhibiting the thrombus formation, it seems that in general, aggregates are formed which finally coalesce. More precisely, it was observed that initially, the protein molecules form clusters (5 min), the size of which increases with time (10 min) and the surface of the thin film is partially covered as it can be concluded by the topography images of AFM, shown in Figure 29.61. The surface of a-C:H

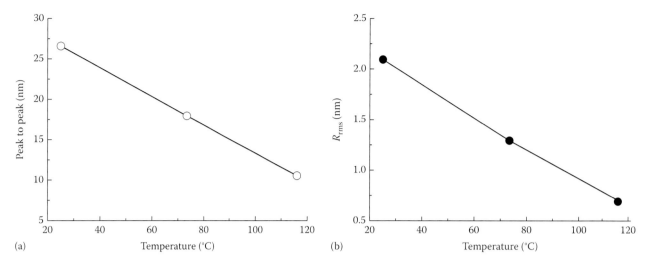

FIGURE 29.60   (a) Peak-to-peak and (b) root-mean-square roughness R_{rms} vs. temperature as derived from AFM semicontact scanning mode measurements.

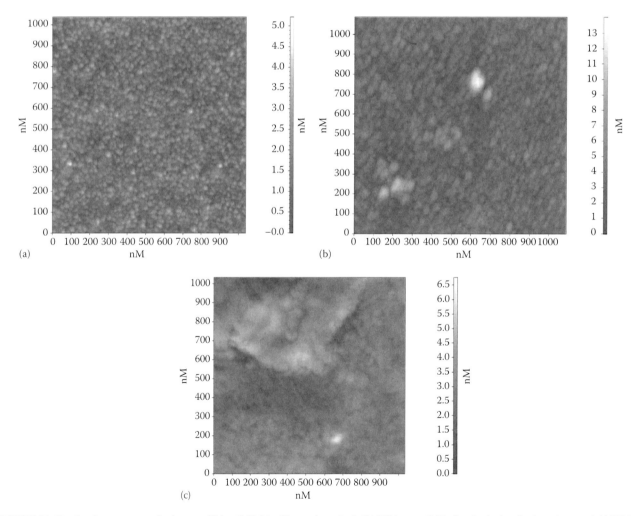

FIGURE 29.61 $1 \times 1\,\mu m$ topography image of (a) a-C:H thin film as deposited, (b) HSA on a-C:H after 5 min incubation time, and (c) HSA on a-C:H after 10 min incubation time.

film without proteins appears to have grain-like surface features, the size of which is around 20–30 nm (Figure 29.61a). In Figure 29.61b (protein incubation time 5 min), it is clearly seen that protein aggregates exist. Typical height is around 8 nm (once or twice the height of one HSA molecule). The protein molecules tend to spread laterally, rather than forming "hills." Figure 29.61c shows that in 10 min of incubation time, there are larger protein clusters and the surface is partially covered. Finally the protein islands coalesce, and the surface of the thin films is totally covered by the proteins (Logothetidis 2007, Mitsakakis et al. 2007).

29.2.4.3 Investigation of Platelets Morphology and Activation by AFM

29.2.4.3.1 Observation of Single Platelet Activation

Platelets play an important role in blood-material interactions as their activation leads to thrombus formation and onto their surfaces take place essential steps of coagulation cascade. Platelets undergo a change in their shape, which exposes a phospholipid surface for those coagulation factors that require a surface and also release agonist compounds of dense and a-granules to attract and activate additional platelets and leucocytes promoting the growth of thrombus. During their activation, there is an increased membrane expression of receptors such as glycoprotein GPIIbIIIa, which is the receptor for Fib and von Willebrand factor, leading to the linkage of adjacent platelets via Fib.

In a first approximation, the morphological characteristics of platelets during their adhesion on the examined films and activation were observed by AFM. More precisely, platelets from their resting form, which is characterized by a discoid shape without pseudopodia, when they come in contact with a surface and activate, present fully spread hyaloplasm with extended, flattening pseudopodia (Figure 29.62a and b). More precisely, in Figure 29.62a, one single platelet is observed, the developing pseudopodia (Karagkiozaki et al. 2009a,b). This is observed in an early stage of activation. On the contrary, in Figure 29.62b, after 2 h of incubation, six pseudopodia are observed around the platelet, with three more pseudopodia growing. The dimensions of the topography image of Figure 29.62b are 15 × 15 μm, so that the full length of the pseudopodia is presented. One stage of platelets activation

includes the formation of pseudonucleus (egg-like type), made by their granules gathered at the center of the cell. These structural changes indicant of the increased platelet activation, are necessary for their spread onto surfaces and aggregation via mainly of Fib. Additionally, platelets release chemical compounds of their granules to facilitate other platelets and leucocytes, resulting in a clot formation (Karagkiozaki et al. 2008).

29.2.4.3.2 Evaluation of Thrombus Formation Potential on a Surface

In the following figures, platelet aggregation and activation on a-C:H films, which have been found to be nonhemocompatible (Karagkiozaki et al. 2008), is presented. In Figure 29.63a, it can be easily noticed that after 1 h of incubation, platelets appear with a "pseudonucleus" in their center, being a index of activation, and after 2 h they aggregate, forming a cluster look like an "island" with a height of approximately 700 nm (Figure 29.63b). Performing X-section at 19,000 nm in Figure 29.28b, it can be deduced that the platelets indicated by the green arrows, have the height of 100–150 nm and length of 2000–2500 nm and at some areas they aggregate (blue arrows) and between them, clusters of platelet-rich plasma proteins having the size of 50–100 nm and height of 20–50 nm, prevail, as shown by purple arrows (Figure 29.63c). The conclusion that the areas indicated by the purple arrows in Figure 29.63c correspond to plasma protein clusters is supported by their size (diameter/length and height), which is quite smaller from the one of the platelets. On the other hand, the green arrows indicate some platelets on the surface of the a-C:H, with much larger diameter and height.

The differences of platelet activation and aggregation on hemocompatible and nonhemocompatible surfaces are pronounced in Figures 29.64 and 29.65. 3D AFM topography images of platelets on a-C:H films after 1 h (Figures 29.64a and 29.65a) and 2 h (Figures 29.64b and 29.65b) of incubation are presented. In Figure 29.64a, the circles indicate activated platelets having the "egg-like" type structure, which remains the same after 2 h of incubation (Figure 29.64b), confirming the good hemocompatibility properties of the film. On the contrary, in Figure 29.65, "egg-like" type platelets (Figure 29.65a) are aggregated and transformed to clusters after 2 h incubation time (Figure 29.65b) (Karagkiozaki et al. 2009).

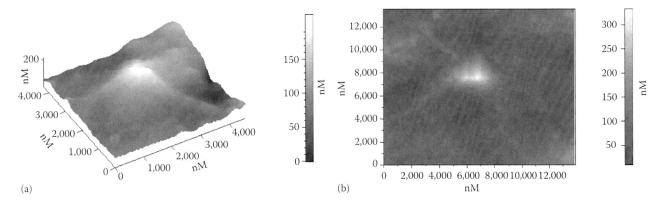

(a)   (b)

FIGURE 29.62 AFM topography image of (a) a single platelet at an early stage of activation when it starts to develop pseudopodia and (b) a highly activated platelet with pseudopodia and increased size on a-C:H film after 2 h of incubation.

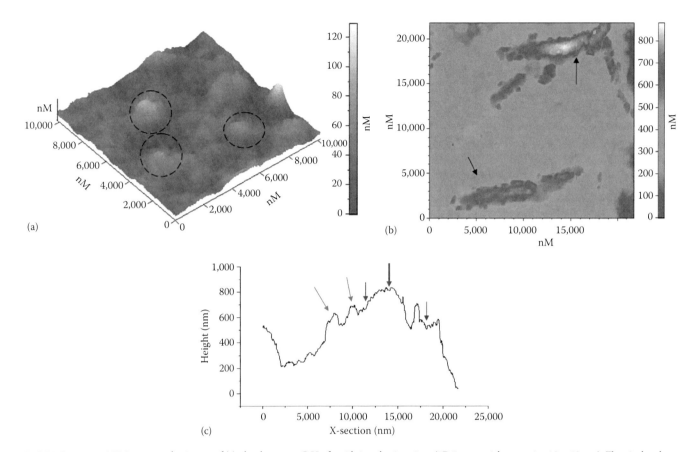

(a)

(b)

(c)

FIGURE 29.63 AFM topography image of (a) platelets on a-C:H after 1 h incubation time (3D image with scan size $10 \times 10 \mu m$). The circles show the activated "egglike" type platelets. (b) Platelets on a-C:H after 2 h incubation time (scan size $21 \times 21 \mu m$). The arrows indicate the platelets aggregation and the formation of clusters. (c) Diagram of X-section at 19,000 nm in Figure 29.52b. The first two left arrows show the platelets (height of 100–150 nm and length of 2000–2500 nm), the fourth arrow from the left indicate platelets clusters whereas clusters of PRP proteins with size of 50–100 nm, are shown by the third and fifth arrow from the left.

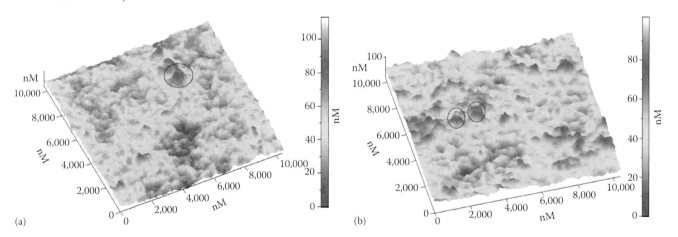

(a)

(b)

FIGURE 29.64 3D AFM topography image of (a) platelets on a-C:H (grown without the application of bias voltage on the substrate during their development) after 1 h incubation time (scan size $10 \times 10 \mu m$). The circle indicates an activated platelet having the "egg-like" type structure (b) Platelets on a-C: H after 2 h incubation time ($10 \times 10 \mu m$). The circles denote the "egg-like type" activated platelets.

These AFM observations are in line with the estimation of surface roughness via the measurements mean values of peak-to-peak and root mean square roughness (R_{rms}) parameters. In Table 29.4, the mean values of peak-to-peak distance and R_{rms} roughness parameters with their standard deviations, for 10 randomly scanned AFM areas are presented, for platelets' incubation times 0 min (bare substrate), 1 h, and 2 h. Thus, AFM can be used for high-resolution real-time studies of dynamic changes in cells, in order for revolutionizing our understanding of biological specimen–surface interactions (Karagkiozaki et al. 2008).

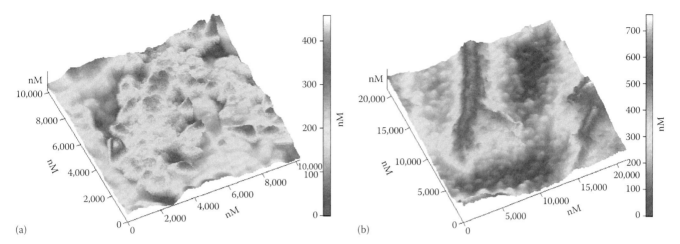

(a)

(b)

FIGURE 29.65 3D AFM topography image of (a) platelets on a-C:H (grown with the application of bias voltage on the substrate during their development), after 1 h incubation (scan size 10 × 10 μm). (b) Platelets on the a-C:H after 2 h incubation time (scan size 20 × 20 μm). The arrows indicate the platelets aggregation and the formation of a cluster like an island.

TABLE 29.4 Film Types, Incubation Times of Platelets (PLTs), and Morphology Parameters of Surfaces, Including the Mean Values of Peak-to-Peak Distance and RMS Roughness and Their Standard Deviation Values, as Measured for Ten Randomly Scanned AFM Areas

Film Type	Incubation Time (h)	Peak-to-Peak (nm)	RMS Roughness, R_{rms} (nm)
a-C:H	0	3.2 ± 0.1	0.28 ± 0.05
PLTs /a-C:H	1	129 ± 6	13.9 ± 1.7
PLTs /a-C:H	2	458 ± 53	58.4 ± 4.4

29.2.4.4 Scanning Near Field Optical Microscopy

Scanning near-field optical microscopy (SNOM) is a relatively new SPM technique with great potential giving optical images of surfaces with subdiffraction resolution. The main idea behind SNOM is to break the Abbe diffraction limit, which governs the resolution in conventional optical microscopy. According to the Rayleigh criteria, two point light sources, which are in distance l_{min}, can be resolved if the distance between the centers of the Airy disks, formed due to diffraction, equals the radius of one of the two Airy disks:

$$l_{min} = 0.61 \frac{\lambda}{n \cdot \sin \alpha}, \qquad (29.47)$$

where n is the refraction index of the medium, where the light source lies, and 2α is the aperture. Usually, $n \sin \alpha < 1.5$ and $l_{min} \sim 0.4\lambda$. Thus, in the conventional optical microscopy we cannot distinguish two objects if the distance between them is lower than 0.4λ.

In SNOM, this limit can be overcome by taking the advantage of the near-surface position an SPM probe (usually a shear force tuning fork) and gluing on it a fiber, through which the light source (laser beam) illuminates the sample surface from ~10 nm distance.

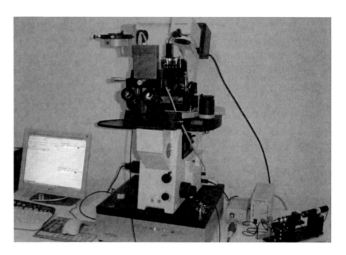

FIGURE 29.66 A SNOM apparatus mounted on an inverted optical microscope. The system is placed on an electronic antivibration table to avoid mechanical vibrations. The laser emits green light (λ = 532 nm). (Courtesy of Lab for Thin Films Nanosystems and Nanometrology, Physics Department, Aristotle University of Thessaloniki, Thessaloniki, Greece.)

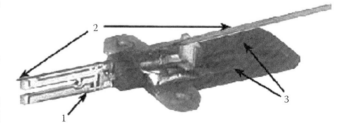

FIGURE 29.67 The SNOM probe. 1. The quartz resonator tuning fork. 2. The single-mode optical fiber and 3. The base with the metallic contacts. (From NT-MDT Co., *Instructions Manual*, NTEGRA Solaris Probe NanoLaboratory.)

A SNOM apparatus is shown in Figure 29.66. The SNOM probe (Figure 29.67) is a single-mode fiber glued on the one "leg" of the U-shape quartz resonator. The fiber is glued in such a way that its end exceeds 0.5–1 mm for the U-shaped quartz. Also, the radius of the optical fiber end is very crucial for resolution purposes and it is sharpened by chemical etching and coated with a metal, in order to achieve a very sharp tip (roundness 50–100 nm). The spatial resolution of SNOM is defined by this tip-surface distance and the aperture diameter of the tip, which is less than 100 nm. During the SNOM operation, a piezodriver is used to excite oscillations of the system tuning fork-optical fiber parallel to the sample's surface. Similar to the AFM semicontact mode, when the tip approaches the sample surface, a reduction of the oscillations amplitude is induced. A feedback system then is utilized to keep this amplitude constant during scanning. In this way, topographic shear force images can be obtained.

SNOM transmission images can be obtained from transparent samples, while SNOM reflection images can be obtained from opaque ones. Transmitted or reflected light from near-field is collected through an objective lens and finally is recorded from a photomultiplier. Fluorescence from the sample can be obtained at both transmission or reflection mode by using a notch filter to cut the excitation light and drive the fluorescence light to a spectrometer or a photomultiplier. SNOM is mostly used to study polymeric and biological systems. Among its various possibilities, the study of fluorescence from single molecules is of great importance.

Information about the topography and transmission of cells can be obtained by the use of SNOM technique. An example is presented in Figure 29.68, in a $45 \times 45\,\mu m^2$ image of platelet-rich

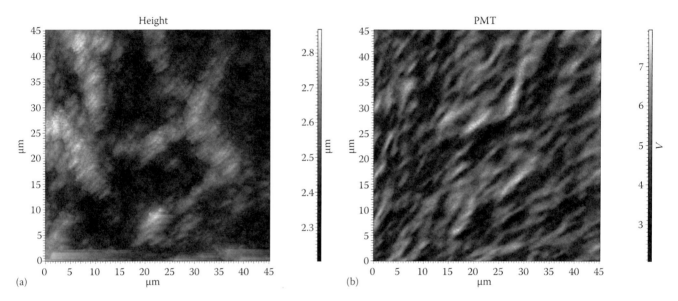

FIGURE 29.68 SNOM topography (a) and transmission image (b) of platelet rich plasma on a-C:H thin film deposited on glass.

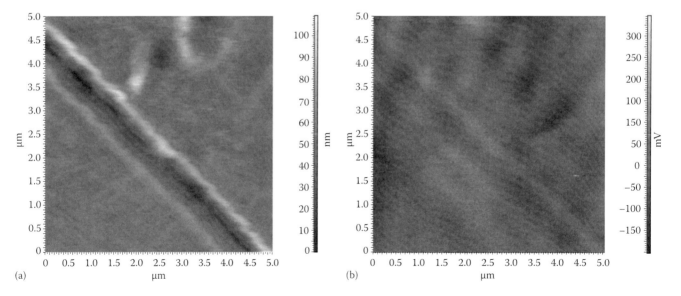

FIGURE 29.69 SNOM topography (a) and transmission image (b) of ZnO deposited on flexible PET substrate coated with hybrid laminate. Scan size is $5 \times 5\,\mu m^2$.

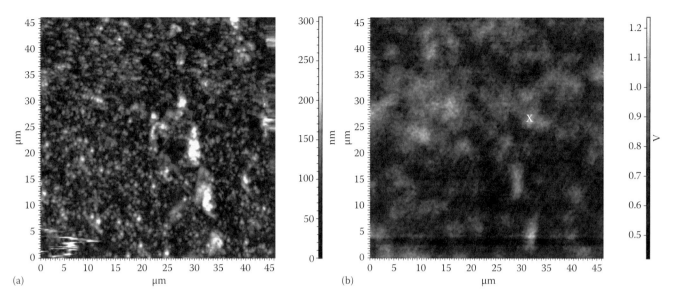

FIGURE 29.70 (a) SNOM topography and (b) fluorescence transmission image.

plasma (PRP). Figure 29.55a presents the topography of the PRP on a-C:H thin film deposited on glass. The platelets of this sample have a very large concentration and form large aggregates of large height values (about 3 μm), although areas of smaller height are also observed (Figure 29.68a). On the other hand, in Figure 29.68b, the image of the same area provides information about light transmission through the platelets. More precisely, the areas with a lighter color (larger transmission) seem to correspond to the topography areas where the height of the platelet aggregates is smaller. This kind of results is the first step showing the potential implementation of SNOM in the medical field.

In Figure 29.56a and b, the effect of surface irregularities to the transmission of the laser light is presented. The sample is a ZnO nanocoating deposited by *dc magnetron sputtering* on a flexible PET substrate coated with a hybrid polymer (organic–inorganic film, described in Section 29.2.1.4) ~2 μm thick laminate. The ZnO nanocoating is a candidate material to replace the brittle indium tin oxide as an electrode in flexible optoelectronic devices. Thus any surface features or irregularities can the affect the device functionality. In the topographic image shown in Figure 29.69a, a large crack on the surface of ZnO can be observed. The same crack is also clearly observed in the transmission image (Figure 29.69b). An example of SNOM florescence in present in Figures 29.70 and 29.71. A green laser ($\lambda = 532$ nm) was used to illuminate latex spheres stained with rhodamine. In Figure 29.70a and b, the SNOM topography and the transmission image from the same surface area are presented. In the transmission image, the whiter regions are possible places of fluorescence emission from the rhodamine, a fact that was verified by the received fluorescence spectra (Figure 29.71). The fluorescence spectra was taken for the position marked with (X) in the transmission image and the characteristic peak of rhodamine fluorescence is presented.

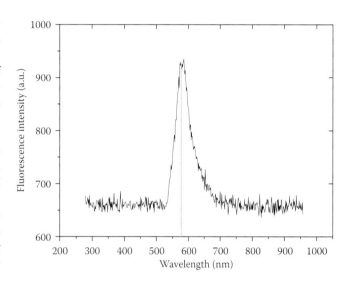

FIGURE 29.71 Fluorescence spectrum of rhodamine. Excitation laser was at 532 nm.

29.3 Summary

Nanotechnology is an emerging technology with applications in several scientific and research fields, such as information and communication technology, biology, medical technology, etc. Novel nano- and biomaterials, and nanodevices are fabricated and controlled by nanotechnology tools and techniques, which investigate and tune the properties, responses, and functions of living and nonliving matter, at sizes below 100 nm. In order to investigate in depth the complex nanosystems, highly sophisticated nanoscale precision metrology tools are required. The advances in nanomaterials necessitate parallel progress of the nanometrology tools and techniques. Examples of important nanometrology tools as they have been discussed in this chapter include in-situ monitoring techniques that allow the monitoring and evaluation of building block assembly and growth; ellipsometry, an optical nondestructive technique, with the capability of measuring in air and liquid environment (e.g., protein solution), highly focused x-ray sources and related techniques that provide detailed molecular structural information by directly probing the atomic arrangement of atoms; nanoindentation allow the quantitative determination of mechanical properties of thin films and small volumes of material; scanning probe microscopy (scanning tunneling microscopy, AFM, etc.) that allow three-dimensional topographical atomic and molecular views or optical responses (SNOM) of nanoscale structures. The above-described nanometrology methods contribute toward the understanding of several aspects of the state-of-the-art nanomaterials in terms of their optical, structural, and nanomechanical properties. The nanoscale precision and the detailed investigation that these nanometrology techniques offer give them an enormous potential for even more advanced applications for the improvement of the quality of research and of everyday life.

29.4 Future Perspective

The advances in fundamental nanosciences, the design of new nanomaterials, and ultimately the manufacturing of new nanoscale products and devices all depend to some degree on the ability to accurately and reproducibly measure their properties and performance at the nanoscale. Therefore, nanometrology tools and techniques are both integral to the emerging nanotechnology enterprise and are two of the main areas critical to the success of nanotechnology. Decades of nanoscience research have led to remarkable progress in nanotechnology as well as an evolution of instrumentation and metrology suitable for some nanoscale measurements. Consequently, today's suite of metrology tools has been designed to meet the needs of exploratory nanoscale research. New techniques, tools, instruments, and infrastructure will be needed to support a successful nanomanufacturing industry.

The currently available metrology tools are also beginning to reach the limits of resolution and accuracy and are not expected to meet future requirements for nanotechnology or nanomanufacturing. Novel methods and combinations, such as the TERS technique that has been described in the Introduction (Section 29.1), achieve much higher resolution values since provides a significant increase of the Raman signal and of the lateral resolution by up to nine orders of magnitude. This combination overcomes the difficulties that originate from low signal since the Raman systems have limit in lateral resolution of 300 μm and require high laser power for surface investigation because the measured Raman intensity is six orders of magnitude lower than the excitation power. Thus, TERS is a promising technique and in the near future, it could be used for probing the chemical analysis of very small areas and for the imaging of nanostructures and biomolecules such as proteins.

However, clever new approaches need to be developed. For this, the fundamental mechanisms by which the probes of the nanometrology measuring systems interact with the materials and objects that are being measured need to be understood. Also, it is important to develop standard samples and to construct standardized procedures for measurements at the nanometer scale, which enable the transfer of the properties and response of the unit from the nanometer to macroscopic scale without any appreciable loss of accuracy for certifying, calibrating, and checking nanometrology instruments. Finally, even with the vast array of current tools available, the important question is whether or not they are providing the required information or reams of inconsequential data. Revolutionary approaches to the nanometrology needed may be required in the near future and therefore, revolutionary and not just evolutionary instrumentation and metrology are needed.

Acknowledgments

The author would like to thank the stuff of the Lab for Thin Films, Nanosystems and Nanometrology (LTFN) (http://ltfn.physics.auth.gr) for their support and contribution.

References

Alford T.L., Feldman L.C., and Mayer J.W. 2007. *Fundamentals of Nanoscale Film Analysis*. Springer, Berlin, Germany.

Arwin H. 2000. Ellipsometry on thin organic layers of biological interest: Characterization and applications, *Thin Solid Films* 48: 377–378.

Azzam R.M.A. and Bashara N.M. 1977. *Ellipsometry and Polarized Light*. North Holland Publishing, Amsterdam, the Netherlands.

Bhattacharya A.K. and Nix W.D. 1988. Analysis of elastic and plastic deformation associated with indentation testing of thin films on substrates. *Int. J. Solids Struct.* 24: 1287–1298.

Bhushan B. 2004. *Handbook of Nanotechnology*. Springer, Berlin, Germany.

Binning G., Gerber Ch., Stoll E. et al. 1987. Atomic resolution with atomic force microscope. *Europhys. Lett.* 3: 1281–1286.

Bruggeman D.A.G. 1935. Berechnung verschiedener physikalischer konstanten von Substanzen. I. Dielektrizitatskonstanten und Leitfahigkeiten der Mischkorper aus isotropen Substanzen. *Ann. Phys. (Leipzig)* 24: 636.

Chen C.T., Song Y.C., Yu G.P. et al. 1998. Microstructure and hardness of hollow cathode discharge ion-plated titanium nitride film. *J. Mater. Eng. Perform.* 7: 324–28.

Cole K.C., Guevremont J., Ajji A. et al. 1994. Characterization of surface orientation in poly(ethylene terephthalate) by front-surface reflection infrared spectroscopy. *Appl. Spectrosc.* 48: 1513–1521.

Cole K.C., Ajji A., and Pellerin E. 1998. Fourier transform spectroscopy, in J.A. de Haseth (Ed.) *Proceedings in 11th International Conference*, Woodbury, NY.

Cuypers P.A., Corsel J.W., Janssen M.P. et al. 1983. The adsorption of prothrombin to phosphatidylserine multilayers quantitated by ellipsometry. *J. Biol. Chem.* 258: 2426–2431.

Daillant J. and Gibaud A. 2009. *X-Ray and Neutron Reflectivity.* pp. 85–131. Springer, Berlin/Heidelberg, Germany.

Daldosso N., Luppi M., Ossicini S. et al. 2003. Role of the interface region on the optoelectronic properties of silicon nanocrystals embedded in SiO_2. *Phys. Rev. B* 68: 085327-1–085327-8.

Fix W., Ullmann A., Ficker J. et al. 2002. Fast polymer integrated circuits. *Appl. Phys. Lett.* 81: 1735–1737.

Forrest S.R. 2004. The path to ubiquitous and low cost organic electronic appliances on plastic. *Nature* 428: 911.

Gioti M., Logothetidis S., Patsalas P. et al. 2000. Magnetron sputtered carbon nitride: Composition and chemical bonding of as-grown and post-annealed films studied with real-time & in-situ diagnostic techniques. *Surf. Coat. Technol.* 125: 289–294.

Gioti M., Laskarakis A., and Logothetidis S. 2004. IR–FUV ellipsometry studies on the optical, electronic and vibrational properties of polymeric membranes. *Thin Solid Films* 283: 455–456.

Gioti M., Logothetidis S., Schroeder J. et al. 2009. Real-time evaluation of thickness, optical properties and stoichiometry of SiO_x gas barrier coatings on polymers. *Thin Solid Films,* 517: 6230–6233.

Gould S.A.C., Schiraldi D.A., and Occelli M.L. 1997. Analysis of poly(ethylene terephthalate) (PET) films by atomic force microscopy. *J. Appl. Polym. Sci.* 65: 1237–1243.

Gravalidis C. and Logothetidis S. 2006. X-ray diffuse scattering investigation of polytetrafluorethylene surfaces. *Polimery/Polymers* 51: 359–364.

Gravalidis C., Gioti M., Laskarakis A. et al. 2004. Real time monitoring of silicon oxide deposition processes. *Surf. Coat. Technol.* 180–181: 655–658.

Gravalidis C., Logothetidis S., Hatziaras N. et al. 2006. Characterization of Si nanocrystals into SiO_2 matrix. *Appl. Surf. Sci.* 253: 385–388.

Hamers R.J. 2001. Flexible electronic futures. *Nature* 412: 489–490.

Herzinger C.M., Johs B., McGahan W.A. et al. 1998. Ellipsometric determination of optical constants for silicon and thermally grown silicon dioxide via a multi-sample, multi-wavelength, multi-angle investigation. *J. Appl. Phys.* 83: 3323–3336.

http://www.dupont.com/releasesystems/en/products/dryfilm_dispersion.html

http://www.ntmdt.com

Hultman L., Shinn M., Mirkarimi P.B. et al. 1994. Characterization of misfit dislocations in epitaxial (001)-oriented TiN, NbN, VN, and (Ti,Nb)N film heterostructures by transmission electron microscopy. *J. Cryst. Growth* 135: 309–317.

Ioannou-Sougleridis V., Nassiopoulou A.G., and Travlos A. 2003. Effect of high temperature annealing on the charge trapping characteristics of silicon nanocrystals embedded within SiO_2. *Nanotechnology* 14: 1174–1179.

Jellison G.E. and Modine F.A. 1996. Parameterization of the optical functions of amorphous materials in the interband region. *Appl. Phys. Lett.* 69: 371–373.

Kalfagiannis N., Zyganitidis I., Kassavetis S. et al. 2009. Microstructural characteristics, nanomechanical properties of Ti/TiB_2 multi-layered metal ceramic composites. To be submitted.

Karagkiozaki V., Logothetidis S., Laskarakis A. et al. 2008. AFM study of the thrombogenicity of carbon-based coatings for cardiovascular applications. *Mater. Sci. Eng.: B* 152: 16–21.

Karagkiozaki V., Logothetidis S., Kalfagiannis N. et al. 2009. Atomic force microscopy probing platelet activation behavior on titanium nitride nanocoatings for biomedical applications. *Nanomed. Nanotechnol. Biol. Med.* 5: 64–72.

Kassavetis S., Mitsakakis K., and Logothetidis S. 2007. Nanoscale patterning and deformation of soft matter by scanning probe microscopy. *Mater. Sci. Eng. C* 27: 1456–60.

Keldysh L.V., Kirzhnits D.A., and Maradudin A.A. 1989. *The Dielectric Function of Condensed Systems*, North-Holland Publishing, Amsterdam, the Netherlands.

Kitano Y., Kinoshita Y., and Ashida T. 1995. Morphology and crystal-structure of an a axis oriented, highly crystalline poly(ethylene-terephthalate). *Polymer* 36: 1947–1955.

Koidis C., Logothetidis S., and Georgiou D. 2008. Growth, optical and nanostructural properties of magnetron sputtered ZnO thin films deposited on polymeric substrates. *Phys. Status Solidi A* 205: 1988–1992.

Koidis C., Logothetidis S., Laskarakis A. et al. 2009. Thin film and interface properties during ZnO deposition onto high-barrier hybrid/PET flexible substrates. *Micron* 40: 130–134.

Krug T.G. 1990. Transparent barriers for food-packaging. In *Proceeding of the 41st Society of Vacuum Coaters Annual Conference*, Boston, MA.

Laskarakis A. and Logothetidis S. 2006. On the optical anisotropy of poly(ethylene terephthalate) and poly(ethylene naphthalate) polymeric films by spectroscopic ellipsometry from visible-far ultraviolet to infrared spectral regions. *J. Appl. Phys.* 99: 066101-1–066101-3.

Laskarakis A. and Logothetidis S. 2007. Study of the electronic and vibrational properties of poly(ethylene terephthalate) and poly(ethylene naphthalate) films. *J. Appl. Phys.* 101: 053503-1–053503-9.

Laskarakis A., Logothetidis S., and Gioti M. 2001. Study of the bonding structure of carbon nitride films by IR spectroscopic ellipsometry. *Phys. Rev. B.* 64: 125419-1–125419-15.

Laskarakis A., Gioti M., and Pavlopoulou E. 2004. A complementary spectroscopic ellipsometry study of PET membranes from IR to Vis-FUV. *Macromol. Symp.* 205: 95–104.

Laskarakis A., Georgiou D., and Logothetidis S. 2010. Real-time optical modelling and investigation of inorganic nano-layer growth onto flexible polymeric substrates. *Mater. Sci. Eng. B,* 166: 7–13.

Laskarakis A., Georgiou D., Logothetidis S. et al. 2009. Study of the optical response of hybrid polymers with embedded inorganic nanoparticles for encapsulation of flexible organic electronics. *Mater. Chem. Phys.* 115: 269–274.

Logothetidis S. 2001. *Thin Films Handbook: Processing, Characterization and Properties.* Academic Press, New York.

Logothetidis S. 2005. Polymeric substrates and encapsulation for flexible electronics: Bonding structure, surface modification and functional nanolayer growth. *Rev. Adv. Mater. Sci.* 10: 387–397.

Logothetidis S. 2007. Haemocompatibility of carbon based thin films. *Diam. Relat. Mater.* 16:1847–1857.

Logothetidis S. and Laskarakis A. 2008. Organic against inorganic electrodes grown onto polymer substrates for flexible organic electronics applications. *Thin Solid Films* 518: 1245–1249.

Logothetidis S., Kassavetis S., Charitidis C. et al. 2004. Nanoindentation studies of multilayer amorphous carbon films. *Carbon* 42: 1133–1136.

Logothetidis S., Gioti M., Lousinian S. et al. 2005. Haemocompatibility studies on carbon-based thin films by ellipsometry. *Thin Solid Films* 482: 126–132.

Lousinian S. and Logothetidis S. 2008. In-situ and real-time protein adsorption study by spectroscopic ellipsometry. *Thin Solid Films* 516: 8002–8008.

Lousinian S., Logothetidis S., Laskarakis A. et al. 2007a. Haemocompatibility of amorphous hydrogenated carbon thin films, optical properties and adsorption mechanisms of blood plasma proteins. *Biomol. Eng.* 24: 107–12.

Lousinian S., Kassavetis S., and Logothetidis S. 2007b. Surface and temperature effect on fibrinogen adsorption to amorphous hydrogenated carbon thin films. *Diam. Relat. Mater.* 16: 1868–1874.

Lousinian S., Kalfagiannis N., and Logothetidis S. 2008. Albumin and fibrinogen adsorption on boron nitride and carbon-based thin films. *Mater. Sci. Eng.: B* 152: 12–15.

Lucas B.N., Oliver W.C., and Swindeman J.E. 1998. Dynamics of frequency-specific, depth-sensing indentation testing. *Mater. Res. Soc. Symp.—Proc.* 522: 3–14.

Martínez-Antón J.C. 2002. Measurement of spectral refractive indices and thickness for biaxial weakly absorbing films. *Opt. Mater.* 19: 335–341.

Mathioudakis C., Kelires P.C., Panagiotatos Y. et al. 2002. Nanomechanical properties of multilayered amorphous carbon structures. *Phys. Rev. B* 65: 205203-1–205203-14.

Miller C.E. and Eichinger B.E. 1990. Analysis of rigid polyurethane foams by near-infrared diffuse reflectance spectroscopy. *Appl. Spectrosc.* 44: 496–504.

Mitsakakis K., Lousinian S., and Logothetidis S. 2007. Early stages of human plasma proteins adsorption probed by atomic force microscope. *Biomol. Eng.* 24: 119–124.

Motohashi M., Hayazawa N., Tarun A. et al. 2008. Depolarization effect in reflection-mode tip-enhanced Raman scattering for Raman active crystals. *J. Appl. Phys.* 103: 034309-1–034309-9.

Nardes A.M., Kemerink M., and Janssen R.A.J. 2007. Anisotropic hopping conduction in spin-coated PEDOT: PSS thin films. *Phys. Rev. B* 76: 085208 1–0852081 7.

Nevot L. and Croce P. 1980. Characterization of surfaces by grazing x-ray reflection—Application to study of polishing of some silicate-glasses. *Rev. Phys. Appl.* 15: 761–779.

Niklasson G.A. and Granqvist C.G. 1984. Optical properties and solar selectivity of coevaporated Co-Al$_2$O$_3$ composite films. *J. Appl. Phys.* 55: 3382–3410.

Nomura K., Ohta H., Takagi A. et al. 2004. Room-temperature fabrication of transparent flexible thin-film transistors using amorphous oxide semiconductors. *Nature* 432: 488–492.

Oliver W.C. and Pharr G.M. 1992. Improved technique for determining hardness and elastic modulus using load and displacement sensing indentation experiments. *J. Mater. Res.* 7: 1564–1580.

Oliver W.C. and Pharr G.M. 2004. Measurement of hardness and elastic modulus by instrumented indentation: Advances in understanding and refinements to methodology. *J. Mater. Res.* 19: 3–20.

Ouchi I. 1983. Anisotropic absorption and reflection spectra of poly(ethylene terephthalate) films in ultraviolet region. *Polym. J.* (*Tokyo, Jpn.*) 15: 225–243.

Ouchi I., Nakai I., and Kamada M. 2003. Anisotropic absorption spectra of polyester films in the ultraviolet and vacuum ultraviolet regions. *Nucl. Instrum. Methods Phys. Res. B* 199: 270–274.

Pai P.G., Chao S.S., and Takagi Y. 1986. Infrared spectroscopic study of SiO$_x$ films produced by plasma enhanced chemical vapor-deposition. *J. Vac. Sci. Technol. A* 4: 689–694.

Palik E.D. 1991. *Handbook of Optical Constants of Solids II.* Academic Press, Boston, MA.

Parratt L.G. 1954 Surface studies of solids by total reflection of x-rays. *Phys. Rev.* 95: 359–369.

Patsalas P. and Logothetidis S. 2001. Optical, electronic, and transport properties of nanocrystalline titanium nitride thin films. *J. Appl. Phys.* 90: 4725–4734.

Patsalas P. and Logothetidis S. 2003. Interface properties and structural evolution of TiN/Si and TiN/GaN heterostructures. *J. Appl. Phys.* 93: 989–998.

Patsalas P., Logothetidis S., and Kelires P.C. 2005. Surface and interface morphology and structure of amorphous carbon thin and multilayer films. *Diam. Relat. Mater.* 14: 1241–54.

Rabe U., Janser K., and Arnold W. 1996. Vibrations of free and surface-coupled atomic force microscope cantilevers: Theory and experiment. *Rev. Sci. Instrum.* 67: 3281–93.

Sinha S.K., Sirota E.B., and Garoff S. 1988. X-ray and neutron scattering from rough surfaces. *Phys. Rev. B* 38: 2297–2311.

Snaith H.J., Kenrick H., and Chiesa M. 2005. Morphological and electronic consequences of modifications to the polymer anode 'PEDOT:PSS'. *Polymer* 46: 2573–2578.

Sneddon I.N. 1965. The relation between load and penetration in the axisymmetric boussinesq problem for a punch of arbitrary profile. *Int. J. Eng. Sci.* 3: 450–467.

Steidtner J. and Pettinger B. 2008. Tip-enhanced Raman spectroscopy and microscopy on single dye molecules with 15 nm resolution. *Physl. Rev. Lett.* 100: 236101-1–236101-4.

Tabor D. 2000. *The Hardness of Metals.* Oxford University Press, London, U.K.

Tauc J., Grigorovici R., and Vaucu A. 1966. Optical properties and electronic structure of amorphous germanium. *Phys. Status Solidi B* 15: 627–637.

Thompson C., Palasantzas G., and Feng Y.P. 1994. X-ray-reflectivity study of the growth kinetics of vapor-deposited silver films. *Phys. Rev. B* 49: 4902–4907.

Tompkins H.G. and Irene E.A. 2005. *Handbook of Ellipsometry.* William Andrew Publishing, New York.

Tonelli A.E. 2002. PET versus PEN: What difference can a ring make? *Polymer* 43: 637–642.

Wangand Y. and Lehmann S. 1999. Interpretation of the structure of poly(ethylene terephthalate) by dynamic FT-IR spectra. *Appl. Spectrosc.* 53: 914–918.

Whitehouse D.J. 2002. *The Handbook of Surface and Nanometrology.* Taylor & Francis Publishing, London, U.K.

Wormeester H., Kooij E.S., and Mege A. 2004. Ellipsometric characterisation of heterogeneous 2D layers. *Thin Solid Films* 455–456: 323–334.

Yanaka M., Henry B.M., Roberts A.P. et al. 2001. How cracks in SiO_x-coated polyester films affect gas permeation. *Thin Solid Films* 397: 176–185.

Yashar P.C. and Sproul W.D. 1999. Nanometer scale multilayered hard coatings. *Vacuum* 55: 179–90.

Yeo, B.-S., Schmid, T., Zhang, W. et al. 2007. Towards rapid nanoscale chemical analysis using tip-enhanced Raman spectroscopy with Ag-coated dielectric tips. *Analyt. Bioanalyt. Chem.* 387: 2655–2662.

Yi L.X., Heitmann J., and Scholz R. 2002. Si rings, Si clusters, and Si nanocrystals—Different states of ultrathin SiO_x layers. *Appl. Phys. Lett.* 81: 4248–4250.

Yi L.X., Heitmann J., and Scholz R. 2003. Phase separation of thin SiO layers in amorphous SiO/SiO_2 superlattices during annealing. *J. Phys.: Condens. Matter* 15: S2887–2895.

Yi K.J., He X.N., Zhou Y.S. et al. 2008. Tip-enhanced near-field Raman spectroscopy with a scanning tunneling microscope and side-illumination optics. *Rev. Sci. Instrum.* 79: 073706-1–073706-8.

Zhang W., Yeo B.S., Schmid T. et al. 2007. Single molecule tip-enhanced Raman spectroscopy with silver tips. *J. Phys. Chem. C* 111: 1733–1738.

Aerosol Methods for Nanoparticle Synthesis and Characterization

Andreas Schmidt-Ott
Delft University of Technology

30.1 Introduction

Nanoparticles are essential building blocks in the fabrication of nanostructured materials. Much of the progress in creating novel nanoassemblies or nanocomposites depends on the progress in producing inorganic nanoparticles, including atomic clusters, nanotubes, and nanowires. Particle production processes are needed that are flexible with respect to size, structure, and composition to enable tailoring of the product properties in view of the application. Systems in which the particles retain their special size-determined properties are of special interest. This may, for example, be achieved by coating, so that the distances and therewith the interactions between the particle cores are controlled. Major long-term research goals in this area are to develop ways of designing and producing new nanoscale materials of chosen properties in application domains including electronic, semiconductor, and optical properties as well as selective catalytic behavior, unusual strength and lightweight, resistance to corrosion, and fast kinetics in hydrogen and lithium storage.

Particle production from the gas phase, the aerosol route, has inherent advantages such as purity, the absence of liquid wastes, and feasibility of continuous processes as opposed to the colloid route, where usually only batch processes are possible. Versatility and flexibility of aerosol methods with respect to particle material and size and structure represent additional advantages. In addition, various methods of online characterization and classification of the particulate product are provided by the state of the art aerosol technology. The lack of monodispersity has been seen as a major drawback of gas-phase nanoparticle production, but this difficulty can be overcome, as shown below. For these reasons, vapor phase nanoparticle production is becoming the dominant nanoparticle production route, although liquid-phase methods had a head start of 100 years.

Gas-phase synthesis of nanoparticles for applications has been reviewed by Kruis et al. [1], Swihart [2], and Biskos et al. [3]. Hahn [4] presented a useful overview of gas-phase synthesis of nanocrystalline materials. This chapter is partly inspired by these reviews and focuses on "round" inorganic particles smaller than 100 nm in diameter and down to the atomic cluster size range. The term "atomic cluster" is used for the subnanometer range here. The production of fullerenes, carbon nanotubes, and related materials has been treated by many authors and is outside the scope of this chapter. However, well-defined particles are the key to nanotube and nanofiber production [5].

30.2 Online Characterization and Classification of Aerosol Nanoparticles

Online characterization of nanoparticles in gas suspension with respect to specific properties is important in connection with gas-phase production in order to control particle formation to give the desired product. Characterization is usually based on effects that also classify or separate particles with respect to certain properties, so that classification and characterization can be treated simultaneously. Separation methods are used to obtain particles that are pure with respect to a certain property. This chapter does not treat techniques that are customarily used for experimental atomic cluster research like time-of-flight mass spectrometry and drift cell mobility analyzers. Techniques that require particle sampling for inspection, for example, by microscopy, are also be left aside. The present section is restricted to size, charge, and concentration analysis in aerosols, but this is the key to assessing any property that induces a change in one of these quantities in a suitable experiment [6,7]. For example, charge measurement combined with aerosol photoelectron emission leads to information on the electronic structure [8]. Photoemission from particles in gas suspension can also be used to measure the adsorption of gases to particles suspended in inert gas [9], because it reveals the work function, which is the energy required to emit an electron from a solid. The work function sensitively depends on molecules adsorbed on the particle surface and can be used to monitor adsorption [9]. Size and concentration change quantify collisional growth of particles and can be used to estimate forces between them [10]. The concentration of aerosol particles before and after passing a magnetic filter yields the particle magnetic moment [11]. The reaction rate of particles with any specific species added to the aerosol or adsorption behavior can be observed via size change. For example, a dual size analyzer comparing size before and after adding water vapor is being applied to measure hygroscopicity of airborne particles. These methods have emerged in the field of aerosol science and technology and have a great potential in characterization and study of nanoparticles and clusters for nanotechnology.

The fact that in the aerosol state particles can easily be separated with respect to their size is also a great advantage for nanoparticle production from the gas phase. It is useful firstly to determine the size distribution of the particles produced and secondly to select particles of a certain narrow size range. This is necessary for basic studies on the particles produced, for example, the size dependence of catalytical properties or for obtaining a product that requires a certain particle size. Most aerosol size classifiers are basically filters that transmit a certain particle size. Of course, such filters could also be designed to transmit a specific size distribution desired for a product. Existing flexible principles for size separation of nanoparticles in gas suspension that cover the whole range down to atomic clusters are inertial impaction and electrical mobility classification [12]. The former method uses the effect that the acceleration of particles in an accelerated gas stream depends on their mass and mobility. The mobility b is defined as

$$b = \frac{v_{\mathrm{rel}}}{F_{\mathrm{d}}}, \tag{30.1}$$

the ratio of the particle velocity relative to the gas and the resulting drag force on the particle. The quantity in terms of which the particles can be separated is, for example, the stopping distance of a particle in an abruptly stopped gas flow

$$S = bmv_0. \tag{30.2}$$

where

  m is the particle mass
  v_0 the initial gas flow velocity

The mass is related to the particle diameter D_p through

$$m = \frac{\pi}{6} D_p^3 \rho \tag{30.3}$$

with the particle density ρ, assuming spherical particles. The mobility b is related to the particle diameter D_p via

$$b = \frac{3}{2p(D_p + d)^2} \frac{(kT/2\pi\mu)^{1/2}}{1 + \pi\alpha/8} \tag{30.4}$$

and

$$\mu = \frac{mm_p}{m + m_p} \tag{30.5}$$

[12] with the gas pressure p, the effective gas-particle collision diameter d, Boltzmann's constant k, temperature T, and the momentum accommodation coefficient α for the gas-particle collisions. Equation 30.4 has been derived by Tammet [13] and experimentally verified by Fernandez de la Mora et al. [12].

The most important device that determines S is the so-called impactor, the principle of which is illustrated in Figure 30.1. The aerosol is forced through a nozzle, behind which a plate causes an abrupt change in the flow direction. The vertical velocity component is reduced to zero, and particles with a stopping distance that exceeds a value in the order of the nozzle-plate distance are impacted onto the plate. They stick there due to van der Waals forces. The device thus separates particles smaller than a specific size D_{po}, which follow the flow, from those that are larger than D_{po}, which adhere to the plate. By varying the pressure downstream of the nozzle, D_{po} can be varied [12,14]. By recording either the current (number per unit time) of impacted particles or the current of nonimpacted particles as function of the pressure, the size distribution can be determined. If a known fraction of the particles carry an elementary charge, the current of the impacted particles is measurable by connecting an electrometer to the plate, and the current of the nonimpacted particles can be determined by an aerosol electrometer (see below). In principle, a combination of impactor-like devices could also be designed to transfer only a narrow size interval. However, the alternative method of electrical mobility classification has been developed into more practical devices. The electric field E exerts the force $F_{el} = qE$ on the particles. Size and charge q are determined from v_{rel} via Equations 30.1 and 30.4 considering the force balance $F_{el} = F_{d}$.

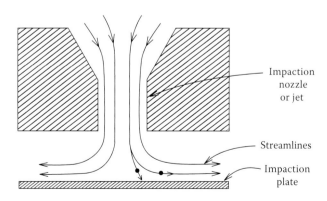

FIGURE 30.1 Impactor (principle).

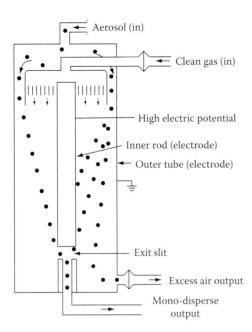

Aerosol (in)

Clean gas (in)

High electric potential

Inner rod (electrode)

Outer tube (electrode)

Exit slit

Excess air output

Mono-disperse output

FIGURE 30.2 Differential mobility analyzer (principle).

The commonly used designs of electrical mobility classifiers are referred to as differential mobility analyzers (DMAs) [15,16]. These instruments usually consist of two concentric cylindrical electrodes that are connected to a voltage source (see Figure 30.2). The polydisperse aerosol enters the DMA in a perimetric flow close to the outer electrode surrounding a particle-free sheath flow. Because of the electric field produced by the applied voltage, charged particles of the right polarity obtain a radial drift velocity toward the central rod, which is superimposed onto the laminar flow. Particles within a narrow range of mobilities exit the DMA through a circular slit in the centre rod, and this flow of gas carrying equally sized particles is available for further studies or processing of these particles while they remain in gas suspension. Alternatively, the particles can be collected on a filter or applied to coat a surface, by using electrostatic or inertial forces. The highest particle concentration obtained behind a DMA is 10^6 cm^{-3} if a hot wire is used as the source of charged particles [17,18] (see Section 30.3.2.3). With the highest aerosol flow rate presently achievable with commercial systems of 100 L/min, this is equivalent to a maximum production rate of about 2×10^9 s^{-1}.

For size analysis, the monodisperse flow is transferred to a detector, where the particle number concentration is measured. Convenient detectors are condensation nucleus counters [19] that saturate the aerosol with a vapor (usually a hydrocarbon) in a continuous flow arrangement. In a cooled duct, the vapor becomes supersaturated and condenses on the particles, magnifying them to an optically detectable size in the micron range. Optical counting at a controlled flow rate yields the particle concentration in the aerosol. By recording this concentration as a function of the DMA voltage, a mobility distribution is obtained, which can be converted into a size distribution via Equation 30.4, if the charge per particle is known for all the sizes

present. Condensation nucleus counters do not detect particles smaller than a few nanometers (typically 3 nm). This is because the supersaturation S required for condensation on very small particles is not far from the threshold of homogeneous nucleation, S_{hom}. In a confinement, the supersaturation can hardly to be made uniform in space, which means that S_{hom} is locally reached if S is near S_{hom}, and additional particles are formed, distorting the particle count.

An alternative method to "count" the particles applicable to all sizes consists in measuring the electric current corresponding to the charged particle flow in the DMA output. This is done by the means of a so-called aerosol electrometer [20], consisting of a particle filter in a metal housing, serving as a Faraday cup. The charged particles are trapped in the filter, and the current flowing from the housing to ground corresponds to the particle current. Divided by the elementary charge, it represents the number of particles per unit time. The currents measurable this way are about 10^{-15} A. For flow rates of a few liters per minute, this corresponds to particle concentrations of 10^2–10^3 particles/cm^3, each carrying an elementary charge.

Most DMAs used in recent studies are based on the improved version of the Hewitt mobility analyzer proposed by Knutson and Whitby (1975) [15]. This design is unsuitable for particles smaller than 10 nm because a large fraction of particles is lost through diffusion to the ducts through which they flow. Devices that can be used down to the molecular range have been developed by the group of Fernandez de la Mora [21]. The DMA model shown in Figure 30.2 [22] can be operated with aerosol flow rates that limit the time of the particles in the deflection zone to about 1 ms. It is important to reduce this time as much as possible, because during deflection by the electric field the particles diffuse, performing Brownian motion. This has the consequence that the mobility classification becomes unsharp. The shorter the time t in the deflection zone, the smaller is the broadening in terms of the diffusional mean square displacement according to Einstein's formula

$$\overline{x^2} = 4Dt, \tag{30.6}$$

where D is the diffusion coefficient. The resolution of the device described in [21,22] is sufficient to separate atomic clusters differing in one atom for a number of atoms $n < 10$ [23] and possibly for larger n. The relative standard deviation of a peak in the size spectrum referring to a cluster is around 1% [23]. A practical problem for particles smaller than 5 nm in most DMA concepts is the change in potential generally required between the aerosol inlet and the outlet lines. As both lines often need to be grounded, the output flow must be passed through an insulating tube, where substantial electrophoretic losses occur. This problem has been overcome by the isopotential nano-DMA of Labowsky and Fernández de la Mora [24], where both the inlet and outlet aerosol slits are at ground potential.

DMA technology and particle current measurement rely on electric charging of the particles. Various designs of aerosol chargers have been applied during the last decades [25,26].

Most of them use diffusive transport of ions to the particles and are referred to as diffusion chargers. The charging efficiency depends on the image force between the ion and the particle, but this material influence is so weak that diffusion charging can be regarded as material independent. Ion production is usually done by a corona discharge or by radioactive gas ionization. If the polarities are not separated in the latter case, the aerosol obtains a bipolar charge distribution. Although the charged fraction is then small, this technique is often used, because an equilibrium charge distribution (Boltzman distribution [27]) is approximately reached [28], which is rather independent of the conditions in the charging device and thus well defined. In photoelectric charging [29], the aerosol is exposed to UV radiation of a photon energy above the work function of the particles and below the ionization energy of the gas molecules. This is an efficient way of charging particles positively. The charging efficiency is material dependent, which can be used for material separation [30] but is not desired in all cases. For most charging principles, nanoparticles below 10 nm in diameter seldomly obtain more than a single elementary charge, so that there is a one-to-one relation between size and mobility. For larger particles, special care has to be taken to avoid multiple charging [31]. Uniform charging is not required for size analysis, but the size-dependent charge distribution has to be known to calculate the size distribution.

30.3 Nanoparticle Generation

30.3.1 Particle Synthesis by Electrohydrodynamic Atomization

Electrohydrodynamic atomization (EHDA), or electrospraying, can be used for producing equally sized liquid droplets in the nanometer size range using electrical forces. This is of interest for solid nanoparticle production, because a solution can be sprayed, and evaporation of the liquid of each aerosol droplet leads to crystallization of the solute. Besides electrospray-drying, electrospray

pyrolysis is an established method [32] (e.g., Messing et al. 1993). In the latter case, the spray aerosol is heated, the solvent evaporates, and the residual particles decompose to form the product particles.

The electrospray process has been described in many publications, see for example, [33–36]. By applying an electric field to a droplet at the end of a capillary tube, the droplet is deformed to a cone (Taylor cone [37]). Under the right conditions, in the so-called "cone jet mode," a jet emanates from the cone tip, which decays into small droplets (see Figure 30.3) [33,34]. These droplets are equally sized and their diameters reach from tens of nanometers up to several microns.

Figure 30.3 also shows a photo of an electrospraying liquid. The liquid is pumped through a nozzle at flow rates between 1 L/h and 1 mL/h. A high voltage (HV) applied between the capillary and an electrode establishes an electric field. The electrode can be ring shaped to allow the spray to escape through it. Electrospray is a versatile way of producing well-defined nanoparticles of various compositions [39–41]. For metals or metal oxide particle production, the precursors can be salts of these metals. For example, if a mixture of metal nitrates is dissolved in the spray liquid, spray pyrolysis in an oxidizing atmosphere produces mixed oxide particles. The droplet aerosol is passed through a tube oven to induce decomposition of the salt, leading to solid metal particles or their oxides. Catalyst particles can be produced by this electrospray pyrolysis method and electrostatically precipitated to a surface [41]. Decomposition may also be induced on that surface instead of in the aerosol state. Salts that only produce gaseous products besides the metal, such as chlorides or nitrates, have been widely used. Spray pyrolysis has been applied for the production of complex mixed oxide particles in the micrometer range [42], for example, for high-temperature superconductors. In analogy, electrospray pyrolysis could be used to produce nanoparticles of a large variety of mixed substances.

The droplets produced in electrospray are highly charged. This has the consequence that agglomeration is avoided. However,

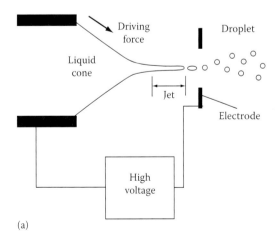

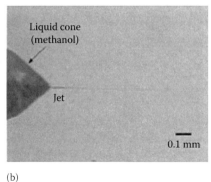

(a)   (b)

FIGURE 30.3 (a) Formation of a liquid cone at the end of a capillary tube and exerted jet and droplets in electrospray. The voltage applied between the nozzle and the liquid is typically around 4 kV for a nozzle-electrode distance of 1 cm. (b) Photo of a methanol cone at the capillary tip in a stable cone-jet mode with a fine jet a few micrometers in diameter. (From Okuyama, K., and Lenggoro, I.W., *Chem. Eng. Sci.*, 58, 537, 2003.)

the high space charge of the spray leads to dispersion of the spray cloud, which is associated with loss of particles to the container wall in practice. The charge level in an electrospray droplet is close to the so-called Rayleigh limit [43], where Coulomb repulsion leads to fragmentation. Under the influence of evaporation, that limit is reached, the droplet disintegrates into smaller droplets, and these evaporate and fragment again, and so on. This chain of Coulomb explosions is undesired, if equally sized particles are to be produced. Partial charge compensation by the introduction of ions avoids this problem [44]. Weak radioactive sources or corona discharges are applied for this purpose. The remaining charge can be applied for effective deposition of the particles onto a surface to produce a particulate film [45].

Another application of electrospray consists in bringing nanoparticles or large molecules from colloidal solution into gas suspension. Coulomb fragmentation is desired here because if there is more than one particle in the droplet, these are separated. As any liquid contains dissolved contaminants, this mechanism also separates particles from these, at least to a certain degree. From gas suspension, large molecules can easily be transferred into a vacuum system, where they are accessible to mass spectrometry [46]. John Fenn received the Nobel Prize for this analytical technique and its application to biomolecules in 2003.

Quantitative understanding of EHDA is difficult, but a lot of progress has been made in this subject. The forces contributing are indicated in Figure 30.4. If the liquid has some conductivity, the field draws carriers of one polarity to the surface by the normal component of the electric stress. The tangential component pulls the charge carriers towards the electrode. This causes a surface flow in that direction. Within a narrow range of the external electric field, the flow converges into a jet (cone jet mode). The cone and jet are shaped by the balance of the electrical, viscous, and surface tension stresses. The jet breaks up due to axisymmetric instabilities, also called varicose instabilities,

or due to varicose and lateral instabilities [43,48]. Finally, the highly charged particles form a plume, which rapidly expands due to its space charge.

A complete model of EHDA should firstly calculate the shape of the liquid cone and jet, the electric fields inside and outside the cone, and the surface charge density on the cone and jet. Further it should estimate the liquid velocity at the liquid surface [47]. The second part should describe the jet breaking up into droplets due to instabilities and the third part should treat droplet motion in the spray plume emitted. These three parts are coupled, and a complete physical model of this kind has never been presented due to the complexity. However, helpful scaling relations giving the droplet radius and the current have been derived by several authors. A scaling law for the droplet diameter is given by

$$d = \alpha \frac{Q^{a_1} \varepsilon_0^{a_2} \rho^{a_3}}{\sigma^{a_4} \gamma^{a_5}}, \qquad (30.7)$$

where

Q is the liquid volume flow rate
γ is the liquid bulk conductivity
ε_0 is the permittivity of free space
ρ is the mass density of the liquid
σ is the surface tension of the liquid
α depends on the liquid permittivity

The exponents proposed in different studies are indicated in Table 30.1 [49]. We see from Equation 30.7 that the droplet size can be decreased by decreasing the flow rate or by increasing the conductivity.

The minimum flow rate at which the cone-jet mode can operate in the steady state was determined by Barrero and Loscertales [53] as

$$Q_{min} = \frac{\sigma \varepsilon_0 \varepsilon_r}{\rho \gamma}. \qquad (30.8)$$

This relation implies a practical limit of about 10 nm for the smallest droplets to be produced by electrohydrodynamic atomization. Of course, the final particle size from spray-drying or spray pyrolysis of solutions is smaller than the droplet size and can further be varied downward by decreasing the solution concentration.

Industrial application of EHDA produced particles is limited by the low mass production rate. Extremely small flow rates are required to generate particles in the size range of a few

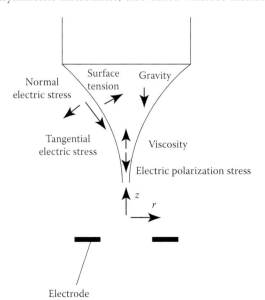

FIGURE 30.4 Forces involved in electrospray. (From Hartman, R.P.A. et al., *J. Aerosol Sci.*, 30, 823, 1999.)

TABLE 30.1 Exponents in Equation 30.1 Derived by Different Authors

Authors	a_1	a_2	a_3	a_4	a_5
Fernandez de la Mora and Loscertales [48,50]	1/3	1/3	0	0	1/3
Ganan Calvo and Ganan Calvo et al. [51,52]	1/2	1/6	1/6	1/6	1/6
Hartman et al. [43]	1/2	1/6	1/6	1/6	1/6

nanometers. To obtain droplets in the micron size range, flow rates are typically less than 1 mL/h. According to Equation 30.7, less than 1 µL/h is then needed for nanosized droplets with the mass production rate that is accordingly small. It is impossible to increase the throughput of material production from a single nozzle, so numbering up the injection ports is the only way to increase the production rate. A number of efforts have been made to scale out EHDA including the use of an array of capillaries and an array of holes in combination with nonwetting material. Other promising approaches use self-organization of Taylor cones forming on a liquid in a serration or groove. MEMS-based manufacturing can be used for this purpose, as summarized by Deng and Gomez [54].

30.3.2 Particle Synthesis from the Vapor Phase

30.3.2.1 General Features

The most straightforward way of producing nanoparticles is evaporation from a surface followed by nucleation and growth in an inert gas. This is a widely used approach for generating particles of well-defined chemical composition from the size range of atomic clusters to the micron range. Compared with methods where nanoparticles are formed in a liquid, where surfactants are usually needed, particle synthesis from the gas phase achieves much higher purity. This is also because noble gases are easily obtained in a very pure state in contrast to water or other solvents. While liquid-phase processes usually have to be performed batch-wise, it is easy to set up continuous processes for particle production from condensing vapors. The production rate is scalable and can be adapted to the application. In contrast to liquid-phase production, which basically requires a new recipe for each nanoparticulate product, gas-phase synthesis is usually very flexible regarding size and material.

Furnace reactors and glowing wires use electrical energy to heat the material to be evaporated and so do methods applying an arc or spark discharge. Flame reactors make use of the heat released in an exothermal reaction. In plasma reactors, the gas temperature may be rather low, whereas hot electrons induce decomposition of a precursor, and the product condenses to form particles. In infrared laser pyrolysis, radiation is absorbed by the gas that provides the energy for precursor decomposition. In laser ablation, a laser beam hits a target, heating and evaporating it.

All evaporation–condensation methods have in common that the vapor produced becomes supersaturated, and condensation takes place under controlled conditions. The saturation ratio is defined as

$$S = \frac{p}{p_{\text{sat}}}, \qquad (30.9)$$

where
   p is the partial pressure of the component to be condensed
   p_{sat} is the saturation pressure under a given temperature

A vapor can condense on a flat surface if $S > 1$, but it cannot condense to form small droplets or particles unless $S > S_{\text{hom}}$, the saturation ratio that allows spontaneous formation of particles in the volume of a vapor. This effect is usually referred to as homogeneous nucleation. It is the initial phase of condensation, where atoms or molecules collide and reversibly stick to form metastable clusters. At a given value of S (>1), there is a critical cluster size, above which growth occurs. In a macroscopic thermodynamic model, this is the size of a droplet, the vapor pressure of which balances the surrounding partial pressure. The vapor pressure of a droplet is given by the Kelvin equation [55], and the smaller the diameter, the higher is its vapor pressure. Simple hydrodynamic models like the classical nucleation theory are unsatisfactory, though [56]. The phenomenon of homogeneous nucleation (see, e.g., [57] and references therein), is very complex in reality. Various efforts of modeling using density functional theory and molecular dynamics have led to improved results, but there is no established general approach that reproduces experimental results. In addition, experiments are very difficult because atoms cannot be observed with the necessary resolution in time and space. To overcome this difficulty, atoms have been replaced by easily observable colloid particles in model experiments for noble gas condensation [57]. Much of the theoretical difficulty comes from the fact that the initial stage of nucleation depends very much on the specific atomic or molecular properties with respect to cluster formation. Homogeneous nucleation requires high saturation ratios, and values exceeding 10^6 have been used. Supersaturation is reached either by rapid cooling of the vapor or by the decomposition of a precursor into a product that has a much smaller saturation pressure than the precursor at the process temperature.

Particle synthesis from the gas phase has mainly been used for generating single-component particles. For multiple-component particles, the difference in condensation behavior of the constituents induces demixing, and layered structures are formed. In the case of spark discharge, quenching and the resulting supersaturation are so extreme that all atoms or molecules more or less stick at the initial few collisions with the critical cluster size being in the range of the atomic size. If this is so for all vapor components, mixed nanoparticles showing the absence of layers or even homogeneous mixtures can be produced [58,59].

The diagrams in Figure 30.5 illustrate growth kinetics of particles from a vapor. Two extreme cases can be distinguished. In Figure 30.5a, the vapor is quickly used up after the nucleation stage. The particles formed keep colliding with each other and coalesce (i.e., merge into new round particles) to form larger particles. This mechanism has been called cluster–cluster growth or coagulation and is governed by Smoluchowski's theory (e.g., [27]). Usually it is assumed that every collision sticks. This is certainly justified for particles of a few nanometers in size or larger, because these particles also stick to surfaces irreversibly [60]. Even if only van der Waals forces are considered, the binding energy at contact is much higher than thermal energy, and the impact energy is readily removed by the surrounding gas. For smaller particles, the assumption of sticking is also justified

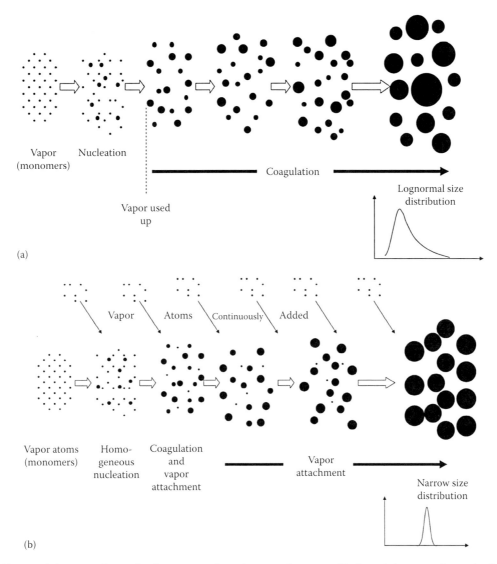

Vapor
(monomers)　Nucleation

Vapor used
up

Coagulation

Lognormal size
distribution

(a)

Vapor　Atoms　Continuously　Added

Vapor atoms
(monomers)　Homo-
geneous
nucleation　Coagulation
and
vapor
attachment

Vapor
attachment

Narrow size
distribution

(b)

FIGURE 30.5 (a) Growth kinetics of particles from vapor: Coagulation is dominant. (b) Growth kinetics of particles from vapor: Vapor attachment is dominant.

where coalescence occurs, as Figure 30.5 suggests. Coagulation leads to size distributions that can usually be approximated well by a lognormal function (Figure 30.5a). The tail of this distribution on the large particle end is undesired, where uniform size is advantageous.

Figure 30.5b demonstrates that uniform size distributions can also be produced, although this has not been achieved often in practice. Size distributions become narrow, if the growth process is not dominated by particle–particle collisions but by particle–atom (or molecule) collisions referred to as vapor attachment. After nucleation, the saturation ratio should be below the homogeneous nucleation threshold and above saturation. This can be achieved, for example, if supersaturated vapor is constantly added to the growing aerosol as indicated in Figure 30.5b, compensating the vapor lost by attachment. The corresponding qualitative time dependence of the vapor concentration n_v and particle concentration n_p is shown in Figure 30.6. n_v initially rises according to the vapor feed rate.

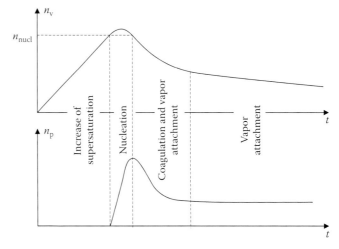

FIGURE 30.6 Particle formation under steady addition of vapor: Qualitative time dependence of vapor concentration n_v and particle concentration n_p.

When homogeneous nucleation occurs, particles are formed. They continue forming, until so much vapor is consumed by this process as well as by vapor attachment to the newly formed particles that S falls below S_{hom}, and homogeneous nucleation stops. From here, particles grow by further vapor condensation and by particle–particle collisions. The latter process rapidly reduces the particle concentration, as the particle–particle collision rate goes with the square of the concentration. Vapor attachment then becomes the dominant growth mechanism because vapor is continuously provided. The particle concentration hardly changes from here.

In the free molecular range, where particle size is smaller than the mean free path of the vapor atoms (e.g., 66 nm at normal conditions in N_2), the growth rate by vapor attachment is proportional to the particle cross section, so

$$\frac{dN}{dt} \propto \frac{d(R^3)}{dt} \propto R^2 n_{\text{v}}. \tag{30.10}$$

Here N is the number of atoms a particle consists of, which can be expressed through the particle radius R, and n_{v} is the vapor concentration. It follows

$$\frac{dR}{dt} \propto n_{\text{v}}, \tag{30.11}$$

which means that the particle growth rate is independent of R. This implies that from the point where vapor attachment is dominant, all particles grow by the same amount ΔR in a given time, regardless of differences in the "initial" size R_0, which arise from different nucleation times and particle–particle collisions. If $\Delta R \gg R_0$, we get a size distribution with a small relative standard deviation.

We encounter the case of Figure 30.5a in processes where a vapor is suddenly cooled, which induces supersaturation, homogeneous nucleation, and subsequent coagulation. The case of Figure 30.5b corresponds to droplet growth in a cloud chamber, for example, where equally sized droplets are formed. Here, no new vapor is added, but a small degree of supersaturation is maintained by a process of adiabatic expansion cooling. More and more vapors become condensable during that process, which compensates vapor depletion by attachment to the particles. Furthermore, the conditions of Figure 30.5b can, in principle, be achieved in reactors, where a precursor is continuously decomposed, so that the condensable vapor is continuously added to the growing particles. For example, this may happen in a nonequilibrium plasma reactor. Indeed, a stunning example for size uniformity has been given by Vollath et al. [61] for particles generated in such a plasma (see Section 30.3.2.6.4.1). It is likely to be the dominance of vapor attachment that leads to the monodispersity observed according to the considerations above. In any case, processes according to Figure 30.5b have a great potential in producing pure and equally sized particles in a gas-phase process. In practice, homogeneity of the conditions is an important prerequisite for a uniform size distribution. For example, temperature or concentration gradients with growing

distance from the walls of the confinement counteract a uniform size distribution of the product. Size limiting effects would be helpful but they have hardly been observed or pursued in gas-phase systems. Vollath et al. [61] point out that charges play an important role in a plasma, and Coulomb forces may have an important influence on the growth kinetics and suppress cluster–cluster aggregation, because particles above a certain size tend to carry the same charge polarity. This explanation is further elaborated on in connection with nonequilibrium plasma particle production in Section 3.2.6.4.1).

The models qualitatively illustrated by Figure 30.5 assume that particle–particle collisions lead to round particles again because coalescence occurs. For the atomic cluster size range, this practically happens under any conditions, and even 5 nm gold particles have been found to coalesce at room temperature, if their surfaces are completely clean [62]. As the melting point is strongly reduced with particle size [63] and particles a few nanometers in size have a liquid-like surface, coalescence is frequently observed and made use of in nanoparticle production. For a given particle size, coalescence stops when the temperature drops below a certain point. Fractal-like [64] irregular structures are then formed, the round units of which (primary particles) may only be bound by van der Waals forces. In processes that produce high particle concentrations, such agglomerated fractal-like structures frequently form. As small particles show a higher tendency toward coalescence, the observed primary particle size is often determined by coalescence. It is the largest size showing coalescence under the relevant process conditions. Incomplete coalescence or sintering at the contact point may form "hard" aggregates. Note that the term "agglomerate" has been established for loosely (van der Waals) bound primary particles and the term "aggregate" indicates strong (e.g., metallic bond) forces between them. In applications, where size effects are used that vanish, when contact with other particles occurs, individual, nonagglomerated nanoparticles are desired. For example, advanced, nanoparticulate ceramics require powders, where interparticulate forces are small, so agglomeration is allowed, while aggregation is undesired because it hinders homogeneous compaction of the powder. In other applications, for example, where high surface-to-volume ratios are required (catalysts, battery electrodes), aggregation may be very favorable. Sintering and coalescence strongly depend on the state of the particle surface concerning adsorbed molecules. A thin layer of oxide hinders particle growth by coalescence and avoids sintering [62,65].

Agglomerates that have once formed can hardly be fractionated, even if the forces involved are only van der Waals type. If the particles are transferred into a liquid, this reduces van der Waals forces, and ultrasound treatment induces some deglomeration. Agglomerates with primary particle sizes of a few nanometers can hardly be split down to sizes below 50 nm, though. The same applies to splitting nanoparticulate agglomerates in a polymer–particle composite by applying shear forces. Where the properties of individual nonagglomerated primary particles are to be applied, avoidance of coagulation directly behind the formation process is the only option. Fast dilution is a way to achieve this, and Coulomb forces between particles with equal polarity has been described as a favorable effect in nonequilibrium plasma production routes [61], with the possibility of adding a coating step that guarantees a distance between the particle cores to retain the size effect. If round particles are desired, coalescence can be induced by heating of the aerosol by passing it through an additional tube oven. Metal particles can usually be sintered easily by heating to moderate temperatures [66].

Among the methods reviewed below, only laser ablation and the hot wire generator method have been demonstrated to effectively produce subnanometer particles, but it is clear that in principle, any method based on the condensation of vapor could be modified to stop growth at this early stage. The reason for the very limited activities in the atomic cluster field using aerosol methods lies in the extremely high-purity requirements because every atom counts. The hot wire method is of special interest where the purity standards do not correspond to ultrahigh vacuum because the wire "emits" clusters as well as K^+ atoms, so that these selectively charge the clusters and not the gas contaminants. Only charged particles are selected or detected. In more sophisticated setups, ultrahigh vacuum-like conditions with respect to contaminants can be reached in one atmosphere of noble gas [9], and this should make aerosol systems competitive with respect to vacuum experiments on atomic clusters on a longer run.

The considerations above illustrate how complex nanoparticle formation in the gas phase is. The result of such a process sensitively depends on a large set of thermodynamic and electronic properties of the particle-forming species as well as the surrounding gas composition including impurities down to the parts per billion range. For example, oxygen impurities chemisorb to the surface of coagulating particles, hindering coalescence. A large number of attempts in modeling nanoparticle formation in various condensation-based processes have been made, which exceed the scope of the present chapter. The perfect model would require coupling of chemical reaction kinetics with computational fluid dynamics simulations with high time resolution and consideration of the inhomogeneity of most processes, with spatially varying concentrations and particle-size distributions. A complete description of the critical step of nucleation would involve first principle calculations of the atomic cluster growth. While empirical approaches still play a dominant role in nanoparticle production research, improvements in simulation methodologies and advances in computing power are making the discrepancies between theory and experiment smaller and increasingly enable useful predictions from theory.

The following sections outline some of the most commonly used evaporation–condensation methods for the synthesis of nanoparticles and atomic clusters in the gas phase.

30.3.2.2 Furnace Generators

The materials of interest are heated in a tubular flow reactor (Figure 30.5) such that the partial pressure of the vapor is high enough to provide the supersaturation necessary for homogeneous

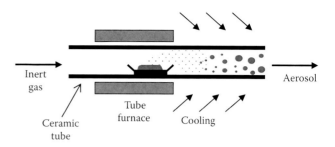

FIGURE 30.7 Tube furnace aerosol generator.

nucleation when cooled back to room temperature. A gas stream is passed over the evaporating material to carry the vapor away from the heated section. As the vapor cools, nucleation occurs and subsequent growth by particle–particle collisions (coagulation) determines the final particle-size distribution according to Figure 30.5a. Accordingly, the resulting size distributions of the primary particles or agglomerates produced this way are usually log-normal. A diluting stream may be introduced to reduce or avoid agglomerate formation. The residence times, cooling rates, and mixing characteristics allow some control over particle size and morphology. Furnace generators (Figure 30.7) have been used to generate nanoparticles of various substances [67,68]. They deliver a continuous and constant output and are frequently combined with a differential mobility analyzer to produce equally sized particles.

Of course, the material to be evaporated should have a much lower melting point than the inner walls of the furnace. Otherwise, vapors from the furnace walls lead to contamination. This is intolerable where surface contamination influences basic particle properties, and it becomes more critical the smaller the particles are. Jung et al. [69] proposed the use of a small ceramic heater and demonstrated that the high-purity silver particles can be generated in this way. Where purity is even more crucial, for example, for atomic clusters, elegant ways to avoid surface contamination by heating only the material to be evaporated have been developed by a resistively heated wire or spark discharge method (see Sections 30.3.2.3 and 30.3.2.4).

30.3.2.3 Hot Wire Generator

The hot wire method has been introduced with the first gas-phase experiments on basic properties of small particles in gas suspension by Schmidt-Ott et al. [70] and subsequently applied in a number of cases for research purposes (e.g., [9,12,71,72]). The particle purity is given by the wire purity. Using wires of high purity, production of neutral adsorbate-free particles was demonstrated by Müller et al. [9]. Helium evaporating from the surface of liquid helium was used to obtain an impurity partial pressure corresponding to ultrahigh vacuum. The amount of material produced hardly reaches 1 µg/h, so that possible applications are restricted to those that require only small amounts of nanoparticulate material such as gas sensors [73].

The material is evaporated by resistive heating of a metal wire subjected to a flowing inert gas (Figure 30.8). The vapor is quenched by diffusional mixing with the gas, which induces

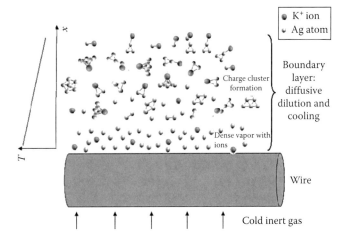

FIGURE 30.8 Formation of charged atomic clusters in the vicinity of a hot wire containing Ag and K as the most volatile components.

nucleation. A minimum evaporation rate is required for particles to be formed by this process which is reached below the melting point of most metals. A list of these metals for which the method is applicable is given by Peineke et al. [17]. The method can be extended toward many more materials by coating wires of high melting point with the material to be evaporated. This has been demonstrated for Au on W [74], and myriad other combinations, including nonconducting and semiconducting materials, should be possible. The conditions are that the surface material to form the particles should wet the electrically heated substrate and that alloying of the two materials is negligible.

A significant fraction of the particles formed by a hot wire carries an elementary charge, and this is of special interest, because the aerosol is directly applicable for electrostatic size classification, avoiding an additional charging step that usually introduces contaminants, as shown by Fernandez de la Mora et al. [12]. The origin of negative particle charge is thermoemission of electrons by the wire. These electrons attach to vapor atoms or the particles forming from them. The positive polarity is also observed and has been explained by cation emission from the wire surface. This effect is governed by the Saha-Langmuir equation [18,75]. It is very effective if the wire contains trace components of low ionization potential like alkali atoms. For example, a K atom sitting on an Ag surface has a high probability of being emitted as a K^+ ion because the energy required to bring the valence electron to the Fermi level of the metal can be provided thermally below the melting point of Ag.

The voltage V across the wire is usually kept constant because in this mode, the heating power P reduces if the resistance R rises due to the diameter reduction through evaporation loss. This is because the power is given by $P = V^2/R$ under these conditions. Wire breakage is avoided this way. The wire diameter changes very slowly, resulting in a change in the current $I = V/R$ of some 10 mA (<0.1%) per hour. The constant voltage mode delivers a sufficiently constant aerosol output for many purposes. Experiments with control circuits using temperature or power control have been carried out, but their performance

did not show any advantage with respect to the constant voltage mode, because the shape and heat distribution along the wire change in the course of time.

Electrical mobility measurements show that the size distributions of the particles generated by hot wires is log normal, as expected according to Figure 30.5a. The self-charging phenomenon allows production of high concentrations (10^5–10^6 cm^{-3}) of monodisperse particles at a few liters per minute by means of an electrostatic size classifier [17]. By biasing the wire potential with respect to the surrounding metal housing, the yield of charged particles can be increased further because particles of the desired polarity are separated from the other polarity, reducing charge recombination.

30.3.2.3.1 Production of Atomic Clusters by a Hot Wire

An extremely interesting feature of the hot wire method consists in the fact that it allows effective production of atomic clusters, that is, particles in the subnanometer size range. Every atom counts in the cluster regime, as stability and all other properties critically depend on the number and sort of atoms they are composed of. A major challenge in experimental studies on atomic clusters is to produce arbitrary pure and well-defined samples. In view of cluster assembled materials, increase of the production rate with respect to existing vacuum technology is required, and aerosol technology is very promising with respect to this requirement. Ag_nK^+ clusters were obtained from a hot wire containing silver, with traces of potassium (<0.1%) [23]. Figure 30.8 illustrates the process of cluster formation in the vicinity of a heated wire. In a boundary layer around the wire, the concentration of the clusters forming is high, and the K^+ ions, released from the same source, rapidly attach to these. Inside of the boundary layer, the clusters are diluted and cooled by heat conduction (diffusion) and outside of it convective mixing with the noble gas contributes, avoiding further growth. On the basis of size, they can be classified continuously with atomic resolution by means of a differential mobility analyzer, specially designed for high-mobility species. The gas contaminants do not obtain any charge because all the K^+ ions are used up in the high-concentration zone. They are therefore not selected in the subsequent electrostatic separation. Figure 30.9 shows the relative abundances determined as a function of the inverse electrical mobility $(be)^{-1} = Z^{-1}$. The quantity measured is the current of an aerosol electrometer representing the cluster number per unit time coming from a DMA as a function of the mobility Z selected by the DMA. An Ag wire with traces of K and a Pd wire with traces of Ag and K as the most volatile constituents were used. The relative abundances agree well with first principles calculations of the stability of Ag_n in terms of the removal energy of Ag [23]. The calculated Ag_nK^+ cluster properties are similar with respect to the pure Ag_n clusters in terms of energetic and electronic stability and cluster structure. Thus, K^+ attachment is an ideal, noninvasive way of charging for mobility classification. The equally sized clusters are available for reactivity, coalescence, and deposition studies. As the bond between the Ag clusters and K^+ is weak, K^+ could easily be removed by heating, which would enable the study of neutral clusters.

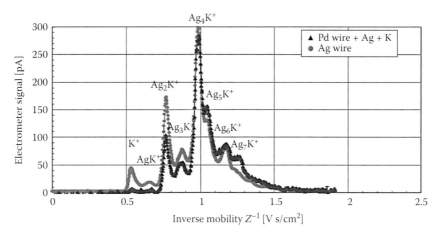

FIGURE 30.9 Cluster abundance in terms of electrometer current versus inverse mobility.

The production rate of size-separated clusters is orders of magnitude higher than conventional vacuum-based sources.

30.3.2.4 Spark Discharge Generator

The method of producing nanoparticles by spark discharge was introduced by Schwyn et al. [76] and has frequently been applied since then. It is one of the most versatile techniques for generating nanoparticles in the gas phase. The technique has been applied for the production of carbon [77], metal [62,76], and metal oxide nanoparticles [78,79] The particles are very similar to those obtained by laser ablation [80] (see below). With respect to this frequently used technique, spark discharge has the advantage of applying only inexpensive components. It can easily be scaled up by operating many sparks in parallel. The method is explained in detail in [62]. Only the material to form nanoparticles is heated, which enables high purity, as is done in the wire method. It can be applied to any conductive material including doped semiconductors and is unique for mixing materials due to the extremely fast quenching of the vapor (10^6–10^7 K/s). Mixing was studied in detail by Tabrizi et al., who observed mixing on an atomic scale and on a nanoscale on metals that are miscible [59] and immiscible [58] in the bulk. These new possibilities of mixing on a nanoscale are of great interest, for example, in preparing advanced catalysts. Coming from a plasma, a fraction of the particles produced carry an electric charge, which implies the same advantage as the wire with respect to electrostatic classification.

Figure 30.10 shows the principle of a spark discharge generator. A capacitor is continuously charged with constant current by an HV source. When the breakdown voltage is reached, the capacitor is rapidly discharged via the spark. Gas breakdown forms a conducting plasma channel. The rapid discharge consists in a current associated with a high temperature (typically larger than 10,000 K). Electrode material is evaporated in the vicinity of the spark by the high plasma temperature and through ion bombardment of the cathode. The vapor then cools rapidly, initially by adiabatic expansion and radiation. Below the evaporation temperature, cooling is dominated by thermal conduction of the gas. The cooling period below the boiling point

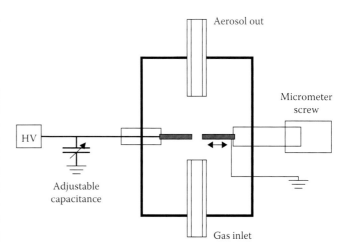

FIGURE 30.10 Spark discharge generator (principle).

is relatively fast because the dimensions of the vapor cloud are in the micron range and thus small compared with other evaporation–condensation methods. The consequence is very high a cooling rate of 10^6–10^7 K^{-1}, inducing a very high supersaturation, which leads to a high concentration of very small particles. Spark generators typically consist of a chamber about 300 cm^3 in volume, in which two opposing cylindrical electrodes a few millimeter in diameter are mounted at an adjustable distance of around 1 mm. The electrodes are connected to an HV source parallel to a capacitance of some tens of nanoFarad. The voltage source delivers a constant current, periodically recharging the capacitor after discharge has occurred at the breakdown voltage. The leads of the electric circuit represent an inductance, which forms an oscillator together with the capacitance, damped by the resistance of the spark. The damped oscillatory discharge typically takes a few microseconds [62]. Because the current is temporarily reversed during this oscillation, both electrodes are ablated to a similar extent. This can be used for mixing the material of one electrode with the material of the other one [58,59]. The spark repetition rate can usually be adjusted up to 1 kHz. Above that the discharge tends to become continuous. Such a continuous arc process produces particles tens of nanometer in diameter [81] (see Section 30.3.2.6.4.2), whereas microsecond sparks lead to nanoparticles smaller than 10 nm in diameter. The mean particle size can be controlled via the energy per spark, which is given by the capacitance and the distance between the electrodes. The particle mass produced per unit time is proportional to the spark frequency. Separated, unagglomerated particles a few nanometers in size can be obtained, if the inert gas flow through the generator is high enough with respect to the spark repetition frequency [62]. The nanoparticle mass production rate is typically 5 g/kWh. Taking gold as an example, it was shown that the sintering of agglomerated particles occurs at room temperature, leading to branched solid structures [62]. This indicates that the particle surfaces are adsorbate free. The mean primary particle sizes and mass production rate mainly depend on thermal conductivity, evaporation enthalpy, and the boiling point of the material. Knowing these properties, an estimate for the mass production rate can be obtained [62].

30.3.2.5 Laser Ablation

Laser ablation is a reproducible way of generating nanoparticles from the gas phase on a laboratory scale. Usually, a pulsed laser beam is focused onto a solid target, which partially evaporates with each pulse. Rotating the target guarantees the same initial condition for each event. Usually, excimer lasers are applied. Upon interaction of the laser beam with the target material, the surface is heated far above its boiling point. The vapor becomes superheated and ionized, forming a plasma. Carrier gas flowing over the heated surface cools the vapor and carries it away from the ablation zone. The particles produced are similar to those formed by spark discharge. An advantage of laser ablation with respect to that technique consists in the fact that also nonconducting materials can be used. The production rate is similar to the spark generator and of the order 100 ng per pulse, and control over the ablated mass is achieved by varying pulse energy and frequency of the laser. In contrast to spark discharge, scaling up by using multiple sources is not feasible because each source would require a laser.

Laser ablation techniques give access to a range of metal and metal oxide nanoparticles [82]. For the oxide, a metal oxide target can be ablated or, alternatively, the pure metal in an oxidizing gas. Laser ablation of aluminum in an oxidizing atmosphere produces extremely pure alumina nanoparticles [83]. Several studies have focused on the operating parameters that control particle size and morphology in laser ablation systems. Basically, increased dilution, for example, by increasing the flow rate, reduces size, whereas increased laser energy per pulse increases the vapor concentration and therewith particle size. Several studies have empirically examined the influence of operating conditions on the particle characteristics [84–86]. At normal pressure, the size of the primary particles has been shown to be restricted to a narrow size range below 15 nm independent of the materials used [80]. Reducing the pressure and/or the time allowed for growth in pulsed laser setups (see, e.g., [87]) leads to formation of atomic clusters. The time of growth is usually limited in such

systems by expanding the condensing vapor into vacuum. Thus, such cluster sources are usually referred to as laser vaporization in the atomic cluster literature. They have been used exclusively in vacuum-based experiments, which are not treated in the present chapter.

30.3.2.6 Chemical Vapor Synthesis

Chemical vapor synthesis (CVS) refers to a process in which a vapor is decomposed in a reactor, forming a nonvolatile inorganic product. This inorganic vapor is highly supersaturated as it forms and homogeneous nucleation and collisional growth lead to particle formation. Nanoparticles of metals, metal oxides, Si, and more have been produced this way. The precursor must fulfill the condition of having a much higher vapor pressure than the product. A large variety of precursors, mainly metalorganic compounds, are available in the market. The method can take advantage of a large number of precursors that have been developed for chemical vapor deposition (CVD) processes, and huge database of precursor chemistries is available. To induce the reaction, heated wall reactors as well as plasmas and flames are being used. Other methods such as laser photolysis or laser pyrolysis use absorption of infrared radiation by the gas to decompose the precursor.

30.3.2.6.1 Hot Wall Reactors

Such a reactor is usually a heated tube, as the furnace generator of Figure 30.7. Instead of evaporating the substance, a precursor is introduced together with an inert gas flow. Inside the tube, the precursor is thermally decomposed in the heated section. For example, copper and copper oxide particles have been produced from copper acetylacetonate [88] and tungsten particles by the decomposition of tungsten hexacarbonyle [89]. In a two-stage reactor, oxide-coated silicon nanoparticles were produced for high-density memory devices [90]. Here particle formation occurs in the first stage and oxidation in the second one. As decomposition of the precursor leads to an enormous supersaturation, any product atom virtually sticks to any other product atom, and this gives the opportunity of doping or mixing by CVS. Silicon nanoparticles doped with erbium [91] and zirconia particles doped with alumina [92] have been prepared using disilane and organometallic precursors. By adding precursors sequentially, coating nanoparticles of one material with another one is possible. Ehrman et al. [93] produced, among others, NaCl-encapsulated Si particles by reacting SiCl with sodium vapor. The salt coating avoids agglomeration of the Si particles, acting as a spacer. In aqueous solution, the salt is dissolved, and a dispersion of individual, round Si particles is produced [2]

30.3.2.6.2 Flame Synthesis

Flames are widely applied in industry to produce carbon black and ceramic particles (e.g., TiO_2 or SiO_2) for use as material reinforcing agents and catalysts [94,95]. Flame reactors for nanoparticle generation use either vapor or sprayed liquid precursors. At flame temperatures reaching up to 2500 K, the precursor is in the gas phase and decomposes. The reaction product has a much smaller vapor pressure than the precursor and rapidly condenses at the flame temperature or during cooling behind the combustion zone. Various particle compositions and mean sizes can be generated by selecting the appropriate conditions and precursors. The composition and crystal modification of the particles can be influenced by adding dopants into the feed stream and by controlling their temperature history [96]. Electric fields have also been shown to alter the physical and chemical characteristics of the particles [97,98]. Carbon blacks are produced using hydrocarbons as precursors [99]. Metal oxide nanoparticles can be made by oxidation of metal halide vapors in hydrocarbon-supported flames [100–102] Metal alloy nanoparticles can be prepared by mixing precursors [103], but this is only feasible in special cases with inert metals. Oxides are formed automatically for all oxidizable metals.

Figure 30.11 shows a flame spray pyrolysis setup applied by Mueller et al. [104]. A precursor fuel (ethanol) is mixed with the precursor [hexamethyldisiloxane (HMDSO)] and fed into an external mixing gas-assisted atomizer. The atomizer is composed of a capillary nozzle in the axis of a tube, through which the precursor mixed with the fuel flows. The annular gap between capillary and tube transports the dispersion gas, which aids break-up of the liquid droplets at the exit. The dispersion gas tube is surrounded by a number of coaxial tubes, which bring CH_4 and oxygen to the flame region. The CH_4 burns to form a supporting flame, which stabilizes the main flame fueled by the spray. The silica nanoparticles formed are collected in a baghouse filter.

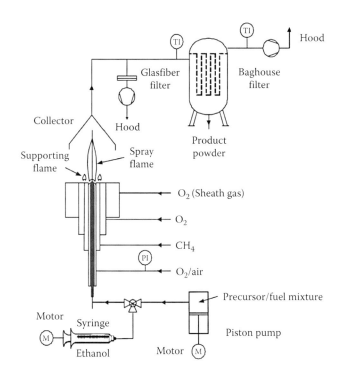

FIGURE 30.11 Experimental setup for fumed silica nanoparticles synthesis at high production rates by flame spray pyrolysis using a commercially available air-assisted stainless-steel nozzle and an annular supporting CH_4/O_2 diffusion flame. (Adapted from Mueller, R. et al., *Chem. Eng. Sci.*, 58, 1969, 2003.)

The material with the largest yearly nanoparticle production besides carbon black is probably titanium dioxide. Here flame synthesis has completely replaced earlier liquid-phase techniques. Reasons for this transition were environmental issues (no solvents are needed in gas-phase production) and the fact that the process is continuous. Both modifications, rutile and anatase, can be produced, depending on the conditions. Rutile is commonly used in pigments, whereas as anatase is preferred in photocatalysis and as an effective UV absorber [105].

Formation of hard aggregates is a frequent phenomenon in flame particle synthesis. This may happen if conditions are such that primary particles rapidly collide but do not coalesce, while at the same time, the flame reaction produces more condensing material that coats the nonspherical agglomerates. As mentioned above, hard aggregates are often undesired. Flames are usually not chosen for synthesizing multiple-component particles, because of the relatively slow quenching process. Materials behave differently concerning nucleation and growth that means bad mixing. The great advantage of flame synthesis is that a high particle production rate up to a few hundreds of grams per hour can be achieved at low cost.

Modeling flame synthesis is extremely complex because chemical reaction and flow phenomena interact with each other, and the conditions are usually strongly inhomogeneous. Janzen and Roth [106] have presented a detailed study of flame synthesis of γ-Fe_2O_3 nanoparticles and compared their results to a theoretical model.

30.3.2.6.3 Laser Photolysis

In laser photolysis, also called laser pyrolysis or photothermal synthesis, a gas containing a precursor is heated by absorption of infrared laser radiation and thermal decomposition of the precursor. The gas must contain molecules that have a high absorption cross section for the laser energy used. Such gases, for example, SF_6, are called photosensitizers. This function may be fulfilled by the precursor itself. Laser pyrolysis is routinely used, usually applying a CO_2 laser. It has advantages compared with methods using furnaces, since only the gas is heated and this is feasible in a highly localized and homogeneous manner, enabling rapid cooling. Condensation of the product on the walls of the confinement is minimized. Nanoparticles of a large variety of materials have been made using this method. These include silicon [107,108], MoS_2 [109], and SiC [110]. Pulsing of the laser radiation shortens the reaction time and leads to smaller particles [107,108]. The nanopowder mass production rates that have been reported are between 100 and 1000 g/h.

30.3.2.6.4 Plasma Synthesis

A plasma can be defined as a highly ionized gas, where electrons, excited molecular or atomic states, and/or a high gas temperature bear energy. This energy can be transferred to a precursor, decomposing it. If the reaction product has a low enough vapor pressure, it is supersaturated under the given conditions, and if the supersaturation exceeds the value required for homogeneous nucleation, particles are formed. In a hot plasma or thermodynamic equilibrium plasma, electrons and ions as well as the unionized gas have the same temperature, typically some thousands of Kelvins. In a nonequilibrium plasma, the electrons have a much higher temperature than the other species, typically several electron volts, so that they can induce chemical reactions. The gas temperature may be close to ambient.

30.3.2.6.4.1 Nonequilibrium Plasma Generators Particle production with nonequilibrium or cold plasmas requires a volatile precursor such as a hydrocarbon, a metal halide, or a metal organic compound. The molecules of the precursor are dissociated by the energetic species in the plasma, resulting, for example, in metal vapor that rapidly forms particles. Hydrocarbons introduced into a nonequilibrium plasma may polymerize under the influence of the active species, forming particles or coating existing particles. Vollath and Szabo used a nonequilibrium microwave plasma at a pressure of several milibar to produce metal oxide nanoparticles [61] (Figure 30.12). The temperature is between 400 and 750 K, and particles leave the reaction zone carrying charges of equal (positive) polarity. This greatly reduces agglomeration, as identically charged particles repel each other. They also showed that the absence of agglomeration enables coating of individual nanoparticles. They added a coating step, where monomethylmetacrylate vapor (monomers) [111] condenses on the particles and is polymerized by UV radiation from the plasma to form a PMMA (Perspex) coating, see Figure 30.12. Superparamagnetic particles were produced by this plasma method, and the coating avoids contact of the magnetic cores, retaining the superparamagnetic property of the powder.

The nonequilibrium microwave plasma method of Vollath and Szabo produced particles remarkably uniform in size. This may have to do with the fact that condensable vapor is constantly added to the process because more and more precursors decompose as illustrated in Figure 30.5b and explained above. The interplay of mechanisms in the plasma that control the charge of the growing particles includes secondary electron emission by hot electron impact, UV photoelectron emission, and ion attachment. There is evidently a growing probability for positive particle charge as a particle grows, because electron impact and UV charging both have cross sections that grow with the geometric particle cross sections. Coulomb repulsion thus forbids

particle–particle collisions above a certain size. Above that size, growth by attachment of vapor or very small particles dominates over coagulation according to Figure 30.5b because not only the vapor is continuously produced but also the coagulation is limited by Coulomb repulsion. The latter effect is pointed out by the authors [61].

Stable nonequilibrium plasmas are feasible between 5 and 100 mbar. Above this range, the plasma tends to become thermal. At normal pressure, the production rate would be higher, and no vacuum pumps would be required. Some successful attempts of using atmospheric pressure cold plasmas have been made using two approaches, namely, a dielectric barrier discharge (DBD) and a so-called plasma jet. In a DBD, an oscillating electric field is applied between a pair of electrodes. At least one of these electrodes is coated by a nonconductive material, like alumina or borosilicate glass, the dielectric barrier. It limits the current and prevents the transition into a hot arc plasma [113]. DBSs are simple setups that can be operated at atmospheric pressure using nitrogen, argon, or helium. DBD plasmas are used on a large scale for ozone generation from air for water purification. In small-scale experiments, iron and iron oxide particles were successfully produced from ferrocene, carbon particles from acetylene, silicon particles from HMDSO [114], and titania particles from titaniumisopropoxide [115,116].

A plasma jet at atmospheric pressure was used for the production of iron and carbon nanoparticles by Barankin et al. [117]. A radiofrequency plasma is maintained between parallel grids by applying an HV in the respective frequency range. Ignition of a hot arc discharge is avoided by fast enough gas exchange. An He flow perpendicular to the grids leads to a plasma flare behind the double grid, which induces decomposition of the precursor. Iron oxide particles of rather narrow size distribution were produced, which points to dominance of vapor attachment over coagulation (Figure 30.5), possibly enhanced by the charge effect described above.

30.3.2.6.4.2 Equilibrium Plasma Generators Particle synthesis in equilibrium or hot plasmas (thermal plasmas) allows for a large variety of feedstocks besides volatile precursors. Liquids and solids can be fed into the plasma. The high temperature of a thermal plasma can be used to evaporate micron-sized solid particles. Heberlein et al. [118] have applied such methods for the production

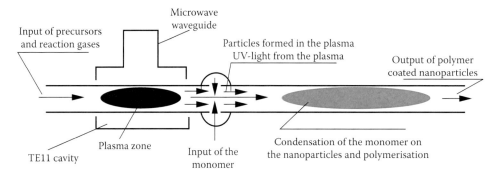

FIGURE 30.12 Microwave plasma process according to Ref. [112]. Particles are formed in the plasma zone, and a monomer is added that condenses on the particles and polymerizes under the influence of UV radiation coming from the plasma. (From Vollath, D. and Szabo, D.V., *Innovative Processing of Films and Nanocrystalline Powders*, eds. K.L. Choy, Imperial College Press, London, U.K., 2002. With permission.)

of SiC and TiC nanoparticles for hard coatings. In a continuous arc discharge between two electrodes, temperatures high enough to evaporate any metal are easily reached. Often the discharge takes place between a carbon anode and a metal cathode, the latter being the feedstock to be evaporated [116,119]. If the gas contains a hydrocarbon [120], this induces carbon coating on the particles. Such a coating is useful for passivation or as a spacer, if size effects of the cores are to be exploited such as superparamagnetism. To achieve the quenching and dilution necessary for nanoparticle formation, cold gas is mixed to the vapor plume. Effective cooling is reached if a cathode rod is surrounded by a coaxial anode tube with the gas flowing through the tube. Such configurations are called arc jets or plasma torches. Precursors are decomposed this way, the products condensing downstream of the gap. Gaseous precursors like SiCl$_4$ and CH$_4$ are applied [121,122]. The electrodes are not meant to be consumed, but they do show wear with resulting contamination of the product.

Electrodeless plasmas overcome this problem. These usually apply the principle of the inductively coupled plasma, where a coil fed by a radiofrequency source surrounds a glass or ceramic tube, through which the gas and the precursor flows [123–125]. Frequencies up to the microwave range have been used [126]. If a plasma has once been ignited within the tube, it is further sustained by currents induced into this conducting cloud by the alternating magnetic field of the coil. The plasma continuously loses energy by radiation. Ignition is done by means of a spark discharge. Quench gas is often injected behind the plasma to reach the supersaturations sufficient for nanoparticle formation. Again, the precursor can be fed into the inductively coupled plasma as a gas, a suspended solid powder, or a spray [127,128]. The plasma is usually hot enough to decompose any compound, and the remaining elements form nanoparticles.

Hot plasmas have high production rates, which makes them important for industrial application. The particle concentrations directly after particle formation are so high that rapid agglomeration and/or aggregation takes place. Individual coating of nanoparticles by a separate process is therefore hardly feasible. However, coating may be achieved in special cases by feeding the coating precursor into the process with the particle core precursor as mentioned for the *C* coatings. Flexibility with respect to feed stocks is a major advantage of hot plasmas. The particle sizes they deliver are generally larger than 30 nm and thus significantly larger than those typically produced by the spark generator, where small vapor clouds are produced that enable extremely fast cooling (see Section 30.3.2.4). The particle-size distributions obtained in hot plasma systems are broad because particles grow as in Figure 30.5a. As the temperatures used are generally much higher than the boiling point of the precursor and as all the gas molecules are heated, the energy consumption per gram of nanoparticulate material is much higher than in nonequilibrium plasmas, where only the electrons are hot.

30.3.2.7 Spray Pyrolysis

Rather than delivering the nanoparticle precursors into a hot reactor as a vapor, one can use a nebulizer to inject the precursor

in a liquid form as described for flame pyrolysis in Section 30.3.2.6.2. The precursor can be the liquid itself or a substance dissolved in the liquid. Basically, any of the CVS methods described in Section 30.3.2.6 can be used, provided the temperature is high enough to evaporate the liquid precursor or to evaporate the solute, if present. If the solute is to be decomposed to produce the particulate material desired, the conditions in the reactor must enable this either thermally or by electron impact (nonequilibrium plasma) or by absorption of radiation. The reaction may also (partly) take place in the droplets, followed by solvent evaporation. Such methods have, for example, been used for synthesis of TiO$_2$ and copper nanoparticles by Ahonen et al. [129] and Kim et al. [130], respectively.

30.4 Conclusions

Gas-phase methods for nanoparticle production, or aerosol methods, are promising and relatively new with respect to liquid-phase methods that range back to the beginning of the last century. Nanoscience and technology is increasingly profiting from aerosol routes that easily allow continuous flow arrangements, and many of the existing nanoparticle production methods that can be used for research purposes on a laboratory scale have been scaled up or have the potential to be scaled up as far as applications require this. There is great flexibility with respect to the particulate material and particle size. Methods like spark discharge enable mixing of materials on an atomic scale up to the scale of nanometers. It is usually simple to add a step of particle coating based on precursor decomposition or polymerization. Gas-phase methods produce no liquid waste. Because of the possible toxicity of the nanoparticles, the processes must be run in closed vessels.

Aerosol routes can be divided into those based on liquid atomization (spray) technology and those based on vapor condensation. The only spray principle presently reaching down to nanometric droplet sizes is EHDA (electrospray). This technique produces equally sized particles. Presently, the quantities produced per unit time are small, but efforts of scaling up are promising. Basically, spraying methods suffer from the same problem concerning purity as colloid methods, since liquids cannot be purified easily. Modeling of electrospray is complex, but helpful scaling laws have been derived. There is a practical lower size limit for the droplets of some tens of nanometers. The residual particles may be smaller, depending on the precursor concentration chosen, but the purity problem increases with decreasing particle size.

Those aerosol methods based on condensation of a vapor have the advantage of arbitrarily high purity. There is a large variety of methods to choose from that are based either on precursor or on evaporation of the same material or material mixture to form the particles. Modeling of condensation methods is very complex. Two mechanisms of particle growth, namely, particle–particle collisions (coagulation) and vapor attachment have qualitatively been described above. They can be seen as an orientation for qualitative understanding of most vapor phase nanoparticle production methods and a starting point for detailed modeling. The second mechanism (Figure 30.5b) illustrates that gas-phase production can, in principle, produce equally sized particles and aids a proper design of the process applied.

Gas-phase methods of nanoparticle production profit from a wealth of methods for online characterization and classification that have been developed in the field of aerosol technology. Where relatively small quantities are required (below the mg range), a specific size can be selected from a broad size distribution, and this size can easily be varied and tuned to the application. The strongest technique presently available for this is mobility classification. Size analysis with this method also gives access to a large variety of properties that result in a change in size, charge, or concentration in a suitable experiment. This can be used for optimization of the desired properties at the particle source. Aerosol methods give access to the full size range from 100 nm down to atomic cluster size. The purity problem with atomic clusters has limited activities in this field, but new approaches described above are leading to increased interest in aerosol methods, mainly because of the larger feasible material quantities. These are useful for research purposes, giving access to more analytical methods and in view of cluster assembled materials.

Particles from aerosols are conveniently collected in filters or brought to surfaces for devices and functional assemblies like chemical sensors, catalysts, solar cells, battery electrodes, and so on. This chapter does not address assembly methods for particles from the gas phase, but very interesting approaches are presently emerging. These include formation of nanoporous layers by impaction and electrostatic focusing of nanoparticle beams to create arbitrary nanostructures and charge-controlled deposition with a resolution in the 10 nm range. Self-assembly involving supramolecular structures or biomolecules requires liquid suspension. Methods that effectively transfer nanoparticles into liquids are needed for this purpose, and approaches based on bubbling or spray scavenging are available.

References

1. F.E. Kruis, H. Fissan, A. Peled, Synthesis of nanoparticles in the gas phase for electronic, optical, and magnetic applications—A review, *J. Aerosol Sci.*, 29, 11–35 (1998).
2. M.T. Swihart, Vapor-phase synthesis of nanoparticles, *Curr. Opin. Colloid Interface Sci.*, 8, 127–133 (2003).
3. G. Biskos, V.A. Vons, C.U. Yurteri, A. Schmidt-Ott, Generation and sizing of particles for aerosol based nanotechnology, *Kona: Powder Sci. Technol. Jap.*, 26, 13–35 (2008).
4. H. Hahn, Gas phase synthesis of nanocrystalline materials, *Nanostruct. Mater.*, 9, 3–12 (1997).
5. K. Bayer, K.A. Dick, T.J. Krinke, K. Deppert, Targeted deposition of Au aerosol nanoparticles on vertical nanowires for the creation of nanotrees, *J. Nanopart. Res.*, 9, 1211–1216 (2007).
6. A. Schmidt-Ott, Aerosol methods in small article research—A review, *J. Aerosol Res.*, 6, 208–216 (1991).
7. W. Fendel, Th. Kauffeldt, A. Schmidt-Ott, Measuring properties of nanoparticles by aerosol methods, *Nanostruct. Mater.*, 6, 655 (1995).
8. U. Müller, H. Burtscher, A. Schmidt-Ott, Photoelectric quantum yield of free particles near threshold, *Z. Physik B-Condens. Matter*, 73, 103–106 (1988).
9. U. Müller, A. Schmidt-Ott, H. Burtscher, First measurement of gas adsorption to free ultrafine particles: O_2 on Ag, *Phys. Rev. Lett.*, 58, 1684 (1987).
10. H. Burtscher, A. Schmidt-Ott, Enormous enhancement of van der Waals forces between small silver particles, *Phys. Rev. Lett.*, 1734 (1982).
11. Th. Kauffeldt, H. Kleinwechter, A. Schmidt-Ott, Absolute on-line measurement of the magnetic moment of aerosol particles, *Chem. Eng. Commun.*, 151, 169–185 (1996).
12. J. Fernández de la Mora, L. de Juan, K. Liedtke, A. Schmidt-Ott, Mass and size determination of nanometer particles by means of mobility analysis and focused impaction, *J. Aerosol Sci.*, 34, 79 (2003).
13. H. Tammet, Size and mobility of nanometer particles, clusters and ions, *J. Aerosol Sci.*, 26, 459–475 (1995).
14. J. Fernandez de la Mora, Drastic improvements on the resolution of aerosol size spectrometers via aerodynamic focusing: The case of variable-pressure impactors, *Chem. Eng. Commun.*, 151, 101–124 (1996).
15. E.O. Knutson, K.T. Whitby, Aerosol classification by electric mobility: Apparatus, theory, and applications, *J. Aerosol Sci.*, 6, 443–451 (1975).
16. W. Winklmayr, G.P. Reischl, A.O. Lindner, A. Berner, A new electromobility spectrometer for the measurement of aerosol size distributions in the size range from 1 to 1000 nm, *J. Aerosol Sci.*, 22, 289–296 (1991).
17. C. Peineke, M. Attoui, A. Schmidt-Ott, Using a glowing wire generator for the production of charged, uniformly sized nanoparticles at high concentrations, *J. Aerosol Sci.*, 37, 1651–1661 (2006).
18. C. Peineke, A. Schmidt-Ott, Explanation of charged nanoparticle production from hot surfaces, *J. Aerosol Sci.*, 39, 244–252 (2008).
19. B.Y.H. Liu, D.Y.H. Pui, A Submicron aerosol standard and the primary, absolute calibration of the condensation nuclei counter, *J. Colloid Interface Sci.*, 47, 155–171 (1974).
20. M.L. He, P. Marzocca, S. Dhaniyala, A new high performance battery-operated electrometer, *Rev. Sci. Instrum.*, 78, 105103 (2007).
21. J. Fernandez de la Mora, L. de Juan, T. Eichler, J. Rosell, Differential mobility analysis of molecular ions and nanometer particles, *Trends Anal. Chem.*, 17, 328–339 (1998).
22. J. Fernandez de la Mora, M. Attoui, A DMA covering the 1–100 nm particle size range with high resolution down to 1 nm, in *Proceedings of the European Aerosol Conference*, Salzburg, Austria, (2007).
23. C. Peineke, M. Attoui, R. Robles, A.C. Reber, S.N. Khanna, A. Schmidt-Ott, Production of equal size atomic clusters by a hot wire, *J. Aerosol Sci.*, 40, 423–430 (2009).

24. M. Labowsky, J. Fernández de la Mora, Novel ion mobility analyzers and filters, *J. Aerosol Sci.*, 37, 340–362 (2006).

25. J. Jiang, C.J. Hogan Jr., D.R. Chen, P. Biswas, Aerosol charging and capture in the nanoparticle size range 6–15 nm by direct photoionization and diffusion mechanisms, *J. Appl. Phys.*, 102, 034904 (2007).

26. H. Burtscher, L. Scherrer, H.C. Siegmann, A. Schmidt-Ott, B. Federer, Probing aerosol by photoelectric charging, *J. Appl. Phys.*, 53, 3787 (1982).

27. W.C. Hinds, *Aerosol Technology: Properties, Behaviour and Measurement of Aerosol Particles*, New York: Wiley (1999).

28. M.M. Vivas, E. Hontañón, A. Schmidt-Ott, Design and evaluation of a low-level (0.24 μ Ci) radioactive aerosol charger based on ^{241}Am, *J. Aerosol Sci.*, 39, 191–210 (2008).

29. T. Jung, H. Burtscher, A. Schmidt-Ott, Multiple charging of ultrafine aerosol particles by aerosol photoemission, *J. Aerosol Sci.*, 19, 485 (1988).

30. H. Kirsch, A. Schmidt-Ott, Separation of nanoparticles according to their photoelectric properties, in *Proceedings of PARTEC 2001*, Published by Nurnberg Messe GmbH, Nuremberg, Germany, March 27–29, 2001.

31. M.M. Vivas, E. Hontanon, A. Schmidt-Ott, Reducing multiple charging of submicron aerosols in a corona diffusion charger, *Aerosol Sci. Techn.*, 42, 97–109 (2008).

32. G.L. Messing, S.C. Zhang, G.V. Jayanthi, Ceramic powder synthesis by spray-pyrolysis, *J. Am. Ceramic Soc.*, 76, 2707–2726 (1993).

33. J. Zeleny, The electrical discharge from liquid points, and a hydrostatic method of measuring the electric intensity at their surfaces, *Phys. Rev.*, 3, 69–91 (1914).

34. J. Zeleny, Instability of electrified liquid surfaces, *Phys. Rev.*, 10, 1–6 (1917).

35. M. Cloupeau, B. Prunetfoch, Electrohydrodynamic spraying functioning modes—A critical review, *J. Aerosol Sci.*, 25, 1021–1036 (1994).

36. J.M. Grace, J.C.M. Marijnissen, A review of liquid atomization by electrical means, *J. Aerosol Sci.*, 25, 1005–1019 (1994).

37. G. Taylor, Disintegration of water drops in electric field, *Proc. R. Soc. Lond. Ser. A-Math. Phys. Sci.*, 280, 383 (1964).

38. K. Okuyama, I.W. Lenggoro, Preparation of nanoparticles via spray route, *Chem. Eng. Sci.*, 58, 537–547 (2003).

39. J. Fernandez de la Mora, J. Navascues, F. Fernandez, J. Rosell Llompart, Generation of submicron monodisperse aerosols in electrosprays, *J. Aerosol Sci.*, 21, S673–S676 (1990).

40. I.W. Lenggoro, K. Okuyama, J.F. Fernandez de la Mora, N. Tohge, Preparation of ZnS nanoparticles by electrospray pyrolysis, *J. Aerosol Sci.*, 31, 121–136 (2000).

41. J. van Erven, R. Moerman, J.C.M. Marijnissen, Platinum nanoparticle production by EHDA, *Aerosol Sci. Technol.*, 39, 941–946 (2005).

42. N. Tohge, M. Tatsumisago, T. Minami, K. Okuyama, K. Arai, Y. Inada, Y. Kousaka, Preparation of superconducting fine particles in the Bi-(Pb)-Ca-Sr-Cu-O system using the spray-pyrolysis method, *J. Mater. Sci.: Mater. Electron.*, 1, 46–48 (1990).

43. R.P.A. Hartman, D.J. Brunner, D.M.A. Camelot, J.C.M. Marijnissen, B. Scarlett, Jet break-up in electrohydrodynamic atomization in the cone-jet mode, *J. Aerosol Sci.*, 31, 65–95 (2000).

44. J.N. Smith, R.C. Flagan, J.L. Beauchamp, Droplet evaporation and discharge dynamics in electrospray ionization, *J. Phys. Chem. A*, 106, 9957–9967 (2002).

45. A.A. Van Zomeren, E.M. Kelder, J.C.M. Marijnissen, J. Schoonman, The production of thin films of $LiMn_2O_4$ by electrospraying, *J. Aerosol Sci.*, 25, 1229–1235 (1994).

46. J.B. Fenn, M. Mann, C.K. Meng, S.F. Wong, C. Whitehouse, Electrospray ionization for mass spectrometry of large biomolecules, *Science*, 246, 64–71 (1989).

47. R.P.A. Hartman, D.J. Brunner, D.M.A. Camelot, J.C.M. Marijnissen, B. Scarlett, Electrohydrodynamic atomization in the cone-jet mode—Physical modelling of the liquid cone jet, *J. Aerosol Sci.*, 30, 823 (1999).

48. J. Fernandez de la Mora, I.G. Loscertales, The current emitted by highly conducting Taylor cones, *J. Fluid Mech.*, 260, 155–184 (1994).

49. A. Jaworek, A.T. Sobczyk, Electrospraying route to nanotechnology: An overview, *J. Electrostatics*, 66, 197–219 (2008).

50. J. Fernandez de la Mora, The fluid dynamics of Taylor cones, *Annu. Rev. Fluid Mech.*, 39, 217–243 (2007).

51. A.M. Ganan-Calvo, New microfluidic technologies to generate respirable aerosols for medical applications, *J. Aerosol Sci.*, 30 (Suppl. 1), 541–542 (1999).

52. A.M. Ganan-Calvo, J. Davila, A. Barrero, Current and droplet size in the electrospraying of liquids. Scaling laws, *J. Aerosol Sci.*, 28(2), 249–275 (1997).

53. A. Barrero, I.G. Loscertales, Micro- and nanoparticles via capillary flows, *Annu. Rev. Fluid Mech.*, 39, 89–106 (2007).

54. W. Deng, A. Gomez, Influence of space charge on the scale-up of multiplexed electrosprays, *J. Aerosol Sci.*, 38, 1062–1078 (2007).

55. W.T. Thomson, On the equilibrium of vapour at a curved surface of liquid, *Phil. Mag.*, 42, 448 (1871).

56. S. Girshick, C.-P. Chiu, Kinetic nucleation theory: A new expression for the rate of homogeneous nucleation from an ideal supersaturated vapor, *J. Chem. Phys.*, 93, 1273 (1990).

57. T.H. Zhang, X.Y. Liu, Nucleation: What happens at the initial stage?, *Angew. Chem. Int. Ed.*, 48, 1308–1312 (2009).

58. N.S. Tabrizi, Q. Xu, N.M. van der Pers, A. Schmidt-Ott, Generation of mixed metallic nanoparticles from immiscible metals by spark discharge, *J. Nanopart. Res.*, 12, 247–259 (2010).

59. N.S. Tabrizi, Q. Xu, N.M. van der Pers, U. Lafont, A. Schmidt-Ott, Synthesis of mixed metallic nanoparticles by spark discharge, *J. Nanopart. Res.*, 11, 1209–1218 (2009).

60. C. van Gulijk, E. Bal, A. Schmidt-Ott, Experimental evidence of reduced sticking of nanoparticles on a metal grid, *Aerosol Sci.*, 40, 362–369 (2009).

61. D. Vollath, D.V. Szabo, The microwave plasma process—A versatile process to synthesize nanoparticulate materials, *J. Nanopart. Res.*, 8, 417–428 (2006).

62. N.S. Tabrizi, M. Ullmann, V.A. Vons, U. Lafont, A. Schmidt-Ott, Generation of nanoparticles by spark discharge, *J. Nanopart. Res.*, 11, 315–332 (2009).

63. Ph. Buffat, J.P. Borel, Size effect on the melting temperature of gold particles, *Phys. Rev. A*, 13, 2287–2298 (1976).

64. A. Schmidt-Ott, In situ measurement of the fractal dimensionality of ultrafine aerosol particles, *Appl. Phys. Lett.*, 52, 954 (1988).

65. H. Kleinwechter, A. Schmidt-Ott, P. Roth, Monitoring carbon particle oxidation by on-line thermograms, in *Proceedings of PARTEC 2001*, Published by Nurnberg Messe GmbH, Nuremberg, Germany, March 27–29, 2001.

66. M. Shimada, T. Seto, K. Okuyama, Size change of very fine silver agglomerates by sintering in a heated flow, *J. Chem. Eng. Jap.*, 27, 795–802 (1994).

67. H.G. Scheibel, J. Porstendorfer, Generation of monodisperse Ag- and NaCl-aerosols with particle diameters between 2 and 300 nm, *J. Aerosol Sci.*, 14, 113 (1983).

68. A.S. Gurav, T.T. Kodas, L.M. Wang, E.I. Kauppinen, J. Joutsensaari,. Generation of nanometer-size particles via vapor condensation, *Chem. Phys. Lett.*, 218, 304–308 (1994).

69. J.H. Jung, H. Cheol Oh, H. Soo Noh, J.H. Ji, S. Soo Kim, Metal nanoparticle generation using a small ceramic heater with a local heating area, *J. Serosol Sci.*, 37, 1662 (2006).

70. A. Schmidt-Ott, P. Schurtenberger, H.C. Siegmann, Enormous yield of photoelectrons from small particles, *Phys. Rev. Lett.*, 45, 1284 (1980).

71. A. Schmidt-Ott, B. Marsen, K. Sattler, Characterizing nanoparticles by scanning tunneling microscopy and scanning tunnelling spectroscopy, *J. Aerosol Sci.*, 28, S729–S730 (1997).

72. A.G. Nasibulin, A. Moisala, D.P. Brown, H. Jiang, E.I. Kauppinen, A novel aerosol method for single walled carbon nanotube synthesis, *Chem. Phys. Lett.*, 402, 227–232 (2005).

73. F. Favier, E.C. Walter, M.P. Zach, T. Benter, R.M. Penner. Hydrogen sensors and switches from electrodeposited palladium mesowire arrays, *Science*, 293, 2227–2231 (2001).

74. H. Burtscher, A. Schmidt-Ott, H.C. Siegmann, Photoelectron yield of small silver and gold particles sin a gas up to a photon energy of 10 eV, *Z. Phys. B.*, 56, 197 (1984).

75. I. Langmuir, K. Kingdon, Thermionic effects caused by vapours of alkali metals, *Proc. R. Soc. Lond.: Ser. A*, 107, 61–79 (1925).

76. S. Schwyn, E. Garwin, A. Schmidt-Ott, Aerosol generation by spark discharge, *J. Aerosol Sci.*, 19, 639–642 (1988).

77. C. Helsper, W. Molter, Investigation of a new aerosol generator for the production of carbon aggregate particles, *Atmos. Environ.*, 27A, 1271–1275 (1993).

78. J.-T. Kim, J.-S. Chang, Generation of metal oxide aerosol particles by a pulsed spark discharge technique, *J. Electrostatics*, 63, 911–916 (2005), doi:10.1016/j.elstat.2005.03.066

79. H. Oh, J. Ji, J. Jung, S. Kim, Synthesis of titania nanoparticles via spark discharge method using air as a carrier, *Eco-Mater. Process. Des.* VIII, 544–545, 143–146 (2007).

80. M. Ullmann, S.K. Friedlander, A. Schmidt-Ott, Nanoparticles formation by laser ablation, *J. Nanopart. Res.*, 4, 499–509 (2002).

81. J.-P. Borra, Nucleation and aerosol processing in atmospheric pressure electrical discharges: Powders production, coatings and filtration, *J. Phys. D: Appl. Phys.*, 39, R19–R54 (2006), doi:10.1088/0022-3727/39/2/R01

82. M. Kato, Preparation of ultrafine particles of refractory oxides by gas evaporation method, *Jap. J. Appl. Phys.*, 15, 757–760 (1976).

83. G.P. Johnston, R. Muenchausen, D.M. Smith, W. Fahrenholtz, S. Foltyn, Reactive laser ablation synthesis of nanosize alumina powder, *J. Am. Ceram. Soc.*, 75, 3293–3298 (1992).

84. T. Sasaki, S. Terauchi, N. Koshizaki, H. Umehara, The preparation of iron complex oxide nanoparticles by pulsed-laser ablation, *Appl. Surf. Sci.*, 129, 398–402 (1998).

85. R.P. Camata, M. Hirasawa, K. Okuyama, K. Takeuchi, Observation of aerosol formation during laser ablation using a low-pressure differential mobility analyzer, *J. Aerosol Sci.*, 31, 391–401 (2000).

86. K. Ogawa, T. Vogt, M. Ullmann, S. Johnson, S.K. Friedlander, Elastic properties of nanoparticle chain aggregates of TiO_2, Al_2O_3 and Fe_2O_3 generated by laser ablation, *J. Appl. Phys.*, 87, 63–73 (2000).

87. D.E. Powers, Supersonic metal cluster beams: Laser photoionization studies of copper cluster (Cu_2), *J. Phys. Chem.*, 86, 2556 (1982).

88. A.G. Nasibulin, O. Richard, E.I. Kauppinen, D.P. Brown, J.K. Jokiniemi, I.S. Altman, Nanoparticle synthesis by copper(II) acetylacetonate vapor decomposition in the presence of oxygen, *Aerosol Sci. Techn.*, 36, 899–911 (2002).

89. M.H. Magnusson, K. Deppert, J.-O. Malm, Single-crystalline tungsten nanoparticles produced by thermal decomposition of tungsten hexacarbonyle, *J. Mater. Res.*, 15, 1564–1569 (2000).

90. M.L. Ostraat, J.W. De Blauwe, M.L. Green, L.D. Bell, H.A. Atwater, R.C. Flagan, Ultraclean two-stage aerosol reactor for production of oxide-passivated silicon nanoparticles for novel memory devices, *J. Electrochem. Soc.*, 148, G265–G270 (2001).

91. R.A. Senter, Y. Chen, J.L. Coffer, L.R. Tessler, Synthesis of silicon nanocrystals with erbium-rich surface layers, *Nano Lett.*, 1, 383–386 (2001).

92. V.V. Srdic, M. Winterer, A. Moller, G. Miehe, H. Hahn, Nanocrystalline zirconia surface-doped with alumina: Chemical vapor synthesis, characterization, and properties, *J. Am. Ceram. Soc.*, 84, 2771–2776 (2001).

93. S.H. Ehrman, M.I. Aquino-Class, M.R. Zachariah, Effect of temperature and vapor-phase encapsulation on particle growth and morphology, *J. Mater. Res.*, 14, 1664–1671 (1999).

94. G.D. Ulrich, Flame synthesis of fine particles, *Chem. Eng. News*, 62, 22–29 (1984).

95. S.E. Pratsinis, Flame aerosol synthesis of ceramic powders, *Prog. Energy Combust. Sci.*, 24, 197–219 (1998).

96. J. Jiang, D.R. Chen, P. Biswas, Synthesis of nanoparticles in a flame aerosol reactor with independent and strict control of their size, crystal phase and morphology, *Nanotechnology*, 18, 285603 (2007).

97. H.K. Kammler, R. Jossen, P.W. Morrison, S.E. Pratsinis, G. Beaucage, The effect of external electric fields during flame synthesis of titania, *Powder Technol.*, 135, 310–320 (2003).

98. H.K. Kammler, S.E. Pratsinis, Electrically-assisted flame aerosol synthesis of fumed silica at high production rates, *Chem. Eng. Proc.*, 39, 219–227 (2000).

99. J.-B. Donnet, R.C. Bansal, M.J. Wang, *Carbon Black*, New York: Marcel Decker (1993).

100. J.R. Jensen, T. Johannessen, S. Wedel, H. Livbjerg, Preparation of ZnO-Al2O3 particles in a premixed flame, *J. Nanopart. Res.*, 2, 363–373 (2000).

101. J.R. Jensen, T. Johannessen, S. Wedel, H. Livbjerg, A study of $Cu/ZnO/Al_2O_3$ methanol catalysts prepared by flame combustion synthesis, *J. Catal.*, 218, 67–77 (2003).

102. O.I. Arabi-Katbi, K. Wegner, S.E. Pratsinis, Aerosol synthesis of titania nanoparticles: Effect of flame orientation and configuration, *Ann. Chim. Sci. Mater.*, 27, 37–46 (2002).

103. W.J. Stark, K. Wegner, S.E. Pratsinis, A. Baiker, Flame aerosol synthesis of vanadia-titania nanoparticles: Structural and catalytic properties in the selective catalytic reduction of NO by NH_3, *J. Catal.*, 197, 182–191 (2001).

104. R. Mueller, L. Mädler, S.E. Pratsinis, Nanoparticle synthesis at high production rates by flame spray pyrolysis, *Chem. Eng. Sci.*, 58, 1969–1976 (2003).

105. R. Strobel, A. Baiker, S.E. Pratsinis, Aerosol flame synthesis of catalysts, *Adv. Powder Technol.*, 17, 457–480 (2006).

106. C. Janzen, P. Roth, Formation of Fe_2O_3 nano-particles in doped low-pressure $H_2/O_2/Ar$ flames, *Combust. Flame*, 125, 1150–11561 (2001).

107. G. Ledoux, J. Gong, F. Huisken, O. Guillois, C. Reynaud, Photoluminescence of size-separated silicon nanocrystals: Confirmation of quantum confinement. *Appl. Phys. Lett.*, 80, 4834–4836, (2002).

108. G. Ledoux, D. Amans, J. Gong, F. Huisken, F. Cichos, J. Martin, Nanostructured films composed of silicon nanocrystals, *Mater. Sci. Eng. C*, 19, 215–218 (2002).

109. E. Borsella, S. Botti, M.C. Cesile, S. Martelli, A. Nesterenko, P.G. Zappelli, MoS_2 nanoparticles produced by laser induced synthesis from gaseous precursors, *J. Mater. Sci. Lett.*, 20, 187–191 (2001).

110. Y. Kamlag, A. Goossens, I. Colbeck, J. Schoonman, Laser CVD of cubic SiC nanocrystals, *Appl. Surf. Sci.*, 184, 118–122 (2001).

111. D. Vollath, D.V. Szabo, Coated Nanoparticles: A new way to improved nanocomposites, *J. Nanopart. Res.*, 1, 235–242 (1999).

112. D. Vollath, D.V. Szabo, in K.L. Choy, ed., *Innovative Processing of Films and Nanocrystalline Powders*, London, U.K.: Imperial College Press (2002).

113. E.E. Kunhardt, Generation of large volume, atmospheric pressure, non-equilibrium plasmas, *IEEE Trans. Plasma Sci.*, 28, 189–200 (2000).

114. V. Vons, Y. Creyghton, A. Schmidt-Ott, Nanoparticle production using atmospheric pressure cold plasma, *J. Nanopart. Res.*, 8, 721–728 (2006).

115. H.L. Bai, C.C. Chen, C.S. Lin, W. Den, C.L. Chang, Monodisperse nanoparticle synthesis by an atmospheric pressure plasma process: An example of a visible light photocatalyst, *Ind. Eng. Chem. Res.*, 43, 7200–7203 (2004).

116. C.C. Chen, H.L. Bai, H.M. Chein, T.M. Chen, Continuous generation of TiO_2 nanoparticles by an atmospheric pressure plasma-enhanced process, *Aerosol Sci. Technol.*, 41, 1018–1028 (2007).

117. M.D. Barankin, Y. Creyghton, A. Schmidt-Ott, Synthesis of nanoparticles in an atmospheric pressure glow discharge, *J. Nanopart. Res.*, 8, 511–517 (2006).

118. J. Heberlein, O. Postel, S.L. Girshick et al., Thermal plasma deposition of nanophase hard coatings, *Surf. Coat. Technol.*, 142–144, 265–271 (2001).

119. W. Mahoney, A.P. Andres, Aerosol synthesis of nanoscale clusters using atmospheric arc evaporation, *Mater. Sci. Eng. A*, 204, 160–164 (1995).

120. X.L. Dong, Z.D. Zhang, Q.F. Xiao, X.G. Zhao, Y.C. Chuang, S.R. Jin, W.M. Sun, Z.J. Li, Z.X. Zheng, H. Yang, Characterization of ultrafine γ-Fe(C), α-Fe(C) and Fe_3C particles synthesized by arc-discharge in methane, *J. Mater. Sci.*, 33, 1915–1919 (1998).

121. N. Rao, S. Girshick, J. Heberlein, P. McMurry, S. Jones, D. Hansen, B. Micheel, Nanoparticle formation using a plasma expansion process, *Plasma Chem. Plasma Proc.*, 15, 581–606 (1995).

122. J. Hafiz, R. Mukherjee, X. Wang, J.V.R. Heberlein, P.H. McMurry, S.L. Girshick, Analysis of nanostructured coatings synthesized by ballistic impaction of nanoparticles, *Thin Solid Films*, 515, 1147–1151 (2006).

123. A. Gutsch, M. Kramer, G. Michael, H. Muhlenweg, M. Pridohl, G. Zimmerman, Gas phase production of nanoparticles, *KONA Powder Technol.*, 20, 24–37 (2002).

124. B.M. Goortani, N. Mendoza, P. Proulx, Synthesis of SiO_2 nanoparticles in RF plasma reactors: Effect of feed rate and quench gas injection, *J. Chem. Eng.*, 4, A33 (2006).

125. R. Ye, J.-G. Li, T. Ishigaki, Controlled synthesis of alumina nanoparticles using inductively coupled thermal plasma with enhanced quenching, *Thin Solid Films*, 515, 4251–4257 (2007).

126. J.C. Weigle, C.C. Luhrs, C.K. Chen, W.L. Perry, J.T. Mang, M.B. Nemer, G.P. Lopez, J. Phillips, Generation of aluminum nanoparticles using an atmospheric pressure plasma torch, *J. Phys. Chem. B*, 108, 18601–18607 (2004).

127. M. Suzuki, M. Kagawa, Y. Syono, T. Hirai, Synthesis of ultra-fine single-component oxide particles by the spray-ICP technique, *J. Maert. Sci.*, 27, 679–684 (1992).

128. T. Ishigaki, S.-M. Oh, J.-G. Li, D.-W. Park, Controlling the synthesis of TaC nanopowders by injecting liquid precursor into RF induction plasma, *Sci. Technol. Adv. Mater.*, 6, 111–118 (2005).

129. P.P. Ahonen, J. Joutsensaari, O. Richard et al., Mobility size development and the crystallization path during aerosol decomposition synthesis of TiO_2 particles, *J. Aerosol Sci.*, 32, 615–630 (2001).

130. J.H. Kim, T.A. Germer, G.W. Mulholland, S.H. Ehrman, Size-monodisperse metal nanoparticles via hydrogen-free spray pyrolysis, *Adv. Mater.*, 14, 518–521 (2002).

Tomography of Nanostructures

Günter Möbus
University of Sheffield

Zineb Saghi
University of Sheffield

31.1 Introduction and Background

An important point in technology development is the demand for progress from an application point of view. The driving force for three-dimensional (3D) nanotomography would be weak, if there were no 3D nanoscale objects of timely interest to the research community. While the bio-/medical research field immediately realized the potential of electron tomography for virus particles, cell organelles, and macromolecular complexes, which are all 3D nanoobjects or microobjects with nanoscale internal details, the research in materials science concentrated for decades on planar cross-sections, which lack the need for tomography. Only the steep rise of (inorganic) nanotechnology, including the fabrication of novel nanoparticles, nanowires, and nanocomposites, finally proved that development of a nanoscale tomography technique has become indispensable. Materials are nowadays more than ever determined by 3D morphology on the nanoscale. The surface-per-volume ratio is exceptionally large for nanoobjects, and even subsurface regions have enhanced functional importance, e.g., in the study of distinct *low-dimensionality* effects, such as quantum confinement.

31.1.1 General 3D Imaging

Scientific imaging in 3D is all about generating 3D data fields describing an object's shape and inner microstructure. This is more than just generating a 3D visual impression in the brain (the field of stereography), or triangulating positions of an object consisting of multiple particles (the field of stereology), both normally just in need of two viewing directions. To recover the interior of materials in 3D, two major techniques have emerged, which we call "cut-and-image" and "tilt-and-project."

- *Cut-and-image* (Figure 31.1) is resolving the aspect of non-visibility of the materials interior by opening it up and making it accessible to surface-sensitive radiation, e.g., visible light or scanning electron microscopy. The technique turns 3D by cutting multiple sections. The approach is therefore often referred to as sectioning, and only part of the literature would classify it as tomography. By planar cuts through an object, multiple 2D images $I(x,y)$ can be acquired by imaging all of the freshly cut surfaces. Reconstruction into a 3D object $I(x,y,z)$ requires tracking, alignment, and interpolation of the 2D images. Essential parameters are the depth of iteration (the step width) in the z-direction and its ratio to the lateral resolution of the imaging mode used in 2D (Figure 31.1a).

- *Tilt-and-project* (Figure 31.2) relies on making opaque materials transparent by exchanging the type of radiation, e.g., using x-rays. Imaging aims at generating projection views (integration of the inner density) of the object. It only becomes 3D by observing the object in various viewing directions. This group of techniques includes computed tomography according to the majority of textbook definitions. *Tomo-graphy* (old greek) stands for a combination of section (tomo-) and draw (graph). The "sections" referred to in the tilt-and-project approach are cross-sections perpendicular to the tomographic rotation axis. In the early days of tomography, no array-valued detectors existed and both source and detector had to be scanned "section-by-section" across the object.

Frequently, this basic classification of techniques is denoted as "destructive" and "nondestructive," which is nearly equivalent to cut-and-image and tilt-and-project, except for the cases of virtual (optical) sectioning of transparent objects. Such a focal

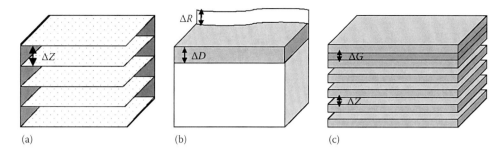

(a)   (b)   (c)

FIGURE 31.1 Cut-and-image: Reconstruction from a series of sections under tomography in a wider sense (a), either by imaging freshly freed surfaces (b) in iteration, or by cutting the entire block into many slices which are then imaged in transmission (c).

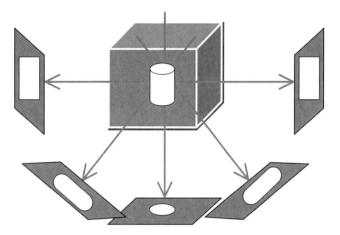

FIGURE 31.2 Projection of a 3D object onto 2D images in various tilt directions.

plane sectioning using low depth of focus does neither involve mechanical cutting nor tilting. This latter field of "confocal" microscopy is considered outside the scope of the chapter, as the wavelengths used in laser scan microscopy do not provide nanoscale lateral resolution. Its application to electron microscopy is briefly covered at the very end in Section 31.3.5.

31.1.2 3D Imaging at the Nanoscale

The list of tomography techniques with state-of-the-art performance below micrometer resolution was long limited to transmission electron tomography (in TEM), which provided nanometer resolution since its invention in 1968 (De Rosier and Klug, 1968; Hoppe et al., 1968), and is therefore the oldest nanotomography technique. Over the last 10 years (Figure 31.3), the diversity of nanotomography techniques has significantly expanded. Within the tilt-and-project approach, we now find for the submicron range x-ray microtomography (Section 31.2.3), while for the sub-10 nm resolution range various novel electron tomography modes have been added to the original bright-field TEM mode (Section 31.2.2). Nanoscale cut-and-image techniques include 3D atom probe (a field-ion microscopy-derived technique, Section 31.3.4.), and sectioning instruments in combination with surface imaging or surface chemistry mapping such as focused-ion-beam microscopy (FIB, Section 31.3.2), atomic force microscopy (AFM, Section

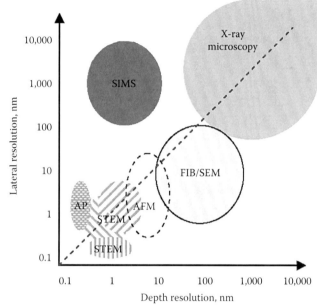

FIGURE 31.3 Length scales of operation of nanoscale tomography techniques. Lateral versus depth resolution ranges for anisotropic or isotropic techniques. Electron projection tomography (S/TEM) is shaded. AP, atom probe tomography; STEM, STEM focal sectioning. (Modified after Möbus et al., *Ultramicroscopy*, 96, 433, 2003.)

31.3.3), and (outside the scope of this chapter) so-called *nano-secondary ion mass spectrometry* (SIMS) and related techniques (Hutter et al., 1993; Zaitsev et al., 2001).

Some of the methods are isotropic in nature and therefore appear on the diagonal of a graph separating depth versus lateral resolution. Some methods operate along a specific (z-) direction, and the resolution differs parallel and perpendicular to this direction (Figure 31.3, see also Möbus et al., 2003).

31.1.3 Alignment and Reconstruction Algorithms

31.1.3.1 Alignment Techniques

For both reconstructions from a series of projections and from a series of sections, the input data has to be first aligned to the same rotation axis or cut axis, respectively. There are two ways of achieving the alignment (Frank, 2007):

1. By assuming that two neighboring images show comparable features, one could apply a cross-correlation technique to study the similarity of the two images. If the maximum of the cross-correlation function is at the centre, it means that the two images are perfectly aligned. If not, the maximum position gives us the relative shift values which are then used to realign the second image with respect to the first (reference) image. As this approach is performed on pairs of neighboring images, it may induce some cumulative errors which can be corrected for by manual refinement.

2. A second approach consists in introducing special markers to the sample and tracking them throughout the projections or sections by cross-correlation or least-square fitting routines. In sectioning techniques, it is common to add a straight thin hole or cross-bar running all through the sample, while in electron tomography it is a standard practice to use gold nanoparticles evenly distributed in the sample.

31.1.3.2 Reconstruction from a Series of Projections

The basic task of tomography is to nondestructively reconstruct 2D cross-sections of a 3D object from several 2D projections taken at different tilt angles (Herman, 1979). For simplicity, the technique is illustrated here by reconstructing a 2D object from its 1D projections.

Tomography is based on the Radon transform, which maps the image domain (x, y) to the projection domain (θ, r) $(\theta \in [0, \pi[)$, with x and y as image coordinates on an infinite grid, and θ, r the polar coordinate transformation of it. The Radon transform $Rf(\theta, r)$ of a 2D function $f(x, y)$ at the angle θ and distance r from the origin is the integral of $f(x, y)$ along the straight line $L_{r,\theta}$ defined by its distance r from the origin and its angle of inclination θ (Herman, 1979). $Rf(\theta, r)$ is defined by Equation 31.1, where the function δ restricts the integral to the points (x, y), which belong to the line $L_{r,\theta}$:

$$Rf(\theta, r) = \int_{-\infty}^{\infty} \int_{-\infty}^{\infty} f(x, y)\delta(x\cos\theta + y\sin\theta - r)\mathrm{d}x\,\mathrm{d}y \quad (31.1)$$

An off-centered point in the image domain is mapped into a sinusoid in the projection domain, hence the name "sinogram" commonly used to describe the projection data. In Figure 31.4, the Radon transform is illustrated on a 2D simulated image formed of two discs with different gray values. The 1D projections at 0°, 90°, and 135° are shown on Figure 31.4a. Figure 31.4b is the sinogram of the simulated object, which corresponds to its projections at all angles from 0° to 180°. The task of tomography is then to retrieve the unknown object (Figure 31.4a) from its sinogram (Figure 31.4b). In 1956, Bracewell introduced a theorem that links the projection of a function to its Fourier transform. This fundamental *projection–slice* theorem states that the 1D Fourier transform $G(\theta, u)$ of the projection $Rf(\theta, r)$ in the θ direction is a central slice of the 2D Fourier transform of the unknown object $f(x, y)$, noted $F(u_1, u_2)$, at the same angle θ:

$$G(\theta, u) = F(u\cos\theta, u\sin\theta) \quad (31.2)$$

Intuitively, if one acquired projections at all angles and filled in the Fourier space of the object, an inverse Fourier transform would be sufficient to retrieve it correctly. Unfortunately, as we fill in the Fourier space in a radial fashion, the high frequencies are undersampled in comparison with the low frequencies, which results in a blurred reconstruction, as shown in Figure 31.4c. To correct for this effect, a filtering of the projections is often performed before reconstruction. This is the basic of the well-known filtered backprojection algorithm (FBP), which was introduced by Bracewell and Riddle in 1967 (Bracewell and Riddle, 1967). From the projection–slice theorem, the more projections we acquire the closer the reconstructed volume is to the original object. However, in many applications, only a limited number of projections can be acquired. This is the case in electron tomography where, due to hardware limitations and beam irradiation damage, the tilt increment and range are often chosen between 2° and 5°, and −60°:+60° to −70°:+70° with state-of-art tomography holders, respectively. These acquisition limitations result in missing directions in the Fourier space of the object. Figure 31.5 shows FBP reconstructions from different number of projections, over the entire angular range. With

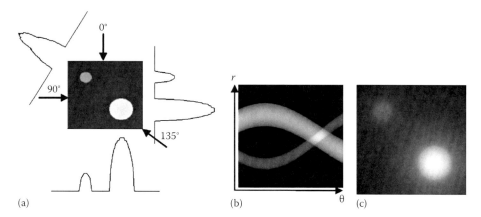

(a) (b) (c)

FIGURE 31.4 (a) 0°, 90°, and 135° projections of a simulated test object, (b) corresponding sinogram, and (c) smearing back the projections results in a blurred version of the object.

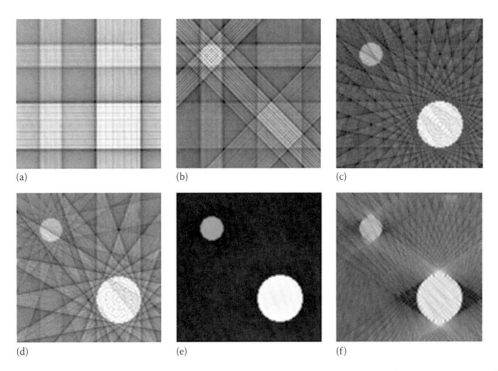

FIGURE 31.5 FBP reconstruction from 2 (a), 4 (b), 9 (c), 18 (d), and 180 (e) projections, over the angular range −90°: +90°. (f) Combination of limited tilt range and increment: −60: + 5: + 60.

1° increment (Figure 31.5e), the reconstruction looks artifact-free, but as the tilt increment increases, star-shaped patterns start appearing.

The limited tilt range has more critical consequences than the limited tilt increment, as it results in a whole missing wedge in the Fourier space, and an elongation of the reconstructed object with anisotropic resolution. Figure 31.5f shows the combination of these two effects on the FBP reconstruction from −60° to +60° with 5° increment, with the two simulated discs appearing now as ellipses elongated in the missing wedge direction.

In addition to filtered backprojection, other algorithms can be found in the literature, such as the iterative direct space methods of algebraic reconstruction technique (ART) (Gordon et al., 1970) or simultaneous iterative reconstruction technique (SIRT) (Gilbert, 1972), which rely on optimizing the reconstruction by iterative comparison between projections of the reconstruction and the original projections from the acquired tilt series. These techniques can incorporate a priori knowledge about the object and perform better than the FBP algorithm under noisy and limited-data conditions (Rangayyan et al., 1985). Recently, Batenburg and Sijbers (Batenburg and Sijbers, 2007; Bals et al., 2007) proposed a modified form of ART, called discrete ART (DART), which reconstructs iteratively a finite number of intensity levels (gray values) in the object. For the case of a sample known a priori to consist of one or a few constant density and constant chemistry phases, this technique seems to be more robust under limited angle conditions than the other techniques.

31.1.3.3 Reconstruction from a Series of Sections

The reconstruction of the original object from a series of sections is mathematically simpler than with projection based tomography. A mere 3D assembly of data rather than post-processing of data would be sufficient were there not unavoidably mis-registration errors (lateral errors) and missing data due to gaps (vertical errors), as illustrated in Figure 31.1.

Two major cutting strategies need to be distinguished:

1. The material can either be iteratively milled or grinded and the top surface of the single remaining block is imaged. The z-resolution is then given by the largest of the quantities of the attempted depth of iteration Δz, the possible fluctuations of the fresh surfaces (roughness, or long-range non-flatness) ΔR, and the possible contribution of subsurface layers of materials to the image formation (depth mixing), ΔD, depending on what radiation is used (Figure 31.1b).

2. Another variant of 3D sectioning is the cutting of the entire object of interest into separate slices of thickness Δz. Each slice is then imaged by means of a projection. Resolution in this mode again comprises contributions from the projection depth Δz, the non-flatness ΔR, and possibly from missing abrasive material due to imperfect cutting, a finite gap ΔG (blue on Figure 31.1c).

To compensate for the missing volume in z-direction, two algorithms serve two complementary purposes:

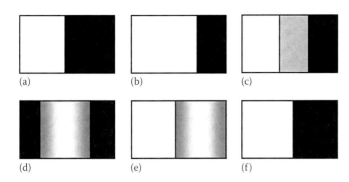

FIGURE 31.6 Interpolation of binary intensity sections for "cut-and-image" reconstruction: (a–c) Interpolation of an inclined contour leading to unwanted shadow images. (d–f) Interpolation after wedge-transformation of edges and final rebinarization at the combined wedge-maximum.

1. *Density-interpolation*: The data field can be a gray-level-sensitive description of variations in object density I ~ Δ (x, y; n) for the nth slice. Then the filling of missing voxels using a selection of bilinear, cubic, or spline interpolation to arrive at a continuous dense data cube $I(x, y, z)$ is simply a question of computing time and size of the slice interval relative to how rapid the data change from slice to slice.
2. *Contour-shape interpolation*: If the aim is to reconstruct shapes of volumes of data with constant interior, the first step of post-processing would conveniently be a binarisation of the data into a black/white data field. Straight interpolation as of (a) would then lead to an undesirable reintroduction of gray values due to superimposition of images with moved edges (Figure 31.6a through c). In this case edge contours of binary images should be converted back into a floating point data array, in which the gray value indicates the distance of a pixel from the sharp edge (wedge pattern). Interpolation along the sectioning direction z then superimposes the two wedge patterns which, upon rebinarization, results in a new sharp edge, this time in the middle between the edge positions of the two neighboring slices (Figure 31.6d through f). Overall step-free smooth surface contours for the binary object result.

31.2 Projection Tomography Techniques on the Nanoscale

31.2.1 Background

There are obviously two possible routes to nanoscale tomography. Either an existing 2D imaging technique with nanoscale resolution can be developed into a tomographic mode. Or existing micro- or macro-resolution tomography techniques can be pushed in resolution into the nanoscale. The two basic techniques covered in this section, x-ray (Section 31.2.3) and electron (Section 31.2.2) tomography, developed differently: Electron tomography started life as a 2D nanoscale imaging technique, which only needed to be turned to 3D by enlarging the angular tilt range of the specimen holders. X-ray tomography originated on the macro- and microscales and it needed x-ray-focusing techniques (e.g., Fresnel lenses) to be invented to reach the sub-millimeter resolution level. Other radiations for projection imaging of materials, such as neutrons, have not reached nanoscale.

Another kind of tomography are emission tomographies, where the radiation emitted from a materials cell (corresponding to a voxel in the reconstruction) is either spontaneous or induced by some other radiation, and is mapped into an image array, corresponding (more or less accurately) to a projection. Early developments in this topical area are presented in Section 31.2.4.

31.2.2 Electron Tomography

TEM tomography became a reality even before x-ray tomography was experimentally established. This is amazing as electrons are not normally considered a radiation for which most kind of matter would classify as transparent. As charged particles, the interaction of electrons with material is so strong that no radiation can reach the detector above an object thickness of a few microns, depending on density and atomic number. The limit of opaqueness is reached, as soon as the signal-to-noise ratio drops to the level of detector noise. Even before that, the linearity of projection (or monotony which can be linearized), that means the intensity being proportional to the thickness times density, might fail. This case was least of a problem as long as electron tomography dealt with its first field of applications, amorphous materials made of carbon or carbohydrate. This includes all the living world, and opened vast biomedical interest. Bone tissue, entire virus particles, cell organelles are just a few examples from this field. For inorganic crystalline materials applications, however, the high coherence of electrons leading to interference patterns through Bragg scattering can easily destroy tomography-interpretable contrast.

The angular range of the tilt series is a crucial point. The closer the full tilt range of ±90° is approached, the less anisotropy artifacts will occur. TEMs for biological applications always had a large specimen holder chamber ("pole piece gap" of the objective lens), as biological specimens do not require ultimate resolution below 0.2 nm. When the extended interest in electron tomography in materials science arouse since 2000, tilt ranges were a major problem, as most TEMs for inorganic crystalline specimens are designed to maximize resolution, and minimize the pole piece gap (to lower spherical aberration). Novel specimen holders needed to be designed (Midgley et al., 2001; Möbus et al., 2004) to fit into those narrow pole piece gaps (2–4 mm) for modern high-resolution TEMs, in order to increase the tilt limit from previously typically ±15°–40° to over ±60°–80°.

In order to overcome the mentioned problems, various new TEM tomography subdisciplines have been developed since 2000. These could be classified by the imaging mode used (TEM/STEM), the detector and aperture setup used (bright field/ dark field) or the energy of electrons detected (elastic/inelastic).

We outline here a classification based on "purpose" of the tomography experiment, which first of all discriminates three cases:

- Density-sensitive methods (Section 31.2.2.1): A typical question for this case would be: How are density fluctuations or particles in a matrix distributed within the materials volume of the sample?
- Shape-sensitive methods (Section 31.2.2.2): Questions for this case include: How anisotropic is the object (e.g., a nanoparticle), are the surfaces round or flat and what symmetry does it show?
- Elemental mapping methods (Section 31.2.2.3): How are selected chemical elements distributed in a chemically nonhomogeneous medium (even if density might be nearly constant).

31.2.2.1 Density Mapping—Quantitative Projections

Since neutrons do not contribute to electron scattering, density always refers to *positional density* of scatterers (depending on atomic number), rather than mass density. In contrast to x-rays, the scattering cross-section is not directly proportional to electron density but to the electrostatic potential in the sample (both quantities are, however, related by Poisson's equation). For constant chemistry, the result is pure positional density of atoms. Complementary, for constant positional density, a quantity which is a monotonous function of the average atomic number per voxel is reconstructed. This can give these techniques a certain chemical sensitivity (Z-contrast) with ability for distinguishing and mapping known chemical phases in a sample, however, not to be confused with true elemental mapping as of Section 31.2.2.3 and Figure 31.7d and e (distributions of one single chemical element). Only in the case of (speculative future) atomic resolution tomography, density mapping and elemental mapping would become a merged technique, if no more than one atom fits into one voxel (Möbus and Inkson, 2003).

The applicability of electron tomography has to account for three major object classes, distinguishing crystalline versus amorphous materials, low versus high atomic number, and maximum thickness (Frank, 2007). Also classes of TEM modes (coherent/incoherent; absorptive vs. scattering contrasts; bright-field vs. dark-field, elastic vs. inelastic) need to be considered: It should be noted, that unlike for the case of photon irradiation, *absorption of electrons* denotes—to great majority—the case

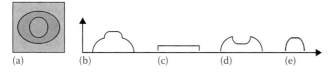

(a)   (b)   (c)   (d)   (e)

FIGURE 31.7 Classification of major TEM projection modes: (a) An elliptical object with atomic number Z_1 enclosing a spherical object with $Z_2 > Z_1$. (b) Quantitative projection (e.g., Z-contrast) for computed tomography. (c) Binary projection for geometric tomography. (d, e) Element specific spectroscopic tomography tuned to element Z_1 and Z_2 respectively. (Modified after Möbus, G. and Inkson, B.J., *Microsc. Microanal.*, 9, 176, 2003.)

where the electrons do leave the sample, however, without reaching the detector, such as for high angle scattering. Absorption and scattering are therefore inseparable.

Figure 31.8 introduces a basic ray diagram for electron tomographies. Using elastic bright-field electrons (Figure 31.8a) leads to the object appearing darker and thicker on a bright background following a Lambert–Beer absorption law, due to the objective aperture filtering out highly scattered electrons. In contrast to this, two cases where the signal appears bright on dark background and is proportional to object thickness include inelastic imaging (Figure 31.8b, Section 31.2.2.3) and emission tomography (Figure 31.8c, Section 31.2.4).

1. For ultrathin samples, low or medium atomic number and high beam coherence, the maximum shift of the electron wave front inside the material relative to the vacuum remains small against a full phase shift of π. This case is aptly named the *weak-phase object approximation* (WPOA). The dependency of image intensity with thickness and atomic number becomes monotonous within the range of validity (a kind of Z-contrast):

$$I(x, y) = I_0(1 + 2\sigma_{ic}\phi_p(x, y) * FFT^{-1}(\sin\chi(u,v)));$$

$$\sigma_{ic}\phi_p(x, y) \ll 1 \qquad (31.3)$$

where
   I is the image intensity
   I_0 is the intensity in a specimen hole
   σ_{ic} is the (wavelength dependent) interaction constant
   ϕ_p is the projected electrostatic specimen potential

This latter quantity is the target of the tomography experiment. The convolution (*) with an inversely Fourier transformed sine function of the wave aberration χ accounts for the oscillating contrast transfer function of the imaging lens, although this can be omitted near Scherzer focus for low partial coherence. The absorption is small, as for such thin objects scattering is very forward-oriented and passes the aperture.

2. For objects thicker than a few nanometers and/or heavier elements, but now restricted to amorphous structure as limiting condition, a Beer-law exponential absorption contrast in bright-field TEM becomes dominating, which can be easily linearized, and conveniently inverted to get brighter signals for higher thickness. Contrast transfer corrections still apply, however, would become less important for lower magnification ranges which are often adopted along with increasing object thickness.

$$I(x, y) = I_0 e^{-\sigma(\alpha)\rho_p(x, y)} \qquad (31.4)$$

Here σ is (different from the interaction constant above) an objective aperture-dependent scattering cross-section, and ρ_p is the product of density and thickness, the projected atomic number density.

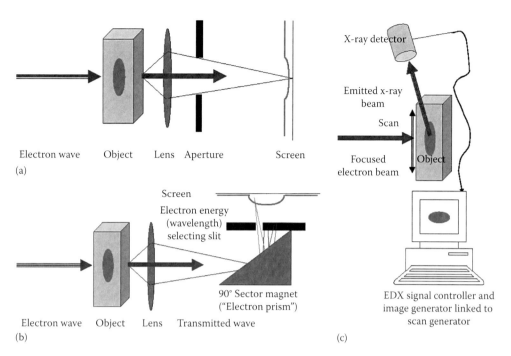

FIGURE 31.8 Some basic ray diagrams for electron tomography. (a) Bright-field coherent TEM with objective aperture induced scattering contrast (dark object on bright background). (b) Energy filtered tomography (EFTEM) with electron prism and energy selection slit to generate an inelastic incoherent image (bright object on dark background). (c) Electron induced x-ray emission tomography (incoherent scattering, scanning TEM, bright object on dark background).

The dominant field of illumination mode of class (2) objects has been for decades the biological and medical bright-field electron tomography, where carbohydrate-based organic matter assures both low atomic number and amorphous structure. A few nonbiological applications of carbon materials, such as carbon-black particles (Hunt and Herd, 2001) and polymer nanostructures (Spontak et al., 1996) have also emerged. For non-carbon inorganic applications the impact of the conditions of class (2) become more severe, and explains why more than 30 years passed between the introduction of electron tomography and its usage for nanoscale meta-structures, e.g., for meso-porous silicates commonly selected for catalyst supports (Koster et al., 2000).

3. For crystalline objects beyond the WPOA, projected image intensity is dominated by Bragg scattering contrast, especially if an angle limiting object aperture is inserted. As intensity is not proportional and not even monotonously dependent on the object density or thickness, this imaging mode is not directly applicable for tomography. However, for the last 8 years it has been realized that incoherent imaging modes provide a solution for this third class of objects. This kind of incoherent imaging has been developed originally for the purpose of 2D analytical electron microscopy (chemical mapping) and finds here a new field of use within tomography.

The major two incoherent modes to generate an image which is a projection of a specimen density (as above) are *energy filtered imaging* (EFTEM) and *annular dark-field STEM* (ADF) (Midgley et al., 2001; Midgley and Weyland, 2003; Möbus and Inkson, 2001; Möbus et al., 2003). Both eliminate the coherent Bragg scattering artifacts in projection images to a variable degree. EFTEM images are the less coherent, the higher the energy loss selected by the energy filter slit. ADF images lose their coherence gradually, the higher the dark-field angle the better, with high-angle detectors (HAADF) being considered a good approximation to incoherency with intensity proportional to Z^α, with α between 1.7 and 2.

Figures 31.9 and 31.10 illustrate a comparison of various imaging modes applied to one object. A comparison of bright-field TEM, EFTEM, and ADF-STEM for a geometrically perfect nanoparticle of CeO_2 (Figure 31.9) illustrates the degree of coherency of the input images, as can be judged by the amount of scattering artifacts (e.g., dark patches at thin object regions, which should be bright), seen dominantly in the BF-series (top row). The incoherent images (EFTEM, middle row) are artifact-free, though slightly blurred. They can be recorded nearly in parallel by inserting the energy filter slit alternately. The ADF-STEM reconstruction (bottom row), obtained on a separate but equivalent particle, is also artifact-free. A BF-STEM tilt series could be recorded in parallel to an ADF-STEM tilt series on some microscopes. The reason why for this particular particle the artifacts in the input images do not translate into any artifacts in the result will be the topic of Section 31.2.2.2.

A comparison of BF TEM, ADF-STEM, and HAADF-STEM in Figure 31.10, applied to platinum particles inside a carbon support, illustrates advantages in contrast and homogeneity

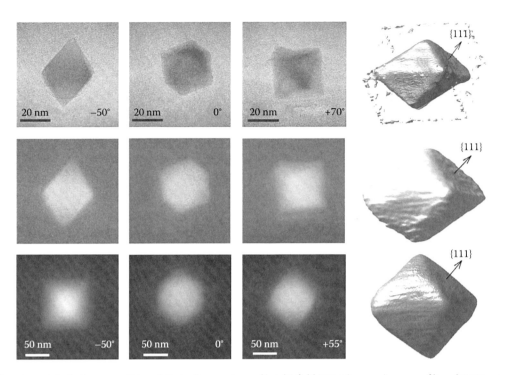

FIGURE 31.9 Regular octahedral nanoparticles of Ceria. Comparison of bright-field TEM (top row), energy filtered TEM using 130 eV energy loss electrons (middle row), and annular dark-field STEM (bottom row). Each imaging mode is represented by three tilt angles out of the tilt series and an iso-surface 3D reconstruction with indexing of facets. (From Xu, X. et al., *Nanotechnology*, 18, 225501, 2007. With permission.)

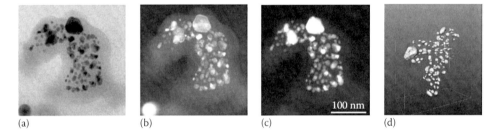

FIGURE 31.10 Nanoscale platinum particles for catalyst purpose embedded in a carbon matrix. Comparison of bright-field TEM (a), annular dark-field STEM (b), high-angle annular dark-field STEM (c). 3D reconstruction using a threshold technique to isolate the high-Z nanoparticles (d). (From Cervera Gontard, L. et al., *J. Phys. Conf. Ser.*, 26, 203, 2006. With permission.)

of intensity inside the particles in favor of the HAADF mode. The large atomic number difference between Pt and C is fully exploited here.

31.2.2.2 Shape Mapping—Geometric Tomography

The scenario of reconstructing a general, multiphase or composite, 3D object greatly simplifies if we introduce conditions based on a priori knowledge: If the atomic number density of the object is constant and if the topology is single-surfaced and connected (an assumption that will be relaxed later), then the only problem is to find the surface or contours of an object. This type of tomography is called "geometric tomography" and is commonly used by the computer vision and pattern recognition communities for real time virtual rendering of scenes, motion tracking, and obstacle avoidance for mobile robots.

For this simplified type of tomography, the projections can be binary images, denoted as shadow or silhouette (Figure 31.7c), as the only information required is the support of the object. Due to this binarisation, any intensity artifacts, such as Bragg scattering, become irrelevant as the only function of the electron wave is to probe whether in a particular viewing direction the rays representing pixels (x, y) do hit the object or not. This then leads to a rehabilitation of coherent imaging modes, such as bright-field TEM as the simplest form of electron tomography, which had to be excluded for most crystalline objects in Section 31.2.1.2.

To extract the 3D shape from a series of shadows, one could use the shape-from-silhouette algorithm (SFS), which consists in backprojecting the shadows from their respective viewing directions, and extracting their intersection. If the object is convex, then the SFS is capable of reconstructing the exact shape. But if the object has concavities, the SFS reconstruction will only be the smallest convex volume containing the object. A special case is when a convex structure has holes (or pores) and that the shadow with the viewing direction parallel to the holes is available. In this case, the condition of single-surfaced and connected topology of objects can be relaxed. Detailed conditions are complex and need to be referred to the literature (Saghi et al., 2008).

An example for this technique (apart from Figure 31.9, top row) is presented for a tungsten tip as used for scanning tunneling microscopy (STM) or for mechanical testing by

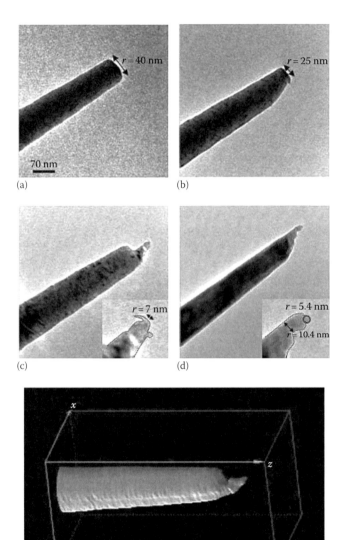

(a) (b)

(c) (d)

(e)

FIGURE 31.11 Bright-field TEM tomography of crystalline Ni material based on the geometric tomography principle, combined with *tomographic nanofabrication*. (a–d) Electron beam-induced atom ablation for tip sharpening in two viewing directions. (e) Tomographic reconstruction of sharpened nanostructure. (From Saghi, Z. et al., *Appl. Phys. Lett.*, 93, 153102, 2008. With permission.)

nanoindentation (Figure 31.11). The tip has been sharpened in situ in the TEM using the atomic ablation power of the focused electron beam. As the cross-sections are convex, binarized projections which are insensitive to Bragg scattering from the crystalline tungsten core, are sufficient to retrieve the 3D shape of this tip.

The combination of shape modification and shape reconstruction within the same electron microscopy session opens up a new field of research, "tomographic nanofabrication," as discussed in more detail in (Saghi et al., 2008).

31.2.2.3 Chemical Mapping—Spectroscopic Tomography

Using spectroscopic imaging modes, which allow the tuning of a spectrometer to pass electrons characteristic for one chemical element only, we can achieve true spectroscopic electron tomography (Möbus and Inkson, 2001). If applied to entire spectra or consecutively to various energy windows, this technique returns a 4D data volume with $(x, y, z, \delta E)$ as 4D voxel and δE as spectroscopic energy. Each tomogram per energy δE contains the density of one element per voxel only, suppressing the existence of other elements. Energy filtered TEM-imaging with a magnetic prism spectrometer (either in the projector lens system or below the viewing screen) has been found useful for this spectroscopic electron tomography (Figure 31.8b). A first EFTEM tomography example has been presented as part of Section 31.2.2.1 and Figure 31.9, where the purpose of energy filtering was merely to achieve incoherence, rather than elemental uniqueness. The EFTEM reconstruction of Figure 31.9, bottom row, is artifact-free in its projection images, unlike BF-TEM. However, the benefit from a simultaneous acquisition of a BF and EFTEM tilt series has to be emphasized, as it provides both sharp contour information and spectroscopic information. In this case, a single energy window including the Ce–N-edge at around 130 eV has been chosen as true elemental mapping is redundant here due to constant inner composition and density of the particle.

For a chemically pure signal, at least two or three spectroscopic images need to be recorded before and after an inner shell energy edge. Only the difference (the jump) in the signal is proportional to the element's local density; therefore, any background which originates from a mixture of all elements present somewhere along the path of view through the sample needs to be subtracted. Figure 31.12 shows the separation of oxide nanoparticles from their iron–aluminide matrix in which they are embedded using iron *L*-edge signals acquired as a tilt series and reconstructed to map the distribution of the particles (Figure 31.12, Möbus et al., 2003).

31.2.2.4 Applications of Electron Tomography for Nanostructures

The list of nanostructures reconstructed by electron tomography for the nonbiological research fields has become quite extensive

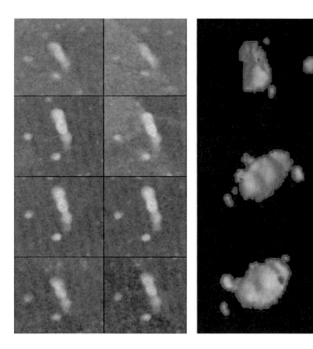

FIGURE 31.12 Spectroscopic tomography using energy filtered TEM. Yttria-based oxide-dispersion particles in FeAl intermetallic alloy. Left: Eight viewing directions of the tilt series using Fe-L-edge inelastic imaging. Inverted contrast (Fe deficiency mapping). Field of view of 220 nm. Right: Three views of reconstructed iso-surface rendered particle assembly. (Modified from Möbus, G. and Inkson, B.J., *Microsc. Microanal.*, 9, 176, 2003.)

in just 8 years. A possible classification sorts the applications into the following:

1. *Free-standing nanoobjects*, such as zero-dimensional or 1D particles and wires. Examples include: bundles of carbon nanotubes, magnetite crystals of biological origin (Koguchi et al., 2001; Friedrich et al., 2005), metallic particles such as Pd (Park et al., 2007) and Cu (Oikawa et al., 2006), and ceramic particles, CeO_2 in (Xu et al., 2007a) (examples only), and Si nanowires (Arbiol et al., 2004), and W-tips for nanoindentation or STM imaging (Xu et al., 2007c).
2. *Nanoporous materials* as, e.g., used for catalysis and electrochemical applications have been successfully reconstructed with and without heavy metal catalyst particle (Ziese et al., 2004) on low-Z or porous substrates. HAADF-STEM (Cervera Gontard et al., 2006) is particularly beneficial due to the large Z-difference, see also Wikander et al. (2007).
3. *Nanocomposites and multiphase devices*: These objects are a logical extension of the nanoporous materials with the filler particles firmly embedded in a dissimilar matrix. Examples include nanoparticle precipitates in glasses (glass–ceramic composites) (Xu et al., 2007b; Yang et al., 2007), and oxide nanoparticles in intermetallic alloys or specialist steels, either volume dispersed or aggregated at grain boundaries (Midgley and Weyland, 2003; Möbus et al., 2003). Within the field of metallurgy, also 3D

distribution mapping of dislocations by dark-field TEM has been realized by tomography. This task has traditionally been solved by stereology, however, a full tilt series allows tomographic reconstruction without the need for dislocation line tracking (Barnard et al., 2006). An example from bone-mineralization research reconstructs collagen-fibrils in 3D (Baldock et al., 2002). Also 3D engineered components of semiconductor device and interconnect nanostructures belong to the class of nanocomposites in the wider sense. Examples of reconstructions by ADF-STEM are found in (Stegmann et al., 2003; Kübel et al., 2005).

31.2.3 X-Ray Nanotomography

X-rays are the "default" choice of radiation when it comes to applications in tomography. Tomography without any prefix always refers to x-ray instrumentation. The popularity and suitability of x-rays for tomography goes directly back to their first application at the time of discovery, the observation of bones in a human hand by transmission. That means the absorption of x-rays by matter is weak enough to allow centimeter-thick objects to be transparent. On the other hand, interaction is strong enough to produce contrast between areas of different product of density times atomic number. Soon the second major breakthrough application for x-rays became apparent: Due to their wavelength-matching atomic distances in crystals, x-ray diffraction could be used to reveal symmetry and lattice constants and herewith pave the way to quantitative crystal structure determination. The small wavelength, however, does not at all supply "resolution."

X-ray micro- and nanotomography has been reviewed recently in Larson and Lengeler (2004), Withers (2007), and Banhart (2008). To judge the suitability of x-ray tomography for the nanoscale at least three setups must be distinguished:

1. Projection imaging by rays from a point source through the object in a parallel beam or in cone geometry is the most common setup for microtomography. This lens-less imaging case follows most closely the initial medical application of x-rays. Resolution is not limited by optics, but merely by the size of the source, the size of detector elements, and the source-object-detector distances. To reach submicron-scale resolution without lenses is the subject of ongoing research.
2. Lens-based x-ray microscopy (transmission or scanning transmission setup) is the furthest developed technique for x-ray tomography below micrometer resolution. The two cases of transmission or scanning setup are most equivalent to the duality between TEM and STEM using electrons; however, the complicated development of beam-converging condenser lenses for x-rays caused this technique to mature much later than STEM. The focusing lens in the condenser part (before the object) or imaging part (after the object) can be made from curved mirrors, (weakly) refractive lenses (Schroer et al., 2005), or from Fresnel zone plates (FZP) as used in Uesugi et al. (2006), Toda et al. (2006), and Yin et al.

(2006). The latter allow more compact setups and are mostly preferred over refractive lenses due to the refractive index of most materials hardly deviating from unity at important x-ray wavelengths. FZPs consist of concentric rings with decreasing fringe distance fabricated by techniques used to pattern semiconductor devices (e.g., lithography). The resolution of the projection image is given by the lens aberrations (full-field geometry) or the beam diameter (scanning geometry) which depends on the (outermost) fringe spacing of the Fresnel lens, its manufacturing precision, the coherence, and wavelength of the radiation. This type of microscopy is capable of sub-50 nm resolution imaging in 2D, and can reach 100 nm under optimum circumstances in a tomogram. Ongoing developments concern the usage of high beam coherence in synchrotrons not only for magnifying optics but also for coherent object penetration to obtain phase contrast instead of absorption contrast similar to high-resolution electron microscopy. Another active research field is two or multiple-wavelength x-ray microscopy to tackle a major artifact source, the variation of absorption with wavelength, but also to exploit absorption-spectroscopy-based imaging modes, equivalent to the EELS mode in electron tomography. The latter allows 3D chemical mapping.

3. Finally, the two major fields of x-ray applications in the last century, *projection imaging with tomography* and *crystallography by diffraction,* have been merged into the technique of coherent x-ray diffraction tomography (Miao et al., 1999; Williams et al., 2003; Robinson and Miao, 2004). Lenses again become obsolete as the scattering of x-rays by the object's local atomic arrangements encodes information about strain, chemical phases and atomic order of the object. The directly transmitted x-ray beam used in the above setups, (1) and (2), is blocked in this method, so that the detector only picks up x-rays scattered into a selected angular range. As no image is generated, the collection of multiple diffraction patterns upon rotation of the sample has to be fed into computer reconstruction software to iteratively deduce the 3D distribution of scattering power in the object (Gerchberg and Saxton, 1972; Fienup, 1978). Sophisticated algorithms succeed in converging against a unique solution. Currently resolution has been pushed to around 100 nm. The object quantity to be mapped by diffraction tomography is not mainly the density but the local atomic lattice plane distance. In other words, while absorption tomography would be blind to changes in lattice orientation (e.g., grain boundaries) or lattice strain, diffraction tomography is conversely of lower sensitivity to changes in atomic number in case of epitaxial orientation relationship.

Amongst the vast amount of materials research into the application of x-ray tomography two examples are presented here to illustrate setups (2) and (3), which have great potential to achieve nanoscale resolution:

Fresnel lens scanning projection tomography as of (2) in its spectroscopic variant has been used to reconstruct the inner density distribution of a cosmic dust fragment (Figure 31.13, Schroer et al., 2003; 2004). A synchrotron x-ray beam has been focused with a nanofocusing lens and scanned in 50 steps of 600 nm pitch across the particle. The number of tilt angles for tomography was 70. Although resolution of around 1 μm does not fully reveal the latest capabilities of x-ray nanotomography, this example is particularly instructive as the selection of a large number of energy channels for the absorption edge of elements in this sample was achieved. This results in images mapping the buried materials density for each chemical element separately probing the homogeneity of elements inside the cosmic particle nondestructively. Therefore, a similar elemental mapping capability (although on the microscale) than with spectroscopic electron tomography as of Section 31.2.2.3 is achieved.

The other example, using diffraction tomography as of (3) reconstructs a gold particle of micron size with a resolution of 75 nm (Figure 31.14, Williams et al., 2003). In addition to evaluating the external shape, which for this particle turns out highly anisotropic, this research also revealed insight into internal twinning of the arrangement of gold lattice planes, a feature that directly exploits the coherent and diffractive nature of this setup.

An innovative further setup, which conceptually refers to projection cone beam imaging as of (1), but which radically improves resolution, has been proposed using the chamber of a scanning electron microscope. The focused electron beam is not used to illuminate the object of interest (as would be done in ET of Section 31.2.2 and hybrid tomography of Section 31.2.4) but is focused on a first sample, the "target" (typically a pure homogenous piece of, e.g., Cu), which emits x-rays, which then penetrate the field-of-view on a second nearby sample, the object of study. The advantage of this SEM setup is the proximity of x-ray generation, object penetration and x-ray detection, which compensates via its high detection efficiency the overall low counts. It currently provides the only low-cost desktop laboratory setup for true nm-scale x-ray tomography (Mayo et al., 2002).

31.2.4 Hybrid Electron/X-Ray Nanotomography

An interesting cross-over between the methods of Sections 31.2.2 and 31.2.3 can be designed by using a focused electron beam to illuminate the specimen in scanning mode (Möbus et al., 2003; Saghi et al., 2007b).

Due to inelastic scattering events, the sample will emit characteristic x-rays typical for the particular elements present at the position of the beam. For example, an excitation of an s-electron out of an oxygen atom (its K-shell to be precise) leads to another electron from the L-shell to fill the gap in the K-shell. The energy difference between L- and K-shells is used to emit an x-ray photon at the wavelength corresponding to this level difference. Using a large condenser aperture and no objective aperture, a large penetration depth of electrons through the volume of interest of over 1000 nm can be achieved. Different from brightfield electron tomography as of Section 31.2.2.1, absorption of the electron beam is not the signal. However, this time we do not collect the electron beam at all, as it emerges from the bottom

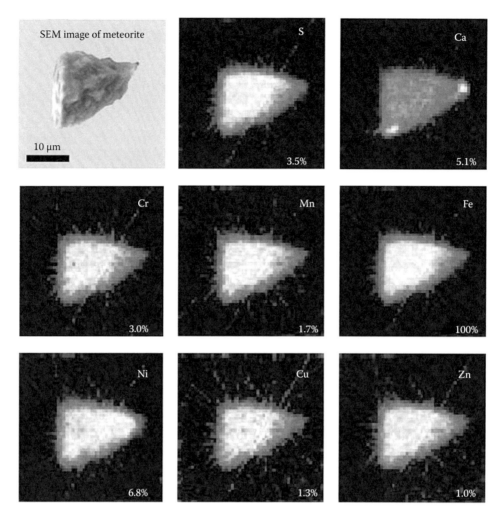

FIGURE 31.13 X-ray microtomograms of a cosmic dust particle (micrometeorite L2036H18 from NASA Cosmic Dust Catalog 15). Fluorescence x-ray images mapping the distribution of eight chemical elements. The b/w image top left is an SEM image for comparison and scaling. All chemical maps are recorded using the Kα line of the annotated element. (Modified after Schroer, C.G. et al., *MRS Bull.*, 29, 157, 2004.)

of the sample. Instead the x-ray signals emitted from the top of the sample are collected in the energy-dispersive x-ray detector sitting one-sided slightly above the specimen plane. We therefore deal with an "emission tomography" method rather than an absorption tomography. The latter aspect is also the major difference to x-ray tomography as of Section 31.2.3. For every beam position, one pixel value for each energy is recorded ("counted") by the detector: $I = I(x, y; E)$. If this procedure is repeated at a number of sample tilts, a 4D data space is collected: $I = I(x, y; \Theta; E)$. It is important that specimen shape and specimen holder are specially modified for this type of tomography. Not only do we have to guarantee that the electron beam passes unobstructed through the specimen to avoid the missing wedge problem as of Section 31.1.3.2, this time we also have to make sure that the emitted x-rays can reach the detector.

This new technique of energy-dispersive x-ray emission tomography in the electron microscope has several key-advantages:

- The x-rays are generated right at the region of interest and collected at close distance by the EDX detector.

- The energy-dispersive detection allows chemically sensitive tomography at high count rates. In x-ray tomography, absorption spectroscopy can give similar chemical identification, however, at the price of very long exposure times due to low counts.

- The resolution is determined by the electron beam, not by any x-ray optics or detector geometries. With down to 1 nm beam size the resolution range is two orders of magnitude beyond common x-ray tomographies.

- In comparison to the other electron tomography methods, advantages include the higher penetration depth, allowing larger objects to be reconstructed before onset of contrast inversion. In comparison to EFTEM tomography, the usual concept holds, where EDX is more suitable for heavy elements and pre-screening of many elements in parallel, while EELS is better for light elements, needs shorter exposure, and avoids nonlocal artifacts.

In Figure 31.8c, the concept of ED x-ray emission tomography is outlined. As a scanning TEM (STEM) technique, the

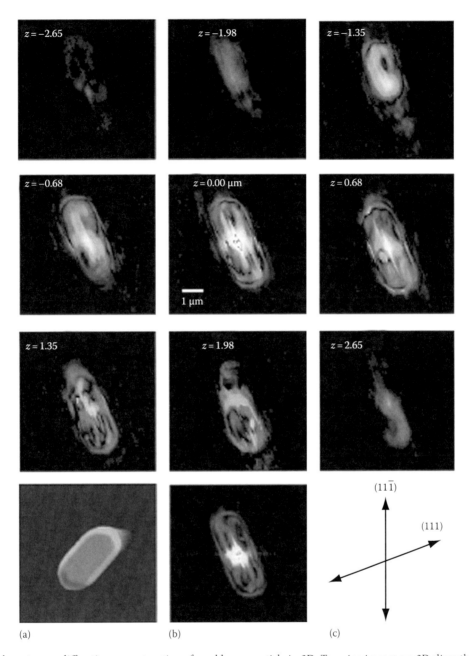

FIGURE 31.14 Coherent x-ray diffraction reconstruction of a gold nanoparticle in 3D: Top nine images are 2D slices through the particle at z-heights as labeled (a) is an SEM image to be compared with (b) the central slice of the x-ray reconstruction, (c) gives the crystallographic axes of the single crystal. (Reprinted from Williams, G.J. et al., *Phys. Rev. Lett.*, 90, 175501-1, 2003. With permission.)

projection image at each tilt angle is collected by EDX control software to record counts per beam position and energy channel. All energy channels (corresponding to all elements in the sample) can be acquired in parallel without any time penalty. Due to the long exposure times for EDX mapping, only large tilt increments of, e.g., 10° are affordable normally, unless the specimen is very stable. A choice of two geometries is obvious: If a 2D chemical EDX map is recorded at every tilt angle with an acquisition time of, e.g., 30 min., a total tilt series over, e.g., 140° requires 14 × 30 min, i.e., 7 h of total experimental acquisition. The result is then a 4D chemical map $I = I(x, y, z; E)$,

as introduced above. While experimentally demonstrated (Möbus et al., 2003), the demand on specimen and drift stability is high. Often, however, a situation arises where the interior of a material needs to be explored at one typical cross-section. In this case, the 2D EDX map per tilt angle can be replaced by a line scan, a 1D EDX plot. The result of the tomography experiment becomes a data field $I = I(y, z; E, x_0)$, which is 3D as x_0 is now a fixed parameter indicating the position of the line scan along an axis parallel to the tomographic rotation axis. Repeating the line scan at various x_0 would finally merge into the 4D technique described at first.

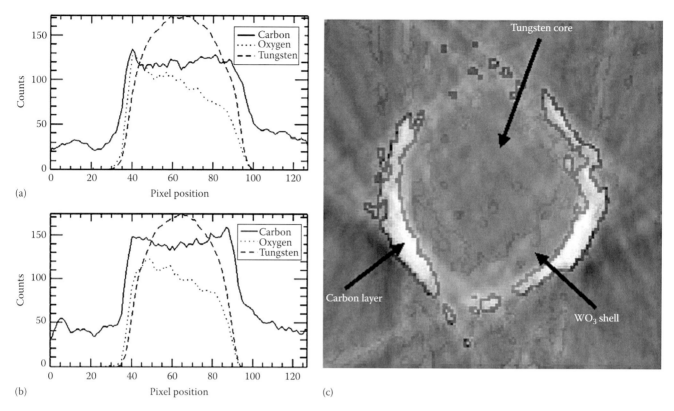

FIGURE 31.15 **(See color insert following page 25-16.)** Energy-dispersive x-ray (EDX) spectroscopic line scan across a carbon and oxide coated tungsten tip. Two members of a tilt series of line scans using C-K, O-K, and W-Lα signals are shown at (a) −70° and (b) +50° (c) RGB-map of the tomographic reconstruction of the chemical distribution of W-Lα (blue), O-K (red), and C-K (green). (Modified from Saghi, Z. et al., *Microsc. Microanal.*, 13, 438, 2007.)

As an application example (Saghi et al., 2007b), Figure 31.15 shows a cross-sectional EDX tomography analysis of an etched tungsten wire in shape of a conical tip. The crystalline W core is surrounded by a shell of oxide (WO_3) and on the surface a layer of carbon contamination has accumulated. The x-ray detection tuned to the energies of C, O, and W reveals the relative distribution of elements through a typical cross-section of the tungsten tip.

31.3 Sectioning Techniques ("Cut-and-Image")

31.3.1 Background

Sectioning techniques are a very heterogeneous area of 3D reconstruction and the diversity of both the cutting and imaging part is large:

Examples of cutting tools used include

1. Sawing of a block into multiple segments
2. Grinding/polishing in iteration (especially where a relief-free surface is required, e.g., for SEM secondary electron imaging)
3. Cutting of small samples into TEM transparent membranes by means of a diamond knife (ultramicrotomy).

4. Etching of a block in liquid phase or plasma with intermittent imaging of fresh surfaces
5. FIB cutting in an FIB microscope (ion beam parallel to fresh surfaces) or using nanobeam-SIMS (ion beam perpendicular to fresh surfaces)
6. Field-ion microscope ablation of a small and sharply curved surface region atom-by-atom

Examples of imaging techniques used to view the sections:

a. Light microscopy
b. Scanning electron microscopy
c. Scanning ion beam secondary electron microscopy using FIB
d. TEM
e. AFM
f. Mass spectrometer coupled to SIMS scanner as of (5) above
g. Position-sensitive atom probe (POSAP) coupled to field-ion ablation as of (6) above

The combination of cutting (1) through (6) and imaging tools (a) through (g) into pairs is flexible as long as both operate on the same length scale (McDowall et al., 1983; Bron et al., 1990; Martinez and Pinoli, 1996; Li et al., 1998; Alkemper and Voorhees, 2001) .

Mechanical cutting with a microwire saw can be ideally combined with light microscopy (2 + a), while ultramicrotomy is best for TEM (3 + d). Some pairs are exclusive, e.g., the field-ion ablation with atom probe imaging system. Of those pairs that qualify for "nanoscale" 3D reconstruction techniques, we select FIB tomography (5 + b) to be presented in most detail, while the combination of nanoscale thinning with AFM (3 and 4 + e), and the 3D atom probe (6 + g) are summarized in brief.

The two lists above do not include virtual optical sectioning where the cutting is not performed destructively but by means of scanning the focal setting of a variable focus microscope through a sample parallel to the beam axis. Our chapter will conclude with an outlook on such a technique.

31.3.2 Focused-Ion Beam Tomography

The resolution of cutting techniques was at any time in the past (and still is) lacking behind the resolution of imaging techniques. This leads at best to a 3D sampling technique far off-axis in the diagram of Figure 31.3. A cutting technique that at least approaches the SEM imaging resolution in a systematic and programmable way became available with the invention of the FIB instrument. Also known as a scanning ion microscope (SIM) it can be used both as a nanoscale knife and for ion-beam-induced secondary electron imaging (Orloff et al., 2003).

FIB tomography requires the following steps if applied to sub-volume of interest of a larger sample: the surface of incidence of the ion beam is protected by a (e.g., Pt) sputter-resistant metal coating to achieve sharp edges in the subsequent cutting steps. A trough is then milled once at the beginning to a depth of interest Δx which makes possible to image an initial cross-section of interest ($\Delta x \cdot \Delta y$) by SEM or SIM. The milling of a volume Δz (as of Figure 31.1) to free up the next cross-sectional surface marks then the first cycle of an iteration. Each iteration consumes a volume $\Delta z \cdot \Delta y \cdot \Delta x$ (Sakamoto et al., 1998; Tomiyasu et al., 1998; Dunn and Hull 1999; Inkson et al., 2001a; Uchic et al., 2007).

The imaging step is performed by secondary electron imaging, and depending on the type of instrument this can be a SEM scan (in case of a dual-beam instrument with ion and electron column) or a low-energy ion beam (SIM) scan (case of single-beam instruments).

- SIM has the advantage of generating crisper images due to the higher secondary electron yield and also grain orientation-sensitive images due to ion channeling into grains oriented near a zone axis, which allows 3D grain reconstruction (Inkson et al., 2001b). The main disadvantage is that the sample must be tilted for every iteration step from the cutting orientation to the imaging orientation by an angle of about 45° which is a slow and inaccurate process, and also the ion beam imaging scan will introduce a small amount of ablation during imaging.
- SEM imaging allows to conveniently share the cutting and imaging procedures between ion and electron beams in a dual-beam instrument. The throughput is much higher and alignment more precise. Many dual-beam instruments are nowadays equipped with automated acquisition

control software. The best voxel resolution achieved by dual-beam FIB tomography has reached to below 50 nm for Δz (Holzer et al., 2004).

- A third approach uses SIMS as the imaging device, with the benefit of spectroscopic imaging, while sacrificing resolution compared to the SEM/SIM imaging modes (Tomiyasu et al., 1998; Dunn and Hull, 1999).

The lateral resolution of FIB tomography is determined by the beam diameter convoluted with a cross-section for secondary electron emission, reaching 1 nm for field-emission electron beams. The cutting resolution (achievable Δz) is more complex to predict with the ion beam diameter being the major factor. Also issues of materials redeposition, uneven milling of composite materials, and depth-dependent lack of flatness of the fresh surfaces will contribute to the ultimate limit, which is strongly dependent on the milling rate (speed), leading to a range of resolutions around 10–100 nm.

Major application areas of nanoscale FIB tomography include semiconductor devices (Inkson et al., 2003; Yeoh et al., 2006), nanocapsules (Tomiyasu et al., 1998), 3D grain mapping in polycrystals (Inkson et al., 2001b), and nanoporous materials (Holzer et al., 2004). Mapping of nanocracks formed after scratch loading was achieved in ceramics. Chemical phase mapping has been performed in various nanocomposites, e.g., superalloys (Uchic et al., 2006) and semiconductors (Dunn and Hull, 1999).

Response of materials to mechanical testing, such as deformation by indentation or after cyclic loading is another major field of application (Motoyashiki et al., 2007).

Figure 31.16 presents a Cu–Al multilayer structure grown by MBE on Al_2O_3 substrate after pyramidal nanoindentation (Inkson et al., 2001a; Steer et al., 2002). The figure shows the

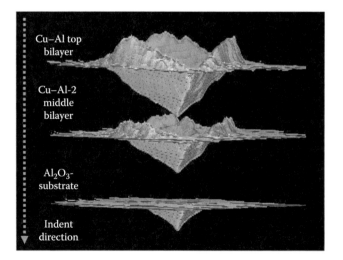

FIGURE 31.16 Application of focused ion beam (FIB) tomography for the 3D reconstruction of a metal-multilayer thin film structure on sapphire after nanoindentation. The three reconstructed "internal" surfaces are in x–z orientation perpendicular to the milled surfaces imaged by ion induced secondary electrons in x–y direction. Field of view = 8000 nm. (Modified from Steer, T.J. et al., *Thin Solid Films*, 413, 147, 2002.)

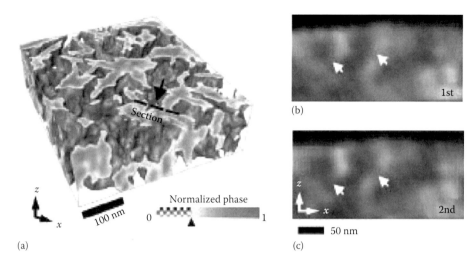

(a)   (c)

FIGURE 31.17 (See color insert following page 25-16.) 3D reconstruction of a volume (512 × 512 × 14 voxels) of elastomeric polypropylene (ePP). (a) Isosurface representation of the boundary between two intensity levels corresponding to amorphous and crystalline polymer phase. (b) and (c) are cross-sections in x–z-orientation with the AFM-height and AFM cantilever phase signal respectively. (From Dietz, C. et al., *Appl. Phys. Lett.*, 92, 143107, 2008. With permission.)

two Cu surfaces of this four-layer Cu–Al–Cu–Al thin film with the upper surface exceeding the lower one in both indentation depth and pile-up amplitude. On the hard substrate surface, no pile-up and only a small residual plastic indent has occurred. With the exception of the top Cu surface, each of these virtual surfaces which are located inside the material could not be imaged by AFM.

The most recent trends in FIB tomography attempt to include further imaging signals other than secondary electrons: EDX chemical mapping (Kotula et al., 2006, Schaffer et al., 2007) has been successfully combined with FIB for the chemical identification of phases in composites leading to a element-specific 3D reconstruction. Another typical SEM signal is electron back scatter diffraction (EBSD) which can be applied as a diffraction map for grain orientation crystallography (Groeber et al., 2006).

31.3.3 AFM Sectioning

Secondary electron imaging by SEM alongside milling with FIB (Section 31.3.1) requires expensive equipment and image specimen surfaces with a contrast mechanism that contains some (often unwanted) subsurface information. It could be advantageous therefore to use scanning probe microscopy (SPM) as it is exactly surface-specific and less complex, e.g., not in need of vacuum. The challenge is then to cut the material next to the sensitive probe of an SPM instrument without mounting the entire SPM inside an FIB. For this purpose, AFM sectioning (Magerle, 2000) could be successfully demonstrated to work in a reactive environment. Application of an etching medium (gas or liquid) over a short well-controlled time-interval δt will remove a layer of height $\Delta z = v \times \mathrm{d}t$ from the material's surface (v being the etchant speed). Imaging then is applied iteratively benefiting from the excellent lateral resolution of AFM of better than 1 nm under optimal conditions. At first, plasma-etching was used to remove 7.5 nm thick layers from a styrene-block-butadiene-

block-styrene (SBS) sample (Magerle, 2000). The applicability of liquid cell etching was soon afterwards confirmed for a bone sample (Dietz et al., 2007).

As shown in Figure 31.17, the 3D volume of a polymer which consists of regions of amorphous and crystalline phases can be mapped. The AFM cantilever is oscillated and (1) the scan height as well as (2) the phase of the oscillation will be influenced by the object density and surface topography. One tomogram can be reconstructed for each of the signals (1) and (2) to enhance sensitivity and resolution (Dietz et al., 2008).

The method of AFM sectioning requires for every material considered to test appropriate etchants and concentrations, which not only should etch at the right speed, but also provide homogeneity of etching speed across a multiphase block of material without introducing cavities or curvature of the surfaces. For benign cases of nonflatness, correction procedures can be applied (Dietz et al., 2007).

Etching in the gas or liquid phase can also be replaced by a combination of ultramicrotome knife with AFM to achieve iterative sectioning (Efimov et al., 2007).

31.3.4 3D Atom Probe

While ion beams are used to sputter atoms at high rates, etchant liquids or plasmas to remove materials layers, or microtome knifes for mechanical sectioning, none of these techniques can nearly approach atomic resolution as too much material is removed at once. Sectioning a material from its surface atom-by-atom (or monolayer-by monolayer) is exclusively possible through the field-ion microscope. A large electrostatic force ablates single atoms as ions from a sharp, tip-shaped, surface due to the strong field by a localized applied voltage. The lateral position where the ion came from can be detected by a position-sensitive detector, which allows calculation back from detector to specimen space due to the known geometry. The depth coordinate of the atoms is traced via the recording of the time of ablation. Together the

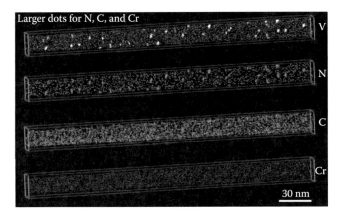

FIGURE 31.18 **(See color insert following page 25-16.)** Sub-15 nm particles for strengthening of high-strength low-alloy steels. 3D reconstruction of a nanoscale volume from atom probe data. Vanadium and nitrogen are mainly partitioned into the particles, while carbon and chromium have a minor enrichment in the particles. (From Craven, A.J. et al., *Mater. Sci. Technol.*, 24, 641, 2008. With permission.)

lateral coordinates and depth coordinates will build up a 3D scatter plot, in which every dot corresponds to one detected atom. Resolution is anisotropic as the depth sensing is more accurate than the lateral sensing. Under optimum circumstances and orientation, atomic lattice fringes can be resolved. The technique of 3D atom probe (also tomographic atom probe) has recently been extensively reviewed (Miller, 2000; Cerezo et al., 2007; Kelly and Miller, 2007). An application in metallurgy by Craven et al. (2008) is presented in Figure 31.18 and shows an example from steel microstructure research on the nanoscale. The enrichment of alloying elements, such as vanadium and nitrogen, into finely dispersed nanoparticles within high-strength low-alloy steel, is mapped in 3D with elemental sensitivity. The achieved nanometer resolution in all three dimensions by atom probe is confirmed by comparison with 2D TEM analysis of a thin specimen in both HAADF-STEM and EELS spectrum imaging mode.

31.3.5 Optical Sectioning in STEM

For the two mentioned tomography techniques using STEM (such as ADF-STEM and energy-dispersive x-ray mapping), it was essential that the beam focused onto the specimen passes the specimen approximately in parallel. Any change of beam diameter due to the cone shape of a convergent beam would reduce resolution and falsify the projections. Normally the depth-of-focus of STEM beams is larger than the specimen thickness, as commonly a small enough condenser aperture is selected, and also the change of height of the specimen within the field of view is small enough for the entire view to be in focus simultaneously. To the other extreme, if the convergence angle of the beam is deliberately maximized, e.g., by choosing the largest condenser aperture, only a part of the specimen would be imaged in focus. This is equivalent to assuming that the free-space propagation of a conical beam from a convergent beam via a cross-over to a divergent beam can also be realized inside material, in spite of diffraction, channeling, and de-channeling effects (Möbus and Nufer, 2003; Möbus et al.,

2003). By controlling and iterating a series of condenser focus selections, a depth scan can be achieved in analogy but not restricted to the principle of confocal laser scan microscopes. Examples for this technique shown so far comprise the location of heavy single atoms inside a thin foil (Van Benthem et al., 2005) using aberration corrected STEM. For the case of nanoparticles, depth sensing could be demonstrated for the particles being placed on a largely inclined carrier film (Möbus et al., 2006), or embedded into a matrix (Borisevich et al., 2006). The selection of the focused and suppression of the unfocused parts can be achieved by high-pass Fourier filtering. Alternatively, also a true confocal setup has been proposed (Nellist et al., 2006).

31.4 Discussion and Conclusions

The field of tomography seems to experience one of its most rapid periods of developments and diversification since its beginnings. The main reason for this step change is due to being identified as a highly promising tool for nanoscience and nanotechnology. Most aspects of nanoobjects are linked to their 3D nature, and therefore the idea of characterization of "typical cross-sections," which has dominated materials science for decades, is no longer available or appropriate.

The two available types of radiation on the nanoscale, electron and x-rays, have been found to be so perfectly complementary, that one could hardly have invented a better division of tasks between these tools. X-rays cover the upper range of the nanoworld length scale with resolutions not better than ~50 nm, but tolerating field of views (which in tomography is equivalent to object thickness) of many tens of micrometers to millimeters. Electrons, while pushing resolution to ~1 nm, suffer from the stronger and nonlinear interaction with matter, and therefore the upper limit in tolerable thickness hardly exceeds a few micrometers at best. The coexistence of both techniques is therefore expected to stay. In both fields, some common lines or trends of recent developments can be identified:

1. The transformation from a density-sensitive to a chemistry-sensitive technique, e.g., by exploiting absorption edges (or even their fine structure) in x-ray spectroscopy or in electron energy loss spectroscopy.
2. The deliberate exploitation of coherent scattering, often seen as a disturbance in classical density mapping, whether by using diffractive x-ray scattering in the synchrotron, or in the TEM by combining atomic resolution imaging or diffraction data with traditional morphological tomography.
3. Flexibility in pairing data acquisition and data reconstruction algorithms: The most suitable technique for a particular purpose can be selected from a range of projective or nonprojective imaging techniques with sensitivity for atomic number, elemental distribution, atomic surface contours, or internal porosity, and with a variety of tilt increments and tilt ranges. Similarly a wide range of reconstruction techniques can be matched to suit best the available (and possibly incomplete) data to be processed.

While the nondestructive techniques mentioned so far have their obvious justification against destructive techniques for unique specimens, the applicability of destructive sectioning for nanomaterials is mostly no obstacle, as often thousands of copies of the object of interest exist from typical synthesis methods. The two sectioning techniques, FIB microscopy and 3D atom probe, therefore, play a similarly complementary role in terms of field of view and resolution than x-rays and TEM for transmissive tomography.

Valuable insight is gained by applying a pair of techniques to the same sample: at first a nondestructive projection tomography followed by a sectioning approach (do not try it the other way round). For the combination of TEM and atom probe, a dual-mode 3D reconstruction has recently been presented by Arslan et al. (2008) and Gorman et al. (2008).

Applicability and usefulness of a 3D reconstruction technique are often linked to key features of the instrumentation involved, such as cost, size, ease of use, and restrictions to materials classes. Nanoscale x-ray tomography, if applied in a synchrotron, is certainly at the top of the price scale, although this would not appear as a cost for the individual user's laboratory, as synchrotrons are always national facilities. Some lab-scale nanoscale x-ray systems are also being marketed, with costs in the range of a TEM. Electron tomography and dual-beam FIB sectioning, as well as 3D atom probe, all need high-capital equipment, while 3D AFM sectioning is probably the lowest-cost technique covered in this chapter. Most important to be noted are recent step changes in applicability: TEM has been a materials characterization technique notoriously difficult in the preparation of specimens. This has radically changed with nanotechnology specimens which require merely mounting and dispersing rather than any thinning to electron transparency. Even more so, 3D atom probe used to suffer from the challenging preparation of a sharp and conductive tip, which has seen recent improvements through FIB preparation of the tip, and also the recent inclusion of semiconductive samples via laser-assisted ablation.

The driving force for pushing tomography to higher and higher resolution will probably continue until true and unambiguous atomic resolution is achieved in all directions of space. 3D atom probe, 3D TEM/STEM projections, and STEM optical sectioning are the three contenders with different and possibly complementary capabilities with respect to atomic position measurement and elemental chemistry identification. It will be fascinating to follow the further development of these techniques in the years to come.

References

Alkemper, J. and Voorhees, P.W. 2001. Three-dimensional image analysis of irregularly sized and shaped microscopic pores in aluminum alloys. *Journal of Microscopy* 201: 388–394.

Arbiol, J., Rossinyol, E., Cabot, A. et al. 2004. TEM 3D tomography of noble metal nanowires growth inside SiO_2 mesoporous aggregates. *Material Research Society Symposium Proceedings* 818: M4.5.1–M4.5.5.

Arslan, I., Marquis, E.A., Homer, M., Hekmaty, M.A., and Bartelt, N.C. 2008. Towards better 3-D reconstructions by combining electron tomography and atom-probe tomography. *Ultramicroscopy* 108: 1579–1585.

Baldock, C., Gilpin, C.J., Koster, A.J. et al. 2002. Three-dimensional reconstructions of extracellular matrix polymers using automated electron tomography. *Journal of Structural Biology* 138: 130–136.

Bals, S., Batenburg, K.J., Verbeeck, J., Sijbers, J., and Van Tandeloo, G. 2007. Quantitative three-dimensional reconstruction of catalyst particles for bamboo-like carbon nanotubes. *Nano Letters* 7: 3669–3674.

Banhart, J. 2008. *Advanced Tomographic Methods in Materials Research and Engineering*. Oxford University Press, Oxford, U.K.

Barnard, J.S., Sharp, J., Tong, J.R., and Midgley, P.A. 2006. Weak-beam dark-field electron tomography of dislocations in GaN. *Journal of Physics: Conference Series* 26(1): 247–250.

Batenburg, J. and Sijbers, J. 2007. DART: A fast heuristic algebraic reconstruction algorithm for discrete tomography. *Proceedings of the IEEE International Conference on Image Processing (ICIP)*, San Antonio, TX, IV: 133–136.

Borisevich, A.Y., Lupini, A.R., and Pennycook, S.J. 2006. Depth sectioning with the aberration-corrected scanning transmission electron microscope. *Proceedings of the National Acadamy of Sciences* 103: 3044–3048.

Bracewell, R.N. and Riddle, A.C. 1967. Inversion of fan-beam scans in radio astronomy. *Astrophysics Journal* 150: 427–434.

Bron, C., Gremillet, P., Launay, D. et al. 1990. Scanning transmission and computer aided volumic electron microscopy: 3D modelling of entire cells by electronic imaging. *Proceedings of SPIE- Biomedical Image Processing* 1245: 61–67.

Cerezo, A., Clifton, P.H., Galtrey, M.J. et al. 2007. Atom probe tomography today. *Materials Today* 10: 36–42.

Cervera Gontard, L., Dunin-Borkowski, R.E., Chong, R.K.K., Ozkaya, D., and Midgley, P.A. 2006. Electron tomography of Pt nanocatalyst particles and their carbon support. *Journal of Physics: Conference Series* 26: 203–206.

Craven, A.J., MacKenzie, M., Cerezo, A., Godfrey, T., and Clifton, P.H. 2008. Spectrum imaging and three-dimensional atom probe studies of fine particles in a vanadium micro-alloyed steel. *Material Science and Technology* 24: 641–650.

de Rosier, D.J. and Klug, A. 1968. Reconstruction of three dimensional structures from electron micrographs. *Nature* 217: 130–134.

Dietz, C., S. Röper, S. Scherdel, A. et al. 2007. Automatization of nanotomography. *Review of Scientific Instruments* 78: 053703.

Dietz, C., Zerson, M., Riesch, C. et al. 2008. Nanotomography with enhanced resolution using bimodal atomic force microscopy. *Applied Physics Letters* 92: 143107-1–143107-3.

Dunn, D.N. and Hull, R. 1999. Reconstruction of three-dimensional chemistry and geometry using focused ion beam microscopy. *Applied Physics Letters* 75(21): 3414–3416.

Efimov, A.E., Tonevitsky, A.G., Dittrich, M., and Matsko, N.B. 2007. Atomic force microscope (AFM) combined with the ultra-microtome: A novel device for the serial section tomography and AFM/TEM complementary structural analysis of biological and polymer samples. *Journal of Microscopy* 226(3): 207–216.

Fienup, J.R. 1978. Reconstruction of an object from the modulus of its Fourier transform. *Optics Letters* 3: 27–29.

Frank, J. 2007. *Electron Tomography: Three-Dimensional Imaging with the Transmission Electron Microscope.* 2nd edn. Plenum, New York.

Friedrich, H., McCartney, M.R., and Buseck, P.R. 2005. Comparison of intensity distributions in tomograms from BF TEM, ADF STEM, HAADF STEM, and calculated tilt series. *Ultramicroscopy* 106(1): 18–27.

Gerchberg, R.W. and Saxton, W.O. 1972. A practical algorithm for the determination of phase from image and diffraction plane pictures. *Optik* 35: 237–246.

Gilbert, P. 1972. Iterative methods for the three-dimensional reconstruction of an object from projections. *Journal of Theoretical Biology* 36: 105–117.

Gordon, R., Bender, R., and Herman, G.T. 1970. Algebraic reconstruction techniques (ART) for three-dimensional electron microscopy and x-ray photography. *Journal of Theoretical Biology* 29: 471–482.

Gorman, B.P., Diercks, D.R., and Jaeger, D. 2008. 3-D cross-correlation of atom probe and STEM tomography. *Microscopy and Microanalysis* 14(Suppl 2): 1042–1043.

Groeber, M.A., Haley, B.K., Uchic, M.D., Dimiduk, D.M., and Ghosh, S. 2006. 3D reconstruction and characterization of polycrystalline microstructures using a FIB-SEM system. *Materials Characterization* 57(4–5): 259–273.

Herman, G.T. 1979. Image reconstruction from projections: Implementation and applications, *Topics in Applied Physics* vol. 32. Springer, Berlin, Germany.

Holzer, L., Indutnyi, F., Gasser, P., Münch B., and Wegmann, M. 2004. Three-dimensional analysis of porous $BaTiO_3$ ceramics using FIB nanotomography. *Journal of Microscopy* 216(1): 84–95.

Hoppe, W., Langer, R., Knesch, G., and Poppe, C. 1968. Protein-Kristallstrukturanalyse mit Elektronenstrahlen. *Die Naturwissenschaften* 55(7): 333–336.

Hunt, E.M. and Herd, C.R. 2001. Tomography of carbon black for surface area measurement. *Microscopy and Microanalysis* 7(Suppl 2): 82.

Hutter, H., Wilhartlitz, P., and Grasserbauer, M. 1993. Topochemical characterisation of materials using 3D-SIMS. *Fresenius' Journal of Analytical Chemistry* 346: 66–68.

Inkson, B.J., Steer, T., Möbus G., and Wagner, T. 2001a. Subsurface nanoindentation deformation of Cu-Al multilayers mapped in 3D by focused ion beam microscopy. *Journal of Microscopy* 201(2): 256–269.

Inkson, B.J., Mulvihill, M., and Möbus, G. 2001b. 3D determination of grain shape in a FeAl-based nanocomposite by 3D FIB tomography. *Scripta Materialia* 45(7): 753–758.

Inkson, B.J., Olsen, S., Norris, D.J., O'Neill, A.G., and Möbus, G. 2003. 3D determination of a MOSFET gate morphology by FIB tomography. *Institute of Physics: Conference Series* 180: 611–616.

Kelly, T.F. and Miller, M.K. 2007. Invited review article: Atom probe tomography. *Review of Scientific Instruments* 78: 031101.

Koguchi, M., Kakibayashi, H., Tsuneta, R. et al. 2001. Three-dimensional STEM for observing nanostructures. *Journal of Electron Microscopy* 50(3): 235–241.

Koster, A.J., Ziese, U., Verkleij, A.J., Janssen, A.H., and De Jong, K.P. 2000. Three-dimensional transmission electron microscopy: A novel imaging and characterization technique with nanometer scale resolution for materials science. *Journal of Physical Chemistry B* 104(40): 9368–9370.

Kotula, P.G., Keenan, M.R., and Michael, J.R. 2006. Tomographic spectral imaging with multivariate statistical analysis: Comprehensive 3D microanalysis. *Microscopy and Microanalysis* 12(1): 36–48.

Kübel, C., Voigt, A., Schoenmakers, R. et al. 2005. Recent advances in electron tomography: TEM and HAADF-STEM tomography for materials science and semiconductor applications. *Microscopy and Microanalysis* 11(5): 378–400.

Larson, B.C. and Lengeler, B. 2004. High-resolution three-dimensional x-ray microscopy. *MRS Bulletin* 29(3): 152–156.

Li, M., Ghosh, S., Rouns, T.N. et al. 1998. Serial sectioning method in the construction of 3-D microstructures for particle-reinforced MMCs. *Materials Characterization* 41(2–3): 81–95.

Magerle, R. 2000. Nanotomography. *Physical Review Letters* 85: 2749–2752.

Martinez, S.J. and Pinoli, J.-C. 1996. Three-dimensional image analysis of irregularly sized and shaped microscopic pores in aluminum alloys. *Journal of Computer-Assisted Microscopy* 8(3): 131–144.

Mayo, S.C., Miller, P.R., Wilkins, S.W. et al. 2002. Quantitative x-ray projection microscopy: Phase-contrast and multi-spectral imaging. *Journal of Microscopy* 207: 79–96.

McDowall, A.W., Chang, J.J., and Freeman, R. 1983. Electron microscopy of frozen hydrated sections of vitreous ice and vitrified biological samples. *Journal of Microscopy* 131(1): 1–9.

Miao, J., Charalambous, P., Kirz, J., and Sayre, D. 1999. Extending the methodology of x-ray crystallography to allow imaging of micrometre-sized non-crystalline specimens. *Nature* 400: 342–344.

Midgley, P.A. and Weyland, M. 2003. 3D electron microscopy in the physical sciences: The development of Z-contrast and EFTEM tomography. *Ultramicroscopy* 96: 413–431.

Midgley, P.A., Weyland, M., Meurig Thomas, J., and Johnson, B.F.G. 2001. Z-contrast tomography: A technique in three-dimensional nanostructural analysis based on Rutherford scattering. *Chemical Communications* 10: 907–908.

Miller, M.K. 2000. *Atom Probe Tomography*, Plenum Press, New York.

Möbus, G. and Inkson, B.J. 2001. 3-Dimensional reconstruction of buried nanoparticles by element-sensitive tomography based on inelastically scattered electrons. *Applied Physics Letters* 79(9): 1369–1371.

Möbus, G. and Inkson, B.J. 2003. Novel nanoscale tomography modes in materials science. *Microscopy and Microanalysis* 9(Suppl 02): 176–177.

Möbus, G. and Nufer, S. 2003. Nanobeam propagation and imaging in a FEGTEM/STEM. *Ultramicroscopy* 96: 285–298.

Möbus, G., Doole, R., and Inkson, B.J. 2003. Spectroscopic electron tomography. *Ultramicroscopy* 96: 433–451.

Möbus, G., Inkson, B.J., Ross, I.M., and Morrison, R. 2004. Unlimited tilt for ultra-narrow lenses: Tomography at highest resolution? *Microscopy and Microanalysis* 10: 1196–1197.

Möbus, G., Al-Bermani, S., Xu, X. et al. 2006. 3D reconstruction of nanostructures from incomplete data. *Proceedings of the 16th International Microscopy Congress*, Sapporo, Japan, 2: 949.

Motoyashiki, Y., Brückner-Foit, A., and Sugeta, A. 2007. Investigation of small crack behaviour under cyclic loading in a dual phase steel with an FIB tomography technique. *Fatigue and Fracture of Engineering Materials and Structures* 30(6): 556–564.

Nellist, P.D., Behan, G., Kirkland, A.I., and Hetherington, C.J.D. 2006. Confocal operation of a transmission electron microscope with two aberration correctors. *Applied Physics Letters* 89: 124105.

Oikawa, T., Langlois, C., Mottet, C., and Ricolleau, C. 2006. Shapes and atomic arrangements of pure Cu-nanoparticles grown in high vacuum and in non-stress condition observed by 3D-tomography in a TEM. *Microscopy and Microanalysis* 12(Suppl 02): 616–617.

Orloff, J., Utlaut, M., and Swanson, L. 2003. *High Resolution Focused Ion Beams: FIB and Its Applications*. Kluwer, New York.

Park, J.-B., Lee, J.H., and Choi, H.-R. 2007. Three-dimensional imaging of stacked Pd nanoparticles by electron tomography. *Applied Physics Letters* 90(9): 093111-1–093111-3.

Rangayyan, R., Dhawan, A.P., and Gordon, R. 1985. Algorithms for limited-view computed tomography: An annotated bibliography and a challenge. *Applied Optics* 24: 4000–4012.

Robinson, I.K. and Miao, J. 2004. Three-dimensional coherent x-ray diffraction microscopy. *MRS Bulletin* 29(3): 177–181.

Saghi, Z., Xu, X., Peng, Y., Inkson, B.J., and Möbus, G. 2007a. Chemical analysis of tungsten tips by EDX line scan tomography. *Microscopy and Microanalysis* 13(Suppl 3): 438–439.

Saghi, Z., Xu, X., Peng, Y., Inkson, B.J., and Möbus, G. 2007b. Three-dimensional chemical analysis of tungsten probes by energy dispersive x-ray nanotomography. *Applied Physics Letters* 91: 251906.

Saghi, Z., Gnanavel, T., Peng, Y. et al. 2008. Tomographic nanofabrication of ultrasharp three-dimensional nanostructures. *Applied Physics Letters* 93: 153102.

Sakamoto, T., Cheng, Z., Takahashi, M., Owari, M., and Nihei, Y. 1998. Development of an ion and electron dual focused beam apparatus for three-dimensional microanalysis. *Japanese Journal of Applied Physics, Part 1: Regular Papers and Short Notes and Review Papers* 37(4 Suppl A): 2051–2056.

Schaffer, M., Wagner, J., Schaffer, B., Schmied, M., and Mulders, H. 2007. Automated three-dimensional x-ray analysis using a dual-beam FIB. *Ultramicroscopy* 107(8): 587–597.

Schroer, C.G., Kuhlmann, M., Hunger, U.T. et al. 2003. Nanofocusing parabolic refractive x-ray lenses. *Applied Physics Letters* 82: 1485–1487.

Schroer, C.G., Cloetens, P., Rivers, M. et al. 2004. High-resolution 3D imaging microscopy using hard x-rays. *MRS Bulletin* 29(3): 157–165.

Schroer, C.G., Kurapova, O., Patommel, J. et al. 2005. Hard x-ray nanoprobe based on refractive x-ray lenses. *Applied Physics Letters* 87: 124103-1–124103-3.

Spontak, R.J., Fung, J.C., Braunfeld, M.B. et al. 1996. Architecture-induced phase immiscibility in a diblock/multiblock copolymer blend. *Macromolecules* 29(8): 2850–2856.

Steer, T.J., Möbus, G., Kraft, O., Wagner, T., and Inkson, B.J. 2002. 3-D focused ion beam mapping of nanoindentation zones in a Cu-Ti multilayered coating. *Thin Solid Films* 413(1–2): 147–154.

Stegmann, H., Engelmann, H.-J., and Zschech, E. 2003. Characterization of barrier/seed layer stacks of Cu interconnects by electron tomographic three-dimensional object reconstruction. *Microelectronic Engineering* 65(1–2): 171–183.

Toda, H. Minami, K, Kobayashi, M., Uesugi, K., and Kobayashi, T. 2006. Observation of precipitates in aluminium alloys by sub-micrometer resolution tomography using Fresnel zone plate. *Materials Science Forum* 519: 1361.

Tomiyasu, B., Fukuju, I., Komatsubara, H., Owari, M., and Nihei, Y. 1998. High spatial resolution 3D analysis of materials using gallium focused ion beam secondary ion mass spectrometry (FIB SIMS). *Nuclear Instruments and Methods in Physics Research Section B* 136(1–4): 1028–1033.

Uchic, M.D., Groeber, M.A., Dimiduk, D.M., and Simmons, J.P. 2006. 3D microstructural characterization of nickel superalloys via serial-sectioning using a dual beam FIB-SEM. *Scripta Materialia* 55(1): 23–28.

Uchic, M.D., Holzer, L., Inkson, B.J., Principe, E.L., and Munroe, P. 2007. Three-dimensional microstructural characterization using focused ion beam tomography. *MRS Bulletin* 32(5): 408–416.

Uesugi, K., Takeuchi, A., and Suzuki, Y. 2006. Development of micro-tomography systems with Fresnel zone plate optics at SpRing-8. *Progress in Biomedical Optics and Imaging* 7(38): 63181F-1–63181F-9.

Van Benthem, K., Lupini, A.R., Kim, M. et al. 2005. Three-dimensional imaging of individual hafnium atoms inside a semiconductor device. *Applied Physics Letters* 87: 034104-1–034104-3.

Williams, G.J., Pfeifer, M.A., Vartanyants, I.A., and Robinson, I.K. 2003. Three-dimensional imaging of microstructure in Au nanocrystals. *Physical Review Letters* 90: 175501-1–175501-4.

Wikander, K., Hungria, A.B., Midgley, P.A. et al. 2007. Incorporation of platinum nanoparticles in ordered mesoporous carbon. *Journal of Colloid and Interface Science* 305(1): 204–208.

Withers, P. 2007. X-ray nanotomography. *Materials Today* 10: 26–34.

Xu, X., Saghi, Z., Gay, R., and Möbus, G. 2007a. Reconstruction of 3D morphology of polyhedral nanoparticles. *Nanotechnology* 18(22): 225501–225508.

Xu, X., Saghi, Z., Yang, G. et al. 2007b. Electron tomography of SPM probes, nanoparticles and precipitates. *Material Research Society Symposium Proceedings* 928E: 0982-KK02-04.

Xu, X., Peng, Y., Saghi, Z. et al. 2007c. 3D reconstruction of SPM probes by electron tomography. *Journal of Physics: Conference Series* 61: 810–814.

Yang, G., Saghi, Z., Xu, X., Hand, R., and Möbus, G. 2007. EELS spectrum imaging and tomography studies of simulated nuclear waste glasses. *Material Research Society Symposium Proceedings* 985: 0985-NN06-01.

Yeoh, T.S., Ives, N.A., Presser, N. et al. 2006. Analysis of successive focused ion beam slices by scanning electron imaging and 3D reconstruction. *Material Research Society Symposium Proceedings* 908E: 0908-OO05-11.1.

Yin, G.C., Tang, M.T., Song, Y.F. et al. 2006. Energy-tunable transmission x-ray microscope for differential contrast imaging with near 60 nm resolution tomography. *Applied Physics Letters* 88: 241115-1–241115-3.

Zaitsev, V., Seo, Y., Shin, K. et al. 2001. ToF-SIMS study of polymer nanocomposites. *Material Research Society Symposium Proceedings* K.K.6.2

Ziese, U., de Jong, K.P., and Koster, A.J. 2004. Electron tomography: A tool for 3D structural probing of heterogeneous catalysts at the nanometer scale. *Applied Catalysis A: General* 260: 71–74.

Local Probes: Pushing the Limits of Detection and Interaction

Adam Z. Stieg
California NanoSystems Institute

and

National Institute for Materials Science

James K. Gimzewski
California NanoSystems Institute

and

University of California

and

National Institute for Materials Science

32.1 Introduction

At the core of all science is mankind's innate curiosity toward understanding the observable universe. While a multitude of phenomena can be perceived through the basic physiological senses, such macroscopic processes are commonly subject to governance by physical laws operating at the microscopic scale and beyond. Experimental approaches toward understanding the behavior of matter at increasingly smaller dimensions require the capacity to carry out increasingly sensitive measurement associated with the concomitant decrease in signal, to push the limits of detection. Tremendous progress has been made in the detection of almost every physical observable related to matter and energy (Welland and Gimzewski, 1995). Detection alone is a limiting case. Through the addition of time and spatial resolution as viable experimental variables, the power to truly probe matter at ever-decreasing scales becomes reality.

With the dawn of the technological age, scientists around the world began to search for means to gain further insight into the laws of nature through the application of newly realized breakthroughs in engineering, information technology, and precision manufacturing. It was in the year 1590 that Zacharias Janssen invented the first optical microscope in an attempt to visualize material objects too small to be resolved by the naked eye, thereby enabling the discovery of the nucleus

and Brownian motion, which promulgated new understanding in biology and atomic physics (Croft, 2004). While optical microscopy continues to be a powerful tool for visualizing the microscopic world, inherent resolution limits stem from the use of light as the measurement probe. The desire to observe matter on ever-smaller scales led Ernst Ruska to create the first electron microscope in 1932, enabling humankind to generate images of matter on a length scale never before seen (Ruska, 1987). While electron microscopy opened the doors to a whole new microscopic world, the question still remained, could even smaller particles, such as individual molecules, atoms, or even electrons be "seen"?

The experimental approach toward understanding the behavior of matter at the atomic and molecular scale (Gimzewski and Joachim, 1999) has made tremendous strides due to the invention of scanning probe microscopy (SPM) (Binnig and Rohrer, 1982; Binnig et al., 1986). Drawing inspiration from the insightful paper *Scanning Tunneling Microscopy—Methods and Variations* (Rohrer, 1990), this chapter seeks to provide an updated perspective on the subsequent history, fundamentals, and current state of local probe methods. While it is not intended to be a comprehensive review of the many achievements and advances in the field, both critical concepts and elegant examples of the power of local probe at ultimate limits of resolution and measurement are presented.

32.2 Local Probes

The relatively brief history of nanotechnology has closely followed developments of local probe methods, specifically in the field SPM. This ever-increasing array of techniques are readily distinguished from other experimental approaches through their use of a local probe, defined here as a mechanical signal transducer whose position is physically adjustable in multiple dimensions. In contrast to other microscopic techniques, these methods enable the direct detection and manipulation of fundamental physiochemical interactions that are spatially controlled with nanoscale sensitivity and resolution. The ability to carry out controlled measurements at this scale using a local probe can be viewed as "the lab on a tip" (Meyer and Bennewitz, 2004).

Reasonable and intuitive starting points for conceptualizing a local probe measurement can be found through analogy, such as considering the probe as a phonograph needle whose size and interaction are drastically reduced or as a stethoscope of a medical practitioner that enables localized probing of motion through the detection of acoustic waves at a resolution of λ/100. This resolution is achieved by probing the near field, also called the evanescent field, where separation of the aperture to the object is well below λ. Local probe analogues of the stethoscope are the near field optical, scanning tunneling, and scanning force microscopies, all of which limit the interaction distance and area to below the λ of the photons or electrons being probed (Wiesendanger, 1994). The local probe may also be considered as an extension of the human sense of touch. Not only can tactile sensing enable the generation of complex spatial representations objects in three dimensions, it also allows for direct, localized detection and interaction. While it is certainly true that nanoscale phenomena were measured prior to the invention of SPM, the available spatial resolution and detection sensitivity of these methods continue to push the limits of experimental science to the level of individual atoms, molecules, or clusters thereof.

32.2.1 The Origin: Scanning Tunneling Microscopy

In 1928, Synge suggested the use of a tiny aperture at the apex of a nontransparent cone to "extend microscopic resolution into the ultra-microscopic region" (Synge, 1928). The idea was reborn by Ash and Nicholls (1972) who experimentally demonstrated a scanned probe microscope operating in the microwave region of the spectrum with a resolution of λ/60. Russell Young invented a scanned probe microscope called the "topografiner," a measurement system that operated mainly in the field emission mode through the application of an electrical bias typically above the work function of a metal [≥5 V] between a sharp, metallic tip in close proximity to a conductive surface (Young et al., 1972), G. Binnig and H. Rohrer successfully measured a controlled tunneling current at voltages below the work function whose local behavior was analyzed to generate the first topographic profiles of surface electronic local density of states (LDOS) with atomic scale spatial resolution (Binnig and Rohrer, 1982). As a

result of this tremendous accomplishment and its broad-ranging implications, the field of scanning tunneling microscopy (STM) has since received a great deal of attention from the scientific community, including the Nobel Prize in Physics in 1986. Since then, the STM became one of the most powerful tools available not only for analysis of surfaces, but also in the visualization and manipulation of matter at the scale of individual atoms and molecules. During its early developmental stages, the STM was employed at cryogenic temperature in an ultrahigh vacuum environment to provide a clean, controlled surface, and because it was believed that atomic vibrations would prevent its operation. However, it was with the successful realization of STM in multiple environments, including air, liquid, vacuum, and high or low temperatures, that launched the STM into the forefront of not only surface science but also electrochemical (Siegenthaler, 1992) and biological studies (Guckenberger et al., 1992).

The essential property of matter and energy required for the function of an STM is the phenomenon of electron tunneling. Prior to Louis de Broglie's suggestion that electrons had wave-like properties (de Broglie, 1924), the scientific community treated electrons as miniscule particles governed by classical mechanics. With the development of quantum mechanics, stemming from the quantum model of the atom (Bohr, 1914) and proof of the mathematical equivalence of the wave and matrix formulations of quantum theory (Schrödinger, 1926), scientists began to realize the true nature of electrons and their wave-like properties. While classical mechanics predicts a zero probability for the crossing of a potential barrier by a particle whose kinetic energy is less than the energy of the barrier itself, quantum mechanics shows us that there is indeed a nonzero probability value for this process (Hund, 1927). As a result, it was realized that an electron has the capability to tunnel through a potential energy barrier between two classically allowed states (Oppenheimer, 1928). In addition, an exponential dependence of the tunnel current, I, on barrier distance was discovered, and is represented, in a vacuum, by

$$I \propto e^{-2\kappa d} \tag{32.1}$$

Here, κ is a constant that behaves as

$$\kappa^2 = \frac{2m(V_B - E)}{\hbar^2}. \tag{32.2}$$

where

E is the energy of the state
V_B is the potential in the barrier
$\hbar$ is Planck's constant divided by 2π

By applying common work functions for most metallic systems, between 4 and 5 eV, to Equation 32.2, it is seen that $2\kappa \sim 2$ Å⁻¹. Consequently, the tunneling current decreases approximately one order of magnitude per Angstrom of distance between the two electrodes. This was an extremely important discovery, as this distance dependence requires maintenance of extremely small separations in a highly stable manner for tunneling to be observed. At very close proximity, the tunneling current saturates at a quantized resistance of

$$R_\kappa = \frac{h}{e^2} \approx 25 \text{ k}\Omega \qquad (32.3)$$

In the STM application, the tunneling effect is applied using a two-electrode system, the first being the atomically sharp tunneling tip and the second being a metallic or conductive surface of interest. Bringing the probe tip within an ~5–10 Å of the surface results in nonzero overlap of the electronic wave functions of both electrodes. Application of a small bias voltage between the tip and the surface induces a flow of tunneling electrons from the Fermi level of one electrode into unoccupied orbitals of the other electrode. Numerous studies of the tip-surface tunneling current interaction have been performed. In general, it is agreed that when the tip-surface distance is on the order of a several Angstroms and a small bias voltage is applied, the tunneling event can be described by the tunnel Hamiltonian method (Tersoff and Hamann, 1985). However, there has been a vast amount of theoretical work carried out in an attempt to model the true nature of the interaction between the probe tip and the surface (Chen, 1990). It is the tunneling current's exponential dependence on tip–sample distance that is the key to the extreme level of resolution made possible by the STM. As a result, control of the tip–sample separation, and thereby the tunneling current, is an essential component of how the STM functions to give topographical images of surface structures.

32.2.2 The Next Generation: Atomic Force Microscopy

A major limitation of the STM technique is the intrinsic requirement of a conductive substrate for tunneling events to occur. Although much of surface science involves elucidating the details of interactions at metallic and semiconductor surfaces, interest in extending SPM to nonconductive surfaces and larger biological systems inspired further innovation. In addition to the measurement of tunneling events, early experiments in STM speculated on the presence of forces interacting between the probe tip and the surface and were later experimentally confirmed (Coombs and Pethica, 1986; Durig et al., 1986). This observation led to a conceptualization of using these forces as the imaging signal for a new local probe, the atomic force microscope (AFM) by Binnig, Gerber, and Quate. Soon thereafter, concept became reality with the first experimental demonstration of AFM in surface imaging using a crushed diamond tip glued to a gold foil cantilever where the force was detected by an integrated tunneling sensor (Binnig et al., 1986).

Force microscopy facilitates investigations of surface topography, local mechanics, as well as a variety of other physical properties. The true test for AFM rested on its capacity to meet or exceed the remarkable resolution achieved by STM measurements, namely, atomic resolution imaging. It took only one year for the first atomic resolution images on both conducting (Binnig et al., 1987) and insulating surfaces (Albrecht and Quate, 1987).

Applications in a liquid (Marti et al., 1987) and electrochemical (Manne et al., 1991) imaging environments soon followed with great success. It is essential to note, however, the fundamental difference between "true atomic resolution" and "atomic resolution" images as the presence of defects in addition to periodic structure. Despite their success in reproducing the expected periodic surface structures visualized by STM, early AFM studies did not strongly reveal surface defects such as step edges or vacancies with atomic resolution which suggested an image corrugation arising from a large tip moving over a periodic surface. With continued diligence and development, true atomic resolution by AFM was ultimately claimed in both vacuum (Giessibl and Binnig, 1992) and liquid environments (Ohnesorge and Binnig, 1993). The final hurdle set forth and achieved by the scientific community was the visualization of a reactive surface that would strongly interact with the tip, namely, the Si(111) 7 × 7 surface (Giessibl, 1995).

In the original conception, it was proposed that the AFM could function in a variety of modes. A critical criteria, generally independent of the mode of operation, is the desire to maintain a constant, minimized force interaction between the probe and the surface. This approach serves to reduce both sample perturbation and tip wear. In the most basic permutation, called "contact mode," a constant probe deflection is used as the control variable. It is reasonable to consider the cantilever as a simple spring whose mechanics can be described by Hooke's law,

$$F = -k \cdot x \qquad (32.4)$$

where
F is the force
k is the spring constant or stiffness of the probe
x is the displacement, commonly referred to as deflection, of the probe from its equilibrium position

The application of a controlled, constant force is then directly related to deflection, as the probe spring constant (k) depends on its overall dimensions and material properties as represented in following equation for a rectangular cantilever:

$$k = \frac{E_Y w t^3}{4 l^3} \qquad (32.5)$$

where
t is the thickness
l is the length
w is the width
E_Y is the materials Young's modulus

For instance, with $k = 1$ N/m, a deflection of 1 nm would correspond to a force of 1 nN. While contact mode measurements have a demonstrated ability to carry out atomic and molecular resolution imaging, there are disadvantages due to the nature of the interaction including: imaging compliant and adhesive surfaces such as polymers and biological materials remains challenging, constant probe–sample contact involves intrinsic frictional effects that result in tip wear and loss of resolution over time, and operating in the repulsive force regime limits the capacity to examine more subtle forces.

A drive to reduce the tip–sample force resulted in an increased focus on "non-contact" modes of imaging. It is, of course interesting to again consider the exact nature of how "contact" itself is defined in such cases. In what have become known as dynamic AFM Applications, the probe is intentionally oscillated at or near its resonance frequency. Force interactions between the tip and the surface directly affect this oscillatory motion as energy provided to and stored in the probe is dissipated by the surface. Here attractive and repulsive forces result in a downward or upward shift in frequency, respectively. Transduction of these effects is carried out through observation of either amplitude, phase, or frequency modulation of the oscillating body. Operation at constant amplitude modulation (AM-AFM), phase modulation (PM-AFM), or frequency modulation AFM (FM-AFM) have all enabled true atomic resolution even on reactive surfaces.

Recent advances in FM-AFM have demonstrated the capacity to achieve atomic resolution imaging in liquid (Fukuma et al., 2005). An attractive feature of this method is the quantitative nature of the images in addition to the lack of necessity for a conductive surface. In contrast to the more conventional AM-AFM technique, commonly known as "tapping" mode, where the generated image is composed of mainly topographic information, the FM-AFM method produces images that can be considered to be force curves at every data point (Sader and Jarvis, 2004). The intrinsic sensitivity of the frequency-modulated detection scheme (Albrecht et al., 1991) also allows for measurements to be carried out within a single force regime, either attractive or repulsive, by the use of small amplitude oscillations of the probe tip. The advantages of such small amplitude oscillation in FM-AFM, <1 nm, were demonstrated to attenuate the effective contribution of long-range interactions on the tip–sample interaction (Giessibl, 2003). It is fundamentally desirable for the imaging signal to be composed solely of force interactions between the tip apex and nearby surface atoms. Both contact and tapping mode imaging are subject to such van der Waals forces that ultimately limit resolution (Hartmann, 1991).

32.2.3 Resolution

The most common perspective presented regarding SPM methods typically involves their capacity to generate spatial maps of topography and/or physiochemical properties at atomic scale resolution. It is essential to note, however, that resolution in a broader sense does not solely refer to the vertical and lateral dimensions in imaging applications. Both imaging and localized spectroscopic measurements carried out with the probe tip, aptly called scanning probe spectroscopy, rely heavily on the ability to detect and resolve extraordinarily small signals. The performance of any local probe technique, SPM included, depends most directly on the nature of the probe itself and the stability/control of its interaction with a sample. As a result, a great deal of focus has been applied toward improvements in materials, instrumental design, and measurement electronics over the last 25 years.

Given this consideration, the resolution can generally be broken down into two aspects. The vertical resolution is solely determined by the stability of the probe–sample separation and is therefore directly dependent on the mechanical stability of the measurement system. Lateral resolution also intrinsically depends on mechanical stability; however, a number of other factors play critical roles. Most obvious is the effective size of the probe itself, a fact well known in microscopy. Local probe methods make this limitation glaringly obvious to anyone who has learned that using tweezers rather than their fingers, and thereby reducing the effective size of the probe, greatly enhances ones ability to interact on smaller length scales. It is commonplace for the consideration of resolution to begin and end with probe radius, and an immense amount of research has focused on the development of ever-smaller probe tips as seen in Figure 32.1, which shows a selection of a variety of microfabricated tips developed to probe different signal interactions and/or nanostructures.

A thorough assessment of resolution requires the consideration of additional factors. As the probe–sample interaction generally involves both local variability and distance-dependent effects, the decay length of a given interaction plays an essential role in determining the fundamental resolution limits. For example, electron tunneling events employed in STM exhibit

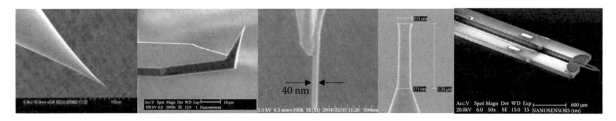

FIGURE 32.1 SEM micrographs of probes used in SPM applications. From *left*, an electrochemically etched tungsten wire for STM, a specialty A-Tec probe for accurate position (Image courtesy of NanoWorld AG.), a magnetically coated carbon nanotube probe for high resolution magnetic force microscopy (Reprinted from Kuramochi, H. et al., *Nanotechnology*, 16, 24, 2005. With permission.), a critical dimension probe (Image courtesy of Veeco Instruments, Inc.), and the Akiyama probe, combining the advantageous features of tuning fork and cantilever technologies. (Image courtesy of NanoWorld AG.)

a strong exponential dependence with respect to distance. This behavior serves to localize the interaction radius to the region between the object of interest and the atomic apex of the probe itself, where the last apex atom wins. On the contrary, interactions that exhibit long decay lengths such as electrostatic, magnetic, and van der Waals forces serve to increase the effective interaction radius and limit the resolution. It becomes clear that resolution limits become quite complex and are extremely dependent on measurement type. In many instances, the intrinsic properties of the probes themselves continue to be viewed as a limiting factor in overall performance. While it is clear that the probe does impart inherent restrictions, a thoughtful assessment of properties related to both the probe and interaction of interest provides insight into possible avenues for advancement of these measurement technologies.

32.2.4 Separation of Control and Measurement Variables

While it is quite common to find descriptions of local probe methods simply in terms of a given distance-dependent interaction, a thorough treatment critically distinguishes between the control variable and the measurement variable. It is true that in many SPM methods, the probe–sample distance, termed here a "control variable," is strictly regulated whereas the distance dependence of a given property of interest, termed here a "measurement variable," provides the dataset to be collected and plotted. It is certainly advantageous for both the control and the measurement to exploit the same particular interaction. However, there are many situations in which inhomogeneous surfaces provide interactions that are spatially variable, composed of multiple interactions or some combination thereof. Progress in method development, especially in spectroscopic applications, has produced a number of powerful techniques, where the measurement variable is not strictly linked to the control variable. As such, it is essential to illuminate the subtle yet distinct difference between the two.

As a general rule, all SPM systems operate using the same underlying principles to position the probe tip relative to the sample. SPM applications are commonly carried out using one of two control methods, namely, constant-interaction or constant-height mode as seen in Figure 32.2, both of which employ piezoelectric actuators that control the tip position.

In constant-interaction mode, an electronic feedback loop is used to maintain a constant interaction magnitude as the tip is raster scanned in the X- and Y-directions (Figure 32.2a). A few simple examples of feedback loops and their control variables encountered in everyday life include a thermostat (temperature) and the cruise control in an automobile (speed). The voltage signal applied to the scanner piezo in the Z-direction dynamically controls the tip–sample separation in order to maintain a constant interaction. For example, in STM, the distance-dependent tunneling current serves as the control variable whose magnitude is maintained throughout the scan. Recording the voltage applied to the Z-component of a calibrated scanner results in a plot of vertical tip position at each data point during the scan. Therefore, we consider distance, or more commonly height, to be the measurement variable. These plots can then be translated into a topographical image of the surface if the voltage sensitivity of the piezoelectric scanner is known. Similarly, in AFM, the control variable is based on a distance-dependent force interaction between the probe tip and surface. Maintenance of this variable also provides height as a measurement variable.

In an attempt to achieve faster scan speeds, constant-height mode was introduced for STM measurements (Bryant et al., 1986). This mode essentially operates with the sensitivity of the feedback loop drastically reduced to omit frequency response. Hence, there is no control variable in this case and the feedback error signal is recorded (Figure 32.2b). By maintaining a nearly constant tip-surface separation and monitoring changes in the tunneling current, surface topographical information can be obtained due to the distance dependence of the tunneling current. Plotting of this information versus the tip position produces images of electronic surface structure based on the transmission coefficient of tunneling electrons. STM images acquired in constant height mode also provide information on variations in the local surface work function.

Both modes offer their own inherent advantages and disadvantages. While may seem, at first glance, that the constant-height mode would be the mode of choice, there are numerous disadvantages to this mode of imaging. The most obvious difference between the two modes is the ability of the constant-interaction mode to operate on surfaces that are not atomically flat. If the constant-height mode were applied to a surface with extensive step edges, surface corrugations, or other surface defects, the probe tip would commonly collide with the surface resulting in

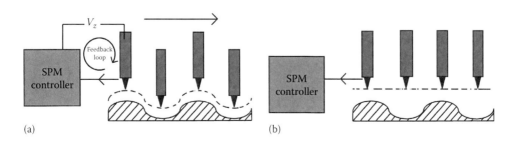

FIGURE 32.2 Graphical representation of the two modes of operation in SPM, constant interaction mode (a) and constant height mode (b). V_z represents the feedback voltage controlling the tip–sample separation in the z-direction.

damage to the probe tip and a loss of resolution during imaging. Also, the translation of the variations in the measurement variable into quantitative values is quite complicated without discrete knowledge of the distance dependence of the said interaction. The disadvantage of imaging in the constant-interaction mode is the limited response time of the piezo elements, and thereby the feedback loop, to changes in applied voltage in the vertical direction. As a result, scan speed is commonly slow. Nonetheless, these two modes of operation, in conjunction with numerous other modes such as differential microscopy and spectroscopic applications, provide researches with an enormous amount of versatility in the analysis of surfaces and interfaces (Wiesendanger, 1994).

32.2.5 Other Local Probe Methods

In addition to the STM and AFM, local probes methods include a vast array of experimental techniques aimed at characterization of new signal domains pertaining to physiochemical properties at the nanoscale. In total, these techniques constitute the field of SPM that can be broken down into subcategories based on the type of property of interest or on the nature of the resolution criterion. For example, Table 32.1 shows those methods aimed at the measurement of forces including atomic (AFM), frictional/lateral (FFM/LFM), magnetic (MFM), acoustic (SAM), electrostatic (EFM), and surface potential or Kelvin Probe (KPFM) while those that characterize local electrical properties include tunneling (STM), electrochemical (SEcM), ion conductance (SICM), capacitance (SCM), and so on. With a brief glance at this list of the techniques, it becomes obvious that there are areas of overlap between individual methods. However, through an assessment of the factors that limit measurement resolution, it can be seen that two cases stand out. Methods that exploit control interactions based on near-field effects tend to exhibit the highest spatial resolution, whereas those whose resolution is limited by spatially confinement of the interaction, typically through optimization of the probe geometry, tend toward lower resolution. Table 32.1 additionally provides an overview of the various techniques, their respective control and measurement variables, the determining factors for resolution limits, and the applicable experimental conditions that range from vacuum,

atmospheric, liquid and electrochemical environments as well as variable temperature ranging from milliKelvin to thousands of degrees Celsius.

32.2.6 Image Convolution

While it is normally adequate to interpret SPM images as representations of surface topography or some other measured property, the exact nature of the image can be complicated by many factors. Ideally, the probe would be a source of controlled interaction operating through a single point. This would not only result in maximum resolution, but also minimize any complex interaction between the tip and the surface. For example, it is agreed that the STM topograph approximately represents a contour of the constant density of charge of the surface being investigated. The STM probe is exponentially biased to tunneling events that occur at or near the Fermi level of the surface (E_F). However, all electrons, whether at, near, or below Fermi level, contribute to the overall charge density of the surface. Norton Lang also demonstrated that atoms and molecules adsorbed on surfaces may generate virtual electron tunneling processes where energetically higher states can generate a nonzero weight at E_F (Eigler et al., 1991).

If the tunneling tip were considered to have an atomically sharp point as the source of current, the equation representing the tunneling current can then be simplified to represent proportionality where current is a function of the LDOS of electrons at the Fermi level. While this may seem to be an extremely simplistic approach, this approximation has been shown to hold as long as the interaction between the tip and surface can be approximated by those for an s-wave tip wave function, which is not always the case as will be discussed below. Thus, the treatment of STM images as an actual surface topograph is limited at best, especially in the region of an adsorbate.

While it has also been shown that at small tip–sample separations, the electronic coupling between the probe tips and the surface tends to be very weak, allowing their wave functions to be treated separately (Noguera, 1989), both short- and long-range forces are always present. Attempts to minimize these effects have included compensation of the contact potential difference between tip and sample, operation of the measurement within

TABLE 32.1 Description of SPM Methods Based on the Nature of the Control/Measurement Variable and the Effect This Has on Resolution

Method	Control Variable	Measurement Variable	Resolution Criterion	Environment
STM	Tunneling current	LDOS/distance	Near-field	(v), (a), (l), (ec)
AFM (contact and tapping)	Short and long range force	Distance, phase	Near-field	(v), (a), (l), (ec)
AFM (non-contact)	Short range force	Distance, adhesion, stiffness	Near-field	(v), (l)
FFM/LFM	Short range, repulsive force	Lateral force	Near-field	(v), (a), (l)
MFM	Distance	Magnetic force	Probe geometry	(v), (a)
EFM/KPFM	Distance	Electrostatic force, work function	Probe geometry	(v), (a)
SECM/SICM	Electrical/ion current	Distance	Probe geometry/aperture	(l), (ec)
NFOM/NSOM	Force	Optical properties	Probe geometry/aperture	(v), (a), (l)

Note: Experimental environments are also presented as vacuum (v), ambient (a), liquid (l), and electrochemical (EC) conditions.

a single force regime through, and the fabrication of sharpened, high-aspect ratio probes. This final approach has resulted in a growing field of specialty probe production, a few examples of which are shown in Figure 32.1, that employ techniques such as oxide etching, ion beam milling, or the use of nanotubes/wires attached to the probe apex.

32.3 Local Probe Physics

The complex physics underlying the operable interactions of local probe methods can by no means be fully described here. However, the fundamental aspects that both enable and set apart these techniques from other experimental approaches can be introduced. By using beams of photons or charged/neutral particles to gather spatial information, traditional microscopic techniques limit their interaction volume, and thereby ultimate resolution, to the wavelength or de Broglie wavelength of the incident probe. Local probe techniques exploit the capacity to position a detector in close proximity to a sample (on the order of angstroms) in a controlled fashion. Flexibility in positional control of the probe allows for the study of both fundamental short- and long-range interactions, shown in Figure 32.3, through the use of probes sensitive to a given physical phenomenon or with a specific geometry. Those that employ short-range interactions, including both attractive (A) and repulsive (R), or even equilibrium (E) chemical bonding forces, often exhibit an exponential dependence on distance that serves to reduce the effective aperture of interaction to subatomic dimensions, thereby enabling imaging and spectroscopy at the atomic and molecular scale. Numerous other local probe techniques rely on a short-range control interaction to facilitate the measurement of a long-range interaction such as magnetic or electrostatic

properties. In the simplest case, measurements can rely solely on the probe positioning system ability to provide spatially resolved measurements with nanoscale precision. In all cases, the nature of both the interaction and the probe itself are deterministic of performance.

32.3.1 Short-Range Interactions

The study of matter at the scale of single atoms and molecules through quantum mechanical treatments quickly becomes analytically intractable. As a result, the use of numerical methods and approximations are commonplace (Heisenberg, 1925). The most well-known molecular system that can be solved analytically in the context of local probes is the H_2^+ ion, composed of two protons and one electron (Chen, 1993). With only one electron, there are no electron–electron interactions to consider, and the only source of repulsion is nuclear. Despite this simplicity, the H_2^+ system allows insight into a system with purely short-range interactions stemming from wave function overlap, namely, Coulomb repulsion and the chemical bond. The well-known Morse potential very accurately describes the potential energy function for this diatomic system and provides a sense of scale when comparing short- and long-range interactions.

$$V(r) = D_e(1 - e^{-a(r-r_e)})^2 - D_e \quad \text{and} \quad a = \sqrt{\left(\frac{k_e}{2D_e}\right)} \quad (32.6)$$

where

D_e is the potential minimum
k_e the bond force constant
r the distance between the atomic nuclei
r_e the equilibrium internuclear distance

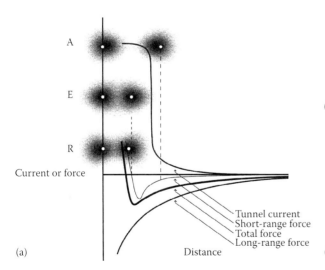

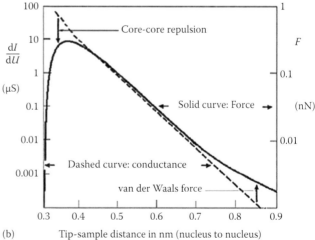

FIGURE 32.3 (a) Schematic potential energy diagram of two atoms as a function of internuclear separation. Above are cartoons of the electron density and nuclear position. In E, the atoms are at their equilibrium bond distance that corresponds to a minimum in the energy. In A, the atoms are in a short-range attractive regime, and in R, they are in a repulsive regime. nc-AFM may operate in regime A, E, R, or a combination of all three. In general, STM operates at distances corresponding to regime A. However, at very close distance, the tunneling barrier collapses and plateaus at the quantized conductance $2e^2/h$ provided no atomic rearrangement occurs. (b) A graph to correlate the raw data of tunneling conductance and the raw data of force. (Reprinted from Chen, C.J., *Nanotechnology*, 16, S27, 2005. With permission.)

The introduction of additional electrons into the system complicates matters significantly from an analytical perspective as both electron–electron and Pauli-repulsions become important. In addition to Coulombic components, the Pauli-repulsion principal prevents two electrons with the same four quantum numbers from sharing space. While not actually a force, the Pauli-repulsion principal is a quantum mechanical requirement that is responsible for the majority of repulsive interactions that gives atoms their volume-occupying characteristic.

The probe apex can be positioned such that the proximity of the apex atoms to surface atoms gives rise to an equilibrium energetic minimum in the short-range overlap of their respective electron density, as represented by the diffuse structures in the upper and lower portions of Figure 32.3, respectively, corresponding to the formation of a single atomic chemical bond. At even shorter distances, electron–electron repulsion sets in, whereas at distances slightly larger, short-range attraction will occur. The magnitude of these distances coincides with those used in both STM and noncontact AFM applications that utilize exponential interaction distance dependencies. Further, the overlap integrals of both electron tunneling and chemical bonding have been shown to be mathematically equivalent, with variations only occurring due to short-range repulsive and long-range attractive effect as seen in Figure 32.3b (Chen, 2005). Thus, through the use of sub-nanometer vertical modulations in the position of the probe, one can effectively and reversibly make and break partial chemical bonds. This isolation and consideration of solely short-range interactions represents the essence and power of local probes methods.

32.3.2 Long-Range Interactions

Even for neutral atoms separated by relatively large interaction distances (>1 nm), additional electrons introduce the possibility of instantaneous oscillating dipoles caused by the random shifting of the electron cloud, polarization via charged species, or interaction between instantaneous dipoles. These attractive electrostatic interactions are called van der Waals forces, and like traditional electrostatics are highly dependent on geometry and charge.

The van der Waals force is a slow decay represented by

$$F_{vdW} = \frac{-A_H R}{6z^2} \qquad (32.7)$$

where
  R is the microscopic, not nanoscopic, tip radius
  A_H is the Hammaker constant
  z is the tip–sample separation

Likewise, electrostatic forces scale also as $1/z$.

$$F_e = -\pi\varepsilon_0 R \frac{U^2}{z} \qquad (32.8)$$

where
  ε_0 is the dielectric permittivity of vacuum
  R is the radius of the probe
  U is the voltage between probe tip and sample

Electrostatic forces may also result from the difference in work functions, ϕ, of probe tip and sample, which are commonly balanced through the application of an appropriate bias between the probe tip and the sample,

$$V_{bias} = \phi_{Sample} - \phi_{Tip} \qquad (32.9)$$

or may result from trapped electrical charges. Recent progress in nc-AFM involves using smaller subnanometer modulation and the use of high aspect ratio tips to minimize long-range interactions (Giessibl, 2003).

32.3.3 Cumulative Interactions

The collective forces at play between the atomic apex of a local probe and sample can be considered as the summation of all the nearby long- and short-range forces, which is shown in the total force curve of Figure 32.3. In the case of a real tip positioned in the short-range interaction regime, other atoms of the probe apex and surface will also contribute to the measurement through long-range interactions. It is therefore possible for the cantilever to simultaneously experience an overall attractive force, whereas the actual apex atom experiences a repulsive interaction (Loppacher et al., 2003) that depends on the microscopic geometry of the tip and surface. In order to directly probe specific short-range interactions, especially in AFM-based methods, it is advantageous to minimize long-range contributions to the measured interaction. In the STM method, the exponential decay of the tunneling current with respect to distance intrinsically eliminates contributions to the measured tunneling current from atoms at larger separations.

As introduced above, calculations of even the simplest intermolecular interactions require the use of numerical methods and approximations. One of the most accessible methods to approximate the intermolecular potential ($U(r)$) is the use of a pair of potentials, as used in the Mie potential, which are of the form

$$U(r) = -\frac{a}{r^m} + \frac{b}{r^n} \qquad (32.10)$$

In this way, all the repulsive and attractive forces are each reduced to two opposing terms. The most well-known version of the Mie potential is the Lennard-Jones potential, which uses $m = 6$ and $n = 12$ to simulate both the long-range van der Waals attractive regime and exponential decay of the short-range repulsive regime for two neutral, monatomic particles. Despite the approximations and assumptions employed in calculating the actual potential felt by a local probe, experimentally determined energy versus distance curves appear remarkably similar to the Lennard-Jones potential, represented as the total force in Figure 32.3.

32.4 Probing the Ultimate Limits

There are a considerable number of spatially confined physiochemical properties that can be readily examined through the use of local probes, a comprehensive review of which is not presented here. The following examples represent a few selected applications of local probes techniques toward the study of long-standing areas of scientific intrigue and serve to demonstrate the power of such techniques as experimental tools.

32.4.1 Spin and Magnetism

The quantum mechanical property known as "spin," though mysterious and abstract to many at a fundamental level, manifests itself ubiquitously in everyday life. While its most obvious expression, namely, magnetism, has been known and studied since the sixth century BC, the theory of spin and its roots in quantum mechanics was only been thoroughly described less than a century ago (Dirac, 1928a,b). What followed was a series of groundbreaking discoveries including Maxwell's equations of electrodynamics, Einstein's theory of special relativity, and the Standard Model. Clearly a topic of paramount importance in understanding the nature of the physical universe, validation, and direct measurements of spin have been intensely studied since the Stern-Gerlach experiment first demonstrated the quantized nature of intrinsic spin (Gerlach and Stern, 1922). However, very few microscopic methods exist that can probe magnetic properties in a scalable fashion.

With regard to local probe methods, the majority of research related to spin resides in the field of magnetism and employs techniques such as magnetic force microscopy (MFM) and spin-polarized scanning tunneling microscopy (SP-STM). The MFM has the capacity to characterize the magnetic properties of a given material with nanoscale precision. However, as it relies on the long-range magnetic interaction between the surface and a magnetically coated tip, its resolution limits are difficult to overcome. On the contrary, SP-STM, despite being far more experimentally complex, has a demonstrated capacity to examine, as seen here in Figure 32.4, and manipulate the magnetic properties of individual atoms by using the spin-dependent tunneling current through a magnetic probe tip placed in an external magnetic field (Heinrich et al., 2004). Considering the ongoing drive for increased density in magnetic storage devices and the emerging field of "spintronics," the value of local probe methods in the study of spin and magnetism once again becomes clear.

32.4.2 Imaging Orbitals

Real-space visualization of atomic and molecular orbitals represents a fundamental limit in the study of bonding interactions. As the nature of these interactions depends on the relative occupation of orbital shells, namely, which electronic orbitals are involved in bond formation, the ability to characterize and directly study them is indispensable. Theories related to covalent

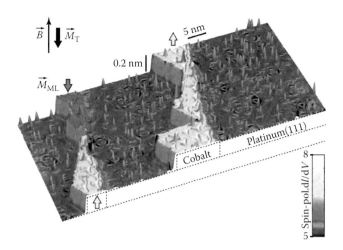

FIGURE 32.4 **(See color insert following page 25-16.)** Using magnetic probe tips and spin-polarized tunneling effect has enabled the detection of magnetization in individual atoms. A spin-polarized dI/dV map of cobalt atoms adsorbed onto a nonmagnetic platinum surface, Pt (111), shown here in blue, demonstrates a dependence on the external magnetic field. (Reprinted from Meier, F. et al., *Science*, 320, 82, 2008. With permission.)

bonding have evolved immensely over the last century, from the inception of the covalent bond to valence bond theory, the linear combination of atomic orbitals (LCAO), hybridization (Pauling), molecular orbital theory, and quantum chemistry. When considering covalent bonding in a surface or surface-bound adsorbate, the local distribution of charge indicates the specific covalent interactions involved. Variations in this charge distribution exist for all systems exhibiting bonding through the higher p and d angular momentum states.

Local probe methods are an ideal experimental system for the study of charge density. As submolecular resolution STM images are composed, to a reasonable approximation, of the spatially resolved bias-dependent tunneling probabilities of individual molecular orbitals, it has been possible to examine the nature of bonding within individual molecules for some time. It is important to note, however, that unless the molecule in question is decoupled electronically from the metallic substrate, for instance, by using a nanoscale thick layer of insulation, the STM image cannot be solely represented as stemming from the molecule itself. Likewise, the exact makeup of the probe apex, in all SPM measurements, is deterministic of the specific interaction that is probed as well as the appearance of the resulting image. The true nature of image contrast continues to be a topic of vigorous research. Extending on the ability of SPM methods to position individual atoms or molecules in extremely close proximity has recently enabled both STM (Chaika and Myagkov, 2008) and AFM (Giessibl et al., 2001) experiments to demonstrate subatomic resolution of atomic orbitals, the latter of which is provided here in Figure 32.5, where a silicon atom was imaged showing a distinct lobed structure, which was assigned to two dangling bonds on the apex of a silicon tip.

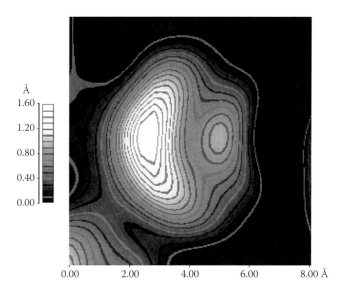

Å

1.60

1.20

0.80

0.40

0.00

0.00   2.00   4.00   6.00   8.00 Å

FIGURE 32.5 **(See color insert following page 25-16.)** The current limits of resolution by local probe methods are personified by this image, the first experiment to resolve subatomic features. In this case, two lobes of charge density are detected within the single atom located at the apex of the probe tip through the angular dependence of tip–sample forces resulting in real-space characterization of a single atomic orbital. (Reprinted from Giessibl, F.J. et al., *Annalen Der Physik*, 10, 887, 2001. With permission.)

32.4.3 Force Spectroscopy

In addition to imaging applications, increased measurement accuracy has extended the use of force spectroscopy toward direct, localized measurement of nanoscale mechanical properties as well as interactions between individual chemical species. The varieties of physical properties that have been examined through traditional indentation studies (Li and Bhushan, 2002) and other force measurement techniques (Isrealachvili, 1992) are extended by the AFM to include spatial resolution and increased sensitivity. Unfortunately, deconvolution of the force from the tip–sample interaction is not always straightforward. Methods aimed at acquisition of quantitative values for local mechanical properties are under constant development, especially in relation to the "soft" surfaces encountered in the study of biological systems (Zlatanova et al., 2000).

Moving beyond local nanomechanics are applications directed at the measurement of inter- and intramolecular interactions. This vast field of research involves work in physics, materials science, and chemistry but has received special attention from the life science community. As many biological interactions are governed by single molecular binding events, all with a related force interaction, force spectroscopy provides a fantastic method for the study of such interactions at the single molecule level. While other spectroscopic techniques are indeed capable of similar examinations, they typically involve the extreme dilution or isolation of

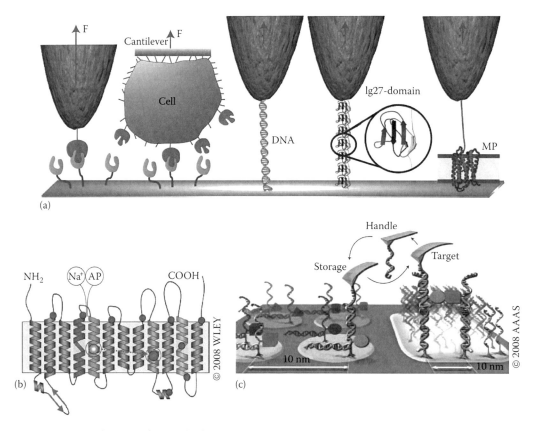

FIGURE 32.6 The numerous applications of molecular force spectroscopy are represented in this schematic, including probing of force interactions such as binding and unfolding in isolated systems (a) or in more complex systems such as biological membranes (b) as well as mechanically assembly (c) at the single molecular level. (Reprinted from Muller, D.J. and Dufrene, Y.F., *Nat. Nanotechnol.*, 3, 261, 2008. With permission.)

the system of interest in order to achieve single molecular events. Applications of local probe spectroscopy, as represented schematically in Figure 32.6, are readily capable of isolating a variety of interactions: single molecule recognition, binding, and unwinding events to name a few, even in complex environments.

32.4.4 Chemical Identification

Since the early days of SPM, dreams of characterizing both the position and the identity of atomic and molecular species have been ubiquitous. Outstanding progress in the field was made quite rapidly in the structural characterization of surfaces and adsorbates. However, chemical identification at such length scales proved much more challenging. A variety of approaches have been taken to this end, most of which involve inelastic processes related to electron tunneling spectroscopy. These methods have provided a direct means to address the LDOS for metallic surfaces and vibrational properties of adsorbed molecules and carry out optical spectroscopy at the molecular scale via tunneling-induced photon emission (Hoffmann et al., 2002). Such techniques have provided some truly astounding results but contain inherent limitations due to the requirements of cryogenic temperatures and conductive substrates.

Common approaches toward atomic and molecular recognition and mapping have involved chemical modifications of the tip surface. Recently, advances in the sensitivity of detection in SPM have allowed the forces associated with covalent bond formation to be used as a characterization tool, even on

reactive and insulating surfaces. These methods employ the small amplitude, FM-AFM approach to carry out quantitative force spectroscopic measurements along with atomic resolution imaging. The ultimate extension of this force detection method, shown in Figure 32.7, has been recently demonstrated in the chemical identification of atomic species at the surface multicomponent system at room temperature (Morita et al., 2005; Sugimoto et al., 2007). These astounding results provide a proof-of-principle for a capacity to chemically identify complex systems at the limits of resolution.

32.4.5 Tribology

Elucidating the nature of interactions between surfaces in relative motion and the laws that govern them have been a subject of intense interest since the late sixteenth century work of Leonardo da Vinci on friction. Subsequent classical representations of friction (Amonton and Coulomb) described frictional forces with respect to three discrete variables. Specifically, their three laws stated friction to be proportional to normal force and independent of both contact surface area and velocity. It was not until the pioneering work of Bowden and Tabor (1939) that distinct postulates and observed contradictions to Amonton's second law could be explained by the "true" microscopic contact area (Tomlinson, 1929). What followed was a series of developments in the use of sensitive force detection schemes to develop a modern description of friction at the atomic scale (Bhushan et al., 1995). The first local probe measurement of friction carried out

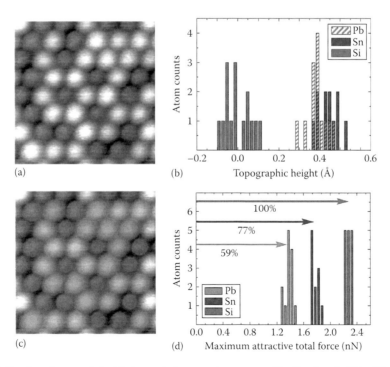

FIGURE 32.7 Atomic resolution imaging coupled with dynamic force spectroscopy enables the chemical identification of individual atoms in a mixed alloy system. Despite the lack of topographic contrast between lead (Pb) and antimony (Sn), measurement of the attractive force between the probe tip and individual atoms during the onset of covalent bond formation allows chemical contrast to be obtained. (Reprinted from Sugimoto, Y. et al., *Nature*, 446, 64, 2007. With permission.)

using the AFM directly contradicted Amonton's first law by demonstrating friction to be independent of normal force (Mate et al., 1987). Further advances in SPM methods have produced a revised perspective that shows frictional force to be proportional to the microscopic contact area, dissipative normal forces, and sliding speed of the contact interface as well as lattice vibrations, electronic friction, and other dissipative forces.

Normally referred to as lateral or frictional force microscopy, operation of the AFM in this area involves the measurement of forces acting on the probe parallel to the scan direction. Again, contact mode is the most straightforward method to detect and characterize local friction. Scanning perpendicular to the longitudinal axis of the probe produces transverse, torsional forces that are readily detected as lateral deflections of the probe itself. Variations in local friction coefficients cause shifts in magnitude of the measured lateral deflection signal allowing for direct spatial comparison and characterization, which is in some cases quantitative. The use of dynamic AFM modes has resulted in improved performance by oscillating the probe parallel to the surface rather than perpendicular, and in some cases have yielded true atomic resolution (Giessibl et al., 2002). Figure 32.8 demonstrates the capacity of these techniques to elucidate unexpected phenomena, such as the effects of subsurface structure and composition, and provide fundamental insight into the nature of friction at the atomic scale.

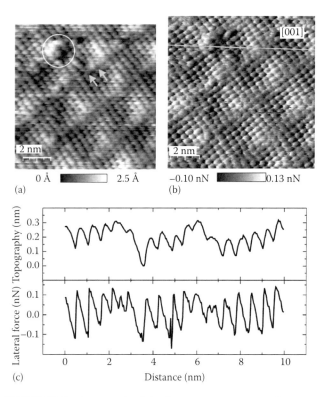

FIGURE 32.8 An example of simultaneous measurement of topography and lateral forces with atomic scale resolution on an insulating surface. This work demonstrated a marked modulation of frictional effects due to the underlying topographic superstructure of a buried interface. (Reprinted from Maier, S. et al., *Phys. Rev. B*, 78, 045432, 2008. With permission.)

32.4.6 Controlling Matter

Beyond detection lies interaction, which leads to control. While quantum theory states that all observations include an intrinsic interaction (Heisenberg, 1925), observation of the interaction itself and harnessing the control it provides has proven to be both intriguing and elusive. Schrödinger (1952a,b) himself postulated man's inability to directly interact with a particle of matter in a controlled fashion, be it electrons, atoms, or molecules. While this perception of impossibility held strong for many years, science would prove reality to be otherwise with the first controlled manipulations (Stroscio and Eigler, 1991) and spectroscopic investigations of individual atoms (Binnig et al., 1983). While extremely challenging experimentally, the former case demonstrated the realistic possibility of "bottom up" construction and the limits of structural fabrication, on which the vision of nanotechnology drew its inspiration (Feynman, 1960). Thousands of years of human history passed before mankind settled on bricks as the basic building blocks for what are now timeless structures and thousands more to construct the first arch. In a sense, researchers in this field are like these builders and engineers of old, seeking to understand not only how to design and fabricate a given nanoscale architecture but also to first begin by assessing what the nature of the building blocks, how they interact, how to arrange them effectively, and even possibly how to have them arrange themselves. Since these initial experiments, much work has focused on more efficient approaches to fabrication such as self- and directed-assembly. However, atomic and molecular manipulation provides a vehicle to gain essential insight into this new paradigm in fabrication (Sugimoto et al., 2008).

Stemming from spectroscopic investigations and a variety of observed inelastic effects, use of the SPM probe as a means to provide energetic stimulus directly to a surface or adsorbate also became a fascinating prospect. Exciting results have been achieved toward inducing chemical reactions at the single molecular scale such as the tip-induced self-propagating polymerization (Okawa and Aono, 2001) or single molecular hydrogen tautomerization reaction (Liljeroth et al., 2007). Clearly an intriguing prospect for chemists, this work also provides a means to address the limitations of controlled manipulation as a fabrication strategy. In addition, controlled stimulus of this type has also garnered attention as a possible avenue toward atomic and molecular electronic devices as well as quantum computation, an elegant example of which is given in Figure 32.9. As a result, controlling matter with local probes continues to be a vibrant, yet challenging, field of fundamental research that is critically important to bridging the gap between nanoscale science and technology.

32.5 Outlook and Perspectives

Since bursting onto the scene of the scientific research community some 25 years ago, local probe methods have seen a precipitous rise in profile as experimental methods in the study of matter and energy at the nanoscale. From their initial applications in

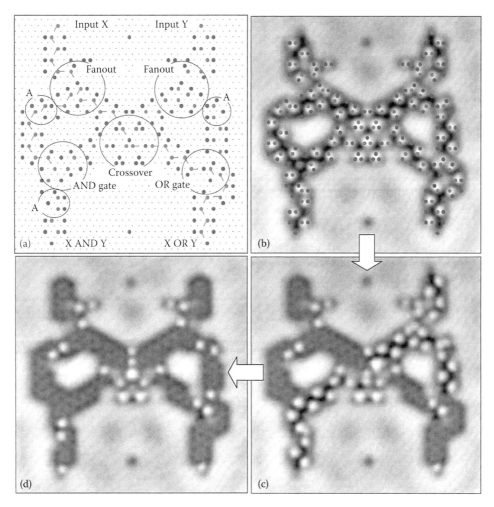

FIGURE 32.9 **(See color insert following page 25-16.)** An elegant example of design-oriented molecular manipulation for computational logic applications. Fabrication of a two-input sorter was carried out through STM manipulation of individual carbon monoxide molecules into predetermined, atomically specific locations. "Molecule cascades" capable of computing the AND/OR logic were produced by inducing motion of one molecule with the STM tip. (Reprinted from Heinrich, A.J. et al., *Science*, 298, 1381, 2002. With permission.)

fundamental physics and surface science to more recent adoption by the life science community, developments in instrumental methods, materials, and control have fostered a period of dynamic discovery. In order to continue along this trajectory, it becomes clear that method development should focus on flexibility and the detection of ever-smaller signals while pursuing higher speed without a loss of resolution. Despite all its advantages, local probe measurements remain painstakingly slow in most cases. Enabling time to become a viable experimental variable will drastically enhance the applicability of these methods to dynamic process in all fields of experimental research.

A common perspective on the balance of speed and resolution is that an increase in speed leads to a decrease in resolution. Given the current state of instrumentation, this may be true in a number of cases. However, these models are shortsighted in accepting the notion that an increase in measurement bandwidth necessarily leads to a decrease in sensitivity and resolution. In fact, with intelligent hardware design and high-speed control, speed and resolution can be seen to scale together. Ambient and

in-situ SPM methods provide obvious examples of this fact. At room temperature, thermal effects such as mechanical drift are limiting factors toward high-resolution imaging and local spectroscopy. Only through increasing measurement speeds can experimental fluctuations be reduced, ultimately leading to better sensitivity and resolution. The creation of local probe instrumentation capable of data collection at higher bandwidths, for example, video rate imaging (>60 and even up to 1000 frames/s), is a challenge currently being addressed in the SPM design community (Humphris et al., 2005; Ando et al., 2006; Picco et al., 2007; Howard-Knight and Hobbs, 2008).

The current limitations on high-speed scanning stem from three criteria. First, the mechanical design of the scanning element itself plays a major role. Inherent to any structural design are mechanical resonances that, if excited, cause detrimental effects. As such, high-speed SPM instrumentation necessitates extremely high structural rigidity. Next, the probe response and signal transduction needs to operate at higher frequency. Methods to achieve these criteria include employment of high-frequency oscillation of the measurement probe in combination with low-noise, high-bandwidth signal transduction. A variety of approaches have been taken toward the prior point, including cantilevers (Yamashita et al., 2007; Nishida et al., 2008), membranes (Degertekin et al., 2005), and tuning forks (Heike and Hashizume, 2003). Finally, as SPM measurements rely on one or more dynamic feedback loops, the issue of control and signal processing becomes an issue. Analog systems of control enable very high speeds but are less user-friendly and inhibit total system integration. Traditional approaches to digital signal processing (DSP) operate in the 10–100 kHz range. Operation of an SPM at video rate requires control electronics capable of generating feedback signals of 5 MHz or more. Use of field programmable gate arrays is bridging the DSP gap by operating at more than 500 MHz and enabling multichannel signal collection and feedback.

The research community has demonstrated tremendous ingenuity in pushing the limits of detection and interaction through developments in SPM. Further advances in materials science, electrical engineering, and control theory will undoubtedly continue to allow the field of SPM to flourish. However, to move toward ever-higher speeds, reduction in the size of mechanical elements will eventually be essential. In particular, despite the use of micromechanics in sensor and tip fabrication, the use of electromechanical systems has been mainly limited to actuation and detection perpendicular to the surface. The simultaneous use of multiple cantilevers, more than 1000 in the case of the IBM Millipede, was an exciting advance in this direction (Lutwyche et al., 1999; Despont et al., 2000; Vettiger et al., 2000). However, eventually, the complete integration of X, Y, and Z motion and sensing in NEMS could re-revolutionize the force sensitivity speed and range of applications of AFM. As early as 1995, members of the Hitachi ARL published a prototype, totally micromechanical STM (Lutwyche and Wada, 1995). Around the same time, IBM also developed a functioning two-axis AFM (Chui et al., 1998), which is shown in Figure 32.10. It is therefore

surprising that all SPM systems remain so structurally large. We predict the merging of nanoscale and microscale fabrication and processing as the key to the future of nanoscience.

Acknowledgments

It is our pleasure to acknowledge Heinrich Rohrer for inspiring this chapter and for many fruitful discussions. Discussions with Franz Giessibl, Christoph Gerber, Gerd Binnig, Don Eigler, Hermann Gaub, Masakazu Aono, and Mark Welland over a long period of time were inspirational in writing this manuscript. We also wish to thank MANA for partial financial support.

References

Albrecht, T. R. and Quate, C. F. (1987) Atomic resolution imaging of a nonconductor by atomic force microscopy. *Journal of Applied Physics*, 62, 2599–2602.

Albrecht, T. R., Grutter, P., Horne, D., and Rugar, D. (1991) Frequency-modulation detection using high-q cantilevers for enhanced force microscope sensitivity. *Journal of Applied Physics*, 69, 668–673.

Ando, T., Uchihashi, T., Kodera, N., Miyagi, A., Nakakita, R., Yamashita, H., and Sakashita, M. (2006) High-speed atomic force microscopy for studying the dynamic behavior of protein molecules at work. *Japanese Journal of Applied Physics Part 1-Regular Papers Brief Communications and Review Papers*, 45, 1897–1903.

Ash, E. A. and Nicholls, G. (1972) Super-resolution aperture scanning microscope. *Nature*, 237, 510–512.

Bhushan, B., Israelachvili, J. N., and Landman, U. (1995) Nanotribology—Friction, wear and lubrication at the atomic-scale. *Nature*, 374, 607–616.

Binnig, G. and Rohrer, H. (1982) Vacuum tunnel microscope. *Helvetica Physica Acta*, 55, 128–128.

Binnig, G., Rohrer, H., Gerber, C., and Weibel, E. (1983) 7×7 Reconstruction on Si(111) resolved in real space. *Physical Review Letters*, 50, 120–123.

Binnig, G., Quate, C. F., and Gerber, C. (1986) Atomic force microscope. *Physical Review Letters*, 56, 930–933.

Binnig, G., Gerber, C., Stoll, E., Albrecht, T. R., and Quate, C. F. (1987) Atomic resolution with atomic force microscope. *Surface Science*, 189, 1–6.

Bohr, N. (1914) Atomic models and x-ray spectra. *Nature*, 92, 553–554.

Bowden, F. P. and Tabor, D. (1939) The area of contact between stationary and between moving surfaces. *Proceedings of the Royal Society of London Series A-Mathematical and Physical Sciences*, 169, 0391–0413.

Bryant, A., Smith, D. P. E., and Quate, C. F. (1986) Imaging in real-time with the tunneling microscope. *Applied Physics Letters*, 48, 832–834.

Chaika, A. N. and Myagkov, A. N. (2008) Imaging atomic orbitals in STM experiments on a Si(111)-(7×7) surface. *Chemical Physics Letters*, 453, 217–221.

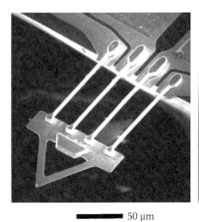

■■■ 50 μm   ■■■ 2 μm

FIGURE 32.10 Scanning electron micrographs of a custom, microfabricated device capable of motion along two axes. Attached to the micromechanical device is a triangular probe whose tip apex is seen at left. (Reprinted from Chui, B.W. et al., *Appl. Phys. Lett.*, 72, 1388, 1998. With permission.)

Chen, C. J. (1990) Tunneling matrix-elements in 3-dimensional space—The derivative rule and the sum-rule. *Physical Review B*, 42, 8841–8857.

Chen, C. J. (2005) A universal relation in NC-AFM, STM, and atom manipulation. *Nanotechnology*, 16, S27–S34.

Chen, J. C. (1993) *Introduction to Scanning Tunneling Microscopy*. Oxford, U.K.: Oxford University Press.

Chui, B. W., Kenny, T. W., Mamin, H. J., Terris, B. D., and Rugar, D. (1998) Independent detection of vertical and lateral forces with a sidewall-implanted dual-axis piezoresistive cantilever. *Applied Physics Letters*, 72, 1388–1390.

Coombs, J. H. and Pethica, J. B. (1986) Properties of vacuum tunneling currents-anomalous barrier heights. *IBM Journal of Research and Development*, 30, 455–459.

Croft, W. J. (2004) *Under the Microscope: A Brief History of Microscopy*. Singapore: World Scientific Publishing Company.

de Broglie, L. (1924) Recherches sur la théorie des quanta. *Physics*. Paris, France: University of Paris.

Degertekin, F. L., Onaran, A. G., Balantekin, M., Lee, W., Hall, N. A., and Quate, C. F. (2005) Sensor for direct measurement of interaction forces in probe microscopy. *Applied Physics Letters*, 87, 213109.

Despont, M., Brugger, J., Drechsler, U., Durig, U., Haberle, W., Lutwyche, M., Rothuizen, H. et al. (2000) VLSI-NEMS chip for parallel AFM data storage. *Sensors and Actuators A-Physical*, 80, 100–107.

Dirac, P. A. M. (1928a) The quantum theory of the electron. *Proceedings of the Royal Society of London Series A-Containing Papers of a Mathematical and Physical Character*, 117, 610–624.

Dirac, P. A. M. (1928b) The quantum theory of the electron—Part II. *Proceedings of the Royal Society of London Series A-Containing Papers of a Mathematical and Physical Character*, 118, 351–361.

Durig, U., Gimzewski, J. K., and Pohl, D. W. (1986) Experimental-observation of forces acting during scanning tunneling microscopy. *Physical Review Letters*, 57, 2403–2406.

Eigler, D. M., Weiss, P. S., Schweizer, E. K., and Lang, N. D. (1991) Imaging Xe with a low-temperature scanning tunneling microscope. *Physical Review Letters*, 66, 1189–1192.

Feynman, R. P. (1960) There's plenty of room and the bottom. *Engineering and Science Magazine*, 23, 22–28.

Fukuma, T., Kobayashi, K., Matsushige, K., and Yamada, H. (2005) True atomic resolution in liquid by frequency-modulation atomic force microscopy. *Applied Physics Letters*, 87, 034101.

Gerlach, W. and Stern, O. (1922) The experimental evidence of direction quantisation in the magnetic field. *Zeitschrift Fur Physik*, 9, 349–352.

Giessibl, F. J. (1995) Atomic-resolution of the silicon (111)-(7X7) surface by atomic-force microscopy. *Science*, 267, 68–71.

Giessibl, F. J. (2003) Advances in atomic force microscopy. *Reviews of Modern Physics*, 75, 949–983.

Giessibl, F. J. and Binnig, G. (1992) Investigation of the (001) cleavage plane of potassium-bromide with an atomic force microscope at 4.2-K in ultra-high vacuum. *Ultramicroscopy*, 42, 281–289.

Giessibl, F. J., Bielefeldt, H., Hembacher, S., and Mannhart, J. (2001) Imaging of atomic orbitals with the Atomic Force Microscope-experiments and simulations. *Annalen Der Physik*, 10, 887–910.

Giessibl, F. J., Herz, M., and Mannhart, J. (2002) Friction traced to the single atom. *Proceedings of the National Academy of Sciences of the United States of America*, 99, 12006–12010.

Gimzewski, J. K. and Joachim, C. (1999) Nanoscale science of single molecules using local probes. *Science*, 283, 1683–1688.

Guckenberger, R., Hartmann, T., Wiegrabe, W., and Baumeister, W. (1992) The scanning tunneling microscope in biology. In *Scanning Tunneling Microscopy II*. Wisendanger, R. and Guntherodt, H. J. (Eds.). New York: Springer Verlag.

Hartmann, U. (1991) van der Waals interactions between sharp probes and flat sample surfaces. *Physical Review B*, 43, 2404–2407.

Heike, S. and Hashizume, T. (2003) Atomic resolution noncontact atomic force/scanning tunneling microscopy using a 1 MHz quartz resonator. *Applied Physics Letters*, 83, 3620–3622.

Heinrich, A. J., Lutz, C. P., Gupta, J. A., and Eigler, D. M. (2002) Molecule cascades. *Science*, 298, 1381–1387.

Heinrich, A. J., Gupta, J. A., Lutz, C. P., and Eigler, D. M. (2004) Single-atom spin-flip spectroscopy. *Science*, 306, 466–469.

Heisenberg, W. (1925) Quantum-theoretical reinterpretation of kinematic and mechanical connections. *Zeitschrift Fur Physik*, 33, 879–893.

Hoffmann, G., Libioulle, L., and Berndt, R. (2002) Tunneling-induced luminescence from adsorbed organic molecules with submolecular lateral resolution. *Physical Review B*, 65, 212107.

Howard-Knight, J. P. and Hobbs, J. K. (2008) Video rate atomic force microscopy using low stiffness, low resonant frequency cantilevers. *Applied Physics Letters*, 93, 104101.

Humphris, A. D. L., Miles, M. J., and Hobbs, J. K. (2005) A mechanical microscope: High-speed atomic force microscopy. *Applied Physics Letters*, 86, 034106.

Hund, F. (1927) On the explanation of molecular spectra I. *Zeitschrift Fur Physik*, 40, 742–764.

Isrealachvili, J. N. (1992) *Intermolecular and Surface Forces*. New York: Academic Press.

Kuramochi, H., Uzumaki, T., Yasutake, M., Tanaka, A., Akinaga, H. and Yokoyama, H. (2005) A magnetic force microscope using CoFe-coated carbon nanotube probes. *Nanotechnology*, 16, 24–27.

Li, X. D. and Bhushan, B. (2002) A review of nanoindentation continuous stiffness measurement technique and its applications. *Materials Characterization*, 48, 11–36.

Liljeroth, P., Repp, J., and Meyer, G. (2007) Current-induced hydrogen tautomerization and conductance switching of naphthalocyanine molecules. *Science*, 317, 1203–1206.

Loppacher, C., Guggisberg, M., Pfeiffer, O., Meyer, E., Bammerlin, M., Luthi, R., Schlittler, R., Gimzewski, J. K., Tang, H., and Joachim, C. (2003) Direct determination of the energy required to operate a single molecule switch. *Physical Review Letters*, 90, 066107.

Lutwyche, M. I. and Wada, Y. (1995) Manufacture of micromechanical scanning tunneling microscopes for observation of the tip apex in a transmission electron-microscope. *Sensors and Actuators A-Physical*, 48, 127–136.

Lutwyche, M., Andreoli, C., Binnig, G., Brugger, J., Drechsler, U., Haberle, W., Rohrer, H. et al. (1999) 5×5 2D AFM cantilever arrays a first step towards a Terabit storage device. *Sensors and Actuators A-Physical*, 73, 89–94.

Maier, S., Gnecco, E., Baratoff, A., Bennewitz, R., and Meyer, E. (2008) Atomic-scale friction modulated by a buried interface: Combined atomic and friction force microscopy experiments. *Physical Review B*, 78, 045432.

Manne, S., Hansma, P. K., Massie, J., Elings, V. B., and Gewirth, A. A. (1991) Atomic-resolution electrochemistry with the atomic force microscope-copper deposition on gold. *Science*, 251, 183–186.

Marti, O., Drake, B., and Hansma, P. K. (1987) Atomic force microscopy of liquid-covered surfaces-atomic resolution images. *Applied Physics Letters*, 51, 484–486.

Mate, C. M., McClelland, G. M., Erlandsson, R., and Chiang, S. (1987) Atomic-scale friction of a tungsten tip on a graphite surface. *Physical Review Letters*, 59, 1942–1945.

Meier, F., Zhou, L. H., Wiebe, J., and Wiesendanger, R. (2008) Revealing magnetic interactions from single-atom magnetization curves. *Science*, 320, 82–86.

Meyer, E. and Bennewitz, R. (2004) *Scanning Probe Microscopy: The Lab on a Tip*. New York: Springer.

Morita, S., YI, I., Sugimoto, Y., Oyabu, N., Nishi, R., Custance, O., and Abe, M. (2005) Mechanical distinction and manipulation of atoms based on noncontact atomic force microscopy. *Applied Surface Science*, 241, 2–8.

Muller, D. J. and Dufrene, Y. F. (2008) Atomic force microscopy as a multifunctional molecular toolbox in nanobiotechnology. *Nature Nanotechnology*, 3, 261–269.

Nishida, S., Kobayashi, D., Sakurada, T., Nakazawa, T., Hoshi, Y., and Kawakatsu, H. (2008) Photothermal excitation and laser Doppler velocimetry of higher cantilever vibration modes for dynamic atomic force microscopy in liquid. *Review of Scientific Instruments*, 79, 123703.

Noguera, C. (1989) Validity of the transfer Hamiltonian approach-Application to the STM spectroscopic mode. *Journal De Physique*, 50, 2587–2599.

Ohnesorge, F. and Binnig, G. (1993) True atomic-resolution by atomic force microscopy through repulsive and attractive forces. *Science*, 260, 1451–1456.

Okawa, Y. and Aono, M. (2001) Materials science-Nanoscale control of chain polymerization. *Nature*, 409, 683–684.

Oppenheimer, J. R. (1928) Three notes on the quantum theory of aperiodic effects. *Physical Review*, 31, 66–81.

Picco, L. M., Bozec, L., Ulcinas, A., Engledew, D. J., Antognozzi, M., Horton, M. A., and Miles, M. J. (2007) Breaking the speed limit with atomic force microscopy. *Nanotechnology*, 18, 044030.

Rohrer, H. (1990) Scanning tunneling microscopy—Methods and variations. In *Scanning Tunneling Microscopy and Related Methods*. Behm, R. J., Garcia, N., and Rohrer, H. (Eds.). Dordrecht, The Netherlands: Kluwer.

Ruska, E. (1987) The development of the electron-microscope and of electron-microscopy. *Reviews of Modern Physics*, 59, 627–638.

Sader, J. E. and Jarvis, S. P. (2004) Accurate formulas for interaction force and energy in frequency modulation force spectroscopy. *Applied Physics Letters*, 84, 1801–1803.

Schrödinger, E. (1926) Quantisation as an eigen value problem. *Annalen Der Physik*, 79, 361–376.

Schrödinger, E. (1952a) Are there quantum jumps, Part I. *The British Journal for the Philosophy of Science*, 3, 109–123.

Schrödinger, E. (1952b) Are there quantum jumps, Part II. *The British Journal for the Philosophy of Science*, 3, 233–242.

Siegenthaler, H. (1992) STM in electrochemistry. In *Scanning Tunneling Microscopy II*. Wisendanger, R. and Guntherodt, H. J. (Eds.). New York: Springer-Verlag.

Stroscio, J. A. and Eigler, D. M. (1991) Atomic and molecular manipulation with the scanning tunneling microscope. *Science*, 254, 1319–1326.

Sugimoto, Y., Pou, P., Abe, M., Jelinek, P., Perez, R., Morita, S., and Custance, O. (2007) Chemical identification of individual surface atoms by atomic force microscopy. *Nature*, 446, 64–67.

Sugimoto, Y., Pou, P., Custance, O., Jelinek, P., Abe, M., Perez, R., and Morita, S. (2008) Complex patterning by vertical interchange atom manipulation using atomic force microscopy. *Science*, 322, 413–417.

Synge, E. H. (1928) A suggested method for extending microscopic resolution into the ultra-microscopic region. *Philosophical Magazine*, 6, 356–362.

Tersoff, J. and Hamann, D. R. (1985) Theory of the scanning tunneling microscope. *Physical Review B*, 31, 805–813.

Tomlinson, G. A. (1929) A molecular theory of friction. *Philosophical Magazine*, 7, 905–939.

Vettiger, P., Despont, M., Drechsler, U., Durig, U., Haberle, W., Lutwyche, M. I., Rothuizen, H. E., Stutz, R., Widmer, R., and Binnig, G. K. (2000) The "Millipede"—More than one thousand tips for future AFM data storage. *IBM Journal of Research and Development*, 44, 323–340.

Welland, M. E. and Gimzewski, J. K. (1995) *Ultimate Limits of Fabrication and Measurement*. Boston, MA: Kluwer Academic Publishing.

Wiesendanger, R. (1994) *Scanning Probe Microscopy and Spectroscopy: Methods and Applications.* Cambridge, U.K.: Cambridge University Press.

Yamashita, H., Kodera, N., Miyagi, A., Uchihashi, T., Yamamoto, D., and Ando, T. (2007) Tip-sample distance control using photo-thermal actuation of a small cantilever for high-speed atomic force microscopy. *Review of Scientific Instruments*, 78, 083702.

Young, R., Ward, J., and Scire, F. (1972) Topografiner-Instrument for measuring surface microtopography. *Review of Scientific Instruments*, 43, 999–1011.

Zlatanova, J., Lindsay, S. M., and Leuba, S. H. (2000) Single molecule force spectroscopy in biology using the atomic force microscope. *Progress in Biophysics & Molecular Biology*, 74, 37–61.

Quantitative Dynamic Atomic Force Microscopy

Robert W. Stark
Ludwig-Maximilians-
Universität München

Martin Stark
Ludwig-Maximilians-
Universität München

33.1 Introduction

Unquestioned, the atomic force microscope (AFM) (Binnig et al. 1986) is capable of surface imaging at very high resolution. However, at its very base, the AFM is an instrument to measure—and to exert—small forces. Typical forces range from one piconewton to several micronewtons. Various measurement modes are used to address specific questions in analytic materials science, physics, chemistry, and life sciences.

A force measurement by AFM includes approaching the tip to the sample surface, touching and deforming the sample, and finally retracting the tip. The forces measured during such a cycle give insights into the details of the interaction between tip and sample. This includes dissipative contributions such as viscous deformation or adhesion, elastic material properties, and magnetic or electrical properties. The test cycle is referred to as "approach–retract curve," "force–distance curve," or simply "nanoindentation." It is a typical AFM experiment because it relates measured forces with distance.

Consecutive force measurements can be used for imaging where the maps of surface properties such as elasticity or adhesion are generated from the approach–retract data. Standard force–distance cycles are performed at repetition frequencies of several Hertz. The main drawback of this approach is the limited data acquisition rate of only 10^{-1} s/pixel. At this data acquisition rate, it takes about 7 h to measure a high-resolution image with 512×512 pixel^2 size. For comparison, let us consider a measurement, where the frequency of the approach–retract cycles coincides with the resonance of the cantilever (few hundred kilo Hertz in air or vacuum). For the acquisition of each data pixel,

one now has to wait until the oscillator equilibrates. The equilibration time thus depends on the resonance frequency and the quality of the oscillator (typically $Q = 400$) and the bandwidth of feedback regulation (typically 1 kHz). In practical applications, the data rate is mainly limited by the controller bandwidth. Thus, about 500 data points per second can be measured. The acquisition time for a 512×512 pixel^2 map is reduced to about 10 min. This measurement principle is an add-on to imaging in a dynamic mode such as tapping or intermittent contact mode.

From a physical point of view, the distinction between the quasi-static force curve and dynamic nanoindentation is artificial because both modes only differ by their operation frequency. From a mathematical point of view, the analysis of the quasi-static force curve is rather simple because the system is operated far away from its resonances. The frequency dependency of the measurement process thus can be neglected. The quantitative analysis of dynamic AFM, however, requires a sophisticated mathematical model that accounts for the system resonances in order to extract material parameters. Nevertheless, the quasi-static case can be understood as a special case of the dynamic nanoindentation.

With this notion in mind, we can compare the concept of quantitative dynamic AFM with other AFM methods as summarized in Table 33.1 and Figure 33.1. The choice of measurement mode depends on the specific application. As is discussed in the following sections, quantitative dynamic AFM provides access to transient events at a time resolution of microseconds. The interaction forces between tip and sample can be measured during routine imaging. By contrast, force volume mapping (Radmacher et al. 1994), which can be understood as sequential nanoindentation, provides full information on

TABLE 33.1 Comparison of Several Force Microscopy Techniques That Can Be Used for Surface Mapping of Elastic Properties

	Mapping Speed	Small Forces	Time Resolution	Transient Events
Quantitative dynamic force microscopy	+ +	+ +	+	+ + (−)[a]
Nanoindentation by force–distance spectroscopy	− −	+[b]	+ +	+ +
Nanoindentation pulsed force mode	+	+[b]	+ + (−)[a]	+ + (+)[a]
Force modulation	+ +	+[b]	−	−

Note: Mapping speed "+ +" means that the sample can be characterized during imaging (typical measurement time 5–10 min for 512 × 512 points), "+" indicates a reduced speed as compared to standard imaging (10–30 min), and "− −" a dramatically reduced image rate (1 h+). The column "small forces" indicates how well the interaction forces can be controlled ("+ +": no difference to standard imaging, "+" slightly increased as compared to imaging), "time resolution" ranks temporal resolution ("+" microseconds, "+ +" time resolution only depends on signal acquisition rate), and "transient events" how well single events such as ruptures can be acquired.

[a] Only if the full signal data is stored. Otherwise time resolution is lost and only specific points of the transients are recorded.

[b] Strongly depends on the choice of cantilever.

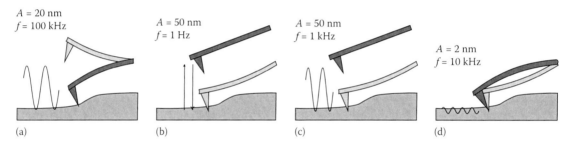

$A = 20$ nm
$f = 100$ kHz

$A = 50$ nm
$f = 1$ Hz

$A = 50$ nm
$f = 1$ kHz

$A = 2$ nm
$f = 10$ kHz

(a) (b) (c) (d)

FIGURE 33.1 Measurement of elastic properties by atomic force microscopy. Typical tip oscillation amplitudes and frequencies are indicated. (a) Dynamic AFM. Surface properties can be extracted from the harmonic distortion of the tip trace. (b) Nanoindentation force curve. The tip is approached to and retracted from the surface to measure the compression of the material. The tip velocity is constant during approach and retract. (c) Pulsed force mode. The tip oscillates sinusoidally to perform nanoindentation. (d) Force modulation. The surface is scanned, while the tip is in permanent contact. The loading force is modulated that leads to a varying indentation into the sample.

the tip-sample interaction at high force resolution but is much slower. Nanoindentation mapping by pulsed force mode (Rosa-Zeiser et al. 1997; de Pablo et al. 1998) is faster than force volume imaging but requires additional electronic circuits. Force modulation microscopy is a variant of contact mode AFM and provides information on the local viscoelasticity (Maivald et al. 1991; Radmacher et al. 1993). Both quantitative dynamic AFM and force modulation are measurement modes, which do not affect the scan speed during imaging because they are an extension of an established imaging mode (tapping or contact mode). By contrast, force distance spectroscopy and pulsed force mode are dedicated force mapping measurement modes.

The following paragraphs discuss a generalized mathematical model for dynamic AFM, which provides a description that includes several eigenmodes. It can serve as a generic model to derive a description of the precise interpretation of sensor data. The approach can also be used to measure time-dependent forces. The concepts outlined in this chapter illustrate how a spectral analysis makes dynamic AFM a time-resolving technique and that it opens the field to measure transient surface forces.

33.2 Modeling Multiple Degrees of Freedom

33.2.1 Overview

In response to the nonlinear interaction forces, the cantilever can undergo a variety of oscillatory motions related to the flexural, torsional, and bending modes (Drobek et al. 1999, 2001; Scherer et al. 1999; Stark et al. 1999, 2001; Reinstaedtler et al. 2003). Obviously, single-degree-of-freedom models are not capable to cover the broadband dynamics of these higher eigenmodes that are relevant to capture time-dependent (transient and high-frequency) contributions, as they occur, for example, in multifrequency techniques (Cuberes et al. 2000; Rodríguez and García 2004; Martinez et al. 2006; Proksch 2006; Platz et al. 2008). These multimodal AFM techniques take advantage of the various cantilever resonances (Kikukawa et al. 1996; Glatzel et al. 2003; Rodríguez and García 2004). For both applications, nano-analytics and nanomanipulation, a numerically efficient model is needed where the cantilever is treated as an extended flexible structure with a sufficiently large number of eigenmodes (Butt and Jaschke 1995; Rabe et al. 1996; Salapaka et al. 1997; Zypman and Eppell 1998; Drobek et al. 1999, 2001; Scherer et al. 1999; Stark et al. 1999, 2001; Reinstaedtler et al. 2003; Raman et al. 2008). Two key elements can be identified that affect the cantilever dynamics and lead to deviations from a one-dimensional harmonic oscillator: first, the nonlinearity of the tip sample contact and second, higher eigenmodes.

Despite of the highly nonlinear interaction in dynamic AFM, the tip oscillation remains periodic under typical imaging conditions (Salapaka et al. 2000). Weak chaos may occur at the transition between attractive and repulsive regime due to a grazing bifurcation (Hu and Raman 2006) or at small tip sample separations (Jamitzky et al. 2006). As an important consequence of the nonlinearities, higher harmonics of the fundamental frequency appear in the spectrum (Cleveland et al. 1998; Dürig 1999, 2000; Hillenbrand et al. 2000; Stark and Heckl 2000; Stark et al. 2000; Sahin and Atalar 2001; Balantekin and Atalar 2003; Hembacher et al. 2004). Also energy dissipation causes a hysteresis of the tip motion and introduces higher frequency components in the spectrum (Dürig 2000; Tamayo et al. 2001; Balantekin and Atalar 2003). Thus, the signal amplitudes of the higher harmonics of a periodic signal generated by a nonlinear system can be used to infer the characteristics of the nonlinearity (elastic and dissipative) by balancing the harmonics of the system input and output (Sebastian et al. 1999, 2001; Wei and Turner 2001; Hu et al. 2004).

33.2.2 Cantilever Model

Usually, only the fundamental mode is taken into account for a theoretical modeling of dynamic AFM. This single-degree-of-freedom or harmonic oscillator approximation allows for the prediction of basic features of the nonlinear dynamics such as the existence of different oscillatory states (Gleyzes et al. 1991; García and San Paulo 2000), the phase lag between driving force and system response (Cleveland et al. 1998), or the pressure at the contact area (Behrend et al. 1998). The cantilever is characterized by its angular resonant frequency ω_0, spring constant k, and the quality factor Q in the harmonic oscillator model. With external force $F(t)$ and the state variables tip displacement x and velocity $\dot{x}$, the equation of motion is

$$\ddot{x}(t) + \frac{\omega_0}{Q}\dot{x}(t) + \omega_0^2 x(t) = \frac{\omega_0^2 F(t)}{k}. \tag{33.1}$$

The effective mass m can be calculated by $m = k/\omega_0^2$. The quality factor Q is related to the damping γ by $Q = 1/(2\gamma)$. In the state-space form, Equation 33.1 can be rewritten as

$$\dot{\mathbf{x}} = \underbrace{\begin{bmatrix} 0 & 1 \\ -\omega_0^2 & -\omega_0/Q \end{bmatrix}}_{\mathbf{A}} \mathbf{x} + \underbrace{\begin{bmatrix} 0 \\ 1 \end{bmatrix}}_{\mathbf{b}} \frac{\omega_0^2}{k} F(t), \tag{33.2}$$

by introducing the state vector $\mathbf{x} = [x, \dot{x}]^T$, the system matrix $\mathbf{A}$, and the input vector $\mathbf{b}$.

The harmonic oscillator approximation holds as long as there are only small contributions to the system dynamics at frequencies above the fundamental resonance (Rodríguez and García 2002). This approximation completely fails, however, to predict the high-frequency response such as transients, generation of higher harmonics, or details of a chaotic response.

Several strategies exist to include higher order modes of the cantilever in the model. A straightforward approach is a finite element (Hirsekorn 1998; Chudoba et al. 1999; Yaralioglu and Atalar 1999; Stark et al. 2001; Arinero and Leveque 2003) or a finite difference analysis (Turner et al. 1997). The dynamics can be directly solved, or cantilever parameters (such as resonant frequencies, spring constant, or damping) can be extracted by calculating eigenvectors and eigenvalues of the finite element model. The modal eigenvalues and vectors can be used as an input for a higher order state-space model.

As an analytic example, we discuss the case of a rectangular cantilever beam, which is used as an AFM sensor. The equation of motion for the flexural vibrations of a freely vibrating and undamped cantilever beam is approximated by the Euler–Bernoulli equation (Clough and Penzien 1993):

$$EI\frac{\partial^4 z(\xi,t)}{\partial \xi^4} + m\frac{\partial^2 z(\xi,t)}{\partial t^2} = 0, \tag{33.3}$$

where

$z(\xi, t)$ is the vertical displacement
$\xi \in [0,1]$ is the normalized position along the cantilever
t is the time
EI is the flexural stiffness
m is the constant mass per unit length

The resonant frequency of the nth eigenmode ω_n is related to the respective eigenvalues k_n by

$$\omega_n^2 = \frac{(k_n)^4 EI}{m}. \tag{33.4}$$

The boundary conditions determine the eigenvectors (modal shapes) $\varphi(\xi)$ of the cantilever; at $\xi = 0$, the beam is clamped and it is free at the other end. The boundary conditions are $\varphi(0) = 0$ for the displacement and $\varphi'(0) = 0$ for the deflection slope. At $\xi = 1$, no external torques or shear forces act, leading to $\varphi''(1) = 0$ and $\varphi'''(1) = 0$. The eigenvalues are calculated by solving the characteristic equation

$$\cos k_n \cosh k_n = -1. \tag{33.5}$$

The asymptotic solution $k_n^{(a)} = (n - (1/2))\pi$ of Equation 33.5 is a good approximation for higher modes ($n > 3$). The eigenvectors of the free cantilever are

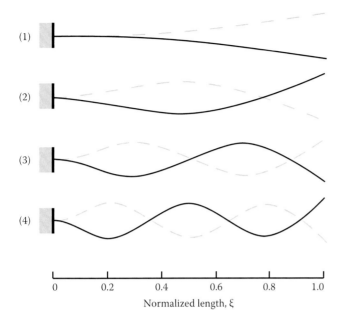

FIGURE 33.2 Modal shapes of the first four flexural eigenmodes of a freely vibrating rectangular cantilever beam.

$$\varphi_n(\xi) = \cos k_n\xi - \cosh k_n\xi - \frac{\cos k_n + \cosh k_n}{\sin k_n + \sinh k_n}(\sin k_n\xi - \sinh k_n\xi).$$

(33.6)

For illustration, the shapes of the first four flexural modes are shown in Figure 33.2. To further simplify the problem, we assume that the tip is massless and that it is located at the free end ($\xi = 1$). We further assume that the laser for the light-lever detection is focused on the free end and that all forces, such as actuation and tip-sample forces, act on the free end. From the modal-bending shape, we obtain the modal displacement $\varphi_n(\xi)$ and the modal deflection angle $\varphi'_n(\xi)$, which corresponds to an idealized light-lever sensor with an infinitely small laser spot. For realistic laser spot geometries, the calculations are more tedious (Stark 2004b; Schäffer and Fuchs 2005).

In order to investigate the system dynamics, we conceive the cantilever as a linear and time invariant (LTI) system, which is subject to a nonlinear output feedback due to the tip-sample interaction (Figure 33.3). The feedback perspective (Sebastian et al. 2001; Stark et al. 2002) allows for a numerically efficient investigation of the system dynamics (Stark 2004a; Stark et al. 2004). The system can be accessed by applying forces $u(t)$ directly as a distributed force (input 1). The other input (2) corresponds to a force directly acting on the tip. Experimentally, driving forces at input (2) can be realized by a magnetic actuation of the cantilever. We will not consider further possible system inputs such as the displacement of the sample or base excitation of the cantilever. Output (1) corresponds to the light lever read-out of the system, that is, the signal that is typically measured and output (2) is the deflection of the free end of the cantilever. The tip deflection $z_{tip}(t)$ (output 2) determines the tip-sample interaction force. This output often cannot be observed because

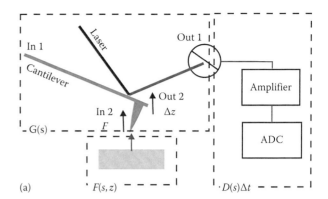

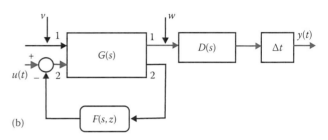

FIGURE 33.3 Schematics of dynamic atomic force microscopy. (a) Inputs and outputs of an AFM. (b) Graphical representation of the dynamic system. The cantilever is represented as an LTI submodel $G(s)$. The tip-sample interaction force is a nonlinear output feedback $F(s, z)$, which also depends on the tip position. The dynamic system of the data acquisition (preamplifier, amplifier, and digitization) can be described by an additional LTI system $D(s)$ and a time delay Δt. (Reproduced from Stark, M. et al., *J. Appl. Phys.*, 98, 114904, 2005. With permission.)

the angle of the cantilever deflection is measured using optical lever detection (output 1).

The state-space form for an N-dimensional system is as given in following equation:

$$\dot{\mathbf{x}} = \mathbf{A}\mathbf{x} + \mathbf{b}u,$$
$$\mathbf{y} = \mathbf{C}\mathbf{x},$$

(33.7)

which is similar to the harmonic oscillator formulation in Equation 33.2. The time-dependent state-vector $\mathbf{x} = (x_1, x_2, \ldots) = (x_{n=1}, \partial_t x_{n=1}, \ldots)$ contains the generalized displacements and velocities of the modes. The $2N \times 2N$ matrix $\mathbf{A}$ is the system matrix, $\mathbf{b}$ the input vector, and scalar u the force input. In the case of multiple inputs, vector $\mathbf{b}$ and scalar u transform into a matrix and a vector, respectively. The output vector $\mathbf{y}$ consists of the tip displacement output y_1, the photo diode signal output y_2, and the tip velocity v_1, which are linear combinations of the system states x_i. The corresponding weights are given by the output matrix $\mathbf{C}$. Note that it is now necessary to introduce an output matrix that distinguishes between different outputs. There is no feed-through between the input and the outputs.

The submatrices of the system matrix $\mathbf{A}$ are constructed using the resonance frequencies $\hat{\omega}_n = \omega_n/\omega_1$ and the modal quality factors Q_n. The dynamics of the *n*th mode is thus described by the modal state-vector $\mathbf{x}_n$, system matrix $\mathbf{A}_n$, and input vector $\mathbf{b}_n$ by

$$\mathbf{x}_n = \begin{bmatrix} x_n \\ \dot{x}_n \end{bmatrix},$$

$$\mathbf{A}_n = \begin{bmatrix} 0 & 1 \\ -\hat{\omega}_n^2 & -\hat{\omega}_n/Q_n \end{bmatrix}, \qquad (33.8)$$

$$\mathbf{b}_n = \begin{bmatrix} 0 \\ \varphi_n(1)/M_n \end{bmatrix}.$$

The modal state vector simply consists of the modal displacement and velocity. The modal system matrix corresponds to that of a harmonic oscillator. The first component of the input vector $\mathbf{b}_n$ is zero, and the second component describes the coupling of an input force to the eigenmode. It is given by the respective modal displacement φ_n at the tip position $\xi_{tip} = 1$, weighted with the respective generalized modal masses $M_n = \int_0^1 m\varphi_n(\xi)^2 \, d\xi$. Here, we have normalized the mass by $M_i = m = 1$.

The dynamics of the *N*-degree-of-freedom cantilever is described by

$$\begin{bmatrix} \dot{\mathbf{x}}_1 \\ \dot{\mathbf{x}}_2 \\ \vdots \\ \dot{\mathbf{x}}_N \end{bmatrix} = \begin{bmatrix} \mathbf{A}_1 & 0 & \cdots & 0 \\ 0 & \mathbf{A}_2 & \cdots & 0 \\ \vdots & \vdots & \ddots & \vdots \\ 0 & 0 & & \mathbf{A}_N \end{bmatrix} \begin{bmatrix} \mathbf{x}_1 \\ \mathbf{x}_2 \\ \vdots \\ \mathbf{x}_N \end{bmatrix} + \begin{bmatrix} \mathbf{b}_1 \\ \mathbf{b}_2 \\ \vdots \\ \mathbf{b}_N \end{bmatrix} u. \qquad (33.9)$$

The output matrix $\mathbf{C}$ combines the system states to the output vector with the tip displacement y_1 and deflection readout y_2. The third channel is the instantaneous velocity of the tip v. The modal contribution to the consolidated output is

$$\mathbf{C}_n = \begin{bmatrix} \varphi_n(\xi_{tip}) & 0 \\ \varphi_n'(\xi_{sens}) & 0 \\ 0 & \varphi_n(\xi_{tip}) \end{bmatrix}. \qquad (33.10)$$

Here, $\xi_{tip} = 1$ and $\xi_{sens} = 1$ are the positions of the tip and detection laser along the cantilever, respectively. The modal bending shapes are given by Equation 33.6. This leads to the system output

$$\begin{bmatrix} y_1 \\ y_2 \\ \dot{y}_1 \end{bmatrix} = \begin{bmatrix} \mathbf{C}_1 & \mathbf{C}_2 & \cdots & \mathbf{C}_N \end{bmatrix} \begin{bmatrix} \mathbf{x}_1 \\ \mathbf{x}_2 \\ \vdots \\ \mathbf{x}_N \end{bmatrix}. \qquad (33.11)$$

The nonlinear interaction between tip and sample can be modeled as an output feedback, where the tip displacement is fed back to input (33.1) through the interaction force $F_{ts}(y_1)$.

33.2.3 Forces between Tip and Sample

The interaction between tip and sample is determined by surface forces, which depend on the distance. The distance between tip and sample is $D = z_s + z$. The scalar z is the tip deflection and the scalar z_s the distance between the undeflected cantilever and the sample. van der Waals forces dominate the interaction in the attractive regime ($D \geq a_0$). In the repulsive regime ($D < a_0$), the tip-sample forces are calculated from the Derjaguin–Muller–Toporov model (Derjaguin et al. 1975) that describes the mechanical interaction between a compressible sphere and a compressible plane. The energy dissipation caused by the tip-sample contact is neglected. To avoid numerical divergence, the parameter a_0 is introduced (García and San Paulo 2000). The tip sample forces are given by

$$F_{ts}(z) = \begin{cases} -HR/[6(z_s + z)^2], & D \geq a_0, \\ -HR/6a_0^2 + \frac{4}{3}E^*\sqrt{R}(a_0 - z_s - z)^{3/2}, & D < a_0, \end{cases} \qquad (33.12)$$

where

H is the Hamaker constant
R is the radius of the tip

The effective contact stiffness is calculated from $E^* = [(1 - v_t^2)/E_t + (1 - v_s^2)/E_s]^{-1}$, where E_t and E_s are the respective elastic moduli and v_t and v_s are the Poisson ratios of tip and sample, respectively. A viscoelastic term

$$F_{vis}(z) = -\eta \dot{z} \sqrt{R(a_0 - z_s - z_0)} \qquad (33.13)$$

can be added in the contact regime in order to account for energy loss due to viscous sample properties.

For very small oscillations around the equilibrium position z_0, Equation 33.12 can be linearized (Rabe et al. 1996)

$$k_{ts}^* = -\frac{\partial}{\partial z} F_{ts}(z) \bigg|_{z=z_0} = \begin{cases} -HR/[3(z_s + z_0)^3] & D \geq a_0 \\ 2E^*\sqrt{R}(a_0 - z_s - z_0)^{1/2} & D < a_0 \end{cases}. \qquad (33.14)$$

Here, the contact stiffness k_{ts}^* is normalized to the cantilever spring constant k by $\hat{k}_{ts} = k_{ts}^*/k$.

33.3 Dynamics of AFM

33.3.1 Linearized Tip-Sample Interaction

Attractive and repulsive interaction forces induce frequency shifts of the modes. It is very instructive to discuss the case of very small oscillations, which allows one to use the linearized Equation 33.14. In this case, the output feedback is directly proportional to the system's position output (2). The elastic surface properties can thus be conceived as a proportional feedback with

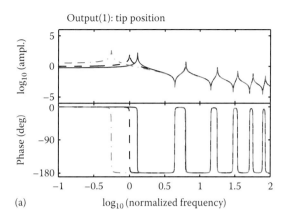
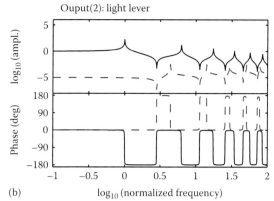

(a) $\log_{10}$ (normalized frequency)   (b) $\log_{10}$ (normalized frequency)

FIGURE 33.4 Frequency response (Bode plot) of a rectangular cantilever beam. (a) Position output (deflection). Moderately attractive (dash dot, $k_{ts} = -0,7$) or repulsive (solid, $k_{ts} = 0,7$) interaction forces shift the resonances to lower or higher frequencies, respectively. (b) Idealized light lever readout. The resonances can even be measured for a strong surface coupling (dashed, $k_{ts} = 10^5$). (Adapted from Stark, R.W. et al., *Phys. Rev. B*, 69, 085412, 2004. With permission.)

the gain parameter $\hat{k}_{ts}$. The influence of the tip-sample interaction on amplitude and phase response is illustrated by the Bode plots in Figure 33.4a. The linearized van der Waals interaction corresponds to a spring with a negative spring constant. The dynamic system softens and the resonant frequencies (dash-dotted) shift to lower frequencies. By contrast, repulsive tip-sample interaction forces correspond to springs with positive force constants, the interaction hardens the system (solid). The frequency shift depends on the stiffness ratio $\hat{k}_{ts}/m\hat{\omega}_n^2$, which rapidly decreases with an increasing mode number n. The resonant frequencies depend on the gain factor of the output feedback, but they are not affected by the choice of the output channel. By contrast, the frequencies of the transmission minima are different for the position and the light lever output.

For a very stiff tip-sample contact with $\hat{k}_{ts} = 10^5$, the system corresponds to a beam with a pinned end (Figure 33.4). The resonant frequencies of the pinned system are the frequencies of the transmission minima of the free system as measured by the displacement output. The pinned resonances still can be measured in the light lever output. In the case of a pinned tip, the tip does not move but the cantilever itself oscillates, which leads to a varying slope at the end of the beam.

This effect illustrates that more generally, the poles (resonances) of a constrained system correspond to the zeros (antiresonances) of the free system (Miu 1993). In the constrained system, poles and zeros cancel for the position output, but not for the deflection angle. A physical interpretation is straightforward: the cantilever was modeled as a beam that is actuated by a force at the free end. However, there is now an additional stiff spring

attached to the free end, which directly counteracts an external driving force.

These examples illustrate that a very precise definition of the system inputs and outputs is essential in order to obtain a realistic description of the system dynamics. This means for the experimentalist that the system transfer characteristics depend on how the tip displacement is measured and how the forces couple to the sensor.

33.3.2 Nonlinear Dynamics

For larger oscillations, the nonlinearity in the tip-sample contact leads to a more complex dynamics. This includes the generation of higher harmonics, the existence of different regimes (touching vs nontouching), or chaos. An in-depth discussion of the nonlinear dynamics is beyond the scope of this chapter. In the following paragraphs, we will highlight only a few consequences of the nonlinearity: the existence of two distinct oscillation regimes and the role of noise.

For the numerical simulations, a rectangular cantilever interacting with the surface as described by the nonlinear force law of Equation 33.12 was assumed (Stark et al. 2004). The simulations were based on Equations 33.8 through 33.14 and were implemented in MATLAB® Release 2008a and Simulink® (The Mathworks, Inc., Natick, Massachusetts). The model parameters are summarized in Table 33.2. The interaction of a silicon tip with a glass surface was simulated numerically. In the repulsive regime, a small energy loss was assumed. For the numerical simulation of the nonlinear dynamics, a model with $N = 3$ modes was used. The

TABLE 33.2 Model Parameters

Resonant Frequency	Spring Constant	Modal Quality Factor	Young's Modulus Tip/Sample	Poisson Ratio Tip/Sample	Tip Radius	Hamaker Constant	Parameter	Viscosity
ω_0	k	Q_n	E_t	ν_t	R	H	a_0	η
1	10 N/m	200	129 GPa, 70 GPa	0.28/0.17	20 nm	6.4 e–20 J	0.166 nm	500 Pa s

Note: For the numerical simulation, typical parameters for tapping mode with a silicon cantilever were chosen and a uniform modal damping Q was used for all modes. Sample parameters for fused silica SiO_2 were assumed.

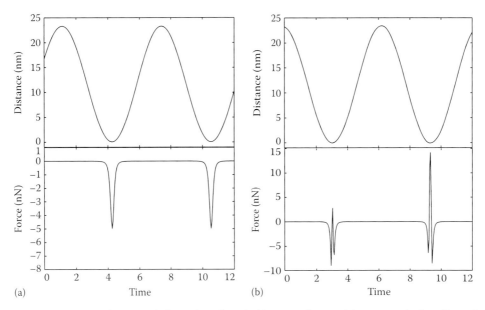

FIGURE 33.5 Time domain tip trace and tip-sample forces (simulation). (a) Tip oscillation of the unperturbed oscillator. Only attractive (negative) forces occur. (b) Tip oscillation with additional noise. Although the noise cannot be seen directly in the time traces, the tip sample forces are strongly affected. The forces are repulsive (positive sign) and fluctuate.

time was normalized to the fundamental resonance. One oscillatory cycle thus corresponds to $T = 2\pi$. The resonance frequencies were $\omega_1 = 1.0$, $\omega_2 = 6.2669$, and $\omega_3 = 17.5475$.

Two simulations were carried out. In the first, a perfect experiment without noise was simulated; in the second, an additional stimulus by a random force was assumed. Such a random force may be caused by thermomechanical forcing (Brownian motion), noise in the electronic circuits, or mechanical vibrations. For the simulation of noise, a band-limited white noise was assumed. The time trace of the oscillation and the tip sample interaction are plotted in Figure 33.5. The solutions without noise (Figure 33.5a) and with noise (Figure 33.5b) differ in the interaction forces. Without noise, only negative (attractive) forces occur. By contrast, in the presence of noise also positive (repulsive) forces occur. The noise cannot be seen in the time traces of the tip position, but it seriously affects the tip sample interaction. The interaction is not periodic and the interaction forces fluctuate. Figure 33.6a shows the simulated evolution of amplitude and phase of the fundamental oscillation. Without noise, the system remains in the nontouching regime and no energy is dissipated. With the additional noise, amplitude and phase show the well-known transition to the repulsive regime that is accompanied by energy dissipation (Schirmeisen et al. 2003). The transition can well be identified by the phase jump and the onset of energy dissipation. The physical reason for this transition is that two stable solutions can coexist, one in the noncontact (small amplitude) and the other in the repulsive regime (large amplitude) (San-Paulo and García 2002). In the case of a bistable dynamics, it depends on the initial conditions, which of both solutions is realized. During approach, the basins of attraction for both solutions change and more and more of the phase space coordinates belong to the repulsive regime (García and San Paulo 2000). This

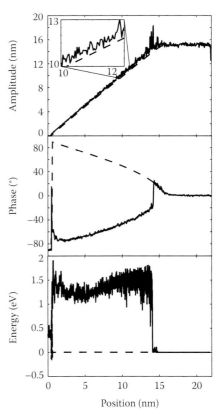

FIGURE 33.6 Amplitude and phase of the fundamental mode and average energy dissipated per oscillatory cycle (simulation). The results of two numerical simulations are plotted: without (dashed) and with noise (solid line). The inset in the amplitude curve illustrates the offset between the noncontact and the contacting solution. Note that the phase shift is given with respect to the phase of the undisturbed oscillation.

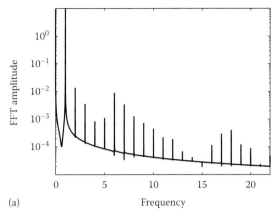
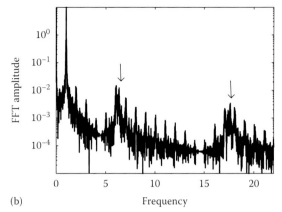

(a)   Frequency   (b)   Frequency

FIGURE 33.7 Fast Fourier transform (FFT) of the position output at a set point of $A/A_0 = 80\%$. (a) Without and (b) with additional noise. The arrows indicate the second and third flexural eigenmodes, which were excited by the white noise.

means that the nontouching solution becomes less and less stable against perturbations. At a certain point, the variations in position and amplitude due to the additional noise are sufficiently large to induce a transition from one solution to the other.

The difference between both oscillatory states can also be seen in the FFT spectra in Figure 33.7. In a noise-free numerical experiment, the noise level was small and well-defined higher harmonics prevailed. The additional noise led to an increased background. Moreover, the second and third modes were excited to a random oscillation and now occur in the spectrum. The higher order harmonics were much stronger in the repulsive regime (for the 21st harmonic nearly one order of magnitude). This difference was caused by hard repulsive interaction forces, which lead to a significant high-frequency response.

The numerical forward simulation shows that the interaction force is encoded in the spectral response. For the experimentalist, also the inverse problem is of high interest. The spectral response of the oscillating cantilever is given and the interaction force is to be reconstructed from the signal. We shall discuss the solution of the inverse problem in Section 33.4.

Recently, VEDA (http://www.nanohub.org/learningmodules/veda as of October 7, 2008), a web-based simulator for dynamic AFM was released (Melcher et al. 2008). VEDA is capable of treating several higher modes. The simulation can include experimentally relevant parameters such as viscoelastic and hysteretic energy dissipation and a liquid environment just to mention a few. We recommend that the readers program an example in MATLAB (or similar software) themselves or use the platform VEDA to run their own simulations in order to further explore the nonlinear dynamics of AFM.

33.4 Reconstruction of the Interaction Forces

33.4.1 Overview

Time-dependent forces mediate adsorption, ordering phenomena, and viscoelasticity. These forces are important in rheology and tribology, as well as in biology and catalysis. The importance

of dynamic aspects becomes obvious when looking at the viscoelastic properties of polymers (Wilhelm 2002). Even on the level of a single biomolecule under external stress, velocity dependence can be observed: the stability of the molecule increases with the applied force rate (Evans and Ritchie 1997). Dynamic forces occurring during mechanical contact of surfaces, however, are experimentally not easily accessible.

In this context, various modes of dynamic AFM offer a large potential to mechanically investigate local material behavior at (sub-) microsecond timescales. As current techniques only allow for the measurement of effective forces as average quantities, the time course has to be estimated from models (Giessibl 1997; Hölscher et al. 2000; Hölscher and Anczykowski 2005). As discussed in the preceding sections, it is exactly the information encoded in the anharmonic contributions that contains the duration and the strength of the interaction. The nonlinear interaction generates higher harmonics of the fundamental oscillation, which are resonantly enhanced to significant signal contributions by higher eigenmode excitation. In the following paragraphs, we will discuss the inverse problem: the reconstruction of the effective force at the tip from measured data without a priori assumptions regarding the interaction forces.

33.4.2 Signal Formation in AFM

System theory provides a convenient formalism to describe the relation between the force input and the measured signal output. In the Laplace domain, with the Laplace variable $s = \sigma + j\omega$, the continuous transfer function of the force sensor is defined by

$$G(s) = \frac{y(s)}{u(s)}. \tag{33.15}$$

This characterizes the relation between the force sensor output, $u(s)$, and the input force acting on the tip, $y(s)$. The transfer function can be approximated theoretically by various methods: an infinite dimensional model (Spector and Flashner 1990; Yuan and Liu 2003), a truncated model (Rabe et al. 1996; Stark et al. 2004), or a discrete approximation (Arinero and Leveque 2003).

The naive application of theoretical models as discussed in the preceding chapters is hampered by significant deviations of the actual cantilever geometry from the idealized geometry and difficulties due to the finite spot size of the detection laser (Stark 2004b; Schäffer 2005; Schäffer and Fuchs 2005). This means that the transfer function has to be estimated from experimental data, since theoretical models may fail to provide sufficient accuracy.

The signal flow diagram of dynamic AFM exhibits two parts, as indicated in Figure 33.3b. The nonlinear circuit (lower part) represents the interaction $F(s, z)$ that couples back on the force–distribution input $u(t)$. The measurement of the cantilever motion relies on a second, linear path, wherein the linear operator D represents the detection (i.e., the photodiode and the electronics). It converts the bending angle $\varphi'(x_{tip}, t)$ into the signal $y(t)$. Here, x' denotes the position of measurement along the cantilever.

For both operators, G and D, linearity is an appropriate assumption because in a typical AFM-experiment cantilever, deflections from the equilibrium are small and only elastic waves with wavelengths larger than $0.1l$ (cantilever length l) have to be considered. Under these conditions, the linear equation

$$P(s) = G(s)D(s)\exp(-s\Delta t) \qquad (33.16)$$

describes the signal formation path in dynamic AFM. The relation between system input and output is given by

$$P(s) = \frac{y(s)}{u(s)}. \qquad (33.17)$$

Thus, the operator $P(s)$ maps the time trace of the force into the time trace of the signal. Specifically, one physical value is mapped on exactly one signal value and vice versa, which is an important requirement for a sensor. Linearity of the operator $P(s)$ allows for the description of dynamic AFM in the framework of linear response theory.

To estimate the system input $u(t)$ from the system output $y(t)$, the transfer function has to be inverted. Although tempting, such an inversion $P^{-1}(s)$ is not straightforward. First, the time delay has to be separated from the transfer function. Second, the bounded input bounded output criterion (BIBO stability) has to be fulfilled. Third, for a real world implementation, the resulting transfer function has to be causal; that is, the system response has to follow the stimulus. In simple words, BIBO stability means that as long as the input is a stable signal, we are guaranteed to have a stable output. Mathematically, this means that a system C with an impulse response $h(t)$ that fulfills the BIBO criterion is absolutely integrable

$$\int_{-\infty}^{\infty} |h(t)|\mathrm{d}t < \infty. \qquad (33.18)$$

For the rational transfer function $C(s)$, this is only the case if both poles and zeros are located in the left half plane, because

due to the inversion, the roles of poles and zeros are exchanged. An AFM force sensor, however, can also exhibit nonminimum phase response, which means that there are zeros in the right half plane (Vazquez et al. 2007).

The presence of noise poses an additional problem. The transfer characteristics exhibit weakly damped antiresonances that transform into resonances of the inverted system. In these frequency bands, small model errors or additional noise lead to large measurement errors. The same is true for signals beyond the cut-off frequency of the low-pass characteristics of $D(s)$. Thus, the bandwidth available for data analysis is limited.

33.4.3 Identification of the Transfer Function

To reconstruct the time trace of the force from the time series of the deflection signal obtained in the experiment, the transfer function $P(s)$ is essential. $P(s)$ can be determined by system identification procedures that rely on the analysis of the system response to a well-defined stimulus (Ljung 1999). The force input has to act on the tip of the free cantilever, and should contain all frequencies. A full spectral characterization of the AFM system allows for a virtual real-time representation of AFM experiments (Couturier et al. 2001), for fast and efficient dynamic control (Sulchek et al. 2000; Sahoo et al. 2003), for nanorobotics (Guthold et al. 2000; Stark et al. 2003), and for time-resolved force measurements (Stark et al. 2002; Todd and Eppell 2003).

In a typical calibration procedure, the static relation between the cantilever deflection and the photo diode signal has to be determined. The sensitivity is usually estimated from data of force-curve experiments on hard substrates. The spring constant is the second key parameter that has to be determined. For this purpose, various methods can be used. Among the most common techniques are the thermal noise method (Hutter and Bechhoefer 1993; Butt and Jaschke 1995; Burnham et al. 2003; Proksch et al. 2004), the added mass method (Cleveland et al. 1993), the Sader method (Sader et al. 1999), and calibration procedures employing a reference spring (Gibson et al. 1996; Torii et al. 1996; Cumpson et al. 2004). So far, system resonances (Rabe et al. 1996) as well as the mechanical impedance in the frequency range below 100 kHz (Scherer et al. 2000) were determined for a dynamic characterization of the cantilever by external excitation.

The experimental estimation of the full transfer function of a force sensor used in AFM goes far beyond these calibration procedures. The need for a priori knowledge is reduced by system intrinsic measurements, which are based on the analysis of discontinuous events in standard force–displacement curves.

Figure 33.3 sketches the AFM cantilever in interaction with the sample surface and the corresponding representation as a dynamic system with an output feedback (Sebastian et al. 2001; Stark et al. 2002, 2004). While the forces between tip and sample play the role of a nonlinear output feedback (lower branch), the signal forming path is modeled as a linear time invariant system (upper branch). The experimentally relevant signal path in

Figure 33.3 leads from the tip–sample forces to the sensor read-out, described by the transfer function of the entire signal path

$$P(s) = G_{21}(s)D(s)\exp(-s\Delta t), \tag{33.19}$$

which also includes a time delay $\exp(-s\Delta t)$. A time delay can be introduced by analog to digital conversion or may be caused by a finite traveling time of a signal from an input to a distant output. Here, input 2 (located at the tip) and output 1 (located at the laser spot) are collocated on the cantilever. Thus, the traveling time of the mechanical wave from input 2 to output 1 can be neglected. For convenience, the transfer function $P(s)$ is split into two components: the rational transfer function

$$G_D(s) = G(s)D(s) \tag{33.20}$$

and the time delay $\exp(-s\Delta t)$. The transfer function $G_D(s)$ describes the dynamic characteristics of the microcantilever together with the detection system. The time delay is treated as a separate parameter.

In the Laplace domain, the forces between tip and sample $u(s)$ translate into the signal $y(s)$ by

$$y(s) = \frac{k_{ols}}{k_c} G_D(s)\exp(-s\Delta t)u(s). \tag{33.21}$$

This equation contains the parameters that have to be determined for a full dynamic calibration: the static optical lever sensitivity k_{ols}, which is usually obtained from quasi-static force curve data, the spring constant k_c as obtained by standard methods, the normalized transfer function $G_D(s)$, and the time delay Δt.

The theoretical background for estimation of an empirical transfer function estimate (ETFE) is only briefly recalled here, a detailed treatment can be found in Chapter 6 by Ljung (1999). The basic idea of this parameter-free estimation procedure is the direct application of Equation 33.15 in order to calculate $G(s)$ without further physical assumptions. Because fast Fourier transformation provides an efficient numerical tool for this purpose, the transfer function is calculated in the Fourier domain. The Fourier transformed of the discrete system output is

$$Y_N(\omega) = \frac{1}{\sqrt{N}} \sum_{t=1}^{N} y(t)e^{-i\omega t}, \tag{33.22}$$

with the time series data, $y(t)$, of length N. The system input is given by

$$U_N(\omega) = \frac{1}{\sqrt{N}} \sum_{t=1}^{N} u(t)e^{-i\omega t}. \tag{33.23}$$

In the following, we determine the ETFE, defined by

$$G_N(e^{i\omega}) = \frac{Y_N(\omega)}{U_N(\omega)}. \tag{33.24}$$

This definition assumes that $U_N(\omega) \neq 0$ for all $\omega < \omega_{bw}$. Thus, the ETFE is not defined for frequencies where $U_N(\omega) = 0$. The estimate for the transfer function is referred to as empirical because the only assumption is the linearity of the dynamic system.

For the experimental estimation of the transfer function $G_D(s)$, it is essential to apply a force stimulus to the tip. In this case, forces between tip and sample acting close to the free end couple with nearly equal efficiency to the cantilever. Other stimulating forces may not act directly on the tip or may even be distributed over the structure. Such distributed loads can be produced, for example, by inertial excitation or by electrostatic fields.

The jump-out-of-contact response of the AFM cantilever $y(t)$ provides a signal that was used to estimate the transfer characteristics of the cantilever (Figures 33.8 and 33.9). The advantage of this concept is that it dispenses with an external excitation, where the response of the external transducer (e.g., the driving piezo) and its coupling to the cantilever would also have to be determined. In the jump-out-of-contact, the system input, $u(t)$, essentially corresponds to a step function with a negative force, $F = -F_{adh}$, as long as the cantilever is attached to the specimen and $F = 0$ after the rupture event. Thus, the cantilever is loaded by a point force acting on the tip before the cantilever is released. At the rupture event, the force load at the tip is zero. Thus, the system intrinsic feedback in Figure 33.8b is switched off for system identification. Ensembles of several snap-off contact events were extracted from the data by iterative correlation averaging. An average rupture event was calculated and served as basis for the subsequent analysis. The rupture event of the snap-off contact was considered an input force by extrapolating the signal without force load long after the rupture to the moment of rupture represented zero force. Extrapolating the linear slope before the event, that is, the adhesion keeping the cantilever at the surface, to the moment of rupture allowed for estimation of the quasi-static force load culminating at −330 nN at the moment of rupture. The resulting time trace of the force load was estimated combining both extrapolations (Figure 33.8d). The rupture of the polymer in the tip–sample contact at the jump-out-of-contact time t_0 could not be measured directly. Because theoretical models do not predict a time delay in the mechanical part of the collocated system (Spector and Flashner 1990), the time t_0 was adjusted to the largest possible value compatible with a causal transfer function of the identified system $G_D(s)$. An uncertainty regarding the time delay Δt of about five samples ($\pm 0.5\,\mu s$) remained.

As first step of the identification procedure, the periodograms of the input $U(\omega)$ and the output $Y(\omega)$, which are estimates of the respective power spectral density, were calculated (data not shown). The ETFE G_N was calculated from the windowed data from Equation 33.24 (Tukey window). The dimensionless magnitude was normalized to unity for a quasi-static input. The corresponding Bode plot is shown in Figure 33.9.

The ETFE is highly reliable at the resonances, while the anti-resonances are difficult to detect due to the low signal level at those frequencies. Appropriate local smoothing routines improve the quality of the ETFE. Here, we applied local polynomial smoothing to the real and imaginary parts of the ETFE.

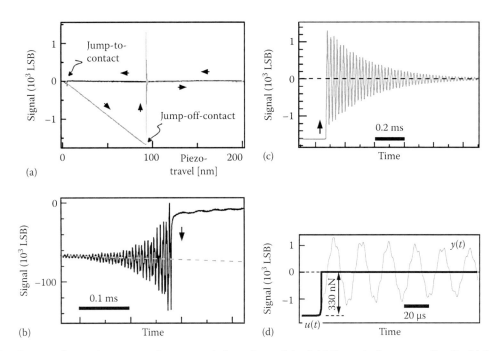

FIGURE 33.8 Cantilever oscillations in an approach retract cycle (experimental data). (a) The entire force curve. Details: (b) Oscillations induced by the snap to contact. (c) Oscillations of the free cantilever after snap-off from the surface. (d) The averaged oscillations after snap-off (thin line) and the estimate for the tip force (thick line); (1 LSB ≈ 0.977 mV). (Reproduced from Stark, M. et al., *J. Appl. Phys.*, 98, 114904, 2005. With permission.)

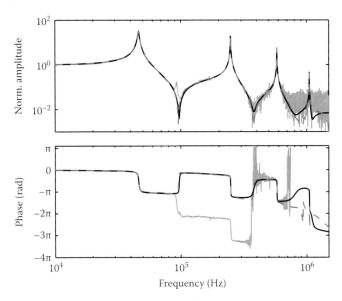

FIGURE 33.9 ETFE (gray), smoothed ETFE (gray, dashed), and an 11th order parametric estimate (black) of the transfer function of a free v-shaped cantilever. (Reprinted from Stark, M. et al., *J. Appl. Phys.*, 98, 114904, 2005. With permission.)

Along with this smoothing procedure, the peak at 10^5 Hz was removed from the data. The Bode plot of the smoothed ETFE is displayed in Figure 33.9. For comparison, the transfer function as obtained by a parametric estimation procedure is also shown (Stark et al. 2005).

The transfer function of the free cantilever is a sensor description that is independent of the interaction force. The nonlinear

interaction force is considered as a system-external feedback; that is, the force acts as input into the system. Thus, describing the surface-coupled cantilever (contact mode) by a linear transfer function means approximating the nonlinear feedback by a linear system. Such a simplification is only valid in the case of very small oscillatory amplitudes. In other cases, the system identification problem of the nonlinear closed loop system (cantilever and sample) requires a much more elaborate mathematical treatment.

33.4.4 Signal Inversion

In the final step of the analysis, the signal is subject to a purely linear transformation as defined by the transfer function. The time trace of the measured AFM signal $s(t)$ was split into consecutive windows with a length compatible to FFT. For each window, the force $F_N(\omega)$ was calculated in the Fourier domain by

$$F_N(\omega) = \frac{S_N(\omega)}{G_N(\omega)}. \tag{33.25}$$

Here, $G_N(\omega)$ is the ETFE as discussed in the preceding section and $S_N(\omega)$ is the FFT of the AFM sensor signal measured during imaging. The time trace of the force $f(t)$ was determined by inverse FFT (Figure 33.10).

Noise degraded the signal and thus the analysis had to exclude frequency bands with unreliable information. In critical frequency bands, $G_N(e^{i\omega}) \ll 1$. In these frequency bands, the signal became small and could fall below the noise level. Accordingly, this information was considered as unreliable and excluded from

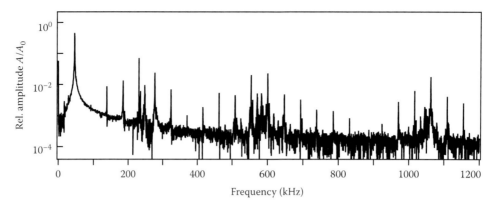

FIGURE 33.10 FFT of a time series signal measured on silicon. The amplitude is normalized to the free oscillation ($A/A_0 = 44.6\%$). Higher harmonics occurred due to the nonlinear tip-sample interaction. (Reprinted from Stark, M. et al., *Proc. Natl. Acad. Sci. USA*, 99, 8473, 2002. With permission.)

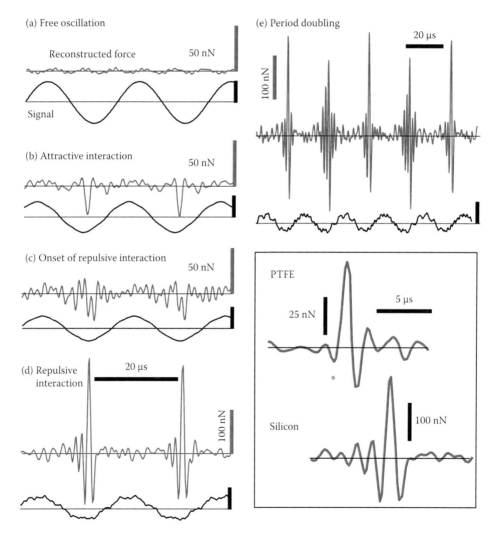

FIGURE 33.11 Time traces of force and signal. The signal (scale bar: amplitude of the free oscillation) is given beneath the corresponding force graph. Various oscillatory regimes can be identified: (a) free oscillation, (b) attractive interaction, (c) onset of repulsive interaction, and (d) dominant repulsive interaction. During approach, the strength of the attractive forces increases together with the contact time (3.5 µs). (e) Period doubling occurs at a small tip sample gap. (Inset) Force pulse on silicon with energy dissipation due to capillary forces and on PTFE with viscous energy dissipation. (Adapted from Stark, M. et al., *Proc. Natl. Acad. Sci. USA*, 99, 8473, 2002. With permission.)

the signal inversion, even though these frequencies contributed to the original input. Nevertheless, the reconstructed time trace of the force *f*(*t*) is a good estimate for the true time trace because the missing information can be extrapolated from adjacent data.

Although average forces are small—in this example, the average force remains smaller than 2 nN—impact events may reach large peak forces. Figure 33.11 shows force *f*(*t*) and signal *s*(*t*) of typical oscillation regimes. Data are representative for: (a) free oscillation, (b) attractive interaction, (c) onset of mechanical contact, (d) dominantly repulsive interaction, and (e) period doubling. All traces are corrected for the external excitation, which resulted in a sinusoidal force with 5 nN amplitude at the tip. Because of the limited bandwidth, one underestimates peak heights, and fast events may be smoothed.

Comparing the impact events on silicon and PTFE (Teflon™) under similar conditions demonstrates differences in response to the material (Figure 33.11b). The asymmetry due to viscous energy dissipation on the polymer sample (PTFE) is evident (see asterisk). Additionally, the repulsive forces are significantly smaller on PTFE, while the force pulse is broader. This is caused by the elastic response of the polymer. The force pulse on silicon indicates a pronounced influence of wetting by a water film due to humidity. Basically, the signature of the impacts reflects the nonlinear interaction similar to quasi-static force curves.

33.5 Conclusion

In this chapter, we have discussed a multiple-degree-of-freedom model of dynamic AFM. Such a model is essential for the investigation of the nonlinear dynamics and to understand the signal formation in tapping mode (amplitude modulation) and other dynamic measurement modi. The basic idea of the approach is a feedback perspective on the AFM. The cantilever is described as an LTI system, which is subject to a nonlinear output feedback due to the interaction between tip and sample. This concept can also be used to solve the inverse problem—the estimation of the tip-sample forces from the signal. Such an approach, however, requires a mathematical model for the transfer characteristics of the linear system components. This goal can be accomplished by system identification procedures, which yield parameter-free or parametric estimates. With such a dynamic calibration, a full spectral analysis of the signal makes dynamic AFM a time-resolving technique. Surface forces, both attractive and repulsive ones, can be measured with high temporal resolution.

So far, we have discussed the reconstruction of forces measured with a standard cantilever. For routine application of this method, the transmission minima of the cantilever transfer function pose a mathematical challenge because small measurement errors may lead to large errors in the estimated force. To overcome this limitation, the torsional harmonic cantilever has been introduced recently (Sahin et al. 2007, 2008; Sahin 2008). The basic idea of this cantilever design is to translate the tip sample forces mechanically from the flexural vibrations into a small torsional vibration before digital signal processing is carried out. Because the resonant frequency of a torsional vibration is much

larger than that of a flexural vibration (typically one order of magnitude), 20 or more harmonics of the fundamental flexural oscillation can be measured in the torsional signal without signal suppression due to transmission minima. This makes the system identification procedure much more tolerant against measurement errors and much more easy to use for the experimentalist. We thus expect that the approach discussed in this chapter will be widely adopted. Fast nanoscale mapping of elastic sample properties will allow for new insights into the surface properties of thin polymer films, naoparticles, and molecules.

References

Arinero, R. and G. Leveque. 2003. Vibration of the cantilever in force modulation microscopy analysis by a finite element model. *Rev. Sci. Instrum.* 74: 104–111.

Balantekin, A. and A. Atalar. 2003. Power dissipation analysis in tapping-mode atomic force microscopy. *Phys. Rev. B* 67: 193404.

Behrend, O.P., F. Oulevey, D. Gourdon et al. 1998. Intermittent contact: Tapping or hammering? *Appl. Phys. A* 66: S219–S221.

Binnig, G., C.F. Quate, and C. Gerber. 1986. Atomic-force microscope. *Phys. Rev. Lett.* 56: 930–933.

Burnham, N.A., X. Chen, C.S. Hodges et al. 2003. Comparison of calibration methods for atomic-force microscopy cantilevers. *Nanotechnology.* 14: 1–6.

Butt, H.J. and M. Jaschke. 1995. Calculation of thermal noise in atomic-force microscopy. *Nanotechnology.* 6: 1–7.

Chudoba, T., N. Schwarzer, and F. Richter. 1999. New possibilities of mechanical surface characterization with spherical indenters by comparison of experimental and theoretical results. *Thin Solid Films* 355–356: 284–289.

Cleveland, J.P., S. Manne, D. Bocek, and P.K. Hansma. 1993. A nondestructive method for determining the spring constant of cantilevers for scanning force microscopy. *Rev. Sci. Instrum.* 64: 403–405.

Cleveland, J.P., B. Anczykowski, A.E. Schmid, and V.B. Elings. 1998. Energy dissipation in tapping-mode atomic-force microscopy. *Appl. Phys. Lett.* 72: 2613–2615.

Clough, R.W. and J. Penzien. 1993. *Dynamics of Structures.* Singapore: McGraw-Hill.

Couturier, G., J.P. Aime, J. Salardenne, and R. Boisgard. 2001. A virtual non contact-atomic force microscope (nc-afm): Simulation and comparison with analytical models. *Eur. Phys. J. Appl. Phys.* 15: 141–147.

Cuberes, M.T., H.E. Assender, G.A.D. Briggs, and O.V. Kolosov. 2000. Heterodyne force microscopy of pmma/rubber nanocomposites: Nanomapping of viscoelastic response at ultrasonic frequencies. *J. Phys. D Appl. Phys.* 33: 2347–2355.

Cumpson, P.J., P. Zhdan, and J. Hedley. 2004. Calibration of afm cantilever stiffness: A microfabricated array of reflective springs. *Ultramicroscopy* 100: 241–251.

de Pablo, P.J., J. Colchero, J. Gomez-Herrero, and A.M. Baro. 1998. Jumping mode scanning force microscopy. *Appl. Phys. Lett.* 73: 3300–3302.

Derjaguin, B.V., V.M. Muller, and P. Toporov Yu. 1975. Effect of contact deformations on the adhesion of particles. *J. Coll. Interf. Sci.* 53: 314–326.

Drobek, T., R.W. Stark, M. Gräber, and W.M. Heckl. 1999. Overtone atomic-force microscopy studies of decagonal quasicrystal surfaces. *New J. Phys.* 1: 15.

Drobek, T., R.W. Stark, and W.M. Heckl. 2001. Determination of shear stiffness based on thermal noise analysis in atomic force microscopy: Passive overtone microscopy. *Phys. Rev. B* 64: 045401.

Dürig, U. 1999. Relation between interaction force and frequency shift in large-amplitude dynamic force microscopy. *Appl. Phys. Lett.* 75: 433–435.

Dürig, U. 2000. Interaction sensing in dynamic force microscopy. *New J. Phys.* 2: 5.

Evans, E. and K. Ritchie. 1997. Dynamic strength of molecular adhesion bonds. *Biophys. J.* 72: 1541–1555.

García, R. and A. San Paulo. 2000. Dynamics of a vibrating tip near or in intermittent contact with a surface. *Phys. Rev. B* 61: R13381–R13384.

Gibson, C.T., G.S. Watson, and S. Myhra. 1996. Determination of the spring constants of probes for force microscopy/spectroscopy. *Nanotechnology.* 7: 259–262.

Giessibl, F. 1997. Forces and frequency shifts in atomic-resolution dynamic-force microscopy. *Phys. Rev. B* 56: 16010–16015.

Glatzel, T., S. Sadewasser, and M.C. Lux-Steiner. 2003. Amplitude or frequency modulation-detection in kelvin probe force microscopy. *Appl. Surf. Sci.* 210: 84–89.

Gleyzes, P., P.K. Kuo, and A.C. Boccara. 1991. Bistable behavior of a vibrating tip near a solid surface. *Appl. Phys. Lett.* 58: 2989–2991.

Guthold, M., M.R. Falvo, W.G. Matthews et al. 2000. Controlled manipulation of molecular samples with the nanomanipulator. *IEEE/ASME Trans. Mech.* 5: 189–198.

Hembacher, S., F.J. Giessibl, and J. Mannhart. 2004. Force microscopy with light-atom probes. *Science.* 305: 380–383.

Hillenbrand, R., M. Stark, and R. Guckenberger. 2000. Higher harmonics generation in tapping-mode atomic-force microscopy: Insight into tip sample interaction. *Appl. Phys. Lett.* 76: 3478–3480.

Hirsekorn, S. 1998. Transfer of mechanical vibrations from a sample to an afm-cantilever-a theoretical description. *Appl. Phys. A* 66: S249–S254.

Hölscher, H. and B. Anczykowski. 2005. Quantitative measurement of tip-sample forces by dynamic force spectroscopy in ambient conditions. *Surf. Sci.* 579: 21–26.

Hölscher, H., A. Schwarz, W. Allers, U. Schwarz, and R. Wiesendanger. 2000. Quantitative analysis of dynamic-force-spectroscopy data on graphite(0001) in the contact and noncontact regimes. *Phys. Rev. B* 61: 12678–12681.

Hu, S.Q. and A. Raman. 2006. Chaos in atomic force microscopy. *Phys. Rev. Lett.* 96: 036107.

Hu, S.Q., S. Howell, A. Raman, R. Reifenberger, and M. Franchek. 2004. Frequency domain identification of tip-sample van der waals interactions in resonant atomic force microcantilevers. *J. Vib. Acoust.* 126: 343–351.

Hutter, J.L. and J. Bechhoefer. 1993. Calibration of atomic-force microscope tips. *Rev. Sci. Instrum.* 64: 1868–1873.

Jamitzky, F., M. Stark, W. Bunk, W.M. Heckl, and R.W. Stark. 2006. Chaos in dynamic atomic force microscopy. *Nanotechnology*: S213–S220.

Kikukawa, A., S. Hosaka, and R. Imura. 1996. Vacuum compatible high-sensitive kelvin probe force microscopy. *Rev. Sci. Instrum.* 67: 1463–1467.

Ljung, L. 1999. *System Identification—Theory for the User.* Upper Saddle River, NJ: PTR Prentice Hall.

Maivald, P., H.J. Butt, S.A.C. Gould et al. 1991. Using force modulation to image surface elasticities with the atomic force microscope. *Nanotechnology* 2: 103–106.

Martinez, N.F., S. Patil, J.R. Lozano, and R. Garcia. 2006. Enhanced compositional sensitivity in atomic force microscopy by the excitation of the first two flexural modes. *Appl. Phys. Lett.* 89: 153115.

Melcher, J., S.Q. Hu, and A. Raman. 2008. Invited article: Veda: A web-based virtual environment for dynamic atomic force microscopy. *Rev. Sci. Instrum.* 79: 061301.

Miu, D.K. 1993. *Mechatronics.* Heidelberg, Germany: Springer.

Platz, D., E.A. Tholen, D. Pesen, and D.B. Haviland. 2008. Intermodulation atomic force microscopy. *Appl. Phys. Lett.* 92: 153106.

Proksch, R. 2006. Multifrequency, repulsive-mode amplitude-modulated atomic force microscopy. *Appl. Phys. Lett.* 89: 113121.

Proksch, R., T.E. Schäffer, J.P. Cleveland, R.C. Callahan, and M.B. Viani. 2004. Finite optical spot size and position corrections in thermal spring constant calibration. *Nanotechnology* 15: 1344–1350.

Rabe, U., K. Janser, and W. Arnold. 1996. Vibrations of free and surface-coupled atomic-force microscope cantilevers—theory and experiment. *Rev. Sci. Instrum.* 67: 3281–3293.

Radmacher, M., R.W. Tilmann, and H.E. Gaub. 1993. Imaging viscoelasticity by force modulation with the atomic-force microscope. *Biophys. J.* 64: 735–742.

Radmacher, M., J.P. Cleveland, M. Fritz, H.G. Hansma, and P.K. Hansma. 1994. Mapping interaction forces with the atomic-force microscope. *Biophys. J.* 66: 2159–2165.

Raman, A., J. Melcher, and R. Tung. 2008. Cantilever dynamics in atomic force microscopy. *Nano Today* 3: 20–27.

Reinstaedtler, M., U. Rabe, V. Scherer, J.A. Turner, and W. Arnold. 2003. Imaging of flexural and torsional resonance modes of afm cantilevers using optical interferometry. *Surf. Sci.* 532–535: 1152–1158.

Rodríguez, T.R. and R. García. 2002. Tip motion in amplitude modulation (tapping-mode) atomic-force microscopy: Comparison between continuous and point-mass models. *Appl. Phys. Lett.* 80: 1646–1648.

Rodríguez, T.R. and R. García. 2004. Compositional mapping of surfaces in atomic force microscopy by excitation of the second normal mode of the microcantilever. *Appl. Phys. Lett.* 84: 449–451.

Rosa-Zeiser, A., E. Weilandt, S. Hild, and O. Marti. 1997. The simultaneous measurement of elastic, electrostatic and adhesive properties by scanning force microscopy: Pulsed-force mode operation. *Meas. Sci. Technol.* 8: 1333–1338.

Sader, J.E., J.W.M. Chon, and P. Mulvaney. 1999. Calibration of rectangular atomic force microscope cantilevers. *Rev. Sci. Instrum.* 70: 3967–3969.

Sahin, O. 2008. Time-varying tip-sample force measurements and steady-state dynamics in tapping-mode atomic force microscopy. *Phys. Rev. B* 77: 115405.

Sahin, O. and A. Atalar. 2001. Simulation of higher harmonics generation in tapping-mode atomic force microscopy. *Appl. Phys. Lett.* 79: 4455–4457.

Sahin, O., S. Magonov, C. Su, C.F. Quate, and O. Solgaard. 2007. An atomic force microscope tip designed to measure time-varying nanomechanical forces. *Nat. Nanotechnol.* 2: 507–514.

Sahin, O., O. Uzun, M. Sopicka-Lizer, H. Gocmez, and U. Kolemen. 2008. Dynamic hardness and elastic modulus calculation of porous sialon ceramics using depth-sensing indentation technique. *J. Eur. Ceram. Soc.* 28: 1235–1242.

Sahoo, D.R., A. Sebastian, and M.V. Salapaka. 2003. Transient-signal-based sample-detection in atomic force microscopy. *Appl. Phys. Lett.* 83: 5521–5523.

Salapaka, M.V., H.S. Bergh, J. Lai, A. Majumdar, and E. McFarland. 1997. Multi-mode noise analysis of cantilevers for scanning probe microscopy. *J. Appl. Phys.* 81: 2480–2487.

Salapaka, M.V., D.J. Chen, and J.P. Cleveland. 2000. Linearity of amplitude and phase in tapping-mode atomic-force microscopy. *Phys. Rev. B* 61: 1106–1115.

San-Paulo, A. and R. García. 2002. Unifying theory of tapping-mode atomic-force microscopy. *Phys. Rev. B* 66: 041406.

Schäffer, T.E. 2005. Calculation of thermal noise in an atomic force microscope with a finite optical spot size. *Nanotechnology* 16: 664–670.

Schäffer, T.E. and H. Fuchs. 2005. Optimized detection of normal vibration modes of atomic force microscope cantilevers with the optical beam deflection method. *J. Appl. Phys.* 97: 083524.

Scherer, V., W. Arnold, and B. Bhushan. 1999. Lateral force microscopy using acoustic friction force microscopy. *Surf. Interf. Anal.* 27: 578–587.

Scherer, M.P., G. Frank, and A.W. Gummer. 2000. Experimental determination of the mechanical impedance of atomic force microscopy cantilevers in fluids up to 70 kHz. *J. Appl. Phys.* 88: 2912–2920.

Schirmeisen, A., B. Anczykowski, and H. Fuchs. 2003. Dynamic force microscopy. In *Applied Scanning Probe Methods*. B. Bhushan, H. Fuchs, and S. Hosaka, editors. Berlin, Heidelberg, New York: Springer.

Sebastian, A., M.V. Salapaka, D.J. Chen, and J.P. Cleveland. 1999. Harmonic analysis based modeling of tapping-mode afm. In *Amer. Control Conf. (ACC)*. Vol. 1. Piscataway, NJ/San Diego, CA: IEEE. pp. 232–236.

Sebastian, A., M.V. Salapaka, D.J. Chen, and J.P. Cleveland. 2001. Harmonic and power balance tools for tapping-mode atomic force microscope. *J. Appl. Phys.* 89: 6473–6480.

Spector, V.A. and H. Flashner. 1990. Modeling and design implications of noncollocated control in flexible systems. *J. Dyn. Syst.-Trans. ASME* 112: 186–193.

Stark, R.W. 2004a. Spectroscopy of higher harmonics in dynamic atomic force microscopy. *Nanotechnology* 15: 347–51.

Stark, R.W. 2004b. Optical lever detection in higher eigenmode dynamic atomic force microscopy. *Rev. Sci. Instrum.* 75: 5053–5055.

Stark, R.W. and W.M. Heckl. 2000. Fourier transformed atomic force microscopy: Tapping mode atomic force microscopy beyond the hookian approximation. *Surf. Sci.* 457: 219–228.

Stark, R.W., T. Drobek, and W.M. Heckl. 1999. Tapping-mode atomic-force microscopy and phase-imaging in higher eigenmodes. *Appl. Phys. Lett.* 74: 3296–3298.

Stark, M., R.W. Stark, W.M. Heckl, and R. Guckenberger. 2000. Spectroscopy of the anharmonic cantilever oscillations in tapping-mode atomic-force microscopy. *Appl. Phys. Lett.* 77: 3293–3295.

Stark, R.W., T. Drobek, and W.M. Heckl. 2001. Thermomechanical noise of a free v-shaped cantilever for atomic-force microscopy. *Ultramicroscopy* 86: 207–15.

Stark, M., R.W. Stark, W.M. Heckl, and R. Guckenberger. 2002. Inverting dynamic force microscopy: From signals to time-resolved interaction forces. *Proc. Natl. Acad. Sci. USA* 99: 8473–8478.

Stark, R.W., F.J. Rubio-Sierra, S. Thalhammer, and W.M. Heckl. 2003. Combined nanomanipulation by atomic force microscopy and uv-laser ablation for chromosomal dissection. *Eur. Biophys. J.* 32: 33–39.

Stark, R.W., G. Schitter, M. Stark, R. Guckenberger, and A. Stemmer. 2004. State-space model of freely vibrating and surface-coupled cantilever dynamics in atomic force microscopy. *Phys. Rev. B* 69: 085412.

Stark, M., R. Guckenberger, A. Stemmer, and R.W. Stark. 2005. Estimating the transfer function of the cantilever in atomic force microscopy: A system identification approach. *J. Appl. Phys.* 98: 114904.

Sulchek, T., R. Hsieh, J.D. Adams et al. 2000. High-speed tapping mode imaging with active q control for atomic force microscopy. *Appl. Phys. Lett.* 76: 1473–1475.

Tamayo, J., A.D.L. Humphris, R.J. Owen, and M.J. Miles. 2001. High-q dynamic force microscopy in liquid and its application to living cells. *Biophys. J.* 81: 526–537.

Todd, B.A. and S.J. Eppell. 2003. Inverse problem of scanning force microscope force measurements. *J. Appl. Phys.* 94: 3563–3572.

Torii, A., M. Sasaki, K. Hane, and S. Okuma. 1996. A method for determining the spring constant of cantilevers for atomic force microscopy. *Meas. Sci. Technol.* 7: 179–184.

Turner, J.A., S. Hirsekorn, U. Rabe, and W. Arnold. 1997. High-frequency response of atomic-force microscope cantilevers. *J. Appl. Phys.* 82: 966–979.

Vazquez, R., F.J. Rubio-Sierra, and R.W. Stark. 2007. Multimodal analysis of force spectroscopy based on a transfer function study of micro-cantilevers. *Nanotechnology* 18: 185504.

Wei, B. and J.A. Turner. 2001. Nonlinear vibrations of atomic force microscope probes in hertzian contact. *AIP Conf. Proc.* 557B: 1658–1665.

Wilhelm, M. 2002. Fourier-transform rheology. *Macromol. Mater. Eng.* 287: 83–105.

Yaralioglu, G.G. and A. Atalar. 1999. Noise analysis of geometrically complex mechanical structures using the analogy between electrical circuits and mechanical systems. *Rev. Sci. Instrum.* 70: 2379–2383.

Yuan, K. and L.Y. Liu. 2003. On the stability of zero dynamics of a single-link flexible manipulator for a class of parametrized outputs. *J. Robotic Sys.* 20: 581–586.

Zypman, F.R. and S.J. Eppell. 1998. Analysis of scanning force microscope force-distance data beyond the hookian approximation. *J. Vac. Sci. Technol. B* 16: 2099–2101.

34

STM-Based Techniques Combined with Optics

Hidemi Shigekawa
University of Tsukuba

Osamu Takeuchi
University of Tsukuba

Yasuhiko Terada
University of Tsukuba

Shoji Yoshida
University of Tsukuba

34.1 Introduction

Through the efforts of the researchers devoted over the last decade, unprecedented advances have been made in developing nanoscale materials and devices with novel functions. Through nanoscale fabrications, elemental blocks of various characteristics are integrated and organized on a designed stage to produce desired or new functions in a system on the macroscopic scale (Figure 34.1). However, with the reduction in the size of structures, differences in electronic properties, for example, that due to the structural nonuniformity in each element, has an even more crucial effect on macroscopic functions.

Figure 34.2a and b shows a scanning tunneling microscopy (STM) image of an Si nanoparticle of ~3 nm diameter on a graphite surface and the cross sections of the same region obtained at different bias voltages. Inhomogeneous structures are observed in the cross sections. Since STM provides information on the electronic structures at the observed bias voltage (Wiesendanger 1994, Sakurai and Watanabe 1999), the result indicates the presence of complex electronic structures even at the single-molecular level. The miniaturization of semiconductor devices into a 10 nm scale brings about another example. Local structures, such as atomic defects and dopant materials, inevitably affect and govern the characteristic properties of macroscopic functions (Figure 34.3).

Therefore, the direct observation of the characteristic properties of nanostructures, which provides us with the basis for

the macroscopic analysis of results, is of great importance. Thus, for further advancement in the development of new functional materials and devices, a method of exploring the transient dynamics of the local quantum functions in organized small structures is desired.

In STM, a sharp tip is placed close to a target material (Figure 34.4), and information on the region underneath the STM tip is obtained using a tunneling current. In general, the local density of states of the material under an equilibrium condition is measured with a certain bias voltage applied between the STM tip and the sample. However, when some perturbation is added from outside, we can analyze the dynamics of the system by observing its responses to the changes in conditions.

When optical modulation is adopted, the dynamics of transient electronic states and structural changes induced in materials by photoexcitation, providing a variety in functional properties of materials and devices (Figure 34.5), may be analyzed by STM. Therefore, in order to take full advantage of functions of materials in miniaturized devices, it is of great importance to develop new microscopy techniques that enable us to study photoinduced phenomena on the nanoscale.

This chapter describes the STM and related techniques combined with optical technologies in detail. Section 34.2 presents the specific issues brought by the combination with optical technologies and the methods of treating them. Section 34.3 shows the basic principles and techniques for probing the carrier dynamics

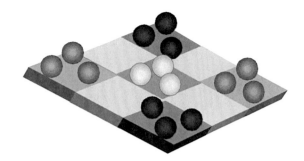

FIGURE 34.1 Schematic model of nanoscale system.

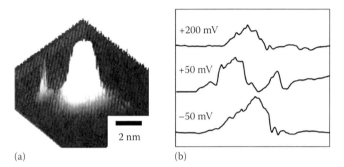

(a)   (b)

FIGURE 34.2 (a) STM current image of Si nanoparticle on graphite surface and (b) cross sections of same region obtained at different bias voltages.

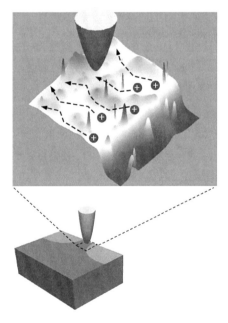

FIGURE 34.3 Schematic of carrier dynamics modulated in nanoscale potential landscape.

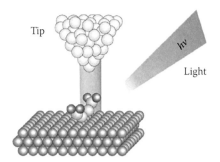

FIGURE 34.4 STM setup with additional modulation from outside (light modulation in this case).

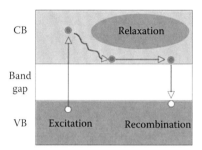

FIGURE 34.5 An example presenting excitation and relaxation processes for photoexcited carriers in semiconductor.

of nanoscale semiconductor structures using laser-combined STM. In Section 34.4, some other cases are introduced. Although it is difficult to cover all the related topics, the materials discussed in this chapter are presented in a form that is instructive to non-specialists as well as specialists. This makes it possible to develop the ability of handling the issues we face in the measurement using the new microscopy techniques. With ingenuity, a new method can be developed.

34.2 Specific Issues in Techniques Combined with Optical Technologies

A super-band gap photoillumination of a semiconductor sample, for example, produces electron–hole pairs, inducing a photocurrent and the surface photovoltage (SPV) effect in STM measurement, which can be probed as a signal and used for analyzing the carrier dynamics. However, since optical excitation produces other various phenomena, we should carefully develop experimental and analytical methods to obtain meaningful information using the new STM and related techniques. Typical issues are discussed in this section.

34.2.1 Thermal Expansion Effect

The most obvious and critical issue is the thermal expansion of the STM tip and sample, due to the heat brought by photo-illumination. In most cases, the effect of tip expansion is dominant because of the shape of the tip. When the light intensity is constant, the temperature and thereby the length of the STM tip gradually increase until they reach equilibrium under illumination. Once an equilibrium condition is achieved, STM observation can be performed stably. In contrast, when the light intensity oscillates, the tip repeatedly expands and shrinks. Since the tunneling current is extremely sensitive to the tip-sample distance,

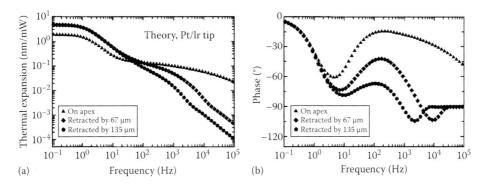

FIGURE 34.6 (a) Chopping frequency dependence of thermal expansion effect on Pt/Ir STM tip (theory). (b) Three series correspond to the results obtained at three different illumination spot positions. (From Grafström, S., *J. Appl. Phys.*, 83, 3453, 1998. With permission.)

this phenomenon affects the STM measurement. In the worst-case scenario, the tip crashes into the sample surface.

The modulation of the light intensity is caused, for example, by the use of an optical chopper or a train of a pulse laser, and the noise due to the fluctuation in the light source. To reduce the effect of heat without decreasing the light density, it is important to reduce the light spot size. When pulsed illumination is used, increasing its frequency is effective for reducing the effect of heat. As shown by the response characteristics of tip length against the chopping frequency of photoillumination in Figure 34.6, the thermal expansion effect is smaller at higher frequencies (Grafström 1998). However, note that the finite effect still exists even when the modulation frequency is as high as 100 kHz, which is already above the bandwidth of the STM tunnel current preamplifier. A superior method that is effective for the use of a short-pulse laser with STM is explained in Section 34.3.3.

34.2.2 Photoelectron Emission

Photoillumination with energy above the work function causes photoelectron emission. For a pulse laser with a high peak intensity, the multiple photon absorption effect may additionally cause photoemission even when the photon energy is below the gap energy. Thus, the photon energy and peak intensity must be chosen to suppress this effect (Jersch 1999). Since the tip–sample distance dependence of the photoelectron current is small, its effect on the measurement can be distinguished from the tunnel current by increasing the tip–sample separation up to ~10 nm, where the tunnel current significantly diminishes. The photoelectron current is smaller in air because the mean free path becomes shorter. The use of photocurrent is introduced in Section 34.4.5.

34.2.3 Displacement Current

Another possible source of the unfavorable current for the measurement of tunnel current is the displacement current from the stray capacitance at the tunnel junction. The nanoscale metal–insulator–semiconductor (MIS) junction, consisting of the STM tip, tunnel gap, and semiconductor sample (Figure 34.9 in Section 34.3.1), works as a capacitor that accumulates a certain amount of charge as a function of the applied bias voltage, similarly to the

gate capacitor of a field effect transistor. When the gap is illuminated and the SPV is induced, the effective bias voltage changes, resulting in the flow of the displacement current from the capacitor. This current appears only when the light intensity changes.

The stray capacitance has a smaller dependence on the tip–sample distance, similarly to the photocurrent. Thus, this effect is also distinguished from the tunnel current by increasing the tip–sample separation up to ~10 nm.

Although the spatial resolution is limited to ~1 μm, the measurement of displacement current enabled the analysis of carrier dynamics in a semiconductor with the temporal resolution of ~10 ns (Hamers and Cahill 1991). An example for the use of the displacement current is shown in Section 34.4.3.

34.2.4 Tip Shape Effect

The shadowing of the photoillumination by the STM tip is an issue that must be considered. This effect can be reduced by STM tip sharpening. Figure 34.7 shows an example of a sharp tungsten tip electrochemically etched in NaOH solution (Williams 2008). Such a sharp tip has an apex with a curvature radius smaller than the optical wavelength. Thus, it does not drop its shadow on the sample surface. Its effect is rather scattering than shadowing, allowing us to illuminate the tunnel gap despite the existence of the STM tip.

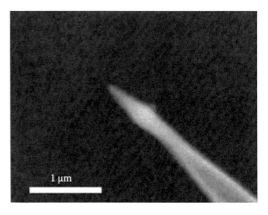

FIGURE 34.7 Example of sharp tungsten tip electrochemically etched in NaOH solution.

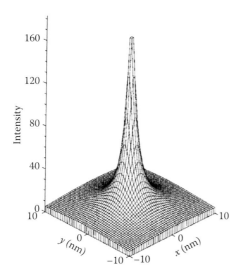

FIGURE 34.8 Calculated profile of plasmon resonance enhancement. (From Martin, O.J.F. and Girard, C., *Appl. Phys. Lett.*, 70, 705, 1997. With permission.)

In contrast to the shadowing effect, in some cases, such a well-sharpened metal tip strongly increases the effective light intensity beneath the STM tip. This is known as the field enhancement effect; the local electric field at the tunnel gap is increased by the plasmon resonance in the metallic tip, resulting from the physical shape of its nanometer-scale apex (Figure 34.8; Martin 1997). According to the literature, this effect can result in a light intensity magnification of as high as ×1000 under carefully prepared conditions. Since a light-induced STM signal is also enhanced, this effect is preferable for measurement. In addition, this also allows us to selectively excite a local area (~10 nm) of the sample, whose size is much smaller than the wavelength. Therefore, this effect increases the spatial resolution of light-induced STM.

34.2.5 Stray Light

The handling of stray light, as well as the stability of light source, is important for a photorelated measurement. A dark environment, for example, is the basis for the accurate measurement of STM-induced light emission (Section 34.4.6). In this respect, atomic force microscopy (AFM) that uses an optical system to measure the deflection of a lever is not suitable for a photorelated measurement. Some fraction of laser beams used inevitably illuminates the sample surface during observation and affects the measurement. Furthermore, the external illumination of the sample may affect the force detection of an optical lever. To solve these problems, a piezoresistive cantilever (Tortonese 1991), which can sense its deformation by itself without the aid of laser illumination, is used for a photorelated AFM measurement. Using a contact-mode AFM with a self-detecting cantilever, photocurrent mapping was performed over InAs nanowires grown at step edges on a GaAs substrate (Masuda 2005). In the experiment, the sample was intermittently illuminated with a variable-wavelength laser, and the nanoscale variation in photocurrent on the wires caused by the inhomogeneity of mechanical stress, composition, and defect density in the wires was observed.

34.3 Probing Carrier Dynamics in Semiconductors

This section describes the basic principles and techniques for probing carrier dynamics in nanoscale semiconductor structures.

34.3.1 Tip-Induced Band Bending and Surface Photovoltage

When STM is performed on a semiconductor sample, the (STM tip)–(tunneling gap)–(semiconductor sample) configuration forms an MIS structure, as shown in Figure 34.9. Here, a p-type semiconductor is used as a sample and a negative bias voltage is applied to the sample. In an STM/STS measurement performed under a dark condition at a certain bias voltage, tip-induced band bending (TIBB) appears owing to the leakage of the applied electric field into the sample, as illustrated in Figure 34.9a (Prins et al. 1996, Sommerhalter et al. 1997, Yoshida et al. 2006). When the sample below the STM tip is photoilluminated with a sufficient intensity, the redistribution of photoexcited carriers reduces band bending, resulting in the flat-band condition illustrated in Figure 34.9b. This change observed in the surface potential, change in the band bending, due to photoillumination is defined as SPV (Kronik and Shapira 1999). Since SPV is related to the local carrier dynamics induced by photoillumination, the analysis of SPV provides us information on the carrier dynamics in the sample.

When SPV is induced by photoexcitation, as shown in Figure 34.9, the barrier height (BH) for tunnel current decreases. Since the tunneling current is determined by the local density of states of the tip and sample and the transient probability that depends on BH, the reduction in BH due to SPV results in an increase in tunneling current. The difference in BH depends on the TIBB under the dark condition and is determined by the local carrier density. If a flat-band condition is achieved under the illuminated condition, the SPV spectrum represents TIBB. Therefore, by measuring the change in tunneling current due to photoillumination, we can observe the carrier density through the analysis of SPV.

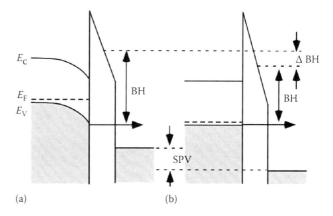

FIGURE 34.9 Schematic illustrations of metal–insulator–semiconductor (MIS) structures in cases (a) without and (b) with photoillumination (BH local barrier height).

Sections 34.3.2 and 34.3.3 show two laser-combined techniques that enable us to visualize the carrier dynamics in semiconductors on the nanoscale on the basis of the mechanisms of TIBB and SPV, namely, light-modulated scanning tunneling spectroscopy (LM-STS) and femtosecond time-resolved STM.

34.3.2 Light-Modulated Scanning Tunneling Spectroscopy

34.3.2.1 Basic Principle

A measurement technique in STM, STS, enables us to probe the local density of electronic states of materials on an atomic scale. In STS, the tunneling current versus sample bias voltage (I–V) curve is measured under a fixed tip-sample distance. The concept of LM-STS is to combine this technique with optical excitation. An I–V curve is measured under a chopped laser illumination (Figure 34.10a), which, for example, provides information on photocurrent generation and local carrier dynamics in a semiconductor sample (Takeuchi et al. 2004a,b).

Figure 34.10b and c shows an I–V curve obtained under a chopped laser illumination on a cleaved clean n-GaAs(110) surface and its magnification, respectively. The tunneling current oscillates at a chopping frequency of laser illumination, and two virtual I–V curves obtained from the envelopes, which correspond to those under dark (solid circles) and illuminated

(open circles) conditions, can be simultaneously obtained. The SPV spectrum shown in Figure 34.10d is obtained by calculating the lateral shift of the two I–V curves with respect to the bias voltage for the I–V curve under the dark condition. For an n-type semiconductor, upward TIBB is induced under a positive sample bias voltage because no surface states exist within the band gap, and the tip can bend the bands until one of the band edges intersects with the Fermi level. In contrast, only a small downward band bending is induced at a negative sample bias voltage due to the Fermi-level pinning by the conduction band edge. If a flat-band condition is achieved under the illuminated condition, the SPV spectrum (Figure 34.10d) represents the bias voltage dependence of TIBB. Therefore, a large SPV is observed for a positive sample bias voltage. For the measurement on a p-type semiconductor, the opposite bias voltage dependence occurs.

34.3.2.2 Measurement on p–n Junction

As explained in Section 34.3.1, the magnitude of TIBB depends on the carrier density, and the minority-carrier dynamics in a semiconductor, which changes local carrier density, can be obtained with a nanoscale spatial resolution through the measurement of SPV by LT-STS (Yoshida et al. 2007).

The experimental setup of LM-STS measurement is illustrated in Figure 34.11. Here, a p–n junction was prepared by growing n-type (Si-doped, $2.0 \times 10^{18}\,\mathrm{cm^{-3}}$, 500 nm) and p-type (Be-doped,

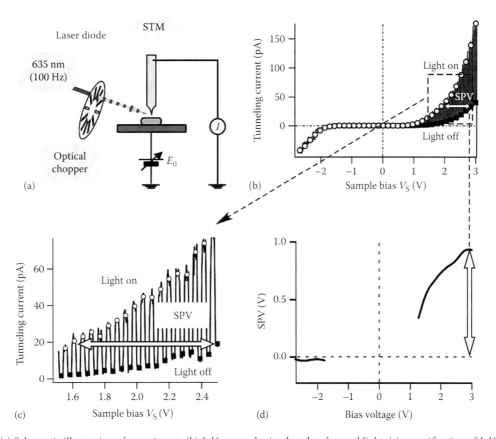

FIGURE 34.10 (a) Schematic illustration of experiment, (b) I–V curve obtained under chopped light, (c) magnification of I–V curve in (b), and (d) SPV (shift of I–V curves indicated by arrow in (b) plotted against bias voltage with respect to bias voltage for I–V curve under dark condition).

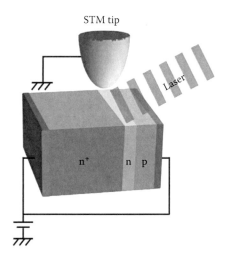

FIGURE 34.11   Schematic of LM-STS measurement setup.

2.0×10^{18} cm^{-3}, 500 nm) GaAs layers on an n-GaAs(001) substrate (Si-doped, 8.3×10^{18} cm^{-3}) using molecular beam epitaxy. Au–Zn Ohmic-contact electrodes were formed on both sides of the samples. LM-STS measurements were performed on a cleaved clean (110) surface at room temperature in ultrahigh vacuum ($<1 \times 10^{-8}$ Pa).

Spatially resolved LM-STS measurements were performed simultaneously with topographic imaging. The feedback loop of STM was opened at equally spaced measurement points during topographic imaging to fix the tip position, and an *I–V* curve was acquired under chopped light at each point. SPV was calculated

from the *I–V* curves for the chosen bias voltages and presented with a gray scale.

Figure 34.12a shows the SPV images obtained at the p–n junction interface for positive (+2.5 V, up) and negative (−2.5 V, bottom) sample bias voltages. As described above, the magnitude of TIBB under a positive V_S is large (negligible) for the n-type (p-type) region, and the change in SPV across the p–n junction is thus clearly observed. In these images, a large SPV area at a positive (negative) V_S corresponds to the n-type region (p-type region).

When a forward bias voltage is applied to a p–n junction, minority holes (electrons) flow from the p-type (n-type) region to the n-type (p-type) region (Figure 34.12b), which decreases the magnitude of TIBB depending on the number of flowing minority carriers, as is explained in Section 34.3.1. Therefore, such dynamics of minority carriers can also be visualized by measuring SPV using LM-STS.

Figure 34.12c shows SPV images of a forward-biased p–n junction ($V_F = 0.5–0.9$ V). Since the decrease in SPV is related to the decrease in the magnitude of TIBB, the observed change is directly related to the change in the number of excess carriers below the STM tip. Using the logarithmic relationship of SPV with the minority-carrier density, the spatial distribution of minority carriers can be calculated from the SPV images (Figure 34.12d). The number of flowing minority carriers decreases with the distance from the p–n junction interface, owing to recombination with the majority carriers, as expected.

The inhomogeneous flow of minority carriers in a forward-biased p–n junction, shown in Figure 34.12, is considered to be

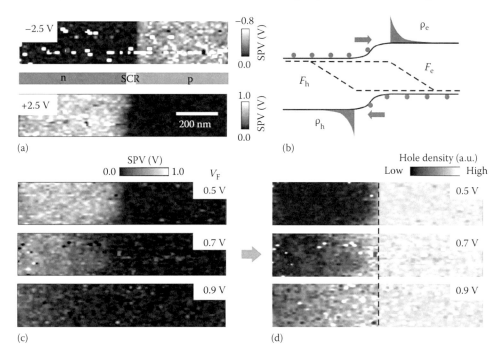

FIGURE 34.12   (a) SPV mappings of zero-bias ($V_F = 0$) p–n junction obtained at $V_S = -2.5$ V (*top*) and $V_S = +2.5$ V (*bottom*) (1000×200 nm, 80×20 LM-STS spectra). *SCR*: space charge region. (b) Band diagram of p–n junction with distribution of minority electrons (ρ_e) and holes (ρ_h) under forward-biased voltage condition. F_h and F_e denote the Fermi levels for holes and electrons, respectively. (c) SPV images of p–n junction with forward bias voltages. (d) Images of minority carrier (hole) flow calculated from (c).

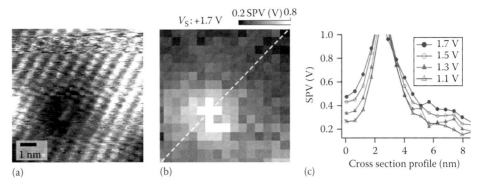

FIGURE 34.13 (a) Filled-state topographic image of Ga vacancy, (b) SPV image simultaneously obtained with topographic image in (a) (V_S = +1.7 V, 17 × 17 points), and (c) cross sections of SPV maps obtained at various bias voltages, along white line in (b).

caused by the nanoscale electrostatic potential fluctuation due to charged impurities, such as dopants and defects. The result indicates the importance of the method of evaluating the potential landscape on an atomic scale. In Section 34.3.2.3, a method of providing information on the modulation of the potential landscape around a single charged defect is presented (Yoshida et al. 2008).

34.3.2.3 Measurement on Charged Atomic Defect

Figure 34.13a shows a filled-state topographic image of a Ga vacancy on a GaAs (110) surface. The Ga vacancy exhibits a contrast of a bright elevation at the center of the defect location due to the dangling bond existing at the Ga vacancy that appears bright in a filled-state image. The surrounded dark area is caused by the depression region, originating from the screened Coulomb potential (SCP) around the negatively charged defect. Figure 34.13b shows the SPV image obtained for the same area in Figure 34.13a (V_S = +1.7 V). Since photoillumination reduces the magnitude of band bending around a charged defect with the same mechanism for reducing the magnitude of TIBB, the spatial variation in SPV represents the spatial variation in SCP. Figure 34.13c shows the cross-sectional profiles acquired at different bias voltages. Because of the effect of SCP, SPV, which has a maximum value at the center of the Ga vacancy, exponentially decreases with the distance from the center.

The magnitude of SPV increases with bias voltage; however, the spatial variation in cross-sectional profile remains unchanged. Since the shape of SCP is not affected by the bias voltage, the bias-dependent component of SPV originates from TIBB that shifts only the entire SPV profile along with bias voltage. Therefore, the spatial distribution of SPV directly represents SCP, and the charge state of the Ga vacancy can be determined by fitting the SPV profile with the SCP equation (McEllistrem et al. 1993).

34.3.3 Femtosecond Time-Resolved STM

34.3.3.1 History of Time-Resolved STM Development

"Smaller" and "faster" are the key factors in nanoscale science and technology. Indeed, important and interesting phenomena in various systems, such as the phenomena observed in functional materials and electronic devices, signal transfer in biosystems, and chemical reactions, are observed in the space range from several tens of nanometers to the single molecule and in the time range from several tens of picoseconds to the subpicosecond. However, it is extremely difficult to obtain spatial and temporal resolutions simultaneously on this scale. STM has an atomic-scale spatial resolution, but its temporal resolution is limited to less than 100 kHz owing to the circuit bandwidth, and the ultrafast carrier dynamics has been beyond the field of vision. In contrast, in leading-edge technology fields of quantum optics, wideband spectroscopy has been established, for example, using an ultrashort-pulse laser. However, the spatial resolution is generally limited to the wavelength. Therefore, since the invention of STM in 1982, combining ultrashort-pulsed-laser technology with STM has been one of the most attractive targets (Hamers and Markert 1990, Grafström 2002, Shigekawa et al. 2005, 2008, Yamashita et al. 2005).

An optical pump-probe technique is a prominent method, in which a sample is illuminated with a sequence of paired laser pulses with a certain delay time t_d. A pump pulse with a high peak intensity excites the sample, and a subsequently arriving probe pulse with a low peak intensity tracks the temporal evolution of the excited stated induced by the first pump pulse (Othonos 1998, Shah 1999). For example, in optical pump-probe reflectivity (OPPR) measurement (Figure 34.14a), the reflectivity of the probe pulse R is measured as a function of delay time t_d between two pulses, which is often performed by recording the reflected light intensity using a photodiode. When the pump-probe pair repeatedly illuminates the surface at a repetition of 1 k–100 MHz, the photodiode detector does not need to resolve each temporal profile of the reflected light intensity, but it only needs to detect the time-averaged intensity. The dependence of reflectivity R on delay time gives a measure of the relaxation of the excited states induced by a pump. The temporal resolution attainable in such experiments is limited only by the pulse width, which is generally in the femtosecond range.

If the tunneling caused by optical excitation can be measured by STM, high temporal and spatial sensitivities can be simultaneously achieved with the atomic-scale resolution of STM (Shigekawa et al. 2005, 2008). A pulse-pair-excited STM (PPX-STM)

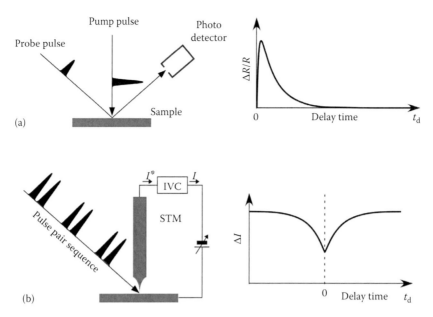

(a)

(b)

FIGURE 34.14 Principle of (a) OPPR and (b) PPX-STM measurements.

system has been a promising setup, in which, in analogy with pump-probe experiments, a sequence of paired laser pulses with a certain delay time t_d illuminates a sample surface beneath the STM tip, and the tunneling current I is measured as a function of t_d (Figure 34.14b). However, efforts toward its realization failed because of experimental difficulties. In most cases, the delay-dependent current component is so faint ($\sim10^{-3}$ of the average current) that lock-in detection with a modulation technique is required for observation. In purely optical pump-probe experiments, the laser intensity is generally modulated; however, this technique is not directly applicable to PPX-STM because it induces the thermal expansion and shrinking of the STM tip in synchronization with modulation, and consequently results in a large change in tunneling current. To suppress this thermal effect, retracting the STM tip far away from the surface has been attempted by many researchers; however, in this case, the detected signal was not a tunneling current but a displacement current or photoelectron current, and thus, no spatial resolution higher than 1 μm has been attained (Hamers and Markert 1990).

34.3.3.2 Shaken-Pulse-Pair-Excited STM

A more applicable technique is shaken-pulse-pair-excited STM (SPPX-STM) (Takeuchi et al. 2004a,b, Terada et al. 2007), which has been developed using the principle of PPX-STM. As described above, in PPX-STM, in analogy with the pump-probe technique, a sequence of paired laser pulses with a certain delay time t_d illuminates a sample surface beneath the STM tip, and the tunneling current I is measured as a function of t_d (Figure 34.14b). When the paired optical pulses arrive at the sample, they generate current pulses in the raw tunneling current I^*, reflecting the excitation and relaxation of the sample (Figure 34.15a). If these current pulses decay faster than the time scale of the STM preamplifier bandwidth, they are temporally averaged in the preamplifier and cannot be detected directly in the signal I. Even in this case, the relaxation

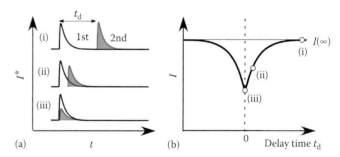

(a)   (b)

FIGURE 34.15 (a) Relationship between raw tunneling current I^* and delay time t_d. (b) Measured tunneling current I as function of delay time, where delay times of (i), (ii), and (iii) correspond to those of (i), (ii), and (iii) in (a).

dynamics can be probed through the t_d dependence of I, as shown in Figure 34.15.

When t_d is small, that is, the second probe arrives at the sample excited by the first pump and not relaxed, the second tunneling-current pulse changes under the strong effect of excited states (Figure 34.15-iii), as opposed to the case in which the probe arrives when the states excited by the pump are completely relaxed (Figure 34.15-i). As t_d increases, the effect of the pump becomes smaller (Figure 34.15-ii), and eventually, at a large t_d, the tunneling current I saturates to a value $I(\infty)$ (Figure 34.15-i). Then, $\Delta I(t_d) \equiv I(t_d) - I(\infty)$ provides information on excited states at the delay time t_d after pump excitation. This technique has a temporal resolution of pulse width equivalent to that of ultrafast spectroscopy (a few femtoseconds) and a spatial resolution of tunneling current equivalent to that of STM (subangstrom).

In SPPX-STM, the delay time is modulated instead of the laser intensity for lock-in detection. To date, two optical setups have been used for delay time control. One is a mechanically controlled setup, which is used in conventional interferometers.

In this setup, a corner-cube mirror is inserted into a pump or probe optical line and is mechanically shaken to periodically modulate the optical lengths. The other is an electro-optically controlled setup, in which the polarization plane of a part of a pulse sequence is electrically rotated with an electro-optic modulator, so that pulses are "picked" by a polarizing beam splitter to generate a desired pulse sequence. The delay time is periodically modulated by changing the timing of picking pulses from the pump and probe pulse sequences. The latter method is superior to the former in performance characteristics, such as the wide delay time range (up to a few microseconds), the high signal-to-noise ratio, the high signal level, and the stability of the optical line, and therefore, it is more suitable for SPPX-STM measurement.

34.3.3.3 Basic Principle of SPPX-STM Measurement on Semiconductors

Although the basic principle of PPX-STM is simple, the underlying physics is not straightforward. Here, an SPPX-STM measurement is assumed to be performed under the condition in which an MIS diode formed by a sample and a tip is reversely biased. As in the case of LM-STS measurements, TIBB occurs in the surface region in the dark owing to the leakage of the electric field into the sample (Figure 34.16a). With optical illumination (Figure 34.16b), the charge separation of photocarriers occurs owing to the electric field, where majority carriers flow into the bulk side and minority carriers are trapped at the surface. This carrier redistribution reduces the electric field and changes the surface potential, which is called SPV (Section 34.3.1), and increases the effective bias voltage applied to the tunnel junction (Figure 34.16c). Consequently, the illumination increases the raw tunneling current I^*.

The excited state subsequently relaxes to the original state through two processes. One is the decay of the photocarriers on the bulk side (bulk-side decay) via recombination, drift, and diffusion (Figure 34.16d). The other is the decay of the carriers trapped at the surface (surface-side decay) via thermionic emission and recombination with counterpart carriers tunneling from the tip (Figure 34.16e). When the tunneling current is low, few counterpart carriers exist near the surface, and hence, the surface-side decay constant is longer than the bulk-side decay constant. Both of these decay processes appear in the SPPX-STM signal as follows.

The bulk-side decay is probed through the mechanism of absorption bleaching, which also provides a basis of OPPR measurement in some cases. If the bulk-side carriers remain when the second optical pulse arrives, the absorption of the second optical pulse is suppressed, which depends on the bulk-side carrier density at the moment. The corresponding decrease in I^* is detected through the decrease in I; hence, $\Delta I(t_d)$ is a measure of the bulk-side carrier density at the delay time t_d after the excitation by the first optical pulse. On the other hand, the surface-side decay is probed through the mechanism related to the charge-separation efficiency instead of photocarrier generation. If the surface carriers remain when the second optical pulse arrives, the electric field in the surface region remains low. Therefore, the excited photocarriers are less efficiently trapped at the surface. Thus, the SPV generated by the second optical pulse decreases, and the height of the second current pulse in I^* also decreases. Accordingly, $\Delta I(t_d)$ is a measure of the surface-side carrier density at the delay time t_d as well as the bulk-side carrier density. The decay constants for both components are derived from the fitting of $\Delta I(t_d)$ by a double exponential function; the

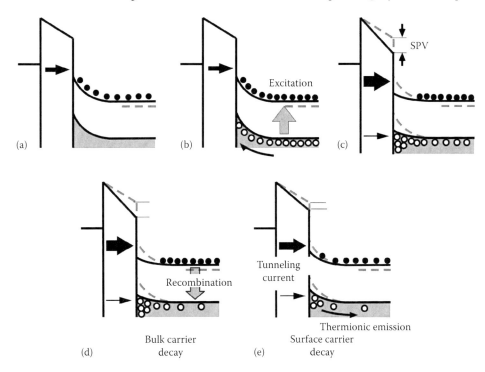

FIGURE 34.16 Time evolution of band diagram for n-type semiconductor due to laser pulse excitation under reversely biased condition: (a) in dark, (b) arrival of laser pulse, (c) after redistribution of excited carriers, (d) bulk-side carrier decay, and (e) surface-side carrier decay.

shorter time constant τ_b is attributed to the bulk-side decay, and the longer time constant τ_b to the surface-side decay.

34.3.3.4 Carrier Dynamics Probed by SPPX-STM

As an example of SPPX-STM measurement results, carrier decay spectra via recombination in bulk (τ_b component) for a $GaN_{0.36}As_{0.64}$ sample are shown in Figure 34.17. The same sample was measured by both SPPX-STM (Figure 34.17a) and OPPR (Figure 34.17b), using a pulse laser with a wavelength of 800 nm, a pulse width of 120 fs, and a repetition frequency of 90 MHz. The time-resolved signal $\Delta I(t_d)$ has a minimum value when $t_d = 0$ and exponentially decays toward zero as t_d increases. The negative sign of the $\Delta I(t_d)$ signal, which corresponds to the decrease in I, agrees with an absorption bleaching mechanism. The decay constant was obtained from the exponential fitting of $\Delta I(t_d)$ to be 444 ps, in good agreement with the bulk-recombination lifetime determined by the temporal evolution of the reflectivity change $\Delta R/R$ (406 ps).

Another example of SPPX measurement results is shown by two-dimensional (2D) maps of SPPX-STM signals obtained for a low-temperature-grown GaAs (LT-GaAs)/$Al_{0.5}Ga_{0.5}As$/GaAs sample. LT-GaAs contains high-density defects that act as recombination sites for carriers. Thus, its carrier lifetime is much shorter than that of GaAs. The AlGaAs barrier layer has a band gap larger than the optical excitation. As expected, the LT-GaAs region exhibits an ultrafast decay component with a time constant of 4 ps, whereas the GaAs region exhibits a time constant of 4.8 ns. These values are consistent with the recombination lifetimes determined by OPPR measurement, namely, 1.5 ps and 2.7 ns for the LT-GaAs and GaAs samples, respectively.

For the spatial mapping of $\Delta I(t_d)$, two modes are possible. One is a constant delay-time mode, where the values of $\Delta I(t_d)$ are recorded for each t_d while the STM tip is scanning the surface. The other is a grid mode. In this mode, the scan area is divided by scan grid points, similarly to that in the case of STS. During the STM scan, the tip is fixed at each grid point and $\Delta I(t_d)$ is measured as a function of t_d.

Figure 34.18 shows $\Delta I(t_d)$ maps for $t_d = 2$, 31 ps, 11, and 200 ns obtained using the constant delay-time mode, together with the STM topographic image (top). At 2 ps, a clear contrast is observed across the AlGaAs/LT-GaAs boundary, which becomes smaller as t_d increases. As the magnitude of $\Delta I(t_d)$ corresponds to that of absorption bleaching, Figure 34.18 shows the spatial distribution of photoexcited carrier density at each t_d after excitation. In this manner, $\Delta I(t_d)$ maps for different t_d values provide spatial information on carrier density at different times. Combining the STM topography with the atomic-scale spatial resolution makes it possible to investigate the fundamental processes of the carrier dynamics.

By combining an advanced ultrashort-pulsed-laser technology with STM, a new microscopy technique is realized that enables the probing and direct analysis of the various carrier dynamics modulated in a nanometer-scale potential landscape (Terada et al. 2010).

34.4 Other Techniques

This section introduces some other cases.

34.4.1 With Optical Manipulation

The simplest photorelated technique is a comparison of STM/AFM images obtained before and during, or after illumination. Figure 34.19 shows an example of such a technique: a derivatized azobenzene molecule, a typical photoactive isomer, adsorbed on a gold substrate was alternately illuminated by visible and ultraviolet lights, and the corresponding change in molecular shape was observed by STM, showing the change in intramolecular electronic structure due to photoisomerization (Yasuda et al. 2003). The structural change in molecules due to photoexcitation is also detectable using AFM by measuring the force induced by the change in the length of molecules bridged between the tip and the substrate (Hugel et al. 2002).

Figure 34.20 shows another example, that is, the desorption of adatoms on a Si(111)-7×7 surface using ultrashort-pulsed-laser irradiation (Futaba et al. 2003). It was shown that one of the four types of adatoms could selectively be desorbed by tuning the temporal separation of successive pulse excitations to the molecular vibration periodicity of the target adatom.

The metal–insulator transition (MIT) of low-dimensional materials is one of the most attractive phenomena observed in the research studies of phase transitions. One-dimensional (1D) nanowires are nanoscale building blocks that can efficiently transport electrical charges that constitute information on nanoelectronics, and an In/Si nanowire system, for example, has

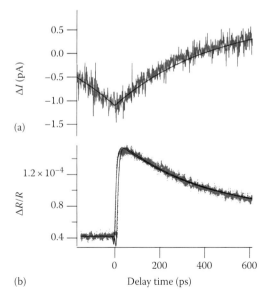

FIGURE 34.17 Carrier decay spectra via recombination in bulk (τ_b component) for $GaN_{0.36}As_{0.64}$ sample measured by (a) SPPX-STM and (b) OPPR.

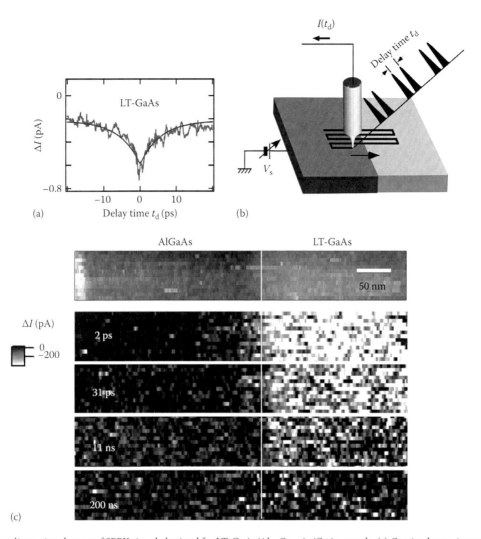

FIGURE 34.18   Two-dimensional maps of SPPX signal obtained for LT-GaAs/Al$_{0.5}$Ga$_{0.5}$As/GaAs sample. (a) Carrier decay via recombination in bulk for LT-GaAs region. (b) Schematic of constant delay-time mode. (c) Two-dimensional maps of SPPX signal obtained for sequential delay times.

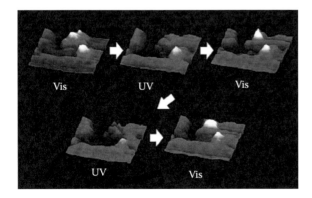

FIGURE 34.19   STM observation of photoactivated isomerization of azobenzen derivative molecule.

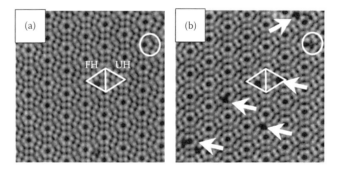

FIGURE 34.20   Si(111) clean surface (a) before and (b) after adatom desorption due to femtosecond pulsed-laser irradiation. (From Futaba, D. N. et al., *Appl. Phys. Lett.,* 83, 2333, 2003. With permission.)

extensively been studied. This system has a metal/semiconductor structure, and surface band bending, caused by charge transfer at the interface, was found to prevent MIT even below the critical temperature owing to the shift in Fermi level from the half-filled state (Terada et al. 2008). Using laser-combined STM, a reversible

MIT due to the band filling of the 1D surface state by the optical doping of carriers was observed (Figure 34.21).

These examples indicate that the laser-combined STM provides useful information on the nanoscale by observing the structural and electronic changes manipulated by photoexcitation.

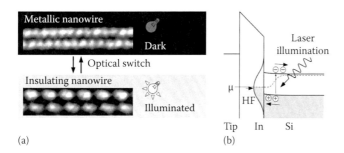

(a)

(b)

FIGURE 34.21 Schematic illustrations of metal–insulator transition of In/Si nanowire by optical doping (a) and band-filling mechanism (b). SPV reduces TIBB (Section 34.3.1), resulting in the change in the magnitude of band filling.

34.4.2 Optical Modulation of Bias Voltage

When STM is combined with photoillumination, a photo-induced change in tunnel current, other than the structural changes represented in Section 34.4.3, reflects various important sample properties when a suitable measurement condition is adopted (Grafström 2002). When a sample is illuminated with a p-polarization geometry, the oscillating electric field of the light induces a small high-frequency modulation of the bias voltage at the tunnel gap. This ac modulation in turn induces a small change in dc current when the I–V characteristics of the tunnel junction are nonlinear, as explained below. If the bias voltage is modulated as $V_0 + \Delta V \sin \omega t$, the tunnel current responds to it as $I(V_0 + \Delta V \sin \omega t)$. Then, its dc component can be approximated as $I(V_0) + (\partial^2 I/\partial V^2)\Delta V^2/2$ (Bragas 1997). Thus, the photoinduced current becomes proportional to $\partial^2 I/\partial V^2$ and the square of the modulation amplitude. With this concept, the ultrafast propagation of voltage pulses along a gold transmission line was detected by pulse-laser-combined time-resolved STM with a time resolution of 10 ps (Khusnatdinov et al. 2000). Although the origin of the spatially resolved contrast was the variation in dI/dV characteristics rather than the transient phenomenon itself, a spatial resolution of 1 nm was achieved with the temporal resolution.

34.4.3 With Variation in Photoexcitation Wavelength

In this method, tunnel current or displacement current is measured as a function of photoexcitation wavelength. With the modulation of the electric field applied to a sample, interband transition energies in semiconductors is analyzed on the nanoscale by detecting the change in the generation of the STM current reflecting the change in optical absorption coefficient depending on the wavelength (Hida et al. 2001). The chopping frequency should be chosen high enough to remove the thermal expansion effect (Section 34.2.1). An advanced rapid acquisition of reliable photoabsorption spectra has been realized by Fourier transform (FT) photoabsorption spectroscopy (Naruse et al. 2007), in which samples are illuminated with a multiplexed light from an FT interferometer, and the corresponding change in tunneling current is measured using an interferogram to obtain an FT spectrum.

34.4.4 Spin-Polarized Excitation

STM observation with an optically pumped GaAs tip allows us to conduct spin-sensitive STM measurements (Suzuki 1997). When the tip is illuminated by a circularly polarized laser, a spin-polarized population of electrons in the conduction band is produced. This is due to the spin-orbit splitting of the p-like valence band and the selection rules for optical transitions (Pierce and Meier 1976). The tunnel current from such a tip is sensitive to the magnetization of the sample surface. Magnetic domains of several hundred nanometer size in a Co thin film were imaged using this technique.

34.4.5 With Synchrotron Radiation

The combination of STM with synchrotron radiation has been studied to realize atom-resolved elemental identification (Saito et al. 2006), in which sample was intermittently illuminated by a mechanically chopped hard x-ray (~11 keV) from a synchrotron under a total reflection condition. The photoinduced current was detected by a lock-in amplifier and spatially mapped simultaneously with the STM topography, as shown in Figure 34.22. While no contrast in photoinduced current was obtained for x-rays with energies below the absorption edge of the inner core electron of a Ge atom, a clear contrast appeared with a spatial variation of a few nanometers when the energy was above the absorption edge, as shown in Figure 34.22b. Although the physical origin of the photoinduced current is not fully understood yet, this technique has the potential of realizing x-ray absorption spectroscopy on a single atom. To obtain a high resolution, a coating STM tip has been considered to block photoinduced electrons impinging on the sidewall of the tip and extract photoelectron current from a small area below the tip apex (Akiyama et al. 2005, Eguchi et al. 2006).

34.4.6 STM-Induced Light Emission

STM luminescence spectroscopy, which is also called photon STM measurement, is used to investigate local photoluminescence properties of conductive samples. It measures the inverse

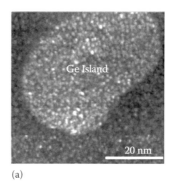
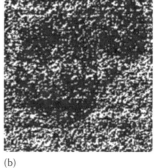

(a)   (b)

FIGURE 34.22 Ge island on Si(111) substrate observed by synchrotron-radiation-based STM. (a) STM topography and (b) photoinduced current by x-ray with energy above Ga adsorption edge. (From Saito, A. et al., *J. Synchrotron Rad.* 13, 216, 2006. With permission.)

photoelectric effect stimulated by the electrons or holes injected from an STM tip to the sample during STM measurement. The injected carriers cause photon emission via various physical phenomena that depend on the sample material, namely, surface plasmon polariton excitation in metallic samples, LUMO-HOMO transition in molecular samples (Qiu 2003), and interband transition in semiconductor samples (Sakurai 2004). The emitted photons are collected using lenses or concave mirrors into an optical fiber and guided to a spectrometer. Under a moderate condition, photon intensities up to 1000 count per second (cps) can be obtained.

Figure 34.23 shows STM luminescence spectra from a ZnEtioI molecule adsorbed on a thin alumina film grown on an NiAl substrate. The intramolecular position of the tunnel current injection was precisely selected on one of the four submolecular

lobes, as shown in the inset. After a suitable normalization, the three spectra measured using three different STM tips were observed to be identical, showing the fluorescence spectrum due to the LUMO-HOMO transition of electrons, with equally spaced (spacing: 40 ± 2 meV) small peaks marked by vertical lines in Figure 34.23a. These peaks correspond to molecular vibration excitations in the fluorescence process, as also observed in macroscopic fluorescence spectroscopy. These results suggest that STM luminescence realizes the vibrational spectroscopy of molecules with a submolecular resolution.

34.4.7 With Near-Field Optics

Near-field scanning optical microscopy (NSOM), a SPM that uses light as a probe, was developed in 1984 (Pohl 1984, Kim and Song 2007). A typical setup is illustrated in Figure 34.24. A sample surface is scanned using an optical fiber with a small aperture to investigate the local optical properties of the sample. The fiber is used either for illuminating the local area of the sample or for collecting photons from the sample, or for both of them. Far-field optical access to the sample may also be adopted, using a conventional optical microscope or a simple lens. The high spatial resolution of NSOM is based on the geometrical setup, where a subwavelength-scale light source or light detector (the small aperture in Figure 34.24) is placed at a subwavelength distance from the sample. Under such conditions, the electromagnetic field around the probe contains a large fraction of nonpropagating, evanescent (near-field) waves that decay exponentially as a function of the distance from the probe. Consequently, the area sensitively probed is determined by the probe size (aperture size), not by the light wavelength. The NSOM probe with a small aperture is made, for example, by pulling a heated optical fiber until it is elongated, thinned, and eventually broken. Successive chemical etching and

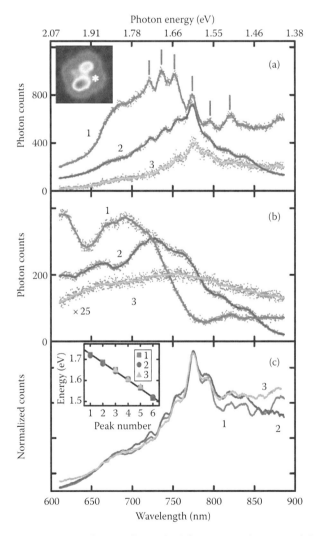

FIGURE 34.23 Vibrationally resolved fluorescence from one of the four submolecular lobes of ZnEtioI molecule (inset). (a) Raw fluorescence spectra from molecule with three different STM tips: Ag tips for spectra 1 and 2 and W tip for spectrum 3. (b) Raw spectra from NiAl substrate with same tips. (c) Smoothed molecular spectra (a) were divided by substrate spectra (b) for normalization. (From Qiu, X. H. et al., *Science*, 299, 542, 2003. With permission.)

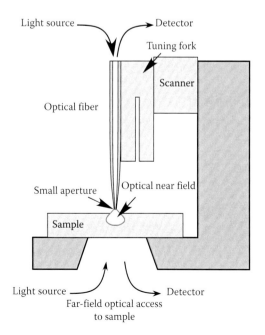

FIGURE 34.24 Typical NSOM setup.

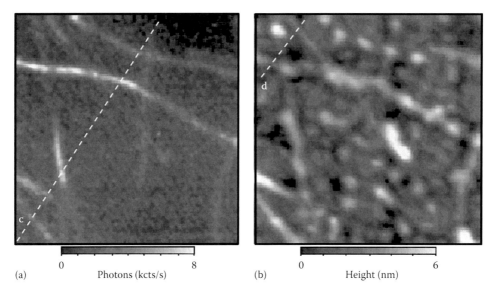

FIGURE 34.25 AFM-enhanced Raman microscopy of single-walled carbon nanotubes. (a) Raman intensity mapping and (b) AFM topography of same area. The scan area is $1 \times 1\,\mu m^2$. (From Hartschuh, A. et al., *Phys. Rev. Lett.*, 90, 095503, 2003. With permission.)

metal coating are also carried out to reduce the aperture diameter and to maximize the throughput of the aperture, respectively. The resolution achieved by such an aperture is ~30 nm. On the other hand, an apertureless NSOM has a higher resolution, ~10 nm (Kawata and Inouye 1995, Kawata et al. 1999, Novotny et al. 1995). An example is STM-enhanced Raman spectroscopy (Hartschuh et al. 2003), where the intensity of the far-field illumination with a spot size of a few micrometers is locally increased by the field enhancement effect by the AFM sharp metal tip. Figure 34.25 shows the high-resolution near-field Raman microscopy of single-walled carbon nanotubes conducted with AFM-enhanced Raman microscopy (Hartschuh et al. 2003).

To achieve a high spatial resolution, the regulation of the probe-sample distance at a certain subwavelength value (typically 1–50 nm) is important. When both the sample and the probe are conductive, an STM regulation mode, where the tunnel current between the probe and the sample is used to measure the distance, can be employed. For nonconductive samples, the distance is often controlled by an AFM regulation mode. For example, as shown in Figure 34.19, the optical fiber is mounted on a quartz tuning fork and oscillated parallel to the sample surface by a self-oscillator circuit. When the probe apex approaches the sample surface, the resonant frequency of the tuning fork shifts slightly, which gives the measure of the probe-sample distance.

Time-resolved NSOM is also a promising method for observing the carrier dynamics in materials and has been developed to provide the spatial and temporal resolutions of ~150 nm and ~150 fs, respectively (Levy et al. 1996, Nechay et al. 1999, Nagahara et al. 2004).

34.4.8 Fabrication

Laser-combined STM and related techniques can be used not only for the investigation of material properties but also for the fabrication of nanostructures (Jersch and Dickmann 1996, Lu 2000, Chimmalgi et al. 2005, Wang 2005). As described in Section 34.2, photo-irradiation on nanoprobes of STM and AFM, for example, causes the thermal expansion of the nanoprobes. Therefore, when probe position and pulsed-laser irradiation are designed, the mechanical indentation of sample surface is performed in a controlled manner. On the other hand, the field enhancement effect induced by p-polarized laser illumination is known to generate extremely localized stress accumulation in the subwavelength area beneath the nanoprobes, which mechanism is applicable for fabrication. This method is superior to the mechanical indentation because of the less damage on the probe apex. A demonstrative study shows that nanotrenches as narrow as 10 nm can be reproductively fabricated on an Au thin film with such a technique (Chimmalgi et al. 2005).

34.5 Summary

In this chapter, new microscopy techniques realized by combining STM and related techniques with optical technologies have been described in detail. The methods with high potentials that, for example, enable the investigation and real-space imaging of nanoscale electronic and structural dynamics induced by optical excitation have been explained. The direct view of such dynamics enables further advancement in nanoscale science and technology. Although it is difficult to cover all related topics, the materials discussed in this chapter are presented in a form that is instructive to nonspecialists as well as specialists. Understanding the basic mechanisms of the new microscopy techniques makes it possible to develop the ability of handling the issues we face in the measurement using the new microscopy techniques. With ingenuity, a new method can be developed.

References

Akiyama, K., Eguchi, T., An, T., and Hasegawa, Y. 2005. Fabrication of a glass-coated metal tip for synchrotron-radiation-light-irradiated scanning tunneling microscopy. *Rev. Sci. Instrum.* 78, 8: 083711.

Bragas, A.V. 1997. Spectroscopic response of photoinduced currents in a laser-assisted scanning tunneling microscope. *J. Appl. Phys.* 82: 4153–4158.

Chimmalgi, A., Grigoropoulos, C.P., and Komvopoulos, K. 2005. Surface nanostructuring by nano-/femtosecond laser-assisted scanning force microscopy. *J. Appl. Phys.* 97: 104319-1–104319-12.

Eguchi, T., Okuda, T., Matsushima, T., Kataoka, A., Harasawa, A., Akiyama, K., Kinoshita, T. et al. 2006. Element specific imaging by scanning tunneling microscopy combined with synchrotron radiation light. *Appl. Phys. Lett.* 89:243119.

Futaba, D.N., Morita, R., Yamashita, M., Tomiyama, S., and Shigekawa, H. 2003. Site-selective silicon adatom desorption using femtosecond laser pulse pairs and scanning tunneling microscopy. *Appl. Phys. Lett.* 83: 2333–2335.

Grafström, S. 1998. Thermal expansion of scanning tunneling microscopy tips under laser illumination. *J. Appl. Phys.* 83: 3453–3460.

Grafström, S. 2002. Photoassisted scanning tunneling microscopy. *Appl. Phys. Rev.* 91: 1717–1753.

Hamers, R.J. and Cahill, G.D. 1991. Ultrafast time resolution in scanned probe microscopies: Surface photovoltage on Si(111)-7x7. *J. Vac. Sci. Technol.* B9: 514–518.

Hamers, R.J. and Markert, K. 1990. Atomically resolved carrier recombination at Si(111)-7×7 surfaces. *Phys. Rev. Lett.* 64: 1051–1054.

Hartschuh, A., Sánchez, E.J., Xie, X.S., and Novotny, L. 2003. High-resolution near-field Raman microscopy of single-walled carbon nanotubes. *Phys. Rev. Lett.* 90: 095503-1–095503-4.

Hida, A., Mera, Y., and Maeda, K. 2001. Electric field modulation spectroscopy by scanning tunneling microscopy with a nanometer-scale resolution. *Appl. Phys. Lett.* 78: 3029–3031.

Hugel, T., Holland, N.B., Cattani, A., Moroder, L., Seitz, M., and Gaub, H.E. 2002. Single-molecule optomechanical cycle. *Science* 296: 1103–1106.

Jersch, J. 1999. Time-resolved current response of a nanosecond laser pulse illuminated STM tip. *Appl. Phys. A* 68: 637–641.

Jersch, J. and Dickmann, K. 1996. Nanostructure fabrication using laser field enhancement in the near field of a scanning tunneling microscope tip. *Appl. Phys. Lett.* 68: 868–870.

Kawata, S. and Inouye, Y. 1995. Scanning probe optical microscopy using a metallic probe tip. *Ultramicroscopy* 57: 313–317.

Kawata, Y., Xu, C., and Denk, W. 1999. Feasibility of molecular-resolution fluorescence near-field microscopy using multi-photon absorption and field enhancement near a sharp tip. *J. Appl. Phys.* 85: 1294–1301.

Khusnatdinov, N.N., Nagle, T.J., and Nunes, G. 2000. Ultrafast scanning tunneling microscopy with 1 nm resolution. *Appl. Phys. Lett.* 77: 4434–4436.

Kim, J.H. and Song, K.B. 2007. Recent progress of nano-technology with NSOM. *Micron* 38: 409–426.

Kronik, L. and Shapira, Y. 1999. Surface photovoltage phenomena: Theory, experiment, and applications. *Surf. Sci. Rep.* 37: 1–206

Levy, J., Nikitin, V., Kikkawa, J.M., Cohen, A., Samarth, N., Garcia, R., and Awschalom, D.D., 1996. Spatiotemporal near-field spin microscopy in patterned magnetic heterostructures. *Phys. Rev. Lett.* 76: 1948–1951.

Lewis, A. 1984 Development of a 500 Å spatial resolution light microscope. *Ultramicroscopy* 13: 227–231.

Lu, Y.F., Mai, Z.H., Zheng, Y.W., and Song W.D. 2000. Nanostructure fabrication using pulsed lasers in combination with a scanning tunneling microscope: Mechanism investigation. *Appl. Phys. Lett.* 76: 1200–1202.

Martin, O.J.F. and Girard, C. 1997. Controlling and tuning strong optical field gradients at a local probe microscope tip apex. *Appl. Phys. Lett.* 70: 705–707.

Masuda, H. 2005. Local photocurrent detection on InAs wires by conductive AFM. *Ultramicroscopy* 105: 137–142.

McEllistrem, M., Haase, G., Chen, D., and Hamers, R.J. 1993. Electrostatic sample-tip interactions in the scanning tunneling microscopy. *Phys. Rev. Lett.* 70: 2471–2474.

Nagahara, T., Imura, K., and Okamoto. H. 2004. Time-resolved scanning near-field optical microscopy with supercontinuum light pulses generated in microstructure fiber. *Rev. Sci. Instrum.* 75: 4528–4533.

Naruse, N., Mera, Y., Fukuzawa, Y., Nakamura, Y., Ichikawa, M., and Maeda, K. 2007. Fourier transform photoabsorption spectroscopy based on scanning tunneling microscopy. *J. Appl. Phys.* 102: 114301.

Nechay, B.A., Siegner, U., Achermann, M., Bielefeldt, H., and Keller, U. 1999. Femtosecond pump-probe near-field optical microscopy. *Rev. Sci. Instrum.* 70: 2758–2764.

Novotny, L., Pohl, D., and Hecht, B. 1995. Scanning near-field optical probe with ultrasmall spot size. *Opt. Lett.* 20: 970–972.

Othonos, A. 1998. Probing ultrafast carrier and phonon dynamics in semiconductors. *J. Appl. Phys.* 83: 1789–1830.

Pierce, D.T. and Meier, F. 1976. Photoemission of spin-polarized electrons from GaAs. *Phys. Rev. B* 13: 5484–5500.

Pohl, D.W. 1984. Optical stethoscopy: Image recording with resolution λ/20. *Appl. Phys. Lett.* 44: 651–653.

Prins, M.W.J., Jansen, R., Groeneveld, R.H.M., van Gelder, A.P., and van Kempen, H. 1996. Photoelectrical properties of semiconductor tips in scanning tunneling microscopy. *Phys. Rev. B* 53: 8090–8104.

Qiu, X.H. 2003. Vibrationally resolved fluorescence excited with submolecular precision. *Science* 299: 542–546.

Saito, A., Maruyama, J., Manabe, K., Kitamoto, K., Takahashi, K., Takami, K., Yabashi, M., and Tanaka, Y. 2006. Development of a scanning tunneling microscope for in situ experiments with a synchrotron radiation hard-x-ray microbeam. *J. Synchrotron Rad.* 13: 216–20.

Sakurai, M. 2004. Optical selection rules in light emission from the scanning tunneling microscope. *Phys. Rev. Lett.* 93: 046102-1–4.

Sakurai, T. and Watanabe, Y. 1999. *Advances in Scanning Probe Microscopy*. Springer, New York.

Shah, J. 1999. *Ultrafast Spectroscopy of Semiconductors and Semiconductor Nanostructures*. Springer, Berlin, Germany.

Shigekawa, H., Takeuchi, O., and Aoyama, M. 2005. Development of femtosecond time-resolved scanning tunneling microscopy for nanoscale science and technology. *Sci. Technol. Adv. Mater.* 6: 582–588.

Shigekawa, H., Yoshida, S., Takeuchi, O., and Terada, Y. 2008. Nanoscale dynamics probed by laser-combined scanning tunneling microscopy. *Thin Solid Films* 516: 2348–2357.

Sommerhalter, C., Matthes, T.W., and Boneberg, J. 1997. Tunneling spectroscopy on semiconductors with a low surface state density. *J. Vac. Sci. Technol. B* 15: 1876–1883.

Suzuki, Y. 1997. Magnetic domains of cobalt ultrathin films observed with a scanning tunneling microscope using optically pumped GaAs tips. *Appl. Phys. Lett.* 71: 3153–3155.

Takeuchi, O., Yoshida, S., and Shigekawa, H. 2004a. Light-modulated scanning tunneling spectroscopy for nanoscale imaging of surface photovoltage. *Appl. Phys. Lett.* 84: 3645–3647.

Takeuchi, O., Aoyama, M., Oshima, R., Okada, Y., Oigawa, H., Sano, N., Morita, R. et al. 2004b. Probing subpicosecond dynamics using pulsed laser combined scanning tunneling microscopy. *Appl. Phys. Lett.* 85: 3268–3270.

Terada, Y., Takeuchi, O., Shigekawa, H., Aoyama, M., Kondo, H., and Taninaka, A. 2007. Ultrafast photoinduced carrier dynamics in GaNAs probed using femtosecond time-resolved scanning tunneling microscopy. *Nanotechnology* 18: 044028.

Terada, Y., Yoshida, S., Okubo, A., Kanazawa, K., Xu, M., Takeuchi, O., and Shigekawa, H. 2008. Optical doping: Active control of metal–insulator transition in nanowire. *Nano Lett.* 8: 3577–3581.

Terada, Y., Yoshida, S., Takeuchi, O., Shigekawa, H. et al. 2010. Imaging ultrafast carrier dynamics in nanoscale potential landscapes by femtosecond time-resolved scanning tunneling microscopy, in press.

Tortonese, M. 1991. Atomic force microscopy using a piezoresistive cantilever. *International Conference on Solid-State Sensors and Actuators*, Pennington, NJ, IEEE Publication 91CH2817-5, pp. 448–451.

Wang, X. 2005. Large-scale molecular dynamics simulation of surface nanostructuring with a laser-assisted scanning tunnelling microscope. *J. Phys. D: Appl. Phys.* 38: 1805–1823.

Wiesendanger, R. 1994. *Scanning Probe Microscopy and Spectroscopy*. Cambridge University Press, Cambridge, U.K.

Williams, C. 2008. Fabrication of gold tips suitable for tip-enhanced Raman spectroscopy. *J. Vac. Sci. Technol. B* 26: 1761–1764.

Yamashita, M., Shigekawa, H., and Morita, R. 2005. *Mono-Cycle Photonics and Optical Scanning Tunneling Microscopy—Route to Femtosecond Angstrom Technology*. Springer, Berlin, Germany.

Yasuda, S., Nakamura, T., Matsumoto, M., and Shigekawa, H. 2003. Phase switching of a single isomeric molecule and associated characteristic rectification. *J. Am. Chem. Soc.* 125: 16430–16433.

Yoshida, S., Takeuchi, O., Shigekawa, H., Kikuchi, J., Kanitani, Y., and Oigawa, H. 2006. Tip-induced band bending and its effect on local barrier height measurement studied by light modulated scanning tunneling spectroscopy. *Surf. Sci. Technol.* 4: 192–196 (e-journal).

Yoshida, S., Takeuchi, O., Shigekawa, H., Kanitani, Y., Oshima, R., and Okada, Y. 2007. Microscopic basis for the mechanism of carrier dynamics in an operating p-n junction examined by using light-modulated scanning tunneling spectroscopy. *Phys. Rev. Lett.* 98: 026802–026805.

Yoshida, S., Kanitani, Y., Takeuchi, O., and Shigekawa, H. 2008. Probing nanoscale potential modulation by defect-induced gap states on GaAs(110) using light-modulated scanning tunneling spectroscopy. *Appl. Phys. Lett.* 92: 102105–102107.

Contact Experiments with a Scanning Tunneling Microscope

Jörg Kröger
Christian-Albrechts-
Universität zu Kiel

35.1 Introduction

From quantum mechanics, we know that properties of materials at the atomic scale differ from those of their macroscopic counterparts. One of the properties is the electrical conductance, which is discussed in detail in this chapter. For macroscopic conductors, Ohm's law holds and states that the conductance of the conductor is inversely proportional to its length. Intuitively, we can understand Ohm's law if we consider electrons as particles, which, on their way from one electrode to the other, suffer scattering in the conductor. However, intuition fails if dimensions reach the Fermi wavelength of electrons. Electrons are now transported ballistically through the microscopic conductor and the character of conductance changes conceptually. Currently, many experimental and theoretical investigations are devoted to an understanding of how electrons are transported through atoms, atomic wires, and molecules. The reason for this increased interest may be found in the ongoing miniaturization in the microelectronic industry. Although we are far from replacing silicon technology by the so-called molecular electronics, it is important to unravel the quantum properties of conductors with nanometer dimensions and to discover new challenges, which must be faced upon reducing the size.

This chapter mainly addresses ballistic electron transport through constrictions with atomic dimensions. Therefore, it is useful to characterize this conductance regime. While for macroscopic systems Ohm's law is applicable, in atomic-sized conductors such simple approaches are no longer valid. Atomic-sized conductors are a limiting case of mesoscopic systems whose transport properties are influenced by quantum coherence. A measure of the preservation of quantum coherence is the phase coherence length, L_Φ, which for a mesoscopic conductor is larger than its length L. We can further compare the length of the conductor and the elastic mean free path, ℓ, of the electrons. If $\ell \ll L$, then the conductance is referred to as diffusive: starting from one electrode, the momentum of the electrons changes a lot of times along its trajectory to the other electrode (Figure 35.1a). For $\ell > L$, the electron momentum is constant and the electron motion is limited only by scattering at the boundary of the conductor. This conductance regime is referred to as ballistic (Figure 35.1b).

Conductance of a point contact in the diffusive regime was studied theoretically already in 1873 by Maxwell (1954). By solving Laplace's equation for an orifice with radius R and resistivity ϱ, Maxwell found a conductance of $G = 2R/\varrho$. This is a classical result reflecting Ohm's law. The situation changes when the ballistic regime of transport is considered. A large potential gradient near the contact causes the electrons to accelerate within short distances. In a semiclassical approach, the conductance was given by Sharvin in 1965 according to $G = e^2 k_F^2 R^2/(2h)$ with k_F the Fermi wave vector, $-e$ the electron charge, and h Planck's constant (Sharvin, 1965). The conductance of a so-called quantum point contact, that is, a contact whose width is comparable to the Fermi wavelength λ_F, was calculated by Landauer (1957):

$$G = G_0 \sum_{i,j=1}^{n} |\tau_{ij}|^2, \qquad (35.1)$$

where

$G_0 = 2e^2/h$ is the quantum of conductance

$|\tau_{ij}|^2$ specifies the transmission probability of an electron incident on the contact mode i and transmitted into mode j

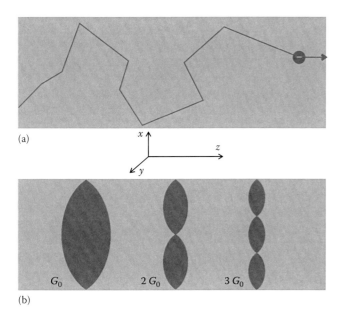

(a)

(b)

FIGURE 35.1 (a) Diffusive transport. The mean free path of electrons is considerably smaller than the length of the conductor. Many scattering events of the electrons take place on their trajectory from one electrode to the other. (b) Ballistic electron transport and conductance quantization in an atomic-sized conductor between two electrodes. The electron motion is confined by hard walls at the boundary of the constriction, whereas the potential along the conductor (*z*-direction) is assumed to be flat. The first three eigenmodes are illustrated as standing waves with 1, 2, 3 antinodes giving rise to 1, 2, 3 conductance channels each contributing a quantum of conductance G_0.

Büttiker (1988) has pointed out that it is possible to define another set of modes that are referred to as eigenchannels and do not mix in the scattering region of the quantum point contact. As a consequence, in Equation 35.1, we set $|\tau_{ij}|^2 = T_i \delta_{ij}$ with T_i the transmission probability of the *i*th eigenchannel. A recipe for the transformation of an arbitrary set of channels to eigenchannels has been given by Brandbyge and Jacobsen (1997). If the cross section of the quantum point contact varies slowly along the constriction, the transmission probability T_i will be either 1 or 0 (Bogachek et al., 1990; Torres et al., 1994; Brandbyge et al., 1995, 1997). Neglecting further the smearing due to finite temperatures as well as any effects caused by internal disorder and inelastic impurity scattering, the sum of transmission probabilities will attain integer values. In this case, the conductance will be quantized in units of G_0. Landauer's concept relates the electrical conductance of atomic-sized conductors to the transmission probability of an electron wave incident on one side of the conductor. This idea, as we will see in a variety of examples in this chapter, applies to electron transport through atoms and molecules. The Landauer approach to the electrical conductance through molecules has been employed for carbon nanotubes (Dekker, 1999) as well as for biologically relevant molecules, for example, DNA (Fink and Schönenberger, 1999; Porath et al., 2000; Kasumov et al., 2001).

Below we illustrate the quantization of the conductance in a quantum point contact in a simplified picture (Bürgi, 1999).

Figure 35.1 shows the conductance quantization as a consequence of quantized transverse electron momentum in the contact constriction. Assuming that the potential is flat along the conductor (*z*-direction in Figure 35.1) and confines the electron motion in transverse directions (*x* and *y* in Figure 35.1), the electron energy may be written as

$$E = E_i + \frac{\hbar^2 k^2}{2 m_e}, \tag{35.2}$$

where
E_i denotes the transverse mode energy
k the wave vector along *z*
m_e the free electron mass

Coupling this conductor to two reservoirs with potential difference V gives rise to a net current of (Datta, 1995)

$$I_i = e^2 V \varrho_i(E_F) \nu_i(E_F) \tag{35.3}$$

in the *i*th channel provided that $E_i < E_F$. The density of states, ϱ_i, as well as the group velocity, ν_i, are evaluated at the Fermi energy, E_F, since only electrons in a narrow region around this energy contribute to the current. Every mode *i* accommodates a one-dimensional electron gas with $\varrho_i = 2/(h\nu_i)$ leading to a total conductance of

$$G = \frac{1}{V} \sum_i I_i = G_0 \sum_i \Theta(E_F - E_i) = G_0 n \tag{35.4}$$

with *n* the number of open transverse modes or conductance channels and Θ the Heaviside step function. In this simple picture, changing the energy of ballistically transferred electrons gives rise to standing waves with *n* antinodes. Each antinode leads to an additional conductance of G_0. This simplified picture of conductance quantization is in agreement with the above-mentioned special case of the multichannel Landauer expression (Landauer, 1957, 1970, 1981; Büttiker et al., 1985; Büttiker, 1988).

The quantization of conductance was first shown experimentally in GaAs–AlGaAs heterostructures by two groups (van Wees et al., 1988; Wharam et al., 1988). In these experiments, the width of the point contact was controlled by an applied gate voltage. Upon varying the width of the point contact, van Wees et al. (1988) and Wharam et al. (1988) observed a change of the conductance in steps of G_0 with intermediate conductance plateaus. Interestingly, Montie et al. (1991) observed the optical analogue of quantized conductance of a point contact. By two-dimensional diffuse illumination of a slit, they found a stepwise increase of the transmission cross section whenever the slit width passed through integer multiples of half of the light wavelength. The diffuse illumination corresponds to the isotropic velocity distribution of incoherent electron waves incident on the point contact.

The first report of a metallic contact of atomic dimension fabricated with the tip of a scanning tunneling microscope (STM) was given by Gimzewski and Möller (1987). Moving an iridium tip toward a thick silver layer adsorbed on Si(111), Gimzewski and Möller (1987) observed a sharp transition from tunneling to contact. Relating the resistance of the tip-surface junction to the Sharvin resistance (Sharvin, 1965; Jansen et al., 1980), Gimzewski and Möller (1987) concluded that the contact radius was ≈1.5 Å and thus corresponded to the electrical resistance of one or two atoms in close contact. García and Escapa then suggested using a STM to observe oscillatory conductance caused by the confinement of electrons (García and Escapa, 1989). Experimental data obtained by Pascual et al. (1993) were interpreted along these lines.

With the development of the mechanically controlled break junction technique (Moreland and Ekin, 1985; Muller et al., 1992b), conductance steps for niobium and platinum electrodes were reported in 1992 (Muller et al., 1992a). Striving for an objective data analysis, statistics on a wealth of conductance curves must be performed. Typically, results are presented in the form of histograms (Krans et al., 1995b). Interestingly, jumps in the integer conductance plateaus were observed and thus distracted from a picture of simply quantized conductance. In fact, these observations inflamed a debate about the origin of conductance steps (see, e.g., Olesen et al., 1994; Krans et al., 1995a; Olesen et al., 1995). A scenario based on mechanical instabilities through which the contact area evolves discontinuously upon breaking or forming the contact could explain conductance steps of the order of G_0 even for imperfect conductors (Landman et al., 1990; Todorov and Sutton, 1993). However, this scenario could not account for the pronounced peaks in the conductance histograms at integer multiples of the quantum of conductance. Combination of atomic force microscopy and STM (Agraït et al., 1995; Rubio et al., 1996) led to the picture of conductance through atomic-sized conductors we have today: integer values of G_0 indicate quantized conductance due to laterally confined quantum modes in the conductor, whereas a jump between the different values is caused by abrupt atomic rearrangements.

By using the mechanically controlled break junction technique, Scheer et al. (1998) presented current–voltage characteristics of superconducting single-atom contacts, from which the number and the transmission probability of transport channels could be estimated. The basis of this experiment is provided by so-called Andreev reflections, which leave their spectroscopic fingerprint in the region of voltages below $2\Delta/e$, where Δ denotes the superconducting gap energy (Klapwijk et al., 1982; Arnold, 1987). The experimental results reported by Scheer et al. (1998) are in agreement with the following simple picture provided by Cuevas et al. (1998): the number of conductance channels of a single-atom contact is determined by the number of valence orbitals of the atom. For instance, noble metal atoms with a single s valence orbital give rise to a single transport channel, whereas transition metal atoms with d valence orbitals provide five transport channels. The conductance of the respective atom then still depends on the transmission probability of these channels.

An extensive review article on quantum properties of atomic-sized conductors as measured mainly by the mechanically controlled break junction technique was given by Agraït et al. (2003). The reader who is interested in experimental aspects of this technique and in a thorough overview of experimental results and theoretical models is referred to this article.

We notice that table-top experiments performed under ambient conditions likewise revealed conductance quantization. For instance, quantum point contacts were formed by placing two macroscopic metallic wires close to each other and inducing vibrations in the wires by tapping the table top on which the experiment was set up. The wires moved in and out of contact and just before the loss of the contact quantized conductance was observed (Costa-Krämer et al., 1995). A slightly different experimental setup based on the same idea consisted of hanging a pin very close to a gold-plated wafer. Small vibrations cause the pin to go in and out of contact (Landman et al., 1996). Hansen et al. (1997) showed that quantized conductance could also be observed in the breaking contact of commercial electromechanical relays. In a fully automated setup, point contacts were formed and broken several thousand times and the conductance was measured.

A variety of experiments using the tip of a STM to study the conductance of atomic-sized contacts were performed (Agraït et al., 1993; Pascual et al., 1993; Olesen et al., 1994; Gai et al., 1996). The main characteristics of these experiments consisted in moving the tip toward the surface, forming and stretching the contact, and simultaneously recording the current at a fixed junction voltage. A step forward to more control of the contact junction was provided by the combination of STM with transmission electron microscopes whose focal point was set at the tip apex. With this technique, Ohnishi et al. (1998) were able to relate the quantized conductance through a gold chain directly to the number of gold atoms in the contact. Erts et al. (2000) used a similar approach to study the conductance of gold point contacts as a function of the contact radius. Thus, a direct comparison of the diffusive and ballistic transport regimes was performed and the predictions by Maxwell (1954) and Sharvin (1965) were verified.

Recent STM experiments (Joachim et al., 1995; Yazdani et al., 1996; Bürgi, 1999; Moresco et al., 2001; Limot et al., 2005; Jensen et al., 2007; Kröger et al., 2007; Néel et al., 2007a,b,c, 2008a,b,c; Kröger, 2008; Schulze et al., 2008; Temirov et al., 2008) make use of the imaging capability of the instrument with atomic resolution. Contacts to flat surfaces, adsorbed atoms, or molecules can therefore be performed controllably and reproducibly without deterioration of neither the tip nor the sample status. These experiments thus allow to characterize the electrodes (tip and substrate surface) and the contact constriction (atom or molecule) prior to and after the contact experiment. Therefore, specific atoms or molecular orbitals can be addressed and conductances measured. Because of the detailed characterization of the contact, the results of the experiments may serve as precise input for model calculations.

35.2 Experiment

Measurements presented in this chapter were performed using a custom-made STM operated at 8 K and in ultrahigh vacuum with a base pressure of 10^{-9} Pa. Sample surfaces and chemically etched tips were cleaned by argon ion bombardment and annealing. *In vacuo*, the tip was gently indented into the substrate until single adatoms appeared with nearly circular shape in constant-current STM images. As a consequence of this tip preparation, we expect the tip apex to be covered with substrate material. Single atoms originating from substrate material were deposited onto the surface by controlled tip-surface contacts as reported in Limot et al. (2005) and reviewed in Section 35.3.1. Deposition of cobalt atoms was performed using an electron beam evaporator whereas C_{60} molecules were sublimed from a heated tantalum crucible. Conductance curves were acquired such as to cover tunneling to contact regimes. To this end, the tip was displaced by 5–6 Å toward the surface and simultaneously the current ranging from the order of 1 nA to the order of 10 µA was recorded. A linear voltage ramp was applied to the piezoelectric tube, which carries the tip. As a result, the tip was approached to or retracted from the sample with velocities of 30–50 Å/s. Figure 35.2 illustrates the principle of the experimental setup. The tunneling junction consists of a sharp metal tip, which is attached to a piezoelectric tube, and a conducting sample. The tunneling current, I, is fed into a transimpedance amplifier, which converts the current into a voltage. The tunneling voltage, V, is applied to the sample. Zero displacement of the tip is defined for a given pair of current and voltage, at which the feedback loop is opened prior to conductance measurement. The differential conductance (dI/dV) was measured by superimposing a sinusoidal voltage onto the sample voltage (root-mean-square amplitude between 1 and 5 mV, frequency of 10 kHz) and by detecting the

current response with a lock-in amplifier. Different gains of the transimpedance amplifier had to be set to measure the current that varied over four to five orders of magnitude. The input impedance of the amplifier is 50 Ω at the gain setting, which is relevant for the single-atom contacts. Further, cables that carry tunneling voltage and current exhibit a resistance of 200 Ω. Consequently, in single-atom contacts with a quantized resistance of $R_0 = G_0^{-1} \approx$ 12.9 kΩ, the voltage drop at peripherical resistances is less than 2% of the voltage applied to the junction. Therefore, the associated error, which occurs by calculating the contact conductance via $G = I/V$, is smaller than the precision of the method by which this value is determined from conductance curves.

35.3 Contact to Surfaces and Adsorbed Atoms

35.3.1 Surfaces

A decrease of the tip-surface distance is related, for instance, to a shift of the binding energy of Shockley-type surface states on noble metal (111) surfaces (Limot et al., 2003; Kröger et al., 2004). For a certain displacement interval, the binding energy shifts almost linearly to higher values. Then an accelerated shift sets in, which cannot be explained by a junction with constant tip-surface geometry. Rather, due to structural relaxations of the tip as well as of the surface, the distance between tip and surface decreases faster than that given by the tip displacement (Limot et al., 2003; Kröger et al., 2004). Increasing the current further typically leads to a controlled contact between the tip and the surface via a single atom. This scenario can be monitored by a conductance-versus-displacement curve as shown in Figure 35.3. The conductance curve is divided into three parts (denoted as 1, 2, and 3 in Figure 35.3).

In Part 1, the conductance curve appears as a straight line on a logarithmic scale, that is, the current varies exponentially with the displacement that is characteristic for a tunneling current. Therefore, Part 1 of the conductance curve is referred to as the tunneling regime.

Part 2 of the conductance curve signals an abrupt change of the exponential behavior. With the time resolution of data acquisition (≈0.1 ms), a discontinuous jump of the conductance is observed (Figure 35.3b). This part of the conductance-versus-displacement characteristics is referred to as the transition regime since it shows the transition from tunneling to contact. The conductance right after the jump may be defined as the contact conductance G_c and is, for the particular case of Cu(111) presented in Figure 35.3, given by $G_c \approx 0.8 \ G_0$ (see Table 35.1 for a comparison of further contact conductances). We notice, however, that in Limot et al. (2005), the contact conductance was defined by the intersection of a linear extrapolation of the tunneling conductance with the conductance curve in the contact regime. Consequently, contact conductance values taken from Limot et al. (2005) are larger than those obtained by the present definition. The tip-surface contact formation exhibited a random character with respect to the contact displacement as well as to the contact conductance (see also the error margins for G_c in Table 35.1). This behavior can be understood in the light of recent

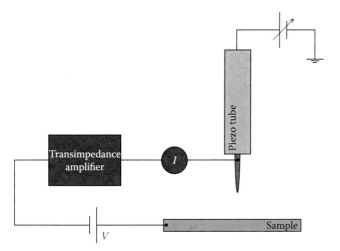

FIGURE 35.2 Illustration of a tunneling junction circuit. The piezoelectric tube carries the tip. Applying a voltage ramp to its inner segment moves the tip along the surface normal. The tunneling current, I, is fed into a variable gain transimpedance amplifier, which converts the current into a voltage. In all experiments presented in this chapter, the tunneling voltage, V, was applied to the sample.

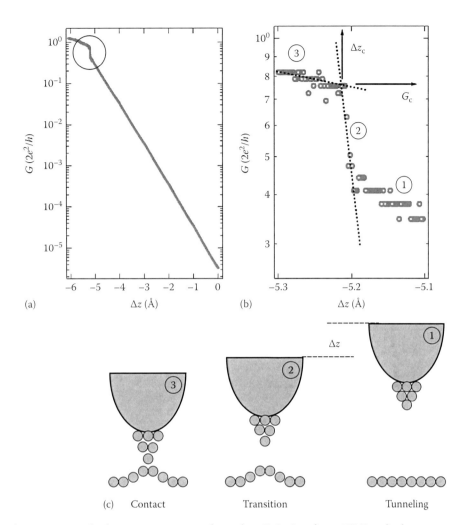

FIGURE 35.3 (a) Conductance-versus-displacement curve acquired on a clean Cu(111) surface at 8 K. Zero displacement corresponds to a tunneling gap set by feedback loop parameters of 0.1 nA and 0.4 V prior to data acquisition. The conductance is plotted in units of the quantum of conductance on a logarithmic scale. (b) Close-up view of transition from tunneling to contact including the graphical definition of contact conductance, G_c, and displacement, Δz_c. Three regimes are discernible: tunneling (1), transition (2), and contact (3). (c) Illustration of contact formation between tip and flat surface. In the transition regime (2), adhesive forces between the tip and surface lead to relaxations of the tip and surface crystal structure. The tip-surface distance does not change according to the tip displacement Δz. The contact regime (3) reflects ballistic electron transport through a single atom.

simulations (Hofer et al., 2001). The jump to contact depends on where the approach is performed on the surface. Regardless of its chemical nature, when the tip is positioned on the top of a surface atom, the jump should be detected at tip–surface distances higher by ≈0.5 Å earlier compared with a threefold hollow position, whereas all other positions lead to a jump to contact within this range. Since surface atoms of compact (111) surfaces are not usually resolved, the conductance measurements were performed at random locations of the surface and the jumps occur randomly within a finite displacement interval of the tip around 0.5 Å, which is in agreement with the above simulation. Therefore, contact conductances obtained for flat surfaces usually exhibit larger standard deviations [≈0.5 G_0 (Limot et al., 2005)] than those obtained for single adatoms [≈0.05 G_0 (Limot et al., 2005)].

In the contact regime (Part 3), the conductance of the tip-surface junction varies slowly with tip displacement. Retracting the tip after this discontinuous jump and imaging the contacted

surface area in ≈70%–80% of the cases, a single adatom was found. This observation together with simulations (Limot et al., 2005) put forward a scenario, in which tip and surface are connected via a single atom during contact.

Why is a jump to contact observed? The jump occurs when chemical bonds between the surface and the tip apex start to weaken the adhesion of the atom to the tip structure. In this case and over a relatively small distance variation of less than 0.1 Å, the atom will be transferred from the tip to the surface. We notice in this context that from continuum models and atomistic simulations, an attractive force between two clean metal surfaces was predicted that leads to an intrinsic instability at a distance of 1–3 Å and causes the surfaces to snap together on a timescale of the order of 100 fs (Pethica and Sutton, 1988; Smith et al., 1989).

In all cases, we observed a contact conductance close to G_0 that together with the observation of a single adatom after contact is evidence for a single-atom contact. Monovalent atoms

TABLE 35.1 Contact Conductance, G_c, for Various Surfaces and Adatoms

Substrate	Tip Apex	$G_c(G_0)$	Reference
Ag(111)	Ag	1.5 ± 0.6	Limot et al. (2005)
Au(111)	Au	1.0 ± 0.4	Kröger et al. (2007)
Cu(111)	Cu	1.1 ± 0.3	Limot et al. (2005)
Ag(111)–Ag	Ag	0.93 ± 0.05	Limot et al. (2005)
Au(111)–Au	Au	0.9	Kröger et al. (2007)
Cu(111)–Cu	Cu	0.98 ± 0.06	Limot et al. (2005)
Cu(111)–Co	Cu	1.0	Néel et al. (2009)
Cu(100)–Co	Cu	1.0	Néel et al. (2007c)
Cu(100)–Gd	Au	0.52 ± 0.10	Bürgi (1999)
Cu(100)–Mn	Au	0.87 ± 0.07	Bürgi (1999)
Cu(100)–Mn	Mn	0.94 ± 0.04	Bürgi (1999)
Ni(110)–Xe	W	0.2	Yazdani et al. (1996)
Ni(110)–Xe	Xe	0.001	Yazdani et al. (1996)

Note: In most cases, the tip apex is covered with substrate material due to the *in vacuo* treatment of the tip. For Cu(110)–Mn, –Gd, an Au tip was used, which was modified by picking up an Mn adatom. The W tip used for contact measurements on Ni(110)–Xe was terminated by a single Xe atom.

like Ag, Au, and Cu are expected to provide a single channel for electron transport (Scheer et al., 1998). In particular, the transmission probability for a single Au atom contact was calculated to give ≈1 for the sp_z orbital that exhibits the highest density of states at the Fermi level (Cuevas et al., 1998).

35.3.2 Adatoms

The controlled contact to individual adsorbed atoms using the tip of an STM was first investigated by Yazdani et al. (1996). The conductance of individual xenon atoms adsorbed on a

nickel surface was studied using a bare tungsten tip and a tip terminated by a xenon atom. Contact conductances of ≈0.2 G_0 and of ≈0.001 G_0 were reported for the single xenon atom and the two-atom chain, respectively. In particular, no jump to contact was observed. In Bürgi (1999), contact experiments were reported for single Mn and Gd atoms adsorbed on a Cu(100) surface. A clean Au tip was used to contact these atoms giving rise to contact conductances of (0.87 ± 0.07) G_0 and (0.52 ± 0.10) G_0 for Mn and Gd atoms, respectively. A contact junction consisting of two Mn atoms bridging the gold tip and the copper surface was fabricated by contacting an adsorbed Mn atom with a Mn-terminated Au tip. The resulting contact conductance was (0.95 ± 0.04) G_0 (Bürgi, 1999). All conductance-versus-displacement curves exhibited a smooth transition from tunneling to contact rather than a jump to contact.

Contact to metallic adatoms using the tip of an STM were then reported by Limot et al. (2005), Néel et al. (2007a,c). In accordance with contact experiments reported by Yazdani et al. (1996) and Bürgi (1999), no discontinuous transition from tunneling to contact was observed. Rather, for the adatoms investigated, that is, Ag on Ag(111) and Cu on Cu(111) (Limot et al., 2005), and Co on Cu(100) (Néel et al., 2007a,c), a continuous transition takes place. Additionally, in contrast to the tip-surface contact, no material is transferred to the surface. Moreover, the tip-adatom contact does not exhibit the random character as observed for the tip-surface contact (see also the significantly lower standard deviation of G_c in Table 35.1). In Figure 35.4, we present a conductance curve for Cu(111)–Cu. Similar to the tip-surface junction, three regions of the conductance characteristics are discernible. In the tunneling regime, the current varies exponentially with the displacement. As a marked difference, the transition between tunneling and

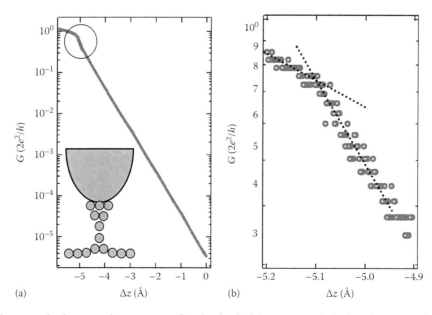

(a)   (b)

FIGURE 35.4 (a) Conductance–displacement characteristics of single adsorbed Cu atom on a Cu(111) surface acquired at 8 K. Zero displacement corresponds to a tunneling gap set by feedback loop parameters of 0.1 nA and 0.4 V prior to data acquisition. Inset: illustration of tip-adatom contact. (b) Close-up view of transition between tunneling and contact regime.

contact is now continuous at the time resolution of data acquisition (Figure 35.4b). Using density functional methods it was argued (Limot et al., 2005) that in case of the tip-adatom contact the tip–adatom interaction does not surmount the value, which is required for a tip fracture. This may be understood in terms of the tip apex atom being farther away from the surface at the tip-adatom contact than at the tip–surface contact. Moreover, elastic constants of the adatom were shown (Limot et al., 2005) to be almost twice as large as those found for the flat surfaces. This increase of the adatom bonding was traced to the creation of a surface dipole due to Smoluchowski effect. Consequently, ever when the tip apex atom is very close to the adatom relaxations are small compared to the tip–surface contact.

Noble metal atoms are monovalent and are thus expected to exhibit a single conductance channel. If the transmission probability equals 1, then the contact conductance should be exactly 1 G_0. There are possible reasons why the ideal contact conductance of a monovalent atom is not observed in the experiment. Equation 35.1 gives the contact conductance as a sum over conductance channels that exhibit a certain transmission probability. In case of a monovalent atom, Equation 35.1 simply reads $G = \tau\, G_0$. A deviation of the conductance from 1 G_0 may therefore be related to a deviation of the transmission probability τ from 1. In fact, the transmission probability for a given atomic conductance channel depends on the connection of the atom to neighboring atoms in the metallic leads. Only in the absence of defects in the leads close to the central atom and in the absence of excitations of other degrees of freedom, we would measure a transmission probability of one. Any partial reflection of the electron wave as a result of, for instance, the mismatch of the waves at both sides of the constriction may alter the current and thus the conductance. Therefore, scattering centers near the contact give rise to a number of corrections, the most obvious of which is a reduction of the total conductance.

The conductance of a single Co atom adsorbed on Cu(100) was determined to be 1.03 G_0 (Néel et al., 2007c). According to Scheer et al. (1998), transition metals give rise to five conductance channels. Thus, completely open channels would lead to a contact conductance of 5 G_0, rather than ≈ 1 G_0 as observed for the Co atom. Assuming that all channels contribute equally to the total conductance, each conduction channel would exhibit a transmission probability of ≈ 0.2. Using the mechanically controlled break junction technique (Scheer et al., 1998; Ludoph et al., 2000), Nb atomic contacts were investigated yielding conductances between 1.5 and 2.5 G_0. Here five conduction channels with transmission probabilities ranging from ≈ 0.02 to ≈ 0.73 were reported.

35.4 Contact to Single Adsorbed Molecules

35.4.1 Conductance and Local Heating of a C_{60} Molecule

The first STM experiment addressing a tip-molecule contact was reported by Joachim et al. (1995) for C_{60} adsorbed on Au(110). They studied the conductance of the tip-molecule junction as a function of the tip displacement at room temperature. Below we review the results of more recent experiments (Jensen et al., 2007; Néel et al., 2007a,b, 2008a; Kröger, 2008), addressing the controlled contact to a C_{60} molecule on Cu(100) at low temperatures.

Figure 35.5a shows an STM image of Cu(100)–C_{60} in a submonolayer coverage regime. The bright and dim rows correspond to chains of C_{60} molecules along the indicated crystallographic directions. By combining STM and x-ray photoelectron diffraction data, Abel et al. (2003) concluded that these stripe patterns were due to a C_{60}-induced missing-row reconstruction of the underlying substrate surface. The close-up view in Figure 35.5b displays individual C_{60} molecules within the bright and dim stripes with submolecular resolution. We attribute this pattern exposed to vacuum to the spatial distribution of the local density of states of the second-to-lowest unoccupied molecular orbital (LUMO + 1) in agreement with Hou et al. (1999) and Lu et al. (2003). Figure 35.5c identifies the orientation of C_{60} molecules encircled by dashed lines in Figure 35.5b. Molecules **1** and **2** adsorb on a single missing row of the Cu(100) surface, whereas molecules **3** and **4** reside in a double missing row. Molecule **1** exhibits a carbon–carbon bond between a pentagon and a hexagon at the top, molecule **2** shows a carbon–carbon bond between two adjacent carbon hexagons, and molecules **3** and **4** expose a pentagon and a hexagon at the top, respectively. This section focuses the discussion on molecules of type **1**.

How does the conductance of an individual molecule change with decreasing distance between the tip and the molecule? Figure 35.6 shows the conductance–displacement characteristics of molecules with orientation **1**. The conductance curves for the single molecule bear resemblance to the conductance curve of a single adatom. In particular, a continuous transition between tunneling and contact is again observed. A marked difference, however, concerns the contact conductance which is only ≈ 0.25 G_0. Figure 35.6 contains calculated data depicted as squares. A detailed description of the applied model can be found in Néel et al. (2007b). The most intriguing observation is that theory is able to model the sharp increase of the conductance as well as a contact conductance below G_0. The rapid rise of the conductance in the displacement interval between ≈ -1.6 and ≈ -2.0 Å can be understood from the relaxed tip-molecule geometries on which the model calculations are based. As the electrode separation is reduced by only 0.05 Å, the tip-molecule distance decreases by 0.84 Å. This strong reduction of the tip-molecule distance corresponds to the formation of a bond between the tip apex and the C_{60} that hence effectively closes the tunneling gap.

In this transition regime between tunneling and contact, only small energy differences discriminate between the configurations with or without the tip-molecule bond (Néel et al., 2007c). At finite voltages and under nonequilibrium conditions imposed by the bias voltage, it is thus likely that the junction will fluctuate between these different configurations. With G_t and G_c, the conductances just before and just after the contact, respectively, a thermally averaged conductance can be calculated according to

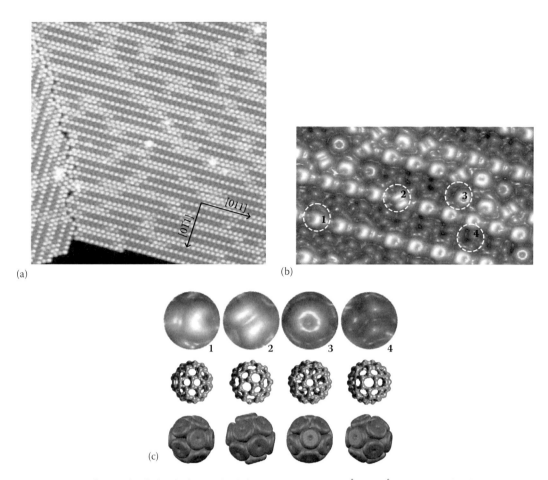

(a)   (b)

(c)

FIGURE 35.5 (a) STM image of a C_{60} island adsorbed on Cu(100) ($V = 1.5\,V$, $I = 1\,nA$, $400\,\text{Å} \times 400\,\text{Å}$, $T = 8\,K$). Molecule chains appear as bright and dim rows due to adsorption-induced surface missing-row reconstruction. (b) Close-up view of (a) showing individual C_{60} molecules with intramolecular structure ($130\,\text{Å} \times 75\,\text{Å}$). (c) Molecule orientations **1–4** indicated in (b) by dashed circles on missing-row reconstructed Cu(100). Molecules with orientations **1** (C bond between C pentagon and C hexagon at top), **2** (C bond between two C hexagons at top) reside on a single missing row, whereas molecules with orientations **3** (C pentagon at top) and **4** (C hexagon at top) are adsorbed in a double missing row. The bottom panel illustrates the spatial distribution of the second-to-lowest unoccupied molecular orbital, which is centered on the C pentagons of the molecule.

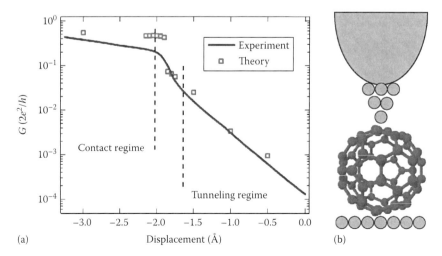

(a)   (b)

FIGURE 35.6 (a) Conductance-versus-displacement curve for a C_{60} molecule adsorbed on Cu(100) with orientation **1**. Owing to the high number of data points, experimental data appear as a line. Calculated data are depicted as squares. Zero displacement corresponds to a tip-molecule distance defined by feedback loop parameters of $3\,nA$ and $0.3\,V$. (b) Illustration of the single-molecule contact.

$$\bar{G}(\Delta z) = \frac{G_t(\Delta z)\exp(-\beta E_t(\Delta z)) + G_c(\Delta z)\exp(-\beta E_c(\Delta z))}{\exp(-\beta E_t(\Delta z)) + \exp(-\beta E_c(\Delta z))} \quad (35.5)$$

with $\beta = (k_B T)^{-1}$. E_t and E_c denote the total energies of a tunneling and a contact configuration, respectively. In Figure 35.7, we focus on the conductance curve in a displacement interval around the instability point. Increasing the effective temperature of the tip-molecule junction from 0 K (squares in Figure 35.7a) via 8 K (thin dashed line) up to 400 K (thick dashed line) leads, first, to an increase of the width of the transition regime and, second, to a shift of the instability displacement to more positive values. At an effective temperature of 400 K, the width of the calculated conductance curve matches well the width of the experimental curve. Because of this elevated temperature, we propose a scenario of local heating of the molecule in contact with the tip. This suggestion is further corroborated by the experimental observation of a voltage-dependent width of the transition regime. The corresponding effective temperatures according to Equation 35.5 are indicated in Figure 35.7b.

The dependence of the effective temperature on the voltage was modeled for atomic-sized contacts by assuming a bulk thermal conduction mechanism (Todorov, 1998; Todorov et al., 2002; Chen et al., 2003). From these investigations, it follows that the effective temperature for low ambient temperatures is proportional to the square root of the voltage. However, there are examples where the effective temperature rises more rapidly than $\sqrt{V}$, for instance, in the case of bias-induced local heating of Zn atomic-sized contacts (Tsutsui et al., 2007). For octanedithiol, the $\sqrt{V}$ law for the effective temperatures seems to approximately hold (Huang et al., 2006), whereas in our case of C_{60}, more experimental data and theoretical analysis would be needed to estimate the voltage dependence of the effective temperature.

The first experiment using the mechanically controlled break junction technique to investigate single molecule conductance

was reported in 1997 (Reed et al., 1997). Gold electrodes interacted with a solution of benzene-1,4-dithiol forming a self-assembled monolayer on the electrodes. The junction was then closed and opened a number of times and current–voltage characteristics were recorded at a position just before the contact was lost. A fairly large energy gap of 2 eV was attributed to a metal–molecule–metal junction. Further similar experiments were performed by Kergueris et al. (1999) and Reichert et al. (2002, 2003). Smit et al. (2002) observed that although a clean Pt single-atom contact exhibited a conductance of $(1.5 \pm 0.2)\,G_0$, hydrogen-covered electrodes lead to a conductance peak near $1\,G_0$. To date, the molecular arrangement responsible for the conductance of $1\,G_0$ is unclear. In a recent work by Böhler et al. (2007), a mechanical break junction with gold electrodes was used to measure the conductance of adsorbed C_{60} molecules. Inferring the presence of the molecule from characteristic vibrational spectra, a somewhat smaller contact conductance of $0.1\,G_0$ was reported.

In the following paragraph, contact experiments on differently oriented C_{60} molecules using the tip of a STM are introduced. These experiments owing to imaging the molecules prior to and after conductance measurements provide precise information about the contact geometry.

35.4.2 Conductance of Oriented Molecular Orbitals

Recent work demonstrated that molecular conformations (Mujical, 1999; Moresco et al., 2001; Dulić et al., 2003; Martin et al., 2006; Venkataraman et al., 2006) and orientations (Néel et al., 2008b) affect the conductance. Despite this progress, the atomic arrangement at the molecule–metal interface usually remains unknown. Theoretical work, for instance (Xue and Ratner, 2003a,b), and references therein, as well as experiments on metallic single atom contacts (Agraït et al., 2003; Untiedt et al.,

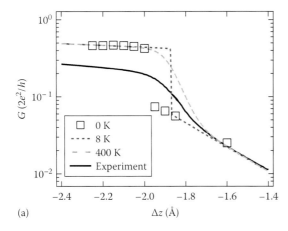

(a)

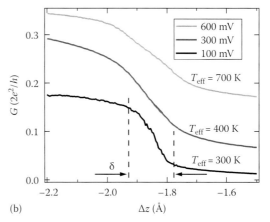

(b)

FIGURE 35.7 (a) Width of transition regime dependent on the effective temperature of the junction. Experimental data taken at 8 K appear as a line, whereas calculated data are depicted as squares for $T_{eff} = 0$ K, as a thin dashed line for $T_{eff} = 8$ K, and as a thick dashed line for $T_{eff} = 400$ K. (b) Comparison of conductance curves acquired at indicated voltages. The width, δ, of the displacement interval for the transition is shown for the curve taken at 100 mV. Conductance curves at 300 and 600 mV are vertically offset. Effective temperatures are indicated.

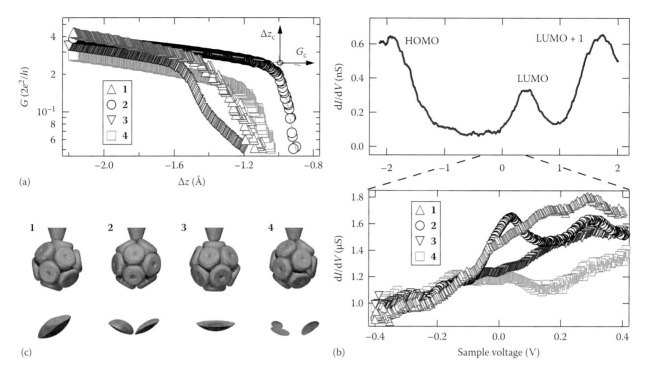

FIGURE 35.8 (a) Conductance curves of C_{60} orientations **1–4** in the transition and contact regimes. Dashed lines indicate the graphical defini-tion of contact displacement and conductance. Zero displacement is defined by feedback loop parameters of $1\,\mu A$ and $0.4\,V$. (b) Top: spectrum of the differential conductance (dI/dV) acquired on a single C_{60} molecule at $8\,K$ in the tunneling regime. The spectrum is fairly insensitive to the molecular orientation. Peaks are attributed to highest occupied molecular orbital (HOMO), lowest unoccupied molecular orbital (LUMO), and the second-to-lowest molecular orbital (LUMO + 1). Bottom: dI/dV spectra at contact for the different molecular orientations. For comparison, spectra were normalized to the conductance at $-0.4\,V$. (c) Top row: illustration of tip contact to molecules **1–4**. Molecules are identified via characteristic orientation of the LUMO. Bottom row: intersection of a spherical Cu $4s$ orbital with toroidal LUMO at $5.4\,\text{Å}$ distance. Volumes decrease from left to right in the sequence 100:57:33:8.

2007), however, highlight the importance of the details of this interface. Therefore, an ideal experiment would allow to address specifically the object to be contacted. As a consequence, the contact geometry would be unambiguously determined and may find its way into calculations that otherwise would have to assume the contact configuration. This paragraph reviews the results obtained for a prototypical molecular contact of a single C_{60} molecule attached to Cu contacts in different orientations (Néel et al., 2008b). A remarkable influence of the contact geometry on the individual molecule conductance is found. We concentrate on the contact conductances of molecules with orientations **1–4** introduced in the preceding paragraph (see Figure 35.5).

While conductance curves for all C_{60} orientations exhibit the same general characteristics discussed for molecule **1** in the pre-ceding paragraph, there are marked and important differences to be discussed next. Since these differences concern the transi-tion and contact regimes, Figure 35.8a focusses on conductance curves of the four C_{60} species in the corresponding displacement interval. In particular, only a small fraction of the tunneling regime is visible for molecule **3**. Two results are obvious: (1) the displacement for contact formation, Δz_c, as well as (2) the contact conductance, G_c, differ for the molecule orientations. We notice that conductance values depend on the tip shape. Blunt tips giv-ing rise to STM images of C_{60} without submolecular resolution typically resulted in contact conductance values between ≈ 0.5

and $0.7\,G_0$. To obtain reproducible conductance curves as shown in Figure 35.8, the tip had to be prepared *in vacuo* such as to give sharp STM images with submolecular resolution.

The different contact displacement may be explained by the spatial extension of the LUMO into vacuum. The conductance curves shown in Figure 35.8a were acquired after opening the feedback loop at a voltage of $0.4\,V$. At this voltage, electrons can tunnel resonantly into the LUMO [top panel of Figure 35.8b (Schull et al., 2008)]. Depending on the local density of states of the LUMO with specific orientation probed by the tip, the tip-molecule distances differ. As a consequence, the tip starts its trajectory toward the molecule at different heights above the molecule and therefore must be displaced differently until contact is formed. The differences in contact displace-ments (Table 35.2) agree very well with differences in apparent

TABLE 35.2 Comparison of Contact Displacements, Δz_c, and Conductances, G_c, of Molecules Exhibiting Orientations **1–4**

Orientation	Δz_c (Å)	G_c (G_0)
1	−1.39	0.26
2	−0.98	0.25
3	−1.57	0.26
4	−1.18	0.17

heights inferred from constant-current STM images acquired at 0.4 V (not shown).

The second marked difference of the conductance curves concerns the contact conductances. While contact conductances of molecules with orientations **1–3** are similar, molecules with orientation **4** with a hexagon pointing toward the tip clearly are less conducting (see Table 35.2 for a comparison). We therefore conclude that electron transport through the same molecule depends crucially on the orientation the molecule adopts between the contacting electrodes. Our observation that contact conductances differ for all C_{60} orientations (and are below $1\,G_0$) may be related to the number of transport channels and their transmission probability. Preliminary calculations reveal that the main transport channel of the tip is provided by the $4s$ orbital of the Cu tip apex atom, whereas the secondary channel given by the $4p$ orbital is conducting an order of magnitude less. For C_{60} due to a lack of calculations, we assume that the LUMO contributes considerably to electron transport through the molecule. To test the validity of this assumption to some extent, we performed additional investigations.

First, we looked for the energy of the LUMO in a spectrum of dI/dV acquired in the tunneling regime above a C_{60} molecule (Figure 35.8b). At a sample voltage of 0.4 V—at which the conductance curves presented in Figure 35.8a were acquired—the spectrum is dominated by the spectral signature of the LUMO. Peaks related to the highest occupied molecular orbital and the LUMO + 1 contribute to the spectrum at widely different energies. We notice that these assignments of peaks in dI/dV spectra to molecular orbitals are not based on calculations but on a qualitative comparison with findings reported for the similar system Ag(100)–C_{60} (Lu et al., 2003). In the tunneling regime, these spectra are similar for all molecule orientations (Schull et al., 2008). *A priori*, it is not at all clear that the molecular orbitals remain unchanged at contact. For instance, previous work on metal contacts has revealed shifts of electronic states in the increasing electric field between tip and sample for decreasing tip-sample distances (Limot et al., 2003; Kröger et al., 2004). Therefore, second, we recorded spectra of the differential conductance at contact. The molecular junctions are stable enough to enable this type of measurement. The bottom panel of Figure 35.8b presents the results. While dI/dV spectra in the tunneling regime vary little even at elevated currents, significantly altered characteristics are observed at contact. For instance, molecule **1** exhibits a steady increase of its dI/dV signal, whereas molecule **4** shows an almost flat dI/dV curve. From these spectra, we infer that the signature of the LUMO is still present for all molecules, but is significantly modified depending on the molecular orientation. In Figure 35.8c, we combine the available information and estimate the overlap of the Cu $4s$ orbital at the tip apex with the C_{60} LUMO for the different orientations by calculating their volume of intersection. For the Cu $4s$ orbital, we assumed a sphere with a diameter of 2.5 Å that corresponds to the nearest-neighbor distance in the copper crystal. For the LUMO, we assumed a toroidal shape as reported by Lu et al. (2003). We further assumed a distance between the center of the spherical Cu

$4s$ orbital and the center of the C_{60} cage of 5.4 Å. This contact distance relies on calculations performed in by Néel et al. (2007b). As an approximation, we applied the same contact distance for all molecules. Within this simplified picture of tip-molecule contact (top row of Figure 35.8c), we calculated the volume of intersection between the Cu $4s$ and the toroidal molecule orbital (bottom row of Figure 35.8c) and took it as a measure for the degree of overlap. Consistent with experimentally measured G_c, the overlap between tip and molecule orbital is largest for type **1** and smallest for type **4**. While yielding a correct sequence of conductances, the model is clearly too simple for reliable predictions. Nevertheless, it provides further indication of the relevance of conductance through the LUMO.

35.4.3 Orientation Change in a Single-Molecule Contact

In this paragraph, we report on a controlled and reversible change of C_{60} adsorption orientations on Cu(100) when the molecule is in contact with the STM tip (Néel et al., 2008c). Currently, considerable interest is devoted to employ small ensembles or even individual atoms or molecules as building blocks in electronic circuits. For instance, bistable switches were realized by the motion of single xenon atoms between the tip of an STM and a nickel surface (Eigler et al., 1991), by reversibly switching between two stable orientational configurations (Comstock et al., 2005) and by conformational changes of molecules induced by tunneling electrons (Moresco et al., 2001; Qiu et al., 2004; Choi et al., 2006; Henzl et al., 2006) or photons (Comstock et al., 2007). Further examples are the regulation of single-molecule conductivity by an electrostatic field (Piva et al., 2006), the direct modification of dynamical properties of an individual molecule by single atom manipulation (Martin et al., 2006), and the control of a complex sequence of coupled rotational and translational dynamics by rolling of a single C_{60} molecule on a silicon surface (Keeling et al., 2005). Despite these pioneering works, the specific addressing of individual molecules to modify reversibly mechanical or electronic properties remains scarce. In particular, experiments providing control of the contact geometry as well as of the contacting electrodes may give complementary information to the widely used mechanically controlled break junction technique.

Controlled and reversible switching of a C_{60} adsorption occurs during contact with the tip. We observed that displacing the tip beyond a threshold excursion can cause the molecule to rotate. The threshold is located at displacements in the contact regime, $\approx$1–1.5 Å away from the contact displacement. In this region, reproducibility of the measurements strongly depends on the tip shape as well as on the position of the contact over the molecule. Different slopes and shapes of this conductance rise were observed. However, for all measurements, this continuous rise was observed to lead to a—on the 100 μs timescale of data acquisition—discontinuous jump of the conductance at higher tip displacements. The discontinuous jump of the conductance may be attributed to a modified geometry of the contact atomic structure (Néel et al., 2007b). It is reasonable to assume that during the continuous rise of the conductance preceding the rearrangement of the contact atomic structure, the molecule adsorption geometry is already slightly modified. Once the tip is retracted, the molecule returns to one of its stable configurations. The rotation occurred when this final configuration differs from the initial one before the contact with the tip.

Rotations of the molecule are illustrated by the STM images of Figure 35.9. The indicated molecules in Figure 35.9 were contacted by the tip of the microscope and by increasing the tip displacement above a threshold at which the molecules switched from one to the other molecule type (Figure 35.9, from **1** to **2**)

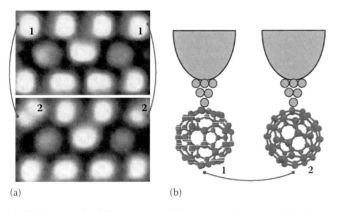

(a)   (b)

FIGURE 35.9 (a) STM images of the same surface area of Cu(100) covered with a single layer of C_{60} ($V = 1.7\,V$, $I = 0.1\,nA$, $35\,\text{Å} \times 20\,\text{Å}$, $T = 8\,K$). Orientation change of molecules in contact with the tip. Top: prior to contact, indicated molecules exhibit orientation **1**. Bottom: after contact, indicated molecules are in orientation **2**. (b) Illustration of switching.

and then back. Modifying the tip apex shape by the indentation of the tip into the substrate surface led to the same observation, indicating that this phenomenon is rather tip-independent.

As shown in the previous paragraph, the adsorption configuration of the C_{60} molecule directly determines the conductance characteristics. In particular, given that feedback loop parameters are the same prior to performing contact measurements, the contact displacement of the tip is different for molecules **1–4**. Here we restrict the discussion to **1** and **2** orientations. It is possible to observe the rotation via the conductance–displacement characteristics as illustrated in Figure 35.10a.

The single (nonaveraged) conductance curves are assigned to a molecule with orientation **1** (black) and **2** (gray). The black curve was acquired just before an orientation switch, whereas the gray curve was taken directly after. Both curves exhibit fluctuations in the transition region as well as in the contact region when the conductance starts to rise again. Fluctuations in the transition region have been interpreted in terms of local heating of the molecule (see Jensen et al., 2007; Néel et al., 2007a,b; Kröger, 2008; and Section 35.4.1). These fluctuations do not lead to rotation of the molecule. Displacing the tip beyond excursions where the conductance rises again, however, can cause the molecule to rotate. Thus, a threshold excursion of the tip must be reached before a change of the adsorption configuration occurs (Néel et al., 2007b).

To investigate the probability of switching for different tip displacements, we used the fact that the conductance curve is characteristic of the adsorption configuration. For each value of the displacement with same initial conditions, 500 conductance measurements were performed and the number of switching

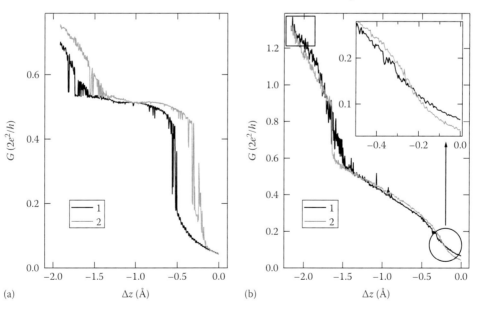

(a) Δz (Å)   (b) Δz (Å)

FIGURE 35.10 (a) Single conductance-versus-displacement curve of molecule **1** (black) together with single conductance–displacement characteristics of molecule with orientation **2** (gray). The black curve was acquired prior to and the gray curve just after rotation. (b) Conductance curves acquired during tip approach (gray) and tip retraction (black) from a C_{60} molecule. During the tip displacement, the molecule changed from **2** to **1**, most probably at a tip displacement exceeding −2.1 Å where conductance curves coincide (square). Feedback loop settings prior to data acquisition: $1\,\mu A$, $0.3\,mV$. Inset: close-up view of conductance traces close to zero tip displacement showing a conductance difference of $\approx 0.02\,G_0$, which is related to a different apparent height of the molecules.

events was counted. The switching probability rises sharply above a threshold displacement of ≈ -1.9 Å to $\approx 30\%$ rather independent of the tip displacement.

To explore whether hysteresis-like effects may be observable, in Figure 35.10b, we present a conductance curve comprising a full cycle for the tip displacement—the feedback loop is opened at a voltage of 300 mV and a current of 1 μA, and $\Delta z = 0$ is fixed to these feedback loop settings. The approach is performed over a molecule of **2** orientation (gray), but upon retraction this orientation has changed to **1** (black), as revealed by STM images (not shown). The rotation from **2** to **1** most probably occurred at displacements exceeding ≈ -2.1 Å since both conductance curves coincide for displacements less than -2.1 Å (see square in Figure 35.10b). On tip retraction, conductance curves deviate from each other for displacements more than -2.1 Å. At zero displacement, the conductance of molecule **1** is slightly higher than that of molecule **2** (see inset of Figure 35.10b). With $G_1 \approx 0.06\,G_0$, $G_2 \approx 0.04\,G_0$, and an apparent barrier height of $\Phi \approx 10$ eV [see above and Néel et al. (2007a)], we find a corresponding height difference of ≈ 0.15 Å that is close to the height difference inferred from cross-sectional profiles in constant-current STM images acquired at 300 mV, that is, at the voltage when the conductance traces in Figure 35.10b were acquired. We notice that the displacement of ≈ -0.23 Å, at which both conductance curves intersect (see circle in Figure 35.10b) likewise reflects the difference in the apparent height of the molecules.

Below we argue that the molecular rotation is mechanically induced, whereas local heating plays a minor role in exciting rotations. An analysis of conductance curves on C_{60} at 300 mV (Néel et al., 2007b) showed that the energy dissipation in the tip-molecule junction leads to an effective heating that could cause the rotation of the molecule. However, in the present experiments, we varied the total power dissipated by a factor of 40. The probability for molecule rotation was found to be insensitive to the dissipated power. Although only a small fraction of the total power is dissipated directly at the molecule, this finding suggests that thermal excitation alone is not the driving force for rotation. Moreover, the tunneling current was not observed to be decisive for inducing rotation. We therefore suggest that mechanical contact with the tip causes C_{60} to rotate.

Mechanically induced rotation explains that certain rotation angles are less frequently observed. For instance, switching a C_{60} from **1** to **2** or vice versa requires a rotation by 20.6°. The apparent rotation by 90° in the surface plane can also be achieved by an out-of-plane rotation of 19.2° and 36.0° for the **1** and **2** orientations, respectively. The large angle needed for the apparent in-plane rotation of **2** molecules is consistent with the observed low frequency of this switching event. In the case of orientation **1**, the apparent in-plane rotation results in a different adsorption geometry on the copper surface with now the carbon–carbon bond between the hexagon and the pentagon of the molecule parallel to the copper missing row. This specific orientation of molecules with orientation **1** is rarely observed, indicating a less favorable adsorption energy. This explains why this specific rotation is rare despite the relatively small angle needed to induce an apparent in-plane rotation of 90°. Switching between **3** and **4**

orientations would require a relatively large rotation angle of 37.4°. In addition, their adsorption on the top of the two missing copper rows leads to a stronger bonding of these molecules with the surface. The bonding is not limited to only the top most copper atoms of the surface as for the three other configurations but also occurs with the copper atoms at the bottom of the missing row. The large angle needed for the rotation and the higher coordination of these configurations with the surface can then explain the absence of switching events between **3** and **4** configurations on elevated tip displacements.

35.5 Conclusion and Outlook

State-of-the-art experiments addressing single-atom and single-molecule contact with the tip of a STM were addressed. Owing to the imaging capability of the instrument, the contacted objects as well as the tip can be characterized prior to and after contact. Therefore, precise and unambiguous information about the contact geometry is provided and the status of the contacting electrodes can be monitored. The objects to be contacted can moreover be chosen independently from the hosting substrate. The reviewed results showed that controlled and reproducible contact to surfaces, adsorbed atoms, and molecules including spectroscopy in contact are feasible.

Future experiments will provide spatially resolved contact conductance maps of complex molecules. The manipulation of structures at the nanometer scale and in contact with the tip will be addressed as well as the ballistic transport of spin-polarized electrons through magnetic structures (Néel et al., 2009). Inelastically transferred electrons giving rise to phonon or photon (Berndt et al., 1993; Aizpurua et al., 2002) excitation will be analyzed by contact spectroscopy.

Acknowledgments

Discussions and cooperation with K. Palotas, W. Hofer (University of Liverpool), A. Garcia-Lekue, Th. Frederiksen (Fundación DIPC, San Sebastián), M. Brandbyge (University of Denmark), L. Limot (University of Strasbourg), and N. Néel, R. Berndt (University of Kiel) are gratefully acknowledged. I thank C. Cepek (Laboratorio Nazionale TASC) for providing clean C_{60} molecules. Technical assistance from Ch. Hamann (University of Kiel) for designing illustrative figures is acknowledged. Image processing was performed by Nanotec WSxM (Horcas et al., 2007). Funding by the Deutsche Forschungsgemeinschaft through SFB 668 and SFB 677 is acknowledged.

References

Abel, M., Dmitriev, A., Fasel, R., Liu, N., Barth, J. V., Kern, K., 2003. Scanning tunneling microscopy and x-ray photoelectron diffraction investigation of C_{60} films on Cu(100). *Phys. Rev. B* 67, 245407.

Agraït, N., Rodrigo, J. G., Vieira, S., 1993. Conductance steps and quantization in atomic-size contacts. *Phys. Rev. B* 47, 12345.

Agraït, N., Rubio, G., Vieira, S., 1995. Plastic deformation of nanometer-scale gold connective necks. *Phys. Rev. Lett.* 74, 3995.

Agraït, N., Yeyati, A. L., van Ruitenbeek, J. M., 2003. Quantum properties of atomic-sized conductors. *Phys. Rep.* 377, 81.

Aizpurua, J., Hoffmann, G., Apell, S. P., Berndt, R., 2002. Electromagnetic coupling on an atomic scale. *Phys. Rev. Lett.* 89, 156830.

Arnold, G. B., 1987. Superconducting tunneling without the tunneling Hamiltonian. II. Subgap harmonic structure. *J. Low Temp. Phys.* 68, 1.

Berndt, R., Gimzewski, J. K., Johansson, P., 1993. Electromagnetic interactions of metallic objects in nanometer proximity. *Phys. Rev. Lett.* 71, 3493.

Bogachek, E. N., Zagoskin, A. N., Kulik, I. O., 1990. Conductance jumps and magnetic flux quantization in ballistic point contacts. *Sov. J. Low. Temp. Phys.* 16, 796.

Böhler, T., Edtbauer, A., Scheer, E., 2007. Conductance of individual C_{60} molecules measured with controllable gold electrodes. *Phys. Rev. B* 76, 125432.

Brandbyge, M., Jacobsen, K. W., 1997. *Proceedings of the NATO Advanced Research Workshop on Nanowires*. NATO Advanced Studies Institute, Vol. 340 Series E: Applied Sciences. Kluwer, Dordrecht, the Netherlands.

Brandbyge, M., Schiøtz, J., Sørensen, M. R., Stoltze, P., Jacobsen, K. W., Nørskov, J. K., Olesen, L. et al., 1995. Quantized conductance in atom-sized wires between two metals. *Phys. Rev. B.* 52, 8499.

Brandbyge, M., Jacobsen, K. W., Nørskov, J. K., 1997. Scattering and conductance quantization in three-dimensional metal nanocontacts. *Phys. Rev. B* 55, 2637.

Bürgi, L., 1999. Scanning tunneling microscopy as local probe of electron density, dynamics, and transport at metal surfaces. PhD thesis, Ecole Polytechnique Fédérale de Lausanne, Lausanne Switzerland.

Büttiker, M., 1988. Coherent and sequential tunneling in series barriers. *IBM J. Res. Dev.* 32, 63.

Büttiker, M., Imry, Y., Landauer, R., Pinhas, S., 1985. Generalized many-channel conductance formula with application to small rings. *Phys. Rev. B* 31, 6207.

Chen, Y. C., Zwolak, M., DiVentra, M., 2003. Local heating in nanoscale conductors. *Nano Lett.* 3, 1691.

Choi, B.-Y., Kahng, S.-J., Kim, S., Kim, H. W., Song, Y. J., Ihm, J., Kuk, Y., 2006. Conformational molecular switch of the azobenzene molecule: A scanning tunneling microscopy study. *Phys. Rev. Lett.* 96, 156106.

Comstock, M. J., Cho, J., Kirakosian, A., Crommie, M. F., 2005. Manipulation of azobenzene molecules on Au(111) using scanning tunneling microscopy. *Phys. Rev. B* 72, 153414.

Comstock, M. J., Levy, N., Kirakosian, A., Cho, J., Lauterwasser, F., Harvey, J. H., Strubbe, D. A. et al., 2007. Reversible photomechanical switching of individual engineered molecules at a metallic surface. *Phys. Rev. Lett.* 99, 038301.

Costa-Krämer, J. L., García, N., García-Mochales, P., Serena, P. A., 1995. Nanowire formation in macroscopic metallic contacts: Quantum mechanical conductance tapping a table top. *Surf. Sci.* 342, L1144.

Cuevas, J. C., Yeyati, A. L., Martín-Rodero, A., Bollinger, G. R., Untiedt, C., Agraït, N., 1998. Evolution of conducting channels in metallic atomic contacts under elastic deformation. *Phys. Rev. Lett.* 81, 2990.

Datta, S., 1995. *Electronic Transport in Mesoscopic Systems*. Cambridge University Press, Cambridge, U.K.

Dekker, C., 1999. Carbon nanotubes as molecular quantum wires. *Phys. Today*, 52, 22.

Dulić, D., van der Molen, S. J., Koudernac, T., Jonkman, H. T., de Jong, J. J. D., Bowden, T. N., van Esch, J., Feringa, B. L., van Wees, B. J., 2003. One-way optoelectronic switching of photochromic molecules on gold. *Phys. Rev. Lett.* 91, 207402.

Eigler, D. M., Lutz, C. P., Rudge, W. E., 1991. An atomic switch realized with the scanning tunnelling microscope. *Nature* 352, 600.

Erts, D., Olin, H., Ryen, L., Olsson, E., Thölén, A., 2000. Maxwell and Sharvin conductance in gold point contacts investigated using TEM-STM. *Phys. Rev. B* 61, 12725.

Fink, H.-W., Schönenberger, C., 1999. Electrical conduction through DNA molecules. *Nature* 398, 407.

Gai, Z., He, Y., Yu, H., Yang, W. S., 1996. Observation of conductance quantization of ballistic metallic point contacts at room temperature. *Phys. Rev. B* 53, 1042.

García, N., Escapa, L., 1989. Elastic oscillatory resistances of small contacts. *Appl. Phys. Lett.* 54, 1418.

Gimzewski, J. K., Möller, R., 1987. Transition from the tunneling regime to point contact studied using scanning tunneling microscopy. *Physica B* 36, 1284.

Hansen, K., Lægsgaard, E., Stensgaard, I., Besenbacher, F., 1997. Quantized conductance in relays. *Phys. Rev. B* 56, 2208.

Henzl, J., Mehlhorn, M., Gawronski, H., Rieder, K.-H., Morgenstern, K., 2006. Reversible cis-trans-Isomerisierung eines einzelnen Azobenzol-Moleküls. *Angew. Chem.* 118, 617.

Hofer, W. A., Fisher, A. J., Wolkow, R. A., Grütter, P., 2001. Surface relaxations, current enhancements, and absolute distances in high resolution scanning tunneling microscopy. *Phys. Rev. Lett.* 87, 236104.

Hofer, W. A., Garcia-Lekue, A., Brune, H., 2004. The role of surface elasticity in giant corrugations observed by scanning tunneling microscopes. *Chem. Phys. Lett.* 397, 354.

Horcas, I., Fernández, R., Gómez-Rodríguez, J. M., Colchero, J., Gómez-Herrero, J., Baro, A. M., 2007. A software for scanning probe microscopy and a tool for nanotechnology. *Rev. Sci. Instrum.* 78, 013705.

Hou, J. G., Yang, J., Wang, H., Li, Q., Zeng, C., Lin, H., Bing, W., Chen, D. M., Zhu, Q., 1999. Identifying molecular orientation of individual C_{60} on a Si(111)-(7 × 7) surface. *Phys. Rev. Lett.* 83, 3001.

Huang, Z., Xu, B., Chen, Y., DiVentra, M., Tao, N., 2006. Measurement of current-induced local heating in a single molecule junction. *Nano Lett.* 6, 1240.

Jansen, A. G. M., van Gelder, A. P., Wyder, P., 1980. Point-contact spectroscopy in metals. *J. Phys. C* 13, 6073.

Jensen, H., Kröger, J., Néel, N., Berndt, R., 2007. Silver oligomer and single fullerene electronic properties revealed by a scanning tunnelling microscope. *Eur. Phys. J. D* 45, 465.

Joachim, C., Gimzewski, J. K., Schlittler, R. R., Chavy, C., 1995. Electronic transparence of a single C_{60} molecule. *Phys. Rev. Lett.* 74, 2102.

Kasumov, A. Y., Kociak, M., Guéron, S., Reulet, B., Volkov, V. T., Klinov, D. V., Bouchiat, H., 2001. Direct measurement of electrical transport through DNA molecules. *Science* 291, 280.

Keeling, D. L., Humphry, M. J., Fawcett, R. H. J., Beton, P. H., Hobbs, C., Kantorovich, L., 2005. Bond breaking coupled with translation in rolling of covalently bound molecules. *Phys. Rev. Lett.* 94, 146104.

Kergueris, C., Bourgoin, J.-P., Palacin, S., Esteve, D., Urbina, C., Magoga, M., Joachim, C., 1999. Electron transport through a metal-molecule-metal junction. *Phys. Rev. B* 59, 12505.

Klapwijk, T. M., Blonder, G. E., Tinkham, M., 1982. Explanation of subharmonic energy gap structure in superconducting contacts. *Physica B* 109–110, 1657.

Krans, J. M., Muller, C. J., van der Post, N., Postma, F. R., Sutton, A. P., Todorov, T. N., van Ruitenbeek, J. M., 1995a. Comment on 'Quantized conductance in atom-sized point contact'. *Phys. Rev. Lett.* 74, 2146.

Krans, J. M., van Ruitenbeek, J. M., Fisun, V. V., Yanson, I. K., de Jongh, L. J., 1995b. The signature of conductance quantization in metallic point contacts. *Nature* 375, 767.

Kröger, J., 2008. Nonadiabatic effects on surfaces: Kohn anomaly, electronic damping of adsorbate vibrations, and local heating of single molecules. *J. Phys.: Condens. Matter* 20, 224015.

Kröger, J., Limot, L., Jensen, H., Berndt, R., Johansson, P., 2004. Stark effect in Au(111) and Cu(111) surface states. *Phys. Rev. B* 70, 033401.

Kröger, J., Jensen, H., Berndt, R., 2007. Conductance of tip-surface and tip-atom junction on Au(111) explored by a scanning tunnelling microscope. *New J. Phys.* 9, 153.

Landauer, R., 1957. Spatial variation of currents and fields due to localized scatterers in metallic conduction. *IBM J. Res. Dev.* 1, 223.

Landauer, R., 1970. Electrical resistance of disordered one-dimensional lattices. *Philos. Mag.* 21, 863.

Landauer, R., 1981. Can a length of perfect conductor have a resistance? *Phys. Lett. A* 85, 91.

Landman, U., Luedtke, W. D., Burnham, N. A., Colton, R. J., 1990. Atomistic mechanisms and dynamics of adhesion, nanoindentation, and fracture. *Science* 248, 454.

Landman, U., Luedtke, W. D., Salisbury, B. E., Whetten, R. L., 1996. Reversible manipulations of room temperature mechanical and quantum transport properties in nanowire junctions. *Phys. Rev. Lett.* 77, 1362.

Limot, L., Maroutian, T., Johansson, P., Berndt, R., 2003. Surface-state Stark shift in a scanning tunneling microsope. *Phys. Rev. Lett.* 91, 196801.

Limot, L., Kröger, J., Berndt, R., Garcia-Lekue, A., Hofer, W. A., 2005. Atom transfer and single-adatom contact. *Phys. Rev. Lett.* 94, 126102.

Lu, X., Grobis, M., Khoo, K., Crommie, M. F., Louie, S. G., 2003. Spatially mapping the spectral density of a single C_{60} molecule. *Phys. Rev. Lett.* 90, 096802.

Ludoph, B., van der Post, N., Bratus', E. N., Bezuglyi, E. V., Shumeiko, V. S., Wendin, G., van Ruitenbeek, J. M., 2000. Multiple Andreev reflection in single atom niobium junctions. *Phys. Rev. B* 61, 8561.

Martin, M., Lastapis, M., Riedel, D., Dujardin, G., Mamatkulov, M., Stauffer, L., Sonnet, P., 2006. Mastering the molecular dynamics of a bistable molecule by single atom manipulation. *Phys. Rev. Lett.* 97, 216103.

Maxwell, J. C., 1954. *A Treatise on Electricity and Magnetism*. Dover Publication, New York.

Montie, E. A., Cosman, E. C., 't Hooft, G. W., van der Mark, M. B., Beenakker, W. J., 1991. Observation of the optical analogue of quantized conductance of a point contact. *Nature* 350, 594.

Moreland, J., Ekin, J. W., 1985. Electron tunneling experiments using Nb-Sn 'break' junctions. *J. Appl. Phys.* 58, 3888.

Moresco, F., Meyer, G., Rieder, K.-H., Tang, H., Gourdon, A., Joachim, C., 2001. Conformational changes of single molecules induced by scanning tunneling microscopy manipulation: A route to molecular switching. *Phys. Rev. Lett.* 86, 672.

Mujical, V., 1999. Electron transfer in molecules and molecular wires: Geometry dependence, coherent transfer, and control. *Adv. Chem. Phys.* 107, 403.

Muller, C. J., van Ruitenbeek, J. M., de Jongh, L. J., 1992a. Conductance and supercurrent discontinuities in atomic-scale metallic constrictions of variable width. *Phys. Rev. Lett.* 69, 140.

Muller, C. J., van Ruitenbeek, J. M., de Jongh, L. J., 1992b. Experimental observation of the transition from weak link to tunnel junction. *Physica C* 191, 485.

Néel, N., Kröger, J., Limot, L., Berndt, R., 2007a. Conductance of single atoms and molecules studied with a scanning tunnelling microscope. *Nanotechnology* 18, 044027.

Néel, N., Kröger, J., Limot, L., Frederiksen, T., Brandbyge, M., Berndt, R., 2007b. Controlled contact to a C_{60} molecule. *Phys. Rev. Lett.* 98, 065502.

Néel, N., Kröger, J., Limot, L., Palotas, K., Hofer, W. A., Berndt, R., 2007c. Conductance and Kondo effect in a controlled single-atom contact. *Phys. Rev. Lett.* 98, 016801.

Néel, N., Kröger, J., Limot, L., Berndt, R., 2008a. Conductance of single atoms and molecules studied with a scanning tunnelling microscope. *J. Scann. Probe Microsc.* 3, 9.

Néel, N., Kröger, J., Limot, L., Berndt, R., 2008b. Probing the conductance of oriented molecules. *Nano Lett.* 8, 1291.

Néel, N., Limot, L., Kröger, J., Berndt, R., 2008c. Rotation of C_{60} in a single-molecule contact. *Phys. Rev. B* 77, 125431.

Néel, N., Kröger, J., Berndt, R., 2009. Quantized conductance of a single magnetic atom. *Phys. Rev. Lett.* 101, 102, 086805.

Ohnishi, H., Kondo, Y., Takayanagi, K., 1998. Quantized conductance through individual rows of suspended gold atoms. *Nature* 395, 780.

Olesen, L., Lægsgaard, E., Stensgaard, I., Besenbacher, F., Schiøtz, J., Stoltze, P., Jacobsen, K. W., Nørskov, J. K., 1994. Quantized conductance in an atom-sized point contact. *Phys. Rev. Lett.* 72, 2251.

Olesen, L., Lægsgaard, E., Stensgaard, I., Besenbacher, F., Schiøtz, J., Stoltze, P., Jacobsen, K. W., Nørskov, J. K., 1995. Reply to Comment on 'Quantized conductance in atom-sized point contact'. *Phys. Rev. Lett.* 74, 2147.

Pascual, J. I., Méndez, J., Gómez-Herrero, J., Baró, A. M., García, N., Binh, V. T., 1993. Quantum contact in gold nanostructures by scanning tunneling microscopy. *Phys. Rev. Lett.* 71, 1852.

Pethica, J. B., Sutton, A. P., 1988. On the stability of a tip and flat at very small separation. *J. Vac. Sci. Technol. A* 6, 2490.

Piva, P. G., DiLabio, G. A., Pitters, J. L., Zikovsky, J., Rezeq, M., Dogel, S., Hofer, W. A., Wolkow, R. A., 2006. Field regulation of single-molecule conductivity by a charged surface atom. *Nature* 435, 658.

Porath, D., Bezryadin, A., deVries, S., Dekker, C., 2000. Direct measurement of electrical transport through DNA molecules. *Nature* 403, 635.

Qiu, X. H., Nazin, G. V., Ho, W., 2004. Mechanisms of reversible conformational transitions in a single molecule. *Phys. Rev. Lett.* 93, 196806.

Reed, M. A., Zhou, C., Muller, C. J., Burgin, T. P., Tour, J. M., 1997. Conductance of a molecular junction. *Science* 278, 252.

Reichert, J., Ochs, R., Beckmann, D., Weber, H. B., Mayor, M., Löneysen, H., 2002. Driving current through single organic molecules. *Phys. Rev. Lett.* 88, 176804.

Reichert, J., Weber, H. B., Mayor, M., Löneysen, H., 2003. Low-temperature conductance measurements on single molecules. *Appl. Phys. Lett.* 82, 4137.

Rubio, G., Agraït, N., Vieira, S., 1996. Atomic-sized metallic contacts: Mechanical properties and electronic transport. *Phys. Rev. Lett.* 76, 2302.

Scheer, E., Agraït, N., Cuevas, J. C., Yeyati, A. L., Ludoph, B., Martín-Rodero, A., Bollinger, G. R., van Ruitenbeek, J. M., Urbina, C., 1998. The signature of chemical valence in the electrical conduction through a single-atom contact. *Nature* 394, 154.

Schull, G., Néel, N., Becker, M., Kröger, J., Berndt, R., 2008. Spatially resolved conductance of oriented C_{60}. *New J. Phys.* 10, 065012.

Schulze, G., Franke, K. J., Gagliardi, A., Romano, G., Lin, C. S., Rosa, A. L., Niehaus, T. A. et al., 2008. Resonant electron heating and molecular phonon cooling in single C_{60} junctions. *Phys. Rev. Lett.* 100, 136801.

Sharvin, Yu. V., 1965. A possible method for studying Fermi surfaces. *Sov. Phys.-JETP* 21, 655.

Smit, R. H. M., Noat, Y., Untiedt, C., Lang, N. D., van Hemert, M. C., van Ruitenbeek, J. M., 2002. Measurement of the conductance of a hydrogen molecule. *Nature* 419, 906.

Smith, J. R., Bozzolo, G., Banerjea, A., Ferrante, J., 1989. Avalanche in adhesion. *Phys. Rev. Lett.* 63, 1269.

Temirov, R., Lassise, A., Anders, F. B., Tautz, F. S., 2008. Kondo effect by controlled cleavage of a single-molecule contact. *Nanotechnology* 19, 065401.

Todorov, T. N., 1998. Local heating in ballistic atomic-scale contacts. *Philos. Mag. B* 77, 965.

Todorov, T. N., Sutton, A. P., 1993. Jumps in electronic conductance due to mechanical instabilities. *Phys. Rev. Lett.* 70, 2138.

Todorov, T. N., Hoekstra, J., Sutton, A. P., 2002. Current-induced embrittlement of atomic wires. *Phys. Rev. Lett.* 86, 3606.

Torres, J. A., Pascual, J. L., Sáenz, J. J., 1994. Theory of conduction through narrow constrictions in a three-dimensional electron gas. *Phys. Rev. B* 49, 16581.

Tsutsui, M., Kurokawa, S., Sakai, A., 2007. Bias-induced local heating in atom-sized metal contacts at 77 K. *Appl. Phys. Lett.* 90, 133121.

Untiedt, C., Caturla, M. J., Calvo, M. R., Palacios, J. J., Segers, R. C., van Ruitenbeek, J. M., 2007. Formation of a metallic contact: Jump to contact revisited. *Phys. Rev. Lett.* 98, 206801.

Venkataraman, L., Klare, J. E., Nuckolls, C., Hybertsen, M. S., Steigerwald, M. L., 2006. Dependence of single-molecule junction conductance on molecular conformation. *Nature* 442, 904.

van Wees, B. J., van Houten, H. H., Beenakker, C. W. J., Wiliamson, J. G., Kouwenhoven, L. P., van der Marel, D., Foxon, C. T., 1988. Quantized conductance of point contacts in a two-dimensional electron gas. *Phys. Rev. Lett.* 60, 848.

Wharam, D. A., Thornton, T. J., Newbury, R., Pepper, M., Ahmed, H., Frost, J. E. F., Hosko, D. G., Peacock, D. C., Ritchie, D. A., Jones, G. A. C., 1988. One-dimensional transport and the quantisation of the ballistic resistance. *J. Phys. C* 21, L209.

Xue, Y., Ratner, M. A., 2003a. Microscopic study of electrical transport through individual molecules with metallic contacts. I. Band lineup, voltage drop, and high-field transport. *Phys. Rev. B* 68, 115406.

Xue, Y., Ratner, M. A., 2003b. Microscopic study of electrical transport through individual molecules with metallic contacts. II. Effect of the interface structure. *Phys. Rev. B* 68, 115407.

Yazdani, A., Eigler, D. M., Lang, N. D., 1996. Off-resonance conduction through atomic wires. *Science* 272, 1921.

36

Fundamental Process of Near-Field Interaction

Hirokazu Hori
University of Yamanashi

Tetsuya Inoue
Yamanashi Industrial Technology College

36.1 Introduction

One of the most important issues of nanophysics is near-field electromagnetic interaction between nanometer-sized electronic systems, which generates correlation between electronic systems and governs excitation transfer and the resulting functions of nanodevices and nanosystems (Hori 2001). As the nature of strongly coupled fields plus matter system, near-field interactions generate complicated space–time correlation depending on the shape, size, and layout of material systems and bring us with a wide variety of novel functions useful for the exploration of nanoscience and nanotechnology. The strong space–time correlation also reveals basic problems involved in electromagnetic interactions in nanometer scales, such as difficulties in defining electromagnetic properties of matter in mean-field regime, in implementing unidirectional signal transport processes, and in considering electromagnetic causality related to near-field interactions. The key study to investigate these problems lies in the fields of near-field optics and nanophotonics, because relatively large optical wavelengths enable us to identify near-field phenomena from those related to light wave propagation (Ohtsu and Hori 1999). A rapid progress in near-field optics and nanophotonics has been made in the last two decades based on the developments of nano-fabrication techniques, scanning probe microscopy, and precision laser spectroscopy, and brought us with both the theoretical and the experimental backgrounds enough to study the dissipation processes and hierarchy properties of near-field interactions in relation to innovation in functional devices and systems.

In this chapter, as the basis to understand and handle these issues involved in general near-field problems, we study theoretical treatment of optical near-fields and near-field optical interactions on the basis of both the classical and the quantum optical theories of half-space problems described by using angular spectrum representation of scattered field. The excitation energy transfer through the assumed boundary plane via near-field optical interactions is described in a similar manner to the electronic tunneling current, in which the importance of the final state density can be clearly understood in terms of Fermi's golden rule. This point is related to another important issue to realize novel near-field optical functions in nanophotonics devices and systems.

36.2 Optical Near-Field and Near-Field Optical Interaction

36.2.1 Optical Near-Field

The interaction of electronic systems with optical fields is one of the most important issues of nanophysics. In nanometer space, optical field behaves as a coupled state of electromagnetic fields with electromagnetic excitations in material environment. The optical fields, therefore, are not free but coupled with polarizations or surface charges of matter lying in sub-wavelength vicinity of the observation point. The surface charges are produced due to the abrupt discontinuation of induced electric polarizations at the surface of irradiated material object, so optical fields near matter involve electric field lines connected to the surface charges. In particular, in a sub-optical-wavelength sized electronic system, no retardation effect of electromagnetic phenomenon is significant, so surface charges dominate the behavior of optical fields near matter. Such a quasi-static and short-ranged nature of optical fields in a sub-optical-wavelength vicinity of material environment is described in terms of "optical near-field." The strong matter-field coupling in nanometer space alters the nature of optical fields and extends the significance of optical processes in nanophysics.

The surface charge distribution consistent with optical field distribution is determined by polarization dynamics and electric current driven in irradiated material system and is, therefore, strongly dependent on the material shape. Another important issue is spatial composition of material systems consisting of isolated electronic systems, for which optical near-field distribution is determined according to the consistency of electromagnetic fields and material polarizations and associated surface charges.

36.2.2 Near-Field Optical Interaction

The generation and elimination of optical near-field, or creation and annihilation of photons related to interaction with optical near-fields, therefore, are not only associated with excitation and relaxation of the electronic systems under consideration but also relevant to those processes in the environmental material systems lying in the sub-optical-wavelength vicinity. In fact, when two isolated electronic systems interact with each other via optical near-fields, the physical quantities such as momentum and angular momentum exchanged between these systems in the course of optical energy transfer are quite different from those quantities exchanged via interaction with the optical far-fields of propagating nature in an approximately free space. This means that optical near-fields exhibit off-shell properties with respect to photon dispersion relation in free space. Such an interaction of isolated electronic systems via optical near-field is referred to as "near-field optical interaction." Indeed, optical near-fields are associated with the optical far-field of propagating nature but play the significant role in nanophysics.

36.2.3 Near-Field Interaction and Half-Space Problems

The near-field optical interaction has its significance provided that it produces a certain modification of signal relevant to our observation system. Here, the signal means scattered optical fields, electronic signals, modifications of material structure, all of which are available in nanophysics. The optical near-field interaction, especially, is significant in optics and photonics of nanometer scale where the optical processes are described in terms of half-space problems. The half-space problems are those in which an entire space under consideration can be separated by an assumed planar boundary into two spaces, say "left half-space" and "right half-space," and optical sources and detectors are correlated individually to the optical fields of either right

half-space or left half-space. In the half-space problems, we can clearly evaluate the optical energy transfer between sources and detectors belonging to one of the half-spaces with those in the other half-space. When interaction of an optical source in one of the half-spaces and an optical detector in the other half-space is only through near-field optical interactions at the assumed boundary plane, even a minute modification of optical response in the sub-optical-wavelength vicinity of the boundary plane results in an observable effect in the signal transfer between the source and detector under consideration.

36.2.4 Evanescent Waves and Angular Spectrum in Half-Space Problems

As it will be seen later in the theoretical treatments of half-space problems, the near-field optical interactions for each monochromatic component are described by electromagnetic interactions via evanescent waves propagating along the boundary plane and exhibiting exponential amplitude decay in the direction perpendicular to the boundary plane. Due to the nature of exponential amplitude decay, the evanescent waves are sustainable only in a narrow vicinity of an irradiated or radiating material object. The spatial range where the near-field optical interactions are effective is referred to as the penetration depth of evanescent waves involved in the interaction. Because of the dispersion relation of optical fields, the penetration depth is directly related to the wavelength of evanescent wave propagating along the assumed boundary plane, so the locality of near-field optical interaction taking place at the boundary plane is described in terms of a coherent superposition of monochromatic evanescent waves with different wavelengths. In other words, the range and locality of the near-field optical interaction are described by a spatial frequency spectrum of monochromatic evanescent waves involved in the near-field optical interaction under consideration, which is referred to as "angular spectrum representation" of optical near-field (Wolf and Niet-Vesperinas 1985, Inoue and Hori 1996, Inoue et al. 1998, Ohtsu and Hori 1999).

36.2.5 Hierarchical Nature of Near-Field Interactions

In nanophysics, the size of matter relevant to the phenomena under consideration is much smaller than optical wavelengths in free space, so the optical processes related to near-field optical interactions are clearly identified from those related to propagating light waves. This makes theoretical description of optical processes in nanophysics rather simple and comprehensible compared with those of nanometer scale electronic phenomena in which electronic wavelengths lies in nanometer range and near-field phenomena cannot be separated from the propagation of electronic waves in the entire system. On the other hand, the vector nature of electromagnetic fields introduces complexity and wide variety of near-field optical phenomena in nanophysics. For instance, one can introduce hierarchical properties on the basis of the hierarchy in optical response of material systems to fabricate the microscopic to macroscopic interconnection of near-field optical processes as the basis of nanophotonics functions.

36.2.6 Properties of Optical Near-Fields

It is noted that the near-field optical interactions can be strongly enhanced, when an optical near-field under consideration is coupled to an elementary excitation of matter in a resonant way. In such cases, the optical near-field is attributed to "polariton" coupled to the elementary excitation by employing the terminology and theoretical background of condensed matter physics. For example, plasmonic resonances are available for extremely strong optical near-field enhancement and observation of non-linear near-field optical processes, and, on the other hand, excitonic resonances are useful for realizing quantum optical phenomena in nanophotonics systems. The nature of coupled state of optical fields with material environment provides us with manifold of novel applications of optical processes beyond the diffraction limit and photon dispersion relation.

36.3 Half-Space Problem and Evanescent Waves

Near-field optical interactions are represented in terms of tunneling of electromagnetic excitation via evanescent waves. Tunneling picture is applicable to general electromagnetic interactions by describing Poynting vector based on angular spectrum representation of scattered fields, in which energy transfer of tunneling regime is described in terms of an overlap integral of evanescent waves with corresponding penetration depth and pseudo-momentum involved in the angular spectrum. This provides us with useful idea of "size resonance," which is commonly acknowledged for sample-probe interactions in scanning near-field optical microscopy. The tunneling picture is especially important in the theoretical evaluation of near-field optical interactions in general nanometer-sized optoelectronic devices exerting functions based on the transport of electromagnetic excitation.

Let us consider a basic configuration of half-space problem described in Figure 36.1, in which a planar boundary separates the entire space into two half-spaces, and assume the left half-space being filled with a nonmagnetic, transparent, homogeneous, and isotropic dielectric medium of refractive index n ($z < 0$), and the right half-space vacuum ($z \geq 0$).

When a monochromatic light source is placed in the right half-space at a distance Z away from the boundary, an optical detector placed in the same half-space at a position far from the boundary observes direct radiation from the source and light reflected from the boundary. For the source to boundary distance Z larger than the optical wavelength λ, these two components interfere with each other and result in modification of observed intensity according to the change in Z. Such an optical field of propagating nature is referred to as "homogeneous wave."

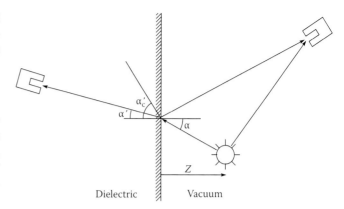

FIGURE 36.1 Schematic diagram showing the half space problems under consideration. The space, half of which is filled with a nonmagnetic, transparent, homogeneous, and isotropic dielectric medium of refractive index n (the left half-space, $z < 0$), and the other half is vacuum (the right half-space, $z \geq 0$).

When an optical detector is placed in the dielectric side far away from the boundary, the continuity relation of optical fields at the boundary known as Snell's law

$$n\sin\alpha' = \sin\alpha \qquad (36.1)$$

tells us that for the incident angle α taking values between 0 and $\pi/2$, the optical detector observes nonzero refracted light propagating only in the direction with an angle of refraction α' taking the value between 0 and α'_c, where $\sin \alpha'_c = 1/n$ is the angle of total internal reflection.

The situation is, however, quite different when an optical source of sub-wavelength size is placed in a sub-wavelength vicinity of the boundary $Z < \lambda$ as shown in Figure 36.2, since the radiation fields described on the basis of half-space configuration always involve both homogeneous waves of propagating nature and evanescent waves exhibiting exponential amplitude decay according to the distance from the optical source. For the evanescent waves, the nature of amplitude decay is described mathematically in terms of the complex incident angle of radiation to the boundary α taking the values in the range

$$\alpha = (\pi/2) - i\gamma, \quad (0 \leq \gamma < \infty). \qquad (36.2)$$

The physical counter part of the amplitude decay in the direction normal to the boundary corresponds to the wavenumber of field propagation in the direction along the boundary, $k_\parallel$, taking values larger than the wavenumber of light propagating in free space $k_0 = \omega/c$. Snell's law tells us that for γ between 0 and γ_c, where $\sin[(\pi/2) - i\gamma_c] = n$, the angle of refraction α' becomes a real number between α'_c and $\pi/2$, and the refracted propagating optical field arises in the direction out of the angle of total internal reflection. In this case, the optical field reflected back into the right half-space is also an evanescent wave, so the pair of incident and reflected evanescent waves produces an optical energy transfer corresponding to the refracted light observed by

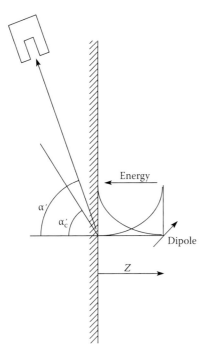

FIGURE 36.2 The tunneling energy corresponds to the energy into the medium in directions without the critical angle α'_c.

a detector placed in far-field of the dielectric side. As will be discussed later, this corresponds to an optical excitation tunneling via evanescent waves from the optical source placed in the right half-space to the dielectrics in the left half-space.

For γ larger than γ_c, the refracted wave corresponds also to an evanescent wave, so that there is no propagating light carrying optical energy into the detector placed in far-field. In this sense, γ_c corresponds to another critical angle of total reflection. In this case, no optical energy transfer is driven by the pair of incident and reflected pair of evanescent waves in the right half-space. As will be discussed later, the probability of optical excitation tunneling is determined not only by the transition amplitude but also by the density of corresponding final state, as is known by Fermi's golden rule. According to this, when an optical detector of sub-wavelength size is placed to observe the evanescent wave refracted into the left half-space, the optical excitation tunneling arises to carry optical energy corresponding to the optical energy dissipated by the detector. The situation is similar when an optical scatterer is placed in a subwavelength vicinity of the boundary to generate some propagating optical fields detectable in far-field. These situations correspond to the basic process of optical near-field measurements. As mentioned above, the placement of an optical detector or a scatterer in the near-field of left half-space alters the excitation tunneling described by the incident and reflected evanescent waves in the right half-space, so one should note that the process of near-field detection corresponds to a destructive measurement of optical near-fields, as is general in any near-field observation processes.

In the followings of this section, we study a general theoretical treatment of optical near-fields and near-field optical

interactions based on the half-space configuration and evanescent waves. Since both evanescent and homogeneous waves are described in terms of plane waves with complex angles of propagation, the expansion of radiated or scattered field into these components is referred to as "angular spectrum representation" (Wolf and Niet-Vesperinas 1985, Inoue and Hori 1996, Inoue et al. 1998, Ohtsu and Hori 1999). Based on angular spectrum representation, we can establish a quantum theory of radiation in half-space problems, which is briefly reviewed in the following sections.

Before proceeding to theoretical treatments, it is instructive to note that the physical significance or reality of evanescent waves have been demonstrated by means of high-resolution nonlinear laser spectroscopy of atoms interacting with evanescent waves produced near a planar dielectric surface. The short penetration depth of evanescent waves and the associated large momentum transfer at resonant absorption of atoms as well as spontaneous emission of optical evanescent waves have been demonstrated experimentally (Matsudo et al. 1997, 1998). Further, angular momentum transfer between evanescent waves and sub-wavelength-sized matter has been demonstrated, which reveals the property of local circular polarization important in spin-related optical phenomena near material surface (Ohdaira et al. 2001, 2008).

36.4 Angular Spectrum Representation of Optical Near-Field

Here, we consider monochromatic radiation fields from a point-like oscillating electric dipole, $P(t)$, with oscillation frequency K located in vacuum at R. Throughout this chapter, we employ the unit in which the light velocity is taken to be unity, $c = 1$, so that the oscillation frequency K is identical to the wavenumber of radiation in free space. It is convenient to introduce the following complex amplitude of electric dipole, P, defined by

$$P(t) = P \exp(-iKt) + P^\star \exp(+iKt). \tag{36.3}$$

The electromagnetic field amplitudes corresponding to the electric dipole radiation is given by

$$E(r) = \left(\frac{iK^3}{4\pi\varepsilon_0} \right) \left\{ -\frac{1}{3} \left[P - 3(n \cdot P)n \right] h_2^{(1)} \left(K|r - R| \right) \right.$$

$$\left. + \frac{2}{3} P h_0^{(1)} \left(K|r - R| \right) \right\}, \tag{36.4}$$

where the unit vector $n = (r - R)/|r - R|$ represents the direction of an observation point from the dipole and $h_0^{(1)} \left(K|r - R| \right)$ and $h_2^{(1)} \left(K|r - R| \right)$ are, respectively, the zeroth and second order spherical Hankel functions of the first kind.

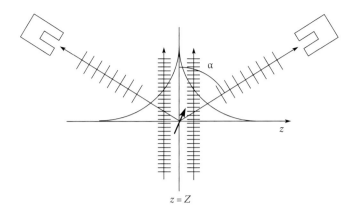

FIGURE 36.3 Angular spectrum representation of scattered fields.

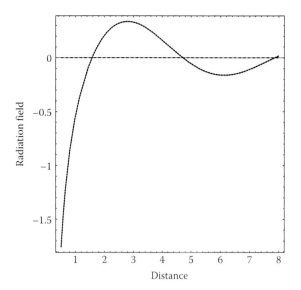

FIGURE 36.4 The imaginary part of $h_0^{(1)}(Kr)$ versus Kr.

As the basis of angular spectrum representation, one can introduce an arbitrary plane as an assumed boundary plane that separates entire space under consideration into right half-space and left half-space. Here, the existence of physical planar boundary is not necessary. When an assumed boundary plane is placed infinitesimally close to the electric dipole at Z, as shown in Figure 36.3, the angular spectrum representation of dipole radiation field is obtained.

For an example, the angular spectrum representation of outgoing scalar spherical wave corresponding to $h_0^{(1)}(K|r - R|)$ is given by (Weyl representation)

$$h_0^{(1)}(K|r - R|) = \frac{1}{2\pi}\int_0^{2\pi}\mathrm{d}\beta\int_C \sin\alpha\,\mathrm{d}\alpha\exp[iKs^{(\pm)}\cdot(r - R)], \quad (36.5)$$

where $s^{(\pm)}$ stand for unit wavevectors with Cartesian components

$$s^{(\pm)} = (\sin\alpha\cos\beta,\ \sin\alpha\sin\beta,\ \pm\cos\alpha) \quad (36.6)$$

representing the plane waves with wavevector $Ks^{(+)}$ propagating in the right half-space, $z > Z$, and that with $Ks^{(-)}$ in the left half-space for $z < Z$. Nature of each plane wave component is described by the contour of integration with respect to the complex angle α. Compared with the radial field amplitude corresponding to the imaginary part of $h_0^{(1)}(K|r - R|)$ indicated for $K|r - R|$ in Figure 36.4, the propagating nature of outgoing field is represented by homogeneous waves involved in the angular spectrum representation with the real values of α ($0 \to \pi/2$). On the other hand, the strong local field of quasi-static nature near the source ($K|r - R| < 1$ in Figure 36.4) is represented by evanescent waves with imaginary values of α ($\pi/2 \to (\pi/2) - i\infty$) propagating parallel to the boundary plane ($z = Z$) and exhibiting exponential decay in the direction normal to the assumed boundary plane. In other words, as it is widely recognized, far-field behavior and near-field behavior of radiated or scattered fields are dominated, respectively, by homogeneous plane waves and evanescent waves involved in the angular spectrum representation of the fields.

Due to the symmetry of half-space problems under parallel displacement along the assumed boundary plane, the wavevector along the boundary plane becomes a conserved quantity, so it is convenient to introduce the projection of the unit wavevector onto the assumed boundary plane, $s_{\parallel} = \sin\alpha$, which is referred to as "spatial frequency" along the boundary plane, and to employ another form of the angular spectrum representation given by (Sommerfeld integral)

$$h_0^{(1)}(K|r - R|) = \frac{1}{2\pi}\int_0^{2\pi}\mathrm{d}\beta\int_0^{\infty}\mathrm{d}s_{\parallel}\left(\frac{s_{\parallel}}{s_z}\right)\exp[iKs^{(\pm)}\cdot(r - R)], \quad (36.7)$$

where

$$s^{(\pm)} = (s_{\parallel}\cos\beta, s_{\parallel}\sin\beta, \pm s_z),\quad s_z = i\sqrt{s_{\parallel}^2 - 1}. \quad (36.8)$$

The contour of the integration with respect to $s_{\parallel}$ is taken in the Rieman plane on which the real parts of $\sqrt{s_{\parallel}^2 - 1}$ involved in s_z take positive values to avoid ambiguity and singularity introduced in the $s_{\parallel}$ representation. In fact, s_z is represented by

$$s_z = \begin{cases} \sqrt{1 - s_{\parallel}^2} & \text{for } 0 \le s_{\parallel} < 1\ (\text{homogeneous mode}) \\ i\sqrt{s_{\parallel}^2 - 1} & \text{for } 1 \le s_{\parallel} < +\infty\ (\text{evanescent mode}) \end{cases}. \quad (36.9)$$

The angular spectrum representation of complex amplitude, E, of electric dipole radiation is given by

$$E(r) = \left(\frac{iK^3}{8\pi^2\varepsilon_0}\right)\int_0^{2\pi}\mathrm{d}\beta\int_0^{\infty}\mathrm{d}s_{\parallel}\left(\frac{s_{\parallel}}{s_z}\right)\left[P - (s^{(\pm)}\cdot P)s^{(\pm)}\right]\exp\left[iKs^{(\pm)}\cdot(r - R)\right].$$

$$(36.10)$$

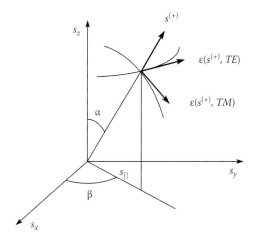

FIGURE 36.5 The polarization vectors, $\varepsilon(s^{(+)}, TE)$ and $\varepsilon(s^{(+)}, TM)$, corresponding, respectively, to TE and TM waves, defined with respect to the unit wavevector $s^{(+)}$.

It is convenient to introduce a set of polarization vectors $\varepsilon(s^{(\pm)}, TE)$ and $\varepsilon(s^{(\pm)}, TM)$ with respect to the unit wavevector $s^{(\pm)}$ defined, respectively, by

$$\varepsilon\left(s^{(\pm)}, TE\right) = e_z \times \frac{s_{\parallel}}{s_{\parallel}} = (-\sin\beta, \cos\beta, 0), \qquad (36.11)$$

$$\varepsilon\left(s^{(\pm)}, TM\right) = \varepsilon\left(s^{(\pm)}, TE\right) \times s^{(\pm)} = (\pm s_z \cos\beta, \pm s_z \sin\beta, -s_{\parallel}), \qquad (36.12)$$

which correspond to two orthogonal polarization vectors transverse to the unit wavevector $s^{(\pm)}$ introduced according to the rotational transform of TE and TM polarization vectors with respect to the propagation vector directing in z-axis by extending the angle of rotation α into complex values, as schematically shown in Figure 36.5.

The TE-polarization vector $\varepsilon(s^{(\pm)}, TE)$ is a real unit vector for both homogeneous and evanescent waves, since it does not involve complex component of wavevector s_z. On the other hand, $\varepsilon(s^{(\pm)}, TM)$ is a real unit vector for $0 \le s_{\parallel} < 1$, and a complex vector for $1 \le s_{\parallel} < +\infty$. The unit wavevector and the polarization vectors satisfy the following orthogonality relations:

$$\varepsilon\left(s^{(\pm)}, \mu\right) \cdot \varepsilon\left(s^{(\pm)}, \mu'\right) = \delta_{\mu,\mu'}, \quad \varepsilon\left(s^{(\pm)}, \mu\right) \cdot s^{(\pm)} = 0, \qquad (36.13)$$

where μ and μ' take TE and TM. With regard to these orthogonality relations, E is represented by

$$E(r) = \left(\frac{iK^3}{8\pi^2\varepsilon_0}\right) \sum_{\mu=TE}^{TM} \int_0^{2\pi} d\beta \int_0^{\infty} ds_{\parallel} \left(\frac{s_{\parallel}}{s_z}\right) \left[\varepsilon\left(s^{(\pm)}, \mu\right) \cdot P\right] \varepsilon(s^{(\pm)}, \mu)$$

$$\times \exp\left[iKs^{(\pm)} \cdot (r - R)\right]. \qquad (36.14)$$

36.5 Electric Dipole Radiation Near a Dielectric Surface

As an example of half-space problem based on the angular spectrum representation, let us evaluate electric dipole radiation near a dielectric surface. Let us consider a basic configuration of half-space problem described in Figure 36.1, in which a planar boundary separates the entire space into two half-spaces, and assume the left half-space being filled with a non-magnetic, transparent, homogeneous, and isotropic dielectric medium of refractive index n ($z < 0$), and the right half-space vacuum ($z \ge 0$).

It is convenient to introduce unit wave vectors of incoming waves in the right half-space, $s^{(-)}$, and that of outgoing waves in the left half-space, $\kappa^{(-)}$, defined, respectively, by

$$s^{(\pm)} = (s_{\parallel}\cos\beta, s_{\parallel}\sin\beta, \pm s_z), \quad s_z = i\sqrt{s_{\parallel}^2 - 1}, \qquad (36.15)$$

$$n\kappa^{(-)} = (s_{\parallel}\cos\beta, s_{\parallel}\sin\beta, -n\kappa_z), \quad n\kappa_z = i\sqrt{s_{\parallel}^2 - n^2}. \qquad (36.16)$$

According to the discussion in Section 36.3, the fundamental processes of the near-field optical interactions of electric-dipole with dielectric-medium involve three characteristic components described in Figure 36.6, where each process is related to the contour of integration with respect to the complex angle α in the angular spectrum representation of electric dipolar radiation.

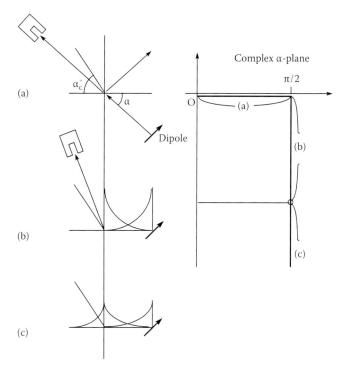

FIGURE 36.6 Three characteristic scattering processes of dipole field described in terms of angular spectrum representation.

(a) $0 \le s_\| < 1$ (α: $0 \to \pi/2$): $s^{(-)}$, $s^{(+)}$, $\kappa^{(-)}$; real
(b) $1 \le s_\| < n$ (α: $\pi/2 \to (\pi/2) - i\gamma_c$, $\sin[(\pi/2) - i\gamma_c] = n$): $s^{(-)}$, $s^{(+)}$; complex, $\kappa^{(-)}$; real
(c) $n \le s_\| < +\infty$ (α: $(\pi/2) - i\gamma_c \to (\pi/2) - i\infty$): $s^{(-)}$, $s^{(+)}$, $\kappa^{(-)}$; complex

The optical energy transferred via near-field optical interaction of the electric dipole with the dielectric medium is calculated by integrating Poynting vector consisted of the cross products of the incident and reflected evanescent waves over a plane placed at z lying between the dipole and the dielectric surface $(0 < z < Z)$

$$I = 2\varepsilon_0 \mathfrak{Re} \left\{ \int\int_{-\infty}^{+\infty} dx\, dy\, (-\boldsymbol{e}_z) \cdot \left[\boldsymbol{E}(\boldsymbol{r}) \times \boldsymbol{B}^{(R)\,*}(\boldsymbol{r}) + \boldsymbol{E}^{(R)}(\boldsymbol{r}) \times \boldsymbol{B}^*(\boldsymbol{r}) \right] \right\}.$$

(36.17)

The angular spectrum representation of the reflected electric dipole field from the dielectric surface is obtained according to the Fresnel's reflection coefficient $R_{21}(s_\|, \mu)$ for each wave component with conserved spatial frequency $K s_\|$ and polarization μ. Substituting the electromagnetic fields in the Poynting vector by their angular spectrum representations, we obtain the angular spectrum representation of Poynting Vector as

$$I = \left(\frac{K^4}{4\pi^2\varepsilon_0} \right) \mathfrak{Re} \left\{ \sum_{\mu=TE}^{TM} \int_0^{2\pi} d\beta \int_1^n \frac{s_\| ds_\|}{i\sqrt{s_\|^2 - 1}} \left[\boldsymbol{P}^\star \cdot \varepsilon \left(\boldsymbol{s}^{(+)}, \mu \right) \right] \left[\varepsilon (\boldsymbol{s}^{(-)}, \mu) \cdot \boldsymbol{P} \right] \right.$$

$$\left. \times R_{21} \left(s_\|, \mu \right) \exp \left(-2KZ \sqrt{s_\|^2 - 1} \right) \right\}.$$

(36.18)

Each component in the integral is interpreted as the optical excitation tunneling between the electric dipole and the dielectrics since it is described as a cross product of evanescent waves that exhibit exponential decay in the opposite direction to each other with respect to z. It is noted that Poynting vector vanishes when evanescent waves are not connected to outgoing fields or optical detector in near-field region in the left half-space.

As a numerical example of angular spectrum representation, let us consider optical near-field of a z-oriented oscillating electric dipole, $\boldsymbol{P} = |\boldsymbol{P}|\boldsymbol{e}_z$, with frequency K placed at a distance Z from an assumed boundary plane at $z = 0$ (Figure 36.7). Here, we consider a near-field regime and assume Z shorter than the optical wavelength in vacuum. For an observation point placed at $\boldsymbol{r} = r_\|\boldsymbol{e}_\|$ on the assumed boundary plane the z-component of the electric field, E_z, corresponding to the near-field of dipole radiation, $1 \le s_\|$, is given in terms of angular spectrum representation by

$$\left[E_z(r_\|\boldsymbol{e}_\|) \right]_{eva} = \left(\frac{K^3|\boldsymbol{P}|}{4\pi\varepsilon_0} \right) \int_1^\infty \frac{s_\| ds_\|}{\sqrt{s_\|^2 - 1}} s_\|^2 J_0(Kr_\|s_\|) \exp\left(-KZ\sqrt{s_\|^2 - 1} \right).$$

(36.19)

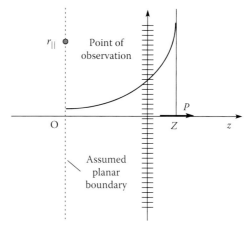

FIGURE 36.7 Angular spectrum representation of dipole field at a point of observation on the assumed boundary plane.

Figures 36.8 and 36.9 indicate the numerical values of the angular spectra of near-field regime, $[E_z(r_\|\boldsymbol{e}_\|)]_{eva}$, normalized by the factor $K^3|\boldsymbol{P}|/(4\pi\varepsilon_0)$,

$$s_\|^2 J_0(Kr_\|s_\|) \exp\left(-KZ\sqrt{s_\|^2 - 1} \right).$$

The angular spectra show the amplitudes of evanescent waves as the function of the spatial frequency, $s_\|$, normalized by the wavenumber of light in free space K.

Figure 36.8 shows the angular spectrum calculated at $\boldsymbol{r} = \boldsymbol{0}$, as is indicated in the inset of the figure for several values of the dipole-boundary distance Z. It is clearly seen that the angular spectra take their maxima at the spatial frequencies corresponding to $\sim 1/KZ$ and have the spectral widths $\sim 1/KZ$. These results provide a very useful criterion to identify the spatial-frequency components dominating the near-field optical interactions when

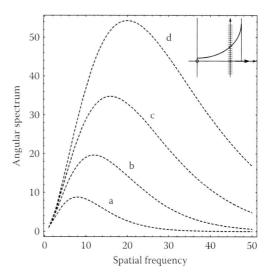

FIGURE 36.8 The angular spectrum of evanescent modes for dipole radiation field versus the spatial frequency $s_\|$ at a point of observation $R_\| = 0$; (a) for $KZ = 1/4$, (b) for $KZ = 1/6$, (c) for $KZ = 1/8$, and (d) for $KZ = 1/10$.

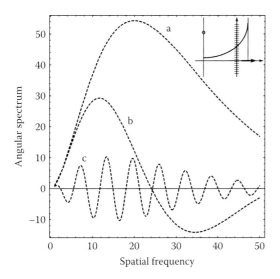

FIGURE 36.9 The angular spectrum of evanescent modes for dipole radiation field versus the spatial frequency $s_\parallel$ at a distance $KZ = 1/10$; (a) for $Kr_\parallel = 1/4$, (b) for $Kr_\parallel = 1$, (c) for $Kr_\parallel = 1/10$, and (d) for $Kr_\parallel = 0$.

a scatterer or a detector is placed at the observation point. It should be noted that the principal part of the angular spectra around their maxima, i.e., the spectral region within the full width at the half maximum, corresponds to the radial dependence of the optical near-field with third power of the inverse distance $1/KZ$ involved in the Hankel function $h_2^{(1)}$, which is the origin of the field singularity in the near-field limit.

Figure 36.9 shows the angular spectrum calculated for a fixed value of the dipole-boundary distance Z at several different observation points r on the assumed boundary plane. The angular spectrum shows an oscillating property for $r_\parallel$ larger than Z, so the overall contribution of evanescent waves to the electric field vanishes as the observation point moves away from the dipole. This property shows how the locality of optical near-fields is represented in angular spectrum representation.

As is discussed in the previous sections, the angular spectrum representation provides a very useful tool to evaluate locality and range of near-field optical interactions as well as optical energy transfer of tunneling regime based on theoretical treatment of half-space problem. For a system with complicated material distribution of near-field regime, one can utilize angular spectrum representation for evaluation of spatial correlation or mode matching with respect to an assumed boundary plane, which characterizes the problem under consideration.

36.6 Quantum Theory of Optical Near-Fields

Field quantization including evanescent waves have been established by Carniglia and Mandel based on the so-called triplet modes (Carniglia and Mandel 1971) that consist of a set of normal modes fit for half-space problems. Each of the triplet modes is composed of a set of incident, reflected, and transmitted waves connected at the planar boundary via Fresnel's relations based

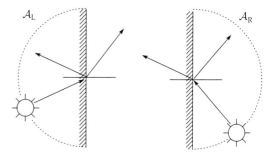

FIGURE 36.10 The triplet modes. R-triplet mode couples to a single optical-source on the hemisphere $\mathcal{A}_R$. L-triplet mode couples to that on the hemisphere $\mathcal{A}_L$.

on the symmetry under parallel spatial displacement along the boundary plane (Figure 36.10). The triplet mode involves a single incident wave and provides a convenient basis for the theoretical description of photon absorption processes in optical near-field problems. On the other hand, the triplet mode describes photon emission processes based on a correlation measurement between two outgoing branches, as is similar to beam splitter problems. One can also establish field quantization including evanescent waves, employing the so-called detector modes as a convenient basis to study photon emission processes in half-space problems (Inoue and Hori 2001, 2005). Each of the detector modes is composed of a set of two incident waves and one outgoing wave connected at the planar boundary via Fresnel's relations, so it includes an interference process between reflected and transmitted waves resulting in a single outgoing wave (Figure 36.11). The detector mode description is quite useful for the evaluation of radiative lifetime of optically excited systems near matter since the final state density of radiation is well specified. In this section, we present a brief review of field quantization in half-space problems based on the triplet and detector modes.

Let us consider the basic configuration of half-space problem described in Figure 36.1 and assume, for completeness, that the entire system is enclosed in a sphere with approximately infinite radius. Therefore, a planar boundary separates the entire space into two hemispheres, the left $\mathcal{A}_L$ and the right $\mathcal{A}_R$ hemispheres of radius $r_\pm$ in far-field, $kr_\pm$, $Kr_\pm \gg 1$, on which optical sources and sinks are considered to be placed. Here, we assume that the left hemisphere $\mathcal{A}_L$ is filled with a nonmagnetic, transparent,

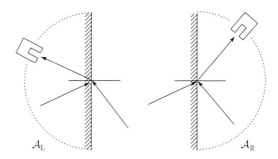

FIGURE 36.11 The detector modes. R detector mode couples to a single optical-sink on the hemisphere $\mathcal{A}_R$. L detector mode couples to that on $\mathcal{A}_L$.

homogeneous, and isotropic dielectric medium of refractive index n ($z < 0$).

Let us consider monochromatic fields of frequency K corresponding to Fourier component of electric field $\mathbf{E}(\mathbf{r})\exp(-iKt)$, under the unit in which the light velocity is taken to be unity, $c = 1$. The complex amplitude $\mathbf{E}(\mathbf{r})$ of electric field satisfies the Helmholtz equation

$$\left[\nabla^2 + K^2 n^2(\mathbf{r}) \right] \mathbf{E}(\mathbf{r}) = \mathbf{0}, \tag{36.20}$$

with the refractive index function defined by $n(\mathbf{r}) = n$ for $z < 0$, and $n(\mathbf{r}) = 1$ for $z \geq 0$, where n is assumed to be a real number.

As the basis to introduce normal modes in half-space problems, we introduce the unit wavevectors of incoming waves from the right of the boundary plane as $\mathbf{s}^{(-)} = \mathbf{K}^{(-)}/K = (s_x, s_y, -s_z)$, and those of outgoing fields to the left of the boundary plane as $\kappa^{(-)} = \mathbf{k}^{(-)}/(nK) = (\kappa_x, \kappa_y, -\kappa_z)$, which satisfy the relations in Equation 36.16. We also introduce the unit wave vectors of incoming waves from the left as $\kappa^{(+)} = \mathbf{k}^{(+)}/nK = (\kappa_x, \kappa_y, \kappa_z)$ and those of outgoing field to the right as $\mathbf{s}^{(+)} = \mathbf{K}^{(+)}/K = (s_x, s_y, s_z)$. The projection of wavevector onto the boundary plane is conserved because of the spatial translational symmetry in half-space problems, so we introduce $nK\kappa_\parallel$ and $Ks_\parallel$ defined, respectively, by $\kappa_\parallel = \sqrt{\kappa_x^2 + \kappa_y^2}$ and $s_\parallel = \sqrt{s_x^2 + s_y^2}$.

When we consider a problem with a single optical source, it is convenient to employ the triplet modes, shown in Figure 36.10, composed of one incoming and two outgoing plane waves being connected by Fresnel's relations at the boundary plane. The incoming waves can be connected, respectively, to the optical sources on the right hemisphere, $\mathcal{A}_R$, and left hemisphere, $\mathcal{A}_L$, and the two outgoing waves exhibit quantum correlation. Based on the orthogonality relations and completeness confirmed by Carniglia and Mandel (1971), we can introduce annihilation operator, $\hat{a}_R(\mathbf{K}^{(-)}, \mu)$, and creation operator, $\hat{a}_R^\dagger\left(\mathbf{K}^{(-)}, \mu\right)$, of photon in the R-triplet mode specified by the wavenumber $\mathbf{K}^{(-)}$ and polarization μ, and annihilation operator, $\hat{a}_L(\mathbf{k}^{(+)}, \mu)$, and creation operator, $\hat{a}_L^\dagger\left(\mathbf{k}^{(+)}, \mu\right)$, of photon in the L-triplet mode specified by the wavenumber $\mathbf{k}^{(+)}$ and polarization μ, which satisfy the following commutation relations:

$$\left[\hat{a}_R(\mathbf{K}^{(-)},\mu), \hat{a}_R^\dagger\left(\mathbf{K}'^{(-)},\mu'\right) \right] = \delta_{\mu,\mu'}(2\pi)^3 \delta\left(\mathbf{K}^{(-)} - \mathbf{K}'^{(-)}\right), \tag{36.21}$$

$$\left[\hat{a}_L(\mathbf{k}^{(+)},\mu), \hat{a}_L^\dagger\left(\mathbf{k}'^{(+)},\mu'\right) \right] = \delta_{\mu,\mu'}(2\pi)^3 \delta\left(\mathbf{k}^{(+)} - \mathbf{k}'^{(+)}\right), \tag{36.22}$$

$$\left[\hat{a}_R(\mathbf{K}^{(-)},\mu), \hat{a}_R\left(\mathbf{K}'^{(-)},\mu'\right) \right] = 0, \quad \left[\hat{a}_L(\mathbf{k}^{(+)},\mu), \hat{a}_L\left(\mathbf{k}'^{(+)},\mu'\right) \right] = 0, \tag{36.23}$$

$$\left[\hat{a}_R(\mathbf{K}^{(-)},\mu), \hat{a}_L\left(\mathbf{k}'^{(+)},\mu'\right) \right] = 0, \quad \left[\hat{a}_R\left(\mathbf{K}^{(-)},\mu\right), \hat{a}_L^\dagger\left(\mathbf{k}'^{(+)},\mu'\right) \right] = 0. \tag{36.24}$$

It is noted that the tree wave components involved in each of the triplet modes are created or annihilated at once in entire space by one of these operations. The electric field operator is represented in terms of the triplet-mode operators by

$$\hat{\mathbf{E}}(\mathbf{r},t) = \frac{1}{(2\pi)^3} \int\limits_{-K_z < 0} \mathrm{d}^3 K^{(-)} \sum_{\mu=TE}^{TM} \left(\frac{\hbar K}{\varepsilon_0} \right)^{1/2}$$
$$\times \left[\hat{a}_R(\mathbf{K}^{(-)}, \mu) \mathbf{E}_R(\mathbf{K}^{(-)}, \mu, \mathbf{r})\exp(-iKt) + \text{H.c.} \right]$$
$$+ \frac{1}{(2\pi)^3} \int\limits_{+k_z > 0} \mathrm{d}^3 k^{(+)} \sum_{\mu=TE}^{TM} \left(\frac{\hbar K}{\varepsilon_0} \right)^{1/2}$$
$$\times \left[\hat{a}_L(\mathbf{k}^{(+)}, \mu) \mathbf{E}_L(\mathbf{k}^{(+)}, \mu, \mathbf{r})\exp(-iKt) + \text{H.c.} \right] \tag{36.25}$$

As the complementary description of the half-space problems, we can introduce the so-called detector modes, shown in Figure 36.11, composed of a single outgoing and two incident plane waves being connected by Fresnel's relations at the boundary plane. The detector-mode description is especially useful in investigations of radiation properties of photonic sources, since one of the single outgoing waves can be connected to a detector placed on the right hemisphere, $\mathcal{A}_R$, and left hemisphere, $\mathcal{A}_L$, in far-field as a well defined final state related to the radiation process. Based on the orthogonality relations and completeness confirmed by Inoue and Hori (2001), we can introduce annihilation operator, $\hat{a}_{DR}(\mathbf{K}^{(+)}, \mu)$, and creation operator, $\hat{a}_{DR}^\dagger(\mathbf{K}^{(+)}, \mu)$, of photon in the R-detector mode specified by the wavenumber, $\mathbf{K}^{(+)}$, and polarization, μ, and annihilation operator, $\hat{a}_{DL}(\mathbf{k}^{(-)}, \mu)$, and creation operator, $\hat{a}_{DL}^\dagger(\mathbf{k}^{(-)}, \mu)$, of photon in the L-detector mode specified by the wavenumber, $\mathbf{k}^{(-)}$, and polarization, μ, which satisfy the following commutation relations:

$$\left[\hat{a}_{DR}(\mathbf{K}^{(+)},\mu), \hat{a}_{DR}^\dagger(\mathbf{K}'^{(+)},\mu') \right] = \delta_{\mu,\mu'}(2\pi)^3 \delta(\mathbf{K}^{(+)} - \mathbf{K}'^{(+)}), \tag{36.26}$$

$$\left[\hat{a}_{DL}(\mathbf{k}^{(-)},\mu), \hat{a}_{DL}^\dagger(\mathbf{k}'^{(-)},\mu') \right] = \delta_{\mu,\mu'}(2\pi)^3 \delta(\mathbf{k}^{(-)} - \mathbf{k}'^{(-)}), \tag{36.27}$$

$$\left[\hat{a}_{DR}(\mathbf{K}^{(+)},\mu), \hat{a}_{DR}(\mathbf{K}'^{(+)},\mu') \right] = 0, \quad \left[\hat{a}_{DL}(\mathbf{k}^{(-)},\mu), \hat{a}_{DL}\left(\mathbf{k}'^{(-)},\mu'\right) \right] = 0, \tag{36.28}$$

$$\left[\hat{a}_{DR}(\mathbf{K}^{(+)},\mu), \hat{a}_{DL}\left(\mathbf{k}'^{(-)},\mu'\right) \right] = 0, \quad \left[\hat{a}_{DR}\left(\mathbf{K}^{(+)},\mu\right), \hat{a}_{DL}^\dagger\left(\mathbf{k}'^{(-)},\mu'\right) \right] = 0. \tag{36.29}$$

It is noted that the tree wave components involved in each of the detector modes are created or annihilated at once in entire space by one of these operations. The electric field operator is represented in terms of the detector-mode operators by

$$\hat{E}(r,t) = \frac{1}{(2\pi)^3} \int_{K_z>0} d^3 K^{(+)} \sum_{\mu=TE}^{TM} \left(\frac{\hbar K}{\varepsilon_0}\right)^{1/2}$$

$$\times \left[\hat{a}_{DR}(\boldsymbol{K}^{(+)},\mu)\boldsymbol{E}_{DR}(\boldsymbol{K}^{(+)},\mu,\boldsymbol{r})\exp(-iKt) + \text{H.c.}\right]$$

$$+ \frac{1}{(2\pi^3)} \int_{-k_z<0} d^3 k^{(-)} \sum_{\mu=TE}^{TM} \left(\frac{\hbar K}{\varepsilon_0}\right)^{1/2}$$

$$\times \left[\hat{a}_{DL}(\boldsymbol{k}^{(-)},\mu)\boldsymbol{E}_{DL}(\boldsymbol{k}^{(-)},\mu,\boldsymbol{r})\exp(-iKt) + \text{H.c.}\right] \qquad (36.30)$$

Based on the triplet-mode and detector-mode formalisms, we can evaluate quantum optical processes in half-space problems. As in general, for the theoretical treatment of radiation problems related to material two-level systems in near-resonant regime with monochromatic optical fields, Hamiltonian for electromagnetic interactions is given by

$$\hat{V}_I(t) = -e \int d^3 r \hat{\boldsymbol{J}}(r,t) \cdot \hat{\boldsymbol{A}}(r,t), \qquad (36.31)$$

where $\hat{\boldsymbol{J}}(r,t)$ is the probability current density defined for the material field $\hat{\Psi}(r,t)$ by

$$\hat{\boldsymbol{J}}(r,t) = \frac{i\hbar}{2m}\left[\left\{\nabla\hat{\Psi}^{\dagger}(r,t)\right\}\hat{\Psi}(r,t) - \hat{\Psi}^{\dagger}(r,t)\left\{\nabla\hat{\Psi}(r,t)\right\}\right], \qquad (36.32)$$

and $\hat{\boldsymbol{A}}(r,t)$ is the vector potential operator defined by the operator relation derived from Maxwell's equation

$$\hat{\boldsymbol{E}}(r,t) = -\frac{\partial}{\partial t}\hat{\boldsymbol{A}}(r,t) \qquad (36.33)$$

from the corresponding electric-field operators of the triplet and detector modes.

When we consider a quantum optical process in which the entire system composed of the fields and the two-level material system exerts a transition because of the electromagnetic interaction from an initial state $|i\rangle$ to a final state $|f\rangle$, the transition probability $d\Gamma$ is given, according to Fermi's golden rule, by

$$d\Gamma = \frac{2\pi}{\hbar^2}\left|\langle f|V_I(0)|i\rangle\right|^2 d\rho(K), \qquad (36.34)$$

where $d\rho(K)$ indicates the final state density of the electromagnetic fields plus material two-level system.

It should be stressed that the density of the final state governs the transition process regardless of the dynamics of the quantum mechanical coupled matter-field system described by the transition amplitude. Since the density of final state is determined by interaction processes with environmental systems described in terms of dissipation, the dynamics of the overall transition process between the initial and final states of the coupled matter-field system depends strongly on the dynamics of the environmental systems. In our current problem, the environmental

system consists of the optical sources and detectors placed on each hemisphere, $\mathcal{A}_R$ and $\mathcal{A}_L$. Here, the difference between the triplet and detector modes becomes significant with respect to the final state, since the triplet modes involves two outgoing waves in contrast to the detector modes, each of which has a single outgoing wave in its final state. We should discuss further on this problem since it is related to one of the most important issues in the study of functional systems in nanoscience and nanotechnology including nanophotonics.

Before proceeding to the study of these issues, we should clearly identify the two different aspects of these problems. One is the problem in which we consider absorption and emission of optical energy by the material two-level system inside the two hemispheres, which is coupled with the approximately free optical fields described in terms of the triplet or detector modes. The other is the problem in which we consider optical energy transfer through the assumed boundary plane in the half-space problem under consideration. The latter is related to the function of nanophotonics devices and systems, which is based on optical energy transport via near-field optical interactions.

Firstly, let us consider the former problem related to absorption and emission processes of the material two-level system inside the hemisphere. When we employ the triplet-mode description, we should be careful about the correlation between the observation processes at the two detectors, since each of the triplet modes involves two outgoing waves. This aspect involved in the triplet-mode description is similar to those found in beam-splitter problems in quantum optics. That is, when we consider emission properties of a single excited material two-level system by using an uncorrelated detector pair on the $\mathcal{A}_R$ and $\mathcal{A}_L$ hemispheres, we can observe the antibunching property in photon detection processes, which results in a single photon detection by only one of the two detectors within the coherence time of the emission process. In contrast, when we consider a correlated measurement of the outgoing waves by introducing a certain coherence property between two detectors, such as interferometry observation, the correlated detectors dissipate the corresponding higher-order coherence involved in the photon emission process. In conclusion, the photon emission properties of the material two-level system, and therefore the transition probability, depend on the correlation between the detectors. In contrast to the above, when we employ the detector mode description, the evaluation of the final state density is straight forward for the photon emission problem under consideration, since each of the detector modes involves only a single outgoing wave. The coherence properties are described as the correlation between the two incoming waves involved in each of the detector modes, which is established during the internal interaction of the coupled matter-field system in the two hemispheres. For instance, if we consider that the left half-space is filled with dielectrics and that an atomic two-level system resonant with optical field is put in the optical near-field of the dielectric surface in the right half-space, the incoming wave in the dielectric side represents the overall response of dielectrics to the radiation from atomic dipole and, therefore, corresponds to image dipole picture appeared in the treatment of classical electromagnetic boundary problems. Here, we restrict our discussion within notification of these aspects. For a detailed study on the optical near-field photon emission and absorption, one can refer to the articles by the authors (Inoue and Hori 2001, 2005, Inoue et al. 2005).

Secondly, let us consider the latter aspect related to energy transport between the sources and detectors placed, respectively, on the left and right hemispheres. According to authentic treatments of scattering problems based on field theory, all the photons injected in the system from the source are considered to be annihilated in the system, and the photons emitted out of the system are those created in the system. The energy transfer through the boundary plane should be described in terms of the creation of photon in the left-triplet mode by the photonic source on the left hemisphere and succeeding annihilation of the photon by the material systems inside the hemispheres followed by creation of photon in the right-detector mode by the material systems inside and succeeding annihilation of the photon by the detector on the right hemisphere. The connection between the annihilation and creation processes of photons inside the system is governed by the coupling of the system with an internal environmental system or an additional reservoir connected to the system within the two hemispheres. This corresponds to the generally recognized aspect for directional transport processes that the dynamics of the entire system becomes irreversible due to dissipation. The issues regarding the dissipation of transport processes in a nanometer scale or, in other words, the study of transport processes in a local system, remain unsolved and are under extensive study at present. These issues, however, are one of the key issues toward innovation in functional devices and systems in nanotechnology including nanophotonics since signal transfer processes, in general, are based on the electromagnetic energy transport driven by electromagnetic interactions between electronic systems (Hori 2001).

References

Carniglia C. K. and Mandel L. 1971, Quantization of evanescent electromagnetic waves, *Phys. Rev. D* 3: 280–296.

Hori H. 2001, Electronic and electromagnetic properties in nanometer scales, in *Optical and Electronic Process of Nano-Matters*, ed. M. Ohtsu, pp. 1–55, KTK Scientific Publishers, Tokyo Japan.

Inoue T. and Hori H. 1996, Representations and transforms of vector field as the basis of near-field optics, *Opt. Rev.* 3: 458–462.

Inoue T. and Hori H. 2001, Quantization of evanescent electromagnetic waves based on detector modes, *Phys. Rev. A* 63: 063805-1–063805-16.

Inoue T. and Hori H. 2005, Quantum theory of radiation in optical near-field based on quantization of evanescent electromagnetic waves using detector mode, in *Progress in Nano-Electro Optics*, Vol. 4, ed. M. Ohtsu, pp. 127–199, Springer-Verlag, Berlin, Germany.

Inoue T., Banno I., and Hori H. 1998, Theoretical treatment of electric and magnetic multipole radiation near a planar dielectric surface based on angular spectrum representation, *Opt. Rev.* 5: 295–302.

Inoue T., Ohdaira Y., and Hori H. 2005, Theory of transmission and dissipation of radiation near a metallic slab based on angular spectrum representation, *IEICE Trans. Electron.* E88-C: 1836–1844.

Matsudo T., Inoue T., Inoue Y., Hori H., and Sakurai T. 1997, Direct detection of evanescent electromagnetic waves at a planar dielectric surface by laser atomic spectroscopy, *Phys. Rev. A* 55: 2406–2412.

Matsudo T., Takahara Y., Hori H., and Sakurai T. 1998, Pseudomomentum transfer from evanescent waves to atoms measured by saturated absorption spectroscopy, *Opt. Comm.* 145: 64–68.

Ohdaira Y., Kijima K., Terasawa K., Kawai M., Hori H., and Kitahara K. 2001, State-selective optical near-field resonant ionization spectroscopy of atoms near a dielectric surface, *J. Microsc.* 202: 255–260.

Ohdaira Y., Inoue T., Hori H., and Kitahara K. 2008, Local circular polarization observed in surface vortices of optical near-fields, *Opt. Express* 16: 2915.

Ohtsu M. and Hori H. 1999, *Near-Field Nano-Optics*: Kluwer Academic/Plenum Publisher, New York.

Wolf E. and Niet-Vesperinas M. 1985, Analyticity of the angular spectrum amplitude of scattered fields and some of its consequence, *J. Opt. Soc. Am. A* 2: 886–890.

37

Near-Field Photopolymerization and Photoisomerization

Renaud Bachelot
Université de Technologie de Troyes

Jérôme Plain
Université de Technologie de Troyes

Olivier Soppera
Centre national de la recherche scientifique

37.1 Introduction

An important domain of nanotechnology is the nanostructuration that involves numerous scientific and economic challenges (Gentili et al. 1994; Bucknall 2005). In particular, the innovative trend of modern technology lies in smaller, cheaper, faster, and better performances. The industry must improve yield by increasing smaller instruments. For instance, cars, cameras, and wireless telephone have combined many functions in a small box. Nanotechnology is an exact example of this trend representing complex technology in commodities. During the last few years, novel structures, phenomena, and processes have been observed at the nanoscale from a fraction of nanometer to about 100 nm, and new experimental, theoretical, and simulation tools have been developed for investigating them. These advances provide fresh opportunities for scientific and technological developments in nanoparticles, nanostructured materials, nanodevices, and nanosystems. From the practical effect, the miniaturization of integration circuit and systems means the reduction of raw materials and energy waste. The smaller products are conducive to transportation and utilization, which proves to have more advantages of miniaturization over traditional products. In order to fit into the development of modern technology, the advances of nanotechnology are in urgent needs for its benefits. So the development of nanotechnology is driven by both science itself and market. This context explains why nanolithography and matter nanostructuration are important branches of nanotechnology (Gentili et al. 1994), envisioning various applications

and research fields including ultrahigh density storage, nanoelectronics, nanomechanics, and nanobiotechnology.

The current techniques of nanolithography are various and numerous (Plain et al. 2006, Sotomayor Torres 2003, Vieu et al. 2000). An exhaustive description and classification of these techniques is beyond the scope of this chapter. Some of them (among the most important ones) are illustrated in Figure 37.1. Most of the lithographic techniques involve far-field illumination of the material to be modified. They are diffraction limited, and the spatial resolution is theoretically not better than $\lambda/2n$, where λ is the wavelength of the source used and n is the refractive index of the medium. Among these techniques, let us cite x-ray and electron beam lithographies that allow for resolution in the 30–50 nm range.

Lithography using light (wavelengths included in the near-UV to near-IR domain) is particularly appreciated for several reasons of costs and simplicity. Presently, mask far-field optical lithography is the most widely used technique for pattern mass production in various fields such as microelectronics and microoptics (Sheats and Smith 1998). This technology is optimized and easy to implement compared with x-ray or ion-beam lithography. The photopolymers used are various, well known, and have been optimized for several applications (Chochos et al. 2008). However, the main limit of the optical lithography is its diffraction-limited spatial resolution. The typical resolution permitted by UV sources is currently about 100 nm. As shown in Figure 37.1, the current trend is to develop and use new small wavelength deep and EUV sources that would enable a resolution better than 50 nm in the near future (Wagner et al. 2000,

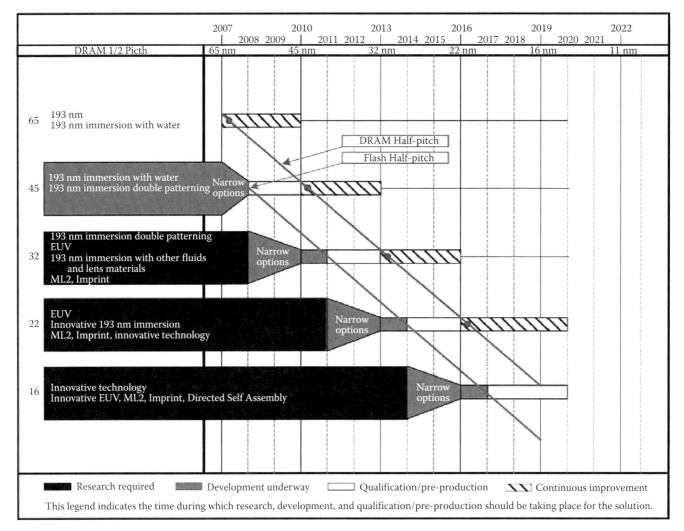

FIGURE 37.1 Far-field lithographies. (From IRTS roadmap 2005, www.itrs.net.)

Lin 2006). However, this approach requires expensive development of new technologies: new sources, new optics, and new photopolymers. As an example, the typical price for a 193 nm exposure tool is approximately 15 MDollars; the price of 157 nm exposure tools is as high as 25 MDollars and extreme ultraviolet (EUV) exposure tools may cost as much as 30 MDollars. Another approach consists in increasing the refraction index: immersion photolithography (using liquid), double patterning, and solid immersion lens lithography (using a scanning microlens) allow resolution to be improved by a factor 2, mainly limited by the refractive index of the available materials (Ghislain and Elings 1998; and Sheats and Smith 1998).

The near-field optical lithography and manipulation (NFOLM) is an alternative and elegant method of improving the resolution (Bachelot 2007, Inao et al. 2007, Tseng 2007). NFOLM relies on the use of spatially confined evanescent fields as optical sources. In the case of near-field illumination, the spatial confinement of the light–matter interaction is not limited by the light wavelength but rather by both source size and source-to-matter distance (Kawata 2001; Courjon 2003; Prasad 2004).

The advantage of NFOLM compared with the other kinds of lithography is thus to be an optics-based technique without any $\lambda/2n$ resolution limit. As it will be seen in Section 37.2, NFOLM actually relies on the use of lateral (parallel to material surface) wave vectors $k_{//}$ that are superior to k, the wave vector in medium n. Such high lateral wave vectors can be obtained by either total internal reflection (TIR) or diffraction by spatial frequencies $>2n/\lambda$ (Goodman 1996). High lateral wave vectors involve evanescent waves and nanometer scale light confinement. The control, use, and study of such electromagnetic waves constitute the near-field optics that has aroused large interest and efforts over the last two decades (Kawata 2001; Courjon 2003; Prasad 2004, Novotny and Hecht 2006). The advent of this science has opened a new field, and appreciation, for the control and manipulation of light at the nanoscale.

In this chapter, we focus our attention on the use of local optical near-fields for high-resolution optical lithography and matter manipulation on photopolymerizable and photoisomerizable systems. Through some examples, we show that this domain not only enabled production of nanostructures using light but also opened the door to nano photochemistry based on the use of evanescent waves. This chapter is divided into the following three sections. In Section 37.2, general considerations on nanooptics are given. The physical effects that are related to specific optical nano sources used and studied for NFOLM are described. Obviously, a complete review of this area of science is beyond the scope of this chapter. The reader is referred to recent reviews in comprehensive books on nanooptics (e.g., Novotny and Hecht 2006). Here we just remind and summarize some important phenomena. This reminder will be necessary for both commenting and appraising reported experiments. Section 37.3 is devoted to nanoscale photopolymerization. In Section 37.4, principles and experiments of controlled nanoscale photoisomerization are presented. Finally, we conclude and evoke some promising routes.

37.2 Some Concepts on Nanooptics: Examples of Optical Nanosources

37.2.1 Principles

Figure 37.2 illustrates the properties of the optical near-field. An illuminated planar object (x, y plane) diffracts the light though an angular spectrum of plane waves:

$$E(x,y,z) = \iint \tilde{E}(u,v,z) \exp\left[i2\pi(ux+vy)\right] du\, dv \qquad (37.1)$$

This expression is a result of the Fourier theory of diffraction, a consequence of the Huygens–Fresnel principle, described, for example, by Goodman (1996). In Equation 37.1, the amplitude of each plane wave is given by

$$\tilde{E}(x,y,z) = \tilde{A}_o(u,v) \exp(ik_z z)$$

$$= \tilde{A}_o(u,v) \exp\left(\frac{i2\pi z}{\lambda}(1 - \lambda^2 u^2 - \lambda^2 v^2)^{1/2}\right) \qquad (37.2)$$

where
  u and v are spatial frequencies of the Fourier (reciprocal) space
  $\tilde{A}_o$ is the spatial Fourier function of A_o, the object transmittance

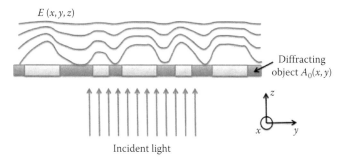

$E(x,y,z)$

Diffracting object $A_0(x,y)$

z

x y

Incident light

FIGURE 37.2 Principles of near-field optics.

Two cases have to be considered:

1. Object features of typical size $>\lambda$ are associated to low spatial frequencies and keep positive the term in root square in Equation 37.2. In that case, the k_z component is real and the waves propagate along z: they can be far-field detected. Obviously, the higher the lateral spatial frequency, the higher the $k_{x,y}$, the weaker the k_z, the more tilted the axis of wave propagation, and the higher the needed numerical aperture of the objective lens used to detect this propagating wave.

2. Object features of typical size $<\lambda$ are associated to high spatial frequencies and make negative the term in root square in Equation 37.2. In that case, the k_z component gets purely imaginary, and the wave becomes evanescent along z: it cannot be far-field detected, and its amplitude is proportional to $\exp(-\beta z)$, where β is a positive real characteristic to the considered spatial frequency.

As an example, a planar object whose transmittance is sinusoidal with a period of 20 nm has three spatial frequencies: 50 μm^{-1}, 0, and −50 μm^{-1}. These high spatial frequencies lead to evanescent waves that can not be detected in the far-field. As a result, only the specular field (corresponding to 0 spatial frequency) can be detected, and the object looks homogenous. Another example is that the angular spectrum of a diffracting single object of 20 nm diameter can be only partially detected: only the propagating part of the spectrum can be far field detected. As a result, using the best immersion objective lenses, the object will be seen in the visible with an apparent size of $\lambda/2n$, corresponding to about 200 nm. In other words, far-field detection is equivalent to a low pass filter for the spatial frequencies of the object to be analyzed. This diffraction limit, called the Abbe barrier, cannot be overcome unless the evanescent part of the angular spectrum is also detected. That requires a detector to be put in the near-field at a very small distance from the object surface. This is the principle of the scanning near-field optical microscopy (Novotny and Hecht 2006).

The above principle can be generalized for 3D nanoscale structures: diffracted field is evanescent in all directions perpendicular to an object plane containing high spatial frequencies. In that way, nanoscale optical sources can be generated. It should be pointed that the Eisenberg principle of uncertainty predicts well this relationship between high wave vectors distribution Δk and spatial precision/uncertainly Δx: $\Delta k\, \Delta x > 2\pi$.

37.2.2 Examples of Optical Nanosources

Relying on the principles introduced in Section 37.2.1, large efforts have been dedicated to the development of efficient optical nano sources over the past 20 years. Approaches of nano sources are numerous. Most of them have been developed at the extremity of probes for illumination-mode scanning near-field optical microscopy (Novotny and Hecht 2006). Figure 37.3 illustrates four examples of evanescent optical confinement at a scale smaller than the

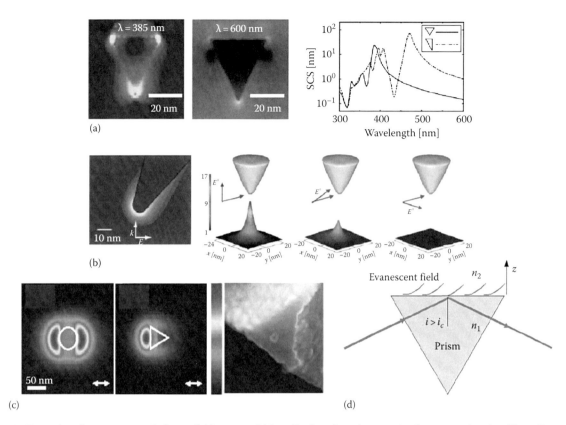

FIGURE 37.3 Examples of evanescent optical near-field sources. (a) Localized surface plasmons in silver nanotriangles. (From Kottmann, J. P. et al., *J. Microsc.*, 202, 60, 2001.) (b) Tip-enhanced optical near-field. (Left: From Sánchez, E. J. et al., *Phys. Rev. Lett.*, 82, 4014, 1999.) (c) Diffraction by a nanoaperture. (From Kottmann, J. P. et al., *J. Micros.*, 202, 60, 2001 and Molenda, D. et al., *Opt. Express*, 13, 10688, 2005.) (d) Total internal reflection. See text for details.

light wavelength. These four examples have been selected because they correspond to optical nanosources that have given rise to many studies and application over the past 20 years, especially in the context of NFOLM. Figure 37.3a shows an optical nanosource supported by localized surface plasmon at the surface of a noble metal nanostructure (MNS). Surface plasmons are quanta of plasma oscillation at a metallic surface (Kreibig and Vollmer 1996). Plasmon modes, corresponding to surface electronic resonance, exist in a number of geometries and in various metals, with the strongest responses in noble metals such as gold and silver. Localized surface plasmons can be coupled to light through evanescent waves diffracted by the metal nanostructures themselves. In confined geometries such as gold or silver nanoparticles, not only does the plasmon resonance depend on materials parameters (conductivity, electron effective mass and charge,…), but it also is affected by the shape of the object and its local environment as well as the condition of illumination (Hutter and Fendler 2004). In particular, for simple geometries (spheres, oblate, prolate, etc.), plasmon properties can be predicted by analytical theoretical models issued from the Mie Theory. Specifically, the polarizability α of a spheroid takes the well-known following form (Bohren and Hoffmann 1983):

$$\alpha(\omega) = \varepsilon_o V \frac{\varepsilon(\omega) - \varepsilon_m}{\varepsilon_m + L_i \left[\varepsilon(\omega) - \varepsilon_m \right]} \qquad (37.3)$$

where $\varepsilon(\omega)$, ε_o, and ε_m are the complex dielectric functions of the MNS, the vacuum, and the surrounding medium, respectively. Metal dispersion function $\varepsilon(\omega)$ can be correctly described through sophisticated Drude–Lorentz models taking into account both intraband and interband electronic transitions (Vial et al. 2005). V is the MNS volume, and L_i describes the spheroid geometry along the axes i ($i = x$, y, or z). For example, for a sphere, $L_x = L_y = L_z = 1/3$. Resonance of the MNS corresponds to a minimization of the denominator in Equation 37.3. This condition depends notably on the particle geometry, that is to say L_i, as well as on the dielectric environment, that is to say ε_m. In particular, in the case of a metallic sphere in air, the resonance condition is $\varepsilon(\omega) = -2$, which occurs typically in the visible region of the spectrum for gold and silver particles. In the case of non-regularly shaped MNS, numerical calculations are needed, as illustrated by Figure 37.3a that shows the numerically calculated optical properties of silver nanotriangles (Kottmann et al. 2001). At the right side of Figure 37.3a, the scattering cross section is calculated for two types of triangles as a function of the excitation wavelength, revealing the sensitivity of the plasmons resonance to the particle geometry. The two left-side images of Figure 37.3a show the corresponding near-field intensity distribution in the case of resonance and off resonance. For plasmon resonance ($\lambda = 385$ nm), the local field intensity is much higher than that for off resonance excitation

($\lambda = 600\,nm$). It is interesting to note that in both cases, the field tends to be confined at the corners of the structure where radius of curvature of the object is small. This phenomenon, illustrated in Figure 37.3b, is an off-resonance optical effect similar to the well-known electrostatic lighting rod effect. The term lightning-rod effect refers to an electrostatic phenomenon in which the electric charges on the surface of a conductive material are spatially confined by the shape of the structure. For a conductor with a nonspherical shape, surface charge density σ varies from point to point along the shape. In the regions of high curvature, σ is locally increased, resulting in a large electric field just outside the material. Although this effect originates from macroscopic electrostatic considerations, a similar phenomenon can be observed in nanometer-sized metallic structures excited by an electromagnetic radiation. This point of view was discussed in detail by Van Bladel (1995). The free electrons of the metal react to an electromagnetic excitation by inducing oscillating surface charges. When the surface presents a geometrical singularity such as a tip apex, the local surface charge density is drastically increased in this region. As a consequence, the electromagnetic field outside the tip is not only locally enhanced over the driving field but also highly confined around the tip apex. Figure 37.3b, left, shows an example of calculated enhanced confined field, acting as an optical nanosource, at the extremity of a gold tip illuminated at $\lambda = 830\,nm$ (Sánchez et al. 1999). The tip is illuminated by a significant field component parallel to the tip axis, i.e., by an incident p-polarization (i.e., incident field parallel to the incident plane). p-Polarization is actually suitable for exciting an electromagnetic singularity in the vicinity of metal tips, as shown at the right side of Figure 37.3b that represents field distribution at the extremity of a tungsten tip when illuminated by different polarization from p to s (incident field parallel to the incident plane). It should be stressed that this polarization effect is a consequence of the metal-dielectric boundary condition of the electric field, as reminded in Equations 37.4 and 37.5:

$$n \times (E_{met} - E_{m}) = 0 \qquad (37.4)$$

$$n \cdot (D_{met} - D_{m}) = \sigma \qquad (37.5)$$

where
  E_{met}, E_{m}, D_{met}, and D_{m}, are, respectively, the electric field within the metal, the electric field within the dielectric medium, the electric displacement within the metal, and the electric displacement within the dielectric medium, all at the metal surface
  n is the unit vector perpendicular to the metal surface of charge density σ

Let us remember that the electric displacement is proportional to the electric field through the respective permittivity ε of the media (ε is a scalar for isotopic media and a tensor for anisotropic ones). Equation 37.4 says that the tangential field component is continuous at the metal surface: the field vanishes (through the skin depth) both inside and outside the metal. Equation 37.5

says that the normal field component is discontinuous at the metal surface as a consequence of the Gauss Theorem: this field component vanishes inside the metal but can be high outside depending on σ. In the case of incident p polarization, the field is mainly parallel to the tip axis and normal to the metal surface just beneath the foremost tip's end, resulting in a local strong field. In the case of s polarization, field is tangential beneath the tip's end, which remains uncharged. With incident s-polarization, only the tip's edges can present significant (but nonlocalized) fields because fields are mainly normal at the edges. As it will be seen in Sections 37.3 and 37.4, such sources have been efficiently used for optical near-field nano-manipulation of photopolymers.

Figure 37.3c shows optical nanosources that are generated at a nanoaperture surrounded by a metallic screen. This effect can be described by the Bethe–Bouwkamp model of diffraction by nanoholes (Bethe 1944, Bouwkamp 1950). For such a source, the light is squeezed at the aperture by the surrounded metal. Aperture scanning near-field microscopy has been relying on this effect for 25 years (Novotny and Hecht 2006). In general, Aperture-based nano sources are developed at the extremity of tapered optical fibers. The development of these sources has given rise to many scientific and technological challenges, whose detailed description (Novotny and Hecht 2006) is beyond the scope of this chapter. The example presented in Figure 37.3c shows calculated field distribution (left) as a function of the geometry of an aperture integrated at the extremity of a tip (see an example of realization at the right side of Figure 37.3c), illustrating the high sensitivity of the near-field features to the local geometry.

Figure 37.3d illustrates a total internal reflection resulting in a light nanoscale confinement along one direction. When an incident light beam emerges from a high refractive index (n_1) medium into a lower index (n_2) medium with an angle greater than the critical value i_c defined by

$$i_c = \arcsin(n_2/n_1) \qquad (37.6)$$

an evanescent electromagnetic field is generated from the interface into the medium of refractive index $n_2 < n_1$. The beam intensity at a distance z into the film $I(z)$ is given by Equation 37.7, where I_0 is the intensity of the incident beam. The decay rate γ, of the exponential term depends on n_1 and n_2, the wavelength of the light beam λ, and the angle of incidence i (Equation 37.8).

$$I(z) = I_0 \cdot \exp(-2\gamma z) \qquad (37.7)$$

$$\gamma = \frac{2\pi}{\lambda}\sqrt{(n_1 \cdot \sin(i))^2 - n_2^2} \qquad (37.8)$$

$1/2\gamma$ has the dimension of a length, and it is called the characteristic penetration depth. As explained in Section 37.2.1, this evanescence is enabled by the high wave vector lateral component that is continuous at the n_1/n_2 interface. As it will be seen in the next section, such a nanoscale field penetration depth was used to produce polymer nanofilms.

37.3 Nanoscale Photopolymerization

Photopolymers are chemical systems of high interest. The general concept of crosslinking photopolymerization is to transform and vitrify irreversibly a fluid monomer or oligomers into a macromolecule structure by a light-induced chain reaction (Selli and Bellobono 1993). Among many advantages such as low temperature conditions and solvent-free formulations (Decker 1993), the interest of light-induced polymerization is to allow a control of the polymer structure via selected monomers, wavelength irradiation, and light intensity. If light is focused onto a limited area, the extent of the polymerization can be advantageously spatially controlled and confined to create complex objects with various functions. This possibility makes them attractive for many applications such as holographic recording (Lougnot 1993), optical data storage (Guattari et al. 2007), and diffractive optical elements fabrication (Carré et al. 2002). Moreover, since their elementary building blocks (monomers or oligomers) are of nanometric scale, these materials present a high potential for nanotechnology applications and, in particular, for nanophotolithography. Like in thermal polymerization, several mechanisms can be involved: free-radical, cationic, and anionic. An example of free-radical system sensitive in the green is given in Figure 37.4. This figure also details the pathway that leads to the photocrosslinking of an acrylate-based monomer by a free radical process. It entails four main steps:

1. Absorption of the photon by the dye
2. Production of reactive species from the dye excited states
3. Polymerization
4. Termination of polymerization by inhibition or other phenomena

The photoinitiator is the central element in this mechanism. Many different systems have been developed to fit the different families of monomer, different excitation wavelength and irradiation conditions (atmosphere, presence of interfering chemical, etc...). An overview of photoinitiators can be found in ref (Fouassier 1993).

The system depicted in Figure 37.4 is a mixture of three components: a xanthenic dye sensitizer (Eosin Y), a co-initiator (methyldiethanolamine, 8 wt.%), and a triacrylic monomer base (pentaerythritol triacrylate—PETA). Due to the presence of Eosin, the system is particularly sensitive to visible wavelength (spectral range from 450 to 550 nm), and thus it fits well with the main green emission ray of the argon laser (514 nm). The photopolymerization reaction is initiated through a free-radical process (Lougnot and Turck 1992). Two components are involved in the radical formation: the dye that absorbs the visible radiation and the amine that can be oxidized by the triplet state of the dye. Once the radicals trigger the polymerization, the monomers polymerize, developing into a three dimensional network. The unreacted monomers are eliminated by rinsing with ethanol.

One specific characteristic of this formulation is related to the existence of a threshold of polymerization that allows a sharp control of the polymerized area. This effect is mainly due to the well-known effect of free-radical reaction quenching by oxygen (Croutxe-Barghorn et al. 2000). In addition, the PETA contains

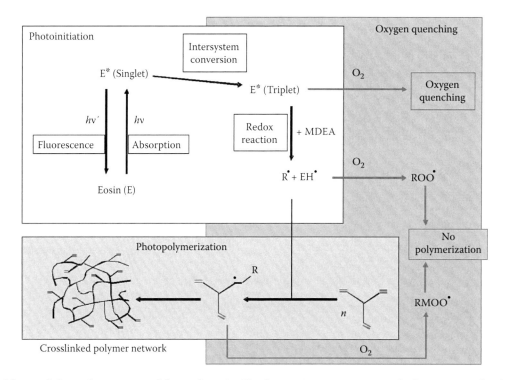

FIGURE 37.4 Scheme of photopolymerization of the acrylic resin. The three main steps are presented: photoinitiation by absorption of light by the photoinitiating system (top left), polymerization of the acrylic monomer (bottom left), and competitive process of quenching by oxygen (right).

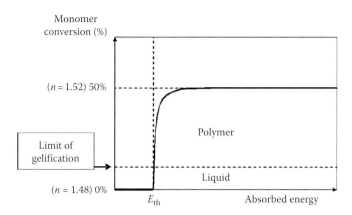

FIGURE 37.5 Typical response of a photocrosslinkable resin, showing the typical reticulation rate and associated refraction index as a function of absorbed energy density. E_{th} is the threshold energy.

300 ppm of thermal polymerization inhibitors added to stabilize the monomer, e.g., to avoid any unlike thermal polymerization. The effect of these compounds is to inhibit the polymerization process as long as the absorbed light dose remains lower than a threshold value. Consequently, the sensitivity of the system is characterized by a curve that shows the degree of cross-linking as a function of the received energy E_r. This curve, presented in Figure 37.5, corresponds to the typical behavior of a formulation that can be polymerized following a radical process. One can note that polymerization starts only when the absorbed energy is greater than a threshold value E_{th}. The formulation is thus characterized by a non linear threshold behavior, allowing for high-resolution patterning.

This resin has already proved its interest for applications in holographic data storage (Jradi et al. 2008a), optical microdevices fabrication (Jradi et al. 2008b). To demonstrate its potential in the frame of nanofabrication, two configurations involving optical near-field photopolymerization were evaluated: evanescent wave created by total internal reflection and MNS plasmon excitation.

37.3.1 Photopolymerization Using Evanescent Waves

As explained in Section 37.2, evanescent waves can be generated by total internal reflection at an interface between two materials with different refractive indexes (see typical configuration at the top of Figure 37.6). In our case, we used a high refractive index prism in contact with a photosensitive resin. When light is totally reflected internally at the interface between media of high and low refractive indexes, evanescent waves are created along the interface with highly limited penetration in the second medium. If this medium is composed of a photosensitive resin, the evanescent waves can be advantageously used to induce the photoinduced modification of the resin. This approach was first developed by Ecoffet et al. (1998), and it allowed one to demonstrate that optical near-field can be used to trigger free-radical polymerization. The penetration depth defined in Section 37.2

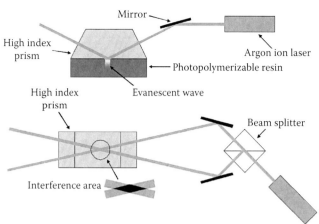

FIGURE 37.6 Experimental setups for evanescent wave photopolymerization. Top: One beam configuration. Bottom: Interferometric configuration used to generate 1D periodical patterns (top view).

can be interpreted here as the thickness of the layer in which actinic light is confined. The thickness and/or the shape of the polymer parts obtained by this technique can be correlated with the photonic properties of the laser beam. According to the properties of evanescent waves, the energy received by the material decays exponentially with the distance from the interface and is proportional to the exposure time and laser power. One can assume that the material solidifies as soon as the energy received exceeds a threshold value E_{th}. Under this assumption, the relationship described in Equation 37.9 predicts the thicknesses of the polymerized layer e, as a function of the intensity of the incident beam at the interface I_0, the exposure time t_e, and the decay rate γ of the evanescent wave. The validity of this theoretical relationship was checked by studying the thickness of planar polymer films as a function of the photonic and optical parameters (Espanet et al. 1999a).

$$e = \frac{1}{2\gamma} \ln\left(\frac{I_0 t_e}{E_{th}}\right) \tag{37.9}$$

In Equation 37.9, the incident intensity at the interface I_0 can be homogeneous along the irradiated area. In this case, a polymer film with constant thickness is created. I_0 can also be a function of planar spatial coordinates (x, y). If a sinusoidal light repartition generated by interference between two coherent laser beams is used, the thickness of the polymer film is then a direct image of the sinusoidal incident field. This configuration was used to generate periodical structures of submicronic height.

Figure 37.7 shows several objects obtained by evanescent wave photopolymerization. One of the interests of this process is its simplicity and versatility since the dimensions of the objects can be easily tuned by adjusting optical or photonic parameters. Moreover, such configuration was used to study the photochemistry at nanoscale and the impact of nonhomogeneous irradiation on photopolymerization. In this context, the role of oxygen diffusion is underlined since the typical pattern dimensions and

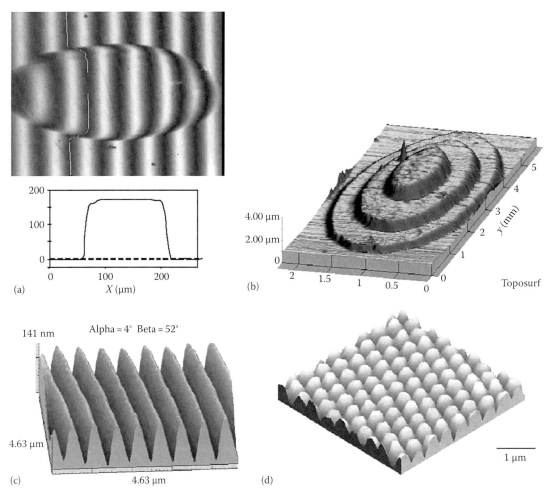

FIGURE 37.7 Example of microstructures obtained by evanescent wave photopolymerization. (a) Example of microspot (thickness of 160 nm) (From Soppera, O. et al., *Proc. SPIE*, 6647, 66470I, 2007. With permission.), (b) 3D microstructure obtained by multiple irradiation (Ecoffet, C. et al., *Adv. Mater.*, 10(5), 411, 1998. With permission.), (c) 1D periodic structures obtained by interferometry (pitch of 500 nm, relief amplitude of 140 nm), and (d) 2D periodic structure obtained by multiple exposure interferometry (pitch of 500 nm in both directions and relief amplitude of 120 nm).

irradiation time are compatible with the diffusion parameters of Oxygen within the acrylate matrix.

37.3.2 Plasmon-Induced Photopolymerization

Recently, the possibilities of near-field nanofabrication were extended to plasmon-induced polymerization. This recently introduced approach is based on controlled nanoscale photopolymerization triggered by local enhanced electromagnetic fields of MNS (Ibn El Ahrach et al. 2007). Its principle is depicted in Figure 37.8. A drop of liquid photopolymerizable formulation with the same composition as in the previous part is deposited on MNSs made by electron beam lithography. In a first approximation, we simplified the photonic response of the photopolymerizable resin to a binary function: polymerization is supposed to be completely ineffective under E_{th} and complete after E_{th}. After formulation deposition, the sample is illuminated ($\lambda = 514$ nm) in normal incidence by a linearly polarized plane wave. The incident energy is below the threshold so that polymerization occurs only around MNS where local near-field is enhanced by

surface plasmon resonance. Silver is chosen as a particle material to achieve mutual spectral overlapping between photopolymer absorption and surface plasmon resonance of metal particles embedded in liquid polymer. After exposure, the sample is washed out with ethanol to remove any unpolymerized material, dried with nitrogen and UV post-irradiated to complete and stabilize the polymerization, and finally characterized by both atomic force microscopy (AFM) and polarized extinction spectroscopy. Knowledge of the threshold value ensures control of the procedure and is therefore of prime importance. This threshold was determined as to be 10 mJ/cm^2 by using two-beam interference pattern as a reference intensity distribution. The characterized formulation was used for near-field photochemical interaction with the MNS. The MNSs covered by the formulation were illuminated with an incident energy density *four times* weaker than the threshold of polymerization. Figure 37.9a and b shows the result of the experiment as imaged by AFM. Two symmetric polymer lobes built up close to the particles can be observed, resulting in metal/polymer hybrid particles. The two lobes originate from the excitation of MNS's dipolar. Surface

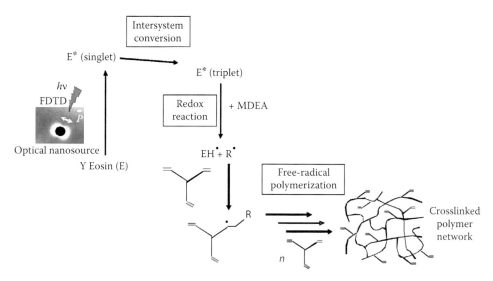

FIGURE 37.8 Principle of nanoscale near-field free-radical photopolymerization. The first step of the process is the local absorption of light by eosin. (From Ibn El Ahrach, H. et al., *Phys. Rev. Lett.*, 98, 107402, 2007. With permission.)

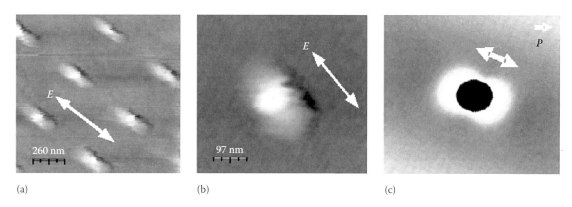

(a)   (b)   (c)

FIGURE 37.9 Near-field nanoscale photopolymerization in the vicinity of silver nanostructures. (a and b) AFM images recorded after irradiation and developing of the silver nanoparticles arrays covered with the photopolymerizable formulation. (c) Intensity distribution in the vicinity of an Ag particle embedded in the formulation as calculated by FDTD method ($\lambda = 514$ nm). The white arrows represent the incident polarization used for exposure. (From Ibn El Ahrach, H. et al., *Phys. Rev. Lett.*, 98, 107402, 2007. With permission.)

plasmon resonance (SPR), as numerically illustrated in Figure 37.9c obtained through a finite difference time domain (FDTD) calculation (Taflove and Hagness 2000). The field distribution associated with the resonance is enhanced in a two-lobe region oriented with the incident polarization. The localized nanoscale photopolymerization is the result of the inhomogeneous field distribution showed in Figure 37.9c. The two lobes can be viewed as a three-dimensional polymer molding of the locally enhanced optical fields. Figure 37.9a and b shows that it is possible to control nanoscale photopolymerization in the visible region of the spectrum by using the near-field of resonant metal nanoparticles. This control was made possible by precise knowledge of the polymerization threshold and results from the abilities of the confined optical near-field of MNS to quickly consume dissolved oxygen at the nanometer scale (oxygen acts as an inhibitor of polymerization) (Espanet et al. 1999b). Figure 37.9 also shows that the intrinsic resolution of the material is very high. This property is one intrinsic characteristic of the negative tone resin

that was used to hybridize the metal nanoparticles: the elementary building blocks are of molecular size, and in addition, a fast transition from liquid to gel under light excitation allows obtaining a well-defined border between reacted and unreacted parts of the photopolymerizable material. On the other hand, our approach constitutes a unique way of quantifying experimentally the field enhancement associated with localized surface plasmon resonance. This approach relies on the precise knowledge of a value characteristic of the photosensitive material: the threshold energy. In the present case, we learn that the intensity enhancement factor is greater than four because the polymerization threshold was locally exceeded. This result is in agreement with the calculated enhancement factor of 12 (not shown). We performed the same exposure using gold particles instead of silver particles. No local polymerization was observed. This is certainly due to the fact that resonance enhancement factor is not superior to 4. This point was confirmed by FDTD calculation that predicted an intensity enhancement factor of about

3.5 for embedded gold particles at $\lambda = 514$ nm. Further studies based on multiple exposures will allow us to quantify precisely the enhancement factors involved in SPR in the near future.

Figure 37.9 clearly shows that polymerization was not isotropic due to the inhomogeneous nature of the actinic field. This suggests that the modification of the Ag particle's plasmon resonance due to local change of the medium is not isotropic. This is confirmed by Figure 37.10. Extinction spectra of the array taken under different conditions are shown at the top of Figure 37.10. Spectrum (a) is the initial spectrum of the particles deposited on glass exposed in air. Spectrum (b) shows a 50 nm red shift in the

liquid polymer (just before exposure). For spectra (a) and (b), the SPR shows an isotropic response to the polarization, within the sample plane, due to the circular symmetry of the particles. Spectra of the hybrid particles were measured for two extreme polarization angles. Spectrum (d) was measured for a polarization parallel to the major axis of the hybrid particle. Compared to spectrum (a), it shows a 28 nm red shift in the resonance. Spectrum (c) was obtained for a polarization perpendicular to the minor axis of the hybrid particle, and compared to spectrum (a), a 8 nm red shift is measured. The anisotropy of the medium surrounding the particle is the reason for these two different red shifts, which are the indicators of a spectral degeneracy breaking. Before local photopolymerization, metallic nanoparticles are characterized by a $C\infty v$ symmetry corresponding to the rotation Cv axis (Tinkham 1964). After polymerization, the two polymerized lobes induce a new lower symmetry: $C2v$, for which any pattern is reproduced by π in-plane rotation. This new symmetry induces the breakdown of the SPR spectral degeneracy. At the bottom of Figure 37.10 is a polar diagram of SPR peaks obtained by measuring 50 spectra of the hybrid particles for different angles of in-plane linear polarization. The two polymerized lobes clearly induced a quasi-continuously tunable SPR in the 508–522 nm range. The local polymerization leads to two plasmon eigenmodes that are centered at 508 and 522 nm, respectively. For any polarization angle, a linear combination of the two eigenmodes is excited, with respective weights depending on the polarization direction. We conclude that the apparent continuous plasmon tuning is the result of a shift in position of the barycenter of the spectral linear combination. This was confirmed by analysis of the full width at half maximum (FWHM) of the spectra acquired. The FWHM was found to be maximal for a polarization at 45° relative to the axis of the hybrid particle, where both eigenmodes are equally excited. These results confirm the importance of symmetry in a nanoparticle in nanophotonics.

The data of Figure 37.10 can be discussed in terms of nanoscale effective index distribution n_{eff} that expresses the effect of the respective weights of the eigenmodes. n_{eff} is equal to $n_{\text{m}} + \Delta n_{\text{m}}$, where n_{m} is the initial refraction index of an external medium taken as a reference, and Δn_{m} is the polymerization-induced shift in the refractive index. n_{m} was chosen to be 1.48 (silver particles embedded in photopolymer formulation before exposure). Δn_{m} was deduced from following Equation 37.10 that results from differentiating the denominator of the particle polarizability expressed in Equation 37.3:

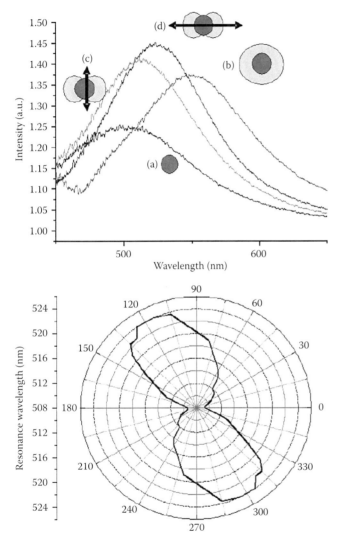

FIGURE 37.10 Spectral properties of the hybrid metal/polymer obtained through near-field photopolymerization. Top: the extinction spectra from a silver nanoparticles array (a) array in air on a glass substrate, (b) array in initial liquid polymer (before exposure), (c) on hybrid particles excited along the minor axis, and (d) on hybrid particles excited along the major axis. Bottom: Polar diagram showing the particle plasmon resonance peak as a function of the polarization angle around the hybrid nanoparticles. 120° and 30° correspond to the particle major axis and minor axis, respectively. (From Ibn El Ahrach, H. et al., *Phys. Rev. Lett.*, 98, 107402, 2007. With permission.)

$$\Delta\lambda = -4n_{\text{m}}\Delta n_{\text{m}} \left(\frac{d\varepsilon}{d\lambda} \right)^{-1} \tag{37.10}$$

where
  ε is the dielectric constant of silver, whose dispersion is known from Palik (1985)
  $\Delta\lambda$ is the measured shift of SPR peak relative to the reference spectrum (see bottom of Figure 37.10)

The derivation was performed for spherical particles, which is a good approximation for in-plane measurements. For Ag particles in air deposited on a glass substrate, n_{eff} is found to be 1.06 as a result of the glass/air interface modifying the SPR. For the hybrid particles, an effective index of 1.15 was found for an excitation along minor axis (suggesting the presence of thin polymer layer along this axis), while the major axis is associated to a 1.3 index, as a consequence of a thicker polymer region along this axis. Note that the bulk polymer index is 1.52. The difference between the effective index associated with the long axis and the bulk value can be attributed to the spatial extent of polymerization, which is limited by the threshold value. The field surrounding the hybrid particle extends beyond the area defined by

the two polymerized lobes, resulting in a lower effective index. Between the two extreme values, a continuous variation of n_{eff} was deduced. The method thus allowed the controlled production of a dielectric encapsulant that can be viewed as an artificial nanometric refractive-index ellipsoid.

The approach presented in this section has unique and numerous advantages compared to the standard approaches based on control of the geometry of the metal particles. In particular, several properties and processes involved in polymer science (Mark 2001) can be coupled to (or assisted by) MNS at the nanoscale. They include nonlinear/electro-optical properties, possible doping with luminescent (non)organic materials, and chemical control of the refractive index. This approach has been recently used in another experimental configuration, using different metal nano-objects and a commercial SU-8 resin with two-photon polymerization (Figure 37.11) (Ueno et al. 2008). This example demonstrates the high versatility of this concept in the field of nanophotochemistry.

37.4 Nanoscale Photoisomerization

Over the past decade, many groups have studied self-developing photopolymers that can spontaneously develop surface topography under optical illumination. It means that the matter moves under the optical illumination, thereby inducing holes and bumps at its surface. One of the most used self-developing photopolymers consists of azobenzene-type Disperse Red 1 (DR1) molecules grafted on the poly(methylmethacrylate) (PMMA). The movement of this PMMA-DR1 copolymer has been shown to result from optically induced isomerization of the azo-dyes. Details on this material can be found in comprehensive review papers such as that by Natansohn and Rochon (Natansohn and Rochon 2002). Here we just briefly recollect the photochemistry of this photopolymer that is illustrated in Figure 37.12. Figure 37.12 shows azobenzene group grafted to a polymer matrix of Polymethylmethacrylate (PMMA). The absorption band of the azo dye is centered in the visible (see absorption spectrum in the inset of the figure). The absorption peak is typically situated around $\lambda = 500\,nm$. This is why the dye is usually named DR1 (Dispersed Red One). Figure 37.12b shows schematically the process of isomerization. The stable state of the azobenzene molecule presented on the left side is the trans-isomeric configuration. The absorption in the visible range of a photon induces the transition to the cis-isomer. This state is metastable, and the reverse transition to the trans-state takes place through either thermal activation or optical absorption. Therefore, a molecule absorbing a photon undergoes a complete *trans-cis-trans* isomerization cycle. Provided that there is a nonzero component of the light polarization vector along the intensity gradient, this transition induces a motion of the molecule and a related deformation of the matrix to which the molecule is grafted. The polarization selectivity lies in the fact that the azobenzene moieties in their *trans*-state are strongly anisotropic and preferentially absorb the light polarized along their main axis. The required illumination conditions to observe any photoinduced topography are actually

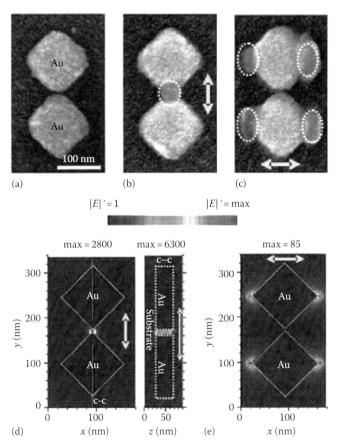

FIGURE 37.11 (a) SEM image of a pair of gold nanoblocks measuring $100 \times 100 \times 40\,nm$ and separated by a 5.6 nm wide nanogap before irradiation by an attenuated femtosecond laser beam. (b) SEM image of other nanoblockpairs after 0.01 s exposure to the laser beam polarized linearly along the long axis of the pair. (c) SEM image of another pair after 100 s exposure to the laser beam polarized in the perpendicular direction. (d and e) Theoretically calculated near-field patterns at selected planes for the excitation conditions of the samples shown in (b) and (c), respectively. In (d), the field pattern is shown on the *x-y* plane bisecting the nanoblocks at half of their height (i.e., 20 nm above the substrate), and in (e), the field is calculated on the plane coincident with the line c-c shown in (d). The field intensity is normalized to that of the incident wave and therefore represents the intensity enhancement factor. (From Ueno, K. et al., *J. Am. Chem. Soc.*, 130(22), 6928, 2008. With permission.)

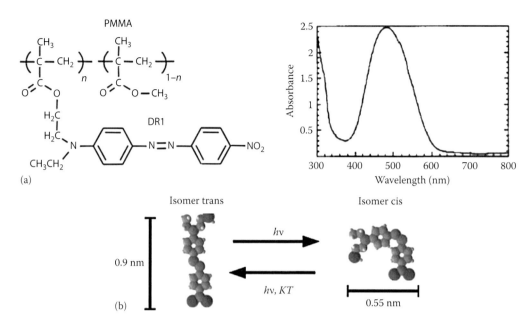

FIGURE 37.12 Photochemistry of the PMMA-DR1. (a) Structure. (b) Photoizomerization. Inset: absorption spectrum of DR1.

intensity gradient and a nonzero component of the polarization vector along the intensity gradient. As a consequence, the polymer is self-developing, and after illumination, its surface presents a topography related to the incident intensity distribution. In a simple way, we can imagine that the polymer is pulled (or pushed) by these photoactivated molecular engines. Nevertheless, the understanding of the motion process at the molecular level is still debated. The simplest proposed model to imagine this molecular motion is the notion of worm-like displacement introduced by Lefin et al. (1998a,b). They proposed that the DR1 molecule acts as a worm that pulls (or pushes) the polymer chain. This hypothesis implies that the displacement is parallel to the long axis of the molecule (and the permanent dipole), as schematized in Figure 37.12.

In this section, we present significant results of structuration at the nanoscale of a PMMA-DR1 film induced by different confined optical sources.

37.4.1 Tip-Enhanced Near-Field Photoisomerization

After many years of interest of the response of a PMMA-DR1 film under a far field illumination in different configuration of polarization and shape (interferences, gaussian beam...) (Natansohn and Rochon 2002), Davy et al. have shown for the first time the possibility to modify a PMMA-DR1 film using the confined electromagnetic field at the end of an aperture NSOM probe similar to that shown in Figure 37.3c (Davy and Spajer 1996). Using the same approach, Bachelot et al. have demonstrated the possibility to investigate the field enhancement effect (such as that illustrated in Figure 37.3b) at the extremity of an apertureless metal probe (Bachelot et al. 2003, this example will be described in details below). Then, Andre et al. have shown for the first time the possibility to modify the topography of a

PMMA-DR1 film under near-field illumination using the optical response of copper colloids deposited on a PMMA-DR1 film (Andre et al. 2002). P. Andre and coworkers deposited copper colloids on a PMMA-DR1 film and then illuminated the sample with a laser and subsequently recorded, through self-induced topography, a fingerprint of the diffraction of the light by the colloids. They showed the dipolar response of the near-field induced. Nevertheless, the preparation of the sample was pretty tricky, and the fact that the nanoparticles are deposited on the film induces a breaking symmetry of the refraction index (i.e., nanoparticles are situated at the PMMA-DR1 - air interface), thus inducing complexity in the results interpretations. These preliminary results somehow opened the door to chemical photoimaging of the electromagnetic near-field associated. In 2003, Bachelot et al. have shown the possibility to image the complete response (i.e., both the far- and the near-field components) of the confined EM field at the extremity of a metal tip under illumination. PMMA-DR1 was illuminated with the presence of a metallized Atomic Force Microscopy (AFM) tip at its surface. After illumination, the optically induced topography was characterized *in situ* by AFM using the same tip. This approach enabled a parametric analysis, leading to valuable information on tip-field enhancement (TFE) (Bachelot et al. 2003, Royer et al. 2004). This physical effect has been described in Section 37.2. In particular, TFE was evaluated as a function of the illumination condition (state of polarization, angle of incidence, etc.) and tip's features (radius of curvature, material, etc.). As a significant example, Figure 37.13 shows the influence of the illumination geometry. Figure 37.13a shows the AFM image of the PMMA-DR1 surface obtained after the p-polarization reflection mode illumination of a platinum-coated Si tip in interaction with the polymer surface. The centre of the image corresponds to the tip position during the exposure. The figure exhibits two different fabricated patterns: a far-field-type fringes system, which corresponds to

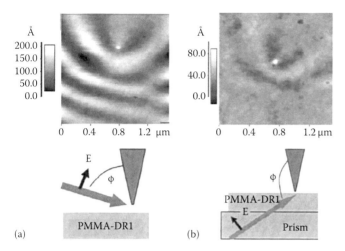

FIGURE 37.13 Tip-enhanced lithography on PMMA-DR1 films under p polarization. Influence of the illumination geometry. (a) Tapping mode AFM image obtained after the reflection mode illumination ($\varphi = 80°$, see configuration below). (b) Tapping mode AFM images obtained using total internal reflection ($\varphi = 130°$, see configuration below). (From Bachelot, R. et al., *J. Appl. Phys.*, 94, 2060, 2003.)

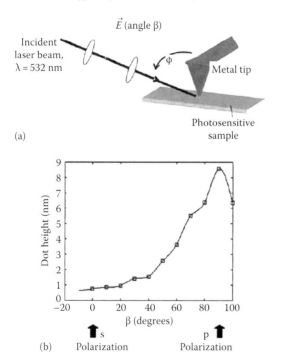

FIGURE 37.14 Tip-enhanced lithography on PMMA–DR1 films. Influence of the incident polarization (a) Schematic diagram of the experimental configuration. $\varphi = 80°$. (b) Obtained dot height as a function of polarization angle β. (From Bachelot, R. et al., *J. Appl. Phys.*, 94, 2060, 2003.)

diffraction by the tip cone, and a central nanometric dot due to the local enhancement of the electromagnetic field below the metallic tip. The far-field contrast (the fringes) vanishes if the tip is illuminated by total internal reflection (see Figure 37.13b), providing valuable information about the suitable way of illuminating a metal tip for near-field optical lithography. The height of

the central near-field spot is very sensitive to the incident angle of polarization (Figure 37.14), confirming that the intensity of the local tip field is enhanced gradually from s-polarization to p-polarization. The above cited near-field investigations lead to preliminary 30 nm resolution nanopatterning based on TFE (see Figure 37.15 as an illustration).

All these experiments stimulated interesting discussions on the nature of the optical response of the molecule. It was specifically shown that while lateral field components (parallel to the polymer surface) tend to make escape molecules from light along the polarization, longitudinal components (perpendicular to the polymer surface) lift the matter vertically, as a consequence of free space requirement from the molecule. This point of view was confirmed by the investigation of surface deformations in azo-polymers using tightly focused higher-order laser beams (Gilbert et al. 2006) and surface plasmons interference (Derouard et al. 2007) as vectorial sources. Considering that the field involved in tip-enhanced NFOLM is mainly longitudinal, contrasts observed in Figures 37.12, 37.13, and 37.15 can be explained by the above described polarization sensitivity.

37.4.2 Plasmon-Based Near-Field Photoisomerization

Hubert et al. have shown the possibility to photoimage directly the complex near-field of plamonic nanostructures using the above approach (Hubert et al. 2005). In particular, they showed the possibility to get all the components of the near-field. It means that using the PMMA-DR1, it is possible to get a photography of the three different components of the field (Hubert et al. 2008). This photochemical optical near-field imaging consists of three steps. Silver 50 nm high nanostructures are first fabricated by electron-beam lithography, typically through the lift-off method. The second step is the deposition of the PMMA-DR1. In our case, the thickness of the polymer film is equal to 80 nm, which is sufficient to fully cover the structures and thin enough to be sensitive to the optical near-field of the particles. No drying was performed after spin coating. The third step consists of illuminating the sample, in this case, at normal incidence. To overlap the 400–600 nm absorption band of the DR1 molecule (see Figure 37.12a, inset), we used the 514 or 532 nm lines of an argon-ion laser or a frequency doubled, diode-pumped Nd:YAG laser, respectively. The polarization and irradiation intensity were carefully controlled and correlated with the detected topographic features. Following the illumination process (step 3), the "imaging" of the optically induced topography is performed through atomic force microscopy. We rigorously calculate the near-field optical intensities using the finite-difference time-domain (FDTD) method (Royer et al. 2004) and show that the negative image of the computed near-field intensities can be correlated with the observed photoinduced topographies. This negative image illustrates the fact that the matter escapes from high-intensity regions to low intensity regions and that the involved field is mainly lateral. The FDTD calculations were fully three-dimensional with appropriate periodic and absorbing boundary conditions. The metals were described

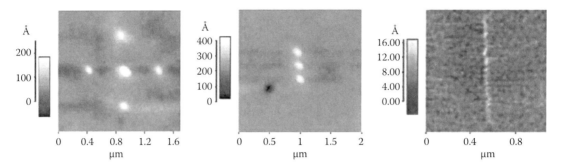

FIGURE 37.15 Tip-enhanced lithography on PMMA-DR1 films under p-polarized total internal reflection mode illumination. Examples of dotted/continuous produced nanostructures with λ = 532 nm (AFM images). (From Bachelot, R. et al., *J. Appl. Phys.*, 94, 2060, 2003.)

by Drude models with parameters fitting the experimental dielectric constant data for the wavelengths of interest, as described by Gray and Kupka (2003). The glass substrate and PMMA-DR1 were also included in the calculations, with dielectric constants of 2.25 and 2.89, respectively. Fourier transformations of the time-domain fields on the wavelengths of interest then yield steady-state fields and thus field intensities. As a first example of the near-field imaging capability of this method, Figure 37.16 shows the results obtained with an array of silver nanoparticles. The extinction spectra of the arrays performed after spin-coating show a typical maximum near 540 nm. A 532 nm irradiation wavelength was used to overlap this resonance plasmon resonance (see Section 37.2). In the case of linear incident polarization, after irradiation, two holes can be observed (top of Figure 37.16a) in the polymer that are close to the particles and are oriented with the incident light polarization. The depressions correlate remarkably well with

the expected dipolar near-field spatial profile. Numerical calculations of the electric field intensity distribution around silver nanoparticles covered with PMMA-DR1 and irradiated with a linearly polarized laser beam show that intensity maxima are located at the same position as the holes observed in the topographic image. The theoretical negative of the electric field intensity is displayed at the bottom of Figure 37.16a; it agrees qualitatively with its experimental counterpart. From these observations, we can argue that topographic modifications observed after irradiation are due to a mass transport phenomenon photoinduced by the optical near-field of the metallic silver particles produced by dipolar plasmon resonance. Figure 37.16b shows an example of the ability of this method to spatially resolve complex fields. It shows the result obtained with silver nanoparticles covered with PMMA-DR1 but now irradiated with a circularly polarized laser beam. The silver particles are 50 nm in height, 100 nm in diameter, and have a periodicity of 1 μm. In this case, large topographic modifications are again observed at the polymer film surface. The AFM image at the top of Figure 37.16b shows that an inner array of lobes around each particle as well as an outer array of lobes, whose periodicity is equal to the lattice spacing, can be distinguished. These outer lobes probably result from interferences of diffraction orders. Quite encouragingly, the negative computed field intensity (bottom of Figure 37.16b) also shows inner and outer high relief features around the particles, although they are not as structured as the experimental result. This probably originates from the fact that theoretical calculations do not actually take into account the diffusion of azodye molecules (and the resulting local change in effective refraction index) and thus cannot perfectly reproduce their behavior under illumination with very intense localized near-fields. Figure 37.17 shows a few other selected results that confirm the high potential of the method. It shows different examples of near-field optical imaging of silver nanostructures illuminated by a linearly polarized green light. In each figure, the white bar represents the light wavelength (532 nm), while the black arrows represent the direction of the incident polarization. Figure 37.17a shows optical near-field around gold ellipsoidal particles under light polarization perpendicular to the long axis of the ellipsoids. The ellipsoids are 50 nm in height, with long and short axis lengths of 1000 nm and 60 nm, respectively. Particle-to-particle distances are 800 nm in the long axis direction. Extinction spectra

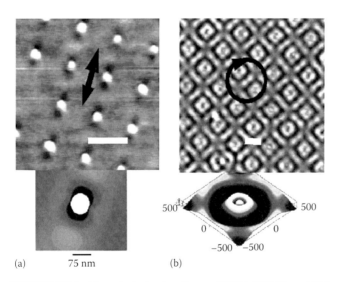

FIGURE 37.16 Nanophotostructuration on PMMA-DR1 films. AFM images (top) of array of silver particle taken after illumination with either linear polarization (a) or circular polarization (b). Irradiation wavelength, time, and intensity were equal to 532 nm, 30 min, and 100 mW/cm², respectively. Theoretical images (bottom) represent the negative of the intensity. The black arrows depict the incident polarization. The white bars represent 500 nm. (From Hubert, C. et al., *Nano Lett.*, 5, 615, 2005.)

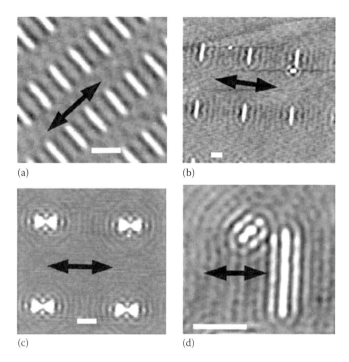

FIGURE 37.17 NFOLM on PMMA-DR1 films. AFM images of different silver nanostructures taken after linearly polarized exposure (normal incidence). Black arrows depict the incident polarization. The white bars represent the incident wavelength of 532 nm. (a) Ellipses, (b) longer ellipses, (c) bowtie antenna, and (d) coupled rod and circular particles.

performed after spin coating of the azo-polymer layer indicate that for polarization parallel to the long axis, the 514 nm irradiation wavelength used is outside the resonance plasmon band. On the other hand, in the case of a polarization direction perpendicular to the long axis, the 514 nm irradiation wavelength is resonant with the plasmon band. This agrees well with the experimental observation. In Figure 37.17a, dips can be observed along the long axis of the particles. These dips are located at the same position as the optical near-field intensity maxima (not shown here) around ellipsoids for such a polarization. This is confirmed by looking at the corresponding calculated negative image of the field intensity around the particles for this illumination condition. Hence, Figure 37.17a is an observation of the resonant charge density

excitation at the edges of the nanorods. For polarization parallel to the rod, an off-resonant electromagnetic singularity is excited, and dips are observed at the extremities of the rods (Hubert et al. 2005). Partial interpretation of Figure 37.17b through d has been proposed (Hubert et al. 2008). It includes the polarization sensitivity evoked in Section 37.4.1. These images are believed to give insight into interesting effects and behaviors of MNSs: the excitation of quadripolar plasmon modes in elongated gold particles (Figure 37.17b), near-field diagram of bow-tie optical antenna (Figure 37.17c), and near-field coupling between two close different nanostructures (Figure 37.17d).

37.4.3 Model of Optical Matter Migration

Different models have been proposed to allow a better understanding of the matter migration of PMMA-DR1 under illumination (Barrett et al. 1996, 1998, Pedersen and Johansen 1997, Lefin et al. 1998a,b, Viswanathan et al. 1999a,b, Barada et al. 2004, 2005, 2006). Nevertheless, the proposed models do not allow to envisage the photoinduced topography at the nanoscale. Recently, a new model based on two hypotheses (Juan et al. 2008) was introduced. First, the probability of absorption by the DR1 is given by

$$P_{abs} \propto E^2 \cos^2 \varphi \tag{37.11}$$

with

E the electric field amplitude

φ the angle between the dipole of the molecule (corresponding to the axis of the molecule) and the polarization of the electric field

Second, the movement of the dye is assumed to occur along the axis of the molecule (identified as the direction of the molecular dipole) as described by the inchworm translation model proposed by Lefin et al. (Lefin et al. 1998a). Using a Monte Carlo approach, matter migration under various far-/near-field illumination conditions was numerically studied. As an example, Figure 37.18 shows the two typical surface relief gratings (i.e., realized with p-polarized and s-polarized interferometric light beam, respectively) calculated using our model. The topography obtained through

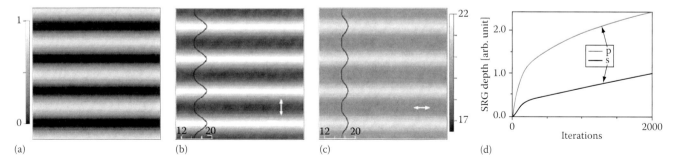

FIGURE 37.18 Simulation of grating formation on PMMA-DR1 films with the stochastic model. (a) The assumed optical interferometric electric field intensity distribution. (b and c) The resulting photoinduced topography after 2000 iterations. The axis refers to the height of the topography (arbitrary units, 0 corresponds to the substrate on which the polymer is deposited). (b) p polarization, (c) s polarization. White arrows represent the direction of the light field polarization. (d) Time evolution of the SRG depth. (From Juan, M.L. et al., *Appl. Phys. Lett.*, 93, 153304, 2008. With permission.)

the calculation is fully comparable to the experimental results (Cojocariu and Rochon 2004) and turns out to describe at the same time the dynamics of molecular displacement and associated local molecular orientation. This model could constitute a valuable tool for developing optical molecular nanomotors based on the use of azobenzene molecules under polarized laser illumination.

37.5 Conclusions

This chapter presented recent exploration of optical interaction between confined optical near-fields and photosensitive organic materials. The selected examples of near-field photopolymerization and photoisomerization have shown the ability of the optical near-field to induce physical and chemical processes at the molecular scale. These experiments not only allowed for the production of nanostructures using visible light but also have opened the door to nanophotochemistry and nonophotophysics based on the use of optical nanosources. Moreover, they lead to progress in the development of new nanosources, in both experimental and theoretical points of view.

Taking into account the associated economical and scientific challenges, the future of the near-field optical matter manipulation based is likely to be successful. Regarding the tip enhanced approach, it is now possible to control the array of tips working simultaneously (Despont 2007), making the concept of high-density multiprobes optical data storage a priori relevant. Regarding the mask-based NFOLM approach (using a planar structure), the control of the mask-photosensitive material distance turns out to be the key parameter. This issue will certainly take advantage of the superlens concept that enables the access of near-field information through negative refraction (Pendry 2000, Fang et al. 2005). Near-field optical lithography and manipulation are not intended to compete with the objectives defined by the International Technology Roadmap for Semiconductors (ITRS, see www.itrs.net) that relies on the decrease of the wavelength. Rather, near-field approaches should be viewed as complementary tools that are appreciated because of their low cost and easy procedures. In particular, the use of visible light and the possibility of taking advantage of the polarization state of the light represent clear assets. Furthermore, the domain of application of NFOLM is far beyond that of nanolithography for microelectronics since it can involve, for example, high data storage and molecular manipulation. NFOLM can also be used as one step of a more complex procedure of nanolithography. For example, the mold used in nanoimprint (Chou et al. 1996) can be fabricated by NFOLM. Finally, this technology shall take advantage of the large variety of powerful physical near-field effects in nanometals that have been investigated recently, such as second harmonic generation (Hubert et al. 2007), photoluminescence (Bouhelier et al. 2005), strong near-field coupling (Atay et al. 2004), and multipole excitation (Krenn et al. 2000). All of these effects will permit higher resolution and better control of intensity and wavelength of the actinic light with regard to photosensitive materials.

Acknowledgments

The author would like to both cite and thank current and former co-workers for this subject of research (in alphabetical order): D. Barchiesi, A. Bouhelier, S. Chang, M. Derouard, C. Ecoffet, R. Fikri, Y. Gilbert, S. K. Gray, F. H'Dhili, C. Hubert, H. Ibn-El-Ahrach, M. Juan, S. Kostcheev, N. Landraud, G. Lerondel, D.J. Lougnot, J. Peretti, P. Royer, A. Rumyantseva, C. Triger, A. Vial, G.P. Wiederrecht, and G. Wurtz.

References

Andre P., Charra F., Chollet P. A. et al. 2002. Dipolar response of metallic copper nanocrystal islands, studied by two-step near-field microscopy. *Adv. Mater.* 14: 601–603.

Atay T., Song J.-H., and Nurmikko A. V. 2004. Strongly interacting plasmon nanoparticle pairs: From dipole-dipole interaction to conductively coupled regime. *Nano Lett.* 4: 1627–1631.

Bachelot R. 2007. Near-field optical structuring and manipulation based on local field enhancement in the vicinity of metal nanostructures. In *Advances in Nano-Optics and Nanophotonics—Tip Enhancement*, pp. 205–234, S. Kawata and V. M. Shalaev (eds.), Amsterdam, the Netherlands: Elsevier.

Bachelot R., H'Dhili F., Barchiesi D. et al. 2003. Apertureless near-field optical microscopy: A study of the local tip field enhancement using photosensitive azobenzene-containing films. *J. Appl. Phys.* 94: 2060–2072.

Barada D., Itoh M., and Yatagai T. 2004. Numerical analysis of photoinduced surface relief grating formation by particle method. *J. Appl. Phys.* 96: 4204–4210.

Barada D., Fukuda T., Itoh M. et al. 2005. Numerical analysis of photoinduced surface relief grating formation by particle method. *Opt. Rev.* 12: 217–273.

Barada D., Fukuda T., Itoh M. et al. 2006. Photoinduced chirality in an azobenzene amorphous copolymer bearing large birefringent moiety. *Jpn. J. Appl. Phys.* 45: 6730–6737.

Barrett C., Natansohn A. L., and Rochon P. L. 1996. Mechanism of optically inscribed high-efficiency diffraction gratings in azo polymer films. *J. Phys. Chem.* 100: 8836–8842.

Barrett C., Rochon P. L., and Natansohn A. L. 1998. Model of laser-driven mass transport in thin films of dye-functionalized polymers. *J. Chem. Phys.* 109: 1505–1516.

Bethe H. A. 1944. Theory of diffraction by small holes. *Phys. Rev.* 66: 163–182.

Bohren C. F. and Hoffmann D. R. 1983. *Absorption and Scattering of Light by Small Particles*, New York: Wiley.

Bouhelier A., Bachelot R., Lerondel G. et al. 2005. Surface plasmon characteristics of tunable photoluminescence in single gold nanorods. *Phys. Rev. Lett.* 95: 1–4.

Bouwkamp C. J. 1950. On the diffraction of electromagnetic waves by small circular disks and holes. *Philips Res. Rep.* 5: 401–422.

Bucknall D. J. (ed.) 2005. *Nanolithography and Patterning Techniques in Microelectronics*, Boca Raton, FL: CRC.

Carré C., Saint-Georges P., Lenaerts C., and Renotte Y. 2002. Customization of a self-processing polymer for obtaining specific diffractive optical elements. *Synth. Met.* 127(1–3): 291–294.

Chochos C. L., Ismailova E., Brochon C. et al. 2008. Hyperbranched polymers for photolithographic applications–Towards understanding the relationship between chemical structure of polymer resin and lithographic performances. *Adv. Mater.* 20: 1–5.

Chou S. Y., Krauss P. R., and Renstrom P. J. 1996. Imprint lithography with 25-nanometer resolution. *Science* 272: 85–87.

Cojocariu C. and Rochon P. 2004. Light-induced motions in azobenzene-containing polymers. *Pure Appl. Chem.* 76: 1479–1497 (and references therein).

Courjon D. 2003. *Near-Field Microscopy and Near-Field Optics*, London, U.K.: Imperial College Press.

Croutxe-Barghorn C., Soppera O., Simonin L. et al. 2000. On the unexpected rôle of oxygen in the generation of microlens arrays with self-developing photopolymers. *Adv. Mater. Opt. Electron.* 10: 25–38.

Davy S. and Spajer M. 1996. Near field optics: Snapshot of the field emitted by a nanosource using a photosensitive polymer. *Appl. Phys. Lett.* 69: 3306–3308.

Decker C. 1993. New developments in UV-curable acrylic monomers. In *Radiation Curing in Polymer Science and Technology. Vol. III. Polymerisation Mechanisms*, pp. 33–64, J. P. Fouassier and J. F. Rabek (eds.), London, U.K. and New York: Elsevier.

Derouard M., Hazart J., Lérondel G. et al. 2007. Polarization-sensitive printing of surface plasmon interferences. *Opt. Express* 15: 4238–4246.

Despont M. 2007. *Millipede Probe-Based Storage*, IBM Corp., SPIE Advanced Lithography, February 25–March 2, 2007, United States.

Ecoffet C., Espanet A., and Lougnot D. J. 1998. Photopolymerization by evanescent waves: a new method to obtain nanoparts. *Adv. Mater.* 10(5): 411–414.

Espanet A., Dos santos G., Ecoffet C., and Lougnot D. J. 1999a. Photopolymerization by evanescent waves: Characterization of photopolymerizable formulation for photolithography with nanometric resolution. *Appl. Surf. Sci.* 87: 138–139.

Espanet A., Ecoffet C., and Lougnot D. J. 1999b. PEW: Photopolymerization by evanescent waves. II - Revealing dramatic inhibiting effects of oxygen at submicrometer scale. *J. Polym. Sci.: Part A: Polym. Chem.* 37: 2075.

Fang N., Lee H., Sun C. et al. 2005. Sub-diffraction-limited optical imaging with a silver superlens. *Science* 308: 534–537.

Fouassier J. P. 1993. *Radiation Curing in Polymer Science and Technology*, J. P. Fouassier and J. F. Rabek (eds.), London, U.K. and New York: Elsevier.

Gentili M., Giovannella C., and Selci S. (eds.) 1994. *Nanolithography: A Borderland between STM, EB, IB, and X-Ray Lithographies*, NATO Science Series E, Dordrecht, the Netherlands: Kluwer Academic Publishers.

Ghislain L. P. and Elings V. B. 1998. Near-field scanning solid immersion microscope. *Appl. Phys. Lett.* 72: 2779–2781.

Gilbert Y., Bachelot R., Royer P. et al. 2006. Longitudinal anisotropy of the photoinduced molecular migration in azobenzene polymer films. *Opt. Lett.* 31: 613–615.

Goodman J. W. 1996. *Introduction to Fourier Optics*, New York: McGrawHill.

Gray S. K. and Kupka T. 2003. Propagation of light in metallic nanowire arrays: Finite-difference time domain of silver cylinders. *Phys. Rev. B* 68: 045415-1–045415-11.

Guattari F., Maire G., Contreras K. et al. 2007. Balanced homodyne detection of Bragg microholograms in photopolymer for data storage. *Opt. Express* 15: 2234–2243.

Hubert C., Rumyantseva A., Lerondel G. et al. 2005. Near-field photochemical imaging of noble metal nanostructures. *Nano Lett.* 5: 615–619.

Hubert C., Billot L., Adam P.-M. et al. 2007. Role of surface plasmon in second harmonic generation from gold nanorods. *Appl. Phys. Lett.* 90: 181105.

Hubert C., Bachelot R., Plain J. et al. 2008. Near-field polarization effects in molecular-motion-induced photochemical imaging. *J. Phys. Chem. C* 112: 4111–4116.

Hutter E. and Fendler J. H. 2004. Exploitation of localized surface plasmon resonance. *Adv. Mater.* 16: 1685.

Ibn El Ahrach H., Bachelot R., Vial A. et al. 2007. Spectral degeneracy breaking of the plasmon resonance of single metal nanoparticles by nanoscale near-field photopolymerization. *Phys. Rev. Lett.* 98: 107402.

Inao Y., Nakasato S., Kuroda R., and Ohtsu M. 2007. Near-field lithography as prototype nano-fabrication tool. *Microelectron. Eng.* 84: 705–710.

Jradi S., Soppera O., and Lougnot D. J. 2008a. Analysis of photopolymerized acrylic films by AFM in pulsed force mode. *J. Microsc.* 229: 151–161.

Jradi S., Soppera O., and Lougnot D. J. 2008b. Fabrication of polymer waveguides between two optical fibers using spatially controlled light-induced polymerization. *Appl. Opt.* 47(22): 3987–3993.

Juan M. L., Plain J., Bachelot R. et al. 2008. Stochastic model for photoinduced surface relief grating formation through molecular transport in polymer films. *Appl. Phys. Lett.* 93: 153304.

Kawata S. (ed.) 2001. *Near-Field Optics and Surface Plasmon Polaritons*, Heidelberg, Germany: Springer-Verlag.

Kottmann J. P., Martin O. J. F., Smith D. R. et al. 2001. Non-regularly shaped plasmon resonant nanoparticle as localized light source for near-field microscopy. *J. Microsc.* 202: 60–65.

Kreibig U. and Vollmer M. 1996. *Optical Properties of Metal Structures*, Vol. 25. Springer Series in Materials Science, Berlin, Germany: Springer.

Krenn J. R., Schider G., Rechberger W. et al. 2000. Design of multipolar plasmon excitations in silver nanoparticles. *Appl. Phys. Lett.* 77: 3379–3381.

Lefin P., Fiorini C., and Nunzi J. 1998a. Anisotropy of the photo-induced translation diffusion of azobenzene dyes in polymer matrices. *Pure Appl. Opt.* 7: 71–82.

Lefin P., Fiorini C., and Nunzi J. 1998b. Anisotropy of the photoinduced translation diffusion of azo-dyes. *Opt. Mater.* 9: 323–328.

Lin B. J. 2006. The ending of optical lithography and the prospects of its successors. *Microelectron. Eng.* 83: 604–613.

Lougnot D. J. 1993. Photopolymers and holography, in *Radiation Curing in Polymer Science and Technology*, Vol. III. *Polymerisation Mechanisms*, J. P. Fouassier and J. F. Rabek (eds.), pp. 65–100. London, U.K. and New York: Elsevier.

Lougnot D. J. and Turck C. 1992. Photopolymers for holographic recording: II—Self developing materials for real-time interferometry. *Pure Appl. Opt.* 1: 251.

Mark J. E. 2001. *Physical Properties of Polymers Handbook*. New York: Springer-Verlag.

Molenda D., Colas des Francs G., Fischer U. C., Rau N., and Naber A. 2005. High-resolution mapping of the optical near-field components at a triangular nano-aperture. *Opt. Express* 13: 10688–10696.

Natansohn A. and Rochon P. 2002. Photoinduced motions in Azo-containg polymers. *Chem. Rev.* 102: 4139–4175.

Novotny L. and Hecht B. 2006. *Principles in Nano-Optics*, Cambridge, U.K.: Cambridge University press.

Palik E. D. 1985. *Handbook of Optical Constants of Solids*, Orlando, FL: Academic Press.

Pedersen T. G. and Johansen P. M. 1997. Mean-field theory of photoinduced molecular reorientation in azobenzene liquid crystalline side-chain polymers. *Phys. Rev. Lett.* 79: 2470–2473.

Pendry J. B. 2000. Negative refraction makes a perfect lens. *Phys. Rev. Lett.* 85: 3966–3969.

Plain J., Pallandre A., Nysten B. et al. 2006. Nanotemplated crystallization of organic molecules. *Small* 2: 892–897.

Prasad P. N. 2004. *Nanophotonics*, Hoboken, NJ: John Wiley & Sons, Inc.

Royer P., Barchiesi D., Lérondel G. et al. 2004. Near-field optical patterning and structuring based on local-field enhancement at the extremity of a metal tip, *Philos. Trans. R. Soc. Lond., Ser. A* 362: 821–842.

Sánchez E. J., Novotny L., and Xie X. S. 1999. Near-field fluorescence microscopy based on two-photon excitation with metal tips. *Phys. Rev. Lett.* 82: 4014–4017.

Selli E. and Bellobono I. R. 1993. Photopolymerization of multifunctional monomers: Kinetic aspects, in *Radiation Curing in Polymer Science and Technology*, Vol. III. *Polymerisation Mechanisms*, J. P. Fouassier and J. F. Rabek (eds.), pp. 1–32, London, U.K. and New York: Elsevier.

Sheats J. R. and Smith B. W. (eds.) 1998. *Microlithography Science and Technology*, New York: Marcel Decker Inc.

Soppera O., Jradi S., Ecoffet C., and Lougnot D. J. 2007. Optical near-field patterning of photopolymer. *Proceedings of SPIE* 6647: 66470I.

Sotomayor Torres C. M. 2003. *Alternative Lithography: Unleasing Potentials of Nanotechnology; Nanostructure Science and Technology*, New York: Kluwer Academic/Plenum Publishers.

Taflove A. and Hagness S. C. 2000. *Computational Electrodynamics: The Finite-Difference Time-Domain Method*, 2nd edn, Norwood, MA: Artech House.

Tinkham H. 1964. *Group Theory and Quantum Mechanics*, New York: McGraw-Hill.

Tseng A. 2007. Recent developments in nanofabrication using scanning near field optical microscope lithography. *Opt. Laser Technol.* 39: 514–526.

Ueno K., Juodkazis S., Shibuya T., Yokota Y., Mizeikis V., Sasaki K., and Misawa H. 2008. Nanoparticle plasmon-assisted two-photon polymerization induced by incoherent excitation source. *J. Am. Chem. Soc.* 130(22): 6928–6929.

Van Bladel J. 1995. *Singular Electromagnetic Field and Sources*, Oxford, U.K.: IEEE.

Vial A., Grimault A.-S., Macías D., Barchiesi D., and Lamy de la Chapelle M. 2005. Improved analytical fit of gold dispersion: Application to the modelling of extinction spectra with the FDTD method. *Phys. Rev. B* 71(8): 085416–085422.

Vieu C., Carcenac F., Pepin A. et al. 2000. Electron beam lithography: Resolution limits and applications. *Appl. Surf. Sci.* 164: 111–117.

Viswanathan N. K., Balasubramanian S., Li L. et al. 1999a. A detailed investigation of the polarization-dependent surface-relief-grating formation process on azo polymer films. *Jpn. J. Appl. Phys.* 38: 5928–5937.

Viswanathan N. K., Kim D. Y., Bian S. et al. 1999b. Surface relief structures on azo polymer films. *J. Mater. Chem.* 9: 1941–1955.

Wagner C., Kaiser W., and Mulkens J. 2000. Advanced technology for extending optical lithography. *Proc. SPIE* 4000: 344–357.

38

Soft X-Ray Holography for Nanostructure Imaging

Andreas Scherz
SLAC National Accelerator
Laboratory

38.1 Introduction

With the advent of third generation synchrotrons, new techniques became feasible such as lensless x-ray microscopy and x-ray photon correlation spectroscopy (XPCS) based on the coherent properties of the x-ray beam. In recent years, the development of these scientific areas have gained tremendous momentum due to the prospect of upcoming x-ray free-electron lasers (XFEL's) which will provide transversely coherent, intense x-ray pulses on femtosecond timescale. The investigation of ultrafast processes has so far been the domain of optical techniques. Because of the larger wavelength, however, the details of the underlying processes remain unresolved. With the development of new accelerator-based sources, coherent x-rays will provide the spatial resolution to follow ultrafast phenomena on the relevant timescales of atomic motions. Scientists dream of taking snapshots or movies of transient phenomena in materials like ultrafast phase transitions, melting and nucleation effects, nonlinear interaction of x-rays with matter, etc. X-ray holography is one of the key x-ray microscopy techniques for high-resolution imaging which is compatible with full-field single-shot imaging in coherent beams and is the focus of this chapter.

The understanding of the building blocks of materials and macromolecules is of fundamental importance in physics, material science, biology, and chemistry. For example, x-ray crystallography is widely used for structure determination. An important prerequisite is the periodicity of the material. This approach excludes a wide range of substances which cannot be grown as crystals, such as amorphous and disordered structures, macromolecules, polymers, magnetic nanostructures, as well as non-periodic electronic and magnetic phases in condensed matter. On the other hand, the fabrication of x-ray lenses for x-ray microscopy is challenging. Building an x-ray microscope based on holography was thought of in the early days (Baez 1952a), since its invention in 1947 by Gabor (1948). This work, for which Gabor received the Nobel Prize in physics in 1971, was focused on improving the resolution of electron microscopes. At that time, the aberrations in electron optics prevented the microscopes from reaching atomic resolution. Gabor proposed a radical, lensless route to high-resolution imaging as a "new microscopic principle."

The precursor of the lensless x-ray microscope was given by Bragg (1939, 1942). This concept recovers an optical image of the atomic structure from experimental diffraction data. An array of holes drilled in a plate mimicked the amplitudes and the positions of a recorded x-ray diffraction pattern. Under coherent illumination, the array emanated secondary wavelets forming an image of the atomic lattice on a screen in the far field. This two-step imaging process rests on Abbe's diffraction theory of image formation by a lens. Since the phases of the wavelets could not be correctly reproduced, an image of the atomic lattice could only be obtained for real-valued diffraction amplitudes where the phase ambiguity of π (amplitudes can either be positive or negative) could be lifted when the diffraction pattern was positively biased, e.g., by a central heavy atom. Gabor followed this direction and envisioned a lensless approach where a coherently diffracted electron wave from the sample interferes with the undiffracted wave. In this way, the phase information can be recorded in the form of a hologram on a photographic plate. The image reconstruction could then be performed as an analogue to Bragg's x-ray microscope by illuminating the hologram with a suitable-scaled optical replica of the undiffracted wave. An all-optical demonstration of this new principle was published by Gabor (1948). Its full potential was recognized more than a decade later when the first optical lasers became available. X-ray holography appeared impracticable at that time due to the lack of sufficiently temporally and spatially coherent x-ray sources, as well as of suitable x-ray optics and detectors (besides one successful image reconstruction of a thin wire from an x-ray diffraction pattern using visible light by El-Sum and Kirkpatrick (1952))

Meanwhile, with the advent of optical lasers with longer temporal coherence, Leith and Upnatnieks (1962, 1964) could circumvent the twin-image problem, which inflicted noise on the reconstruction in the Gabor geometry, by shifting the reference beam off-axis at an angle to the object. They further demonstrated three-dimensional reconstructions. The lower spatial frequencies in these holograms could be recorded with low resolution detectors. Based on the theoretical foundations and applications in Fourier optics and communication theory, VanderLugt (1964) generated Fourier transform holograms as masks for use in coherent optical processors. Fourier transform holography with x-rays was finally proposed by Stroke and Falconer (1964) to attain high-resolution imaging with x-rays. Simultaneously, Winthrop and Worthington (1965) outlined the principles of lensless x-ray microscopy base on successive Fourier transformation. Winthrop and Worthington (1966), Stroke et al. (1965a) demonstrated the feasibility of the above approach, and the implication of using a point source as a reference in FTH was addressed by Stroke et al. (1965b,c).

Fourier transform holography was first demonstrated at x-ray wavelengths by Aoki et al. (1972). In the late 1980s and early 1990s, x-ray holography surpassed for the first time, the optical resolutions. Howells et al. (1987) recorded Gabor holograms using photoresist detectors at a resolution of 40 nm. Shortly thereafter, McNulty et al. (1992) succeeded with Fourier transform holography-resolving features down to 60 nm length scales using a zone plate to generate a bright and a small reference source. In both experiments, the object reconstruction was performed with a computer. X-ray holography at atomic resolutions was demonstrated by Tegze and Faigel (1996) based on the inside-source concept at hard x-ray energies. This approach, however, is difficult to extend to larger length scales. With wavelengths of 5–0.5 nm, soft x-rays fill this gap to the optical regime although they are currently limited by the available coherent flux at third generation synchrotrons. With the birth of XFEL's, these techniques will likely close the resolution gap and reach ≪ 10 nm resolution. This is comparable to transmission and scanning x-ray microscopes which achieve resolutions down to 15 nm using Fresnel zone plates Chao et al. (2005).

Finally, holography is one technique among others that exploit the coherence properties for x-ray microscopy. Coherent diffractive imaging based on iterative phase reconstruction has

rapidly evolved in parallel, since its first experimental demonstration by Miao et al. (1999). Based on the Shannon–Whittaker sampling theorem, it is possible to iteratively recover the phases of diffraction patterns (Gerchberg and Saxton 1972, Fienup 1978, Marchesini et al. 2003) when recording the intensities at higher than the Nyquist frequency (oversampling). Applications range from the imaging of nanocrystals by Robinson et al. (2001), bacteria and cells by Miao et al. (2003) and Shapiro et al. (2005) to x-ray ptychography on extended samples with zone plates by Thibault et al. (2008). It is beyond the scope of this article to describe those techniques and the reader shall be referred to detailed reviews on this subject, e.g., by Fienup (1982), van der Veen and Pfeiffer (2004), Miao et al. (2008), and Marchesini et al. (2008).

38.2 Principles of Holography

Holography is based on the interference of two waves: the object wave and a reference wave which is ideally a plane or a spherical wave. A sufficient degree of coherence is required in order to have a constant phase relation in time between the two waves and therefore to observe an interference pattern or hologram on the detector. Holography or "whole writing" reflects the capability to encode both the amplitude *and* the phase information of the object with a reference beam prior to its recording. The original wave front is reconstructed in a second step by illuminating the hologram with the reference beam. The back-propagation to the image plane can also be performed by computational methods when the reference information is known. A key element of holography is the absence of a lens in the hologram formation and object reconstruction. Since interference is a general phenomenon, holography can be applied to all forms of waves, e.g., to particle waves and acoustic waves, as long as their degree of coherence is sufficient enough.

Let the wave, $S = |S|\exp(i\phi_S)$, be emanating from the object and superposed with a reference wave, $R = |R|\exp(i\phi_R)$. Then the interference pattern recorded on a suitable detector yields

$$I = |S + R|^2 = SS^\star + RR^\star + SR^\star + RS^\star$$

$$= |S|^2 + |R|^2 + 2|S||R|\cos(\Delta\phi), \quad (38.1)$$

where the first two terms contain the self-interference of object and reference. Obviously, the phase information is lost in the self-interference terms while the cross-terms contain both the object and the reference phase. The phase information is encoded in the relative or holographic phase, $\Delta\phi = \phi_S - \phi_R$. The hologram recording and the wave reconstruction for Gabor, off-axis, and Fourier transform holography are illustrated in Figure 38.1.

The image reconstruction is performed by re-illumination of the hologram with the reference wave. The following wave fields are generated:

$$RI = R|S|^2 + R|R|^2 + |R|^2 S + R^2 S^\star. \quad (38.2)$$

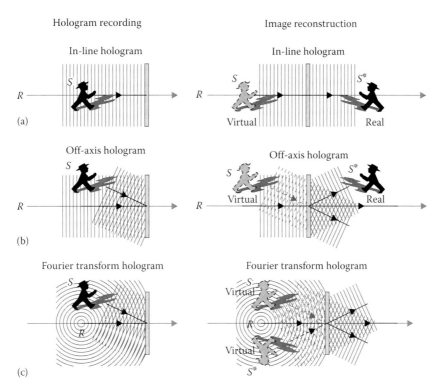

FIGURE 38.1 (a) In-line hologram recording (Gabor 1948) using a coherent reference beam, R, illuminating the object S. The interference of the object and the reference wave is recorded in the holographic plane. Image reconstruction is performed by re-illumination of the hologram with the reference. Back-propagation of the diffracted wave R^2S gives the image of the object. Because of in-line geometry, reconstruction is prone to noise from the zero-order beam and the twin image. (b) Hologram recording in off-axis geometry spatially separates the zero-order beam and the twin images S and $S^\star$ after Leith and Upnatnieks (1964). (c) In Fourier transform holography, the reference is placed next to the object, generating Young's double slit-like fringe pattern. Recording the hologram in the far field of the object simplifies back-propagation by Fourier inversion.

Here, the zero-order beam becomes modulated by the first two terms, $|S|^2$ and $|R|^2$, and would give upon recording speckle patterns but no image of the object or reference. The third term, $S|R|^2$, is the reconstructed wave field with its origin where the object was positioned. This *virtual image* comes with its twin, which is the last term, $R^2S^\star$ in Equation 38.2. This *real* image is the conjugate image of the object. The spatial separation of the zero-order beam and the twin images in an off-axis geometry, Leith and Upnatnieks (1964), which avoids background noise from the other waves reconstructed under illumination of the hologram (see Figure 38.1b).

38.2.1 Wave Propagation and Diffraction

In the following, we formulate an expression relating an incident wave subjected to an object disturbance to the diffracted wave field. For this, we consider that the object consists of point scatterer distributed over a finite region, as illustrated in Figure 38.2, and is described by a scattering potential $t(\mathbf{r})$. Let the incident plane wave $\psi_0(\mathbf{r}) = A_0\exp(ikz)$ with wave number $k = 2\pi/\lambda$, and amplitude A_0 propagate along the optical axis $z \equiv z'$ and penetrate the disturbance such that the total wave field emanating into the half space behind the object is formed by a superposition of the incident wave and the diffracted wave. The propagation of the total wave is a formal solution of the inhomogeneous wave equation,

where the incident and the diffracted wave are the solutions of the homogeneous and inhomogeneous parts respectively. We make use of Green's function in order to obtain an integral equation for the total wave field:

$$\psi(\mathbf{r}') = A_0\exp(ikz) + \frac{ik}{2\pi}\iiint G(\mathbf{r}-\mathbf{r}')t(\mathbf{r})\psi(\mathbf{r})\,\mathrm{d}\mathbf{r}. \qquad (38.3)$$

The retarded Green's function of the wave equation is given by $G(\mathbf{r}-\mathbf{r}') = \exp(ik|\mathbf{r}-\mathbf{r}'|)/|\mathbf{r}-\mathbf{r}'|$, i.e., an outgoing spherical wave emitted from the inhomogeneity located at $\mathbf{r}$. The amplitude of the emitted wave is proportional to the local scattering potential $t(\mathbf{r})$ and the local amplitude of the wave field $\psi(\mathbf{r})$ at a given point. The diffracted wave is therefore the integral over all emitted spherical waves emanating from the object's region.

The formulation of the wave field (38.3) requires the knowledge of the local fields in the scattering region. A more manageable formulation can be found using the Born approximation where the impact of the local disturbance onto the incident wave field is negligible and $\psi(\mathbf{r}) \simeq \psi_0(\mathbf{r})$. This condition is valid for large wave numbers k or weak-scattering potentials as is the case for x-rays. Because each point of the object sees the same incident wave field, the approximation further ignores multiple scattering scenarios and allows the projection of the scattering potential onto a single diffraction plane at $z = 0$ normal to the propagation direction of the

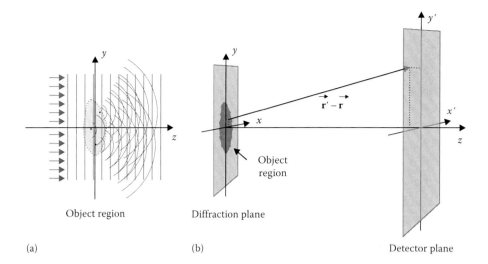

FIGURE 38.2 (a) Schematic of the diffraction of an incident wave field by an object. (Forward scattering is here considered.) (b) Geometry of the wave propagation.

undiffracted wave such that $t(\mathbf{r}) \rightarrow t(x, y)$ is the transmittance of the object. The integral expression for the total wave field becomes

$$\psi(\mathbf{r}') = A_0 \exp(ikz) + \frac{ik}{2\pi} \iiint \frac{\exp(ik\,|\,\mathbf{r} - \mathbf{r}'\,|)}{|\,\mathbf{r} - \mathbf{r}'\,|} A_0 t(x, y) \mathrm{d}\mathbf{r}. \quad (38.4)$$

Here, $\psi_e(x, y) = A_0 t(x, y)$ can be viewed as the wave front that exits the object at $z = 0$. According to the Huygen–Fresnel principle, this new wave front is formed by a superposition of wavelets given by the object's transmittance.

Considering only forward and small-angle scattering in Figure 38.2b, (38.4) can be further simplified deriving the Fresnel and Fraunhofer approximations. For this, we approximate

$$|\,\mathbf{r} - \mathbf{r}'\,| = z\sqrt{1 + \frac{x^2 + y^2}{z^2} + \frac{x'^2 + y'^2}{z^2} - 2\frac{xx' + yy'}{z^2}}$$

$$\approx z\left(1 + \frac{x^2 + y^2}{2z^2} + \frac{x'^2 + y'^2}{2z^2} - \frac{xx' + yy'}{z^2}\right), \quad (38.5)$$

and omit the unscattered wave ("direct beam") to obtain the diffracted wave front

$$\psi(x', y', z) = A(x', y', z)$$

$$\times \iint t(x, y) \exp\left(ik\frac{x^2 + y^2}{2z}\right) \exp\left(-ik\frac{xx' + yy'}{z}\right) \mathrm{d}x \mathrm{d}y, \quad (38.6)$$

where the prefactor is given as

$$A(x', y', z) = A_0 \frac{ik \exp(ikz)}{2\pi z} \exp\left(ik\frac{x'^2 + y'^2}{2z}\right). \quad (38.7)$$

The Fresnel integral (38.6) can be further simplified with $k = 2\pi/\lambda$ and by introducing spatial frequencies:

$$u_x = \frac{x'}{\lambda z} \quad \text{and} \quad u_y = \frac{y'}{\lambda z}. \quad (38.8)$$

We find for the near field

$$\psi_{\text{near}}(u_x, u_y, z) = A(\lambda z u_x, \lambda z u_y, z)$$

$$\times \iint t(x, y) \exp\left(i\pi\frac{x^2 + y^2}{\lambda z}\right) \exp\left(-2\pi i\left\{u_x x + u_y y\right\}\right) \mathrm{d}x \mathrm{d}y, \quad (38.9)$$

where the diffracted wave in the plane at a given distance z is simply given by the Fourier transform of the exit wave multiplied by a phasor representing a finite curvature of the superposed wavelets. As a consequence, the interference pattern is rapidly evolving as a function of distance in the Fresnel regime. In the far field of a small lateral object, the phasor becomes unity because $z\lambda \gg (x^2 + y^2)$. The diffracted wave then reduces to a simple Fourier transform of the exit wave:

$$\psi_{\text{far}}(u_x, u_y, z) = A(\lambda z u_x, \lambda z u_y, z)$$

$$\times \iint t(x, y) \exp(-2\pi i\{u_x x + u_y y\}) \mathrm{d}x$$

$$\propto \mathcal{F}\{t(x, y)\}, \quad (38.10)$$

where $\mathcal{F}\{\}$ is the Fourier operator. In the Fraunhofer regime, the diffracted wave is a composition of plane waves emanating at small diffraction angles to the propagation direction.

38.2.2 Optical Constants

The transmittance of the object depends on its index of refraction:

$$n(\omega) = 1 - \delta(\omega) + i\beta(\omega) \quad (38.11)$$

where β represents the absorptive and δ the refractive properties of the material. In the x-ray regime, the atomic scattering factors $f_1 = Z + f'(\omega)$ and $f_2 = f''(\omega)$ in units of number of electrons are commonly used to describe the interaction of x-rays with matter. Neglecting polarization effects and magnetic scattering, there exists a simple relation between the optical constants and the atomic scattering factors, see e.g., Attwood (1999):

$$\delta(\omega) = \frac{r_0\lambda^2}{2\pi}\rho f_1(\omega) \quad \text{and} \quad \beta(\omega) = \frac{r_0\lambda^2}{2\pi}f_2(\omega), \qquad (38.12)$$

where

- Z is the number of atomic electrons
- r_0 is the classical-electron radius
- λ is the x-ray wavelength
- ρ is the atomic number density

For simplicity, the notation of optical constants is used in the following.

In the soft x-ray regime, the optical constants δ and β are of the order of $\simeq 10^{-3}$, while in the hard x-ray regime the refractive part $\delta \approx 10^{-6}$ is several magnitudes larger than the absorption. Hence, the index of refraction is $n \approx 1$. Assuming that the object has a total thickness d along the propagation direction of the incoming wave in Equation 38.6, the transmittance can be expressed as

$$t(x,y) = \exp\left(-k\int_0^d \left\{i\delta(x,y,z) + \beta(x,y,z)\right\}dz\right). \qquad (38.13)$$

Since we considered the Born approximation to be valid and the optical constants to be $\delta, \beta \ll 1$, the expression in Equation 38.13 simplifies to

$$t(x,y) \simeq 1 - i\phi(x,y) - \mu(x,y), \qquad (38.14)$$

where $\phi = (2\pi d/\lambda)\delta$ and $\mu = (2\pi d/\lambda)\beta$ are the phase-shifting and the attenuating sample properties for the incoming wave field, respectively. In this weak contrast limit of the sample, the transmittance depends linearly on the optical constants.

38.2.3 Gabor X-Ray Holography

Rogers (1950) draws an analogy between Gabor holograms and generalized zone plates. Suppose a reference point source R, with wavelength λ, is placed at a distance q to a holographic plane, cf. Figure 38.3a. Diffraction of the reference wave from a point object at the position S on the optical axis creates a rotationally symmetric hologram, where the radial distribution is given by the interference $2|S||R|\cos(\Delta\phi)$ (Equation 38.1). The phase difference, $\Delta\phi$, is given by the different path lengths the wavelet has to travel before reaching point H, i.e., $\Delta\phi = (2\pi/\lambda)\Delta l$. By comparing the two distances $R \rightarrow H$ and $R \rightarrow S \rightarrow H$ this path difference yields

$$\Delta l = \sqrt{r^2 + q^2} - \left(q - p + \sqrt{r^2 + p^2}\right) \approx \frac{r^2}{2}\left(\frac{1}{q} - \frac{1}{p}\right). \qquad (38.15)$$

Constructive interference can be found when $\cos(\Delta\phi) = 1$ and the path difference equals $n\lambda$, where $n = 1, 2, 3,\ldots$. Using Equation 38.15, the "point" hologram results in a concentric ring pattern with the radii

$$r_n = \sqrt{2nf\lambda} \quad \text{where} \quad \frac{1}{f} = \frac{1}{q} - \frac{1}{p}, \qquad (38.16)$$

which follows the construction of a zone plate with the only difference that the radial fringes have a sinusoidal envelope while a Fresnel zone plate is a binary or "digital hologram" of a point object. The sinusoidal intensity distribution of the hologram therefore diffracts the reference beam only up to first orders resembling the virtual image S and its conjugate image S^* at a distance f which is analogous to zone plates an effective focal length in the lens Equation 38.16.

The hologram of a more complex object is an agglomerate of "point" holograms formed from the secondary wavelets

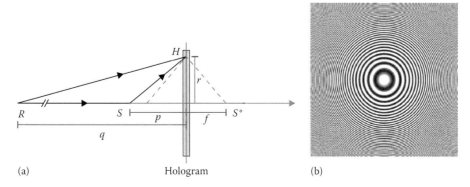

FIGURE 38.3 (a) Illustration of the Gabor hologram formation of a point scatterer. (b) The resulting hologram has a close relation to a Fresnel zone plate with focal length f.

emanated from the object. The resolution of the reconstructed image is determined by the smallest fringe spacings that can be sampled. High-resolution photoresist detectors can provide a resolution of about $\Delta r \approx 10\,nm$. The optimum working distance for a Gabor hologram is determined by the image resolution (ideally the detector resolution) and the coherence properties of the source (Jacobsen et al. 1990). The working distance equals the focal length of the hologram, when the object is coherently illuminated ($q \rightarrow \infty$). Using Equation 38.16, we find for two adjacent rings:

$$r_n^2 - r_{n-1}^2 = 2f\lambda$$

$$= r_n^2 - (r_n - \Delta r)^2 \simeq 2r_n\Delta r, \qquad (38.17)$$

where

$r_n = d/2$ is the outermost zone to be captured from each point of the object

$\Delta r = r_n - r_{n-1}$ is the smallest fringe-spacing to be resolved by the detector

The optimum working distance yields

$$f_{opt} = \frac{r_n \Delta r}{\lambda}, \qquad (38.18)$$

where the soft x-ray beam has to be at least transversely coherent over r_n. With a typical transverse coherence of $\xi_t \lesssim 50\,nm$, the working distance of the sample to the detector becomes $f = 250\,\mu m$ with a 2 nm wavelength for achieving 10 nm resolution, which lies in the near field or Fresnel regime. In addition, the x-ray beam has to be monochromatic, such that there are as many longitudinal coherent waves as zones (optical path difference between adjacent zones is λ, see above). For the determined working distance, it follows from Equation 38.16 that at least $n = 2500$ coherent wavelengths, which is achievable with soft x-ray monochromators.

Under similar conditions, Howells et al. (1987) were able to recorded Gabor holograms of zymogen granules with a resolution of 40 nm using polymethyl methacrylate (PMMA) photoresist in Figure 38.4. The photoresist was processed with a solvent which dissolves the exposed areas at a higher rate than the unexposed ones leaving a surface relief pattern behind. Instead of using an optical laser light shined on a properly magnified version of the hologram (which had little success in previous attempts), the image reconstruction was done numerically. An electron micrograph of the holographic pattern was recorded, and the magnified hologram was then digitized and reconstructed by numerically back-propagating to the object plane using the Fresnel approximation Equation 38.6.

Baez (1952b) suggested producing Fresnel zone plates from Gabor point holograms and using them for x-ray focusing or lens-based x-ray microscopy. Modern zone plates are fabricated by means of electron-beam lithography allowing patterns below 10 nm to be written. The highest resolution achieved with a zone plate in a full-field transmission x-ray microscope (TXM) is

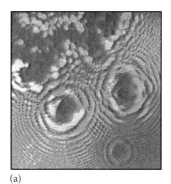 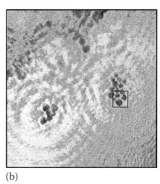

(a) (b)

FIGURE 38.4 (a) Gabor hologram of zymogen granules recorded with a high resolution photoresist detector. (b) Image reconstruction from the digitized version of the developed hologram. (Reprinted from Jacobsen, C. et al., *J. Opt. Soc. Am. A*, 7, 1847, 1990. With permission.)

15 nm and better at ~2 nm soft x-ray wavelengths (Chao et al. 2005). Since the high-resolution photoresist detectors rely on the same principles, zone-plate-based microscopy prevailed against Gabor x-ray holography in avoiding the complex steps of hologram processing and reconstruction which may further degrade the photoresist detector resolution.

38.2.4 Fourier Transform X-Ray Holography

Fourier transform x-ray holography was examined by Stroke and Falconer (1964), Stroke (1965) and Winthrop and Worthington (1965, 1966) for attaining high-resolution x-ray holography. The concept avoids the use of high-resolution detectors. In Fourier transform holography (FTH), a spherical reference is positioned next to the object as shown in Figure 38.1c. The reference-object plane is perpendicular to the optical axis and parallel to the detector plane. When coherently illuminated, both the reference and object waves interfere analogously to Young's double-slit experiment. The latter provides a simple picture of the FTH principle. Let the object S and a point reference R be arranged in the diffraction plane such that the transmittance of the holographic mask is

$$t(x, y) = s(x, y) + \delta(x - x_0, y - y_0). \qquad (38.19)$$

Considering the diffracted wave in the far field of the holographic mask using Equations 38.10 and 38.19, a simple Fourier transform relation between the hologram and the diffraction plane exists

$$\mathcal{F}\{t(x, y)\} = S(u_x, u_y) + \exp\left(-2\pi i \left\{u_x x_0 + u_y y_0\right\}\right), \qquad (38.20)$$

where we used the shift-theorem (Goodman 2005) and the fact that $\mathcal{F}\{\delta(x, y)\} = 1$. The formed hologram follows readily from Equation 38.1,

$$I(u_x, u_y) = |S^2| + 1 + |S| \cos\left(2\pi \left\{u_x x_0 + u_y y_0\right\} + \phi_s\right) \qquad (38.21)$$

using the definitions of the spatial frequencies, Equation 38.8, in the detector plane. In its simplest form, i.e., $S = \delta(x, y)$, we obtain

Fourier transform hologram Autocorrelation

 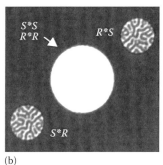

(a) (b)

FIGURE 38.5 (a) Fourier transform holography of a magnetic-domain pattern recorded with circular polarization: object and reference interference leads to the diagonal fringes encoding the object's phase information. (b) Autocorrelation given by a simple inverse Fourier transformation. The correlations of the object and the reference with themselves appear in the center while the cross-correlation terms between the object and the reference appear off-axis. (Reprinted from Eisebitt, S. et al., *Nature*, 432, 885, 2004. With permission.)

sinusoidal interference fringes perpendicularly oriented to the connecting line between the object and the reference (cf., Figure 38.5). The spacing of the fringes is $d/(\lambda z)$, where the distance between the object and the reference is $d = \sqrt{x_0^2 + y_0^2}$. For small object-reference spacings d and long propagation distances z at the given wavelength λ, the object is sampled at low spatial frequencies which can be easily recorded with detectors. This has a major advantage for imaging at x-ray wavelengths. The limiting factor in Gabor holography is the detector resolution required to resolve the fine-fringe patterns. On the contrary, FTH attains high spatial resolution, since the achievable resolution is only determined by the detector size or numerical aperture. The sampling requirements of the holographic mask are discussed in the following section.

The object reconstruction from a FTH hologram can be explained by considering Young's fringes as a diffraction grating. Since the grating does not contain higher Fourier coefficients, light is diffracted only up to the first order forming two virtual images of the object as illustrated in Figure 38.1c. Numerically back-propagating the hologram to the image plane is a simple and immediate step and can be performed with an inverse Fourier transformation of the recorded intensity pattern. Defining the autocorrelation as

$$g(x, y) \otimes g(x, y) = \int\!\!\!\int_{-\infty}^{\infty} g(x', y') g^*(x'-x, y'-y) dx' dy' \quad (38.22)$$

and using the autocorrelation theorem,

$$\mathcal{F}\{s(x,y) \otimes s(x,y)\} = \mathcal{F}\{s(x,y)\} \mathcal{F}\{s(x,y)\}^* = |S(u_x, u_y)|^2, \quad (38.23)$$

the inverse Fourier transform of the intensity pattern, Equation 38.21, yields

$$\mathcal{F}^{-1}\left\{I(u_x, y_y)\right\} = s(x, y) \otimes s(x, y) + \delta(x, y)$$

$$+ s^*(-x+x_0, -y+y_0) + s(x+x_0, y+y_0)$$

$$\equiv t(x, y) \otimes t(x, y). \quad (38.24)$$

In other words, the Fourier transformation of the hologram gives the autocorrelation of the holographic mask as in Equation 38.19. It contains the autocorrelation of the sample and the reference in the center, and the two cross-correlations of the sample and reference located at (x_0, y_0) and $(-x_0, -y_0)$. The reference point source is represented by a δ-function providing a uniform spectral range with a constant phase. In practice, a suitable "point-like" source will have a finite size. Accordingly, the uniformity of the spectral range and the phase of the reference will limit the achievable spatial resolution of the reconstruction. Concepts to compensate for extended sources are given in the next section.

The first demonstration of FTH with a significant improvement in a resolution below optical wavelengths was given by McNulty et al. (1992). The hologram was recorded from a sample placed near the first-order focal point of a Fresnel zone plate (cf., Figure 38.3). The latter served as a bright point-like reference wave while other orders were blocked except for illuminating the sample. The hologram was captured with a charge-coupled device (CCD) camera which allows for fast numerical reconstruction by Fourier inversion after readout. The resolution of the test patterns was better than 60 nm using 3.4 nm soft x-rays. By milling a holographic mask into an opaque gold film on a silicon nitride membrane and by placing a sample in the front, the experimental setup for FTH was greatly simplified. Eisebitt et al. (2004) tuned the soft x-rays to an absorption resonance and recorded a magnetic-domain pattern by exploiting the x-ray magnetic circular dichroism effect. Depending on the orientation of the magnetization along the propagation direction of the x-rays, the optical constants vary, providing a detectable scattering contrast. The charge scattering from the circular object aperture is superposed with magnetic speckles as shown in Figure 38.5. The diagonal fringes represent the holographic encoding of the object phases by a pinhole reference of a 100 nm in diameter. After a Fourier inversion of the hologram, the images of the magnetic-domain pattern appear off-center in the autocorrelation with a spatial resolution of 50 nm.

Because of the technical limitations in fabricating small and bright references for Fourier transform holography, a diffraction-limited resolution can only be achieved by "numerically" compensating for an extended source. We will discuss the possible concepts in the following sections.

38.3 Practical X-Ray Holography for Nanostructure Imaging

38.3.1 Holographic Mask Fabrication

Silicon nitride (Si_3N_4) membranes of ~100 nm thickness, fabricated on silicon wafers, are typically used to support e.g., nanoparticles and biological specimens. Nanostructures, such as point

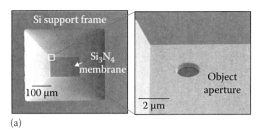

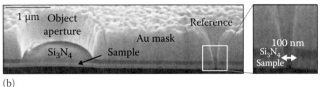

FIGURE 38.6 (a) Holographic samples are prepared on Si_3N_4 membranes with an object aperture milled through the soft x-ray opaque Au film (~1 µm) in the corner using a focused ion beam. (b) Cross section of a holographic mask: The object aperture stops at the membrane leaving the magnetic Co/Pt multilayer intact. The reference is drilled through the entire structure.

references and coded apertures can serve as references for holography Marchesini et al. (2008). These are prepared with electron lithography on the membranes using the same methods to produce x-ray zone plates. Alternatively, references can be milled directly into the silicon nitride using a focused ion beam (FIB). For continuous structures such as magnetic films and polymers that exceed the spatial-coherence length of the source, an object aperture is required that confines the sample area under investigation. This can be realized by depositing an opaque Au film on the rear of the membrane. The Au films of 600–1200 nm thickness are sufficiently opaque at soft x-ray wavelengths. A holographic mask consisting of an object aperture and a reference can be patterned into the Au film (Schlotter 2007). Due to the different milling rates of Au and silicon nitride, the milling process can be controlled to stop at the membrane shown in Figure 38.6. This allows a good definition of the object apertures which leave the sample on the other side intact. Reference holes are drilled through the entire structure. Since the ion beams can be focused down to ~10 nm with a focal depth of about ~100 nm, references with large aspect ratios can be fabricated. This is important to produce small reference structures with a strong scattering contrast for high-resolution imaging.

38.3.2 Detector and Sampling

An array detector samples the hologram at N discrete values along one dimension at intervals of $\Delta u = \Delta x'/(\lambda z)$, where $\Delta x'$ is the detector pixel size. Here, we use the definition of the spatial frequencies in the detector plane given by Equation 38.8. The shortest spatial frequency that can be resolved with the detector requires at least two pixels in order to observe maxima and minima of the fringes. This corresponds to half the length of the autocorrelation which is sampled at an interval Δx over a length of $L = 1/\Delta u = N\Delta x$. As a result, to resolve the largest length-scale $d = L/2$ between the object and the reference

requires the sampling with the detector at finer intervals of at least $\Delta u = 1/2d$ satisfying the Shannon–Whittaker sampling theorem (Goodman 2005). Therefore, the highest spatial resolution achievable with Fourier transform holography is achieved with an optimum sampling of

$$\Delta x = 2 \frac{\lambda z}{\Delta x' N}. \tag{38.25}$$

38.3.3 High-Resolution Imaging

In order to achieve a diffraction-limited resolutions, the reference size must be of the same order as the wavelength. At the same time, the scattering of the reference should be comparable to the object for optimum-fringe visibility in the interference pattern. Hence, point-like scatterers such as nanocubes and nanospheres have a suitable spectral range for high-resolution imaging but also have small cross sections. Compared to the object, the reference wave is faint and consequently, the interference term is small resulting in a weak image reconstruction. Using Equation 38.1, the signal-to-noise ratio (SNR) of a weak-reference hologram can be expressed as

$$\text{SNR} = \frac{|S||R|}{\sqrt{|S|^2 + |R|^2}}. \tag{38.26}$$

Assuming a sphere of radius a with a weak-scattering contrast $\alpha \ll 1$, which is defined as the difference in the optical constants of the sphere with the surroundings, the number of scattered photons by the sphere becomes $|R|^2 \propto \sigma_s F \propto a^4 \alpha^2 F \tau$ (London et al. 1989), where σ_s is the total cross section, F the flux density, and τ the exposure time. In the weak-reference limit $|S|^2 \gg |R|^2$, where the object dominates the hologram, the SNR strongly depends on the size of the reference scatterer:

$$\text{SNR} \approx |R| \propto \sqrt{\sigma_s F} \propto a^2 \alpha \sqrt{F \tau}. \tag{38.27}$$

Therefore, to improve the resolution of the reconstructed image by a factor of 2 while sustaining the image quality, the exposure time will increase by a factor of 16. The latter sets limits for high-resolution imaging of radiation-sensitive samples which are difficult to overcome. Using point-like references in flux-limited experiments, such as single-shot imaging with pulsed x-ray sources, will produce similar problems. It is therefore, advantageous to use extended references to improve the SNR. Utilizing source-compensation methods, high-resolution imaging becomes possible from "low-resolution" holograms.

38.3.4 Compensation for Extended References

A prerequisite for high-resolution imaging with extended references is that the reference structure must be known and should provide a suitable spectrum over which the object and the reference can be deconvolved. When the reference and the object have a large-enough spacing as in FTH, the term R^*S in Equation 38.1 can be spatially filtered using the autocorrelation of the hologram. The blurred image can be expressed as

$$b(x, y) = \mathcal{F}^{-1}\left\{B(u_x, u_y)\right\}$$
$$= \mathcal{F}^{-1}\left\{R^*(u_x, u_y)S(u_x, u_y)\right\} = r(-x, -y) \star s(x, y), \quad (38.28)$$

where the convolution is defined as

$$g(x, y) \star h(x, y) = \int\int_{-\infty}^{\infty} g(x', y')h(x - x', y - y')\mathrm{d}x'\mathrm{d}y'. \quad (38.29)$$

In Equation 38.28, the reference $r(x, y)$ represents the impulse-response function, which describes the spread of a point due to the imaging process using an extended reference source. In order to deblur the image $b(x, y)$, a suitable filter F must be found such that the object

$$s(x, y) = \mathcal{F}^{-1}\left\{FB\right\} \quad (38.30)$$

is fully recovered. From Equation 38.28, it follows that $F = 1/R^*$. Because the power spectrum of an extended reference R^* will have in general many zero crossings, a Wiener deconvolution filter is used (Howells et al. 2001):

$$s(x, y) = \mathcal{F}^{-1}\left\{\frac{|R|^2}{|R|^2 + \Phi}\frac{B}{R^*}\right\}. \quad (38.31)$$

The remaining problem is how to obtain the power spectrum of the reference. This can be done, either by having an high-resolution image of the reference, e.g., from a scanning electron micrograph, or by iteratively reconstructing the reference with phase retrieval algorithms (He et al. 2004).

38.3.5 Multiple References and Coded Apertures

An increase of the SNR can also be achieved by spatial multiplexing without increasing the radiation exposure. The technique allows for using many point-like references without the need of further image processing. Schlotter et al. (2006) demonstrated that the detection limit of high-resolution FTH can be extended by introducing multiple references. Each of the n references will generate a unique image of the object upon reconstruction. A composite image can then be obtained by adding the n images resulting in a composite SNR of

$$\mathrm{SNR}_{\mathrm{composite}} = \sqrt{N}\,\mathrm{SNR}_1, \quad (38.32)$$

where $\mathrm{SNR}_1 = \bar{s}/\sigma_b$ of a single reconstructed image with $\bar{s}$, the mean value of intensity over the image area, and σ_b the standard deviation in the background surrounding the sample. As mentioned

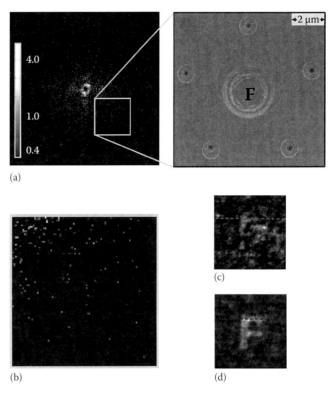

(a)

(b)

(c)

(d)

FIGURE 38.7 (a) Hologram of the "F" sample with multiple references in the weak illumination limit, i.e., 2500 detected photons. (area up to $q = \pm0.09\,\mathrm{nm}^{-1}$ is shown). (b) The magnification reveals single-photon events over a wide range of momentum transfers. (c) The reconstruction from a single reference (SNR ~ 3) is not recognizable. (d) The composite provides a clear image of the sample ($\mathrm{SNR}_5 \sim 9$). (Reprinted from Schlotter, W. F. et al., *Appl. Phys. Lett.*, 89, 163 112, 2006. With permission.)

above, the exposure time or the number of photons may be limited and the image quality is then affected by the photon noise in the hologram. The performance of multiple references in the weak-illumination limit is shown in Figure 38.7. A letter "F" and five reference holes were written into a 1 μm thick Au mask using a focused ion beam. The hologram was recorded with a short exposure of the mask containing only 2500 photons. Over a wide area of scattering angles, the hologram reveals single-photon counts. The letter "F" is not recognizable upon image reconstruction, while the letter is clearly visible in the composite image. The boost in the SNR has its limit in the number of reference holes that can be placed, such that upon reconstruction, the images do not overlap in the autocorrelation.

The spatial multiplexing can be further exploited when one allows overlap of the reconstructed images. The most compact way of producing many replicas of the object is by utilizing a uniform redundant array (URA) (Fenimore and Cannon 1978). The application of URAs ranges from astronomy to medical science. The URA consists of a mosaic pattern of pinholes, such that its power spectrum becomes uniformly flat, see Figure 38.8. This reduces noise caused by the deconvolution procedure to reconstruct the image. URAs maintain the high resolution provided by a single pinhole but reach the high SNR ratios

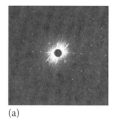

(a)   (b)

FIGURE 38.8 (a) Hologram of a spiroplasma cell using a 17×19 twin-prime URA: The diffraction pattern was taken with a single pulse of a VUV FEL(1×10^{12} photons, 15 fs, 13.5 nm). (b) Reconstruction using the Hadamard transform and extending the resolution by iterative phase-retrieval methods (resolution 75 nm). (Reprinted from Marchesini, S. et al., *Nat. Photon.*, 2, 560, 2008. With permission.)

of large apertures. The number of pinholes in such arrays can be orders of magnitude and dramatically improves the SNR according to Equation 38.32. The efficiency of this technique is of value for single-shot imaging with VUV (Ackermann et al. 2007) and upcoming x-ray free-electron lasers or even table-top x-ray sources with lower photon counts per pulse (Sandberg et al. 2008, Wang et al. 2008). Coded aperture imaging at 13 nm VUV radiation from the FLASH free-electron laser has been demonstrated by Marchesini et al. (2008). Figure 38.8 shows a 17×19 twin-prime URA with 162 pinholes of 150 nm in diameter utilized to image a bacterium with a resolution of 75 nm after improving resolution beyond the pinhole diameter by iterative phase retrieval algorithms.

38.3.6 Spectroscopy and Multiple-Wavelength Anomalous Diffraction

So far we have not addressed resonant phenomena, but instead have considered the optical constants to be smoothly varying as functions of wavelength. The soft x-ray regime contains important resonances of 3d-transition metals (L edges) and rare earth (M edges) as well as C, N, and O absorption K edges providing elemental insight into the electronic and magnetic properties of matter (Stöhr and Siegmann 2006). X-ray spectroscopy techniques are therefore of fundamental importance for material science and biology. Here, x-ray spectro-microscopy is in particular important for investigations in nanoscience. Lensless spectro-holography of magnetic nanostructures was first demonstrated by Eisebitt et al. (2004) exploiting x-ray magnetic circular dichroism (Schütz et al. 1987) as a magnetic contrast mechanism. This effect has been used to investigate magnetic domain structures on sub-100 nm length scales, see e.g., Günther et al. (2008).

Resonant scattering inherently increases absorption that may limit the exposure time for radiation-sensitive samples and shorten the attenuation length of the x-rays making it difficult to image buried structures. Both effects can be mitigated by using resonant phase-imaging with x-ray holography (Scherz et al. 2007). The direct relation between the optical constants and the image reconstruction is shown in Figure 38.9. The technique is based on the fact that phase and absorption contrast have their resonant-scattering maxima at slightly different energies. According to Equation 38.14,

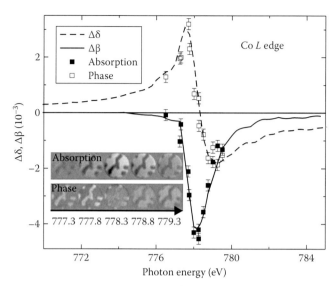

FIGURE 38.9 Optical constants $\Delta\delta$ and $\Delta\beta$ of the magnetic circular dichroism (difference between left and right circularly polarized x-rays) across the cobalt L_3 edge. The scaled phase and the absorption contrast (data points) are taken from the reconstructed complex image. (Inset) Top row is the real part with absorption contrast. The bottom row shows the imaginary part with dispersive-like phase contrast.

the phase component is the imaginary part and the absorption is the real part of the sample transmittance. The phase-dominated image can therefore be recorded before the absorption edge, where the autocorrelation of the phase hologram has mainly an imaginary component. An absorption hologram can be recorded at the resonance maximum where the phase-shift is zero and the autocorrelation only has a real component. The resonant-scattering contrast can therefore be optimized in terms of exposure time between the two energies to mitigate sample damage and to improve image quality at higher resolutions.

In the following, we show that resonant scattering can also serve as a holographic reference for structure determination. The intricacy of structure determination due to lost-phase information in diffraction measurements can be downsized to a manageable problem by labeling the unknown specimen with a few heavy atoms that serve as reference scatterers. In this way, multiple-wavelength anomalous diffraction (MAD) phasing has revolutionized macromolecular structure determination on atomic length scales and has become a well-established technique in x-ray crystallography with dedicated synchrotron beamlines (Hendrickson 1991). An important prerequisite for conventional MAD is the periodicity of the sample which is required for assembling the biological molecules into a crystal.

Scherz et al. (2008) extended the methodology of MAD to nonperiodic structures using coherent x-rays. By utilizing the strong energy-dependence of the optical constants in the vicinity of the resonance to define a reference wave, the resonant wave, ψ_R, emerges from one part of the sample and interferes with the nonresonant wave, ψ_N, emanated from other parts of the sample to form a hologram. With this generalization of the reference wave, no prior information of its structure is required. While the optical

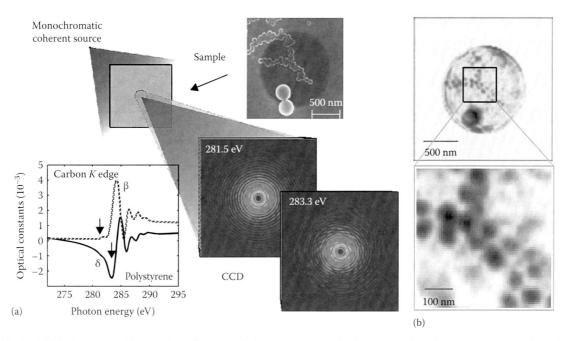

FIGURE 38.10 (a) MAD imaging: The sample is illuminated with monochromatized and spatially coherent x-rays across the carbon *K edge* (the plot shows the optical constants derived from Kramers-Kronig conversion of the experimental absorption of the 90 and 300 nm polysterene spheres). Resonantly- (dark gray) and nonresonantly- (light gray) scattered photons interfere on the detector and form an energy-dependent MAD hologram. (b) Image reconstruction of the polysterene spheres with a ~25 nm resolution limited by photon statistics at a high momentum transfer. (Reprinted from Scherz, A. et al., *Phys. Rev. Lett.*, 101, 076101, 2008. With permission.)

constants of the material in resonance strongly changes within a short energy interval, the nonresonant constituents are approximately energy-independent. Let the nonresonant transmittance of the sample be described by $s(x, y)$ and the resonant transmittance be more explicitly written as $(\delta_R + \beta_R)r(x, y)$. We assume a weak-resonant contrast and linearize the object transmittance, see Equation 38.14. The MAD hologram in the far-field of the sample can be expressed as

$$I(E, u_x, u_y) = |S(u_x, u_y) + (\delta_R(E) + \beta_R(E))R(u_x, u_y)|^2, \quad (38.33)$$

where S and R are the Fourier transforms of the spatial distributions.

For the reconstruction of the object, we must know the resonant reference wave. Given the optical constants and by recording holograms at multiple resonant energies $E = E_1, E_2, \ldots$, a set of linear equations can be obtained:

$$I(E_i) = |S|^2 + (\delta_R^2(E_i) + \beta_R^2(E_i))|R|^2$$
$$- 2|S||R|(\delta_R(E_i)\sin\Delta\phi + \beta_R(E_i)\cos\Delta\phi). \quad (38.34)$$

Three energies are enough to analytically solve the three unknowns: These are the amplitudes of the resonant wave R, and the nonresonant-wave S, as well as their relative phase $\Delta\phi = \phi_S - \phi_R$. An immediate reconstruction of the object is possible if one of the two phases is known. In practice, this is typically not the case. The phases are therefore iteratively recovered where S, R, and $\Delta\phi$ serve as reciprocal-space constraints. The latter dramatically improves standard-phase retrieval algorithms in terms of uniqueness and rapid convergence. It is beyond the scope of this text to describe the details of iterative phase retrieval. The concepts are described in great detail, for example, by Fienup (1982), van der Veen and Pfeiffer (2004), Miao et al. (2008), Marchesini (2007).

The experimental realization is shown in Figure 38.10. The holograms were recorded from dispersed polysterene spheres on a silicon nitride membrane with a gold mask. The interference patterns significantly changed within 2 eV of the incoming x-rays according to the optical constants at the carbon *K* edge. The optical constants have been determined by measuring the absorption of the sample and applying the Kramer-Kronig conversion, see e.g., Attwood (1999). By applying iterative methods, the resonant image of the polysterene spheres and the nonresonant image of the aperture, are simultaneously reconstructed in less than 30 iterations with a resolution of ~25 nm. The inherent resonant aspect provides sensitivity to the elemental, chemical, and magnetic state of the sample that further renders lensless MAD imaging widely applicable in nanoscience. The resolution of MAD "holography" is neither limited by the detector resolution nor by the size of the reference structure in comparison to Gabor or Fourier transform holography, but is, in principle, only limited by the wavelength.

38.4 Ultrafast X-Ray Holography

The investigation of ultrafast processes has so far been the domain of optical techniques. However, due to longer wavelengths, the details of the underlying processes remained unresolved. With the development of new accelerator-based (Ackermann et al. 2007) and even table-top sources (Sandberg et al. 2008, Wang et al. 2008), holographic techniques will provide high-spatial resolution combined with femtosecond time resolution which are both relevant to capture complex dynamics on the nanoscale. Pioneering experiments by Chapman et al. (2007) and Barty et al. (2008) explore the possibilities of this novel scientific field. X-rays allow for exploring transient phenomena in materials such as ultrafast phase transitions, melting, and nucleation effects, as well as nonlinear interaction of x-rays with matter.

38.4.1 Toward Ultrafast Panoramic Imaging

We have discussed techniques to improve SNR in holographic imaging. Panoramic imaging in real space can either be performed by zooming to a larger field of view and losing spatial resolution, or by creating a mosaic of images at a higher resolution. Fourier transform holography allows for panoramic imaging without compromising spatial resolution (Goodman and Gaskill 1969) by using multiple objects and references. For each region-of-interest (ROI) on the sample, an individual reference is placed at a distance smaller than the field-of-view of the detector. The arrangement is such that the ROI images do not overlap but tile up in the autocorrelation upon inversion of the hologram. The concept is illustrated in Figure 38.11.

The ROIs have to be captured by the x-ray beam but in principle can be separated over large distances. Schlotter et al. (2007) proposed using panoramic imaging to capture ultrafast temporal evolutions spanned across the ROIs and transforming

the high-spatial resolution into a high-temporal resolution. Such a configuration can be realized by a cross-beam geometry as illustrated in Figure 38.11. Suppose an ultrafast optical or near-optical pump pulse excites the sample under an angle α over a distance d establishing a timeline

$$\Delta t = t_{max} - t_0 = \frac{d \cos \alpha}{c}, \qquad (38.35)$$

where $c \approx 300$ nm fs^{-1} is the speed of light. The excited system along the ROIs can be simultaneously recorded with a subsequent single-shot probe pulse. This removes the time jitter between the pump and the probe pulse. A dense positioning of ROIs further reduces the uncertainty of finding $t_0 = 0$. In the experimental proof-of-principle, the ultrafast stopwatch could stretch over a distance of 180 μm which corresponds, at an incident angle of $\alpha = 30°$, to a timeline of $\Delta t = 520$ fs. The pixel resolution in the autocorrelation of the hologram, Equation 38.25, sets the ultimate time resolution and therefore may reach even sub-fs time resolutions. Hence, the time resolution is limited by the femtosecond pulse durations achievable with upcoming x-ray lasers.

38.4.2 Time-Delay X-Ray Holography

A femtosecond short, intense x-ray pulse can also be used to trigger dynamics in the sample. By reflecting the x-ray pulse from a mirror at a distance l back onto the sample, Figure 38.12a, the beam will diffract from the sample another time with a delay of $\Delta t = 2l/c$. Both diffracted waves, the prompt and the delayed one, interfere and form a time-delay hologram of the sample. Chapman et al. (2007) used this geometry to investigate the explosion of polysterene nanospheres by a focused VUV FEL pulse at a 32 nm wavelength. The particles of 140 nm in diameter were dispersed on a 20 nm thick silicon nitride membrane. The multilayer mirror was slightly tilted against an array of membranes providing variable distances $l = 30$–1200 μm which correspond to delays of 200 fs − 8 ps. Here, the time-delay and the structural evolution of the particles after the pumping are encoded in the hologram of Figure 38.12b. They can be quantitatively determined from the observed Newton fringes (Chapman et al. 2007). Spatial information is contained in the speckles that are visible in the fringe pattern and can be unfolded using phase-retrieval methods.

An expansion of the spheres can be observed from the narrowing of the intensity envelop at longer time-delays, see Figure 38.12b. Furthermore, the expansion and the consequent explosion of the particles change its refractive index toward unity, which causes measurable phase shifts in the hologram. From the observed phase shifts at 350 fs time delay in Figure 38.12b, Chapman et al. (2007) were able to conclude that the nanospheres expand no more than 6 nm, while after several picoseconds, the size of the particles increased by more than 40%. These first results with exceptional time and spatial resolution serve as benchmarks for soft x-ray microscopy of cells with ultrashort intense pulses to overcome radiation-damage limits (Neutze et al. 2000),

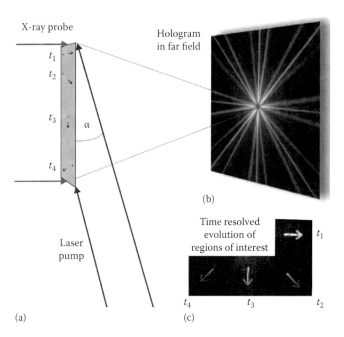

FIGURE 38.11 (a) Ultrafast time-resolved imaging in a single shot using a cross-beam geometry. The pumped region extends over a large area spanning different states of the evolving sample which (b) are captured in a single hologram. (c) By strategically placing the reference holes, the individual states can be rearranged in the autocorrelation without compromising the spatial resolution. (Reprinted from Schlotter, W. F. et al., *Opt. Lett.*, 32, 3110, 2007. With permission.)

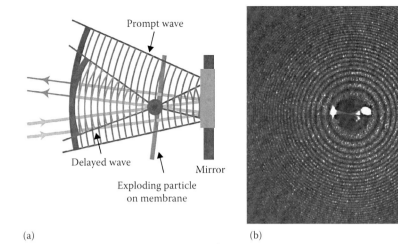

(a)   (b)

FIGURE 38.12   Time-delay ultrafast x-ray holography: (a) A polysterene sphere is pumped with a focused x-ray pulse ($(0.5 \pm 0.2) \times 10^{14}$ W cm^{-2}, 25 fs). The pulse is reflected back onto the exploding particle at a delay time $t = 2l/c$. (b) Both, delayed and prompt diffracted waves propagate together, forming Newton fringes on the detector (time delay 348 ± 1 fs, $q = 4.5\,\mu$m^{-1}). The rings reveal speckles because of the local arrangement of the dispersed spheres on the silicon nitride membrane. (Reprinted from Chapman, H. N. et al., *Nature*, 448, 676, 2007. With permission.)

see also Chapter 39. The resolution of such structure determination depends on how far the atoms of the sample drift during the ultrashort x-ray pulse. With an expansion of 0.4 nm during a 25 fs x-ray pulse, this goal appears feasible in the future. Since biological specimens contain mainly low-Z elements, the scattering contrast at hard x-rays energies is low and achieving atomic resolution is challenging. A holographic approach could provide a path to efficiently enhance the signal by linking high-Z nanostructures like Au nanospheres as strong reference scatterers to macromolecules (Shintake 2008).

38.5 Outlook

The development of x-ray holography for nanostructure imaging was the focus of this chapter. In parallel, coherent diffractive imaging based on iterative phase reconstruction has rapidly evolved since its first experimental demonstration by Miao et al. (1999). Both x-ray holography and coherent diffractive imaging complement one another. As the resolution in x-ray holography is limited by the fabrication of suitable reference structures, phase retrieval algorithms can extend the resolution closer to the diffraction limit. The algorithms themselves greatly benefit from the unique and the robust holographic image reconstruction which can be used as an initial input. The resolution of both techniques is currently limited by the available coherent x-ray flux and it is anticipated that with the increased brilliance of next-generation x-ray sources, spatial resolution well below 10 nm will be achievable. However, not all systems can be prepared on x-ray-transparent membranes and so holography in reflection geometry is desired. Ultrafast coherent x-ray pulses from VUV and x-ray FELs (Ackermann et al. 2007), as well as novel table-top x-ray lasers (Sandberg et al. 2008, Wang et al. 2008) will revolutionize this field, providing both the time and the spatial resolution to elucidate complex dynamic processes. Other techniques like x-ray photon-correlation spectroscopy

will progress toward much shorter time scales compared to third generation synchrotrons (Grübel et al. 2007). X-ray microscopy of cells and macromolecules with focused, intense x-ray pulses may enable scientists to resolve structures that are prone to radiation damage and cannot be otherwise crystallized. Finally, the capabilities of these techniques depend on the development of fast array detectors to provide a larger dynamic range, faster readout, and gating.

Acknowledgments

The author is grateful for the contributions of William F. Schlotter, Diling Zhu, Benny Wu, Catherine Graves, Jan Lüning, Stefan Eisebitt, Ramon Rick, Kang Chen, Christian Günther, Olav Hellwig, Ian McNulty, Sujoy Roy, and Joachim Stöhr.

References

Ackermann, W., Asova, G., Ayvazyan, V. et al. (2007), Operation of a free-electron laser from the extreme ultraviolet to the water window, *Nat. Photon.* **1**, 336.

Aoki, S., Ichihara, Y., and Kikuta, S. (1972), X-ray hologram obtained by using synchrotron radiation, *Jpn. J. Appl. Phys.* **11**, 1857.

Attwood, D. (1999), *Soft X-Ray and Extreme Ultraviolet Radiation: Principles and Applications*, Cambridge University Press, Cambridge, U.K.

Baez, A. V. (1952a), Resolving power in diffraction microscopy with special reference to x-rays, *Nature* **169**, 963.

Baez, A. V. (1952b), A study in diffraction microscopy with special reference to x-rays, *J. Opt. Soc. Am.* **42**, 756.

Barty, A., Boutet, S., Bogan, M. et al. (2008), Ultrafast single-shot diffraction imaging of nanoscale dynamics, *Nat. Photon.* **2**, 415.

Bragg, W. L. (1939), A new type of 'x-ray microscope, *Nature* **143**, 678.

Bragg, W. L. (1942), The x-ray microscope, *Nature* **149**, 470.

Chao, W., Harteneck, B. D., Liddle, J. A., Anderson, E. H., and Attwood, D. T. (2005), Soft x-ray microscopy at a spatial resolution better than 15 nm, *Nature* **435**, 1210.

Chapman, H. N., Hau-Riege, S. P., Bogan, M. et al. (2007), Femtosecond time-delay x-ray holography, *Nature* **448**, 676.

Eisebitt, S., Lüning, J., Schlotter, W. F. et al. (2004), Lensless imaging of magnetic nanostructures by x-ray spectro-holography, *Nature* **432**, 885.

El-Sum, H. M. A. and Kirkpatrick, P. (1952), Microscopy by reconstructed wavefronts, *Phys. Rev.* **85**, 763.

Fenimore, E. E. and Cannon, T. M. (1978), Coded aperture imaging with uniformly redundant arrays, *Appl. Opt.* **17**, 337.

Fienup, J. R. (1978), Reconstruction of an object from the modulus of its Fourier transform, *Opt. Lett.* **3**, 27.

Fienup, J. R. (1982), Phase retrieval algorithms: A comparison, *Appl. Opt.* **21**, 2758.

Gabor, D. (1948), A new microscopy principle, *Nature* **161**, 777.

Gerchberg, R. W. and Saxton, W. O. (1972), A practical algorithm for the determination of the phase from image and diffraction plane pictures, *Optik* **35**, 237.

Goodman, J. W. (2005), *Introduction to Fourier Optics*, 3rd edn., Roberts & Company Publishers, Englewood, CO.

Goodman, J. W. and Gaskill, J. D. (1969), Use of multiple reference sources to increase the effective field of view in lenseless Fourier transform holography, *Proc. IEEE* **57**, 823.

Grübel, G., Stephenson, G. B., Gutt, C., Sinn, H., and Tschentscher, T. (2007), Xpcs at the European x-ray free electron laser facility, *Nucl. Instrum. Meth. B* **262**, 357.

Günther, C. M., Radu, F., Menzel, A. et al. (2008), Steplike versus continuous domain propagation in Co/Pd multilayer films, *Appl. Phys. Lett.* **93**, 072505.

He, H., Weierstall, U., Spence, J. C. H., Howells, M., Padmore, H. A., Marchesini, S., and Chapman, H. N. (2004), Use of extended and prepared reference objects in experimental Fourier transform x-ray holography, *Appl. Phys. Lett.* **85**(13), 2454–2456.

Hendrickson, W. A. (1991), Determination of macromolecular structures from anomalous diffraction of synchrotron radiation, *Science* **254**, 51.

Howells, M., Jacobsen, C., Kirz, J., Feder, R., McQuaid, K., and Rothman, S. (1987), X-ray holograms at improved resolution: A study of zymogen granules, *Science* **238**, 514.

Howells, M. R., Jacobsen, C. J., Marchesini, S., Miller, S., Spence, J. C. H., and Weierstall, U. (2001), Toward a practical x-ray Fourier holography at high resolution, *Nucl. Instrum. Meth. Phys. Res. A* **467**, 864.

Jacobsen, C., Howells, M., Kirz, J., and Rothman, S. (1990), X-ray holographic microscopy using photoresists, *J. Opt. Soc. Am. A* **7**, 1847.

Leith, E. N. and Upnatnieks, J. (1962), Reconstructed wavefronts and communication theory, *J. Opt. Soc. Am.* **52**, 1123.

Leith, E. N. and Upnatnieks, J. (1964), Wavefront reconstruction with diffused illumination and three-dimensional objects, *J. Opt. Soc. Am.* **54**, 1295.

London, R. A., Rosen, M. D., and Trebes, J. E. (1989), Wavelength choice for soft x-ray laser holography of biological samples, *Appl. Opt.* **28**, 3397.

Marchesini, S. (2007), A unified evaluation of iterative projection algorithms for phase retrieval, *Rev. Sci. Instrum.* **78**, 011301.

Marchesini, S., He, H., Chapman, H. N., Hau-Riege, S. P. et al. (2003), X-ray image reconstruction from a diffraction pattern alone, *Phys. Rev. B* **68**, 140101(R).

Marchesini, S., Boutet, S., Sakdinawat, A. E. et al. (2008), Massively parallel x-ray holography, *Nat. Photon.* **2**, 560.

McNulty, I., Kirz, J., Jacobsen, C., Anderson, E. H., Howells, M. R., and Kern, D. P. (1992), High-resolution imaging by Fourier transform x-ray holography, *Science* **256**, 1009.

Miao, J., Charalambous, P., Kirz, J., and Sayre, D. (1999), Extending the methodology of x-ray crystallography to allow imaging of micrometre-sized non-crystalline specimens, *Nature* **400**, 342.

Miao, J., Hodgson, K. O., Ishikawa, T., Larabell, C. A., LeGros, M. A., and Nishino, Y. (2003), Imaging whole *Escherichia coli* bacteria by using single-particle x-ray diffraction, *PNAS* **100**, 110.

Miao, J., Ishikawa, T., Shen, Q., and Earnest, T. (2008), Extending x-ray crystallography to allow the imaging of noncrystalline materials, cells, and single protein complexes, *Annu. Rev. Phys. Chem.* **59**, 387.

Neutze, R., Wouts, R., Spoel, D., Weckert, E., and Hajdu, J. (2000), Potential for biomolecular imaging with femtosecond x-ray pulses, *Nature* **406**, 752.

Robinson, I. K., Vartanyants, I. A., Williams, G. J., Pfeifer, M. A., and Pitney, J. A. (2001), Reconstruction of the shapes of gold nanocrystals using coherent x-ray diffraction, *Phys. Rev. Lett.* **87**(19), 195505.

Rogers, G. L. (1950), Gabor diffraction microscopy: The hologram as a generalized zone-plate, *Nature* **166**, 237.

Sandberg, R. L., Song, C., Wachulak, P. W. et al. (2008), High numerical aperture tabletop soft x-ray diffraction microscopy with 70-nm resolution, *PNAS* **105**, 24.

Scherz, A., Schlotter, W. F., Chen, K. et al. (2007), Phase imaging of magnetic nanostructures using resonant soft x-ray holography, *Phys. Rev. B* **76**, 214410.

Scherz, A., Zhu, D., Rick, R. et al. (2008), Nanoscale imaging with resonant coherent x rays: Extension of multiple-wavelength anomalous diffraction to nonperiodic structures, *Phys. Rev. Lett.* **101**(7), 076101.

Schlotter, W. F. (2007), Lensless Fourier transform holography with soft x-rays, PhD thesis, Stanford University, Menlo Park, CA.

Schlotter, W. F., Rick, R., Chen, K. et al. (2006), Multiple reference Fourier transform holography with soft x rays, *Appl. Phys. Lett.* **89**, 163112.

Schlotter, W. F., Lüning, J., Rick, R. et al. (2007), Extended field of view soft x-ray Fourier transform holography: Toward imaging ultrafast evolution in a single shot, *Opt. Lett.* **32**, 3110.

Schütz, G., Wagner, W., Wilhelm, W. et al. (1987), Absorption of circularly polarized x-rays in iron, *Phys. Rev. Lett.* **58**, 737.

Shapiro, D., Thibault, P., Beetz, T. et al. (2005), Biological imaging by soft x-ray diffraction microscopy, *PNAS* **102**, 15343.

Shintake, T. (2008), Possibility of single biomolecule imaging with coherent amplification of weak scattering x-ray photons, *Phys. Rev. E* **78**, 041906.

Stöhr, J. and Siegmann, H. C. (2006), *Magnetism*, Springer-Verlag, Berlin Heidelberg, Germany/New York.

Stroke, G. W. (1965), Lensless Fourier transform method for optical holography, *Appl. Phys. Lett.* **6**, 201.

Stroke, G. W. and Falconer, D. G. (1964), Attainment of high resolutions in wavefront-reconstruction imaging, *Phys. Lett.* **13**, 306.

Stroke, G. W., Brumm, D., and Funkhouser, A. (1965a), Three-dimensional holography with "lensless" Fourier-transform holograms and coarse p/n polaroid film, *J. Opt. Soc. Am.* **55**, 1327.

Stroke, G. W., Restrick, R., Funkhouser, A., and Brumm, D. (1965b), Resolution-retrieving compensation of source effects by correlation reconstruction in high resolution holography, *Phys. Lett.* **18**, 274.

Stroke, G. W., Restrick, R., Funkhouser, A., and Brumm, D. (1965c), Resolution-retrieving source-effect compensation in holography with extended sources, *Appl. Phys. Lett.* **7**, 178.

Tegze, M. and Faigel, G. (1996), X-ray holography with atomic resolution, *Nature* **380**, 49.

Thibault, P., Dierolf, M., Menzel, A., Bunk, O., David, C., and Pfeiffer, F. (2008), High-resolution scanning x-ray diffraction microscopy, *Science* **321**, 379.

van der Veen, F. and Pfeiffer, F. (2004), Coherent x-ray scattering, *J. Phys.: Condens. Matter* **16**, 5003.

VanderLugt, A. (1964), Signal-detection by complex spatial-filtering, *IEEE Trans. Inform. Theor.* **10**, 139.

Wang, Y., Granados, E., Pedaci, F. et al. (2008), Phase-coherent, injection-seeded, table-top soft-x-ray lasers at 18.9 nm and 13.9 nm, *Nat. Photon.* **2**, 94.

Winthrop, J. T. and Worthington, C. R. (1965), X-ray microscopy by successive Fourier transformation, *Phys. Lett.* **15**, 124.

Winthrop, J. T. and Worthington, C. R. (1966), X-ray microscopy by successive Fourier transformation ii. An optical analogue experiment, *Phys. Lett.* **21**, 413.

Single-Biomolecule Imaging

Tsumoru Shintake
RIKEN SPring-8 Center

39.1 Overview of Single-Biomolecule X-Ray Diffraction Imaging

When you take a photo of your friend using a digital camera, you obtain a 2D projection image of a 3D object. Figure 39.1 shows a photo of wire art. From this projection image, it is rather hard to understand the overlapping structure. To obtain a complete 3D structure information, we have to take multiple photos from different directions as shown in Figure 39.2. Combining those photos, we may construct a 3D structure information, that is, a three-dimensional coordinate of all winding wires. Proteins are linear polymers built from 20 different L-α-amino acids, and thus we may consider them a sort of wire art. However, proteins are nanometers in size and much more complicated.

In this chapter, we aim to determine the 3D structure of a biomolecule, whose physical size is about a few nm to a few 100 nm, with a resolution in a few Å (0.1–0.3 nm). Biomolecules are nucleic acids, proteins, carbohydrates, and lipids. A structural study of a virus is also our scope, whose size is 10–300 nm. We use a light wave to observe the structure, but its wavelength

is around 1 Å and it is called x-ray, which is required to meet atomic resolution. There are big technical challenges; there is no optical lens or mirror for x-ray with a sufficient deflecting angle to satisfy Å resolution optics, and biomaterials are fairly transparent against x-ray since they mostly consist of light atoms; hydrogen, carbon, nitrogen, and oxygen. Thus the interaction of a single biomolecule with x-ray is fairly weak. As you know, in hospitals, x-ray has been used in radiography to see inside one's body, most commonly to view bones. Mussels, made by biomolecules, are fairly transparent. Therefore, the available x-ray signal from the biomolecule is very weak (Figure 39.3).

In order to overcome these difficulties, the crystallography technique has been applied to analyze the 3D structure of a biomolecule, as schematically shown in Figure 39.4. Crystal structures of proteins began to be solved in the late 1950s, beginning with the structure of myoglobin by Max Perutz and Sir John Cowdery Kendrew, for which they were awarded the Nobel Prize in Chemistry in 1962. Using crystal, a large number of identical molecules of the same orientation are arranged at regular spaces. Thus, weak diffraction power from each molecule is dramatically

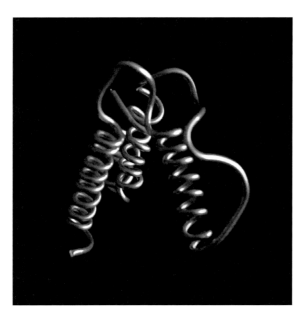

FIGURE 39.1   Wire art. This is an example of a 2D projection image of a 3D object taken by a digital camera.

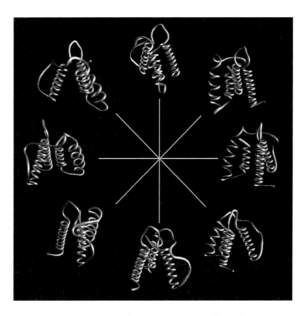

FIGURE 39.2   By rotating the wire art, you will understand the 3D structure.

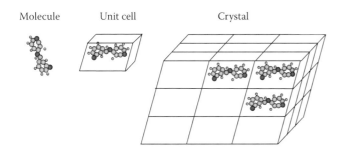

FIGURE 39.3   Crystal structure of a molecule. A crystal arranges a large number of molecules in the same orientation.

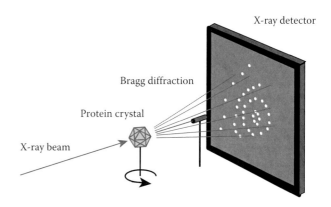

FIGURE 39.4   Setup for taking x-ray diffraction data in x-ray crystallography.

amplified by interference effect. This is called Braggs diffraction that can be easily observed as many intense spots in a 2D x-ray detector. By analyzing the position and intensity of these Bragg's diffraction spots and by applying an additional process to determine the relative phase of these spots, we can reconstruct the image of a target molecule. By rotating crystal, we can collect many projection images, from which we can construct a 3D structure of a sample in Å resolution. The rotation of crystal is also required to satisfy all Bragg's condition on Ewald sphere and to fill diffractions in reciprocal space (see later sections).

Figure 39.5a shows an example of Bragg's diffraction spots from a protein crystal. Figure 39.5b shows an example of the solved 3D structure of a protein. It is named "rhodopsin," and plays a pivotal role in viewing eye retina ranging from invertebrates to mammals (Palczewski et al., 2000). The structure is well determined by x-ray crystallography: the typical seven-transmembrane helices with many bends and eight successive short helices (bottom). The crystal structure of a bovine rhodopsin first revealed these typical features, including the photo-absorption pigment retinal (balls at the center) as a Vitamin E derivative. The photo-sensitivity of rhodopsin is very high and it has been thought to detect only five photons or less in a retina. The rhodopsin has great physiological importance: it is a member

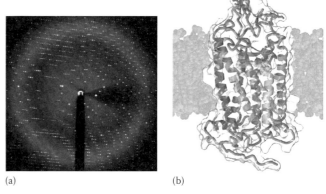

(a)   (b)

FIGURE 39.5   (a) An example of Bragg diffraction spots from a protein crystal. (b) An Example of solved rhodopsin by x-ray crystallography. (Courtesy of Dr. Masashi Miyano, RIKEN SPring-8 Center.)

of the largest protein family, namely, GPCR (G-protein coupled receptor), which is one of the most important components of cellular signaling, and which is also a therapeutic target for half of all prescribed drugs because of its diversified signal acceptation from hormone to virus like the HIV virus in humans.

As seen in this example, very detailed structure information is required in physiological study. Even if we can take the image of a single biomolecule at several Å resolutions but cannot create a 3D structure, there will not be much useful information available from this work. Our ultimate target is to determine the 3D coordinate of each atom in a single biomolecule in Å resolution.

Protein crystallography has advanced through the use of high-quality x-ray beams generated by synchrotron light sources (Dauter, 2006), and especially with the development of higher-brightness x-ray beams produced by an undulator device placed in low-emittance electron storage rings (Nave, 1999). The increase in the brightness of the x-ray beams has made it possible to increase the flux incident on small protein crystals. Currently, crystals with a diameter of 100 μm are being routinely analyzed, and the analysis of 10 μm diameter crystals is under investigation. It is comparatively easier to form high-quality small crystals than larger crystals with a lesser number of defects.

Solved structures are usually deposited in the protein data bank (PDB), a freely available resource from which structural data about thousands of proteins can be obtained in the form of Cartesian coordinates for each atom in the protein. The data are submitted by biologists and biochemists from around the world, and released into the public domain. The PDB database is updated weekly, and its statistics can be found in the PDB holding list (http://www.rcsb.org/pdb/statistics/holdings.do). Most structure data are determined by x-ray crystallography, and about 15% of structures are determined by protein NMR spectroscopy. The number of proteins in the PDB has grown exponentially; 7263 structures were added in 2007.

Membrane proteins form a large component of the genomes and include several proteins of great physiological importance, such as ion channels and receptors (Lundstrom, 2006). However, the crystallization of membrane proteins is extremely challenging; therefore, in PDB databases, membrane proteins of known 3D structure are of only 174 kinds.

A structural analysis of GPCR is the most important target for drug development. If we can detect a single biomolecule GPCR without crystallization, it will be an extraordinary breakthrough in this field. To do this, we need an extremely strong x-ray beam to illuminate a single molecule and also enough intensity of x-ray scattered from a single molecule. Recently, the practical applications of single-pass free-electron lasers (FELs) (Bonifacio et al., 1984; Ackermann et al., 2007) at short wavelengths have increased as a result of long-term research and development of important devices and technologies such as low-emittance electron sources, accurate undulator devices and high-precision electron-beam control techniques. Currently, three large x-ray FELs emitting 1 Å x-ray laser beams are under construction and will be ready in the next few years (Arthur et al., 2002; Tanaka and Shintake, 2005; Altarelli et al., 2006).

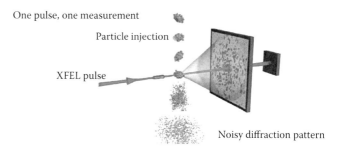

FIGURE 39.6 Single molecule imaging with XFEL. Injecting a molecular beam into a vacuum, and illuminating with an x-ray beam from a normal direction. With ultra-short XFEL pulse, the diffraction image of the molecule can be obtained before the Coulomb explosion. (From Janos, K., *Nat. Phys.*, 2, 799, 2006. With permission.)

Since x-ray FELs produce extremely high peak power with a short-pulse duration, the potential use of these devices to determine the structure of a single biomolecule or individual cells from scattered photons at one shot has been studied (Neutze et al., 2000). The idea is shown in Figure 39.6. We inject a beam of single molecules into a vacuum, followed by illumination with an intense x-ray beam. With an ultrashort XFEL pulse, the diffraction image of the molecule can be obtained before the Coulomb explosion. By numerical process of the so-called phase retrieval and Fourier transform (see later sections), we obtain a projection image of a single molecule. By repeating this process for a large number of molecules, adjusting the position and the rotation by least-mean-square process, and averaging data on reciprocal space, we finally reconstruct the 3D structure of the molecule.

39.2 Resolution Limit of Observation with Light

39.2.1 Optical Microscope

In observing small objects using microscopes, the spatial resolution is limited by imperfections in the optical system, i.e., misalignment and defect in lens. Even in a perfect optical system, there is a fundamental maximum to the resolution that is due to diffraction of light. The resolution (minimum distance between two adjacent objects which can be distinguished) of a given instrument is proportional to the aperture of the objective lens, and inversely proportional to the wavelength of the light being observed.

$$\delta \approx \frac{\lambda}{2n\sin\alpha} = \frac{\lambda}{2NA} \tag{39.1}$$

where
λ is the wavelength of light
n is the diffractive index of the media around the object
α is the half angle of the maximum cone of light that can pass the lens
NA is the numerical aperture of the objective lens: $NA = n \sin \alpha$ (refer Figure 39.7)

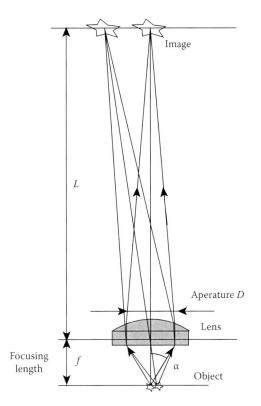

FIGURE 39.7 Resolution is defined by wavelength and the numerical aperture of the lens (opening angle) in an optical microscope.

Due to the limitations of the values α, λ, and n, the resolution limit of a light microscope using visible light is about 240 nm in air. This is because: α for the best lens is about 70° (sin α = 0.94), and the shortest wavelength of visible light is blue (λ = 450 nm). Extending L gives a higher image-magnification factor, $M = L/f$, while the resolution remains the same as defined by Equation 39.1.

Using shorter wavelength radiation, we obtain higher resolution in principle. However, the radiation wavelength falls to below 100 nm, the photo-absorption process in air and water dominates, and it becomes hard to realize optical microscopes at this wavelength.

39.2.2 X-Ray Diffraction

At even shorter wavelengths, i.e., λ < a few nm, the electromagnetic radiation becomes again transparent, being able to propagate in air without substantial attenuation.

Scientists have been using x-ray at 1.54 Å (a line spectral of characteristic x-ray at a K_α line of copper atom from an x-ray tube using copper anode). Ways of observing small objects in optical light and at x-ray wavelengths are different. However, on resolution, the same principle works. Using 1.5 Å x-ray, and an aperture angle of 30° (NA = sin α = 0.5), from Equation 39.1, we have the spatial resolution $\delta \sim \lambda$ = 1.5 Å. This is the comparable distance between C–C in ethane. The resolution of 1.5 Å is enough for a 3D structural analysis for most proteins, and other macromolecules.

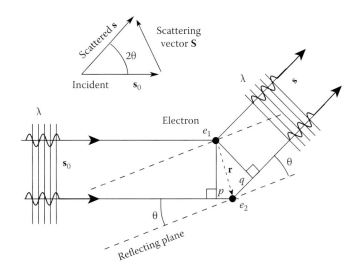

FIGURE 39.8 Scattering vector **S** in a two electron system.

Figure 39.8 shows the scattering of x-ray by two electrons: e_1 and another electron e_2 at position **r** with respect to e_1. The phase difference between the two beams scattering each electron is

$$\Delta\phi = -\frac{2\pi(p+q)}{\lambda} = -2\pi \cdot \mathbf{r} \cdot (\mathbf{s}_0 - \mathbf{s}) = 2\pi \mathbf{r} \cdot \mathbf{S} \qquad (39.2)$$

where **S** is the scattering vector: $\mathbf{S} = \mathbf{s} - \mathbf{s}_0$, whose length is

$$S = |\mathbf{S}| = \frac{2(\sin\theta)}{\lambda} \qquad (39.3)$$

Note that the inverse of Equation 39.3 is equal to Equation 39.1. Both equations explain that the resolution limit is defined by the maximum diffraction angle of light that can be captured by the detector, or by an observer in the ordinal microscope.

39.3 Lens-Less Diffraction Microscopy

39.3.1 From the Optical Microscope to the Lens-Less Diffraction Microscope

Figure 39.9 shows a comparison of the optical microscope and the lens-less x-ray diffraction microscope. In the optical microscope, the object is usually illuminated by white light, which is incoherent. The reflected wave from the object surface, or the transmitted wave through the transparent object, reaches the lens. Because lens is made of glass, whose refractive index is higher than in air, a refraction on rays is caused, resulting in a focus on the real object.

According to the Huygens–Fresnel principle, the propagating wave can be formulated by a series of wave-fronts, where a fresh new wave-front is created by adding many wavelets from each point right behind the old wave-front. At each step, the wavelets interfere and constructively add for the same phase and cancel between wavelets with a π-phase difference. When it arrives at the lens surface, the wavefront at the interface starts

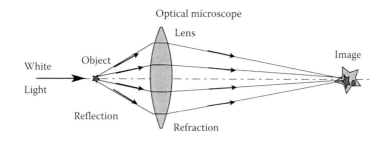

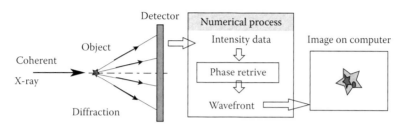

FIGURE 39.9 Optical microscope and lens-less x-ray diffraction microscope.

to bend because the wave speed is slower inside the glass. As a result, it causes refraction, and the rays are focused. Hence, the effective functioning of the focusing lens lies in its application of the phase shift on the wavelet according to its radial position (Schnars and Jueptner, 2005),

$$\Psi_2(r) = \Psi_1(r) \exp\left[i\frac{\pi r^2}{\lambda f}\right] \qquad (39.4)$$

where
   r is radial position on the lens
   f is its focusing length
   $\Psi_1(r)$, $\Psi_2(r)$ are the wave functions at entrance and exit surfaces on the lens

After passing this phase shifter, the wavefront starts to focus and creates the image at the downstream. This equation teaches that once we have the incoming wave information (amplitude and phase), we may obtain the output wavefront by Equation 39.5, followed by focusing the wave on the object.

This may be done through the numerical process as shown in Figure 39.9, which is called lens-less diffraction microcopy. Unfortunately, there is no wave detector to determine the amplitude and the phase. Instead, we obtain only two-dimensional amplitude data from a detector, such as the CCD cell. Therefore, we need to retrieve phase information from the amplitude data. To do this, iterative routines have been developed, which is discussed in the later sections. In reality, we do not use Equation 39.4, but we use the Fourier transform to reconstruct images.

The phase-retrieval process requires the illuminating light to be monochromatic and coherent, and thus we need laser. At the moment, there is no phase-retrieval routine for white-light illumination.

39.3.2 Diffraction from a Small Object

We learned about the two-electron system in Section 39.2.2. If we repeat the same calculations for many electrons, we may compute the diffraction from a complicated system. X-ray diffraction from a single molecule is given by integrating the scattering contributions from all electrons in the molecule by taking into account the phase difference as follows (Drenth, 1994):

$$F(\mathbf{S}) = \int \rho(\mathbf{r}) \cdot \exp[2\pi i \mathbf{r} \cdot \mathbf{S}] \cdot d\mathbf{r}$$

$$= \iiint \rho(x, y, z) \cdot \exp[2\pi i (x \cdot S_x + y \cdot S_y + z \cdot S_z)] \cdot dx\, dy\, dz$$

$$(39.5)$$

This equation describes that the diffraction power into the direction of $\mathbf{S}$ is given by integrating the electron charge-density by taking account the phase term. If you look carefully at this equation, you will find that the formula is identical to the Fourier transform (three-dimensional Fourier transform with a complex number). This Fourier transform computes the spatial components on the electron distribution including phase (since F is a complex number). According to crystallography theory, the three-dimensional space, (S_x, S_y, S_z), is called reciprocal space, since it is reciprocal to the real space and has a dimension of the inverse distance: Å^{-1}.

By multiplying the phase term of $\exp[-2\pi i \mathbf{r} \cdot \mathbf{S}]$ from both sides on Equation 39.5, and by integrating on $\mathbf{S}$, we find,

$$\rho(x, y, z) = \int F(\mathbf{S}) \cdot \exp[-2\pi i \mathbf{r} \cdot \mathbf{S}] \cdot d\mathbf{S}$$

$$= \iiint F(S_x, S_y, S_z) \cdot \exp[-2\pi i (x \cdot S_x + y \cdot S_y + z \cdot S_z)] \cdot dS_x dS_y dS_z$$

$$(39.6)$$

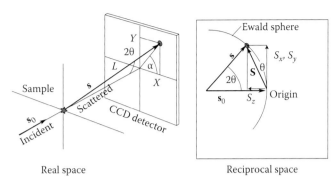

Real space   Reciprocal space

FIGURE 39.10 Coordinate in diffraction experiment and reciprocal space.

This equation tells us that in order to obtain a three-dimensional charge-density profile, we need to know three-dimensional spatial components, $F(S_x, S_y, S_z)$, on the reciprocal space. If we obtain $F(S_x, S_y, S_z)$ from diffraction experiments, a simple Fourier transform gives the charge density.

Figure 39.10 shows the coordinate in the diffraction experiment.

$$\mathbf{S} = \mathbf{s} - \mathbf{s}_0 = S\left[\mathbf{e}_x \cos\theta \cdot \cos\alpha + \mathbf{e}_y \cos\theta \cdot \sin\alpha - \mathbf{e}_z \sin\theta\right] \quad (39.7)$$

The relation between the pixel position of the CCD detector and the scattering angle is given by

$$X = L\sin 2\theta \cdot \cos\alpha$$
$$Y = L\sin 2\theta \cdot \sin\alpha \quad (39.8)$$

where L is the distance from the sample to the CCD detector.

The scattering angle is given by

$$2\theta = \sin^{-1}\left(\frac{\sqrt{X^2 + Y^2}}{L}\right)$$
$$\alpha = \tan^{-1}\left(\frac{Y}{X}\right) \quad (39.9)$$

We obtain a diffraction image on a CCD detector, which is the 2D intensity map: $I(X, Y)$, followed by the application of the phase retrieval process (described later), through which we obtain the Fourier component $F(S_x, S_y, S_z)$. We have to note that the two-dimensional data set $I(X, Y)$ is converted into the two-dimensional data set on a sphere as shown in Figure 39.10, on the right side, which is called Ewald sphere. The relation between the pixel coordinate $I(X, Y)$ and the reciprocal space (S_x, S_y, S_z) is given by Equations 39.7 through 39.9. To fill up all data points in the reciprocal space, we need to rotate sample azimuth and also altitude angles, as we did in Figure 39.2.

39.4 Phase Retrieval

In the previous section, we learned that the electron density is obtained by the Fourier transformation of the Fourier component, $F(S_x, S_y, S_z)$, which is a complex number, that is, the

amplitude and the phase. However, what we obtain in the diffraction experiment is $I = |F|^2$, where the phase information is missing. Therefore, we need to retrieve the phase from the intensity data.

39.4.1 History (Robinson and Miao, 2004)

It was first suggested by Sayre in 1952 that possessing additional measurements of the Fourier magnitudes for a crystal between the Bragg peaks might provide phase information (Sayre, 1952). In 1982, using the autocorrelation function of object, Bates concluded that the retrieval of the phases from the diffraction intensities required twice finer over-sampling of the intensities than the spatial frequency in the diffraction pattern in each dimension (Bates, 1982). Millane generalized the over-sampling criterion into three dimensions (Millane, 1996). Miao et al. found that both Bates' and Millane's criteria were overly restrictive and proposed a different justification whereby the product of the over-sampling ratio in three dimensions should be greater than two (Miao et al., 1998).

Figure 39.11 shows the two-dimensional fast Fourier transformation (FFT). Applying the FFT on a large object as in Figure 39.11a, the Fourier component (diffraction) shows a high frequency speckle pattern, whose phase information is fairly complicated. When the object size becomes smaller as in Figure 39.11b, the Fourier amplitude and its phase become smoother. Object shapes are the same in both cases, but their Fourier components are totally different. The oversampling criterion is satisfied in Figure 39.11b, but not in Figure 39.11a.

The uniqueness of the phase solution was discussed theoretically by Bruck and Sodin (1979), where it has been shown that while there are multiple solutions for 1D objects, multiple solutions are rare for 2D and 3D objects. However, no analytical solutions have yet been developed.

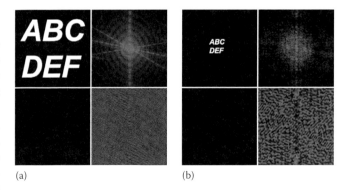

(a)   (b)

FIGURE 39.11 Four images are in one set: where two images on the left half represent the real and the imaginary parts of the object, and in the right half, the two images are the amplitude and the phase of Fourier components (diffraction wave). (a) When the object size is large, the Fourier component shows a very complicated structure. (b) When it becomes smaller, surrounded by empty space, the Fourier components become simple and smooth. Each image size is 256×256 pixels.

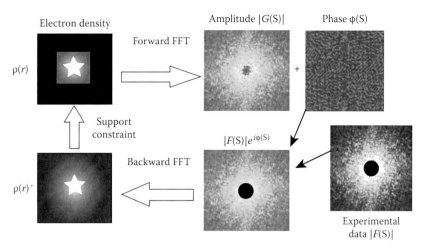

FIGURE 39.12 Iterative phase-retrieval process.

39.4.2 Iterative Phase Retrieval

At the moment, the most effective way is the iterative algorithms, which was developed by Fienup in 1978 improving on the method of Gerchberg and Saxton (Gerchberg and Saxton, 1972; Fienup, 1978). The algorithm iterates data back and forth between the real and the reciprocal spaces as shown in Figure 39.12. The process consists of the following four steps: (1) Prepare a complex array (reciprocal-space), using the square root of experimental-diffraction data as the Fourier amplitude. For the initial cycle, a random-phase array is used. (2) Apply the FFT. The electron-density distribution is obtained. (3) Apply a "support constraint" on the previous electron density, that is, the electron density outside the support, and the negative-electron density inside the support is pushed to zero. (4) Apply the FFT on the electron-density array. A complex reciprocal array: amplitude and phase is obtained. Keep the phase array (with the phase on the central pixel at zero), and replace the amplitude with the experimental data. Repeat the process above until it reaches conversion. The hole in the center is the x-ray block for the direct beam.

The shape of the support is fairly important to this process. It is essential to determine the size and the shape of the support, which can be obtained from external sources or may have to be "learned" using algorithms (Marchesini et al., 2003).

39.4.3 First Experiment on X-Ray Diffraction Microscopy

The first experimental demonstration of x-ray diffraction microscopy was carried out by Miao et al. in 1999, where the phase for a diffraction pattern of a nano-crystalline sample was determined by the iterative algorithm (Miao et al., 1999). Since then, the method has been successfully applied to image a variety of samples and nanocrystals. Figure 39.13 shows one of the works in the early stages of development (Miao et al., 2002). A double-layered sample was prepared, fabricated in Ni, by electron-beam lithography on a Si_3N_4 membrane, where

the two layers of the pattern had the same structure but were separated by a 1 μm thick polymer film and rotated 65° relative to each other. On observation through a scanning electron microscope (SEM), it was seen that the image of the sample (Figure 39.13b) showed only the top pattern, while the second pattern was blurred. The sample was illuminated with a coherent x-ray provided by a third generation synchrotron beam line at SPring-8, and diffraction data was collected. Using the iterative algorithm, the phases were retrieved from the oversampled intensities, and the image was reconstructed as shown in Figure 39.13b. The resolution of the image was ~8 nm, which was determined by the upper range cutoff of the diffraction pattern. Due to the longer penetration length of x-rays compared to the electrons, both the top and the bottom patterns in the sample were clearly visible.

In order to determine the 3D structure of the sample, 31 diffraction patterns by rotating the sample from −75° to +75° in 5° increments were taken. The 3D structure was reconstructed by a 3D FFT on the assembled 3D diffraction data as shown in Figure 39.13d. This experiment clearly demonstrated the efficiency of the x-ray diffraction microscope in carrying out a 3D structural study on a nanoscale sample.

39.5 Ultrafast Diffraction Imaging with X-Ray FELs

From this section, we learn about the usage of the new technology x-ray FEL, and the associated problems: sample damage, Coulomb explosion and a possible solution by ultrafast imaging.

39.5.1 Interaction of X-Ray with Matter

In order to clearly understand the later discussions, here we learn basic physics on the interactions of x-ray with matter. X-ray is electromagnetic radiation, whose wavelength is in the range of 10–0.01 nm. When an x-ray beam interacts with matter, it is

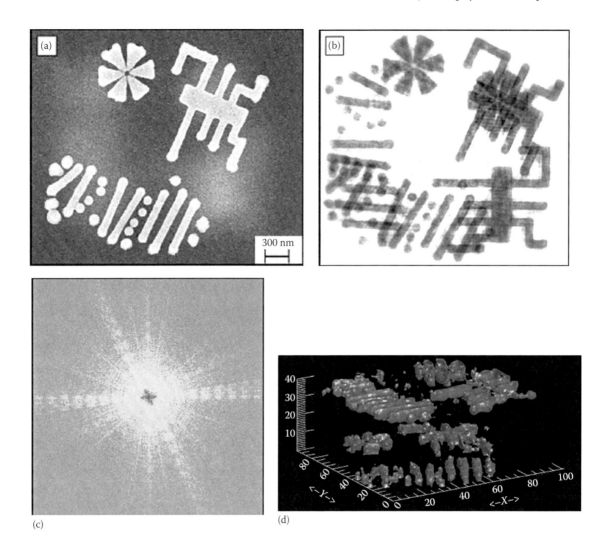

FIGURE 39.13 The x-ray diffraction microscope experiment on a double-layered nanostructure. (a) SEM image, (b) reconstructed x-ray image, (c) the diffraction recorded, (d) reconstructed 3D structure of a Ni sample. (From Miao, J. et al., *Phys. Rev. Lett.*, 89, 88303, 2002. With permission.)

reflected, diffracted, or absorbed. Figure 39.14 shows the dependence of these interactions in a carbon atom (Janos, 2001). When we discuss the interactions of an electromagnetic wave with matter, we use photon energy in an eV unit to specify its frequency, instead of using its wavelength as a parameter. This is because, the transfer of energy from the electromagnetic wave to a matter is quantized at photon energy, which is simply given by the Planck relation

$$E_p = h \cdot \upsilon = \frac{hc}{\lambda} \qquad (39.10)$$

where

h is the Plank constant: $h = 6.626 \times 10^{-34}$ J s $= 4.135 \times 10^{-15}$ eV s
υ is the frequency of the electromagnetic wave
c is the speed of light

X-ray has rather high photon energy; it becomes 1.2–12 keV at the wavelength of 1–0.1 nm. Therefore, the quantum effect plays an important role.

The vertical axis in Figure 39.14 is expressed in the *barn* unit. The incident x-ray beam can be treated as a flow of photons, where the classical meaning of amplitude of the electromagnetic wave is described as the photon number density. The degree of interaction is measured by the crosssection: likelihood of interaction between particles (photon and atoms). The cross section σ is a hypothetical area of atoms facing the flow of photons. A number of scattered or absorbed photons per second are then given by simply multiplying the scattering or absorbing cross section with photon density in the incident x-ray beam. The unit: *barn* of the cross section is $1\ b = 10^{-24}$ cm².

In Figure 39.14, at a lower photon energy (longer wavelength), the photo-absorption dominates (top curve), which decreases as the photon energy increases. A discontinuity at a few 100 eV is the absorption edge, which corresponds to the excitation energy of the inner-shell electron. The coherent-scattering coefficient (Thomson scattering) stays constant at a lower energy but decreases due to competition with the Compton scattering. At the moment, x-ray crystallography is done at a wavelength

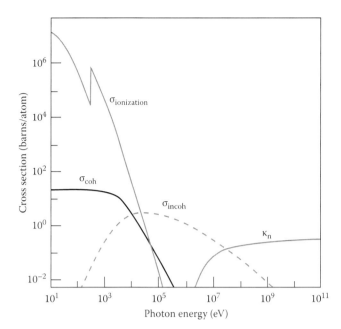

FIGURE 39.14 Interaction of x-ray with carbon. (From Hubbell, J.H. et al., *J. Phys. Chem.*, 9, 1023, 1980.)

of 1–1.5 Å regions, where absorption and Compton scattering problems are moderated, and also importantly the wavelength matches the desired atomic resolution.

Here, we need to know the wave–particle duality of quantum particles; the concept that all matter and energy exhibit both wave-like and particle-like properties. When we describe a small object, quantum mechanics addresses this duality. X-ray plays both the wave-like property as the electro-magnetic wave, and the particle-like property as the photon. Thomson scattering is the elastic scattering of photons by a charged particle. It is sometimes called coherent diffraction as there is no energy loss Thus, the frequencies of incident and the scattered waves are the same, and the phase is kept constant through the interaction. Importantly, in this process, there is no energy transfer from the wave to the matter. Thus, the process is not quantized, and may be included under the classical electromagnetic theory. On the other hand, in photo absorption and Compton scattering the quantum effect has to be taken into account, as it also plays an important role in observing photons on the CCD detector. This is the origin of the statistical fluctuations in our measurement.

The radiation damage problem is originated from the fact that the photo-absorption coefficient is almost ten times higher than the coherent scattering around 10 keV. Only the coherent scattering provides useful information for imaging, while ten times of x-ray photons will be absorbed in the sample that causes the major difficulty in single biomolecule imaging.

39.5.2 Thomson Scattering

When electromagnetic wave incidents to an electron in free space, it accelerates the electron in sinuous motion along the direction of the oscillating electric field, which produces electromagnetic

FIGURE 39.15 Dipole radiation pattern. Snap shot of Radiation2D. (From Shintake, T., New mathematical method for radiation field of moving charge. In *Proceedings EPAC2002*, Paris, France, 2002.)

dipole radiation. Figure 39.15 shows a snapshot of the dipole radiation simulated by a real-time simulator Radiation2D (Shintake, 2002). This software is available at http://ShintakeLab.com, free of charge, which gives a graphical impression of the radiation field from a moving charge and helps to understand various physical concepts: dipole radiation, synchrotron, undulator, thermal radiations, etc.

In dipole radiation (i.e., the Thomson scattering),

1. The radiation power takes the maximum value in a direction perpendicular to motion of the electron, i.e., the direction of the electric field of incident wave.
2. The radiated field is polarized along the direction of the electron motion.
3. The radiation field is proportional to the acceleration on the charged particle, but changes its polarity, thus causing a phase shift by π.

The total radiated power is given by integrating the power flow in all directions, from which the Thomson scattering cross section is calculated

$$\sigma_T = \frac{8\pi}{3}\left(\frac{e^2}{4\pi\varepsilon_0 m_e c^2}\right)^2 = 6.625 \times 10^{-25} \text{ cm}^2 \quad (39.11)$$

where $e^2/(4\pi\varepsilon_0 m_e c^2) = 2.8 \times 10^{-15}$ m is called the *classical electron radius*. We have to note that there is no wavelength term or photon energy in this equation; therefore, it takes a constant value for different wavelengths at lower-photon energies, until it starts competing with Compton's effect as shown in Figure 39.14.

Incidentally, the proton has 1700 times larger mass than an electron and possesses the same amount of charge as the electron of the opposite sign. The Thomson scattering cross section for a proton becomes $(1/1700)^2 \approx 10^{-6}$ times smaller than that of an electron. Therefore, we may neglect the contribution of protons and consider only the interaction with the electron. This is the reason why a discussion, to determine "electron density distributions" through x-ray diffraction experiments, is needed.

39.5.3 Elastic Scattering Factor of Atom

The size of atoms (the diameter of an electron cloud around nuclei) is comparable to an x-ray wavelength. Thus, in order to calculate diffraction from an atom, we need to take into account the electron distribution within the atom. The probability of finding an electron in each orbit is defined by wave function and is localized in a specific orientation according to its spin number. For a rough estimate (Drenth, 1994), we may assume that the radial-density profile dominates the scattering factor and the electron cloud is centro-symmetric around the origin; $\rho(\mathbf{r}) = \rho(-\mathbf{r})$. The atomic scattering factor f becomes

$$
\begin{aligned}
f(S) &= \int_r \rho(\mathbf{r}) \exp[2\pi i \mathbf{r} \cdot \mathbf{S}] d\mathbf{r} \\
&= 2 \int_r \rho(\mathbf{r}) \cos[2\pi \mathbf{r} \cdot \mathbf{S}] d\mathbf{r}
\end{aligned}
\tag{39.12}
$$

As an example, the atomic scattering factor f for a carbon atom is shown in Figure 39.16 as a function of the length of the scattering vector $2(\sin\theta)/\lambda$. A unit of the scattering factor is usually represented by an equivalent number of electrons; thus, at a small scattering angle (forward scattering), it approaches the atomic number Z of carbon. The scattering factor f represents the amplitude of the scattered x-ray with a normalized incident field. Thus the scattering power scales as Z^2. The scattering factor for other atoms and the related parameter can be found elsewhere (Chantler et al., 1995).

As we learned from Equation 39.1, in order to achieve a higher spatial resolution, we need to collect scattered x-ray for a wide angle of θ, i.e., at a larger scattering vector. However, as seen in Figure 39.16, the scattering factor from an atom decays quickly at a larger scattering angle. This is also one obstacle to achieving an atomic resolution in single bio-molecule imaging.

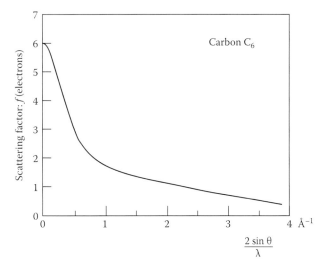

FIGURE 39.16 X-ray scattering factor of a carbon atom.

39.6 Single Biomolecule Imaging with X-Ray FEL

In this section, we discuss the possibility of single biomolecule imaging with x-ray FEL. Because of the extremely high x-ray flux, we will be able to image the molecule through a lens-less diffraction microscope. However, there are two major problems. One is the radiation damage on the sample, and two is the limited number of available photons on the detector.

39.6.1 Sample Damage Problem

The potential for biomolecular imaging with femtosecond x-ray pulses from a future XFEL was first studied by Neutze, Wouts, and Spoel in 2000. Since the elastic scattering x-ray with single molecules is incredibly small, only by using an extremely strong x-ray pulse from the future XFEL would it seem feasible to obtain enough x-ray signal to perform structure analysis. However, the in-elastic scattering crosssection for light material, for example, a carbon, at 12 keV (1 Å wavelength), is ten times higher (refer Figure 39.14), and thus the sample damage by x-ray becomes a serious problem. Analyses of the dynamics of radiation damage formation suggest that the conventional damage barrier may be extended at a very high dose-rate by introducing very short exposure times. Through a computer simulation, which considers in detail the damage associated with photoabsorption, and Auger electron emission and their various effects, it has been demonstrated that coherently scattered photons from a single biomolecule can be collected using femtosecond pulses before atoms start drifting due to Coulomb forces.

Figure 39.17 shows a simulation of the Coulomb explosion of T4 lysozyme under an XFEL beam (Neutze et al., 2000). The integrated x-ray intensity was 3×10^{12} (12 keV) photons per 100 nm diameter spot ($=3.8 \times 10^6$ photons per Å^2). If the XFEL pulse is shorter than 10 fs, x-ray diffraction will be completed before the atoms move. In another simulation of the same integral intensity, with a 50 fs pulse duration, it became difficult to obtain the correct structure data since all atoms moved.

It has been predicted that ultrashort pulses, as low as 20 fs, at least through computer simulations, will be available from x-ray FELs based on today's accelerator technology. A few fs pulses will

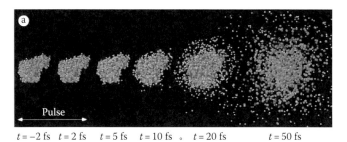

$t = -2\,\text{fs}$ $t = 2\,\text{fs}$ $t = 5\,\text{fs}$ $t = 10\,\text{fs}$ $t = 20\,\text{fs}$ $t = 50\,\text{fs}$

FIGURE 39.17 Explosion of T4 lysozyme induced by x-ray radiation damage. The x-ray pulse length is 2 fs FWHM. The integrated x-ray intensity was 3×10^{12} photons per 100 nm diameter spot. (From Neutze, R. et al., *Nature*, 406, 752, 2000. With permission.)

be challenging, and will require research and development on electron-beam dynamics including x-ray seeding technology.

39.6.2 Limited Number of Coherent Scattering Photons

In order to obtain the diffraction image of a single biomolecule, we have to collect an adequate number of photons on the CCD detector. For example, we need at least an average of a 100 photons per pixel on a 512 × 512 pixel detector; thus the order of 10^7 photons will be required. However, there exits a principal limitation; the number of scattering photons is still very small even if we use x-ray FELs.

The physical origin of this problem is the very small elastic scattering crosssection of an electron. As we learned in the previous section, it was $\sigma_T = 6.625 \times 10^{-25}$ cm^2. By counting the contribution from each electron in one atom, the atomic scattering crosssection becomes larger in forward scattering, $\sigma_A = Z^2\sigma_T$. However, it is still small, for the carbon atom $\sigma_A = 2 \times 10^{-23}$ cm^2. For example, a case of lysozyme, which has a mass of 14.4 kDa, contains about 1000 carbon-like atoms, and thus the total cross section becomes $\sigma_M = N_C Z^2 \sigma_T = 2 \times 10^{-20}$ cm^2.

On the other hand, the photon flux from x-ray FEL is extremely high. The instantaneous peak power reaches to about 100 GW at a 1 Å wavelength, or a 12 keV photon energy, for a 10 fs pulse duration. The accumulated energy in one pulse becomes 1 mJ. Therefore, the number of photons in this pulse is given by

$$N_p = \frac{\tau P}{h\nu} = \frac{W/e}{h\nu/e} = \frac{1 \text{ mJ}/1.6 \times 10^{-19} C}{12 \times 10^3 \text{ eV}} = 5 \times 10^{11} \text{ photons} \quad (39.13)$$

To increase the probability of collision, we focus the FEL beam into a spot of a 100 nm diameter using a spherical mirror in the beam line, and thus the number of scattering photons from the single lysozyme becomes

$$\sigma_s = \frac{N_p \times \sigma_M}{\pi a^2} \approx \frac{5 \times 10^{11} \times 2 \times 10^{-20} \text{ cm}^2}{\pi (50 \text{ nm})^2}$$
$$= 127 \text{ photons} \quad (39.14)$$

This is an extremely small number. If we use a CCD detector of 256 × 256 pixels, the average photon number in one pixel becomes only 0.01, which means that almost all the pixels will be empty (i.e., it is dark).

According to numerical simulation of photon flux available from a single T4 lysozyme under an irradiation of 3×10^{12} (12 keV) photons per 100 nm diameter spot (Neutze et al., 2000), there are high peaks at the center (on the axis), related to the forward diffraction that gives only a total cross section of the sample and the gross shape, while there are no photons found at the higher-diffraction angle, which is related to a higher-spatial frequency and contains a detailed structure inside the lysozyme. Therefore, we cannot obtain useful information from this x-ray diffraction data.

In order to overcome this difficulty, various schemes have been proposed, most of which basically average the diffraction data. Recently, a new scheme using the x-ray heterodyne technique has been proposed (Shintake, 2008), in which the number of x-ray photons can by amplified by attaching a gold particle to a single biomolecule. This idea is described in detail, later.

39.6.3 Demonstration of Ultrafast Diffractive Imaging

Ultrafast diffractive imaging with an intense x-ray pulse from an x-ray FEL was first demonstrated by H. N. Chapman and his co-workers using the FLASH soft-x-ray free-electron laser in 2006 (Chapman et al., 2006).

As discussed in the previous sections, due to the strong photo-absorption and the associated photo-ionization, the sample will cause Coulomb explosion or will turn into plasma under the XFEL pulse. The key issue is how to capture the diffraction image before the atoms start to move toward a Coulomb explosion. A demonstration of the principle was performed using the FLASH soft-x-ray free-electron laser, whose experimental setup is shown in Figure 39.18. An intense 25 fs, 4×10^{13} W/cm^{-2} pulse, containing 10^{12} photons at a 32 nm wavelength, produced a coherent diffraction pattern from a nanostructured, nonperiodic object. The sample is made by a 20 nm thick silicon nitride membrane with a picture milled through it and mounted on the sample plate. The direct beam passes through the sample window and exits through a hole in a graded multilayer mirror. The diffracted light from the sample reflects from this mirror onto a CCD detector. The numerical aperture of the detector is 0.25, and thus the theoretical resolution limit is about 60 nm. The reconstructed image, obtained directly from the coherent pattern by phase retrieval through oversampling, shows no measurable damage, and is reconstructed at the diffraction-limited resolution. In the second shot, a different diffraction pattern was obtained, which indicated that the test pattern was destroyed under FEL power in the first shot (Figure 39.19).

This experiment demonstrated the capability of ultrafast diffraction imaging with XFELs. One can obtain diffraction data right before the explosion using the ultrashort x-ray FEL pulse.

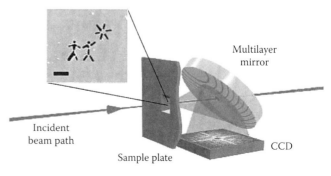

FIGURE 39.18 Schematic diagram of the experimental apparatus of ultrafast diffractive imaging. (From Chapman, H.N. et al., *Nat. Phys.*, 2, 839, 2006. With permission.)

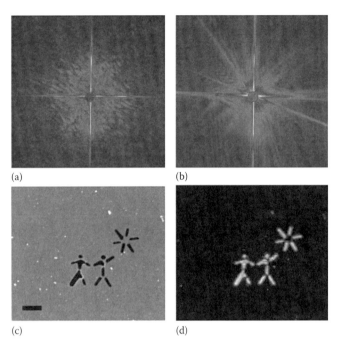

(a)   (b)

(c)   (d)

FIGURE 39.19 (a) Diffraction pattern recorded with a single FEL pulse from a test object (b) The diffraction pattern recorded with a second FEL pulse, showing diffraction from the hole in the sample created by the first pulse power. (c) Scanning electron microscope image of the test object. The scale bar denotes 1 μm. (d) The image reconstructed from the single-shot diffraction pattern shown in (a). (From Chapman, H.N. et al., *Nat. Phys.*, 2, 839, 2006.)

39.6.4 Reconstructing 3D Structure

As predicted by the numerical simulation on x-ray diffraction on a single biomolecule, the available x-ray photons from a single-shot diffraction event is incredibly small.

In order to improve the photon-number statistics and also to fill up the reciprocal space with different orientations, a large number of diffraction data will have to be collected and averaged on the reciprocal space. A 3D structure analysis of a single biomolecule was studied through computer simulation by Miao, Hodgson, and Sayre in 2001 (Miao et al., 2001). It was shown, that a simulated molecular-diffraction pattern at a 2.5 Å resolution, accumulated from multiple copies of single rubisco biomolecules, each generated by a femtosecond level x-ray FEL pulse, can be successfully phased and transformed into an accurate electron density map. In this study, the orientation of the molecule was assumed to be known.

However, in reality, to do so, we need information on molecule orientation. Huldt et al. (2003) presented a possible scheme for classifying the diffractions and determining the orientation from the correlation technique. Figure 39.20 shows a 3D arrangement of the diffractions after classifications. Accumulating many diffraction data on 3D diffraction space and applying possible averaging, followed by 3D FFT, the electron-density map of a single biomolecule can be obtained.

FIGURE 39.20 Three diffraction images that intersect in diffraction space. Each diffraction has a spherical section (Ewalds sphere), shares the same origin, but different angles according to the molecule orientation. (From Huldt, G. et al., *J. Struct. Biol.*, 144, 219, 2003. With permission.)

This technique will be applicable to a large macromolecule and virus, both of which produce a sufficient number of photons, i.e., one photon per pixel on an average at the rim of the CCD. It will be difficult to apply this technique for smaller single-biomolecule imaging, since most of all pixels will be empty.

Classification of diffraction data from a flying molecule has been studied through computer simulation (Fung et al., 2008). By using the appropriate polar coordinate and interpolating the photon number, an efficient classification was demonstrated even with a photon density as low as the 10^{-2} level.

39.6.5 Demonstration of Diffractive Imaging of Nanoscale Specimen in Free Flight

As we learned in the demonstration of the FFT (Figure 39.11), to satisfy the oversampling condition for iterative phase retrieval, it is better to take diffraction from an isolated object in free space. When the specimen size becomes smaller, especially in the single biomolecule case, it will be hard to perform the imaging of single molecules on a substrate membrane. The scattered photons coming out from the substrate will dominate the signal, causing a high background noise, and the phase retrieval will not work. Therefore, technology development of the free flight-imaging technique is the most important issue.

The diffractive x-ray imaging of a nanoscale specimen in free flight was first demonstrated by Bogan et al. in 2008. Figure 39.21 shows the imaging system. (1) The FEL pulses are incident from the right, and pass through an aperture designed to block stray light. (2) They are focused to a 20 μm spot that defines the interaction region. (3) Electrospray-generated aerosols are charge-neutralized via an α–source and delivered into the vacuum through a differential-pumping interface equipped with an aerodynamic focusing system. (4) The pressure in the three pumping regions is set to optimize the focusing of the

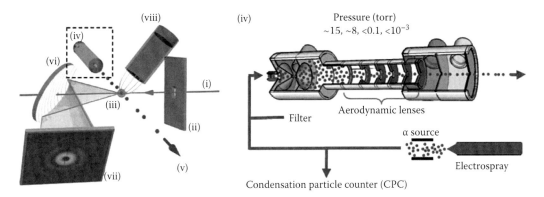

FIGURE 39.21 Experimental apparatus of diffractive x-ray imaging of a nanoscale specimen in free flight. (From Bogan, M.J. et al., *Nano Lett.*, 8, 310, 2008. With permission.)

nanoparticles through an aerodynamic lens stack (green). The resultant continuous stream of nanoparticles travels at ~150 m/s and is confined to a beam of ~250 μm diameter. (5) The nanoparticles are steered into the interaction region using a mechanical translation of the entire differential-pumping interface. (6) X-rays diffracted from the interception of single nanoparticles by the FEL reflect from a graded multilayer planar mirror onto the single photon sensitive x-ray CCD. (7) The direct beam passes through a hole in the center of the mirror. (8) Ions created from the FEL interactions are detected by a miniature TOFMS: time-of-flight measurement system.

Bogan et al. successfully obtained diffractions. Some of them showed a periodic interference pattern, whose image was retrieved clearly and showed two particles attached. This success indicated the feasibility of single particle x-ray diffractive imaging of single biomolecules, viruses or cells in free flight.

TOF: time-of-flight also provided useful information corresponding to the ionization of a particle by XFEL light. A possible combination of the ion-reaction microscope to provide molecule orientation from the knowledge of ion momentum after Coulomb explosion is under development (Ullrich et al., 2003).

39.6.6 Amplification of Weak X-Ray Signal by 2D Crystal Structure

In order to overcome the difficulty associated with the weak scattering power of the single biomolecule, scientists are trying to make a 2D array of the nanostructure, i.e, an artificial 2D-crystal, which will provide strong Bragg's diffractions, from which the structure information in one unit cell will be solved. Mancuso et al. (2009) have demonstrated 2D crystallography using a free-electron laser at an 8 nm wavelength. The finite crystal sample was manufactured by milling holes (a pair of large and small holes) using a focused ion beam (FIB) on a 100 nm thick Si_3N_4 membrane coated with 600 nm of Au and 200 nm Pd. The sample was illuminated with a soft-x-ray beam from the FLSH facility at DESY. The diffraction data showed Bragg peaks corresponding to the 2D array size (1 μm × 1 μm pitch). In between

those peaks, oscillations were observed which corresponded to the internal structure of each cell.

This scheme has a great scope in improving the signal-to-noise ratio of the x-ray diffraction data, as there are a number of copies of the same samples. To realize single biomolecule imaging, future research and development will be required to align all single biomolecules on the membrane in the regular spacing and in the same direction within the accuracy to guarantee the atomic resolution, i.e., a 1 Å displacement and a 1 mrad angle (1 Å for 100 nm full size).

39.7 Coherent Amplification and X-Ray Heterodyne Detection of a Single Biomolecule

39.7.1 Introduction

As we learned in the previous sections, the major difficulty in single biomolecule imaging is the limited number of photons available from coherent scattering with even intense x-ray FEL lights. There were several theoretical considerations to improve the S/N ration by introducing, (1) an artificial structure as a reference wave in holography, (2) multiple holes as the coherent illumination source, (3) averaging diffraction on a large number of events in reciprocal space.

Another new approach using x-ray heterodyne detection was proposed (Shintake, 2008), in which the weak diffraction power from the single biomolecule is coherently amplified and detected by the heterodyne technique using an intense reference wave from a nano-sized gold particle linked to the single molecule. Figure 39.22 shows the schematic setup. The biomolecule, linked with a nano-sized gold particle will be injected into the x-ray beam line from the molecular-beam generator. After illuminating with x-ray FEL pulse, the resulting interference pattern between the intense-reference wave and the weak-object wave is recorded as a hologram on the CCD detector. Since there are 10^3 times more photons than that from a single molecule alone, it is fairly easy to apply the iterative phase-retrieval process and recover the image.

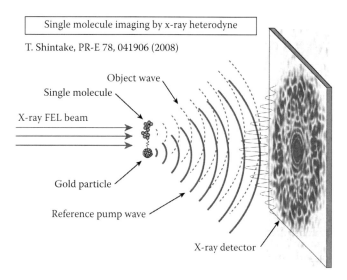

FIGURE 39.22 Single biomolecule imaging with an x-ray heterodyne detection scheme. A nano-sized gold particle provides an intense reference pump wave, which amplifies the weak scattering wave from the single molecule by the x-ray heterodyne scheme.

There are various discussion on this scheme:

1. The atomic structure of the gold particle will vary in each sample, thus making it difficult to classify diffraction images, and to re-align and average them.
2. Linking to a gold particle will possibly affect the conformation of the single molecule.
3. How much can the S/N be improved?

However, the available photon number dramatically increases in this scheme. Thus, any kind of data processing becomes easier. This situation is much better than that without using a gold particle, where most of all the CCD pixels are empty.

The scattering diffraction data from a nano-sized particle was analyzed, and the amplification effect was observed experimentally at FLSH FEL, which strongly suggests that the scattering object will support the increase of the weak signal (Boutet et al., 2008).

39.7.2 Wave Particle Duality

As is well known, light possesses both wave and particle features. This is called wave-particle duality. Light propagates as electromagnetic wave and also runs as a particle of photon. According to Copenhagen interpretation, when a particle is running in free space, the motion can be described by a propagation of the probability wave. When we try observe it, the probability wave suddenly collapses at a point and its energy is transferred into matter by a discrete amount, i.e., by the photon energy. If the wave intensity is low, this process truncates most of all the detailed information carried from a single molecule.

Incidentally, the inelastic scattering or the photo-ionization event is a particle-like phenomenon, where the photon energy plays an important role. On the other hand, the coherent x-ray diffraction, or the elastic scattering can be considered as a wave-like phenomena since there is no energy loss in this process. The wave diffraction can be treated within the classical electromagnetic theory, i.e., Maxwell's equations, without using the quantum mechanical treatment. The amplitude and the phase of the diffraction wave are functions of the incoming x-ray wave multiplied by the electron density distribution in the sample, and importantly they are linear in function, and not a stepwise function as is the case with photon energy. Even though the diffraction wave carries detailed information from the single biomolecule sample, once it is quantized in the photo-ionization event, the detail is lost.

39.7.3 Heterodyne Detection

In this section, we consider how to amplify wave information before quantization in a photo detector. Figure 39.23 shows a modified version of Young's double-slit experiment. The width of slit S_2 is considerably less than that of S_1, such that the intensity (and photon flux) of light passing through S_2 is considerably lower than that passing through S_1. The wave functions of the two beams arriving at the screen are expressed as follows:

$$\psi_1 = A_1 \cdot \exp i(\omega t - kz + \phi_1) \tag{39.15a}$$

$$\psi_2 = A_2 \cdot \exp i(\omega t - kz + \phi_2) \tag{39.15b}$$

These amplitudes are normalized such that the conjugate product $\psi^* \psi = |\psi|^2$ is equal to the photon flux of each beam. Therefore, the probability of finding a photon in the beams is proportional to $I_1 = \psi_1^* \psi_1$ for beam −1 and $I_2 = \psi_2^* \psi_2$ for beam −2.

The two waves interfere and the resulting interference fringes, are observed on the screen, and can be represented by the linear summation of the two wave functions, $\psi = \psi_1 + \psi_2$. The probability of finding a photon on the screen is given by

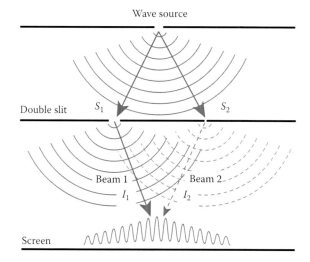

FIGURE 39.23 Modified double-slit experiment. (From Shintake, T., *Phys. Rev. E*, 78, 041906, 2008. With permission.)

$$I = |\psi|^2 = \psi^\star \psi = (\psi_1^\star + \psi_2^\star) \cdot (\psi_1 + \psi_2)$$

$$= |\psi_1|^2 + 2|\psi_1||\psi_2|\cos(\phi_1 - \phi_2) + |\psi_2|^2$$

$$= I_1 + 2\sqrt{I_1 I_2}\cos(\phi_1 - \phi_2) + I_2 \qquad (39.16)$$

In the x-ray diffraction experiment, ψ_1 represents the reference pump wave and ψ_2 represents the diffracted wave from the single biomolecule. The second term in Equation 39.16 represents the interference fringes that contain information about the second slit, i.e., its width, the intensity of the light beam passing through it, and its distance from the first slit.

Next, we discuss information that can be retrieved from the fringe contrast. From the maximum and the minimum amplitudes given in Equation 39.16, we determine the modulation depth as follows:

$$M = \frac{I_{max} - I_{min}}{I_{max} + I_{min}} = \frac{2\sqrt{I_2/I_1}}{1 + I_2/I_1} \qquad (39.17)$$

The modulation depth has a maximum value of 1 when the intensities are equal, i.e., $I_2 = I_1$. By decreasing I_2, the modulation depth decreases, although the rate of this decrease is slower than a first-order dependence, which indicates that the modulation depth or fringe contrast is visible even for very small values of I_2. For example, letting $I_2/I_1 = 0.01$, we find $M = 0.2$, which means only 1% of the second beam creates a 20% modulation amplitude on the main beam.

This phenomenon has been widely used to detect a weak signal in a radio frequency communication system and in an optical laser system; it is called the heterodyne detection method or the homodyne detection method when the frequencies of the two beams are identical. The high-intensity beam is a local oscillator in the case of radio frequency applications or a reference pump beam in the optical laser system.

39.7.4 Gold Nanoparticle Attached to Single Biomolecule

Figure 39.24 shows an example case of a single lysozyme molecule linked with a gold particle. Gold is chosen because it is a heavy atom, the atomic number Z is high, and thus provides an intense reference wave. The gold particle and the single biomolecule correspond to the wide and narrow slits, respectively, in the double-slit experiment described previously. The scattered waves from these two objects interfere and form an interference pattern on the planar detector.

Now, we roughly estimate the required number of atoms in the gold particle. The number of scattering photons under the same flux becomes

$$\frac{n_{p.Au}}{n_{p.C}} = \frac{N_{Au}\sigma_{Au}}{N_C\sigma_C} = \frac{N_{Au}Z_{Au}^2}{N_C Z_C^2}. \qquad (39.18)$$

We approximate the lysozyme molecule as a 1000-atom carbon cluster, denoted by the subscript C. Using $Z_C = 6$, $N_C = 1000$, $Z_{Au} = 79$, and the same number of Au atoms as those in the lysozyme, i.e., $N_{Au} = 1000$, we find the intensity ratio $n_{p.Au}/n_{p.C} \approx 200$. A gold particle in the metallic phase with a diameter of less than 3 nm contains 1000 atoms. Using the latest advanced gold-labeling technology (Hainfeld and Powell, 2000), we can bind a nano-sized gold particle to various types of biomolecules such as proteins, lipids, or ATPs. Products such as Nanogold are manufactured using a well-established technology and are commercially available with diameters ranging from 1.4 to 40 nm. The linking arm of the gold particle can be specifically made to react with thiols; thus, the location of the link to the biomolecule is well-defined.

When an intense x-ray beam with a flux of 3×10^{12} photons per pulse in a 100 nm diameter spot is incident on the gold particle linked to a single lysozyme molecule, we obtain 2×10^4 photons in the scattering angle for a 2–30 Å resolution on the x-ray detector. In order to provide a uniform illumination of the reference pump wave within the planar x-ray detector, a non-crystal structure will be suitable for the gold particle. Intensive research will be required to optimize its metallic structure (amorphous gold, quasi-crystal gold, glassy alloy, etc.).

39.7.5 Demonstration of Coherent Amplification

In order to demonstrate coherent amplification phenomena, a simple experiment using optical laser is shown here. HeNe-laser beam was split into two beams and overlapped again on a CCD detector on a digital camera. Beam-2 was attenuated 200 times to simulate the signal from a single biomolecule.

Figure 39.25 shows the results of CCD images. (a) Beam-1 and (b) Beam-2 are obvious. In (c), the Beam-2 image brightness was enhanced 200 times on the image process software, where a faint image of Beam-2 can be seen while it is buried under the CCD noise. (d) Beam-1 and Beam-2 were overlapped, which created an interference fringe, whose visibility was 15% as expected from Equation 39.17. Thus, Beam-2 information was amplified and recorded as the interference fringe.

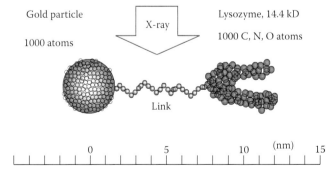

FIGURE 39.24 Example case of a lysozyme linked with a gold particle. Both have 1000 atoms, and provide 200 times more photons. (From Shintake, T., *Phys. Rev. E*, 78, 041906, 2008. With permission.)

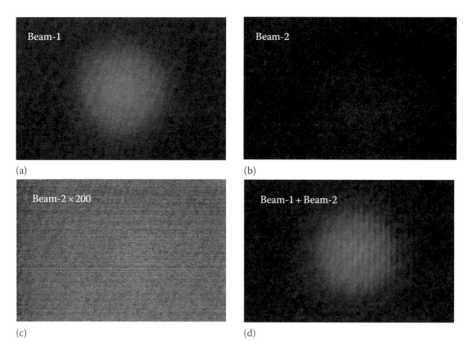

FIGURE 39.25 Images captured on the CCD detector. (a) Beam-1 only, without Beam-2. (b) Beam-2 only. It is 200-times lower than Beam-1, and thus it is dark. (c) Beam-2, with its brightness is 200-times enhanced on computer. (d) Beam-1 + Beam-2 overlapped, clearly generated an interference fringe.

Since an x-ray FEL will provide a coherent x-ray beam, whose coherent length is longer than the expected path-length difference, the same phenomena will happen with a HeNe laser, and we can utilize the coherent amplification phenomena to detect weak scattering x-ray photons.

Figure 39.26a shows the test pattern simulating a single molecule linked with a gold particle. The diffraction image was in Figure 39.26b. The complicated speckle is related to the random atom position in the gold particle, and the fine vertical fringe is the amplified wave from the single molecule. Applying the iterative-phase retrieval, we obtain the original image as shown in Figure 39.26c. Note that, the signal is amplified as the interference fringe and is recorded in diffraction data, but it does not mean that the retrieved molecule image becomes brighter. The amplification phenomena help to enhance the image quality (better resolution). Without the gold particle, we cannot retrieve the molecule image as in Figure 39.26c.

39.7.6 Three-Dimensional Structure Reconstruction

In actual experiments, a large number of biomolecules with gold labeling must be injected into a beam line formed by a spraying technique and illuminated by an x-ray FEL. The orientation of the biomolecules will be random. The experiment is carried out as follows (refer Figure 39.27).

1. Select diffraction images in which the gold particle and the biomolecule lie in the *xy*-plane. This selection can be done by simple FFT on diffraction intensity data.
2. Apply an iterative phase retrieve and the FFT to obtain the projection image.
3. Rotate the image on the *xy*-plane to align it on the *y*-axis.
4. Assemble many images according to the azimuth angle. To find the orientation, the ion-reaction microscope will be a good candidate.
5. Create the 3D structure by 3D FFT.

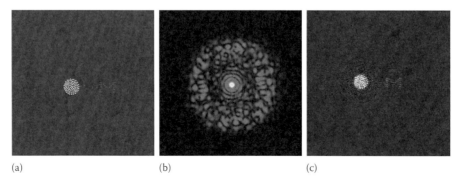

FIGURE 39.26 Demonstration of coherent amplification on test pattern.

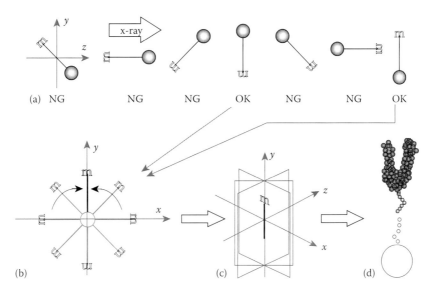

FIGURE 39.27 Reconstruction of a 3D structure from randomly oriented molecules.

This method has the high potential to realize single biomolecule imaging, because we have a large number of photons on the detector in one shot. Extensive research and development will be required on (1) the development of a stable link-structure to the biomolecule, (2) the optimization of the nano-sized gold particle (crystal or amorphous), (3) optimum data-processing routines to handle background noise, and (4) intelligent data averaging in reciprocal space. For better determination of molecule orientation, the ion-reaction microscope has to be implemented.

39.8 X-Ray Free-Electron Laser

In this section, the newly developed x-ray FEL technology is explained. Till March 2009, there were three x-ray FEL projects under construction; LCLS at SLAC Stanford (Arthur et al., 2002), XFEL/SPring-8 at Hyogo Japan (Tanaka and Shintake, 2005), and Euro-XFEL at DESY Hamburg (Altarelli et al., 2006). These projects aim at generating a coherent, intense, x-ray beam at a Å wavelength. The peak brightness of such x-ray FELs reaches 1×10^{33} photons/mm²/mrad²/0.1%BW, which is 10^{10} times higher than that of third generation light sources.

Figure 39.28 shows the schematic diagram of x-ray free-electron laser. The basic configuration of x-ray free-electron laser is quite simple; the electron source generates a low emittance (i.e., low divergence) electron beam, followed by a linear accelerator to raise the energy in the GeV energy range, and then a beam is sent into the undulator line.

We use an undulator device to generate a coherent x-ray. Figure 39.29a shows the schematic arrangement of the undulator device, where series of permanent magnet arrays create a periodic transverse magnetic field of alternating polarity. Running an electron through this device excurses transverse oscillation, and generates a series of spherical waves at each curve on the sinuous trajectory. Since the electron is high energy and its velocity is very close to the speed of the light velocity, the wavelength (distance between the series of the spherical wave fronts) becomes very small (this is the Doppler effect). Figure 39.29b shows a snapshot image of the radiation field made by the real-time-simulator Raidtion2D. After traveling a distance, the field pattern becomes a plane wave with nice periodic and narrow spectrum. This is called "undulator radiation." Today's light sources are equipped with many undulator devices providing monochromatic x-ray beams for important scientific research, i.e., crystallography, spectroscopy, etc.

The wavelength of the generated undulator radiation is related to the undulator period and the electron beam energy as follows:

$$\lambda_X = \frac{\lambda_u}{2\gamma^2}\left(1 + \frac{K^2}{2}\right) \tag{39.19}$$

where

λ_u is the longitudinal period of the undulator field
K is the normalized field strength in the undulator
γ is the electron energy in the rest-mass energy unit: $\gamma = 1 + E/m_0 c^2$

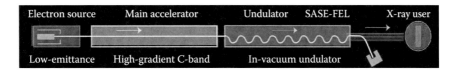

FIGURE 39.28 Basic configuration of an XFEL machine. The drawing is an example of the SCSS, which consists of a low-emittance electron source, the main C-band accelerator and the in-vacuum undulator. One or two bunch compression systems are included in the main accelerator.

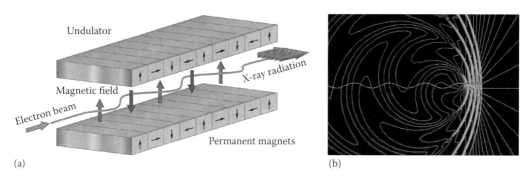

FIGURE 39.29 (a) Undulator of permanent magnet type. (b) Undulator radiation simulated by Radiation2D. (From Shintake, T., New mathematical method for radiation field of moving charge. In *Proceedings EPAC2002*, Paris, France, 2002.)

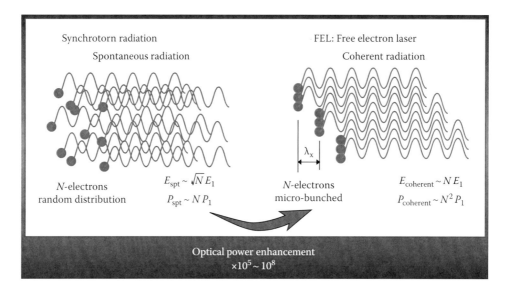

FIGURE 39.30 In an ordinal synchrotron, many electrons radiate in random phases as spontaneous radiation. In the FEL mode, electrons are aligned at the wavelength. The peak power is enhanced dramatically with an interference effect.

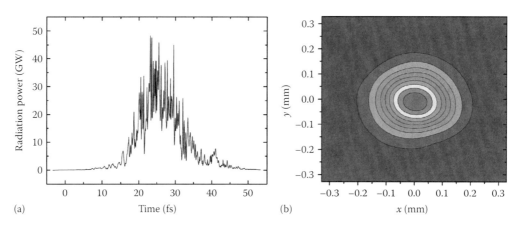

FIGURE 39.31 (a) Temporal profile of XFEL pulse. (b) Transverse beam profile at 50 m downstream.

Using a practical number, the undulator period λ_u of 18 mm, $K = 1.5$, and the target wavelength of x-ray as 1 Å, $\lambda_X = 1 \times 10^{-10}$, we find the required electron beam energy $\gamma = 1330$, or $E = 6.8$ GeV. This energy is rather high; therefore it requires a large scale accelerator and the x-ray FEL becomes a large-scale facility.

In the accelerators, there are a large numbers of electrons; typically 10^9–10^{10} electrons run together in one bunch. Since each electron possesses a random position within the bunch, the radiation field also becomes random, and the total electric field intensity becomes the statistical deviation of the particle group.

If we use a long undulator line, and a high quality electron beam (a small transverse emittance and a small energy spread), the electron, when passing through the undulator, starts to form micro-bunching (i.e., periodic density modulation) at the x-ray wavelength. This is created by a transverse kick from the radiation field, followed by a momentum conversion (from transverse to longitudinal) along with a sinuous trajectory in the undulator. Once the density modulation is created, the radiation field is enhanced by the interference effect, which further accelerates the longitudinal modulation on the electron bunch, resulting in the amplification of the radiation field in terms of exponential growth. This process starts from a shot noise on the incoming electron bunch. It is called the SASE-mode FEL: Self-Amplified Spontaneous-Emission type Free-Electron Laser. It does not require a mirror-resonator, and there is no wavelength limitation associated with radiation damage or reflection loss on the mirror at a shorter wavelength. Thus, it can operate at any wavelength, ultimately in x-ray wavelength (Figure 39.30).

Figure 39.31 shows the temporal structure of the XFEL beam and the transverse-beam profile at 50 m downstream. SASE-FEL starts from the spontaneous noise, and thus its shows remained a noise feature. The longitudinal coherent length is principally limited due to the random phase in the noise source. A typical coherent length is a few 100 nm. However, this is enough length for a single molecule imaging, because the expected maximum path difference in the diffraction imaging on a single molecule is less than 100 nm. The transverse mode is fairly uniform, and it becomes even better after focusing the beam into a 100 nm spot through focusing optics (higher frequency transverse modes escape quickly from the Gaussian beam core).

TABLE 39.1 Typical Beam Parameter of X-Ray FEL (XFEL/SPring-8)

Operational wavelength	<1 Å
Peak output power	~20 GW
Spot size electric field intensity@50 m downstream (after focusing 100 nm spot)	0.2 mm FWHM, 14 GV/m = 1.4 V/Å) (100 nm FWHM, 30 TV/m = 3000 V/Å)
X-ray pulse length	~20 fs FWHM
Optical pulse energy@1 Å	0.4 mJ (photon energy 12 keV)
Photon number in one pulse@1 Å	2×10^{11} photons/pulse
Maximum peak brightness@1 Å	1×10^{33} photons/mm^2/mrad^2/0.1%BW
Maximum pulse repetition	10–60 pps
Electron beam energy	8 GeV
Electron charge per bunch	0.3 nC/bunch
Normalized slice emittance (projected)	0.8 πmm · mrad (1.2 πmm · mrad)
Slice energy spread (projected)	0.8×10^{-4} (1×10^{-3})
Undulator period, K-parameter	18 mm, $K = 1.9$
Undulator length	90 m

Table 39.1 summaries the typical x-ray beam parameter available from XFEL (example is, XFEL/SPring-8). For single biomolecule imaging, the most important parameter is the available photon number in one pulse, and the short pulse duration. In the example case, a 2×10^{11} photons/pulse is available for a 20 fs pulse duration, which is almost compatible to the required x-ray beam for the imaging of a single lysozyme with a x-ray heterodyne technique using a gold particle.

Figure 39.32 shows an aerial view of the XFEL/SPring-8 under construction. The beam operation is scheduled from FY 2011.

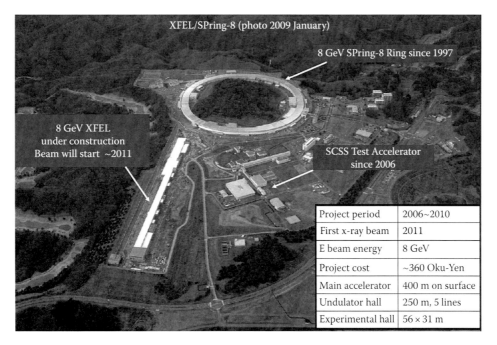

FIGURE 39.32 Aerial view of XFEL/SPring-8. It consists of a 400 m long main accelerator and a 250 m long undulator area.

39.9 Summary

The world's first x-ray FEL (LCLS) will be in operation in 2009, followed by the XFEL/SPring-8 in 2011. Then Euro-XFEL will be ready in 5 years. In these facilities, the highlighted experiment will be diffraction x-ray microscopy on nanostructure study, and bio samples. To achieve the ultimate goal of single biomolecule imaging, we need continuous research and development on various aspects, and the trial and error experience using these new facilities.

It seems all the remaining problems are technical issues and will be solved. The author believes that single biomolecule imaging will be realized in the near future.

References

Ackermann, W., G. Asova, V. Ayvazyan et al. 2007. Operation of a free-electron laser from the extreme ultraviolet to the water window. *Nat. Photon.* **1**: 336–342.

Altarelli, A., R. Brinkmann, M. Chergui et al. (Eds.) 2006. *The European X-Ray Free-Electron Laser, Technical Design Report*, Preprint DESY 2006-097, DESY, Hamburg, Germany.

Arthur, J., P. Anfinrud, P. Audebert et al. 2002. *Linac Coherent Light Source (LCLS) Conceptual Design Report, SLAC-R593*, Stanford, CA.

Bates, R.H.T. 1982. Fourier phase problems are uniquely solvable in more than one dimension. I: Underlying theory. *Optik* **61**: 247.

Bogan, M.J., W.H. Benner, S. Boutet et al. 2008. Single particle x-ray diffractive imaging. *Nano Lett.* **8**: 310–316.

Bonifacio, R., C. Pellegrini, and L.M. Narducci. 1984. Collective instabilities and high-gain regime in a free electron laser. *Opt. Commun.* **50**: 373–377.

Boutet, S., M.J. Bogan, and A. Barty. 2008. Ultrafast soft X-ray scattering and reference-enchanced diffractive imaging of weakly scattering nanoparticles. *J. Electron. Spectros. Relat. Phenom.* **166–167**: 65–73.

Bruck, Y.M. and L.G. Sodin. 1979. On the ambiguity of the image reconstruction problem. *Opt. Commun.* **30**: 304.

Chantler, C.T., K. Olsen, R.A. Dragoset et al. 1995. X-ray form factor, attenuation, and scattering tables, in *Physical Reference Data*, NIST National Institute of Standards and Technology (http://physics.nist.gov/PhysRefData).

Chapman, H.N., A. Barty, M. Bogan et al. 2006. Femtosecond diffractive imaging with a soft-X-ray free-electron laser. *Nat. Phys.* **2**: 839–843.

Dauter, D. 2006. Current state and prospects of macromolecular crystallography. *Acta Crystallogr.* **D62**: 1.

Drenth, J. 1994. *Principles of Protein X-ray Crystallography*, 2nd edn., Springer Verlag, New York.

Fienup, J.R. 1978. Reconstruction of an object from the modulus of its Fourier transform. *Opt. Lett.* **3**: 27–29.

Fung, R., V. Shneerson, D.K. Saldin, and A. Ourmazd. 2008. Structure from fleeting illumination of faint spinning objects in flight. *Nat. Phys.* **5**: 64–67.

Gerchberg, R.W. and W.O. Saxton. 1972. A practical algorithm for the determination of phase from image and diffraction pictures. *Optik* **35**: 237.

Hainfeld, J.F. and R.D. Powell, 2000. New frontiers in gold labeling. *J. Histochem. Cytochem.* **48**: 471–480.

Hubbell, J. H., H. A. Gimm, and I. Fverf. 1980. Pair, triplet, and total atomic cross sections (and mass attenuation coefficients) for 1 MeV-100 GeV photons in elements Z: 1 to 100. *J. Phys. Chem.* **9**: 1023.

Huldt, G., A. Szoke, and J. Hajdu. 2003. Diffraction imaging of single particle and biomolecules. *J. Struct. Biol.* **144**: 219–227.

Janos, K. 2001. Scattering of X-rays from electrons and atoms, in *X-Ray Data Booklet*, Thompson, A.C., D.T. Attwood, E.M. Gullikson et al. (Eds.) 341. Lawrence Berkeley National Laboratory, Berkeley, CA.

Janos, K. 2006. Free electron lasers, FLASH microscopy. *Nat. Phys.* **2**: 799.

Lundstrom, K. 2006. Structural genomics for membrane proteins. *Cell Mol. Life Sci.* **63**: 2597.

Mancuso, A.P., A. Schropp, B. Reime et al. 2009. Coherent-pulse 2D crystallography using a free-electron laser x-ray source. *Phys. Rev. Lett.* **102**: 035502/1–5.

Marchesini, S., H. He, H.N. Chapman et al. 2003. X-ray image reconstruction from a diffraction pattern alone. *Phys. Rev. B* **68**: 140101.

Miao, J., D. Sayre, and H.N. Chapman. 1998. Phase retrieval from the magnitude of the Fourier transforms of non-periodic objects. *J. Opt. Soc. Am. A* **15**: 1662.

Miao, J., P. Charalambous, J. Kirz et al. 1999. Extending the methodology of X-ray crystallography to allow imaging of micrometre-sized non crystalline specimens. *Nature* **400**: 342.

Miao, J., K.O. Hodgson, and D. Sayre. 2001. An approach to three-dimensional structures of biomolecules by using single-molecule diffraction images. *Proc. Natl. Acad. Sci. USA* **98**: 6641–6645.

Miao, J., T. Ishikawa, B. Johnson et al. 2002. High resolution 3D X-ray diffraction microscopy. *Phys. Rev. Lett.* **89**: 88303.

Millane, R.P. 1996. Multidimensional phase problems. *J. Opt. Soc. Am. A* **13**: 725.

Nave, C. 1999. Matching X-ray source, optics and detectors to protein crystallography requirements. *Acta Crystallogr.* **D55**: 1663.

Neutze, R., R. Wouts, D. van der Spoel et al. 2000. Potential for biomolecular imaging with femtosecond X-ray pulses. *Nature* **406**: 752.

Palczewski, K., T. Kumasaka, T. Hori et al. 2000. Crystal structure of rhodopsin: A G protein–coupled receptor. *Science* **289**: 739–745.

Robinson, I.K. and J. Miao. 2004. Three-dimensional coherent x-ray diffraction microscopy. *MRS Bull*. March: 177–181.

Sayre, D. 1952. Some implications of a theorem due to Shannon. *Acta Crystallogr*. **5**: 843.

Schnars, U. and W. Jueptner. 2005. Phase transformation of a spherical lens, in *Digital Holography,* Springer, Berlin, pp. 145–149.

Shintake, T. 2002. New mathematical method for radiation field of moving charge. In *Proceedings EPAC2002*, Paris, France.

Shintake, T. 2008. Possibility of single biomolecular imaging with coherent amplification of weak scattering x-ray photons. *Phys. Rev. E* **78**: 041906.

Tanaka, T. and T. Shintake (Eds.). 2005. *SCSS X-FEL Conceptual Design Report*, RIKEN Harima Institute, Hyogo, Japan.

Ullrich, J., R. Moshammer, A. Dorn et al. 2003. Recoil-ion and electron momentum spectroscopy: Reaction-microscopes. *Rep. Prog. Phys.* **66**: 1463–1545.

Amplified Single-Molecule Detection

Ida Grundberg
Uppsala University

Irene Weibrecht
Uppsala University

Ulf Landegren
Uppsala University

40.1 Introduction

Biomolecules such as nucleic acids and proteins provide valuable building material for nanoengineering complex structures with atomic precision. In particular, DNA and/or protein constructs combined with enzymes that covalently modify these have proven useful as probes for high-performance molecular detection reaction in biomedical research and diagnostics. In this chapter, we describe some mechanisms in use in biomolecular detection reactions and, in particular, focus on some approaches being explored in our laboratory for detection at the ultimate level of single molecules in biological samples.

The recent availability of total information about the genomes of humans and other species, and about the RNA molecules and proteins encoded therein, has motivated the development of new generations of techniques to investigate this wide array of molecules in health and disease—their levels, locations, and interactions in complexes. This challenge requires the development of powerful molecular tools, and we discuss herein some trends for the design and production of molecular-detection reagents suitable for these tasks. But first, a general introduction to DNA-based nanoengineering is given.

40.2 DNA as a Material for Molecular Design

Among materials available for nanometer-scale construction work, DNA molecules present a number of striking opportunities: (1) The chemical synthesis of DNA strands composed of the four nucleotide building blocks of DNA: A, C, G, and T, in any sequence, has been brought to such perfection that strands of 100 or more nucleotides can be built with ease, and the equivalent of a bacterial genome of a million nucleotides can be fabricated by a parallel synthesis at modest expense. (2) The simple rules of base pairing of complementary DNA strands—where A basepairs to T and G to C—allow a convenient design of duplex DNA molecules at will. The biophysics of these interactions is well-characterized with respect to thermal stability, sensitivity to mismatches, influence of dangling ends, the absence of an internucleotide bond (stacked hybridization), etc. (3) Yet another advantage over other nanoscale construction materials is afforded by the availability of a wide repertoire of DNA-specific enzymes including polymerases that copy strands of DNA, ligases that join shorter DNA strands into longer ones, and nucleases that digest DNA from the ends (exonucleases) or internally (endonucleases). (4) It is also of importance that nucleic acids can be replicated several times over. (5) This process preserves any information embodied in the sequence of the amplified molecules.

40.3 Building Nanoscale DNA Structures

Extreme examples of nanoscale DNA construction have been described by Zhang and Seeman (1994) and Rothemund (2006), where highly complex structures that include truncated octahedra and nanometer-sized objects in the shape of "smileys" have been produced, respectively (Figure 40.1). Short synthetic DNA strands are used to create these constructions, sometimes in combination with longer strands produced by bacteria, by using the rules of basepairing together with some further construction features. The structures are stabilized by duplexes of DNA strands to undergo branching, and the strands are covalently sealed using DNA ligases. This type of sequence-directed folding and joining of DNA strands into larger predefined structures

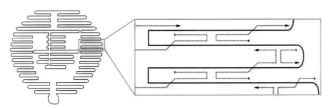

FIGURE 40.1 Nanofabricated DNA design. A nanofabricated structure in the shape of a smiley was constructed using a long single-stranded DNA molecule derived from bacteriophage M13 (solid line). The strand was twisted into a predetermined structure by annealing shorter oligonucleotides (dashed lines) and sealing gaps between some of these through DNA ligation. The structure induced by the complementary DNA sequences from part of the construction is shown in a magnified box. The diameter of the whole structure is approximately 100 nm. (Adapted from Rothemund, P.W., *Nature*, 440, 297, 2006.)

has been referred to as molecular origami and it is promising as a means of engineering highly complex structures at an unprecedented level of precision.

40.4 Clonal Amplification of Nucleic Acid Molecules

The two- and three-dimensional DNA structures described above are different from anything observed in nature, and the topological links introduced for stability preclude replication of the DNA strands by polymerases. This ability of normal nucleic acids to be replicated, generating copies of itself, is of course one of the most salient properties of this class of molecules both in biology and biotechnology. The *in vivo* expansion of single DNA molecules carried by bacteria, plants, and other organisms into clones with large numbers of identical members is a well-known phenomenon in nature. The advent of the molecular biology revolution brought the ability to expand any DNA sequences at will after recombination into a carrier DNA molecule (a vector) and introduction in bacteria or other host species. However, other techniques have been available since quite some time for amplifying single DNA molecules. For example, Chetverin et al. demonstrated that single RNA molecules could be grown as colonies in agarose gels with the aid of the enzyme Qβ replicase (Chetverina and Chetverin 1993, Chetverina et al. 2002). This enzyme is known to copy certain RNA molecules into complementary strands that, in turn, can be copied and so on, producing large numbers of replicates of the RNA molecules. The currently dominant method for *in vitro* amplification of DNA is through the polymerase chain reaction (PCR) (Saiki et al. 1985) (Figure 40.2). This method

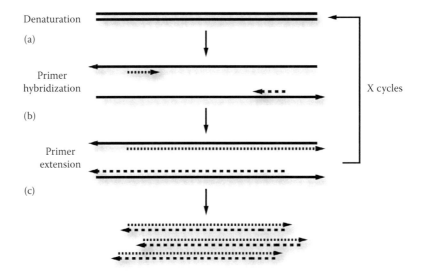

FIGURE 40.2 PCR. Amplification of target DNA sequences by the PCR. (a) The double-stranded DNA target molecule is denatured by heating. (b) Two PCR primers shown in light and dark gray anneal to their target sequences when the temperature is decreased. (c) Next, the temperature is increased to allow a DNA polymerase to extend the primers. These steps are repeated several times in every cycle, doubling the amount of copies of the targeted sequence thus leading to an exponential amplification of the sequence of interest.

has proven useful for an exponential amplification of small numbers or even single DNA molecules, allowing individual molecules to be grown as molecular clones, and this is now one of the fundamental technologies of DNA manipulation and in DNA diagnostics. The method allows specific DNA sequences to be cyclically replicated, each cycle doubling the amount of DNA targeted by the reaction.

Another method for DNA amplification is that of rolling-circle amplification (RCA) (Baner et al. 1998, Fire and Xu 1995, Liu et al. 1996, Lizardi et al. 1998) (Figure 40.3). In this method, a circular DNA strand is continuously copied by a polymerase, producing a long DNA strand composed of repeated complements of the DNA circle. Unlike PCR, this reaction does not proceed in an exponential fashion but only accumulates copies at a rate that is linear over time. However, the fact that the product is a single contiguous DNA strand that tends to bundle up in a random coil renders this amplification method highly suitable for localized detection reactions. It is also possible to achieve a more rapid amplification if the concatemeric RCA products are cleaved into monomers, directed by an oligonucleotide. These monomers can then be turned into DNA circles in ligation reactions templated by the same oligonucleotide. Next, the circles that form, complementary to the original ones, are now ready to be replicated in a second RCA reaction, primed by the same oligonucleotide. This process is referred to as circle-to-circle amplification (C2CA) and it can be repeated until the desired amount of product is obtained (Dahl et al. 2004) (Figure 40.4).

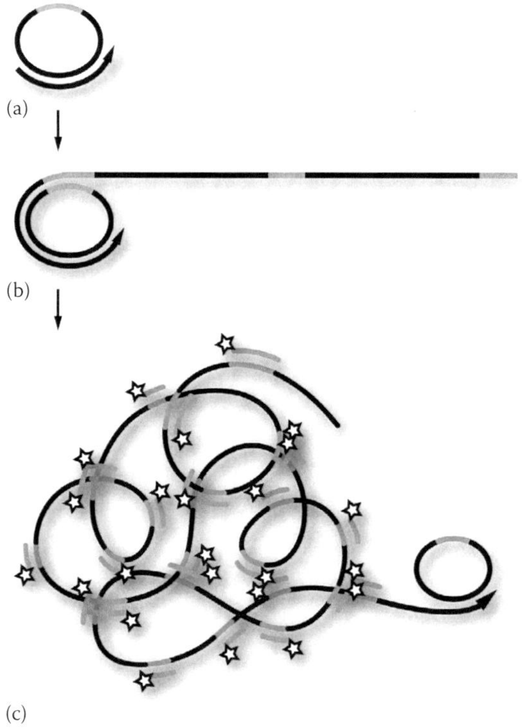

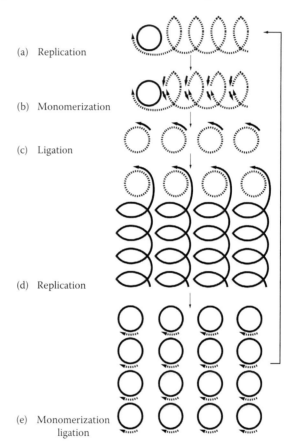

(a) Replication

(b) Monomerization

(c) Ligation

(d) Replication

(e) Monomerization ligation

FIGURE 40.3 RCA. Rolling circle amplification is a linear amplification reaction that results in a localized reaction product. (a) First, a primer oligonucleotide hybridizes to a circular single-stranded DNA molecule to be amplified. The circular molecule contains a site for detection probe hybridization (gray). (b) Upon addition of a suitable DNA polymerase (e.g., phi29 polymerase), the primer is extended. Because of the capacity of the polymerase to perform strand displacement, the extension reaction can continue many turns around the DNA circle. The reaction produces a complementary DNA molecule consisting of contiguous concatemeric copies of the target DNA circle. (c) The RCA typically yields a product of ~1000 copies of a 100 nucleotide DNA circle after a one-hour amplification. This reaction product collapses into a DNA-bundle that can be detected by e.g., hybridization of fluorescence-labeled detection oligonucleotides (stars), resulting in a ~1 μm brightly fluorescent spot that can be conveniently visualized through a fluorescence microscope with minimal background from remaining un-hybridized detection probes.

FIGURE 40.4 C2CA. Amplification of circular DNA strands by C2CA. (a) A circular DNA strand is used to template a first round of RCA, generating a complementary RCA product, shown as dashed line. (b) Next, an oligonucleotide is hybridized to the product, directing its cleavage into monomers by a restriction enzyme. (c) After denaturing the enzyme and dissociating hybridized oligonucleotide fragments by elevating the temperature, the temperature is lowered again. This allows the remaining undigested oligonucleotides to hybridize to the monomeric RCA products, directing their joining into short DNA circles by a DNA ligase. (d) These circles, complementary to the one initiating the reaction, are next used to template another generation of RCA, primed by the same oligonucleotide. (e) The reactions can be repeated to achieve greater amplification, and they can be used to produce circular or linear monomers or long single-stranded concatemers of the desired polarity.

40.5 Molecules That Have Both a Genotype and a Phenotype

DNA molecules and, in the case of some virus, RNA molecules carry the genotypic information that allows organisms to propagate with preservation and evolution of properties through generations. It was noted early in the era of molecular biology that nucleic acid molecules like DNA or RNA, in addition to having a genotype, could also carry a phenotype that has observable and selectable properties without first being translated to proteins (Kacian et al. 1972). For example, the rate of replication of RNA molecules by the Qβ replicase exhibits sensitivity to inhibition by chemicals that bind to RNA molecules, or enzyme that degrade RNA. Because of the frequent replication errors by the replicase, it proved possible to evolve *in vitro* mutant RNA molecules that exhibited increased resistance to inhibitory agents by including these in the replication reactions.

In 1990, three reports appeared that demonstrated that repeated cycles of selection from large libraries of nucleic acids allowed the isolation of nucleic acid molecules with properties such as the ability to bind specific target molecules or even exert enzymatic activities (Ellington and Szostak 1990, Robertson and Joyce 1990, Tuerk and Gold 1990). RNA or DNA molecules isolated by such procedures and shown to bind specific target molecules are referred to as aptamers. Related methods have been developed to isolate nucleic acids that encode proteins with selectable properties without cell-based cloning by creating covalent or noncovalent complexes of nucleic acids and the proteins they encode. The technology of using repeated cycles of selection and amplification to isolate binding molecules from libraries of molecules that include both an RNA genotype and a protein-based phenotype is referred to as an RNA display (Liang and Pardee 1992, Welsh et al. 1992). The technique is of interest in the context of this chapter both because these molecules represent instances of advanced nanoengineered hybrid nucleic acid-protein molecules, and also because the methods provide a valuable means of selecting reagents that bind target protein molecules with high affinity and specificity.

40.6 Reagents for Molecular Detection

We are now ready to discuss reagents used for detecting macromolecules. Detection of specific nucleic acid sequences or protein molecules is most often accomplished using affinity probes that bind the target molecules. In the case of nucleic acids, this is fairly easily done by taking into account the simple rules of nucleic acid complementarity to produce probe nucleic acid strands, often prepared as short synthetic DNA strands or oligonucleotides that bind target nucleic acid sequences. For proteins, the task is more complex, however, as suitable binding reagents cannot be designed by following any set of theoretical rules, but instead they must be selected from large libraries of molecules by processes of trial and error. This can be accomplished either using the immune systems of immunized animals to produce antibodies directed against a protein of choice, or by taking advantage of *in vitro* selection methods such as the aptamer selection or the RNA display methods mentioned above. Nonetheless, efforts are now underway both in academia and industry to develop resources of affinity reagents against all human proteins (www.proteomebinders.org/, www.proteinatlas.org/).

In the case of nucleic acid probes, a single 20-nucleotide sequence from the human genome is sufficiently long that its sequence is unlikely to occur more than once in a genome the size of the human genome, since the 4^{20} possible probes are more plentiful than the 6×10^9 nucleotides present on both strands of the human genome. Nonetheless, because of the subtle difference in stability between a perfectly matched probe-target pair and the many slightly mismatched probe-target hybrids that can form, it is generally not possible to uniquely hybridize such probes to their correct target sequences while avoiding hybridization to any irrelevant sequences. The PCR technique solves this problem of specific detection by enlisting two mechanisms. First, in order to amplify a DNA sequence by PCR, two oligonucleotides are needed, one hybridizing to each of the two complementary strands of the targeted sequence. This requirement for detection by not just one but two oligonucleotides greatly reduces the risk that incorrect target molecules will be detected. Second, in order for polymerases to extend the two oligonucleotides, thus priming replication, it is particularly important that the ends to be extended are correctly base paired, further reducing the risk of amplifying irrelevant sequences.

In the case of protein detection by antibodies or other affinity probes, a similar situation exists. The affinities of protein-binding reagents such as antibodies for their target proteins can vary widely, and they regularly also show affinity for related or unrelated proteins other than those they were raised against. If these other target proteins are present at higher concentrations than the targeted protein in a biological sample, then specific detection can be difficult or impossible. Here, again matters can be improved by using assays where two binding reactions are required. In sandwich immune assays, proteins to be analyzed in a sample are first bound to immobilized antibodies. After washes, antibodies that are capable of binding other determinants on the bound target proteins are added and recorded via detectable groups after washes (Wide et al. 1967). This classical sandwich-immune assay has been slightly modified in recent years, for example, by using oligonucleotide-coated gold nanoparticles as detectable moieties (Nam et al. 2003). As an alternative, an amplified detection signal can be obtained from bound antibodies by attaching DNA strands to the antibodies added in the solution for detection via PCR (immuno-PCR) (Sano et al. 1992), or by priming the RCA of added DNA circles (immuno-RCA) (Schweitzer et al. 2002).

40.7 Need for Improved Molecular Detection

The development of new molecular detection reactions proceeds along several trajectories. As described above, a central goal is to improve the specificity and the sensitivity of detection. These two properties are linked: if the specificity is poor, no amount of signal amplification can improve the sensitivity. In complex samples, such as total genomes or among all proteins present in serum, the problem of specificity is of paramount importance, and it is likely that much of the proteome is currently unavailable for detection by standard methods because of insufficient sensitivity of detection. Other desirable properties are the ability to interrogate many molecules in parallel, and also to efficiently handle many samples, in particular, for research applications. Clinically, there is a requirement to develop tests that produce results in less than 30 min or faster still, and the cost and the precision are other important factors to consider in both research and routine.

40.8 Pushing Detection to the Limit of Single Molecules

An emerging requirement in molecular testing is the need to study individual cells and even individual molecules. Current biological assays typically record average results across many cells in a tissue or other samples, and for millions or more molecules. However, in order to appreciate the heterogeneity and the division of labor within a tissue, individual cells should ideally be evaluated separately. Moreover, to investigate the composition of the molecular complexes that execute cellular functions, it is desirable to observe individual complexes, but this requires better specificity and sensitivity of detection than what has been available. The advent of green fluorescent protein (GFP) technology allows the detection of few and even single protein molecules that have been genetically manipulated to include a fluorescent domain (Chalfie et al. 1994). In the case of biological specimens where genetic manipulation is impossible, such as for patient

samples, one has to depend on molecular-probing methods for detection purposes. As already discussed, molecular amplification can serve to improve detectability and also to enhance the specificity of detection provided the amplification reaction is only initiated upon highly specific detection, as is the case in the PCR technique. Some methods for amplified single-molecule detection will be discussed next.

A number of recent techniques for parallel sequencing of DNA fragments require that many single DNA molecules are locally amplified to larger numbers, in order to next determine the sequence of each of the hundreds of thousands or more molecular colonies in parallel. The methods used for these procedures include the amplification of single-target molecules and trapping on a particle, all contained in individual aqueous droplets in an emulsion (Diehl et al. 2005). In a variant of this procedure, surfactant-coated droplets can be precisely formed in a microfluidic device and made to merge or divide, or be subjected to thermocycling for amplification of individual molecules (www.raindancetechnologies.com/). Alternatively, PCR primers can be immobilized on a planar surface, such that the products initiated from single targets accumulate in little isolated patches on the surface, ready for sequencing (www.illumina.com/) (Figure 40.5). Finally, yet one more method to obtain a cluster of copies of a DNA molecule for sequencing purposes is to employ the RCA procedure to generate concatemers of sequences complementary to circularized DNA fragments (Goransson et al. 2008, Pihlak et al. 2008). All these single-molecule amplification techniques start from pure strands of DNA, but biological samples present considerably greater challenges. They require powerful probing strategies involving molecular-probe designs that provide the necessary specificity and sensitivity. Next, we will describe approaches developed in our lab for the detection of first, nucleic acids and then proteins.

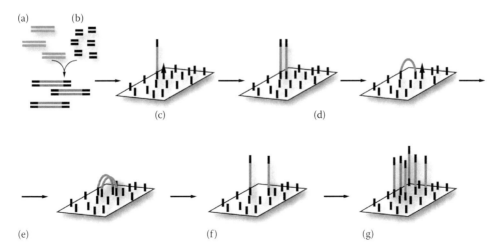

FIGURE 40.5 Bridge amplification. Generation of clusters of identical DNA molecules on a solid surface by bridge amplification. (a) Adapter sequences shown in black are attached to DNA target fragments (gray). (b) Individual target molecules are randomly distributed on a solid support covered with primers for amplification (black) and (c) copied by extension by a DNA polymerase from one of the primers. (d) After denaturation, the extension product bridges over to a nearby complementary primer. (e) Following the renewed primer extension, a double-stranded bridge molecule is created and after a second denaturation. (f) Two single-stranded molecules have formed and are ready to hybridize to complementary oligonucleotides. (g) After several rounds of replication dense clusters of copies of individual target molecules are formed.

40.9 Padlock Probes for Parallel and Localized Detection of Nucleic Acids

Padlock probes are oligonucleotides of around 100 nucleotides that can be used for the detection and the distinction of nucleic acid sequences (Nilsson et al. 1994) (Figure 40.6a). The probes have one target-complementary sequence at each end. Upon hybridization to a target sequence, the end sequences are juxtaposed. If the probes are correctly hybridized, the padlock probe can be ligated to a closed circle by a DNA ligase. Just as for PCR, the requirement for hybridization by two probe segments ensures sufficient specificity to identify unique sequences in large genomes. Moreover, the action of the ligase depends on an accurate hybridization by the ends to be joined, permitting a convenient distinction of single-nucleotide target sequence variants.

The dual-recognition mechanism of these probes has allowed them to be successfully combined for parallel genotyping ten thousand genomic loci (Hardenbol et al. 2005). By contrast, PCRs cannot be combined in similar numbers because of the many opportunities for cross-reactions between different primers, producing incorrect target sequences and decreasing the efficiency of amplification.

The act of circularizing the probes causes them to become topologically linked around long target molecules, simplifying localized detection of bound probes by microscopy. As an alternative, the target-dependent probe circularization can be combined with RCA to generate large, easily detected bundles of DNA where the probes have become bound (Christian et al. 2001, Lizardi et al. 1998). Probes that have become circularized, but not unreacted linear probe molecules, can be copied in an RCA reaction where the target sequence serves as a primer allowing a polymerase to spin out the amplified DNA strand, which remains anchored to the target sequence during detection steps for locating the target sequence (Larsson et al. 2004) (Figure 40.7a through d). More recently, the technique has been adapted for the detection of RNA sequences in a manner that can permit an analysis of the differential tissues expression of variants of RNA molecules from genes encoded on paternally or maternally inherited chromosomes (Larsson et al. submitted) (Figure 40.7e through i). The combination of highly specific detection with a localized amplification reaction allows single-molecule detection with sufficient precision to distinguish single nucleotide sequence variants in the genome or among transcribed genes.

40.10 Proximity Ligation for Advanced Protein Analyses

Regrets are often heard that proteins cannot be amplified in the same manner that PCR amplifies target DNA sequences. While methods for protein amplification remain unavailable, the detection of proteins can be enhanced via DNA amplification reactions using immuno-PCR or immuno-RCA. Proximity ligation is a more recent technique that allows the specific detection of target protein molecules to be recorded via amplification of DNA strands that can only form by ligation when two or more probes have been brought in proximity, such as by being bound to the same target molecule. The proximity probes (Figure 40.6b) used in these reactions are chimeric molecules with one part, normally an antibody, selected to specifically recognize a protein, and another part that is one or more oligonucleotides. Upon the proximal binding of two proximity probes to the same target molecules, the oligonucleotides can be joined by ligation, and the ligation product then serves as a reporter molecule that can be amplified for sensitive detection (Fredriksson et al. 2002) (Figure 40.8).

The mechanism of proximity ligation can be varied in many ways. For example, it is possible to use a set of three proximity probes so that an amplifiable DNA strand can only form when all three bind the target protein, or protein complex (Schallmeiner et al. 2007). This variant of the assay further increases the specificity since now three target epitopes have to be recognized, and nonspecific reactions are even more unlikely. Proximity ligation has been shown to permit sensitive detection of a wide range of targets, including protein complexes or proteins with posttranslational modifications such as phosphorylated sites, protein–DNA complexes, and also exceedingly small numbers of virus or bacteria (Gustafsdottir et al. 2006, 2007, Jarvius et al. 2007). The method has furthermore been shown to permit the parallel detection of different proteins, since probes can be designed to avoid cross-reactions between different proximity probes, in contrast to the situation for sandwich-immune assays that are more difficult to perform in multiplex (Fredriksson et al. 2007).

Like the padlock probes described earlier, proximity ligation assays (PLAs) can also be adapted for a localized detection of target molecules in samples prepared for microscopy. For this purpose, the oligonucleotides attached to antibodies that have bound in proximity direct the formation of a circular DNA molecule. Next, one of the antibody-bound oligonucleotides primes an RCA reaction, giving rise to easily detectable bundles of DNA at sites where

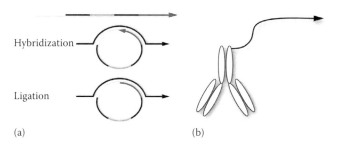

Hybridization

Ligation

(a)   (b)

FIGURE 40.6 Affinity probes. Two of the affinity probes discussed in this chapter. (a) A linear padlock probe, which upon hybridization to its target strand can be ligated into a circular molecule upon juxtaposition of the target-complementary end sequences, shown in white and dark gray, to the target molecule (black). The hybridization site for a detection reagent in the padlock probe is depicted in light gray. (b) A proximity probe consists of a protein-binder, typically an antibody, shown in white, and an attached oligonucleotide, shown in black.

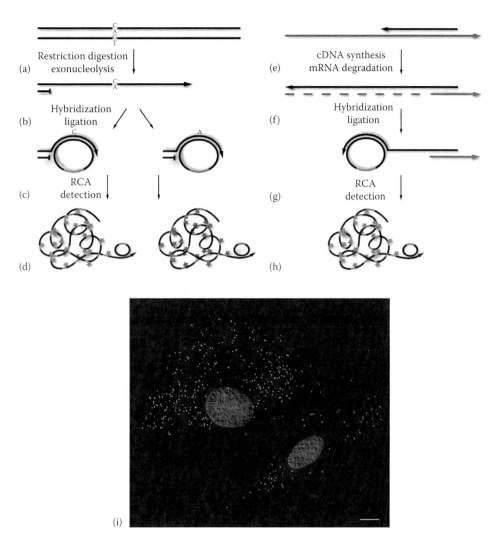

FIGURE 40.7 **(See color insert following page 25-16.)** Padlock probes *in situ*. Detection and genotyping of DNA and mRNA *in situ* using padlock probes. (a) Double-stranded DNA to be investigated for the presence of a C or an A residue in a particular position is made single-stranded (b) to permit probe hybridization by restriction digestion and exonucleolysis to remove the non-target strand. (c) Two padlock probes, designed to have the appropriate G or T nucleotides at one end in order to recognize the two genotypes, and distinct tag sequences (green or red) are hybridized to their relative targets and the open circles are closed upon ligation. (d) Localized amplification products containing numerous copies of the original padlock probes are produced by RCA and detected by the hybridization of fluorescence-labeled oligonucleotides (green and red stars) for detection in a fluorescence microscope. (e) mRNA molecules can be detected in a similar manner by first annealing a primer to the target mRNA (gray) and generating a cDNA molecule (black) by reverse transcription. (f) Next, the copied mRNA sequence is degraded by treatment with RNase H. (g) A padlock probe is allowed to hybridize to the new cDNA strand. Upon ligation, the probe is subjected to RCA creating a bundle of DNA representing a concatemer of complements of the original padlock probe. (h) Finally, the amplification product is visualized in a fluorescence microscope by hybridizing fluorescence-labeled detection probes. (i) Detection of β-actin transcripts in human and mouse fibroblasts using padlock probes. The target sequences are similar except for a single-nucleotide difference, giving rise to red RCA products in the human fibroblasts and green signals in the mouse cells. The DNA in the nuclei is stained in blue using DAPI. The scale bar represents 10 μm.

the two antibodies have bound together (Soderberg et al. 2006) (Figure 40.9). The requirement for dual, proximal binding ensures high detection specificity, and it also allows an analysis of interactions among pairs of proteins being recognized by separate antibodies. In this manner, the chimeric protein-DNA reagent in combination with enzymatic reactions steps involving both ligation and polymerization together for the first time enable analyses of signal transmission via protein complexes in patient tissues, furnishing entirely new diagnostic opportunities.

40.11 Read-Out of Molecular Detection Reactions

We have described two nanoengineered probing systems, padlock probes for nucleic acid analyses and proximity probes for investigating proteins and other macromolecules, and the means of locating the target molecules or molecular complexes by microscopy. Next, we will discuss some alternative read-out options for high-throughput, high-precision measurement

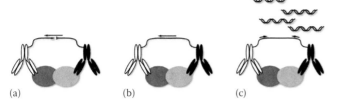

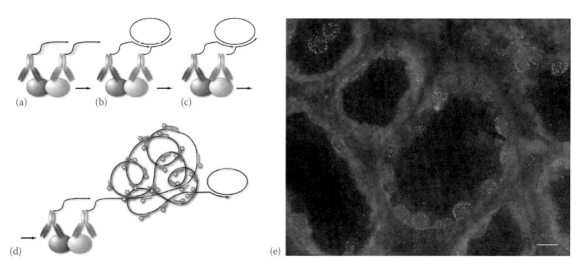

(a)   (b)   (c)

FIGURE 40.8 PLA. Detection of proteins or protein complexes via the PLA. (a) Upon binding by two proximity probes to the same protein complex, the oligonucleotides carried by the proximity probes are brought in proximity and can hybridize to an added connector oligonucleotide. (b) The oligonucleotides are next joined by ligation, creating a new DNA sequence. (c) This new DNA sequence can then be amplified and the accumulation of amplification products during the PCR is recorded as a measure of ligation products, representing the numbers of detected protein complexes.

of probe-reaction products. The general aim is to preserve the specificity of the probing systems by ensuring that only reacted probes can be recorded, and to obtain high quantitative precision. The optimal achievable precision is that which results by counting individual reacted probes.

RCA products from either of the two probing methods can be distributed on planar surfaces and interrogated by methods used for parallel DNA sequencing, resulting in digital measures of numbers of reaction products for all probes combined in a reaction. In particular, RCA products on planar surfaces have been interrogated by a sequence of hybridization reactions

to determine their identities, providing digital counts of their prevalence for precise parallel measurements of different reaction products (Goransson et al. 2008, Pihlak et al. 2008) (Figure 40.10a).

Another means to enumerate reacted probes is by hybridizing fluorescent probes to solution-phase RCA products and pumping these through a microfluidic channel, and digitally recording and identifying all single-molecule amplification products as they travel through the device. This approach was shown to provide very rapid and precise digital measures of specific target DNA or protein molecules detected by padlock or proximity probes, respectively (Jarvius et al. 2006) (Figure 40.10b).

Finally, a dual-tag microarray platform has been developed to further speed-up assay read-out for recording reaction products of large sets of probes (Ericsson et al. 2008) (Figure 40.11). This technique allows highly specific measurements of reaction products from different probes at separate locations on a planar array. Unlike commonly used microarrays, the recognition depends on the juxtaposition of two-tag sequences, located at either ends of amplified reporter strands. The reporter strands are produced by PCR or RCA of reacted probes that include two separate tag-sequence elements, incorporated in padlock probes or in pairs of proximity probes that become joined by proximity-dependent ligation. After ligating the two tags at either ends of the reporter molecules to form DNA circles, RCA reactions are initiated on the arrays to further amplify detection. If required, individual amplification products can be detected and enumerated in individual features of the arrays

FIGURE 40.9 **(See color insert following page 25-16.)** *In situ* PLA. Visualization of protein interactions using the *in situ* PLA. (a) Upon binding of two proximity probes to the same target molecule complex, the oligonucleotides of the proximity probes are brought into proximity. (b) Next, two oligonucleotides are added that can hybridize to the antibody-bound oligonucleotides. (c) The added oligonucleotides are ligated into a circular DNA-molecule, guided by the antibody-bound oligonucleotides. (d) Subsequently, one of the oligonucleotides on the proximity probes can prime an RCA, allowing amplification products to be detected, e.g., through hybridization of fluorescence-labeled oligonucleotides (red stars). Since the RCA product remains attached to one of the proximity probes, each amplification product gives rise to a distinct localized signal per detected protein–protein interaction where the antibodies bound their target complex. (e) An example of the visualization of the interaction of the oncoprotein c-Myc and its dimerization partner Max as red fluorescent dots by *in situ* PLA in fresh frozen human colon tissue. The cytoplasms were counterstained using a FITC-labeled anti-actin antibody (green) and the DNA in the nuclei was stained with Hoechst 33342 (blue). The scale bar represents 10 μm.

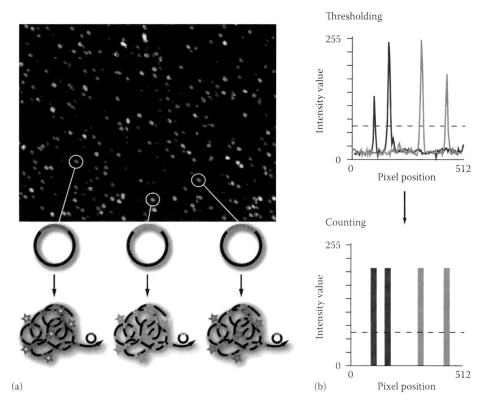

Thresholding

Counting

(a)

(b)

FIGURE 40.10 **(See color insert following page 25-16.)** Multiplexed detection. Two alternative read-out methods for an amplified detection of single DNA molecules using padlock probes. (a) Random array: Padlock probes are hybridized to complementary DNA molecular targets, and after RCA, the amplification products are randomly distributed on microscope slides. Individual amplification products are visualized in a fluorescence microscope as bright objects by hybridization of short oligonucleotide detection probes carrying different fluorophores (blue, green and red stars). (b) Detection in microfabricated channels: Individual RCA products are pumped through a microchannel and detected using a confocal fluorescence microscope in line-scan mode. Amplification products representing individual-detected target molecules are easily separated from the background by applying a fluorescence threshold. Individual amplification products are recorded as digital objects.

by microscopy for increased quantitative precision. Because of its ability to decode pairs of tag sequences, the method can be used to detect and quantitate any pairwise interactions in a set of investigated target molecules, allowing an efficient measurement of interactions among DNA, RNA, or protein molecules in biological specimens.

40.12 Conclusions and Future Perspectives

As illustrated herein, there are great opportunities for measurements of molecular states in cells, tissues, and in bodily fluids. Improved methods can provide access to new classes of diagnostic targets for early detection of disease and for monitoring responses to therapy. Advanced probes, such as the protein-DNA hybrids described herein, and their combination with carefully adjusted DNA hybridization and enzymatic modifications by ligases and polymerases provide excellent specificity and a potential to amplify molecules with little or no background, enabling single-molecule detection. A major

area of growth in diagnostics is at the point-of-care, that is, using simple devices capable of delivering accurate responses to unskilled users. This is also an area where a new generation of high-performance molecular tools can greatly extend the classes of molecules that can be targeted for analysis, and also permit parallel analyses of many more biomolecules of different classes.

Finally, while there is need for further development of detection schemes and of methods to produce the reagents, there are also new areas where improved detection methods will be required. For example, the techniques we have focused on herein are suitable to obtain snap shots of biological processes by observing molecular states at the time of analysis. This is appropriate for diagnostic analyses, however, for a profound understanding of biological processes in experimental samples, similar requirements for sensitivity, specificity, and parallel detection are seen in real-time analyses within living cells. Such probes for dynamic molecular-detection reactions could also lend themselves to imaging disease processes in patients, for example, locating tumor cells or sites of cell death by methods like positron emission tomography (PET).

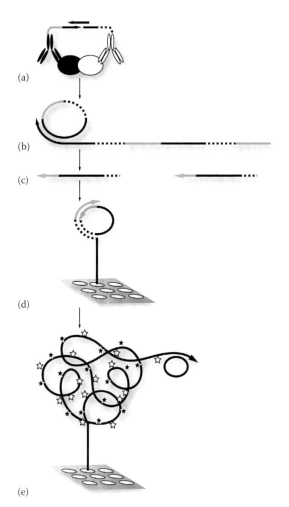

FIGURE 40.11 Dual-tag microarray read-out. Detection of protein interaction via proximity ligation with dual-tag microarray read-out. (a) A proximity ligation assay reaction joins two oligonucleotides on antibodies having bound interacting protein molecules. The two oligonucleotides include tag sequences, shown as solid gray and dashed black lines, that represent the specificity of the antibodies they are attached to. (b) Next, the ligated oligonucleotides are cleaved from the antibodies and circularized (not shown), the DNA circle templates an RCA. (c) The RCA product is digested into monomers, leaving tag sequences at either end. (d) The monomers, resulting from cleaving all RCA products from a multiplex proximity ligation reaction, are next added to a DNA microarray with oligonucleotides containing different combinations of the tag sequences. Hybridization to the appropriate combination of tags allows the monomerized RCA products to be ligated into DNA circles. (e) Finally, another generation of RCA is performed on the microarray, primed from the immobilized oligonucleotides and the amplification products are detected by hybridizing fluorescent oligonucleotides (stars).

Acknowledgments

We would like to thank Tim Conze, Jenny Göransson, Chatarina Larsson, and Ola Söderberg for providing the images included in this chapter. The work in our group is supported by grants from the Wallenberg Foundation, the EU FP6, and FP7, the Swedish Research Councils for Medicine and for Natural Sciences and Technology, VINNOVA/SSF, and by Uppsala-Bio.

References

Baner, J., Nilsson, M., Mendel-Hartvig, M., and Landegren, U. 1998. Signal amplification of padlock probes by rolling circle replication. *Nucleic Acids Res.* 26(22): 5073–5078.

Chalfie, M., Tu, Y., Euskirchen, G., Ward, W. W., and Prasher, D. C. 1994. Green fluorescent protein as a marker for gene expression. *Science* 263(5148): 802–805.

Chetverina, H. V. and Chetverin, A. B. 1993. Cloning of RNA molecules in vitro. *Nucleic Acids Res.* 21(10): 2349–2353.

Chetverina, H. V., Samatov, T. R., Ugarov, V. I., and Chetverin, A. B. 2002. Molecular colony diagnostics: Detection and quantitation of viral nucleic acids by in-gel PCR. *Biotechniques* 33(1): 150–152, 154, 156.

Christian, A. T., Pattee, M. S., Attix, C. M., Reed, B. E., Sorensen, K. J., and Tucker, J. D. 2001. Detection of DNA point mutations and mRNA expression levels by rolling circle amplification in individual cells. *Proc. Natl. Acad. Sci. U S A* 98(25): 14238–14243.

Dahl, F., Baner, J., Gullberg, M., Mendel-Hartvig, M., Landegren, U., and Nilsson, M. 2004. Circle-to-circle amplification for precise and sensitive DNA analysis. *Proc. Natl. Acad. Sci. U S A* 101(13): 4548–4553.

Diehl, F., Li, M., Dressman, D. et al. 2005. Detection and quantification of mutations in the plasma of patients with colorectal tumors. *Proc. Natl. Acad. Sci. U S A* 102(45): 16368–16373.

Ellington, A. D. and Szostak, J. W. 1990. In vitro selection of RNA molecules that bind specific ligands. *Nature* 346(6287): 818–822.

Ericsson, O., Jarvius, J., Schallmeiner, E. et al. 2008. A dual-tag microarray platform for high-performance nucleic acid and protein analyses. *Nucleic Acids Res.* 36(8): e45.

Fire, A. and Xu, S. Q. 1995. Rolling replication of short DNA circles. *Proc. Natl. Acad. Sci. U S A* 92(10): 4641–4645.

Fredriksson, S., Gullberg, M., Jarvius, J. et al. 2002. Protein detection using proximity-dependent DNA ligation assays. *Nat. Biotechnol.* 20(5): 473–477.

Fredriksson, S., Dixon, W., Ji, H., Koong, A. C., Mindrinos, M., and Davis, R. W. 2007. Multiplexed protein detection by proximity ligation for cancer biomarker validation. *Nat. Methods* 4(4): 327–329.

Goransson, J., Wahlby, C., Isaksson, M., Howell, W. M., Jarvius, J., and Nilsson, M. 2008. A single molecule array for digital targeted molecular analyses. *Nucleic Acids Res.* 36(8):e45.

Gustafsdottir, S. M., Nordengrahn, A., Fredriksson, S. et al. 2006. Detection of individual microbial pathogens by proximity ligation. *Clin. Chem.* 52(6): 1152–1160.

Gustafsdottir, S. M., Schlingemann, J., Rada-Iglesias, A. et al. 2007. In vitro analysis of DNA-protein interactions by proximity ligation. *Proc. Natl. Acad. Sci. U S A* 104(9): 3067–3072.

Hardenbol, P., Yu, F., Belmont, J. et al. 2005. Highly multiplexed molecular inversion probe genotyping: Over 10,000 targeted SNPs genotyped in a single tube assay. *Genome Res.* 15(2): 269–275.

Jarvius, J., Melin, J., Goransson, J. et al. 2006. Digital quantification using amplified single-molecule detection. *Nat. Methods* 3(9): 725–727.

Jarvius, M., Paulsson, J., Weibrecht, I. et al. 2007. In situ detection of phosphorylated platelet-derived growth factor receptor beta using a generalized proximity ligation method. *Mol. Cell Proteomics* 6(9): 1500–1509.

Kacian, D. L., Mills, D. R., Kramer, F. R., and Spiegelman, S. 1972. A replicating RNA molecule suitable for a detailed analysis of extracellular evolution and replication. *Proc. Natl. Acad. Sci. U S A* 69(10): 3038–3042.

Larsson, C., Koch, J., Nygren, A. et al. 2004. In situ genotyping individual DNA molecules by target-primed rolling-circle amplification of padlock probes. *Nat. Methods* 1(3): 227–232.

Liang, P. and Pardee, A. B. 1992. Differential display of eukaryotic messenger RNA by means of the polymerase chain reaction. *Science* 257(5072): 967–971.

Liu, D. Y., Daubendiek, S. L., Zillman, M. A., Ryan, K., and Kool, E. T. 1996. Rolling circle DNA synthesis: Small circular oligonucleotides as efficient templates for DNA polymerases. *J. Am. Chem. Soc.* 118(7): 1587–1594.

Lizardi, P. M., Huang, X., Zhu, Z., Bray-Ward, P., Thomas, D. C., and Ward, D. C. 1998. Mutation detection and single-molecule counting using isothermal rolling-circle amplification. *Nat. Genet.* 19(3): 225–232.

Nam, J. M., Thaxton, C. S., and Mirkin, C. A. 2003. Nanoparticle-based bio-bar codes for the ultrasensitive detection of proteins. *Science* 301(5641): 1884–1886.

Nilsson, M., Malmgren, H., Samiotaki, M., Kwiatkowski, M., Chowdhary, B. P., and Landegren, U. 1994. Padlock probes: Circularizing oligonucleotides for localized DNA detection. *Science* 265(5181): 2085–2088.

Pihlak, A., Bauren, G., Hersoug, E., Lonnerberg, P., Metsis, A., and Linnarsson, S. 2008. Rapid genome sequencing with short universal tiling probes. *Nat. Biotechnol.* 26(6): 676–684.

Robertson, D. L. and Joyce, G. F. 1990. Selection in vitro of an RNA enzyme that specifically cleaves single-stranded DNA. *Nature* 344(6265): 467–468.

Rothemund, P. W. 2006. Folding DNA to create nanoscale shapes and patterns. *Nature* 440(7082): 297–302.

Saiki, R. K., Scharf, S., Faloona, F. et al. 1985. Enzymatic amplification of beta-globin genomic sequences and restriction site analysis for diagnosis of sickle cell anemia. *Science* 230(4732): 1350–1354.

Sano, T., Smith, C. L., and Cantor, C. R. 1992. Immuno-PCR: Very sensitive antigen detection by means of specific antibody-DNA conjugates. *Science* 258(5079): 120–122.

Schallmeiner, E., Oksanen, E., Ericsson, O. et al. 2007. Sensitive protein detection via triple-binder proximity ligation assays. *Nat. Methods* 4(2): 135–137.

Schweitzer, B., Roberts, S., Grimwade, B. et al. 2002. Multiplexed protein profiling on microarrays by rolling-circle amplification. *Nat. Biotechnol.* 20(4): 359–365.

Soderberg, O., Gullberg, M., Jarvius, M. et al. 2006. Direct observation of individual endogenous protein complexes in situ by proximity ligation. *Nat. Methods* 3(12): 995–1000.

Tuerk, C. and Gold, L. 1990. Systematic evolution of ligands by exponential enrichment: RNA ligands to bacteriophage T4 DNA polymerase. *Science* 249(4968): 505–510.

Welsh, J., Chada, K., Dalal, S. S., Cheng, R., Ralph, D., and McClelland, M. 1992. Arbitrarily primed PCR fingerprinting of RNA. *Nucleic Acids Res.* 20(19): 4965–4970.

Wide, L., Bennich, H., and Johansson, S. G. 1967. Diagnosis of allergy by an in-vitro test for allergen antibodies. *Lancet* 2(7526): 1105–1107.

Zhang, Y. W. and Seeman, N. C. 1994. Construction of a DNA-Truncated Octahedron. *J. Am. Chem. Soc.* 116(5): 1661–1669.

Index

Printed and bound by CPI Group (UK) Ltd, Croydon, CR0 4YY

26/10/2024

01779567-0001